MARINE POLLUTION AND SEA LIFE

LA POLLUTION DES MERS ET LES RESSOURCES BIOLOGIQUES

LA CONTAMINACIÓN DEL MAR Y LOS RECURSOS VIVOS

Rédacteur en chef: **General Editor:**   Redactor Jefe:

MARIO RUIVO

Director, Fishery Resources Division, FAO
Directeur de la Division des Ressources halieutiques, FAO
Director de Recursos Pesqueros, FAO

Published by arrangement with the Food and Agriculture Organization of the United Nations
Publié en accord avec l'Organisation des Nations Unies pour l'alimentation et l'agriculture
Editado de acuerdo con la Organización de las Naciones Unidas para la Agricultura y la Alimentación

FISHING NEWS (BOOKS) LTD.

23 ROSEMOUNT AVENUE, WEST BYFLEET, SURREY, ENGLAND
and
110 FLEET STREET, LONDON EC4A 2JL ENGLAND

© FAO 1972

ISBN 0 85238 021 6

Editorial Team (FAO)		Equipe de rédaction (FAO)	Equipo de redacción (FAO)
M. RUIVO	General Editor	Rédacteur en chef	Redactor jefe
D. SAHRHAGE	Assistant Editor	Rédacteur adjoint	Redactor adjunto
L. ANDREN	Technical Editor	Rédacteur technique	Redactor técnico
V. ANGELESCU	Technical Editor	Rédacteur technique	Redactor técnico
G. TOMCZAK	Technical Editor	Rédacteur technique	Redactor técnico

Editorial assistance was also received from :

Ont également collaboré à la mise au point rédactionnelle:

Tambien han colaborado en los trabajos editoriales :

R. B. Clark, R. I. Clutter, S. B. Olsen and J. E. Portmann.

MARINE POLLUTION AND SEA LIFE

Introduction

AS human populations multiply and industrialization increases and diversifies, the problem of the pollution of the environment becomes more critical. Not least of the environmental problems is water pollution. The seas and oceans, which cover 70 per cent of the world's surface and offer one of man's great hopes for future food supplies, are not exempt from the menace of pollution. Pollution problems mount as populations move to the coasts seeking the amenities and recreational opportunities of the sea shore, as well as the convenience and advantages to be found there for certain kinds of industry. With the growing use of sea lanes for commerce, the ever-increasing size and variety of cargo ships and tankers and the use of the sea-bed for mineral extraction, the threat of pollution to the marine environment from deliberate or accidental release of noxious materials from ships and cargoes becomes more acute every day. The sea is also polluted by fall-out from the atmosphere and large amounts of pollutants and wastes reach the oceans through the rivers and run-off from the land.

Authorities are reacting to public pressures to maintain clean inland and coastal waters and are moving wastes further offshore by pipelines, tankers and barges. However, because of the enclosed character of some seas, the wastes may not be rapidly diluted and dispersed by natural processes. Depending on the nature of the waste, and on the transport by currents and winds, it may drift to the coasts of other nations. Drifting out to sea, it may even adversely affect the open sea environment for plants and animals.

During the last two decades considerable efforts have been made internationally against marine pollution. Since 1954 there has already been in existence an International Convention for the Prevention of Pollution of the Sea by Oil which has been amended subsequently to improve its efficiency. Also in 1971 a Convention for the Prevention of Marine Pollution by Dumping from Ships and Aircraft has been concluded for the northeastern Atlantic, including the North Sea, (the so-called "Oslo Convention"), and international negotiations are under way with a view to establishing a dumping convention on a world-wide basis. Generally, however, there is still little, if any, control over the discharge of wastes into international waters.

The menace of pollution to the marine environment has been brought into sharp focus by a number of events in recent years. Fall-out from nuclear tests and the discharge of radioactive wastes from nuclear reactors have brought increasing protest from a radioactivity-sensitive public. Various tanker disasters, the Santa Barbara oilwell seepage, and many other accidents have demonstrated the devastation that can arise in coastal waters and on beaches, and of its impact on living resources, from the uncontrolled entry of oil into the sea. The detection, far from any obvious sources, of pesticides in the marine environment and in the marine organisms has brought to light the devious ways in which dangerous materials can be transmitted through the marine food chain and transported long distances by physical and biological processes. The synthesis of an increased number of new compounds for a variety of purposes and their release into the environment before they are fully tested on aquatic organisms, presents potential harm to living marine resources. Moreover, the possible synergistic effects of this multiplicity of toxic substances can be far more severe than single-factor bio-assays would suggest.

Pollution is a growing problem and we should be preparing ourselves for events to come. We must assemble "baseline" information on essentially unpolluted areas of the marine environment against which we can measure the effects of pollution. Subtle changes can only be recognized by means of a long series of painstaking observations. The exploration and exploitation of non-living resources, both advancing rapidly in many coastal seas, are bound to have certain undesirable effects—at least locally—on water quality and bottom characteristics. There is a particular need to obtain data on existing conditions in unpolluted tropical waters, about which so little is known, where exploitation is planned or is proceeding.

It should also be pointed out that there are certain developments on the positive side. There is a tendency in industry to develop means of recovering valuable elements and compounds from wastes before these are discharged. Sometimes it is also possible, as in the case of "pollution" by warm water discharges, to turn the "waste" product to advantage. At least some countries are adopting practices which might reduce the immediately harmful effect of dumping (by putting wastes further offshore, and in deeper waters or in more effective containers) and minimize the risk of pollution by accidents. All these factors, however, require further scientific and economic study, development, and controlled application. This is particularly true where, as in dumping and deep pipeline construction, we are concerned essentially with balancing one use of the ocean resource against its other uses.

Among these other uses must now be counted not only traditional fishing activities, but also the great potential for aquaculture in coastal waters. In the development of this potential, the maintenance of "water quality" may be of paramount importance.

Apart from the effects on the living resources, fishing operations are affected in a variety of ways by pollutants. Trawlers may tear their nets on discarded automobile bodies and on other large, heavy objects—even on abandoned cables and gears discarded by other fishermen. They may "catch" in their nets explosives and containers of dangerous chemicals. Nets can be fouled by oil and by other similar substances. The means of controlling these nuisances and hazards to fishermen are a vital requirement for the continuing exploitation of the sea for commercial and sport fisheries.

Lastly, pollution—and the public's fear of pollution—can have other adverse effects on the economy of the fishery industries. Marine products may be tainted or poisonous, and hence unsaleable, or quality impaired and prices depressed. To regain lost quality, processing and treatment costs may be added, and the price to the consumer is correspondingly increased. Public knowledge that some fish and marine products are affected by pollution can lead to a buyer's reaction against all similar products.

It was in this context that FAO decided to organize a Technical Conference on Marine Pollution and its Effects on Living Resources and Fishing. The programme of this Conference was prepared with the assistance of the FAO Advisory Committee on Marine Resources Research (ACMRR) and of the IMCO/FAO/Unesco/WMO/WHO/IAEA Joint Group of Experts on the Scientific Aspects of Marine Pollution (GESAMP).

Notice to the Reader

THIS volume contains edited contributions and summaries of discussions at the FAO Technical Conference on Marine Pollution and its Effects on Living Resources and Fishing, organized by the Department of Fisheries, Fishery Resources Division, of FAO and held at FAO Headquarters in Rome from 9 to 18 December 1970. A summary report on this Conference has been issued as an unpriced document *FAO Fisheries Reports* No. 99.

A Seminar on Methods of Detection, Measurement and Monitoring of Pollutants in the Marine Environment, organized by FAO and cosponsored by Unesco, IAEA, SCOR and WMO, was held in conjunction with the Conference, from the 4 to 10 December 1970. A report on the Seminar has been similarly issued as *FAO Fisheries Reports* No. 99 Suppl. 1, and in an expanded version as a book entitled "A Guide to Marine Pollution" and published by Gordon and Breach Science Publishers, New York, London, Paris.

Preface

CONCERN about the degree to which man is tapping natural resources available to him, the way in which he does so, and the effects of his actions on the environment, has intensified in recent years. It has reached a point beyond that of academic interest and, in many societies, is a subject of constant public concern.

The concern represents a crisis of economic and social values. It questions the value of the technology on which man has based his activity, by wanting to examine results besides the products expected. The concern is universal. In societies relying on a high degree of technology, the questions regard the feasibility of repairing damage. In others, which have so far been spared damage, the questions are in the nature of a paradox: is it necessary to choose between development and damage to the environment? Is it possible to avoid past pitfalls and to achieve rational development?

No one government of any or its constituent departments has all the answers. It is the same with all international agencies. This is therefore the time for analysing the symptoms and formulating man's strategy vis-à-vis the natural resources on which he has to rely. It is the time for each institution concerned with a sector of man's activity to examine the experience of that sector's development, the environmental problems generated or suffered, the likely pattern of future growth and the policies to be followed if that is to be rational.

This reasoning led FAO to organize its Technical Conference on Marine Pollution and its Effects on Living Resources and Fishing, held in Rome in December 1970. The Conference provided the first world forum where experts from all backgrounds and disciplines concerned examined the problems of pollution of the sea in relation to sea life and fisheries. It contributed to identifying and advising on scientific and technical problems requiring further study or action, It also helped to relate the problems of marine pollution to the wider issues of the environment. Its timing ensured that its findings could contribute to the United Nations Conference on the Human Environment, held in Stockholm, in June 1972. They will also help in the preparation of the next United Nations Conference on the Law of the Sea. These United Nations Conferences are designed to lead to global understanding, policies and plans for action requiring a substantial international effort. FAO, with its mandate for promoting the development of world fisheries, and at the same time conserving natural resources, is ready to assume its responsibilities under this expanded effort.

I should like to take this opportunity to express my appreciation and sincere thanks to all those persons who contributed to the success of the Technical Conference and to assisting FAO to serve its member nations more fully and competently. I trust that the publication of this book will help to spread the findings of the Technical Conference among a wider audience, and create the climate of action required to deal with the problems of marine pollution.

F. E. Popper
Assistant Director-General (Fisheries)
Food and Agriculture Organization of the
United Nations

List of Contributors

[xi]

[xiii]

[xv]

Contents

Section 5

TECHNICAL ASPECTS OF MINIMISING POLLUTION AND COUNTERING ITS EFFECTS

Section 6

EFFECTS OF POLLUTANTS ON QUALITY OF MARINE PRODUCTS AND EFFECTS ON FISHING

[xxiii]

Errata

P 189. Authorship of the paper listed as commencing on p 238 should read: *G. Suárez, M. E. Ramiro y J. J. Tapanes.*

P 189. The title of the paper listed to commence on p 289 should read in the final line "aerobic and anaerobic conditions".

Section 1

MARINE POLLUTION IN THE WORLD TODAY

Summary of Discussion

Special requirements of sensitive ecosystems

Discussion started on the development of Arctic areas, with particular reference to the oilfield development in Arctic North America, and to the apparent lack of information on which adequate control measures designed to protect natural communities of the region could be based. Some facets of Arctic communities which make these regions sensitive were mentioned, e.g. the absence of planktonic dispersal phases in Arctic bottom animals. The effects of oil on coral reef systems also merit special attention. Preliminary indications are that where the reef is permanently submerged the risks of damage are slight. However, in the presence of a continuous thick film of oil, interruption of the photosynthetic cycle of the zooxanthsellae could occur and lead to further secondary effects. Occasional observations of coral reefs continuously submerged in Persian Gulf waters which were exposed to refinery effluents also showed no immediate effects. Reference was made to damage to other components of the reef ecosystems including fish kills in the Pacific Islands. Excess of nutrients may cause changes in the reef ecosystem. As an example, it was noted that a reef in Hawaii was being destroyed through blanketing green bubble algae, as a result of fertilization of coastal waters. It was pointed out that funds devoted to tropical research in general, and coral reefs in particular, are insufficient.

The vulnerability to pollution of lagoons, and of other coastal embayments, with a growing need for their use for mariculture was emphasized. Some support was expressed for the concept of classification of coastal areas in respect of their proposed utilization. Great care would be required in framing and administering pollution legislation based on such an approach, owing to the combination of both political and technical problems inherent in such schemes. In particular, it might be politically undesirable to designate areas for waste disposal in perpetuity.

Regional trends in marine pollution and monitoring schemes

The impact of atmospheric and river-borne pollutants was emphasized, and the consequent need for a co-ordinated approach and regionally based studies was made apparent. Some international organizations, e.g. FAO and WHO, and regional bodies, e.g. ICES, have been supporting these activities and fruitful cooperation is already established in the North Sea, the Baltic and the Mediterranean. Critical reviews of other parts of the world have not yet been made owing to the lack of adequate machinery. Such machinery is an essential prerequisite for the development of regional monitoring schemes.

The concept of monitoring was discussed and the views of the Seminar on Methods of Detection, Measurement and Monitoring of Pollutants in the Marine Environment were given by the Convener of the Panel on the design of a global marine monitoring system. Mention was also made of the steps being taken by the Intergovernmental Oceanographic Commission to elaborate the technical basis for monitoring within the framework of the IOC Long-term and Expanded Programme of Oceanic Research (LEPOR). A global marine monitoring system must be broad enough to include a variety of the components of the ecosystem and should be developed on the basis of national and regional systems. This presents major problems in standardization of methods; there is a need for immediate action in this direction by international organizations. Monitoring is required to understand the chronic effects of pollution and to forecast future risks.

Some discussion was devoted to the unsatisfactory nature of the definition of "pollution" adopted for the conference; but the consensus favoured retaining a general description related to protection of the environment for the beneficial use by humanity. Within such a generalized definition, the environmental quality requirements must nevertheless be rigorously defined for each particular situation. The use of regional units, based on biophysical or ecological criteria, allowing a flexible approach to pollution management was supported.

Establishment of ecological baseline studies, nationally and regionally

Discussion occurred as to whether ecological baseline studies were essential for effective pollution control at this stage of knowledge. The conference recognized the difficulties of establishing useful baselines in view of the inherent variability of ecological systems, but nevertheless emphasized the need for such studies, particularly in the light of the chronic effects of long-term exposure to persistent pollutants. Physiological criteria were recognized as essential components of baseline studies. If the use of ecologically defined management regions is adopted as an administrative concept in pollution control, then such baseline studies are an integral part of that approach. Irrespective of how sophisticated chemical monitoring techniques become, they will not provide information on the status of the biological components of an ecosystem.

Speakers expressed concern lest such biological monitoring schemes concern themselves exclusively with commercially important species and pointed out that the early recognition of the far-reaching effects of certain pesticides depended on observations on reproduction in birds. The conference accepted the need for broad, multiple-index studies, without diverting attention from the legislative and technical control of the sources of pollution.

The use of ecological baseline studies on an international or regional basis raised the problem of standardizing methods. Some discussion took place regarding the usefulness of marine pollution information centres, and it was considered that these should be developed nationally and regionally, and lead toward an international service. The role of FAO in the development of reference systems, particularly to assist developing countries, was also reviewed, and its expansion to cover marine pollution adequately was encouraged.

National organization for marine pollution research

The prospects of waste-recycling techniques were mentioned. Recycling, where feasible, could not only reduce or prevent a pollution problem, but could also facilitate a more rational conservation of natural resources. It was felt that more research should be encouraged on these problems, both to solve the technical difficulties and to overcome the socio-economic problems of implementation. The position of responsible industrial management, and its attitude to funding research, was discussed. In Europe, North America and Australia, joint or complete funding of mission-orientated research by industry is common and this appeared to be a positive response to public opinion. A conservation-educated public would exert a more rational influence on policy makers. Attention was drawn to the role that international organizations such as FAO, as well as non-governmental scientific organizations, could play in encouraging governments to include environmental management in general educational curricula.

Offshore dumping in the high seas

In view of the recovery by trawlers of drums containing toxic chemicals from the North Sea the necessity for providing a sound scientific basis for the control of deep-sea dumping of noxious materials was noted as requiring urgent attention. Mention was made of the suggestion that such containers be dumped into the deep trenches where they will eventually be buried by geo-physical in-folding. The legislative position regarding control of extra-territorial dumping by nationals appears ill-defined in many countries, and this presents a potential international problem.

Concluding points

The Convener, in summing up, said pollution problems fell into three groups:

(i) Those requiring *local action*, e.g. by placing restraints on industry or development. This is by far the largest group. Such problems lay within national jurisdiction, in estuaries and coastal waters. These areas were important, not merely as the main production areas for shellfish but often also as nursery grounds for young fish. Their potential importance for mariculture was also recognized. These were also the areas of greatest importance for recreational purposes.

(ii) Those requiring *regional cooperative action*, e.g. (a) in enclosed or semi-enclosed seas such as the Baltic or Mediterranean; or (b) at national boundaries, such as in the Gulf of Finland; or (c) where large river systems carry wastes derived from several countries, e.g. the Rhine and the St. Lawrence.

(iii) Those requiring *global international action* of which oil is perhaps the best known example. There are, however, other categories of pollutants included here: (a) persistent and wide-spreading materials such as chlorinated hydrocarbons; (b) materials transported in substantial quantities through the air, such as mercury and lead; and (c) materials of all kinds dumped in international waters.

Such a classification, while not perhaps fully comprehensive, would serve as a basis for waste management and the development of pollution control.

North Sea Pollution

*H. A. Cole**

La pollution en mer du Nord

En 1968, le Conseil international pour l'exploration de la mer a terminé une étude pratique de la pollution dans la mer du Nord et les eaux adjacentes (publiée en 1969 sous le titre "Cooperative Research Report, Series A, No. 13"). La Norvège, la Suède, le Danemark, la République fédérale d'Allemagne, les Pays-Bas, la Belgique, la France et le Royaume-Uni ont fourni des renseignements en vue de ces travaux. Le rapport contient un sommaire de la législation adoptée pour lutter contre la pollution dans les divers pays situés sur le pourtour de la mer du Nord et l'on y note que cette législation, en voie de développement, porte notamment sur les problèmes suivants : (i) la nécessité de lutter contre la pollution des mers par les grandes voies d'eau qui arrosent plusieurs pays, telles que le Rhin; (ii) l'intérêt qu'il y aurait à intégrer les divers éléments de la législation pour lutter contre la pollution sous toutes ses formes; et (iii) la nécessité d'apporter certaines restrictions au déversement de substances toxiques dans les eaux internationales situées au-delà des limites territoriales.

Des renseignements pratiques concernant les agents polluants ont été regroupés sous les rubriques suivantes: pollution par les eaux d'égouts, pollution industrielle, pesticides, hydrocarbures et détergents. Pour chaque catégorie, on a résumé la situation dans chacun des Etats Membres. Les méthodes et les résultats des études sur la toxicité ont fait l'objet d'un examen et l'on a envisagé la nécessité de compléter les tests destinés à mesurer la toxicité aiguë par des études à long terme sur les effets sublétaux des polluants.

On a étudié les mécanismes de transport et de diffusion qui jouent un rôle important dans la propagation des polluants; cette

La contaminación en el mar del Norte

En 1968 el Consejo Internacional para la Exploración del Mar llevó a cabo una encuesta circunstanciada sobre la contaminación que llegaba al Mar del Norte y aguas adyacentes (publicada en 1969 como Cooperative Research Report, Series A, No. 13). Facilitaron datos Noruega, Suecia, Dinamarca, la República Federal de Alemania, Países Bajos, Bélgica, Francia y el Reino Unido. Se llevó a cabo un resumen de la legislación en materia de lucha contra la contaminación en los diversos países que rodean el Mar del Norte y se observó que aquélla se encontraba en un estado de desarrollo y que se dedicaba atención particular a: (i) la necesidad de combatir la contaminación que llegaba al mar desde los ríos principales que atravesaban varios países, como el Rhin, por ejemplo; (ii) la conveniencia de integrar la legislación a fin de combatir la contaminación de todas las procedencias; y (iii) la necesidad de poner limitaciones a la descarga de materiales tóxicos en aguas internacionales fuera de los límites territoriales.

La información de datos efectivos referentes a los contaminantes se agrupó bajo los siguientes subtítulos: contaminación por aguas residuales, contaminación industrial, plaguicidas e hidrocarburos y detergentes. Para cada categoría de contaminación se llevó a cabo un resumen de la situación en cada uno de los países miembros. Se examinaron los métodos y resultados de los estudios de toxicidad y se consideró la conveniencia de apoyar los ensayos de toxicidad aguda con estudios más prolongados de los efectos subletales.

Igualmente se examinaron los mecanismos de transporte y difusión en relación con la dispersión de los contaminantes, siendo esta sección de gran valor en cuanto a indicar el tipo de investigaciones ulteriores necesarias.

*Ministry of Agriculture, Fisheries and Food, Fisheries Laboratory, Lowestoft, Suffolk, England.

section est particulièrement utile pour préciser la nature des recherches nouvelles qu'il convient d'entreprendre.

Le rapport se termine par une bibliographie complète compilée au moyen de documents provenant des Etats Membres. Il donne en annexe une série de cartes où sont indiquées les principales sources de pollution en Norvège, en Suède et au Royaume-Uni, des données relatives aux tests de toxicité, et le texte de l'accord international fixant les modalités de la coopération en matière de lutte contre la pollution des eaux de la mer du Nord par les hydrocarbures. Cet accord peut servir de base à une coopération régionale élargie pour lutter contre d'autres agents polluants.

Depuis la publication du rapport du CIEM en 1969, la recherche a connu un développement considérable en plusieurs des pays situés sur le pourtour de la mer du Nord; il en est résulté une prise de conscience plus aiguë de la nécessité de limiter le déversement d'effluents toxiques et rémanents, tels que les pesticides organochlorés, les biphényles polychlorés et les métaux lourds comme le mercure, le plomb et le cuivre. Le rôle que peut jouer la coopération internationale pour mettre sur pied des normes servant au contrôle des déversements d'agents polluants est devenu plus évident.

Les effets catastrophiques d'un déversement de thiodan (endosulfan) dans le Rhin, en 1969, ont mis en lumière les dangers qui peuvent résulter des grandes voies d'eau dont le cours traverse plusieurs pays.

Termina el informe con una amplia bibliografía recopilada de las fuentes nacionales de los distintos países miembros. En los apéndices figuran una serie de mapas en los que se señalan las fuentes principales de contaminación en Noruega, Suecia y el Reino Unido, los datos de las pruebas de toxicidad y el texto del acuerdo internacional que establece la cooperación para combatir la contaminación por hidrocarburos en el Mar del Norte. Este acuerdo puede formar la base para una cooperación regional más amplia para ocuparse de otros contaminantes.

Desde la publicación del informe del CIEM en 1969 ha avanzado considerablemente la investigación en varios países limítrofes del Mar del Norte lo que ha llevado a un reconocimiento más agudo de la necesidad de limitar las descargas de desechos tóxicos y persistentes, tales como los plaguicidas órganoclorados, los bifenilos policlorinados y los metales pesados tales como el mercurio, el plomo y el cobre. Se ha puesto de manifiesto en forma más evidente la importancia de la cooperación internacional en cuanto a formular normas para controlar las descargas de contaminantes.

Los peligros que pueden derivarse de los principales sistemas fluviales que pasan a través de varios países quedaron bien de manifiesto por los efectos catastróficos de una descarga accidental de tiodan (endosulfan) en el Rhin, en 1969.

THE International Council for the Exploration of the Sea decided in 1967 to assemble factual data regarding pollution reaching the North Sea, and to summarize the national legislation governing the discharge or dumping of wastes into this sea and the relevant research in progress in the countries abutting on to it. Full information was obtained from Norway, Sweden, the Federal Republic of Germany, Denmark, the Netherlands, Belgium and the United Kingdom, with some contribution from France. A report was issued (ICES, 1969) containing an extensive bibliography, and has been widely distributed, being followed in 1970 by a similar study of the Baltic (ICES, 1970).

This action by ICES was extremely timely and coincided with an upsurge of concern regarding the possible effects of polluting the sea on its biological resources, which although generated initially by accidents to oil tankers involving spectacular releases (see for example: Smith, 1968; Simpson, 1968; and Straughan, 1969) embraced also many other kinds of short- and long-term pollution effects. Attention was drawn especially to the highly persistent and toxic organochlorine pesticide residues, to heavy metals such as mercury and lead, and to radioactive materials. While public reaction has been most evident in the highly industrialized countries of the northern hemisphere there has been a rapid growth of public opinion on a world-wide scale which may be expressed in simple terms as a wider awareness of the fact that man has but one environment and should be more careful than he has been so far to ensure that this is not so damaged that the quality of human life is degraded. This concern is possibly more evident in regard to the sea than to the land or the air, resulting mainly, perhaps, from a widespread mystical approach to the oceans, a reflection of their vastness and power and of our relative ignorance of undersea conditions, rather than a balanced appreciation of the relative merits and disadvantages of sea and land disposal of human wastes. To the author there are no good ethical or economic reasons why man should deny himself the use of the oceans for waste disposal but we need to know, more precisely than we do at present, the effects of what we are doing on the biological productivity of the oceans and their living resources. The alternative to using the oceans is very frequently damage to the land environment or pollution

of the air. We must seek to preserve the quality of all three but it should be appreciated that in many places we are already running short of suitable land while vast areas of ocean are at present unused except as a means of communication.

There is no intention in this paper to summarize the data contained in the ICES Report, or to repeat its comprehensive list of references; instead a commentary is provided on the main issues highlighted by the Report, with some indication of priorities for further action needed to conserve living resources.

Control of North Sea pollution

So far there is international agreement only on the control of oil pollution from ships under the provisions of the International Convention for the Prevention of the Pollution of the Sea by Oil, London, 1954, and its subsequent revisions, and regionally under the Agreement for Cooperation in Dealing with Pollution of the North Sea by Oil, signed in 1969 by all countries with coastlines in this area. Currently, close consideration is being given to the choice of international or regional bodies in which to promote further consideration of the need for international or regional agreements on the control of persistent and widespreading pollutants such as heavy metals and pesticide residues: there seem to be strong arguments in favour of a regional approach based on principles which could later be extended to a wider area or even adopted internationally. Not unnaturally, industrialists in any one country do not wish to be placed under more onerous restrictions regarding waste disposal than their commercial competitors overseas. It is highly desirable therefore that pollution-prevention practices in countries around the North Sea should have the same general goals. Moreover, looking at the matter from the point of view of protection of fisheries and other living resources, it must be remembered that the North Sea stocks are for the most part a common resource, exploited by the ships of many nations, including a few which have no North Sea coastlines. Regulation of these fisheries is a function of the North-East Atlantic Fisheries Commission, but this body has not concerned itself with the effects of pollution.

The ICES Report showed that, in approximate terms, the control of pollution in countries surrounding the North Sea was based on similar principles and in no country

[4]

could it be said to be fully adequate to protect the living resources from at least local damage. All countries contain estuarine and coastal areas where pollution has severely damaged the resources and, indeed, the fish-less estuary is a feature of all heavily industrialized areas. Even more widespread are rivers which formerly carried migratory fish but which now do not do so because certain reaches, especially the upper estuaries, are impassable on account of pollution.

While all countries have some statutory control of pollution discharged from rivers or pipelines to the sea (and many improvements have taken place since the ICES Report was completed in 1968), the control of dumping of wastes from ships, outside the narrow band of territorial waters, is usually on a voluntary basis. Industry makes proposals for sea disposal, which are scientifically evaluated in government laboratories, followed by the specification of conditions, including containers, and the designation of a disposal site chosen according to the toxicity and persistence of the wastes involved. As a general rule wastes which are both toxic and persistent are taken out of the North Sea and deposited in very deep water in the Atlantic Ocean well beyond the end of the continental shelf. In the North Sea solid non-toxic wastes are deposited in areas of little or no value for fishing, and liquid wastes or sludges are dispersed from moving ships in areas of good water movement so as to ensure rapid dilution and dispersion.

Potential hazards to the environment from the carriage of cargoes of toxic materials by ships are being carefully studied by IMCO and are taken into account in framing regulations and codes of practice, in addition to possible risks to the ship, its crew, and passengers. Both the variety and the total amount of chemicals carried in bulk are increasing. In some cases, e.g. tetraethyl lead, specially designed ships are used, constructed in such a way as to minimize the chances of an escape of toxic material. Very few cases of collision followed by escape of toxic materials have occurred in recent years, but of course the volume of such traffic is very small compared with the traffic in oil and oil products. The scale of the damage which could occur in estuarine or shallow-water coastal areas is evident from the effects of the accidental release of the pesticide thiodan (endosulfan) in the Rhine in 1969.

Sewage

Very large quantities of sewage reach the North Sea, much of it untreated except for comminution. There is a growing tendency in several countries, e.g. the United Kingdom and the Netherlands, to consider the discharge of untreated sewage from coastal towns by long pipelines carrying the waste well away from holiday beaches as an alternative to providing additional treatment works. Moreover, certain schemes are being developed for providing trunk sewers, collecting sewage from conurbations sited around heavily industrialized estuaries, for discharge directly to the sea. One of the aims of such schemes is to reduce the pollution of such estuaries, which are often overloaded. In the United Kingdom there are schemes of this kind related to the discharges to the rivers Tyne and Thames. These tendencies seem likely to increase rather than diminish the polluting load received by the North Sea and must give rise to some

anxiety because of the large amounts of phosphorus and nitrogen discharged. Such pipelines can also be used, rather too conveniently perhaps, for toxic wastes which could not be passed through a sewage works without interfering with the processes of purification. Very often determination to avoid contamination of estuaries and tourist beaches may result in the discharge offshore, possibly directly on to fish nursery areas or spawning grounds, of very large volumes of wastes with a high BOD and containing appreciable quantities of persistent metallic wastes. Such schemes need to be considered very carefully at the planning stage, taking into account their future growth and ultimate capacity, in relation to the mixing characteristics of the receiving waters and the coastal resources of fish and shellfish and, especially, the distribution of grounds on which fish spawn or on which the young stages live during their first few months of bottom life. It is already established that such juvenile stages are generally more sensitive to pollution than adult fish. Unfortunately, even in such a well-studied area as the North Sea, the precise positions and extent of all spawning grounds and nursery areas have not been determined. However, their importance in maintaining production of sea fish cannot be disputed; in this connection recent Dutch evidence regarding the Wadden Zee (Zijlstra, in press) is of great interest.

Although reliable techniques are now available for the purification from bacterial contamination of filter-feeding molluscan shellfish which may become polluted by sewage (e.g. see Wood, 1961), these techniques do not necessarily remove risk of contamination with virus (e.g. of hepatitis). It has been suggested that suitability for the production of molluscan shellfish for direct sale would provide a good criterion for the quality of coastal waters, since the requirements for bathing are less stringent; indeed investigations in the United Kingdom (Moore, 1959) suggested that unless conditions were aesthetically unpleasant there was unlikely to be any direct health risk from bathing. The areas studied during this investigation included some North Sea beaches. From the production standpoint there are good reasons for giving special protection from pollution damage to those estuarine and coastal areas which are specially suitable for shellfish production.

A secondary problem of some importance associated with sewage purification is the disposal of sludge. Use of this on land presents difficulties and disposal at sea from coastal works is fairly common and the scale is increasing (see, for example, "Taken for Granted", Anon., 1970). Sludge from London treatment works has been dumped in the Outer Thames Estuary for 80 years and now exceeds 5 million tons per annum, but recent investigations reported by Shelton at the 1970 Meeting of ICES (Doc. C.M. 1970/E:8) suggest that neither oxygen content of the overlying sea water nor the character of the bottom communities has been materially affected. The area is one of rapid movement and mixing and there is no substantial local build-up of organic material, in contrast to conditions in the area used by New York for the same purpose (as described by Dr J. Pierce of the Sandy Hook Laboratory of the U.S. Bureau of Commercial Fisheries). Such sewage sludge may contain substantial quantities of adsorbed metals, e.g. zinc, lead and copper, in addition to organochlorine pesticides and polychlorinated biphenyls,

The recycling of these through food webs needs to be carefully studied.

The direct effects of the added nutrients entering the North Sea as sewage can only be clearly seen over a wide area off such large river systems as the Rhine and the Thames. Otherwise enrichment is only evident in close proximity to outfalls. The importance of such enrichment on fish production has not been determined. The relationship between sewage pollution and the development of adverse biological conditions in Oslofjord, including the occurrence of toxicity in mussels, is well established (Braarud, 1955 and subsequent studies). Such conditions occur, however, in only a few localities along the North Sea coasts, but there have been suggestions that there may be some connection between the build-up of nutrients due to sewage and the occurrence since 1968 of toxic blooms of dinoflagellates off the north-east coast of Britain (McCollum, Pearson, Ingham, Wood and Dewar, 1968; Ingham, Mason and Wood, 1968) and the west coast of Denmark (Hansen and Sarma, 1969). There is, however, no definite evidence that these events are connected.

Pesticides

Since the ICES survey was made in 1968, the usage by countries draining into the North Sea of persistent and highly toxic organochlorine pesticides such as dieldrin, aldrin, endrin and DDT has almost certainly declined. This follows action taken in many countries to protect wildlife and to reduce the contamination of streams. The manufacture of pesticides containing these compounds may not, however, have decreased to the same extent, or at all, since there is an extensive export trade in formulated products for use outside Europe where conditions may require a different assessment of the advantages and disadvantages.

Although levels of organochlorine pesticide residues in some North Sea fish are rather high (Portmann, 1968, 1970; Holden and Portmann, 1970) they are not increasing. Substantially higher figures have, however, been obtained from seals (Holden and Marsden, 1967; Holden, (see Section 3) and seabirds (Presst, 1970).

Generally speaking, where schemes exist for screening pesticides against selected animals and plants prior to their release for use in agriculture or horticulture, no marine forms are included; yet tests have shown (Portmann, 1968) that crustacea, for example shrimps and prawns, are highly susceptible. Such crustacea often live and breed in shallow estuarine or coastal waters where they are particularly liable to be exposed to pollution. In the North Sea no damage to shrimp fisheries by organochlorine pesticide residues has been established but it has been stated that brown shrimps, *Crangon crangon*, no longer breed very successfully close in to the coast in the Netherlands, while off the Thames and in the Wash the fishery for pink shrimps, *Pandalus montagui*, shows very great fluctuations in abundance with an overall tendency to decline. There is in most countries abutting on to the North Sea a drive towards the elimination of non-biodegradable pesticides which may accumulate in biological materials, but there is still very little work on their effects on marine fish and shellfish and their food. There is a particular need to look for effects on physio-

logy, behaviour and breeding success which may be produced by pesticide concentration levels an order of magnitude or more below those showing acute toxicity in normal laboratory tests (Anderson and Prins, 1970; Anderson, 1970, in press). There is a particularly urgent need to develop convenient methods of bio-assay for the quantification of these effects. It is now generally accepted that acute toxicity tests should be regarded as the first step only in the assessment of the effects of pollutants.

Industrial waste

There is probably a larger concentration of industry along the North Sea coasts and in the river basins discharging into it than around any other sea area of comparable size. This industry is of almost infinite variety and its wastes cover the whole spectrum of human activities. Moreover, most of the North Sea countries are heavily populated, with no extensive land areas which can be made available for waste disposal. To these countries sea disposal may be indispensable. This contrasts sharply with the situation in, say, North America or the U.S.S.R., where vast unproductive land areas exist which can, if necessary, be used for waste disposal. Fortunately, the mixing characteristics of the North Sea are very favourable for the dispersion and dilution of pollutants and there is a good interchange of water with the Atlantic through the English Channel and around the north of Scotland.

The influences of most of the wastes discharged are quite local, affecting principally estuarine and coastal fisheries within territorial waters but sometimes causing serious local problems. Bulky organic wastes, such as those from pulp and paper mills, may give rise to deoxygenation and the destruction of bottom animals by smothering. Marked changes in the character of the bottom may also arise from inert materials in suspension, such as coal dust, china clay (Portmann, 1970), fly ash from power stations, "red mud" from bauxite reduction and gypsum. These changes may not be entirely harmful and in certain areas may lead to an increased production of soft bottom animals such as small light-shelled molluscs, worms and crustacea very suitable as fish food (Howell and Shelton, 1970). On the other hand materials in suspension cut down light penetration and so may affect the growth of bottom algae and reduce phytoplankton production. Such conditions are also objectionable from an amenity standpoint, leading to turbid discoloured water and deposits on beaches. The discharge of such inert materials as a slurry through long pipelines or by barges is increasingly favoured in countries around the North Sea and is usually to be preferred, on both fishery and amenity grounds, to discharge into estuaries where local effects may be acute, and include siltation of navigable channels. Such effects, for example, may be occurring in the Humber estuary in the United Kingdom which receives heavy industrial discharges including gypsum.

Some plastics wastes are virtually indestructible and are already accumulating at an alarming rate on sea verges. Disposable polythene containers are a particular problem and their enforced replacement by materials which either disintegrate or are biodegradable is a matter of urgency if coastal amenities are to be saved. Some of the soluble wastes of plastics production, e.g. alkyl chlorinated compounds, may be accumulated by biological materials

[6]

and their toxicity and persistence is only now coming to light. Dumping of these materials has been alleged to cause death of plankton animals and 0-group fish living near the surface. The bewildering array of organic materials in commerce makes it increasingly likely that wastes are being produced and discharged whose toxicity to marine animals and plants has not been assessed. The only practicable way of preventing damage arising in this way seems to be to place on industry the responsibility for establishing that the wastes it wishes to discharge are of low toxicity.

Oil

Oil pollution is a constant threat in the North Sea, especially perhaps in the Southern Bight because of the concentration of shipping at the approaches to the major Continental ports and in the Thames Estuary and the Straits of Dover. So far no major spillage has occurred in this area but there is a steady succession of small "incidents" causing local damage to amenities and the death of seabirds.

Since the publication of the ICES Report there have been notable advances in methods of dealing with oil spills, including the development of relatively non-toxic dispersants of wide application (mainly through the efforts of the oil companies) and the testing of a promising new method of oil-sinking using a slurry of amine-treated sand. Special organizations have been set up in many countries to deal with oil spills both near the shore and in open water, and North Sea countries have agreed to cooperate in surveying, reporting and tackling any major spillages. However, from an examination of the causes of major disasters, from the *Torrey Canyon* to the recent accident to the *Pacific Glory* off the Isle of Wight, one can only conclude that the risks of collision and stranding will not be materially reduced by additional navigation aids and traffic lanes unless some means can also be found of reducing the effects of human frailties, especially in relation to navigation and ship handling. Meanwhile, IMCO is pressing ahead with its avowed aim of eliminating pollution by oil discharges from ships during normal operation, but such improvements as are achieved by reduction in discharges per vessel tend to be counterbalanced by the steady growth of tanker tonnage.

From the point of view of protection of fisheries and other living resources, current policies for dealing with oil spills are in the main soundly conceived. Although there is still some division of opinion on when to use dispersants, it is generally accepted that, in open water away from fish nursery areas, spawning grounds or shellfish beds, treatment with the newer non-toxic dispersants will not cause any material damage to fisheries. The new methods of oil-sinking have yet to be tried on a large scale, but there is some evidence that sunken oil may move about on the bottom and possibly affect bottom animals. The conditions in which sunken oil may foul fishing gear and taint the catch are being re-investigated; the conclusions will be of considerable importance in relation to the future use of this method. Sinking should certainly not be done in the vicinity of shellfish beds or herring spawning grounds. Oil stranded on the shore is unlikely to do lasting damage to fisheries but may cause local tainting of molluscs and crustacea. Such stranded

oil, except on sandy beaches, may however create serious clearance problems. In emergencies, when popular tourist resorts are threatened, the need to preserve amenities from massive oil contamination is likely to override all other considerations. For this reason, there is value in having immediately available to both central government and local authorities information and maps showing the areas of greatest sensitivity from a fisheries standpoint, i.e. shellfish beds, young-fish nursery grounds and productive estuaries, where some priority should be given to their protection.

No measures have been suggested to reduce materially the toll of seabirds, especially auks, and almost total destruction by oil pollution of breeding colonies near main shipping lines seems a possibility.

There is considerable speculation regarding the effects of oil ingested by fish and shellfish and their food organisms, for example, plankton and the possible accumulation of carcinogens has received some attention; there is, however, little evidence of value and much more systematic long-term studies are required.

Radioactive waste

In countries surrounding the North Sea the disposal of radioactive waste is subject to strict controls based generally on the principle that there shall be minimal risk to man and living resources. In the United Kingdom, which has relatively large research and nuclear energy programmes, discharges to the North Sea are of a minor character and represent but a small fraction of discharges to the Irish Sea. Although the latter has a relatively enclosed circulation it has been firmly established by comprehensive monitoring that exposure to man is well within the limits recommended by the International Commission on Radiological Protection and there is no harm to fisheries or other living resources (Mitchell, 1969).

There is no significant dumping of solid radioactive waste within the North Sea and any such waste produced by European countries which is disposed of at sea is taken to a site in the North Atlantic well beyond the edge of the continental shelf and approved by the European Nuclear Energy Agency, The International Atomic Energy Agency has made a study of the disposal of radioactive waste at sea and the recommendations of the Brynielsson Report have now been generally accepted. It is understood that this study is being updated following analysis of the results of disposal and consideration of the associated research.

General condition of fisheries and living resources

Although it is easy to point to estuaries in the North Sea where fisheries for shellfish and migratory species of fish have been lost, and to establish that certain toxic pollutants are present in easily measurable concentrations in coastal waters adjoining industrial areas, it is not possible to demonstrate that the North Sea as a whole is less productive than it used to be. The question is, of course, a complicated one because the production achieved from fisheries depends upon the patterns of recruitment and exploitation. As is well known, overfishing, if carried far enough, will produce not only a reduction in catch per unit of effort, but also in total catch. Moreover, it is becoming increasingly evident that if stocks are reduced to a very low level by heavy fishing the production and survival of young to replace the stock may fail unless the environmental conditions are exceptionally favourable during the early stages of life, probably during the first few months in the case of many fish and shellfish (Cushing, in press; Hancock, in press). In these circumstances, to separate the effects of pollution from the influence of climatic trends and the effects of short-term fluctuation in natural environmental factors is an almost impossible task. This is true even when there is a long series of data such as we possess for various North Sea fisheries. The records of the Edinburgh Plankton Recorder Surveys, already maintained over 25 years, do show certain downward trends in the abundance of particular planktonic organisms (SMBA, 1968), including some important components of fish food, and the suggestion has been made (Colebrook, in press) that these changes are similar to those which might be produced by pollution. This, however, is very debatable: in the first place the changes seem more likely to be the result of climatic trends similar to those described by Lee in a special lecture "Long-term environmental changes in the North Atlantic" delivered at the 1970 Statutory Meeting of the International Council for the Exploration of the Sea and, secondly, pollution often results not in a gradual reduction in abundance of species but in a marked reduction in the variety of species without an accompanying reduction in biomass; indeed, among bottom organisms an increase in biomass may sometimes be observed.

In fact, the production of the North Sea fisheries has reached an all-time maximum in recent years (Table 1), due mainly to the occurrence of some exceptionally rich year-classes (e.g. of haddock and cod) following the very severe winter of 1963. However, figures of total fish catches may be misleading, since landings of some species have declined (e.g. of turbot) and those of others fluctuate markedly over a short time cycle (e.g. herring, Norway pout, sand eels and sprats) due to variable year-classes and changing effort. The greatly increased exploitation of mackerel in recent years—the fishing off of an accumulated stock—has also had a marked effect on total landings.

It may, indeed, be impossible to recognize pollution effects by their direct influence upon production, unless this is very severe. Moreover, caution needs to be exercised in thinking that so-called "baselines" can easily be established for littoral organisms which may be thought to be useful as indicators of pollution: a different approach may be preferable, by which the effects of potentially-toxic pollutants on processes of growth, physiology, behaviour and breeding are established in controlled laboratory experiments and are then looked for in the field. Where gradients in the concentration of pollutants in the sea can be found, as, for example, off the mouths of rivers carrying sizable quantities of easily recognizable materials such as metallic wastes, it should be possible to establish in sedentary animals and plants a comparable gradation in pollution effects, for example in growth rates, behaviour, maturation of gonads, etc. From such a combination of precise laboratory experiments and field studies of polluted situations it should be possible to build up an appreciation of the ecological consequences of pollution and so reach sound judgements regarding the capacity of the marine environment to accept waste without damage to living resources or productivity. The North Sea is a particularly suitable venue for work of this kind because it necessarily receives a great variety of wastes which, in practical terms, cannot be diverted elsewhere (although particularly objectionable elements can certainly be reduced or eliminated), while at the same time supporting some of the most important and productive fisheries of the world.

Programme for the future

Further population increases and the development of even larger industrial complexes along the fringes of the North Sea, and on the drainage basins discharging there, are inevitable. At present facilities for the treatment of

TABLE 1. NORTH SEA FISH CATCHES, IN METRIC TONS (SOURCE: BULLETIN STATISTIQUE DES PECHES MARITIMES)

Year	Total catch of all species by all countries	Cod	Haddock	Plaice	Herring	Whiting	Mackerel	Norway pout	Sand eel
1950	1,529,745	66,823	56,429	67,380	1,107,467	45,441	31,834		
1951	1,751,734	61,381	56,478	66,533	1,220,216	73,446	48,421		
1952	1,673,287	76,371	52,372	70,778	1,145,095	73,340	39,661		
1953	1,773,754	81,090	60,380	78,883	1,211,390	62,531	41,273		
1954	1,818,964	80,571	70,135	66,965	1,297,296	64,815	45,325		
1955	2,018,180	83,448	87,656	63,315	1,411,220	72,436	53,224	NOT	
1956	2,013,569	80,267	93,917	63,881	1,364,126	74,943	43,666	RECORDED	
1957	1,819,383	94,981	105,304	69,272	1,047,834	84,314	68,934	SEPARATELY	
1958	1,585,998	103,733	96,191	72,429	804,643	77,484	68,774		
1959	1,798,766	109,467	79,670	78,324	903,654	80,491	70,055		
1960	1,553,106	104,399	66,424	86,289	786,950	53,123	72,927		
1961	1,465,698	105,811	67,238	85,783	689,778	83,289	85,826	33,834	83,716
1962	1,570,297	89,558	52,419	87,419	678,515	68,967	66,285	156,959	110,041
1963	1,919,713	105,921	59,398	107,062	805,301	98,653	55,402	166,827	162,134
1964	2,146,298	121,550	198,706	110,368	932,046	91,528	79,390	82,669	128,501
1965	2,597,043	179,469	221,700	96,927	1,230,315	106,694	151,721	59,342	130,802
1966	2,920,300	219,702	268,958	100,130	1,038,851	155,153	505,134	52,737	161,110
1967	3,064,596	249,803	167,408	100,646	819,324	91,245	909,879	180,173	188,795
1968	3,361,058	285,314	139,469	108,838	850,127	144,920	808,578	468,713	194,225
1969	3,236,689	199,035	639,175	121,652	724,853	199,029	713,866	134,549	112,953
1970*	3,173,007	224,742	671,831	130,334	748,750	181,506	297,488	273,628	187,801

*provisional figures

[8]

sewage and industrial wastes are barely keeping pace with development, and pollution is increasing despite greater public awareness of the hazards to the environment. In all the European countries concerned serious consideration is being given to the need for tighter legislative control of the discharge of wastes to rivers, and standards for effluents are rising as knowledge of their effects on living resources and wildlife are more clearly established. The major obstacle in the way of a spectacular improvement in the condition of rivers and estuaries is the great cost of treatment works or of trunk sewerage schemes diverting effluent from the estuaries to the open coasts. At the coast discharge by long pipelines is being increasingly favoured as an alternative to expensive treatment which demands suitably placed sites in areas of great value for development or amenity purposes. This provides the opportunity to include more industrial effluents in these pipeline discharges, including toxic materials which would be unacceptable in a treatment works. While the discharge of wastes by suitably designed pipeline systems in deep water well away from the shore, in carefully chosen situations favourable for dispersion, is likely to minimize their effects on fisheries this is only true if persistent and toxic wastes are present in great dilution or are excluded altogether. It may be necessary to apply new standards to such pipe-line discharges and to think particularly in terms of persistence, toxicity and accumulation in biological materials. Releases of such potentially dangerous materials as mercury, cadmium and organochlorine pesticides should be minimal, and increasingly the onus should be transferred to industry to establish that the wastes it wishes to discharge will present no threat to fisheries or other marine resources. Toxicities need to be established to a wide range of marine animals and plants, including zooplankton and unicellular algae, in a form which takes account of longer-term exposure to low levels of pollutants as well as acute effects. Possible synergistic action between pollutants needs to be taken into consideration. Progressive elimination of the more persistent and toxic materials from discharges, especially those accumulating in biota or adsorbed on to bottom sediments, might well be an agreed aim of all countries and this needs to be allied with a policy designed to limit their release to the air and subsequent deposition with rain. It is clearly established that certain pesticides and heavy metals have been widely distributed beyond national boundaries through aerial transport.

Dumping at sea in such a relatively shallow area as the North Sea needs to be controlled according to the same principles as those governing coastal and river discharges. There is no real division between these activities on the basis of their effects except that dumped discharges may be more easily and rapidly dispersed and there are fewer constraints in the selection of dumping areas of relative unimportance for fisheries.

Because industry exists in a competitive situation and commercial advantage can result from cheap waste disposal practices, even though harmful to the environment, there would be great value in regional agreements among countries surrounding the North Sea and exploiting its fisheries in common. Such agreements could perhaps be framed in the form of progressive exclusion of a series of particularly toxic wastes from sea and river discharges within a stated period, backed by an agreed system of monitoring based on standard methods of analysis, with regular reporting through an appropriate international body.

References

ANDERSON, J M Assessment of the effects of pollutants on
1970 physiology and behaviour. *Phil. Trans. R. Soc.*

ANDERSON, J M and PRINS, H B Effects of sublethal DDT on a
1970 simple reflex in brook trout. *J. Fish. Res. Bd Can.*, 27:331–4.

BRAARUD, T The effect of pollution by sewage upon the waters
1955 of the Oslofjord. *Proc. int. Ass. theor. appl. Limnol.*,
12:811–3.

COLEBROOK, J M Changes in the distribution and abundance of
1971 plankton in the North Sea 1948–69. *In* Proceedings of the
Symposium on Conservation and Productivity of Natural
Waters, 22–23 October 1970. London, Zoological Society
(in press).

CUSHING, D H The influence of planktonic production cycles on
1971 the recruitment of young fish. *In* Proceedings of the
Symposium on Conservation and Productivity of Natural
Waters, 22–23 October 1970, London, Zoological Society
(in press).

HANCOCK, D A The role of predators and parasites in a fishery
1971 for the mollusc *Cardium edule* L. *In* Proceedings of the
Advanced Study Institute on Dynamics of Numbers in
Populations, Arnhem, Holland, 6–19 September 1970 (in
press).

HANSEN, V K and SARMA, A H W On a *Gymnodinium* red water
1969 in the eastern North Sea during autumn 1968 and accompanying fish mortality, with notes on the oceanographic
conditions. ICES, C. M. 1969, Doc. L:21 (mimeo).

HOLDEN, A V and MARSDEN, K Organochlorine pesticides in
1967 seals and porpoises. *Nature, Lond.*, 216(5122):1274–6.

HOLDEN, A V and PORTMANN, J E Organo-chlorine residue monitoring
1970 of fish and shellfish. *Mar. Pollut. Bull.*, 1(3):41–2.

HOWELL, B R and SHELTON, R G J The effect of china clay on the
1970 bottom fauna of St. Austell and Mevagissey Bays. *J. mar.
biol. Ass. U.K.*, 50:593–607.

IAEA Radioactive waste disposal into the sea. *Saf. Ser. int.
1961 atom. Energy Ag.*, (5): 174 p.

ICES Report of the ICES Working Group on Pollution of the
1969 North Sea. *Coop. Res. Rep. int. Coun. Explor. Sea (A)*,
(13): 61 p.

ICES Report of the ICES Working Group on Pollution of the
1970 Baltic Sea. *Coop. Res. Rep. int. Coun. Explor. Sea (A)*,
(15): 86 p.

INGHAM, H R, MASON, J and WOOD, P C The distribution of toxin
1968 in molluscan shellfish following the occurrence of mussel
toxicity in north-east England. *Nature, Lond.*, 220:25–7.

McCOLLUM, J P K, *et al.* An epidemic of mussel poisoning in
1968 north-east England. *Lancet*, (7571):766–70.

MITCHELL, N T Radioactivity in surface and coastal waters of the
1969 British Isles 1968. *Tech. Rep. Fish. radiobiol. Lab., Lowestoft*
(FRL 5): 39 p.

MOORE, B Sewage contamination of coastal bathing waters in
1959 England and Wales. A bacteriological and epidemiological
study. *J. Hyg., Camb.*, 57(4):435–72.

PORTMANN, J E Progress report on a programme of insecticide
1968 analysis and toxicity-testing in relation to the marine
environment. *Helgoländer wiss. Meeresunters.*, 17(1–4):
247–56.

PORTMANN, J E Monitoring of organo-chlorine residues in fish
1970a from around England and Wales, with special reference to
polychlorinated biphenyls (PCBs). ICES, C. M. 1970, Doc.
E:9 (mimeo).

PORTMANN, J E The effects of china clay on the sediments of St.
1970b Austell and Mevagissey Bays. *J. mar. biol. Ass. U.K.*,
50:577–92.

PRESST, I Techniques for assessment of pollution effects on
1970 seabirds. *Phil. Trans. R. Soc.*

SHELTON, R G J The effects of the dumping of sewage sludge on
1970 the fauna of the outer Thames Estuary. ICES, C. M. 1970,
Doc. E:8, 9 p. (mimeo).

SIMPSON, A C The "Torrey Canyon" disaster and fisheries. *Lab.
1968 Leafl. Minist. Agric. Fish. Fd*, (18): 43 p.

SMBA Annual report of the Scottish Marine Biological Association,
1968 1967–68. *Rep. Scott. mar. biol. Ass.*, (1967–68): 69 p.

SMITH, J E (Ed.) "Torrey Canyon" pollution and marine life:
1968 a report by the Plymouth Laboratory of the Marine
Biological Association of the U.K. Cambridge University
Press, 196 p.

STRAUGHAN, D A Santa Barbara study. *In* Proceedings of the
1969   Joint API and SWPCA Conference on Prevention and
        Control of Oil Spills, New York, 17 December 1969, pp.
        309–12.
WOOD, P C Principles of water sterilisation by ultra-violet light,
1961   and their application in the purification of oysters. *Fishery
        Invest., Lond.* (2), 23(6): 48 p.

ZIJLSTRA, J J On the importance of the WaddenSea as a nursery
1971   area in relation to the conservation of southern North Sea
        fishery resources. *In* Proceedings of the Symposium on
        Conservation and Productivity of Natural Waters, 22–23
        October 1970, London, Zoological Society (in press).
ANON.   Taken for granted. Report of the Working Party on
1970   Sewage Disposal. London, HMSO.

A Pollution Survey of the Trondheim Fjord as Influenced by Sewage and the Pulp Mill Industry

*G. Berge, R. Ljøen and K. H. Palmork**

Enquête sur la pollution du Tronheimfjord: système influencé par les eaux usées et les effluents des fabriques de pulpe de bois

Les auteurs se sont efforcés de combiner plusieurs techniques océanographiques standard pour déterminer le type de pollution que l'on trouve dans les systèmes des fjords et en mesurer le degré. Le fjord choisi était soumis à divers modes de pollution: déchets domestiques d'origine urbaine, eau de drainage de l'agriculture et de l'industrie minière, et effluents industriels rejetés par les fabriques de pulpe de bois. La démonstration porte sur trois types d'effets et sur leurs sources; eutrophisation, pollution particulaire et pollution par le mercure. Les auteurs examinent l'influence de l'hydrographie du fjord sur la répartition et l'accumulation des polluants.

Estudios sobre la contaminación del fiordo de Trondheim poraguas residuales y desechos de la industria de fabricación de pulpa de papel

Se ha tratado de combinar diferentes técnicas oceanográficas para determinar la clase y el grado de contaminación de los fiordos. El fiordo elegido está sometido a diversos tipos de contaminación: aguas residuales procedentes de las ciudades, desagües de la agricultura y de la industria minera, y aguas residuales industriales de las fábricas de pulpa. Se describen tres clases de efectos y sus orígenes: eutroficación, contaminación por partículas y mercurio. Se examina el efecto de la hidrografía de los fiordos en la distribución y acumulación de contaminantes.

TRONDHEIM FJORD in midwest Norway, between N 63° 30′ and 64° 00′ and E 9° 30′ and 11° 30′ (fig 1, I) is one of the larger fjord systems of Norway with a total length of about 100 km. It is surrounded by agricultural country, mining industry, pulp mill and paper factories and the district is rather densely populated. Important fisheries, especially on sprat, are developing. Only occasional observations on pollution have been made but for some time the effect of pollutants from cities and industry on the fjord system have been discussed. Release of fibres from the pulp mill factories, with the associated mercury fungicides, have been of special concern to fisheries, as both fibres and mercury could affect the quality and quantity of available fish.

An *ad hoc* study was undertaken with the R/V *Johan Hjort* during a three-day survey in May 1970.

Hydrography

The shallowest sill in the fjord area (fig 1, area F), limits the layer of free communication with coast water throughout the whole fjord to the upper 60–100 m. A typical estuarine circulation appears in this layer. The thickness of the brackish water flowing out of the fjord varies between 2–10 m. The residual current reaches a velocity of more than 50 cm/sec in periods of swelling, but is generally 10 cm/sec (Wendelbo 1970). Less is known about the ingoing compensating current just below this layer as regards velocity and thickness.

No stagnant water has been found in any of the deep basins of the fjord. A complete renewal of water in the outer basin normally occurs three to four times a year and in the inner basin one to two times a year (Wendelbo 1970). At the time of this observation there was a significant inflow of relatively cold coast water with rather low salinity ($t < 6.5°C$ and $S°/_{00} = 34.6$–34.7) to the outer basin (fig 2, I, II). This inflow replaced part of the old water of the outer fjord and the upper layer of the inner part was lifted and overflowed the shallow threshold into the next basin and to a great extent exchanged the waters there also. The result of the exchanges is demonstrated by the intermediate temperature maximum and the minimum in oxygen saturation, and also by the salinity and density in the inner basin water being equal to that near the sill depth outside the threshold (fig 2, I–IV). The compensating outgoing current from the inner basin was in the layer of temperature maximum.

This circulation continued at least until mid-July (fig 2, VII, VIII), as indicated by the two moored, continuously recording current and temperature meters at the entrance of the fjord. The flow near the bottom (326 m) was dominantly eastward, whereas that at 126 m was mainly westward. At this locality the true east and west current components represented the water transport in and out of the fjord respectively. The temperature of the inflowing water increased, indicating a contribution of Atlantic water. The maximum outflow was associated with temperatures clearly above 6.5°C, which showed that during the period 10–30 June a great part of the oldest water of the fjord was replaced.

Vertical diffusion is significant in the entire fjord, and in the outer basin (Wendelbo 1970).

Particle distribution

Figure 1, II is drawn on the basis of continuous recordings at the 5 m level of relative values of particle concentration (Berge 1963). Vertical distributions at selected sections are inserted. High concentrations of particles were usually located in the southern side of the fjord and towards the westward and northward opening bights. This feature is probably related to the common water circulation in the upper layer of the fjord and to heavier runoff and contributions from the southern side (fig 1, II, C, D and E). The heaviest concentrations were located to area E and I. The highest concentrations were always confined to the uppermost surface layer, decreas-

* Institute of Marine Research, Bergen, Norway.

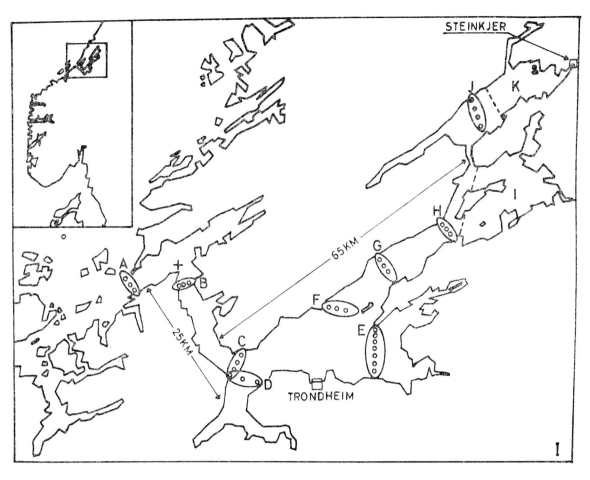

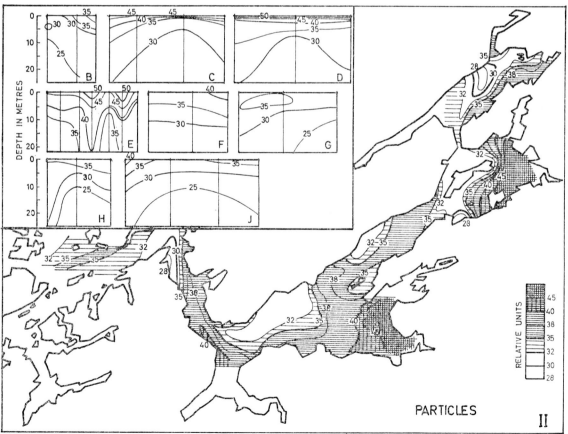

Fig 1. I. Trondheim fjord and the investigated areas with stations. II. Particle distributions at 5 m level. Inserted are vertical distributions at selected areas. Ultimo May 1970

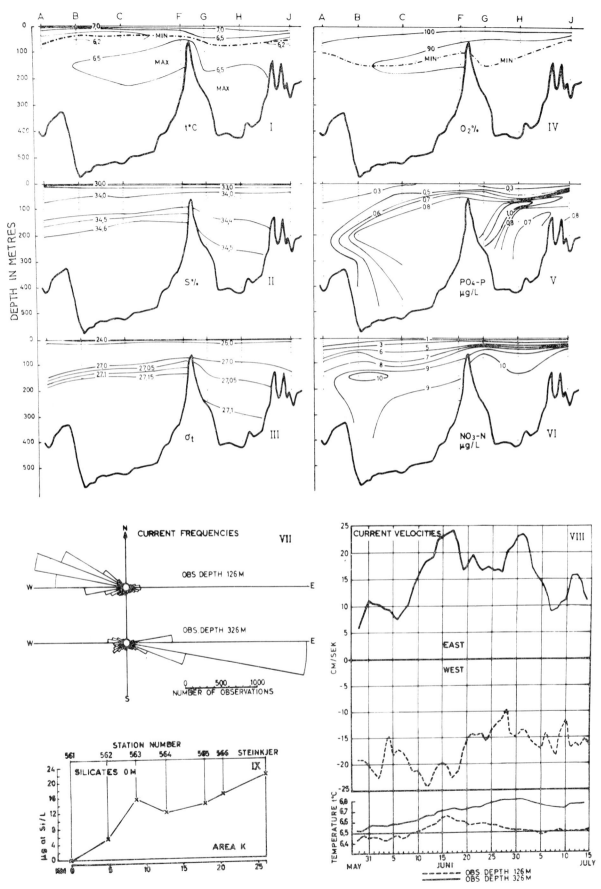

Fig 2. I–VI and IX. Hydrographic and chemical conditions of Trondheim fjord. Ultimo May 1970. VII and VIII. Current frequencies, velocities and temperatures for the period 30 May to 15 July 1970. Moored buoy system

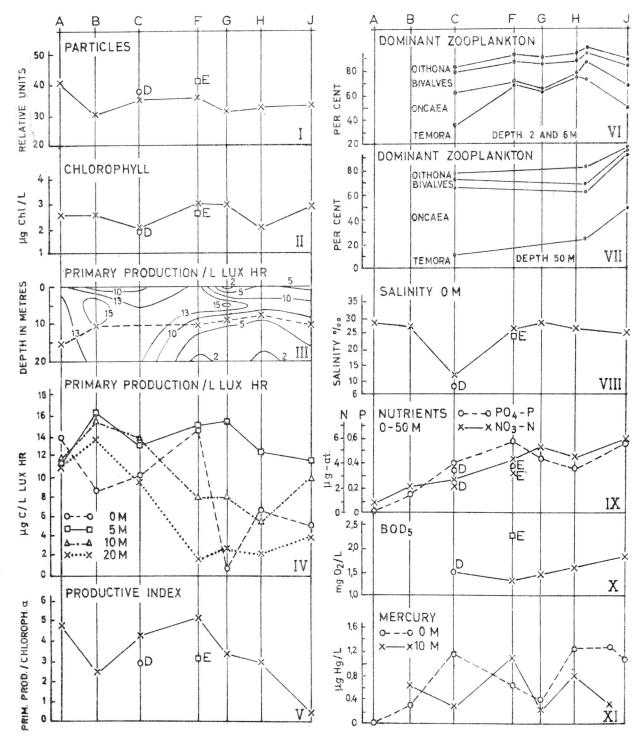

Fig 3. I–XI. *Longitudinal sections through the observed areas A–J showing averaged values of the parameters observed. The euphotic zone is indicated by a dashed line on III*

ing more or less rapidly with depth, but still significant even at the maximum measured depth of 50 m. The concentrations of particles in the fjord were negatively correlated to the chlorophyll content (r = − 0.4) indicating that they did not significantly originate in eutrophic plankton. Net samples at 2 m, 6 m and 50 m revealed that wood fibres were present in all the sampled localities and contributed to the particle concentrations as far out as area C.

Primary production

The rates of primary production per unit of light in the central areas were generally high for the time of the year (fig 3, III, IV). The range was similar to that of open coast waters during spring bloom. However, the spring bloom of phytoplankton usually occurs much earlier in the fjord, and a typical summer "situation" was probably established. Using the average values for each locality, a general trend of decreasing rates of primary production

[13]

towards the inner fjord was evident (fig 3, IV). A depression from this trend was observed in area D (0 m) attributable to the water of very low salinity from the Orkla fjord branch (fig 3, VIII). This depression, showing production rates of 4.4 μg C/L·h was also visible in area B, where a rate of 0.8 μg C/L·h at the northernmost station in this section was found. Exceptionally high rates were measured at one station in area E (average of three depths = 190 μg C/L·h).

The primary production at light saturation/chlorophyll-a (Strickland 1960) was calculated from the average 0 m values in each area (fig 3, V). The indices show a rapid decrease towards the inner fjord as well as towards the branch Orkla fjord (area D). Low values were also measured in areas B and E.

Nutrients

In the euphotic zone nutrients were exhausted at the surface of the outermost area A (fig 2, V, VI). Inward, significant amounts were present, increasing towards the inner fjord. A trend of strongly increasing nutrients in the zone above 50 m (temperature minimum layer) towards the inner fjord was also evident (fig 3, IX), with the average concentrations of phosphate and nitrate increasing from nearly 0–0.5 and 6.0 μg-at respectively.

Similarly, an increase is found in nutrients in deep water (fig 2, V, VI). Inflowing water is much lower in phosphate (0.45 μg-at P/L) than the older water that is lifted into the inner basin (average >0.7 μg-at P/L).

Figure 3, II is based on average surface values for each locality. The higher values correspond with those observed in coast waters during the spring bloom. A slight increase towards the inner fjord is evident.

Bottom sediments

Sampling was conducted at areas F, H, J and K. The combustible organic matter was high in all samples, averaging 5.4, 5.5, 10.7 and 6.8 per cent respectively. In area J, which had the maximum organic matter, wood particles and fibres were visible in the water. These fibres were traced by echosounders from the outlet of a pulp mill factory as a continuous beam until they settled 500 m from the river mouth at a depth of 200 m. A sediment layer 8 m thick had formed here. Bottom sediments in and around this area showed fractionated sedimentation of wood and bark particles.

Zooplankton and wood fibres*

Horizontal hauls at 2.6 and 50 m (90 μ nets) were used to estimate zooplankton and floating wood fibres. Figure 3, VI and VII shows the percentage distribution of the four dominant zooplankton species recorded. No typical indication of pollution effects can be found. The species' composition and quantities were normal for that time of the year. The average combustible organic matter in the samples from areas C, D, E and J were 57, 70, 250, 169 and 183 mg/m³ respectively. The contribution from wood fibres varied considerably with general values of 1–10 per cent of the total volume. An extreme value of 80 per cent was recorded in area J. The wood fibres fraction was 10 per cent as far out as area C.

The levels recorded (fig 3, X) were all considerably

*The zooplankton samples were analysed by Herman Bjoerke, Institute of Marine Research, Bergen.

above the values of natural sea water, which seldom exceed 0.7 ml/l. The levels increased from the central part of the fjord inward, and extremely high figures were recorded in area E (2.2 ml O_2/l) and area J (1.8 ml O_2/l).

Silicates—In area D silicate analyses made on samples from all standard depths from surface to bottom revealed very high values in the surface samples (22 μg-at/l) as compared with the lower depths (10 m = 0.5 μg-at). Analyses were also made in areas H, J and K where the surface values revealed a strong increase towards the innermost bottom of the fjord (fig 2, IX). Increasing values are attributable to the outflow from the river systems in areas D and K.

Mercury—Normal values (<0.03 μg Hg/l) were measured at the surface in the outer area A. In the main parts of the fjord mercury contamination of the surface water was evident at all the observed localities, and was especially high in areas C, H, I and J (fig 3, XI). The average in the upper 10 m showed an increase towards areas H, I and J. Two of the bigger pulp mills and paper factories are situated in these areas. Filtered water samples from the main sewage outlet of the pulp mill in area I contained 42 μg Hg/l. Samples from the main sewage outlet of the pulp mill in area J, measured 82 μg Hg/l. The mercury content of the river water was less than 0.03 μg Hg/l above the factory and 5.0 μg Hg/l at the river mouth. Mercury analysis of sediments sampled 500 m from the outlet revealed that up to 15.6·10³ μg Hg/l of sludge was released through ultrasonic treatment of the sediment in acidified water.

Discussion

Particle concentrations were relatively high throughout the fjord, and in several areas were higher than those experienced during heavy spring blooms in the coast waters. The particles affected the light penetration. The euphotic zone was usually less than 10 m (fig 3, III), thus limiting the primary production and consequently the oxygen supply to a rather thin surface layer. Due to good exchange in the upper waters the latter effect seemed to be of minor importance since the resulting oxygen depression was less than 20 per cent. The high particle concentrations in areas D and K coincided with the very high silicate contents of the upper waters indicating that material from the rivers emptying into these fjord branches contributed significantly. Rivers might also be responsible for the high percentage of the particles in areas E and I. However, in area E a very high BOD_5 figure was recorded together with high values both of chlorophyll, primary productivity and the particulate organic matter (biomass) of the net samples. Possibly important parts of the particles originated in organic waste and sewage from Trondheim city.

In late April the productivity indices of open coast waters were found to average 4.6 (Berge, personal communication). This fits in well with the values observed in areas A, C and F. Except for these localities, the indices were low, indicating less favourable ecological conditions which increased towards the inner part of the fjord, as well as towards areas D and E. This condition was probably not due to nutrient deficiency since the nutrients showed increasing amounts in the mentioned directions. Possibly some other inhibitors were affecting area J especially.

[14]

The increase of nutrients towards the inner fjord might be due to long-term additions from the drainage area and increasing dilutions outward by the exchanging deep water. Both processes were considered significant in explaining nutrient increase. The integrated phosphate content was calculated to 4,370 t. At the last period of exchange no observations were available on the basic phosphates content in the fjord water and calculations could not be made. When the average values of nutrients in the inner basin were compared to those of similar fjord systems (Saelen 1962), the increases observed did not seem to have a major effect. The maximum value measured was about 1.2 μg-at P/L which is less than half the value of 2.8 μg-at P/L recorded in the central part of the heavily polluted Oslo fjord (Føyn, 1970).

All the BOD_5 measurements were made on samples from 5 m. The particle distributions indicated that the surface values would be significantly higher. In any case, the recorded levels were generally so high that the waters were considered overloaded. For comparison, the levels in the Oslo fjord in October 1967 were about 1.5 ml O_2/l (Andersen 1968). In spite of the good water exchange in the fjord the increased levels of BOD_5 should be a matter of concern in relation to any planned increase in the use of the fjord as a waste recipient.

The observations of mercury in the waters have indicated that the pulp mills in areas J and I are sources of mercury pollution. Additions are still evident in absorbed mercury compounds in the sewage systems, and in the huge quantities of fibre materials settled in the bottom sediments. Although the use of mercury as a fungicide ceased more than half a year ago, the contamination from the previously settled material must be considered a problem pollutant for some time. The rather high levels of mercury, combined with the ability of certain organisms to concentrate mercury about 3,000 times, suggests that the stationary organisms can also be significantly contaminated, a condition which should be monitored.

Conclusions

The methods applied show that standard oceanographic techniques can provide information on the state of pollution of the fjord, including the organic load on the water, the degree of eutrophication, the particulate contamination, and the eventual toxic agents. Combined use of BOD_5 primary production, productive indices and transparency measurements appear to be adequate techniques. Moored hydrographic buoys at the entrance of the fjord and preferably at each threshold between the basins, combined with nutrient measurements throughout the fjord, can be used to estimate accumulations and to calculate the capacity of the basin as a waste recipient.

It was concluded from the measurements made that the situation in the Trondheim fjord at the time of investigation was one of slight eutrophication in the central part of high organic load on the surface waters all over, and marked particulate contaminations in special areas. Toxic agents originated primarily from the inner fjord, but probably also from the Orkla branch of the fjord. Special analyses revealed serious contamination throughout the fjord from accumulated mercury in the sediments, rivers, and sewage systems of pulp mills.

References

ANDERSEN, T A whole year (1966–1967) chemical-hydrographical
1968   investigation of a polluted basin in the inner Oslo fjord. Thesis, University of Oslo (in Norwegian).

BERGE, G A recording transparency meter for oceanic plankton
1963   estimation. *Fisk Dir.Skr. (Havunders.)*, 13(6):95–105.

FØYN, L Nitrogen and phosphorus exchanges in the inner Oslo
1970   fjord. Thesis, University of Oslo (in Norwegian).

SAELEN, O H The natural history of the Hardanger Fjord. 3. The
1962   hydrographical observations 1955–1956. Tables of observations and longitudinal sections. *Sarsia*, (6):1–25.

STRICKLAND, J D H Measuring the production of marine phyto-
1960   plankton. *Bull. Fish. Res. Bd Can.*, (122):172 p.

WENDELBO, P S The hydrographic conditions of the Trondheim
1970   fjord 1963–1966. Thesis, University of Oslo (in Norwegian).

Pollution in the Baltic

B. I. Dybern*

La pollution dans la Baltique

Les conditions caractéristiques des fjords qui règnent dans la Baltique—eaux de surface de faible salinité et couches profondes de plus forte salinité—rendent les eaux du fond plus ou moins stagnantes, notamment dans la Baltique même et dans les zones adjacentes. Le léger accroissement de salinité de la couche profonde survenu lors des dernières décennies s'est accompagné d'une diminution de la teneur en oxygène. De ce fait, les grands fonds présentent souvent des déficits d'oxygène. La baisse la plus marquée des taux de ce gaz s'est produite ces dernières années et coïncide avec un accroissement de la teneur des eaux en phosphore.

Les pays situés sur le pourtour de la Baltique déversent de plus en plus des quantités considérables d'eaux d'égouts et d'effluents industriels dans cette mer. Le traitement des eaux usées est très insuffisant. Aucun des établissements de traitement ne prévoit la précipitation chimique du phosphore.

Nombreuses sont les eaux côtières fortement polluées. Le déficit d'oxygène actuel des bassins profonds du large est certainement attribuable à la stagnation des eaux, qu'accentua l'augmentation récente de la salinité. Toutefois, ce déficit en eau profonde peut être causé indirectement par une augmentation de la production de matières organiques dans les couches supérieures à la suite d'un apport accru en nutriments d'origine terrestre. L'apport annuel net de phosphore à toute la région se situerait d'après les

La contaminación en el mar Báltico

En el Báltico y zonas adyacentes de fiordos con una capa superficial de agua de baja salinidad y otra profunda de salinidad más elevada, las aguas profundas están más o menos estancadas. En los últimos decenios se ha registrado un ligero aumento de la salinidad en las capas profundas, acompañado por una disminución del contenido en oxígeno. En la actualidad, es frecuente encontrar falta de oxígeno en muchas zonas de aguas profundas. La disminución más pronunciada del oxígeno se ha registrado en los últimos años, coincidiendo con un aumento del contenido en fósforo.

Todos los países circunvecinos descargan en el Báltico cantidades considerables, y siempre mayores, de aguas negras e industriales. El tratamiento a que se someten dichas aguas residuales está lejos de ser aceptable. En ninguna planta se procede a la precipitación química del fósforo.

Muchas aguas costeras están gravemente contaminadas. La causa principal de la falta actual de oxígeno en las zonas profundas del mar Báltico es, sin duda alguna, el estancamiento, acentuado por el reciente aumento de salinidad. Pero el déficit de oxígeno de las zonas de aguas profundas puede deberse, también, indirectamente, al aumento de la producción orgánica en las capas superficiales derivado del aumento de nutrientes procedentes de tierra. Anualmente se descargan en todo el Báltico entre 7.000 y 20.000

*Institute of Marine Research, Lysekil, Sweden.

estimations entre 7.000 et 20.000 tonnes; la plus grande partie de cet apport fait sentir son influence sur la zone située au sud du golfe de Botnie et du golfe de Finlande.

La haute teneur des composés de DDT et de PCB chez les poissons et autres organismes de la haute mer constitue l'un des aspects préoccupants de la pollution en mer Baltique. Les concentrations de ces substances dépassent celles que l'on trouve chez les animaux de la mer du Nord. La pollution mercurique a atteint certaines régions côtières. Il semble que des substances résistantes et toxiques tendent à s'accumuler dans la Baltique, du fait de la durée relativement prolongée de leur séjour dans ses eaux.

La pollution par les hydrocarbures constitue un problème grave, mais la pollution radio-active est soumise à un strict contrôle. D'autres formes de pollution résultant du déversement d'eaux chaudes, de l'extraction du sable et du dépôt de boue et de vase, sont limitées à des régions locales où, toutefois, elles peuvent causer des difficultés aux opérations de pêche.

La législation de lutte contre la pollution des eaux en vigueur dans les divers pays est généralement applicable également aux eaux territoriales, mais le contrôle exercé sur les eaux côtières est d'ordinaire moins strict que celui exercé sur les eaux intérieures.

La coopération internationale se développe pour faire face aux problèmes de pollution dans la Baltique. Depuis plus d'un an des navires de recherche appartenant aux divers pays contrôlent de manière plus ou moins permanente un réseau de stations fixes nstallées en haute mer.

toneladas de fósforo; la mayor parte de esta cantidad afecta a la zona situada al sur de los golfos de Botnia y de Finlandia.

Una grave característica de la contaminación del Báltico son los muchos compuestos de DDT y bifenilos policlorados (BPC) que contienen el pescado y otros organismos de alta mar, con concentraciones superiores a las que se encuentran en los ejemplares procedentes del Mar del Norte. La contaminación de mercurio ha afectado a algunas zonas costeras. Parece que las sustancias tóxicas resistentes tienden a acumularse en el Báltico, debido al escaso movimiento de sus aguas. La contaminación con hidrocarburos constituye un serio problema, mientras la contaminación radiactiva está estrictamente vigilada. Otros tipos de contaminación, debida a descargas de agua caliente, operaciones de succión de arena, y acumulación de lodo y cieno, etc., se limitan a algunas zonas determinadas, en las que pueden crear problemas para las faenas de pesca.

Las leyes contra la contaminación de las aguas en vigor en los diversos países son válidas también, generalmente, para las aguas costeras territoriales, pero en ellas la vigilancia es de ordinario menos estricta que en las continentales. La cooperación internacional en torno a los problemas planteados por la contaminación del Báltico va en aumento. Barcos de investigación de los diversos países se encargan en forma más o menos continua desde hace más de un año de recoger datos en una serie de estaciones fijas en alta mar.

THE countries around the Baltic Sea have long been aware of local coastal pollution problems; in recent years increased attention has been paid to the situation in the open sea.

In 1968 the International Council for the Exploration of the Sea (ICES) established a Working Group on Pollution of the Baltic. Its report of September-October 1969, was published in February 1970 (ICES Cooperative Research Report, Series A, No. 15). This paper is based mainly on this Report, with additions.

The Baltic Sea includes the Bights of Kiel and Mecklenburg, the Baltic proper, the Gulf of Finland, the Bothnian Sea and the Bothnian Bay (fig. 1). The Danish Belt Sea and the Kattegat are also considered in this paper though they are not part of the Baltic proper.

Hydrographical conditions in relation to pollution

The Baltic is a large fjord with a split-up and narrow entrance. The sill (Sill of Darss) depth is 17–18 m. Inside the entrance there are a series of basins, sometimes, as in the Landsort Deep (459 m; F78 in fig. 1), of considerable depth. For depths, areas and volumes of the different parts of the Baltic see Table 1.

An excess of fresh water enters the Baltic and the outflow is about twice the amount of inflowing saline water:

$$\text{Runoff} + \text{Precipitation} + \text{Inflow} = \text{Evaporation} + \text{Outflow}$$

$$470 \text{ km}^3 \quad 200 \text{ km}^3 \quad 430 \text{ km}^3 \quad 180 \text{ km}^3 \quad 920 \text{ km}^3$$

The mean "residence time" for Baltic water has been estimated as about 21 years.

The coastal configurations vary. Along the Finnish and Swedish coasts there are several widespread archipelagos which reduce the water exchange with the open sea. The German, Polish and Soviet coasts are mainly open and have better water exchange.

Since much fresh water enters the Baltic, the surface layers have low salinity. They are separated from deeper, more saline layers by a permanent halocline which in the central Baltic proper lies at 50–70 m and in the Gulf of Finland and the Bothnian Sea between 15 and 30 m. Average salinity (‰) of surface and bottom layers in different parts are:

Area	Surface	Bottom
Southern Baltic proper	7–8	15–17
Central Baltic proper	6–8	11–13
Bothnian Sea	3–4	4–5
Bothnian Bay	1–4	3–4
Bight of Kiel	16–17	21–22
Central Kattegat	21–22	33–34

The permanent halocline, especially in the Baltic proper and westwards, is rather sharp and hampers exchange between surface and deep water. During the summer a thermocline, which further restrains the exchange, occurs at 15–20 m.

Since the beginning of the 20th century the salinity of the Baltic water has increased by about 1–2‰. This has somewhat raised the halocline and increased its stability.

Oxygen conditions

The breakdown of sinking organic material reduces oxygen in the deeper water layers. The halocline (and to some extent the temporary thermocline) restricts the transfer of new oxygen from the surface to the deeper water layers. The oxygen content of the deeper water layers (except in the shallow Bothnian Bay) has decreased during recent decades. This decrease has been most pronounced in the last few years. (Figure 2; Landsort Deep, F78 in fig. 1). In recent times a total oxygen deficit and the production of H_2S has occurred in some areas. Figure 3 shows the conditions in September 1968 when large parts of the deep bottom of the Baltic proper were covered by H_2S water (shaded areas).

There are two principal theories as to the oxygen decrease:

1. The increased salinity particularly of the deeper water layers, and the accompanying strengthening of the halocline have made the bottom waters more stagnant.
2. Increased nutrient supply to the water, presumably from pollution, has increased the biological oxygen demand in deeper waters.

Other causes proposed include the lower oxygen values in inflowing water and the slight increase of temperature

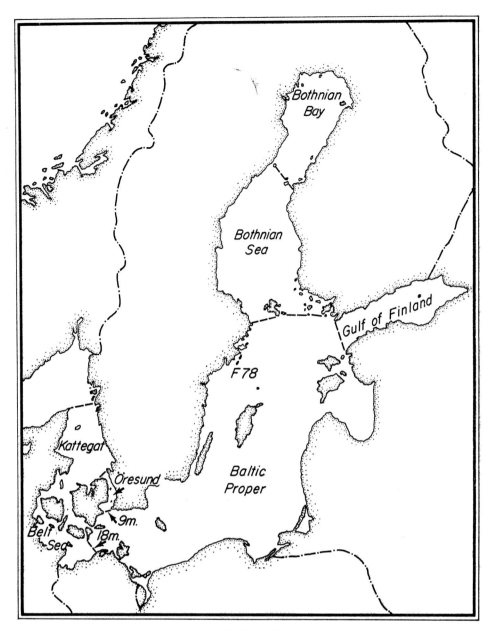

Fig 1. The Baltic Sea area

(1–2°C) that has occurred during recent decades in the deep water of the southern and central Baltic.

The nutrient content of the Baltic has always been fairly low compared with the North Sea. But the phosphate content in the deeper water layers of the Landsort Deep has increased about threefold in the last 15 years (fig. 4). Similar conditions occur in other parts of the Baltic. This may indicate an increase in the nutrient content of the Baltic followed by increased production in the upper water layers, and increased organic breakdown and oxygen consumption in the deeper layers.

The increase of phosphate values roughly coincides with an increased outflow of nutrients from land, especially since World War II, caused by increased human population in coastal areas, the use of washing powders containing phosphorus, and, to some extent, the use of artificial fertilizers. This increase of phosphate values has seriously polluted many coastal areas.

It cannot be denied that the increased salinity and the strengthening of the halocline has a strong effect on the oxygen exchange between the surface and deeper water layers; but an increased breakdown of organic material, as indicated by higher phosphate values, may have contributed to decreased oxygen.

During 1969 and early 1970 the oxygen conditions became somewhat better and most of the H_2S water was replaced by oxygen-containing water from the entrance sounds. Oxygen values of this water are, however, very low, often less than 1 ml/l, and many scientists expect it to soon become H_2S water.

Short accounts of the Baltic position during the last two years have been given by Francke and Nehring (1970) and Carlberg (1970).

Coastal waters

There is no doubt that the coastal waters of all countries are influenced by the increased outflow of sewage and industrial wastes. Polluted areas lie like a string of pearls

TABLE 1. Areas, volume and depth conditions of the sub-areas

Name of Basin or Deep	Area km²	Volume km³	Max. depth m	Mean depth m	Sill depth m
Kattegat	23,000	580		25	
Belt Sea	21,000				
Baltic proper	202,000	13,600	459	67	17
1. Arkona basin (below 30 m)		70	55		17
2. Bornholm basin (below 60 m)		160	105		45
3. Central basin (below 60 m)		4,100	459		60
A. Eastern Gotland basin (below 100 m)		920	249		60
a Gdańsk basin (below 100 m)		10	116		88
b Gotland Deep (below 150 m)		196	249		60
c Fårö Deep (below 150 m)		25	205		140
B. Northern Central basin (below 100 m)		558	459		115
a Northern basin (below 100 m)		228	219		115
b Landsort Deep (below 100 m)		270	459		138
C. Western Gotland basin (below 100 m)		101	205		100
a Norrköping Deep			205		100
b Karlsö Deep			112		101
4. Riga Bay	16,700	460	51	28	c.20
Gulf of Finland	29,500	1,125	100	38	
Baltic proper + Gulf of Finland (0–60 m)	249,000	9,500			17
Baltic proper + Gulf of Finland (total)	249,000	15,190	459	61	17
Åland Sea	5,200	405	301	77	
Archipelago Sea	8,300	195		23	
Bothnian Sea	66,600	4,595	293	69	40 (70?)
Bothnian Bay	37,000	1,540	126	42	25
Baltic Sea (total)	366,000	21,960	459	60	17

TABLE 2. Approximate BOD$_5$ in tons per year from sewage and industrial waste waters

Area	Million inhabitants		Sewage BOD$_5$ tons/year		Industrial waste waters BOD$_5$ tons/year	BOD$_5$ tons/year
	Direct	Indirect	Direct	Indirect	Direct + Indirect	Total
Bothnian Bay	0.230	0.167	5,600	4,500	159,000	169,000
Bothnian Sea	0.565	0.495	13,600	12,200	384,000	410,000
Gulf of Finland	4.868	0.797	121,400	19,200	115,000	256,000
Baltic proper	3.335	2.065	60,900	30,000	~ 110,000	~ 200,000
Sum	8.998	3.524	201,500	65,900	~ 760,000	~ 1,025,000
Belt Sea	1.195	2.600	40,100	21,000	< 10,000	70,000
Öresund	1.770	0.310	40,600	2,600	< 10,000	53,000
Kattegat	0.625	0.360	16,000	5,000	~ 25,000	46,000
Total	12.588	6.794	298,200	94,500	~ 800 000	~ 1,200,000

TABLE 3. Approximate BOD$_5$ in tons per year from the different countries

Area	Denmark	Finland	Poland	U.S.S.R.	Sweden	Fed. Rep. of Germany	German. Dem. Republic
Bothnian Bay		148,000			21,000		
Bothnian Sea		50,000			361,000		
Gulf of Finland		83,000		170,000			
Baltic proper	> 2,000		>10,000	96,000	79,000		> 10,000
Sum	> 2,000	281,000	> 10,000	266,000	461,000		> 10,000
Belt Sea	> 45,000					10,000	> 10,000
Öresund	> 40,000				6,000		
Kattegat	> 6,000				37,000		
Total	> 93,000	281,000	> 10,000	266,000	503,000	10,000	> 20,000

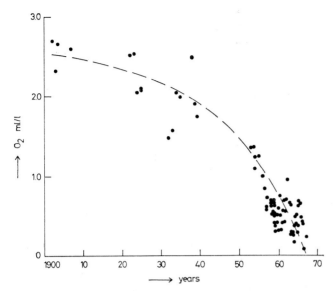

Fig 2. Mean values of O_2 (ml/l) below the halocline in Landsort Deep (fig. 1, F 78), 1902–1967

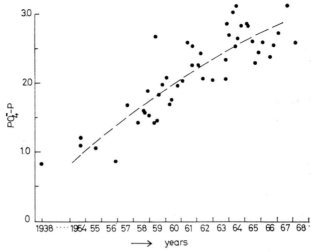

Fig 4. Mean values of $PO_4 - P$ at 100, 200, 300 and 400 m, in Landsort Deep (fig. 1, F78), 1938 and 1954–1968

Pollution in Terms of BOD_5

The ICES Working Group estimated the pollution of the Baltic by sewage and industrial wastes in terms of BOD_5 because most available BOD figures are based on the old BOD_5 method. This method has a series of limitations, and indicates only part of the total organic load in waste waters. There is a considerable difference in the rate of breakdown between the organic matter in sewage water and the waste waters from such industries as paper and pulp. Estimation of BOD_{21}, or estimations for even longer times, would in many cases have been appropriate.

In making BOD_5 estimates it was assumed that the organic load is equivalent to about 25 kg O_2 per person per year. For industrial wastes the estimates were based on different information in different countries, and in some cases estimates were adjusted in relation to the degree of treatment or local circumstances.

Although the figures mentioned below are very rough, especially when indirect and certain industrial discharges are considered, they nevertheless indicate the degree of organic pollution in the Baltic.

Estimated pollution

Table II summarizes the pollution in terms of BOD_5. In the Bothnian Bay and the Bothnian Sea, industrial pollution outweighs sewage pollution by about 15 to 1 because of the sparse human population and the heavy paper and pulp industries. In the Baltic proper and Gulf of Finland the ratio is about 1 to 1. Here there are large human populations in Leningrad, Helsinki, Tallinn, Stockholm, and the Riga and Gdańsk regions.

In the Belt Sea and the Öresund, sewage pollution is about 5 times as high as industrial pollution. The coastal areas are densely populated and industries are fairly small and scattered. In the Kattegat the ratio is again about 1 to 1.

For such small sea areas, the BOD_5 figures are relatively high. This is especially true of the Bothnian Sea and the Gulf of Finland, largely because of the cellulose industries. Since fibres from these industries are broken down slowly, the values would probably be higher if long-term BOD methods had been used for the calculations. Although the archipelagos prevent much of the

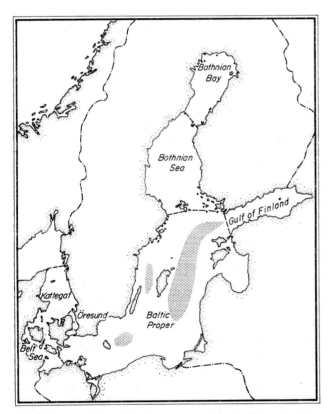

[Fig 3. Approximate distribution of H_2S-containing bottom waters in September 1968

along the Swedish Baltic coast, especially where water exchange is hampered by embayments and archipelagos. Similar conditions are sometimes found along the Finnish coast. Along the more open southeastern and southern coasts polluted areas are most common outside the larger cities and towns.

In a sense the archipelagos of Sweden and Finland and such semi-enclosed embayments as the Bight of Riga act as filters for many pollutants. What would conditions in the open sea have been without this hindrance?

[19]

pollutants from reaching the open sea, coastal waters are entirely destroyed in some places.

In Table 3 Finland, Sweden and the U.S.S.R. are the main producers of organic wastes. The figures for Sweden may be overestimated and those for U.S.S.R. underestimated, but the main trend is shown.

TABLE 4. APPROXIMATE PHOSPHORUS DISCHARGED INTO THE SEA IN TONS PER YEAR

Area	Direct	Indirect	Total
Bothnian Bay	290	220	510
Bothnian Sea	780	600	1,380
Gulf of Finland	5,050	870	5,920
Baltic proper	3,880	2,370	6,250
Sum	10,000	4,060	14,060
Belt Sea	1,240	2,800	4,040
Öresund	1,960	350	2,310
Kattegat	890	400	290
Total	14,090	7,610	21,700

TABLE 5. APPROXIMATE PHOSPHORUS DISCHARGE IN TONS PER YEAR FROM FINLAND, SWEDEN AND THE U.S.S.R.

Area	Finland	U.S.S.R.	Sweden	Others
Bothnian Bay	280		230	
Bothnian Sea	800		580	
Gulf of Finland	1,470	4,550		
Baltic proper		1,840	2,700	1,700
Total	2,550	6,390	3,510	1,700

TABLE 6. ESTIMATES OF THE NET SUPPLY OF PHOSPHORUS IN TONS PER YEAR

	P tons/year	
	Highest values	Lowest values
Supply:		
Sewage	15,000	15,000
Air	3,000	1,000
Natural	3,000	500
The Sounds	6,000	0
Total	27,000	17,000
Outflow:	↙ ↘	↙ ↘
The Sounds	7,000 10,000	7,000 10,000
	—↓— —↓—	—↓— —↓—
Net supply:	20,000 17,000	10,000 7,000

Phosphorus

Phosphorus in different compounds stimulates the production of living organic material in the sea. An increased phosphorus supply increases the number of organisms which, when they decompose, increases oxygen consumption. This may be an advantage in areas where the oxygen consumed is easily replaced, but not in many Baltic coastal waters and deep basins. Because of its great interest the ICES Working Group tried to estimate the phosphorus supply to the Baltic.

The estimates assumed that the discharge is equivalent to about 1 kg per person per year. The figures obtained include industrial phosphorus, which constitutes only a minor fraction of the whole sum (probably between 10 and 30%). In some cases the figures were adjusted in relation to local circumstances.

See Tables 4 and 5.

The phosphorus supply is heaviest outside the most populated coastal areas such as those around Leningrad, Helsinki, Stockholm, Riga, Gdynia-Gdańsk, and Malmö-Copenhagen.

Estimated net supply

The phosphate values from the deep water of the Baltic proper have been increasing during the last 15 years (cf. fig. 4). During this time the coastal human population has increased, especially in the southern Baltic area. The use of phosphorus compounds in, for instance, synthetic washing powders, also has increased.

In Sweden it is estimated roughly that sewage water 10–15 years ago contained about 1.5 g P per person per year compared with about 4 g today. There is reason to believe that the general trend is similar for the other Baltic countries.

When estimating the net supply we must take into account the quantities coming from sources other than sewage and the quantities going out through the entrance sounds. Table 6 shows a calculation of the net supply of phosphorus. According to the calculations, the net supply at present is somewhere between 7,000 and 20,000 tons per year. This is a fairly small amount compared with, for instance, the 300,000 to 400,000 tons of P in the deep waters of the Baltic proper. But some scientists consider the estimated net supply to be too small and furthermore, as suggested earlier, the phosphorus is distributed mainly outside the most densely populated regions. It is therefore at present very difficult to evaluate the influence of the increased phosphorus outflow during the last few years on the oxygen conditions in the Baltic deep basins, although it seems quite certain that many coastal waters are negatively influenced.

There is an urgent need for detailed studies on the phosphorus balance in the Baltic. Since the Baltic must be regarded as a large fjord where the water has a long residence time, phosphorus is likely to accumulate.

Treatment of sewage and industrial waste

Considerable amounts of sewage enter the Baltic and most of it is not treated or undergoes only primary (mechanical) treatment. Less than approximately 20 per cent undergoes partial or complete biological treatment. No treatment plants for the chemical precipitation of phosphorus are known in the area. In all countries treatment of sewage inland tends to be better than in coastal areas. Unless more treatment plants are constructed, sewage pollution, especially in coastal waters, will continue to increase. Only if chemical precipitation methods are included in the treatment plants will phosphorus discharge be brought under control.

Most industrial waste waters in the area come from the cellulose industries. In both Finland and Sweden, treatment of waste waters is improving and the peak of pollution is considered to have been passed. The control of wastes from food-processing industries is also improving but there is need for a greater reduction of the organic load in industrial waste waters.

Organic pollution and marine life

Pollution by sewage and by industrial waste waters that contain a heavy organic load including phosphorus compounds reduces the oxygen content of the water. This occurs extensively in inshore waters along the coast of Sweden and similar situations occur in, e.g., the fjords of Schleswig-Holstein, the Bay of Lübeck, and the archipelagos of Turku and Helsinki. The fauna is influenced in various ways. Generally the species composition becomes poorer.

In many places the dark-coloured liquor from pulp industries greatly hampers the production of plants; in other places heavy phytoplankton often blooms. Local fisheries are often influenced and sometimes discontinued.

High contents of coliform bacteria have been reported from several polluted areas, and living *Salmonella* virus has been found in the Bight of Gdańsk and in the Öresund. Adeno virus from Stockholm sewage caused an epidemic among bathers some years ago.

In the areas of oxygen deficit in the deep basins of the Baltic proper, bottom life is depleted but it is difficult to say to what extent these conditions are caused by organic pollutants from land. In the upper water layers organic pollutants may have been beneficial for production because the open waters of the Baltic are relatively poor in nutrients.

Toxic substances

Many substances in sewage, industrial waste waters, and wastes from agriculture are in some way toxic to marine organisms. They often accumulate in organisms or bottom sediments from which they can be taken up by bacteria and bottom-living animals.

Mobile organisms sometimes react directly to some of these substances and can move away from them. Other substances, no doubt most of them, however, act as slow poisons. They accumulate in the food chains and, on reaching certain threshold values, become harmful to the organisms and even to man through fishes. Many known toxic substances are heavy metal compounds and chlorinated hydrocarbons.

Mercury, DDT and PCB

Mercury compounds have been used as pesticides in most countries. In Finland and Sweden especially they have been used as fungicides and slimecides in the paper and pulp industry.

High mercury contents have been found in fish from many inland and coastal waters in Denmark, Finland and Sweden. The use of mercury is now very restricted but it will remain in nature for a considerable time. In Sweden, fishing in waters where the fish have a content higher than 1 mg/kg wet weight has been prohibited except by the fisherman, for himself, at his own risk.

Mercury pollution seems to have affected only the inland waters and coastal zones. High values in fish from the open sea have not been found.

The content of DDT, DDD, DDE and PCB (polychlorinated biphenyls) in fish, seals and fish-eating birds from the Baltic and the Swedish west coast is under extensive investigation. DDT and its breakdown products poison organisms, and PCB compounds seem to have a similar or perhaps stronger effect. The PCB compounds derive from many industries and are also a component in many paints.

The results of the Swedish investigations show that DDT and PCB have accumulated in considerable quanti-

TABLE 7A. CONCENTRATIONS OF ORGANOCHLORINE COMPOUNDS IN SWEDISH MARINE ORGANISMS 1965–1968
(after Jensen, Johnels, Olsson & Otterlind, 1969)

	No. in Sample	ΣDDT*	ppm in fat DDT	PCB	ΣDDT*	ppm in fresh tissue DDT	PCB
Swedish West Coast							
Mussel	17	1	0.6	2	0.02	0.007	0.084
Oct. 1966, Dec. 1967		(0.4–5)	(0.3–1.3)	(0.5–7.0)	(0.005–0.04)	(0.002–0.03)	(0.011–0.33)
Plaice	3	1		5	0.006	0.004	0.021
Sept. 1966		(0.9–2)	n.e.	(0.4–14)	(0.003–0.009)	(trace–0.006)	(0.002–0.056)
Cod	4	1		7.3	0.005	0.003	0.019
Sept. 1967		(0.6–2)	n.e.	(1.8–16)	(0.001–0.006)	(n.d.–0.006)	(0.006–0.030)
Picked Dogfish	7	1.5	0.91	1.5	0.15	0.091	0.15
Aug. 1968		(0.29–3.9)	(0.15–2.3)	(0.81–2.4)	(0.028–0.33)	(0.015–0.21)	(0.054–0.30)
Fish Oil	3	2.1	1.2	0.74			
Oct. 1968		(1.5–2.6)	(0.83–1.4)	(0.54–1.0)			
Baltic Sea proper incl. the Sound							
Mussel	40	6	1.8	4.3	0.03	0.02	0.03
Oct. 1966, Dec. 1966, Dec. 1967, Jan. 1968		(0.9–10)	(0.5–2.9)	(1.9–8.6)	(0.009–0.07)	(0.003–0.023)	(0.008–0.057)
Herring	18	17	9.7	6.8	0.68	0.40	0.27
Apr./Sept. 1966–68		(4.1–37)	(1.5–21)	(0.5–23)	(0.093–2.3)	(0.012–1.3)	(0.009–1.0)
Plaice	6	2.7	2.1	2.7	0.018	0.013	0.017
Sept. 1967		(1.4–7.8)	(0.6–7.2)	(1.7–4.8)	(0.006–0.036)	(0.003–0.029)	(0.010–0.032)
Cod	5	19	9.8	11	0.063	0.032	0.033
Sept. 1967		(12–31)	(3.5–19)	(3.2–20)	(0.027–0.11)	(0.008–0.058)	(0.012–0.57)
Salmon*	11	31	14	2.9	3.4	1.5	0.30
Autumn 1968		(20–53)	(7.7–20)	(1.1–8.2)	(0.26–7.1)	(0.095–3.1)	(0.014–0.54)
Fish Oil	1	16	7.3	3.5			
Oct. 1968							
Seal (grey) liver	1	96	41	44	3.9	1.7	1.8
Seal (common and grey)	2	130	62	30	66	32	15
Sept., Nov. 1968		(110–150)	(57–66)	(16–43)	(58–74)	(31–32)	(8.5–21)
Eggs from Guillemot	9	570	20	250	40	1.2	16
May 1968		(300–790)	(7.5–38)	(140–360)	(20–51)	(0.7–2.3)	(7.9–21)

*ΣDDT stands for DDT + DDE + DDD. For salmon and seal there were respectively 41% DDD in ΣDDT, the mean figure being 17%.
n.e. = not estimated. n.d. = not detected.

TABLE 7B. CONCENTRATIONS OF ORGANOCHLORINE COMPOUNDS IN SWEDISH MARINE ORGANISMS 1965–1968
(after Jensen, Johnels, Olsson & Otterlind, 1969)

	No. in Sample	ΣDDT*	ppm in fat DDT	PCB	ΣDDT*	ppm in fresh tissue DDT	PCB
The Archipelago of Stockholm							
Mussel	15	3	1	5.2	0.04	0.02	0.37
Oct. 1966, Dec. 1967		(1–4.7)	(1–1.8)	(3.4–7.0)	(0.01–0.061)	(0.01–0.024)	(0.032–0.044)
Herring	4	7.7	3.9	5.1	0.23	0.11	0.17
May 1965		(4.3–11)	(2.0–5.3)	(3.3–8.5)	(0.094–0.30)	(0.044–0.15)	(0.073–0.23)
Seal (grey)*	3	170	17	30	36	4.2	6.1
May 1968		(97–310)	(11–21)	(16–56)	(35–36)	(2.4–6.6)	(5.7–6.4)
White-tailed eagle							
March–June 1965-66	4	25,000		14,000	330		190
Pectoral muscle		(16,000–36,000)	n.d.	(8,400–17,000)	(290–400)	n.d.	(150–240)
Brain	3	1,900		910	100		47
		(1,700–2,100)	n.d.	(490–1,500)	(99–110)	n.d.	(29–70)
Eggs from white-tailed eagle	5	1,000		540			
May–June 1966		(610–1,600)	n.d.	(250–800)	n.e.	n.e.	n.e.
Heron							
April 1967	1	14,000	n.d.	9,400	71	n.d.	48
Gulf of Bothnia							
Herring	4	6.2	3.5	1.5	0.26	0.14	0.065
		(5.2–8.1)	(2.9–4.8)	(0.93–2.0)	(0.15–0.42)	(0.091–0.21)	(0.026–0.091)
Seal (ringed)	2	120	56	13	63	30	6.8
May–Oct. 1968		(110–130)	(54–57)	(9.7–16)	(58–68)	(28–31)	(5.0–8.5)
Gulf of Finland							
Seal pup (ringed)	2	42	23	6.5	25	14	3.9
March 1968		(41–43)	(22–23)	(6.0–7.0)	(24–26)	(13–14)	(3.4–4.4)
Seal milk							
March 1968	1	36	21	4.5	11	6.5	1.4

*ΣDDT **stands** for DDT + DDE + DDD. For salmon and seal there were respectively 41% DDD in ΣDDT, the mean figure being 17%
n.e. = not estimated. n.d. = not detected.

ties in animals from the Baltic proper and the Öresund (cf. Tables 7A and 7B; after Nature **224**, 1969, p. 249, Table 1, somewhat abbreviated). High values have been found especially in white-tailed eagles, Guillemot eggs, and seals. These animals eat fishes of different kinds.

Fish from the Baltic generally has a much higher content of DDT and PCB than fish from the Swedish west coast. It therefore seems that the water of the Baltic contains more of these toxic substances. Baltic fishes and Guillemot eggs sampled far away from the mainland show the same trend towards high concentrations.

General

There are certainly many toxic compounds other than the DDT and PCB complexes found in Baltic marine organisms. Since many of them break down very slowly they will occur in animals, plants and bacteria for a long time, even if some of them, such as DDT, aldrine, endrine and lindane, are now, or will be, entirely or partially forbidden in several countries. The occurrence in the open sea suggests that considerable quantities are brought by air, and the sources may be far from the sea area concerned.

Another source of toxic substances is dumping of both industrial wastes and war material from ships. This has often occurred in the Baltic. Outside the Swedish east coast there is, for instance, dumped material containing compounds of arsenic, mercury, chromium and various acids. In the Bornholm Deep, war material, including mustard gas from World War II, has caused considerable trouble to fishing.

Many scientists consider toxic substances to be the most conspicuous and dangerous feature of Baltic pollution today.

Oil pollution

Most Baltic tanker routes go to Danish, Finnish and Swedish harbours and these countries possess refineries. The size of tankers is increasing.

There is a marked tendency for oil pollution to increase in all areas. Several accidents have occurred through collision or running aground. There has been much discussion about the sensitivity of the Baltic waters to oil pollution. As yet it has been impossible, however, to reach any international agreement about these problems.

Other kinds of pollution

Radioactive pollution is strictly controlled in all Baltic countries and as yet is not a special problem.

Warm water pollution is at present very restricted but may be more important when new electrical power stations are constructed. Sweden, for instance, plans to build about ten nuclear power stations during the next 25 years. Each will emit about 150 m³/sec of heated water, which corresponds to the outflow of ten average Swedish rivers.

The ship traffic is considerable and wastes are discharged from ships on many routes. Some sea bottoms have been so filled with discarded objects that trawling has become difficult or impossible.

Legislation controlling water pollution

In most Baltic countries laws against inland water pollution are also valid for their coastal territorial waters. The U.S.S.R. has special regulations. The degree of control varies, but generally marine control has been much less strict than for inland waters. The present pollution situation of the Baltic has intensified the control of coastal waters in all countries.

[22]

Discharge of harmful wastes in the open sea outside the territorial border is regulated by law only in Finland. In the Federal Republic of Germany such discharge is subject to voluntary control.

Most countries have ratified the London Oil Convention. The German Democratic Republic and the U.S.S.R. have rules very similar to the stipulations of the Convention.

All Baltic countries are active in pollution research. A large-scale international investigation, "The Baltic Year", initiated by the Conference of Baltic Oceanographers, with members from all countries, was carried out in 1969–1970. The investigation involved hydrographical surveys and the sampling of plankton and bottom animals at a number of fixed stations, mainly in the Baltic proper. In this survey, research ships from the different countries succeeded each other in order to keep the survey continuous.

Bilateral cooperation began in 1968 between Finland and the U.S.S.R. concerning the pollution of the Gulf of Finland. Similar cooperation has existed since 1955 between Denmark and Sweden concerning the Öresund. Attempts are being made at present to get more bilateral or multilateral agreements on Baltic pollution problems.

References

Most basic facts, as well as the figures and some of the tables in this paper, are derived from the Report of the ICES Working Group on Pollution of the Baltic Sea (ICES Cooperative Research Report, Series A, No. 15, February 1970) (ICES Secretariat, Charlottenlund Slot, Charlottenlund, Denmark). The report contains comprehensive lists of literature concerning work done in the Baltic waters. References to some more recent work are:

CARLBERG, S (Ed.) Compiled cruise-reports from the Baltic Year. 1970 *Meddn Havsfiskelab., Lysekil*, (87).

FONSELIUS, S H Om Östersjön och svavelvätet. *Meddn Havsfiskelab., Lysekil*, (78). 1970

FRANCKE, E and NEHRING, D First results on a new inflow of high saline water into the Baltic during February 1969. *In* Abstracts. 7th Conference of the Baltic Oceanographers, Helsinki, 11–15 May 1970. 1970

NEHRING, D, FRANCKE, E and BROSIN, H J Observations on the oceanological variations in the Gotland Deep during the turnover in October, 1969. *In* Abstracts. 7th Conference of the Baltic Oceanographers, Helsinki, 11–15 May 1970. 1970

On Eutrophication and Pollution in the Baltic Sea

S. H. Fonselius*

De l'eutrophisation et de la pollution de la Baltique

L'auteur décrit le cycle de fertilisation en bassins semi-stagnants et examine les effets de la pollution sur les conditions locales. Il expose plus en détail les conditions propres à la Baltique, ainsi que la diminution de la teneur en oxygène des eaux profondes de 1900 à 1970. Les raisons possibles de ce phénomène sont discutées. Quelques estimations sont fournies de la charge polluante de la Baltique. L'auteur décrit l'effet de l'enrichissement de l'eau en nutriments, en particulier par l'intermédiaire du phosphore. Il discute de l'accumulation de matières organiques qui s'oxyde lentement; à ce propos, il cite comme exemple l'accumulation et les effets des hydrocarbures chlorés. Il traite de l'intoxication des organismes animaux par le mercure et des problèmes qui en découlent. Enfin, il examine la pollution par les produits pétroliers et ses effets éventuels.

Sobre la eutroficación y contaminación en el Mar Báltico

En la presente contribución, se describe y se discuta el ciclo de fertilización en una cuenca de aguas semi-estancadas con los consiguentes efectos de la contaminación.

Se detallan algunos datos sobre las condiciones hidrográficas del Mar Báltico, comprobándose una disminución del contenido en oxígeno de las capas profundas durante los años del periódo 1900–1970. Se discutan luego las posibles causas de este fenómeno, y se hacen algunas estimaciones de las descargas de contaminantes en el Báltico. Además, se examina el efecto de la fertilización del agua por los nutrientes, especialmente el fósforo. También se discuta la acumulación paulatina de materias orgánicas oxidables, como igualmente la acumulación y los efectos de los hidrocarburos clorinados. Se insiste en el proceso del envenenamiento de los organismos animales por mercurio y en los problemas derivados del mismo en el Báltico. Por último, se discuta la contaminación por petróleo y sus posibles efectos.

THE problem of pollution in the marine environment becomes ever more serious. Intracontinental mediterranean seas are easily exposed to pollution because of restricted water exchange and long shore lines. The Baltic is the best example, and is among the most contaminated sea areas in the world. Surrounded by seven countries, all with highly developed industries and dense populations, it has a very restricted water exchange with the open ocean via the North Sea.

Some 20 million people travel by ship every year between Sweden and Denmark. In summer some 500 passenger ships cruise in the Baltic area, and there are regular daily ferry tours across the Baltic. From this traffic between 250 and 400 m³ of rubbish is dumped into the sea every day, making fishing trawlers reluctant to trawl across ferry routes because of getting the trawl clogged with empty bottles, beer cans, paper, plastic articles, etc.

Fertilization cycle in stagnant seas

Areas with stagnant or partly stagnant conditions are especially sensitive to pollution due to slow water exchange. Such areas generally have a positive water balance, which causes formation of a light surface water layer which isolates the heavier deep water from contact with the atmosphere. A shallow sill at the entrance to such a basin restricts the horizontal water exchange. Examples are fjords in Norway, Greenland and Canada, and the Black Sea, the Baltic, etc. Stagnant basins act as nutrient traps. Nutrients are removed from the surface water through uptake by micro-organisms. When these organisms die they, or parts of them, sink through the halocline into the deep water. Through bacterial oxidation processes the nutrients of these organisms are transformed back into inorganic form and are dissolved. Due to the slow water exchange through the halocline large amounts of nutrient salts are thus accumulated in the deep water basin. In spite of this high nutrient concentration there, the surface water may be on the starving limit regarding nutrient salts. Figures 1 and 2 show the vertical distribution of total P, inorg. P and org. P in the Baltic; this distribution is typical for stagnant basins (Fonselius 1969).

There is of course always a certain exchange of water

---

*Fishery Board of Sweden, Göteborg 4, Sweden.

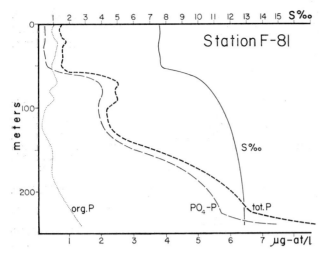

Fig 1. Vertical distribution of phosphorus at Station F–81, 8.2.1968, Central Baltic Sea

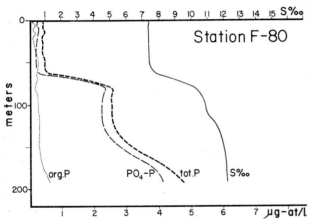

Fig 2. Vertical distribution of phosphorus at Station F–80, 12.2.1968, Central Baltic Sea

through the halocline and therefore nutrient rich water is mixed up into the surface water, but this amount is small.

Organic wastes from communities and industries will cause an oxygen reduction process in the water. Oxygen is utilized for oxidation of the organic matter and nutrients bound in the matter will be released. This process will cause an increase of the primary plankton production in the area. This again will increase the oxygen utilization. A secondary oxygen reduction process will begin, where the new organic matter is oxidized. This secondary oxygen reduction may require two to five times more oxygen than the primary process. This secondary process is not accounted for in the conventional BOD techniques.

In coastal areas with restricted water exchange, e.g. in archipelagos, gulfs and bays, this discharge may cause eutrophication of the water. The primary production increases until the light penetration limits it. A thick layer of green algae is formed along the shores and in the surface water. Anyone who has had an aquarium has certainly experienced this. Signs of such eutrophication can be observed in the Baltic area around all large cities, especially around Stockholm and Helsinki. Many fresh water lakes in industrial areas also show signs of eutrophication, e.g. the Boden See in Central Europe,

Mälaren, Vänern and Vättern in Sweden, Baikal in the U.S.S.R. and the Great Lakes in the U.S.A.

In sea areas with a low nutrient content a release of organic wastes may cause eutrophication of the surface water in the whole area. The increased biological production may be beneficial to the surface water by increasing the fish yield, but it can have serious effects on conditions in the deep water. Increased decaying organic matter there may lead to total oxygen deficiency. Hydrogen sulphide will form and the area be transformed into a dead oceanic desert—bottom fauna being completely destroyed. When hydrogen sulphide is formed, the environment changes from oxidizing to reducing. During reducing conditions nutrient salts will be dissolved from the bottom sediments and increase high nutrient concentration still more.

Fertilization of the surface water through renewal from the deep water may cause an enormous plankton bloom. This may start a "vicious circle" of fertilizations. During every fertilization large amounts of organic matter are produced. Therefore more organic matter is brought down into the deep water. The oxygen there will fast be utilized, hydrogen sulphide will again be formed and more nutrients will be released from the sediments. The next renewal will again cause a large plankton bloom, producing new organic matter which will sink down into the deep water, etc. It seems to be very difficult for nature to restore oxidizing conditions when such a fertilization cycle has started.

Effects of pollutants in the Baltic Sea

The Baltic is an example of such a partly stagnant basin. Its surface water has a very low nutrient content and large amounts of nutrients are accumulated in the deep water. Phosphorus is considered the limiting nutrient for primary production in the Baltic. As long as oxygen measurements have been carried out, the deep water has had a low oxygen concentration. Hydrogen sulphide formation has occasionally occurred in some isolated deep areas. The long series of oxygen measurements made since the 1890s on the international deep stations show that the oxygen concentration of the deep water during this century has decreased from about 3 ml/l to values

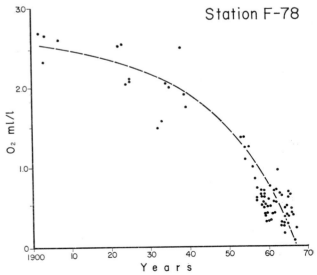

Fig 3. Dissolved oxygen content below the halocline during the period **1902–1967**

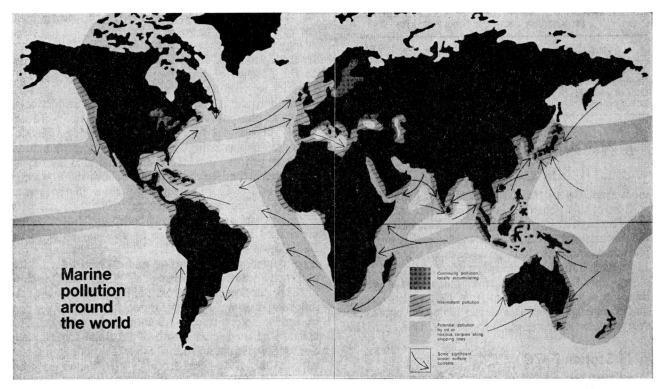

Marine pollution around the world

This map illustrates present and potential pollution round the world in relation to main surface currents indicated by arrows. Heavy shading and dots show continuing pollution locally accumulating; light shading with sloping strokes indicates intermittent pollution; and the light clear shading the potential pollution by oil or noxious cargoes along shipping lines.
Source: M. Waldichuk, Fisheries Research Board of Canada, and L. Andrén, FAO.—FAO copyright.

close to zero (fig. 3). Hydrogen sulphide has several times appeared in the bottom water of the central basin. Increasingly larger areas have been poisoned. The last stagnation period produced hydrogen sulphide in the whole deep area of the Baltic proper. The deep water was renewed in 1969 and all hydrogen sulphide had disappeared in the winter of 1969–1970. Recent measurements show, however, that the oxygen values are again fast decreasing and that hydrogen sulphide formation has again begun in the northern parts of the central basin.

It is known from sediment cores that stagnation periods have occurred earlier in the Baltic's geological history. It is also known that natural factors, e.g. increased salinity and temperature, may have diminished the water exchange and increased the oxygen utilization, but it is quite obvious that the enormous load of urban and industrial waste water discharged into the Baltic has enforced oxygen utilization.

An international working group established by the ICES has published a report on pollution of the Baltic (ICES 1970). It is estimated that organic matter equal to a biological oxygen demand (BOD_5) of 1.2 million tons per year is discharged into the Baltic by industries and in sewage water. The basic assumption was that the organic load of discharged sewage is equivalent to 25 kg O_2/person/year. (See Dybern, Tables 7A and 7B on pages 21-22 this book). It was agreed that BOD_5 values are not a good way to express oxygen demand, but that it was difficult to find a generally acceptable better way. When multiplying the BOD_5 value 1.2 million tons by 3 and considering that the total volume of the Baltic is about 22,000 km³, we get a total BOD of 0.45 mg O_2/m³/day.

The oxidation of this material causes a secondary oxygen demand which is two to five times the primary demand. Therefore the actual oxygen demand caused by this waste discharge may be 1 to 2 mg O_2/m³/day. This is true only if the wastes were evenly distributed in the whole water mass. All waste discharge is made at the shores and therefore the actual oxygen demand in the neighbourhood of large cities and industries may be several times higher.

The working group also estimated that about 14,000 tons of phosphorus are annually discharged into the Baltic, most originating from sewage water. The basic assumption was that sewage water contains 1 kg P/person/year. Swedish investigations have shown that the phosphorus concentration in sewage water has increased from 1.5 g/person/day to more than 4 g/person/day during the last 15 years. This phosphorus increase comes mainly from the increasing domestic use of synthetic

TABLE 1. ESTIMATED AMOUNT OF PHOSPHORUS IN DISCHARGED WASTES (TONS P/YEAR)*

Area	Direct	Indirect	Total
Bothnian Bay	290	220	510
Bothnian Sea	780	600	1,380
Gulf of Finland	5,050	870	5,920
Baltic proper	3,880	2,370	6,250
Baltic Sea	10,000	4,060	14,060
Belt Sea	1,240	2,800	4,040
Öresund	1,960	350	2,310
Kattegat	890	400	1,290
Total	14,090	7,610	21,700

TABLE 2. ESTIMATED AMOUNT OF PHOSPHORUS DISCHARGED IN WASTES BY DIFFERENT COUNTRIES (TONS P/YEAR)*

Area	Finland	U.S.S.R.	Sweden	Others
Bothnian Bay	280		230	
Bothnian Sea	800		580	
Gulf of Finland	1,470	4,550		
Baltic proper		1,840	2,700	1,700
Total	2,550	6,390	3,510	1,700

*Data from ICES 1970.

detergents. Several popular washing powders contain more than 30 per cent phosphate (Tables 1 and 2).

Most of this phosphate is certainly removed from the water through biological filtering in shallow water close to the discharge area but there is evidence that the Baltic phosphate concentrations have increased during the last 15 years, in both surface and deep water. A critical review of all phosphate measurements made in the central Baltic in depths of 0 to 10 m from 1951 to 1970, shows that phosphate values due to the biological production every summer decrease down to values close to zero, but

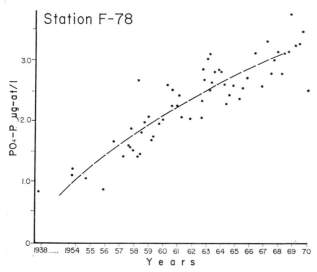

Fig 4. *Mean values of phosphate in deep water layers (100, 200, 300 and 400 m) in the Central Baltic Sea during the years 1938 and 1954–1970*

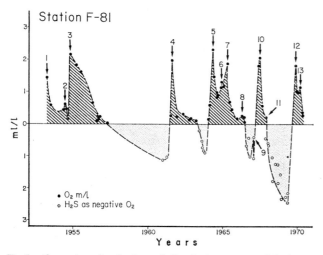

Fig 5. *Alternating distribution of dissolved oxygen and hydrogen sulphide at Station F-81 (Central Baltic Sea) during the period 1955–1970*

that there is a clear tendency for the surface values to increase from year to year.

Figure 4 shows the phosphate increase in the deep water from 1954 to 1970. The tendency is here much clearer due to the absence of primary production. The hydrogen sulphide formation causes dissolution of phosphate from the sediments but there must of course be an increasing supply of phosphorus from the surface water in order to supply phosphates to the sediments during oxidation periods. The increasing fertilization of the surface water increases primary production there and also the oxygen utilization in the deep water. Figure 5 shows the alternating oxygen and hydrogen sulphide periods in the central Baltic (Fonselius and Rattanasen 1970).

This phosphorus increase is better understood if it is considered that the residence time of the water in the Baltic is, on average, 24 years. The actual exchange rate is, of course, faster in surface and slower in deep water and such a value is a very rough simplification. One-third of the water flowing out from the Baltic returns with the inflowing deep water. Therefore only about 2.85 per cent of, for example, a tracer, will annually be removed. This means that an accumulation of 35 times the annually discharged amount will occur. Such an accumulation may happen only if an ideal tracer is used which does not decompose, sediment or participate in the biological life cycle.

DDT and poly-chlorinated biphenyls

Ordinary sewage waste will therefore not accumulate in such a degree, but there are some compounds which are only slowly decomposed, e.g. pesticides such as DDT, and some industrial products such as PCB (poly-chlorinated biphenyls) which is used in paints, transformer oils, etc. PCB is almost indestructible and a Swedish investigation has shown that DDT and PCB accumulate in marine organisms and that the concentration in marine species from the Baltic may contain up to 100 times more of these products than species from the Swedish west coast or the North Sea. Figure 6 shows some of these results, quoted in Jensen *et al.* (1969). The

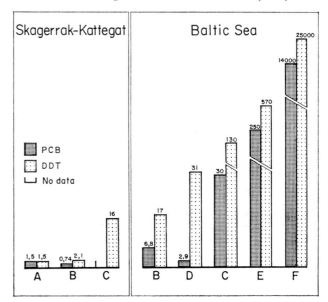

Fig 6. *PCB and DDT contents in different marine animals: A = picked dogfish, B = herring, C = seals, D = salmon, E = eggs of guillemot, F = sea eagle (after Jensen et al. 1969)*

[26]

concentrations are expressed as mg/kg fatty tissues. Analyses of tissues of organisms belonging to the marine food chain, e.g. mussels, herring, plaice, picked dogfish, cod, salmon, grey seal, ringed seal, guillemot, white-tailed eagle and heron, have shown that organisms from the Baltic contain about 8–10 times higher concentrations of PCB and DDT than organisms from the Swedish west coast (Kattegat) or from England or Canada. The eagles (all found dead) contained 100 times higher concentrations of these compounds than eagles caught in Northern Sweden (Lapland). These latter eagles generally build their nests on the Norwegian north Atlantic coast. The investigation in the Baltic shows that the concentration of DDT and PCB increases in the food chain so that the animals at the top of it have the highest concentrations. Since many of the investigated animals were caught in the open sea, there must be transport there of pesticides and other toxic substances. The reason for the seemingly ten times higher concentration of chlorinated hydro-carbons in the Baltic than in the Kattegat and Skagerrak is obviously the Baltic's restricted water exchange.

DDT is probably the most used pesticide in the world. Many investigations have shown the dangers of its use in large amounts. It is found in the air and has been reported in fatty tissues of penguins in the Antarctic, far away from all DDT sources. The thin surface film on natural waters consists mainly of fats and fatty de-composition products from dead plankton organisms. DDT dissolves in this film and is therefore enriched there. Through wave action and turbulence it is spread downwards. It may affect primary production by inhibiting the assimilation process of phytoplankton—one of the first links in the food chain. The annihilation or even a considerable decrease of phytoplankton may have very serious effects on marine life and may ultimately be a danger for mankind by decreasing the amount of food in the oceans.

Mercury

Much of the mercury found in the marine environment is spread from the air. Mercury is brought into the air in smoke and fumes from factories and by paper-burning in garbage disposal plants. It may also be added to the water through direct discharge from factories, e.g. paper and pulp industry, chlor-alkali plants, dental clinics and through agricultural drainage from mercury compounds used as seed disinfectants. In the paper and pulp industry and in agriculture organic mercury compounds are used. The pulp industry in Sweden has recently replaced mercury compounds used as slimecides and fungicides with another, unfortunately very poisonous compound, penta-chlor-phenol.

Metallic mercury has generally a low toxicity as has also the inorganic mercury ion, if used in low concentrations. Mercury ions and fumes may cause mercury poisoning, which injures skin and mucous membranes and may accumulate in and injure the kidneys. Inorganic mercury compounds are, however, removed from the body by the help of the kidneys and the victim recovers if the exposure ceases.

Organic mercury compounds are considerably more dangerous. Organic mercury is accumulated in the body and attacks the nerve system. The organic part of the molecule, to which the mercury is attached, makes it easily soluble in fats. Therefore it may dissolve in the thin fat layer surrounding the nerve cords. When symptoms of poisoning by organic mercury compounds are observed there is no recovery and severe cases may lead to death. This was recorded in Japan some years ago.

There seem to be large differences in the toxicity of organic mercury compounds. The most dangerous compounds are the alkyl mercury compounds, e.g. methyl-mercury-dicyan-diamine. It is believed that less poisonous organic mercury compounds may be transferred into alkyl-mercury compounds, e.g. in the bottom sediments.

No fatal case of mercury poisoning, caused by high mercury content of fish or meat, is known to have occurred in Sweden (Tables 3 and 4). It has, however, been found that fish from certain areas along the Swedish coast and especially from inland waters may contain up to 5 mg mercury/kg fish meat. It has also been found that mercury concentrates in the food chain so that animals at the top have the highest concentrations. The high concentrations seem to be locally restricted and cod caught in the open Baltic have, according to German investigations, a normal mercury concentration.

TABLE 3. ANALYSIS OF FISH FOR MERCURY
(DATA FROM FOLKHÄLSOINSTITUTET, PUBLISHED IN VÅR FÖDA, 1967:1)

Area	Range of Hg concentration mg Hg/kg	Mean value	Type of fish
Certain lakes and coastal areas	0.13–3.95	1.30	Perch
	0.15–5.20	1.09	Pike
	0.17–2.55	1.00	Pike-perch
	0.12–1.40	0.36	Vendace

TABLE 4. MERCURY CONTENTS IN MUSCLES OF FISH
(DATA PUBLISHED IN LÄKARTIDNINGEN 64 (37), 1967)

Area	Range of Hg concentration mg Hg/kg	Type of fish
The West Coast	0.03 – 0.2	Pike, perch, pike-perch, vendace
The Baltic	0.02 – 2.5	Pike, perch, pike-perch, vendace
Lakes	0.05 –10.0	Only pike
Off the coast in the Baltic and the Atlantic	0.016– 0.110	Various salt water fishes

The publication of reports on contamination of fish often has an economically damaging effect on fishermen through the panic caused in the consumer. In Sweden, commercial fishing is now prohibited in the most polluted areas, but fishermen have encountered difficulties in selling uncontaminated fish caught in the open sea far away from polluted areas. During the first few months after the prohibition was announced the sale of fish in Gothenburg decreased by more than 20 per cent, although most of the fish marketed there is caught far out in the Skagerrak and the North Sea.

Oil

In the Baltic the 17 m deep entrance sill at Darss puts an upper limit of some 120,000 tons on the size of tankers. The control of such large ships is rigorous and there is no risk of oil dumping. However, accidents may happen. The handling of such ships in narrow waters is difficult and the North Sea and the Baltic are not free from mines outside the narrow swept channels.

Oil dumping from small tankers, cargo ships and trawlers occurs almost daily and even if the amount dumped by each one is small, the total amount is certainly a serious threat to marine life and beaches. Almost every week drifting oil masses are reported and most originate from illegal tank flushing. In 1968, 125 discharges of oil from ships are known to have occurred in Swedish coastal areas.

The situation is especially serious in the Baltic as some countries in the area are short of fresh water and plans have been made to distil or dialyse Baltic sea water which is especially suitable due to low salinity. If the water is contaminated by oil, difficulties arise.

The effect of oil on marine life was until recently not considered serious except with sea birds who cannot survive even light contamination. An oil film on the sea surface may affect plankton and marine algae are known to be very sensitive to oil and oil products (Mironov and Lanskaia, 1966). Developing eggs of *Rhombus maeoticus* and of cod and herring were found to be very sensitive to oil derivates and fish flesh may be tainted (Kühnold 1969).

The most serious damage by oil accidents is from the solvent detergents used. The oil compound may sink to the bottom, destroy fishing grounds, and eventually float to the surface again. The solvent part of the detergent may destroy marine life. Only mechanical methods should be used to remove the oil from the sea surface, but considering the heavy damage caused by crude oil on beaches, most authorities prefer to use detergents.

Conclusion

In the Baltic region there is very good cooperation in pollution research. All countries are aware of the problems and intensive research is carried out to estimate the degree of pollution and prevent it as far as possible.

International cooperation here is a good example of how these problems should be attacked. Steps by single nations are not enough to cope with the present situation. It is now necessary to take up the problem on an international level, for example in the United Nations. We need international laws and international control of pollution before it is too late.

References

ACKEFORS, H   A survey of the mercury problem in Sweden with
1968   special reference to fish. ICES C.M. 1968/E:3.
ACKEFORS, H A short report of the mercury problem in Sweden
1968a   *Meddn. Havfiskelab. Lysekil (Biol.)*, (47).
FONSELIUS, S H   Hydrography of the Baltic Deep basins. 3. *Rep.*
1969   *Fishery Bd Sweden (Ser.Hydrogr.)*, (23).
FONSELIUS, S H and RATTANASEN, C   On the water renewals in the
1970   eastern Gotland Basin after World War II. ICES C.M.
    1970/C:8.
ICES   Report of the ICES working group on pollution of the
1970   Baltic Sea. *Coop. Res. Rep. Int. Coun. Explor. Sea(A)*, (15).
JENSEN, S *et al.*   DDT and PCB in the marine environment. *Fauna*
1969   *Flora, Upps.*, 64:4 (in Swedish).
KÜHNOLD, W W   Effect of water soluble substances of crude oil on
1969   eggs and larvae of cod and herring. ICES C.M. 1969/E:17.
MIRONOV, O G and LANSKAIA, L A   The influence of oil on the
1966   development of marine phytoplankton. Paper presented at
    second International Oceanographic Congress, Moscow.

Review on the State of Pollution in the Mediterranean Sea

*GFCM/ICSEM Group of Experts on Marine Pollution**

Etude sur l'état de la pollution dans la mer Méditerranée

Seize des 21 pays riverains de la Méditerranée (y compris la mer de Marmara et la mer Noire) ont envoyé au CGPM des informations sur l'état de la pollution sur leurs côtes. Ces données ont été utilisées pour établir une étude sur l'état de la pollution dans la région.

Dans tous les pays, des lois contre la pollution des eaux intérieures et des eaux côtières territoriales sont en vigueur. Cependant, leur mise en application est insuffisante dans les zones côtières. Les effluents domestiques sont déversés dans la mer en grandes quantités, soit directement sur la côte, soit par pipelines. Dans presque tous les cas ces eaux ne sont pas traitées. Elles subissent parfois un traitement primaire mais il semble que les installations équipées dans ce but ne fonctionnement pas correctement. Quelques zones, en particulier les côtes européennes de la Méditerranée occidentale, sont très fortement polluées. Ceci est particulièrement vrai au cours de la saison des bains de mer alors que le chiffre de la population dépasse dans quelques régions de 2 à 10 fois la normale.

La pollution industrielle concerne principalement les zones dans lesquelles la pollution domestique représente déjà un risque. Il existe des points cruciaux, particulièrement le long des côtes espagnoles et françaises (Barcelone, côte entre Sète et Toulon) et dans le golfe de Venise. La pollution par les hydrocarbures est un problème important dans les eaux côtières des pays qui produisent et traitent le pétrole, particulièrement dans la Méditerranée orientale. Des résidus de pétrole brut, qui se présentent sous forme de masses d'apparence goudronneuse, ont été trouvés en grandes quantités à la surface de toute la Méditerranée. Les pesticides sont utilisés de

Estudio sobre el estado de la contaminación en el Mediterráneo

Dieciseis de los 21 países que bordean el Mediterráneo, el Mar de Mármara y el Mar Negro han facilitado al Consejo General de Pesca del Mediterráneo detalles sobre el estado de la contaminación costera. Estas informaciones se han aprovechado para redactar un estudio sobre el estado de la contaminación en el Mediterráneo.

En todos los países existen leyes contra la contaminación de las aguas dulces y marinas dentro del límite de las aguas territoriales. Sin embargo, la mayoría de las veces su aplicación es insuficiente en las regiones costeras.

Las aguas de las alcantarillas van a parar al mar en grandes cantidades, sea directamente sobre la costa, o sea a través de tuberías. En casi todos los casos las aguas de desecho no se someten a ningún tratamiento. En ciertas circunstancias se someten a un tratamiento previo pero, según se comunica, dicho tratamiento no es satisfactorio. Algunas regiones, especialmente las costas europeas del Mediterráneo occidental, están muy contaminadas. Esto ocurre particularmente durante el período del veraneo, en el que la población excede en algunas regiones de 2 a 10 veces la población normal.

La contaminación industrial ocurre principalmente en las mismas regiones donde los desechos domésticos significan un riesgo. Los puntos cruciales se encuentran especialmente a lo largo de las costas españolas y francesas (Barcelona, y de Sète a Toulon) y en el Golfo de Venecia.

La contaminación por petróleo constituye un problema grave en las regiones costeras de los países que producen y tratan el petróleo, principalmente en el Mediterráneo oriental; en efecto, se han

---

* Nominated by the General Fisheries Council for the Mediterranean and the International Commission for the Scientific Exploration of the Mediterranean Sea.

THIS paper summarizes the most important chapters of the final report being presented March 1972 to the General Fisheries Council for the Mediterranean following the preliminary report given at the FAO Conference December 1970. It represents the work of a joint group of experts.

Hydrography

The Mediterranean's enclosed sea has only limited water exchange with the open ocean so its continuous receipt of waste which cannot be naturally processed to become harmless to living resources will, sooner or later, become a real danger. The currents are generally not strong. Only under the influence of strong winds, or when there are favourable tidal streams (as in the Strait of Messina, Euboea and parts of the Adriatic), or near the entrances of channels or straits (Strait of Gibraltar, Dardanelles, Bosporus) are there horizontal movements strong enough to favour quick dilution or dispersion of waste matter. Main areas of deep-reaching vertical mixing water masses are in the Ligurian Sea, the northern Adriatic and the northern Levant Basin between Rhodes, Crete and Cyprus. However, this mixing occurs only during autumn and winter when surface water masses are cooled and so initiate mixing.

Equally, stratification by thermocline can only be observed during summer, but then it is well established in most parts of the Mediterranean and may be used for introducing domestic sewage below this layer so keeping surface water safe for recreational purposes.

General oxygen conditions vary considerably from area to area and season to season. The vertical mixing of winter provides deep water with a new oxygen supply. The heating of the water during summer and high biological activity in the euphotic zones diminishes the oxygen content of surface layers.

Domestic sewage

Information from all countries bordering the north and east Mediterranean, and from Tunisia in the south, gives quite a clear picture of the organic load of the coastal zones. In all these countries the main urban centres still discharge sewage directly into the sea or into rivers at very short distances from the sea. More pipelines are being used, but only exceptionally is the waste discharged far enough from the coast and/or at sufficient depths to avoid beaches being affected.

Generally, discharged sewage is not treated. There are some purification stations but it is well known that their operation leaves much to be desired and their capacity is too small compared with population.

The situation is particularly severe in the north-western basin—the coast from the river Ebro in Spain to the river Arno in Italy, in the North Adriatic and along the coasts of Lebanon and Israel.

Projects and improvements are under way as everywhere people are becoming increasingly concerned, but because of high construction costs of adequate treatment plants, improvement will be slow. Israel and Cyprus, for instance, are planning that within five years no more polluted water will be discharged into the sea.

TABLE 1

ORGANIC LOAD OF DOMESTIC SEWAGE DISCHARGED DIRECTLY INTO THE SEA OR THROUGH RIVERS

	BOD_5[a] (tons/year)	P[a] (tons/year)	BOD_5/km coastline (tons/year)	P/km coastline (tons/year)
Spain	130,000	5,900	60	2.7
North-western Basin	360,000	16,000	336	15.0
Italy	400,000	18,000	61	2.7
Yugoslavia	17,800	800	27	1.2
Malta	8,000	320	67	2.7
Greece	100,000	4,500	37	1.7
Turkey[b]	100,000	4,500	36	1.6
Cyprus	9,600	430	15	0.7
Lebanon	31,250	1,250	149	6.0
Israel	32,000	1,400	145	6.5

[a] The BOD_5 has been calculated on the assumption that the organic load of domestic sewage equals 20–25 kg/person/year.
[b] The Black Sea coast of Turkey has not been included.

Industrial waste

Industrial pollution is not restricted to the coastal zones; industrial wastes are often dumped in the open sea and, due to their persistency, are spread over wide sea areas. Of the organic load of industrial effluents, not enough is known yet about the adverse effects of effluents due to their toxic components or solid substances. The level of organic load and the impact of toxic substances on the water in the eastern Mediterranean do not seem to be so dangerous at present as in the western Mediterranean where large concentrations of industries seriously affect the water masses nearby. However, in some areas industrial wastes are becoming an increasing risk to aquatic life, e.g. in the Sea of Marmara and some parts of the Aegean Sea and the Levant Basin.

Most of the factories now situated along the coasts and in river estuaries discharge waste waters untreated or with insufficient treatment. There is a clear indication of further rapid industrial development in nearly all Mediterranean countries involving numerous new factories along the coasts. This may mean further pollution hazards.

Some facts follow:

(a) **Mahgreb—Balearic region** (south-western Mediterranean). Most industrial activity along the Spanish coast comes from chemical production and textile and leather works. Ironworks are under construction. As a spectacular example of discharge of waste, a lead mine near Cartagena discharges 2,000–3,000 tons of mineral powder daily into a bay. This bay has not been used for fishery or tourism in the past but has already been half filled up. The important activity on the African coast is food production, but large phosphate and zinc factories are planned here and it is disturbing that, for the zinc factory, no plans for treatment of waste are envisaged.

(b) **North-western Basin** (coasts of Catalonia, Gulf of Lions, Provence and Gulf of Genoa). This is the most populated and industrialized region in the Mediterranean with some 50,000 industrial enterprises. The organic load of the industrial waste equals that of more than 23 million inhabitants, i.e. to a BOD_5 of 350,000 tons/year, which is nearly the same as the organic load of the domestic sewage in that area.

A well-known example of discharge in the area is the "boues rouges", which is the waste of a bauxite processing factory near Marseille. This contains a high content of ferric oxide and is reddish-brown in colour. From 2,000–3,000 tons of this waste are discharged daily by pipeline into a depth of 350 m at a distance of 2,000 m from the coast. From there the waste sinks into a submarine canyon 2,000 m deep.

(c) **Tyrrhenian Sea and Gulf of Taranto.** There are some large industrial centres in the area (Naples, Leghorn, the marble quarries near Trapani, and mining areas in Sardinia). Serious effects have been observed, especially through the wastes of the mines near Iglesias (Sardinia) and the marble industry near Trapani (Sicily). These have adversely affected tuna migration and so damaged fisheries.

(d) **Adriatic Sea.** This is semi-enclosed and for this reason is especially susceptible to pollution. In particular from the northern part (shallow, not deeper than 200 m) and having, along its coasts, such industrial centres as Venice, Trieste, Rijeka and Split, serious effects are already reported.

Waste waters from 76 industrial plants are discharged into the Venice lagoon and 10,000 out of its 69,000 hectares are deemed seriously affected. The lagoon of Merano has at times the characteristics of partially purified waste water instead of brackish water. Finally, it is most disturbing that concentrations of some heavy metals in the water of the open sea and in edible marine organisms is only at, or even slightly above, the acceptable safety level.

(e) **Aegean Sea and Sea of Crete.** Although the total number of factories is relatively small in this area, some parts of the coastal zone, mainly bays and gulfs, must be classified as heavily polluted—causing bad smells and with adverse effects on fisheries (Sea of Marmara, İzmir Bay, estuary of the river Strimon).

(f) **Levant Basin.** In the inner part of the region (Cyprus, Gulf of Iskenderun, Beirut, and Haifa/Tel Aviv) there are important industrial centres. Adverse effects on marine life have been observed in Morphou Bay where the number of fish species has decreased and the yield of fish catches has been reduced to only 20 to 30 per cent of the previous level. The discharge of effluents into the sea (including the rivers) along the coast of Israel amounts to 1,700,000 m³/year from only the largest factories. Fish kills caused by large-scale release of acid wastes have sometimes been observed.

Oil pollution

In some countries bordering the Mediterranean, and mainly those situated in the eastern part, there are high capacity oil wells and pipelines. In 1970 about 300 million tons of oil were loaded from their harbours and terminals. Of this, about 165 million tons were destined for harbours within the Mediterranean involving such short hauls that the so-called "load-on-top" system for reducing the oil residues could not be used.

Only five out of 14 harbours have facilities for receiving oil residues from tankers. Therefore, most ships make use of the two zones in the Mediterranean in which discharge of oil residues is still allowed. It is calculated that the amount of oil released into the area is 300,000 tons annually, and this strongly suggests that the entire Mediterranean Sea should be classified as a prohibited zone for the release of any oil from ships.

In addition to this figure another 20,000 tons of oil per year flows into the Mediterranean with refinery effluents—estimated on the assumption that 1.1 m³ water per ton of oil is needed for processing the oil and that this water on leaving the factories again contains about 100 ppm of oil. The uptake of oil by the cooling water has not been taken into account and the real figure of the pollution load through refineries will therefore be more than 20,000 tons/year.

In view of the possibility of tanker accidents in the area, various countries are taking measures to reduce pollution. Such precautions seem insufficient in some countries where mechanical cleaning of the beaches has been considered. Oilsink means have been envisaged by one country; most countries will use emulsifying solvents. Spain, France and Italy have agreed to coordinate their measures in the case of an accident off their coasts following the example given by North Sea States which, in 1970, signed a treaty for technical cooperation to combat oil pollution caused by tanker accidents.

Most Mediterranean beaches are heavily polluted by oil—up to 80 per cent of the coastline; in many places tar lumps are not only found on beaches but also in great quantities in the open sea (mainly in the eastern Mediterranean); damage to fishing gear is becoming more frequent. It is becoming increasingly difficult to sell (except at very reduced prices) various species of fish caught in some areas, for instance the Sea of Marmara, the northern Adriatic, the Tyrrhenian Sea and the French and Spanish coasts. Finally, ecological damage has been reported at places in the Tyrrhenian Sea (Naples, Cagliari), in the Adriatic Sea (mainly the lagoons in the northern part), and in the Sea of Marmara. In these areas a clear decrease in the number of commercial species has been observed.

Pesticides

In all areas of high agricultural production large amounts of pesticides are distributed. If these areas are close to the coast, for example in the plains of Lebanon, Israel and Egypt, the coastal waters and even the open sea receive appreciable amounts through wind transport.

All commercial pesticides are used. The quantities vary with the product and the country. There appears to be a general tendency to increase the use of organophosphorus and carbamate insecticides substituting the persistent chlorinated products.

Although in general the concentrations of pesticide residues in the open water body are considerably lowered by dilution, especially due to the efficient mechanism for vertical circulation in the Mediterranean, adverse effects have been found in various coastal zones.

In the western Mediterranean the intensive use of pesticides for agricultural purposes has, in some areas, been found responsible for the death of marine organisms at all levels (destruction of the fauna of the Etang de l'Olivier; damage at Martigues due to arseniates, and on the coast of Languedoc following the fight against mosquitos).

In the eastern part, pollution by pesticides is still a local problem. Several cases of fish mortality through indiscriminate use of pesticides have occurred. In Israel, Cyprus and the Nile delta where intensive agriculture provides some probability for runoff or, at least, wind transport of pesticides to coastal waters, the problem will become more serious and extend over larger areas.

Other types of pollutants

Pollution by radioactive substances, heat, floating materials, suspended matter and polychlorinated biphenyls (PCBs) does not seem to have seriously affected the water quality in the Mediterranean yet.

Pollution by radioactive substances originating from peaceful uses of atomic energy is at present well under control. As more nuclear power reactors are under construction and in the planning stage, further attention has to be given to this point.

Regarding thermal pollution problems, it has been found that the number of power stations and industrial plants releasing cooling waters into the sea is increasing. Six countries have reported that hot water from 70 power stations, generally about 6°–8°C warmer than the recipient, is discharged. Some 40 other stations are planned.

In the case of floating materials, persistent synthetic materials such as plastic foils represent the biggest problems, hindering navigation of smaller vessels and decreasing the recreational value of beaches.

Suspended matter and bottom deposits interfere with marine ecosystems in several areas. Mackerels, for example, have practically disappeared from Turkish waters although they were once of major economic importance. This is probably due to the increased turbidity in the Sea of Marmara where the mackerels' spawning ground is situated. The deposit of suspended matter coming from dredging has been harmful to spawning grounds on the French coast.

Legislation

Less attention is given to the problem of pollution control in most Mediterranean countries than in countries of northern Europe. For instance, the countries bordering the North Sea have agreed upon pollution control under a convention drafted at Oslo in October 1971. This convention could be adapted to the special needs of the Mediterranean area.

It is suggested that the Mediterranean countries should follow the recommendations of the European Inter-Parliamentary Seminar, held in Rome in September 1971, and achieve cooperation among themselves to solve the specific problems in this region. In this connection,

attention should be drawn to the joint statement made by representatives of Cyprus, France, Italy, Malta, Morocco, Turkey and Yugoslavia during the first meeting of the Inter-Governmental Working Group on Marine Pollution in preparation for the UN Conference on the Human Environment to be held in Stockholm in June 1972. This statement proposes common planning for establishing purification stations, for scientific and technical co-operation on pollution control, the definition of standards and the establishment of a monitoring network.

References

American Petroleum Institute, Manual on disposal of refinery
1969   wastes. Liquid wastes. American Petroleum Institute.
ARMAGAU, C   Tentatives d'utilisation des détergents anioniques
1967   comme traceurs de pollution fécale. Expériences réalisées
       sur l'émissaire de Lez à Palavas. Revue Trav. Inst. (scient.
       tech.) Pêch. marit. 31(4):417–24.
AUBERT, M, AUBERT, J and DANIEL, S Côtes de France. Inventaire
1968   national de la pollution bactérienne des eaux littorales.
       Méditerranée. Cartes et graphiques. Revue int. Océanogr.
       méd., Suppl. Tome 4:167 p.
AUBERT, M et al.   Etat actuel des pollutions bactériennes le long
1968a   des côtes méditerranéennes françaises. Revue int. Océanogr.
        méd., 9:45–72.
AUBERT, M, et al. Côtes de France. Etude générale des pollutions
1969   chimiques rejetées en mer. Méditerranée. Revue int.
       Océanogr. méd., Suppl. Tome 2:135 p.
BELLAN, G   Pollution et peuplements benthiques sur substrat
1967   meuble dans la région de Marseille. Première partie. Le
       secteur de Cortiou. Revue int. Océanogr. méd., 6–7:53–87.
BELLAN, G Pollution and peuplements benthiques sur substrat meuble
1967a   dans la région de Marseille. Deuxième partie. L'ensemble
        portuaire marseillais. Revue int. Océanogr. méd., 8:51–95.
BELLAN, G and PÉRÈS, J M   Etat général des pollutions sur les côtes
1969   méditerranéennes de France. In Symposium FAST,
       "Acqua per il Domani", Milan, April 1969.
BELLAN-SANTINI, D   Contribution à l'étude des peuplements infra-
1967   littoraux sur substrat rocheux (études qualitative et
       quantitative de la frange supérieure). Thèse Fac. Sciences
       Marseille (A-O CNRS 1857), 2 vols:395 p. Also issued as
       Recl Trav. Stn mar. Endoume, (47–63).
BEYCHOK, M R   Aqueous wastes from petroleum and petro-
1967   chemical plants. London, J. Wiley and Sons, 380 p.
BIANCHI, A and MARQUET, R Etude de la pollution du golfe de
1965   Marseille. 1. Note préliminaire: étude de la diffusion des
       eaux polluées en fonction de la répartition de certains
       germes intestinaux. In Symposium sur les pollutions
       marines par les microorganismes et les produits pétroliers,
       avril 1964, Monaco. C.I.E.S.M.M., pp. 59–66.
BOURCIER, M   Ecoulement des "boues rouges" dans le Canyon de
1969   la Cassidaine (décembre 1968). Tethys, 1:3.
BRISOU, J   La pollution microbienne virale et parasitaire des eaux
1968   littorales. Conséquences pour la santé publique. Bull. Wld
       Hlth Org., 38–79.
C.I.E.S.M.M.,   Symposium international des pollutions marines
1969   par les microorganismes et les produits pétroliers, Monaco,
       1964. Rapports et procès verbaux de la Commission inter-
       nationale pour l'exploration scientifique de la mer Médi-
       terranée, 384 p.
CUSMAI, R   Preliminary results of an inquiry into the state of pol-
1969   lution of Italian coastal waters and attempts at a study of
       the various aspects of the problem. In Report of proceedings
       International Conference on Oil Pollution of the Sea,
       7–9 October 1968 at Rome, pp. 330–48.
DANALI, S D   Functional arrangement and organisational pro-
1967   blems of a laboratory for monitoring radioactivity in the
       biosphere. Rep. Univ. Calif. radioact. Lab., (1356) July.
EMIG, C C   Essai d'étude de la teneur en gaz d'hydrocarbures dans
1966   le milieu marin. Recl Trav. Stn mar. Endoume, (41–57):9–16.
EMIG, C C Etude de la teneur en gaz d'hydrocarbure des sédiments.
1969   2. Etang de Berre. Recl Trav. Stn mar. Endoume, (45–61).
FAUVEL, Y   La pollution dans le canal de Sète au Rhône. XXIe
1968   Ass. plein., Monaco, C.I.E.S.M.M.
GOLUBIC, S   Effect of organic pollution on benthic communities.
1970   Mar. Pollut. Bull., 1:4.
GOUGENHEIM, A   Une élimination rationnelle de déchets industriels.
1970   Cah. océanogr., 22(3):213–7.
HATZIKAKIDIS, A D   Chemical and microbiological study of
1950   marine waters. Praktika 'ell. 'udrobiol. Inst., 4(1):103–20.
HATZIKAKIDIS, A D   Rougissement périodique des eaux. Contri-
1952   bution à l'étude des sulfabactères. Praktika 'ell. 'udrobiol.
       Inst., 27:492–501.

HATZIKAKIDIS, A D Pollution of the sea shore waters. *Praktika*
1964 *'ell. 'udrobiol. Inst.*, 9(4):1–47.
IOC, Publications et références sur la pollution par les hydro-
1965 carbures à Monaco. *Rapp. Commn inter-gouv. Océanogr.*
JUILLAN, M *et al.* Etude bactériologique des eaux du port et de la
1962 baie d'Alger. *Archs Inst. Pasteur Alger*, 40(1):33.
KALOPISSIS, J *et al.* L'expérience de la lutte contre le Dacus à
1953 Kirra (Itea), Athènes.
KOCH, P Les modifications de la vie marine au débouché des
1968 émissaires d'eaux résiduaires urbaines. *Revue int. Océanogr.
méd.*, 10:11–26.
LACROIX, A *et al.* Evacuation des eaux usées des hôpitaux de
1952 l'agglomération algéroise. *Alger. méd.*, 581–5.
MAJORI L Research experiments on the pollution of the waters of
1969 the upper Adriatic. *In* Report and Proceedings. International
Conference on Oil Pollution of the Sea, 7–9 October 1968
at Rome, pp. 349–81.
MALLET L Pollution par les hydrocarbures en particulier du type
1965 benzo 4.4 pyrène des rivages méditerranéens français, et
plus spécialement de la baie de Villefranche. *In* Symposium
sur les pollutions marines par les microorganismes et les
produits pétroliers, 1964, Monaco, C.I.E.S.M.M., pp.
325–30.
MICALLEF, H and BANNISTER, W H On a dinoflagellate bloom
1969 (*Plectodinium nucleovalvatum Biech*) causing "Red Water"
in Pietà Creek (Malta). *Experientia*, 25:655.
MONTOUSIS, C and PAPAVASSILIOU, J The bacteriological examina-
1957 tion of sea water in swimming places near to Athens.
Delt. 'ell. mikrobiol. 'ug. 'Etair, 2:25.
OREN, O H Tar pollutes the Levant Basin. *Mar. Pollut. Bull.*,
1970 1(10):149–50.
ORPHANIDIS, P S and TSAKMAKIS, A A Recherches expérimentales
1957/58 sur la phytotoxicité du Malathion. *Annls Inst. Phytopath.
Benaki*, 1:300–5.

ORPHANIDIS, P S and SOULTANOPOULOS, C Observations pré-
1962 liminaires sur les courbes de densité de la population de
certains insectes vivant dans les oliveraies en 1961. *Annls
Inst. Phytopath. Benaki*, 4(2):148–54.
ORPHANIDIS, P S *et al.* Résidus de l'insecticide organophosphore
1968 Lebaycid dans l'huile et les olives. *Annls Inst. Phytopath.
Benaki*, 8(3):93–101.
ORPHANIDIS, P S *et al.* Les insecticides carboniques par comparaison
1969 aux insecticides organophosphorés dans les appâts
d'hydrolysats de protéines pour la lutte contre la mouche
d'olive. *Z. angew. Entomol.*, 63:389–405.
PANELLA, S Preliminary report on the pollution of Italian coastal
1969 waters in relation to biological environment and fisheries.
In Report and proceedings. International Conference on
Oil Pollution of the Sea, 7–9 October 1968 at Rome,
pp. 50–62.
RAMIREZ, J Sur le dosage par Spectrophotométrie des hydro-
1965 carbures de l'eau de mer et des organismes marins. D.E.S.,
Marseille Fac. Sci., Juin.
SITTIG, M Water pollution control and solid waste disposal.
1969 *Chem. Process Rev. Lond.*, (32).
TORRES, L Contribution à l'étude des lipides et biocides dissous
1969 dans l'eau de mer. Thèse, Fac. sc. Marseille A-O CNRS
3811.
TYSSET, C and BRISOU, J Considérations sur les flores bactériennes
1954 et commensales des fruits de mer consommés dans la région
algéroise. *Annls Inst. Pasteur, Lille*, 15:193.
ANON. A survey on the legislative position on refinery liquid
1968 effluents, and average refinery performances. London,
Caltex, Process Engineering.
ANON. Water—1969: chemical engineering progress. *Symp. ser.*
1968 *Am. inst. Chem. Engnrs*, 1968: 65–97.

La Pollution dans le Bassin Méditerranéen (Quelques Aspects en Méditerranée Nord-occidentale et en Haute Adriatique: Leurs Enseignements)

G. Bellan*
et J.-M. Pérès*

Pollution in the Mediterranean basin (some aspects in the north-western Mediterranean and the upper Adriatic: their lessons)

From the time of Marion (1883) the bibliography of pollution in the Mediterranean includes some hundreds of references. One of the latest works, that produced by the FAST (Federation of Scientific and Technical Associations of Milan) at the symposium on the subject of sea protection held in April 1969, endeavours to summarize the body of knowledge on the consequences of pollution in the Mediterranean. Most countries bordering on this sea are concerned about these problems.

Data are available on chemical, bacteriological and radioactive pollution and its effects on marine organisms and public health. There are plenty of bacteriological data. However, only the regions of Marseilles (France) and Trieste (Italy) seem to have been studied in a more general way. The conclusions reached by the authors are extremely close. Bacteriological and chemical data (such as on detergents or dissolved oxygen) tally. Homologous plant and animal populations are very similar even in detail. It would be easy to draw up a balance sheet of the "current status of pollution" in most of the Mediterranean, with the use of certain widely scattered perennial benthic populations. Some pollution factors such as detergents and coliforms can be measured by routine analysis, which would increase our knowledge.

Since most waste enters the sea untreated, proper sanitary engineering particularly adapted to the specific problems of the marine environment must be developed. The "sea-cleaning power" of sea water cannot be regarded as sufficient. The Mediterranean, which is a closed sea surrounded by industrial countries in which mass tourism is increasing, is rushing towards complete pollution. Fragmented efforts by one or more bordering countries will not improve the position and we must combine our energies and know-ledge, when necessary, under the wing of some international body.

La contaminación en el Mediterráneo (algunos aspectos que presentan el noroeste del Mediterráneo y el norte del Adriático: lo que nos enseñan)

Desde los días de Marion (1883), la bibliografía de la contaminación del Mediterráneo recoge varios centenares de referencias. En uno de los últimos trabajos, el de la FAST (Simposio sobre la "Difesa del Mare", Milán, abril de 1969) se tendió a resumir el conjunto de nuestros conocimientos sobre las consecuencias de la contaminación del Mediterráneo. Estos problemas inquietan a la mayor parte de los países ribereños.

Se tienen datos sobre la contaminación química, bacteriológica y radiactiva, y sus efectos sobre los organismos marinos y sobre la higiene pública. Abundan los datos bacteriológicos. No obstante, solamente las regiones de Marsella (Francia) y Trieste (Italia), parecen haber sido estudiadas de una manera más general. Las conclusiones a las que han llegado los autores son muy parecidas. Los datos bacteriológicos y químicos (como el oxígeno disuelto o los detergentes) son similares. Hasta en los detalles se parecen mucho las poblaciones de vegetales y animales homólogos. Utilizando algunas poblaciones bentónicas ampliamente repartidas, será fácil preparar un balance del "estado actual de la contaminación" en la mayor parte del Mediterráneo. Hay ciertos factores de la contaminación tales como los detergentes y los coliformes que pueden medirse efectuando análisis de rutina, y que incrementarán nuestros conocimientos.

La mayor parte de los desechos que llegan al mar no han sido tratados, por lo que hay que crear una verdadera ingeniería sanitaria adaptada a los problemas que plantea el medio marino. No basta creer sólo en el "poder autodepurador" del agua del mar. El Mediterráneo es un mar cerrado rodeado de países industrializados en el que aumenta el turismo de masa y que está en vías de una contaminación completa y acelerada. Un esfuerzo parcial de uno de los países ribereños no mejorará la situación por lo que hay que combinar las energías y conocimientos, de ser menester, dentro de una organización internacional.

EN 1876, Stossich vantait la variété et l'abondance de la faune et de la flore marines du "Vallone di Muggia", en haute Adriatique, à proximité des

Trieste, zone, qui, comme on le verra plus loin, est aujourd'hui très polluée. Quelques années plus tard, Marion (1883) publiait les résultats de ses recherches sur

* Station Marine d'Endoume et Centre d'Océanographie, Faculté des Sciences de Marseille, 13 Marseille, France.

peuplements marins de la région de Marseille, en Méditerranée nord-occidentale, mettant en évidence leur richesse naturelle; il signalait, par contre, que l'impureté des eaux des ports de Marseille provoquait un appauvrissement considérable de la faune. C'était le premier cri d'alarme lancé contre les méfaits de la pollution en mer.

Depuis, l'étude des pollutions dans le bassin méditerranéen a donné lieu à de multiples travaux. Le récent symposium "Difesa del Mare" qui s'est tenu à Milan (avril 1969) sous l'égide de la "Federazione delle Associazioni Scientifiche e Tecniche" a résumé l'ensemble de nos connaissances sur les causes et le degré de pollution d'un certain nombre d'aires de la Méditerranée. On peut dire que la plupart des pays riverains s'inquiètent de ces problèmes, l'Italie et la France étant, pour le moment, au premier rang de ceux-ci. On possède des données sur les pollutions chimiques, bactériologiques et radioactives, leurs effets sur les organismes marins et dans le domaine de l'hygiène publique. Ce sont, toutefois, les données bactériologiques (et leurs conséquences éventuelles dans le domaine sanitaire) qui sont les plus abondantes. On en possède, notamment, pour l'Espagne, la France, l'Italie, la Yougoslavie, le Liban, Israël.

Les régions de Marseille et de Trieste, ont été, jusqu'à présent, les seules étudiées sur un plan plus général. La présence de différents polluants a donné lieu à des recherches et leur action sur la flore et la faune marines a été mise en évidence. C'est en confrontant les résultats obtenus dans ces deux secteurs géographiques, que nous nous efforcerons de discerner les conséquences de la pollution en Méditerranée et d'en tirer quelques conclusions d'ordre pratique de portée plus générale.

La pollution dans la région de Trieste

La Haute Adriatique est soumise à un phénomène de marées (jusqu'à 1 m) qui provoquent, de par leur caractère régulier, un mouvement des eaux et la formation de courants de marées. Ces courants de marées peuvent être perturbés par l'action des vents qui sont dominants dans les secteurs: N-E d'une part et E-S-E d'autre part.

Les pollutions sont d'origine variée mais classique: domestiques, pluviales, industrielles: (raffineries et des chantiers navals, etc.). Majori (1968) a étudié de nombreux indices chimiques et bactériologiques de cette pollution; il s'est plus spécialement penché sur la présence et l'abondance d'*Escherichia coli*, sur la "Demande biologique en oxygène" (D.B.O.) et sur la mesure des détergents anioniques, classant les eaux, étudiées en un lieu donné et en un temps donné, en très pure, peu polluée, modérément polluée, polluée et très polluée, en fonction de différents niveaux numériques atteints par les trois indices essentiels étudiés.

Pollution par les bactéries

Selon Majori (1968), l'ensemble du "Vallone di Muggia", sur le flanc nord duquel s'étend la ville de Trieste et qui est bordé à l'est et au sud par la zone industrielle, peut être considéré comme pollué à très pollué. Le plus souvent, on note plus de 30 000 *E. coli* au litre. *Streptococcus fecalis* est plus rare, mais tout aussi constant; *Salmonella typhi* a été reconnue. Des concentrations de 1 000 à 5 000 *E. coli* au litre se retrouvent à plusieurs milles au large de Trieste.

La D.B.O. est considérable (supérieure à 6 ppm) dans le port de Trieste et ne descend guère en dessous de 4 ppm dans la baie de Muggia. L'oxygène dissous en baie de Muggia atteint en surface 60 à 70 pour cent de la saturation, mais tombe à 30–40 pour cent en profondeur.

Selon Majori (1968), le port de Trieste serait très pollué par les détergents anioniques (jusqu'à 2 ppm, généralement plus de 0,8 ppm). Dans la baie de Muggia, la pollution par les détergents serait moindre (0,1 à 0,8 ppm).

Les divers tests utilisés par Majori se superposent étroitement. Les Salmonelles sont récoltées dans les secteurs où les tests colimétriques indiquent un danger bactérien sérieux. La D.B.O. se superpose, à quelques détails près, aux résultats obtenus par les tests bactériologiques. Les cartes de répartition des détergents anioniques recouvrent exactement celles fournies par la bactériologie et le calcul de la D.B.O. Majori *et al.* (1968) écrivent à ce sujet: "Le détergent peut témoigner, soit d'un état de pollution dans lequel sont actuellement présents les autres éléments polluants, soit d'une condition de diminution de pollution lorsque les autres indices sont déjà affaiblis ou même totalement disparus."

Les conséquences de la pollution sur les peuplements de la baie de Muggia

Nous emprunterons à Ghirardelli *et al.* (1968) leurs résultats.

113 espèces végétales ont été récoltées. Le nombre d'espèces recueillies diminue au fur et à mesure que l'on s'éloigne de la mer ouverte pour pénétrer plus avant vers le fond de la baie et les zones industrielles polluées. On passe de 50 espèces par prélèvement à 4–6. Les Phaeophycées disparaissent rapidement puis les Rhodophycées. L'importance relative des Chlorophycées s'accroît en raison inverse.

Les auteurs ont récolté 40 espèces sur la centaine signalée dans l'ensemble de la baie de Trieste. Certaines des espèces "absentes" sont retrouvées à l'état subfossile sous le sédiment récent.

Ils distinguent trois secteurs dans la baie:

—Sur la côte sud, Ghirardelli *et al.* (1968) signalent, dans le fond de la baie, une zone à *Cladophora* et une à *Balanes* qui s'enrichit au fur et à mesure qu'on se rapproche de la mer ouverte où l'on note le peuplement classique de haute Adriatique à *Fucus*, *Mytilus*, etc.

—Sur la côte nord, une bande de 100 à 500 m de large où le fond est azoïque. On n'a qu'une vase noire, fétide, avec des débris divers. Les quais des ports sont très faiblement peuplés, les animaux caractéristiques des étages supra et médio-littoral ont disparu; le peuplement à *Mytilus* ne débute qu'à 10 cm en dessous du niveau moyen de la mer.

—Le troisième secteur, central, correspond à des fonds de 13 à 21 m de profondeur. Le peuplement, riche en *Corbula gibba*, *Apporhais pespelicani*, *Turitella communis*, *Tapes aureus*, *Upogebbia littoralis*, etc., se rapproche d'un aspect très appauvri de la biocénose "*Schizaster chiajei*" de Vatova, biocénose qui occupe en mer ouverte des fonds homologues. L'affinité avec cette biocénose *S. chiajei* n'est détectable que grâce aux mollusques, les Echinodermes (éléments les plus caractéristiques de la biocénose) *Schizaster canaliferus* et *Amphiura chiajei* ayant disparu.

La biomasse (poids formolé) est inférieure à ce qu'elle est dans l'ensemble du golfe de Trieste (52,40 g/m² au lieu de 150,06 g/m²) et à Rovinj (262,70 g/m²). L'épifaune est particulièrement appauvrie.

Ainsi donc la pollution provoque un appauvrissement de la baie de Muggia, aussi bien sur le plan qualitatif que quantitatif.

La pollution dans la région de Marseille

Dans la région de Marseille, la marée luni-solaire est faible et son effet nul. Les différences extrêmes du niveau de l'eau sont liées aux marées barométriques renforcées par les vents (Bellan *et al.*, 1967). Les mouvements des masses d'eau superficielles (celles qui nous intéressent) sont liés directement aux courants, lesquels sont, pour l'essentiel, sous l'influence des vents. Les vents dominants sont ceux de N-W (Mistral) et ceux du secteur Est.

Ce sont les mêmes qu'à Trieste, comme dans n'importe quelle grande agglomération avec ports et industries. Les nuisances se concentrent principalement dans deux zones: à l'est de l'agglomération, au débouché en mer du grand émissaire d'égouts de la ville (pollutions domestiques, pluviales, industrielles) et à l'ouest au voisinage de l'ensemble portuaire.

Son origine est domestique. Nous n'insisterons que sur les coliformes utilisés comme indice de pollution bactérienne. De très nombreuses bactéries pathogènes ont été signalées. Selon Bianchi *et al.* (1965), la zone polluée dans le secteur du débouché des égouts a, par vent d'est, une surface double de celle qu'elle présente par vent d'ouest. Le même phénomène se reproduit dans la baie de Marseille. Le long de la côte on dénombre généralement plus de 20 000 coliformes par litre et toujours plus de 5 000. Dans les ports et à proximité de l'émissaire, on note couramment plus de 1 000 000 col/l. A 2 milles au large de l'émissaire d'égouts, on trouve le plus souvent plus de 5 000 col/l. Les sédiments marins sont encore pollués par des germes intestinaux à plus de 100 m du rivage (Bianchi *et al.*, 1967).

Il n'a été étudié que dans les ports de commerce où Minas (1960) trouve une eau sous-saturée surtout en profondeur et dans le vieux port où la sous-saturation e.t encore plus nette et, où, sporadiquement par période de temps calme, il peut y avoir absence d'oxygène en profondeur (Leung Tack, comm. pers.).

Entre 50–100 m du débouché de l'émissaire d'égouts, on note jusqu'à 10 ppm de détergents et 0,5 ppm à 2 milles; 0,1 ppm à 3–4 milles. Les détergents sont pratiquement localisés en surface. Dans la baie de Marseille et les ports, on a des chiffres similaires (Arnoux, comm. pers.).

Torres (1969) a étudié différents pesticides présents dans la baie de Marseille. Au débouché d'un petit fleuve côtier, égout à ciel ouvert, il trouve (en γ/l d'eau de mer): DDT op: 21, DDT pp': 18, Lindane: 17, Aldrine: 16; à un mille à l'ouest les concentrations ont légèrement fléchi et à 3 milles au large, il signale 16 γ/l de DDT op et 12 γ/l de DDT pp'.

Dans le vieux port, on a relevé de 0,84 à 16,12 ppm d'hydrocarbures (Ramirez, 1965). On a trouvé 8,5 µg de

benzo (3, 4) pyrènes par 100 g de sédiment dans un sable fin de la baie de Marseille (Mallet, 1965).

Les données en notre possession, lorsqu'elles sont suffisamment complètes, se recoupent. C'est le cas pour les coliformes et les détergents anioniques. Ces derniers, dans le cas des émissaires urbains, peuvent parfaitement servir de tests de pollution à défaut d'être des traceurs de pollution fécale comme le voulait Paoletti (1966). Les données que l'on possède sur les teneurs en oxygène dissous et les hydrocarbures sont des indices précieux et concordants.

Les conséquences de la pollution sur les peuplements marins dans la région de Marseille

Bellan-Santini (1969) a signalé la disparition de certaines espèces dans les zones polluées. C'est, en particulier, le cas de *Cystoseira stricta* et de *Petroglossum nicaeense* qui, en eau pure, forment l'élément de base de communautés (ou faciès). D'autres espèces, telles *Ulva lactuca*, des entéromorphes, sont, au contraire, favorisées par la pollution.

On distingue 2 grands ensembles de peuplements: ceux établis sur substrat solide et ceux établis sur substrat meuble. Dans les deux cas, on note une zone de pollution maximale dépourvue de toute vie macroscopique, là où la pollution est particulièrement intense, et, ailleurs, une diminution générale du nombre d'espèces.

Sur substrat dur, Bellan-Santini (1969) note la succession suivante en fonction de la pollution croissante (pour des niveaux superficiels en mode agité): pelouse à *Cystoseira stricta;* zone de disparition de *C. stricta;* zone sans *C. stricta* (peuplement à *Mytilus gallo-provincialis* et à *Corallina cf. officinalis* dominantes); zone azoïque.

Sur substrat meuble, Bellan (1967) décrit une zone de peuplement normal (compte tenu de la granulométrie du sédiment, de la profondeur, etc.); une zone subnormale, stade de dégradation de la précédente dont le peuplement, très riche en polychètes vasicoles, est dominé par *Corbula gibba* et *Thyasira flexuosa;* une zone polluée à *Capitella capitata* et *Scolelepis fuliginosa* (accompagnée, ou non, de *Nereis caudata* et *Staurocephalus rudolphii*); enfin, une zone de pollution maximale azoïque.

L'accroissement apparent de la biomasse, sur substrat solide, est lié à l'augmentation du poids de calcaire organique et non de la matière organique sèche qui, elle, décroît avec la pollution.

Sur substrat meuble, les faits sont moins nets et varient d'un point à l'autre. Il pourrait y avoir une légère augmentation de la matière organique (exprimée en poids sec). On a: zone de pollution maximale: 0 g/m²; zone polluée: 1 à 5,5 g/m²; zone subnormale: 6,5 à (exceptionnellement) 61 g/m²; zone normale 3 à 10 g/m².

Discussion

Il n'est point besoin d'une longue discussion pour montrer combien sont similaires, à Trieste et à Marseille, les causes, les niveaux et les conséquences de la pollution.

Les causes (domestiques, industrielles, portuaires) sont les mêmes.

Les colimétries, les teneurs en oxygène dissous et en détergents apparaissent respectivement bien proches les unes des autres, dans des milieux comparables et à des distances analogues des sources essentielles de pollution. On retrouve les mêmes germes pathogènes. Mais plus spectaculaires sont peut-être les analogies, voire les identités, frappantes dans les peuplements marins benthiques. On note, par exemple, une régression qualitative et quantitative des peuplements.

Pour s'en tenir aux peuplements établis sur substrat meuble, la parenté des peuplements de la zone subnormale dominée par *Corbula gibba* est évidente. Il est aussi curieux de constater que *C. gibba* a son maximum d'abondance, à Trieste, dans une zone draguée. Bellan (1967) à Marseille, la cite d'une zone récemment draguée du Vieux Port de Marseille, comme la première (et seule) espèce établie. Si, à Trieste, la zone polluée à *C. capitata* et *S. fuliginosa* n'a pas été reconnue, Bellan (1967a) note qu'une telle zone peut être absente de la succession normale des peuplements, en particulier dans des enceintes portuaires protégées par des digues (c'est le cas à Trieste où la baie de Muggia ressemble fort à la partie nord des ports de Marseille). Les similitudes sont donc remarquables, jusques et y compris dans les détails.

Les données que nous avons rassemblées pour Marseille et Trieste ne sont pas limitées à ces deux ports. Les études déjà publiées, les renseignements que nous possédons de sources diverses, les observations que nous avons pu faire personnellement, montrent que les mêmes causes et les mêmes effets se retrouvent au voisinage de n'importe quelle grande agglomération bordant la Méditerranée.

Il apparaît donc facile de dresser un inventaire, un bilan de l'état actuel de la pollution dans la plus grande partie de la Méditerranée, en utilisant certains peuplements benthiques judicieusement sélectionnés. Ces peuplements pérennants, soustraits aux variations continuelles des eaux qui les baignent, intègrent l'ensemble des facteurs du milieu, reflétant ainsi les conditions non seulement au moment de la récolte, mais celles qui prévalaient auparavant. L'assemblage des espèces indique le degré de pollution. Ces mêmes assemblages peuvent coïncider avec des taux déterminés de polluants (cf. Bellan-Santini, qui a pu superposer une carte de peuplements benthiques et une carte de distribution de coliformes). De l'étude de ces peuplements on peut encore prévoir l'évolution de la pollution dans un secteur déterminé et proposer des moyens de lutte pour empêcher ou réduire l'extension de la pollution (Bellan, 1967a).

Toutefois ces peuplements ne sont pas spécifiques de tel ou tel type de pollution. Ils intègrent trop bien l'ensemble des facteurs du milieu et il faudra toujours, lorsqu'on voudra reconnaître la présence de tel polluant, effectuer des recherches spécifiques (à moins que la connaissance des peuplements planctoniques, moins stables que les peuplements benthiques, mais dont l'étude est à peine ébauchée, nous donne une réponse satisfaisante). L'expérience nous a montré que les recherches particulières sur la présence d'un certain polluant devront être faites régulièrement après étude préliminaire de différents facteurs du milieu et ne sauraient se limiter à quelques prélèvements sporadiques. Une fois ces principes définis et les modalités d'exécution clairement établies, ces recherches pourront être réalisées de façon routinière.

Conclusion

De l'ensemble de ces recherches pourrait découler un véritable "engineering" bien adapté aux problèmes propres au milieu marin, particulièrement en Méditerranée. On ne saurait se contenter du seul "pouvoir antibiotique" (notion encore controversée) de l'eau de mer, ni de solutions de facilité (tel le broyage des plus gros éléments de pollution dans un but purement "visuel"), encore moins de considérer la mer, si vaste et si profonde soit-elle, comme un réceptacle idéal des déchets de toutes sortes.

La Méditerranée, mer fermée, entourée de pays fortement industrialisés ou en voie d'industrialisation, vouée à un tourisme de masse en voie de développement explosif, nous paraît en cours de pollution accélérée. Un effort parcellaire d'un ou de quelques pays n'améliorera pas la situation et nous devons regrouper nos énergies et nos connaissances, au besoin dans le cadre de quelque organisme international, étant entendu que les pays les plus riches et qui par conséquent sont à l'origine des pollutions les plus intenses se doivent de faire un effort à la mesure de leurs possibilités et des nuisances qu'ils engendrent.

Bibliographie

BELLAN, G Pollution et peuplements benthiques sur substrat
1967 meuble dans la région de Marseille. Première partie. Le secteur de Cortiou. *Revue int. Océanogr. méd.*, 6–7:53–87.
BELLAN, G Pollution et peuplements benthiques sur substrat meuble
1967a dans la région de Marseille. Deuxième partie. L'ensemble portuaire marseillais. *Revue int. Océanogr. méd.*, 8:51–95.
BELLAN, G *et al.* Etat général des pollutions sur les côtes méditer-
1967 ranéennes de France. Document présenté au Congrès international "Acqua per il Domani" IV inchiesta internazionale "La Difesa del Mare", Milan.
BELLAN-SANTINI, D Contribution à l'étude des peuplements
1969 infralittoraux sur substrat rocheux (Etude qualitative et quantitative de la frange supérieure). *Recl. Trav. Stn mar. Endoume*, (63–47):5–294.
BIANCHI, A *et al.* Etude de la pollution du golfe de Marseille. 1.
1965 Note préliminaire; étude de la diffusion des eaux polluées en fonction de la répartition de certains germes intestinaux. *In* C.I.E.S.M.M. Symposium sur les pollutions marines par les microorganismes et les produits pétroliers, Monaco, avril 1964, pp. 59–66.
BIANCHI, A *et al.* Etude de la pollution du golfe de Marseille. La
1967 pollution des sables. *Rapp. P. V. Réun. Commn int. Explor. Mer Méditerr.*, 18(3):599–602.
GHIRARDELLI, E *et al.* Conséquences de la pollution sur les
1968 peuplements du "Vallone de Muggia" près de Trieste. *Revue int. Océanogr. méd.*, 10:111.
MAJORI, L Research experiments on the pollution of the waters of
1968 the upper Adriatic. Paper presented at the International Conference on oil pollution of the sea, Rome, 7–9 October 1968.
MAJORI, L *et al.* Recherches sur la pollution des eaux de mer dans
1968 le golfe de Trieste. *Rev. int. Océanogr. Méd.*, 9:83–95.
MALLET, L Pollution par les hydrocarbures en particulier du type
1965 benzo 3,4 pyrène des rivages méditerranéens français, et plus spécialement de la baie de Villefranche. *In* C.I.E.S.M.M. Symposium sur les pollutions marines par les microorganismes et les produits pétroliers, Monaco, avril 1964, 325–30.
PAOLETTI, A Les détergents, nouvel indice chimique de la pollution
1966 fécale des eaux de surface. *Revue int. Océanogr. méd.*, 3:5–10.
RAMIREZ, J Sur le dosage par spectrophotométrie des hydro-
1965 carbures de l'eau de mer et des organismes marins. D.E.S., Fac. Sci. Marseille, juin 1965.
STOSSICH, A Breve sunto sulle produzioni marine del golfo di
1876 Trieste. *Boll. Soc. adriat. Sci. nat.*, 3:349–71.
TORRES, J P Contribution à l'étude des lipides et biocides dissous
1969 dans l'eau de mer. Thèse, Fac. Sci. Marseille 19/12/69 et A.O. C.N.R.S. 3811.

Etat Actuel de la Pollution Bactérienne au Large des Côtes Françaises

M. Aubert,* J. Aubert,*
S. Daniel* et N. Desirotte*

Current bacterial pollution of French offshore waters

The authors present a synthesis of the results of this inventory of fecal pollution germs in all French coastal waters. These bacteriological assays were performed on the Atlantic and English Channel coasts at rising and falling tide and in the Mediterranean, under various weather conditions. The sampling sites were chosen an average of 3 000 m apart and were used for one sampling near the shore and another further offshore perpendicular to the −5 bathymetric line. Analysis methods included the investigation and enumeration of coliform bacilli, *Escherichia coli* and fecal streptococci, as well as a total germ count.

The results of these bacteriological assays are presented as variables—first for each site and second for the length of the shoreline—under various hydrological and meteorological conditions. At the same time, a conspectus is given of the dynamics of coastal waters which cause the dissemination of pollutants and which enable the breakdown of such telluric pollution to be explained in terms of contamination sources (waste water outlets, estuaries, port complexes, etc).

In the Mediterranean a study has also been made of the distribution of water masses by the investigation of surface isotherms and isohalines. On the basis of these various data, it is possible to determine general laws relating to the diffusion of bacterial pollution, and construct a mathematical model allowing forward studies to be performed on the control of this type of pollution.

Estado actual de la contaminación bacteriana a lo largo de las costas francesas

Los autores presentan una síntesis de los resultados del inventario de los gérmenes de contaminación fecal, que han realizado en las aguas costeras francesas próximas a las desembocaduras de los ríos. Se han realizado mediciones bacteriológicas en las costas atlántica y de la Mancha, con marea ascendente y descendente, y en la costa mediterránea, en condiciones meteorológicas diferentes. Las estaciones para la obtención de muestras se eligieron a la distancia de unos 3 000 m por término medio y se tomaron dos muestras: una cerca de la orilla y otra mar afuera, a lo largo de la línea batimétrica − 5. Como métodos de análisis se han empleado la investigación y recuento de los coliformes, los *Escherichia coli* y los streptococci fecales y el recuento total de gérmenes.

Los resultados de estas mediciones bacteriológicas se presentan, primero, estación por estación, y luego siguiendo la línea costera, en diversas condiciones hidrológicas y meteorológicas. Al mismo tiempo se presenta un panorama de la dinámica de las aguas costeras, de la que depende la difusión de los agentes contaminantes, y que permite explicar la distribución de las contaminaciones telúricas en función de las fuentes de contaminación (desagües de alcantarillas, estuarios, zonas portuarias, etc).

En el Mediterráneo se ha estudiado, además, la distribución de las masas hidrológicas, determinando las curvas isotermas e isohalinas de la superficie. A partir de todos esos datos, es posible obtener leyes generales sobre la difusión de la contaminación bacteriana y un modelo matemático que permite realizar estudios preventivos en la lucha contra ese tipo de contaminación.

AU cours de l'année précédente, nous avons achevé la réalisation des mesures bactériologiques, hydrologiques et courantologiques poursuivies par notre laboratoire sur l'ensemble des zones marines bordant le littoral français.

Cette vaste enquête a permis d'apporter, ainsi, une information directe sur l'état sanitaire des rivages mais, de plus, elle nous a fourni un ample matériel permettant une analyse des phénomènes de dispersion bactérienne et des mécanismes de contamination du milieu marin. Dans ce but, elle a été poursuivie non seulement par la mesure des taux bactériens d'échantillons prélevés selon un rythme variable de 2 à 5 km de développement linéaire de côte en fonction des vents d'afflux et des vents d'amont ou selon les conditions diverses de marées, mais également elle a été doublée d'une étude simultanée de l'évolution des courants côtiers et de la répartition des masses hydrologiques superficielles qui véhiculent les micro-organismes telluriques.

Méthodes d'étude

Cet ensemble de données a été recueilli de la manière suivante:

Les prélèvements bactériens ont été faits, d'une part à la laisse de basse mer ou de haute mer et d'autre part, en surface à quelques dizaines ou quelques centaines de mètres au large, selon la profondeur. Ils ont été exécutés au moyen de flacons stériles placés immédiatement dans la glace en containers étanches et ramenés au laboratoire par voiture ou transport aérien. L'analyse a porté sur la numération totale des germes et sur les germes-tests de contamination fécale: coliformes, *Escherichia coli*, streptocoques fécaux, selon les techniques classiques définies par Buttiaux (1958).

L'étude des courants et de la diffusion des apports terrigènes a été faite par courantomètres à flotteurs, ainsi que par rejets d'un grand nombre de cartes-flotteurs dont l'évolution est suivie en mer sous diverses conditions météorologiques par nos différents bateaux océanographiques à bord desquels ont été notées les positions successives par relèvement sur des amers. Les points d'atterrissage des cartes flotteurs situent les zones de pollution issues des rejets polluants existants ou envisagés dans l'urbanisation future des zones côtières. Ces mesures courantométriques recueillies par nos soins ont été complétées pour certaines zones par des mesures fournies par des documents du Service Hydrographique de la Marine.

L'étude de la répartition et de l'évolution des masses d'eau véhiculant les apports des fleuves ou des émissaires urbains a été effectuée principalement en Méditerranée par la détermination des courbes isothermes et isohalines dans les eaux de surface pratiquement seules intéressées par la diffusion bactérienne au large. Chaque zone de travail a été l'objet d'un quadrillage serré de mailles de 1 mille de côté, jusqu'à 5 milles au large, dont chaque point est étudié du point de vue salinité et température.

Il est à noter que les méthodes que nous avons utilisées pour nos études hydrologiques et courantométriques ne visent pas une recherche fondamentale dans ce domaine, mais ont simplement pour but de détecter les influences générales ou locales apportées par les rejets en mer d'eaux résiduaires actuels ou envisagés pour l'avenir.

On obtient ainsi par cette méthodologie une image assez fidèle de la diffusion bactérienne autour du territoire national et des phénomènes hydrodynamiques qui la conditionnent, cet état étant valable pour la période actuelle de réalisation. L'absence de tout document d'ensemble antérieur ne permet pas pour le moment de connaître son évolution.

La publication de l'ensemble de ces résultats est faite

* Centre d'Etudes et de Recherches de Biologie et d'Océanographie Médicale, 06 Nice, France.

sous la forme de quatre tomes: le tome I comprend l'exposé des méthodes d'étude employées et des considérations d'ordre théorique sur la diffusion de ces pollutions et les conséquences qu'elles impliquent. Le tome II donne un tableau aussi complet que possible de l'état bactériologique des zones marines bordant les côtes de la Manche et des courants qui assurent la dispersion de ces agents microbiens. Les tomes III et IV traitent des mêmes sujets en ce qui concerne les côtes atlantiques et les côtes méditerranéennes.

Il est bien évident que les chiffres que nous citons donnent une image de la pollution bactérienne valable pour la présente période pendant laquelle cet inventaire a été réalisé. Au cours des années à venir, une évolution doit apparaître qui peut modifier sensiblement ces taux en micro-organismes selon les variations démographiques constantes ou saisonnières liées à l'évolution économique des régions. Un inventaire est valable à un instant donné; il doit être suivi et rajeuni; c'est à partir de cette évolution que pourront être mises en évidence les lois encore cachées qui président à la transformation biologique du milieu en fonction de l'activité humaine.

Nous avons présenté l'ensemble de nos données d'ordre biologique ou physique sur un fond de cartes marines classiques; car, par cet artifice apparaissent d'une manière plus objective les liaisons existant entre les phénomènes étudiés et la morphologie exacte des rivages et des fonds marins.

L'ensemble de ces résultats n'a pu être acquis que grâce à des moyens logistiques dont il ne faut pas se dissimuler l'importance: équipes effectuant les prélèvements le long des rivages; équipes à bord de petits navires, chargées des observations en mer; équipes en laboratoire réalisant les mesures bactériologiques ou chimiques sur les prélèvements; équipes de cartographie; enfin, équipes de chercheurs et d'ingénieurs qui se sont efforcés de réaliser la synthèse de cette dynamique des pollutions des rivages et des zones marines avoisinantes et de déterminer les rythmes de reproductibilité des résultats permettant ainsi de dégager les lois sur les phénomènes de dispersion bactérienne en mer.

La réalisation de cet Inventaire National a porté sur l'ensemble du littoral des côtes françaises et a comporté l'étude de 2.064 points de prélèvements. Leur répartition se fait de la manière suivante:

—En mer du Nord et en Manche: 267 stations ont été étudiées suivant deux conditions de marée différentes, soit 534 points.

—En océan Atlantique: 171 stations ont été étudiées également pour deux conditions de marée, soit 342 points.

—Enfin en Méditerranée: 297 stations ont fait l'objet d'études pour deux conditions de vents différentes et respectivement pour le rivage et pour le large, soit 1.188 points de prélèvements.

Résultats

Les résultats de nos études portent essentiellement sur les données bactériologiques, mettant ainsi en évidence le bilan général des zones marines côtières explorées. Nous les résumerons rapidement:

On constate une pollution avoisinant généralement 1.000 coliformes/100 ml, mais dont les points de pollution maximale se situent au niveau des grandes villes: Dunkerque, Gravelines, Calais où le nombre total des germes est supérieur à 150.000 germes/100 ml et surtout Boulogne où le taux des coliformes atteint 3.000.000/100 ml. A l'approche de l'estuaire de la Seine, à Fécamp, la pollution s'accroît d'une manière très sensible pour atteindre au Havre un taux très important (10.000.000 de coliformes/100 ml).

Pour l'étude des côtes du Calvados, réalisée en collaboration avec le Service du Pr Jacquet et pour laquelle les prélèvements ont été effectués au cours de deux saisons différentes (la première période d'afflux touristique et de climat estival, et la deuxième période après la saison estivale et par mer agitée), nous avons dégagé les données suivantes:

En ce qui concerne la saison d'afflux touristique, les pollutions bien que très importantes si on les compare à celles trouvées le long des côtes méditerranéennes, se sont avérées moins élevées que lors de la deuxième saison.

Dans l'ensemble les pollutions sont plus marquées à marée montante qu'à marée descendante: dans certaines communes le taux de coliformes atteint 850.000/100 ml (Langrune-sur-Mer); entre l'Orne et la Seine, on trouve des taux dépassant 500.000 coliformes/100 ml dans les zones proches de débouchés d'émissaires.

En ce qui concerne la deuxième saison, les pollutions sont beaucoup plus importantes entre la Vire et l'Orne, et au contraire, relativement moins sensibles entre l'Orne et la Seine: ainsi à marée montante, tout le littoral compris entre Arromanches-les-Bains (coliformes 742.000/100 ml) et Luc-sur-Mer (coliformes: 310.000/100 ml) est largement contaminé; à marée descendante, la région comprise entre Grandcamp-les-Bains et Saint-Come-de-Fresne est la plus polluée.

En conséquence, on peut penser qu'en ce qui concerne les résultats des côtes normandes, l'apport continu des bactéries terrigènes par les grands fleuves côtiers de cette région ne justifie pas, à lui seul, les importantes pollutions rencontrées le long de ce littoral. Celles-ci sont dues à la présence constante d'émissaires plus ou moins épurés et de longueur nettement insuffisante tout au long des plages de Normandie.

Les côtes du Cotentin, notamment sur la face est, sont moins polluées; néanmoins, il importe de souligner que la région de Cherbourg est le siège d'une très forte pollution, approchant 5.000.000 de coliformes/100 ml par marée descendante.

La côte ouest du Cotentin, de Granville au Mont Saint-Michel est nettement plus contaminée, notamment la baie du Mont Saint-Michel (de 20.000 à 60.000 coliformes/100 ml). Ce fait semble dû aux effets du flux et du reflux particulièrement manifestes dans cette zone.

Les rivages des côtes du nord de la Bretagne présentent une pollution moyenne où le taux de coliformes est rarement supérieur à 10.000/100 ml.

Les côtes de l'océan Atlantique

Par leur topographie et leur morphologie, elles conditionnent en quelque sorte la pollution des rivages et pourront être différenciées en deux parties:

—d'une part, le littoral qui s'étend de Brest à St-Nazaire.

Celui-ci, très découpé, présente des endentures nombreuses, délimitant des baies plus ou moins fermées, dont les eaux non renouvelées sont parfois le siège de pollution importante. Ainsi dans la rade de Brest, le taux de coliformes se situe entre 10.000 et 30.000 et le taux de streptocoques fécaux atteint 3.000/100 ml, ou encore à Douarnenez et Audierne où le taux de coliformes est voisin de 10.000/100 ml; par contre, la baie de Bénodet, plus ouverte sur le large, est peu contaminée (moins de 1.000 coliformes/100 ml alors que des taux plus élevés se rencontrent près de Concarneau (50.000 coliformes/ 100 ml).

Si ensuite, l'ensemble des rivages est moins pollué même aux abords du port de Lorient, on retrouve de nouveau dans la baie de Quiberon (par ex. à St-Pierre) plus de 20.000 coliformes/100 ml, et dans le golfe du Morbihan 800.000 germes aérobies/100 ml.

Puis, sur un fond de pollution presque constant oscillant entre 100 à 1.000 coliformes/100 ml, on constate des pics plus marqués, soit au niveau des agglomérations à caractère estival comme le Croisic et la Baule, soit surtout à St-Nazaire et à l'embouchure de la Loire où l'on note de 10.000 à 30.000 coliformes/100 ml.

—d'autre part, le littoral d'aspect linéaire qui s'étend de St-Nazaire à la frontière espagnole.

Celui-ci présente une contamination faible: ainsi de St-Nazaire aux Sables d'Olonne, les eaux sont relativement propres (faible taux de coliformes, absence fréquente des autres germes-tests), par contre la présence de l'agglomération des Sables d'Olonne se signale par un taux de coliformes voisin de 10.000/100 ml, puis de nouveau la côte est propre (moins de 1.000 coliformes/ 100 ml) jusqu'à l'embouchure de la Gironde où même la présence d'un port comme La Rochelle ou de stations balnéaires importantes comme Royan, modifient peu les taux bactériens (entre 1.000 et 5.000 coliformes/100 ml).

La côte des Landes, déserte, présente très peu de contamination; par contre, le bassin d'Arcachon est assez pollué, le maximum apparaissant au voisinage d'Andernos et d'Arcachon (de 1.000 à 10.000 coliformes/ 100 ml). Au sud du bassin et jusqu'à Biscarosse, la côte est de nouveau propre, pauvre en germes coliformes, puis on retrouve quelques contaminations à Mimizan, à la plage de Contis et au Vieux Boucau. Du Boucau à la frontière espagnole, les plages de Biarritz, Guétary, St-Jean-de-Luz accusent encore en saison estivale 1.000 coliformes/100 ml.

En conclusion, le littoral atlantique, comparativement aux côtes de la Manche, est dans l'ensemble peu pollué, que ce soit à marée montante ou à marée descendante où l'on retrouve en moyenne des taux bactériens du même ordre de grandeur. Seuls, les points situés au voisinage immédiat des agglomérations sises en général dans les baies ou encore à l'embouchure des fleuves côtiers, montrent une pollution élevée supérieure à 10.000 coliformes/100 ml.

Les côtes de la Méditerranée

L'étude des zones marines situées au large des rivages du Languedoc et du Roussillon montre que les phénomènes hydrologiques et les mouvements des masses d'eau sont en corrélation intime avec les vents dominants. Les résultats des analyses bactériologiques montrent, dans l'ensemble, un faible taux de germes-tests de contamination fécale. Même la présence d'une grande ville comme Sète modifie peu le taux de contamination de cette zone.

Cependant, les pollutions les plus importantes corres-pondent, en général, aux prélèvements effectués aux embouchures des fleuves ou rivières le long desquels sont situées les grandes villes de l'intérieur; aux embouchures du Tet, du Tech, de l'Aude, on trouve 95.000 coliformes/ 100 ml; aux embouchures de l'Orb, de l'Hérault, on trouve 25.000 coliformes/100 ml; les canaux portuaires de Port-La-Nouvelle, de Palavas, de Carnon et du Grau-du-Roi, sont également le siège d'une pollution très élevée (800.000 coliformes/100 ml au Grau-du-Roi).

Il faut également signaler la partie littorale située au sud-ouest du Cap d'Agde, entre l'Hérault et l'Aude—et particulièrement entre ce dernier fleuve et l'Orb—dont la côte se trouve elle-même polluée sur quelques kilomètres. Ce fait est dû aux courants d'afflux qui viennent buter sur le Cap d'Agde, créant ainsi une zone tourbillonnaire par vent d'ouest et rabattant vers le rivage les eaux de l'Hérault, l'Aude et l'Orb.

Le long des rivages des Bouches-du-Rhône, les pollutions et leurs extensions sont grandement influencées par les deux conditions de vents principales: secteur est et secteur ouest.

La zone de rejet du grand collecteur de Marseille, entre l'île Riou et la côte, la zone située au niveau de l'embouchure de l'Huveaune, sont particulièrement polluées (taux de coliformes de 100.000 à 900.000/100 ml). Mais, de part et d'autre de cette vaste agglomération, les pollutions des rivages sont faibles: dans la baie de la Ciotat, les taux de coliformes sont inférieurs à 100/ 100 ml avec absence d'*Escherichia coli* et de strepto-coques fécaux; la région de Cassis et des premières Calanques est également peu polluée. Par contre, à l'approche de l'agglomération marseillaise, la pollution apparaît de plus en plus importante: Morgiou et surtout Sormiou (coliformes: 1.200/100 ml; *Escherichia coli*: 20/100 ml; streptocoques fécaux: 90/100 ml; numération totale des germes: 6.000/100 ml).

La côte bordant le massif de l'Estaque est au contraire très peu contaminée, car les courants superficiels diffusent largement vers le sud sous l'influence des vents de secteur nord-ouest: en effet, à l'Estaque, Port Niolon, Carry et Carro, les taux de pollution sont très faibles (100 coliformes/100 ml, peu ou pas d'*Escherichia coli* ni de streptocoques fécaux). Dès que l'on atteint la région située à l'ouest du Cap Couronne, on retrouve un certain degré de pollution provenant des agglomérations de Port de Bouc et de celles situées aux abords du canal de Caronte et de l'étang de Berre.

Le littoral varois, est, en général, moins pollué—exception faite pour certaines baies soumises à l'accroissement démographique estival (comme la baie de Cavalaire, la baie du Lavandou, la grande rade de Toulon au nord et à l'ouest (1.000 coliformes et 100 streptocoques fécaux/100 ml).

Sur le littoral des Alpes-Maritimes, on remarque la présence de germes polluants d'une manière continue au taux de 10 germes/ml, taux qui s'élève considérablement à proximité des grandes agglomérations pouvant atteindre plus de 100.000 germes/ml. Parmi les zones les plus polluées, il est à noter la région située au droit de la Principauté de Monaco, la partie médiane de la Baie des Anges, la zone ouest de la rade de Golfe-Juan et la partie médiane de la rade de Cannes.

En ce qui concerne les eaux littorales de la côte corse, la pollution bactérienne est, en général, très faible sauf à la sortie immédiate des émissaires de certaines villes littorales situées dans le fond des baies, comme le montrent des prélèvements d'eau de mer au nord d'Ile Rousse (1.000.000 de coliformes/100 ml, 40.000 *E. coli*/100 ml,) ou encore ceux de la baie d'Ajaccio (50.000 coliformes/100 ml), de Bonifacio et Porto-Vecchio où l'on trouve encore 5.000 coliformes/100 ml. La dispersion de ces pollutions est très rapide, favorisée par la prédominance des vents forts qui balayent ces côtes. Le plus constant de ces vents est celui du secteur sud-ouest qui accélère le courant général de direction sud-nord, venant buter sur les caps avancés de la face ouest, formant ainsi des zones tourbillonnaires le long du golfe de Valinco, de la baie d'Ajaccio et de la rade de Calvi.

Sur la côte est, le long des rivages de cette côte basse, il existe des courants de direction sud-est dispersant rapidement vers le large les pollutions bactériennes issues de quelques bourgades côtières. Bien que gênés par vents de secteur est, ils restent encore nets et facilitent la dispersion des eaux résiduaires.

En résumé, l'état sanitaire de la côte corse est actuellement très satisfaisant, tant au niveau des plages qu'au niveau des eaux côtières, cependant, quelques points situés dans le fond des baies relativement peuplées traduisent la distance actuellement beaucoup trop proche des rivages des rejets d'eaux résiduaires.

D'une manière générale, l'ensemble de ces analyses montre que :

— les eaux marines bordant les grandes villes présentent une contamination entérique très importante dont l'extension est évaluée dans les cas maximaux à environ 2 à 3 milles marins. De même, les zones situées dans la proximité des émissaires courts desservant des petites agglomérations à population saisonnière et les localités situées à l'embouchure des fleuves côtiers, sont le siège d'une pollution élevée ;

— la zone de diffusion bactérienne se situe dans le sens du courant, lui-même soumis à la direction des vents et à la répartition des masses d'eau déterminée par l'étude du complexe température-salinité ;

— en général, la comparaison des pollutions de la zone marine et du rivage lui-même correspondant, montre une contamination souvent plus élevée au niveau de ce dernier.

Cette recherche présente un intérêt particulier pour la Santé Publique, étant donné l'influence de ces pollutions sur la morbidité et les conséquences qu'elles peuvent avoir sur les implantations littorales urbaines, industrielles ou encore touristiques. D'autre part, elles sont susceptibles de favoriser une infection microbienne des animaux marins ou des parcs coquillers se trouvant à proximité des zones de déversement.

Application aux lois de la diffusion bactérienne en mer

L'étude systématique des taux bactériens à des distances plus importantes des rejets, la confrontation de ces résultats avec l'étude conjointe de traceurs d'ordre physique, soit naturels, soit artificiels, nous ont conduits à essayer de dégager les lois mathématiques qui président à la diffusion en mer des micro-organismes telluriques et qui permettent de prévoir, avec une approximation suffisante, l'extension des zones de contamination qui proviennent d'un rejet existant ou, ce qui est plus utile,

qui découleront d'un futur rejet d'émissaire prévu dans une région côtière et urbanisée.

Les éléments initiaux de ce calcul prévisionnel sont tirés des expériences de diffusion en mer réalisées par le C.E.R.B.O.M. conjointement avec le C.E.A., en étudiant comparativement la dispersion de traceurs radioactifs et traceurs biologiques, expériences rapportées en détail par l'un de nous (Aubert, 1966).

Les équations de départ sont les suivantes :

$$\text{Log } \frac{B_o}{B_t} = 1,8 \text{ Log } \frac{A_o}{A_t} \quad \text{et} \quad \text{Log } \frac{B_o}{B_t} \Big/ \frac{A_o}{A_t} = \frac{t}{70}$$

avec : A_t : concentration en un point quelconque de l'espace du traceur physique au temps t,

B_t : concentration au même point du traceur biologique au temps t,

A_o et B_o : les concentrations de départ,

t : temps de contact (exprimé en minutes).

Des deux équations précédentes, nous tirons une troisième qui nous donne directement la variation de la concentration en bactéries en fonction du temps ou encore en fonction de la distance de l'atterrissage sur le rivage et de la vitesse de courant de transfert :

$$\text{Log } \frac{B_o}{B_t} = \frac{t}{70 \times 0,444}.$$

Nos études de 1967 et des années précédentes avaient abouti à la mise en évidence de courbes définissant la dispersion bactérienne, calculée à partir d'un rejet ponctuel, c'est-à-dire en "régime transitoire". Au cours de l'année 1968, nous avons étudié cette diffusion en "régime continu" qui correspond aux cas concrets des rejets en mer, nous permettant ainsi d'établir une loi de décroissance bactérienne, globalement obtenue à partir des effets cumulés de la dilution mécanique et de l'action bactéricide de l'eau de mer.

C'est pourquoi, après de nombreuses expérimentations réalisées en "régime continu" sur un émissaire type, nous avons abouti à une courbe expérimentale donnée par l'équation suivante :

$$\log_{10}C = (\log_{10}C_o + \alpha)\frac{\beta \gamma}{1,1}t$$

dont les paramètres sont ainsi définis :

C_o : concentration en bactéries des eaux résiduaires avant leur émission sur le fond marin,

C : concentration en bactéries au temps t,

t : temps de contact (exprimé en heures),

α : coefficient lié à la profondeur du rejet,

β : coefficient de débit,

γ : coefficient d'action bactéricide de l'eau de mer.

Il est à remarquer que les prélèvements d'eau sur lesquels furent réalisés les comptages bactériens ont été effectués sur l'axe principal de dérive, par conséquent les concentrations que nous donne l'équation ci-dessus sont des concentrations maximales puisque ce sont celles qui ont été enregistrées le long de l'axe préférentiel du courant.

Les coefficients intervenant dans cette formule demandent à être précisés :

α : ce coefficient de profondeur tient compte de l'influence de la hauteur du cône de rejet qui a pour effet de faire varier la concentration initiale, c'est pourquoi il intervient dans l'ordonnée à l'origine de la courbe établie pour un émissaire-type (hauteur = −10 m) et pour lequel nous posons le coefficient: $\alpha = -1$. Dans l'état actuel de nos connaissances, on peut admettre que le coefficient α varie comme suit: $\alpha = -\log_{10}h$.

h: hauteur du cône de rejet (exprimée en mètres).

β: ce coefficient tient compte de l'influence du débit en modifiant la pente de la courbe obtenue pour un émissaire-type (Q = 1 m³/sec.) et pour lequel nous posons le coefficient: $\beta = 1$.

La variation de ce facteur se fera selon une loi de décroissance angulaire de raison 3/4 pour une variation de débit de 1/2.

γ: ce coefficient tient compte de l'intensité de l'action bactéricide de l'eau de mer qui agit sur la vitesse de disparition des bactéries; c'est donc un coefficient de même type que le précédent, c'est pourquoi il intervient également dans la pente de la courbe établie pour un émissaire-type en Méditerranée pour laquelle nous posons: $\gamma = 1$.

Ce coefficient γ a été mesuré comparativement en Méditerranée, en Atlantique et en Manche, pour des périodes différentes de l'année. Il existe des variations quelquefois importantes de sa valeur. Il n'est donc pas sans intérêt de le faire figurer parmi les variables qui conditionnent cette diffusion bactérienne.

L'ensemble de cette vaste enquête bactériologique que nous avons présentée assez sommairement permet de se rendre compte de l'état sanitaire des eaux littorales françaises, mettant en évidence des zones d'importantes pollutions localisées en général aux abords des grandes villes et aux embouchures de fleuves. Ces pollutions peuvent présenter un danger pour la santé publique, en entraînant de graves infections tant au niveau des organismes marins que des parcs coquilliers.

En outre, l'étude systématique de ces taux bactériens a permis de dégager une loi mathématique prévisionnelle des phénomènes de diffusion des eaux résiduaires dans le milieu marin; son utilisation permettra de résoudre les problèmes d'aménagement des territoires côtiers d'une manière plus efficace et plus conforme à l'intérêt de l'homme.

Bibliographie

AUBERT, M Le comportement des bactéries terrigènes en mer.
1966   Relations avec le phytoplancton. Thèse. Université Aix-Marseille, 285 p.

AUBERT, M et DESIROTTE, N Théorie formalisée de la diffusion
1968   bactérienne. *Revue int. Océanogr. méd.*, 12:5–48.

AUBERT, M, AUBERT, J et DANIEL, S Côtes de France. Inventaire
1968   national de la pollution bactérienne des eaux littorales. Atlantique. Cartes et graphiques. *Revue int. Océanogr. méd.*, 3, Suppl.:103 p.

AUBERT, M, AUBERT, J et DANIEL, S Côtes de France. Inventaire
1968a   national de la pollution bactérienne des eaux littorales. Méditerranée. Cartes et graphiques. *Revue int. Océanogr. méd.*, 4, Suppl.:167 p.

AUBERT, M, AUBERT, J et GAMBAROTTA, J P Côtes de France.
1968   Inventaire national de la pollution bactérienne des eaux littorales. *Revue int. Océanogr. méd.*, 1, Suppl.:73.

AUBERT, M, AUBERT, J et GAMBAROTTA, J P Côtes de France.
1968a   Inventaire national de la pollution bactérienne des eaux littorales. Mer du Nord et Manche. Cartes et graphiques. *Revue int. Océanogr. méd.*, 2, Suppl.: 122 p.

AUBERT, M *et al.* Etat actuel des pollutions bactériennes le long
1968   des côtes méditerranéennes françaises. Compte rendu IIIee Coll. Internationale Océanographie Médicale, Nice. *Revue int. Océanogr. méd.*, 9:45–72.

BUTTIAUX, R Surveillance et contrôle des eaux d'alimentation. 3.
1958   La standardisation des méthodes d'analyse bactériologique de l'eau. *Revue Hyg. Méd. soc.*, 6(2):170–92.

HARRAMOES, P Prediction of pollution from planted sewage outlets.
1964   *In* C.I.E.S.M.M. Symposium sur les pollutions marines par les micro-organismes et les produits pétroliers, Monaco, avril 1964, pp. 11–7.

KETCHUM B H, AVERS, J C and VACCARO, R F Processes contribut-
1962   ing to the decrease of coliform bacteria in a tidal estuary. *Ecology*, 33(2):247–58.

ORLOB, G T Viability of sewage bacteria in sea-water. *Sewage ind.*
1966   *wastes*, 28:1147.

Estuarine and Coastal Pollution in the United States

*T. A. Wastler and L. C. Wastler**

La pollution estuarine et côtière aux Etats-Unis

La zone côtière des Etats-Unis est divisée en dix régions sur la base d'analogies physiques et biologiques. Ces régions portent les désignations suivantes: Atlantique Nord, Atlantique Moyen, Baie de Chesapeake, Atlantique Sud, Zone des Caraïbes, Golfe du Mexique, Pacifique Sud-Ouest, Pacifique Nord-Ouest, Alaska et Iles du Pacifique.

L'auteur discute les problèmes de pollution côtière et estuarine dans ces régions en soulignant leur influence sur les ressources biologiques des régions côtières. La localisation des problèmes de pollution dans la zone côtière des Etats-Unis résulte de l'évolution historique des populations et de la création de centres industriels dans ces régions, mais l'influence qu'ils exercent sur les ressources biologiques est étroitement liée aux caractéristiques du milieu naturel.

Les modifications constantes apportées au milieu par de vastes opérations de dragage et de comblement des zones côtières sont encore plus dommageables que ne le sont les déversements d'effluents. Ces changements ont souvent abouti à détruire des zones estuaires d'alevinage pour les crevettes et pour d'autres espèces du biota, importantes du point de vue commercial, et ont provoqué

La contaminación en los estuarios y en las costas de los Estados Unidos

La zona costera de los Estados Unidos se divide en diez regiones basadas en semejanzas biológicas y físicas. Estas regiones son el Atlántico Norte, Atlántico Medio, Bahía de Chesapeake, Atlántico Sur, Caribe, Golfo de México, Pacífico Sudoccidental, Pacífico Noroccidental, Alaska e Islas del Pacífico.

Se examinan los problemas de la contaminación en la costa y en los estuarios en cada una de estas regiones, insistiendo en sus consecuencias sobre los recursos vivos de las zonas costeras. Los emplazamientos de los problemas de la contaminación en la zona costera de los Estados Unidos han resultado del desarrollo histórico de los centros de población e industriales en ellas, pero sus efectos sobre los recursos vivos están estrechamente relacionados con las características del ambiente natural de las regiones.

Incluso más perjudiciales que las descargas de los desechos son los cambios permanentes de tipo ambiental que están siendo causados por el dragado y el relleno de las zonas costeras. Frecuentemente, los cambios han dado lugar a la destrucción de los criaderos de camarones y de otra biota comercial importante

---

* Technical Assistance Branch, Federal Water Quality Administration, Washington, DC 20242, U.S.A.

une altération des mouvements aquatiques qui pourrait avoir des répercussions globales sur certains systèmes d'eaux estuarines.

L'étude sur la pollution estuaire nationale, qui vient d'être achevée, a recommandé la mise en oeuvre d'un programme national destiné à aménager les ressources côtières aux fins d'utilisation optimale tout en maintenant leur valeur pour l'ensemble du pays par des mesures de conservation et de protection. Dans le cadre du rôle qu'il exerce dans ce programme, le Gouvernement fédéral a mis sur pied un système de traitement automatique de l'information, connu sous le nom d'"Inventaire national des eaux estuarines", pour aider les Etats en leur ouvrant un accès commode à la multitude de renseignements dont ils ont besoin pour prendre des décisions rationnelles dans le domaine de l'aménagement.

existentes en los estuarios así como a la alteración de los movimientos de las aguas que pueden afectar a sistemas enteros de estuarios.

El recién acabado Estudio sobre la Contaminación de los Estuarios Nacionales recomendó un programa nacional de tipo general para manejar los recursos costeros con vistas a un aprovechamiento óptimo, conservando y protegiendo a la vez su valor para toda la nación. Como parte de su función en este programa, el Gobierno Federal ha creado un sistema automático de información, que se conoce con el nombre de Inventario de los Estuarios Nacionales, para ayudar a los Estados proporcionándoles un acceso rápido a la gran variedad de información que hace falta para adoptar decisiones racionales de explotación.

FOR over 300 years the coastal areas of the United States have provided the harbors through which a growing nation's commerce moved and around which great centers of population and industry developed. The many embayments and marshes along the coast are not only the habitat for a great variety of wildlife and seafood, but also the nursery grounds for many fish harvested from the oceans. In recent decades, increasing development and pollution of coastal areas have caused damage to the living marine resources of the United States and have begun to limit the use of coastal areas for other purposes.

In 1966 the United States Congress directed that a comprehensive study of coastal pollution be undertaken and recommendations be made for a national coastal zone management program to preserve these areas for the future of the nation. The Federal Water Quality Administration of the Department of the Interior submitted to Congress in November 1969 the first definitive report of the National Estuarine Pollution Study.

THE COASTAL ZONE OF THE UNITED STATES

The 40,595 km of United States ocean coastline lie along the Atlantic Ocean, the Pacific Ocean, and the Gulf of Mexico. Within this zone, there are 144,281 km of tidal shoreline, or about 3.5 km of shoreline per kilometer of ocean coastline. While there are river entrances, embayments, and coastal marshes along all major coastlines the numbers of each and their abilities to assimilate wastes vary tremendously because of differences in dominating environmental conditions.

The seven major variable environmental factors are:

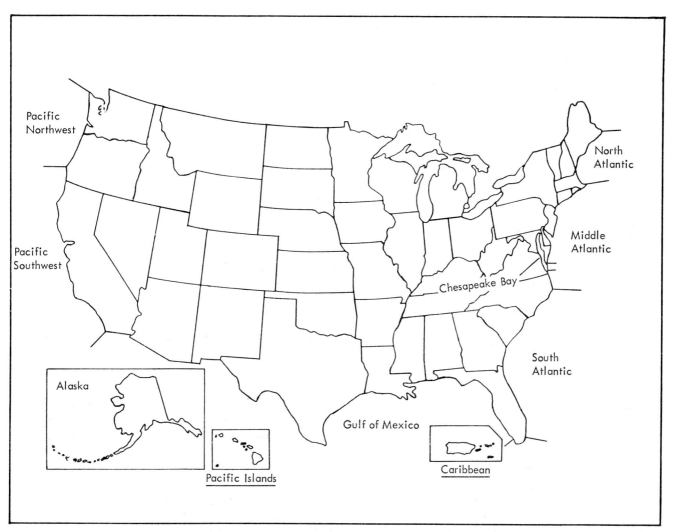

Fig. 1. Biophysical Regions of the United States.

[41]

TABLE 1. DOMINATING ENVIRONMENTAL FACTORS OF THE ESTUARINE ZONE U.S.A.

Environmental factor	North Atlantic	Middle Atlantic	Chesapeake Bay	South Atlantic	Caribbean	Gulf of Mexico	Pacific Southwest	Pacific Northwest	Alaska (1)	Pacific Island
Continental Shelf Width Range (Kilometers) Bottom Type	240–320 Irregular, Rocky	80–160 Smoothly Sloping, Lagoon off MJR	Not Applicable	50–110 Smoothly Sloping	5–15 E<FLA Puerto Rico, Virgin Islands 190–N of Florida Keys	105–225 Smoothly Sloping	3–32 Average about 10	15–60, with Indentations on the Outer Edge	48–240 on SE & S Coast 400' on W Coast	0, Volcanic Island Rising From S
Ocean Currents (2) Influenced By:	Labrador Current	Labrador Current	None	Gulf Stream	Gulf Stream	Waters Forming Gulf Stream	California Current	California-Aleutian Current	Alaska Current	Mid-Pacific Circulation
Temperature (°C) Mean	8	12	42	22	27	23	16	13	6	25
Summer	18	23	27	31	31	31	21	19	13	27
Winter	0	2	33	10	22	12	13	7	–1	23
Salinity (%) Mean	3.03	3.15	1.37	3.55	3.57	3.23	3.36	3.08	3.19	3
Dry Season	3.1	3.18	1.60	3.60	2.63	3.70	3.38	3.25	3.21	3
Wet Season	2.9	3.13	1.10	3.25	3.49	3.03	3.33	2.85	3.12	3
Coastline Structure	Rocky, Very Irregular, Many Embay- ments	Smooth, Many Large Embayments	Very Irregular Extensive Marshes on East Side	Smooth, Low Lying, Extensive Marshes	Irregular Mangroves, Coral & Rocks	Smooth, Low Lying with Barrier Islands, Marshes	High Land Close to Shore, Bluffs & Beaches	High Land Close to Shore, Bluffs & Beaches	All Glaciated, Irregular Except in NW	Steep Irregular with Beaches
River Flow Runoff Per Kilometer of Tidal Shoreline (Cubic meters per second)	.28	.27	.27	.28	.05	.92	.48	1.4	No Data	No Data
Number of Major River Basins	5	6	6	12	0	21	8	9	16	0
CBX.02832 = CMS Total Runoff (CFS)	2040	3,000	2,260	4,360	330	22,600	2,360	10,400	No Data	No Data
Sedimentation Quantity of Suspended Load (1,000 tons/yr)	No Data: Sedimentation not Severe Problem Great Amount of Rock	15,300	8,640	58,100	No Data	362,600	24,000	127,000	No Data	No Data
Climate (3) Temperature (°C) Mean	7	12	4	19	24	21	16	13	10	24
Summer	27	29	31	33	32	33	24	26	24	28
Winter	–11	–4	–1	4	18	7	12	8	–15	26
Precipitation (cm) Total in × 2.540 = cm	109	107	112	127	117	140	30	102	142	61
Snow, Ice	183	41	58	0	0	0	0	23	227	0
Tide (3) Type	Equal Semidiurnal	Equal Semidiurnal	Equal Semidiurnal	Equal Semidiurnal	Equal Semidiurnal	Diurnal	Unequal Semidiurnal	Unequal Semidiurnal	Unequal Semidiurnal	Unequal Semidiurnal
Mean Range (meters)	3.0	1.5	0.7	1.5	0.7	0.7	1.5	2.3	7.0	0.7

(1) Data are given for Southeast and South Coasts only
(2) Data are Typical Near-Coastal Values for the Region, except for Chesapeake Bay, where Data are for the Middle of the Bay
(3) Data are for a Point Typical of the Region
Reference: The National Estuarine Inventory
Data Source: U.S. Coast and Geodetic Survey, U.S. Geological Survey, U.S. Weather Bureau

[42]

continental shelf width, ocean currents, coastline structure, river flow, sedimentation, climate, and tidal regime. A careful analysis of variations in these factors revealed that the coastal zone could be subdivided naturally into 10 biophysical regions, each dominated by a unique combination of these seven factors. The geographical limits of each biophysical region are shown in fig 1, and typical regional characteristics for each environmental factor are presented in Table 1.

Each biophysical region has its characteristic circulation patterns and water quality conditions dependent upon the shape and size of the coastal features and the general environmental factors. In the coastal zone there are seven major types of forms: (1) smooth shoreline, (2) indented shoreline, (3) marshy shoreline, (4) unrestricted river entrance, (5) embayment with only coastal drainage, (6) embayment with continuous upland river flow, and (7) fjord. Except for fjords, each morphological type exists in all biophysical regions.

Alaska, Maine, and Washington, were once subjected to massive glaciation, and have deeply indented shorelines and many long, deep embayments. Penobscot Bay and Puget Sound are classic examples. The typical estuary along the rest of the Atlantic and Gulf coasts is the drowned river valley. Sedimentation has evolved many of these into elongated shallow embayments separated from the ocean by barrier beaches, or, further, into coastal marshes.

The drowned river valleys of the Middle Atlantic biophysical region, for example Delaware Bay, are in an early state of evolution. The South Atlantic biophysical region contains extensive coastal marsh areas, particularly along the South Carolina and Georgia coasts. The Gulf of Mexico biophysical region exhibits estuarine systems such as Matagorda Bay and Mobile Bay. These are separated from the sea by barrier beaches and embayments which receive only local fresh-water runoff.

Near the Pacific shoreline large mountain ranges form barriers to the drainage from the interior. A typical configuration is that of the Columbia River, which cuts through the western slopes of the mountain barrier and discharges directly into the sea.

Each of the three major coastlines, however, has one major estuarine system which is unique. Pamlico Sound, on the Atlantic coast, was formed by the development of offshore barrier islands at the juncture of the major northward and southward flows of ocean circulation. Cape Hatteras and Diamond Shoals are still zones of sediment deposition and present a major hazard to navigation. On the Gulf coast the Mississippi River drains 41 per cent of the continental United States and has built a delta of extensive marshlands all the way across the continental shelf. On the Pacific coast, San Francisco Bay was apparently formed within the coastal mountain range by a seismic disturbance.

Living resources

The natural living resources of the coastal zone include not only the life forms which spend their entire existence in a single estuarine system, but also those which use the estuaries as passages between sea and river and those which use the estuaries as nursery areas but spend their mature existence in the open sea.

Of those creatures which spend all of their lives in a small area, the only animals of economic significance are oysters, clams, crabs, and lobsters. The estuarine zone, however, also plays a direct role in the life cycles of shrimp, salmon, menhaden and many other species which are of great economic importance.

The Atlantic oyster, *Crassostrea virginica*, the primary commercial oyster in the United States, is an animal well adapted to fluctuations in salinity and temperature. Its range is from the Gulf of St. Lawrence to the Mexican coast. It is most abundant in estuarine systems characterized by considerable inflows of fresh water, constant water movement, and fluctuating salinities within the range of 5 and 30 parts per thousand. Among the most productive areas for the Atlantic oyster are the Chesapeake Bay and those Louisiana bays and sounds affected by the great flow of the Mississippi River.

The three species of commercially important shrimp are concentrated along the South Atlantic and Gulf coasts. The species are: brown (*Penaeus aztecus*), white (*P. setiferus*), and pink (*P. duorarum*). The pink shrimp spawns in offshore waters at depths of 30 to 45 m, where salinity is between 36.1 and 37.7 parts per thousand, and temperatures between 18 and 25°C. After 13 or 14 hours the eggs hatch and larval shrimp begin a series of developmental stages while drifting towards the mainland about 160 km distant. Movement to the estuary probably takes about three to five weeks. During this time postlarval shrimp have developed from planktonic to benthic feeders and have developed a wide tolerance to varying salinity and temperature conditions. For about two to nine months juvenile shrimp grow rapidly and attain commercial size before returning to the sea and completing their life cycle. The life cycles of the three main commercial species differ primarily in the species' penetration of the estuary and their utilization of the estuarine environment after adult stage is attained. The brown shrimp spawns in waters 45 to 70 m in depth and remains a relatively short time in the estuary. The white shrimp is rarely found in waters deeper than 30 m and possesses a greater affinity for fresh water than the other two species.

The Atlantic salmon (*Salmo salar*) has all but disappeared as a major commercial fish in the U.S.A. On the Pacific coast, however, there are five major commercial species of salmon. The spawning range of these species is from Monterey Bay in California to the northwest tip of Alaska. Only the King Salmon (*Oncorhynchus tschawytscha*) occupies spawning streams throughout the geographic range. The Silver Salmon (*O. kisutch*) has the next longest range extending from the Sacramento River to the Bering Strait. The Red (*O. nerka*), Chum (*O. keta*), and Pink (*O. gorbuscha*) salmon range from the Columbia River to the Bering Sea.

The distance upstream that the Pacific salmon migrates to spawn varies from species to species, as well as within species. Some spawn in the headwaters 2,400 km from the estuarine zone; others spawn within a few miles of the estuary. Both the young and the adult salmon of all species pass through the estuarine zone. During the passage through the brackish estuary the adult, but not the juvenile, salmon ceases feeding. Young Silver salmon may remain for extended periods within the highly nutritious estuarine portions of their natal stream and adult Silver salmon are caught there throughout the year. Pink salmon enter the brackish estuarine waters soon after hatching in spring and are known to remain until August.

These few examples indicate the essential role of the estuarine zone in life cycles of important commercial forms of marine life; yet these waters are far more important than a few examples can show. From the continent's interior rivers bring to coastal areas organic and inorganic nutrients which form the food supply of a vast variety of microscopic life forms which themselves are food for marine life, some of which never enter the estuarine zone. The most valuable of the living resources of any estuarine zone are probably these tiny plants and animals which form the foundation of the energy conversion chain.

Non-living resources

The continental shelf of the United States contains great deposits of sand and gravel, phosphate rock, sulfur, oil, and other valuable minerals. These offshore deposits are becoming increasingly important to the national economy.

Sand and gravel abound in the estuarine zone and are generally readily accessible. In the Gulf region are also large deposits of oyster shell used for some forms of construction and road building.

Phosphate rock deposits underlie large areas in the South Atlantic region and the eastern part of the Gulf region. At present, well over half of the United States supply of phosphate comes from central Florida. Other important deposits are beneath Pamlico Sound in North Carolina and along the Georgia coast. Deposits of nearly pure sulfur lie off Texas and the bulk of the U.S. supply is produced there.

Petroleum deposits occur along the Gulf coast, the southern California coast, and the Cook Inlet in Alaska. Great new deposits of petroleum have also been discovered along Alaska's arctic coast. These deposits are all being commercially exploited and new deposits are constantly being found.

SOCIAL AND ECONOMIC DEVELOPMENT

Urban and agricultural development

Table 2, which shows the 1960 population and agricultural development in the coastal zone of the United States, illustrates very clearly the existence of several distinct environments in the estuarine zone. Population and agricultural data exist in political subdivision groupings, while the Standard Metropolitan Statistical Areas (SMSA) cross State and county boundaries to present unified economic groupings. The classification by biophysical regions cuts across the boundaries of some political subdivisions, but is compatible with the SMSA economic units.

The coastal counties contain only 15 per cent of the U.S. land areas, but within this area is concentrated 33 per cent of the nation's population, with about four-fifths of that living in primarily urban areas which form about 10 per cent of the total estuarine zone area. Another 13 per cent of the estuarine land area is farmland, but this accounts for only 4 per cent of the total agricultural land of the nation. The estuarine zone is nearly twice as

densely populated as the rest of the country, and supports only one-fourth as much agriculture per unit area.

The magnitude of population and agricultural development in the estuarine zone is shown in Table 3 by densities in terms of tidal shoreline. The few estuarine areas in the Pacific Southwest show the greatest shoreline development for both living and farming. This is shown by a population density of 2,474 persons per km of tidal shoreline and a farmland density of 1.2 ha per km. The Middle Atlantic region, in contrast, has a very high population density and a low farmland density.

The differences between the east coast and the west coast in estuarine land use development probably results from the difference in rainfall. The low rainfall in the Pacific Southwest required the intensive farming of all land amenable to irrigation, of which a major part was near the mouths of rivers. Plentiful rainfall in the Middle Atlantic region, however, permitted the farming of land away from the estuarine zone.

In those regions between Cape Hatteras and Canada, as well as in the Pacific Southwest, over 90 per cent of the population lives in urban areas; over much of the Atlantic estuarine zone stretches the great Northeastern megalopolis with population densities averaging over 385 persons per square kilometer. The remainder of the estuarine zone of the United States exhibits a pattern of major centers of population clustered around natural harbors and separated by stretches of coastline which are either empty and inaccessible or are beginning to be sprinkled with private residences and resort communities in the vicinities of population centers.

Industrial development

Table 3 gives a general picture of the extent of industrial development in the estuarine zone of the United States. The coastal countries have within their borders 40 per cent of all manufacturing plants in the United States. The mixture of manufacturing types in the estuarine zone is the same as the national composition with only minor exceptions. Distribution of manufacturing types among the biophysical regions shows regional differences related to historical development as well as raw material and market availability.

Over half of all plants in the coastal counties and one-fifth of all manufacturing plants in the United States are located in the Middle Atlantic biophysical region. The Pacific Southwest is the major industrial center of the Pacific coast, and its shoreline has now the same intensity of development as that seen in the Middle Atlantic region. Leather product plants are clustered in the North Atlantic region, and lumber manufacturing plants are most plentiful in the Pacific Northwest. Food processing plants follow closely the population distribution.

While much of the industrial development located in coastal counties affects the estuarine zone indirectly through use of adjacent land, some of the water-using industries have an impact on the estuarine zone far beyond their numbers. The paper, chemical, petroleum, and primary metal industries are the major water users and these plants are widely distributed throughout the estuarine zone. Brackish estuarine waters may become an increasingly important source of water supply for industries and for municipalities as desalting technology improves.

TABLE 2. POPULATION AND AGRICULTURE IN THE ESTUARINE ZONE, 1960

Bio-physical Regions and States	State Population Density (Persons per sq km)	Coastal* Counties Population	Coastal Urban Areas Population	Coastal Counties Population Density on (Persons per sq km)	Coastal County Land in Farms (Per cent)	Coastal County Tidal Shoreline Density — Population Density (Persons per sq km)	Coastal County Tidal Shoreline Density — Farmland (sq km/km)	Population in SMSAs (Per cent)	Total Coastal Counties Population (per cent)
North Atlantic		3,258,798	3,541,000	113	17.6	286	.39	109	5.6
Maine	12			24	21.6				
New Hampshire	26			58	22.7				
**Massachusetts	254			346	10.7				
Middle Atlantic		22,387,123	20,852,000	450	28.1	1,081	.26	93	38.1
**Massachusetts	254			278	16.2				
Rhode Island	315			315	17.9				
Connecticut	239			270	17.0				
New York	136			1,934	4.5				
New Jersey	331			371	17.1				
Pennsylvania	97			807	26.4				
Delaware	87			87	54.8				
**Maryland	121			19	45.2				
**Virginia	39			128					
**North Carolina	36			15	38.6				
Chesapeake Bay		5,127,824	4,956,000	143	38.0	363	.37	97	8.8
**Maryland	121			153	51.5				
**Virginia	39			100	32.6				
D. of Columbia	4,804			4,804	0.0				
South Atlantic		2,202,669	1,659,000	34	31.5	87	.31	75	3.8
**North Carolina	36			15	27.5				
South Carolina	3			26	30.6				
Georgia	26			36	15.9				
**Florida	36			49	42.6				
Caribbean		3,682,667	935,000			413		25	6.3
**Florida	36			81	11.7				
Puerto Rico	265			265					
Virgin Islands	51			51	52.2				
Gulf of Mexico		5,833,149	3,109,000	47	49.1	147	.59	53	10.0
**Florida	35			37	34.8				
Alabama	25			50					
Mississippi	18			41					
Louisana	28			46					
Texas	14			56					
Pacific Southwest		12,198,082	10,991,000	151	48.1	1,537	1.89	90	20.7
**California	39			151	48.1				
Pacific Northwest		3,126,000	2,414,000	11	15.1	251	.52	77	5.3
**California	39			10	20.0				
Oregon	7			22	18.5				
Washington	17			37	15.5				
Alaska	0.1	168,721	85,531	0.2	<1.0	2	.03	50	0.3
Pacific Islands		632,772	500,000		64.7	185	1.30	79	1.1
Hawaii	38			38	64.7				
Guam	122			122	24.0				
American Samoa	102			102					

* Based on Standard Metropolitan Statistical Areas (SMSA), Except for Alaska, which are those communities with a population Density of Over 1000 Persons Per Square Mile.

** States with area in more than one bio-physical region.

Reference: The National Estuarine Inventory.

Source: U.S. Dept. of Commerce, Bureau of the Census, U.S. Coast and Geodetic Survey.

EXTENT OF POLLUTION

The population and industrial development outlined above has, in numerous places, caused a severe and damaging pollution of the estuarine and coastal resources.

There are five kinds of pollution. First, there is bacterial contamination associated with the discharge of untreated human sewage and with stormwater runoff from cities. This contamination can cause epidemics of such waterborne diseases as typhoid fever and infectious hepatitis.

Second, there are discharges of decomposable organic materials from human sewage and some industrial wastes which deplete dissolved oxygen resources. Third, there are toxic materials from industrial wastes and from land runoff, pesticides, herbicides, and the wastes from a variety of chemical manufacturing plants. These wastes may kill directly or cause damage to reproduction capability. Fourth, there are materials which act as fertilizers and tend to stimulate growth of some life forms

TABLE 3. EXTENT OF INDUSTRIAL DEVELOPMENT IN THE ESTUARINE ZONE

Bio-physical Regions	Total Number of Plants	Number of Plants with 20 Employees					Major Industries in Region (Plants with 20 Employees)				Density of Ind. Devel. (Number of Plants/Mile of Tidal Shoreline)
		Total	Major Water Use Industries				Industry Group	Number of Plants	Per cent of Total Plants in Region	Per cent of Total Industry Group in Estuarine Zone	
			Paper & Allied Prods.	Chemical & Allied Prods.	Petroleum & Allied Prods.	Primary Metals Ind.					
North Atlantic	8,617	2,933	132	88	13	60	Apparel & other Textile Products	368	12.5	4.4	0.7
							Leather & Leather Products	342	11.7	34.0	
							Fabricated Metal Products	225	7.8	6.7	
							Machinery & Electrical Equipment	483	16.5	10.0	
							Food & Kindred Products	325	11.1	7.8	
Middle Atlantic	65,000	21,847	761	840	78	532	Apparel & other Textile Products	6,547	30.0	79.4	2.7
							Printing & Publishing	1,701	7.8	56.5	
							Fabricated Metal Products	1,677	7.7	49.6	
							Machinery & Electrical Equipment	2,353	10.8	48.1	
							Textile Mill Products	1,280	5.9	79.4	
							Food & Kindred Products	1,413	6.5	33.2	
Chesapeake Bay	5,186	2,064	86	127	13	41	Food & Kindred Products	510	24.7	12.1	0.4
							Printing & Publishing	240	11.6	8.0	
							Fabricated Metal Products	158	7.7	4.7	
							Apparel & other Textile Products	191	9.3	2.3	
							Lumber & Wood Products	142	6.9	8.7	
							Stone, Clay, & Glass Products	133	6.4	9.8	
South Atlantic	2,695	693	40	59	12	10	Food & Kindred Products	164	23.7	3.9	0.07
							Lumber & Wood Products	97	14.0	6.0	
							Stone, Glass, & Clay Products	67	9.7	4.9	
							*Chemicals & Allied Products	59	8.5	3.4	
							Printing & Publishing	51	7.4	1.7	
Caribbean	2,554	654	18	16	5	9	Apparel & other Textile Products	123	18.8	1.5	0.2
							Fabricated Metal Products	75	11.5	2.2	0.2
							Furniture & Fixtures	56	8.6	4.8	
							Printing & Publishing	54	8.3	1.8	
							Food & Kindred Products	81	12.4	1.9	
Gulf of Mexico	6,980	2,013	70	192	62	61	Food & Kindred Products	493	24.5	11.7	0.1
							Fabricated Metal Products	216	10.7	6.4	
							*Chemicals & Allied Products	192	9.5	11.2	
							Machinery, except Electrical	161	8.0	6.5	
							Stone, Clay, & Glass Products	168	8.3	12.3	
							Lumber & Wood Products	123	6.1	7.6	
Pacific Southwest	27,508	7,633	240	332	54	286	Food & Kindred Products	846	11.1	20.1	2.5
							Apparel & other Textile Products	846	11.1	10.3	
							Printing & Publishing	545	7.1	18.1	
							Electrical Equipment	664	8.7	28.9	
							Fabricated Metal Products	867	11.4	25.6	
							Transportation Equipment	429	5.6	39.1	
Pacific Northwest	7,584	1,804	76	55	14	47	Food & Kindred Products	292	16.2	40.8	0.4
							Lumber & Wood Products	662	36.7	40.8	0.4
							Fabricated Metal Products	115	6.4	3.4	
							Machinery, except Electrical	97	5.4	3.9	
							Printing & Publishing	93	5.2	3.1	
							*Paper & Allied Products	76	4.2	5.2	
Alaska	303						Data Not Available For Coastal Area Only				
Pacific Islands	672	194	26	8	No Data	No Data	Food & Kindred Products	75	37.8	1.8	0.1
							Textile Mill Products	37	19.1	2.3	
							*Paper & Allied Products	26	13.4	1.8	
							Lumber & Wood Products	19	9.8	1.2	
Total Estuarine Zone	126,796	39,835	1,449	1,717	251	1,046	Apparel & other Textile Products	8,239	20.7		
							Food & Kindred Products	4,199	10.5		
							Fabricated Metal Products	3,381	8.5		
							Printing & Publishing	3,011	7.6		
							Machinery, except Electrical	2,476	6.2		
							Electrical Equipment	2,315	5.8		
							*Chemicals & Allied Products	1,717	4.3		
							Lumber & Allied Products	1,623	4.1		
							Textile Mill Products	1,613	4.0		
							*Paper & Allied Products	1,449	3.6		
							Stone, Clay, & Glass Products	1,361	3.4		
							Furniture & Fixtures	1,175	2.9		
Total United States	306,619	99,355	3,552	3,985	689	3,585	Food & Kindred Products	14,113	14.1		
							Apparel & other Textile Products	13,011	13.0		
							Fabricated Metal Products	212	9.2		
							Machinery, except Electrical	8,426	8.4		
							Printing & Publishing	7,215	7.2		
							Lumber & Wood Products	5,765	5.8		
							Electrical Equipment	4,722	4.7		
							Stone, Clay, & Glass Products	4,655	4.7		
							Textile Mill Products	4,367	4.4		
							*Chemicals & Allied Products	3,985	4.0		
							*Primary Metal Industries	3,585	3.6		
							*Paper & Allied Products	3,552	3.6		

*Major Water Use Industries
References: U.S. Census of Manufacturers, 1964
National Estuarine Inventory

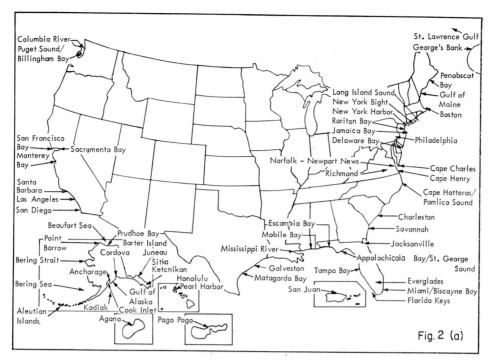

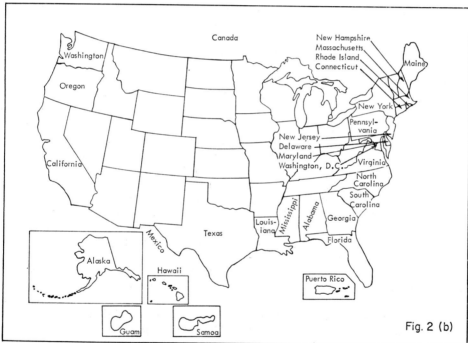

Fig. 2. Geographic Locator: Points mentioned in text

at the expense of others. Fifth, there are inert materials, such as sediments, which fill in valuable estuarine areas and smother benthic life forms.

The impact of these pollutants depends not only on their quantity but also upon the assimilative characteristics of the biophysical region and the individual estuarine and coastal areas.

North Atlantic region

The region extends from the Canadian border to Cape Cod. Cool, fertile waters with a large tidal range strike a steep indented coast with deep water close inshore. The region is protected from the full force of the ocean waves by a wide continental shelf. Moderate precipitation with heavy snow fall leads to heavy spring river runoff which dominates local circulation. Natural erosion and sedimentation are not severe problems. The drowned river valley estuaries are in an early stage of their progression toward becoming coastal marshes.

The heavily populated and industrialized Boston area has severe pollution problems from municipal and industrial wastes, particularly in the harbor area, where dissolved oxygen depletion is common during the late summer months. Nuisance algae growths occur in some areas, and shellfish harvesting is restricted because of bacterial contamination.

North of the Boston area localized pollution problems center around industrial operations and relatively small communities. One of the major problem areas is Penobscot Bay, a deep, natural embayment a maximum of 13 km wide with depths of 82 m near the ocean and averaging 9 to 18 m in its upper parts. The Penobscot River, with an average summer flow greater than 140 m^3 per sec is the major source of fresh water inflow. The tide is semi-diurnal with a range of about 3.5 m. Where the Penobscot River enters the upper part of the Bay some horizontal salinity stratification is found.

Both sewage and industrial wastes are discharged to the tidal portion of Penobscot River and to Penobscot Bay from numerous small towns and a variety of industrial operations. Sewage is the principal source of bacterial pollution in the upper Bay area, although poultry processing plants in the town of Belfast contribute more than 11 per cent of the 70,035 bacterial population equivalents discharged into the system. Over 99 per cent of the sewage discharged locally receives no treatment, and the two towns of Bangor and Brewer account for about 65 per cent of the bacterial load which has forced the closing of highly productive clamming areas.

A variety of industrial activities add suspended solids, decomposable organic wastes, and toxic wastes to the Bay. Total suspended solids equivalent to that in the sewage of 633,000 people and biochemical oxygen demand (BOD) equivalent to that from 1,191,000 people are discharged into the Bay from all sources. Over 90 per cent of these waste loads are from industrial plants, and nearly 85 per cent of the total comes from only two pulp and paper manufacturing plants, which also discharge waste liquors from sulfite pulp manufacture. These wastes have had severe impact on water quality and the living marine resources of the system. Sludge beds have covered much of the bottom of the tidal Penobscot River and parts of the Bay as well. These sludge deposits are believed responsible for decreasing lobster and scallop harvests and have also damaged some clam beds. During the summer the large decomposable organic loads have completely depleted dissolved oxygen in several kilometers of the river with resulting damage to the anadromous fisheries.

Sulfite waste liquor causes discoloration of water, has a nauseating smell, and has been shown toxic to shellfish. The presence of these wastes has reduced the recreational use of Penobscot Bay and has had unknown effects on living marine resources.

The polluted condition of Penobscot Bay and River has, however, resulted in strong action by the State of Maine and by the United States government.

Middle Atlantic region

This region extends from Cape Cod to Cape Hatteras, exclusive of Chesapeake Bay. A wide gently sloping continental shelf with a smooth shoreline is cut by several major river systems carrying moderate amounts of sediments. The same cool, fertile waters as in the North Atlantic estuarine region wash this coastline, but with a smaller tidal range.

The region has a very high coastal concentration of population and industry and includes the major population centers of Philadelphia and New York metropolitan areas. Coastal waters near all of the major population centers are degraded in some degree, but off New York even ocean waters are adversely affected.

The New York Harbor area is an intricate network of channels, canals, and embayments dominated by the flow of the Hudson River (annual mean = 263 m^3 per sec) and a tide range of 1–2 m. The eastern end of the New York Harbor is connected with Long Island Sound and the different tidal heights and phases in the two systems make the zone where they meet an area of extreme turbulence and vicious currents. In spite of increasing efforts to provide more effective sewage treatment, uncontrolled storm and surface runoff from a city of 8,000,000 persons, and the effluents from the many sewage treatment plants, cause severe damage to water quality in the New York Harbor system. Bacteria counts are high and dissolved oxygen values are low, particularly during late summer and early fall when temperatures are highest and river flows lowest. New York Harbor is primarily a commercial sea port, a use which is only minimally affected by pollution. The marine resources of adjacent areas, however, have been damaged by pollution. The runs of American shad (*Alosa sapidissima*) up the Hudson River have been severely curtailed, and many formerly productive shellfishing areas in Raritan Bay and Jamaica Bay have been closed because of bacterial contamination or have been damaged by sedimentation or toxic waste discharges.

Partially treated sewage sludge is taken by barge from the New York metropolitan area and dumped in the New York Bight about 11 km offshore. Such dumping began about 1940 and has increased progressively over the years until an average of over 8,400 m^3 per day was dumped in this area during 1964–65; the amounts dumped have increased since then. In addition acid wastes from chemical plants, mud, and other materials are dumped in several areas near the sludge dumping grounds.

Studies of the effects of these massive dumping operations on marine resources are under way. There is evidence that there are some adverse effects of these wastes on the fish population, since some fish caught near these areas have some forms of fin rot and show other evidence of disease. Benthic populations are smothered or are otherwise severely affected. This biophysical region has an area affected by a unique pollution problem. Along the southern shore of Long Island "Long Island Duckling", a poultry delicacy of some repute, is a major commercial enterprise. The drainage of fecal matter from duck farms has caused severe damage to the water quality of Moriches and Great South Bays, two shallow coastal embayments on the south shore of Long Island. At the present time, the Duck Farmers Cooperative is investigating methods of phosphate removal and many previously closed shellfish beds have been reopened. However, the inputs of nitrogen and phosphorus compounds from the runoff of duck farms continue to over-fertilize parts of these embayments and stimulate blooms of *Nannochloris*, a form of plankton which is not an acceptable food material for the clams and oysters which were once a major commercial product here. Runoff from septic tanks of suburban developments has also contributed to the over-enrichment of the waters of these Bays. Similar pollution problems also occur in many embayments along the New Jersey coast.

[48]

Chesapeake Bay region

The region includes all Chesapeake Bay systems inland from Cape Charles and Cape Henry. This is the drowned mouth of the Susquehanna River. It is characterized by isolation from direct oceanic effects in the numerous estuaries tributary to the Chesapeake Bay. The reduced salinities in the tributary estuarine systems produce circulation patterns which are possibly unique to the Chesapeake Bay.

Industrial and population development in the region are concentrated on the tributary estuarine systems, in many cases at the head of navigable waters; for example Richmond, Virginia, Washington, D.C., and Baltimore, Maryland. Thus, in the Chesapeake Bay, pollution problems are associated with population and industrial activities located up the tributaries of the Bay, and the obvious and directly damaging effects of these activities appear primarily in the tidal portions of the tributary rivers rather than in the Bay itself. Nevertheless, a buildup of inorganic nutrients and other persistent pollutants is being observed in the Bay, and recent severe plankton blooms have been attributed to these pollutants.

The area is one of the finest and best-known regions in the United States for large-scale commercial harvesting of oysters and crabs. It also supports a very large recreational industry centered around the great variety of year-round sport fishing available in vast areas of sheltered waters suitable for sailing and cruising.

The Susquehanna River discharges a mean flow of 990 m^3 per sec into the head of the Chesapeake. Its effects on salinity are observable two-thirds of the way down the Bay. The waters of the Susquehanna contain acid mine waters and other forms of pollution accumulated in its passage through a large region of coal mines and heavily industrialized and populated commercial centers. The seasonal variation in the flow of the Susquehanna and other major rivers entering the Bay, combined with generally reduced salinities over most of the Bay and a seasonal influx of denser water near the bottom, cause unusual circulation patterns and rapid water renewal in some of the tributaries. In Baltimore Harbor, for example, there is a three-layered circulation pattern in which water enters the Harbor near the surface and the bottom and leaves the Harbor at mid-depth.

Heavy manufacturing industry is concentrated in the upper part of the Bay, particularly in the Baltimore area, and the waters of Baltimore Harbor, even with excellent flushing conditions in much of this system, have reduced dissolved oxygen and cause high bacterial counts. In addition, these waters show the effects of industrial waste effluents from metal industries, i.e., reduced pH, increased color, hardness, and metal concentration. One large steel mill uses the secondary municipal sewage treatment plant effluent as cooling water for some processes and then discharges the effluent to the Bay.

Below the head of the Bay the western and eastern shores are different in physical and ecological character and have different pollution problems. The western shoreline is higher than the eastern, has many rivers that drain the mountainous interior, and has greater population concentration than the eastern shore. The low-lying land of the eastern shore has small rivers, tidal through much of their length, extensive marsh areas, and scattered small communities and industrial operations.

On the western shore the James River has Richmond at the head of tidewater and Norfolk and Newport News at its mouth, both of which are major ports and population centers. The dissolved oxygen depletion and high bacterial counts characteristic of such centers are found in this area.

Also on the western shore, at the head of tidewater on the Potomac River is Washington, D.C. This is a unique community, having little industrial or commercial development. Yet solely from municipal sewage from Washington metropolitan area, the Potomac River below Washington becomes a livid green from the heavy algal blooms attributed to nutrients from these sources. Low dissolved oxygen values and high bacterial counts are typical during summer months.

The comparatively sparsely populated eastern shore does not yet have pollution problems to compare with those across the Bay, but even here there are problems of localized water quality degradation in the vicinities of the small communities and the numerous food processing plants which are the basis of the eastern shore economy.

The effects of pollution on the living resources of Chesapeake Bay have not been accurately determined. The side effects of municipal and industrial pollution are over-fertilization or the addition of trace metals or pesticides to fish and shellfish, making them toxic to consumers. These seem to be of greater ecological and commercial importance than the easily identifiable fish kills and noxious odors which often appear near a waste discharge. High concentrations of trace metals have been found in Chesapeake Bay oysters and there is a gradual increase of inorganic nutrients in Chesapeake Bay waters. The effects of heated waste discharges on living resources are being studied.

Also of concern are the effects wrought by changes in flow regimes or by channel deepening or widening. In the Potomac River, the oyster drill is killed each spring by the rapid salinity reduction caused by high spring river flows. This periodic elimination of a major predator permits commercial oyster production in areas which could not otherwise support it. Plans to reduce the flow below the level that kills the predator would change this position. Hydraulic model studies have shown that projects to widen and deepen the James River would so affect circulation patterns that the oyster larvae could be carried beyond the bars most favorable for their development. The oyster harvest would thus be reduced.

The effects of such artificial changes in flow patterns and conditions are so poorly known that the Congress has authorized the building of a hydraulic model of Chesapeake Bay to study a variety of problems. This model will cover more than 22,250 m^2 of concrete and to collect the data needed for verification will require several years of intensive study. The problems of the Chesapeake Bay biophysical region are so complex that only a massive effort such as this can hope to achieve an acceptable solution.

South Atlantic region

This region extends from Cape Hatteras to Fort Lauderdale, Florida. It is characterized by a wide continental shelf brushed by the warm water of the Gulf Stream. The low-lying coastal plain terminates in barrier islands and marshes in which large amounts of sediments are continually being deposited by moderate-sized rivers fed by heavy summer rainfall. Many of the drowned river valley estuaries have evolved to the final stage and have become coastal marshes. Tidal ranges are small to moderate depending on local conditions.

Over much of this region population and industrial development are centered in metropolitan areas located around natural harbors. Between these metropolitan centers are long stretches of sandy coastline or marsh, thinly populated or devoted to recreational development and resort communities.

Pollution problems in this region are associated with the discharge of large quantities of municipal and industrial wastes especially from pulp and paper mills into confined harbor areas. The harbors of Charleston, Savannah, and Jacksonville all exhibit depressed dissolved oxygen levels due in part to pulp and paper mill wastes. The waters along much of this coast receive partly degraded wood fibers from the drainage of extensive marshes. They have, therefore, a naturally high content of organic material. This, combined with the high natural temperatures, reduces the natural dissolved oxygen content so that the waste assimilation capacities of these estuaries are less than those of estuaries of similar size to the north.

This region contains also a unique example of damage done by modifying an estuarine system. The harbor of Charleston was once a deep-water bay with depths of about 15 m over a large area. There was a fresh-water inflow of about 6 m³ per sec from the surrounding coastal plains. In 1940, as part of a hydroelectric project, a river with an average daily flow of 425 m³ per sec was diverted into the harbor. Heavy siltation began within a year, and in the following five years annual dredging costs rose from about $18,000 to $2,000,000. Many docks filled with silt and had to be abandoned. Recent studies showed that the increase in river flow caused stratification to develop and this further increased the settling of sediment. A project to redivert the river flow around the harbor is now under way.

There is increasing pressure to fill in or otherwise destroy the extensive marshlands in this region. The discovery of large deposits of phosphate rock underlying Pamlico Sound in North Carolina and off the coastal marshes of Georgia has aroused interest in their commercial exploitation. The increasing need for more waterfront residential areas in Florida has brought pressure to fill in marshes and shallow estuarine areas for this purpose.

The coastal marshes of this region play a major role in the life cycle of shrimp and menhaden, both major commercial fishery products. Oysters and crabs also abound in these waters and are of great commercial value. Damage to these resources is as yet highly localized, but the increasing industrial development and population pressure which this region is experiencing cause alarm at what may happen in the future.

Caribbean region

This region extends around the tip of Florida, from Fort Lauderdale to Cape Romano, and the oceanic areas including Puerto Rico and the Virgin Islands. High temperatures, heavy rainfall and warm ocean currents along practically non-existent continental shelves result

in tropical marine environments throughout this region. Coral reefs and mangrove swamps are typical coastal features, and the islands are fringed with coral reefs and beaches. Tidal ranges are small. Except for major metropolitan areas of Miami, and San Juan, Puerto Rico, the coastal zone is not heavily populated. Southern Florida is composed mainly of vast marshes of the Everglades and the coral islands of the Florida Keys. Rivers of Puerto Rico and the Virgin Islands are small, embayments are few, and much coastal development is lacking.

Along the southeast coast of Florida the continental shelf is narrow and with deep water adjacent. The Gulf Stream nearly touches the coast and with these favorable conditions many communities, including Miami and Miami Beach, use ocean outfalls to dispose of their municipal sewage. This form of waste disposal is increasing and studies are under way to determine the effects on living resources.

The increasing need for electric power in the Miami metropolitan area has caused the construction of more and larger power plants with prospects of damage to living resources due to heated effluents of power plants. Measurements in Biscayne Bay, near Miami, already show temperatures consistently in excess of 32°C, and occasionally above 38°C. Some damage to attached algae and benthic biota has been shown but the effects on fishery resources have not yet been demonstrated.

San Juan Harbor has a very small fresh-water inflow, small tidal range, and high natural temperatures. Large amounts of municipal and industrial wastes from the surrounding metropolitan area have lowered dissolved oxygen levels and raised coliform bacteria counts within the Harbor.

Gulf of Mexico region

This region extends from Cape Romano to the Mexican border. A wide continental shelf extends all round this large embayment in which warm tropical waters are moved gently by weak currents and small tidal ranges. Heavy rainfall brings sediments from the broad coastal plains. Barrier islands are extensive and have large shallow bays behind them. The Mississippi River forms one of the major deltas of the world. It is unique among the estuarine systems of the United States, both in its size and in the extent it has built out over the continental shelf.

Population and industrial development here are similar to that of the South Atlantic region. People and industry are concentrated into metropolitan areas centered around harbors but separated from each other by relatively undeveloped marsh or beach areas. Accessible beaches are rapidly turning into thriving and expanding resort communities of high seasonal population density and little industry.

In the Gulf region, however, coastal water pollution problems are caused more by chemical manufacturing industries than the pulp and paper plants, although these are also important. The coasts of Texas and Louisiana are massive producers of petroleum. The nearshore waters abound in oil well drilling platforms. Huge amounts of pure sulfur are also mined from subterranean and submarine deposits, and the area around Tampa Bay, Florida, produces over half of the nation's supply of phosphate rock. Oil, sulfur, and phosphates are major raw materials and there is a wide variety of chemical producers. The waste products of such processes often include materials whose effects on living creatures are not fully known. The overall impact of such a great variety of subtly damaging wastes has not yet been measured.

The Gulf coast marshes and coastal embayments are nursery areas for shrimp and they also play an essential role in the life cycles of other important commercial and sports fish.

As the Mississippi system drains 41 per cent of the land area of the United States, including much of the agricultural heart-land, the pesticides and fertilizers draining from it flow into the Mississippi and then into the Gulf of Mexico. Fish kills in the Mississippi and the disappearance of the brown pelican along the Gulf coast have been traced to excessive pesticide concentrations. The Mississippi delta has an intricate system of channels and marshes. Its salt water/fresh water balance has led to the commercial production of furbearing mammals such as muskrat and nutria, a major industry in Louisiana, and to the development of an extensive oyster and shrimp harvest in adjacent areas. Canals to move oil drilling equipment about the delta threaten to upset the current ecological balance and are a major pollution problem in the area.

The Gulf region is rich in natural resources and has a pleasant climate. Increasing industrial development and associated population pressure will undoubtedly cause increasing strain on the living resources. Widespread off-shore oil well facilities constitute a continuing danger of catastrophic damage to living resources from accidents involving spills.

Pacific Southwest region

This area extends from the Mexican border to Cape Mendocino. Because of the narrow continental shelf there is a periodic upwelling of cool fertile water as winds during several months of the year force the California current offshore. Coastal population and commercial development is centered in the San Francisco Bay area and, along the southern part of the coast, in the Los Angeles and San Diego metropolitan areas. Much of the coastline in this region is mountainous and un-suitable for metropolitan or industrial development; it does, however, receive much recreational use.

Ocean outfalls are used for municipal waste disposal at Los Angeles, San Diego, and other smaller coastal cities. No damage to living resources due to these discharges has been demonstrated. In fact studies on the area surrounding one of the Los Angeles outfalls suggest that the fisheries are actually enhanced by these discharges.

The major problem in the southern part of this region, particularly in the vicinity of Los Angeles and Santa Barbara, is the extensive development of offshore oil drilling. This is occurring in a region historically subject to earthquakes and to that extent adds a new danger.

In January 1969, such a mishap did occur off Santa Barbara, releasing a large quantity of oil which polluted the beaches and caused damage to coastal wildlife. The leak was contained 12 days later; but, subsequently, oil began leaking through the ocean floor. This region is subject to many natural oil emissions from the bottom, and the natural quantity leaked is sometimes so great that the odor of petroleum is noticable over large areas. It is difficult to assess the amount of ecological damage resulting from this one spill; however, to clean the beaches and other property damaged by the oil cost about $5,000,000. New restrictions and additional safety measures have been imposed to prevent a recurrence.

In the San Francisco Bay area extensive municipal and industrial development, combined with the need for additional agricultural land near the Bay, caused the filling in of much marshland and many estuarine shallows. Over two-thirds of the marshes have been destroyed. Total damage assessment has not been made, but there has been a drastic decline, and possibly the elimination, of crab and shrimp fishing. This loss of vast quantities of marshes and shallow estuarine areas in San Francisco Bay, formerly one of the largest areas suitable for nursery grounds, must have had an impact on the ocean fisheries off the Pacific coast. There are not, however, the historical records necessary to measure this impact.

Recognizing the damages inherent in unrestrained filling and pollution of the Bay, the State and local governments appointed a commission to recommend solutions. This definitive study culminated in the establishment of a permanent agency with far-reaching authority to zone shoreline and Bay areas and to regulate future development.

The San Diego metropolitan area is a classic example of how positive community action can correct pollution problems, enhance the environment, and aid the economy at the same time. During the 1950s and early 1960s there was extensive pollution of San Diego Bay as shown by the low dissolved oxygen and high coliform bacteria counts, and quantities of suspended and floating solids which reduced clarity. Water contact activity was not permitted in much of the Bay. The disposal of untreated sewage into the Bay was the principal cause of pollution, but other problems included storm water runoff, agricultural water runoff, vessel wastes, oil discharges, and heated effluents.

In 1963 the San Diego Metropolitan Sewage System was initiated to treat sewage and divert it from the Bay to the ocean. Bay water quality has been consistently compatible with water-contact ever since. This has been coupled with an overall awareness of the importance of the resources of the coastal zone in the San Diego area. Mission Bay for example has been developed from a tidal mudflat into an intensively used estuarine recreational resource. The success of the San Diego community in cleaning up and effectively using its estuarine resource is a major example of the value of proper management.

Pacific Northwest region

Here the region extends from Cape Mendocino to the Canadian border. The continental shelf and coastal configurations are similar to those of the Pacific Southwest, but ocean water temperatures are lower. The movement of the California current away from the coast is not as pronounced as it is farther south. Major rivers through the coastal mountains form deeply embayed estuarine systems. Extensive erosion and sedimentation have made wide tidal flats, bars, and shoals. The straits of Juan de Fuca and Puget Sound, glacier-formed, do not show such severe sedimentation, and have retained much of their original configuration.

Two major coastal features in this region are Puget Sound and the mouth of the Columbia River. The tidal river reaches and headwaters of many of the streams of the Northwest region are the spawning grounds for the several species of salmon. The environmental requirements of these fish during their spawning cycle determine the major pollutional concerns. Salmon require cool, clean water with a small temperature range for reproduction, and, like so many living creatures, they are especially, sensitive to toxic materials and other environmental anomalies during the breeding season. Salmon eggs require dissolved oxygen levels near saturation for successful hatching.

The Columbia River has an average flow of 7,000 m³ per sec and drains 667,000 km² in the United States and Canada. Numerous high dams have created large impoundments across the watershed. Fish ladders are used to assist salmon past these dams, but the increase in time required for migration and the change in temperature regime associated with such dams have caused considerable concern about their ultimate effect on the salmon fishery. There is increasing industrial and power plant development throughout the watershed, and concern for the impact of such development on the future of the salmon fishery has caused the Federal Water Quality Administration (FWQA) to initiate a massive study on the thermal problems of the Columbia. This study is now under way, but the authorities are already opposed to any waste heat discharges entering streams of the Columbia River Basin.

Industrial wastes, discharged near the mouth of the Columbia primarily from pulp mills, stimulated the growth of *Sphaerotilus* to the extent of fouling the gill nets of commercial fishermen. Recent improvements in waste treatment show signs of alleviating this problem.

Farther north, in Puget Sound, the wastes of pulp mills have a more subtle and far-reaching effect. There are several sulfite process pulp mills in the northern area and their untreated wastes have reduced dissolved oxygen, created sludge deposits, and increased suspended solids concentrations in several embayments. An intensive study completed in 1967 of the damage caused by these discharges found that these obvious changes in water quality affected the salmon and shellfishery resources of Bellingham Bay and other parts of Puget Sound. The study also found that sulfite liquor wastes, even in relatively dilute concentrations (5 to 15 ppm sulfite waste liquor), was damaging to immature forms of indigenous fish and shellfish.

The FWQA study observed damage from sulfite waste liquor to oyster larvae and to English sole eggs. The damage included both lethal effects and the development of abnormal juveniles. The recognition of the teratogenic effect of sulfite waste liquor and its potential for subtle, far-reaching effects on the food chain as well as on the commercial fishery resources is one of the most important results of this intensive study, which is now resulting in a great improvement in the water quality of Puget Sound. This proven cause-effect relationship among industrial wastes, water quality, and genetic damage to living resources is bringing about new approaches toward the study and control of industrial wastes.

Alaska region

This region includes all of the Alaskan coastline and includes the Aleutian and Bering Sea Islands. The dominant factors in this region are temperature and precipitation. Water temperatures are near freezing and much of the precipitation falls as snow. The continental shelf is wide throughout the region and the tide range is very great. There is active glaciation on the southeast and south coasts and the area is primarily glacier-cut embayments and fjords. The west and north coasts are much flatter and have been modified by sediments including glacial silt eroded from the interior and by the grinding action of pack ice during winter.

Alaska and its neighboring ocean areas are rich in a great variety of natural resources—now beginning to be exploited on a massive scale. The southeastern coast is rich in timber which is cut for lumber and for pulpwood. Pulp and paper mills near Ketchikan and Sitka have caused some water quality degradation, but pollution problems have thus far been isolated.

The mill on Silver Bay (Sitka) discharges sulfite waste liquor to the surface of the water. Water quality sampling of the Bay demonstrated extensive degradation of the surface water stratum as indicated by depressed dissolved oxygen concentrations, changes in pH, and an increase in turbidity. Vertical profiles of these water quality parameters indicated that the waste materials remained on or near the surface in a low-density layer in concentrations great enough to be toxic to many of the natural food chain organisms and to cause the same abnormalities found in the Puget Sound study. The lack of vertical mixing and dispersion is characteristic of waste disposal problems where heated plant effluents are discharged on the surface of cold, dense arctic waters. The lack of mixing serves to concentrate the effects of such waste discharges near the surface and to increase the probability that they will move as patches of pollution driven by the wind. While these conditions are not unique to the Arctic, they are intensified in higher latitudes.

The southern, southwestern, and northern coasts of Alaska are rich in oil and minerals. These are now under intensive exploration and, in some areas, are already being exploited. In spite of the problems of winter ice, Cook Inlet (an embayment as large as the Chesapeake Bay) contains numerous oil drilling platforms and its shores are being developed as sites for refinery, petrochemical, and other industrial operations. Anchorage, the largest city in Alaska, lies at the head of Cook Inlet. Even though industrial development is in its infancy, there are already frequent instances of oil pollution in Cook Inlet and at least one significant loss of waterfowl has occurred. The heavy glacial silt load of Cook Inlet and similar embayments causes concern that oil and toxic materials from other waste sources may become entrained in the sediments, sink to the bottom and present a potential threat to bottom-feeding organisms of the future.

The development of oil reserves in the Beaufort Sea, and on the northern slope and coast of Alaska present a potential pollution problem of unknown magnitude. Investigations are being made to determine baseline ecology before major development begins. In September 1970, a combined FWQA-Coast Guard investigating team, supported by other Federal agencies, carried out an ecological baseline study of the nearshore areas of the Beaufort Sea. This study is being followed by others as soon as ice conditions next summer permit. These studies are centered around Prudhoe Bay, but also cover the area from Point Barrow to Barter Island.

Exploitation of the living resources of Alaskan waters has been increasing for over 50 years. The most famous fishery products are salmon, and king crab. Yet even the harvesting of these living resources leads to pollution. Salmon canneries abound and their organic debris pollutes the waters around many towns; Ketchikan, Juneau, Cordova, and Kodiak are examples.

Alaska is a region in which the drive to obtain the most economic value from natural resources comes into direct conflict with the need of conservation for future generations.

Pacific Islands region

This region includes the Hawaiian Islands, Guam, and American Samoa. These are all tropical ocean islands of volcanic origin. Dominating factors are the lack of a continental shelf, full exposure to oceanic conditions, and pleasantly warm temperatures. There are few major rivers, but there are small streams with mean flows of 0.05–0.5 m³per sec.

The economy of the region is centered on agriculture, food processing, commercial fishing, and tourism. Guam and American Samoa are underdeveloped compared to Hawaii, which has a rapidly expanding population and economy. The Hawaiian Islands are the shore base for a major tuna fleet and the coastal embayments are the spawning, nursery, and rearing grounds for the nehu (*Stolephorus purpureus*), the principal bait fish for tuna.

On American Samoa and on Guam, the flow of streams is steady but small. In the Hawaiian Islands many streams are diverted to irrigate sugar cane, pineapple, and other crops. The fresh-water flow into many Hawaiian estuaries is therefore erratic and often contains fertilizer and pesticide residues. Where metropolitan centers are located there is generally water quality damage even though ocean outfalls are used for waste disposal. Pearl Harbor and Honolulu Harbor are examples.

Pearl Harbor is Hawaii's major source area for nehu. This fish supplies one third of the total local supply. Formerly the Harbor was the site of intensive bait fishing, but now there are numerous fish ponds which supply most of the nehu. The shoreline is heavily developed for military, residential, and industrial use and the primary sources of pollution are sanitary sewage from the city, sugar mill waste discharges, ships, storm sewer discharges, cesspool seepage, and heated power plant effluents. These wastes cause depressed dissolved oxygen values and high bacterial counts in large areas of Pearl Harbor. Coliform counts of over 1,000,000 MPN/100 ml have been found in oysters, and fish kills have been attributed to domestic sewage and to pesticides.

Honolulu Harbor is the principal commercial port of the islands. The harbor was originally a natural channel in the reef, but it has now been artificially expanded. Sand Island, created by dredging and filling, now occupies the area of the original reef. Other reef and lowland areas have also been filled.

Pollution enters Honolulu Harbor from many sources. Most municipal wastes are discharged directly to the ocean, and a large drainage canal entering the Harbor carries not only agricultural residues but also wastes from three large pineapple packing companies and cooling waters from another plant. Additional pollution sources to the Harbor are an oil refinery, a power plant, storm drainage from the surrounding business and industrial areas, and the ships in the harbor.

Coliforms bacteria counts are generally low. In the drainage canal, however, and the harbor area adjacent to it, counts of 1,000,000 MPN/100 ml are found. Three moderate to heavy fish kills have occurred in the drainage canal during the past few years; at least two of these kills were attributed to food processing wastes. Harbor waters are generally oily. Oil contamination is occasionally noted but rarely traced to specific sources.

On the north coast of Hawaii are six sugar cane processing plants which discharge their wastes directly to the ocean. These mills are on an inaccessible shoreline with steep cliffs. The alongshore currents push the wastes long distances along the shore and then out into the ocean. The principal effect of the sugar cane wastes has been the shading of coral by the highly turbid waters, the occurrence of high phosphorus and coliform concentrations, and the lowering of fish diversity and productivity. The slope of the nearshore ocean floor is steep and the dilution capacity of the deep water minimizes these effects within a short distance offshore.

One of the distinguishing characteristics of a tropical coast is the large quantity of coral. In the sugar mill waste disposal area at Honokaa, the coral has been completely covered with sludge (composed mainly of bagasse and settleable solids) within a radius of one-quarter mile on each side of the sludge deposit. For more than 1 km down current from the outfall, the coral coverage is between 10 and 55 per cent. The coral coverage on the down current area does not reach normal density until about 1.5 km from the outfall, where coverage is about 55 per cent (considered normal for comparable areas). There is little doubt that the reduced coral density is a result of the increased turbidity; coral relies upon light for formation and maintenance.

Many tropical fish are dependent upon the coral reef for protection from predators and for food. Since the coral in this area was destroyed, it is reasonable to expect that the fish population also deteriorated. The diversity of fishes in the outfall area has decreased to 16 species, as compared to a normal 60 found 3 km away. The biomass of fish was also reduced near the waste disposal area to 30 kg per hectare during the sugar cane grinding season, compared to 110 kg per hectare 2 mi away.

CONTROL OF POLLUTION THROUGH RESOURCE MANAGEMENT

The consequence of damage to the biophysical environment is the immediate or future loss of its usefulness to man. Institutional management must exercise responsibility and authority in achieving the maximum multiple use of the estuarine resource. The primary objective of technical management is to achieve the best possible combination of uses while protecting, preserving, and enhancing the biophysical environment for continuing benefit.

Nearly all estuarine uses involve both land and water,

[53]

either directly or indirectly. For example, the construction of a manufacturing plant on the shore of an estuarine system may not involve any direct use of the water (even for waste disposal), yet it limits access by its occupation of the shoreline and so may interfere with other uses. Conversely, the disposal of liquid wastes into the water may not use any appreciable space but may make the shoreline unusable for recreation and make the water unsafe.

The impact of one estuarine use on another may be either "prohibitive" or "restrictive" depending on the kind of use or the manner in which it is carried out.

Prohibitive impacts involve permanent changes in the environment and prohibit all uses that cannot be adapted to such changes. Examples include: dredging and filling, solid waste disposal, construction of bridges, dikes, jetties and other structures, shoreline development, mining from the estuarine bottom, and flow regulation.

Some estuarine uses may restrict use for other purposes but do not automatically exclude them. These do not require a permanent modification of the estuarine system. They generally involve the estuarine waters and other renewable resources. Restrictive impacts may involve damage to water quality, living organisms, or aesthetic quality, and they may also result from the exclusive appropriation of space. The key feature of restrictive uses is that they may, with proper management, be carried out simultaneously with other uses. The major restrictive use is, of course, waste disposal.

Where there is conflict one activity must be substituted for another. This may cause cases of uncompensated damage but these damages are difficult to prove. While commercial enterprises, such as commercial fishing, can be quantified in terms of the economic loss, the intangible values of recreation and estuarine habitat are very difficult to measure. Recreational loss would have to be measured in terms of how many people *do not* swim or go boating because the water is polluted. The value of estuarine habitat is just as difficult to establish. There are now only about 2.1 million km² of important estuarine marsh and wetland remaining in the coastal zones of the United States and each area is irreplaceable.

Some aspects of technical management

Damage to the estuarine and coastal zones is not a necessary feature of our civilization, but use-conflicts will continue to exist as more and more demands are made on the natural environment. The ability of any management authority to prevent use-damage and to resolve use-conflicts depends not only upon its institutional composition and legal authority, but also upon social, economic, and biophysical characteristics of the specific estuarine or management unit over which it has authority. Effective control of the social and economic demands that are made on the estuarine environment can be maintained only if both the major sources of damages and the geographic range of their influences are subject to unified control.

An estuarine or coastal management unit should therefore include not only the waters, bottoms, and associated marshlands, but also all of the shoreline and as much of the adjoining land as is necessary to regulate the discharge of wastes into estuarine waters.

Allocation of part of the estuarine resource for a

single use is a necessity of estuarine management. The shoreline must be used for both shipping docks and for swimming beaches, but they cannot occupy the same place. Similarly, dredged channels and oyster beds cannot occupy the same place. Resolution of such conflicts can be achieved by the allocation of adequate space to each use by adequate institutional mechanisms.

The evaluation of the effects of prohibitive uses of the estuarine environment is probably the most difficult problem currently facing technical management. The immediate and obvious effects on the habitat can be measured and described fairly easily, but the ultimate results of the modification of water movement patterns and flushing characteristics can be estimated only in general terms. Of primary concern is the effect of any physical modifications on the estuarine ecosystem and the limitations of knowledge mentioned above present a critical problem in present efforts to resolve prohibitive use-conflicts.

More difficult problems arise when a massive dredge or fill operation is proposed. When such modifications are necessary it may be possible to create new, equivalent habitats in a different part of the management unit equivalent to the one destroyed.

Disposal of liquid wastes to the estuarine environment is the major restrictive use. Technology exists to provide thorough treatment for nearly every kind of municipal and industrial waste, and there is no reason not to provide treatment sufficient to protect the environment from damage and to permit other uses. Treatment requirements for different wastes may vary according to local conditions, but damage to the environment can always be prevented.

Water quality standards have been set and are now being implemented in all the coastal states. These standards are the foundation upon which the effective control of estuarine pollution rests, and they provide the framework within which technical management can effectively operate. The long-range water quality goals of estuarine management should be to keep all waters safe for direct contact by humans and usable as a fish and wildlife habitat.

FUTURE POLLUTIONAL TRENDS AND EFFECTS

The amounts and impact of wastes generated by man's activities are a function of population growth, urbanization, industrial and commercial development, changing technologies, and consumption of goods and services. The effects of man's activities extend well beyond the defined area of the estuarine zone, both landward and seaward.

Fresh-water inflows

Many of the sources that determine estuarine water quality are external to the zone itself and the quantity and quality of fresh-water inflows are largely determined by upstream water use. Water diversion for irrigation, impoundment for flood control, and a host of other uses tend to cut the natural stream flow necessary for the successful assimilation and diffusion of both natural and man-made wastes.

Pressures for increased upstream diversion and use of fresh water are certain to increase in all biophysical regions, but the relatively arid and rapidly developing Western Gulf and Southwest Pacific coasts are projected to experience the greatest pressures on estuarine systems for at least three main reasons:

(1) Much of the upstream water supports irrigation with accompanying loss of water to the inflow systems by evaporation, transpiration, absorption, and mineralization through leaching.
(2) Rainfall and snow pack are highly variable and often result in extended periods of flooding or drought in these regions.
(3) Consumption of water other than for irrigation will increase at a high rate in response to an expected population growth well above the national average.

It should be noted, however, that these diversion projects may also allow an increased control of water inflows and this could be beneficial to the maintenance of existing estuarine productivity.

Municipal wastes

Municipal waste water disposal is the most frequently cited cause of water quality degradation. The extent of future municipal water pollution is indicated by the projection that even if secondary treatment were provided for all urban population served by sewers, the amount of residual wastes reaching the nation's waters would be about the same as it is today when much of this population is not served by secondary treatment facilities. From approximate coefficients developed by the Federal Water Quality Administration for municipal wastes generated in areas served by sewers, the following estimate (Table 4) can be made:

TABLE 4 APPROXIMATE MUNICIPAL WASTES GENERATED YEARLY BY THE ESTUARINE ZONE POPULATION OF THE UNITED STATES, 1960–1980

	1960	1970	1980	Numerical increase 1960–1980
Waste water, cu m	6.10	7.20	8.06	1.96
Standard BOD, millions of kg	1,011.3	1,193.6	1,336.91	325.2
Settleable and Suspended Solids, millions of kg	1,218.4	1,438.1	1,610.7	392.3

These projections are based on formulae found in the FWPCA publication, *The Cost of Clean Water*, Vol. II, "Detailed Analysis" (Washington, D.C.: U.S. Government Printing Office, 1968), p. 68

Although these figures are approximate, and understate the magnitude of the municipal waste load, they indicate the tremendous pressure increasing population will place on the water quality of the estuarine and coastal zone in the future. They do not take into account an increasing use of such appliances as washing machines, dish-washers, and garbage disposal units.

These figures are reasonable statements of pressures from urban populations, but suburban and rural populations presently not served by sewers will undoubtedly contribute further wastes to the estuaries. Beach front and estuarine communities, particularly resort-oriented developments, depend mostly on septic tank disposal. Waste seepage from septic treatment has been noted at

Long Island, the Florida communities, and the Delaware-Maryland-Virginia shoreline. Furthermore, many coastal communities are sewered with primary treatment facilities that often discharge directly into shallow back bays. These are no longer adequate, especially in the critical summer months.

The heavily populated estuarine-associated States such as California, New York, New Jersey, and Florida will require nearly two-thirds of the total expected $5.5 billion expenditure for waste treatment required in the near future. The estuarine portions of the marine States located in the Middle Atlantic biophysical region (New York to Delaware) will account for nearly 44 per cent of the total for these areas. These and other urban areas will require the fullest possible resources of technology, and planning, if water quality is to be enhanced or even maintained.

Industrial wastes

Although municipal wastes are a major source of pollution, manufacturing is the principal source of controllable water-borne wastes. The Federal Water Quality Administration estimates that manufacturing establishments are responsible for approximately three times more waste loading than that caused by the nation's population. Moreover, the volume of industrial production which gives rise to industrial wastes is increasing at about 4.5 per cent a year, or three times as fast as the population growth rate.

Approximately 85 per cent of the 14.2 trillion gal of water used by manufacturing plants in 1964 was accounted for by four major industry groups: Primary Metal Industries, Chemical and Allied Products, Paper and Allied Products, and Petroleum and Allied Products. Most of the increase in manufacturing water demands between 1954 and 1964 may be attributed to these four groups. This increase in demand may be expected to continue.

Estuarine areas with significant concentrations of high water use industries are: (1) Chemicals and allied products: New York-Northeast New Jersey, Philadelphia-New Jersey-Delaware and the Texas North Gulf; (2) Petroleum refining: Philadelphia-New Jersey-Delaware, Louisiana, Texas North Gulf and Texas South Gulf, and California; (3) Paper and allied products: Maine, South Carolina, Georgia-Eastern Florida, Central Florida Gulf, Mississippi-Alabama-West Florida, Oregon and Washington. All these industries have high growth potential.

Other high water use industries of importance to individual estuarine areas are: (1) Textiles: Massachusetts-Rhode Island, New York-Northeast New Jersey, North Carolina, and Mississippi-Alabama-West Florida; (2) Primary metals: Connecticut, Maryland-Virginia and the Texas North and South Gulfs; (3) Food and kindred products: Philadelphia-New Jersey-Delaware, North Carolina, Southern Florida Gulf, Central Florida Gulf, Louisiana, California and Oregon and Washington.

Thermal wastes

Although heated effluents come from many sources, electric power generation is estimated to produce 81 per cent of the total. Demand and supply has doubled every 10 years this century. Power requirements in 1980 will triple those of 1963. Areas of particularly rapid growth include Florida, parts of the Gulf Coast, Texas, and Puerto Rico.

Modern plants will be larger in unit size, and it is estimated that by 1975 about half of generating capacity will be nuclear fueled.

This last growth is significant because nuclear power plants require about 50 per cent more condenser water than fossil fuel plants of equal size. Twenty-one major projects affecting estuarine waters are planned to produce 15,187 megawatts with consequential heavy use of cooling water.

Solid wastes

Solid wastes, particularly those associated with urban areas and concentrations of industry, must be recognized as major hazards. The disposal of solid wastes becomes particularly acute as land surrounding cities is developed. The amount of land necessary to store and/or process solid wastes for ultimate disposal will nearly double from 1966 to 1976.

A New York study showed that residential wastes in 1965 amounted to 11 million tons and business wastes were 6½ million tons. By the year 2000 the combined total would be about 22 million tons. This trend is being repeated throughout the nation, particularly in metropolitan areas associated with estuaries.

Not only will solid wastes increase at a rate substantially exceeding population growth but their volume and character will alter through the use of non-degradable packaging materials such as plastics, adding greatly to total bulk. A great commitment of money, manpower, and technology will clearly be required to alleviate the ill effects of current practices and to prevent future damage.

The commercial fishery

The commercial fishing industry in the United States has grown relatively slowly over the years. From 1925 through 1966, the quantity of catch increased by only 60 per cent. During the same period, the amount paid to fishermen for their catch increased by something less than 100 per cent. Since 1964 the average annual catch per fisherman has remained below the 1957–1959 average.

Industrial uses of commercial fish, rather than human consumption uses, have accounted for most of the increase in tonnage, particularly in the period 1961–1966. Industrial uses of marine fish are primarily for fish oils, fish solubles, and fish meal (used for industrial processing), pet food, agricultural feed, and fertilizers.

The primary industrial fish is the estuarine-dependent menhaden. Its productive areas have been the Middle Atlantic, Chesapeake, South Atlantic and Gulf of Mexico biophysical regions. Production in the Middle Atlantic region has decreased markedly in recent years, and the catch in the Chesapeake Bay has fluctuated. Fishing pressure for menhaden in all regions has intensified, and stocks may have been over-fished in some areas.

The increased harvesting of industrial fish is ultimately dependent on renewable supplies of the resource. Although sizeable stocks of under utilized species exist, such as the thread herring in the Gulf of Mexico, other stocks may be over-fished or will be in the future. Further degradation or destruction of estuarine nursery grounds for menhaden could well reduce or eliminate this major source.

Penaeid shrimp, a valuable resource, are dependent upon the estuary for nursery grounds and are harvested in coastal shelf waters principally in the southern South Atlantic and Gulf of Mexico biophysical regions. Estuarine economic areas that support this fishery, and allied processing, are the Georgia-Eastern Florida Coast, the Louisiana Coast, the Mississippi-Alabama-West Florida Coast and the Texas North and South Gulf Coasts.

The shallow water shrimp fishery is estimated to be fully utilized and perhaps over-fished in the traditional South Atlantic and Gulf of Mexico grounds. Deep water shrimp supplies are estimated to be large and are relatively untapped, but there are considerable technological problems in locating and harvesting them.

It is uncertain whether shrimp stocks can meet the growing demand for them. Recent declines in shrimp landings have been noted in estuarine areas of relatively little industrial and population pressure, as well as in areas of considerable development. In Florida's Apalachicola Bay, the shrimp fishery experienced a dramatic decrease between 1964 and 1967 and nearby St. George Sound experienced a similar decline.

Galveston Bay has been a prime nursing ground for shrimp and a major area of shrimp harvesting and processing. These activities are threatened by industrial and municipal pollution, dredging and filling, and decreases in quantity and quality of fresh-water inflows. The total Galveston Bay catch declined drastically from 4,192,900 lb in 1962 to 1,941,000 lb in 1966.

The record of the oyster industry shows a loss in quantity and a decline in the productivity of the beds. Declines have taken place in nearly all estuarine areas that naturally supported oyster populations. Depletion has occurred for many reasons, both natural and man-induced.

Natural catastrophies have depleted the oyster beds in some areas. In Narragansett Bay, for example, the hurricane of 1954 is considered the prime factor in the destruction of beds and the decline of the processing industry. By 1956 the oyster harvest from Narragansett Bay had declined to 31,000 lb, from 252,000 lb in 1953. In 1957 the last oyster dealer went out of business.

Most of the reduction in domestic oyster production, however, can be attributed to man's activities. Examples are many. New Jersey's Raritan Bay, outstanding producer in the 19th century, is now almost barren, mainly due to municipal and industrial waste discharge. Many other areas of oyster production are closed because municipal wastes contaminate oysters with bacteria.

Silting due to dredge operations has appreciably diminished the quality of many oyster-producing areas. The silt may either smother the beds, or so seriously disturb the estuary floor as to cause deleterious effects from lowered levels of dissolved oxygen.

Perhaps the most difficult management problem is the legal labyrinth surrounding the ownership and use of oyster beds. The management and sound economic use of the oyster resource is almost impossible under present institutional constraints. These range from public ownership in Massachusetts to a tangle of leasing and private ownership in the Georgia coast area, the Chesapeake Bay, and James River estuaries.

The future of a viable oyster industry, and the continued availability of this nutritious food, is thus linked not only to the quality of the biophysical environment, but to the workings of the economic and institutional environment as well.

Landings of anadromous fish, particularly those of economic importance such as the salmon and shad, have steadily declined in recent years.

The diminution of the continental salmon fishery is a classic example of damage by modification of the environment. As dam building, lumbering, and other activities increased, the once-abundant salmon catches declined. The Atlantic salmon has almost completely disappeared from the east coast. On the west coast, reduction in the quality and quantity of fresh water, sedimentation in spawning areas, pollution of the transitional zones in estuaries, and heavy fishing pressure by both sport and commercial fishermen, have combined to reduce the once-flourishing salmon industry.

Future prospects

Examples of the decline and projected increasing pressures on the domestic commercial fishery are many. It is the conclusion of many experts that a harsh choice must be made in the near future: either the management of the nation's estuarine resources will be substantially strengthened, institutional constraints relieved, and the trend toward degradation of the estuarine environment stemmed, or the supply of commercially valuable finfish and shellfish will diminish.

Mariculture, the manipulation of the estuarine or marine environment to increase production of commercial species, is often cited as a method to overcome the depletion of natural stocks and fill increasing market demands. The ability of artificial culture to significantly increase yields has been proven in countries such as Japan, where shrimp, oyster, and certain finfish are raised on a profitable basis. However, the economic use of mariculture is in its infancy in the United States. The ultimate impact of mariculture would appear great. Yields may increase five to as much as twenty times. However, the present economic and social climate would seem to indicate that the impact of mariculture will be relatively slight in future decades. When other ancillary values are added, it would appear that proper management of the natural estuarine environment is a preferable course of action both to preserve and perhaps enhance the production of fish and to maintain the quality of this unique environmental resource.

INFORMATION FOR MANAGEMENT

In the 1966 mandate by Congress there was a directive that all existing pertinent information on the estuaries and estuarine zones of the United States be assembled, organized and coordinated. The key word in this instruction was the word "pertinent". In considering marine pollution, the coastal zone and the estuaries and estuarine zone, what is "pertinent" information? As a general classification of the kinds of information and data that are pertinent to understanding marine pollution problems, the total environment may be divided into three components.

The **biophysical environment** is the world of nature. Its description includes the natural state of the land, water, and life, and the changes they have undergone at the hand of man. The parameters of description include:—

land—its composition, configuration, vegetation, the uses to which it is put, and the changes and damages which it has suffered;

water—its quality, quantity, movement, the uses to which it is put, the factors which change or damage it, and the extent to which this change or damage has occurred;

life—its member species, their relative abundance, their use of an area, their importance to the total ecology, and the changes and damages they have undergone; and

climate—the normal range and extremes of temperature, humidity, precipitation, wind speed and direction, and the effect of the nearby ocean currents.

The **socio-economic environment** is the world of man and commerce. Its description includes uses of the environment for commercial enterprise and as a place to live and enjoy. The parameters of description include:

primary economic factors—commercial shipping, fishing, and mining; resort, residential, and industrial development; employment and directly water-related gains;

secondary and tertiary water-related economic gains— services to directly water-oriented economic elements and industries which would not be present in the absence of water such as seafood processing and shipping establishments;

recreational use—commercially oriented elements such as hotel-motel-marina development, charter boats, fishing piers, etc., and non-commercial recreation such as public access beaches, camping sites, and water-oriented public and private sports facilities; and

use as a habitat for man—population density, aesthetic values, safety of contact with this environment; and the changes and damages brought about by the reciprocal impacts of the social and economic segments.

The **institutional environment** is the world of law and regulation. Its descriptors include the identification of all entities which have authority to control any segment of the water-related environment, their scopes of interest, authority, missions, functions and activities; the types and sizes of the areas controlled or managed; and the administration of their management. These entities include all levels of government: International, Federal, interstate, State, intra-state, and local, as well as private entities and interested national organizations.

There is, of course, overlap among these three environments. For example, a game or bird refuge, part of the *biophysical* sphere is also a part of the *institutional* environment since it is established and managed by law. It also may restrict exploitation of mineral resources, which is a *socio-economic* element.

Obviously, within these three broad environmental components and their basic descriptors, are a multitude of individual parameters. These parameters must be chosen with some care, for the purpose of such a vast undertaking would be defeated if one component of the environment were accented to the detriment of our understanding of the other two. Yet, since these three environments are relatively separate, defining them offers a mechanism for data and information classification which is natural and simple.

The national estuarine inventory

During the National Estuarine Pollution Study (1966–1969), some 200 million individual items of coastal-related data were assembled and automated on the broad base of the three environment classifications outlined above. The data bank is called the National Estuarine Inventory (NEI), although the word "estuarine" is misleading. Actually the data encompassed in this system include land areas as far inland as the effect of the tides are noted, the quality of inflowing streams, and the ocean seaward to the 100-fathom contour.

The data and information assembled are directed towards one objective: management. While much of the data bank consists of detailed information which can be retrieved as needed, emphasis is placed on summary materials oriented for use by the manager, rather than by the technical analyst.

The basic storage and retrieval programs are unsophisticated with the data manipulation programs entirely external. This method serves a dual purpose. First and most important is flexibility. There are 26 coastal governments to serve as well as a multiplicity of Federal agencies and inter-state, regional and local governments with coastal interests, each of which has unique information requirements.

The second purpose is to save computer time. Simple retrievals do not have to include checking for manipulation options and search time is considerably shortened.

Definition of data gaps

The first five sections of this paper have illustrated the types of information gathered into the NEI. While the data bank has enabled a fuller description of the United States coastal zone than would otherwise have been possible, it has also served to delineate where data and information are lacking. Vast geographic areas have never been sampled for background ecological or water chemistry data, and many subjects vital to understanding the coastal environment have never been approached by research.

Basically, data and information are available in those areas where problems exist. The parameters studied are those of immediate concern only. For example, it is almost impossible to document damage to a non-commercial species of biota, no matter how important it may be to the food chain. Recreation and aesthetic damage are also very difficult to document.

There is also much missing information which exists but is not available for inclusion. This data gap consists of records held proprietary, files which require massive amounts of manpower to gather and organize before any attempt at automation can be made and, perhaps most important, information which is available in readily procurable forms such as navigation charts, microfilm files, and published reports. To extract data from this last category requires knowledgeable manpower, a commodity always in short supply. Nonetheless, efforts continue to assimilate these records and files of existing data into the National Estuarine Inventory.

Future of estuarine inventories

There is nothing new about storage and retrieval systems. They abound in both the scientific and business worlds. In the case of the National Estuarine Inventory, however, an old concept has been used for a new purpose: to attempt to gather the full spectrum of information needed to manage the complete coastal resource of a nation.

The available pertinent information on a wide variety of subjects has been gathered, organized, and coordinated. This compilation is vast and reveals many areas in which information is poor or lacking. Some can be obtained by careful, routine monitoring of the estuarine environment, but the acquisition of other knowledge requires an integrated, multi-disciplinary research and study program.

The most important knowledge to be gained by the use of such a system is a sufficient understanding of the estuarine environment to permit the recognition and interpretation of inter-relationships which, in turn, provide the capability to predict the effects of natural and human activities.

Many countries are involved in both institutional and technical management planning, plan implementation, and research in the coastal zone. An inventory automation system such as discussed is capable of supplying interested groups with data pertinent to their own needs with these two advantages: first, available information can be acquired from a single national source, providing a rapid-access baseline of usable information for the planner; secondly, knowledge gaps are indentifiable, making it possible for the manager, the scientist or the technician to concentrate in areas of true ignorance, and directing efforts to new or complementary, rather than duplicative, activities.

CONCLUSION

The ever-increasing and often conflicting social and economic demands of modern human civilization are placing significant pressures on the U.S. limited estuarine resources. The delicately balanced natural ecology of the estuarine zone has been subjected to over three hundred years of exploitation and alteration. Objective analysis of the results of this use and misuse shows that positive action is needed to preserve, conserve, and enhance the finite resources of the coastal zone.

It is the value of the estuarine zone as a fish and wildlife habitat, a recreational resource, and an aesthetic attraction that make the coastline a unique feature of the human environment, yet it is these very values that have been generally ignored in satisfying the immediate social and economic needs of civilization. The value of the estuarine and coastal zone for these uses is probably greater than its value for commercial exploitation. Unfortunately, we have not yet developed the ability to adequately express these social and humanistic values in quantitative terms. The value of the estuarine resource lies more in the multiple purposes it can serve than in the economic worth of a single use. Population and economic pressures are increasing more rapidly now than they have in the past, and continuation of present attitudes and approaches toward coastal zones can bring only an increasing rate of damage to its ecology and resources.

Properly supported and managed research to increase present knowledge and information can contribute greatly to effective management of estuaries and coastal areas. Over and above this must be added a stronger and better institutional environment to provide the integrated and comprehensive planning needed to convert the processes of loss and damage to actions leading to enhanced and broadened values.

References

Battelle Memorial Institute, Pacific Northwest Laboratories, The economic and social importance of estuaries. Report prepared under F.W.P.C.A., Contract No.14-12-115, for the National Estuarine Pollution Study, 219 p. (in press).

Bendix Marine Advisors, Inc., A case study of estuarine sedimenta-
1969   tion in Mission Bay-San Diego Bay, California. Report prepared by marine advisers for F.W.P.C.A. under Contract No. 14-12-425, 200 p.

Butler, P A   Pesticide residues in estuarine mollusks. In Proceed-
1967   ings of the National Symposium on Estuarine Pollution, Stanford University, Stanford, California, Vol. 1: 107–21.

Colberg, M R   The social and economic values of Apalachicola
1968   Bay, Florida. Report prepared by the University of Florida under F.W.P.C.A., Contact No. 14-12-117, for the National Estuarine Pollution Study, 58 p.

Gulf Universities Research Corporation, Case studies of estuarine
1969   sedimentation and its relation to pollution of the estuarine environment. Report prepared by Gulf Universities Research Corporation under F.W.C.P.A., Contract No. 14-12-445, for the National Estuarine Pollution Study, 280 p.

Hargis, W J   Final report on results of Operation James River.
1966   Spec. scient. Rep. Va Inst. mar. Sci., (7):73 p.

Harvey, H W   The chemistry and fertility of sea waters. Cam-
1963   bridge, Cambridge University Press, 240 p.

Kuenen, P H   Marine geology. New York, John Wiley and Sons,
1950   568 p.

Odum, H T, (Ed.)   Coastal ecological systems of the United States.
1969   Report prepared by the University of North Carolina under F.W.P.C.A., Contract No. 14-12-429, for the National Estuarine Pollution Study, 1878 p.

Pickard, G L   Descriptive physical oceanography. New York,
1963   MacMillan, 199 p.

Ralph Stone and Company, Estuarine-oriented community
1969   planning for San Diego Bay. Report prepared under F.W.P.C.A., Contract No. 14-12-189, for the National Estuarine Pollution Study, 178 p.

Rorholm, N A   socio-economic study of Narragensett Bay.
1969   Report prepared by the University of Rhode Island under F.W.P.C.A., Contract No. 14-12-93, for the National Estuarine Pollution Study, 200 p.

San Francisco Bay Conservation and Development Commission,
1968   The San Francisco Bay plan. San Francisco, California, U.S.A.

Stevens, M D   Solid waste disposal and San Francisco Bay. San
1966   Francisco, California, San Francisco Bay Conservation and Development Commission, 6 p.

Stommel, H   The Gulf stream, A physical and dynamical study.
1958   Berkeley, California, University of California Press, 202 p.

Sverdrup, H V, Johnson, M W and Fleming, R H   The oceans.
1942   Englewood Cliffs, New Jersey, Prentice-Hall, 1087 p.

U.S. Army Corps of Engineers, Survey report on Cooper River,
1967   South Carolina. Shoaling in Charleston Harbor. Charleston, S.C., U.S. Army Corps of Engineers.

U.S. Department of Commerce, Bureau of the Census, Statistical
1967   abstract of the United States. U.S. Department of Commerce, Bureau of the Census, 1050 p.

U. S. Department of Health, Education and Welfare, A study of
1962   water circulation in parts of Great South Bay, Long Island. Cincinnati, Ohio, U.S. Public Health Service, (Unpubl. Rep.) 25 p.

U.S. Department of the Interior, Federal Water Pollution Control
1966   Administration, Report on water pollution caused by the operation of vessels. Washington, D.C., U.S. Department of the Interior, F.W.P.C.A. 20 p.

U.S. Department of the Interior, Federal Water Pollution Control
1967   Administration, Proceedings of the conference on the pollution of Raritan Bay and adjacent waters. Boston, Massachusetts, U.S. Department of the Interior, F.W.P.C.A., Northeast Region, 448 p.

U.S. Department of the Interior, Federal Water Pollution Control
1968   Administration, Water quality criteria. Report of the National Technical Advisory Committee to the Secretary of the Interior. Washington, D.C., U.S. Government Printing Office, 234 p.

U.S. Department of the Interior, Federal Water Pollution Control
1969   Administration, National estuarine pollution study. Washington, D.C., Government Printing Office, 3 vols: pag. var.

U.S. Department of the Interior, Federal Water Pollution Control
1969   Administration, Report on pollution of the Potomac River in the Washington, D.C., Metropolitan Area. Charlottesville, Virginia, U.S. Department of the Interior, F.W.P.C.A. Middle Atlantic Region, 150 p.

University of Georgia, Advisory Committee on Mineral Leasing,
1968   A report on proposed leasing of State-owned lands for phosphate mining in Chatham County, Georgia. Athens, Georgia, Advisory Committee on Mineral Leasing, University of Georgia. 1:C-22.

Wastler, T A and de Guerrero, L C   National estuarine invent-
1968   ory handbook of descriptors. U. S. Department of the Interior, Federal Water Pollution Control Administration, 77 p.

The Gulf of St. Lawrence from a Pollution Viewpoint

R. W. Trites*

Etude du Golfe du Saint-Laurent du point de vue de la pollution

Le Golfe du Saint-Laurent, qui couvre une superficie d'environ 156.000 km² et est presque entièrement entouré de terres, peut être considéré comme une vaste zone estuarine. Le réseau fluvial du Saint-Laurent, dont le cours arrose sur quelque 3.200 km d'importantes zones urbaines et industrialisées, fournit au Golfe des matériaux de drainage provenant d'une zone qui couvre environ 1.300.000 km².

On dispose de renseignements sur les polluants spécifiques de quelques estuaires et baies situés au pourtour du Golfe, où des fabriques de pulpe de bois, des exploitations minières ou d'autres activités humaines ont eu visiblement des effets locaux. On a peu de données sur la charge que représentent ces polluants, mais il est possible que leurs effets se fassent sentir dans toute l'étendue du Golfe. Dans le passé, de grandes quantités de DDT ont été pulvérisées sur les forêts du Nouveau-Brunswick et on a observé que la vie aquatique avait souffert de graves dégâts dans les cours d'eau qui drainent ces régions. Bien que l'on n'ait pas encore eu confirmation de l'existence de concentrations appréciables d'insec-

El Golfo de San Lorenzo desde el punto de vista de la contaminación

El Golfo de San Lorenzo que tiene una superficie de cerca de 60.000 millas cuadradas y que casi se halla rodeado de tierra, se puede considerar como un gran complejo estuarino. Desde la red fluvial del San Lorenzo, que se extiende hacia el interior unas 2.000 millas hacia las principales zonas urbanas e industrializadas, el Golfo recibe las aguas procedentes de medio millón de millas cuadradas aproximadamente.

Se dispone de información sobre contaminantes específicos de algunas de las ensenadas del estuario y costeras situadas alrededor del Golfo, en donde han ejercido claros efectos locales las fábricas de pasta para pepel, la minería y otras actividades humanas. Se sabe poco de la extensión de estos contaminantes, pero es posible que en gran parte del Golfo puedan experimentarse efectos detectables. En el pasado se han usado grandes cantidades de DDT en los bosques de Nueva Brunswick, habiéndose observado graves daños en la vida acuática de los ríos que recorren esta zona. Aunque todavía no se ha confirmado la existencia de apreciables

---

* Marine Ecology Laboratory, Fisheries Research Board of Canada, Bedford Institute, Dartmouth, Nova Scotia, Canada.

ticide dans la chaîne trophodynamique, on a constaté qu'au moins une colonie de fous de Bassan parmi celles vivant dans cette région se ressent de ses effets comme le prouve la ponte d'oeufs à coquille mince. D'après des renseignements préliminaires, les taux de concentration du mercure sont relativement faibles chez les poissons capturés dans le Golfe.

L'auteur passe en revue les caractéristiques océanographiques générales communes du Golfe et, avec ces renseignements de base, formule des observations et des suggestions en vue de recherches éventuelles intéressant la pollution.

concentraciones en la cadena alimentaria, al menos una colonia de alcatraces del Golfo pone huevos de cáscara delgada y débil que le pueden causar perjuicios. La información preliminar sobre la concentración de mercurio en los peces capturados del Golfo indica que esta es relativamente baja.

Se examinan los rasgos oceanográficos generales del Golfo y con esta información se hacen comentarios y sugerencias para efectuar posibles investigaciones futuras desde el punto de vista de la contaminación.

WHILE we currently have but scanty information on the distribution, life history and action of most contaminants in the sea, it is clear that as long as we continue to "get rid of" wastes by discharging them into natural waterways we can expect some of them to show significant accumulations in large marine basins. Large basins adjacent to major river systems are of special concern and the Gulf of St. Lawrence in eastern Canada is one such area.

This Gulf plays an important role either directly or indirectly in the livelihood and recreation of a large number of people. About one quarter of all Canadian fishermen catch nearly 25 per cent of the total Canadian commercial fish production there; it is an important avenue of transportation even though the service is impeded or interrupted by winter ice; in summer, it is used extensively for recreational activities; it is the site for active oil exploration; predictably, it also provides a waste receptacle for many millions of people and much of the industrial wastes of eastern and central United States and Canada.

To date most pollutant measurements and pollution research in the Gulf of St. Lawrence have been confined

to areas relatively near to urban areas or industrial operations where man-made activities have had obvious local effects. It has been generally assumed that the capacity of the main body of the Gulf to accept and dispose of wastes is sufficient to keep concentrations in the environment, as well as in organisms, below unacceptable levels. In view, however, of the situation in such areas as the Baltic (ICES, 1970), it is obvious that we must place the Gulf of St. Lawrence under more careful scrutiny, since it receives the wastes from nearly 30 million people and their industrialized activities. The character, stability and behaviour of the pollutants will determine what proportion of these waste products will eventually reach the Gulf, and beyond. This paper will review what is known about the physical oceanographic features of the Gulf so as to judge its physical capacity to "dilute" the various pollutants discharged into it.

PHYSICAL FEATURES

Hydrography and physiography

The Gulf of St. Lawrence has an area of 214×10^3 km² (Forrester and Vandall, 1968). The principal connection

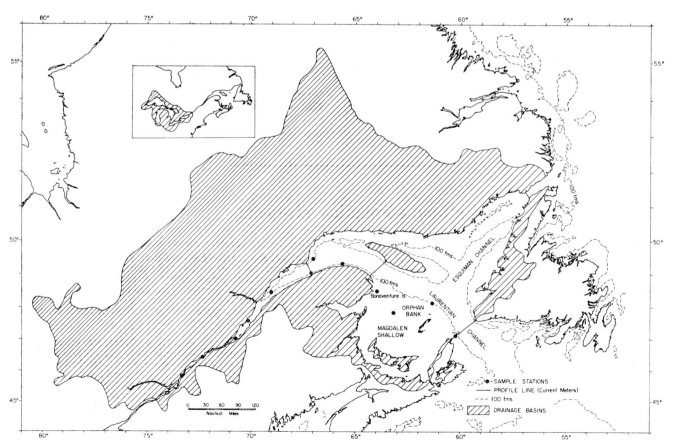

Fig. 1. Map showing drainage basin of Gulf of St. Lawrence, location of sampling stations taken in July 1970, and cross-section locations where extensive current measurements have been taken

with the Atlantic Ocean is through Cabot Strait (figs 1, 2) with a width of 104 km, a maximum depth of 480 m and a cross-sectional area of 35×10^6 m². A more restricted connection is through the Strait of Belle Isle, with a minimum width of 16 km, a maximum depth of 60 m, and a minimum cross-sectional area of 10^6 m². A dominant bathymetric feature of the Gulf is the Laurentian Channel, a deep trench with a maximum depth of 500 m, extending from the continental shelf to the mouth of the Saguenay River (fig. 1). The Esquiman Channel branches off the Laurentian Channel and extends from the central part of the Gulf northeastwards towards the Strait of Belle Isle. A second major feature is the large shallow area south of the Laurentian Channel known as the Magdalen Shallow, where depths are mostly less than 50–75 m. One quarter of the Gulf is shallower than 50 m, while less than one fifth is deeper than 300 m (Lauzier *et al.*, 1957).

Fresh-water discharge

The St. Lawrence River, which has a drainage basin extending inland approximately 2,000 mi, constitutes the largest single source of fresh water. According to Parde (1948), the mean annual discharge at Quebec City is 1.04×10^4 m³/sec. Figure 1 shows that a very significant proportion of the Gulf of St. Lawrence drainage basin lies seaward of Quebec City. Although the watershed areas in New Brunswick, Nova Scotia, Prince Edward Island, and Newfoundland are relatively small, the rivers discharging into the Gulf along the northern shores are relatively large and numerous.

In an attempt to determine the total fresh-water input to the Gulf, data have been extracted from the published Water Resource Papers (Canada) and extrapolated to cover unmetered drainage basins. In addition, precipitation figures have been derived from the Gulf area by extrapolating from shore-based stations. A mean evaporation rate of approximately 75 cm/year (assumed constant throughout the year) has been used to compute the net influx of fresh water. Lauzier (1962) has carried out some preliminary calculations on the precipitation and evaporation for the Gulf on a seasonal basis. The computations of precipitation, P, minus evaporation, E, (i.e. P—E) revealed marked variations seasonally. Moreover, large differences were found between the northern and southern part of the Gulf, yielding a mean annual value of P—E of + 6.1 in and − 1.5 in for the northern and southern sectors respectively. Lauzier's results indicate that including time and space variability in P—E would provide a significant refinement in computing the fresh-water budget of the Gulf.

A summary of the fresh-water budget is given in Table 1. These computations indicate that the net total fresh-water input varies from a mean monthly minimum of approximately 8.8×10^3 m³/sec in March to a maximum of approximately 28.4×10^3 m³/sec during May. A comparison of the discharge at Quebec City to the mean annual total input of 15.8×10^3 m³/sec indicates that the St. Lawrence River supplies approximately two-thirds of the total input to the Gulf.

Water masses

In terms of vertical temperature distribution, the Gulf in summer can be considered as a three-layer system. At

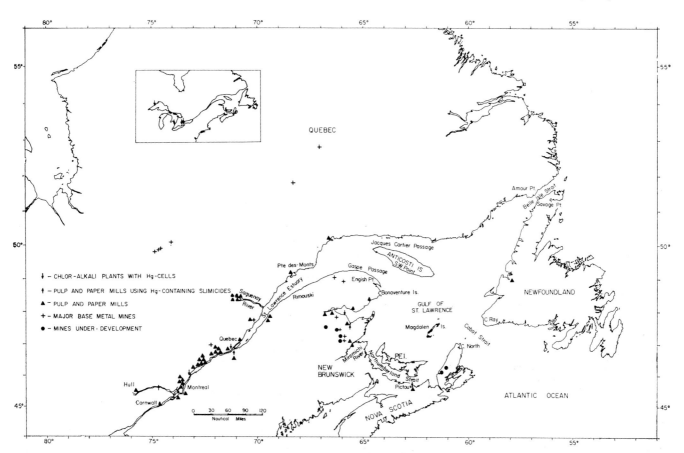

Fig. 2. Map showing location of some of the possible pollution sources for Gulf of St. Lawrence

[61]

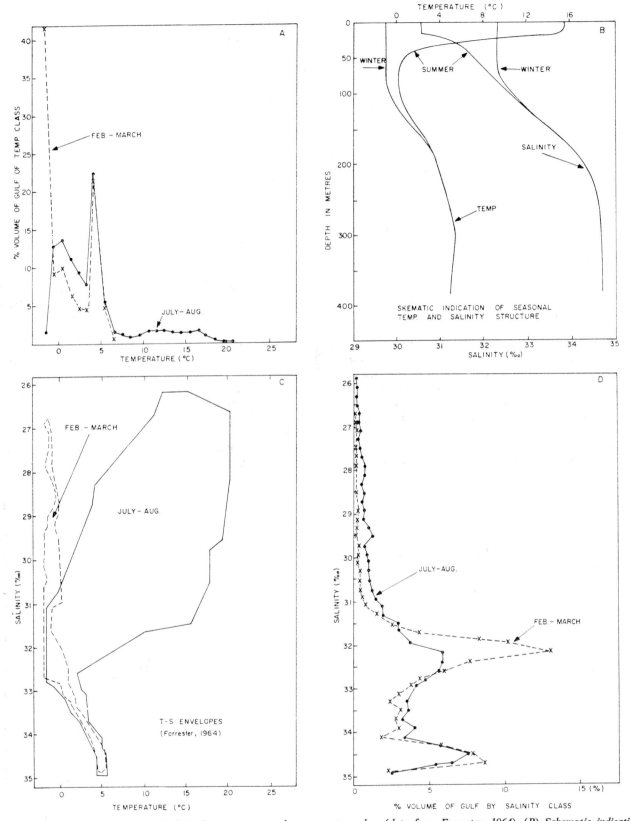

Fig. 3. (A) *Percent volume of Gulf of St. Lawrence waters by temperature class (data from Forrester, 1964), (B) Schematic indication of seasonal temperature and salinity structure, (C) T-S envelopes for February-March and July-August in Gulf of St. Lawrence (after Forrester, 1964), (D) Percent volume of Gulf of St. Lawrence waters by salinity class (data from Forrester, 1964)*

this time, a warm surface layer, 10–20 m thick, overlays an intermediate cold layer that is usually less than 1°C, and a deep warm layer (4–6°C) with a temperature maximum at 200–300 m (fig. 3B). The two upper layers undergo seasonal variations and become one, thermally,

during winter. The origin of the intermediate cold layer has been the subject of considerable speculation (Tremblay and Lauzier, 1940; Lauzier, 1958; Forrester, 1964; Banks, 1966). Earlier it was thought that this cold water was formed outside the Gulf as its T–S characteristics were very similar to Labrador Current Water. With the acquisition of additional data, it became evident that the formation of the cold-water layer is related to the cooling of the upper layers of the Gulf and to the production of the mixed layer in winter. The current thinking is that the cold layer is formed almost entirely locally during the winter.

TABLE I. CALCULATED FRESH-WATER BUDGET FOR THE GULF OF ST. LAWRENCE

Month	Total river input (m^3/sec)	Total precipitation over Gulf (m^3/sec)	Precipitation less evaporation (m^3/sec)	Net total fresh-water input (m^3/sec)
Jan.	8,116	9,485	3,380	11,496
Feb.	7,333	8,587	2,482	9,815
Mar.	7,779	7,161	1,056	8,835
Apr.	14,124	6,258	153	14,277
May	29,268	5,275	−830	28,438
June	22,791	9,421	3,316	26,107
July	14,459	6,340	235	14,694
Aug.	13,198	10,073	3,968	17,166
Sept.	10,749	8,646	2,541	13,290
Oct.	11,268	9,788	3,683	14,951
Nov.	12,042	10,508	4,403	16,445
Dec.	10,192	10,121	4,016	14,208
Mean	13,443	8,472	2,367	15,810

Forrester (1964) has summarized concisely and quantitatively the T–S composition of the Gulf. His analysis permits one to readily see the major water masses, what proportion of the total volume of the Gulf they occupy, and how they vary from summer to winter. He determined the total volume of water falling inside T–S boxes of size 1°C by 0.2‰. Figure 3(C) shows an extract from his analysis. The envelopes that are shown encompass most of the points on the T–S scatter diagram. Figures 3(A) and 3(D) also show the volume of water by salinity and temperature class respectively for 0.2‰ and 1°C intervals (the points have been joined for visual convenience). On the T–S diagram, the most striking change from summer to winter is the shrinkage in area representing the water in the upper 75 m. This prominent part of the envelope, however, represents less than half the volume of the Gulf. Examination of figs 3(A) and 3(D) indicates a very strong T–S mode centred at 5°C and 34.6‰ both in summer and winter. About 17 per cent of the Gulf water falls within ± 0.5°C and ± 0.3‰ of this point. In winter, a strong mode centred at − 1.5°C and 32.2‰ is evident. About 26 per cent of the water in the Gulf lies within ± 0.5°C and ± 0.30‰ of this point. The residue of this mode is still present in summer at 0.0°C and 32.4‰, slightly shifted from its winter position. About 8 per cent of the water in the Gulf falls within ± 0.5°C and ± 0.3‰ of this point. On the basis of the T–S diagrams, the summer-winter differences in water with salinity > 33‰ appear insignificant. This value indicates that approximately 45 per cent of the volume of water in the Gulf is little affected by local seasonal changes in heat budget and fresh-water inflow.

The temperature maximum of the deep layer may vary seasonally from about 4°C to 6°C, and the salinity at the maximum seldom departs from 34.6‰ by more than 0.2‰. The source of this water is outside the Gulf (Lauzier and Bailey, 1957) and its volume correlates with its maximum temperature, which in turn can be correlated with the temperature of Labrador Water (Lauzier and Trites, 1958). Observations indicate that the maximum temperature varied from a low of about 4°C in the 1920's to a high of nearly 6°C in the 1950's, accompanied by an increase in volume of the deep layer.

Ice

For several months each year, ice in varying concentrations is present in the Gulf. It arises from three sources (El-Sabh, 1969): (a) Labrador ice that enters through the Strait of Belle Isle; (b) ice from the St. Lawrence River and Estuary; and (c) ice fields which are locally formed in the Gulf. Based on five years of ice cover data from 1961 to 1965 published by the Meteorological Branch, Matheson (1967) has compiled mean ice concentration maps at fortnightly intervals during the ice season. Five of these have been selected to show the general features. Ice starts to form in December in sheltered areas. During January, ice increases rapidly, although west of Newfoundland remains unfrozen due to the influx of warmer surface water through Cabot Strait. By the end of January, the southwestern and central parts of the Gulf are covered by heavy ice originating in the St. Lawrence River and Estuary. As winter progresses, ice moves seaward through Cabot Strait. Major ice concentrations usually persist until April when a break-up commences.

Tides and tidal currents

The semidiurnal and diurnal tides from the North Atlantic Ocean are both propagated through Cabot Strait (Farquharson, 1962) and are illustrated in figs 4A and 4B, which show the semidiurnal lunar tidal constituent M_2 and the lunisolar diurnal tidal constituent K_1 respectively. There are two amphidromic points for the M_2 constituent—one near the Magdalen Islands and a second near the western end of Northumberland Strait. In most areas of the Gulf, the semidiurnal constituent dominates. Tidal range increases rapidly towards the St. Lawrence River with a mean range of about 13 ft near Quebec City.

Except in the St. Lawrence Estuary, Cabot, Belle Isle and Northumberland Straits, and other locally confined regions, tidal currents seldom exceed 0.5 kn. In Cabot Strait, tidal streams are typically of the order of a knot. In some areas, the phase of the tidal stream varies significantly with depth. Forrester (1970) has found evidence of internal tides in the St. Lawrence Estuary seaward of the Saguenay River entrance. It is possible that internal tides exist throughout much of the Gulf, but sufficient data are lacking at present to clarify the situation.

Circulation

Current measurements have been taken across Cabot Strait (1959, 1966), Gaspé Passage (1962), St. Lawrence Estuary at Pte. des Monts (1963), Belle Isle Strait (1963), and near Rimouski (1965), for periods of approximately one month (Farquharson, 1962, 1966; Farquharson and Bailey, 1966; Forrester, 1967). Self-recording current meters were moored, usually at three depths, at selected sites along a cross-section. In addition, extensive current measurements were undertaken in Northumberland

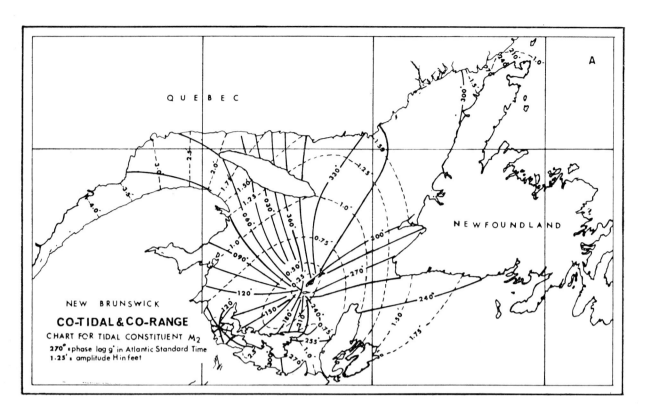

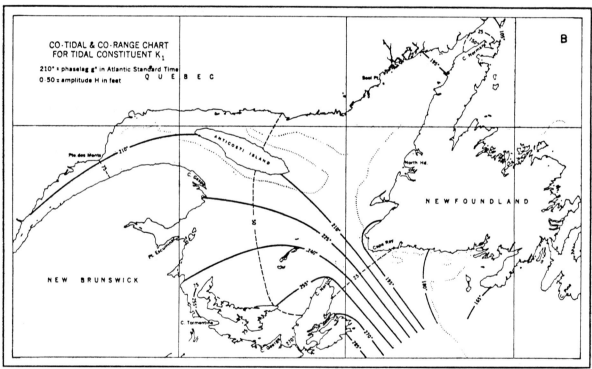

Fig. 4. (A) Semi-diurnal lunar tidal constituent, M_2 (after Farquharson, 1962), (B) Lunisolar diurnal tidal constituent, K_1 (after Farquharson, 1962)

Strait in connection with a study of a proposed causeway (Farquharson, 1959). Single station moorings have been placed at other selected sites in the Gulf by staff of the Bedford Institute and operated for periods of about one month.

Data from four sections are shown in fig. 5. The St. Lawrence Estuary and Gaspé Passage sections clearly show the outflow of surface water along the Gaspé coast. This outflow is mostly confined to the upper 25–50 m.

An equally prominent feature is the upstream current immediately below the seaward-moving surface current, with its core at about 100 m depth. The flow through Cabot and Belle Isle Straits appears to be somewhat differently structured. The outflow through Cabot Strait is similar to that in the Gaspé section in that the strongest currents are associated with the brackish seaward-moving layer, but the upstream current appears less well defined and occupies a proportionately larger area

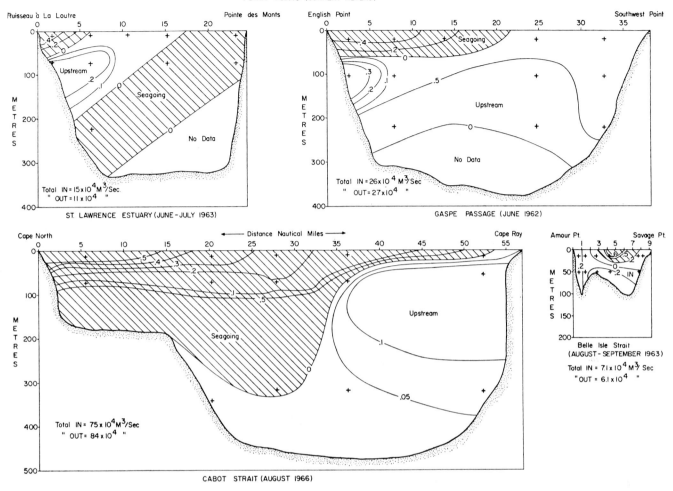

Fig. 5. *Residual currents through four sections in Gulf of St. Lawrence, as determined by direct current measurements*

of the section. Fluctuations in the daily residual flow occur in all sections and at times it may be unidirectional throughout the entire section, or reversed.

Circulation patterns have also been studied in the Gulf using Lagrangian techniques. Drift bottles and sea bed drifters have been used by Bumpus and Lauzier (1965) and Lauzier (1967), and parachute drogues have been employed by other investigators (Blackford, 1965, 1967; Trites, 1968). Most of the observations of currents have been taken during summer months. Very few measurements have been made when the Gulf is ice covered. However, the information referred to above, together with geostrophic computations and the recent drift-bottle observations of Lauzier (personal communication), permits one to sketch a typical summer surface circulation pattern (fig. 6) and indicates the daily drift in selected areas. The dominant features are: the general two-way flow in both entrance Straits; the counterclockwise circulation in the interior part of the Gulf; and the Gaspé current, which begins to develop in the Rimouski-Pte. des Monts areas and extends throughout the entire length of the Gaspé coast. Highest speeds are found in the Gaspé current and in the outflow through Cabot Strait where values of 10–20 mi/day are reached.

Data on subsurface currents in the Gulf are sparse and it is not feasible at present to form a picture for the entire Gulf at any season. Sea bed drifters released by Lauzier

(1967 and 1970, personal communication) in the southern and central part of the Gulf show in general a well-marked seaward movement along the 50–100 fath contour along the southeastern border of the Laurentian Channel, that is, along the edge of the Magdalen Shallows. An inward flow usually is present along the 50–100 fath contours on the northwestern side of the Laurentian Channel. A rather complex pattern emerges for the southwestern Gulf, although a large area ($\sim$7,000 mi^2) surrounding the Magdalen Islands shows a general convergence towards the Islands. Residual bottom currents based on sea bed drifter experiments appear to be mostly in the range of 0.3–0.7 mi/day.

Theoretical models based on simplified forms of the equations of motion have been used to study the gross features of the circulation in the Gulf (Blackford, 1965, 1966; Murty and Taylor, 1969). These models considered wind a primary driving force; in general they have revealed a pattern rather similar to that of fig. 6. It seems clear from the success of these models that the wind, as well as fresh-water discharge, plays an important role in producing the large-scale surface layer circulation in the Gulf.

Flushing time

It is important to have some knowledge of how rapidly a pollutant may be carried from its point of discharge,

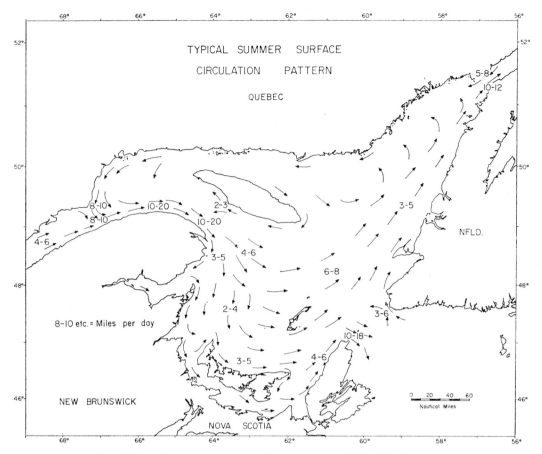

TYPICAL SUMMER SURFACE
CIRCULATION PATTERN

QUEBEC

NFLD.

8-10 etc.= Miles per day

NEW BRUNSWICK

NOVA SCOTIA

Fig. 6. Typical summer surface circulation pattern in Gulf of St. Lawrence

and at what rate it is dispersed. When detailed current and diffusion measurements are not available, it is sometimes helpful to estimate flushing times from very simple models, using parameters more readily measured than currents and diffusion. The following elementary considerations can be useful:

If S is the salinity of water at any point in an estuarine system and S_0 is the salinity of the sea water outside the system, and the source of water used for mixing, the fresh-water fraction at that point is:

$$f = \frac{S_0 - S}{S_0}$$

The total quantity of fresh water within the system is given by:

$$F = \int_{vol} f \, d(vol)$$

where the integration is carried out over the entire volume of the estuary.

If D represents the rate of inflow of fresh water to the system and a steady state is assumed, then the rate of removal of fresh water must also equal D. The flushing time t may then be defined as the time taken to remove the equivalent of the accumulated volume of fresh water present at a given instant at this rate and is given by:

$$t = \frac{F}{D}$$

Half of this time is sometimes referred to as the mixing half-life of the system. Data for making such a calculation of flushing time are available for the Gulf.

From a cursory examination of approximately 20 oceanographic sections taken at various times of the year across Cabot Strait, the average salinity on the southwest or Cape Breton side was approximately 29.1‰ at the surface and 31.2‰ at 50 m depth. On the northeast or Newfoundland side, the salinity averaged about 31.6‰ at the surface, 32.2‰ at 50 m, and the 33‰ isohaline was, on an average, at about 130 m depth. From the current measurements across Cabot Strait, and from geostrophic computations carried out by MacGregor (1956), the core of inflowing water appears to be centred on the Newfoundland side of the Strait and to be mostly confined to the upper 150 m. If it is assumed that the core of the inflowing water is at 130 m, and S_0 taken as 33‰, and the quantitative salinity computations of Forrester (1964) are utilized, then the total quantity of fresh water accumulated in the Gulf is equal to 637 km³ in February-March and 975 km³ in July-August. It has been calculated that the total fresh-water input (D) to the Gulf varies from a minimum of 8.8 × 10³ m³/sec in March to a maximum of 28.4 × 10³ m³/sec in May. The yearly mean is 15.8 × 10³ m³/sec (Table 1). If one estimates the mean volume of fresh water in the Gulf by averaging the February-March and July-August salinity data, then the mean flushing time based on mixture with sea water of 33‰ is approximately 510 days. This is likely to be near the upper limit of possibilities. Alternatively, if it is assumed that the bulk of the inflowing water is confined to the upper 75 m on the Newfoundland side of the Strait, then S_0 might better be chosen at about 32‰. Under these assumptions, the

[66]

mean flushing time would be approximately 220 days.

There are other oceanographic observations which permit one to make some refinements in the choice between these alternatives. For example, MacGregor (1956) concluded that the seaward transport through Cabot Strait is approximately 10^6 m³/sec which is about 60 times the fresh-water discharge. Similarly, the current meter data of 1966 (fig. 5) gave a total seaward transport of 0.84×10^6 m³/sec which is about 50 times the mean fresh-water discharge. If it is assumed that the bulk of the seaward-moving water through the Strait is in the shape of a wedge 50 m thick on the Cape Breton side and thinning to zero on the Newfoundland side, the mean salinity of this layer would appear to be approximately 31.3‰. Combining this value with the above transports implies that the diluting water must on average be in the neighbourhood of 32‰. Thus it would appear that the mean flushing time of the fresh water in the Gulf is more likely to be in the neighbourhood of 6–8 months.

Figure 6 gives an idea of how long it would take a surface particle in the mean flow to move through the Gulf from the Saguenay to Cabot Strait. A particle trajectory in the general direction of the arrows indicates that in summer months the particle would reach Cabot Strait in just under three months.

POLLUTION CONCERNS

Pesticides

Quebec and the Atlantic Provinces are the last North American stronghold for Atlantic salmon and eastern brook trout. Both species may attain most of their growth in the sea, but at least two or three years of their early life must be spent in fresh water. The Miramichi system in New Brunswick (fig. 2) is the greatest producer of Atlantic salmon in the world. Unfortunately, this area is a major producer of timber and a huge aerial forest spraying programme has been mounted against the spruce budworm. Fifty-six per cent of the province's 24,000 mi² of forest has been sprayed with DDT since the insect control programme began in 1952, and as much as 37 per cent of the area has been sprayed in a single year (Sprague and Ruggles, 1967). During the first years, effects on fish were particularly serious, with mortality of young salmon in sprayed areas varying from 50 per cent to 98 per cent depending on the size of the fish and the manner of spraying. Organophosphate insecticides, such as Sumithion, are now being used in place of DDT. These appear to be much less damaging to fish.

The post-application fate of organochlorine pesticides is relatively unpredictable. Numerous studies have shown, however, that the disappearance of the pesticide from the site of application is usually due to transport rather than its degradation to harmless substances. Although highly variable, the half-life of DDT is measured in decades. Portions of the enormous total quantity of DDT applied to areas of New Brunswick may therefore yet be in transit via hydrological or meteorological processes.

While forest spraying operations in New Brunswick were the largest single use of organochlorine compounds in the immediate vicinity of the Gulf of St. Lawrence, there is reason to expect that other significant sources of input continue to exist. More than half of all fresh water reaching the Gulf is contributed by the St. Lawrence River which drains extensive farmlands. U.S. studies show that runoff from agricultural watersheds is the greatest single source of pesticides (Nicholson, 1970). Although improvements in pesticide regulation continue to be made both in the U.S.A. and Canada, the use of a wide spectrum of persistent organochlorine compounds is still permitted. There is ample reason to believe that the Gulf of St. Lawrence is subject to appreciable waterborne organochlorine contamination and to airborne input from the predominantly continental weather patterns affecting the Gulf.

The rate of input of polychlorinated biphenyls (PCB's), plasticizing agents used in many industrial processes, is unknown at this time but they may be appreciable due to heavy industrialization of parts of the lower Great Lakes drainage basin and St. Lawrence River valley. It is possible that PCB contamination of the Gulf could be a major problem in the near future.

While the direct effects of DDT and other organochlorine compounds on salmon and trout have been well established, little is known about the accumulation of pesticide residues in the marine food chains of the Gulf of St. Lawrence. However, the gannet colony on Bonaventure Island off the Gaspé coast is suffering from the production of thin-shelled eggs. The primary food of gannets is medium-sized fish such as mackerel and herring. Brown, of the Canadian Wildlife Service (personal communication), believes that the major feeding areas for this colony are: (a) the Gaspé Passage area; (b) the Magdalen Shallow (Orphan Bank); and (c) near the mouth of Chaleur Bay. Analyses of mackerel from the vicinity of Prince Edward Island (Duffy and O'Connell, 1968) and Miramichi (Sprague et al., 1969) indicate that they would contribute in the order of 0.1 to 1.0 ppm of DDT and its metabolites to predators. If the mackerel and herring in the gannet feeding areas exhibit similar levels they are sufficient, together with the food chain concentration effect usually found to occur with persistent organochlorine compounds, to account for the observed impairment of gannet reproduction.

Limited analyses suggest that, other than mackerel, a number of commercial species of fishes and shellfish from the Canadian Atlantic possess relatively low level DDT (Sprague et al., 1969). Mackerel, however, are normally present in the Gulf for only a few months of the year and migrate seasonally over considerable distances. Although their migration and habits are not fully known, mackerel present in the Gulf in summer may have been along the American seaboard and as far south as Long Island earlier in the year (Sette, 1950). It is possible, therefore, that an appreciable fraction of the DDT in mackerel caught in the Gulf has been accumulated prior to their entry into Gulf waters.

Pulp and paper industry wastes

There are approximately 50 pulp and paper mills in Quebec and the Atlantic Provinces which discharge their effluents into the drainage basin of the Gulf of St. Lawrence. The bulk of these are along the St. Lawrence River, and in the harbours and estuaries surrounding the Gulf (fig. 2). The effluents, which may be toxic, place an increased demand on the available oxygen supply, and are aesthetically displeasing. To date, measurable effects which could be directly attributed to these wastes have been confined to the area of discharge. However, on a recent oceanographic cruise, spot sampling from Cabot Strait to Montreal revealed that dissolved oxygen values in the river above Quebec City (figs 1, 2) are only about 85 per cent of saturation in a water column that was completely mixed from top to bottom with currents of 3–5 kn. This lowered oxygen concentration may suggest that a significant oxygen demand is being made by domestic and pulp mill wastes.

A comprehensive survey is in progress on the Miramichi Estuary to investigate the effects of pollution from pulp mills and other sources on the migratory movements of salmon. It has been found that salmon commonly delay entering or avoid areas of industrial pollution (Mann and Sprague, 1970).

Effluent from a 500 t/day bleached kraft (sulphate) pulp mill recently established near Pictou, Nova Scotia, has received considerable study. The wastes are held in a natural lagoon before they reach the sea. The retention time appears to be short and the escaping effluent still has a considerable oxygen demand. Moreover, the pond is relatively ineffective in removing the smaller fractions of suspended matter, and these flocculate rapidly when the effluent is discharged into salt water (Bewers and Pearson, personal communication). The resulting pollution of the shoreline near the outlet has damaged recreational facilities. This area is adjacent to an important lobster fishery and concern for the welfare of the lobster stocks is understandable. Bioassays with effluent material have indicated that lobster larvae are moderately resistant to bleached kraft pulp mill waste (Sprague and McLeese, 1968). Field survey work on the distribution and abundance of adult lobster and their larvae has failed to demonstrate any direct effect on them. However, some of the filter-feeding organisms in the area show evidence of an altered chemical environment. This has been established in the course of chemical studies at the Bedford Institute employing neutron-activation techniques. The concentration of some 25–30 isotopes has been determined in samples of clams and foraminifera taken at Pictou, N.S., not far from the discharge of the pulp mill. Samples from two stations about a mile apart have been analysed and show significant variation in chemical composition (Bewers and Pearson, personal communication). Clams near the outlet showed a significantly higher concentration of some 10 isotopes. Iron, bromine, scandium, and several rare earth elements showed large differences. By comparison, calcarious foraminifera showed much less variation in composition, but the concentration of mercury was significantly higher than in the clams at both stations. As adequate background data from other areas are as yet unavailable, it would be premature to judge the overall implications of these preliminary results.

Mercury

Concern about mercury pollution has risen rapidly in the past two years. Fimreite (1969) has reported on mercury uses in Canada and the possible hazards of mercury contamination. Mercury is used for a variety of purposes but the Chlor-alkali industry is undoubtedly the most important consumer in Canada. Altogether these plants have an inventory of approximately 2,000,000 lb of mercury and require about 2,000,000 lb annually to make

up for mercury losses. Most of these losses find their way to streams and lakes (Fimreite, 1969). In fig. 2, the distribution of Chlor-alkali plants using mercury cells is shown for the Gulf of St. Lawrence drainage basin. As all but two of the Canadian plants using mercury cells are within the Gulf of St. Lawrence drainage basin, nearly 200,000 lb must be released into the watershed annually. The locality of the sites shown in fig. 2, indicates that risks of mercury contamination are greater in the St. Lawrence River and Estuary than in the Gulf. Lesser amounts of mercury may reach the Gulf from other sources, including pulp and paper mills. In 1968, five mills situated in the Gulf of St. Lawrence drainage basin were reported to be using mercury compounds as slimicides. The total annual usage was probably of the order of only 3,000 lb of phenylmercury acetate (Fimreite, 1969). Small quantities of mercury can also be expected to be present in the sodium hydroxide made in Chlor-alkali plants employing mercury electrodes. In pulp mills using NaOH, some mercury is likely to be present in the waste waters. For a mill consuming 20–30 tons of NaOH per day, roughly 100 g of Hg is involved. It is possible that the mercury detected by Bewers and Pearson (personal communication) in suspended sediments taken near the effluent discharge of the pulp mill near Pictou came through this route.

Mercury compounds are also employed as seed-dressings to control soil and seed-borne diseases, in horticulture as foliar sprays, in turf fungicides, paints, pharmaceuticals, dental preparations, and in scientific laboratories. The total loss of mercury to the watershed of the Gulf through these uses probably does not exceed 20,000 lb annually.

The recent discovery of relatively high mercury concentrations in commercial fish species in several areas of Canada has prompted an accelerated sampling and analysis programme (Bligh, 1970). High concentrations have been measured in some species of fish taken within the Great Lakes system and in the St. Lawrence River and Estuary as far seaward as (and including) the Saguenay Estuary. In the Gulf of St. Lawrence, analyses have been undertaken on clams, herring, flounder, mussels, lobsters, crabs, oysters and shrimps. Preliminary results indicate that apart from a few samples taken near known mercury sources, levels have not exceeded the 0.5 ppm (wet weight) guideline of the Department of Health and Welfare (Bligh, personal communication). Relatively few analyses have been undertaken on the biota from the St. Lawrence Estuary seaward of the Saguenay. From the distribution of known mercury sources, it is likely that values in this area will be somewhat higher than in other parts of the Gulf.

The foregoing results suggest that there is no gross mercury pollution in the Gulf at present. However, the current situation is unlikely to represent an equilibrium level.

Petroleum

The immediate, short-term effects of oil pollution are in many respects obvious. Coastal fouling and the damage to bird populations have been documented abundantly. Catastrophic spills, such as the Torrey Canyon, Santa Barbara, and most recently the Arrow on the Atlantic Coast of Nova Scotia, bring dramatic aspects to the forefront, but these accidents contribute only a very small fraction of the oil entering the marine environment (Hoult, 1969). The greater ecological impact may result from the day-to-day spills or discharges which take place during "normal" shipping operations and which, with present regulations, are difficult or impossible to control.

Surveys undertaken after the *Arrow* disaster in February 1970, revealed that oil particles as small as a few microns were present in the water column, not only in the vicinity of the wreck but for distances of at least 200 km from the source. Particles were observed to depths of at least 80 m. Within a 10–20 km radius of the wreck, typical concentrations in the water column below the sea surface were in the neighbourhood of 20 ppb (Forrester, 1970).

Absorption and fluorescence spectrophotometric methods for measuring the concentration of Bunker C in sea water were developed by Levy (1970), and were used to estimate the oil present in the water column at a number of stations off the Canadian Atlantic Seaboard. To obtain a preliminary survey of the possible extent of oil pollution from Bunker C or similar oils in the Gulf of St. Lawrence, samples were collected in July 1970 at a series of stations extending from Cabot Strait to Montreal (fig. 1). Concentrations of a few parts per billion were found at all stations and all depths. While these levels are near the limits of reliability of the method, Levy is of the opinion that there is a low background level of oil which has resulted from man-made activities. Based on these preliminary observations, there are not sufficient grounds for either grave concern or complacency. Additional sampling is required to determine the geographic extent of this pollution and repetitive sampling is needed to establish possible rates of change.

Other pollutants

Most domestic sewage from the urban centres of Quebec and the Atlantic Provinces is discharged untreated into the nearest waterway. This practice makes it hazardous to use areas for recreational purposes, especially in waters of low salinity. Furthermore, a high proportion (>25 per cent for the Maritime Provinces as a whole) of the potential inshore shellfish areas are closed because of contamination by faecal bacteria. Fish processing plants use large quantities of water and some of these experience difficulty in obtaining clean sea water. Standards of cleanliness for processing water are stringent, and traces of marine pollution force plants to develop, often at great expense, alternate fresh-water supplies.

In addition to the hazards discussed earlier, the salmon stocks in the Miramichi have on occasion been threatened by effluents from base metal mining activities in New Brunswick (fig. 2). In fresh water, copper and zinc are extremely toxic to fish and can be effective barriers to migratory species (Sprague, Elson and Saunders, 1965). The concentrations of metals are so low, however, that they produce no measurable effects in the estuaries or Gulf.

Circulation and man-made changes

Although not commonly thought of as pollution, the regulation and diversion of fresh water entering the sea may produce important changes in circulation patterns and ecology. Today's technology can deliberately alter systems as large as the Gulf of St. Lawrence and engineering works have already significantly affected the system unintentionally.

Fresh water from the St. Lawrence River drainage system plays a prominent role in the generation of currents and large-scale movements and mixing processes in the Gulf. The amount of subsurface sea water brought into circulation during different seasons is significantly influenced by the volume of fresh water discharged. Neu (1968 and personal communication) has made a preliminary investigation of the effect this regulation may have on the physical oceanographic conditions in the Estuary. He has hypothesized that the currents along the Gaspé coast should decrease during summer months and increase during winter when greater flow regulation is introduced. Under natural conditions, before any regulation of discharge was employed, he has estimated that the ratio between the summer and winter flow of the St. Lawrence River was 1.7 to 1. Since then, this ratio has been modified to approximately 1.4 to 1, a change of 30 per cent in the seasonal outflow volume. There is a continuing effort to achieve the 1 to 1 ratio which will indicate optimum electrical power production. Although there are insufficient oceanographic data from which to measure the precise effects of the seasonal discharge on water-flows or temperatures, meteorological data from Rimouski support Neu's hypothesis.

Although the current temperature and circulation changes produced by the present regulation of the St. Lawrence River system may not be great for the Gulf as a whole, there is a continued and steady increase in the control of this system. Furthermore hydroelectric plants are being constructed on the many large, formerly untouched streams which enter along the north shore of the Gulf. In view of the present variability in ice formation, it appears that a relatively minor change in the heat budget might be sufficient to eliminate ice cover and possibly lead to marked climatic effects over a substantial region of the country. Undoubtedly, many of the effects would be generally considered "good". They could equally be detrimental. What is of concern here is that the changes may be unintentional, and while the consequences are unknown, it is highly likely that they are, from a practical viewpoint, irreversible. The magnitude and the irreversible nature of the changes are likely to be similar for most of the world's major river systems. The study of the mechanisms involved and the consequences of continuing indiscriminate regulation and alteration of fresh water flowing into major gulfs are therefore at least equal in priority with the more immediate but interacting problems of pollution by addition of contaminants.

DISCUSSION

It is convenient to think of the Gulf of St. Lawrence as a large complex estuary, with physical oceanographic features determined by a spectrum of parameters, including precipitation, fresh-water discharges, wind, topography, heat transfer, tides and tidal currents. The fresh water is largely confined to the mixed surface layer which varies in thickness from approximately 10 m in July to about 100 m in March.

A much stronger thermocline, halocline and pycnocline is present during summer than in winter. The mechanisms determining the vertical structure are not particularly well known. The very stable summer conditions are brought about through increased fresh-water discharge in spring, increased solar heating and decreased wind action. Winter conditions are the result of decreased fresh-water discharge, rapid heat loss through the sea surface, and strong winds. Mixing energy, supplied by the tides, appears to be of lesser importance, except in the St. Lawrence Estuary and in restricted straits and passages.

If it is assumed that pollutants behave similarly to fresh water, then the bulk of material added to Gulf waters is confined to the upper 100 m. During winter months, the "age" of the pollutant is probably similar throughout the upper 100 m of the water column at any given position. In the summer, however, pollutants in the surface layer will be much "younger" than those below the pycnocline. It is unlikely that the differences in "age" exceed one year.

If pollutants behave similarly physically to the fresh water, then they are present in the immediate environment of the bulk of the organisms in the food chain, including benthic organisms. If the concentration of a particular pollutant is assigned a value of unity in the fresh-water input to the Gulf, and if it were further assumed to be confined to the upper 40 m, and that the undiluted sea water used as a mixing source has a salinity of 32‰, then the mean pollutant concentration in the Gulf would be in the neighbourhood of 0.05. Thus, if wastes are discharged relatively uniformly into the fresh water entering the Gulf, one cannot expect a dilution of as much as two orders of magnitude with sea water during its residence in the Gulf.

There is often no reason, however, to assume that the behaviour of a pollutant is similar to a "parcel" of fresh water. A pollutant in suspension may be assimilated by marine organisms or deposited on the bottom without undergoing any chemical change. Once the pollutant leaves the fluid part of the environment, physical oceanographic processes cease to play a direct role, although they do continue to act indirectly to influence the pollutant's pathway. It is evident that the situation in the Gulf of St. Lawrence demands a far more detailed knowledge of the behaviour of specific pollutants and the spatial and time scales involved.

Earlier the general ice conditions in the Gulf were reviewed. In many harbours and coastal embayments ice may be shorefast for several months. During this period, the exchange of oxygen across the sea surface is substantially reduced. Consequently, if these areas also coincide with sites where pollutants with a heavy oxygen demand are being discharged, one can expect a lower dissolved oxygen concentration in the water during the ice cover period. Little is known about the behaviour and fate of oil released into ice-covered waters. In the case of the Gulf, a major spill in winter months could be particularly detrimental to the large seal herds which enter the Gulf and pup during the late winter months. Although ice might temporarily contain the oil, experience gained from the *Arrow* disaster suggests that, for residual petroleum products at least, oil trapped in ice remains undispersed until the ice breaks up or melts. Thus it appears that regardless of the time of year, the oceanographical and meteorological conditions in the Gulf are such that a good proportion of any major oil spill will probably be found along the shoreline during some phase of its life.

In July 1970, the Bedford Institute sampled the water column, the bottom, and the lower and intermediate trophic levels of the food chain at widely spaced stations between Cabot Strait and Montreal (fig. 1), to gain some insight into the distribution of some of the pollutants entering the Gulf. The largest single source was expected to be the St. Lawrence River. Therefore, the intent of the survey was to sample along the general path of the fresh water as it traverses the Gulf from the Saguenay to Cabot Strait. Oceanographic parameters, such as temperature, salinity, dissolved oxygen, nutrients and fluorides in the water column, were measured, and samples of the sediment, plankton and selected larger organisms were taken for mercury and organochlorine analysis. Suspended particle counts with a Coulter Counter have been undertaken and analyses for the presence of residual petroleum products in the water column are being made. Most analyses of these samples are still incomplete, but the preliminary results suggest that pollutants may be widespread at low concentration levels. Trends in concentration in relation to the circulation patterns may aid in early appreciation of the rates of accumulation or disappearance of various pollutants.

CONCLUSIONS AND RECOMMENDATIONS

The flushing time for fresh water in the Gulf appears to be on the average somewhat less than one year. However, the internal circulation pattern is such that the bulk of the fresh water does not move directly through the Gulf but probably makes one or more "circuits" before exiting through Cabot or Belle Isle Straits.

Knowledge of the physical oceanography of the Gulf is inadequate to satisfactorily describe the physical behaviour of a particular pollutant. Much more information is required on water circulation at all times of the year. For the winter season especially there is a paucity in all oceanographic measurements.

Pollution is known to be a problem in a number of local areas within the Gulf and the rivers that discharge into it. The pollutants include pesticides, mercury, effluent from base metal mines, pulp and paper mill wastes, and domestic sewage. Preliminary results indicate relatively low levels of mercury, pesticide residues, and residual petroleum products for the Gulf as a whole. However, there is no evidence that this represents an approach to equilibrium with respect to present supply sources.

The source of pesticide residues in the gannet colony of Bonaventure Island has not yet been established. It is possible that the source in the Gulf is mackerel and its supporting food chain, but the migrations of both gannets and mackerel to areas well beyond the Gulf, make other sources for contamination possible. Further specific sampling of the marine food chain in the gannet feeding areas west of Anticosti Island and Orphan Bank may shed light on the pollutants' pathway.

Mercury levels in samples of commercial fish have also been lower than was originally expected. More extensive sampling in the St. Lawrence Estuary-Gaspé Passage area and at all trophic levels are needed to establish more clearly the mercury budget and pathways.

With acceleration in the advent of new chemicals, many of them stable or toxic at very low levels, it is increasingly important that techniques be developed for rapidly analysing the physical, biological and geological environment for a large number of substances. Analyses employing techniques such as neutron-activation of marine organisms that are particularly sensitive to changes in the chemical environment can make a major contribution to monitoring environmental quality. Additional methods must be developed. Without some improvement in this field, we shall continue to experience "panics" as, more or less accidentally, it is discovered that some particular pollutant is turning up in unexpected places and in unanticipated concentrations.

In view of the many uses of the Gulf of St. Lawrence, both present and potential, we can ill afford to continue to assume that there are no important pollution problems on a Gulf-size scale. Most pollution problems that receive widespread public attention and concern are the *Torrey Canyons*, the "Long Harbours" and the "Minamata diseases" where dramatic effects are readily in evidence. However, in many instances, it is not the direct toxic effects that are particularly important, but the long-term indirect and sublethal effects. Some of these are undoubtedly irreversible. Our knowledge and understanding of pollutant pathways in the marine environment are at present too rudimentary to do more than conjecture about man's future successes in the manipulation of his environment to achieve continuing maximum benefit.

References

BANKS, R E   The cold layer in the Gulf of St. Lawrence. *J. geophys.*
1966   *Res.*, 71(6):1603–10.

BLACKFORD, B L   A simple two-dimensional electrical analog
1965   model for wind-driven circulation in the Gulf of St. Lawrence. *Manuscr. Rep. oceanogr. limnol. Fish. Res. Bd Can.*, (185):48 p.

BLACKFORD, B L   Some oceanographic observations in the south-
1965a   eastern Gulf of St. Lawrence—summer 1964. *Manuscr. Rep. oceanogr. limnol. Fish. Res. Bd Can.*, (190):22 p.

BLACKFORD, B L   Results from a current meter moored in Kouchi-
1965b   bougaac Bay, June and July 1964. *Manuscr. Rep. oceanogr. limnol. Fish. Res. Bd Can.*, (196):8 p.

BLACKFORD, B L   A simple two-dimensional electrical analog
1966   model for wind-driven circulation in the Gulf of St. Lawrence. *J. Fish. Res. Bd Can.*, 23(9):1411–38.

BLACKFORD, B L   Some oceanographic observations in the south-
1967   ern Gulf of St. Lawrence—summer 1965. *Tech. Rep. Fish. Res. Bd Can.*, (26):34 p.

BLIGH, E G   Mercury and the contamination of freshwater fish.
1970   *Manuscr. Rep. Ser. Fish. Res. Bd Can.*, (1088):27 p.

BUMPUS, D F and LAUZIER, L M   Surface circulation on the
1965   continental shelf off Eastern North America between Newfoundland and Florida. *Ser. Atlas mar. Envir.*, Folio 7, 8 plates.

Canada, Department of Northern Affairs and National Resources, Water Resources Branch, Surface water supply of Canada. *Wat. Resourc. Pap.*, Ottawa, (126) (129) (130) (133) (134) (137) (140) (143) (144).

DUFFY, J R and O'CONNELL, D O   DDT residues and metabolites
1968   in Canadian Atlantic coast fish. *J. Fish. Res. Bd Can.*, 25(1):189–95.

EL-SABH, M I   Bibliography and some aspects of Physical Ocean-
1969   ography in the Gulf of St. Lawrence. *Manuscr. Rep. mar. Sci. Cent. McGill Univ.*, (14).

FARQUHARSON, W I   Causeway investigation Northumberland
1959   Strait, Report on tidal survey 1958. Canada, Department of Mines and Technical Surveys, Surveys and Mapping Branch, 137 p.

FARQUHARSON, W I   Tides, tidal streams and currents in the Gulf
1962   of St. Lawrence. Canada, Department of Mines and Technical Surveys, Marine Sciences Branch, 76 p.

FARQUHARSON, W I   St. Lawrence Estuary current surveys.
1966   *Manuscr. Rep. Bedford Inst. Oceanogr.*, (66–6):84 p. (Unpubl.).

FARQUHARSON, W I and BAILEY, W B   Oceanographic study of
1966   Belle Isle Strait, 1963. *Manuscr. Rep. Bedford Inst. Oceanogr.*, (66–9):78 p. (Unpubl.).

FIMREITE, N   Mercury uses in Canada and their possible hazards
1969   on sources of mercury contamination. *Manuscr. Rep. Ser. Can. Wildl. Serv. Pestic. Serv.*, (17):19 p.

FORRESTER, W D   A quantitative temperature-salinity study of the
1964   Gulf of St. Lawrence. *Manuscr. Rep. Bedford Inst. Oceanogr.*, (64–11):16 p. (Unpubl.).

FORRESTER, W D   Currents and geostrophic currents in the St.
1967   Lawrence Estuary. *Rep. Bedford Inst. Oceanogr.*, (67–5).

FORRESTER, W D   Geostrophic approximation in the St. Lawrence
1970   Estuary. *Tellus*, 22(1):53–65.

FORRESTER, W D   Distribution of suspended oil particles following wreck of tanker *Arrow*. (In press).

FORRESTER, W D and VANDALL, P E Jr,   Ice volumes in the Gulf
1968   of St. Lawrence. *Manuscr. Rep. Bedford Inst. Atlant. Oceanogr. Lab.*, (68–7):16 p. (Unpubl.).

HOULT, D P and CRAVEN, J P (Eds.)   Oil on the sea. Proceedings of
1969   a symposium on the scientific and engineering aspects of oil on the sea held in Cambridge, Massachusetts, 16th May, 1969. New York, Plenum Press, 114 p.

ICES   Report of the ICES Working Group on Pollution of the
1970   Baltic Sea. *Coop. Res. Rep. int. Coun. Explor. Sea (A)*, (15) 86 p.

LAUZIER, L M   Some aspects of oceanographic conditions in the
1958   Gulf of St. Lawrence from autumn 1956 to spring 1957. *Manuscr. Rep. oceanogr. limnol. Fish. Res. Bd Can.*, (9): 43 p. (Unpubl.).

LAUZIER, L M   Evaporation in the Gulf of St. Lawrence. *Rep. biol.*
1962   *Stn. St Andrews Fish. Res. Bd Can.*, 1961–2:213–4

LAUZIER, L M   Bottom residual drift in the continental shelf area of
1967   the Canadian Atlantic coast. *J. Fish. Res. Bd Can.*, 24(9): 1845–59.

LAUZIER, L M and BAILEY, W B   Features of the deeper waters of
1957   the Gulf of St. Lawrence. *Bull. Fish. Res. Bd Can.*, (111): 213–50.

LAUZIER, L M and TRITES, R W   The deep waters in the Laurentian
1958   Channel. *J. Fish. Res. Bd Can.*, 15(6):1247–57.

LAUZIER, L M, TRITES, R W and HACHEY, H B   Some features of
1957   the surface layer of the Gulf of St. Lawrence. *Bull. Fish. Res. Bd Can.*, (111):195–212.

LEVY, E M   A shipboard method for the estimation of Bunker C
1970   in sea water. Paper presented to Symposium, Atlantic Section of Chemical Institute of Canada, Charlottetown, P.E.I., August.

MacGREGOR, D G   Currents and transport in Cabot Strait. *J. Fish.*
1956   *Res. Bd Can.*, 13(3):435–48.

MANN, K H and SPRAGUE, J B   Combating Pollution on the East
1970   Coast of Canada. *Mar. Pollut. Bull.*, 1(5):75–7.

MATHESON, K M   The meteorological effect on ice in the Gulf of
1967   St. Lawrence. *Manuscr. Rep. mar. Sci. Cent. McGill Univ.*, (3):110 p.

MURTY, T S and TAYLOR, J D   A numerical calculation of the wind-
1969   driven circulation in the Gulf of St. Lawrence. Paper presented at the 50th annual meeting of the AGU, Washington, D.C., April 21–25.

NEU, H J A   Man-made changes: the potential benefits and liabili-
1968   ties. Dartmouth, N.S. Canada. Bedford Inst. Oceanogr., Gulf of St. Lawrence Workshop, 86 p.

NICHOLSON, H P   Occurrence and significance of pesticide residues
1970   in water. *J. Wash. Acad. Sci.*, 59(4–5):77–84.

PARDE, M   Hydrologie du fleuve Saint Laurent. *Revue can. Géogr.*,
1948   (2):35–83.

SETTE, O E   Biology of the Atlantic mackerel (*Scomber scombrus*)
1950   of North America. Part 2. Migrations and habits. *Fishery Bull. U.S. Fish Wildl. Serv.*, 51 (49): 251–358.

SPRAGUE, J B and McLEESE, D W   Toxicity of kraft pulp mill
1968   effluent for larval adult lobsters and juvenile salmon. *Wat. Res.*, 2:753–60.

SPRAGUE, J B and RUGGLES, C P   Impact of water pollution on
1967   fisheries in the Atlantic provinces. *Can. Fish. Rep.*, (9):11– 15.

SPRAGUE, J B, DUFFY, J R and CARSON, W G   A measurement of
1969   DDT residues in marine commercial fish and shellfish collected in Atlantic Canada in 1967. *Manuscr. Rep. Ser. Fish. Res. Bd Can.*, (1061).

SPRAGUE, J B, ELSON, P F and SAUNDERS, R L   Sublethal copper-
1965   zinc pollution in a salmon river—a field and laboratory study. *Int. J. Air Wat. Pollut.*, 9:531–43. Also issued as *Adv. Wat. Pollut. Res.*, 2(1):61–82, 99–102.

TREMBLAY, J and LAUZIER, L M L'origine de la nappe d'eau froide
1940   dans l'estuaire du St. Laurent. *Naturaliste Can.*, 67(11):5–
        23.
TRITES, R W Are gyres a common feature of the Gulf? Dartmouth,
1968   N.S., Canada, Bedford Inst. of Oceanogr., Gulf of St.
        Lawrence Workshop, 86 p.

Acknowledgement

The author expresses his thanks to L. M. Dickie and S. R. Kerr
and to J. M. Bewers, E. G. Bligh, W. D. Forrester, L. M. Lauzier,
E. M. Levy, J. H. A. Neu and J. B. Sprague for providing information and comments from as yet unpublished material.

Pollution Problems in the Strait of Georgia

*T. R. Parsons**

Les problèmes de pollution dans le détroit de Georgie

Le détroit de Georgie constituant un milieu marin semi-fermé
pourrait être susceptible d'une pollution à grande échelle. A l'heure
actuelle les principales sources de pollution sont les usines de pâte
à papier et les égouts. Cependant, l'étendue de la pollution due à
ces sources ainsi qu'à d'autres est actuellement bien localisée et une
étude, portant sur 5 années, des eaux libres du détroit n'a indiqué
aucune pollution générale de cet environnement. Une action
unifiée gouvernementale ou internationale est nécessaire pour
assurer la préservation de cet environnement qui représente une
région importante pour les pêches industrielles et sportives.

Problemas de contaminación en el Estrecho de Georgia

El Estrecho de Georgia, considerado como un ambiente marino
semicerrado, puede ser muy afectado por la expansión de la
contaminación. Por ahora, los causantes principales de la contaminación son los desechos de las factorías para pasta de papel y las
aguas cloacales. Sin embargo, la extensión de la contaminación de
una o de otra fuente es hasta el presente de carácter enteramente local
y los estudios que se han hecho durante cinco años de las aguas
abiertas del Estrecho de Georgia no señalan alguna contaminación
general de este ambiente. Se requiere una acción unificada, gubernamental o internacional, para asegurar la preservación de este
ambiente como una región importante para la pesca comercial y
deportiva.

THE Strait of Georgia (fig. 1), is located on the coast
of British Columbia between the mainland of
Canada and Vancouver Island. It is 220 km long,
has an average width of 33 km and a mean depth of 156 m.
A major river, the Fraser, flows into the Strait near
the southern end. South of 49°N, the international
boundary between Canada and the United States has
been drawn through the "centre" of the Strait. The area
is studded with numerous islands and inlets (fjords), the
latter indenting the coast by as much as 90 km.

The primary factors influencing the hydrographic
conditions are tide, wind, insolation, and fresh-water
runoff; other forces involved are Coriolis, centrifugal,
and topographical. Tully and Dodimead (1957) have
described three tidal regions. The northern and southern
regions are in the immediate vicinity of access to the
ocean where rapid and turbulent tidal currents cause
strong mixing in the island passages. These regions advance 10 to 20 km into the Strait during the flood, and on
the ebb tide retreat into the passages. In a large central
region of the Strait, tidal currents are believed to be
generally less than 1 kn except where river discharge
influences tidal flow. It appears that the flooding tide is
stronger and of longer duration than the ebb on the
eastern (B.C. mainland) side of the Strait. The reverse is
true on the western side. The tendency toward a counterclockwise (surface) circulation is believed to be enhanced
by a counterclockwise wind pattern considered to exist in
the southern part of the Strait during at least part of the
year.

The effects of heating and of fresh-water runoff have
been described by Waldichuck (1957). In the immediate
vicinity of the Fraser River estuary, high stability is
maintained throughout the year by the direct effect of
low salinity surface layers. In the southern part of the
Strait and over a large area north of the Fraser River,
stability is reduced to nearly zero during the winter
months but is re-established during the summer by
insolation and by run-off from the Fraser River, which
reaches a maximum discharge from May to July.

Salinities in the region of the Fraser River range from 0
to 20‰ in the surface layers and from 28 to 31‰ in
deeper water. Over the rest of the Strait, surface salinities
vary from 24 to 28‰, except in the areas which are
dominated by tidal mixing and are virtually homogeneous—with salinities of 31 to 32‰—from top to bottom.
Seasonal temperatures in the area range from about
7° to 22°C at the surface but only from 8° to 9°C in
deeper water.

Levels of biological production in the pelagic environment of the Strait have been summarized by Parsons,
LeBrasseur and Barraclough (1970). The annual primary
production of phytoplankton is approximately 120 g
C/m^2. In addition to this, the annual river discharge
introduces about the same amount of organic carbon,
as allochthonous detritus, from the land. The annual
primary production of near-shore areas is generally
appreciably higher than the above figure for the pelagic
environment. This is in part due to the continual re-
plenishment of nutrients through near-shore tidal mixing
as well as to the production of attached macrophytes. An
impressive number of commercially important fish spend
their early life history stages in the Strait before going
offshore to feed. The most important fish in this group
are the herring and the four species of North American
salmon. As an indication of the numbers involved, the
average annual output of juvenile salmon from the
Fraser River alone can be estimated as 97×10^6 sockeye
(*O. nerka*), 234×10^6 pink (*O. gorbuscha*) and 70×10^6
chum salmon (*O. keta*). The resident adult fish populations
of interest to fisheries are herring (*Clupea pallasii*), ling
cod (*Ophidon elongatus*), cod (Gadidae), dog fish (*Squalus
suckleyi*) and flatfish.

Agencies and methods in pollution research

The three principal Canadian agencies which have been
involved in pollution research in the Strait of Georgia are
the federal government's Department of Fisheries and the
Fisheries Research Board, and the provincial government's Department of Recreation and Conservation.

* Environmental Research Group, Fisheries Research Board of Canada, Nanaimo, B.C., Canada.

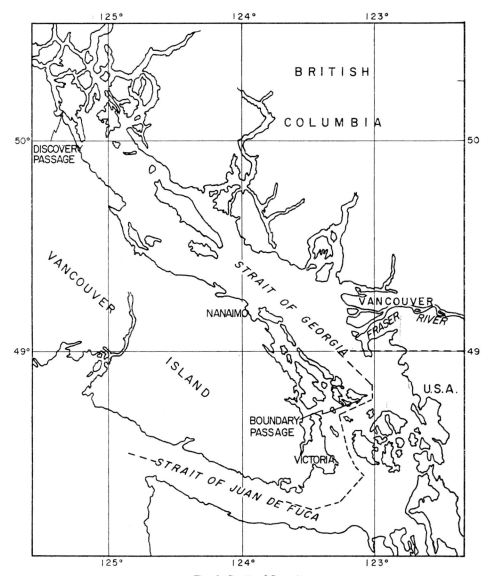

Fig. 1. Strait of Georgia

Research by the latter agency has been concerned primarily with pollution in fresh water entering the Strait, while the two fisheries agencies of the federal government have carried out research in the marine environment. More recently, the federal Department of Energy, Mines and Resources has started a pollution research programme on the physical and chemical environment of the Strait. In addition, pollution research has been carried out by the International Pacific Salmon Fisheries Commission (IPSFC), the University of British Columbia, the Department of National Health and Welfare, the British Columbia Research Council, and private companies working for industry or under contract for municipalities. The question of jurisdiction in controlling pollution is at present largely related to the type of pollution and to the particular environment in which pollution occurs. Thus contamination of oysters from sewage concerns the departments of National Health and Fisheries, while IPSFC and the federal Department of Fisheries are concerned with types of pollution which may affect migration and survival of salmon.

Methods for examining polluted environments have been developed to meet specific situations. Generally, colorimetric analyses have been used for all nutrients, oxygen is determined by the Winkler method, and sewage contamination is assessed from coliform plate counts. Heavy metals are now being determined by newer techniques, such as atomic absorption spectrometry, while pesticides, phenols and other organic poisons are determined by solvent extraction followed by gas chromatography and IR or UV spectrophotometry. Some routine monitoring is carried out in the vicinity of known sources of pollution, but no routine sampling is carried out in the open waters of the Strait.

Sources of pollution

The Report of the Task Force on West Coast Water Pollution Research Programming (Anon. 1969) lists a number of causes of water pollution in British Columbia; nine of these are summarized in Table 1 as being applicable to the Strait of Georgia.

The most widespread and severe pollution is due to pulp mill effluent, which is harmful to the marine environment for three principal reasons. It has a high Biological Oxygen Demand (BOD), it is directly toxic to marine life, and it contains pulp fibres which sink and

TABLE 1 SOURCES OF POLLUTANTS IN THE STRAIT OF GEORGIA

Source	Type of pollutant	Comment
1. Forest industry, including pulp and paper, logging and lumber	Toxic and O$_2$ consuming compounds	Most severe form of pollution in the Strait
2. Domestic sewage	Infectious organisms and O$_2$-consuming compounds	Contamination of bathing beaches, shellfish and fish-processing water
3. Mineral and Mineral Processing	Toxic compounds, sediments and dusts	Medium to low local severity
4. Agriculture and Forestry	Pesticides and nutrients	Medium to low severity
5. Food Processing	Organic wastes	Moderate local pollution from fish-processing plants
6. Industrial Organic and Inorganic Wastes (other than 1, 3 and 5 above	Specific substances depending on industry	Moderate local pollution
7. Petroleum Industry	Petroleum products	Sporadic and low severity
8. Thermal Pollution	Heat	Not assessed
9. Radioactive Pollution	Radioactive materials	Largely potential

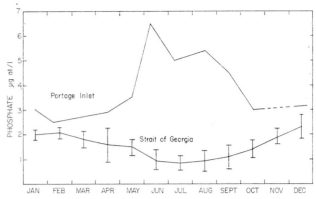

Fig. 2. *Phosphate concentration in the Strait of Georgia (lower graph) and in Portage Inlet (upper graph). Average for the Strait, and one standard deviation shown for the years 1965–1968, from Parsons et al. (1970); data for Portage Inlet, 1965–1966, from Waldichuck (1969)*

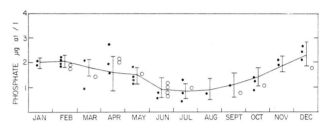

Fig. 3. *Average phosphate concentration in the Strait of Georgia (data as in Fig. 2). Phosphate concentration 5 mi (○) and 25 mi (●) from Vancouver City outfall (data from Fulton et al., 1967 and 1968)*

accumulate as sediment. The gradual accumulation and decay of the fibres may lead to the smothering of benthic fauna and flora as well as to the production of an anaerobic-H$_2$S zone on the bottom. The high BOD of pulp mill effluent causes an oxygen gradient in estuarine areas which may be the principal cause of fish avoiding pulp mill polluted waters. However, the direct toxicity of the effluent in the immediate vicinity of a pulp mill appears to eliminate all except the lowest forms of life such as fungi and bacteria. The lethal effects of pulp mill effluent have been studied by a number of authors. Alderdice and Brett (1957) showed, for example, that a concentration of 4.8‰ full-bleach kraft mill effluent (BKME) in sea water of salinity 20‰ and of temperature 17.8°C was a limiting level for toxicity, below which survival of sockeye salmon (*O. nerka*) underyearlings was complete. Conclusions of this type have to be repeated, however, for a much longer study period. Oyster larvae have been shown to be particularly sensitive to sulphite waste liquor (Woelke, 1960); concentrations of 2 ppm and greater have been shown to affect larval development. Experiments carried out in other laboratories (e.g. Sprague and McLeese, 1968) have indicated that more than one component is involved in the toxicity of BKME. These compounds appear to be polychlorinated polyphenols and quinones (e.g. tetrachlorobenzoquinone and tetrachlorocatechol). Storage may decrease toxicity in some cases, but the most effective lessening of toxicity appears to be by means of chemical oxidation in aeration towers or, more recently, through biological oxidation (Werner, 1962; Howard and Walden, 1965, Servizi, *et al.*, 1966). Pulp mills are now beginning to use the latter process, and biological digestion ponds, to which inorganic nitrogen and phosphorus are added, are now being installed on an experimental basis.

In marine environments close to urban areas, pollution from domestic sewage has been responsible for the sporadic closing of several swimming beaches and the fouling of fishermen's nets, particularly in the vicinity of the Fraser River estuary. Among marine products, oyster production has suffered most from domestic sewage pollution, and certain areas have been closed permanently to commercial use. Tanks in which harvested oysters are allowed to cleanse themselves in unpolluted water have been installed and operated successfully on a small scale. Nutrient enrichment from sewage pollution has been detected in isolated inlets bordering on urban areas. The example shown in fig. 2 is for phosphate concentration in Portage Inlet, an area which has been studied extensively (Waldichuck, 1969). The inlet is 15 km long and surrounded by residences and small industries. It is actually located outside the Strait of Georgia (near Victoria), but the data serve as a means of comparing local values from an urban inlet undergoing eutrophication with seasonal changes in the open waters of the Strait of Georgia. High phosphate concentrations throughout the year are characteristic of sewage pollution, the most probable cause of a summer maximum in concentration being due to decreased drainage volume coupled with a steady input of nutrients from the urban population.

The use of pesticides in British Columbia has been widespread, particularly in agriculture and for the control of forest insects by means of DDT. The extent to which sprays used on land may have entered the marine environment of the Strait of Georgia is not known, but, in addition to forest spraying, direct application of pesticides has been made both to rivers entering the Strait and to marine inlets. In this connection DDT has been used for mosquito control in river valleys, benzene hexa-

[74]

chloride is routinely used to spray log booms for Ambrosia beetle control, while sodium arsenite is employed for Teredo control for docks. In all cases the application of these materials has been studied with respect to their immediate effects on marine life, especially on young salmonids. The results of these tests have generally permitted the continued use of pesticides based on 48- and 96-hour mortality experiments. However, because of the latest knowledge on long-term detrimental effects of pesticides such as DDT, readily degradable substitutes are now beginning to replace the use of the accumulative poisons.

Potential and actual sources of pollution from the mining industry in the Strait of Georgia include the release of metals of known toxicity, such as copper, lead, and zinc, as well as the accumulative effect of mine tailings in smothering benthic fauna and flora in certain local areas. Further, recent construction of a large coal port outside Vancouver has raised the potential threat that coal dust on sea surfaces may cause sufficient light attenuation to destroy the littoral macrophytes (*Zostera*, *Fucus*, and *Sargassum*) on which herring spawn.

Industrial organic and inorganic wastes have been responsible for sporadic fish and waterfowl kills in isolated parts of the Strait. In the immediate vicinity of Vancouver, heavy metals, cyanides, chlorine, phenol, formaldehyde, and caustic soda in effluents are probably among the most lethal contaminants to be diluted and flushed into the sea. Generally, the short-term harmful effects of these substances are only apparent when an accident or faulty operation allows the sudden release of very large quantities. Two such fish and bird kills have recently been reported in the Strait; in these cases the poisons were identified as cyanide and chlorine, respectively. On the other hand, where a small sustained discharge of mercury (20 lb per day) had been escaping into a large inlet over a period of several years, analyses of edible marine animals in the area failed to show general levels of mercury considered harmful to humans; an exception to this was a high level of mercury in commercial crabs and a number of other benthic feeders obtained from the immediate vicinity of the industrial outfall. Pollution from oil is not widespread in the Strait although the concentration of phenols in the waters of the port of Vancouver is generally about 10 ppb and may be greater than 2 ppm in oil slicks associated with a petroleum refinery (Waldichuck and Werner, unpublished data).

Of other sources of pollution listed in Table 1, the effects of pollution from fish canneries and reduction plants were at one time very apparent in the Fraser River delta, but allowance for proper flushing of these areas has alleviated the situation. The possible effects of industrial thermal pollution have not been assessed but cannot at present be considered serious.

Extent of pollution

Six pulp mills are located in the Strait of Georgia, and various effects of the disposal of effluents have been discussed by Waldichuck (1962, 1968). One mill (on the edge of Discovery Passage) has been in operation for nearly 20 years, and normal dissolved oxygen levels are encountered outside a radius of about 100 m from the outfall. Strong tidal currents rapidly dilute the BKME

and keep the bottom free of any deposits of particulate materials. In contrast, the other five mills all impart some degree of pollution to the environment. In total, however, it must be emphasized that the pollution is localized and not apparent in the open waters of the Strait. The only obvious exception is the frequent presence of large logs which are a considerable hazard to small craft in the open waters.

Sewage and urban runoff from Vancouver are collected in primary treatment ponds on Iona Island in the Fraser River delta. The flow of effluent from the ponds is between 30 and 60 million gallons per day, depending on the season. The concentration of effluent is decreased by three to four orders of magnitude by dilution associated with Fraser River waters, so that, although the annual output of phosphate from sewage could amount to between 0.1 and 3 million pounds per year, the effect of this enrichment on the quantity of phosphate already present in the Strait would be insignificant. This is borne out by data shown in fig. 3 in which the phosphate concentration at points in the Strait approximately 5 and 25 mi west and northwest of the sewage outfall is compared throughout the year with the average concentration of phosphate in the surface waters of the Strait. From these results it may be seen that there is no apparent phosphate enrichment of the marine environment in the vicinity of the Fraser River. The worst effect of sewage disposal in the Strait of Georgia is in fact only apparent in local situations, and these are indicated primarily by high coliform counts in near-shore waters and in shellfish. It may be concluded, in fact, that both pulp mill and sewage pollution, together with the other types of pollution discussed, are all of "local" concern and that the general level of productivity of the open waters of the Strait does not reflect pollution of this environment, either by poisons or by enrichments. This is borne out in Table 2, which shows the mean and standard deviation for various materials in the marine environment of the Strait of Georgia, as determined from ten stations over a period of four years. From these data it is apparent that the growth of phytoplankton and the decrease in nutrients follow a normal pattern which is determined largely by the increase in solar radiation, March to June (see also Parsons *et al.*, 1969 for further discussion). The only departure from this sequence is that silicate, which reaches a rate limiting concentration for diatoms in April, suddenly increases again in May. This has been observed previously by Tully and Dodimead (1957) and is largely due to the high concentration of silicate in Fraser River water; the effect of Fraser River water is most apparent in the Strait during May when the volume of water entering the Strait is five times greater than in April. The maximum level of primary production reached in May is normal for coastal temperate waters, and levels of particulate materials are not excessive. If this total system was under the stress of a pollutant which could affect the lower trophic levels (e.g. by the sustained introduction of light attenuating materials, poisons or over-enrichment from nutrients), an abnormal cycle of events would have been apparent in Table 2, such as has already been demonstrated in fig. 2. As well as demonstrating a lack of pollution, however, the data are particularly important in that they serve as a baseline from which to judge future effects of a growing population.

[75]

TABLE 2. BASELINE DATA FOR THE SURFACE WATERS OF THE STRAIT OF GEORGIA EXPRESSED AS AVERAGE VALUES FOR EACH MONTH DURING THE PERIOD 1965–1968 (Parsons *et al.*, 1970)

	Jan.	Feb.	Mar.	Apr.	May	Jun.	Jul.	Aug.	Sep.	Oct.	Nov.	Dec.	Max.	Min.
Irradiance [a] (l ys/month)	75	166	273	412	562	599	606	503	362	171	95	60	701	57
	(9)	(34)	(57)	(30)	(54)	(122)	(37)	(71)	(29)	(19)	(14)	(5)		
Secchi depth (m)	9.5	8.5	9.5	7.5	4.0	3.8	5.0	5.8	11.6	8.2	8.4	9.3	16	1
	(3.0)	(1.7)	(3.2)	(3.8)	(2.9)	(2.9)	(4.8)	(2.6)	(2.6)	(2.6)	(2.0)	(6.3)		
Silicate [b] (μg.at./l)	*	56	50	22	48	33	24	25	26	38	44	52	100	8
		(5)	(13)	(10)	(20)	(9)	(10)	(15)	(11)	(13)	(4)			
Nitrate [b] (μg.at./l)	22	23	21	15	14	7	8	9	12	17	23	24	27	<1
	(0.8)	(0.8)	(2.7)	(3.3)	(2.9)	(2.8)	(4.2)	(4.4)	(4.9)	(6.0)	(2.3)	(1.9)		
Phosphate [b] (μg.at./l)	2.0	2.1	1.8	1.6	1.5	0.9	0.8	0.9	1.1	1.4	1.9	2.3	3.2	0.3
	(.18)	(.19)	(.29)	(.69)	(.32)	(.41)	(.29)	(.37)	(.48)	(.41)	(.23)	(.53)		
Seston [b] (mg/l)	3.3	1.0	2.0	2.5	4.2	4.8	2.7	2.7	1.8	2.0	1.8	*	8.8	0.2
	(2.6)	(0.7)	(2.1)	(1.6)	(3.0)	(2.6)	(1.4)	(1.7)	(1.0)	(1.7)	(1.3)			
Particulate C [2] (μg/l)	250	200	210	220	320	410	390	410	280	230	170	180	740	50
	(64)	(57)	(57)	(110)	(130)	(95)	(130)	(160)	(130)	(86)	(43)	(87)		
Particulate N [2] (μg/l)	15	16	21	32	32	53	50	40	34	32	15	14	115	9
	(3.4)	(13)	(5.3)	(11)	(26)	(13)	(17)	(19)	(15)	(11)	(2.9)	(5.9)		
Chlorophyll a [2] (μg/l)	0.9	1.0	2.1	3.0	3.2	4.3	3.0	2.4	2.0	2.0	0.8	1.0	10	<0.5
	(0.3)	(0.5)	(1.6)	(2.1)	(2.4)	(1.9)	(1.6)	(1.4)	(1.1)	(1.3)	(0.5)	(0.4)		
Soluble C [2] (mg/l)	*	1.5	2.0	2.5	2.8	2.9	3.0	3.2	3.0	2.0	1.2	1.2	8.5	0.3
		(0.6)	(1.1)	(1.1)	(0.9)	(1.4)	(1.9)	(1.9)	(1.0)	(1.0)	(1.0)	(0.1)	(*)	
Primary productivity (mgC/m²/day)	20	250	470	550	1200	320	270	310	200	230	120	*	2562	10
	(*)	(122)	(240)	(450)	(820)	(300)	(130)	(280)	(120)	(85)	(48)			
Total herbivorous zooplankton [c] (mg. wet wt./m³)	50	60	450	950	1420	700	300	120	100	80	50	50	5600	<50

[a] Measured at Departure Bay
[b] Average value for 0 to 20 m
[c] 1965–1967 average value for 20 m vertical net haul at eight stations using a 350 μ Hensen net

* Insufficient data to report a value; value in brackets, e.g., (29), is one standard deviation

☐ Maximum and ⌐⌐⌐ minimum monthly means

In addition to the data shown in Table 2, other data have been accumulated (Parsons *et al.*, 1970) on the occurrence of larval and juvenile fish. Since these life forms are most easily affected by pollution, future waste disposals should be located (in the case of cities), or timed (in the case of a single dumping of material), so as not to pose a special threat to the young of marine species. There is also a clear need for both the systematic collection of data on all industrial and domestic materials which are now entering the Strait and for a continual monitoring of their effects, especially the long-term results of their residence time in the marine food chain. Jurisdictional differences between local, provincial and federal governments should not be allowed to obscure the urgency of carrying out these tasks if the Strait of Georgia is to be preserved as a relatively unpolluted environment. The formation of an international agency, similar to the International Joint Commission for the Great Lakes, would possibly serve the purpose of coordinating all anti-pollution activities in the Strait.

References

ALDERDICE, D F and BRETT, J R   Some effects of kraft mill effluent
1957   on young Pacific salmon. *J. Fish. Res. Bd Can.*, 14:783–95.

FULTON, D J, *et al.*   Data record. Physical, chemical and biological
1967   data, Strait of Georgia, 1966. *Manuscr. Ser. Fish. Res. Bd Can.*, (915):1–145.
FULTON, D J   Data record. Physical, chemical and biological data,
1968   Strait of Georgia, 1967. *Manuscr. Ser. Fish. Res. Bd Can.*, (968):1–197.
HOWARD, T E and WALDEN, C C   Pollution and toxicity character-
1965   istics of kraft pulp mill effluents. *TAPPI*, 48:136–41.
PARSONS, T R, LEBRASSEUR, R J and BARRACLOUGH, W E   Levels of
1970   production in the pelagic environment of the Strait of Georgia, British Columbia: a review. *J. Fish. Res. Bd Can.*, 27(7):1251–64.
PARSONS, T R, STEPHENS, K and LEBRASSEUR, R J   Production
1969   studies in the Strait of Georgia. Part 1. Primary production under the Fraser River plume, February to May 1967. *J. exp. mar. Biol. Ecol.*, 3(1):27–38.
SERVIZI, J A, STONE, E T and GORDON, R W   Toxicity and treatment
1966   of kraft pulp bleach plant waste. *Progr. Rep. int. Pacif. Salm. Fish. Commn*, (13):34 p.
SPRAGUE, J B and MCLEESE, D W   Different toxic mechanisms in
1968   kraft mill effluent for two aquatic animals. *Wat. Res.*, 2:761–5.
TULLY, J P and DODIMEAD, A J   Properties of the water in the
1957   Strait of Georgia, British Columbia, and influencing factors. *J. Fish. Res. Bd. Can.*, 14(3):241–319.
WALDICHUCK, M   Physical oceanography of the Strait of Georgia,
1957   British Columbia. *J. Fish. Res. Bd Can.*, 14(3):321–486.
WALDICHUCK, M   Water pollution in British Columbia. *A. Rev.*
1962   *Fish. Coun. Can.*, (17):26–9, 31–3.
WALDICHUCK, M   Waste disposal in relation to the physical
1968   environment—oceanographic aspects. *In* Seminar "You and Your Environment". Victoria, B.C., Adult Education Division, Victoria School Board, 22 p.

WALDICHUCK, M   Eutrophication studies in a shallow inlet on
1969   Vancouver Island. *J. Wat. Pollut. Control Fed.*, 41:745–64.

WERNER, A E   Sulphur compounds in kraft mill effluents. *Pulp*
1962   *Pap. Can.*, 16:35–43.

WOELKE, C E   Effects of sulphite waste liquor on the normal
1960   development of Pacific oyster (*Crassostrea gigas*) larvae.
Res. Bull. Wash. St. Dep. Fish., (6):150–61.

ANON.   Report of the Task Force on West Coast Water Pollution
1969   Research Programming. Vancouver, B.C. Department of
Fisheries and Forestries (MS) 68 p.

Acknowledgement

The author acknowledges the assistance of Mr A. Werner in having
read and commented on the draft of this presentation.

Marine Pollution in Japan

*T. Nitta**

La pollutión des eaux de la mer au Japón

Le présent document rend compte de l'étude de plusieurs cas
relatifs à chaque agent de pollution et d'enquêtes générales portant
sur des zones pertinentes.

Les études portent sur les cas suivants: verdissement des huîtres
dans la baie de Nobeoka et sur la côte de Nagoya et d'Hitachi;
altération des huîtres et des laitues marines (algues comestibles)
causée par les effluents d'industries alimentaires dans la baie de
Matsushima; carcinogénèse des laitues marines et coloration des
bivalves par boue charbonneuse dans la baie d'Ohmuta; dégâts
causés aux fonds de pêche de la baie d'Ariake par les biphényles
chlorés, contamination des poissons par des odeurs provenant
d'effluents des industries pétrochimiques de Yokkaichi et d'Iwakuni;
perturbation de la formation des couches de l'huître perlière par
des résidus du dragage dans le détroit de Bisan; fréquentes marées
rouges résultant de l'eutrophisation des baies de Tokuyama et
d'Ohmura.

Les enquêtes sur le milieu ont comporté des observations générales et spéciales sur la dispersion de divers effluents et la dégradation
du fond marin et de la faune benthique provoquée par des résidus
de nature organique et inorganique. Il est possible de prédire
l'influence des déversements côtiers par évaluation du degré de
dispersion. On a procédé à des études comparatives de la pollution
dans les principales baies de Tokyo, d'Ise et d'Osaka par examen
de la faune benthique. Des enquêtes systématiques ont été entreprises et l'on a dressé des cartes de l'état actuel de la pollution dans
diverses régions.

La contaminación de las aguas del mar en el Japón

Se exponen los estudios de casos de cada contaminante y los
reconocimientos generales de las zonas pertinentes.

Los casos estudiados comprenden: ostras frescas en la Bahía de
Nobeoka y en la costa de Nagoya e Hitachi; corrupción de ostras
y de algas comestibles debido a los desechos procedentes de la
elaboración de alimentos en la Bahía de Matsushima; carcinogenesis de ovas y pudrición de bivalvos por el fango del carbón en
la Bahía de Ohmuta; daños producidos en los ostiones causados por el
PCP en la Bahía de Ariake; contaminación de peces a causa de los
olores procedentes de las descargas de las fábricas petroquímicas
en Yokkaichi e Iwakuni; formación interrumpida de capas de
madreperla en las ostras a causa del dragado en los estrechos
de Bisan; y frecuentes floraciones letales de mareas rojos debidas a
la eutroficación en Tokuyama y Bahía de Ohmura.

Los reconocimientos del ambiente han incluido observaciones
generales y particulares sobre la dispersión de las distintas aguas
negras y la deterioración del fondo del mar y de la fauna béntica
ocasionada por desechos orgánicos e inorgánicos. Se puede predecir
la influencia de las descargas en la costa calculando la amplitud de
la dispersión. Examinando la fauna béntica se han realizado
estudios comparativos de la contaminación en las principales
entradas a las Bahías de Tokio, Ise y Osaka. Se han inicidao
investigaciones sistemáticas y se están preparando cartas de la
actual situación de la contaminación en las distintas zonas.

THE coast of Japan has many embayments and the
coastal waters are liable to become polluted because
the rapid development of industries has increased the
flow of wastes. Water pollution is also pronounced in rivers
because their waters are used for many different purposes,
including the discharge of sewage and industrial effluents.
All rivers in the developed regions are being polluted.

In contrast to inland pollution, marine pollution is
observed only in limited areas along the coast but it is
increasing. The decrease of transparency in the embayments along the coast from Tokyo Bay to the Seto Inland
Sea is prominent.

Two projects are being carried out to examine the
present status of water pollution. One is the application of
environmental water quality criteria to specific regions,
and the other is the compilation of a map of polluted areas.
These operations are still incomplete. Polluted areas are
recognized through: (1) investigations in connection with
specific pollution problems; and (2) general observations
on areas polluted by individual sources. In the Tokyo and
Ise bays, where industrial plants stand side by side, single
synthesized pollution areas are formed.

The accumulation of pollutants.

A dark green coloration in oysters, caused by a high
content of copper and zinc, is known as green oyster. It
occurs in waters where the concentrations of copper or
zinc are raised to about 0.01 ppm and 0.1 ppm respectively.

These elements harm oysters above these concentrations. The effluents of copper refineries (Hitachi, Hiihama,
and Takehara are examples) and bemberg plants
(Nobeoka Bay) have caused green oyster. In Nagoya
Harbour green oyster was caused by the outdoor storage
of raw ores.

In Nagoya, Takehara, and Niihama, green oyster is
observed only in limited areas near the outlets of effluents. In Nobeoka Bay, where effluents flow into the bay
in river water, green oyster is found 3 km from the mouth
of the river. The effluent of the Hitachi plants flows to the
sea in river water. Oysters are not found within about
1 km of the river mouth because of the toxicity of the water,
and green oyster is found 4 to 5 km away. Other areas are
affected in the same way. Green oyster is also caused by
high concentrations of zinc without particularly high
concentrations of copper. The soft parts of green oysters
contain copper and zinc in concentrations as much as 100
times greater than normal.

The copper content of industrial wastes is usually removed by precipitation. The bemberg plant of Nobeoka
reduces the concentration of copper in raw wastes from
45 ppm to 0.2 ppm by treatment with ion exchange
resins. Despite this treatment the oysters in Nobeoka Bay
still have a green colour.

It has been established that oysters become green in 2
to 3 weeks. This is longer than the time required for fish
and shellfish to become tainted by odours from mineral

* Tokai Regional Fisheries Research Laboratory, Tokyo, Japan.

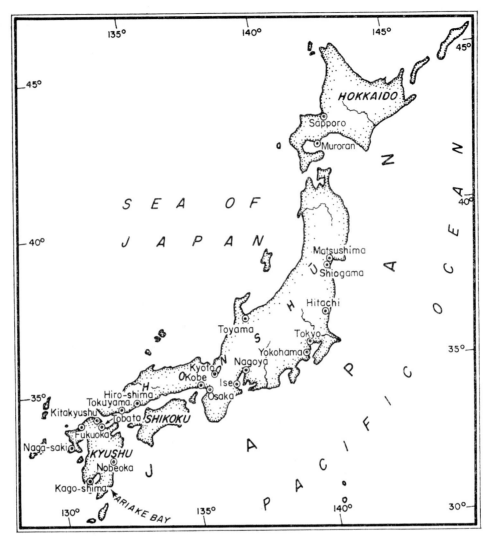

Fig. 1

oils or accumulate methyl mercury. It is difficult therefore, to be certain that green oyster is caused only by the uptake of copper. Laboratory experiments also show that oysters lose some of their green colour during periods of active metabolism.

Minamata disease

Minamata disease broke out in 1953 at Minamata in Kumamoto Prefecture. There were 89 cases and 40 per cent proved lethal. The disease was a toxicosis of more than 100 ppm of mercury caused by eating fish and shellfish from Minamata Bay. The mercury was derived from methyl mercury in the effluent from a chemical plant. It was presumed that the insufficient replacement of water in the bay had accelerated the poisoning of organisms there. After the incident the plant provided facilities for treating its wastes.

In 1964 Minamata disease broke out again in the basin of the River Agano in Niigata Prefecture; 29 cases and five deaths. Since these accidents a regulation that methyl mercury must not be detectable in industrial effluents has been enforced. The use of agricultural chemicals that contain methyl mercury has been also limited. It is believed that the danger of Minamata disease has been removed.

The addition of this type of mercury to Minamata Bay has been stopped and the mercury content of the organisms in the bay has decreased.

Seafoods with foreign odours

Fish become tainted with foreign odours in harbours and in the vicinity of engineering and petroleum plants. Osaka Harbour has this problem. Although there is no oil refinery or other plant producing oily wastes near the harbour, the oil content of the bottom mud is high from the disposal of oil by ships. Fish kept in fish preserves in the harbour become unmarketable.

Iwakuni Airport is an example of the problems caused by engineering plants. The effluent from the airport deposits oil on the bottom of the reservoir through which the effluent flows out to the sea. Fish caught in the vicinity of the reservoir, particularly grey mullet, have the odour of oil. There are petroleum and pulp plants in the vicinity of Iwakuni and their effluents combine.

The most intensive studies on fish spoilage due to effluents of petroleum plants were made at Yokkaichi where all fish caught within 2 km of the harbour, half of the fish within 2.5 to 4 km, and several species of animals, including sea eel, squilla and flat fish, within 4 to 15 km had offensive odours.

[78]

Fish may take up odours from both sea water and the sea bed. The oils in sea water are derived mainly from the effluents of oil refineries and petro-chemical plants. Oils in the sea bed are introduced by precipitation from these effluents and by the dumping of industrial wastes.

It has been reported that oysters, laver (edible seaweed) and fish take on odour from sea water containing 1.5 ppm and 1.7 ppm of oil respectively. Experimental studies done at Yokkaichi in 1963 showed that the minimum concentration of oil in sea water necessary to impart an odour to fish was 0.01 ppm. Saurel kept in sea water containing 0.01 ppm of oil take on a slight odour in 24 hours. Mackerel takes on odour in sea water containing about 0.05 ppm of oil. The maximum acceptable concentration of oil in sea water for fish for human consumption is 0.01 ppm.

The treatment of wastes with activated sludge reduces its smell as well as its oil concentration. The minimum concentration of treated oil that gives an odour to fish was found to be 0.1 ppm. On the basis of these results, permissible oil concentration has been set at less than 1 ppm for effluents treated with activated sludge. In Yokkaichi and Iwakuni, however, the effect of this ruling has not yet been ascertained.

It was difficult to give fish offensive odour by oral administration and it is likely that the oil in bottom mud is absorbed through the gills of fish that touch the bottom. By experiment mud containing 0.2 per cent of oil (dry basis) gave fish a foreign odour. This odour in fish can be removed by clean sea water.

The ambient temperature of water in the harbours of Yokkaichi and Mizushima has been increased by warm water effluents from power plants. Therefore fish gather there in winter, and odour in fish is more intensified then. So oil effluents should not be discharged into the harbour.

Problems of disease and injury

Agricultural chemicals until recently had caused problems only in fresh waters. Rice culture in paddy fields is the nucleus of Japanese agriculture. Fresh-water animals such as loach, carp, and shrimp, are produced in paddy fields and other waters. While fresh-water animals had been injured by pesticides, there had been few problems with coastal animals because the inflow of these chemicals into coastal waters usually had not been sufficiently concentrated to injure sealife. In the Ariake Bay, however, there was a large-scale death of mysid shrimps in 1953 and of soft clams in 1962.

Along the coast of the Ariake Bay an agricultural plain covers the basin of the Chikugo River. Since it was estimated that the death of the mysid shrimps accounted for 80 per cent of the total damage in Japan by agricultural chemicals in 1953, attention was paid to the use of pesticides. The studies begun after this accident amounted to no more than biological assay of the TL_m values of pesticides.

The use of PCP as an herbicide increased rapidly in 1962, and it was spread over the paddy fields in the basin of the Chikugo River because of the unusual weather that summer. It rained heavily over the area immediately after that, and the PCP flowed quickly into the Ariake Bay. This resulted in heavy damage to soft clams in tidelands all around the Ariake Sea. PCP caused large-scale death among fish in Lake Biwa at the same time. The harmful effect of PCP drew deep concern. Since then, its use has been regulated, and the development of other herbicides accelerated.

Death of oysters in Matsushima Bay

Matsushima Bay has been famous for its oyster culture but large-scale deaths have taken place every summer since 1961. Heavy pollution was caused by the effluents from fish processing plants in Shiogama Harbour. It was found that in the bottom mud organic materials had increased from 13 to 41 per cent and total sulphides increased 27 per cent over the values of 10 years before. Moreover, water temperature becomes high during the summer because of the shallowness and poor water circulation in the bay.

Bacterial injury to oysters was once suggested as a reason for oyster deaths in Hiroshima Bay. But bacterial studies showed no distinct evidence of such injury. On the basis of patho-histological studies, it seemed rather that mortality might be caused by overgrowth and over-maturity from high water temperature and eutrophication.

Carcinomatous disease of laver, *Porphyra tenera*

A carcinomatous disease of laver is found in culture grounds where industrial effluents are discharged, and where there are bacterial diseases caused by the excess of organic substances. The cells of laver affected by this disease divide irregularly, and its surface becomes rough, something like millet grain, and the leaves wrinkle. These symptoms are caused experimentally by cyanide, phenol and the other chemicals.

The disease was found first in the culture ground near Nagoya Harbour and later in Tokyo Bay. It can now be observed within 1 km of the mouth of the River Ohmuta.

Bitter taste of soft clams

There are many coal plants and other industries in the basin of the Ohmuta River. The river is heavily polluted by their effluents, and soft clams in its estuary have a bitter taste. These clams contain aromatic compounds, and it is thought that the bitter taste is produced by an abnormal metabolism that accompanies the accumulation of these compounds. Quantities of aromatic nitro compounds transported by river water occur in the bottom mud of the estuary. The distribution of these compounds coincides with the occurrence of bitter tasting clams. A survey in 1963 showed 10 per cent of the clams within 5 km of the outlet of the river were bitter.

Red tides, "blooms" of phytoplankton

The coastal waters of Japan have been made eutrophic by the inflow of urban effluents. In recent years there have been frequent occurrences of red tides. These explosive blooms of phytoplankton are caused, under certain conditions, by a high concentration of nitrogen and phosphorus.

Red tides occur throughout the summer in many of Japan's bays, including those of Tokyo, Ise, Osaka, and Hakata. These bays are supplied with a large quantity of nitrogen and phosphorus from sewage. It is thought that red tides are produced in waters where COD values are about 1 ppm. In most rivers the COD values are higher than 1 ppm, and red tides develop within the area where the river waters are somewhat diluted. In Tokyo Bay red tides occur locally even in the winter.

Since 1957 red tides have occurred frequently and for fairly long periods in Tokuyama Bay. Here nitrogen is supplied by the industrial effluents and the limiting factor is phosphorus. In 1957 the bay was dredged. Phosphorus from the deoxidized bottom mud was released into the water column and a red tide resulted. In Ohmura Bay the effluents from numerous industries producing starch probably contribute to the eutrophication of the bay and there is little exchange between bay water and the open sea. Red tides occur frequently throughout the summer. Apparently the deoxidization of bottom water releases nutrients from the bottom mud. Plankton bloom near the bottom and then rise to the surface to form a red tide.

Effects of dumping sewage

Sewage is now dumped beyond the prohibited area established by the Clean Law, but sometimes it drifts ashore. In Osaka Bay sewage is dumped 10 km offshore and the dumped materials usually sink in 10 to 60 min. Occasionally small trawl nets are fouled and spoil the catch.

There have been 24 oil releases that have resulted in heavy casualties over the past 10 years—through collisions, 12 strandings, and deliberate disposals. The amounts of oil released approximated 40,000 tons

Solid materials

In the Tokyo and Ise bays solid wastes from petroleum plants have frequently caused foreign odours in fish and damaged fishing gear. In Tokyo Bay an operation to remove these materials was once made. It is estimated that the mean quantity of such solids as plastics in fishing grounds in Osaka Bay is as much as 3 1/100 m², and the total amount in the bay is 46,000 m³. Drifting solids such as pieces of wood, plastic, cloth and paper are deposited in specific areas by tidal currents. Bottles, cans and ropes are distributed throughout the harbours and along the fairway.

In 1965, the pearl oysters cultured near Bisan Channel were badly damaged and it was concluded that the cause was suspended solids produced by dredging the fairway. Suspended solids in sea water usually amount to 25 to 100 ppm at the dredging point and from 15 to 20 ppm 2 km away, but in some cases the suspended solids are moved as far as 15 km by tidal currents. The area that is affected by dredging operations also depends on the technical conditions. In Toyama Bay dredging spoil was dumped in deep water outside the harbour, and it was difficult to find the remains after 1 year.

Spatial aspects of pollution problems

Because of their low density, fresh-water effluents and the cooling water from power plants spread over sea surfaces, but effluents of the same density as sea water, such as those produced by dredging, disperse in another pattern.

Many field surveys, particularly for pulp mills, have been made on the dispersion of the effluents in terms of the extent of spread, speed of dispersion, and the diffusion process. The diffusion of fresh-water effluents has been clarified to some extent. A rapid slope current flows outward when a large quantity of effluent is discharged, and then spreads on the sea surface. Sea water under the effluent rises up and dilutes the effluent from 60 to 100 times at the edge of spreading. Since a stream occurs along the outside edge an obvious "shione" is sometimes formed. The thickness of effluents is from 1 to 2 m inside the edge and it takes from 1 to 2 h for the effluent to travel from outlet to edge. While the pollutant effluent is concentrated in proportion to its dilution inside the edge, there is scarcely any pollutant beyond the edge.

The relationship between the quantity (X) and the dispersion area (Y) of the effluent has been determined experimentally:

$$\log Y = 1.23(\log X) + 0.086$$

Fresh-water effluents spread on the surface of the sea until they are diluted about 100 times. Warm sea water effluents dilute until their temperature is decreased to about 1°C higher than ambient. Therefore, the dispersion area of warmed coolant is far less than that of fresh water. Sea water containing suspended materials disperses by eddy diffusion until the materials are precipitated.

Deterioration of the sea bed

Although the sea bed is deteriorated by the sedimentation of suspended materials from effluents, the area affected is usually far less than the water area affected by the effluent. Heavy deterioration takes place where organic suspended materials are discharged, such as pulp mill effluents. The sea bottom within 500 m off the coast has been affected at Mishima and Kawanoe, where there are a number of pulp plants. Deterioration of the sea bed near embayments depends on such local conditions as the width of mouth, water depth, and the quantity of fresh-water inflow that affects the subsurface inflow of sea water into the bay. When fish and shell fish are cultured in these embayments their waste products will accelerate deterioration.

Sediment COD values of more than 30 mg/g, on a dry weight, are abnormal. The ratio of COD value to ignition loss is useful for following the deterioration process. Sulphide contents greater than 0.2 mg/g also are abnormal; in these cases methane gas bubbles out at low tide. Furthermore, when the deoxigenation of deep water occurs, nutrients and metallic salts in the sediments are dissolved.

About 0.2 per cent of fats and oils (not mineral oils) often exists in normal bottom sediments, but the presence of more than 0.2 per cent mineral oil in mud produces fish with a foreign odour.

Alteration of fauna

The alteration of fauna by pollution usually occurs among the benthic animals in tidal regions and in bottom sediments. Since the effluents spread on the sea surface animals in the tidal regions are first affected. The animals are reduced in number in proportion to the degree of pollution. There are also physiological effects such as the occurrence of green oyster and the accumulation of methyl mercury by shellfish. Slight eutrophication makes benthic fauna more rich. As pollution proceeds polychaetes prevail, particularly *Capitella capitata japonica* which has a strong tolerance to pollution. In some cases only *Capitella* exists at the stage before all animals are killed.

Characteristics of some polluted areas

The effluent from pulp mills, because of its lignin and large volume, pollutes wide areas in coastal waters, and deteriorates the sea to some extent by precipitating suspended materials. Pulp mill wastes are obvious in Yatsushiro, Nichinan and Saheki on Kyushu Island, Mishima, Kawanoe and Iwakuni-Ohtake on Seto Inland Sea, Gotsu, Yonago, the outlet of the River Zintsu, Akita on the coast of the Japan Sea in Honshu, Tomoegawa, Yoshihara, Takahagi, Ishinomaki, Hachinoe at the Pacific coast in Honshu, Shiraoi, Tomakomai, Yufutsu and Kushiro in Hokkaido.

Pollution by the effluent of fermentation plants (e.g. Nobeoka and Mitajiri Bays) is remarkable because the effluent is coloured and is of great volume. At Mitajiri, deterioration of the sea bed by the sedimentation of suspended materials is serious. Fish processing plants also cause problems in Shiogama, Hachinoe, Nemuro and Wakkanai and in many other places, although the scale is small. The chemical fibre plants in Mitajiri, Ohtake, Iwakuni and Mihara, discharge acidic effluents of fairly large volume which make the sea water turbid.

Steel works discharge effluents that turn the sea water red in the harbours of Tobata, Kure and Hirohata. Titanium refinery plants also cause coloured areas at Ube, Okayama and Onahama. The effluents from petroleum plants give a foreign odour to fish and shellfish at Yokkaichi and Mizushima, and also, on a smaller scale, at Matsuyama, Tokuyama, Iwakuni and Shimotsu.

Power plants discharge warmed coolant, flyash and sometimes chlorine. Recently, warmed coolants of large volume have become a serious problem for fishery resources and fishing operations.

Effects of effluents over a wide area

In Tokyo, Osaka and Ise bays, where human population and economic activities are centralized, large volumes of sewage and industrial effluents are discharged, and form an extensive polluted area. The rivers that flow into these bays have become drainage canals for wastes; the rivers Sumida and Tama are examples. Furthermore, a number of industrial effluents flow directly into these bays. Most of Tokyo Bay is covered with polluted water to a depth of 4 to 5 m. It flows out of the bay at ebb tide.

Deterioration of the sea bed is most serious in Tokyo Bay. Although there is an inflow of oceanic water through the east side of the mouth of the bay, at depth water substitution is not great and the sea bed is heavily polluted by sediments, particularly the inner part. Osaka Bay is not as polluted as Tokyo Bay, and heavy pollution is limited to its inner part. In Ise Bay, eutrophication of surface water is caused by effluent from Nagoya Harbour. Some eutrophication occurs in the west part of the bay but considerable deterioration of sea bed has not yet developed.

Outline of the polluted area

The effluents disperse around outlets and form polluted areas proportional to their volumes but modified by local conditions, such as an inflow of river water, and geographical characteristics.

The pulp mill at Gotsu has a single effluent to the open sea that forms a coloured area 3 km in radius. At Mishima and Kawanoe, a group of pulp mills discharges effluents on the coast forming an area of coloured water 5 km in width 1 km off the coast. The sea bed is deteriorated to 500 m off the coast. A pulp mill at Yonaga discharges its effluent to the sea via river water that spreads over a radius of 1.5 km. However, coloured water is limited to one-fifth of the area. In Nobeoka, effluents of bemberg, amino acid fermentation, and many pollutants are discharged into the sea via a river. The water, which contains copper and ammonia and is coloured brown, spreads over the sea 2 km off the coast. It causes green oyster and sometimes drifts into the bay 3 km away and injures fish.

The Ohmuta River receives the effluents of many chemical plants related to the coal industry. Its colour varies. Though there is little colour at its outlet, laver are infected with carcinomatous disease within 1 km of the outlet, and soft clams within 3 km have a bitter taste.

In Mitajiri Bay geographical conditions have helped develop pollution. In Mitajiri Harbour, the inner part of the bay, there are two fermentation and one chemical fibre plants. The effluents of the fermentation plants have much suspended material, and the sea bed is heavily damaged. At ebb tide polluted water covers the bay and sometimes causes the death of fish.

Matsushima Bay is an example of pollution in a closed bay. Shiogama Harbour, in the western part of the bay, has a number of fish processing plants that discharge much organic waste; most of Shiogama Harbour is heavily polluted. Although oysters have been cultured extensively in the bay, they have been replaced by laver culture because large-scale death of oysters has taken place every year. Recently it has become eutrophic and the laver suffers too; its culture ground is being limited to areas away from the harbour.

Present status of marine pollution

A project has been established to apply environmental water quality criteria to individual aquatic regions. In this project regions are classified into three groups. The limited area around waste outlets is graded as class C and the areas away from the outlets are classified as A or B according to their use for fisheries and other purposes. The operation has begun in fresh-water regions and is in the planning stage for the coastal regions. In 1968 the Fishery Agency started a 5-year plan for economic and scientific research on the principal aquatic regions. The scientific research is intended to make clear the present status of pollution by the examination of water and sea bed quality, dispersion of wastes, changes of flora and fauna, and the identification of harmful effects from wastes.

Marine Pollution and Research in the Philippines

*R. M. Lesaca**

Pollution des mers et recherches aux Philippines

Pour le moment la pollution des mers ne semble pas poser de problème majeur aux Philippines excepté dans quelques estuaires proches de centres industrialisés des Philippines. La plupart de ces cas de pollution se rapportent à la pollution de l'eau due à des déchets domestiques et des résidus d'exploitations minières. Le problème est grave dans les régions près de baies côtières et les régions côtières près des estuaires de rivières qui traversent ces régions urbaines. Une loi sur le contrôle de la pollution a été promulguée et des recherches sur les problèmes de la pollution sont en cours.

Contaminación marina e investigaciones en la República de Filipinas

Hasta el tiempo presente la contaminación marina no aparece en la República de Filipinas como un problema de mayor importancia, con excepción de pocas regiones estuarinas que bordean algunos pocos centros industrializados de las Filipinas. En la mayoría de los casos de contaminación se trata de aguas de desperdicios domésticos y residuos minerales de las explotaciones mineras.

El problema es grave solamente en las regiones cerradas de bahías playares y en aquellas de la boca de los ríos que traspasan áreas urbanizadas. Ha sido promulgada una ley del control de la contaminación y se han comenzado las investigaciones sobre los problemas de contaminación.

THE Philippines are largely agricultural, but since 1955 an industrialization programme has established numerous industries. These were attracted to specific areas by the presence of raw materials, cheap power, availability of water, and accessible transportation.

Marine pollution in the Philippines is at present limited to a few specific areas.

Manila Bay is a major shipping centre and its waters are also used for recreation and commercial fishing. A fairly detailed study of the pollution of Manila Bay was made in 1969. The study was financed by the United Nations and was part of a more general study to investigate the possibility of an improved sewerage system for metropolitan Manila. It was found that the dissolved oxygen levels in Manila Bay are depressed below the minimum level of five parts per million defined by the National Water and Air Pollution Control Commission. Coliform counts obtained from monthly shore sampling very often gave values above 1,000 per 100 ml. It was estimated that at least 265 tons of 5-day BOD is contributed daily by the pollution from the Manila area alone.

There are various sources of industrial pollution in the Bay. The Pasig River, which flows through the city of Manila, receives the effluents from numerous industries. Fortunately the currents in the bay are such that the river waters are carried rapidly out into the China Sea. The many wharfs and piers in the bay cause a light but perennial oil slick along the eastern shore. On the northern shore are an oil refinery and a fertilizer plant. A pulp and paper plant on the northeastern shore contributes a small but significant amount of pollution. Manila Bay is industrializing rapidly and pollution from industrial effluents is expected to increase.

Pollution from mining operations

A principal pollutant is the silt that is discharged into rivers by plants processing the metal ores, particularly copper, mined in the mountains. The quantities of silt transported to the sea by rivers per year are summarized in Table 1.

Other sources of industrial pollution

Other sources of pollution include: A sugar processing plant that releases fermenting molasses and distillery wastes into the Palico River, three coconut processing plants that release fermenting coconut water into Tayabas Bay, a fertilizer plant whose effluents go into the Sapang Daku River, and logging operations that pollute fish

TABLE 1 POLLUTION BY SILT FROM MINING OPERATIONS

Recipient of discharge	River	Mines	Silt/year m. tons
China Sea	Abra	2	2
Lingayen Gulf	Agno	6	6
Sulu Sea	Sipalay-Taoangan	1	2
Tañong Strait	Sapang Daku	1	10
Philippine Sea	Taft	1	0.3

ponds in Tagabuli Bay with sawdust and bark. Subic Bay is polluted by oil and oily wastes from ships and docking facilities of the United States naval base. The Agus River on Mindanao Island receives effluents from chemical plants, pulp and paper mills, lumber operations and the only integrated steel mill in the country. In several places the Agus River has been harnessed to provide hydroelectric power. The area around the Agus River is industrializing rapidly and pollution is increasing.

Marine research in the Philippines

Marine pollution research in the Philippines is coordinated by the National Committee on Marine Sciences, which was created by the National Science and Development Board (NSDB). The principal function of this Committee is to coordinate national activities in the marine sciences, including marine pollution and research. This body is made up of representatives from approximately 15 government agencies including the National Water and Air Pollution Control Commission.

Most research in the marine sciences pertains to physical and chemical oceanography, marine geodesy and geology, and studies on ocean currents in relation to commercial fishing. There is very little research on marine pollution as such, probably because there are relatively few industrial complexes that discharge effluents into the ocean. But the process of industrialization is growing. At present, only two major marine pollution studies are being undertaken. The first is on the pollution of Manila Bay and the second on the oceanography and general marine biology of Macajalar Bay in Northern Mindanao.

The Manila Bay study was undertaken by an American group with UNDP technical assistance. It concerns the possibility of using the bay as the recipient of low-level wastes discharged from metropolitan areas which are to be provided with a sanitary sewerage system. The study included determining the direction of currents in the bay, the dissolved oxygen content and the present pollution load received. The study also included characterizing the

* National Water and Air Pollution Control Commission, Manila, Philippines.

biological and physical quality of Manila Bay water. Preliminary results indicate that there is relatively less pollution during the dry season than during the rainy season when surface runoff transports pollution from the metropolis into the bay.

The Macajalar Bay study aims to gather information on the marine fauna and flora of the area and the physical and chemical characteristics of the waters. The study has received financial assistance from the NSDB.

The pollution control law

Although the country started industrializing in 1955, it was only in 1964 that the Congress passed a law creating the National Water and Air Pollution Control Commission and defined its functions, duties and responsibilities. The Commission started to function in early 1966, and has been in continuous operation since 1968. The law defines pollution as any "alteration of the physical, chemical and/or biological properties of any water and/or atmospheric air of the Philippines, or any such discharge of any liquid, gaseous or solid substance into any of the waters and/or atmospheric air of the country as will or is likely to create or render such waters and/or atmospheric air harmful or detrimental or injurious to public health, safety or welfare, or to domestic, commercial, industrial, agricultural, recreational or other legitimate uses, or to livestock, wild animals, birds, fish or other aquatic life". The Commission's most important duty is to determine the existence of pollution in the waters or air within the territorial limits. It is also authorized to conduct public hearings to determine the nature and extent of any pollution and to order its remedy.

The Philippine Congress has adopted a national policy of maintaining reasonable standards of purity for the waters and air of the Philippines and, in the implementation of this policy, the Commission has always worked on the philosophy of prevention and control rather than correction. It also realizes the contribution of industry to the economic development of the country and has always sought cooperation and voluntary compliance rather than compulsion. Most of the cases brought before the Commission dealt with inland waters and river pollution rather than marine pollution.

Marine Pollution Studies at Hong Kong and Singapore   H. R. Oakley and T. Cripps*

Etudes sur la pollution des mers à Hong Kong et à Singapour
Le document décrit les événements qui ont mené à ces travaux, les problèmes auxquels se heurtent les gouvernements intéressés et leurs ingénieurs-conseils, ainsi que les objectifs et les méthodes qui ont guidé le travail de prospection.

L'étude menée à Hong Kong avait pour objet d'examiner quantitativement et qualitativement les effets des déversements actuels des eaux d'égout non traitées dans les ports de Victoria et de Tolo, et d'évaluer l'aptitude de ces ports et des eaux adjacentes à assimiler de futurs déversements d'eaux d'égout traitées ou non sans qu'il en résulte des effets préjudiciables.

A Singapour, les travaux avaient pour but de déterminer si les déchets ou les boues des égouts peuvent être déversés dans la mer sans dommage et d'évaluer le danger de pollution par les hydrocarbures provenant des raffineries locales et du mouvement des pétroliers.

Estudios sobre la contaminación marina efectuados en Hong Kong y Singapur
En esta contribución se describen los hechos que dieron lugar a las investigaciones, los problemas con que tropiezan los gobiernos interesados y sus ingenieros consultores, y los objetivos y métodos empleados en la labor de reconocimiento.

El estudio de Hong Kong tenía por objeto examinar el efecto de las descargas de aguas residuales sin tratar en los puertos de Victoria y Tolo, y de determinar la capacidad de éstas y de sus aguas adyacentes para asimilar las futuras descargas de aguas residuales crudas o tratadas sin que se produzcan efectos dañosos.

En Singapur, la investigación se encaminó a determinar si era posible descargar con seguridad las aguas residuales o los sedimentos en el mar y evaluar el peligro de la contaminación producida por hidrocarburos procedentes de las refinerías locales y de los petroleros.

HONG KONG is one of the most densely populated areas in the world. Of a total estimated population of about 4 million, over 80 per cent are in the urban areas of Hong Kong Island (7,000 ha) and Kowloon (970 ha). The present total population of the New Territories and the islands is about ½ million concentrated mainly in four principal centres but with a number of small isolated village communities. In 1966 it was estimated that the total population of the Colony would increase by 3 million people by 1986 and the bulk of this increase would be in the undeveloped areas of the New Territories (fig 1).

Marine fish is one of the Colony's main products. Over 9,000 fishing vessels are worked by over 75,000 people. Approximately 60 per cent are inshore vessels mainly operating south of the Colony inside the 20 fm line. Edible oyster industries and pearl oyster culture farms are located around the Colony. Within and without the urban areas there are important heavy industries and many light industries, farming and fishing are the main occupations in the New Territories.

Sewage effluent from water-borne systems is disposed of, after screening in some instances, through a number of sea outfalls (fig 1). Recognizing that population increase in the New Territories would cause new problems, the Government commissioned in 1965 an examination of the problems of sewage disposal in those inland areas. That report (Watson and Watson, 1965) gives a comprehensive review and it was supplemented by another report (Watson and Watson, 1966). This recommended more detailed studies of the consequences of the marine disposal of sewage in Victoria Harbour and in Tolo Harbour, the sheltered inlet in the New Territories. This led to a Marine Survey here considered.

Hong Kong Island lies on the coastal shelf of the mainland and the adjacent estuary of the Pearl or Canton River is important in local hydrology. Three types of water are present: fresh water from the river system, surface shelf water and deep shelf water. The first has low salinity and wide annual temperature range, the second high salinity and variable temperature and the third high salinity and relatively constant temperature.

*Partners, J. D. and D. M. Watson, Westminster, London S.W.1

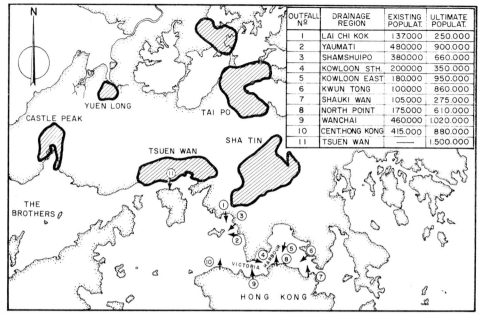

OUTFALL N⁰	DRAINAGE REGION	EXISTING POPULAT.	ULTIMATE POPULAT.
1	LAI CHI KOK	137.000	250.000
2	YAUMATI	480.000	900.000
3	SHAMSHUIPO	380.000	660.000
4	KOWLOON STH.	200.000	350.000
5	KOWLOON EAST	180.000	950.000
6	KWUN TONG	100.000	860.000
7	SHAUKI WAN	105.000	275.000
8	NORTH POINT	175.000	610.000
9	WANCHAI	460.000	1.020.000
10	CENT.HONG KONG	415.000	880.000
11	TSUEN WAN	—	1.500.000

Fig. 1. Hong Kong: areas of prospective development (shaded) and principal sea outfalls

The deep shelf water occurs throughout the year over the 50 m stratum about 96 km off Hong Kong. It moves shorewards in the summer along the shelf bottom in conjunction with upwelling, resulting from the seaward movement of the surface water discharged from the Pearl River (maximum strength June/August). The fresh water spreads over the surface layers of the sea and in the summer its influence may be detected as far as 160 km south of Hong Kong, but in the winter sea water invades the estuary and extends upwards for an unknown distance. Consequently, local waters around Hong Kong and the New Territories show varying annual salinity and temperature characteristics.

The climate is determined by the monsoonal system. In the Victoria Harbour area the predominant wind direction is east-north-east from October to December but during the weak south-west monsoon in June to August the winds are variable. The surface currents of the China Sea flow in accordance with the wind so that the main currents are generated by the monsoons, and they reverse direction with the change of monsoon.

The tidal range at Hong Kong does not normally exceed 2.5 m and the tidal pattern is complicated as both diurnal and semi-diurnal winds occur. Tidal currents are of moderate strength on the west shores of the New Territories increasing in strength as the channel of the New Territories and Lantao Island narrows (fig 2).

Survey objectives

The aims were:

(1) to suggest water quality standards from health, aesthetic and ecological viewpoints for the waters concerned;

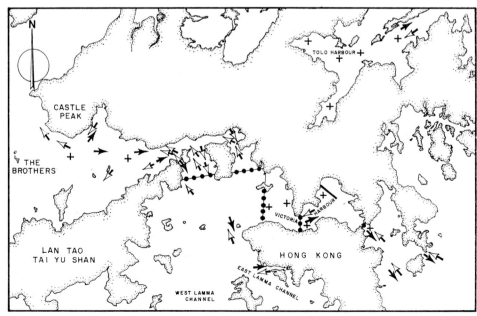

Fig. 2. Hong Kong: main tidal streams (ebb current→ ; flood current ←+) and Phase II fixed stations positions (+) and traverse lines (-•-•-•-)

(2) to assess the pollution load these bodies of water can take yet maintaining suggested standards. The assessment for Victoria Harbour and its associated tidal streams is based on the treatment and disposal method presently employed in Hong Kong. For Tolo Harbour, the assessment is based on different degrees and types of treatment before disposal;

(3) to recommend an improvement programme with particular reference to when and where action should be taken if it is anticipated that the assessed capacity of the water may be exceeded;

(4) to recommend monitoring methods and programmes to ensure that suggested standards are not exceeded and/or to report on the pollution level when incorporation of improvement programmes becomes necessary.

Differences in objectives and circumstances dictated different approaches in the surveys of Tolo Harbour and Victoria Harbour. In Tolo Harbour the existing pollution is slight and the aim was to determine the best location of future outfalls and the degree of necessary treatment. Since the sheltered waters permitted unrestricted survey, float tracking was one of the most useful methods. In Victoria Harbour, busy with shipping, float tracking was not possible and current meter observations and other instrumental evidence such as dye or bacterial tracers were used.

Experimental work was conducted in two phases. The first, over a three-months period, was aimed at general water movements and the selection of key points. The second and longer phase spanned the two principal monsoon periods and lasted approximately twelve months. Three surveying teams were established for phase I, using a suitable boat and a full set of survey gear including direct reading directional current meters and salinity, dissolved oxygen, temperature and turbidity meters. Two teams were sufficient for phase II. Most of the observations were carried out over a 25 h period. The selected fixed-station positions and traverse lines are shown in fig 2. In addition, sampling at bathing beach locations was arranged.

Float tracking was carried out using canvas covered drogues set at different depths. For sampling for bacterial analysis below the surface, a simple and easily sterilized unit was devised; membrane filtration was used for bacterial counts. Grab and core sampling apparatus was used for bottom sediments. Boat location relied on visual methods. Visual methods were also used for float tracking; experiments were carried out using radio beacons but these proved unreliable.

Aerial observation was also used on the visible extent of the sewage field. Given suitable weather conditions this was effective but aerial photography proved disappointing and limited by surface conditions.

Use of bacterial tracers

The species *Serratia indica* was used in 10 experiments as a bacterial tracer and the procedures for culturing and use generally followed those recommended by Ormerod (1964) and Pike, Bufton and Gould (1968). A culture with a concentration of about 10^9 per ml was dosed at a rate of about 1 litre per hour per 10,000 population served into selected foul sewage outfalls for a period of $12\frac{1}{2}$ hours to cover one tidal cycle.

The large volumes of culture required some changes to procedure, as follows:

(1) all cultures were incubated in 50 l plastic containers in the laboratory and transported to site as required;

(2) it was not practicable to maintain the full recommended precautions for sterile equipment;

(3) the sucrose concentration was maintained at 0.3 per cent instead of the 3 per cent recommended by Ormerod;

(4) constant temperature control during dosing was not practicable, but times were chosen when the ambient temperature was within the range 25–30°C.

During the dosing period and for $12\frac{1}{2}$ hours after, (25 hours in all), water samples were taken in 500 ml sterile bottles at surface and mid-water from selected points up to a maximum of 120 samples in each experiment which was as much as the laboratory could handle. Membrane filtration techniques were used in analysis.

In general the experiments were very successful. However, one main difficulty was that pigmentation of the colonies, although satisfactory at the beginning of the experiments, became less and less intense and by the end very few red colonies were developed. The colourless colonies were similar to the red ones; they were all circular, low convex, translucent with smooth surfaces and entire edges. They could be counted. The other difficulty was that antibiotic medium was not as effective in inhibiting the growth of other bacteria present in the sea water samples as had been hoped.

Summary of results

Chau and Abesser (1958) have published previous work in the area and have discussed temperature, salinity and dissolved oxygen content of the coastal waters.

The conditions they reported were generally confirmed. Water temperature closely follows ambient air temperature and there is no pronounced vertical stratification except during the relatively infrequent periods of heavy rainfall, when the surface salinity may fall to relatively low values. Additionally, the salinity of the southern waters drops dramatically during the spring floods from the Pearl River, reaching lows of 5 ‰ in the western sector. Water quality was usually high in Tolo Harbour and less satisfactory in the southern waters where contents at or below 50 per cent saturation are common (fig 3).

The low levels in Victoria Harbour and nearby are largely a function of sewage discharges, particularly in the winter months when surface water run off is low. In the Castle Peak area, however, dissolved oxygen drops only after the spring floods from the Pearl River. At these times the river's effect is visible for a hundred miles offshore. During the floods algae bloom abundantly and there is evidence of the resulting photosynthetic activity raising dissolved oxygen into the supersaturated range before algal decay combines with the pollution base load from the river to deplete the reserves. The Pearl River effect on Victoria Harbour is less marked but still contributes to lowering DO to 50 per cent saturation in the spring compared with a winter minimum of about 60 per cent.

As expected, water behaviour in Castle Peak Bay and Victoria Harbour, which is open to tidal flushing, is quite different to that in land-locked Tolo Harbour.

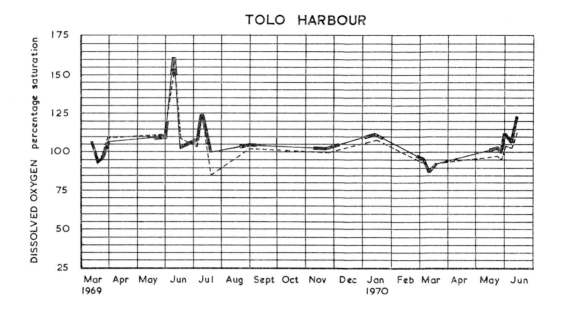

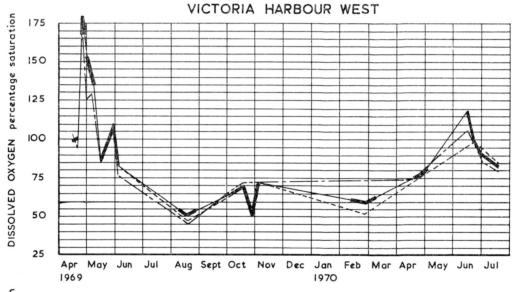

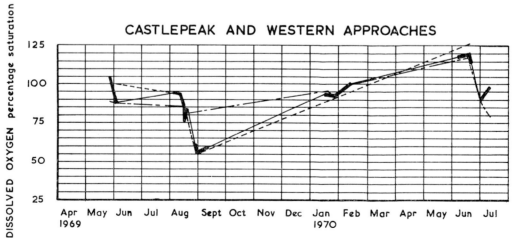

LEGEND·——·Depths : Surface ———— 7½ feet ------- 15 feet —·—·—

Fig. 3. Dissolved oxygen findings in the regions of Tolo Harbour, Victoria Harbour West, Castlepeak and Western Approaches

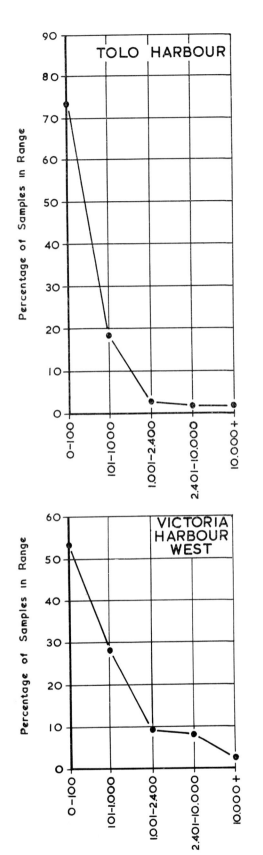

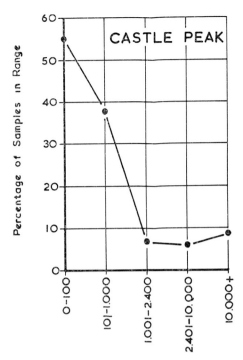

Fig. 4. *Bacterial counts in the regions of Tolo Harbour, Victoria Harbour West and Castle Peak*

Within Victoria Harbour the ebb is predominant and net movement is easterly. The estimated mass transport ranges from 6.5 to 50 × 10⁶ m³ according to tide range and wind direction. Castle Peak Bay waters are also replenished largely from the west with removal eastwards and southwards into the West Lamma Channel. To the west and northwest of the Kowloon peninsula, movement

is less satisfactory; during spring tides, ebb currents carry polluted waters into the strong east-going current through Victoria Harbour, but the area tends to be isolated during neap periods, particularly with south-west winds when there is reciprocating movement along the coast and little evidence of net removal. In the southerly section of the western approaches escape of polluted water is both easterly and westerly as the tide ebbs and flows.

In Tolo Harbour the predominant north-easterly and easterly winds blowing along Tolo Channel have a marked effect on surface water movement. Generally surface water spreads landwards across the harbour and is compensated by outward flowing deeper water or by surface water on lee shores. Net movements are small and are therefore difficult to quantify by path tracing or current measuring methods. Examination of average salinities measured, and reference to average rainfall and the "base" salinity of the open waters beyond Tolo Channel, enabled an assessment of average residence time of fresh water run off to be made. The method adopted was that described by MacKay (1969), and residence time was estimated at 35 days.

Bacterial counts in Victoria Harbour were generally very low in spite of the large sewage discharges (fig 4); 6 per cent to 8 per cent of coliform counts were above 2,400 per 100 ml and only 1 per cent to 2 per cent above 10,000 per 100 ml. In Tolo Harbour and Tolo Channel the counts were still lower. Coliform counts in the Castle Peak area were higher, often over 10,000 per 100 ml, with maxima in excess of 10⁶ per 100 ml. The high readings were all coincident with Pearl River flooding.

In the southern waters, therefore, the spring floods from the Pearl River have a major influence both on dissolved oxygen and coliform count.

In Tolo Harbour, visual evidence was of healthy water conditions, except close to the developed areas at Sha Tin and Tai Po. In Victoria Harbour, sewage solids were observed in some localities and there is smell nuisance on some occasions, particularly to the west and north-west of Kowloon. A number of outfalls in this area are sea-wall discharges with consequent poor mixing potential.

Bottom muds in Tolo Harbour are thick, contain up to 47 per cent volatile matter and have 5 day B.O.D. of between 150 and 700 mg/l. Coliform counts in the mud reached as high as 107 per 100 ml at the western end of Tolo Channel, near Sha Tin and near Tai Po. Comparison of echo soundings with seventy-year old charts suggests a build-up of bottom deposits, as much as 7 ft in the south of the harbour and 15 ft in the deeper water at the western end of Tolo Channel.

In Victoria Harbour muds predominated again but the bottom in the main channel was relatively clean, consisting of sands and gravels. Volatile solids ranged up to 66 per cent in mud areas, with associated BOD from 40 to 700 mg/l. Comparison of depth with old charts indicates some accretion.

Conclusions

(a) Recreational use of the Colony waters does not form a severe health hazard, but amenity should be preserved by ensuring that recognizable sewage particles and sewage slicks do not occur at bathing or other water contact sport areas. An initial dilution of 1 in 200 over diffusers for screened sewage is desirable; floating matter and solids above 10 mm in size should be removed; and not less than three hours should elapse before polluted receiving water reaches bathing beaches or water contact sport areas.

(b) A minimum dissolved oxygen level of 50 per cent of saturation is desirable, both to maintain fish life and to give ample reserve against anaerobic conditions and associated nuisance.

(c) Because of the influence of the Pearl River and in view of the general difficulty of applying and monitoring bacterial standards the setting of such standards is impracticable, but routine counts may be of value for comparison and for design purposes.

(d) Special regard should be paid to ecological changes arising from pollution and affecting prawn and oyster production. Historical evidence on general ecology is insufficient to gauge the present effects of sewage discharges but long-term observations should be commenced so that changes can be assessed.

(e) Toxic discharges do not pose problems at present, but control of industrial waste discharges should be instituted.

(f) The measures required for aesthetic reasons will be effective in avoiding the fouling of ships' intakes and bottoms.

(g) Legislation is necessary with regard to the discharge of polluting matter by ships in the harbour, and with regard to the behaviour and navigation of oil tankers.

Proposals for the future are based on planning and water behaviour predictions and are made subject to continuing monitoring indicating the need for further improvements from time to time. The provision of screening and long outfall facilities on all discharges goes a long way to satisfying aesthetic objection and ensuring efficient dilution, and will probably be adequate for discharges in the Castle Peak Bay and Victoria Harbour areas. Discharges to poorly flushed areas like Tolo Harbour and North West Kowloon will need pretreatment by sedimentation and in some instances biological methods. Monitoring of dissolved oxygen, algal activity and nutrient loads is recommended, together with periodic examination of bottom deposits.

SINGAPORE STUDY

Singapore island covers an area of some 58,000 ha and has a population of about 2.25 million.

The developed areas are served by a water-borne system of sewers and drains to two principal treatment works sited away from the coast and discharging treated sewage effluent to separate estuaries (fig 5). In 1969, due to new development, the Government explored the possibility of sewage disposal, after treatment, direct to coastal waters.

To the north the island is separated from the mainland of West Malaysia by the Johore Strait which is sectioned by the Causeway from the island to the mainland. The little water movement and its shallowness preclude any significant marine discharges. The water around the remainder of the island is also shallow, seldom exceeding 22 fathoms, and at low water of spring tides the sea bed is exposed up to 2.4 km from the shore and shoals extend for 13 km on the eastern tip of Johore at only 1 or 2 fathoms. The island is slightly affected by monsoon conditions and water usually moves westward from the South China Sea through the Straits of Malacca into the Indian Ocean. North-easterly and south-westerly winds may predominate during the monsoon periods and tides follow the modified diurnal pattern with main floods interrupted by tides having usually less than half the range of the main tides.

Inshore water fisheries have progressively decreased in annual catch with palisade traps; 1969 being 40 per cent lower than in 1954. Although inshore fishing accounts for only a small proportion of total catch, it employs about 80 per cent of total fishermen.

Intertidal prawn ponds are also well established but, becoming less economic to operate, in some instances are being drained and reclaimed for development. There is little local shellfish production although green mussels are gathered from the east end of Johore Straits.

Singapore waters are not an important breeding ground, but large numbers of fry and immature fish move with the prevailing ocean current from the South China Sea into Singapore waters where they remain to feed and reach marketable size. The plankton count in local waters is similar to that in the English Channel, but few fish eggs or fry are found and the bulk of the plankton is found at more than 2 m depth. Therefore the disposal of sewage into the inshore waters would probably not have as high an impact on plankton as in temperate zones and might even be beneficial by adding to the nutrient salt concentration. There is, however, a danger that pollution would cause incoming fish to divert away from the coast.

Survey objectives

(1) The investigation was to determine if sewage wastes or sludges, either fully or partially treated, could be safely discharged to sea from new proposed developments. Consideration was also to be given to pollution by oil and oil wastes.

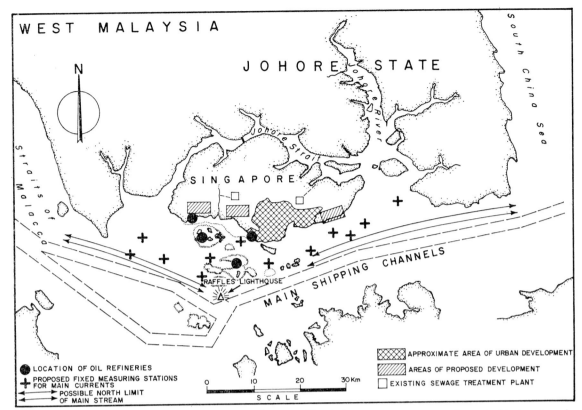

Fig. 5. Singapore: areas of existing and proposed development with existing sewage treatment works, main shipping channels, and Phase II fixed measurement stations

(2) Investigation of marine conditions was broadly spread along the south coast of the Republic with special attention to possible discharge from the proposed developments of Jurong, Telok Blangah and Bedok.

(3) Consideration was given to the present polluting discharges and the effect of further discharges on amenities and the public health situation.

Method

The survey was divided into two phases: the first lasting four months included the collection and assessment of available information concerning weather, tides, tidal currents, bottom profiles, land reclamation, fishing and shipping activities, etc., and enabled planning to be done for the second phase lasting nine months covering the monsoon periods and including field measurements of ocean, tidal and wind induced currents, measurement of water quantity and quality, and the study of sediment movement. The final phase will consist of analysis of experimental results and evaluation of sewage disposal possibilities and oil pollution protection methods.

Assisting in the first phase, the firm of Marine Consultants, Captain C. W. McMullen & Associates, made preliminary offsite investigations and submitted two reports, one a preliminary report on sea disposal facilities and the other discussing the general problems of oil pollution. In both, shipping considerations are of particular importance.

Fixed stations for current and other measurements were chosen in relation to the main tidal movements and with consideration of the main shipping channels (fig 5).

The two survey teams were each equipped with direct reading directional current meters, salinity and dis-

solved oxygen meters and sediment samplers. Because of the shipping, drogue tracking was used as little as possible, preference being given to current meter readings and other fixed station observations. When prospective outfall positions were located, dispersion characteristics were assessed using Bromine 82 as a tracer.

Oil pollution

The risk of oil pollution in Singapore is very great. Oil pollution can occur from collision or grounding of ships, or spillages from ocean-going and coastal vessels, in transit or in port, or spillages from refineries, submerged pipelines and bunkering points, or spillages in docks and shipyards.

In 1969, eight reports of major discharges were lodged. Ships of every type pass through Singapore, including tankers. Due to the restricted waters in the Straits, the inadequacy of navigational aids and the present lack of any routing system, collisions occur. This unsatisfactory situation has been recognized and is being studied, and steps taken internationally to reduce the hazards of ships.

Four major refineries exist to the west of Singapore (fig 5), which also must be regarded as potential sources of oil spillages. Steps have been taken by refinery companies and the Port Authority to reduce the danger of accidental spillages and to contain any spillages which occur. Measures have also been taken at Government level to establish the necessary control, identification and treatment of oil spillages.

Marine disposal of sewage

Full analysis of the results has not yet been made, but

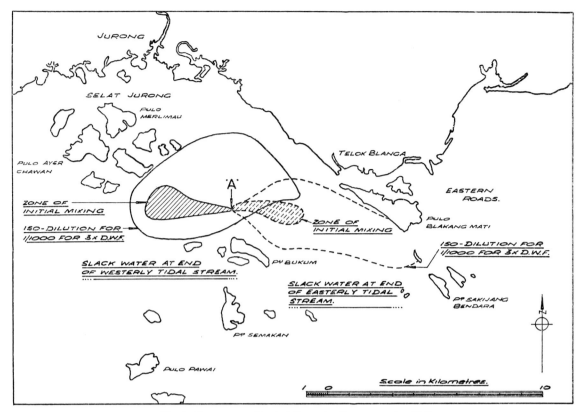

Fig. 6. Singapore: estimated dispersion for proposed outfall

preliminary consideration has been given to the possibilities of marine disposal of sewage from the Telok Blangah development (fig 6).

The current measurements were analysed to eliminate wind effects and to separate the net drift due to ocean current movements from tidal currents. Winds off Singapore are generally light and variable and the wind induced current was usually negligible.

Analysis of the observed currents at point A indicated the relationships:

$$v_{maj} = 0.92 \text{ m} \sin \frac{4\Pi t}{T} + \text{m} \sin \frac{2\Pi t}{T} + d_{maj}$$

and

$$v_{min} = 0.02 \text{ n} \sin \frac{\Pi t}{T} - 0.04 \text{ n} \frac{4\Pi t}{T} + d_{min}$$

where v_{maj} = major component of velocity

v_{min} = minor component of velocity

m & n = constants

d_{maj} = overall drift component along major axis over one tidal cycle

d_{min} = overall drift component along minor axis over one tidal cycle

t = time elapsed after higher high water

T = Period of tidal cycle

Minimum tidal cycles are critical from pollution aspects, and for these conditions and near-surface currents good correlation was obtained for values of:

m = 0.306   n = 1.00

d_{maj} = + 0.09 m/sec d_{min} = + 0.01 m/sec

Using this relationship an assessment was made of water movement and dispersion.

It was concluded that a discharge of untreated sewage at A would adversely affect waters around some of the islands in the vicinity where tourist development with swimming and pleasure boating is contemplated, and that both floating and settlable solids should be removed before discharge.

Conclusions

The water pollution control problems in many Asian coastal towns and cities follow patterns that are evident in Hong Kong and Singapore. Industrial activity offers a means of raising the standards of life but at an increased risk of damage to the environment. Provided that proposals for marine disposal are responsibly designed on the basis of adequate preliminary studies of this kind, developing coastal communities in Asia should be able to continue to have recourse to the sea as a means of assimilating their wastes, without detriment.

The early steps taken in Hong Kong and Singapore for the provision of water-borne sanitation and the means provided for sewage disposal have contributed effectively towards the well-being of these communities, and have up to this time provided environmental safeguards. In Hong Kong it was considered that no great environmental harm would result from the discharge of untreated sewage (except for screening on some outfalls) through short outfalls into the deep waters of Victoria Harbour. In Singapore it was recognized that the relatively shallow waters around the island could not effectively assimilate untreated discharges and it was necessary to provide inland treatment plants and to discharge the treated or partially treated effluents to river or estuarial outfalls.

Rapid population increases, industrialization and

general development, considered in conjunction with improved methods of water pollution control, both in Hong Kong and Singapore, have resulted in the need to review existing arrangements and their environmental effects, in order to plan effectively for the future.

In Hong Kong, the Victoria Harbour study checks the present and continued effectiveness of existing provisions, in relation to the marine environment and amenity considerations. The Tolo Harbour study provides data for the formulation of original proposals to serve extensive new development. In Singapore the main purpose is to learn, in relation to modern outfall construction techniques, if it is possible, with economic advantage and without additional risk to the environment, to diversify the means of sewage and industrial effluent disposal.

References

CHAU, Y K and ABESSER, R   A preliminary study of the hydrology
1958   of Hong Kong territorial waters. *Hong Kong Univ. Fish. J.*, (2):43–57.

MACKAY, D M   Control of pollution in the sea: the Holy Loch.
1969   *J. Inst. Wat. Pollut. Control*, 68(1):118–26.

ORMEROD, J G   *Serratia indica* as bacterial tracer for water move-
1964   ments. Oslo, Norway, Norwegian Institution for Water Research.

PIKE, B E, BUFTON, A W J and GOULD, D J   The use of *Serratia*
1968   *indica* and *Bacillus sabtilis* var. niger spores for tracing sewage dispersion in the sea. Stevenage, England, Water Pollution Research Laboratory.

WATSON, J D and WATSON, D M   Report to Government of Hong
1965   Kong on sewerage and sewage disposal in northern New Territories.

WATSON, J D and WATSON, D M   Supplementary Report to Govern-
1966   ment of Hong Kong on sewerage and sewage disposal in northern New Territories.

Marine Pollution in Australia: A Review of the Present Situation Regarding Problems, Research Investigations and Management Techniques

A. J. Gilmour*

La pollution des mers en Australie, examen de la situation actuelle: problèmes, recherches et techniques d'aménagement

Dans le présent document, on examine les caractéristiques physiques des plaines côtières, le ruissellement et les courants sur le plateau continental de l'Australie. Le climat et les connaissances actuelles sur le milieu marin du plateau continental sont brièvement esquissés. Les renseignements manquent pour une grande partie des mers adjacentes.

Les ressources halieutiques et leur exploitation font l'objet d'un bref exposé qui démontre leur dépendance à l'égard des stocks du littoral voisin et des zones estuarines. La notion de contrôle de la pollution des eaux, dans le cadre de l'aménagement des ressources, est discutée et la définition de la pollution formulée par la Conférence est acceptée.

Un exposé du cadre législatif et institutionnel du Commonwealth australien, de ses six Etats fédérés et de ses territoires, est accompagné d'observations sur la situation du point de vue des normes appliquées par chaque Etat pour le déversement des effluents dans le milieu marin.

Le document esquisse les sources potentielles de pollution sur les côtes australiennes compte tenu de la répartition des centres urbains, de l'ampleur du traitement des eaux usées et des déversements d'effluents dans le milieu marin. Il indique la répartition des ports et des installations portuaires destinés à recevoir les pétroliers, ainsi que des industries primaires et secondaires y compris l'industrie énergétique. Ces dangers potentiels sont ensuite évalués sur la base des agressions polluantes connues. L'on souligne le manque de données quantitatives pour l'ensemble de l'Australie.

L'auteur passe en revue les travaux de recherche et les programmes d'aménagement en cours et ceux qui sont prévus dans le proche avenir pour chaque Etat et pour le territoire du Nord. Ce répertoire est limité dans l'espace compte tenu de la longueur du littoral australien, bien qu'au stade actuel, les situations critiques soient concentrées dans quelques zones. Toutefois, l'exécution d'enquêtes sur le milieu de l'ampleur et de la portée voulues dans ces zones mêmes est limitée par l'insuffisance des crédits dont dispose l'ensemble de la recherche marine.

On fait valoir la nécessité d'enquêtes écologiques complètes, telle celle menée dans la baie de Port Phillip (Victoria), et celle recommandée pour la baie de Westernport dans le même Etat, pour chacun des bras de mer sur lesquels sont situées les capitales d'Etat. Ces zones devraient être reliées par une chaîne de stations de surveillance du milieu situées dans les eaux du plateau continental en bordure de la côte. La nécessité d'enquêtes écologiques sur les côtes est également soulignée. Le document discute la possibilité de créer des situations se prêtant à de multiples utilisations, notamment en ce qui concerne l'emploi des effluents thermiques en mariculture.

La contaminación marina en Australia: examen de la situación actual con respecto a los problemas, investigaciones y técnicas de ordenación

En este estudio se examinan las características físicas de las llanuras costeras, la naturaleza del avenamiento de la superficie y la plataforma continental del continente australiano. Se describen brevemente el clima y el conocimiento que actualmente se posee del ambiente marino de la zona continental. Falta información sobre una gran extensión de los mares contiguos.

Se detallan brevemente los recursos pesqueros y su explotación, y con ello se demuestra la dependencia que hay de las poblaciones ícticas próximas a la costa y de los estuarios. Se examina el concepto de la lucha contra la contaminación de las aguas, que constituye una parte de la ordenación de los recursos; se acepta la definición de contaminación dada por la Conferencia.

Se describe la estructura legislativa e institucional de la Commonwealth de Australia, sus seis Estados federados y los territorios, junto con los comentarios sobre la situación en lo que respecta a las normas para la descarga de aguas residuales en el ambiente marino de cada Estado.

Se indican las fuentes potenciales de contaminación en torno al litoral australiano, considerando la distribución de los centros urbanos, el grado de tratamiento de las aguas residuales y los desagües de éstas en el ambiente marino. Se indica la distribución de los puertos y terminales del petróleo, y de las industrias primariae y secundarias, incluida la industria de la energía eléctrica. Se evalúan estas amenazas potenciales basándose en las situaconss de contaminación conocidas. Se pone de relieve la falta de datos cuantitativos para toda Australia.

Se examinan con respecto a cada Estado y al Territorio del Norte los activos programas de investigación y ordenación que se efectúan actualmente y los que se proyectan comenzar en el inmediato futuro. Esta lista es de amplitud limitada por lo que se refiere a la longitud de la costa, pero no obstante, en esta fase las situaciones críticas se concentran en un número pequeño de zonas. Sin embargo, la provisión de estudios de la magnitud y alcance necesarios sobre el medio ambiente, incluso en dichas zonas, se halla limitado por los fondos disponibles para todas las investigaciones marinas.

Se expone la necesidad de llevar a cabo, para cada una de las ensenadas en las que se hallan situadas las capitales de los Estados, estudios en gran escala sobre el ambiente, tales como los efectuados sobre la Bahía de Port Phillip en Victoria, y los recomendados para la Bahía de Westernport en este mismo Estado. Estas zonas se deberían unir con una cadena de estaciones de inspección del medio ambiente en las aguas de la plataforma en torno a la costa. Se subraya también la necesidad de efectuar reconocimientos ecológicos a lo largo de las costas. Se discuten las posibilidades de desarrollar situaciones de uso múltiple, particularmente con respecto al empleo en la maricultura de aguas residuales calentadas.

* Fisheries and Wildlife Dept., Melbourne, Victoria, Australia.

AUSTRALIA lies between longitudes 113° E and 153° E, and between the parallels of 10° S and 43° S latitude. It has a continental area of 7.69 million sq km.

Climatically the continent north of the Tropic of Capricorn comes under the influence of the wet north-west monsoon in summer and the south-east trades in winter. The southern portion is influenced by the moist westerly air streams in winter while in summer the sub-tropical high pressure belt moves south and brings drier air-flows. Some 70 per cent of this, the second driest continent to Antarctica, receives an average of less than 50 cm of rainfall a year. Less than 7 per cent receives more than 100 cm per annum and these areas are virtually restricted to the coastal belt.

Most of Australia's rivers are unreliable in flow and nearly half of the run-off is carried by the Tasmanian and northern Queensland rivers. The estimated total average annual discharge is 34,537 million cu m of which 12,322 million cu m is measured discharge. Only 4.92 million sq km are regarded as contributing to stream flow.

The marine environment

The East Australian Current is a dominating feature of northern East Coast conditions from 30° S (Wyrtki, 1960)

to Eden (N.S.W.) at 37° S. Hamon (1961) has pointed out that under certain conditions the southerly run can be found further offshore, at the latitude of Sydney and to the south. It is replaced by a cold water mass from the south. Brandon (1970) has discussed the oceanography of the Great Barrier Reef area. The sea conditions generally, and currents in particular, are poorly delineated for most of northern and western Australia.

Fishery resources and exploitation

There has been little fishing in Australian waters since first settlement. The earlier settlers exploited immediately available mollusc and scale fish stocks, and it is only in comparatively recent years that open ocean waters have been exploited. One exception to this has been the whaling and sealing industries that helped open up the continent.

Of 1969–70's catch of fish—90,353 metric tons—44,452 came from the continental shelf and 45,903 from coastal and estuarine waters—a quite significant proportion. The 1966 census revealed that the number of persons classifiable as engaged in fishing was 8,021; this was 1.56 per cent of persons involved in all primary industries and 0.17 per cent of the total work force (Commonwealth Year Book 1969). The survey of this

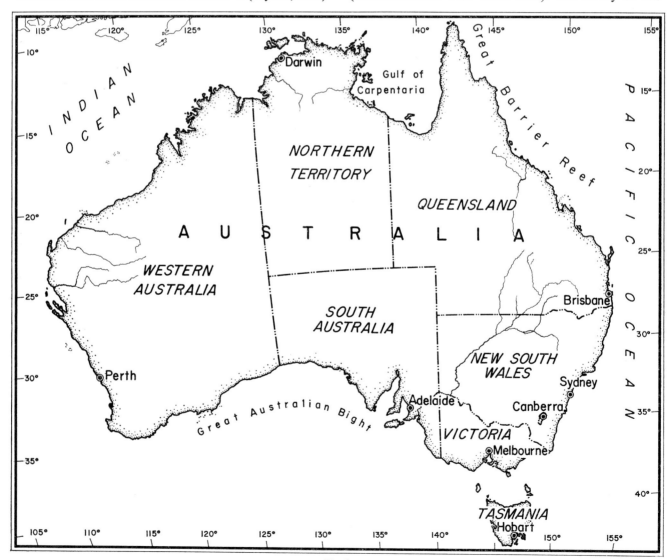

Fig. 1

industry for 1967–68 showed that 90.3 per cent of all registered fishing boats were less than 15 m in length and only 0.5 per cent exceeded 25 m. While the inshore catches will continue to satisfy some part of the demand for fresh fish it is probable that mariculture will offer substantial possibilities for increasing production.

Institutional framework

Australia was proclaimed a Federation of six States on 1st January, 1901. The Commonwealth assumes exclusive powers in certain areas including foreign affairs, excise duties and overseas trade, while some powers are jointly administered and others remain with the States. These arrangements can only be modified on the approval of a two-thirds majority in a national referendum. Many areas of Commonwealth-States relations are therefore subject to some debate leading to compromise solutions.

The State Parliaments enact legislation on all aspects of government related to intra-State matters. An increasingly important feature of many areas of government in recent years is the complementary legislation brought down in all State Parliaments and in the Commonwealth Houses. Many State agencies and semi-government instrumentalities, such as water supply and sewerage authorities, are given powers to promulgate regulations and schedules which permit variations in detail but not in intent.

Traditionally the limits of seaward jurisdiction of the States have been the "three mile limit" but recently it has been argued that Commonwealth jurisdiction commences at low water (Spender, 1969).

Pollution concept

The conference definition of pollution is a variant of the concept of impairment of beneficial use that has been adopted by one fisheries agency (Fisheries and Wildlife, Victoria, 1968) in Australia. At present there is no legislation in force in the Commonwealth that uses such a conceptual basis.

Linked with this type of definition is the concept of "total pollution load" which requires the prescription of receiving water standards with respect to the designated "beneficial use" of that water. In this regard there is a lack of knowledge on both background or baseline levels and on the requirements of the biological communities in all marine waters of Australia. There is also a lack of appreciation on the part of non-biologists of the dynamic nature of these ecological baselines.

Due to the relatively short history of industrialization there has been little pressure to develop strong anti-pollution measures and there are no laws whose primary purpose is to protect the marine environment.

Commonwealth powers at present relate to the territories and to the high seas beyond the "three mile limit". Only three Federal acts have any application: on oil pollution from shipping (Pollution of the Sea by Oil Act, 1960–65); offshore oil exploration and development (Petroleum Submerged Lands Act, 1967–68); and regulation of the discharge at sea of materials injurious to health, and the disposal of ships by sinking in territorial waters (Beaches, Fishing Grounds and Sea Routes Protection Act, 1932–66).

There are at present no statutory requirements by the State Health Departments for the discharge of sewage or trade wastes into marine waters. In Port Phillip Bay,

Victoria, however, the Government must approve the method of sewage treatment prior to discharge into the bay. The Departments responsible for fisheries and for ports in each State have powers which enable them to control discharges, but these are used on a discretionary basis. The Maritime Services Board of N.S.W. has standards relating to discharges to navigable waters, but no other authorities have such regulations. The list of authorities that are in some way responsible for pollution problems in Port Phillip Bay, a body of saline water of some 1,820 sq km on which Melbourne (pop. 2.1 million) stands, can be taken as typical of many places in Australia. The Melbourne and Metropolitan Board of Works is the city sewerage and water supply authority and controls discharges to creeks and drains in the capital and carries out foreshore work around most of the Bay; the Melbourne Harbour Trust in its statutory area astride the lower reaches of the Yarra River, upon which Melbourne stands, exercises its control on discharge of waste to the port; the State Rivers and Water Supply Commission controls discharge to streams which enter Port Phillip Bay elsewhere other than those in that area set aside to the Geelong Harbour Trust; the Port Phillip Bay Authority controls 20 m each side of the high water mark; the Ports and Harbours Branch of the Public Works Department is responsible for safe navigation through the bay and for enforcing the State oil pollution regulations; the Fisheries and Wildlife Department is responsible for the fish and other living resources in the bay.

Local or municipal councils, which are responsible to Departments of Local Government, have limited powers to control pollution through bye-laws.

Potential sources of pollution

Only 17 per cent of Australia's population is classified as rural, and 58 per cent lives in capital city metropolitan areas. The number of cities and towns with populations in excess of 10,000 in areas draining to the coasts in the various States is: N.S. Wales 9, Victoria 8, Queensland 11, South Australia 5, West Australia 4, Tasmania 4, Northern Territory 1.

Sewerage

The public utilities responsible for sewage collection differ in constitution and nature. Thistlethwayte (1969) has reviewed the situation from an engineering standpoint.

New South Wales: The Sydney Metropolitan Water, Sewerage and Drainage Board (SMWSDB) has in, and adjacent to, the metropolitan area, including the Bellambi-Wollangong-Port Kembla (South Coast) region, 20 separate sewerage systems. Ten of these discharge to the ocean, six undergoing some form of primary treatment, and nine discharge through outfalls on the 19 km of coastline around Sydney. Three sewerage systems discharge through outfalls on the south coast. Three further systems proposed for construction by 1975 will discharge effluent from high-rate activated sludge plants to the ocean.

Newcastle: Second largest city in the State, is served by the Hunter District Water Board system. The board has five outfalls discharging raw or primary sewage to the Pacific Ocean, and only one meets the requirements of the Health Department (Anon, 1969). Proposals are current for the progressive introduction of deep water outfalls followed by an activated sludge plant. Some unspecified plants are operated by the N.S.W. Public Works Department, and all except the one at Coffs Harbour are "fully treated" and chlorinated before discharge. The Coffs Harbour plant discharges to the Pacific Ocean from bio-filters without further treatment.

Currently a proposal to discharge sewage effluent into Brisbane Water, a seaward branch of the Hawksbury River Estuary (Broken Bay) has met adverse public reaction.

Victoria: The capital city is served by the Melbourne and Metropolitan Board of Works which has one major system and eight subsidiary systems. Some 97 per cent of the sewage flow from the city is treated by land filtration, grass filtration or ponding at the Board's Werribee Farm which lies on the shores of Port Phillip Bay 26 km south-west of the city. The farm has an area of 10,854 ha and 63 per cent of it is used for the treatment process. Discharge into the bay is through four outlets. The present flow to Werribee is estimated at 454,586 cu m per day with a contributing population of 1.8 million. Construction work has started on a second major sewerage collection system to divert flows and to accommodate expansion in the eastern and south eastern districts of Melbourne. The south eastern system is based on a modern activated plant, designed to accommodate 290,000 cu m per day, to be built at Carrum (some 32 km south-east of the city), which will produce a highly purified effluent for disposal into Bass Strait.

The Geelong Water Works and Sewerage Trust serves 100,000 people in Victoria's second city. Discharge to the Southern Ocean 14.5 km to the south, is via screens and comminutors which open directly into the sea. The wastes of certain plants are not accepted and these are causing problems in areas of Port Phillip Bay. There are some five local sewerage authorities whose systems discharge raw sewage to Bass Strait (Southern Ocean). Only one is from a town with a population exceeding 10,000.

Queensland: The sewage treatment works of Brisbane City Council, which serves 340,000 people, discharges raw sewage to the mouth of the Brisbane River and thence into Moreton Bay. Brisbane City Council also has 23 small systems designed to serve between 2,000 and 36,000 people, in various areas around the city. Of these, 11 are biological filtration plants and six are based on the activated sludge principle. There are some 53 cities or towns which have sewerage systems in the extra-metropolitan areas of the State. Gold Coast City is a mixture of established and post-1945 holiday development. It covers 122 sq km along 32 km of Pacific Ocean beach front some 48 km south of Brisbane. Extensive studies have been made to establish design criteria for ocean outfalls.

South Australia: The State Government Engineering and Water Supply Department is responsible for water and sewerage systems throughout the State. The capital, Adelaide, has four treatment plants which drain into St. Vincent's Gulf. The largest of these is the new Bolivar Treatment works, commissioned first in 1965 and in the final stage in 1969. It is a high-rate stabilization lagoon plant with a design capacity for 1.3 million people, of which 700,000 is the population equivalent for the industrial component. Discharge is via an open channel to the sea,

and provision has been made for the use of part of the effluent for irrigation (Lewis, 1969). The Glenelg system provides for 210,000 persons and an anticipated 350,000. The other metropolitan plant is the Port Adelaide works which discharges a high quality effluent from an activated sludge plant serving 130,000 persons. Adelaide has the largest proportion of residents supplied with water on a sewerage system of any Australian capital city. There are coastal raw sewage outfalls from Port Lincoln and a few other localities. Some modern plants for sewage treatment are under construction.

Western Australia: The capital, Perth, is serviced by the Metropolitan Water Supply Sewerage and Drainage Board. There are three treatment plants which discharge through two submarine ocean outfalls. The Swanbourne outfall terminates 1,967 m offshore in 11 m water. It is connected to the Subiaco plant, a high rate (modified) activated sludge plant of 54,500 cu m per day dry weather flow, and the Swanbourne plant which has primary treatment with separate sludge digestion on a flow of 5,500 cu m per day. Effluent from the latter is fine screened and chlorinated while the former is chlorinated only during the summer. The second outfall terminates 1,829 m offshore in 17 m of water. The associated treatment works (Woodman Point) has a flow of 15,000 cu m per day and provides primary treatment, separate sludge digestion, and fine screening. Chlorination is not currently being provided. Four new plants are planned, three for discharge to the ocean.

Tasmania: The population of this State is decentralized and so there are a large number of small schemes. The capital, Hobart (131,070), is served by seven separate small sewerage schemes.

The second largest urban area is Launceston (37,217) on the River Tamar and the city discharges raw sewage to the river. Three of the municipalities peripheral to the city of Launceston have primary treatment plants with digesters or sedimentation and plans are in hand by the Launceston Council to provide a treatment works by 1973. Burnie (15,806) is served by treatment works that provide clarifiers, digesters and drying beds. Nevertheless, notices prohibit bathing on the town front beach where one of the ocean outfalls terminates at low water level.

Northern Territory: Darwin (pop. 21,205) installed its first communal facilities for the collection and disposal of sewage for four separated areas in 1950. There are fourteen discharge points from communitors or from Imhoff tanks. Two of the existing outfalls have recently been extended to 1,073 m offshore. Currently plans are under review (Anon, 1969a) for augmenting the city sewerage systems.

Ports

In all ports the relevant Port Authorities are responsible for pollution control. Until recently Port Authorities have been concerned only with the safety of the port but with the increasing interest in pollution some authorities have commenced using their general provisions to control pollution. There are no provisions limiting discharge of ships' sewage, but incinerators are provided at major ports to dispose of garbage from all shipping.

Major Australian ports handled 106,792,795 metric tons of cargo with 28,330 ship movements during 1968–69. Of this 48.1 per cent was handled through the ports associated with the capital cities, and 64.9 per cent was bulk cargo. These bulk terminals have a dust problem and those that load a wetted product may discharge the conveyor-belt washing into the sea, causing a reduction in transparency and blanketing of the sea bed.

Principal ports for the shipment of ores are: Groote Eylandt (manganese), Gove (bauxite), and Darwin (iron) in the Northern Territory; Yampi Sound (iron), Port Hedland (iron and manganese), and Dampier (iron) in Western Australia; Whyalla (iron) in South Australia; Port Latta (iron) in Tasmania; Sydney, Wollongong-Port Kembla, Lake Macquarie and Newcastle (coal) in N.S.W.; and Gladstone (coal) and Weipa (bauxite) in Queensland. Newcastle and Wollongong-Port Kembla are the two major importers of iron ore, iron, and mineral fluxes while Bell Bay (Tasmania) takes in bauxite and manganese. In such areas there may be a dust problem similar to that associated with the unloading of gypsum, elemental sulphur and phosphate rock.

Oil and petroleum products moved through Australian ports totalled 6.7 million metric tons in 1967–68. In each State the Government responsible for marine matters and the relevant port authority have jurisdiction within the "three mile limit" and a declared port respectively. In waters beyond the "three mile limit" the Commonwealth Department of Shipping and Transport is the responsible authority. Each authority has its own plan for dealing with oil spills and these are in the main based on the use of "detergents". Details have yet to be announced of any integrated plan that could make provision for minimizing detrimental effects of oil spills on the ecology of coastal areas.

The majority of Australia's thermal power stations are located on the N.S.W. coast.

Pesticides and fertilizers

The coastal plans are fairly narrow, of the order of 80 km, particularly in those areas well supplied with rainfall. In these areas agriculture and forestry are practised. To date there is little data available to indicate the extent to which pesticides, fertilizers, and general land utilization techniques have affected the marine environment around Australia. Bacher (1970) has quoted some figures for levels of seven organochlorines in sixteen specimens (five birds, three fish, and some molluscs of one species and a whale). This work is in its preliminary stages and further data will be required before a definitive statement can be made.

Thomson (personal communication) working on the estuaries of the Mary and Brisbane rivers considers that flood run-off, particularly in sugar areas, may carry sufficient pesticides to cause fish kills. Similarly the northern New South Wales rivers may also become contaminated from agricultural areas (Scribner, personal communication). Spraying for mosquito control, particularly in northern areas, and for weed control on river banks, road sides, and in irrigation and drainage canals also appears to be contributing to such problems. Tranter (personal communication) considers that a build-up of persistent pesticides in the waters landward of the Great Barrier Reef may be occurring with a consequent threat to the Reef System.

Thistlethwayte and Robbins (1970) give values for nitrate-nitrogen of 10 and 25 mg/l obtained from two wells in a citrus producing area. No data are available for a marine situation that can be directly attributed to agriculture.

Mineral and oil extraction

The only major pollution problems associated with mining, apart from the beneficiation of ores at ports are in the beach sand mining operations and in the recovery of gravels, sands and shell grit. Beach sand mining in Queensland, New South Wales and Western Australia produced during 1968 in excess of 80 per cent of the free world's rutile and zircon production, and some 20 per cent of the ilmenite production. Concern has been expressed by conservationists in Queensland and New South Wales that inadequate coastal areas have been set aside from mining, and in particular that certain high dune formations, areas of rain forest and other areas of special interest require complete protection. Government inquiries are under way in some cases. The mining companies are required to restore their lease areas to the satisfaction of local authorities and State Mining Departments, but rarely, to date, have ecologists been consulted. Figures provided by the Rutile and Zircon Development Association Limited for two regions in N.S.W. indicate that rehabilitation costs are about 2.5 per cent of the gross production value.

An application to mine limestone from Ellison Reef, a small reef in the north Central Region (Maxwell, 1969) of the Great Barrier Reef, was refused at a Queensland Mining Warden's hearing. The reasons for mining on the reef are outlined by Carruthers (1969). Applications for gravel, sand and shell-grit removal, whether on beaches or the sea bed, appear to be referred to Fishery Departments in all States and to date the working of successful licensee areas appears to present few problems.

The offshore production of natural gas and petroleum commenced in Australia at Barrow Island on the north Western Australian coast in 1967 and in the Bass Strait area in 1969. The latter is operated at present from five platforms and these will probably be increased. The crudes from the Bass Strait fields have the following API gravities: Barracuda, 43 degrees; Halibut, 42 degrees; and Kingfish, 46 degrees. The API gravity of the oil from Barrow Island (W.A.) is 38 degrees. Exploration is currently being undertaken on a number of potential oil-bearing basins on the continental shelf of Australia. Geophysical surveys are being carried out off the southern N.S.W. coast, in the Great Australian Bight, off the northern coast of W.A. and in the Timor Sea off the Northern Territory coast. Drilling began in June 1970 in the Gippsland basin in Bass Strait; in two basins, the Carnarvon and Canning, off the northern W.A. coast, and in Bonaparte Gulf, Northern Territory. Only the Gippsland and Barrow Island fields are currently producing gas.

Two gas-blows have occurred. The one at the Marlin rig in Bass Strait was quickly brought under control, and visual surveys suggest that it did not cause any damage. The second, in the Bonaparte Gulf, has been blowing since 1969.

Reclamation

The development of marsh, mangrove swamps and mud-flats as sports fields, public reserves, housing developments, marinas, and car parks has long been considered good practice by local government. In all States co-operation between the controlling authority, usually the Department of Lands, the Port Authority or a local Council, and the Department of Fisheries is being developed. The closest co-ordination is achieved in New South Wales where the Fisheries Department has two biologists involved in this work.

Reported sites of pollution

New South Wales: Population and industry concentration in the Sydney, Wollongong and Newcastle areas ensures that it is in these areas that pollution is most evident. Sydney and Newcastle discharge raw or primary treated sewage to the ocean and at times it is blown back on to the beaches. Overflows in wet weather and sullage contribute to extremely high bacterial counts in many streams (Thistlethwayte & Robbins, 1970) (see **Research and management investigations**).

A report by Barton (1970) on the problem of waste disposal in the metropolitan areas of Sydney Harbour and Georges River shows that values greater than 60 per cent saturation exist over most of the navigable areas. Values between 0 and 20 per cent have typically been recorded in the upper reaches of Sydney Harbour, that is, parts of the Parramatta and Duck rivers. The presence of a marine borer, *Martesia sp.*, further up the inner reaches of Sydney Harbour is suggested to be an indication of improved water quality (Moore, 1967).

A report by the Department of Health identified 13 discharges to the Parramatta and Duck rivers, while on the Georges River the location of some 50 discharges have been established to date but it is estimated that this constitutes only half of the probable number. The cause of the gross pollution of Mill Creek, a tributary of the Georges River that is typically 0 to 20 per cent saturated with dissolved oxygen (Barton, 1970), has not been determined, but a contributing cause may be the leaching of grease and other materials from waste disposal sites. Many such sites are causing problems in the Sydney area. Botany Bay has 12 major pollution sources identified at present, while the Cooks River, and particularly a tributary of it called the Alexandra Canal, contributes a significant pollution load. Some 40 separate sources have been identified discharging to the canal. Oil pollution affects the bay from time to time as there are two major oil facilities located near its entrance.

Outside the Sydney area Newcastle and Port Kembla are major concentrations of industry. In the latter there are a steel works, two non-ferrous and two ferrous metal works, and a fertilizer plant. In Newcastle 92 drains or storm water channels discharge from mainly industrial areas to the port. High levels of zinc have been measured that appear to be derived from a viscose and acetate yarn plant upstream, and from metal finishing plants in the area. The steel mill and its associated plants, the largest complex in the port area, currently discharges some 26 million cu m of salt and fresh water per month with various contaminants. An extensive survey has been carried out by the firm which has provided an excellent picture of its pollution loadings. The company concerned, the Broken Hill Proprietary Co. Ltd., is considering extensive recycling of fresh water and the adaptation of heat ex-

changers, and this will assist in mitigating the situation.

Victoria: The lower reaches of the Yarra River, the Maribyrnong River and Stony Creek that run into it, and the Kororoit Creek all suffer from the discharge of industrial wastes and seepage of sullage and septic tank effluents of the city of Melbourne. Results of investigations conducted as part of the Environmental Study of Port Phillip Bay show that the Yarra, in contrast with the waters of Bass Strait, is high in ortho-phosphate (Reinsch & Hardy, 1970). Corio Bay, an appendix of Port Phillip Bay, receives a thermal load from the sea water cooling system of a refinery and the hydrocarbon content varies between 2.5 ppm and 9.5 ppm. The through-put of sea water can be up to 20,300 metric tons per hour. Discharges of pickling waste effluent from a nearby wire works occur periodically and stain the water red-brown for some distance. Samples of substrate taken in the area show evidence of an "oily" film and have a high content of iron sulphide. Superphosphate dust is blown and washed into the bay from a bulk loading dock to the south. The refinery in Corio Bay was established in 1953 and within a year fish caught in the bay were condemned at the fish market due to a "kerosene flavour" when cooked. Some six species are involved, two of which are herbivorous and the others carnivorous (Butcher, Wharton & Pearce, 1969).

Westernport Bay to the east of Port Phillip is being developed as an industrial port but has no major pollution loadings at present other than those attributable to agricultural runoff. It is an area that will require most stringent controls if the quality of the environment is not to deteriorate. A further area of the State that will require close attention is the Gippsland Lakes, a system of five coastal lakes and six rivers in one of the favourite water sports areas of the State. In the summer this area receives runoff contaminated by household sullage and septic tank effluent. A further threat is the diversion of the headwaters of some of the rivers to provide water supplies for Melbourne. The effects of industry, including heated effluents from power plants, land-use practice, and possible sewage seepage from the Latrobe Valley Water and Sewerage Board's disposal farm, have combined to produce cultural eutrophication in the westernmost body of water, Lake Wellington.

Queensland: Pollution orientated surveys of the Brisbane River have indicated that oxygen depletion has occurred (Henry, 1969). Some 700 drains etc. discharge into the city reaches of the river. Henry (1966) has reviewed the waste problems of the sugar and meat industries in Queensland in two most comprehensive reports. He notes that some fish kills have been attributed to sugar mill wastes and that some thermal pollution from sugar mills occurs.

South Australia: Marine pollution has not been a serious problem to date except in the lower reaches of the Port River which flows through Port Adelaide. This is the driest state on the continent and consequently there has been much preoccupation with the pollution of water supplies. Until the commencement of operations of the first stage of the new Bolivar treatment works in 1965 problems had been encountered with heavy pollution loadings on the lower reaches of the river. The condition of the river has improved although discharges from two large industries are still causing pollution. One of these, an alkali works, discharges some 100,000 metric tons of sulphates and carbonates, together with grit, per annum into the river. Remedial dredging is charged to the company which is contemplating a $1 million treatment plant. A major steel works at Whyalla discharges blast furnace flue washings contaminated with oil and dust. No data are available on the quality or quantity of this effluent nor on its effects on the environment. The same qualifications apply to the discharges to a mangrove swamp at Port Pirie of a lead and zinc works.

Western Australia: The Swan River Conservation Board considers that although there is some pollution from discharges of industrial wastes, septic tank effluents and sullage, land runoff, discharges from water craft, misuse of pesticides and rotting vegetation in the river, the situation is being contained. (Courtney, in evidence to the Senate Select Committee on Water Pollution, 1970). Although the Board is a part-time committee with wide representation it succeeds in its objectives and it is agreed that the Swan River is in much better condition now than when the Board was formed in 1958.

The Cockburn Sound is controlled by the Fremantle Port Authority (FPA) and there are recent proposals to develop it as a major naval port with modifications to its topography. There are a number of major industries on the eastern shore (Chittleborough, 1970).

A pollution problem exists at Bunbury on the south west of the State where a titanium oxide plant has been established. Pumping the effluent to the open ocean failed and settling ponds were installed. These were built in 1968 and are considered a short term solution. Long term solutions are currently under investigation.

Tasmania: Sewage pollution is a significant problem in Tasmania (Senate Select Committee Report, 1970). In Hobart, on the Derwent, a confectionery factory may occasionally discharge milk wastes, while a textile firm, a glass works, a zinc and fertilizer plant, a food processing plant and a brewery, also contribute discharges. Food canneries at Devonport, Ulverston and Smithton discharge untreated wastes into estuarine waters (Senate Select Committee Report, 1970). Both the pulp paper and hardboard mill and the titanium dioxide plant at Burnie have aesthetic problems with their discharges (see **Research and management investigations**) and evidence is lacking on the effects on marine life.

A sulphuric acid plant at Burnie dumps 238,000 metric tons per annum of calcine wastes wetted with scrubbing effluent containing calcium arsenite and silico-fluorides in 37 m of water 6 km north east of Burnie in Bass Strait, Discussions are still in progress although dumping has commenced.

Northern Territory: Darwin has a large number of sewage discharges, and data give evidence of sewerage pollution. Industrialization is limited but there is evidence of oil pollution in mangrove swamps close to the major oil installations. The Gove bauxite plant is being developed and extensive areas of mangrove swamp have been set aside for reclamation by settling ponds for "red mud". On Groote Eylandt the manganese mine and port does not provide any pollution problems, but a prawn processing plant pumps its waste into a bay some 20 m from the plant. The horticulturist at the mining company is to experiment with the use of prawn heads, possibly in conjunction with sewage sludge, as a soil conditioner.

Research and management investigations

Investigations in Australia have been essentially on an *ad hoc* basis until the last five years. More extensive studies of particular aspects have been developed by a few research workers virtually in isolation. There has been no comprehensive study of any area until a broad scale environmental study of Port Phillip Bay, Victoria, was commenced in 1968.

New South Wales: Some studies of oceanographic conditions have been carried out by the Sydney Metropolitan Water Sewage and Drainage Board (SMWSDB) on its major ocean outfalls and of the physico-chemical characteristics of Sydney Harbour. These are being used to assess priorities for the installation of treatment works and to enable better siting of new outfalls. Officers of the Board have also investigated sewage pollution and sea bathing on Sydney beaches (Flynn & Thistlethwayte, 1965). The results indicate that during the survey period coliform counts exceeded 1,000 per 100 ml MPN almost everywhere along the 32 km of beaches. Estimates of 10,000 per 100 ml MPN were obtained on beaches within 4 to 6 km of outfalls, while the highest figure obtained was 150,000 per 100 ml MPN. Grease extracts at high levels were found on some beaches. A three month study (Wells & Edwards, 1969) of bacterial contamination of Sydney Rock Oyster (*Crassostrea commercialis*) from four localities in the Georges River, which flows into Botany Bay through the southern suburbs of Sydney, showed that significant contamination occurred at all sites from time to time. At one site oysters were unacceptably polluted during the entire test period. The study suggested that there were four main contributing causes: large unsewered residential areas with septic tank installations, overflow from sewers, pumping stations, and leaching from a local shire tip (Wells & Edwards, 1969). The river is the subject of a joint physico-chemical investigation by the Board and the State Fisheries Department. The latter also have two scientists responsible for work on the control of estuarine development, and one on pollution surveys.

The Maritime Services Board of New South Wales has carried out surveys in the Newcastle area of the Hunter River, in Botany Bay and to a limited degree in Port Kembla. These studies, directed at solving particular problems, commenced in 1948 on Lane Cove, 1950 on Sydney Harbour, 1957 on Hunter River, and some sporadic work commenced on the Georges River in 1955. There are no pollution data available or published reports. Two committees have been set up to co-ordinate activities of government and semi-government agencies involved in anti-pollution work.

The State Electricity Commission has funded studies on the effects of the discharge of heated effluents on aquatic weeds (Sydney University), on the biology of mussels and jellyfish (University of N.S.W.), and on the sea disposal of flyash (Thomson, 1963). A pollution orientated research programme on the sediments of Botany Bay is based at the Department of Geology and Geophysics at the University of Sydney. Two Hydraulics Research laboratories at Manly Vale have carried out model studies of significance to pollution investigations (University of N.S.W. Water Research Laboratory, 1969, and Lucas, 1969).

Victoria: The only programmes of marine pollution

research being carried out are based on the two major bays, Westernport and Port Phillip, on the south coast of the State.

The Fisheries and Wildlife Department has conducted marine research in Westernport Bay since 1963. The first programme was a biological study of the King George Whiting (*Sillaginodes punctatus*) (C & V) of which a major part was an evaluation of the fishing success of amateur fishermen and the development of baseline data on catch per unit of effort for the species.

A Port Phillip Bay Environmental Study Co-ordinating Committee was established in 1967 with representation from the Melbourne and Metropolitan Board of Works, the Fisheries and Wildlife Department and three other agencies. A full description is given by Gilmour (1970). Early in 1969 the Government of Victoria considered a proposal to discharge 290,000 cu m per day of secondary treated sewage effluent to Port Phillip Bay. The Government directed that the discharge should be made to Bass Strait but adopted a recommendation that the broad scale, multi-disciplinary environmental research programme be continued.

Queensland: Considerable work has been carried out by the Department of Local Government (DLG) (Henry, 1969a) in river and ocean surveys. The physico-chemical characteristics of up to 50 river systems have been examined with varying degrees of repeat sampling, and this is continuing. A continuous monitoring station has been established on the Brisbane River. An ocean survey, under joint direction of the Co-ordinator General's Department and the DLG has been conducted off the beaches of the Gold Coast areas in connection with beach erosion studies (McGrath, 1967) and major sewer outfall design (Henry & Barton, 1968). The Hydraulic Laboratory of Delft, Holland, has reported on a comprehensive programme of investigation. Limited biological work for this study has been carried out by divers of the Queensland Littoral Society (Henry & Barton, 1968) and this is continuing.

A benthonic survey of the mouth of the Brisbane River, operated from the Department of Zoology, University of Queensland, was concluded during 1970 and is being written up (Hailstone, MS). This area is subject to heavy pollution loads from the city of Brisbane, particularly through the discharge of raw sewage at Luggage Point. Two recent reports on tainting of Sea Mullet (*Mugil cephalus*, Linneaus) have discussed the magnitude and possible causes (Grant, 1969) and the analysis of contaminated fish (Vale *et al.* 1970, Skipton *et al.* 1970). A study of the physical regime of Moreton Bay has been completed (Newell, 1970).

Although limited biological work has been carried out on the Great Barrier Reef the paucity of knowledge has been underlined rather sharply recently when three potential threats to the stability of the reef system were recognized. The first is the Crown of Thorns starfish (*Acanthaster planci*), the second was a proposal to mine the reef for lime, and the third is proposals to drill in the area for oil. A Royal Commission is currently evaluating the potential for oil pollution while a Committee has examined the starfish problem (Australian Academy of Sciences, 1970). A joint Queensland–Commonwealth Government Committee is also expected to report in the last quarter of 1970 on the latter.

South Australia: No major studies related to pollution control of marine waters have been carried out.

Western Australia: Pollution of the Swan River has been the subject of debate since 1870. The most notable investigation was by a series of local experts (Swan River Reference Committee, 1955) and this led to the setting up, in 1958, of the Swan River Conservation Board, a step first recommended in 1922.

Routine weekly bacteriological tests are made by the State Public Health Department on all ocean beaches in the metropolitan area and twice weekly on beaches adjacent to the two ocean outfalls. Studies by staff and students on marine topics in the Departments of Botany, Zoology, Geography, Civil Engineering, and Geology at the University of Western Australia will have a bearing on any future bench-mark studies that may be set up.

Tasmania: After the establishment of a sulphuric acid plant at Burnie, on the north coast, and the proposed dumping of calcine wastes from it into Bass Strait, some limited field sampling was carried out. However, knowledge of the physico-chemical characteristics of Bass Strait is so limited, and the lack of staff is such that it would appear that little useful work can be accomplished.

Australian Paper Manufacturers Ltd. undertook, in conjunction with the Australian Atomic Energy Commission, a tracer study prior to the establishment of their pulp mill at Port Huon. An untreated effluent from this neutral sulphite semi-chemical pulp mill, based on hardwoods (eucalypt), is discharged to the Huon Estuary. The company, in conjunction with the CSIRO Division of Fisheries carried out a follow-up survey in 1967 (Newell, 1969).

A second paper company, Australian Pulp and Paper Manufacturing Co. Ltd., have been investigating at their Burnie Plant, improvements in the economic recovery of heat and chemicals from their soda pulp mill liquor. Current practice is to discharge about 82,000 cu m of cooling water containing some residual materials such as fibre, clay, bleach, water solubles, and black liquor to low water in Bass Strait. The topography of the coastline and prevailing winter winds (northerly) are such that nuisance conditions occur. The same company, also in conjuction with the CSIRO Division of Fisheries and Oceanography, has carried out a current study in the lower reaches of the Tamar prior to the establishment of a new plant (Newell, 1969a).

Also at Burnie is the Australian Titan Products titanium dioxide plant. The effluent from this plant, a mixture of ferrous sulphate, dilute sulphuric acid and process water, was discharged to low water in Bass Strait. Recently the effluent line was extended to 1.5 km and now discharges in 14 m of water, but a problem can still exist at times. The company is carrying out a series of experiments using "Woodhead drifters" to study the effects of various combinations of wind and tide with a view to extending the pipeline further.

Conclusions

The legislative solution in Australia has been demonstrated to be inadequate in that it does not provide for full control of the marine environment. All States are currently reviewing their capabilities and three have announced their intention to provide new water pollution legislation. Western Australia and Victoria have announced proposals to provide for all forms of environmental contamination, while South Australia has established a Committee of Inquiry to advise the Government on this broader situation. The Commonwealth Government has received reports from the Senate Select Committees of Inquiry on Air and Water Pollution. The latter urged an integrated approach to the problem of water resource management.

While there is sufficient subjective information to conclude that acute pollution, albeit in varying degrees and with respect to various beneficial uses, exists in the waters close to the 12 major population-industrial centres there is a lack of quantitative data. Data does exist in a number of government agencies in each State on some of the commoner criteria, such as dissolved oxygen values and bacterial counts, but there is a complete lack of systematic data on such pollutants as heavy metals and pesticides. Thus there is little opportunity, and no basic data, to cope with the problems of evaluating chronic pollution of the marine environment. Linked with this lack of basic long-term physico-chemical data is a dearth of biological knowledge, and even in fisheries research compatible Commonwealth-wide statistics are only now becoming a reality.

Industry is no better provided with the necessary data on the quantity and quality of its discharges to the environment. With some exceptions, monitoring, even on an intermittent basis, is lacking. Thus wastes of unknown quality and quantity are being discharged into an environment which is ill defined.

A three-tiered approach on the part of industry, State, and Commonwealth Governments to this problem must be forthcoming in the immediate future, The definition of current and future problems should not be delayed any further.

State authorities should develop research programmes on estuaries and bays, selected on an Australia wide basis, to establish the necessary bench-mark data. This would mesh with Commonwealth-State operated programmes in onshore waters and Commonwealth activity in oceanic waters all around the continent. State legislation should require all discharges, whether direct or indirect, to be monitored on a basis acceptable to a centralized State environmental agency. The impact of land utilization and the distribution of urban and industrial centres on the near shore environment must be recognized, and adequate provision made to measure it. Such an all-embracing approach is implied in the definition of pollution adopted, and the acceptance of the concept of "total pollution loading".

There is a good reason, in the Australian situation at least, to postulate that such a centralized State environmental agency, which should be provided with economists and engineers in addition to ecologists, should be responsible for co-ordination of research and management in such matters. These State agencies should co-ordinate their activities through a regular Commonwealth-States Conference.

References

Australia, Commonwealth of, Fisheries Branch, Australian fisheries
1968   a review on the occasion of the 13th Session, IPFC, Brisbane. Canberra, Department of Primary Industry, 73 p.
Australia, Commonwealth of, Official year book of the Commonwealth of Australia. *Off. Yb. Commonw. Aust.*, (54):959.
1968

Australia, Commonwealth of, Official year book of the Commonwealth of Australia. *Off. Yb. Commonw. Aust.*, (55):980.
1969

Australia, Commonwealth of, Senate Select Committee, Water
1970   pollution in Australia. *In* Report of the Parliament of the Commonwealth of Australia, p. 214.

Australian Academy of Science, *Acanthaster planci* (Crown of
1970   Thorns starfish) and the Great Barrier Reef. *Rep. Aust. Acad. Sci.*, (11): 20 p.

BACHER, G J Pesticides in the marine environment. Paper presented to Australian Marine Science Association Symposium, Melbourne (mimeo).
1970

BARTON, A E Report on investigations into the problem of waste
1970   disposal in the metropolitan area. New South Wales, Sydney, 38 p.

BENNETT, I and POPE, E C Intertidal zonation of the exposed
1953   rocky shores of Victoria, together with a rearrangement of the biogeographic provinces of temperate Australian shores. *Aust. J. mar. freshwat. Res.*, 4:105–59.

BENNETT, I and POPE, E C Intertidal zonation of the exposed
1960   rocky shores of Tasmania and its relationship with the rest of Australia. *Aust. J. mar. freshwat. Res.*, 11(2):182–221.

BRANDON, D E Oceanography of the Great Barrier Reef and the
1970   Queensland continental shelf, Australia. Thesis, University of Michigan, Ann Arbor.

BRODIE, R and RADOK, R Tides, weather and surface drift in
1970   Bass Strait. *Res. Pap. Flinders Univ. Sth Aust.*, (35): 9 p.

BUTCHER, A D, WHARTON, J C F and PEARCE, T S Evidence to
1969   Senate Select Committee on water pollution, Melbourne and Geelong.

CARRUTHERS, D S Limestone mining. *Spec. Publs. Aust. Conserv.*
1969   *Fdn*, (3):47–50.

CHITTLEBOROUGH, R G Conservation of Cockburn Sound
1970   (Western Australia). *Spec. Publs. Aust. Conserv. Fdn*, (5): 27 p.

FLYNN, M J and THISTLETHWAYTE, D K B Sewage pollution and
1965   sea bathing. *Int. J. Air. Wat. Pollut.*, 9:641–653.

GILMOUR, A J The implications of industrial development on the
1965   ecology of a marine estuary. *Fish. Contr. Victoria*, (20): 15 p.

GILMOUR, A J Marine pollution research in Victoria, Australia.
1970   *Mar. Pollut. Bull.*, 1:120–7.

GRANT, E M "Kerosene" taint in sea mullet (*Mugil cephalus*,
1969   Linnaeus). *Fish. Notes Dep. Harb. Mar. Qd*, (3):1–13.

HAMON, B V Structure of the east Australian current. *Tech. Pap.*
1961   *Div. Fish. Oceanogr. C.S.I.R.O. Aust.*, (11): 11 p.

HAMON, B V The east Australian current, 1960–1964. *Deep-sea*
1965   *Res.*, 12(6):899–921.

HENRY, L Sugar industry wastes in Queensland. *Rep. Dep. loc.*
1966   *Govt, Brisbane*, 1966:33 (mimeo).

HENRY, L Meat industry wastes in Queensland. *Rep. Dep. loc.*
1966a   *Govt, Brisbane*, 1966:26 (mimeo).

HENRY, L Water pollution control. *Rep. Dep. loc. Govt, Brisbane*,
1969   (3):3 (mimeo).

HENRY, L Changes in water pollution control in Queensland.
1969a   *Div. Tech. Pap. Instn Engrs Aust.*, 10 (16):1–15.

HENRY, L and BARTON, C Ocean outfall investigations with
1968   particular reference to diffusion and dispersion. *Tech. Pap. Aust. Wat. wastewat. Ass. Fed. Conv.*, (3):19–44.

HIGHLEY, E Oceanic circulation patterns off the east coast of
1967   Australia. *Tech. Pap. Div. Fish. Oceanogr. C.S.I.R.O. Aust.*, (23).

HIGHLEY, E The International Indian Ocean Expedition: Australia's contribution. *Tech. Pap. Div. Fish. Oceanogr. C.S.I.R.O. Aust.*, (28).
1968

KNOX, G A Littoral ecology and biogeography of the Southern
1960   oceans. *Proc. Soc. (B)*, 152:577–624.

KNOX, G A The biogeography and intertidal ecology of the
1963   Australasian coasts. *Oceanogr. mar. biol.*, 1:341–404.

LEWIS, K W Water pollution control in South Australia. *In*
1969   Proceedings of the 41st ANZAAS Congress, Adelaide (mimeo).

LUCAS, A H Diffusion of salinity in an estuary with and without
1969   a pneumatic barrier. *Proc. int. Conf. Ass. Hydraul. Res.*

McGRATH, B L Erosion of Gold Coast beaches, 1967. *J. Instn*
1967   *Engrs Aust.*, 40:155–66.

MacINTYRE, R J Oxygen depletion in Lake Macquarie. *Aust. J.*
1968   *mar. freshwat. Res.*, 19:53–6.

MAXWELL, W G Physical geology and oceanography. *Spec.*
1969   *Publs. Aust. Conserv. Fdn*, (3):5–14.

MOORE, D D Problems in the control of water pollution. *Aust.*
1967   *Chem. Process. Engng J.* 20:32–6.

NEWELL, B S Hydrology of south-east Australian waters. *Tech.*
1961   *Pap. Div. Fish. Oceanogr. C.S.I.R.O. Aust.*, (10).

NEWELL, B S Dispersal of pulp mill effluent in Hospital Bay,
1969   Tasmania. *Rep. Div. Fish. Oceanogr. C.S.I.R.O. Aust.*

NEWELL, B S Total transport and flushing times in the Lower
1969a   Tamar River. *Rep. Div. Fish. Oceanogr. C.S.I.R.O. Aust.*, (45).

NEWELL, B S   A preliminary description of the physical environ-
1970   ment of Moreton Bay, Queensland, 1967–68. *Tech. Pap.
Div. Fish. Oceanogr. C.S.I.R.O. Aust.*

PEARSON, R G and ENDEAN, R   A preliminary study of the Coral
1969   Predator *Acanthaster planci* (L) Asteroidea on the Great
Barrier Reef. *Fish Notes Dep. Harb. Mar. Qd*, 3(1):27–69.

PHIPPS, C V G   Topography and sedimentation of the Continental
1963   shelf and slope between Sydney and Montague Island,
N.S.W. *Australas. Oil Gas J.*, 10:40–6.

PHIPPS, C V G   The character and evolution of the Australian
1967   Continental Shelf. *Aust. Petrol. Explor. Ass. J.*, 7(2):44–9.

REINSCH, D A and HARDY, M J   Physico-chemical studies of Port
1970   Phillip Bay. Paper presented to Australian Marine Science
Association Annual Conference, Melbourne, p. 11
(mimeo).

ROCHFORD, D J   Port Phillip Survey 1957–63. Hydrology. *Mem.
1966   natn. Mus. Vict.*, 27:107–18.

ROCHFORD, D J   Seasonal interchange of high and low salinity
1969   surface waters off south-west Australia. *Tech. Pap. Div.
Fish. Oceanogr. C.S.I.R.O. Aust.*, (29).

SHIRLEY, J   An investigation of the sediments on the Continental
1964   shelf of New South Wales, Australia. *J. geol. Soc. Aust.*,
11:331–41.

SKIPTON, J, VALE, G and MURRAY, K E   Studies of a kerosene-like
1970   taint in mullet (*Mugil cephalus*). Pt. 1. *J. Sci. Fd Agric.*,
21:433–6.

SPENCER, R S   Studies in Australian estuarine hydrology. *Aust. J.
1956   mar. freshwat. Res.*, 7(2): 193–253.

SPENDER, P C   The Great Barrier Reef: Legal aspects. *Spec. Publs
1969   Aust. Conserv. Fdn*, 3:25–41.

Swan River Reference Committee, Pollution of the Swan River.
1965   *Rep. Swan River Ref. Comm.*, Perth, 1965:153.

THISTLETHWAYTE, D K B   Water and waste water and water
1969   pollution control in Australia. *Wat. Pollut. Control*,
68(3):256–74.

THISTLETHWAYTE, D K B and ROBBINS, D M   Water pollution in
1970   New South Wales. *Tech. Pap. Aust. Wat. wastewat. Ass.
Fed. Conv.*, (4):1–24.

THOMSON, J M   Mortality thresholds of fish in fly ash suspension.
1963   *Aust. J. Sci.*, 25(9):414–5.

University of New South Wales, Water Research Laboratory. *Res.
1969   Rep. Univ. N.S.W.*, 1967–69.

VALE, G L, et al.   Studies of a kerosene-like taint in mullet (*Mugil
1970   cephalus*). Pt. 2. *J. Sci. Fd Agric.*, 21:429–32.

VAN DER BOSCH, C C, CONOLLY, J R and DIETZ, R S   Sedimenta-
1970   tion and structure of the continental margin in the vicinity
of the Otway Basin, Southern Australia. *Mar. Geol.*,
8:59–83.

VAUX, D   Surface temperature and salinity for Australian waters,
1970   1961–65. *Atlas, Div. Fish. Oceanogr. C.S.I.R.O. Aust.*
(1): 198 p.

WELLS, G C and EDWARDS, R A   A survey of sewage pollution in
1969   Georges River oysters. *Fd Technol. Aust.*, 21:616–9.

WILSON, B R   Survival and reproduction of the mussel *Xenastrobus
1968   secnris* in Western Australian estuary. *J. nat. Hist.*, 2:
307–28.

WOMERSLEY, H B S and EDMONDS, V A general account of the
1958   intertidal ecology of South Australian coasts. *Aust. J. mar.
freshwat. Res.*, 9:217–60.

WYRTKI, K   The surface circulation in the Coral and Tasman
1960   seas. *Tech. Pap. Div. Fish. Oceanogr. C.S.I.R.O. Aust.* (8).

ANON.   Standards and requirements of sewage effluent for disposal
1969   into ocean and inland waters. Paper presented at the 14th
Conference of Engineers (Local Government), Brisbane.

ANON.   Augmentation of sewerage services, Darwin, Northern
1969a   Territory. *Parliament. Pap. Commonw. Aust.*, Canberra, (13).

ACKNOWLEDGEMENT

Appreciation is extended to the Chief Secretary for permitting an
around-Australia familiarization trip prior to preparing this paper,
and to many officers and scientists from government agencies and
industries around Australia; to colleagues in Melbourne, and
particularly to Mr. A. Dunbavin Butcher, Director of Fisheries and
Wildlife, Victoria, for cooperation and support.

Panorama General de la Contaminación de las Aguas en México

*Juan Luis Cifuentes Lemus, Rene Rodriguez Castro y Amin Zarur Menez**

General review of water pollution in Mexico

This document reviews types of pollutants of the marine and continental waters of Mexico; these include oil, sulphur, sugar mill
effluent and agricultural wastes. The effects of sewage waters and
heat are also described. The consequences of such pollution, in the
form of ecological changes affecting marine and brackish water,
fauna and flora, as well as the grave social consequences are
reported.

There follows a discussion of measures being taken or that can be
taken to prevent and eliminate such contamination, the main
possible solutions and the legal aspects of this problem. A detailed
description of the present situation is given, with short and long-term
projections.

Attention is also paid to lagoons and coastal estuaries of such
great importance to the economy of Mexico, as these are the
breeding grounds and habitat for many species which particularly
suffer from the effects of contamination. Mention is made of the
problems arising from the tapping of rivers which flow into the
areas, whether it be for agricultural purposes or otherwise, of great
significance in connection with ecological changes in these areas
and stocks of certain exploitable marine species.

Examen général de la pollution dans les eaux mexicaines

Dans le présent document, on examine les types de déchets qui
polluent les eaux marines et continentales du Mexique; on cite
notamment le pétrole, le soufre, les effluents de l'industrie sucrière et
les résidus d'origine agricole; les auteurs décrivent également les
effets des eaux d'égouts et de la chaleur. Ils mentionnent les
conséquences qu'entraînent ces effluents en apportant à l'écosystème des modifications qui affectent les organismes marins et
d'eaux saumâtres et qui ont des conséquences importantes dans le
domaine social.

Le document passe en revue les moyens utilisés et ceux que l'on
pourrait appliquer pour prévenir cette pollution ou l'éliminer; il
examine les principales solutions possibles ainsi que les aspects
juridiques du problème. La situation actuelle est décrite en détail
et fait l'objet de projections à court et à long termes.

Les auteurs concentrent leur attention sur les lagunes et les zones
marécageuses littorales qui ont une grande importance pour
l'économie du Mexique car elles constituent des sites favorables à
la reproduction et à la croissance de diverses espèces; ces zones sont
d'ailleurs particulièrement exposées aux effets de la contamination.
On signale les problèmes posés par le captage des rivières qui
alimentent ces régions en eau, à des fins agricoles et autres, car ces
problèmes sont importants du fait des modifications écologiques
qui en découlent et des disponibilités offertes par certaines espèces
exploitables.

SE considera que el agua es de vital importancia
para el desarrollo de las actividades humanas, y
cuyo uso se puede clasificar en los siguientes aspectos: doméstico, agrícola, energético, pesquero, recreativo,
industrial y de transporte; por lo tanto, se hace indispensable la conservación de la calidad del agua, así
como su uso racional. La contaminación se puede

definir como la adición de algún material o cualquier
acción o condición que interfiera, degrade o impida
alguna propiedad útil del agua. La acción del hombre
para combatir la contaminación debe enfocarse a mantener el agua con una calidad semejante a la que se
encuentra en su condición natural.

México, por el desarrollo industrial en continuo

---

* Dirección General de Pesca e Industrias Conexas, México.

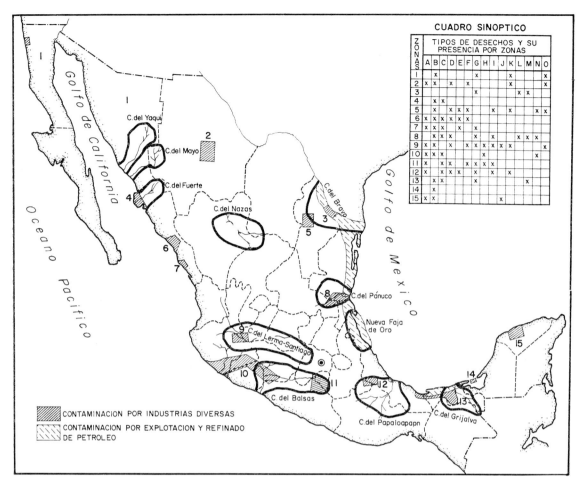

Fig. 1. Cuencas hidrológicas de México y zonas de mayor contaminación industrial; las regiones de contaminación de petróleo se refieren principalmente a petróleo crudo, mezclas químicas de perforación y sólidos en suspensión (para la explicación del cuadro sinóptico en el mapa, ver Cuadro I del texto)

CUADRO I. PRINCIPALES TIPOS DE DESECHOS INDUSTRIALES Y DOMÉSTICOS VERTIDOS EN AGUAS DE MÉXICO

Industrias y otras fuentes	Productos contaminantes: orgánicos (Or), inorgánicos (In) y mixtos (Mi)	Referencias al Cuadro Sinóptico de la fig. 1
Cervecerías y destilerías	(Or) Materia orgánica; alcohol etílico; aguas de lavado de envases: detergentes y sosa cáustica	A
Alimentaria	(Or) Materia orgánica; aguas de lavado de maquinaria: detergentes y sosa cáustica	B
Azucarera	(Or) Materia orgánica; residuos de blanqueadores: hipocloritos y cal (CaO); aguas de lavado de maquinaria: detergentes y sosa cáustica	C
Farmacéutica y química	(Mi) Diversos desechos y residuos químicos	D
Tenerías	(Mi) Materia orgánica; taninos; sales de cromo y sulfuro de sodio	E
Metalúrgica	(In) Aceites (empleados en cortes y enfriamiento); soluciones desoxidadoras ácidas y alcalinas, óxidos metálicos; cianuros y diversas sales metálicas	F
Aguas negras	(Or) Materia orgánica de origen doméstico, etc.	G
Fibras sintéticas	(Mi) Hemicelulosa; glucosa; sosa cáustica; compuestos de sodio; polisulfuros; ácidos débiles; jabones	H
Celulosa y papel	(Mi) Líquidos digeridos; líquidos de descortezado y desmenuzado; fibrillas de madera y papel; compuestos químicos diversos: ácido arsénico, licores sulfatados, sales de lignina, azúcar de madera	I
Conservación de madera	(Or) Materia orgánica; creosota	J
Textil	(Mi) Fibrillas de lana y algodón; colorantes	K
Refinerías de petróleo	(Or) Petróleo crudo; sales de plomo; jabones; fenoles; ácidos y álcalis	L
Petroquímica	(Or) Derivados del petróleo: benceno, etilbenceno, tolueno, acetaldehído, aromáticos pesados, dodecilbenceno y 20 derivados más	M
Minería	(In) Extracción de metales: ácido sulfúrico, sales de diversos metales; extracción de carbón: cieno, arena suelta, carbón fino	N
Pesticidas	(Mi) Residuos de DDT, aldrín, endrín, lindano, etc	O

ascenso y la intensificación de la agricultura, está en el momento oportuno de realizar una campaña proteccionista de sus aguas nacionales, antes de llegar a situaciones irreversibles; asimismo deberá, en la medida de sus posibilidades, colaborar en programas inter-nacionales para evitar la contaminación de las aguas de los mares. Lo anterior cobra mayor importancia por haberse intensificado su programa nacional para el desarrollo de la industria pesquera y por haber establecido programas para incrementar sus recursos, tanto en

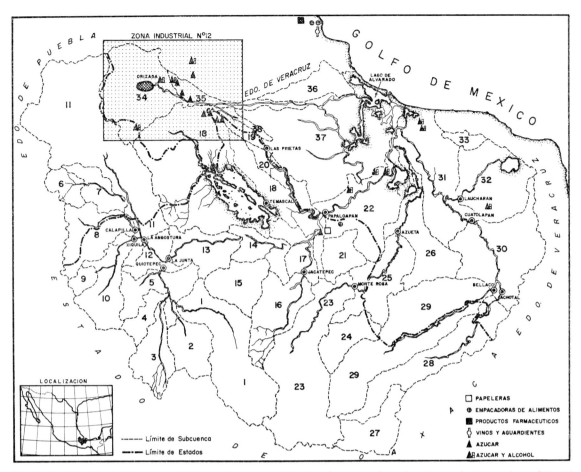

Fig. 2. Distribución de diferentes industrias en la cuenca del Río Papaloapan, en la cual se destaca la Zona industrial No. 12
(ver también fig 5)

aguas continentales como salobres, y al mismo tiempo, por estar aplicando una política que le permita descentralizar su industria hacia las zonas alejadas de la capital.

En este trabajo, se expone la situación actual sobre la contaminación de las aguas mexicanas, indicando además, las fuentes de contaminación, los alcances de ésta, la situación legal y administrativa, los programas de investigación actual, así como algunas recomendaciones para tomar las medidas necesarias y evitar en un futuro problemas que afecten los recursos y a la industria pesquera del país.

Fuentes de contaminación

Se pueden señalar como principales fuentes de contaminación del agua en México, a la industria del petróleo y sus derivados; a las aguas negras; a los desechos de las industrias cervecera, azucarera, alimentaria; a los derivados de la minería, metalurgia, tenerías; asimismo a los productos farmacéuticos y químicos y a los de las fibras artificiales, la celulosa, los hilados y tejidos y, por último, a los que se desprenden de la agricultura, actividad que cada día presenta un mayor desarrollo en el país. En el Cuadro I se presentan en forma general los componentes de estos desechos y en las figs 1 y 2, se ubican las zonas de mayor concentración de fuentes de desechos vertidos en aguas nacionales. Como se puede observar, algunas áreas de la costa este de México, se encuentran más expuestas a los efectos de la contaminación industrial, siendo la industria petrolera, la fuente de mayor importancia.

En la industria petroquímica operan 32 plantas donde se elaboran 26 productos diferentes; su producción total en el año 1969 fue de 1 567 112 toneladas. En la fig 1, se puede observar que 19 de estas plantas se encuentran localizadas en las zonas 8 y 13 (figs 3 y 4), industrias subsidiarias de 2 refinerías de petróleo, en las cuales se producen 22 subproductos petroquímicos; estas plantas tienen una capacidad nominal de 1 586 165 toneladas métricas al año y casi todos sus desechos van directamente a la zona litoral marina o áreas adyacentes.

Otra fuente de contaminación importante de aguas interiores, cuyos efectos llegan hasta el Golfo de México, es la que deriva de la industria azucarera. En la actualidad, se encuentran en operación 92 ingenios azucareros en todo el país, cuya producción fue para 1969 de 2 363 219 toneladas. De estos ingenios, 37 se encuentran en los estados costeros del Golfo de México y de éstos, 28 están ubicados en las cuencas del Río Grijalva y Río Papaloapan (figs 2 y 4). La cuenca del Río Papaloapan es quizá, en lo que se refiere a contaminación por desechos industriales, una de las más expuestas, tanto por la variedad de desechos como por la cantidad de algunos de ellos (figs 2 y 5). Para tener una idea de la importancia industrial y, por lo tanto de las posibilidades de contaminación, se dan a continuación los siguientes datos indicativos: 16 de los 17 ingenios que están dentro de esta cuenca producen el 33,2 % del azúcar nacional;

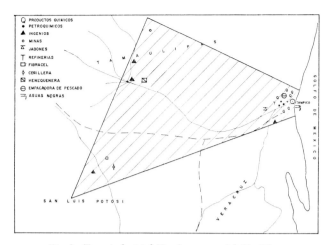

Fig. 3. Zona industrial No. 8, cuenca del Río Pánuco

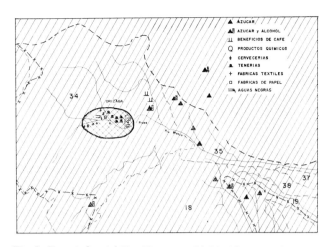

Fig. 5. Zona industrial No. 12, cuenca del Río Blanco, en la cual se destaca la región de Orizaba; ver también fig 2

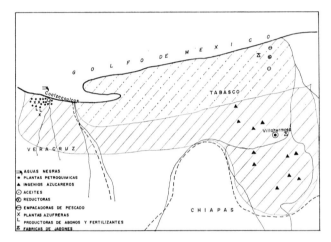

Fig. 4. Zona industrial No. 13, cuencas de losríos Coatzacoalcos Grijalva, etc.

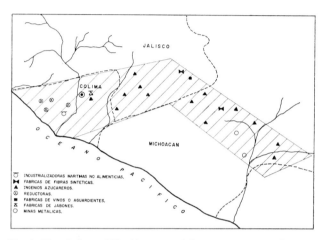

Fig. 6. Zona industrial No. 10, cuencas de los ríos Armería Cuahuayana y Balsas

para este mismo año, 9 destilerías ubicadas en esta cuenca, produjeron 15 169 687 litros de alcohol de caña, que representa el 34 por ciento de la producción nacional. En la subcuenca No. 34 del Río Papaloapan, con un área de 1 731,6 km², se encuentran instaladas 6 tenerías, 5 industrias textiles, 2 cervecerías, 1 ingenio azucarero, 2 fábricas de papel y celulosa y 2 fábricas de productos químicos; en la subcuenca 35, con un área de 449,8 km², se encuentran 4 ingenios azucareros y 1 destilería de alcohol. Además de las 23 industrias mencionadas, es necesario mencionar los desechos urbanos de 3 poblaciones de cierta importancia, que acentúan aún más el problema.

Sobre la costa oeste de México, la acción de contaminación es en general de menor intensidad; no obstante, ella está en continuo aumento debido al desarrollo demográfico e industrial, como por ejemplo se comprueba en la zona 10 (figs 1 y 6).

En cuanto a las aguas negras, podemos señalar que, además del problema que ocasiona este tipo de desechos en la zona del altiplano de México, en las ciudades costeras se hace más grave el problema, en virtud de la práctica que anteriormente se seguía de arrojar los desechos al mar, sin el más mínimo análisis del sitio de descarga, con la única confianza en la enorme capacidad de dilución, dispersión y difusión de los desechos en el mar.

La mayoría de las ciudades costeras de cierta importancia, como es el caso de Mazatlán, Sin., Tampico, Tamps, y Coatzacoalcos, Ver., no cuentan con sistemas adecuados para el tratamiento o descarga de los desechos en la zona costera marina.

Sin embargo, en la actualidad se está realizando un programa de obras tendientes a descargar las aguas negras en áreas alejadas de la población, como es el caso de Acapulco, Gro., en otros, como el caso de Veracruz, donde se encuentra en proceso la instalación de una planta de tratamiento de ese tipo de desechos. Es indudable que se hace necesario un programa integral para el control de las aguas negras en México, en el cual deben participar dependencias estatales, del sector público, federal y municipal, a efecto de reducir al mínimo los daños ocasionados por dichos desechos.

Influencia de pesticidas

Ha sido en las pesquerías de las costas de la región noroeste del país, donde se ha supuesto la mayor influencia de pesticidas, provenientes de las cuencas de los ríos Mayo, Yaqui y Fuerte (fig 1). De los 30 000 000 de hectáreas cultivables en México, 125 349 hectáreas se cultivan por sistema de riego y de éstas, el 30,7 por ciento corresponde a aquella región. La distribución de las zonas de riego del potencial total de labor en el país es la siguiente:

riego de temporal	20 000 000 ha
riego, zona noroeste	961 164 ,,
riego, zona complementaria	2 163 836 ,,
medio riego por gravedad y bombeo	5 875 000 ,,
riego con necesidad de saneamiento agrícola	1 000 000 ,,

En las áreas cultivadas por sistema de riego, se practica una agricultura muy tecnificada, aplicando fertilizantes y pesticidas. De las cantidades totales anuales de pesticidas clorados, aplicados en toda el área cultivable que a continuación se dan, gran proporción corresponde a aquella región.

Pesticidas	Toneladas	Hectáreas
Aldrín 100%	836,9	418 929
Endrín 100%	13 667,2	273 344
Lindano 100%	57,7	464 222
DDT 100%	1 325,6	313 836
Totales	15 886,4	1 470 331

ALCANCES DE LA CONTAMINACION Y EFECTOS SOBRE RECURSOS PESQUEROS

Algunos ejemplos de los casos de contaminación de las aguas de que se tiene noticia, son los siguientes:

(a) En el Río Blanco en el Estado de Veracruz, en cuyo recorrido de 156 km, prácticamente se ha reducido al mínimo la explotación de especies pesqueras durante todo el año; además de los problemas que presenta la calidad del agua para uso doméstico y como abrevaderos de ganado, los efectos de la contaminación del río se dejan sentir en la Laguna de Alvarado, lugar hasta donde desemboca.

(b) En el Lago de Chapala, el más grande de México, desde hace algunos años se ha reducido la captura del pescado blanco (*Chirostoma* spp.), sus causas se desconocen, pero se considera que la influencia de los desechos urbanos y el gran número de industrias establecidas en los ríos alimentadores de este lugar (Ríos Lerma y Santiago), son factores importantes en la disminución de las poblaciones de esta especie.

(c) Se han presentado casos de mortandad de peces, originados por los desechos de ingenios azucareros y, aunque no se tiene un dato preciso sobre la concentración de dichos desechos, se presentan los datos registrados en el Cuadro II.

(d) En las zonas estuarinas de San Blas y el Latero, de la región noroeste del país, se han recibido informes de mortandad de peces atribuidas por los pescadores, a residuos de herbicidas aplicados en los canales de irrigación que desembocan en esos estuarios.

(e) En las aguas estuarinas de la costa oriental de México, se ha observado mortandad de ostiones en forma masiva y cuyo origen es atribuido por los pescadores, a los desechos de perforación de pozos petroleros. Estos residuos han provocado problemas de mortandad masiva en la Laguna de Tamiahua, Ver. y Laguna Machona, de Tabasco. Cuando se han presentado los casos de mortandad, se han realizado algunos análisis y aunque los agentes dañinos de los desechos no se han registrado en concentraciones tóxicas, se presume que el impacto mayor se debe a la densidad con que los desechos de los pozos petroleros se acumulan en las poblaciones ostrícolas. En la Laguna Machona antes mencionada, se ha observado una disminución paulatina en las existencias de especies de peces y se tienen registros de que en la época de desove de estas especies, el Río Santa Ana, principal afluente de esta área, acarrea una gran cantidad de desechos provenientes de un ingenio azucarero (Cuadro II), además de otras

CUADRO II. Casos de mortandad de peces y crustáceos producidos por desechos de la industria azucarera

Año	Ingenio	Lecho acuático	Efectos	Intensidad	Evidencia*
1963	Independencia	Río Nautla	Mortandad de peces	?	Acumulación de materia orgánica (desechos de caña)
1965	Independencia	Río Nautla	Mortandad de peces	?	Acumulación de materia orgánica (desechos de caña)
1965	El Modelo	Río Hediondo	Mortandad de peces	2 toneladas	Coincidencia con lavado de maquinaria
1965	La Gloria	Río San Carlos	Mortandad de peces	3 toneladas	Coincidencia con lavado de maquinaria
1970	La Gloria	Río San Carlos	Mortandad de peces	?	Coincidencia con lavado de maquinaria
1968	Mahuixtlán	Río Pixquiac	Mortandad de peces	?	Acumulación de materia orgánica (desechos de caña)
1963	San Cristóbal	Río El Salado	Mortandad de peces y crustáceos	?	Bioensayos en sus aguas de desecho
1964	San Cristóbal	Río El Salado	Mortandad de peces y crustáceos	?	Coincidencia con lavado de maquinaria
1970	San Cristóbal	Río El Salado	Mortandad de peces y crustáceos	?	Acumulación de materia orgánica de H_2S en concentraciones tóxicas en sus desechos y en la laguna
	Sta. Rosalía	Ríos San Vicente y Sta. Ana	Alteración de las condiciones naturales	7 km	p.p.m. en agua: 4,0 H_2S; 380 de sólidos totales; 2,0 de oxígeno disuelto. Olor fétido

* En ningún caso se ha encontrado otra fuente de contaminación; emplean detergentes, sosa cáustica y ácido muriático; se acumulan los desechos por la escasez de lluvias durante la producción de azúcar.

razones de tipo ecológico a las que pudiera atribuirse dicha disminución.

Situación legal y administrativa

A continuación, se dan en forma concisa, los diferentes aspectos de los distintos ordenamientos legales existentes en México para controlar la calidad del agua de propiedad nacional.

(a) **Ley de pesca.** Prohíbe a los pescadores, dueños de embarcaciones y terceros, usar explosivos para pescar y dejar correr o verter sustancias que dañen a las especies acuáticas. El organismo ejecutivo es la Secretaría de Industria y Comercio.

(b) **Ley de aguas de propiedad nacional y su reglamento.** Establece que para otorgar permiso de usar el agua con fines industriales, el usuario deberá regresar el agua a su cauce con la misma calidad con que fue tomada; además prohíbe arrojar a los cauces de propiedad nacional desechos de diversa naturaleza que contaminen las aguas, haciéndolas dañosas para la salud de personas o animales, para la pesca, agricultura o la industria. El organismo ejecutivo es la Secretaría de Recursos Hidráulicos.

(c) **Codificación sanitaria mexicana.** En ella se establecen las normas de calidad de agua potable; se indica los mecanismos para el control de desechos municipales y se prohíbe arrojar en los lechos acuáticos productos que alteren las propiedades del agua para los siguientes usos: de pesca, recreativos, agrícolas, domésticos y que dañen la salud de las personas o sus bienes. El organismo ejecutivo es la Secretaría de Salubridad y Asistencia.

(d) **Otras medidas administrativas.** Existe también la Comisión Técnica Mexicana para la Prevención de la Contaminación de las Aguas del Mar por Hidrocarburos, creada en base a la Conferencia Internacional sobre Contaminación Aceitosa del Mar, celebrada en Londres, Inglaterra, en 1954. Sus funciones son de estudiar la contaminación aceitosa y recomendar las medidas prácticas para prevenirla.

No obstante estos instrumentos legales, se ha observado una renuencia por parte de la mayoría de los industriales para acatar sus disposiciones, aduciendo para ello incosteabilidad en la instalación y operación de los sistemas adecuados para el tratamiento de los desechos. Recientemente, la Dirección General de Pesca, ha adoptado la medida administrativa de otorgar permisos a las industrias nuevas para que viertan sus desechos en aguas nacionales, condicionado a que cumplan con las normas de calidad preestablecidas que van fundamentalmente enfocadas a evitar alteraciones de las condiciones naturales del agua. Además, para estudiar y tratar de resolver el problema en conjunto, se han hecho esfuerzos para crear una Comisión Intersecretarial contra la Contaminación de las Aguas de Propiedad Nacional y del Subsuelo, que se integraría con las distintas dependencias gubernamentales relacionadas con el manejo de las aguas.

Programa de investigación actual

El programa que actualmente se sigue en la Dirección General de Pesca, a través de su Departamento Químico, está enfocado en su primera etapa a completar el inventario de las industrias situadas en las proximidades de los lechos acuáticos y cuyos desechos son descargados en ellos; para tal efecto, se ha dividido el país en zonas, atendiendo al número de industrias y a la importancia de las cuencas hidrológicas. La segunda etapa del programa, consiste en conocer el grado de contaminación de cada área establecida; para tal efecto, se realizan inspecciones a las industrias y se llevan a cabo muestreos de sus desechos en las corrientes acuáticas en donde se descargan. En dichos muestreos se practican análisis físicos y químicos de acuerdo con la composición de los desechos. En virtud de la extensión del territorio nacional, los muestreos se han realizado aplicando cierta jerarquía de acuerdo con la magnitud del problema, y la frecuencia de las inspecciones ha dependido de la intensidad en la descarga de los agentes contaminantes que afectan a la vida acuática.

La Secretaría de Recursos Hidráulicos, a través de su Departamento de Control y Prevención de la Contaminación de las Aguas, ha iniciado los trabajos de análisis químicos de aguas del Río Lerma, cuya longitud es de 746 km. La Escuela Nacional de Ciencias Biológicas del Instituto Politécnico Nacional, en colaboración con la Comisión Federal de Electricidad, está realizando el estudio de las comunidades marinas en una zona del Golfo de México, donde se piensa instalar una planta nuclear para producir energía eléctrica; por computadoras, se simularán los efectos que producirá el cambio térmico en esa zona y se realizarán bioensayos con los radio-isótopos que se verterán; en el caso de que se observen efectos de radiactividad en los organismos acuáticos, se evitará arrojar los desechos radiactivos en el mar, solidificándolos en la planta.

Avances de la investigación

De los trabajos realizados hasta la fecha, se tiene conocimiento de 567 industrias que vierten sus desechos en aguas nacionales: en 77 de ellas se ha comprobado que originan problemas de contaminación, 15 de las cuales han adoptado medidas para evitarlo y 20 más están en la etapa de estudio para tal fin. De estas últimas, 18 industrias que se encuentran próximas entre sí, situadas en las márgenes del Río Blanco (figs 2 y 5), han acordado instalar una planta de tratamiento común, con participación proporcional a sus volúmenes y tipos de desecho. También la Empresa Petróleos Mexicanos, después de experiencias tenidas en años anteriores, ha decidido no verter sus desechos de perforación en el seno de lagunas costeras, en donde se han provocado algunos problemas, sino que actualmente se han descargado los contaminantes en la zona marina litoral. Además, cuando se realizan exploraciones sismológicas en el fondo marino, esa empresa ha substituido la dinamita por la energía sónica para sus trabajos.

También existen casos en los que a través del establecimiento de normas de calidad de los desechos antes de verterse a las aguas, y especialmente aplicadas a industrias de nueva instalación, han dado por resultado experiencias positivas. Un ejemplo de ello es el caso de la

Empresa Fertilizantes Fosfatados de México, la cual construyó grandes presas de neutralización, precipitación y decantación de sus desechos, constituidos fundamentalmente por ácidos fosfóricos, hidrofluorurosílico y sulfato de calcio; los sobrenadantes, después de algún tiempo, se vierten a la zona marina en donde sus efectos son mínimos para las especies pesqueras de la región. Otra empresa química llamada Cyanaquim, S. A., que desechará gran número de solventes orgánicos y colorantes, está construyendo presas de neutralización con el mismo fin.

La Empresa Bayer de México, S. A., que tenía intenciones de instalarse en la cercanía del mar y a la cual se le negó el permiso para descargar 36 toneladas diarias de sólidos, compuestos de cromatos, dicromatos, óxidos de cromo y otros metales, tuvo que instalarse en el interior de la República, con el objeto de evitar los daños que se habían previsto por la descarga de estos desechos. En el caso de los ingenios azucareros, 8 de ellos han puesto en práctica la aplicación de sus desechos a los campos de cultivo, los cuales obran como fertilizantes; esto ha originado que no tan sólo se evite la contaminación, sino que la producción de caña se ha mejorado hasta en un 35 por ciento sólo con esa medida; también emplean el bagazo como combustible de las calderas y parte de este subproducto se destina a la producción de celulosa; la ceniza de la combustión del bagazo se emplea para neutralizar la acidez en los campos de cultivo producida por la degradación de la materia orgánica.

Consideraciones finales

Además de presentar una breve síntesis de la situación actual en México respecto a la contaminación de sus aguas continentales y marinas, es necesario señalar algunas observaciones y recomendaciones que podrían servir para la planificación de actividades nacionales e internacionales relacionadas con este problema, tal como sigue:

(1) Al tener en cuenta que el ciclo vital de muchas especies de importancia pesquera se cumple tanto en aguas dulce y marina o marina y de estuario y que en muchos países para aumentar la agricultura se realizan obras para derivar el agua dulce a las tierras de cultivo, indudablemente se alterarán las características de los esteros y lagunas litorales y de las condiciones de vida de los organismos que los habitan y también de algunos de las aguas marinas adyacentes. Por tanto, como varias especies marinas que se encuentran en esos casos son explotadas comercialmente por los países vecinos, es necesario que se consideren en conjunto estos puntos de vista para estudiar las causas de las alteraciones de la vida de algunas áreas marinas de interés común. Como un ejemplo concreto, se pueden mencionar los programas que se desarrollan en México para lograr mediante el uso racional de las aguas de ríos y lagunas litorales el mejoramiento de las condiciones ecológicas, principalmente para las especies de camarón, con el fin de aumentar su producción. En los Estados de Sonora, Sinalosa, Nayarit y Oaxaca, se están realizando obras de conexión entre ríos y lagunas para aprovechar parte del agua dulce en la agricultura y parte en las pesquerías; a la vez, se dragan las desembocaduras y se conectan los esteros con el mar.

(2) De ningún modo, la contaminación del mar se puede considerar separadamente de la contaminación de las aguas continentales. Mientras que unos tipos de contaminación son endémicos en el mar, por ejemplo la contaminación por petróleo, la mayoría de los tipos se origina de las actividades humanas en tierra, siendo la contaminación del mar sólo una etapa dentro de una amplia secuencia. Los efectos de la contaminación en tierra no deben ser descuidados, y la estrategia de control de la contaminación debe empezarse ahí mismo. Además, muchas especies pasan algunas fases de su vida en habitats de aguas continentales así como en aguas marinas. En consecuencia, la estrategia respecto a la contaminación debe ser orientada de tal manera, al considerar el habitat del hombre en su totalidad.

(3) Ni la demostración ni el control de las fuentes de contaminación, como tampoco la aplicación coordinada de estas dos medidas, pueden ser suficientes como estrategia contra la contaminación. Debe tenerse en cuenta el principio general de que en la naturaleza, los desechos de una población o de una comunidad son retirados o usados por otras especies, o son depositados en sumideros o bien se acumulan, resultando de todo esto un proceso de sucesión ecológica con desalojamiento y quizás con supresión de una o varias especies. Entonces es responsabilidad del hombre de manejar el ambiente con sus recursos explotables, diseñar prácticas para el uso de sus desechos, etc. No es suficiente dirigir acusaciones contra los empresarios respecto a los desechos de sus fábricas, porque esas personas son agentes de la comunidad, y la comunidad debe compartir con ellos la tarea de descubrir la práctica para el aprovechamiento de los desechos o encontrar la modalidad de almacenarlos en lugar seguro.

Contaminantes Potenciales que Pueden Afectar a los Organismos del Ambiente Marino a lo Largo de la Costa del Perú
L. Chang Reyes*

Potential pollutants which could affect living organisms in the marine environment along the coast of Peru

The present work is an evaluation of possible pollution problems resulting from the discharge of industrial sewage into the sea by the major industries along the Peruvian coast. This refuse, as well as the domestic sewage from the population, presents aesthetic pro-

Polluants potentiels qui pourraient nuire aux organismes vivant dans le milieu marin le long de la côte du Pérou

Ce travail présente une évaluation des problèmes de pollution possibles résultant des décharges industrielles au large des côtes du Pérou. Celles-ci, ainsi que les décharges domestiques, posent actuellement certains problèmes d'esthétique sur les plages; ils ne

* Universidad Nacional de Ingeniería Sanitaria, Lima, Perú.

blems on the beaches—problems, however, which could be resolved by adequate placing of submarine outlets.

Due to lack of study there is no precise knowledge concerning biological changes in the flora and fauna of the Peruvian coast. In specific cases there is no evidence of change in the normal behaviour of marine organisms; however, it would be necessary to carry out surveys and studies under the guidance of experts and international organizations in order to preserve the normal conditions of this, one of the richest regions in the world in marine life, and taking into consideration future programmes of mining and industrial expansion, especially of exploitation of the sea.

Therefore it is extremely important that international organizations intervene to forbid completely atomic experiments and the dumping of radioactive residues and toxic elements into the Pacific Ocean, considering that this could alter the life of marine populations, which could result in the destruction of the most important future food resources for man and which could also damage the economic position of the countries depending on these resources.

pourront être résolus que par une localisation appropriée des émissaires sous-marins.

Par manque d'études on connaît mal les effets biologiques que cette pollution entraîne sur la faune et la flore de la côte péruvienne. Pour quelques cas particuliers il n'apparaît pas que le comportement normal des organismes marins soient modifiés. Les conditions normales d'une des régions les plus riches du monde en organismes marins ne seront préservées que si des études sont entreprises sous la supervision d'organisations et d'experts internationaux. Les programmes futurs de développement industriel et minier devront être pris en considération.

Il est donc extrêmement important que les organisations internationales interviennent afin d'interdire totalement dans l'Océan Pacifique les expériences atomiques et la décharge de résidus radioactifs et d'éléments toxiques, car ceux-ci peuvent réellement nuire à la vie des organismes marins, et par conséquent détruire la plus importante source alimentaire future de l'humanité, tout en ayant une répercussion sur l'économie des pays qui dépendent de telles ressources.

EL Perú está situado en la costa occidental de Sudamérica, con una superficie territorial de 1 285 2156 km² y una longitud de costa de 2 864 km. Básicamente está dividido en cuatro regiones geográficas: la marítima, la costa, la sierra y la selva.

La región de la costa a pesar de su latitud no tiene un clima tropical debido a la corriente fría de Humboldt que corre paralela a la costa en dirección norte. La presencia de esta corriente y la proximidad de la cordillera de los Andes hacen que en la costa peruana no llueva.

La actividad económica del país gira alrededor de tres renglones principales: la minería, la pesca y la agricultura. La explotación de estas tres actividades, origina desechos industriales que son arrojados a los cursos de agua, los que finalmente se vierten al mar, sobre el lado occidental de la cordillera andina.

Contaminación por desechos mineros

La explotación minera está basada principalmente en la extracción de minerales como: plata, plomo, hierro, cobre, cinc, oro, antimonio y otros; tiene lugar en la región de la sierra y cuando se realiza en la región occidental andina, los desechos líquidos resultantes se

vierten en los ríos de la costa. La región costera tiene una anchura que varía entre 40 y 80 km, siendo la diferencia de altitud de 4 500 m sobre el nivel del mar. Esto significa que los ríos tienen una gran velocidad y alcanzan rápidamente el Océano Pacífico. Debido a esta situación muy especial, existe un peligro potencial de contaminación a lo largo de la costa peruana por los elementos tóxicos minerales que pueden causar una alteración en la fauna marina.

Aunque aún no se han notado síntomas de estas alteraciones en la zona marítima peruana, es de mencionar que algunos ríos sí han sido alterados biológicamente. Siendo el Perú un país que tiene planes futuros de expansión minera, especialmente de refinerías de metales que serán situadas en la costa, este problema merece estudios que revisarán y mejorarán los procesos de explotación minera así como estudios propios en los cursos de agua que acarrean los desechos mineros y que finalmente vierten sus aguas en el océano.

Contaminación por el uso de pesticidas en la agricultura

El Perú es un país eminentemente agrícola ya que el 44,5 por ciento de la población económicamente activa

CUADRO 1. PESCA TOTAL DEL PERÚ (TONELADAS MÉTRICAS)

Especies	1962	1963	1964	1965	1966	1967	1968
Totales mundiales	47 000 000	48 200 000	52 700 000	53 500 000	57 200 000	60 700 000	64 000 000
Totales del Perú	6 464 986,8	6 609 756,6	9 037 899,4	7 391 200,1	8 712 358,3	10 712 358,3	10 440 402,0
Anchoveta	6 274 624,5	6 423 243,9	8 863 366,9	7 242 394,0	8 529 820,8	9 529 820,8	10 262 661,0
Atún	10 901,7	11 230,8	9 490,2	3 187,5	5 461,0	3 359,5	3 447,3
Barrilete	12 722,8	16 911,3	7 114,9	8 517,2	6 387,8	13 777,9	6 528,4
Bonito	86 525,0	88 179,3	76 405,0	62 308,7	71 430,3	63 574,4	54 273,6
Caballa	13 275,2	7 911,3	2 048,2	3 807,5	7 559,3	13 432,4	7 186,5
Lorna	8 712,9	7 814,3	2 146,3	1 713,1	2 229,7	4 351,4	4 331,1
Machete	10 883,0	7 863,0	13 948,0	7 060,1	13 419,8	18 415,5	11 880,1
Otras especies	47 341,7	46 602,7	63 379,9	62 212,0	72 690,6	92 572,8	90 094,0

Notas: Las discrepancias existentes con las cifras de las anteriores Memorias de la Sociedad Nacional de Pesquería se deben a que las cifras han sido variadas por la fuente estadística.
La pesca de anchoveta para 1969 asciende a 8 960 460 t según el Instituto del Mar del Perú.

Fuente: Totales mundiales: Anuario Estadístico de Pesca, FAO, Vol. 26 1968.
Estadística: Totales anchoveta: Instituto del Mar del Perú.
Resto de especies: 1962/63, Servicio de Pesquería del Ministerio de Agricultura, Perú.
1964–68, Instituto del Mar del Perú.

trabaja en los campos. A pesar de que la costa del Perú en su mayor parte es arenosa y árida, existen algunos valles fértiles, los que se dedican en su mayor parte a la explotación de algodón, arroz, maíz, caña de azúcar, etc. Esta explotación agrícola necesita del uso de pesticidas los que muchas veces se emplean en forma incontrolada. Sin embargo, la contaminación del mar por pesticidas es bastante remota ya que debido a las escasas lluvias en la costa peruana no hay posibilidad de lavado de la tierra y del consecuente acarreo de estos productos a los cursos de agua ni al mar.

Contaminación por desechos derivados de la explotación pesquera del mar

La industria pesquera peruana se ha desarrollado en forma tan violenta en los últimos años que en la actualidad ha situado al país a la cabeza de la producción de harina y aceite de pescado del mundo.

En el año 1955 se produjeron aproximadamente 20 000 toneladas de harina habiéndose capturado 1 000 000 de toneladas de anchoveta (*Engraulis ringens*). En el año 1969 se han producido 1 610 800 toneladas de harina de pescado con una captura de 8 960 460 toneladas de anchoveta. Es decir, que la producción y extracción es cien veces mayor que hace 15 años. En el Cuadro I se muestran las variaciones de captura de anchoveta en estos últimos años. Esta extracción del mar ha sido acompañada de un aumento del número de fábricas de industrialización del pescado, así como de embarcaciones dedicadas a la pesca. Igualmente se han desarrollado más puertos pesqueros, contándose en la actualidad con 21 puertos con instalaciones de fábricas de harina de pescado y sus derivados.

Proceso de fabricación de la harina de pescado

La fabricación de harina de pescado es una industria relativamente sencilla: en forma esquemática ella consiste en las siguientes operaciones: captura; transporte de la materia prima; almacenamiento; cocción del pescado; prensado; secado; molienda y envasado de la harina.

Como línea colateral debe agregarse el centrifugado del aceite y, a veces, la concentración de las aguas residuales del prensado. A continuación, se describen las distintas operaciones con la finalidad de indicar los residuos que ellas son capaces de producir, así como la significación sanitaria de dichos residuos (fig 1).

Captura. Se efectúa a cierta distancia de la costa, por lo que no produce residuos capaces de representar un problema, en sentido de la contaminación del mar.

Transporte de la materia prima. El pescado se extrae de los barcos habitualmente mediante bombas de succión. Con este objeto, se agrega al contenido de las bodegas agua de mar en proporción similar a la cantidad del pescado capturado, formándose de esta manera un fluido susceptible de ser aspirado mediante una bomba. El efluente llega a un tamiz que separa el sólido devolviendo el líquido al mar. Al paso por la bomba y la cañería, que a veces llega a tener 500 o más metros de longitud,

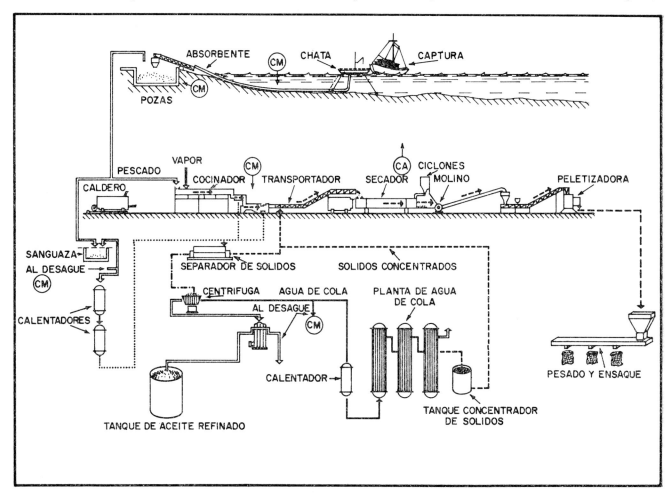

Fig. 1. Proceso esquemático de la fabricación de harina de pescado, indicándose dentro de círculos las fuentes principales de contaminación de mar (CM) y del aire (CA)

una parte del pescado se rompe perdiendo sangre y trozos pequeños de sólidos que son arrastrados al mar. Dependiendo de la longitud de las cañerías, se pierde 3 a 6 y aún 8 por ciento del total de la materia orgánica recogida, lo que significa que teniendo en cuenta un promedio de 4 por ciento en pérdidas, en 1959 se devolvieron al mar 358 418 t de materia orgánica. Si la descarga se efectúa en una bahía cerrada como la mayor parte de las bahías de la costa del Perú, sin corrientes de importancia, se produce la descomposición del agua por el aporte de sangre y residuos orgánicos. Calculando la población equivalente en base a la DBO (Demanda Bioquímica de Oxígeno), este residuo que tiene una DBO de $\pm$ 30 000 mg/l representa para el año 1969 una población equivalente diaria de 18 230 000 habitantes.

Almacenamiento. La materia prima se recibe en grandes pozas desde las que pasa a la elaboración. Por la presión que sufre el pescado al quedar uno sobre otro, pierde cierta cantidad de sangre y algo de agua que queda en el fondo de las pozas. Estas son utilizadas en forma alternativa para evitar que parte de la materia prima sea almacenada durante mucho tiempo y son lavadas casi diariamente. Las aguas de lavado arrastran sangre y trozos de pescado que pueden representar una carga orgánica importante en el proceso de la contaminación del mar.

Un problema adicional de especial gravedad se puede producir durante el almacenamiento; la anchoveta utilizada es una materia de fácil descomposición sobre todo cuando está comprimida y amontonada y, si no es utilizada dentro de pocas horas después de su recepción, se inicia la putrefacción al convertirse en una masa gelatinosa, difícil de manipular y casi imposible de elaborar. El problema sanitario que pueden representar algunas toneladas de materia orgánica en putrefacción es fácil de imaginar.

Cocción del pescado. Se efectúa al vapor, en cocederos continuos. Si éstos no son herméticos caerá agua y trozos de pescado que pueden ser llevados al alcantarillado durante las operaciones de limpieza.

Prensado. El pescado cocido pasa a filtros prensas que le extraen una parte importante del agua y casi todo el aceite. Queda una torta con aproximadamente 50 por ciento de agua y 1 a 2 por ciento de grasa. Las aguas del filtrado tienen en suspensión el aceite y en disolución un 5 a 6 por ciento de proteínas solubles, que representan además aproximadamente el 10 por ciento del total de las proteínas del pescado.

Secado. Las tortas obtenidas del prensado se secan en hornos rotatorios, mediante aire caliente o vapor indirecto, para eliminar el agua que contienen. No se producen residuos líquidos durante esta operación. Los gases del secado, sin embargo, representan un importante problema de contaminación del aire que debe ser controlado. Se utilizan con este objeto sistemas en caliente, mediante una incineración de la materia orgánica contenida en sus gases, o por vía húmeda, lavado, enfriamiento y condensación.

Molienda y envasado de la harina. No se presentan problemas de residuos. Las aguas extraídas durante el prensado son sometidas a una centrifugación para separar las grasas suspendidas. Queda un líquido lechoso con 5 a 6 por ciento de materia orgánica. Dado que este residuo está constituido casi exclusivamente por proteínas solubles de excelente calidad, las fábricas más modernas tratan de recuperarlas, lo que mejora su rendimiento y la calidad del producto final elaborado. Numerosas industrias, lamentablemente, se limitan todavía a arrojarlas a los desagües y finalmente al mar. Considerando que sólo un 30 por ciento de este líquido llamado agua de cola es procesado en las fábricas, el resto, arrojado al mar durante el año 1969 constituye un volumen de alrededor de 2 822 545 t, lo que representa en sólidos aproximadamente 169 352,7 t. Si se calcula la población equivalente en base a la DBO, este residuo que tiene una DBO del orden de 70 000 mg/l, representa para el año 1969 una población equivalente diaria de 13 440 000 habitantes.

Significación sanitaria de los residuos derivados de la fabricación de harina de pescado

Según se ha visto en los distintos procesos de fabricación, la población equivalente total diaria está representada por más de 31 millones de habitantes; esta población es 2,3 veces mayor a la población total del Perú.

Los residuos, hasta hace poco arrojados directamente al mar, por la facilidad de descomposición de la materia orgánica, han originado una inutilización total de las playas frente a las fábricas. Sin embargo, la costa del Perú goza del privilegio de tener una corriente hacia el norte que transporta y diluye estos residuos, no constituyendo un peligro tal que pueda afectar la fauna y flora marina, especialmente por estar constituidos por materias orgánicas del propio pescado que sirven de nutrientes a la producción primaria en el mar. En la actualidad, se están efectuando obras de saneamiento en la mayor parte de los puertos pesqueros con el fin de recolectar los desagües de las fábricas en cada puerto y disponerlos en un punto alejado en el mar a través de emisores submarinos, con el propósito de evitar la contaminación de playas. Estos trabajos se vienen efectuando con el propósito de dar cumplimiento a los reglamentos de preservación de las aguas marítimas de la nueva Ley General de Aguas que se mencionan más adelante.

Esfuerzos de la industria de harina y aceite de pescado por mejorar los procesos de producción. Conforme se ha visto en los distintos procesos, existe una gran cantidad de materia orgánica que no es utilizada. En la actualidad se viene realizando una serie de estudios y experimentos para aprovechar la totalidad de la materia prima de manera tal que no se arroje nada al mar, evitando consecuentemente la contaminación de las costas peruanas.

Proyectos futuros de explotación pesquera del mar. El Gobierno del Perú está actualmente desarrollando una serie de proyectos de construcción de puertos pesqueros a lo largo del litoral peruano, con el fin de lograr una mejor explotación de los recursos marinos, especialmente para la pesca de consumo y de envasado de pescado. En todos los proyectos se prevé la disposición adecuada de los residuos derivados de esta explotación, de manera que se evite la contaminación de las aguas costeras.

Reglamento peruano para la preservación de las aguas marítimas

Este reglamento ha entrado en vigor en el mes de julio

de 1969 y clasifica a las zonas costeras del país de acuerdo con las siguientes clases:

Clase I. Las aguas marítimas comprendidas dentro de esta zona, por las características físico-químicas y bacteriológicas, podrán ser destinadas a fines de recreación y para agua potable; sólo podrán recibir descargas, con o sin tratamiento, que no alteren las características de ellas, en la línea de playa incluyendo su temperatura normal.

A este efecto se consideran las siguientes características de las aguas en la línea de playa:

1. Color	Ausente
2. Sustancias que causen olor o sabor	ausentes
3. Sólidos sedimentables	ausentes
4. Aceites y grasas	ausentes
5. Fenoles	menos de 0,001 mg/l
6. Sustancias tóxicas o potencialmente tóxicas	en cantidades no mayores a las indicadas:

	mg/l
Plomo	0,10
Flúor	1,50
Arsénico	0,20
Selenio	0,05
Cromo hexavalente	0,05
Cianuro	0,01
Bario	0,10
Cadmio	0,01
Plata	0,05
Hierro	0,30
Manganeso	0,10
Cobre	1,00
Cinc	5,00

7. Número más probable de bacilos coli	de 50 a 500:100 ml
8. Demanda bioquímica de oxígeno a 5 días y 20°C	máximo 1 mg/l
9. Oxígeno disuelto en cualquier día	6 mg/l
10. pH	entre 6 y 9

Clase II. Las aguas marítimas comprendidas en esta zona por sus características físico-químicas y bacteriológicas podrán ser destinadas para la conservación de la fauna ictiológica y con fines de agua potable; sólo podrán recibir desechos, con o sin tratamiento, que no alteren las características de ellas en la línea de playa, con excepción de la temperatura que podrá alterarse en 2,5°C.

A este efecto, se consideran las siguientes características de las aguas marinas vecinas a la línea de playa:

1. Color	Máximo 5 unida des
2. Sustancias que causen olor o sabor	ausentes
3. Sólidos sedimentables	ausentes
4. Aceites y grasas	ausentes
5. Fenoles	menos de 0,001mg/l
6. Sustancias tóxicas o potencialmente tóxicas	iguales a las de la Clase I
7. Número más probable de bacilos coli	de 500 a 3 000 c/100 ml
8. Demanda bioquímica de oxígeno a 5 días y 20°C	máximo 2 mg/l
9. Oxígeno disuelto en cualquier día	mínimo 5 mg/l
10. pH	entre 6 y 9

Clase III. Las aguas marítimas comprendidas en esta zona, por las características físico-químicas y bacteriológicas pueden ser utilizadas con fines industriales; sólo podrán recibir desechos que, con o sin tratamiento, no alteren las características de ellas en la línea de playa, con excepción de lo que respecta a temperatura en la que podrán admitirse alteraciones mayores a 2,5°C previo estudio en cada caso específico.

A este efecto, se consideran las siguientes características de las aguas marinas vecinas a la línea de playa:

1. Color	Máximo 20 unidades
2. Sustancias químicas que causen olor o sabor	ausentes
3. Sólidos sedimentables	ausentes
4. Aceites y grasas	ausentes
5. Fenoles	máximo 0,002 mg/l
6. Sustancias tóxicas o potencialmente tóxicas	cantidades no mayores a las que se indican:

	mg/l
Aluminio	1,00
Arsénico	1,00
Bario	0,50
Cromo	5,00
Cobalto	0,20
Cobre	3,00
Cinc	15,00
Vanadio	10,00
Selenio	0,05
Níquel	0,50
Litio	5,00

7. Número más probable de bacilos coli	de 3 000 a 20 000 c/100 ml
8. Demanda bioquímica de oxígeno	máx. 25 mg/l; mín. 4 mg/l
9. Oxígeno disuelto en cualquier día y a cualquier tiempo	entre 6 y 9

Clase IV. Las aguas marítimas comprendidas en estas zonas, podrán ser destinadas para fines de navegación y puertos; sólo podrán recibir desechos que, con o sin tratamiento, no alteren las características de ellas en la línea de playa, con excepción de lo que respecta a temperatura, en que podrán admitirse alteraciones mayores a 2,4°C previo estudio en cada caso específico.

A este efecto, se consideran las siguientes características de las aguas:

1. Color	Máximo 30 unidades
2. Sustancias que causen olor o sabor	ausentes
3. Sólidos sedimentables	ausentes
4. Sólidos suspendidos	máximo 100 mg/l
5. Aceites y grasas	ausentes
6. Fenoles	máximo 0,01 mg/l
7. Sustancias tóxicas o potencialmente tóxicas	cantidades no mayores a las que se indican:

	mg/l
Aluminio	1,00
Arsénico	1,00
Bario	0,50
Cromo	5,00
Cobalto	0,20
Cobre	3,00
Cinc	15,00
Vanadio	10,00
Selenio	0,05
Níquel	0,50
Litio	5,00

8. Número más probable de bacilos coli	de 20 000 c/100 ml
9. D.B.O. a 5 días y 20°C	máximo 50 mg/l
10. Oxígeno disuelto	mínimo 3 mg/l
11. pH	entre 6 y 8,5

Clase V. Las aguas marinas de esta clase podrán ser destinadas a otros usos no previstos en las clases anteriores; sólo podrán recibir desechos, con o sin tratamiento, que no alteren las características de ellas en la línea de playa, con excepción de la temperatura en que podrán admitirse alteraciones mayores a 2,5°C previo estudio en cada caso específico.

A este efecto se consideran características de calidad inferior que las que corresponden a la clase IV.

Consideraciones finales

La posibilidad de contaminación del mar, en la costa del Perú, por las actividades derivadas de sus pobladores y de la industria, no representa en la actualidad ningún peligro. Sin embargo, las autoridades de la sanidad vienen ejerciendo un control estricto a través de una reglamentación adecuada. Esto ha sido comprendido por los industriales, quienes realizan esfuerzos para mejorar sus procesos industriales y de esta manera, evitar la contaminación de las aguas del mar.

Se espera que en un futuro cercano, los desagües de las poblaciones e industrias se dispongan adecuadamente en al costa evitando así el problema sanitario y estético de contaminación de playas y, en cualquier caso, prohibir totalmente la disposición de desagües que contengan sustancias tóxicas capaces de destruir la flora y fauna marinas.

Además, debe tenerse en cuenta la contaminación radiactiva del mar y exigir a través de los organismos internacionales la prohibición de ensayos atómicos y de la disposición de residuos radiactivos tóxicos en el Océano Pacífico. Este tipo de contaminación podría afectar una de las fuentes más importante de alimentos obtenidos por la pesca y de la cual depende la economía de muchos países litorales del Océano Pacífico.

Section 2

BEHAVIOUR AND FATE OF POLLUTANTS IN THE MARINE ENVIRONMENT

Summary of Discussion

Physical behaviour

Pollutants are transported and distributed in the marine environment by currents, waves and turbulent mixing processes. These will differ for different physical states of the pollutant. At the present time, there is a considerable amount of data available concerning the physical behaviour of pollutants on discharge to the sea. The information essentially refers only to the surface waters. Although the way in which diffusion processes operate are not fully known, the way in which pollutants will behave in the marine environment under the influence of various wind and turbulence conditions away from immediate coastal effects can now be predicted with a certain degree of accuracy. A number of mathematical models have been constructed to enable such predictions to be made. These have been tested by comparison with the observed behaviour of a pollutant in field experiments, using a variety of tracer techniques, e.g. Rhodamine-B and radioactive tracers. These tests indicate that under many conditions, there is good agreement between the predicted and observed dispersion rates in surface waters.

There is, however, a suggestion that under certain conditions the rate of diffusion does not increase steadily with size of the system but that as the diffusion field increases beyond a certain size, there is a marked reduction in the dependence of the rate of diffusion on further increase in size of system. This indicates the dangers in drawing conclusions from small-scale short-term experiments. When the boundaries are grossly irregular, as in coastal waters, the mathematical models require modification to allow for these factors. A number of proposals have been made on ways of overcoming this problem, and one of these is to make use of a hydro-dynamical model in which coastal irregularities are included in the boundary conditions. It is, therefore, important when coastal areas are involved to obtain as much data as possible on the various local conditions such as runoff, topography, locally induced currents, etc.

The Conference considered that the behaviour of a pollutant on discharge from an outfall was a somewhat specialized case which required rather more sophisticated models than those necessary for open sea problems. Although the problems involved do seem to be relatively well understood, since so many factors are involved, e.g. density of plume, size of outlet, etc., prediction remains very difficult. It is possible, nevertheless, with the exception of particularly refractory materials, to design outfalls by a combination of theory and empirical data, which will allow pollution prevention on beaches. In special cases where a thermocline exists, it is possible to design an outfall so that the plume does not rise to the surface. In the special case of a sewage discharge, factors can be introduced into the models to allow for the die-off of bacteria. Thus, predictions can be extended to probable bacterial distribution at some distance from the outfall, e.g. on bathing beaches.

Physical behaviour in combination with chemical and biological behaviour

The chemical behaviour of many pollutants and the way in which they may be altered or taken up by marine organisms is in many cases at least partially understood. Mathematical models can be constructed which represent such behaviour. However, in order to be able to predict the full fate of a pollutant, with a view to protecting the environment, the problem is to try to combine the various models in order to describe and predict the combined physical, chemical and biological behaviour of the pollutant.

Some workers are already beginning to investigate these possibilities. One preliminary approach considers that the biological factor will behave as a particle with a vertical migration pattern. The Conference considered that, although limited at present by the working speed of present computers, it would be possible as bigger and more efficient computers become available to introduce many more factors. Models could be constructed to take account of all the various parameters, but extensive testing of the model with appropriate chemical and biological data would be necessary. Although much information is available, there remains the problem, particularly with the biological data, of selecting the best set of data for testing the models. The Conference concluded that there was a need for collaboration of biologists, chemists and physicists with mathematicians in solving these problems.

Deep-water processes

The discussions indicated that at the present time it is not possible to accurately predict either the physical behaviour or the chemical or biological fate of a pollutant in the deep waters of the ocean. It has to be proved that the present mathematical models for dispersion can be used for the prediction of dispersion in deep waters and how in such areas the constants may have to be changed. Furthermore, it is known from work in the field of marine radioactivity that chemical behaviour is different in deep waters. Knowledge of bottom currents was considered inadequate and demands further investigations.

For technical reasons present tracer methods are not readily applicable to deep waters. There appears to be a complete lack of information regarding deep-sea organisms which might concentrate certain chemicals. It was realized that there would always be a need for waste disposal and that we have to prove carefully the possible capacity of the ocean to accept wastes of different kinds. However, in order to be able to predict the possible consequences and to calculate the amount which could safely be disposed of in deep waters, it is especially necessary to have a better knowledge of the physical, chemical and biological conditions near the bottom of the oceans.

Sedimentation processes

Many pollutants, particularly metals and halogenated hydrocarbons, are readily adsorbed onto suspended particulate material. If the particulate material settles on the bottom, these materials may become locked in the sediments. There is evidence to suggest that into certain cases as much as half of the nutrients discharged in rivers may become bound up in the sediments. Under certain circumstances, these scavenging processes may be considered to be beneficial, since potentially biologically hazardous materials may become inactivated in the sediments.

On the other hand, sedimentation of a pollutant may lead to long-term problems in an area, since the benthic population may be drastically altered. Alternatively, biochemical decomposition may take place, with subsequent release of noxious materials such as sulphides. In certain special instances, e.g. mercury, the pollutant may even become altered to a more biologically hazardous form. When quantities of organic matter decompose on the sea bed, the pH of the sediments is liable to be reduced, although in the absence of oxygen, various processes tend to counteract this.

Many materials may not be permanently bound in a sediment, and recent investigations indicate that up to 75 per cent of certain metals may be released from the upper layers of a sediment. The Conference also noted that with certain substances, e.g. mercury, the suspended material may contain considerably higher proportions than the bottom sediments, as a result of different size of particles and matrix type. On the other hand, some organic materials, e.g. DDT and PCB's, tend to be more firmly bound, although some recycling may be possible. The amount of material in suspension depends very much on the degree of turbulence, and in shallow water areas recycling will be increased in times of storms with even small changes in wind strengths, bringing about considerable resuspension of sedimentary materials. It was clear that knowledge on the role of sediments in the fate of a pollutant is at present inadequate and that more studies will be necessary before the effects of suspended and sedimentary particulates on pollution are fully understood.

Degradation

Many pollutants, including components of crude oil, will be degraded by the action of bacteria yeasts and other microflora either in the water or in the bottom sediments. The rate of degradation is related to the presence of other microflora, additional nutrients, the chemophysical conditions and, in certain instances the presence of other microflora. The Conference noted that in general the rate of bacterial degradation of a pollutant is dependent not only on the bacterial flora present, but also upon the amount of organic substrate available. Thus, in deep waters, bacterial degradation may be very slow owing to limiting amounts of substrate even though bacteria are present. (Bacteria have been found in large numbers even in very deep water, but little is known of their *in situ* activity.) However, even in the presence of adequate concentrations of substrate and bacteria, degradation in deep waters has been shown to be slow. The Conference considered that these low rates of decomposition of organic pollutants would be particularly important in considerations of quantities and materials which can be safely dumped in deep-water areas.

Radioactivity considered

Although not a special concern of the Conference, the field of radioactivity was considered in the way that it could offer valuable experience in both scientific and organizational fields.

The Chairman referred to the activities of IAEA in the field of radioactivity. The IAEA has the statutory mandate to establish or adopt standards of safety for the protection of health and minimization of the danger arising from the peaceful uses of atomic energy. It is establishing a register of radioactive waste disposal to the sea. With the assistance of the Government of Monaco and the Musée Océanographique, the Laboratory for Marine Radioactivity is presently working on a review of current knowledge in the field of radioactive waste disposal in the sea, and the intercalibration and standardization of measuring methods. The coordinated intercalibration programme involves more than 60 laboratories distributed around the world.

The Conference noted that when a radioactive material is to be discharged to the sea studies are made of dispersion before and after discharge. It also noted that different organisms have different rates of uptake for different radioactive nuclides and this led to a discussion on the possible use of indicator species which, because they concentrate a particular chemical, might be useful in the detection of pollutants. This would be particularly useful for low-concentration pollutants, for instance, in the study of the fate of highly toxic compounds.

The fate of oil

The behaviour and fate of oil in the marine environment was considered in some detail. Automatic sensing and possible monitoring methods were discussed. It was recognized that many areas are severely contaminated by oil and the suggestion was made that this should be quantified by analysis, if possible. There is considerable evidence available today to suggest bacterial and other routes of decay. It was noted, however, that certain polynuclear aromatics some of which are recognized as being commonly present in the marine environment, are only degradable under certain conditions. For example, benzpyrene decomposition is dependent particularly on the available light, but also on temperature and oxygen levels. Thus, in deep waters, degradation of such substances would be extremely slow.

The fate of mercury

The fate of mercury in the marine environment is complicated by the fact that not only is it liable to be concentrated by marine organisms but there is considerable evidence of biochemical alteration to the more toxic methyl mercury. Aerobic methylation of mercury is known to occur; and there is now some indication that ionic mercury can be methylated under anaerobic conditions. The uptake of mercury by plants is considered to be largely a surface absorption effect, but fish appear to take up mercury both via their gills and food. There is considerable evidence of cycling of mercury within the bio-geochemical system, and although all mechanisms are not fully understood, various cycling processes have now been suggested.

Enteric organisms

The fate of coliforms and other enteric organisms in the sea was briefly considered. Coliforms are known to have a fairly rapid die-off in sea water and since the rates of die-off have been measured it is possible to use counts of coliforms in order to estimate the rate of dispersion of sewage plumes. The rate of die-off or inactivation of viruses, on the other hand, is a relatively new field of study. The information available to date suggests that viral inactivation processes may be considerably slower than those for coliforms. The rate of inactivation appears to be increased with increase in bacterial numbers. There appears also to be an antagonism between naturally occurring marine viruses and enteric viruses. There may be some measure of chemical inactivation as well. If the sea water is rich in colloidal suspended material, then the virus may be protected by adsorption. Coliphages, which are occasionally suggested as a means of measuring viruses, die more rapidly and fluctuate in numbers in a way which does not always reflect viral numbers.

Concluding points of Convener

The task for the scientists at the present time is to prove the possible capacity of the oceans to accept wastes of different kinds. We have to look for representative parameters, their critical values and for methods of measuring and controlling pollutants in the different oceanic areas. The governments and the competent national and international organizations need our help for preparing regional or global regulations to avoid dangerous pollution of the sea. This required us to make predictions. One of our efforts should therefore be to develop mathematical or other models not only for the physical but also for the biological and chemical processes which influence the behaviour and fate of pollutants in a marine environment. This would facilitate very much our task of prediction especially of dangerous concentrations. It is a difficult but surely not hopeless problem through patient work in laboratories and in the ocean itself to determine with increasing accuracy the necessary physical, biological and chemical information which might be put in a computer programme. This task obviously requires broad and intensive international collaboration. That this is possible has, however, already been demonstrated by the international coordinated research programme in marine radioactivity studies of the IAEA.

Dilution and Dispersion of Pollutants by Physical Processes

H. Weidemann and H. Sendner*

La dilution et la dispersion des polluants par des procédés physiques

Le présent document traite en premier lieu de l'influence des propriétés physiques sur la persistance des polluants et sur les sites où ils se déposent dans la mer. Les matériaux peuvent être solides ou liquides, ils peuvent flotter, être dissous ou submergés. La deuxième partie est consacrée à l'influence du milieu physique sur les polluants. Les facteurs dynamiques les plus importants, c'est-à-dire la turbulence et les courants, provoquent le mélange ou la diffusion par tourbillons et le transport des matériaux. Les auteurs envisagent également le rôle de paramètres statiques tels que la température, la densité et la viscosité.

Les différentes chaînes de transformation énergétique qui dépendent des propriétés physiques du milieu sont examinées, ainsi que leurs effets sur les agents de pollution. Parmi celles-ci figurent le réchauffement et le refroidissement de la température de surface dus au rayonnement, d'où proviennent des courants de convection ou de couches indépendantes, des vagues de surface d'origine éolienne, des courants de dérive et d'autres courants. L'influence éventuelle d'autres sources d'énergie d'origine biologique, chimique ou provenant de procédés techniques artificiels est d'ordinaire négligeable. Enfin, la troisième partie du document est consacrée à l'examen des méthodes et fournit des exemples d'études systématiques de ces techniques, avec des calculs théoriques et pratiques.

La conclusion regroupe les résultats d'ensemble et indique les questions qui doivent faire l'objet de recherches à l'avenir.

Dilución y dispersión de contaminantes mediante procedimientos físicos

La primera parte trata de la influencia de las propiedades físicas en la duración de la permanencia de los contaminantes en el mar y en los lugares en que se depositan. Los materiales pueden ser sólidos o flúidos y hallarse flotando disueltos o sumergidos.

La parte siguiente estudia la influencia del medio ambiente físico sobre los contaminantes. Los factores dinámicos más importantes, turbulencia y corrientes, dan lugar a mezclas o a la difusión y transporte en remolinos. Se estudian también parámetros estáticos tales como la temperatura, densidad y viscosidad.

Se examinan las distintas cadenas de transformación de la energía relacionadas con estas propiedades del ambiente físico y sus consecuencias sobre los contaminantes. Comprenden éstas el calentamiento y enfriamiento de la superficie por la radiación, que ocasionan corrientes de convección y la formación de capas, olas superficiales debidas al viento y corrientes en general. Tiene poca importancia la influencia de otras posibles fuentes de energía tales como las biológicas, químicas o técnicas artificiales. La parte última examina los métodos y da ejemplos de estudios sistemáticos de estos procedimientos, incluyendo mediciones teóricas y prácticas.

Las conclusiones resumen los resultados generales e indican las cuestiones que quedan todavía por investigar en el futuro.

THE behaviour of pollutants in the marine environment is strongly dependent on their nature and their density.

Insoluble substances with lower density will float on the surface. They may be solids such as plastics or wood, or liquids such as mineral oil. If the substance is neutrally buoyant it will float between surface and sea-bed. Many organic particles and colloidal suspensions belong to this group. If the substance is heavier than sea water it will sink with a velocity depending on its density, shape and size. Most inorganic materials and some insoluble liquids are of that kind.

Soluble substances will at first show the same behaviour, but sooner or later, depending on their solubility rate, will be completely dissolved. This also applies to all substances mentioned above. Gases may dissolve completely or partly in sea water; their origin will usually be in the atmosphere, but can also be in chemical or biochemical processes in the sea.

Influence of density and solubility on pollutants

The main difference between insoluble and soluble substances becomes evident by their duration of stay in the sea and location of their deposition. Floating insoluble

---

* Deutsches Hydrographisches Institut, Hamburg, Germany (FRG)

substances will usually sooner or later be deposited on shore from which they may again be removed by waves, surf or currents.

Neutrally buoyant and soluble substances however, may stay in the sea for an almost unlimited period. The best examples are the natural sea salts dissolved over millions of years from the earth. A few natural physical processes may remove small fractions of dissolved substances. Freezing processes in arctic or antarctic regions separate salt from fresh water, but the salt usually remains in the sea water. Separation of sea salts occurs in lagoons when they are cut off from the ocean and if their location is in an arid climate but this process obviously had a greater importance during the early stages of the earth's history.

Dissolved gases, however, are different from other soluble substances because the vapour pressure always balances that of the atmosphere.

Environmental physical properties and processes

Static properties mainly interact with the properties of the pollutants. Temperature and salinity influence the rates of solubility or precipitation of soluble pollutants or of coagulation of dispersed material. The density of sea water is mainly determined by these two factors particularly when the density of the pollutants is very near to that of the water. Viscosity, which depends on temperature and salinity may also have an indirect influence, e.g. by determining the sinking velocity of particles. Surface tension is a very important factor in its interaction with oil films, etc.

The most important physical influences on the distribution of pollutants however, are exerted by motions of any kind.

The spectrum of motions in the sea extends from molecular dimensions to oceanwide current systems. Molecular diffusion may be ignored in this context as it is inefficient compared with the larger scale motions defined as turbulence. The irregular, multi-directional motion of turbulence causes mixing processes very similar to that of molecular diffusion but with much greater efficiency. This effect is usually known as "eddy diffusion" and it differs from molecular diffusion in the variability of its coefficients which are not so clearly defined functions of the temperature as those of the molecular processes.

Distinction between turbulence and currents

Eddy diffusion is responsible for mixing and dilution of pollutants. Transport is only possible by currents which have directional motion but which may also be turbulent. It is sometimes difficult to differentiate between "turbulence" and "current", the difference depending on the time and spatial scales under consideration. An example of this is a tidal current which is uni-directional but has real transport capability only if considered for one half of a tidal cycle. If averaged over many tidal cycles or if considered within a greater current system e.g. the Gulf Stream, the tidal ellipses have the character of turbulence elements. According to Heisenberg and Weizsäcker a continuous spectrum of eddies, up to the dimensions of the ocean itself has to be observed. Thus even the main North Atlantic current system is a single large eddy.

Sources and chains of energy

The direct effect of solar radiation is production of thermal stratification in the near-surface layers of the sea.

This process may change the static properties, but also influences the vertical components of turbulent motions. The thermal layering—in extreme cases forming a sharp thermocline with strong vertical gradients—increases the stability of the density stratification. If vertical components of turbulent motions, e.g. owing to wind, waves, try to transport surface water downwards the potential energy of the stable layering has to be overcome by sacrificing part of the motion energy.

Negative radiation by night or during the fall and winter seasons reduces the stability of the surface layers. When this occurs surface water becomes denser than lower layers, vertical convection starts and initiates additional turbulence, even in the absence of other turbulent motions. If present, however, the vertical mass exchange due to their vertical components is not impeded by the potential energy of the masses. Similar effects can be produced by cold or warm air masses mainly when they meet with meridional components, thus having temperatures differing from those of the sea surface.

Wind waves

Probably the most important and most effective source of energy is the wind which produces wind waves. Because surface waves are not merely periodical displacements of the mean water level but also have deep-reaching orbital motions below the surface, their influence on mixing and dispersion of pollutants is considerable. There is a difference, however, between wind or storm waves and swell from distant wind fields, the latter being much less turbulent than the wind waves. The breaking of waves means the transformation of a great deal of their energy into small-scale turbulence, which rapidly disperses floating pollutants. However, an oil film on the sea surface has a considerable effect in preventing the breaking of waves. The range of the vertically mixed layer owing to wind waves is usually monitored by a homothermal layer above the thermocline.

Ekman currents

The second important wind effect on the surface is wind stress, resulting first in a surface drift and, secondly, by internal friction in a drift current decreasing with depth as theoretically investigated by Ekman. Both of these effects are important for pollutants; the surface drift mainly for oil films and other floating substances which "sail" with the wind with a velocity higher than that of the water. The vertical shear of the Ekman current speed and direction combined with the orbital wave motions may be the origin of very complex mixing patterns for dissolved substances. As Ekman has described, the scheme of drift currents becomes still more complicated in the presence of bottom friction, near coasts, etc. The piling-up of water by onshore-winds or the upwelling effect with offshore-winds results in additional vertical components and thus in increased vertical mass exchange. The mass balance will usually result in a compensating current near the bottom or in intermediate depths directed opposite to the surface current. In this way it is possible for pollutants deposited on the bottom at some distance from the shore to be transported back to the shore.

Other sources of energy are currents set up due to density gradients, e.g. as a result of tidal currents, intrusions of fresh water, salinity differences between different sea regions (Mediterranean and Atlantic). Although density currents are usually clearly directed, typical tidal currents periodically change their speed and direction. The effect of tidal currents should, therefore, be that of a turbulence element without essential net transport; however, most tidal currents, when integrated over longer periods, yield rest currents which may reflect the mean density current in that region or the effect of wind drift during the integration interval. It is to be expected, therefore, that the eddy diffusion coefficients are higher here than in tideless regions, assuming the same weather conditions.

Other energy sources

Compared with the energy sources already mentioned, other sources are of only secondary or local importance. Biochemical or chemical processes may deliver some heat energy on very small scales, and the motion of animals may produce some turbulence, but this will have hardly any measurable effects on the distribution of pollutants. The transport of pollutants via the food chain, especially when actively moving animals are involved, is more important because they combine the ability to accumulate with that of moving.

In special regions there are other energy sources able to produce additional turbulent motions, e.g. volcanic heat in regions like the Red Sea, or thermal pollution from the outlets of coastal power stations. Heavy ship traffic, may also increase the degree of turbulence, but the effects are usually small and restricted.

Investigating physical processes in pollution

The direct measurement of turbulence is very difficult, partly because of the problem of defining limits between turbulence and currents and partly on account of missing sensors. An ideal current meter, capable of recording the whole energy spectrum with uniform precision in speed as well as in direction, would also enable us to derive necessary information on turbulence. Unfortunately, most known instruments contain moving components and, therefore, have systematic errors owing to their inertia. A method applicable in the atmosphere uses hot wires or films cooled according to the intensity of turbulent motion; in water, however, difficulties arise due to generation of vapour bubbles, corrosion and fouling, etc.

Diffusion measurements and investigations

To obtain information on the effect of turbulence on the rate of diffusion it is better to determine the diffusion rate directly. There are many methods for this and literature on the subject is extensive. Most describe research on actual pollution problems in inlets (Bowden and Sharaf El Din, 1966; Gade, 1968; Gyllander, 1966; Munk, 1949), estuaries (Alsaffar, 1966; Francis et al., 1953, Pritchard, 1969), or other coastal regions (Carter and Okubo, 1965; Hulburt, 1968; Kullenberg, 1968; Meersburg, 1970), but there are also many general papers (Okubo, 1962; Ozmidov et al., 1969). The methods fall into three main groups.

Natural Tracers—Examples of natural tracers are the salinity distribution near estuaries (Gade, 1968; Visser,

1968) or in straits between seas of different salinity like the Straits of Gibraltar where the intrusion of the high salinity Mediterranean water can be traced well out into the Atlantic (Defant, 1956; Düing and Schwill, 1967; Wüst, 1935). It is also possible to evaluate turbidity distribution measurements (Joseph, 1958; Voitov, 1964) or distribution of certain biological matter, e.g. bacteria, species of plankton, fish eggs (Aubert et al, 1966; Gunnerson, 1959; Shuval, 1968). It must be remembered that these materials are particulate and are, therefore, subject to those influences already mentioned.

Artificial Tracers—The use of soluble dyes is very common since their distribution and dispersion can be observed visually or photographically (Carter and Okubo, 1965; Joseph et al, 1964). If fluorescent dyes are used, e.g. Fluoresceine (Rosset, 1962) or more commonly Rhodamine B (Carter and Okubo, 1965; Joseph et al, 1964; Meersburg, 1969) the tracer concentration can be measured down to much lower values (1 part in 10). Rhodamine B has several advantages over other tracers, it is non-toxic, it is decomposed by natural processes and its fluorescence lines are not present in natural waters (Pritchard and Carpenter, 1960).

Radioactive tracers have also been used but suffer disadvantages due to the radiation hazard involved in using the large quantities necessary for sea experiments and the selection of isotopes with optimal life and an energy spectrum sufficiently different from that of natural sea salt components (Folsom and Vine, 1957; Guizerix, 1965; Nelepo, 1962; Prospero, 1965). The release of drift cards or similar floating particles (Hanzawa, 1953; Neumann, 1966; Tomczak, 1964) permits the derivation of diffusion coefficients from the frequency distribution of measured distances between single cards (Richardson and Stommel, 1948; Stommel, 1949). This method is particularly suited for oil pollution research.

Diffusion theories

Theoretical considerations on the problem of turbulent diffusion aims at results of more general validity than those obtained from local experiments (Ichiye, 1959; Joseph and Sendner, 1962; Okubo, 1962; Ozmidov, 1958). Practically all authors start with a functional equation known in probability theory as the "Markov Process" or from differential equations derived from it (Okubo, 1962; Schönfeld, 1962). Differences exist in certain assumptions or suppositions, e.g. on homogeneity, isotropy, or in definitions of special parameters like "diffusion constant", diffusion velocity", (Joseph and Sendner, 1958), "energy dissipation parameter" (Ozmidov, 1960a; Okubo, 1962) etc. The resulting formulae are usually applied to horizontal processes only, trying to prove the validity by means of the consistency of the formula with the existing observations and measurements (Okubo, 1962, 1962a, 1970). However, the precision of the measurements is usually not sufficient to make a clear decision between the different formulae which often yield only slightly different results. It seems doubtful that it is appropriate or even possible to set up a universally acceptable formula because the parameters are too complex and moreover, vary with time and circumstances.

The theories are however useful for interpreting and comparing the results of experimental measurements and for making roughly quantitative predictions for pollution cases or accidents. In this connection it is possible to predict situations with "continuous sources" of pollution such as sewage inlets from "point source" experiments with which a single release of tracer material is investigated (Pritchard and Carter, 1965). These experiments represent the situation which arises, e.g. with accidents, the release of toxic or radioactive substances, and with single dumping of soluble wastes.

Current measurements

The dispersion of pollutants is not only determined by the rate of eddy diffusion but also by movement due to currents of different origins. The literature on this subject, however, is at least as extensive as that on diffusion, and papers have therefore only been embodied in the references when they are directly concerned with the pollution problem.

Two principal methods are used for current investigations: the Eulerian with flow measurements at a fixed point, and the Lagrangian tracing the trajectories of water particles. The trajectory method is better suited for pollution studies because the question is always where will the pollutants be carried in a given time interval. Measurements by means of the flow method, using any of the many types of current meters at the inlet point, are not conclusive because the current pattern in nearshore waters is usually complex and irregular; hence a single point measurement represents only its immediate surroundings.

Unfortunately there are only very limited possibilities to practise the trajectory method which is sometimes tedious and expensive. The problem is to get identifiable marks which are transported precisely with the water mass itself such as drift cards. If the cards can be traced for a longer period, e.g. by vessel or aircraft, or by recalculating their way after recovery ashore, good results representative of surface film drifts may be obtained. It becomes much more difficult, however, when the transport of suspended or dissolved material investigated. The best way is, as with diffusion studies, to use natural tracers such as salinity ("Kernschicht"-method) or turbidity, or of dye tracers again, in this case observing their path rather than their dilution. Another method, developed by Swallow, uses neutrally buoyant floats which sink to a predetermined depth according to their density and drift with the current. The floats transmit acoustic signals which enable a surface vessel to trace their route. This at present seems to be the only reliable method of measuring subsurface trajectories. Large drogues connected to a surface float or buoy with a thin wire, give faulty results because of the drag exerted on the wire by the integrated effect of the currents throughout the water column and of the wind wave effect on the surface buoy. This system allows the use of microwave or other wireless ranging methods for locating the buoy whilst with the Swallow method the vessel must maintain contact with the float throughout the experiment.

Summary of results and outlook

From the results of experimental and theoretical work which are of more general application some statements for the diffusion problem and the problem of transportation can be made.

The results of practically all experiments show that the turbulence responsible for the observed diffusion is not isotropic. The dye patches originally nearly circular, become elongated, crescent-shaped or irregular with time. Systematic variations of this behaviour induced by tidal currents, wind influences etc. can be presumed but not yet proved.

In spite of the irregularity of shapes it is possible to compare results of different experiments with each other and with theories, when the areas of the various concentration contours are determined and represented by an equivalent or mean radius (Joseph and Sendner, 1958). Their values usually fit well into calculations and diagrams.

Diagrams which can be drawn using the measured values of concentration and radius are especially useful in obtaining answers to questions such as:

How long a certain critical concentration could be observed?
How long at a certain distance from the centre?
What was the largest area filled with critical concentrations?
How did the maximum concentration at the centre decrease with time? (Okubo, 1962).

From some of these diagrams it is possible to read general results, e.g. the decrease in maximum concentration has a tendency to reach an asymptotic with the third power of time. If experiments last for a longer period with changing weather conditions it is sometimes possible to interpret variations in the fundamental relations with special weather events, etc. In general it appears that at different stages of development different relations apply. Whether this reflects a dependence on the scale of the experiment or simply the time of the experiment cannot yet be stated.

With only a few exceptions (Schuert 1970) all of the experiments have been made in the surface or near surface layers. Since many pollutants are dumped at the bottom, often in deep sea, it is important that reliable measuring techniques be developed for deep sea experiments with a tracer released at the bottom. Even in shallow waters the available techniques do not yet allow a survey with equivalent precision in all three dimensions.

As mentioned earlier it is difficult to repeat and vary the sites, weather situations and season of experiments with a view to obtaining a clearer impression of the influence of diffusion parameters. Methods of simulating turbulence and turbulent diffusion by computer techniques would allow the parameters to vary systematically, compared to the random chances with field experiments. The results would probably lead to a better understanding of the connection between turbulence theory and diffusion.

Since recording current meters have been moored in chains in close vertical and horizontal distances, it is known that the current field in the sea is usually much more complex than was hitherto assumed. Moreover, it has been found that the energy spectrum is spread over a wide band of periods; the interpretation of the energy chain, however, is usually not easy. The calculation of rest currents may suffer from systematic errors owing to the behaviour of the instruments, or the mooring system, etc.

Current measurements by the trajectory method are much less numerous. Sometimes they reveal very strange current patterns, e.g. rest currents in certain depths going in one direction for one or two weeks, then suddenly bending sharply or even reversing for a similarly long period. The correlation of such behaviour with oceanographic or atmospheric events of similar periods has not yet been successful. The results of drift card experiments, however, usually showed good correlations with wind conditions between release and recovery time. Computer methods were used successfully to calculate the most probable effective wind drift factor.

To obtain satisfactory results for pollution transport problems it would be necessary to use closely spaced recording current meters. This is expensive, and the evaluation of the results of such a 3-dimensional network would involve further high costs because computers, plotters etc. have to be used.

The use of trajectory methods is more reliable and from the instrumental point of view, less expensive, but much more tedious and also quite expensive in terms of ship's time. The situation could be improved if techniques were developed which would enable a single ship or even a shore station to trace many floats or markers simultaneously.

A problem which awaits technical solution is the precise measurement of bottom current profiles without disturbing or falsifying the current field too much by the sensors. The solution of this problem would improve the means of investigating diffusion and transport problems with pollutants deposited at the bottom and also for geological research work on sedimentation and migration of sediments.

Conclusion

It should be remembered that the physical processes discussed, especially those responsible for turbulent diffusion, are statistical in their nature. This means that the equations, laws or rules which can be found to describe these processes are only valid on an average of a great number of similar processes. It will generally not be possible to predict single events precisely, e.g. traces of single particles.

References

ALSAFFAR, A M Lateral diffusion in a tidal estuary. *J. geophys. Res.*,
1966 71(24): 5837–41.

ARTHUR, R S Horizontal diffusion from a radioactive source.
1959 University of California, Scripps Institution of Oceanography, 5 p.

AUBERT, M, *et al.* Etude de la dispersion bactérienne d'eaux
1966 résiduaires en mer au moyen de traceurs radioactifs. *Revue int. Océanogr. méd.*, 1: 56–91.

BODVARSSON, G, BERG, J W and MESECAR, R S Vertical temperature
1967 gradient and eddy diffusivity above the ocean floor in an area west of the coast of Oregon. *J. geophys. Res.*, 72(10): 2693.

BOGDANOVA, A K (On the mixing of water mass in the upper and
1967 lower currents in the Bosporus). *In* Dinamika vod i voprosy gidrochimii cernogo morja, Kiev, p. 14.

BOGDANOVA, A K (Intensity variation of the turbulent exchange
1967a with depth and time in the Black Sea). *In* Dinamika vod i voprosy gidrochimii cernogo morja, Kiev, p. 3.

BOGUSLAVSKII, S G and PARANICHEV, G L (Vertical turbulent ex-
1966 change in the area of the Faroe-Shetland-Ridge). *Ekspress Inf. Mosk. gidrofiz. Inst.*, (3): 29.

BOURRET, R and BROIDA, S Turbulent diffusion in the sea. *Bull.*
1960 *Mar. Sci. Gulf Caribb.*, 10(3): 354–66.

BOWDEN, K F Horizontal mixing in the sea due to a shearing
1965 current. *J. Fluid Mech.*, 21(1): 83.

BOWDEN, K F Currents and mixing in the ocean. *In* Chemical
1965a oceanography, edited by J. P. Riley and G. Skirrow, London, Academic Press, vol. 1:43.

BOWDEN, K F Stability effects on turbulent mixing in tidal currents.
1967 Boundary layers and turbulence. *Physics Fluids*, 1967 Suppl.:278.

BOWDEN, K F and SHARAF EL DIN, S H Circulation and mixing
1966 processes in the Liverpool Bay area of Irish Sea. *Geophys. Jl R. astr. Soc.*, 11(3): 279–92.

BROECKER, W S Radioisotopes and the rate of mixing across the
1966 main thermoclines of the ocean. *J. geophys. Res.*, 71(24): 5827–36.

BROOKS, N H Diffusion of sewage effluent in an ocean current. *In*
1960 Waste disposal in the marine environment, edited by E. A. Pearson, New York, Pergamon Press, pp. 247.

BRYANT, G T and SEYER, C T The travel time of radioactive wastes
1958 in natural waters. *Trans. Am. geophys. Un.*, 39(3): 440.

BUBNOV, V A Vertikal'nyi turbulentnyi obmeni transformatsiia
1967 schediyemnomorskikh vodv Atlantischeskom okeane (Vertical turbulent exchange and transformation of the Mediterranean water in the Atlantic Ocean). *Okeanologiia*, 7(4): 586–92.

BURKE, C J Horizontal diffusion of dyes in the ocean, *Oceanogrl*
1946 *Rep. Scripps Instn Oceanogr.*, (2): 18 p.

CARPENTER, J H Tracer from circulation and mixing in natural
1960 waters. *Publs Chesapeake Bay Inst.*, (91): 110. Also issued as *Coll. Repr. Chesapeake Bay Inst.*, (5) (1963).

CARTER, H H The distribution of excess temperature from a heated
1968 discharge in an estuary. *Tech. Rep. Chesapeake Bay Inst.*, (44): 39 p.

CARTER, H H and OKUBO, A A Study of the physical processes of
1965 movement and dispersion in the Cape Kennedy area. *Tech. Rep. Chesapeake Bay Inst.*, (65–2). Also issued as *Rep. U.S. atom. Energy Commn*, (NYO-29731).

CHECHOTILLO, K A (To the calculation of coefficients of horizontal
1960 turbulent diffusion at sea). *Trudy Inst. Okeanol.*, 39: 130.

COSTIN, M, Davis, P and Gerard, R Dye diffusion experiments in
1963 the New York Bight,٦ *Tech. Rep. U.S. atom Energy Commn*, (CU-2-62): 18 p.

DAUBERT, A and LEBRETOW, J C Effet de la houle sinusoidale sur la
1965 diffusion entre liquides de salinités différentes. *Houille blanche*, 20(1): 45.

DEFANT, A Turbulenz und Vermischung im Meer. *Dt. hydrogr. Z.*,
1954 7: 2.

DEFANT, A Die Ausbreitung des Mittelmeerwassers im Nord-
1956 atlantischen Ozean. *Deep-Sea Res.*, 3:465.

DENNER, W W, GREEN, T and SNYDER, W H Large scale oceanic
1968 drogue diffusion. *Nature, Lond.*, 219(5152): 361.

DOWLING, G D and OLSON, F C W Diffusion in oceans and fresh
1964 waters. *Science, N.Y.*, 146(3650): 1492–3.

DÜING, W and SCHWILL, W D Ausbreitung und Vermischung des
1967 salzreichen Wassers aus dem Roten Meer und aus dem Persischen Golf. *Meteor Forsch. Ergebn. (A)*, (3): 44.

FISHER, L J Field dye dispersion studies. *In* Proceedings of the First
1964 U.S. Navy Symposium on Military Oceanography, Washington, p. 281.

FISHER, L J Preliminary results and comparison of dye tracer
1967 studies conducted in harbors, estuaries and coastal waters. *Proc. Conf. ost. Engng*, 10(2): 1481.

FOLSOM, T R and VINE, A C On the tagging of water masses for the
1957 study of physical processes in the oceans. *In* The effects of atomic radiation on oceanography and fisheries. *Publs natn. Res. Coun., Wash.*, (551): 121–32.

FORD, W L The distribution of the Merrimack River effluent in
1947 Ipswich Bay. *Tech. Rep. Woods Hole oceanogr. Instn*, (3): 23 p.

FORD, W L Radiological and salinity relationship in the water at
1949 Bikini Atoll. *Trans. Am. geophys. Un.*, 30(1):46–54.

FRANCIS, J R D, *et al.*, Observations of turbulent mixing processes in
1953 a tidal area. Woods Hole, Oceanography Institute, (53–22): 20 p.

GADE, H Horizontal and vertical exchange and diffusion in the
1968 water masses of the Oslo-fjord. *Helgoländer wiss. Meeresunters.*, 17: 462.

GEZENTSVEI, A N O gorizontal'nom makroturbulentnom obmene v
1961 Chernom more (Large-scale horizontal turbulent exchange in the Black Sea). *Trudy Inst. Okeanol.*, 52: 115–32.

GLINSKY, N T (The influence of internal waves on the vertical
1963 exchange in the sea). *Izv. Akad. Nauk SSSR (Geofiz).*, (10): 1554.

GRAY, E and POCHAPSKY, T E Surface dispersion experiments and
1964 Richardson's diffusion equation. *J. geophys. Res.*, 69(24): 5155.

GUIZERIX, J Etude à l'aide de traceurs radioactifs de la distribution
1965 des eaux usées dans un projet de rejet. *Cah. C.E.R.B.O.M.*, 17: 109 1–23.

GUIZERIX, J Etude des phénomènes de dispersion d'eaux usées
1968 par traceurs radioactifs et colorés. *Revue int. Océanogr. méd.*, 9:135.

GUNNERSON, C G Sewage disposal in Santa Monica Bay. *Trans. Am.*
1959 *Soc. civ. Engnrs*, 124: 823.

GYLLANDER, C Water exchange and diffusion processes in Tvaeren,
1966 a Baltic bay. *In* Disposal of radioactive wastes into seas, oceans and surface waters, Vienna, IAEA, p. 207.

HANSEN, D W Currents and mixing in the Columbia River estuary.
1965 *Ocean Sci.*, 2: 943.

HANZAWA, M On the eddy diffusion of pumices ejected from
1953 Myojin Reef in the southern Sea of Japan. *Oceanogrl Mag.* 4(4): 119–42.

HARREMOËS, P Tracer studies on jet diffusion and stratified disper-
1967 sion. *Adv. wat. Pollut. Res.*, 3(3): 65.

HECHT, A On the turbulent diffusion of the water of the Nile
1964 floods in the Mediterranean Sea. *Bull. Sea Fish. Res. Stn, Haifa*, (36): 24 p.

HELA, I Vertical eddy diffusivity of waters in Baltic Sea. *Geophysica*,
1966 *Helsinki*, 9(3): 219–34.

HERRERA, L E Un experimento sobre difusión turbulenta. *Boln*
1967 *Inst. Oceanogr. Univ. Oriente*, 6(2): 163–85.

HIRANO, T On the dilution area of effluent in the sea. *Spec. Contr.*
1966 *geophys. Inst. Kyoto Univ.*, 6.

HULBURT, E M Stratification and mixing in coastal waters of the
1968 western Gulf of Maine during summer. *J. Fish. Res. Bd Can.*, 25(12): 2609–21.

LAEA Radioactive waste disposal into the sea. Report of the Ad
1961 Hoc Panel convened by the International Atomic Energy Agency under the chairmanship of Henry Brynielsson in 1960. *Safety Ser. int. atom. Energy Ag.*, (5).

ICHIYE, T A note on horizontal diffusion of dye in the ocean. *J.*
1959 *oceanogr. Soc. Japan*, 15 (4): 171–6.

ICHIYE, T On a dye diffusion experiment off Long Island. *Tech.*
1964 *Rep. Off. nav. Res. Lamont geol. Obsns*, (CU-9-74). Also issued as *Tech. Rep. atom. Energy Commn*, (CU-10-64): 19 p.

ICHIYE, T Diffusion experiments in coastal waters using dye
1965 techniques. *In* Symposium on diffusion in oceans and fresh waters, edited by T. Ichiye, Palisades, N.Y., Lamont Geological Observatory of Columbia University, p. 54.

ICHIYE, T Turbulent diffusion of suspended particles near the
1966 ocean bottom. *Deep-Sea Res.*, 13(4):679–85.

ICHIYE, T and OLSON, F C W Über die "neighbour diffusivity" im
1960 Ozean. *Dt. hydrogr. Z.*, 13(1): 13–23.

INOUE, E On the turbulent diffusion phenomena in the ocean. *J.*
1951 *oceanogr. Soc. Japan*, 7 (1): 1–8.

ISAEVA, L S (Horizontal turbulent diffusion in the sea). *Trudy*
1963 *morsk, gidrofiz. Inst.*, 28: 36.

ITO, N On the small-scale horizontal diffusion near the coast. *J.*
1964 *oceanogr. Soc. Japan*, 19(4): 182–6.

ITO, N, FUKUDA, M and TANIGAWA, Y Small-scale horizontal
1966 diffusion near the coast. *In* Disposal of radioactive wastes into the seas, oceans and surface waters, Vienna, IAEA, p. 471.

IVANOFF, A Quelques résultats concernant les propriétés diffusantes
1961 des eaux de mer. *Monogr. int. Un. Geod. Geophys.*, (10): 45.

JORDAAN, J M The contribution of scale-model studies to the
1964 determination of mixing and dispersion characteristics of the surf zone. *C.I.S.R. Res. Rev., Pretoria*, (222): 27.

JOSEPH, J Die Trübungsverhältnisse in der Irminger See im Juni
1958 1955 und ihre Hydrographischen Ursachen. *Ber. dt. wiss. Kommn Meeresforsch.*, 15: 238.

JOSEPH, J and SENDNER, H Über die horizontale Diffusion im Meere.
1958 *Dt. hydrogr. Z.*, 11(2): 49–77.

JOSEPH, J and SENDNER, H On the spectrum of the mean diffusion
1962 velocities in the ocean. *J. geophys. Res.*, 67(8):3201–6.

JOSEPH, J, SENDNER, H and WEIDEMANN, H Untersuchungen über
1964 die horizontale Diffusion in der Nordsee. *Dt. hydrogr. Z.*, 17(2): 57–75.

KAMENKOVICH, V M (On the coefficients of eddy diffusion and
1967 viscosity in large-scale oceanic and atmospheric motions). *Izv. Akad. Nauk SSSR. (Fiz. Atmosf. Okean.)*, 3(12): 1326.

KARABASHEV, G S (A new device for the diffusion measurements in
1966 the sea). *Izv. Akad. Nauk SSSR. Atmosf. Okean.*), 2(5): 548.

KARABASHEV, G S (New data on the admixture diffusion in the
1969 sea). *Izv. Akad. Nauk SSSR. (Fiz. Atmosf. Okean.)*, 5(11): 1191.

KARABASHEV, G S (Some results of the mixture diffusion investi-
1969a gation in the sea). *Okeanologiia*, 9(1):82.

KARABASHEV, G S and OZMIDOV, R V (A study of turbulent diffu-
1965 sion in the sea with the help of fluorescent dye). *Izv. Akad. Nauk SSSR. (Fiz. Atmosf. Okean.)*, 1(11): 1178.

KENT, R E Turbulent diffusion in a sectionally homogeneous
1958 estuary. *Tech. Rep. Chesapeake Bay Inst.*, (16): 86 p.

KETCHUM, B H Hydrographic factor involved in the dispersion of
1950 pollutants introduced into tidal waters. *J. Boston Soc. civ. Engnrs*, 37(3): 296.

KETCHUM, B H and FORD, W F Rate of dispersion in the wake of a
1952 barge at sea. *Trans. Am. geophys. Un.*, 33(5) Pt. 1: 680–4.

KHLOPOV, V V (The variation of the diffusion coefficient according
1958 to Black Sea observations). *Izv. Akad. Nauk SSSR (Geofiz.)*, (2): 129.

KHLOPOV, V V (Variations of the coefficient of vertical turbulent
1969 exchange with depth and time in some regions of the North Atlantic). *Trudy gos. okeanogr. Inst.*, 96: 127.

KITAIGORODSKII, S A (On the theory of turbulent mixing in the
1961 sea and computation of the thickness of the upper homo-geneous layer). *Trudy Inst. Okeanol.*, 52: 87–96.

KITAIGORODSKII, S A and MIROPOLLSKII, Iu Z (On the theory of
1967 turbulent mixing in the upper ocean). *Izv. Akad. Nauk SSSR (Fiz. Atmosf. Okean.)*, 3(11): 1196.

KOLESNIKOV, A G The vertical turbulent exchange under conditions
1959 of stable sea stratification. *In* Preprints of abstracts of papers presented at the International Oceanographic Congress, edited by M. Sears, Washington, D.C., AAAS, pp. 404–8.

KOLESNIKOV, A G, PANTELEV, N A and PISAREV, V D (Direct
1964 determination data on the intensity of turbulent exchange in the depths of the Atlantic Ocean). *Dokl. Akad. Nauk SSSR*, 155(4): 788.

KULLENBERG, G Measurements of horizontal and vertical diffusion
1968 in coastal waters. *Rep. Inst. fys. Oceanogr. Københ. Univ.*, (3): 50 p.

LAIKHTMAN, D L and DORONIN, Iu P (The coefficient of turbulent
1959 exchange in the sea and an estimate of the heat flow from ocean waters). *Trudy arkt. antarkt. nauchno-issled Inst.*, 226: 61–5.

MALONEY, W E and CLINE, C H A study of diffusion in an estuary.
1961 *Proc. Conf. cst. Engng*, 7(2): 536.

MCEWEN, G F A statistical model of instantaneous point and disk
1950 sources with applications to oceanographic observations. *Trans. Am. geophys. Un.*, 31(1): 35.

MEERSBURG, A J Diffusie in Zee. *Zee*, 90(8), 352.
1969

MEERSBURG, A J Een diffusie experiment voor de Nederlandse
1970 kust med behulp von Rhodamin B. *Wet. Rapp. K. ned. met. Inst.*, (1):13 p.

MIYAKE, Y and SARUHASHI, K Distribution of man-made radio-
1958 activity in the North-Pacific through summer 1955. *J. mar. Res.*, 17: 383–9.

MOON, JR, F L, BRETSCHNEIDER, C L and HOOD, D W A method for
1957 measuring eddy diffusion in coastal embayments. *Publs Inst mar. Sci. Univ. Tex.*, 4(2): 14–21. Also issued as *Cont. Oceanogr. Met. agric. mech. Coll. Tex.*, 3(84): 325.

MUNK, W H, EWING, G C and REVELLE, R R Diffusion in Bikini
1949 Lagoon. *Trans. Am. geophys. Un.*, 30(1): 59–66.

NANNITI, T On the Austausch coefficient in the sea. *Oceanogrl Mag.*,
1951 3(1): 49–52.

NELEPO, B A Détermination des coefficients de diffusion turbulente
1962 dans la mer, à l'aide de traceurs radioactifs. CEA Transl No. R1469, unpag. 16 p. Also issued as *Vest. mosk. gos. Univ. (Fiz. Astr.)*, 4: 64.

NEMCHENKO, V I (A study of horizontal turbulent diffusion in the
1964 Atlantic Ocean). *Okeanologiia*, 4(5): 805.

NEUMANN, H Die Beziehung zwischen Wind und Oberflächen-
1966 strömung auf Grund von Triftkartenuntersuchungen. *Dt. hydrogr. Z.*, 19: 253–66.

OKUBO, A Horizontal diffusion from an instantaneous point-source
1962 due to oceanic turbulence. *Tech. Rep. Chesapeake Bay Inst.*, (32): 124.

OKUBO, A A review of theoretical models of turbulent diffusion in
1962a the sea. *Tech. Rep. Chesapeake Bay Inst.*, (30): 113 p.

OKUBO, A A note on horizontal diffusion from an instantaneous
1966 source in a nonuniform flow. *J. oceanogr. Soc. Japan*, 22(1):35–40.

OKUBO, A The effect of shear in an oscillatory current on hori-
1967 zontal diffusion from an instantaneous source. *Int. J. Oceanol. Limnol.*, 1(3):194–204.

OKUBO, A A new set of oceanic diffusion diagrams. *Tech. Rep*
1968 *Chesapeake Bay Inst.*, 38 (Ref. 68-6):35 p.

OKUBO, A Oceanic mixing. *Tech. Rep. Chesapeake Bay Inst.*, 62
1970 (Ref. 70-1): 119.

OKUBO, A and CARTER, H H An extremely simplified model of the
1966 "shear effect" on horizontal mixing in a bounded sea. *J. geophys. Res.*, 71(22): 5267

OKUBO, A and KARWEIT, M J Diffusion from a continuous source
1969 in an uniform shear flow. *Limnol. Oceanogr.*, 14(4): 514–520. Also issued as *Contr. Chesapeake Bay Inst.*, (139).

OKUBO, A, *et al.*, (Report of the observation concerning the diffu-
1957 sion of dye patch in the sea off the coast off Tokai-mura). *Res. Pap. Japan atom. Energy Res. Inst.*, (2): 17.

OLSON, F C W A simple derivation of the noble equation for the
1964   diffusion of a circular dye patch. *J. geophys. Res.*, 69(2): 366.
ORLOB, G T Eddy diffusion in homogeneous turbulence. *J. Hydraul.*
1959   *Div. Am. Soc. civ. Engrs*, 85: 75.
OZMIDOV, R V Eksperimental'noi issledovanie gorlizontal'noi
1957   turbulentnoi diffuzii v more i iskusstvannom vodoeme
       nebol'shoi glublny. (Experimental investigation of hori-
       zontal diffusion in the sea and in shallow artificial reser-
       voir). *Izv. Akad. Nauk SSSR (Geofiz.)*, (6): 756–64.
OZMIDOV, R V (Calculation of horizontal turbulent diffusion of
1958   patchy admixtures in the sea). *Dokl. Akad. Nauk SSSR*,
       120:761–3.
OZMIDOV, R V (On the importance of eddies of different dimen-
1958a sions in the diffusion process). *Izv. Akad. Nauk SSSR*
       (Geofiz.), (2):149.
OZMIDOV, R V (Horizontal turbulent diffusion of the pollutant
1960   patches in the sea). *Trudy Inst. Okeanol.*, 37:164.
OZMIDOV, R V (On the rate of dissipation of turbulent energy in
1960a sea-currents and on the dimensionless universal constant in
       the "4/3-power law"). *Izv. Akad. Nauk SSSR (Geofiz.)*,
       (8):821.
OZMIDOV, R V Statisticheskie kharakteristiki gorizontal'noi
1962   nakroturbulentnosti v Chernon more. (Statistical charac-
       teristics of the Black Sea horizontal large-scale turbulence.).
       Trudy Inst. Okeanol., 60:114–9.
OZMIDOV, R V (On the turbulent exchange in the stable stratified
1965   ocean). *Izv. Akad. Nauk SSSR (Fiz. Atmosf. Okean.)*, (8):
       853.
OZMIDOV, R V Turbulent diffusion of conservative and non-
1966   conservative admixtures. *In* Radioactive contamination of
       the sea, edited by V. I. Baranar and L. M. Khitov, Jerusa-
       lem, IPST cat. no.(1666):161–9.
OZMIDOV, R V Turbulentnosti i turbulentnoe peremeshivanie v
1967   okeane. Kratkara istoniia issledovanii v SSSR za 50 let.
       (Turbulence and turbulent exchange in the ocean). *Okean-
       ologiia*, 7(5):860–7.
OZMIDOV, R V (On the dependence of the lateral turbulent diffu-
1968   sivity Kl in the ocean of the phenomenon scale). *Ivz.
       Akad. Nauk SSSR (Fiz. Atmosf. Okean.)*, 4(11):1224.
OZMIDOV, R V and GEZENTSVEI, A N (New data on the admixture
1969   diffusion in the sea). *Izv. Akad. Nauk SSSR (Fiz. Atmosf.
       Okean.)*, 5(11): 1191.
OZMIDOV, R V, GEZENTSVEI, A N and KARABASHEV, T S Nekotorye
1969   rezul'taty issledovaniia diffuzii primesei v more. (Some
       results of the mixture diffusion investigation in the sea).
       Okeanologiia, 9(1): 82–6.
PREDDY, W S and WEBBER, R The calculation of pollution of the
1963   Thames estuary by a theory of quantized mixing. *Int. J. Air
       Wat. Pollut.*, 7(8): 829.
PRITCHARD, D W The application of existing oceanographic
1960   knowledge to the problem of radioactive waste disposal into
       the sea. *In* Disposal of radioactive wastes into seas, oceans
       and surface waters, Vienna, IAEA, pp. 229–48.
PRITCHARD, D W Dispersion and flushing of pollutants. *In* Report
1965   of the Corps of Engineers U.S. Army Committee on Tidal
       Hydraulics, No. 3 (Chapt.8):1 p.
PRITCHARD, D W Dispersion and flushing of pollutants in
1969   estuaries. *J. Hydraul. Div. Am. Soc. cib. Engnrs*, 95 (HY 1):
       115.
PRITCHARD, D W and CARPENTER, J H Measurements of turbulent
1960   diffusion in estuarine and inshore waters. *Bull. int. Ass.
       scient. Hydrol.*, (20): 37–50.
PRITCHARD, D W and CARTER, H H On the prediction of the
1965   distribution of excess temperature from a heated discharge
       in an estuary. *Tech. Rep. Chesapeake Bay Inst.*, 33(65–1):
       45 p.
PRITCHARD, D W, OKUBO, A and CARTER, H H Observations and
1966   theory of eddy movement and diffusion of an introduced
       tracer-material in the surface layers of the sea. *In* Disposal
       of radioactive wastes into seas, oceans and surface waters,
       Vienna, IAEA, p. 397. Also issued as *Contr. Chesapeake
       Bay Inst.*, (89).
PROSPERO, M The determination of mixing and circulation in the
1965   ocean with radioactive tracers. *Tech. Rep. Inst. mar. Sci.
       Univ. Miami*, (65-1): 13.
REINERT, R L Near surface oceanic diffusion from a continuous
1965   point source. Symposium on diffusion in oceans and fresh
       waters 1964. *Contr. geophys. Sci. Lab., New York Univ.*, (32).
RICHARDSON, L F and STOMMEL, H Note on eddy diffusion in the
1948   sea. *J. met.*, 5: 238.
ROCHFORD, D J Mixing trajectories of intermediate depth waters of
1963   the south-east Indian Ocean as determined by a salinity
       frequency method. *Aust. J. mar. freshwat. Res.*, 14(1): 1–23.
ROSSET, M Expérience de dispersion d'un effluent dans la mer.
1962   *Cah. océanogr.*, 14(2): 103–19.

ROSSOV, V V (The magnitude problem of the coefficient of vertical
1967   turbulent exchange for dissolved oxygen in the sea.) *In*
       Novje modeli i rezul'taty rasceta tecenij v baroklinom
       okeane, A.S. Sarkijan, Sevastopol, p. 56.
SAINT-GUILY, B On vertical heat convection and diffusion in the
1963   South-Atlantic. *Dt. hydrogr. Z.*, 16: 263.
SAINT-GUILY, B Diffusion thermique turbulente au niveau de la
1965   pycnoline. *Rapp. P.-v. Reun. Commn int. Explor. scient. Mer
       Mediterr.*, 18(31):865.
SAINT-GUILY, B Diffusion verticale au niveau de l'eau intermédiaire.
1966   *Bull Inst. océonogr., Monaco*, 66(1367): 12p.
SAINT-GUILY, B Diffusion verticale dans les eaux superficielles de
1968   l'ouest de la Méditerranée et du sud de la Mer Rouge. *Vie
       Milieu (B)*, (2-B):225.
SCHÖNFELD, J C Integral diffusivity. *J. geophys. Res.*, 67(8):3187.
1962
SCHOTT, F and EHRHARDT, M On fluctuations and mean relations
1970   of chemical parameters in the northern North Sea. *Kieler
       Meeresforsch.*, 26(2): 272.
SCHUERT, E A Turbulent diffusion in the intermediate waters of the
1970   North Pacific Ocean. *J. geophys. Res.*, 75(3): 673.
SHAPKINA, V F (Turbulent water mixing in presence of tidal
1964   currents). *Izv. Akad. Nauk SSSR (Geofiz.)*, (8): 1223.
SHUVAL, H J, COHEN, N and YOSHPE PURER, Y The dispersion of
1968   bacterial pollution along the Tel-Aviv Shore. *Revue int.
       Océanogr. méd.*, 9: 107.
STERN, M E Lateral mixing of water masses. *Deep-Sea Res.*, 14(6):
1967   747.
STOMMEL, H Horizontal diffusion due to oceanic turbulence. *J. mar.*
1949   *Res.*, 8(3): 199–225.
STOMMEL, H Computation of pollution in a vertically mixed
1953   estuary. *Sewage ind. Wastes*, 25(9):1065.
SVERDRUP, H U Speculations on horizontal diffusion in the ocean.
1946   *Oceanogrl Rep. Scripps Instn Hydrog. Off.*, (1) Project No.
       4901: 3 p. (Prelim. rep.).
TOMCZAK, G Investigations with drift cards to determine the
1964   influence of the wind on surface currents. *In* Studies in
       oceanography, by Hidake Jubilee Committee, Tokyo,
       Geophysical Institute, University of Tokyo, pp. 129–39.
VAN DAM, G C and DAVIS, J A S Radioactive waste disposal and
1966   investigations on turbulent diffusion in the Netherlands
       coastal areas. *In* Disposal of radioactive wastes into seas,
       oceans and surface waters, Vienna, IAEA, p. 233.
VERWEY, C J and MCMURRAY, W R Tracers for the study of mixing
1964   in the surf. *C.I.S.R. Res. Rev., Pretoria*, (222): 45.
VISSER, M P Note on the estimation of eddy diffusivity from salinity
1966   and current observations. *Neth. J. Sea Res.*, 3(1): 21–7.
VLADIMIRTSEV, IU A K voprosu ob izuchenii protsessov peremeshi-
1961   vaniia v Chernom more. (On the problem of studying mixing
       processes in the Black Sea). *Okeanologiia*, 1(6): 976–82.
       Also issued as *Deep-Sea Res.*, 10(5): 659 (1963).
VOITOV, V I (The optical characteristics of water masses regarded
1964   as indices of turbulent mixing processes at sea.) *Okeano-
       logiia*, 4(3): 386.
WALDICHUK, M Vertical diffusion in the sea from a radioactive
1963   point source. *Manuscr. Rep. Fish. Res. Bd Can.*, (155).
WALDICHUK, M Dispersion of kraft-mill effluent from a submarine
1964   diffuser in Stuart Channel in British Columbia. *J. Fish. Res.
       Bd Can.*, 21(5):1289.
WEICHART, G Berechnung der Vertikaldiffusion von natürlichen
1966   Stoffen und Abfallstoffen in der Iberischen Tiefsee-Ebene
       aus der vertikalen Konzentrationsverteilung der natür-
       lichen Stoffe über dem Meeresboden. *Dt. hydrogr. Z.*,
       19(6): 266–84.
WEIDEMANN, H Untersuchungen des DHI zum Problem der
1968   Vermischung und Ausbreitung gelöster Fremdstoffe im
       Meer. *Fischerblatt*, 16(12): 294.
WÜST, G Die Stratosphäre des Atlantischen Ozeans. *Wiss. Ergebu-
1935   dt. atlant. Exped. "Meteor"* 1925–27, 6(1).
WYRTKI, K and BENETT, E B Vertical eddy viscosity in the Pacific
1963   equatorial undercurrent. *Deep-Sea Res.*, 10(4): 449.
ZATS, V I (On the problem of horizontal turbulence of diffusion in
1964   the near-coastal-zone of the Black Sea). *Okeanologiia*,
       4(2): 249.
ZATS, V I (Estimation of horizontal turbulent exchange according
1967   to current speed observations in the Black Sea). *In* Dinamika
       vod i voprosy gidrochimii cernogo morja, Kiev, p. 26.
ZATS, V I (Experimental investigations of horizontal turbulent
1967a diffusion in the near-coastal zone of the Black Sea.) *In*
       Gidrofiziceskie i gidrochimiceskie issledovanija v cernom
       more, A.G. Kolesnikov, Moskva, p. 24.
ZATS, V I (Time dependence of the diffusion of due admixtures at
1967b the sea surface.) *In* Dinamika vod i voprosy gidrochimii
       cernogo morja, Kiev, p. 34.

The Influence of Tides, Waves and other Factors on Diffusion Rates in Marine and Coastal Situations

*J. W. Talbot**

Influence des marées, des vagues et d'autres facteurs sur les taux de diffusion dans les milieux marins et côtiers

L'auteur décrit la mesure des taux de diffusion dans les milieux marins et côtiers, qui sont ensuite comparés. Chaque enquête côtière a porté sur un problème donné de dispersion des polluants et ses effets éventuels sur les pêches locales. Pour mesurer les taux de diffusion, on a utilisé un indicateur colorant à base de Rhodamine-B et, dans la plupart des cas, un dispositif de courantomètres sur amarres. Du point de vue géographique, les milieux étudiés variaient depuis la mer du Nord au grand large, à plus de 100 milles de la côte la plus proche, jusqu'à 1 mille environ du rivage. La gamme des états du vent et de la mer variait également entre vent fort et grosses lames.

L'auteur examine les résultats de la diffusion à la lumière des taux de turbulence indiqués par l'analyse spectrale des courantogrammes et en fonction de l'influence relative des marées, de la houle et d'autres facteurs sur les taux de diffusion.

Efecto de las mareas, olas y otros factores en la velocidad de difusión en situaciones marítimas y costeras dadas

Se describen y comparan las medidas de las velocidades de difusión en situaciones dadas de alta mar y cerca de la costa. Cada uno de los estudios costeros estaba relacionado con un problema determinado de dispersión de contaminantes y de su posible efecto en la pesca local. Las medidas de las velocidades de difusión se efectuaron con el colorante Rodamina-B y en casi todos los casos también se empleó un grupo de correntómetros registradores fondeados. Geográficamente, los lugares estudiados se encuentran en puntos del mar del Norte, a más de 100 millas de la tierra más próxima, y en otros a menos de una de la costa. Las determinaciones se hicieron en diversos estados del viento y el mar, hasta temporales y mar gruesa.

Los resultados de la difusión se examinan teniendo en cuenta la intensidad de la turbulencia indicada por el análisis espectral de los registros de los correntómetros y el efecto relativo de la marea, la marejada y otros factores que influyen en la velocidad de difusión.

A GROWING public concern for protecting the environment is reflected in increasing attention to examining proposals for discharging effluent to the sea. In examining such proposals a quantitative assessment of the contribution of diffusion to dispersal is usually important. This involves using a diffusion equation containing a parameter representing the local rate of diffusion. The most fundamental of the parameters so used is the coefficient of eddy diffusion (K) which, for the one dimensional case, may be defined by the equation

$$\frac{\delta S}{\delta t} = \frac{\delta}{\delta x}\left(K\frac{\delta S}{\delta x}\right)$$

where S represents the concentration of pollutant in the water,

x is distance along the axis of the distribution of pollutant,

and t is time.

Many workers have used equations involving one or more eddy diffusion coefficients and it has been recognized that a very wide range of values may be associated with this parameter. The assumption that there is an inertial subrange in the spectrum of turbulence has been shown by Kolmogoroff (1941) and Batchelor (1950) to lead to an expression for the eddy diffusion coefficient of the form

$$K = A\,G^{\frac{1}{3}}\,L^{\frac{4}{3}} \qquad (1)$$

where A is a constant,

G is the average rate of energy dissipation per unit mass,

and L is a measure of the mean size (scale) of the eddies participating in the diffusion process.

The energy dissipation parameter G will depend on the level of turbulence and will therefore be affected by such factors as local velocity shear, surface wind and wave action. The size of the diffusing system is also important because it determines the range of eddy sizes that contribute to diffusion. Bigger eddies increase diffusion and the appropriate value of the diffusion coefficient may increase. This dependence of the diffusion coefficient on size may be of considerable practical importance, but there is no general agreement on the

mathematical relationship involved for diffusion in the sea.

In addition to the direct action of turbulence represented in Eq. (1) it has been shown that a relatively steady velocity shear may have considerable effect on diffusion (Bowden, 1965; Okubo, 1967). Such a steady velocity shear is often associated with tidal flow but surface wind could have a similar effect.

PRACTICAL MEASUREMENTS OF DIFFUSION RATES

In this paper the results of a number of measurements of rates of diffusion by the Fisheries Laboratory, Lowestoft, are reviewed with particular reference to the relevant physical conditions. The situations covered range from a sheltered estuary—the Swale in Kent—to the open North Sea, and ranged from flat calm to a full gale.

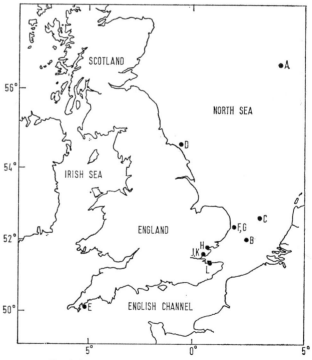

Fig 1. Positions of the diffusion experiments

*Ministry of Agriculture, Fisheries and Food, Fisheries Laboratory, Lowestoft, Suffolk, England

TABLE 1. DETAILS OF DIFFUSION EXPERIMENTS

Location	Reference letter	Time	Duration	Approximate sea state
Central North Sea	A	Aug.–Sept. 1965	23 days	1–7
Central southern North Sea	B	Jan.–Feb. 1969	9 days	4–6
Central southern North Sea	C	Sept.–Oct. 1969	3½ days	0–1
Yorkshire coast	D	Jan.–Feb. 1970	1¼ days	4
South Cornish coast	E	Feb. 1970	1¼ days	3–4
Suffolk coast	F	Feb. 1968	7 hours	2
Suffolk coast	G	Mar. 1968	7 hours	4
Blackwater Estuary	H	June 1968	8 days	2–3
Crouch Estuary	J	June 1969	1¼ days	—
Roach Estuary	K	June 1969	1¼ days	—
Swale Estuary	L	Sept. 1969	2 days	—

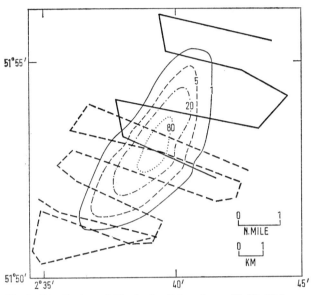

Fig 2. Distribution of Rhodamine-B (as g/cc × 10⁻¹¹) 68 hours after a release in the southern North Sea (experiment B)

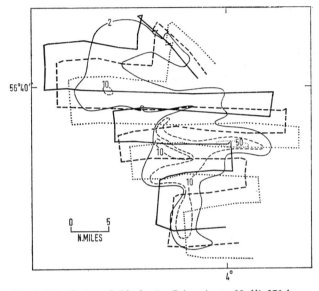

Fig 4. Distribution of Rhodamine-B (as g/cc × 10⁻¹¹) 270 hours after a release in the central North Sea (experiment A)

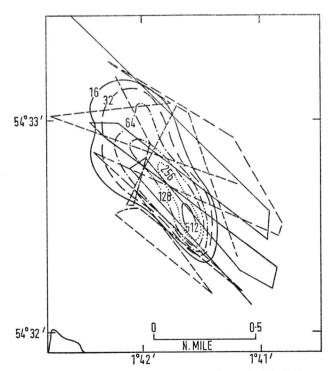

Fig 3. Distribution of Rhodamine-B (as g/cc × 10⁻¹¹) 6½ hours after a release off the Yorkshire coast (experiment D)

Diffusion measurements involved using dye tracer Rhodamine-B and in most experiments water velocities were measured, either by direct-reading current meters suspended from the research ship or by an array of moored recording current meters (see fig 1 and Table 1).

Experiment A was an international experiment under the auspices of ICES; it was designed to provide information on large-scale diffusion needed to predict the consequences of an accidental release of pollutant after a mishap. Experiments B and C were smaller than A but also provide information concerning open sea diffusion but in addition they were intended to provide data for plaice larvae studies.

Experiments D and E were carried out when investigating the dispersal of effluent from potash extraction and china clay works respectively.

Experiments F, G and L aimed to investigate dispersal of sewage from proposed outfalls off the Suffolk coast and in the Swale estuary.

Experiment H was part of a programme of investigations into the effect of the discharge from the Bradwell nuclear power station on the Blackwater estuary.

Experiments J and K were concerned with studies on the dispersal of oyster larvae, but their results also relate to dispersal of effluent in narrow unstratified estuaries.

[123]

Results on diffusion

A full account of these experiments cannot be given here, but the results relating to the influence of physical conditions on diffusion rates will be discussed.

The results fall under two main headings: the effect of the local spectrum of turbulence, and the effect of steady velocity shear, although these two effects are not entirely unrelated. Wave action is also briefly discussed.

It follows from the concept of the spectrum of eddies contributing to turbulence that the effect on a particular pollutant distribution will depend on the size of the eddies relative to the distribution. Those eddies that are appreciably smaller than a distribution will produce the spreading of the distribution that we call diffusion. Large eddies, on the other hand, will produce lateral motion of the distribution as a whole, and eddies of similar size to the distribution itself will produce changes in shape which may however result in an increased effectiveness of the small eddies in the diffusion process. It is therefore the medium- and small-scale eddies which are most important in diffusion, and in the following discussion the effect of large eddies will be ignored.

Distributions measured after approximately point releases of Rhodamine-B are illustrated in figs 2, 3, 4 and 5. The first two figures are based on experiments B and D respectively. In both cases the regular shape of the dye patch suggests that turbulence in the areas was predominantly of small scale. These results contrast with those shown in figs 4 and 5 which demonstrate appreciable distortion of the overall shape of the distributions, suggesting the existence of irregular water movements whose size was similar to the distributions. Figure 4 illustrates a dye distribution 270 hours after a release in the central North Sea and fig 5 is based on measurements following a release close to the south Cornish coast. The latter distribution divided into two parts a few hours after the measurements illustrated were obtained. In such cases it is difficult to make exact predictions of diffusion rates, since the action of the water movements in distorting the shape of the distribution will produce relatively greater areas where small-scale turbulence will act on significant local concentration gradients and may appreciably increase the rate of dispersal.

In fact, the value of the diffusion coefficient estimated for the Cornish coast survey was somewhat greater than that deduced for the Yorkshire coast, although in the latter area higher velocities of water movement were measured. It seems that in situations where moderate-scale irregularities in water movement are known to occur, comparatively high diffusion rates are to be expected, but direct experimental measurement of these may be necessary. The shape of a coastline may give a general indication that such an effect will occur, and there may be a significant increase in diffusion rates in those areas where there are gross irregularities such as are found on the Cornish coast. Similar results may be expected in an area where the circulation is complicated by the confluence of major current systems or when storms have produced irregularities of appropriate size in the pattern of water movement.

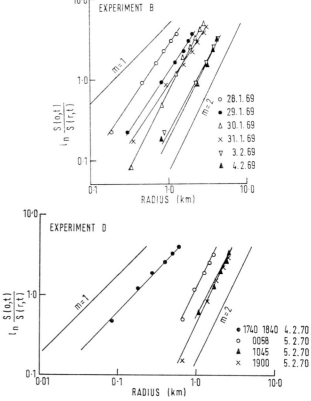

Fig 6. Graphs illustrating the diffusion behaviour during experiments B and D

As a distribution increases in size so does the range of eddy sizes that are smaller than the distribution, and this produces the well-known tendency for the diffusion coefficient to increase. The diffusion of a distribution in space and time should therefore depend on the distribution of turbulent energy as a function of eddy size, the turbulence spectrum. Ozmidov (1965) has remarked that in the open sea there are a restricted number of wave number regions through which energy may enter the marine system and suggests that these correspond to linear scales of the order 1,000 km, 10 km and 10 m. In estuarine and coastal regions tidal flow over a rough bottom or past a coastline may provide an important contribution to local turbulence. Such local contributions will normally be of comparatively small scale and in these areas it is also to be expected that proximity to the coast will limit the formation of large-scale turbulence. Hence in many

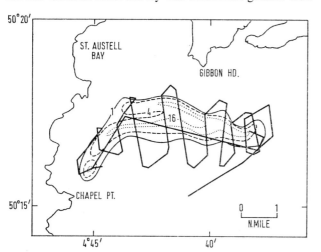

Fig 5. Distribution of Rhodamine-B (as g/cc × 10^{-11}) 20 hours after a release off the south Cornish coast (experiment E)

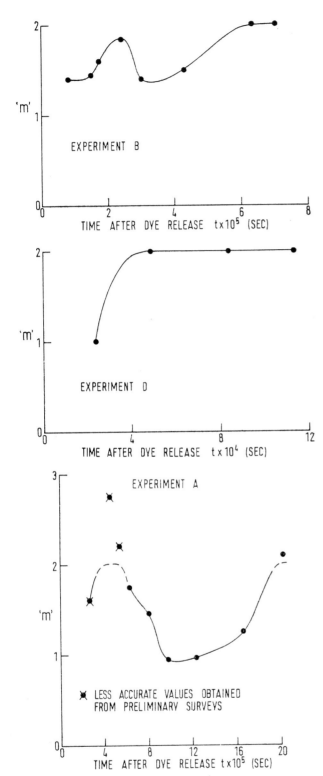

Fig 7. *Variation of the parameter "m" during experiments B, D and A*

some persistent increase in the diffusion coefficient with scale or time.

The general diffusion equation for radially symmetric diffusion in two dimensions may be written

$$\frac{\delta S}{\delta t} = \frac{1}{r} \frac{\delta}{\delta r} \left(K r \frac{\delta S}{\delta r} \right) \tag{2}$$

where r is distance from the centre of the distribution.

Okubo (1962) reviewed the solutions of this equation proposed up to that time and pointed out that if a dependence of the diffusion coefficient (K) on scale (r) and time (t) is assumed of the form

$$K = a \, r^{2-m} \, t^n \tag{3}$$

where a is a constant, and m and n are constants such that $0 < m \leqslant 2$ and $0 \leqslant n \leqslant 2$
then a solution of (2) is

$$S(r,t) = \frac{mM}{2\pi \, m^{4/m} \, \Gamma\left(\frac{2}{m}\right) a^{2/m} \{\psi(t)\}^{2/m}} e^{-\frac{r^m}{m^2 \, a \Psi(t)}} \tag{4}$$

where $\psi(t) = \dfrac{t^{(n+1)}}{n+1}$, $M =$ mass of material/unit depth, and $S(r,t) =$ concentration of material in the water.

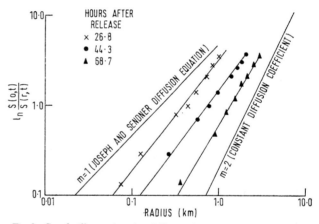

Fig 8. *Graph illustrating the diffusion behaviour during a North Sea experiment reported by Joseph, Sendner and Weidemann (1964).*

Equation (4) may, for the present purpose, be simplified to

$$S(r,t) = S(o,t)e^{-br^m} \tag{5}$$

where $S(o,t)$ represents the maximum concentration at time t in a diffusing system,

and $b = \dfrac{1}{m^2 a \psi(t)}$ is constant at a particular time.

Variation of diffusion coefficient with scale is dependent on m which may be determined by rewriting (5) in the form

$$1_n \frac{S(o,t)}{S(r,t)} = br^m \tag{6}$$

Values of m less than 2 indicate a tendency for the diffusion coefficient to increase with scale according to Eq. (3), and m is equal to 1 in the case of the Joseph and Sendner (1958) equation where the diffusion coefficient is directly proportional to r. Graphs of $1_n \dfrac{S(o,t)}{S(r,t)}$ against r for the results of experiments B and D are shown in

coastal regions turbulence will be of predominantly small scale, with the practical consequence for pollution dispersal that a diffusing system may be expected to reach a critical size beyond which little further increase in value of the diffusion coefficient is to be expected. As a result measurements made on patches smaller than the critical size could be quite misleading if used to predict diffusion at larger scales, since all solutions of the diffusion equation except the Fickian equation assume

[125]

fig 6; the graphs are approximately linear, confirming that these results may reasonably be represented by eqn. 4. It is also notable that in these cases there is a general tendency for the slope (m) to increase as the patch ages and therefore becomes larger. This feature is illustrated in fig 7, where m is plotted against time for experiments B and D and for experiment A. Diffusion during the latter experiment was considered to have been influenced appreciably by turbulence of moderate scale (Sendner and Talbot, in press); this work was done during a period of increasingly stormy weather which probably produced changes in the diffusion behaviour, represented by the value of m falling as low as 1. Some 80 hours after the dye release in experiment B a period of gales also developed and these are thought to have contributed to the minimum of m exhibited in fig 7. However, in this experiment the decrease in the value of m was appreciably less than in experiment A and the gales appear to have produced little distortion of the shape of the patch.

The general variation of the parameter m discussed above suggests that at first the diffusion coefficient increased as the distribution spread over a larger area but that a critical size was reached beyond which further increase in diffusion rate was small, at least for the range of sizes measured. Clearly, where there is appreciable variation of diffusion rate with size it is important that the diffusion coefficient behaviour should be known for distributions covering a sufficient range of dimensions. Okubo and Pritchard (1969) comment that although the Joseph and Sendner (1958) equation, which corresponds to a linear increase of diffusion coefficient with size, is widely applied in practical problems, a Gaussian distribution ($m = 2$) often gives a better fit with experimental observations. It is interesting to compare the results obtained by Joseph, Sendner and Weidemann (1964) with these ideas; thus fig 8 shows results reported by these authors in the same form as those described above.

The tendency for the slope of the graph to approach $m = 2$ is apparent, again suggesting that the diffusion coefficient became less dependent on scale as the diffusing system increased in size. This behaviour is summarized in fig 9 which shows the relation estimated for diffusion coefficient and scale, represented by radius of the distribution, for experiments B and D.

The approach to conditions corresponding to an almost constant diffusion coefficient with regard to radius of a distribution does not discount variation of the diffusion coefficient with time. Okubo and Pritchard's equation (Pritchard 1960) was in fact of this type. However, whilst the spectrum of eddy turbulence provides an acceptable explanation for variation of diffusion rate with radius, it is less easy to see why a persistent variation with time should take place. The Okubo-Pritchard equation corresponds to a diffusion coefficient varying linearly with time but always independent of position. Physically there would appear to be no direct justification for this assumption, and the use of such an equation may be determined largely by mathematical convenience. Thus, if, in some situations the diffusion coefficient does tend to a limiting value, it would be better to determine this value rather than use an equation which may overestimate diffusion rates at large radii or times.

Importance of the level of turbulence

In addition to the consequences of the distribution of turbulent energy with respect to size, the absolute level of turbulence, represented by the parameter G in Eq. (1), will have a considerable effect on diffusion rates. An estimate of the level of turbulent water movement may be obtained from current meter records. Where water velocity can be represented by a steady, or relatively slowly varying, component, combined with random fluctuations about this, it is convenient to write

$$U = \overline{U} + u$$

where the actual velocity is represented by U, $\overline{U}$ represents the mean (or slowly varying component) and u the fluctuating component.

Following this terminology the level of turbulence may be denoted $\overline{u^2}$. On dimensional grounds the diffusion coefficient K can then be written

$$K = A\sqrt{\overline{u^2}} \times L \qquad (7)$$

where L has the dimensions of length and should give a measure of the size of the eddies contributing to $\overline{u^2}$.

Equations 1 and 7 are equivalent if $\overline{u^2} = G^{2/3}L^{2/3}$, but Eq. (7), does not restrict $\overline{u^2}$ to such a dependence on L. Practical application of Eq. (7) requires evaluation of the contribution to $\overline{u^2}$ arising from eddies of the appropriate size. It is a simple matter to resolve the variance $\overline{u^2}$ into components as a function of periodic time t by normal spectral analysis techniques, but the evaluation of a spectrum as a function of a linear dimension representing eddy size is very much more difficult. Analysis of cospectra produced by an array of current meters may lead to information of this type and such analyses are being carried out, but obtaining this kind of data is rather time-consuming and in this paper the problem is approached more simply by examining the results obtained from a single meter record. Some theoretical rigour is lost but there is a considerable gain in practical convenience, since it may often be possible to use one meter where an array would be impracticable.

The relation between diffusion coefficient and the level of turbulence will be most clearly revealed where other factors have a minimum influence on the diffusion rate. This is likely to be the case in situations where diffusion depends chiefly on the action of small-scale turbulence. Two experiments where this appeared to be so were experiments B, in January 1969 in the southern North Sea, and D, in January 1970 in the Yorkshire coastal area. The spectra for one current meter in each of these two cases are shown in fig 10 A and B respectively. These spectra were produced by an analysis carried out in two parts; the lower frequency part was based on the whole length of record available ($\sim$ 250 hours), but the higher frequencies were investigated using a length of only 30 hours' duration obtained during the respective dye release experiments. This procedure allows investigation of the high frequency part of the spectrum for various sections of the original meter record with acceptable resolution. The low frequency parts of these spectra show the main tidal component and harmonics of this but no significant structure is evident at higher frequencies. In fig 10 A the higher frequency spectrum (up to 6 cycles per hour) is shown for two periods during the diffusion experiment,

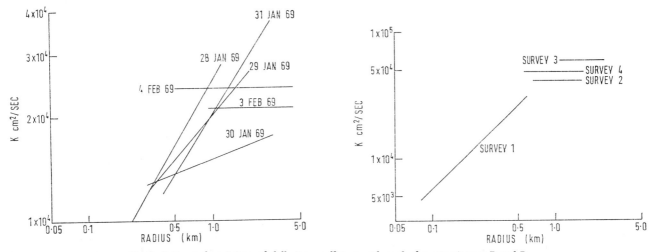

Fig 9. *Estimated variation of diffusion coefficient with scale for experiments B and D*

before and after the storm which is thought to have influenced the diffusion process; the spectra show a considerable increase of the higher frequency components in the latter period. Spectra of current meter records obtained during experiment A also showed increased levels during a period of severe gales (Talbot and Talbot, 1969).

These spectra may be compared with an analysis of a record obtained during experiment C in October 1969 (fig 10 C). During this period the sea was particularly calm and although the low frequency spectrum is similar to that obtained during the rougher conditions of experiment B the spectral energies at higher frequencies

are lower. The results so far obtained are insufficient to deduce a quantitative relationship between spectral energies and diffusion rates, and further records are being examined.

Relation between mean velocity shear and diffusion coefficient

The results discussed suggest that in some cases much of the energy of water turbulence will be derived from local instabilities of the mean flow past a local boundary, especially in coastal or estuarine situations. When considering flow in a channel Bowden (1963) pointed out that if the bottom stress is assumed to be propor-

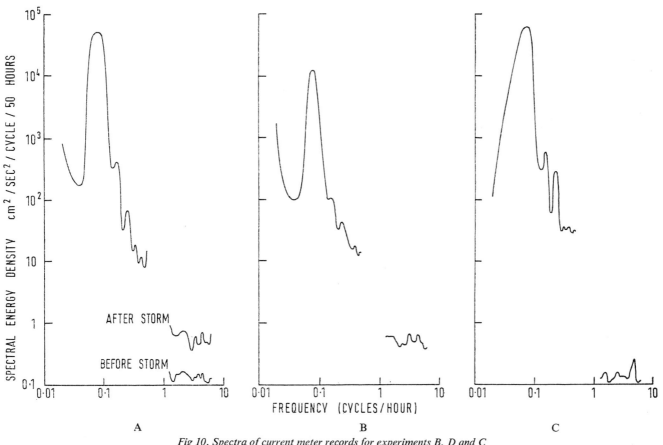

Fig 10. *Spectra of current meter records for experiments B, D and C*

tional to the square of the depth mean velocity (U_D) then the diffusion coefficient along the direction of flow (K_x) is given by

$$K_x = 0.295\, U_D H \qquad (8)$$

where H is the depth of water.

Further consideration of diffusion in vertical shear flow (Bowden, 1965) led to the equation

$$K_x = \frac{U_0^2 H^2}{{}_M K_z}\, \phi\!\left(\frac{z}{H}\right) \qquad (9)$$

where ${}_M K_z$ is the maximum value of the coefficient of vertical diffusion, U_0 is the velocity at the water surface, z is distance from the bottom, and $\phi\!\left(\dfrac{z}{H}\right)$ depends on the vertical distribution of K_z and velocity, U.

Equation 9 demonstrates the importance of vertical mixing in contributing to horizontal diffusion. In practice the magnitude of this shear diffusion effect is such that in many cases it will be the dominant mechanism for diffusion. In deriving Eq. (9) and introducing the expression $\phi(z/H)$ allowance was made for a variety of vertical profiles of K_z and U, but the derivation was restricted to the condition that the horizontal concentration gradient was independent of z, a condition that is unlikely to apply near the centre of a diffusing system, particularly shortly after a small-volume release. This condition is more likely to be valid at appreciable distances from the centre of a distribution and in the important practical case of the diffusion along the length of an unstratified estuary.

Bowden also considered the effect of an oscillatory current on vertical shear diffusion and showed that where vertical mixing is essentially complete within the period of the oscillation there will be a similar contribution to horizontal diffusion but of approximately half the magnitude given by Eq. (9). Okubo (1967) reached similar conclusions after considering the problem from a different point of view.

Since in Eq. (9) the coefficient of horizontal diffusion is inversely proportional to the coefficient of vertical diffusion the vertical shear diffusion effect may be expected to be most evident where vertical stability or lack of turbulence result in a low degree of vertical mixing. The shelter of an estuary may favour such conditions, or calm weather on the open sea. Experiments C and F illustrate the effect and fig 11 shows a typical distribution measured $4\frac{1}{2}$ hours after a point release less than a mile off the Suffolk coast.

Okubo's (1967) treatment of vertical shear diffusion was less general than Bowden's in that it was restricted to a linear velocity distribution in the vertical direction. The vertical current profile is of some importance in vertical shear diffusion, and a number of measurements in Dutch estuaries and the Straits of Dover by van Veen (1938) have suggested a distribution of the form

$$U = U_0 \left(\frac{z}{H}\right)^{\beta}$$

where $\beta \simeq \dfrac{1}{5.2}$.

Bowden showed that for a van Veen velocity profile the vertical distribution of K_z was given by an expression which may be written

$$K_z = 8.72 \left(\frac{z}{H}\right)^{\frac{21}{26}} \left(1 - \frac{z}{H}\right) U_D H 10^{-3} \qquad (10)$$

where U_D is the depth mean velocity.

In this case Eq. (9) reduces to

$$K_x = 1.11\, U_D H \text{ for a steady flow} \qquad (11)$$

$$\text{and } K_x = 0.56\, U_M H \text{ for an oscillatory flow} \qquad (12)$$

where U_M is the maximum depth mean tidal velocity.

However, Bowden found that measured values of K_x were an order of magnitude or more greater than those given by this expression for the Severn, Thames and Mersey estuaries, and was only partly successful in explaining this difference as the result of vertical stability of the water column. A number of the diffusion experiments mentioned in this paper were in situations where validity of Eq. (12) might have been expected. The values of the horizontal diffusion coefficient derived for these cases, together with corresponding values of the depth and water velocity, are given in Table 2.

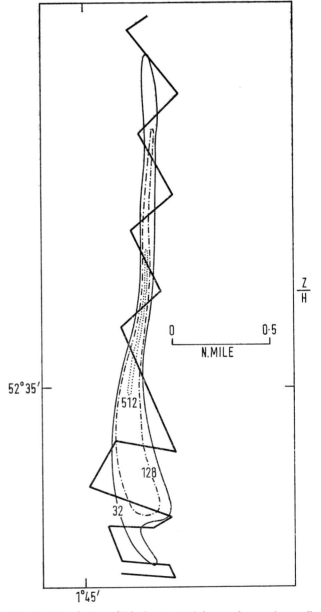

Fig. 11. Distribution of Rhodamine-B $4\frac{1}{2}$ hours after a release off the Suffolk coast (experiment F)

[128]

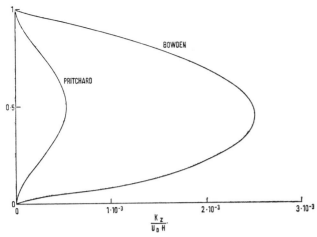

Fig 12. Comparison of Bowden and Pritchard formulae for vertical mixing coefficient

Of these results, those based on river flow data for the Crouch and Roach are probably least accurate; omitting these cases the mean result becomes

$$K_z = 3.6 U_M H \qquad (13)$$

or about seven times the value predicted by Eq. (12).

It seems possible that the differences between the values of diffusion coefficient predicted by Bowden and our experimental measurements may result partly from an over-estimation of the vertical mixing coefficient K_z. Using an empirical approach based on the Prandtl (1925) mixing length hypothesis, Pritchard (1959) obtained an expression for K_z which, in the case of a van Veen vertical velocity profile, becomes

$$K_z = 8.97 \left(\frac{z}{H}\right)^2 \left(1 - \frac{z}{H}\right)^2 U_D H 10^{-3}. \qquad (14)$$

Figure 12 illustrates the vertical distribution of K_z predicted by this equation in comparison with that given by Eq. (10), and shows that Pritchard's expression gives values that on average are smaller by a factor of about 6. Since the diffusion coefficients given by Eqs. (12) and (13) differ by a factor of about 7, the results reported in this paper are generally consistent with Bowden's derivation of the coefficient of longitudinal diffusion for a van Veen vertical velocity profile, if the empirical expression obtained by Pritchard is used for the vertical mixing coefficient.

Effect of wave action on diffusion

It has been shown above that in many shallow water situations vertical current shear combined with vertical mixing may produce a marked horizontal diffusion in the direction of flow. It follows, from the part played in this process by vertical diffusion, that wave action, by increasing local turbulence levels, should lead to a reduction of the velocity shear diffusion effect along the direction of mean flow. Wave-induced turbulence would clearly tend also to increase that part of the diffusion process produced directly by horizontal turbulence. One effect of increased wave action on coastal diffusion should therefore be to make the diffusion process behave in a more nearly horizontally isotropic fashion. A comparison of the results of experiments F and D illustrates this; in F, carried out in comparatively calm conditions, the effect of vertical shear diffusion is clearly

evident (fig 11), but much more nearly horizontally isotropic behaviour was observed in D (fig 3). During the latter experiment there was a long swell in the survey area, although the water depth was also greater, so that it may no longer have been an important factor limiting the development of vertical turbulence. A comparison of the results of experiments carried out in the central Southern Bight of the North Sea leads to similar conclusions. The distributions measured under calm conditions (experiment C) exhibited greater elliptical eccentricity than those obtained during a period of rough seas (experiment B), and the rate of diffusion along the major axis of the distribution during calm conditions was about twice that derived for a rough sea.

TABLE 2 HORIZONTAL DIFFUSION COEFFICIENTS FOR SOME CASES WHERE VERTICAL DIFFUSION IS SIGNIFICANT

Region and experiment	U_M (Maximum depth mean velocity, cm/s.)	H (Depth cm)	K_z (cm²/sec10⁵)	$\frac{K_z}{(U_M H)}$
Lowestoft				
G	119	1,650	12.0	6.1
F	94	1,280	2.3	1.9
Crouch				
J	49	750	2.3	6.3
J (Royal Corinthian Y.C.)*	55	800	4.6	10.5
Roach				
K	56	840	2.0 / 1.25	4.3 / 2.7
K (Whitehouse Hole)*	62	1,170	6.3	8.7
Blackwater				
H	52	950	1.8	3.6
H	98	1,100	1.4	
H*	100	975	2.9	3.0
H (Nuclear Power Station)*	55	975	1.8	3.4
Swale				
L	60	360	0.26	1.2

*Values based on river flow data.

Conclusions

In some areas eddies of moderate size may produce a complex local flow pattern. This may occur near an irregular coastline, over uneven bottom topography or where major current systems meet; storms may produce a similar effect. The increase in effective diffusion rate due to these factors is difficult to predict on purely theoretical grounds and direct investigation will generally be necessary.

In other areas, where the main contributions to local turbulence are of comparatively small scale, experimental evidence suggests that, although the diffusion coefficient may initially increase appreciably with the size of the diffusing system, this increase may become much less marked after the system has reached a certain critical size. In such cases it may be important to determine the diffusion coefficient behaviour over a sufficient range of dimensions.

Measurements of diffusion rates in a number of situations where vertical shear diffusion was important suggest that the longitudinal diffusion coefficient K_z is related to the maximum depth mean tidal velocity U_M and depth H, by the equation

$$K_z = 3.6 U_M H.$$

Increases in the level of turbulence produced by gales may appreciably affect diffusion behaviour, and increased spectral levels for current meter records during such conditions have been obtained, particularly for the highest frequencies analysed (up to 6 cycles per hour).

During rough sea states increased rates of vertical mixing may result in a reduction of the vertical shear diffusion effect, causing the pattern of diffusion to exhibit less elliptical eccentricity.

References

BATCHELOR, G K The application of the similarity theory of
1950   turbulence to atmospheric diffusion. *Q. Jl R. met. Soc.*, 76:133–46.

BOWDEN, K F The mixing processes in a tidal estuary. *Int. J. Air*
1963   *Wat. Pollut.*, 7:343–56.

BOWDEN, K F Horizontal mixing in the sea due to a shearing
1965   current. *J. Fluid Mech.*, 21(1):83–95.

JOSEPH, J and SENDNER, H Über die horizontale Diffusion im
1958   Meere. *Dt. hydrogr. Z.*, 11(2):49–77.

JOSEPH, J. SENDNER H and WEIDEMANN, H Untersuchungen über
1964   die horizontale Diffusion in der Nordsee. *Dt. hydrogr. Z.*, 17(2):57–75.

KOLMOGOROFF, A N The local structure of turbulence in incom-
1941   pressible viscous fluid for very large Reynolds numbers. *Dokl. Akad. Nauk SSSR*, 30:301–5.

OKUBO, A A review of theoretical models for turbulent diffusion
1962   in the sea. *J. oceanogrl Soc. Japan*, 20th Anniv. Vol.:286–320.

OKUBO, A The effect of shear in an oscillatory current on hori-
1967   zontal diffusion from an instantaneous source. *Int. J. Oceanol. Limnol.*, 1(3):194–204.

OKUBO, A and PRITCHARD, D W Summary of our present knowl-
1969   edge of the physical processes of mixing in the ocean and coastal waters, and a set of practical guidelines for the application of existing diffusion equation etc. *Rep. Chesapeake Bay Inst.*, (NYO 3109.40/69–1

OZMIDOV, R V Energy distribution between oceanic motions of
1965   different scales. *Izv. Akad. Nauk SSSR* (*Atmos. Ocean. Phys.*), 1(4):257–61.

PRANDTL, L Bericht über Untersuchungen zur ausgebildeten
1925   Turbulenz. *Z. angew. Math. Mech.*, 5:136–9.

PRITCHARD, D W The movement and mixing of contaminants in
1959   tidal estuaries. *In* Proceedings of the First International Conference on Waste Disposal in the Marine Environment pp. 512–25.

PRITCHARD, D W The application of existing oceanographic
1960   knowledge to the problem of radioactive waste disposal into the sea. *In* Disposal of radioactive wastes, Vienna, IAEA, Vol. 2:229–53.

SENDNER, H and TALBOT, J W Consideration of the horizontal diffusion process during the RHENO experiment. *Rapp. P. -v. Réun. Cons. perm. int. Explor. Mer.* in press.

TALBOT, J W and TALBOT, G A Some aspects of the variation of
1969   turbulent energy in the sea during the RHENO experiment. ICES, C. M. 1969, Doc. C:16 (mimeo).

VAN VEEN, J Water movements in the Straits of Dover. *J. Cons.*
1938   *perm. int. Explor. Mer*, 13:7–36.

Experimental Studies of Horizontal Diffusion in the Black Sea Coastal Zone

V. I. Zats*

Études expérimentales sur la diffusion horizontale dans la zone côtière de la mer Noire

Les recherches sur la diffusion par turbulence comportent diverses études touchant à l'auto-épuration de la mer et l'analyse de la transformation des zones d'épandage ou des déversements d'eaux usées. Des travaux complexes ont été entrepris pour estimer la diffusion horizontale et le taux de dilution dans la mer.

1. Études sur la diffusion selon la méthode de la "diffusivité voisine" (méthode de Richardson). Elles indiquent que la "loi des 4/3" n'est pas universelle et qu'elle se vérifie rarement dans les zones côtières. La valeur de la diffusivité tourbillonnaire K dépend de l'échelle du phénomène de diffusion l conformément à la relation $K \sim l^n$. La puissance n pour l'échelle du phénomènes jusqu'à 10 km est inférieure à 4/3 dans la plupart des cas. Toutefois, la valeur de n pour les "mésoprocessus" (échelle 1–10 km) est 1,5 à 2 fois plus élevée que pour les petits processus (quelques centaines de mètres).

   L'auteur examine aussi les rapports existant entre K et d'autres facteurs déterminants—vitesse du courant et stratification en fonction de la densité de l'eau.

   En appliquant la méthode de diffusion de Richardson, on a estimé les paramètres ci-après:
   (i) l'ampleur de l'expansion relative d'une aire de diffusion non stationnaire dépendant de la durée de la diffusion et de la valeur de n;
   (ii) l'ampleur de l'expansion relative d'une aire de diffusion dans des conditions stationnaires dépendant de la valeur de n et d'autres facteurs.

2. Etudes d'aires de diffusion horizontale par aérophotographie. Les expériences ont été faites en 1968–1969. L'analyse des données obtenues par aérophotographie sur les aires de diffusion en conditions non stationnaires a permis de déterminer: (1) les rapports existant entre la superficie des aires S et le temps t, que l'on peut rendre approximativement par l'expression $S \sim t^m$; (2) la vitesse probable de diffusion (P) d'après la méthode de Joseph et Sendner; (3) les valeurs des coefficients de diffusion à petite échelle. Les données disponibles montrent que la "loi des 4/3" ne se vérifie pas. L'auteur examine les résultats obtenus et les compare avec ceux d'autres chercheurs.

Estudios experimentales de la difusión horizontal en la zona costera del mar Negro

La labor de investigación sobre la difusión turbulenta la constituyen diversos aspectos de estudios sobre la manera en que el mar se autopurifica de la contaminación y comprende el análisis de la transformación de zonas o de aportes de aguas residuales. Se efectúaron complejas investigaciones para estimar la mezcla horizontal y la velocidad de dilución en el mar.

1. Los estudios de la difusión por el método de "difusibilidad vecina" (método de Richardson) indican que la "ley 4/3" no es universal y raramente ocurre en la zona costera. El índice de difusibilidad de remolinos K depende de la escala del fenómeno de difusión l según la relación $K \sim l^n$. En la mayor parte de los casos el exponente n de la escala del fenómeno hasta 10 km es inferior a 4/3, pero el índice para el "mesoproceso" (escala de 1–10 km) es de 1,5 a 2 veces mayor que para los procesos pequeños (cientos de metros).

   También se examina la relación entre K y otros factores determinantes, como velocidad de la corriente y estratificación en función de la densidad del agua. En función de la "difusión Richardson" se estimaron los parámetros siguientes:

   (i) amplitud de la expansión relativa de una mancha de difusión no estacionaria según la duración de la difusión y el exponente n;

   (ii) amplitud de la expansión relativa de una mancha de difusión estacionaria, según el valor de n y otros factores.

2. Los estudios de las manchas de difusión horizontal mediante aerofotografía. Los experimentos se realizaron en 1968 y 1969. El análisis de los datos de las manchas de difusión no estacionaria obtenidos mediante la aerofotografía permitieron determinar: (1) la relación entre el área de la mancha S y el tiempo t que se obtiene una aproximación por la expresión $S \sim t^m$; (2) probable velocidad de difusión (P) según Joseph y Sendner; (3) valores de los coeficientes de difusión a pequeña escala. Los datos recogidos indican que la "ley 4/3" no se cumple. Se examinan los resultados obtenidos y se comparan con los logrados por otros investigadores.

* Institute of Biology of the Southern Seas, Academy of Sciences of the Ukrainian Republic, Sevastopol, U.S.S.R.

IN studies of processes affecting the reduction of the contaminant admixture, attention has been paid mainly to the physical oceanographic factors. This is an imperfect reflection of the whole phenomenon. According to the data of many investigators, the transport, diffusion and mixing processes that depend on oceanographic factors are most important in all stages of distribution and transformation of sewage fields in the sea. We have carried out complex research work on turbulent diffusion to estimate the intensity of horizontal mixing.

Study of "Richardson" diffusion

Long-term experiments on "Richardson" diffusion were carried out by means of the "neighbour diffusivity" method (Richardson and Stommel, 1948) on the coasts of the Black Sea (Crimea and Caucasus). Drift floats were used and were tracked from the shore or ship at anchor, or by following them with vessels (Zats, 1964, 1967, 1970). The values of horizontal diffusion coefficients were determined, and their dependence on scale was demonstrated for small and mesoprocesses (Zats, 1964, 1967, 1970). The scale length covered ranges from tens of meters to 500–1,000 m and 1–10 km. The relationship between the diffusion coefficients K and the scale l is approximately

$$K(l) = cl^n \qquad (1)$$

The long-term experiments indicated that in most cases n was less than 4/3 confirming that the "4/3 law" is not universal (Okubo, 1962; Foxworthy, Barsom and Tibby, 1966; Zats, 1967, 1970; Ozmidov, 1968). For some near-shore releases n varied between 0.21 and 1.30

(fig 1/1, 2, 3). Chiefly within distances 2.5–5 mi offshore, the "4/3 law" conforms to observations (fig 1/2). The value of n for the mesoprocesses of diffusion is however 1.2–2 times the value for the small scale processes (to 1 km) fig 1 (4–7). According to Ozmidov (1968) the validity of the function $K(l)$ with $n = o$; 1:4/3 for idealized mid-ocean conditions can be related to the distribution of kinetic energy density relative to different-scale motions in the ocean. The function $K(l)$ in the complex near shore conditions seems variable and it is difficult at present to predict its variability.

The "Richardson" diffusion experiments allowed the dependence of the coefficient K on current speed to be determined. It approximates in general form to the power function $K(V) \sim V_l^R$ for the small and mesoprocesses (fig 2). For small-scale phenomena (150–250 m) and current speeds from 4 to 42 cm/sec in the South Crimea coastal zone, $K(V)$ was linear (Zats, 1964). But for most cases where the scales are increased from 0.5–1.0 to 3–5 km, it approximates the power function (fig 2).

The curves in fig 2 are of great interest because current velocities are determined under almost all oceanographic conditions. From these one can estimate the value of the diffusion coefficients. Within the framework of "Richardson" diffusion an estimate can be made of some kinematic parameters of horizontal diffusion, such as: (a) the extent of relative expansion of a diffused patch of particles (Zats, 1967, 1967a) depending on the duration of diffusion and on n; (b) the extent of expansion of the flow of particles from the linear stationary source (Brooks, 1960; Zats, 1967) depending on the source length B, the current speed V and n.

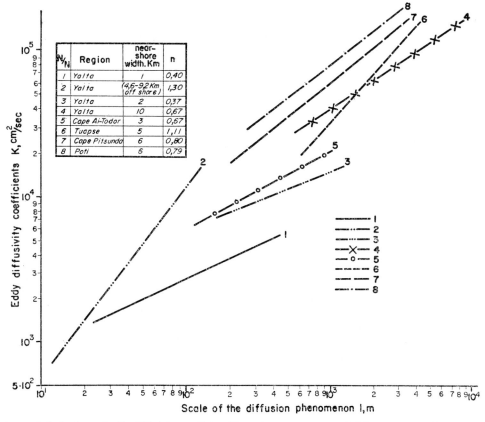

Fig 1. The dependency of the horizontal eddy diffusivity coefficient K on the phenomenon scale l for some coastal zones of the Black Sea. 1–4 Yalta, 5–Cape Ai-Todor, 6–Tuapse, 7–Cape Pitsunda, 8–Poti vicinity. Logarithmic scales

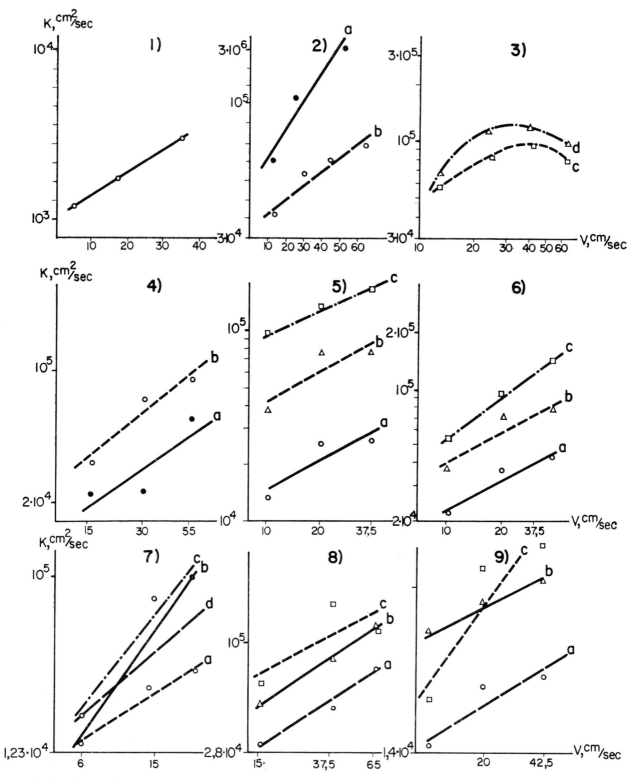

Fig 2. The dependency of the eddy diffusivity K *on the current speeds* V *for the definite phenomenon scales* l *for some regions of the Black Sea*

1. *Yalta: 1150–250 m*
2. *Yalta: (a)–1,000–2,000 m, (b)–4,000–5,000 m*
3. *Yalta: (c)–2,000–3,000 m, (d)–3,000–4,000 m*
4. *Tuapse: (a) 500–1,500 m; (b) 1,500–3,000 m*
5. *Pitsunda—mild period: (a) 150–500; (b) 500–2,000; (c) 2,000–6,000*
6. *Pitsunda—cold period: (a) 150–700; (b) 700–1,500; (c) 1,500–3,000*
7. *Sukhumi-Bathmi: (a) 1,000–1,500 m; (b) 1,500–2,500; (c) 2,500–4,000*
   (d) 4,000–7,000
8. *Poti—mild period:* ⎱
9. *Poti—cold period:* ⎰ *(a) 200–700; (b) 700–1,500; (c) 1,500–3,000*
   1–2 linear scales; 3–9 logarithmic scales

Here the relationships are

$$K(l) = cl^n \text{ or } K/K_0 = (l/l_0)^n \quad (2)$$

(where K_0 is the initial diffusion coefficient at scale l_0) and

$$K = \frac{1}{2}\frac{d(\delta^2)}{dt} \quad (3)$$

This is the expression for the diffusion coefficient K in differential form, where δ^2 is the mean value of the derivation quadrate of particles from the average distribution.

Proceeding from (2) and (3) the relation between diffusion parameters of the non-stationary patch of particles t, l, l_0, K_0, n, (where $0 < M \leqslant 2$) is given by

$$t = l_o^n(l^{2-n} - l_0^{2-n})/12 K_0(2 - n) \text{ or}$$
$$(l/l_0)^{2-n} = 12 tK_0(2 - n)/(l_0^2 + 1) \quad (4)$$

l_0 and l can be interpreted as the diameters of the horizontal patch at the initial and next moment. The other conditions being equal, l and t depend essentially on n, which determines the form of the function $K(l)$.

The problem of the relative expansion of the stationary flow of particles from the linear source due to diffusion processes was solved first by Brooks (1960) for some particular cases ($n = 4/3$; $n = 1$, $n = 0$). We have summarized it for variable n ($0 < n \leqslant 2$).

The following ratio was obtained

$$l/B = \left[1 + (2 - n)\beta\frac{x}{B}\right]^{\frac{1}{2-n}} \quad (5)$$

where B is the length of the linear source of particles; l is the width of the particle flow at a distance x downstream from the source; $\beta = 12K_0/BV$; K_0 is the diffusion coefficient for the initial width of the flows VB; is the current speed. The value l/B estimated by Eq. (5) agrees with that obtained from the experiments with floats.

The study of "Richardson" diffusion enables us to have a more valid understanding about n in Eq. (1) and to use data about n for solving some practical problems concerning admixture diffusion on the sea surface.

Study of dye patches in the sea by aerial methods

Aerial photography of dye patches from an "instantaneous point source" at definite intervals permits some important parameters of turbulent diffusion to be determined. The experiments were carried out in the Black Sea in 1968-1969, 5-10 mi offshore. Uranine and Rhodamine-C were used as indicators. "The instantaneous point source" for the initial moment had an area of 10-100 m²; during the diffusion process the patch area increased to 1×20^5-4×10^5 m². This test lasted over a period of 2-8 h. The areas of the diffused patches were found by assuming (Zats and Andrutshenko, 1969) that the visible patch area has an isoline contour of definite concentration. The area variability of the diffused patches S within time t is approximated by the function (fig 3)

$$S = at^m \quad (6)$$

The value of m varied over the range 1.29-1.92 during the experiments of 1968, and over a range of 0.63-1.29 during the experiments of 1969 for the patch radii that changed from 20 to 350 m. The patch radius

variability R relative to time t is given for most experiments by a linear function

$$R = at + c \quad (7)$$

The difference between the average values of m for experiments with uranine and Rhodamine-C is negligible: $m_{ur} = 1.05$, $m_{rhod} = 0.93$. Therefore the function $S(t)$, when averaged for all releases, is similar to the linear function.

In the work of Zats and Andrutshenko (1969) the values of m are listed for the test data of Nanniti, Okubo, Isaev and Isaeva: $m = 0.45 + 1.50$. We attempted to find m from the data of additional investigators. According to the works of Fukuda, Ito, Sakagishi (1965) m varies from 1.04 to 1.96 in the nearshore zone and according to the tests of Fujimoto and Tanaka (1968) m was within the limits 0.93-1.25 (patch radii from 40 to 378 m). Therefore, from our test data and those of the Japanese investigators, for most releases in the near shore zone m will be within the limits 0 to 2 for patch radii to 500-1,000 m (taking the radii as the characteristic phenomenon scales). The data given refer to the stages of patch increase from $t = 0$ to the time when S_{max} is reached, at other stages and for other scales the function $S(t)$ will be more complex.

When R_{max} is determined at time t_{max} it is possible to determine the probable velocity of diffusion according to the Joseph-Sendner theory (1958)

$$R_{max} = 2Pt_{max} \quad (8)$$

Values of P determined for two experiments were 0.87 and 0.97 cm/sec, i.e. of the same order as those found in our experiments with drift floats (1 cm/sec) in the near shore zone (Zats, 1967).

Aerial photography also permits the coefficients of horizontal diffusion to be determined by the method of Richardson and Stommel (1948). Assuming that one may define the patch size variability by its effective diameter changes D (where $D = 2\sqrt{S/\pi}$), it is possible to define the diffusion coefficient

$$K_D = \frac{\overline{(\Delta D)^2}}{2\Delta t} \quad (9)$$

For the 1968-1969 experiments the average values for each of 4 releases ranged from 10^3 to 10^4 cm²/sec for patch diameters of tens of metres to 500-700 m. There was close agreement between values of K_D and K_l as obtained from free float spar buoys for phenomenon scales up to 1,000 m; K_l varied between 2.2×10^4 and 7.0×10^4 cm²/sec, and K_D varied from 1.6×10^4 to 5.4×10^4 cm²/sec. According to a relationship that is analagous to that given by Zats and Andrutshenko (1969) K_D depends on the phenomenon scale

$$K_D = cD^{\frac{(2m-1)}{m}} \quad (10)$$

From this relationship it is apparent that D does not exceed 1.0 for the values m mentioned above. Hence, according to the data from aerial photography of dye patches the "4/3-law" is not valid for small-scale diffusion processes and for current speeds from 5-10 to 60-72 cm/sec.

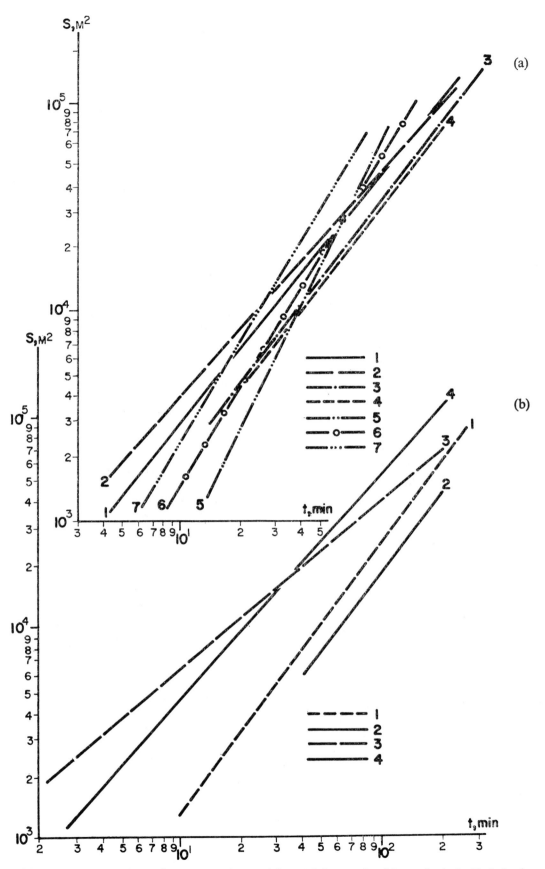

Fig 3. Diagrams of the function S(t) *according to the experiments by use of the aerial photography of dye patches in the Black Sea* (experiments 1968–1969)

(a) *Uranine patches:*
1–8.VIII.1969; 2–12.VIII.1969; 3–15.VIII.1969; 4–9.VIII.1969;5–4.VII.1968; 6–5. VIII.1968; 7–30.VIII.1968
(b) *Rhodamine-C patches:*
1–4.VIII.1968; 2–8.VIII.1969; 39.VIII.1969; 4–12.VIII.1969

References

BROOKS, N H Diffusion of sewage effluent in an ocean-current.
1960 *In* Proceedings of the First International Conference on
waste disposal in the marine environment, University of
California, Berkeley, 22–25 July 1959, pp. 246–67.

FOXWORTHY, J E, BARSOM, G M and TIBBY, R B On eddy diffusi-
1966 vity and the fourthirds low in highly stratified nearshore
coastal waters. *Trans. Am. geophys. Un.*, 47(3):478.

FUJIMOTO, M and TANAKA, K Dye diffusion experiments in the
1968 ocean by use of aerial photography. *J. Japan Soc. Photo-
gramm.*, 7(1):8–20.

FUKUDA, M, ITO, N and SAKAGISHI, S Diffusion phenomena in
1965 coastal areas. *Adv. Wat. Pollut. Res.*, 2(3):193–204.

JOSEPH, J and SENDNER, H Über die horizontale Diffusion im Meere.
1958 *Dt. hydrogr. Z.*, 11(2):49–77.

OKUBO, A A review of theoretical models for turbulent diffusion
1962 in the sea. *J. Oceanogrl Soc. Japan*, 20th Anniversary vol.:
286–320. Also issued as *Tech. Rep. Chesapeake Bay Inst.*,
(30):113 pp.

OZMIDOV, R V About coefficient dependency of the horizontal
1968 turbulent exchange in the ocean on the phenomenon scale.
Izv. Akad. Nauk SSSR (Fiz. Atmos. Okean.), 4(11);1224–5
(in Russian).

RICHARDSON, L F and STOMMEL, H Note on eddy diffusion in the
1948 sea. *J. Met.*, 5:238.

ZATS, V I To the question of the horizontal eddy diffusion in the
1964 near-shore of the Black Sea. *Okeanologiia*, 4(2):249–57
(in Russian).

ZATS, V I The oceanographic aspects of the problem of sewage
1967 discharge in the near-shore of the Black Sea. *Vop. bio-
okeanograph. Trudy Inst. biol. jug. morej AN USSR*, pp.
22–30 (in Russian).

ZATS, V I About the dependency of the distribution extent of the
1967a admixture patches on the sea surface. *Dynamica wod i
voprosy hydrochim. Chernoho morja. Trudy Inst. biol. jug.
morej AN USSR*, pp. 34–40 (in Russian).

ZATS, V I Characteristic of the mesoprocesses of the horizontal
1970 turbulent diffusion in the near-shore. *Okeanograph.
aspekty samoochitsh. morja ot zagr. Mater. konfer. Trudy
Inst. biol. jug. morej AN USSR*, pp. 71–87 (in Russian).

ZATS, V I and ANDRUTSHENKO, B F The study of the horizontal
1969 turbulent diffusion processes of the admixture patches in
the sea by use of aerial photography. *Materialy III Vses.
Symp. po vopr. samoochitshen. vodoemov i smesh. stoch.
wod. Tallin*, 2:56–65 (in Russian).

On the Predictability of Waste Concentrations

G. Abraham* and G. C. van Dam†

De la possibilité de prévoir les concentrations de déchets

On détermine souvent l'effet de l'évacuation des déchets sur le milieu marin au moyen d'études reposant sur l'usage d'indicateurs. Les expériences se font généralement par observation de la diffusion de colorants fluorescents ou de traceurs radio-actifs émis en tant que sources ponctuelles instantanées dans la zone considérée. La diffusion des substances continuellement émises dans l'océan à partir d'une source donnée est déterminée ensuite par superposition des répartitions de concentration résultant des diverses émissions instantanées qui se sont succédées. Dans cette intégration, il est possible d'envisager les effets de la dégradation et de l'advection.

Etant donné la nature du phénomène de diffusion en cause, il est extrêmement difficile de faire des prévisions précises sur la réparti-tion des concentrations de déchets, notamment à de grandes distances du point de décharge. Ceci est dû aux causes suivantes:

1. Dans les expériences d'émission instantanée, il faut faire un effort considérable pour effectuer un nombre suffisant d'observations simultanées afin d'obtenir des courbes faibles d'égale concentration.
2. L'intégration visant à déterminer les effets d'une émission continue ne peut se faire que par l'intermédiaire de formules mathématiques servant à représenter les courbes d'égale concentration telles qu'elles ont été constatées lors des expériences d'émission instantanée. Ces formules mathé-matiques doivent correspondre aux mesures des répartitions de concentration, c'est-à-dire que l'on choisit les valeurs les plus probables des paramètres numériques dans les formules mathématiques.
3. La durée des expériences d'émission instantanée peut être brève par comparaison au temps nécessaire pour obtenir des conditions d'équilibre avec émission continue, en partant d'une mer propre comme condition initiale de l'intégration. L'intégration suppose alors l'extrapolation des données relatives à l'émission instantanée au-delà de la période couverte par les expériences.
4. Par ailleurs, on peut se demander si les conditions qui prévalent pendant la période nécessaire pour faire un nombre limité de mesures au moyen des indicateurs sont représenta-tives des conditions moyennes.

Dans le présent document, on montre que la variation des formules mathématiques dans les proportions imputables aux limites naturelles de la technique de mesure a parfois un effet considérable sur les taux de concentration calculés par intégration pour des conditions d'émission continue. Il en est ainsi en particulier lorsqu'on se trouve à de grandes distances de la source et en cas d'émission de substances qui ne se conservent pas. Lorsqu'on interprète les résultats des expériences d'émission instantanée avec intégration, il faut tenir compte des facteurs mentionnés ci-dessus et s'efforcer d'en évaluer leur influence. On peut ainsi mesurer l'importance des inexactitudes des taux de concentration prévus pour des conditions données d'émission continue.

Sobre la posibilidad de prever las concentraciones de desechos

El efecto que la evacuación de desechos tiene en el medio marino se determina con frecuencia basándose en estudios de elementos trazadores. Generalmente estos experimentos se realizan obser-vando la difusión de colorantes fluorescentes o trazadores radiacti-vos que se sueltan como fuentes puntiformes instantáneas en el lugar que se estudia. La difusión de la sustancia, que se suelta constantemente en el océano desde un punto dado, se determina más tarde mediante la superposición de las distribuciones de las concentraciones que resultan de las distintas liberaciones instan-táneas o posteriores. En esta integración se pueden examinar los efectos de la descomposición y del transporte advectivo.

A causa de la naturaleza del fenómeno de difusión es muy difícil hacer pronósticos exactos de la distribución de la concentración de desechos, particularmente a mucha distancia de la fuente. Esto lo motivan las circunstancias siguientes:

1. En los experimentos de suelta instantánea, hay que esforzarse mucho para realizar simultáneamente un número de obser-vaciones que sea suficiente para lograr líneas fidedignas de la misma concentración.
2. La integración para determinar el efecto de una liberación continua sólo puede hacerse introduciendo ciertas expresiones matemáticas que describen el desarrollo de las líneas de igual concentración, como se han determinado en los experimentos de suelta instantánea. Las expresiones matemáticas se tienen que adaptar a las distribuciones de las concentraciones medidas, por ejemplo, seleccionando los valores más probables de los parámetros numéricos dentro de las expresiones matemáticas.
3. Es posible que la duración de los experimentos de suelta instantánea sea corta con respecto al tiempo necesario para obtener condiciones de equilibrio con una suelta continua que comienza en un mar limpio, como condición inicial de inte-gración. En este caso, la integración implica extrapolación de los datos de suelta instantánea en un período de tiempo superior al comprendido en los experimentos.
4. La incertidumbre que se presenta de si las condiciones pre-valentes durante el número limitado de mediciones indicadoras sean representativas de las condiciones normales.

Se muestra en este trabajo que en algunos casos, la variación de las expresiones matemáticas que quedan dentro de los márgenes permitidos por las limitaciones naturales de la técnica de medida, tiene un enorme efecto en los índices de la concentración calculados por medio de la integración para condiciones de suelta continua. Esto ocurre particularmente cuando a grandes distancias de la fuente se sueltan sustancias que no se conservan. Al interpretar mediante la integración los resultados de los experimentos de suelta instantánea, se han de tener presentes los factores men-cionados y tratar de determinar su influencia. Es posible que esto haga comprender la magnitud de las inexactitudes en los índices de concentración pronosticados cuando se trata de sueltas continuas.

* Delft Hydraulics Laboratory, Delft, Netherlands
† State Public Works Mathematics and Physics Division, Directorate for Water Management and Water Research, The Hague, Netherlands

WHEN designing a sea outfall one has to select such an outfall-length that harmful concentrations of effluent do not reach the coast. The prediction of the concentration at the shore for different outfall-lengths is often done by observing the spread of fluorescent dyes or radioactive tracers released at a chosen point in the proposed disposal area. The concentration at the shore of the effluent is subsequently determined by superposition of the results from different successive releases; this method also permits consideration of the effects of decay and advective transport.

This discussion is restricted to the case of a continuous release of waste effluent into a sea with tidal currents approximately parallel to a nearly straight coast line. The effluent source is some distance from the coast and small compared with that distance. The discussion is further restricted to soluble wastes or suspensions that disperse like solutions. The waste water is considered to have a density such that after initial mixing no buoyancy effects interfere. Initial mixing will not be discussed.

Superposition method

The dispersion of dissolved or suspended materials continuously released into a stream from a small source, especially if the stream is unsteady, is much more difficult to visualize than the dispersion of an instantaneously released single patch. Therefore it is useful to relate the case of continuous release to the simpler case of instantaneous release by considering the continuous injection as a series of many instantaneous injections, equally spaced in time at such short intervals ($\triangle t$) that it approximates to a continuous introduction. It is necessary that $m\triangle t$ units of material are always released per instantaneous injection, where m is the quantity released per unit of time in a continuous release.

The simplest case is injection in a constant uniform stream where the centres of the individual patches will be equally spaced along a straight line (see fig 1a). Since the stream is stationary, each patch may be expected to behave identically. Therefore, in order to find the result of a continuous release by summation of the successive patches of various ages and with centres at different positions, it is sufficient to know the rate of development of a single patch. Mathematically this can sometimes be done very accurately by letting $\triangle t$ go to zero, i.e. by

integration. If the behaviour of one patch is not known as an analytical function, the summation procedure must be applied, preferably by computer in view of the number of terms.

A continuous release into an (alternating) tidal current may also be considered by superposition of a series of elementary patches. The convective transport of the different patches by the tidal currents may imply that the central parts of very "old" and large patches (released say ten tides ago) cover very "young" ones (released say one tide ago) (see fig 1b). However, only when considering a continuous release into a steady current can the behaviour of one single patch be taken as representative of the behaviour of all patches since the development of a single patch released during tidal slack could be different from that of a patch released during maximum velocities.

The essential thesis of the above is that the individual patches of a long series behave in precisely the same way as they would if they were alone. In other words, all patches develop independently and may be superposed or added linearly. Although, in this presentation, the proposition sounds quite acceptable, it should be mentioned that it is not self evident and not always valid, as explained by Abraham (1966) for the release of large quantities of domestic sewage.

For a continuous release into a steady current, the superposition method was described by Harremoes (1966). A more mathematically developed version, including the effect of a periodically changing though uniform stream, was presented by Schönfeld (1964). Computer programmes based upon the work of Schönfeld have been used in the Netherlands since 1962.

To explain the application of the superposition method consider the case of a continuous constant release of effluent. Assume the release to be started at time $t = 0$, into an initially clean sea. We want to know the concentration of the effluent in a given point near the source at a time $t = t_1$. In accordance with the above considerations this problem can be solved as follows:

(a) Separate the continuous release into successive instantaneous releases, the first release being at time $t \approx 0$ and the last release being at time $t \approx t_1$.
(b) Determine the location of the centre of each patch at time $t = t_1$. This requires substantial informa-

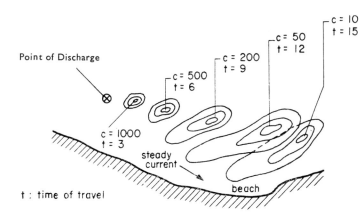

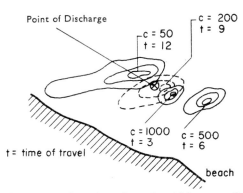

Fig. 1a. Continuous release into steady current divided into separate patches (after Harremoes, 1966)

Fig 1b. Continuous release into alternating tidal current divided into separate patches

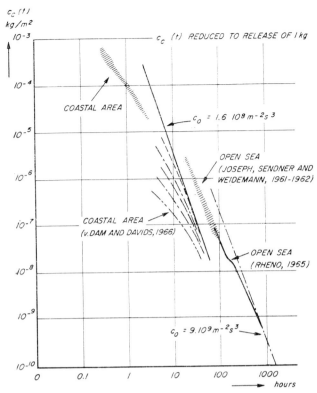

c_c (t)
kg/m²

10^{-3}

c_C (t) REDUCED TO RELEASE OF 1kg

10^{-4}

COASTAL AREA

c_0 = 1.6 $10^8 m^{-2} s^3$

10^{-5}

OPEN SEA
(JOSEPH, SENDNER AND
WEIDEMANN, 1961-1962)

10^{-6}

COASTAL AREA
(v. DAM AND DAVIDS, 1966)

10^{-7}

OPEN SEA
(RHENO, 1965)

10^{-8}

10^{-9}

c_0 = $9.10^9 m^{-2} s^3$

10^{-10}

0   0.1   1   10   100   1000
                                                hours

Fig 2. Instantaneous release experiments in the North Sea (van Dam, 1968)

tion on the current field in the projected disposal area.

(c) Determine the concentration distribution within each patch at time $t = t_1$. This information can be obtained only by observing the spread of some tracer released as an instantaneous point source.

(d) Determine the concentration, which at time $t = t_1$ will be found at a given point, as a result of the continuous release, by adding the contribution of each separate patch to the concentration at the point considered. This can be done on the basis of the information indicated under (b) and (c).

In general, when designing an outfall for a given effluent, one has to obtain the information indicated under (c) from experiments. Usually these experiments are done with tracers that have a specific decay time. This decay time may be different from that of the effluent to be discharged (e.g. Rhodamine B in experiments on the design of an outfall of domestic sewage, containing coli bacteria as a pilot component). The superposition method makes it possible to take the difference in decay time into account when making predictions for continuous release.

The superposition method further allows consideration of the effect of variations in the displacement curves of the separate patches. Possible causes of these variations are listed in the following section.

Complicating physical factors

In the procedure thus explained, one must know the displacement curve of each separate patch and the development of the concentration distribution within the patch. This implies that one must describe a complicated physical phenomenon. The complications may be divided into two main groups: systematic effects and irregular effects. This division corresponds to an essential difference with respect to the mathematical formulation of the superposition method. The first group only causes the mathematical formulation to be somewhat more complex, the second requires an essentially different approach. However, in specific cases it may be possible to simulate irregular effects by using regular (periodical) terms in the mathematical formulation.

Systematic effects that influence displacement curves are tidal currents, and slow, regular meteorological variations like seasons. Effects that influence the development of concentrations within separate patches are changes in dispersion rates caused by periodic changes in the regular current components. Other systematic effects include: regular but non-uniform current fields; non-uniform water depth; curved coastline; and non-uniform vertical distribution of injected material caused by stratification or by slow vertical diffusion by small diffusion coefficients and/or great water depth.

Irregular effects that influence displacement curves include: turbulent components of the current vector that cause irregularities in the shape of the curve along which the individual patch centres are situated, and irregularities in their mutual separation distance; current variations by small and rapid changes in meteorological conditions (wind) having the same effect. Irregular effects that influence the development of concentration within separate patches cause deviations from the average development of the individual patches, resulting in irregular shapes as well as deviations from the average size development.

Complicating factors require schematizing

The displacement curve of each separate patch and especially the development of the concentration distribution within each patch must be determined by observing the spread of tracers, but this raises several problems:

The simultaneous measurement of the concentration at enough locations to permit reliable lines of equal concentration to be drawn to describe the patch development is not easy. The mathematical representation must be adapted to the actual (irregular) measured concentration by selection of suitable values for the mathematical coefficients. The duration of the instantaneous release experiments may be short compared with the time required to get equilibrium conditions when beginning to release continuously into a clean sea, i.e. the initial condition of the superposition method. This means extrapolation of the instantaneous data beyond the time period covered in these experiments. The factors mentioned previously require that one determines whether the conditions prevailing during the restricted experimental period are representative of average or other conditions.

Frequency distributions of shoreline concentrations are difficult to predict and can only be made for shoreline concentrations when the results of the systematic and irregular effects listed are known. Generally speaking this requires such extensions of the empirical knowledge about the coastal waters in the projected disposal area that predictions of the frequency distributions are difficult to make. Because of the factors explained in the following paragraph, this especially holds for non-conservative (decaying) effluents. It is difficult to apply standards for beach water quality which ensure that concentrations of an effluent will not exceed an acceptable value for relatively short periods.

Decay reinforces the effects of the complicating factors previously listed which lead to a certain degree of inaccuracy in the information required for the super-position proper (step (d) of the procedure given). These inaccuracies, amongst others, lead to errors in the estimates of the time that it will take the effluent to travel from the outfall site to the shore. This means that predictions of concentrations of non-conservative (decaying) effluents at shore are especially affected by the inaccuracies.

This is of special interest when considering the disposal of domestic sewage because its quality is commonly assessed by coliform bacteria counts. Coliform bacteria in the waste water gradually diminish after discharge into the receiving water. As a measure of the rate of decay the parameter T_{90} is used. This is the time required for the die-off of 90 per cent of the organisms. T_{90} is known only within certain limits of accuracy which is unfortunate because the inaccuracies caused by the factors previously listed and the inaccuracies in the T_{90} values reinforce each other in their effect on predictions of concentrations at shore.

When considering the behaviour of a separate patch on the basis of a fully developed cascade of turbulent energy, maximum concentrations within the patch are proportional to t^{-3} (injection is at time $t = 0$). The spatial distribution of concentration is Gaussian with respect to the centre of the patch if the classical diffusion concept (with time dependent diffusion coefficients) is applied. More fundamental generalizations of the diffusion concept, e.g. by Schönfeld (1962), are not yet quite satisfactory but demonstrate that spatial distributions of non-Gaussian distribution must be expected. Both the Netherlands' data and the other theoretical distributions mentioned by Schönfeld (1962) are represented best by a distribution steeper than a Gaussian one.

Substantial information about oceanographic diffusion confirms that the maximum concentration within a patch is proportional to t^{-3} (Okubo, 1968; van Dam, 1968; van Dam and Davids, 1966). This may be expressed as:

$$c_c(t) = c_0 M t^{-3}$$

where c_c is the quantity of considered substance per unit area, determined at the centre of the patch by integration over the depth. M is the quantity of substance released, e.g. in the case of a series representing a "continuous" release of a quantity m per unit time, $M = m.\triangle t$. The coefficient c_0 has dimensions length^{-2} time^3. In general for coastal waters the information given by Okubo (1968) and van Dam (1968) (see fig 2) may be summarized by indicating that, for coastal waters, values of c_0 were found to vary from 10^8 to 10^9 m^{-2} sec^3.

The above-mentioned experimental information shows that in first approximation the lines of equal concentrations within the patch can be described as ellipses. In coastal waters, values of a/b—the ratio of the long axis over the short axis of the ellipses—were found to vary from 2 to 10, the long axis being parallel with the shoreline (van Dam and Davids, 1966).

The parameters c_0 and a/b contain sufficient informa-

tion to describe the development of the concentration distribution within separate patches if one assumes the concentration distribution to be Gaussian. Doing so, one can perform calculations to show how the development of the concentration distribution within separate patches, as characterized by the parameters c_0 and a/b, affects the predictions of the concentration at shore.

In the case of a continuous release into a sea that is stagnant except for the turbulent motion, figs 3, 4, 5 and 6 show how the relationship between the maximum concentration at shore, C(A), and the outfall-length, A, is influenced by variations of the parameters c_0, a/b and T_{90} within the range of values found in literature. Figures 3, 4, 5 and 6 are determined for the case of a continuous release of a quantity Q_A per unit time into coastal water of constant depth h, assuming the concentration distribution within the patch to be Gaussian.

As a first approximation a continuous release from a source which moves periodically through a stagnant sea in a direction parallel to the shore may be considered to be equivalent to a continuous release from a fixed source into a sea with tidal currents in a direction parallel to the shoreline with no net flow of sea water. Hence in first approximation the latter case may be treated by superposition of a number of continuous releases into a stagnant sea, provided that the different sources are situated in such a way with respect to each other that the tidal motion is represented properly. Thus in first approximation the effect of variations of the parameters c_0, a/b and T_{90} on the concentration at shore found for the hypothetical case of a stagnant sea can be applied for a sea with tidal motion and no net flow of ocean water. The ratio between the concentrations found at shore for two different combinations of c_0, a/b and T_{90} will be the same with and without tidal action, provided there is no net flow of ocean water.

For tidal conditions with no net flow of sea water figs 3, 4, 5 and 6 show that variation of the decay time T_{90} has a strong effect on the shoreline concentration. Moreover these figures show the effect of variations of the parameters c_0 and a/b to be increasing with decreasing values of the decay time. Therefore notwithstanding the complicating factors listed previously, accurate information must be available about the development of the concentration distribution within the patches. This especially holds for the prediction of the concentration of continuously released non-conservative effluents.

Figures 3, 4, 5 and 6 are derived by assuming a Gaussian distribution of concentration within the separate patches. In this connection it is of interest to mention that variations of the shape of the distribution of concentration within the separate patches also may have a significant effect on the concentrations at shore.

Conclusions

For the case of continuous release of effluents into a sea with tidal currents in a direction roughly parallel to an approximately straight coastline, the following conclusions were derived; the conclusions hold when the net flow of sea water along the shore is nil or small:

(1) Predictions of concentrations at shore can be made by the superposition method.

(2) In applying the superposition method one must

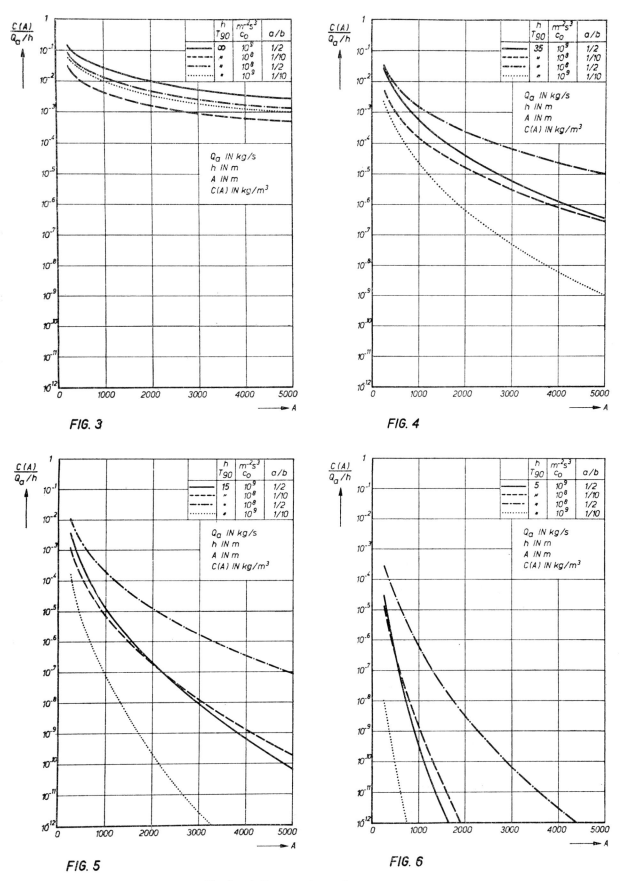

Figs 3 to 6. Concentration at shore, stagnant sea

have information about the development of the concentration of a tracer within separate patches of tracer material. This information can be obtained only from instantaneous release tracer studies.

(3) Because of the complicating factors listed, inaccuracies must be accepted in the results of the experiments referred to in conclusion (2). These inaccuracies may lead to considerable inaccuracies in the predicted shoreline concentrations, especially for non-conservative effluents (see figs 3, 4, 5 and 6). Accurate predictions of shoreline concentrations can be made only if a large number of instantaneous release tracer studies have been performed. (The programme of instantaneous release tracer studies performed along the Netherlands coast covers a period of several years.)

(4) Considerable inaccuracies in the predicted shoreline concentration must be accepted if the accurate value of the decay time T_{90} is not known (as holds for, e.g. coli bacteria, the index organisms for domestic sewage) (see figs 3, 4, 5 and 6). These inaccuracies can be eliminated only by accurate determinations of the T_{90} value.

(5) It is a common practice to select an outfall-length such that shoreline concentrations do not exceed a certain value. If an insufficient number of instantaneous release tracer studies are performed and if no accurate T_{90} values are available it follows from conclusions (3) and (4) that the outfall-length determined by this procedure may be substantially inaccurate.

(6) Finally one can state that the superposition method is the only method available for determining the required outfall-length. However, due to the complex nature of the considered phenomenon a considerable experimental effort is required in order to obtain accurate predictions. Even so, experience in the Netherlands shows that fairly accurate predictions can be obtained.

References

ABRAHAM, G   Formal discussion of paper of P. Harremoes. *In*
1966   Third International Conference on Water Pollution Research, Munich, Germany. Water Pollution Control Federation, Washington, Pap. III–4.

HARREMOES, P   Prediction of pollution from planned wastewater
1966   outfalls. *J. Wat. Pollut. Contr. Fedn*, 38(8):1323–33.

OKUBO, A   A new set of oceanographic diagrams. *Tech. Rep.*
1968   *Chesapeake Bay Inst.*, (38).

SCHÖNFELD, J C   Integral diffusivity. *J. geophys. Res.*, 67(8): 3187.
1962

SCHÖNFELD, J C   Lozing van unit een bron in zee. *Notulen Math.*
1964   *Fys. Afd. Rijkwaterstaat*, (6411) (in Dutch).

VAN DAM, G C   Dispersie van opgeloste en zwevende stoffen in zee
1968   gebracht op 3 km uit de Nederlandse kust ter hoogte van Wijk aan Zee. *Notulen Math. Fys. Afd. Rijkwaterstaat*, (6812) (in Dutch, English summary).

VAN DAM, G C and DAVIDS, J A G   Radioactive waste disposal
1966   and investigations on turbulent diffusion in the Netherlands' coastal areas. *In* Disposal of radio active wastes into seas, oceans and surface waters, Vienna, IAEA, pp. 233–48.

State of the Art for Simulation of Pollution Problems and Controls in Estuaries

*D. J. Baumgartner and R. J. Callaway**

La simulation des problèmes de pollution et de son contrôle en estuaire—état de la question

Ce document passe en revue les applications des techniques de simulation aux problèmes de pollution dans les estuaires et examine les lacunes que présente la méthodologie et qui requièrent des études dans le domaine de la recherche et du développement. Les auteurs considèrent les types d'application suivants:

1) l'utilisation des techniques de simulation à la détermination des sites les plus appropriés au déversement dans les estuaires des effluents domestiques et industriels et l'importance du traitement auquel ces effluents doivent être préalablement soumis;

2) les effets d'une modification physique sur les composants de la qualité de l'eau dans un estuaire;

3) l'évaluation de diverses mesures proposées pour lutter contre les éléments qui nuisent à la qualité de l'eau dans les estuaires et que l'on peut rattacher à un certain nombre de sources de pollution.

Le document décrit les avantages et les limites des modèles mathématiques mono, bi et tri-dimensionnels basés sur l'emploi de diverses valeurs moyennes et de solutions en état d'équilibre.

Les modèles de la qualité de l'eau sont envisagés comme des applications de la description hydrodynamique fondamentale du transfert de masse et de la dispersion dans les estuaires, compte tenu de la multitude d'interactions physiques, chimiques et biologiques qui peuvent agir sur la forme et la concentration des polluants.

Un chapitre porte sur les diverses méthodes qui permettent de résoudre les équations différentielles et sur les conditions qui définissent le modèle mathématique. Les avantages et les limitations des ordinateurs analogiques et digitaux sont comparés.

Dans certain cas, l'utilisation de maquettes hydrauliques se révèle supérieure à celle des modèles mathématiques; ces cas sont décrits dans un bilan sur les limitations présentées par les deux méthodes.

Técnica de la simulación de los problemas de la contaminación y vigilancia en los estuarios

Los autores reseñan la aplicación de las técnicas de la simulación en el estudio de los problemas de contaminación de los estuarios y examinan los vacíos que existen en la metodología y las investigaciones que hay que efectuar. Las aplicaciones que se examinan son:

1) el empleo de las técnicas de simulación para determinar el lugar exacto del estuario en el que se descargan los desechos municipales e industriales e indicar el grado del tratamiento necesario;

2) los efectos de la modificación física en los constituyentes de la calidad del agua de un estuario;

3) la evaluación de otras medidas propuestas para suprimir los constituyentes indeseables de la calidad del agua de los estuarios relacionados con diversas fuentes de contaminación.

Se describen las ventajas y limitaciones de los modelos matemáticos de una, dos y tres dimensiones, que emplean diversos valores medios y soluciones en estado de equilibrio.

Los modelos de calidad del agua se consideran como ampliaciones de la descripción hidrodinámica básica del transporte y dispersión de masas en los estuarios, reconociendo que en la forma y concentración de los contaminantes pueden influir gran número de acciones recíprocas físicas, químaicas y biológicas.

Se dedica una sección a los diversos métodos para solucionar las ecuaciones diferenciales y las condiciones límite de modelo matemático. Se evalúan las ventajas e inconvenientes relativos de las computadoras análogas y digitales.

En algunos casos conviene más emplear modelos hidráulicos físicos que matemáticos, presentándose algunos para equilibrar el estudio de los límites.

---

* National Coastal Pollution Research Program, Federal Water Quality Administration, Environmental Protection Agency, Corvallis, Oregon, U.S.A.

Les auteurs examinent les aspects généraux de la dynamique des populations vivantes du point de vue de la méthodologie du contrôle de la pollution dans les estuaires.

Les éléments qui ont servi de base à cette étude figurent dans un rapport préparé par la TRACOR Corporation d'Austin, Texas, en coopération avec le personnel de la Federal Water Quality Administration et un groupe d'experts-conseils de renom.

Se exponen los aspectos generales de la dinámica de la población biológica con respecto a la metodología de la vigilancia contra la contaminación en los estuarios.

La información examinada para preparar el documento está contenida en un informe preparado por la TRACOR Corporation de Austin, Texas, en cooperación con el personal de la Federal Water Quality Administration y un grupo de consultores distinguidos.

THIS paper reviews the use of modelling to evaluate proposed waste discharges and physical modifications on estuarine water quality, and to assess the relative merits of alternative procedures which might be recommended to control pollution in estuaries.

Much of the material was compiled by the TRACOR Corporation of Austin, Texas, under a research contract with the Federal Water Quality Administration. Their report (TRACOR, 1970) contains chapters on similar subject matter. Additional material and concepts concerning laboratory modelling are also included.

The ultimate objective in building a model is to be able to predict system responses in advance of a real system.

Kinds of simulation techniques

Simulation of complex estuarine systems can be provided either mathematically or physically. Physical models are generally built on empirical relationships, whereas mathematical models are either entirely empirical, entirely deterministic, or a mixture of both.

Models of system behaviour which are based on statistical relationships among observed phenomena are called phenomenological, statistical, or empirical models. The phytoplankton standing crop (M), for example, is known to be influenced by concentrations of predators (C_p), nutrients (C_N), various toxicants (C_T), salinity (S), temperature (T), depth of the photic zone (H), and so on. Conceivably, an entirely acceptable empirical model could be constructed to predict the influence of various levels of one or more of the control factors on the standing crop, viz.;

$$M = AC_p^a + BC_N^b + DC_T^d + ES^e + FT^f + GH^g + \ldots \quad (1)$$

The functional relationship can take a number of forms and can be made more or less extensive in the number of parameters included. The coefficients and exponents (A, a, B, b ...) are evaluated on the basis of observed data in the prototype, and some measure of variability, or confidence, is provided for use of the model. However, there is no guarantee that the relationship will hold for values ranging outside those observed, and little insight is provided as to the mechanisms of the processes which take place within the system.

A significant advance in the structuring of empirical models was provided with the introduction of dimensional analysis, which, through the use of non-dimensional quantities, allowed reduction in the number of necessary parameters (Kline, 1965). This, for example, has led to the development of hydraulic models of estuarine systems where similitude between model and prototype is guaranteed by equality of two dimensionless quantities, a length scale and the Froude number. The same principle is involved in a number of mathematical models for mass and momentum transport mechanisms.

Deterministic models

It is scientifically more sound but not always possible or practical to describe a process in terms of the physical mechanisms within the process rather than by some empirical relationship between cause and effect. If the governing differential equations can be written in sufficient detail and completeness to represent the physical process, or processes, in the region of interest then the equations along with the boundary conditions constitute a model of the system. This type of mathematical model is described as "deterministic". Unfortunately, not all of the processes within an estuarine system have been defined with equal degrees of certainty and even the most advanced deterministic models contain terms, or coefficients, or both, which are less well founded than others. For example, the dependence of phytoplankton mass on the concentration of a critical nutrient associated with a waste discharge previously described by Equation (1) might be more rationally described by:

$$\frac{dM}{dt} = \frac{C_N R}{k_m + C_N}, \quad (2)$$

where R = maximum rate
$k_m = C_N$ at $R/2$.

This equation, the so-called Michaelis-Menten mechanism, is an extension of qualitative relationships observed for enzyme reaction with a single substrate. Extension of this model to represent the total mass of phytoplankton in a complex natural environment should still be considered with scepticism; however, it represents an advance in that it proposes an explanation of nutrient utilization by the organism. The dependence of M on C_p, the predator concentration, might be expressed as a simple function of phytoplankton mass, as a function of both predator mass and phytoplankton mass, or as an even more complex mechanism. The choice depends on how much is known regarding the process and on how much detail is required for practical analysis. Although not explicitly indicated in Equation (1), terms might be added representing transport and dispersion processes. In these, uncertainty exists in evaluating coefficients but there is general acceptance of the form of mechanism.

The processes which must be included in either a deterministic or empirical model depend to some degree on the nature of the problem of interest; however, achievement of the desired prediction might require inclusion of other terms. For example, a pollution control agency might want to know the temperature distribution in an estuary resulting from a power plant cooling water discharge. Even though they have no direct interest in currents or tides, it may be necessary to include them in the system model to determine the temperature patterns to the precision required.

Transport and dispersion

A pollutant discharged to an estuary will move with the mean velocity of the system and disperse according to the turbulence of the system. Well known equations exist for computing the velocities for one- and two-dimensional cases; turbulence is estimated with coefficients of eddy diffusion derived from the distribution of dissolved substances in the estuary or from its physical dimensions.

The relationship describing the time change of a property at a given point is known as the advection–diffusion (A–D) equation:

$$\underset{[1]}{\frac{\partial s}{\partial t}} + \underset{[2]}{u\frac{\partial s}{\partial x}} + \underset{[3]}{v\frac{\partial s}{\partial y}} + \underset{[4]}{w\frac{\partial s}{\partial z}} - \underset{[5]}{\frac{\partial}{\partial x}\left(K_x\frac{\partial s}{\partial x}\right)} -$$

$$- \underset{[6]}{\frac{\partial}{\partial y}\left(K_y\frac{\partial s}{\partial y}\right)} - \underset{[7]}{\frac{\partial}{\partial z}\left(K_z\frac{\partial s}{\partial z}\right)} = \underset{[8]}{\Sigma S} \quad (3)$$

where

S = concentration of the pollutant;

u,v,w = velocities in the x,y,z directions;

K_x, K_y, K_z = eddy diffusion terms in the x,y,z directions;

ΣS = sum of sources and sinks.

In Equation (3) the time change of concentration of a substance at a fixed point [1] is computed from a knowledge of the velocity field (the advective transport terms [2], [3] and [4]), the level of turbulence (turbulent transport terms [5], [6] and [7]) and the rates of discharge and/or uptake of the pollutant, [8].

The A–D equation was applied in estuaries by Stommel (1953) using a steady-state assumption for a well-mixed system, i.e. negligible variations in the vertical and lateral directions as compared with those along the axis of the estuary. He then utilized terms [2], [5], and [8] only. The velocity term was estimated from river runoff and estuary cross sectional areas; the diffusion term from runoff, area and salinity gradient in the x-direction.

O'Connor (1960) and Thomann (1963) extended Stommel's formulation to the non-steady case by including term [1]. Velocity was determined as in Stommel's case; dispersion was computed via an exchange coefficient or from data extrapolated from dye releases.

Shubinski *et al.* (1965) solved the A–D equation for a branched network of interconnecting channels by first calculating, via the equations of motion and continuity, velocities with time in each channel; dispersion was computed from the scale (depth) of each channel according to relationships suggested for estuaries by Orlob (1959). The terms employed by Shubinski *et al.* were [1], [2], [5] and [8]. This was a new approach since an independent calculation of velocities was made prior to insertion in Equation (1) and because the dispersion term was based somewhat less on empiricism.

Leendertse (1970) has recently extended the solution of Equation (3) to include terms [1], [2], [3], [5], [6] and [8]. A well-mixed system in the vertical was again postulated and the horizontal velocity terms, u, v were computed independently from the equations of motion and continuity. Dispersion terms were extrapolated from coefficients derived from laboratory flume tests.

Before the three-dimensional case can be broached it remains to satisfy the most difficult problem in estuarine systems, namely solutions for those systems which cannot be treated as well-mixed. Difficulties exist in solving Equation (3) for given substances since there may be feedback or feedforward interactions which require simultaneous solution of several sets of equations. It is doubtful if a single diffusion coefficient will suffice to accurately describe the turbulent transport terms for each individual constituent in a system. In addition, functional expression of the terms is rather an art form while field estimates are hard to come by and rather shakily extrapolated. An extensive discussion of this problem is included in the TRACOR report (1970). A further limitation in treating the three-dimensional and laterally averaged two-dimensional case arises because the velocity and turbulence terms are usually not known *a priori* and must, ultimately, be computed as an integral part of the simulation procedure.

This discussion has been mainly concerned with soluble pollutants which move passively with the current system. Active transport of nutrients, trace elements, pollutants, etc., is possible via plankton and higher trophic levels. The contribution by such mechanisms is by no means trivial and presents complications which appear capable of solution only by extensive laboratory and field investigations. It appears at this time that attempts to include such processes are best relegated to "box model" approximations which sacrifice precision in the name of the realities associated with such difficult undertakings.

Physical, chemical, and biological interactions

The concentration of a pollutant in an estuary is influenced by the chemical, biological, and physical interactions within the system. For example, heat as a pollutant may be lost due to boundary transfer mechanisms, and toxic chemicals may be altered to non-toxic end products. Studies of the biochemical degradation of dissolved organics in liquid systems having been started in the early 1920's, the fundamental foundation for expression of mechanisms has not been as well established as for the hydrodynamic principles which have been rigorously studied and applied for about 200 years. While significant efforts in the last decade have been made to elucidate multi-component mechanisms, the expression in use almost exclusively is that of first order decay, as shown in Equation (4).

$$\frac{\mathrm{d}c}{\mathrm{d}t} = -K_1 c \quad (4)$$

where K_1 = reaction velocity (decay) constant;

c = concentration of degradable constituent.

This expression has been found useful in most estuarine pollution applications where the concern has been on the effect of the dissolved oxygen resource of the estuary. Modelling of the dissolved oxygen resource has been limited to the one-dimensional case and the quasi two-dimensional case by extension of the work of Streeter and Phelps (1925), by O'Connor (1960) and Thomann (1963). This has been accomplished by including a first order expression for atmospheric oxygen transfer to the estuarine system coupled with the oxygen utilization of the dissolved organics. Few attempts have been made to describe the vertical distribution of dissolved oxygen or dissolved organics because detailed specification of the vertical transfer properties has not been established.

Although less work has been done on temperature, the state of the art appears to be as far or further advanced as the dissolved oxygen–dissolved organics subsystem (Callaway *et al.*, 1969; TRACOR, 1970). There are, for example, several situations for which vertical distributions of temperature can be estimated.

[142]

Thomann *et al.* (1970) contributed a major advancement to estuarine modelling by describing a system of nitrogen compound oxidation and reduction plus algal utilization and release for steady-state one-dimensional systems. All the mechanisms are approximated by first order kinetics. To date, simulation has been attempted on only a few estuarine systems using reaction velocity constants computed from distributions of the various nitrogen components within the same system. Although this is a very recent contribution the mere presence of the simulation technique will undoubtedly speed its use and refinement. However, as the range of estuarine system models includes more and more biological processes, nagging doubts remain about the validity of first order kinetics. Conceivably, systems for phosphorus, iron, sulphur, cobalt, and many trace materials could be constructed with the single proviso that empirical data from natural systems or physical models be available.

DiToro *et al.* (1970) presented a dynamic model of phytoplankton production based on Michaelis-Menten kinetics, light limiting terms and predator-prey relationships. Application to an estuarine system was made by employing a non-steady state, one reservoir, box model approach. The time average used was such as to minimize the contribution of dispersion and the model actually considered only unidirectional advective transport.

There are no operational models to describe the turbidity in an estuary resulting from one or more waste discharges and the effects of sedimentation and re-suspension. The chemical and physical processes which influence precipitation, flocculation, and sedimentation in a system where vertical turbulent fluctuations and density distributions exist, have not yet been specified. Similarly, no effective method is available to compute the distribution of floatables and immiscible liquid films although considerable work is being recorded in the scientific literature on oil films. The contribution of materials contained in tidal flats on water quality is largely unknown, although recent work has been initiated on this problem (Bella *et al.*, 1970).

The myriad interactions which influence the distribution of coliform organisms from waste discharges, has traditionally been grouped in a single first order decay expression. Foxworthy and Kneeling (1969), although experimenting with oceanic rather than estuarine waters, described a model which can be applied to the estuarine system. The mechanism for die-off was written as the sum of two simultaneous first order reactions, and distribution was determined for steady-state lateral and vertical dispersion. Extensive field studies were required in order to obtain enough data for a simultaneous determination of the dispersion and die-off coefficients which were found to be extremely variable. The populations of other bacteria and viruses from waste discharges in estuarine systems must be assessed in relation to protection of marine water quality for living resources and food production. At present there are no models available for animal and plant pathogenic microorganisms, due primarily to the lack of a quantitative technique for identification in natural systems.

Effects

In many cases it is not feasible to represent the mechanism of pollution effects independently of the fate mechanisms.

Where there is no physical coupling between pollutant and a component of the system, the effect relationship can be independently established. For instance, turbidity caused by waste discharge can impede the growth of attached plants by interfering with light. Mathematical models can be written for light intensity on plant growth. The same is true for, say, fish survival or reproductive success as a function of parameters such as temperature and the concentration of a stress agent (Alderdice, 1963). This does not mean that all that needs to be done in quantitatively describing the effects of pollutants on marine resources is on the verge of being announced. For instance much of the stress work reported in literature is related to lethal manifestations while the management of estuarine resources requires the consideration of long-term effects on marine populations (indeed, pollution control technology is such that short-term acute effects can now be avoided). Also, much experimental work has been done under static conditions, whereas environmental simulation requires knowledge of transient effects. Although not covered in the TRACOR report details on effects models may be obtained from Stark (1966), Paulik (1967), Box and Youle (1955), Sjolseth (1968), and Paulik (TRACOR, 1970).

Mathematical solutions via machine methods

Researchers and resource managers must have access to easy means of solving questions once the questions have been properly framed. Hopefully, equation solving will be a minor part of the total effort in the overall pollution problem. For instance, a solution of Equation (4) is given by

$$C = C_i e^{-K_1 t} \qquad (5)$$

where C_i = initial concentration.

If the formulation of the problem is found to suffice through use of Equation (4), then the solution simply consists of table lookup or use of a slide rule. Real-life formulations and solutions of estuarine problems are rarely so easily dispatched; recourse must be made to machine computation which can handle complex geometries and non-linear equations.

Electrical analogue computers

For simulation of estuarine systems electrical analogues provide the greatest advantage in simulation when there are only a few dependent variables in the system and demands for accuracy are not high (TRACOR, 1970). A principal advantage has been that the results of manipulation of one or more of the parameters in the system provide immediate output in a graphical form thus allowing the analyst considerable flexibility in simulating and verification. Although large systems have been built (Riley *et al.*, 1966) for complex river basin systems the use of analogue simulation has not kept pace with the use of digital computer for simulation (TRACOR, 1970).

Digital computers

If the velocity terms are not known throughout the estuarine system they must be calculated as a function of winds, tides, and fresh water inflow responsible for water movement within the system. For a two-dimensional system with vertical uniformity the equations which must be solved to provide the necessary data, as written by Leendertse (TRACOR, 1970) are shown in equations (6), (7), and (8).

$$\frac{\partial u}{\partial t} + u\frac{\partial u}{\partial x} + v\frac{\partial u}{\partial y} + g\frac{\partial \eta}{\partial x} - fv + R(x) = F(x) \qquad (6)$$

$$\frac{\partial v}{\partial t} + u\frac{\partial v}{\partial x} + v\frac{\partial v}{\partial y} + g\frac{\partial \eta}{\partial y} + fu + R(y) = F(y) \qquad (7)$$

$$\frac{\partial \eta}{\partial t} + \frac{\partial[(h+\eta)u]}{\partial x} + \frac{\partial[(h+\eta)v]}{\partial y} + S = 0 \qquad (8)$$

where η = water level referred to a datum;
h = depth of channel referred to a datum;
g = gravitational acceleration;
f = Coriolis parameter;
R = boundary roughness term;
F = forcing function for wind, etc.;
S = lateral inflow or outflow.

The solution of these equations must then be incorporated in terms [2] and [3] of Equation (3) as stated above. The computational time for this can be quite large and usually exceeds or is equal to that for a solution of Equation (3).

For some simplified cases approximations can be made and the number of equations and their complexity reduced so that exact analytical solutions can be written. For most cases, however, the simultaneous solution of the three equations requires the speed and storage capacity of the digital computer. Although the computer time required for the numerical solution of the differential equations can become quite large, it will be only a fraction of the total effort expended in setting up the solution, schematizing the estuary, etc. The availability, access to, and capacity of the computer have to be considered at the outset. The time and space requirements for solution of the differential equations for medium and large systems approach the practical limit for utilization of digital computers generally available in 1969.

Aside from the hydrodynamic computations, the expressions for complex chemical, biological and physical interactions of the pollutants in the system frequently require simultaneous solution of a group of partial differential equations, again requiring a digital computer. For some cases there may also be coupling between the pollutant and the hydrodynamic property making the solution even more complex.

The real time simulation and graphical display advantages previously enjoyed by analogue systems are rapidly being eroded by advancements in digital computer peripheral equipment such as cathode ray tubes and the built-in computer programmes for rapidly converting vast arrays of digits to graphical form and vice versa (TRACOR, 1970).

Hybrid computers

Computing systems made up of dual operations on analogue and digital computers were designed to provide economies in storage on digital computers by performing some operations on the analogue system in parallel with digital storage and/or to take advantage of the graphical display and real time coefficient adjustability offered by analogues. For certain routine operations, this arrangement remains technically and economically attractive.

However, advances in peripheral systems for digital computers are beginning to offer the same attraction (TRACOR, 1970).

Physical models

This section reviews the use of physical models employing hydraulic models, rather than a broad review of all physical analogues.

Estuarine models scaled to the proportions of the prototype system and also to the gravitational forces through use of the Froude number provide the greatest utility for investigation of three-dimensional phenomena of transport and dispersion. Thus, the effects of suspended solids on sediment build-up, the effects of harbour development on distribution patterns and the distribution of wastes from one or more discharges can be observed and measured in scale. Since viscous forces are not guaranteed to be similar in the model and prototype, and the horizontal and vertical scales are frequently not the same, there are situations where the dispersive character of the model is not similar to the prototype (TRACOR, 1970). Consequently, considerable personal skill is involved in conducting simulation and interpreting the results.

The models for major estuarine systems are relatively expensive to construct and operate and usually require a rather large facility for housing the model and the control and measurement devices.

Physical, chemical and biological interaction

Physical models for measurement and visualization of transport and dispersive functions are usually not employed for simulation of other processes, primarily because the time scale is distorted. Most physical, chemical, and biological interactions cannot be time adjusted. Consequently, separate hydraulic models are utilized: in many the mixing processes are not modelled but are provided by pumps or stirrers. Apparatus ranging from simple tanks to elaborate environmentally controlled flow-through channels have been used—a recent example is given by DeBen (1970).

There are two questions which remain unsatisfactorily answered by tank simulation. One is the scale effect and the other is the effect of physical interaction of dispersive properties on the biochemical interactions. For instance, the rate of growth of phytoplankton is dependent on vertical light integration and distribution of phytoplankton mass in the water column.

Perhaps the greatest physical model utility is provided through a study of portions of the natural environment. Naturally occurring or artificially constructed enclosures of estuaries, bays or other coastal environments have been used to study both interactions and effects. The obvious problems are lack of control capability for some parameters and the logistics of operating the system. The major disadvantages are the limited range and limited frequency of parameters.

Economic impact of damages

Estuarine management agencies are primarily motivated to clean up pollution problems. Mathematical simulation techniques for economic analysis have been available for some time (Dorfman et al., 1958). However, the state of the art for simulation techniques advanced more quickly than appropriate quantitative economic data could be supplied regarding the effects of pollution on marine water uses (Kneese, 1962). The economic value of the Delaware estuary fishery resource was considered and evaluated with regard to costs required to improve dissolved oxygen concentrations (Federal Water Pollution Control Administration, 1966). The difficulties are basically two-fold. First, the economist needs input from the marine biologist on what quantitative effects various water quality levels will have on marine organisms, and second, he must determine how individuals and social institutions value various levels of the marine resource. Although not related specifically to the estuaries, a good review of simulation techniques for systems analysis including technological, economic, and management inputs is provided by Maass et al. (1966). More recently, Kneese and Bower (1968) have emphasized the difficulties in determining economic benefits associated with technical alternatives.

The only physical models of the estuarine economic processes are those which are segments of the prototype. The same segment of a given economy can be effective in providing estimates of what might happen to other economies under various sets of technological influences. The implications of the statistical theory of sampling are evident.

The National Estuarine Pollution Study report (U.S. Department of the Interior, 1970) concludes that economic research together with management and legal controls is perhaps the most important area in need of research.

Model verification

The generality of estuarine systems models depends upon the range of situations for which they have been verified and is related to the number and types of processes which are meant to be included and the scales of space and time.

Verification of distorted hydraulic models is a highly refined art. In the main, it consists of adding roughness elements to the bed of the model to provide distribution patterns similar to the prototype, whereas in the mathematical model the same result is achieved by numerical adjustment of the roughness coefficient. The process is usually conducted over a wide range of conditions since the setting of the roughness elements for a given set of tidal conditions does not guarantee the proper dispersion for another set of tidal conditions. In general, such models cannot be applied to the investigation of transport and dispersion process in any other prototype estuary, whereas mathematical models can usually accommodate different sets of input data and hence can easily represent different estuaries.

Laboratory scaled hydraulic models of chemical, physical and biological interactions and effects may be adjusted in much the same way to produce a degree of similarity between model and prototype. Arbitrary adjustments of bottles, mixers, light intensity, and scale factors seem to be all that is available. For some physical-chemical mass transfer operations which can be scaled on the basis of non-dimensional groups such as the Schmidt number, verification is hampered due to the difficulty of measuring the same kind of phenomena in the prototype and the laboratory model. We are not aware of a comprehensive review on this topic and are reluctant to offer an opinion on its general applicability.

For complete system models there is the question of general applicability within the system. It is not unusual for some parameters to be evaluated not by external measurements but by numerical adjustment so that the simulated output for one or more components of the system match what has been observed in the prototype. For example, the roughness coefficient in the equations of motion may be adjusted to assure matching of the tidal heights and tidal velocities, and the first order decay coefficient in the advection-diffusion equation for dissolved organics may be adjusted to assure matching of the model output and observations of the dissolved organics in the prototype. This is an acceptable and very useful procedure if observations are taken over a sufficiently long and variable range of natural conditions.

Technological gaps and research needs

As it stands now, the state-of-the-art is satisfactory, because of accumulated knowledge or the ability to collect data in a short time period.

But as we attempt to build long-range ecological response models, deficiencies in our understanding of the chemical and biological interaction terms will become more apparent. Reactions and reaction rates which on the short term appear to be insignificant may be vital in the long run. The best experimental approach to modelling future needs regarding DDT and PCB appears to be the use of large-scale environmental simulators closely coupled with prototype verification of pathways and reservoirs.

Intermediate between acute toxicity and long-term ecological problems is that of excessive undesirable growth, of either phytoplankton or attached plants. Advanced kinetic models are required to assess the impact of waste discharges on this measure of estuarine water quality. It is at once apparent that two-dimensional horizontal transport models are not sufficient to be compatible with step advances in the biological modelling process. The major need here is for the development of two-dimensional models which can accommodate vertical non-uniformities.

Some transport and dispersion models are not satisfactory for description of time and space distribution of pollutants very close to the discharge point. The major research need is for models describing the distribution of the transition region between present plume and estuary models.

References

ALDERDICE, D F Some effects on simultaneous variation in salinity,
1963   temperature and dissolved oxygen on the resistance of young coho salmon to toxic sub-substances, *J. Fish. Res. Bd Can.*, 20(2):525–50.

BAUMGARTNER, D J and TRENT, D S Ocean outfall design. Part 1,
1970   Literature review and theoretical development. Corvallis, U.S. Department of the Interior, Federal Water Pollution Control Administration, 129 p.

BELLA, D A, RAMM, A and PETERSON, P Effects of mud flats on
1970   estuarine water quality. Paper presented at Annual Meeting Pacific NW Pollution Control Association, Victoria, B. C. October, 1970.

BOX, G E P and YOULE, P V The exploration and exploitation of
1955   response surfaces: an example of the link between the fitted surface and the basic mechanism of the system. *Biometrics*, 11:287–323.

CALLAWAY, R J, BYRAM, K V and DITSWORTH, G R Mathematical
1969   model of the Columbia River from the Pacific Ocean to Bonneville Dam. Corvallis, U.S. Department of the Interior, Federal Water Pollution Control Administration, 155 p.

DeBen, W A Design and construction of a saltwater environ-
1970 mental simulator. Corvallis, U.S. Department of the Interior, Federal Water Quality Administration, Working Pap. (71):30 p.

DiToro, D M, O'Connor, D J and Thomann, R V Modeling of
1970 the nitrogen and algal cycles in estuaries. Paper presented at Fifth International Water Pollution Research Conference, San Francisco, California.

Dorfman, R, Samuelson, P A and Solow, R M Linear programm-
1958 ing and economic analysis. New York, McGraw-Hill, 527 p.

Federal Water Pollution Control Administration, Delaware estuary
1966 comprehensive study, preliminary report and findings. Philadelphia, U.S. Department of the Interior, 94 p.

Foxworthy, J E and Kneeling, H R Eddy diffusion and bacterial
1969 reduction in the ocean. Los Angeles, Allan Hancock Foundation, University of Southern California, 176 p.

Kline, S J Similitude and approximation theory. New York,
1965 McGraw-Hill, 229 p.

Kneese, A V Water pollution: economic aspects and research
1962 needs. Washington, Resources for the Future, 107 p.

Kneese, A V and Bower, B T Managing water quality: economics,
1968 technology, tuitions. Baltimore, Johns Hopkins Press, 328 p.

Leendertse, J J A water quality simulation model for well-mixed
1970 estuaries and coastal seas. Principles of computation. Santa Monica, The Rand Corporation, vol. 1:71 p.

Maass, A M et al. Design of water-resource systems. Cambridge,
1966 Harvard University Press, 620 p.

National Academy of Sciences, Committee on Oceanography,
1970 National Academy of Engineering, Committee on Ocean Engineering, Waste management concepts for the coastal zone—requirements for research and investigation. Washington, 126 p.

Orlob, G T Eddy diffusion in homogeneous turbulence. J.
1959 Hydraul. Div. Am. Soc. civ. Engrs, 85(HYD):75–101.

O'Connor, D J Oxygen balance of an estuary. J. sanit. Engng
1960 Div. Am. Soc. civ. Engrs, 86(SA3):35–55.

Paulik, G J Digital simulation of natural animal communities. In
1967 Pollution and marine ecology, edited by T. A. Olson and F. J. Burgess. New York, Interscience Publishers, 364 p.

Riley, J P, Chadwick, D G and Bagley, J M Application of
1966 electric analog computer to solution of hydrologic and river basin planning problems: Utah simulation model. Logan, Utah State University, 129 p.

Sjolseth, D E A study of some factors possibly affecting the
1968 distribution of salmon and plankton in Bellingham Bay, Seattle, Fisheries Research Institute, University of Washington, 16 p.

Shubinski, R P, McCarty, J C and Lindorf, M R Computer
1965 simulation of estuarial networks, American Society of Civil Engineers Water Resources Engineering Conference, Mobile, Alabama, Conf. Prepr., (168):28 p.

Stommel, H Computation of pollution in a vertically mixed
1953 estuary. Sewage Ind. Wastes, 25:1065–71.

Streeter, H W and Phelps, E B. A study of the pollution and
1925 natural purification of the Ohio River. Bull. U.S. publ. Hlth. Serv., (146):75 p.

Stark, R W The organization and analytical procedures required
1966 by a large ecological systems study. In Systems analysis in ecology, edited by K. E. F. Watt, New York, Academic Press.

Thomann, R V Mathematical model for dissolved oxygen, J.
1963 sanit. Engng Div. Am. Soc. civ. Engrs., 89(SA5):30 p.

Thomann, R V, O'Connor, D J and DiToro, D M Modeling of
1970 the nitrogen and algal cycles in estuaries. Paper presented at Fifth International Water Pollution Research Conference, San Francisco, California.

Tracor Estuarine water quality modeling. An assessment of
1970 pollution control capabilities. Work performed under contract 14–12–551 for the Federal Water Quality Administration, U.S. Department of the Interior, 3 vols.

U.S. Department of the Interior, The National estuarine pollution
1970 study, Report of the Secretary of the Interior to the United States Congress, Senate Doc. (No. 91–88) 25 March, 1970.

Investigation and Prediction of Dispersion of Pollutants in the Sea with Hydrodynamical Numerical (HN) Models

P. M. Wolff,* W. Hansen† and J. Joseph‡

Étude et prévision de la dispersion des polluants dans la mer avec modèles numériques hydrodynamiques (HN)

L'étude expérimentale du mouvement et de la diffusion des polluants dans les estuaires et les eaux côtières coûte cher et prend du temps. Cependant, à l'aide de grands ordinateurs, on peut appliquer des modèles numériques hydrodynamiques aux problèmes de la pollution en tout emplacement dont on connaît avec précision la distribution des profondeurs et dans des conditions limites bien définies (marées, arrivée d'eau douce, vent et source de pollution).

Les auteurs donnent les formules hydrodynamiques de base, la formule des différences finies de W. Hansen et le programme d'ordinateur permettant de calculer la répartition synoptique et le régime des courants, et la diffusion des polluants dans les estuaires et les eaux côtières. La grille utilisée dans les exemples donnés varie de 1 à 18 km, et la période de référence de 12 à 300 secondes. On calcule le mouvement aux emplacements souhaités en se fondant sur la mesure nette des courants de marée et des courants aériens. On peut également, le cas échéant, tenir compte de la composante thermohaline des courants. Certains problèmes spéciaux de délimitation et le traitement qui leur est apporté sont décrits dans le document.

Le modèle comporte le schéma de diffusion de Joseph-Sendner sous une forme modifiée, ce qui permet d'établir la répartition zonale de la concentration des polluants à divers moments du processus de production. Il est prouvé que l'on peut également traiter sous forme numérique, dans le même modèle, la dégradation chimique, radiologique et biologique des polluants.

Les auteurs donnent quelques exemples et vérifications de l'application du modèle. Ils font également une brève description d'un modèle multi-couches plus élaboré, accompagnée d'exemples des résultats qu'il permet d'obtenir.

Investigación y predicción de la dispersión de contaminantes en el mar, con modelos hidrodinámicos numéricos (HN)

La investigación experimental de la dispersión y difusión de contaminantes de los estuarios y en las aguas costeras es costosa y requiere mucho tiempo. Sin embargo, gracias al empleo de grandes computadoras es posible aplicar modelos hidrodinámicos numéricos a los problemas de la contaminación en cualquier zona, especificando convenientemente la distribución batimétrica y las condiciones del contorno (mareas, afluencia de agua dulce, viento y fuente de contaminación).

Se dan las fórmulas hidrodinámicas fundamentales, el modelo de diferencia finita de W. Hansen y el programa de computadora que sirven para computar la distribución sinóptica y el comportamiento de las corrientes y su difusión en los estuarios y aguas costeras. En los ejemplos presentados, las cuadrículas de extensión geográfica varían de 1 a 18 km y los intervalos de tiempo de 12 a 300 segundos. La dispersión se computa, en las zonas deseadas, teniendo en cuenta solamente las mareas y las corrientes de aire. Si se desea, se puede tener también en cuenta la influencia de la convección termohalina. Se describen algunos problemas especiales y la forma de abordarlos.

En el modelo se incluye en forma modificada el método de Joseph-Sendner para establecer la difusión, que permite determinar la distribución, por zonas, de la concentración en los momentos deseados. Se muestra cómo es posible tratar en forma numérica, con el mismo modelo, la degradación química, radiológica y biológica de los contaminantes. Se dan algunos ejemplos de la forma de aplicar el modelo y se presentan algunas verificaciones. Sigue una breve descripción de un modelo poliestratificado más avanzado y algunos ejemplos de los resultados obtenidos.

PROBLEMS of diffusion and transport of substances in the sea have received considerable theoretical and experimental attention in recent decades— especially radioactive pollutants (Joseph and Sendner 1958; Okubo 1970; and Schönfeld and Groen 1961). Most attempts to find accurate, universal solutions to dispersion formula have encountered the same difficulties:

* Fleet Numerical Weather Central, Monterey, California, U.S.A.
† Institut für Meereskunde, Universität Hamburg, Germany (FRG).
‡ IAEA Laboratory of Marine Radioactivity, Monaco, Principality of Monaco.

(a) Numerical solutions require simultaneous treatment of many variables (diffusion, transport, decay, etc.) all of which vary with time and are interdependent.

(b) Specification of the velocity field has been incomplete. Fluctuations in transport and turbulence vectors were not known. Averaging over extended periods introduced errors because the processes are non-linear.

Two recent developments show promise in handling these complex problems through (a) application of Hydrodynamical Numerical models which include diffusion formulae in finite difference form; and (b) use of large computers for synoptic analysis/prediction in the oceans.

The combined effect of these developments permits simultaneous, numerical solution of complex formulations in short time steps. The solutions are essentially Lagrangian in form although certain Eulerian treatments will be indicated where possible.

HN models and their verifications

The Hydrodynamical Numerical model is based on the following hydrodynamical formulas:

$$\frac{\partial u}{\partial t} - fv - v\Delta u + \frac{\gamma}{H}u\sqrt{u^2 + v^2} + g\frac{\partial \zeta}{\partial x} = X + \frac{\tau^{(x)}}{H}$$

$$\frac{\partial v}{\partial t} + fu - v\Delta v - \frac{\gamma}{H}v\sqrt{u^2 + v^2} + g\frac{\partial \zeta}{\partial y} = Y + \frac{\tau^{(y)}}{H}$$

$$\frac{\partial \zeta}{\partial t} + \frac{\partial}{\partial x}(Hu) + \frac{\partial}{\partial y}(Hv) = 0$$

where ζ is surface elevation, H—depth, γ—friction coefficient, v—coefficient of horizontal eddy viscosity, τ—components of wind stress, X, Y—components of external force and other symbols are conventional for hydrodynamics.

These equations are solved numerically in finite difference form. For presentation of these formulae see Hansen 1956, 1966 (for an operational description of Hansen's model, see Laevastu and Stevens, 1969). The model requires the use of a medium or large computer, as many arrays must be available in core memory, and the time step is short (determined by Friedrich-Courrant-Lewis criterion and varies from 6 sec to 10 min, depending upon grid size and maximum depth).

An example of the grid used is in fig 1. The model includes bottom friction, surface wind stress and atmospheric pressure effects. The tides are introduced at open boundaries or at the coast with as many constituents as are available and/or significant. Fresh-water inflow can be included. Recently, treatment of three open boundaries has been studied and two slightly different two-layer HN models are operational (Hansen, unpublished).

Verifications of results from these HN models have been done at the Oceanographic Institute, University of Hamburg and at Fleet Numerical Weather Central in Monterey, California. In general, it has been found that the model reproduces sea level changes (tides and storm surges) quite well. Although little synoptic information on currents is available, the few observations provided are in good agreement with results from the HN model.

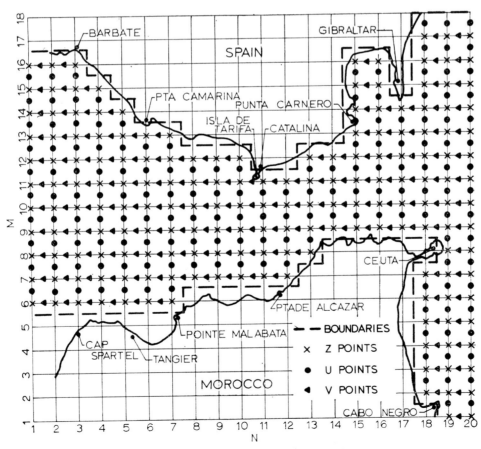

Fig 1. Example of a grid net for HN model

[147]

Dispersion equations

As HN models provide a current vector at each grid point at small time steps one is tempted to solve the interacting advection and diffusion problem using time-stepping finite difference method. In this paper, we treat the dispersion of substances which can be readily mixed in shallow water or in a surface mixed layer. If these problems can be solved by the HN model, no difficulties are foreseen in problems such as dispersion of oil, brackish water, etc. The following were considered in designing the dispersion portion of the model:

(a) Vertical diffusion should be neglected (can be added to multi-layer models).

(b) A basic Lagrangian approach should be used because of its application to computer solutions for grid arrays in finite-difference time stepping.

(c) Eulerian modifications should be used for instantaneous releases where the size of the blob is small in comparison with the mesh length.

(d) The grid should be fine enough to permit neglect of nonlinear terms.

(e) Fickian diffusion with constant diffusivity is used since advection and diffusion equations are solved separately in short, interlocking time steps (similar to the "two-step method" of Schönfeld and Groen, 1961).

(f) The Austausch coefficient for eddy diffusion *per se* must be related to grid size.

Diffusion in water bodies has been presented by many formulae, a few of which are presented here (neglecting vertical diffusion).

The general diffusion formula is:

$$\frac{\partial^2 S}{\partial x^2} + \frac{\partial^2 S}{\partial y^2} - \frac{1}{A}\frac{\partial S}{\partial t} = 0$$

The basic dispersion formula:

$$\frac{\partial S}{\partial t} = Y - \frac{S}{n} - \frac{\partial}{\partial x}(u_x + P_x^s) - \frac{\partial}{\partial y}(u_y + P_y^s)$$

The Fickian equation:

$$\frac{\partial S}{\partial t} = Y - \frac{S}{n} - K\nabla^2 S - \frac{\partial}{\partial x}(S_u u) - \frac{\partial}{\partial y}(S_v v)$$

where:

Y—addition (release), n—decay, S—concentration, $P_{x,y}^s$ concentration velocity component, t—time $K = \beta a v_p$; a — depth, $V_r = \sqrt{u^2 + v^2}$, $\beta = 0.003$ A—diffusion coefficient.

The Lagrangian approach of diffusion we have adopted is given by:

$$S_{n,m}^{t+\tau} = S_{n,m}^t\left(1 - \frac{4\tau A}{l^2}\right) + \frac{A\tau}{l^2}(S_{n-1,m}^t + S_{n+1,m}^t + S_{n,m-1}^t + S_{n,m+1}^t - 4S_{n.m}^t)$$

where:

t is time, τ—time step, l—grid size, S—concentration, A—diffusion coefficient, n and m are grid coordinates and u and v are velocity components.

The above finite equation deviates from the usual finite difference diffusion formula only in the addition of the last term ($-4S$), which makes the solution similar to the solution of the "Laplacian" ($K\nabla^2 S$). Secondly, the

advection is computed linearly in finite difference form:

$$S_{n,m}^{t+\tau} = S_{n,m}^t - \tau|u_{n,m}^{t+\tau}|\left(\frac{S_{n,m\pm1}^{tt} - S_{n,m\pm1}^t}{l}\right) - \tau|V_{n,m}^{t+\tau}|\left(\frac{S_{n,m}^t - S_{n\pm1,m}^t}{l}\right)$$

where $Sn, m-1$ or $n, m+1$ (respectively $n-1, n+1$) are used, depends on the direction (sign) of U and V.

Due to the short space and time steps in the HN model, it was found that secondary terms such as the following can be neglected:

$$1/2\,(U_{n,m} - U_{n,m-1})(S_{n,m} - S_{n,m-1}) + 1/2\,(U_{n,m+1} - U_{n,m})(S_{n,m+1} - S_{n,m})$$

If the grid size is large in relation to the size of the dispersing cloud (or the time step is long), linear interpolation of gradients introduces errors. In this case, an additional term (empirical) has been added to the transport components:

$$-C\left(\frac{S_{n,m} - S_{n,m\pm1}}{|u,v|\tau l}\right)$$

Numerical experiment shows, however, that the last term is relatively unimportant.

If the diffusion formula alone is used, A can be determined from the relations given by Joseph and Sendner (1958).

$$A = \frac{Pl}{2}$$

where $P \simeq 1.5$ cm/sec. This formula requires a diffusion coefficient of 1.5×10^5 for our numerical runs for the Strait of Gibraltar. If diffusion and advection are treated separately, a much smaller A is needed ($A = 3.0 \times 10^3$).

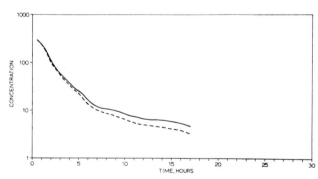

Fig 2. Example of dispersion from an instantaneous release of 300 units in Strait of Gibraltar. (Change of maximum concentration with time; full line: $A = 3.0 \times 10^4$; dashed line: $A = 1.4 \times 10^5$).

Preliminary results

Figures 2 to 4 present some preliminary results of dispersion computations in the Strait of Gibraltar. Figure 2 shows the dispersion from an instantaneous release of 300 arbitrary units. If we compare the results with an experiment conducted by Hela and Voipio (1960) and express the maximum concentration change with time with their formula:

$$\log C = \log C_1 - m \log t,$$

we find that the coefficient m in our computation is 1.5 with $A = 3.0 \times 10^4$ and 1.65 with $A = 1.4 \times 10^5$ as compared to 2.4 in Hela and Voipio's experiment. The difference is apparently because (a) our computation did

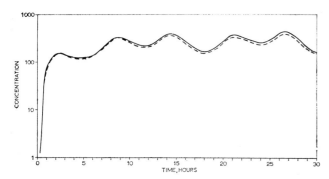

Fig 3. Example of dispersion from a continuous release of 5 units per minute in Strait of Gibraltar. (Change of maximum concentration with time; full line: $A = 3.0 \times 10^4$; dashed line: $A = 1.4 \times 10^5$)

not include decay and (b) there was some eddy diffusion in the vertical in the Hela-Voipio experiment.

Figure 3 shows the change of maximum concentration with time from a continuous release of five units per minute. Note that the maximum concentration is reached asymptotically and fluctuates in a tidal rhythm. With this model it is thus possible to compute either the maximum concentration from a given continuous release or the release level if a maximum permissible concentration is prescribed. The areal distribution of the concentration after 30 hours of continuous release is given in fig 4, which also shows the computation of dispersion (diffusion) without transport for comparison.

If the dimensions of a blob are small in relation to grid size (in case of an instantaneous release), it is preferable

to use an Eulerian approach initially. In this treatment the centre of the blob is advected by the currents, and its position is recorded in every time step. The concentrations at neighbouring grid points and the change of central concentration are computed from the Joseph Sendner (1958) formula. This is repeated until a number of surrounding grid points have concentrations above a predetermined minimum level, whereafter the Lagrangian treatment is continued.

Decay (uptake by biota, remineralization, etc.) was not included in the tests reported above; however, inclusion would not require special programming. Furthermore, "thermal pollution" can be treated by including the heat exchange terms, as done synoptically in FNWC (Laevastu and Hubert, 1965). The model must be fully tested and tuned before it could be universally applied.

Summary

(a) The Hydrodynamical Numerical (HN) models provide the velocity field in small time steps and are suited for inclusion of dispersion formulae in definite difference form.

(b) The HN models reproduce the changes of sea levels and currents very well. As the treatment of open boundaries as well as multiple layers has been successfully solved recently, these models have wide application.

(c) The dispersion formulae (including diffusion, transport and decay) have been programmed

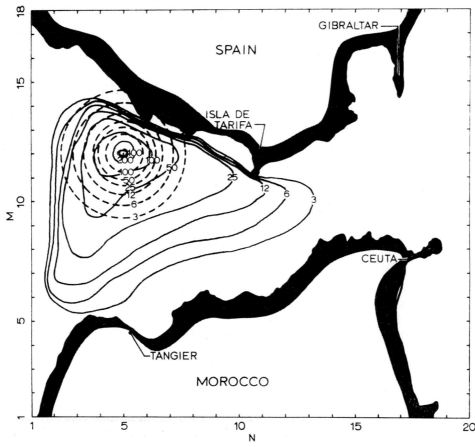

Fig. 4. Example of areal distribution in Strait of Gibraltar after 30 h from a continuous release of 5 units per minute. (Full line, diffusion and transport; dashed line, diffusion only)

into HN models in Lagrangian form and solved in small time steps. They can also be used in Eulerian form when the pollutant field is small initially in relation to grid size.

(d) Test results show that this model can be used for computation of pollutant distribution at any given time and determination of the level of release if a maximum permissible concentration is given.

References

HANSEN, W Theorie zur Errechnung des Wasserstandes und der
1956 Strömungen in Randmeeren nebst Anwendungen. *Tellus*,
8(3):287–300.

HANSEN, W The reproduction of the motion in the sea by means
1966 of hydrodynamical numerical methods. *Mitt. Inst. Meeres.
Univ. Hamburg*, 5: 57 p.

HELA, I and VOIPIO, A Tracer dyes as means of studying turbulent
1960 diffusion in the sea. *Suomal. Tiedeakat. Toim. (F.)*, 69A:9 p.

JOSEPH, J and SENDNER, H Über die horizontale Diffusion im
1958 Meer. *Dt. hydrogr. Z.*, 11(2):49–77.

LAEVASTU, T and HUBERT, W E Analysis and prediction of the
1965 depth of the thermocline and near-surface thermal structure.
Tech. Notes Fleet numer. Weath. Facil., Monterey, (10):
90 p.

LAEVASTU, T and STEVENS, P Applications of hydrodynamical-
1969 numerical models in ocean analysis/forecasting. Part 1. The
single-layer models of Walter Hansen. *Tech. Notes Fleet
numer. Weath. Facil., Monterey*, (51):45 p.

OKUBO, A Oceanic mixing. *Rep. Chesapeake Bay Inst.*, 62(70–1):
1970 119 p.

SCHÖNFELD, J C and GROEN, P Mixing and exchange processes.
1961 Radioactive waste disposal into the sea. *Saf. Ser. int. atom.
Energy Ag.*, (5):100–32.

Studies on Biodegradation of Organic Materials in the Deep Sea*

*H. W. Jannasch** and K. Eimhjellen†*

Études sur la biodégradation des matières organiques dans les grands fonds

Des expériences de dégradation microbienne récemment effectuées *in situ* comportaient le dépôt d'échantillons de matières organiques simples et complexes sur le fond ou près du fond de l'océan, à des profondeurs comprises entre 2 000 et 5 000 m. Certains des récipients contenant les échantillons étaient pourvus d'un mécanisme permettant l'ensemencement *in situ* par pression hydrostatique. D'autres échantillons avaient été ensemencés au préalable soit avec des colonies bactériennes mixtes prélevées dans les eaux de surface, soit avec des cultures pures aux propriétés métaboliques connues, y compris des souches psychrophiles de bactéries isolées précédemment dans les fonds marins. Les échantillons ont été exposés aux conditions du milieu des grandes profondeurs pendant une période de 2 à 6 mois. On a mesuré les activités microbiennes, notamment la croissance bactérienne, la production d'ammoniaque, l'incorporation de radio-carbone organique marqué et la formation de gaz carbonique marqué.

Ces expériences montrent que les taux d'activité microbienne sont étonnamment faibles à grande profondeur si on les compare avec des témoins soumis à des températures d'incubation analogues. On ne connaît pas d'explication simple pour le phénomène observé, si ce n'est le rôle de la pression joint à une moindre aptitude d'adaptation des colonies microbiennes naturelles. Cela présente un intérêt pratique considérable si l'on envisage le rôle préservateur du milieu des grandes profondeurs sur le potentiel de biodégradation des matières organiques. L'accumulation de déchets organiques dans ce milieu associée à l'action lente, mais réelle, des courants de fond pourraient donner naissance à des situations dangereuses. Des études détaillées sont en cours pour obtenir des évaluations plus précises des taux de dégradation et pour estimer leurs effets potentiels à long terme sur la multiplication microbienne dans les profondeurs marines.

Estudios sobre la biodegradación de materias orgánicas en alta mar

Se han efectuado recientes experimentos *in situ* sobre degradación microbiana depositando muestras de sustratos complejos y orgánicos en los fondos marinos o cerca de ellos, a profundidades de 2 000 a 5 000 m. Algunos de los recipientes que contenían las muestras se equiparon con un instrumento para la inoculación *in situ* mediante presión hidrostática. Otras muestras se inocularon previamente, ya con poblaciones bacteriales mixtas obtenidas de aguas superficiales o con cultivos puros de propiedades metabólicas conocidas, comprendiendo razas sicrófilas de bacterias aisladas previamente en alta mar. Las muestras objeto de estudio se expusieron a las condiciones ambientales de alta mar de dos a seis meses. La medición de las actividades microbianas comprendía proliferación, producción de amoníaco, incorporación de carbono orgánico radiactivo marcado y formación de dióxido de carbono marcado.

Han sorprendido estos experimentos porque indican que en alta mar son lentas las actividades microbianas en comparación con los testigos incubados a iguales temperaturas. No existe una explicación fácil del fenómeno observado excepto por lo que se refiere a los efectos de la presión en combinación con las limitadas capacidades de adaptación de la población natural microbiana. Considerado como un efecto conservador de las condiciones ambientales de aguas profundas sobre la biodegradabilidad de las materias orgánicas, el fenómeno es del mayor interés práctico. La acumulación de materias orgánicas de desecho en alta mar, en relación con las lentas pero constantes corrientes del fondo, podría crear situaciones peligrosas. Se efectúan estudios detallados para obtener estimaciones más precisas de las tasas de degradación y los posibles efectos a largo plazo sobre los aumentos de microbios en aguas profundas.

INCREASING quantities of wastes produced in highly populated coastal areas create serious disposal problems. Biological, chemical, and physical oceanographers, health officials, and engineers study the effects of coastal pollution and the possibilities of designing offshore and deep-sea disposal operations as an alternative to expanding land-operated treatment plants.

The efficiency of such disposal operations will ultimately depend on the rates and completeness of biodegradation in the particular deep-water environment. Information has been gathered for several decades on the corresponding processes in soil, lakes, and rivers. Study of the various parameters of life in the deep sea, however, is relatively recent, and information on rates of bio-degradation of organic matter in deep waters is virtually absent. Research in this laboratory has been directed toward qualitative and quantitative measurements of microbial growth and biochemical transformations in surface sea water.

On 16th October, 1968, the research submersible *Alvin* of the Woods Hole Oceanographic Institution sank in about 1,500 m of water 135 mi southeast of Woods Hole, Massachusetts. On 1st September 1969, when

* Contribution No. 2552 from the Woods Hole Oceanographic Institution, supported by Grant No. 20956 of the National Science Foundation.
** Woods Hole Oceanographic Institution, Woods Hole, Massachusetts 02543, U.S.A.
† Department of Biochemistry, The Technical University, Trondheim, Norway.

[150]

the *Alvin* was brought to the surface, one of the items recovered from the water-filled pressure hull was the crew's lunch. From general appearance, taste, smell, certain bacteriological and biochemical evidence, the food materials were in a strikingly well-preserved condition. Brought under refrigeration at similar temperature to the deep sea (ca 3°C), the starchy and proteinaceous materials submersed in sea water spoiled in a few weeks.

Possible implications of this unexpected finding caused us to make some additional observations. There was no evidence of reducing conditions or a noticeable lack of dissolved oxygen in the pressure hull of the vessel nor in the box containing the food materials. In addition, there was no evidence of the presence or the possible leakage of a soluble material that could have acted as a preservative.

When pieces of bread were streaked on sea-water agar, bacteria and moulds grew profusely. Placed in tubes with sterile sea water and kept at 3°C, the bread decayed with slight gas production (floating to the surface) within 6 weeks. The slices of meat (boloney) were greyish on the outside but still pink in the centre. Submerged in sterile sea water, the meat spoiled with a putrefactive smell within 4 weeks at 3°C and within 5 days at 30°C. Two apples found in the lunch box had a pickled appearance but showed no sign of obvious decay. The soup, originally prepared with hot (not boiling) water from canned meat extract, was perfectly palatable in hot and cold condition. Samples of this broth showed a maximum turbidity caused by bacterial growth in 22 days when incubated at 3°C and in 5 days when incubated at 30°C. Sporeforming bacteria were observed while the majority of bacteria were represented by gram-negative rods that grew well on sea-water media.

In short, the food materials recovered from the *Alvin* exhibited a degree of preservation that, in the case of fruit, equalled that of careful storage and, in the case of starch and proteinaceous materials, appeared to surpass that of normal refrigeration.

Experiments

With the co-operation of the Dept. of Physical Oceanography, WHOI, several sample containers were lowered from deep-sea moorings to depths of 4,300 and 5,300 m and exposed to deep-sea conditions for periods of 2 to 5 months. Control samples were kept in the laboratory at 3°C.

Dissolved substrates were submersed in pressure-equalizing bottles and in syringes. The materials included peptone, yeast extract, a variety of carbohydrates and amino acids. For determining degradation, chemical analyses were employed when the initial concentrations of the particular substrate exceeded 10 $\mu g/ml$. At lower concentrations, liquid scintillation counting of C^{14}-labelled compounds was used. A more detailed description of techniques, methods and results will be published elsewhere (Jannasch *et al.* 1971).

The results of one representative set of experiments are shown in Table 1. The exposure time of the samples was 8 weeks at a depth of 5,300 m (location 33° 58′N, 70°W); the controls were kept at 3°C for 6 weeks. The substrates used were uniformly labelled with C^{14} and the recovery of C^{14} was excellent. The data given in columns 2 and 3 of the table are re-converted in μg substrate per total volume of the sample (120 ml). The relative degradation in the sample, corrected for exposure time, and expressed as a percentage of the controls is shown in column 4.

In the controls the ratio of the amount of C^{14}-labelled CO_2 to the amount of particulate C^{14}-labelled carbon ranged from 1.5 to 3.4. In the deep-sea samples, however, the amount of C^{14}-labelled CO_2 was too small for significant measurements and very low relative to the amount of C^{14}-labelled carbon in the particulate fraction. For this reason, only the conversion of substrate into particulate carbon is given in Table 1. It is conceivable that dissolved products other than CO_2 had been formed, although in none of the experiments was oxygen a limiting factor.

The percentage of the substrate degraded in the sample relative to that in the controls ranged from 0.15 to 12.9, i.e. the degradation in the deep-sea samples was 10–100 times slower than in the controls. With exception of the casamino acids, the percentage appears to decrease with decreasing concentration of the particular substrate. In addition, based on the turnover of organic carbon, the nitrogenous substrates degraded 2 to 4 times as fast as the carbohydrates.

TABLE 1 MICROBIAL DEGRADATION OF SOME ORGANIC SUBSTRATES IN THE DEEP SEA

Substrate	μg substrate added	μg substrate in particulate fraction		Per cent degradation relative to controls
		Control	Sample	
Acetate	3,600	88.9	3.58	3.0
,,	1,200	146.1	2.08	1.07
,,	600	138.7	0.29	0.15
,,	240	16.2	0.073	0.34
Mannitol	3,600	166.6	3.45	1.55
,,	1,200	46.0	1.06	1.7
,,	600	41.1	0.6	1.1
,,	240	40.1	0.13	0.24
Na-Glutamate	3,600	252.5	6.50	1.9
,,	1,200	130.6	1.66	0.95
,,	600	59.6	2.20	2.77
,,	240	43.8	0.50	0.86
Casamino Acids	3,600	406.7	48.80	9.0
,, ,,	1,200	336.0	14.40	3.2
,, ,,	600	123.6	17.10	10.4
,, ,,	240	49.9	8.60	12.9

The results of other experiments showed that there was no perceptible quantitative difference between rates of degradation per cell number when a bacteria-rich inshore surface water and an offshore surface water containing few bacteria were used as inoculum. In pure culture experiments, a selection of mesophilic and psychrophilic bacteria isolated from various depths in the open ocean was used. Only an obligate psychrophilic strain showed appreciable biochemical activity in the deep-sea samples. In no instance did any of the liquid media exposed to deep-sea conditions give rise to turbid cell suspensions.

Conclusions

Although the experiments described were carried out at greater depth they substantiate the observations of unexpectedly well-preserved food materials recovered from the *Alvin*. The surprisingly large difference between rates of microbial activity in samples exposed to deep-sea conditions and those in the appropriate controls must be considered as real. The common findings of waterlogged wood materials (Bruun and Wolff, 1961) in deep-sea dredgings seem to support this fact. No obvious explanation offers itself except for some clues that may be derived from an apparent temperature-pressure relationship indicated by some of our data.

Psychrophilic bacteria (exhibiting growth optima at relatively low temperatures) were found throughout the water column, but in the strains isolated, K. Eimhjellen (unpublished) found under laboratory conditions (1 atm)

a distinct decrease of the minimal growth temperature with the depth of isolation. Data available in the literature (Zobell, 1968; Morita and Haight, 1962) indicate an increase of maximal temperatures for growth and enzymatic activities with pressure. The relatively low rates of microbial activity found under deep-water conditions indicate that deep-sea waste disposal would be extremely inefficient with possible transport of waste materials, or intermediate decomposition products, by bottom currents. Our data are not sufficient to entirely rule out the possibility of the formation of an enriched microbial deep-sea population that might develop slightly higher rates of biochemical activities. The ratio between the rate of oxygen supply and microbial degradation activities will determine whether deep-sea bottom fauna will participate in the enrichment or whether anaerobic conditions will arise eliminating benthic populations of non-microbial fauna. Long-term experiments on this problem have been started.

References

Bruun, A F and Wolff, T   Abyssal benthic organisms: nature,
1961   origin, and influence on sedimentation. *In* Oceanography, edited by M. Sears, Washington, American Association for the Advancement of Science, pp. 391–7.

Jannasch, H W, *et al.*   Microbial degradation of organic matter
1971   in the deep-sea. *Science N.Y.*, 171:672–5.

Morita, R Y and Haight, R D   Malic dehydrogenase activity at
1962   101° C under hydrostatic pressure. *J. Bact.*, 83:1314–6.

Zobell, C E   Bacterial life in the deep sea. *Bull. Mar. Stn Kyoto*
1968   *Univ.*, (12):77–96.

Waste Removal and Recycling by Sedimentary Processes   *M. G. Gross**

L'élimination par les techniques de décantation des déchets transportés par les eaux

Les particules sédimentaires présentent une forte capacité d'élimination des matériaux dissous, y compris divers types d'effluents, dans les eaux de l'océan. Cela est particulièrement important dans les eaux côtières, où les concentrations de particules sont généralement élevées. Ces particules proviennent des rivières, de l'affouillement des dépôts sédimentaires par les courants, des retombées aériennes et du déversement par chalands d'effluents des villes côtières. Elles se déposent pour la plupart à proximité du point de déversement, bien que les matériaux de fine granulométrie ou de faible densité puissent être transportés sur de longues distances. Les déchets dissous peuvent être éliminés par des réactions d'échange ionique, par précipitation et par action mécanique sous l'effet de la pesanteur ou par l'action de processus biologiques. Une fois au fond, les déchets déposés peuvent échapper à de nouvelles interactions avec les eaux de l'océan et les organismes marins. De ce fait, la décantation peut constituer une solution élégante pour l'élimination des déchets. Le déversement d'effluents dans l'océan peut également causer des problèmes. Les zones utilisées pour de vastes opérations d'évacuation des déchets sont parfois exemptes d'organismes benthiques normaux. Mais les déchets, une fois dissous, peuvent être incorporés aux organismes par ingestion des particules. Les dépôts de déchets peuvent également constituer des problèmes régionaux à long terme s'ils dégagent des nutriments ou des substances délétères lorsque les dépôts de fine granulométrie sont soumis à des pressions qui provoquent l'expulsion des eaux interstitielles. Le déversement des eaux souterraines à proximité des côtes et le brassage des sédiments par les bulles des dégagements gazeux peuvent également provoquer l'apparition de matériaux dissous provenant de dépôts de déchets.

Supresión de los desechos hidrotransportados por sedimentación

Las partículas de sedimentos tienen bastante capacidad para eliminar los materiales disueltos, comprendidas diversas clases de desechos, de las aguas del océano. Esto reviste especial importancia en las aguas costeras en las que tienden a ser elevadas las concentraciones de partículas que proceden de los ríos, la socavación por la corriente de depósitos de sedimentos, la caída de partículas de la atmósfera, y las basuras de ciudades costeras que descargan las barcazas. La mayor parte de las partículas se depositan cerca del lugar de descarga, aunque las de grano fino y poca densidad pueden ser transportadas a largas distancias. Los desechos disueltos pueden suprimirse mediante reacciones de intercambio de iones, por precipitación y gravedad o por procesos biológicos. Una vez están en el fondo, los depósitos de desechos pueden aislarse del agua y de los organismos del océano, en cuyo caso, esto puede ser una manera interesante de eliminar desechos. El tirar los desechos al mar plantea problemas. Los lugares en los que se efectúan grandes descargas de desechos se ven privados de sus organismos normales del bentos. Los desechos disueltos originalmente pueden quedar al alcance de los organismos que se les incorporan por las partículas que ingieren. Los depósitos de desechos tambièn pueden constituir problemas regionales de largo plazo si liberan nutrients o sustancias nocivas como resultado de la consolidación de los de granos finos que expulsan las aguas de los intersticios. Las descargas de aguas terrestres en la costa y la agitación de los depósitos por la formación de gases y burbujas también pueden liberar materiales disueltos provenientes de los depósitos de desechos.

---

* Marine Sciences Research Centre, State University of New York, Stony Brook, New York 11790, U.S.A.

[152]

WHEN attempts are made to solve marine pollution problems, those processes whereby materials introduced to the ocean are removed and concentrated to be either deposited on the bottom or assimilated by organisms are frequently ignored. However, some specific examples can be given which indicate the significance of these processes.

Importance of interfaces

Most chemical reactions occur in the ocean at the air-sea interface or at the sediment-water interface or at the boundaries between the water and particles (living or dead) that are suspended in the water (Horne, 1969).

Some materials are concentrated at the air-water interface, where they are subjected to oxidation, ultra-violet radiation, evaporation of the more volatile constituents, and polymerization. If not destroyed by the process, these materials become heavier and eventually sink to be incorporated in sediment deposits. Petroleum products, chlorinated hydrocarbon pesticides, and many other materials can be concentrated in such surface films. Substances with appreciable vapour pressures, such as chlorinated hydrocarbon pesticides (Wurster, 1969) or methyl mercury (Lofroth, 1969) may evaporate to the vapour phase. Bursting bubbles can also inject surface-associated materials into the atmosphere (Horne, 1969).

Many hydrophobic compounds become associated with particle surfaces. This has been little studied in marine sediments, but is well known from studies of organic materials in soils (Mortensen and Himes, 1964).

Sources of particles

Particles in sea water come from river-borne detritus, atmospheric fallout, biological activity, chemical reactions, and resuspension of sediment from the ocean bottom. Estimates of the annual sediment load that reaches the ocean generally range from 20 to 36 × 10⁹ tons per year (Holeman, 1968), coming primarily from the Asian continent. More than half of a river's sediment load usually is deposited near the river mouth (Strakhov, 1967). Estuaries and deltas also form effective sediment traps that prevent sediments in certain regions from depositing on the continental shelf (Meade, 1969). If present in large amounts, the sediment load may exceed the ocean's capacity to transport the sediment away from the river mouth, and a delta forms. When the sediment load is not large enough to form a delta, the sediment frequently moves along the continental shelf approximately parallel to the coast (Gross, 1966).

The estuarine-like circulation that prevails along most of the world's coastline further inhibits sediment movement across continental shelves. Subsurface currents are generally directed landward. Thus materials settling from surface layers may move landward and into estuaries. Considering the rate of sediment supply to the ocean and the apparent rate of accumulation in deep ocean basins, it seems likely that 90 per cent or more of the river-borne particles are deposited on continental shelves (Arrhenius, 1963).

Most sediment consists of rock and mineral fragments which have been substantially altered during weathering (Jackson, 1964; Marshall, 1964). Particle size is reduced during weathering, and ions in the silicate lattice are removed leaving a stripped lattice with H⁺, H₂O, or various organic molecules associated with the grains. Quartz, feldspat, and various micas (including clay minerals) are common constituents of river-borne particulate matter. There is little evidence of major alteration of silicate materials upon entering sea water (Strakhov, 1967) and mineral changes that may take place, apparently require long periods of time at the relatively low temperatures and pressures of the earth's surface.

Particle concentrations in open ocean surface waters typically range from 0.1 to 1 mg/litre (Manheim *et al.*, 1970) and from about 1 to 30 mg/l in surface water within a few tens of kilometre of the coast (Conomos, 1968; Manheim *et al.*, 1970). Particle concentrations near the bottom are generally much higher and variable owing to resuspension of previously deposited sediment (Conomos, 1968). In coastal waters, organic matter typically makes up 20 to 40 per cent of the particles. The proportion increases to more than 50 per cent in offshore waters and may exceed 80 per cent of the suspended particles in open ocean waters (Manheim *et al.*, 1970).

There is often a subsidiary maximum of particles near the pycnocline (Jerlov, 1959), ascribed to retarded settling velocities of particles by the strong density variation. Large concentrations of clay particles at mid-depths near the Columbia River mouth are thought to occur because the settling velocity of clay particles is nearly equalled by the upward velocity of waters moving into the surface layer (Conomos, 1968).

Particles larger than 0.5 microns have a residence time in the ocean water column of less than 100 years; for smaller particles the residence time is more than 200 years, but substantially less than 600 years (Arrhenius, 1963). This rapid removal is probably caused by filter-feeding organisms which remove particles and incorporate them in faecal pellets which sink fairly quickly to the bottom. Coagulation does not appear to play a significant role except possibly in waters with high sediment concentrations.

Prior to the industrial revolution, the major sources of air-borne dust were deserts, mountain areas devoid of vegetation, and volcanic eruptions. These wind-borne particles (2–10 microns diameter), were eventually deposited over large parts of the ocean bottom where they can be recognized by their mineral composition and characteristic surface features. Atmospheric particle transport is most conspicuous in the South Pacific where there are few other sources of particles (Rex and Goldberg, 1962).

Since the industrial revolution, industrial and agricultural activities contribute particles to the atmosphere, many of them carried long distances (Delaney *et al.*, 1967). Such particles include fly ash from coal, lead aerosols from internal combustion engines, aerosols of industrial minerals, and carbon particles from incomplete oil combustion. Thus, both chemical and biological interactions of these particles may be markedly different from naturally occurring materials.

Tidal currents, waves, and surges resuspend sediments, causing large particle concentrations in near-bottom coastal waters (Conomos, 1968) and deep ocean waters (Ewing and Thorndike, 1965). In shallow continental shelf areas and marginal seas like the North Sea, this is an important source of suspended particles. Resuspended particles commonly include rock and mineral fragments as well as lesser amounts of biogenous detritus. Near the bottom, particles tend to be coarser than in the surface layers.

Organic particulate matter frequently exceeds the amount of inorganic matter in the surface waters of the open ocean. Below the photic zone this organic matter usually consists of diatom frustules, and skeletal and faecal pellet fragments of zooplankton; wood fibres and pollen grains are also common near shore. Where large amounts of sewage plant effluents and sewage solids are discharged into coastal rivers or ocean waters, cellulose fibres also occur (Conomos, 1968; Manheim et al., 1970).

Also in the ocean are delicate plate-like aggregates of organic matter, ranging from a few microns to several millimetres in diameter (Riley, 1963). They are pale yellowish or brownish, apparently amorphous matrices of organic matter usually including inorganic particles, bacteria and sometimes plankton. They apparently form, at the air-sea interface or on bubbles, from dissolved organic substances secreted by the primary producers (Baylor et al., 1962). Pollutants may also be concentrated by these aggregates, providing a potentially important route for dissolved materials to be removed from sea water and assimilated by organisms.

Particulate wastes are also discharged directly into the coastal ocean. The largest amount of such wastes comes from the disposal of dredge spoils, removed from navigation channels and slips. This material is largely "natural" sediment mixed with wastes. In many coastal areas, navigation channels form relatively efficient sediment traps. In many coastal areas, the disposal of dredged wastes exceeds the natural discharge of riverborne sediment by several orders of magnitude (Gross, 1970). Other types of materials are also barged to the coastal ocean for disposal, including building and demolition rubble, the solids in waste chemicals, coal ash, sewage solids from treatment facilities, wrecks, and defective products.

Removal of particles and wastes

Most particulate matter brought to the ocean settles out rapidly owing to relatively large grains or their high density, relative to water. Because of the importance of settling, both for the initial removal and subsequent clearing of the system once a waste is stirred up, it is important to determine settling velocities for the common types of wastes.

Next to gravitational settling for removing materials from sea water, ion exchange is probably the most important factor. In its simplest form, ion exchange reactions can be visualized as follows:

$$Na^+RK^+ = Na^+ + K^+R^-$$

where R is a solid material such as a sediment particle, Na^+ and K^+ are ions (Van Olphen, 1963). In this reaction the particle reacts with its surrounding, through the replacement of one ion with another. The ion attached to the particle is then removed from the solution, the other ion moves with the water.

Sediment particles tend to be small so they expose a large surface area per unit weight. Ion-exchange reactions require contact between water and particles; therefore, small grain size favours such reactions. Many sediment particles have been extensively altered during weathering and structural defects cause unsatisfied changes, many of them on the grain surface. Clays not only have exchange sites on the surface of the grains, but also exchangeable ions in positions between lattice sheets. This tends to cause a much larger ion-exchange capacity for such clays than would be predicted based on particle size alone. Among the factors that determine the ion exchange capacity of solids are: accessibility of exchange sites; their number, distribution and nature; and physical-chemical conditions in solutions surrounding the particle.

Ion exchange reactions are sensitive to the type and charge of ion involved. For example, many cations with charges of +1 or +2 occur in sea water, but virtually none with charges greater than +2. The higher the charge the more tightly are ions bonded to an ion exchange medium, and thus it seems probable that ion exchange reactions play a role in removing charged cations. If the charge is greater than +2, the cation will most probably become associated with particulate matter and be incorporated in sediment deposits. A few highly charged ions such as uranium do not obey this generalization because they form strong complexes with one of the common constituents in sea water. Anions apparently behave differently.

Chemical models of sea water

Successful equilibrium models of the chemical processes in the world ocean have been formulated by using available equilibrium constants determined in the laboratory (Sillen, 1961). The results have been most useful for elements whose residence time (calculated by dividing the reservoir of an element in the ocean by the annual flux) exceeds 100,000 years (fig. 1). Considering that the characteristic time for ocean circulation is a few thousand years, elements with long residence time tend to be well mixed in sea water. Concentration changes normally result from dilution, evaporation, or mixing with different water masses. Equilibrium models have been developed for sea water and various fresh water systems (Strumm, 1967).

Equilibrium models have been less successful in predicting the behaviour of those elements whose residence time in the ocean is less than about 10,000 years. In general these elements occur in low concentrations in sea water; many are involved in biological processes or complex ion exchange or solid solution reactions with solid phases. Furthermore, it is difficult to detect the minute amounts of these elements in sea water and to determine which chemical species are present (Baker, 1968).

While equilibrium models have been useful for understanding the evolution of sea water during the billions of years of earth history, they have not been used for predictions of waste behaviour in the ocean. It is imperative, however, that adequate models be developed to permit useful predictions to be made for waste discharges and their effects on the coastal ocean. There is especially a need for kinetic models suitable for use with transient phenomena as many discharges tend to be.

Particle-associated radionuclides

Detailed information about the behaviour of wastes in the ocean has come from studies of radionuclides in the

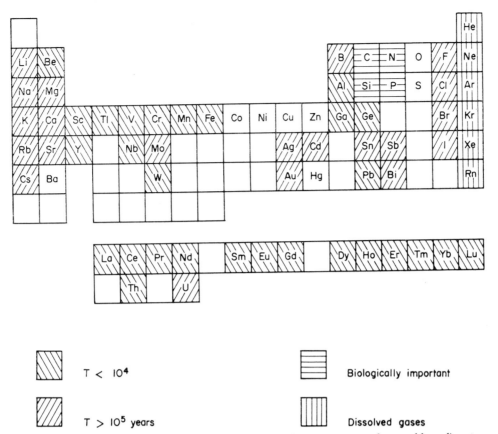

Fig. 1. *Grouping of elements by residence times as calculated from the apparent rate of removal by sedimentary processes*

ocean (IAEA, 1966; Seymour, 1970). Studies of natural sediment-water indicate that quantitative predictions are not yet possible, but it is possible to indicate qualitatively some of the interactions to be expected (Duursma and Gross, 1970). For instance, there is a tendency for fine-grained sediments to absorb more radionuclides than coarse-grained deposits. On the other hand, laboratory experiments with water-sediment systems have shown little correlation between median-grain size and the amounts of radionuclide taken up. In general, it seems that sediment bulk properties (such as grain size, mineral composition, or organic carbon content) are poor predictors of chemical behaviour. The amount of a given radionuclide taken up by sediments during an experiment was commonly affected by the conditions (amount of stirring, manner of nuclide addition) of the sediment-water system immediately before and after introduction of the radionuclide (Duursma and Gross, 1970).

Even after radioactive wastes are associated with particles and incorporated in sediment deposits, they can still be exposed to the overlying water and possibly recycled. Vertical transport in sediment is caused by biological activity, and stirring of the deposits by gas movement (Reesburgh, 1968). Sediments down to a depth of 2 cm below the interface are probably recycled annually; recycling of the upper few millimetres may take place daily (Rhoads, 1967). This has been observed to control concentrations of sediment associated radionuclides below the water-sediment interface (Barnes and Gross, 1966).

Recycling of surficial sediment is aided by burrowing organisms which distribute wastes to depths of several centimetres below the water-sediments interface. This reduces the concentration at the surface by diluting newly deposited material through mixing with previously deposited sediment. Assuming the typical depth of burrowing to be about 5 cm (Arrhenius, 1963) and the typical rate of sediment accumulation on those continental shelves receiving modern sediment to be about 15 cm/1000 years (Gross, 1966), a single layer of wastes will be almost completely buried within 500 to 1,000 years. On the deep ocean bottom, the time required to bury such a layer is much longer, because rates of sediment accumulation range from 0.1 to 1 cm/1,000 years. Using the same assumptions as above, it would take from 20,000 to 200,000 years to completely bury a single thin layer of wastes.

Water deposited with the sediment is gradually expelled as the sediment compacts. This movement of interstitial water carries materials through the sediment and into the overlying waters. Elements soluble in oxygen-free environment, typical of sediments, are likely to be dissolved off the grains and returned to the overlying water (Anikouchine, 1967).

Where ground water discharges through the ocean bottom on the continental shelf (Jenne and Wahlberg, 1968; Manheim, 1967), such water movements may also flush materials out of overlying deposits. This would be especially important for wastes deposited in shallow coastal areas.

Studies of radionuclide uptake in the Clinch River, Tennessee (Jenne and Wahlberg, 1968) showed that ^{90}Sr was apparently associated with carbonates, probably precipitated in the river, whereas ^{60}Co was associated with Fe and Mn coatings on mineral grains. Substitution

[155]

of ^{137}Cs, probably substituting in the crystal structure for K, was the dominant means of uptake for that nuclide. In Columbia River sediment, only a minor amount of the [65]Zn sorbed was involved in ion exchange reactions; most was apparently bound by more complex and specific interactions, perhaps with organic matter (Osterberg et al., 1966). Substances changing the valence state of ^{51}Cr in these sediments were found to have an important influence.

Soils and weathered rock fragments are commonly coated with organic matter and various metallic oxide coatings (Jenne, 1968). It seems highly probable that these coatings affect the chemical behaviour of particles in water systems (Chave, 1965; Jenne, 1968). For instance, hydrous (possible amorphous) Mn and Fe oxide coatings on grains may cause sorption of Co, Cu, Ni, and Zn by particles. A small amount of these oxides as surface coatings would have chemical effects far exceeding that predicted from the bulk chemical composition of the sediment. Among the processes affected by such coatings are oxidation of Mn^{2+} to Mn^{4+} and removal of metals (such as Zn) (Jenne, 1968). Such coatings may inhibit or even prevent the establishment of chemical equilibrium between the solid and liquid phases.

Organic coatings on the particles may serve to concentrate other wastes from the liquid phase. A striking example is the concentration of DDT on particles exposed to repeated oil pollution (Hartung and Klinger, 1970). In this way even neutral or non-polar molecules can be taken up by sediments and concentrated as much as 10^6 fold.

It may well be that these coatings include or perhaps are even formed by bacteria. For instance, particles in sea water may (1) provide a substrate for bacteria; (2) concentrate substances from the water and hold them on the substrate so that they may be attacked by bacteria; (3) retard the diffusion of bacterial enzymes away from the bacterium (Harvey, 1955). Mineral composition of the particulate matter may also be important in the ocean as it apparently is in certain soils (Stotzky, 1966). This process is commonly ignored in constructing models of chemical behaviour of sediment-water systems.

New York Metropolitan Region

Although marine disposal of wastes is common (Føyn, 1965) there is generally little detailed information available for specific areas. Data are available, however, for the New York Metropolitan Region (Gross, 1970) indicating both the amount and types of solids barged to sea for disposal.

The average amount discharged per year for the period of 1964 to 1968 was 10^7 tons, and is probably larger than for other coastal metropolitan areas (Brown and Smith, 1969). About 20 per cent of this went into Long Island Sound, the remainder was dumped into the coastal ocean near the entrance to New York Harbour (Gross, 1970).

Solids dredged from navigation channels and slips accounts for 7 million tons per year. Although is commonly considered to be clean sand and gravel, it is usually mixed with substantial amounts of waste solids. The simplest assumption is that dredged materials consist of a mixture of sands, silts and wastes to form the surficial deposits. Minor element concentrations in

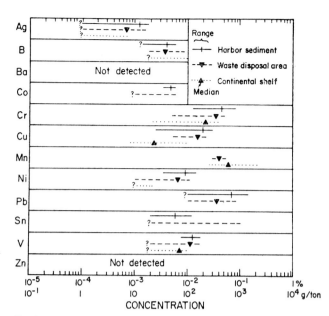

Fig 2. *Minor element concentration of harbour sediments and dredged wastes as compared to continental shelf sediment*

harbour sediments and dredged wastes are given in fig 2 and compared to continental shelf sediment.

Large volumes of liquid wastes are barged to sea for disposal. These materials come from many sources and are difficult to characterize. The solid content is highly variable, but typically about 5 per cent by weight. One of the wastes dumped in large volumes in the region is an iron-rich waste from TiO_2 production (Redfield and Walford, 1951). Sewage sludges, the solids removed during sewage treatment, are barged to sea as liquids (about 5 per cent solid) and discharged. The sewage solids consist of about 55 per cent organic matter, about 45 per cent aluminosilicates, chemically similar to shale. The solids are enriched by at least ten-fold in Cr, Cu, Pb, and Sn (relative to living organisms, fig 3).

Beneath the sewage sludge and dredged waste disposal sites, there are areas of 14 and 7 mi² respectively devoid

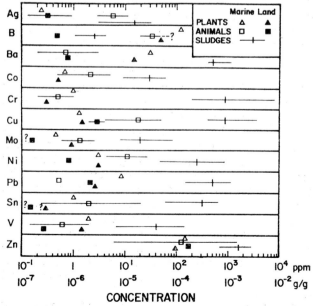

Fig 3. *Minor element concentrations in sewage sludge and solids as compared to organisms*

of "normal benthic organisms" (Pearce, 1969). Among the possible causes of these observed effects are (1) low dissolved oxygen concentrations (2) toxicity of metals in the sludges (3) pathogenic effects of the wastes (4) physical effects of the change in substrate and (5) greatly increased rate of solid deposition.

Needed research

Monitoring will continue to rely on widely accepted operational criteria, such as bacteriological indices and experience factors (USFWPCA, 1968). Work should be undertaken, however, to determine whether particulate matter in the water or removed by organisms, should also be monitored, either for possible deleterious effects on organisms (and food stocks) or as an indicator of hazardous wastes.

In dealing with cumulative poisons or possibly carcinogenic agents, particles and sediment deposits are likely to be extremely important. Originally low concentrations of dangerous substances may be concentrated by particles and stored for long periods. The problem may be further aggravated by chemical processes taking place in the sediment deposits which are rich in microorganisms and have chemical and physical characteristics not commonly found elsewhere in the earth's surface environment.

Here we have little knowledge. For example, Cu, Cr, Sn, and Pb occur in industrial wastes and often in mixed domestic wastes (Gross, 1970). They are known to be toxic to varying degrees, carcinogenic or cumulative poisons (Epstein *et al.*, 1966). Mercury is another example of a cumulative poison, but its possible association with sediments is essentially unknown (Lofroth, 1969).

Organic compounds, especially polynuclear aromatics commonly associated with airborne particles, are known to be carcinogenic agents, but their behaviour in sediment-water systems apparently has not been studied. Chlorinated hydrocarbon pesticides have been more extensively studied and transport by particles has been shown to be important (Wurster, 1969). Similar studies of the widespread petroleum products (Horn *et al.*, 1970) in surface ocean waters are needed. Future work should also be focused on common industrial materials such as Cu, Cr, Hg, Sn, and Pb, which are likely to pose health hazards if taken into human food.

References

ANIKOUCHINE, W A   Dissolved chemical substances in compacting
1967   marine sediments. *J. geophys. Res.*, 72(2):505–9.
ARRHENIUS, G   Pelagic sediments. *In* The Sea, edited by M. N. Hill,
1963   New York, Interscience, Vol. 3:655–727.
BAKER, R A   Trace inorganic in water. Washington, D.C., Ameri-
1968   can Chemical Society, 396 p.
BARNES, C A and M G GROSS   The distribution at sea of Columbia
1966   River water and its load of radionuclides. *In* Disposal of
   radioactive wastes into seas, oceans, and surface waters,
   Vienna, I.A.E.A., pp. 291–302.
BAYLOR, E R, SUTCLIFFE, W H and HIRSCHFELD, D H   Adsorption
1962   of phosphate onto bubbles. *Deep-Sea Res.*, 9:120–4.
BOWEN, H J M   Trace elements in biochemistry. New York,
1966   Academic Press, 241 p.
BROWN, R P and SMITH, D D   Marine disposal of solid wastes: an
1969   interim report. *Mar. Pollut. Bull.*, (18):12–6.
CHAVE, K E   Carbonates: association with organic matter in
1965   surface sea water. *Science, N.Y.*, 148:723–4.
CONOMOS, T J   Processes affecting suspended particulate matter in
1968   the Columbia River effluent system. Thesis, Seattle,
   University of Washington, 141 p. (Unpubl.)
DELANEY, A C *et al.*   Airborne dust collected at Barbados. *Geochim.*
1967   *Cosmochim. Acta*, 31:885–909.

DUURSMA, E K and GROSS, M G   Marine sediments and radio-
1970   activity. *In* Radioactivity in the marine environment,
   edited by A H Seymour, Washington, D.C., National
   Academy of Sciences.
EPSTEIN, S S *et al.*   Carcinogenicity of organic particulate pollutants
1966   in urban air after administration of trace quantities to neo-
   natal mice. *Nature, Lond.*, 212:1305–7.
EWING, M and THORNDIKE, E M   Suspended matter in deep ocean
1965   water. *Science, N.Y.*, 147(3663):1291–4.
FØYN, E   Disposal of wastes in the marine environment and the
1965   pollution of the sea. *Oceanogr. mar. Biol.*, 3:95–114.
GOLDBERG, E D   Minor elements in sea water. *In* Chemical oceano-
1965   graphy, edited by J P Riley and G Skirrow, New York,
   Academic Press, Vol. 1:163–96.
GROSS, M G   Distribution of radioactive marine sediment from
1966   the Columbia River. *J. geo-phys. Res.*, 71(8):2017–21.
GROSS, M G   New York Metropolitan region—a major source of
1970   sediment. *Wat. Resour. Res.*, 6.
HARTUNG, R and KLINGLER, G W   Concentration of DDT by
1970   sedimented polluting oils. *Envir. Sci. Technol.*, 4:407–10.
HARVEY, H W   Chemistry and fertility of sea water. Cambridge,
1955   England, Cambridge University Press, 224 p.
HOLEMAN, J N   Sediment yield of major rivers of the world. *Wat.*
1968   *Resour. Res.*, 4:737–47.
HORN, M H, TEAL, J M and BACKUS, R H   Petroleum lumps on the
1970   surface of the sea. *Science, N.Y.*, 168(3928):245–6.
HORNE, R A   Marine chemistry. The structure of water and the
1969   chemistry of the hydrosphere. New York, Wiley-Inter-
   science, 568 p.
International Atomic Energy Agency. Disposal of radioactive
1966   wastes into seas, oceans and surface waters. Vienna,
   I.A.E.A., 898 p.
JACKSON, M L   Chemical composition. *In* Chemistry of the soil,
1964   edited by F. E. Bear, New York, Reinhold, pp. 71–141.
JENNE, E A   Controls on Mn, Fe, Co, Ni, Cu, and Zn concentration
1968   in soils and water. The significant role of hydrous Mn and
   Fe oxides. *In* Trace inorganics in water, edited by R. A.
   Baker, Washington, D.C., American Chemical Society,
   pp. 337–87.
JENNE, E A and WAHLBERG, J S   Role of certain stream-sediment
1968   components in radioion sorption. *Prof. Pap. U.S. Geol.*
   Surv., (443-F):16 p.
JERLOV, N G   Maxima in the vertical distribution of particles in
1959   the sea. *Deep-Sea Res.*, 5:173–84.
LOFROTH, G   Methylmercury. *Bull. ecol. Res. Comm.*, Stockholm,
1969   (4):35 p.
MANHEIM, F T   Evidence for submarine discharge of water on the
1967   Atlantic continental slope of the southern United States and
   suggestions for further search. *N.Y. Acad. Sci.*, 29:839–53.
MANHEIM, F T, MEADE, R H and BOND, G C   Suspended matter in
1970   surface waters of the Atlantic continental margin from
   Cape Cod to the Florida Keys. *Science, N.Y.*, 167(3917):
   371–6.
MARSHALL, C E   The physical chemistry and mineralogy of soils.
1964   New York, Wiley, Vol. 1: 388 p.
MEADE, R H   Landward transport of bottom sediment in estuaries
1969   of the Atlantic coastal plain. *J. sedim. Petrol.*, 39:222–34.
MORTENSEN, J L and HIMES, F L   Soil organic matter. *In* Chemistry
1964   of the soil, edited by F. E. Bear, New York, Rheinhold,
   pp. 206–41.
OSTERBERG, C L *et al.*   Some non-biological aspects of Columbia
1966   River radioactivity. *In* Disposal of radioactivity into seas,
   oceans, and surface waters. Vienna, I.A.E.A., pp. 321–35.
PEARCE, J B   The effects of waste disposal in the New York Bight.
1969   Interim report for January 1, 1970. Sandy Hook, N.J.,
   U.S. Bureau of Sport Fisheries and Wildlife, Sandy Hook
   Marine Laboratory, 100 p.
REDFIELD, A C and WALFORD, L A   A study of the disposal of
1951   chemical waste at sea. Washington, D.C., National Academy
   of Sciences, 49 p.
REESBURGH, W S   Determination of gases in sediments. *Envir.*
1968   *Sci. Technol.*, 2:140–1.
REX, R W and GOLDBERG, E D   Insolubles. *In* The Sea, edited by
1962   M. N. Hill, New York, Interscience, vol. 1:295–304.
RHOADS, D C   Biogenic reworking of intertidal and subtidal
1967   sediments in Barnstable Harbor and Buzzard's Bay.
   J. Geol., 75:461–76.
RILEY, G A   Organic aggregates in sea water and the dynamics of
1963   their formation and utilization. *Limnol. Oceanogr.*, 8(4):
   372–81.
SEYMOUR, A H   Radioactivity in the marine environment.
1970   Washington, D.C., National Academy of Sciences.
SILLEN, L G   The physical chemistry of sea water. *In* Oceanography,
1961   edited by M. Sears, Washington, D.C., American Association for the Advancement of Science, pp. 549–82.
STOTZKY, G   Influence of clay minerals on microorganisms. *Can. J.*
1966   *Microbiol.*, 12:831–48.

STRAKHOV, M N   Principles of lithogenesis. New York, Consul-
1967   tants Bureau, vol. 1:245 p.

STRUMM, W   Equilibrium concepts in natural water systems.
1967   Washington, D.C. American Chemical Society, 344 p.

U.S. Federal Water Pollution Control Administration. Report of
1968   Committees on water quality criteria. 234 p.

VAN OLPHEN, H   Clay colloid chemistry for clay technologists,
1963   geologists and soil scientists. New York, Interscience
        publishers, 301 p.

WURSTER, C F   Chlorinated hydrocarbon insecticides and avian
1969   reproduction: how are they related? In Chemical fallout,
        edited by M. W. Miller and G. C. Berg, Springfield,
        Illinois, Charles C. Thomas, pp. 368–89.

Water Sediment Exchange and Recycling of Pollutants through Biogeochemical Processes

*E. Olausson**

L'échange de sédiments aquatiques et le recyclage des polluants par des procédés biogéochimiques

L'influence de l'homme sur les sédiments marins se manifeste de deux manières : (1) accumulation accrue de matières carbonées avec modification du pH et de l'Eh du milieu, et (2) pénétration des polluants dans les sédiments. Le premier effet peut modifier les réactions dans les sédiments, ainsi que celles qui se produisent entre l'eau et ces derniers. Il en résulte des perturbations dans la distribution des végétaux et des animaux. L'auteur examine deux groupes parmi les agents de pollution les plus importants : les métaux lourds et les pesticides. Il fait une place particulière aux problèmes de la reconcentration par des procédés biogéochimiques.

Intercambio de sedimentos en el agua y reciclado de contaminantes por medio de procesos biogeoquímicos

La influencia del hombre sobre los sedimentos marinos se registra de dos modos: (1) mediante una creciente acumulación de materia carbonada con un cambio en el pH y Eh del medio ambiente, y (2) mediante el enterramiento de los contaminantes en los sedimentos. El primer efecto puede cambiar las reacciones en los sedimentos y entre el agua y aquéllos. Esto da lugar a disturbios en la distribución de las plantas y de los animales. Se examinan dos grupos de los contaminantes más importantes: metales pesados y plaguicidas. Se presta especial atención al problema de la reconcentración por procesos biogeoquímicos.

MAN'S activities are leading to an increased accumulation of carbonaceous matter with a change in pH and Eh of the environment, and in the burial of pollutants in sediments. A significant portion of the heavy metals (25 to 75 per cent) seems to escape from the sediments into the water (organisms). The ratio $PCB/\Sigma DDT$ changes rapidly with depth in sediments suggesting that PCB is not recycled while most DDT is recycled/decomposed rapidly.

MARINE ENVIRONMENTS IN TERMS OF pH AND Eh

pH. The oxidation of organic matter can be described by the equation

$$CH_2O + O_2 \rightarrow CO_2 + H_2O \qquad (1)$$

where CH_2O is carbohydrate. In order to oxidize a mol of carbon (12 g) there is a theoretical need for 22.4 1 O_2. The oxygen content of sea water is $\angle 8$ ml/1 which is sufficient to oxidize $\angle 4$ mg carbon. This amount of carbon (as living or dead material) is found in several regions of the oceans, which suggests that oxygen deficiency can occur in basins with restricted water exchange.

It is considered that the pH of sea water is fixed by the H_2CO_3-$CaCO_3$-system (Buch, 1939). The pH of a sediment is a function of the activity of its acid (H_2CO_3) and alkaline products (e.g. NH_3) with $CaCO_3$ as a buffer. If we take $C_{106}N_{16}P$ as an average composition for plankton their oxidation can be illustrated by

$$(CH_2O)_{106}(NH_3)_{16}H_3PO_4 + 106\ O_2 \rightarrow$$
$$106\ CO_2 + 16\ NH_3 + H_3PO_4 + 106\ H_2O \qquad (2)$$

which shows that the breakdown of organic material produces about 6 times more acid products than alkaline ones. However, the amount of O_2 available in the sediments limits this reaction. When O_2 is used up

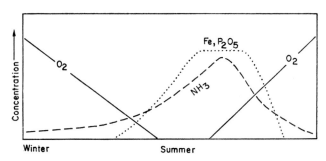

Fig. 1. Concentration of oxygen just above the mud surface of dissolved substances during summer stagnation (simplified from Hayes, 1964, Fig. 6)

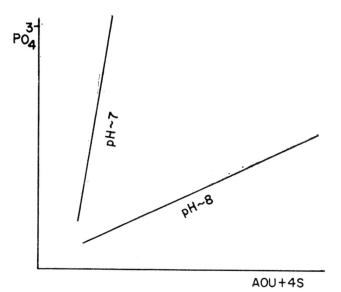

Fig. 2. Total redox reactants in stagnant basins in relation to the pH of the environment. Data from various authors and theoretical calculations (see Olausson, 1967)

*Marine Geological Laboratory, University of Göteborg, Göteborg, Sweden.

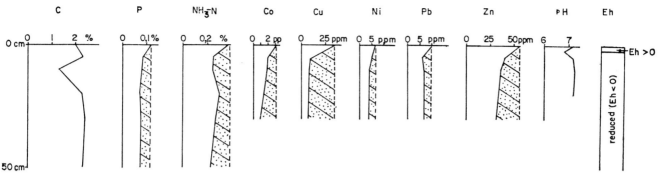

Fig. 3. Core 141 (Askeröfjorden, 11°47′E, 58°5.3′N). Distribution C, P, N, and some heavy metals in sediment samples. Judging from the pH measurements and the colour of the sediments, the redox discontinuity appears at a depth of 3 cm below the depositional interface. The highest concentration of acids corresponds to the pH minimum (see further Wood, 1965). If steady-state conditions have existed during the time of deposition of the uppermost 10 (-30) cm, the dotted-hatched areas of the curves indicate the amount of each element which must have escaped from the sediments into the overlying water.

$SO_4^=$ is used by anaerobic bacteria and nitrogen regeneration leads to the formation of NH_3, N_2 and N_2O (Vaccaro, 1965). The reaction

$$SO_4^= + 2\,H^+ + 2\,C \rightarrow H_2S + 2\,CO_2 \qquad (3)$$

will increase the pH of the environment (a strong acid is changed into a weak base; see fig 3).

While calcareous matter is present it acts as a buffer and the acid products can be used to dissolve carbonates:

$$CaCO_3 + H_2CO_3 \rightarrow Ca^{2+} + 2\,HC\bar{}_3 \qquad (4)$$

and apatite

$$Ca_5(PO_4)_3OH + 4\,H^+ \leftrightharpoons$$
$$5\,Ca^{2+} + 3\,HPO_4^= + H_2O \qquad (5)$$

If we only consider the molar (m) concentration of the three components mentioned, four theoretical possibilities exist:

		pH	Sediment
$^mCH_2O > {}^mO_2$	$^mCaCO_3 > {}^mCH_2O$	~8	Calcareous mud
	$^mCaCO_3 < {}^mCH_2O$	~6–7	Inorganic mud (Shelf deposits)
$^mCH_2O < {}^mO_2$	$^mCaCO_3 > {}^mCH_2O$	~8	Calcareous ooze
	$^mCaCO_3 < {}^mCH_2O$	~6–7	Red clay (Deep-sea deposits)

In many shelf areas the content of particulate and dissolved organic matter has increased due to a greater supply of nutrients and raised quantities of organic matter from rivers and sewage. In areas of low carbonate accumulation (chiefly at higher latitudes) this may lead to a lower pH of the environment (Eq. 2).

Eh. A simplified redox potential equation of a system can be expressed as

$$Eh = E^0 + \frac{k.}{n}\,\log\frac{(ox)}{(red)}$$

where Eh is the oxidation potential, E^0 is the voltage of the reaction when all substances involved are at unit activity, and (ox) and (red) stand for the oxidizing and reducing agents. For a sediment, a heterogenous system which is slowly brought into equilibrium, we may use the relation

$$Eh \sim k + \log(ox) - \log(red)$$

Since oxygen is the most important oxidizing substance and organic matter is the compound which consumes oxygen we may illustrate Eh of a sediment as

$$Eh \sim {}^mO_2 - {}^mCH_2O$$

If $^mO_2 > {}^mCH_2O$ (Eh>0) the environment is oxidized but if $^mO_2 < {}^mCH_2O$ (Eh<0) the pollution is reduced.

The oxygen consumption over marine sediments off Woods Hole is 250 ml/m²/day (Kanwisher, 1962) with a maximum of 330 in summer and a decrease to 150 or less in winter. An increase in the supply of carbonaceous matter (CH_2O) to the bottom sediment results in an upward migration of the redox discontinuity zone in the sediment and, if the renewal of the bottom water is not continuous, can create euxinic conditions in the bottom water. However, the absence of H_2S in the sediments is exceptional and many basins with comparatively slow exchange of water contain H_2S in their bottom water even when unpolluted. Climatic change can alter this water exchange and it is not known whether the presence of H_2S in a few deep holes in the Baltic is natural (Olausson, 1970) or whether it is due to pollution. The Swedish west coast Byfjorden contains H_2S in the water below the sill depth due to contamination from the town of Uddevalla.

When oxygen is removed from the water, as during summer stagnation, the redox discontinuity moves up and reduced solutes make their appearance in the water. This is illustrated in fig 1.

RECYCLING OF POLLUTANTS

Phosphate

The regeneration of organic compounds is dependent on the Eh of the environment. Nazarkin (1960) found that 5 to 7 per cent of the organic matter produced yearly was lost in mud. Judging from the classic study by Rittenberg *et al.* (1955), 1/3 to 7/8 of the fallout may return to the water in a dissolved state.

Regarding phosphate regeneration, it can be stated from Eq. 5 that the lower the pH of the environment, the more rapidly apatite grains are dissolved (fig. 2). If the pH of the sediment is lowered due to pollution (see above), the return of phosphate from the sediment is increased. The regeneration of phosphate bound into soft organic materials is influenced by the Eh of the environment (see Eq. 2): it may be complete in oxidized conditions but incomplete in reduced ones.

[159]

Heavy metals

The analytical treatment of sediments includes freeze-drying, oxidation in a low-temperature asher, followed by leaching with HCl (pH about 2.6) and APDC-MIBK-extraction. A Unicam SP 90 Serie 2 atomic absorption unit was used. This process only extracts metals bound to organic matter not those in minerals.

Oxidized, polluted marine sediments

A profile from Askeröfjorden (50 km N Göteborg, Sweden), was analysed to determine the content of "non-minerogenic" Co, Cu, Ni, Pb, and Zn. The distribution of these elements together with carbon, phosphate and nitrogen (NH_3), pH and Eh are given in fig. 3. The uppermost 3 cm was oxidized below which reduced conditions prevailed. The probable rate of accumulation was 3 cm/year.

Askeröfjorden receives much sewage water from the industrial area at Stenungsund. The pollution was intensified in the mid-1960's but was regulated by a decision of the water right court. It seems justified to assume a steady-state condition for the last 3 to 4 years.

The carbon curve is irregular but probably depends on inhomogeneity of the sediments. However, the phosphate content drops from the oxidized ("actual") layer into the reduced ("historical") sequence where it is constant. Theoretically this represents a 25 per cent return of phosphate to the water. One third or more of the N-containing material is decomposed in the oxidized zone and the solutes are returned to the water.

Figure 3 shows that most of the copper and about half the content of Zn may escape from the sediments during early diagenetic changes. An appreciable portion of Co, Ni, and Pb (chiefly from gasoline) return to the water/biosphere. As a result recycling noticeably increases the content of metals available to organisms beyond the level fixed by the water exchange and sewage discharge.

Reduced marine sediments

Sediment cores from two euxinic fjords (Byfjorden and Koljefjorden) have been analysed. Both fjords are euxinic due to human activity. The chemistry of their sediments deviates from the analyses showed in fig. 3 chiefly by the uniform distribution of PO_4^{3-} and NH_3-N in the uppermost 60 to 100 cm.

There are similar drops in the distribution of heavy metals in the upper 10 cm of the cores from the euxinic fjords (e.g. core 141). The decrease in the content of the heavy metals of core 141 occurred in a weakly oxidized zone but it occurred in strongly reduced environments in the other cases. It therefore appears that the recycling of Co, Zn, Pb, Ni, and Cu is independent of Eh.

Chlorinated hydrocarbons

Pesticides may enter the marine environment as a result of dust storms over agricultural and urban areas and application of pesticides by aircraft (Risebrough et al. 1968; Seba and Corcoran, 1969). The main source is probably the transport of soil particles to which the residues are attached (Lichtenstein, 1966). The quantity of pesticides in marine water is extremely small ($<10^{-9}$) and the solubility of PCB (polychlorinated biphenyls) in water is $<10^{-8}$ (Höhne, lecture 22 September, 1970). Many studies have monitored organochlorine pesticide residues in estuaries (e.g. Butler, 1968; Keith and Hunt, 1966; Odum et al., 1969). However, little attention has been given to their persistence in marine muds.

In a Long Island New York estuary, Woodwell, Wurster and Isaacson (1967) found 0.04 ppm of DDT in plankton and $\sim$0.02 ppm in organic debris at the bottom but the concentration was considerably higher in benthic organisms (0.2 to 0.4 ppm). Apparently a portion of the DDT returns to the biosphere through biogeochemical processes.

At the author's laboratory a study of the concentration of chlorinated hydrocarbons in muds and in different organisms, and the processes of biological decomposition of pesticides is in progress (Report no. 1 from the Marine Geological Laboratory, University of Göteborg, 1970). In surface sediments from the estuary of Göta älv, the PCB concentration is 40 to 160 ppb. At a depth of 10 to 25 cm it is usually 10–20 ppb lower but it can be higher suggesting a very slow rate of decomposition and/or recycling. In other areas off the Swedish west coast the concentration in surface muds is less (<40 ppb).

Dieldrin is present in small amounts (1 to 4 ppb) and rather uniformly distributed along the coast. The amount decreases by half from the depositional interface down to 5 cm. The DDT (and DDE) concentration in the Göta älv estuary is up to 14 ppb. In the Kungsbacka fjord (30 km S Göteborg) and in samples taken north of Göteborg this concentration is reduced by half. This suggests that within 1 to 3 years half the DDT is either decomposed or returned to the water/organisms. The comparatively high concentration of DDT in mud-feeding organisms found in the Long Island study and in a few analyses of *Mytilus edulis* from the Swedish west coast suggests an important recycling of pesticides in marine environments.

According to Jensen et al. (1969), the ratio PCB/Σ DDT in fish from the Swedish west coast is generally about 0.4. If 0.4 to 1 is taken as an average ratio between PCB/Σ DDT of living planktonic organisms for the area in question, the change from 0.4 to a ratio of 6 to 10 can give a rough idea of the amount of DDT recycled (or decomposed) in marine muds.

References

BUCH, K Kohlensäure in Atmosphäre und Meer an der Grenze
1939 zum Arktikum. *Acta Acad. Åbo.*, (*Math. phys.*), 11(12).

BUTLER, P A Pesticides in the estuary. *In* Proceedings of the Marsh
1968 and Estuary Management Symposium, Louisiana State University, July 1967, pp. 120–4.

EMERY, K O and RITTENBERG, S C Early diagenesis of California
1952 basin sediments in relation to origin of oil. *Bull. Am. Ass. Petrol. Geol.*, 36:735–806.

HAYES, F R The mud-water interface. *Oceanogr. mar. Biol.*,
1964 2:121–45.

JENSEN, S *et al.* DDT and PCB in marine animals from Swedish
1969 waters. *Nature, Lond.*, 224:247–50.

KANWISHER, J Gas exchange of shallow marine sediments. *Occ.*
1962 *Publs Narragansett mar. Lab.*, (1):13–9.

KEITH, J O and HUNT, E G Levels of insecticide residues in fish
1966 and wildlife in California. *Trans. N. Am. Wildl. Nat. Res. Conf.*, (31):150–77.

LICHTENSTEIN, E P *et al.* Toxicity and fate of insecticide residue
1966 in water. *Archs envir. Hlth*, 12:199–212.

NAZARKIN, L A On the role played by the rate of sedimentation
1960 in the accumulation of absolute masses of organic matter in the precipitate. *Dokl. Akad. Nauk SSSR*, 130:868–70.

ODUM, W E, WOODWELL, G M and WURSTER, C F DDT residues
1969 absorbed from organic detritus by fiddler crabs. *Science, N.Y.*, 164, (3879):576–7.

OLAUSSON, E Climatological, geoeconomical and paleooceano-
1967 graphical aspects on carbonate deposition. *Progr. Oceanogr.*, 4:245–65.

OLAUSSON, E Är Östersjön förorenad (Is the Baltic polluted?)
1970 *Svenska Dagbladet*, Stockholm, 21 May.

OLAUSSON, E and OLSSON, I U Varve stratigraphy in a core from
1969 the Gulf of Aden. *Palaeogeog. Palaeoclimat. Palaeoecol.*,
6:87–103.

RISEBROUGH, R W, *et al.* Pesticides: transatlantic movements in
1968 the northeast trades. *Science, N.Y.*, 159(3820):1233–5.

RITTENBERG, S C, EMERY, K O and ORR, W L Regeneration of
1955 nutrients in sediments of marine basins. *Deep-Sea Res.*,
3(1):23–45.

SEBA, D B and CORCORAN, E F Pesticides in water: surface slicks
1969 as concentrators of pesticides in the marine environment.
Pestic. monitg J., 3(3):190–3.

VACCARO, R F Inorganic nitrogen in sea water. *In* Chemical
1965 oceanography edited by J. P. Riley and G. Skirrow, London, Academic Press, vol. 1:365–408.

WOOD, E J F Marine microbial ecology. London, Chapman and
1965 Hall, 320 pp.

WOODWELL, G M, WURSTER, C F JR. and ISAACSON, P A DDT
1967 residues in an east coast estuary: a case of biological concentration of a persistent insecticide. *Science, N.Y.*, 156:
821–3.

Acknowledgement

The author is indebted to the Swedish Natural Research Council
and to the Bank of Sweden Tercentenary Fund for financial support.
The chemical analyses were carried out by the chemists at the Marine
Geological Laboratory. Documentation service was given by Dr.
Rutger Irgens at the Medical Library, University of Göteborg.

Dissolved Oxygen as an Indicator of Water Pollution in Egyptian Brackish-water Lakes

*M. A. H. Saad**

L'oxygène dissous, indicateur de la pollution des eaux dans les lacs égyptiens d'eau saumâtre

Les variations saisonnières de la teneur en oxygène dissous du lac Mariout, lagune d'eau saumâtre sur la côte méditerranéenne à proximité d'Alexandrie (Egypte), oscillent entre 0 et 20,31 mg/litre. La teneur en oxygène dissous de l'hydroaérodrome de Nozha, qui a été séparé artificiellement du lac Mariout, varie entre 5,58 et 10,79 mg/litre. Ce lac "secondaire" n'est alimenté que par l'eau du Nil non modifiée, alors que le lac Mariout reçoit des effluents industriels et des eaux usées ménagères; aussi sa teneur en matières organiques est-elle plus élevée que celle de l'hydroaérodrome.

Dans le lac Mariout, l'oxygène dissous sert principalement à oxyder ces grandes quantités de matières organiques. Pendant toute la durée de l'étude (janvier 1969–janvier 1970), la consommation d'oxygène a été très forte à certaines stations, où sa disparition était totale et où l'on notait une forte odeur d'hydrogène sulfuré. Les déchets industriels apportent aussi dans le lac différentes substances toxiques qui exercent un effet délétère sur les poissons. La production totale de poissons est tombée de 9.977.815 kg en 1961 à 1.868.600 kg en 1967. Cette diminution est essentiellement due à la pollution du lac.

Divers écoulements alimentent aussi le lac Mariout. Les pompes Mex déversent dans la mer l'excédent d'eau du lac afin de le maintenir à environ 2,8 mètres au-dessous du niveau de la mer. Le taux de mélange de l'eau polluée du lac avec l'eau de la mer et les effets de la pollution marine résultante sur la côte méditerranéenne près d'Alexandrie feront l'objet d'autres études.

El oxígeno disuelto como indicador de la contaminación de los lagos de agua salobre de Egipto

Oscila entre 0 y 20,31 mg/l la variación estacional del oxígeno disuelto en el Lago Mariut, una cuenca somera de agua salobre junto a la costa del Mediterráneo cerca de Alejandría, Egipto. El oxígeno disuelto en el hidroaeródromo de Nozha, que se separó artificialmente del Lago Mariut, varía de 5,58 a 10,79 mg/l. Este lago "apéndice" recibe tan sólo agua del Nilo sin modificar, en tanto que el Lago Mariut recibe desechos domésticos e industriales que hacen que contenga más materia orgánica que el hidroaeródromo.

En el Lago Mariut el oxígeno disuelto se consume principalmente por oxidación de estas enormes cantidades de materia orgánica. Durante el período que duró la investigación, de enero de 1969 a enero de 1970, el consumo de oxígeno fue muy elevado en algunas estaciones, en las que se observó su completo agotamiento y un olor constante a sulfuro de hidrógeno. Además, los desechos industriales aportan al lago diferentes sustancias tóxicas perjudiciales para los peces. La producción pesquera total ha disminuido de 9.977.815 kg en 1961 a 1.868.600 kg en 1967, lo que se debe principalmente a la contaminación del agua.

Al Lago Mariut descargan diversas cloacas. Las bombas Mex envían el exceso de agua del lago al mar, con objeto de mantener su nivel en 2,8 m por debajo del mar. La velocidad de mezcla del agua contaminada del lago con la del mar y el efecto de la contaminación de esta última resultante en el Mediterráneo cerca de Alejandría es otro problema que se tendrá que estudiar más a fondo.

L AKE MARIUT is the smallest of four shallow, brackish lakes, adjoining the Mediterranean coast of Egypt. The total lake area has been greatly decreased, due to continuous land reclamation for agriculture and is now artificially divided into four parts (fig 1). In the past Lake Mariut was a productive aquatic habitat but owing to increased population and industry the problem of pollution is increasing and fish production is now greatly reduced.

One of the most important single chemical analyses to determine water quality is the DO concentration. Foehrenbach (1969) concluded that the oxygen content can be a reflection of organic loading, nutrient input, and biological activity. This investigation deals with the influence of pollution on the concentration and distribution of dissolved oxygen (DO) in the lake. Six stations were chosen to represent conditions within the lake (fig 1).

The Umum Drain supplies drainage water from Beheira Province. Water from the Alexandria drainage system, mixed with sewage and industrial wastes, enters through Moharrem Bey Bridge. Mexpumps discharge the surplus water from the lake into the sea to maintain the lake water level at about 2.8 m below sea level. Repeated fish mortalities observed in the sea near the area of connection showed that the polluted lake water is toxic.

Results

The DO shows irregular concentration and distribution in the lake water (fig 2). Both high and low values of DO were found throughout the period of investigation. The seasonal variations of DO in the lake proper range between 0 and 20.31 mg/l. Complete depletion of DO was observed at some stations (I, II and III), during certain months. Station V was always completely depleted of DO, except in March, October and November 1969, when the DO values were 32.73, 8.84 and 0.40 mg/l respectively. Samples from Umum Drain gave DO values of 8.03 and 7.32 mg/l in December 1969 and January 1970 respectively.

* Oceanography Department, Alexandria University, Alexandria, A.R.E.

DO at stations I and II reached high values of 13.88 and 12.46 mg/l in October and April 1969 respectively. Low values of DO at these two stations (2.02 and 4.86 mg/l) were recorded in July and March 1969. Station III had the highest value of DO found in the lake proper during the investigation period. This value (20.31 mg/l) was recorded in October 1969. The lowest value of DO at this station (1.70 mg/l) was in February 1969. Stations IV and VI had the highest values of DO (16.69 and 18.87 mg/l), in June and October 1969 respectively. The lowest values of DO at these two stations (0.48 and 2.42 mg/l) were recorded in August 1969 and January 1970.

Discussion of results

The DO varied greatly during the study period. Such irregularities in concentration and distribution were not previously found in most of the Egyptian lakes. The irregular concentration of DO may be caused mainly by the decomposition of variable amounts of dissolved organic matter introduced into the lake in sewage and industrial wastes. The fraction of organic matter in these wastes is high. Under certain conditions decomposition of organic matter proceeds so fast that complete depletion of DO occurs.

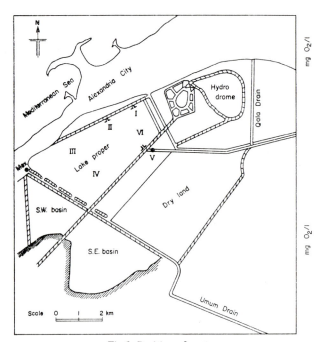

Fig 1. Position of stations

After complete depletion of DO further decomposition of organic matter occurs anaerobically. This produces hydrogen sulphide which can be smelt at great distances particularly in summer when high temperature favours decomposition.

Especially in summer a thick layer of green phytoplankton covers the surface of the lake proper. Similar phytoplankton abundance was observed by Ohle (1954) in some German lakes. These phytoplankton blooms may also contribute to the irregular distribution of DO in the lake. After death, the added autochthonous organic materials remove oxygen from water during decomposition.

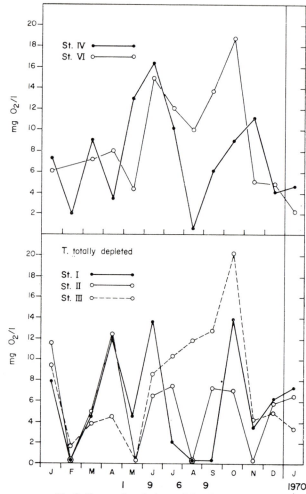

Fig 2. Seasonal variations of dissolved oxygen

The higher values of DO observed during the investigation may be attributed to these factors:
1. Decrease in the amount of organic load introduced into the lake.
2. Photosynthesis, particularly when the photosynthetic zone is deep enough to allow the oxygen produced to exceed the oxygen consumed. Water transparency may be a limiting factor for oxygen production.
3. Absorption from the atmosphere, especially in absence of the thick surface layer of green algae. Agitation is important for such a process. The reaeration is proportional to the existing oxygen deficit (Phelps 1944).
4. Low temperature which enables the lake to dissolve more oxygen (Ruttner 1953).

The effect of organic pollution on the distribution of DO in Lake Mariut is demonstrated by comparison with similar data for Nozka-Hydrodrome which is separated from Lake Mariut and receives only unmodified Nile water. Generally DO concentration and distribution in the Hydrodrome are normal and complete depletion was never observed during the study period. As observed by Naguib (1958) the DO in summer months was lower than in other months. The lowest DO value (5.58 mg/l) was observed in June 1969 and the highest (10.78 mg/l) in March 1969.

Effect of pollution on fish

The total fish production in Lake Mariut decreased from 9,977,815 kg in 1961 to 1,868,600 kg in 1967. This

[162]

great decrease is due mainly to lake pollution, although land reclamation must also be considered. In recent years both populations and industries have increased around the lake. Accordingly, the amount of sewage and industrial wastes has been increased. It was concluded that small concentrations of pollutants killed the fish even when other conditions were favourable. Dead fish were observed.

It is not the organic matter itself that kills the fish, but the low DO concentration resulting from the biochemical activities. The effect of low oxygen on the survival and distribution of fish was extensively studied, both under natural and laboratory conditions. The amount of DO required by fish is variable. DO deficiency in water may cause death or cause the fish to avoid the area of low DO.

References

FOEHRENBACH, J Pollution and eutrophication problems of great
1969 south bay, Long Island, New York. *J. Wat. Pollut. Control Fed.*, 41:1456–66.
NAGUIB, M Studies on the ecology of Lake Qarûn (Faiyum,
1958 Egypt). Part 1. *Kieler Meeresforsch.*, 14(2):187–222.
OHLE, W Die zivilisatorische Schädigung der Holsteinischen Seen.
1954 *Z. Städtehyg.*, 9:1–5.
PHELPS, B Stream sanitation. New York, John Wiley & Son.
1944
RUTTNER, F (D G Frey and F E J Fry, Transl.) Fundamentals of
1953 limnology. Toronto, University Press, 242 p.

The Discharge of Nutrients from Estuaries and their Effect on Primary Productivity

*A. James and P. C. Head**

Le déversement de nutriments par les estuaires et leurs effets sur la productivité primaire

Les estuaires constituent les principaux canaux par lesquels s'effectue le transport des nutriments vers la mer, mais les enquêtes concernant les effets de ce transport et les quantités sur lesquelles il porte sont rares. Les effets du déversement de nutriments dans la mer du Nord sont examinés ici en liaison avec les enquêtes détaillées menées dans la rivière Tyne et les calculs relatifs aux quantités de nutriments fournis par les principaux estuaires de Grande-Bretagne et d'Europe septentrionale.

Sur un apport quotidien de 9,2 tonnes d'azote dans l'estuaire de la Tyne, 5 tonnes seulement sont déversées chaque jour dans la mer, le reste s'accumulant surtout parmi les dépôts du fond. Entre les hautes eaux et les basses eaux, les matériaux déversés par l'estuaire constituent une couche d'eau enrichie de nutriments et de faible salinité, qui se prolonge jusqu'à 10 km de la côte. On peut la représenter sous la forme d'une jet flottant qui se déverserait dans un liquide de plus forte densité, et dont la salinité et les nutriments se repartiraient conformément à la théorie. A marée montante cette couche est déplacée vers le sud, ce qui accentue sa dispersion. Cette couche est déplacée vers le sud qui accentue sa dispersion. En été, l'utilisation biologique des nutriments aboutit à l'élimination quasi totale de ces derniers des eaux côtières superficielles. Dans la couche de déversement, des quantités appréciables de nutriments se maintienenent, bien que les peuplements de phytoplancton y soient d'une plus forte densité par comparaison avec les eaux côtières.

Les effets du déversement dans la rivière Tyne, à la fois sur le plan local et dans l'ensemble de la mer du Nord, sont faibles, mais les calculs portant sur les quantités totales de nutriments pénétrant dans cette mer indiquent qu'il s'y produit une lente accumulation.

Le descarga de sales nutrientes de los estuarios y sus efectos sobre la productividad primaria

Los estuarios son las principales vías de entrada de nutrientes al mar, pero existen pocas investigaciones sobre los efectos de tal incorporación o de las cantidades de aquéllos. Se examinan los efectos de la adición de nutrientes al mar del Norte en relación con estudios detallados efectuados en el río Tyne y los cálculos de las cantidades que llevan los principales estuarios de Gran Bretaña y Norte de Europa.

De las 9,2 toneladas diarias de nitrógeno que recibe el estuario de Tyne, sólo descargen en el mar cinco toneladas diarias, acumulándose el resto principalmente en los depósitos del fondo. Entre la pleamar y la bajamar el material procedente del estuario, en una extensión que llega hasta los 10 kilómetros de la costa, se puede describir como un chorro flotante que descarga en un líquido más denso, pudiéndose predecir las distribuciones de salinidad y nutrientes. Durante la marea creciente el penacho se desplaza hacia el sur y después se dispersa. Durante el verano, la utilización biológica de los nutrientes da lugar a que éstos desaparezcan casi por completo de las aguas superficiales costeras. Dentro del penacho producido por la descarga se mantienen apreciables cantidades de nutrientes, a pesar de los mayores efectivos de fitoplancton existentes en comparación con las aguas costeras.

El efecto de la descarga del río Tyne, lo mismo localmente que en el mar del Norte en su conjunto, es pequeño pero los cálculos de las cantidades totales de nutrientes que penetran en el mar del Norte sugieren que está teniendo lugar una lenta acumulación de ellos.

THIS paper describes an investigation into the amounts of nutrients being discharged from the estuary of the River Tyne and discusses the effect of estuarine discharges in the North Sea.

The River Tyne has an average flow of 50 m³/sec. Untreated domestic and industrial effluents can comprise up to one-third of the dry weather flow. Calculations of the amounts of nitrogen entering the estuary (Table 1) indicate that considerably more nitrogen enters the estuary than is in fact discharged in solution. This excess nitrogen is deposited in the bottom muds, some of which are ultimately removed to the sea by dredging. The discharge of phosphorus amounts to 0.3 tons per day.

Seaward extent of the discharge

During the period between high and low water the water discharged from the river builds up as a plume of low salinity nutrient enriched water overlying the indigenous sea water. The salinity of the coastal water is usually in the range 34.40 to 34.50⁰/₀₀ although lower values do occur inshore. The 34.30⁰/₀₀ surface isohaline of the discharge extends up to 8 km from the river mouth. The thickness of the plume is between 5 and 10 m at the river mouth. Higher concentrations of nitrate, ammonia, phosphate and silicate are found within 5 km of the river mouth. Typical winter concentration of these substances in the coastal water are:

TABLE 1. NITROGEN BALANCE IN THE RIVER TYNE ESTUARY

Gains	tons/day	Losses	tons/day
From upstream	2.4	To deposition and dredging	4.1
From domestic and		To atmosphere	0.1
industrial wastes	6.8	To sea	5.0
Total	9.2	Total	9.2

*Department of Civil Engineering, University of Newcastle-upon-Tyne, Newcastle-upon-Tyne, U.K.

10 μg-at NO_3N litre, 2–3 μg-at NH_3-N/litre, 0.75 μg-at PO_4P/litre and 8 μg-at SiO_3Si/litre.

The dispersion pattern of the water leaving the estuary on the ebb tide has been modelled as a buoyant jet discharging into a denser fluid by using a modification of the approach used by Tamai (1969). Salinity differences have been used to delineate the shape and dimensions of the discharge plume, and it has been possible to obtain predictions from the model which are in reasonable agreement with the observed pattern. Since the concentrations of nutrients in the area depend not only on this dispersion but also on rates of uptake and loss by biota it is not possible to use the model directly to generate patterns of nutrient concentrations.

Effect on primary production

During the winter period the standing crop of phytoplankton is low in both the coastal water and the discharge plume (chlorophyll-a values near 0.10 μg/litre). During the spring bloom, which usually occurs in March or April, the chlorophyll-a concentrations increase to between 7.0 and 10.0 μg/1. This is accompanied by a rapid reduction in the levels of phosphate, nitrate and silicate from the surface waters. The chlorophyll-a concentration of the coastal waters falls to about 1.0 μg/1 and is maintained at this level during the summer. Within the discharge plume the levels do not drop to the same extent, and concentrations of between 3 and 6 μg/1 are found associated with higher levels of nutrients throughout the summer. Within 1 km of the river mouth concentrations of 0.45 μg-at P/1, 1.6 μg-at NO_3N/1, 5.00 μg-at NH_3N/1 and 2.0 μg-at Si/1 were found in July. These concentrations decrease with distance from the shore to typical coastal values of about 0.15 μg-at P/1, 0.1 μg-at NO_3N/1, 0.5 μg-at NH_3N/1 and 1.0 μg-at Si/1. The build-up of the standing crop in the discharge plume is probably the result of increased growth of algae discharged from the river and species entrained in the discharge by the mixing processes.

The effect of the discharge from the River Tyne is to increase the planktonic production of the inshore waters near to its mouth. The beneficial effect of the extra supply of nutrients, especially nitrogen, outweighs any detrimental effects caused by increased turbidity or the presence of toxic substances.

Nutrient budget of the North Sea

Investigations into the currents and salinity distribution of the North Sea (Carruthers 1925, 1926; Tait, 1937; Goedecke et al., 1967; Lee and Ramster, 1968) have shown that the main circulation is anticlockwise. Water entering the sea from the North Atlantic and English Channel leaves in a current near the Norwegian coast. Mixing of the North Atlantic inflow and the Channel

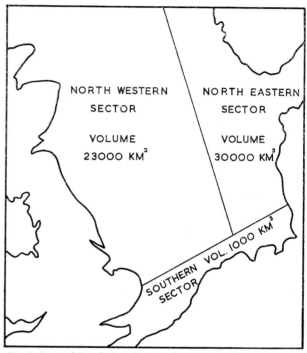

Fig 1. Map of North Sea showing division into three sectors for establishing a nutrient budget

TABLE 2. SUPPLY OF NUTRIENTS TO THE NORTH SEA

	Volume km³/day	Concn. of N μg-at/1	Concn. of P μg-at/1	Amount of N tons/day	Amount of P tons/day
N.W. Sector					
1. North Atlantic	54.8 [a]	10.0 [b]	0.70 [b]	7,700	1,230
2. Runoff	0.06	350.0	26.00	305	50
3. Precipitation	0.3 [e]	28.6 [e]	1.29 [d]	120	12
Total	55.16	10.5	0.75	8,125	1,292
S. Sector					
4. Channel	4.9 [a]	6.0 [b]	0.50 [b]	400	76
5. Runoff	0.3	372.0	21.00	1,389	194
6. Precipitation	0.08	28.6	1.29	32	3
Total	5.28	25.5	1.70	1,821	273
N.E. Sector					
7. From N.W. Sector	55.16	10.5	0.75	8,125	1,292
8. From S. Sector	5.28	25.5	1.70	1,821	273
9. Baltic	1.3 [e]	10.0 [e]	0.25 [e]	180	10
10. Runoff	—	—	—	54	12
11. Precipitation	0.3	28.6	1.29	120	12
Total	62.04	12.0	0.84	10,300	1,599

[a] Laevastu (1963)   [b] Johnston & Jones (1965)   [e] Vaccaro (1965)   [d] Fruh (1968)   [e] ICES (1970)

TABLE 3. NUTRIENT CONTRIBUTIONS TO THE NORTH SEA FROM SEWAGE AND RUNOFF

Country	Nitrogen tons/day	Phosphorus tons/day	Source of information
France (discharge to N. Sea)	4	1	Volume of sewage discharged[a]
Belgium	100	24	Volume of sewage discharged[a]
Netherlands	100	24	Population[b]
Germany, via R. Rhine	650	40	River concentrations
via Rivers Ems, Weser & Elbe	450	100	Population[c]
Denmark	27	6	Population[b]
Norway	27	5	Population[b]
United Kingdom	150	35	Volume of sewage discharged[a]
U.K. via major rivers	240	20	River concentrations
Total	1,548	256	

[a] ICES (1969)
[b] 9 g N and 1 g P per person per day
[c] May be low as no estimate of industrial pollution has been included

inflow occurs predominantly in an area between the Dogger Bank and the Danish Coast. On the basis of this type of circulation and the nutrient distributions given by Johnston and Jones (1965) the North Sea has been divided into three sectors (fig 1).

The supply of nutrients to each of the sectors has been estimated on the basis of the sea water inflow, the fresh water runoff and the precipitation (Table 2). A fuller treatment of the runoff and sewage estimates is given in Table 3.

The data given in Table 2 represent the equilibrium conditions existing in each of the three sectors at the moment. If each of the sectors is considered as a fully stirred tank, changes in concentration with time can be calculated from the relationship

$$Cm = C_1 (C_0 - C_1) e^{-\frac{ut}{v}}$$

where Cm is the mixed concentration, C_1 is the concentration of inflow, C_0 is the concentration of outflow, u is the volume of the inflow, t is time and v is the volume of the tank. For all three sectors the time required for the mixed concentration to equal that of the inflow, is found to be fairly short, approximately three years for the two northerly sectors, and less than one year for the southern.

The contribution of nutrients from sewage and runoff to the Northwestern Sector is small compared with the supply from the Northern Atlantic. Thus very large increases in this supply would be required to raise the concentration of the sector as a whole. In the Southern Sector the contribution from runoff is much greater. Increases in the concentration of the runoff would quickly result in an overall increase in the nutrient concentration of the sector. Changes in the Southern Sector would be reflected more slowly in the Northeastern Sector.

The concentration of nutrients in the North Sea at the present is fairly typical of northern temperate waters except for the shallow Southern Sector which receives most of the drainage of Northwestern Europe and a considerable portion of that from the United Kingdom. It is only in this sector that any serious build-up of nutrients could occur as a result of increases in the supply from land runoff. Build-up in the other sectors

would occur much more slowly and it is unlikely that concentrations would ever reach a level where detrimental effects would occur.

Summary and conclusions

The supply of nutrients from a polluted estuary has a beneficial effect on the standing crop of phytoplankton in the immediate vicinity.

Land runoff into the North Sea contributes about 17 per cent of the total nutrient supply from other sources. About 80 per cent of the runoff contribution is discharged into the shallow Southern Sector. Mixing of this nutrient enriched water with the water of the Northeastern Sector quickly reduces the concentration to levels typical of the North Sea as a whole.

References

CARRUTHERS, J N The water movements in the southern North
1925 Sea. Part 1. The surface drift. Fishery Invest., Lond. (2), 8(2):1–119.
CARRUTHERS, J N The water movements in the southern North
1926 Sea. Part 2. The bottom currents. Fishery Invest., Lond. (2), 9(3):1–114.
FRUH, E G Biological responses to nutrients—eutrophication:
1968 problems in fresh water. Adv. Wat. Qual. Improv., 1:49–64.
GOEDECKE, E, SMED, J and TOMCZAK, G Monatskarten des
1967 Salzgehaltes der Nordsee, dargestellt für verschiedene Tiefenhorizonte. Dt. hydrog. Z., 9 Suppl. B(4):13 p.
ICES, Report of the ICES working group on pollution of the North
1969 Sea. Coop. Res. Rep. int. Coun. Explor. Sea, (A),(13):61p.
ICES, Report of the ICES working group on pollution of the North
1970 Sea. Coop. Res. Rep. int. Coun. Explor. Sea, (A), (15):86 p.
JOHNSTON, R and JONES, P G W Inorganic nutrient in the North
1965 Sea. Ser. Atlas mar. Envir., Folio 11:10 plates.
KALLE, K Der Einfluss des englischen Küstenwassers auf den
1952 Chemismus der Wasserkörper in der südlichen Nordsee. Ber. dt. wiss. Kommn Meeresforsch., 13:130–5.
LAEVESTU, T Surface water types of the North Sea and their
1963 characteristics. Ser. Atlas mar. Envir., Folio 4:2 plates.
LEE, A and RAMSTER, J The hydrography of the North Sea. A
1968 review of our knowledge in relation to pollution problems. Helgoländer wiss. Meeresunters., 17(1–4): 44–63.
TAIT, J B Surface water drift in the northern and middle areas of
1937 the North Sea and Faeroe-Shetland Channel. Scient. Invest. Fishery Bd Scotl., (1):66 p.
TAMAI, N Diffusion of horizontal buoyant jet discharged at water
1969 surface. In Proceedings of the 13th Congress of the International Association for Hydraulic Research, Kyoto, Science Council of Japan, 3 Subject C, vol.1:215–22.
VACCARO, R F Inorganic nitrogen in sea water. In Chemical
1965 oceanography, edited by J. P. Riley and G. Skirrow, London, Academic Press, vol.1:365–403.

Effects of Effluent Discharge on Concentration of Nutrients in the Saronikos Gulf

R. C. Dugdale, J. C. Kelley*
and T. Becacos-Kontos†

Les effets du déversement d'effluents sur la concentration en nutriments dans le golfe de Saronique

Lors d'une croisière effectuée en Méditerranée par le R/V *Thompson*, en mars 1970, on a étudié le schéma de concentration des nutriments dans la partie supérieure du Golfe de Saronique au moyen d'un dispositif sous-marin d'acquisition des données. L'eau de mer, aspirée à la pompe d'une profondeur de 3 mètres, a fait l'objet d'une analyse continue visant à déterminer les taux de concentration de nitrate, d'ammonium, de silicate, de phosphate et de chlorophylle. Les données recueillies ont été codées et portées automatiquement sur bandes perforées à intervalles d'une minute, à mesure que le navire suivait l'itinéraire prescrit. Une fois cet itinéraire couvert, les données ont été lues à l'ordinateur marin IBM 1130; un programme de relevés par calculateur a permis d'obtenir sur marqueur Calcomp une carte pour chacune des propriétés mesurées. Avec deux séries de cartes, on a pu étudier les régions directement affectées par la décharge d'effluents de Karatsini, située immédiatement à l'ouest de l'entrée du Pirée. On a constaté l'existence de deux situations antithétiques: (1) une "plume" de concentration dense en nutriments et en chlorophylle à proximité de la côte de l'île de Salamine; et (2) une couche concentrique riche en nutriments et en chlorophylle centrée au voisinage de la décharge. Les taux de concentration des nutriments présentaient des écarts considérables. Il semble que lorsque les vents sont orientés dans certaines directions, les effluents soient chassés tout d'abord dans le golfe de Salamine où l'activité du phytoplancton modifie la teneur en nutriments. La "plume" visible à proximité de l'île de Salamine résultait apparemment de ce "traitement" partiel des eaux du golfe. Les taux exceptionnels des concentrations en nutriments de cette "plume" ont permis de mener plusieurs expériences destinées à vérifier des hypothèses sur les processus d'absorption des nutriments et sur la limitation de la croissance du phytoplancton sous leur action dans la mer Méditerranée. Le document donne les résultats de ces expériences et en fait l'évaluation.

Efectos producidos por la descarga de aguas residuales sobre la concentración de nutrientes en el Golfo de Saronikos

Durante un crucero realizado en el Mediterráneo, en marzo de 1970 por el buque de investigación *Thompson*, se estudió el tipo de concentración de nutrientes en la parte alta del Golfo de Saronikos empleando un sistema en movimiento de adquisición de datos. Se analizó continuamente agua de mar bombeada desde una profundidad de tres metros con objeto de determinar las concentraciones de nitrato, amonio, silicato, fosfato y clorofila. Los datos se memorizaron y perforaron automáticamente en cinta de papel a intervalos de un minuto, a medida que la embarcación navegaba siguiendo el rumbo prescrito. Al final del crucero, se leyeron los datos a bordo del barco en una computadora IBM 1130, y en una trazadora Calcomp se levantó un mapa de cada una de las propiedades medidas en un programa del perfil del terreno. Valiéndose de dos series de estos mapas se han definido las regiones directamente afectadas por las descargas de aguas residuales en Keratsini, al oeste de la entrada al puerto del Pireo. Se observaron dos condiciones contrastantes: (1) un sector en forma de penacho de elevada concentración de nutrientes y de clorofila cerca de la costa de la Isla Salamis; y (2) una zona concéntrica de alta concentración de nutrientes y de clorofila que tiene su centro en las cercanías de la salida de las aguas de desecho. Se observó una considerable diferencia en las proporciones de los nutrientes. Al parecer, cuando el viento sopla en una dirección determinada, las aguas residuales pueden ir a parar, en primer lugar, al Golfo de Salamis en donde la actividad del fitoplancton modifica el contenido de nutrientes. A lo largo de la Isla Salamis, el penacho se produjo, al parecer, por esta agua parcialmente "elaborada" del Golfo de Salamis. Las proporciones de nutrientes poco corrientes que hay en este penacho hizo posible realizar varios experimentos para evaluar las hipótesis sobre los procesos de absorción de nutrientes y de limitación de éstos en el desarrollo del fitoplancton en el Mar Mediterráneo. Se presentan y evalúan los resultados de dichos experimentos.

ALTHOUGH the effect of sewage outfalls on the distribution of bacteria is the most common cause of concern, in the Mediterranean Sea and other oligotrophic regions, the role of added nutrients in bringing about phytoplankton blooms also requires study. The impact of a particular outfall is difficult to assess because of the small areas involved and the rapidity with which changes in the sewage-sea water system occur. A computer model of a marine sewage outfall describing the spatial distribution of ammonium and phytoplankton has been constructed and when expanded should provide predictive capacity (Dugdale and Whitledge, 1970). An automated data acquisition system installed on board the R V *Thomas G. Thompson* was used to study the distribution of nutrients and chlorophyll in the vicinity of the Keratsini sewage outfall near Piraeus, Greece, in March 1970.

Methods

Water for analysis was obtained from an intake in the engine room at a depth of 3 m below the surface. The sea water was first pumped into a small tank with temperature and salinity sensors and then to the chemical laboratory where continuous samples for nutrient and chlorophyll analyses were taken. The nutrient analyses were made continuously by Autoanalyzer (R) techniques by methods basically given by Hager, Gordon and Park (1968) for nitrate, nitrite, reactive silicate, and phosphate. Ammonium analyses were made using the method of MacIsaac and Olund (1971). Chlorophyll was measured with the flow technique of Lorenzen (1966) using a Turner Model 111 fluorometer. The analogue outputs of

the Autoanalyzer colorimeters and of the fluorometer were connected to a Hewlett-Packard scanning digitizer and punched onto paper tape at predetermined intervals as the ship steamed over a pre-assigned path.

The data from two passes over the northern part of the Saronikos Gulf have been contoured automatically on the shipboard IBM 1130 computer. The contouring technique involves interpolating from the raw data to a regular grid. The interpolation is linear, based on the five data points closest to the grid intersection of interest. Once the grid is calculated, a non-linear contouring method is employed to position the contour lines, which are drawn on a Calcomp plotter. The position of the cruise track is shown as the dashed line. Data points

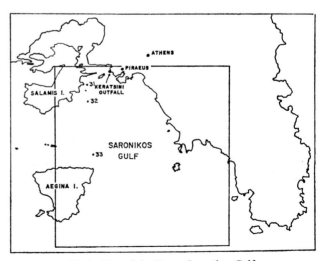

Fig 1. Map of the Upper Saronikos Gulf

*Department of Oceanography, University of Washington, Seattle, Washington, U.S.A.
†Democritos Nuclear Research Centre, Greek Atomic Energy Commission, Athens. Greece.

were recorded every minute and three minute averages were used to develop the contoured maps.

Athens and Piraeus lie at the north end of the Saronikos Gulf. The major portion of sewage collected is discharged after rudimentary settling through the Keratsini outfall just north of Piraeus Harbour (fig 1). The boil from the outfall is clearly visible and primary production is visibly enhanced. Studies of primary production and nutrient concentration in the unpolluted Saronikos Gulf which, in common with much of the eastern Mediterranean, is low in nutrients and production, are described by Becacos-Kontos (1968) and Ignatiades and Becacos-Kontos (1969).

Results

The set of surface maps made on 7th March 1970 is given in figs 2a–e. The track was made from south to north over a period of about 12 hours. The winds had been calm for the preceding two weeks. The enhancement of the phytoplankton population is readily apparent with the highest concentration appearing in the northwest and noticeable increases as far south as the island of Aegina. However, on Pass 5, the outfall was not approached sufficiently close to measure the presence of the boil and main plume. The pattern of phosphate and nitrate at 3 m depth (figs 2b and d) are similar to that of chlorophyll. The pattern of silicate distribution, fig 2c, is strikingly

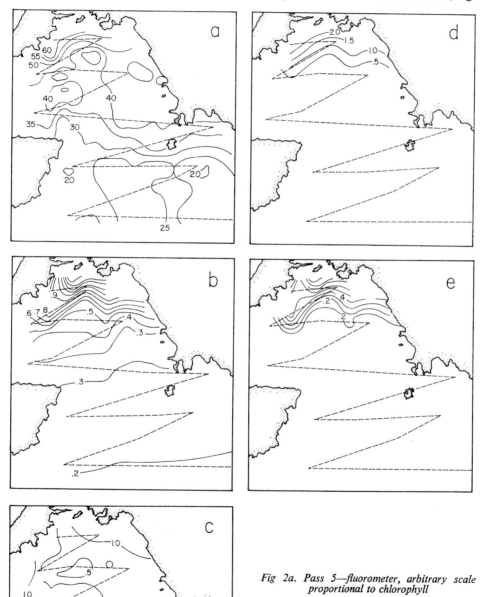

Fig 2a. Pass 5—fluorometer, arbitrary scale proportional to chlorophyll

Fig 2b. Pass 5—phosphate, µg at/litre

Fig 2c. Pass 5—silicate, µg at/litre

Fig 2d. Pass 5–nitrate, µg at/litre

Fig 2e. Pass 5—nitrate µg at/litre

[167]

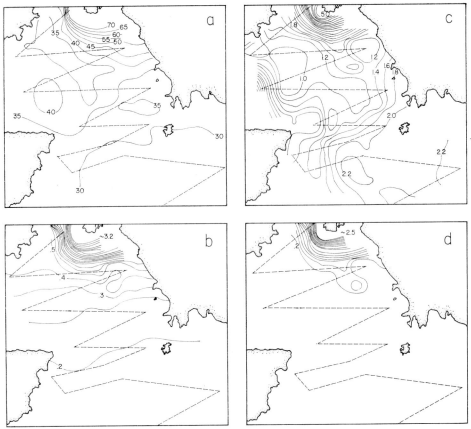

Fig 3a. Pass 15 fluorometer, arbitrary scale proportional to chlorophyll
Fig 3b. Pass 15—phosphate, μg at/litre
Fig 3c. Pass 15—silicate, μg at/litre
Fig 3d. Pass 15—nitrate, μg at/litre

different, the concentration decreasing from an ambient value of about 2 μg at/litre in the south to as low as 0.5 μg at/litre at the north where the highest phytoplankton populations were found.

The surface maps obtained on 14th March 1970 are given in figs 3a–d. The light southerly winds prevailing at that time appear to hold the plume of phytoplankton (fig 3a) in the northeastern bay and the distribution of phosphate (fig 3b) and nitrate (fig 3d) parallel that of chlorophyll very well. Silicate (fig 3c) shows a more complex distribution, with both an area of low concentration corresponding to a chlorophyll high to the west, and one of high silicate concentration around the outfall. The source of high silicate to the west along Salamis Island is unknown and appears to be unrelated to the other nutrients or to chlorophyll. Again the effects of the outfall can easily be discerned as far as Aegina.

Although some of the obvious differences between the two sets of maps can be ascribed to differences in the cruise track it appears that the patterns of nutrient distribution can vary widely at different times, probably primarily in response to the wind-induced currents. For example, the nutrient and chlorophyll distribution observed on Pass 5 may be the result of older phytoplankton populations originating in the Bay of Salamis and moved southward by a current along Salamis Island.

Three hydrographic stations to 25 m were worked immediately after Pass 5 to assess the depths to which the added nutrients could be detected. A section for

ammonium is shown in fig 4. The maximum effects appear to be confined to the upper 20 m.

The ready availability of these surface maps enhance our capability of using the marine environment as a natural laboratory for the physiological experiments necessary to obtain critical rate constants for insertion into simulation models. It was possible to carry out a

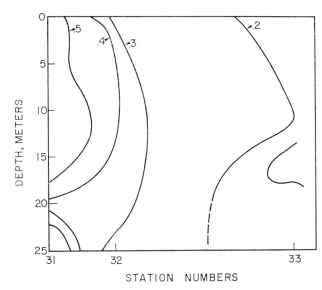

Fig 4. Vertical section for ammonium from Salamis Island (Sta.31) to the north end of Aegina Island (Sta.33)

number of experiments following the construction of the maps. These experiments confirmed laboratory experiments showing that ammonium, at the low concentrations occurring in the Upper Saronikos Gulf, strongly inhibits the uptake of nitrate by phytoplankton. Inhibition of the intracellular enzyme nitrate reductase as a function of naturally occurring ammonium was also documented (Packard, et al, 1971).

Future plans

A research programme to acquire the necessary inputs to a computer simulation model of the nutrient-phytoplankton relationships in the Upper Saronikos Gulf has been established on a co-operative basis between the Democritos Nuclear Research Centre in Athens and the Upwelling Program of the United States IBP. Arrays of moored current meters, coupled with meteorological observations and dye diffusion studies will provide the rate constants for the phytoplankton-nutrient model.

This research should develop quickly enough to provide useful advice to the local authorities charged with planning for sewage treatment and discharge.

References

BECACOS-KONTOS, T   The annual cycle of primary production in
1968   the Saronikos Gulf (Aegean Sea) for the period November
1963–October 1964. *Limnol. Oceanogr.*, 13:485–9.

DUGDALE, R C and WHITLEDGE, T   Computer simulation of phyto-
1970   plankton growth near a marine sewage outfall. *Revue int.*
Océanogr. méd., 17:201–10.

HAGER, S W, GORDON, L I and PARK, P K   A practical manual for
1968   use of the Technicon Autoanalyzer (R) in seawater nutrient
analyses. *Tech. Rep. Dep. Oceanogr. Oregon St. Univ.*, (68–
33):1–31.

IGNATIADES, L and BECACOS-KONTOS, T   Nutrient investigations in
1969   Lower Saronikos Bay, Aegean Sea. *Vie Milieu (Océanogr.)*,
20:51–62.

LORENZEN, C J   A method for the continuous measurement of *in*
1966   *vivo* chlorophyll concentration. *Deep-Sea Res.*, 13(2):223–7.

MACISAAC, J J and OLUND, R K   An automated extraction pro-
1971   cedure for the determination of ammonia in seawater.
Investigación pesq. 35(1):221–32.

PACKARD, T T, *et al.*   Variations of nitrate reductase activity in
1971   marine phytoplankton. *Investigación pesq.* 35(1):209–19.

Trace Elements, Radionuclides and Pesticide Residues in the Hudson River

T. J. Kneip,* G. P. Howells*
M. E. Wrenn*

Traceurs, radionuclides et résidus de pesticides dans le fleuve Hudson

L'Institute of Environmental Medicine de l'Université de New York procède depuis 1964 à des études sur les oligo-contaminants dans l'Hudson et sur leur bioconcentration dans divers biota. La distribution relative des radionuclides dans l'eau, des sédiments et des biota, est semblable à celle que l'on a observée dans d'autres zones pour les radionuclides et leurs éléments stables associés. Les radio-nuclides naturels potassium-40, radium-226 et radium-228 représentent de 80 à 90 pour cent de l'activité du fleuve; la plus grande partie de cette activité est imputable au potassium-40. Exception faite du tritium, la radioactivité provenant des activités humaines dans l'eau de l'Hudson est inférieure à 1/100 000 des limites réglementaires de l'AEC. On a étudié la bioaccumulation de plusieurs radionuclides artificiels. Parmi les nuclides retenus (manganèse-54, cobalt-60, fer-55, cérium-144 et strontium-90), seuls le manganèse-54 et le cobalt se sont reconcentrés en proportions mesurables. L'activité accumulée n'augmente pas avec le niveau trophique dans le cycle alimentaire, mais est régie par d'autres facteurs.

En ce qui concerne les résidus de pesticides du type des hydro-carbures chlorés, il apparaît que leur concentration augmente avec le niveau trophique dans de nombreux systèmes aquatiques. On l'a vérifié dans le cas des biota de l'Hudson: les poissons prédateurs et les oiseaux au sommet du cycle alimentaire présentent les concentrations les plus élevées, la proportion de certains résidus atteignant 1 à 3 microgrammes par gramme d'échantillon. Cela sous-entend une amplification biologique supérieure à 100 000 par rapport aux concentrations maximales observées dans l'eau, à savoir quelques nanogrammes par litre. Bien que l'on ne possède aucune preuve d'effets biologiques sur les espèces ou populations observées, le facteur de proportionnalité entre les niveaux relevés et les concentrations signalées ailleurs comme étant létales pour quelques espèces est inférieur à dix. Chez deux espèces au moins des poissons étudiés, la quantité de résidus de pesticides dépassait les concentrations actuellement fixées à titre indicatif par la Food and Drug Administration pour les portions comestibles de poisson.

Elementos de traza, radionúclidos y residuos de pesticidas en el río Hudson

El Institute of Environmental Medicine (Universidad de Nueva York) estudia desde 1964 los oligo-contaminantes y su bioconcentración en biotas seleccionadas en el río Hudson. La distribución relativa de radionúclidos en el agua, sedimentos y biota es similar a la que se observó en otras zonas en cuanto a los radionúclidos y sus elementos estables asociados. A los radio-núclidos naturales, potasio-40 y radio-226 y 228 corresponde del 80 al 90 por ciento de la actividad en el río; la mayor parte de ésta se atribuye a potasio-40. Excepto en el caso del tritio la radiactividad originada por el hombre en el agua del río es inferior a 1/100 000 que es el límite reglamentario de la AEC. Se ha estudiado la bioacumulación de varios radionúclidos originados por el hombre; de los seleccionados (manganeso-54, cobalto-60, hierro-55, cerio-144, estroncio-90) sólo el manganeso-54 y el cobalto se encontraron reconcentrados en cantidades mensurables. La actividad acumulada no aumento con el nivel trófico en la cadena alimentaria, sino que está gobernada por otros factores.

Se ha demostrado que los residuos de plaguicidas a base de hidratos de carbono clorurados tienen por norma aumentar en concentración con el nivel trófico en muchos sistemas acuáticos. Tal es el caso de la biota del río Hudson con los peces y las aves predadores en el vértice de la cadena alimentaria, mostrando las mayores concentraciones con residuos que alcanzan de 1 a 3 microgramos/gramo de muestra. Esto representa un aumento biológico de más de 100 000 a partir de las concentraciones máximas observadas de unos pocos nanogramos/litro de agua. Aunque no se han encontrado datos que indiquen efectos biológicos en las especies o poblaciones observadas, existe un factor de menos de 10 entre las concentraciones observadas y las comunicadas de otras partes como letales para ciertas especies. Por lo menos en el caso de dos especies de peces examinadas, los residuos de plaguicida exceden las actuales concentrationes citadas a modo de orientación por la Food and Drug Administration para les partes comestibles del pescado.

WHEN studying trace contaminants the goals of the programme must be clearly defined before initiating the project because these establish the methods that will be used. In routinely determining the quality of a body of water with reference to standards set by regulatory bodies, values designated "less than" or "not detected" suffice when concentrations are below regulatory limits. However, for research studies, it is necessary to use methods sensitive enough to measure absolute concentrations. This has become obvious in the light of present evidence of unexpected reconcentration in biological systems. Thus, we find

---

* Institute of Environmental Studies, New York University Medical Center, New York 10016, U.S.A.

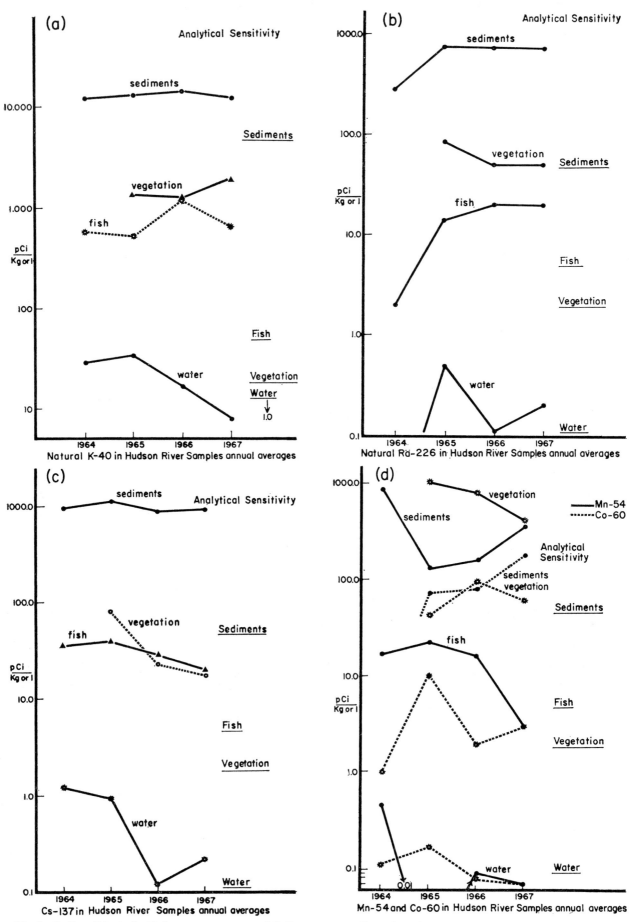

Fig. 1a. Radionuclides in the Hudson ecosystem: natural K-40
Fig. 1c. Radionuclides in the Hudson ecosystem: Cs-137

Fig. 1b. Radionuclides in the Hudson ecosystem: natural Ra-226
Fig. 1d. Radionuclides in the Hudson ecosystem: Mn-54 and Co-60

ourselves working harder and harder to discover less and less of any given contaminant, in an attempt to define the capacity of an ecosystem to absorb contaminants without significant effects.

The concentrations of most substances tested are usually so low in environmental samples that, in general, concentration and/or separation from interfering substances is required before quantitative analysis can be made.

Radionuclide data have been obtained by standard γ-spectrometric methods including computer conversion of raw sample spectra to numerical nuclide concentrations by comparison with selected standards. Sample handling is confined to evaporation of water samples and drying at 120° and ashing at 425°C for other samples.

Trace metals have been determined by atomic absorption. Extraction and concentration is done by liquid extraction using ammonium pyrolidine dithiocarbamate or oxine (manganese only) complexing agent. Standards are submitted to an identical procedure to correct for all yield factors.

Pesticide residues have been separated and concentrated by standard methods (U.S. Dept. Health, Educ. & Welfare, 1968) using hexane extraction, and concentrations determined by gas liquid chromatography using an electron capture detector. Occasional samples have been examined by thin layer chromatography to obtain a qualitative confirmation of the presence of the residues.

Results of survey

Trace element concentrations in fresh water are important in relation to suitability of the water for drinking or industrial use, as well as their potential toxicity to the natural fauna and flora either directly or indirectly (F.W.P.C.A., 1968). Trace element distribution in water and biota can be used to interpret the distribution of radionuclides in the same aquatic system.

TABLE 1. OCCURRENCE OF TRACE CONSTITUENTS IN HUDSON RIVER AT POUGHKEEPSIE, OCTOBER 1962–JUNE 1967

| | Per cent occurrence | Concentration of[a] positive samples | | Minimum level of detection[b] $\mu g/l$ |
		Average $\mu g/l$	Range $\mu g/l$	
Aluminium	44[c]	33	3–54	3–29
Arsenic	0	—	—	10–53
Barium	100	35	11–52	
Beryllium	0	—	—	0.03–0.07
Boron	100	46	21–89	—
Cadmium	0	—	—	1–17
Chromium	60	11	2–33	2–7
Cobalt	0	—	—	1–15
Copper	90	19	5–72	4
Iron	90	43	4–154	4
Lead	30	14	5–23	3–29
Manganese	40	3.3	0.6–7.0	0.6–4.3
Molybdenum	20	32	9–55	2–29
Nickel	0	—	—	1–15
Phosphorus	60	72	20–232	0.3–1.5
Strontium	100	99	63–155	—
Vanadium	0	0	—	2–29
Zinc	90	52	8–93	8–15

[a] Raw data obtained from Storet system of the United States Department of the Interior, Federal Water Pollution Control Administration
[b] Reported detection limits vary from year to year. The range given is that reported for all negative samples
[c] 1962 sample not reported

Cumulated data for a 5-year period (Table 1) provide estimates of the "normal" concentration of trace metals in the Hudson River at a single point in the river above the "salt front". The data in Table 2 (Institute of Environmental Medicine) demonstrate the considerable variability in the concentrations at different locations along the river; variability also occurs seasonally due to the changes in fresh-water flow and the moving "salt front". In a tidal estuary the degree of intrusion of salt water is important in relation to trace element concentrations. At Poughkeepsie, chloride is usually <100 mg/1, and the water is practically "fresh", but south of this, the water varies from fresh to 20 per cent of sea water in the reach of the river studied, and the trace element composition is comparable to sea water dilutions.

Concentrations are usually well below standards set for raw waters to be used for drinking, but some metals approach or exceed the limits for this use in isolated instances (see Table 2).

TABLE 2. OCCURRENCE OF MINOR CONSTITUENTS IN HUDSON RIVER WATER (27–100 MILES NORTH OF BATTERY PARK) COMPARED TO FEDERAL DRINKING WATER STANDARDS[d]

	Per cent occurrence	Concentration of positive samples range $\mu g/l$[a, b]	Minimum level of detection $\mu g/l$[c]	Standard $\mu g/l$[d]
Cadmium	34	3.0–140	2.0–6.0	10
Copper	67	1.7–260	—	1,000
Iron	80	2.0–280	2.0–10.0	—
Lead	88	4.0–63	—	50
Manganese	48	1.0–38	1.0–4.0	300
Nickel	100	0.5–44	10	—
Strontium	100	15–570	—	—
Zinc	85	3.0–310	2.0	5,000

[a] Raw data from Howells and Bath, 1969; McCrone, 1966 and 1967. July and August 1965, November 1966, March through June 1967, April through December 1968. Sampling sites were variable from year to year
[b] Average not calculated because of the variability in time and location of sampling. In some cases extremes of range represent fresh water–salt water contrast
[c] Range reported for all negative results
[d] References F.W.P.C.A., 1968

The concentrations of potentially toxic metals are of interest also, in that many organisms are known to be sensitive to them (Doudoroff and Katz, 1953; Pickering and Henderson, 1966; Sprague, 1968) or to reconcentrate them (Bowen, 1966; Brooks and Rumsby, 1965; Polikarpov, 1966; Pringle et al., 1968; Vinogradov, 1953). An example of bioconcentration in Hudson River biota is the selective accumulation of manganese by Potomogeton spp. (to 2.3 mg/g wet weight) and Chara sp. (to 6.0 mg/g), while water concentrations are only about 3 mg/1.

Radionuclides

There are three sources of the radionuclides in the Hudson River; a major part comes from natural sources such as washout from soils, the rest being man-made nuclides from atmospheric fallout, or from industry or other institutions. The nuclear reactor at Indian Point is the most significant source of man-made nuclides in the Hudson.

Since estimates of low water concentrations of radionuclides are often at the limit of accuracy, accumulation in the biota allows a better evaluation of releases into the river. In general, radionuclides will follow the same path in metabolism and distribution as the stable element, so that stable element data can be used to interpret or predict radionuclide data. The comparison has been of value in studying stable manganese and manganese-54 in the ecosystem.

The distribution of selected radionuclides at Indian Point is shown in fig 1 a–d. In each case, the analytical sensitivity is adequate to measure the concentration present as shown by the levels at the right of each figure. Potassium-40 and radium-226 represent the natural radionuclides. Levels in water are lowest of all samples while sediments are high in concentration (on an equivalent "wet weight" basis), while vegetation and fish samples are intermediate. Note that radium-226 levels are generally about 10 times lower than potassium-40. Radium-228 has essentially the same distribution and concentrations as radium-226.

Cesium-137 is an isotope derived principally from fallout due to weapons testing. The source was practically eliminated several years ago, and sediments and soils appear to have reached an equilibrium. Decreasing concentrations in water still follow the decreasing input from fallout. Concentration values for components of the ecosystem range from 0.1 to 1,000 pCi/kg or litre. Fish and vegetation samples range from 10 to 100 pCi/kg or litre, similar to the range for radium and about 1/10 that for potassium-40.

Manganese-54 and cobalt-60, derived both from fallout and a fission source, are subject to biological concentration, as seen in fig 1 d. The highest radioactivity for manganese-54 in vegetation is about equal to that for potassium-40 but manganese and cobalt activities in fish are about 100 times less than that of potassium-40 in the same samples. In general, total radioactivity from artificial nuclides is 10 to 10,000 times less than that from radionuclides from natural sources. For fish, most of the radioactivity is attributable to potassium-40.

Measured radionuclide concentrations in water and the established regulatory limits may be compared (Table 3). In addition, concentrations in locally caught fish, and the quantity of that fish which could be eaten in a year before the acceptable dose would be exceeded, are given. The average per capita consumption of fish (all varieties) in the United States of America is about 8 kg/year (U.S. Dept. Agric., 1969).

For the radionuclides shown, only natural radium-228 approaches the regulatory limit for water, and even here a factor of more than a 100 still exists between the observed concentration and the permissible limit. It is also obvious from the calculated quantities of fish which

TABLE 4. CHLORINATED HYDROCARBON PESTICIDES IN THE HUDSON RIVER WATER (NG/L)

	PHS Fall–1964	Inshore water March–1967	Channel Mar.–Oct. 1968
Dieldrin	8	2	8
DDT	ND	200	T
DDE	ND	50	2
DDT(TDE)	4	180	T
Aldrin	P	ND (I)	ND (I)
Heptachlor	ND	ND	ND
Heptachlor Epoxide	ND	ND (I)	ND (I)
Lindane	ND	6	2

U.S. Dept. Health, Education and Welfare, 1965. I = Interference; ND = Not Detected; P = Probably Present.

TABLE 5. PESTICIDE RESIDUES (DDT METABOLITES AND DIELDRIN) IN HUDSON RIVER BIOTA

Trophic level	Pesticide concentration range (ng/g)	Concentration factor
Water	0–20 × 10⁻³ ng/ml	—
Plankton	20–150	1,000–7,500
Algae	30–100	1,500–5,000
Bivalves	30–100	1,500–5,000
Some fish	30–100	1,500–5,000
Some fish	100–1,000	5,000–50,000
Birds	1,000–3,000	50,000–150,000

would have to be eaten before acceptable levels are exceeded, that none of the radionuclides studied presents a realistic hazard. Nor do the measured activities in any of the biota indicate that any biological effects could be observed when the radioactivity due to artificial radionuclides is compared to that normally present from potassium and radium.

Pesticide residues

Typical levels for several chlorinated hydrocarbon pesticides in Hudson River water are shown in Table 4. Samples taken from shallow backwaters along the river in March 1967 showed significantly higher levels for several pesticides, presumably reflecting a localized seasonal input. The bulk of water present in the main channel of the river shows consistently low residue concentrations throughout the year.

In contrast to stable and radioactive nuclides which show a complex distribution pattern in the ecosystem, governed by their chemical and physiological behaviour, pesticides show a consistent increase in concentration, progressing from the lower to the higher trophic levels. This results basically from the insolubility of the chlorinated hydrocarbons in water and their high

TABLE 3. COMPARISON OF MEASURED RADIONUCLIDE CONCENTRATIONS IN THE HUDSON RIVER AT INDIAN POINT IN 1968 AND REGULATORY LIMITS FOR WATERS

	Ra-228	Nuclide Cs-137	Co-60	Mn-54
Concentration in water (pCi/1)	0.12	0.01	ND	0.37
Maximum permissible concentration in water (pCi/1)ᵃ	30	20,000	50,000	100,000
Fraction of MPC	4 × 10⁻³	5 × 10⁻⁷	NA	4 × 10⁻⁶
Concentration in fish (pCi/kg)	10	31	5	27
Quantity of fish containing yearly permissible dose (kg)	2,400	516,000	8,000,000	2,960,000

ᵃ U.S. Atomic Energy Commission 1967. ND = Not detected. NA = Not applicable.

solubility in oils or fats. Thus, concentrations of pesticide residues are higher at the upper levels of the food web than in the aquatic medium in which the biota live, and are higher in animals or tissues which have higher fat content. However, a quantitative interpretation of pesticide concentrations in the food web and the relation to water concentration requires an understanding of the routes and dynamics of uptake.

The pesticide concentrations in sediments are many times higher than those in water. They have been found to be very variable in relation to both season and sampling site. Concentrations in sediments are nearly always detectable, but may vary over an order of magnitude from one month to the next at a single location.

TABLE 6. COMPARISON OF PESTICIDE RESIDUES IN FISH FLESH TO FOOD AND DRUG ADMINISTRATION GUIDELINE UG/G (PPM)[a]

	FDA limit	Striped Bass	Perch	Summer Herring	Sturgeon eggs
DDT & Metabolites	5	0.4	1.9	0.6	1.6
Dieldrin	0.3	0.1	0.8	0.1	0.9

[a] Duggan, personal communication.

In Table 5, the general pattern of concentration levels in the biota from the Hudson River is shown both in terms of pesticide concentrations observed and in terms of a "concentration factor" (CF) calculated from the concentration in the wet biological sample and the concentration in water. Although the concentrations and concentration factors shown are for DDT metabolites and dieldrin, the concentration factors are similar to other published values despite differences in the pesticide measured, different locations, species, or exposure route, or even of the actual concentration in the ambient water system (Godsil and Johnson, 1968; Terriere et al., 1966; Woodwell, Wurster and Isaacson, 1967).

The Food and Drug Administration has established Administrative Guidelines for pesticide residues in edible fish portions (Duggan, pers. comm.). Tolerance limits, taking account of the dietary pattern, have not yet been established, but if a sample exceeds the administrative guideline limit, it may not be marketed as food. Fish flesh and sturgeon eggs (taken from the Hudson in 1969) are compared with FDA guidelines for DDT and metabolites, and for dieldrin in Table 6: concentrations range from one-tenth to three times the guideline limits even though water concentrations were about 10 ng/l or less.

CONCLUSIONS

Artificial radionuclides found in the Hudson River water are at least 250 thousand times less concentrated than the levels permitted by existing regulations; while radium, a naturally occurring nuclide, is within a factor of 250 of the permitted concentration.

Both trace metals and pesticide residues occasionally exceed the concentrations allowed by State or Federal regulations. Sporadic high level inputs of trace metals could be the cause of localized, elevated concentrations

seen in water samples. Pesticides, on the other hand, while mostly found in negligible concentration in water, are so concentrated through build-up in the biological food web that some fish may be too contaminated for commercial use.

A continuing need exists for improved definition of the fate of radionuclides in the ecosystem in order to make safe predictions of the effect of future nuclear installations along the river. Source studies relating to trace metals should be pursued to evaluate the potential biological significance of occasional high concentrations. Organic contaminants, among them pesticide residues, should be examined on a broader scale to determine the identities and concentrations of the compounds present, and their potential hazards both in food sources and to species survival. Control of organic contaminants may be of the greatest importance in improving the utility of the river both for recreation and water supplies.

References

BOWEN, J H M Trace elements in biochemistry. London, Acad-
1966   emic Press.
BROOKS, R R and RUMSBY, M G The biogeochemistry of trace
1965   element takeup by some New Zealand bivalves. Limnol.
         Oceanogr., 10:521–7.
DOUDOROFF, P and KATZ, M Critical review of literature on the
1953   toxicity of industrial wastes and their components to fish.
         The metals as salts. Sewage Ind. Wastes, 25:1380–97.
DUGGAN, R Deputy Associate Commissioner for Compliance,
         Food and Drug Administration, personal communication.
Federal Water Pollution Control Administration, Water quality
1968   criteria. Report of the National Technical Advisory Com-
         mittee to the Secretary of the Interior, Washington, D.C.
GODSIL, P J and JOHNSON, W C Residues in fish, wildlife and
1968   estuaries. Pestic. Monitg J., 1(4):21–6.
HOWELLS, G P and BATH, D Trace elements in the Hudson River
1969   Development of a biological monitoring system and a
         survey of trace metals, radionuclides and pesticide residues
         in the Hudson River. New York University Medical Center,
         Institute of Environmental Medicine.
McCRONE, A W The Hudson River estuary, sediments and
1966   pollution. Geogrl Rev., 52:175–89.
McCRONE, A W The Hudson River estuary, sedimentary and
1967   geochemical properties between Kingston and Haverstraw.
         J. sedim. Petrol., 37:475–86.
PICKERING, H and HENDERSON, C The acute toxicity of some
1966   heavy metals to different species of warm water fishes. Int.
         J. Air. Wat. Pollut., 10(6/7):453–63.
POLIKARPOV, G G Radioecology of aquatic organisms. Amster-
1966   dam, North-Holland Publishing Co., 320 pp.
PRINGLE, B H, et al. Trace metal accumulation by estuarine
1968   mollusks. J. sanit. Engng Div. Am. Soc. Civ. Engrs, 94:455–
         75.
SPRAGUE, J B Avoidance reaction of salmonid fish to representa-
1968   tive pollutants. Wat. Res., 2:23–4.
TERRIERE, L C, et al. The persistence of toxaphene in lake water
1966   and its uptake by aquatic plants and animals. J. agric. Fd
         Chem., 14(1):66–9.
U.S. Atomic Energy Commission, Rules and Regulations, Title 10
1967   Code of Federal Regulations Chapter I, Part 20. Standards
         for protection against radiation. Washington, D.C., United
         States Government Printing Office.
U.S. Department of Agriculture, Food consumption of households
1969   in U.S.A. (ARS 62–76).
U.S. Department of Health, Education and Welfare, Food and Drug
1968   Administration, Pesticide analytical manual, Vol. 1, Revised.
U.S. Department of Health, Education and Welfare, Public
1965   Health Service Report on pollution of the Hudson River
         and its tributaries. New York, New York Division of
         Water Supply and Pollution Control Region II.
VINOGRADOV, A P The elementary chemical composition of marine
1953   organisms. Mem. Sears Fdn Mar. Res., (2):648 p.
WOODWELL, G M, WURSTER, C F, JR. and ISAACSON, P A DDT
1967   residues in an east coast estuary: A case of biological con-
         centration of a persistent insecticide. Science N.Y., 156:
         821–4.

Etude de la Pollution Marine par les Détergents Anioniques provenant des Eaux d'Egouts de Marseille

*A. Arnoux et F. Caruelle**

Study of marine pollution by anionic detergents from sewage of the city of Marseilles

Each day the dumping by the City of Marseilles of 500 000 m³ of sewage into the sea constitutes an input of anionic detergents of the order of 30 t. A short study has supplied useful information on variations in the detergent concentration in sewage. Our study has essentially covered the investigation of these detergents in the sea and their systematic quantitative assaying. On the basis of large grouped series of samplings, maps have been compiled which show, in quantitative terms, the dispersion and persistency of these detergents from the sewage outlet to the open sea. The heavy fluctuations in surface distribution made it possible to show the movements of the mass of polluted water under different weather conditions and to gain a better idea of areas of maximum pollution. This study, which has been performed mainly at the surface, has been amplified by one on distribution in depth which brings out certain stratification anomalies.

Estudio de la contaminación de las aguas del mar causada por los detergentes aniónicos procedentes del alcantarillado de Marsella

Los 500 000 m³ de aguas del alcantarillado de Marsella que se vierten diariamente en el mar aportan cerca de 30 toneladas de detergentes aniónicos. Un breve estudio nos ha permitido reunir algunas informaciones útiles sobre las variaciones de la concentración de dichos detergentes en la red de alcantarillado. Nuestro estudio se ha centrado esencialmente sobre la investigación y la medición sistemática de las concentraciones de esos detergentes en el mar. Procediendo a series agrupadas de muestreos hemos podido preparar algunos mapas que permiten apreciar la dispersión y la persistencia de los detergentes aniónicos desde que salen del alcantarillado hasta que se adentran en el mar. Las notables fluctuaciones de la distribución de los detergentes en la superficie nos han permitido comprobar los desplazamientos de la masa de agua contaminada en función de las condiciones meteorológicas y conocer mejor las zonas de contaminación máxima. Aunque el estudio se ha realizado primordialmente en la superficie, se ha completado con análisis de la distribución de los detergentes en profundidad, que han revelado algunas anomalías de estratificación.

L'UTILISATION intensive, aussi bien domestique qu'industrielle, des détergents en fait un des éléments les plus fréquents des effluents urbains. Parmi les différents constituants entrant dans la composition des produits commerciaux, une place prépondérante revient aux agents tensio-actifs dont les effets sont particulièrement néfastes au maintien de l'équilibre biologique dans les eaux.

Il ressort des données publiées en France ces dernières années que 75 à 80 pour cent de ces produits appartiennent au type anionique (alkyl-sulfates, alkyl-sulfonates, alkyl-aryl-sulfonates). A ce titre, ils nous ont paru suffisamment représentatifs de la pollution par les détergents et en particulier de celle de l'environnement marin de la ville de Marseille.

Situation géographique

L'agglomération marseillaise est assainie par un réseau de plus de 650 kilomètres de galeries convergeant vers un grand émissaire qui débite chaque jour environ 500.000 m³ d'eaux usées. Le rejet se fait à la surface de la mer, dans l'anse de Cortiou, située au sud de Marseille, en un point désertique de la côte rocheuse du massif de Marseille–Veyre. Vers le large, ce secteur est limité par la succession, du sud-est au nord-ouest, des îles Riou, Plane, Jarre et Maire (fig 1a).

La topographie du fond dans cette zone est schématisée sur la fig 1 par une série de profils dont la direction est repérée sur la carte générale. La coupe longitudinale A suivant approximativement l'axe de ce secteur (fig 1b), illustre le relèvement du fond qui passe de 70 m, au droit de l'émissaire, à moins de 7 m au plateau des Chèvres, entre la côte et l'île Jarre, et redescend ensuite à une cinquantaine de mètres entre Jarre et Maire. Les coupes transversales B, C et D établies entre la côte et les îles (fig 1c), sont réunies sur un même graphique afin de mettre en évidence le rétrécissement accompagnant le relèvement du fond.

Prélèvements

L'étude entreprise était essentiellement comparative puisqu'elle avait pour but d'apprécier, à partir d'un point de rejet, les modalités de la pollution par les détergents dans le milieu marin. Les modifications pouvant être assez rapides, il s'avérait indispensable d'établir des situations générales en un instant donné dans l'ensemble ou une partie du secteur de Cortiou. Comme il était matériellement impossible d'effectuer tous les prélèvements en même temps, nous recueillions à chaque sortie et au cours d'une même matinée 40 à 60 échantillons qui étaient transportés sans tarder au laboratoire et analysés dans les heures suivantes.

La méthode de dosage utilisée est dérivée de celle de Longwell *et al.* (1955) adaptée à nos conditions de travail en tenant compte des indications d'Abbott (1962) et d'Armangau (1967). Le principe repose sur la formation en milieu alcalin (tampon borate) d'un complexe bleu de méthylène-détergent anionique extractible par le chloroforme. La spécificité du dosage est améliorée par lavage de l'extrait chloroformique avec une solution aqueuse acide de bleu de méthylène. Le complexe bleu ainsi obtenu se prête à une mesure spectrophotométrique à 655 nm.

En veillant tout particulièrement à la qualité des réactifs (bleu de méthylène pour mesures Redox, chloroforme distillé extemporanément) et à la reproductibilité exacte des conditions opératoires, nous disposions d'une méthode suffisamment sensible et fidèle pour pouvoir mettre en évidence des variations très faibles de concentrations.

Résultats

L'intensité de la coloration du complexe dépendant de la longueur de la chaîne alkyl, elle peut varier largement suivant la nature des produits ou leur degré de dégradation. Le dosage dans les eaux représente ainsi une détermination globale dont la valeur doit être rapportée à un corps de référence, tel que le dioctylsulfosuccinate de sodium (Manoxol OT), dont la réactivité moyenne peut être considérée comme représentative d'un mélange de détergents anioniques.

C'est donc par rapport au Manoxol OT que nous avons apprécié la fidélité du dosage et la sensibilité de

---

*Faculté de Médecine et de Pharmacie, Marseille, France.

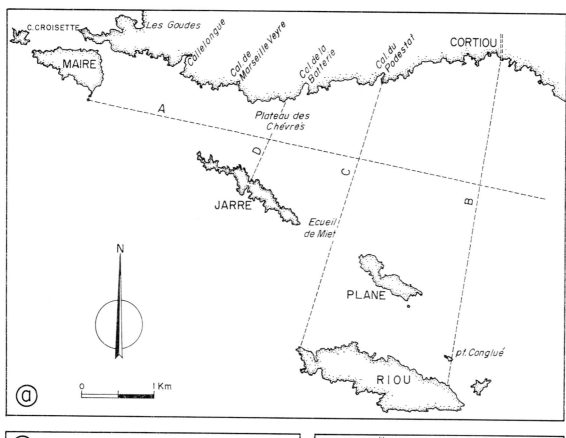

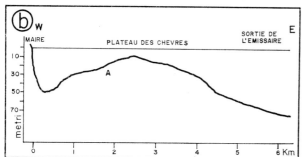

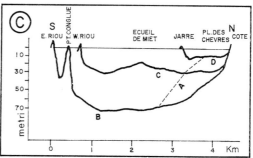

Fig. 1. Secteur de Cortiou: a) Situation géographique; b) Topographie du fond, coupe longitudinale; c) Topographie du fond, coupes transversales

la réaction. Ainsi, pour une prise d'essai de 100 ml de solution à 1 mg par litre, extraction par 25 ml de chloroforme et lecture en cuve de 1 cm, nous avons calculé un écart type de 0,86.

La limite de détection dans ces conditions est de 5 µg/l. Nous avons pu l'abaisser à moins de 2 µg en utilisant des cuves de 4 cm.

Apport des égouts en détergents anioniques

La composition d'un effluent urbain peut varier au cours d'une même journée dans des proportions très importantes. Pour illustrer ce fait nous avons procédé à des prélèvements toutes les heures dans la station de relèvement du Prohibé située dans le quartier central du Vieux Port de Marseille.

Le graphique supérieur de la fig 2 (A) est parfaitement représentatif des fluctuations rapides de la concentration en détergents, qui reflètent assez bien l'activité de ce quartier à vocation commerçante et touristique. Nous en retiendrons essentiellement le maximum à 17 mg/l vers 15 heures et l'augmentation marquée vers une

heure du matin que l'on peut attribuer à l'activité nocturne de ce quartier.

Le graphique inférieur de la fig 2 (B) concerne les eaux du grand émissaire dans sa partie terminale. Malheureusement l'accès de l'émissaire étant interdit après 20 heures, il nous a été impossible de mener notre enquête sur 24 heures. Cependant il nous donne une indication intéressante sur l'amplitude des variations de la concentration entre les premières heures de la matinée (1,2 mg/l à 7 heures) et l'après-midi (20,4 mg/l à 15 heures).

Le tracé est ici beaucoup plus régulier que dans le cas précédent, car il représente l'apport d'une population de plus de 800.000 habitants aux activités les plus diverses. Il en résulte un amortissement des fluctuations passagères locales.

Connaissant le débit dans cette partie de l'émissaire, il nous a été possible de superposer à la courbe des variations de concentration des détergents, celle des quantités correspondantes exprimées en kg/heure. Par simple cumulation, nous avons calculé que l'apport en 13 heures serait de 2.840 kilos. Une extrapolation mini-

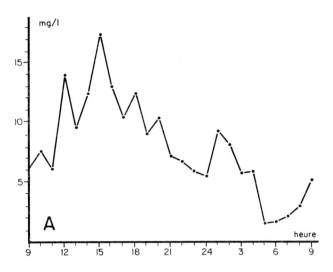

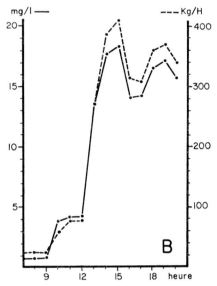

Fig. 2. Variations horaires des détergents anioniques dans les eaux d'égouts: A: station du Prohibé; B: station du grand émissaire.

male nous donnerait pour 24 heures un chiffre de l'ordre de 4 à 5 tonnes. Evidemment, il ne s'agit là que d'une approximation grossière, mais elle nous donne une idée de l'importance de la pollution subie par ce secteur du littoral.

Dispersion en mer

Le rejet en mer des eaux de l'émissaire se fait en surface et, du fait de leur plus faible densité, elles y persistent longtemps. C'est donc à ce niveau que l'étude de la dispersion sera la plus significative. Cependant nous l'avons complétée par quelques séries de prélèvements en profondeur de manière à préciser leur distribution au-dessous de la surface.

Variations dans le temps à proximité de l'émissaire

Nous avons vu dans un paragraphe précédent l'importance de l'amplitude des variations de la concentration des eaux d'égout au cours d'une période de 24 heures. Il nous a semblé utile d'apprécier les répercussions de ces fluctuations dans l'environnement du point de rejet.

Pour cela nous avons effectué pendant une même journée deux séries de prélèvements horaires, l'une en un

point situé à 50 m environ de la sortie de l'émissaire, l'autre à 500 m au sud. Les résultats sont représentés par les deux courbes de la fig 3. Le graphique A correspondant au point le plus proche, se caractérise d'abord par une diminution brutale de la concentration: le maximum de 15 heures atteint ici 2 mg/l alors qu'il était égal à 20 mg/l dans l'émissaire. Par contre l'amplitude des variations est comparable. Le graphique B présente un aspect sensiblement identique. Cependant nous constatons que, malgré l'éloignement, la dilution est beaucoup plus faible que celle subie par les eaux d'égout dès leur arrivée en mer. Par ailleurs les deux courbes ne sont pas exactement parallèles et se chevauchent même. Ce phénomène trouve son explication dans les variations du débit de l'émissaire qui provoquent aux heures de pointe une progression rapide de la nappe de pollution vers le large.

Par contre en période de faible débit, le contrecoup se manifeste soit par une stagnation d'eaux plus chargées en détergents, soit par un rapprochement d'eau moins polluée du large vers la côte. Nous pourrions comparer ce phénomène à une marée dont la périodicité et l'amplitude seraient conditionnées par le débit de l'égout.

Nous retiendrons enfin de cette étude, la lenteur de l'amortissement des variations de concentration depuis le point de rejet. Cela peut expliquer quelques anomalies dans les cartes de dispersion que nous décrirons par la suite. A leur propos, nous devons rappeler que les séries de prélèvements ayant presque toujours été effectuées au cours de la matinée, les résultats obtenus ne reflètent souvent qu'un état minimum de la pollution.

Influence des vents dominants sur la dispersion en surface

Dans le secteur de Cortiou, les eaux de surface sont soumises à l'influence de vents dominants d'est et de nord-ouest (mistral). Il peut s'ensuivre des transferts parfois très importants de la masse des eaux polluées dans des directions différentes avec dans certains cas des accumulations dans des zones plus exposées.

Pour illustrer cette évolution, nous avons établi une série de 5 cartes dont nous ne ferons ici qu'un bref com-

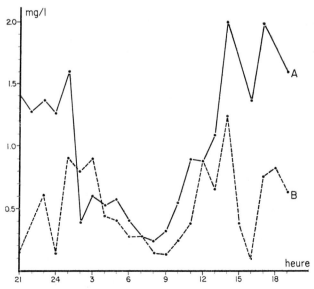

Fig. 3. Variations horaires des détergents anioniques dans l'eau de mer, station de Cortiou: A: 50 m de la sortie de l'émissaire; B: 500 m de la sortie de l'émissaire.

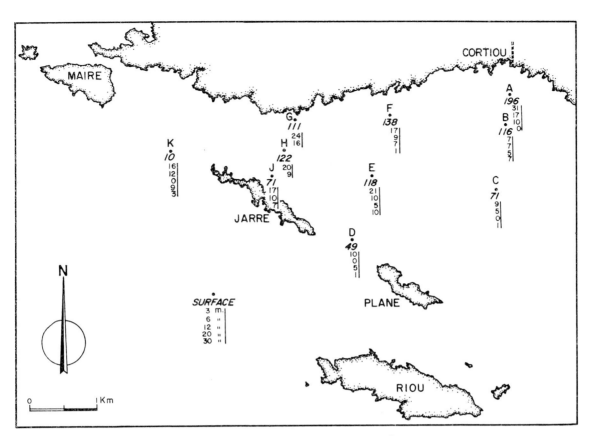

Fig. 4. Dispersion des détergents par temps calme.

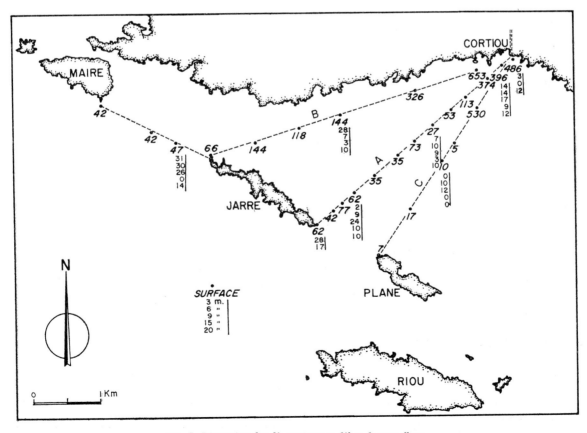

Fig. 5. Dispersion des détergents par début de vent d'est.

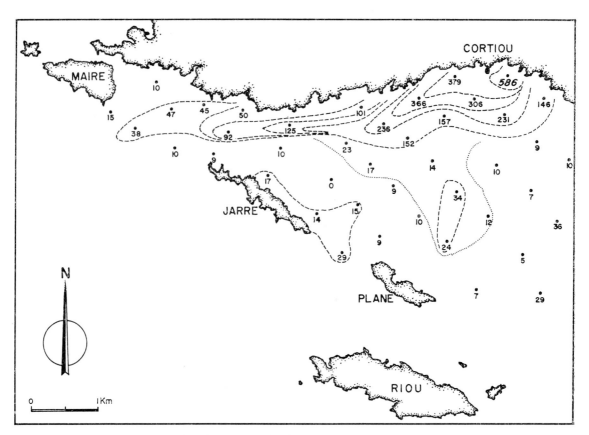

Fig. 6. Dispersion des détergents par vent d'est.

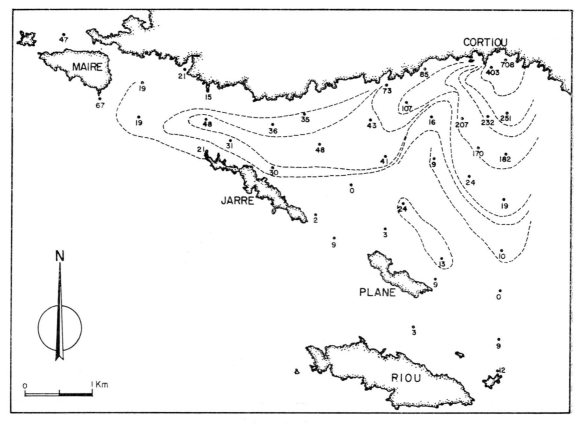

Fig. 7. Dispersion des détergents après vent d est.

mentaire. Toutes les valeurs inscrites correspondent à des concentrations exprimées en microgrammes de Manoxol OT par litre, le premier chiffre se rapportant toujours au prélèvement en surface.

La première carte de la fig 4 présente la situation par temps calme. Elle se caractérise par un envahissement de tout le secteur sans que l'on puisse mettre en évidence des zones bien distinctes mais seulement une diminution très progressive des concentrations de l'ordre de 200 à 100 μg/l depuis la côte jusqu'aux îles Plane et Jarre. Les trois cartes suivantes nous renseignent sur les variations en période de vent d'est (fig 5, 6 et 7).

La fig 5 correspond à l'évolution de la nappe de pollution lorsque le vent se lève. Nous avons distingué les trois radiales de prélèvements, car seule la A a été faite le matin tandis que les deux autres ont été réalisées en début d'après-midi. Ce décalage se manifeste, comme on pouvait le prévoir, par des concentrations relativement plus faibles sur A à proximité de l'émissaire. Malgré cela, nous pouvons constater un rabattement vers l'ouest comme le montrent la diminution des concentrations entre Cortiou et Plane, et le maintien de valeurs aussi élevées que précédemment entre Jarre et la côte.

La fig 6 illustre de façon très nette l'état créé par une période de plusieurs jours de vent d'est. La nappe de pollution est ramenée à une bande étroite plaquée le long de la côte et se détachant franchement des eaux à faible concentration qui la bordent au sud. Cependant il faut noter la persistance de faibles quantités de détergents dans des zones très limitées comme, par exemple, en bordure de l'île Jarre.

La fig 7 est un essai de représentation des modifications de la situation précédente lorsque tombe le vent d'est.

La propagation depuis l'émissaire se fait surtout vers le sud en auréoles régulières. Par contre à l'ouest, nous assistons à une uniformisation progressive se traduisant par une homogénéisation des concentrations entre la côte et Jarre.

Ainsi s'installe peu à peu le "faciès de temps calme" avec dispersion des détergents dans toutes les directions.

Enfin pour terminer cette étude nous devions envisager l'influence d'un autre vent dominant soufflant du nord-ouest: le mistral. Bien que le secteur de Cortiou soit relativement à l'abri de ce vent, nous y avons effectué une série de prélèvements après une période de mistral. Les résultats en sont rassemblés sur la fig 8. Si la déformation vers l'est depuis l'émissaire et le rabattement au nord de Plane sont manifestes, il ne semble pas qu'ils atteignent l'ampleur de celui observé par vent d'est dans une direction opposée.

Distribution en profondeur

Si la plus grande partie des détergents demeure en surface, il doit cependant se réaliser à la longue une dilution avec les eaux sous-jacentes plus denses.

Pour étudier cette distribution nous avons effectué à l'aide de bouteilles Nansen, des séries de prélèvements à diverses profondeurs, en général ne dépassant pas 20 à 30 m. La fig 4 nous montre que, par temps calme, la diminution au-dessous de la surface est toujours très rapide et régulière. Cependant la concentration vers 3 et 6 m a tendance à augmenter au fur et à mesure que l'on se rapproche du plateau des Chèvres. La topographie du fond plus que la simple diffusion doit en être responsable. Une place particulière doit être faite aux distributions aux points E et D. Nous relevons dans les

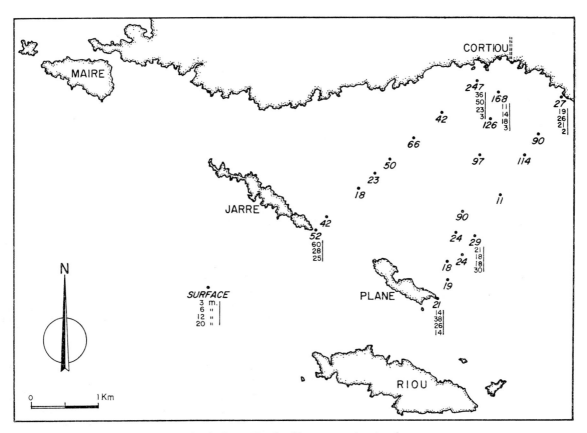

Fig. 8. Dispersion des détergents après mistral.

[179]

deux cas un minimum de concentration à 12 m pour E et à 9 m pour D. Il s'agit en fait de différences très faibles, de l'ordre de quelques microgrammes. Nous les attribuons à des stratifications créées par des mouvements de flux et de reflux liées aux variations de débit de l'émissaire.

Par contre, pendant les périodes de rabattement des zones de pollution sous l'influence des vents dominants, les anomalies de distribution en profondeur deviennent plus complexes. Ainsi par début de vent d'est (fig 5) nous observons un phénomène identique à celui décrit précédemment pour les séries de la radiale B, c'est-à-dire dans une zone soumise à l'influence directe des eaux d'égout.

Il en va différemment au niveau des radiales A et C situées dans la zone de rabattement. La poussée d'est en ouest d'eaux moins polluées peut entraîner un enfoncement par recouvrement d'eaux relativement plus riches en détergents. Il en résulte, à l'inverse du cas précédent, une stratification comportant un maximum de concentration vers 6 ou 12 m.

Une mention spéciale doit être faite pour les deux séries effectuées aux extrémités de l'île Jarre: elles se caractérisent par l'importance relative des concentrations en profondeur par rapport à celles de la surface, aboutissant à une distribution pratiquement homogène des détergents dans les 15 premiers mètres.

Une situation analogue à celle créée par le vent d'est va se manifester, dans une direction opposée, au cours du rabattement de la nappe de pollution par le mistral (fig 8). Là encore, nous observons des concentrations plus élevées au-dessous de la surface. Ce phénomène est surtout marqué aux pointes sud-est de Jarre et de Plane, et au nord de cette île. Cependant, ici l'agitation importante de l'eau de mer peut être rendue responsable de la pénétration à des profondeurs importantes de quantités parfois notables de détergents.

En conclusion, l'étude de la dispersion en mer des détergents anioniques rejetés par le grand émissaire de Marseille montre les difficultés de l'appréciation de cette pollution. Elle se trouve sous la dépendance directe des variations de composition et de débit des effluents et sous l'influence des vents dominants d'est et de nord-ouest. Ceux-ci amènent des remaniements fréquents de la distribution en surface et en profondeur pouvant aboutir à des accumulations passagères dans des zones limitées où l'effet nocif des polluants pourra davantage s'exercer. Enfin cette étude démontre, si besoin était, que toute détermination isolée et éloignée dans le temps ou dans l'espace est dépourvue de signification et que seules des séries nombreuses et groupées peuvent nous apporter des indications précises sur les modalités et l'importance d'une pollution.

Bibliographie

ABBOTT, D C The determination of traces of anionic surface-
1962   active materials in water. *Analyst, Lond.*, 87:286.
ARMANGAU, C Tentative d'utilisation des détergents anioniques
1967   comme traceurs de pollution fécale. *Revue Trav. Inst. (scient. tech.) Pêch. marit.*, 31:4.
LONGWELL, J, *et al.* Determination of anionic detergents in
1955   sewage, sewage effluents, and river water. *Analyst, Lond.*, 80:167.

Etude d'un Site Marin dans la Perspective de Rejets d'Effluents Radioactifs

*J. Ancellin, M. Avargues et P. Bovard**

Study of a marine site with respect to discharges of radioactive effluents

The discharge of radioactive effluents in the marine environment causes a potential risk of secondary irradiation of the populations due to the contamination of marine products and of their utilization for alimentary or other purposes. Definition of the limits of the discharges has to be taken from the studies and research concerning the transfers of radio nucleides from their starting point until reaching man, through the components of the marine environment: water, sediments, living organisms. Research should also cover the eventual action of the radioactive substances on the biological equilibrium of the environment. Such studies have been undertaken by the Atomic Energy Commissariat, or under its direction, on the occasion of the opening at La Hague of a factory for treatment of irradiated combustibles. The authors describe the fields explored and the principal results obtained. They stress the importance of the physico-chemical state of the radionucleides in their evolution and their transfer to the marine environment.

Research which has been developed for the study of radioactive pollution of the marine environment could be extended, using an identical methodology and techniques, to the study of marine contamination due to non-radioactive chemical pollutants.

Estudio de un lugar marino considerando el arrojo en el de efluvios radiactivos

El arrojo de efluvios radiactivos en ambiente marino lleva consigo un peligro potencial de irradiación secundaria de las poblaciones a consecuencia de la contaminación de los productos marinos y de su utilización para fines alimenticios y otros. La definición de los límites y de las modalidades de estas expulsiones debe basarse en estudios e investigaciones sobre los movimientos de los radionucleidos desde su punto de emisión hasta el hombre, a través de los componentes del medio marino: agua, sedimentos, organismos vivos. Las investigaciones deben cubrir igualmente la posible acción de las sustancias radiactivas sobre el equilibrio biológico del ambiente.

Tales trabajos han sido llevados a cabo por la Comisaría para la Energía Atómica, o bajo su dirección, con ocasión de la puesta en marcha en La Hague de una fábrica de tratamiento de combustibles irradiados. Los autores exponen cuáles han sido los dominios explorados e informan de los principales resultados obtenidos. Ponen en evidencia, especialmente, la importancia del estado físico-químico de los radionucleidos en su evolución y en sus movimientos en ambiente marino.

Las investigaciones ampliadas del estudio de la polución radiactiva del ambiente marino podrían extenderse, según una metodología idéntica y de técnicas parecidas, al estudio de las contaminaciones marinas debidas a las poluciones químicas no radiactivas.

UNE des solutions adoptées pour se débarrasser des déchets radioactifs résultant de l'utilisation de l'énergie nucléaire consiste à rejeter ceux-ci dans le milieu marin. En raison de la dispersion à laquelle on peut s'attendre dans une énorme masse d'eau mouvante et des processus d'épuration secondaire appelés à se produire sur des fonds hors d'atteinte de l'homme, l'océan paraît, à première vue, offrir des garanties

---

* Département de la Protection sanitaire, Commissariat à l'Energie atomique, 92 Fontenay-aux-Roses, France.

optimales pour ce genre de rejets. Cependant un examen plus attentif permet de se rendre compte que l'on doit aborder avec circonspection la question de l'élimination d'effluents radioactifs dans le milieu marin.

Lors de l'implantation près de Cherbourg, dans le Nord-Cotentin, d'une usine de traitement chimique de combustibles nucléaires irradiés, la France s'est trouvée placée pour la première fois devant le problème de l'élimination en mer d'effluents radioactifs et le Commissariat à l'Energie atomique a dès lors entrepris, sur le plan de la protection sanitaire, les recherches à moyen et long terme que motivait cette entreprise.

En matière de radioactivité, les doses d'irradiation maximales admissibles définies pour l'homme constituent la barrière qu'il importe de ne pas s'exposer à franchir par suite de l'utilisation du milieu marin et de ses ressources vivantes, sièges de redistribution et de concentration éventuelles des radioéléments. Dans ce domaine, à côté des paramètres biologiques interviennent des paramètres physiques et chimiques qui conditionnent également la répartition des radioéléments dans le milieu et dans les organismes. Dans un premier temps il s'agissait de dégager un certain nombre de données permettant de cerner de plus près les valeurs admissibles pour les rejets sur le site de La Hague. Ceci comportait tout d'abord des enquêtes socio-économiques sur les conditions d'utilisation du littoral et d'exploitation des ressources marines locales, des observations sur le régime océanographique du secteur et, enfin, des expériences de contamination en laboratoire destinées à évaluer les facteurs de concentration globaux d'un certain nombre de radioéléments par diverses espèces.

Cette première étape franchie on s'est attaché à élucider les mécanismes mêmes de la contamination du milieu—solide et liquide—et des espèces qui y vivent. Ces études fournissent de nouveaux éléments concernant le bilan de contamination et la "capacité d'acceptation nucléaire" d'un site marin donné.

A ce stade il convient de rechercher quelles sont les incidences sur la contamination de facteurs propres:

(a) aux radioéléments
(b) aux organismes
(c) au milieu physique.

Les aspects qualitatif, quantitatif et évolutif de la contamination sont à considérer en fonction de l'état physico-chimique des radioéléments, du comportement des espèces et en particulier de leur physiologie et des interactions qui s'établissent entre le polluant radioactif et les constituants du milieu.

Parallèlement les observations in situ entreprises dès le début des rejets industriels de l'usine de La Hague sont à poursuivre puisqu'elles ont le double intérêt de permettre des comparaisons avec les expériences réalisées en laboratoire et de fournir des renseignements sur les espèces plus particulièrement "indicatrices" de la contamination du milieu.

Océanographie physique: bionomie et radioactivité littorales

L'étude des paramètres fondamentaux que représentent la dispersion physique et le transport des effluents par les courants marins a été abordée par plusieurs voies. Tout d'abord des études hydrographiques ont été poursuivies sur des modèles réduits (Laboratoire national hydraulique de Chatou, Laboratoire de mécanique des fluides de Grenoble). Les essais menés sur une de ces maquettes fondée sur le principe d'une plaque tournante reconstituant la force de Coriolis, ont porté notamment sur une étude qualitative du phénomène de dispersion par injection d'une solution de $K Mn O_4$ et de solutions très diluées de fluorescéine suivies par photographie et cinématographie à la lumière de Wood.

En second lieu une série d'expérimentations in situ a consisté à larguer, pour l'étude des courants locaux, des bouées lestées au niveau du point d'aboutissement prévu de l'émissaire, pendant la période favorable définie par les études antérieures sur maquette, et des cartes-témoins (neuf lâchers au total correspondant à 1 730 cartes) destinées à apporter des informations complémentaires sur le transport sur de longues distances des masses d'eau ayant reçu l'effluent. Ces essais ont été complétés par une série d'expériences d'injection en mer de colorant (rhodamine B). Les mesures sont effectuées soit par lecture directe du prélèvement, soit au laboratoire, après extraction du colorant par l'alcool isoamylique. Les lectures sont faites au spectrophotomètre et la sensibilité de la méthode, au laboratoire, est de l'ordre de 5.10^{-11} g/cm^3, le seuil de détection se situant à une concentration de 10^{-11} g/cm^3. Les concentrations des solutions de rhodamine B utilisée dans les expérimentations ont été de l'ordre de 10^{-3} g/cm^3.

Cinq opérations de rejets de colorant ont eu lieu entre 1962 et 1964, la dernière s'étant étendue sur une période de 15 jours et ayant comporté 30 rejets successifs de 90 m^3 de solution à 3,3 g/l aux moments favorables des courants de marée.

Ces expériences ont confirmé que la région du Raz Blanchart, entre l'île d'Aurigny et le Cap de La Hague, était tout à fait favorable à la dispersion des effluents en raison des courants intenses qui règnent dans ce secteur. D'autre part nos connaissances se sont précisées en ce qui concerne le trajet de l'effluent pour différents coefficients de marée et les niveaux d'accumulation qui, d'après la dernière expérience effectuée, se stabilisent dans le Raz à environ 7.10^{-11} g/cm^3, après le 10e jour de rejet; enfin il a été possible d'établir des gradients de dispersion à partir du point de rejet et de vérifier la loi de diffusion du colorant en fonction de la distance: la décroissance de la concentration s'avère proportionnelle à l'inverse de la distance au point de rejet (Ausset et Farges, 1968; Avargues et Jammet, 1966).

En même temps, dès 1961 l'étude systématique de la radioactivité naturelle et artificielle des organismes marins, animaux et végétaux a été entreprise dans le but d'établir le niveau de leur radioactivité avant les rejets.

Une étude bionomique et écologique systématique de la faune et de la flore dans la zone de balancement des marées sur le littoral du nord-ouest du Contentin a permis de dresser un inventaire des principales espèces accompagné de données sur leur répartition. Cette étude a été également orientée vers l'estimation des biomasses des espèces et l'observation de leurs fluctuations naturelles en l'absence de toute contamination radioactive (Ancellin et al., 1969).

Enquêtes socio-économiques

Enfin, un ensemble d'enquêtes économiques et humaines (cf. Colloque de Radioécologie Marine, Cherbourg 22—25 avril 1964, Université de Caen, Commissariat à l'Energie atomique) a porté sur les points suivants:

conditions de travail en mer des pêcheurs et particulièrement manipulation des engins de pêche;
temps de séjour des habitants de la région et des estivants sur les plages;
nature, importance et distribution commerciale des produits alimentaires d'origine marine;
alimentation des populations dans le Nord-Cotentin, et plus spécialement en ce qui concerne les produits marins (nature, quantité, provenance).

Ces enquêtes économiques et humaines ont permis d'évaluer plus sûrement l'importance relative des risques divers encourus par les populations et de définir le groupe qualifié de critique qui doit être surveillé avec le plus d'attention.

Détermination expérimentale de facteurs de concentration par diverses espèces marines

La détermination expérimentale des facteurs globaux de concentration de radioéléments a porté sur les espèces utilisées par l'homme à des fins diverses et notamment pour son alimentation. Elle a également eu pour but de reconnaître les espèces qui, par leur pouvoir élevé de rétention des radionucléides, pouvaient constituer des indicateurs biologiques de contamination en vue de contrôles systématiques ultérieurs. Ces résultats ont pu être comparés par la suite à ceux des observations *in situ*. Ces expériences sont actuellement approfondies par l'analyse des mécanismes de contamination conditionnés par des facteurs ou des ensembles de facteurs tels que l'état physico-chimique des radioéléments, les constituants du milieu, la "réponse" physiologique des espèces soumises à l'action du contaminant (Ancellin *et al.*, 1965; Ancellin et Vilquin, 1966, 1968; Avargues *et al.*, 1968; Vilquin, 1969).

Conditions expérimentales, espèces étudiées, prélèvements et mesures

Les expériences sont réalisées dans une salle climatisée à 14°–16°C: c'est la température moyenne de l'eau de mer dans la région en dehors de la période hivernale.

On utilise des aquariums en résine armée de fibres de verre et dont le volume utile est de 40 l à 200 l. Une face verticale de plexiglass permet l'observation. Un couvercle, également en résine armée, réduit l'évaporation. Ces aquariums fonctionnent en circuit fermé: ils sont pourvus d'un "microfiltre" constitué par un conduit en matière plastique, perforé, disposé horizontalement sur le fond de l'aquarium. L'aspiration est obtenue par injection d'air à la base d'une colonne verticale placée à une extrémité du conduit horizontal: l'oxygénation, la circulation et la filtration de l'eau se trouvent ainsi assurées. (On peut améliorer la filtration en recouvrant de laine de verre et de sédiments le tube horizontal perforé.)

Toutefois, dans le cas de contamination par des radionucléides sous forme particulaire nous avons été amenés à supprimer le dispositif de filtration qui retient les particules; seules subsistent l'aération par injection d'air à travers des diffuseurs poreux et l'agitation au moyen de palettes tournantes. Pour une durée d'expérimentation ne dépassant pas 3 à 5 semaines et pour des peuplements peu importants des aquariums cette technique s'est révélée satisfaisante. Les expériences ont été réalisées avec les produits de fission suivants: césium 137, ruthénium 106, cérium 144, zirconium 95, en équilibre avec leurs descendants radioactifs respectifs. Il s'agit soit de formes chimiques simples (sulfate, trichlorure), soit de formes complexes, soit de mélanges de radioéléments préparés en laboratoire ou bien encore d'extraits d'effluents industriels. Enfin, à la suite des travaux de physico-chimie effectués au laboratoire sur le ruthénium, nous avons pu réaliser des expériences de contamination par des formes solubles et insolubles dans l'eau de mer de ce radionucléide.

En principe l'aquarium est contaminé par apport de ces radioéléments en solution ou en suspension, de manière à obtenir une radioactivité initiale de l'eau de 1 à 10 microcuries par litre. Après une chute de l'activité de l'eau observée pendant les premiers jours, un équilibre relatif s'établit. Le plus souvent les espèces n'ont été introduites dans les aquariums qu'une fois cet équilibre atteint. Toutefois, étant donné l'évolution initiale très rapide et difficile à contrôler des contaminants sous forme particulaire (Ru et Zr par ex.), l'attente de cette stabilisation ne se justifie plus: dans ce cas l'introduction des espèces dans les aquariums précède la contamination de l'eau. Les aliments ajoutés en cours d'expérimentation —à raison d'une ou deux fois par semaine—ne sont pas contaminés (sauf bien entendu s'il s'agit d'une étude de la contamination par voie alimentaire).

Nos recherches ont surtout porté sur des espèces marines du littoral du Cotentin; on peut citer entre autres: Algues: *Fucus serratus, Laminaria flexicaulis, Chondrus crispus, Corallina officinalis;* Cnidaires: *Anemonia sulcata, Actinia equina;* Mollusques: *Mytilus edulis, Chlamys opercularis, Cardium edule, Gryphaea angulata, Tapes* sp., *Nucella lapillus;* Crustacés: *Balanus* sp., *Leander serratus, Homarus vulgaris, Portunus puber;* Ascidie: *Dendrodoa grossularia;* Poissons: *Blennius pholis, Mugil* sp., *Pleuronectes platessa, Solea solea.*

On a tenté dans un certain nombre d'expériences de respecter les équilibres naturels en reconstituant approximativement dans les aquariums les ensembles biologiques et les groupements d'espèces observés sur le littoral lui-même. A des intervalles de temps réguliers, des individus de chaque espèce sont prélevés et sacrifiés pour la mesure de la radioactivité. Cependant certaines mesures sont faites sur les animaux vivants afin de suivre la dynamique de leur contamination et de leur décontamination.

Toutes les mesures sont faites en comptage gamma (sélecteur monocanal ou multicanaux) avec un détecteur à scintillation. Les résultats sont donnés sous forme de facteurs de concentration, exprimant le rapport:

$$\frac{\text{Activité spécifique de l'échantillon biologique (imp./min./g, poids frais)}}{\text{Activité spécifique de l'eau de mer (imp./min./g)}}$$

La manipulation des produits de la mer provenant d'une zone contaminée, leur commercialisation, leur utilisation industrielle, leur consommation font, dans bien des cas, intervenir des risques—s'ils existent—dès que ces produits sont sortis de l'eau. C'est pourquoi nous nous sommes surtout attachés, dans un premier temps, à évaluer la résultante globale de la contamination chez différentes espèces.

Résultats généraux

Sans donner les résultats obtenus par espèce nous pouvons indiquer les valeurs moyennes suivantes (facteurs de concentration ou d'accumulation après équilibre entre l'activité du milieu et celle des espèces) observées pour trois radioéléments:

Césium 137	Algues:	40
	Invertébrés:	40
	Poissons:	40
Cérium 144	Algues:	2 000 à 10 000
Ruthénium 106	Algues:	100 à 200
(trichlorure, forme	Invertébrés:	20 à 50
présumée soluble)	Poissons:	1

La comparaison de ces résultats fait apparaître les points suivants:

Pour le césium 137 l'accumulation se produit surtout dans la chair de certaines espèces: crevettes, blennies; elle est par contre très faible au niveau de la coquille des mollusques.

L'algue *Corallina officinalis* et l'ascidie *Dendrodoa grossularia* figurent parmi les espèces qui concentrent le plus le ruthénium 106.

Pour le cérium 144 l'algue *Corallina officinalis* est remarquable par son facteur de concentration élevé (de l'ordre de 20 000); les spongiaires et les ascidies sont parmi les invertébrés qui concentrent le plus ce radio-élément.

Certains des organismes mentionnés (corallines, ascidies) peuvent, en raison de leur pouvoir élevé de concentration des radioéléments, jouer le rôle d'espèces "indicatrices".

En ce qui concerne les espèces commerciales et pour deux radioéléments considérés (ruthénium et cérium), il apparaît que l'algue *Chondrus crispus* (appelée vulgairement lichen carragaheen, elle a divers usages industriels et alimentaires) présente un pouvoir d'accumulation plus élevé que les mollusques et les crustacés tels que le vanneau, la coque, la moule et la crevette rose.

L'accumulation des radioéléments chez les espèces étudiées paraît varier, en général, en sens inverse du degré d'évolution des organismes.

Comparaisons avec d'autres données

Il n'est pas sans intérêt de comparer ces résultats à ceux déjà publiés sur des expériences comparables aux nôtres, ou sur les mesures directes dans un milieu naturel contaminé (Bryan, 1966; Keckes *et al.*, 1966; Mauchline, 1963; Polikarpov, 1966; Preston et Jefferies, 1969).

Nos valeurs expérimentales sont souvent du même ordre que celles publiées pour des expériences réalisées sur des espèces identiques ou voisines. Par contre les comparaisons de certaines données expérimentales avec celles provenant de mesures faites dans un milieu naturel contaminé font apparaître—notamment en ce qui concerne le ruthénium—des différences assez marquées.

Il faut considérer d'une part que les contaminations (faisant suite aux premières contaminations "globales") réalisées expérimentalement à partir de formes particulaires du ruthénium déterminent des facteurs de concentration 5 à 10 fois plus élevés que ceux obtenus à partir de formes solubles (contaminants utilisés: ruthénium extrait des effluents; séparation des formes sur résine cationique). D'autre part si l'on rapproche ces résultats expérimentaux des observations *in situ* on constate que ces dernières fournissent des valeurs comparables à celles obtenues à la suite de contaminations expérimentales par formes particulaires (Ancellin et Bovard, 1969; Ancellin *et al.*, 1967; Vilquin, 1969).

Ces constatations laissent présumer l'importance de la forme physico-chimique du radioélément dans certains processus de contamination. L'exemple de l'algue coralline donné dans le tableau suivant est, à ce sujet, particulièrement démonstratif:

Corallines—Facteurs de concentration (Ruthénium 106)

	Résultats expérimentaux		Observations in situ
	Formes solubles	Formes insolubles	
Facteurs de concentration	100–200	600–1 000	1 000–1 300

Les observations *in situ*

Il a paru intéressant de procéder à des mesures de radioactivité d'un certain nombre d'espèces vivant dans un secteur déterminé du littoral soumis à l'influence des rejets. Ce travail devait permettre des comparaisons avec certains résultats expérimentaux concernant les facteurs de concentration des radioéléments étudiés. On se trouvait d'autre part conduit à réaliser, dans le secteur même des rejets futurs, une expérience permettant de vérifier *in situ* le comportement et le transfert des radioéléments rejetés en mer.

Enfin il était important, sur le plan de la surveillance, d'obtenir à cette occasion de nouvelles données sur les espèces indicatrices susceptibles, en raison de leur pouvoir d'accumulation élevé, de donner un reflet plus sensible de la contamination du site.

Etant donné les moyens à mettre en œuvre pour obtenir un aperçu de la contamination dans un milieu où la radioactivité due aux rejets est toujours demeurée très faible nous avons limité nos observations à la Baie d'Ecalgrain, lieu de référence situé à 3–4 kilomètres au nord-est du point de rejets et où l'influence de ceux-ci, sans être faussée par la proximité de l'émissaire, reste néanmoins significative. Les observations ont porté sur une période de deux années (1966–68); des résultats antérieurs au fonctionnement de l'usine peuvent servir de termes de comparaison (Bovard, 1968; Guegueniat, 1967; Guegueniat et Lucas, 1969; Guegueniat *et al.*, 1969).

Méthode de détermination de la radioactivité dans l'eau de mer et les espèces vivantes

Radioactivité de l'eau de mer et des espèces.

Les mesures concernent les trois produits de fission suivants: cérium 144, ruthénium 106, zirconium 95, qui constituent, avec le césium 137, la part prépondérante des radionucléides rejetés en mer.

La radioactivité de l'eau de mer a été déterminée par une méthode d'adsorption au moyen de bioxyde de manganèse colloïdal. En effectuant des prélèvements quotidiens de 40 à 80 litres sur le littoral, la concentration mensuelle de l'eau de mer en Ru, Ce, Zr a été mesurée (on a tenu compte dans les calculs de la décroissance radioactive, du rendement d'adsorption et des retombées radioactives).

A titre d'exemple nous avons obtenu pour le ruthénium:

Avant rejet: $0,20 \pm 0,13$ p.Ci/litre
en décembre 1966: $0,50 \pm 0,32$,,   ,,
en février 1967: $1,25 \pm 0,55$,,   ,,
en juin 1968: $0,65 \pm 0,22$,,   ,,

En ce qui concerne les espèces, les échantillons prélevés ont été desséchés à l'étuve puis broyés et comptés directement à l'aide d'un analyseur gamma S A 40 B. Voici pour l'algue *Corallina officinalis* quelques valeurs exprimées en p.Ci/kg frais:

mars   1967: $1,080 \pm 360$
juin   1967: 500 ± 200
mars   1968: $1,020 \pm 120$
octobre 1968: 750 ± 80

Facteurs de concentration.

La contamination variable de l'eau de mer entraîne des fluctuations dans la radioactivité des espèces prélevées; tout dépend de l'évolution de l'équilibre:

contamination de l'espèce $\rightleftarrows$ contamination du milieu,

lorsque la contamination du milieu varie. Nous avons essayé de tenir compte de cette dynamique en faisant intervenir plusieurs hypothèses relatives à la vitesse des échanges entre les espèces et le milieu et en tenant compte de l'activité maximale et minimale de l'eau de mer enregistrée pendant une période donnée.

Résultats

On trouvera ci-dessous quelques résultats exprimés sous formes de facteurs de concentration (ordres de grandeur) concernant les principales espèces étudiées et deux des radioéléments:

Ruthénium 106

Corallina officinalis	1,000–1,300
Porphyra umbilicalis	200–1,500
Chondrus crispus	800– 900
Laminaria sp.	300– 400
Fucus sp.	400– 500
Halichondria panicea	900–1,200
Pachymatisma johnstoni	5,000–6,000
Dendrodoa grossularia	1,500–1,600

Zirconium 95

Corallina officinalis	4,300
Chondrus crispus	1,000
Halichondria panicea	2,000
Dendrodoa grossularia	2,800

Ces résultats peuvent être rapprochés de ceux obtenus à partir d'observations *in situ* réalisées en Grande-Bretagne dans le secteur de Windscale en 1965–66 (Preston et Jefferies, 1969): pour le ruthénium 106, les ordres de grandeur suivants ont été obtenus pour les facteurs de concentration chez ces groupes d'organismes:

Algues:	10^2–2.10^3
Mollusques:	10^3–2.10^3
Crustacés:	10^3
Poissons:	10

Les résultats de nos mesures en Baie d'Ecalgrain donnent pour les algues: 10^2–10^3, pour les spongiaires: 10^3–10^4, pour les ascidies: 10^3.

Comportement physico-chimique du ruthénium dans l'eau de mer

Comme nous l'avons déjà indiqué les comparaisons établies entre les facteurs de concentration résultant de contaminations *in situ* et ceux obtenus expérimentalement font apparaître pour certains radioéléments des écarts, parfois importants. Ceci est particulièrement le cas du ruthénium et il s'avérait nécessaire pour mieux comprendre les mécanismes de la contamination par ce radioélément d'approfondir nos connaissances sur son comportement physico-chimique dans le milieu eau de mer. Des formes solubles (anioniques, neutres, cationiques, organiques) et insolubles (colloïdales, particulaires) du ruthénium apparaissent au contact de l'eau de mer; nous avons essayé de les isoler sans les détruire afin de les utiliser séparément dans des expériences de contamination (Guegueniat, 1970; Guegueniat et Lucas, 1969; Guegueniat *et al.*, 1969).

Dès à présent on doit considérer que les données acquises vont permettre d'étendre les techniques mises au point à l'étude de la physico-chimie d'autres radioéléments tels que: cobalt, fer, césium, zirconium, cérium.

Propriétés générales des résines cationiques en milieu électrolytique

Il convient de rappeler brièvement les propriétés générales des résines cationiques utilisées pour séparer les diverses formes de ruthénium.

Ainsi que le souligne Schubert (1948, 1950) cette utilisation d'échangeurs cationiques n'est applicable qu'à des traceurs en milieu fortement électrolytique, ce qui est le cas de l'eau de mer.

Nous avons employé la résine Dowex 50×8 et nous pouvons considérer que les propriétés mises en jeu sont les suivantes:
—Les anions et les substances neutres ne sont pas fixés non plus que les radioéléments complexés par des composés organiques.
—La fixation d'un cation décroît lorsque la concentration d'autres cations augmente dans la solution.
—L'adsorption d'un radioélément par un échangeur de cations met quelquefois en jeu des mécanismes plus complexes que le simple échange de cations. Dans un milieu électrolytique les colloïdes sont fixés sur la résine par adsorption, l'échangeur offrant des sites favorables à une adsorption superficielle; ainsi les échangeurs de cations pourront fournir de nombreuses informations en ce qui concerne les colloïdes.
—Contrairement à ce qui se passe dans le cas d'échange cationique l'adsorption d'un radiocolloïde par la résine s'accroît lorsque la concentration des électrolytes augmente dans le milieu, ces derniers favorisant la coagulation des colloïdes.
—Au point de vue élution lorsqu'il y a fixation par échange cationique, l'élément fixé pourra être déplacé lorsque la résine sera mise en présence de nouvelles solutions électrolytiques débarrassées du cation considéré.

—Les acides permettent dans certains cas de resolubiliser les colloïdes.

—Certaines substances peuvent jouer le rôle d'agents complexants à l'égard de formes de radioéléments insolubles dans l'eau de mer et provoquent une notable diminution de l'adsorption. C'est ainsi que nous avons pu mettre en évidence la complexation du ruthénium, du cérium et du zirconium par des composés organiques.

Application au cas du ruthénium dans l'eau de mer

Les propriétés des résines que nous venons de rappeler peuvent être appliquées à l'étude de divers radioéléments. Nous examinerons le cas particulier du ruthénium.

Formes solubles, anioniques, neutres et cationiques

Nous avons dans un premier temps isolé l'ensemble des formes solubles. La résine, mise en contact avec la solution à analyser, fixe les espèces colloïdales et une partie des formes cationiques: par conséquent, restent en solution l'autre partie cationique, les formes anioniques et neutres. Les espèces cationiques fixées seront déplacées par mises en contact successives de la résine avec des eaux de mer non contaminées. Il sera ainsi relativement facile de récupérer la totalité des espèces solubles sans les détruire si la résine a été au préalable régénérée par contact avec l'eau de mer: en effet, lors de la fixation et lors de l'élution, les réactions d'échanges ne mettront en jeu que les ions de l'eau de mer; il faut proscrire une régénération par des ions H^+ qui, d'une part, resolubiliseraient les formes colloïdales fixées sur la résine, et qui, d'autre part, risqueraient fort de changer la nature des espèces solubles très sensibles à des variations de pH.

Formes insolubles (particulaires, colloïdales)

On peut s'assurer de l'existence de ces formes de ruthénium par une étude comparée de la fixation par la résine du césium et du ruthénium introduits dans des solutions dont on fait varier la salinité (mélange eau de mer—eau distillée dans des proportions différentes).

Dans une série de petits béchers pourvus d'un système de brassage, nous avons introduit des solutions eau de mer—eau distillée (500 cm³) de salinité variable d'un bécher à un autre, la résine (5 g), puis 250 µl de la solution-mère de ruthénium* et du césium 137. Les pourcentages de césium et de ruthénium fixés par la résine après quelques heures de brassage ont été reportés dans le tableau ci-dessous:

% d'eau de mer dans le mélange eau de mer—eau distillée	% sorption Ru	% sorption Cs
1	38	70
10	34	35
20	64	23
40	61	15
60	67	14
80	70	10
90	60	7
100	65	8

* La solution-mère de ruthénium est préparée en introduisant 0,1 mCi de la source choisie (en milieu acide) dans 100 cm³ d'eau de mer préalablement filtrée (millipore, 0,4 µ); le pH de la solution est ensuite ajusté à celui de l'eau de mer (pH 8). Dans le cas présent la solution est utilisée un mois après sa préparation.

La fixation du césium présent dans l'eau de mer sous forme cationique est importante dans le cas de solution de faible salinité puis diminue au fur et à mesure que la salinité augmente, les cations de l'eau de mer déplaçant par échanges ioniques les espèces fixées (compétition).

Par contre la fixation du ruthénium est plus importante dans le cas de solution à forte salinité. L'adsorption sur la résine de formes colloïdales du ruthénium est en effet favorisée par la présence des cations du milieu eau de mer (rôle coagulant de l'eau de mer). Il est à noter que si l'adsorption n'est pas de 100 pour cent c'est que le ruthénium se présente également dans l'eau de mer sous forme anionique et neutre non retenue par la résine.

Récupération des formes colloïdales—Utilisation pour des contaminations expérimentales

La méthode employée pour la désorption et la différentiation des formes particulaires (colloïdales), plus ou moins énergiquement fixées sur la résine, consiste à soumettre celle-ci—après disparition des formes solubles—au contact de solutions successives d'acide de normalités différentes. Après neutralisation de l'acide ces formes recueillies en milieu acide sont ensuite utilisées pour les contaminations expérimentales; nous avons pu d'autre part nous assurer que le traitement à l'acide n'altérait pas leur nature particulaire.

Nous avons vu que les facteurs de concentration obtenus expérimentalement, à partir des formes solubles, sont nettement moins élevés que ceux obtenus avec les formes particulaires. Ces derniers sont par contre voisins de ceux observés *in situ*. On peut penser dans ces conditions que les formes particulaires du ruthénium seraient les principales responsables de la contamination globale des espèces marines étudiées—sans préjuger de la part revenant au métabolisme et de celle due à des processus d'adsorption superficielle.

Remarques générales et conclusion

On doit attendre des études de radioécologie marine une meilleure connaissance des processus de contamination du milieu marin. Ces recherches font appel, outre celles ayant trait à la radioactivité, à des disciplines variées: biologie, chimie, océanologie, etc.; il a été d'autre part nécessaire de procéder à la mise au point de techniques appropriées dans ces différents domaines, notamment en matière d'aquariologie, de séparations chimiques, de mesures de radioactivité et d'autoradiographie. Il ne paraît pas inutile de souligner pour terminer, que ces moyens en personnel et en matériel se trouveraient sans difficulté adaptés pour recherches portant sur les pollutions chimiques et les nuisances non radioactives observées en mer et sur le littoral: le Laboratoire de La Hague représente à cet égard une structure d'accueil et constitue un outil potentiel pouvant répondre à des besoins plus étendus concernant la protection du milieu océanique.

Bibliographie

ANCELLIN, J et BOVARD, P Observations concernant les con-
1969 taminations expérimentales et les contaminations *in situ* d'espèces marines par le ruthénium 106. Document présenté au IVᵉ Colloque International d'Océanographie Médicale, Naples, 2–5 octobre 1969.

ANCELLIN, J et VILQUIN, A Contaminations expérimentales
1966   d'espèces marines par le cérium 144, le ruthénium 106 et le
       zirconium 95. *In* Disposal of radioactive wastes into seas,
       oceans and surface waters. Vienne, A.I.E.A., pp. 583–604.

ANCELLIN, J, et VILQUIN, A Nouvelles études de contamina-
1968   tions expérimentales d'espèces marines par le césium 137,
       le ruthénium 106 et le cérium 144. *Radioprotection*,
       3(3):185–213.

ANCELLIN, J, BOVARD, P et VILQUIN, A Nouvelles études de
1967   contaminations expérimentales d'espèces marines par le
       ruthénium 106. *In* Actes du Congrès International sur la
       radioprotection du milieu, Société Française de Radio-
       protection, Toulouse, 14–16 mars 1967, pp. 213–34.

ANCELLIN, J, MICHON, G et VILQUIN, A Contaminations expéri-
1965   mentales de crevettes roses par le césium 137. *Rapp. CEA*,
       (R2818):14 p.

ANCELLIN, J, *et al.* Observations sur la distribution de la faune
1969   et de la flore dans la zone de balancement des marées le
       long du littoral du nord-ouest du Cotentin. *Mém. Soc.
       natn. Sci. nat. math. Cherbourg*, 52:139–99.

AUSSET, R et FARGES, L Utilisation de la rhodamine B dans
1968   l'étude des caractéristiques hydrologiques d'un site marin
       de rejets d'effluents radioactifs. *Revue int. Océanogr.
       méd.*, 9:167–89.

AVARGUES, M et JAMMET, H P Etude du site marin de La Hague
1966   en relation avec le rejet d'effluents radioactifs. *In* Disposal
       of radioactive wastes into seas, oceans and surface waters.
       Vienne, A.I.E.A., pp. 787–95.

AVARGUES, M, ANCELLIN, J et VILQUIN, A Recherches
1968   expérimentales sur l'accumulation des radionucléides par
       les organismes marins. *Revue int. Océanogr. méd.*, 9:87–99.

BOVARD, P Radioécologie marine—recherches *in situ* (études
1968   effectuées en France). *In* Compte rendu du colloque sur la
       radioécologie marine organisé par l'Agence Européenne
       pour l'Energie Nucléaire (OCDE), Cherbourg, 3–6
       décembre 1968.

BRYAN, G W Accumulation of radionuclides by aquatic organisms
1966   of economic importance in the United Kingdom. *In*
       Disposal of radioactive wastes into seas, oceans and
       surface waters, Vienna, I.A.E.A., pp. 623–37.

Commissariat à l'Energie Atomique, Colloque de radioécologie
       marine, Cherbourg, 22–25 avril 1964. Université de Caen.

GUEGUENIAT, P Détermination de la radioactivité de l'eau de
1967   mer en ruthénium, cérium, zirconium par entraînement et
       adsorption au moyen du bioxyde de manganèse. *Rapp.
       CEA*, (3284):30 p.

GUEGUENIAT, P Etude sur la physico-chimie du ruthénium dans
1970   l'eau de mer.

GUEGUENIAT, P et LUCAS, Y Observations sur la contamination
1969   *in situ* de quelques espèces marines. *Note CEA*, (N-1185):
       22 p.

GUEGUENIAT, P, BOVARD, P et ANCELLIN, J Influence de la forme
1969   physico-chimique du ruthénium sur la contamination des
       organismes marins. *C. r. hebd. séanc. Acad. Sci., Paris*,
       (D), 268:967–9.

KECKES, S, PUCAR, Z et MARAZOVIC, L The influence of the
1966   physico-chemical form of ruthenium 106 on its uptake by
       mussels from sea water. Radioecological concentration
       processes. *In* Annual report on research contract,
       Institute "Ruder Boskovic", July 1965–June 1966.

MAUCHLINE, J The biological and geographical distribution in
1963   the Irish Sea of radioactive effluent from Windscale Works,
       1959 to 1960. Harwell, U.K.A.E.A. *A.H.S.B. Rep.*,
       (RP)R27.

POLIKARPOV, G G Radioecology of aquatic organisms. Amster-
1966   dam, North Holland Publishing Company, 320 p.

PRESTON, A et JEFFERIES, D F Aquatic aspects in chronic and acute
1969   contamination situations. *In* Seminar on agricultural and
       public health aspects of environmental contamination by
       radioactive materials, Vienna, I.A.E.A., pp. 183–211.

SCHUBERT, J The use of ion-exchangers for the determination of
1948   physio-chemical properties of substances particularly
       radiotracers in solution. *Int. J. Phys. Colloid. Chem.*,
       52:340–50.

SCHUBERT, J et RICHTER, J W The use of ion exchangers for the
1950   determination of physico-chemical properties of substances,
       particularly radiotracers. 3. The radiocolloids of zirconium
       and niobium. *J. Colloid Sci.*, 5:376–85.

VILQUIN, A Contaminations expérimentales d'espèces marines
1969   par des formes de ruthénium 106 solubles et insolubles
       dans l'eau de mer. *Radioprotection*, 4(3):185–92.

Chemical and Physical Investigations on Marine Pollution by Wastes of a Titanium Dioxide Factory

G. Weichart*

Enquêtes chimiques et physiques sur la pollution marine provoquée par les effluents d'une usine de bioxyde de titane

Depuis le mois de mai 1969, les effluents de TiO₂ d'une usine située à proximité de Bremerhaven sont rejetés dans la baie d'Heligoland. La zone de déversement est un rectangle de 31 km² dont le centre se trouve à 14 milles nautiques au nord-ouest d'Heligoland. Dans cette zone, les fonds sont de 25 à 28 mètres.

Les eaux usées de l'usine de TiO₂ contiennent 10 pour cent d'acide sulfurique et environ 14 pour cent de sulfate ferreux, les autres éléments constituants ayant moins d'importance. Chaque jour, 1 800 tonnes de déchets sont ainsi rejetées dans la zone de déversement par un bateau-citerne affecté à cette tâche. Afin d'obtenir une dilution rapide, les déchets sont déversés dans le sillage du navire qui marche à une vitesse de 8 noeuds environ. Le déchargement du bateau prend de 1 à 2 heures.

Le navire de recherche allemand "Gauss" a étudié en mai et en octobre/novembre 1969 les processus physiques et les modifications chimiques les plus importantes de l'eau de mer et du fond. Les principales constatations ont été les suivantes:

1. Le taux de dilution initial des effluents par l'hélice du bateau-citerne était environ de 1:1 000.
2. La première phase du processus de dilution secondaire de l'eau du sillage par l'eau de mer non polluée était très rapide. Une heure ou deux après le déversement, le facteur de dilution atteignait environ 10. Par la suite, il était beaucoup plus faible.
3. Le pH atteignait sa valeur minimale immédiatement après le déversement des déchets. Quatre minutes après cette opération, le pH au milieu de l'eau agitée par l'hélice était d'environ 6,0. Le déficit en O₂ au milieu de cette eau, n'a dépassé à aucun moment 30 pour cent. Quelques heures après le déversement, la dilution atteignait un tel degré que le pH

Investigaciones químicas y físicas sobre la contaminación de las aguas del mar causadas por los desechos procedentes de una fabrica de bióxido de titanio

Desde mayo de 1969 se depositan en el mar del Norte los desechos de una fábrica de TiO₂, situada cerca de Bremerhaven. La zona de evacuación es un rectángulo de 12 millas cuadradas con el centro situado a 14 millas náuticas al noroeste de Heligoland. En dicha zona la profundidad del agua es de 25 a 28 metros.

Las aguas residuales de la fábrica de TiO₂ contienen el 10 por ciento de ácido sulfúrico y alrededor de un 14 por ciento de sulfato ferroso. Los demás componentes son de poca importancia. Un barco cisterna especial descarga a diario 1 800 toneladas de estos desechos en la zona de evacuación, y para facilitar su disolución rápida, se lanzan en la corriente creada por la hélice del barco que navega a unos ocho nudos. La operación dura de una a dos horas.

El buque de investigación alemán "Gauss" estudió en mayo y octubre/noviembre de 1969 los procesos físicos y las alteraciones químicas más importantes en el agua y en el fondo del mar. Los principales resultados obtenidos fueron:

1. La disolución inicial de agua de desechos por la hélice del barco era de cerca 1:1 000.
2. La primera fase de la disolución secundaria del agua de la hélice por agua de mar no contaminada procedía muy rápidamente, y de una a dos horas después de la descarga el coeficiente de disolución era de alrededor de 10. Posteriormente tal coeficiente era mucho más pequeño.
3. El pH alcanzaba su valor mínimo inmediatamente después de evacuar los desechos. Cuatro minutos después, el pH en el centro del agua de la hélice era de cerca de 6,0. El déficit de O₂ en el centro de la corriente de agua de la hélice no excedió en ningún momento de 30 por ciento. Algunas horas después de la descarga, la disolución había procedido hasta tal punto que

* Deutsches Hydrographisches Institut, Hamburg, Federal Republic of Germany

et la concentration en O_2 avaient presque retrouvé leur valeur normale.

4. Pendant les premières heures suivant le déversement, les effluents ne pénétraient pas dans le fond de la mer. Après que le pH et la concentration en O_2 avaient retrouvé approximativement leur valeur normale quelques heures après le déversement, on n'observait à aucun moment de diminution notable de ces valeurs à proximité du fond de la mer.

5. Les analyses d'échantillons prélevés sur le fond en novembre 1969, c'est-à-dire 5 mois $\frac{1}{2}$ après le début du déversement, ne sont pas encore terminées. Il semble toutefois que le taux de concentration du fer dans la couche supérieure de sédiment ait marqué un certain accroissement.

6. Cinq mois $\frac{1}{2}$ après le début du déversement des déchets, la quantité d'hydroxyde ferrique en suspension était relativement faible. La zone où l'on a constaté un accroissement notable de la concentration en fer couvrait une superficie d'environ 500 km². La concentration maximale trouvée dans cette région atteignait 500 μg de Fe par litre.

el pH y la concentración de O_2 habían recuperado valores casi normales.

4. Durante las horas siguientes a la descarga, los desechos no llegaban al fondo del mar. Como el pH y la concentración de O_2 habían casi alcanzado sus valores normales a las pocas horas después de la descarga, en ningún momento había cerca del fondo disminuciones importantes de los valores de aquéllos.

5. Los análisis de las muestras del fondo tomadas en noviembre de 1969, es decir, 5 meses y medio después de iniciarse la evacuación de los desechos, no se han completado todavía, pero es casi seguro que se encontrará un aumento de la concentración de Fe en la capa superior del sedimento.

6. Cinco meses y medio después de la evacuación de los desechos la cantidad de hidróxido férrico en suspensión era relativamente pequeña. La superficie de la zona en la que había aumentado de manera notable la concentración de Fe tenía una extensión de unos 500 km². La mayor concentración encontrada en esta zona era de 500 μg Fe por litro.

THE wastes from a TiO_2 factory near Bremerhaven, have been disposed of in the German Bight since May 1969 in a rectangular area of 12 n mi² with the centre 14 n mi NW of Heligoland. In this area the water is 25 to 28 m deep. The wastes are mainly 10 per cent sulphuric acid and about 14 per cent ferrous sulphate in water. Special tankers discharge 1,800 t per day of these wastes in the disposal area. In order to achieve a quick dilution and reduce the negative physiological effect of the wastes on marine organisms the wastes are introduced into the propeller water of the tankers underway at about 8 kn over a period of 1 to 2 h. The physical and chemical changes in the sea water and on the sea bed were studied by the German research vessel *Gauss* in May and October/November 1969.

Initial and secondary dilution in propeller water

From the Fe concentration in the undiluted wastes and in the propeller water the initial dilution was calculated to be 1:1,000. The secondary dilution was followed by measuring the Fe concentration in samples from the centre of the propeller water at different distances behind the tanker. In the first hour after the discharge the secondary dilution factor was about 10, in the second hour about 2. This result was obtained under medium turbulance conditions at a little less than 1 m wave height and at medium current velocity.

Changes of pH, Fe and O_2 in propeller water

The most important changes were a reduction in the pH of the sea water (normally 8.1 to 8.2) and marked increase in the CO_2 partial pressure. The ferrous sulphate increases the iron concentration, and the Fe^{2+} is oxidized to Fe^{3+} and precipitated as hydroxide. This reaction causes an O_2 deficit in the sea water.

Water samples were taken at different distances behind the tanker at 4 m depth in the centre of the propeller water. The pH in the samples was measured by a glass electrode. The Fe concentration was determined by gravimetry and photometry and the O_2 concentration by titration. Four minutes after the release of the wastes, the pH in the centre of the wake was about 6 and the CO_2 partial pressure was extremely high. At the same time the Fe concentration was about 40 mg/l, that is 1,000 to 2,000 times more than the normal value. In the following hours the pH increased and the Fe concentration decreased rapidly. Two hours after the release

the pH was about 8 and the Fe concentration was 2 mg/l.

Penetration of wastes to sea bed

The density of the wastes is about 1.25 g/cm³ and it would be expected that part of the wastes would sink to the sea bed immediately after the release. In which case the bottom flora and fauna would be seriously affected by the rather concentrated wastes. To study this very important question three different techniques were applied:

(a) Water samples were taken by plastic Nansen bottles from different depths at different distances behind the tanker and the pH, the Fe concentration, and the O_2 concentration were measured. During the first hours after release no pollution was detected below 15 m depth.

(b) About 20 cm above the sea bed the pH was recorded continuously by a specially constructed apparatus consisting of a cage containing a watertight case in which a pH recorder, an accumulator, and a glass electrode for high pressure (up to 6 atmospheres) were installed. Before the cage was lowered into the sea the interior of the electrode was pressurized to a pressure a little higher than that at the sea bed. By means of this apparatus the pH could be recorded continuously for about 4 weeks. Although the tanker passed over the instrument several times while discharging wastes, no change of the pH 20 cm above the sea bed was detected.

(c) Water was pumped from the sea bed into the laboratory of the RV *Gauss* while the tanker was passing. The pH of the water was recorded automatically but no change was detected.

These results show that the wastes did not reach the sea bed during the first hours after release. Since within the same period the pH and O_2 content had almost returned to normal at no time could there have been a significant decrease in pH or O_2 near the bottom. However, the Fe concentration in the bottom layer increased several hours after the release.

Accumulation of pollutants in sea water

Five and a half months after the beginning of the waste disposal the possible accumulation of wastes was investigated. Water samples taken 5 m below the surface were examined for pH and Fe concentration. The area polluted by ferric hydroxide had an elliptical shape with the long axis in the SSW-NNE direction and was centred

[187]

some nautical miles NNE of the disposal area. The size of the area of increased Fe concentration was about 150 n mi².

The highest Fe concentration in the sea water 5 m below the surface (0.5 mg/1) was found at the northern edge of the disposal area. Similar concentrations were observed between Heligoland and the coast of Schleswig-Holstein where the water is polluted by the rivers Elbe and Weser. No change of the pH was detected.

Accumulation of ferric hydroxide and oxide in sediment

The ferric hydroxide formed by reaction of the wastes with sea water sinks to the sea bed. The possible accumulation of ferric hydroxide and oxide in the sediment was studied $5\frac{1}{2}$ and 16 months after the beginning of the waste disposal. Compared with the samples that were taken before the beginning of the waste disposal no appreciable increase of the Fe content in the upper layers of the sediment could be found.

Section 3

EFFECTS OF POLLUTANTS ON THE BIOLOGY AND LIFE CYCLE OF MARINE ORGANISMS

Summary of Discussion

The discussion clearly reflected the considerable complexity of this problem. One of the main difficulties appears to be the paucity of knowledge of biological parameters concerning the large majority of species which may be used as test organisms. It was pointed out that there is still a tendency to limit experimental work to species which are easy to culture, and that often these species are not adequate for toxicity studies, in view of their resistance to pollution.

The criteria for choosing test animals appear to vary largely with the types of experiments and their objectives. The importance of using a wide spectrum of organisms, as far as possible representing different trophic levels, was emphasized. The criterion of ecological significance of the organism should always be taken into account, but unfortunately, not all ecologically significant species may be suitable test organisms.

Indicator species should show the effects of waste releases, and thus give warning of unwanted changes. Such organisms must fulfil certain criteria: adequate numbers in the area affected; stationary; sufficiently long-lived, and affected by pollution. It seems preferable that such species also be large enough to provide material for the study of histological changes and for chemical analysis on the content of various accumulated substances.

Observe the food chain

The food chain was noted as a means of concentration of pollutants at higher trophic levels, and in this connection, phagocytosis of bacteria by cells of coelenterates and other lower invertebrates was believed to be of particular importance. Several speakers pointed out the desirability of using test organisms from non-polluted areas, as they would not have adapted to the pollutant used in the experiment. Studies of the mechanisms of action of pollutants on specific physiological functions may require different kinds of organisms in each case. In this connection, the paucity of information on the physiology of marine organisms and on chronobiological mechanisms was pointed out. As a result, an important criterion for choosing test organisms was considered to be the amount of information available on a species. Several speakers mentioned the desirability of choosing species of economic importance as test animals, but it was considered that the choice among such organisms would be rather limited. It was noted, however, that the effects of pollution in many cases might be reflected earlier in non-commercial benthic organisms than in economically important species which are generally migratory and belong to higher trophic levels.

Practical experiences

On the usefulness of various taxae as test organisms, several speakers referred to their practical experiences. Especially noted were certain benthic invertebrates, especially polychaetes, which are relatively stationary and reflect not only the conditions at the time of sampling, but also those which have existed before. Such species as *Capitella capitata* are useful as indicators of polluted areas, being cosmopolitan and permitting comparisons between many regions of the world. The advantages were pointed out of comparing the relative densities of two species, such as *Capitella capitata* and *Nereis grubei*, which react differently to pollution.

Other useful test organisms are: marine bacteria, in view of their easy culture and sensitivity to pollutants; larvae of some fish species which, besides being economically important, are biologically better known and—as with all young stages of organisms—are more sensitive to pollutants than the adults; microscopic algae, as they are easy to cultivate, and because pollutant effects on such cultures can easily be measured by the influence on their reproduction rate; hydrozoa in which the effects of pollutants could be measured by counting the number of polyps in the culture; macroalgae which would also be particularly well suited as indicators; and molluscs which are of special interest in view of their capacity to accumulate pollutants, and sensitivity to the mechanical action of certain pollutants. It was noted that mariculture projects could provide an excellent opportunity for the study in the field as well as in the laboratory of the action of pollutants.

Greater range needed

It was concluded that the range of organisms presently used for toxicity tests should be considerably expanded. Moreover, research should be focused on certain species of organisms, which would satisfy the various basic objectives and requirements of experimental research, such as abundance of the species, sedentarism, minimum life-span, high sensitivity to pollutants and a size sufficiently large for making biochemical and histological analyses. In this connection, reference was made to the work of the Panel on Monitoring Organisms of the Seminar on Methods of Detection, Measurement and Monitoring of Pollutants in the Marine Environment, which has already identified a number of criteria for the selection of organisms for such purposes.

The Conference recommended that a working group on the choice of test organisms be established and that a special effort be made by FAO to make available up-to-date information on the biology of such species.

Tests of acute toxicity

The criterion first and currently most employed when testing the effects of pollution is acute toxicity. The results are often expressed as the minimum dose or concentration required to kill 50 per cent of the test population in a given time, e.g. 48 or 96 hours.

Discussion concentrated largely on the considerable variations in the toxicity of a pollutant, depending on: (1) the physico-chemical conditions of the water used in the experiments (e.g. presence of chelating agents, degree of ionization, hardness, and salinity, which is negatively correlated with surface tension, etc.); (2) the physico-chemical condition and behaviour of the pollutant (which may occur in different forms such as ionic, particulate, organically bound, emulsified or as a surface film, and may be chelated or adsorbed on suspended organic matter); and (3) the condition and behaviour of the test organisms (i.e. the amount of toxic material actually penetrating into the organism, the degree of their retention and rate of excretion, the adaptation which might have been developed by test animals at an earlier stage to the pollutants tested, etc).

On the basis of these considerations, several methodological questions were raised. The need was emphasized for providing, together with the results of the tests, precise data on the physico-chemical environment and on the condition of the organisms tested. Most of the presently available test results were considered useless, in fact, for the purpose of predictions, because they were not substantiated by this kind of information. It was found desirable to extend the duration of tests of lethal toxicity, as mortality may sometimes occur only after the termination of the test. The active concentration or the toxic agent in the water during the test, as well as the amount actually penetrating into the organisms, was considered to be of vital importance and should be measured.

Advantageous tests

Several speakers pointed out the advantages of continuous flow-tests over batch culture-tests. The former facilitate maintenance of more stable conditions in the chemical environment and a better control—at least in the case of micro-algal cultures—of the population size. There may be some limitations of this method, however, in cases where test organisms are used which are sensitive to water flow. It was suggested that not only the LD 50 but also the LD 0 and the LD 100 application factors be given with the result of the tests. This would be particularly important in the case where advice on permissible levels of waste dumping is required. Evidence was given of the great variation in radio-sensitivity of certain test organisms. This showed the broad spectrum of species that must be examined in order to find out which of them should be particularly monitored in a biocenosis exposed to radioactive pollution.

The need was expressed for a better comparability of test results, and reference made to the problem of standardization of methods.

The Chairman of European Inland Fisheries Advisory Commission (EIFAC) informed the meeting that his organization was aware of the multiplicity of fresh-water toxicity terminology and is preparing a multilingual glossary of terms used in this field which could be made comprehensive for fresh and marine waters.

Concerning the usefulness of lethal toxicity tests, some speakers pointed out the large variability of application factors for such tests, according to the nature of the pollutant. Quantitative toxicity tests should be aimed at measuring biological parameters, such as reproduction rate, rather than mortality. Thus lethal toxicity tests may be used primarily as a basis for establishing a warning system of acute toxicity accidents, rather than for determining biological significance.

Knowledge of the biological action of toxicants required for the elaboration of sublethal toxicity tests

In most cases, the limits for permissible concentrations of pollutants in the marine environment must be set on grounds other than their acute toxic effects on the organisms. Of primary importance in this connection are the chronic and sublethal effects of the pollutants, including avoidance reactions of fishes. Many substances are detected and avoided by fish at very low concentrations. Other sublethal effects may result, either from the gradual accumulation of substances to levels where they have biological effects on organisms, or from the continued stress exerted by pollutants. Among long-term effects, it is also necessary to consider those on genetics of marine organisms. However, there is very little information on this subject.

Numerous forms of organic pollution lower the oxygen concentration. If we consider only the limit which allows marine organisms to survive for a short period, we generally obtain a low lethal threshold for oxygen; however, for normal nutrition, growth and reproduction, a higher oxygen content than the minimum lethal threshold is needed. Only long-term physiological tests, for example, on respiratory function, nutritional behaviour, or fecundity, can show the critical level of oxygen concentration at which the organism ceases to be normal.

The need for extending the duration of sublethal tests was also emphasized, first because of data such as those showing the slow action of pollutants and the possibility of their accumulation in the organisms, and secondly from a chronotoxicological point of view: the variation in sensitivity of test organisms according to time rhythm. These rhythms seem to exist at all levels of organization (living organisms as a whole, as well as tissues and cells), and they are an expression of what has been called the "biological clock". The effect of pollution on this biological clock deserves special consideration for future studies.

To make the assessment of sublethal effects more accurate, it was recommended that not only biochemical but also histological control analysis be applied whenever possible. The desirability of establishing standard measures for the effect of pollutants on certain species was voiced, and in this connection, the assessment of water quality by the degree of opening of oyster shells was mentioned as an example.

Effects on bacteria and plants

It is well known that a number of pollutants, e.g. of food and paper industries, favour the proliferation of bacteria and lower fungi, while others inhibit them. Some studies demonstrate the stimulating effect of viruses on the growth of marine bacteria. In this connection, the positive effect of oil on the development of certain populations of oil-degrading marine bacteria was mentioned.

As regards the studies of sublethal effects on algae, attention was drawn to the action of numerous pollutants, notably DDT at concentrations as low as 4 parts per billion, on the photosynthetic activity of phytoplankton. It has been demonstrated that DDT also inhibits cellular division in marine phytoplankton. Examples were given of several planktonic algae in which the reproduction is stopped or altered by minor doses of DDT, dieldrin or endrin, but sensitivity varies greatly between species. The large specific variations in resistance towards radio-nuclides was stressed, particularly with *Nitzchia* cells showing high resistance to Cs—137. On the other hand this alga is very sensitive toward organomercurials used as fungicides. Organic mercury compounds are more toxic than inorganic ones. Bacteria play a prominent role in conversion of the inorganic form to the toxic methyl mercury. Crude oil constituents have a retarding effect on cellular division of phytoplankton. Still more toxic are some detergents used for cleaning the beaches after a massive pollution. Certain metallic elements, such as copper and molybdenum, and organic materials, like humic acids and vitamin complexes, can either stimulate or inhibit algal development.

Also mentioned were the effects of hydrocarbons and detergents on the cellular membrane of algae, leading to a penetration of the pollutant into the cell and to the extrusion of cellular fluid. A high sensitivity of some macroalgae, like *Enteromorpha*, to copper at low concentrations was reported from Sweden; but at higher concentrations, a chelating effect was observed which protects the algae from the pollutant.

With regard to methodological aspects, the advantages and disadvantages were examined of using pure clone-cultures or a genetically mixed material as test organisms in flow-culture of microalgae. Although genetically homogeneous, clone-cultures may involve the risk of not representing the average sensitivity of the species to the tested pollutant.

The extensive work in the U.K. on cultivation of macroalgae which has led to the development of a new practical method of using seaweeds as test organisms in continuous flow-culture was mentioned.

Effects on animals in relation to nutrition

It was noted that a large number of pollutants can act on different nutritional phases by: reducing the food resources of the environment; altering certain sensorial mechanisms which change the abilities of the animal to detect its prey; acting on the behaviour of the consumer, including its appetite; or bringing about increased catabolism and excessive loss of substance. A number of examples were given of effects of this kind, brought about, single or combined, by various types of pollution (i.e. malnutrition caused in mussels as a result of reduced food consumption and production of mucous faeces or pseudo faeces following contamination by flocculated iron hydroxide; action of pesticides on the food consumption and the feeding period of fishes; action of several pollutants on the taste papillae of fishes, and hence, on their nutritional behaviour, etc.).

The effect of certain pollutants is a function of the nutritional condition of the organisms. Thus, the compounds PCB and DDT accumulated in the blubber of seals, when they are liberated and released into the blood stream during starvation may give rise to serious physiological effects. On the other hand, it was also pointed out that bacterial pollution of urban origin can favourably affect the nutrition of certain filtering organisms and/or of detritus feeders.

Numerous pollutants probably act on the intermediate metabolism, and at least in certain cases through the neuroendocrine system, but this mechanism has not been sufficiently studied in marine animals. The intermediate metabolism is not only interesting in connection with the physiopathology of the animals concerned, but also in view of its general role in the biocenosis. The possibility, under certain circumstances, of bacterial methylation of inorganic mercury has been demonstrated. Fish also seem capable of methylating the several forms of mercury introduced into the organism.

As shown by several experiments with mussels and fish the degree of toxicity, and the rates of accumulation and excretion of certain pollutants, appear to be dependent on their physico-chemical condition, concentration, environmental factors, and on the pathway through which they are introduced into the organisms.

The importance of studying the metabolism of toxic agents throughout the food chain was emphasized—the results of some experiments suggest that marine polychaetous annelids may play an important role as a food chain link for entry of DDT into fish.

Effects on animal respiration

It was shown that pollutants can act on the respiratory functions of organisms directly and indirectly through a variety of mechanisms. These include: clogging of the gills through a purely mechanical action; modification of the branchial cells (by certain metals), sometimes causing histological changes in the gill tissue; and diminution of the oxyphoretic capacity of the blood. The latter may happen through a reduction of concentration of the respiratory pigment in the blood, of the number of red blood cells particularly, damage to the haematopoietic tissue and possibly even through qualitative modification of the blood respiratory pigment (i.e. oxygen dissociation curve). Further research is required.

Examples of the effects of low oxygen concentrations on the physiology and behaviour of some organisms were given, such as disturbances of the reproductive function and migration patterns. Oxygen consumption and efficiency of oxygen uptake were shown to be reduced in the Pacific salmon under such conditions. It was reported that salmon has been found to survive at rather low oxygen concentrations for limited periods, by resorting to anoxybiotic metabolism as a provisional source of energy. However, this was accompanied by deleterious metabolic effects (accumulation of lactic acid in the blood, excretion of this lactic acid through urine, loss of energy, depletion of glycogen reserves in the muscles, etc.).

Attention was drawn to the need for studying particularly the effect of fluctuations in temperature on respiration and on other physiological functions. With regard to values of oxygen tolerance determined under experimental conditions, it was pointed out that they would tend to be too low, if the whole life cycle of an organism were taken into consideration. On the other hand, experimental values might be too high as a result of increased metabolism due to the stress of experimental conditions.

Effects on osmoregulation and ionic regulation

Some pollutants, such as methylparathion, endrin and methoxychlor, were shown to alter the concentrations of various cations in the fish *Sphaeroides maculatus*, indicating an effect on the hydromineral regulation. These alterations probably have partly a cellular base. Certain metals, such as cadmium, copper, mercury, lead, can also interfere by combining with the cellular membrane. As the osmoregulatory capacity in fish and invertebrates varies with temperature, thermal pollution may interfere with the osmoregulation and ionic regulation of marine animals. This is particularly of great practical importance in tropical regions.

The osmo- and ionic regulatory processes are also of prime importance in the triggering and the development of migratory phenomena. The inherited gradual modifications of osmoregulatory mechanisms during the life cycle may be an essential factor in this connection. Disturbances of these mechanisms may bring about developmental malfunctions, as well as important modifications of migratory behaviour, and even death.

It was of great interest, therefore, to study the effects of pollutants on the osmoregulatory mechanisms of different marine species. On the other hand, modifications of osmoregulation, either as a function of a change of the natural environment (for example, salinity change), or under the action of a given pollutant, will have an effect on absorption and excretion of other pollutants. Finally, the effects of changes in chemical composition of the environment on a homeo-osmotic marine animal may provide a basis for a suitable sublethal test.

Effects on reproduction, development and growth

A number of examples were given of the effect of various pollutants on those physiological functions which are particularly vulnerable to modifications of environmental conditions. Heat discharge, although it sometimes has advantageous effects on the establishment or the culture of certain resources in specific areas, was considered in most cases to be deleterious. It particularly adversely affects the processes of maturation and fertilization, as well as the survival of eggs and larvae, in the latter case the effect being either direct or indirect, e.g. through the inhibition of feeding. Changes in salinity, turbidity as well as reduced dissolved oxygen concentration of the water, resulting from pollution, can also be harmful to the reproductive functions of marine organisms.

Fish eggs and advanced yolk-sac larvae of fishes, especially at the beginning of feeding, were shown to be particularly sensitive to insecticides and to oil combined with detergents. Pesticides were also reported to prevent the hatching of oyster and mussel eggs. Larval development of molluscs was found to be seriously affected by a number of pesticides. Degenerative changes in testes of crabs were observed as a result of the combined effect of oils and detergents.

In birds, numerous observations have shown the disastrous effect of pesticides on reproduction, which can be explained on the basis of the concentrations of these substances in higher trophic levels. The reproduction of sea birds is reduced in proportion to the amount of pesticides in the parents and in the eggs. One of the most striking effects on the eggs is the thinning of the shell.

Observations on the inhibition of the settling mechanism in oyster larvae in a culture, otherwise completely normal and where no pollutant could be traced chemically, suggest that such low doses in some substances can affect specific stages of larval development which could not be detected by available chemical means.

Effects on neuro-muscular system and behaviour

A number of substances were reported to have an effect on the nervous system and on the behaviour of marine organisms. Thus, organophosphate pesticides inhibit the acetylcholinesterase activity in the brain of fresh water fish. Oils may inhibit the development of the byssus in mussels, probably through a narcotic action on the muscular activity of the foot or by affecting the secretion of byssus-collagen. With regard to methodological aspects, a technique was referred to for measurement of the action potential within the lateral nerve of fish, which would allow the detection of the neurotoxic action of a pollutant. Ionizing radiation may modify the behaviour of certain aquatic organisms. Zinc affects the migration patterns of salmon. Numerous chemical substances alter the schooling behaviour of some fish.

The ability of animals (generally those with well developed olfactory sense) to detect toxic substances appears to be of special importance, as it may lead to flight reactions. Although these reactions protect the organisms from harm by the pollutant, they may in turn affect their nutritional and reproductive behaviour. Some given examples, showed that this ability of some animals by no means applies to all toxic substances. Thus, certain fish capable of avoiding acids and sulphides seem to be attracted to other toxic substances, such as copper, sulphates and ammonia. Some invertebrates (*Crangon crangon*) can detect and avoid oil dispersants at low concentrations, but when non-aromatic solvents are used together with the dispersants, the ability of detection is reduced.

The effects of pollutants on the mechanisms of secretion, uptake and retention of ectocrine substances were considered to be of particular interest. It was suggested that this problem should be the object of some sublethal tests.

Effects on cellular physiology

Recent observations have shown that the degree of resistance of various organisms to certain pollutants is closely correlated with the resistance of their tissues. It was considered essential, in order to understand the action of pollutants, to utilize the resources of biochemistry and biophysics and to apply them to the tissues and cells.

The influence of pollutants on the endocrine and enzymatic systems of marine organisms was illustrated by several examples. Such action can be stimulating, as was shown by the effects of sublethal doses of hydrocarbons on the glycogenesis in mussels (*Mercenaria mercenaria*); by the effect of chlorinated hydrocarbon pesticides on birds, which may cause the induction of hepatic enzymes that metabolize steroids, and by the drop in the glycogen level within the liver caused by kerosene. On the other hand, inhibiting action occurs, as exemplified by methyl-parathion on the esterase activity in the blood serum. The presence of pesticides may also result in considerable changes of biochemical characteristics of certain organs, e.g. of the liver of some fish (*Sphaeroides maculatus*).

Varied and extensive alterations of cellular activities were reported in oysters (*Ostrea gigas*) subjected to waste waters from pulp and paper mills, and in this connection, a reduction of the quantity of RNA in the cells and a decrease of calcium absorption were especially pointed

[193]

out. Differences in toxicity of pesticides on different species of fish could be explained by specific differences in the enzymatic acitvities which catabolize such compounds.

The potential risks of cancer induction by several types of pollutants (such as the polynuclear aromatic compounds entering the marine environment with oil) and by biosynthesis (probably by anaerobic bacteria) of carcinogenic substances were discussed. It appears that this risk cannot be clearly and quantitatively determined, as we do not have any information on critical thresholds of action of these substances.

It was considered that present development of techniques for culturing tissues of marine fish will provide a possible means of gaining a deeper insight into the mechanism of action of pollutants. The more refined the techniques for studying sublethal effects become, the more the experimental conditions must be accurately kept under control. This would include the important aspects of chronotoxicology.

Biological effects of pollution in the natural environment

The need for better complementary programmes, including laboratory tests and studies in the natural environment was especially emphasized. The discussion concentrated around the complex interactions of abiotic and biotic variables and their effects on the responses of marine organisms to poisonous substances in a natural environment. These interactions render predictions of their effects difficult. The properties of many of these variables are still not well understood, and it was considered necessary to gain a better knowledge of the multiparametric responses occurring in natural conditions, in order to correctly assess the ecological effect of different types of pollution. The need for ecological toxicology was expressed, meaning by this, the toxicology of an ecological system which might be closely reproduced under experimental laboratory conditions. On the other hand, attempts have been made already to determine the specific susceptibility levels of some marine invertebrates, namely polychaetes, to certain types of pollutant. Suggestions were made to consider the establishment of a biological test scale for marine waters, comparable to the saprobial system presently used in fresh waters. In this connection, the intensification of ecological studies in unpolluted waters, especially on benthic communities, appears to be of prime importance.

The development of a satisfactory methodology for the study of the effects of pollutants on organisms in the natural marine environment appears to be still in a rather preliminary stage. However, the development of some new interesting techniques was reported, such as for example, the use in the U.K. of transducers attached to molluscs under natural circumstances, which made it possible to measure certain physiological functions like the cardiac rhythm. These results were correlated with those obtained in the laboratory. Another reported technique was the exposure of benthic invertebrates (polychaetes) in cages to pollution in the surroundings of outlets of waste effluents.

Finally, the side effects of some types of pollution were considered. For example, the danger was noted of proliferation of certain pathogenic bacteria, which might affect schools of commercially important fish, often tending to concentrate around outfalls discharging certain types of organic wastes. The correlation between the releases of certain wastes and fish diseases reported for some areas could be explained partly by deterioration of the physiological condition of the fish, as a result of the direct action of a pollutant.

L'Action des Polluants sur la Vie Marine.
Du Choix de Critères Expérimentaux

*M. Fontaine**

Effects of pollutants on marine life—choice of experimental criteria
To study and monitor the effects of pollutants on living resources, sublethal tests which are more sensitive than the lethal ones should be employed. Several possible tests are identified from a review of the main known or supposed effects of pollution on the structure of marine organisms or communities as well as on the various biological functions of these organisms. It is suggested that a special and standardized methodology should be established in order to allow exact comparisons between experiments and to reduce dispersion of efforts. This implies that a series of tests should be selected according to the nature of the test species, their possible utilization by man, and the nature of pollutants and their effects.

El efecto de los contaminantes sobre la vida marina: selección de criterios experimentales
Para estudiar y vigilar los efectos de los contaminantes en la vida marina hay que emplear pruebas subletales más sensibles que las letales. El examen de las principales acciones conocidas o supuestas de los contaminantes sobre la estructura de los organismos o de las poblaciones marinas, así como sobre las diversas funciones biológicas de estos organismos, permite identificar algunas de estas pruebas. El perfeccionamiento de una metodología especial y normalizada que permitiría efectuar comparaciones exactas y reducir la dispersión de los esfuerzos, implica una selección de las pruebas elegidas en función de la naturaleza de las especies testigo (ellas deben ser ubicuas, sensibles y con su biología bien conocida) de su empleo eventual por el hombre, y de la naturaleza de los contaminantes y sus efectos de contaminación.

L E vaste problème de l'action des polluants sur la vie marine peut se poser de différentes façons. Tantôt, un accident spectaculaire (le naufrage d'un pétrolier, ou un événement imprévu dans un forage sous-marin par exemple, déversant une masse importante d'hydrocarbures dans le milieu marin) peut entraîner une pollution, dont la nature est incontestable, et dont certains effets immédiats apparaissent évidents. Mais les effets à long terme ou géographiquement lointains, étant donné les déplacements du polluant, qui peuvent d'ailleurs s'effectuer aussi bien par voie aérienne que par voie aquatique, sont souvent plus difficiles à préciser et appellent, pour une interprétation sérieuse, non seulement l'observation, mais l'expérimentation.

---

* Muséum National d'Histoire Naturelle, Physiologie Générale et Comparée, Paris, France.

Tantôt, une modification des biocœnoses marines se produit dans une région où les pollutions sont d'origines variées, par exemple pollution par les eaux d'un fleuve qui reçoit des volumes importants d'eau d'usage domestique et les déchets d'usines de types différents, et alors la recherche de l'élément ou des éléments en cause dans telles ou telles modifications observées, exige l'expérimentation pour dissocier ce qui, dans les effets néfastes constatés, revient à tel ou tel polluant. C'est là une recherche essentielle pour connaître à quelle source doit s'exercer d'abord et surtout la prévention. Parfois, des modifications d'un écosystème se produisent sans que l'on puisse trouver dans son environnement la cause possible d'une pollution. Dès lors, il s'agit de rechercher si l'eau de ce biotope exerce certaines influences néfastes sur l'écosystème considéré—les fluctuations observées pouvant être dues à des rythmes internes, à certains cycles dont nous avons maints exemples dans la vie sauvage—et, dans le cas d'une réponse positive, encore faut-il rechercher si cette influence néfaste est due à une auto-pollution (pollution endogène), ou à une exo-pollution (pollution exogène) (Fontaine, 1966).

De toute façon, se pose la question des méthodes biologiques à employer pour rechercher si l'eau de mer est nocive pour certains organismes marins. La méthode la plus courante en toxicologie consiste à rechercher la toxicité léthale de l'eau suspectée vis-à-vis d'un pourcentage donné d'un lot homogène d'organismes marins touchés par une pollution éventuelle. Si le polluant est identifié, on détermine la 48 ou 96 LC 50 ou TL 50, c'est-à-dire la concentration de ce polluant dans l'eau de mer, qui entraîne en 48 ou 96 heures, une mortalité de 50 pour cent des individus. Mais il faut souligner que cette toxicité peut varier beaucoup selon les conditions mécaniques, physiques, physico-chimiques et chimiques du milieu et selon la forme sous laquelle se trouve le polluant. Par exemple, selon que l'eau est agitée ou calme, des produits volatils, tels ceux se trouvant dans une pollution par les pétroles, étant plus ou moins rapidement éliminés dans l'atmosphère, apparaîtront plus ou moins toxiques (Lacaze, 1967; Ryland, 1966). La plupart des polluants ont une toxicité qui varie selon la température et la salinité de l'eau. Quant à la condition sous laquelle se trouve le polluant, citons l'exemple du pétrole qui, sous la forme d'émulsions avec l'eau de mer est plus toxique qu'un film de surface (Mironov, 1970), probablement parce que sous forme d'émulsion, les fines gouttelettes de pétrole exercent sur des surfaces d'échanges essentielles, notamment sur les branchies, en plus de leur action toxique, une action mécanique nocive. La dose léthale varie beaucoup selon l'espèce test et, pour un organisme donné, est souvent très différente selon le stade de développement de celui-ci. De nombreux travaux ont montré la sensibilité particulière à divers polluants des oeufs et des larves de poissons aussi bien que celle des nauplii de crustacés. Mais ces tests létaux, quel que soit leur intérêt qui est incontestable, surtout quand ils sont soigneusement standardisés (Portmann, 1970), ne nous informent que très incomplètement sur la nocivité des polluants, car un polluant donné peut être très dangereux pour un individu, souvent plus encore pour une population, sans être mortel dans les 48 ou 96 heures.

Le choix de l'espèce test peut répondre à des motivations diverses; celle-ci peut être sélectionnée en fonction du rôle important qu'elle joue comme maillon d'une biocœnose que l'on juge menacée et selon que l'on a affaire à une zone déjà polluée ou à une zone d'eau pure, on aura probablement à choisir des espèces différentes, par exemple dans le premier cas on prendra vraisemblablement, en ce qui concerne les annélides, une espèce typique des zones polluées comme *Capitella capitata*, alors que dans d'autres cas, on aura à prendre une annélide caractéristique d'une zone saine, *Nereis grubei* (Reish, 1970). On peut aussi choisir une espèce en raison de sa sensibilité supposée au (ou aux) polluant(s) considéré(s), de son mode d'alimentation (par exemple, un animal se nourrissant par filtration étant particulièrement intéressant si le filtre est assez fin pour retenir un polluant en suspension), en raison des caractères particuliers qu'elle offre pour des investigations ultérieures plus fines (présence de cellules géantes par exemple *Valonia*, *Loligo*), en raison du pouvoir de concentration d'une substance toxique ou radioactive spécialement étudiée, en raison du rythme rapide de reproduction si l'on veut étudier les effets génétiques, en raison aussi de la valeur économique des espèces, de leur importance pour l'alimentation humaine et de la nécessité particulièrement flagrante qui apparaît de les protéger. Aucune de ces motivations ne paraît contestable, mais il est évident que pour éviter une trop grande dispersion des efforts et autant que les caractères biogéographiques des espèces le permettent, il serait souhaitable de concentrer un certain nombre de travaux sur quelques organismes types—répondant à certaines de ces motivations—assez largement répandus et représentant divers maillons de la chaîne alimentaire. Le choix de certaines de ces espèces, autant que possible ubiquistes, étant fait, une institution comme la FAO rendrait un grand service en rassemblant sur celles-ci une documentation mondiale. Il serait sans doute souhaitable que, parmi celles retenues, se trouvent des espèces d'élevage (comestibles et espèces végétales ou animales servant de nourriture à celles-ci, particulièrement aux premiers stades de développement) car ces espèces ou du moins les genres correspondants sont en général largement répandues dans le monde et tout leur cycle de développement bien connu peut faciliter la recherche.

Quant aux critères de toxicité sublétale, ils peuvent porter sur des structures et des fonctions variées qui auront d'autant plus d'intérêt que celles-ci jouent un rôle plus important dans la physiologie normale de l'organisme considéré.

Cellules

Envisageons d'abord les organismes unicellulaires et les cellules. On a utilisé avec succès comme tests sublétaux le freinage ou l'inhibition du développement et de la croissance de cultures d'algues ou de flagellés et les variations de la photosynthèse (dosages d'oxygène—fixation du ^{14}C—dosage des chlorophylles, etc. . . .). De nombreux polluants altèrent l'activité photosynthétique, notamment le DDT à des concentrations infimes, de l'ordre de quelques milliardièmes.

On peut aussi étudier l'action des polluants soit sur des gamètes, soit sur des cellules isolées de tissus d'organismes multicellulaires, soit sur des cultures de cellules. Ces recherches devraient nous permettre de mieux connaître le mécanisme intime de l'action des polluants en décelant les sites d'action de ces toxiques. On a pu ainsi observer des altérations de structures mitochondriales sous l'effet de certains polluants et ces altérations apparaissent souvent pour des seuils beaucoup plus bas que ceux déterminés par la mort sur l'animal total (Öberg, 1961, 1964, 1967, 1967a). Elles témoignent pourtant d'une atteinte déjà sérieuse de l'organisme par le polluant. Il sera très important, en particulier du point de vue génétique, de développer la recherche des actions éventuelles des polluants sur les diverses structures subcellulaires jusqu'au niveau moléculaire et notamment sur les acides nucléiques. Il y aurait sans doute aussi intérêt à utiliser les méthodes électrophysiologiques. En effet, la membrane des cellules vivantes est électriquement polarisée, cette polarisation et ses variations jouent un rôle fondamental dans le fonctionnement cellulaire, en particulier dans le passage à travers les structures membranaires des substances actives. Je pense que la plupart des polluants ne peuvent exercer leur action nocive que par une action primaire sur les membranes cellulaires et les méthodes électrophysiologiques modernes doivent nous permettre de déceler les premières atteintes des polluants sur la perméabilité cellulaire.

C'est parce que l'atteinte des polluants se manifeste fondamentalement par une action sur certaines cellules que, à mon sens, les premières expériences visant à dépister une pollution marine devraient porter sur le métabolisme et les structures cellulaires. Mais certaines cellules spécialisées des métazoaires peuvent être plus sensibles à tel ou tel polluant que les cellules des organismes unicellulaires. De plus, elles peuvent transmettre une information, olfactive ou gustative par exemple, qui conduise l'animal à modifier son comportement dans un sens défavorable pour l'individu ou l'espèce sans que la cellule subisse un dommage. C'est pourquoi, il est nécessaire aussi d'étudier l'action des polluants sur certaines fonctions du métazoaire.

Nutrition

Considérons par exemple la nutrition: les polluants peuvent agir sur elle par des voies très variées, soit en diminuant les ressources alimentaires du milieu, soit en altérant certains mécanismes sensoriels qui permettent à l'animal de détecter sa proie, soit en agissant sur le comportement du prédateur et sur son aptitude à saisir sa victime ou encore sur son appétit, soit en se substituant à la nourriture habituellement consommée, soit en entraînant un catabolisme accru et une perte excessive de substance, ces divers mécanismes ou certains d'entre eux pouvant coexister dans un type donné de pollution. C'est ainsi que Winter (1970) montre comment les floculats d'hydroxyde de fer provenant de certaines industries peuvent entraîner chez les moules un état grave de dénutrition par suite de la diminution de la nourriture consommée et de la perte de substance que représente une sécrétion abondante de mucus pour constituer, soit des pseudofèces, soit des fèces en grande partie composées d'hydroxyde de fer. Il y a chute de poids des parties molles de l'animal et mort en quelques mois, l'un et l'autre phénomène étant fonction de l'intensité de la pollution. Beaucoup d'autres exemples des voies variées par lesquelles les polluants peuvent intervenir

sur l'alimentation ont été donnés, par exemple la concentration des eaux en oxygène, car à partir de certains seuils, divers animaux marins ne s'alimentent plus. Halstead (1970) signale que des hydrocarbures saturés à point d'ébullition élevée peuvent interférer avec la nutrition. Hiatt *et al.* (1953) notent que les pesticides sont susceptibles d'inhiber le comportement alimentaire des poissons, Eisler (1970) que les *Spheroïdes maculatus* exposés au méthylparathion seul, insecticide organophosphoré, ou au méthylparathion en combinaison avec le méthoxyclor, insecticide organochloré, ne s'alimentent plus. De nombreuses observations sont décrites dans la littérature, mais les mécanismes de l'action du polluant sont probablement complexes et trop souvent ne sont pas encore éclaircis. Notons toutefois que des travaux comme ceux de Bardach *et al.* (1965) sur *Ictalurus natalis* montrent à quelle faible concentration (0,5 partie par million dans l'eau ambiante) certains détergents lèsent les bourgeons du goût de ce poisson—on constate à l'examen histologique une érosion de ceux-ci—et peuvent entraîner des troubles du comportement alimentaire. Ces lésions sont sévères puisque 6 semaines après avoir été remis dans l'eau pure, les poissons n'ont pas recouvré leur comportement normal. Il est probable que de nombreux polluants agissent sur le métabolisme intermédiaire et dans certains cas au moins, par l'intermédiaire du système neuroendocrinien, mais cette action des polluants sur l'activité de ce système n'a pas été suffisamment étudiée chez les animaux marins. Il doit être rappelé aussi que la toxicité de certains polluants est fonction de l'état nutritionnel de l'organisme. Holden (1970) le note très opportunément en citant le cas des phoques amaigris chez lesquels l'utilisation des lipides de réserve qui avaient stocké les résidus de pesticide a pour conséquence de libérer dans le courant circulatoire ces résidus qui vont exercer une action toxique sur le système nerveux et le foie. De plus, la faible quantité subsistant de lipides de réserve ne permet plus d'offrir aux apports nouveaux de pesticide une voie de garage où ils pourraient être stockés.

Signalons enfin que la pollution bactérienne d'origine urbaine, néfaste à certains égards et qui mérite donc bien alors le nom de pollution, peut intervenir favorablement dans la nutrition de certains organismes filtrants ou limivores (Paoletti, 1970).

Les études sur le métabolisme intermédiaire des polluants présentent souvent un grand intérêt car elles peuvent rendre compte de leur mode d'action et des fluctuations de leur toxicité. Ainsi selon Miettinen *et al.* (1970), la demi-vie de la plus grande partie du méthylmercure est plus longue que celle observée pour le bichlorure de mercure, caractère qui, joint à sa résorption facile, à sa grande stabilité chimique, fait du méthylmercure un polluant particulièrement toxique. Il serait fort intéressant de connaître les conditions exactes de méthylation du mercure car il semble que cette réaction puisse être effectuée par des bactéries de la vase marine dans certains milieux (Jensen et Jernelöv, 1969), pas dans d'autres (Rissanen *et al.*, 1970). Il apparaît probable que les poissons soient capables de méthyler le mercure introduit sous des formes diverses dans l'organisme (Noren et Westöö, 1967). C'est parce que les composés organomercuriels sont lipotropes qu'ils tendent à s'accumuler dans le système nerveux, y causant des altérations histopathologiques et fonctionnelles qui peuvent être sévères. Au cours des recherches sur la maladie de Minamata, on a décrit chez le poisson touché, des changements histologiques dans les neurones, témoignant de leur dégénérescence et une disparition des cellules granuleuses dans le cerebellum.

Le mécanisme d'action des insecticides organophosphorés a été mieux compris après qu'il eût été montré que ces substances exerçaient par phosphorylation de l'enzyme une action inhibitrice sur l'activité acétylcholinestérasique du cerveau de poisson, activité dont on sait qu'elle est essentielle pour la conduction de l'influx nerveux. Il a été signalé que la mort survient quand cette activité tombe au-dessous de 19 pour cent de l'activité cholinestérasique normale. De telles déterminations utilisées pour des critères sublétaux présentent toutefois l'inconvénient d'exiger le sacrifice de l'animal (Holland *et al.*, 1967). Tel n'est point le cas des techniques où l'on suit l'influence des polluants sur certaines activités enzymatiques du sérum, telle par exemple l'inhibition de l'activité estérasique signalée par Eisler (1970) sous l'influence du méthylparathion. Grâce aux techniques maintenant bien au point de cathétérisme chez les poissons, on peut suivre dans le temps les variations non seulement des actions enzymatiques, mais des éléments figurés, de l'équilibre hydrominéral (reflétant les variations de l'ionorégulation) des produits du catabolisme, des protéines sanguines, etc., sous l'influence de certains polluants et déceler ainsi les seuils d'action de ceux-ci. On trouvera ici l'exemple d'un dispositif utilisé par Smith et Bell (1964) et permettant des prélèvements de sang artériel répétés sur un saumon nageant librement (fig. 1). Les polluants peuvent tantôt

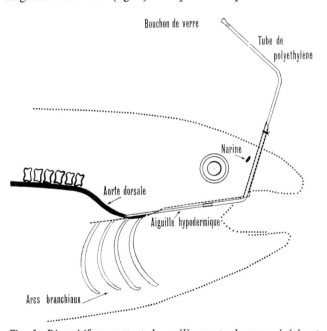

Bouchon de verre

Tube de polyethylène

Narine

Aorte dorsale

Aiguille hypodermique

Arcs branchiaux

Fig. 1. Dispositif permettant des prélèvements de sang répétés et pendant une durée prolongée sur un poisson en nage libre (d'après Smith et Bell, 1964).

diminuer les actions enzymatiques, tantôt les exalter. C'est ainsi que Engel *et al.* (1970) montrent que l'administration chronique de doses sublétales d'hydrocarbures chlorés peuvent conduire à des altérations métaboliques dans le sens d'un accroissement de la glycogénolyse et de

l'inhibition de la gluconéogenèse chez *Mercenaria mercenaria*. Chez le mulet, Sidhu *et al.* (1970) soulignent que le traitement prolongé au kérosène entraîne une chute du glycogène du foie par diminution de l'activité des enzymes synthétisant le glycogène. Au contraire, selon Wurster (1969) les pesticides chlorés provoquent, chez les oiseaux, l'induction des enzymes hépatiques qui métabolisent les stéroïdes. Buhler et Rasmusson (1968) suggèrent que les différences observées dans les toxicités des pesticides chez des espèces variées de poissons peuvent résulter de différences dans leur action sur les enzymes catabolisant la drogue. Un prétraitement par le DDT ou le phénylbutazone peut entraîner, du moins chez certaines espèces, une induction sélective des enzymes métabolisant la drogue considérée. Mais il faut signaler que les résultats peuvent être fort différents selon l'espèce considérée dans la série des vertébrés. Ainsi l'activité spécifique de la glucose-6-P déshydrogènase hépatique diminue de façon très significative chez le rat recevant un régime contenant du DDT alors qu'aucun effet n'a été observé chez le saumon coho (*Oncorhynchus kisutch*) recevant un régime semblable (Buhler et Benville, 1969). Il n'est pas exclu que des différences spécifiques existent au sein du groupe si hétérogène des poissons et *a fortiori* quand on passe des vertébrés aux invertébrés. Mais il est sûr que dans certains cas, l'étude des variations d'activités enzymatiques sous l'action d'un polluant puisse constituer un test sublétal intéressant.

Respiration

La fonction respiratoire peut faire aussi l'objet de tests d'une large utilisation, car de nombreux polluants agissent sur la fonction respiratoire par des mécanismes variés: par exemple en abaissant la teneur en oxygène du milieu, soit que la matière organique polluante soumise à des processus d'oxydation chimique entraîne une chute d'oxygène des eaux, soit qu'une augmentation de la turbidité des eaux détermine une diminution de la photosynthèse, soit qu'un développement excessif de la biomasse chlorophyllienne faisant suite à un apport de matières organiques nutritives n'entraîne, au temps et aux lieux où la photosynthèse ne peut s'exercer et du fait de sa respiration, une chute de la teneur en oxygène des eaux dangereuse pour les autres éléments de la biocœnose. Mais ils peuvent aussi agir directement sur la fonction respiratoire même de l'organisme par des mécanismes variés, par exemple par une action purement physique en colmatant les branchies. Certains métaux précipitent et adhèrent à l'épithélium branchial, certains hydroxydes métalliques résultant de diverses industries se collent aux filaments branchiaux. Le polluant peut agir sur le mécanisme physiologique intime de la cellule branchiale: Öberg (1965, 1967) a constaté que la roténone cause une inhibition de la consommation d'oxygène des branchies isolées et il a pu montrer en étudiant les structures subcellulaires, notamment les mitochondries, que le point d'attaque de la roténone se situait dans la chaîne respiratoire transporteuse d'électrons, dans la région d'intervention de la flavoprotéine. Le polluant peut agir sur les mécanismes de transport des gaz de la branchie aux cellules et vice versa. Jusqu'ici, on a surtout décrit chez diverses espèces de poissons une diminution du nombre d'hématies à la suite de l'action de divers polluants et chez la crevette, une chute des protéines sanguines par suite de l'action du DDT entraînant vraisemblablement une chute de l'hémocyanine. Une plus large investigation montrera sans doute que des polluants peuvent agir sur les caractères de la courbe de dissociation du pigment respiratoire, action qui sera surtout néfaste pour les animaux actifs à métabolisme élevé, exigeant un pigment respiratoire parfaitement adapté, alors que pour des animaux lents, ledit pigment apparaît souvent comme un mécanisme non indispensable. D'autres polluants (sel de cuivre par exemple) agissent sur le tissu hématopoïétique. Ainsi, des techniques très variées pourront être mises en œuvre pour dépister l'action d'une eau polluée sur le métabolisme respiratoire: histologie et cytologie des tissus hématopoïétique et branchial—numération des cellules à pigment respiratoire dans le milieu intérieur—étude des protéines respiratoires et enfin mesures du métabolisme respiratoire soit du tissu branchial isolé, soit de l'organisme entier. Ces mesures sont souvent exprimées par le terme de métabolisme de base, expression qui a un sens bien précis chez les homéothermes mais qui ne peut avoir en elle-même une signification aussi précise chez les poïkilothermes. En effet, chez les homéothermes, le métabolisme de base est le métabolisme minimal de l'organisme nécessaire pour assurer le déroulement normal des mécanismes vitaux essentiels à la vie dans une zone de température bien particulière: la température de neutralité thermique. Et si, partant d'une température basse, on trace la courbe du métabolisme (ordonnée) de l'homéotherme au repos et à jeun en fonction d'un accroissement de température (abscisse), on obtient une courbe d'abord descendante, passant par un minimum, puis ascendante. On a donc bien une valeur minimale du métabolisme de base pour laquelle les mécanismes thermorégulateurs n'ont pas à intervenir. Mais pour la majorité des poïkilothermes* qui ne présente pas de température fixe et où les mécanismes de thermorégulation sont extrêmement réduits, la température corporelle étant très proche de la température ambiante, il n'existe pas de valeur minimale du métabolisme de base si ce n'est celle qu'on peut observer au voisinage de la température inférieure létale alors que l'animal est immobile et au repos digestif. En effet, chez un poïkilotherme, le métabolisme respiratoire augmente avec la température jusqu'à une valeur de celle-ci peu éloignée de la température létale. On évoque donc chez les poïkilothermes sous l'expression de métabolisme de base, la notion d'une série de valeurs du métabolisme indispensable au fonctionnement physiologique minimal de l'organisme, valeurs essentiellement variables selon la température et l'activité musculaire. C'est pourquoi les auteurs qui utilisent et utiliseront ce test (mesure du métabolisme respiratoire) qui présente un réel intérêt, ne doivent pas se laisser abuser par ce terme de métabolisme de base qui ne présente nullement le caractère relativement précis qu'il possède chez les homéothermes et doivent donner avec le plus grand soin les conditions exactes de l'expérimentation relatives au milieu et à l'organisme (température, activité motrice, état nutritionnel, etc.), ainsi que les conditions physiques et physico-chimiques du biotope dans lesquelles

vivait, avant l'expérience, l'organisme considéré, conditions dont l'expérimentation doit tenter de se rapprocher le plus possible et dont dépendent *pro parte* les résultats obtenus. C'est seulement sous ces conditions que pourront être établies des comparaisons intéressantes d'un expérimentateur à l'autre. Signalons parmi les nombreuses techniques proposées pour la mesure du métabolisme respiratoire l'une des plus récentes et intéressantes (Smith et Newcomb, 1970).

Osmorégulation et ionorégulation

Divers auteurs ont signalé que les phénomènes d'osmo et d'ionorégulation peuvent être altérés par des polluants variés et notamment par la pollution thermique. Le plus souvent ce sont des modifications de composition ionique du milieu intérieur qui en témoignent—Eisler (1970) par exemple a proposé une représentation très frappante dite "stress profiles" des principaux effets d'un polluant donné, aussi bien sur le sang d'ailleurs que sur les tissus—ou encore des modifications de structure des cellules branchiales sécrétrices de chlorures (Conte, 1965; Öberg, 1961, 1964, 1967, 1967a). Mais maintenant que les techniques permettant de mesurer les flux d'ions (flux nets, flux entrants, flux sortants) sont bien élaborées (Maetz, 1970; Motais, 1970), il serait fort intéressant de connaître l'action de certains polluants sur ces flux s'exerçant essentiellement au niveau de la

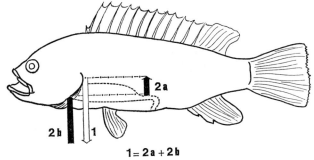

Fig. 2. *Représentation schématique des flux unidirectionnels de sodium chez un téléostéen marin, d'après Maetz (1970) et Motais (1970).*
2a: flux entrant intestinal 2b: flux entrant branchial
1: flux sortant branchial

branchie, du tube digestif et de l'appareil urinaire—ce dernier surtout efficace dans l'élimination des ions bivalents—et de telles mesures pourraient constituer un type de test sublétal probablement de grande valeur, car les phénomènes d'ionorégulation et d'osmorégulation jouent un rôle important dans la vie et en particulier dans le comportement migratoire des animaux marins (fig. 2). Depuis qu'il a été prouvé que la grande lamproie marine capturée au début de sa migration reproductrice en eau douce n'était plus capable de retourner dans l'océan qu'elle venait de quitter et cela parce que ses mécanismes osmorégulateurs vis-à-vis d'un milieu hypertonique étaient perturbés (Fontaine, 1930, 1930a), de nombreux chercheurs ont montré que l'évolution normale (au cours du cycle annuel ou pluriannuel) des mécanismes osmorégulateurs peut être l'un des facteurs essentiels du comportement sédentaire ou migratoire. Il apparaît donc qu'il serait particulièrement intéressant de développer les études de l'influence des polluants sur les mécanismes osmorégulateurs de tous les animaux marins

---

* Nous écrivons la majorité car il semble que pour certains gros poissons—le thon rouge, certains squales—il existe une certaine aptitude à la thermorégulation comme nous le verrons plus loin.

parce que leur perturbation peut déterminer des troubles de développement et la mort, mais aussi parce qu'ils peuvent entraîner des modifications, très importantes pour la pêche, du comportement migratoire.

Reproduction—Développement—Croissance

Les fonctions que nous examinons maintenant sont très sensibles aux variations de température des eaux. C'est pourquoi nous envisagerons d'abord la pollution thermique. L'élévation de la température des eaux dans certaines zones par suite du déversement d'eaux chaudes (centrales thermiques par exemple—centrales thermonucléaires) peut, dans des régions nordiques, exercer un effet favorable en permettant à des espèces de peupler des zones qu'elles n'auraient point habitées sans cet apport d'eaux chaudes ou en accélérant la maturation génitale et la croissance d'espèces comestibles. Ce phénomène est utilisé en mariculture. Mais inversement, et c'est alors qu'apparaît la notion de pollution, tout particulièrement dans les régions tropicales et subtropicales, où la température normale des eaux est souvent très proche de la température supérieure admissible pour certaines espèces, l'apport d'eaux chaudes peut perturber la fonction de reproduction en intervenant défavorablement par exemple soit sur la maturation des organes génitaux, soit sur la fécondation, soit sur la survie des œufs, car on sait que toutes ces étapes ne peuvent se dérouler que dans des marges bien définies de température. De même, l'apport par l'industrie d'effluents d'eau douce à des bassins d'eau salée, ou inversement de détournement d'eau douce qui arrivait normalement dans des aires d'eau saumâtre peut perturber la reproduction. Les interactions température salinité sont aussi très importantes à considérer (Kinne, 1967). Toutes les pollutions qui diminuent la teneur en oxygène des eaux et l'intensité du métabolisme respiratoire peuvent être néfastes pour la reproduction.

De nombreux travaux montrent l'action nocive de divers polluants sur les gamètes, la fécondation, l'éclosion, le développement et la croissance. Il serait fastidieux de les énumérer ici et nous renvoyons le lecteur à l'excellente revue de Davis (1970). Nous signalerons le fait que plusieurs de ces résultats suggèrent l'établissement de tests sublétaux, par exemple:

(a) *La division cellulaire de certaines espèces de phytoplancton* marin qu'il est facile de cultiver et dont la reproduction peut être bloquée ou altérée par des doses faibles de polluants. Il faut toutefois noter que le choix de l'espèce est très important car il existe de grandes différences spécifiques vis-à-vis d'une classe de polluants ou d'un polluant donné (*Dunaliella tertioleata* par exemple est très résistante aux pesticides).

(b) *Le développement des œufs fécondés et des larves* notamment d'échinodermes, de mollusques, de crustacés et de poissons qui se révèlent d'après les données acquises comme des réactifs extrêmement sensibles de certains polluants, soit par l'action exercée par ceux-ci sur la vitesse de développement embryonnaire, soit par les anomalies provoquées chez les embryons, soit par les phénomènes tératologiques observés et le pourcentage de survie des alevins, soit par l'action exercée sur certaines métamorphoses (voir fig. 3 de Wilson, 1970)*. Ainsi, une concentration de 1 mg par litre d'insecticides Sevid

* Publié aussi dans ce volume, Section III.

n'affecte pas l'éclosion des œufs du crabe *Cancer magister*, mais bloque la mue du stade prézoé ou stade zoé. En général, c'est à la fin de la résorption du sac vitellin et au début de la phase d'alimentation que les jeunes alevins sont le plus sensibles aux polluants (Wilson, 1970).

(c) *La croissance des mollusques, crustacés, poissons d'élevage, leur fécondité et les conséquences génétiques éventuelles d'une pollution*. C'est dans ces derniers domaines que doit apparaître particulièrement féconde la coopération avec les stations de mariculture.

Système neuromusculaire et comportement

Nous avons déjà vu comment l'action inhibitrice des insecticides organophosphorés sur l'activité acétylcholinestérasique retentissait sur la conduction de l'influx nerveux. Les insecticides organochlorés peuvent stimuler le système nerveux jusqu'à la convulsion en produisant une élévation importante de l'acétylcholine par libération d'une forme liée. Les dosages d'acétylcholine et d'activité cholinestérasique peuvent donc être des tests intéressants de l'action de certains polluants. Mais plus généralement et plus rapidement, la technique de mesure du potentiel du courant d'action de certains nerfs—notamment du nerf latéral (Halsband, 1970)— permet de déceler l'effet neurotoxique de certains polluants. Ceux-ci sont capables d'agir directement sur le système musculaire, mais il est probable que plusieurs aussi agissent par une libération d'histamine ou de catécholamines. L'étude des variations des potentiels de repos et d'action de certains muscles sous l'influence de certains polluants, peut être instructive. Rappelons ici que les pétroles exercent une action narcotique sur le système musculaire des mollusques et inhibent la formation du byssus chez la moule (Griffith, 1970). Signalons aussi l'utilisation possible de l'électrocardiographie (Halsband, 1970).

Sans doute y aurait-il intérêt à rechercher selon la technique décrite par Hara (1967) et mise en œuvre notamment chez des *Oncorhynchus* (Hara *et al.*, 1965), quel peut être le seuil d'action de certains polluants sur la réponse (potentiel évoqué) induite dans le bulbe olfactif, soit par l'eau des frayères quand il s'agit d'un poisson amphihalin en migration anadrome, soit par une solution saline pour un poisson amphihalin en migration catadrome, nous informant ainsi de l'action que les polluants peuvent exercer sur une réponse impliquée dans le comportement migratoire de ces espèces.

Ainsi sommes-nous conduits aux problèmes très importants concernant l'action des polluants sur les comportements qui jouent un rôle capital dans la biologie des espèces.

Les pollutions peuvent agir sur le comportement par un abaissement de la tension d'oxygène des eaux. Smith *et al.* (1970) montrent que l'*Oncorhynchus*, lors de sa migration anadrome dans un estuaire pollué aux eaux sous-oxygénées peut faire appel, pendant un temps limité, au métabolisme anaérobique comme source d'énergie mais non sans inconvénient (accumulation d'acide lactique dans le sang, baisse du pH et modification de la capacité respiratoire du sang, excrétion de lactate par l'urine d'où perte d'énergie, baisse des réserves de glycogène musculaire), mais cette étape ne peut être que transitoire et les auteurs envisagent les diverses altérations que cette pollution est susceptible d'entraîner dans le comportement migratoire et la reproduction. Effectivement, travaillant dans la nature et sur le *Salmo salar* dans la rivière Miramichi, Elson *et al.* (1970) montrent que l'abondance de matières organiques en suspension ou en solution dans l'eau ralentit la vitesse de migration.

Il a été montré que nombreux sont les polluants qui aux doses sublétales diminuent l'activité motrice sous l'influence des stimuli lumineux* et d'une façon générale sont susceptibles d'altérer la vitesse de nage et les caractères de celle-ci. Cette action peut être particulièrement grave pour l'avenir de certaines espèces chez lesquelles le comportement natatoire reproducteur des deux sexes apparaît comme admirablement réglé pour que les œufs soient fécondés dans les délais les plus rapides. Toute altération de ce comportement peut être néfaste pour la reproduction en retardant ou en empêchant la rencontre des gamètes mâles et femelles.

Certains animaux marins présentent un comportement de fuite quand ils détectent des polluants qui leur sont particulièrement nocifs. Ainsi Portmann (1970a) rapporte les expériences de Shelbon, selon lesquelles *Crangon crangon* peut éviter des dispersants des pétroles à des concentrations égales au 1/100 de la 48 DL 50 cependant que Sprague et ses collaborateurs (Sprague et Drury, 1969; Sprague *et al.*, 1965) observant que les saumons peuvent modifier leur migration à la suite d'une pollution causée par une exploitation de minerai et expérimentant au laboratoire ont constaté que les jeunes saumons évitaient des concentrations de métaux égales à 9 pour cent du seuil fatal pour le zinc, à 5 pour cent pour le cuivre et à 2 pour cent pour le mélange des deux. Ces résultats pourraient inciter à conclure que chez les animaux marins, les substances toxiques sont détectées à une faible concentration très inférieure à la concentration dangereuse, et qu'elles entraîneraient une réaction d'évitement avec ce que celle-ci peut présenter d'inconvénients pour divers comportements intéressant notamment les fonctions de nutrition et de reproduction, mais aussi avec l'avantage de soustraire l'individu à l'action nocive de ces polluants. Ce n'est cependant pas le cas général et nombreux sont les cas cités de polluants qui attirent telle ou telle espèce marine et ceux auxquels l'organisme apparaît indifférent, peut-être d'ailleurs parce que, selon l'hypothèse de Kühnhold (1970), les chémorécepteurs sont bloqués très rapidement dès le premier contact. On peut envisager que les polluants puissent perturber le fonctionnement de bien d'autres récepteurs sensoriels tels les thermo et électrorécepteurs qui jouent très probablement un rôle important dans la répartition de nombreux animaux marins.

Nous avons vu plus haut comment certains polluants peuvent agir sur le comportement par une action sur les mécanismes osmo et ionorégulateurs et tout spécialement chez les poissons amphihalins, c'est-à-dire dont le cycle de migration comporte des déplacements d'eaux de forte salinité à des eaux de faible salinité, mais ils doivent être susceptibles d'agir aussi chez certaines espèces par une intervention sur la thermorégulation qui se manifeste

* Voir notamment le travail de Blaxter (1968) sur les larves de harengs, dans lequel est décrite une technique de mesure intéressante.

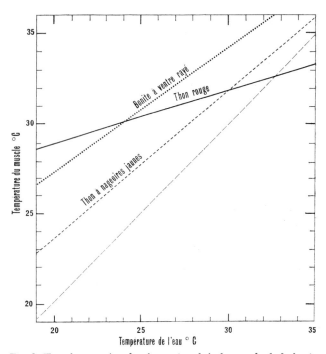

Fig. 3. Températures (en degrés centigrades) du muscle de la bonite à ventre rayé (Katsuwonus pelamis), du thon a nageoires jaunes (Thunnus [Neothunnus] albacora) et du thon rouge (Thunnus thynnus) capturés dans des eaux de températures différentes (d'après les données de Barrett et Hester, Jones et Mather, voir Carey et Teal, 1969).

à des degrés divers chez des espèces holohalines mais avec une intensité particulière chez le thon rouge (*Thunnus thynnus*) qui peut présenter une température corporelle de plus de 20° supérieure à celle de l'eau ambiante (fig. 3). Certains auteurs avaient autrefois attribué de telles différences de température entre le milieu et le poisson à l'activité musculaire, mais les travaux de Carey et Teal (1969a) ne permettent plus d'en douter: ce thon possède bien de véritables capacités thermorégulatrices, attribuables, au moins *pro parte*, à certaines particularités circulatoires et notamment à l'importance des dispositifs artérioveineux dits à contre-courant. A notre sens, c'est en raison de ces capacités que ce thon peut parcourir dans un temps relativement court de longs trajets dans des eaux de températures très différentes: ainsi les marquages ont-ils montré qu'il peut se rendre en quelques semaines des eaux des Bahamas d'une température proche de 30° à celles des eaux des côtes norvégiennes voisines de 6°. Il est probable que cette fonction thermorégulatrice n'est pas de même intensité pendant toute la durée du cycle vital de ce thon et que ces fluctuations sont impliquées dans la recherche ou l'abandon de certaines eaux de niveaux thermiques différents. Dès lors on peut se demander si certaines pollutions susceptibles d'apparaître en haute mer (insecticides, mercure, hydrocarbures, etc.) ne sont pas capables d'altérer la fonction thermorégulatrice de cette espèce, par suite ses migrations et par conséquent le rendement des activités halieutiques. Nous avons pris comme exemple le thon rouge en raison de son intérêt économique et du fait qu'il apparaît, dans l'état actuel de nos connaissances, comme le poisson présentant la thermorégulation la plus efficace, mais d'autres espèces sont aptes à une certaine thermorégulation (par exemple les élasmobranches, *Isurus oxyrhyncus* et *Lamna nassus*) (Carey et Teal, 1969); certains polluants en touchant cette fonction notamment par la voie de vaso-dilatation ou vasoconstriction des vaisseaux péri-phériques, de la vitesse et du volume du sang les parcourant, peuvent modifier leur thermopréférendum. Les progrès réalisés récemment en télémétrie doivent permettre de progresser dans ce domaine.

Nous avons réuni dans ce paragraphe système neuro-musculaire et comportement car l'intégrité de celui-là est indispensable au déroulement normal de celui-ci. Mais en fait, il faudrait faire figurer les formations endocrines parmi les fonctions conditionnant le comportement (Fontaine, 1970). Il est regrettable que nous soyons si pauvres en données relatives à l'action des polluants sur celles-ci. En toxicologie, on a cependant étudié depuis longtemps les réponses de diverses glandes et notamment de la surrénale aux agressions toxiques. Nous avons maintenant des données qui nous permettent d'estimer l'activité des tissus des poissons homologues de ceux de la surrénale des mammifères (interrénal et tissu chromaffine) et d'autres glandes dont l'assimilation au système surrénalien est discutable, mais qui semblent cependant jouer un rôle non négligeable dans la défense de l'organisme des téléostéens, les corpuscules de Stannius (Lopez et Fontaine, 1967). Il serait intéressant de rechercher chez les poissons les seuils de réponse de ces glandes aux polluants. Peut-être pourraient-ils constituer les critères d'une activité sublétale ? Signalons enfin que les polluants peuvent, soit par action sur le système nerveux, soit par intervention sur la sécrétion des hormones hypophysaires agissant sur le système pig-mentaire cutané, altérer les rapports entre individus et populations.

Conclusions

Cette revue des actions démontrées ou supposées vraisemblables des polluants sur le monde vivant et qui peuvent faire l'objet de tests sublétaux, ne prétend pas être exhaustive. Telle qu'elle est présentée avec ses certitudes, ses lacunes et ses hypothèses, nous espérons qu'elle pourra déjà susciter certaines initiatives et suggérer certains choix, choix nécessaires pour ne pas aboutir à une trop grande dispersion des efforts. Il faut toutefois souligner que divers tests doivent être sélec-tionnés en fonction des caractères particuliers d'une biocœnose donnée ou des organismes les plus menacés, ou les plus menaçants pour l'homme dans le cas où ils sont comestibles et pollués d'une façon dangereuse pour l'espèce humaine, en fonction aussi de la nature des polluants connus ou supposés et de leur mécanisme d'action et que, par suite, il faut laisser ouverte aux chercheurs la possibilité d'une sélection de tests hors programme mais adaptés à certains problèmes par-ticuliers. Une institution comme la FAO peut être très efficace dans le choix des tests d'un large spectre d'application et dans la définition précise des techniques qui doivent être appliquées aux matériels vivants et inanimés retenus. Mon opinion toutefois est que cet éventail de tests doit aller du plus bas niveau trophique au plus haut, du plus simple au plus complexe, c'est-à-dire de la cellule et des structures subcellulaires aux tissus, aux organes, aux organismes entiers, aux biocœnoses expérimentales ou naturelles, lesquelles permettent de déceler les interactions de leurs divers

constituants et les effets d'une accumulation de polluants tout au long de la chaîne alimentaire. Les expérimentateurs, après avoir considéré isolément l'action des polluants, doivent aussi se préoccuper des effets synergiques ou antagonistes de certains d'entre eux qui, dans la nature, peuvent se trouver en contact dans un même milieu marin et de l'interdépendance de l'action toxique de ces polluants et des caractères physico-chimiques du milieu. Ces tests doivent comporter toutes les informations nécessaires pour une reproduc-tion exacte de l'expérience. On est frappé de la relativement bonne concordance des données écolo-giques qui, bien qu'acquises en divers points du globe, permettent fréquemment une synthèse satisfaisante alors que les données expérimentales sont assez souvent contradictoires et dans bien des cas, il apparaît que c'est parce que les conditions des expériences ne sont pas suffisamment précisées. L'étude des pollutions, qui devrait être une véritable science avec sa propre méthodologie, n'a pas encore atteint ce stade et cette promotion devrait être l'un des buts de la FAO. Ainsi les auteurs ne devraient pas ignorer, comme tel semble être trop souvent le cas, les grands progrès de la chronotoxicologie et de la biométéorologie. Par exemple il est sûr que les expérimentateurs devraient tenir compte de l'existence probable de rythmes de périodes variées, circadiens, circatrigentiens, saisonniers, de la sensibilité du vivant aux polluants. De tels rythmes paraissent susceptibles d'exister à tous les niveaux d'organisation, aussi bien à celui de l'organisme dans son ensemble qu'au niveau subcellulaire et cependant il est rare que dans les protocoles expérimentaux, les renseignements temporels soient donnés qui permettent de les mettre en évidence et, par là, d'expliquer certaines divergences dans les résultats obtenus, comme l'ont permis souvent dans leur domaine la chronotoxicologie et la chrono-pharmacodynamie. D'ailleurs le mécanisme horloge biologique apparaît aujourd'hui essentiel dans le com-portement de nombreux animaux marins. Comment les polluants peuvent-ils ou non dérégler cette horloge ? Il est actuellement bien difficile de donner une réponse à cette question.

Il est très probable de plus que certaines substances nutritives, dissoutes dans les eaux de mer, jouent un rôle important dans le développement des organismes cellulaires, des œufs en voie de développement et des stades larvaires. Ainsi d'après Bass *et al.* (1969), des larves d'annélides polychètes peuvent capter en 1 heure, 0,05 pour cent de leur poids frais de leucine. Il est vraisemblable que d'autres acides aminés et d'autres molécules organiques peuvent être ainsi captés à partir de solutions très diluées et jouer un rôle dans la croissance des êtres marins, surtout aux premiers stades de leur développement. Il est probable que des pollutions libèrent, soit directement, soit indirectement, des substances de cette nature dont les concentrations respectives interviennent dans la reconstitution d'une biocœnose altérée ou au contraire inhibent la sécrétion naturelle ou encore, par combinaison physique ou chimique, suppriment toute action biologique de ces substances ectocrines. Elles peuvent évidemment altérer aussi bien les phénomènes de captation que ceux de sécrétion et très probablement les modifications observées par Schafer (1968) de la composition en acides aminés des moules, des ormeaux et des crabes, selon qu'ils sont en eau pure ou en eau polluée, sont liées à des mécanismes de cette nature influençant la teneur des eaux marines en substances ectocrines. Il me paraît opportun aussi de souligner que l'action d'un polluant peut, avec le temps, s'exercer dans le sens d'une adaptation d'une population à ce polluant, adaptation qui peut s'effectuer par voie enzymatique. Mais il ne faut pas sous-estimer les dangers que fait courir à l'homme cette adaptation, non seulement parce que le polluant peut s'accumuler en plus grandes quantités dans la chair de l'animal marin comestible, mais parce que la présence de ce polluant dans l'organisme de l'animal marin peut modifier certains métabolismes et par voie de conséquence, les qualités nutritionnelles et organoleptiques de sa chair. D'autre part, l'utilisation d'organismes marins issus d'une zone souillée par un polluant déterminé comme agents de tests sublétaux vis-à-vis de ce polluant pourrait conduire à des résultats erronés, soit que l'organisme ayant déjà accumulé une certaine quantité de polluant ou ayant été sensibilisé à l'égard de ce polluant, donne lieu à la détermination d'un seuil inférieur au seuil normal, soit, ce qui est plus grave, qu'une certaine adaptation de l'organisme conduise à la détermination d'un seuil trop élevé.

J'aimerais souligner que la mise en œuvre de ces tests sublétaux trouvera certainement avantage à être associée aux travaux d'une station de mariculture chaque fois que les organismes vivants choisis comme sujet d'expérience seront élevés dans ces stations soit en tant qu'animaux comestibles, soit en tant que nourriture de ceux-ci, car la connaissance approfondie de toutes les étapes de leur cycle par les zootechniciens qui leur sont attachés facilitera beaucoup la détection des effets sublétaux liminaires. De plus la surveillance continue pendant une longue durée d'une race bien définie, choisie pour l'élevage, permettra de suivre des effets génétiques éventuels ou des conséquences pathologiques à long terme (cancérisation par exemple) d'un polluant avec une sûreté toute particulière. Il faut enfin rappeler que, dans la nature, nous ne nous trouvons pas bien souvent en présence d'un seul polluant, mais de plusieurs qui peuvent exercer des actions simplement additives ou antagonistes ou synergiques. L'expérimentateur ne doit pas négliger la mise en évidence de telles interactions.

Enfin, émettons l'espoir qu'une large connaissance de l'action de certains polluants sur des espèces variées permette, en mettant en évidence des seuils de toxicité très différents selon les espèces, de déceler, au-dessous de ceux-ci, des concentrations favorables de certains d'entre eux vis-à-vis de certaines espèces. La notion de pollution est en effet très subjective. Un hydrocarbure par exemple est un polluant pour l'oiseau ou le baigneur qui en est souillé sur les plages, c'est un aliment pour certaines bactéries. Une plus vaste connaissance de l'action biologique des polluants doit permettre, en orientant leur destinée, de retirer à certains d'entre eux, dans des conditions particulières, le qualificatif de polluant et c'est là évidemment, la solution la plus satisfaisante. Il n'en reste pas moins que jusqu'à l'obtention de ce résultat encore bien lointain pour la plupart d'entre eux, il nous faut tenter de maintenir ou mieux encore de réduire la concentration des polluants dans les eaux marines au-dessous du seuil correspondant à leurs premiers effets sublétaux sur les organismes les plus sensibles, directement ou indirectement, par l'intermédiaire de la chaîne alimentaire notamment. L'élaboration et la mise en application des tests envisagés ci-dessus devraient nous y aider.

Bibliographie*

BARDACH, J E, FUJIYA, M and HOLL, A   Detergents: effects on the
1965   chemical senses of the fish, *Ictalurus natalis* Le Sueur. *Science N.Y.*, 148 (3677): 1605–7.

BASS, N, CHAPMAN, G and CHAPMAN, J M   Uptake of leucine by
1969   larvae and adults of *Nereis. Nature, Lond.*, 221:476–7.

BLAXTER, J H S   Visual thresholds and special sensitivity of
1968   herring larvae. *J. exp. Biol.*, 48:39–53.

BUHLER, D R and BENVILLE, P   Effect of feeding and of DDT on
1969   the activity of hepatic glucose 6 phosphate deshydrogenase in two salmonids. *J. Fish. Res. Bd Can.*, 26:3209–16.

BUHLER, D R and RASMUSSON, M E   The oxydation of drugs by
1968   fishes. *Comp. Biochem. Physiol.*, 25:223–39.

CAREY, F G and TEAL, J M   Mako and Porbeagle: warm-bodied
1969   sharks. *Comp. Biochem. Physiol.*, 28:199–204.

CAREY, F G and TEAL, J M   Regulation of body temperature by
1969a   the blue fin tuna. *Comp. Biochem. Physiol.*, 28:199–204.

CONTE, F P   Effect of ionizing radiation on osmoregulation in fish
1965   *Oncorhynchus kisutch. Comp. Biochem. Physiol.*, 15:293–302.

DAVIS, C C   The effects of pollutants on the reproduction of
1970   marine organisms. Paper presented to the FAO Technical Conference on Marine Pollution and its effects on Living Resources and Fishing, Rome, Italy, 9–18 December 1970, FIR:MP/70/R-2:8 p.

EISLER, R   Pesticide-induced stress profiles. Paper presented to
1970   FAO Technical Conference on Marine Pollution and its effects on Living Resources and Fishing, Rome, Italy, 9–18 December 1970, FIR:MP/70/E-12:9 p.

ELSON, P F, LAUZIER, L M and ZITKO, V   A preliminary study
1970   of salmon movements in a polluted estuary. Paper presented to the FAO Technical Conference on Marine Pollution and its effects on Living Resources and Fishing, 9–18 December 1970, FIR:MP/70/E-81:11 p.

ENGEL, R H, NEAT, M J and HILLMAN, R E   Sublethal chronic
1970   effects of DDT lindane on glycolytic and gluconeogenetic enzymes of the quahog, *Mercenaria mercenaria*. Paper presented to the FAO Technical Conference on Marine Pollution and its effects on Living Resources and Fishing, Rome, Italy, 9–18 December 1970, FIR:MP/70/E-13:6 p.

FONTAINE, M   Modifications du milieu intérieur des poissons
1930   potamotoques au cours de la reproduction. *C. r. hebd. Séanc. Acad. Sci.*, Paris, 191:736.

FONTAINE, M   Recherches sur le milieu intérieur de la lamproie
1930a   marine. *C. r. hebd. Séanc. Acad. Sci., Paris*, 191:680.

FONTAINE, M   Les pollutions des océans et leurs répercussions sur
1966   les grands problèmes de la faim et de la soif dans le monde. *Ind. natn.*, 1966 (2):1–15.

FONTAINE, M   Glandes endocrines et comportement chez les
1970   poissons. *In* Hormones et comportement. Problèmes actuels d'endocrinologie et de nutrition. *Expans. Rech. scient.*, 1970:335–48.

GRIFFITH, D de G   Toxicity of crude oil and detergents to two
1970   species of edible molluscs under artificial tidal conditions. Paper presented to FAO Technical Conference on Marine Pollution and its effects on Living Resources and Fishing, Rome, Italy, 9–18 December 1970, FIR:MP/70/E-16:12 p.

HALSBAND, E   Etude physiologique du degré de toxicité de
1970   différentes substances contenues dans l'eau de mer. Document présenté à la Conférence Technique de la FAO sur la pollution des mers et ses effets sur les ressources biologiques et la pêche, Rome, Italie, 9–18 décembre 1970, FIR:MP/70/E-17:13 p.

HALSTEAD, B W   Toxicity of marine organisms caused by pollu-
1970   tants. Paper presented to FAO Technical Conference on Marine Pollution and its effects on Living Resources and Fishing, Rome, Italy, 9–18 December 1970, FIR:MP/70/R-6:21 p.

HARA, T J   Electrophysiological studies of the olfactory system
1967   of the goldfish *Carassius auratus* L. 2. Response patterns of the olfactory bulb neurones to chemical stimulation and their centrifugal control. *Comp. Biochem. Physiol.*, 22:199–208.

HARA, T J, UEDA, K and GORBMAN, A   Electroencephalographic
1965   studies of homing salmon. *Science, N.Y.*, 149(3686):884–5.

* Il ne pouvait être envisagé de tenter de donner une bibliographie complète d'un domaine aussi vaste.

HIATT, R W, NAUGHTON, J J and MATTHEWS, D C Effects of
1953   chemicals on a schooling fish, *Kuhlia sandvicensis*. *Biol. Bull. mar. biol. Lab.*, *Woods Hole*, 174:28–44.

HOLDEN, A V Monitoring organochlorine contamination of the
1970   marine environment by the analysis of residues in seals. Paper presented to FAO Technical Conference on Marine Pollution and its effects on Living Resources and Fishing, Rome, Italy, 9–18 December 1970, FIR:MP/70/E-63:14 p.

HOLLAND, H T, COPPAGE, D L and BUTLER, P A Use of fish brain
1967   acetylcholinesterase to monitor pollution by organophosphorus pesticides. *Bull. Environ. Contamin. Toxicol.*, 2(3):152–62.

JENSEN, S and JERNELOV, A Biological methylation of mercury
1969   in aquatic organisms. *Nature*, *Lond.*, 223(5207):753–4.

KINNE, O Physiology of estuarine organisms with special reference
1967   to salinity and temperature: general aspects. *In* Estuaries, edited by G. H. Lauff. *Publs Am. Ass. Adrmt Sci.*, 83: 525–710.

KÜHNHOLD, W W The influence of crude oils on fish fry. Paper
1970   presented to FAO Technical Conference on Marine Pollution and its effects on Living Resources and Fishing, Rome, Italy, 9–18 December 1970, FIR:MP/70/E-64: 10 p.

LACAZE, J C Etude de la croissance d'une algue planctonique en
1967   présence d'un détergent utilisé pour la destruction des nappes de pétrole en mer. *C. r. hebd. Séanc. Acad. Sci., Paris*, 265:1489–91.

LOPEZ, E ET M FONTAINE Réponse des corpuscules de Stannius de
1967   l'anguille (*Anguilla anguilla* L.) à des blessures expérimentales. *C. r. Séanc. Soc. Biol.*, 161(1):36–9.

MAETZ, J Les mécanismes des échanges ioniques branchiaux
1970   chez les Téléostéens. Etudes de cinétique isotopique. *Bull. Infs. scient. tech. Commt Energ. atom.*, (145):3–33.

MIETTINEN, V, *et al.* Preliminary study on the distribution and
1970   effects of two chemical forms of methyl mercury in pike and rainbow trout. Paper presented to FAO Technical Conference on Marine Pollution and its effects on Living Resources and Fishing, Rome, Italy, 9–18 December 1970, FIR:MP/70/E-91:12 p.

MIRONOV, O G The effect of oil pollution on flora and fauna of
1970   the Black Sea. Paper presented to FAO Technical Conference on Marine Pollution and its effects on Living Resources and Fishing, Rome, Italy, 9–18 December 1970, FIR:MP/70/E-92:4 p.

MOTAIS, R Les mécanismes branchiaux des échanges ioniques
1970   chez les téléostéens en rapport avec la salinité: aspect biochimique. *Bull. Infs scient. tech. Commt. Energ. atom.*, (146):3–19.

NOREN, K and WESTÖÖ, G Methylmercury in fish. *Var. Föda*,
1967   19:13–24.

ÖBERG, K E The reversibility of the respiratory inhibition in
1961   gills and the ultrastructural changes in chloride cells from the rotenone poisoned marine teleost, *Gadus callarias* L. *Expl Cell. Res.*, 24:163.

ÖBERG, K E The reversibility of the respiratory inhibition in gills
1964   and the ultrastructural changes in chloride cells from the rotenone poisoned marine teleost, *Gadus callarias* L. *Expl Cell Res.*, 36:407.

ÖBERG, K E On the principal way of attack of rotenone on fish.
1965   *Ark. Zool.* 18(11):217–20.

ÖBERG, K E On the mode of action of rotenone upon the cell
1967   respiration and the structural pattern of the chloride cells on the gills of the rotenone poisoned marine teleost, *Gadus callarias* L. *Acta Univ. uppsal.*, 85:1–17.

ÖBERG, K E The reversibility of the respiratory inhibition in gills
1967a and the ultrastructural changes in chloride cells from the rotenone poisoned marine teleost, *Gadus callarias* L. *Expl Cell. Res.*, 45:590–602.

PAOLETTI, A Les "filter-feeders" dans le milieu marin: leur
1970   influence positive et négative sur la salubrité des plages polluées. Document présenté à la Conférence Technique de la FAO sur la pollution des mers et ses effets sur les ressources biologiques et la pêche, Rome, Italie, 9–18 décembre 1970, FIR:MP/70/E-27:7 p.

PORTMANN, J E Discussion of the results of acute toxicity tests with
1970   marine organisms using a standard method. Paper presented to FAO Technical Conference on Marine Pollution and its effects on Living Resources and Fishing, Rome, Italy, 9–18 December 1970, FIR:MP/70/E-31:13 p.

PORTMANN, J E Toxicity-testing with particular reference to oil-
1970a removing materials and heavy metals. Paper presented to FAO Technical Conference on Marine Pollution and its effects on Living Resources and Fishing, Rome, Italy, 9–18 December 1970, FIR:MP/70/E-33:7 p.

REISH, J A critical review of the use of marine invertebrates as
1970   indicators of varying degrees of marine pollution. Paper presented to FAO Technical Conference on Marine Pollution and its effects on Living Resources and Fishing, Rome, Italy, 9–18 December 1970, FIR:MP/70/R-9:11 p.

RISSANEN, K, ERKAMA, J and MIETTINEN, J K Experiments on
1970   microbiological methylation of mercury (2+) ion by mud and sludge in anaerobic conditions. Paper presented to FAO Technical Conference on Marine Pollution and its effects on Living Resources and Fishing, Rome, Italy, 9–18 December 1970, FIR:MP/70/E-61:4 p.

RYLAND, J S Observations on the development of larvae of
1966   *Pleuronectes platessa* L. in acquaria. *J. Cons. perm. int. Explor. Mer*, 30(2):177–95.

SCHAFER, R Variations in free amino acid composition in some
1968   invertebrates subjected to specific types of pollutants. *Revue int. Océanogr. Méd.*, 10:55–67.

SIDHU, G S, *et al.* Nature and effects of a kerosene-like taint in
1970   mullet (*Mugil cephalus*). Paper presented to FAO Technical Conference on Marine Pollution and its effects on Living Resources and Fishing, Rome, Italy, 9–18 December 1970, FIR:MP/70/E-39:9 p.

SMITH, L S and BELL, G R A technique for prolonged blood
1964   sampling in free swimming salmon. *J. Fish. Res. Bd Can.*, 21(4):711–7.

SMITH, L S and NEWCOMB, T W A modified version of the Blazka
1970   respirometer and exercise chamber for large fish. *J. Fish. Res. Bd Can.*, 27(7):1321–4.

SMITH, L S, *et al.* Physiological changes experienced by Pacific
1970   salmon migrating through a polluted urban estuary. Paper presented to FAO Technical Conference on Marine Pollution and its effects on Living Resources and Fishing, Rome, Italy, 9–18 December 1970, FIR:MP/70/E-40:7 p.

SPRAGUE, J B and DRURY, D E Avoidance reactions of salmonid
1969   fish to representative pollutants. *Adv. Wat. Pollut. Res.*, 4(1):1–11.

SPRAGUE, J B, ELSON, P F and SAUNDERS, R L Sublethal copper-
1965   zinc pollution in a salmon river: a field and laboratory study. *Int. J. Air. Wat. Pollut.*, 9:531–43.

WILSON, K W The toxicity of oil spill dispersants to the embryos
1970   and larvae of some marine fish. Paper presented to FAO Technical Conference on Marine Pollution and its effects on Living Resources and Fishing, Rome, Italy, 9–18 December 1970, FIR:MP/70/E-45:7 p.

WINTER, J E Long-term laboratory experiments on the influence
1970   of ferric hydroxide flakes on the filter-feeding behaviour, growth, iron content in soft parts and mortality in *Mytilus edulis* L. Paper presented to FAO Technical Conference on Marine Pollution and its effects on Living Resources and Fishing, Rome, Italy, 9–18 December 1970, FIR:MP/ 70/E-112.

WURSTER, C F Chlorinated hydrocarbon insecticides and avian
1969   reproduction: how are they related. *In* Chemical Fallout, edited by M. W. Miller and G. G. Berg, Springfield, Ill., C. C. Thomas, pp. 368–89.

The Use of Marine Invertebrates as Indicators of Varying Degrees of Marine Pollution

D. J. Reish

L'utilisation des invertébrés marins comme indicateurs des divers degrés de pollution des mers

C'est Wilhelmi qui, le premier, appliqua au milieu marin en 1916 l'utilisation des invertébrés comme indicateurs de la pollution en s'inspirant des découvertes faites précédemment par Kolkwitz et Mansson en 1908 en ce qui concerne la pollution des aux douces. La définition et l'application aux eaux de mer de la notion d'organis-

Uso de invertebrados marinos como indicadores de los distintos grados de contaminación marina

El uso de invertebrados como indicadores de la contaminación fue aplicado por primera vez al ambiente marino por Wilhelmi en 1916. Siguió los conceptos descubiertos anteriormente por Kolkwitz y Mansson, en 1908, con respecto a las aguas dulces contaminadas. La amplia descripción y aplicación del concepto de organismo

mes indicateurs se poursuivant de nos jours depuis une quinzaine d'années. On a donné la préférence aux espèces d'invertébrés benthiques car elles reflètent non seulement les conditions existant au moment de l'échantillonnage, mais aussi celles qui existaient quelque temps auparavant. Trois à quatre zones ont été choisies pour indiquer divers degrés de pollution: (1) une zone salubre où les peuplements sont normaux et ne se ressentent pas de déversements d'effluents, (2) une zone semi-salubre ou intermédiaire, où l'on note une réduction de la diversité des espèces, (3) une zone polluée dans laquelle deux ou trois espéces seulement sont présentes, l'une d'entre elles étant généralement le polychète *Capitella capitata*, et (4) une zone très polluée chaque fois que les conditions sont si défavorables qu'elles excluent toute vie d'animaux macroscopiques. On a constaté que le polychète *Capitella capitata* prolifère chaque fois qu'il y a une accumulation de matériaux organiques, comme c'est le cas au voisinage des décharges d'égouts. Cette espèce s'est révélée utile comme indicateur de pollution des eaux en Californie, au Japon, en France et en Afrique du Sud.

L'utilisation des invertébrés comme indicateurs des divers degrés de pollution des mers fait l'objet d'une évaluation critique et l'auteur discute son application aux études présentes et futures.

indicador de las aguas marinas empezó hace unos 15 años y ha continuado hasta la fecha. Las especies de invertebrados bénticos han sido favorecidas como indicadores, ya que señalan no sólo las condiciones cuando se recogen las muestras sino también algún tiempo antes. Se han empleado de tres a cuatro zonas para indicar los distintos grados de contaminación: (1) una zona sana en donde la población es normal y no está afectada por descargas de desechos; (2) una zona semi-sana, o intermedia, en la que se observa una reducción de diversidad de especies; (3) una zona contaminada en la que sólo existen dos o tres especies, siendo generalmente una de ellas el poliqueto *Capitella capitata*, y (4) una zona muy contaminada cuando lo está tanto que excluye toda vida macroscópica animal. El poliqueto *Capitella capitata* abunda siempre que se acumula material orgánico, como en las vecindades de las salidas de aguas de alcantarilla. Esta especie se ha usado como indicador en las aguas contaminadas de California, Japón, Francia y Sudáfrica.

El empleo de invertebrados como indicadores de los distintos grados de contaminación marina se evalúa críticamente, examinándose su aplicación a los estudios actuales o futuros.

THE use of marine invertebrates or populations as indicators of pollution dates from the work of Wilhelmi (1916), who found that the polychaete *Capitella capitata* plays a similar role in marine waters of Germany as the oligochaete *Tubifex* does in fresh water. Blegvad (1932) studied benthic communities near outfall sewers in Copenhagen harbour and divided the area into a number of zones. These zones, which have since also been identified in other parts of the world, are: a polluted zone characterized by two or three species (or, if the pollution is severe, a total lack of macroscopic animals), a semi-polluted zone, and an unpolluted zone.

Studies during the past decade have not only extended the previous general findings to other geographical areas, but also included experimental studies dealing with the physiological, biochemical, and pathological effects of pollutants on marine organisms (Fujiya 1965, Reish 1966; Reish and Barnard, 1960).

Invertebrates as indicators

The use of organisms for this purpose is based on the belief that natural, unpolluted environments are characterized by balanced biological conditions and contain a great diversity of plant and animal life with no one species dominating. A natural environment is composed of animals at all trophic levels. Pollutants kill the more sensitive organisms, thus eliminating the enemies of the more tolerant species which in turn increase in numbers. If the amount of waste discharge is increased, a further reduction of species occurs with the surviving species increasing further. If the discharge is great, the accumulated effect may kill the surviving species, leaving an area totally devoid of macroscopic life. Therefore, knowledge of what species are present is of paramount importance in evaluating the effect of waste discharge.

The food cycle is altered by adding waste discharges. Plants are killed and cannot repopulate because of direct or indirect toxic effect of the discharge, the elimination of suitable substrates through silting, the decreasing availability of sunlight from increased water turbidity, or the combination of these factors. With elimination of plant life, herbivores can no longer exist. Carnivores may be eliminated directly and also by the absence of herbivores. Further increases in the amount of waste discharges eliminate the filter feeders leaving only the detritus feeders. Finally, if pollution is severe all detritus feeders are killed. The degree of pollution can be evaluated by knowing the food habits of the particular population.

Principles have been learned by correlating different populations of organisms to degrees of pollution in fresh-water environment chiefly through such workers as Kolkwitz and Mansson (1908), Forbes and Richardson (1913), Patrick (1949) and by Gaufin and Tarzwell (1952).

Some case studies

Filice (1954, 1954a) sampled the benthos in the Castro Creek region of San Pablo Bay, part of the San Francisco Bay system, which received industrial and domestic wastes. It was possible to separate groups of stations into (1) normal, (2) near industrial outfalls, (3) at industrial outfalls, (4) near domestic outfalls, and (5) at domestic outfalls. With all types of waste discharge he was able to divide the benthos into three zones on the basis of distance from the outfall. The natural environment was characterized by 39 species—an average of 6.5/10 l of dredged material. The marginal or semi-polluted zone near industrial outfalls had 12 species with an average of 3.0/10 l, whereas the same zone near domestic outfalls had 30 species and averaged 11.5 species/10 l. The polluted zone had three species at industrial outfalls and averaged 1/10 l and at the domestic outfall five species and an average of 2/10 l. On the basis of abundance, the crab *Rhithropanopeus harrisii* indicates the polluted zone both at the industrial and domestic outfalls and the polychaete *Capitella capitata* at only the domestic outfalls.

A biological, chemical, and physical survey was made at 54 stations in Los Angeles–Long Beach Harbours, California, three times during 1954 (Reish 1959). These harbours constitute one body of water oceanographically. There are about 235 outfalls in the harbours carrying wastes from industries, domestic, and storm drains. It was possible to divide the harbours into five zones based on bottom conditions, species composition, and water characteristics. These data are summarized in Table 1. Proceeding from healthy or normal bottom to very polluted bottom there is reduction and eventual exclusion of polychaetes and other invertebrates, the dissolved oxygen concentration decreases, sulphide odour in the substrate becomes more prevalent, and the organic carbon content of the substrate increases. Each of the zones is characterized by one or more polychaete species except at the very polluted stations which lack macroscopic animal life. The polychaete *Capitella capitata* is found especially in the vicinity of either fish cannery or domestic outfall sewers.

TABLE 1. SUMMARY OF BIOLOGICAL, CHEMICAL, AND PHYSICAL CHARACTERISTICS OF FIVE ECOLOGICAL AREAS OF LOS ANGELES—LONG BEACH HARBOURS

Characteristic	Healthy bottom	Semi-healthy bottom *I*	Semi-healthy bottom *II*	Polluted bottom	Very polluted bottom
Dominant species of polychaete	*Tharyx parvus*, *Cossura candida*, *Nereis procera*	*Polydora paucibranchiata*, *Dorvillea articulata*	*Cirriformia luxuriosa*	*Capitella capitata*	No animals
Number of animal species (average)					
Polychaetes	7	5	5	3	0
Non-polychaetes	3	2	2	2	0
Dissolved oxygen (ppm) (median)					
Surface	6.0	2.5	2.5	3.5	1.6
20-ft depth	6.0	3.2	3.2	3.5	2.2
pH (median)					
Surface	7.8	7.3	7.4	7.6	7.5
20-ft depth	7.8	7.4	7.6	7.6	7.5
Substrate	7.2	7.2	7.2	7.3	7.1
Nature of substrate (in order of importance)	grey mud, black mud, black sulphide mud	black sulphide mud, grey clay, sand, and mud, black mud	black sulphide mud, grey clay, black mud	black sulphide mud	black sulphide mud
Organic carbon of substrate (%) (median)	2.5	2.0	2.7	2.7	3.4

A study of possible biological indicators of marine pollution in Los Angeles Harbour was approached differently by Crippen and Reish (1969). They reported on a 15-month quantitative study of fouling organisms attached to floating boat docks. Emphasis was placed on the polychaetous annelids at five selected stations from the outer to inner harbour area. The distribution of these species was correlated with dissolved oxygen, turbidity, temperature, chlorinity, nitrates, nitrites, and phosphates. Four of the stations were characterized by the presence of at least one unique polychaete species; polychaetes were absent from the fifth station which was in a highly polluted area. The number of polychaete species, the concentration of dissolved oxygen, nitrates, nitrites, and chlorinity decreased as the degree of pollution increased. Phosphates and turbidity of the water increased with an increase in pollution. *Capitella capitata* comprised an increasing percentage of the total polychaete population from the outer harbour to the inner harbour.

The use of fouling organisms from floating boat docks as indicators of pollution offers certain advantages in the study of marine pollution in protected waters. It greatly facilitates sampling with respect to time, equipment, and manpower.

Kitamori and his associates (1958, 1959, 1960, 1963) sampled benthic communities in coastal waters to study the extent of pollution. In all regions (Fukuyama Inlet, Mihara Bay, Osaka Bay, Kanzaki River, and the Seto Island Sea Area) the polychaete *Capitella capitata* was the dominant species. In general they found the number of species decreased, but the total of the remaining species increased with increased pollution. In some instances it was possible to distinguish four zones (1) characterized by the eel grass *Zostera*, (2) by the alga *Caulerpa*, (3) typical of off-shore waters, and (4) by *Capitella capitata*.

Pollution at Marseille, was studied by Gilet (1960) and Bellan (1967). Four zones were described by Bellan from Marseille Harbour: (1) maximum zone lacking macroscopic animals, (2) a polluted zone with the polychaete *Capitella capitata* and *Scolelepis fuliginosa* dominating with *Nereis caudata*, *Staurocephalus rudolphii*, and *Audounia* sp. present locally, (3) a subnormal zone with

the molluscans *Corbula gibba*, *Thysira flexuosa* and a rich polychaete fauna dominating, and (4) a clear water zone.

Validity of such indicators of pollution

Benthic organisms reflect not only conditions at time of sampling but also conditions for some time previously. Since many are either attached forms or burrowers in the substrate and hence relatively immobile, they are therefore subjected to the same conditions of pollution. This means that their condition indicates the community structure at a particular locality over several sampling periods. It would therefore be of value to know the detailed benthos conditions prior to the onset of pollution but this knowledge is generally inadequate.

Experience in the East Basin–Consolidated region of Los Angeles Harbour (Anon 1952, Reish 1957, 1959) illustrates what occurred when the bottom sediments of a heavily polluted region were dredged and accumulated sludge removed. This shows the rapidity of deterioration. Oil refinery wastes were emptied into the Dominguez Channel, flowed into Consolidated Slip, then into the East Basin before entering the Main Channel of Los Angeles Harbour. Benthic animals were present only in the outer limits of East Basin prior to dredging. Dredging to provide a deeper channel removed the accumulated oil refinery wastes from the bottom thus exposing uncontaminated sediments. A benthic survey taken four months after completion revealed 10 species where previously were only five. The very polluted bottom, which lacked animals, was restricted to that region which had not been dredged. Ten months later the number had decreased to five and a year later had fallen to two.

An attack of red-tide in Alamitos Bay reduced the number of species taken at 10 stations from 56 before the outbreak to 29 afterwards and changed the substrate from grey to black as well as creating a sulphide odour. The die-off lowered dissolved oxygen concentration from the range of 5.0 to 7.5 down to 0.1 ppm. Recovery to former conditions took some six months. The polychaete *Capitella capitata* (present formerly in small numbers) suddenly appeared in a peak of 1200 specimens two months later and then fell to its previous level 8 months after the outbreak as recovery to former conditions proceeded.

Before the removal of a sewage outfall in estuarine Raritan River (N.J.) (Dean and Haskin 1964) 17 marine species of invertebrates were collected and no fresh-water forms. Afterwards the number increased to 28 and fresh-water forms increased to 14.

The number of case studies available for analysis is limited primarily because we do not know what organisms were present before discharge began. The indications are, however, that change in the marine environment is reflected in the species composition and population of the bottom-dwelling animals. The presence and number of *Capitella capitata* serves as a clue to indicate changes are occurring.

Effects of environmental variables on indicator species

Oxygen is required by all animals and aquatic animals depend upon dissolved oxygen. The minimum concentration to support fish and aquatic life is about 5.0 ppm. Dissolved oxygen concentration is probably the most universally applied water quality criterion.

While each species has its own specific requirements for oxygen and other necessities, it can tolerate some departure from them. Since tolerance limits are known for only a few animals, we cannot tell when waste is added to the environment which members of the local fauna will be able to tolerate the changes.

Laboratory studies are necessary, but it is uncertain how far the results of such studies can be applied to the extremely complicated situation that exists in the ocean. Further complication is introduced by seasonal and other variations in the physiological responses of animals and variation in the quality of the seawater itself. Nevertheless, physiological investigations are vital although results must be interpreted cautiously.

The effects of varying concentrations of dissolved oxygen on four species of polychaetes which have been used as indicators of varying degrees of pollution were determined by Reish (1966). *Nereis grubei* was used as an indicator of normal or healthy conditions, *Neanthes arenaceodentata* and *Dorvillea articulata* as indicators of semi-polluted or semi-healthy conditions, and *Capitella capitata* as an indicator of polluted conditions. The amount of dissolved oxygen in 100 ml of sea water in a 500 ml sealed Erlenmeyer flask was varied by flushing the overlying air with N_2 gas. Each experiment ran for 28 days, and the results were expressed as 28-day TLm values. The 28 TLm was 2.95 ppm for *Nereis grubei*, an indicator of healthy zone, 0.90 and 0.65 ppm for *Neanthes arenaceodentata* and *Dorvillea articulata*, respectively, the indicators for semi-healthy zone, and 1.5 ppm for *Capitella capitata*, the indicator for the polluted zone. In general, these data indicate a relationship between the type of indicator and the concentration. However, in both instances the semi-healthy species were able to tolerate lower dissolved oxygen concentrations than the indicator of the polluted zone. It is possible that some factors other than dissolved oxygen limit the distribution of the semi-healthy indicators.

By using the same methods of dissolved oxygen control, Raps (1970) was able to demonstrate haemoglobin compensation experimentally in a laboratory population of the polychaete *Neanthes arenaceodentata*. The total haemoglobin content remained unchanged at dissolved oxygen levels of 7.3 to 4.2 ppm; however, below 4.2 ppm haemoglobin compensation occurred and the amount produced was inversely related to the dissolved oxygen concentration. This species has a 28-day TLm of 0.95 ppm dissolved oxygen and a concentration of about 4.2 ppm represents its normal dissolved oxygen environment (Reish 1966).

Since dissolved oxygen is essential its relationship to reproduction has been studied in two varieties of indicator species (Reish and Barnard 1960, Davis 1969). *Capitella capitata* were placed in plastic traps and suspended for 54 days in Los Angeles Harbour in areas of varying degrees of dissolved oxygen. A male and female were placed in each trap which was covered with a mesh cloth. The animals were fed periodically and the specimens were examined for survival, feeding, and reproduction. Oxygen levels were less than 2.5 ppm for survival, 2.9 ppm for feeding, and over 3.5 ppm for reproduction.

A laboratory study on the effects of reduced dissolved oxygen levels on egg production in *Neanthes arenaceodentata* was done by using the same technique of dissolved oxygen control (Davis 1969). The number of eggs produced by a female was the same at dissolved oxygen concentrations from 5.9 to 3.0 ppm. It was 50 per cent at 2.0 ppm, 35 per cent at 1.5 ppm, and 10 per cent at 1.0 ppm. Egg growth was slower between 1.5 and 3.0 ppm, and the eggs failed to develop between 1.0 and 1.5 ppm.

Studies of Kraft pulp mill wastes on the cytochemistry of the Japanese oyster *Ostrea gigas* by Fujiya (1965) showed that cellular changes induced in the tissue of the oyster were related to the concentration of the waste discharge as follows: (1) a decrease in calcium absorption as indicated by Ca^{45}, (2) death of cells in the stomach, intestine, and gills, (3) a decrease in the amount of RNA (ribose nucleic acid) in the stomach, intestine, and digestive glands, (4) and an increase in mucus secretion by the digestive system and gills.

Schafer (1961) detected changes in the free amino acid composition of species of mussels, abalones and crabs collected from polluted environments. The significance of these changes is not known but they may indicate a disturbance of normal metabolism.

Use of monitoring

Most information gathered on the effects of pollutants on marine invertebrates has been academic, fragmentary, and of little direct use to public officials concerned with pollution control and abatement. Since one primary purpose in pollution control is to protect and improve the environment, the most useful application of the indicator organism concept is monitoring a region or an outfall by determining the distribution of animal associations or perhaps more subtly by noting physiological or biochemical changes to the organisms.

Although most natural unpolluted waters have a diverse fauna without a single dominant species and a polluted environment has few species individually in great abundance, lack of information about the normal fauna in our area often hampers monitoring of the effects of discharged wastes. Because of this, pollution biologists have favoured the indicator organism concept. This has its difficulties, however, for the indicator species may be found in abundance in its own natural environment in the absence of pollution (Barnard 1970). Thus *Capitella capitata* is present with few accompanying species in the upper parts of non-polluted Alamitos Bay (Stone and Reish 1965). Large numbers of *Capitella* is not an infallible guide to pollution damage but merely warrants further examination.

Whenever a new outfall sewer is contemplated a prior study of the biological, chemical and physical characteristics of the region should be undertaken as part of the construction cost. The operators should then be required to monitor the area periodically after commencement of operations as an integral cost of plant operation. Since this practice has been initiated only in recent years, the supervision and critical evaluation of its merit, frequency of sampling, data collected, etc., should be under the jurisdiction of a state or federal agency. The benthic communities around a new outfall sewer should not be allowed to deteriorate. Initially some data may be of little or no value; but since we really have very little background experience in monitoring marine areas we should consider many parameters, then eliminate those of minimal value after a period of time. The possible deterioration of the benthic community at Point Loma is an example to consider. *Capitella capitata* was found for the first time in the entire Point Loma region in 1965, or less than two years after the sewage treatment plant began to discharge into the ocean. Unfortunately, this area has not been sampled since 1965. The chief value of marine invertebrates as indicators is in serving as a danger signal.

There are many additional measures of the effects of waste discharges besides survival. These include feeding, reproduction, haemoglobin concentration, cytochemical and biochemical changes. An additional technique for monitoring an outfall which would give us an earlier warning of a possible problem would be to measure physiological changes in the organisms being stressed by it. But it is imperative that whatever species of organism is used, identification must be precise.

Summary

It may not be possible to use the same species of organisms as indicators of polluted waters throughout the world because many species have either limited geographical or ecological distribution. However, the most widely used marine indicator *Capitella capitata*, is cosmopolitan in distribution. Furthermore, two of the polychaetes used in the Los Angeles–Long Beach Harbours study were also used in Marseille. The fact that the same assemblage of species would not necessarily be a universal indicator of polluted or semi-polluted conditions should not be a deterrent to the indicator concept being used.

References

BARNARD, J L Benthic ecology of Bahia de San Quintin, Baja
1970   California. *Smithson. Contr. Zool.*, (44): 60 p.
BELLAN, G Pollution et peuplements benthiques sur substrat
1967   meuble dans la région de Marseille. Deuxième partie. L'ensemble portuaire marseillais. *Revue int. Océanogr. méd*, 8: 51–95.
BLEGVAD, H Investigations of the bottom fauna at outfalls of drains
1932   in the Sound. *Rep. Dan biol. Stn.*, (37): 1–20.
CRIPPEN, R W and REISH, D J. An ecological study of the poly-
1969   chaetous annelids associated with fouling material in Los Angeles Harbor with special reference to pollution. *Bull. Sth. Calif. Acad. Sci.*, 68(3): 170–87.

DAVIS, W R Oogenesis and its relationship to dissolved oxygen
1969   suppression in *Neanthes arenaceodentata* (Polychaeta: Annelida). Thesis, California State College, Long Beach, California, 62 p.
DEAN, D and HASKIN, H H Benthic repopulation of the Raritan
1964   River estuary following pollution abatement. *Limnol. Oceanogr.*, 9(4): 551–63.
FILICE, F P An ecological survey of the Castro Creek area in San
1954   Pablo Bay. *Wasmann J. Biol.*, 12(1): 1–24.
FILICE, F P A study of some factors affecting the bottom fauna of
1954a   a portion of the San Francisco Bay estuary. *Wasmann J. Biol.*, 12(3):257–92.
FORBES, S A and RICHARDSON R E Studies on the biology of the
1913   upper Illinois River. *Bull. Ill. St. Lab. Nat. Hist.*, (9): 481–574.
FUJIYA, M Physiological estimation on the effects of pollutants
1965   upon aquatic organisms. *Adv. Wat. Pollut. Res.*, 2(3):315–31.
GAUFIN, A R and TARZWELL, C M Aquatic invertebrates as indica-
1952   tors of stream pollution. *Publ. Hlth Rep., Wash.*, 67: 57–64.
GILET, R Water pollution in Marseilles and its relation with flora
1960   and fauna. *In* Waste disposal in the marine environment, edited by E. A. Pearson, London, Pergamon Press, pp. 39–56.
KITAMORI, R Studies on the benthos communities of littoral areas
1963   in the Seto-Inland Sea and the adjacent waters. *Bull. Naikai reg. Fish. Res. Lab.*, (21): 90 p.
KITAMORI, R and FUNAE, K The benthic community in polluted
1960   coastal waters. 5. Hiro Bay. *Bull. Naikai reg. Fish. Res. Lab.*, (13):11–8.
KITAMORI, R and KOBAYASHI, S The benthic community in polluted
1958   coastal waters. 1. Fukuyama Inlet. *Bull. Naikai reg. Fish. Res. Lab.*, (11): 6 p.
KITAMORI, R, KOBAYASHI, S and NAGATA, K The benthic commu-
1959   nity in polluted coastal waters. 2. Mikara Bay. *Bull. Naikai reg. Fish. Res. Lab.*, (12): 201–14.
KITAMORI, R and FUNAE, K The benthic community in polluted
1959   coastal waters. 3. Osaka Bay. *Bull. Naikai reg. Fish. Res. Lab.*, (12): 215–22.
KITAMORI, R and KOBE, Z The benthic community in polluted
1959   coastal waters. Kanzaki River. *Bull. Naikai reg. Fish. Res. Lab.*, (12): 223–226.
KOLKWITZ, R and MANSSON, M Oekologie der pflanzlichen Sapro-
1908   bien. *Ber. dt. bot. Ges.*, 26: 505–19.
PATRICK, R A proposed biological measure of stream conditions
1949   based on a survey of the Conestoga Basin, Lancaster County, Pennsylvania. *Proc. Acad. Nat. Sci., Philad.*, 101: 227–341.
RAPS, M E The effects of dissolved oxygen on the hemoglobin
1970   levels of the polychaetous annelid, *Neanthes arenaceodentata* (Moore). Thesis, California State College, Long Beach, California, 58 p.
Reish, D J An ecological study of lower San Gabriel River,
1956   California, with special reference to pollution. *Calif. Fish Game*, 42(2): 51–61.
REISH, D J Effect of pollution on marine life. *Ind. Wastes*, 2:
1957   114–8.
REISH, D J An ecological study of pollution in Los Angeles–Long
1959   Beach Harbors, California. *Occ. Pap. Allan Hancock Fdn.*, (22): 119 p.
REISH, D J The use of the sediment bottle collector for monitoring
1961   polluted marine waters. *Calif. Fish Game*, 47(3):261–72.
REISH, D J Mass mortality of marine organisms attributed to the
1963   "red tide" in southern California. *Calif. Fish Game*, 49(4): 265–70.
REISH, D J Relationship of polychaetes to varying dissolved
1966   oxygen concentrations. Paper presented at Third International Conference Water Pollution Research, Munich, Sect. 3, paper 10:10 p.
REISH, D J and BARNARD, J L Field toxicity tests in marine waters
1960   utilizing the polychaetous annelid *Capitella capitata* (Fabricius). *Pacif. Nat.* 1(21–22): 1–8.
SCHAFER, R Effects of pollution on the free amino acid content of
1961   two marine invertebrates. *Pacif. Sci.*, 15(1): 49–55.
STONE, A N and REISH, D J The effect of freshwater run off on a
1965   population of estuarine polychaetous annelids *Bull. 8th Calif. Acad. Sci.* 64(3): 111–9.
TURNER, C H and STRACHAN, A R The marine environment in the
1969   vicinity of the San Gabriel River mouth. *Calif. Fish Game*, 55(1): 53–68.
TURNER, C H, EBERT, E E and GIVEN R R The marine environment
1968   offshore from Point Loma, San Diego County. *Fish Bull. Calif.*, (140): 85 p.
WILHELMI, J Übersicht über die biologische Beurteilung des Was-
1916   sers. *Ges. naturf. Freunde Berl.*,: 297–306.
ANON. Los Angeles–Long Beach Harbors pollution survey. *Rep.
1952   Calif. reg. Wat. Pollut. Control Bd*, (4): 43 p.

Utilisation de la Chaine Trophodynamique dans l'Etude de la Toxicité des Rejets d'Eaux Chimiquement Polluées

M. Aubert, J. Aubert,
B. Donnier et M. Barelli*

Utilization of the trophodynamic chain in the study of the toxicity of chemically-polluted waste water

The authors have performed a general study of chemically-polluted waste along the coasts of France. They examined the effects of such waste on a marine food chain (phytoplankton, zooplankton, fish and the higher mammals) and in this way determined, first the toxicity thresholds (lethal doses) of each type of polluted water with respect to the constituents of a marine biological chain and second by following the successive passages between the various levels of this biological chain, the transmitted or induced toxicity due to nutrient concentration factors. By this method, which is employed for purposes of comparison with conventional techniques for the chemical measurement of the constituents of these waste waters, it is possible to identify more readily the influence of such pollution on the elements of the marine environment and the toxic effects which it may have on the final consumer in the trophodynamic chain.

The studies covered over a hundred types of industrial wastes and included testing effluent from the petrochemical industry, wood chemistry, the metallurgical, plastics and ceramics industries, from cement works, canneries, etc. The results allow a more realistic approach to the dangers to the marine environment and its nutritional utilization as a result of pollution due to chemically-polluted water.

Utilización de la cadena trofodinámica para el estudio de la toxicidad de las aguas de descarga contaminadas químicamente

Los autores han realizado un estudio general de las aguas de descarga químicamente contaminadas a lo largo de las costas francesas, estudiando sus efectos en una cadena trófica marina (fitoplancton, zooplancton, peces, mamíferos superiores) y determinando, por un lado, los umbrales de toxicidad (dosis letales) de cada una de esas aguas contaminadas para los componentes de una cadena biológica marina, y, por otro, la toxicidad transmitida o inducida (siguiendo los pasos sucesivos por los diversos escalones de la cadena biológica) debida a los factores de concentración nutricional. Esta metodología, que se compara con las técnicas clásicas de medición química de los elementos constitutivos de las aguas residuales, permite apreciar mejor la influencia de la contaminación química en los diversos elementos del medio marino y las consecuencias tóxicas que puede tener para el consumidor final de la cadena trofodinámica.

Se han estudiado más de un centenar de tipos de aguas de descarga industriales, examinando aguas procedentes de la industria petroquímica, xiloquímica, metalúrgica, plástica, del cemento, cerámica, conservera, etc. En conjunto, los resultados obtenidos permiten un enfoque más realista de los peligros que corre el medio marino y del riesgo que representa su aprovechamiento para la nutrición, por razón de la contaminación derivada de las aguas químicamente contaminadas.

UNE rapide vue d'ensemble des techniques d'analyse nous a permis de constater que, hormis quelques cas bien précis, l'analyse chimique des polluants, dont souvent on ignore *a priori* la nature, est une tâche difficile donnant la plupart du temps des résultats imparfaits; de plus, l'utilisation de méthodologies parfois très lourdes interdit leur emploi en analyse de routine. Lorsqu'il existe un procédé simple d'analyse, les mesures effectuées *in situ* ou au laboratoire permettent d'évaluer avec précision l'étendue et l'importance de la pollution. Cependant, pour ce type de mesures il est souvent préférable d'utiliser des traceurs colorés ou radio-actifs à vie très courte—comme cela a été fait dans le cas de pollution bactérienne (Aubert, 1966)—car l'on obtient alors une plus grande précision et une meilleure sélectivité dans les mesures. Mais l'imperfection majeure de l'étude des pollutions par l'analyse chimique réside dans le fait que si les résultats permettent de mesurer la dilution et la diffusion de l'agent chimique *in situ*, ils n'apportent aucun renseignement sur la toxicité effective de ces corps, c'est-à-dire sur l'aspect biologique du problème et sur les conséquences effectives de cette pollution tant pour l'homme que pour les animaux et les végétaux marins.

C'est dans cet ordre d'idées que nous avons orienté et affiné nos recherches beaucoup plus vers l'aspect biologique que vers celui purement chimique des pollutions. Dans ce but, nous avons réalisé dans notre Institut une section particulière que nous appelons "compteur biologique", dont une importante partie est orientée vers l'étude de la pollution chimique des eaux marines; notre travail a consisté essentiellement à étudier la pollution des rivages français par les effluents industriels et les conséquences de tels rejets sur la faune et la flore marines et sur le consommateur terrestre.

Pour ce faire nous avons réalisé des élevages d'organismes faciles à cultiver en masse si cela est nécessaire et pouvant représenter une chaîne alimentaire, qui, partant des micro-organismes marins, aboutit aux mammifères supérieurs.

Description de la chaîne alimentaire-test

Le premier échelon est représenté par des organismes autotrophes. Comme élément phytoplanctonique, nous avons choisi une diatomée néritique dont nous avons étudié en détail depuis plusieurs années les courbes de croissance et les conditions d'élevage: *Asterionella japonica* (Cleve). Cette micro-algue fréquemment rencontrée sur nos côtes pendant la saison froide se cultive aisément en milieu de type Provasoli. Ce milieu de culture est préparé à partir d'eau de mer naturelle vieillie à l'obscurité et stérilisée, qui reçoit un enrichissement stérile à la fois minéral et organique par l'adjonction de mélange E. S. Provasoli à 2 pour cent.

Les cultures sont effectuées en erlenmeyers de 250 à 500 ml dans une salle de cultures maintenue à une température constante ($18° + 1$) et éclairée selon un rythme nycthéméral (12 heures de lumière—12 heures d'obscurité).

Cette diatomée s'est révélée impropre à la nutrition de certains stades larvaires de l'échelon zooplanctonique; nous y avons adjoint dans ce but précis une micro-algue unicellulaire de la famille de coccomyxacées: *Diogenes* sp. (Pennington) isolée à partir de pêches effectuées dans nos eaux (Aubert et ses collaborateurs).

Ces micro-algues sont douées d'une grande vitalité et d'une forte capacité de reproduction; dans la plupart des cas, des cultures effectuées en tubes à essais sont suffisantes pour obtenir une masse de ces organismes compatibles avec les besoins de la chaîne alimentaire. Ces cultures ont été réalisées sur milieu S. T. P. de Provasoli dans lequel leur rendement est supérieur au milieu E.S. à 2 pour cent. Les diverses espèces constituant la collection de notre laboratoire sont cultivées en permanence d'une

*Centre d'Etudes et de Recherches de Biologie et d'Océanographie Médicale, 06 Nice, France.

part telles qu'elles ont été isolées et d'autre part en souche parfaitement axénique, ce qui permet, lorsqu'il existe des phénomènes expérimentaux de dégradation des toxiques, de séparer ce qui est occasionné par les micro-algues elles-mêmes de ce qui incombe aux bactéries marines (Aubert, 1968; Aubert et ses collaborateurs, 1968). Les souches de micro-algues sont rendues axéniques par la technique des lavages en cascade de Provasoli; pour les crustacés planctoniques ce résultat est obtenu par trempage des œufs dans diverses solutions antibiotiques. Dans tous les cas, le résultat doit être vérifié par des tests de stérilité effectués sur milieu de Oppenheimer-ZoBell. Les expérimentations de routine sont effectuées sur des souches non axéniques afin de se placer dans des conditions voisines de celles existant en mer.

Notre deuxième échelon, c'est-à-dire l'échelon zooplanctonique, est constitué par un crustacé résistant d'élevage facile: *Artemia salina* (branchiopode anostracé). Les œufs d'*Artemia salina* sont introduits dans une ampoule à décanter contenant de l'eau de mer; une certaine agitation initiale est nécessaire pour dissocier les amas d'œufs; seuls les œufs descendant au fond du récipient seront utilisés dans la suite des expériences, car ceux qui sont plus légers que l'eau sont morts et doivent être rejetés d'emblée. Parmi les œufs restants le pourcentage d'éclosion est de 60 à 70 pour cent dans les conditions optimales d'élevage. Dans l'étude des phénomènes de toxicité, cet inconvénient est levé par la comparaison des essais avec leurs témoins homologues. Dans les premiers stades de croissance, ce crustacé est alimenté par les nutrilites présents dans le milieu S. T. P. de Provasoli, puis il se nourrit de micro-algues. Enfin, lorsqu'il atteint l'âge adulte il est capable d'ingérer des éléments phytoplanctoniques de taille plus importante, tel *Asterionella japonica*. La culture s'effectue à 25°C en erlenmeyers ou en tubes à essais éclairés selon un rythme nycthéméral. Ce crustacé résiste très bien aux variations de salinité, ce qui permet de se rapprocher des conditions naturelles de dilution des eaux de mer par les eaux douces polluantes.

Le troisième échelon de la chaîne est constitué par un cyprinidé *Carassius auratus*, capable de se nourrir à partir de l'échelon zooplanctonique. L'élevage des poissons se fait dans des bacs en verre de 5 l; l'aération du milieu se fait par barbotage selon la technique classique de l'aquariologie.

Le stade ultime de notre chaîne biologique est constitué par de petits mammifères nourris à partir de l'échelon directement inférieur (poissons ou coquillages selon le cas); il s'agit de souris de laboratoire, *Mus musculus*.

Méthodologie d'emploi pour les produits toxiques de composition connue

La chaîne biologique test a été utilisée en premier lieu pour la recherche de la toxicité due à des produits chimiques purs tels que:

—des détergents commerciaux d'usage courant: détergents anioniques à base d'alkyl aryl sulfonate de soude, détergents anioniques polyvalents à base de sulfonate, détergents anioniques de laboratoire à base de lauryl sulfate de soude,
—des insecticides divers,
—des hydrocarbures.

Nous n'exposerons pas les résultats de ces expériences qui ont déjà été publiés antérieurement (Aubert *et al.*, 1969a).

Application de la méthode à l'étude de la toxicité des rejets d'eaux chimiquement polluées

Les échantillons d'eaux usées ont été prélevés dans des rejets se déversant en mer, soit directement, soit par l'intermédiaire du réseau d'égouts urbains, soit par l'intermédiaire de ruisseaux, qui, compte tenu de leur surcharge en eaux résiduaires, n'ont plus qu'un caractère d'égout, tel l'Huveaune qui débouche dans la rade de Marseille après avoir collecté les effluents de nombreuses industries. Quelques prélèvements effectués plus à l'intérieur des terres ont été destinés à caractériser certains types d'industrie. Les prélèvements ont été effectués soit à l'intérieur, soit à l'extérieur des enceintes industrielles, au débouché des émissaires.

Pour rendre compte de la toxicité globale d'eaux rejetées au même endroit par plusieurs usines, nous avons fait des prélèvements sur différents cours d'eau. Pour les plus importants, un échantillon moyen a été réalisé sur 10 ou 24 heures. Pour chaque étude, nous prélevons 10 l d'eau dans des flacons en polyéthylène. Ceux-ci sont stockés à −40°C de façon à arrêter les réactions chimiques ou bactériennes, puis ils sont dégelés au moment de l'utilisation.

Notre but étant essentiellement une étude de la toxicité et les analyses d'eaux résiduaires nécessitant des méthodes et des appareillages complexes, nous nous sommes contentés d'analyses succinctes au moment de l'utilisation des eaux:

—pH mesuré à l'aide d'un pH mètre,
—densité à 20°C, mesurée à l'aide de densimètres gradués au 1/1000,
—extrait sec obtenu comme suit: l'eau est homogénéisée à l'aide d'un broyeur à ultra-sons, puis concentrée sous vide dans un évaporateur Buchi; le résidu est séché dans une étuve à 105°C, et pesé.

Recherche des seuils de toxicité

Nous adopterons la définition suivante: le seuil de toxicité est la concentration de produits chimiques testée pour laquelle l'organisme-test reste à l'état stationnaire, c'est-à-dire dont le développement est arrêté sans pour autant qu'apparaisse la destruction complète de la culture ou de l'élevage.

Toxicité vis-à-vis du phytoplancton: A partir de l'eau résiduaire à tester, nous effectuons des dilutions successives de 1/2, 1/4, 1/8 etc., jusqu'à 1/256, au moyen d'eau de mer préalablement stérilisée, puis enrichie au taux de 2 pour cent avec le milieu E. S. Provasoli et réensemencée en bactéries marines avec quelques gouttes d'eau de mer filtrée sur filet à plancton. Pour chaque concentration, nous ensemençons séparément trois tubes à essais avec 0,5 ml d'inoculum phytoplanctonique provenant d'une culture en pleine croissance d'*Asterionella japonica* (soit environ 100 000 cellules) dans 10 ml de milieu. La même opération est pratiquée sur trois tubes témoins sans eau chimiquement polluée.

Les diatomées atteintes par l'eau résiduaire subissent une lyse totale, elles sont ainsi facilement identifiables au microscope. Des comptages sont effectués, les 1er, 4e et 9e jours, ces intervalles ayant été choisis en fonction des variations observées dans la courbe de croissance de la diatomée. Ainsi, par comparaison avec les comptages effectués sur les témoins, nous pouvons mettre en évidence le seuil de toxicité.

Toxicité vis-à-vis du zooplancton: Comme précédemment, on réalise les différentes dilutions de l'eau toxique. Pour chacune d'elles, 3 tubes à essai contenant chacun 10 ml reçoivent une vingtaine de larves d'*Artemia salina* prises au stade métanauplius et 1 ml d'une culture de micro-algue destinée à leur servir de nourriture. Ces tubes sont alors placés en salle de cultures et on effectue un comptage journalier pendant 9 jours. Le seuil toxique est calculé par rapport aux 3 tubes témoins.

Toxicité vis-à-vis des poissons: L'élevage des poissons (*Carassius auratus*) est réalisé dans des bacs en verre contenant chacun 4 l d'eau toxique, diluée au taux désiré avec un mélange 1/4 eau de mer 3/4 eau douce. Les bacs sont soumis à une aération continue et reçoivent chacun trois poissons. Le seuil toxique est calculé, comme précédemment, par rapport aux poissons d'un bac témoin contenant uniquement de l'eau de mer diluée au quart.

Etude de la transmission de la toxicité

Transmission de la toxicité par chaîne courte:

Afin de rechercher la transmission possible de la toxicité des différents rejets directement en fin de chaîne, nous avons alimenté le dernier échelon de la chaîne biologique, constitué par des mammifères, avec des poissons intoxiqués.

Les poissons ayant servi à la recherche des seuils toxiques sont tués le dernier jour de l'expérience et mis au congélateur. Ils sont ensuite donnés comme nourriture aux souris à raison d'un poisson par jour et par souris, en commençant par les poissons ayant séjourné dans les eaux toxiques les plus fortement diluées. Pour que cette nourriture soit bien assimilée, les souris sont habituées à ce régime quelques jours avant leur mise en expérience par une alimentation en poisson non intoxiqué additionnée d'un minimum de graines.

Quand une souris meurt pendant l'expérience, elle est remplacée, la souris suivante absorbant les dilutions restantes. Les souris mortes sont stockées au congélateur de façon à ce qu'une autopsie puisse être pratiquée ultérieurement. On reproduit ainsi les possibilités de transmission de toxicité apportées au consommateur final par le simple séjour *in situ* du dernier échelon marin de la chaîne.

Induction de la toxicité sur la totalité de la chaîne biologique:

La 2e phase de notre expérimentation a porté sur la recherche de la toxicité induite par les rejets industriels sur la totalité de la chaîne biologique. Ayant déterminé les seuils de toxicité directe d'un effluent, nous choisissons une concentration inférieure de moitié à la dilution la plus forte ayant entraîné la disparition de l'un des échelons; de cette façon la substance toxique pourra être transmise d'un échelon à l'autre et concentrée au cours des différentes étapes sans que la chaîne biologique soit rompue par la mort d'un des intermédiaires. Si au cours de l'expérimentation, on a observé une atteinte de l'un des échelons inférieurs, l'expérience est reprise à

une dilution plus faible. La dose de départ étant établie, l'expérience consiste à alimenter les échelons supérieurs de la chaîne à l'aide des échelons inférieurs ayant subi une intoxication préalable de 4 jours par les eaux chimiquement polluées.

Notre chaîne biologique va comporter 4 échelons:

—une espèce phytoplanctonique, la micro-algue: *Diogenes* sp. (Pennington), ce choix a été guidé par le fait que les diatomées ont une taille trop importante pour être absorbées par l'échelon immédiatement supérieur,
—une espèce zooplanctonique: *Artemia salina*, prise au stade métanauplius,
—un poisson: *Carassius auratus*,
—un petit mammifère, la souris blanche: *Mus musculus*.

Détail expérimental

Dans un erlenmeyer, nous introduisons environ 200 ml d'une culture stérile de micro-algues et nous ajoutons la quantité d'eau résiduaire nécessaire pour obtenir la dilution désirée. Ce milieu étant mis en salle de cultures, au bout de 4 jours on rajoute le même volume d'eau résiduaire qu'au départ de façon à compenser la quantité absorbée par la micro-algue et éventuellement celle qui s'est dégradée. Le mélange est ensuite réparti dans 24 tubes à essais et on introduit dans chacun d'eux une centaine de larves d'*Artemia salina*, prises au stade métanauplius.

Au bout de 4 jours, 18 poissons sont répartis dans 3 bacs contenant l'eau résiduaire à la dilution choisie pour la chaîne induite. Ils sont alors nourris exclusivement avec les larves intoxiquées d'*Artemia salina* à raison de 2 tubes par bac et par jour. Au bout du quatrième jour, les poissons sont tués et donnés comme nourriture à 3 souris, à raison d'un poisson par souris et par jour.

On reproduit ainsi les conditions dans lesquelles peuvent avoir vécu *in situ*, dans une biomasse polluée, tous les éléments successifs d'une chaîne biologique marine et on évalue ainsi l'induction éventuelle de la toxicité apportée d'une extrémité à l'autre de la chaîne.

Résultats

Le Tableau 1 illustre quelques exemples pratiques de toxicité expérimentale choisis parmi les plus importants et parmi les rejets industriels effectués le long des côtes françaises. Ils sont classés par type d'effluent.

Il est possible de tirer quelques observations de ces résultats:

Pour un certain nombre de rejets, il n'est pas apparu de phénomènes de toxicité directe vis-à-vis des organismes testés; ceci ne signifie pas que ces corps soient inoffensifs car il peut exister des phénomènes de toxicité induite ou des phénomènes de transmission de l'effet toxique par les stades intermédiaires de la chaîne alimentaire. Ce phénomène est observé nettement sur les rejets en provenance des industries de la céramique et sur certains rejets en provenance des raffineries de pétrole.

De même, la plupart des collecteurs généraux testés, en raison de la dilution déjà importante des effluents industriels, n'incommodent pas notablement les différents échelons qui peuvent ainsi accumuler une certaine quantité de polluant. La toxicité n'apparaît alors qu'au

TABLEAU 1. EXEMPLES PRATIQUES DE TOXICITÉ EXPÉRIMENTALE CLASSÉS PAR TYPES DE POLLUANTS

Rejets	Seuils toxiques (dilution en eau de mer)			Toxicité transmise aux souris	
	Asterionella japonica	Artemia salina	Poissons	Transmission directe Dose moyenne d'apparition de toxicité (ml)	Transmission par induction Dose moyenne d'apparition de toxicité (ml)
Parfums synthétiques	1/2	>1/2	>1/2	328	315
Céramiques et sels minéraux	>1/2	>1/2	>1/2	259	315
Céramiques et sels minéraux	1/2	>1/2	>1/2	1 328	301
Céramiques et sels minéraux	1/2	1/8	1/8	663	86
Céramiques et sels minéraux	>1/2	>1/2	>1/2	15,6	172
Sels et Oxydes métalliques	1/4	>1/2	>1/2	36	430
Produits chimiques organiques:					
Acétylène	1/64	1/8	1/16	1 328	31,8
Acide tartrique et divers	1/2	1/8	1/8	>1 328	>516
Polymères synthétiques	1/256	1/128	1/32	36,4	14,7
Produits organiques et minéraux divers	1/16	1/8	1/32	>1 328	>64
Dérivé de l'éthylène	>1/2	>1/2	1/32	>1 328	>64
Dérivés de l'éthylène	>1/2	>1/2	1/8	>1 328	>257
Carbon black	>1/2	>1/2	>1/2	>1 328	>516
Carbon black	>1/2	>1/2	>1/2	1 328	>516
Fongicides—Insecticides	>1/2	>1/2	1/16	>1 328	>128
Engrais	1/4	>1/2	1/8	413	257
Engrais	1/4	>1/2	>1/2	>1 328	>1 028
Engrais	1/16	1/8	1/8	>1 328	>257
Dérivés résiniques	>1/2	>1/2	>1/2	415	>516
Raffineries de pétrole	>1/2	1/8	1/16	>1 328	>64
Raffineries de pétrole	>1/2	1/2	>1/2	134	172
Raffineries de pétrole	>1/2	1/2	>1/2	334	>516
Raffineries de pétrole et Pétrochimie	>1/2	>1/2	>1/2	>1 328	>516
Raffineries de soufre	1/2	>1/2	>1/2	64	>516
Raffineries de soufre	>1/2	>1/2	>1/2	36	430
Raffineries de soufre	>1/2	>1/2	>1/2	>1 328	>516
Soufre et Pigments colorés	>1/2	>1/2	1/2	>1 328	516
Soufre et Industries diverses	>1/2	>1/2	>1/2	>1 328	172
Métallurgie	1/16	1/2	>1/2	>1 328	>128
Métallurgie	>1/2	>1/2	>1/2	>1 328	>1 028
Métallurgie	>1/2	>1/2	>1/2	454	>1 028
Cimenteries	1/4	1/4	1/8	78	128
Cimenteries	1/2	1/8	>1/2	370	287
Papeterie	1/4	>1/2	1/4	>1 328	268
Blanchisserie	1/16	1/4	1/16	78	129
Conserverie	>1/2	>1/2	1/4	797	128
Collecteurs généraux	>1/2	>1/2	>1/2	>1 328	>516
Collecteurs généraux	1/4	>1/2	>1/2	328	172
Collecteurs généraux	1/2	>1/2	>1/2	412	229
Collecteurs généraux	>1/2	>1/2	1/4	672	172
Collecteurs généraux	>1/2	>1/2	1/2	119,6	343
Collecteurs généraux	>1/2	>1/2	1/16	78	128
Collecteurs généraux	>1/2	>1/2	1/4	>1 328	215
Collecteurs généraux	1/2	1/8	1/8	663	86

niveau du dernier consommateur en raison de la concentration effectuée par les échelons précédents.

Les concentrations létales pour une même eau résiduaire sont variables d'une espèce à l'autre.

Dans l'ensemble de la transmission de la toxicité, les résultats observés pour la transmission directe et pour la transmission induite sont comparables. Quelquefois, lorsque la toxicité directe est trop forte, la concentration choisie pour la chaîne induite est trop faible pour qu'une éventuelle transmission de la toxicité puisse apparaître pendant la durée de l'expérience.

Bibliographie

AUBERT, M   Le comportement des bactéries terrigènes en mer.
1966   Relations avec le phytoplancton. Thèse Université Aix-Marseille, 285 p.
AUBERT, M   Etude des effets des pollutions chimiques sur le phy-
1968   toplancton. Revue int. Océanogr. méd., 10:81–91.

AUBERT, M, AUBERT, J et DANIEL, S   Côtes de France. Inventaire
1968   national de la pollution bactérienne des eaux littorales. Atlantique. Cartes et graphiques. Revue int. Océanogr. méd., 3, Suppl.:103 p.
AUBERT, M, AUBERT, J et DANIEL, S   Côtes de France, Inventaire
1968a   national de la pollution bactérienne des eaux littorales. Méditerranée. Cartes et graphiques. Revue int. Océanogr. méd., 4 Suppl.:167 p.
AUBERT, M, AUBERT, J et GAMBAROTTA, J P   Côtes de France.
1968   Inventaire national de la pollution bactérienne des eaux littorales. Revue int. Océanogr. méd., 1, Suppl.:73 p.
AUBERT, M, AUBERT, J et GAMBAROTTA, J P   Côtes de France.
1968a   Inventaire national de la pollution bactérienne des eaux littorales. Mer du Nord et Manche. Cartes et graphiques. Revue int. Océanogr. méd., 2, Suppl.:122 p.
AUBERT, M, et al.   Etude des effets des pollutions chimiques sur le
1969   plancton. Dégradabilité du fuel par les micro-organismes telluriques et marins. Revue int. Océanogr. méd., 13–14:107–24.
AUBERT, M, et al.   Etude de la toxicité des produits chimiques
1969a   vis-à-vis de la chaîne biologique marine. Revue int. Océanogr. méd., 13–14:45–72.

BELLAN, G Pollution et peuplements benthiques sur substrat
1967 meuble dans la région de Marseille. lre partie. Le secteur
      de Cortiou. *Revue int. Océanogr. méd.*, 6–7: 53–87.
REISH, D J The use of marine invertebrates as indicator of water
1960 quality. *In* Proceedings of the International conference on
      waste disposal in the marine environment, edited by
      E. A. Pearson, Oxford, Pergamon Press, pp. 92–103.

REISH, D J and BARNARD, J L Field toxicity tests in marin waters
1960 utilizing the polychaetous Annelid *Capitella capitata*
      (Fabricius). *Pacif. Natst*, 1(21–22): 1–8.

UI, J Minamata disease and water pollution by industrial waste.
1969 *Revue int. Océanogr. méd.*, 13–14: 37–44.

Results of Acute Toxicity Tests with Marine Organisms, using a Standard Method

*J. E. Portmann**

Examen des résultats des tests de toxicité aiguë effectués avec des organismes marins selon une méthode standard

L'auteur expose les résultats de tests de toxicité portant sur diverses substances et effectués selon une méthode standard. Les diverses classes de composés étudiés comportent des insecticides, des herbicides, des métaux, des détergents, et des dispersants d'hydrocarbures. On examine les résultats et leur exactitude en fonction de la nature chimique et physique des matériaux étudiés et de la méthode suivie. Aux fins de comparaison, le document donne certains résultats d'essais effectués sur des formes larvaires et juvéniles de crustacés, mollusques et poissons ainsi que sur des espèces divers d'algues planctoniques. Quelques valeurs de la CL₅₀ sont indiquées et envisagées à la lumière des résultats d'un certain nombre de réactions d'évitement faisant suite à des tests de comportement.

Discusión de los resultados de las pruebas de toxicidad intensa con organismos marinos empleando un método patrón

El autor presenta en este trabajo los resultados de las pruebas de toxicidad con diversas sustancias, empleando un método patrón. Los compuestos examinados comprenden insecticidas, herbicidas, metales, detergentes y dispersantes del petróleo. Los resultados y su exactitud se examinan en función de la naturaleza química y física de los materiales ensayados y del método de prueba. Para poder hacer comparaciones se incluyen resultados de pruebas efectuadas con formas larvales y juveniles de crustáceos, moluscos y peces y varias especies de algas planctónicas. Unos pocos de los valores de LC₅₀ dados, también se examinan a la vista de los resultados de pruebas de comportamiento para evitar la reacción.

ALTHOUGH we have a great deal of information on fresh water toxicity (Applegate *et al.*, 1957; McKee and Wolf, 1963), there is a very limited amount of corresponding data for the marine environment. Since it is unlikely that the toxicity of a given material would be the same to, e.g. fresh water and marine fish, separate tests are needed for most chemicals. Since early in 1966 we have tested over 160 different "pollutants" against a variety of marine organisms, and some results are here given.

The method used in the tests involves the use of a series of perspex tanks, each containing 10 l of sea water, to which varying additions of the pollutant under test are made (Portmann, 1968; Portmann and Connor, 1968). All tests are carried out at 15°C in a constant-temperature environment and duplicate tanks are used. The tests last 48 h and, depending upon the species, up to 25 animals are used per tank. A variety of animals has been used; but all the pollutants were tested with the European brown shrimp (*Crangon crangon*) and the European cockle (*Cardium edule*) and frequently also with at least one species of fish. *Crangon* was selected as being the most sensitive species available, and *Cardium* was chosen as a representative bivalve which was unable to avoid contact with toxin owing to its serrated valve rims.

Usually dilutions based on a half-log scale are used and the LC₅₀ (the concentration required to kill 50 per cent of the animals in 48 h) is then expressed as being within a half-log range, e.g. 10–33 ppm, although in some of our earlier work closer definition of the LC₅₀ by means of probit analysis was attempted. This method of testing has a number of weaknesses which are discussed in detail below; basically these stem from the fact that the water has to be aerated and from the fact that, the water being static, no allowances are made for any losses of the pollutant under test. However, the method is simple to

use and large numbers of experiments can be conducted fairly readily.

The method used for testing with unialgal cultures of flagellates is of necessity somewhat different from that used for testing animals. Stock cultures of algae are grown in Erd-Schreiber medium under artificial light and controlled temperature conditions in 1 l Erlenmeyer flasks. Each culture is gently aerated. When an experiment is set up a portion of the stock culture is filtered through

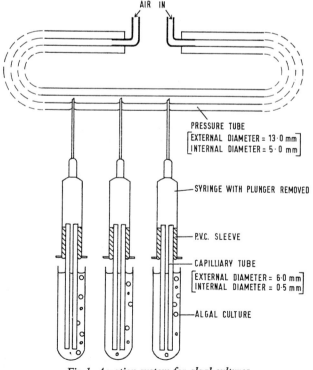

Fig 1. Aeration system for algal cultures

* Ministry of Agriculture, Fisheries and Food, Fisheries Laboratory, Burnham-on-Crouch, Essex, England.

loosely packed glass wool and diluted with Erd-Schreiber medium to an appropriate cell density, as estimated by optical density measurements; 50 ml aliquots are then measured into culture tubes to which 1 ml of a pollutant is added. All pollutant additions are of 1 ml of appropriate strength solutions and the controls receive 1 ml of distilled water. The tubes are placed in racks in a light box at a constant 15°C, and aerated by the system shown in fig 1; this aids growth and prevents clumping. Growth is estimated at, for example, 1, 2, 4 and 6 days, etc. after the start of the experiments, by measurement of optical density of 5 ml aliquots from each tube. The culture is killed before measurement by 1 drop of formalin solution. Results are then expressed in terms of the concentration which totally prevents, or which just inhibits, growth.

Results

Results of some tests with a number of metals, detergents, solvent emulsifiers, insecticides, herbicides and miscellaneous other chemicals are given under appropriate sub-headings in Tables 1–6. Of the materials tested the most toxic were the insecticides, particularly to the Crustacea. With exception of mercury and copper, metals were not particularly toxic to the species used. The oil dispersants had a wide range of toxicities, as did the miscellaneous group, but in general newer oil dispersants were considerably less toxic than earlier products.

Metals

Six metals, all commonly occurring in industrial effluents, were tested against *Crangon* and *Cardium*, and the three most toxic were also tested with fish (Table 1). Mercury and copper (in that order) proved the most toxic, confirming earlier work by Wisely and Blick (1967) who used larvae of marine animals. The remaining four metals proved rather less toxic to the species we used.

Of various disadvantages, the principal one was that it was not possible to ensure a constant concentration of metal throughout. Losses due to uptake by animals and by adsorption on the vessel walls may have occurred, although it is unlikely that these accounted for a significant proportion of the metal ions added, and in this sense the LC_{50} values quoted are unlikely to be radically different from any obtained with a continuous flow constant dosing method.

Most metals precipitated out to a noticeable extent (e.g. a visible precipitate of copper occurred at the 10 ppm level and above) and it is perhaps somewhat unrealistic to describe the LC_{50} in terms of added concentration when this was not actually the concentration in the water. However, precipitates were to a large extent kept in suspension by aeration and the movement of test animals. Brown *et al.* (1968) show that metals act largely on gill surfaces where they cause thickening of the epithelial walls and ultimately death by suffocation as oxygen transfer is impeded. It was felt that suspended metals precipitates would come in contact with gill surfaces, thus causing similar damage; also if such damage were not caused, sufficient precipitate would possibly gather around the gills and ultimately cause suffocation.

In 1968 the author reported that young brown shrimps were more susceptible to mercury than adults and that less than half the quantity of mercury was needed to kill 50 per cent of small young shrimps compared with that needed for larger animals. Using adult oysters (*Ostrea edulis*) the 96 h LC_{50} for zinc was found to be greater than 100 ppm, but P. R. Walne (personal communication) has found that the equivalent figure for larval oysters was about 1/100th of this**.

TABLE 1. TOXICITY DATA FOR METAL IONS

Metal ion tested	Valency	48 h LC_{50} in ppm, w/v		
		Crangon crangon	*Cardium edule*	Fish*
Copper	2+	10–33	**1**	1–3.3 (F)
Chromium	6+	**100**	100–330	33–100 (A)
Iron	3+	33–100	100–330	
Mercury	2+	3.3–10	3.3–**10**	**3.3** (F)
Nickel	2+	100–330	>330	
Zinc	2+	100–330	100–330	

* (A) = *Agonus cataphractus*; (F) = *Pleuronectes flesus*
NB. Bold figures in all tables indicate an LC_{50} very close to this number

Experiments were conducted with copper, zinc and mercury to test whether or not *Crangon* would avoid concentrations of these metals which were potentially lethal. The apparatus used was essentially similar to that described by Sprague (1965). Shrimps were found to have a positive avoidance reaction to copper; the minimum concentration causing significant avoidance was between 0.33 and 0.5 ppm, compared with 48 h LC_{50} of 10–33 ppm. There was no avoidance reaction to mercury up to 100 ppm or with zinc up to 33 ppm; on the contrary, at low concentrations, i.e. with 0.1–1 ppm of mercury and 3.3–33 ppm of zinc, there was an attraction effect. Since the 48 h LC_{50} for mercury lies between 3.3 and 10 ppm an attraction effect at concentrations only just below this could have serious consequences, particularly if the more susceptible larvae exhibit the same behaviour†.

The effect of two metals, mercury and copper, on the flagellate *Dunaliella tertiolecta* was also examined. The experiment with mercury lasted 18 days, since no aeration was used and growth rate was therefore slower. The concentrations of mercury used were in the range 0.033–3.3 ppm; only the two highest, 1 and 3.3 ppm, caused significant reduction in growth at the end of this period. At 3.3 ppm growth was completely suppressed and at 1 ppm it was noticeably reduced compared with controls. Concentrations of mercury less than 1 ppm had no effect. In an experiment with copper only one concentration, 10 ppm, was used and this caused no reduction in growth rate over a period of 3 weeks.

Detergents

Of the detergents tested, three: octyl phenol 11 ethylene oxide, nonyl phenol 12 ethylene oxide and coco monoethanolamide, were non-ionics; the remainder were anionics (Table 2). Although none proved to be particularly toxic to the animals used, octyl phenol 11 ethylene oxide was marginally more toxic than the other surfac-

** All the tests with larval oysters mentioned in this paper were conducted at Conway oyster hatchery (Ministry of Agriculture, Fisheries and Food), unless otherwise stated.

† My colleague Mr P. C. Wood gives this and subsequent similar information quoted in this paper.

tants. As a group the non-ionics appeared to be slightly more toxic than the anionics. According to Marchetti (1965) the toxicity of the non-ionics decreases with increase in the number of ethylene oxide groups incorporated in the molecule, and this might explain the difference in toxicity between octyl phenol 11 ethylene oxide and nonyl phenol 12 ethylene oxide.

TABLE 2. TOXICITY DATA FOR DETERGENTS

	48 h LC_{50} in ppm, w/v		
Material tested	Crangon crangon	Cardium edule	Fish*
Octyl phenol 11 ethylene oxide	33–100	10–33	33–100 (F)
Nonyl phenol 12 ethylene oxide	33–100	33–100	
Coco monoethanolamide	>100	>100	
Dobs JN	>100	>100	
Dobs 055	>100	33–100	10–33 (F)
Lauryl ether sulphonate 3 ethylene oxide	>100	10–33	

* (F) = *Pleuronectes flesus*

Testing of detergents at concentrations above about 20 ppm by the method described here is somewhat unsatisfactory since foaming occurs at the aerators, thus leading to losses of detergents from the sea water solution. According to Moss (1958) much higher concentrations of detergents are found in the foam than in the mother liquor and it is likely that the toxicity determined by the method described here was an underestimate. Other losses, for example by biodegradation, were probably small by comparison in view of the relatively short experimental period. Symptoms of distress shown by fish suggested that respiration was impaired by the detergents, agreeing with Brown *et al.* (1968) who demonstrated that detergents affected gill surfaces similar to metals.

Above results suggest that detergents discharged to the marine environment are relatively harmless, since concentrations in excess of 20 ppm are unlikely even in raw sewage (U.K., Ministry of Technology, 1966). Many workers have however suggested that it is the phytoplankton and the larval stages of marine life which are particularly vulnerable to surface-active agents. Calabrese and Davis (1967) showed that at only 0.025 ppm a linear alkyl sulphonate detergent decreased the number of oyster eggs hatching normally and that at 0.5 to 1 ppm growth rate and survival were also affected. P. R. Walne (personal communication) found that the LC_{50} for Dobs 055 to oyster larvae was 0.1 to 0.5 ppm, whereas we found that adults were unaffected by 100 ppm.

Avoidance reaction experiments with brown shrimps showed that they could detect and actively avoid concentrations of Dobs 055 in excess of 0.3 to 1 ppm, whereas in the 48 h LC_{50} test it was found that concentrations up to 100 ppm were not lethal.

Ukeless (1965) worked with a variety of detergents and several species of unicellular algae and was able to detect inhibition of growth by 1 ppm of certain detergents. Similar work at our laboratory showed that growth of *Dunaliella tertiolecta* was significantly reduced over a period of 6 weeks by 10 ppm of octyl phenol 11 ethylene oxide. Similarly, 10 ppm of Dobs 055 completely suppressed growth of the same organism for 1 week, after

which recovery was rapid and growth normal. The same detergent, Dobs 055, suppressed growth of *Isochrysis galbana* and *Phaeodactylum tricornutum* in concentrations above 1 ppm, although at 0.33 ppm there was no effect; *Tetraselmis suecica* was unaffected by concentrations of up to 3.3 ppm but at 10 ppm growth was considerably enhanced. Coco monoethanolamide did not affect *Dunaliella tertiolecta* at up to 10 ppm.

The apparent enhancement of growth in the case of *Tetraselmis suecica* in Dobs 055 is not understood but might have been caused by mortality of competing bacteria or removal of other growth-inhibiting factors. Where growth was initially suppressed and then recovery occurred it is possible that biodegradation of the detergent took place, followed by growth of the culture which although not growing had not been killed.

Oil dispersants

Results in Table 3 are for a selected range of oil dispersant mixtures which include the most toxic and the least toxic encountered. In general, oil dispersants consist of a mixture of detergent or emulsifier and a solvent to aid penetration of the oil mass. There is now little doubt that the high toxicity of materials such as BP 1002 is largely due to the solvent fraction of the mixture (Corner *et al.*, 1968; Portmann and Connor, 1968). In view of this it seems highly likely that the figures quoted in Table 3 are underestimates of toxicity for this type of material, since, due to aeration of sea water, loss of the solvent fraction must be quite rapid. Preliminary tests with *Crangon crangon* suggest that the static system did produce an underestimate of the toxicity of BP 1002, but only by a factor of 3–5 times.

TABLE 3. TOXICITY DATA FOR OIL DISPERSANTS

	48 h LC_{50} in ppm, v/v		
Material tested	Crangon crangon	Cardium edule	Fish*
BP 1002	3.3–10	33–100	10–33 (L)
Slickgone	3.3–10	10–33	
Gamlen OSR	3.3–10	10–33	
Polycomplex A	100–330	33–100	33–100 (L)
Aquaclene	100–330	33–100	
Foilzoil	330–1,000	33–100	
Ridzlick	330–1,000		
Corexit 7664	3,300–10,000	3,300–10,000	1,000–3,300 (L)
BP 1100	>3,300	1,000–3,300	1,000–3,300 (A)
Dispersol OS	3,300–10,000	330–1,000	
Corexit 8666	>3,300		

* (A) = *Agonus cataphractus;* (L) = *Pleuronectes limanda*

Since it became known that the toxicity of earlier oil dispersants was caused largely by the solvent fraction, (particularly the aromatic components), there has been a change to less toxic non-aromatic solvents. In view of this lower toxicity of the newer dispersants, such as BP 1100 and Dispersol, the static system used for testing such materials probably produces even smaller inaccuracies than those for older materials. This is now being investigated, using continuous flow method.

Tests with larvae of *Crangon crangon* suggested that they were likely to be susceptible to concentrations of solvent emulsifiers between 3 and 10 times lower than those which affect the adults (Portmann and Connor,

1968). These tests were restricted to the more toxic materials and there is a lack of similar data for the newer and less toxic formulations. However, P. R. Walne (personal communication) has shown that 50 per cent of oyster larvae were killed in 96 h in 10 ppm of Dispersol OS. Although tests have been conducted with adult oysters they would probably not be affected by concentrations less than 1,000 ppm, since they are likely to be at least as resistant as *Cardium edule*.

Although animals on a beach treated with oil dispersants could have little chance of escape, animals in the water, provided they can detect an oil dispersant, may be able to avoid potentially lethal concentrations. Investigations by divers during the *Torrey Canyon* incident showed very few dead animals in the littoral zone off treated beaches (Simpson, 1968); this suggested normal populations had detected the dispersants and moved away. Recent avoidance reaction tests at our laboratory, using *Crangon crangon*, confirm that marine creatures do avoid oil dispersants. The minimum concentrations causing significant avoidance for BP 1002, Dispersol OS and Corexit 7664 were 0.03–0.05 ppm, 1,000–3,300 and 330–1,000 respectively. Shrimps were able to detect much lower concentrations of BP 1002 than of the other two materials and, more importantly, at levels which gave them a much greater safety margin. It seems probable that the reason for this difference can be found in the different solvents used in the three materials.

Tests with algal cultures and BP 1002 showed that growth of *Phaeodactylum tricornutum* was inhibited by a concentration of 3.3 ppm and was completely stopped at 33 ppm and above. Growth of *Tetraselmis suecica* was not prevented by concentrations of BP 1002 up to 100 ppm, although in those above 1 ppm the rate of growth was noticeably reduced. Dispersol OS and Corexit were very similar to BP 1002 in their effect on algal cultures. Dispersol OS inhibited growth of *Isochrysis galbana* and *Phaeodactylum tricornutum* at 10 and 3.3 ppm respectively. Corexit reduced the growth rate of both *Tetraselmis suecica* and *Phaeodactylum tricornutum* at 3.3 ppm. However, concentrations of up to 100 ppm of either Dispersol OS or Corexit failed to prevent completely growth of any of the cultures used. It therefore appears that it is the emulsifier which does the damage with algae and that the type of solvent is of less importance than it is with animals.

Insecticides

The insecticides tested were of two types, organochlorine and organophosphorus, and were by far the most toxic class of compounds tested. There was little difference between the groups as a whole, although the most toxic single compound tested was an organophosphorus compound. The most striking feature of these materials was that they were relatively much more toxic to crustacea than to molluscs; perhaps not surprising, since they are designed to kill creatures very closely related to crustacea. The insecticides also proved fairly highly toxic to fish but were apparently much less toxic to *Cardium edule*, at least short term (Table 4).

All the insecticides were added in acetone solution—minimum volume. Data on the solubility of insecticides in water are rather scanty but most of the compounds tested are acknowledged to have very low solubilities (Martin, 1967). It is therefore somewhat doubtful that the concentrations quoted in Table 4 represent the concentrations in solution. However, at the low concentrations used the non-soluble material must have been present as very finely divided particles which it is felt must be available to the test animals. There was certainly nothing in the detailed experimental results to suggest inactivation by precipitation at any concentration. Although all the insecticides have very low volatility, aeration of the sea water solutions must lead to some losses and it is probable that the results quoted slightly underestimate the toxicity. The quoted results were in fact in general slightly higher than those obtained by Butler (1963) for brown shrimps (*Penaeus aztecus*) using continuous flow methods, e.g. 0.0004 ppm for BHC, 0.001 ppm for DDT and 0.006 ppm for dieldrin.

TABLE 4. TOXICITY DATA FOR INSECTICIDES

| Material tested | 48 h LC_{50} in ppm | | |
	Crangon crangon	Cardium edule	Fish*
DDT	0.0033–0.01	>10	0.3–1 (P)
γ-BHC	0.001–0.0033	10	
Dieldrin	0.01–0.033	>10	
Endosulfan	0.01	>10	0.033–0.1 (A)
Parathion	0.0033–0.01	3.3–10	
Malathion	0.33–1.0	3.3–10	
Azinphos-methyl	0.00033–0.001	>10	0.01–0.03 (L)
Morphothion	1–3.3	>10	

* (A) = *Agonus cataphractus*; (P) = *Pleuronectes platessa*; (L) = *Pleuronectes limanda*

Although organochlorine insecticides are now generally dispersed throughout the marine environment it would be of advantage to mobile marine animals if they could detect the more concentrated sources. This would also apply to the less persistent organophosphorous compounds. Unfortunately this does not appear to be the case, at least for azinphos-methyl and DDT which were tested with *Crangon crangon*. This species was apparently unable to detect either 0.1 ppm of DDT or 1 ppm of azinphos-methyl, concentrations which are far in excess of the 48 h LC_{50} values. Since both types of insecticide are believed to act on the nervous system it is possible that any response mechanism was blocked by their action.

The toxicity of insecticides to the larvae of marine animals does not appear to have received as much attention as it deserves. Davis (1961) reported the effects of various pesticides on bivalve larvae and showed they were in general more susceptible than adults. Certain insecticides appeared to cause little or no mortality and BHC at up to 5 ppm increased the growth rate of clam larvae, a fact attributed to possible bactericidal properties of BHC. Insecticides in low concentrations do not, in general, affect terrestrial plants and the same might be expected for marine plants. Preliminary experiments with DDT at concentrations as low as 0.0001 ppm and *Tetraselmis suecica* and *Isochrysis galbana* do not bear out this view and conflict with the findings of Ukeless (1962) who showed that 1 ppm of DDT and BHC did not affect a variety of unialgal phytoplankton cultures. It has also been reported recently that photosynthesis rates of marine phytoplankton are reduced by concentrations of DDT as low as a few parts per thousand million (Wurster, 1968).

Herbicides

No herbicides tested were as toxic as the insecticides and only carbyne and maneb gave LC_{50} values for *Crangon crangon* of less than 10 ppm (Table 5). As with the insecticides, *Cardium edule* was much less susceptible than *Crangon* in these short-term tests and fish were only slightly more resistant than *Crangon*.

TABLE 5. TOXICITY DATA FOR HERBICIDES

Material tested	48 h LC_{50} in ppm			
	Crangon crangon	Cardium edule	Fish*	
Carbyne	3.3–10	**100**	3.3–10	(A)
Maneb	3.3–10	100–330	0.33–1	(A)
Paraquat	>10	>10		
Diquat	>10	>10		
Simazine	>100	>100		
Atrazine	10–33	>100		
Dalapon	>100	>100	100	(F)
Aminotriazole	**1,000**–3,300			
Fentin acetate	>33			

* (A) = *Agonus cataphractus*; (F) = *Pleuronectes flesus*

With the exception of dalapon, atrazine and simazine, which were added as suspensions of wettable powders, all herbicides were added in solution in water. None appeared to precipitate and most compounds used were probably truly present in solution. As with insecticides small losses probably occurred due to blow-off at the aerators, and results recorded are probably underestimates rather than overestimates. The results agree well with those reported by other workers; Butler (1965) found that 1 ppm of paraquat had no effect and 1 ppm of atrazine killed 30 per cent of *Penaeus aztecus* in 48 h.

As with adults, the toxicity of herbicides to larvae does not appear particularly high. Davis (1961) found that urea herbicides in concentrations of up to 0.5 ppm slightly impaired the growth rate of bivalve larvae and that at 1 ppm adverse effects were only just detectable. Concentrations of herbicides are unlikely to reach such levels in the marine environment. The behavioural test with *Crangon crangon* was used with only two herbicides, aminotriazole and atrazine; the results suggest that, as with insecticides, *Crangon* are not very sensitive to herbicides. The shrimps failed to detect 10 ppm of atrazine and 1,000 ppm of aminotriazole.

Since herbicides are designed to kill terrestrial plants and not animals their comparatively low toxicity to marine animals might be expected. More reasonably, they might be expected to have an effect on marine algae, and Ukeless (1962) showed that the urea herbicides were certainly more toxic to marine algae than were insecticides. Experiments at our laboratory using *Dunaliella tertiolecta* showed that 1 ppm of both atrazine and simazine markedly reduced growth and that 10 ppm virtually prevented growth. Even more striking was the effect of aminotriazole, which did not affect *Crangon crangon* at 3,300 ppm, whereas at only 0.1 ppm the growth rate of *Phaeodactylum tricornutum* and *Isochrysis galbana* was reduced by almost 50 per cent.

Miscellaneous chemicals

The toxicity of a selected group of chemicals which fall in no category hitherto mentioned is given in Table 6. This table is not comprehensive but does include a number of chemicals which occur commonly in effluents and all of the more toxic ones tested. Also included are data on the toxicity of a number of polychlorinated biphenyl compounds (PCBs) recently of considerable interest.

TABLE 6. TOXICITY DATA FOR MISCELLANEOUS CHEMICALS

Materials tested	48 h LC_{50} in ppm			
	Crangon crangon	Cardium edule	Fish*	
PCBs				
Clophen A30	0.3–1	3	>10	(A)
Clophen A60	>10	>10	>10	(A)
Aroclor 1248	0.3–**1**	>10	>10	(A)
Aroclor 1254	3–10	>10	>10	(A)
Aroclor 1260	>10	>10	>10	(A)
Phenol	10–33	>330	33–100	(F)
Cresol	10–33	>100	10–33	(A)
Sulphuric acid	33–100	100–330	100–330	(F)
Nitric acid		330–1,000	100–330	(A)
Hydrochloric acid	100–330	330–1,000	100–330	(A)
Sodium hydroxide	33–100	330–1,000	33–100	(A)
Vinylacetate	10–100		>100	(F)
Allyl alcohol	1–10	>100		
Acrylonitrile	10–33*			
Formaldehyde	330–1,000		100–330	(F)

* 24 h experiment. (A) = *Agonus cataphractus*; (F) = *Pleuronectes flesus*

As can be seen none of the chemicals tested was as highly toxic as the insecticides. The PCBs show a wide variation in toxicity, the highest apparently being associated with the compound containing the least amount of substituted chlorine (the last two figures of each formulation name indicate the approximate percentage of chlorine substitution). As with the majority of other chemicals tested fish were, in general, less susceptible than *Crangon crangon* and more so than *Cardium edule*.

Some of the materials tested were highly volatile, for instance acetaldehyde and acrylonitrile, and in these cases the figures quoted for LC_{50} are certainly higher than those which would be obtained using a continuous flow method. On the other hand the alkalis, acid and PCBs all have very low volatility, and for this type of material the static method used gives a result which is probably not different from that which would be obtained by continuous flow methods.

Very little work on behavioural testing has been done with this group of compounds and the only results available are for sulphuric acid and phenol. *Crangon crangon* were able to detect and moved away from 3.3–10 ppm of sulphuric acid (a pH of 6.0–5.2) and 10–33 ppm of phenol. In neither case was the concentration avoided much less than the 48 h LC_{50}.

Conclusions

For some of the chemicals tested it appears likely that the static method used for assessing toxicity has produced an LC_{50} somewhat higher than that which would be obtained if a continuous flow method were used. This was mainly due to losses arising from aeration. It is clear however that the static method has enabled us to assess the toxicity of a wide variety of materials while a continuous flow method was being developed.

In all the experiments described and with all the materials tested, *Crangon crangon* was the most susceptible species; this was particularly noticeable with the insecticides. Fish were in general less susceptible than *Crangon* and more so than *Cardium edule*, which seemed to be moderately resistant to most chemicals. It must however be borne in mind that the tests were all short term and that no observations were made of the animals in clean sea water after the 48 h testing. Many animals which had been exposed to the material tested were moribund, but definitely not dead, when the experiment ended. There is therefore a strong possibility of considerable mortalities afterwards—particularly to cockles.

The behavioural tests mentioned are particularly interesting. With only a few exceptions the test animal *Crangon crangon* could apparently detect and avoid concentrations below the 48 h LC_{50}, although two metals at intermediate concentrations did show attraction effects. The margin of safety varied considerably but was greatest with materials such as copper and sulphuric acid, where ions are present in sea water. With completely alien materials such as DDT and atrazine there was no safety margin. However, with oil dispersants, detergents and phenol there was a small safety margin, and it can be argued that these materials are not dissimilar to chemicals which are found naturally in sea water. It should be pointed out here that, although shrimps reacted to some chemicals in laboratory experiments at concentrations below the LC_{50}, we cannot yet assess the significance of this in relation to avoidance of toxic concentrations under natural sea conditions.

The few tests carried out with larvae and the published data avilable all confirm the view that larvae are more sensitive than their parents, the LC_{50}s differing by factors of 3 to 100 times. Phytoplankton species also appear to be more susceptible than adult marine animals to most of the chemicals tested; in general they appear to be as sensitive as larvae. The methods for testing with phytoplankton are simple and these organisms appear to be useful for routine experiments. This aspect of our testing programme is at present being extended. However, I do not wish to imply that tests with phytoplankton can take the place of tests with animals; they are simply an additional test that is necessary to complete the toxicity picture.

References

APPLEGATE, V C, HOWELL, J H and HALL, A E Jr. Toxicity of
1957   4,346 chemicals to larval lampreys and fishes. *Spec. scient. Rep. U.S. Fish Wildl. Serv. (Fish.)*, (207):157 p.
BROWN, V M, MITROVIC, V V and STARK, G T C Effects of
1968   chronic exposure to zinc on toxicity of a mixture of detergent and zinc. *Wat. Res.*, 2:255–63.
BUTLER, P A Commercial fisheries investigations. *In* Pesticide-
1963   wildlife studies. A review of Fish and Wildlife Service investigations during 1961 and 1962, edited by J L George. *Circ. U.S. Fish. Wildl. Serv.*, (167):11–25.
BUTLER, P A Commercial fishery investigations. *In* The effects of
1965   pesticides on fish and wildlife. *Circ. U.S. Fish Wildl. Serv.*, (226):65–77.
CALABRESE, A and DAVIS, H C Effects of "soft" detergents on
1967   embryos and larvae of the American oyster (*Crassostrea virginica*). *Proc. natn Shellfish. Ass.*, 57(1966):11–6.
CORNER, E D S, SOUTHWARD, A J and SOUTHWARD, E C Toxicity
1968   of oil-spill removers ("detergents") to marine life: an assessment using the intertidal barnacle *Elminius modestus*. *J. mar. biol. Ass. U.K.*, 48(1):29–47.
DAVIS, H C Effects of some pesticides on eggs and larvae of
1961   oysters (*Crassostrea virginica*) and clams (*Venus mercenaria*). *Comm. Fish. Rev.*, 23(12):8–23.
MARCHETTI, R Critical review of the effects of synthetic detergents
1965   on aquatic life. *Stud. Rev. gen. Fish. Coun. Mediterr.*, (26):32 p.
MARTIN, H (Ed.) Pesticide manual. British Crop Protection Coun-
1967   cil, 464 p.
McKEE, J E and WOLF, H W (Eds.) Water quality criteria. 2nd
1963   edition. Sacramento, California, Resources Agency of California State Water Quality Control Board, Publication (3A):548 p.
MOSS, H V Synthetic detergents. Continuing research related to
1958   water and sewage treatment. *Wat. Wks Engng*, 61(745):116–9.
PORTMANN, J E Progress report on a programme of insecticide
1968   analysis and toxicity-testing in relation to the marine environment. *Helgoländer wiss. Meeresunters.*, 17(1–4):247–56.
PORTMANN, J E and CONNOR, P M The toxicity of several oil-spill
1968   removers to some species of fish and shellfish. *Mar. Biol.*, 1(4):322–9.
SIMPSON, A C The "TORREY CANYON" disaster and fisheries.
1968   *Lab. Leafl. Minist. Fish. Fd*, (18):43 p.
SPRAGUE, J B Apparatus used for studying avoidance of pollutants
1965   by young Atlantic salmon. *In* Biological problems in water pollution, edited by C M Tarzwell. *Publ. Hlth Serv. Publs. Wash.*, (999-WP 25):315.
UKELESS, R Growth of pure cultures of marine phytoplankton
1962   in the presence of toxicants. *Appl. Microbiol.*, 10:532–7.
UKELESS, R Inhibition of unicellular algae by synthetic surface-
1965   active agents. *J. Phycol.*, 1(3):102–10.
U.K., Ministry of Technology, Synthetic detergents. II. Non-ionic
1966   detergents. *Notes Wat. Pollut.*, (34):4 p.
WISELY, B and BLICK, R A P Mortality of marine invertebrate
1967   larvae in mercury, copper, and zinc solutions. *Aust. J. mar. freshwat. Res.*, 18(1):63–72.
WURSTER, C F DDT reduces photosynthesis by marine phyto-
1968   plankton. *Science, N.Y.*, 159(3822):1474–5.

Toxicity-Testing with Particular Reference to Oil-Removing Materials and Heavy Metals

*J. E. Portmann**

Détermination de la toxicité, notamment dans le cas des substances servant à éliminer les hydrocarbures et des métaux lourds

Ce document sert d'introduction aux problèmes posés par la détermination de la toxicité. On y évoque notamment la mesure de la toxicité des métaux lourds et des produits chimiques qui servent à éliminer les hydrocarbures, et on y discute brièvement de leurs voies d'arrivée dans la mer. Un examen plus détaillé porte sur les types d'appareils qui peuvent être utilisés pour les essais généraux de toxicité menés sur les animaux marins; les avantages et les inconvénients des diverses méthodes sont discutés et des suggestions son faites concernant la valeur de certaines méthodes pour des

Prueba de toxicidad con especial referencia a los metales pesados y las sustancias para la remoción del petróleo

Se intenta en este documento presentar los problemas que plantea la prueba de toxicidad. Se presta especial atención a la de metales pesados y sustancias químicas que eliminen el petróleo y se exponen sucintamente las diversas maneras en que llegan al mar. Se mencionan con detalle los aparatos que pueden emplearse en las pruebas generales de toxicidad sobre animales marinos. Se mencionan las ventajas y los inconvenientes de los diversos métodos y se hacen propuestas sobre la conveniencia de métodos especiales para materiales determinados. También se menciona la importancia de

* Ministry of Agriculture, Fisheries and Food, Fisheries Laboratory, Burnham-on-Crouch, Essex, England.

[217]

produits déterminés. L'auteur examine également l'intérêt d'utiliser toute une gamme d'organismes et les écarts de sensibilité manifestés par certains animaux à divers stades de leur développement.

emplear diversos organismos y las diferencias en la susceptibilidad de animales manifestadas en fases diversas de su desarrollo.

IN oil pollution incidents the oil spilled has most often been either crude or heavy fuel oil. Neither type of oil is particularly toxic to marine life unless the pollution is so extensive and thick that it literally smothers and suffocates any animals affected (Simpson, 1968). The toxic components of crude oils are, in the main, highly volatile and are rapidly lost by evaporation. As a result, even in the littoral zone, oil alone causes little damage to marine flora and fauna with the exception of sea birds, though it can render fish and shellfish un-saleable due to tainting. It is partly on account of the sea birds, and partly because oil pollution is unsightly, smelly and renders beaches unsuitable for recreation, that such efforts are made to remove it.

Various methods for dealing with oil-spills have been proposed (Wardley-Smith, 1968). They include physical removal and subsequent burial or burning; sinking at sea by some sinking agent; recovery from the sea surface by various means; and dispersion either at sea or on the sea-shore by an oil-dispersing chemical. Since only the last method has potentially serious effects on marine life, the other methods of dealing with oil pollution have been disregarded.

The oil-dispersing chemicals are usually mixtures of a solvent, used to penetrate the oil, and a detergent, whose purpose is to disperse the oil into fine droplets and prevent them from recombining while physical dispersion occurs, either by tidal action or by artificial means. In this way unsightly oil pollution is removed and at the same time a larger surface area is made available for bacterial attack. Many chemicals for this purpose have been marketed; unfortunately many have proved much more damaging to marine life than the oil would have been if left alone (Smith, 1968). Until recently oil dispersants were designed for use both at sea and on beaches, but it is now recognized that these two uses pose different problems from a toxicity point of view and that a material suitable for use at sea may be too toxic for use on a beach or in an enclosed area such as an estuary. This is be-cause when an oil dispersant is used at sea the chemical is rapidly diluted so that relatively few animals are exposed to high concentrations of it. Also, a depth of water being available means that mobile animals can escape, assuming they can detect it and have an avoidance reaction to it. On a beach, exposure will be to an undiluted chemical, at least until this is diluted by hosing or tidal action. These factors need to be borne in mind when toxicity-testing is considered.

The term "heavy metals" is a loose one chemically. Normally this group includes the transition metals chromium, cobalt, nickel, copper, zinc, cadmium and mercury, together with lead and occasionally arsenic, antimony and bismuth. Their main feature in common is that they are almost all relatively toxic and readily concentrated by animals, in comparison with other metals. They may all be expected to find their way into the marine environment as a result of mine drainage, mineral ore separations or smelting operations, and in a few instances following aerial pollution by burning of fossil fuels containing them. Industrial discharges are, however, generally considered to be their largest single

source. The toxicity-testing of heavy metals does not pose quite the same sort of problem as do oil-dispersant mixtures, since almost certainly the concentrations involved will be low and heavy metals are non-volatile. The main factors to be considered are their solubility and their likely chemical and/or physical form in an effluent or sea water.

Methods of testing

In the testing of oil-dispersant chemicals it is important to decide whether the tests are to simulate conditions in the field or whether the tests are to be of a standard type. As the animals in a given species do not all have the same susceptibility it is usual to employ a number of animals for any one exposure concentration and to define the results in terms of Tolerance Limits; for example, TL_{10} would be the concentration which would cause death in 10 per cent of test animals.

For routine screening of oil-dispersant chemicals for toxicity a standard test is generally used, with the object of determining a specified tolerance limit. These limits can be determined by a static or continuous flow type of test. Normally the static type of test is only considered suitable for testing materials which are comparatively non-volatile and stable and without a significant oxygen demand. If the oxygen demand is appreciable this can be allowed for either by controlled artificial oxygenation or by changing the test solution at regular intervals. The static type of test is not really suitable for materials such as oil dispersants, many of which contain volatile toxic components.

To determine the tolerance limit for an oil-dispersant mixture it is preferable to use a continuous flow test. These, almost by definition, provide well-oxygenated test solutions, eliminate problems arising from unstable test materials, provide for the removal of metabolic waste matter, and permit extended exposure times and chronic toxicity-testing. Straightforward screening tests of the type usually envisaged for oil dispersants are normally of the acute type and last 24, 48 or 96 h. It has been shown that for many materials the acute toxic effect is lost after a period of about 4 days (Sprague, 1969) and work at the author's laboratory confirms that this applies to oil dispersants (Portmann and Connor, 1968). The tolerance limit defined is usually the median or 50 per cent limit (TL_m or TL_{50}). Sometimes the term LC_{50} is used in this context, particularly in European work, and this is equivalent to TL_m. A further expression—the LD_{50}—or lethal dose to 50 per cent of the animals is also occasionally, but wrongly, used by some workers. Good reviews on the subject of toxicity-testing and the important factors involved have been given by Doudoroff *et al.* (1951), Tarzwell (1965) and Sprague (1969). The most important features are summarized below.

For any type of toxicity test care is necessary in selecting the species to be used; it must be of a type adaptable to laboratory conditions and this to some extent implies that it should be hardy, but at the same time not so hardy as to be insensitive. The species selected should be available in adequate numbers and from the same place. In freshwater testing this frequently means hatchery-reared specimens can be used, but in marine work generally the best arrangement is a collection of test specimens from the same area. Ideally they should be small enough to allow 10-25 individuals in each test tank. It is essential the animals should be representative of those likely to be exposed to the chemical in question and that as wide a variety of species as possible should be used. Moreover, the range in size should be such that the largest is not more than 1.5 times the size of the smallest. Once in the laboratory the selected animals should be kept in water of quality and quantity to maintain them in good condition. They should be acclimatized to conditions, particularly the temperature, and fed if necessary, though not during the two days preceding the test.

As species of different biological groups show different susceptibilities to toxins, the major groups should be represented and particularly commercially useful species, i.e. fish, molluscs and crustaceans. It is also important that phytoplankton be included, and a readily reared flagellate such as *Tetraselmis suecica* is recommended. At Burnham-on-Crouch the species normally used for routine testing are the small fish *Agonus cataphractus*, the cockle *Cardium edule*, the brown shrimp *Crangon crangon*, the flagellates *Tetraselmis suecica* and *Isochrysis galbana*, and the diatom *Phaeodactylum tricornutum*.

For an experimental period of up to 96 h it is neither necessary nor desirable to feed the test animals. When transferring them to the test solution from the holding tank great care should be taken to avoid harming them—particularly important with fish and small crustaceans such as shrimps. The concentrations used for the tests may be selected according to individual choice, but in screening work it is normal to use a log, half-log or quarter-log scale. The number of test animals used per concentration should not be less than ten, although in preliminary work smaller numbers may be permissible. Controls are of course essential, and if a solvent is used for initial solution of the material under test the controls should be subjected to the solvent at a concentration equivalent to the highest used in the test. For testing oil dispersants, although a solvent is involved, it is an essential part of the mixture and is therefore treated as such, the control experiments being conducted in clean sea water.

At the author's laboratory, duplicate tanks are used for each concentration and for controls, and a half-log concentration scale with concentrations in the range 1–10,000 ppm is used for experiments with oil dispersants. With this system it is usually possible in a single experiment to give the TL_m within a half-log interval, e.g. 100–330 ppm, and to reproduce this result at any future date if required. Experience has shown that, particularly with oil dispersants, it is very difficult to give a reproducible result with greater accuracy, since formulations vary from batch to batch. Thorough cleaning of the test tanks is essential following a test; we normally use 10 l perspex tanks, which are reasonably robust and stand up to most solvents. Where large numbers of tests are being carried out Hesselberg and Burress (1967) have described various labour-saving devices for cleaning which may be helpful. The number of animals used depends upon the species under test, for instance 25 for a small crustacean and 8 for fish. As increased tempera-

ture typically results in high toxicity (Portmann, 1968) the temperature chosen should be approximately that of inshore waters in summer; 15°C has been adopted at Burnham-on-Crouch and is widely used in the United Kingdom. Whichever temperature is selected it is essential that all tests for comparative purposes be conducted at the same temperature. Some form of thermostatic control is clearly necessary and suitable facilities have been described in detail by Henderson and Tarzwell (1957) and Lennon and Walker (1964).

As discussed briefly above, provided that certain limitations are accepted and the absolute TL_m is not required, a static test may be acceptable for testing oil-dispersants even though these contain volatile components. In a static test aeration is essential and it is advisable to do this with the minimum of agitation. Pure oxygen in the form of large bubbles at the rate of 30–180 per min is a suitable method. If it is particularly desirable that the presence of volatile substances should be maintained as long as possible, then controlled oxygenation as described by Doudoroff *et al.* (1951), Hart *et al.* (1948) or Carter (1963) may be adopted. As a poor substitute for continuous flow methods it is possible to change the test solutions at regular intervals.

The basic essential required for continuous-flow testing is a supply of fully oxygenated sea water at a constant temperature and with an equal flow rate to each test tank. To this must be added a supply of toxin solution at a measured rate so as to produce the desired test concentration. A wide variety of methods has been described which achieve these objectives. Jackson and Brungs (1966) have described a method suitable for materials of low toxicity. For more toxic materials a Mariotte bottle, which supplies a small volume slowly but accurately, may be useful (Surber and Thatcher, 1963; Burrows, 1949). Metering pumps (Lemke and Mount, 1963; Symons, 1963) may be used, but these suffer the disadvantage that toxin may continue to be metered when the water supply has failed. Special serial dilution types of apparatus which overcome all these failure difficulties have also been described—Mount and Warner (1965) or Mount and Brungs (1967). The apparatus described by Mount and Brungs is so designed that up to five dilutions, plus water for a control, can be supplied with flow rates of up to 400 ml/min and dilution factors of between 0.75 and 0.50. Extra metering cells can be constructed to give dilutions outside this range.

At our laboratory a constant head tank supplies aerated sea water at a thermostatically controlled temperature via rotameter-type flow meters. This sea water is mixed in a small sealed mixing chamber, with a concentrated solution of the toxicant supplied via an accurate metering pump. The outflow from the mixing chamber supplies the test tank and this overflows to waste. The flow rates are such that a complete water change is supplied to each 10 l tank every hour. The system is so arranged that each constant head tank can supply a maximum of 20 test tanks. Unless appreciable volumes of toxin are supplied to each tank no thermostatic control of toxin temperature is needed, since the whole arrangement is housed in a constant temperature room. A cut-out device is fitted to turn off the metering pump supply should the seawater supply fail.

Although it is normal in short-term toxicity-testing

to describe one's results in terms of lethal concentration (LC_{50} or TL_m), the actual definition of death is not always easy. With fish and shrimps, changes in colour may be observed and the animals usually turn over prior to death, providing good indication of pending mortality. However, particularly with crustaceans, recovery of balance once the animal is placed in clean sea water can be very rapid and any result based on this criterion should be described as an effective concentration (EC). At our laboratory, failure to respond to touch or, in the case of molluscs, a gentle tap, is the normal criterion used to define death.

In testing flagellates, growth can be estimated by cell counting or more rapidly by measurement of optical density (absorbance) of an aliquot of culture. The aliquot of culture should be killed by addition of one drop of formalin prior to measurement. Results can then be expressed in terms of the concentration which either inhibits growth to a stated degree or completely prevents it.

Since the young stages of most marine animals are generally more susceptible than the adults, tests with larvae are extremely important. Using the larvae of *Crangon crangon* and *Carcinus maenas* it has been demonstrated that, with solvent emulsifiers, in only 3 h the LC_{50} was at least 3 times and up to 10 times lower than that for adults of the same species (Portmann and Connor, 1968). Similarly, juveniles of the same species appear to be more susceptible than fully grown adults (Portmann, 1968).

The above briefly describes the various methods available, and the many precautions that must be taken, when conducting standard bioassay tests. The methods used in the author's laboratory are described in some detail, but the one chosen by any investigator will depend upon his particular requirement. This may mean a continuous-flow or static test. The static test used by the author, though not ideal, has proved useful for the task posed. We have, in addition to screening many of the marketed materials, conducted tests on development samples and components which have led to the marketing of two effective and low-toxicity oil dispersants.

Simulated field tests

Although a standard toxicity test of the type described clearly has its uses it is also very important to know precisely what the effect of an oil dispersant will be in the field. In order to simulate these conditions, test animals can be dipped into undiluted chemical for a short period, representing the time when the spray is applied, followed by a period of thorough rinsing and then by a suitable period in clean sea water. Tests of this nature were first carried out at the author's laboratory some 10 years ago and have been described by Shelton (1970). It was found that bivalve molluscs such as oysters (*Ostrea edulis* and *Crassostrea angulata*), and mussels (*Mytilus edulis*) were completely unaffected by such treatment. Winkles (*Littorina* sp.) were however affected to varying degrees by the different chemicals, although over 80 per cent of them had recovered within 7 days.

In an effort to simulate conditions in an enclosed body of water or immediately offshore from a beach, tests were also carried out in which animals were exposed to varying concentrations of oil dispersant in the water but for limited periods. They were then placed in clean sea water for a period of 24 h. During exposure to the oil dispersant the water was kept well aerated, and although this undoubtedly led to loss of the more volatile constituents it was considered that the losses were not appreciably greater than would have occurred as a result of wind and water movements.

In the first series of experiments (Shelton, 1970) oysters (*Crassostrea angulata* and *Ostrea edulis*), mussels (*Mytilus edulis*) and cockles (*Cardium edule*) were used and the exposure period was 24 h. It was found that very few of the oysters or mussels suffered any damage but that the cockles were severely affected. This suggested that the oysters and mussels could close their shells and avoid contact with the chemical for most of the exposure period. The cockles, however, were unable to do this, owing to the indented edges of their shells. In the second series of tests (Portmann and Connor, 1968), a wider variety of species was used including fish, lobsters and shrimps. The exposure period was varied in an attempt to obtain the minimum concentration which had any effect on the animals. It was found that with the majority of the materials tested the toxicity was much reduced after the first few hours and further experiments revealed that this was due to the loss of the more toxic aromatic components of the solvent, a view later confirmed by other workers (Corner *et al.*, 1968; Wilson, 1968). One interesting discovery made in the course of this work was that crustaceans which had been affected to the extent that they suffered loss of balance recovered rapidly when placed in clean sea water (Portmann and Connor, 1968). A similar phenomenon was noted by Wilson (1968) using *Sabellaria spinulosa* larvae, although, after a much longer period, it became evident that the larvae had suffered permanent damage and failed to metamorphose.

Testing of metals

The heavy metals differ from the oil-dispersant chemicals in one major respect, i.e. they are relatively much more stable and for practical purposes non-volatile. A further difference is that they are in general much less soluble in sea water; a practical limit of 10 ppm applies to most of the metals in the group. In addition they are generally more toxic. In most instances it is possible to use a static bioassay method without serious disadvantages being involved. Losses which may occur are adsorption on the container walls and uptake by the test animals, but in the course of experiments lasting up to 96 h losses due to the latter cause are unlikely to be important. The toxicity of the metal will depend very much upon the valency state, and it is important to select the correct one either for the effluent in question or for the sea water. If a lower valency state to that in sea water is expected there may be an appreciable oxygen demand and aeration must therefore be adequate to cope with this. Thorough cleaning of the tanks after use is particularly important and a rinse with acid to remove adsorbed metal ions is advisable. Apart from these minor differences the precautions which must be taken are the same as those for oil dispersants.

[220]

Discussion

At present very little is known about sub-lethal effects of oil dispersants. Exposed surfaces such as the gills would be expected to show first signs of tissue damage, and a careful study of their fine structure before and after exposure to the oil dispersant and also after a "recovery period" is likely to be rewarding. Some work of this nature has been done with metals and fish, but mostly with freshwater species (Kuhn and Koecke, 1956; Cairns and Scheier, 1964; Brown *et al.*, 1968).

An important factor in pollution is the extent to which it can be detected by the sensory systems of animals involved. If detected in time it is possible for mobile species to swim away and avoid damage. With oil dispersants, experiments at the author's laboratory (Shelton, 1970) using a Sprague-type trough (Sprague, 1962) have shown that the shrimp *Crangon crangon* can detect and avoid some oil dispersants at concentrations of one hundredth of the 48 h TL_m. With oil dispersants using non-aromatic solvents the detection level is only a seventh to a tenth of the 48 h TL_m. Much more work of this type has been done with metals, although many of the experiments were made with freshwater species (Sprague, 1964, 1968). Similar work at our own laboratory has shown that certain metals, e.g. zinc and mercury, at particular concentrations have an attracting rather than a repelling effect. Effects such as this may well have serious ecological repercussions and tests are continuing.

Other criteria which may be used in short-term sub-lethal toxicity-testing are such things as loss of balance for fish, etc., and rate of shell growth. This latter technique is particularly well suited for testing with newly settled bivalve spat, which grow rapidly for the first few days after settlement. It has also been used by Butler (1966) with adult oysters. The technique in this instance involves filing the shell edge back to the mantle and observing the rate of shell regrowth relative to controls. In all experiments on sub-lethal effects, the results are expressed as effective concentrations; for example EC_{50} would be the concentration which affected 50 per cent of the animals in a given way.

Tainting

Although toxic effects must always be very important in assessing the effects of oil-dispersant usage or of effluents containing metals, in many cases (particularly with oil pollution) greater economic harm is caused by tainting than by toxic effects. Tainting by oil alone occurs with littoral molluscs such as mussels where oil settles on the shell and taints the flesh during cooking. During stormy weather, or following an oil dispersant, the oil may become well dispersed and the small globules formed can be ingested by filter-feeding molluscs. These animals are also very susceptible to tainting by metals such as copper, since they are extremely efficient in concentrating some metal ions from the sea water; for example, in the Fal in the United Kingdom oysters are unpalatable due to high copper concentration (Cole, 1956). Fish may be affected particularly by oil tainting, and reports of inedible fish caught in and around harbours are not uncommon. Isolated instances of tainting of fish (plaice, sea-trout and lobsters) were recorded after the *Torrey Canyon* spill (Simpson, 1968). The tainting of lobsters was of fairly short duration, about 3 weeks, and the paraffin-like taste

of the affected animals suggested that the aromatic solvent of an oil dispersant was the most likely cause.

Recently it has been suggested that, although tainting may not be a long-term problem, many of the constituents of oil are very similar to compounds found in flesh and that they might therefore be assimilated and retained by fish in oil-polluted areas. Although the compounds in question do not taint the fish, some are known carcinogens and may present a hazard both to the fish and to man (Blumer, 1969). Blumer therefore questions the wisdom of using dispersants on oil, because this renders them easier to assimilate.

References

BLUMER, M   Oil pollution of the ocean. *Oceanus*, 15(2):2–7.
1969
BROWN, V M, MITROVIC, V V and STARK, G T C   Effects of
1968   chronic exposure to zinc on toxicity of a mixture of detergent and zinc. *Wat. Res.*, 2:255–63.
BURROWS, R E   Prophylactic treatment for control of fungus
1949   (*Saprolegnia parasitica*) on salmon eggs. *Progve Fish Cult.*, 11:97–103.
BUTLER, P A   The problem of pesticides in estuaries. *Spec. Publs
1966   Am. Fish. Soc.*, (3):110–5.
CAIRNS, J and SCHEIER, A   The effects of sub-lethal levels of zinc
1964   and high temperature upon the toxicity of a detergent to sunfish (*Lepomis gibbosus*). *Notul. Nat.*, (367):4 p.
CARTER, L   Toxicity of trade wastes to fish. *Effluent Wat. Treat. J.*,
1963   3:206–8.
COLE, H A   Oyster cultivation in Britain: a manual of current
1956   practice. London, HMSO, 43 p.
CORNER, E D S, SOUTHWARD, A J and SOUTHWARD, E C   Toxicity
1968   of oil spill removers (detergents) to marine life: an assessment using the intertidal barnacle *Elminius modestus*. *J. mar. biol. Ass. U.K.*, 48(1):29–47.
DOUDOROFF, P, *et al.*   Bioassay methods for the evaluation of acute
1951   toxicity of industrial wastes to fish. *Sewage ind. Wastes*, 23:1380–97.
HART, W B, WESTON, R F and DeMANN, J G   An apparatus for
1948   oxygenating test solutions in which fish are used as test animals for evaluating toxicity. *Trans. Am. Fish. Soc.*, 75:225.
HENDERSON, C and TARZWELL, C M   Bioassays for control of
1957   industrial effluents. *Sewage ind. Wastes*, 29:1002–17.
HESSELBERG, R J and BURRESS, R M   Investigations in fish control.
1967   21. Labor-saving devices for bioassay laboratories. *Resour. Publs Bur. Sport Fish. Wildl.*, (38):8 p.
JACKSON, H W and BRUNGS, W A   Biomonitoring of industrial
1966   effluents. *Proc. 21st ind. Waste Conf. Purdue Univ. Engng Ext. Bull.*, 21(121):117.
KUHN, O and KOECKE, H U   Histologische und cytologische
1956   Veränderungen der Fischkieme nach Einwirkung in Wasser enthaltener schädigender Substanzen. 2. *Zellforsch. mikrosk. anat. Forsch.*, 43:611–43.
LEMKE, A E and MOUNT, D I   Some effects of alkyl benzene
1963   sulfonate on the bluegill, *Lepomis macrochirus*. *Trans. Am. Fish. Soc.*, 92:372–8.
LENNON, R E and WALKER, C R   Investigations in fish control.
1964   1. Laboratories and methods for screening fish control chemicals. *Circ. Bur. Sport Fish. Wildl.*, (185):5 p.
MOUNT, D I and BRUNGS, W A   A simplified dosing apparatus for
1967   fish toxicology studies. *Wat. Res.*, 1:21–9.
MOUNT, D I and WARNER, R E   A serial dilution apparatus for
1965   continuous delivery of various concentrations of materials in water. *Publ. Hlth Serv. Publs, Wash.*, (999-WP-23):16 p.
PORTMANN, J E   Progress report on a programme of insecticide
1968   analysis and toxicity-testing in relation to the marine environment. *Helgoländer wiss. Meeresunters.*, 17(1–4):247–56.
PORTMANN, J E and CONNOR, P M   The toxicity of several oil-spill
1968   removers to some species of fish and shellfish. *Mar. Biol.*, 1(4):322–9.
SHELTON, R G J   Dispersant toxicity-test procedures. *In* Pro-
1970   ceedings of the Joint API/FWPCA Conference on Prevention and Control of Oil Spillages.
SIMPSON, A C   The *Torrey Canyon* disaster and fisheries. *Lab.
1968   Leafl. Minist. Agric. Fish. Fd*, (18): 43 p.
SMITH, J E, (Ed.)   *Torrey Canyon* pollution and marine life.
1968   Report by the Plymouth Laboratory of the Marine Biological Association of the U.K., Cambridge University Press, 196 p.

SPRAGUE, J B Apparatus used for studying avoidance of pollutants
1962   by young Atlantic salmon. *Publ. Hlth Publs, Wash.*, (999-WP-25):315.

SPRAGUE, J B Avoidance of copper-zinc solutions by young
1964   salmon in the laboratory. *J. Wat. Pollut. Control Fed.*, 36(8):990–1004.

SPRAGUE, J B Avoidance reactions of rainbow trout to zinc
1968   sulphate solutions. *Wat. Res.*, 2(5):367–72.

SPRAGUE, J B Measurement of pollutant toxicity to fish. 1. Bio-
1969   assay methods for acute toxicity. *Wat. Res.*, 3:793–821.

SURBER, E W and THATCHER, T O Laboratory studies of the effects
1963   of alkyl benzene sulfonate (ABS) on aquatic invertebrates. *Trans. Am. Fish. Soc.*, 92:152–60.

SYMONS, J M Simple continuous-flow low and variable rate
1963   pump. *J. Wat. Pollut. Control Fed.*, 35:1480.

TARZWELL, C M Bioassay methods for the evaluation of acute
1965   toxicity of industrial waste waters and other substances to fish. *In* Standard methods for examination of water and waste water including bottom sediments and sludges, New York, American Public Health Association, pp. 545–63.

WARDLEY-SMITH, J W Recommended methods for dealing with
1968   oil pollution. *Rep. Warren Spring Lab.*, (LR 79 (E 15)).

WILSON, D P Long-term effects of low concentrations of an oil
1968   spill remover (detergent). Studies with the larvae of *Sabellaria spinulosa*. *J. mar. biol. Ass. U.K.*, 48(1):177–82.

Effect of Oil Pollution on Flora and Fauna of the Black Sea

O. G. Mironov*

Effets de la contamination par les hydrocarbures sur la flore et la faune de la mer Noire

Des expériences ont été faites avec des organismes phytoplanctoniques et zooplanctoniques ainsi qu'avec diverses espèces de poissons dans de l'eau de mer contaminée par différentes concentrations d'hydrocarbures. Les espèces holoplanctoniques se sont révélées beaucoup plus sensibles à l'action des contaminants que les espèces mésoplanctoniques et benthiques; en particulier, les algues monocellulaires ont manifesté une vulnérabilité 100 à 1.000 fois supérieure en ce qui concerne les indices de division cellulaire et les taux de mortalité.

Dans de l'eau de mer contenant de 0,05 à 0,1 ml de contaminants par litre, les organismes zooplanctoniques ont succombé le premier jour de l'expérience. Avec des concentrations de 0,001 ml/l, le taux de mortalité a augmenté par rapport à celui des organismes témoins en eau non contaminée.

Des spécimens de poissons jeunes ont fait preuve d'une certaine résistance à la contamination par le pétrole, demeurant actifs pendant quelques jours à des concentrations de 0,1 ml/l; parmi les espèces étudiées, la plus résistante a été *Mugil saliens*. Chez les poissons, le taux de mortalité a considérablement augmenté lorsque l'eau était contaminée par du pétrole sous forme d'émulsion; ce fait est dû à l'action mécanique provoquée par les gouttelettes microscopiques du contaminant sur les branchies.

Le pétrole exerce aussi un effet toxique sur les organismes benthiques, mais dans une mesure très variable, en fonction du degré de réaction des espèces et de leur stade de développement. Les stades larvaires sont les plus sensibles. Les larves de crustacés benthiques n'ont pas survécu dans de l'eau de mer à des concentrations de pétrole de 0,1–0,01 ml/l, alors que les adultes ont survécu à des concentrations 2 à 3 fois plus fortes. Les larves de poisson en cours de croissance sont en particulier extrêmement sensibles á cette action; à des concentrations de l'ordre de 10^{-5} ml/l, le nombre des larves anormales de *Rhombus maeoticus* a été plusieurs fois plus élevé que chez les témoins.

Efectos de la contaminación por petróleo sobre la flora y fauna del Mar Negro

Se han realizado experimentos con organismos del fitoplacton y zooplancton y especies de peces en agua de mar contaminada a diferentes concentraciones de petróleo.

Las especies del holoplancton han sido mucho más sensibles a la acción del contaminante que las del mesoplancton y bentos; particularmente, las algas unicelulares han evidenciado una sensibilidad en un grado de 100 a 1.000 veces mayor en lo referente a las tasas de división de las células y de la mortalidad.

En agua de mar conconcentraciones de 0,05 a 0,1 ml/l, los organismos del zooplancton han muerto en el primer día del experimento. Ya a concentraciones de 0,001 ml/l, la tasa de mortalidad ha aumentado en relación con la de los organismos testigo mantenidos en agua no contaminada.

Ejemplares de peces jóvenes han demostrado una cierta resistencia a la contaminación por petróleo quedando activos por algunos días a concentraciones de 0,1 ml/l; de las especies estudiadas, la más resistente ha sido *Mugil saliens*. La tasa de mortalidad en los peces aumentó considerablemente caundo se experimentó con petróleo en estado de emulsión; este hecho se debe al efecto mecánico producido por las gotas microscópicas del contaminante sobre las branquias.

También el petróleo tiene un efecto tóxico sobre los organismos del bentos, pero con una gran diferencia, según el grado de reacción de las especies y su estado de desarrollo. Los estadios larvales son los más sensibles. Las larvas de crustáceas bentónicos han muerto en agua de mar con concentraciones de petróleo de 0,1–0,01 ml/l, mientras que los adultos han sobrevivido en concentraciones de 2 a 3 veces mayores. Especialemnte las larvas de peces en curso de desarrollo son muy sensibles a esta acción; en concentraciones del orden de 10^{-5} ml/l, el número de larvas anormales aumentó para la especie *Rhombus maeoticus* varias veces con respecto al número de ejemplares testigo.

THE Black Sea flora and fauna are subjected to oil pollution in different ways (Mironov, 1967) but experimental work on the effect of oil on Black Sea organisms is limited (Aliakrinskaia, 1966, 1966a; Milovidova, 1967; Kryshtyn, 1968). Experiments (Mironov and Lanskaia, 1967, 1968) indicate that oil products are toxic to phytoplankton. Differences in sensitivity occur between species (Table 1). This is most

TABLE 1. REACTION OF ALGAE TO DIFFERENT CONCENTRATIONS OF OIL POLLUTION (ml/l) DURING 5 DAYS

Algae	Death of cells 100% concentration	No cell division or delayed cell division		Not different from the control
Glenodinium foliaceum	1.0–0.1	0.1 –0.01	0.001	–0.0001
Chaetoceros curvisetus	1.0–0.1	0.01	0.001	–0.0001
Gymnodinium wulffii	1.0–0.1	0.01 –0.0001		–
Ditylum brightwellii	0.1–0.0001	–		–
Gymnodinium kovalevskii	1.0–0.001	0.001–0.0001	0.00001	
Prorocentrum trochoideum	1.0	0.1 –0.00001		–
Licmorpha chrenbergii	1.0	0.1 –0.001	0.0001	–0.00001
Platiminas viridis	1.0	0.1 –0.001	0.0001	
Coscinodiscus granii	1.0	1.0 –0.1	0.01	–0.0001
Peridinium trochoideum	1.0	1.0	0.1	–0.00001
Melosira moniliformis	1.0	1.0 –0.1	0.01	

* Institute of Biology of the Southern Seas, Academy of Sciences of the Ukrainian Republic, Sevastopol, U.S.S.R.

clearly seen in the case of *Ditylum brightwellii* and *Melosira moniliformis* which differed in sensitivity by 3–4 orders of magnitude.

We found that oil and oil products at 0.001 ml/l accelerated the death of zooplankton (Table 2) (Mironov, 1969a, 1969c) but that generally the reduction in the survival times of these marine animals was less than 20 per cent. Some deviations occur in *Oithona nana* when oil and mazout are used, but, in general, zooplankton are apparently not sensitive to oil at concentrations of 0.001 ml/l. At 0.1 ml/l, however, the organisms died during the first 24 h.

Information on the effect of oil pollution on fish is limited. Data obtained have shown that young *Sargus annularis* and *Crenilabrus tinca* remained active in sea water for several days in concentrations of 0.25 ml/l. *Mugil saliens* developed in a normal way for several months in clean water following the experiments in oil concentrations of 0.25 ml/l for many days.

TABLE 2. PERCENTAGE MORTALITY (COMPARED WITH CONTROLS) OF ANIMALS IN 0.001 ml/l OIL IN SEA WATER

	Oil products					
	Oil		Mazout		Diesel oil	
	50	100	50	100	50	100
Acartia clausi	69	89	79	89	78	78
Paracalanus parvus	64	70	64	70	72	70
Panilia averostris	80	67	80	67	80	67
Centropages ponticus	83	87	83	84	83	95
Oithona nana	—	57	—	42	—	71

The survival of fishes depends mainly on the way the oil is introduced. When oil products were emulsified in sea water, the damage was much greater than with oil films on the surface. It appears that besides being toxic the mechanical action of tiny droplets of oil on the gill apparatus is important. The survival of marine organisms in relatively high concentrations on the first day, does not necessarily indicate that they resist oil pollution.

The data on the effect of oil on benthic organisms, particularly molluscs, are contradictory. The material was obtained from adult molluscs *R. euxinica*, *B. reticulatum*, *G. divaricata*, which are abundant in the coastal zone and on which fishes feed (Mironov, 1967b). The results showed that oil produces a clear toxic effect and that the sensitivity to oil pollution varied between species. Of the three species examined, *R. euxinica* was the most susceptible to oil and oil products. A large percentage of molluscs and other species remained active for 10 to 15 days in sea water containing oil in concentrations of 1.0 ml/l.

Diogenus pugilator were quite sensitive and died in oil concentrations of 0.01 ml/l. The external manifestation of hydrocarbon toxicity in most organisms was exhibited by delayed activity. Some *Diogenus pugilator* abandoned the shell before death. Oil in a concentration of 1.0 g per 1 kg of bottom sediments, accelerated the death of *Nereis diversicolor* but *Pachygrapsus marmoratus* was relatively resistant to oil pollution. Individuals weighing 2 to 2.5 g remained active for 15 days in sea water with mazout concentrations of 1.0 and 0.1 ml/l, and their behaviour was similar to that of the control organisms. The crabs continually passed through mazout films while crawling

onto the rocks above water and remained for long periods on rocks contaminated by oil. This species is sometimes found in large numbers on the sea coast and remains in areas even when polluted. *Balanus* sp. were also active in the oil film remaining there for several days.

Oil moves rapidly with currents and winds. This suggests the possibility of a short contact time between oil and marine organisms. A 5-minute contact between *Ditylum brightwellii* and sea water containing 1.0 ml/l mazout caused a statistically significant delay of development upon being returned to clean water. If contact time was increased to 1 hr, death followed three days after return to clean water.

Five, 30 and 60 min exposures to diesel oil (concentrations of 1 ml/l) on zooplankton shortened the life of experimental organisms. Similar results were obtained with planktonic stages of benthic organisms. The damaging effect here was due to soluble and emulsified components of oil pollution, as no contact with the oil film was observed. Larval forms of benthic organisms and fishes were more resistant to oil pollution than plankton.

Developing fish eggs were found to be highly sensitive (Mironov, 1967). Eggs of *Rhombus maeoticus* died on the second day in sea water containing oil and oil products in concentrations of 10^{-4} and 10^{-3} ml/l. At other concentrations (10^{-4} and 10^{-5} ml/l) only 55 to 89 per cent of eggs actually hatched. Hatching was irregular and extended over a rather long time period whereas over 90 per cent of the control larvae hatched on the fourth day. In other concentrations the hatching rate varied from 60 to 100 per cent. In a number of cases there was a delay in the hatching of larvae due to effects of oil and oil products.

Most larvae hatched in sea water containing oil and oil products were abnormal (body distortions, often in many parts) and proved inactive. In concentrations of 10^{-4} ml/l, practically all hatched larvae had defects and died the following day. In oil concentrations of 10^{-3} ml/l the number of abnormal larvae was 23 to 40 per cent while in the controls it did not exceed 7 to 10 per cent. Investigations sought the influence of oil and oil products on fish larvae. Several hours after hatching the most active individuals were selected. All died on the second day of exposure to oil concentrations of 0.1 ml/l and mazout in concentrations of 0.01 ml/l. There were no significant deaths of control organisms. No difference was noticed between the survival of experimental and control larvae in oil extracts. Larvae placed in the main extraction immediately sank and became motionless and their reaction to touch was weak. However, in some 2 to 3 h they regained their activity and subsequently did not differ from the control. This may be accounted for by the rapid destruction of toxic components in oil extractions which we have observed in the case of marine organisms. These findings suggest that the larvae are more resistant to oil pollution than developing eggs.

The data indicate that oil and oil products in the sea are highly toxic to developing fish eggs and cause their destruction in concentrations of 10^{-3} and 10^{-4} ml/l, and also in some cases at 10^{-5} ml/l.

Similar data have been obtained regarding the developing eggs of Black Sea fishes (Mironov, 1967). Some nauplii of plankton and benthic crustacea are also highly sensitive to oil pollution (Mironov, 1968, 1969a, 1969c). The observations in this and most other papers are concerned with the ability of marine organisms to survive concentrations of harmful agent. This is an important and necessary stage which enables us to set up guide lines of toxicity of some substances, to obtain initial information about their danger to sea life, and in particular to find the order of sensitivity of different groups of marine organisms. Longer term investigations are required.

References

ALIAKRINSKAIA, I O   O povedenii i fil'tratsionnoi sposobnosti
1966   Chernomorskikh midii *Mitylus galloprovinciale* v vode, zagriaznennoi neft'iu. *Zool. Zh.*, 45(7).

ALIAKRINSKAIA, I O   Effect of oil on the survival and rate of growth
1966a   of young (Black Sea) mullets. *Ryb. Khoz.*, 42(3):16–8.

KRYSHTYN, E G   Nekotorye voprosy biologii razmnozheniia
1968   khamsy, barabuli i stavridy v Novorossiiskoi bukhte. Biologicheskie issledovaniia Chernogo moria i ego promyslovykh resursov. Moskva, Nauka, pp. 204–7.

MILOVIDOVA, N Iu   Zoobentos bukht severo-vostochnoi chasti
1967   Chernogo moria. Avtoreferat dissertatsii na soiskanie uchenoi stepeni kandidata biologicheskikh nauk. Rostovna-Donu.

MIRONOV, O G   Effect of low concentrations of oil and oil products
1967 · on the developing eggs of the Black Sea flatfish. *Vop. Ikhtiol.*, 7(3):577–80 (in Russian).

MIRONOV, O G   K voprosu o zagriaznenii vod Chernogo moria
1967a   nefteproduktami. Dinamika vod i voprosy gidrokhimii Chernogo moria. Kiev, Naukova dumka.

MIRONOV, O G   Deistvie nefti i nefteproduktov na nekotorykh
1967b   molliuskov pribrezhnoi zony Chernogo moria. *Zool. Zh.*

MIRONOV, O G   Hydrocarbon pollution of the sea and its influence
1968   on marine organisms. *Helgoländer wiss. Meeresunters.*, 17(1–4):335–9.

MIRONOV, O G   Development of some Black Sea fishes in oil-
1969   polluted sea water. *Vop. Ikhtiol.*, 9(6):1136–9 (in Russian).

MIRONOV, O G   The effect of oil pollution upon some representa-
1969a   tives of the Black Sea zooplankton. *Zool. Zh.*, 48(7):980–4.

MIRONOV, O G   Microorganisms of the Black Sea, growing on
1969b   hydrocarbons. *Mikrobiologiia*, 38:728–31 (in Russian, English summary).

MIRONOV, O G   Viability of larvae of some crustacea in seawater
1969c   polluted with oil products. *Zool. Zh.*, 48(7):1734–7 (in Russian, English summary).

MORONOV, O G   Vyzhivaemost' lichinok nekotorykh rakoo-
1969d   braznykh v morskoi vode, zagriazennnoi nefteproduktami. *Zool. Zh.*, 48(11).

MIRONOV, O G   Biological aspects of pollution of the seas with oil
1970   and oil products. Synopsis of thesis, Moscow State University, 26 p. (in Russian).

MIRONOV, O G and LANSKAIA, L A   Razvitie nekotorykh dia-
1967   tomovykh vodoroslei v morskoi vode, zagriaznennoi nefteproduktami. Biologiia i raspredelenie planktona iuzhnykh morei. Kiev.

MIRONOV, O G and LANSKAIA, L A   The capacity of survival in sea
1968   water polluted with oil products inherent in some marine planktonic and benthoplanktonic algae. *Bot. Zh.*, 53:661–9 (in Russian, English summary).

Toxicity of Crude Oil and Detergents to Two Species of Edible Molluscs under Artificial Tidal Conditions

D. de G. Griffith*

Toxicité du pétrole brut et des détergents pour deux espèces de mollusques comestibles dans des conditions de marée artificielle

Une enquête sur les toxicités relatives du pétrole brut léger d'Arabie en partie évaporé et du même produit accompagné de deux détergents—Corexit 7664 et Dispersol OS—a été exécutée dans des conditions de marée artificielle, avec comme sujets le bigorneau (*Littorina littorea*) et la moule (*Mytilus edulis*). Dans chaque expérience, le régime de marée a été suivi par une période de rétablissement dans des réservoirs propres qui sont restés remplis d'eau. Des expériences parallèles (avec également des phases de marée et de rétablissement) ont été effectuées à des températures contrôlées de 4,6°C et 11°C.

On a constaté que, dans le régime de marée, les mélanges pétrole/Corexit et pétrole/Dispersol ont été immédiatement toxiques pour *L. littorea* (28 sur 30 des animaux expérimentaux sont morts dans l'heure). Le pétrole utilisé seul était moins toxique; $DT_{50} = 2$–12 heures selon la température. Après transfert dans des réservoirs d'eau claire non soumis à la marée, *L. littorea* s'est rétabli plus ou moins complètement soit que l'expérience ait été menée avec pétrole/détergent ou avec pétrole seul, principalement en fonction de la température, la température plus basse (4,6°C) étant plus favorable.

Le pétrole par lui-même n'a exercé aucune toxicité apparente sur *M. edulis* dans le régime de marée. On a constaté un effet particulier dans le cas du mélange pétrole/Dispersol à 4,6°C: une haute toxicité initiale (21 animaux tués sur 30) a été suivie d'un rétablissement partiel dans le régime de marée. Une mortalité progressive normale a fait suite à cette phase. On n'a pas observé un tel schéma à 11°C ou dans le cas d'utilisation du mélange pétrole/Corexit à diverses températures; on pense que cette particularité résulte de la persistance de la fraction isopropanol dans le Dispersol. Après transfert dans des réservoirs d'eau propre, *M. edulis* n'a manifesté aucun rétablissement après avoir été soumis à la pollution par pétrole/détergent. Le document examine les résultats sous leurs aspects écologiques.

Toxicidad del petróleo bruto y de los detergentes para dos especies de moluscos comestibles sometidos a un régimen artificial de mareas

Con bígaros (*Littorina littorea*) y mejillones (*Mytilus edulis*) sometidos a un régimen artificial de mareas se ha estudiado la toxicidad relativa de petróleo bruto ligero de Arabia, parcialmente evaporado, solo y combinado sucesivamente con dos detergentes: Corexit 7664 y Dispersol OS. El período de régimen de mareas fue seguido, en cada experimento, por un período de recuperación en estanques limpios, constantemente llenos de agua. Se realizaron experimentos paralelos (en las dos fases: de mareas y de recuperación) a temperatura controlada de 4,6°C y 11°C.

Se encontró que, en el régimen de mareas, tanto la mezcla de petróleo y Corexit como la de petróleo y Dispersol tenían efectos tóxicos inmediatos en *L. littorea* (28–30 de cada 30 animales experimentales resultaron "muertos" al cabo de una hora). El petróleo solo resultó menos tóxico: $DT_{50} = 2$–12 horas, según la temperatura. Cuando se trasladaron los ejemplares de *L. littorea* a los tanques de agua limpia no sujetos a régimen de mareas, los animales se recuperaron gradualmente de los efectos causados por la mezcla de petróleo y detergente y por el petróleo solo; uno de los factores que más influyeron en la recuperación fue la temperatura, siendo la más favorable la más baja (4,6°C).

El petróleo solo no tuvo efectos tóxicos aparentes en *M. edulis* durante el período de régimen de mareas. Con la mezcla de petróleo y Dispersol se observó un efecto peculiar a la temperatura de 4,6°C: se produjo una elevada toxicidad inicial (21 de 30 animales), seguida de una recuperación parcial dentro del régimen de mareas, que dio lugar más tarde a una mortalidad progresiva normal. Esto no se repitió a la temperatura de 11°C con la misma mezcla ni a ninguna de las dos temperaturas utilizando la mezcla de petróleo y Corexit, suponiéndose que la causa fue la persistencia del Isopropanol presente en el Dispersol. Al pasar los ejemplares de *M. edulis* sometidos a la contaminación de petróleo y detergente a los tanques limpios, no se observó recuperación. Se examinan los aspectos ecológicos de los resultados.

TOXICITY tests are generally done with straight dilutions of pollutants. Such tests may not accurately reflect the reactions of animals that live in the intertidal zone because the effects of alternate exposure and immersion during a tidal cycle cannot be taken into account. Therefore, the investigations reported here were done under artificial tidal conditions.

Materials and method

Aramco crude (Arabian light) oil was used throughout.

* Department of Agriculture and Fisheries, Fisheries Division, Dublin, Ireland.

Before the start of the experiment, 2 1 of oil were stirred for 24 h in a large (3 1) beaker with a mechanical stirrer to allow evaporation. This was done at room temperature. It caused the loss of 11–14 per cent of the original volume of the oil. This procedure minimized changes in toxicity that would have taken place during the experimental period due to continuous evaporation of these more volatile fractions. Secondly, it was an attempt to simulate the conditions of an actual oil spill at sea, in which most of the volatiles would have been lost by the time the oil reached the shore. The oil dispersants used were Corexit 7664 and Dispersol OS. Corexit 7664 (Standard Oil Company of New Jersey, U.S.A.), is "a non-ionic surfactant" soluble in fresh water in 5 per cent NaCl solution and in isopropanol, and dispersible in fuel and crude oils. It is said to contain no organic halides or heavy metals and its specific gravity at 15.6°C is given as 1.004. The recommended working dilution is 1 part Corexit to 10 parts oil.

Dispersol OS (Imperial Chemical Industries Ltd.) is "a solution of biodegradable, non-ionic emulsifying agent in isopropanol", having a density of 0.796 g/ml at 20°C. The active agent is insoluble in water. When Dispersol is added directly to water the active agent precipitates out and floats, dissolving in any oil it contacts, and the isopropanol present dissolves. Dispersol is completely soluble in oil and the recommended dosage (personal communication from the manufacturers) is 1 part Dispersol to 10–15 parts oil.

The animals used in the experiments were the periwinkle (*Littorina littorea*) and the mussel (*Mytilus edulis*); both were freshly collected for each experiment. Only one species and one dispersant was investigated at a time. Parallel experiments were done simultaneously at two controlled temperatures, 4.6°C and 11°C.

For each experiment three circular polythene basins (35.5 cm diameter), each containing 5 1 clean sea water were allowed to reach the control temperature. Thirty specimens of a single species were then added to each tank and allowed to acclimatize for 18 h under aeration through a porous block. After acclimatization, aeration was removed and approximately 435 ml evaporated oil was poured onto the surface of each of two of the basins and to one of them about 43.5 ml dispersant was also added. The surface layers were then agitated for about 60 sec in turn with a glass rod. The third tank served as a control. All the vessels were then drained through a stopcock low down on the side, so that the sludge of oil or oil/dispersant settled onto the animals in the polluted tanks.

Seven hours after draining the tanks were refilled with 5 1 of clean sea water at the correct temperature and the aeration was resumed; the oily layer floated on the surface of Tanks I and II; 6 h after refilling the tanks were drained again. This "tidal" régime was maintained for 31 h (2½ cycles) with *L. littorea* and for 120 h (4 cycles) with *M. edulis*. At all times the tanks were kept loosely covered. The animals were examined at intervals throughout these cycles.

Following the tidal régime the animals from the polluted tanks were rinsed vigorously in sea water, which caused most of the oil adhering to them to be transferred to the sides of the washing vessel. They and the animals from the control tanks were then put into

separate tanks of clean sea water and maintained at the same temperature as before. These tanks were kept covered and aerated but they were not drained, and the animals were kept under observation for 90 h (*L. littorea*) and 48 h (*M. edulis*).

Results

The physical condition of *Littorina littorea* was assessed by their ability to adhere to the wall or floor of the tank. If they were attached to the substratum they were classed as healthy, and if they were movable when touched lightly with a glass rod they were considered moribund or dead. When an animal that was touched immediately withdrew into the shell and became detached from the tank it was considered healthy on that occasion and was not classed as moribund unless it was seen to be unattached at the next examination. The same criterion was used in assessing the condition of the animals when they had been transferred to the clean recovery tanks. The numbers of healthy animals at different times of the tidal régime and recovery period are shown in Table 1.

Both Corexit-and-oil and Dispersol-and-oil were

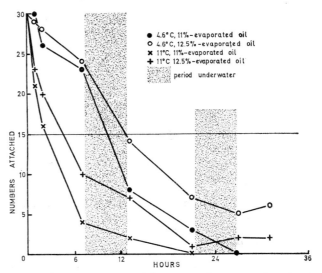

Fig. 1. Littorina littorea—*tidal regime with evaporated crude oil.*

immediately toxic to almost all the test animals—no more than two were healthy at the end of the first hour of low tide conditions, and no recovery took place during the rest of the tidal régime.

A slower but progressive toxic effect was seen among the *Littorina* subjected to crude oil alone; the TD_{50} (time to 50 per cent deaths) ranged from 2–12 h. The TD_{50} at 4.6°C was 10 h for 11 per cent evaporated oil and 12 h for 12½ per cent evaporated oil, but only 2 h and 4 h respectively at 11°C (Fig 1.) Throughout the tidal régime, animals that had become unattached littered the floor of the tank, some within the shell and others extended.

During the recovery period in clean tanks not only the *Littorina* that had been subjected to crude oil alone but also those subjected to the dispersant/oil mixtures could become re-attached to the substratum. Recovery from both dispersant/oil mixtures was more rapid at 4.6°C, where the TR_{50} (time to 50 per cent recovery) was 20 h for Corexit with 11 per cent evaporated oil and 35 h for Dispersol with 12½ per cent evaporated oil,

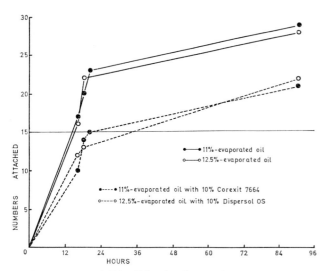

Fig. 2. Recovery at 4.6°C of Littorina littorea *from evaporated crude oil, with and without dispersants.*

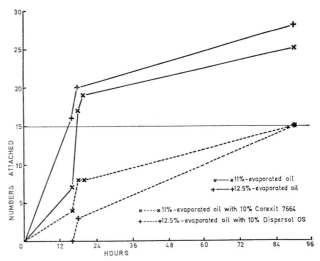

Fig. 3. Recovery at 11°C of Littorina littorea *from evaporated crude oil, with and without dispersants.*

than at 11°C, where the TR_{50} was 90 h for both mixtures (Figs. 2 and 3).

Neither the temperature at which the animals were kept, nor the volume of oil lost by prior evaporation appeared to affect significantly the TR_{50} for oil alone. The TR_{50} at 4.6°C was 14 h for 11 per cent evaporated oil and 15 h for the $12\frac{1}{2}$ per cent evaporated oil, while at 11°C the respective figures were 18 h and 15 h (Table 2).

With *Mytilus edulis* it was more difficult to assess the condition of the animals under test. This was done by tapping them hard on the shell with a glass rod. If the shell closed when handled, and gave a solid "thump" sound when tapped, the animal was classed as healthy. If the shell was agape, or if it appeared to be closed but

gave a hollow sound when tapped, the tapping was continued until the solid "thump" was heard. Where a "gaper" failed to respond to this treatment the soft inner parts were prodded in an attempt to stimulate closure. Only when these methods failed to produce a firmly closed shell, which did not give a hollow noise when tapped, was the animal considered moribund or dead. It was felt that this was the only way to ensure objectivity.

No toxic effects were seen among the animals subjected to a tidal régime of crude oil without dispersants.

Dispersal-and-oil, however, showed an initial toxicity at 4.6°C, which was followed by partial recovery and in turn by progressive mortality. This initial effect was not

TABLE 1. *Littorina littorea*—NUMBERS ATTACHED TO SUBSTRATUM DURING TIDAL AND RECOVERY PERIODS

| | Oil 11% evaporated | | | | | | Oil 12.5% evaporated | | | | | | Total hours |
| | Corexit and oil | | Oil | | Control | | Dispersol and oil | | Oil | | Control | | |
TIDAL REGIME	4.6°C	11°C	4.6°C	11°C	4.6°C	11°C	4.6°C	11°C	4.6°C	11°C	4.6°C	11°C	
1 hour after draining	1	0	30	21	28	29	1	2	29	23	30	28	1
2 hours after draining	0	0	26	16	28	29	0	0	28	20	30	30	2
7 hours after draining	0	0	23	4	28	30	0	0	24	10	30	30	7
6 hours underwater													
On draining	0	0	8	2	30	30	0	0	14	7	29	29	13
8 hours after draining	0	0	3	0	30	30	0	0	7	1	29	30	21
6 hours underwater													
On draining	0	0	0	0	30	30	0	0	5	2	29	30	27
4 hours after draining	0	0	0	0	30	30	0	0	6	2	29	30	31
CLEAN TANKS													
After 16 hours	10	4	17	7	30	29	12	0	16	16	28	27	16
After 18 hours	14	8	20	17	30	30	13	3	22	20	29	30	18
After 20 hours	15	8	23	19	30	30	—	—	—	—	—	—	20
After 90 hours	21	15	29	25	30	30	22	15	28	28	27	28	90

TABLE 2. *Littorina littorea*: TD_{50} DURING TIDAL REGIME, TR_{50} DURING RECOVERY PERIOD

	Temperature	*11% evap. oil*	*$12\frac{1}{2}$% evap. oil*	*Corexit + 11% evap. oil*	*Dispersol + $12\frac{1}{2}$% evap. oil*
TD_{50}	4.6°C	10	12	1	1
(hours)	11°C	2	4	1	1
TR_{50}	4.6°C	14	15	20	35
(hours)	11°C	18	15	90	90

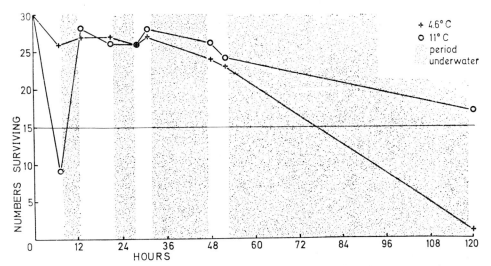

Fig. 4. Mytilus edulis—*tidal regime with crude oil (14% evaporated) and 10% Dispersol OS.*

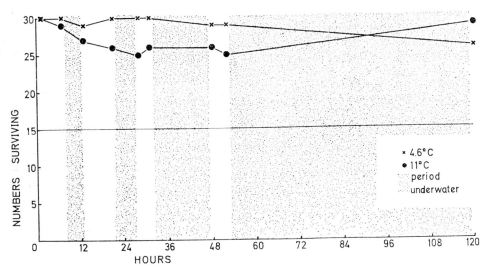

Fig. 5. Mytilus edulis—*tidal regime with crude oil (13% evaporated) and 10% Corexit 7664.*

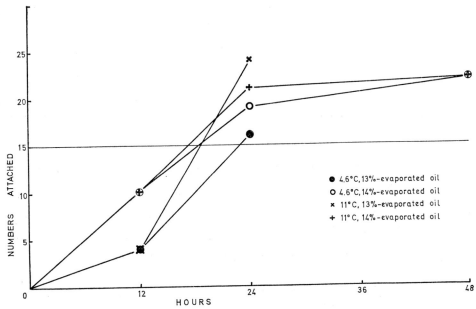

Fig. 6. *Recovery of* Mytilus edulis *from evaporated crude oil.*

[227]

seen at 11°C, although after 120 h of the tidal régime there were fewer survivors at 11°C than at 4.6°C (Figs 4 to 6)·

Corexit-and-oil was less toxic at either temperature than Dispersol-and-oil. Out of a total of 30 animals no more than 5 individuals at any time showed any adverse effects. These results are summarized in Table 3.

When the *Mytilus* were placed in clean tanks, recovery was assessed by inspecting for byssus attachment. If one byssus thread was observed the animal was considered to be attached. Among those which had been subjected to either of the oil/dispersant mixtures there was virtually no attachment.

The animals from the tanks that had contained only crude oil showed progressive attachment with a TR_{50} of about 18 h—those from the 13 per cent evaporated oil becoming attached somewhat more slowly, at least in the earlier part of the recovery period, than those from the 14 per cent evaporated oil.

The TR_{50} for the control *Mytilus* was less than 12 h.

Discussion

A higher degree of toxicity has been seen to be associated with the presence of the lower boiling range fractions of crude oil (Kühnhold 1969), and freshly spilled diesel oil was seen to have serious effects on a range of animal and plant species (North *et al.* 1965).

propanol at the lower experimental temperature. It was not seen at 11°C, where this solvent fraction of the detergent would not have persisted for so long. Portmann and Connor (1968) and Wilson (1968) have demonstrated a loss of toxicity of BP 1002 due to evaporation; Wilson showed that *Sabellaria spinulosa* larvae apparently recovered from 1 ppm BP 1002 after 2 days, but 4 weeks later they were moribund on the bottom of the tank. The *Mytilus* in Dispersol at 4.6°C behaved in a fashion similar to Wilson's *Sabellaria* larvae (although over a much shorter period). In both instances one may conclude that the initial toxicity is caused by the volatile solvent of the dispersant. The effect was not seen with Corexit-and-oil at either temperature. Corexit is a non-solvent emulsifier.

Following experiments with cockles (*Cardium edule*) in shallow open-air tanks, Portmann and Connor (1968) concluded that "evaporation caused by gentle aeration was effectively similar to that caused by a breeze blowing over the surface of the water". The effects observed here with Dispersol and *Mytilus edulis* could have significance for use in the field at temperatures around freezing point. At these temperatures the lighter crude oils should still be sufficiently fluid to make dispersant treatment worth while; the Arabian light crude used in these experiments has a pour point of − 26°C.

TABLE 3. *Mytilus edulis*—NUMBERS OF SURVIVORS DURING TIDAL REGIME; NUMBERS ATTACHED IN RECOVERY PERIOD

| TIDAL REGIME | Oil 13% evaporated | | | | | | Oil 14% evaporated | | | | | | Total hours |
| | Corexit and oil | | Oil | | Control | | Dispersol and oil | | Oil | | Control | | |
	4.6°C	11°C	4.6°C	11°C	4.6°C	11°C	4.6°C	11°C	4.6°C	11°C	4.6°C	11°C	
1 hour after draining	30	30	30	30	30	30	—	—	—	—	—	—	1
2 hours after draining	30	30	30	30	30	30	—	—	—	—	—	—	2
7 hours after draining	29	30	30	30	30	30	9	26	30	30	30	30	7
6 hours underwater													
On draining	27	29	30	30	30	30	28	27	30	30	30	30	13
8 hours after draining	26	30	30	30	30	30	26	27	30	30	30	30	21
7 hours underwater													
On draining	25	30	30	30	30	30	26	26	29	30	30	30	28
3 hours after draining	26	30	30	30	30	30	28	27	30	30	30	30	31
17 hours underwater													
On draining	26	29	30	30	30	30	26	24	30	30	30	30	48
4 hours after draining	25	29	30	30	30	30	24	23	30	30	30	30	52
68 hours underwater													
On draining	29	26	30	30	30	30	16	1	30	30	30	30	120
CLEAN TANKS													
After 12 hours	0	0	4	4	20	29	0	0	10	10	21	26	12
After 24 hours	2	2	16	24	30	29	0	0	19	21	28	30	24
After 48 hours	—	—	—	—	—	—	0	0	22	22	28	30	48

In these investigations the small difference in *Littorina* TD_{50} between crude oil from which 11 per cent had been lost by evaporation and that which had lost $12\frac{1}{2}$ per cent—a difference in TD_{50} of 2 h at both 4.6°C and 11°C—is hardly significant, particularly because there was only a $1\frac{1}{2}$ per cent difference in volume loss on the original crude. The 8 h difference in *Littorina* TD_{50} at 4.6°C and 11°C is more significant. A temperature-dependent effect has been reported by Portmann (1968) for heavy metal toxicities as well.

The acute but (at least initially) reversible effect on *Mytilus* of Dispersol-and-oil at 4.6°C is thought to have been the result of a slower evaporational loss of iso-

When the *Littorina* were transferred to clean tanks it immediately became apparent that the effects that had been seen during the tidal régime were sub lethal. Although the *Littorina* subjected to oil/dispersant and those subjected to oil alone both demonstrated this ability to recover, those from oil/dispersant took longer to do so. Temperature did not affect the recovery rate of *Littorina* from oil alone, but the difference of up to 70 h between the TR_{50} from oil/dispersant at 4.6°C and that at 11°C is interesting because although a distinct temperature effect was observed in the TD_{50}s for oil it was not seen in the TD_{50}s for oil/dispersant. Presumably the toxicity (or narcotic effect) of the oil/dispersant was too acute for any such temperature dependence to show up. The difference between the 4.6°C-TR$_{50}$ for Corexit-and-oil and that for Dispersol-and-oil (20 h compared with 35 h) is not considered to be significant; the Dispersol-and-oil curve shows that a considerable lowering of the TR$_{50}$ could be brought about by only a small increase in the numbers of animals recovered during the first 24 h of this recovery period.

With *Mytilus*, however, it appears that Dispersol-and-oil is toxic. At the end of the recovery period at 4.6°C and 11°C, 24 and 30 animals respectively were "gapers". The six individuals at 4.6°C were able to close tightly, but had still not made any byssus attachment however (these data do not appear in the tables). There is insufficient information on which to base similar remarks concerning Corexit-and-oil.

These conclusions support the field observations (O'Sullivan and Richardson 1967)—that the effects on shore-dwelling animals of crude oil treated with a dispersant are more severe than the effects of untreated oil.

It is considered that the results of these investigations are relevant to the use of these oil dispersants in the field. Even if an effect is only sub-lethal under laboratory conditions, a narcotized animal in an intertidal habitat is at an obvious disadvantage with regard to predators and/or wave action. The inhibition of byssus formation in adult *Mytilus edulis* may have serious implications for successful spat settlement of this species.

References

KÜHNHOLD, W W   Effect of water soluble substances of crude oil
1969   on eggs and larvae of cod and herring. I.C.E.S. CM1969/ E:18 (mimeo).
NORTH, W J, NEUSHUL, M and CLENDENNING, K A   Symposium
1965   sur les pollutions marines par les microorganismes et les produits pétroliers, avril 1964 Monaco. Paris, C.I.E.S. M.M., 384 pp.
O'SULLIVAN, A J and RICHARDSON, A J   The TORREY CANYON
1967   disaster and intertidal marine life. *Nature, Lond.*, 214(5087): 448, 541–2.
PORTMANN, J E   Progress report on a programme of insecticide
1968   analysis and toxicity testing in relation to the marine environment. *Helgoländer wiss. Meeresunters.*, 17(1–4): 247–56.
PORTMANN, J E and CONNOR, P M   The toxicity of several oil-spill
1968   removers to some species of fish and shellfish. *Mar. Biol.*, 1(4):322–9.
WILSON, D P   Long-term effects of low concentrations of an oil-
1968   spill remover ("detergent"): studies with the larvae of *Sabellaria spinulosa*. *J. mar. biol. Ass. U.K.*, 48(1):177–82

Pesticide-Induced Stress Profiles

*R. Eisler**

Les profils de tension engendrés par les pesticides

Les résultats des dosages de forte toxicité, pratiqués avec certains insecticides organochlorés et organophosphorés et diverses espèces de poissons, crustacés et mollusques marins, sont résumés dans le présent document; on y trouve une liste des facteurs physico-chimiques connus pour influer sur les régimes de mortalité. Une technique de profil est mise au point pour évaluer les effets des pesticides en concentration sublétale sur la faune marine; elle est fondée sur l'hypothèse en vertu de laquelle la probabilité de détection des troubles métaboliques est infiniment plus grande si l'on isole un certain nombre de paramètres pour en suivre le tracé simultanément. A titre d'exemple, la tension engendrée par les pesticides chez certaines espèces marines est profilée au moyen des paramètres suivants: hémoglobine, protéines sériques, estérases et autres composants du sang, taux de concentration du zinc, du calcium et d'autres cations dans certains tissus et dans l'animal entier. Les schémas visuels résultant de ces profils fournissent un indice qui sert à l'identification de conditions écologiques défavorables, avant que ne se produisent des modifications morphologiques ou physiologiques visibles chez le sujet. En fin de compte, il devrait être possible d'établir une corrélation entre des profils spécifiques et des agents toxiques déterminés.

Perfiles de los efectos negativos de los plaguicidas

Se resumen los resultados de los bioensayos de toxicidad aguda realizados con algunos insecticidas a base de cloruro orgánico y fósforo orgánico sobre varias especies de teleósteos marinos, crustáceos y moluscos y se enumeran los factores físico-químicos que ciertamente influyen en la mortalidad. Se presenta una técnica de trazado de perfiles para evaluar los efectos de concentraciones subletales de plaguicidas en la fauna marina, basada en la hipótesis de que es mucho más probable detectar los trastornos metabólicos, si se determinan y combinan simultáneamente varios parámetros. Se presentan algunos ejemplos de efectos negativos de plaguicidas en algunas especies marinas, utilizando como parámetros la hemoglobina, las proteínas del suero, las esterasas y otros elementos constitutivos de la sangre, y la concentración de cinc, calcio y otros cationes en determinados tejidos y en todo el animal. Los esquemas visuales que resultan de estos perfiles facilitan obtener un índice útil para la identificación de las condiciones ambientales desfavorables antes de que se produzcan cambios morfológicos o fisiológicos obvios. En último término, tal vez llegue a ser posible poner en relación determinados perfiles específicos con los agentes tóxicos de manera individual.

SYNTHETIC pesticides, especially organochlorine and organophosphorus insecticides, have become increasingly important additions to chemical wastes polluting natural aquatic communities. Many of these insecticides are considered hazardous because of their ability to kill or immobilize fresh-water, estuarine, and marine organisms at extremely low concentrations, i.e. at levels substantially below 10 μg/l (Tarzwell, 1963; Cope, 1965; Eisler, 1969). Some insecticides constitute a potential hazard because of their high resistance to physical and microbial degradation over a period of several years (Wilkinson *et al.*, 1964; Croker and Wilson, 1965) or because they are occasionally found hundreds of kilometers from the initial point of application (George,

1965)—presumably transported there by winds, underground water, or biological vectors. Particularly insidious is the inclination for several nondegradable insecticides to accumulate in lipid deposits of desirable species; afflicted animals exhibit lowered resistance to disease, to exertion, and to seasonal starvation (Johnson, 1968).

Fish, other aquatic species, and wildlife that frequent inshore coastal areas are especially vulnerable to chemical insecticides because of the tendency of these compounds to disseminate in drainage systems and to accumulate in estuaries (Butler, 1966, 1968). In addition to pesticides the coastal zone receives other types of chemical wastes. These include herbicides, petrochemicals, detergents, radioactive materials, effluents from electroplaters and

*Environmental Protection Agency, Water Quality Office, National Marine Water Quality Laboratory, West Kingston, Rhode Island 02892, U.S.A.

paper mill manufacturers, as well as domestic discharges. It is almost certain that some of these compounds will display synergism or antagonism, thereby directly or indirectly affecting the biocidal properties of insecticides, the susceptibility of endemic fauna, or both. As a consequence, results of acute toxicity bioassays conducted in the laboratory with a single compound may be of limited usefulness under field conditions. This account briefly summarizes information on the major biological effects of pesticides to marine organisms and also expands on the multiparametric approach suggested by Stokinger (1962) and Stokinger *et al.* (1966) for detecting toxicant-induced metabolic stress before obvious morphological or behavioural changes occur.

Biological effects

Results of recent studies with insecticides and marine organisms demonstrate that concentrations which are not sufficient to control many species of pestiferous insects, including several species of salt-marsh mosquitos, nevertheless can inhibit the productivity of phytoplankton populations (Butler and Springer, 1963); kill or immobilize crustaceans, fishes, and molluscs (Eisler, 1969, 1970a, 1970b); kill eggs and larvae of bivalve molluscs (Davis, 1961); induce deleterious changes in tissue composition of molluscs (Eisler and Weinstein, 1967) and teleosts (Eisler and Edmunds, 1966; Eisler, 1967); disrupt the schooling and feeding behaviour of fishes (Hiatt *et al.*, 1953); and interfere with ovary development in molluscs (Eisler, 1970b) and teleosts (Boyd, 1964). Indirect harm to higher animals can result by concentration of chemicals in their passage through food chains or by reduction in numbers of important food organisms. Clams and oysters, for example, can concentrate pesticides from the medium by factors of 70,000 and greater (Butler, 1966). Advanced forms, such as fishes, can also concentrate appreciable quantities of insecticide directly from the medium in less than 5 min (Premdas and Anderson, 1963) and retain most of it for at least 4 months (Croker and Wilson, 1965). Marine species, unlike some freshwater and terrestrial groups, are not only unable to acquire resistance to any pesticide, but are sensitized during sublethal exposures (Lowe, 1964). Killifishes which survived short-term immersion in high concentrations of organochlorine insecticides, and their F_1 progeny, experienced heavy mortality after exposure to relatively low pesticide levels (Holland *et al.*, 1966).

Acute toxicities of 7 organochlorine and 5 organo-phosphorus insecticides under controlled environmental conditions for 7 species of marine teleosts, 3 species of decapod crustaceans, and 2 species of molluscs are shown in Table 1. In general, organochlorine insecticides are more toxic to marine teleosts and crustaceans than other agricultural, industrial, and domestic wastes—including organophosphorus insecticides, soaps and synthetic detergents, slimicides, heavy metals (with the possible exception of some salts of silver, gold, and mercury), crude and refined oils, chemical oil dispersants, and aziridinyl insect sterilants (Eisler, unpublished). However, comparatively small variations in the regimens of temperature, pH, or salinity can shift the LC–50 values in Table 1 by at least one order of magnitude. Several organophosphorus insecticides, for instance, produce greatest kill ratios under conditions of high temperature,

high salinity, or low pH, whereas some organochlorine insecticides are most toxic at intermediate temperatures, i.e. 20°C, when pH is >9 or <7, over a wide range of salinities (Eisler, 1970).

TABLE 1. SUMMARY ON RANGE IN CONCENTRATIONS OF ORGANO-CHLORINE AND ORGANOPHOSPHORUS INSECTICIDES FATAL TO 50 PER CENT OF INDIVIDUAL SPECIES IN 96 HOURS AT 24 PER THOUSAND SALINITY, pH 8.0, AND 20°C. ALL VALUES ARE IN μG/L ACTIVE INGREDIENTS

	Teleosts[a]		Crustaceans[b]		Molluscs[c]
Organochlorines					
endrin	0.05–	3.1	1.7–	12.0	>10,000
p,p′–DDT	0.4 –	89.0	0.6–	6.0	>10,000
heptachlor	0.8 –	188.0	8	–440	>10,000
dieldrin	0.9 –	34.0	7	– 50	>10,000
aldrin	5 –	100.0	8	– 33	>10,000
lindane	9 –	66	5	– 10	>10,000
methoxychlor	12 –	150	4	– 12	>10,000
Organophosphates					
dioxathion	6 –	75	38	–285	>25,000
malathion	27 –	3,250	33	– 83	>25,000
Phosdrin (R)	65 –	800	11	– 69	>25,000
DDVP	200 –	2,330	4	– 45	>25,000
methyl parathion	5,200 –	75,000	2	– 7	>25,000

(a) from Eisler (1970a): species include puffer, *Sphaeroides maculatus*; killifishes, *Fundulus heteroclitus, F. majalis*; mullet, *Mugil cephalus*; eel, *Anguilla rostrata*; silverside, *Menidia menidia*; and bluehead, *Thalassoma bifasciatum*.

(b) from Eisler (1969): species tested were grass shrimp, *Palaemonetes vulgaris*; sand shrimp, *Crangon septemspinosa*; and hermit crab, *Pagurus longicarpus*.

(c) from Eisler (1970b): species tested were quahaug clam, *Mercenaria mercenaria* and mud snail, *Nassa obsoleta*.

Stress profiles

Industrial toxicologists, physicians, and others concerned with the health of persons who routinely work with dangerous or hazardous materials have developed a profile technique based on the acceptable hypothesis that metabolic disturbances have a far greater probability of detection if a number of parameters are simultaneously determined and plotted. Stokinger (1962) stated that the pattern or profile obtained by radially plotting the activity of a number of variables was a natural extension and growth of the use of enzymes and other systems as indicators of response to poisons. This concept has been tested in individuals poisoned by a number of toxic chemicals and with a number of disease conditions. Profiles have been plotted of poisonings from trichloro-ethylene, carbon tetrachloride, carbon monoxide, quinidine, phosphorus, benzene, and other compounds (Frajola, 1960). Radial profiles appear to possess great potential for detecting metabolic derangement and early disease states. The method apparently is limited only by the ingenuity of the investigator in selecting the variables most likely to furnish an insight on the condition resulting from intoxication.

Application to marine species

As one example of a profile technique consider the influence of endrin, a chlorinated hydrocarbon insecticide, on some blood and tissue constituents of the tetraodont fish, *Sphaeroides maculatus* (Table 2). These data are also shown as Fig. 1 in the form of a series of bar graphs arranged radially. It appears that high sublethal concentrations of endrin are associated with reductions in

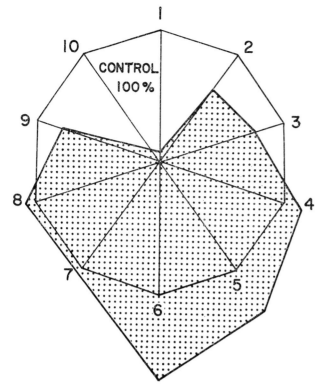

Fig. 1. Profile on effect of 1.0 μg/l of endrin on selected blood and tissue constituents of northern puffer, Sphaeroides maculatus, *surviving exposure for 96 h.*

1 = Liver Ca; 2 = Gill Zn; 3 = Liver Na; 4 = Hemoglobin; 5 = Serum K; 6 = Serum cholesterol; 7 = Serum Na; 8 = Serum Ca; 9 = Liver K; 10 = Liver Zn.

TABLE 2. EFFECT OF ENDRIN ON VARIOUS BLOOD AND TISSUE CON-STITUENTS OF NORTHERN PUFFER, *Sphaeroides maculatus*, SURVIVING 96-H EXPOSURE TO 1.0 μG/L, (FROM EISLER AND EDMUNDS, 1966)

Constituent		Controls	1.0 μg/l endrin	Percentage deviation from controls
Liver Ca	g/kg ash	127.5	10.0	−92
Liver Zn	g/kg ash	37.5	4.5	−88
Gill Zn	g/kg ash	4.4	3.0	−32
Liver K	g/kg ash	165.0	130.0	−21
Liver Na	g/kg ash	137.0	120.0	−12
Hemoglobin	g/100 ml	7.86	8.39	+ 7
Serum Ca	mg/100 ml	13.8	14.8	+ 7
Serum Na	mg/100 ml	225.4	246.1	+ 9
Serum K	mg/100 ml	66.8	91.4	+37
Serum cholesterol	mg/100 ml	269.4	440.8	+64

from the tissues, and other changes (Fig. 2). With the possible exception of serum esterases, no single variable measured is a useful indicator of response to methyl parathion, a known anti-cholinesterase agent. Nevertheless, it is apparent that the heart-shaped profile shown in Fig. 2 for methyl parathion is significantly different from those of controls or from methoxychlor—a DDT analogue—(Fig. 3), as indeed it is statistically (except for variable 6, Zone V per cent serum proteins). On the other hand, the methoxychlor profile (Fig. 3) is not significantly different from controls at the 0.05 level except for items 6, 9 (hematocrit), and 11 (gill filament Zn), and none of those were significant at the 0.01 level.

In another study, quahaug clams, *Mercenaria mercenaria*, were exposed for 96 h to various concentra-

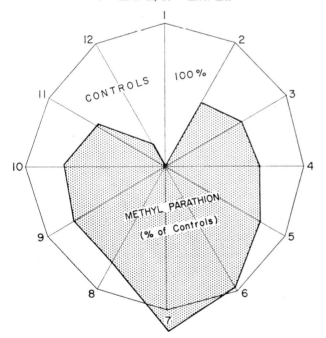

Fig. 2. Profile on effect of methyl parathion (10,100 μg/l and higher) on various blood and tissue constituents of puffers (from Eisler, 1967)

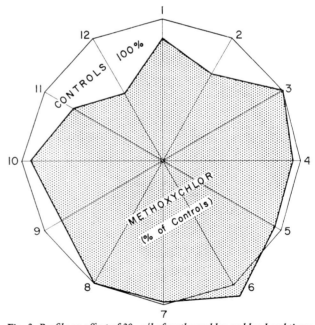

Fig. 3. Profile on effect of 30 μg/l of methoxychlor on blood and tissue constituents of puffers (from Eisler, 1967)

1 = Serum esterase; 2 = Zinc in liver; 3 = Zinc in gill arch; 4 = Magnesium in liver; 5 = Hemoglobin; 6 = Zone V % serum proteins; 7 = Zone 1 % serum proteins; 8 = Erythrocytes; 9 = Hematocrit; 10 = Total serum protein; 11 = Zinc in gill filament; 12 = Zinc in serum.

levels of Na, K, and Ca in liver, and increases in levels o Na, K, Ca, and cholesterol in serum. Lower concentrations of endrin had less effect on each of the 10 variables measured; higher concentrations caused death.

Puffers, *S. maculatus*, that survive exposure for 40 days to comparatively high levels of methyl parathion, an organophosphorus compound, exhibit complete inhibition of serum esterase activity, a generalized loss of zinc

tions of methoxychlor or malathion. Obvious changes occurred in their profiles when the variables were tissue cations. Patterns induced by both compounds were similar; only those for methoxychlor are shown (Fig. 4). Disproportionate changes in Ca and Zn levels in mantle were considered sufficiently important to warrant further study. Quahaug clams subjected to all levels of methoxychlor tend to concentrate the pesticide from the medium.

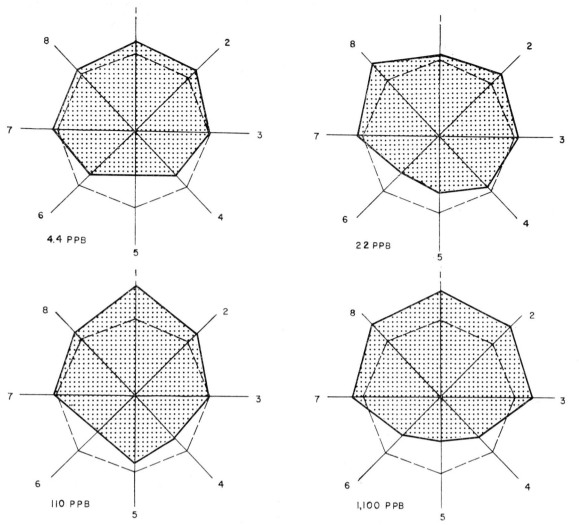

Fig. 4. *Profiles on effect of 4.4, 22, 110 and 1,100 µg/l of methoxychlor on various metal constituents of quahaug clams after exposure for 96 h (from Eisler and Weinstein, 1967). Area enclosed by the broken line represents normal (100 per cent) control values.*

1 = Mantle Zn; 2 = Muscle K; 3 = Muscle Mg; 4 = Whole clam Ca; 5 = Muscle Ca; 6 = Mantle Ca; 7 = Mantle K; 8 = Muscle Na.

At 1100 µg/l for example, methoxychlor residues on a wet weight basis in gill, mantle, and whole animal were about 7700, 2500, and 1600 µg/l respectively. The normal outward appearance and pumping rate of quahaug clams containing whole body residues of 1600 µg/l, and the apparently normal growth, survival, and behaviour during a protracted post-treatment observation period (133 days) in uncontaminated media suggests that this species can not detect methoxychlor at those concentrations.

A note of caution: substantial variability may occur in many of the indices shown in Figs. 1–4 according to results of recent unpublished studies at the National Marine Water Quality Laboratory in Rhode Island, U.S.A. For example, whole body levels of Na, K, Ca, Mg, Zn, and Sr of small marine fishes are functions of the age of the animal, salinity of the medium, length of acclimatization period to different temperature regimens, and conceivably other factors. As data on selected biological indicator species continue to accumulate it should become possible to establish the range in activity for a wide number of metabolic systems under different environmental conditions. Deviations from ranges established for individual physiological or chemical variables that appear substantially before obvious morphological or behavioral changes occur will be useful

in constructing profiles for one or several pollutants. Eventually, it could become possible and practical to correlate specific profiles with individual toxic agents.

References

BOYD, C E   Insecticides cause mosquitofish to abort. *Progve Fish*
1964   *Cult.*, 26:138.
BUTLER, P A   The problem of pesticides in estuaries. *Spec. Publs*
1966   *Am. Fish. Soc.*, (3):110–5.
BUTLER, P A   Pesticides in the estuary. *In* Proceedings of Marsh and
1968   Estuary Management Symposium, edited by J. D. Newsom. Baton Rouge, La, Louisiana State University, pp. 120–4.
BUTLER, P A and SPRINGER, P F   Pesticides—a new factor in coastal
1963   environments. *Trans. N. Am. Wildl. Conf.*, 28:378–90.
COPE, O B   Agricultural chemicals and fresh water ecological
1965   systems. *In* Research in pesticides, edited by C. O. Chichester, New York, Academic Press, pp. 115–27.
CROKER, R A and WILSON, A J   Kinetics and effects of DDT in a
1965   tidal marsh ditch. *Trans. Am. Fish. Soc.*, 94(2):152–9.
DAVIS, H C   Effects of some pesticides on eggs and larvae of
1961   oysters (*Crassostrea virginica*) and clams (*Venus mercenaria*). *Comml Fish. Rev.*, 23(12):8–23.
EISLER, R   Tissue changes in puffers exposed to methoxychlor and
1967   methyl parathion. *Tech. Pap. U.S. Fish Wildl. Serv.*, (17):1–15.
EISLER, R   Acute toxicities of insecticides to marine decapod
1969   crustaceans. *Crustaceana*, 16:302–10.
EISLER, R   Factors affecting pesticide-induced toxicity in an
1970   estuarine fish. *Tech. Pap. U.S. Fish Wildl. Serv.*, (45):1–20.
EISLER, R   Acute toxicities of organochlorine and organo-
1970a   phosphorus insecticides to estuarine fishes. *Tech. Pap. U.S. Fish Wildl. Serv.*, (46):1–12.

EISLER, R   Latent effects of insecticide intoxication to marine
1970b   molluscs. *Hydrobiologia*, 36(3–4):345–52.
EISLER, R and EDMUNDS, P H   Effects of endrin on blood and
1966   tissue chemistry of a marine fish. *Trans. Am. Fish. Soc.*,
          95(2):153–9.
EISLER, R and WEINSTEIN, M P   Changes in metal composition of
1967   the quahaug clam, *Mercenaria mercenaria*, after exposure to
          insecticides. *Chesapeake Sci.*, 8(4):253–8.
FRAJOLA, W T   Serum enzyme patterns. Part 1. *Fedn. Proc. Fedn.*
1960   *Am. Socs. exp. Biol.*, 19(1).
GEORGE, J L   DDT reportedly discovered in Antarctic seals and
1965   fish. *Agric. Chem.* 20:59.
HIATT, R W, NAUGHTON, J J and MATTHEWS, D C   Effects of
1953   chemicals on a schooling fish, *Kuhlia sandvicensis. Biol. Bull.*
          mar. biol. Lab., *Woods Hole*, 104(1):28–44.
HOLLAND, H T, COPPAGE, D L and BUTLER, P A   Increased
1966   sensitivity to pesticides in sheepshead minnows. *Trans. Am.*
          Fish. Soc., 95:110–2.

JOHNSON, D W   Pesticides and fishes—a review of selected
1968   literature. *Trans. Am. Fish. Soc.*, 97(4):398–424.
LOWE, J I   Chronic exposure of spot, *Leiostomus xanthurus*, to
1964   sublethal concentrations of toxaphene in seawater. *Trans.*
          Am. Fish. Soc., 94:396–9.
PREMDAS, F H and ANDERSON, J M   The uptake and detoxification
1963   of C¹⁴-labelled DDT in Atlantic salmon, *Salmo salar.*
          J. Fish. Res. Bd Can., 20:827–37.
STOKINGER, H E   New concepts and future trends in toxicology.
1962   *Am. ind. Hyg. Ass. J.*, 23:8–19.
STOKINGER, H E MOUNTAIN, J T and DIXON, J R   Newer
1966   toxicologic methodology. *Archs. envir. Hlth*, 13:296–306.
TARZWELL, C M   Hazards of pesticides to fishes and the aquatic
1963   environment. In Pesticides—their use and effect. *In* New
          York State Joint Legislative Committee on Natural
          Resources, Albany, N.Y., U.S.A., pp. 30–41.
WILKINSON, A T S, FINLAYSON, P G and MORLEY, H V   Toxic
1964   residues in soil 9 years after treatment with aldrin and
          heptachlor. *Science, N.Y.*, 143:681–2.

Etude Physiologique du Degré de Toxicité de Différentes Substances Contenues dans l'Eau de Mer

E. Halsband*

Physiological studies on the toxicity of various constituents of sea water

In the framework of research being done on the pollution of coastal regions by industrial effluents, a study has been made of the various effects produced on the physiology and metabolism of fish. These effects can be used as a yardstick to measure precisely the degree of pollution of the water. Research has been done on:

the basal metabolism of plaice, eels and shrimp measured in terms of oxygen consumption in the entire body and of respiration through gill tissues;

the modification in the composition of the blood by the toxins in polluted waters;

the variations in the conductibility of the body of the fish, caused by different polluted waters;

the functional potential of the *nervus lateralis*. These measurements are particularly instructive when the material to be examined contains a "neurotoxin", for example, a herbicide;

the functioning of the heart by electrocardiogram;

the K and Na content of the entire animal and of the serum;

histological changes in the gills.

Estudios fisiológicos sobre el grado de toxicidad de diversas sustancias contenidas en las aguas del mar

En el marco de las investigaciones sobre la contaminación de las regiones costeras por aguas residuales industriales, se han examinado los diversos efectos producidos sobre la fisiología de los peces, como también en el metabolismo.

Estos efectos pueden servir a medir de una manera muy precisa el grado de contaminación de las aguas. Se han efectuado investigaciones sobre:

el metabolismo basal de la platija, la anguila y el camarón, medido a partir del consumo de oxígeno del animal entero y de la respiración del tejido branquial;

la modificación de la fórmula sanguínea por las toxinas de las aguas contaminadas;

las variaciones de la conductibilidad del cuerpo del pez causadas por diferentes aguas contaminadas;

la potencia de acción del *nervus lateralis*. Estas mediciones son especialmente instructivas en los casos en que la sustancia que ha de examinarse contiene "neurotoxina", como por ejemplo un herbicida;

el funcionamiento del corazón, registrado mediante electro-cardiogramas;

el contenido de K y Na del animal entero y del suero;

los cambios histológicos de las branquias.

D ANS le cadre de cette étude sur la pollution des eaux côtières par les eaux industrielles, les recherches ont porté sur des eaux polluées par diverses industries chimiques ainsi que par certaines substances chimiques. On a utilisé des méthodes d'étude ayant fait leurs preuves lors de recherches antérieures sur la toxicité des eaux douces polluées. Les méthodes adoptées ont donc été utilisées en eau saumâtre et en eau de mer.

Les valeurs obtenues peuvent servir de critère d'évaluation du degré de pollution de l'eau, car, d'une part, l'influence nocive des eaux polluées est perçue de façon plus exacte par les réactions physiologiques de l'organisme et d'autre part, le comportement métabolico-physiologique et les réactions neurophysiologiques des animaux aquatiques sont soumis aux conditions existantes dans le milieu extérieur. Les recherches ont porté sur:

—le métabolisme basal des plies, des anguilles et des crevettes: il était mesuré d'après la consommation en oxygène de l'animal et d'après la respiration du tissu branchial. De même, les mouvements respiratoires des plies et des anguilles étaient déterminés à l'aide d'une méthode électronique;

—la modification de la formule sanguine sous l'effet des toxines des eaux polluées: à cette fin, on a étudié le nombre et la surface des érythrocytes et la valeur hématocrite;

—la conductibilité du corps du poisson: afin d'enregistrer les modifications neurophysiologiques des poissons, causées par différentes eaux polluées, les réactions des poissons dans le champ électrique ont été examinées à l'aide d'un appareil spécial. La méthode, la réalisation des mesures et les calculs mathématiques ont été décrits par l'auteur dans ses publications antérieures;

—le potentiel du courant d'action du *nervus lateralis*: ces mesures sont spécialement instructives dans le cas où le produit responsable contient une "neuro-toxine", comme par exemple un herbicide;

* Bundesforschungsanstalt für Fischerei, Institut für Küsten- u.Binnenfischerei, Hamburg, Federal Republic of Germany.

—le fonctionnement du cœur à l'aide de l'électro-cardiogramme;

—la teneur en "K" et en "Na" de l'animal entier et du sérum;

—les modifications histologiques: ces observations histologiques servent seulement à vérifier indirectement l'existence d'une modification physiologique dans le poisson. Une modification individuelle des organes et des structures du tissu entraîne, en général, une modification du fonctionnement de l'organe en question. Les préparations histologiques sont, selon la nature de l'objet, fixées, coupées et teintées suivant une méthode appropriée.

Méthodes d'étude et résultats

Quelques exemples pratiques concernant les sept types d'études effectuées par l'auteur sont décrits ci-dessous.

(1) Métabolisme basal

Pour mesurer le métabolisme basal de l'animal entier, on peut distinguer:

—le seuil létal au-delà duquel l'animal ne peut vivre et meurt rapidement;

—le seuil de perturbation où apparaissent des perturbations physiologiques fortes, lesquelles ne sont pas perceptibles superficiellement et n'entraînent pas immédiatement un effet létal. L'animal peut survivre quelque temps, et, perturbé par l'altération de l'eau, il s'échappe des eaux polluées s'il en a la possibilité. Si le poisson ne peut émigrer dans des zones biologiquement plus saines, les modifications physiologiques conduisent à la longue à la mort. Le seuil de perturbation est particulièrement important en ce qui concerne l'appréciation, car il sépare le secteur biologiquement normal du secteur pathologique. La concentration en polluant d'une eau encore tolérable doit être, dans tous les cas, inférieure à la concentration correspondant à ce seuil de perturbation;

—le seuil de tolérance en deçà duquel des fluctuations seulement minimes peuvent encore être compensées.

Sur la fig 1 A les fluctuations du métabolisme peuvent être distinguées. Il s'agit ici d'une concentration de 2,5 mg/l de CuSO$_4$. Les animaux témoins étaient des plies (*Pleuronectes platessa*). Cette concentration représente, par conséquent, le seuil de perturbation.

Les fluctuations enregistrées sont nettement inférieures pour une concentration de 1,5 mg/l de CuSO$_4$; l'animal peut compenser les fluctuations en peu de temps (fig 1 B). Dans ce cas aussi les animaux expérimentaux étaient des plies. La concentration de 1,5 mg/l de CuSO$_4$ représente, pour le métabolisme basal des plies, la tolérance limite.

La série d'expériences a également porté sur les anguilles adaptées à une longue période en eau de mer puis placées dans de l'eau contenant du CuSO$_4$. Dans ce cas, le seuil de perturbation s'observe pour une concentration de 2,5 mg/l (fig 1 C) et la tolérance limite pour une concentration de 1,5 mg/l (fig 1 D).

Les recherches, effectuées à l'aide d'un appareil de Warburg-Longator, sur la consommation en oxygène du tissu branchial des plies et des anguilles montraient également qu'une concentration de 1,5 mg/l de CuSO$_4$ constituait la tolérance limite.

(2) Modifications de la formule sanguine par des toxines d'eau polluée

Pour déterminer dans quelle mesure la formule sanguine des poissons est modifiée par les toxines contenues dans une eau polluée, on a mesuré le nombre et la surface des globules rouges et la valeur hématocrite. La surface est mesurée sous microscope, au grossissement 800, et calculée d'après la formule $(\times a \times b) \times 2$, a représentant la longueur et b, la largeur des érythrocytes. Le comptage des globules rouges est effectué à l'aide du compteur de Thoma: le sang à examiner est prélevé directement dans le cœur, par une canule; on compte pour chaque poisson au moins 200 carrés.

La valeur hématocrite est déterminée d'après la méthode de Hedin-Gärtner. Le volume (%) des globules sanguins et du plasma est calculé d'après la hauteur du culot sédimentaire des composants solides par rapport à celle du sérum. On a d'abord examiné le sang de poissons témoins adaptés à l'eau normale. Puis d'autres poissons ont été placés dans différentes solutions d'eau polluée et les déviations (en pourcentage) par rapport aux valeurs obtenues pour les poissons témoins ont été déterminées.

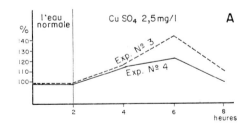
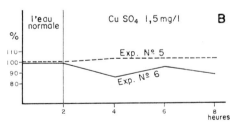
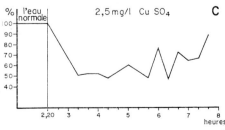
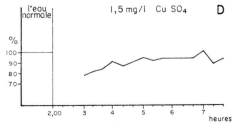

Fig 1.

TABLEAU 1. NOMBRE D'ÉRYTHROCYTES $\left(\text{POUR } \dfrac{1}{400} \text{ mm}^2\right)$ CHEZ UNE ANGUILLE ADAPTÉE DEPUIS LONGTEMPS À L'EAU DE MER

8	6	6	5	5	4	4	4	3	4	4	5	4	4	3	5	4	5	5	5
3	4	5	4	3	5	4	5	3	4	4	5	4	5	2	4	4	4	4	4
4	3	4	4	6	1	3	1	4	4	4	4	4	6	4	4	4	5	4	4
4	0	4	4	3	4	2	5	5	3	4	4	5	4	4	5	4	2	5	4
4	2	4	5	4	6	5	3	5	4	4	4	4	4	4	4	2	6	4	6
6	3	4	4	4	4	4	3	5	4	4	3	5	4	4	2	4	5	5	3
4	6	3	4	3	4	4	4	4	4	4	4	5	4	3	5	3	4	4	3
3	4	5	5	4	5	4	5	2	2	5	4	4	3	4	3	4	6	4	4
6	5	0	5	4	5	4	4	1	2	4	6	6	4	5	4	3	4	0	5
4	4	3	4	5	3	3	4	3	5	5	5	4	4	4	4	5	5	5	6
4	5	4	3	3	3	4	5	1	4	6	5	3	4	4	4	4	6	5	4
4	5	4	2	1	3	6	4	4	3	5	5	3	4	3	3	3	4	4	4
5	4	5	2	5	4	5	2	6	3	5	4	2	5	2	5	4	4	6	4
6	4	6	5	5	6	3	5	3	3	3	4	2	4	2	3	3	5	5	5
3	4	5	1	6	5	3	5	3	3	4	3	5	4	2	4	5	4	4	4
3	5	4	2	5	5	5	4	4	4	5	2	5	4	3	3	5	4	5	3
4	6	4	5	4	3	3	4	4	4	1	3	4	4	4	3	5	3	3	4
6	6	5	4	4	4	5	4	1	3	4	4	6	3	4	4	6	4	4	4
4	2	3	3	4	4	4	4	2	3	4	2	3	0	6	5	6	4	4	4
3	5	4	3	4	4	4	5	5	5	5	4	1	4	4	6	6	5	5	5

Valeur moyenne: 4,0025 = 1 601 000 érythrocytes par mm³

TABLEAU 2. SURFACE DES ÉRYTHROCYTES D'UNE ANGUILLE ADAPTÉE DEPUIS LONGTEMPS À L'EAU DE MER

No	a	b	F in μ²	No	a	b	F in μ²
1	8,0	5,0	161	26	6,5	5,0	131
2	7,0	5,0	141	27	6.2	5,0	125
3	6,5	5,0	131	28	7,0	4,8	135
4	6,5	5,5	144	29	7,0	4,8	135
5	8,0	4,5	148	30	6,0	5,2	125
6	6,5	5,2	136	31	6,0	5,2	125
7	6,8	5,0	137	32	6,5	5,2	136
8	6,5	5,2	136	33	6,0	5,5	133
9	6,5	5,2	136	34	7,0	4,5	127
10	6,5	5,2	136	35	6,0	5,0	121
11	6,0	5,5	133	36	6,5	5,0	131
12	7,0	4,8	135	37	6,5	5,2	136
13	7,2	4,8	139	38	6,5	5,2	136
14	6,5	5,2	136	39	8,0	4,0	129
15	8,0	4,2	135	40	6,5	5,2	136
16	6,2	5,5	137	41	6,0	5,0	121
17	6,5	5,5	144	42	6,0	5,0	121
18	6,5	5,2	136	43	7,0	4,2	118
19	6,5	5,2	136	44	6,2	5,0	125
20	6,5	5,2	136	45	6,0	5,5	133
21	6,5	5,5	144	46	6,5	5,2	136
22	6,2	5,5	137	47	6,2	5,0	125
23	6,5	5,2	136	48	6,0	5,2	125
24	6,5	5,5	144	49	6,5	5,0	131
25	6,5	5,2	136	50	6,5	5,2	136

Valeur moyenne: 134 μ²

TABLEAU 3. NOMBRE D'ÉRYTHROCYTES $\left(\text{POUR } \dfrac{1}{400} \text{ mm}^2\right)$ CHEZ UNE ANGUILLE MARINE PLACÉE PENDANT 24 h DANS UNE EAU CONTENANT 50 g/l DE BOUE ROUGE

2	0	0	0	1	1	2	3	1	1	2	2
2	2	1	1	1	2	3	2	2	4	0	3
2	5	1	4	2	4	1	2	4	2	0	1
2	0	1	3	3	4	2	2	1	1	5	0
3	2	1	1	1	3	3	2	3	2	2	2
1	1	4	0	2	3	4	3	2	2	2	4
2	0	2	3	2	1	2	1	3	3	3	1
2	0	1	1	3	3	1	2	3	4	1	2
2	1	2	1	3	2	1	1	1	3	4	1
4	2	3	2	4	3	2	2	2	2	4	5
2	4	2	1	3	2	2	3	2	1	2	3
3	1	2	0	3	3	2	3	3	2	0	2
3	4	2	1	2	3	1	3	2	3	1	2
1	3	2	2	2	2	1	2	1	0	2	3
1	4	0	3	3	1	2	0	2	1	2	3
2	1	0	1	2	2	3	1	6	0	4	3
2	3	1	2	0	2	4	1	3	0	2	2
4	1	1	2	1	3	2	3	4	2	0	2
2	2	1	2	3	4	2	1	2	1	1	1
3	3	2	4	2	3	2	1	2	3	3	0

Valeur moyenne: 2,038 = 815 000 érythrocytes par mm³

No	a	b	F in μ^2	No	a	b	F in μ^2
1	7,0	5,0	141	26	6,5	5,2	136
2	7,0	5,0	141	27	7,0	5,0	141
3	6,5	5,0	131	28	7,0	5,0	141
4	6,5	5,2	136	29	7,0	5,0	141
5	6,5	5,2	136	30	6,5	5,2	136
6	8,0	4,5	145	31	6,5	5,2	136
7	8,0	4,2	135	32	6,5	5,2	136
8	6,5	5,2	136	33	6,5	5,5	144
9	6,5	5,2	136	34	6,5	5,2	136
10	6,5	5,2	136	35	7,0	4,8	135
11	6,0	5,8	140	36	7,0	4,8	135
12	6,5	5,2	136	37	6,5	5,2	136
13	6,5	5,2	136	38	7,0	5,0	141
14	6,5	5,2	136	39	6,5	5,2	136
15	6,0	5,8	140	40	6,2	5,2	130
16	6,5	5,2	136	41	7,0	5,0	141
17	6,5	5,2	136	42	7,0	5,0	141
18	6,5	5,2	136	43	7,0	4,8	135
19	6,5	5,2	136	44	8,0	4,2	135
20	5,8	5,8	135	45	6,5	5,2	136
21	6,5	5,5	144	46	6,5	5,0	131
22	6,5	5,2	136	47	6,2	5,5	137
23	6,5	5,2	136	48	6,2	5,5	137
24	6,5	5,2	136	49	6,5	5,0	131
25	6,0	5,8	140	50	7,0	5,0	141

Valeur moyenne: 137 μ^2

Les Tableaux 1 et 2 montrent le nombre et la surface des érythrocytes d'anguilles, adaptées depuis longtemps à l'eau de mer. Leur nombre s'élève en moyenne à 1 600 000/mm³ et leur surface à 134 μ^2.

Les Tableaux 3 et 4 représentent les nombres et les surfaces des érythrocytes d'anguilles placées pendant 24 h dans une eau contenant 50 g/l d'argile.

Pour cette démonstration, j'ai choisi des concentrations où les réductions du nombre d'érythrocytes sont très marquées: elles sont presque de 50 pour cent inférieures à la norme. Je décrirai plus loin les électrocardiogrammes enregistrés pour les mêmes concentrations.

(3) Conductibilité du corps du poisson

La résistance, plus exactement la conductibilité du corps du poisson, dépend, on le sait, de la consistance de la masse interne du poisson et de la nature de la surface de son corps. La consistance du corps du poisson peut être influencée par la composition du milieu extérieur. Comme les réactions aux excitations du milieu extérieur, les modifications de la résistance du poisson sont, soit plus élevées, soit réduites, selon la nature de la substance qui l'affecte. Les mesures de la résistance peuvent être effectuées, soit sur l'animal entier, soit sur les muscles du poisson dépouillés ou non. Les mesures chez l'animal entier sont réalisées de la façon suivante: le poisson est placé dans un tube de verre rempli d'eau de mer, le diamètre du tube correspondant à l'épaisseur moyenne du poisson. Les électrodes sont constituées de feuilles d'argent arrondies du même diamètre que les tubes. Elles sont disposées de telle manière que la distance qui les sépare puisse être adaptée à la longueur du poisson placé directement entre les deux électrodes, l'une des électrodes se trouvant à l'extrémité de la tête, l'autre à l'extrémité de la queue.

La résistance est mesurée à l'aide d'un oscillographe suivant le principe de la perte de potentiel. Il est préférable d'utiliser un oscillographe à double faisceau et un courant alternatif de 0,5 V de tension.

(4) Modifications du potentiel du courant d'action du *nervus lateralis*

Ces mesures sont spécialement instructives lorsque l'on examine des substances contenant une "neurotoxine".

Sur la fig 2, qui représente la méthode de mesure du potentiel de repos et de la chronaxie chez une truite, la paire des électrodes côté droit est placée pour la mesure des potentiels de repos et d'action. Les électrodes sont constituées de fil d'argent chloré, l'électrode neutre ayant un diamètre de 0,8 mm, l'électrode sous tension de 0,1 mm. Le *nervus lateralis*, préparé sous turbocurarin, est mis sur l'électrode neutre et l'électrode sous tension est introduite dans le nerf. Les potentiels peuvent être observés à l'aide d'un oscillographe, et être photographiés ou filmés.

La fig 3 A présente un exemple du potentiel de repos

Fig 2. Dispositif de montage pour la mesure du potentiel de repos et de la chronaxie chez la truite.

[236]

chez une plie en eau normale (réglage de l'oscillographe: 1 mV/cm; 5 ms/cm); puis la plie a été placée durant 18 h dans une eau contenant 200 mg/l d'aminotriazol. Pour cette concentration aucune fluctuation appréciable du métabolisme ou de la physiologie (ni du métabolisme basal, ni du métabolisme des tissus) n'apparaît; ainsi cette concentration serait inférieure au seuil de tolérance limite. Pourtant le potentiel du courant d'action montre à la fois une augmentation d'amplitude et une réduction

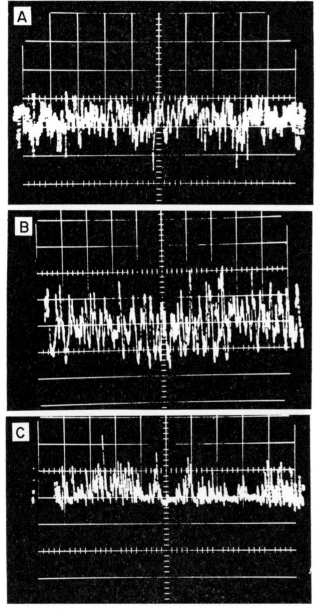

Fig 3.

de fréquence (fig 3 B). Le réglage de l'oscillographe était identique à celui utilisé pour l'enregistrement présenté à la fig 3 A. C'est seulement pour une concentration de 100 mg/l d'aminotriazol que l'on peut constater la normalisation du potentiel du courant d'action (fig 3 C). On a voulu démontrer par ces exemples que l'examen du *nervus lateralis* est nécessaire pour déceler la tolérance limite pour quelques substances ayant physiologiquement un effet excitant.

(5) Electrocardiographie

L'électrocardiographie permet, de taçon tres élégante et surtout rapide, de mettre en évidence la nocivité de certaines substances. Les plus faibles fluctuations de l'oxygénation, causées par exemple par le dépôt d'une couche sur les branchies, sont décelées par électrocardiographie. Il est seulement important de choisir toujours pour la dérivation la même position du cœur, par exemple base-tête, et de s'assurer d'une mise sous écran suffisante.

La fig 4 A montre l'électrocardiogramme d'une anguille adaptée depuis longtemps à l'eau de mer: "P" et "T" sont dans la zone positive, la ligne isoélectrique se déroule régulièrement. Ensuite la même anguille a été placée pendant 24 h dans une solution à 50 g/l d'argile (rebuts d'une usine d'aluminium); l'électrocardiogramme correspondant est représenté sur la fig 4 B: on observe, à côté de la pointe négative "P", le déplacement de la ligne iso-

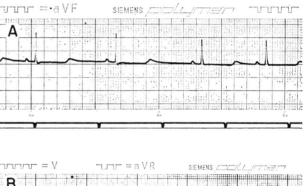

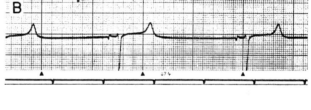

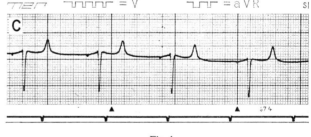

Fig 4.

électrique, la modification de la pointe "T" et avant tout un ralentissement du rythme cardiaque d'environ 50 pour cent. Cette réduction provient, selon l'auteur, de la couche qui se dépose sur les branchies. Après un séjour de 24 h dans une solution à 20 g/l d'argile, le rythme cardiaque reste encore sub-normal (fig 4 C), mais les autres caractères de l'électrocardiogramme restent identiques et le déplacement de la ligne isoélectrique s'est même encore accentué par rapport à l'expérience précédente.

(6) Teneur en K et en Na de l'animal entier et du sérum

Les ions K et Na sont dosés au photomètre de flamme. Pour la mesure du sérum, le sang est centrifugé et le sérum obtenu est dilué. Les diverses solutions sont pulvérisées et dosées au photomètre.

(7) Modifications histologiques du poisson

Surtout chez les poissons benthiques, les sédiments ou les floculats causés par des transformations chimiques, et qui se déposent sur les branchies, ont un effet sur la circulation et l'oxygénation des tissus (fig 4 B, 4 C). Cette sédimentation sur les branchies peut être prouvée histologiquement sans difficulté. En certains cas la couche de sédiment est même si épaisse qu'elle est visible macroscopiquement. Mais dans la plupart des cas une vérification histologique est nécessaire, puisque les sédiments se déposent et s'agglutinent entre les lamelles branchiales. Le traitement préalable des coupes dépend de la composition chimique de l'eau polluée. A titre d'exemple, les sédiments déposés sur les branchies d'une anguille placée depuis 18 h dans une eau contenant 50 g/l d'argile, sont représentés sur la fig 4 C.

Bibliographie

HALSBAND, E Eine neue Methodik der Atemfrequenzmessung
1955   zur Untersuchung der Schädlichkeit von Abwassergiften. *Arch. Hydrobiol.*, 23(3/4):323–8.

HALSBAND, E Entwicklung eines elektrobiologischen Testes zur
1960   Bestimmung der Abwasserlast von Flüssen. *Arch. Fisch-Wiss.*, 11(1):48–60.

HALSBAND, E Eine elektrische Messmethode zur Bestimmung des
1965   Körperwiderstandes von Fischen. *Elektro-Med.*, 10(3): 126–9.

HALSBAND, E Physiologische Untersuchungsmethoden zur Bestim-
1968   mung des Schädlichkeitsgrades von Abwassergiften in Süss-, Brack- und Salzwasser. *Helgoländer wiss. Meeresunters.*, 17(1–4):224–46.

HALSBAND, E Die schädigende Wirkung des Kaltreinigers C 4/65
1968   für Fische. *Fischwirt.*, 9:232–6.

HALSBAND, E und HALSBAND, I Untersuchungen über die
1954   Störungsschwellen im Stoffwechsel der Fische und Fischnährtiere nach Einwirkung verschiedener Abwassergifte. *Arch. FischWiss.*, 5(3/4):119–32.

HALSBAND, E und HALSBAND, I Veränderungen des Blutbildes von
1963   Fischen infolge toxischer Schäden. *Arch. FischWiss.*, 14(1–2):64–84.

HALSBAND, E und HALSBAND, I Untersuchungen über die elek-
1965   trische Leitfähigkeit des Fischkörpers. *Arch. FischWiss.*, 16(1):21–32.

HALSBAND, E und HALSBAND, I Die Leitfähigkeit des Fischkörpers
1967   nach Einwirkung verschiedener Aussenfaktoren. *Arch. FischWiss.*, 17(2):134–40.

HALSBAND, E und HALSBAND, I Eine Apparatur zur Messung der
1968   Stoffwechselintensität von Fischen und Fischnährtieren. *Arch. FischWiss.*, 19(1):78–82.

HALSBAND, E und HALSBAND, I Am Beispiel eines eisensul-
1968a   fathaltigen Abwassers: Physiologische Methoden einer Untersuchung von Abwässern in Süss- und Seewasser. *Wass. Luft Betr.*, 12(3):140–3.

Influencia del Dimecrón en la Supervivencia de la Langosta *Panulirus argus* en Relación con la Circulación de Agua Sobre la Plataforma de Cuba

Gerardo Suarez, A., María Elena Ramiro, G. y Juan J. Tápanes, D. *

Effects of Dimecron on survival of the spiny lobster, *Panulirus argus*, in relation to the water circulation on the Cuban continental shelf

Through direct bio-assays with Dimecron, a systemic insecticide used largely for rice culture in Cuba, the proportion of DL50 in the spiny lobster *Panulirus argus*, after 24, 48, 72 and 96 hours was found to be 1.70, 0.57, 0.42 and 0.30 ppm respectively.

The "boundary" value of this pollutant was between 0.2 and 0.3 ppm, which shows the necessity of monitoring the effluents which carry it into the coastal waters.

The working methods used are set out in detail with the results of the bio-assays; the possible circulation of the pollutant on the Cuban continental shelf in the southwestern region is also considered. The concentrations at present used in agriculture in Cuba do not produce lethal accumulations of this pollutant under the natural conditions in the fishing grounds of this species.

Effets de l'insecticide "Dimecron" sur la survie de la langouste *Panulirus argus*, en relation avec la circulation des eaux continentales à Cuba

Le Dimecron, insecticide qui agit sur les systèmes physiologiques des insectes, est largement utilisé dans les rizières cubaines. Par des expériences de survie réalisées sur la langouste *Panulirus argus*, on a déterminé que les doses létales de ce produit étaient respectivement de 1,70, 0,57, 0,42 et 0,30 ppm pour des temps d'expérience de 24, 48, 72 et 96 heures. Comme la valeur limite de concentration létale se situe entre 0,2 et 0,3 ppm, il est nécessaire de surveiller les effluents qui drainent ce produit dans les eaux côtières.

Ce travail décrit de façon détaillée les méthodes employées dans les expériences de survie et analyse les possibilités d'un drainage de ce polluant au-dessus du plateau continental dans la région sud occidentale de Cuba. Cette étude permet de conclure que les quantités actuellement utilisées à Cuba ne risquent pas d'entraîner des concentrations supérieures au seuil létal dans les zones de pêche de la langouste.

ESTE trabajo se realizó por la importancia prestada al clamor mundial de cuidar los recursos marinos y permanecer en actitud vigilante, hoy en día que el desarrollo de un país implica la mecanización e industrialización, y hace que en la agricultura se empleen sustancias químicas contra las plagas dañinas a los cultivos. Estas sustancias en diferentes concentraciones, pueden provocar como consecuencia, al llegar al mar por medio de los escurrimientos o efluentes desde la tierra, el envenenamiento de los animales marinos. El plaguicida estudiado en este trabajo, "Dimecrón 50", está ampliamente usado para los cultivos de arroz, en áreas agrícolas cercanas a las costas; por ende es evidente la importancia de conocer los efectos sobre la langosta *Panulirus argus*, especie comercial de gran valor económico en Cuba.

Se determinaron los valores de las dosis letales del

---

* Centro de Investigaciones Pesqueras, Playa Habana, Bauta, Cuba.

plaguicida Dimecrón, por medio de ensayos directos, para el 50% de langostas (DL$_{50}$) en períodos de 24, 48, 72 y 96 horas.

Método

Se empleó para los experimentos con la langosta *Panulirus argus*, el producto comercial "Dimecron 50", conocido vulgarmente como "Fosfamidón" y usado ampliamente para combatir las plagas de los cultivos de arroz en Cuba. Como método básico de experimentación se eligió el de bioensayos; estos se llevaron a cabo de acuerdo a las instrucciones dadas por la American Public Health Association (A.P.H.A.) en 1963. Las temperaturas de las soluciones de ensayos fluctuaron entre 21,08 y 24,80°C, con un valor medio de 23,7°C, manteniéndose las variaciones dentro de los límites recomendados por la A.P.H.A. Para cada ensayo se usaron de 6 a 12 ejemplares de langosta, los que se mantuvieron en recipientes plásticos de 60 litros, en forma aislada para evitar un posible efecto de grupo. Los ejemplares fueron colectados en una misma localidad del Golfo de Batabanó y transportados a los tanques de aclimatación, donde se mantuvieron con circulación constante de agua a razón de 75 a 80 litros por hora. La talla de los ejemplares de experimentación osciló entre 260 y 385 mm (longitud total); la longitud del ejemplar más grande no excedió en 1,5 veces la del ejemplar más pequeño, garantizándose de este modo un resultado estándard de los experimentos. Se aclimataron las langostas durante una semana en condiciones similares a las de los bioensayos, suspendiéndose la alimentación 48 horas antes del comienzo del ensayo. La mortalidad en los tanques de aclimatación fué del 5%. Se mantuvieron todos los ejemplares en las mismas condiciones de alimentación para evitar que el ayuno variado produzca cambios en la susceptibilidad

de las langostas a los plaguicidas, ya que Portmann (1968) señaló que el ayuno intensifica la acción del producto químico. El agua para los experimentos fué obtenida en una región costera que se supuso no contaminada. El plaguicida se diluyó aereándose el agua durante 30 minutos; el contenido en oxígeno disuelto (OD) de las soluciones de prueba nunca descendió de 1,2 ml O_2, siendo ésta una concentración no perjudicial para la langosta; en efecto, Suárez y Ramiro (MS) señalaron como concentración letal para esta especie la de 0,7 ml por litro a temperatura de 24°C.

A través de bioensayos exploratorios se determinó que las concentraciones de "Dimecrón 50" requeridas se encontraban entre 2,5 y 0,1 ppm; se experimentó enseguida con una serie de 4 a 6 concentraciones, llegándose finalmente a 8 concentraciones diferentes (6,3; 2,5; 1,0; 0,7; 0,5; 0,35; 0,2; y 0,1 ppm). Culley y Ferguson (1969) usaron tres concentraciones diferentes como número mínimo, para determinar el efecto de 28 plaguicidas en *Gambusia affinis*, expresado como DL$_{50}$.

El valor de la DL$_{50}$ se determinó graficamente por el método de interpolación gráfica de lineas rectas. El valor umbral, o sea, la concentración mínima a la cual un ejemplar está fuera de peligro, se determinó por un gráfico, en el cual en la ordenada se colocan las horas de supervivencia y en la abcisa, las concentraciones del plaguicida.

Resultados

Las observaciones de la supervivencia a 24; 48; 72; 96 horas (Cuadro 1), permitieron construir los gráficos de interpolación (fig 1), por los cuales se calcularon las DL$_{50}$. Estos valores para el Dimecrón están comprendidos entre 0,3–1,7 ppm; para las ratas, la DL$_{50}$ es de 1–7,5 ppm (CIBA, 1967). Una permanencia para *Panulirus argus* de 96 horas en este plaguicida hace que la potencia (Finney, 1964) de la DL$_{50}$ sea 5,7 veces mayor que la de 24 horas.

Se realizó un análisis de variancia (Dixon y Massey, 1966) para conocer si había diferencia entre los ejemplares que morían en las primeras 24 horas, estudiándose si la talla (entre 260–385 mm) influía o no en la muerte. Se observó que no hay diferencia significativa para un $x = 0,05$ entre las tallas, tampoco alguna diferencia con respecto al sexo.

El valor umbral para *Panulirus argus*, en concentraciones de ensayo de Dimecrón, está comprendido entre 0,2 y 0,3 ppm (fig 2) o sea bajo esta concentración, las langostas están fuera de peligro de efectos agudos. En el pez *Phoxinus laevis*, el valor umbral es de 100 ppm y en ratas de 1 ppm (CIBA, 1967).

Este valor tiene importancia, pues permitirá predecir si se producirá mortalidad en los ejemplares de un área dada, cuando habría una descarga al mar que arrastre concentraciones de Dimecrón superiores a estos valores. Por lo tanto, se impone mantener un control sistemático de las aguas de la plataforma, para descubrir cualquier aumento de este plaguicida en los efluentes que los arrastran a la costa. Debe tenerse en cuenta que las langostas envenenadas pasan antes de morir por un período en el cual pierden su equilibrio y, bien se acuestan en posición completamente dorsal, o se colocan verticalmente con el cefalotórax hacia el fondo y el abdomen hacia la superficie. Estas circunstancias pueden

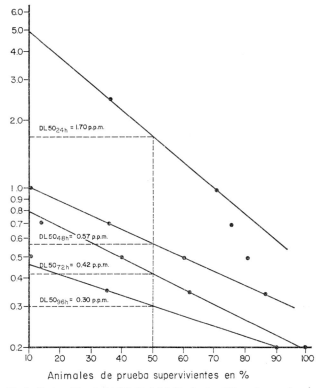

Fig 1. Estimación de la DL50 del "Dimecrón 50" en langostas por interpolación gráfica de línea recta para: 24, 48, 72 y 96 horas

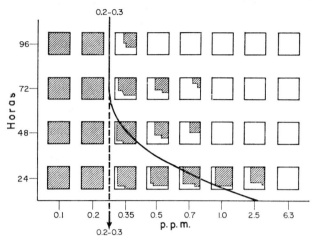

Fig. 2 Determinación del valor umbral en ppm para el Dimecrón; el área sombrada en los cuadrados representa el porcentaje de supervivencia de las langostas; el valor umbral encontrado es de 0.2-0.3 ppm

ser catastróficas ya que, si un efluente trae el plaguicida y provoca este fenómeno y, si la concentración aunque no letal continúa persistiendo en el área, las langostas recibirán nuevas dosis que se acumularán en su cuerpo y, por la pérdida de equilibrio, terminarán muriendo.

En los experimentos de laboratorio se comprobó que la pérdida de equilibrio ocurre en las langostas a una dosis de Dimecrón de 0,7 ppm para un período de 24 horas, con un 75% de animales sobrevivientes.

Características hidrográficas de la plataforma cubana

Actualmente, parece existir gran preocupación para los problemas biológico-pesqueros originados por los diferentes contaminantes vertidos por algunas industrias y actividades agrícolas (Morales, 1968, 1968a, 1969). Sin embargo, es también necesario, como paso previo antes de tomar una decisión, realizar trabajos de investigación sobre las características hidrográficas de las aguas supuestas o realmente contaminadas de las costas de Cuba (Emilsson, Tápanes y Godoy, 1969; Tápanes, 1969, 1970), con el propósito de conocer el régimen hidrológico de las mismas.

La plataforma que rodea las costas de Cuba (fig 3) constituye una región de considerable extensión, 42 000 km² aproximadamente, en relación con la superficie del territorio nacional (14 000 km²) y su profundidad media es muy reducida: entre 6 y 14 m.

La marea en la plataforma es bastante débil con una variación máxima de aproximadamente 50 cm en la región suroriental y de 25 cm en la sudoccidental; sin embargo, a pesar de su escasa amplitud, constituye un factor importante para el intercambio de aguas entre la plataforma y el mar exterior. El movimiento de las aguas sobre la plataforma cubana, además de su importancia hidrográfica, ejerce un papel decisivo en la traslación de las posibles masas de aguas contaminadas.

Con excepción de algunas botellas de deriva lanzadas y recuperadas en las aguas de la plataforma (Tápanes, 1963), no se dispone de datos conclusivos de observa-

CUADRO 1. DATOS PARA LA ESTIMACIÓN DE LAS MEDIANAS DE LOS LÍMITES DE TOLERANCIA POR INTERPOLACIÓN GRÁFICA DE LÍNEA RECTA

Concentración en ppm	Número de langostas de prueba	Langostas sobrevivientes en %			
		24 h	48 h	72 h	96 h
6.30	6	0	0	0	0
2.50	8	43	0	0	0
1.00	10	70	10	0	0
0.70	8	75	37	13	0
0.50	10	80	60	40	10
0.35	8	87	87	62	35
0.20	10	100	100	100	100
0.10	2	100	100	100	100
0.00	6	100	100	100	100

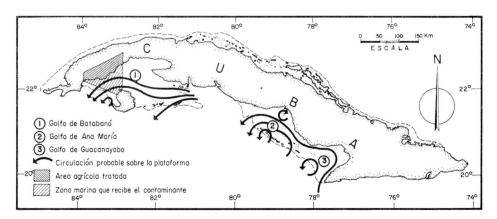

Fig 3. Esquema general de la circulación del agua sobre la plataforma de Cuba, región sur y sud oeste (se indica la zona de contaminación)

[240]

ciones directas sobre la circulación costera. El Instituto de Oceanología de la Academia de Ciencias de Cuba está trabajando activamente en el estudio del sistema de corrientes proprio al Golfo de Batabanó (región sud-occidental) por métodos eulerianos (Blázquez, comunicación personal); también se han hecho estudios de carácter teórico (Tápanes, 1970a) pero el conocimiento que se tiene no es de alcance total.

En razón de que faltan observaciones directas, se intenta presentar a continuación, en base a datos meteorológicos, topográficos e hidrográficos, un esquema general de la circulación de las aguas en la región de la plataforma cubana.

Bajo las condiciones de viento encontradas en la región y dada la escasa profundidad del talud, se presume que las aguas sobre la plataforma deben desplazarse hacia el oeste, con una velocidad promedio de 5 millas náuticas por día. Esto significa que una masa de agua contaminada puede atravesar el Golfo de Batabanó (si el contaminante no se deposita sobre el fondo), de este a oeste, en 30 días, y el Golfo de Ana María, en el mismo sentido y bajo las mismas condiciones, en unos 15 días. En la fig 3 se indica el sentido de desplazamiento eólico promedio del agua sobre la plataforma, suponiendo un viento predominante de ENE, con una velocidad de seis metros por segundo. Este esquema general pretende indicar solamente, la dirección y sentido del movimiento promedio del agua que cubre la región, excluyendo cualquier información sobre su intensidad. No obstante, la topografía irregular que caracteriza la mayor parte de la región perturba, en gran medida, el flujo de agua, dando origen a una distribución muy variada de la velocidad de la corriente.

Influencia de la circulación marina sobre la distribución del contaminante

Sobre la base del esquema general presentado, es posible predecir la ruta que seguirá una masa de agua contaminada sobre la plataforma, y determinar las zonas que se verán afectadas, si se sabe donde se encuentran las fuentes de contaminación. También debido al lento movimiento de traslación de la masa de contaminante, éste puede ejercer su efecto nocivo durante un período de tiempo prolongado (de 15 días a un mes), sobre las especies que habitan la plataforma.

Partiendo del dato de que el contaminante estudiado se riega en una proporción de 16 galones por caballería (1 caballería = 13 hectáreas), es posible calcular la cantidad del mismo por metro cuadrado en una zona dada. Por ejemplo, para una caballería que cubre un área de 130 000 m^2 y en la que el plaguicida se emplea a una concentración de 500 gm/l, es fácil deducir la cantidad por metro cuadrado, la que será de 0.28 gm/m^2, aproximadamente.

Ahora bién, suponiendo que se riegue un área total de 10 000 caballerías (fig 3) esto hace que la cantidad de plaguicida empleada se eleve hasta un total teórico aproximado de 390 toneladas para la zona considerada. Suponiendo además, que sólo un 5% del total del plaguicida regado llegue al mar (por medio de los gastos de los ríos de la región, escurrimiento de agua dulce, etc.), y siendo el volumen de agua que recibe este total de $2,4 \times 10^{10}$ m^3 se estima pues, que la concentración de

plaguicida será de 0,00008 ppm, en base a los valores teóricos anteriormente considerados.

Si se admite una dilución total y homogénea de toda la masa del contaminante considerado, la concentración del mismo (0,00008 ppm) se encuentra muy por debajo de los valores letales determinados en los presentes ensayos para *Panulirus argus;* esto quiere decir que la zona marina en cuestión absorbería por su amplitud hasta mil veces más cantidad de plaguicida, lo cual llevaría la concentración del mismo hasta un nivel de 0,08 ppm, una cantidad todavia no peligrosa para la vida de las langostas.

Conclusiones

De los resultados de los ensayos realizados en laboratorio y de las correlaciones establecidas con los presumibles movimientos del agua costera y distribución del contaminante, se puede llegar a las siguientes conclusiones:

1. La DL_{50} para langostas *Panulirus argus* entre 260 y 385 mm de longitud total mantenidas en agua con temperatura entre 21,0° a 24,8°C durante lapsos de 24, 48, 72 y 96 horas, empleando el insecticida fósforo orgánico "Dimecrón 50", fue de 1,70, 0,57, 0,42 y 0,30 ppm, respectivamente.

2. La potencia de la DL_{50} del Dimecrón, en 96 horas, es 5,7 veces mayor que la para un lapso de 24 horas.

3. El valor umbral está comprendido entre 0,2 y 0,3 ppm.

4. A una concentración de 0,7 ppm y en un lapso de 24 horas, se produce la pérdida de equilibrio para las langostas de ensayos.

5. No hay diferencia con respecto a talla y sexo entre los ejemplares que murieron durante las primeras 24 horas, para el rango de las tallas estudiadas.

6. Bajo las condiciones de viento encontradas sobre la plataforma cubana, las aguas deben desplazarse hacia el oeste, con una velocidad promedio de 5 millas náuticas por día.

7. Sobre la base del esquema hidrográfico general considerado, es posible predecir la ruta que seguirá una masa de agua contaminada y delimitar las zonas que serán afectadas, si se sabe donde se encuentran las fuentes de la contaminación.

8. Con las concentraciones del "Dimecrón 50" actualmente empleadas en Cuba (16 galones por caballería, o sea 0,28 gm/m^2), y siempre que el área de la masa de agua receptora sea lo suficientemente grande con respecto al área de terreno agrícola tratado (3 veces mayor), se producirá una dilución total y homogénea del contaminante sin acumulaciones peligrosas en ciertas áreas críticas y no habrá peligro de consecuencias fatales en las agua de la región de la plataforma, a pesar de su escasa profundidad.

Bibliografía

CIBA   Dimecrón Insecticida sistémico. CIBA, Suiza, Deparmento
1967   agroquimico, 8 p.
CULLEY, D D Jr. and FERGUSON, D E Patterns of insecticide
1969   resistance in the mosquito fish *Gambusia affinis. J. Fish. Res. Bd Can.*, 26:2395-401.
DIXON, W J y MASSEY, F J Introducción al análisis estadístico.
1966   Madrid, Ediciones del Castillo, 489 p.

EMILSSON, I, TÁPANES J. J. and GODOY, J Contribution to the hydro-
1969   graphy of the insular shelf off the Cuban south coast.
          FAO Fish. Rep., (71:1):108 (abstract).

FINNEY, D J Statistical method on biological assay. London,
1964   Charles Griffin & Co., 668 p.

MORALES, J J Desarrollo industrial, pero sin destruir la riqueza
1968   pesquera. *Téc. Pesq.*, 1(1):14–8.

MORALES, J J Contaminación, incapacidad o negligéncia punible.
1968a   *Téc. Pesq.*, 1(3):13–6.

MORALES, J J Contaminación en Orizaba. *Tec. Pesq.*, 2(21):15–6.
1969

PORTMANN, J E Progress report on a programme of insecticide
1968   analysis and toxicity-testing in relation to the marine
          environment. *Helgolander wiss. Meeresunters.*, 17:247–56.

SUÁREZ,, A G y RAMIRO, M E Instrucciones para la manipulación
          y embarque de langostas *Panulirus argus*. Latreille 1804.
          (MS).

TÁPANÈS, J J Afloramientos y corrientes cercanas a Cuba.
1963   *Contrnes Cent. Invest. pesq. Habana*, (17):29 p.

TÁPANES, J J Consideraciones sobre la hidrografía de la plata-
1969   forma de Cuba. *En* Informe a la Primera Reunión sobre
          Planes de Trabajo, Sept. 1969. La Habana, Centro de
          Investigaciones Pesqueras, 20 p.

TÁPANES, J J Contaminantes y su tratamiento. *En* Informe a la
1970   Segunda Reunión sobre Planes de Trabajo, Sept. 1969.
          La Habana, Centro de Investigaciones Pesqueras, 25 p.

TÁPANES, J J Empleo de modelos de flujo planetario estacionario
1970a   en cuencas limitadas, y su aplicación al Golfo da Batabanó,
          Cuba. *En* Informe a la Segunda Reunión sobre Planes de
          Trabajo. La Habana, Centro de Investigaciones Pesqueras,
          8 p.

Chlorinated By-products from Vinyl Chloride Production: A New Source of Marine Pollution

S. Jensen,* A. Jernelov,†
R. Lange‡ and K. H. Palmork§

Les sous-produits chlorés de la production de chlorique de vinyle:
une nouvelle source de pollution des mers

Lors de la chloration des hydrocarbures aliphatiques, de nombreux
composés aliphatiques chlorés (Cl–C) indésirables se forment. Ces
sous-produits, dont le montant s'élève à 75.000 tonnes au moins par
an pour la seule Europe du Nord, sont couramment brûlés dans des
vaisseaux spéciaux ou rejetés à la mer par des bateaux-citernes. On a
étudié le taux d'accumulation des composés de Cl–C dans les organismes biologiques, ainsi que sa dégradation et sa solubilité dans
l'eau de mer. On a procédé à des estimations des concentrations des
composés de Cl–C dans des échantillons d'eau de mer et dans
des organismes marins recueillis lors de croisières et par des stations
à terre, dans la mer du Nord, le long de la côte norvégienne et dans
la mer de Barents. Afin de recueillir des données utilisables en cas
d'évaluation des effets nocifs éventuels des composés de Cl–C,
plusieurs expériences de laboratoire ont été consacrées aux effets
graves de ces substances sur les algues, les invertébrés et les poissons.

Les données préliminaires indiquent que la plus forte concentration de Cl–C trouvée en haute mer n'atteignait qu'environ un
dizième de la valeur la plus faible connue pour provoquer de façon
certaine un effet biologique aiguë.

Los subproductos clorados procedentes de la producción de cloruro de
vinilo, nueva fuente de contaminación marina

Cuando los hidrocarburos alifáticos adquieren cloro se forman
numerosos alifáticos clorados (Cl–C) no deseados. Estos subproductos, cuya cantidad asciende a 75.000 toneladas anuales, por
lo menos, en el norte de Europa únicamente, se queman por lo
general en el mar en barcos especiales, o se arrojan al mar desde los
petroleros. Se ha estudiado el ritmo de acumulación de Cl–C en los
organismos vivos, así como su degradación y solubilidad en el agua
del mar. Se han hecho estimaciones del Cl–C en muestras de agua
de mar y en organismos marinos recogidos durante los cruceros
efectuados, y desde estaciones en tierra en el Mar del Norte, a lo
largo de la costa noruega y en el Mar de Barents.

En los intentos de establecer datos que sirvan en las determinaciones de los posibles efectos perjudiciales del Cl–C se ha llevado
a cabo algunos experimentos de laboratorio sobre los efectos agudos
de estas sustancias en las algas, invertebrados y peces.

Los datos preliminares indican que la concentración más alta de
Cl–C encontrada en alta mar es una décima parte, aproximadamente,
del valor más bajo que produce un inequívoco y agudo efecto
biológico.

UNWANTED chlorinated aliphatic hydrocarbons
(C–Cl) are formed as by-products in vinyl
chloride production. Reports in May 1970 on
dumping of large quantities of C–Cl into the sea,
initiated a joint Norwegian/Swedish research programme.

The first report from the research vessel *Johan Hjort*,
indicated the problem was even more serious than was
originally assumed. It stated: ". . . high densities of
particles were observed, and within some areas the sea
was coloured red-white. . . . Fish in bad condition were
observed and the particles could be seen down to a depth
of about 2 to 3 m. They looked like dead plankton."

Materials and methods

Samples were taken on various cruises in the North
Atlantic from 18 June to 28 September 1970. Most of the
biological material was sampled by other means along
the Norwegian coast. Fifty species of animals were
collected (invertebrates, elasmobranchs, teleosts, birds
and grey seal). Samples of about 10 g were collected,
wrapped in Al-foil, quickly deep frozen and stored
ready for analysis.

Toxicity tests

Primary production was measured at room temperature
according to the method by Steemann-Nielsen (1952)
using C^{14}-bicarbonate (Isotope Laboratory, Charlottenlund Slot, Denmark. 1 μCi/ampulla). Aqueous solutions
of the by-products (C–Cl) were prepared by mixing the
tar with synthetic sea water (Andersen and Føyn 1969)
for 3 h. This solution (80 μl/l SW) was added to the
reaction vessels (125 ml) in amounts which gave the
desired final concentration. One ml of an algae suspension
was then added and pre-incubated for 1, 2, 3 and 4 h,
respectively, before one ampulla C^{14}-bicarbonate was
added to each vessel. A pure culture of *Dunaliella* sp. was
used in most experiments, whereas in others, the algae
(*Chlorella stigmatophora* and *Porphyridium violaceum* +
some dead diatoms) present in surface water (S =22.22
per thousand, t = 17.8°C) from the mouth of the Oslo
fjord were used.

Tests on the acute toxicity were performed on cod
(*Gadus morrhua*), plaice (*Pleuronectes platessa*) and the
invertebrates *Leander adspercus* (Crustacea), *Mytilus
edulis* (Mollusca) and *Ophiura texturata* (Echinodermata)

* Environmental Protection Board, Uppsala 7, Sweden.
† Swedish Water and Air Pollution Research Laboratory, Stockholm, Sweden.
‡ Institute of Marine Biology, University of Oslo, Oslo, Norway.
§ Institute of Marine Research, Directorate of Fisheries, Bergen, Norway.

according to methods described by Litchfield and Wilcoxon (1949). The test populations were also analysed for heterogenity regarding sensitivity to C–Cl by χ^2 (chi-square)-test in accordance with the observed values to straight dose/effect line. The tests were made in 60 l glass aquaria at 13 to 18°C with a salinity of 30 to 35 per thousand. The LC_{50}-values obtained were plotted against time (fig. 2). *Mytilus* exhibited a pronounced latent period before reacting to C–Cl. They were exposed for a maximum of 96 h and then transferred to pure sea water and observed for 11 days.

Chemical analysis of C–Cl

Isolation of C–Cl was achieved by co-distillation with water and cyclohexane. The procedure for biological samples was: 10 g of homogenized tissue in 250 ml distilled water, 3 ml cyclohexane and 1 ml antifoaming agent were set up for downward distillation in a 1-litre round bottomed flask. The distillation was ended when the 3 ml of cyclohexane and 25 ml of water had passed over and into a 25 ml volumetric flask. The procedure for sea water samples was: one litre unfiltered sea water was distilled with 3 ml cyclohexane. The separation of the C–Cl components, now quantitatively dissolved in the cyclohexane layer, was achieved by gas liquid chromatography. Ten μl of the cyclohexane distillate were injected on a gas chromatograph fitted with an electron capture detector. The chromatograms were then compared to those obtained using a standard solution made from samples of the by-product of vinyl chloride production by the oxychlorination method. (The composition of the by-product is given in Table 1). A chromatogram (fig 1B) is a positive identification if all retention times of all peaks are in agreement with those of the standard (fig 1 A). If the intensities of the peaks also agree, a rough quantitation is possible (fig 1 A and B). In other cases

(fig 1 E) only some of the retention times correspond with those of the standard (fig 1 D). The result is then only indicative and quantitation is impossible at present. In still other cases peaks appear which are absent in the standard solution of the by-product. These components may be by-products obtained during chlorination of other types of aliphatic hydrocarbons, but as the electron capture detector used is not specific to chlorinated hydrocarbons, they may not contain chlorine at all.

Results

C–Cl has a wide distribution but in the Atlantic Ocean and the Barents Sea only surface water samples have been analysed. In some cases, i.e. Skagerrak and along the Norwegian coast, midwater and bottom water have also been taken. These samples show that C–Cl is not confined to the surface water and that the peak pattern of the chromatograms changes somewhat with depth. This may be because C–Cl undergoes changes in the eutrophic zone comparable to those experimentally induced by exposure to ultraviolet light, in addition to the changes caused by the biological activity within this layer.

Relatively few biological samples have been analysed (about 80 samples from 18 different species). Although C–Cl is absent in some specimens tested and unfiltered water samples analysed, the data strongly indicate that C–Cl originates from industries on both sides of the Atlantic and is widely distributed.

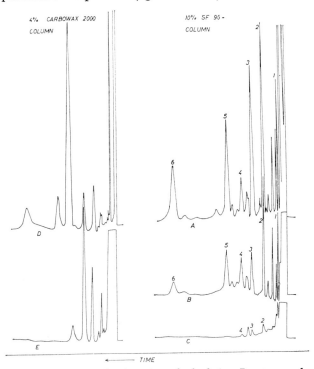

Fig 1 Chromatograms showing: A standard solution; B water sample, equal to 0.6 ng/g ppb, taken close to Iceland; C distilled water (1 litre); D Standard solutions as A; E 10 g fish sample (Saithe sampled at NB 62°20', EL 10°25')

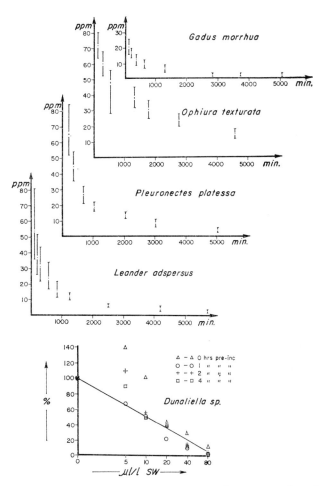

Fig 2 Toxicity curves for C–Cl on four species of marine animals and the effect on the photosynthetic activity of Dunaliella sp.

TABLE 1. COMPOSITION OF BY-PRODUCTS FROM VINYL CHLORIDE PRODUCTION USING THE OXYCHLORINATION PROCESS (ANALYSIS MADE BY A PETROCHEMICAL FACTORY AT LAKE CHARLES, U.S.A.)

		Mv	Bp	Weight %
1.2–Dichloroethane (EDC)	$CH_2Cl–CH_2Cl$	99.0	83.6°C	20
Trichloroethane	$CHCl = CCl_2$	131.4	87	1
1.1.2–Trichloroethane	$CHCl_2–CHCl_2$	133.4	113.5	67
1.2–Dichlorobutane	$CH_3–CH_2–CHCl–CH_2Cl$	127.0	124	2
1.3–Dichlorobutane	$CH_3–CHCl–CH_2CH_2Cl$	127.0	131	3
asym. Tetrachloroethane	ClC_3CH_2Cl	167.9	130.5	1
Monochlorobenzene	C_6H_5Cl	112.6	132	
1.3–Dichloro–2–butene, cis and trans	$CH_3–CCl = CHCH_2Cl$	125	128–30	0.5
sym. Tetrachloroethane	$CHCl_2CHCl_2$	167.9	146.3	
1.4–Dichloro–2–butene, cis and trans	$CH_2ClCH = CH–CH_2Cl$	125	152.5	2
1.4–Dichlorobutane	$CH_2ClCH_2CH_2–CH_2Cl$	127	161–63	
2.3.4–Trichloro–1–butene	$CH_2ClCHClCCl = CH_2$	159.4	$60^{20\ mm}$	0.2
1.1.2–Trichlorobutane	$CCl_2–CHClCH_2CH_3$	161.5		0.3
1.3.4–Trichloro–1–butene, cis and trans	$CH_2ClCHClCH = CHCl$	159.4		0.3
1.2.3–Trichlorobutane	$CH_3CHClCHClCH_2Cl$	161.5	165–67	
bis (2-Chloroethylether)	$(CH_2ClCH_2)_2O$	143	178	0.3
Pentachloroethane	$CHCl_2CCl_3$	203.3	162	1
1.2.4–Trichlorobutane	$CH_2ClCH_2CHClCH_2Cl$	161.5		0.3
1.2.4–Trichloro–2–butene	$CH_2ClCH = CCl–CH_2Cl$	159.4	67^{10}	1.1
				100.0 %

Toxicity tests

It appears from fig 2 that the concentrations of C–Cl compatible with the "indefinite survival time" range from 2 to 20 ppm. Cod and plaice, are the most susceptible, followed by *Leander adspercus* and *Ophiura texturata*. No heterogenity was observed in any of these populations. Due to the latent period between exposure and death in the common mussel, a complete toxicity curve cannot be drawn. The concentration of C–Cl which gives the "indefinite survival time", however, appears to be about 20 ppm.

As for C–Cl's effect on photosynthesis, it appears (fig 2) that about 13 ppm (10 μl by-product/l SW) causes an immediate 50 per cent reduction. In some cases the lowest concentrations tested stimulated the photosynthesis of *Dunaliella* sp. Other experiments, using algae samples in Oslo fjord, confirmed this.

Although concentration figures have been omitted in this paper it can be stated that a water sample from one station showed about 0.2 ppm C–Cl which is only one order of magnitude lower than the lowest concentration giving rise to an immediate effect. That figure illustrates the possible risk but it should be stressed that this was the highest observed so far.

Degradation and accumulation

1.4 ppm of C–Cl was added to 1 m³ of sea water sampled at 71°NB, 20°EL and held at 9°C. The loss of the main components (2 to 6 in fig 1A) was followed for 9 days.

On the second day a glass bottle filled with 1 litre of water and 70 mg of *Calanus finmarchicus* was added. After 2 days at 4°C the plankton was separated into dead and live specimens and analysed. The accumulation coefficient for components 2, 3, 4, 5, and 6 was 23, 0, 102, 94, and 164 respectively for living *Calanus* and 0, 0, 0, 8, and 24 for dead specimens.

TABLE 2. THE DISAPPEARANCE OF C–Cl DISSOLVED IN WATER

Incubation time	Amounts of componds 2–6 (in % of theoretical)				
	2	3	4	5	6
0 days	90	97	82	96	95
3 ,,	19	85	74	94	86
5 ,,	8	92	54	82	75
7 ,,		72	55	91	78
9 ,,		64	42	67	68

Conclusions

The data secured demonstrate the omnipresence of C–Cl in North Atlantic waters, and also, that C–Cl is a danger to marine organisms. This paper is however based on a very short period of research (July to 29 October 1970), and because of that and because the products made from chlorinated hydrocarbons are so widespread that some sources of error may have been overlooked and samples with small peaks on the chromatograms corresponding to levels from 0.1 to 0.6 ng/g (ppb) have been regarded as negative, we feel further investigation is needed.

Although the degradation experiments showed that the various C–Cl components disappeared at faster or slower rates it was also found that in water with no biological activity, C–Cl remained unchanged up to 27 days at room temperature.

References

ANDERSEN, A T and FØYN, L Common methods for analysis of sea
1969 water. *In* Chemical oceanography: an introduction, edited by R. Lange, Oslo, Universitetsforlaget, pp. 111–47.
LITCHFIELD Jr., J T and WILCOXON, F A simplified method of
1949 evaluating dose effect experiments. *J. Pharmac. exp. Ther.*, 96:99.
STEEMANN NIELSEN, E The use of radioactive carbon (C¹⁴) for
1952 measuring organic production in the sea. *J. Cons. perm. int. Explor. Mer.* 18(2):117–40.

Action *in vitro* de Détergents sur Quelques Espèces Marines

G. Bellan, J.-P. Foret, P. Foret-Montardo et R. A. Kaim-Malka*

Action *in vitro* of detergents on some marine species

This study covers the action of 20 non-ionic detergents and 15 anionic base product detergents on *Scolelepis fuliginosa* and *Capitella capitata* (polychetes), *Mytilus galloprovincialis* (mollusc), *Sphaeroma serratum* (crustacean). The 12 concentrations examined fall between 0.1 and 800 mg/litre. The following have been defined:

1. For each concentration, the time required to kill an individual or all individuals subjected to test.
2. The median tolerance limit after 48 h and 96 h. The value and limits of each of these data are discussed. The greatest frequency of $DL50_{48h}$ for anionic detergents lies between:

  10–25 mg/l for *Scolelepis*
  1–10 mg/l for *Capitella*
  over 800 mg/l for *Mytilus* and *Sphaeroma*

For $DL50_{96h}$ the greatest frequencies are:

  0.1– 1 mg/l and 10–25 mg/l for *Scolelepis*
  1–5 mg/l for *Capitella*
  10–25 mg/l and over 800 mg/l for *Sphaeroma*
  5–25 mg/l for *Mytilus* collected in polluted water
  5·25 mg/l for *Mytilus* collected in pure water.

For non-ionic detergents, the greatest frequencies of $DL50_{48h}$ and $DL50_{96h}$ are lower:

Scolelepis $DL50_{48h}$: 0.5– 5 mg/l
Capitella $DL50_{48h}$: 1– 5 mg/l $\Big\}$ $DL50_{96h}$: 0.1–2.5 mg/l
Mytilus $DL50_{48h}$: 1–25 mg/l
  $DL50_{96h}$: 0.5– 5 mg/l
Sphaeroma $DL50_{48h}$: 10–100 mg/l
  $DL50_{96h}$: 5– 50 mg/l

Non-ionic detergents are from 1.5 to 2 times more active than anionic detergents. Polyoxyethylenated straight-chain alcohols are more dangerous than alkyl aryl sulphanates. The four species may be classified in order of increasing sensitivity: *Sphaeroma*, *Mytilus* collected in polluted water, *Mytilus* collected in pure water, *Capitella* and *Scolelepis*.

Acción *in vitro* de los detergentes sobre algunas especies marinas

Se examina en este estudio la acción de 20 detergentes no-iónicos y de 15 aniónicos (productos de base) sobre *Scolelepis fuliginosa* y *Capitella capitata* (poliquetos), *Mytilus galloprovincialis* (molusco), *Sphaeroma serratum* (crustáceo). Las 12 concentraciones estudiadas están comprendidas entre 0,1 y 800 mg/l. Se han definido:

1. Para cada concentración, el tiempo necesario para matar un ejemplar, o todos los ejemplares del experimento.
2. La tolerancia límite media al cabo de 48 horas y de 96 horas. Se examinan el valor y los límites de cada uno de estos datos. La frecuencia más grande de la $DL50_{48h}$ para los detergentes aniónicos se sitúa entre:

  10–25 mg/l para *Scolelepis*
  1–10 mg/l para *Capitella*
  superior a 800 mg/l para *Mytilus* y *Sphaeroma*.

Para la $DL50_{96h}$ las frecuencias más grandes son:

  0,1– 1 mg/l y 10–25 mg/l para *Scolelepis*
  1– 5 mg/l para *Capitella*
  10–25 mg/l y más de 800 mg/l para *Sphaeroma*
  5–25 mg/l para *Mytilus* recogidos en agua contaminada
  5–25 mg/l para *Mytilus* recogidos en agua pura.

Para los no-iónicos, las frecuencias más grandes de $DL50_{48h}$ y $DL50_{96h}$ son reducidas:

Scolelepis $DL50_{48h}$: 0,5– 5 mg/l
Capitella $DL50_{48h}$: 1– 5 mg/l $\Big\}$ $DL50_{96h}$: 0,1–2,5 mg/l
Mytilus $DL50_{48h}$: 1– 25 mg/l
  $DL50_{96h}$: 0,5– 5 mg/l
Sphaeroma $DL50_{48h}$: 10–100 mg/l
  $DL50_{96h}$: 5– 50 mg/l

Los detergentes no iónicos son de 1,5 a 2 veces más activos que los aniónicos. Los alcoholes en cadena normal polioxietilenos son más peligrosos que los sulfonatos alkyl-arylos. Las cuatro especies se pueden clasificar por orden de sensibilidad creciente: *Sphaeroma*, *Mytilus* recogidos en agua contaminada, *Mytilus* recogidos en agua pura, *Capitella*, *Scolelepis*.

BELLAN-SANTINI (1964, 1969) a montré que, parmi les facteurs de pollution jouant un rôle très important dans la destruction des populations naturelles, figurait l'action des détergents issus de la pétroléochimie. Nous avons donc été amenés à étudier expérimentalement cette action sur divers organismes marins caractéristiques de peuplements soumis à l'influence de pollutions variables en nature et en intensité (voire d'absence de pollution). Cette étude a donné lieu à une note préliminaire dans laquelle est exposée l'action des détergents sur la polychète *Scolelepis fuliginosa* (Bellan *et al.*, 1969).

Extrait d'un travail concernant l'action de 50 détergents sur 9 espèces animales, nous présentons, ici, l'action *in vitro* de 35 détergents sur 4 espèces animales. Il s'agit de produits de base et non de préparations commerciales (les produits commercialisés sont constitués, outre le produit de base, par un solvant et divers composants plus ou moins neutres). Les préparations commerciales seront étudiées ultérieurement, quand l'action des composants essentiels sera mieux connue. Les 35 détergents se répartissent en 15 anioniques et 20 non-ioniques. Parmi les espèces étudiées, nous avons choisi pour le présent travail:

Polychètes: *Scolelepis fuliginosa, Capitella capitata*.
Mollusques: *Mytilus galloprovincialis* provenant de populations vivant en eau pure,
  Mytilus galloprovincialis provenant de populations vivant en eau polluée.
Crustacé: *Sphaeroma serratum*.

Dans le milieu marin, l'action des détergents est la résultante de l'action particulière de chacun des types de détergents rejetés. Aussi, si nous avons procédé à l'étude *in vitro* de l'action de chaque produit, on peut présenter, sous une forme globale, synthétique, ces résultats obtenus *in vitro*. Cette synthèse représente mieux ce qui se passe dans le milieu naturel.

Processus expérimental

Les animaux ont été récoltés au sein d'une même population. Les concentrations choisies sont (en mg/l d'eau de mer): 0,1; 0,5; 1; 2,5; 5; 10; 25; 50; 100; 200; 400; 800. Par comparaison, nous signalerons que les doses (anioniques seuls) sont, au voisinage immédiat d'un émissaire d'eaux usées, de 50 à 100 mg/l (exceptionnellement, 400 mg/l) et, à 5 milles de celui-ci, de 0,1 mg/l, dose limite dosable avec certitude. Dix individus sont mis en expérience pour chaque concentration et chaque produit. Des témoins sont utilisés. Ainsi, nous avons pu préciser:

—Pour chaque concentration, le temps nécessaire pour tuer tous les individus mis en expérience. Nous avons préféré envisager le facteur temps, plutôt que les notions de dose minimale mortelle (DMM) et dose létale 100 pour cent (DL100);

—La tolérance limite médiane au bout de 48 h et de 96 h, qui est la concentration qui tue 50 pour cent des individus mis en expérience en 48 h et en 96 h (=DL50). En fait, ce que nous désignons par DL50 correspond beaucoup plus à un intervalle de concentrations dans lequel se situe la DL50 effective de chacun des produits

---

*Station Marine d'Endoume, 13 Marseille, France.

testés. L'interpolation usuelle de la DL50 nous paraît délicate. Cette notion ne peut être appliquée d'une manière rigoureuse: ainsi que nous l'indiquerons plus loin, il peut y avoir, pour un même produit, un même organisme, un même laps de temps, au cours d'une même série d'expériences, deux DL50 comprises, chacune, entre deux intervalles de deux concentrations bien distinctes. Ceci peut s'expliquer par le fait que les détergents auraient des actions différentes selon les concentrations (faibles ou fortes).

Etude de la mortalité totale

Détergents anioniques

Nous envisagerons l'action des produits pour une période de 24 h, 48 h, 96 h et au-delà. Nous nous contenterons d'indiquer, pour chacune de ces périodes, la "gamme de concentrations agissantes", c'est-à-dire les concentrations maximale et minimale ayant tué tous les animaux mis en expérience dans le laps de temps considéré. Nous indiquerons, de même, le pourcentage de détergents ayant, pour ces concentrations et le laps de temps considéré, agi, c'est-à-dire ayant tué tous les animaux mis en expérience; nous les appellerons "produits exterminateurs".

Pour 24 h

Espèces	Gamme des concentrations agissantes	Pourcentage de produits exterminateurs
Capitella	10–800 mg/l	21–100
Scolelepis	10–800 mg/l	21–100
Sphaeroma	100–800 mg/l	7

Mytilus (eau pure et eau polluée): aucun produit exterminateur

Pour un laps de temps relativement court, les Mytilus résistent bien aux détergents anioniques; il en est de même pour Sphaeroma (qui est cependant un peu plus sensible). Les polychètes sont sensibles aux détergents, avec une action un peu plus marquée pour Scolelepis.

Pour 48 h

Espèces	Gamme des concentrations agissantes	Pourcentage de produits exterminateurs
Scolelepis	1– 50 mg/l	7–95
Capitella	1–100 mg/l	7–93
Mytilus (eau polluée)	50–800 mg/l	7–13
Sphaeroma	25–800 mg/l	7–20

Aucun produit n'est exterminateur sur Mytilus en eau pure. Nous pouvons faire sur ces résultats les mêmes remarques que précédemment.

Pour 96 h

Espèces	Gamme des concentrations agissantes	Pourcentage de produits exterminateurs	dont entre 10–0,1 mg/l
Scolelepis	0,1– 10 mg/l	7–62	7–62
Capitella	0,1– 25 mg/l	20–87	20–69
Mytilus (eau pure)	25 –800 mg/l	20–53	
Mytilus (eau polluée)	25 –800 mg/l	13–53	
Sphaeroma	2,5–800 mg/l	7–47	0–7

Capitella se révèle être un peu plus sensible que Scolelepis; pour les autres espèces, les résultats sont conformes à ce qui a déjà été observé.

Au-delà de 96 h

Espèces	Gamme des concentrations agissantes	Pourcentage de produits exterminateurs
Scolelepis	0,1– 50 mg/l	7–95
Capitella	0,1–100 mg/l	20–93
Mytilus (eau pure et eau polluée)	0,1–800 mg/l	0–53
Sphaeroma	0,1–800 mg/l	0–46

Pour un laps de temps relativement long, parmi les polychètes, Capitella est un peu plus sensible aux détergents que Scolelepis. Les Mytilus sont moins résistantes que Sphaeroma.

DL50

Nous indiquerons les concentrations minimale et maximale qui, en fonction de l'ensemble des détergents testés, ont provoqué les DL50 au bout de 48 h et de 96 h. Le détergent le plus toxique provoquera la DL50 à la concentration la plus faible, le moins toxique à la concentration la plus forte.

Espèces	$DL50_{48h}$	$DL50_{96h}$
Capitella	1–10 mg/l	1 – 5 mg/l
Scolelepis	10–25 mg/l	0,1– 1 mg/l et 10–25 mg/l
Mytilus (eau pure)	>800 mg/l	5 –25 mg/l
Mytilus (eau polluée)	>800 mg/l	5 –25 mg/l
Sphaeroma	>800 mg/l	10 –25 mg/l et 800 mg/l

Ces résultats confirment les données précédentes.

Détergents non ioniques

Pour 24 h

Espèces	Gamme des concentrations agissantes	Pourcentage de produits exterminateurs
Capitella	5–800 mg/l	20–100
Scolelepis	5–800 mg/l	25–95
Mytilus (eau pure)	200–800 mg/l	5–15
Mytilus (eau polluée)	25–800 mg/l	5–15
Sphaeroma	25–800 mg/l	5–55

La sensibilité des animaux, vis-à-vis des non ioniques, présente des modalités proches de celles des anioniques.

Pour 48 h

Espèces	Gamme des concentrations agissantes	Pourcentage de produits exterminateurs
Capitella	0,1–200 mg/l	0–90
Scolelepis	0,1–800 mg/l	5–100
Mytilus (eau pure)	10 –800 mg/l	15–60
Mytilus (eau polluée)	25 –800 mg/l	20–35
Sphaeroma	10 –800 mg/l	5–75

Il est important de remarquer que les produits non ioniques agissent sur *Mytilus*, ce qui n'était pas le cas des anioniques pour des durées d'action identiques.

Pour 96 h

Espèces	Gamme des concentrations agissantes	Pourcentage de produits exterminateurs
Capitella	0,1–100 mg/l	15–90
Scolelepis	0,1–200 mg/l	15–95
Mytilus (eau pure)	1 –800 mg/l	5–95
Mytilus (eau polluée)	2,5–800 mg/l	5–90
Sphaeroma	0,1–800 mg/l	5–80

Au-delà de 96 h

Espèces	Gamme des concentrations agissantes	Pourcentage de produits exterminateurs
Capitella	0,1–100 mg/l	15–90
Scolelepis	0,1–200 mg/l	15–95
Mytilus (eau pure)	0,1–800 mg/l	0–95
Mytilus (eau polluée)	0,1–800 mg/l	0–90
Sphaeroma	0,1–800 mg/l	5–75

Pour les trois dernières espèces citées, les produits sont très peu exterminateurs entre 0,1 et 2,5 mg/l.

DL50

Espèces	$DL50_{48h}$	$DL50_{96h}$
Capitella	1 – 5 mg/l	0,1–2,5 mg/l
Scolelepis	0,5– 5 mg/l	0,1–2,5 mg/l
Mytilus (eau pure)	1 –25 mg/l	0,5–5 mg/l
Mytilus (eau polluée)	1 –25 mg/l	0,5–5 mg/l
Sphaeroma	10 –100 mg/l	5 –50 mg/l

Comparaisons

De l'ensemble des résultats obtenus, nous pouvons déduire que nous avons dans l'ordre de résistance croissante: polychètes, *Mytilus*, *Sphaeroma*. Les polychètes ont à peu près la même sensibilité vis-à-vis des détergents. Cette sensibilité varie légèrement selon la concentration, le laps de temps considéré et la nature du détergent (anionique ou non ionique). Il en est de même pour les *Mytilus* provenant d'eau pure ou d' eau polluée.

Si l'on compare les résultats obtenus pour les anioniques et les non ioniques, nous observons que les non ioniques sont plus actifs que les anioniques, en particulier les pourcentages des produits exterminateurs sont plus élevés parmi les anioniques. Par ailleurs, pour 24 h et 48 h, l'action des anioniques est faible ou nulle, tandis que, pour les mêmes périodes, les non ioniques ont un effet plus important. De surcroît, les DL50 par 48 h et 96 h sont abaissées dans le cas des non ioniques. Il en ressort que les détergents non ioniques sont de 1,5 à 2 fois plus nocifs vis-à-vis des animaux testés que les anioniques et que leur action est plus rapide. Une étude actuellement en cours sur différents groupes d'animaux permettra de préciser l'action des différentes familles chimiques de détergents à l'intérieur des deux grandes catégories (anioniques et non ioniques).

Conclusion

Les produits détergents non ioniques (produits de base) s'avèrent être plus dangereux que les anioniques. Pourtant les non ioniques tendent à remplacer, légalement, ces derniers parce qu'ils seraient biodégradables. Or des expériences préliminaires ne nous ont pas montré de diminution de toxicité avec le temps, pour les détergents réputés biodégradables. Ceci risque d'avoir des conséquences graves pour l'équilibre du milieu naturel. Il convient de pousser plus avant les recherches, en particulier sur les produits nouveaux qui apparaissent chaque jour sur le marché, produits qui se veulent tous plus puissants et plus efficaces, mais dont les effets risquent d'être irrémédiables sur le milieu naturel. Il convient surtout de rechercher, dans le milieu marin, des espèces à partir desquelles les résultats expérimentaux obtenus auront une valeur générale et il faudra, aussi, adapter au milieu marin les méthodes expérimentales utilisées en milieu terrestre et dans les eaux continentales.

Bibliographie

Bellan, G, *et al.* Contribution à l'étude de différents facteurs
1969 physicochimiques polluants sur les organismes marins. 1. Action des détergents sur la polychète *Scolelepis fuliginosa* (note préliminaire). *Téthys*, 1(2):367–74.

Bellan-Santini, D Influence de la pollution sur quelques
1965 peuplements superficiels de substrats rocheux. *In* Symposium sur les pollutions marines par les micro-organismes et les produits pétroliers, C.I.E.S.M.M., Monaco, avril 1964, 123–6.

Bellan-Santini, D Contribution à l'étude des peuplements
1969 infralittoraux sur substrat rocheux. (Etude qualitative et quantitative de la frange supérieure). *Recl Trav. Stn mar. Endoume*, (63–47): 5–294.

Foret-Montardo, P Etude de l'action des produits de base
1970 entrant dans la composition des détergents issus de la pétroléochimie, vis-à-vis de quelques espèces animales marines. Thèse de spécialité, mention Océanographie biologique U.E.R. "Sciences de la mer", Université Aix-Marseille II, 30 juin 1970. Aussi publié dans *Téthys*, 2(3) et (4).

ANNEXE 1

Détergents anioniques (15 produits)

Produits contenant un groupement aryl et un groupement alkyl:

+Acides: E6 et E7
+Sels: Phosphates: Celanol PS19
         Sulfonates: Hexaryl L30, Perolène SPZ
         Sulfates: Celanol 251.

Produits dérivés de l'alcool laurique:

+Phosphates: Laural LA, Laural LP4A
+Sulfates: Laural 729, Laural LS, Mélanol L90, Neopon Lam.

En marge de ces groupes chimiques, il faut placer 3 produits:

MSK, composition inconnue, Syntaryl A990 et Lensex LE40, des mélanges de détergents.

Détergents non ioniques (20 produits)

Alkylaryls polyoxyéthylénés:

Lensex TA01, Cémulsol NPT9, Syntopon C et NP 936 et OP 1062.

Produits contenant une fonction alcool:

Plurafac RA 30, Plurafac RA43, Cémulsol DB312, Cémulsol 870, Alcool Oxo 431 Nam.

Produits correspondant à une condensation de l'oxyde d'éthylène et de l'oxyde de propylène sur un alcool:
   Pluronic L61 R, Pluronic L62.
Acide oléique condensé sur 14 molécules d'oxyde d'éthylène: E2.
Produits possédant une fonction éther:
   El, Simulsol 330M.

Ester oxyéthyléné: Cémulsol B.
Amine grasse oxyéthylénée: Ethomeen C25.
Composition inconnue: DO 60, OP 9, Oxane Z.

Les détails sur la composition chimique des détergents seront présentés dans une thèse actuellement en cours de publication (Foret-Montardo, 1970).

Toxicity and Degradation of Tensides in Sea Water

H. G. W. Mann*

Toxicité et dégradation des détergents dans l'eau de mer
Entre 1950 et 1960, les détergents synthétiques ont pris la place des savons "naturels" jusqu'alors utilisés aussi bien au foyer que dans l'industrie. Les détergents arrivent dans l'eau par l'intermédiaire des réseaux d'égouts et passent ensuite dans les zones côtières et dans les eaux salées et saumâtres. Il faudrait donc entreprendre des études pour déterminer si la salinité influe sur la toxicité des détergents. Un certain nombre d'expériences ont été faites sur les anguilles et d'autres organismes présents dans les eaux côtières. Toutes ces expériences ont montré que la toxicité des détergents augmentait proportionnellement à la salinité. Selon les recherches effectuées ultérieurement, la tension superficielle décroît en même temps que la salinité. Des travaux réalisés en eau douce nous ont appris qu'une tension superficielle inférieure à 50 dynes/cm est dangereuse pour le poisson. Aussi estime-t-on que l'abaissement de la tension superficielle de l'eau salée est la cause de l'accroissement de la toxicité des détergents dans les eaux saumâtres et salées.

En Allemagne, la loi de 1969 sur les détergents n'autorise la commercialisation que des détergents biodégradables ("doux"). Le tétrapropylène-benzolsulfonate à chaîne alcoylée ramifiée, que l'on utilisait autrefois, se décompose à 30 pour cent alors que les nouveaux produits à chaîne linéaire se dégradent à 90 pour cent, perdant ainsi leurs propriétés nocives. Toutes ces observations concernent les eaux douces. Des recherches ont été faites sur la dégradation des détergents "doux" dans une installation modèle de traitement biologique des eaux usées. Les expériences avec des eaux présentant divers degrés de salinité ont montré que les détergents "doux" introduits se dégradaient à 80–90 pour cent.

Toxicidad y degradación de los agentes tensoactivos en el agua del mar
Entre 1950 y 1960 los agentes tensoactivos sintéticos han sustituido a los jabones "naturales" empleados hasta entonces en el hogar y la industria. Los agentes tensoactivos llegan a los ríos a través del alcantarillado, y entran en las zonas costeras, donde encuentran aguas saladas y salobres. Es necesario, pues, estudiar si la toxicidad de los agentes tensoactivos resulta afectada por la salinidad. Se han realizado varios experimentos con anguilas y otros animales presentes en las aguas costeras, encontrando en todos ellos que la toxicidad de los agentes tensoactivos aumenta con la salinidad.

Las investigaciones han mostrado que la tensión superficial disminuye al aumentar la salinidad. Por los experimentos realizados en aguas dulces se sabe que una tensión superficial inferior a 50 dinas/cm es peligrosa para los peces. Se cree, por tanto, que la disminución de la tensión superficial de las aguas saladas es el factor causante de la mayor toxicidad de los agentes tensoactivos en aguas salobres y saladas. En Alemania se decidió, en relación con la ley sobre los detergentes de 1969, que sólo se admitirían en el comercio los agentes tensoactivos biodegradables (blandos). En la práctica, se descomponía el 30 por ciento tetrapropileno-benzolsulfonato con una cadena alquílica ramificada utilizado precedentemente, mientras que los nuevos productos, con una cadena normal, se descomponen el 90 por ciento, con lo que desaparecen las características insatisfactorias. Todas estas observaciones se refieren a las aguas dulces. Se han realizado investigaciones sobre la degradación de los agentes tensoactivos "blandos" en una planta modelo para el tratamiento biológico de aguas residuales. Durante los experimentos realizados con salinidades diferentes, se descomponía del 80 al 90 por ciento de los detergentes "blandos" introducidos.

I N the last decades, synthetic detergents have replaced soaps in many industrial applications as well as in households. Many undesirable secondary effects occurred which were eliminated by the introduction of biodegradable (soft) tensides (Bock and Wickbold 1963; Bucksteeg 1966). Although the non-biodegradable (hard) tensides are less toxic than the soft tensides (Mann 1965), the threat of ill effects on fish and other aquatic organisms has, nevertheless, been diminished by the introduction of soft tensides. Today the tenside concentration in surface waters is far below the critical values that are toxic for fish and other aquatic life (Bock and Wickbold 1966). The toxicity limits of hard tensides range between 10 and 20 mg/l. A concentration of about 4 mg/l of soft tensides is toxic for sensitive species, such as rainbow trout.

These statements refer to fresh-water organisms (Lüdemann and Kayser 1963; Mann 1965). Tensides are carried with household and industrial waste water by way of rivers into coastal waters. Therefore, the question is whether toxicity values for fresh water also apply to brackish and sea water.

Toxicity tests
During our tests, hard tensides were used as the only way of keeping desired tenside concentration constant.

The animals used were eels (*Anguilla anguilla* L., 8–10 cm long), adult *Gammarus tigrinus* Sexton, and larvae of *Artemia salina* L. These three species tolerate considerable fluctuations in salinity. The tests were done at water temperatures between 19 and 22°C. The varying salinities were obtained by mixing sea and tap water. All tests were from 8 to 24 h. The concentrations of tetra-propylene-benzene-sulphonate that were tested ranged between 5 and 35 mg/l (Bock and Mann 1969).

All tests showed that the toxicity of the tenside increased with increasing salinity. The effect of the

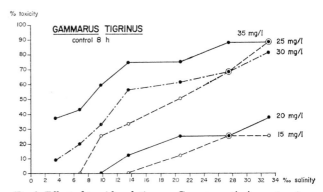

Fig 1. Effect of tenside solutions on Gammarus tigrinus *at various salt concentrations (8 h test) (according to Bock and Mann 1969)*

* Bundesforschungsanstalt für Fischerei, Institut für Küsten-und Binnenfischerei, Hamburg, Federal Republic of Germany.

tenside solution on *Gammarus tigrinus* is an example. During a period of 8 h, the highest concentration of tenside, 35 mg/l at a salt content of $2^0/_{00}$, had a toxicity of 40 per cent; this concentration has a toxicity of almost 100 per cent when the salt content reaches $34^0/_{00}$ (fig 1). Previous tests (Schmid and Mann 1962) showed that the epithelium of the gills of fish is destroyed by the tensides. There is a close relationship between this and the decrease in surface tension caused by the tensides (Bock 1966).

Since all tests were done in the same manner, it was safe to assume that increasing toxicity was due to a change in surface tension under the influence of the salt content. The decrease in the surface tension of the various tenside solutions was, therefore, tested at salt-water concentrations of 3 to $26^0/_{00}$. It appears (fig 2) that the surface tension decreases as the salt content increases.

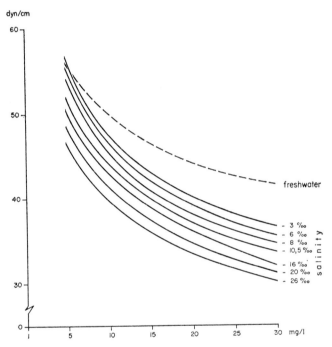

Fig 2. *Decrease in surface tension of tenside solutions at various salt concentrations (according to Bock and Mann 1969)*

Under our test conditions, the concentration becomes lethal for the various species of marine life when the surface tension is reduced to 40 dyn/cm. This corresponds to tenside concentrations of 5 to 10 mg/l. This means that tensides will be harmful only at outlets of concentrated and biologically undegraded waste water.

Degradation of tensides in sea water

All degradation tests with tensides have so far been done only in fresh water (Bock and Wickbold 1963). It was necessary, therefore, to do some tests in brackish and salt water. A 10 mg/l tenside solution was prepared with natural sea water in a 60 l plastic basin, and at intervals of 3 days the content of active detergent (AD) and the surface tension were measured. The tensides degraded within 7 to 10 days. After 2 weeks, the lower limit of traceability was reached. Within the same period, the decrease in surface tension was reversed because after about 2 weeks the critical values of 40 to

50 dyn/cm that are dangerous for fish and other marine species were exceeded (fig 3). Similar results were obtained in a large-scale test whereby the degradation of 20 mg/l AD in a concrete basin containing 120 m³ of sea water was examined under natural conditions. The original amount of tensides degraded after about 10 days to such an extent that only about 2 mg/l could still be traced. After another 4 days the tenside content had dropped to 1 mg/l.

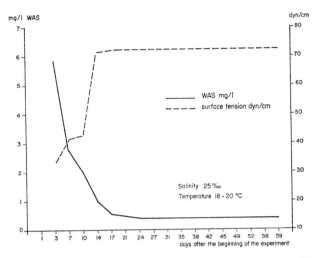

Fig 3. *Degradation of tensides in sea water (initial content 10 mg/l, salt content $25^0/_{00}$, temperature 18–20°C)*

The tests have shown that in sea water the toxic effect of tensides of the same concentration increases with increasing salt content. The cause is the considerable decrease of the surface tension in brackish and sea water. According to our test results, the toxic limit under reduced surface tension lies between 40 and 50 dyn/cm. From the tests on the degradation of tensides in sea water, it is apparent that after a certain time of adaptation the degradation of the tensides takes place in the same manner as in fresh water.

Tests concerning the content of tensides in coastal areas near Cuxhaven (mouth of the River Elbe), near Emden (mouth of the River Ems), and near Tönning (mouth of the River Eider) have shown that the tensides reached a maximum concentration of 0.3 mg/l only in the immediate proximity of the discharge outlet of household and industrial waste water. All other values remained below the limit of exact traceability (0.1 mg/l). This means that, if easily degradable (soft) detergent raw materials are being used that meet the requirements of the law on detergents, the degradation of tensides in coastal areas does not pose a major problem.

References

Bock, K J Uber die Wirkung von Waschrohstoffen auf Fische.
1966 *Arch. FischWiss.*, 17(1):68–77.

Bock, K J and Mann, H Die Bedeutung des Salzgehaltes für die
1969 Wirkung von Tensiden. *Ber. Dt. wiss. Kommn Meeresforsch.*, 20:278–81.

Bock, K J and Wickbold, R Untersuchungen an biologisch
1963 abbaubaren Waschrohstoffen. *Seifen-öle-Fette-Wachse*, 26(12):870–2.

Bock, K J and Wickbold, R Auswirkungen der Umstellung auf
1966 leicht abbaubare Waschrohstoffe in einer grosstechnischen Kläranlage und im Vorfluter. *Vom Wass.*, 33:242–53.

BUCKSTEEG, W Detergentienabbau in Kläranlagen und Vorflutern
1966   vor und nach der Umstellung auf weiche Waschrohstoffe.
        Wass. Luft Betr., 1966:28–31.
LÜDEMANN, D and KAYSER, H Beiträge zur Toxizität von grenz-
1963   flächenaktiven Substanzen (Detergentien) für Fische. *Z.
        angew. Zool.*, 50:229–38.

MANN, H   Die Bedeutung der Waschmittel für die Fischerei. *Fette-
1965   Seifen AnstrMittel*, 67(12):977–80.

SCHMID, O J and MANN, H   Die Einwirkung von Dodecylbenzol
1962   sulfonat auf die Kiemen von Forellen. *Arch. FischWiss.*
        13(1/2):41–51.

Acute Effects of Heated Effluents on the Copepod *Acartia tonsa* from a Sub-Tropical Bay and Some Problems of Assessment

M. R. Reeve and E. Cosper*

Les effets aigus des effluents chauds sur le copépode *Acartia tonsa* dans une baie sub-tropicale et quelques problèmes concernant leur évaluation

La plupart des renseignements concernant les effets des effluents chauds sur les organismes marins proviennent d'expériences faites dans des eaux tempérées septentrionales. Dans les baies et estuaires peu profonds et bordés de mangroves de la Floride méridionale (et de bien d'autres zones semblables du reste du monde), les températures maximales de l'été, à savoir 31–34°C, sont déjà dangereusement proches du seuil létal supérieur pour de nombreuses espèces. Comme ces températures peuvent persister plusieurs mois et s'accompagner de valeurs extrêmes en ce qui concerne la variation de la salinité, les populations naturelles se trouvent déjà exposées à de fortes tensions. De plus, dans les régions hautement développées, c'est en été que la demande d'électricité atteint son maximum, ce qui contribue à renforcer le déversement d'effluents dans le milieu naturel.

Les auteurs décrivent les effets de températures et de salinités extrêmes associées sur une grande variété de zooplancton en recourant à des tests types de 48 et 96 heures, ainsi que la mesure dans laquelle ces effets évoluent avec l'acclimatation saisonnière. Ils étudient également les effets thermiques à court terme de l'entraînement de l'eau à travers les installations de la centrale électrique: élévation rapide de la température, maintien à température élevée pendant 3 heures au cours du passage dans un canal d'évacuation, puis refroidissement rapide par mélange avec l'eau de la baie. A pleine capacité, la centrale provoquera tous les 14 jours l'entraînement du volume total du système fermé de la baie à côté de laquelle elle est située. Les auteurs en étudient les incidences sur le rôle du zooplancton considéré comme maillon de la chaîne alimentaire aboutissant aux invertébrés et poissons commerciaux et comme partie intégrante du cycle biologique de la plupart de ces organismes.

L'accent est mis sur le copépode *Acartia tonsa* qui représente peut-être la plus importante de toutes les espèces zooplanctoniques en termes de biomasse dans les eaux côtières de l'est des Etats-Unis.

Efectos agudos de los efluentes calentados sobre el copépodo *Acartia tonsa* en una bahiá subtropical y algunos problemas de su evaluación

Casi toda la información relativa a los efectos que los efluentes calentados tienen en los organismos marinos procede de las experiencias adquiridas en aguas templadas del hemisferio norte. En el sur de Florida (y en muchos otros lugares análogos de todo el mundo) hay bahías y estuarios someros y rodeados de manglares, cuya temperatura ambiente máxima en el verano, de 31° a 34°C, está ya peligrosamente cercana a la letal para muchas especies. Como estas temperaturas pueden durar varios meses e ir acompañadas de variaciones de la salinidad extremas, las poblaciones naturales se encuentran ya sometidas a graves tensiones. Además, en zonas muy habitadas y desarrolladas, las demandas de electricidad son mayores en el verano que en el invierno, con lo cual aumenta todavía más el efecto de los efluentes en el medio.

Los autores describen los efectos de las combinaciones de temperaturas y salinidades extremas para diversas especies de zooplancton, en ensayos normales de 48 y 96 horas, y el grado en que éstos cambian con la aclimatación estacional. También se investigaron los efectos térmicos de corta duración del paso del agua por la central eléctrica, con el consecuente rápido aumento de la temperatura, el mantenimiento de ésta durante las tres horas en que ésta en un canal de descarga y el enfriamiento rápido posterior por mezcla con el agua exterior. A plena capacidad, la central tarda días en usar toda el agua del sistema de la bahía cerrada adyacente en cuyo borde se encuentra. Se estudian las repercusiones que esto tiene tanto en la función del zooplancton como un eslabón en la cadena alimentaria hasta llegar a los invertebrados y peces comerciales y como parte integrante de la biología de casi todos estos organismos.

Se hace especial mención del copépodo *Acartia tonsa*, que posiblemente es la especie del zooplancton más importante en lo que se refiere a la biomasa de las aguas costeras del este de los Estados Unidos.

IN the shallow mangrove-fringed bays and estuaries of south Florida ambient summer maxima in the lower thirties (°C) are already dangerously close to the upper lethal level for many species, so the natural populations are already under stress. When operating to projected capacity, the power plant in south Biscayne Bay (Miami) will pass 7.4×10^7 m³/week through the condensers from a basin having a mean volume of 56.6×10^7 m³ (Michel, 1970). At present, entrained water is elevated in seconds to a temperature averaging 5°C above ambient, and held within a degree of this for approximately 3 hours, while it passes down an outflow canal before mixing takes place in the bay.

The copepod *Acartia tonsa* Dana was chosen as the assay animal (Reeve, 1970a). Heinle (1969) showed that northern (Patuxent) populations tolerated poorly temperatures above 30°C. This paper describes the results of subjecting *A. tonsa* to temperature cycles simulating entrainment.

Plankton were collected between March and July 1970. Approximately 100 copepods were transferred by pipette into each of several glass tubes 30 cm long containing 250 ml of 1.2 μ-filtered sea water at ambient temperature. Apart from the ambient controls, 2 tubes were placed in each of several water baths at temperatures between 32 and 37°C (±0.05) for periods up to 6 hours, then returned to an ambient water bath. The following morning animals in each tube were recovered and total numbers and numbers dead counted with aid of a microscope. Dates and ambient water temperatures were (A) 16th March, 21°C; (B) 6th April, 26°C; (C) 9th June, 27°C; (D) 7–17th July, 30°C.

Figure 1 A–D graphs the results of each experiment in terms of percentage survival at each temperature for periods up to 6 hours. Experiment A took place at end of winter when ambient temperatures had been below 21°C for several months. Shock temperatures above 32°C produced over 50 per cent mortality within 3 hours. By 6th April, the bay water had warmed rapidly and the

*Rosenstiel School of Marine and Atmospheric Sciences, University of Miami, Miami, Florida 33149, U.S.A.

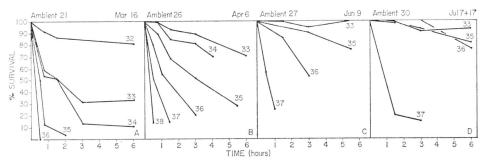

Fig 1. *The percentage survival of populations of* Acartia tonsa *held at elevated temperatures for up to 6 h on four occasions between March and July*

population appeared to have acclimated, yielding higher experimental survival rates than before. The bay water stabilized in this temperature range for over 2 months, during which period further acclimation occurred, except at 37°C. By mid-July, acclimation had proceeded to the point where no more than 25 per cent mortality occurred over 6 hours at 36°C. There was, however, still no acclimation to 37°C which appears to be beyond the limit of tolerance of *Acartia tonsa*, whatever the season.

Problems in assessment of data

The experimental warm-up time (5 min) was much less rapid than in the condensers. Although at certain critical temperatures this could reduce mortality slightly if measured within minutes, overnight mortalities were found to be independent of warm-up rates over a range below 1 min up to 30 min. Cool-down by remixing in the bay may have been considerably slower than in these experiments which was approximately 5 min. A check of cool-down times from 1 to 15 minutes did not, however, reveal any effect on survival. No attempt was made to simulate mechanical damage which might occur on passing through pumps, down condenser tubes, and at discharge.

Critical thermal maximum and death

Under certain conditions (e.g. experiment A, at 34°C after 1½ hours) a large percentage of animals would lose their power to swim effectively, and would sink to the bottom—their critical thermal maximum temperature, defined by Mihursky and Kennedy (1967) being exceeded. Since, however, there were intermediates between active swimming near the surface and ineffective twitching on the bottom, actual death appeared a preferable criterion. From Fig. 2 (A) it can be seen that immediate analysis of the live/dead ratio yielded a much higher apparent survival than after 24 hours, suggesting that loss of swimming ability was largely irreversible.

Effects due to the experimental technique

Oxygen depletion was considered as a factor in the experimental tubes, but several trials using a small oxygen electrode at the end of experiments suggested no systematic variation between tubes, although an average decrease to 80 per cent saturation was usual. Provision of single-bubble aeration using pasteur pipettes produced

higher overall and very erratic mortalities presumed to be due to mechanical damage.

This laboratory is situated some 25 miles from the power plant, so it could be argued that differences in both water and copepods could produce different results. On several occasions, this was tested by using both water and copepods taken from the site, and transported to the laboratory within 2 hours. The results were essentially similar to those of Fig. 1.

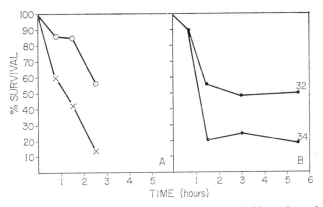

Fig. 2. *(A) Percentage survival of* Acartia tonsa *held at elevated temperatures and analyzed immediately (circles) and after 24 h following return to ambient temperature (crosses); (B) Percentage survival of populations of* Acartia tonsa *from the power plant site held at elevated temperatures at the same time of year as in Fig 1A*

In all the experiments presented in graphic form, mortality in control tubes was less than 20 per cent over 24 hours. Up to mid-May, mortalities of controls averaged 10 per cent. After this, control mortalities often exceeded 30 and sometimes 50 per cent. It was assumed that the "quality" of the water at these high ambient temperatures was such that it rapidly became unable to support animals in small volumes in the laboratory. Over this period unfiltered water always produced lower mortalities than that filtered through 1.2 μ or 0.45 μ filters at ambient temperatures, but in experimental warm-ups these differences disappeared. There were no discrepancies in dissolved oxygen or pH, and additions of 300 mg/l EDTA or combination of penicillin and streptomycin in a range of concentrations up to 15 and 1.5 mg/l respectively did not reduce mortality in filtered water. Additions of the alga *Isochrysis* to give a faint coloration to the water, as well as a process combining filtration with passage through activated charcoal, reduced mortalities but never below those in the corresponding unfiltered water. Mortalities in unfiltered water fluctuated daily and sometimes in water treated in the standard way were reduced.

Conclusions

As pointed out by Mihursky and Kennedy (1967) no single temperature can be applied by a regulatory agency as that beyond which water may not be raised, unless it is that chosen on the basis of winter conditions. Otherwise, prohibited temperatures must be closely correlated with the patterns of seasonal change of the environment water temperature. Acclimation to progressively higher temperatures occurs naturally in *Acartia tonsa*, and probably continues to some extent even after stabilization following environmental temperature change. Even so there is an ultimate upper limit beyond which acclimation will not proceed.

From Fig. 1 LD 50 temperatures may be derived for this copepod under the conditions specified, but it must be emphasized that these are likely to be higher than the values obtained under operational conditions. Under conditions of temperature stress, other stress factors not considered here might assume importance—such as mechanical effects of entrainment, and the effects of low salinity. A second major source of error tending to under-estimate the adverse effects of any specific temperature is in the 24-hour death criterion.

It is clear from experience in this laboratory, that in experimental procedures involving biological systems at ambient temperatures in the upper twenties and thirties (°C), we are dealing with situations highly sensitive to even the slightest (and probably unwitting) modification, including removal and isolation from the original system. A better understanding of the factors working to stabilize the natural system may be the most important requirement in laboratory studies of tropical environments.

References

HEINLE, D R   Temperature and zooplankton. *Chesapeake Sci.*,
1969   10:186–209.
MICHEL, J F   An analysis of the physical effects of the discharge of
1970   cooling water into Card Sound by the Turkey Point Plant
          of Florida Power and Light Company. University of Miami,
          Rosenstiel School of Marine and Atmospheric Science,
          (Unpubl. rep.).
MIHURSKY, J A and KENNEDY, V S   Water temperature criteria to
1967   protect aquatic life. *Spec. Publs Am. Fish. Soc.*, (4):20–32.
REEVE, M R   The growth of *Sagitta hispida* from egg to maturity
1970   under a variety of experimental conditions. Paper presented
          at the Thirty-Third Annual Meeting American Society of
          Limnology and Oceanography.
REEVE, M R Seasonal changes in the zooplankton of South Biscayne
1970a   Bay and some problems of assessing the effects on the zoo-
          plankton of natural and artificial thermal and other
          fluctuations. *Bull. mar. Sci.*, 20(4):894–921.

Sublethal Effects of Pollutants on Fish

V. V. Mitrović*

Les effets sublétaux des polluants sur le poisson

Il convient de modifier certains des critères sur lesquels reposent les taux de concentration admissibles des substances toxiques dans l'eau, puisque ces derniers ont été établis sur la base des résultats de dosages biologiques à court terme où l'on avait pris la mortalité du poisson comme mesure de la toxicité. Il ressort des résultats de dosages biologiques à long terme et de certaines observations pratiques que des quantités de substances toxiques très inférieures aux doses létales pourraient soit causer des troubles fonctionnels et morphologiques chez le poisson, soit modifier les conditions de l'écosystème aquatique à un degré tel que l'existence des populations naturelles y serait en danger. De plus, la pêche pourrait être compromise dans une région lorsque le goût du poisson est dénaturé (par exemple, le cas du poisson "phénolé"), ou lorsque des substances toxiques pour l'homme sont accumulées dans les tissus du poisson (par exemple, dans le cas des pesticides).

L'auteur du présent document s'efforce de définir la notion d'effets sublétaux et passe en revue la littérature relative aux conséquences d'une longue exposition du poisson à plusieurs groupes de substances toxiques (phénols, détergents, métaux lourds et pesticides).

Efectos subletales de los contaminantes en los peces

Se tendrán que modificar algunos de los criterios de las concentraciones permisibles de sustancias tóxicas en el agua porque se derivaban de los resultados de bioensayos de corta duración en los que la mortalidad de los peces se consideraba como una medida de toxicidad. Los resultados de los bioensayos de larga duración y algunas observaciones hechas en la práctica indican que cantidades muy inferiores a las letales podrían causar disturbios funcionales o morfológicos en los peces o alterar de tal manera las condiciones ambientales en el ecosistema acuático que estarían en peligro las poblaciones naturales. Además, pueden perder valor algunos caladeros si los peces que se capturan en ellos tienen mal sabor (como ocurre en el caso del fenol) o en sus tejidos se acumulan sustancias tóxicas para el hombre (como plaguicidas).

El autor trata de definir el concepto de efectos subletales en el pescado y examina la literatura sobre los efectos de una larga exposición a varios grupos de sustancias tóxicas (fenoles, detergentes, metales pesados y plaguicidas).

THE effect on fish of small quantities of various toxic materials, although less apparent than sudden fish kills may be even more harmful; therefore it is difficult to speak about a maximum safe concentration of a toxic substance for fish.

To demonstrate the effect of sublethal concentrations it is necessary to define a very vague concept of sublethal effects. In our opinion it covers the effects of all those concentrations which are not necessarily lethal for individuals, even at prolonged exposures, but increase the population mortality, decrease its size or change its composition. In this case the concept includes a group of effects which are reflected on the growth rate, metabolism, reproduction or which impair the defence mechanism of fish, to the deterioration of the environmental conditions in a water eco-system by the changes of physical, chemical and biotic factors.

In this short review only sublethal effects on fish will be discussed. Various parameters are used, which conditionally, according to the method of approach, could be listed under three categories:

(1) Study of the changes of the most important biologic characteristics, such as growth rate and pattern, feeding, maturation, capability of fertilization and development of eggs, survival of fry etc.—

---

* Faculty of Agriculture, University of Belgrade, Belgrade-Zemun, Yugoslavia.

Mount and Stephan's concept of "laboratory production index".

(2) Study of impaired functions (pathophysiologic study) by physiologic and biochemical methods (most often haematologic, metabolic rate measurements, study of immunobiologic and enzymic activities or behaviour observations).

(3) Study of pathomorphologic changes which include observations of all changes connected with fish morphology from the external shape to histologic and cytologic disturbances.

Sublethal effects of heavy metals

Death of fish in toxic solutions of heavy metals was attributed to asphyxiation (Lloyd, 1960) or precipitation of mucus over the gill surface (Carpenter, 1927, 1930).

After a 3 months exposure of *Lebistes reticulatus* to sublethal concentrations of zinc sulphate and lead nitrate (1.15 mg Zn/l and 1.25 mg Pb/l) Crandall and Goodnight (1963) found no changes in respiratory epithelium or mucus accumulation over the gill surface. They did observe some other pathohistologic changes, reduction of renal lymphoid tissue, dilatation of renal tubules, heart damage, destruction of blood cells and retardation of gonadal development. They also recorded decreased growth rate and retarded maturation. Brown *et al.* (1968) found thinning of secondary lamellae in the gills of rainbow trout that had lived for 100 days in 0.8 mg Zn/l and Baker (1969) in a detailed histological and electron microscopical study on copper poisoning in *Pseudopleuronectes americanus* found degenerative and necrotic changes in liver, kidneys, haemopoetic tissues and gills of fish exposed for 2 weeks to copper sulphate (0.18–3.2 mg/l as Cu) in sea water. While higher concentrations tested induced necrotic changes, lower concentrations induced lesions of haemopoetic tissue leading to haemolitic anaemia and distrophic changes in epithelial and mucous cells in the gills. It appears that copper had a coagulative effect upon the epithelium of gills.

Brungs (1969) reported a reduced number of eggs with the female of *Pimephales promelas* held for 10 months in sublethal concentrations of zinc (1.3; 0.66 and 0.18 mg/l), so that their reproduction was almost completely inhibited by concentrations which did not influence their survival, growth or maturation. In the author's opinion, the maximum acceptable zinc concentration in this case could only be based on the spawning results and must be lower than the lowest concentration tested. Similarly, Mount and Stephan (1969) in the study of chronic toxicity of copper on *Pimephales promelas* found retarded growth and maturation and inhibition of reproduction. They concluded that maximum permissible concentration of copper in soft water must be between 0.13 and 0.22 of 96 h TL_m.

Results obtained by Sprague *et al.* (1965) indicated that avoidance reaction of fish to sublethal concentrations of copper and zinc might influence their distribution. Weir and Hine (1970) described the symptoms of poisoning with different heavy metals (Pb, Hg, As, Se) in goldfish and studied the behaviour of fish in different concentrations. The strongest response of fish was towards mercury, apparent even in concentration of 0.0003 mg/l.

Jones (1947, 1948) using reaction tube apparatus showed that fish, *Pygosteus pungitus*, *Phoxinus phoxinus* and others, can recognize solutions of heavy metal salts and avoid them but that the avoidance reaction differs with the fish species. The stickleback seemed able to detect and avoid heavy metal salts in this order: lead, mercury, zinc and copper.

Sublethal effects of detergents

Acute poisoning with hard and soft detergents impairs respiration and salt balance. Depending on concentration and exposure different degrees of damage to the gill tissue were described (Schmid and Mann 1962; Bock 1965; Brown *et al.*, 1968).

Several authors found that sublethal concentrations of various detergents also cause changes of the respiratory epithelium of the fish gills—in some cases thinning and elongation of respiratory epithelium cells and the thickening and fusing of secondary lamellae caused by proliferation of epithelium tissue on the lamellae tips and between adjacent lamellae (Mount, 1963; Lemke and Mount, 1963; Cairns and Scheier, 1962, 1964).

Such changes are not restricted to detergents. Similar changes were found in trout exposed to high concentrations of suspended solids (Herbert and Merkens, 1961; Slanina, 1962) or as a reaction to various gill parasites. These morphological changes did not necessarily lead to death but the respiration function of the gills was impaired, probably for a long period, since Scheier and Cairns (1963) reported that 24 h exposure of *Lepomis gibbosus* to nonlethal concentrations of alkyl benzene sulphonate (18 mg/l as ABS mixture) resulted in typical gill damage which persisted even after 8 weeks of recovery. This was detrimental for fish in water with low oxygen content.

It was also found (Solon *et al.*, 1969) that resistance of fish (*Pimephales promelas*) to insecticides was lowered by half when the fish were kept in sublethal concentrations of LAS (1.0 mg/l). Mann (1962) found with carp that with the addition of detergents to a phenol solution the accumulation of phenol increased twofold.

Krylov (1969) reported that fish exhibited general pathologic reaction to intoxication by surface active substances, increase in haemoglobin and erythrocyte count with a decrease in leucocyte count which could be interpreted as a protective reaction. Krylov believes that chronic exposure to such substances would ultimately kill the fish.

Lemke and Mount (1963) found detergents retarded growth rate with *Lepomis macrochirus* in sublethal concentrations of alkyl benzene sulfonate (30 days exposure period) and the same was recorded for *Lebistes reticulatus* held for 45 days in 1 mg/l of tetrapropylene benzol sulphonate (Mann and Schmid, 1965) where, at the same time the beginning of liver lesions was noted. Mann and Schmid (1961) studied the effect of dodecyl benzol sulphonate on trout sperm and eggs and found that concentrations between 5 to 10 mg/l arrested movements of sperm. The eggs from these concentrations could be fertilized but most broke afterwards.

In 1965 Bardach *et al.* showed that soft and hard ABS detergents led to the damage of chemoreceptors of *Ictalurus natalis* which hindered its search and intake of food. This damage was still present after 4 to 6 weeks of recovery.

[253]

Sublethal effects of phenols

The majority of data on the pathological effect of phenols on fish referred to acute and subacute poisoning. In acute poisoning phenol acts as a nerve poison (Havelka and Effenberger, 1957; Luk'ianenko, 1967) causing death by paralysis. In subacute poisoning a general intoxication of fish organism occurs followed by inflammatory and necrotic changes of vital organs, gills, circulatory system, brain, liver, kidneys, spleen and gonads (Kuhn and Koecke, 1956; Lammering and Burbank, 1960; Vischnevetsky, 1962; Waluga, 1966; Mitrović et al., 1968; Andres and Kurazhovskaia, 1969). In a detailed study with *Abramis brama* held for 7 days in 9 mg/l phenol, Waluga (1966a) found erithropenia, leucopenia and necrobiotic changes of the blood cells.

Data on sublethal effects of phenol are scarce. Halsband and Halsband (1963) described a 25 per cent reduction in the number of erythrocytes and a 26 per cent swelling in trout exposed for 28 h to sublethal concentration of phenol (1.5 mg/l). Reichenbach-Klinke (1965, 1966) recorded decreased numbers of the erythrocytes in fish from phenol polluted waters (6 to 8 mg/l).

Mikriakov (1969) found the amount of serum proteins decreased and formation of immunoglobulins was inhibited in carp held for 2 months in 12.5 mg/l phenol. Weight loss was also recorded.

Reichenbach-Klinke (1965) studied the effect of low phenol concentrations (0.02 to 0.07 mg/l) on 14 fish species caught alive in the Rhine and Elbe. He found serious lesions in the gills of fish, inflammation, loss of mucous cells, and in some cases, complete destruction of the secondary lamellae walls. Inflammatory and degenerative changes were found in skin, intestine, muscles and liver.

Stepanov and Flerov (1969) reported that *Lebistes reticulatus* kept for a year in a rather high phenol concentration (12.5 mg/l) first spawned at the age of 5 months while the control fish spawned when 10 months old. In the course of the year there were 4 to 5 spawns both in test and control fish. They found also that the weight increases of the two fish were similar. Although *Lebistes reticulatus* is considered to be a fairly resistant fish species and we have no comparable data, reports of damage to gonads with the destruction of oocytes vitelus in subacute phenol poisoning (Reichenbach-Klinke, 1965; Mitrović et al., 1968; Waluga, 1966) do not concur with those results.

Avoidance reactions of fish to phenol were studied by several authors but controversial results were obtained. As early as 1935, Kalabina, from the results of a field study, concluded that fish would not inhabit parts of the river with phenol content higher than 0.2 mg/l. Ishio (1969) described an unbalanced fish distribution in the Onga river (Japan), where parts of the river were polluted by coal washing wastes containing phenol in concentrations 0.024–0.1 mg/l. Only four fish were caught per 150 m² from polluted parts as compared with 413 per 150 m² from the upper stream and abundant fish fauna at the mouth of a small tributary. Avoidance tests in tanks with *Zacco platypus* (dominant fish species from the river) proved that most fish avoided water with phenol concentration higher than 0.1 mg/l. Conversely, Sprague and Drury (1969) did not find avoidance reactions in rainbow trout exposed to different phenol concentrations

(0.001–10 mg/l). Jones (1951), using higher concentrations of phenol (4–400 mg/l) and the reaction tube apparatus concluded that *Phoxinus phoxinus* did not show any ability to recognize the solution. On the other hand Hasler and Wisby (1949) trained *Hyborhynchus notatus* to recognize phenol in concentration of 0.01 mg/l and some of them even as low as 0.0005 mg/l which is much lower than the human odour threshold.

These data about the avoidance reactions of fish to phenol are not easily comparable because different devices were used and a wide range of phenol concentrations were tested.

Sublethal effects of pesticides

Several authors have reported pathological changes in various fish organs induced by sublethal amounts of some pesticides, but in many cases results are inconclusive. Pathohistological changes usually are not specific for the action of a certain pesticide but are similar to those found for other toxic substances. Lowe (1964) found that the gill lamellae thickened when exposed to sublethal concentrations of toxaphene. Schultz (1970) reported necrotic changes in epithelial cells and increase of mucous cells in the gills of carp exposed to different concentrations of Na–Ta (natrium trichlor acetate). Rudd and Genelly (1956) noticed degenerative and necrotic changes of the liver with BHC and chronic nephritis in goldfish exposed to DDT. Gilderhus (1966) described histopathologic changes of heart, liver, gills and degeneration of oogonies and oocytes with bluegill kept for 16 weeks in sublethal concentrations of sodium arsenite. Christie and Battle (1963) found increased mucous secretion in the gills of larval lampreys and trout and vasodilatation in the liver and cloaca of larval lampreys after a prolonged exposure to TFM.

Eller (1969) reported recovery of fish which had been seriously affected by one treatment with a herbicide. He found serious lesions of gills, liver, testes and blood of *Lepomis microlophus* after a single application of Hyrothol 191 to earthen ponds in concentrations between 0.03 and 0.3 mg/l. Interesting cytological aberrations of testicular tissue were recorded with the appearance of ova-like hypertrophic testicular cells. Then the damage began receding. Lesions in gills began to disappear after 14 days, in testes after 28 and in liver after 56 days. The appearance of the fish organs became similar to the control after 100 days in the 0.3 mg/l ponds. Similar results were obtained by Cope et al. (1970). Twenty-four hours to 14 days after the treatment of experimental ponds with 2,4–D (0.1–10 mg/l), they found pathologic lesions of liver, vascular system and brain of *Lepomis macrochirus*. After 14 days these changes gradually regressed. After 28 days they were visible only in the highest concentration tested and after 84 days they disappeared.

Among pathological changes induced by pesticides were reduction of haemoglobin without a change in the erythrocyte count after a prolonged exposure to DDT (Rudd and Genelly, 1956) and haematologic changes after exposure of *Ictalurus punctatus* to apholate (Dodgen and Sullivan, 1969). Microhematocrit readings of pesticides were often done on fish from contaminated water but consistent changes in this value cannot be as yet connected with the action of a definite pesticide.

Allison *et al.* (1963) found no relation between the hematet value and the exposure to malathion. Cope *et al.* (1970) recorded elevated hematocrit readings 1 to 3 days after application of 2,4–D (10 mg/l) but after 7 days the readings did not differ from controls.

Reduction in cholinesterase activity in fish exposed to sublethal concentrations of organic phosphate insecticides was reported by several authors. Henderson (1966) referred to by Johnson (1968) found a 40 per cent reduction in brain cholinesterase activity in fish captured from a malathion contaminated stream and Weiss (1961) reported that normal levels of cholinesterase activity were not restored even after 60 days following exposure of fathead minnow to parathion.

In several instances, changes in growth rate and reproduction were correlated with the effect of sublethal concentrations of pesticides. Andrews *et al.* (1966) found a reduced growth rate of bluegills receiving heptachlor with food and Gilderhus (1966) found the same with bluegills exposed to sublethal concentrations of sodium arsenite. Growth rate and reproductive performance of *Poecilia latipinna* were adversely affected by very low concentrations of dieldrin (0.0015 to 0.00075 mg/l) (Lane and Livingston, 1970) and a 10 per cent decrease in weight of carp kept for 63 days in Na–Ta was recorded by Schultz (1970). However, Cairns *et al.* (1967) found better weight increase in guppies exposed to sublethal concentrations of dieldrin. Macek (1968) reported that brook trout fed 2 mg DDT/kg body weight per week gained more weight than fish on DDT free diet, but had higher mortality when starved. Increased growth rate of *Lepomis macrochirus* after the treatment of ponds with 2,4–D was explained (Cope *et al.*, 1970) by the thinning of fish population and the control of weed in higher concentrations of 2,4–D.

Cairns *et al.* (1967) observed reduced reproduction of guppies held in sublethal concentrations of dieldrin. Mount and Stephan (1967) exposed *Pimephales promelas* for 10 months to sublethal concentrations of malathion and 2,4–D. From the effect of those concentrations on growth and reproduction of fish they concluded that only concentrations as low as 1/45 and 1/19 respectively of their TL_m could be considered safe for this fish.

Solly and Ritchie (1969) showed that egg viability of trout could be directly correlated with DDT residue level. It was found (Allison *et al.*, 1963) that the exposure of cutthroat trout to 3 mg of DDT/kg of body weight did not reduce the number and volume of eggs but did increase mortality of sac fry. Adverse effects of some insect chemosterilants on fish reproduction was also reported. Eisler (1966) found that ovarian development of *Fundulus majalis* was affected by apholate and Stock and Cope (1969) showed changes in testicular tissue of *Poecilia reticulata* 134 days after bathing in sublethal levels of TEPA (25 mg/l).

Mount (1962) reported a 25 per cent increase in oxygen consumption and hypersensitivity to stimuli in bluntnose minnows in sublethal concentrations of endrin and Cairns and Scheier (1964) found increased oxygen consumption in *Lepomis gibbosus* exposed for 12 weeks to 0.00168 mg/l of dieldrin. Rongsriyam *et al.* (1968) observed decreased feeding activity in guppies exposed to several insecticides (lindane, malathion, dieldrin, etc.) Warner *et al.* (1966) attempted to establish the minimal

concentration of malathion which would lead to significant changes in fish behaviour.

In addition to influencing the biologic characteristics of fish and increasing their susceptability to unfavourable environmental factors (e.g. low oxygen content), functional disturbances also impair the self-defence mechanism of fish and open the way for the action of secondary factors, such as various infectious and parasite diseases of which fungi are the most frequent. Luk'ianenko (1967) noted that fish which had survived phenol poisoning often died later of saprolegniosis. Herbert and Merkens (1961) found fin rot in the trout kept in high concentrations of suspended solids, while Schoental (Cope, 1965) observed increased susceptibility to furunculosis in salmonids where the accumulation of DDT in the body was recorded.

Questions of fish adaptation to toxic substances are important theoretically and practically. Some information suggests the possibilities of adaptation to certain substances with stress effect (Luk'ianenko, 1967; Mount, 1962), existence of more tolerant individuals in a group of fish, or formation of resistant populations in pesticide contaminated streams (Ferguson *et al.*, 1964). Such resistant fish can accumulate much higher amounts of toxic compounds in their bodies which could be a potential hazard for their consumers and for man (Ferguson, 1967, in Johnson, 1968).

Phenomena like adaptation and acclimation, although comprehensive from the ecological standpoint, have not been studied or even defined in a wider biological sense and we know very little about mechanisms involved in such phenomena. They can be explained only by proper physiological and genetic studies.

The sublethal effects of a few other substances (ammonia, suspended matter, etc.) have been studied, but in general this field of work has been neglected until recently. Our knowledge about longterm effects of pollutants is fragmentary and far from conclusive. Field data are scarce and very often can not be correlated with laboratory data. Nevertheless, data about the sublethal effects have been much help in assessing safe concentrations. This is essential for such substances which penetrate unnoticeably into the water, accumulate in the fish body and whose concentration in the water cannot be taken as a measure of the fish's toxicity.

We do not yet know enough about the true ecological significance of the damage found in fish and about the questions of greatest importance to the fishery of a certain area, e.g. which concentration of a certain toxic substance will be in the long run fatal for a certain fish species and other questions. The answers will be obtained by detailed studies.

References

ALLISON, D, KALLIMAN, B J and VAN VALIN, C C Insecticides: 1963   effects on cutthroat trout of repeated exposure to DDT. *Science, N.Y.*, 142(3594):958–61.

ANDRES, A G and KURAZHOVSKAIA, T N Gistopatologicheskie 1969   izmenenia u leshcha (*Abramis brama* L.) pri ostrom otravlenii fenolom v eksperimente. *Trudy Inst. Biol. Vodokhran.*, 19(20):73–86.

ANDREWS, A K, VAN VALIN, C C and STEBBINGS, B E Some effects 1966   of heptachlor on bluegill (*Lepomis macrochirus*). *Trans. Am. Fish. Soc.*, 95(3):297–309.

BAKER, J T P Histological and electron-microscopical observa- 1969   tions on copper poisoning in the winter flounder (*Pseudopleuronectes americanus*). *J. Fish. Res. Bd Can.*, 26(11):2785–93.

BARDACH, J E, FUJIYA, M and HOLL, A   Detergents: effects on the
1965   chemical senses of fish *Ictalurus natalis* (Le Sueur). *Science,
N.Y.*, 148(3677)1605–7.

BOCK, K J   Über die Wirkung von Waschstoffen auf Fische. *Arch.
1965   FischWiss.*, 17(1):68–76.

BROWN, V M, MITROVIĆ, V V and STARK, G T C   Effects of chronic
1968   exposure to zinc on toxicity of a mixture of detergent and
zinc. *Wat. Wks.*, 2:225–63.

BRUNGS, W A   Chronic toxicity of zinc to the fathead minnow
1969   (*Pimephales promelas* Rafinesque). *Trans. Am. Fish. Soc.*,
98(2):272–9.

CAIRNS, J and SCHEIER, A   The acute and chronic effects of
1962   standard sodium alkylbenzene sulphonate upon the pumpkinseed sunfish, *Lepomis gibbosus* (Linn.) and the blue-gill
sunfish, *L. macrochirus* (Raf.). *Proc. ind. Waste Conf. Purdue
Univ.*, 17(112):14–28.

CAIRNS, J and SCHEIER, A   The effect upon the pumpkinseed sun-
1964   fish, *Lepomis gibbosus* (Linn.) of chronic exposure to lethal
and sublethal concentrations of dieldrin. *Notul. Nat.*, (70):
1–10.

CAIRNS, J, SCHEIER, A and LOOS, J J   Changes in guppy population
1968   resulting from exposure to dieldrin. *Progve Fish Cult.*, 28(4):
220–6.

CAIRNS, J et al.   Effects of sublethal concentrations of dieldrin on
1967   laboratory populations of guppies. *Proc. Acad. nat. Sci.
Philad.*, 119(3):75–91.

CARPENTER, K E   The lethal action of soluble metallic salts on
1927   fishes. *J. exp. Biol.*, 4:378–90.

CARPENTER, K E   Further research on the action of metallic salts on
1930   fishes. *J. exp. Zool.*, 56:407–22.

CHRISTIE, R M and BATTLE, H J   Histological effect of 3-trifluor-
1963   methyl-4-nitrophenol (TFM) on larval lamprey and trout.
Can. J. Zool., 41(1):51–61.

COPE, O B   Agricultural chemicals and freshwater ecological
1965   systems. *In* Research in pesticides, New York, Academic
Press, pp. 115–28.

COPE, O B, WOOD, E and WALLEN, G   Some chronic effects of
1970   2,4–D on the bluegill (*Lepomis macrochirus*). *Trans. Am. Fish.
Soc.*, 99(1):1–13.

CRANDALL, C A and GOODNIGHT, C J   The effect of sublethal con-
1963   centrations of several toxicants to the common guppy,
Lebistes reticulatus. *Trans. Am. microscop. Soc.*, 82:59–73.

DODGEN, L and SULLIVAN, S   Hematological effects of apholate on
1969   Channel catfish (*Ictalurus punctatus*). *Proc. Soc. exp. Biol.
Med.*, 131(1):124–6.

EISLER, R   Effects of apholate, an insect sterilant, on an estuarine
1966   fish, shrimp and gastropod. *Progve Fish Cult.*, 28:154–158.

ELLER, L L   Pathology in redear sunfish exposed to Hydrothol 191.
1969   *Trans. Am. Fish. Soc.*, 98(1):52–9.

FERGUSON, D E et al.   Resistance to chlorinated hydrocarbon
1964   insecticides in three species of freshwater fish. *Bioscience*,
14:43–4.

GILDERHUS, P A   Some effects of sublethal concentrations of
1966   sodium arsenite on bluegills and the aquatic environment.
Trans. Am. Fish. Soc., 95(3):289–96.

HALSBAND, E and HALSBAND, I   Veränderungen des Blutbildes von
1963   Fischen infolge toxischer Schäden. *Arch. Fisch Wiss.*, 14(1–
2):68–84.

HASLER, A and WISBY, W   Use of fish for the olfactory assay of
1949   pollutants (phenols) in water. *Trans. Am. Fish. Soc.*, 79:64–
70.

HAVELKA, J and EFFENBERGER, M   Symptoms of phenol poisoning
1957   of fish. *Sb. col. Acad. Zemed. Ved.*, 30:421–4. Also issued as
Ann. Acad. tchecosl. Agric., 2(5):421–4.

HERBERT, M V D and MERKENS, C J   The effect of suspended
1961   mineral solids on the survival of trout. *Int. J. Air Wat. Pollut.*,
5(1):46–55.

ISHIO, S   Formal Discussion, Section I, paper 10/Sprague. *Adv.
1969   wat. Pollut. Res.*, 4.

JONES, J R E   The reactions of *Pygosteus pungitus* L. to toxic solu-
1947   tions. *J. exp. Biol.*, 24:110–22.

JONES, J R E   A further study of the reaction of fish to toxic solu-
1948   tions. *J. exp Biol.*, 25:22–34.

JONES, J R E   The reactions of the minnow, *Phoxinus phoxinus* (L)
1951   to solutions of phenol. ortho-cresol and para-cresol. *J. exp.
Biol.*, 28:261–70.

JOHNSON, D W   Pesticides and fishes—a review of selected litera-
1968   ture. *Trans. Am. Fish. Soc.*, 97(4):398–424.

KALABINA, M M   Der Phenolzerfall in Fliess- und Staugewässern.
1935   *Z. Fisch.*, 33:295–315.

KUHN, O and KOECKE, H V   Histologische und cytologische
1956   Veränderungen der Fischkieme nach Einwirkung im Wasser
enthaltener schädigender Substanzen. *Z. Zellforsch.
mikrosk. Anat.*, 43:611–643.

KRYLOV, O N   Vlianie poverhnostno aktyvanyh veshestv na ryb.
1969   Sb. "Biol. produktivnost vodoemov Sibiri" M. "Nauka",
pp. 276–9.

LAMMERING, M V and BURBANK, N C   The toxicity of phenol, o-
1960   chlorophenol and o-nitrophenol to bluegill sunfish. *Proc.
ind. Waste Conf. Purdue Univ.*, 15(106):541–55.

LANE, C E and LIVINGSTON, J R   Some acute and chronic effects of
1970   dieldrin on the Sailfin Molly, *Poecilia latipinna*. *Trans. Am.
Fish. Soc.*, 99(3):489–95.

LEMKE, A and MOUNT, D I   Some effects of alkyl benzene sulfonate
1963   on the bluegill, *Lepomis macrochirus*. *Trans. Am. Fish. Soc.*,
92(4):372–8.

LLOYD, R   The toxicity of zinc sulphate to rainbow trout. *Ann. appl.
1960   Biol.*, 48(1):84–94.

LLOYD, R   Factors that effect the tolerance of fish to heavy metal
1965   poisoning. *Publs. publ. Hlth Serv., Wash.*, (999–WP–25):
181–7.

LOWE, J I   Chronic exposure of spot, *Leiostomus xanthrus* to sub-
1964   lethal concentrations of toxaphene in sea water. *Trans. Am.
Fish. Soc.*, 93(4):396–9.

LUK'IANENKO, V I   Tosikologiia ryb. Moskva, 216 pp.
1967

MACEK, K J   Growth and resistance to stress in brook trout fed
1968   sublethal levels of DDT. *J. Fish. Res. Bd Can.*, 25(11):2443–
51.

MANN, H.   Die Förderung der Geschmacksbeeinflussung bei
1962   Fischen durch Detergentien. *Fischwirt*, 12:237–40.

MANN, H and SCHMID, O J   Der Einfluss von Detergentien auf
1961   Sperma, Befruchtung und Entwicklung bei der Forelle. *Int.
Rev. ges. Hydrobiol. Hydrogr.*, 46:419–26.

MANN, H and SCHMID, O J   Der Einfluss subletaler Mengen von
1965   Detergentien (Tetrapropylenbenzolsulfonat) auf das Wachstum von *Lebistes reticulatus*. *Arch. FischWiss.*, 16(1):16–20.

MIKRIAKOV, V R   Vlianie fenola na kolichestvo belka v syvorotki
1969   karpov (*Cyprinus carpio* L.) v usloviah hronicheskogo
eksperimenta. *Trudy Inst. Biol. Vodokhran.*, 19:70–2.

MITROVIĆ, V V, BROWN, V M and SHURBEN, D G   Some patho-
1968   logical effects of sub-acute and acute poisoning of rainbow
trout by phenol in hard water. *Wat. Res.*, 2:249–54.

MOUNT, D I   Chronic effects of endrin on bluntnose minnow and
1962   guppies. *Res. Rep. U.S. Fish Wildl. Serv.*, (58):38 p.

MOUNT, D I   Some effects of metals and detergents on fishes. *Proc.
1963   a. Conf. Air Wat. Pollut. Univ. Mo.*, 8(Bull.64):24.

MOUNT, D I   Chronic toxicity of copper to fathead minnows
1968   (*Pimephales promelas*, Rafinisque). *Wat. Res.*, 2:215–23.

MOUNT, D I and STEPHAN, C E   Chronic toxicity of copper to the
1969   fathead minnow (*Pimephales promelas*) in soft water. *J. Fish.
Res. Bd Can.*, 26(9):2449–57.

REICHENBACH-KLINKE, H H   Der Phenolgehalt des Wassers in
1965   seiner Auswirkung auf den Fischorganismus. *Arch. Fisch-
Wiss.*, 16(1):1–16.

REICHENBACH-KLINKE, H H   The blood components of fish with
1966   relation to parasites, infections and water pollution. *Bull.
Off. int. Epizoot.*, (65):1039–54.

RONGSRIYAM, Y, PROWNEBON, S and HIRAKOSO, S   Effects of
1968   insecticides on the feeding activity of the guppy, a mosquito-
eating fish in Thailand. *Bull. Wld Hlth Org.*, 39(6):977–80.

RUDD, R I and GENELLY, E R   Pesticides: their use and toxicity in
1956   relation to wildlife. *Caif. Fish Game Bull.*, (7):209 p.

SCHEIER, A and CAIRNS, J   Persistance of gill damage in *Lepomis
1963   gibbosus* following a brief exposure to alkylbenzene
sulphonate. *Notul. Nat.*, (39):1–7.

SCHMID, O J and MANN, H   Die Einwirkung von Dodecylbenzol-
1962   sulfonat auf die Kiemen von Forellen. *Arch. Fisch Wiss.*,
13(1/2):41–51.

SLANINA, K   Beitrag zur Wirkung mineralischer Suspensionen auf
1962   Fische. *Wass. Abwass.*, 1962:1–10.

SCHULTZ, D   Studien über Nebenwirkungen des Herbizids Na–Ta
1970   (Na-Trichloracetat) auf Karpfen. I Mitt. Kiemen. *Zentbl.
VetMed.*, 17(3):230–51.

SOLON, J M, LINGER, J L and NAIR, J H   The effect of sublethal
1969   concentrations of LAS on the acute toxicity of various
insecticides to the fathead minnow (*Pimephales promelas*
Rafinisque). *Wat. Res.*, 3:767–75.

SOLLY, S R B and RITCHIE, A R   DDT in trout and its possible
1969   effect on reproductive potential. *N.Z. Jl mar. freshwat. Res.*,
3:220–9.

SPRAGUE, J B, ELSON, P F and SAUNDERS, R L   Sublethal copper-
1965   zinc pollution in a salmonid river—a field and laboratory
study. *Adv. Wat. Pollut. Res.*, 1:61–82, 99–102.

SPRAGUE, B J and DRURY, E D   Avoidance reaction of salmonid
1969   fish to representative pollutants. *Adv. Wat. Pollut. Res.*,
4:169–79.

STEPANOV, V S and FLEROV, B A   Vlianie subtoksicheskih kon-
1969   centracii fenola na kolicestvo i kacestvo potomstva u
Lebistes reticulatus. *Trudy Inst. Biol. Vodokhran.*, 19(22):
60–1.

STOCK, J N and COPE, O B Some effects of TEPA, an insect
1969 chemosterilant on the guppy *Poecilia reticulata. Trans. Am.
Fish. Soc.*, 98(2):280–7.

VISCHNEVETSKY, F E Patomorfologia otravelnia ryb fenolom i
1962 vodnorastvorymimi komponentami syrei nefti, kamen-
nougholnoi smoli i mazuta. *Trudy astrakh. gos. Zapov. Ved.*,
5:350–2.

WALUGA, D Zmiany anatomo-histologiczne u loszeza pod
1966 wplywem fenolu. (Phenol effects on the anatomo-histo-
pathological changes in bream (*Abramis brama* L.). *Acta
hydrobiol.*, Krakow, 8(1):55–78.

WALUGA, D Zmiany we krwi obwodewj leszezy pod wplywem
1966a fenoly. (Phenol induced changes in the peripheral blood of
the breams (*Abramis brama* L.). *Acta hydrobiol.*, Krakow,
8(2):87–95.

WARNER, R E, PETERSON, K K and BORGMAN, L Behavioural
1966 pathology in fish: a quantitative study of sublethal pesticide
toxication. *J. appl. Ecol.*, 3(Suppl.):223–47.

WEIR, P and HINE, CH Effects of various metals on behaviour of
1970 conditioned goldfish. *Archs envir. Hlth*, 20(1):45–51.

WEISS, C V Physiological effect of organic phosphorus insecticides
1961 on several species of fish. *Trans. Am. Fish. Soc.*, 90(2):143–
52.

Sublethal Chronic Effects of DDT and Lindane on Glycolytic and Gluconeogenic Enzymes of the Quahog, *Mercenaria mercenaria*

R. H. Engel, M. J. Neat and R. E. Hillman*

Effets chroniques sublétaux du DDT et du lindane sur les enzymes glycolytiques et gluconéogéniques de la praire, *Mercenaria mercenaria*

Des populations distinctes de *M. mercenaria* ont été exposées à une concentration de 2 µg/1 de DDT et de lindane dans une eau de mer courante pendant une période de 30 semaines, à l'issue de laquelle on a dosé plusieurs enzymes glycolytiques et gluconéogéniques essentielles dans les tissus du pied, des branchies et du manteau des mollusques. On a comparé les taux d'enzymes constatés dans les praires exposées au DDT et au lindane avec ceux des praires maintenues dans une eau témoin. On a également déterminé les activités enzymatiques dans un groupe de praires prélevées directe-ment sur le milieu marin.

On a constaté que le tissu normal des praires contient les enzymes suivantes: hexikinase, glucose-6-phosphate déhydrogénase, phos-phofructokinase, fructose disphosphatase, pyruvate kinase, phos-phoénolpyruvate carboxytransphosphorylase et déhydrogénase lactique. Bien que les activités catalytiques aient été extrêmement faibles (approximativement 5 pour cent sur la base de foie de rat en poids de tissu), lors d'une comparaison des taux d'hexikinase dans chaque tissu, le rapport général des taux d'activité s'est révélé analogue entre le tissu des mollusques et celui des mammifères. Après réduction de la ration alimentaire, on a constaté chez les praires témoins une diminution des réserves de glycogène, une baisse du taux de glucose-6-phosphate déhydrogénase et une élévation de l'activité du fructose diphosphatase, ce qui indique le fonctionnement de mécanismes régulatoires analogues à ceux du tissu des mammifères.

Outre ces modifications, les praires exposées au DDT ont accusé les phénomènes suivants: diminution du taux de fructose diphos-phatase (40 pour cent) dans le pied, de celui de glucose-6-phosphate-déhydrogénase (100 pour cent) dans les branchies et accroissement de celui de phosphofructokinase dans les branchies (70 pour cent). L'exposition au lindane a provoqué une diminution analogue du taux de fructose diphosphatase (30 pour cent) dans le pied. Ces données indiquent que l'administration continue de doses sub-létales d'hydrocarbures chlorés aux praires pourrait provoquer des altérations enzymatiques indicatrices d'un accroissement du taux de dégradation glycolytique du glucose et d'une suppression de la gluconéogenèse.

Efectos subletales crónicos del DDT y el lindane en las enzimas glucolíticas y gluconeogénicas del verigüeto, *Mercenaria mercenaria*

Se conservaron varias poblaciones separadas de *M. mercenaria* en agua marina corriente, con una concentración de DDT y lindane de 2 µg/1, durante 30 semanas, examinándose al final de dicho período la cantidad de enzimas glucolíticas y gluconeogénicas fundamentales contenidas en los tejidos del pie, las branquias y el manto. El contenido de enzimas de los verigüetos expuestos a la acción del DDT y el lindane se comparó con el de verigüetos mantenidos en una capa freática de control. Se determinó también la actividad enzimática de un grupo de verigüetos tomados directamente del medio ambiente marino.

Se encontró que los tejidos de verigüetos normales contenían las enzimas siguientes: hexoquinasa, glucosa-6-fosfato deshidrogenasa, fosfofructoquinasa fructosa difosfatasa, piruvato quinasa, fosfo-fenolpiruvato carboxitransfosforilasa, y deshidrogenasa láctica. Aunque las actividades catalíticas resultaron muy bajas (un 5 por ciento, aproximadamente, de las del hígado de rata, basándose en el peso de los tejidos) cuando se comparan con los contenidos relativos de hexoquinasa en cada tejido, aparece una similitud general de actividad proporcional entre el tejido de los moluscos y el de los mamíferos. Restringiendo la ingestión de alimentos, se observó en los verigüetos empleados como testigo una depaupera-ción de las reservas de glucógeno, una disminución de la glucosa-6-fosfato deshidrogenasa y un aumento de la actividad de la fructosa difosfatasa, lo que indica la entrada en funcionamiento de mecanismos reguladores análogos a los de los tejidos de los mamíferos.

Además de estos cambios, los verigüetos expuestos a la acción del DDT revelaron una disminución de la fructosa difosfatasa en los tejidos del pie (40 por ciento) y de la glucosa-6-fosfato deshidro-genasa en los tejidos de la branquia (100 por ciento) y un aumento de la fosfofructoquinasa en los tejidos de la branquia (70 por ciento). La exposición al lindane se tradujo en una reducción semejante de fructosa difosfatasa en los tejidos del pie (30 por ciento). Estos datos indican que la administración crónica de dosis subletales de hidrocarburos clorados al verigüeto puede producir alteraciones enzimáticas, que revelan un aumento del índice de degradación glucolítica de la glucosa y la desaparición de la gluconeogénesis.

CHRONIC exposure to environmental contamin-ants can have sublethal effects on marine organ-isms. Significant reductions in the estuarine productivity of a number of fish and shellfish are occur-ring that are related to pesticide concentrations which do not necessarily kill significant numbers of individual organisms (Miller and Berg 1968). This phenomenon may be caused by subtle alterations in metabolic processes that direct cellular mechanisms such as gonad development or shell deposition and could lead to the slow attrition of a particular species in a polluted area—to become evident only after several years.

Evidence is accumulating that a number of important enzymes, among them succinic dehydrogenase, cyto-chrome oxidase, and carbonic anhydrase, are inhabited by the chlorinated hydrocarbons (O'Brien 1967). The present work was done to determine whether sublethal chronic doses of DDT and lindane administered to a marine mollusc, the quahog, *Mercenaria mercenaria*, at concentrations currently existing in estuarine waters could alter the activity of key regulatory enzymes and thereby affect intermediary metabolism.

In mammalian tissue several key enzymes have been identified which regulate the flow of glycolytic inter-mediated in the direction of glycolysis or gluconeogenesis (Scrutton and Utter 1968), and are usually present in rate-limiting concentrations and catalyse essentially irreversible reactions. Their relative activities at any given time are a function of the concentrations of various allosteric effectors such as ATP and NADH and the

---

* William F. Clapp Laboratories, Duxbury, Massachusetts 02332, U.S.A.

circulating levels of various hormones such as insulin and glucocorticoids which can initiate or repress the synthesis of specific enzyme proteins.

In contrast to mammalian tissue, the intermediary metabolism of marine molluscs has received little attention in the past, particularly regarding the glycolytic pathway. However, a number of hexose and triose phosphate glycolytic intermediates have been identified (Usaki and Okamura 1956) as well as associated enzymes such as phosphoenolpyruvate carboxytransphosphory-lase, malic enzyme (Simpson and Awapara 1964) and pyruvate carboxylase (Jodrey and Wilbur 1955). Therefore, it seems reasonable to presume that glucose metabolism in molluscan tissue proceeds via the usual glycolytic pathway. Thus, the enzymes selected for study were chosen on the basis of their demonstrated importance in mammalian tissue.

Methods

Quahogs were obtained from mudflats within 2 km of the laboratory and placed in running sea water in fibreglass sea water tables containing an operating volume of 200 1. Following an acclimatization period of one week, the quahogs were covered with 7 to 10 cm of mud and then exposed to DDT and lindane. Stock solutions of the pesticides were prepared weekly in 38 per cent acetone and delivered via a peristaltic pump to the water tables at a final concentration of 2 μg/1. A control table received 38 per cent acetone at a final concentration of 19 mg/1.

The extreme insolubility of DDT in natural waters makes its exact solubility difficult to quantify. Although the most reliable values are probably derived from radioactive tracer studies, the reported values vary from as low as 1.2 μg/1 (Bowman *et al.* 1960) to 37 μg/1 (Babers 1955). The pesticide level chosen in this study, 2 μg/1, was selected as a high value likely to be reached fairly frequently in the average estuary (Hammerstrom *et al.* 1967). Excess DDT which did not remain in solution could be expected to have been deposited in the mud substrate. This process, which normally occurs in the marine environment, leads to a natural accumulation

of DDT in bottom sediments in excess of the surface water and was the primary reason for maintaining the quahogs in a mud substrate. Lindane, with a water solubility of 7.3 μg/1 (Richardson and Miller 1960) entered the water table in solution; however, subsequent adsorption on the mud surface cannot be discounted.

Exposure to DDT and lindane was continued for 30 weeks, then samples of the mud substrate were examined for interstitial meiofauna. In each table there were substantial numbers of nematodes, copepods and other biotic forms normally present in the upper mud layers, indicating that acute toxic conditions were not present during the study.

Results and discussion

Histological examination of the gonads revealed that in each water table gonad development had proceeded to the stage at which spawning could have occurred. Most of the gonads contained mature gametes but we could not definitely establish whether the gonads were being reabsorbed after spawning or following a period during which spawning could have occurred but did not. The relatively few mature gametes compared with other cell types associated with gonad development indicate that spawning did occur and the gonads were undergoing normal reabsorption.

Enzyme studies were done on 10 per cent homogenates of quahog foot, gill and mantle tissue prepared individually in 0.9 per cent saline. Six to ten organisms were used per group. Following centrifugation, the 6000xg supernatant was decanted and assayed for enzymic activity and protein. All enzyme systems were coupled to the oxidation of NADH or the reduction of NADP.

Substrates were added at zero time and activity at 27°C was measured against identical mixtures lacking substrate. Catalytic rates were calculated from the extinction coefficients of NADH and NADP and a minimum acceptable 3 minute linear absorbance change of 0.04. Mixtures containing boiled enzyme showed absorbance changes not exceeding 0.005.

Enzyme levels were also determined on a group of quahogs taken directly from the marine environment. The data in Table 1 compare the enzyme activities found in quahog tissue with the maximum catalytic rates reported for rat liver (Scrutton and Utter 1968). With the exception of glucose-6-phosphatase, all of the enzymes assayed were present.

Enzyme activities in quahog tissue are extremely low compared with rat liver tissue (approximately 5 per cent). Although the delineation of optimum assay conditions for the quahog may increase these rates somewhat, it seems clear that the overall catalytic capacity of *M. mercenaria* is well below that of mammalian tissue.

Quahogs maintained in the control water table receiving 19 mg acetone/l showed a 62 per cent reduction in foot glycogen. A similar reduction in foot glycogen of 75 per cent occurred in a group of quahogs maintained in running, filtered sea water receiving no acetone. It is therefore likely that these changes reflect decreased metabolism as a result of partial starvation, the food source in each table being limited to the plankton entering at the standard flow rate of 3 l/min. Histological examination of the gut epithelium revealed a slight metaplasia which would also indicate partial starvation.

TABLE 1.* CATALYTIC RATES OF GLYCOLYTIC AND GLUCONEOGENIC ENZYMES IN RAT LIVER AND QUAHOG TISSUES

Enzyme	Rat[a] liver	Quahog tissues		
		Foot	Gill	Mantle
Hexokinase	4.3 (100)+[b]	0.16 (100)	0.21 (100)	0.09 (100)
Glucose-6-phosphate dehydrogenase	6.7 (156)	0.22 (137)	0.71 (338)	0.30 (333)
Phosphofructokinase	3.3 (77)	0.15 (125)	0.13 (62)	0.19 (212)
Fructose diphosphatase	15 (349)	0.18 (150)	0.45 (214)	0.30 (333)
Pyruvate kinase	50 (1162)	8.42 (5260)	5.43 (2580)	2.34 (2600
Phosphoenolpyruvate Carboxytransphosphorylase	6.7 (156)	0.25 (156)	— —	— —
Lactic dehydrogenase	230 (5350)	1.07 (668)	0.35 (168)	0.49 (466)

[a] Maximum catalytic rates reported (Scrutton and Utter, 1968).
[b] Values are expressed as μMoles/min/g wet weight. Numbers in parentheses are per cent of hexokinase activity in each tissue.
* Reproduced from *Comp. Biochem. Physiol.*, 37:397–403 (1970).

TABLE 2. ENZYME ACTIVITIES IN DDT AND LINDANE-EXPOSED QUAHOG TISSUE (PER CENT ACTIVITY RELATIVE TO CONTROL)

Enzyme	Insecticide	Foot		Gill		Mantle	
		Wet weight	Protein	Wet weight	Protein	Wet weight	Protein
Hexokinase	DDT	—	—	—	—	—	—
	Lindane	—	—	—	—	—	—
Glucose-6-phosphate dehydrogenase	DDT	—	—	0	0	—	—
	Lindane	159 (0.05)	214 (0.005)	—	—	—	—
Phosphofructokinase	DDT	—	—	169 (0.1)	—	—	—
	Lindane	—	—	—	—	—	—
Fructose diphosphatase	DDT	66 (0.001)	69 (0.05)	70 (0.1)	64 (0.05)	50 (0.1)	—
	Lindane	71 (0.005)	69 (0.05)	57 (0.005)	68 (0.1)	—	—
Pyruvate kinase	DDT	77 (0.1)	83 (0.05)	—	—	—	114 (0.1)
	Lindane	—	88 (0.05)	—	—	—	—
Lactic dehydrogenase	DDT	—	—	—	—	—	174 (0.05)
	Lindane	—	—	—	—	—	—

Numbers in parentheses represent values of "P" calculated from "Students" t test.

These changes in glycogen levels and histology were accompanied by alterations in the activity of a number of enzymes. Although glucose-6-phosphate dehydrogenase remained constant in the foot it decreased more than 70 per cent in the gill and fell to negligible levels in the mantle. This response has also been noted in the rat following severe starvation (McDonald and Johnson 1965). Glucose-6-phosphate dehydrogenase is severely depressed along with hexokinase and pyruvate kinase. Upon refeeding, normal enzyme levels are established within a few days. No significant change was observed in hexokinase. Pyruvate kinase levels rose significantly in the foot (45 per cent) in relation to both tissue weight and protein.

Fructose diphosphatase was doubled in the foot. It has been observed that in the fasting rat, fructose diphosphatase levels remain high during the fasting period, and, in terms of cell protein, show an increase (Weber 1959). The increase in fructose diphosphatase noted in the control group may thus reflect stimulation of gluconeogenesis. Thus with respect to both glucose-6-phosphate dehydrogenase and fructose diphosphatase, the quahog appears to respond to partial starvation with essentially a mammalian response. This may indicate that a number of the same metabolic control mechanisms operate in mammalian and molluscan tissue.

Following 30 weeks of exposure to DDT and lindane, additional alterations in enzyme activities were noted (Table 2). DDT reduced the glucose-6-phosphate dehydrogenase content of gill tissue to negligible levels. Similar inhibition of this enzyme by DDT has been reported for rat liver under in vitro (Tinsley 1964) and in vivo conditions (Tinsley 1965). However, exposure to lindane resulted in a significant increase in the foot glucose-6-phosphate dehydrogenase activity of 59 per cent on a tissue weight basis and 114 per cent in terms of protein.

The most striking effect of DDT and lindane appears to be a consistent decrease in fructose diphosphatase activity compared to control quahogs. This would indicate that chlorinated hydrocarbons may interfere with gluconeogenesis. The quahogs used in this study can be assumed to be under a degree of metabolic stress as the result of partial starvation. Normal variations in the marine environment which result in the development of similar natural stress situations might be expected to be exacerbated by the presence of DDT or lindane.

Although the significance did not reach P<0.1 except in the case of the DDT-exposed gill, there was a trend toward higher levels of phosphofructokinase. This and fructose diphosphate catalyse the phosphorylation of fructose-6-phosphate and the dephosphorylation of fructose-1, 6-diphosphate, respectively, and have been shown to be under subtle allosteric control by substrate, product and a number of cofactors. These exert opposite influences to ensure that both enzymes will not operate simultaneously, the result of which would be the hydrolysis of ATP. A concomitant increase in phosphofructokinase and decrease in fructose-diphosphatase in the presence of DDT raises the possibility of DDT interference with these critical allosteric interactions, which could conceivably increase glucose use and suppress gluconeogenesis. Stimulation of glucose catabolism by DDT has been reported by Silva et al. (1959).

Exposure to both DDT and lindane resulted in a slight but significant decrease in foot pyruvate kinase and a 75 per cent increase in mantle lactic dehydrogenase.

The results support the view that glycolysis is an operative metabolic pathway in molluscan tissue and that it responds to stress conditions such as starvation with compensating mechanisms analogous to those identified in mammalian tissue. The data further suggest that the chronic administration of sublethal doses of chlorinated hydrocarbons may lead to an increase in glucose degradation and suppression of gluconeogenesis.

References

BABERS, F H   Solubility of DDT in water, determined radiometrically. J. Am. chem. Soc., 77:4666.
1955

BOWMAN, M C, ACREE, F, JR. and CORBETT, M K   Solubility of C14 DDT in water. J. agric. Fd Chem., 8:406–8.
1960

HAMMERSTROM, R S, et al.   Study of pesticides in shellfish and estuarine areas of Louisiana. Publ. Hlth Serv. Publs Wash., (999):1–26.
1967

JODREY, L and WILBUR, K   Studies on shell formation. 4 The respiratory metabolism of the oyster mantle. Biol. Bull. Mar. biol. Lab., Woods Hole, (108):346–58.
1955

McDONALD, B E and JOHNSON, B C   Metabolic response to realimentation following starvation in adult male rat. J. Nutr., 87:161–7.
1965

MILLER, M W and BERG, G G   Chemical fallout. Springfield, Ill.. Charles C. Thomas, 532 pp.
1969

O'BRIEN, R D   Insecticides, action and metabolism. New York, Academic Press, 319 pp.
1967

RICHARDSON, L T and MILLER, D M   Fungitoxicity of chlorinated hydrocarbon insecticides in relation to water solubility and vapor pressure. Can. J. Bot., 38:163–75.
1960

SCRUTTON, M C and UTTER, M   The regulation of glycolysis and gluconeogenesis in animal tissue. Ann. Rev. Biochem., 37:249–302.
1968

SILVA, G M, DOYLE, W P and WANG, C H   Glucose catabolism in
1959   the DDT treated American cockroach (*Periplaneta ameri-
cana L.) Archos port. Bioquim.*, 3:298–304.
SIMPSON, J and AWAPARA, J   Phosphoenolpyruvate carboxykinase
1964   activity in invertebrates. *Comp. Biochem. Physiol.*,
12:457–64.
TINSLEY, I J   Ingestion of DDT and liver glucose-6-phosphate
1964   dehydrogenase activity. *Nature, Lond.*, 202:1113–4.

TINSLEY, I J   DDT ingestion and liver glucose-6-phosphate dehy-
1965   drogenase activity. 2. *Biochem. Pharmac.*, 14:847–51.

USAKI, I and OKAMURA, N   Glycolytic intermediates in the oyster
1956   gill. *Scient. Rep. Tohoku Univ.*, 22:225–32.

WEBER, G   Pathology of glucose-6-phosphate metabolism. *Revue
1959   can. Biol.*, 18:245–82.

Accumulation and Metabolism of DDT-¹⁴C (Dichloro-Diphenyl-Trichloro-Ethane) in Marine Organisms

*W. Ernst**

Accumulation et métabolisme du DDT-¹⁴C (Dichloro-diphényl-trichloro-éthane) dans les organismes marins

Ces études à court terme avaient pour objet de mesurer l'accumulation du DDT dans les vers marins à partir de solutions très diluées administrées par voie orale. Etant donné que les vers utilisés sont des organismes entrant dans l'alimentation des poissons plats, qui sont consommés par l'homme, on a étudié la transmission du DDT aux poissons dans des concentrations corrélatives, simulant ainsi une partie de la chaîne alimentaire.

Les animaux d'expérience provenaient de la mer du Nord et de l'estuaire de la Weser, où on les trouve en grand nombre.

1. Accumulation du DDT par les vers marins:

Nereis diversicolor et *Lanice conchilega*. Ces animaux sont gardés dans des solutions d'eau de mer additionnée de DDT, aux concentrations suivantes:

0,01   0,05   0,10   1,0   2,0   10 parties par milliards (*Lanice conchilega*)

0,3   0,75   3,0   30,0   parties par milliards (*Nereis diversicolor*)

On a en outre administré du DDT oralement à *Nereis diversicolor* en doses uniques et renouvelées de 1 à 5 µg/animal. Les diverses durées d'exposition et de rétablissement, les concentrations de DDT dans les vers et dans le liquide coelomique de *Lanice conchilega* étaient dans la gamme des parties per million.

2. Transmission et accumulation du DDT dans les poissons plats:

Platichthys flesus et *Solea solea*. Des doses de DDT ont été administrées par voie orale et on a mesuré la répartition du pesticide dans divers tissus (cerveau, chair, viscères, foie, rein), ainsi que dans les excréments. On a déterminé le taux d'absorption de l'appareil gastro-intestinal en fonction du temps et l'incidence du DDT dans le cerveau. Les niveaux de DDT étaient dans la gamme des parties per million.

Dans toutes les expériences, l'examen des produits métaboliques du DDT par radiochromatographie a révélé la dégradation vers le DDE, le DDD et d'autres composés polaires. Les données expérimentales sont discutées compte tenu des tolérances résiduelles de ce pesticide dans d'autres denrées alimentaires. Les procédures suivies lors des expériences seront examinées en détail.

Acumulación y metabolismo de DDT-¹⁴C (Dicloro-difenil-tricloro-etano) en organismos marinos

El objetivo de estos estudios era medir la acumulación de DDT en poliquetos inmersos en soluciones muy diluidas de ese producto y sometidos a tratamiento oral con el mismo, en investigaciones a corto plazo. Teniendo en cuenta que los poliquetos empleados sirven de alimento a los peces planos, consumidos luego por el hombre, se estudió la transmisión del DDT a los peces, con concentraciones correlativas, simulando así una parte de la cadena de alimentación. Los animales empleados en los experimentos procedían del Mar del Norte y del estuario de Weser, donde son muy abundantes.

1. Acumulación de DDT en poliquetos:

Nereis diversicolor y *Lanice conchilega*. Se mantuvieron en soluciones de DDT en agua marina a las concentraciones siguientes:

0,01   0,05   0,10   1,0   2,0   10 partes per billón (*Lanice conchilega*)

0,3   0,75   3,0   30,0   partes per billón (*Nereis diversicolor*)

Además, a los *Nereis diversicolor* se les suministró DDT por vía oral en dosis únicas y repetidas de 1 a 5 µg por animal. Variando el tiempo de exposición y de recuperación, las concentraciones de DDT en los poliquetos y en el fluido celómico de *Lanice conchilega*, resultaron del orden de partes por millón.

2. Transmisión y acumulación de DDT en peces planos:

Platichthys flesus y *Solea solea*. Se administraron dosis de DDT oralmente y se midió la distribución del plaguicida en diversos tejidos (cerebro, carne, vísceras, hígado, riñón) y en los excrementos. Se determinó la absorción de DDT en el tubo gastroentérico en función del tiempo, y la presencia de DDT en el cerebro. Las concentraciones de DDT resultaron del orden de partes por millón.

En todos los experimentos se examinaron radiocromatográficamente los productos metabólicos del DDT, encontrándose una degradación hacia DDE, DDD y otros compuestos más polares. Se examinan los datos experimentales teniendo en cuenta las tolerancias de los residuos de este plaguicida en otros alimentos. Se examinarán en detalle los procedimientos aplicados en la experimentación.

VERY low concentrations of persistent pesticides in water may produce harmful effects in the marine environment by accumulating in marine organisms and becoming distributed in the food chain. Degradation of pesticides to lesser toxic compounds by various mechanisms in marine organisms is therefore important to the survival of marine animals.

We did experiments on the metabolism and some aspects of accumulation of DDT-¹⁴C at very low concentrations in the marine polychaetes *Lanice conchilega* and *Nereis diversicolor* (Ernst 1969) and in the flatfishes *Platichthys flesus* and *Solea solea*, which feed partly on these polychaetes. Laboratory studies of the absorption and detoxification of DDT have been done by other workers (Greer and Paim 1968; Premdas and Anderson 1963; Cherrington *et al.* 1969) on other species and with other concentrations.

Methods and results

The animals were kept in glass tanks in sea water or diluted sea water with salinities and temperatures corresponding with natural conditions. The polychaetes were sampled from the Weser River and conditioned for 2–4 weeks preceding the experiments. *Nereis diversicolor* were kept in glass tubes which were placed vertically in the glass tank during the experiments. *Lanice conchilega* were allowed to prepare their tubes from washed sea sand 1–2 days before the experiment. Flatfishes were dredged in the Weser estuary and were kept under the experimental conditions. Before the administration of DDT they were trained to feed immediately on *Lanice conchilega* and *Nereis diversicolor*. DDT-¹⁴C was given orally on food particles or as an ethanol in water solution from which it was absorbed. The experimental conditions and results for *Lanice conchilega* and *Nereis diversicolor* are summarized in Table 1.

After exposure to DDT-¹⁴C the animals were extracted by grinding them with quartz sand in acetone, evaporating the acetone extract, and dissolving the residue in n-heptane. Quantitative measurements of radioactivity

* Institut für Meeresforschung, Bremerhaven, Federal Republic of Germany.

TABLE 1. METABOLIC PRODUCTS AND ACCUMULATION OF DDT–^{14}C BY *Lanice conchilega* AND *Nereis diversicolor* FROM AQUEOUS SOLUTIONS

Organisms	ppb DDT in water	Time of exposure	Metabolites	ppb DDT in animals	Experimental conditions
Lanice[a] *conchilega*	0.009	67 hours	DDT	2.1	Single dose of DDT–^{14}C in 4 l water, 27°/$_{00}$ S, 10° C
	0.06	67 hours	DDT	12.5	
	0.11	67 hours	DDT	30	
	1.12	2 days	DDT	84	Single dose of DDT–^{14} in 0.5 l water, 27°/$_{00}$ S, 10° C
	2.24	2 days	DDT	187	
	11.2	2 days	DDT	788	
	2.24	31 days	DDE, trace DDD, polar compounds	5,120 10,400[b]	20 repeated doses of DDT–^{14}C in 0.5 l water, 27°/$_{00}$ S
Nereis[c] *diversicolor*	30	2 h	DDT	5,400	Single dose of DDT–^{14}C in 0.5 l water, 10°/$_{00}$ S, 10° C
	30	7 h	DDT	8,500	
	30	24 h	DDT	14,700	
	0.3	5 days	DDT	610	Daily repeated doses of DDT–^{14}C in 0.5 l water
	0.75	5 days	DDT	1,240	
	3.0	5 days	DDT	4,200	

[a] Wet weight 5 g, for repeated doses 1.9 g for each concentration.
[b] Coelomic fluid
[c] Wet weight 0.25 g for each concentration.

were made by liquid scintillation counting. In order to detect any metabolites, the concentrated extracts were separated by thin layer chromatography on silica gel with subsequent radioscanning.

Unchanged DDT was the only radioactive compound in *Nereis*-extracts except in the case of oral application of 5 μg DDT-^{14}C with a waiting period of 5 days, where traces of polar metabolites were detected. After repeated application of DDT in *Lanice*, we detected DDE and traces of DDD and polar compounds in the extracts from the coelomic fluid and the tubes of the polychaetes. The extract from tubes that were exposed to DDT in absence of *Lanice* showed similar metabolic products, which are evidently formed by bacterial degradation. The nature of the more polar compounds, which were also present in the water of the tanks, has not yet been elucidated. In order to check removal of DDT from *Lanice* with a DDT-content of 230 ppb, the animals were placed in a glass vessel which was kept in a bigger glass tank that contained untreated *Lanice* under aeration. After 5 weeks the loss of activity from the treated animals was about 6 per cent.

In the experiments with flatfishes the animals were kept single in glass tanks of 10°/$_{00}$ salinity for *Platichthys flesus* and 20°/$_{00}$ for *Solea solea*. During the experimental period the food contained known amounts of DDT-^{14}C. A total of 5.4 μg DDT-^{14}C was absorbed from the gastrointestinal tract of *Platichthys* within 6 h. After 6 h the DDT levels in brain tissue were 20–260 ppb and in liver tissue 30–80 ppb; the wet weight of the animals was between 6 g and 42 g. The distribution of DDT-^{14}C in various tissues at higher doses and over longer periods is shown in Table 2. After a waiting period of 28 days no remarkable decrease of DDT levels was observed in the tissues of *Platichthys*.

The DDT content of flesh was 0.2–0.5 ppm and 2.0–3.8 ppm in the brain of *Platichthys* (Table 2). DDE and DDD could be detected after the 28-day waiting period in the liver, the gastrointestinal tract and in the

TABLE 2. DISTRIBUTION OF RADIOACTIVITY OVER VARIOUS TISSUES IN FLATFISHES AFTER ADMINISTRATION OF DDT–^{14}C

Experimental animal	*Platichthys flesus*			*Solea solea*		
Body weight (g)	13.0	13.4	28.2	25.8	30.8	34.7
DDT–^{14}C (μg) total	14.8	16.3	24.2	1.8	5.4	9.0
Time of application (days)	14	14	14	1	1	1
Waiting period after the last application (days)	28	—	—	8	8	8
% radioactivity related to initial dose:						
brain	0.85	1.20	0.40	0.80	0.49	0.59
kidney	0.68	0.59	0.31			
gastrointestinal tract	3.3	3.8	0.9			
flesh	13.30	14.25	22.0			
total body	1.85	0.41	2.31	55.0	58.1	61.1

flesh. The proportions were DDD/DDE/DDT = 1:2:12 in the flesh. More unidentified polar compounds were found in the gastrointestinal tract and in faeces.

Conclusions

The animals in these experiments are part of the bottom fauna of the Weser River and the coastal area of the North Sea. The possibility of accumulating DDT from very low environmental concentrations by marine polychaetes seems to be an important factor for the introduction of DDT and presumably other persistent pesticides into the marine food chain, e.g. polychaetes to flatfishes.

The absorbed DDT is metabolized in both polychaetes and flatfishes only after longer periods of exposure or higher doses. Accumulation data from these experiments are preliminary for the polychaetes because of the short experimental period and the lack of constant DDT concentrations in most cases. But it is clear that very low concentrations of 0.01 ppb led to a level of about 2 ppb in the *Lanice conchilega* within a short time. The concentration in water listed in Table 1 is calculated from the applied dose and the volume of water. Since some DDT absorbs to the glass walls of the tanks, the true concentration in water is lower than this.

References

CHERRINGTON, A D, PAIM, U and PAGE, O T *In vitro* degradation
1969   of DDT by intestinal contents of Atlantic salmon (*Salmo salar*). *J. Fish. Res. Bd Can.*, 26(1):47–54.

ERNST, W Stoffwechsel von Pesticiden in marinen Organismen.
1969   1. Vorläufige Untersuchungen über die Umwandlung und Akumulation von DDT–¹⁴C durch den Polychaeten *Nereis diversicolor. Veröff. Inst. Meeresforsch. Bremerh.*, 11(2):327–32.

ERNST, W Stoffwechsel von Pesticiden in marinen Organismen.
1970   2. Biotransformation und Akumulation von DDT–¹⁴C in Plattfischen, *Platichthys flesus. Veröff. Inst. Meeresforsch. Bremerh.*, 12:353–60.

ERNST, W Stoffwechsel von Pesticiden in marinen Organismen·
1970   3. Abbau und Speicherung von DDT–¹⁴C in Plattfischen, *Solea solea*, im Kurzzeitversuch. *Veröff. Inst. Meeresforsch., Bremerh.*, 12:361–4.

GREER, G L and PAIM, U Degradation of DDT in Atlantic salmon
1968   (*Salmo salar*). *J. Fish. Res. Bd Can.*, 25(11):2321–6.

PREMDAS, F H and ANDERSON, J M The uptake and detoxification
1963   of C14-labelled DDT in Atlantic salmon (*Salmo salar*). *J. Fish. Res. Bd Can.*, 20(3):827–37.

The Association of DDT Residues with Losses in Marine Productivity

Philip A. Butler, *Ray Childress†* *and Alfred J. Wilson‡*

Relation entre les résidues de DDT et la baisse de productivité des eaux marines

Dans le cadre de son programme coopératif de surveillance de la pollution marine, le Bureau of Commercial Fisheries du Fish and Wildlife Service des USA a suivi au cours des cinq dernières années l'écosystème des estuaires de la côte du Texas. Tous les mois des échantillons d'huitres et de poissons sont prélevés à des stations permanentes; les teneurs pour 10 pesticides à base d'hydrocarbures chlorés couramment rencontrés, ainsi que celles correspondant à certains composés biphénylchlorés sont dosés par chromatographie avec une précision d'environ 10 µg/kg. Des échantillons recueillis dans un estuaire par le Parks and Wildlife Department du Texas contenaient des concentrations relativement élevées de résidus de DDT et de ses métabolites pendant toute la durée du programme de contrôle. Les auteurs communiquent les résultats des analyses de ces échantillons en y joignant des données corrélatives sur l'utilisation du DDT et la présence de résidus de pesticides dans les voies d'eau et les sédiments fluviaux et formulent l'hypothèse d'un rapport causatif entre l'usage à des fins agricoles de DDT dans le biota des eaux estuarines.

La baisse régionale de la production du sciaenide (*Cynoscion nebulosus*) dans les estuaires, démontrée par des études fondées sur le recensement annuel des poissons, est attribuée à l'action inhibitrice des résidus de DDT contenus dans le jaune de l'oeuf sur le développement du frai dans les sacs embryonnaires.

Le document fournit des données à l'appui de l'hypothèse selon laquelle les zones estuarines et les deltas des rivières constituent les principaux réceptacles des pesticides chimiques rémanents qui ont été appliqués à l'échelle mondiale lors des 25 dernières années.

La asociación entre los residuos de DDT y las pérdidas en la productividad marina

Dentro del programa cooperativo del Bureau of Commercial Fisheries del Fish and Wildlife Service de los Estados Unidos para la vigilancia de la contaminación del mar, se realizaron durante los últimos cinco años estudios sobre el ecosistema de los estuarios de la costa de Texas. Mensualmente se recogieron muestras de ostras y peces en estaciones permanentes y se utilizaron métodos analíticos cromatográficos para determinar, con una sensibilidad de cerca de 10 µ/kg, los 10 plaguicidas de hidrocarburos clorinados contenidos comunmente asi como algunos de los compuestos de bifenilos clorinados. Las muestras recogidas por el Department of Parks and Wildlife de Texas contenian residuos relativamente abundantes de DDT y metabolitos en todo el curso del programa de vililancia. Los resultados analíticos de estas muestras más los datos correlativos sobre la utilización de DDT y residuos de pesticidas en las aguas y sedimentos de los ríos tributaríos se presentan en este trabajo, para demostrar la supuesta relación causal entre el empleo para fines agrícolas del DDT en las cuencas adyacentes de drenaje y los residuos de plaguicidas en la biota de los estuarios.

Los descensos regionales y bien documentados de la producción de la corvina (*Cynoscion nebulosus*) en los estuarios, comprobados por los estudios de los censos anuales de los peces, se atribuyen a la inhibición del desarrollo de los embriones vesiculados a causa de los residuos de DDT en el vitelo.

Se presentan datos que apoyan la hipótesis de que las zonas de estuario y deltas de los ríos son probablemente los depositarios mayores de los plaguicidas químicos persistentes que se han aplicado en todo el mundo en los últimos 25 años.

IN the past decade much evidence has been accumulated demonstrating the persistence of DDT, its metabolites, and other chlorinated hydrocarbon compounds in various sectors of the biosphere. The occurrence of these pollutants in aerially transported dust (Risebrough, *et al.*, 1968), rain water (Cohen and Pinkerton, 1966), and the fauna of fresh and salt water environments in polar as well as temperate climates shows clearly the ubiquity of these man-made organics. In most instances, reports of these residues have resulted from essentially random surveys. The mechanisms for the transport of DDT from soils, its movement into surface and ground water, and its volatilization into the atmosphere are still poorly understood. Terriere (1966) presents evidence to show that 40% of the DDT applied to two orchards remained *in situ* after 20 years. About 50% was presumed lost to the atmosphere through volatilization, and less than 5% of the DDT had been leached into surface and ground water. Van Middelem (1966) has reviewed concisely the physical and metabolic factors influencing the persistence and fate of organic pesticides in the environment.

In 1965, the Water Resources Division (U.S. Geological Survey) expanded its programme of monitoring the quality of irrigation water to include analyses for 12 of the more common insecticides and herbicides (Brown and Nishioka, 1967; Manigold and Schulze, 1969). In 1965, the Bureau of Commercial Fisheries initiated a nationwide programme to determine the levels of persistent pesticides in estuarine molluscs (Butler, 1966). A preliminary summary of data showed widespread occurrence of DDT in the nation's estuaries (Butler, 1969).

The Coastal Fisheries Division of Texas Parks and Wildlife Department conducted annual surveys designed

---

*Gulf Breeze Biological Laboratory, U.S. Department of the Interior, Gulf Breeze, Florida 32561, U.S.A.
†Coastal Fisheries, Texas Parks and Wildlife Department, Seadrift, Texas 77893, U.S.A.
‡Gulf Breeze Biological Laboratory, U.S. Department of the Interior, Gulf Breeze, Florida 32561, U.S.A.

to show the strength of the different year classes and data obtained from these surveillance programmes form the basis of this report.

Study objective

The surveys of finfish populations in the Laguna Madre indicated a progressive decline in the numbers of juvenile spotted seatrout, *Cynoscion nebulosus*—important in the sport fishery. Typically, seatrout enter the estuaries and bays in early spring and move offshore into deeper water in the fall after spawning but do not migrate very far along the coasts (Idyll and Fahy, 1970).

Since the seatrout population decline was restricted to juvenile fish, it was apparent that over-fishing was not a factor. The oyster monitoring programme had shown that DDT residues in Laguna Madre oysters, *Crassostrea virginica*, were routinely higher than in oysters elsewhere. The work of Burdick, *et al.*, (1964) had shown that DDT residues in the eggs of fresh water fish caused increased mortality. We therefore initiated a project to determine the kinetics and accumulation of DDT and its metabolites in seatrout and their food organisms in Texas estuaries.

The lower Laguna Madre on the south Texas coast is a usually hypersaline estuary measuring about 80 by 25 kilometres separated from the Gulf of Mexico by a narrow sand barrier, Padre Island, and tidal circulation is restricted. Water depth varies from about 0.3 to 4.0 metres. Despite wide fluctuations in salinity due to river discharge and solar evaporation, the area supports a mixed fauna of commercially valuable fish and crustaceans. Childress (1965) has estimated that in 1965 the associated watershed had been treated with about 950 tons of DDT to control pests on cotton and other crops. The agricultural use of DDT was terminated in 1966–67.

The Arroyo Colorado, a natural but formerly intermittent river draining into the Laguna Madre, has been dredged for a distance of about 30 kilometres to Port Harlingen and now supports a varied estuarine fauna.

Methods

Throughout this report, reference to DDT include its metabolites DDD (TDE) and DDE. Samples of 12 oysters were collected at monthly intervals beginning in July 1965. There are two permanent stations in each of the six major Texas estuaries. Each group of 12 oysters is homogenized and mixed with the desiccant sodium sulphate plus 10% by weight of Quso, a micro fine precipitated silica. The resultant free-flowing granular homogenate can be held for at least 30 days at room temperatures without degradation of any chlorinated hydrocarbon pesticide residues. This technique makes it possible to send samples by ordinary mail to the Gulf Breeze Laboratory for analysis. Fish samples were collected with trawls and cast nets. At least 15 individuals of uniform size and presumably the same year class comprised each sample.

Oyster and fish homogenates were extracted for 4 hours with petroleum ether in a Soxhlet apparatus. Extracts were concentrated and partitioned with acetonitrile. The acetonitrile was evaporated to dryness and the residue eluted from a Florisil column (Mills, Onley and Gaither, 1963).

The sample was then analysed by electron capture gas chromatography. Three columns of different polarity (DC-200, QF-1, mixed DC-200, QF-1) were used to confirm identification. Operating parameters on Varian Aerograph 610D gas chromatographs were as follows: Column; 5-foot, ⅛-inch (O.D.) Pyrex glass; 3% DC-200, 5% QF-1 and a 1:1 ratio of 3% DC-200 and 5% QF-1; all on 80/100 mesh Gas Chrom Q. Temperature; detector 210°C, injector 210°C, oven 190°C. Carrier Gas; pre-purified nitrogen at a flow rate of 40 ml/min. In a few samples, thin layer chromatography was employed. The lower limit of detectability was 0.01 mg/kg. Recovery rates were: p,p'-DDE, 80–85%; p,p'-DDD (TDE), 92–95%; p,p'-DDT, 91–95%. Data in this report have not been corrected for percentage recovery.

Seatrout study

A lined trawl was used to collect fish census samples. From 5 to 40 hectares were trawled depending on the density of fish. In the period 1964–69, seatrout populations in the major Texas estuaries were reasonably constant except in the lower Laguna Madre where the number of juvenile seatrout per hectare declined dramatically as follows: 1964—74; 1965—62; 1966—30; 1967—no sample; 1968—6; and 1969 only 0.4. Field observations showed, however, that seatrout continued to spawn.

Mature seatrout were collected at monthly intervals beginning in March 1968. Samples of 10 to 20 fish, fork length 300 to 670 mm, were dissected and either the eggs or entire ovaries were composited and analysed. DDT in the ovaries reached a peak concentration of about 8 mg/kg in September prior to spawning. The lower residue levels in October suggest that the spawned eggs contained a major portion of the DDT residues. DDT residues in the ovaries of seatrout collected in other estuaries along

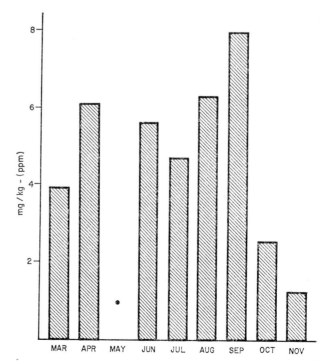

Fig 1. DDT residues in the ovaries of mature seatrout collected in the Arroyo Colorado in 1968; spawning is completed by October (•No sample in May)

[263]

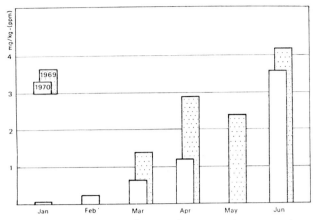

Fig 2. DDT residues in the eggs of seatrout prior to spawning in the Arroyo Colorado

the Texas coast in the summer of 1968 did not exceed 0.5 mg/kg (fig 1).

In the spring of 1969 and 1970, residue analyses were made primarily on the seatrout eggs (fig 1). The gradual accumulation of DDT in the developing ova is apparent. It is clear also that in 1970 DDT residues were lower than in 1969 (fig 2).

Oyster residues

Oysters accumulate chlorinated hydrocarbon pesticides when the concentration of the pollutants is as low as 0.01 μg/kg in the surrounding water. When the pollution ceases, the residues are flushed out of the oyster at regular rates (Butler, 1966). Oysters and some other molluscs, therefore, are excellent indicator organisms to monitor pesticide pollution in estuaries. In the first three years of the Bureau's nationwide estuarine monitoring programme, more than 5,000 samples were analysed (Butler, 1969). These show that DDT residues are consistently higher in oysters from estuaries draining areas where there is intensive farming. Oysters from the Arroyo Colorado contain consistently higher DDT residues than oysters from any other state.

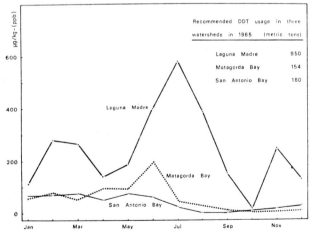

Fig 3. DDT residues in oysters collected in three representative Texas estuaries in 1966

The casual relationship between DDT residues in oysters and the agricultural use of DDT is implied by the results of analyses of oyster samples from three estuaries with different watersheds (fig 3). Childress (1965) esti-

mated that about five times more DDT was used in the Laguna Madre area than in the Matagorda Bay and San Antonio watersheds. A year later, 1966, DDT residues in oysters showed a similar discrepancy in the three estuaries.

Although the use of DDT for agricultural purposes in Texas was restricted in 1966–67, there was no significant change in average DDT residues in oysters in the Laguna Madre area in 1969 (fig 4). The percentage of samples containing DDT actually increased from 75 to

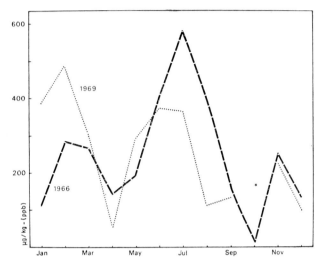

Fig 4. DDT residues in Arroyo Colorado oysters (*No sample in October 1966)

100%. In the five other major bays on the Texas coast, however, the percentage of samples positive for DDT residues decreased about 20% since 1967, and in three of the five bays the average levels of residues in oysters were lower in 1969 than in 1966. The three seasonal peaks in DDT residues in oysters from the Laguna Madre area probably reflect the multiple annual harvest possible in this semi-tropical climate.

Water and sediment residues

In 1967 and 1968, 20 river stations were monitored at monthly intervals (Manigold and Schulze, 1969) and DDT was detected on numerous occasions in depth integrated samples, i.e., water collected in a narrow-mouth bottle as it was lowered slowly from the surface to the river bottom. In most instances, samples collected at the three stations on Texas rivers contained DDT residues in the range of 0.09 μg/l or less; the maximum observed was 0.18 μg/l in the Colorado River at Wharton, Texas in January 1968. There was no evidence of correlation between residue levels and discharge rates or season of the year. River water samples and bottom sediments collected at 29 stations by the U.S. Geological Survey in 1969 near or in Texas estuaries contained the highest levels of DDT (Childress, ms). The maximum level of DDT observed in estuarine sediments was 99.0 μg/kg and in estuarine water it was 1.02 μg/l. This value approximates the solubility of DDT in water but we assume that most of the DDT was probably sorbed on silt. Although the estuarine samples were collected only once or twice a year, analyses indicate that DDT residues were highest at the junction of river and bay, i.e., where the decrease in water current permitted silt containing DDT to drop to the bottom. DDT residues were uniformly low or absent in bottom sediments at the junction zone of estuarine water with the Gulf of Mexico. Maximum DDT residues in water and sediment of the five bays sampled were similar and were generally not correlated with residues in local oyster samples.

Menhaden residues

Juvenile menhaden, *Brevoortia patronus* and *B. gunteri*, about 60 mm in fork length, were sampled monthly in the Arroyo Colorado. Whole body residues of DDT ranged from about 3 to 8mg/kg. in 1968 but were appreciably lower in 1969 (fig. 5). These DDT residue

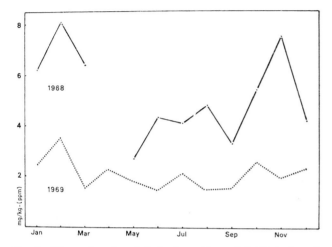

Fig 5. Fluctuations in whole body residues of DDT in juvenile menhaden collected at Port Harlingen (No sample in April 1968)

levels are significantly higher than in menhaden from other Texas estuaries where the range was 0.04 to 0.6 mg/kg. In 1970 menhaden from the Atlantic coast had DDT residues of about 0.2 mg/kg.

Discussion

The correlation between levels of DDT in oysters and the reported amounts of DDT used on agricultural land in the associated drainage basins in Texas is quite striking. This is supported by data collected from the Bureau's monitoring programme which indicate very low DDT residues in marine molluscs in Maine and Washington where relatively insignificant amounts of DDT are used in the coastal drainage basins.

Seatrout are widely distributed from Delaware to Mexico but characteristically the populations are localized and coastal movements are very restricted (Idyll and Fahy, 1970). Pearson (1928) estimated that about 80% of the food of adult trout consists of shrimp. However in the Arroyo Colorado and the lower Laguna Madre spotted seatrout feed primarily on menhaden. Since shrimp are usually killed by DDT before they build up significant residues of the pesticide (Butler, 1969a), whereas menhaden build up relatively large deposits of DDT without apparent effect on their mortality, the seatrout's diet in different estuaries will largely control its DDT residues. Mature seatrout entering the Laguna Madre in the early spring feed largely on menhaden whose DDT deposits will be metabolically incorporated into the developing eggs of the seatrout. In other estuaries,

seatrout feed on shrimp and DDT residues are negligible. It would be of interest to analyse seatrout from Lake Pontchartrain in Louisiana, where their diet is reported to be almost exclusively fish (Idyll and Fahy, 1970).

Menhaden feed on plankton. The fact that menhaden in the Arroyo Colorado have high DDT residues compared to menhaden collected in other estuaries indicates the levels of contamination of the plankton and infers the relatively high DDT contamination of water in the Arroyo Colorado. The higher contamination with DDT of Arroyo Colorado plankton is also reflected in the high DDT residues found in oysters there.

Farm lands around the six major Texas estuaries were estimated in 1965 to be using amounts of DDT which ranged from 100 to 950 metric tons annually. Yet, in the 1970 survey, DDT residues in estuarine sediments in the different estuaries were on average less than an order of magnitude. It is established that biological systems may become saturated with a given pollutant when the environmental loading is relatively constant. We suggest a similar situation for DDT deposits in estuarine sediments. The solubility of DDT in water is about 1.0 $\mu g/l$; a level that would not be exceeded normally in drainage waters. DDT residues are sorbed onto bottom deposits until a plateau is reached for the type of sediments. Excess deposits of DDT are lost by partitioning when the overlying tidal waters are not saturated with the pesticide.

Occasional high DDT residues in other Texas oyster populations present an anomalous situation. In Tres Palacios Bay, for example, DDT residues in oysters were uniformly low at a level of only about 0.1 mg/kg in 1966 and the first half of 1967. This is despite the fact that bottom deposits in the bay contain as much as 1.0 mg/kg of DDT. Then late in 1967, DDT residues in oysters suddenly increased 5 to 7 times and remained high for about 10 months before returning to their former low levels. We suggest that this was due to the tropical storm which passed through the area in September resuspending DDT residue-laden bottom sediments; torrential rains leached nutrient chemicals from the drainage basin and a period of superproductivity of plankton was initiated. Plankton took up the newly available DDT and the recycling of these residues, which had perhaps been immobilized for several years in bottom deposits, was reflected for the next 10 months in the oysters which fed upon the plankton.

Average DDT residues in juvenile menhaden collected in the Arroyo Colorado in 1969 were less than half the levels present in 1968. Since we believe that the DDT contamination of drainage water is the result of its agricultural use, we attribute the decline in menhaden DDT residues in 1969 to the restrictions on the agricultural use of DDT in 1966–67.

Although some seatrout mature in their first and second years, they live a relatively long time and their reproductive capacity increases with age. Our results suggest that the DDT residues were directly responsible for the failure of seatrout eggs to develop even though the spawning processes were normal. Since DDT residues in menhaden, the seatrout's chief food in the Arroyo Colorado area, are declining, we anticipate that DDT residues in seatrout eggs will begin to decline and that the reproductive capacity of the seatrout in this locality will soon return to its former norm.

Conclusions

1. Agricultural use of DDT is the chief source of DDT contamination of the estuarine environment in Texas.

2. Trophic magnification of DDT residues in the estuarine food web resulted in the reproductive failure of seatrout populations in the lower Laguna Madre, Texas in 1969.

3. Seatrout populations in other Texas estuaries were not harmed because of different food chain interactions.

4. Data suggest that estuarine sediments build up to a plateau of DDT residues over a period of years and these residues do not reflect the seasonal levels of waterborne pesticide pollution in the environment.

5. Sedimentary residues of persistent DDT may be resuspended physically by storms and recycled in the biota.

6. Restrictions on the agricultural use of DDT were reflected by decreased residues in estuarine biota in adjacent estuarine areas within three years.

7. The reproductive capacity of long-lived fish populations damaged by DDT residues may be restored by prohibiting the use of DDT in adjacent drainage basins.

References

BROWN, E and NISHIOKA, Y A   Pesticides in selected western
1967   streams—a contribution to the national program. *Pestic. Montig J.*, 1(2): 38–46.

BURDICK, G E *et al.*   The accumulation of DDT in lake trout and
1964   the effect on reproduction. *Trans. Am. Fish. Soc.*, 93(2): 127–36.

BUTLER, P A   Pesticides in the marine environment. *J. appl. Ecol.*,
1966   3(Suppl.):253–9.

BUTLER, P A   The significance of DDT residue in estuarine fauna.
1969   *In* Chemical fallout, edited by M. W. Miller and G. G. Berg, Springfield, Ill., Charles C. Thomas, pp. 205–20.

CHILDRESS, V R   A determination of source, amount, and areas of
1965   pesticide pollution in some Texas bays. Texas Parks and Wildlife, Progress Report Coastal Fisheries, pp. 245–55.

COHEN, J H and PINKERTON, C   Widespread translocation of pesti-
1966   cides by air transport and rain-out. *Adv. Chem. Ser.*, (60): 163–76.

IDYLL, C P and FAHY W E   Spotted seatrout. *Leafl. Atlant. St. mar.*
1970   *Fish Commn.* (13): 4–p.

MANIGOLD, D B and SCHULZE, J A   Pesticides in water. *Pestic.*
1969   *Monitg J.*, 3(2):124–35.

MILLS, P A, ONLEY, J H and GAITHER, R A   Rapid method for
1963   chlorinated pesticide residues in non-fatty foods. *J. Ass. off. agric. Chem.*, 46(2):186–91.

PEARSON, J C   Natural history and conservation of the redfish and
1928   other commercial sciaenids on the Texas coast. *Fishery Bull. U.S. Fish Wildl. Serv.*, 64:129–214.

RISEBROUGH, R W *et al.*   Pesticides: transatlantic movements in
1968   the northeast trades. *Science, N.Y.*, 159(3820):1233–5.

TERRIERE, L C *et al.*   Persistence of pesticides in orchards and
1966   orchard soils. *Adv. Chem. Ser.*, (60): 263–70.

VAN MIDDELEM, C H   Fate and persistence of organic pesticides in
1966   the environment. *Adv. Chem. Ser.*, (60): 228–49.

Monitoring Organochlorine Contamination of the Marine Environment by the Analysis of Residues in Seals

*A. V. Holden**

La surveillance de la contamination du milieu marin par l'analyse des residus organo-chlorés trouvés chez les phoques

Les composés organo-chlorés s'accumulent dans les tissus adipeux en concentrations croissantes à mesure de leur passage dans la chaîne trophodynamique animale, et des chercheurs ont déjà signalé l'existence de niveaux relativement élevés de contamination par les pesticides organo-chlorés dans les échantillons de lard de phoque provenant des deux côtés de l'Atlantique et de la Baltique. On s'est servi des analyses de ces échantillons pour comparer les niveaux de contamination du milieu marin à la fois par les pesticides organo-chlorés et par les biphényles polychlorés (PCB), composés utilisés largement dans l'industrie. Grâce à une technique d'analyse améliorée, on a pu faire séparément le dosage des deux groupes de composés, en évitant les réactions parasites causées précédemment par les PCB lors des analyses des résidus de pesticides.

Le document examine les différences des niveaux de contamination par la dieldrine, le DDT et ses produits de dégradation, et par les composés de PCB dans les diverses zones marines, en fonction de leur utilisation dans les zones terrestres adjacentes. Le problème posé par l'évaluation précise des concentrations de PCB est esquissé; et les proportions des isomères de PCB dans les différentes zones font l'objet d'une comparaison. Selon l'auteur, il se produit sans doute quelque dégradation de certains isomères de PCB dans les préparations commerciales, mais la dégradation du DDT semble être moindre que dans de nombreux autres types d'échantillons fauniques. Les concentrations de dieldrine sont généralement très inférieures à celles du groupe DDT, qui atteignent près de 100 ppm sur la côte canadienne; on trouve des concentrations de PCB supérieures à 100 ppm dans des échantillons provenant des côtes britanniques.

Determinación de la contaminación del ambiente marino por cloro orgánico, analizando los residuos en las focas

Los compuestos de cloro orgánico se acumulan en los tejidos grasos en concentraciones crecientes al pasar por las cadenas alimentarias de los animales. En trabajos anteriores se daba cuenta de una contaminación bastante fuerte por pesticidas a base de cloro orgánico en muestras de grasa de focas de ambos lados del Atlántico y del Báltico. Se han empleado análisis de tales muestras para comparar la intensidad de la contaminación del ambiente marino por los dos plaguicidas a base de cloro orgánico y los bifenilos policlorados (PCB) tan empleados en la industria. Con una técnica analítica mejorada se ha podido separar la determinación de los dos grupos de compuestos, evitando la interferencia que previamente causaba el PCB en el análisis de los residuos de plaguicidas.

El autor examina la diferencia de la intensidad de la contaminación por dieldrín, DDT y los productos de su degradación y los compuestos de PCB en diversos lugares del mar, con respecto a su empleo en zonas terrestres adyacentes. Se esboza el problema de la evaluación exacta de las concentraciones de PCB y se comparan las proporciones relativas de sus isómeros en diversos lugares. Se indica que tiene que ocurrir una cierta degradación de algunos isómeros del PCB en los productos del comercio, pero la degradación del DDT parece ser menor que en otras muchas clases de muestras de vida silvestre. Generalmente, las concentraciones de dieldrín son mucho menores que las del grupo del DDT, que se aproxima a 100 ppm en la costa canadiense. En muestras de las costas de la Gran Bretaña se encuentran concentraciones de PCB que exceden las 100 ppm.

THE occurrence of organochlorine pesticide residues in a wide range of living organisms in all parts of the global environment is due in part to their relatively high stability as chemical compounds and in part to their lipophilic properties. By virtue of their ability to be stored in body fats, they can be transferred progressively through a food chain and can accumulate in increasingly high concentrations. At each stage, digestion of the food of a predator may result in partial metabolism of at least some of the residues contained in

* Freshwater Fisheries Laboratory, Pitlochry, Scotland, U.K.

the food, and some loss by excretion is possible, nevertheless it is found that the concentrations of residues in predators are generally higher than in their prey, usually by about one order of magnitude.

In the marine environment, the fish-feeding aquatic mammals such as seals and porpoises are the final (or penultimate) stages in long food chains and could be expected to contain relatively high organochlorine residue levels. Furthermore they carry a high proportion of their body weight as subcutaneous fat or blubber. These species are therefore worthy of study as possible indicators of the levels of contamination of marine environments by organochlorine compounds. Organochlorine pesticide residues have been reported in various marine species in the past few years, and more recently other organochlorine compounds, the poly-chlorinated biphenyls (PCBs) have been identified by several workers (Risebrough et al., 1968; Koeman et al., 1969; Jensen et al., 1969). Two earlier papers by the author (Holden and Marsden, 1967; Holden, 1969) have described residue levels found in seals from the coasts of Great Britain, Canada, Norway and Sweden, and other workers (George and Frear, 1966; Sladen et al., 1966; Anas and Wilson, 1970) have reported analyses from several other areas.

The PCB residues occur as groups of individual polychlorinated biphenyl isomers of differing chlorine arrangements, and interfere with the identification and determination of pesticide residues unless precautions are taken to separate them from the pesticide residues. They are sometimes found in relatively high concentrations, and neglect of such precautions can result in substantial errors in the estimation of pesticide residues. A method of separation of the two types of residues has been developed, and many of the samples described in an earlier paper have been re-analysed by the new method to assess the extent of such errors. In addition, many more recent samples have been analysed, and the data obtained by the modified method are described in this paper.

Sources of samples

The earlier samples of seal blubber, many of which have been re-analysed, were obtained from the east coast of England, the east, north and west coasts of Scotland, the Canadian Arctic and the Gulf of St. Lawrence, the Norwegian Arctic coast and the Gulf of Bothnia (north Baltic Sea). The later samples include seals from the north-east and south-west coasts of England, the Welsh coast, and the Scottish coasts.

The species of seals examined were as follows:

Ringed seal (Arctic and Baltic) *Pusa hispida*
Harp seal (Canada) *Pagophilus groenlandicus*
Hooded seal (Canada) *Cystophora cristata*
Grey seal (Canada, U.K.) *Halichoerus grypus*
Common seal (Canada, U.K.) *Phoca vitulina*

Scottish samples have been deep-frozen at −20°C until required for analysis. Samples from other areas have been mainly preserved in 5 per cent aqueous formalin for transport to the laboratory, and subsequently deep-frozen at −20°C. Where comparative tests have been made of the effects of formalin preservation, no major differences from deep frozen samples have been detected.

Analysis practice

Aliquots of seal blubber (approximately 5g) were accurately weighed, ground with approximately five times their weight of anhydrous sodium sulphate to a fairly free-flowing powder and extracted in a rapid-action Soxhlet apparatus with n-hexane for 30 min (approximately twenty siphon-extractions). This has been found to remove virtually all the detectable residues. Both the sodium sulphate and the hexane were routinely tested to be free of residues interfering in analysis. The extracts were made up to 100ml and 50ml evaporated at 100°C in tared dishes to determine the percentage of extractable residues (primarily fat).

Appropriate aliquots of the extracts (usually 1ml) were cleaned-up to remove fats by alumina column chromatography, and PCB residues separated from organochlorine pesticide residues on silica columns as described by Holden and Marsden (1969). Both the alumina and silica were specially prepared, and tested to be free of organochlorine residues, and each batch of silica was checked to ensure that the intended residue separation was achieved. To be certain that no pp′–DDT was eluted in the first (n-hexane) eluate, about 5–7 per cent of the pp′–DDE was allowed to elute in the second eluate (10 per cent ether in hexane). This DDE is determined easily in the GLC analysis.

GLC analysis

The two eluates from the silica column were analysed by gas-liquid chromatography (GLC) using two different types of column (DC-200 and DC-200/QF-1) as described by Holden and Marsden (1967). The first eluate (PCBs − pp′–DDE) was injected on to the DC-200 column and the second eluate (mainly dieldrin, pp′–TDE and pp′–DDT) on to the DC-200/QF-1 column. The small amount of pp′–DDE referred to above can also be determined on this column without confusion. The relative retention times (Rx) of the PCB peaks were calculated with reference to dieldrin = 100 on the DC-200 column.

Confirmation of residues

Confirmation of the residues was provided by: (1) the fraction in which they were eluted from the silica column, (2) their retention times on the GLC columns with reference to standards, (3) alcoholic alkaline hydrolysis at 50°C under reflux, which converts pp′–DDT to pp′–DDE and pp′–TDE to pp′–MDE, subsequent GLC analysis confirming the conversion, (4) oxidation of pp′–DDE to pp′–dichlorobenzophenone (confirmed by GLC) with chromic oxide in glacial acetic acid, and (5) destruction of dieldrin by shaking extracts with concentrated sulphuric acid (confirmed by GLC). PCB isomers are unaffected by reactions (3), (4) and (5).

Calibration graphs using appropriate standard solutions were produced daily, the commercial PCB formulation Aroclor 1254 (made by Monsanto Ltd.) being used to quantify the PCB residues. The pp′–DDE residues elute with the PCBs, but the PCB residue having the same GLC retention time was small in most samples, and the peak was measured entirely in terms of pp′–DDE. No fully accurate method of determining the concentration of PCB residues is yet available, and the results obtained using commercial formulations as standards depend on the choice of formulation used. The peak patterns obtained from seal blubber extracts are not identical in proportions with those in commercial formulations, although the relative retention times of the peaks occurring agree with those in the commercial products. The peaks of shorter retention time (lower chlorine number) are, however, absent or almost so. The analyses of PCBs given in this paper were obtained by comparing the sum of the peak heights of the four major PCBs (exluding that at the pp′–DDE position) with the sum of the heights of the corresponding peaks of Aroclor 1254, using the DC-200 column for GLC analyses. (The large peak at Rx = 122 in Aroclor 1254 was small in all seal samples.)

Calculation of residue concentrations is normally based on wet weight of tissue, but for some purposes comparison of samples is more appropriately made with calculations based on extractable lipids. The concentrations of residues

in different organs of an individual seal are usually closely similar when the latter form of calculation is used.

Results

The results of the analyses obtained for the various samples, in terms of wet weight of blubber (or the sample supplied as blubber) are given in Tables 1 (Great Britain) and 2 (elsewhere). The mean values and ranges are given, together with the means and ranges of the percentage of extractable lipids found in the samples. Most specimens contain 80–90 per cent of extractable lipids, and thus calculation on the basis of lipid content only increases the residue concentration by some 11–25 per cent. In a few instances lipid values of 40–60 per cent have been found, usually in samples from dead or poorly-conditioned seals. Where much lower values have been found as in certain seal pups or emaciated adults, the sample supplied as blubber was in fact sub-

TABLE 1. ORGANOCHLORINE RESIDUES IN SEAL BLUBBER SAMPLES FROM GREAT BRITAIN
(means and ranges in parts per million tissue)

Source	Species	No.	% Fat	Dieldrin	DDE	TDE	DDT	PCB
Scotland, N.								
Shetland	Common (pups)	4	80 (69–88)	0.06 (0.06–0.07)	1.3 (0.7–1.7)	0.13 (0.08–0.18)	1.2 (0.8–1.7)	4 (2–6)
Orkney	Grey	8	69 (43–90)	0.18 (0.06–0.31)	6.4 (1.4–12.9)	0.60 (0.17–0.97)	6.0 (1.3–11.0)	18 (3–30)
Scotland, W.								
Hebrides	Grey	3	83 (78–87)	0.29 (0.24–0.32)	8.5 (5.6–12.6)	0.65 (0.41–1.1)	5.9 (4.0–9.5)	30 (19–40)
Summer Is.	Grey	4	55 (42–73)	0.38 (0.07–1.1)	3.9 (1.3–5.8)	0.54 (0.15–1.2)	5.4 (2.6–11.1)	16 (11–19)
Mull	Common (March 1969)	12	77 (54–86)	0.15 (0.09–0.22)	2.9 (0.9–4.6)	0.25 (0.20–0.36)	2.4 (0.72–3.5)	12 (5–19)
	Common (Dec. 1969)	5	65 (59–71)	0.12 (0.07–0.25)	2.5 (1.8–4.8)	0.32 (0.12–0.61)	2.2 (1.2–3.4)	12 (6–20)
Clyde	Common/Grey	3	87 (79–95)	1.3 (1.2–1.5)	13.2 (11.5–14.5)	1.2 (0.96–1.48)	10.8 (7.4–15.4)	75 (58–99)
Scotland, E.								
Aberdeen–Montrose	Grey	16	77 (45–91)	0.83 (0.46–1.7)	9.7 (4.2–19.1)	0.93 (0.54–1.5)	9.5 (3.8–15.7)	38 (12–88)
England, E.								
Farne Is.	Grey (pups)	5	81 (67–87)	0.46 (0.20–0.59)	6.6 (3.3–10.3)	0.86 (0.46–1.17)	5.6 (2.6–7.6)	40 (25–50)
Wash	Common (pups)	12		0.33 (0.16–0.66)	2.8 (1.4–4.1)	0.44 (0.28–0.73)	3.3 (1.8–4.9)	15 (7–24)
Scroby	Grey	2	79 (76–82)	2.3 (1.8–2.8)	16.4 (10.1–22.6)	2.6 (1.3–3.8)	20.7 (15.7–25.7)	123 (100–146)
	Common	3	83 (74–89)	0.23 (0.19–0.26)	14.0 (7.3–23.2)	0.83 (0.75–0.89)	9.1 (7.6–10.3)	131 (93–185)
England/Wales, W.								
Wales	Grey	3	48 (40–57)	0.58 (0.20–1.2)	11.0 (10.7–11.2)	1.2 (0.46–2.2)	5.3 (3.1–7.5)	212 (200–235)
Cornwall	Grey (pups)	3	17 (4–37)	0.25 (0.08–0.44)	6.7 (3.2–8.7)	0.69 (0.27–1.1)	3.7 (1.7–5.0)	160 (118–187)

TABLE 2. ORGANOCHLORINE RESIDUES IN SEAL BLUBBER SAMPLES OTHER THAN FROM GREAT BRITAIN
(means and ranges in parts per million tissue)

Source	Species	No.	% Fat	Dieldrin	DDE	TDE	DDT	PCB
Arctic (Canada)	Ringed	3	91 (all 91)	0.13 (0.09–0.18)	1.1 (0.4–1.6)	0.18 (0.14–0.24)	1.4 (0.8–2.1)	3 (2–4)
Arctic (Norway)	Ringed	2	84 (75–92)	0.18 (0.15–0.20)	0.8 (0.7–0.8)	0.24 (0.18–0.30)	1.4 (0.9–1.8)	1.5 (1–2)
Baltic (N. Sweden)	Ringed	1 (tail)	33	0.14	14.4	8.9	0.5	22
Gulf of St. Lawrence (Canada)								
Sable Is.	Grey	5	83 (81–84)	0.25 (0.12–0.49)	24.9 (6.9–50.6)	2.0 (0.8–3.1)	18.5 (7.2–31.6)	27 (12–65)
Basque Is.	Grey	2	85 (84–86)	0.10 (0.08–0.13)	24.6 (18.2–31.0)	1.9 (0.8–2.9)	23.4 (17.1–29.7)	32 (17–46)
Magdalen Is.	Hood	1	80	0.09	3.5	0.55	6.2	3

cutaneous muscle with no appreciable layer of fat beneath the skin. In such instances the calculation of residue levels on the basis of extractable lipid becomes of more practical significance.

Samples from Great Britain

Examining first the data from the British samples in Table 1, the lowest levels of all four pesticides and PCBs were found in the common seal pups taken in the Shetland Isles some 250 km north of the Scottish mainland. Although pups are generally found to contain lower concentrations than their mothers (see later), these levels still imply that the adult population was the least contaminated of the British stocks and only slightly more contaminated than Arctic samples of ringed seals (Table 2). The Orkney specimens of grey seals contained similar levels of dieldrin, but significantly more of the DDT group and of PCBs. To the north-west of Scotland, samples from the Hebrides (grey seals) contained slightly higher dieldrin concentrations, similar total DDT residues, but more PCBs than the Orkney stocks. Farther south, the Summer Isles population contained similar levels of dieldrin, less DDE but similar TDE and DDT levels, and less PCB than the Hebridean stocks. Still farther south off the Scottish west coast the common seals on the island of Mull were even less contaminated than those of the previous two groups, but here the change of species may have some relevance. Two separate groups of samples from Mull, taken nine months apart, showed remarkably little difference in either means or ranges of the five residues. The highest concentrations of organochlorine residues in Scottish seals have been found in the Firth of Clyde area of south-west Scotland. PCB concentrations were particularly high, and it is known that the sewage sludge dumped in the area contains unusually high concentrations of PCBs (Holden, 1970). The fish in this area also contain higher PCB levels than those from other Scottish waters.

On the east coast of Scotland a larger sample of grey seals contained significantly more dieldrin than any other Scottish samples, as well as slightly more of the DDT group and of PCBs than other Scottish stocks, and PCB levels up to 88 ppm have been found. Farther south, off the north-east coast of England, grey seal pups from the Farne Islands contained slightly less than those of the east Scottish adult stock, and it is probable that the Farnes adults have higher levels than the east Scottish stocks. A complication here is that a proportion of the seals taken off Scotland is known to originate from the Farne Islands. To the south, common seal pups taken on the east coast of England, in the Wash, have a generally lower level of contamination than the grey seal pups of the Farnes. This may in part be due to a species difference but it is probably also in part due to a lower level of industrial and sewage discharge in this area as compared with the densely populated and industrialized north-east coast. Finally, both grey and common seal stocks farther south on the East Anglian coast, north of the Thames Estuary, contained the highest levels of the DDT group and of the PCB group found in east coast samples. The grey seals also contained much more dieldrin than has been found anywhere else.

The only other samples so far examined have been from two areas on the west coast, one in south-west England and the other on the Welsh coast. The sample from south-west England was of grey seal pups, which contained similar levels of dieldrin and the DDT group to those sampled from the Faroes and Orkney Islands. The PCB levels were very much higher than in these other samples and similar to those in the East Anglian adults. The pups from the south-west coast, however, were in poor condition, with a very low percentage of extractable lipid, and on a lipid basis these specimens (which were found dead or dying) contained the highest PCB residue levels in any seals so far examined from all sources. Table 3 compares the residue levels (on a lipid basis) of these pups with those from north-east England. One other grey seal pup, found dead on the Welsh coast at the same time, had similar levels. Adult grey seals on the same coast, while having similar levels of dieldrin and DDT group residues to those of eastern Scotland, had much higher PCB group concentrations. As the blubber samples from these specimens contained only 40–60 per cent of extractable fat the PCB concentrations in the lipids were about twice as high, exceeding considerably the corresponding levels in the same species on the east coast of Great Britain. These high levels, in conjunction with the very high levels in pups already referred to, indicate a higher PCB contamination level on the west coast of England and Wales than anywhere else around the British Isles.

TABLE 3. MEAN RESIDUE LEVELS IN GREY SEAL PUPS ON BASIS OF EXTRACTABLE LIPID (parts per million)

Area	Tissue	Dieldrin	DDE	TDE	DDT	PCB
S.W. England	Liver	1.6	40	c. 10	c. 5	1020
	Blubber	1.9	65	6	37	1800
N.E. England	Liver	0.8	8.4	1.4	7.7	56
	Blubber	0.6	8.4	1.1	7.1	50

Samples from other areas

The other samples examined have been obtained from Canada, Norway and Sweden (Table 2). Ringed seals from the Canadian Arctic and the Norwegian Arctic coast contained very low levels of all three groups of residues, about 0.1 ppm of dieldrin and 2 ppm each of the DDT group and PCBs. These seals were the least contaminated of all specimens so far examined. A ringed seal from the Gulf of Bothnia (N. Sweden) had a similar dieldrin level but both the DDT group and PCB concentrations were ten times greater than in the Arctic specimens. A ringed seal from southern Sweden examined earlier had DDT and PCB concentrations about sixty times the levels found in the Arctic specimens, the DDT values being among the highest found in all the samples examined.

The remaining samples were all obtained from the Gulf of St. Lawrence (E. Canada). Here the hood seals from the Magdalen Isles had the least contamination, although the DDT group of residues totalled about 10 ppm in the specimens examined. Both dieldrin and PCB levels were similar to those found in the Arctic ringed seals. The grey seals from the Basque Isles and Sable Island had very similar concentrations, the dieldrin level again being about 45–50 ppm. One grey seal from Sable Island contained 65 ppm PCBs and 85 ppm of total DDT, the DDT level being similar to those found in Baltic ringed seals.

In a few instances, pups have been culled from populations in association with the females, and analyses of the blubber samples (Table 4) show that the levels in the pups are less than in the parent females. At the time of sampling the pups were still being fed by the females, and the residues were therefore derived from the parent seals only. The figures suggest that only a part of the residues in the lipids of the parent seal is transferred to the pup.

Geographical residue distribution

The data given here confirm the general pattern of distribution of the organochlorine pesticides in seal populations described in an earlier paper (Holden, 1969). The previous estimates of the DDT group of residues were, however, usually too high, due to the interference by PCBs, which were not previously determined. The present data include estimates of PCB concentrations and reveal that high contamination levels occur on both east and south-west coasts of Great Britain. The PCB concentrations are lowest in Arctic samples from both Canada and Norway, but increase in a southerly direction, down the east coast of Canada and both east and west coasts of Great Britain. The highest PCB levels occur in areas where high human population densities, with their associated industrial activities, discharge to the sea. Fairly high PCB levels similar to those of eastern Canada and northern Britain also occur in the southern Baltic, where there is a considerable discharge of polluting effluents, and where the rate of water movement is low (figs 1 and 2).

TABLE 4. MEAN RESIDUE LEVELS IN GREY SEAL PUPS AND ADULT FEMALES (parts per million in blubber)

Area	No. of samples	Dieldrin	DDE	TDE	DDT	PCB
Canada	2 adults	0.37	22	2.1	17	20
(Sable Is.)	2 pups	0.16	5.4	0.7	5.4	12
Scotland	4 adults	0.16	4.7	0.50	5.0	17.0
(Orkney Is.)	4 pups	0.20	3.1	0.40	3.1	7.9

There is a tendency for high PCB and high DDT group residues to occur together, but this is considered to be coincidental; the high DDT levels also being indicative of greater human activity on the adjacent land areas. In the Gulf of St. Lawrence, the total DDT concentrations generally exceed the PCB levels, due probably to the relatively high use of DDT for forest spraying in the past, in areas where industrialization is not so intensive as, for example, in southern England. In the samples from east and south-west England and Wales the PCB concentrations greatly exceed the total DDT concentrations.

As stated earlier, the method of measurement of PCB concentrations is arbitrary, but alternative methods, using fewer or more peaks, or areal integration instead of peak heights on chromatograms do not usually vary the estimate by more than a factor of two, provided the same PCB formulation is used as reference. Formulations of different chlorine content may give estimates differing by as much as threefold.

Other published data

Few analyses of seal tissue have been published, but Koeman and van Genderen (1966) found 9.6–27.4 ppm total DDT (uncorrected for PCBs) in the fat of three common seals in the Netherlands, a further indication of the relatively high DDT levels in the southern part of the North Sea. George and Frear (1966) reported 0.04–0.12 ppm DDT (but no DDE) in fat from Antarctic Weddell seals (*Leptonychotes weddelli*), and Sladen *et al.* (1966) found 0.039 ppm total DDT in fat from a crab-eater seal (*Lobodon carcinophagus*) at Ross Island, Antarctica. Northern fur seals (*Callorhinus ursinus*) from Alaska and the Pacific coast of the U.S.A. were found by Anas and Wilson (1970) to contain mostly less than 1 ppm total DDT and rarely more than 0.01 ppm dieldrin (dry weight) in liver and brain samples, but calculations on a lipid basis were not made. No PCBs were detected.

Wolman and Wilson (1970) examined 23 grey whales (*Eschrichtius robustus*) from the California coast, and found detectable total DDT concentration (all less than 1 ppm) in the blubber of six, and four contained dieldrin (less than 0.1 ppm). Six sperm whales (*Physeter catodon*) contained total DDT residues in blubber from 1.8 to 9.4 ppm, and two contained traces of dieldrin (less than 0.1 ppm), but no PCBs were found. The major residue in all cases was DDE.

The data most comparable with that given in the present paper were reported by Jensen *et al.* (1969) from seals in various parts of the Baltic, and are summarized in Table 5. The DDT group concentrations in extractable fat are higher than most of the Canadian values or those from England, but the PCB levels are lower than in many English seals. The Baltic nevertheless shows evidence of a high degree of contamination.

Relation to food intake

The concentration of organochlorine residues in individual seals will be dependent on the level in food and on the quantity eaten. In general, concentrations in inverte-

TABLE 5. ORGANOCHLORINE RESIDUES IN BLUBBER FROM BALTIC SEALS*
(means in parts per million)

Source	Species	No.	% Fat	DDT	ppm in tissue Total DDT	PCB	DDT	ppm in fat Total DDT	PCB
Gulf of Finland	Grey (pups)	2	60	14	25	3.9	23	42	6.5
Gulf of Bothnia	Ringed	2	54	30	63	6.8	56	120	13
Stockholm Archipelago	Grey	3	27	4.2	36	6.1	17	170	30
Baltic proper	Common and Grey	2	52	32	66	15	62	130	30

* From Jensen *et al.* (1969)

[270]

brates are less than in fish, and thus whales consuming invertebrates may be expected to contain lower residue levels than whales or seals eating fish. Thus grey whales contained very low residues partly owing to feeding on crustacea as well as fish (Mansfield, 1967). The Baltic ringed seals, however, were highly contaminated.

The grey and common seals around the Scottish coast feed mainly on cod (*Gadus morrhua*) and salmon (*Salmo salar*) (Rae, 1960), and organochlorine residues in the muscle of cod from northern Scottish waters are of the order of 0.005 ppm dieldrin, 0.02 ppm total DDT and 0.10 ppm PCB. (Values from cod liver are some 10–30 times greater.) The seal blubber samples from this area contained about fifty times as much dieldrin, and about two hundred times as much total DDT and PCB. In terms of extractable lipid concentrations the difference is usually less than tenfold.

Degradation of organochlorines

One as yet unexplained feature of the residue analyses of seal blubber is that, despite the separation of PCB residues from TDE and DDT, thus ensuring that the estimations of the latter are not in excess due to PCB interference, the DDT levels are often higher than those of DDE, TDE being relatively low. This DDE/DDT proportion is not typical of wildlife and has not been found, e.g., in cod or Atlantic salmon on which grey seals feed. Yet the DDT residues have been confirmed by hydrolysis with alcoholic alkali, which results in complete conversion to DDE. Jensen *et al.* (1969) also found that up to 50 per cent of the total DDT was present as DDT in seals.

The comparison of relative peak heights of individual PCB residues in various Aroclor formulations and typical blubber samples (Table 6) suggest that the isomers of short retention time may be metabolized to a marked degree, but those of longer retention time, if compared with Aroclor 1254, are only slightly reduced in proportion. This formulation appears to represent all blubber residues rather more effectively than the other two, especially in respect of the peaks of longest retention time. The table also indicates that there are only small differences in the residue patterns from Canada, Great Britain and the Baltic.

Physiological effects

There is no conclusive evidence that any physiological effects have resulted in any of the specimens examined. However, the highest concentrations in extractable lipids have been found in seals which were either dead or dying, and in poor condition. When such circumstances arise, body fat is metabolized and organochlorine residues, if not also degraded, are concentrated in the remaining lipid reserves. Circulation of such high-residue lipids may result in damage to the liver or other organs, and the high residue levels in blood could affect the central nervous system. In the presence of other stress factors, high residue levels may thus increase the possibility of physiological damage. Risebrough *et al.* (1968) have shown that dieldrin DDE, DDT and PCBs induce the formation of hepatic enzymes which degrade steroids and interfere with calcium metabolism by causing Vitamin D deficiency in birds. Enzyme induction in mammals may also occur.

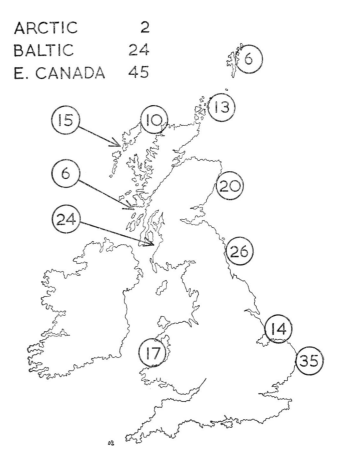

ARCTIC 2
BALTIC 24
E. CANADA 45

Fig. 1. *Mean concentrations of total DDT (ppm) in seal blubber samples from the coasts of Britain, compared with samples from other areas.*

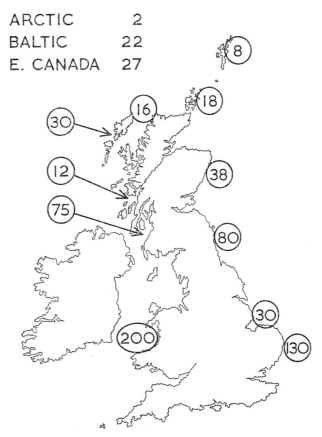

ARCTIC 2
BALTIC 22
E. CANADA 27

Fig. 2. *Mean concentrates of Pc Bs (ppm) in seal blubber samples from the coast of Britain, compared with samples from other areas.*

[271]

TABLE 6. RELATIVE PEAK HEIGHTS OF PCB ISOMERS IN BLUBBER SAMPLES

Sample	Relative retention times of GLC peaks (Dieldrin = 100)						
	81	122	145	171	200	240	283
Aroclor 1254	72	110	100	92	26	12	19
Aroclor 1260	14	30	100	69	70	55	87
Aroclor 1262	8	50*	100	45	105	75	111
Grey seal, E. Scotland	11	8	100	59	25	3	25
Grey seal, E. England	8	8	100	60	36	7	36
Grey seal, S.W. England	9	2	100	59	20	4	26
Grey seal, Wales	7	7	100	51	29	5	36
Ringed seal, Baltic	33	30	100	70	17	3	21
Grey seal, Canada	16	12	100	53	44	9	51

* Double peak

Summary

The analyses of seal blubber have demonstrated that, by comparison with the Arctic, Antarctic and Pacific Oceans, the levels of contamination in the Baltic, North Sea and Irish Sea (west of Great Britain) are very much higher, as judged by the presence of both pesticide and PCB residues (figs 1, 2). Dieldrin and PCB levels are highest around Great Britain and DDT levels highest in the Baltic and Gulf of St. Lawrence. There is no evidence that seals in good condition are affected by these residues, but when fat is mobilized the higher residue concentrations which are produced in the remaining lipids could have a detrimental physiological effect.

References

ANAS, R E and WILSON, A J Organochlorine pesticides in fur
1970 seals. Pestic. Monitg J., 3(4):198–200.
GEORGE, J L and FREAR, D E H Pesticides in the Antarctic.
1966 J. appl. Ecol., 3(suppl.):155–67.
HOLDEN, A V Organochlorine residues in seals. I.C.E.S. C.M.
1969 Fisheries Improvement Committee Doc. (E:22) (mimeo).

HOLDEN, A V A source of polychlorinated biphenyl contamination
1970 in the marine environment. Nature, Lond., 228:1220–1.
HOLDEN, A V and MARSDEN, K Organochlorine pesticides in
1967 seals and porpoises. Nature, Lond., 216(5122):1274–6.
HOLDEN, A V and MARSDEN, K Single-stage clean-up of animal
1969 tissue extracts for organochlorine residue analysis. J.
Chromat., 44:481–492.
JENSEN, S, et al. DDT and PCB in marine animals from Swedish
1969 waters. Nature, Lond., 224:247–50.
KOEMAN, J H and VAN GENDEREN, H Some preliminary notes on
1966 residues of chlorinated hydrocarbon insecticides in birds
and mammals in the Netherlands. J. appl. Ecol., 3(Suppl.):
99–106.
KOEMAN, J H, TEN NOEVER DE BRAUW, M C and DE VOS, R H
1969 Chlorinated biphenyls in fish, mussels and birds from the
river Rhine and the Netherlands coastal area. Nature,
Lond., 221(5186):1126–8.
MANSFIELD, A W Seals of the Arctic and eastern Canada. Bull.
1967 Fish. Res. Bd Can., (137):35 p.
RAE, B B Seals and Scottish fisheries. Mar. Res., 1960(2):39 p.
1960
RISEBROUGH, R W, et al., Polychlorinated biphenyls in the global
1968 ecosystem. Nature, Lond., 220:1098–102.
SLADEN, W J L, MENZIE, C M and REICHEL, W L DDT residues in
1966 Adelie penguins and a crabeater seal from Antarctica.
Nature, Lond., 210–670–3.
WOLMAN, A A and WILSON, A J Occurrence of pesticides in whales
1970 Pestic. Monit J., 4(1):8–10.

Potential Hazards from Radioactive Pollution of the Estuary

T. R. Rice, J. P. Baptist,
F. A. Cross* and T. W. Duke†

Dangers potentiels de la pollution radio-active des estuaires

Parmi les ressources aquatiques, les estuaires sont probablement les lieux les plus exposés aux dégats causés par la pollution résultant de diverses matières usées, notamment la radio-activité. Plus de 70 pour cent des espèces exploitées aux Etats-Unis passent une partie de leur cycle biologique dans les eaux des estuaires, qui peuvent recevoir des matériaux radio-actifs d'origines multiples. Outre la radio-activité provenante des retombées et des réacteurs nucléaires qui se trouvent parfois à proximité des estuaires, il peut y avoir également un apport de radio-activité venant des aires de drainage. Il est toujours possible que des bateaux à propulsion nucléaire, à la suite d'accident, déversent de grandes quantités de matières radio-actives dans les estuaires. Il semble donc que l'homme lui-même expose les ressources aquatiques les plus productives aux dangers les plus graves en raison des progrès de la technologie.

Afin d'évaluer les effets de la radio-activité dans les estuaires, nous avons mené des expériences, en laboratoire et sur le terrain, concernant le cycle élémentaire et les effets biologiques du rayonnement sur les organismes aquatiques. Il ressort de ces expériences qu'une grande partie de la radio-activité qui atteint les estuaires s'associe au biota et aux sédiments. On a suivi au laboratoire le cheminement de certains radionucléides à travers plusieurs niveaux trophiques, du phytoplancton au poisson. On a montré que les sédiments agissent comme réservoirs des isotopes, stables et radio-actifs, de Mn, Fe et

Riesgos potenciales de contaminación radiactiva en los estuarios

Entre los ambientes acuáticos a disposición del hombre, los estuarios son probablemente los más expuestos a posibles daños derivados de la contaminación con diversos materiales de desecho, entre otros los radiactivos. Más del 70 por ciento de las especies pescadas en los Estados Unidos transcurren parte de sus ciclos vitales en aguas de estuarios, que pueden recibir material radiactivo de diversas fuentes. Además de la radiactividad derivada de la lluvia radiactiva y de los reactores nucleares que puedan hallarse en el estuario, la radiactividad puede llegar a través de las cuencas fluviales. Siempre es posible, además, que a causa de averías las naves accionadas por energía nuclear, dejen libres en los estuarios grandes cantidades de radiactividad. Parece, pues, que los ambientes acuáticos más productivos de que el hombre dispone son más amenazados por los progresos de la tecnología.

Para evaluar los efectos de la radiactividad en los estuarios, se han realizado experimentos en laboratorio y en el ambiente contaminado sobre los ciclos elementales y los efectos biológicos de la radiación en los organismos acuáticos. Los resultados obtenidos indican que gran parte de la radiactividad que llega a los estuarios tiende a asociarse a la biota y a los sedimentos. El recorrido de algunos de estos radionúclidos se ha seguido en laboratorio a través de las cadenas tróficas, desde el fitoplancton hasta los peces. Se ha demostrado que

* Center for Estuarine and Menhaden Research, National Marine Fisheries Service, Beaufort, North Carolina 28516, U.S.A.
† Pesticide Field Station, Center for Estuarine and Menhaden Research, National Marine Fisheries Service, Gulf Breeze, Florida, U.S.A.

Zn. Ces éléments, une fois fixés dans l'estuaire, peuvent être accumulés par des organismes marins, qui risquent ainsi de recevoir des doses de rayonnement issues des radionucléides déposés dans l'organisme ainsi que d'autres provenant des radionucléides présents dans le milieu aquatique et les sédiments. Finalement, nous avons examiné les effets biologiques du rayonnement comme facteur du milieu et démontré que les sensibilités aux radiations varient avec la température et la salinité.

A
LTHOUGH the amounts of radioactive waste discharged into estuaries or reaching them from river drainage systems at present are generally small, amounts are likely to increase over the next 20 years.

We have investigated this particular problem by carrying out experiments in the laboratory, in ponds, and in the estuary. This paper summarizes our results.

Cycling of radionuclides

Radionuclides reaching the estuary can remain in solution or in suspension, precipitate and settle on the bottom, or be taken up by plants and animals. Certain factors such as currents and turbulence interact to dilute and disperse these materials, while bioaccumulation and sedimentation simultaneously tend to concentrate them. Each radionuclide tends to take a characteristic route and has its own rate of exchange among the water, biota, and sediments. Although in some instances sediments may initially remove large quantities of radionuclides from the water and thus prevent their immediate uptake by the biota, these radionuclides exchange from the sediment back to the water and again become available for uptake by the biota.

Sediments

The sorption of radionuclides to estuarine sediments is influenced by (1) the chemical and physical states of radionuclides as they enter the estuary, (2) the chemical and physical conditions of the estuary such as pH, Eh, temperature, salinity, tidal prism and geomorphology, (3) the affinity of radionuclides for adsorption on particulate matter, and (4) the rates of sedimentation. Also, the type of sediments and the amount of suspended particulate material may influence the partition of radionuclides within an estuarine ecosystem, particularly in the shallow, coastal plain estuaries along the southeastern coast of the United States. In these estuaries the finer sediments may be re-suspended continuously due to wind turbulence, thus increasing the opportunity to exchange and redistribute many radionuclides.

Of the radionuclides which enter estuaries as waste from nuclear reactors and other sources, the most biologically active elements and, therefore, the ones that would be transferred through food webs, are Mn, Fe and Zn. For this reason, our studies on the cycling of radionuclides in the estuarine environment have dealt mainly with these three elements.

Pond experiments

We followed the movement of ^{65}Zn in two experimental ponds near Beaufort, N.C. One was a closed system since the water in it did not exchange directly with the estuary. The water in the other pond exchanged directly with the estuary. Each pond was stocked with plants and animals representative of our area: marsh plants, seaweeds, oysters, clams, crabs, snails and fish. Ten millicuries of

los sedimentos actúan como depósitos de los isótopos estables y radiactivos de Mn, Fe y Zn. Estos elementos, una vez llegados al estuario, pueden acumularse en los organismos marinos, que de esa forma reciben dosis de radiación de los radionúclidos depositados en su cuerpo y de los presentes en las aguas y los sedimentos. Por fin, se han estudiado los efectos biológicos de la radiación como factor ambiental y se ha demostrado que la sensibilidad a la radiación varía con la temperatura y la salinidad.

^{65}Zn were added to each pond, and samples of water, biota and sediment were removed periodically and analysed for ^{65}Zn and stable Zn. After 100 days, more than 95 per cent of the ^{65}Zn and stable Zn was in the sediment of each pond. Analysis of specific activity (pCi ^{65}Zn/μg Zn) in components of the ecosystem indicated that the exchange of Zn between sediment and water dominated the cycling of this element and controlled the distribution of ^{65}Zn in the ponds.

We measured the concentrations of Mn, Fe and Zn in sediments and water over a 2-year period in the Newport River Estuary near Beaufort, N.C. This is a small (31 km²), shallow (less than 1 m at mean low tide) estuary and thermal or saline stratification rarely occurs. Because of relatively small watershed of the Newport River (about 340 km²), brackish water is carried into the river channel during flood tide where initial mixing of saline and fresh water occurs. Three stations were selected for routine collection of sediment and water: "Tall Pine" (Newport River), "Cross Rock" (upper estuary) and "Newport Bridge" (lower estuary). Median salinities taken monthly near high tide for two years were

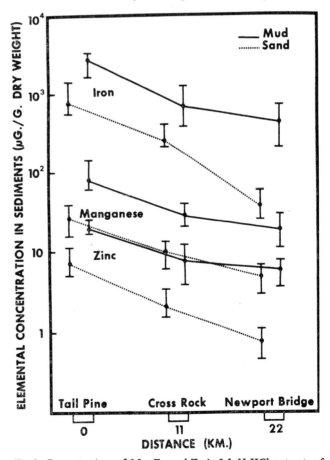

Fig 1 *Concentrations of Mn, Fe, and Zn in 0.1 N HCl extracts of muddy and sandy sediments collected monthly from the Newport River Estuary from October 1966 through September 1968. Values are given in median and 10th–90th percentile ranges and are offset at each station to avoid overlap of data*

less than 0.5 ppt at Tall Pine, 23 ppt at Cross Rock and 32 ppt at Newport Bridge.

Figure 1 gives two years' data for muddy sediments and for sandy sediments. Several facts emerged: muddy sediment samples had consistently higher concentrations than sandy sediments; Fe was the most abundant element at each station, followed by Mn and then Zn; the concentration of each element decreased in a seaward direction.

The concentrations of Mn, Fe and Zn in water from these three stations were similar in the sediment in two respects (Table 1): the order of relative abundance was Fe, Mn and Zn; and concentrations of Mn and Fe also decreased seaward. Concentrations of Zn, however, remained relatively constant at all stations. The water-to-sediment ratios (Table 2) indicate that surface sediments of this estuary serve as potentially greater reservoirs for 0.1 N HCl-extractable Fe and Zn than for Mn.

Biota

Plants and animals, to be of any significance in cycling radionuclides in estuarine environment, must accumulate the radionuclide, retain it, be eaten by another organism and be digestible. The biota can obtain radioactivity by absorption, adsorption and ingestion and lose it by excretion and decomposition.

TABLE 1. CONCENTRATIONS OF MN, FE, AND ZN IN UNFILTERED WATER SAMPLES FROM THE NEWPORT RIVER ESTUARY FROM NOVEMBER 1967 THROUGH SEPTEMBER 1968 (VALUES ARE GIVEN IN MEDIAN AND 10TH–90TH PERCENTILE RANGES)

Element	Median concentrations (μg/l)		
	Tall Pine	Cross Rock	Newport Bridge
Manganese	22 (15–33)	14 (5.4–28)	3.3 (0.5–5.5)
Iron	300 (200–380)	88 (5.0–260)	39 (3.2–81)
Zinc	0.6 (<0.1–2.2)	0.8 (0.1–1.3)	0.6 (<0.1–1.4)

TABLE 2. ESTIMATES OF MN, FE AND ZN IN WATER AND SEDIMENTS OF THE NEWPORT RIVER ESTUARY

	Manganese (kg)	Iron (kg)	Zinc (kg)
Water	2.4×10^2	1.8×10^3	1.9×10^1
Sediments	1.7×10^3	3.8×10^5	4.6×10^3
Water to sediment ratio	1/70	1/210	1/240

Primary producers

In estuarine environment, primary producers can be separated into three broad categories—microscopic floating plants (phytoplankton), macroscopic attached plants (benthic algae) and rooted plants (marsh grass). Phytoplankton concentrate ionic radionuclides by absorption and particulate nuclides by surface adsorption. Because of their tremendous surface area per unit volume, millions of rapidly-dividing phytoplankton cells may accumulate radionuclides from the water within a few hours. Particulate radioisotopes are also rapidly accumulated by several species of phytoplankton, as demonstrated in our laboratory experiments with ^{144}Ce.

Attached algae or seaweeds also accumulate radionuclides from the water and pass them to herbivores. Even dead algae accumulate radionuclides and in one experiment dead *Ulva* took up more ^{65}Zn than live *Ulva*

(Gutknecht, 1961). After death, the algae are consumed by detritus feeders, which in turn accumulate the radionuclides and transfer them to carnivores. A seaweed, *Porphyra umbilicalis*, is eaten by some people in England and accumulates ^{106}Ru to such levels that it can be used as a guide to limit the release of fission products from the Windscale Atomic Works (Dunster *et al.*, 1964).

Marsh grass, *Spartina* sp., detritus may be important in conveying radionuclides of Fe into estuarine food chains near Beaufort, N.C., as concluded by Williams and Murdoch of our staff. *Spartina* is consumed mostly as detritus, and the already high Fe content of the living plant is markedly increased after death by adsorption from the water. The long period required for growth permits considerable decay of short-lived radioisotopes, like ^{59}Fe (half-life 45 days), taken up during growth. Loss due to radioactive decay is, however, compensated by the adsorption of additional amounts of Fe after death. The high iron content of dead *Spartina* may also increase the Fe content of organisms feeding on it (Stevenson and Ufret, 1966) and thus increase the transport of Fe radioisotopes into the animal community.

Primary consumers

In the estuarine environment, probably the most important primary consumers in the cycling of radionuclides are zooplankton and shellfish. To determine the ability of zooplankton to accumulate a variety of radionuclides from water, we measured the rates of accumulation of $^{56, 57, 58}$Co, ^{65}Zn, ^{85}Sr, ^{137}Cs and ^{144}Ce from sea water by the copepod, *Tigriopus californicus* (fig 2). All five radionuclides were concentrated by these zooplankton, although at different rates. Concentration factors varied from less than two for ^{85}Sr to more than 500 for ^{144}Ce after only 48 hours (figs 2 and 3).

We also conducted an experiment to compare the accumulation of a radionuclide by a zooplankter from both food and water. The brine shrimp, *Artemia salina*, concentrated ^{65}Zn from both sources, but food was a more important source of ^{65}Zn (fig 3).

Shellfish (clams, oysters and scallops) may also concentrate radionuclides while filtering radioactive phyto-

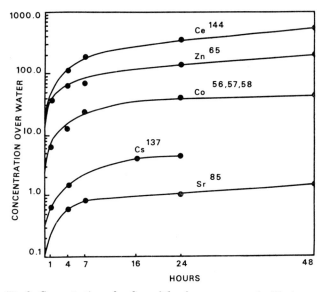

Fig 2 *Concentration of radionuclides from sea water by* Tigriopus californicus

[274]

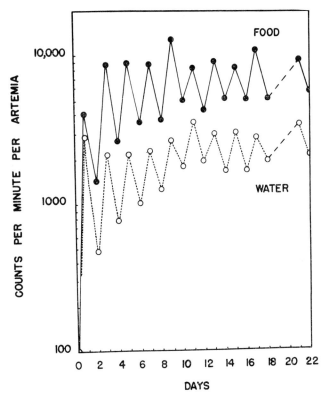

Fig 3 *Comparison of accumulation of* ^{65}Zn *by* Artemia salina *from food and water*

To see how much of a biologically active radionuclide could be concentrated by a carnivorous fish from food, we carried out a four-step food chain experiment in the laboratory. For comparison, we also experimented to measure the uptake of the radionuclide from sea water by each trophic level. The four trophic levels consisted of phytoplankton, *Chlamydomonas*, zooplankton, *Artemia salina*, postlarval croakers, *Micropogon undulatus* and mummichogs, *Fundulus heteroclitus*.

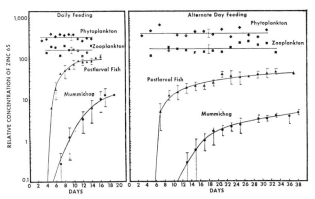

Fig 4 *Transfer of* ^{65}Zn *through an estuarine food chain with daily and alternate-day feeding. Concentrations were based on the initial amount of* ^{65}Zn *(representing a concentration of 1) in the phytoplankton culture*

Zinc–65 was readily transferred through the food chain to the fourth trophic level in both experiments, and the levels of concentration generally declined up the food chain (fig 4). Phytoplankton and zooplankton maintained constant levels of radioactivity because new cultures were used for each feeding. Postlarval fish took up ^{65}Zn rapidly during the first few days but the rate of uptake soon decreased and the ^{65}Zn content in the fish remained more or less constant. Radioactivity in mummichog continued to increase throughout both experiments. With daily feeding, 3.8 per cent of the ^{65}Zn concentration in phytoplankton reached the fourth trophic level; and with alternate-day feeding, 1.1 per cent reached the fourth trophic level. The fourth trophic level ^{65}Zn concentrations were 1.4 and 9.7 per cent, respectively, of the third trophic level concentrations.

plankton or detritus from the water. In laboratory experiments, oysters concentrate Zn to extremely high levels. In estuarine studies at our laboratory, C. L. Schelske found that the scallop, *Aquipecten irradians*, concentrates more ^{54}Mn from atmospheric fallout than other estuarine organisms. Radionuclides not biologically active such as ^{95}Zr–^{95}Nb, ^{106}Ru and ^{144}Ce, which are particulate in sea water, adsorb to the shell and body surfaces of shellfish. In addition, if these radionuclides are ingested, they will remain for some time in the visceral mass with only a small amount being assimilated. Biologically active radionuclides, however, such as ^{65}Zn, ^{54}Mn, ^{59}Fe or ^{60}Co will be readily assimilated by these organisms.

Secondary consumers

Radionuclides accumulated by carnivorous fish are limited to those that can pass through membranes and are either biologically essential or have the same chemical characteristics as essential elements. The concentration reached depends upon the turnover rate and the amount available in the food supply along with various environmental factors—temperature and salinity. Particulate radionuclides such as ^{95}Zr, ^{106}Ru and ^{144}Ce may be assimilated by carnivorous fish in limited quantities from the digestive tracts of food organisms. However, the carnivore rapidly excretes the major portion so that no significant concentration is reached. This was demonstrated at our laboratory in which 99 per cent of the ^{144}Ce pipetted directly into the stomachs of croakers was excreted in 24 hours. Other investigators have shown that ^{95}Zr–^{95}Nb and ^{141}Ce were concentrated by primary producers and herbivores but not by carnivores (Osterberg, *et al.*, 1964).

Biological effects of radiation

At present estuarine organisms are subjected to radiation from radionuclides that occur naturally and from those artificially produced. More than a dozen long-lived, naturally-occurring radionuclides have been identified. Most natural radiation in sea water originates from only three of these radionuclides: ^{40}K, which alone accounts for more than 90 per cent of the natural radiation in the ocean and ^{238}U and ^{232}Th and their daughter products. Since 1945, artificially produced radionuclides have been added to estuaries from fallout, nuclear reactors, medical uses and research.

Because rapidly-dividing cells are more sensitive than inactive cells to radiation, we experimented on the effects of relatively low-level chronic irradiation on the division rate of phytoplankton. *Nitzschia closterium* cells were grown for 56 weeks under conditions of continuous division by subculturing weekly. Each subculture was started with 2 million cells and 14.3 μCi of ^{137}Cs per litre of medium. After 26 weeks, the ^{137}Cs was increased ten-fold to 143 μCi per litre, and subculturing was continued for an additional 30 weeks.

During the 56-week period, cells grown in the radioactive culture medium divided 426 times; and cells grown in the control culture medium, following the same general pattern, divided 427 times (fig 5). We concluded, therefore, that *Nitzschia* cells, known to be quite resistant to acute radiation exposure, are also resistant to long-term chronic exposure at the levels tested.

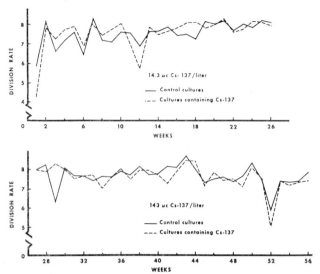

Fig 5 Division rate of Nitzschia closterium *cells grown in medium containing* ^{137}Cs

The comparative radiation sensitivities of estuarine crustacea fall within a wide range. At our laboratory, D. W. Engel determined the LD_{50} values for the blue crab, *Callinectes sapidus*; three species of fiddler crabs, *Uca pugnax*, *U. pugilator* and *U. minax*; and the grass shrimp, *Palaemonetes* sp. The conditions of irradiation were the same, 30 ppt and 20°C. When the LD_{50} 40-day values for these decapod crustaceans were calculated, the LD_{50} for blue crab was 42,000 rads, for the fiddler crabs 9,600–18,000 rads, and for the grass shrimp, 215 rads. Thus, the spread of radiation sensitivities was almost as great as exists among the various phyla of invertebrates, $2\frac{1}{2}$ orders of magnitude. Another interesting point is that an LD_{50} of 215 rads puts the grass shrimp at a slightly lower level of sensitivity than most mammals.

Ionizing radiation in estuaries places an additional stress on organisms living there. Survival of estuarine organisms depends upon the interaction of the environ-mental factors, especially temperature and salinity. These two factors, which can vary rapidly and over wide ranges, are interrelated and effects cannot be considered independently. The effect of the interaction of ionizing radiation, temperature and salinity on the mummichog, *Fundulus heteroclitus*, was tested in our laboratory by J. W. Angelovic, J. C. White and E. M. Davis. Effects of radiation dose and the temperature–salinity interaction were significant at the 5 per cent probability level. The effect of radiation was most apparent at 20°C or where the mortality at any given salinity reached 100 per cent. The temperature–salinity interaction caused an increase in mortality with a decrease in salinity at temperatures below 20°C and an increase in mortality with an increase in salinity at temperatures above 20°C.

Importance of radioactive pollution

The principal source of artificially produced radioactivity in estuaries will be radioactive wastes from nuclear power plants. Although radionuclides in estuaries do not occur in sufficient quantities to damage fishery resources, they could, if permitted to increase without adequate surveillance and discharge limitations, become a threat to fisheries and to man. At present, there are 17 nuclear plants in operation and 100 more planned or under construction in U.S.A. At least 41 will be located on seacoasts, estuaries, or on major rivers. By 1980, nuclear plants will account for 25 per cent of the electric power generating capacity (Seaborg, 1968).

It is impossible to predict with accuracy the effects their radioactive discharges will have on fishery resources. Much more research is needed. The effects of sublethal radiation on the ecosystem must be investigated, including all trophic levels of the food chain. Additional research is needed on the interaction of radiation with other environmental factors, as discussed earlier in this paper.

References

DUNSTER, H J, *et al.*, Environmental monitoring associated with
1964 the discharge of low activity radioactive waste from Windscale works to the Irish Sea. *Hlth Phys.*, 10(5):353–62.
GUTKNECHT, J Mechanism of radioactive zinc uptake by *Ulva*
1961 *lactuca. Limnol. Oceanogr.*, 6(4):426–31.
OSTERBERG, C, PEARCY, W G and CURL, H Jr. Radioactivity and
1964 its relationship to oceanic food chains. *J. mar. Res.*, 22(1):2–12.
SEABORG, G T The nuclear industry—1968 and beyond, remarks at
1968 a Financial Forum on Nuclear Energy, New York. U.S. Atomic Energy Commission No. S–39–68.
STEVENSON, R A and UFRET, S L Iron, manganese and nickel in
1966 skeletons and food of the sea urchins *Tripneustes esculentus* and *Echinometra lucunter. Limnol. Oceanogr.*, 11(1):11–7.

Mercury as a Marine Pollutant

S. Keckes and *J. K. Miettinen*†

Le mercure en tant que polluant marin

Jusqu'à ces derniers temps l'activité de l'homme n'apportait pas de changements considérables dans l'environnement marin, mais avec le développement technologique rapide des dernières décades, le rejet direct ou indirect dans les mers et les océans de divers produits et sous-produits potentiellement dangereux est devenu une réalité. Le mercure et ses composés appartiennent à cette catégorie et l'on estime qu'à présent la moitié de l'apport annuel de mercure dans les mers et océans (environ 5.000 tonnes) est le résultat d'un rejet incontrôlé provenant de l'emploi industriel ou

El mercurio como contaminante de las aguas del mar

Hasta hace poco tiempo, la actividad humana no contribuyó mucho a modificar el ambiente marino, pero con el rápido adelanto tecnológico de los últimos decenios se hizo una realidad la evacuación directa e indirecta de varios productos y subproductos potencialmente perjudiciales en los mares y océanos. Pertenecen a esta categoría el mercurio y sus compuestos, calculándose que en la actualidad la mitad del mercurio que anualmente llega a los mares y océanos (unas 5.000 toneladas) es el resultado de la falta de control con que en la agricultura y en la industria se da salida a los

* Laboratory of Marine Radioactivity, International Atomic Energy Agency, Monaco, Principality of Monaco.
† Department of Radiochemistry, University of Helsinki, Helsinki, Finland.

agricole de composés de mercure. Cet apport a augmenté localement le contenu en mercure de l'eau de mer et, par son intermédiaire, celui des organismes comestibles à un degré tel qu'il a provoqué la mort de plus de 40 personnes au Japon et, dans les pays scandinaves, l'application de mesures strictes pour éviter de pareilles tragédies.

Dans l'eau de mer, la concentration en mercure est très basse, 0,03–0,27 ppb, mais en raison de la réactivité chimique élevée et de la toxicité biologique du mercure, son augmentation a un effet hautement nuisible sur les organismes. La transformation biologique de formes de mercure relativement peu toxiques en formes de haute toxicité et son enrichissement dans les organismes marins—soit directement à partir de l'eau de mer, soit à travers les chaînes alimentaires—provoquent l'accumulation du mercure sous une forme concentrée chez l'homme qui consomme ces organismes. Le mercure ainsi accumulé, mis à part son effet toxique direct, peut aussi avoir des effets somatiques et même génétiques.

Ces aspects du mercure en tant que polluant marin vont être passés en revue et l'état actuel de nos connaissances sur le biogéocycle du mercure et son impact sur les organismes marins vont être discutés.

compuestos del mercurio. Localmente, en el Japón, la cantidad de mercurio que se virtió en las aguas del mar y que pasó a los organismos comestibles elevó su concentración de manera que ocasionó la muerte de más de 40 personas. En los países escandinavos tuvieron que aplicarse disposiciones estrictas para evitar tragedias similares.

La concentración de mercurio en el agua del mar es muy baja, 0,03–0,27 ppm, pero debido a su alta reactividad química y toxicidad biológica su aumento ejerce en los organismos un fuerte efecto deletéreo. La transformación biológica de formas de mercurio relativamente menos tóxicas en otras de más alta toxicidad, y su bioacumulación, directamente a partir de las aguas del mar o a través de las cadenas alimentarias, hace que el mercurio vuelva al hombre en una forma concentrada, lo que aparte de ejercer efectos tóxicos directos, puede tener también una consecuencia de tipo genético.

Volverán a examinarse estos aspectos del mercurio en cuanto contaminante de las aguas del mar, y se analizará la situación actual del conocimiento que poseemos sobre el biogeociclo del mercurio y sus consecuencias en los organismos marinos.

THE oceans are relatively stable and uniform in their chemical composition, and their elemental composition has evolved slowly (Conway, 1942, 1943: Nicholls, 1965). Marine organisms, having developed under such conditions, are very sensitive to sudden changes (Prosser and Brown, 1962).

The detrimental compounds introduced into the seas and oceans cover a wide range of organic and inorganic substances. Among the inorganic substances are the heavy metals, notably mercury and lead, to which organisms are especially sensitive because they interfere with vital biochemical processes.

The high toxicity of metallic mercury and mercury compounds was described in antiquity (Goldwater, 1936; Bidstrup, 1964) but not until the mass poisoning, known as Minamata disease (Kurland et al., 1960; Irukayama, 1967; Study Group of Minamata Disease, 1968; Ui, 1970), did mercury investigations become concerned with the harmful effect of heavy metals released by man into the marine environment.

BIOGEOCHEMISTRY

Properties of mercury

In nature mercury has seven stable isotopes and several artificial radioisotopes with various half-lives, the most useful of them being ^{203}Hg ($T_{\frac{1}{2}} = 47$ days) which emits β and γ-rays (79 per cent at 0.279 MeV).

Mercury is almost insoluble in common solvents: 0.02 ppm in water, but 0.6 ppm in methanol and 2.7 ppm in pentane at 40°C. The solubility in lipids is at ppm level (Hughes, 1957).

Mercurous chloride, Hg_2Cl_2, commonly known as calomel, and mercuric chloride, $HgCl_2$, commonly known as sublimate, are among the most important inorganic mercury compounds.

Occurrence and distribution

Mercury, being highly volatile, is always found in the atmosphere in trace amounts probably as metallic mercury or dimethyl mercury. Mercury concentration in the air is about 0.02 µg/m³ (Saukov, 1953). Williston (1968) found 0.001–0.05 µg of mercury per m³ of air in the San Francisco Bay area; the higher mercury levels coincided with high smog levels. From the atmosphere mercury returns in rain water, which contains 0.02–2 ppb (µg/l) of Hg (Stock and Cucuel, 1934; Saukov, 1953). The storage time of mercury in the atmosphere is estimated at 2 years (Saukov, 1953).

Mercury in the lithosphere behaves as a sulphophile and calcophile element. The mercury content of igneous rocks is 10–100 ppb (Preuss, 1941) and in sedimentary rocks is relatively higher. Klein and Goldberg (1970) give the mercury concentration in dried sediments off the Californian coast as 0.02–1.0 ppm, but in exceptional cases 40 ppm (µg/g) was reported (Lausen, 1936). According to Aidin'Yan and Ozerova (1968) mercury in recent sedimentary rocks was enriched by diagenetic and early epigenetic processes. This agrees with the findings that Pacific marine sediments contain 1–400 ppb Hg (Boström and Fisher, 1969) and that the pattern of Hg distribution indicates a relationship of degassing of the mantle. The deep-sea manganese nodules accumulate up to 800 ppb mercury (Harriss, 1968), especially in areas rich in submarine volcanic exhalations.

The analysis of about 300 natural water samples from central Italy showed 0.01–0.05 ppb of mercury, except in samples from areas containing mercury deposits, where up to 2.2 ppb of Hg was measured (Dall'Aglio, 1968). In 34 ground water samples from different parts of Sweden 0.02–0.07 ppb Hg (mean 0.05 ppb) was reported, while four river water samples contained 0.05–0.07 ppb (Wiklander, 1968). In another study from Sweden, river and lake water samples taken near the bottom gave 0.12–0.29 ppb Hg but two of the seven watercourses studied were heavily polluted by industrial mercury (Jernelöv and Lann, 1970).

Several authors have reported deposition of mercury in metallic form or as HgS from hydrothermal waters (Dickson and Tunell, 1968; Aidin'Yan and Ozerova, 1968; Karzhano, 1969). The mercury content of recent hydrothermal solutions are reported to be below 0.07 ppm. Since part of the element is transported in the gas phase, the hydrothermal vapours have a mercury content several times higher than the atmosphere (Aidin'Yan and Ozerova, 1968).

Ground waters normally contain 0.01–0.07 ppb Hg, lake and river waters 0.08–0.12 ppb, while polluted surface waters may contain 0.1–0.3 ppb Hg or more.

At about pH 6 mercury is present in fresh water as HgOHCl, as $HgCl_2$ at lower and as $Hg(OH)_2$ at higher pH values (Anfält et al., 1968).

The average concentration of mercury in sea water is usually quoted as 0.03 ppb (Stock and Cucuel, 1934; Goldberg, 1963). Hamaguchi et al., (1961) reported values three times higher and Hosohara (1961) found

0.15 ppb in surface waters of Pacific and 0.27 ppb at 3,000 m. It is believed that mercury in sea water is mainly $HgCl_3^-$ and $HgCl_2^{4-}$ (Sillen, 1961; Goldberg, 1963). The residence time of mercury in the sea is estimated as 4.2×10^4 years (Goldberg and Arrhenius, 1958).

Goldschmidt (1937) assumed that mercury in the sea originated primarily from weathering of primary rocks but most of it is no longer in solution. According to Goldberg (in press) present annual input of mercury into oceans is about 10,000 t; half as a result of normal weathering of rocks and half as uncontrolled release from the industrial and agricultural use of mercury compounds.

Due to the great affinity of mercury and organo-mercurials to SH groups and the common occurrence of traces of proteinous materials in natural waters, it can be assumed that mercury in natural waters is usually bound to a matrix-like suspended matter, plankton, higher organisms or bottom material. Even in heavily polluted waters only a fraction of mercury is found free in solution, except close to the point of its discharge (Johnels *et al.*, 1967; Hasselrot, 1968, 1969; Jernelöv and Lann, 1970).

Organisms are able to concentrate mercury but the biological significance of this is unknown. In a work on elementary composition of organisms Webb and Fearon (1937) referred to mercury as a "contaminant". In his extensive monograph Vinogradov (1953) does not list mercury as an element regularly found in organisms, but he reports that it was found in marine algae and fish (Stock and Cucuel, 1934; Raeder and Snekvik, 1941).

Only with the increased interest in mercury as a serious environmental pollutant have the data on its occurrence and distribution in organisms become more abundant. A recent analysis of 81 marine species from the Californian coastal area showed values from 0.4–21 ppm of mercury on a dry weight basis (Klein and Goldberg, 1970).

Circulation and transformation

Even with the lack of sufficient information on the biogeocycle of mercury, Rankama and Sahama (1950) proposed the following general scheme for its circulation:

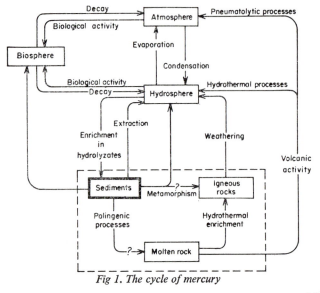

Fig 1. The cycle of mercury

More recently Jernelöv (1969) presented a scheme, considering possible transformations of some industrial and agricultural organomercurials:

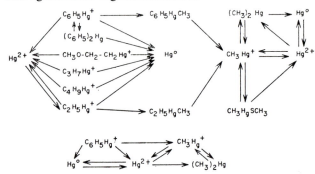

Fig 2. Possible transformations of some organomercurials

The cycling of mercury in nature is facilitated by the volatile character of its metallic and methylated forms and by the easy solubility of mercuric chloride in aqueous media.

Microbial transformation

Volatilization of mercury from sea water (Corner and Rigler, 1957), soil (Booer, 1944; Kimura and Miller, 1964) and from the lungs and body surfaces of mammals (Clarkson and Rothstein, 1964) is established. Most authors mistakenly assume that mercury is always volatilized as metallic mercury (Kimura and Miller, 1964).

Bacterial and viruses can fix mercury (Ruska, 1947; Tonomura *et al.*, 1968a) and transform it biochemically (Furukawa *et al.*, 1969). Within 45 h at 32°C *Pseudomonas fluorescens*, a common marine bacterium, accumulated from peptone-enriched sea water, about three times more mercuric chloride than phenyl mercury chloride (Yoshida *et al.*, 1967). Bacterial activity usually increases the volatilization of ionic mercury, probably because it is biochemically converted to a more volatile compound (Booer, 1944; Corner and Rigler, 1957; Magos *et al.*, 1964; Tonomura *et al.*, 1968; Tonomura and Kanzaki, 1969; Yamada *et al.*, 1969).

When methyl mercury was recognized as the cause of Minamata disease, its microbial formation was suggested (Kurland *et al.*, 1960) but could not be proved. The hypothesis of the formation of mercury in nature was rejected (Takizawa, 1967) after Irukayama *et al.* (1962a) by IR-absorption spectrophotometry detected methyl mercury chloride as a competent of wastes from the Minamata factory. However, Jensen and Jernelöv (1969) and Fageström and Jernelöv (1970) later showed by gas chromatography that under aerobic conditions in the presence of sediments from aquaria, fresh water and coastal waters of Sweden, decaying biological material, inorganic divalent mercury and mercury sulphide were converted into mono- and dimethyl mercury. At the same time Kitamura *et al.* (1969) reported the isolation of mercury-resistant bacteria capable of decomposing phenyl mercury and synthesizing methyl mercury and a small amount of ethyl mercury. Wood *et al.* (1968) showed formation of methyl mercury by cell-free extracts of methanogenic bacteria in the presence of methyl cobalamine and ATP, and in mild-reductive medium the methylation occurred even without the enzyme.

Ackefors (1969), Grant (1969) and Abelson (1970) assumed that microbial methylation of mercury would be enhanced in anaerobic conditions, but numerous tests (Rissanen et al., 1970) with various anaerobic muds and sludges, including the active sludge from a methane factory gave inconclusive results. One explanation of this might be that in anaerobic conditions when H_2S can be formed, the low solubility of HgS removes mercury ions from the reaction field insofar that methylation cannot take place. This explanation conforms with the observations of Booer (1944) and Magos et al. (1964) that microorganisms producing H_2S inhibit the volatilization of mercury from soil and biological material.

Transformation by plants

Phytoplankton accumulates mercury mostly by surface adsorption, as shown by Glooschenko (1969) for the marine diatom, Chaetoceros costatum. Comparing the accumulation of mercury by Chlamydomonas angulosa, Yoshida et al. (1967) obtained removal of 1.3 per cent of mercuric chloride and 6.5 per cent of phenyl mercuric nitrate from solution within 48 h at 26°C. Ünlü et al. (1970), usually found after 4–5 days, more than 10 per cent of mercuric chloride bound to Phaeodactylum tricornutum at about 6×10^8 cells/l. Makhonina and Gileva (1968) showed in aquarium experiments accumulation of ^{203}Hg in fresh-water algae and macrophytes while Hannerz (1968) found mainly surface absorption.

Little is known about the bioaccumulation and biochemical transformation of mercury in higher marine organisms and one must rely largely on extrapolations from related fresh-water and terrestrial organisms.

Transformation by animals

Korringa (1952) suggested mucus which often covers the free surfaces of marine organisms as the agent which collects mercury, as well as some other positive polyvalent ions, from sea water. After incorporation into organisms a typical organomercuric compound of type R-Hg$^+$X (R is the organic radical and X a dissociable anion) may undergo a biotransformation yielding inorganic mercury, Hg^{++}, or another mercury compound with a changed organic radical R. The first type of reaction occurs, for instance, with phenyl mercury (Miller et al., 1960; Gage and Swan, 1961; Gage, 1964) and methoxy ethyl mercury (Daniel and Gage, 1970). The alkyl mercury compounds seem much more stable and once introduced into the body exist unchanged for a long time (Miller et al., 1961; Gage, 1964). These various metabolic pathways are reflected in the biological distribution, accumulation, retention and effect of different mercury compounds.

Irukayama et al. (1962) exposed Venus japonica to 0.3 ppm of various inorganic and organic mercury compounds for 4–12 days. A remarkable concentration ability was observed, higher for alkyl mercury compounds (~70 ppm of dry weight within four days) than for the inorganic compounds (30–40 ppm within 10 days). No organic form of mercury was found in Venus exposed to inorganic forms, and the authors concluded that within the organism of Venus organic mercury cannot be formed

from inorganic but must be accumulated directly from sea water.

Venerupis philippinarum exposed four days to mercuric chloride and phenyl mercuric chloride showed a different distribution of these compounds in various organs and tissues (Yoshida et al., 1967). In both cases the muscles accumulated the least amounts of mercury per gram; the decreasing sequence of mercury's accumulation was the mid-gut gland, Bojanus' organ, intestine, gill for phenyl mercury and the Bojanus' organ, pericardium, gill and mid-gut gland for mercuric chloride.

Miettinen et al. (1969) compared the excretion rates of three different forms of mercury injected into the foot muscle of a fresh-water mussel (Pseudanodonta complanata). Inorganic mercury had the shortest biological half-time, 23 days, phenyl mercury the next, 43 days, and methyl mercury the longest, 86–435 days depending on the age of the animals (the youngest had the shortest half-time).

Ünlü et al. (1970) showed a rapid accumulation of mercuric chloride from sea water by Tapes. More than half of the mercury accumulated in the visceral organs. The retention of the accumulated mercury strongly depended on the means of accumulation: only about 25 per cent of the injected mercury was lost within two months, mostly during the first six days. The retention of mercury accumulated from sea water or through the food chain was weak and had an overall biological half-time of about 10 and five days respectively. The dependence of the biological half-time on the type of mercury compound injected is reported by Miettinen et al. (1970) and Ünlü et al. (1970) for two species of marine molluscs:

Species	Compound	Half-time
Tapes	mercuric chloride	~ 100 days
Tapes	methyl mercury	480 days
Mytilus	methyl mercury	1,000 days

Hannerz (1968) reported in fresh-water snails higher accumulation of methyl, methoxyethyl and phenyl mercury than of mercuric chloride.

In laboratory experiments Yoshida et al. (1967) estimated that the efficiency of the transfer of mercuric chloride through the food chain: sea water → Pseudomonas fluorescens → Artemia salina → Venerupis philippinarum was about 0.1 per cent, while the transfer efficiency through the food chain without Artemia was estimated at 1 per cent.

For Maia high concentrates of mercuric chloride from sea water were found in blood, antennary glands and gills (Corner, 1959). Most of the blood mercury was bound to proteins. When n-amyl mercuric chloride was used instead of mercuric chloride, the level of blood mercury remained very low. In Leander, another marine crustacean, the mercuric chloride and n-amyl mercuric chloride were similarly concentrated predominantly in the gills and antennary glands (Corner and Rigler, 1958). Mercury injected as methyl mercury into the haemolymph of Carcinus had a slow excretion resulting in a biological half-time of about 400 days (Miettinen et al., 1970), which is longer than the 300-day half-time found for the fresh-water Astacus when mercury was injected and 150 days when it was administered per os (Tillander et al., 1969).

Accumulation by fish

Fish apparently can accumulate mercury compounds more than other aquatic organisms, both directly from the sea water and indirectly through the food chain. Earlier analyses (Stock and Cucuel, 1934; Raeder and Snekvik, 1941) indicated 25–155 ppb Hg per wet weight of marine fish. About the same range can be observed today in fish from non-contaminated areas (Johnels et al., 1967, 1968; Westöö, 1967; Häsänen and Sjöblom, 1968; Sjöblom and Häsänen, 1969). In fish from environments with elevated mercury concentration the range runs up to 9.8 ppm (Johnels et al., 1967) which is the highest mercury concentration yet found in fish. Unfortunately the main type of the mercury found in fish from contaminated regions is the most toxic, methyl mercury (Norén and Westöö, 1967; Westöö, 1966, 1967a, 1969a).

By a perfusion method, Hibiya and Oguri (1961) estimated that the absorption rate of mercury as $Hg(NO_3)_2$ through the gills of the eel, Anguilla japonica is $0.3–3.2 \times 10^{-2}$ µg/h. Hannerz (1968) studied the bioaccumulation and distribution of methoxyethyl mercury, methyl mercury and mercuric nitrate in cod. Testing the accumulation of these compounds from sea water (33‰) he observed an increasing rate of mercury accumulation with time in almost all organs. Accumulation rate was highest in the gills, and depended on the size of the fish. There was a higher accumulation rate for methyl mercury and diminished accumulation in cod kept in brackish water. Methyl mercury was more readily accumulated from food than methoxyethyl mercury. The rate of elimination was relatively slow, comparable with the 267 days reported as biological half-time for methyl mercury in Serranus (Miettinen et al., 1970). In brackish water (6‰) long biological half-times were found for the excretion of methyl mercury's slow component: flounder 400 to 700 days, perch 500 days, pike 500–700 days and eel 900 to 1,000 days (Miettinen et al., 1969; Järvenpää et al., 1970). In these species the method of administration, in food or injection, had no noticeable effect on the slow component of excretion. Results from laboratory experiments (Bäckström, 1969; Ohmomo et al., 1969) and experiments done in contaminated environments (Hasselrot, 1969) using fresh water fishes agree with those described by Hannerz (1968).

Mercury not only enters the fish as dissolved mercury compounds. The observed distribution of mercury in different trophic levels (Johnels et al., 1967) indicates that it probably accumulates through the food chains as well.

Transfer route

Fish as an important source of food represent a transfer route of mercury to higher trophic levels (Johnels et al., 1967; Löfroth, 1969; Berlin, 1969; Miettinen, 1970). Berg et al. (1966) and Johnels et al. (1968) found that the osprey, Pandion haliaetus and the crested grebe, Podiceps cristatus feeding predominantly on fish originating from contaminated areas, had the mercury content in their feathers increased from about 4 ppm (probable natural level) to about 20 ppm. In the case of Minamata

disease (Irukayama, 1967) primarily fish with high methyl mercury content provided the fatal concentration. Unfortunately, as well as being able to transfer mercury to higher trophic levels, some aquatic microorganisms seem to be able to transform it into methyl mercury (Norén and Westöö, 1967; Jensen and Jernelöv, 1969), the most toxic form of mercury.

The only data on the mercury content in marine mammals are reported by Henriksson et al. (1969) for the ring seal, Pusa hispida, from the Gulf of Finland and Finnish lakes. The average mercury content in adult specimens, expressed in ppm of total Hg per wet weight was:

Regions	Muscle	Liver	Kidney
Gulf of Finland	0.9	11.8	3.2
Finnish lakes	62.4	137.8	46.3

Probably their diet in these two localities indicates a different mercury content. These values are very high compared to the lethal concentration of methyl mercury in cats: 30 mg/kg of fresh weight (Rissanen, 1969; Skerfving, 1969). The excretion of orally administered methyl mercury in Pusa hispida showed a typical fast and slow component with a biological half-time of about 500 days for a slow component (Tillander et al. 1970).

Berlin and Ullberg (1963, 1963a, 1963b) studied, by autoradiography, the distribution of mercury in mice injected as mercuric chloride, phenyl mercuric acetate and methyl mercuric dicyandiamide. The inorganic and phenyl mercury showed a similar distribution pattern and elimination rate. The methyl mercury, probably because of its slower breakdown and higher solubility in lipids, had a much slower elimination rate from the mice and with time was accumulated in nervous tissues and red blood cells. Among three mercury compounds studied by Berlin and Ullberg only methyl mercury penetrated through the placental barrier and accumulated in the foetus. These results reviewed by Berlin (1963), agree with those of Friberg (1956, 1959), Friberg et al. (1957, 1961), Swensson et al. (1959, 1959a), Suzuki et al. (1962, 1963, 1967), Ulfvarson (1962, 1969, 1969a), Brown et al. (1967), Swensson and Ulfvarson (1968a), Berglund and Berlin (1969), Östlund (1969) and Suzuki (1969) using various other mammals. One should expect that marine mammals would be similarly affected.

"ARTIFICIAL" SOURCES

Most mercury is obtained by oxidation from the ore cinnabar, HgS. Ore is heated in the presence of air or lime and the mercury vapours formed in reactions $HgS + O_2 \rightarrow Hg + SO_2$ or $4HgS + 4CaO \rightarrow 4Hg + 3CaS + CaSO_4$ are readily recovered by condensation (Johns and Bradley, 1966).

Fluctuations in the world mercury production reflects the actual industrial and agricultural uses of mercury and its derivatives. The high wartime production of 9,560 t in 1941 dropped to 3,720 t in 1948 and after 1950 was between 5,600 and 9,000 t/year (Johns and Bradley, 1966). For 1968 production was 9,000 t, but in spring 1970 there was some decrease in demand due to excess surplus at the largest manufacturers (Anon., 1970), but a recent estimate is 9,200 t/year (Goldberg, in press).

About 10 per cent of the world production of mercury is used in agriculture, the bulk being used in industry. According to Grant (1969) more than 80 different industrial processes require mercury as raw material or catalyst. Among the most important industries using mercury are (Bidstrup, 1964; Johns and Bradley, 1966):

> chlorine and alkaline plants for electrolytic production of chlorine and caustic soda;
> cellulose industries for preserving the wet pulp from bacterial and fungal biodeterioration (discontinued in Scandinavia);
> plastic industries for catalytic reactions;
> electric industries for production of relays, switches, batteries, rectifiers, lamps, etc.;
> pharmaceutical industries for production of diuretics, antiseptics, cathartics, some contraceptives and drugs for treatment of congestive heart failure;
> paint industries, mainly for production of anti-corrosive paints;
> metal refinement by amalgamation;
> power plants in special heat engines instead of water vapour;
> plants treating nuclear wastes for electrolytic purification of wastes;
> industries producing industrial and control instruments, e.g. thermometers, barometers, mercury pumps.

Figures for the annual consumption of mercury in United States for the years 1964–68 are published by the Bureau of Mines and Minerals. According to the National Materials Advisory Body in 1968 the total consumption of mercury in the United States was 2,540 t and for 1974–75 the annual use is expected to reach 2,860 t with the following consumption breakdown (in flasks, 1 flask = 34.5 kg):

	1968	1974–75
Agriculture	3,430	2,650
Amalgams	259	250
Catalysts		
urethanes	800	1,560
vinyl chlorine monomer	500	250
anthraguinone derivatives	175	220
miscellaneous	236	340
Dental applications	3,500	3,800
Electrical apparatus	17,200	22,700
Electrolytic preparation of chlorine		
and caustic soda	17,424	22,864
General laboratory use	2,075	2,075
Industrial and control instruments		
switches and relays	2,500	2,650
other instruments	6,400	6,600
Paints	10,566	10,725
Paper and pulp	375	250
Pharmaceuticals	600	650
Other*	7,815	5,960
Totals:	73,855	83,544

* Includes start-up of new chlorine cells.

Rough estimates of annual consumption and losses of mercury in two Scandinavian countries have been made by Häsänen (1969):

	Consumption (t/year)	
	Finland 1954–66	Sweden 1967
Chlorine industry, increase of capacity	22	?
Chlorine production, replacement for losses	10	25–38
Paper and pulp industry, as slimicides and fungicides	10	14*
Agriculture, as pesticides	4	5
Other uses	4	~30
Total	50	74–87

	Losses into waters (t/year)	
	Finland 1954–66	Sweden 1967
Chlorine industry	4	8
Paper and pulp industry		
as slimicides	3.6	3.2**
as caustic soda	0.5	1.0
Agriculture, as pesticides	0.3	0.3
Other industries	0.2	1.5
	8.6	14.0

* Now discontinued.
** 1941–68.

The majority of environmental mercury pollution is probably due to the chlorine and cellulose industries. Losses of 150–200 g Hg per ton of chlorine produced are considered common (Smith, 1968; Bouveng and Ullman, 1969), but they will probably be reduced in the U.S.A. by 1975 to about half of the present values.

Studies made

A Swedish study carried out in 1966–67 reported losses of 30–40 g of mercury per ton of chlorine, which at a total annual chlorine production of 250,000 t means losses of 7.5–10 t of mercury. Special measures in old plants could reduce this to 10 g per ton of chlorine (Bouveng and Ullman, 1969) and in new plants the losses can be reduced further. According to Bouveng (1970) it is not difficult today to reduce the quantity of mercury effluents from a plant producing 50,000 t/year of chlorine to below 50 kg/year. The application of efficient external treatment units may further reduce this figure. Losses through ventilation systems are more difficult to control and require further studies. The losses of hydrogen depend on the cooling system; cooling to +5°C with no compression will give about 1 g/t. Filtration of the produced caustic soda reduces this form of mercury loss to 0.5–1.0 g/t.

When mercury is used in pulp industry about 40 g of phenyl mercury acetate per ton (dry weight) of wet mechanical pulp is added as preservative during transport and storage (Bouveng, 1967). Up to 20 per cent of the mercury added to pulp is lost as waste. For Sweden this represents an annual discharge of about 2.5 t of mercury (Bouveng, 1967).

Public concern about the harmful effects of mercury pollution, especially after the Minamata accident, led to significant reductions by some industries in the use of mercury and to the application of more efficient pre-discharge treatments for wastes containing mercury. For example, the use of mercury as a fungicide and slimicide in the U.S.A. paper and pulp industry dropped from 182 t/year in 1959 to 21 t/year in 1966 (Novick, 1969). In Scandinavia this use of mercury was completely discontinued by 1968 (Löfroth, 1969a; Miettinen, 1970).

Agricultural use of mercury compounds as efficient herbicides, pesticides, slimicides and fungicides, mostly in the form of seed dressing, is presently estimated at about 1,000 t/year. In 1958 about 300 t were used (Gayner, 1962), but for 1965–66 Smart (1968) reports a total of 2,100 t/year with the following breakdown:

Country	Tons
U.S.A.	400
Denmark	3.5
Germany	41
Great Britain	20
Bulgaria	5
Finland	1
Italy	26
Poland	9
Austria	4
Norway	0.4
Portugal	0.2
Turkey	22.5
Spain	7.1
Sweden	2
Morocco	1
Israel	0.2
New Zealand	0.5
Japan	1,600 (approx.)

For Finland the figure should really be 4 t/year (Häsänen, 1969) and for Sweden about 5 t/year (Ulfvarson, 1967).

Changes in certain countries

Only alkoxyalkyl mercury compounds are now used in Scandinavian agriculture. They are metabolized more easily and, therefore, considered less dangerous. Methoxyethyl mercury ion is biochemically unstable and rapidly decomposed in soil (Booer, 1944) and in animal tissues (Ulfvarson, 1962; Swensson and Ulfvarson, 1968a; Bäckström, 1969). The liberated mercury ion is believed to be bound by organic matter in humus and thus it is assumed that less than 10 per cent of pesticide mercury reaches the water courses (Häsänen, 1969).

In 1968 the use of mercury in Japan's agriculture was discontinued, except for soil treatment (Fukunaga and Tsukano, 1969). Previously ethyl and phenyl mercury were used in rice cultivation. Ui (1971) reported that in Jinzu river, polluted with mercury from industry and agriculture, a specimen of carp contained up to 20 ppm of total mercury per dry weight, 1.7 ppm methyl mercury and 12.1 ppm ethyl mercury. The latter figure corresponds to about 4 mg of Hg as ethyl mercury per kg fresh fish, probably discharged with drain water from rice fields.

Mercury pollution of certain inland and territorial waters has occurred and can be expected in the future. Present pollution is estimated to be an "artificial" input of about 5,000 t of mercury annually into seas and oceans (Goldberg, in press). In order to reduce harmful effects of this pollution countries where the most grave situations exist, such as Sweden, Finland, Japan, U.S.A. and Canada, have prohibited, discontinued or reduced the use of alkyl and aryl mercury compounds in agriculture and in the cellulose industry, and introduced strong measures to reduce mercury losses from industrial

and other sources. However, as is evident from the facts described, current measures are not sufficient.

BIOLOGICAL EFFECTS

Mercury enters organisms by absorption through free surfaces such as skin (Schamberg, 1918) or gills (Hannerz, 1968), by intake of water or food containing mercury compounds, and in terrestrial organisms by inhalation of mercury vapours. Once in the organism, mercury and its compounds show pronounced and manifold biological effects which serve as a basis for the use of these compounds in medicine (Friedman, 1957; Schoemaker, 1957; Passow et al., 1961), agriculture (Ashworth, 1967; Lihnell, 1967; Walker, 1967; Smart, 1968; Ishiyama, 1969) and in certain industries (Bouveng, 1967; Halldin, 1969). The biological effect of mercury is strongly dependent on its concentration, chemical form and the organism.

The biological effects of inorganic mercury compounds seem to be proportional to their ability to yield active inorganic mercury ions (Rowland, 1952) which probably react with thiols in proteins and enzyme systems forming mercury mercaptide (Hughes, 1950, 1957; Webb, 1966). According to Goodman and Gilman (1955) organo-mercurials are probably reactive in the form of R–Hg$^+$. Weiner and Müller (1955) suggest that in some compounds the active form is R–Hg–R′, where R′ is a cysteine-like complex. This might explain the haemolytic action of some mercurials (Arbuthnott, 1962). Handley and Seibert (1956) and Greif and Pitts (1956) found that some intact organomercurials retain their biological activity. Furthermore some of the organomercurials, for example the alkyl mercury compounds, are lipotropic and tend to accumulate in the nervous system causing severe functional and histopathological changes (Hunter et al., 1940; Hunter and Russel, 1954; Harris et al., 1954; Kurland et al., 1960; Takeuchi et al., 1962; Berlin and Ullberg, 1963b; Bäckström, 1969).

Toxic effects

Most experiments have lasted only a few hours or days and relate to acute toxicity; detailed studies of long exposures to low concentrations are generally missing, hence, reported toxic concentrations may be misleadingly high.

Little is known about the effect of mercury on marine bacteria, but research on pathogenic microorganisms show that the first action site generally is the surface, compounds with higher lipid solubility are more toxic and the toxic effect is mostly irreversible (Meyer, 1964). Passow and Rothstein (1960) showed that physiological damage to yeast cells, at least at low concentrations of mercury ions (0.2 mm), occurs in the surface membrane, which loses its semi-permeability resulting in loss of the K$^+$ ions. At 0.6 mm of Hg the leakage is almost maximal. Mercury is bound very rapidly, the half-time of binding is 2–4 min at pH 3.0 and 26°C. The mechanism is supposed to be formation of S–Hg–S bridges in specific steric positions.

Hoffmann (1950) tested the toxicity of 14 different organic mercury compounds for a phytomonadine. He found that in short-term experiments all except one (oxyphenyl mercuric chloride) showed toxic effects in the range 0.13–0.23 ppm of Hg^{++}. The toxic concentration of the oxyphenyl mercuric chloride was much higher (1.8 ppm of Hg^{++}). There was no correlation made between the toxic effect and the chemical structure of the compounds tested.

Boney et al. (1959) tested the influence of different mercury compounds on the growth and viability of sporelings of the red alga *Plumaria elegans* that were immersed 18 h in toxic medium. The toxicity of organic forms is higher than inorganic, and in a homologous series of primary *n*-alkyl mercuric chlorides the toxicity increased up to n-C_3H_7HgCl and then remained unchanged, with 13 ppb as LD$_{50}$.

Compound	LD$_{50}$ (ppb Hg)
methyl mercuric chloride	44
ethyl mercuric chloride	26
n-propyl mercuric chloride	13
n-butyl mercuric chloride	13
n-amyl mercuric chloride	13
isopropyl mercuric chloride	28
isoamyl mercuric chloride	19
phenyl mercuric chloride	54
phenyl mercuric iodide	104
mercuric iodide	156
mercuric chloride	3,120

Boney and Corner (1959) showed that, apart from the higher sensitivity toward n-C_3H_7HgCl than toward $HgCl_2$, the sporelings of six species of intertidal red algae were affected in proportion to the lipid content of their membranes.

Harriss et al. (1970) reported that some organo-mercurial fungicides (diphenyl mercury, phenyl mercuric acetate, methyl mercury dicyandiamide and *n*-methyl-mercuric tetrahydrodimethanohexachlorophtalimide) at concentrations as low as 0.1 ppb in water reduced photosynthesis and growth in laboratory cultures of marine *Nitzschia delicatissum* and in several fresh water natural phytoplankton communities. This showed that some phytoplankton species are sensitive to much lower concentrations of mercury compounds than those presently accepted for water quality standards.

Detailed study

The poisonous effect of mercuric chloride, mercuric iodide and ethyl mercuric chloride on adults of the marine copepod *Acartia clausi*, on the larvae of the brine shrimp *Artemia salina* and on the larvae of the barnacle *Elminius modestus* were extensively studied by Corner and Sparrow (1956). In all three species the mercuric chloride was the least toxic. LD$_{50}$ values within 2.5 h at 23°C $\pm$ 2°C in sea water were: 0.05 ppm of Hg^{++} for *Acartia*, 0.3 ppm for *Elminius* and even 800 ppm for *Artemia*. The relative toxicity of the other compounds was slightly higher in *Elminius* and *Acartia*, but in *Artemia* it was 24–31 times that of mercuric chloride, probably due to the different penetration rates of the poison. When the toxicities of mercuric chloride, mercuric iodide and methyl, ethyl, propyl, *n*-butyl, *n*-amyl, isopropyl, isoamyl and phenyl mercuric chlorides were tested on larvae of *Artemia* and *Elminius* (Corner and Sparrow, 1957), it was found that in both species mercuric chloride was the least toxic, that primary

alkyl mercuric chlorides were more toxic than the corresponding secondary compounds, and that as the homologous series of primary compounds is ascended, the toxicities increased. *Elminius* was always less resistant than *Artemia* to the same compound.

Barnes and Stanbury (1948) demonstrated with the copepod, *Nitocra spinipes*, the dependence of survival on the concentration of mercury in the medium. After 24 h exposure to sea water containing 0.07–4.40 ppm mercury as $HgCl_2$ the following percentage of copepods died:

ppm of Hg in medium	0.07	0.31	0.40	0.60	0.70	1.50	3.00	4.00
% of deaths	0	1.4	10.1	16.7	50	72	84	100

Clarke (1947) reported that 1 ppm of mercuric chloride killed 90 per cent of adult barnacles (*Balanus balanoides* and *B. eburneus*) within 48 h, but the metamorphosis of cyprides were affected only with about 16.6 ppm of mercuric chloride.

Various zooplankters (copepods, nauplii, larvae of polychaetes, rotatoria) tested with a series of homologous alkyl mercuric chlorides showed clearly decreasing survival time with increased number of C atoms: from 75 min with methyl mercuric chloride to 40 min with i-amyl mercuric chloride at 2 ppm of the compound. In autumn the same species were less sensitive to ethyl mercury than in spring (Hoffman, 1950).

Weiss (1947) compared the tolerance of several common fouling organisms to mercury by observing their attachment to antifouling-paints containing 14–90 per cent mercury (dry weight of paints). The decreasing tolerance sequence was: *Polysiphonia sp.*, *Balanus amphitrite*, *Bugula neritina*, *Balanus improvisus*, *Watersipora cucullata*, *Anomia sp.*, *Enteromorpha sp.*, *Hydroides parvus*, hydroides and tunicates. Only the first five species were frequently found on surfaces coated with paints containing mercury. Wisely and Blick (1967) exposed the larvae of bryozoans, worms, molluscs and brine shrimps to mercuric chloride. At 18°–25°C 50 per cent of the larvae died in 2 h at the following Hg^{++} concentrations: 0.1 ppm (*Watersipora cucullata*), 0.14 ppm (*Spirorbis lamellosa*), 0.2 ppm (*Bugula neritina*), 1.2 ppm (*Galeolaria caespitosa*), 13 ppm (*Mytilus edulis*), 180 ppm (*Crassostrea commercialis*) and 1,800 ppm (*Artemia salina*).

Low tolerance of mercury compounds by molluscs, especially organomercurials, which in short-term experiments had toxic effects at 0.1 ppm, was reported by several investigators (Chandler, 1920; Clarke, 1943; Mozley, 1944; Szumlewicz and Kemp, 1951; McMullen, 1952; Bond and Nolan, 1954; Wisely and Blick, 1967). Irukayama et al. (1962) found that *Venus japonica* exposed to 0.3 ppm of various inorganic mercury compounds survived for 6–10 days. They survived for four days when exposed to the same concentrations of various organic mercury compounds.

Vertebrates are as sensitive to mercury compounds as invertebrates. Earlier literature on the toxicity of industrial wastes, including mercury compounds, to fish, was critically reviewed by Doudoroff and Katz (1953). In fish, mercury ions in their environment, primarily affect the epithelium of the skin and gills (Schweiger, 1957). After peroral or intramuscular administration of mercuric nitrate and phenyl and methyl mercuric nitrate relatively high amounts of mercury were concentrated in the gills (Bäckström, 1969).

The symptoms of acute mercury poisoning in fish are rigidness, widely spread fins, lazy movements and "hanging" on the surface with hind part of the body turned downwards (Boëtius, 1960). These symptoms are followed by loss of balance and finally sinking to the bottom before death occurs. All these symptoms were clearly visible in the fish from Minamata Bay (Takeuchi, 1966), where mercury compounds from industries were released (Irukayama et al., 1962a, 1969; Kiyoura, 1963). Takeuchi (1966) reported cataracts of the fish eye as a typical symptom. Histological examinations of affected fish showed degenerative changes in neurons and decrease or disappearance of the granular cells in the cerebellum (Takeuchi, 1968).

Degree of sensitivity

According to Boëtius (1960) the euryhaline Salmo gairdnerii and Gasterosteus aculeatus are more sensitive than fresh-water species. Phenyl mercuric acetate is more toxic to them than mercuric chloride, and the survival time is decreased at higher temperatures and is longer for heavier fish. This might be indirect evidence that the area of the surface determines the toxic effect. Influence of environmental conditions on the toxicity of mercury was shown by Binet and Nicolle (1940) in experiments with Gasterosteus leiurus. At 18°C the lowest toxicity of 10 ppm of mercuric chloride was in 1.6 per cent NaCl solution. At higher or lower NaCl concentrations the survival time of fish decreased significantly. Similar experiments with Phoxinus phoxinus (Jones, 1940) showed that the toxicity of mercuric chloride decreased as the NaCl concentration increased to about 0.9 per cent. A solution of glucose with the corresponding osmolal value was as effective as NaCl. In explaining their results the authors (Jones, 1940; Binet and Nicolle, 1940) suggest that there may be different ionization of mercuric chloride or the formation of double or complex chlorides of mercury and sodium at different NaCl concentrations.

On young eels, Anguilla japonica, Oshima (1931) compared the relative toxicity of one monovalent and 14 divalent cations in chloride forms. Mercury was the most toxic: 0.02 ppm was the highest concentration tolerated at 20°–22°C for more than 50 h. Similar results were obtained with Notropis (Carpenter, 1927), Orizias latipes (Iwao, 1936) and Gasterosteus aculeatus (Jones, 1938, 1939).

In short-term experiments 1 ppm of Hg^{++} in the form of mercuric chloride was fatal for Gasterosteus aculeatus (Boëtius, 1960) but for prolonged exposures concentrations as low as 0.01–0.02 ppm of Hg^{++} were lethal (Jones, 1940; Uspenskaya, 1946). The product of the survival time and the concentration of mercury in the environment is constant for a given fish and mercury compound. From the figures for mercury concentration in natural waters Boëtius (1960) determined that mercury poisoning could not be the limiting factor for the maximum age of fishes in non-polluted waters, but he noted that Carpenter (1927) demonstrated that the salts of

heavy metals are "infinitely toxic", i.e., there is no theoretical lowest limit for their toxicity.

Amend et al. (1969) studied the influence of water temperature, dissolved oxygen, water hardness (20–256 ppm as $CaCO_3$) and chloride ion concentration (up to 2 mm) on the susceptibility of rainbow trout (Salmo gairdneri) to the acute toxicity of ethyl mercury phosphate after exposing the fish for 1 h to 0.125 ppm of the compound. The death rate increased with the increase of temperature and chloride ion concentration and with the decrease of oxygen. The hardness of water had no effect. At 10–12 ppm of dissolved oxygen no losses occurred, at 7–9 ppm of oxygen 7–8 per cent of the fish died at 16–20°C (no loss at 13°C) but at 4–6 ppm of oxygen deaths ranged from 4 per cent (at 13°C) to 20 per cent (at 20°C). Chlorine ions (2 mm) increased the latter figure to 76 per cent. The toxicity of ionic and protein-bound methyl mercury administered per os to pike (Esox lucius) was compared by Miettinen et al. (1970). The LD_{50} dose in 30 days was about 20 ppm of Hg per fresh body weight in both cases. Rainbow trout (Salmo gairdneri) was more sensitive; the corresponding LD_{50} dose being 10–12 ppm of Hg. Little is known about the toxicity of mercury for marine mammals but it is assumed that mercury intoxication can cause symptoms similar to those of terrestrial mammals (Goldwater, 1957; Bidstrup, 1964; Berglund and Berlin, 1969; Study Group of Minamata Disease, 1968; Swensson, 1969; Löfroth, 1969). The behavioural effects of chronic inorganic mercury poisoning in man are nervousness, increased irritability and rhythmical tremor (erethism). Human exposure to organic mercury compounds, particularly methyl mercury, damaging primarily the central nervous system, leads to generalized ataxia, constriction of the visual field and apathy. In cats a "suicide" type reaction was often observed. Apathy was a marked behavioural effect in a ring seal Pusa hispida (Koivisto, 1970), which showed extremely high mercury concentrations: 197 mg/kg in flesh and 210 mg/kg of liver (Henriksson et al., 1969).

Effect of environment

The toxicity of metallic cations for aquatic organisms depends on the concentration and on the general composition of the environment. Reports on the antagonistic action of some cations are often misleading, because of failure to consider actual changes in the concentration of the toxic compounds, or changes in pH of the solution upon introduction of the "antagonizer" (Doudoroff and Katz, 1953). Although some cations are really antagonistic to more toxic heavy metals (Lloyd 1965), the effect of mixed solutions of heavy metals is usually not only additive, but synergistic. Barnes and Stanbury (1948) demonstrated with harpacticid copepod, Nitocra spinipes, that the toxic effect of mercuric chloride and copper sulphate when applied together is not additive but synergistic, probably because mercury impairs the excretory functions and permits larger concentrations of copper to build up. In an amphipod crustacean, Marinogammarus marinus, copper in very low concentrations has a pronounced effect on animals undergoing mercury poisoning (Hunter, 1949). In Artemia salina and Acartia clausi more than the additive toxic effect was observed when various mercury compounds, such as mercuric iodide, ethyl mercuric chloride or mercuric chloride were used together with copper compounds (Corner and Sparrow, 1956). This supplemental synergism must be seriously considered when discussing the limiting or permissible concentrations of various pollutants in the environment.

"A non-effect level has not been demonstrated for mercury. The level of 0.1 ppm, equivalent to 0.005 mg/kg of body weight per day, produced a slight effect in the rat. Even if this figure were to be adopted as a maximum non-effect level and the customary safety factor applied this would give an acceptable daily intake for man of 0.05 μg/kg of body weight". (FAO/WHO 1967).

Teratogenic effects

Experimental teratogenic effects of mercury as the result of its foetotoxic activity were first reported by Murakami *et al.* (1954, 1955). They found that vaginally applied contraceptive pills containing 0.1 mg phenyl mercuric acetate to pregnant mice resulted in death of foetuses, malformations of the spinal cord and abnormal tails. Miyoshi (1959) reported cataracts and reduction in femur length in foetuses after mercuric chloride was administered to pregnant rats. Cats treated with biethyl mercuric sulphide during pregnancy gave birth to young with ataxia, retarded growth of cerebellum, narrow molecular and granular layer of the brain and disarrangement in Purkinje's cells (Morikawa, 1961; Takeuchi, 1968a). Young rats treated during pregnancy with methyl mercuric sulphide showed diminution and atrophy of cells in cerebral white matter (Tatetsu *et al.*, 1968). In litters of cats and rats treated with methyl mercuric chloride and methyl mercuric sulphide incomplete formation of the cerebral granular layer, changes in the brain cellular architecture and disappearance of the cerebellar granular cells were observed (Harada *et al.*, 1968).

Comparable malformations were partially present in human foetuses of women suffering from Minamata disease (Matsumoto *et al.*, 1965) and in children born from these women (Harada, 1968, 1968a; Murakami, 1969).

Teratogenic and especially foetotoxic effects occur because mercury is able to reach the foetus and provoke histopathological changes including those on genetic material. The foetus is not protected because the placenta is permeable to some mercury compounds (Radaody–Ralarosy, 1938). The ability of various mercury compounds to pass the placental barrier in mice was demonstrated by Berlin and Ullberg (1963, 1963a, 1963b). They found by autoradiography that methyl mercuric dicyandiamide given intravenously, rapidly and easily penetrated the placenta and was evenly distributed in the foetus in concentrations comparable to those found in maternal tissues. Mercury administered intravenously as phenyl mercuric acetate or mercuric chloride was not detected in foetus, although the placenta concentrates large amounts of mercury.

The passage of mercury compounds into reproductive organs and germinal tissues is also common in birds. Bäckström (1967, 1969) reported that the yolk of quails concentrates phenyl mercury and methoxyethyl mercury, while the egg white concentrates only methyl mercury. The yolk also accumulates mercury ion.

The egg-laying frequency in hens fed with corn treated with various mercury compounds decreased and in the group where methyl mercury was used the number of embryos that died during incubation increased (Swensson, 1969), but no malformations were observed in hatched chickens and they developed normally. In the eggs of white leghorn hens fed on wheat treated with methyl mercury, phenyl mercury, methoxyethyl mercury and mercury nitrate, the highest concentration was reached with methyl mercury and then in decreasing order with phenyl mercury, methoxyethyl mercury and mercury nitrate (Kiwimae *et al.*, 1969). In the group with methyl mercury the concentration was higher in the egg white while in all other groups it was higher in the yolk. This agrees with the results reported by Bäckström (1967, 1969) for quails' eggs and by Rissanen and Miettinen (1968) for hens' eggs.

Bäckström (1967, 1969) compared the accumulation of various mercury compounds in the germinal tissues and genital organs of fresh-water trout (*Salvelinus fontinalis*), pike (*Esox lucius*), pike-perch (*Lucioperca lucioperca*), perch (*Perca fluviatilis*) and salmon (*Salmo salar*). After intravenous administration of mercuric nitrate the interstitial tissue of male and female gonads moderately accumulated mercury, and in the testicular lobules no mercury was found. The eggs contained little mercury but the oviduct and the follicular walls showed high mercury concentrations. Intramuscularly injected phenyl mercury was moderately concentrated in the interstitium of gonads but the uptake by germinal cells was very low. Parenterally given methyl mercury showed low concentrations in germinal cells and somewhat higher in gonads. The yolk of fish oocytes did not concentrate any of the mercury compounds used.

Genetic effects

There are no published data on the genetic effect of mercury compounds on marine organisms. The evidence from various plant and animal species, including man, indicates that the genetic effect of mercury might apply to all organisms.

Skerfving *et al.* (1970) reported an increased frequency of chromosome breaks in human lymphocyte cultures. The cultures were prepared from nine persons who had increased mercury levels in their blood from eating fish with high methyl mercury content. The biological significance of these observations is not yet clear.

References

ABELSON, P H, Methyl mercury. *Science, N.Y.*, 169:237.
1970

ACKEFORS, H   Recent advances in mercury investigations with
1969   special reference to fish. ICES, C.M. 1969/E:12:4 p.
   (mimeo).

AIDIN'YAN, N K and OZEROVA, N A   Mercury geochemistry.
1968   *Problemy Geokhim. Kismol.*, 1962:160–5 (in Russian).

AMEND, D F, YASUTAKE, W T and MORGAN, R   Some factors in-
1969   fluencing susceptibility of rainbow trout to the acute
   toxicity of an ethyl mercury formulation (Timsam). *Trans.
   Am. Fish. Soc.*, 98:419–25.

ANFÄLT, T, *et al.*   Chemical state of the twovalence mercury in
1968   natural waters. *Svensk kem.Tidskr.*, 80:340–2 (in Swedish),

ARBUTHNOTT, J P   Haemolytic action of mercurials. *Nature, Lond..*
1962   196:277–8.

ASHWORTH, DE B   Use of organo-mercurials in crop protection in
1967   the United Kingdom. *Oikos*, Suppl. 9:19–20.

BÄCKSTRÖM, J Distribution of mercury compounds in fish and
1967 birds. *Oikos*, Suppl. 9:30–31.

BÄCKSTRÖM, J Distribution studies of mercuric pesticides in
1969 quail and some fresh-water fishes. *Acta pharmac. tox.*, 27,
Suppl. 3:74–92.

BARNES, H and STANBURY, F A The toxic action of copper and
1948 mercury salts both separately and when mixed on the
harpactacid copepod, *Nitocra spinipes* (Boeck). *J. exp.
Biol.*, 25:270–5.

BERG, W, *et al.* Mercury content in feathers of Swedish birds
1966 from the past 100 years. *Oikos*, 17:71–3.

BERGLUND, F and BERLIN, M Risk of methylmercury cumulation
1969 in man and mammals and the relation between body
burden of methylmercury and toxic effects. *In*: Chemical
fallout, edited by M. W. Miller and G. G. Berg,
Springfield, Ill., Ch. C. Thomas, Publ., pp. 258–69.

BERGLUND, F and BERLIN, M Human risk evaluation for various
1969a populations in Sweden due to methylmercury in fish.
In Chemical fallout, edited by M. W. Miller and G. G. Berg,
Springfield, Ill., Ch. C. Thomas Publ., pp. 423–31.

BERLIN, M On estimating threshold limits of mercury in biological
1963 material. *Acta. med. scand.*, 1963 (Suppl.):396.

BERLIN, M Risk of methylmercury cumulation in man with
1969 reference to consumption of contaminated fish by methylmercury. *J. Japan med. Ass.*, 61:1047–50.

BERLIN, M and ULLBERG, S Accumulation and retention of
1963 mercury in the mouse. 1. An autoradiographic study after
a single intravenous injection of mercuric chloride. *Archs.
envir. Hlth*, 6:589–601.

BERLIN, M and ULLBERG, S Accumulation and retention of
1963a mercury in the mouse. 2. An autoradiographic comparison
of phenylmercuric acetate with inorganic mercury. *Archs.
envir. Hlth*, 6:602–9.

BERLIN, M and ULLBERG, S Accumulation and retention of
1963b mercury in the mouse. 3. An autoradiographic comparison
of methylmercuric dicyanamide with inorganic mercury.
Archs. envir. Hlth, 6:610–6.

BIDSTRUP, P L Toxicity of mercury and its compounds. New York,
1964 Elsevier.

BINET, L and NICOLLE, P Influence du degré de salinité du
1940 milieu sur la toxicité du chlorure mercurique pour les
épinoches. *C.r. Séanc. Soc. Biol.*, 134:562–5.

BOËTIUS, J Lethal action of mercuric chloride and phenylmercuric
1960 acetate on fishes. *Meddr. Danm. Fisk, og. Havunders.*,
3(4):93–115.

BOND, H W and NOLAN, N O Results of laboratory screening tests
1954 of chemical compounds for molluscicidal activity.
2. Compounds of mercury. *Am. J. trop. Med. Hyg.*, 3:
187–90.

BONEY, A D and CORNER, E D S Application of toxic agents in
1959 the study of the ecological resistance of intertidal red
algae. *J. mar. biol. Ass. U.K.*, 38:267–75.

BONEY, A D, CORNER, E D S and SPARROW, B W P The effects of
1959 various poisons on the growth and viability of sporelings of
the red alga *Plumaria elegans* (Bonnem.) *Schm. Biochem.
Pharmac.*, 2:37–49.

BOOER, J R The behaviour of mercury compounds in soil. *Ann.
1944 appl. Biol.*, 31:340–58.

BOSTRÖM, K and FISHER, D E Distribution of mercury in East
1969 Pacific sediments. *Geochim. cosmochim. Acta*, 33(6):743–5.

BOUVENG, H O Organo-mercurials in pulp and paper industry.
1967 *Oikos*, Suppl. 9:18.

BOUVENG, H O Personal communication.
1970

BOUVENG, H O and ULLMAN, P Reduction of mercury in waste
1969 waters from chlorine plants. Stockholm. *Rep. Inst. Vatten-
och. Luftvardsforsk., Stockh.*, 1969.

BROWN, J R, JOSE, F R and KULKARNI, M V Studies on the toxicity
1967 and metabolism of mercury and its compounds. *Med.
Servs. J. can.*, 23:1089–1110.

CARPENTER, K E The lethal action of soluble metallic salts on
1927 fishes. *Br. J. exp. Biol.*, 4:378–91.

CHANDLER, A C Control of fluke disease by destruction of the
1920 intermediate host. *J. agric. Res.*, 20:193–208.

CLARKE, G L The effectiveness of various toxics and the course of
1943 poisoning and recovery in barnacles and mussels. *In* Sixth
Report from the Woods Hole Oceanographic Institution
to the Bureau of Ships, May 1, 1943, Paper 11.

CLARKE, G L Poisoning and recovery in barnacles and mussels.
1947 *Biol. Bull. mar. biol. Lab., Woods Hole*, 92(1):73–91.

CLARKSON, T and ROTHSTEIN, A The excretion of volatile mercury
1964 by rats injected with mercuric salts. *Hlth Phys.*, 10:1115–21.

CONWAY, E J Mean geochemical data in relation to oceanic
1942 evolution. *Proc. R. Ir. Acad.*, 48B(8):119.

CONWAY, E J The chemical evolution of the ocean. *Proc. R. Ir.
1943 Acad.*, 48B(9):161.

CORNER, E D S The poisoning of *Maia squinado* (Herbst) by
1959 certain compounds of mercury. *Biochem. Pharmac.*,
2:121–32. Also issued as *J. mar. biol. Ass. U.K.*, 39:414
(1960).

CORNER, E D S and RIGLER, F H The loss of mercury from stored
1957 sea water solutions of mercuric chloride. *J. mar. biol. Ass.
U.K.*, 36:449–58.

CORNER, E D S and RIGLER, F H The mode of action of toxic
1958 agents. 3. Mercuric chloride and N-amylmercuric chloride
on crustaceans. *J. mar. biol. Ass. U.K.*, 37:85–96.

CORNER, E D S and SPARROW, B W The modes of action of toxic
1956 agents. 1. Observations on the poisoning of certain
crustaceans by copper and mercury. *J. mar. biol. Ass. U.K.*,
35:531–48.

CORNER, E D S and SPARROW, B W The modes of action of toxic
1957 agents. 2. Factors influencing the toxicities of mercury
compounds to certain crustaceans. *J. mar. biol. Ass. U.K.*,
36:459–72.

DALL'AGLIO, N The abundance of mercury in 300 natural water
1968 samples from Tuscany and Latium (central Italy). *In*
Origin and distribution of the elements, edited by
L. H. Ahrens, Oxford, Pergamon Press, pp. 1065–81.

DANIEL, J W and GAGE, J C The metabolism of methoxyethyl-
1970 mercury chloride in the rat. *Biochem. J.*

DICKSON, F W and TUNELL, G Mercury and antimony deposits
1968 associated with active hot springs in the western United
States. *In* Ore deposits in the U.S. 1933–1967, edited by
J. D. Ridge, New York, American Institute of Mining,
vol. 2:1673–701.

DOUDOROFF, P and KATZ, M Critical review of literature on the
1953 toxicity of industrial wastes and their components to fish.
2. The metals, as salts. *Sewage ind. Wastes*, 25:802–39.

FAGESTRÖM, T and JERNELÖV, A Formation of methyl mercury
1970 from pure mercuric sulphide in aerobic organic sediment.
5 p. (mimeo).

FAO/WHO Evaluation of some pesticide residues in food, Rome,
1967 FAO, (PL:CP/15):237 p. Also issued as WHO/Food Add/
67.32.

FRIBERG, L Studies on the accumulation, metabolism and
1956 excretion of inorganic mercury (Hg-203) after prolonged
subcutaneous administration to rats. *Acta pharmac. tox.*,
12:411–17.

FRIBERG, L Studies on the metabolism of mercuric chloride and
1959 methyl mercury dicyandiamide. *A.M.A. Archs ind. Hlth*,
20: 42–9.

FRIBERG, L, ODELBLAD, E and FORSSMAN, S Distribution of two
1957 mercury compounds in rabbits after a single subcutaneous
injection. *A.M.A. Archs. ind. Hlth*, 16, 163–8.

FRIBERG, L, SKOG, E and WAHLBERG, J E Resorption of mercuric
1961 chloride and methyl mercury dicyandiamide in guinea-pigs
through skin pre-treated with acetone, alkylaryl-sulphonate
and soap. *Acta derm.-vener., Stockh.*, 41:40–52.

FRIEDMAN, H L Relationship between chemical structure and bio-
1957 logical activity in mercurial compounds. *Ann. N.Y. Acad.
Sci.*, 65:461–73.

FUKUNAGA, F and TSUKANO, Y Pesticide regulations and residue
1969 problems in Japan. *Residue Rev.*, 26:1–16.

FURUKAWA, K, SUZUKI, T and TONOMURA, K Decomposition of
1969 organic mercurial compounds by mercury-resistant bacteria.
Agric. Biol. Chem., 33:128–30.

GAGE, J C Distribution and excretion of methyl and phenyl
1964 mercury salts. *Brit. J. ind. Med.*, 21:197–202.

GAGE, J C and SWAN, A A B The toxicity of alkyl and aryl mercury
1961 salts. *Biochem. Pharmac.*, 8:77.

GAYNER, F C H Fungicides in agriculture and horticulture. Paper
1962 presented at Symposium organized by the Pesticides Group
of the Society of the Chemical Industry, 1962.

GLOOSCHENKO, W A Accumulation of ^{203}Hg by the marine diatom
1969 *Chaetoceros costatum. J. Phycol.*, 5:224–6.

GOLDBERG, E D The oceans as a chemical system. *In* The sea,
1963 edited by M. N. Hill, New York, Interscience Publishers,
vol. 2:3–25.

GOLDBERG, E D Manuscript for the McGraw-Hill Yearbook of
Science and Technology (in press).

GOLDBERG, E D and ARRHENIUS, G O S Chemistry of Pacific
1958 pelagic sediments. *Geochim. Cosmochim. Acta*, 13:153–212.

GOLDSCHMIDT, V M The principles of distribution of chemical
1937 elements in minerals and rocks. *J. chem. Soc.*, 1937: 655–73.

GOLDWATER, L J From Hippocrates to Ramazzini: early history
1936 of industrial medicine. *Ann. med. Hist.*, 8:27.

GOLDWATER, L J The toxicology of inorganic mercury. *Ann. N.Y.
1957 Acad. Sci.*, 65:498–503.

GOODMAN, L and GILMAN, A The pharmacological basis of thera-
1955 peutics. 2nd ed. New York, Macmillan.

GRANT, N Legacy of the mad hatter. *Environment*, 11:43–44.
1969

GREIF, R L and PITTS, R F Distribution of radiomercury ad-
1956 ministered as labelled chlormerodrin in the kidneys of rats
and dogs. *J. clin. Invest.*, 35:38.
HALLDIN, A Industrial sources (of mercury). *In* Proceedings of
1969 the Nordic Symposium on the Problems of Mercury,
Lindigö, Sweden, Oct. 10–11, 1968, pp. 154–9 (in Swedish).
HAMAGUCHI, H, KURODA, R and HOSOHARA, K Photometric deter-
1961 mination of mercury in sea-water. *J. chem. Soc. Japan*,
(Pure Chem. Sect.), 82:347–9. Abstract in *Wat. Pollut.
Abstr.*, 36, (1062)(1963).
HANDLEY, C A and SEIBERT, R A The urinary excretory products
1956 after meraluride administration. *J. Pharm. exp. Ther.*,
116:27.
HANNERZ, L Experimental investigations on the accumulation
1968 of mercury in water organisms. *Rep. Inst. freshwat. Res.
Drottningholm*, 48:120–76 Abstract in *Aquat. Biol. Abstr.*, 1:
Aq661 (1969).
HARADA, Y Infantile Minamata disease. *In* Minamata disease,
1968 edited by the Study Group on Minamata Disease,
Kumamoto, Kumamoto University, Japan, pp. 73–91.
HARADA, Y Congenital of fetal Minamata disease. *In* Minamata
1968a disease, edited by the Study Group on Minamata Disease,
Kumamoto, Kumamoto University, Japan, pp. 93–117.
HARADA, Y, SUHOD, Y and NONAKA, I Congenital Minamata
1968 disease and experimental organic mercury poisoning in
cats and rats. Cited after Murakami (1969).
HARRIS, W H, *et al.* A study of metal ions in the central nervous
1954 system. *J. Neuropath. exp. Neurol.*, 13: 427–34.
HARRISS, R C Mercury content of deep-sea manganese nodules.
1968 *Nature, Lond.*, 219:54.
HARRISS, R C, WHITE, D B and MACFARLANE, R B Mercury
1970 compounds reduce photosynthesis by phytoplankton.
Science, N.Y., 170:736.
HÄSÄNEN, E Determination of the amount of mercury in biological
1969 material by way of activation analysis. *Nord. Hyg. Tidskr.*,
50(2):78.
HÄSÄNEN, E Report of the Finnish Water Protection Commission
1969a and a Lecture given at Department of Radiochemistry,
1969.
HÄSÄNEN, E and SJÖBLOM, V Mercury content of fish in Finland
1968 in 1967. *Suom. Kalatal.*, 36:1–24 (in Finnish).
HASSELROT, T B Report on current field investigations concerning
1968 the mercury content in fish, bottom sediment, and water.
Rep. Inst. freshwat. Res. Drottningholm, 48, 102–11. Also
issued as *Aquat. Biol. Abstr.*, 1:Aq662 (1969).
HASSELROT, T B Which method should be used in tracing
1969 mercurial discharge in water and checking effect of
measures taken. *Nord. Hyg. Tidskr.*, 50(2)160–3 (in
Swedish).
HENRIKSSON, K, HELMINEN, M and KARPPANEN, E The amounts
1969 of mercury in seals from lakes and sea. *In* Proceedings of
the Nordic Symposium on the Problems of Mercury,
Lidingö, Sweden, Oct. 10–11, 1968, pp. 54–9 (in Swedish).
HIBIYA, T and OGURI, M Gill absorption and tissue distribution
1961 of some radionuclides (Cr–51, Hg–203, Zn–65, Ag 110m,
110) in fish. *Bull. Jap. Soc. scient. Fish.*, 27(11):996–1000.
Abstract in *Wat. Pollut. Abstr.*, 37:346 (1964).
HOFFMANN, C Beiträge zur Kenntnis der Wirkung von Giften auf
1950 marine Organismen. *Kieler Meeresforsch.*, 7:38–52.
HOSOHARA, K Mercury content of deep sea water. *J. chem. Soc.
1961 Japan*, 82:1107–8, 1961. Also issued in *Deep-Sea Res.*,
11:319 (1964).
HUGHES, W L Protein mercaptides. *Symp. quant. Biol.*, 14:79–84.
1950
HUGHES, W L A physical rationale for the biological activity of
1957 mercury and its compounds. *Ann. N.Y. Acad. Sci.*, 65:
454–60.
HUNTER, D and RUSSEL, D S Focal cerebral and cerebellar atrophy
1954 in a human subject due to organic mercury compounds.
J. Neurol. Neurosurg. Psychiat., 17:235–41.
HUNTER, D, BOMFORD, R R and RUSSEL, D S Poisoning by methyl
1940 mercury compounds. *Q. Jl Med.*, 9:193–213.
HUNTER, W R The poisoning of *Marinogammarus marinus* by
1949 cupric sulphate and mercuric chloride. *J. exp. Biol.*, 26:
113–24.
IRUKAYAMA, K The pollution of Minamata Bay and Minamata
1967 disease. *Adv. Wat. Pollut. Res.*, 3:153–80.
IRUKAYAMA, K, *et al.* Studies on the origin of the causative agent
1962 of Minamata Disease. 2. Comparison of the mercury
compound in the shellfish from Minamata Bay with
mercury compounds experimentally accumulated in normal
shellfish. *Kumamoto med. J.*, 15:1–12 (in English). Also
issued as *Jap. J. Hyg.*, 16:467–75.
IRUKAYAMA, K, *et al.* Studies on the origin of the causative agent
1962a of Minamata disease. 3. Industrial wastes containing
mercury compounds from Minamata factory. *Kumamoto
med. J.*, 15:57–68. Abstract in *Wat. Pollut. Abstr.*, 38:143
(1965).

IRUKAYAMA, K, *et al.* Mercury pollution in Minamata District
1969 before and after the suspension of the production of
acetaldehyde in Minamata factory. *Kumamoto med. J.*,
43:946–57.
ISHIYAMA, T Fungicidal action of phenylmercuric acetate.
1969 *Residue Rev.*, 25:123–31.
IWAO, T Comparative investigation of the toxicity of various
1936 metals. *Jap. J. exp. Pharmac.*, 10:357–80.
JÄRVENPÄÄ, T, TILLANDER, M and MIETTINEN, J K Methyl-
1970 mercury: halftime of elimination in flounder, pike and eel.
Paper presented to FAO Technical Conference on Marine
Pollution and its Effects on Living Resources and Fishing,
Rome, Italy, 9–18 December 1970, FIR: MP/70/E-66:6 p.
JENSEN, S and JERNELÖV, A Biological methylation of mercury in
1969 aquatic organisms. *Nature, Lond.*, 223(5207): 753–4.
JERNELÖV, A The metabolism of turnover of mercury in nature and
1969 what we can do to affect it. *Nord. Hyg. Tidskr.*, 50(2):
174–8.
JERNELÖV, A and LANN, H Mercury content in bottom sediment,
1970 bottom fauna and water from Oresund, Wänem (outside
Skoghall) and Delangersan—a comparison. *Rep. Inst.
Vatten-och Luftvardsforsk.*, (B.67):18 p.
JOHNELS, A G, OLSSON, M and WESTERMARK, T *Esox lucius* and
1968 some other organisms as indicators of mercury contamination in Swedish lakes and rivers. *Bull. Off. int.
Epizoot.*, 69:1439–52.
JOHNELS, A, *et al.* Pike (*Esox lucius*) and some other aquatic
1967 organisms as indicators of mercury contamination in the
environment. *Oikos*, 18:323–33.
JOHNS, I B and BRADLEY, W Mercury. *In:* Encyclopaedia
1966 Britannica, 15:183–5.
JONES, J R E The relative toxicity of salts of lead, zinc and copper
1938 to the stickleback (*Gasterosteus aculeatus*) and the effect
of calcium on the toxicity of lead and zinc salts. *J. exp.
Biol.*, 15:394–407.
JONES, J R E The relation between the electrolytic solution
1939 pressures of the metals and their toxicity to the stickleback
(*Gasterosteus aculeatus* L.). *J. exp. Biol.*, 16:425–37.
JONES, J R E The toxicity of double chlorides of mercury and
1940 sodium. I. Experiments with the minnow, *Phoxinus
phoxinus* (L.). *J. exp. Biol.*, 17:325–30.
KARZHANO, T K Finding of metallic mercury in central
1969 Kuzulkum thermal waters. *Uzbek. geol. Zh.*, 13:79–80.
KIMURA, Y and MILLER, V L The degradation of organomercury
1964 fungicides in soil. *J. agric. Fd Chem.*, 12:253–7.
KITAMURA, M, TAINA, M and SUMINO, K Synthesis and de-
1969 composition of organic mercury compounds by bacteria.
Jap. J. Hyg., 24:132–3 (in Japanese. Eng. transl.).
KIWIMAE, A, *et al.* Methylmercury compounds in eggs from hens
1969 after oral administration of mercury compounds. *J. agric.
Fd Chem.*, 17:1014.
KIYOURA, R Water pollution and Minamata disease. *Int. J. Air
1963 Wat. Pollut.*, 7:459–70. Abstract in *Wat. Pollut. Abstr.*,
38(1):142 (1965).
KLEIN, D H and GOLDBERG, E D Mercury in the marine environ-
1970 ment. *Envir. Sci. Technol.*, 4:765–8.
KOIVISTO, I Personal communication.
1970
KORRINGA, P Recent advances in oyster biology. *Rev. Biol.*,
1952 27:266–308, 339–65.
KURLAND, L T, FARO, S N and SEIDLER, H Minamata disease.
1960 *Wld Neurol.*, 1:370–95.
LAUSEN, C The occurrence of minute quantities of mercury in
1936 the Chinle shales at Lee Ferry, Arizona. *Econ. Geol.*,
31:610.
LIHNELL, D The use of mercury in Swedish agriculture. *Oikos*,
1967 Suppl. 9:16–7.
LLOYD, R Factors that affect the tolerance of fish to heavy metal
1965 poisoning. *Publ. Hlth Serv. Publs, Wash.*, (999–WP–25):
181–7.
LÖFROTH, G Birds give warning. *Environment.*, 11:10–7.
1969
LÖFROTH, G Methylmercury. A review of health hazards and side
1969a effects associated with the emission of mercury compounds
into natural systems. *Bull. ecol. Res. Comm. nat. Sci. Res.
Coun., Stockh.*, (4):1–29.
MAGOS, L, TUFFERY, A A and CLARKSON, T W Volatilization of
1964 mercury by bacteria. *Br. J. ind. Med.*, 21:294–8.
MAKHONINA, G I and GILEVA, E A Accumulation of Zn–65,
1968 Cd–115 and Hg–203 by freshwater plants and the effect of
EDTA on the accumulation coefficients for these radioisotopes. *Trudy Inst. ekol. Rast. Zhivot.*, 61:72–8.
MATSUMOTO, H G, KOYA, G and TAKEUCHI, T Fetal Minamata
1965 disease: a neuropathological study of two cases of intra-
uterine intoxication by a methyl mercury compound.
J. Neuropath. Neurol., 24:563–74.
MCMULLEN, D B Schistosomiasis and molluscacides. *Am. J.
1952 trop. Med. Hyg.*, 1:671–9.

MEYER, H Der Wirkungsmechanismus von Quecksilberverbin-
1964 dungen bei Mikroorganismen. *Pharm. Zentralhalle Dtl.*, 103:571–6.

MIETTINEN, J K Organic mercurials as food chain problems. *In*
1970 Nuclear techniques for studying pesticide residue problems, Vienna, IAEA, pp. 43–7.

MIETTINEN, J K, HEYRAUD, M and KECKES, S Mercury as hydro-
1970 spheric pollutant. 2. Biological half-time of methyl mercury in four Mediterranean species: a fish, a crab, and two molluscs. Paper presented to FAO Technical Conference on Marine Pollution and its Effects on Living Resources and Fishing, Rome, Italy, 9–18 December 1970, FIR:MP/70/E–90:8p.

MIETTINEN, J K, et al. Distribution and excretion rate of phenyl-
1969 and methylmercuric nitrate in fish, mussels, molluscs and crayfish. *In* Proceedings of the 9th Japan Conference on Radioisotopes, Tokyo, Japan Industrial Forum, Inc., pp. 474–8.

MIETTINEN, V, et al. Preliminary study on the distribution and
1970 effects of two chemical forms of methyl mercury in pike and rainbow trout. Paper presented to FAO Technical Conference on Marine Pollution and its Effects on Living Resources and Fishing, Rome, Italy, 9–18 December 1970, FIR: MP/70/E-91:12 p.

MILLER, V L, KLAVANO, P A and CSONKA, E Absorption,
1960 distribution and excretion of phenylmercuric acetate. *J. Toxicol. appl. Pharmac.*, 2:344–52.

MILLER, V L, et al. Absorption, distribution and excretion of
1961 ethyl mercuric chloride. *J. Toxicol. appl. Pharmac.*, 3:459–68.

MIYOSHI, Y Experimental studies on the effect of toxic
1959 substances on pregnancy. *Med. J. Osaka Univ.*, 8:309–18.

MORIKAWA, N Pathological studies on organic mercury poisoning.
1961 2. Experimental production of congenital cerebellar atrophy by bisethylmercuric sulfide in cats. *Kumamoto med. J.*, 14:87–93.

MOZLEY, A The control of Bilharzia in Southern Rhodesia.
1944 Salisbury, Southern Rhodesia, Rhodesian Printing & Publishing Co., Ltd.

MURAKAMI, U Fetotoxic effect of some organic mercury com-
1969 pounds. *J. Jap. med. Soc.*, 61:1059–72 (in Japanese).

MURAKAMI, U, KAMEYAMA, Y and KATO, T Experiments on the
1954 abnormal formation of embryos by chemical substance (mercury compounds). Effect of contraceptive pills on the embryo and mother. Preliminary report. *Kankio Igaku Kenkyosho Nenpo*, 5:167–8.

MURAKAMI, U, KAMEYAMA, Y and KATO, T Effects of a vaginally
1955 applied contraceptive with phenylmercuric acetate upon developing embryos and their mother animals. *Rep. Res. Inst. Envir. Med. Nagoya Univ.*, 1955:88–99.

NICHOLLS, G D The geochemical history of the oceans. *In*
1965 Chemical Oceanography, edited by J. P. Riley and G. Skirrow, London, Academic Press, vol. 2:277–94.

NORÉN, K and WESTÖÖ, G Methylmercury in fish. *Var Föda*
1967 19:13–24 (in Swedish).

NOVICK, S A new pollution problem. *Environment*, 11:3–9.
1969

ÖSTLUND, K Studies on the metabolism of methylmercury and
1969 dimethyl mercury in mice. *Acta pharmac. tox.*, 27(Suppl. 1).

OHMOMO, Y, et al. Studies on the distribution of ^{203}Hg-labelled
1969 methyl mercury and phenyl mercury in pike. Paper presented at the 5th R.I.S. Symposium, Helsinki.

OSHIMA, S On the toxic action of dissolved salts and their ions
1931 upon young eels (*Anguilla japonica*). *J. imp. Fish. exp. Stn.*, Tokyo, 2:139–93.

PASSOW, H and ROTHSTEIN, A The binding of mercury by the yeast
1960 cell in relation to changes in permeability. *J. gen. Physiol.*, 43:621–33.

PASSOW, H, ROTHSTEIN, A and CLARKSON, T W The general
1961 pharmacology of the heavy metals. *Pharmac. Rev.*, 13: 185–224.

PREUSS, E Beiträge zur spektralanalytischen Methodik. 2. Bestim-
1941 mung von Zn, Cd, Hg, In, Tl, Ge, Sn, Pb, Sb und Bi durch fraktionierte Destillation. *Z. angew. Miner.*, 3:8.

PROSSER, C L and BROWN, F A Comparative animal physiology.
1962 Philadelphia, W. B. Saunders Company, 688 p.

RADAODY-RALAROSY, P Histochemical study of the passage of
1938 arsenic and mercury in the placenta. *Archs Soc. Sci. med. biol. Montpellier*, 19:22–6.

RAEDER, M G and SNEKVIK, E Qecksilbergehalt mariner
1941 Organismen. *K. norske Vidensk. Selsk. Forh.*, 13:169–172

RANKAMA, K and SAHAMA, TH G Geochemistry, Chicago,
1950 University of Chicago Press.

RISSANEN, K Retention and distribution of mercury in cats.
1969 Preliminary communication. *In* Proceedings of the Swedish-Finnish Mercury Symposium held in Helsinki, Nov. 7, pp. 21–7.

RISSANEN, K and MIETTINEN, J K Thin layer chromatography of
1968 alkyl and alkoxy mercury derivatives and location of mercury in the yolk of hens' eggs. *Anns agric. fenn.*, 7(Suppl. 1):22–3.

RISSANEN, K, ERKAMA, J and MIETTINEN, J K Experiments on
1970 microbiological methylation of mercury (2+) ion by mud and sludge in anaerobic conditions. Paper presented to FAO Technical Conference on Marine Pollution and its Effects on Living Resources and Fishing, Rome, Italy, 9–18 December 1970, FIR: MP/70/E-61:4 p.

ROWLAND, R L Mercurial diuretics. 6 Ionization of organic
1952 mercurials. *J. Am. chem. Soc.*, 74:5482.

RUSKA, H Fixation of mercuric chloride on bacterial and virus
1947 *Arch. exp. Path. Pharmak.*, 204:576–85.

SAUKOV, A A Geochemie. Berlin, VEB-Verlag Technik.
1953

SCHAMBERG, J F Experimental study of mode of absorption of
1918 mercury when applied to the skin. *J. Am. med. Ass.*, 70.

SCHOEMAKER, H A The pharmacology of mercury and its com-
1957 pounds. *Ann. N.Y. Acad. Sci.*, 65:504–10.

SCHWEIGER, G Die toxikologische Einwirkung von Schwer-
1957 metallsalzen auf Fische und Fischnährtiere. *Arch. Fisch. Wiss.*, 8(1–2):54–78.

SILLEN, L G The physical chemistry of sea water. *In* Oceanography,
1961 edited by M. Sears, Washington, D.C., American Association for the Advancement of Science, pp. 549–81.

SJÖBLOM, V and HÄSÄNEN, E Mercury content in fish in Finland.
1969 *Nord. Hyg. Tidskr.*, 50(2):37–53 (in Swedish).

SKERFVING, S Comparing toxicity experiments on cats with
1969 methylmercury, biologically accumulated in the Swedish fish and methylmercury hydroxide added into a fish homogenate. Preliminary results. *In* Proceedings of the Swedish-Finnish Mercury Symposium, Nov. 7, Helsinki, pp. 14–7 (in Swedish).

SKERFVING, S, HANSSON, K and LINDSTEN, J Chromosome break-
1970 age in human subjects exposed to methyl mercury through fish consumption. *Archs Envir. Hlth*, 21:133–39.

SMART, N A Use and residues of mercury compounds in agri-
1968 culture. *Residue Rev.*, 23:1–36.

SMITH, W W Comparison of modern mercury and diaphragm type
1968 chlorine cells. *Chem. Engng*, March issue P.CE54.

STOCK, A and CUCUEL, F Die Verbreitung des Quecksilbers.
1934 *Naturwissenschaften*, 22:390–3.

STUDY GROUP ON MINAMATA DISEASE (Ed.), Minamata disease.
1968 Kumamoto, Kumamoto University, Japan, 499 p.

SUZUKI, T Placental transfer of mercuric chloride, phenylmercury
1967 acetate and methylmercury acetate in mice. *Ind. Hlth*, 5:149–55.

SUZUKI, T Neurological symptoms from concentration of mercury
1969 in the brain. *In* Chemical fallout, edited by W. W. Miller and G. G. Berg, Springfield, Ill., C. C. Thomas Publ., pp. 245–72.

SUZUKI, T, MIYAMA, T and KATSUNUMA, H An experimental study
1962 on the accumulation, metabolism and excretion of sublime after repeated subcutaneous administration in rats. *Nisshin Igaku*, 49:745–9 (in Japanese).

SUZUKI, T, MIYAMA, T and KATSUNUMA, H Comparative study of
1963 bodily distribution of mercury in mice after subcutaneous administration of methyl ethyl, and n-propyl-mercury acetates. *Jap. J. exp. Med.*, 33:277–82.

SUZUKI, T, MIYAMA, T and KATSUNUMA, H Mercury in the plasma
1967 after subcutaneous injection of sublimate or mercuric nitrate in rats. *Ind. Hlth. Jap.*, 5:290–2.

SWENSSON, A Comparative toxicity of various organic mercury
1969 compounds. *J. Japan med. Soc.*, 61:1056–9.

SWENSSON, A and ULFVARSON, U Distribution and excretion of
1968 mercury compounds in rats over a long period after a single injection. *Acta pharmac. tox.*, 26:273–83.

SWENSSON, A and ULFVARSON, U Distribution and excretion of
1968a various mercury compounds after single injection in poultry. *Acta pharmac. tox.*, 26:259–72.

SWENSSON, A, LUNDGREN, K D and LINDSTROM, O Distribution and
1959 excretion of mercury compounds after single injection. *Archs. ind. Hlth*, 20:432–43.

SWENSSON, A, LUNDGREN, K D and LINDSTROM, O Retention of
1959a various mercury compounds after subacute administration. *Archs ind. Hlth*, 20:467.

SZUMLEWICZ, A-P and KEMP, H Moluscocidas promissores contra
1951 um caramujo planorbídeo brasileiro. *Revta. bras. Malar. Doenc. trop.*, 3:389–406.

TAKEUCHI, T Minamata disease—a study on the toxic symptoms
1966 by organic mercury, University of Kumamoto, Report of Department of Medical Science.

TAKEUCHI, T Pathology of Minamata disease. *In* Minamata
1968 disease, edited by the Study Group on Minamata Disease, Kumamoto, Kumamoto University, Japan pp. 141–228.

TAKEUCHI, T   Experiments with organic mercury particularly
1968a   with methyl mercury compounds, similarities between experimental poisoning and Minamata disease. *In* Minamata disease, edited by the Study Group on Minamata Disease, Kumamoto, Kumamoto University, Japan, pp. 229–52.

TAKEUCHI, T, *et al.*   A pathological study of Minamata disease in
1962   Japan. *Acta neuropath.*, 2:40–57.

TAKIZAWA, Y   Synthesis of alkylmercury compound in the natural
1967   world. *Niigata med. J.*, 81:45–8 (in Japanese).

TATETSU, M, *et al.*   Experimental development of fetal Minamata
1968   disease. *Seishin Shinkei Gaku Zasshi*, 70:162 (in Japanese).

TILLANDER, M, MIETTINEN, J K and KOIVISTO, I   Excretion rate of
1970   methyl mercury in the seal (*Pusa hispida*). Paper presented to FAO Technical Conference on Marine Pollution and its Effects on Living Resources and Fishing, Rome, Italy, 9–18 December 1970, FIR:MP/70/E–67:4 p.

TILLANDER, M, *et al.*   Excretion of phenyl and methyl mercury
1969   nitrate after oral administration or intramuscular injection in fish, mussel, mollusc and crayfish. *In* Proceedings of the Nordic Symposium on the Problems of Mercury, Lidingö, Sweden, Oct. 10–11, 1968, pp. 181–3 (in Swedish).

TONOMURA, K and KANZAKI, F   The reductive decomposition of
1969   organic mercurials by cell-free extract of a mercury-resistant Pseudomonas. *Biochim. biophys. Acta*, 184:227–9.

TONOMURA, K, *et al.*   Stimulative vaporization of phenylmercuric
1968   acetate by mercury-resistant bacteria. *Nature, Lond.*, 217–644.

TONOMURA, K, *et al.*   Studies on the action of mercury-resistant
1968a   microorganism on mercurials. 1. The isolation of mercury-resistant bacterium and the binding of mercurials to the cells. *J. Ferment. technol., Osaka*, 46:506–12.

UI, J   Mercury pollution of sea and fresh water. Its accumulation
1971   into water biomass. *Revue int. Océanogr. méd.*, 22–23:79–128.

ULFVARSON, U   Distribution and excretion of some mercury com-
1962   pounds after long term exposure. *Int. Arch. Gewerbepath. Gewerbehyg.*, 19:412–22.

ULFVARSON, U   Mercury consumption. *Oikos*, Suppl. 9:22.
1967

ULFVARSON, U   The absorption and distribution of mercury in
1969   rats fed organs from rats injected with various mercury compounds. *Toxic. appl. Pharmac.*, 15:525–31.

ULFVARSON, U   The effect of the size of the dose on the distribution
1969a   and excretion of mercury in rats after single intravenous injection of various mercury compounds. *Toxic. appl. Pharmac.*, 15:517–24.

ÜNLÜ, Y, HEYRAND, M and KECKES, S   Mercury as hydrospheric
1970   pollutant. 1. Accumulation and excretion of ^{203}Hg Cl in *Tapes decussatus* L. Paper presented to FAO Technical Conference on Marine Pollution and its Effects on Living Resources and Fishing, Rome, Italy, 9–18 December 1970, FIR:MP/70/E–68:6 p.

USPENSKAYA, V I   Experimental observations on the influence of
1946   mercury compounds on aquatic organisms. *Gig. Sanit.*, 11:1 (in Russian).

VINOGRADOV, A P   The elementary chemical composition of marine
1953   organisms. *Mem. Sears Fdn mar. Res.*, (2):648 p.

WALKER, K C   Legislative restrictions by federal action in the
1967   United States. *Oikos*, Suppl. 9:21.

WEBB, D A and FEARON, W R   Studies on the ultimate composition
1937   of biological material. 1. Aims, scope and methods. *Scient. Proc. R. Dubl. Soc.*, 21:487–504.

WEBB, J L   Enzyme and metabolic inhibitors. New York, Academic
1966   Press, vol. 2.

WEINER, I M and MÜLLER, O H   Polarographic studies of mersalyl-
1955   thiol complexes and of the excreted products of mersalyl. *J. Pharmac. exp. Ther.*, 113:241.

WEISS, C M   The comparative tolerances of some fouling organisms
1947   to copper and mercury. *Biol. Bull. mar. biol. Lab.*, Woods Hole, 93:56–63.

WESTÖÖ, G   Determination of methylmercury compounds in
1966   foodstuffs. 1. Methylmercury compounds in fish, identification and determination. *Acta chem. scand.*, 20:2131–7.

WESTÖÖ, G   Mercury in fish. *Var Föda*, 19:1–7 (in Swedish).
1967

WESTÖÖ, G   Determination of methylmercury compounds in
1967a   foodstuffs. 2. Determination of methylmercury in fish, egg, meat and liver. *Acta chem. scand.*, 21:1790–1800.

WESTÖÖ, G   Methylmercury compounds in animal foods. *In*
1969   Chemical fallout, edited by M. W. Miller and G. C. Berg. Springfield, Ill., C. C. Thomas Publ., pp. 75–90.

WESTÖÖ, G   Mercury compounds in animal foodstuffs. *Nord.*
1969a   *Hyg. Tidskr.*, 50(2):67–70 (in Swedish).

WIKLANDER, L   Mercury in ground water and river water. *Grund-*
1968   *förbättring*, 21:151–5 (in Swedish).

WILLISTON, S H   Mercury in the atmosphere. *J. geophys. Res.*,
1968   73:7051–5.

WISELY, B and BLICK, R A P   Mortality of marine invertebrate
1967   larvae in mercury, copper and zinc solutions. *Aust. J. freshwat. Res.*, 18(1):63–72. Abstract in *Wat. Pollut. Abstr.*, 41:1343 (1968).

WOOD, J M, KENNEDY, F and ROSEN, C G   Synthesis of methyl
1968   mercury compounds by extract of a methanogenic bacterium. *Nature, Lond.*, 220:173–4.

YAMADA, M, DAZAI, M and TONOMURA, K   Change of mercurial
1969   compounds in activated sludge. *J. Ferment. Technol. Osaka*, 47:155–60.

YOSHIDA, T, KAWABATA, T and MATSUE, Y   Transference
1967   mechanism of mercury in marine environment. *J. Tokyo Univ. Fish.*, 53:73–84.

Anon.,   Moderate growth for mercury. *Metals Week*, April
1970   13:27.

Experiments on Microbiological Methylation of Mercury (2+) Ion by Mud and Sludge under Aerobic and Anaerobic Conditions†

K. Rissanen,* J. Erkama**
and J. K. Miettinen*

Expériences sur la méthylation microbiologique de l'ion mercure (2+) dans des boues et vases placées sous des conditions anaérobies

La méthylation du mercure inorganique dans les boues marines et lacustres et dans les boues résiduelles d'une usine de traitement des eaux d'égout a été étudiée sous des conditions aérobies et anaérobies (sous azote pur), respectivement à 20°C et à 12°C. A 20°C et sous barbottage gazeuse, 1 à 14 pour cent du mercure inorganique introduit se transforme en méthylmercure au bout de deux semaines environ. Pour des échantillons de boues en ampoules scellées, le taux de méthylation est moindre. A 12°C le taux de méthylation est très faible.

Pour la plupart des échantillons de boues, les taux de méthylation sous des conditions aérobies et sous des conditions anaérobies ne diffèrent pas significativement. En général le taux de méthylation est le plus élevé dans les boues riches en matières organiques.

Experimentos sobre la metilación microbiológica del ion de mercurio (2+) mediante barro y fango en condiciones aeróbicas y anaeróbicas

Se efectuaron investigaciones sobre la metilación del mercurio inorgánico en muestras de fangos marino y lacustres y muestras procedentes de una planta de tratamiento de aguas residuales, bajo condiciones aeróbicas y anaeróbicas a temperaturas de 20° y 12°C. Durante un período aproximado de dos semanas, se ha convertido de 1 a 14 por ciento del mercurio inorgánico adicionado en metilmercurio, bajo la agitación del gas a temperatura de 20°C. Muestras contenidas en ampollas cerradas han producido menos metilación. El ritmo de metilación ha sido muy lento a temperatura de 12°C. Para la mayor parte de las muestras de fango no se han encontrado diferencias significativas entre los ritmos de metilación bajo condiciones aeróbicas y anaeróbicas. En general, el ritmo de metilación ha sido más rápido en fangos ricos en materias orgánicas.

MICRO-ORGANISMS, or muds containing them, report Jensen and Jernelöv (1969) and Kitamura (1969), convert phenyl mercury or inorganic mercury to methyl mercury. Although it was not directly stated in either of these publications whether the experiments were carried out under aerobic or anaerobic conditions, it is likely that the conditions were aerobic (Jernelöv, 1968).

However, the methylation of mercury ion in anaerobiosis has been reported by Wood *et al.* (1968), who

* Department of Radiochemistry, and **Department of Biochemistry, University of Helsinki, Helsinki, Finland.
† Investigations supported by International Atomic Energy Agency (Grant no. 702/RB) and Nordforsk.

TABLE 1. MUD AND SLUDGE PREPARATIONS USED IN THE EXPERIMENTS ON AEROBIC AND ANAEROBIC METHYLATION OF MERCURY

Sample No.	Location	Thickness of sediment layer, cm from surface	Sampling depth m	Organic material % of dry weight	Type
Sea mud					
1	Porvoo, Grundvik	0 – 1	0.5	2	sand
2	Tvärminne, Byviken	0 – 5	4	17	gyttja
3	Tvärminne, Storfjärden	0 – 5	36	16	gyttja
4	Porvoo, Suurpellinki	0 – 5	10	15	gyttja
Lake mud					
5*	Pitkänokanlampi	0 – 8	3	77	dy (humus colloid)
6a†	Tuusulanjärvi	0 – 3	10	10	clay-gyttja
6b†	Tuusulanjärvi	3 – 8	10	10	clay-gyttja
7	Kytäjärvi	0 – 5	2	8	clay-gyttja
8	Kytäjä, pond	0 – 5	1	11	clay-gyttja
9a	Kalljärvi	0 –10	0.5	29	gyttja
9b	Kalljärvi	0 –10	1.5	28	gyttja
Treatment plant sludge					
10	"active sludge"				
11	"incubation sludge"				

* Oligotrophic lake
† Polluted lake, almost anaerobic

carried out the reaction with a cell-free extract of a methanogenic bacterium in the presence of hydrogen and ATP. In mild reductive conditions, however, the reaction took place even in the absence of the enzyme. On the basis of the results of Wood et al. (1968), some authors have assumed (Ackefors, 1969; Grant, 1969) that even in nature the methylation occurs primarily by anaerobic organisms. Since this reaction may have a great significance in the future development of polluted waters, the present work compares the methylation of ^{203}Hg-labelled mercuric ion by various bottom muds and sludges under aerobic and anaerobic conditions both in gas-flow and sealed glass-ampul systems.

Samples and aeration procedure

The bottom muds and sludges used in this investigation are presented in Table 1. In all mud-water suspensions pH was approximately 5. Two sludge samples, taken from the sewage treatment plant at Rajasaari, Helsinki, had pH about 5.5 in "active sludge" and 7.5 in "incubation sludge". The dry weight of the samples was determined at 110°C and the ash weight at 450°C.

A mixture of 50 ppm inactive $Hg(NO_3)_2$ (per dry weight of mud) and 1–1.7 μCi of ^{203}Hg($NO_3)_2$ (per 30 ml of suspension) was added to 50–100 g of mud-water suspension. In earlier experimentation, it was found that the optimal concentration for the methylation of mercury, at +20°C and under aerobic agitation, in the systems presently used was 50 ppm Hg (per dry weight of mud).

In the aerobic experiments, the aeration was performed using compressed air bubbled through water. Anaerobic conditions were achieved by using technical N_2 gas (99.9 % N_2) or extra pure N_2 gas (99.999 % N_2) passed through a Cr^{2+} solution (McArthur, 1952) and ascarite. Under both conditions, the pressure of the gas flow to the various mud suspensions was regulated by means of a symmetrical brass distributor and screw clamps. An attempt was made to keep the gas flow as slow as possible (1–2 bubbles per min) in order to permit a natural sedimentation of the suspension.

Before the addition of Hg, the mud suspensions were effectively agitated for 30 min by the gas used in that experiment. The muds in the glass ampuls were also

either saturated with air or else the air was removed by N_2 flow before the addition of Hg and sealing.

Incubation at 19–20°C or at 12°C was continuous and lasted for about two weeks. Volatile mercury compounds which might be formed are carried by the gas flow and absorbed into 2N HCl solution, which thus breaks any dimethyl mercury formed into the mono-methylmercury compound. Mercury compounds not absorbed by the HCl become absorbed by 1 per cent cysteine solution.

Muds in glass ampuls were also kept two weeks at 20°C, with occasional agitation.

The organic mercury compounds formed from inorganic mercury were analysed by a combined extraction, radioactivity measurement, thin layer chromatography and autoradiography procedure.

Analytical extraction procedure

A 30 ml homogenized sample of the air- or N_2-agitated mud suspension, or the content of the glass ampul, was acidified by HCl to approximately 2N, extracted three times by 10 ml benzene in a Griffin flask shaker (5 min each time), and the combined extracts were shaken twice with 2 ml 1 per cent cysteine solution in water, according to Westöö (1967). After acidification, the combined cysteine solutions were extracted with 0.20–0.25 ml benzene, in order to have a concentrated extract suitable for thin layer chromatography.

Inorganic mercury is not soluble in benzene, while most organomercurials are extracted as halide or cyanide complexes. In proteinous materials, mercury and alkyl mercury radicals are firmly bound to SH groups, but can be liberated by acidifying below pH 1.

The extraction procedure was tested for yield by using ^{203}Hg-labelled methyl mercury (50 ppm Hg per dry weight mud), which was incubated for 5 days at 22°C with muds Nos. 7 and 9b and "incubation sludge" No. 11. Two parallel samples were taken of each and aerated for 30 min, one by air and the other by N_2 gas, before the addition of methylmercury. The average yields from the extraction procedures are given in Table 2.

The extraction of methyl mercury from pure water solution was determined, as well as the amount of

TABLE 2. Test extraction of $CH_3{}^{203}HgNO_3$ from bottom mud (No. 9b) containing much organic material, from bottom mud (No. 7) containing less organic material and from "incubation sludge" (No. 11). Modified Westöö (1967) method

| | | ^{203}Hg activity at various stages of extraction as percentage of the $CH_3{}^{203}Hg$ activity added | | |
		No. 9b (28% org. mat.)	No. 7 (8% org. mat.)	No. 11 "incubation sludge"
Extraction:*				
1. Working process	1 benzene extract	26%	37%	10%
	2 benzene extract	26%	23%	20%
	3 benzene extract	15%	14%	19%
	1 + 2 + 3 yield	67%	74%	49%
2. Working process	1 cysteine extract	40%	36%	9%
	2 cysteine extract	14%	22%	7%
	1 + 2 yield	54%	58%	16%
3. Working process	benzene extract	13±1%	18±2%	4±0.1%

* As described in the text.

TABLE 3. Test extraction of $CH_3{}^{203}HgNO_3$ and $^{203}Hg(NO_3)_2$ from pure water. Modified Westöö (1967) method

| | | ^{203}Hg activity at various stages of extraction as percentage of | |
		$CH_3{}^{203}Hg$ activity added	^{203}Hg activity added
Extraction:			
1. Working process	1 benzene extract	63%	0.47%
	2 benzene extract	25%	0.27%
	3 benzene extract	10%	0.14%
	1 + 2 + 3 yield	98%	0.88%
2. Working process	1 cysteine extract	60%	0.47%
	2 cysteine extract	30%	0.12%
	1 + 2 yield	90%	0.59%
3. Working process	benzene extract	21 ± 1%	0.21 ± 0.04%

inorganic mercury that may be extracted or carried in water droplets through the extraction procedure from a pure water solution to the last benzene face. The average yields of duplicate determinations are presented in Table 3.

Certain types of organic material present in some muds (Nos. 6 and 9) and sludges (Nos. 10 and 11) make the extraction procedure according to the Westöö method very difficult due to the formation of interfering emulsions. Centrifugation at 5,000 rpm was not effective in reducing the emulsion in the case of the above samples.

Radioactivity determinations, method sensitivity and identification of methyl mercury

Radioactivity determinations were carried out by counting 10-ml samples in a NaJ(Tl) $1\frac{3}{4}'' \times 2''$ well-type crystal (40.2 mm deep, 25.4 mm in diameter) (Model SCDA 2, Wallac Oy, Turku, Finland). Pulses were directed into a Wallac Model AS-11 one-channel analyzer. The efficiency was 580 cpm per nCi for the photopeak of ^{203}Hg (0.279 MeV), using the energy range 0.230–0.330 MeV. The corresponding background was 300 cpm.

Sensitivity, when defined as equal to the background count, was 0.5 nCi. With 1–1.7 nCi added and about 15 per cent extraction yield, a methylation of about 0.4 per cent of the mercury added could thus have been detected if the separation of inorganic mercury had been complete. However, since 0.2 per cent of the inorganic mercury may be carried through the extractions (Table 3), the true sensitivity is somewhat less. It was concluded that

finding 0.6 per cent or more of the added activity in the final extract is a positive indication of methylation.

In all samples the presence of methyl mercury was also confirmed by thin layer chromatography and autoradiography. This technique has been described elsewhere (Miettinen et al., 1971).

Results and discussion

The results of the methylation experiments, corrected for loss due to the extraction procedure and for small amounts of inorganic mercury carried through the extractions, are presented in Table 4. Mud sterilized by gamma-radiation did not methylate inorganic mercury under similar conditions.

Inorganic mercury was methylated to about the same extent by different bottom muds at +20°C under both aerobic and anaerobic conditions. Only in bottom muds taken from a relatively low depth in the sea (Nos. 1 and 2) was methylation stronger under aerobic conditions, while stronger methylation under anaerobic conditions was found only in the upper sediment layer of the polluted lake (No. 6a). However, all these results are based only on single determinations. The methylation percentage (1–14 per cent in two weeks) for different muds seems to depend on the type of the mud, being usually greater in muds rich in organic substances where more microbes able to methylate mercury are present.

At 12°C, no methylation was found under aerobic conditions, but under anaerobic conditions some methylation was found in muds that were rich in organic

TABLE 4. METHYLATION OF INORGANIC MERCURY (2+) AS PERCENTAGE OF THE ORIGINAL ^{203}Hg ADDED AT DIFFERENT CONDITIONS. RESULTS ARE CORRECTED FOR LOSS DUE TO EXTRACTION PROCEDURE AND FOR SMALL AMOUNT OF INORGANIC MERCURY CARRIED THROUGH EXTRACTIONS

No.	air agitation 19–20°C (10d)	12°C (14d)	N$_2$-agitation 19–20°C (13d)	12°C (14d)	Glass ampuls, saturated before sealing with air 23°C (14d)	N$_2$ 20°C (14d)
1	6%		1%			
2	10%	0	5%	0	2%	2%
3	10%		10%			
4	5%		7%			
5	7%		9%			
6a	4%		9%			
6b	6%		6%			
7	6%	0	6%	0	0.5%	0.4%
8	5%		7%			
9a	11%	0	14%	3%	8%	6%
9b		0		2%	3%	4%
10		0		1%	2%	3%
11		1%		0.6%	5%	1%

material (Nos. 9a and 9b). Therefore, the methylation of mercury in Finnish waters seems to be quite weak since the mean annual temperature of the waters is approximately 5°C.

Formation of volatile organic mercury compounds was not found. This is due to the fact that dimethyl mercury, which may have been formed by micro-organisms, is unstable at the pH of these mud suspensions (pH 5) (Olsson, 1969; Jernelöv, 1969).

Methylation was found to be weaker at the same temperature in the sealed glass ampuls than in the suspensions kept in continuous agitation. Aeration of the mud suspensions by either air or N$_2$ gas before sealing the ampuls did not have any effect on the methylation degree.

The methylation of mercury by active sludge and "incubation sludge" was weak in comparison to that of the natural muds, However, the experimental conditions differed significantly from those of the sewage treatment plant.

The aeration of the mud suspensions by either air or N$_2$ gas before the addition of Hg may have had some effect on the result of this investigation, since a portion of the free H$_2$S may have been carried away by the gas flow. H$_2$S reacts with inorganic mercury to form HgS, which is very insoluble. To a degree, micro-organisms are able to methylate HgS under aerobic conditions (Fagerström and Jernelöv, 1971). According to Jernelöv (1969), however, there is no turnover of HgS to the methyl form under anaerobic conditions.

References

ACKEFORS, H Recent advances in mercury investigations with
1969 special reference to fish. ICES C.M. 1969/E:12:4 p. (mimeo).
FAGERSTRÖM, T and JERNELÖV, A Formation of methyl mercury
1971 from pure mercuric sulphide in aerobic organic sediment. Water Res., 5:121–2.
GRANT, N Legacy of the Mad Hatter. Environment, 11:18–23.
1969
JENSEN, S and JERNELÖV, A Biological methylation of mercury in
1969 aquatic organisms. Nature, Lond., 223(5207):753–4.
JERNELÖV, A Laboratory experiments concerning the turnover of
1968 mercury in various chemical states in the environment. Vatten, 24:53–6 (in Swedish).
JERNELÖV, A Turnover of mercury in nature and what we can do
1969 to influence it. Nord. hyg. Tidskr., 50:174–8 (in Swedish).
KITAMURA, S Synthesis and decomposition of organic mercury
1969 compounds by bacteria. Jap. J. Hyg., 24:132–3 (in Japanese).
MCARTHUR, I A A method for determining low concentrations of
1952 oxygen in gases. J. appl. Chem., 2:91–6.
MIETTINEN, J K, et al. Elimination of ^{203}Hg-methyl mercury in
1971 man. Ann. clin. Res., 3:116–22.
OLSSON, M Mercury evaporation from the lakes. Nord. hyg.
1969 Tidskr., 50:179 (in Swedish).
WESTÖÖ, G Determination of methyl mercury compounds in
1967 foodstuff. 2. Determination of methyl mercury in fish, egg, meat and liver. Acta chem. Scand., 21:1790–800.
WOOD, J M, et al. Synthesis of methyl mercury compounds by
1968 extract of a methanogenic bacterium. Nature, Lond., 220:173–4.

Mercury as a Hydrospheric Pollutant.
I. Accumulation and Excretion of ^{203}HgCl$_2$ in *Tapes decussatus* L.†

*M. Y. Ünlü, M. Heyraud and S. Keckes**

Le mercure en tant que polluant hydrosphérique

I. Accumulation et excrétion du ^{203}HgCl$_2$ chez *Tapes decussatus* L.

Quelques aspects du métabolisme du mercure chez *Tapes decussatus* L. liés au danger potentiel présenté par la pollution de l'environnement marin par le mercure sont décrits. Le mercure ^{203}Hg a été utilisé sous forme de chlorure.

Le mercure est accumulé rapidement par les *Tapes* à partir de l'eau de mer et en un jour les animaux contiennent dix fois plus de mercure par unité de poids que leur environnement. Lorsque l'exposition est plus longue, l'accumulation du mercure est considérablement ralentie. Les organes viscéraux contiennent plus de la moitié du mercure accumulé en 14 jours, tandis que les

El mercurio como contaminante hidrosferico;

1. Acumulación y excreción de ^{203}HgCl$_2$ en *Tapes decussatus* L.

Se describen algunos aspectos del metabolismo del mercurio en *Tapes decussatus* L., en relación con el peligro potencial de la contaminación por el mercurio del medio ambiente marino. El mercurio se empleó como cloruro ^{203}Hg.

El mercurio procedente del agua del mar es acumulado rápidamente por *Tapes* y en un día los animales contienen diez veces más mercurio por peso unitario que su ambiente. Durante exposiciones más largas la acumulación de mercurio disminuye considerablemente. Los órganos viscerales contienen más de la mitad del mercurio acumulado durante 14 días, mientras que los músculos

* Laboratory of Marine Radioactivity, International Atomic Energy Agency, Monaco, Principality of Monaco.
† Investigations supported by International Atomic Energy Agency (Grant no. 702/RB) and Nordforsk.

muscles contiennent moins de 2 pour cent du mercure total accumulé.

La vitesse de perte du mercure accumulé ne dépend pas de la durée d'exposition des animaux à l'élément, mais dépend fortement de son mode de pénétration. La rétention du mercure accumulé est la plus forte après qu'il ait été introduit dans les *Tapes* par injection dans le muscle du pied; environ 25 pour cent seulement du mercure injecté est perdu en deux mois, surtout pendant les six premiers jours. La rétention du mercure accumulé à partir de l'eau de mer ou par l'intermédiaire de la chaîne alimentaire n'est pas très forte et la demi-vie biologique globale est d'environ 10 et 5 jours respectivement.

contienen menos del 2 por ciento de todo el mercurio acumulado.

La tasa de desintegración del mercurio acumulado no depende del tiempo en que el animal esté expuesto a él, sino que depende grandemente de la vía de penetración. La retención del mercurio acumulado es más alta cuando éste se introdujo en *Tapes* mediante inyecciones en el músculo abductor; sólo el 25 por ciento del mercurio inyectado se pierde dentro de un plazo de dos meses, y en su mayor parte en los seis primeros días. La retención del mercurio acumulado procedente del agua del mar o que llega a través de la cadena alimentaria no es demasiado fuerte y posee un período biológico general de semieliminación de unos 10 y 5 días respectivamente.

THE accumulation and excretion of mercury by the bivalve *Tapes* has been studied because it is a commercially available mollusc often used as human food and may therefore contribute to the mercury accumulation in man to a dangerous level.

Materials and methods

The animals, *Tapes decussatus* L., used in these experiments were obtained commercially from La Cooperative Maritime "Le Dauphin", Sète, France, and were kept in the laboratory for several days in running sea water under conditions close to those to which they were subsequently exposed during the experiments. The surface of the animals was cleaned of epibionts by gentle brushing. The animals were about 4.5×3.0 cm, weight about 11.7 g, and their age was estimated as 10–12 months.

The sea water for the experiments, where the accumulation of ^{203}Hg was tested, was collected with all-plastic samplers about 5 mi off the coast of Monaco, stored in polyethylene bottles and filtered through cotton before use. The salinity of this water was 37.5‰ and the temperature was kept at $15° \pm 1°C$. The loss experiments were done in running sea water from the laboratory supply system with temperature variations from 20°C–23°C.

The salinity, pH and temperature of the sea water in the experimental basin was determined routinely (Strickland and Parsons 1968).

In all experiments ^{203}Hg of high specific activity (Hg–203–S–Z, in form of $HgCl_2$, 1.45 mCi/mg),

obtained from the Commissariat à l'Energie Atomique de France through the courtesy of the Centre Scientifique de Monaco, was employed. ^{203}Hg has a relatively short half-life (47 days) and decays to stable ^{203}Tl emitting both beta (0.205 MeV) and gamma (0.279 MeV) radiations (Strominger *et al.*, 1958).

The accumulation of mercury by animals directly from sea water was tested in polyethylene basins filled with 2–5 l of aerated sea water into which ^{203}HgCl$_2$ was added.

The retention of ^{203}Hg in *Tapes* was followed in animals after it was accumulated or introduced into their bodies from sea water, from food or by injection. The results of the radiometric determinations were converted into percentage of the animals' specific activities at the beginning of the loss experiments.

As radioactive food, non-bacteria-free cultures of *Phaeodactylum* were used. In the logarithmic phase of the growth, 10.0–16.6 nCi ^{203}Hg was added per ml of the culture suspension. Four to five days after ^{203}Hg was

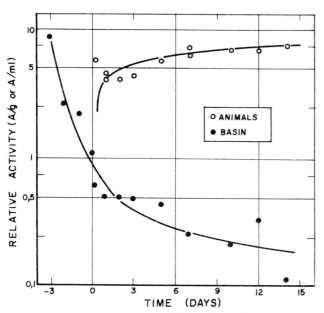

Fig 1. *Accumulation of mercury, added as* ^{203}HgCl$_2$ *to sea water, by eight whole live* Tapes. *Each dot is the mean of eight animals*

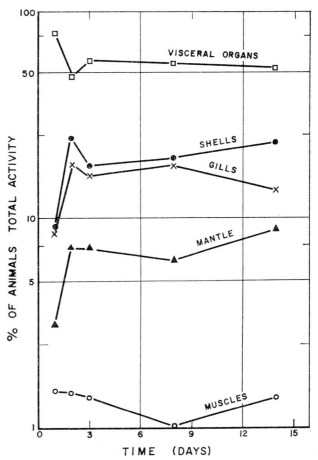

Fig 2. *Relative distribution of mercury in organs and tissues of* Tapes *during its accumulation from sea water containing* ^{203}HgCl$_2$. *Each mark is the mean of six animals*

added to the culture flask, the cells were washed by centrifugation (usually three times) and resuspended in the non-radioactive sea water in to which the animals were introduced. After 4-5 days' exposure more than 10 per cent of the ^{203}Hg was bound to *Phaeodactylum* at about 6×10^8 cells/l. The distribution of ^{203}Hg between the cells, i.e. particles, and the sea water in which they were suspended, was controlled by filtration.

The particulate fraction of ^{203}Hg is defined as the fraction of ^{203}Hg retained specifically on the upper filter after filtering basin samples through double layer Millipore filters. Usually 0.45 μ pore size filters were used.

The injection of ^{203}Hg solution ($\sim$10 nCi in 0.2 ml of sea water or $\sim$20 nCi in 0.1 ml of sea water) was made in the foot muscle of *Tapes* through a fine hypodermic needle.

Radiometric determination of the ^{203}Hg content of the whole live animals, their tissue and organs after dissection, the basin samples and in the particulate fraction, was done with one channel analysers (CEA, Model ECT 32 and SAIP with Mecasserto automatic sample changer) connected to well type NaJ/Tl scintillation probes (Harshaw 3 × 3 in and Quartz et Silice $1\frac{1}{2} \times 1\frac{1}{2}$ in). Using an internal standard, corrections were made for the different counting geometry, decay of ^{203}Hg and sensitivity drift of the instruments during the experiment. The counting error was estimated to be below 5 per cent. The results in the figures are shown as mean values.

Results and discussions

The accumulation of mercury added as ^{203}HgCl$_2$ to the sea water was tested with 6 and 8 animals in two experiments. In both, there was rapid initial uptake and within one day the animals' mercury content per unit weight was about 10 times that in the surrounding water (fig 1). Uptake in subsequent days was much less probably because, although 85 per cent of the mercury was initially in particulate form, there was rapid absorption to the walls of the experimental vessels, loss by evaporation and removal by the animals.

No toxic effects of mercury were observed.

The concentration of mercury in tissues and organs of *Tapes* was studied by dissecting 30 animals at various times from the start of the experiment. Gills and viscera had the highest mercury content (fig 2). The relatively low concentration in structures with a large but inactive surface exposed to the sea water indicates that most of the mercury is not accumulated by passive surface absorption. There is little change in the relative distribution of mercury after the third day.

Loss rate of mercury in the uptake period

The duration of the accumulation can influence the loss rate of some compounds (Keckes *et al.* 1968) and give different values for their biological half-time. To investigate this phenomenon three groups of 5–8 animals each were exposed to ^{203}HgCl$_2$ in the basin for 1, 7 and 14 days and afterwards the loss of the accumulated mercury was tested under similar conditions.

The results did not show significant differences in the three experimental groups. The biological half-time was estimated to be between 7 and 10 days.

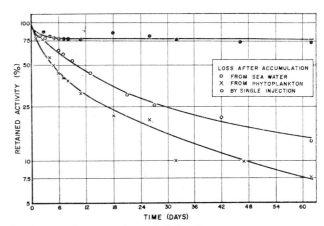

Fig 3. Loss of mercury from whole live Tapes *in running sea water after short uptake from sea water* (○), *phytoplankton* (×) *and after single injection into the foot muscle* (●). *Each mark is the mean of 4–10 animals*

To investigate the influence of the accumulation route of mercury on its retention, the loss of mercury was compared in animals contaminated through the food chain, from sea water and from single injection (fig 3).

The loss of mercury after its accumulation from plankton and sea water showed a similar pattern indicating a continuous loss lasting longer than 2 months. The similarity of the loss curves is understandable because in both groups the mercury was mainly accumulated in particulate form, most probably through the digestive tract.

There is much stronger retention of mercury after it is introduced into the animals by single injection into the foot muscle. Until the sixth day there is a considerable release, but after that the total amount diminishes only slightly.

The overall biological half-time for mercury estimated from these experiments is about 5 days for animals fed with phytoplankton containing mercury, about 10 days for mercury accumulated from sea water and very long for mercury solution injected into the foot muscle. Miettinen *et al.* (1970) give 481 ± 40 days as the biological half-time for the elimination of mercury injected into the foot muscle of *Tapes* as CH$_3$ ^{203}HgNO$_3$ instead of as ^{203}HgCl$_2$, as in these experiments.

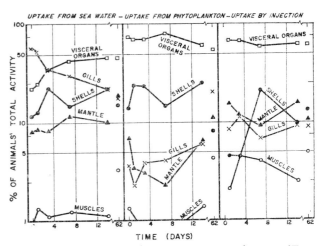

Fig 4. Relative distribution of mercury in organs and tissues of Tapes *during its loss in running sea water after short uptake from sea water, phytoplankton and after single injection into the foot muscle. Each mark is the mean of four animals*

[294]

The distribution of mercury in organs and tissues of *Tapes* during the loss experiment reflects the method of its accumulation also (fig 4).

In all three groups the highest amount of mercury is retained by the visceral organs. The shells of animals injected with mercury show initially only a smaller amount of mercury, but due to its redistribution during the loss experiment after the seventh day, the amount of mercury found in the shells is comparable with that in mantle and gills. Muscles, for obvious reasons, have much more mercury in animals injected with it than in those where mercury was accumulated from sea water or phytoplankton. Relatively higher amounts of mercury were found in the gills and mantle of the animals where it was accumulated from sea water than in those which accumulated it from phytoplankton. This indicates that although in both cases the mercury was most probably accumulated as particulate, the nature of the particles associated with mercury is important.

References

KECKES, S, OZRETIC, B and KRAJNOVIC, M   Loss of Zn^{65} in the
1968   mussel *Mytilus galloprovincialis*. *Malacologia*, 7:1–6.
MIETTINEN, J K, HEYRAUD, M and KECKES, S   Mercury as hydro-
1970   spheric pollutant. II. Biological half-time of methyl mercury in four Mediterranean species: a fish, a crab, and two molluscs. Paper presented to FAO Technical Conference on Marine Pollution and its Effects on Living Resources and Fishing, Rome, Italy, 9–18 December 1970, FIR: MP/70/E-90:8 p.
STRICKLAND, J D H and PARSONS, T R   A practical handbook of
1968   seawater analysis. *Bull. Fish. Res. Bd Can.*, (167):311 p.
STROMINGER, D, HOLLANDER, J M and SEABORG, G T   Table of
1958   isotopes. *Rev. Mod. Phys.*, 30:585–904.

Mercury as a Hydrospheric Pollutant. II. Biological Half-time of Methyl Mercury in Four Mediterranean Species: a Fish, a Crab, and Two Molluscs†

*J. K. Miettinen, M. Heyraud and S. Keckes**

Le mercure en tant que polluant hydrosphérique. II. Demi-vie biologique du méthylmercure dans quatre espèces méditerranéennes: un poisson, un crabe et deux mollusques

L'analyse des vitesse d'excrétion du méthylmercure introduit de différentes façons dans quatre organismes marins (*Serranus scriba* (L.)., *Carcinus maenas* L. *Tapes decussatus* L. et *Mytilus galloprovincialis* Lam.) a montré qu'une importante fraction du composé administré à une vitesse de perte lente, avec une période biologique allant de 267 jours (*Serranus*) à plus de 1 000 jours (*Mytilus*). Ces résultats ont été comparés à ceux obtenus avec d'autres crustacés, mollusques, poissons et un mammifère, soit d'eau douce, d'eau saumâtre ou d'eau de mer. On a conclu que des espèces liées phylogénétiquement ont un mode d'excrétion du méthylmercure similaire, avec une demi-vie biologique qui dépend de la température et de la voie de pénétration du méthylmercure dans les organismes, cette demi-vie étant plus longue après injection intramusculaire qu'après administration pérorale.

El mercurio como contaminante hidrosferico. II. Período biológico del metilmercurio de semieliminación en cuatro especies mediterraneas: un pez, un cangrejo y dos moluscos

El análisis de las tasas de excreción de metilmercurio introducido con distintos métodos en cuatro organismos marinos (*Serranus scriba* (L.)., *Carcinus maenas* L., *Tapes decussatus* L. y *Mytilus galloprovincialis* Lam.) indican que una amplia fracción del compuesto administrado posee una tasa de desintegración con un período biológico de semieliminación de 267 días (*Serranus*) a más de 1 000 días (*Mytilus*). Estos resultados se compararon con los obtenidos en otros cangrejos de aguas dulces, salobres y marinas, en moluscos, peces y en un mamífero. Se llegó a la conclusión de que las especies filogenéticamente afines siguen un tipo similar de excreción de metilmercurio. El período biológico de semieliminación depende de la temperatura y del modo en que el metilmercurio penetra en los organismos, siendo más prolongado después de una inyección intramuscular que después de su administración por vía oral.

MAN's use of mercury in industry and agriculture is increasing. It is estimated that half the mercury used (5,000 tons) reaches the oceans and a similar amount enters the sea as a result of erosion (Goldberg 1970). In at least two areas, Japan and Scandinavia, the industrial release of mercury has resulted in severe pollution of the environment and fatal accidents in two areas, Minamata Bay (Kurland 1960; Irukayama 1967; Study Group on Minamata Disease 1968; Ui 1971) and the Agano river district (Japan 1967; Ui 1969, 1971) in which at least 48 people died and several hundred were injured.

No human casualties have yet been reported in Scandinavia, but fishing is banned in more than 40 Swedish lakes and rivers because the mercury levels in fish, especially pike, perch and burbot, exceed 1 mg/kg fresh weight (Johnels *et al.* 1967). This level is exceeded in at least five large watercourses in Finland (Häsänen and Sjöblom 1968; Sjöblom and Häsänen 1969). In Norway 1 ppm or more is found in pike at least in the country's largest lake, Mjösa (Underdahl 1969) and in cod in some coastal waters (Anon. 1968). Recently (Anon. 1970, 1970a) the Canadian and U.S. authorities banned fishing in Lake Erie, Lake St. Clair and St. Clair River, after it

was found that the fish caught there contained up to 7 mg of Hg/kg of wet weight, far above the U.S. and Canadian standard of 0.5 mg/kg.

In most cases the mercury compound was not the metallic or inorganic form originally released, but the organic methyl mercury form which is more readily absorbed and retained and is more toxic than the inorganic mercury forms (Swensson *et al.* 1959, 1959a; Löfroth 1969). At Minamata about 5 per cent of the mercury released was in methylated form (Sebe *et al.* 1967) but in Scandinavia the industrial releases—from chlorine, alkali, and cellulose industries—have been either in the metallic or inorganic form, or the organic phenyl mercury form (Swedish Royal Commission on Natural Resources 1967). In Sweden methyl mercury was used for seed dressing until 1966, but loss from soil into watercourses can be considered insignificant (Swedish Royal Commission on Natural Resources 1967). The only practical source for methyl mercury in biota is the formation of this compound in nature by methylation of some other forms, as was demonstrated in bottom mud as a result of microbial activity (Jensen and Jernelöv 1969; Kitamura *et al.* 1969).

In spite of many investigations on the mercury prob-

---

* Laboratory of Marine Radioactivity, International Atomic Energy Agency, Monaco, Principality of Monaco.
† Investigations supported by International Atomic Energy Agency (Grant no. 702/RB).

lem, especially in Japan and Scandinavia, very little has been published on the metabolism of mercury and its derivatives in water organisms. The studies of Hannerz (1968), Jernelöv and Martin (1969), Bäckström (1969), Miettinen et al. (1969) and Tillander et al. (1969) are exceptions. The biological half-lives of ^{203}Hg-labelled methyl mercury were determined in fish, crabs and molluscs in fresh water and in fish in brackish water of the Gulf of Finland (salinity about $5^0/_{00}$). It was found that methyl mercury, administered orally in the form of $CH_3^{203}HgNO_3$, followed a two-exponential rate of excretion. A small proportion of the mercury, usually 5 to 15 per cent had a half-life of a few days, whereas the rest was excreted extremely slowly, with a half-life from 50 to 500 days, depending on the species.

For ecological reasons it is important to know whether the excretion of methyl mercury in marine species follows similar routes and rates as in parallel species from brackish and fresh waters. Four analogous species, common in the Mediterranean, were chosen and the excretion rate of methyl mercury was determined using a similar technique as in the earlier studies in Finland (Miettinen et al. 1969; Tillander et al. 1969).

Materials and methods

The experimental animals used were *Serranus scriba* (L.), *Carcinus maenas* L. *Tapes decussatus* L. and *Mytilus galloprovincialis* Lam., all of which are common in the Ligurian Sea. Prior to the experiments the animals were acclimatized to laboratory conditions for several days and marked for identification: fish by clipping their fins, the crabs and molluscs by painting numbers on their surfaces.

The fish were between 7 and 12 cm long and were about 4 years old. The crabs weighed about 60 g, their thickness being 20 to 30 mm. The age of the molluscs was probably 2 years; *Tapes* measured about 30×45 mm and weighed 12 ± 2 g while *Mytilus* measured 30×55 mm.

The compound used for the labelling of fish was $CH_3^{203}HgNO_3$ (spec. activity 0.14 mCi/mg of Hg) dissolved in water containing about 0.4 per cent of agar agar to give 5 μCi/ml. About 0.2 ml of this solution (1 μCi) was introduced *per os* into the stomach of the fish by a small syringe provided with a plastic cannula on the needle. The fish often vomited part of the labelling solution immediately after its administration and were therefore kept for a few minutes in a water container before transfer into the experimental aquarium.

The crabs were labelled by injecting 0.2 ml of 4 per cent sterile NaCl solution containing 1 μCi of $CH_3^{203}HgNO_3$ into the haemolymph through the soft cuticle of the proximal joint of the first leg and were transferred immediately into the experimental aquarium.

Another group of crabs were labelled by feeding for 24 h on *Tapes* which had been made radioactive by keeping them in sea water containing $CH_3^{203}HgNO_3$ for 48 h. During the feeding with radioactive *Tapes* the crabs were kept out of water to avoid accumulation of ^{203}Hg from sources other than food. After feeding the crabs were transferred to an experimental basin with running sea water.

The molluscs were labelled by injecting 0.1 ml of 4 per cent sterile NaCl solution containing 1 μCi of $CH_3^{203}Hg$-NO_3 into the foot muscle between the slightly open shells.

They were then kept on a dry board for 4 h and later transferred into the experimental aquarium.

The labelled animals were kept in open air aquaria protected from direct sunshine and provided with running sea water (3 l/min). The salinity of the sea water used was about $37.5^0/_{00}$ and the temperature roughly followed the temperature of the sea, being about 23°C at the beginning of the experiments (August 1968) and about 17°C at their end (December 1968). The experiment with crabs labelled through food was carried out in June to August 1970 at temperatures from 22°C to 23°C.

Whole body counting of individual live animals was carried out by two separate scintillation counters using the 0.279 MeV gamma energy of ^{203}Hg. The counter used for fish and crabs had an analyser with lower discriminator and a flat 3×4 in NaI(Tl) crystal. The counter used for molluscs had a 3×3 in NaJ(Tl) well-type Harshaw crystal and a CEA, Mod ECT 32 one-channel analyser.

During the counting the fish was held in a plastic box filled with sea water and foam rubber pads which kept the centre of fish 230 ± 5 mm below the surface of the crystal. A standard, 1.5 μCi of ^{203}Hg in 1 ml in a closed glass ampoule in this position, gave 26,700 net cpm/1 μCi. Fish labelled with 1 μCi at the first counting gave an average 22,000 net counts per minute. The individual results varied from 8,000 to 24,000 net cpm, depending on the extent of vomiting and the size of the fish. The background was about 3,700 cpm. The initial counting time used was 1 min. The crabs were counted at a distance of 100 to 103 mm between the midpoint of the animal and the surface of the crystal. The animals were placed without water in a small petri dish with a net cover which did not allow the animals to move more than a few mm. The standard used gave in this closer position 76,000 net cpm/1μCi, while the injected crabs gave initially on average 66,000 net cpm.

The molluscs were counted in plastic tubes inserted in the crystal's 36×65 mm well. A standard, 250 nCi of ^{203}Hg in about 1 ml, gave 240,000 cpm, i.e. the efficiency was 43.4 per cent. Dead time loss at this counting rate was 0.7 per cent. In both counting systems the counting error through the experiments was estimated to remain below ± 2 per cent.

The standards were counted before and after each series of animal countings and the animal's counts were appropriately corrected for decay of the isotope and fluctuations in sensitivity of the counting systems. The intervals between labelling and first counting were for crabs $\frac{1}{2}$ to 1 h, fish 1 to 2 h and molluscs 4 h. A few nonactive controls were introduced into the aquaria and the level of ^{203}Hg in these animals was used for correction of the labelled animal's activity due to ^{203}Hg recirculation within the aquaria. In all experiments only the results for healthy animals were taken into account. The half-lives were evaluated separately for each animal and finally expressed as mean half-life with the corresponding error of the mean.

Results and discussion

In analysing the loss of the ^{203}Hg a rapid initial loss of a small amount of ^{203}Hg can be distinguished (fast component) in all organisms. In *Serranus* and *Carcinus* the "fast component" was only 4 to 6 per cent of the total ^{203}Hg

[296]

Species	Method of administration	Duration of experiment (days)	Number of animals in experiment at the		Slow component Biological half-life (days)	% of the administered dose
			beginning	end		
Serranus scriba	*per os*	60	22	16	267 ± 27	96
Carcinus maenas	injection into haemolymph	35	19	9	400 ± 50	94
Carcinus maenas	*per os*		15			96
Tapes decussatus	injection into foot muscle	99	15	9	481 ± 40	78
Mytilus galloprovincialis	injection into foot muscle	98	14	9	1,000	80

initially present, but in *Tapes* and *Mytilus* it was about 20 per cent of the administered dose. The half-life of the "fast component" was in all animals estimated to be of only a few hours.

Most of the ²⁰³Hg was excreted relatively slowly and it is the rate of this fraction which actually determines the overall biological half-life. For technical reasons the loss rate was not followed for more than 99 days, a fraction of the biological half-life, but the extrapolations appear to be justified by the consistency of the results. Table 1 summarizes the most typical results which are also illustrated in fig 1.

Although the slow component is about 20 per cent of the injected dose in both molluscs used, the biological half-life of ²⁰³Hg in *Mytilus* is double that for *Tapes*. These species belong to two different orders of bivalve molluscs (*Anisomyaria* and *Eulamellibranchiata*) and might therefore have different physiological processes. In earlier studies (Miettinen *et al.* 1969; Tillander *et al.*

1969) where methyl mercury was injected into a freshwater bivalve, *Pseudanodonta complanata* L., the half-life varied from 82 days in small animals to 400 days in larger ones. *Pseudanodonta* and *Tapes* belong to different sub-orders (*Schizodonta* and *Heterodonta*) of the same order (*Eulamellibranchiata*) but their half-lives are quite comparable. In experiments with *Tapes* on the excretion rate of ²⁰³Hg injected as mercury chloride into the foot muscle, about 100 days were estimated as the biological half-life (Ünlü *et al.* 1970). This is a considerably shorter time than that obtained with methyl mercury and could be attributed to the higher lipid solubility of the latter compound.

Comparison of the biological half-life for ²⁰³Hg given *per os* or injected as methyl mercury into the haemolymph of *Carcinus* and given *per os* to *Astacus fluviatilis* Fabr. (Tillander *et al.* 1969; Miettinen *et al.* 1969) shows a much longer half-life for *Carcinus* (400 ± 50 days) than for *Astacus* (144 ± 37 days). By extrapolation from experiments with *Tapes* (Ünlü *et al.* 1970) one can expect that the method of ²⁰³Hg administration favoured its quicker excretion in *Astacus* compared to *Carcinus*. The half-life of about 300 days obtained in only one *Astacus* after intramuscular injection of methyl mercury seems to support this explanation. Although *Carcinus* and *Astacus* belong to different sub-orders of the order (*Decapoda*) this does not seem to have much influence on their excretion rate of methyl mercury.

The biological half-life obtained for *Serranus* (267 ± 27 days) is similar to that for a closely related species, *Perca fluviatilis* L. (110–470 days) in brackish waters of the Gulf of Finland (Tillander *et al.* 1969; Miettinen *et al.* 1969). It seems that neither the salinity (37.5⁰/₀₀ for *Serranus* and 5⁰/₀₀ for *Perca*) nor the relatively small phylogenetic distance (they belong to two families of the same order *Perciformes*) influenced the excretion rate of methyl mercury.

These comparisons indicate that the excretion rates of methyl mercury in evolutionary related fresh water, brackish and marine organisms follow a similar pattern. Generally, most of the administered mercury has a slow loss rate. The biological half-life depends on the method of entry into the organism, being longer after intramuscular injection than after oral administration. The loss is temperature dependent. Recent studies (Tillander *et al.* 1970) on the ringed seal (*Pusa hispida* Schreb.) showed that marine mammals are no exception.

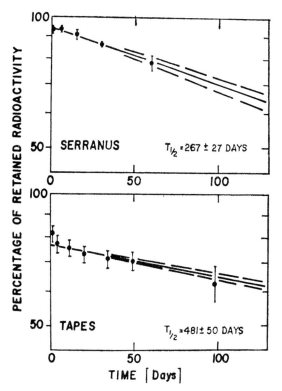

Fig 1. Biological half-life of ²⁰³Hg administered as methyl mercury nitrate to Tapes *by injection into the foot muscle and* per os *to* Serranus. *The dots are the means from 9 (*Tapes*) and 16 (*Serranus*) animals with bars indicating ± 1 standard deviation. The full and broken lines represent the mean loss rate ± 1 standard deviation as determined from individual half-life of each animal separately. All data refer to the slow component of 16°–23°C environmental temperature and are corrected for the physical decay of ²⁰³Hg.*

References

BÄCKSTRÖM, J Distribution studies of mercuric pesticides in quail
1969   and some fresh-water fishes. *Acta pharmac. tox.*, 27, Suppl. 3:1–103.

GOLDBERG, E D *In* McGraw-Hill Yearbook of Science and Tech-
1970 nology 1970, New York, McGraw-Hill Book Co.
HANNERZ, L Experimental investigations on the accumulation of
1968 mercury in water organisms. *Rep. Inst. freshwat. Res.
Drottningholm*, (48):120–76.
HÄSÄNEN, E and SJÖBLOM, V Mercury content of fish in Finland
1968 in 1967. *Suom. Kalatal.*, (36):24 p. (in Finnish, English
summary).
IRUKAYAMA, K The pollution of Minamata Bay and Minamata
1967 disease. *Adv. Wat. Pollut. Res.*, 3:153–80.
Japan, Ministry of Welfare, Tokyo, Special Research Group, Special
1967 report on cases of mercury poisoning in Niigata. (Translated
into Swedish, 1969).
JENSEN, S and JERNELÖV, A Biological methylation of mercury in
1969 aquatic organisms. *Nature, Lond.*, 223(5207):753–4.
JERNELÖV, A and MARTIN, A-L Mercury in sea water. *In* Report
1969 from Institutet för Vatten och Luftvardsforskning, Stockholm, Sweden, (in Swedish).
JOHNELS, A G *et al.* Pike (*Esox lucius* L.) and some other aquatic
1967 organisms in Sweden as indicators of mercury contamination in the environment. *Oikos*, 18:323–33.
KITAMURA, M, SUMINO, K and TAINA, M Synthesis and decompo-
1969 sition of organic mercury compounds by bacteria. *Jap. J.
Hyg.*, 24:132–3 (in Japanese).
KURLAND, L T, FARO, S N and SIEDLER, H Minamata disease. The
1960 outbreak of a neurologic disorder in Minamata, Japan, and
its relationship to the ingestion of seafood contaminated by
mercuric compounds. *Wld Neurol.*, 1:370–91.
LÖFROTH, G Methylmercury. A review of health hazards and side
1969 effects associated with the emission of mercury compounds
into natural systems. *Bull. ecol. Res. Comm., Stockh.*, (4):
38 p.
MIETTINEN, J K *et al.* Distribution and excretion rate of phenyl-
1969 and methylmercury nitrate in fish, mussels, molluscs and
crayfish. *In* Proceedings of the 9th Japan Conference on
Radioisotopes, Tokyo, Japan, Japan Atomic Industrial
Forum, Inc., pp. 474–8.
SEBE, E *et al.* A side reaction accompanied with catalytic hydration
1967 of acetylene. Pts 1 and 2. *Folia pharmac. Jap.*, 63:228–43,
244–60 (in Japanese, English abstracts).

SJÖBLOM, V and HÄSÄNEN, E Mercury content of fish in Finland.
1969 *Nord. Hyg. Tidskr.*, 50(2):37–53 (in Swedish).
Study Group on Minamata Disease, Minamata disease. Kumamoto,
1968 Japan, Kumamoto University, (English ed.).
SWENSSON, A, LUNDGREN, K D and LINDSTRÖM, O Distribution
1959 and excretion of mercury compounds after single injection.
A.M.A. Archs ind. Hlth, 20:432–44.
SWENSSON, A, LUNDGREN, K D and LINDSTRÖM, O Retention of
1959a various mercury compounds after subacute administration.
A.M.A. Archs ind. Hlth, 20:467–72.
Swedish Royal Commission on Natural Resources, The mercury
1967 problem. Symposium concerning mercury in the environ-
ment. Stockholm, 1966. *Oikos*, Suppl. 9:51 p.
TILLANDER, M, MIETTINEN, J K and KOIVISTO, I Excretion rate of
1970 methyl mercury in the seal (*Pusa hispida*). Paper presented
to FAO Technical Conference on Marine Pollution and its
Effects on Living Resources and Fishing, Rome, Italy,
December, 1970.
TILLANDER, M *et al.* The excretion by fish, mussel, mollusc and
1969 crayfish of methyl mercury nitrate and phenyl mercury
nitrate, introduced orally or injected into musculature.
Nord. Hyg. Tidskr., 50(2):181–3 (in Swedish).
Ui, J A short history of Minamata disease research and the present
1969 situation of mercury pollution in Japan. *Nord. Hyg. Tidskr.*,
50(2):139–46.
Ui, J Mercury pollution of sea and fresh water: its accumulation
1971 into water biomass. *Revue int. Océanogr. méd.*, 22-23:79-128.
UNDERDAHL, B Studies of mercury in some food stuffs. *Nord. Hyg.
1969 Tidskr.*, 50(2):60–3 (in Swedish).
ÜNLÜ, Y, HEYRAUD, M and KECKES, S Mercury as a hydrospheric
1970 pollutant. I. Accumulation and excretion of ^{203}Hg Cl_2 in
Tapes decussatus L. Paper presented to FAO Technical
Conference on Marine Pollution and its Effects on Living
Resources and Fishing, Rome, Italy, December, 1970.
ANON. Biocid-Information, 15/2. Nordforsk, Stockholm (in
1968 Swedish).
ANON. Mercury: wiping out an industry. *Chem. Engng News*,
1970 April 13.: 9
ANON. The Mad Hatter's legacy. *Newsweek*, April 20:26–7.
1970a

Preliminary Study on the Distribution and Effects of Two Chemical Forms of Methyl Mercury in Pike and Rainbow Trout

*V. Miettinen, E. Blankenstein, K. Rissanen,
M. Tillander, J. K. Miettinen*
and M. Valtonen†*

Etude préliminaire de la répartition et des effets de deux formes chimiques de méthylmercure chez le brochet et la truite arc-en-ciel

L'expérience dont il est fait mention a pour objet d'établir les éléments suivants:

(1) L'existence éventuelle d'une différence entre la forme ionique "libre" de méthylmercure et celle qui est liée aux protéines dans le cas de toxicité à l'égard du brochet.
(2) La distribution de la radioactivité après administration de ces deux composés de méthylmercure marqués au ^{203}Hg.
(3) Modifications pathologiques et histopathologiques des organes du brochet après administration orale des deux composés de méthylmercure.

La forme ionique "libre" a été appliquée en solution aqueuse de $CH_3{}^{203}HgNO_3$ avec entraîneur (0,5 μCi/mg Hg). La forme liée aux protéines a été préparée par addition d'homogénat de foie de vache frais à la solution aqueuse de $CH_3{}^{203}HgNO_3$ avec entraîneur et par incubation du mélange pendant 24 heures à température ambiante (22°C). Elle a été identifiée comme méthylmercure vrai lié aux protéines par un procédé d'extraction avec ou sans acidification par HCl concentré et extraction au benzène suivie de séparation par chromatographie sur couche mince.

L'administration a été effectuée par voie orale en trois ou quatre petites prises à intervalles de deux à trois jours. Cinq brochets (poids moyen 300 g) ont reçu du MeHg ionique en dose totale moyenne finale de 12,1 mg Hg/kg (retenu) et 5 brochets (poids moyen 340 g) ont reçu du MeHg lié aux protéines en dose totale moyenne finale de 24,7 mg Hg/kg. En outre 6 brochets témoins non radioactifs ont été conservés dans le même aquarium. La longévité moyenne pour les brochets auxquels on avait administré du MeHg ionique a atteint 33 jours; pour ceux auxquels on avait administré du MeHg lié aux protéines, cette longévité atteint 18 jours.

La valeur approximative de la $DL_{50/30j.}$ pour ces deux composés était de 15 ± 3 mg Hg/kg de poids frais. Les symptômes histo-

Estudio preliminar de la distribución y efectos de dos formas químicas de metilmercurio en el lucio y en la trucha arco iris

Este experimento se realizó para establecer:

(1) Cualquier posible diferencia entre la forma iónica "libre" del metilmercurio y la forma del metilmercurio ligado a las proteínas en la toxicidad del lucio.
(2) Distribución de la radiactividad después de administrar estos dos compuestos de metilmercurio marcados con ^{203}Hg.
(3) Cambios patológicos e histopatológicos en los órganos del lucio después de la administración oral de los compuestos de metilmercurio.

La forma iónica "libre" se aplicó como solución en agua de $CH_3{}^{203}HgNO_3$ con transportador (0,5 μCi/mg Hg). La forma ligada a las proteínas se hizo añadiendo homogenizado de hígado fresco de vaca a la solución acuosa de $CH_3{}^{203}HgNO_3$ con transportador e incubando la mezcla durante 24 horas a la temperatura ambiente (22°C). Se caracterizó como auténtico metilmercurio ligado a las proteínas mediante un procedimiento de extracción con o sin acidificación por HCl concentrado y extracción con benceno seguido de una separación efectuada con cromatografía de capas delgadas.

La administración se hizo oralmente, dando de tres a cuatro pequeñas dosis a intervalos de 2–3 días. A cinco lucios (peso medio de 300 g) se les administró MeHg iónico hasta una dosis media definitiva de 12,1 mg Hg/kg (retenido) y a cinco lucios (peso medio 340 g) hasta una dosis media definitiva de 24,7 mg Hg/kg y MeHg ligado a proteínas, respectivamente. Además, en el mismo acuario se tuvieron seis lucios testigos inactivos. La vida media de los lucios a los que se administró MeHg iónico fue de 33 días; a los que se les dio MeHg ligado a proteínas fue de 18 días.

El valor aproximado de $LD_{50/30d.}$ para ambos compuestos fue de 15 ± 3 mg Hg/kg de peso fresco. Los síntomas histopatológicos

---

* Department of Radiochemistry, University of Helsinki.   † State Veterinary Medical Institute, Helsinki, Finland.

pathologiques produits par ces composés de méthylmercure ont été analogues, les organes les plus lésés étant les reins et les pseudobranchies. On n'a constaté aucune différence notable dans le schéma de distribution des deux composés de méthylmercure.

Après administration orale de méthylmercure ionique en dose unique à sept truites arc-en-ciel (220 g), ce qui représente une charge corporelle moyenne de 11,9 mg Hg/kg, tous les sujets sont morts dans un délai de 1½ h à 2 heures. Après administration du même montant de méthylmercure ionique ou lié aux protéines en deux temps, avec un intervalle de 1 à 2 jours, 3 poissons sur 12 seulement sont morts dans un délai variant de 47 à 71 jours. Les autres sujets sont demeurés apparemment en bonne santé pendant au moins 94 jours. Lors de ces expériences, la température était de +18°C.

THE toxicity of organo-mercurials to fish has been investigated less than mercury ions. Its role as an industrial pollutant for fish and man was first indicated in connection with the tragic Minamata accident in Japan (Kurland *et al.*, 1960; Irukayama 1967).

Kitamura (1968) reported 10 to 20 mg Hg/kg fresh weight in disabled fish floating on the sea surface; the maximum value was 24.1 mg Hg/kg in muscle. Cataracts of the eye lense and torpid movements were characteristic symptoms. Additional observations on methyl toxicity to fish are reported by Hannerz (1968).

The distribution of methyl mercury in pike (*Esox lucius* L.) was studied by Ohmomo *et al.* (1969) and Bäckström (1969), and the distribution of ethyl mercury in rainbow trout (*Salmo gairdneri* Richardson) by Rucker and Amend (1969). Amend *et al.* (1970) studied the susceptibility of rainbow trout to the acute toxicity of ethyl mercury phosphate after exposing the fish for 1 h to 0.125 ppm of the compound in water. They found that the death rate increased with an increase in water temperature, a decrease in oxygen, and an increase in chloride ion. Calcium had no effect.

In Scandinavia and the U.S.A. the highest mercury concentrations are reported in pike; in Sweden 9.8 mg/kg fresh weight (Johnels *et al.*, 1967), in the U.S.A. 7.09 mg/kg (Anon., 1970), and in Finland 5.8 mg (Häsänen and Sjöblom, 1968). The amount lethal to pike was unknown, and we therefore carried out preliminary toxicity determinations of methyl mercury on pike and rainbow trout. Salmonoids are usually most sensitive to heavy metal poisoning. The specific purpose of these experiments was:

(1) to determine whether methyl mercury is more toxic in the free ionic form than when bound to SH-groups of proteins, as it is in the food chains in nature,

(2) to determine the approximate dose of methyl mercury that is lethal to pike and rainbow trout,

(3) to find the pathological changes, and the gross distribution of labelled methyl mercury in the fish.

MATERIAL AND METHODS

The fish

The fishes used were pike (*Esox lucius* L.), aged 2–4 years, weight 240–625 g, and rainbow trout (*Salmo gairdneri* Richardson) aged 2 years, weight 200–250 g at the beginning of the experiment. The pike were caught about 6 months before the experiment in Lake Suolijärvi, Kytäjä, about 50 km north of Helsinki, and the rainbow trout were obtained from the Kytäjä hatchery (Hyvinkää, Finland). The pike were fed on small fish, and the rainbow trout on cow liver. The fish acclimatized well to the

de estos compuestos de metilmercurio fueron similares. Los órganos más perjudicados fueron los riñones y las pseudobranquias. No se observó ninguna diferencia digna de mención en el tipo de distribución entre estos dos compuestos de metilmercurio.

Cuando se administró oralmente metilmercurio iónico, como única dosis, a siete truchas arco iris (peso medio de 220 g), representando esta una carga corporal promedia de 11,9 mg Hg/kg, todas murieron al cabo de una hora y media a dos horas. Cuando se administró la misma cantidad de metilmercurio iónico o de metilmercurio ligado a proteínas en dos partes con un intervalo de 1 a 2 días, de 12 peces solamente tres murieron después de los 47 a los 71 días. Los demás se mantuvieron sanos durante 94 días, por lo menos. La temperatura en estos experimentos fue de +18°C.

aquarium conditions, ate well, and no diseases occurred.

The fish were kept in a 2,000 l aquarium provided with good aeration and a continuous throughflow of tap water dechlorinated by active carbon. All experiments were conducted under fully aerated conditions. The tap water was soft; it contained 34.8 mg Ca/l, 11.2 mg Na/l, 4.7 mg K/l and less than 0.08 mg Cl/l. The water temperature was 10°C during the pike experiment and 18°C during the rainbow trout experiment.

Labelling

The methyl mercury compounds were administered orally directly into the stomachs of the fish with a plastic catheter in three or four portions with a 2-day interval. To facilitate handling, the fish were slightly anaesthetized by keeping them for about 1 min in water containing tricaine methanesulphonate (MS-222, Sandoz): 0.008 per cent for trout, 0.056 per cent for pike. Because the fish regurgitated part of the substance, the amount absorbed from each dose was monitored with a whole-body counter (see below). The planned dose of the ionic form was 10 mg Hg/kg fish, and of the protein-bound form 20 mg Hg/kg fish.

Five pike (mean weight 300 g) were given total doses of 9.6–19 mg Hg/kg fish (mean 13.7 mg) of the ionic form, and five pike (mean weight 330 g) were given 15–29.6 mg Hg/kg (mean 24.7 mg) of the protein-bound methyl mercury (absorbed amounts).

Seven rainbow trout (mean weight 220 g) were given 10.0–14.1 mg Hg/kg (mean 11.9 mg) of the ionic form as single doses (Table 3). All these fish died within 1.5–2 h (see below), therefore a second group (170 g) was given 11.0 mg Hg/kg (mean) as two half-doses. Similarly, to five rainbow trout (170 g) 11.0 mg Hg/kg body weight was given as protein-bound methyl mercury in two small doses. (Table 4).

[203]Hg-labelled methyl mercury nitrate (activity 2.8 mCi/ml, specific activity 0.5 mCi/mg Hg; Ab Atomenergi, Studsvik, Sweden), and methyl mercury hydroxide as stable mercury (Casco Ab, Sweden) neutralized with 1-N nitric acid, were used to prepare stock solutions.

The solution of the "free" ionic form of methyl mercury contained 2.23 µCi [203]Hg and 4.45 mg Hg/g of diluted solution and 2 per cent starch to make it slightly viscous. A small amount of natural blueberry dye was added to facilitate the observation of vomiting. The protein-bound methyl mercury was prepared by incubating the stock solution of the ionic form with fresh cow liver suspension for 24 h at +20°C. The combination of methyl mercury with protein and its stability were examined by extraction at pH 1 with benzene (Westöö 1966, 1967) and by separation with thin layer chromatography (Miettinen *et al.*, 1969). Each gram of the suspension of the protein-bound methyl mercury contained 5.4 μCi ^{203}Hg and 10.75 mg Hg.

Radioactivity determinations

The retention of the mercury was measured with a whole-body counter by placing the fish in a plastic box in front of a NaJ (T1) crystal of 3 × 3 in (7.6 × 7.6 cm) (Harshaw) with a distance of 30 cm from the side of the fish to the surface of the crystal. The lead shield was 4 cm thick. The fish were in water, but immobilized in the narrow space of the box. The background of the counter at the 0.279 MeV ^{203}Hg photopeak (0.260–0.300 MeV) was 550 cpm. The effectivity was 5320 ± 70 cpm/μCi ^{203}Hg (γ79 per cent). The pulses were collected in a one channel analyser, model AS-11, by Wallac Oy (Turku, Finland).

Sample preparation

When the first pike died after 10 days (one that had been given protein-bound methyl mercury) one control pike and one that had been given the ionic form were killed in order to compare the pathological changes with the dead one.

When other fish died they were dissected, visual observations were recorded, and the radioactivity of a few organs was measured in a well counter.

The organs taken for the histopathological investigations were skeletal muscle, gills, pseudobranchiae, brains, intestines, liver and kidney (caudal opistonephros). This selection is based upon the earlier results on the distribution of methyl mercury in fish (Bäckström and Ullberg, 1966; Ohmomo et al., 1969). The samples of the organs were fixed with 10 per cent formalin, embedded in paraffin and sections of 5–10 μ thickness stained with Erlich's hematoxylin eosin.

RESULTS

Toxicity of methyl mercury to pike

Pike given the ionic form of methyl mercury retained on average 53 per cent of the doses administered (Table 1). They lived a mean of 33 days and between the last administration and death (about 25 days) they excreted

17 per cent of the retained activity. This gives a mean biological half-time of 94 days for the elimination of the activity originally retained. The retained dose found to be lethal was 14 mg Hg/kg (mean), while at death the fish contained only 12.5 mg Hg/kg.

Pike that received the protein-bound methyl mercury retained on an average 38 per cent of the doses administered and excreted 6 per cent (mean) between the last administration and death (10 days); they lived an average of 18 days (Table 2). The biological half-time of excretion was approximately 110 days. The lethal dose was 24.7 mg Hg/kg, the body burden at death being 23.2 mg Hg/kg.

Because the quantity of this material is small, it is not possible to give a precise LD50/30 days value. It is about 15 ± 3 mg Hg/kg body weight for both forms of methyl mercury at +10°C.

Visual and histopathological observations on pike after death

White muscle: Ionic MeHg, mean conc. 6.4 μg Hg/g; protein-bound MeHg, mean conc. 7.4 μg Hg/g. In both cases the muscle was normal in appearance, but between the muscle segments broken blood vessels were observed, especially along the lateral line, in three pike with ionic Me, that lived 28–40 days, and in three pike with protein-bound MeHg, that lived 14–30 days. No histopathological signs.

Gills: Ionic MeHg, mean conc. 8.5 μg Hg/g; protein-bound MeHg, mean conc. 14.7 μg Hg/g. Necrotic changes were observed in the lamellae of all fish.

Pseudobranchiae; Ionic MeHg, mean conc. 48.7 μg Hg/g; protein-bound MeHg, mean conc. 68.2 μg Hg/g. In earlier water uptake experiments this organ has had the highest amounts of mercury and the second highest amounts in oral administration experiments after the stomach. In this experiment this organ had the third highest amount after stomach and intestine. The organ was inflamed. The pseudobranchiae of the control pike had approximately 1–3 μg Hg/g wet weight due to the mercury the other pike had vomited, although the pike used in the experiment had been kept in another aquarium during the first hour after administration of

TABLE 1. TOXICITY OF CH$_3$Hg TO PIKE (*Esox lucius* L.)

Pike No.	3	2	5	4	1	Mean
Fresh weight, g	260	306	326	242	366	300
Sex	♂	♂	♂	♀	♀	
Given mg Hg/kg	20.0	41.2	17.4	26.4	25.7	
Retained mg Hg/kg	12.5	13.0	9.6	14.2	19.4	12.1
At death mg Hg/kg	12.5	12	8.6	12.3	17	11.1
Lived days	(10) (killed)	26	28	38	41	33

TABLE 2. TOXICITY OF PROTEIN-BOUND METHYL MERCURY TO PIKE (*Esox lucius* L.)

Pike No.	10	9	7	8	6	Mean
Fresh weight, g	250	292	320	372	402	343
Sex	♂	♂	♂	♂	♂	
Given mg Hg/kg	68.4	69.5	69.7	54.5	64.1	
Retained mg Hg/kg	29.6	27.0	29.3	15.0	22.7	24.7
At death mg Hg/kg	29.6	26.1	26.8	13.7	20.0	
Lived days	10	13	14	25	30	18

[300]

the mercury. The pseudobranchiae of the control pike were also slightly inflamed. Degenerated areas were only noticed when the mercury content exceeded 50 μg/g.

Brain: Ionic MeHg, mean conc. 7.4 μg Hg/g; protein-bound MeHg, mean conc. 9.0 μg Hg/g. The brain was normal looking. No pathological changes.

Intestine: Ionic MeHg, mean conc. 41.4 μg Hg/g; protein-bound MeHg, mean conc. 82.5 μg Hg/g. Fish (numbers 2 and 5; lived 26 and 28 days) with ionic MeHg had small inflamed areas in the external wall of the intestine. Perhaps from abnormal action of the gall bladder, the mucous membrane was greenish in the last pike (numbers 5, 4 and 1 lived 28–40 days) that were given ionic MeHg and in the last pike (numbers 8 and 6 lived 25–30 days) that were given protein-bound MeHg. Histopathological examination showed that the mucous membranes were normal, and with the exception of numbers 2 and 5, the external wall was also normal.

Liver: Ionic MeHg, mean conc. 37.0 μg Hg/g; protein-bound MeHg, mean conc. 75.5 μg Hg/g. In all the pike with the ionic MeHg the colour of liver was abnormally dirty brown and in the two pike that lived, 38 and 40 days the edges of the organ were gelatinous and transparent. The last pike had a dirty green area at the end opposite the gall bladder.

In the pike with the protein-bound MeHg the colour of the liver was abnormal in only the two which lived 25 and 30 days. Both had also a dirty green area. Histopathological observations showed that in all cases the green areas of the liver were necrotic. Small degenerated areas in livers of 1, 2, 5, 6, 7 and 8.

Kidney: Ionic MeHg, mean conc. 23.9 μg Hg/g; protein-bound MeHg, mean conc. 30.6 μg Hg/g. The kidneys were swollen in three pike with ionic MeHg (lived 28–40 days) and in three pike with protein-bound MeHg (lived 14–30 days).

In these cases there was oedema in the glomeruli and the tubules. In all the fish the lining cells of some tubules were degenerated.

In the pathological picture of all the investigated organs there was no difference between the ionic and protein-bound forms of methyl mercury.

The amounts of mercury in liver (37 μg and 75.5 μg Hg/g wet weight) were best correlated with the whole body burdens at death, 11.1 and 23.2 mg Hg/kg, respectively.

To summarize, the most damaged organs were the pseudobranchiae, which were severely inflamed, and kidneys, where the tubules were largely degenerated. In the gills parts of some lamellae were necrotic and the liver of several pike contained green necrotic areas. Cataract was not observed. On the surface of the entrails blood vessels were enlarged in several fish. This may be due to a secondary effect, the poison being a stressor factor and liberating histamine, which has caused the vasodilation. The liberation of histamine was, however, not studied.

The only externally visible changes were the necrotic areas of the gills and the broken blood vessels in the ventral skin.

Behavioural changes were observed only shortly before death. The fish began to have difficulties in keeping their balance, their respiration became irregular and they performed sudden fast rushes.

The control pike

To check the amount of reabsorption of mercury that was eliminated, six pike were kept as inactive controls. Three of them were given the same amount of liver suspension as to the experimental pike, but free of mercury. The control pike all took up a small amount of labelled mercury from the water and at the end of the experiment they had 0.5 to 1 mg Hg/kg. The mercury content was below 0.7 μg Hg/g in all other tissues except pseudobranchiae, which contained 1 to 3 μg Hg/g. In some control pike the pseudobranchiae were slightly inflamed, but no changes, visual or microscopic, were observed in other tissues.

Toxicity of methyl mercury to rainbow trout

Fish (numbers 1–7) that received a single dose of ionic methyl mercury corresponding to 4.9 mg Hg/fish all died within 1.5 to 2 h (Table 3). The mean retention was 52 per cent, 11.9 mg Hg/kg body weight. Of the fish (numbers 8–14) that received two doses of 1.47 mg Hg/fish with a 2-day interval, only one fish died after 47 days containing 12.4 mg Hg/kg. On the average these fish

TABLE 3. ACUTE TOXICITY OF THE IONIC FORM OF METHYL MERCURY TO RAINBOW TROUT (*Salmo gairdneri* RICHARDSON)
Fish No. 1–7 received a single dose containing 4.9 mg Hg
Fish No. 8–14 two doses each containing 1.47 mg Hg

Fish No.	Weight g	Retained Hg % of given	Retained Hg mg/kg	Life span
1	228	49	10.6	1.5 h
2	223	54	11.8	1.5
3	220	63	14.1	1.5
4	242	65	13.3	1.5
5	202	41	10.0	2.0
6	210	56	13.1	1.5
7	190	39	10.3	2.0
mean	220	52	11.9	~1.5
8	168	68	12.4	47 d
9	208	48	7.3	>94[a]
10	177	72	12.3	>94
11	132	46	10.5	>94
12	220	43	5.7	>94
13	137	68	13.5	>94
14	141	72	14.9	>94
mean	170	59	11.0	>94

[a] Fish 9 to 14 killed after 95 days when no damage was observed.

contained 11.0 mg Hg/kg but remained healthy for at least 94 days, when they were killed. The importance of the rate of administration of the toxic agent is evident from these results. While 12 mg Hg/kg as ionic methyl mercury was acutely lethal, 11 mg/kg was not observably toxic when given in two smaller doses with a 2-day interval. Mercury concentrations in total body and nine organs at death are given in Table 6. Spleen had the highest concentration in all but one fish (number 4), in which pseudobranchiae had the highest value. Of the five rainbow trout (numbers 15–19, Table 4) that received methyl mercury proteinate in two portions retaining on an average only 26 per cent of the amount given, two fish died after 70 days while no damage was observable in the other three fish when the experiment ended, after 95 days. Mercury distribution in these fish is given in Table 7. Again both forms were about equally toxic, 11 mg Hg/kg fish being lethal to some fish within 3 months at $+18°C$, while most fish lived over 94 days.

Thus, it is evident that it is not possible to give any single $LD_{50/30d}$. value for methyl mercury in rainbow trout because the toxicity of the substance depends greatly on the rate of administration.

TABLE 4. TOXICITY OF PROTEIN-BOUND METHYL MERCURY TO RAINBOW TROUT (*Salmo gairdneri* RICHARDSON)[a]

Fish No.	Weight g	Retained Hg % of given	Retained Hg mg/kg	Life span d
15	168	29	9.6	71
16	182	27	9.1	>95
17	269	26	5.7	>95
18	149	30	11.5	>95
19	98	14	19.4	69
mean	170	26	11.0	70 to over 90

[a] The proteinate was given in two doses with a 1-day interval, the first containing 3.85 mg Hg, the second 1.92 mg Hg.

Visual observations on acute toxicity of methyl mercury to rainbow trout

Fishes (numbers 1–7), which died within 1.5 h, evidently died of suffocation. The highest mercury concentration of the four tissues analysed (Table 5) was in each case found in gills, more than tenfold the average in the whole fish. The fish died with the mouth and gills wide open. No damage of gills was visually observable. The only tissue damage observed visually was a slightly swollen and darker-than-normal spleen and a dark bluish gall bladder. This suggests a damage of the erythrocytes.

The only visually observable symptoms in trout (numbers 8–19) were dark and swollen spleens. Trout (numbers 8–14; ionic form) also had darker than normal gall bladders.

TABLE 5. MERCURY CONTENTS (µg/g) IN TISSUES OF RAINBOW TROUT WHICH DIED WITHIN 2H OF AN ACUTELY TOXIC AMOUNT OF PERORALLY GIVEN IONIC METHYL MERCURY

Fish No.	1	2	4	6
Gill	134.3	140.8	166.4	178.6
Liver	33.2	39.2	57.3	42.1
Kidney	25.8	24.9	9.9	20.1
Heart	—	—	26.4	36.8
Whole body	10.6	11.8	13.3	13.1

Discussion

No difference was found in the present study between the toxicity of the ionic and protein-bound forms of methyl mercury to pike and rainbow trout. With the fast dosing we used, the $LD_{50/30d}$. value is approximately 15 mg Hg/kg for both species. This is evidently less than the fish could tolerate for months or years of chronic intake. The fish were able to defend themselves against the intake of the poison: of the ionic form about 50 per cent, of the protein-bound form 75 per cent was vomited immediately after administration. To what remained absorbed after the first day, an approximate biological half-time of 110 ± 20 days was obtained for pike, which is somewhat shorter than that found by Tillander *et al.* (1969) for pike with non-toxic tracer amounts of the ionic methyl mercury, 140 ± 40 d. within the first 20 days, 600 days after this. Thus, the higher dose of poison does not increase the biological half-time as was found in the case of mouse (Östlund, 1969). Our result for the estimated $LD_{50/30d}$. value, approximately 15 mg Hg/kg, is in excellent conformity with the observations from Mina-

TABLE 6. MERCURY CONCENTRATION (µG/G) IN RAINBOW TROUT AT DEATH, 47–94 DAYS AFTER PERORAL ADMINISTRATION IN 2 DOSES WITH AN INTERVAL OF 1 DAY, OF THE IONIC FORM OF METHYL MERCURY

Fish No.	8	9	10	11	12	13	14	$\bar{x}$
Organs								
Gills	11.1	4.2	12.6	8.2	5.6	9.8	13.8	9.3
Pseudobranchine	29.4	9.1	35.0	34.8	31.5	26.5	40.5	29.5
Stomach	10.5	5.2	11.1	6.7	4.7	9.6	12.9	8.7
Intestine	18.7	6.5	13.2	12.4	2.5	12.1	24.0	12.8
Liver	30.8	16.4	31.6	21.1	15.3	29.1	40.5	26.4
Kidney	28.4	19.1	20.1	22.6	18.0	22.1	31.5	23.1
Brain	11.4	7.9	13.2	9.5	7.1	13.2	16.5	11.3
White muscle	12.7	8.5	16.2	11.1	8.7	16.4	21.5	13.58
Spleen	53.1	27.0	50.5	38.0	15.2	45.8	58.4	41.1
Total body	13.0	7.1	9.6	8.3	3.9	12.4	17.0	10.0
Life span days	47	94[a]	94[a]	94[a]	94[a]	94[a]	94[a]	94

[a] killed

TABLE 7. MERCURY CONCENTRATION (µG/G) IN RAINBOW TROUT AT DEATH, 69–95 DAYS AFTER PERORAL ADMINISTRATION OF PROTEIN-BOUND METHYL MERCURY (2 DOSES WITH A 1-DAY INTERVAL)

Fish No.	15	16	17	18	19	$\bar{x}$
Gills	5.5	9.2	3.8	11.6	14.8	9.0
Pseudobranchine	28.0	26.6	3.9	39.5	21.5	23.9
Stomach	8.5	7.0	3.3	7.7	21.2	9.5
Intestine	12.8	7.0	1.2	12.6	59.4	18.6
Liver	26.2	23.4	13.3	28.5	46.3	27.5
Kidney	25.3	23.6	9.0	24.0	34.5	23.2
Brain	11.3	8.6	6.4	9.5	19.1	11.2
White muscle	11.8	11.3	6.4	13.5	20.5	12.7
Spleen	35.4	37.4	19.2	45.8	39.0	35.3
Total body	9.2	8.3	4.5	8.1	18.2	9.6
Life span days	71	95[a]	95[a]	95[a]	69	70 or over

[a] killed

mata, where the maximal value reported was 59.2 mg Hg/kg dry weight corresponding approximately to 20 mg Hg/kg fresh weight (Kitamura, 1968). There the intake of the poison was probably more chronic.

Our results on distribution and tissue damage are in conformity with those of Bäckström (1969) who made an autoradiographic and scintillometric study on the distribution of mercury, phenyl mercury and methyl mercury in five species, pike, speckled trout (*Salvelinus fontinalis*) and salmon (*Salmo salar*) included. He found no marked difference in the results between these five species. With methyl mercury he found after 10–30 days the highest concentrations in spleen, kidney, blood and liver, then myocardium, pseudobranchiae and gill. By autoradiography Bäckström (1969) also found accumulation of radioactivity in the lens capsule of the eye after intramuscular injection of methyl mercury. This is in conformity with the observations of cataract in Minamata (Minamata Disease Study Group, 1968). Cataract evidently is a more chronic lesion as we could not find it in our experiment. In the kidney of a speckled trout Bäckström (1969) found a heavy accumulation of mercury in the terminal parts of the proximal tubules which precisely corresponds to the greatest damage we found in kidney of pike after oral administration of methyl mercury.

Bäckström (1969) found after the first day maximal concentration of mercury in spleen after intravenous injection of methyl mercury to speckled trout. He refers to Black (1951) who found that piscine erythrocytes are very fragile. Our visual observations on swollen spleen and bluish colour of gall bladder in rainbow trout after acutely toxic amounts of methyl mercury might be explainable for the same reason. Ethyl mercury seems to be less heavily concentrated in spleen, as Rucker and Amend (1969) found after 30-day feeding of fingerlings containing ethyl mercury to 2-year chinook salmon (*Oncorhyncus tshawytscha*). Here the spleen contained 7.4 ppm and was the 8th of the organs studied. The highest concentrations were found in liver, 30.5 ppm, large intestine, 23.6 ppm and posterior kidney, 17.5 ppm.

References

AMEND, D F, YASUTAKE, W T and MORGAN, R Some factors
1970 influencing susceptibility of rainbow trout to the acute toxicity of an ethyl mercury phosphate formulation (Timsan). *Trans. Am. Fish. Soc.*, 98:419–25.
BÄCKSTRÖM, J Distribution studies of mercuric pesticides in
1969 quail and some fresh-water fishes. *Acta pharmac. toxic.*, 27 (Suppl. 3).
BÄCKSTRÖM, J and ULLBERG, S Distribution of different mercury
1966 compounds in organisms. *Svensk Vet. Tidn.*, 10:291–4 (in Swedish).
BOËTIUS, J Lethal action of mercuric chloride and phenylmercuric
1960 acetate on fishes. *Meddr Danm. Fisk.-og. Havunders.*, 3(4):93–115.
DOUDOROFF, P and KATZ, M Critical review of literature on the
1953 toxicity of industrial wastes and their components to fish. 2. The metals, as salts. *Sewage ind. Wastes*, 25:802–39.
HANNERZ, L Experimental investigations on the accumulation of
1968 mercury in water organisms. *Rep. Inst. freshwat. Res.*, Drottningholm, 48: 121–76. Abstract in *Aquat. Biol. Abstr.*, 1:Aq661(1969).
HÄSÄNEN, E and SJÖBLOM V Mercury content of fish in Finland in
1968 1967, *Suom. Kalatal.*, 36:1–24 (in Finnish).
IRUKAYAMA, K The pollution of Minamata Bay and Minamata
1967 disease. *Adv. Wat. Pollut. Res.*, 3:153–83.
JOHNELS, A et al. Pike (*Esox lucius*) and some other aquatic
1967 organisms as indicators of mercury contamination in the environment. *Oikos*, 18:323–33.
KITAMURA, S Determination on mercury content in bodies of
1968 inhabitants, cats, fishes and shells in Minamata District and in the mud of Minamata Bay. *In* Minamata disease, edited by Study Group on Minamata Disease, Kumamoto University, Japan, pp. 257–66 (English ed.).
KURLAND, L T, FARO, S N and SIEDLER, H Minamata disease.
1960 The outbreak of a neurologic disorder in Minamata, Japan, and its relationship to the ingestion of seafood contaminated by mercuric compounds. *Wld Neurol.*, 1:370–91.
LÖFROTH, G Methylmercury. *Bull. ecol. Res. Commn, Stockholm*,
1969 (4):38p.
MIETTINEN, J K et al. Distribution and excretion rate of phenyl-
1969 and methyl-mercury nitrate in fish, mussels, molluscs and crayfish. Paper presented to 9th Japan Conference on Radio-isotopes 13–15 May 1969, Tokyo, Japan, B/II-17.
OHMOMO, Y et al. Studies on the distribution of ^{203}Hg-labelled
1969 methyl mercury and phenyl mercury in pike. Paper presented at the 5th RIS Symposium, Helsinki, May 19–20, 1969, 18 p.
ÖSTLUND, K Studies on the metabolism of methylmercury and
1969 dimethylmercury in mice. *Acta Pharmax. Toxic.*, 27 (Suppl. 1).
RUCKER, R R and AMEND, D F Absorption and retention of
1969 organic mercurials by rainbow trout and chinook and sockeye salmon. *Progve Fish Cultst*, 31:197–201.
SCHWEIGER, G Die toxikologische Einwirkung von Schwermetall-
1957 salzen auf Fische und Fischnährtiere. *Arch. FischWiss.*, 8(1–2):54.
STUDY GROUP ON MINAMATA DISEASE (Ed), Minamata disease,
1968 Kumamoto, Kumamoto University, Japan.
TILLANDER, M et al. Excretion of phenyl and methyl mercury
1969 nitrate after oral administration or intramuscular injection in fish, mussel, mollusc and crayfish. *Nord. hyg. Tidskr.*, 50:181–3 (in Swedish).
UI, J Mercury pollution of sea and fresh water, its accumulation
1971 into water biomass. *Revue int. Océanogr. méd.*, 22-23:79-128
WESTÖÖ, G Determination of methylmercury compounds in
1966 foodstuffs. 1. Methylmercury compounds in fish, identification and determination. *Acta chem. scand.*, 20:2131–7.
WESTÖÖ, G Determination of methylmercury compounds in
1967 foodstuffs. 2. Determination of methylmercury in fish, egg, meat and liver. *Acta chem. scand.*, 21:1790–1800.
ANON. Mercury: wiping out an industry. *Chem. Engng News*,
1970 April 13:9.

Excretion Rate of Methyl Mercury in the Seal (*Pusa hispida*)

M. Tillander, J. K. Miettinen*
and I. Koivisto†

Taux d'excrétion du méthylmercure chez le phoque (*Pusa hispida*)

On a déterminé le taux d'excrétion de méthylmercure chez le phoque (*Pusa hispida*) en administrant oralement du méthylmercure marqué au ^{203}Hg en préparation liée aux protéines et en utilisant la technique de mesure de l'intensité du corps entier. Pendant le premier mois, l'élimination a été masquée par une amélioration des paramètres géométriques, le méthylmercure passant de l'estomac dans les tissus adipeux à la surface de l'animal. Toutefois, entre 100 et 153 jours, on obtient une courbe rectiligne sur une échelle semi-logarithmique donnant pour le

Tasa de excreción de metilmercurio en la foca (*Pusa hispida*)

La tasa de excreción de metilmercurio en la foca (*Pusa hispida*) se determinó usando metilmercurio marcado con ^{203}Hg como preparado ligado a proteínas por vía oral y siguiendo el método de recuento de todo el cuerpo. Durante el primer mes la eliminación quedó enmascarada por un mejoramiento de la geometría de la medición, transfiriéndose el metilmercurio del estómago a los tejidos grasos de la superficie del animal. Sin embargo, entre los 100 y 153 días se obtuvo una línea recta en una gráfica semilogarítmica dando para la componente de excreción

---

* Department of Radiochemistry, University of Helsinki.

† Helsinki Municipal Zoo, Helsinki, Finland.

constituant lent d'excrétion une période de 500 jours. Ce taux d'excrétion vaut pour 45 pour cent de la dose donnée. Entre 30 et 100 jours, un "constituant rapide" se manifeste avec une période d'environ 20 jours; cette valeur est seulement approximative par suite de l'effet masquant de la modification des paramètres géométriques. Cette enquête fait suite à des mesures récentes révélant la toxicité des concentrations mercuriennes dans une petite population de phoques de l'espèce susmentionnée, qui subsistent dans la région du lac Saimaa et figurent sur la liste préparée par l'Union internationale pour la conservation de la nature et de ses ressources des animaux qui nécessitent une protection spéciale.

lenta un período de semieliminación de 500 días. El 45 por ciento, aproximadamente, de la dosis suministrada sigue esta tasa de excreción. Entre 30 y 100 días aparece una "componente rápida" con un período de semieliminación del orden de los 20 días; este valor es sólo aproximado debido al efecto enmascarador del cambio geométrico. Esta investigación se efectuó a causa de que recientemente se midieron niveles tóxicos de mercurio en la pequeña población de focas del lago Saimaa, que figura en la lista, preparada por la Unión Internacional para la Protección de la Naturaleza y sus Recursos, de los animales que necesitan una protección especial.

HIGH levels of mercury have been reported in a relict population of ringed seals *Pusa hispida* living in Lake Saimaa in Eastern Finland (Helminen, Karppanen and Koivisto, 1968). The specimen with the highest concentrations of mercury showed behavioural symptoms of mercury poisoning. The ringed seal also occurs in the Baltic where high contents of organic mercury are found in many water animals. Since the seal eats almost exclusively fish and represents a man-equivalent trophic level, it is a valuable indicator organism.

Nothing is known of the retention and excretion rate of methyl mercury in water mammals and these factors were therefore determined in seals by using ^{203}Hg-labelled methyl mercury and the whole-body counting technique. In order that the resorption of the labelled compound would correspond to natural food, the labelled methyl mercury was given to the animal in protein-bound form, synthesized at the laboratory. A special whole-body counter was constructed to enable measurements to be continued over a five-month period.

The animal

One female ringed seal (*Pusa hispida*) (age about 9 months at the beginning of the study, weight about 40 kg) was

available for the experiment at the Municipal Zoo of Helsinki. It was caught as a pup in the Gulf of Finland and kept at the Zoo in a brackish-water basin where the water was changed regularly.

In this investigation it was desirable to administer the ^{203}Hg-labelled compound as methyl mercury proteinate. The ^{203}Hg (46.7d) has a relatively short physical half-time. Therefore, the preparation of the proteinate by biological incorporation into fish muscle protein, would be difficult. A simplified method was used. The proteinate was prepared as follows: an aqueous solution of $CH_3\,^{203}HgNO_3$ (AB Atomenergi, Studsvik, Sweden), 540 μCi/mg Hg, was mixed with dry milk powder and incubated at room temperature for 24 h. The labelled mercury compound was thus firmly fixed to the sulphydryl groups of milk proteins and could not be removed by prolonged heating (105°C) or by benzene extraction at neutral pH. Extraction with benzene after acidification by HCl to pH 0 to 1, however, removed the labelled compound into the organic phase, where it could be conclusively identified as methyl mercury by thin-layer chromatography.

Ca 1 g of $CH_3\,^{203}$Hg-proteinate containing 6.5 μCi ^{203}Hg and 12 μg stable Hg was packed into gelatine capsules, concealed in two herrings and fed to the animal

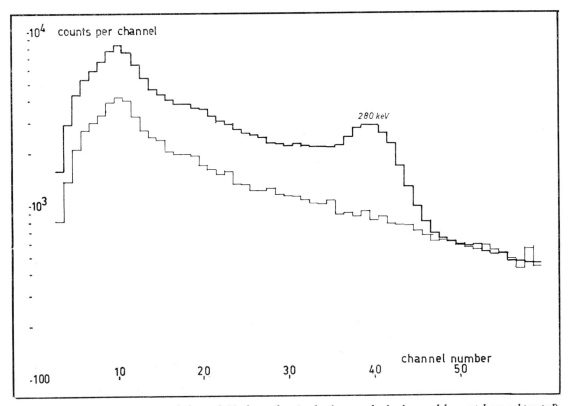

Fig 1. Twenty-minute spectrum of the seal 93 days after intake (upper, the background has not been subtracted) and of the background phantom (lower)

at the beginning of December. The ICRP value of ^{203}Hg for the maximum permissible concentration in the whole body of man as inorganic mercury is 80 μCi. To a 40 kg seal this would correspond to about 45 μCi. The activity given was set below this value which was used as a guide, as no MPL exists for methyl ^{203}Hg. The animal did not suffer from the treatment and remains in excellent health. A third capsule was sealed into a glass tube for use as a standard. Its activity was determined by using an IAEA point source (^{203}Hg No. 29, 21.64 μCi, 1 January 1968).

Whole-body counting

A whole-body counter with a "shadow" type shielding was constructed of 4 cm thick lead bricks in a temperature stabilized storage room near the seal cage. For the measurement the animal was placed in a wooden box with its "axis" 65 cm from the face of the 5×3 in (12.7×7.6 cm) NaJ (Tl) crystal. The animal was always placed with its right side facing the crystal.

The activity of ^{203}Hg was measured from the 279 keV photopeak using a Baird Atomics single channel analyser. The standard was measured and the spectrum checked before and after each actual measurement. On one occasion the spectrum was also measured with a multi-channel analyser (fig 1). The background was measured by using a phantom filled with water to the same mass as the animal, 42.5 kg at the start of the experiment. At least 10,000 counts were collected in each measurement to ensure statistical accuracy. The animal was first measured for 5 min but at the end of the experiment 20 min measurements were necessary. The animal was tranquil during the measurements.

Results and discussion

The measured pulse rates were corrected for phantom background and the net rates for physical decay by comparison with the standard. The corrected results are presented in Table 1.

The ^{203}Hg activity initially shows an apparent increase. This can be attributed to redistribution of methyl mercury in the body of the animal. The substance is evidently translocated into the fat tissue near the body surface. After 27 days the activity decreases and two components can be resolved from the diagram, each of which accounts for about half the activity present 27 days after intake. The longer component has a half-time of about 500 days and

TABLE 1. WHOLE BODY RADIOACTIVITY OF A RINGED SEAL (*Pusa hispida*) AFTER AN ORAL DOSE OF 6.5 μCi OF ^{203}Hg-LABELLED METHYL MERCURY PROTEINATE

Date	Days elapsed	Seal, cpm, net, corrected
Dec. 3	0	4,380
Dec. 4	1	5,920
Dec. 5	2	6,390
Dec. 7	4	6,260
Dec. 9	6	6,270
Dec. 16	13	7,060
Dec. 30	27	7,440
Jan. 27	55	5,800
Mar. 6	93	5,180
Mar. 20	107	4,940
Apr. 9	127	4,800
May 5	153	4,600

the shorter one about 20 days. Because of the apparent initial activity increase the retention can be estimated roughly as 80 to 100 per cent.

This investigation showed that about 45 per cent of the CH_3 ^{203}Hg administered as a single dose of methyl mercury proteinate was excreted with the very long biological half-time of 500 days. Most was excreted rapidly with a half-time of about three weeks. The half-time of the slow component of excretion was of the same order as those found in other sea animals (Tillander *et al.* 1969; Miettinen *et al.* 1970) and much longer than the half-time of 75 days observed in man (Ekman *et al.* 1968; Miettinen *et al.* 1969).

References

EKMAN, L, *et al.* Distribution of mercury-203 after administration
1968 of methylradiomercury-nitrate. *Nord. Med.*, 79:456–8 (in Swedish).
HELMINEN, M, KARPPANEN, E and KOIVISTO, I Mercury content
1968 of the ringed seal of Lake Saimaa. *Suom. LaakLehti*, 74:87–9 (in Finnish).
MIETTINEN, J K, *et al.* Metabolism of radiomercury administered
1969 as methyl mercury protein in man. *In* Fifth R.I.S. Symposium, Helsinki, 19–20 May 1969.
MIETTINEN, V, *et al.* Preliminary study on the distribution and
1970 effects of two chemical forms of methyl mercury in pike and rainbow trout. Paper presented to FAO Technical Conference on Marine Pollution and its Effects on Living Resources and Fishing, Rome, Italy, 9–18 December 1970, FIR:MP/70/E-91:12 p.
TILLANDER, M, *et al.* The excretion by fish, mussel, mollusc and
1969 crayfish of methyl mercury nitrate and phenyl mercury nitrate, introduced orally or injected into musculature. *Nord. hyg. Tidskr.*, 50(2):181–3 (in Swedish).

The Effects of Pollutants on the Reproduction of Marine Organisms

C. C. Davis*

Les effets des polluants sur la reproduction des organismes marins

La pollution thermique modifie les saisons de reproduction et permet à des espèces d'eau chaude de pénétrer et de se reproduire dans des régions qui autrement seraient trop fraîches pour elles. De même, la reproduction d'espèces d'eau froide peut être entravée par des températures trop élevées. De plus, la pollution thermique peut entraîner un accroissement de la production de poissons, de homards, etc. dans les régions plus fraîches.

Efectos de los contaminantes en la reproducción de los organismos marinos

La contaminación térmica altera las temporadas de reproducción y permite a las especies de aguas templadas penetrar y reproducirse en zonas que de otra forma serían demasiado frías para ellas. En cambio, las especies de aguas frías pueden encontrar inconvenientes para reproducirse, debido a las temperaturas demasiado elevadas. Por otra parte, la contaminación térmica puede permitir

*Department of Biology and Marine Sciences Research Laboratory, Memorial University of Newfoundland, St. John's, Newfoundland, Canada.

La diminution de la salinité sous l'effet d'effluents d'eau douce réduit souvent l'activité sexuelle, ainsi que le nombre et la résistance des jeunes.

On a constaté que sous l'action de résidus de DDT, des oiseaux de mer produisaient des oeufs à coquille mince, sans doute par suite des effets de l'insecticide sur l'anhydrisation carbonique dans la glande coquillière. Chez le saumon et la truite, la libération après l'incubation des résidus de DDT accumulés pendant l'absorption du vitellus a provoqué des taux élevés de mortalité.

Dans des conditions expérimentales, des canes auxquelles on avait administré de petites quantités d'hydrocarbures relativement non toxiques ont cessé de pondre. Des oiseaux de mer dont le plumage était enduit d'hydrocarbures ont transmis ces produits à leurs oeufs pendant la couvaison et l'éclosion ne s'est pas produite, bien que les oeufs aient été fertiles à l'origine. De petites quantités d'hydrocarbures ont endommagé des oeufs et des larves de plies en voie de développement et les ont détruits.

Des sulfates ont réduit le pourcentage de fertilisation chez les oeufs de harengs et accru les anomalies structurelles chez les larves à peine écloses. Les déchets sulfateux de fabriques de pulpe de bois ont entravé la reproduction et la fixation des huîtres, mais à l'inverse les déchets de fabriques de papier d'emballage ont stimulé la production de naissain par les moules.

Le développement anormal des oeufs d'oursins et de pélécypodes a fourni un test biologique sensible pour déceler la présence de petites quantités d'ions métal polluants, de phénols, et d'autres polluants.

De petites quantités de radioisotopes influent sur les taux d'amitose et de mitose des chromosomes et ont une incidence sur les difformités que présentent les oeufs des poissons, des polychètes et des oursins.

un aumento de la producción de pescado, langosta, etc., en las regiones más frías. La disminución de la salinidad causada por las descargas de agua dulce reduce frecuentemente la actividad sexual, el número y el vigor de las formas juveniles.

Bajo la acción de los residuos de DDT, las aves marinas pusieron huevos de cáscara muy frágil, debido, probablemente, a la influencia del insecticida en la actividad de la anhidrasa carbónica en la glándula del oviducto que forma la cáscara. En el salmón y la trucha, la liberación de los residuos acumulados de DDT después de la incubación, durante la absorción del saco vitelino, se ha traducido en grados elevados de mortalidad.

Algunos patos alimentados experimentalmente con pequeñas cantidades de hidrocarburos relativamente no tóxicos interrumpieron la puesta. Las aves marinas cuyo plumaje había resultado cubierto de hidrocarburos, transmitieron los hidrocarburos a los huevos durante la formación de los mismos y los huevos no avivaron, aunque inicialmente eran fértiles. Pequeñas cantidades de hidrocarburos produjeron daños en huevos y larvas de platija e incluso la muerte.

Los sulfatos redujeron el porcentaje de fertilización de los huevos de arenque y multiplicaron las anormalidades estructurales de las larvas recién eclosionadas. Los residuos sulfatados de las fábricas de pasta de madera redujeron la reproducción de las ostras, pero, en cambio, los residuos de las fábricas de pasta-kraft estimularon el desove de los mejillones. El desarrollo anormal de los huevos de erizos de mar y pelecípodos se ha utilizado como bioensayo, por su sensibilidad, para determinar la presencia de pequeñas cantidades de iones metálicos, fenoles y otros contaminantes. La presencia de pequeñas cantidades de radioisótopos influye en la segmentación, la fragmentación de los cromosomas e incide en las deformidades de los huevos de peces, poliquetos y erizos de mar.

POLLUTANTS may affect reproduction in many different ways. Teratological development of embryos may result in deformed or malfunctioning larvae which do not survive hatching. Reproduction may be influenced by behavioural changes of the adults during the mating season. Their behaviour, the production of eggs and sperm, the secretion of egg membranes, eggshells and production of egg nutrients, may all be affected by changes of hormone function and enzyme activity. Changes in the ecosystem may influence reproductive success when vitellogenesis is directly influenced by the availability of food.

Mortality of adults obviously prevents reproduction and Chia (1970) has emphasized the importance of this in the Arctic where growth to maturity is slow and most benthic animals lack pelagic larvae. Recovery from the destruction of such a community is very much delayed.

Thermal pollution

There are few direct studies of the effect of heated industrial effluents on the breeding of animals. The recent occurrence and breeding of *Venus mercenaria* in Southampton waters is related to heated effluents and larvae transported by currents from Southampton to adjacent colder waters can settle, but could not breed (Ansell, 1963). Pannell, Johnson and Raymont (1962) reported that gribble prolonged their migration and breeding season in Southampton waters because of heated effluents. At Swansea, South Wales, Stubbings and Houghton (1964) observed that boreal crab, *Carcinus maenas*, could not breed because of higher temperatures caused by a heated effluent. Naylor (1965a) reported that the ascidians *Ciona intestinalis* and *Ascidiella aspersa*, which are summer breeders in Scotland, were able to breed in all seasons at Swansea.

Numerous studies have been made of the temperature tolerances of fertilized fish eggs (Garside, 1959; Kinne and Kinne, 1962, 1962a; Olsen and Foster, 1955). Excessively high temperatures cause failures at gastrulation, the initiation of circulation, and hatching in the

cottid, *Clinocottus analis* (Hubbs, 1966); Worley (1933) showed that cleavage in the mackerel, *Scomber scombrus*, was irregular, resulting in death of embryos at temperatures above 22°C; Swarup (1958, 1959) discussed cleavage abnormalities and twinning caused by high temperatures during embryonic development of the stickleback, *Gasterosteus aculeatus*. Brett (1956, 1960), Kinne (1963), Kuthalingam (1959) and Naylor (1965) concluded that successful reproduction in fish and other marine animals as a rule occurs only within a distinctly narrower range of temperature than the tolerance range for survival and growth (fig. 1). Among invertebrates, Runnström (1928) has shown that embryonic development is abnormal above a certain temperature in the sea urchin, *Strongylocentrotus dröbachiensis* (>11°C), the sea cucumber, *Cucumaria frondosa* (>13°C), the nudibranch, *Dendronotus frondosus* (>14°C), the mussel, *Mytilus edulis* (>16°C) and the urchin, *Echinus esculentus* (>16°C).

It is clear from these studies that increases of temperature as a result of hot water effluents in semi-enclosed bays, estuaries and fjords in boreal or cool-water regions would unfavourably influence reproduction in many forms.

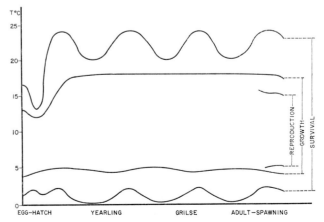

Fig 1. Pacific salmon: a schematic representation of tolerance to temperature at different stages of life. Note reduced tolerance at egg hatching and spawning stages (after Brett 1960)

Colton (1959) observed a mass die-off of pelagic hatching-stage larvae of the yellow-tail flounder (*Limanda ferruginea*) and other boreal species in an intrusion of warm Gulf Stream water just south of Georges Bank. The temperature of water was 18.5°C in contrast to the surrounding water at 8°C. This is a natural example of the type of catastrophe that might be caused by thermal pollution.

Gametogenesis in *Fundulus heteroclitus* (Burger, 1939) and stickleback, *Apeltes quadracus* (Merriman and Schedl, 1941) and spawning in the oyster, *Ostrea virginica* and California sardine, *Sardinops sagax* and many other animals are dependent on, and stimulated by, the attainment of certain minimum water temperatures. If the requisite temperature conditions are attained through warming by thermal effluents at unsuitable seasons, the eggs and/or larvae may then be carried to their destruction in colder waters. Furthermore, the rate of embryonic development is increased greatly by higher temperatures. As has been suggested by Ketchen (cited in Brett, 1956), pelagic eggs and the resultant larvae can develop in this fashion, and attain metamorphosis before they have been carried to their suitable settling grounds.

In contrast to the deleterious effects of thermal pollution, it is possible that the artificial warming of cold waters by thermal effluents can be used to stimulate greater reproductive and growth rates for fish and commercially important invertebrates (Hedgpeth and Gonor, 1969; Mihursky, 1967).

Salinity changes

Addition of dilute aqueous effluents to salt-water basins, or conversely the cutting off of fresh-water supplies to estuaries by diversion of the water can act as pollution by lowering or raising the natural salinity. There have been several observations concerning the effects of salinity on reproduction in natural waters. Colonies of the polyp, *Cordylophora caspia*, grown from planulae, developed gonophores only in salinities between 5 and 16.7 per thousand, but when normal colonies were amputated and allowed to regenerate, gonophores were formed at a wide range of salinities although not in fresh water nor in salinities higher than 30 per thousand (Kinne, 1956). In regenerating colonies, however, eggs ripened only in salinities between 2 and 10 per thousand, and egg size was smaller in unfavourable salinities. Brattström (1941) reported that the starfish, *Asterias rubens*, existed in the low salinities of the eastern Baltic Sea. The species, however, needed to be replenished by recruitment of larvae from elsewhere because they could not breed in the diluted water. Similarly, Schnakenbeck (1940) noted that cod in Finnish waters were unable to breed there. Kinne and Kinne (1962, 1962a) subjected eggs of the pupfish, *Cyprinodon macularius*, to varying salinities from distilled water to 85 per thousand, and found a retardation of development with increasing salinities; there was arrest of development at the extremes.

Insecticides

Persistent insecticides can be carried to the sea by water runoff or as dust in air currents (Cohen and Pinkerton, 1966; Risebrough 1969). Insecticide residues, particularly the chlorinated hydrocarbons, have been reported in the tissues of marine organisms, especially in fat and egg yolk (Moore and Tatton, 1965; Holden and Marsden, 1967; Risebrough *et al.*, 1967; Robinson *et al.*, 1967); some of the reports deal with residues in animals as far from sources of pollution as Antarctica (George and Frear, 1966; Sladen *et al.*, 1966; Brewerton, 1969; Frost, 1969). Several studies have shown that the residues increase in concentration at higher trophic levels (e.g., Woodwell *et al.*, 1967), so that greatest quantities occur in top predators and fish-eating birds (Hickey *et al.*, 1966; Risebrough, 1969). There has been much circumstantial evidence that reproductive success in marine birds is reduced in proportion to the quantity of DDT residues in the parents and in the eggs. For example, Ames (1966), studying osprey reproduction in salt marshes in Long Island Sound, showed that there had been a decline of hatching success from a normal rate of 2.2–2.5 young per nest to 0.1 per nest in 1960 and 0.4 per nest in 1961. DDT residues were very high in the dead eggs. In 1963 a comparison was made with a colony in Maryland where hatching success was more nearly normal; DDT residues in the eggs from Long Island Sound were half again as high as in Maryland.

Keith (1966) examined a population of herring gulls in which there was a very high incidence of broken egg shells, and where egg mortalities were 2.5–5.8 times mortalities reported elsewhere. The dead eggs had higher DDT residues than live eggs. Hickey and Anderson (1968) examined the shell thickness of herring gull eggs from five widely separated areas in the United States (Fig. 2),

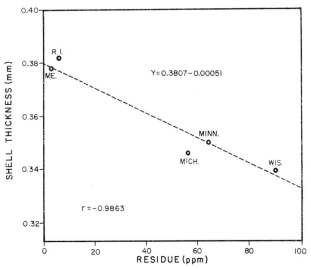

Fig 2. Shell thickness in the eggs of herring gulls from five breeding areas: Rhode Island, Maine, Lake Huron (Michigan), Lake Superior (Minnesota), and Green Bay (Wisconsin) in relation to total DDT residues in the eggs (from Hickey and Anderson, 1968)

and found that the thickness was inversely proportional to the quantity of DDT residues in the various colonies. Wurster and Wingate (1968) investigated the decline of the Bermuda petrel (*Pterodroma cahow*), a rare bird whose breeding success declined drastically from 66.7 per cent in 1958 and 1961 to 28.6 per cent in 1966 and 36.4 per cent in 1967. Although living far from contamination sources, the occurrence of significant quantities of DDT residues in the eggs and chicks convinced the authors that the insecticide caused the decline.

Evidently there have been no experimental studies of

the effects of DDT or its residues on reproduction in marine birds, but there have been a few investigations of terrestrial or fresh-water species. Genelly and Rudd (1956) fed insecticides to pheasants and found that egg laying was decreased, and that the hatchability of those eggs that were produced was reduced, as was survival of hatchlings. Heath, Spann and Kreitzer (1969) fed DDT to mallard ducks; there was thinning and cracking of the egg shells, as well as a marked increase in embryo mortality. Koeman, Oudejans and Huisman (1967), experimenting with chicks, observed an increased mortality from dieldrin at the stage of yolk-sac reabsorption.

As indicated by Moats and Moats (1970), there apparently have been no studies on the effects of pesticides on reproduction in wild fish. On the other hand, Burdick *et al.* (1964) observed a "relatively close relationship" between fry mortality and ppm of DDT in hatchery-raised lake trout. In an experiment, Allison, Kalliman and Van Valin (1963) treated hatchery trout with sublethal doses of DDT. Egg production was not decreased, but mortality among sac fry was greatest in lots treated with the greatest amounts of insecticide. The greater mortality occurred when the yolk sac was absorbed, at which time stored DDT residues would be liberated. Similar results were obtained by Macek (1968).

Two main hypotheses have been presented concerning the mechanism of action of chlorinated hydrocarbons in thinning of egg shells and repressing of reproduction in birds and mammals. Peakall (1967) worked with pigeons and showed that DDT and dieldrin affected the metabolism of testosterone and progesterone. He conjectured that this led to reduced sex activity. Welch, Levin and Conney (1969), using white rats, similarly showed that the metabolism of several steroid sex hormones and related substances was inhibited by chlorthion, whereas DDT and chlordane stimulated it. Wurster (1969) stressed that chlorinated hydrocarbons cause the induction of hepatic enzymes that metabolize the steroids; this in turn influences calcium metabolism in birds, and hence causes the thin and brittle egg shells. Bitman, Cecil and Fries (1970) have presented evidence for a second hypothesis, namely that the DDT inhibits avian shell-gland carbonic anhydrase, and hence causes thinner egg shells.

Evidently there have been no definitive experiments on the behavioural effects of pesticides as this affects sexual activity in fish, although Warner, Peterson and Borgman (1966) have described general changes of behaviour in goldfish resulting from sublethal pesticide toxication. Obviously, in fish where behaviour is critical for sexual responses breeding could be prevented entirely if sublethal concentrations of pesticides were present during the breeding season.

Very little study has been devoted to effects of insecticides on reproduction in marine or other aquatic invertebrates. Loosanoff (1947) reported an investigation of the effects of DDT in preventing metamorphosis of barnacle larvae on oyster beds. Waugh, Hawes and Williams (1952) and Waugh and Ansell (1956) showed that the pesticide was toxic also to settling of oysters, but less so than to barnacles. Subsequent growth of the oyster spat, after an initial retardation, was very good because of reduction of competition from the barnacles.

Loosanoff (1960) subsequently pointed out the toxic effects on oyster larvae. Davis (1961) studied the effects of pesticides on the eggs and larvae of oysters (*Crassostrea virginica*) and clams (*Venus mercenaria*). Lindane and guthion were toxic at certain concentrations, leading to a hatch failure. Grosch (1967) reported that the proportion of resting eggs to non-resting eggs produced by the brine shrimp, *Artemia salina*, was higher when the adults were treated with DDT than when they were not.

Wurster (1968) showed that DDT reduces photosynthesis in marine phytoplankton; this would lead to reduction of reproduction by cell division. Menzel, Anderson and Randtke (1970) have indicated that different phytoplankton species have differing responses. Thus, DDT had no effect on cell division in *Dunaliella tertiolecta*, whereas 100 ppb of DDT blocked cell division in *Skeletonema costatum* after 2–3 divisions. In *Cyclotella nana* cell division was completely blocked by dieldrin and endrin, whereas it continued, albeit more slowly, with DDT.

Organophosphorus insecticides have been investigated far less than the chlorinated hydrocarbons. The only pertinent study discovered was that of Konar (1969), who reported that hatching success for the snake-head (*Channa punctatus*) was lowered, that the hatching time was increased, and that yolk absorption was inhibited by concentrations of DDVP greater than 1.6 mg/l.

TEPA is an insect chemosterilant, not an insecticide; Stock and Cope (1969) found that sublethal doses of TEPA caused testicular damage and reduced male fertility in guppies (*Poecilia reticulata*) although female fertility was less affected.

Oil and detergents

Oils vary considerably in toxic effects—therefore difficult to generalize. The following illustrate how pollutants may act:

Rittinghaus (1956; reported by Hartung, 1965) described an accidental oiling of shore birds (particularly Cabot's terns, *Thalasseus sandvicensis*) at a sea bird sanctuary. The oiled birds transferred oil to the eggs and they failed to hatch. Hartung (1965) thinly oiled mallard eggs with a lubricating oil. Two days after oiling, half the embryos were dead, and only 5 to 24 hatched; nearly all the control eggs hatched. Setting mallard hens were oiled and transferred the oil to their eggs, so none of them hatched.

The effects of oils on yolk-sac stage "embryos" of the toadfish *Opsanus tau* were reported by Chipman and Galtsoff (1949). Mironov (1968) reported that developing eggs of the plaice (*Rhombus maeoticus*) are extremely sensitive to oil products. As little as 10^{-4} to 10^{-5} ml/l of oils produced injuries in the embryos, so that from 40 to 100 per cent of the hatchlings degenerated and died. Mironov emphasizes the dangers of floating oils in the hyponeuston, where many pelagic marine fish eggs occur.

There have been few studies of the direct effect of oils on reproduction in adult animals, as was recognized by Mironov (1968). Hartung (1965) fed adult mallards 2 g/kg of a relatively non-toxic mineral oil by use of a stomach tube; the experimental ducks stopped laying, whereas controls fed water were unaffected.

Mironov and Lanskaya (reported in Mironov, 1968) demonstrated retardation of cell division in phytoplankton submitted to oil contamination. Similarly, Lacaze (1969) tested 1 per cent extracts of crude petroleum of the type spilled in the *Torrey Canyon* disaster on the unicellular marine alga, *Phaeodactylum tricornutum*, obtaining a 10 per cent reduction of colony growth (hence of cell division).

The widespread use of emulsifying "detergents" on the *Torrey Canyon* oil spill led to a rash of publications. Corner, Southward and Southward (1968) tested BP1002, Gamlen, Slipclean and Dasic against newly hatched barnacle nauplii. Higher concentrations killed the nauplii, while sublethal concentrations retarded development. For example, 0.5 ppm of BP1002 retarded development, and prevented the metamorphosis of cyprids. Lacaze (1969) found that Gamosol at 5.6 ppm reduced cell division in *Phaeodactylum tricornutum* by 50 per cent; similar results were attained by using 8 ppm of the surfactant fraction only of Gamosol. Boney (1970) demonstrated that BP1002 was more toxic to the motile reproductive cells than to other stages of the green intertidal alga, *Prasinocladus marinus*.

The detergents Paic, Teepol, and Deter-Baz were tested against eggs of *Artemia salina* by Aubert, Charra and Malara (1969). At lower concentrations hatching was speeded up through increased rate of absorption of water. Paic and Teepol prevented hatching at 0.312 g/l and 0.375 ml/l respectively, but Deter-Baz did not prevent hatching at any strength.

Along a different line of investigation, Bardach, Fujiya and Holl (1965) found that different kinds of detergents adversely affected the taste senses of yellow bullhead, *Ictalurus natalis*, and there was no recovery 6 weeks after treatment. Although this study did not test changes in reproductive behaviour, it is clear that in those fish where reproduction is facilitated by chemical senses, detergent pollution may hinder successful breeding.

Inorganic pollutants

The importance of pulp mill wastes as marine pollutants has led to several studies of these wastes and their components on the reproduction of marine animals. Galtsoff and Chipman (1947) described the reduction of spawning and settling of the oyster, *Crassostrea virginica*, in the upper York River, Virginia. They implied, but did not prove, that the cause was sulphate pulp mill wastes. These authors showed that the wastes reduce the time that the oyster shells are open; hence the time for feeding is reduced, and this could lead to reduced fertility.

Breese, Millemann and Dimick (1963) showed that kraft mill effluent stimulated spawning in mussels, and that the gametes were viable. Local stimulation of spawning could lead to self-stimulation over a much wider area than that immediately affected by the effluent.

Kinne and Rosenthal (1967) studied the effects of $FeSO_4$ and H_2SO_4 on fertilization, embryonic development and larvae of the herring, *Clupea harengus*. The percentage of successful fertilization was reduced, embryonic growth rate was retarded, the embryonic heart frequency was enhanced (stress), duration of incubation was decreased, the percentage of successful hatching was reduced, and the occurrence of structural abnormalities in the hatchlings was increased. There was danger in concentrations as low as 1:32,000. These results were

supported by the findings of Kinne and Schumann (1968).

The use of fertilized eggs of sea urchins and bivalved molluscs as bioassays in the detection of a wide variety of pollutants, mostly inorganic, has been reported by Okubo and Okubo (1962) (summarized also in Føyn, 1965). The sensitivity of the developing eggs was very high; the pollutants caused a disturbance of development and metamorphosis.

Radioactive wastes

Evidently there have been no field studies of the effects of radioactive pollutants on the reproduction of marine animals, but investigators have undertaken laboratory experiments along these lines. Costello, Henley and Kent (1952) showed that early mitotic divisions in the eggs of *Chaetopterus* were markedly affected by ^{32}P levels as low as 0.03 mc/ml. Green and Roth (1955) irradiated *Arbacia* eggs with 0.25 to 5.0 µc/ml doses of ^{32}P, ^{35}S and ^{42}K, subsequently fertilizing them with non-irradiated gametes. Depending on the dose, the first cleavage was accelerated or retarded; at higher doses further development was prevented. As reported by Mauchline and Templeton (1964), Polikarpov and Ivanov (1961) found that 100 µµc of ^{90}Sr retards hatching and causes abnormalities in the eggs of fish. Golovinskaia and Romashov (1958) discussed the Bikini atomic bomb tests and showed numerous pictures of deformities of fish eggs caused by radioactive fallout chemicals.

The effects of continuous sublethal irradiation of cultures of *Daphnia pulex* have been reported by Marshall (1962). The rate of natural increase was inversely proportional to the dose; this was not caused by increased deaths, but by decreased births. Donaldson and Foster (1957) and Bonham and Welander (1963) examined the radioresistance of fish. In both instances it was found that the early stages were far more sensitive than later stages of embryonic development. Donaldson and Foster (1957) showed that LD_{50} for gametes of the rainbow trout was 50–100 r, whereas for the adult it was 1500 r. The LD_{50} for eyed eggs of the chinook salmon was 1000 r, and that for fingerlings was 1250–2500 r.

Eutrophication effects

One result of pollutional enrichment of waters by the ions NO_3^-, PO_4^{3-} and other plant nutrients is eutrophication. This has no direct effect on reproduction of aquatic organisms, but it can have far-reaching indirect effects. Most of these indirect effects, such as oxygen reduction, do not properly belong in this review, but some studies in Lake Constance (Bodensee) are of interest. Karbe (1964; summarized in Stewart and Rohlich, 1967) found that the habitat boundaries of four species of coregonids have become blurred as a result of the advancing eutrophication of the lake; changed amounts of plankton have allowed a change in breeding areas, and this has led to extensive hybridization. Nümann (1964) reported that with the enrichment of the lake the whitefish *Coregonus wartmanni*, grows faster so that at any given size the gonads and the eggs are smaller. Therefore, the existing fishery removes the immatures and prevents adequate spawning. Also, the mortality of the embryos was greater

(50 per cent) than before (20 per cent). In addition, eutrophication has caused the migration of perch from shore into the pelagial, where they probably eat more whitefish eggs and fry than before. Unfortunately, there do not appear to be any such studies of eutrophication in estuaries or the sea.

Research needs

With the exceptions of some work of Kinne (Kinne, 1964; Kinne and Kinne, 1962, 1962a), practically none of the investigations has considered synergistic effects of pollutants, although more often than not industrial and domestic effluents contain more than a single offending ingredient, as for example industrial cooling waters containing chemical wastes. In order to appreciate the potential dangers of actual pollution sources it will be very important to thoroughly examine such matters as the effect of detergents in the presence of various metal ions, the effect of temperature in relation to salinity, BOD, etc.

Behavioural aspects of reproduction as affected by pollutants have hardly been touched by any of the investigators. It will be important for animal behaviourists and ethologists to look closely into this facet of the problem.

The physiology of pollutional effects on reproduction and on embryonic and larval development, including metamorphosis, is little understood. It will be not only of practical value, but of theoretical interest as well, to elucidate these mechanisms.

References

ALLISON, D, KALLIMAN, B J and VAN VALIN, C C Insecticides:
1963 effects on cutthroat trout of repeated exposure to DDT. *Science, N.Y.*, 142(3594):958–61.

AMES, P, DDT residues in the eggs of the osprey in the north-
1966 eastern United States and their relation to nesting success. *J. appl. Ecol.*, 3(Suppl.):87–97.

ANSELL, A D *Venus mercenaria* in Southampton waters. *Ecology*,
1963 44(2):396–7.

AUBERT, M, CHARRA, R and MALARA, G Etude de la toxicité de
1969 produits chimiques vis-à-vis de la chaîne biologique marine. *Revue int. Océanogr. méd.*, 13/14:45–72.

BARDACH, J E, FUJIYA, M and HOLL, A Detergents: effects on the
1965 chemical senses of the fish *Ictalurus natalis* (le Sueur). *Science, N.Y.*, 148(3677):1605–7.

BITMAN, J, CECIL, H and FRIES, F DDT-induced inhibition of
1970 avian shell-gland carbonic anhydrase: a mechanism for thin egg shells. *Science, N.Y.*, 168(3931):594–6.

BONEY, A D, Toxicity studies with an oil-spill emulsifier and the
1970 green alga *Prasinocladus marinus. J. mar. biol. Ass. U.K.*, 50(2):461–73.

BONHAM, K and WELANDER, A D Increase in radioresistance of
1963 fish to lethal doses with advancing embryonic development. *In* Radioecology, edited by V. Schultz and A. W. Klement, New York, Reinhold, pp. 353–8.

BRATTSTRÖM, H *Unters. aus dem Öresund*, 27:329 p. (ref. not seen,
1941 cited in Kinne, 1964).

BREESE, W P, MILLEMANN, R E and DIMICK, R E Stimulation of
1963 spawning in the mussels, *Mytilus edulis* Linnaeus and *Mytilus californianus* Conrad, by Kraft-mill effluent. *Biol. Bull. mar. biol. Lab.*, Woods Hole, 125(2):197–205.

BRETT, J R Some principles in the thermal requirements of fishes.
1956 *Q. Rev. Biol.*, 31(2):75–87.

BRETT, J R Thermal requirement of fish—three decades of study,
1960 1940–1970. *Publ. Hlth Serv. Tech. Rep., Wash.*, (W60–3):110–7.

BREWERTON, H V DDT in fats of Antarctic animals. *N.Z. Jl Sci.*,
1969

BURDICK, G E, *et al.* The accumulations of DDT in lake trout and
1964 the effect on reproduction. *Trans. Am. Fish. Soc.*, 93(2):127–36.

BURGER, J W Some experiments in relation of the external
1939 environment to the spermatogen cycle of *Fundulus hetero-clitus* (L.). *Biol. Bull. mar. biol. Lab.*, Woods Hole, 77(1):96–103.

CHIA, F-S Reproduction of Arctic marine invertebrates. *Mar.*
1970 *Pollut. Bull.*, 1(5):78–9.

CHIPMAN, W A and GALTSOFF, P S Effects of oil mixed with
1949 carbonized sand on aquatic animals. *Spec. scient. Rep. U.S. Fish Wildl. Serv.*, (1):53 p.

COHEN, J M and PINKERTON, C Widespread translocation of
1966 pesticides by air transport and rainout. *Adv. Chem.*, 60:163–76.

COLTON, J B JR. A field observation of mortality of marine fish
1959 larvae due to warming. *Limnol. Oceanogr.*, 4(2):219–22.

CORNER, E D S, SOUTHWARD, A J and SOUTHWARD, E C Toxicity
1968 of oil-spill removers ("detergents") to marine life: an assessment using the intertidal barnacle *Elminius modestus. J. mar. biol. Ass. U.K.*, 48(1):29–47.

COSTELLO, D P, HENLEY, C and KENT, D E Effects of ^{32}P on
1952 mitosis in *Chaetopterus* eggs. *Biol. Bull. mar. biol. Lab.*, Woods Hole, 103:298–99 (Abstract only).

DAVIS, H C Effects of some pesticides on eggs and larvae of
1961 oysters (*Crassostrea virginica*) and clams (*Venus mercenaria*). *Comml Fish. Rev.*, 23(12):8–23.

DE SYLVA, D P Theoretical considerations of the effect of heated
1969 effluents on marine fishes. *In* Biological aspects of thermal pollution, edited by P. A. Kren and F. L. Parker, Nashville, Tenn., Vanderbilt University Press, pp. 229–93.

DONALDSON, L R and FOSTER, R F Effects of radiation on aquatic
1957 organisms. *Publs U.S. natn. Acad. Sci.*, (551):96–102.

FØYN, E Disposal of waste in the marine environment and the
1965 pollution of the sea. *Oceanogr. mar. Biol.*, 3:95–114.

FROST, J Earth, air, water. *Environment*, 11(6):14–29, 31–3.
1969

GALTSOFF, P S and CHIPMAN, W A Ecological and physiological
1947 studies of the effect of sulfate pulp mill wastes on oysters in the York River, Virginia. *Fishery Bull. U.S. Fish Wildl. Serv.*, (43):59–186.

GARSIDE, E T Some effects of oxygen in relation to temperature on
1959 the development of lake trout embryos. *Can. J. Zool.*, 37:689–98.

GENELLY, R E and RUDD, R L Effect of DDT, toxaphene and
1956 dieldrin on pheasant reproduction. *Auk*, 73(3):529–39.

GEORGE, J L and FREAR, D E H Pesticides in the Antarctic.
1966 *J. appl. Ecol.*, 3(Suppl.):155–67.

GOLOVINSKAIA, K A and ROMASHOV, D D Effect of ionizing
1958 radiation on the development and reproduction of fishes. *Vop. Ikthiol.*, 11:16–38 (in Russian).

GREEN, J M and ROTH, J S The effect of radiation from small
1955 amounts of ^{32}P, ^{35}S and ^{42}K on the development of *Arbacia* eggs. *Biol. Bull. mar. biol. Lab.*, Woods Hole, 108(1):21–8.

GROSCH, D S Poisoning with DDT: effect on reproductive
1967 performance of *Artemia. Science*, 155:592–3.

HARTUNG, R Some effects of oiling on reproduction of ducks.
1965 *J. Wildl. Mgmt*, 29(4):872–4.

HEATH, R G, SPANN, J W and KREITZER, J F Marked DDT
1969 impairment of mallard reproduction in controlled studies. *Nature, Lond.*, 244:47–8.

HEDGPETH, J W and GONOR, J J Aspects of the potential effect of
1969 thermal alteration of marine and estuarine benthos. *In* Biological aspects of thermal pollution, edited by P. A. Krenkel and F. L. Parker, Nashville, Tenn., Vanderbilt University Press, pp. 80–132.

HICKEY, J J and ANDERSON, D W Chlorinated hydrocarbons and
1968 eggshell changes in raptorial and fish-eating birds. *Science, N.Y.*, 162:271–3.

HICKEY, J J, KEITH, J A and COON, F B An exploration of
1966 pesticides in a Lake Michigan ecosystem. *J. appl. Ecol.*, 3(Suppl.):141–54.

HOLDEN, A V and MARSDEN, K Organochlorine pesticides in seals
1967 and porpoises. *Nature, Lond.*, 216:1274–6.

HUBBS, C Fertilization, initiation of cleavage and developmental
1966 temperature tolerance of the cottid fish, *Clinocottus analis. Copeia*, 1966(1):29–42.

KARBE, L Die Auswirkung der künstlichen Eutrophierung des
1964 Bodensees auf das Artgefüge seiner Coregonenpopulationen. *Mitt. hamb. zool. Mus. Inst.*, Kosswig Festschrift: 83–90.

KEITH, J A Reproduction in a population of herring gulls (*Larus*
1966 *argentatus*) contaminated by DDT. *J. appl. Ecol.*, 3(Suppl.):57–70.

KINNE, O Über den Einfluss des Salzgehaltes und der Temperatur
1956 auf Wachstum, Form und Vermehrung bei dem Hydroid-polypen *Cordylophora caspia* (Pallas). 1. Mitteilung über den Einfluss des Salzgehaltes auf Wachstum und Entwick-lung mariner, brackischer und limnischer Organismen. *Zool. Jb. (Physiol.)*, 66(4):565–638.

KINNE, O The effects of temperature and salinity on marine and
1963 brackish water animals. 1. Temperature. *Oceanogr. mar. Biol.*, 1:301–40.

KINNE, O The effects of temperature and salinity on marine and
1964 brackish water animals. 2. Salinity and temperature-salinity combinations. *Oceanogr. mar. Biol.*, 2:281–339.

KINNE, O and KINNE, E M Effects of salinity and oxygen on
1962 developmental rates in a cyprinodont fish. *Nature, Lond.*, 193:1097–8.

KINNE, O Rates of development in embryos of a cyprinodont fish
1962a exposed to different temperature–salinity–oxygen com-binations. *Can. J. Zool.*, 40(2):231–253.

KINNE, O and ROSENTHAL, H Effects of sulfuric water pollutants
1967 on fertilization, embryonic development and larvae of the herring, *Clupea harengus*. *Mar. Biol.*, 1(1):65–83.

KINNE, O and SCHUMANN, K-H Biologische Konsequenzen
1968 schwefelsäure-und eisensulfathaltiger Industrieabwässer. Mortalität junger *Gobius pictus* und *Solea solea* (Pisces). *Helgoländer wiss. Meeresunters.*, 17(1–4):141–55.

KOEMAN, J H, OUDEJANS, R C H M and HUISMAN, E A Danger of
1967 chlorinated hydrocarbon insecticides in birds' eggs. *Nature, Lond.*, 215:1094–6.

KONAR, S K Lethal effect of the insecticide DDVP on the eggs and
1969 hatchlings of the snake-head, *Channa punctatus* (Bl.). *Jap. J. Ichthyol.*, 15:130–3.

KUTHALINGAM, M D K Temperature tolerance of the larvae of
1959 ten species of marine fishes. *Curr. Sci.*, 28:75–6 (cited in de Silva, 1969).

LACAZE, J C Effects d'une pollution de type *Torrey Canyon* sur
1969 l'algue unicellulaire marine *Phaeodactylum tricornutum*. *Revue int Océanogr. méd.*, 13–14:157–79.

LOOSANOFF, V L Effects of DDT upon setting, growth and survival
1947 of oysters. *Fishg Gaz. N.Y.*, 64(4):94, 96.

LOOSANOFF, V L Effects of pesticides on marine arthropods and
1960 molluscs. *Publ.Hlth Serv. Tech. Rep., Wash.*, (W60–3):89–93.

MACEK, K J Reproduction in brook trout (*Salvelinus fontinalis*)
1968 fed sub-lethal concentrations of DDT. *J. Fish. Res. Bd Can.*, 25(9):1787–96.

MARSHALL, J S The effects of continuous gamma radiation on the
1962 intrinsic rate of natural increase of *Daphnia pulex*. *Ecology*, 43(4):598–607. Also issued in Radioecology, edited by V. Schultz and A. W. Klement, New York, Reinhold, pp. 363–6 (1963).

MAUCHLINE, J and TEMPLETON, W L Artificial and natural radi-
1964 isotopes in the marine environment. *Oceanogr. mar. Biol.*, 2:229–79.

MENZEL, D W, ANDERSON, J and RANDTKE, A Marine phyto-
1970 plankton vary in their response to chlorinated hydrocarbons. *Science, N.Y.*, 167 (3926):1724–26.

MERRIMAN, D and SCHEDL, H P The effects of light and tem-
1941 perature on gametogenesis in the four-spined stickleback, *Apeltes quadracus* (Mitchill). *J. exp. Zool.*, 83(3):413–46.

MIHURSKY, J A On possible constructive uses of thermal additions
1967 to estuaries. *BioScience*, 17(10):698–702.

MIRONOV, O G Hydrocarbon pollution of the sea and its influence
1968 on marine organisms. *Helgoländer wiss. Meeresunters.*, 17(1–4):335–9.

MOATS, S A and MOATS, W A Toward safer use of pesticides.
1970 *BioScience*, 20(8):459–64.

MOORE, N W and TATTON, J O'G Organochlorine insecticides in
1965 the eggs of sea birds. *Nature, Lond.*, 207:42–3.

NAYLOR E Effects of heated effluents upon marine and estuarine
1965 organisms. *Adv. mar. Biol.*, 3:63–103.

NAYLOR E Biological effects of a heated effluent in docks at
1965a Swansea, S. Wales. *Proc. zool. Soc. Lond.*, 144:253–68.

NÜMANN, W Die Veränderungen im Blaufelchenbestand (*Corego-
1964 nus wartmanni*) und in der Blaufelchenfischerei als Folge der künstlichen Eutrophierung des Bodensees. *Verh. int. Verein. theor. angew. Limnol.*, 15:514–23.

OKUBO, K and OKUBO, T Study on the bioassay method for the
1962 evaluation of water pollution. 2. Use of fertilized eggs of sea urchins and bivalves. *Bull. Tokai reg. Fish. Res. Lab.*, (32):131–40.

OLSEN, P A and FOSTER, R F Temperature tolerance of eggs and
1955 young of Columbia River chinook salmon. *Trans. Am. Fish. Soc.*, 85:203–7.

PANNELL, J P M, JOHNSON, A E and RAYMONT, J E G An investi-
1962 gation into the effects of warmed water from Marchwood Power Station into Southampton water. *Proc. Instn civ. Engrs*, 23:35–62.

PEAKALL, D B Pesticide-induced enzyme breakdown of steroids in
1967 birds. *Nature, Lond.*, 216:505–6.

POLIKARPOV, G G and IVANOV, V N On the action of ^{90}Sr–^{90}Y on
1961 the development of the spawn of the khamsa. *Vop. Ikthiol.*, 1(20):583–90 (in Russian).

RISEBROUGH, R W Chlorinated hydrocarbons in marine eco-
1969 systems. *In* Chemical fallout, edited by M. W. Miller and G. G. Berg, Springfield, Ill., C. C. Thomas, pp. 5–23.

RISEBROUGH, R W *et al.* DDT residues in Pacific sea birds; a
1967 persistent insecticide in marine food chains. *Nature, Lond.*, 216:589–91.

RITTINGHAUS, H Etwas über die "indirekte" Verbreitung der
1956 Ölpest in einem Seevogelschutzgebiet. *Orn. Mitt.*, 8(3):43–6.

ROBINSON, J, *et al.* Organochlorine residues in marine organisms.
1967 *Nature, Lond.*, 214(5095):1307–11.

RUNNSTRÖM, S Über die Thermopathie der Fortpflanzung und
1928 Entwicklung mariner Tiere in Beziehung zu ihrer geo-graphischen Verbreitung. *Bergens Mus. Arb. (Naturv. rekke)* 1927, 2:1–67.

SCHNAKENBECK, W *Tierwelt N.-u. Ostsee*, 1 (Ie):1–48 (ref. not
1940 seen, cited in Kinne, 1964).

SLADEN, W J L, MENZIO, C M and REICHEL, W L DDT residues
1966 in Adelie penguins and a crab eater seal from Antarctica. *Nature, Lond.*, 210:670–3.

STEWART, K M and ROHLICH, G A Eutrophication—a review.
1967 *Publs St. Wat. Qual. Control Bd Calif.*, (34):188 p.

STOCK, J N and COPE, O B Some effects of TEPA, an insect
1969 chemosterilant, on the guppy, *Poecilia reticulata*. Trans. Am. Fish. Soc., 98:280–7.

STUBBINGS, H G and HOUGHTON, D R The ecology of Chichester
1964 Harbour, S. England, with special reference to some fouling species. *Int. Rev. ges. Hydrobiol.*, 49(2):233–80.

SWARUP, H Abnormal development in the temperature-treated
1958 eggs of *Gasterosteus aculeatus* (L.). 1. Cleavage abnormali-ties. *J. zool. Soc. India*, 10:108–13.

SWARUP, H Abnormal development in the temperature-treated
1959 eggs of *Gasterosteus aculeatus* (L.). 2. Gastrulation abnormalities *J. zool. Soc. India*, 11:1–6.

WARNER, R E, PETERSON, K K and BORGMAN, L Behavioural
1966 pathology in fish: a quantitative study of sublethal pesticide toxication. *J. appl. Ecol.*, 3(Suppl.):223–47.

WAUGH, G D and ANSELL, A The effect on oyster spatfall of
1956 controlling barnacle settlement with DDT. *Ann. appl. Biol.*, 44(4):619–25.

WAUGH, G D, HAWES, F B and WILLIAMS, F Insecticides for
1952 preventing barnacle settlement. *Ann. appl. Biol.*, 39:407–15.

WELCH, R M, LEVIN, W and CONNEY, A H Effect of chlorinated
1969 insecticides on steroid metabolism. *In* Chemical fallout, edited by M. W. Miller and G. G. Berg, Springfield, Ill., C. C. Thomas, pp. 390–407.

WOODWELL, G M, WURSTER, C F and ISAACSON, P A DDT
1967 residues in an east coast estuary: a case of biological concentration of a persistent insecticide. *Science, N.Y.*, 156:821–4.

WORLEY, L G Development of the egg of the mackerel at different
1933 temperatures. *J. gen. Physiol.*, 16(5):841–57.

WURSTER, C F DDT reduces photosynthesis in marine phyto-
1968 plankton. *Science, N.Y.*, 159(3822):1474–5.

WURSTER, C F Chlorinated hydrocarbon insecticides and avian
1969 reproduction: how are they related? *In* Chemical fallout, edited by M. W. Miller and G. G. Berg, Springfield, Ill., C. C. Thomas, pp. 368–89.

WURSTER, C F and WINGATE, D B DDT residues and declining
1968 reproduction in Bermuda petrel. *Science, N.Y.*, 159:979–81.

Effects of Low Concentrations of Free Chlorine on Eggs and Larvae of Plaice, *Pleuronectes platessa* L.

R. Alderson*

Incidences de faibles concentrations de chlore libre sur les oeufs et les larves de plie *Pleuronectes platessa* L.

De nombreuses centrales électriques côtières qui utilisent l'eau de mer pour leurs systèmes de refroidissement par condensation additionnent l'eau de chlore au point de capture afin de prévenir l'accumulation de déchets d'origine organique. Lorsque l'eau de refroidissement est déversée, elle contient encore d'ordinaire un faible pourcentage de chlore et, bien que celui-ci puisse être rapidement dispersé par dilution, il est néanmoins assez important de savoir s'il pourrait avoir des effets sur les organismes marins.

Une enquête est actuellement menée sur les incidences de faibles concentrations de chlore dans l'eau de mer aux stades précoces du développement de certains poissons plats et l'on a communiqué les résultats de cette enquête en ce qui concerne les effets aigus et chroniques de ces concentrations sur les oeufs et les larves de plie *Pleuronectes platessa* L. Les effets aigus se mesurent par la teneur en DL_{50} pendant une période de 48 heures; les effets chroniques sur les oeufs se manifestent par les dommages morphologiques infligés aux larves après éclosion, et sur les larves elles-mêmes par les modifications de leur régime alimentaire.

Efectos del cloro, en bajas concentraciones, sobre los huevos y larvas de la solla *Pleuronectes platessa* L.

Son muchas las centrales eléctricas costeras que emplean agua del mar para sus sistemas de refrigeración de los condensadores, y que añaden cloro al agua en el punto de toma, para impedir la proliferación de organismos dañinos. Cuando las aguas del sistema de refrigeración se descargan, contienen aun de ordinario pequeñas cantidades de cloro y, aunque el cloro puede dispersarse rápidamente por dilución, no deja de ser importante saber si puede afectar a los organismos marinos.

Se está realizando un estudio de los efectos del cloro, en bajas concentraciones de agua de mar, sobre las primeras fases de desarrollo de algunos peces planos y en esta contribución se dan los resultados relativos a los efectos agudos y crónicos del cloro sobre los huevos y larvas de la solla *Pleuronectes platessa* L. Los efectos agudos se miden por la DL_{50} tipo en 48 horas; los efectos crónicos se miden: en los huevos, por los daños morfológicos sufridos por las larvas que nacen de ellos, y en las larvas, por los cambios que se registran en la ingestión de los alimentos.

THIS paper enquires into the effects of low levels of chlorine on the survival and growth of flatfish eggs and larvae and results from a programme for exploiting the waste heat of marine power station effluents for the culture of certain flatfish, currently being developed by the White Fish Authority at the Hunterston Nuclear Generating Station in Ayrshire. Chlorine is important as an environmental factor in this programme because it is used in most power stations as an anti-fouling agent. It is injected at the seawater intake and prevents the growth both of mussels in the intake culverts, and of slime bacteria in the condenser cooling systems. Much of the chlorine is removed by reaction with organic material in its passage through the power station, but a detectable chlorine level is usually present in the effluent. The effect of this on the developmental stages of fish, both those under culture in the effluent and wild stock coming into contact with the effluent discharge, is of considerable interest.

In earlier work (Alderson, 1969) chlorine concentrations required for toxicity testing were produced by adding measured volumes of sodium hypochlorite solution to sea water. Tests were then carried out under static conditions. This was unsatisfactory because the level of chlorine in the containers fell during the experiment. This was overcome by a continuous flow of chlorinated sea water thus enabling constant concentrations to be maintained.

Apparatus

The equipment (fig 1) consisted basically of a pair of constant head vessels A_1 A_2 delivering constant flows of fresh and chlorinated sea water into a mixing and holding tank B. The resultant mixture was then passed through a further constant head vessel C into two 34 1 tanks containing test animals. Equivalent flow rates were maintained through both tanks, and for experiments on the egg stage these were 12 1/h. It was possible to reduce the flow to 4 1/h while maintaining a stable chlorine level and these flow rates were used in all subsequent larval experiments. Throughout the apparatus the valves used to adjust flow rates consisted of Hoffman

clips on leached PVC tubing. To produce efficient mixing the chlorinated sea water was introduced through a 2.5 cm diam rigid polythene tube with the ends blocked off and a series of 1.5 mm holes drilled along both sides of its length. Placed on the bottom of the tank, aeration from an air stone at each end provided further water movement and mixing. Five of these systems with varying mixtures of fresh and chlorinated sea water gave the desired range of chlorine concentrations; the control being of fresh water only.

The continuous supply of chlorinated sea water for the

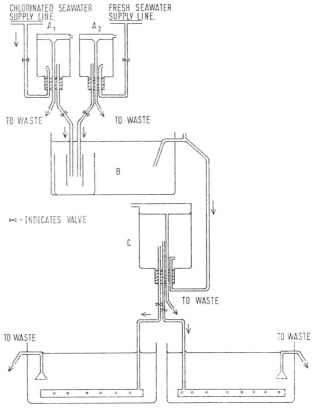

Fig. 1. Semi-diagrammatic representation of system for maintaining constant chlorine concentrations in experimental tanks

---

* Ministry of Agriculture, Fisheries and Food, Marine Hatchery, Port Erin, Isle of Man, U.K.

mixing system was produced by the direct electrolysis of a constant flow of sea water through a Paterson Candy Chlorocel, a constant DC current of about 100 mA at 3.5 V being passed between the electrodes. Under the conditions of this electrolysis the main chemical reactions are given, according to Adamson *et al.*, 1963, by Equations (1) and (2):

At the cathode

$$(1)\ 2H_2O + 2e \to 2OH' + H_2$$

At the anode

$$(2)\ 2Cl' - 2e \to Cl_2$$
$$(3)\ 2OH' - 2e \to \tfrac{1}{2}O_2 + H_2O$$

The other possible anodic reaction, their Equation (3), was unlikely to occur because of the low current density at the electrode surface and the rapid replacement of electrolyte at the flow rates used, which reduced the possibility of depletion of chloride ions. Some bromine may also have been liberated at the anode, but as the concentration of chloride ion in sea water is much higher than that of the bromide ion, chlorine would be the principal halogen discharged. The discharge of hydroxyl ions at the cathode caused some precipitation of magnesium and calcium hydroxides as their solubility products were exceeded. These flocculent precipitates were removed by a glass wool filter.

The chlorine evolved at the cathode would not long remain in molecular form, but would be quickly hydrolysed to hypochlorous acid and hypochlorite ion. Similar bromine compounds, hypobromous acid and the hypobromite ion would also be formed by the reaction of both the chlorine and its hydrolysis products with the bromide ion in sea water (Lewis, 1966; Cavell, personal comment). Further reactions would occur between these halogen compounds and nitrogenous materials in sea water, forming chloramines and bromamines. The relative proportions of the products of these reactions in the final mixture would depend on the pH of the sea water and the amount of nitrogenous material present.

Chlorination of sea water for the cooling systems of power stations is achieved by the direct injection of chlorine gas into the intake water, rather than by electrolysis, but it is probable the reactions occurring in both systems ultimately lead to similar mixtures of halogen derivatives.

Chlorine determinations

Chlorine levels were measured colorimetrically by Palin's DPD method (Palin, 1967) and the acid o-tolidine method (Orland, 1965). The DPD method provides some differentiation between the more active components (hydrochlorous acid, hypochlorite, hypobromous acid, hypobromite and the bromamines) and the less active chloramines, but is not as sensitive as the o-tolidine method. The latter method gives a reliable measure of the total available halogen only, the chloramines being included with the other halogen compounds, but it is possible to determine concentrations in the 0.01–0.1 ppm range to within ±0.002 ppm. Caution has to be observed with o-tolidine, however because it is a potentially carcinogenic compound.

All colorimetric measurements were made with a spectrophotometer with 4 cm cells, transmission being recorded at 555mμ for the DPD method, and 435 mμ or 490 mμ for the o-tolidine. Chlorine concentrations were calculated from previously constructed calibration curves. All measurements are given in terms of ppm chlorine. During each experiment 20 ml samples were removed at least once a day, and chlorine level measured. Some fluctuations in concentration were noticed; for example, in an eight-day experiment at a mean concentration of 0.040 ppm the range was 0.037–0.043 ppm, and at a mean concentration of 0.32 ppm the range was 0.30–0.33 ppm. In experiments of shorter duration the mean levels were more closely maintained.

Experiments

Experiments were done to determine the LD_{50} at several stages in the early development of the plaice, *Pleuronectes platessa* L. Test animals were usually exposed for 96 h, but an assessment was also made of dead individuals after 48 h treatment, so that a 48 h as well as a 96 h LD_{50} result was obtained from some experiments.

After chlorination, dead eggs were those which, after cessation of aeration for 20 min, lay on the bottom. Dead larvae were easier to define by tissue opacity, shrinkage and distortion; these were counted and removed. A count was also made of those larvae which, though still alive, were lying on the bottom, semi-moribund, and incapable of normal swimming movement when stimulated; these, however, were not removed. The mortality data were subjected to probit analysis (Finney, 1952) and the LD_{50} value was obtained by graphical interpretation of results, the straight line being fitted by eye.

Temperatures at which experiments were done differed for each development stage of plaice according to the ascending temperature regime recommended by Shelbourne (1964). The larval stages referred to correspond to the arbitrary division of their morphological development given by Ryland (1966); stage 1c still had the yolk sac present; stage 2a had resorbed the yolk and commenced feeding on *Artemia*, and stages 3b–c were in the late pelagic phase before the onset of metamorphosis.

All experimental animals were obtained from captive

TABLE 1. LD_{50} VALUES FOR PLAICE EGGS AND LARVAE EXPOSED TO A RANGE OF CHLORINE CONCENTRATIONS IN SEA WATER

	Stage at beginning of experiment				
	Eggs		Larvae		
	4 days after spawning	5 days after spawning	1c	2a	3b–c
Duration of exposure (hours)	72	192	96	96	96
LD_{50} at end of exposure (ppm chlorine)	0.7	0.12	0.024	0.028	0.034
Highest concentration of chlorine giving mortality no greater than in the control (ppm)	0.25	0.04	*	0.016	0.020
Number of specimens per tank	500	500	80	80	60
Temperature (°C)	5.4–5.8	5.9–6.5	7.4–7.8	7.5–8.1	8.8–9.3

* Mortalities greater than in the control tanks occurred in all concentrations tested; the mortality at the lowest, 0.016 ppm, was 15%.

[313]

spawners, the eggs and subsequent larvae being held in black polythene tanks (120 × 60 × 30 cm) until required, the larvae being fed daily with *Artemia* nauplii. When established feeding stock was tested, an excess of *Artemia* nauplii was maintained.

Results

Results of experiments to date are in Table 1, together with experimental conditions. Only 96 h mortality figures are given for the larval stages, since the 48 h LD_{50} values, particularly when semi-moribund larvae were included, showed little difference from values obtained for longer exposure. In an experiment on stage 1c larvae, LD_{50} values after 48 and 96 h exposure were as follows:

48 h LD_{50}		96 h LD_{50}	
dead larvae	dead + moribund larvae	dead larvae	dead + moribund larvae
0.032	0.026	0.025	0.024

In contrast experiments in the egg stage showed a considerable difference when exposure was prolonged; LD_{50} values for 72 and 192 h (3 and 8 days) exposures in Table 1.

At the end of the 8-day experiment, eggs from the two lowest concentrations, 0.075 and 0.04 ppm, and the control tanks were allowed to continue their development up to hatching in clean sea water. No significant difference in the percentage hatch was observed, nor was there any evidence of morphological damage to the subsequently hatched larvae. The egg membranes therefore seem to give considerable protection to the embryo, allowing development to continue normally over long periods even in concentrations of chlorine which would be rapidly lethal to hatched larvae.

Eggs from a later spawning were exposed to a range of chlorine concentrations for 3 days at a temperature of 7.5°C. Eggs from all tanks were then transferred to clean sea water and development allowed to continue until yolk sac absorption was completed. Again, there was no evidence of morphological abnormalities due to the chlorine exposure, but there was a difference in the proportion of eggs hatching in the batches exposed to different chlorine concentrations:

Number of eggs exposed	400	200	200	400	400	400
Chlorine concentration (ppm)	0	0.36	0.30	0.14	0.10	0.05
Percentage of eggs failing to hatch	0.75	36	18	15	9	0.75

Unhatched and dead eggs were removed 3 days after hatching had been completed in control tanks. On removal, many eggs which had been exposed to chlorine were found to hatch as they were ejected from the pipette into a glass beaker, and the larvae swam around quite actively. It would appear therefore that in a proportion of the individuals the exposure to chlorine, though not killing the larvae, in some way prevented the hatching mechanism from operating successfully.

Results for larvae in Table 1 show the very low levels of chlorine required to produce a significant mortality. It was also found that the regression of probit on chlorine concentration had a very steep gradient in these experiments, and for all larval stages the level for 5 per cent mortality was found to be only approximately 0.007 ppm less than the level for 50 per cent mortality. Table 1 also shows the increase in LD_{50} as larval development proceeds, but temperature differences between stages could also have had some influence on this result (Brown *et al.*, 1967).

On stage 3b-c larvae the proportion actively feeding at the end of the experiment was also assessed. As the larvae were removed from the experimental tanks they were held in a pipette in front of a bright light and the presence of *Artemia* nauplii in the guts was noted. The numbers of feeding larvae were expressed as a percentage

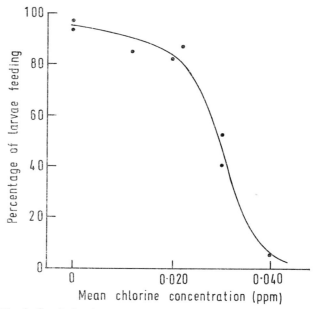

Fig. 2. Graph showing percentage of stage 3b-c larvae which were actively feeding at the end of 96 hours' exposure to a range of chlorine concentrations

of the initial number of larvae exposed, and in fig 2 these percentages are plotted against the chlorine concentrations. The results were also treated by probit analysis and the concentration for 50 per cent feeding was determined as 0.030 ppm. These experiments all show the effects of relatively long-term exposure to chlorine in sea water. A separate series of short-term experiments was carried out on batches of 3b-c larvae to determine the lethal effect of higher concentrations than used in previous experiments.

Batches of 20 larvae were subjected to a particular concentration at a temperature of 9.6-9.9°C for different periods of time, and then transferred to fresh sea water for a further 24 h before mortality was assessed. The percentage mortality on a log scale was plotted against the time of exposure and the time taken to kill 50 per cent of the larvae determined. The results show that even at the relatively low level of 0.1 ppm larvae are rapidly affected:

Chlorine concentration ppm	0.05	0.075	0.10	0.13
Time to 50 per cent kill (min)	460	175	90	70

Discussion

If it can be assumed that electrolytic chlorination is similar in its effect to gas chlorination, the results presented give an indication of the chlorine levels which could be tolerated in power station effluent by plaice eggs or larvae.

Measurements of chlorine concentrations made during 1969 by the White Fish Authority in the effluent at the Hunterston Nuclear Generating Station (Cheetham, personal comment), where continuous low-level gas chlorination is operated, showed that chlorine levels of 0.5 and 0.7 ppm were occasionally reached. These levels would be rapidly lethal to plaice larvae at any of the stages tested. The average range (0.02–0.35) was however lower, but the upper part of this range would also be lethal to plaice larvae. These figures refer to chlorine levels in the effluent before discharge into the open sea, and there are two factors which would probably considerably reduce the chlorine level at the outfall itself. First, the chlorine would probably continue to react with, and be absorbed by, constituents of the sea water in the effluent itself; secondly, mixing of the effluent with fresh sea water at the outfall would result in further rapid absorption of chlorine. In the absence of direct measurements of chlorine in the sea water around power station outfalls, it is not therefore possible to state whether these discharges could represent a real hazard to plaice and perhaps other wild fish larvae.

References

ADAMSON, A F, LEVER, B G and STONES, W F   The production of
1963   hypochlorite by direct electrolysis of sea water: electrode materials and design of cells for the process. *J. appl. Chem., Lond.*, 13:483–95.

ALDERSON, R   The survival of flatfish eggs, larvae and post-larvae
1969   in low concentrations of free chlorine. ICES C.M. 1969 Doc.(F:27).

BROWN, V M, JORDAN, D H M and TILLER, B A   The effect of
1967   temperature on the acute toxicity of phenol to rainbow trout in hard water. *Wat. Resour., Wash.*, 1:587–94.

FINNEY, D J   Probit analysis. Cambridge, Cambridge University
1952   Press.

LEWIS, B G   Chlorination and mussel control 1. The chemistry of
1966   chlorinated sea-water: a review of the literature. Central Electricity Research Laboratories (RD/L/N106/66).

ORLAND, H P (Ed.)   Standard methods for the examination of
1965   water and waste water. New York, American Public Health Association Inc., 769 p.

PALIN, A T   Methods for the determination in water of free and
1967   combined available chlorine, chlorine dioxide and chlorite, bromine, iodine and ozone, using diethyl-p-phenylene diamine (DPD). *J. Instn Wat. Engrs*, 21(6):537–47.

RYLAND, J S   Observations on the development of larvae of the
1966   plaice, *Pleuronectes platessa* L., in aquaria. *J. Cons. perm. int. Explor. Mer*, 30(2):177–95.

SHELBOURNE, J E   The artificial propagation of marine fish. *Adv.*
1964   *mar. Biol.*, 2:1–83.

The Influence of Crude Oils on Fish Fry

*W. W. Kühnhold**

Influence des pétroles bruts sur les alevins

La documentation révèle que la pollution par les hydrocarbures fait peser une menace croissante sur les organismes aquatiques depuis quelque 70 ans. Le présent document passe en revue les renseignements publiés sur les effets des pétroles bruts et de certains de leurs sous-produits.

Des expériences ont été faites récemment pour étudier les effets de divers bruts d'extraction récente (Venezuela, Iran, Libye) sur les oeufs et les larves de harengs, de morues et de plies. Ces poissons ont été exposés à des dispersions de bruts d'extraction récente dans l'eau et à de l'eau contenant des composés de pétrole brut dissous.

Parmi les résultats enregistrés figuraient la mortalité des oeufs, les taux d'éclosion, les anomalies larvaires et la période de survie des larves dans l'eau des essais. Les résultats ont montré que la mortalité des oeufs est proportionnelle au taux de concentration des composés dissous lorsqu'il dépasse une certaine limite. Les oeufs sont le plus fortement atteints pendant les trois premiers jours de leur développement. Les larves sont plus sensibles que les embryons aux mêmes taux de concentration des composés dissous.

Les dispersions ont une toxicité beaucoup plus forte qu'une pellicule d'hydrocarbures d'égal volume répandue à la surface de l'eau. La toxicité des dispersions s'accroît en cas de mélange avec un dispersant, même s'il n'est pas toxique aux concentrations appliquées.

De nouvelles expériences ont montré que les larves semblent incapables de percevoir les dispersions d'hydrocarbures. En effet, les jeunes larves n'évitent pas ces dernières, bien qu'elles aient des effets immédiats sur la nageoire primordiale et l'intégument.

Influencia de los hidrocarburos crudos en los alevines

La literatura existente sobre el tema revela que la contaminación ocasionada por los hidrocarburos ha constituido un creciente y constante peligro para los organismos acuáticos durante los siete pasados decenios. Se analiza la información publicada sobre los efectos de los hidrocarburos crudos y de algunos de sus productos.

Recientemente se realizaron experimentos para estudiar las consecuencias de los hidrocarburos crudos frescos de origen diferente (Venezuela, Irán, Libia) sobre los huevos y larvas del arenque, bacalao y platija. Se expusieron éstos a dispersiones en agua de hidrocarburos crudos frescos y a agua conteniendo compuestos de hidrocarburos crudos disueltos.

El registro incluía la mortalidad de los huevos, ritmo de incubación, anormalidades larvales y período de supervivencia de las larvas en el agua en que se efectuaba la prueba. Los resultados indicaron que la mortalidad de los huevos es proporcional a la concentración de los compuestos disueltos si pasa de un cierto límite. Los huevos se hallan más afectados durante los tres primeros días de su desarrollo. Las larvas son más sensibles a las mismas concentraciones de disueltos compuestos que los embriones.

Las dispersiones tienen una toxicidad mucho mayor que la misma cantidad de una capa de hidrocarburos en la superficie del agua. La toxicidad de las dispersiones aumenta aún más cuando a ellas se mezcla un dispersante, aun cuando éste no sea tóxico en la concentración aplicada.

Otros experimentos demostraron que las larvas no parecen percibir las dispersiones de los hidrocarburos. Aquéllas no evitan las dispersiones que tienen efectos inmediatos en la aleta primordial y en el integumento de las larvas jóvenes.

IN fighting the *Torrey Canyon* disaster contaminated shores and drifting oil patches were treated with enormous quantities of oil spill removers. Field observations and the following tests showed that most of these chemicals were highly toxic to many littoral organisms but damage to fisheries and fish stocks has largely been unassessed (Simpson, 1968).

Many commercially important species have planktonic eggs and larvae which are abundant in the uppermost surface layer and therefore exposed to the influence of oils and dispersants. The effect of floating films of crude oil left untreated or crude oils dispersed by a "non-toxic" agent on fish eggs and larvae has been investigated.

Water extracts of different crude oils showed high

---

* Institut für Meereskunde an der Universität Kiel, Kiel, Federal Republic of Germany.

toxicity on herring eggs when incubated under a film of 10^3 and 2.10^4ppm (Kühnhold, 1969). The mean survival time was 2.5 to 3.5 days. Newly hatched herring larvae were killed in nearly the same time. This high toxicity is mainly caused by low boiling components.

The present paper gives some results of current investigations which will be published later in detail.

Material and methods

Cod eggs of different stages and young larvae of cod (*Gadus morrhua* L.), herring (*Clupea harengus* L.), and plaice (*Pleuronectes platessa* L.), which had been fertilized and reared in laboratories, were contaminated with dissolved and dispersed crude oils of different origin and character: a Venezuelan (Tia Juana) crude belonging to the naphthene-basic type, an Iranian (Agha Jari) and a Libyan (Sarir) crude of paraffin-basic character.

Experiments with dissolved compounds were carried out by extracting the oils in separate 30 1– plastic containers to avoid direct application of oil films in the test jars. Different amounts of oil (10^4, 10^3 10^2ppm) were poured onto the water and left for two days. A calm water circulation in the containers was maintained by means of small pumps, ensuring maximum saturation of soluble compounds. Evaporation, however, could take place at the same time. The clear oil extract was transferred into the test jars and renewed every two days. The amount of crude oil used for preparing the extracts is, however, no criterion for the actual amount and types of dissolved hydrocarbons. Preliminary chemical analyses show that under the described test conditions the amount of hydrocarbons dissolved from 10^4 ppm of crude oil are in the range of 10 ppm.

Oil dispersions were obtained by stirring 10^3 ppm of the Iranian crude oil for one minute at 10,000 rev/min; dilutions were then prepared at once. The experiments were started 1 and 50 hours later.

Results

It is difficult to give the $50LC_{50}$ or $100LC_{50}$ of the eggs as the mortality rate in the controls was 0–30 per cent after 100 hours depending on the spawn quality. Young eggs which were put into the oil extracts 5 to 30 h after fertilization seemed to be most sensitive. After 100 h the extract from 10^4ppm of Venezuelan crude had caused a

40 per cent higher mortality than in the control (fig 1). The same amount of Iranian crude caused a mortality of about 30 per cent above control. An extract from 100ppm still caused 10 to 17 per cent higher mortality. A similar toxic effect was also found with eggs that had finished gastrulation. The Libyan crude oil, on the other hand, was nearly non-toxic, possibly because of the high stockpoint of 18°C or a smaller content of toxic fractions.

Ten-day-old embryos with a fully developed heart were less sensitive to all three concentrations and did not show greater mortality than the controls.

In Iranian crude oil the mortality depended on the duration of exposure and the concentration. One series of eggs was kept in the extract during the whole test period, while another lot was transferred into clean water after 100 h. In this case the mortality rate was obviously lower (Table 1).

TABLE 1. CORRELATION BETWEEN MORTALITY OF COD EGGS AND DURATION OF OIL INFLUENCE (IRAN, AGHA JARI)

Amount of crude oil (ppm)	Test for 100 hours in oil extract	test continued until hatching	
		in oil extract	in clean water
	% dead eggs	% dead eggs	% dead eggs
10^4	30	99	43
10^3	24	63	32
10^2	13	33	18
control	14		21

Embryos contaminated with Venezuelan extract did not recover although they were transferred to clean water after 48 h. The deleterious effect of this oil seems to be manifested much more rapidly.

When eggs were kept under test conditions until hatching, the spectrum of the hatching success given in fig 2 shows more clearly the graduated correlation between oil concentration, sensitivities of the embryos and the different toxicities of the oils. Fifty per cent hatching success was only achieved in some cases.

The biological effects revealed were worse than the numerical scale expresses, because most of the larvae that hatched had deformed bodies or abnormal flexures of the tail so that they were unable to swim normally and most died within one day. The deformation could often be observed from gastrulation onward. In the two

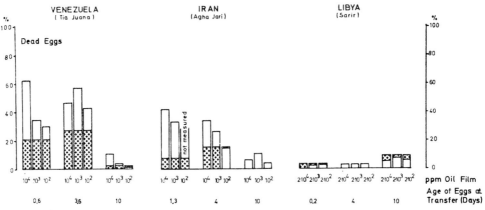

Fig 1. Mortality rates of cod eggs at different ages in water extracts from different crude oils (Venezuela (Tia Juana), Iran (Agha Jari), Libya (Sarir))

▨ = mortality rates in the controls
ppm Oil Film means amount of oil extracted in 30-l containers

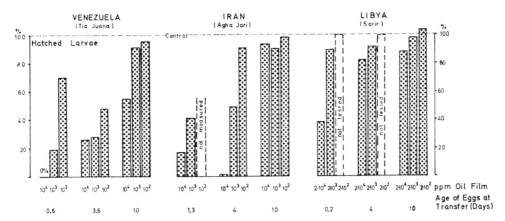

Fig 2. Hatching rates of cod eggs given as percentages of the control (= 100%); ppm Oil Film means amount of oil extracted in 30-l containers

highest concentrations a delay of development was observed beginning 3.5 to 4 days after putting the eggs into the test milieu. In some cases hatching was delayed or did not occur though the embryos looked normally developed.

Effects on larvae

The larvae showed typical behavioural symptoms in oil extracts: increased activity (especially in high concentration) was followed by a reduction of swimming activity, which finally stopped except for sporadic twitches. The larvae then showed signs of beginning narcosis, which slowly deepened until the "critical point" when no responses of the larvae were obtained even by touching or prodding. These observations are similar to those reported by Wilson (1970), who tested the effect of several dispersants with aromatic and kerosene solvents on fish larvae.

After the "critical point" is reached, no recovery from narcosis was observed when the larvae were put back into clean water. This point was regarded as the most important for the judgement of the toxic effect of the oils tested. The "mean critical time", i.e. the mean time until reaching the "critical point", was plotted against concentration of oil.

It was observed that young larvae are much less resistant to the dissolved crude oil compounds than embryos are. The resistance decreased with the advancing resorption of the yolk sac. Cod larvae have a "mean critical time" of 4.2 days when brought into the highest concentration of Iran crude at the age of one day and 0.5 days when 10 days old (Table 2). In the lowest concentration the values were 14 days and 5.5 days respectively, but control larvae reached nearly the same values because they were not fed.

The difference in toxicity between the crude oils was also shown here. The Libyan crude, however, had a more

distinct influence on the larvae than on the embryos. Herring larvae were less and plaice larvae more resistant than cod larvae of the same age.

A very important aspect is the different effects of floating oil and dispersions obtained by mechanical means only or by additional aid of chemicals. For this purpose the relatively non-toxic dispersant Corexit 7664 was used. 10 and 100ppm of Corexit were added when the oil was mixed with the water, representing 1 and 10 per cent of the oil dispersed. (These concentrations lie in the range recommended by the producer). Pure Corexit solutions of the mentioned concentrations were non-toxic to the larvae (fig 3). Oil dispersions with and without Corexit caused "mean critical times" of 3 to 6 h at 10^3ppm of Iran oil and 60 to 100 h at 20ppm (fig 3). After two days the pure oil-water dispersion had lost much of its toxicity while the Corexit oil dispersion containing 100ppm of dispersant had kept or even slightly increased its toxicity at all dilutions tested.

The larval integument was damaged especially in higher concentrations. Typical rows of blisters formed on the primordial fin and the end of the tail looked gnawed. Surprisingly, the larvae did not avoid well defined milky clouds of even highly concentrated oil dispersions. The larvae entered and crossed the clouds though they suffered typical tissue damage, The chemoreceptors seemed to be blocked very quickly at the first contact with oil components.

Conclusions

As these experiments did not simulate natural conditions in the sea the results cannot be applied directly to marine pollution. There are many additional factors which might increase, but probably reduce the toxic limits. But it can be shown that toxic compounds are extracted from oil films, injuring larvae and younger stages of floating eggs, and even if the concentration of dissolved compounds is sublethal to eggs, the embryos may be injured and the hatched larvae can hardly survive.

Similar concentrations of extracted compounds are more toxic to larvae with partly or entirely resorbed yolk sac at the critical phase of beginning feeding.

It is important to note that the observed influences can vary widely depending on the type of crude oil spilled. Therefore more types of crude oil must be tested, and countermeasures against an oil spillage

TABLE 2. "MEAN CRITICAL TIME" (DAYS) OF COD LARVAE IN DISSOLVED CRUDE OIL COMPOUNDS (IRAN, AGHA JARI).

Age of larvae at transfer (days)	Amount of oil extracted (ppm)		
	10^2	10^3	10^4
1	14	8.4	4.2
3	10	7.5	3.5
5	8.2	5.9	2.5
10	5.5	4.5	0.5

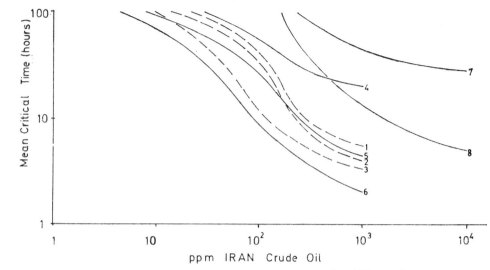

Fig 3. Effect of Iran crude oil and Corexit 7664 on one-day-old herring larvae:

1, 4: dispersions without dispersant
2, 5: dispersions with 10ppm of Corexit 7664
3, 6: dispersions with 100ppm of Corexit 7664
- - - - = test started 1 hour after preparing of dispersions
———— = test started 50 hours after preparing of dispersions
7 : dissolved compounds from oil film of plotted amount
8 : solution of Corexit 7664

should be related to the possible biological effect that the crude oil concerned might have.

Oil dispersions can have a ten to hundredfold higher toxicity to both eggs and larvae.

"Natural" oil dispersions obtained without chemicals gradually lose their toxicity with an oil film being re-established on the water surface. Dispersions stabilized with dispersants keep or increase their deleterious effect for days.

Larvae do not seem to be able to avoid oil contaminated water, especially dispersions, as the chemoreceptors are probably blocked or destroyed rather quickly at the first contact with oil compounds. If larvae remain in oil dispersions they have little chance of survival.

References

KÜHNHOLD, W W The influence of watersoluble compounds of
1969   crude oils and their fraction on the ontogenetic development of herring fry (*Clupea harengus* L.) Ber. dt. Wiss. Kommn Meeresforsch, 20(2):165–71 (in German).
SIMPSON, A C The *Torrey Canyon* disaster and fisheries. *Lab.*
1968   *Leafl. Minist. Agric. Fish. Fd*, (18):43 p.
WILSON, K W Toxicity of oil-spill dispersants to embryos and larvae
1970   of some marine fish. Paper presented to FAO Technical Conference on Marine Pollution and its Effects on Living Resources and Fishing, Rome, 9–18 December 1970, FIR:MP/70/E-45:7 p.

Toxicity of Oil-Spill Dispersants to Embryos and Larvae of Some Marine Fish†

K. W. Wilson*

Toxicité des dispersants des hydrocarbures répandus à la surface de la mer pour les oeufs et les larves de certains poissons marins

L'auteur a examiné les facteurs affectant la toxicité de plusieurs dispersants des hydrocarbures répandus à la surface de la mer pour les embryons et les larves de six espèces de poissons de mer. Les principaux facteurs qui influent sur la toxicité d'un dispersant sont le type et la teneur aromatique du solvant. Le vieillissement des solutions d'essai entraîne une diminution marquée de la toxicité, mais la température et la salinité n'ont guère d'influence sur celle-ci. Pour tous les dispersants, les écarts de sensibilité entre espèces sont moindres que les écarts accusés par divers sujets d'âge différent au sein d'une même espèce. Les larves de toutes espèces manifestent une sensibilité analogue peu après leur éclosion, mais les embryons (au sein du chorion) sont plus résistants. La sensibilité s'accroît au stade du sac vitellin jusqu'à ce que les larves aient atteint le stade de l'autonutrition, la résistance augmente ensuite jusqu'à la métamorphose.

On a conçu des indicateurs plus sensibles que la survie pour déterminer la présence, le degré et la persistance de l'intoxication sublétale. Des larves au stade du sac vitellin ont été élevées jusqu'à la métamorphose après avoir reçu une dose sublétale de produit. Sur les dix paramètres de mesure, on n'a trouvé aucune différence entre

Toxícidad de los dispersantes de petróleo para los huevos y larvas de algunos peces marinos

Se han estudiado los factores que influyen en varios dispersantes de petróleo sobre los embriones y larvas de seis especies de peces marinos. Los principales de aquéllos, son la clase y contenido en aromáticos del solvente. El envejecimiento de las soluciones de ensayo da por resultado una disminución de la toxícidad, pero la temperatura y salinidad ejercen muy poco o ningún efecto. En el caso de todos los dispersantes, las diferencias en la susceptibilidad de diversas especies está menos acusada que las que existen en diferentes grupos de edad de la misma especie. Las larvas de todas las especies tienen análoga susceptibilidad en el momento de la eclosión, pero los embriones (dentro del corión) son más resistentes. La susceptibilidad aumenta durante la fase en que aún queda saco vitelino y hasta que la larva ha comenzado a alimentarse que es cuando crece la resistencia hasta el momento de la metamorfosis.

Se han ideado indicadores más sensibles que la sobrevivencia para determinar la presencia, intensidad y persistencia de la intoxicación subletal. Después de administrarles dosis subletales se criaron larvas con saco vitelino hasta el momento de la metamorfosis. Entre las poblaciones testigo y experimentales no se han

* Zoology Department, University of Aberdeen, Aberdeen, Scotland, U.K.   † This work was supported by a grant from Natural Environment Research Council.

[318]

les populations témoin et celles soumises à l'expérience. Des expériences de nutrition ont montré qu'en présence de substances en concentration sublétale, l'aptitude des larves à capturer les organismes dont elles s'alimentent est compromise, mais qu'après leur retour dans les zones non polluées, elles la recouvrent entièrement. Les mesures de l'activité donnent des résultats analogues, alors que des expériences préliminaires effectuées dans un petit fluvarium indiquent que les larves plus âgées peuvent tout au moins éviter des concentrations létales de dispersants.

encontrado diferencias en ninguno de los diez parámetros medidos. Los experimentos de alimentación han demostrado que en presencia de concentraciones subletales, disminuye la capacidad de las larvas de capturar los organismos de que se alimentan, pero se recuperan por completo cuadno se ponen en agua no contaminada. Arrojan resultados análogos las medidas de la actividad minetras que los experimentos preliminares en un fluvario pueqeño indican que por lo menos, las larvas de más edad, pueden evitar las concentraciones letales de dispersantes.

THE use of chemical dispersants (solvent emulsifiers) has been extremely toxic to many marine animals, due mainly to the toxicity of the aromatic solvent (Corner *et al.*, 1968; Portmann and Connor, 1968; Smith, 1968; Wilson, 1968). This particular criticism of dispersants has largely been nullified by the recent use of solvents which are very much less toxic.

The effects of spraying at sea are restricted to the surface waters, and it is here that deleterious effects on fish stocks will be realized. The majority of commercially important fish species around Britain have planktonic eggs and pelagic larvae, which spend most of their early life feeding on the zooplankton in the surface waters. Most of the flatfishes face an additional hazard as the newly metamorphosed fish move into nursery feeding grounds close inshore where pollution from shore spraying is a distinct possibility.

Recent advances in fish rearing have made it possible for large numbers of larvae to be produced for experimental purposes (Shelbourne, 1964; Blaxter, 1968, 1969). Despite the drawbacks of seasonable availability and difficulties in rearing there are considerable advantages in using larvae as experimental animals. Their small size ensures that large numbers and replicates can be used; the environmental conditions before testing can be closely regulated, and, most important, the developing embryo and larva are the most susceptible stages in life history. Indeed, it is surprising that this particular aspect has received such scant attention from toxicologists.

For details of species used and techniques for their rearing see Table 1.

Measurement of acute toxicity

Stock solutions of 100 ppm were made up from commercial grade dispersants with clean well-aerated sea water (salinity 31–34 ‰). The desired concentrations were prepared by serial dilution of the stock solution; 10 fish were transferred by pipette from their holding tank to plastic containers holding $\frac{1}{2}$ l of test solution. The containers were sealed and the test solutions were renewed every 24 h to avoid discrepancies due to evaporation of the volatile components of the solvent. The period of survival of each fish was recorded as the interval between the time of immersion and the time when the heart ceased beating for $\frac{1}{2}$ min. Dead fish were removed. Experiments were conducted at the rearing

temperature of each species maintained at $\pm 0.5°C$. In experiments on effects of temperature and salinity or susceptibility, the fish were acclimatized over a period of 48 h and the experiment was started 24 h later. All experiments ended after 100 h and the fish were not fed during periods of acclimatization or during the test period. The distributions of survival times were found to be log normal and were analyzed graphically to determine the mean survival time (Bliss, 1937). The lethal concentration at 100 h (100 LC_{50}) was derived by extrapolation of the plots of mean survival time against concentration.

When larvae were added to the higher concentrations they almost immediately became extremely active, with vigorous head-shaking followed by backwards swimming. Narcosis followed, the larvae making only sporadic twitches, and as the narcosis deepened they responded only when touched. Up to this point the effects were largely reversible. When no response could be elicited, even to violent prodding, irreversible changes occurred

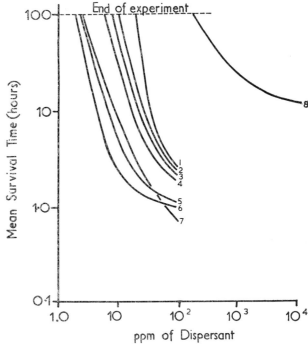

Fig. 1. The toxicity of several oil-spill dispersants to plaice larvae (stage 2b). 1 Atlas, 2 Pentone, 3 Finasol, 4 Houghtosolv, 5 Slipclean, 6 Basol, 7 BP1002, 8 Corexit. 1–4 Kerosene solvent. 5–7 Aromatic hydrocarbon solvent

TABLE 1. DETAILS OF THE SPECIES OF FISH USED AND THE REARING TECHNIQUES

Common name	Scientific name	Rearing temperature	Food (nauplii)	Rearing technique
Herring	*Clupea harengus* L	8°C	*Balanus*	Blaxter 1968a
Pilchard	*Sardina pilchardus* Walbaum	12°C	unfed	Blaxter 1969
Plaice	*Pleuronectes platessa* L	6–12°C	*Artemia*	Shelbourne 1964
Sole	*Solea solea* (L)	12–16°C	*Artemia*	Shelbourne 1964
Lemon Sole	*Microstomus kitt* (Walbaum)	8°C	unfed	as for plaice
Haddock	*Melanogrammus aeglefinus* (L)	8°C	unfed	as for plaice

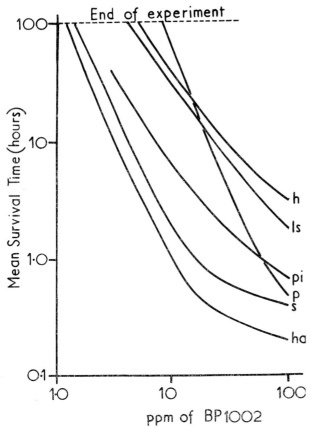

Fig. 2. The toxicity of BP1002 to fish larvae. h-herring, ls–lemon sole, pi–pilchard, p–plaice, s–sole, ha–haddock

and extensive tissue damage, especially to the epidermis of the tail, was common before death finally ensued. Different dispersants and different species exhibited slightly different responses and at lower concentrations the symptoms were less well defined. However, a period of narcosis always preceded death.

The toxicity of several dispersants to plaice larvae (stage 2b, Ryland, 1966) is shown in fig 1. The toxicity of a dispersant depended mainly on the type and aromatic content of its solvent. Their relative toxicities remained the same for all species tested. The toxicity of BP1002 to newly hatched larvae is shown in fig 2. Embryos of different species are not necessarily at a comparable stage of development at hatching but there was no obvious relationship between the susceptibility of a species and

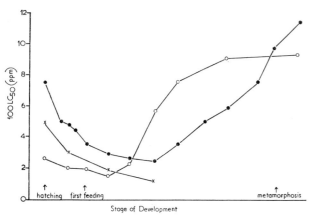

Fig. 3. The effect of the stage of development of larvae on the susceptibility to BP1002. ● *plaice,* ○ *sole,* × *herring*

its size or state of development at hatching. That the subsequent degree of development of a larva is an important factor modifying the toxicity of a dispersant can be seen from the example shown in fig 3. Developing embryos within the chorion are much more resistant than the larvae; the $100 LC_{50}$ was always greater than 20 ppm. The most sensitive stage of the embryo appeared to be before gastrulation and exposure to concentrations of dispersants greater than 20 ppm shortly after fertilization disrupted normal cell division. The degree of protection afforded by the chorion can be seen in the dramatic reduction in $100 LC_{50}$ from in excess of 20 ppm immediately prior to hatching to less than 8 ppm after hatching. The particularly susceptible period during the establishment of feeding conforms to the idea of a "critical stage" during stage 2 as was suggested by Ryland (1966). In rearing tanks with a high density of larvae, swimming larvae at late stage 4 were less tolerant than their demersal contemporaries and a similar reduction of tolerance was achieved by starvation.

Evaporation and/or degradation of the dispersants was the major factor affecting the toxicity. Gentle aeration of the test solutions for 100 h increased $100 LC_{50}$ for BP1002 from 4 ppm to greater than 40 ppm. The temperature of the test solution did not affect the $100 LC_{50}$, but at 100 ppm BP1002 the mean survival time decreased from 450 min at 6°C to 50 min at 20°C. In the range 14–35 ‰ salinity did not affect the $100 LC_{50}$.

Delayed effects of acute exposure to dispersants

Although in acute toxicity testing the embryos were more tolerant than the larvae, growth abnormalities and delays in development were noticed at sublethal concentrations. The effects of sublethal concentrations on plaice embryos and larvae were determined over 96 and 48 h respectively and by transferring the survivors to sea water. The subsequent survival to metamorphosis is shown in Table 2. Hatching success was impaired at all concentrations and many of the larvae that did hatch showed abnormal flexures of the body axis. Kühnhold (1969) describes similar abnormalities in fish embryos treated with crude oil extracts but as such "bent" larvae occur after high incubation temperatures and high bacterial contamination of the egg during development it appears to be a non-specific response possibly due to interference with the gaseous exchange at the chorion. The proportion of "bent" larvae increased with increasing concentration of dispersant and since such larvae were unable to feed successfully this determined the numbers of established feeders. Although after treatment all larvae appeared to recover, many were not able to establish themselves; an

TABLE 2. THE SURVIVAL OF PLAICE EMBRYOS AND LARVAE AFTER TREATMENT WITH BP 1002 FOR 96 AND 48 H RESPECTIVELY

Concn. of BP1002	Treatment as embryos			Treatment as larvae	
	% hatch	% feeding	% metamor-phosed	% feeding	% metamor-phosed
0.0 ppm	83	62	49	83	74
2.5 ppm	62	28	19	70	61
5.0 ppm	47	14	8	52	45
10.0 ppm	5	0	0	9	5
	n = 200			n = 300	

[320]

TABLE 3. POPULATION CHARACTERISTICS OF STAGE 3A PLAICE
LARVAE 30 DAYS AFTER TREATMENT WITH DISPERSANT
BP1002. THE MORTALITY INCURRED DURING TREATMENT IS
NOT CONSIDERED

Population parameter	Control	BP1002 at 7.5 ppm for	
		24 h	48 h
% Survival	99.3	96.3	99.3
% Metamorphosis	94.6	88.0	89.0
% Sinistral	2.0	1.3	3.3
% Abnormally pigmented	46.0	39.3	40.0
% Fin biting	4.0	4.0	9.0
Mean length mm (L)	15.87	15.21	15.90
Mean wet weight mg (W)	36.99	32.30	38.56
Mean condition factor (W/L³)			
n = 150	0.828	0.807	0.871

inability clearly related to dose (Table 2). In all cases once the larvae had established feeding, the mortality rates were the same for all treatments. Treatment of established larvae (stage 3a), even at concentrations which killed over 10 per cent of the population after 48 h, had no effect on the population characteristics of the survivors at metamorphosis, 30 days later (Table 3).

Behavioural changes at sublethal concentrations

In the study of sublethal toxication, behavioural parameters have distinct advantages over physiological and biochemical measurements, since the activities of an animal represent the integration of a diversity of processes. Furthermore, behaviour patterns are highly sensitive to environmental influences and measurements can be made without undue harm to the test organism.

Blaxter (1968a) has shown that herring larvae are positively phototactic at intensities above 10^{-1} m.c. and negatively phototactic down to a threshold of 10^{-5} m.c. This was used to study the responsiveness of larvae after treatment with dispersant. A series of black Perspex troughs, with clear Perspex ends and measuring 17.5 cm × 2.5 cm × 2.5 cm were filled with test solution or sea water and held in a light-proof box. Each trough carried a matched pair of thermistors linked to a pen recorder as shown in fig 4. This is a modification of the activity-recording device described by Heusner and Enright (1966) and arranged so that by setting the pen recorder to the middle of the chart graded deflections to both right and left could be achieved. The troughs could be illuminated from either side by 12 V 36 W car headlamp bulbs which were run from the mains by a transformer. A variable transformer was inserted in the circuit so that the illumination of the troughs could be varied.

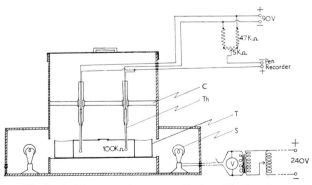

Fig 4. Activity measuring system. C–clasp, Th–thermistor, T–trough, S–light source

After the system had been left for 30 min to stabilize, batches of 5 larvae were transferred to the appropriate troughs and left to dark adapt for 30 min. A total trial lasting 8 min consisted of consecutive recordings of (a) 2 min in total darkness, (b) 2 min of low light intensity (7×10^{-4} lux), (c) 2 min of high light intensity (1.3×10^2 lux), and (d) 2 min of total darkness. The recordings were then scored by measuring the height of each deflection on the pen recording from the baseline and summing these scores over 40 sec intervals. An example of the results is shown in fig 5. Responses were found to be dose dependent in that after an initial increase the response fell below that of the controls. Treated larvae returned to sea water gave high scores which waned over 1–3 days to reach the level of the controls. The sensitivity of the system was such that differences in response were detected after treatment with BP1002 at a level of only 1/50 of the 12 h LC_{50}.

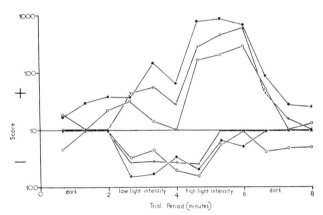

Fig 5. The response of herring larvae to light of different intensities with BP1002. for 12 h. ● 0.5 ppm, ○ 0.1 ppm, × 0.0 ppm

Feeding experiments have proved less sensitive, but in the presence of sublethal concentrations (down to 1/5 100 LC_{50}) the ability of the larvae to capture prey items was significantly impaired. After returning to sea water the larvae quickly recovered and were feeding normally within 48 h.

Conclusions

The dangers of extrapolating the findings of acute toxicity tests, conducted with convenient laboratory animals, to populations in the field cannot be overstated. The stage of development and the condition of the animal is important and the findings of such tests may bear little or no relation to the effects of sublethal toxication. The experiments have shown that larvae are extremely amenable to laboratory experiments and their use in acute toxicity testing and in the study of sublethal toxication may help research.

References

BLAXTER, J H S Rearing herring larvae to metamorphosis and
1968   beyond. *J. mar. biol. Ass. U.K.*, 48(1):17–28.
BLAXTER, J H S Visual thresholds and spectral sensitiviy of
1968a   herring larvae. *J. exp. Biol.*, 48:39–53.
BLAXTER, J H S Experimental rearing of pilchard larvae, *Sardina*
1969   *pilchardus. J. mar. biol. Ass. U.K.*, 49(3):557–75.
BLISS, C I The calculation of the time-mortality curve. *Ann. appl.*
1937   *Biol.*, 24:815–52.

CORNER, E D S, SOUTHWARD, A J and SOUTHWARD, E C Toxicity
1968   of oil-spill removers ("detergents") to marine life: an
        assessment using the intertidal barnacle *Elminus modestus*.
        J. mar. biol. Ass. U.K., 48(1):29–47.
HEUSNER, A A and ENRIGHT, J T Long-term activity in small
1966   aquatic animals. *Science, N.Y.*, 154(3748):523–3.
KÜHNHOLD, W W Der Einfluss wasserlöslicher Bestandteile von
1969   Rohölen und Rohölfraktionen auf die Entwicklung von
        Heringsbrut. *Ber. dt. wiss. Kommn Meeresforsch.*, 20:165–71.
NELSON-SMITH, A Biological consequences of oil pollution and
1968   shore cleansing. *Fld Stud.*, 2(Suppl.):73–80.
PORTMANN, J E and CONNOR, P M The toxicity of several oil-spill
1968   removers to some species of fish and shellfish. *Mar. Biol.*,
        1(4):322–9.

RYLAND, J S Observations on the development of larvae of plaice,
1966   *Pleuronectes platessa* L. in aquaria. *J. Cons. perm. int. Explor.
        Mer*, 30(2):177–95.
SHELBOURNE, J E The artificial propagation of marine fish. *Adv.
1964   mar. Biol.*, 2:1–83.
SMITH, J E (Ed.) "TORREY CANYON" pollution and marine
1968   life. A report by the Plymouth Laboratory of the Marine
        Biological Association of the United Kingdom. Cambridge,
        University Press, 196 p.
WILSON, D P Long-term effects of low concentrations of an oil-
1968   spill remover ("detergent"); studies with the larvae of
        Sabellaria spinulosa. *J. mar. biol. Ass. U.K.*, 48(1):177–82.

Physiological Changes Experienced by Pacific Salmon Migrating through a Polluted Urban Estuary

L. S. Smith, R. D. Cardwell
A. J. Mearns, T. W. Newcomb
*and K. W. Watters, Jr.**

Modifications physiologiques subies par le saumon du Pacifique lors de sa migration dans un estuaire urbain pollué

Le Duwamish Waterway est un estuaire pollué par les égouts qui se trouve sur le territoire de la municipalité de Seattle, Washington (Etats-Unis), et dans lequel le saumon argenté (*Oncorhynchus kisutch*) passe chaque automne pour atteindresa zone de frai, dans le cours supérieur de la Green River, qui se jette dans l'estuaire. L'arrivée du saumon à l'embouchure de l'estuaire coïncide avec les basses eaux de la rivière qui sont caractérisées par de faibles taux d'oxygène dissous à la fin de l'été. D'ordinaire le saumon réussit à franchir l'estuaire lors des premières pluies d'automne.

Avant que les pluies ne commencent, il semble que le saumon remonte l'estuaire aussi loin que possible pour s'arrêter au point où sa puissance de nage se heurte à un courant de force égale. Nous avons étudié l'état physiologique du saumon lors d'expériences menées à bord d'un laboratoire flottant ancré en aval des eaux où l'on trouve le plus faible taux d'oxygène dissous (2–4 ppm) et nous avons constaté ce qui suit:

1. Les taux de lactate sanguin étaient de 2 à 5 fois plus élevés que la normale et les excrétions considérables de lactate représentaient une grave perte d'énergie pour le poisson.

2. La consommation d'oxygène et l'efficacité de processus d'extraction de l'oxygène de l'eau étaient réduites.

3. La puissance de nage était réduite et, même après repos dans une eau aérée, n'était pas suffisante pour permettre au poisson de passer des eaux à faible taux d'oxygène dissous dans la rivière en amont de l'estuaire, avant d'atteindre un degré de fatigue extrême.

4. La concentration d'ammoniaque dans le sang était inférieure à la normale lorsque le taux d'oxygène dissous était faible.

5. De profondes modifications hématologiques se sont produites.

Cambios fisiológicos experimentados por el salmón del Pacífico al emigrar a través de un estuario urbano contaminado

El Duwamish Waterway es un estuario contaminado por aguas de descarga, situado dentro de los límites urbanos de Seattle, Washington (EE.UU.), a través del cual pasa todos los otoños el salmón coho (*Oncorhynchus kisutch*) en su migración hacia los frezaderos situados en la parte alta del río Green. La llegada del salmón a la boca del estuario, a finales del verano, coincide con el estiaje del río y con la presencia en el agua de bajos niveles de oxígeno disuelto. El salmón suele atravesar con éxito el estuario, por lo general, durante las primeras precipitaciones otoñales.

Hasta que lleguen las lluvias, el salmón asciende, al parecer, todo lo que puede por el estuario, deteniéndose cuando su energía natatoria es igual a la velocidad de la corriente. Se estudió el estado fisiológico del salmón, realizando experimentos a bordo de un laboratorio flotante anclado en el extremo inferior, aguas abajo, de la zona de contenido más bajo de oxígeno disuelto (2–4 ppm), hallando que:

1. El contenido de lactatos en la sangre era de 2 a 5 veces superior al normal, eliminándose importantes cantidades de lactatos, lo que representa una gran pérdida de energía para el pez.

2. El consumo de oxígeno del pez y su capacidad de extraer oxígeno del agua eran reducidos.

3. La energía natatoria era reducida e, incluso después de permanecer en aguas aireadas, no bastaba para que el pez atravesara el estuario y entrara en el río, dejando atrás la zona de bajo contenido de oxígeno disuelto, antes de que la fatiga se apoderara de él.

4. La concentración de amoníaco en la sangre durante la permanencia en aguas con bajo contenido de oxígeno disuelto era inferior a la normal.

5. Se producían varios cambios hematológicos.

AS little is known of the estuarine and marine environmental requirements of salmon for survival and reproduction, we undertook a study of several basic physiological functions vital to their transition between fresh and salt water. The estuaries were considered as a point in the salmon's life-long migration where natural stressors† are greatest and mankind's additional stressors would most likely be overwhelming. To study the salmon *in situ* a self-contained laboratory was built aboard a 98.5 ft (30.0 m) steel barge (Smith 1970) which could be anchored in protected waterways along their migration routes.

The Duwamish Waterway

In Autumn 1969, we performed a series of experiments in the Duwamish Waterway, a sewage-polluted estuary

† Stress represents the sum of morphological, physiological, and biochemical changes resulting from the actions of the stressor.

within city limits of Seattle, Washington, U.S.A. Coho salmon (*Oncorhynchus kisutch*) migrate each autumn through the estuary to their spawning grounds in upper Green River. Their arrival coincides with the low river flows of dissolved oxygen (DO) found in late summer. Successful passage through the estuary usually occurs during the first autumn rainfall.

The problems salmon encounter in water of low DO are difficult to predict because there are a large number of considerations and alternatives, some of which may be suitable at one DO level and not another. Our experiments illustrate some of the ways in which salmon select these alternatives.

For these experiments a series of four fish were put in the swimming chamber (Smith and Newcomb 1970):
(1) no catheterizations and no prolonged anaesthesia;
(2) no catheterizations but with prolonged anaesthesia;
(3) dorsal aorta (Smith and Bell 1964) and urethra

---

* Fisheries Research Institute, University of Washington, Seattle, Washington 98105, U.S.A.

catheters; and (4) pre- and post-gill catheterizations (Davis and Watters 1970). The fish were allowed to rest after capture, then anaesthetized, catheterized on an operating table for fish (Smith and Bell 1967), placed in a swimming chamber, and rested overnight.

The turnover of *in situ* estuarine water (salinity 21–25 per cent of sea water and temperature 12.0–13.5°C) in the swimming chamber was 10 l/min. The water was aerated for the first hour, giving DO values ranging from 5.5 to 6.5 mg/l. The aeration was stopped during the second hour and DO decreased to 4.0 to 5.0 mg/l. The current in the chamber was set to provide a velocity of 56 cm/sec. At the end of the 2 h period, the current was reduced. The fish's recovery could then be followed with the fish resting in aerated water.

The experiment with four fish was repeated three times with different fish in ambient estuarine water having a moderately low DO and once in sea water of Elliott Bay (about 2 miles from the first site) where DO was saturated and no pollution was obvious. In Elliott Bay, water flow through the chamber was reduced during the second hour so that the fish's oxygen consumption produced DO levels in the chamber comparable to those observed in the estuary. As an additional control for the coho salmon caught in the estuary and whose swimming stamina might have been affected by pollution agents other than low DO, we brought coho salmon from Hood Canal where pollution contamination was very unlikely and tested them at Elliott Bay on the same schedule as the other salmon.

Results

Some comments on swimming stamina are warranted. Swimming stamina may be an integrated indicator for the

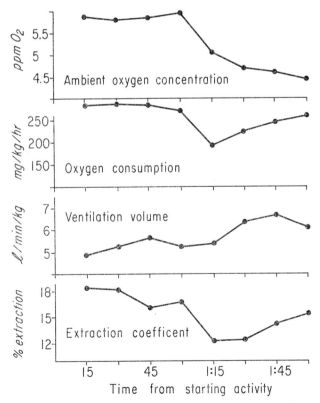

Fig 1. *Oxygen consumption, extraction coefficient, and ventilation volume with change in environmental oxygen concentration in swimming coho salmon. Temperatures ranged from 12° to 15°C, and salinities from 20 to 28 per thousand*

status of a fish's energy transfer systems, but high speed swimming until fatigue probably does not occur in nature. Most fish will not work to complete exhaustion unless forced by unusual circumstances. However, in a group of eight coho tested four fish did fatigue at our moderate test velocity of 56 cm/sec (2 km/h) within the 2 h swimming period. The fatigued coho salmon were exposed to DO levels of 4.0 to 4.5 mg/l while the remaining fish were tested in DO levels over 4.5 mg/l.

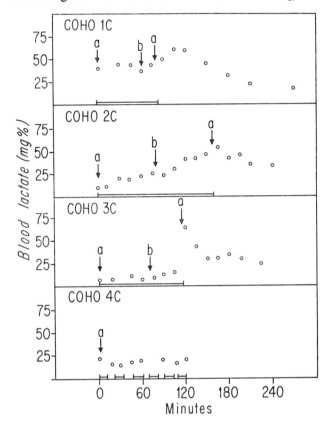

Fig 2. *Changes in blood lactate levels in four coho salmon (Oncorhynchus kisutch) swimming various periods (bar) in (a) elevated (65 per cent saturated), (b) low (50 per cent saturated) dissolved oxygen levels. Temperatures ranged from 12° to 14° C, and salinities from 20 to 28 per thousand.*

The three respiratory variables measured (oxygen consumption, ventilation volume, and extraction coefficient) are summarized in fig. 1. During the first hour of swimming in high DO, the consumption remained nearly constant. The extraction coefficient generally complemented the ventilation volume. As the level of dissolved oxygen was decreased, both oxygen consumption and extraction coefficient decreased almost immediately by about one third and then slowly recovered.

The decreased oxygen consumption occurred during a period of constant or slightly increased activity and the decreased consumption must therefore represent an oxygen debt and lactate accumulation. The ventilation volume increased rather slowly compared with the consumption and extraction, indicating that an increase is more "expensive" physiologically than going into oxygen debt, and that an increase in ventilation volume occurs only when the low DO persists so long that the lactate accumulation begins to become a problem.

Information is presented for individual fish because each was different and each illustrates different points.

The first fish (Fig. 2/1C) began swimming with a moderately high blood lactate level, but it decreased slightly by the end of the first hour. Upon entry into water of low DO, however, the lactate increased rapidly until the fish became fatigued just before the end of the second hour. The recovery period after fatigue was normal and uneventful, declining to the normal 5–20 mg per cent level of a typical rested fish.

The blood lactate of the second fish (Fig. 2/2C) was normal initially but rose during swimming, showing a tendency to stabilize toward the end of the first hour. The adjustments in lactate levels in both this and the first fish probably correspond to a slight decrease in oxygen consumption.

Fig. 2/3C illustrates the effect of erratic behaviour of the fish. The sharp peak in lactate concentration between 105 and 150 min resulted from violent, non-swimming activity. The fish acted as if it would have preferred to turn downstream rather than continue to swim upstream. Fig. 2/4C shows the results for a fish that refused to swim.

Although the differences between the four fish are sufficient to make averages questionable, we believe that the average represents the general blood lactate response to our swimming schedule. In adequate DO there was a minimal resting level of lactate. This rose slightly during increase in activity. When a critical point was reached the blood lactate level increased rapidly until the fish became fatigued and stopped swimming.

The presence of lactate in the urine of salmonids was first reported by Hunn (1969). We have just presented the first evidence that blood lactate levels may increase in proportion to sub-maximal levels of activity. These observations raise the question of how much lactate (energy) a salmon might lose if caused to increase blood lactate while swimming in chronically low DO levels.

Some experiments in progress may answer that question. In a coho salmon whose lactate concentration ranged between 10 and 15 mg per cent, 30–45 μg/kgh were excreted in the urine. If we extrapolate to a blood level of 45 mg per cent and 135 μg/kgh urinary excretion, a 1 kg fish would lose 3.2 mg/day or 97.2 mg/month. A month is not an unusual period for the fish to wait in the estuary before ascending the stream.

Whether this energy loss is significant is impossible to say now, because we do not know how crucial the last few milligrammes of energy stores may be to the fish when it is on the spawning grounds. Also, there may be other routes of lactate loss, such as through the gills. This was suggested by an experiment in which a coho salmon had a very large dose of lactate injected directly into the blood stream. Only 57 per cent of it was recovered in the urine.

Ammonia

Ammonia is the primary nitrogenous metabolite in most teleost fish. In salmonids excretion occurs primarily through the gills. We monitored ammonia excretion from the difference in ammonia content between the water entering and leaving the swimming chamber.

Data from ten fish indicate two patterns of ammonia excretion (Fig. 3). The rate of ammonia excretion increased during activity. Fish that did not fatigue during the second hour (low DO above 4.5 mg/l) characteristically showed a depressed rate of ammonia excretion which was elevated when an adequate DO level was

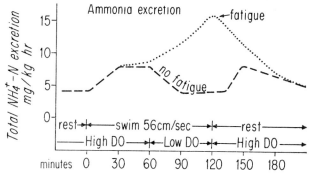

Fig 3. Total ammonia excretion level in swimming adult coho salmon subjected to various levels of dissolved oxygen

restored. Fish that did fatigue during the low DO period had rates of ammonia excretion (dotted line) which remained constant or continued to increase until fatigue occurred.

These data suggest that DO levels above 4.5 mg/l are needed to regulate ammonia excretory metabolism. We suggest that *in vivo* ammonia is normally detoxified and stored by the fish until it can be excreted. This requires energy, which ultimately means the consumption of oxygen. A fish in good condition has energy reserves (e.g. creatine phosphate) which may be used to detoxify *in vivo* ammonia without using oxygen when the fish is exposed to anhypoxic environment. A fish with insufficient energy reserves may thus be forced to excrete *in vivo* ammonia immediately. This demand over-rides the demands of swimming by competing for the meagre oxygen supply. This is shown by fatigue and the total ammonia excretion curve of a fatigued fish (Fig. 3).

Kidney function in marine salmon is concerned primarily with the excretion of divalent ions, particularly magnesium, and organic acids including lactic acid. The major excretory load of sodium, ammonium and chloride ions is excreted by cells in the gills, so that urine volume in marine salmon is only about one eighth that of fresh water salmon (Hickman and Trump 1969). However, any major decrease in the rate of kidney function could cause problems through the toxicity of the accumulated magnesium ions in marine salmon or accumulated water in fresh water salmon.

Our present data indicate an apparent decrease in urine production when the fish was subjected to low levels of DO. If it is important to the fish to excrete lactate during low DO and excretory mechanisms are impaired by low DO, then toxic levels of lactic acid as well as other blood constituents may be reached in the tissues.

Discussion and conclusions

The use of salmon as bioassay organisms for multi-factor studies of sublethal pollution is just beginning. It is clear that no single physiological criterion is sufficient to describe all of the problems faced by salmon even when they are confronted with single factor pollution, such as low dissolved oxygen. The major goal of these studies should be to promote the continuation of the species involved. In the case of sublethal low oxygen levels, reproduction is our concern. The ability of salmon to reproduce effectively is largely a problem of the adult fish having enough energy to reach the spawning grounds and spawn viable eggs and sperm.

Adult coho salmon may fatigue when forced to sustain swimming in estuarine water containing low concentrations of dissolved oxygen (below 4.5 mg/l). Our data suggest that, prior to fatigue, the swimming salmon attempt to acclimate to the hypoxic environment by using anaerobic metabolism as an energy source. Dependence on an anaerobic energy source gives cardiovascular functions (oxygen uptake mechanisms) opportunity to respond and adjust to the lower oxygen level. This also provides temporary energy for continued swimming. Two problems may result, however: (1) increased lactic acid levels in the blood may decrease pH, and therefore affect the oxygen binding capacity of the blood. Lactate may be excreted in the urine resulting in an energy leak; (2) fatigue can occur before the cardiovascular adjustments are complete. This is a direct result of the depletion of anaerobic stores of energy (glycogen) in the muscle (Stevens and Black, 1966).

Our observations were on salmon in a swimming chamber and we would not expect the fish to perform identically in the estuary but they do allow us to postulate alternatives that would be encountered in the estuary:

1. The fish may detect the low DO water and, by random searching, avoid it (Jones, 1964) and find a more suitable environment. This would delay arrival on spawning grounds.

2. Given a strong home-water response, the fish may attempt passage through the low DO environment. If cruising is sustained (2 km/h), partial dependence on anaerobic metabolism and oxygen debt will occur, but progress will be made. The fish in our tests swam between 1 and 2 km while incurring an oxygen debt. Cruising velocity may also be reduced to avoid the occurrence of an oxygen debt. In this case progress toward the spawning grounds will depend more heavily on rates of river flow and tidal change.

These biological alternatives, as well as fluxes in river and tidal flow, help to explain the intermittent presence of a salmon run during the spawning season in the estuary.

Our data also suggest the importance of possible synergism between environmental DO, salinity and metabolic ammonia in relation to kidney function. Continued research may demonstrate that water quality standards should be different for a given waste component at different times and places depending on the other waste components involved.

References

DAVIS, J C and WATTERS, K W, Jr.   Evaluation of several methods
1970   for sampling water expired by fish. *J. Fish. Res. Bd Can.*, 27(9)1627–35.
HICKMAN, C P, Jr. and TRUMP, B F   The kidney. *In* Fish physiology,
1969   edited by W. S. Hoar and D. J. Randall, New York, Academic Press, vol. 1:91–239.
HUNN, J B   Chemical composition of rainbow trout urine following
1969   acute hypoxic stress. *Trans. Am. Fish. Soc.*, 98:20–2.
JONES, J R E   Fish and river pollution. London, Butterworths Publ.,
1964   203 p.
SMITH, L S   Building and operating a floating laboratory. *Lab.*
1970   *Pract.*
SMITH, L S and BELL, R   A technique for prolonged blood sampling
1964   in free-swimming salmon. *J. Fish. Res. Bd Can.*, 21(4):1579–88.
SMITH, L S and BELL R   Anaesthetic and surgical techniques for
1967   Pacific salmon. *J. Fish. Res. Bd Can.*, 24(7):1579–88.
SMITH, L S and NEWCOMB, T W   A modified version of the Blazka
1970   respirometer and exercise chamber. *J. Fish. Res. Bd Can.*, 27(7):1321–4.
STEVENS, D E and BLACK, E C   The effect of intermittent exercise
1966   on carbohydrate metabolism in rainbow trout, *Salmo gairdneri. J. Fish. Res. Bd Can.*, 23(4):471–85.

A Preliminary Study of Salmon Movements in a Polluted Estuary

*P. F. Elson, L. N. Lauzier and V. Zitko**

Les déplacements du saumon dans un estuaire pollué

On a étudié les déplacements du saumon (*Salmo salar* L.) par détection sonique dans l'estuaire du Miramichi, dont le bassin est (en 1969) le plus important producteur de saumons de l'Atlantique au Canada. La pollution due aux effluents industriels et urbains a été mesurée et rapportée à l'hydrographie de l'estuaire et aux déplacements du saumon. L'estuaire du Miramichi est en partie constitué par un mélange d'eaux qui se renouvellent sur une période d'environ dix jours.. Les variations verticales des degrés de salinité dans le cours supérieur de l'estuaire indiquent qu'il existe un état relativement uniforme de faible salinité à la fin des basses eaux (1,3 à 4,3 pour mille de la surface au fond), suivi d'un état stratifié à la fin des hautes eaux (4,2 à 13,8 pour mille).

On a étudié la pollution par spectroscopie d'absorption des ultraviolets. Le pouvoir absorbant des échantillons aquatiques s'accroissait lentement avec la diminution de la longueur d'onde, ce qui indique la présence d'un mélange complexe de composés organiques. Les taux d'absorption A250 s'échelonnaient entre 0,3 et 0,9 dans le cours supérieur, et 0,1–0,5 à l'embouchure de l'estuaire. La corrélation entre la salinité et le pouvoir absorbant était moins nette dans la partie industrialisée de l'estuaire, ce qui indique une répartition hétérogène des polluants. On a suivi pendant plusieurs heures des saumons porteurs de capsules soniques émettant des signaux distincts. La position d'autres poissons a été observée périodiquement. L'expérience s'est prolongée pendant six semaines. On a constaté des déplacements considérables associés au mouvement des marées montantes et descendantes.

Movimientos de los salmones en un estuario contaminado

En 1969 se estudiaron los movimientos del salmón (*Salmo salar* L.) en el estuario del sistema del río Miramichi, principal zona productora de salmón del Atlántico del Canadá, siguiendo acústicamente el recorrido de los peces. Se midió la contaminación causada por las aguas de descarga industriales y urbanas y se puso en relación dicha contaminación con la hidrografía del estuario y los movimientos de los salmones. Las aguas del estuario de Miramichi se mezclan parcialmente y se renuevan en unos diez días, tiempo que tardan los contaminantes en dispersarse en el océano. Las variaciones de los gradientes verticales de salinidad en la parte alta del estuario indican la presencia de condiciones relativamente uniformes de poca salinidad en la bajamar (1,3 a 4,3 por mil, de la superficie hacia el fondo) y condiciones estratificadas en la pleamar (4,2 a 13,8 por mil).

Se estudió la contaminación mediante un espectroscopio de absorción de rayos ultravioletas. La absorción de las muestras de agua aumentó uniformemente al disminuir la longitud de onda, revelando así la presencia de una mezcla compleja de compuestos orgánicos. Los valores de absorbencia A250 variaron de 0,3–0,9 en la parte alta a 0,1–0,5 en la boca del estuario. La correlación entre la salinidad y la absorción resultó menos significativa en la zona industrializada del estuario, indicando así la heterogeneidad de la distribución de los contaminantes. Se siguió durante varias horas a algunos salmones marcados con cápsulas que emitían señales acústicas distintivas y periódicamente se observó la posición de otros salmones igualmente marcados. El experimento duró seis semanas. Los movimientos ascendentes y descendentes fueron numerosos,

* Fisheries Research Board of Canada Biological Station, St. Andrews, New Brunswick, Canada.

Les zones d'arrêt ou de changement de direction du poisson se trouvaient généralement à proximité des points connus de pollution industrielle. Dans la partie industrialisée de l'estuaire du Miramichi, le déplacement vers l'aval était nettement plus lent que dans un estuaire voisin, plus petit et non pollué. Dès lors que le poisson remontait au-delà du secteur industrialisé, il quittait rapidement la zone soumise à l'étude.

coincidiendo con la subida o la bajada de las mareas. Las zonas donde de ordinario se registró una demora en la marcha o un rodeo coincidían con puntos de contaminación industrial bien conocidos. En la zona industrializada del estuario de Miramichi los movimientos aguas arriba fueron notablemente más lentos que en un estuario adyacente más pequeño y sin contaminación. Una vez que los peces superaron la zona industrializada, salieron rápidamente de la zona en estudio.

The Miramichi River is the most important in Canada for Atlantic salmon (*Salmo salar* L.). The Fisheries Research Board of Canada has had a programme of salmon research on the Northwest tributary since 1950. Facilities include a trap 11 km above tidehead, at which migrant fish are counted.

In 1968 the run of adult salmon was the lowest on record. The next two lowest, in 1956 and 1957, resulted from extensive deaths of young from spraying adjoining forests with DDT (Elson, 1967). River discharge in 1968 was exceptionally low, and ascent of rivers by salmon is facilitated by good discharge (Banks, 1969). By comparison with Curventon counts in earlier low-water

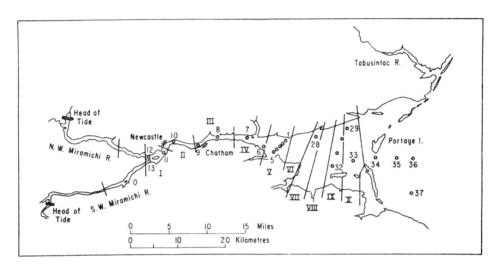

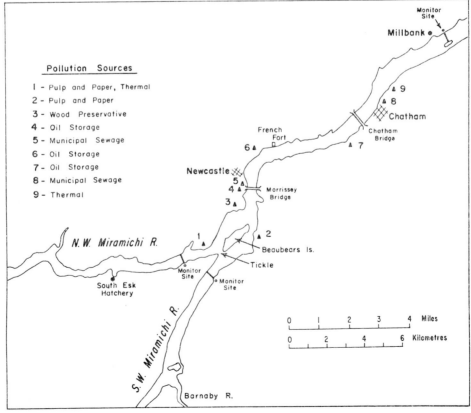

Fig 1. (*Upper*) *Miramichi estuary and bay. Numbered circles = locations of hydrographic stations; lines crossing estuary and bay = segments according to Ketchum (1951).* (*Lower*) *Portion of the Miramichi estuary where movements of Atlantic salmon were followed (1969); principal pollution sources shown.*

years, the 1968 run was puzzlingly low considering that commercial catches in the estuary were not correspondingly depressed.

Experience in Maritime Province streams has shown that pollutants can produce divergence from the normal patterns of salmonid migration (Saunders, 1969; Saunders and Sprague, 1967). In late August 1968, a number of industrial and urban sources were pin-pointed as contributing organic pollution to the Miramichi estuary. Pollutants included substances which young salmon avoid even in great dilution, such as pulp mill effluents (Sprague and Drury, 1969), wood preservatives (Zitko and Carson, 1969), or they may avoid urban effluents and thermal power plant cooling water. Oil spills related to storage depots, industry and shipping occur periodically.

A coordinated study of estuarial mixing, pollution of the estuary and salmon movements was initiated in 1969.

Hydrographic and pollution measurement techniques

Hydrographic and water pollution surveys in the Miramichi estuary were carried out on 17–19 June, 15–17 July and September 16–18, 1969; river discharge was estimated at 283, 167 and 624 m³/sec respectively. Variability of river discharge was substantial during each survey and was largest for September. Temperature and salinity measurements were made *in situ* at all stations with a Beckman salinometer RS-5-3. Water samples were taken at a maximum of 4 depths (surface, bottom and mid-depths) for pollution studies. The stations are shown in fig 1 (upper). A dividing line between the estuary and Miramichi Bay was empirically located near Oak Point, separating segments V and VI.

Ultraviolet (UV) absorption spectra of water samples were used to study the distribution of organic compounds. UV absorption of sea water is due to organic matter, nitrate and bromide (Armstrong and Boalch, 1961, 1961a; Ketchum, 1951) and may serve as an index of organic pollution. The absorbance increases with increasing chemical oxygen demand (COD) and the measurement of absorbance at 250 nm, almost entirely due to organic matter, has been suggested for investigations of highly polluted sea water (Ogura and Hanya, 1966). Beckman DK-2A spectrophotometer and 1 cm cells were used to record UV spectra. Samples were stored at 5°C before analysis and spectra usually recorded within 48 h. Most samples were clear and were not filtered. Large suspended particles occurring in bottom samples were removed by filtration through Pasteur pipets plugged with glass wool.

Segmentation of Miramichi estuary and bay was made according to the Ketchum (1951) method of study exchange of fresh and salt waters in the system and to estimate the flushing rate. Because of sampling restrictions in June and abnormally high river discharge in early September, only July data were used in estimating the flushing rates.

Estuarial mixing and distribution of organic matter

The main feature in longitudinal sections derived from the three surveys is the location of the steep horizontal gradient of salinity, in the downstream half of the estuary (stations 6 to 8) within the upper layers. Within the lower layers such gradient was shifted upstream.

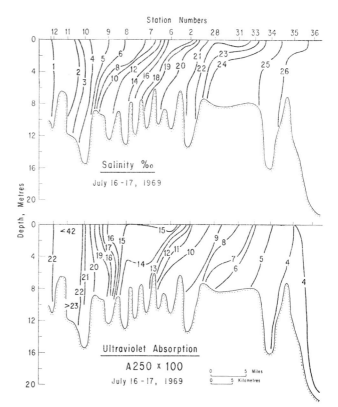

Fig 2. Longitudinal distribution of salinity and ultraviolet absorption in Miramichi estuary and bay

Conditions observed during the July survey (fig 2, upper) are representative.

The vertical salinity gradient in the upper 8 m reached 10.0‰ in June and July in the vicinity of station 7. It was generally less than 5.0‰ at other stations upstream and downstream from that station. During the September survey the vertical gradient was less variable all along the estuary and bay.

These surveys were generally carried out at the end of flooding tide and during most of the ebbing tide.

Observations carried out at different stages of the tide showed greater variability in salinity gradient within the upper estuary, at station 10, than at station 9, 4.8 km downstream. This resulted from the fact that at station 10 relatively uniform conditions of low salinity prevailed at the end of ebb and stratified conditions at the end of flood.

The variations of salinity in relation to the tide also brought about variations of the longitudinal salinity gradient. It was lowest and most uniform within the lower layers and also during or at the end of ebb. The lateral variations of salinity were very small in most of the estuary.

The shape of the UV spectra of all samples was identical with that described in the literature, i.e., absorbance increased with decreasing wavelength. At salinities higher than 19‰ (below juncture of estuary and bay) A250 decreased linearly with increasing salinity. Dilution of organic matter coming from the upper part of the estuary took place in this area. In the middle part of the estuary (salinity 5–18‰) A250 decreased only slightly with increasing salinity (fig 2; upper and lower diagrams, between stations 8 and 2). A250 decreased sharply with salinity increasing from practically fresh water to 5‰ in

[327]

the upper part of the estuary. The correlation between A250 and salinity was better at higher than at lower salinities which indicates that the distribution of organic matter in the middle and upper part of the estuary was less homogeneous than at the mouth of the estuary. The heterogeneity is caused by distribution of the sources of organic matter (run-off, industrial and domestic effluents) in the middle and upper part of the estuary, by partial mixing, tidal movements, and possibly by discontinuity of effluent discharges. With high run-off (September survey), the contribution of organic compounds in the run-off, as determined by A250, overshadowed the contributions of organic compounds from other sources. The slope of the decrease of A250 with increasing salinity was constant in the whole estuary, indicating a rather uniform dilution.

Vertical distribution of organic matter and salinity at several stations during the July survey is presented in fig 2. A250 values decreased with increasing depth and salinity but the gradient of organic matter concentration was lower than the gradient of salinity.

We consider the Miramichi estuary to be a partially mixed estuary for which the segmentation method could be applied for studying the rate of exchange of water within the estuary and the bay.

During the July survey the river discharge was estimated at 167 m³/sec or 744×10^4 m³ per tidal cycle. The segmentation of Miramichi bay and estuary is shown in fig 1 (upper).

Knowing the salinity distribution in the estuary and the bay and assuming that it had reached a steady state, the river water content and exchange ratios in the 10 segments were calculated. A base salinity of 27.00‰ was used for calculating proportion of river water in the bay and estuary.

In computing the accumulation of river water in the estuary and *versus* that in the bay, a more refined estimate was reached for the estuary, 147 to 167×10^6 m³ than for the bay, 119 to 170×10^6 m³. This is due to the fact that small variations of salinity in the bay would result in large discrepancies of river water volume because of the very large column in each segment.

The accumulation of fresh water in the estuary 147 to 167×10^6 m³ was approximately 20 to 22 times the river discharge per tidal cycle, i.e., 10 to 11 days of transport. The segmentation implied that the average tidal incursion was 6.1 km in the estuary and 3.3 km in the bay.

The flushing times were estimated at 20–22 tidal cycles in the estuary and 16–23 tidal cycles in the bay.

Assuming a given level of organic matter calculated from A250, in fresh water, it is possible to calculate the fraction of organic matter, Y, in the estuary and compare this fraction with that of fresh water, X. The content of organic matter in fresh water was taken as the observed level at station 11.

If a simple dilution of organic matter present at station 11 takes place in the estuary, $Y/X = 1$. If additional organic compounds are discharged between two points (downstream) into the estuary, the fraction of organic matter will not decrease as compared to the fraction of fresh water. Then, at a given point $Y/X > 1$. If on the other hand some of the organic matter is precipitated or decomposed in the estuary, the fraction of fresh water, calculated from salinity, would be higher than the frac-

tion of organic matter calculated from A250 and $Y/X < 1$. While in June and September the concentrations of organic compounds at station 11 were fairly uniform, the July data show a significant vertical gradient.

Additional contribution of organic compounds from sources in the estuary below station 11 apparently did occur in June, between stations 6 to 8 and in July between stations 8 and 9. The July data indicate no such effect. This may be caused by the stratification at station 11 at that time and choice of an A250 value not fully representative of the input of organic compounds.

Distribution of organic matter and salinity in the estuary for the July survey is presented in fig 2. There was a steep gradient of organic matter concentration between stations 6 and 28, and another gradient between stations 9 and 11. There was no stratification of organic matter at stations 10 and 12.

Determining movements of salmon

Two ways of learning whether pollutants deter salmon movement in an estuary are to determine the time required to pass from any given position to another farther upstream or to follow the fish. Salmon in an estuary sometimes move back and forth in relation to ebbing and flooding tides (Banks, 1969), so time for upstream progress is best measured over several tidal cycles. Technology involving ultrasonic capsules inserted in stomachs of live fish and then tracking the sound coming from the fish has been developed in recent years (Hasler, 1966).

Tracking equipment (obtained from Smith-Root Inc., Seattle, U.S.A.) consisted of a supply of SR-69 ultra-sonic transmitting capsules, each having a distinctive pulse frequency, an SR-70 directional hydrophone and a TA-60 sonic receiver with a sound-amplifying horn added. This equipment was operated from an outboard-powered boat, with radio-frequency noise suppressors on the motor ignition system.

In addition, 3 stationary monitoring units were borrowed from the U.S. Fish and Wildlife Service, through Dr. W. C. Leggett, then working with the Essex Marine Laboratory in Connecticut, U.S.A., on tracking shad (*Alosa sapidissima* (Wilson)). Each monitoring unit was connected to an upstream and a downstream directional hydrophone. The units recorded the time and pulse frequency of passing sonic transmitters. One monitor was located about 2.4 km below Chatham, another about 2.4 km up the Northwest branch of the estuary, and the third 1.6 km up the Southwest branch (fig 1, lower). Range of the monitors was sufficient to cover the width of the estuary at their locations. Monitors were operative from 09.00 h (Millbank, 14.30 h) on 23 September to 12.00 h on 26 September.

Salmon for tracking were obtained from the Department of Fisheries and Forestry's research net 1.6 km above the Millbank monitor, from the counting fence at Curventon, and by seining in a privately-owned pool about 25km above the mouth of the river, through gracious cooperation of the Tabusintac Salmon Club. Salmon were tracked only in the system in which they were caught.

To record movements, salmon were taken to a pre-determined liberation point by truck or boat, a sonic transmitter (64 mm × 13 mm) was inserted in stomachs, and external numbered tags fixed below the dorsal fin. The fish were placed gently in the water and allowed to swim off. Sometimes the boat with the listening device followed the fish. Signals could be picked up at distances of 0.8 to 2.4 km but were blocked out by wharfs, large rocks, or other obstacles between fish and directional hydrophone. Seven fish were liberated within the fields of the two upstream monitors.

After a number of sonic-tagged fish had been liberated, the tracking boat spent much time cruising about to make spot-checks on their locations.

Most tracking was done in the polluted Miramichi estuary (fig 1, lower), but movements of several fish were followed in the Tabusintac estuary (fig 1, upper). The Tabusintac flows entirely through forest and has no industrial development, hence can be regarded as an unpolluted control system.

Data from monitors and tracking operations were tabulated and plotted on enlarged maps of the estuaries.

Patterns of salmon movement

Useful records were obtained from the movements of 29 salmon liberated in the Miramichi estuary and 7 in the Tabusintac. Twenty-two of the Miramichi fish had to pass through 5–14 km of polluted estuary to reach comparatively clean waters, but only 17 were under observation (time from liberation to time of last record, but not necessarily continuous) for substantially more than a full tidal cycle (minimum 28 h under observation). Only 5 Tabusintac fish were under observation for a full tidal cycle or more (minimum 22 h).

In this study "net upstream progress" is used to indicate the distance between liberation point and point of last record. It takes no account of interim excursions farther upstream and back down again. Thus it is a measure of real progress, within the given time limits, towards fresh water reaches. Net upstream progress was: for 7 fish liberated at Millbank (fig 1, lower) near hydrographic station 8 (fig 2) and under observation for an average of 149 h, 0.10 km/h; for 4 fish liberated at French Fort (near station 10) and under observation for an average of 69 h, −0.05 km/h (i.e., net movement downstream); for 7 fish liberated at Newcastle (just below station 11) and under observation an average of 89 h, 0.11 km/h. The overall mean rate of upstream progress for these 17 Miramichi fish was 0.06 km/h. The corresponding mean for 5 Tabusintac fish under observation an average of 55 h was 0.21 km/h. Observed net upstream progress was more than three times faster in the Tabustinac than in the Miramichi.

Patterns of up- and downstream movement, as indicated by the monitor records, varied depending on position in the estuary and appeared to be related to intensity of pollution and proximity to pollution gradients. For 9 fish recorded by the Millbank monitor, the ratio of first-approach + last-departure movements coinciding with tidal flow to movements against the flow was 2.6 : 1. For 8 fish recorded by Southwest monitor, ratio of first-approach + last-departure with and against tidal flow was 1.5 : 1. For 10 fish recorded at the Northwest monitor the comparable ratio of movements with and against the tide was 1 : 1. Salmon movements were more affected by tidal oscillations at Millbank, *below* the most polluted part of the estuary, than in the comparatively

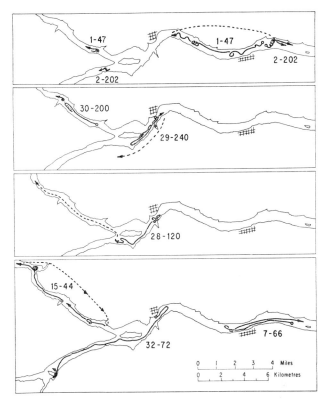

Fig. 3. *Paths taken by salmon with sonic transmitters; flood =rising tide, ebb = falling tide; high = high tide; numbers following dashes are pulses per minute of transmitters. 1–47: up with flood, stopped at high, down with ebb; up 10 miles 10 days later. 2–202: down with ebb; up 10 miles in 10 days. 30–200: Curventon fish; up on 1st flood, then wandered near Hatchery brook. 29–240: crossed estuary on ebb, approached and retreated from Beaubears, then down on flood; caught 20 miles up Southwest in May 1970. 28–120: wandered on ebb, up to and avoided Tickle on flood, then up Northwest in 1½ days. 15–44: taken in Hatchery net above Southwest monitor and liberated in Northwest at Hatchery brook. 32–72: Curventon fish; down on ebb, up Southwest on flood avoiding Tickle at head of Beaubears. 7–66: up on flood, down on ebb.*

clean Southwest branch. Tidal flow was not a dominant factor at the polluted Northwest monitor site.

All the fish recorded in the Millbank monitor were upstream of its location when recording terminated. Of 8 approaching the Southwest monitor, 5 proceeded up and 3 diverted temporarily to the Northwest monitor, then returned and went up the Southwest. Of 10 approaching the Northwest monitor, 4 went up, 6 came back down, of which 4 went up through the Southwest monitor while 2 stayed down.

Paths taken by some of the fish which were followed for several hours by the tracking boat are shown in fig 3.

Discussion

Most of the study of salmon movements in the Miramichi took place in the 3½ weeks from 13 September to 7 October. During this period the combined mean daily discharge of two large Southwest Miramichi tributaries was 98 m³/s and of two large Northwest Miramichi tributaries 44 m³/s. For the three days of the July hydrographic survey the comparable figures were: Southwest, 115 m³/s and Northwest, 39 m³/s (J. E. Peters, Inland Waters Branch, Dept. of Energy, Mines and Resources, personal communication). Hence the distribution of salinity and organic pollution patterns emerging from the July hydrographic survey (fig 2) probably resembles the conditions pertaining during the September-October study of salmon movements.

Movement of salmon liberated at Millbank, illustrated in fig 3: 1–47, 2–202 and 7–66, seems to indicate much back-and-forth movement, somewhat limited at the upper end by the steep pollution gradient between Chatham and French Fort (fig 2, stations 9 and 10). Nos. 1–47 moved up about 8 km between liberation at low tide and the next high tide; Nos. 7–66, liberated about 2 h after low tide, moved up about 5 km and turned downward on the ebb (fig 3). In both instances the downward turn at high tide occurred in the area of rapidly increasing pollution. Range of upstream movement was apparently a little greater than the average tidal incursion of 6.1 km in the estuary.

Fish liberated at French Fort tended to move downstream, hence away from the heavier pollution upstream.

Those liberated at Newcastle and which were followed in the tracking boat swam over to the cleaner, Chatham side of the river before moving up. Although the tracking boat did much searching in the area, at no time were fish located in the channel on the northwest side of Beaubears Island. A marsh at the downstream end of this channel was a reservoir for residual effluent from a wood-preservative plant, which seeped into the river at high tides (Zitko, unpublished data). A pulp and paper plant at the mouth of the Northwest arm (fig 1) also contributes to the organic burden on the northwest side of Beaubears. Some of its effluent flows through the Tickle into the Southwest arm of the estuary (Whitney and Wilson, 1968). It is probable that the paths followed (fig 3) by fish 28–120 and 32–72 reflect avoidance of pulp mill waste (Sprague and Drury, 1969) of Northwest water below Beaubears and at the Tickle above. Although a pulp mill near the foot of Beaubears (fig 1) pours a strongly toxic effluent into Southwest water, it is more diluted than that in the Northwest (ratio of fresh water flows into the two arms is 2 or 3:1).

Finally, behaviour of fish at the monitor sites is also in accord with what might be expected from cleaner water in the Southwest than the Northwest branch. This is indicated in Table 1.

There are no extensive industrial operations on the upper Southwest Miramichi River so that branch has a minimal load of effluents. Whitney and Wilson (1968) describe the effluent plume from the pulp mill on the Northwest, studied by use of fluorescent dye, as "disappearing within 2½ miles" (4 km). But their data indicate that identifiable pollution extended some

TABLE 1. AVERAGE NUMBER OF MOVEMENTS OF INDIVIDUAL FISH INTO AND OUT OF MONITOR FIELDS AND NUMBERS OF FISH GOING UP- OR DOWNSTREAM ON FINAL DEPARTURE, MIRAMICHI ESTUARY, 23–26 SEPT. 1969

Monitor	Into field from		Final departure	
	upstream	downstream	upstream	downstream
Millbank	2.9	2.8	9	0
Southwest	2.0	1.0	8	0
Northwest	3.5	3.6	4	6

distance above the site where our Northwest monitor was located. This pollution probably accounts in large part for the apparent reluctance of salmon to enter the Northwest arm of the estuary.

As shown by our results, nearly all fish traced in the unpolluted Tabusintac estuary moved up comparatively rapidly. Slower upward passage through the Miramichi estuary appears to be attributable to the Miramichi's burden of industrial pollution.

References

ARMSTRONG, F A J and BOALCH, G T The ultra-violet absorption
1961   of sea water. *J. mar. biol. Ass., U.K.*, 41:591–7.
ARMSTRONG, F A J and BOALCH, G T Ultra-violet absorption of
1961a   sea water. *Nature, Lond.*, 192(4805):858–9.
BANKS, J W A review of the literature on upstream migration of
1969   adult salmonids. *J. Fish biol.*, 1:85–136.
ELSON, P F Effects on wild young salmon of spraying DDT over
1967   New Brunswick forests. *J. Fish. Res. Bd Canada*, 24:731–767.
HASLER, A D Underwater guideposts: homing of salmon.
1966   Madison, Milwaukee, University of Wisconsin Press, 167 p.
KETCHUM, B H The exchanges of fresh and salt waters in tidal
1951   estuaries. *J. mar. Res.*, 1(1):18–38.
OGURA, N Ultraviolet absorbance of sea waters of Tokyo Bay,
1965   Sagami Bay and off-shore waters in the western North
Pacific. *J. Oceanogrl Soc. Japan*, 21:(6)237–44.
OGURA, N and HANYA T Nature of ultra-violet absorption of sea
1966   water. *Nature, Lond.*, 212(5063):758.
OGURA, N and HANYA T Ultraviolet absorption of the sea water,
1967   in relation to organic and inorganic matters. *Int. J. Oceanol.
Limnol.*, 1(2):91–102.
OGURA, N and HANYA, T Ultraviolet absorbance as an index of
1968   the pollution of seawater. *J. Wat. Pollut. Control Fed.*,
40:464–7.
SAUNDERS, J W Mass mortalities and behaviour of brook trout
1969   and juvenile Atlantic salmon in a stream polluted by
agricultural pesticides. *J. Fish. Res. Bd Can.*, 26:695–9.
SAUNDERS, R L and SPRAGUE, J B Effects of copper-zinc mining
1967   pollution on a spawning migration of Atlantic salmon.
Wat. Res., 1:419–32.
SPRAGUE, J B and DRURY, D E Avoidance reactions of salmonid
1969   fish to representative pollutants. *Adv. Wat. Pollut. Res.*,
4:169–79.
WHITNEY, W H and WILSON, G G Effluent distribution studies on
1968   the Miramichi and Restigouche estuaries. *Pulp Pap. Mag.*,
69(21):79–85.
ZITKO, V and CARSON, M V Analysis of the effluent from the
1969   Domtar wood preserving plant at Newcastle, N.B.
Manuscr. Rep. Fish. Res. Bd Can., (1024):14 p.

Marine Algae and their Relation to Pollution Problems

W. J. North,* G. C. Stephens
and B. B. North†

Les algues marines et leurs rapports avec les problèmes de la pollution
De nombreuses activités humaines ont des répercussions sur les algues. Les eaux d'égouts et les eaux usées agricoles contiennent des éléments nutritifs et des substances toxiques pour les algues. Les effluents thermiques, le dragage, les épanchements de pétrole et les travaux de construction portuaire peuvent influer sur l'abondance et la composition spécifique des populations d'algues.
Les effets de la pollution sur la croissance des algues vont de

Las algas marinas y sus relaciones con los problemas de la contaminación
Muchas actividades humanas afectan la vida de las algas. Existen nutrientes así como sustancias tóxicas para las algas, en las aguas procedentes de las alcantarillas y de los desechos agrícolas. Pueden influir en la abundancia y composición de las especies de las poblaciones de algas las aguas termales, los derrames de petróleo y los trabaios de dragado y construcción de puertos.

* W. M. Keck Engineering Laboratories, California Institute of Technology, Pasadena, California 91109, U.S.A.
† Developmental and Cell Biology, University of California at Irvine, Irvine, California, U.S.A.

l'inhibition grave à la forte stimulation. Les floraisons d'algues provoquées par la pollution peuvent avoir des effets secondaires indésirables (intoxication des animaux, réduction de la teneur en oxygène dissous pendant la nuit, production d'odeurs, coloration anormale de l'eau, etc.). Des substances toxiques telles que les composés radioactifs et les pesticides peuvent être introduites dans la chaîne alimentaire des animaux par l'intermédiaire des algues. Les mécanismes de stimulation des algues sont donc également sinon plus importants que les effets d'inhibition pour la compréhension des conséquences que comporte l'immersion des déchets.

Les substances connues pour exercer une action toxique ou inhibitrice sur les algues marines comprennent les métaux lourds (Cr, Cu, Hg, Ni, Pb, Zn), des matières organiques simples (benzène, crésol, hexane, phénol, toluène), diverses substances inorganiques (Cl, CN, H₂S), et des déchets complexes (eaux usées industrielles, solution de sulfite, hydrocarbures, détergents). Le dépistage de la toxicité grâce aux algues d'eaux douces donne à penser que de nombreuses autres substances pourraient influer sur les algues marines.

Parmi les substances stimulantes qui se trouvent en concentration élevée dans certains déchets, on peut citer les phosphates, l'azote ammoniacal, les nitrates, l'azote organique et la silice. Les rapports P/N et les concentrations de P et de N sous diverses formes peuvent affecter la composition spécifique ainsi que la productivité des algues.

La comparaison des taux d'absorption de l'azote organique (sous forme d'acides aminés) par les algues de "pollution" (*Ulva, Enteromorpha*) avec d'autres formes (18 macrophytes, 12 microphytes) donne une base expérimentale pour expliquer l'abondance de certaines algues dans des milieux riches en matières organiques. D'autre part, certaines algues secrètent de grandes quantités de matières organiques qui pourraient stimuler la croissance chez certains animaux.

La contaminación influye en el crecimiento de las algas, pudiendo producir desde una inhibición grave hasta un estímulo fuerte. La floración de algas afectadas por la contaminación puede tener efectos colaterales indesables (envenenamiento de la fauna, reducción del oxígeno disuelto durante la noche, producción de olores, descoloración de las aguas, etc.). También pueden introducirse en las cadenas alimentarias de los animales sustancias tóxicas, tales como los compuestos radiactivos y los plaguicidas por medio de su incorporación por las algas. Por ello, los mecanismos de estimulación de las algas son igualmente importantes, si es que no lo son más, que los efectos de inhibición para comprender las consecuencias ocasionadas por los desechos vertidos en el mar.

Las sustancias tóxicas e inhibitorias conocidas de las algas marinas incluyen metales pesados (Cr, Cu, Hg, Ni, Pb, Zn), sustancias orgánicas simples (benceno, cresol, hexano, fenol, tolueno), varias sustancias inorgánicas (Cl, CN, H₂S) y desechos complejos (aguas de descarga industriales, aguas sulfíticas residuales, compuestos de petróleo, detergentes). Las determinaciones de la toxicidad usando algas de agua dulce indican que otras muchas sustancias podrían influir en las algas marinas.

Las sustancias estimulantes que aparecen en ciertas aguas residuales comprenden los fosfatos, nitrógeno amoniacal, nitrato, nitrógeno orgánico y sílice. Las proporciones y concentraciones de P/N de varias formas de P y de N pueden influir en la composición de las especies, así como en la productividad.

Las comparaciones de los índices de absorción de nitrógeno orgánico (en forma de aminoácidos) por las algas "de contaminación" (*Ulva, Enteromorpha*) en relación con otras formas (18 macrofitas, 12 microfitas) ofrecen una base experimental para explicar la resistencia de ciertas algas en ambientes ricos en materias orgánicas. Por el contrario, algunas algas secretan copiosamente sustancias orgánicas que pueden estimular el crecimiento de ciertos animales.

HUMAN wastes commonly entering marine water in significant amounts include domestic and industrial sewage, agricultural drainage, and heat. Accidents liberate petroleum and its derivatives as well as various chemicals. Construction and dredging can affect water quality both during operations and subsequently, due to altered physical conditions. Floral responses range from stimulation through indifference to total devastation. Species may respond uniformly or in diametrically opposite ways. Dependent animal communities are usually profoundly influenced.

Algal blooms cause many undesirable side-effects

TABLE 1. METALLIC SUBSTANCES SERVING AS COMMON ALGAL MICRONUTRIENTS AT LOW CONCENTRATIONS AND AS TOXICANTS AT HIGHER CONCENTRATIONS (WHEN TOXIC LEVELS HAVE NOT BEEN DETERMINED FOR A MARINE ALGA, THRESHOLDS FOR OTHER ORGANISMS ARE GIVEN)

Metallic substance	Concentration in sea water* mg/l	Concentration mg/l	Test organism	Reference
		STIMULATORY LEVELS		
Boron	4.6	0.1–1	*Ulva lactuca*	Suneson (1945)
		0.1	*Nostoc muscorum*	Eyster (1952)
Cobalt	0.0001	0.0004	" "	Holm-Hansen *et al.*, 1954
Copper	0.003	>0.006	*Chlorella* sp. "	Walker (1953)
		0.02	*Monodus subterraneus*	Miller and Fogg (1957)
Iron	0.01	0.0015–0.01	*Chlorella pyrenoides*	Myers (1947)
		6 × 10⁻⁶	*Asterionella japonica*	Goldberg (1952)
Manganese	0.00005	0.005–0.02	*Dunaliella tertiolecta*	Harvey (1947)
		5 × 10⁻⁵–0.05	*Chlorella pyrenoidosa*	Eyster *et al.* (1958)
Molybdenum	0.01	10⁻⁵ (NO₃ red.)	*Nostoc muscorum*	Eyster (1959)
		0.01 (N fix.)		
		>0.0001	*Scenedesmus* "	Ichioka and " Arnon (1955)
		0.001	*Anabaena cylindrica*	Wolfe (1954, 1954a)
Silicon	3	0.5	*Asterionella*	Lund (1951)
		35.0	*Navicula pelliculosa*	Lewin (1955)
Vanadium	0.002	0.1	*Scenedesmus obliquus*	Arnon and Wessel (1953)
Zinc	0.01	>0.0065	*Chlorella pyrenoidosa*	Walker (1954)
		INHIBITORY LEVELS		
Boron		0.5–1	higher plants	Wilcox (1958)
Cobalt		0.5	*Scenedesmus*	Bringmann and Kuhn (1959)
Copper		0.0064	*Chlorella vulgaris*	Greenfield (1942)
		0.1	*Macrocystis pyrifera*	Clendenning (1958)
		1–2	Common concentrations when used as algicide	Krauss (1962)
Iron		not determined		
Manganese		0.005	*Anabaena, Aphanizomenon, Asterionella*	Gusseva (1937)
Molybdenum		54	*Scenedesmus*	Bringmann and Kuhn (1959)
Silicon		not determined		
Vanadium		10–20	higher plants	Warington (1954)
Zinc		0.5–0.7	*Scenedesmus*	Bringmann and Kuhn (1959)
		5	*Macrocystis pyrifera*	Clendenning (1958)

*Concentrations in sea water from Goldberg (1965)

[331]

TABLE 2. INHIBITORY CONCENTRATIONS OF CERTAIN METALLIC IONS

Metallic ion	Concentration in sea water* mg/l	Inhibitory concentration mg/l	Test organism	Reference
Aluminium	0.01	1.5–2.0	Scenedesmus	Bringmann and Kuhn (1959)
Barium	0.03	34	„	„ „ „ „
Cadmium	0.00011	14–140	Navicula pelliculosa	Lewin (1954) „ „ „
		0.1	Scenedesmus	Bringmann and Kuhn (1959)
Cerium	5.2×10^{-6}	0.15–0.20	„	„ „ „ „
Lanthanum	1.2×10^{-5}	0.15	„	„ „ „ „
Lead	0.00003	4.1 (Pb pptd, not inhib.)	Macrocystis pyrifera	Clendenning (1958) „
Mercury	0.00003	0.03	Scenedesmus	Bringmann and Kuhn (1959)
		0.05	Macrocystis pyrifera	Clendenning (1958)
Nickel	0.002	0.09–1.5	Scenedesmus	Bringmann and Kuhn (1959)
		0.001	Macrocystis pyrifera	Clendenning (1958)
Rubidium	0.12	14	Scenedesmus	Bringmann and Kuhn (1959)
Selenium	0.0004	2.5	„	„ „ „ „
Silver	0.00004	0.05	„	„ „ „ „
Thorium	0.00005	0.4–0.8	„	„ „ „ „
Titanium	0.001	2.0	„	„ „ „ „

*Concentrations in sea water from Goldberg (1965)

including odours, dermatitis, interference with filtration apparatus, biotoxicant release, excessive demands on dissolved oxygen, increases in organic loads, rendering human food organisms poisonous, introducing radio-activity or pesticides into animal food chains, mucilage and slime secretion, and affecting water taste. Algal stimulation mechanisms are consequently an important aspect of their relations to pollution problems.

The relevant literature is extensive and the reference list is selective, emphasizing review articles. Where critical information is lacking for marine forms the reference is to fresh water algal studies.

Metallic stimulating and inhibiting substances

Some metallic elements can either stimulate or depress algal growth or metabolism. Effect usually varies with concentration. Other metallic substances apparently lack stimulatory powers but inhibit when above toxic thresholds. Two (iron and silicon) can stimulate but apparently never occur at concentrations inhibiting algae (they can affect animals by physical mechanisms such as gill-clogging). These metallic elements occur as metal cations (Rb^+), as components of anions (BO_3^{---}), as ion pairs ($CdCl^+$), and complexed with inorganic and organic ligands (Goldberg, 1965). Chelation and complexing yields molecules of varying stability. Uptake is frequently highly selective, producing concentrations of 100 to more than a million fold. Specificity may arise from macromolecules within the organism, displaying proper geometry and exposed chemical groups (Bayer, 1964). Goldberg (1965) noted a general correlation between metal-ligand stabilities and relative concentration factors.

Common metallic nutrients occurring in trace quantities in sea water include boron, cobalt, copper, iron, manganese, molybdenum, silicon, and zinc (Table 1). Vanadium may occasionally be required. Physiological functions of these substances were recently reviewed by Wiessner (1962). Stimulation levels frequently lie in the range of parts per billion. As concentrations of parts per million are approached, adverse effects may appear among the micronutrient metals. Many non-nutrient metals are toxic at parts per billion or less (Table 2).

Potentially toxic metallic wastes result from many industrial sources. Examples are the chemical, food, mining, metallurgical, photographic, plating, printing, textile, and tanning industries. Ceramics, dye, electronics,

TABLE 3. INHIBITORY CONCENTRATIONS OF CERTAIN INORGANIC SUBSTANCES

Substance	Inhibitory concentration mg/l or ppm	Test organism	Reference
Ammonia	320–420	Navicula seminulum	
	0.4–0.5	Aphanizomenon	Gusseva (1937)
Chloramines	4	Scenedesmus	Bringmann and Kuhn (1959)
Chlorine	5–10	Macrocystis pyrifera	Clendenning (1958)
	1	phytoplankton control	Fair and Geyer (1961)
$Cr_2O_7^{--}$	0.7	Scenedesmus	Bringmann and Kuhn (1959)
	0.2	Navicula seminulum	
	1.0	Macrocystis pyrifera	Clendenning (1958)
CN^-	26	Scenedesmus	Osterlind (1951)
H^+	>2.5, <8.5	opt. range for algae	McKee and Wolf (1963)
$K_3Fe(CN)_6$	0.25	Scenedesmus	Bringmann and Kuhn (1959)
$K_4Fe(CN)_6$	0.2	„	„ „ „ „
$K(SbO)C_4H_4O_6$	24	„	„ „ „ „
$NaAsO_2$	35–46	„	„ „ „ „
	0.13	Navicula pelliculosa	Lewin (1954) „ „ „
$NaClO_3$	3.8	Scenedesmus	Bringmann and Kuhn (1959)
NaF	95	„	„ „ „ „
Na_2S	97	„	„ „ „ „

and paint manufacture can produce metallic wastes as well as products that eventually may add to this category. Non-industrial sources include agricultural waste waters, deterioration from construction, fertilizers, solid and liquid poisons for rodents and insects, and water treatment processes. Specific sources for each substance are summarized by McKee and Wolf (1963).

Other inhibitory inorganic substances

Man liberates many inorganic substances besides toxic metallics, that can potentially affect algae. These usually do not occur in nature in significant amounts. Except for ammonia, none stimulates algal metabolism. Toxic thresholds are generally higher than cited for metallic substances (Table 3). In addition to the sources listed for metallic substances, this category of inorganics arises in wastes from pulp and textile mills, gas and electric utilities, explosives manufacture, cleaning and laundry wastes, weed control, and water and sewage treatment (McKee and Wolf, 1963).

Inhibitory organic compounds

Organics common in effluents—Toxic thresholds of simple organic compounds tend to be higher than the inorganics (Table 4). Many organics do display high toxicity levels resembling thresholds of inorganics (Krauss, 1962), but fortunately they are uncommon in discharged wastes. Some of the substances of Table 4, such as methylamine, also arise naturally from decomposition. Abundant levels in discharged wastes may arise from the petroleum, food, chemicals, plastics, dye, and pharmaceutical industries. Toxic organics used as fumigants, preservatives, solvents, and in medicine may find their way to natural waters.

Organic pesticides—The older pesticide literature is concerned largely with animal responses. The few algal studies related primarily to algicide uses (Sladeckova and Sladeck, 1968). Recent demonstrations of pesticide concentration via food chains based on algae (Woodwell, 1967) have stimulated interest. The current literature depicts an active field, particularly emphasizing phytoplankton. Because of food chain implications, both nontoxic levels and thresholds are of interest (Table 5). Evidently a wide range of sensitivities exists. Increased inhibition at levels above solubility limits of DDT and dieldrin suggested an influence in non-dissolved states (Menzel et al. 1970).

TABLE 4. INHIBITORY CONCENTRATIONS OF ORGANIC COMPOUNDS FOUND IN DISCHARGED WASTE WATERS

Compound	Inhibitory concentration mg/l	Test organism	Reference
Abietic acid	not detd (most toxic of pulp mill ether-sols.)	*Scenedesmus obliquus*	Kawabe and Tomiyama (1955)
Acetaldehyde	249	*Navicula seminulum*	
Acetic acid	74	,, ,,	
Aniline	10	*Scenedesmus*	Bringmann and Kuhn (1959)
Benzene	>10	*Macrocystis pyrifera*	Clendenning (1960)
Benzyl alcohol	640	*Scenedesmus*	Bringmann and Kuhn (1959)
Benzylamine	6	,,	,, ,, ,, ,,
Butyl acetate	320	,,	,, ,, ,, ,,
Cresol	6–40	,,	,, ,, ,, ,,
	5–10	*Macrocystis pyrifera*	Clendenning (1960a)
Cyclohexane-carboxylic acid	28–29	*Navicula seminulum*	
Dichloropropene	40	*Scenedesmus*	Meinck et al. (1956)
Diethylamine	4	,,	Bringmann and Kuhn (1959)
Diethylphosphate	250	,,	,, ,, ,, ,,
Dinitrocresol	36	,,	,, ,, ,, ,,
Dinitrophenol	40	,,	,, ,, ,, ,,
Ethylamine	10	,,	,, ,, ,, ,,
Ethylenediamine	20	,,	,, ,, ,, ,,
Ethylmercuric phosphate (lignasan)	1	gen. effective algicide	Burrows and Combs (1958)
Formaldehyde	0.3	*Scenedesmus*	Bringmann and Kuhn (1959)
Formic acid	100	,,	,, ,, ,, ,,
Hexane	10	*Macrocystis pyrifera*	Clendenning (1960)
Hydroquinone	4	*Scenedesmus*	Bringmann and Kuhn (1959)
Methylamine	4	,,	,, ,, ,, ,,
Nitrobenzene	40	,,	,, ,, ,, ,,
Nitrophenol	28–72	,,	,, ,, ,, ,,
Phenol	1	*Platymonas*	Hood et al. (1959)
	10	*Macrocystis pyrifera*	Clendenning (1960a)
	40	*Scenedesmus*	Bringmann and Kuhn (1959)
	250	*Navicula seminulum*	
Pyrocatechol	6	*Scenedesmus*	Bringmann and Kuhn (1959)
Pyrogallol	8	,,	,, ,, ,, ,,
Quinhydrone	4	,,	,, ,, ,, ,,
Quinoline	140	,,	,, ,, ,, ,,
Quinone	6	,,	,, ,, ,, ,,
Resorcinol	60	,,	,, ,, ,, ,,
Toluene	120	,,	,, ,, ,, ,,
	10	*Macrocystis pyrifera*	Clendenning (1960a)
Toluidine	8–10	*Scenedesmus*	Bringmann and Kuhn (1959)
Triethylamine	1	,,	,, ,, ,, ,,
Trinitrophenol	240	,,	,, ,, ,, ,,

[333]

TABLE 5. TOXIC AND NON-TOXIC PESTICIDE LEVELS FOR ALGAE

Substance	Concentration, mg/l non-toxic	toxic	Test organism	Reference
Acrolein		1.5–7.5	Cladophora, Spirogyra	
CMU		2	Blue green algae, diatoms	Palmer (1957)
Dichloronaphthoquinone		0.002	Blue green algae	Palmer and Maloney (1955)
Rosin amine D acetate		0.25–2	Algae	Maloney and Palmer (1956)
		0.25	Ulothrix	Johnson (1955)
Toxaphene (60%)		2	Algae	Palmer and Maloney (1955)
Zn dimethyldithiocarbamate		0.004	Microcystis	Palmer (1957)
		0.25	Diatoms	,,
DDT		0.01	Skeletonema, Coccolithus	Menzel et al. (1970)
		0.001	Cyclotella	,, ,, ,,
	1000		Dunaliella	,, ,, ,,
		0.005	Pyramimonas	Wurster (1968)
		0.001	Peridinium	,, ,,
		0.3	Chlorella	Sodergren (1968)
Dieldrin		0.001	Cyclotella	Menzel et al. (1970)
		0.01	Skeletonema, Coccolithus	,, ,, ,,
	1000		Dunaliella	,, ,, ,,
Endrin		0.001–10⁻⁵	Cyclotella	,, ,, ,,
		0.001–0.01	Coccolithus	,, ,, ,,
		0.01	Skeletonema	,, ,, ,,
	1000		Dunaliella	,, ,, ,,
Simazine	200		Chlorella, Scenedesmus, Chlamydomonas	Vance and Smith (1969)
Dacthal	200		,,	,, ,, ,, ,,
24 D	200		,,	,, ,, ,, ,,
245 T	200			,, ,, ,, ,,
Amitrol-T		150	Chlorella, Scenedesmus, Chlamydomonas	Vance and Smith (1969)
Pentachlorphenol		0.1–0.5	Macrocystis pyrifera	Clendenning (1959)
Na pentachlorphenate		0.1–0.5	,, ,,	,, ,,
p-chlorothiophenol		<10	,, ,,	,, ,,
pentachlorobenzenethiol		<10	,, ,,	,, ,,

TABLE 6. TOXIC DETERGENT LEVELS FOR ALGAE

Substance	Toxic concentration mg/l	Test organism	Reference
Methyldodecylbenzyltrimethyl ammonium chloride	2	Cylindrospermum, Gloeocapsa, Scenedesmus, Chlorella, Gomphonema, Nitzschia	Palmer and Maloney (1955)
Dodecylacetamidodimethylbenzyl ammonium chloride	2	,,	,, ,, ,, ,,
Alkyldimethyl ammonium chloride	2	Cylindrospermum, Scenedesmus, Chlorella, Nitzschia	,, ,, ,, ,,
Arquad D, Arquad D	5	Oscillatoria	Williams et al. (1952)
Zephiran chloride	<1	Macrocystis pyrifera	Clendenning (1958)
Na dodecyl sulphate	5–10	,, ,,	,, ,,
Na tetrapropylenebenzene sulphonate (ABS)	0.5	,, ,,	Clendenning (1960a)

Detergents—Detergents appear substantially less toxic than the potent pesticides (Table 6). They typically occur in wash waters that enter sewers and combine with other liquids, enhancing dilution. Significant damage from short-term exposures to algae thus seems unlikely. Long-term exposure effects await investigation (McKee and Wolf, 1963).

Radioactive compounds

Interactions between radionuclides and organisms vary with concentrations, chemical states, and species of nuclides as well as organismic concentration capabilities, transmission pathways, and often other factors. Micronutrients (^{32}P), skeletal components (^{45}Ca), and analogues substituting in such functions (^{90}Sr) frequently display high concentration factors. Following uptake, biological effects will depend on the nuclide's half-life and emission type as well as location within organisms and concentra-

tion or dilution tendencies in food chains. Space precludes even superficial treatment of this complex subject. Fortunately good reviews exist (Boney, 1965; Burton, 1965; Eppley, 1962; McKee and Wolf, 1963; Templeton, 1965).

Recent work indicates *Corallina officinalis* can concentrate ^{144}Ce by 20,000 (Ancellin and Vilquin, 1966), while *Gonyaulax polyedra* in water from a nuclear test site concentrated radioactivity up to almost 7,000-fold (Thomas et al., 1962). ^{65}Zn concentration factors ranged from 1,200 to 13,000 for *Gracilaria*, *Enteromorpha*, and *Sphacelaria* (Bedrosian, 1962). Burkholder (1963) found a decreasing order of concentration tendencies: Rhodophyta < Phaeophyta < Chlorophyta. Zlobin (1968) proposed a model for ^{89}Sr loss and uptake by *Fucus* and *Ascophyllum*. Williams et al. (1965) distinguished between radioactivity from fallout vs effluents within the fucoid belt in Solway Firth. In a survey of Pacific waters receiving 900 curies daily from the Columbia River, Seymour and Lewis (1964) found potentials among diatoms and seaweeds for ^{51}Cr uptake. An analysis of fallout entering Japanese diets suggested certain seaweeds concentrated ^{95}Zr and ^{95}Nb several thousandfold (Yamagata and Iwashima, 1965).

Complex wastes

Domestic and industrial sewage—Although pure domestic wastes are fairly uniform, combined domestic and industrial discharges vary widely in compositions. Not surprisingly, algal responses range from depression to stimulation (Table 7; also see review by Boney, 1965). Profound faunal changes may accompany algal alterations: Oslo Fjord (Baalsrud, 1967); Great South Bay, Long Island (Ryther, 1954; Lackey, 1960); southern California (North, 1964). Artificial reduction in a single sewage nutrient such as NO_3 might not control algal blooms because of differing requirements among species, potential for substituting other compounds, and abilities to economize. Prior history may influence results because algae may store excess nutrients (Baalsrud, 1967; Borchardt and Azad 1968). Eutrophication can create a much greater organic load than contained in the responsible sewage (Baalsrud, 1967). Wilkinson (1964) computed the nitrogen budget of a polluted New Zealand estuary. Sewage input was predominantly ammonia which was converted to organic nitrogen by resident algae (*Euglena, Ulva, Enteromorpha*) and exported to the sea in dissolved and particulate form. Turbid sewage affects water clarity and limits depth of algal colonization (Gilet, 1960). Suspensoids can accumulate at density interfaces in stratified water, reducing oxygen concentrations significantly (Føyn, 1960). ^{14}C productivity measurements near a southern California ocean outfall revealed depressed conditions close to the discharge, changing to significant elevations at distances downstream representing 6 to 10 hours travel time (State Water Quality Control Board, 1965). Katayama and Fujiyama (1957) and North (1966) noted tumorous growths on algae near discharges.

Petroleum spillage—Oil spillages typically liberate heterogeneous mixtures. Compositions differ according to the material's origin, degree of refinement, and nature of additives, if any. Following spillage, weathering can affect composition. Volatiles will evaporate and these are often the most toxic constituents (McKee, 1956). Detergents used for clean-up can seriously damage biota.

Recent well-publicized oil spills off Cornwall, England (wreck of the *Torrey Canyon*), and Santa Barbara, U.S.A. (blowout during offshore drilling), stimulated major research efforts. The English experience established dangers of detergents (Smith, 1968). Preliminary accounts of events at Santa Barbara suggest that relative to Cornwall, damage to biota was light (Holmes, 1969; Jones, 1969). No detergents were used at Santa Barbara.

Oil spillages may affect algae less severely than animals (North *et al.*, 1965; Smith, 1968) but laboratory studies indicate that petroleum and detergents can affect both attached algae and phytoplankton after several days' exposure at levels of parts per million (Lacaze, 1967; Mironov and Lauskaia, 1968; State Water Quality Control Board, 1964).

Heated effluents—Tropical organisms apparently exist closer to their upper heat tolerances than temperate and arctic dwellers. Consequently installations have more leeway in discharging heat the greater their distances from the equator (Mihursky and Pearce, 1969). Extensive algal losses were noted in receiving waters of semitropical Biscayne Bay, Florida (Wood and Zieman, 1969), but the only floral abnormality noted near a temperate-water discharge in southern California by Turner and Strachan (1969) was absence of large phaeophyta. In addition to waste heat, power plants may inject chlorine or copper into cooling waters to reduce fouling. Some operators also occasionally control fouling by increasing temperatures 20° or 30° above ambient levels for a few hours. Such measures must always be considered when evaluating *in situ* conditions.

Temperature influences on metabolic processes are treated extensively in most physiology textbooks. Field results indicate differential sensitivities exist among algal

TABLE 7. Effects of various water samples on photosynthetic rates of *Macrocystis pyrifera*, expressed as percentage of rates from duplicates in unpolluted water*

Source and % conc. of polluted sample, mixed with sea water	% Photosyn. vs control	Remarks
1% chlorinated San Diego sewage	114–142	42–360 h exposures
10% chlorinated San Diego sewage	0–120	" " (disintegrated)
1% unchlorinated San Diego sewage	120–160	" "
10% unchlorinated San Diego sewage	0	" " (disintegrated)
San Diego Bay	98–146	3 day exposure, sample from 1 mile downstream from outfall
San Diego Bay	71–100	3 day exposure, sample from outfall area
San Diego Bay	60–100	plant moored 1 month 1 mile downstream from outfall
2% Los Angeles Co. sewage	87–172	2–10 day exposures
2% refinery waste	0–69	48–144 h exposures
1% refinery waste	0–144	" "
0.5% refinery waste	80–103	" "
2% oil brine, untreated	0–17	96 h exposure
2% oil brine, treated	0–63	"
2% Dominguez Channel	77–168	48–144 h exposures, a tidal slough receiving industrial waste

* Data from State Water Quality Control Board (1964)

[335]

species and adverse effects are not always predictable from laboratory findings. *Macrocystis pyrifera*, for example, displayed temperature optima for photosynthesis between 20° to 25°C (Clendenning, 1957) yet kelp stands deteriorate when ocean temperatures of 18° to 20°C persist for several weeks (Brandt, 1923; North, 1958). Grazing and encrustation damage apparently increase more rapidly than kelp growth rates as temperature rises.

Morgan and Stross (1969) believed that passage through a cooling system of a power plant utilizing Chesapeake Bay water, caused greater phytoplankton inactivation than could be explained by exposure to heat and chlorine alone. They noted that during colder seasons, added heat stimulated productivity but depressions occurred during summer when effluents were substantially warmer than the elevated bay temperatures. North (1969) examined algal distributions near a thermal effluent discharge canal at Morro Bay, California, and concluded normal conditions existed within 500 feet of the canal terminus.

Stimulatory effects of organic compounds

Certain macroscopic and microscopic algae survive and may even flourish in polluted waters. Domestic wastes and run-off in quiet bays and estuaries may encourage growth of seaweeds such as *Ulva* and *Enteromorpha*. *Ulva* blooms can become so luxuriant that the algal decay products are more unpleasant than the sewage itself (Cotton, 1910). Articulated corallines have also been cited as pollution indicators. *Corallina*, *Bossiella* and *Lithothrix* dominate near ocean outfalls in southern California but are inconspicuous elsewhere (State Water Quality Control Board, 1965a). Associations may exist between other macrophytes and domestic wastes (Boney, 1965; Pearson *et al.*, 1960). Pollution may affect phytoplankton populations. *Skeletonema costatum* occurrences and larger diatom standing crops correlate with ocean outfalls in southern California (Marine Advisers, 1965; State Water Quality Control Board, 1965a; Water Resources Engineers, 1967). *S. costatum* is the most common phytoplankter in the polluted waters of Oslo Fjord (Baalsrud, 1967). Natural contamination favours other microalgae, e.g. *Platymonas* (Lewis and Taylor, 1921). Although suggestive, such correlations provide little information about underlying physiological and ecological mechanisms, and pertinent experimental work is scarce. Proposed mechanisms explaining such algal "blooms" include:

(1) Reduced competition rather than growth stimulation might cause algal dominance. Differential resistance to toxicity may exist among algae. If productivity falls in a contaminated area, grazing pressures would spare resistant forms (e.g. calcareous algae).

(2) Waste organics may stimulate growth by increasing micronutrient supply (Provasoli, 1960).

(a) Sewage sludge may contain essential vitamins, particularly vitamin B_{12} (Hoover *et al.*, 1951; Sathyanarayana, 1959). Many algae require B_{12}, thiamin, and biotin (Droop, 1962; Provasoli, 1963). Although sea water usually contains enough B_{12} to support normal crops of phytoplankton (Droop, 1957), localized concentrations may stimulate growth of selected phytoplankters (Daisley, 1957).

(b) Organic chelators may solubilize required trace elements, making them more available to phytoplankton (Barber and Rhyther, 1969; Johnston, 1964). Alternatively, chelator-trace element binding may be so stable that the trace element is unavailable to the alga (Droop, 1962; Eyster, 1968).

(3) Waste organics may provide macronutrients, such as C, P, or N.

(a) Microbes, particularly in sediments, may utilize energy-rich organics to produce inorganic plant nutrients such as NO_3 or NH_3 (Boysen-Jensen, 1914; Newell, 1965; Stewart, 1968).

(b) Algae may utilize organic molecules in wastes directly. Amino acids, sugars, and fatty acids may supply significant quantities of nitrogen and reduced carbon.

Nutritional potential of dissolved organics

The heterotrophic and photoassimilatory abilities of algae have been studied extensively in the laboratory (Danforth, 1962). In batch culture, many phytoplanktors can use a variety of organic compounds. Unfortunately, few of these studies can be applied to natural algal populations because the usual culture technique employs abnormally high concentrations of organic macronutrients, rapidly changing growth conditions, and dense cell concentrations. Valid laboratory models must demonstrate algal utilization of low, natural substrate concentrations. The following points must be investigated before dissolved organics can be implicated as a significant macronutrient source for an alga in the sea:

(1) The alga must be able to accumulate organic compounds from dilute solution. (Organic carbon reaches only a few ppm, even in contaminated waters.)

(2) It must assimilate the accumulated compounds. The molecules must enter synthetic and catabolic pathways to qualify as growth stimulants.

(3) The quantitative relation between substrate concentration, accumulation rate, and assimilation rate must be assessed. For example, assimilation of organic carbon can be compared to total energy metabolism. Uptake of nitrogenous molecules can be related to nitrogen requirements and growth rates. These quantitative comparisons allow an estimate of the importance of dissolved organic material vis-à-vis other nutrient sources.

A study by North and Stephens (1967, 1969) on the uptake of amino acids by *Platymonas* attempted to fulfil these requirements. Rates of uptake from dilute solution were determined with trace quantities of C^{14}-labelled amino acids. Uptake over a range of concentrations followed Michaelis-Menten kinetics. The accumulation system was half-saturated at low concentrations (5-10 $\times$ 10^{-6} moles/l glycine). Leakage of C^{14} was insignificant. Accumulated amino acids entered catabolic pathways (indicated by evolution of $C^{14}O_2$), and C^{14} was assimilated into trichloroacetic-acid insoluble material. Thus,

Platymonas satisfies criteria (1) and (2) listed above. But what is the nutritional significance of amino acid uptake?

Cells in log phase growth (22 hour generation time) accumulate 80 µg glycine/h/100 mg dry weight from a solution containing 1 µmol per litre. When *Platymonas* is grown in continuous culture with a decreased nutrient supply, glycine uptake increases to 410 µg glycine/h/100 mg dry weight. *Platymonas* oxygen consumption is probably in the range, 100 to 500 µl/h/100 mg dry weight. This is equivalent to the complete oxidation of 191 to 955 µg of glycine/h/100 mg dry weight. These calculations indicate that amino acids may be a significant carbon source for *Platymonas*. This argument does not assume that accumulated acids are fully oxidized. The comparison with oxidative rates is merely a way to relate the amino acid input to the total requirements of the organism. Growth and leakage of compounds to the environment may increase metabolic requirements by a factor of two. Nonetheless, dissolved amino acids can provide as much as 50 per cent of the total carbon requirement.

The evaluation of dissolved organics as a nitrogen source is approached in a similar way. Nitrogen content, generation time, and uptake rates can be used to compute the value of amino acids as a nitrogen source. However, each of these parameters varies with growth conditions. As culture nutrient supply decreases, generation time and uptake rate increase, and cell nitrogen decreases. This greatly increases the effective contribution of the amino acid uptake pathway. After growth at high nutrient levels, uptake accounts for 10 per cent of the nitrogen requirement. At lower levels, glycine uptake supplies more than twice the required nitrogen. Even the low amino acid concentrations in nearshore waters (about 1 µmol/l: Chau and Riley, 1966; Degens *et al.*, 1964; Webb and Wood, 1967) can support the nitrogen requirements of an organism such as *Platymonas*. Leakage of nitrogen-rich compounds would increase the nitrogen requirement. Hellebust (1965) measured excretion of organics by 22 species of phytoplankton, including *Platymonas*. Only about 2 per cent of total assimilated carbon is excreted by *Platymonas*. Even assuming that all leaked compounds are nitrogenous, loss of nitrogen is small compared to amino acid uptake.

The ability to utilize low concentrations of dissolved organics differs among algae (Bunt, 1969; Hellebust, 1968; Hellebust and Guillard, 1967; Sloan and Strickland, 1966). *Platymonas* is common in naturally contaminated environments and may be pre-adapted to polluted conditions. Preliminary results for other microalgae suggest that uptake rates are low for cells grown in nitrogen-rich batch culture. However, nitrogen deprivation increases uptake in some cases (Table 8).*

Macroscopic algae may also possess uptake systems. In our laboratory, *Ulva* and *Enteromorpha* show rapid accumulation of amino acids from dilute solution. Assimilation also occurs rapidly. Interestingly, the uptake system does not appear to saturate at low concentrations, as is the case in *Platymonas*. Rather, half-saturation constants are at least 10^{-4} mol/l. Preliminary measurements were also made on other common macro-

* Tables 8 and 9 contain preliminary results to be reported in detail elsewhere.

TABLE 8. AMINO ACID UPTAKE BY PHYTOPLANKTERS GROWN ON TWO CONCENTRATIONS OF KNO_3[a]

Organism	Amino acid tested	% Uptake A	% Uptake B	Growth in culture
Chlorella sp.	glycine	<1	<1	poor
Chlorella sp.	arginine	82	37	good
Platymonas subcordiformis	glycine	49	10	good
Cyclotella nana	alanine	<1	<1	fair
Melosira sp.	arginine	28	15	good
Nitzschia closterium	glycine	5	<1	good
Nitzschia ovalis	glycine	24	1	good
Thalassiosira fluviatilis	glycine	<1	<1	poor
Thalassiosira fluviatilis	arginine	24	4	good

[a] Cells were harvested from cultures with initial concentrations of 2×10^{-3} M KNO_3 (A) or 2×10^{-4} M KNO_3 (B), washed, and suspended in nitrogen-free medium (5×10^5 cells/ml). Trace quantities of a C^{14}-labelled amino acid (corresponding to 100,000 cpm/ml and $1 - 4 \times 10^{-7}$ mol/l) were added. CPM in cells was determined after 30 min. Uptake is expressed as per cent radioactivity removed from the medium by the cells. In separate experiments, the ability of cells to use the amino acid as a nitrogen source in culture was determined. Results are given in the last column. "Good" indicates that growth rates on amino-nitrogen and nitrate-nitrogen were equal

TABLE 9. UPTAKE OF C^{14}-GLYCINE BY MARINE MACROPHYTES (Tissue was exposed to 2×10^{-7} M/l C^{14}-glycine (1000 cpm/ml) for 15 min, washed, and extracted with 80% ETOH. CPM in the alcohol-soluble and alcohol-insoluble fractions were measured)

Organism	Amino acid uptake cpm/mg dry weight* Alcohol-soluble fraction	Alcohol-insoluble fraction
CHLOROPHYTA		
Chaetomorpha	23	29
Codium	2	8
Enteromorpha	360	73
Ulva	347	434
PHAEOPHYTA		
Colpomenia	17	44
Egregia	2	3
Hesperophycus	7	<1
Macrocystis	8	5
Pachydictyon	17	114
Pelvetia	3	<1
RHODOPHYTA		
Bossiella	4	10
Corallina	12	38
Endocladia	41	43
Gelidium	24	35
Gelidium	76	100
Gigartina	1	<1
Gracilariopsis	5	1
Lithothrix	4	14
Porphyra	1	1
Weeksia	4	<1

* Dry weight of corallines divided by 2 to correct for high ash content.

phytes (Table 9). Accumulation (cpm in alcohol-soluble fraction) and assimilation (cpm in alcohol-insoluble fraction) were measured after 15 min exposure to glycine-C^{14}. *Ulva*, *Enteromorpha*, *Gelidium* and *Pachydictyon* show rapid uptake (i.e. at least 100 cpm/mg in a fraction). *Colpomenia*, *Endocladia*, *Corallina*, *Gelidium*, *Lithothrix* and *Chaetomorpha* also demonstrate moderate uptake (at least 10 cpm/mg). The uptake systems in *Ulva*, *Enteromorpha*, *Lithothrix* and *Corallina* may be partially responsible for the success of these algae in polluted environments. Conversely, many of the forms with little or no uptake ability (the fleshy red and brown algae) are conspicuously absent from contaminated waters (State Water Quality Control Board, 1965a).

In the most active forms (*Ulva* and *Enteromorpha*), 10^{-6} mol amino acids could supply significant carbon or nitrogen when growth involves doubling times of four weeks or longer. Doubling times of two to four weeks have been reported for laboratory-cultured *Ulva* (Kale and Krishnamurthy, 1967). However, the concentrations of amino acids in bays and estuaries may even be higher than 10^{-6} mol. Stephens (1963) reports $7\text{-}10 \times 10^{-5}$ mol/l in the interstitial water of mudflats. Other reports on the amino acid content of sediments are available (Degens, 1968) but more information is needed.

Tremendous quantities of soluble organic compounds are discharged into some inshore waters, including amino acids in particulate and supracolloidal form (Hunter and Heukelekian, 1965). The excessive growth of some algae in these regions may be related to direct uptake and utilization of organic wastes.

References

ANCELLIN, J and VILQUIN, A Experimental contamination of
1966 marine species with Ce-144, Rh-106, and Zr-95. *In* Disposal of radioactive wastes into seas, oceans and surface waters, Vienna, IAEA, pp. 583–603.

ARNON, D I and WESSEL, G Vanadium as an essential element
1953 for green plants. *Nature, Lond.*, 172:1039–40.

BAALSRUD, K Influence of nutrient concentrations on primary
1967 production. *In* Pollution and marine ecology, edited by T A Olsen and F J Burgess. New York, Wiley, pp. 159–69.

BARBER, R T and RYTHER, J H Organic chelators: factors affecting
1969 primary productivity in the Cromwell Current upwelling. *J. exp. mar. Biol. Ecol.*, 3(2):191–9.

BAYER, E Structure and specificity of organic chelating agents.
1964 *Angew. Chem.*, (*Int. ed.*), 3(5):325–392.

BEDROSIAN, P H Relation of certain macroscopic marine algae to
1962 Zn⁶⁵. Thesis, University of Florida, University Microfilms No. (61–5510):173 pp. Abstr. in *Dissert. Abstr.*, 22:3146 (1962).

BONEY, A D Aspects of the biology of seaweeds of economic
1965 importance. *Adv. mar. Biol.* 3:105–253.

BORCHARDT, J A and AZAD, H S Biological extraction of nutrients.
1968 *J. Wat. Pollut. Control Fedn*, 40:1739–54.

BOYSEN-JENSEN, P Studies concerning the organic matter of the
1914 sea bottom. *Rep. Dan. biol. Stn*, (22):1–39.

BRANDT, R P Potash from kelp: early development and growth
1923 of the giant kelp *Macrocystis pyrifera*. *U.S. Dept. Agric. Bull.*, (1191):40 p.

BRINGMANN, G and KUHN, R The toxic effects of waste water on
1959 aquatic bacteria, algae, and small crustaceans. *Gesundheitsingenieur*, 80:115–20.

BUNT, J S Observations of photoheterotrophy in a marine
1969 diatom. *J. Phycol.*, 5(1):37–41.

BURKHOLDER, P R Radioactivity in some aquatic plants. *Nature,*
1963 *Lond.*, 198:601–3.

BURROWS, R E and COMBS, B D Lignasan as bactericide and
1958 algaecide. *Progve Fish Cult.*, 20:143–4.

BURTON, J D Radioactive nuclides in sea water, marine sediments,
1965 and marine organisms. *In* Chemical oceanography, edited by J. P. Riley and G. Skirrow, New York, Academic Press, pp. 425–75.

CHAU, Y K and RILEY, J P The determination of amino acids in
1966 sea water. *Deep-Sea Res.*, 13:(b) Pt 1:1115–24.

CLENDENNING, K A Physiology and biochemistry of giant kelp.
1957 1 July–30 Sept. 1957. *Q. Prog. Rep. Inst. mar. Res. Univ. Calif.*, (IMR 57–6):29–35.

CLENDENNING, K A Laboratory investigations. Effects of dis-
1958 charging wastes on kelp, 1957–8. *A. Rep. Inst. mar. Res. Univ. Calif.*, (58–11);27–38.

CLENDENNING, K A Laboratory investigations. Effects of dis-
1959 charging wastes on kelp, 1 Oct.–31 Dec. 1958. *Q. Prog. Rep. Inst. mar. Res. Univ. Calif.*, (IMR-59-4):13 p.

CLENDENNING K A Laboratory investigations. Effects of dis-
1960 charging wastes on kelp, 1 Oct–31 Dec. 1959. *Q. Prog. Rep. Inst. mar. Res. Univ. Calif.*, (IMR-60-10):7–11.

CLENDENNING, K A Laboratory investigations. Effects of dis-
1960a charging wastes on kelp, 1 July–30 Sept. 1959. *Q. Prog. Rep. Inst. mar. Res. Univ. Calif.*, (60–4):44–7.

COTTON, A D On the growth of *Ulva latissima*, L. in water
1910 polluted by sewage. *Bull. misc. Inf. R. bot. Gdns Kew*, 1910:15–9.

DAISLEY, K W Vitamin B₁₂ in marine ecology. *Nature, Lond.*,
1957 180:1042–3.

DANFORTH, W F., Substrate assimilation and heterotrophy. *In*
1962 Physiology and biochemistry of algae, edited by R A Lewin, New York, Academic Press, pp. 99–123.

DEGENS, E T Molecular composition of nitrogenous compounds
1968 in sea water. Woods Hole, Mass, Woods Hole Oceanographic Institution (6–52) 27 p.

DEGENS, E T, REUTER, J H and SHAW, K N F Biochemical
1964 compounds in off-shore California sediments and sea waters. *Geochim. cosmochim. Acta*, 28(1):45–67.

DROOP, M R Vitamin B₁₂ in marine ecology. *Nature, Lond.*,
1957 180:1041–2.

DROOP, M R Organic micronutrients. *In* Physiology and bio-
1962 chemistry of algae, edited by R. A. Lewin, New York, Academic Press, pp. 141–60.

EPPLEY, R W Uptake of radioactive wastes by algae. *In* Physiology
1962 and biochemistry of algae, edited by R. A. Lewin, New York, Academic Press, pp. 839–40.

EYSTER, C Necessity of boron for *Nostoc muscorum*. *Nature,*
1952 *Lond.*, 170:755.

EYSTER, C Mineral requirements of *Nostoc muscorum* for nitrogen
1959 fixation. *Proc. int. bot. Congr.*, 9(2):109.

EYSTER, C Microorganic and microinorganic requirements for
1968 algae. *In* Algae, man, and the environment, edited by D. F. Jackson, New York, Syracuse University Press, pp. 27–36.

EYSTER, C, *et al.* Manganese requirement with respect to growth.
1958 Hill reaction and photosynthesis. *Pl. Physiol.*, 23:235–41.

FAIR, G M and GEYER, J C Water supply and waste-water
1961 disposal. New York, Wiley, 973 p.

FØYN, E Chemical and biological aspects of sewage disposal in
1960 inner Oslofjord. *Proc. int. Conf. Waste Disp. mar. Envir.*, 1:279–84.

GILET, R Water pollution in Marseilles and its relations with
1960 flora and fauna. *Proc. int. Conf. Waste Disp. mar. Envir.*, 1:39–56.

GOLDBERG, E D Iron assimilation by marine diatoms. *Biol.*
1952 *Bull. mar. biol. Lab., Woods Hole*, (102):243–8.

GOLDBERG, E D Minor elements in seawater. *In* Chemical oceano-
1965 graphy, edited by J. P. Riley and G. Skirrow, New York, Academic Press, pp. 163–96.

GREENFIELD, S S Inhibitory effects of inorganic compounds on
1942 photosynthesis in *Chlorella*. *Am. J. Bot.*, 29:121–31.

GUSSEVA, K A Hydrobiology and microbiology of the Uchinskii
1937 Reservoir. 2. Observations of the department of *Anabaena lemmermannii, Aphanizomenon flos aquae*, and *Asterionella formosa*. *Mikrobiologiia*, 6:449–64.

HARVEY, H W Manganese and the growth of phytoplankton.
1947 *J. mar. Biol. Ass. U.K.*, 26(4):562–79.

HELLEBUST, J A Excretion of some organic compounds by
1965 marine phytoplankton. *Limnol. Oceanogr.*, 10(2):192–206.

HELLEBUST, J A The uptake and utilization of organic substances
1968 by phytoplankters. *In* Symposium on organic matter in natural waters, University of Alaska, Sept. 1968.

HELLEBUST, J A and GUILLARD, R R L Uptake specificity for
1967 organic substrates by the marine diatom, *Melosira nummuloides*. *J. Phycol.*, 3(3):132–5.

HOLM-HANSEN, O, GERLOFF, G C and SKOOG, F Cobalt as an
1954 essential element for blue-green algae. *Physiol. Pl.*, 7:665–75.

HOLMES, R W The Santa Barbara oil spill. *In* Oil on the sea,
1969 edited by D. P. Hout, New York, Plenum Press, pp. 15–27.

HOOD, D W, DRAKE, T W and STEVENSON, B Measurement of
1959 toxicity of organic wastes to marine organisms. *Am. Chem. Soc. Abstr.*, 136:3u.

HOOVER, S R, *et al.* Vitamin B₁₂ in activated sewage sludge.
1951 *Science, N.Y.*, 114:213.

HUNTER, J V and HEUKELEKIAN, H The composition of domestic
1965 sewage fractions. *J. Wat. Pollut. Control Fedn*, 37:1142–63.

ICHIOKA, P S and ARNON, D I Molybdenum in relation to
1955 nitrogen metabolism. 2. Assimilation of ammonia and urea without molybdenum by *Scenedesmus*. *Physiol. Pl.*, 8:552–60.

JOHNSON, L D Control of *Ulothrix zonata* in circular ponds.
1955 *Progve Fish Cult.*, 17:126.

JOHNSTON, R Sea water, the natural medium of phytoplankton.
1964 2. Trace metals and chelation and general discussion. *J. mar. Biol. Ass. U.K.*, 44:87–109.

JONES, L G, *et al.* Just how serious was the Santa Barbara oil
1969 spill? *Ocean Ind.*, (June): 53–56.

KALE, S R and KRISHNAMURTHY, V The growth of excised pieces
1967 of thallus of *Ulva lactuca* var. *ridiga* in laboratory cultures. *In* Proceedings of Seminar on Sea, Salt and Plants, edited by V. Krishnamurthy, Bhavnagar (Gujarat), India, Central Salt and Marine Chemicals Research Institute, pp. 234–9.

KATAYAMA, T and FUJIYAMA, T Studies on the nucleic acid of algae
1957 with special reference to the desoxyribonucleic acid contents of crown gall tissues developed on *Porphyra tenera*. Kjellm. *Bull. Jap. Soc. scient. Fish.*, 23:249–54.

KAWABE, K and TOMIYAMA, T Studies on purification of industrial
1955 waste. 5. On the nature of the poisonous substance in alkali pulp waste. *Bull. Jap. Soc. scient. Fish*, 21(1):37–41.

KRAUSS, R W Inhibitors. *In* Physiology and biochemistry of the
1962 algae, edited by R. A. Lewin, New York, Academic Press, pp. 673–85.

LACAZE, M J Etude de la croissance d'une algue planctonique en
1967 présence d'un détergent utilisé pour la destruction des nappes de pétrole en mer. *C. r. hebd. Séanc. Acad. Sci., Paris*(D), 265(20):1489–91.

LACKEY, J B The status of plankton determinations in marine
1960 pollution analyses. *Proc. int. Conf. Waste Disp. mar. Envir.*, 1:404–12.

LEWIN, J C Silicon metabolism in diatoms. 1. Evidence for the
1954 role of reduced sulfur compounds in silicon utilization. *J. gen. Physiol.*, 37:589–99.

LEWIN, J C Silicon metabolism in diatoms. 1. Sources of silicon
1955 for growth of *Navicula pelliculosa*. *Pl. Physiol.*, 30:129–34.

LEWIS, F and TAYLOR, W R Notes from the Woods Hole
1921 Laboratory. *Rhodora*, 23:249–56.

LUND, J W G Recent researches on algae in relation to water
1951 works practice. *J. Instn. Wat. Engrs*, 5:558.

MALONEY, T E and PALMER, C H Toxicity of six chemical
1956 compounds to thirty cultures of algae. *Wat. Sewage Wks*, 103:509.

Marine Advisers Analysis of oceanographic and ecological
1965 monitoring program at the City of San Diego, Pt. Loma outfall. La Jolla, California, Marine Advisers, 39 p.

McKEE, J E Report on oily substances and their effects on the
1956 beneficial uses of water. *Publs St. Wat. Pollut. Control Bd Calif.*, 16:71 p.

McKEE, J E and WOLF, H W Water quality criteria. *Publs. St.*
1963 *Wat. Pollut. Control Bd, Calif.*, (3–A):548 p.

MEINCK, F, STOFF, H and KOHLSCHUTTER, H *In* Industrial waste
1956 waters. 2nd ed. Stuttgart, Fischer Verlag.

MENZEL, D W, ANDERSON, J and RANDTKE, A Marine phyto-
1970 plankton vary in their response to chlorinated hydrocarbons. *Science, N.Y.*, 167(3926):1724–6.

MIHURSKY, J A and PEARCE, J B Introduction. *Chesapeake Sci.*,
1969 10:125–7.

MILLER, J D and FOGG, G E Studies on the growth of Xantho-
1957 phyceae in pure culture. 1. The mineral nutrition of *Monodus subterraneus* Petersen. *Arch. Mikrobiol.*, 28:1–17.

MIRONOV, O G and LAUSKAIA, L A Vyzhivaemost nekotorykh
1968 morskikh planktonnykh i bentoplanktonnykh vodoroslei v morskoi vode, zagriaznennoi nefteproduktami. *Bot. Zh., SSSR*, 53:661–9. Abstr. in *Biol. Abstr.*, 1969:67981.

MORGAN, R P and STROSS, R G Destruction of phytoplankton
1969 in the cooling water supply of a steam electric station. *Chesapeake Sci.*, 10:165–71.

MYERS, J Culture conditions and the development of the photo-
1947 synthetic mechanism. 5. Influence of the composition of the nutrient medium. *Pl Physiol.*, 22:590–7.

NEWELL, R The role of detritus in the nutrition of two marine
1965 deposit feeders, the prosobranch *Hydrobia ulvae* and the bivalve *Macoma baltica*. *Proc. zool. Soc. Lond.*, 144:24–45.

NORTH, B B and STEPHENS, G C Uptake and assimilation of amino
1967 acids by *Platymonas*. *Biol. Bull. mar. biol. Lab. Woods Hole*, 133(2):391–400.

NORTH, B B and STEPHENS, G C Amino acids and *Platymonas*
1969 nutrition. *Proc. int. Seaweed Symp.*, 6:263–73.

NORTH, W J Kelp investigations program. *A. Rep. Inst. mar. Res.*
1958 *Univ. Calif.*, (IMR 58–12):18 p.

NORTH, W J Ecology of the rocky nearshore environment in
1964 southern California and possible influences of discharged wastes. *Proc. int. Conf. Wat. Pollut. Res.*, 1(3):247–437.

NORTH, W J Kelp bed restoration activities. *Rep. Calif. Inst.*
1966 *Tech.*, 1965–66:5–30.

NORTH, W J Biological effects of a heated water discharge at
1969 Morro Bay, California. *Proc. int. Seaweed Symp.*, 6:275–86.

NORTH, W J, NEUSCHUL, M and CLENDENNING, K A Successive
1965 biological changes observed in a marine cove exposed to a large spillage of mineral oil. *In* Proceedings of the C.I.E.S.M.M. Symposium on marine pollution caused by micro-organisms and mineral oils, Monaco, April 1964, pp. 335–54.

OSTERLIND, S Anion absorption by an alga with cyanide
1951 resistant respiration. *Physiol. Pl.*, 4:528–34.

PALMER, C H Evaluation of new algicides for water supply
1957 purposes. *Taste Odor Control J.*, 23:1.

PALMER, C H and MALONEY, T E Preliminary screening for
1955 potential algicides. *Ohio J. Sci.*, 55:1.

PEARSON, E A, POMEROY, R D and McKEE, J E Summary of
1960 marine waste disposal research program in California. *Publs. St. Wat. Pollut. Control Bd, Calif.*, (22):77 p.

PROVASOLI, L Micronutrients and heterotrophy as possible factors
1960 in bloom production in natural waters. *In* Transactions of the Seminar on Algae and Metropolitan wastes, April 27–29. *Publ. Hlth Publs, Wash.*

PROVASOLI, L Organic regulation of phytoplankton fertility.
1963 *In* The Sea, vol. 2., edited by M. N. Hill, New York, Interscience, vol. 2:165–219.

RYTHER, J H The ecology of phytoplankton blooms in Moriches
1954 Bay and Great South Bay, Long Island, New York. *Biol. Bull. mar. biol. Lab.*, Woods Hole, 106(2):198–209.

SATHYANARAYANA, R, *et al.* Vitamin B_{12} in sewage sludges.
1959 *Science, N.Y.*, 129:276.

SEYMOUR, A H and LEWIS, G B Radionuclides of Columbia River
1964 origin in marine organisms, sediments and water collected from coastal and offshore waters of Washington and Oregon, 1961–3. Oak Ridge, Tennessee, U.S. Atomic Energy Commission, Technical Information Service Extension (UWFL–86):73 p.

SLADECKOVA, A and SLADECK, V Algicides—friends or foes? *In*
1968 Algae, man and environment, edited by D. F. Jackson, New York, Syracuse Univ. Press, pp. 441–58.

SLOAN, P R and STRICKLAND, J D H Heterotrophy of four marine
1966 phytoplankters at low substrate concentrations. *J. Phycol.*, 2:29–32.

SMITH, J E *TORREY CANYON* pollution and marine life.
1968 Cambridge. Cambridge Univ. Press, 196 p.

SODERGREN, A Uptake and accumulation of C^{14} DDT by
1968 *Chlorella* sp. *Oikos*, 19:126–38.

State Water Quality Control Board An investigation of the
1964 effects of discharged wastes on kelp. *Publs. St. Wat. Qual. Control Bd, Calif.*, (26):124 p.

State Water Quality Control Board An investigation of the fate
1965 of organic and inorganic wastes discharged into the marine environment and their effects on biological productivity. *Publs St. Wat. Qual. Control Bd*, (29):116 p.

State Water Quality Control Board An oceanographic and
1965a biological survey of the southern California mainland shelf. *Publs St. Wat. Qual. Control Bd, Calif.*, (27).

STEPHENS, G C Uptake of organic material by aquatic inverte-
1963 brates. 2. Accumulation of amino acids by the bamboo worm, *Clymenella torquata*. *Comp. Biochem. Physiol.*, 10:191–202.

STEWART, W D P Nitrogen input into aquatic ecosystems. In
1968 Algae, man, and the environment, edited by D. F. Jackson, New York, Syracuse University Press, pp. 53–72.

SUNESON, S Einige Versuche über den Einfluss des Bors auf die
1945 Entwicklung und Photosynthese der Meeresalgen. *K. fysiograf. Sallsk. Lund Fohr.*, 15:185–97.

TEMPLETON, W L Ecological aspects of the disposal of radioactive
1965 wastes to the sea. *In* Ecology and industrial society, Oxford, Blackwell Science Publications, pp. 65–97.

THOMAS, W H, LEAR, D W and HAXO, F T Oceanographic
1962 studies during operation "Wigwam". Uptake by the marine dinoflagellate *Gonyaulax polyedra* of radioactivity formed during an underwater nuclear test. *Limnol. Oceanogr.*, 7(Suppl.):66–71.

TURNER, S H and STRACHAN, A R The marine environment in
1969 the vicinity of the San Gabriel river mouth. *Calif. Fish Game*, 55(1):53–68.

VANCE, D B and SMITH, D L Effects of 5 herbicides on 3 green
1969 algae. *Tex. J. Sci.*, 20:329–37.

WALKER, J B Inorganic micronutrient requirements of *Chlorella*.
1953 1. Requirements for calcium (or strontium), copper, and molybdenum. *Arch. Biochem. biophys.*, 46:1–11.

WALKER, J B Inorganic micronutrient requirement of *Chlorella*.
1954 2. Quantitative requirements for iron, manganese, and zinc. *Arch. Biochem. biophys.*, 53:1–8.

WARINGTON, K Some interrelationships between manganese,
1954 molybdenum, and vanadium in the nutrition of soybeans, flax, and oats. *Ann. appl. Biol.*, 38:624.

Water Resources Engineers Inc. Effects of ocean discharge of
1967 waste water on the ocean environment near the City of San Diego outfall. Walnut Creek, California, Water Resources Engineers, Inc., 57 p.

WEBB, K L and WOOD, L Improved techniques for analysis of
1967 free amino acids in seawater. *In* Automation in analytical chemistry, Technicon Symposium, 1966, New York, Mediad, pp. 440–4.

WIESSNER, W Inorganic micronutrients. *In* Physiology and bio-
1962 chemistry of algae, edited by R. A. Lewin, New York, Academic Press, pp. 267–86.

WILCOX, L V Determining the quality of irrigation water.
1958 _Agriculture Inf. Bull._, (197):6.

WILKINSON, L Nitrogen transformations in a polluted estuary.
1964 _Proc. Int. Conf. Wat. Pollut. Res._, 1(3):405–20.

WILLIAMS, O B, GRONINGER, C R and ALBRITTON, N F The
1952 algicidal effect of certain quaternary ammonium compounds. _Proc. Mo._, 16:8.

WILLIAMS, B R H, PERKINS, E J and GORMAN, J The biology of
1965 the Solway Firth in relation to the movement and accumulation of radioactive materials. 4. Algae. Productivity Grp Rep. U.K. atom. Energy Auth., (6) (Abstr. in _Chem. Abstr._, 1967:16254h).

WOLFE, M The effect of molybdenum upon the nitrogen meta-
1954 bolism of _Anabaena cylindrica_. 1. A study of the molybdenum requirement for nitrogen fixation and for nitrate and ammonia assimilation. _Ann. Bot._, 18:299–308.

WOLFE, M The effect of molybdenum upon the nitrogen meta-
1954a bolism of _Anabaena cylindrica_. 2. A more detailed study of the action of molybdenum in nitrate assimilation. _Ann. Bot._, 18:309–25.

WOOD, E J F and ZIEMAN, J C The effects of temperature on
1969 estuarine plant communities. _Chesapeake Sci._, 10:172–4.

WOODWELL, G M Toxic substances and ecological cycles.
1967 _Scient. Am._, 216(3):24–31.

WURSTER, C F, Jr DDT reduces photosynthesis by marine
1968 phytoplankton. _Science, N.Y._, 159(3822):1474–5.

YAMAGATA, N and IWASHIMA, K Environmental contamination
1965 with ^{95}Zr–^{95}Nb in Japan. _Annls. Hyg. publ., Tokyo_, 14: 137–47.

ZLOBIN, V S Dinamika nakopleniia radostrontsiia nekotorymi
1968 burymi vodorosliami i vliianie solenosti morskoi vody na koeffitsienty nakopleniia. _Okeanologiia_, 8(1):78–85 (Abstr. in _Biol. Abstr._, 1969:55221).

Nutrient and Pollutant Concentrations as Determinants in Algal Growth Rates

P. J. Hannan and C. Patouillet*

Les concentrations de nutriments et de polluants comme facteurs du taux de croissance des algues

Cette étude a pour objet de déterminer le taux de croissance des algues sur une période de plusieurs jours en présence de divers polluants. Dans des solutions contenant environ une partie par million de cellules on introduit des nutriments et des polluants en diverses concentrations; les solutions sont ensuite versées dans de petites fioles que l'on expose à une lumière constante dont l'intensité est d'environ 5 382 lux. Périodiquement, on passe les fioles au fluoromètre afin de mesurer les modifications de la fluorescence des cellules.

Certains polluants exercent un effet considérablement plus toxique en combinaison que par eux-mêmes, par exemple les hydrocarbures et le DDT. La constatation la plus remarquable en ce qui concerne les effets des polluants dans les eaux naturelles, est que de nombreuses cultures peuvent être fortement inhibées pendant les premiers jours d'exposition aux produits toxiques, mais se rétablissent pendant les jours suivants. Le taux de concentration des nutriments fournis aux cellules détermine largement dans quelle mesure la culture est inhibée ou reprend son développement.

Il faut faire preuve de prudence lorsqu'on évalue les effets des agents tensioactifs sur le taux de croissance des algues, compte tenu du facteur constitué par le taux surface/volume de la fiole d'épreuve. La toxicité est inversement proportionnelle à ce taux surface/volume.

On examine les répercussions des résultats obtenus sur la revitalisation des voies d'eaux polluées.

Las concentraciones de nutrientes y contaminantes y su efecto en los índices de crecimiento de las algas

El objetivo de este estudio es determinar el índice de crecimiento de las algas durante un período de varios días en presencia de varios contaminantes. Se preparan suspensiones con un contenido aproximado de células de una parte por millón y con diversas concentraciones de nutrientes y contaminantes, y se exponen, en pequeñas ampollas, a una intensidad luminosa constante de unas 500 bujías-pie. Las ampollas se introducen periódicamente en un fluorómetro para medir las modificaciones de la fluorescencia de las células.

El efecto tóxico de algunos contaminantes, por ejemplo, los hidrocarburos y el DDT, es considerablemente mayor en combinación que solos. El descubrimiento más significativo, por lo que se refiere a los efectos de los contaminantes en las aguas naturales, es que muchos cultivos pueden experimentar una seria inhibición durante el primer día de exposición a un tóxico para recobrarse después en los días siguientes. El grado de inhibición o recuperación depende en gran parte de las concentraciones de nutrientes a disposición de las células.

Al evaluar los efectos que en los índices de crecimiento de las algas tienen los agentes que modifican la tensión superficial, es preciso tener cuidado, porque un factor que influye es la razón superficie/volumen de la ampolla utilizada en la prueba, ya que la toxicidad está en razón inversa de la relación superficie/volumen.

Se examinan las implicaciones de los resultados obtenidos, en relación con la revitalización de cursos de agua contaminados.

THE purpose of this study was to determine the effect of various additives on the growth rate of two algae: _Chlorella pyrenoidosa sorokiniana_ (fresh water) and _Phaeodactylum tricornutum_ (marine).

Each experiment lasted at least three days, permitting the distinction between immediate and adaptive effects of pollutants on the organisms. Variables included were concentrations of algal cells, nutrients and pollutants, and surface/volume ratios of the glass culture vessels. Measurements of the fluorescence of the test suspensions were used to estimate cell concentrations.

Experiments

Algal cells fluoresce primarily because of their chlorophyll _a_ content (Lorenzen, 1966) and concentrations of cells of less than one part per million can be detected by this method. In this study a Fluoro-Microphotometer equipped with glass filters was used; a log–log plot of its detector output vs concentration was linear over the range of 2–1,500 ppm, the limits used in the calibration. A basic assumption made here is that an increase in fluorescence can be equated, approximately at least, with an increase in growth.

Test suspensions for the assay procedure were illuminated by Cool White fluorescent lights at 500 fc. Cultures were contained in glass vials, 1.0 × 7.5 cm, or in larger tubes and flasks depending on the appropriate surface/volume ratio. Fluorescence measurements were made with the 1.0 × 7.5 cm vials which fit the sample holder of the fluorometer.

Stock cultures of _Chlorella_ were maintained in long test tubes and aerated with 3 per cent CO_2 before illumination; _Phaeodactylum_ was grown in cotton-stoppered flasks and both types of cultures were maintained at about 300 fc constant illumination by incandescent lamps. No attempt was made to maintain the _Chlorella_ bacteria-free, but standard sterile precautions were used in the initial culturing of the _Phaeodactylum_.

Water which had been passed through a Millipore filter previously washed to remove all detergent traces, was used in preparing media. After the addition of cells to the desired fluorescence level (concentration) the suspensions were aerated for 5 minutes with 3 per cent CO_2-in-air and then placed in volumetric flasks for the addition of the various pollutants. Aliquots were dispensed into vials or tubes according to the purpose of

* Ocean Sciences Division, U.S. Naval Research Laboratory, Washington, D.C. 20390, U.S.A.

the experiment, and illuminated for three days or more with fluorescence readings made once each day. Samples were run in triplicate.

All experiments with the *Phaeodactylum* were performed with the culture medium described by Guillard and Ryther (1962); the suspensions were never washed with water or otherwise subjected to osmotic shock. The culture medium used for growing the *Chlorella* was essentially that devised by Dr. Dean Burk, National Institutes of Health, Bethesda, Md. It is a concentrated medium (Hannan and Patouillet, 1963), and in many experiments was diluted by a factor of 10^3–10^5; in such cases the *Chlorella* suspension used as an inoculum was centrifuged and washed three times to remove traces of the original medium.

Results

There was often an initial lag of algal growth rates which made the gain in fluorescence for the first day rather slight, followed by a greater increase on the second day. Growth on the third day tended to be limited, presumably because of nutrient depletion. The variability in the results was about what one would expect; comparative toxicities could be determined in a given experiment, but growth rates in different experiments could not be compared.

The extremely low concentrations of cells in these experiments probably amplified the variations inherent in wall effects. Best results were obtained by acid-cleaning the glass vessels used in the test exposures, then washing thoroughly in water and keeping them in water until use.

Previous research in this laboratory has concerned mass cultures of *Chlorella* and the medium used has been more concentrated than those normally described. For this study the most dilute medium (Burk's medium diluted 10^5) contained 0.017 ppm PO_4 and 0.003 ppm N. In most of the experiments the concentrations were 1.7 and 0.3 ppm, which are considerably greater than those occurring in most natural waters. All *Chlorella* cells used were harvested from undiluted Burk's medium and therefore had at least some reserve of nutrients to draw from in their growth under less favourable conditions.

TABLE 1. INHIBITION OF GROWTH OF *CHLORELLA* AS FUNCTION OF CONCENTRATION OF CULTURE MEDIUM AND POLLUTANT

Dilution of culture medium	Pollutant lead-conc.	Days		
		1	2	3
Undiluted	10.0 ppm	No change	No change	No change
10^3	0.1 ppm	Not measured	32%	63%
10^3	1.0 ppm	Not measured	100%	100%

TABLE 2. EFFECTS OF ANIONIC DETERGENT CONCENTRATION, AND SURFACE/VOLUME RATIO, ON GROWTH OF *PHAEODACTYLUM* (+ INDICATES STIMULATION, − INDICATES INHIBITION)

Concentration of detergent	Surface volume	Days		
		1	2	3
0.1 ppm	4.0	+50%	+6%	+17%
	2.5	+100%	0%	+16%
1.0 ppm	4.0	−37%	−56%	−61%
	2.5	−100%	−82%	−78%
10.0 ppm	4.0	−100%	−100%	−100%
	2.5	−100%	−100%	−100%

In these experiments growth inhibition by a toxicant compared with appropriately treated controls, varied with the concentration of nutrients in the culture as well as with the toxicant concentration. With undiluted Burk's medium cell growth rate was unaffected by a concentration of 10 ppm lead as Pb $(C_2H_3O_2)_2$, but when the medium was diluted a thousandfold the presence of even 0.1 ppm had a significant effect (Table 1). Similar effects were observed with 1 ppm of an anionic detergent added to Burk's medium at dilutions of 10^3–10^5.

DDT (20 ppm) with acetone solvent (1,600 ppm) also had varying effects on *Chlorella*, depending on the nutrient concentration. The *Chlorella* was resistant to the effects of DDT in all these experiments, and the inhibitions were caused principally by the large excess of acetone as solvent.

Surface/Volume effects

In all the early experiments the only culture vessels used were glass vials, 1.0×7.5 cm. When surface-active agents were tested as toxicants, however, it was anticipated that the surface/volume ratio of the culture vessel would affect results, so a variety of shapes of culture vessels were used. The mutual shading of cells was not a factor in these studies because the suspensions were water white.

Varying the culture vessels had an effect on results, the toxicity being more pronounced in vessels with lower surface/volume ratios. This was interpreted to mean that as the opportunity for adsorption on the walls decreased, more of the additive remained in the bulk liquid in contact with the cells. This is illustrated in Table 2 which shows the inhibition of *Phaeodactylum* by an anionic detergent at several concentrations and in culture vessels of different surface/volume ratios. For a given concentration the effect of the additive is greater at the lower surface/volume ratio. The implication to be drawn from this is that in a system of very small surface/volume ratio, such as a natural water system, the effect of a detergent would be more pronounced than in a laboratory test.

It is interesting that traces of 0.1 ppm of detergent cause a noticeable increase in growth.

The dimensions of the culture vessel can have an appreciable effect on the growth rate of the culture, regardless of the presence of organic detergents. There was good growth of *Chlorella* in the 1.0×7.5 cm vials with Burk's medium undiluted, or at dilutions up to 10^3; however, when the medium was diluted 10^4 there was only minimal growth for one day and then the fluorescence of the culture declined. But when larger test tubes or volumetric flasks were used as culture vessels the *Chlorella* grew abundantly in Burk's medium diluted 10^5. This may be a manifestation of loss of phosphate from the medium by adsorption on the walls (Hassenteufel *et al.* 1963).

Toxicant–Time interaction

During three-day exposure to toxicants, algal cells may become more resistant to the toxicant or be progressively inhibited. Once again the result is interpreted as the combined influence of toxicant and nutrient concentrations. The limits of such variables depend on many factors, see Table 1. This shows that for a given DDT concentration, the inhibition decreases with time in the case of the medium diluted by a factor of 10^2, but it increases with time at a tenfold greater dilution.

The same type result is discernible with anionic detergent. The medium diluted 10^3 provides a uniform inhibition over the three-day test, but at greater dilutions the effect becomes more pronounced with time. Cells suffering from nutrient deficiencies are less able to adapt to the presence of toxicants than those which have been well supplied with nutrients.

Short-term changes in fluorescence

Wurster (1968) found that the addition of DDT in the parts per billion range inhibited the uptake of $C^{14}O_2$ by four different algae. The uptake was measured over a 4 h period after the cells had been grown for the preceding 20–24 h in a systematic light-dark regimen. These results differ considerably from our findings, but Wurster's results reflect a change in growth rate over a 4 h period compared with 3 days in our studies.

To test this, hourly readings were taken of the fluorescence of four *Chlorella* cultures representing two concentrations of Burk's medium, with and without DDT, illuminated with 500 fc. In the first hour the fluorescence dropped to approximately 75 per cent of the initial readings of the four cultures, and then increased gradually until at the end of 8 h the fluorescence in each was approximately the same as at the start. After 24 h the cultures had increased in fluorescence to a value roughly twice the original, with slight inhibitions in the DDT-treated samples.

There is no necessary contradiction between our results and those obtained by Wurster.

Discussion

The principal advantages derived from the fluorescence method in studying the effects of pollutants on algae are as follows:

(1) It is convenient to make fluorescence measurements of cell suspensions without filtration or extraction procedures.

(2) The effect of a toxicant can be monitored for several days and this is helpful in determining whether the test suspension can recover from the effects of the toxicant.

A disadvantage of the method is that the results may not be interpreted quantitatively because changes in the chlorophyll *a* content of the cells may not exactly parallel growth. But there was good correlation of growth rate with log chlorophyll *a* in a study made by Eppley and Sloan (1966), which supports the inferences drawn here.

Phaeodactylum is generally more susceptible than *Chlorella* to damage by the pollutants tested. This is particularly evident with high concentrations of detergents (500 ppm) which give an initially high fluorescence reading with the *Phaeodactylum*, after which there is no growth. Growth of *Chlorella* is possible at these detergent levels if undiluted Burk's medium is used.

The relatively slight inhibition of *Chlorella* by DDT found here should not be interpreted as a universal phenomenon; Menzel *et al.* (1970) have shown that there is a diversity of response among the phytoplankton to DDT and some organisms are far more easily inhibited than others. It is possible that an organism unaffected by a pollutant may still act as a concentrator of it and therefore play an important role in the increasing concentrations found along the food chain.

One interesting observation is the apparent increase in growth promoted by trace amounts of surface active agents. In a number of instances there was a significant growth increase over that of the controls when sublethal amounts of surfactants were added. In the case of the very toxic anionic detergent mentioned above, 0.1 ppm promoted growth whereas 1.0 ppm was inhibitory. The same phenomenon occurred with other detergents at different concentration ranges depending on the toxicities of the compounds.

Temperature has not been included as a factor in these studies although it would probably affect the results. Room temperature (21–23°C) is considerably lower than the 38°C temperature optimum of this strain of *Chlorella*, which could account for the generally slow growth. The control cultures of *Chlorella* approximately doubled between the second and third days of culture, whereas under optimum growth conditions this organism will double every two hours.

The variability in the results is characteristic of those obtained when algal cells grown in batch cultures are used in such assays. An algal system is reproducible when operated in continuous culture at constant light intensity, temperature, CO_2 input rate, and dilution rate (Hannan and Patouillet, 1963a) but this does not apply when cells are centrifuged, washed and resuspended in batches to be grown under new conditions. Therefore it is not possible to make a meaningful compilation of growth rates under various conditions. Within a given experiment the comparison of growths obtained as a function of the variables tested is justified, but inferences beyond such direct comparisons are not recommended.

References

EPPLEY, R W and SLOAN, P R Growth rates of marine phyto-
1966   plankton; correlation with light absorption by cell chloro-
        phyll *a*. *Physiologia Pl.*, 19(1):47–59.
GUILLARD, R R and RYTHER, J H Studies on marine planktonic
1962   diatoms I. *Cyclotella nana* Hustedt and *Detonula confer-
        vacea* (Cleve) Gran. *Can. J. Microbiol.*, 8(2):229–39.
HANNAN, P J and PATOUILLET C Gas exchange with mass cultures
1963   of algae. 1. Effects of light intensity and rate of CO_2 input
        on oxygen production. *Appl. Microbiol.*, 11(5):446–9.
HANNAN, P J and PATOUILLET, C Gas exchange with mass cultures
1963a   of algae. 2. Reliability of a photosynthetic gas exchanger.
        Appl. Microbiol., 11(5):450–2.
HASSENTEUFEL, W, JAGITSCH, R and KOCZY, F F Impregnation
1963   (protection) of glass surface against sorption of phosphate
        traces. *Limnol. Oceanogr.*, 8:152–6.
LORENZEN, C J A method for the continuous measurement of *in
1966   vivo* chlorophyll concentration. *Deep-Sea Res.*, 13(1):223–7.
MENZEL, W, ANDERSON, J and RANDTKE, A Marine phytoplank-
1970   ton vary in their response to chlorinated hydrocarbons.
        Science, N.Y., 167(3926):1724–6.
WURSTER, C F, JR. DDT reduces photosynthesis by marine
1968   phytoplankton. *Science, N.Y.* 159(3822):1474–5.

Possible Dangers of Marine Pollution as a Result of Mining Operations for Metal Ores

J. E. Portmann*

Dangers potentiels de la pollution marine à la suite de l'extraction de minerais metalliférés

La technologie moderne ayant un besoin toujours croissant de matières premières de base, telles que sable, ciment et métaux, on exploite maintenant des ressources jugées non rentables auparavant. Les sources de telles matières premières que l'on exploite actuellement comportent déjà des opérations minières et on a exprimé des craintes concernant les risques de pollution marine qui pourraient en résulter. Dans le présent document, l'auteur examine un certain nombre de métaux et les dangers qui pourraient provenir de l'exploitation du minerai au fond de la mer. Chaque métal est considéré individuellement et l'auteur discute les risques qu'il fait courir à la santé des animaux marins et de l'homme.

Posibles peligros de contaminación del agua de mar debido a la extracción de minerales

Aumentan constantemente las demandas que la tecnología moderna hace de materias primas básicas como arena, cemento y metales y esto da por resultado la explotación de recursos que anteriormente se consideraban antieconómicos. Para explotar algunos de ellos es preciso recurrir a la minería y se ha expresado el peligro de contaminación del mar que esto pueda representar. En este trabajo el autor examina los posibles peligros que representan varios metales y la explotación de minerales en el fondo del mar. Cada metal se examina por separado y se discutan los riesgos para el hombre y para los animales marinos.

THE search for new sources of minerals has now extended to the sea, particularly the continental shelf areas. Extraction of minerals from the sea bed is now a recognized source for sand, tin ores and in certain areas raw materials for the manufacture of cement (Mero, 1965). This exploitation of seabed resources will increase in future and this paper indicates some of the hazards which may result.

Recent figures in the United Kingdom suggest that at least 10 per cent of the sand and gravel used in London is derived directly from marine sources. Dredging operations, though not directly toxic, may lead to increased turbidity and reduced primary productivity. In addition important spawning, feeding or nursery grounds may be damaged or lost. The fine material stirred up may affect the migration routes of fishes, and valuable fisheries, e.g. for salmon, may be lost or damaged. Furthermore, marine fish may be driven away. Dredging operations can leave the sea bed unsuitable for fishing by light trawls or seines because of "catchy" ground.

The dumping of large quantities of fine material in the sea can blanket the bottom and kill off all marine life. This does not necessarily happen, as we found off south-west and north-east England, where large quantities of china clay and colliery waste respectively are dumped. In particular off the south-west coast, the dumping had caused some alteration of the fauna but the new fauna was both richer and more abundant than the original, was suitable as a fish food and was being utilized as such (Howell and Shelton, 1970). Certain types of fishing may however be adversely affected, for example crabs, lobsters and spiny lobsters are likely to move away.

Antimony

This element occasionally found in nature, occurs chiefly as stibnite Sb_2S_3, cervanite Sb_2O_4 and valentinite Sb_2O_3. The normal concentration of the element in sea water is 4 µg/l (Portmann and Riley, 1966a). Few of its salts are soluble and in fact antimony tends to precipitate from solution as Sb_2O_3 or Sb_2O_5. It is thus unlikely that large concentrations would occur in sea water close to extraction operations. Direct ingestion by marine animals cannot however be ruled out, and concentration factors of more than 300 have been reported for certain marine animals (Noddack and Noddack, 1940). Since antimony is reported lethal to humans in quantities as small as

97 mg (McKee and Wolf, 1963) any antimony operations will have to be closely monitored.

Arsenic

This element occurs naturally in the elemental form but mainly as arsenides of true metals and as orpiment As S_3. It is commonly found associated with ores of copper, lead, zinc and tin, and pollution due to arsenic must also be allowed for where these ores are involved. The data on toxicity of arsenic to marine animals are somewhat limited but the lethal concentration appears to be of the order of 1–10 ppm in the water. As this is about 1,000 times higher than the normal seawater concentration (Portmann and Riley, 1964), and as arsenic-containing minerals are unlikely to be appreciably soluble in sea water, wide-scale fish mortalities appear unlikely. In man, 100 mg of arsenic causes severe poisoning and 130 mg has proved fatal (Browning, 1961). It must be borne in mind that arsenic is a cumulative poison and symptoms may take years to develop. According to Noddack and Noddack (1940) concentration factors for marine animals can be as high as 3,300. In areas of naturally high arsenic concentration shellfish can contain as much as 100 ppm, therefore any exploitation of arsenic minerals would have to be accompanied by local restrictions on the sale of shellfish and possibly fish.

Beryllium and bismuth

Beryllium is comparatively rare, its only commercial source being the mineral beryl $Be_3Al_2Si_6O_{18}$. The carbonate and hydroxide are virtually insoluble in water and it is unlikely that beryl would dissolve in sea water to an appreciable extent. The risk to humans from eating any marine animal which may concentrate the element must be minimal.

Bismuth occurs as the native metal and a bismuthinite Bi_2S_3 and bismite Bi_3O_3. The principal commercial source is as a by-product in the refining of other metal ores, for example those of copper and lead. Very few bismuth salts are soluble in water and the normal concentration is 0.02 µg/l (Portmann and Riley, 1966). Although high concentrations are unlikely to arise in the event of marine mining operations, marine animals have been reported to concentrate the element by factors of up to 1,000 times (Goldberg, 1957). This could conceivably pose a human health hazard but there is no evidence that bismuth is highly toxic to marine life.

* Ministry of Agriculture, Fisheries and Food, Fisheries Laboratory, Burnham-on-Crouch, Essex, England.

Cadmium

Cadmium occurs as the sulphide ore greenockite Cd S, but the only sources of commercial importance are zinc ores with which it is commonly associated as an impurity. It is difficult to envisage high concentrations of cadmium arising in sea water, where its normal concentration is 0.02 µg/l (Brooks, 1960). Concentration factors of more than 4,500 have been reported for some marine animals (Noddack and Noddack, 1940). The concentration that is toxic to aquatic life depends upon the hardness of the water and the salt in question, and figures in the range 0.01 to 1 ppm are quoted by McKee and Wolf (1963). Cadmium, once absorbed by humans, tends to be concentrated by the liver, kidneys, pancreas and thyroid and is unlikely to be released (Truhaut and Boudene, 1954). In view of this and the possible concentration effect of marine animals, some monitoring of cadmium levels in fish and shellfish may be necessary if cadmium minerals are mined on the sea bed. Risks to marine life would appear small.

Chromium

The principal source of chromium is the mineral chromite $FeO.Cr_2O_3$. Many chromium salts are soluble but hydroxide and carbonate are relatively insoluble. Finely divided chromite ore may dissolve in sea water to a very small extent, but the normal concentration of chromium in sea water is only 0.05 µg/l (Waldichuk, 1961). Lethal concentrations reported in the literature depend upon the valency state of the element and range from 18 to more than 200 ppm (McKee and Wolf, 1963). Work at Burnham-on-Crouch, using chromic oxide, gave a 48 h LC_{50} of about 100 ppm for brown shrimps and fish (*Agonus cataphractus*). Chromium is not particularly toxic to humans.

Cobalt

This element occurs principally in admixture with other elements as an arsenide, e.g. in smaltite $CoAs_2$ and cobaltite CoAsS. The normal concentration in sea water is between 0.3 and 0.7 µg/l (Young *et al.*, 1959). The element has two possible valency states but only the divalent or cobaltous form is likely in sea water. Concentrations of up to 10 ppm do not appear to be toxic to most aquatic organisms, but there is a great shortage of data for marine animals. The concentration factor for certain marine organisms can be as high as 21,000 times (Noddack and Noddack, 1940). Cobalt appears to have a low toxicity to man (Browning, 1961) and it seems unlikely that mining could create much of a hazard. The arsenic content of many cobalt ores may present a health hazard to man.

Copper

Copper occurs naturally as the element in boulder or native copper and in a variety of sulphide (e.g. pyrites $Cu FeS_2$) and oxide (e.g. cuprite Cu_2O) ores. Although the hydroxide and carbonates are insoluble and the normal concentration of copper in sea water is 1–20 µg/l (Riley, 1965) concentrations of up to c. 1 ppm are known in coastal waters near copper deposits. A certain amount of dissolution of copper from finely divided copper ore must be expected. Literature contains a wealth of data

on the toxicity of copper to many aquatic organisms, including marine animals. The 96 h LC_{50} for oysters is reported to be 0.1–0.5 ppm (Galtsoff, 1932), and work at Burnham-on-Crouch has shown that for pink shrimps the 48 h LC_{50} is 0.14 ppm (Portmann, 1968). The concentration factor for marine organisms was reported by Noddack and Noddack (1940) to be 7,500, and oysters are among the creatures which have this capacity; such concentrations in the animal do not appear to affect its well-being. Copper is toxic to man in quantities of about 100 mg (McKee and Wolf, 1963), but copper poisoning as a result of eating copper-contaminated shellfish is unlikely since their taste renders them unpalatable (Cole, 1956; Fujiya, 1960). The taste threshold for water is as low as 5.0–7.5 ppm (Hale, 1942).

Lead

The main sources of lead are the sulphide ore galena PbS and the carbonate ore cerussite $PbCO_3$. The hydroxide and carbonate are relatively insoluble but the concentration found in sea water can be as high as 9 µg/l (Costa and Molins, 1957), and the solubility of lead salts in sea water approaches 1 ppm. The toxicity of lead to marine organisms is not known precisely, but work at Burnham-on-Crouch suggests that at least for short-term exposures it is well in excess of 1 ppm, at which point it precipitates in sea water. Insoluble lead salts do not appear to be readily ingested by or toxic to fish (Wallen *et al.*, 1957) and it is unlikely that marine mining operations would be a hazard to marine life. Lead is a cumulative poison to man and susceptibility varies greatly with individuals. Noddack and Noddack (1940) reported a concentration factor of up to 1,400 for marine animals, and careful monitoring of tissue levels in food fish and shellfish may prove necessary in the vicinity of mining operations.

Mercury

This metal occurs to a limited extent as the free element but is principally found as the sulphide ore cinnabar HgS. Although the hydroxide and carbonate are relatively insoluble and the normal concentration of mercury in sea water is unlikely to exceed 0.15 µg/l (Hosohara, 1961), even small increases in mercury concentration resulting from mining operations may be dangerous. As little as 0.075 ppm was lethal to pink shrimps in 48 h (Portmann, 1968) and even lower concentrations cause the death of bivalve larvae (Woelke, 1966). In recent years it has become apparent that mercury is readily bioaccumulated by fish and shellfish, particularly in the most toxic form, methyl mercury. In some areas fish have been declared unfit for human consumption because of their mercury content (ICES, 1969). A number of deaths and serious illnesses have occurred as a result of people eating shellfish and fish which had high mercury levels (Kiyoura, 1964). The source of mercury in these cases has not been mining, but they must serve as a warning of what might occur if marine sources of mercury-containing ores were exploited. Very careful monitoring of mercury levels would then be necessary.

Nickel

This element occurs in a variety of forms of which the main one is pentlandite or millerite Ni S. It has a very low toxicity to humans and the dangers to human health of marine mining must be very small. Nickel is normally present in sea water in concentrations of up to 6 µg/l (Black and Mitchell, 1952) but higher concentrations are possible. Its toxicity to marine animals does not appear very great. Even small concentrations can injure land plants (Anon, 1948) and marine flora including phytoplankton may suffer similarly.

Selenium

Selenium occurs in a few rare minerals such as crooksite, but at present the principal source is anode mud from copper refining. The normal concentration in sea water is 0.4 µg/l (Chau and Riley, 1965). As most selenium salts are insoluble, high concentrations of dissolved selenium appear unlikely. Acute toxic level appears to be about 2 ppm in fresh water (McKee and Wolf, 1963), though data for marine species are lacking. However, there is some evidence of foodchain concentration, at least in freshwater (Barnhart, 1958), with high concentrations building up in the liver of fishes with fatal results in only a few weeks. The element is highly toxic to man, with the safe level in food considered to be no more than 3 mg/kg (Ohio River Valley Sanitation Commission). If selenium-containing ores were mined on the sea bed investigations into the accumulation of this element by marine life would be necessary both from the marine animal and human standpoints.

Tin

One of the principal ores of tin is cassiterite SnO_2. It is one of the few elements for which important marine deposits are presently being exploited, and intensive dredging operations are in progress off Thailand and Indonesian coasts. Normal concentrations in sea water are of the order of 5 µg/l (Riley, 1965), but mining operations may well increase this. Tin, at least in trace concentrations, may prove beneficial to fish (Finkel and Allee, 1940) and even concentrations as high as 100 ppm do not appear to be toxic. It is therefore unlikely that dredging operations will prove injurious to fish, and since humans tolerate up to 1,000 mg daily (O.R.V.S.C. 1950) any direct danger to man is unlikely.

Uranium

This element occurs in a variety of forms as the oxide, for instance in pitchblende U_3O_8 or uranite UO_2. Uranium is said to be highly toxic purely as an element but in addition it exists as a number of long-lived isotopes which create a radiological hazard. Any attempt to mine uranium ores would certainly be accompanied by careful radiochemical monitoring, and this would be sufficiently stringent to safeguard both human and marine life.

Zinc

Zinc is one of the more abundant toxic metals. It occurs commonly as zinc blende or sphalerite ZnS and in a different crystalline form as wurtzite. Since the carbonate and hydroxide are relatively insoluble it is unlikely that concentrations of zinc will greatly exceed the normal level for sea water of 1–20 µg/l (Riley, 1965). Concentrations up to 0.4 mg/l have been recorded in some estuarine waters and at these concentrations bivalve larvae are killed. The toxic levels for adult fish and shellfish are somewhat higher and are of the order of 10 ppm (Portmann, 1968). There is some evidence to suggest that particulate zinc can be taken up by marine organisms; the concentration factor can be as high as 100,000 for ^{65}Zn (Silker, 1961) and for the stable isotope Noddack and Noddack (1940) quote a figure of 32,500. Zinc is toxic to man only in very large doses, and since zinc in high concentrations imparts a blue-green colour to fish and shellfish (Speer, 1928) it is somewhat unlikely that specimens contaminated to such an extent would be eaten. Zinc mining operations on the sea bed thus appear to present little risk to man but may well adversely affect fish and shellfish, particularly in the young stages.

Phosphorites and manganese nodules

Phosphorite or phosphate rock occurs commonly as a surficial rock on continental shelf areas around the world. Phosphorite concretion or nodules were first reported in the "Challenger" Reports (Murray and Renard, 1891). Abundant deposits occur off California (Mero, 1965) and will undoubtedly prove a valuable source of phosphorite in future. The composition of the concretions varies somewhat but consists largely of calcium and phosphorus, the main mineral being isotropic carbonate fluorapatite.

Manganese nodules were also first discovered on the "Challenger" expedition and are probably economically the most important marine deposits yet found. They occur in a variety of shapes ranging from totally irregular to roughly spherical. As their name suggests they are composed principally of manganese, but other elements such as cobalt, nickel, zinc and copper occur in appreciable quantities. Their mode of formation is by no means certain, but they usually have a nucleus such as a metallic fragment or shark's tooth to which successive layers of manganese dioxide have become adsorbed. As manganese dioxide is an extremely effective scavenger of other elements this explains the presence of cobalt, nickel, etc. Mero (1965) deals with the subject of manganese nodules in great detail and concludes that, in spite of the fact that they are usually found in deep water, they will be dredged on a commercial basis in the not-too-distant future and are likely to become important sources of manganese, cobalt, nickel and copper.

Pollution risks, at least due to the dredging of manganese nodules and phosphorites, do not appear very great, since both are rather hard concretions and are unlikely to become finely divided.

References

BARNHART, R A   Chemical factors affecting the survival of game
1958   fish in a western Colorado reservoir. *Q. Rep. Colo. Co-op. Fish. Res. Unit*, 4:25–8.
BLACK, W A P and MITCHELL, R L   Trace elements in the
1952   common brown algae and in sea water. *J. mar. biol. Ass. U.K.*, 30(3):575–84.
BROOKS, R R   The use of ion exchange enrichment in the deter-
1960   mination of trace elements in sea water. *Analyst*, 85:745–8.
BROWNING, E   Toxicity of industrial metals. London, Butterworths,
1961   325 p.
CHAU, Y K and RILEY, J P   The determination of selenium in sea
1965   water, silicates and marine organisms. *Anal. chim. acta*, 33(1):36–49.
COLE, H A   Oyster cultivation in Great Britain: a manual of
1956   current practice. London, HMSO, 43 p.
COSTA, R L and MOLINS, L R   Determinación colorimetrica del
1957   plomo en el mejillon (*Mytilus edulis*) y en el agua de Mar de la Ria de Vigo. *Boln Inst. esp. Oceanogr.*, (84):13 p.

FINKEL, A J and ALLEE, W C   The effect of traces of tin on the rate
1940   of growth of goldfish. *Am. J. Physiol.*, 130:665–70.

FUJIYA, M   Studies on the effects of copper dissolved in sea
1960   water on oysters. *Bull. Jap. Soc. scient. Fish.*, 26(5):462–8.

GALTSOFF, P S   Life in the ocean, from a biochemical point of
1932   view. *J. Wash. Acad. Sci.*, 22(9):246–57.

GOLDBERG, E D   Biogeochemistry of trace metals. *Mem. geol.*
1957   *Soc. Am.*, 1(67):345–57.

HALE, F E   Relation of copper and brass pipe to health. *Wat. Wks*
1942   *Engng*, 95:84, 139, 187, 240.

HOSOHARA, K   Mercury content of deep sea water. *J. chem. Soc.*
1961   *Japan*, 82:1107–8.

HOWELL, B R and SHELTON, R G J   The effect of china clay on the
1970   bottom fauna of St. Austell and Mevagissey Bays. *J. mar.*
         biol. Ass. U.K., 50(3):593–608.

ICES   Report of the ICES working group on pollution of the
1969   North Sea. *Coop. Res. Rep. int. Coun. Explor. Sea (A)*,
         (13):61 p.

KIYOURA, R   Water pollution and Minamata disease. *Adv. Wat.*
1964   *Pollut. Res.*, 7:291–3.

McKEE, J E and WOLF, H W   Water quality criteria. *Publs Resourc.*
1963   *Ag. Calif. St. Wat. Qual. Control Bd*, (3–A):548 p.

MERO, J L   Mineral resources of the sea. *Elsevier Oceanogr. Ser.*,
1965   (1):312 p.

MURRAY, J and RENARD, A F   Report on deep-sea deposits. *In*
1891   Report on the scientific results of the exploring voyage of
         HMS CHALLENGER 1873–76, edited by C Wyville
         Thomson. London, HMSO, 525 p.

NODDACK, I and NODDACK, W   Die Häufigkeiten der Schwer-
1940   metalle in Meerestieren. *Ark. Zool.*, 32A(4).

Ohio River Valley Sanitation Commission. Subcommittee on
1950   Toxicities, Metal Finishing Industries Action Committee,
         Report (3).

PORTMANN, J E   Progress report on a programme of insecticide
1968   analysis and toxicity testing in relation to the marine
         environment. *Helgoländer wiss. Meeresunters.*, 17(1–4):
         247–56.

PORTMANN, J E and RILEY, J P   Determination of arsenic in sea
1964   water, marine plants and silicate and carbonate sediments.
         Analytica chim. acta, 31:509–19.

PORTMANN, J E and RILEY, J P   The determination of bismuth in
1966   sea and natural waters. *Analytica chim. acta*, 34:201–10.

PORTMANN, J E and RILEY, J P   The determination of antimony
1966a   in natural waters with particular reference to sea water.
         Analytica chim. acta, 35:35–41.

RILEY, J P   Analytical chemistry of sea water. *In* Chemical
1965   oceanography, edited by J P Riley and G Skirrow.
         London, New York, Academic Press, vol. 2:295–424.

SILKER, W B   Separation of radioactive zinc from reactor cooling
1961   water by an isotope exchange method. *Analyt. Chem*,
         33:233–5.

SPEER, C J   Sanitary engineering aspects of shellfish pollution.
1928   *Bull. Md St. Dep. Hlth*, 1(3).

TARZWELL, C M and HENDERSON, C   Toxicity of less common
1960   metals to fishes. *Ind. Wastes J.*, 5(1):12–8.

TRUHAUT, R and BOUDENE, C   Enquiries into the fate of
1954   cadmium in the body during poisoning: of interest to
         industrial medicine. *Arhiv. Hig. Rada*, 5:19–48. Abstract
         in *A.M.A. Archs Ind. Hlth*, 11:179–80.

WALDICHUK, M   Sedimentation of radioactive wastes in the sea.
1961   *Circ. biol. Stn, Nanaimo, Fish. Res. Bd Can.*, (59):24 p.

WALLEN, I E, GREEN, W C and LASATER, R   Toxicity to *Gambusia*
1957   *affinis* of certain pure chemicals in turbid waters. *Sewage*
         ind. Wastes, 29(6):695–711.

WOELKE, C E   Bioassay with bivalve larvae. *Rep. Pacif. mar. Fish.*
1966   *Commn*, 18(1965) app. 2:33–5.

YOUNG, E G, SMITH, D G and LANGILLE, W M   The chemical
1959   composition of sea water in the vicinity of the Atlantic
         provinces of Canada. *J. Fish. Res. Bd Can.*, 16(1):7–12.

ANON   Bibliography of the literature on the minor elements and
1948   their relation to plant and animal nutrition. 4th ed. New
         York, Chilean Nitrate Educational Bureau Inc., vol. 1:
         1037 p.

Section 4

ECOSYSTEM MODIFICATIONS AND EFFECTS ON MARINE COMMUNITIES

Summary of Discussion

THE use of predictive models to explain spatial or seasonal changes of individual species has been of great interest to ecologists, and is of great potential value for management purposes. Accumulating enough information on physiology and population dynamics to allow a sensible assessment of such changes is for the future. However, ecosystem theory does provide us with a means for analysing the causes of large-scale phenomena in ecosystems. For work at the community or ecosystem level, we need to emphasize species similarities, grouping some into new ecological supertaxa or other ecologically definable units of a community. This approach was seriously explored by speakers, with reference to the stress of pollution.

The components of one model include: the Shannon-Wiener diversity index; Margalef's index of stability in the sense of persistence through time of biomass components; measures of the instability of the aquatic medium. Fairly simple models constructed with these components—when placed in the context of what is known about ecological successions, community structure, the dependence of poikilotherm metabolic rates on temperature, and production—provide a general framework. Apparently, such stresses as fisheries exploitation, toxic waste disposal, dumping of non-toxic organic wastes, and nutrient enrichment produce some similar effects. In particular, a stress alone will reduce the stability and diversity of a community by tending to kill the longer-lived, larger organisms, or, in the case of nutrients, by favouring short-lived opportunists. The variable application of stress and the natural fluctuations of abiotic environmental variables accentuate the stress effects on community instability.

It follows that a polluted aquatic area will be dominated by opportunistic, short-lived, prolific, vagile species that will fluctuate markedly in abundance through time and space. Opportunistic species are commonly less valuable than climax species (e.g. large fish) and sometimes harmful (e.g. red tide organisms). Productivity *per se* is not an index of the stability or success of a community, and is not a good indication of its potential usefulness to man as an exploiter.

The marine ecosystems

Because of man's industrial and recreational uses, heretofore untouched ecosystems such as coral reefs, adjacent lagoons and mangrove swamps are in danger. Throughout the tropics of the world, specific cases are recorded of damage by oil spills, sewage, dredging operations and filling. In the ecosystem terminology, most of these tropical ecosystems are characterized by being highly productive and highly diverse; however, in most cases, not enough is known about the biological mechanisms to ascertain the causes or the levels at which pollutants affect (stress) the system to the point of triggering collapse or radical transformation. Some information suggests that mangroves are relatively insensitive to sewage wastes, while small amounts of sewage affect

corals severely. Conversely, small amounts of oil seem to have no short-term effects on corals, but oil damages mangroves.

Because of variability in environmental factors, some of these ecosystems have high diversity indices and some have low indices. Estuaries are good examples of low diversity, presumably because of short-term, severe abiotic fluctuations. Available evidence suggests that these ecosystems are unstable; hence, violent changes in population numbers can occur. Man tends to congregate near estuaries and a large amount of evidence of pollution damage is accumulating. Other littoral ecosystems, beaches and sub-littoral, are characterized by high diversity and stability, at least in tropical and semi-tropical environments. It is suspected that most of these will be found to be extremely sensitive to stress.

The pelagic ocean is a stable ecosystem characterized by high diversity, a great deal of specialization among individuals, and low productivity. Because it is the largest ecosystem on earth, its contribution to total productivity on this planet is large. Typical of stable ecosystems, the patterns of energy flowing through trophic levels is complex.

The only available information suggests that deep ocean benthic communities of organisms have high diversity and high stability. Because of the uniformity of the environment and the genetically fixed specialization of the organisms to this uniformity, these populations can be suspected to be extremely sensitive to stress. Productivity of these communities is low. Prior to a few years ago, it was believed that about 1—10 per cent of the organic material produced in the euphotic zone settled to the bottom and was consumed by the benthic fauna. More recent measurements throughout the water column show that almost all of the organic matter formed in the euphotic zone is decomposed in the upper 200 metres. Speculation as to the mechanism of food supply to deep water organisms includes very rapid settlement of large organic particles, absorption of dissolved organics on inert particles, transfer of shallow water material to deep water by bottom currents, and a successive cycling of organic material by vertically-migrating animals.

In no case do we possess complete knowledge of the flora and fauna of an aquatic community. Over vast areas of the earth even the common, easily obtained forms are poorly known, resulting in a continued need for the descriptive phase of biology. Taxonomic identification of species in communities, and as indicator and test organisms, should be subjected to similar standards of exactitude as are applied to the chemical identification of pollutants.

Dumping in the ocean

The suitability of the deep oceans as a dump site was discussed. Ecological damage caused by dumping highly toxic pollutants in the deep ocean was considered by constructing a simple analytical model from a set of postulated conditions which were considered to be extreme. Some physical, chemical, and biological factors were taken into account in model construction. The most conservative assumptions regarding water movement were to assume that there was no vertical turbulence nor horizontal transport. The hypothetical site was considered to be a flat region. The toxicant was assumed to spread at a constant rate in a thin film from its source in all directions, penetrating to a uniform depth in the underlying sediments.

For a first approximation, the dosage-response curve was assumed to be linear, and the rate of decomposition of the pollutant was assumed to be a linear function of time. Substitution of coefficients into the final working model indicated that the rate of decomposition or detoxication of the material was the most important single factor in determining ecological damage. On the basis of this work, it was suggested that highly toxic (but short-lived) pollutants were less important than those that persisted for long periods of time from the point of view of damage to organisms.

Concluding remarks by Convener

The development of ecosystem theory is an efficient approach to understanding community responses to broad stresses and to the level of sensitivity of ecosystems of our earth and, therefore, of great value for measurement of the environment. This is not seen as a substitute for theory at lower levels of organization, e.g. populations and organisms.

Research at all levels of organization is equally important. The discussions emphasized that to understand the ultimate effects of pollution or other stresses on marine ecosystems, comprehensive multidisciplinary studies are imperative. These should include in particular: structures and processes characteristic of coral reefs and lagoons, mangrove associations, neritic and deep-sea benthos, including interstitial fauna and pelagic systems. The effects of stress on communities must be inferred from replicate studies in a number of communities. A series of long-term base-line studies, perhaps somewhat less intensive and detailed, should be included. Only in the context of such studies would monitoring data become meaningful.

Méthode d'Approche pour l'Evaluation des Niveaux de Pollution Chimique des Milieux Marins et des Chaînes Alimentaires Marines

*R. Bittel et G. Lacourly**

Method of approach for the evaluation of chemical pollution levels in marine environments and marine food chains

A simple mathematical model has been proposed for the evaluation of transfers of radioactive pollution to marine food chains: if D is the dose of radiation delivered to the organisms, Q the ingested quantity of each food vector, F the transfer factor of the radioelement from the environment to the organism and x the water contamination level:

$$D = \Sigma D(Q,F,x)$$

the sum Σ being extended to cover food vectors as a whole and all the radioelements considered. For man, observance of protection standards demands $D \leqslant D_1$.

This study suggests the possibility of proposing a very similar model for transfers of chemical pollution in the different oceanic trophic levels: if D' is the quantity of pollutant incorporated by an organism or one of its organs whose contamination depends on that of the water, Q' the quantity of each food vector of the pollutant, F' the transfer factor and x' the pollution level of the environment, the following relationship will result:

$$D' = \Sigma D'(Q',F',x')$$

$D' \leqslant D'_1$ being required in order that the integrity of the environment, of living organisms and of human health may be safeguarded.

Further, a list can be made of the parameters on which transfers depend. Its application can clarify concepts relative to critical ecological level, permitted concentration in water, foods and critical diet, thus allowing identification of priority studies for the protection of the marine environment and human health.

Método para evaluar los niveles de contaminación química de los medios marinos y de las cadenas alimentarias marinas

A fin de evaluar las transferencias de la contaminación radiactiva a las cadenas alimentarias en el mar, se ha propuesto un sencillo modelo matemático: si D es la dosis de irradiación que pasa al organismo; Q, la cantidad absorbida por cada vector alimentario; F, el factor de transferencia del radioelemento del medio al organismo; x, el nivel de contaminación del agua:

$$D = \Sigma D(Q,F,x)$$

donde la suma Σ abarca el conjunto de los vectores alimentarios y de todos los radioelementos considerados. El respeto de las normas de protección requiere que para el hombre $D \leqslant D_1$.

Este estudio sugiere la posibilidad de proponer un modelo muy parecido para las transferencias de las contaminaciones químicas en los diferentes niveles tróficos oceánicos: si D' es la cantidad de un contaminante incorporado por un organismo o uno de sus órganos cuya contaminación depende de la de las aguas; Q', la cantidad de cada uno de los vectores alimentarios del contaminante; F', el factor de transferencia; x', el nivel de contaminación del medio, se tendrá que:

$$D' = \Sigma D'(Q',F',x')$$

imponiendo el respeto de la integridad del medio, de los organismos vivos y de la salud del hombre $D' \leqslant D'_1$.

Este modelo facilita el inventario de los parámetros de los que dependen las transferencias. Su aplicación es susceptible de precisar las nociones del nivel ecológico crítico, la concentración autorizada en el agua y en los alimentos, el régimen alimentario crítico, y permite con ello determinar los estudios prioritarios sobre la protección del medio marino y la salud del hombre.

UNE partie des produits chimiques que l'homme utilise ou qui constituent les déchets de ses diverses activités parvient tôt ou tard dans le milieu, ce qui peut avoir pour effet de nuire à la santé de l'homme et au maintien des équilibres naturels. On peut distinguer deux types d'action pour les toxiques chimiques présents dans les eaux: d'une part, une toxicité directe ou immédiate sur les organismes, d'autre part, une toxicité induite ou mieux une toxicité par transmission, le polluant parvenant à l'organisme par les aliments (Aubert *et al.*,1969; CERBOM, 1969). Dans chacun de ces cas, la contamination peut être unique et aiguë, ou bien répétée, voire chronique et ses effets peuvent apparaître soit immédiatement, soit à long terme

---

*Commissariat à l'Energie Atomique, Département de la Protection Sanitaire, Fontenay-aux-Roses, France.

(Truhaut, 1966). Les analogies entre ces phénomènes et ceux qui résultent d'une pollution des eaux par des radioéléments semblent donc assez frappantes. Aussi, est-il intéressant de rechercher dans quelle mesure les études entreprises pour évaluer les niveaux de contamination radioactive peuvent servir de base pour proposer une méthode d'approche destinée à apprécier les niveaux de pollution des milieux et des chaînes alimentaires polluées chimiquement. L'objet concret de cette communication est précisément de tenter d'étendre au domaine des transferts des pollutions chimiques des chaînes alimentaires océaniques de l'homme un modèle préalablement proposé dans le cas des pollutions radioactives (Bittel *et al.*, 1969, 1970).

Modèle mathématique pour l'evaluation des niveaux de pollution radioactive des milieux aquatiques

Les substances radioactives parvenant dans les milieux aquatiques peuvent entraîner par des mécanismes divers une irradiation de l'homme. Parmi ces mécanismes, il semble que les transferts de radioéléments par les chaînes alimentaires soient les plus importants.

La Commission Internationale de Protection Radiologique (CIPR) a fixé, pour les personnes professionnellement exposées et le public, des doses limites qui, en principe, ne doivent pas être dépassées. Pour s'assurer que cette dose limite n'est pas atteinte pour les populations intéressées, on cherche à fixer des normes dérivées pour les différents radioéléments susceptibles d'être présents dans les effluents. C'est ainsi qu'on a fixé des concentrations maximales admissibles pour l'eau (CMA) établies à partir de l'hypothèse simplificatrice suivant laquelle l'eau ingérée avec les aliments et l'eau directement consommée comme eau de boisson présentaient le même niveau de contamination et en étaient les uniques vecteurs. Or, l'expérience a montré que les divers aliments présentent des niveaux de contamination très différents. De nouvelles études étaient donc nécessaires pour déterminer les niveaux dérivés limites de contamination, notamment pour les eaux continentales et marines, de façon à s'assurer que l'ingestion par l'homme des produits dont la pollution dépend de celle des eaux n'entraîne, pour le public, aucun dépassement de la dose limite d'irradiation. Ces niveaux dérivés ont été établis sur la base des quantités maximales ingérables quotidiennement.

Le problème a d'abord été traité sur le plan général et théorique par Ledermann (1965), puis, sur un plan plus concret, dans les cas des chaînes alimentaires terrestres, par Lacourly *et al.* (1969) et dans celui des chaînes alimentaires aquatiques par Bittel *et al.* (1969, 1970).

Dans le cas des chaînes alimentaires aquatiques, on a supposé que les radioéléments déposés dans les systèmes aquifères de surface parvenaient à l'homme par l'ensemble des aliments dont la pollution peut dépendre de celle des eaux et par ceux-ci seulement, c'est-à-dire:
— l'eau consommée comme eau de boisson,
— les aliments contaminés par les eaux d'irrigation ("aliments irrigués"),
— les aliments aquatiques d'origine océanique (poissons, crustacés, mollusques et, éventuellement, algues).

Le modèle simple basé sur la notion de facteur de transfert et, en particulier sur celle de facteur de con-

centration, qui a pu être élaboré, va être très brièvement exposé.

On a convenu de désigner par facteur de transfert F le rapport:

$$F = \frac{\text{Dose délivrée du fait de l'ingestion de 1 kg de vecteur alimentaire}}{\text{radioactivité dans 1 kg d'eau}}$$

(F est exprimé en rem par curie ou en sous-multiple de cette unité)

La contribution d'un aliment i à la dose d'irradiation D délivrée par ingestion pendant l'unité de temps est:

$D_i = Q_i d_i$, Q_i étant la quantité ingérée par unité de temps et d_i, la dose délivrée unitaire, c'est-à-dire celle due à l'ingestion de l'unité de poids de l'aliment i. D'après les définitions précédentes:

$d_i = F_i x_i$, x_i étant le niveau de radioactivité de l'eau. On peut donc écrire,

$$D = \Sigma Q_i d_i \qquad (1)$$
$$\text{et } D = \Sigma Q_i F_i x_i \qquad (2)$$

la somme Σ étant étendue à tous les aliments aquatiques et aux divers radioéléments considérés.

A partir de ce modèle, il a été possible de faire une étude théorique des points suivants:
— évaluation des niveaux limites de contamination des eaux,
— étude de l'évolution vers les conditions "critiques", c'est-à-dire vers les conditions entraînant les radio-expositions maximales pour les individus,
— étude de la variabilité des divers paramètres dont dépendent les doses délivrées et les niveaux limites, et étude de l'incidence de cette variabilité sur la dose délivrée du fait de l'ingestion d'aliments dont la pollution dépend de celle des eaux.

Possibilité d'extension du modèle précédent aux cas des pollutions chimiques

Le modèle précédent recourt à trois notions courantes dans le cadre de la protection radiologique des chaînes alimentaires:
— celle de dose délivrée, exprimée en unité d'énergie (le rem),
— celle de niveau de pollution du milieu, exprimée en unité de radioactivité (picocurie),
— celle, enfin, de facteur de transfert, elle-même complexe, puisque les transferts de l'eau à l'homme comprennent les transferts écologiques dans le milieu et les transferts physiologiques intégrant l'ensemble des mécanismes allant de l'ingestion aux dépôts dans les organes. Aux premiers, correspondent des facteurs écologiques de transfert (analogues aux facteurs de concentration) et, aux seconds, des facteurs physiologiques de transfert, exprimés en unités d'énergie (rems) par unité de radioactivité ingérée (picocurie).

L'extension de ce modèle aux pollutions chimiques implique d'abord d'étudier en quelle mesure ces notions sont transposables au domaine des pollutions chimiques.

Notions de dose et de dose limite

Les radioéléments qui sont parvenus dans l'organisme sont nuisibles par suite de l'énergie que les rayonnements émis délivrent aux tissus. Cette énergie est caractéristique du radioélément et ne dépend pas de la forme physico-chimique des combinaisons dans lesquelles il est inclus. D'autre part, les effets biologiques de l'énergie absorbée ne sont spécifiques ni de l'isotope radioactif, ni de la structure de la combinaison chimique où figure cet isotope : la radiotoxicité d'un radioisotope ne dépend que des caractéristiques nucléaires du radionucléide et de la radiosensibilité des différents tissus ou organismes. Enfin, les doses d'irradiation provenant des divers radioéléments présents simultanément dans un tissu ou un organe sont considérées comme additives.

Il a donc été possible, en tenant compte des radiosensibilités des tissus et des organismes, des risques acceptés en fonction notamment de l'importance des groupes de population concernés et des possibilités de contrôle, de définir des doses limites d'irradiation pour les différents tissus ou organes, quel que soit le radionucléide. Ce sont ces limites de dose qui servent de base à l'estimation des niveaux acceptables de contamination du milieu.

En ce qui concerne les pollutions chimiques, la situation est très différente. En effet, dans ce dernier cas, les modalités d'atteinte des organismes sont très variées et le nombre des polluants possibles est énorme, au regard de celui des radioisotopes susceptibles de polluer un milieu donné. D'autre part, faute d'une unité commune, les nuisances ne sont pas additives : il existe en outre des phénomènes de synergie et des phénomènes d'antagonisme encore très incomplètement connus. Enfin, alors que, dans le domaine radioactif, les problèmes de filiation entre radionucléides ou entre radionucléides et nucléides stables sont bien établis, il est loin d'en être de même dans le cas des polluants chimiques. Il semble donc a priori beaucoup plus difficile de fixer des limites de tolérance dans le cas des pollutions chimiques que dans celui des pollutions radioactives, la difficulté étant surtout grande lorsqu'il s'agit de pollutions chroniques ou continues dont les effets se manifestent à long terme. Si, dans le cas des contaminations chimiques aiguës, on dispose de données assez nombreuses relatives aux doses létales ou létales à 50 pour cent, ou bien relatives aux seuils de toxicité (OMS, 1962 et 1968; Truhaut, 1966; Ternisien, 1968; CERBOM, 1969), les données relatives aux pollutions chroniques et aux pollutions transmises par la chaîne alimentaire sont beaucoup plus rares (Aubert *et al.*, 1969; U.S. Department of the Interior, 1968).

Pourtant, on a défini une dose journalière acceptable pour une absorption prolongée exprimée en mg du produit chimique sous la forme qu'il présente dans l'aliment par kg de poids corporel (mg/j/kg) (Ternisien, 1968). Pour un individu de poids moyen, cette dose journalière acceptable (DJA) correspond à ce qui a été appelé quantité maximale ingérable pour les substances radioactives.

Facteurs de transfert

Il est évidemment possible de définir dans le cas des pollutions chimiques, des facteurs de transfert qui s'écriraient :

$$F = \frac{\text{concentration en } \mu\text{g de polluant par kg frais d'organisme}}{\text{concentration en } \mu\text{g de polluant par kg d'eau}}$$

et ici encore, on peut définir des facteurs écologiques et des facteurs physiologiques de transfert, chacun de ces facteurs s'exprimant alors par des nombres sans unité.

Niveau de pollution

Pour les raisons qui viennent d'être citées la détermination des niveaux de pollution des milieux aquatiques soulève des grandes difficultés lorsqu'il s'agit de toxiques chimiques. En ce qui concerne les toxiques minéraux, en particulier les métaux, on doit souligner qu'il importe de connaître à la fois les concentrations dans l'eau et l'évolution des structures chimiques, dans lesquelles ils figurent, en fonction des paramètres de milieu (c'est le cas du mercure en particulier). Le problème est presque identique à celui posé par un certain nombre de radio-isotopes dont la physico-chimie est complexe et dont le comportement dans le milieu dépend de la nature des combinaisons dans lesquelles ces éléments sont inclus (Bittel *et al.*, 1968).

De la dose journalière acceptable (DJA) pour un toxique chimique donné, on peut en principe déduire les niveaux limites admissibles pour l'eau. Ce calcul a été fait en ce qui concerne l'homme, en admettant que l'eau ingérée comme boisson est la seule source de pollution (DGRST, 1966–1967; U.S. Department of the Interior, 1968). Or, il est bien connu que l'eau consommée comme eau de boisson est loin d'être le seul vecteur des polluants, qu'ils soient radioactifs ou chimiques. De toute manière, les valeurs obtenues ne peuvent pas a priori être utilisées pour fixer les niveaux limites de pollution des eaux océaniques qui ne sont pas consommées directement par l'homme. Comme dans le cas des pollutions radioactives, il importe de rechercher des niveaux dérivés qui, compte tenu des diverses utilisations des eaux marines, assurent que les doses acceptables pour une absorption prolongée ne soient pas dépassées.

Modèle relatif aux pollutions chimiques

Les considérations précédentes montrent que, sans sous-estimer les difficultés théoriques et pratiques de mise en oeuvre, un modèle très voisin de celui qui a été élaboré pour les pollutions radioactives peut être proposé pour les pollutions chimiques. On envisage ici le cas de pollutions chimiques chroniques de l'homme par suite de la contamination des chaînes alimentaires aquatiques.

Malgré les différences de base existant entre pollutions radioactives et pollutions chimiques, il est possible d'écrire, en ce qui concerne ces dernières, des relations analogues aux relations (1) et (2), c'est-à-dire :

$$D = \Sigma Q_i d_i, \qquad (3)$$
$$D = \Sigma Q_i F_i x_i, \qquad (4)$$

où D est la DJA pour une absorption prolongée, les Q_i sont les quantités, ingérées quotidiennement, de chacun des vecteurs alimentaires dont la pollution dépend de celle des eaux (en kg/j), les d_i sont ce qu'on a convenu d'appeler doses unitaires, les F_i sont les facteurs de transfert (sans unité), les x_i les niveaux de contamination des eaux continentales ou marines constituant les milieux de dispersion de la pollution (la somme Σ étant étendue à tous les vecteurs alimentaires d'un même polluant).

Evaluation des niveaux limites de pollution des eaux

La relation (4) peut s'expliquer de la façon suivante:

$$D = Q_1 F_1 x_1 + \Sigma_2 Q_2 F_2 x_2 + \Sigma_3 Q_3 F_3 x_3 + \Sigma_4 Q_4 F_4 x_4, \quad (5)$$

les indices 1, 2, 3, 4 étant respectivement réservés à l'eau consommée comme eau de boisson, aux "aliments irrigués", aux poissons des eaux continentales et aux produits de la mer. Dans le cas simple où il n'est consommé qu'un aliment de chaque classe et où on ne considère qu'un polluant déterminé, on peut écrire:

$$D = Q_1 F_1 x_1 + Q_2 F_2 x_2 + Q_3 F_3 x_3 + Q_4 F_4 x_4. \quad (6)$$

A la limite pour D = Dose acceptable D_1, et $x_i = l_i$,

$$D_1 = Q_1 F_1 l_1 + Q_2 F_2 l_2 + Q_3 F_3 l_3 + Q_4 F_4 l_4. \quad (7)$$

La relation (7) montre clairement que les niveaux limites l_i, pour les différents vecteurs primaires 1, 2, 3, 4 ne sont pas indépendants, mais qu'ils sont au contraire liés par une relation linéaire. Cette interdépendance des niveaux correspondant à la dose acceptable D_1 était d'ailleurs évidente a priori: il est, par exemple, clair que le niveau limite pour les eaux océaniques ne peut être explicitement calculé que si on connaît les niveaux de contamination des autres vecteurs aquatiques primaires, ou bien si, dans la dose délivrée limite D_1, on a fait a priori la part du milieu non océanique et du milieu océanique.

Calcul des doses journalières acceptables

Les relations (5) et (6) permettent le calcul de la dose résultant de l'ingestion de l'ensemble des aliments pollués par les eaux et le calcul des doses partielles D_1, D_2, D_3, D_4, correspondant respectivement aux ingestions d'eau de boisson, de produits irrigués, de poissons d'eau douce et de produits de la mer. Ceci implique évidemment qu'on dispose de données suffisantes sur les différents paramètres Q, F, et x qui interviennent dans les relations. S'il en est ainsi, on peut comparer la dose calculée à la dose limite D_1 et évaluer la part relative des divers vecteurs alimentaires aquatiques dans la dose délivrée et ainsi apprécier la voie alimentaire susceptible de contribuer le plus à la contamination des populations critiques.

Evolution vers des conditions critiques

Des conditions seront dites critiques, si elles déterminent la contamination maximale des individus pour un même niveau de contamination du milieu. Soit plusieurs régimes alimentaires R_1, R_2, R_3, où les aliments 1, 2, ... i, ... n contaminés du fait de la pollution des eaux marines apportent la totalité de la contamination. La condition de non-dépassement des doses acceptables impose, pour chaque régime alimentaire,

$$\sum_1^n Q_i d_i \leqslant D_1$$

Dans un espace à $(n + 1)$ dimensions rapporté à des axes de coordonnées d_1, d_2, ... d_i, ... d_n, auxquels on ajoute l'axe des d, doses délivrées par l'unité de poids d'eau de mer si elle était consommée, la relation

$$\sum_1^n Q_i d_i = D_1 \quad (8)$$

représente un "plan" P caractéristique de ce régime alimentaire. Le respect des normes impose que le point figuratif d'une situation donnée soit à l'intérieur ou sur la surface du polyèdre délimité par les plans de coordonnées et le "plan" P. Le respect des normes pour un ensemble de régimes impose donc que le point figuratif d'une situation donnée reste à l'intérieur ou sur la surface du polyèdre convexe minimal limité par les plans de coordonnées et les "plans" P_1, P_2, ..., relatifs à chacun des régimes, la surface extérieure du polyèdre représentant l'ensemble des situations critiques ou "surface critique". Il est commode de rapporter les doses unitaires d_i au paramètre unique d et d'écrire:

$$d_i = F_i d_i \ (i = 1, 2, \ldots, n) \quad (9)$$

L'ensemble des équations (9) caractérise une droite (E) représentant l'évolution des doses unitaires d_i en fonction de la pollution de l'eau et qu'on conviendra d'appeler "droite d'évolution". Si d croît, le point figurant une situation donnée s'éloignera de l'origine des axes de coordonnées vers la "surface critique P". Sur cette surface critique les régimes alimentaires sont représentés chacun par une facette. (E) coupe (P) en un point qui se trouve sur une facette ou à la limite de plusieurs facettes; les régimes correspondants seront les régimes "critiques".

Choix d'études prioritaires

Les considérations théoriques précédentes doivent évidemment déboucher sur des aspects concrets dont le plus important est le choix d'études prioritaires.

Les recherches menées dans le cadre des contaminations radioactives ont fourni un modèle pour l'étude des pollutions chimiques. La mise en œuvre de ce modèle nécessite avant tout qu'on s'accorde sur les valeurs des doses journalières acceptables pour les diverses espèces marines et pour l'homme, dans le cas de pollutions chroniques. Il semble qu'il s'agisse là d'un objectif absolument prioritaire qui, pour être atteint, nécessite des études de longue haleine sur les transferts des pollutions dans les divers échelons trophiques. D'autre part, la détermination des niveaux limites nécessite une connaissance suffisante des régimes alimentaires de l'échelon trophique le plus élevé, ce qui, pour l'homme, peut être obtenu et a été, en fait, obtenu par des enquêtes, et une appréciation fidèle des facteurs de transfert dans chacune des situations possibles. Ce dernier point exige un effort d'autant plus considérable qu'il faut tenir compte des divers paramètres de milieu et qu'on doit en principe envisager le problème pour l'ensemble des polluants chimiques. En vue de tenter de clarifier un problème aussi complexe, il apparaît urgent:

—de réaliser un inventaire des différents paramètres des transferts écologiques et physiologiques des polluants,

—de tenter une classification simplifiée des divers polluants sur des critères précis, physiques, physico-chimiques, biochimiques et biologiques, de manière à n'envisager, dans une première phase des recherches, que des problèmes types.

Bibliographie

AUBERT, M *et al.* Côtes de France. Etudes générales des pol-
1969   lutions chimiques rejetées en mer. Inventaire et études de toxicité. Tome 1. Méthodologie. *Revue int. Océanogr. méd.*, Vol. I, Suppl., 72 p.

BITTEL, R et al. Discussion sur le concept de facteur de concentra-
1968   tion entre les organismes marins et l'eau en vue de l'inter-
        prétation des mesures. Rev. int. Océanogr. méd., 11:107–27.

BITTEL, R et LACOURLY, G Essai d'évaluation des transferts de la
1969   pollution radioactive dans les chaînes alimentaires
        océaniques et marines. Revue int. Océanogr. méd, 21:
        75-83

BITTEL, R Estimation des risques de contamination interne de
1970   l'homme résultant de la pollution radioactive des eaux.
        Document présenté à l'"International Radiation Protection
        Association Congress", Brighton, 3–8 mai 1970.

Centre d'Etudes et de Recherches de biologie et d'océano-
1969   graphie médicale. Rapport d'activité 1969. Rapp. Activ.
        C.E.R.B.O.M., 1969.

Délégation générale à la recherche scientifique et technique
1966–   (DGRST). Les pollutions et nuisances d'origine industrielle
1967   et urbaine. Paris, Documentation française, 2 vols.

LACOURLY, G et al., Evaluation des niveaux de contamination
1969   radioactive du milieu ambiant et de la chaîne alimentaire.
        In Environmental contamination by radioactive materials,
        Vienne, A.I.E.A., pp. 273–91.

LEDERMANN, S Contamination radioactive des denrées alimentaires.
1965   Détermination des niveaux admissibles. Rapp. CEA,
        (R-2707) Also issued as Rapp. EUR, (2177f).

Organisation mondiale de la Santé (OMS). Principes devant régir
1962   la protection de la santé des consommateurs à l'égard des
        résidus de pesticides. Tech. Rep. Ser. Wld. Hlth Org., (240).

Organisation mondiale de la Santé (OMS). Résidus de pesticides.
1968   Tech. Rep. Ser. Wld. Hlth Org., (381).

TERNISIEN, J A Les pollutions et leurs effets. Paris, Presses univer-
1968   sitaires de France.

TRUHAUT, R Problèmes toxicologiques posés par l'emploi des
1966   pesticides en agriculture. Bull. INSERM, 21(6):1063–120.

U.S. Department of the Interior. Water quality criteria. Washington,
1968   U.S. Government Printing Office.

Effects of Radiation in the Marine Ecosystem*

W. L. Templeton, R. E. Nakatani, and E. E. Held†

Les effets des rayonnements sur l'écosystème marin

L'expansion actuelle de l'énergie nucléaire va inévitablement faire peser sur le milieu marin une charge croissante de rayonnements. Nous devons nous préoccuper des quantités et du type de rayonnements que les organismes, les peuplements et l'écosystème peuvent tolérer sans que l'équilibre naturel en soit modifié de manière sensible. Cet équilibre n'est pas statique: il réagit différemment a de multiples facteurs, qu'ils soient naturels ou d'origine humaine. Les rayonnements ne représentent que l'un de ces facteurs.

Le document passe en revue nos connaissances sur le rayonnement naturel de fond, les effects de l'exposition aigue et chronique et les émetteurs internes. On discute l'influence du milieu sur les effets du rayonnement, ainsi que les effets de ce dernier sur le comportement et le métabolisme. Des données pertinentes résultant d'etudes pratiques effectuées sur les terrains d'essais du Pacifique sont examinées dans la mesure où elles relèvent des effets du rayonnement.

Les auteurs discutent les travaux de laboratoire et les études de terrain consacrés à l'influence des rayonnements sur les peuplements notamment en ce qui concerne leurs effets potentiels sur les ressources des pêches et sur l'écosystème.

Efectos de la irradiacion en el ecosistema marino

Dada la actual expansion del empleo de la energía nuclear, es inevitable que el ambiente marino reciba una carga creciente de irradiación. Es preciso interesarse en las cantidades y clases de irradiación que organismos, poblaciones y el ecosistema pueden tolerar sin que se modifique sensiblemente el equilibrio de la naturaleza, equilibrio que no es estático, sino que reacciona de diversas maneras ante muchos factores naturales y artificiales, uno de los cuales es la irradiación.

Este documento reseña nuestros conocimientos de la irradiación natural, los efectos de la exposición externa aguda y crónica y los emisores internos. Se examinan la influencia del medio en los efectos de la irradiación y los de las reacciones y el metabolismo y se dan datos de los estudios prácticos realizados en los terrenos de prueba del Pacífico y en relación con los efectos de la irradiación.

Finalmente, se discutan los estudios de laboratorio y prácticos sobre los efectos en la población, particularmente con relación a las posibles repercusiones en los recursos pesqueros y en el ecosistema.

THE sources of the natural radioactivity in the earth's crust, sea water sediments and biota are the cosmic rays. More than sixty radionuclides have been identified within the marine system and calculations have been made of the dose rates affecting the biota. To those rates of natural radioactivity marine life has long been subject and has acquired or developed various degrees of tolerance. To assess what artificially applied extra rates various organisms can stand tests have been made. It was found that lethal amounts differ widely among organisms, because of biological variations related to species, age, physiological state, body size, etc. In the aquatic environment, these variations are further complicated by the interaction of environmental factors such as temperature, dissolved oxygen, chemical composition, and salinity. Nevertheless, exclusive of the eggs and larvae of invertebrates and fish, most of the freshwater and marine organisms for which data exist are relatively radio-resistant. Marine species differ little in radiation tolerances from fresh water species.

Primitive forms are more resistant than the complex vertebrates, and older organisms are more resistant than the young. Bacteria and algae may tolerate doses of thousands of roentgens; but fresh water fish, the most sensitive group, were affected by considerably lower doses.

Despite research into lethal effects of radiation for over 50 years, surprisingly few LD_{50} values have been determined for marine organisms. For six species of marine adult fish, the $LD_{50}/30$ (lethal dose for 50% mortality in 30 days) ranged from 1050 R to 5550 R, similar to values for fresh water fishes, whilst invertebrates ranged from 2000 R to 110,000 R. Since the levels of radiation required to kill marine organisms are so high, actual kills by radiation in the environment are extremely unlikely. These experimental LD_{50} curves are required, however, to keep in perspective the likelihood of mortality in marine organisms from acute radiation.

A series of long-term experiments involving large numbers of organisms (96,000 to 260,000 fingerlings

---

* This paper of over 9000 words with 87 references to sources was based on Chapter XI "Radiation Effects" in the book "Radioactivity in the Marine Environment" ed. A. H. Seymour, published by the National Academy of Science and the National Research Council, Washington, D.C. in September 1971. A great part of the paper was occupied by detailed record of research into specific aspects. For space reasons it is abbreviated and scientists are referred to the prime source. This summary is given to record the general bearing of radiation on marine life, effects noted and conclusions reached.

† Battelle Memorial Institute, Pacific Northwest Laboratories,' Richland, Washington, U.S.A.

were released per experiment) indicate that irradiation at 0.5 R/day from the fertilization stage to the feeding stage produced no damage to the stock sufficient to reduce the reproductive capability over a period of slightly more than one generation. Although abnormalities in young fish were increased by irradiation, the number of adults returning was not affected. On the contrary, the irradiated stock returned in greater numbers and produced a greater total of viable eggs than the control stock.

The Pacific tests

After outlining various experiments on the effects of radiation on aquatic organisms the paper stated that ultimately concern was more with radiation effects on populations and ecosystems in the marine environment than with individuals.

The first large-scale introduction of man-made radionuclides into a marine environment was at Bikini Atoll in 1946. Two 20 kiloton devices were detonated, the first an air burst and the second the underwater detonation in the 250 mi² lagoon, which has maximum depth of about 60 m. In succeeding years, during and including 1958, Bikini and Eniwetok became the Pacific Proving Ground. During that time, nuclear and thermonuclear devices with a total yield of many megatons were detonated at the atolls. Certainly, these atolls represent the most radioactively contaminated marine environment in the world, as far as is known from public announcements. And yet today, more than 23 years since the initial contamination of the atolls, a statement by Schultz made in 1949 on the observed biological effects still holds true:

> "Undoubtedly, countless animal individuals have perished at Bikini because of the atomic bomb experiments and still others may perish. But, this destruction of life in a large atoll like Bikini amounts to only an extremely small percentage of the total animal life. The overall picture of life on the reefs has changed little because beneath this surface layers, and from extensive adjoining unaffected areas, individuals have come forth to repopulate and occupy the reefs. The pressure of population from all sides into the damaged areas is very great and soon replaces the losses. Thus, nature begins the repopulation cycle, and, if given sufficient time, the wounded reefs will be cleansed of their contamination, biological equilibrium will be reached; and life will establish itself as in past millenniums—similar to that before man released the greatest destructive force in his history."

It is inconceivable that there were no radiation effects at the test sites. Evidently, where the prompt radiation at the moment of detonation is sufficiently intense to produce immediate visible effects, the concomitant effects of blast and heat virtually eliminated the populations. Furthermore, those individuals suffering sufficient injury from the residual radiation to be readily recognized are soon eliminated. In other words, gross radiation injury in marine organisms has not been seen at Bikini and Eniwetok because seriously injured individuals do not survive the natural rigors of the environment and the more subtle injuries are exceedingly difficult to detect.

Bikini and Eniwetok were intensively studied before and following the test series and yet in none of the reports on marine organisms is there reference to anomalous individuals.

The land plants and animals were subjected to greater intensities of radiation both from external sources and from internally deposited radionuclides, but even here, lasting effects of radiation on populations or on the ecosystem are not apparent. The land-dwelling hermit crab and coconut crab are subject to higher levels of chronic radiation from internally deposited radionuclides than any other organism studied at the atolls. The levels of ^{90}Sr and ^{137}Cs were found to remain virtually constant at 4500 pCi of ^{90}Sr per gramme of skeleton and 450 pCi of ^{137}Cs per gramme of muscle in the hermit crab at Eniwetok over a period of two years.

Parallel studies of the coconut crab at Rongelap Atoll showed that the crabs contained more than 700 pCi ^{90}Sr per gramme of skeleton and 100 pCi of ^{137}Cs per gramme of muscle over a period of 10 years, 1954–1964. No gross anomalies were noted among these crabs; and no obvious population changes were noted during this past fallout period; however, population studies as such were not made. The abundance and size of fish and of spiny lobsters and coconut crabs at Bikini Atoll appear to be greater than ever, which presumably results from the absence of predation by man.

Oak Ridge experience

Very few studies have been made of natural populations exposed to chronic radiation higher than background. The salivary chromosomes of the larvae of *Chironomus tentans*, which inhabit the contaminated bottom sediments of White Oak Creek and White Oak Lake, at Oak Ridge National Laboratory, were analysed for 5 years for chromosomal aberrations. Calculations and measurements of the adsorbed dose for the larvae living in the sediments gave values of 230–240 R/yr, or 1000 times the background for that area. More than 130 generations have been exposed to this or greater dose rates over the previous 22 years. The conclusion was that the ionizing radiation from the contaminated environment was increasing the frequency of new chromosomal aberrations in the irradiated population, but that the new aberrations were eliminated by natural selection. Also, the present level of chronic irradiation has not affected the frequency of the endemic inversions.

The fecundity of a natural population of fish, *Gambusia affinis affinis* that had been exposed to chronic irradiation in White Oak Creek for many generations, compared with a control population, was also studied. The calculated dose rate from the bottom sediments was 10.9 R/day. A significantly larger brood size occurred in the irradiated than in the non-irradiated population, although significantly more dead embryos and abnormalities were observed in the irradiated broods. These results suggest that an increased fecundity is a means by which a natural population having a relatively short life cycle and producing a large number of progeny can adjust rapidly to an increased environmental stress caused by radiation.

Effect on resources

The extrapolation of the results of laboratory experiments on effects of radiation into the practical terms of their effects on marine resources must be made with care since without evaluation of the natural variations related to changes in fecundity, mortality, and recruitment, quite erroneous conclusions can be reached.

While it may be valid that the fecundity of both the individual and the population could be reduced by radiation effects, it is extremely doubtful that these would influence stock size beyond the normal range related to environmental changes, except in very heavily exploited stocks. It is certain however, that with the techniques of assessment at present available it would not be possible to obtain an unbiased measure of the effect attributable to radiation alone.

Conclusion

We conclude with a concept that has recurred many times; Man, as an individual, is the critical biological target in predicting the consequences of introducing radioactive materials into marine environments. If the radionuclides are present in concentrations acceptable for man, the individual, then it is difficult to conceive from today's evidence that there will be more than subtle effects on ecosystems-perturbations that would probably be indistinguishable from those due to causes other than radiation. On the other hand, it is not only conceivable, but probable, that with increasing use of atomic energy, accidents will occur that will result in damaging concentrations of radionuclides. In preparation for these contingencies, there is a pressing need to increase the sensitivity of methods for studying the response to radiation of populations and ecosystems in the marine environment.

Biological Effects of Oil Pollution in the Santa Barbara Channel   *D. Straughan**

Effets de la pollution par les hydrocarbures sur la faune et la flore dans le canal de Santa Barbara

Le présent document a été établi d'après une étude visant à déterminer les dégâts causés par l'explosion d'un puits de pétrole survenue en janvier 1969 dans le canal de Santa Barbara. L'étude a porté sur une période de 12 mois afin d'évaluer non seulement les effets initiaux mais aussi certaines des répercussions à long terme de la pollution par le pétrole. Ses résultats donnent aussi à penser que la zone commence à s'assainir. L'épanchement s'est produit dans un secteur contenant de nombreux suintements naturels de pétrole et au cours d'une période de précipitations anormalement abondantes. L'auteur tient compte de ces éléments, ainsi que des facteurs humains, en interprétant les données recueillies. Trois principales conclusions se dégagent:

1. Il est difficile de mettre en évidence *in situ* les dégâts dus à la pollution par le pétrole.
2. Les dommages subis par la faune et la flore ne sont pas aussi importants qu'on le craignait.
3. La zone est en voie d'assainissement.

L'auteur expose les motifs qui l'ont conduit à ces conclusions.

Efectos biológicos de la contaminación por petróleo en el canal de Santa Bárbara

En este documento se dan los resultados de un estudio con el fin de determinar los efectos reales del estallido de un pozo de petróleo en enero de 1969, en el canal de Santa Bárbara. El estudio duró 12 meses para poder determinar los efectos iniciales y algunos más dilatorios de la contaminación por petróleo. También se obtuvo una indicación de la capacidad de recuperación del lugar.

El derrame ocurrió en un lugar en el que se producen naturalmente muchas filtraciones de petróleo y durante un período de lluvias anormalmente intensas, factores, que junto con la intervención por el hombre, se tienen en cuenta al interpretar los datos. Sucintamente, se obtuvieron tres conclusiones principales:

1. Es difícil demostrar que la contaminación por petróleo causara daños en el lugar.
2. Los daños sufridos por la fauna y la flora no fueron tan graves como se creía.
3. El lugar va recuperando la normalidad.

El autor expone las razones de estos resultados.

POINT CONCEPTION at the northern end of the Santa Barbara Channel (fig 1) is usually regarded as the break between northern and southern faunas, but in fact the channel contains a mixture of northern and southern faunal and floral components.

Natural oil seeps have been recorded from the area for over 200 years (Ventura and Wintz, 1971). While their age is unknown, La Brea Tar Pits in the Los Angeles area contain Pleistocene fossils. In a survey of southern California beaches, Merz (1959) found that the average amount of tar deposited is 344 oz per 500 sq ft per day with a maximum recorded at Coal Oil Point of 1,520 oz per 500 sq ft in one day; 69 per cent of the beaches examined contained only traces of tar or none at all. Emery (1960) and Scott (1969) have plotted the position of known offshore seeps and tar mounds in the Santa Barbara Channel. The more recent study showed a seep near Platform A†, but this had become inactive by the time drilling began. Emery had reported inactive tar

† Fixed drilling Platform A is located approximately 9 km south of Santa Barbara, on Federal Lease Parcel No. 402. While the lease is jointly owned by Union Oil, Gulf Oil Corporation, Mobile Oil Corporation, and Texaco, well A-21 is operated by Union Oil.

mounts off Carpinteria but they were seeping in April 1969 (personal communication, W. Stephens). This variable activity makes it difficult to plot the position of natural seeps and estimate the volume of oil they produce. Allen (1969) estimates that an average of 50–70 brl per day was lost in October 1969 from Coal Oil Point, but daily figures range between 11 and 160 brl (1 barrel (brl) = 42 gallons oil).

Drilling began on the A-21 well, the fifth well drilled from Platform A, on 14th January 1969. At mid-morning on 28th January, mud began to flow from the drill pipe and the blowout which followed was not stopped until 11 days later. However, seepage was continuous and by 12th February, had increased and an oil slick formed 30–150 feet wide and 2 miles long. After 23rd February the seepage continued at rates estimated up to 50 brl per day. Seepage rates by April 1970 still varied up to 10 brl per day.

Estimates of quantity varied. Alan A. Allen of General Research Corporation computed that 2.2 million gallons of crude petroleum poured into the Santa Barbara Channel during the first 10 days and that spillage during the spills first year was 3.3 million gallons. Estes and

* Allan Hancock Foundation, University of Southern California, Los Angeles, California 90007, U.S.A.

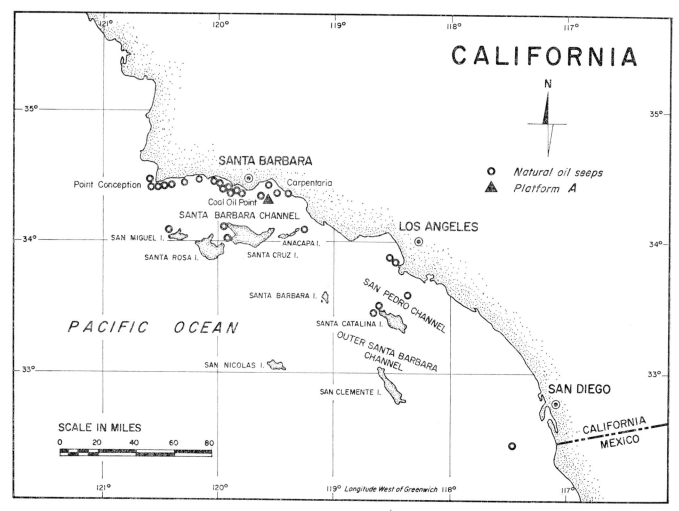

Fig 1. Map of Southern California showing the position of natural oil seeps

Golomb (1970) state that in May 1970, both Allen and Union Oil Company agreed that current flow rates averaged about 30 brl per day although earlier Allen's estimate was only one tenth that of the Union Oil Coy. This extreme diversity accentuates the difficulties of calculating accurately oil spill from an uncontrolled source.

After the spill, oil accumulated at sea under the influence of the prevailing north-northwest winds. A storm on 4th to 7th February shifted the winds to the west so that large quantities of oil entered Santa Barbara Harbor and heavy pollution of the mainland beaches began. Anacapa Island, the eastern end of Santa Cruz Island and the western tip of San Miguel Island received some oil. Oil was washed on and off mainland beaches by the tide and in some cases was buried by sand. Contamination varied almost continuously.

Booms were used to try to control spread but some broke up while others proved ineffective.

Methods used to abate pollution were many and varied. Some rocky shore areas were sandblasted to remove the oil while others were steam cleaned. Straw was the most widely used sorbant. Limited use was made of the dispersants Corexit 7664, Polycomplex A-11 and Ara Chem at sea. A total of 1,367 drums (1 drum = 55 gallons) of dispersant were used outside the one-mile limit to the end of February 1970. On 4th February 1969,

20 drums of Polycomplex A-11 were used to mitigate the fire hazard inside the Santa Barbara Harbor and an equal volume of Corexit 7664 was used in the vicinity of the Sterns Wharf and undersea Garden Aquarium. By and large, however, these particular dispersants were ineffective (Gaines, 1970).

In December 1969, a second spill occurred on Platform A estimated at 400 barrels. After production resumed, seepage from the sand increased due to pressure buildup but no estimate of this loss is available (personal communication, H. Morrison).

Unusually heavy winter rains in January and February 1969 caused flooding in southern California. This resulted in lower salinities in inshore waters. Minimum salinity of 15.26‰ was recorded at the Ventura Marina by the Scripps Institute of Oceanography. In January and February 1968, salinities recorded at the Ventura Marina did not drop below 30‰ while in January–February 1967 there is only one record (23.73‰) below 30‰. There was also an increased rate of sedimentation in the Channel. In sections of the eastern end of the Santa Barbara Channel where the normal annual sedimentation is 0.5 cm, up to 20 cm of sediment were deposited after the winter rains. This sediment is subsequently being re-worked and redistributed in the Channel.

The storm rains occurred soon after the citrus orchards had been sprayed with pesticides. As large numbers of oranges were washed on to Channel shores, it is possible pesticide concentration in the sea increased during this period.

There is a marked lack of background data on marine flora and fauna before the Santa Barbara spill. This fact, combined with the increased fresh water runoff, possible increase in pesticide levels and the inrceased sedimentation in the area, make it extremely difficult to isolate the biological effects of the spill from damage from other causes.

Planktonic investigations

A 12 month study of the zooplankton in the upper 5 m in the eastern end of the Santa Barbara Channel failed to reveal any effects of oil pollution. Peak productivity occurred in January 1970 and minimum productivity in September 1969 (McGinnis, 1971). The chosen indicator species *Labidocera trispinosa* were breeding; blooms of two species, *Doliolum denticulatum* and *Penelia avirostris* were recorded. A survey on 11th February 1969 revealed no significant deviation from the expected kinds and abundance of fish eggs and larvae in the area (U.S. Bureau of Commercial Fisheries, 1969). Similarly the phytoplankton study has not revealed any immediate effects of oil pollution but the productivity pattern parallels that of the nutrient pattern (Oguri and Kanter, 1971; Kolpack, 1971).

Benthic investigations

Background data for this study was obtained from examining the shelf areas of the Santa Barbara region between 1956 and 1959. In the ten years to May 1969, there was a decrease in biomass in the area. The volume of biomass and its distribution did not change significantly between May and October 1969. The major part of the earlier decrease in biomass was due to a decrease in the dominant species in the area, an echuroid worm, *Listriolobus peliodes*. Fauchald (1971) suggests that this decrease in population may be associated with oil exploration and drilling in the Channel. The areas of population decrease, however, are also associated with a fault line and an area of suspected increase in oil seepage over recent years.

Intertidal investigations

A survey of sandy beaches over twelve months following the oil spill failed to pin-point any damage due to oil pollution. Fewer species and fewer animals were recorded during winter than summer. In fact, in January 1970, only one specimen was recorded on Carpinteria Beach, but there was very little sand on the beaches in the winter. A further important fact is that some of the samples were collected during the January rain. As might be expected, few animals were found in these samples (Trask, 1971).

Anderson *et al.* (1969) surveyed intertidal and subtidal stations on the mainland and Channel Islands between 14th and 19th February 1969. They reported that the vast majority of organisms appeared healthy and that moribund and/or dead creatures were so rare that they could be described individually.

Nicholson and Cimberg (1971) resurveyed rocky shores within and to the south of the Santa Barbara Channel that had been surveyed by Dawson between 1956 and 1959, as did Neushal and students in 1967 and Widdowson and students in 1968 and 1969. They reported an overall decrease in number and abundance of algal species. This decrease was not associated with oil pollution but generally with increased human activity on some beaches and increased chronic pollution other than by oil. The California Department of Fish and Game reported that the marine grass *Phyllospadix* spp. and the algae *Hesperophycus harveyanus* were killed in some areas when exposed to heavy oil pollution. These areas had almost recovered by August 1969. *Phyllospadix* is abundant on rocky shores exposed to natural seepage and therefore appears tolerant of continuous exposure to small amounts of oil.

Chthamalus fissus in the upper intertidal areas was smothered by a thick layer of oil in some areas while *Balanus glandula*, a larger barnacle, was able to project above the oil (Nicholson and Cimberg, 1971). Straughan (1971) was unable to detect any effects in breeding in either of these species. *B. glandula* settled on oil from the January 1969 spill prior to 18th March 1969 but *C. fissus* was not recorded settling on oil from the January 1969 spill until the fall of 1969. Breeding in *Pollicipes polymerus*, a stalked barnacle found in lower intertidal areas, was reduced (1) by pollution from the January 1969 oil spill, (2) in areas of natural seepage, and (3) at localities south of the Santa Barbara Channel. These latter localities are near the southern extremes of its distribution. *Pollicipes* larvae, which normally settle on the stalks of adults, did not settle on moderately or heavily oiled stalks. It was recorded on lightly oiled stalks but it was impossible to determine if these stalks were oiled before or after larval settlement.

Fisheries analysis

The survey by the California Department of Fish and Game in 1969 (identical with that in 1965) to determine the abundance and distribution of the important pelagic species showed that all fish appeared healthy and they were unable to find any indications of impairment in the food chains or damage to the fish populations.

The commercial landings of fishery products were lower in each month February–July 1969 than in the years 1965–68 in Santa Barbara Harbor. The greatest decline in landings was during February when the Harbor was closed at times because of the spill. The total commercial fish catch from the Santa Barbara Channel and Channel Islands for the six months in 1969 (2,337,931 pounds) with the same six months in 1968 (2,124,820 pounds) shows no decrease in catch for the area. On a monthly basis there was a decrease in catch for February and March 1969 compared with previous years. Fish spotting data, apart from February 1969 when it was difficult to see because of the oil, does not indicate a decline in the major species of fish in the Channel (Straughan, 1971a).

Bird surveys

In February to March 1969, the living bird population in Santa Barbara Channel was 12,000. In the next two months, due to seasonal migration, the population rose to 85,000. The estimated mortality from all causes, including oiling, was between 3,500 and 4,000 birds (California Department of Fish and Game, 1969).

The majority of birds killed were pelagic and of 432

identified, 253 were grebes and loons. The most abundant species in the area, gulls and terns, did not suffer high mortality (Straughan, 1971b).

Mammal surveys

Grey whales migrated northwards through and to the west of the Santa Barbara Channel following the January 1969 oil spill. These animals do not feed when migrating. Strandings were not significantly higher than in the preceding nine years. In an autopsy on one dead animal, Federal Government Agencies were unable to find any trace of oil (Brownell, 1971).

A small area of the seal colony on San Miguel Island was contaminated by oil about 17th March 1969. This was after the elephant seal pups had stopped suckling. Although many of these animals were completely covered with oil, data from recapture of tagged oiled and unoiled seals twelve months later—indicated no increase in mortality in the oiled animals (Le Boeuf, 1971).

California sea lions gave birth to pups after the area received oil in March, 1969. It is not known if mortality rate was affected by the presence of oil.

Discussion

These data suggest that there was very little mortality in the Santa Barbara Channel due to direct toxic effects of spilled oil. The most toxic components of crude oil evaporate rapidly when exposed to the elements or are rapidly diluted and lost in the body of the ocean. The relative insolubility of oil in general is well known and is clearly illustrated, for example, by the work of Kolpack (1971) who studied the properties of water in natural seep areas.

The marked lack of toxicity of crude oil in open sea conditions contrasts vividly with the effects of spills of refined products on or very close to shore. In such instances these volatile oils do not have a chance to dissipate by evaporation and dilution before reaching the environment of the marine life of the area. Blumer (1969) reports on the high mortality following such a spill.

Plankton might be expected to be affected by crude oil at sea whilst it is still in its toxic condition. That plankton did not appear to be affected by the Santa Barbara spill is probably accounted for by the fact that oil, influenced by wind as well as currents, moves at a different rate than plankton. The latter would be unlikely, therefore, to be exposed to oil for long periods.

Chthamalus fissus was the only marine species badly affected by the oil spill and, in this case, mortality was due to smothering by oil rather than any toxic effects. *Balanus glandula*, which was able to protrude through the oil layer, was not so badly affected.

It is also possible that organisms living in the Santa Barbara Channel that have been continually exposed to small doses of oil over long periods, have a higher tolerance to oil than organisms never exposed to oil. Areas such as Coal Oil Point which are continually receiving oil from natural seeps at sea have abundant intertidal flora and fauna. Fewer species are found in oil seep than non-oil seep areas but these species are more abundant so that intertidal areas are well populated.

Data on *Pollicipes polymerus* suggests that while individuals of these species can survive in the presence of oil their reproductive capacity may be impaired. Con-

sidering the entire range of this particular species (Alaska to northern Mexico), only very small populations of the entire species are exposed to oil pollution, and, in such a situation, while isolated populations may suffer periodically as a result of oil pollution, the species as a whole is not endangered. The situation may be different for a species with a more limited distribution.

While the fresh water flooding probably placed intertidal species under stress by lowering the salinity, it may also have prevented some of the oil from reaching the shore. The increased amount of sediment in the water may have had the effect of a sinking agent on the oil. Geologists are at present investigating the latter.

Following the oil spill many people were disturbed because "nothing moved on the sandy beaches" but Trask's data indicates that during the winter months sandy beach populations are very low. The fact that Trask collected fewer animals in the rain would suggest that the rains in January, February 1969 may have contributed to, if not caused, the low sandy beach populations.

While the commercial fishery landings at Santa Barbara were low during the six months following the January, 1969 spill, much of this can be attributed to the reluctance of fishermen to work in oily waters. The overall fish catch from the Santa Barbara Channel for the same period was similar to that recorded in 1968. This indicates that the fish populations were probably not affected by the oil spill. Similarly, there is no evidence to suggest increased mortality in marine mammal populations in the area. As with the *Torrey Canyon* disaster (Bourne, 1968), high mortality was recorded among pelagic species of birds. These spend most of their time swimming and tend to dive to avoid danger, thus increasing their contamination by oil. It is important to deal with oil as soon as possible at sea to minimize damage to these species. Such action would also incidentally reduce beach pollution and make easier the task of returning the coastline to a clean condition.

References

ALLEN, A A Estimates of surface pollution resulting from sub-
1969   marine oil seeps at platform A and Coal Oil Point. *Tech. Memo. gen. Res. Corpn*, (1230):63 p.
ANDERSON, E K, *et al.* Preliminary report on ecological effects of
1969   the Santa Barbara oil spill. Part of a report submitted to Western Oil and Gas Association.
Battelle Memorial Institute, Review of Santa Barbara Channel
1969   oil pollution incident. *Wat. Pollut. Control Res. Ser.*, (Dust 20).
BLUMER, M Oil pollution of the ocean. *Oceanus*, 2:2–7.
1969
BOURNE, W R P Oil pollution and bird populations. *In* The
1968   biological effects of oil pollution on littoral communities. *Fld Stud.*, 2 (Suppl.):99–122.
BROWNELL, R L Whales, dolphins and oil pollution. *In* Biological
1971   and oceanographical survey of the Santa Barbara Channel oil spill, 1:255–76. Los Angeles, Allan Hancock Foundation.
California Department of Fish and Game Santa Barbara oil leak.
1969   *Interim Rep. Calif. Dep. Fish Game*, (Dec. 15).
EMERY, K O The sea off southern California. New York, J. Wiley
1960   and Sons, 366 p.
ESTES, J E and GOLOMB, B Oil spills: method for measuring their
1970   extent on the sea surface. *Science*, 169:676–8.
FAUCHALD, K The benthic fauna in the Santa Barbara Channel
1971   following the January 1969 oil spill. *In* Biological and oceanographical survey of the Santa Barbara Channel oil spill, 1:61–117. Los Angeles, Allan Hancock Foundation.

[358]

GAINES, T H   Pollution control at a major oil spill. Paper
1970   presented at International Conference on Water Pollution
Research, San Francisco, July 1970.

KOLPACK, R   Biological and oceanographical survey of the Santa
1971   Barbara Channel oil spill, 2. Los Angeles, Allan Hancock
Foundation.

LE BOEUF, B J   Oil contamination and elephant seal mortality: a
1971   "negative" finding. *In* Biological and oceanographical
survey of the Santa Barbara Channel oil spill, 2. Los
Angeles, Allan Hancock Foundation.

McGINNIS, D   Observation on the zooplankton of the eastern
1971   Santa Barbara Channel from May 1969 to March 1970.
In Biological and oceanographical survey of the Santa
Barbara Channel oil spill, 1:49–60. Los Angeles, Allan
Hancock Foundation.

MERZ, R   Determination of oily substances on the beaches and in
1959   nearshore waters. *Publs St. Wat. Pollut. Control Bd Calif.*,
(21).

NICHOLSON, N L and CIMBERG, R L   The Santa Barbara oil spills
1971   of 1969: a post-spill survey of the rocky intertidal. *In*
Biological and oceanographical survey of the Santa
Barbara Channel oil spill, 1:325–400. Los Angeles, Allan
Hancock Foundation.

OGURI, M and KANTER, R   Primary productivity in the Santa
1971   Barbara Channel. *In* Biological and oceanographical survey
of the Santa Barbara Channel oil spill, 1:17–48. Los
Angeles, Allan Hancock Foundation.

SCOTT, C B   A visit to San Miguel Island. *Union Seventy-Six*,
1969   13(5):19–23.

STRAUGHAN, D   Breeding and larval settlement of certain
1971   invertebrates in the Santa Barbara Channel following
pollution by oil. *In* Biological and oceanographical survey
of the Santa Barbara Channel oil spill. 1:223–44.
Los Angeles, Allan Hancock Foundation.

STRAUGHAN, D   Oil pollution and fisheries and the Santa Barbara
1971a   Channel. *In* Biological and oceanographical survey of the
Santa Barbara Channel oil spill, 1:245–54. Los Angeles,
Allan Hancock Foundation.

STRAUGHAN, D   Oil pollution and sea birds. *In* Biological and
1971b   oceanographical survey of the Santa Barbara Channel oil
spill, 1:307–12.

TRASK, T   A study of three sandy beaches in the Santa Barbara,
1971   California area. *In* Biological and oceanographical survey
of the Santa Barbara Channel oil spill, 1:159–78. Los
Angeles, Allan Hancock Foundation.

U.S. Bureau of Commercial Fisheries   Monthly report—Feb-
1969   ruary 1969. La Jolla, California. 92037, Fishery
Oceanography Centre.

VENTURA, C and WINTZ, J   Natural oil seeps: historical back-
1971   ground. *In* Biological and oceanographical survey of the
Santa Barbara Channel oil spill, 1:11–6. Los Angeles,
Allan Hancock Foundation.

The Biological Effects of Oil Pollution and Oil-cleaning Materials on Littoral Communities, Including Salt Marshes†

E. B. Cowell, J. M. Baker and G. B. Crapp*

Les effets biologiques de la pollution par les hydrocarbures et par les matériaux de nettoyage des hydrocarbures sur les communautés littorales, y compris celles des salines

Les auteurs décrivent les méthodes suivies lors des enquêtes hydrographiques menées sur les rivages et les résultats obtenus grâce à ces méthodes.

L'étude des pollutions accidentelles et des pointages expérimentaux ont prouvé que les hydrocarbures, sauf lorsqu'ils sont en grandes quantités, ont peu d'effet sur la faune et la flore intertidales.

Les émulsifiants du type couramment utilisé dans les cas graves de pollution par les hydrocarbures, bien que plus toxiques que les pétroles bruts, ne causent pas une mortalité élevée aux organismes littoraux à moins d'être utilisés en grandes quantités.

Les enquêtes sur le terrain comme celles faites en laboratoire ont démontré que certaines espèces ont une plus forte sensibilité aux émulsifiants. Parmi les végétaux, les plantes annuelles sont particulièrement sensibles, alors que parmi la faune qui fréquente les rivages britanniques, la patelle, *Patella vulgata*, est l'animal qui présente la plus grande sensibilité. On peut utiliser les paramètres simples de la population des patelles pour contrôler le rétablissement de la structure des communautés sur les rivages rocheux.

La sensibilité de la patelle révèle qu'un traitement, même léger, aux émulsifiants (y compris les nouvelles "variétés moins toxiques") causera sans doute autant de dégâts qu'un nettoyage prolongé.

Tant pour les hydrocarbures que pour les émulsifiants l'expérience montre que les dégâts sont plus graves au début du printemps que lors des autres saisons.

Des observations et des expériences ont porté sur les effets de la pollution chronique après que l'on eut observé des dégâts localisés, mais graves au voisinage des décharges de raffineries de pétrole. Dans certains sites, le déversement continu d'une eau contenant des hydrocarbures en faible concentration (10–25 ppm) a provoqué l'accumulation de grandes quantités de ces produits sur les rivages.

Certaines des modifications écologiques ont été reproduites expérimentalement sur la végétation des salines. Il est suggéré que l'on pourrait réduire au minimum ce type de dégât simplement en déplaçant les tuyaux de décharge des effluents vers des eaux profondes où les marées provoquent de forts brassages.

Le document décrit les expériences de laboratoire qui ont porté sur le matériel végétal afin d'étudier le mécanisme des dégâts provoqués par les hydrocarbures et les détergents. Selon les résultats obtenus, il apparaît que la dégradation de la membrane cellulaire est suivie de l'altération des mitochondries, de la dissociation des gênes et de taux de respiration anormaux qui précèdent la mort.

Efectos biológicos de la contaminación con petróleo y con los emulsivos empleados para limpiar las aguas de petróleo, en las comunidades litorales, incluidas las marismas

Se describen algunos métodos para el reconocimiento de las costas y se examinan los resultados conseguidos con ellos. Los estudios realizados en el caso de contaminaciones accidentales y pruebas experimentales han confirmado que el petróleo, a no ser en grandes cantidades, ejerce escaso influjo en la fauna y la flora de la zona que cubra y descubre la marea. Los emulsivos del tipo comúnmente empleado en los casos de graves contaminaciones con petróleo, aunque son más tóxicos que el petróleo crudo, no causan gran mortandad entre los organismos del litoral, a no ser que se empleen en cantidades muy grandes.

Las investigaciones hechas en laboratorio y en el mar han demostrado que algunas especies son más sensibles a los emulsivos. Entre las plantas, las anuales son especialmente sensibles, mientras que de los animales costeros británicos el más sensible es la lapa (*Patella vulgata*). Como indicador de la recuperación de la estructura comunitaria de las costas rocosas pueden emplearse los parámetros de población de las lapas, medidos en forma sencilla. La sensibilidad de la lapa indica que incluso un tratamiento ligero con emulsivos (incluidas las nuevas "variedades menos tóxicas") será con toda probabilidad tan dañoso como un tratamiento prolongado. Los experimentos realizados con petróleo y emulsivos muestran que el daño es mayor a principios de la primavera que en las demás estaciones del año.

Tras observar la existencia de daños localizados pero intensos en las cercanías de los desagües de las refinerías de petróleo, se procedió a hacer algunas observaciones y experimentos sobre los efectos de la contaminación crónica. En algunas zonas, la descarga continua de agua con pequeñas cantidades de petróleo (10–25 ppm) se ha traducido en una exposición de la costa a los efectos de grandes cantidades de petróleo. Algunos de los cambios ecológicos se reprodujeron experimentalmente en la vegetación de marismas. Parece que los daños de este tipo podrían reducirse al mínimo desplazando los tubos de desagüe hacia zonas de aguas más profundas y de mayor turbulencia causada por las mareas. Se describen los experimentos realizados en laboratorio sobre plantas para estudiar los mecanismos de la deterioración producida por el petróleo y los detergentes. Los resultados muestran que se produce una ruptura de la membrana celular, seguida, antes de la muerte, por daños a las mitocondrias, fraccionamiento y anormalidades en la respiración.

* Oil Pollution Research Unit, Field Studies Council, Orielton, Pembroke, Wales, U.K.

† This work was supported by grants from the Institute of Petroleum's Jubilee Fund and the World Wildlife Fund.

MILFORD HAVEN in south-west Wales, although it lies in the newly formed Pembrokeshire Coast National Park, has been developed as Britain's largest oil port since 1960 and in 1968 handled 28 million tons of crude oil. Tankers up to 200,000 tons capacity can be accepted on good tides. Besides marine terminals and crude oil storage tanks, the Haven now has three refineries on its shores.

There have inevitably been frequent spillages of oil. The frequency is correlated with the tonnage of oil handled annually, and is likely to increase with the expansion of the port despite an efficient anti-pollution organization. The oil companies and Milford Haven Conservancy Board (the harbour authority) jointly operate a launch equipped with spray booms and emulsifier, which treats any oil as soon as it is sighted. Responsibility for the spillage is assigned afterwards or, if the culprit is unknown, the cost of clean-up is shared. More than 2,500 gal emulsifier is used monthly on oil spill clean-up.

This paper reports research done by the Field Studies Council's Oil Pollution Research Unit to determine the biological and ecological effects of these and other oil pollution problems. The first of two sections, under J. M. Baker, worked on the effects of oil pollution and shore cleaning on salt-marsh communities; the second, under G. B. Crapp, investigated the effects on rocky shores.

In Britain salt-marsh associations colonize the mudflats of sheltered shore and tidal estuaries between Mean High Water Neap tide levels (MHWN) and Mean High Water Springs (MHWS). They are economically important beeause they can stabilize mudflats, raise their levels, and consequently make land available for sea defence and reclamation (Allen, 1930; Ranwell, 1967). They are important as feeding grounds of many species of wildfowl, waders and swans, and are vital for the maintenance and recovery of the populations of many bird species, some of which are depleted by oil pollution (e.g. Barclay-Smith, 1956; Co-ordinating Advisory Committee on Oil Pollution of the Sea, 1959; Harrison, 1967). They are involved in the energy flow systems of related offshore mudflats and important in maintaining the production of estuarine areas (Odum and Smalley, 1959; Odum, 1961; Teal, 1962). For these reasons it is important to evaluate the damage that occurs to salt-marsh plants from accidental oil pollution.

Methods of research

Following oil spills, experimental oil spraying with an agricultural sprayer was done at different times of the year on a local salt marsh, and experimental chronic pollution has been carried out in three different marsh communities on the north Gower Coast, Wales. Experimental emulsifier cleaning, burning and cutting have been tried as cleaning methods on Milford marshes. The effects of refinery discharge are being investigated in Southampton Water, England. Salt-marsh turves kept in an unheated greenhouse have been used for comparing the effects of different oil fractions and volumes, emulsifiers, and naphthenic acids. Field and greenhouse experiments are usually of randomized block design.

A points frame has been used extensively for estimating how much change in the percentage cover of different

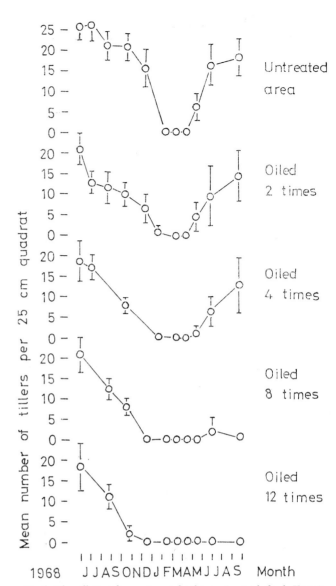

Fig 1. The effects of experimental oil spraying with fresh Kuwait crude oil on Spartina anglica. Plot size 2 × 5 m, dose 4.5 l oil per month, beginning June 1968. The vertical lines show the 95% confidence limits

species is brought about by different treatments (Tiver and Crocker, 1948). Other methods of recording changes in vegetation are visual estimation, photography, density (by counts within quadrats) and dry weight.

Individual plants or rooted tillers have been kept in water cultures for physiological work.

Effects of isolated oil spillage

A short-term effect of oil on a marsh is that the oil adheres firmly to the plants and hardly any is washed off by the tide, except where there are puddles of oil. Leaves may remain green under the oil film for a few days, but eventually they become yellow and die. Plants recover by producing new shoots, a few of which can usually be seen within three weeks of pollution, unless large quantities of oil have soaked into the plant bases and and soil. Seedlings and annuals rarely recover.

In the long term, recovery from oil spillages has been observed many times (e.g. Buck and Harrison, 1967; Ranwell, 1968; Smith, 1968; Stebbings, 1968; Cowell and Baker, 1969). The cases described cover different salt marsh communities, different types, volumes and degrees of weathering of oil, and pollution at different times of year. Vegetative recovery from experimental spraying at different times of year has been observed (Baker, in press). The evidence indicates that marshes recover well from a single oil spillage, or from successive oil spillages provided these are separated by long time intervals.

Experimental chronic pollution on salt marshes in three different marsh communities has shown that recovery from up to four successive monthly oilings can be good, but more than this results in a rapid decline of the vegetation (fig 1) from which recovery is slow (probably years, but this has not yet been measured).

The initial survey of a *Spartina anglica* marsh in Southampton Water, polluted by a refinery waste-water discharge containing up to 25 ppm oil and 2 ppm phenols, showed that approximately 35 ha of *Spartina* had been killed since the discharge started in 1951, and that the marsh surface was generally 15–25 cm lower than that of nearby healthy marsh. The death of the *Spartina* may be due to cumulative oiling from the very low levels of oil in the discharge water, to phenols, to the build up of sulphide in the mud, or to all these and other factors, possibly accentuated by the heat (25–35°C) of the water. The best available evidence suggests that damage is caused by thin oil films rather than other factors (Baker, in press).

Seasonal effects

Since recovery from pollution is by new growth, it appears to be slowest in winter when new growth is slight. But some species die down during winter and perennate in, or under, the soil surface, e.g. *Juncus gerardii* (Mud rush) and *Spartina anglica*. These may not be affected by a winter spillage. Seedlings and annuals rarely recover from oil, and are vulnerable in spring and summer. Spring and summer oiling may also reduce flowering and seed production of many plants; this has been demonstrated in field experiments for *Festuca rubra* (Red fescue), *Plantago maritima* (Sea plantain) and *Spartina anglica*.

The observations and experiments indicate that the long-term susceptibility of salt-marsh species to oil is different from that reported by Cowell (1969), who observed considerable short-term damage in *Puccinellia maritima*, *Festuca rubra*, and *Spartina anglica*. Damage has been observed following experimental oil spraying, but it is now known that recovery by new growth usually occurs within a few months of pollution.

Susceptibility of different species, and reasons for damage

The following grouping of salt marsh plants is based on observations following oil spillages and experimental spraying of different groups.

1—Very susceptible—Shallow rooting plants with no or small food reserves; quickly killed by oil and cannot recover; e.g. *Suaeda maritima* (Seablite); seedlings of all species

2—Susceptible—Shrubby perennials with exposed branch ends which are badly damaged by oil; e.g. *Halimione portulacoides* (Sea purslane)

3—Susceptible—Filamentous green algae. Though filaments are quickly killed, populations can recover rapidly by growth and vegetative reproduction of any unharmed fragments or spores

4—Intermediate—Perennials which usually recover from one spillage or up to four light experimental oilings, but decline rapidly if chronically polluted; e.g. *Spartina anglica*; *Puccinellia maritima* (Sea poa)

5—Resistant—Perennials, usually of rosette form, with large food reserves (e.g. tap roots). Most of them die down in winter. Some have survived twelve experimental oilings; e.g. *Armeria maritima* (Thrift)

6—Very resistant—Perennials of Group 5 type which have in addition a resistance to oils at the cellular level; e.g. members of the Umbelliferae.

These groups of plants vary in their responses to oil because of differences in morphology, anatomy and physiology. The oil may vary in toxicity according to the content of low boiling compounds, unsaturated compounds, aromatics, and acids and oil with higher concentration of these constituents is more toxic.

Emulsifiers

The following emulsifiers have been tested: BP 1002 (a solvent emulsifier commonly used for cleaning oil), test blends X (medium aromatic content) and Y (low aromatic content), and BP 1100 and DS 10239. Small differences have been observed between the toxicities of these emulsifiers to salt marsh grasses. In all cases concentrations below 10 per cent were not permanently damaging and concentrations above 50 per cent usually killed them. The toxicity depends upon dilution rather than the absolute amount of emulsifier; for example 10 ml of undiluted emulsifier kills a salt marsh turf 30 cm × 40 cm, but the same amount of emulsifier has no visible effect if applied as 1,000 ml of 1 per cent solution.

In tests with BP 1002 and its solvent A260, the solvent alone proved as toxic as the whole emulsifier. Any hydrocarbon solvent is liable to penetrate into plants through the lipophilic surfaces, and penetration is crucial in determining toxicity. Once the hydrocarbon is inside the plant it may dissolve in cell membranes and cause loss of cell sap. Penetration of undiluted emulsifiers is very rapid, but the more aqueous the solutions, the longer the penetration time.

Predictions and recommendations

The evidence indicates that isolated pollution incidents, even if severe, do not cause long-term damage to salt-marsh vegetation. Continuous refinery discharge can kill all vegetation, resulting in subsequent erosion of the marsh. Between these two extremes is the case of marshes near oil terminals which are liable to be polluted by small spillages perhaps several times a year. No long-term records are available to show how long a marsh can go on recovering from this type of pollution. At Martinshaven (a Milford Haven salt marsh) recovery has been good after each incident, but it is difficult to predict the effect of, say, two spillages a year for a long time, for example 20 years. A slow decline seems likely. Experimental work shows that successive monthly oil spillages cause a rapid decline after four oilings.

Oiled salt marshes should be left untreated. Emulsifiers cannot decrease the damage to plants, may increase it, and are likely to increase toxic reducing conditions in the soil (since crude oil has high biological oxygen demand) by aiding the penetration of oil. Cutting and burning is less damaging than emulsifier treatments.

Effects of oil and emulsifiers on rocky shores

We studied the effects of oil and emulsifiers on rocky shores by simulating oil spillages in the field, and by monitoring a number of transect sites in the Haven. Most of these sites were first surveyed by Moyse and Nelson-Smith (1963) and Nelson-Smith (1967) in 1961 and 1962, and several have been polluted by oil during the last three years. We observed no biological changes attributable to oil pollution. Records of the fauna and flora of these shores were made using the abundance scale devised by Crisp and Southward (1958), and modified by Ballantine (1961). Although this scale does not give numerical abundance, it has enabled us to determine small changes in the distribution of the barnacles *Balanus balanoides* and *Chthamalus stellatus*, which have been attributed to climatic deterioration.

Field trials showed that most littoral species are resistant to toxic oils, even when spilt at intervals of one month. But some littoral gastropods (*Littorina neritoides*, *L. saxatilis*, and *L. obtusata*) are affected by the thicker oils. A thick layer of oil on the shell increases the volume and mass of the animal, and wave action is more likely to dislodge it from the rock.

Oil left untreated on the shore tends to be swept by wave action to the high water mark habitat of *L. saxatilis* and *L. neritoides*, and these species are likely to be most affected by small but frequent oil spillages.

Effects of cleaning up oil spillages

The chief ways of removing oil from the shore are mechanical removal and washing with solvent emulsifiers. Mechanical removal causes little biological damage but most solvent emulsifiers until recently were highly toxic. We have investigated BP 1002 as an example of widely used toxic detergents.

We applied the emulsifier to the test organisms in the same way as Perkins (1968). The animals were exposed to high concentrations of detergent for 1 h, this was followed by a recovery period in clean sea water. This method is quick, which reduces the evaporation of the toxic fractions. This does not exactly represent field conditions, where initial high concentrations probably decrease slowly.

Most of the laboratory tests have been done with the prosobranch molluscs of the rocky shore. These respond to emulsifier treatment by withdrawing into the shell for a few hours to a few days. Animals are kept until they fully recover or die. The susceptibility of different species to detergent varied with the time of year: the topshell *Monodonta lineata*, a southern, warm water species, is most resistant to detergent treatment in the warmer summer months. The dogwhelk, *Thais* (*Nucella*) *lapillus*, a species which extends into Arctic waters, has greater resistance in December than in July or October (fig 2).

The resistance of various intertidal species to treatment with BP 1002 in September 1969 is shown in fig 3. These figures have been compared with observations made following a spillage in Milford Haven in November 1968, and also with the results of field experiments set up to test the effects of detergent treatment.

The spillage was crude oil that went ashore at Hazel-beach, Milford Haven, a shore which had already been polluted and cleaned in January 1967 when, although many organisms were killed, enough grazing animals survived to prevent an algal flush from covering the shore (Nelson-Smith, 1968a, b). Detergent treatment in November 1968 was light, but the shore community had not fully recovered from the earlier spillage. Three weeks after the cleaning the number of periwinkles and top-shells was drastically reduced, except *Littorina littorea*, but the numbers increased again by January 1969. We believe that these animals behaved the same on the shore as in the laboratory. Exposure to emulsifier was followed by retraction into the shell, during which they were rolled into deep crevices or to the sublittoral zone. Two months later the animals had recovered and regained their normal position on the shore. Numbers were reduced in most cases, particularly amongst those species showing the greatest sensitivity in the laboratory, but *Littorina littorea*, the most resistant species in laboratory tests, appeared to be unaffected by the spillage.

The barnacles *Balanus balanoides* and *Elminius modestus* did not suffer very heavy mortalities, but the density of the very sensitive limpet *Patella vulgata* (fig 3) was reduced from c. 150 per m² to 21 per m². The numbers continued to decline during the spring, and, when young were first found in June 1969, the density had dropped to 5 per m². This density was insufficient to prevent an algal flush from developing, and during the spring the shore was covered by a growth of the green alga *Enteromorpha*. This was replaced in the summer by *Fucus vesiculosus*. The covering of the shore by a belt of *Fucus* has altered its character completely, and the number of barnacles, for instance, is now declining.

The same sequence of events was observed on experimental plots that were treated with detergent. Survival was good amongst many species following light detergent treatment, but limpets and mussels (*Mytilus edulis*) suffered heavy mortalities. The death of these key species was followed by an algal flush, and those species which survived the detergent treatment are now declining.

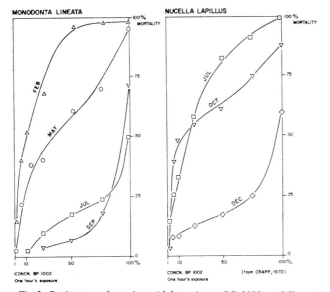

Fig 2. Resistance of two intertidal species to BP 1002 at different times of year

Shores in Devon and Cornwall that suffered heavy detergent treatment following the *Torrey Canyon* and *Fina Norvege* spillages have also been visited. These are now dominated by fucoid algae, but limpets are re-appearing under the algal cover. On these shores few animals were left alive after cleaning, but on a limpet dominated shore it seems likely that the effects of a total kill and a kill of limpets alone will be very similar after a few years have elapsed. The shores resemble the experimental strip at Port Erin described by Jones (1948) and Southward (1964), where limpets were removed intentionally.

Recovery

We expect recovery to be very protracted. The redevelopment of the normal community cannot proceed until the algae disappear. The reappearance of limpets under the canopy is unlikely to accelerate this process. Small limpets may graze on the fucoids to a limited extent, but appear to play only a minor role in the destruction of the plants, although they prevent recruitment to the algal cover. These biological processes have serious consequences in terms of amenity. Shores that were cleaned in order to render them suitable for holiday-makers are now very difficult to travel over, and will remain so for several years.

resistant to treatment with detergents, but when it is removed or destroyed the shore is invaded by *Fucus*, preceded by *Enteromorpha*, as happens on *Patella* dominated shores. We do not yet know how long the *Ascophyllum* takes to re-establish.

Chronic pollution by continuous discharge of oil

When it was found that the continuous discharge of oil in very low concentrations, from a refinery outfall, can have a very marked effect on adjacent salt marsh, we examined rocky shores near to outfalls in Milford Haven to determine whether any such effect could be detected. Near one outfall the shore community has changed from one dominated by *Patella* and barnacles to a *Fucus vesiculosus* type. The extent to which this is due to oily discharge from the outfall pipe is uncertain, as a certain amount of emulsifier cleaning has been done in this area.

Similar effects were noted by Mackin and Hopkins (1962) where oysters were killed by "bleedwater" from the Lake Barre Oil Field in Louisiana, U.S.A. In both Milford Haven installations where pollution gradients appear, the outfall pipes discharge at the lower levels of the beach while a refinery that does not cause measurable effects discharges at the jetty head.

Present results suggest that if oil levels in outfalls cannot be lowered to 10 ppm or less, then damage could

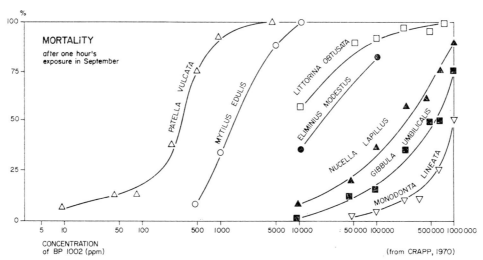

Fig. 3. Resistance of various intertidal species to BP 1002

The solvent emulsifiers used during the *Torrey Canyon* spillage and many subsequent spillages were not designed to be used in the natural environment. Recently new types of detergents have been developed that are much less toxic. We have tested examples of these, and have found that they are much better than those previously available. However, those tested still possess some toxicity to intertidal animals, particularly to the dominant species *Patella vulgata* and *Mytilis edulis*. We therefore predict that heavy use of these, for instance on the scale of post *Torrey Canyon* cleaning, will lead to substantial mortality to the populations of these species. This, in turn, will result in population changes similar to those observed following light treatment with BP 1002.

Most of our investigations have been carried out on shores dominated by limpets; however, many shores are dominated by populations of mussels or of the alga *Ascophyllum nodosum*. *Ascophyllum* appears to be very

be minimized by resiting discharge pipes in areas of deeper water and greater tidal dispersion.

Since it seems likely that technical difficulties will prevent better oil/water separations in the immediate future, we suggest that attention should be given to the design and siting of outfalls.

References

ALLEN, H H *Spartina townsendii*, a valuable grass for reclamation
1930 of tidal mud-flats. 2. Experience in New Zealand. *N.Z. Jl Agric.*, 40:189–96.

BAKER, J M The effects of oil pollution and cleaning on salt marsh
1970 ecology. *In* Second Annual Report, Oil Pollution Research Unit, Field Studies Council.

BAKER, J M Refinery effluent. *In* The ecological effects of oil pollution on littoral communities, edited by E. B. Cowell. Proceedings of a Symposium of the Institute of Petroleum, London, 30 November–1 December, 1970 (in press).

BALLANTINE, W J A biologically defined exposure scale for the
1961 comparative description of rocky shores. *Fld Stud.*, 1(3): 1–19.

BARCLAY-SMITH, P., Oil pollution. *Bird Notes*, 27:81.
1956

BUCK, W and HARRISON, J   Some prolonged effects of oil pollution
1967   on the estuary. *Ann. Rep. Wildfowl Ass.*, 1967: 32–3.

Co-ordinating Advisory Committee on Oil Pollution of the Sea. *In*
1959   Report of proceedings, International Conference on Oil
Pollution of the Sea, 3–4 July, 1959 at Copenhagen, 93 p.

COWELL, E B   The effects of oil pollution on salt marsh communi-
1969   ties in Pembrokeshire and Cornwall. *J. appl. Ecol.*, 6:133–
42.

COWELL, E B and BAKER, J M   Recovery of a salt marsh in Pem-
1969   brokeshire, S.W. Wales, from pollution by crude oil. *Biol.
Conserv.*, 1:291–6.

CRISP, D J and SOUTHWARD, A J   The distribution of intertidal
1958   organisms along the coasts of the English channel. *J. mar.
biol. Ass. U.K.*, 37(1):157–208.

HARRISON, J   Oil pollution fiasco on the Medway estuary. *Birds*,
1967   1:134–6.

JONES, N S   Observations and experiments on the biology of
1948   *Patella vulgata* at Port St. Mary, Isle of Man. *Proc. Lpool
biol. Soc.*, 56:60–77.

MACKIN, J G and HOPKINS, S H   Studies on oyster mortality in
1962   relation to natural environments and to oil fields in Louisiana. *Publs Inst. mar. Sci. Univ. Tex.*, 7:1–131.

MOYSE, J and NELSON-SMITH, A   Zonation of animals and plants on
1963   rocky shores around Dale, Pembrokeshire. *Fld Stud*, 1(5):
1–31.

NELSON-SMITH, A   Marine biology of Milford Haven: the distribu-
1967   tion of littoral plants and animals. *Fld Stud.*, 2(4):435–77.

NELSON-SMITH, A   The effects of oil pollution and emulsifier
1968a   cleansing on shore life in south-west Britain. *J. appl. Ecol.*,
5:97–107.

NELSON-SMITH, A   Biological consequences of oil pollution and
1968b   shore cleansing. *Fld Stud.*, 2(suppl.):73–80.

ODUM, E P   The role of tidal marshes in estuarine production.
1961   *N.Y. St. Conserv.*, June–July.

ODUM, E P and SMALLEY, A E   Comparison of population energy
1959   flow of a herbivorous and deposit feeding invertebrate in a
salt marsh ecosystem. *Proc. natn. Acad. Sci. U.S.A.*, 45:
617–22.

PERKINS, E J   The toxicity of oil emulsifiers to some inshore
1968   fauna. *Fld Stud.*, 2(suppl.):81–90.

RANWELL, D S   World resources of *Spartina townsendii* (*sensu
1967   lato*) and economic use of *Spartina* marshland. *J. appl.
Ecol.*, 4(1):239–56.

RANWELL, D S   Extent of damage to coastal habitats due to the
1968   *Torrey Canyon* incident. *Fld Stud.*, 2(suppl.):39–47.

SMITH, J E (ED.)   *Torrey Canyon* pollution and marine life. A
1968   report by the Plymouth laboratory of the Marine Biological Association of the United Kingdom. Cambridge,
Cambridge U.P., 196 p.

SOUTHWARD, A J   Limpet grazing and the control of vegetation on
1964   rocky shores. *Symp. Br. ecol. Soc.*, 4:265–75.

STEBBINGS, R E   *Torrey Canyon* oil pollution on salt marshes
1968   and a shingle beach 16 months after. Furzebrook, Ware-
ham, Nature Conservancy, 12 p.

TEAL, J M   Energy flow in salt marsh ecosystem of Georgia.
1962   *Ecology*, 43:614–24.

TIVER, N S and CROCKER, R L   Survey methods in grassland
1948   ecology. *J. Br. Grassld Soc.*, 3:1–26.

Coral Reefs and Pollution*

R. E. Johannes†

Récifs coralliens et pollution

Les récifs coralliens constituent l'une des communautés les plus productives du point de vue biologique, les plus diverses sur le plan taxonomique et les plus célèbres pour leur beauté. Ils constituent également la communauté la plus vaste en eaux peu profondes dans le monde entier. En dépit de ces caractéristiques, la dégradation généralisée des communautés de récifs coralliens par suite des activités humaines passe relativement inaperçue. Cela est probablement dû au fait que les récifs coralliens sont par définition invisibles à tous, sauf aux plongeurs sous-marins. La contamination des communautés des récifs par les eaux d'égout et les déchets industriels est extrêmement répandue. L'envasement et la stagnation de l'eau provoqués par le dragage ont donné lieu à de graves problèmes sur certains récifs. La destruction des communautés des récifs par des eaux d'inondation très limoneuses mais de faible salinité, en rapport avec le déboisement, ira sans aucun doute en augmentant. Etant donné que les organismes marins tropicaux ne vivent que dans une très étroite gamme de températures variant de quelques degrés à peine par rapport à leurs limites thermiques supérieures, la pollution thermique par les centrales nucléaires risque de créer un grave problème. La dévastation des récifs coralliens du Pacifique par l'étoile de mer *Acanthaster* est sans doute due à l'action inconsciente de l'homme qui modifie l'équilibre écologique des communautés des récifs.

Bien que les coraux ne représentent qu'une faible fraction de la biomasse totale des récifs, lorsqu'ils meurent toute la communauté des récifs disparaît par suite de la mort ou de l'émigration de la plupart des autres espèces de la faune des récifs. La tolérance écologique de la communauté des récifs n'est donc pas supérieure à celle de son élément constitutif, le corail. Bien que les incidences des activités humaines sur les communautés des récifs dans leur ensemble soient très mal comprises, on dispose d'une masse considérable de renseignements sur la tolérance des diverses espèces coralliennes aux agressions du milieu. Ces renseignements devraient servir de guide pour l'établissement d'un règlement provisoire concernant la conservation des récifs. Si d'autres constituants vitaux de la communauté des récifs présentent des tolérances au milieu moindres que celles des coraux, un tel règlement pourrait être insuffisant; il vaudrait toutefois la peine de commencer par l'établir. De toute évidence, il faut chercher des données sur la réponse à la pollution de l'ensemble des communautés des récifs, mais le temps presse et il faut prendre dès à présent des mesures pour réduire la pollution sur les récifs.

Los arrecifes coralinos y la contaminación

Los arrecifes coralinos son, entre todas las comunidades, los que biológicamente producen más, los más diversos taxonómicamente y los más elogiados desde el punto de vista estético. Constituyen igualmente la comunidad de aguas someras más extensa de la tierra. A pesar de estas características, ha pasado relativamente inobservada para el hombre la extensa degradación sufrida por dichas comunidades. Probablemente esto se debe a que los arrecifes coralinos son invisibles para todas las personas excepto para los buceadores. Es corriente la contaminación de las comunidades coralinas por las aguas negras y los desechos industriales. La sedimentación y estancamiento de las aguas ocasionadas por los dragados han creado graves problemas en algunos arrecifes. Se cree que cada vez será más frecuente la destrucción de las comunidades de arrecifes por las aguas de baja salinidad y cargadas de sedimentos de las inundaciones relacionadas con el desmonte de tierras. Como los organismos tropicales marinos viven a muy pocos grados de sus límites térmicos superiores, la contaminación térmica por las usinas de energía nuclear amenazan convertirse en un grave problema. La devastación de los arrecifes coralinos del Pacífico por la estrella de mar *Acanthaster*, se cree que procede del inconsciente cambio que hace el hombre en el equilibrio ecológico de la comunidad de los arrecifes.

Aunque los corales generalmente representan sólo una pequeña parte de toda la biomasa de las comunidades de los arrecifes, cuando mueren desaparecen todas ellas a causa de la muerte o emigración consiguientes de casi toda la demás fauna que habita en ellos. Así, pues, la tolerancia ambiental de la comunidad indicada no puede superar a la de su componente coral. Aunque los efectos de las actividades del hombre sobre las comunidades de los arrecifes en su conjunto apenas se comprenden, se dispone de considerable información sobre la tolerancia de las distintas especies de coral a las modificaciones ambientales, que debería usarse como orientación para establecer disposiciones provisionales sobre la conservación de los arrecifes. Si otros componentes vitales de tal comunidad demuestran tener tolerancias ambientales menores que los corales, dichas disposiciones podrían ser insuficientes, pero constituirían un comienzo valioso. Debe tratarse de obtener información sobre la reacción de toda la comunidad de los arrecifes a la contaminación, pero no podemos esperar hasta que se recoja para adoptar medidas encaminadas a reducir la contaminación de los arrecifes.

* From Pollution of the Marine Environment. E. J. Ferguson Wood, editor. Amsterdam, Elsevier Publishing Company (In press).

‡ Department of Zoology, University of Georgia, Athens, Georgia 30601, U.S.A. This paper was aided by funds from a U.S. National Science Foundation grant.

CORAL reefs are probably the most extensive shallow marine communities on earth (Goreau, 1961) and among the most biologically productive of all natural communities, marine or terrestrial (Odum, 1971). In the tropics, where man's terrestrial protein sources are often inadequate, reef fishes and shellfish provide high quality protein. Reef communities also abound with species containing a wide range of pharmacologically active compounds (Baslow, 1969). For example, a post-conception birth control pill derived from an extract of a species of Caribbean coral is currently under development.

Coral reefs also provide a buffer against the ocean, preserving about 400 atolls, low tropical islands, and thousands of miles of continental coastlines. Chave, Smith and Roy (1971) estimate the gross production of calcium carbonate by reef communities at between 100/500 tons per acre per year, which builds and maintains both the reef and nearby beaches of marked recreational values.

To foresee the impact of man on the reef community requires knowledge of environmental tolerances of the component organisms. Such studies are few, save investigations on corals themselves. Although these animals do not appear to account for the major reef community biomass (Odum and Odum, 1955), their role is nonetheless pivotal. So central are corals to the integrity of the reef community that when they are selectively killed, migration or death of much of the other reef fauna ensues (Barnes, 1966; Chesher, 1969, 1969a). Therefore the environmental tolerances of the reef community as a whole cannot exceed those of its corals and this fact provides us with convenient preliminary criteria for setting up standards for protecting reef communities from pollution. Where other vital components of the reef community have significantly narrower stress tolerances, such standards would, of course, prove inadequate.

Diversity and pollution tolerance

It is widely believed that a community with "a rich fauna and flora . . . tends to be very stable because of a multiplicity of checks and balances" (Watt, 1964). Since coral reef communities are characterized by great trophic complexity (Wells, 1957) it might therefore be concluded that they are unusually stable (in the sense of ability to withstand stress) and thus unusually resistant to pollution. Observations in this paper suggest that, on the contrary, these communities are quite susceptible to some forms of pollution. Furthermore, shallow mangrove and turtle grass communities seem more stable than coral reef communities at the same latitudes although they are less diverse. Certainly coral communities cannot tolerate the normal extremes of high turbidity, high temperature, or low oxygen encountered routinely by one or both of these other communities, except perhaps the highly modified coral communities in the intertidal zone (Yonge, 1940). Why trophic complexity has not conferred greater stability upon coral reef communities is made apparent by considering at least two processes which determine the degree of stability of a community.

First, individual species develop stress tolerance by natural selection, controlled primarily by the rigours they must survive to colonize the community in question. Secondly, communities as a unit develop a measure of

increased tolerance to stress through an increase in trophic complexity during ecological succession. This is a separate process requiring no increase in the physiological tolerance of the component species any more than the links in a web need be stronger than those in a chain for us to say that the former is stronger than the latter.

The natural environment of a typical subtidal coral reef community does not suffer wide fluctuations in temperature, salinity, sedimentation, or dissolved nutrient levels. The component species have accordingly not developed much resistance to fluctuations in the physical environment; presumably their adaptations are largely to biologically mediated characteristics. Sanders (1968) calls such communities "primarily biologically accommodated" and contrasts them with "primarily physically controlled communities" in which adaptations are largely to physical conditions.

Mangrove and turtle grass communities are examples of primarily physically controlled tropical inshore communities. Their generally shallower waters suffer greater environmental fluctuations than coral communities, so their tolerance is greater.

But it should not be assumed that physically controlled communities are necessarily resistant to types of stresses not normally encountered—we must specify the type of stresses to which we are alluding. For example, mangroves are not very resistant to oil pollution (Rützler and Sterrer, 1970; Spooner, 1969) and are extremely sensitive to defoliants (Boffey, 1971)—two types of stresses not normally encountered by them.

Sedimentation from poor land management

Exposure of reefs to brackish, silt-laden water associated with flood runoff has probably been the single greatest cause of reef destruction historically. Undoubtedly some of this damage would occur naturally, but without doubt bad land management has greatly magnified the problem.

Although corals can cleanse themselves of moderate amounts of falling sediment, most cannot live for long if heavily coated or buried (Mayer, 1918; Edmondson, 1928; Marshall and Orr, 1931), and diversity of coral species decreases as one moves from clear to turbid waters (Roy and Smith, 1971). Coral planulae cannot settle and survive on an unconsolidated substrate. Fine sediments subject to resuspension are inimical to suspension feeding benthos in general (Rhoads and Young, 1970) and bring about the destruction of a variety of reef fauna in addition to corals (Bakus, 1968). In strong currents sand and other coarse suspended materials can inhibit coral growth by abrasion (Bourne, 1888; Crossland, 1927, 1928; Wiens, 1962; Storr, 1964).

High islands and continental lands fringed by reefs frequently have high rainfall and lush vegetation—thus are particularly susceptible to erosion. Kirby (1969) states, "experiments have shown that the greater the rainfall and the thicker the vegetation (up to a point) the greater will be the acceleration of erosion".

Moberly (1963 and personal communication) studied man's activities on reef sedimentation in Hawaii. Ancient Hawaiians cultivated taro in swampy lowlands, and raised pigs, chickens and dogs either penned in the villages or loose in the forests. With arrival of Europeans, virtually all native plants below 3,000 ft were replaced by invaders; sheep, goats and cattle were introduced, and upland soils ploughed for sugar cane and pineapple. Resulting erosion effects on some reefs have been dramatic. On west Molokai for example, ranching started in 1897 and pineapple growing in 1918. Maps of the U.S. Geological Survey and old photographs examined by Moberly show the shoreline has advanced as much as 2 km across the reef flat; in adjacent areas 10 to 70 cm of red-brown silt now overlie the reef.

The coral reefs in Kaneohe Bay, Hawaii, have been increasingly exposed to sedimentation and low salinities associated with flooding. The distribution and abundance of corals on these reefs have greatly decreased concomitantly. The bay watershed is scarred by acres of exposed soil due to land clearing for housing and highway construction despite laws requiring the replanting of land cleared of vegetation. The bay turns red and opaque several times a year after heavy rains. During one rainstorm in 1969 one of ten streams entering the bay was found to be carrying enough sediment in 24 h to cover the entire bay bottom (about 41 km²) with about 1 kg/m² of sediment if spread evenly. Surveys made in 1888 and 1927 yielded essentially the same values for mean bay depth. In 1969 a third survey showed that subsequent sedimentation had reduced the mean depth by more than 1·5 m (Roy, personal communication).

Fine, clay-sized, terrigenous sediments have killed many corals growing below 6–9 m (Maragos, 1971) particularly in the southern end of the bay, where runoff is high and circulation is slowest (Bathen, 1968), and on reefs fringing the shoreline. Sediments here are trapped between coral heads on the reef slopes. Settling of planulae is thus poor or absent, growth of existing corals is slow, and only a few older corals remain alive in many areas (Maragos, 1971).

Fairbridge and Teichert (1948) suggested that because of "colossal soil erosion due to unplanned agriculture" in Australia, continental sedimentation appeared to be gaining on coral growth in the vicinity of the Low Isles, Great Barrier Reef. On the seaward slopes of some of these reefs they reported that much of the coral was dead and covered with a fine coating of terrigenous sediments. Stephenson, Endean and Bennett (1958) note "the widely held opinion that floods have become increasingly severe in North Queensland owing to land clearance". "If so," they state, "a new element—human interference—has entered the ecological complex, and coral growth may never attain its previous luxuriance." Van Eepoel and Grigg (1970) report that in large areas of Lindberg Bay, St. Thomas, Virgin Islands, most corals and other sessile animals have been destroyed and conditions remain unsuitable for their re-establishment due to sedimentation caused by bulldozing, construction and the surfacing of land which drains into the bay. Damage to reef communities due to accelerated terrigenous sedimentation has also been observed in Tanzania by Ray (1968).

Fine sediments carried into coastal waters by flooding are often concentrated initially largely in the upper low salinity layer (Emery, 1962). To some extent this facilitates diffusion and dilution before the sediments settle. Such is not the case when sediments are suspended by dredging, a common but often unrecognized source of reef pollution. Levin (1970) provides a very useful review of the pertinent literature. Brock et al. (1966) give a detailed account of the destruction of corals and reduction of

fish and echinoderm populations at Johnston Island due to siltation brought about by dredging. Over 2,800 ha of reef and lagoon were more or less affected by silty water and 440 ha of reef totally destroyed through dredging and filling.

An airfield was built in Castle Harbour, Bermuda, between 1941 and 1943 by dredging and filling. Visibility was reduced at times to less than one foot. The bay's population of brain corals, *Diploria*, the dominant coral on Bermuda reefs, was destroyed (Burnette-Herkes, personal communication). Dead colonies up to 1 m in diameter can still be seen today. Dredging and filling continues sporadically and no significant recolonization by *Diploria* has occurred.

Coral on Middle Reef, off Cockle Bay near Townsville, Australia, have been destroyed by the dumping of dredge spoil, and "coral gardens" in nearby Nelly Bay, Magnetic Island, face a similar fate (*Townsville Daily Bulletin* Sept. 16, 1970).

Dredging also injured or destroyed some reef communities in Tumon Bay, Guam (Falanruw, 1971), in several bays in American Samoa (Sverdloff, 1971), in Suva Harbour, Fiji, and Kaneohe Bay, Hawaii (personal observations), in Water Bay, St. Thomas, Virgin Islands (Brody et al., 1970), and in the straits near Singapore (Wycherley, 1969).

Sugar mill wastes are another source of reef pollution in Hawaii. The bundles of sugar cane contain soil which, along with some cane leaves, is washed from the cane and often dumped directly into the ocean. After the juice is pressed out, the waste cane pulp, known as "bagasse", is also often dumped into the sea (Burm and Morris, in press). These authors noted one sludge bank which extended for a quarter mile radius from a mill outfall. In the nearby water the numbers and diversity of corals and fishes were reduced and there was an overgrowth of benthic algae. Some sugar companies try to reduce this pollution.

Strong currents in turbid areas help prevent the build-up of sediments on coral surfaces (Marshall and Orr, 1931). Reef communities can thus develop in turbid waters if flushing is adequate (Crossland, 1907; Marshall and Orr, 1931; Verwey, 1931; Motoda, 1940; Roy and Smith, 1971). However the rich and varied biota found in the myriad of caves, tunnels and crevices in the typical reef are undoubtedly more vulnerable to siltation than organisms on the reef surface as sediments tend to fill or plug these cavities. The contribution of this extensive subcommunity to the structure and function of the reef community as a whole, while poorly known, is probably large (Hiatt, 1958). Near Okha in the Gulf of Kutch, India, high tides produce strong currents which continually flush sediments from the raised surfaces of flourishing coral reefs. In the areas I examined the typical cavities characteristic of most coral reefs were not apparent; if they existed at all they were filled with sediments. Di Salvo (1969) found that the internal spaces in dead coral heads in Kaneohe Bay contained unusually large quantities of sediment.

The distribution of reef-building corals is determined in part by light because of the symbiotic algae which live in their tissues and with which they maintain an intimate physiological interdependence. Verwey (1931) for example, found a clear correlation between the maximum depth of reef growth and water clarity in the Bay of Batavia. A chronic increase in turbidity which may not seriously affect the surface biota in shallow water can therefore reduce growth rates of corals and plants in deeper water and cause eventual death.

Low salinities

There are a number of accounts of the destruction of reef biota of fresh-water runoff during floods (Rainford, 1925; Hiatt, 1958; Slack-Smith, 1960; Cooper, 1966; Banner, 1968) and Goreau (1964). This loss of zooxanthellae may be expected to significantly lower the rate of calcification of corals. It may also lower tissue growth rates since dissolved organic compounds released by zooxanthellae are of nutritive value to the coral host (Muscatine and Cernichiari, 1969).

Most species of Indo-Pacific and Atlantic corals that have been tested for tolerance to low salinities survived exposure to 50 per cent sea water for less than two days (Wells, 1932; Vaughn, 1919; Mayer, 1918a; Edmondson, 1928). Eight out of 15 Hawaiian species survived less than a week in sea water diluted by one-quarter with fresh water. Exposure to fresh water for 30 minutes killed most species (Edmondson, 1928). Coral planulae may not settle or grow at reduced salinities (Edmondson, 1929). Fresh water is thus clearly a pollutant when introduced into the reef environment through man's carelessness.

Sewage

Kaneohe Bay, Hawaii, is the best known example of sewage damage. Forty-three years ago Edmondson (1928) described this bay as "one of the most favourable localities—nearly all the reef-forming genera known in Hawaiian waters are represented ... and many species grow luxuriantly".

The south end of the bay is a shallow basin of about 880 ha in surface area and contains extensive patch and fringing reefs. More than 99 per cent of the corals in this basin are dead. Living specimens of the coral species which built the reefs in this area now die within a few weeks when placed there (Maragos, 1971). The specific cause has not been determined, but about three million gallons of sewage receiving primary or secondary treatment flows into this basin daily. In 1967, when sewage output was less than it is today, oxygen levels near the sewer outfalls fell as low as 2·5 mg/l at night (Bathen, 1968). Between 1966 and 1970 dissolved phosphate levels near the sewer outfall increased at an average rate of 0·75 μg–at/l (Caperon and Cattell, in Banner and Bailey, 1970).

Reefs in the bay are being rapidly invaded by *Dictyosphaeria cavernosa* or "green bubble algae." This first colonizes the bases and crevices of coral heads, then extends upward engulfing the entire colony. Finally, it spreads in thick, heavy sheets fusing with other colonies to blanket large areas of reef. Shut off from light, oxygen and food the enveloped corals and many of the associated plants and animals die. Wherever the alga comes into physical contact with corals, the skeleton erodes (Banner and Bailey, 1970) and the reef surface begins to disintegrate. Fish populations become strikingly less abundant.

The accelerating growth of this alga in the bay has paralleled the increasing quantities of nitrogen and phosphorus entering the bay in sewage (Banner and Bailey, 1970). If *D. cavernosa* is grown in bay sea water supplemented with nitrate and phosphate, it grows much faster than it does in bay water alone (Johannes, in Banner and Bailey, 1970). This suggests that further nutrient enrichment of bay waters (inevitable until sewage practices are altered) will magnify this already critical problem. Plans are being made to divert in several years sewage from the bay. In my opinion however, if *Dictyosphaeria* continues to spread at its present rate there will be few viable reef communities left to save by then.

My casual observations between 1959 and 1971 lead me to believe that the numbers and range of a detritus-eating holothurian, *Ophiodesoma spectabilis*, have also increased considerably during this time, possibly because of increased availability of organic matter from sewage.

A recent 18-month multidisciplinary study in Kaneohe Bay led to the publicly expressed opinion that it was not badly polluted and was "in exact balance" (*Honolulu Star Bulletin*, July 19, 1969). This study, focused mainly on water quality, did not hint at the disruption of the reefs that was under way—which is instructive and demonstrates two common shortcomings of pollution surveys.

First, the research was heavily weighted towards public health. It is not always sufficiently recognized by pollution scientists with a public health background, that waters containing levels of pollutants which do not threaten human health can be destructive to aquatic communities.

Secondly, no one surveyed or apparently even looked at bottom communities during the study. The continuous removal and replacement of water and its contents due to tidal flushings, makes water quality an inadequate indicator of pollution in shallow coastal waters. The bottom, in contrast, serves as a reservoir for pollutants, which accumulate there through settling and sorption. The impact of pollution is thus liable to be felt first in bottom communities. It is noteworthy that while reef communities were being destroyed by pollution, only "subtle" changes were observed in the plankton community (Clutter, 1970).

Pollution is likely to first affect populations with rapid turnover times. Thus, benthic microorganisms should provide early warning signs of pollution in coral reef communities as they do elsewhere (Storrs *et al.*, 1969). Predictably, as one moves toward the sewage outfalls in Kaneohe Bay, an expanding anaerobic layer with increasing numbers of sulphate reducing bacteria and decreasing redox potentials are encountered (Di Salvo, 1969; Sorokin, personal communication).

Seepage from a single cesspool in Honaunau Bay, Hawaii, brought about degeneration of a nearby coral community (Doty, 1969). Benthic algal populations are now larger than normal and much coral is dead and encrusted. Both here and in Kaneohe Bay, *Porites compressa* is the coral species most susceptible to sewage effluent (Doty, 1969, and Maragos, personal communication). *P. compressa* is the major reef-building species in shallow water in Hawaii.

Chan (1970) has described how copra farming in the Pacific Islands has led to sewage pollution of reef lagoons. Coconut plantations have forced people, once scattered, to concentrate in villages. Instead of pit latrines, sewage now is generally dumped into the lagoons, which decreases fish and shellfish production and removes already scarce plant nutrients.

Siren and Scheuring (1970) found faecal coliform bacteria in numbers roughly 1,000 times greater than the maximum accepted level for recreational use.

Chan (1970) and Marshall (1969) describe simple, inexpensive sewage treatment units for use on tropical islands. In each case fermentation produces methane which can be bled off and used for cooking or light. When fermentation is complete the residue is sanitary and usable as fertilizer. The effluent, Chan suggests, may be used for aquaculture. A pilot project is in the planning stages at Majuro Atoll.

Recently other proposals have been made for raising various marine animals commercially in atoll lagoons using the entire lagoon rather than creating controlled conditions in restricted areas (Pinchot, 1966, 1970; Inoue, 1968; Isaacs and Schmitt, 1969). This can be expected to produce increases in lagoon nutrient levels either deliberately through fertilization or involuntarily through the increased production of animal excretions containing nutrients—but keep in mind the fate of Kaneohe Bay.

Some of these plans require the closing of lagoon passes to retain the animals being raised but as most corals require fresh, well-circulated waters, this may have serious consequences. Restricting lagoon circulation in this manner at Palmyra Atoll completely destroyed its lagoon reef communities (Dawson, 1959). At Kolonia, Ponape, Caroline Islands, a causeway built in 1970 to an airfield has blocked normal water flow and siltation is destroying the reef communities (Eldredge, personal communication).

In predicting the net value of lagoon aquaculture the possible loss of already productive reef communities must be considered among the costs. In some cases this cost may be acceptable, but this estimate requires careful evaluation.

Constructing special enclosures of limited size in tropical inshore water to utilize the nutrients in sewage and industrial effluents (Chan, 1970; Gundersen and Bienfang, 1970) should produce less ecological dislocation and allow simultaneously for greater control over the species being farmed.

The effect on marine organisms of free residual chlorine in sewage is very poorly known. In view of the well known toxicity of chlorine and its widespread use in sewage treatment plants more research on its effects seems advisable. Unexplained coral mortality in the vicinity of sewage outfalls in Kaneohe Bay might be examined in this context. Effluent from Kaneohe sewage plant contains about 1.1 mg/l of residual chlorine on recent measurements (Young and Chan, 1970). Although chlorine levels diminish rapidly by distance and dilution with sea water, it is worth noting that this concentration is 45 times greater than the lowest level found to be lethal to plaice larvae in the United Kingdom (Alderson, 1970).

Sewage treatment workers are often inadequately trained and sometimes use chlorine well in excess of prescribed levels—particularly when sludge after treatment is discharged and during floods when sewage may bypass normal treatment. The University of Guam is investigating sewage effects.

Thermal pollution

Destruction or alteration of marine communities by heated effluent from power plants and other industrial installations is greatest in the tropics because tropical organisms live at temperatures only a few degrees below their upper lethal limit (Mayer, 1914). Edmondson (1928) showed that 11 of 13 species of Hawaiian corals survived less than 24 h at 32°C, which is only 5 to 6 degrees higher than ambient summer water temperatures. Similarly narrow ranges exist between maximum ambient temperatures and lethal temperatures for corals of the Great Barrier Reef (Mayer, 1918), Caribbean (Mayer, 1918a) and Samoa (Mayer, 1924). Tolerances of high temperatures in these areas appear to be slightly greater than those of Hawaiian corals (although the data are not strictly comparable because of variation in testing methods) but the temperature maxima in these areas are also somewhat higher. Some corals which live naturally in exceptionally warm waters have been reported from the Persian Gulf (Kinsman, 1964) and the Laccadives (Gardiner, 1903). The ranges between maximum environmental temperatures and lethal temperatures are probably even smaller here. The lethal temperatures of gorgonians reported by Cary (1918) were based on experiments lasting only two hours and are therefore of little practical value.

Lethal temperatures determined in the laboratory during short-term experiments must be extrapolated to nature with caution. First, organisms may not be able to tolerate temperatures as high in nature as they do under controlled laboratory conditions (Read, 1967; North, Stephens and North, 1970). Secondly, as Mayer (1914) pointed out, the temperatures where feeding reactions and normal metabolic processes cease are more significant than death-temperatures. Sublethal heat stress is deleterious to corals for a number of reasons. Mayer (1918a) reported that three species of Caribbean corals ceased to feed at temperatures 1.5–3.0°C lower than their lethal temperatures. Edmondson (1928) reported similar results from Hawaii. Yonge and Nicholls (1931) observed that corals extrude their zooxanthellae in strings of mucus at high sublethal temperatures. Coral planulae may not settle at high sublethal temperatures (Edmondson, 1946).

In shallow tropical bays and back-reef areas where water is heated rapidly by the sun, growth of various marine plants and animals may be prohibited by high temperatures even under normal conditions. Mayer (1918) for example, showed that high temperatures within 150 m of shore inhibited coral growth in the Torres Strait. Wells (1952) drew similar conclusions concerning coral growth at Arno Atoll. Obviously any increase in summer water temperatures brought about by thermal effluent released at the water's edge in such environments will expand the area inimical to growth.

Southern Biscayne Bay, Florida, is the best-studied case in point. Under normal conditions some benthic organisms found in the outer bay throughout the year disappear from the warmer inner bay during the summer months (Zieman, 1970). Thermal effluent from a power plant at Turkey Point has killed virtually all plants and killed or greatly reduced animal populations in an area circumscribed closely by a +4°C isotherm (Roessler and Zieman, 1969). Mortalities extend for more than 1.5 km from the outfall. Corals are among the most sensitive species involved, being killed by the effluent at greater distances from the outfall than most other organisms (Anon, 1970).

Colonies of the corals, *Solenastrea hyades* and *Siderastrea siderea*, have been eliminated in the area noticeably affected by heated effluents. In areas of marginal influence many individuals of these species are dead or visibly damaged. Living *S. siderea* specimens are pale and appear to be more susceptible to overgrowth by algae and encrustation by other forms (Purkerson, personal communication).

Thermal output from the Turkey Point plant is greatest during hot summer afternoons when the power demand by Miami air conditioners is highest, thus coinciding with natural peak water temperatures in the bay. In 1968, 120 ha were affected by the effluent. By August, 1969, the damaged area had more than doubled (Anon., 1970). Two additional nuclear powered generating units are scheduled for use by 1972. If efforts to block construction are unsuccessful, thermal output from this station will increase fourfold. There is concern that the ecological effects may extend offshore to the famous underwater park and coral reef preserve at John Pennekamp Coral Reef State Park.

Bays are generally very poor locations for the dumping of pollutants since flushing and dilution are reduced. This is so obvious one might think its mention here superfluous. Yet bays seem to be favoured dumping sites—witness the power plants in Biscayne Bay (hardly a worse spot, ecologically, can be found in south Florida) and in the Virgin Islands in Lindberg Bay. Kaneohe Bay, Hawaii, already seriously damaged by sewage and sediment pollution, is now faced with the construction of a new power plant which will pump heated effluent close to some of the finest remaining reefs. It is hard to believe that the officials responsible for the location of such plants take environmental problems seriously, their earnest public statements to the contrary.

Desalination effluents

Some corals are particularly sensitive to salinities slightly above their ambient range. Nine out of 12 Hawaiian species tested could not tolerate salinities of even 110 per cent of normal for more than two weeks. Three species died within three days. Most species died within 24 h when exposed to salinities 150 per cent of normal (Edmondson, 1928).

Effluents from desalination plants are characterized by elevated levels of not only salinity but also temperature, and toxic metals such as copper, zinc and nickel. They may also contain various chemicals added for pH and corrosion control (Zeitoun, Mandelli and McIlhenny, 1969). These effluents are often sufficiently saline that despite their elevated temperature they sink and flow along or near the bottom where there is no loss of heat directly to the atmosphere and where their potential for damage to benthic communities is greatest (Clarke, Joy and Rosenthal, 1970; van Eepoel and Grigg, 1970).

Toxic metals introduced in the effluent are due to corrosion of metallic surfaces of the distillation system. Zinc and nickel ions, though toxic to marine organisms at elevated concentrations, are much less harmful than copper at equivalent concentrations; copper therefore appears to be the toxic contaminant of greatest concern

in desalination effluents (Zeitoun *et al.*, 1969). The mixture of brine blowdown and cooling water discharged from desalination plants may contain this element at concentrations 6–8 times as high as the recommended maximum concentration of 0.02 mg/l, and 30–40 times the natural copper concentration in coastal waters (Zeitoun, Mandelli and McIlhenny, 1969).

Studies of the effects of desalination effluents are greatly complicated by interactions between different stress factors. For example, zinc and copper can act synergistically in their effect, as can copper and elevated temperature, while elevated salinities may reduce the toxicity of copper due to the presence of more ions to compete with copper for sorption sites (Zeitoun, Mandelli and McIlhenny, 1969).

Desalination effluent in Key West, Florida, appears to be responsible for the reduction or disappearance of some marine species from the outfall area and the increase in some others, including fish and shellfish of several edible species (Clark, Joy and Rosenthal, 1970). The levels of toxic metals in these species were not examined. Gorgonian corals transplanted to the area of the outfall seemed particularly vulnerable. These workers reported that the effluent was detected, due to elevated temperatures and salinity, at greater distances from the discharge point than were gross changes in the biological communities. This suggests, they state, that the levels of effluent dilution necessary to produce minimal environmental modification should be amenable to prediction.

The use of combination power-desalination plants in the vicinity of marine communities may be expected to amplify ecological problems because of the greater temperatures of the effluent. Van Eepoel and Grigg (1970) made preliminary observations near a plant discharging into Lindberg Bay, St. Thomas, Virgin Islands. Salinities and temperatures near the outfall were only slightly higher than ambient on two occasions when surveys were made. On another occasion divers could not measure the effluent temperature because it was above that which they could tolerate. Corals and other invertebrates were killed to a distance of 200 m from the discharge plant. Difficulty in studying the effects of the effluent beyond that point were encountered because the bay is also polluted with oil and soot as well as excessive sedimentation. Power and water output from the plant are expected to double within the next three years.

Desalination plants are proliferating as the demand for fresh water grows and the cost of desalination decreases. It is disturbing therefore to note how little effort seems to have been made to examine the ecological consequences of brine disposal. Of the first 481 research and development reports based on work carried out under contract to the United States Office of Saline Water, only three specifically concern ecological problems attending effluent disposal in marine environments. It is also noteworthy that of the eight desalination plants surveyed by Zeitoun, Mandelli, and McIlhenny (1969) (five of which are located on coastlines in regions of extensive coral reef development) no governmental regulations existed with regard to effluent discharge, and in only one case (Nassau, Bahamas) was an oceanographic or engineering study of the effect of the effluent undertaken before the plant was built.

Oil pollution

There appears to be no conclusive evidence that oil floating above corals damages them. Grant (1970) floated Moonie crude oil in a vessel over specimens of the coral, *Favia speciosa*. The corals showed no visible signs of injury during six days of subsequent observation. Johannes, Coles and Maragos (unpublished) floated five different types of oil over groups of the Hawaiian corals, *Porites compressa*, *Montipora verrucosa* and *Fungia scutaria* for 2.5 h. During 25 days of subsequent observation no visible damage was noted. Lewis (1971) found that four species of coral exposed to crude oil or an oil spill detergent exhibited ruptured tissue and other signs of distress. These experiments were carried out in sealed vessels to prevent escape of volatile oil fractions. Since the harmful effects may have been due to these volatiles it cannot be assumed that the results are representative of what might occur during an oil spill when volatiles would escape to the atmosphere.

Published reports of oil spills near coral reefs also provide no conclusive evidence of injury. In a Panama spill where inshore littoral fauna suffered heavily Rützler and Sterrer (1970) assumed corals escaped readily observable damage because they were continually submerged. Spooner (1969) observed no damage to reef communities in Tarut Bay, Saudi Arabia, in an area of chronic oil pollution.

Although he makes no mention of corals Gooding (1971) reports the extensive destruction of associated reef fauna after the release by a tanker of six million gallons of assorted oils off the harbour entrance at Wake Island. Of the inshore reef fishes killed, 2,500 kg washed ashore. Numerous reef invertebrates were also killed.

In some areas reef surfaces are exposed during low spring tides. This phenomenon, common on Indo-Pacific reefs, was taken into account in an experiment carried out by Johannes, Maragos and Coles (MS). They exposed 22 species of corals, whose upper portions were above water, to Santa Maria crude oil for 1.5 h. Oil adhered to portions of the surfaces of most species. Tissue death ensued in these areas, but not on areas where the oil did not adhere. Branching species of *Acropora* and *Pocillopora* were affected most; species of *Fungia* and *Symphyllia* showed no evidence of obvious damage.

It must also be remembered that not all oil remains afloat. Some oil compounds are soluble and are particularly toxic to marine organisms (Mironov, 1970; Blumer, 1970). Some heavier oil residues sink readily, while lighter oils may quickly absorb to suspend sediment particles (Hawkes, 1961). Commercial species may become tainted and unsaleable (Sidhu *et al.*, 1970). The community may thus be damaged from man's viewpoint even when its structure is unaltered.

Extensive and expanding offshore drilling over large oil deposits occurs in Indonesia and similar activities are afoot for other parts of southeast Asia, the Persian Gulf, the Caribbean, Fiji, Tonga and other reef areas. There is no evidence that serious consideration has been given to possible damage to nearby reefs. Such is not the case in Australia where public opinion forced suspension of oil exploration permits on the Great Barrier Reef until possible damage could be evaluated.

Oil spills associated with shipping have occurred very commonly along the coastal perimeter of the Arab States and Israel for many years. This area appears to offer an excellent opportunity for some badly needed research on the long-term effects of oil on reef communities.

Effects of nuclear weapon tests

The obliteration of coral reef communities in the vicinity of nuclear blasts is of minor significance for man compared with subsequent effects spread over much wider areas. The most obvious immediate biological impact outside the area of complete destruction is that of radiation burns to the biota. Donaldson, cited by Hines (1962), collected living fishes in the vicinity of a recent blast whose skin was missing from one side as if the animals had been "dropped in a hot pan." This also is a comparatively trivial phenomenon. It is the radioactive contamination of the food chain which has by far the gravest consequences.

The geographic distribution of radioactive contaminants has been difficult to predict but it is well known that serious contamination may extend great distances. Because of atmospheric circulation patterns, areas more than 150 mi from a thermonuclear explosion at Bikini Atoll received four times as much radiation as parts of Bikini itself. Parts of Rongelap Atoll 100 mi away received sufficient fallout within 48 h that human survival in the absence of shelter would have been unlikely (Hines, 1962). It was not until four years later that all reef species at Rongelap were declared safe for human consumption (Hines, 1962). A single nuclear blast may thus render fish and shellfish unsafe or unsaleable for thousands of square miles.

Generally it is the lower trophic levels, the algae and herbivores, which become most radioactive immediately after a blast. Later carnivorous reef fishes become more radioactive than omnivores or herbivores (Donaldson, 1960; Hines, 1962). As radioisotope levels in the community subsequently diminish, *Tridacna* and other edible reef clams are among the last reef animals to become safe for human consumption (Hines, 1962; Welander, 1969).

The ecological impact of radioactive contamination on reef communities is almost unknown. Gorbman and James (1963) discovered radiation damage to the thyroid glands of various coral reef fishes after a nuclear test at Eniwetok Atoll. This appears to be the only observed damage attributable to the injurous effects of radioactive contamination of the reef food chain. This may reflect a dearth of relevant research more than lack of damage.

Dying animals on the reef are liable to be consumed by reef scavengers and carnivores before they are observed (Donaldson, 1960). To evaluate the impact of any stress on a reef community population sizes of various species should therefore be measured before and after the stress occurred. This does not seem to have been done with nuclear tests in the Pacific. The statement by Schultz, "the pressure of population from all sides into the damaged area is very great and soon replaces the losses" (cited with approval by Templeton *et al.*, 1970), therefore seems premature. It is to be hoped that the French have made efforts to remedy this situation in connection with their recent tests in the South Pacific.

In January 1971 Maragos and the author had a brief opportunity to examine some shallow patch reef communities which had apparently been destroyed by a nuclear blast at Eniwetok Atoll. The date of destruction could not be determined because several tests were carried out on different dates. The nearest blast created a crater several hundred metres away on the inter-island reef. The last year in which nuclear tests took place here was 1958.

The reefs are located several hundred metres northwest of Yvonne (Runit) Island where the inter-island reef begins to slope toward the lagoon. The tops of the reefs are covered by about 0.5 m of water at low tide. The communities had been dominated structurally by *Acropora corymbosa* ("table acropora"). We could not tell whether its destruction was caused by heat or radiation. Since skeletons of *A. corymbosa* were still standing we assume the communities were not killed by the force of the blast.

Extensive regrowth of *A. corymbosa* had occurred. Some living specimens were nearly 3 m in greatest diameter. Coral diversity was low however. Only four species of corals were common, whereas Maragos counted nine to ten common genera and estimated there were about 15 common species on typical *Acropora*-dominated inner reef communities at Eniwetok. Diversity and abundance of other invertebrates, fishes and algae also appeared low on the affected reefs. Very little grew on the dead skeletons of *A. corymbosa* or the substrate. In short, although regrowth of *A. corymbosa* on this reef was impressive, the community had by no means recovered 13 years after the last nuclear test in the vicinity. Patch reefs farther from the test sites, several hundred yards south of Yvonne Island, showed no readily observable effects of the blasts.

Acanthaster

The destruction of increasing numbers of coral reef communities in the Pacific in the past few years by a coral-eating starfish, *Acanthaster planci*, has received widespread attention. Most hypotheses point the finger at man as the ultimate cause. Chesher (1969, 1969a) suggests that blasting and dredging in reef areas may provide abnormally large areas for settling of larvae of *A. planci* and states that such activities have occurred in many areas where *A. planci* epidemics subsequently developed. Pearson and Endean (1969) offer the hypothesis that shell collectors have seriously reduced the numbers of a predator of *A. planci*, *Tritonia charonis*, the triton or trumpet shell, and thus the starfish population has expanded rapidly. A third hypothesis is that pesticides have attained harmful levels in the tissues of predators on the eggs, larvae or adults of *A. planci* (Chesher, 1969a; Fischer, 1969).

Consistent with the latter hypothesis are the observations of Dr. L. McCloskey (personal communication), who has discovered high concentrations of polychlorinated biphenyls (PCB) in the tissues of *A. planci* taken from Guam, Ifalik and Woleai—three sites of *A. planci* outbreaks in Micronesia. McCloskey is presently investigating the possibility that there is a progressive concentration of this toxicant at each level in the food chain ultimately leading to predators on *A. planci*, so that they can no longer maintain control.

It is also suggested that *Acanthaster* outbreaks may be a natural phenomenon (Yonge, 1966). Recently *Acanthaster* has been found as far east as reefs off western Panama (Glynn, personal communication). There is thus a possibility of passage of *Acanthaster* larvae into the

Caribbean through the Panama Canal in sea water ballast (Chesher, 1968) or through the proposed sea level canal (Thomas, 1970).

Ciguatera

Algae are generally among the first macroscopic organisms to colonize any fresh surface in reef areas (Brock *et al.*, 1966; Dawson, 1959), artificial reefs (Randall, 1963) skeletons of corals eaten by *A. planci* (Chesher, 1969a), sunken ships, abandoned war materials and other human artifacts (Randall, 1958; Halstead, 1970). The apparent connection between these new algal growths and fish poisoning was first noted by Dawson, Aleem and Halstead (1955) and discussed in detail by Randall (1958) and Dawson (1959). These authors pointed to the relationship between the availability of new surfaces in the reef environment, the rapid growth of algae, and the development of toxicity, known as ciguatera, in normally edible species of reef fish in the immediate area. They theorized that toxic algae, possibly blue-greens, growing unusually rapidly on new reef surfaces, are the basis for the toxic food chain leading ultimately to man and causing illness and sometimes death; de Sylva and Hine (1971) suggest that thermal pollution may also be the forerunner of ciguatera since elevated water temperatures favour the growth of toxic blue-green algae.

Randall (1958) points out "The importance of ciguatera goes beyond the purely medical aspects of treatment of patients suffering from the ingestion of toxic fish. Many non-toxic fish are denied to humans or domestic animals as food because of fear of their being poisonous. Sections of reef with a reputation of harbouring poisonous fishes are not fished, whereas non-toxic areas may, as a result, be overfished."

Bagnis (1969) has provided a valuable account of the development of ciguatera in a reef community in the same areas and in the same chronological sequence as disruptive human activities. There are numerous examples of dredged or otherwise disturbed reef areas where ciguatera has not developed. At Johnston Island, for example, where ciguatera has occurred in the past, extensive dredging of reef areas in 1963 has not been followed by any new outbreaks (Banner, personal communication). Neither are there any examples so far of ciguatera developing in the wake of *Acanthaster planci* invasions (Banner, personal communication). Nonetheless, the relationship between the disruption of reefs by man and the subsequent development of ciguatera in the immediate area seems too frequent to be coincidental and research continues on this recalcitrant problem.

Other causes of reef damage

Though often illegal, the use of dynamite and poisons to kill reef fishes is commonplace (Owen, 1969; Powell, 1970; Ramas, 1969; Sverdloff, 1971; Anon., 1952). In the Philippines, according to Ramas (1969) a fisherman may find some large coral formation about the size of a regular house in about 100–150 ft of water. The area after bait is spread is completely wired with dynamite and when the big fish come around in enough numbers, the charge is detonated. So much fish is obtained this way that the smaller fish under five pounds are not even picked up.

[371]

The use of poisons to kill marine fishes in Hawaii was outlawed in 1850. But here, as elsewhere, the practice remains common today.

The danger of lagoon contamination by pesticides (sometimes used in large quantities on atolls) has recently been noted. Fifteen to twenty tons of fish died suddenly in Truk lagoon on April 17, 1970, hospitalizing six people who ate some of them. The flesh was found to contain the pesticide Endrin in the highest concentrations ever recorded in fishes to that date (Bourns, 1970). Barracuda taken on the Flower Gardens reefs (northern Gulf of Mexico) contained high levels of DDT. Significant amounts of Dieldrin and Endrin were also discovered (Bright, personal communication; *Houston Post*, Dec. 12, 1970). People of Cook Islands now use chlorinated hydrocarbon pesticides as a substitute for derris in killing lagoon food fishes (Hambuechen, 1971). High levels of PCBs have been found in *Acanthaster planci* from several Micronesian reefs.

Fate of damaged reef communities

When a reef community is destroyed the ecological conditions that follow cannot be expected to coincide with those preceding. So it cannot be taken for granted that the reef community will ever replace itself. Wood-Jones (1907) reported that 30 years after the destruction of a coral reef by "foul water" released from a volcanic vent, only dead corals remained without sign of regeneration. Rainford (1925) reporting 1918 flood damage on the Great Barrier Reef, doubted that recovery would ever occur. In 1953 the same area showed negligible recolonization (Stephenson, Endean and Bennett, 1958).

In instances where recovery of the reef community does occur, rates differ widely; complete recovery requires decades if damage to corals is extensive (Stephenson, Endean and Bennett, 1958; Chesher, 1969, 1969a; Stoddart, 1969). Some of the larger coral heads killed by starfish in Guam recently were over 200 years old (Goreau, in Chesher, 1969a).

Recovery may be more rapid if only mobile organisms or organisms with shorter lifespans than corals are affected. For example, four years after the fish population was removed from a Bermuda patch reef recolonization was complete (Bardach, personal communication).

When corals die their surfaces are rapidly colonized by algae (Slack-Smith, 1960; Chesher, 1969, 1969a; Weber, 1969; Banner, 1968; Goreau, 1964; Moorhouse, 1936). Benthic algae are inimical to coral regrowth by their presence (Hedley, 1925); they also trap silt and create shifting sandbanks unsuitable for larval settlement (Wood-Jones, 1907).

Subsequent to colonization by algae, dead coral surfaces may become colonized by encrusting animals including zoanthids, hydroids, bivalves and alcyonarians (Slack-Smith, 1960; Banner, 1968; Barnes, 1966; Wood-Jones, 1907; Chesher, 1969a; Weber, 1969). In Guam large numbers of the sea urchin *Echinometra mathaei* appeared and removed most of the growth of filamentous algae on coral recently killed by *A. planci* (Chesher, personal communication).

Unfortunately subsequent stages in succession are almost unknown because no long-term, continual studies of succession on a denuded reef have been published. Such studies are presently under way in Guam and on the Great Barrier Reef in the wake of *A. planci* epidemics.

Although many faunal elements of the community disappear after corals are killed, the standing crop of herbivorous fishes may increase dramatically as a result of this increased food supply (Chesher, 1969a). It has been suggested that the production of harvestable protein under such circumstances might equal or exceed that of the original reef. However, my observations in Fiji and Eniwetok in areas recently infested with *Acanthaster* indicate that an increase in herbivorous fish under such circumstances is not an invariable response. In these two areas herbivorous fishes were conspicuously less numerous than in adjacent areas with abundant living coral, despite a heavy overgrowth of algae on dead coral. Apparently the colonizing algae here were not acceptable. In areas where herbivorous fish populations increase after an *Acanthaster* invasion it is likely that this is just a transient early stage of ecological succession on the recovering reef.

Conclusions

Effective policing of reefs to deter the individual polluter is frequently impossible. Around Truk dynamite to kill reef fishes is extensively used (Fuchs, 1968) as is the widespread use of poisons in other reef communities.

Although the corporate polluter is easier to detect, laws to prevent such pollution are virtually always inadequate and frequently non-existent. Too often the public, including even that segment whose livelihood depends upon the reefs, does not realize the problem exists. The enactment of appropriate laws and public censure of polluters can only be brought about when the public is made aware of their value. The most important step in deterring corporate and individual polluters is thus education.

Numerous customs helped conservation in pre-industrial cultures. Falanruw (1971) quoting former conservation-valued practices stressed that "young people of Micronesia today are not individually conservation minded". This is one of several reasons why the teaching of conservation becomes increasingly imperative as the relentless "westernization" of these cultures proceeds.

To help foreign-born teachers with little knowledge of local fauna and flora, the Bermuda Department of Education engaged scientists to give a series of lectures on the ecology of Bermuda, supplemented by field trips. The information, enthusiastically received, is now being beneficially reflected in their curricula.

A conservation programme was started in Truk in 1967 using daily radio programmes, posters, a newspaper column, adult education classes, etc., to create a desire for conservation and to gain support for a law enforcement programme.

Undoubtedly only a small fraction of the damage man has done to coral reefs has been recognized. Detailed surveys should therefore be made of reef resources, particularly those near populated areas. These surveys should involve not only biologists and geologists, but also economists.

The allocation of funds for coral reef research is very small in relation to the importance of these communities to man, so new or accelerated programmes on some of the problems discussed are clearly needed. But we cannot afford to wait until such research is completed before taking steps to alleviate reef pollution. Recommendations biologists make now regarding reef conservation may be imperfect, but this is no reason to allow the destruction of this important resource to continue unchecked.

References

ALDERSON, R   Effects of low concentrations of free chlorine on
1970   eggs and larvae of plaice, *Pleuronectes platessa* L. Paper presented at FAO Technical Conference on Marine Pollution and its Effects on Living Resources and Fishing, Rome, Italy, 9–18 December 1970, FIR:MP/70/E-3:8 p.

BAGNIS, R   Naissance et développement d'une flambée de
1969   ciguatera dans un atoll de l'archipel des Tuamotu. *Revue Corps Santé Armées*, 10(6):783–95.

BAKUS, G J   Sedimentation and benthic invertebrates of Fanning
1968   Island, Central Pacific. *Mar. Geol.*, 6:45–51.

BANNER, A H   A fresh-water "kill" on the coral reefs of Hawaii.
1968   *Tech. Rep. Hawaii Inst. mar. Biol.*, (15):29 p.

BANNER, A Y and BAILEY, J H   The effects of urban pollution upon
1970   a coral reef system. *Tech. Rep. Hawaii Inst. mar. Biol.*, (25):66 p.

BARNES, J H   The crown of thorns starfish as a destroyer of coral.
1966   *Aust. nat. Hist.*, 15:257–61.

BASLOW, M H   Marine pharmacology. Baltimore, Williams and
1969   Wilkins Co., 286 p.

BATHEN, K H   A descriptive study of the physical oceanography of
1968   Kaneohe Bay, Oahu, Hawaii. *Tech. Rep. Hawaii Inst. mar. Biol.*, (14):353 p.

BLUMER, M   Oil contamination and the living resources of the sea.
1970   Paper presented to the FAO Technical Conference on Marine Pollution and its Effects on Living Resources and Fishing, Rome, Italy, 9–18 December 1970, FIR:MP/70/R-1:11 p.

BOFFEY, P M   Herbicides in Vietnam: AAAS study finds wide-
1971   spread devastation. *Science, N.Y.*, 171:43–7.

BOURNE, G C   The atoll Diego Garcia and the coral formations in
1888   the Indian Ocean. *Proc. R. Soc. Lond.*, 43:440–61.

BOURNS, C T   Truk Island fish kill, April 1970. Water Quality
1970   Contingency Report. U.S. Department of the Interior, Federal Water Quality Administration Pacific Southwest Region, 13 p.

BROCK, V E, VAN HEUKELEM, W and HELFRICH, P   An ecological
1966   reconnaissance of Johnston Island and the effects of dredging. *Tech. Rep. Hawaii Inst. mar. Biol.*, (11):56 p.

BRODY, R W, *et al.*   A study of the waters, sediments and biota of
1970   Chocolate Hole, St. John, with comparison to Cruz Bay, St. John. Report of the Caribbean Research Institute to Government of Virgin Islands, Department of Health, Division of Environmental Health, 20 p.

BURM, R J and MORRIS, D E   The effect of turbid, high carbo-
hydrate, sugar processing waste on tropical open sea. *Adv. Wat. Pollut. Res.*, 5:14 p. (in press).

CARY, L R   A study of respiration in Alcyonaria. *Pap. Dep. mar.*
1918   *Biol. Carnegie Instn. Wash.*, (12):185–91.

CHAN, G L   Use of potential lagoon pollutants to produce protein
1970   in the South Pacific. Paper presented to FAO Technical Conference on Marine Pollution and its Effects on Living Resources and Fishing, Rome, Italy, 9–18 December 1970, FIR:MP/70/E-10:4 p.

CHAVE, K E, SMITH, S V and ROY, K   Carbonate production by
1971   coral reefs

CHESHER, R H   Transport of marine plankton through the Panama
1968   Canal. *Limnol. Oceanog.*, 13:383–8.

CHESHER, R H   Destruction of Pacific corals by the sea star *Acan-
1969   thaster planci. Science, N.Y.*, 165:280–3.

CHESHER, R H   *Acanthaster planci* impact on Pacific coral reefs.
1969a   Westinghouse Research Laboratories, Report to U.S. Department of the Interior, 152 p.

CLARKE, W D, JOY, J W and ROSENTHAL, R J   Study of Key West
1970   desalting plant effluent. Final Report to U.S. Department of the Interior. Westinghouse Research Laboratories, No. 14–12–470, 73 p.

CLUTTER, R I   Subtle effects of pollution on inshore tropical
1970   plankton. Paper presented to FAO Technical Conference on Marine Pollution and its Effects on Living Resources and Fishing, Rome, Italy, 9–18 December 1970, FIR:MP /70/E-52:11 p.

COOPER, M J   Destruction of marine flora and fauna in Fiji caused
1966   by the hurricane of February 1965. *Pacif. Sci.*, 20(1):137–41.

CROSSLAND, C   Reports on the marine biology of the Sudanese
1907   Red Sea. *J. Linn. Soc. Lond. (Zool.)*, 31:3–30.

CROSSLAND, C   Marine ecology and coral formation in the Panama
1927   region, Galapagos, and Marquesas Islands, and the atoll of Napuka. *Trans. R. Soc. Edinb.*, 56:531–54.

CROSSLAND, C   Coral reefs of Tahiti. *J. Linn. Soc. Lond. (Zool.)*, 36:
1928   577–620.

DANA, T F   *Acanthaster*: a rarity in the past? *Science, N.Y.*,
1970   169:894.

DAWSON, E Y   Changes in Palmyra Atoll and its vegetation
1959   through the activities of man 1913–1958. *Pacif. Nat.*, 1(2):1–51.

DAWSON, E Y, ALEEM, A A and HALSTEAD, B W   Marine algae
1955   from Palmyra Island with special reference to the feeding habits and toxicology of reef fishes. *Occas. Pap. Allan Hancock Found.*, (17):39 p.

DE SILVA, D P and HINE, A E   Ciguatera—marine fish poisoning—
1970   a possible consequence of thermal pollution in tropical seas? MS accepted by FAO Technical Conference on Marine Pollution and its Effects on Living Resources and Fishing, Rome, Italy, 9–18 December 1970.

DI SALVO, L H   Thesis, University of North Carolina.
1969

DONALDSON, L R   Radiobiological studies at the Eniwetok test
1960   site and adjacent areas of the western Pacific. *In* Transactions of the 2nd Seminar Biological Problems in Water Pollution, U.S. Department of Health, Education, Welfare, pp. 1–7.

DOTY, M S   The ecology of Honaunau Bay, Hawaii. *Hawaii Bot.*
1969   *Sci. Pap.*, (14):211 p.

EDMONDSON, C H   The ecology of an Hawaiian coral reef. *Bull.*
1928   *Bernice P. Bishop Mus.*, (45):64 p.

EDMONDSON, C H   Growth of Hawaiian corals. *Bull. Bernice P.*
1929   *Bishop Mus.*, (58):38 p.

EDMONDSON, C H   Behaviour of coral planulae under altered saline
1946   and thermal conditions. *Occas. Pap. Bernice P. Bishop Mus.*, (18):283–304

EMERY, K O   Marine geology of Guam. *Prof. Pap. geol. Surv.*,
1962   (403B):76 p.

FAIRBRIDGE, R W and TEICHERT, C   The Low Isles of the Great
1948   Barrier Reef: a new analysis. *Geogrl. J.* 111: 67–88.

FALANRUW, M V C   Conservation in Micronesia. *Atoll Res. Bull.*,
1971   (148):18–20 p.

FISCHER, J L   Starfish infestation: hypothesis. *Science, N.Y.*,
1969   165:645.

FUCHS, E H   Conservation problems in Truk. Paper prepared for
1968   South Pacific Fisheries Technical Meeting, Palau, 1968.

GARDINER, J S   The fauna and geography of the Maldive and
1903   Laccadive Archipelagos. Cambridge, vol. 1:423 p.

GOODING, R M   Oil pollution on Wake Island from the tanker
1971   *R.C. Stoner. Spec. Sci. Rep. Natl Oceanic Atmos. Admin.*, (636)1:10 p.

GORBMAN, A M and JAMES, M S   An exploratory study of radiation
1963   damage in the thyroids of coral fishes from Eniwetok Atoll. *In* Radioecology, edited by V. Schultz and A. W. Klement, Washington, D.C., Rheinhold and AIBS, pp. 385–99.

GORDON, M S and KELLY, H M   Primary productivity of an
1962   Hawaiian coral reef: a critique of flow respirometry in turbulent waters. *Ecology*, 43:473–80.

GOREAU, T F   Problems of growth and calcium deposition in reef
1961   corals. *Endeavour*, 20:32–40.

GOREAU, T F   Mass expulsion of zooxanthellae from Jamaican reef
1964   communities after Hurricane Flora. *Science, N.Y.*, 145:383–6.

GRANT, E M   Notes on an experiment upon the effect of crude oil
1970   on live corals. *Fish. Notes Dep. Primary Ind., Brisbane*, (1):1–13.

GUNDERSEN, K and BIENFANG, P   Thermal pollution: use of deep,
1970   cold, nutrient rich sea water for power plant cooling and subsequent aquaculture in Hawaii. Paper presented to FAO Technical Conference on Marine Pollution and its Effects on Living Resources and Fishing, Rome, Italy, 9–18 December 1970, FIR:MP/70/E-84:9 p.

HALSTEAD, B W   Toxicity of marine organisms caused by pol-
1970   lutants. Paper presented to FAO Technical Conference on Marine Pollution and its Effects on Living Resources and Fishing, Rome, Italy, 9–18 December 1970, FIR:MP/70/R-6:21 p.

HAMBUECHEN, W H   Pesticides in the Cook Islands. Paper
1971   presented to the South Pacific Commission Regional Symposium on Conservation of Nature—Reef and Lagoons, SPC/RSCN/wP.1:6 p.

HAWKES, A L   A review of the nature and extent of damage
1961   caused by oil pollution at sea. *Trans. N. A. Wildl. Conf.*, 26:343–355.

HEDLEY, C   The natural destruction of a coral reef. Report of the
1925   Great Barrier Reef Commission. *Trans. R. geogrl Soc. Australas. Qd*, 1:35–40.

HIATT, R W   Factors influencing the distribution of corals on the
1958   reefs of Arno Atoll, Marshall Islands. *Proc. Pacif. Sci. Congr.*, 8:929–70.

HINES, N O   Proving ground. Seattle, University of Washington
1962   Press, 366 p.

INOUE, M   Cited in *National Fisherman*, Aug. (1968), p. 22B.
1968

ISAACS, J D and SCHMITT, W R   Stimulation of marine productivity
1969   with waste heat and mechanical power. *J. Cons. Int. Explor. Mer*, 33:20–9.

KINSMAN, D J J   Reef coral tolerance of high temperatures and
1964   salinities. *Nature, Lond.*, 202:1280–2.

KIRBY, M J   Erosion by water on hillslopes. *In* Water, earth and
1969   man, edited by R J Chorley, London, Methuen, pp. 229–38.

LEVIN, J   A literature review of the effects of sand removal on a
1970   coral reef community. *Publs Univ. Hawaii Dept. Ocean Engng.*, 78 p.

LEWIS, J B   The effect of crude oil and an oil spill detergent upon
1971   reef corals. *Mar. Pollut. Bull.*, (2):59–62.

MARAGOS, J   Thesis, Univ. of Hawaii.
1971

MARSHALL, K   A new method for sewage treatment on coral
1969   atolls. *Atoll Res. Bull.*, (126):10–1.

MARSHALL, S M and ORR, A P   Sedimentation on Low Isles Reef
1931   and its relation to coral growth. *Scient. Rep. Gt Barrier Reef Exped. 1928–29*, 1(5):94–133.

MAYER, A G   The effects of temperature upon tropical marine
1914   animals. *Publs Carnegie Instn, Wash.*, (183):24 p.

MAYER, A G   Ecology of the Murray Island coral reef. *Publs*
1918   *Carnegie Instn, Wash.*, (213):48 p.

MAYER, A G   Toxic effects due to high temperature. *Publs Carnegie*
1918a   *Instn, Wash.*, (252):175–8.

MAYER, A G   Structure and ecology of Samoan reefs. *Publs*
1924   *Carnegie Instn, Wash.*, (340):1–25.

MIRONOV, O G   The effect of oil pollution on flora and fauna of
1970   the Black Sea. Paper presented to FAO Technical Conference on Marine Pollution and its Effects on Living Resources and Fishing, Rome, Italy, 9–18 December 1970, FIR:MP/70/E-92:4 p.

MOBERLY, R, JR   Coastal geology of Hawaii. *Rep. Hawaii Inst.*
1963   *Geophys.*, (41):216 p.

MOORHOUSE, F W   The cyclone of 1934 and its effect on the Low
1936   Isles, with special observations on Porites. *Rep. Gt Barrier Reef Commn Brisbane*, 4:37–44.

MOTODA, S   Comparison of the conditions of water in bay, lagoon
1940   and open sea in Palao. *Stud. Palao Trop. biol. Stn*, (2):41–8.

MUSCATINE, L and CERNICHIARI, E   Assimilation of photo-
1969   synthetic products of zooxanthellae by a reef coral. *Biol. Bull. mar. biol. Lab. Woods Hole*, (137):506–23. Also issued as *Am. Zool.*, 8:771(1968).

NORTH, W J, STEPHENS, G C and NORTH, B B   Marine algae and
1970   their relations to pollution problems. Paper presented to the FAO Technical Conference on Marine Pollution and its Effects on Living Resources and Fishing, Rome, Italy, 9–18 December 1970, FIR:MP/70/R-8:22 p.

ODUM, E P   Fundamentals of ecology. Philadelphia, W. B.
1971   Saunders, 3rd ed., 574 p.

ODUM, H T and ODUM, E P   Trophic structure and productivity of
1955   a windward coral reef community on Eniwetok Atoll. *Ecol. Monogr.*, 25:291–320.

OWEN, R P   The status of conservation in the Trust Territory of
1969   the Pacific Islands. *Micronesica*, 5(2).

PEARSON, R G and ENDEAN, R   A preliminary study of the coral
1969   predator *Acanthaster planci* (L.) (*Asteroidea*) on the Great Barrier Reef. *Fish. Notes Dept. Harbours Mar. Qd*, 3(1):27–68.

PINCHOT, G B   Whale culture—a proposal. *Perspect. biol. Med.*,
1966   10:33–43

PINCHOT, G B   Marine farming. *Scient. Am.*, 223(6):15–21.
1970

POWELL, R   Conservation problems in the Truk Lagoon. (Unpubl.
1970   MS):7 p.

RAINFORD, E H   Destruction of the Whitsunday Group fringing
1925   reefs. *Aust. Mus. Mag.*, 2:175–7.

RAMAS, G C   Effects of blast fishing. *Underwat. Natst*, 6(2):31–3.
1969

RANDALL, J E   A review of ciguatera, tropical fish poisoning, with
1958   a tentative explanation of its cause. *Bull. mar. Sci. Gulf Caribb.*, 8(3):236–67.

RANDALL, J E   Overgrazing of algae by herbivorous marine fishes.
1961   *Ecology*, 42(4):812.

RANDALL, J E   An analysis of the fish populations of artificial and
1963   natural reefs in the Virgin Islands. *Caribb. J. Sci.*, 3:31–48.

RAY, C   Marine parks for Tanzania. Washington, D.C., Conserva-
1968   tion Foundation, 47 p.

READ, K R H   Thermal tolerance of the bivalve mollusc *Lima*
1967   *scabra* Born, in relation to environmental temperature. *Proc. malac. Soc. Lond.*, 37:233–41.

RHOADS, D C and YOUNG, D K   The influence of deposit-feeding
1970   organisms on sediment stability and community trophic structure. *J. mar. Res.*, 28:150–78.

ROESSLER, M S and ZIEMAN, J C, JR   The effects of thermal
1969   additions on the biota of southern Biscayne Bay, Florida. *Proc. Gulf Caribb. Fish. Inst.*, 22:136–45.

ROY, K J and SMITH, S V   Sedimentation and coral reef develop-
1971   ment in turbid water: Fanning Lagoon. *Pacif. Sci.*, 25:234–48.

RÜTZLER, K and STERRER, W   Oil pollution damage observed in
1970   tropical communities along the Atlantic seaboard of Panama. *BioScience*, 20:222–4.

SANDERS, H L   Marine benthic diversity: a comparative study.
1968   *Am. Natst*, 102:243–82.

SIDHU, G S, *et al.*   Nature and effects of a kerosene-like taint in
1970   mullet (*Mugil cephalus*). Paper presented to the FAO Technical Conference on Marine Pollution and its Effects on Living Resources and Fishing, Rome, Italy, 9–18 December 1970, FIR:MP/70/E-39:9 p.

SIREN, N and SCHEURING, D L   Murky waters of Micronesia.
1970   *Publs Trust Territ. Pacif. Islands Hlth Serv.*, 45 p.

SLACK-SMITH, R J   An investigation of coral deaths at Peel Island,
1960   Moreton Bay, in early 1956. *Pap. Dep. Zool. Univ. Qd*, (1):211–22.

SPOONER, M   Oil spill in Tarut Bay, Saudi Arabia. *Mar. Pollut.*,
1969   *Bull.*, (1):166–7.

STEPHENSON, W, ENDEAN, R and BENNETT, I   An ecological survey
1958   of the marine fauna of Low Isles, Queensland. *Aust. J. mar. freshwat. Res.*, 9:261–318.

STODDART, D R   Post-hurricane changes on the British Honduras
1969   reefs and cays; re-survey of 1965. *Atoll Res. Bull.*, 131:25 p.

STORR, J F   Ecology and oceanography of the coral reef tract,
1964   Abaco Island, Bahamas. *Spec. Publs geol. Soc. Am.*, (79).

STORRS, P N, *et al.*   Estuarine water quality and biologic population
1969   indices. *Adv. Wat. Pollut. Res.*, 4:901–10.

SVERDLOFF, S N   The status of marine conservation in American
1971   Samoa. South Pacific Commission Regional Symposium on Conservation of Nature—Reef and Lagoons, SPC/RSCN/-wP.4:4 p.

TEMPLETON, W L, NAKATANI, R E and HELD, E   Effects of
1970   radiation in the marine ecosystem. Paper presented to FAO Technical Conference on Marine Pollution and its Effects on Living Resources and Fishing, Rome, Italy, 9–18 December 1970, FIR:MP/70/R-10:19 p.

THOMAS, L P   Another *Acanthaster* disaster. *Nature, Lond.*,
1970   225:1269–70.

VAN EEPOEL, R P and GRIGG, D I   Survey of the ecology and water
1970   quality of Lindberg Bay, St. Thomas. *Wat. Pollut. Rep. Caribb. Res. Inst.*, Jan. issue:6 p.

VAUGHN, T W   Corals and the formation of coral reefs. *Rep.*
1919   *Smithsonian Instn.*, (17):189–238.

VERWEY, J   Coral reef studies: geomorphological notes—the coral
1931   reefs of Batavia Bay. *Treubia*, 13:199–215.

VERWEY, J   Coral reef studies. The depth of coral reefs in relation to
1931a   their oxygen consumption and the penetration of light in the water. *Treubia*, 13:169–98.

WATT, K E F   Comments on fluctuations of animal populations
1964   and measures of community stability. *Can. Entomolst*, 96:1434–42.

WEBER, J N   Disaster at Green Island—other Pacific islands may
1969   share its fate. *Earth Mineral Sci.*, 38(5):37–44.

WELANDER, A D   Distribution of radionuclides in the environment
1969   of Eniwetok and Bikini Atolls, August 1964. *Proc. Conf. Radioecol.*, 2(AEC CONF–670503):474–82.

WELLS, J W   Study of the reef corals of the Tortugas. *Ybk*
1932   *Carnegie Instn, Wash.*, (31):290–1.

WELLS, J W   The coral reefs of Arno Atoll, Marshall Islands. *Atoll*
1952   *Res. Bull.*, (9):14 p.

WELLS, J W   Coral reefs. *Mem. geol. Soc. Am.*, 67(1):609–31.
1957

WHITE, L, JR.   The historical roots of our ecological crisis,
1969   *Science, N.Y.*, 155:1203–7.

WIENS, H J   Atoll environment and ecology. New Haven Yale
1962   University Press, 532 p.

WOOD-JONES, F   On the growth-forms and supposed species in
1907   corals. *Proc. zool. Soc. Lond.*, 1907:518–56.

WOODWELL, G M   Effects of pollution on the structure and
1970   physiology of ecosystems. *Science, N.Y.*, 168:429–33.

WYCHERLEY, P R   Conservation of coral reefs in West Malaysia.
1969   *Biol. Conserv.*, 1:259–60.

YAMAUCHI, H   An economic evaluation of the values of a sub-
   tropical bay from the standpoint of water quality management. *Adv. Wat. Pollut. Res.*, 5:12 p. (in press).

YONGE, C M   The biology of reef-building corals. *Scient. Rep. Gt*
1940   *Barrier Reef Exped.*, (1):353–91.

YONGE, C M   Introduction to the Great Barrier Reef. *Aust nat.*
1966   *Hist.*, 15:233–6.

YONGE, C M and NICHOLLS, A G   Studies on the physiology of
1931   corals. 4. The structure, distribution and physiology of the zooxanthellae. *Scient. Rep. Gt Barrier Reef Exped.*, (1):135–76.

YOUNG, R H F and CHAN, P L   Oahu wastewater treatment plant
1970   efficiency. *J. Wat. Pollut. Control*, 42:2052–9.

ZEITOUN, M A, MANDELLI, E F and McILHENNY, W F Disposal of
1969   the effluents from desalination plants into estuarine waters.
        Rep. Off. Saline Wat. Res. Develop., (415):140 p.

ZEITOUN, M A, *et al.*   Disposal of effluents from desalination plants:
1969   the effects of copper content, heat and salinity. *Rep. Off.
        Saline Wat. Res. Develop.*, (437):192 p.

ZIEMAN, J C, JR.   The effects of a thermal effluent stress on the sea-
1970   grasses and macro-algae in the vicinity of Turkey Point,
        Biscayne Bay, Florida. Thesis, University of Miami.
ANON.   Fisheries in the Caribbean. Caribbean Commission Central
1952   Secretariat, 170 p.
ANON.   Report on thermal pollution of intrastate waters Biscayne
1970   Bay, Florida. Fort Lauderdale, Florida, Southeast Water
        Laboratory, 44 p.

Apercu sur l'Influence des Pollutions sur les Peuplements Benthiques

J.-M. Pérès et G. Bellan*

A review of the influence of pollution on benthic populations

A general summary of the effects of different types of pollution on
the benthic populations has been made from the data contained in
the literature and through observations. In general, previous authors
placed more emphasis on the action of the pollutants on the different
species than on the population itself (biocoenoses and communities).

Four main types of pollution are envisaged:
1. Chemically dissolved pollution of urban and industrial origin.
   On both soft substrates and solid substrates one can delimit
   the zones where the composition of populations, rather similar
   all over the world, includes a certain pollution level.
   Quantitative and qualitative aspects are considered. The action
   of pesticides and herbicides, studied *in vitro*, is known only for
   some isolated species.
2. Thermal pollution. Because it acts in a particular manner on
   each species, no general plan can be given. The macrophyte
   populations seem to be more sensitive than the animal
   populations.
3. Pollution by solid residues. Generally devoid of so-called
   toxicity, these muds influence the populations by hyper-
   sedimentation.
4. Petroleum pollution becomes apparent in instances of disaster.
   Fuels appear to be more dangerous to macrobenthic populations than crude oil. They have a lasting effect on the equilibrium of the populations. The consequences of pollution by
   lighter products, which are difficult to discriminate from those
   of group 1, are practically unknown, at least at the population
   level, if not by particular species. Accidental pollution due to
   organochloride and organophosphoric compounds may be
   very serious if massive.

Revista de la influencia de la contaminación en poblaciones bentónicas

Una revista general somera de la acción de los diversos tipos de
contaminación sobre las poblaciones bentónicas ha sido llevado a
cabo, tomando datos de publicaciones y por medio de observaciones
propias. Generalmente, los autores han considerado más atenta-
mente la acción de los contaminantes sobre las diversas especie que
sobre las poblaciones propiamente dichas (biocoenoses y
comunidades).

Cuatro grandes tipos de contaminación han sido considerados:
1. Contaminaciones químicas disueltas de origen urbano e
   industrial. Tanto en subestratos blandos como sólidos se
   pueden delimitar las zonas en las que la composición de las
   poblaciones, bastante parecidas en todo el mundo, incluyen
   un cierto nivel de contaminación. Los aspectos cualitativos y
   cuantitativos están considerados. La acción de los pesticidas
   y herbicidas estudiados "in vitro" se conoce solamente en
   especies aisladas.
2. Contaminación térmica. Actúa de diversas maneras en las
   distintas especies por lo que no se puede proporcionar un
   esquema general. Las poblaciones de macrófitos parecen ser
   más sensibles que las de animales.
3. Contaminación por restos sólidos. Generalmente desprovisto
   de toxicidad propia, el lodo resultante actúa sobre las
   poblaciones por medio de la hipersedimentación.
4. Contaminación petrolífera. Se la conoce principalmente cuando
   es accidental y en grandes cantidades; los combustibles parecen
   ser más peligrosos para las poblaciones macrobentónicas que
   el petróleo bruto; los primeros alteran durante más tiempo el
   equilibrio de las poblaciones. Las consecuencias de la
   contaminación de grupo 1, son prácticamente desconocidas,
   al menos a nivel de poblaciones, ya que lo que se sabe es sobre
   especies en particular. La contaminación accidental de
   organocloros y organofósforos puede ser grave si es masiva.

DEPUIS plusieurs années d'assez nombreux
chercheurs de divers pays (Etats-Unis, France,
Grande-Bretagne, Italie, pays scandinaves, essentiellement) ont été amenés à considérer que l'étude de la
distribution de diverses espèces animales et végétales
benthiques peut permettre de caractériser des niveaux
différents de pollution en milieu marin (estuaires
compris).

L'étude des altérations des communautés ou, d'une
façon plus générale, des unités de peuplement, sous
l'influence des agents polluants peut être utilisée dans le
même but, les différences d'assemblage des espèces au
sein de celles-ci permettant une meilleure approche du
problème, car il apparaît que chaque assemblage
particulier possède une signification du point de vue du
niveau général de la pollution.

Les unités de peuplement traduisent non seulement les
facteurs ambiants climatiques et édaphiques agissant au
moment du prélèvement, mais représentent également une
véritable intégration des conditions ayant prévalu
pendant les semaines ou les mois précédents, compte tenu

de la longévité moyenne des espèces constitutives et des
conditions de leur multiplication. Encore que ce soit là
une appréciation relativement subjective, il ne semble pas
que, sauf exceptions dont certaines seront évoquées plus
loin, les facteurs biotiques, c'est-à-dire les actions
mutuelles des espèces interviennent de façon majeure
(notamment sur les substrats meubles) dans la composition des peuplements de zones polluées. La connaissance
de peuplements liés à la pollution, la possibilité de les
délimiter, l'utilisation d'espèces indicatrices devraient
permettre, non seulement de juger du niveau de pollution
en une station donnée, mais aussi des probabilités de voir
s'installer en un lieu donné tel ou tel type de peuplement,
autrement dit d'apprécier l'évolution probable des
peuplements benthiques d'une aire donnée, en fonction
de l'accroissement prévisible de la pollution liée à
l'expansion urbaine ou industrielle, et compte tenu de
l'action des autres facteurs ambiants.

L'étude du benthos et notamment celle, plus avancée
actuellement, du macrobenthos, est donc de nature à
révéler certains effets cumulatifs de la pollution sur le bios

---

* Station Marine d'Endoume, 13 Marseille, France.

marin, notamment en raison de l'intégration, par des populations d'êtres fixés ou sédentaires, des effets à long terme d'une ambiance défavorable à la vie benthique en général. Les changements d'ordre qualitatif et la prédominance d'une ou d'un petit nombre d'espèces dans des unités de peuplement données sont liés au fait que l'ensemble des facteurs abiotiques n'est plus toléré que par un nombre de plus en plus restreint d'espèces au fur et à mesure que croît la pollution. Ceci permet, parfois, l'accroissement des populations de celles qui ont réussi à subsister et qui supplantent les espèces moins résistantes qui cohabitaient avec elles, ou même laissent la place à des espèces nouvelles venues. Ces espèces à résistance maximale demeurent éventuellement seules, et finissent à leur tour par disparaître lorsque la pollution atteint un niveau tel que la vie du macrobenthos n'est plus possible. Les passages d'un aspect d'un peuplement à un autre, dans des conditions de pollution croissante, se manifestent le plus souvent, tant sur le plan temporel que sur le plan spatial, comme de véritables discontinuités.

L'action de la pollution globale a été surtout étudiée: sur les peuplements de substrat meuble; sur substrats solides dans les zones portuaires. En ce qui concerne ces dernières la plupart des auteurs ne font pas la distinction entre les peuplements de salissures, lesquels peuvent apparaître sur des substrats solides artificiels en eau pure, et les peuplements réellement baignés d'eau polluée. Ceci n'a rien d'étonnant, les recherches sur les salissures ayant été le plus souvent faites en milieu portuaire.

La pollution globale comprend trois aspects essentiels:

(1) La pollution industrielle et domestique (à dominance de rejets organiques ou chimiques dissous); c'est la plus courante, la mieux connue et, apparemment, la plus grave;
(2) La pollution thermique (souvent associée à la précédente) mais qui peut être envisagée isolément pour son action propre;
(3) La pollution par produits solides, apparemment non toxiques en eux-mêmes mais susceptibles d'altérer les peuplements en modifiant les conditions de sédimentation.

Dans les pages qui suivent, les peuplements pris en considération sont essentiellement ceux qui sont constamment immergés et qui, en fait, appartiennent surtout à l'étage infralittoral (Pérès et Picard, 1964), l'étage circalittoral sous-jacent étant relativement peu intéressé par les pollutions sauf cas particuliers: plateau continental fortement incliné ou déversement des polluants à une profondeur notable. Il est assez étonnant de constater que les peuplements des horizons superficiels (étages médiolittoral et infralittoral) sur les côtes intéressées par des marées d'amplitude supérieure à 1 m–1,50 m en vive-eau n'ont pratiquement fait l'objet d'aucune investigation, à l'exception des excellents travaux de Fischer-Piette (1960) et S. Lefévère (1965).

Pollution industrielle et domestique
Substrats meubles
Que la pollution soit domestique, industrielle, chimique (notamment produits organiques), et généralement plus ou moins associée à l'action de produits pétroliers légers,

on observe typiquement à partir du point de rejet et en s'éloignant de celui-ci une succession de zones qui ont été définies notamment par Bellan (1967):

(A) Une zone "morte" où le macrobenthos (et le plus souvent le microbenthos) a disparu; c'est la zone de pollution maximale.

(B) Une zone plus ou moins complexe, dite "polluée" par Bellan, dont le peuplement animal est très *appauvri qualitativement*. On y trouve à peu près toujours *Capitella capitata* et il peut s'y ajouter quelques autres espèces: soit à distribution géographique particulière (*Scolelepis fuliginosa* dans les eaux de l'Europe occidentale); soit relativement tolérantes à la pollution, quoique à un moindre degré que *C. capitata*: *Nereis caudata*, *Staurocephalus rudolphii*, *Audouinia tentaculata*. Ces espèces sont les seules bien représentées quantitativement. On observe parfois des aires d'"explosion" numérique (plusieurs dizaines de milliers d'individus au m²) mais il y a aussi des aires où la population est très dispersée. Il ne semble pas que cet ensemble d'espèces puisse être considéré comme constituant une biocoenose individualisée, ou une communauté (Bellan, 1967). Leur coexistence provient simplement de ce qu'elles ont des marges de tolérance comparables et rien ne permet, dans l'état actuel des connaissances, d'avancer l'existence d'interactions biotiques. Ces espèces se retrouvent plus ou moins isolées dans des biocoenoses situées hors de la zone polluée, y compris en eau pure, notamment lorsque l'évolution naturelle d'un facteur édaphique reproduit des conditions comparables à celles d'un début de pollution, par exemple lors de la dégradation de quantités importantes de matériel organique végétal. Ainsi la chute saisonnière des feuilles de certaines zostéracées, ou de portions de thalles d'algues, peut amener une pullulation locale et temporaire de certaines des polychètes mentionnées plus haut. Ces espèces ne sont donc pas des indicatrices absolues d'une pollution d'origine humaine, mais leur présence est liée à une teneur élevée en matériel organique déposé, quelle qu'en soit l'origine; l'influence sur elles d'une pollution chimique dissoute ne provenant pas de la dégradation de matériel organique reste à étudier.

(C) Une zone appelée "subnormale" par Bellan, dans laquelle on rencontre, en s'éloignant de l'origine des polluants, un peuplement beaucoup plus riche en espèces; la composition de ce peuplement varie en fonction de la situation biogéographique et en fonction des conditions édaphiques locales (nature du sédiment, hydrodynamisme, profondeur, etc.). Les peuplements benthiques de cette zone subnormale présentent des analogies évidentes avec ceux que l'on trouverait dans la station concernée si la pollution ne s'y faisait pas sentir. Toutefois, ces peuplements se caractérisent, par rapport à la biocoenose que l'on pourrait considérer comme normale, c'est-à-dire indemne de toute pollution, par les particularités suivantes:

(a) Absence quasi-totale des espèces caractéristiques exclusives de la biocoenose normale, espèces qui sont apparemment les plus sensibles à la modification de la qualité de l'environnement. Par exemple, dans la région marseillaise, *Leiocapitella dollfusi* et *Tellina serrata* disparaissent de la biocoenose du détritique envasé (Picard, 1965) lorsque celle-ci empiète sur la zone subnormale; de même *Ova canalifera* et *Amphiura chiajei* n'ont pas été retrouvées par Ghirardelli *et al.* (1968) dans la zoocénose de *Schizaster chiajei* étudiée antérieurement par Vatova (1949).

(b) Prolifération des espèces les moins sensibles à l'altération de l'ambiance ou caractéristiques de tel ou tel facteur particulier du milieu; il s'agit le plus souvent, d'espèces à larges potentialités écologiques, par exemple, en Méditerranée nord-occidentale, d'espèces classées par Picard (1965) parmi les minuticoles ou les vasicoles tolérantes: *Nematonereis unicornis, Lumbriconereis latreillei, Heteromastus filiformis, Corbula gibba, Thyasira flexuosa*, etc. (Bellan, 1965; Leppäkoski, 1968; Ghirardelli *et al.*, 1968; Bagge, 1969; etc.).

McNulty (1961, 1970) décrit des communautés équivalentes en Baie de Biscayne. De plus, il faut souligner que les espèces citées ci-dessus (par. 2) sont en principe absentes dans les fonds de la zone subnormale.

Le nombre d'espèces (n) est, en général, important; le nombre d'individus (N) peut l'être également, mais il y a typiquement un rapport n/N $\left(\text{ou } \dfrac{n-1}{\text{Log } N}\right)$ nettement différent de celui qu'on trouve dans la zone polluée. En règle générale, le rapport n/N est beaucoup plus élevé que dans la zone polluée. A titre d'exemple, nous donnerons quelques résultats calculés à partir des données de deux auteurs (Bellan, 1967, 1967a, 1968; Leppäkoski, 1968).

Les données de Bellan sont des moyennes établies à partir des prélèvements de 1/10 m² dans des biotopes variés:

Débouché de l'émissaire urbain de Marseille

Zone polluée	Zone subnormale	Zone d'eau pure
0,0045	0,0607	0,0576

Ensemble portuaire marseillais

Zone polluée	Zone subnormale superficielle	Zone subnormale profonde
0,0972	0,1129	0,2431

Les données de Leppäkoski (établies à partir du nombre d'individus par m²) nous montrent l'évolution du rapport n/N à partir du point d'origine de la pollution dans le sens des numéros croissants des stations (données extraites de la figure 4 de l'auteur):

Stations:	6	9	10	11	13	15	17	19	20
n/N:	0	0	0	0,00077	0,01111	0,01087	0,02240	0,02667	0,02895

exemple, dans les estuaires et d'une façon générale, dans les zones à salinité abaissée (Tulkki, 1960), la zone **B** n'est pas représentée; de même il y a parfois télescopage entre les zones A et C, la zone B étant complètement écrasée, peut-être pour des raisons d'hydrodynamisme empêchant une sédimentation suffisante (Filice, 1959). Une autre anomalie par rapport au schéma classique résulte du fait que beaucoup de travaux (Reish, 1959 par exemple) ont été exécutés dans des aires portuaires où la zone normale d'eau pure (D) ne saurait exister. Le Tableau 1 résume l'ensemble de ces diverses modalités des peuplements sur substrat meuble.

Substrats solides

Les études de l'action des pollutions dissoutes industrielles et domestiques sur les substrats solides laissent, en général, beaucoup à désirer, et ceci pour plusieurs raisons. Tout d'abord elles ont été faites essentiellement dans des zones portuaires, où seuls les peuplements d'eau calme sont représentés et où l'impact humain peut prendre d'autres formes que celles des polluants chimiques (décapages de surfaces, par exemple); le plus souvent, ces études ont été faites en liaison avec les recherches sur les salissures, dont les espèces, justement, sont favorisées par le décapage. L'utilisation de plaques artificielles immergées pour juger de la pollution, méthode proposée par Relini (1969) est un exemple de cette fâcheuse interférence entre les études de deux problèmes distincts, celui de la pollution et celui des salissures. De plus, très peu de travaux ont été exécutés en mer ouverte à l'exception de celui de Bellan-Santini (1969) et de celui de Golubic (1968) lequel est malheureusement limité aux phytocoenoses. Or, il apparaît indispensable que toutes les études de ce genre portent sur un examen approfondi des peuplements suivant les deux gradients, de pollution d'une part et d'agitation des eaux d'autre part.

Caractères généraux

Sur les substrats rocheux il semble que le tableau général des zones à partir de la source de pollution soit le suivant:

(a) Une zone dépourvue de macrobenthontes tant végétaux qu'animaux;

(b) Une zone de peuplements oligospécifiques dans laquelle domine toujours un mytilidé (en Europe occidentale, *Mytilus edulis* ou *M. galloprovincialis* suivant les régions biogéographiques) accompagné d'un cortège

(D) Au-delà de la zone subnormale on trouve des peuplements qu'on peut considérer comme d'eaux pures, compte tenu des conditions ambiantes, l'influence de la source de pollution pouvant être alors considérée comme négligeable.

L'existence de quatre zones distinctes, de degré de pollution décroissant, telles qu'elles viennent d'être décrites, correspond à une disposition qu'on peut considérer comme classique. Une telle disposition n'a pas souvent été décrite; parfois une zone peut manquer et c'est alors généralement la zone polluée (zone B). Par

d'espèces parmi lesquelles les plus constantes et les plus caractéristiques paraissent être: des polychètes du g. *Polydora* (*P. ligni* en Californie, *P. flava* et *P.* cf. *ciliata* sur les côtes d'Europe occidentale); des amphipodes (*Corophium, Elasmopus, Jassa*). Ces amphipodes sont d'ailleurs beaucoup plus abondants dans les modes calmes (Reish, 1964; 1964a), où, d'autre part, les chlorophycées du g. *Ulva* deviennent généralement assez abondantes;

(c) Une zone d'enrichissement graduel en espèces.

Parmi les algues, qui apparaissent comme les organismes les plus sensibles à la pollution chimique,

	BLEGVAD 1932	TULKKI 1960	FILICE 1959	REISH 1959	BELLAN 1967	BAGGE 1969
	Copenhague	*Finlande*	*San Francisco*	*Long Beach–Los Angeles*	*Ports et environs de Marseille*	*Saltkällefjord*
Zone interne azoïque	"dead area"	*présente*	*présente*	"very polluted bottom"	"zone de pollution maximum"	*présente*
Zone moyenne soumise à la pollution	Zone à *Scoloplos armiger* avec *Nereis, Mya, Macoma,* enrichie par Gastéropodes et autres espèces (peut manquer)	*Nereis Macoma Chironomidae* puis apport de *Tubifex Mya Corophium Mesidothea*	"marginal zone" à *Capitella, Neanthes succinea, Streblospio benedicti,* *Mya arenaria Macoma inconspicua*	"polluted bottom" à *Capitella capitata* dominant "semi-healthy bottom 11" à *Cirriformia luxuriosa* dominant, *Capitella* présente "semi-healthy bottom 1" à *Dorvillea articulata* dominant, *C. capitata* présente "healthy bottom" (enrichissement et changement de la faune) Polychètes nombreuses (*Tharyx parvus, Cossura candida, Nereis procera* dominantes)	Zone polluée à *Capitella capitata, Scolelepis fuliginosa* s'enrichissant avec *Nereis caudata* puis *Audouinia tentaculata* et (ou) *Stautocephalus rudolphii* "zone subnormale" disparition des espèces précédentes; apparition de nombreuses espèces de Polychètes et Mollusques à large potentialité écologique dans les biotopes de substrat meuble (sédiment non grossier)	*Capitella fuliginosa* community *Amphiura filiformis— A. chiajei* community appauvrie
				Une de ces zones peut manquer		
Zone externe des peuplements en eau pure	peuplements classiques du Sound	enrichissement en *Harmothoe, Cardium*, etc.	"healthy zone"	non reconnue avec certitude	Peuplements classiques en Méditerranée nord-occidentale et en fonction des conditions de milieu (nature du sédiment, hydrodynamisme, profondeur, etc.)	*Maldane filiformis* community et *Melesina tenu tenuis* community

l'enrichissement, lorsqu'on va vers les eaux plus pures, porte essentiellement sur les phéophycées et rhodophycées tandis que décroît l'importance relative des chlorophycées. Ghirardelli *et al.* (1968) ont bien mis en évidence ce phénomène dans la baie de Trieste: de l'intérieur (eau polluée) vers l'extérieur (eau pure) le pourcentage de phéophycées présentes dans les stations étudiées passe de 0 à 21,4, celui des rhodophycées de 0 à 51,9, celui des chlorophycées de 75 à 19, celui des cyanophycées de 25 à 7; les espèces d'algues apparaissant graduellement au fur et à mesure que la pollution diminue sont très différentes selon les régions et ce n'est qu'à titre d'exemples que nous citerons les cystoseires pour la Méditerranée, et les fucacées pour les côtes atlantiques de l'Europe.

Les espèces animales se multiplient de même, mais il apparaît malaisé de mettre en évidence des variations similaires au niveau des grands groupes systématiques. Ceux-ci sont toujours représentés et c'est plutôt l'apparition de nouvelles espèces qu'on peut mettre en corrélation avec la diminution de la pollution.

Nous donnerons, à titre d'exemple de ce qui précède, d'une part, le nombre d'espèces animales présentes dans quatre peuplements (deux présents en eau pure et deux—homologues—en eau polluée) et les variations (en pourcentage) des groupes systématiques les plus importants (données extraites de Bellan-Santini, 1969).

Nombre d'espèces (pour 10 prélèvements de 1/25 m²)	*Mytilus* en eau pure 98	*Mytilus* en eau polluée 76	Corallines en eau pure 127	Corallines en eau polluée 108
Composition des peuplements (%)				
Polychètes	30,49	32,83	36,87	33,05
Crustacés	29,27	28,35	28,36	26,44
Mollusques	20,73	20,89	17,02	22,31
Cnidaires	6,09	2,98	5,67	6,61
Bryozoaires	2,43	1,49	3,54	1,65
Pycnogonides	3,65	1,49	0,70	1,65
Groupes divers	7,31	11,94	7,80	9,09

Il ne semble pas que la zone à *Capitella* (zone B sur les substrats meubles) ait un équivalent exact, sur les substrats solides, quoique l'existence—peut-être seulement saisonnière—de zones à peuplement apparemment exclusif de cyanophycées soit possible et mérite de nouvelles investigations (Golubic, 1960); il ne serait pas étonnant *a priori*, qu'un peuplement exclusivement végétal fût l'homologue, sur des substrats solides bien éclairés, du peuplement à *Capitella capitata* des vases polluées. Il faut signaler aussi que, dans certains ports (Los Angeles) où la sédimentation est très importante, certaines espèces peuvent se contenter de la pellicule de substrat meuble qui s'établit sur substrat solide du fait de cette hypersédimentation; on voit alors se réinstaller les espèces citées dans la zone polluée des substrats meubles (Crippen *et al.*, 1969). Le plus souvent ces espèces (*Nereis caudata, S. rudolphii, A. tentaculata*) se trouvent également dans le peuplement à *Mytilus*, par exemple.

Milieux portuaires

Dans les aires franchement portuaires, considérées d'un point de vue plus général que le cas particulier évoqué ci-dessus, le peuplement des surfaces solides en milieu pollué est complexe; il paraît comprendre les compartiments suivants:

(a) Un stock d'espèces de substrats solides baignés d'eaux pures particulièrement tolérantes à la pollution (*Mytilus, Corallina, Platynereis*);

(b) Le petit stock d'espèces liées à la pollution et aux substrats durs, citées précédemment en B, augmenté, dans les conditions d'hypersédimentation, des espèces de substrat meuble citées à propos du port de Los Angeles;

(c) Un stock d'espèces des salissures, espèces pionnières que l'on rencontre aussi en eau pure lorsqu'on y immerge des substrats artificiels neufs: *Hydroides elegans, Bougainvillea ramosa, Bugula neritina, B. stolonifera, Caprella aequilibra, Balanus amphitrite, Styela partita,* etc.;

(d) Un stock d'espèces sciaphiles qui peuvent effectuer dans ces eaux de turbidité élevée, une remontée très importante: *Amphitrite rubra, Antedon mediterraneum, Ophiothrix fragilis, Rissoa semistriata, Turris pulchellus, Tritonalia blainvillei*, etc.; des espèces normalement endobiontes (voire perforantes) comme *Cirratulus cirratus, Polydora ciliata*, deviennent épibiontes (Ledoyer, 1968).

Peuplements de mer ouverte

Nous entendons par peuplements de mer ouverte ceux qui sont situés hors des enceintes portuaires mais dans des eaux auxquelles, par exemple, le voisinage d'une concentration urbaine plus ou moins importante confère un certain degré de pollution globale. Dans les stations de ce type, l'influence du mode, négligeable dans les enceintes portuaires toujours abritées, joue un rôle relativement important. Si dans les milieux d'eau pure, l'hydrodynamisme est un facteur important de la détermination des faciès, au voisinage d'une source de pollution son rôle est un peu différent; les mouvements de l'eau, tant sous l'influence des courants locaux que par l'action, au voisinage du rivage, des houles et des vagues, tendent à distribuer les polluants horizontalement et verticalement, dans un volume beaucoup plus vaste, ce qui du même coup dilue les diverses substances; l'intensité globale de la pollution reste la même, mais son intensité à une distance donnée de l'émissaire, et notamment dans les horizons superficiels, est moindre que dans un milieu abrité.

La pollution, dans les milieux de mer ouverte, apparaît comme un facteur édaphique particulier, de caractère artificiel, dont, à partir d'une intensité relativement faible, l'action devient prépondérante par rapport à celle des facteurs climatiques ou édaphiques naturels; il en résulte une certaine uniformité verticale et horizontale des peuplements de substrat dur dans toute la zone intéressée par cette pollution. Cette uniformité relative, concomitante d'ailleurs d'un certain appauvrissement, procède de la conjonction de trois éléments:

(1) La prédominance, comme, par exemple, sur tous les autres, du facteur édaphique pollution provoque la suppression des faciès, (en Méditerranée, disparition du faciès de mode battu et forte insolation à *Cystoseira stricta*; disparition des faciès de mode battu et éclairement atténué à *Plocamium coccineum* et *Petroglossum nicaeense*, etc. (Bellan-Santini, 1969; Gamulin-Brida *et al.*, 1969).

(2) Le niveau global de pollution est suffisant pour que soient éliminées de nombreuses espèces appartenant aux peuplements d'eau pure; ainsi, dans la région de Marseille (Bellan-Santini, 1969), les peuplements à *Mytilus* montrent-ils seulement 76 espèces en eau polluée contre 98 en eau pure et les peuplements à *Corallina* 108 en eau polluée contre 127 en eau pure.

(3) Les espèces citées dans les peuplements oligospécifiques d'eau polluée (zone B) sont rares ou absentes dans les peuplements de substrat dur influencés par la pollution en mer ouverte; la disparition de ces espèces ne peut être obligatoirement imputée à un niveau insuffisant de pollution, car beaucoup d'autres facteurs (et notamment le calme des eaux, la sédimentation, l'abaissement de la teneur en oxygène dissous, les fluctuations extrêmes de la température et salinité, etc.) agissent dans les milieux portuaires où ce peuplement oligospécifique est particulièrement florissant.

Stabilité et instabilité des peuplements benthiques influencés par les sources de pollution

Les recherches sur l'influence des pollutions industrielles et domestiques sur les peuplements sont, pour la plupart, trop récentes ou n'ont été conduites par chaque auteur que pendant une période trop brève, pour qu'on puisse véritablement apprécier l'action à long terme de ces pollutions. Toutefois, dans la région de Marseille, des observations à peu près régulières sont faites depuis 1965 sur les substrats meubles. En ce qui concerne les substrats durs, ceux de mer ouverte ont été suivis assez attentivement depuis 1960; ceux de caractère portuaire (Vieux-Port de Marseille) ont fait l'objet d'observations superficielles de 1940 à 1943, et à partir de 1948 et, de façon méthodique, depuis la fin de 1968.

Provisoirement, il semble possible, à titre d'hypothèse de travail, d'avancer l'existence d'une nette différence de stabilité à long terme en fonction du degré de pollution.

Les peuplements de la zone subnormale sur substrats meubles (zone C) et les peuplements d'eau polluée en mer ouverte sur substrats solides ne paraissent présenter que des variations à long terme insignifiantes et du même ordre que celles qu'on observe dans les peuplements d'eau pure.

En revanche, les peuplements soumis à une pollution intense, ou tout au moins certains groupes d'espèces qui en font partie, paraissent sujets à des altérations brutales lorsque certains facteurs ambiants présentent temporairement une intensité anormale. Ainsi, une période prolongée de calme et d'élévation de la température de l'air et donc de celle des eaux, a provoqué en juillet 1969, au large de l'émissaire urbain de Marseille, la disparition à peu près totale du peuplement de la zone polluée (zone B à *Capitella capitata*, etc.) décrite ci-dessus, peuplement qui n'a retrouvé sa densité initiale qu'environ six mois plus tard. De même, le peuplement portuaire à *Mytilus—Polydora—*amphipodes paraît présenter certaines variations discontinues; celles-ci étaient particulièrement nettes avant l'extension, à partir de 1955–60, de l'emploi des détergents; l'augmentation massive de la teneur des eaux en ces produits a provoqué un net appauvrissement du peuplement qui tend à masquer les mortalités brutales que l'on observait antérieurement, soit pendant des périodes prolongées de grande chaleur, soit au moment où de fortes précipitations orageuses (généralement en automne) entraînaient l'apparition de températures ou de salinités anormales dans les couches superficielles (mortalité massive de *Ciona intestinalis*, de divers bryozoaires, etc.).

Pollution urbaine et eutrophisation

On admet généralement que la pollution peut entraîner une augmentation de la biomasse et de la production, par rapport aux peuplements d'eau pure.

Cette hypothèse, relativement subjective, d'une eutrophisation repose essentiellement sur l'observation, dans certains cas, d'un accroissement spectaculaire du nombre d'individus, ou de la grande taille atteinte par les individus de certaines espèces, notamment sur les substrats durs (par exemple des *Mytilus galloprovincialis* dépassant 15 cm). En fait si, en général, on observe une biomasse élevée dans des peuplements benthiques moyennement pollués, l'eutrophisation est loin d'être démontrée. Nous présentons ci-après à ce sujet quelques remarques:

(a) Il existe des zones dépourvues de vie macroscopique où la biomasse est infime; on peut évidemment arguer que la pollution y est trop intense pour que le macrobenthos s'y développe.

(b) Les zones polluées apparemment riches sont souvent en équilibre instable sur le plan écologique, ce qui peut entraîner des mortalités massives et, corrélativement, un gaspillage de la production potentielle qui était la leur.

(c) Les biocoenoses voisines en eau pure ont souvent des teneurs en matière organique sèche du même ordre que celles d'eau polluée apparemment eutrophique.

C'est ainsi que l'un de nous (G. Bellan, 1970) signale, pour la zone subnormale profonde (15–35 m) d'un port de commerce, des biomasses moyennes (exprimées en poids de matière organique sèche) de 6,560 g/m², alors qu'à des profondeurs semblables dans des zones voisines non polluées on a de 6,390 à 9,830 g/m² (Reys, 1968). Le même auteur signale que l'accroissement de la biomasse limité aux zones polluées et subnormales, au large d'un émissaire urbain ne se manifeste plus guère au-delà de 1 500 m du point de rejet.

(d) Des cas de diminution de la biomasse avec la pollution croissante ont été constatés: ainsi Bellan-Santini (1969) indique, sur les substrats durs du port de Marseille (en poids de matière organique sèche), 929 g/m² à l'entrée du port et 378 g/m² dans les zones internes plus polluées; Ghirardelli *et al.* (1968) indiquent un appauvrissement quantitatif (biomasse exprimée en poids du matériel formolé) de la zoocoenose *Schizaster chiajei*, dans la baie polluée de Muggia (52,40 g/m²) par rapport à cette zoocoenose dans des aires non polluées du golfe de Trieste (262,70 g/m²).

(e) Enfin, il ne faut pas oublier que l'éventuelle augmentation de la biomasse dans des aires polluées peut n'être due qu'à l'accroissement du calcaire organique. Bellan-Santini (1969) s'est longuement étendu sur ce phénomène et cite de multiples exemples parmi lesquels nous ne retiendrons que les suivants:

Rapport $\dfrac{\text{P. calcaire}}{\text{P. sec}}$	eau pure	eau moyennement polluée
Moulières	10,81	15,95
Faciès des corallines	9,81	17,94

De même, une moulière en eau moyennement polluée fournit 10 fois plus de calcaire organique et deux fois moins de matière organique sèche que le peuplement d'eau pure à *Cystoseira stricta* auquel elle se substitue.

Pesticides, herbicides, fongicides

Quoique les produits organiques utilisés en agriculture (pesticides, herbicides, fongicides) ne puissent évidemment être rangés dans la catégorie des polluants d'origine urbaine, le fait qu'ils soient présents dans l'eau de mer sous forme dissoute, comme les détergents, nous a conduits à les mentionner à la fin de ce chapitre. Comme pour les détergents, l'action de ces produits n'est connue qu'à court terme et sur des espèces prises en particulier (Butler, 1966). On ne sait rien de leur action sur les peuplements benthiques et les études de ce genre seront certainement très difficiles, car les teneurs les plus importantes en ces divers produits se rencontrent naturellement dans les estuaires de fleuves ou rivières

draînant de vastes territoires cultivés, mais qui amènent également à la mer—hors l'eau douce qui perturbe par elle-même les peuplements marins—les pollutions chimiques urbaines des territoires traversés. D'après les études *in vitro*, il semble que les herbicides soient moins toxiques que les insecticides proprement dits. D'après les observations faites sur certains peuplements d'eau douce, on peut supposer que les biocides exerceront à long terme une action néfaste graduelle, et sans doute cumulative, sur les peuplements benthiques et la vie maritime en général.

Pollution thermique

La pollution thermique résulte de l'utilisation de l'eau de mer pour le refroidissement de diverses installations industrielles (centrales thermiques ou nucléaires, raffineries de pétrole, etc.); l'élévation de température de l'eau rejetée par rapport à la valeur normale pour l'eau de pompage est en général de 7 à 8°C mais peut dépasser 11°C (North, 1969).

Les conséquences de la pollution thermique sont beaucoup plus difficiles à évaluer que celles de la pollution urbaine précédemment étudiée, et ceci pour trois raisons. Tout d'abord, d'une façon générale, les données dont on dispose dans la littérature sont beaucoup moins nombreuses et, de plus, ne concernent souvent que les métaphytes. Ensuite, bon nombre des observations relatives à la pollution thermique ont été faites dans des estuaires; la différence qui existe dans l'espace, et dans le temps lorsque ces estuaires sont affectés par des courants de marée, entre le point de pompage et le point de rejet, peut entraîner des fluctuations plus ou moins importantes de la salinité au niveau de ce dernier, et ceci d'autant plus que le réchauffement des eaux de l'émissaire en diminue la densité et favorise donc la stratification; de plus, l'eau de l'émissaire peut renfermer des substances dissoutes, en suspension, ou en émulsion qui peuvent agir par elles-mêmes (cas des résidus pétroliers, par exemple). Enfin la plupart des publications s'attachent davantage à la résistance propre de telle ou telle espèce à l'élévation de température qu'aux modifications apportées à la composition des peuplements.

Toutefois, il nous apparaît possible de dégager, principalement d'après la littérature, un certain nombre de caractéristiques générales de l'action de la pollution thermique:

(a) Plus la température de l'eau rejetée est élevée, plus nocif est son effet sur le peuplement benthique.

(b) La sensibilité des espèces à l'eau réchauffée de l'émissaire s'accroît au fur et à mesure que celle-ci se rapproche de la température maximale annuelle correspondant aux conditions naturelles, et *a fortiori*, lorsqu'elle dépasse cette dernière; il en résulte que l'altération de l'ambiance est plus grande en été qu'en hiver; un certain nombre d'espèces présentes en hiver disparaissent donc en été, ce qui diminue l'indice de diversité du peuplement (Warriner *et al.*, 1966; Pearce, 1969; Nauman *et al.*, 1969; etc.). Les espèces d'un peuplement donné qui sont originaires de latitudes élevées résistent mieux tout au long de l'année (même en été) que celles qui ont un caractère plus méridional (Mihursky *et al.*, 1969; Ferguson-Wood *et al.*, 1969).

(c) L'équilibre entre les espèces existant normalement dans le peuplement peut être altéré par leurs différences respectives de tolérance à l'échauffement de l'eau. Ainsi les stades jeunes de *Mytilus edulis* sont relativement peu affectés par celui-ci; alors que l'activité de prédateurs tels que *Asterias forbesi*, *Carcinus maenas* et *Thais lapillus* est fortement diminuée (Pearce, 1969). Dans certains cas on observe une pullulation plus ou moins marquée d'espèces dont l'élévation de température allonge la période de reproduction et améliore les conditions de fixation des larves, par exemple, dans le port de Swansea, *Hydroides incrustans*, *Corophium acherusicum*, *Ciona intestinalis*, *Ascidiella aspersa*, etc. (Naylor, 1965). Enfin l'apport de larves d'espèces exotiques thermophiles par les navires— soit du fait de la présence d'adultes d'espèces sessiles en période de reproduction, fixés sur leur coque, soit au cours d'opérations de ballastage—pourrait amener l'installation de telles espèces au voisinage de l'émissaire d'eaux réchauffées; Naylor (1965) cite, pour le Queen's Dock de Swansea: *Aiptasiomorpha luciae*, *Bugula neritina*, *Mercierella enigmatica*, *Balanus amphitrite*, *Atherina boyeri*, etc.; ces espèces favorisées par l'élévation de température pourraient d'ailleurs supplanter des espèces indigènes d'écologie et éthologie analogues et Naylor avance à ce sujet l'exemple de *Balanus amphitrite* qui aurait remplacé les autres espèces de Balanes. Si l'hypothèse de Naylor paraît recevable pour *M. enigmatica* et *A. boyeri*, ainsi que pour *B. neritina* (encore que cette dernière soit une espèce des salissures), nous pensons que les cas d'introduction d'espèces subtropicales ou tropicales dans les zones de pollution thermique de mers plus froides, doivent être examinés de façon critique, en tenant compte de l'habitat de ces espèces et de leur aire d'expatriation végétative normale. Le cas de *B. amphitrite* prête également à discussion, car on peut supposer que cette espèce, qui est une caractéristique des peuplements des salissures; ne s'est installée que parce que les autres balanes ont été éliminées par la pollution thermique.

Substrats meubles

Sur les substrats meubles on observe typiquement, à partir du point de rejet, la succession des zones suivantes:

(A) Une zone "morte" dépourvue de macrobenthontes, zone qui peut manquer si l'élévation de la température de l'eau, et sans doute aussi le débit, ne sont pas trop importants.

(B) Une zone où le nombre des espèces est fortement réduit (surtout en été), mais où il ne semble pas que les espèces subsistant présentent un accroissement du nombre des individus ou de la biomasse.

(C) Une large zone où se rétablit plus ou moins graduellement un peuplement correspondant à celui qui existe dans les biotopes comparables du voisinage non soumis à la pollution thermique (Cory *et al.*, 1969; Ferguson-Wood *et al.*, 1969; Warinner *et al.*, 1966; etc.).

On doit souligner que la zone polluée (zone B)—définie précédemment à propos des pollutions chimiques d'origine urbaine, à *Capitella*, *Scolelepis*, *Nereis*, etc.— paraît caractéristique de celles-ci et n'apparaît pas dans la zone d'action d'une pollution thermique. Il faut signaler également des cas d'accélération de la croissance chez certaines espèces dans des fonds sableux baignés d'eau réchauffée (*Tellina tenuis* et *T. fabula* au voisinage d'Hunterston (Ecosse) d'après Barnett *et al.*, 1969).

Substrats solides

Sur les substrats solides le schéma général est assez analogue à celui qui vient d'être décrit à propos des substrats meubles, à savoir, à partir du point de rejet: une zone morte, une zone de peuplement relativement pauvre en espèces, une zone de passage graduel aux peuplements sessiles normaux pour l'aire considérée. Toutes les conséquences générales de la pollution thermique énumérées précédemment sont particulièrement marquées dans les peuplements de substrats durs, beaucoup plus souvent étudiés, d'ailleurs, que ceux des fonds meubles.

Toutefois, une remarque nous paraît devoir être faite. Dans la zone d'appauvrissement qualitatif sur substrats solides, la pauvreté quantitative n'est pas de règle, comme elle paraît l'être sur les fonds meubles, tout au moins pour les espèces animales; en effet, si les métaphytes sont toujours chétifs et dispersés (Ferguson-Wood *et al.*, 1969; North, 1969), il semble que, dans certains cas, les espèces animales résistant à la pollution thermique présentent un développement exubérant, notamment en raison de certaines altérations de la composition du peuplement évoquées ci-dessus. A la limite, il semble que puissent exister dans cette zone des peuplements totalement dépourvus d'algues multicellulaires. Il est possible que cette dominance plus ou moins marquée des espèces animales soit en rapport avec le fait que, chez les végétaux, l'accroissement de température augmente davantage la respiration que la fonction chlorophyllienne.

Pollution par rejets solides

Nous ne prendrons en considération ici que les rejets solides dépourvus de toxicité propre. On néglige habituellement les effets de la pollution des fonds par les rejets solides non industriels et, notamment, par ceux qui découlent du rôle de plus en plus important joué par les matières plastiques dans la vie de l'homme moderne. Les emballages de produits alimentaires, d'entretien, de beauté, les vêtements en plastique usagés, etc., rejetés à la mer s'accumulent sur certains fonds très proches du littoral ou sur le rivage même et paraissent pratiquement indestructibles. A titre d'exemple, nous indiquerons que l'un de nous a pu observer au voisinage de Beyrouth des fonds densément couverts de débris de ce genre et qui étaient rigoureusement dépourvus de faune et de flore, en même temps qu'un voile bactérien blanchâtre, pratiquement continu s'y était développé. De plus, il faut souligner que ces déchets plastiques flottants tendent à fixer les hydrocarbures; en raison de leur légèreté ils ont tendance à la fois à étendre horizontalement la pollution chronique par ces hydrocarbures du fait qu'ils peuvent dériver sur de longues distances et à polluer les horizons les plus superficiels (étages supra- et médiolittoral), notamment sur les plages.

La pollution solide d'origine industrielle, elle, se traduit par le rejet d'un matériel pulvérulent généralement mêlé à de l'eau douce. Il est certain que, lorsque le rejet est pratiqué au voisinage même du rivage, donc à faible profondeur, l'accroissement de turbidité qui en résulte peut diminuer fortement la pénétration de la lumière et, de ce fait, altérer la composition des peuplements benthiques: diminution ou disparition de la fraction végétale de ces peuplements, remontée dans l'étage infralittoral d'espèces de l'étage circalittoral, etc.

Les exemples de pollution par du matériel solide finement divisé paraissent peu nombreux; nous n'en avons guère trouvé de mention dans la littérature, et nous ferons état principalement des deux cas dont nous avons une connaissance personnelle. Dans l'un et l'autre, le produit rejeté ne présente aucun caractère de toxicité et l'influence du rejet sur les peuplements découle uniquement de l'hypersédimentation.

Immédiatement à l'est de Marseille, les résidus de fabrication de l'alumine à partir de la bauxite sont rejetés, à 350 m de profondeur sur une pente de l'ordre de 40 pour cent, dans le canyon de Cassis, depuis le printemps de 1967. Ces résidus vulgairement désignés sous le nom de boues rouges, d'une densité voisine de 2, sont constitués d'une pâte alcaline d'oxydes de fer et d'aluminium, et de silicates; le rejet se fait à la cadence de 85 m³/h. Les conséquences du rejet sont suivies régulièrement (Bourcier, 1969). On distingue deux zones en forme de cônes de déjections étirés dans le sens de la pente: l'une axiale, absolument azoïque, d'écoulement continu des boues, où l'on retrouve les tests morts des espèces qui peuplaient antérieurement la vase bathyale et où l'épaisseur de boue paraissait être, vers le milieu de l'année 1969, d'une vingtaine de centimètres; l'autre, périphérique, appelée zone de dépôt par Bourcier, où sédimentent plus lentement les particules qui ont été mises en suspension à la sortie de l'émissaire; le peuplement de cette zone de dépôt a déjà été altéré, puisqu'on y récolte, à côté d'espèces caractéristiques de la biocoenose des vases bathyales (*Abra longicallus*, *Nicomache*, *Golfingia minuta*) diverses espèces vasicoles (*Onuphis lepta*, *Lumbriconereis fragilis*, *Nucula sulcata*), et diverses espèces sans exigences écologiques précisées; les espèces limivores ingèrent sans dommage le sédiment mélangé de boues rouges et diverses polychètes en édifient leur tube. La zone de dépôt paraît se poursuivre jusqu'au-delà de 2 000 m de profondeur.

L'autre exemple que nous pouvons traiter sommairement ici est celui de la mine d'amiante de Canari (Corse) dont l'exploitation a d'ailleurs cessé depuis une quinzaine d'années. D'après J. Picard (communication personnelle) qui avait observé les conditions de pollution lorsque la mine fonctionnait encore, le rejet consistait en minerai impropre au traitement, formant une sorte de bouillie floconneuse d'un gris bleuâtre, ressemblant à de la pâte à carton. L'action de ce dépôt était purement d'hypersédimentation, sans aucune toxicité, comme le prouve le fait que les feuilles vertes des posidonies dépassant d'un épais dépôt de cette boue portaient leurs épibiontes normaux: foraminifères, hydroïdes, bryozoaires. L'importance de la fraction fine avait amené dans l'axe du golfe de St. Florent, l'extinction des peuplements appartenant à la biocoenose du détritique côtier (Picard, 1965) et son remplacement graduel, après une phase d'instabilité caractérisée par l'abondance de *Leda pella*, *Lucina borealis*, *Dosinia lupina*, etc. (Pérès et Picard, 1964), par un peuplement apparenté à la biocoenose de la vase terrigène côtière. Il semble aussi que l'abondance de la fraction fine ait légèrement altéré certains peuplements coralligènes du golfe de St. Florent et diminué le rendement de la pêche à la langouste. L'évolution de la situation n'a pas été étudiée depuis l'arrêt de l'exploitation.

En ce qui concerne les résidus de traitement du minerai de nickel de Nouvelle Calédonie, une enquête faite sur place n'a révélé aucune altération importante des peuplements benthiques, encore peu étudiés.

Les cas des rejets à la mer des sédiments dragués à l'occasion des travaux d'extension de zones portuaires posent, quant à leurs conséquences, des problèmes assez particuliers. Lorsque les sédiments sont enlevés d'aires encore non ou faiblement polluées—comme ce fut le cas peu après la seconde guerre mondiale pour l'extension de la zone portuaire de Port-de-Bouc—Lavera, au nord-ouest de Marseille—et déversés dans la mer, ils sont assez rapidement colonisés par les espèces du peuplement environnant, pour autant que les caractéristiques granulométriques du matériel déversé ne soient pas trop différentes de celles des sédiments en place. Dans ce cas, on ne saurait considérer qu'il y a pollution, d'autant que les rejets cessent avec l'achèvement des travaux. Les conséquences de rejets périodiques de nature comparable mais effectués régulièrement s'étendant sur une très longue période pourraient sans doute être marquées, en raison de la perturbation également périodique du peuplement en cours d'installation dans l'intervalle de deux rejets. Toutefois, la nocivité de ces rejets pour l'ensemble de l'écosystème benthique reste à démontrer. On peut citer dans cet ordre d'idées l'exemple publié tout récemment par Howell *et al.* (1970) relatif au déversement dans les baies de St. Austell et Mevagissey (Manche) et celui d'un cas tout à fait opposé (Lower Bay, près de New York) cité dans une communication personnelle du Dr. Pearce à Bellan.

Pollutions pétrolières accidentelles et autres

Nous traitons à part, intentionnellement, des pollutions pétrolières, généralement de caractère accidentel, ainsi que des autres cas de pollution qu'il nous a paru impossible de faire entrer dans la catégorie des pollutions chimiques dissoutes d'origine urbaine ou industrielle, pollutions qu'on peut considérer comme ayant un caractère chronique.

A vrai dire, les exemples de ces pollutions fortuites sont peu nombreux. Tout le monde connaît le cas de la pollution par un dérivé mercuriel survenu à Minamata (Japon) dont les conséquences pour la santé publique ont été graves mais dont l'action sur les peuplements marins ne paraît pas avoir été envisagée. Un autre exemple qui nous a été communiqué par J. Picard est le naufrage, en 1968, en rade de Tuléar (Madagascar) d'un chaland chargé d'organophosphorés; le peuplement endogé de la basse plage (portion exondable de l'étage infralittoral) paraît avoir été entièrement détruit; tandis que les crustacés et les petits apodes fouisseurs, ainsi que les entéropneustes sortaient du sédiment et étaient trouvés morts à la surface de celui-ci, les polychètes et les mollusques paraissent avoir été tués au sein même de ce sédiment; le dommage ayant eu un caractère très localisé, le repeuplement a dû s'effectuer à partir des aires voisines restées indemnes, mais ce repeuplement n'a pu être surveillé, eu égard aux difficultés qu'aurait présentées l'étude d'un peuplement totalement endogé, en un site peu favorable à une étude suivie.

Lorsqu'on parle de pollution par les hydrocarbures on pense essentiellement aux grands accidents spectaculaires qui ont été assez nombreux ces dernières années, et qui ont fait l'objet d'études parfois très poussées, malheureusement, ici encore, conçues sous l'angle de la

[382]

mortalité puis de la réapparition, des diverses espèces prises en particulier et non sous l'angle des communautés. En fait, il existe également une pollution, de type chronique, par les hydrocarbures légers, mais la connaissance que l'on en a est pratiquement nulle car elle se manifeste essentiellement dans les aires marines situées au voisinage immédiat des grandes implantations urbaines ou industrielles et l'action propre de ces produits légers, notamment par le jeu de possibles concentrations au niveau de certains maillons de la chaîne alimentaire (Blumer, 1969), probable à long terme, se confond avec celle des détritus organiques et des diverses substances chimiques étudiés dans une partie de la présente note; toutefois, au voisinage même de grandes implantations d'industries pétrolières, telles celles du port de Fos-sur-Mer à l'ouest de Marseille, nous avons pu constater nous-mêmes la prédominance de l'élément hydrocarbure sur les autres facteurs de pollution, notamment par l'altération des caractères organoleptiques de certaines espèces de poissons. Nous nous bornerons donc à rappeler quelques-unes des catastrophes pétrolières les plus marquantes et les plus étudiées: celle du *Tampico-Maru* au Mexique dont North *et al.* (1965) ont étudié en détail les conséquences sur la faune et sur la flore; celle du *Witwater* dans la zone atlantique du canal de Panama en 1968 (voir Rützler *et al.*, 1970); celle du *Torrey Canyon* dans la Manche en 1967 (voir O'Sullivan et Richardson, 1967 et J. E. Smith, 1968); enfin, celle du puits sous-marin de Santa Barbara (Californie) en 1969.

De ces diverses catastrophes pétrolières on peut tirer un certain nombre de conclusions générales provisoires que nous résumerons comme suit:

(a) Les déversements massifs de pétrole brut paraissent beaucoup moins dangereux que ceux de produits partiellement raffinés comme les fuels.

(b) Tous les facteurs susceptibles de diviser la masse d'hydrocarbures liquides et en particulier de l'émulsionner accroissent considérablement les dommages; certains de ces facteurs, comme l'agitation de l'eau au voisinage de la ligne de rivage sont difficiles à contrôler; en revanche, l'utilisation des détergents est à proscrire, car elle a pour résultat d'ajouter aux effets nocifs liés à leur pouvoir émulsionnant, ceux qui découlent de leur toxicité propre. Quant aux produits dits dispersants (et non détergents) dont l'emploi a été récemment proposé par divers groupes industriels, et qui sont réputés non toxiques, leur effet bénéfique sur les nappes d'hydrocarbures liquides reste à démontrer; la dispersion des nappes ainsi traitées, si elle est certainement propice à une dégradation plus rapide des hydrocarbures par les micro-organismes marins, peut aussi contribuer à une certaine extension de la pollution sur des surfaces importantes (Blumer, 1969).

(c) Dans la plupart des cas les moyens physiques et mécaniques de lutte paraissent devoir être préférés aux moyens chimiques.

(d) L'action des fuels, aussi bien que celle du pétrole brut traité par les détergents, amène un déséquilibre profond et durable des peuplements; les algues apparemment moins sensibles que les animaux, connaissent une phase de prospérité entre six mois et un an au moins après le rejet, prospérité qui paraît généralement imputable à la lenteur avec laquelle se reconstituent les populations d'invertébrés herbivores.

Discussion

La nature des agents de pollution, la diversité et l'intensité de leurs modalités d'action, le fait que la pollution frappe les peuplements les plus variés des mers froides aux mers tropicales, sur les substrats meubles comme sur les substrats durs (naturels ou artificiels), ne permet guère évidemment de tirer des conclusions générales sur l'influence de ce phénomène sur la composition et la structure des peuplements benthiques. Tout au plus peut-on ébaucher une discussion des aspects les plus généraux du problème de la pollution en ne retenant que les faits les plus marquants:

(a) Dans un biotope baigné d'eaux pures et peuplé par une biocoenose donnée, la pollution commence toujours par éliminer les espèces les plus caractéristiques de la biocoenose, ce qui est normal puisque ce sont celles dont les tolérances sont le plus étroitement ajustées sur une certaine intensité des divers facteurs climatiques et édaphiques.

(b) Lorsque la pollution croît ou lorsqu'elle se prolonge, d'autres espèces sont progressivement éliminées; les dernières formes macrobenthiques qui persistent sont toujours des espèces ayant de larges tolérances à l'action des divers facteurs ambiants et qui, de ce fait, sont plus ou moins ubiquistes; ceci entraîne une monotonie certaine de toute la zone influencée par une pollution d'égale intensité, puisque seules les espèces les plus résistantes, qui sont celles qui étaient communes à des biotopes originels assez divers, ont réussi à se maintenir. Beaucoup de ces espèces étant non seulement ubiquistes mais cosmopolites (souvent à l'exception des mers polaires), il en résulte aussi une relative uniformité des zones polluées dans les diverses régions et provinces biogéographiques; sur le plan des pollutions urbaines et chimiques dissoutes, on peut, dans une certaine mesure, considérer qu'il y a un petit stock paucispécifique qui a une distribution mondiale.

(c) Lorsque le taux de pollution est particulièrement élevé, le macrobenthos disparaît totalement et il n'y a plus que des bactéries.

(d) La diminution du nombre des espèces dans les peuplements benthiques soumis à une pollution est la marque d'un profond déséquilibre qui peut se traduire par des conséquences de trois ordres:

(i) Si les sources de nutrition sont suffisamment abondantes (sels minéraux pour les algues, matériel organique en suspension ou dissous pour les invertébrés microphages), il peut y avoir pullulation de certaines espèces, mais le caractère oligospécifique du peuplement fait qu'on n'observe généralement pas une chaîne alimentaire aboutissant à des espèces commercialisables, de sorte que la production, aux échelons primaire et secondaire est en général inutilisée par les échelons supérieurs.

(ii) Les peuplements benthiques de zones polluées sont exposés, de par leur instabilité, à des mortalités massives, soit lorsque l'intensité de la pollution augmente brusquement, soit lorsque certains facteurs abiotiques prennent temporairement une intensité anormale; il y a donc un gaspillage de biomasse, un gaspillage d'une richesse d'ailleurs plus apparente que réelle.

(iii) Lorsque l'ensemble des facteurs ambiants (pollution comprise) revient à un niveau comparable à celui qui a précédé la phase de destruction totale du peuplement considéré, la reconstitution de celui-ci est toujours lente.

(e) En cas d'arrêt définitif de la source de pollution, il semble que les peuplements benthiques d'eaux pures qui existaient dans les biotopes de l'aire concernée avant le début de la pollution puissent se reconstituer, sauf dans le cas où les caractères granulométriques du sédiment ont été modifiés de façon définitive par un polluant solide.

Bibliographie

La liste qui suit ne prétend pas être exhaustive. Nous nous sommes limités aux travaux répondant le plus strictement à notre but. De nombreuses références n'ont pas été citées dans le texte de ce rapport pour ne point l'alourdir, mais les conclusions auxquelles nous arrivons n'en ont pas moins été inspirées par ces travaux.

ARÍAS, E, *et al.* Ecología del puerto de Barcelona y desarrollo de
1963 adherencias organicas sobre embarcaciones. *Investigación pesq.*, (24):139–63.

ARTHUR, D R *Torrey Canyon*: different kinds of death. *New*
1968 *Scient.*, 37(589):625–7.

BAGGE, P Effects of pollution on estuarine ecosystems. 1. Effects
1969 of effluents from wood processing industries on the hydrography, bottom and fauna of Saltkällefjord (W. Sweden). 2. The succession of the bottom fauna communities in polluted estuarine habitats in the Baltic–Skagerrak region. *Meer. jukl., Suomi*, (228):130 p.

BARNARD, J L, *et al.* Ecology of Amphipoda and Polychaeta at
1959 Newport Bay, California. *Occ. Pap. Allan Hancock Fdn.*, (21):106.

BARNARD, J L, *et al.* Field toxicity test in marine waters utilizing
1960 the polychaetous annelid. *Capitella capitata. Pacif. Natst.*, 1:21–22.

BARNETT, P R O, *et al.* The effects of temperature on the benthos
1969 near the Hunterston Generating Station, Scotland. *Chesapeake Sci.*, 10(3–4):255–6.

BASYE, D E Santa Barbara sparking in wake of clean-up job.
1969 *Oil Gas J.*, 67(34):33–8.

Battelle Memorial Institute, Pacific Northwest Laboratories. Oil
1967 spillage study literature search and critical evaluation for selection of promising techniques to control and prevent damage. Richland, Washington. Battelle Memorial Institute, pag. var.

BELLAMY, D J, *et al.* Effects of pollution from the *Torrey Canyon*
1967 on littoral and sub-littoral ecosystems. *Nature, Lond.*, 216(5121):1170–3.

BELLAMY, D J, *et al.* Problem in the assessment of the effects of
1968 pollution on inshore marine ecosystems dominated by attached macrophytes. *Fld Stud.*, 2:49–54.

BELLAN, G Contribution à l'étude systématique, bionomique et
1964 écologique des annélides polychètes de la Méditerranée. *Recl. Trav. Stn. mar. Endoume*, 49(33):1–372.

BELLAN, G Pollution et peuplements benthiques sur substrat
1967 meuble dans la région de Marseille. Première partie. Le secteur de Cortiou. *Revue int. Océanogr. méd.*, 6–7:53–87.

BELLAN, G Pollution et peuplements benthiques sur substrat
1967a meuble dans la région de Marseille.2ème—partie. L'ensemble portuaire marseillais. *Revue int. Océanogr. méd.*, 8:51–95.

BELLAN, G Contribution à la connaissance des peuplements de
1968 substrat meuble établis dans les zones polluées de la région de Marseille. *Rapp. P.-v. Réun. Commn int. Explor. scient. Mer Méditerr.*, 19(2):91–2.

BELLAN, G, *et al.* Etat général des pollutions sur les côtes méditer-
1969 ranéennes de France. *In* Congresso internazionale. Acqua per il domani. La difesa del Mare, Milan.

BELLAN-SANTINI, D Influence de la pollution sur quelques peuple-
1964 ments superficiels de substrats rocheux. *In* Symposium sur les pollutions marines par microorganismes et produits pétroliers, C.I.E.S.M.M., Monaco, pp. 127–37.

BELLAN-SANTINI, D Influence de la pollution sur les peuplements
1968 benthiques. *Revue int. Océanogr. méd.*, 10:27–53.

BELLAN-SANTINI, D Contribution à l'étude des peuplements infra-
1969 littoraux sur substrat rocheux (Etude qualitative et quantitative de la frange supérieure) *Recl Trav. Stn. mar. Endoume*, 63(47):5–294.

BEYER, F Zooplankton, zoobenthos and bottom sediments as
1968 related to pollution and water exchange in the Oslofjord. *Helgoländer wiss. Meeresunters.*, 17:496–509.

BICK, H Untersuchungen zur Verträglichkeit von Meer und
1968 Brackwasser für Ciliaten des Saprobiensystems der Wassergütebeurteilung. *Helgoländer wiss. Meeresunters.* 17(1–4):257–68.

BLEGVAD, H Investigations of the bottom fauna at outfall of
1932 drains in the Sound. *Rep. Dan. biol. Stn*, (37):1–20.

BLUMER, M Oil pollution of the ocean. *Oceanus*, 15(2):2–7.
1969

BOURCIER, M Ecoulement des "boues rouges" dans le Canyon de
1969 la Cassidaigne (décembre 1968). *Tethys*, 1(3):779–82.

BROWN, R P Marine disposal of solid wastes: an interim summary.
1969 *Mar. Pollut. Bull.*, (18):12.

BUTLER, P A Pesticides in the marine environment. *J. appl. Ecol.*,
1966 3(Suppl.):253–9.

BUTLER, P A Pesticides in the estuary. *In* Proceedings of the Marsh
1968 and Estuary Management Symposium, 1967, Baton Rouge, La, pp. 120–4.

CADWALLADER, L W Rational approach to thermal pollution.
1969 *Pwr Engr.*, 73(11):26–9.

CASPERS, H Der Einfluss der Elbe auf die Verunreinigung der
1968 Nordsee. *Helgoländer wiss. Meeresunters.*, 17(1–4):422–34.

CASTENHOLTZ, R W Stability and stresses in intertidal populations.
1967 *In* Pollution and marine ecology, edited by T.A. Olson and F.J. Burgess, New York, Interscience Publishers, pp. 15–28.

CHAMROUX, S, *et al.* La marée noire sur la côte nord du Finistère.
1967 *Penn Bed*, 6(50)3:99–106.

CHASSE, Cl, HOLMES, M T H et PERROT, Y Esquisse d'un bilan des
1967 pertes biologiques provoquées par le mazout du *Torrey Canyon* sur le littoral du Trégor. *Penn Bed*, 6(50)3:107–12.

COOPER, L H N Scientific consequences of the wreck of the *Torrey*
1968 *Canyon*. *Helgoländer wiss. Meeresunters.*, 17(1–4):340–55.

COPELAND, B J Effects of industrial waste on the marine environ-
1966 ment. *J. Wat. Pollut. Control. Fed.*, 38(6):1000–10.

COPELAND, B J, *et al.* Community metabolism in ecosystems
1964 receiving oil refinery effluents. *Limnol. Oceanogr.*, 9:431–47.

CORY, R L, *et al.* Epifauna and thermal additions in the upper
1969 Patuxent River estuary. *Chesapeake Sci.*, 10(3–4):210–217.

CRIPPEN, R W, *et al.* An ecological study of the polychaetous
1969 annelids associated with fouling material in Los Angeles harbor with special reference to pollution. *Bull. Sth Calif. Acad. Sci.*, 68(3):169–87.

DEAN, D, *et al.* Benthic repopulation of the Raritan river estuary
1964 following pollution abatement. *Limnol. Oceanogr.*, 9:551–63.

FERGUSSON-WOOD, E J, *et al.* The effects of temperature on
1969 estuarine plant communities. *Chesapeake Sci.*, 10(3–4): 172–4.

FILICE, F P Invertebrates from the estuarine portions of San
1958 Francisco Bay and some factors influencing their distributions. *Wasmann J. Biol.*, 16(2):159–211.

FILICE, F P The effect of wastes on the distribution of bottom
1959 invertebrates in the San Francisco Bay estuary. *Wasmann J. Biol.*, 17(1):1–17.

FISCHER-PIETTE, E Le bios intercotidal d'une côte battue devant
1960 une usine métallurgique. *Bull. Inst. océanogr. Monaco*, 1184:1–17.

FLEMER, D A, *et al.* Biological effects of spoil disposal in
1968 Chesapeake Bay. *J. sanit. Engng Div. Am. Soc. civ. Engrs*, 94(SA 4):683–706.

GAMULIN-BRIDA, H, *et al.* Contribution aux études des bio-
1967 coenoses subtidales. *Helgoländer wiss. Meeresunters.*, 15:429–44.

GHIRARDELLI, E, *et al.* Conséquence de la pollution sur les
1968 peuplements du "Vallone de Muggia" près de Trieste. *Revue int. Océanogr. méd.*, 10:111–22.

GILET, R Water pollution in Marseilles and its relation with flora
1960 and fauna. 1. *In* Waste disposal in the marine environment, edited by E. A. Pearson, New York, Pergamon Press, pp. 39–56.

GILMOUR, A J Biology of marine pollution. *Aust. Fish*, 28(6):26–9.
1969

GOLUBIC, S Vegetecija cijanofita u lukama sjevenog Jadrana.
1960 *Thalassia jugosl.*, 2(2):5–36.

GOLUBIC, S Die Verteilung der Algenvegetation in der Umgebung
1968 von Rovinj (Istrien) unter dem Einfluss häuslicher und industrieller Abwässer. *Wass. Abwass. Forsch.*, 3:87–97.

GOLUBIC, S Effect of organic pollution on benthic communities.
1970 *Mar. Pollut. Bull.*, 1(4):56–7.

HADERLIE, E C Influence of pesticide run-off in Monterey Bay.
1970 *Mar. Pollut. Bull.* 1(3):42–3.

HEDGPETH, J W Marine biogeography. *In* Treatise on marine
1957 ecology and paleoecology. *Mem. geol. Soc. Am.*, 1(67): 359–82.

HENRIKSSON, R The bottom fauna in polluted areas of the Sound.
1968 *Oikos*, 19(1):111–25.

HENRIKSSON, R Influence of pollution on the bottom fauna of the
1969 Sound (Öresund), *Oikos*, 20(2):507–24.

HOOD, D W, STEVENSON, B W and JEFFREY, L M Deep sea
1958 disposal of industrial wastes. *Ind. Engng Chem.*, 50(6):885–8.

HOPKINS, S H   Biological and physiological basis of indicator
1967   organisms and communities. *In* Pollution and marine
ecology, edited by T. A. Olson and F. J. Burgess, New York,
Interscience Publishers, 291–2.

HOULT, D P and CRAVEN, J P (Eds.)   Oil on the sea. Proceedings
1969   of a Symposium on the scientific and engineering aspects of
oil on the sea, held in Cambridge, Massachusetts, May 16,
1969, New York Plenum Press, 114 p.

HOWELL, B R, *et al.*   The effect of china clay on the bottom fauna
1970   of St. Austell and Mevagissey Bays. *J. mar. biol. Ass. U.K.*,
50(3):593–608.

ITO, T, *et al.*   Aquatic communities in polluted streams with
1964   industrial and mining wastes. 1. Effect of the papermill
waste on the benthic invertebrates. *Rep. Noto mar. Lab.*,
(4):33–43.

JONES, L G, *et al.*   Just how serious was the Santa Barbara oil
1969   spill? *Ocean Ind.*, 4(6):53–6.

JOUIN, C   Est-il inévitable que l'humanité soit asphyxiée par ses
1967   déchets. *Penn Bed*, 6(50)3:150–2.

JOW, T   Trawling survey of Santa Barbara oil spill. *Cruise Rep.*
1969   *Calif. Fish Game mar. Resour. Opl Fish. Lab.*, (69–4):5.

KINNE, O, *et al.*   Biologische Konsequenzen schwefelsäure- und
1968   eisensulfathaltiger Industrialabwässer. Mortalität junger
Gobius pictus und *Solea solea* (Pisces). *Helgoländer wiss.
Meeresunters.*, 17(1–4):141–55.

KITAMORI, R, *et al.*   The benthic community in polluted coastal
1958   waters. 1. Fukuyama inlet. *Bull. Island Sea Res. biol. Stn*
(11):1–16.

KITAMORI, R, *et al.*   The benthic community in polluted coastal
1959   waters. 2. Nukara Bay. *Bull. Island Sea Res. biol. Stn*,
(12):201–14.

KITAMORI, R, *et al.*   The benthic community in polluted coastal
1959a   waters. 3. Osaka Bay. *Bull. Island Sea Res. biol. Stn*,
(12):215–22.

KITAMORI, R, *et al.*   The benthic community in polluted coastal
1959b   waters. 4. Kanzaki River. *Bull. Island Sea Res. biol. Stn*,
(12):223–6.

KOCH, P   Les modifications de la vie marine au débouché des
1968   émissaires d'eaux résiduaires urbaines. *Revue int. Océanogr.
méd.*, 10:11–25.

KOCH, P   Le rejet en mer des déchets littoraux. *Eau*, 5:209–16.
1968a

KOEMAN, J H, *et al.*   Residues of chlorinated hydrocarbon insecti-
1968   cides in the North Sea environment. *Helgoländer wiss.
Meeresunters.*, 17(1–4):375–80.

LEDOYER, M   Ecologie de la faune vagile des biotopes méditer-
1968   ranéens accessibles en scaphandre autonome (Région de
Marseille principalement). 4. Synthèse de l'étude écologique.
Recl. Trav. Stn mar. Endoume, 60(44):125–295.

LEFEUNE, P H   Pollution of marine waters. Damages to the flora
1958   and fauna of Rio de Janeiro. *Mems Inst. Oswaldo Cruz*,
50(1):39–79.

LEFÉVÈRE, S   Le recouvrement biogène le long de la côte belge.
1965   *Bull. Inst. r. Sci. nat. Belg.*, 41(26):1–10.

LEFÉVÈRE, S, LELOUP, E et VAN MEEL, L   Observations biologiques
1956   dans le port d'Ostende. *Mem. Inst. R. Sci. nat. Belg.*,
(133):1–157.

LEIGHTON, D L, *et al.*   Ecological relationships between giant kelp
1966   and sea urchins in southern California. *Proc. int. Seaweed
Symp.*, 5:143–53.

LELOUP, E, *et al.*   Observations sur la salissure dans le port
1967   d'Ostende. *Bull. Inst. r. Sci. nat. Belg.*, 42(23):14 p.

LEPPAKOSKI, E   Some effects of pollution on the benthic environ-
1968   ment of the Gullmarsfjord. *Helgoländer wiss. Meeresunters.*,
17(1–4):291–301.

LEPPAKOSKI, E   Transitory return of the benthic fauna of the
1969   Bornholm basin after extermination by oxygen insufficiency.
Cah. Biol. mar., 10(2):163–72.

L'HARDY, J P   La pollution des océans par les hydrocarbures et
1967   ses conséquences biologiques. *Penn Bed*, 6(50)3:123–38.

LUCAS, A   Menaces sur un milieu vivant. *Penn Bed*, 6(50)3:77–8.
1967

LÜDEMANN, D   Gewässerverschmutzung durch Aussenbordmoto-
1968   ren und deren Wirkung auf Fauna und Flora. *Helgoländer
wiss. Meeresunters.*, 17(1–4):356–69.

MARION, A F   Esquisse d'une topographie zoologique du golfe de
1883   Marseille. *Ann. Mus. Hist. nat. Marseille*, 1(1)

McKEE, J E   Report on oily substances and their effects on the
1956   beneficial uses of water. *Publs Calif. St. Wat. Pollut. Control
Bd.*, (16):72 p.

McNULTY, J K   Ecological effects of sewage pollution in Biscayne
1961   Bay, Florida: sediments and the distribution of benthic and
fouling macroorganisms. *Bull. mar. Sci. Gulf Carrib.*,
11(3):394–447.

McNULTY, J K   Effects of abatement of domestic sewage pollution
1970   on the benthos. Volumes of zooplankton and the fouling
organisms of Biscayne Bay. *Florida Stud. trop. Oceanogr.*,
(9):107 p.

MICHANECK, G   Quantitative sampling of benthic organisms by
1967   diving on the Swedish west coast. *Helgoländer wiss.
Meeresunters.*, 15:455–9.

MIHURSKY, J A   On possible constructive uses of thermal additions
1967   to estuaries. *Bioscience*, 17(10):698–702.

MIHURSKY, J A, *et al.*   Introduction (Second thermal workshop of
1969   the U.S.I.B.P.) *Chesapeake Sci.*, 10(3–4):125–7.

MILEIKOVSKY, S A   The influence of human activities on breeding
1968   and spawning of littoral marine bottom invertebrates.
Helgoländer wiss. Meeresunters., 17(1–4):200–8.

MIRONOV, O G   Hydrocarbon pollution of the sea and its influence
1968   on marine organisms. *Helgoländer wiss. Meeresunters.*,
17(1–4):335–9.

MOO PING CHOW   The pollution of a harbour in a subtropical area.
1964   *Adv. Wat. Pollut. Res.*, 2(3):65–84.

NAUMAN, J W, *et al.*   Thermal additions and epifaunal organisms
1969   at Chalk Point, Maryland. *Chesapeake Sci.*, 10(3–4):
218–226.

NAYLOR, E   Biological effects of a heated effluent in docks at
1965   Swansea, S. Wales. *Proc. zool. Soc. Lond.*, 144:253–68.

NELSON-SMITH, A   Biological consequences of oil pollution and
1968   shore cleaning. *In* Biological effects of oil pollution on
littoral communities, edited by J. D. Carthy and D. R.
Arthur, London, Field Studies Council, pp. 73–80.

NELSON-SMITH, A   The effects of oil pollution and emulsifier
1968a   cleansing on shore life in south-west Britain. *J. appl. Ecol.*,
5(1):97–107.

NEUSHUL, M   A preliminary study of oil spill damage in the
1969   intertidal regions of Santa Barbara and Ventura Counties,
California (U.S.A.). *Mar. Pollut. Bull.*, (16):12.

NICHOLSON, N   Intertidal flora and fauna. *In* Biological and
1969   oceanographic effects of oil spillages in the Santa Barbara
Channel following the 1969 blowout. *Mar. Pollut. Bull.*,
(13):4.

NORTH, W J   Ecology of the rocky nearshore environment in
1964   Southern California and possible influences of discharged
wastes. *Adv. Wat. Pollut. Res.*, 2:247–74.

NORTH, W J   An investigation of the effects of discharged wastes on
1964a   kelp. *Rep. St. Wat. Qual. Control Bd*, Finland (26):123 p.

NORTH, W J   Biological effects of heated water discharge at Moro
1969   Bay, California. *Proc. int. Seaweed Symp.*, 6:275–86.

NORTH, W J, *et al.*   Successive biological changes observed in a
1965   marine cove exposed to a large spillage of mineral oil.
In C.I.E.S.M.M. Symposium sur les pollutions marines par
microorganismes et produits pétroliers, Monaco, 335–54.

OGLESBY, R and JAMISON, A   Intertidal communities as monitors
1968   of pollution. *J. sanit. Engng Div. Am. Soc. civ. Engnrs*,
94,5A3, pap. No. 542–50.

OLIFF, W D, *et al.*   The ecology and chemistry of sandy beaches
1967   and nearshore submarine sediments of Natal. 1. Pollution
criteria for sandy beaches in Natal. *Wat. Res.*, 1:115–29.

OLIFF, W D, *et al.*   The ecology and chemistry of sandy beaches and
1967   nearshore submarine sediments of Natal. 2. Pollution criteria
for nearshore sediments of the Natal coast. *Wat. Res.*,
1:131–46.

OLSON, T A and BURGESS, F J (Eds.)   Pollution and marine ecology,
1967   New York, Interscience Publishers, 364 p.

O'SULLIVAN, A J and RICHARDSON, A J   The *Torrey Canyon*
1967   disaster and intertidal marine life. *Nature, Lond.*, 214(5087):
448, 541–2.

PEARCE, S B   Thermal addition and the benthos: Cap Cod Canal.
1969   *Chesapeake Sci.*, 10(3–4):227–33.

PÉRÈS, J M   Océanographie biologique et biologie marine. 1. La vie
1961   benthique. Paris, Presses Univ. France, 541 p.

PÉRÈS, J M and PICARD, J   Nouveau manuel de bionomie benthique
1964   de la mer Méditerranée. *Recl. Trav. Stn mar. Endoume*,
47(31):1–137.

PERKINS, E S   The toxicity of oil emulsifiers to some inshore fauna.
1968   *Fld. Stud.* 2:81–90.

PERSOONE, G and DE PAUW, N   Pollution in the harbour of Ostend
1968   (Belgium): biological and hydrographical consequences.
Helgoländer wiss. Meeresunters., 17(1–4):302–20.

PICARD, J   Recherches qualitatives sur les biocoenoses marines des
1965   substrats meubles dragables de la région marseillaise.
Recl Trav. Stn mar. Endoume, 52(36):5–160.

PIGNATTI, S and DE CRISTINI, P   Associazioni di alghe marine come
1967   indicatori di inquinamenti delle acque nel Vallone di
Muggia presso Trieste. *Archo. Oceanogr. Limnol.*, 15 Suppl.:
185–91.

PORTMANN, J E   Progress report on a programme of insecticide
1968   analysis and toxicity-testing in relation to the marine
environment. *Helgoländer wiss. Meeresunters.*, 17(1–4):
247–56.

RANKIN, JR, J S   Immediate gain vs. long loss through marsh
1964   destruction. *IUCN Publs New Ser.*, 3(1):80–6.

REISH, D J   The relations of polychaetous annelids to harbor
1955   pollution. *Publ. Hlth Rep. Wash.*, 70:1168–74.

REISH, D J   An ecological study of lower San Gabriel River,
1956   California, with special reference to pollution. *Calif. Fish Game*, 42(1):51–61.

REISH, D J   An ecological study of pollution in Los Angeles—Long
1959   Beach Harbors, California. *Occ. Pap. Allan Hancock Fdn*, (22):119 p.

REISH, D J   The uses of marine invertebrates as indicators of water
1960   quality. *In* Waste disposal in the marine environment, edited by E. A. Pearson, New York, Pergamon Press, pp. 92–103.

REISH, D J   A study of benthic fauna in a recently constructed boat
1961   harbor in southern California. *Ecology*, 42(1):84–91.

REISH, D J   Further studies on the benthic fauna in a recently con-
1963   structed boat harbor in southern California. *Bull. Soc. Calif. Acad. Sci.* 62(1):23–32.

REISH, D J   Studies on the *Mytilus edulis* community in Alamitos
1964   Bay, California. 1. Development and destruction of the community. *Veliger*, 6(3):124–31.

REISH, D J   Studies on *Mytilus edulis* community in Alamitos Bay,
1964a   California. 2. Population variations and discussion of the associated organisms. *Veliger*, 6(4):202–7.

REISH, D J   Discussion of the *Mytilus californianus* community on
1964b   newly constructed rock jetties in southern California. Veliger, 7(2):95–101.

REISH, D J   The effect of oil refinery wastes on benthic marine
1965   animals in Los Angeles harbor, California. *In* Symposium sur les pollutions marines par micro-organismes et produits pétroliers, C.I.E.S.M.M., Monaco, pp. 355–61.

REISH, D J and WINTER, H A   The ecology of Alamitos Bay,
1954   California, with special reference to pollution. *Calif. Fish Game*, 40(2):105–21.

RELINI, G   La comunità dominante nel "fouling" portuale di
1966   Genova. *Natura, Milano*, 57(2):136–56.

RELINI, G, *et al.*   Possibilité d'étudier les effets de la pollution sur
1970   les organismes benthiques en employant des panneaux immergés. *Revue int. Océanogr. méd.*, 17:189–99.

REYS, J P   Quelques données quantitatives sur les biocoenoses
1968   benthiques du golfe de Marseille. *Rapp. P.v. Reun. Commn int. Explor. Mer Méditerr.*, 19(2):121–3.

RÜTZLER, K, *et al.*   Oil pollution damage observed in tropical
1970   communities along the Atlantic seaboard of Panama. *Bio Science*, 20(4):222–4.

SANDERS, H L   Benthic studies in Buzzard's Bay. 3. The structure
1960   of the soft bottom community. *Limnol. Oceanogr.*, 52:138–53.

SICSIC, M, *et al.*   L'intérêt biologique des milieux portuaires.
1966   *Annls. Soc. Sci. nat. Toulon*, (Var 18): 141–56.

SMITH, J C   The fauna of a polluted shore in the Firth of Forth.
1968   *Helgoländer wiss. Meeresunters.*, 17(1–4):216–23.

SMITH, J E (Ed.)   *Torrey Canyon* pollution and marine life.
1968   Cambridge, Cambridge University Press, 196 p.

SMITH, J W   Problems in dealing with oil pollution on sea and land.
1968   *J. Inst. Petrol.*, 54(539):358–66.

SOUDAN, F   Incidence de la pollution sur la vie marine. *Sci. Pêche*,
1968   169:1–10.

SWEDMARK, B   Pollution along the west coast of Sweden. *Mar.*
1970   *Pollut. Bull.*, 1(1):10–1.

TARZWELL, C M   Water quality criteria for aquatic life. *In* Bio-
1957   logical problems in water pollution, edited by C. M. Tarzwell, U.S. Department of Health, Education and Welfare, Robert A. Taft, Sanitary Engineering Center, pp. 246–72.

TENDRON, G   La pollution des mers par les hydrocarbures et la
1962   contamination de la flore et de la faune marines. *Penn Bed*, 3(29), 2:173–82.

TULKKI, P   Studies on the bottom fauna of the Finnish south-
1960   western archipelago. 1. Bottom fauna of the Airisto Sound. *Soumal. elain-ja kasvit. Seur. van. kasvit. Julk.*, 21(3):1–26.

TULKKI, P   Studies on the bottom fauna of the Finnish south-
1960a   western archipelago. 2. Bottom fauna of the polluted harbour area of Turku. *Soumal. elain-ja kasvit. Seur. Van. kasvit. Julk.*, 18(3):169–88.

TULKKI, P   Disappearance of the benthic fauna from the Basin of
1965   Bornholm (Southern Baltic), due to oxygen deficiency. *Cah. Biol. mar.*, 6:455–63.

TULKKI, P   Effect of pollution on the benthos off Gothenburg.
1968   *Helgoländer wiss. Meeresunters.*, 17(1–4):209–15.

TURNER, C H   Inshore survey of Santa Barbara oil spill. *Cruise*
1969   *Rep. Calif. Fish Game mar. Resour. Opl. Fish. Lab.*, 69A(2):3 p.

VATOVA, A   La fauna bentonica dell'alto e medio Adriatico. *Nova*
1949   *Thalassia*, 1(3).

WALDICHUK, H   Effects of pollutants on marine organisms;
1969   improving methodology of evaluation—a review of the literature. *J. Wat. Pollut. Control Fed.*, 41:1586–601.

WARINNER, J E, *et al.*   The effects of thermal effluents on marine
1966   organisms. Industrial Waste Conference. *Int. J. Air Wat. Pollut.*, 10:277–89.

WASS, M L   Biological and physiological basis of indicator
1967   organisms and communities. 2. Indicators of pollution. *In* Pollution and marine ecology, edited by T. A. Olson and F. J. Burgess, New York, Interscience Publishers, pp. 271–83.

ZOBELL, C E   The occurrence, effects and fate of oil polluting the
1949   sea. *Adv. Wat. Pollut. Res.*, 2(3):85–109.

Monitoring of Creeping Over-Fertilization in the Baltic

I. Haahtela *

Contrôle de la surfertilisation progressive dans la Baltiqu

Les études sur la pollution des mers sont généralement effectuées dans des zones où le biome naturel a déjà subi des changements notables. Dans les bassins fermés, il est nécessaire de contrôler la productivité des biomes naturels et donc d'obtenir, à un stade aussi précoce que possible, des indications concernant la surfertilisation et la pollution, ce qui permet en outre de s'opposer à une évolution fâcheuse. A l'heure actuelle, la pollution dans la mer Baltique est devenue préoccupante. Il est généralement reconnu que cette aire d'eau saumâtre a une certaine limite de tolérance. En Finlande, on estime que l'industrie de la pâte à papier est la principale responsable de l'altération des conditions naturelles, mais récemment l'attention s'est également portée sur la pollution par les eaux d'égout et les hydrocarbures.

On a soigneusement étudié la biologie de plusieurs organismes, mais l'on sait peu de chose de leur productivité ou des collectivités des eaux saumâtres, notamment sur le benthos littoral. Un programme de recherche PBI, entrepris au titre du projet national PBI-PM à l'entrée du golfe de Finlande et dans le golfe de Bothnie, vise à établir un paramètre de la productivité des zones non polluées de la Baltique.

Le benthos littoral a été étudié au moyen de nouvelles méthodes d'échantillonnage quantitatif mises au point par le Groupe finlandais PBI-PM. On a étudié la productivité primaire et le phytoplancton, de même que la zone profonde, à l'aide de méthodes classiques.

Vigilancia de la sobrefertilización lenta del Báltico

En general, los estudios de la contaminación del agua del mar se efectúan en lugares en los que el bioma natural ha experimentado ya cambios notables. En las cuencas cerradas es necesario vigilar la productividad de los biomas naturales y obtener lo antes posible señales de sobrefertilización y contaminación, además de tener la posibilidad de comprobar una evolución indeseada. La contaminación del mar Báltico ha pasado a constituir en el momento actual una cuestión interesante. Se acepta de manera casi general que esta masa de agua salobre tiene un cierto límite de tolerancia. En Finlandia se achaca a la industria de la pasta para papel al ser la principal perturbadora de las condiciones naturales, pero últimamente se presta también atención a las aguas cloacales y a la amenaza de la contaminación por petróleo.

Se ha estudiado a fondo la biología de diversos organismos, pero se sabe muy poco de su productividad o de las comunidades de agua salobre, particularmente en la zona bentónica litoral. Un programa de investigaciones nacionales iniciado por el PBI en la boca del Golfo de Finlandia y en el de Botnia tiene por objeto establecer un parámetro de la productividad de las zonas del Báltico no contaminadas.

El bentos litoral se estudia con nuevos métodos de muestreo cuantitativo preparados por el Grupo finlandés del PBI-PM. La productividad primaria y el fitoplancton así como la zona profunda

* The Finnish IBP-PM Group, Tvärminne Zoological Station, Tvärminne, Finland.

Chaque année, on effectue de trois à cinq périodes d' échantillonnage de deux semaines chacune dans les deux zones à l'étude. Le matériel animal et végétal prélevé est identifié au niveau de l'espèce; on relève le poids à l'état frais, le poids à l'état sec et le poids de cendres, anisi que les teneurs en pigments d'assimilation. Il est prévu d'exercer un contrôle continu sur les données les plus importantes.

se estudian empleando métodos tradicionales. Anualmente se efectúan de tres a cinco períodos de muestreo de dos semanas de duracion en ambas zonas de estudio. Se identifican, hasta el nivel de las especies, los animales y plantas obtenidos en las muestras y se registra el peso en fresco, en seco, el de la ceniza y el contenido de los pigmentos de asimilación. Se proyecta una vigilancia continua de los datos más esenciales.

ALTHOUGH localized research into bottom fauna of the coastal waters of Finland had been made by Bagge et al. (1965), Bagge and Voipio (1967), Dahlström and Sormunen (1965), Laakso (1965), Särkkä (1969) Tulkki (1964) and while plankton and primary production had been studied by Bagge and Lehmusluoto (1970), it was felt that detailed but comprehensive investigation should be made of the Finnish littoral as part of the International Biological Programme Project. Accordingly, a project started in May 1968 was divided into three parts: (1) studies of the littoral biocenoses; (2) benthic macrofauna; and (3) phytoplankton and primary production. It soon became clear this was important not only in relation to domestic pollution but also in relation to oil damage and the advent of power stations, nuclear and others.

Two principal stations

The Tvärminne Zoological Station (University of Helsinki) on the Gulf of Finland and the Krunnit Biological Station (University of Oulu) on the Bothnian Bay were chosen. Both are in areas considered to be unpolluted. The tidal range is unnoticeable.

Rantakokko, personal communication). Areas around the largest cities and towns by the Bothnian Bay are heavily polluted. A remarkable feature is that about one third of the whole water mass of Bothnian Bay is renewed yearly by an inflow of more saline water from the Bothnian Sea (Palosuo, 1964).

The littoral project

This project forms the largest part of IBP-PM research in Finland. Sampling is carried out in periods 10 to 20 days long four to five times a year at Tvärminne and three to four times a year at Krunnit, beginning shortly after the ice has broken up and ending before the ice forms again. In March 1970 sampling was also carried out under the ice in Tvärminne.

The four principal biocenoses sampled, and the communities in them, are:

A. Rocky or stony bottom by the open sea.
  1 Filamentous algae at the water's edge.
  2 Dense and short bladder wrack, *Fucus vesiculosus*, depth about 1 m.
  3 Filamentous algae, depth 3–5 m.

TABLE 1. THE MAIN CHARACTERISTICS OF AREAS AT TVÄRMINNE AND AT KRUNNIT ARE AS FOLLOWS

	Archipelago zone	Bottom	Depth range	t°C at 0 m <3°/>10°	Ice cover	S‰ at 0m	Colour transp. max.	Nutrients
Tvärminne	mainly outer	smooth bedrock, ooze, sand; steeply sloping	0–40 m	Dec.-April/ June-Sept	(Dec.) Jan-April	6	Green 9 m	Abundant
Krunnit	solitary isles	stones, sand; gently sloping	0–40 m	Dec.-May/ July-Sept	Dec.-May	2–2.5	Brown 7 m	Poor

Both domestic and industrial wastes are discharged in increasing amounts into the area north of Tvärminne, but the effects have not yet proved to be seriously harmful. Intensive shipping occurs in the study area. In December 1969 an oil disaster threatened Tvärminne (Haahtela, 1970). The weather conditions especially at the end of the year are very unfavourable for shipping. Dumping from ships has some harmful effect on the shores, but not on the water.

The Krunnit is quite different from Tvärminne. No archipelago exists, but there are solitary, exposed isles about 20km west of the mainland. The sandy bottom slopes very gently, stones and boulders occur everywhere, and the bedrock lies deep and is not exposed. Seasonal changes in temperature are even more drastic than in Tvärminne. During the winter, fresh water from the mainland forms a layer as much as 4 m in thickness between the ice and the brackish water lying below (T. Valtonen, personal communication). Humus colloids colour the water brownish. The area is typically poor in nutrients; a surplus has thus far been registered only in 1969 due to a discharge of industrial waste water rich in nitrogen and phosphorus from the river Oulu (Anneli

  4 Large, solitary *Fucus vesiculosus*, depth about 5m.
  5 Underwater rock, depth 5–10 m.

B. Rocky or stony bottom in the outer archipelago.
  1–4 as above.

C. Sheltered sediment bottom of the archipelago.
  1 Soft bottom, depth 0–0.5 m.
  2 Pondweed, *Potamogeton perfoliatus*, depth 2–4 m.
  3 Detached *Fucus vesiculosus*, depth 2–3 m.
  4 Loose-lying Leafweed *Phyllophora*, depth 10 m.
  5 Eel-grass, *Zostera marina*, depth 4–5 m.

D. Sheltered skerry area in the outer archipelago.
  1 Large, solitary *Fucus vesiculosus*, depth 2 m.

Stations A1, A3, A5, B1, C1 and C2 are sampled in both study areas, the remainder only in Tvärminne. Stony bottoms are, however, sampled at Krunnit, because no bedrock occurs. The fresh water moss *Fontinalis antipyretica* is irregularly sampled at Krunnit as a substitute for the *Fucus* (A4) community of Tvärminne.

The biomes form a mosaic pattern in the archipelago

TABLE 2a. THE MAIN SPECIES OF PLANTS AND ANIMALS OF THE BIOCENOSES AT TVÄRMINNE

Species/Biocenose	A1	A2	A3	A4	B1	B2	B3	B4	C1	C2	C5	D1
Cyanophycae spp.										*		
Ulothrix subflaccida					*							
Cladophora glomerata	***	*	*	(*)	***	*				*		*
Chara aspera										***		
Sphacelaria arctica	(*)	(*)	*	*		(*)	**				(*)	(*)
Pilayella littoralis	*	*	**	*	*	*	**	*	*	(*)	(*)	*
Ectocarpus confervoides			**				*	*		(*)		
Stictyosiphon tortilis	**	*	*	**	*		**	*	(*)	*	*	*
Dictyosiphon foeniculaceus	**	*	*	**	*	*		*				
Chorda filum	(*)			(*)			*	*		(*)		*
Fucus vesiculosus	**1	****	**	***	*1	****	**	***		(**)	***2	*****3
Furcellaria fastigiata			**	**		*	*	*				
Phyllophora spp.			*	*								
Ceramium tenuicorne	(*)	*	**	*		*	*	*	(*)	*	(*)	*
Potamogeton perfoliatus										***		
Zostera marina											***	
Laomedea loveni						*		**				**
Procerodes ulvae			*	**		*	*	**				
Dendrocoelum lacteum			*	*	(*)	*		**			*	**
Prostoma obscurum			*		(*)	*	*	*	*		*	**
Theodoxus fluviatilis	**		*	***		**	**			*	*	***
Hydrobia ulvae		(*)	***	**	(*)		****	*		*	**	*
Potamopyrgus jenkinsi				*			*	*		*	*	
Lymnaea pereger	**		*		(*)	*	*	**	*			*
Mytilus edulis		*	*****	***		*	**	****		(*)	**	***
Cardium lamarcki		*					*	*			*	**
Macoma baltica			*				*		*	**	**	
Nereis diversicolor			(*)							*	**	
Pygospio elegans			(*)								**	
Oligochaeta spp.			*				*	*	*	**	*	**
Praunus inermis		*	*		*		*	**				**
Idotea baltica		*	*	*		**	(*)	*			*	***
Jaera albifrons spp.		*	***	*	**	**	**	****	*		*	***
Gammarus oceanicus	*	**	*	**	**	**	(*)	**			*	*
Gammarus salinus	*		**	*		*		***		(*)	**	***
Gammarus zaddachi	*		*							*	*	
Pontoporeia affinis										**	*	
Corophium volutator										**	*	
Trichoptera larvae								**	*	*	*	**
Chironomidae larvae	***	*	*	*	***	*	***	***	***	**	**	***

TABLE 2b. THE MAIN SPECIES OF PLANTS AND ANIMALS AT KRUNNIT

Species/Biocenose	A1	A2	A3	A4	B1	B2	B3	B4	C1	C2	C5	D1
Cyanophycae spp.	(*)				(*)				*			
Ulothrix subflaccida	**				**							
Ulothrix zonata	***				***					(*)		
Cladophora aegagropila			***	**						(*)		
Cladophora glomerata	***		*		**					(*)		
Chara aspera									*	(*)		
Fontinalis antipyretica				**								
Zannichellia palustris									*			
Potamogeton perfoliatus										***		
Potamogeton pusillus									**	(*)		
Potamogeton filiformis	*				(*)				*			
Eleocharis acicularis									***			
Ephydatia fluviatilis			*	*								
Hydridae spp.					*					***		
Cordylophora caspia			*	*								
Prostoma obscurum	*		*		**				*	*		
Theodoxus fluviatilis			**	**	*					*		
Potamopyrgus jenkinsi									*	*		
Lymnaea pereger	**		**	*	***				**	**		
Oligochaeta spp.	*		*	**	*				***	**		
Mesidotea entomon			*	*						*		
Gammarus duebeni	***				*					*		
Gammarus zaddachi	*		*	*	**				*	**		
Ostracoda spp.									**	**		
Coleoptera larvae					*				*	*		
Ephemeroptera larvae	*		*		*				*			
Trichoptera larvae	*		*						**	*		
Chironomidae larvae	***		***	**	***				***	**		
Other Diptera larvae	*				*				**			

Explanations:

The biocenoses   A = Rocky or stony bottom by the open sea. B = Rocky or stony bottom in the outer archipelago. C = Sheltered sediment bottom of the archipelago. D = Sheltered skerry area in the outer archipelago (Details in the text)

Frequency   (*) = seasonal or occasional, * = rare, single specimens, ** = common, *** = abundant, **** = very abundant

Footnotes   1 = sporlings, 2 = mf. nana, 3 = with a rich growth of the Bryozoan, *Electra crustulenta* especially in D1, but also in other biocenoses.

[388]

of the Baltic. Therefore sampling is not carried out at random, but at selected sites. When planning the programme it became clear that there was no suitable equipment which would sample the biocenoses quantitatively and include both animals and plants in the same sample. Partially or totally new equipment was developed, and was built and described by the Finnish IBP-PM Group (1969). It consists of a plexiglass frame for sampling communities A1 and B1, a bag for A2, A4, B2, B4 and D1, a suction sampler for A3, A5 and B3, and a metal cylinder (the Tvärminne sampler) for C1–C5. Half the research team are trained SCUBA divers.

The fauna and flora in Tvärminne are characterized by marine or brackish water species, which often occur in large quantities, e.g. *Fucus vesiculosus*, filamentous algae, molluscs and crustaceans (Tables 2a and 2b). Some limnic species, however, produce one or more generations yearly, and have a very important role in the production studies. These are, for example, *Potamogeton perfoliatus* and chironomid larvae. At Krunnit the limnic element prevails: fresh water vascular plants and insects dominate in many biotopes. It must be emphasized that Tables 2a and 2b present only preliminary and summarized biological data of the study areas.

The material sampled was immediately sorted and the main components identified. The samples were treated very accurately. The sampling was planned to be finished in 1970, and the data were to be compiled during 1971. Final results were to be published after completion.

Phytoplankton and primary production

Tvärminne—A very intensive study has been carried out since 1967 throughout the year, from the open sea to the innermost part of the Pojo Bay. The composition of the phytoplankton, its biomass and chlorophyll-a content are analysed. Primary production is studied by the ^{14}C method. Temperature, salinity, oxygen, nutrients (nitrate, nitrite, ammonium, phosphates, total phosphorus and silicate) are measured, as well as light intensity and transparency, turbidity and the colour of water. Some of these are also measured at the littoral group sampling stations.

Krunnit—In 1968 and 1969 a less detailed investigation than the one described above was carried out during the ice free period near the mainland northwest of Oulu and near Krunnit. The results are almost completed and will be published as soon as possible. This study is to be continued on a smaller scale from 1970 onward.

Conclusions

Good progress in the study of the hydrography and chemistry of the Baltic has been made during the last few years due to improved methods and intensified study (Fonselius 1969, 1970, Koroleff 1969, Voipio 1969, The Baltic Oceanographers 1970, 7th Conference of the Baltic Oceanographers in Helsinki 1970). The bottom water of the central Baltic basin, which had stagnated for a long period, was renewed in 1969 by an influx of more saline North Sea water (The Baltic Oceanographers 1970, Fonselius 1970). Simonov and Justchak (1970) show that pollution by oil-products contaminate the Baltic water to a depth of 100 m. In

Sweden the National Nature Conservancy Office (Statens Naturvårdsverk 1969) has developed a plan for investigations into the pollution of the Baltic. Corresponding investigations are planned in Finland and apparently also in other Baltic countries. A study similar to that of the Finnish IBP-PM littoral group has been done by Hiddensee, DDR (von Oertzen 1968) and one has been started in Sweden by Askö (Wulff 1970). Although the present IBP-PM study in Finland mainly records the biomass and its seasonal changes, it is valuable in establishing the parameters of the productivity of unpolluted Baltic areas. A study on primary and secondary production should be started as soon as possible during the MAB (Man and Biosphere) programme, a continuation for the IBP. This kind of objective and independent continuous recording with simple methods and a comparatively small investment is necessary for the future welfare of the Baltic. Such monitoring should be carried out in several localities covering larger areas than the present.

References

BAGGE, P, *et al.* Bottom fauna of the Finnish southwestern
1965   archipelago. 3. The Lohm area. *Annls zool. fenn.*, 2:38–52.

BAGGE, P and LEHMUSLUOTO, P O   Primary production in Finnish
1970   coastal waters in relation to pollution. *In* Proceedings of the 7th Conference of the Baltic Oceanographers, Helsinki, Institute of Marine Research. Abstracts. (mimeo).

BAGGE, P and SALO, A   Biological detectors of radioactive con-
1967   tamination in the Baltic. Helsinki, Institute of Radiation Physics, Report SFL-A9:1–29 (44).

BAGGE, P and VOIPIO, P   Disturbed bottom and hydrographic
1967   conditions in some coastal areas of Finland. 1. Loviisa. *Merentutkimuslait. Julk.*, 223:3–12.

The Baltic Oceanographers, The Baltic year 1969–70: cruise
1970   reports. Göteborg, 59 p. (mimeo).

Conference of the Baltic Oceanographers, 7th, Helsinki, Finland,
1970   May 11 to 15, 1970. Abstracts. Helsinki, Finland, Institute of Marine Research, 89 p. (mimeo).

DAHLSTRÖM, H   Tutkimus Kymijoen itäisten suuhaarojen
1964   kalastusoloista ja kalastuksesta. Helsinki, Kalataloussäätiön tutkimuslaitos, 78 p. (mimeo).

DAHLSTRÖM, H and SORMUNEN, T   Tutkimus Oulun edustan
1965   merialueen kaloista ja kalastuksesta. Helsinki, Kalataloussäätiön tutkimuslaitos, 108 p. (mimeo).

The Finnish IBP-PM Group, Quantitative sampling equipment for
1969   the littoral benthos. *Int. Revue ges. Hydrobiol. Hydrogr.*, 54:185–93.

FONSELIUS, S H   Hydrography of the Baltic deep basins. 3. *Rep.*
1969   *Fishery Bd Swed. (Hydrog.)*, (23):1–97.

FONSELIUS, S H   Om Östersjön och svavelvätet (Föredrag för
1970   riksdagsmän och forskare den 25.2.1970). *Meddn Havsfiskelab. Lysekil,*. (78):1–8.

HAAHTELA, I   Oil spills off Finland. *Mar. Pollut. Bull.*, (1):19–20.
1970

KOROLEFF, F   En översikt av de kemiska förhållandena i Östersjön.
1969   *Nordenskiöld-Samf. Tidskr.*, 29:50–61.

LAAKSO, M   The bottom fauna in the surroundings of Helsinki.
1965   *Annls zool. fenn.*, 2:18–37.

LEMMETYINEN, R   Jäteöljyn vesilinnuille aiheuttamista tuhoista
1966   Itämeren alueella. *Suom. Riista*, 19:63–71.

MICHANEK, G   Marinekologin tar puls på framtiden. *Svensk*
1968   *Natur.*, 1968:219–25.

PALOSUO, E   A description of the seasonal variations of water
1964   exchange between the Baltic proper and the Gulf of Bothnia. *Merentutkimuslait. Julk.*, 215:1–32.

SEGERSTRÅLE, S G   Studien über die Bodentierwelt in südfinn-
1933   ländischen Küstengewässern. Untersuchungsgebiete, Methodik und Material. *Acta Soc. Sci. fenn. (Comment. Biol.)*, 4(8):1–62.

SEGERSTRÅLE, S G   Studien über die Bodentierwelt in südfinn-
1933a   ländischen Küstengewässern. 2. Übersicht über die Bodentierwelt, mit Berücksichtigung der Produktionsverhältnisse. *Acta Soc. Sci. fenn. (Comment. Biol.)*, 4(9):1–77.

SIMONOV, A I and JUSTCHAK, A A   Recent hydrochemistry
1970   variations in the Baltic Sea. *In* Proceedings of the 7th Conference of the Baltic Oceanographers. Abstracts. Helsinki, Institute of Marine Research, 5 p. (mimeo).

Statens Naturvårdsverk, Coordinated plan for investigations into
1969   the pollution of the Baltic. Stockholm, 6 p. (mimeo).
SÄRKKÄ, J The bottom fauna at the mouth of the river Kokemäen-
1969   joki, southwestern Finland. *Annls zool. fenn.*, 6:275–88.
TULKKI, P Studies on the bottom fauna of the Finnish south-
1960   western archipelago. 1. Bottom fauna of the Airisto Sound.
  Annls zool. Soc. "Vanamo", 21(3):1–26.
TULKKI, P Studies on the bottom fauna of the Finnish south-
1964   western archipelago. 2. Bottom fauna of the polluted harbour
  area of Turku. *Archs Soc. "Vanamo"*, 18:175–88.

VOIPIO, A On the cycle and the balance of phosphorus in the
1969   Baltic Sea. *Acta Chem. fenn.*, A42:48–53.
VON OERTZEN, J-A Untersuchungen über die Besiedlung der
1968   Fucusvegetation der Gewässer um Hiddensee. *Z. Fisch.*,
  16:253–77.
WULFF, F Studies on the effects of water pollution within the
1970   Trosa archipelago in the Northern Baltic. 2. Benthic
  macrofauna. *In* Proceedings of the 7th Conference of the
  Baltic Oceanographers. Abstracts. Helsinki, Institute of
  Marine Research (mimeo).

On the Influence of Industrial Waste Containing H₂SO₄ and FeSO₄ on the Bottom Fauna off Helgoland (German Bight)

E. Rachor*

Influence des effluents industriels contenant H₂SO₄ et FeSO₄ sur la faune benthique au large d'Heligoland (baie d'Heligoland)

Depuis le mois de mai 1969, on déverse quotidiennement par bateaux-citernes spéciaux quelque 1.600 tonnes de déchets acides (contenant 10 pour cent d'H₂SO₄ et 14 pour cent de FeSO₄ additionnés de minéraux) dans une zone située à 11,5 milles au nord-ouest d'Heligoland. On a prouvé expérimentalement que les déchets acides ont des effets nocifs, même à des taux de dilution élevés, sur divers organismes marins. Plusieurs instituts de recherche allemands ont effectué des travaux sur les effets écologiques de ces déchets acides. Les recherches de l'auteur concernent l'influence exercée par ces effluents sur les peuplements benthiques (en particulier la macrofaune). On dispose de renseignements satisfaisants sur les conditions qui régnaient auparavant parmi la faune de la zone de recherche et de la zone proche située au sud. Cette zone fait partie d'une région de sables moyens occupée par un peuplement de Venus-gallina de faible biomasse.

Depuis le mois d'avril 1969, on mène régulièrement des enquêtes mensuelles en prélevant des échantillons par van-Veen-grab dans la zone polluée ainsi que dans des zones témoins. La macrofaune est identifiée et fait l'objet de relevés quantitatifs; on calcule la croissance et la répartition par âges des peuplements par la mesure de la taille ou du poids. Les dragages et la pêche nous fournissent des renseignements qualitatifs supplémentaires, en particulier en ce qui concerne la faune vagile. A la suite d'une année de recherche, on a constaté peu d'altérations dans l'écosystème. Des enquêtes à long terme sont nécessaires pour mettre en lumière une action significative des effluents acides sur les peuplements benthiques.

Sobre la influencia de los desechos industriales que contienen H₂SO₄ y FeSO₄ sobre la fauna de fondo en aguas de Heligoland (bahia de Heligoland)

Desde mayo de 1969 tanques especiales han descargado diariamente en una zona a 11,5 millas al noroeste de Heligoland, una cantidad de unas 1.600 toneladas de residuos de ácidos (conteniendo 10 por ciento de H₂SO₄ y 14 por ciento de FeSO₄ más minerales). Se han comprobado experimentalmente sobre varios organismos marinos los efectos perjudiciales de los residuos de ácidos, incluso muy diluidos. Diferentes institutos de investigación alemanes están llevando a cabo investigaciones sobre los efectos ecológicos de estos ácidos residuales. Las investigaciones del autor se refieren a las influencias sobre las comunidades bénticas (particularmente la macrofauna). Existen abundantes datos acerca de las condiciones previas de la fauna en la zona investigada y en las proximidades meridionales. La zona forma parte de una región arenosa media ocupada por una comunidad Venus-gallina de escasa biomasa.

Desde abril de 1969 se llevan a cabo investigaciones mensuales tomando muestras con una draga van-Veen en la zona contaminada, así como en otras zonas de comparación. Se registra y determina cuantitativamente la macrofauna; se efectúa un cálculo del crecimiento y la distribución por edad de la población mediante mediciones de talla o de peso. Mediante el dragado y la pesca se obtiene adicional información cualitativa, especialmente acerca de la fauna vagil. Los resultados de un año de investigaciones indican escasa alteración del ecosistema. Para comprobar influencias significativas de los residuos ácidos sobre la comunidad béntica son precisas investigaciones más prolongadas.

SINCE May 1960, about 1,750 t of waste acid left by the production of TiO₂-pigment has been discharged daily by special tankers in an area of 2.5 × 5 n mi 11.5 mi northwest of Helgoland. This waste contains about 10 per cent H₂SO₄, 14 per cent FeSO₄ and also mineral pollutants (MgSO₄, TiOSO₄, Mn⁻, V⁻ and other salts) in low quantities.

This waste acid has been shown to have harmful effects even in great dilution under experimental conditions using various marine organisms (Kinne and Rosenthal, 1967; Kinne and Schumann, 1968; Halsband, 1968; Kayser, 1969), but little is known about its ecological effects (Redfield and Walford, 1951; Hickel, 1969). Investigations are therefore being carried out in the discharge area by German research institutes. My investigations are concerned with influences on the benthic macrofauna community.

Waste is discharged from moving tankers into an area influenced by tide currents. Thus, the waste is diluted, the acid being neutralized and the Fe(II) oxidized and precipitated as Fe(III)-oxide-hydrate.

There is good information about the previous faunistic and sedimentary conditions in the research area (Stripp and Gerlach, 1969; Hickel, 1969) and in the southern neighbourhood (Stripp, 1969, 1969a). The discharge area is part of a fine to medium sand region with small patches of coarse sand indicating influences of bottom currents. A clay layer found by Hickel in 15 cm depth at one station indicates variations in the sedimentation conditions. The water depth ranges from 23.5 to 27.5 m.

This area is occupied by a Venus-gallina community of low biomass (wet weight 16 g/m²); and the boundary of a more productive Echinocardium-cordatum-Amphiura-filiformis community is found about 1 n mi southwest. The residual tide currents are mainly towards northeast.

Material and methods

Since April 1969 regular monthly samples (using a van-Veen-grab) have been taken in the polluted area and in comparison areas. The sediment is sieved (mesh width 1 mm) and the macrofauna recorded quantitatively and identified to species level. By measuring size or weighing (wet weight: methods in Stripp, 1969) calculations are made of biomass and growth and age distribution in the population. Dredging and fishing

* Institut für Meeresforschung, Bremerhaven, Federal Republic of Germany.

provide additional qualitative information especially about the vagil fauna.

The results in this report were obtained from samples taken in the discharge area or its immediate neighbourhood. At all the stations the sediment consisted of fine to medium sand. The iron content of sediment samples was determined regularly by a KSCN-method (described by Lüneburg, 1966). This method provided information about the "mobile" iron, but not about that which is a part of the mineral grains. Since summer 1970 we have been measuring the iron content of some bottom animals by a modified KSCN-method after wet oxydation of the soft parts.

Results

Since autumn 1969, bottom samples have shown a loose mass of Fe(III)-oxide-hydrate flakes floating above the sediment. The iron content of the fine sandy sediment itself was about 2 mg/g dry weight, but did not rise during the months February to September 1970. Little or no incorporation of the Fe-oxide-hydrate flakes has been found in the sediment of the discharge area itself, and no crust has been formed.

Animals producing mucous substances, such as tubicolous polychaetes, show Fe-flakes affixed to the mucus and some tests of the gut content of polychaetes indicate a high proportion of iron shares. Some preliminary results of measurements of the Fe-content in the soft parts of the filter-feeding bivalve Venus gallina L. have been surprising because the Venus from a station six mi east of the disposal area show Fe-values which are about 50 per cent higher than those from the disposal area itself or from a station 7.5 mi southeast of the area. There are no symptoms of intoxication in these bivalves.

Some results of the faunistic survey at the central station of the discharge area are shown in Table 1. These results can be compared with the list presented by Stripp

and Gerlach (1969) for a sample taken in June 1967 (first column).

In April 1969, just before the start of TiO_2 manufacture and the disposal of the waste acid, we observed an increase in numbers of individuals as well as in numbers of species. At least eight species were not only new for this central station, but for the whole disposal area. (Cumacea and Amphipoda, except Ampelisca brevicornis, are not taken into consideration, as they are not determined to species level in Stripp and Gerlach).

Comparing these data of the central station with an evaluation of May 1970, no significant change in individual or species numbers was found. But, at least five additional species are new at the station, and others have vanished.

These facts can be verified when comparing the whole disposal area and its environs during 1967–68 and May 1970 (see columns 5 and 6). Thirty-five of the 47 species determined by Stripp and Gerlach were also found in May 1970. Those found in more than 30 per cent of all the samples during 1967–68 were found again. In addition at least 26 species were new in the area, ten of which were present in more than 30 per cent of the samples.

Samples taken in October 1970 revealed little specific alterations (two new species), but they were distinguished from the earlier samples by a considerable increase in individual numbers, caused mainly by an increase in abundance of a few polychaete species (Spio filicornis and Magelone papillicornis). It must be stressed that during the autumns of 1967–68 and 1969 no comparable fluctuations in abundance occurred.

Discussion

My investigations show that there has been little or no permanent deposition of iron-oxide-hydrate in the sediment of the disposal area up to autumn 1970,

TABLE 1. INDIVIDUAL AND SPECIES NUMBERS AND BIOMASS (PER m²) OF THE BOTTOM MACROFAUNA IN THE WASTE DISPOSAL AREA NORTHWEST OFF HELGOLAND

Species	Central station				Total area	
	June 1967	April 1969	May 1970	Oct. 1970	1967–1968	May 1970
Owenia fusiformis Delle Chiaje	70	255	80	188	32	21
Ophiura albida Forbes	40	106	293	180	19	223
Tellina fabula Gmelin	30	15	3	30	60	40
Edwardsia spec.	25	25	45	48	5	64
Cultellus pellucidus (Pennant)	20	10	3	24	15	1
Nephtys hombergi Audouin & M.-E.	15	35	18	14	17	19
Scoloplos armiger (Müller)	10	5	70	18	26	47
Echinocardium cordatum (Pennant)	10	3	5	6	20	9
Venus gallina Linnaeus	5	2	10	—	9	6
Ophelia limacina (Rathke)	5	3	5	—	14	16
Magelone papillicornis Müller	—	25	13	524	21	16
Ampelisca brevicornis (A. Costa)	—	14	3	74	12	13
Goniada maculata (Oersted)	—	5	3	4	11	6
Chaetozone setosa Malmgren	—	45	10	44	—	10
Spiophanes bombyx (Claparède)	—	185	148	112	—	117
Phoronis spec.	—	15	3	242	1	1
Lanice conchilega (Pallas)	—	5	3	38	—	1
Spio filicornis (Müller)	—	—	—	1,106	—	1
Other species: individual number	100	82	88	125	114	266
(species number)	(11)	(20)	(21)	(23)	(33)	(57)
Total number of species except Cumacea and Amphipoda	21	33	30	37	47	61
Number of additional new species	—	12	5	2	—	26
Wet biomass (g)	4.0	2.8	4.4	10.1	11.0	8.4

(O. albida, E. cordatum, V. gallina excluded)

although there is considerable accumulation above the sediment. It seems very probable that the iron flakes are transported by currents out of the area, perhaps mainly in a northeasterly direction. This assumption may explain the differences found in the iron content of the bivalve *Venus gallina*.

It is impossible at this time to prove any harmful effect to the macrofauna by the wastes because the research area seems to be an area of inconstancy. During the past three to four years we can observe an increase in individual and species numbers, but it must be stressed that this trend of increase was visible before waste disposal began.

In spite of the considerable accumulation of Fe-oxide-hydrate flakes directly above the sediment, the settlement of new species has not been hindered, nor was the settlement of larvae of the polychaete and echinoderm species found in the area before. It is not known if this statement is valid for all the filter-feeding bivalves. A settlement of young individuals can be reported only for *Cultellus pellucidus* and *Tellina fabula*.

The observations made in October 1970 are very remarkable. Some macrofauna species which used to be scarce in the research area now appear in great numbers and make up more than 50 per cent of all the individuals there. Further investigations will show whether these fluctuations are natural or whether they have been favoured by the waste disposal. There may be an improvement of the nutritive qualities of the sediment or of the seas above the sediment, and thus certain feeding types will be favoured as was found in the disposal area of china clay waste (Howell and Shelton, 1970). But, equally, the observations made in October 1970 may be the first symptoms of injury to the eco-system; if this is true, the macrofauna community will not only be altered but also damaged or impoverished.

References

HALSBAND, E Physiologische Untersuchungsmethoden zur Be-
1968 stimmung des Schädlichkeitsgrades von Abwassergiften in Süss-, Brack- und Salzwasser. *Helgoländer wiss.Meeresunters.*, 17(1–4):224–46.

HICKEL, W Sedimentbeschaffenheit und Bakteriengehalt im
1969 Sediment eines zukünftigen Verklappungsgebietes von Industrieabwässern nordwestlich Helgolands. *Helgoländer wiss.Meeresunters.*, 19(1):1–20.

HOWELL, B R and SHELTON, R G J The effect of china clay on the
1970 bottom fauna of St. Austell and Mevagissey Bays. *J. mar. biol. Ass. U.K.*, 50:593–607.

KAYSER, H Züchtungsexperimente an zwei marinen Flagellaten
1969 (Dinophyta) und ihre Anwendung im toxikologischen Abwassertest. *Helgoländer wiss.Meeresunters.*, 19(1):21–44.

KETCHUM, B H, YENTSCH, C S and CORWIN, N Some studies of
1958 the disposal of iron wastes at sea. 1. The distribution of plankton in relation to the circulation and chemistry of the water. Woods Hole, Oceanographic Institution, (58–55): 1–35 (Unpubl. MS).

KINNE, O and ROSENTHAL, H Effects of sulfuric water pollutants
1967 on fertilization, embryonic development and larvae of the herring, *Clupea harengus*. *Mar. Biol.*, 1(1): 65–83.

KINNE, O and SCHUMANN, K-H Biologische Konsequenzen
1968 schwefelsäure- und eisensulfathaltiger Industrieabwässer. Mortalität junger *Gobius pictus* und *Solea solea* (Pisces). *Helgoländer wiss.Meeresunters.*, 17(1-4): 141–155.

LÜNEBURG, H Ursprung und Transport der Sedimente in den
1966 Wattrinnen des Grådyb-Systems bei Esbjerg. *Folia geogr. dan.*, 10(2):1–33.

REDFIELD, A C and WALFORD, L A A study of the disposal of
1951 chemical waste at sea. *Publs natn. Res. Coun., Wash.*, (201): 1–49.

STRIPP, K Jahreszeitliche Fluktuationen von Makrofauna und
1969 Meiofauna in der Helgoländer Bucht. *Veröff.Inst.Meeresforsch.Bremerh.*, 12(2):65–94.

STRIPP, K Die Assoziationen des Benthos in der Helgoländer
1969a Bucht. *Veröff. Inst.Meeresforsch.Bremerh.*, 12:95–142.

STRIPP, K and GERLACH, S A Die Bodenfauna im Verklappungs-
1969 gebiet von Industrieabwässern nordwestlich von Helgoland. *Veröff. Inst.Meeresforsch.Bremerh.*, 12:149–56.

Long-Term Laboratory Experiments on the Influence of Ferric Hydroxide Flakes on the Filter-Feeding Behaviour, Growth, Iron Content and Mortality in *Mytilus edulis* L.

*J. E. Winter**

Expériences de longue durée en laboratoire concernant l'influence des flocons d'hydroxide ferrique sur la filtration de l'eau, la croissance, la teneur en fer et la mortalité de *Mytilus edulis* L.

Pendant cinq mois on a étudié chez 3.400 individus l'influence des flocons d'hydroxide ferrique sur le comportement de *Mytilus edulis* —filtration de l'eau, croissance, teneur en fer des parties molles et mortalité. Tous les animaux ont reçu la même quantité d'aliment de qualité égale, tandis que pour chaque groupe la quantité de flocons d'hydroxide ferrique ajoutée à la nourriture variait entre 0,4 et 4,0 mg de Fe/l.

On a constaté une corrélation distincte entre la filtration de l'eau et les divers taux de concentration des flocons d'hydroxide ferrique. La production de pseudo-fèces augmente proportionellement à la quantité de flocons ajoutés, alors que l'ingestion des aliments diminue. De plus, parallèlement à la production accrue de pseudo-fèces, il se produit une perte croissante de matière organique sous forme de mucus. Il en résulte une réduction du poids des parties molles pendant la période expérimentale.

La teneur en fer de ces mêmes parties augmente proportionnellement à l'adjonction des flocons d'hydroxide ferrique. Pour un taux d'adjonction donné, cette teneur marque un accroissement considér-

Experimentos de larga duración efectuados en el laboratorio sobre el efecto de los flóculos de hidróxido férrico en la alimentación por filtración, el crecimiento, el contenido en hierro y la mortalidad de *Mytilus edulis* L.

Se estudió el efecto de diferentes concentraciones de flóculos de hidróxido férrico sobre el comportamiento de la alimentación por filtración, el crecimiento, el contenido en hierro de las partes blandas y la mortalidad en 3.400 ejemplares de *Mytilus edulis* durante un período de más de cinco meses. Todos los animales recibieron la misma cantidad de alimento de igual calidad, pero para cada grupo de experimentos, la cantidad adicionada de flóculos de hidróxido férrico varió de 0,4 a 4,0 mg Fe/l.

La alimentación por filtración del agua està relacionada de manéra diferente con la diversas cantidades de flóculos de hidróxido férrico empleadas. La eliminación de pseudo-feces aumentó con la cantidad de flóculos, en tanto que disminuyó la ingestión de alimento. Además, al aumentar la producción de pseudo-feces, también aumentó la pérdida de materia orgánica en forma de mucosidad. Por esta razón, el peso de las partes blandas disminuyó durante el período de los experimentos.

El contenido en hierro de las partes blandas aumentó proporcionalmente con la cantidad de flóculos de hidróxido férrico. Existe un notable incremento en el contenido de hierro en los tejidos, cual-

* Institut für Meeresforschung, Bremerhaven, Federal Republic of Germany.

able en fonction du temps (processus d'accumulation). L'accumulation du fer présent dans les tissus est due à l'assimilation de 0,36 pour mille au maximum de la quantité totale de fer ajoutée au cours de l'expérience.

Des concentrations fortes et moyennes de flocons d'hydroxide ferrique provoquent une mortalité de 75 pour cent en l'espace de trois mois; concentrations peu élevées provoquent une mortalité d'environ 40 pour cent en l'espace de cinq mois. Il existe une corrélation entre le taux de mortalité et la quantité de flocons d'hydroxide ferrique ingérés dans l'estomac. A part les effets mécaniques, il n'existe aucune preuve que le fer ait des effets toxiques aigus sur les moules.

quiera que sea la cantidad aplicada, como función del tiempo (acumulación). Es suficiente para que se acumule hierro en los tejidos cuando la asimilición no exceda de 0,36 por mil de la cantidad total de hierro utilizada durante los experimentos.

Cantidades grandes y medianas de flóculos de hidróxido férrico causan una mortalidad del 75 por ciento en un periodo de 3 meses; en cambio, la aplicacion en pequeña cantidad causa una mortalidad de cerca del 40 por ciento dentro de un período de 5 meses. La mortalidad guarda relación con la cantidad de flóculos de hidróxido férrico que llega al estómago. Si se excluyen los efectos mecánicos, no hay pruebas de que el hierro produzca una toxicidad aguda en los mejillones.

IRON pollutants from the production of titanium dioxide pigment are released in high amounts into the North Sea near Helgoland. The ferrous sulphate is transformed to ferric hydroxide flakes, causing a high degree of turbidity (Kinne and Schumann, 1968; Rachor, 1970; Weichart, 1970). The question is in what way these ferric hydroxide flakes could influence the filter-feeding behaviour of suspension-feeding organisms, since the studies of Loosanoff *et al.* (1942, 1947, 1948, 1962), Davis (1960) and Davids (1964) showed a detrimental influence of turbidity-creating substances on various bivalves and their larvae.

Material and methods

Mussels (*Mytilus edulis* L.) were collected from buoys in the German Weser Estuary at a salinity of 25 per thousand in March 1970. At the beginning of the experiments, the mussels were about 15 mm long, had an average volume of 0.29 cm³ and average dry weight of 147.4 mg (shell, 138.0 mg; soft parts, 9.4 mg).

The water was changed every other day, was well aerated and maintained at 12°C, a salinity of 25 per thousand and a pH range of 7.9–8.1. The mussels were adapted to these conditions for one month and then placed in 34 glass aquaria each containing 100 individuals in 2.222 l of sea water. Ten aquaria contained control animals while the remaining 24 aquaria were divided into 6 groups (Table 1).

TABLE 1. EXPERIMENTAL CONDITIONS FOR THE VARIOUS GROUPS OF MUSSELS, INDICATING THE AMOUNT OF ALGAE AND IRON FLAKES (DRY WEIGHT) PER APPLICATION OR PER 1 L SEA WATER CONTAINING 45 MUSSELS. THERE WERE THREE APPLICATIONS PER DAY.

Group	Fe	Ferric hydroxide flakes	Standard food	Total suspended matter
0 (control)	—	—	1.59 mg	1.59 mg
0 (no food)	—	—	—	—
0.4	0.4 mg	0.74 mg	1.59 mg	2.33 mg
1.0	1.0 mg	1.84 mg	1.59 mg	3.43 mg
2.0	2.0 mg	3.68 mg	1.59 mg	5.27 mg
2.0 (no food)	2.0 mg	3.68 mg	—	3.68 mg
4.0	4.0 mg	7.36 mg	1.59 mg	8.95 mg

The mussels were fed three times a day with a suspension of *Dunaliella marina* (size: $7.5 \times 5.0\,\mu m$) and *Saccharomyces cerivisiae* (5–6 μm in diameter) to produce in the aquaria initial densities of 40×10^6 cells/l. The pure ferric hydroxide flakes (0.6–5.0 μm in diameter) were stored in sea water for at least 2 days before being given to the mussels mixed with the standard food.

The dead animals, which were counted for mortality rate determinations, were removed before the mussels were fed for the first time during the day. The amount of

sea water, of standard food and of iron flakes were then adjusted to the number of mussels left, so each mussel had at any time the chance to filter an equal amount of food and iron flakes from the water.

Using subsamples, measurements of shell length, total volume, dry weight of soft parts and iron content of soft parts were made every 30 days. The iron content of soft parts was determined by measuring the optical density of its coloured complex with KSCN. The iron content of the separated faeces and pseudo-faeces was determined by the same method.

Filter-feeding behaviour

At low iron applications (0.4 Fe mg/l) and in controls, no production of pseudo-faeces took place (fig 1). All ferric hydroxide flakes and standard food was transported

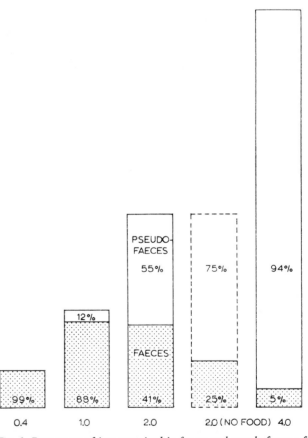

Fig. 1. *Percentage of iron contained in faeces and pseudo-faeces of* Mytilus edulis *in relation to the amount of iron given. These determinations were carried out during the second month of the experimental period, long enough for the mussels to be adapted to the various experimental conditions. In some cases, the amount of iron determined in faeces and pseudo-faeces is not equal to the total amount of iron given. Any discrepancy should be attributed to a small amount of iron that may be still within the intestine (the mussels were kept in filtered sea water for 2 days to eliminate undigested material), sticking to the shells of the mussels, or to the walls of the aquaria, and not to the small amount of iron accumulated in the tissues (0.36%).*

into the stomach. Thus, no selection of ferric hydroxide flakes and algal cells occurred on the mussels' gills and palps. At higher applications, a considerable production of pseudo-faeces took place. At the highest application (4.0 mg Fe/l), almost the total amount of ferric hydroxide flakes and algal cells was disposed of as pseudo-faeces. At higher applications the shell movements were clearly associated with frequent ejections of large quantities of pseudo-faeces (fig 1).

In the experiments where 2.0 mg Fe/l were applied, the percentage of iron disposed of as pseudo-faeces was higher in animals which did not get algal food.

Growth of soft parts

The quality and quantity of microorganisms given as "standard food" to the mussels were sufficient to bring about an increase of body weight of about 10 per cent over 5 months (see controls). On the other hand, the dry weight of soft parts (fig 2) decreases with increasing applications of ferric hydroxide flakes. An application of ferric hydroxide flakes causes a decrease down to 38 per cent after a period of 3 months. Mussels which were

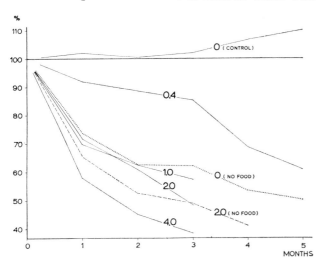

Fig. 2. Decrease in dry weight of soft parts of Mytilus edulis in relation to the amount of ferric hydroxide flakes given and to the length of experimental time. The initial dry weight of soft parts (average value: 9.38 mg) is designated as 100.

never given food nor iron distinctly demonstrate the negative influence of ferric hydroxide flakes. After 3 months the dry weight of these animals was higher than the dry weight of those fed with food and more than 1.0 mg Fe/l per application.

Iron content of soft parts

After the period of adaptation, the iron content of soft parts was 0.23 µg Fe/mg dry weight or 2.2 µg Fe/animal. In controls (fig 3), the iron content was reduced during the experiment. But in all experiments, where ferric hydroxide flakes were given, the iron content increased with increasing applications. At the highest application (4.0 mg Fe/l), the iron content per dry weight of soft parts increased ninefold within 3 months. During the same period, the initial dry weight of soft parts in this group was reduced to 38 per cent (see fig 2), so that the increase of the iron content per animal was only three times more than the initial iron content.

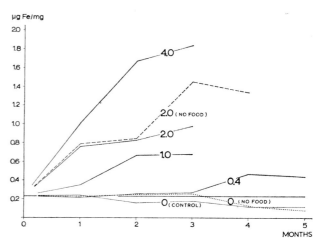

Fig. 3. Iron content in soft parts (per mg dry weight) of Mytilus edulis in relation to the amount of ferric hydroxide flakes given and to the length of experimental time.

It can be calculated that the amount of iron accumulated within the first three months did not exceed 0.36 per thousand of the amount given. However, from fig 4 it is clear that there exists a positive correlation between the amount of iron applied and the average amount of iron in the tissues.

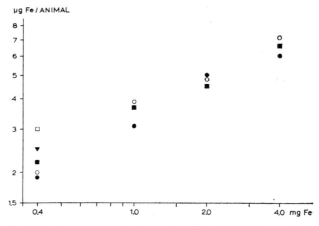

Fig. 4. The iron content of soft parts (per animal) of Mytilus edulis in relation to the amount of ferric hydroxide flakes given and to the length of experimental time. The vertical ranges indicate the respective lowest and highest values of iron in tissues from the end of the first month onward. During the first month, no determinations of the iron content were conducted, to allow the mussels to adapt to the experimental conditions.

● Values at the end of the first month
○ Values at the end of the second month
■ Values at the end of the third month
□ Values at the end of the fourth month
▼ Values at the end of the fifth month

Mortality

High and medium amounts of ferric hydroxide flakes caused a mortality of 75 per cent within 3 months (fig 5). Low applications caused a mortality about 40 per cent within 5 months, whereas the control animals had a mortality of about 20 per cent during the same period. The low rate of mortality during the first month at the high application of 4.0 mg/l deserves special attention. Apparently, very high applications can be tolerated for a short time because almost the total amount is refused and returned as pseudo-faeces and only a small amount of iron enters the digestive tract.

[394]

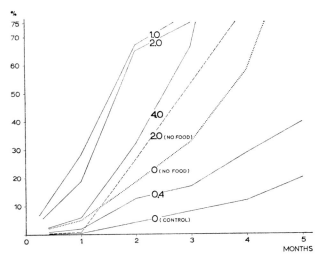

Fig. 5. Mortality of Mytilus edulis *in relation to the amount of ferric hydroxide flakes given and to the length of experimental time. The computation of mortality was made in such a way that the number of animals dead plus the number of animals alive at any time sum up to 100.*

Discussion

The initial iron content of soft parts was, in my experiments, 0.23 µg Fe/mg dry weight. This amount agrees well with the findings of Bellamy *et al.* (1970) for *M. edulis* from unpolluted areas and lies within the range of the iron content of *M. edulis* from Southampton determined by Hobden (1967).

The iron content in controls and starving animals goes down to 0.08 µg Fe/mg dry weight, but most values lie within the range of the "permanent store" (Hobden, 1967) of iron. The experiments with 0.4 mg Fe/l show that about this amount of iron is necessary to retain the initial iron content.

The highest amount of iron in tissues obtained in my experiments was 1.84 µg Fe/mg dry weight, similar to the results obtained by Bellamy *et al.* (1970) regarding the iron content of *M. edulis* of the same size from polluted areas.

My experiments show a distinct correlation between the amount of iron accumulated and the amount present in suspension. Hobden (1969), however, did not find such a correlation.

Mortality

Granted that ferric hydroxide flakes do not bring about acute toxicity, one has still to admit that there is a big decrease in body weight and a relatively high mortality rate among the mussels fed with ferric hydroxide flakes. An addition of iron flakes to the standard food can result in three different reactions:

First, an increased pseudo-faeces production brings about an increase in the formation of mucus which in turn signifies a loss of organic substance, especially when this occurs for a long period of time.

Second, with an increase in the production of pseudo-faeces, there is a corresponding increase in the amount of algal cells entangled in the pseudo-faeces (Winter, 1969, 1970) and these algal cells are no longer available as food to the animals.

Third, indigestable products and waste material left by the process of digestion in the phagocytosing cells

are returned to the mid-gut within large "excretory spheres" (fig 6), resulting in a further loss of organic material (List, 1902; Owen, 1955, 1956, 1970; Vonk, 1924; Yonge, 1923, 1926, 1926a). The substantial loss of body weight at low applications (0.4 mg Fe/l), when no pseudo-faeces are produced is of special interest because this loss must be attributed mostly to these processes taking place in the gut.

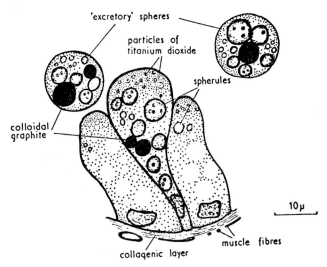

Fig. 6. Formation of excretory spheres by the cells of *a tubule of the digestive diverticula of* Cardium edule *6 h after feeding with a mixture of titanium dioxide and colloidal graphite (after Owen, 1955, Fig. 8).*

It can be assumed that the higher the amount of ferric hydroxide flakes transported through the digestive tract, the higher will be the amount phagocytosed and the higher the number of "excretory spheres" produced. The highest mortality was observed in mussels which were fed only medium quantities of ferric hydroxide flakes along with algae. This may be due to the fact that the amount of ferric hydroxide flakes transported through the digestive system (fig 1) was highest in this group. At high concentrations of ferric hydroxide flakes only a very small fraction passes through the gut.

The ultimate function of the iron stored is still a matter of conjecture. The "temporary store" (Hobden, 1967, 1970) seems to be of little significance for metabolic processes since it is so easily lost. Furthermore, the question whether iron has a real toxic effect on mussels, aside from mechanical influences, is open for further research. In my experiments, there is no evidence of any acute toxicity.

References

BELLAMY, D J *et al.* The place of ecological monitoring in the study
1970 of the marine environment. Paper presented to FAO Technical Conference on Marine Pollution and its Effects on Living Resources and Fishing, Rome, Italy, December 9–18 1970, FIR:MP/70/E–65:12 p.

DAVIDS, C The influence of suspensions of microorganisms of
1964 different concentrations on the pumping and retention of food by the mussel (*Mytilus edulis* L.). *Neth. J. Sea Res.,* 2:233–49.

DAVIS, H C Effects of turbidity-producing materials in sea water
1960 on eggs and larvae of the clam (*Venus (Mercenaria) mercenaria*). *Biol. Bull. Mar. biol. Lab., Woods Hole,* (118): 48–54.

HOBDEN, D J Iron metabolism in *Mytilus edulis*. 1. Variation
1967 content and distribution. *J. mar. biol. Ass. U.K.,* 47(3):597–606.

HOBDEN, D J   Iron metabolism in *Mytilus edulis*. 2. Uptake and
1969   distribution of radioactive iron. *J. mar. biol. Ass. U.K.*, 49(3): 661–8.

HOBDEN, D J   Aspects of iron metabolism in freshwater mussels.
1970   *Can. J. Zool.*, 48:83–6.

KINNE, O and SCHUMANN, K H   Biologische Konsequenzen schwe-
1968   felsäure- und eisensulfathaltiger Industrieabwässer. Mortalität junger *Gobius pictus* und *Soleu solea* (Pisces). *Helgoländer wiss. Meeresunters.*, 17(1–4):141–55.

LIST, T   Die Mytiliden. *Fauna Flora Golf. Neapel.*, 27:312 p.
1902

LOOSANOFF, V L   Effects of turbidity on some larval and adult
1962   bivalves. *Proc. Gulf Caribb. Fish. Inst.*, 14:80–94.

LOOSANOFF, V L and ENGLE, J B   Effects of different concentrations
1942   of plankton forms upon shell movements, rate of water
pumping and feeding and fattening of oysters. *Anat. Rec.*,
84:86–7.

LOOSANOFF, V L and ENGLE, J B   Effect of different concentrations
1947   of micro-organisms on the feeding of oysters. *Fishery Bull.
U.S. Fish. Wildl. Serv.*, 51(42):31–47.

LOOSANOFF, V L and TOMMERS, F D   Effect of suspended silt and
1948   other substances on the rate of feeding of oysters. *Science,
N.Y.*, 107(2768):69–70.

OWEN, G   Observations on the stomach and digestive diverticula of
1955   the Lamellibranchia. 1. The Anisomyaria and Eulamelli-
branchia. *Q. Jl microsc. Sci.*, 96:517–37.

OWEN, G   Observations on the stomach and digestive diverticula of
1956   the Lamellibranchia. 2. The Nuculidae. *Q. Jl microsc. Sci.*,
97(4):541–67.

OWEN, G   The fine structure of the digestive tubules of the marine
1970   bivalve *Cardium edule. Phil. Trans. R. Soc.*, 258:245–60.

RACHOR, E   On the influence of industrial waste containing H_2SO_4
1970   and $FeSO_4$ on the bottom fauna off Helgoland (German
Bight). Paper presented to FAO Technical Conference on
Marine Pollution and its Effects on Living Resources and
Fishing, Rome, Italy, December 9–18 1970, FIR:MP/70/E–
101:5 p.

VONK, H J   Verdauungsphagocytose bei den Austern. *Z. vergl.
1924   Physiol.*, 1:607–23.

WEICHART, G   Chemical and physical investigations in the German
1970   Bight on marine pollution caused by wastes of a TiO_2 factory.
Paper presented to FAO Technical Conference on Marine
Pollution and its Effects on Living Resources and Fishing,
Rome, Italy, December 9–18 1970, FIR:MP/70/E–44:2 p.

WINTER, J E   Über den Einfluss der Nahrungskonzentration und
1969   anderer Faktoren auf Filtrierleistung und Nahrungs-
ausnutzung der Muscheln *Arctica islandica* und *Modiolus
modiolus*. (On the influence of food concentration and other
factors on filtration rate and food utilization in the mussels
Arctica islandica and *Modiolus modiolus*). *Mar. Biol.*, 4:87–
135.

WINTER, J E   Filter feeding and food utilization in *Arctica islandica
1970   and Modiolus modiolus* at different food concentrations. In
Marine food chains, edited by J. Steele, Edinburgh, Oliver &
Boyd, pp. 106–206.

YONGE, C M   The mechanism of feeding, digestion and assimila-
1923   tion in *Mya. Br. J. exp. Biol.*, 1:15–64.

YONGE, C M   The digestive diverticula in the lamellibranchs. *Trans.
1926   R. Soc., Edinb.*, 54:703–18.

YONGE, C M   Structure and physiology of the organs of feeding
1926a   and digestion in *Ostrea edulis. J. mar. biol. Ass. U.K.*, 14:
295–386.

Influence de la Pollution sur les Peuplements Marins de la Région de Marseille

G. Bellan et D. Bellan-Santini*

Influence of pollution on marine populations in the Marseilles region

1. *Modifications in specific population composition*

   Solid substrata

Pure water	Polluted water
	populations composed of
Mytilus galloprovincialis (slopes)	*M. galloprovincialis*
Cystoseira stricta (shelves)	
Corallina cf. *mediterranea*	*Corallina* cf. *officinalis*
Cystoseira crinita	
Halopteris scoparia	*Ulva lactuca*
Padina pavonia	

Soil factors (in particular, hydrodynamics) play an important part.

   Loose substrata

Starting from the pollution source, a succession of concentric zones is observed: *maximum pollution*, azoic; *polluted*, poor in species (5, cosmopolitan); *subnormal* (alteration of the natural population through disappearance of sensitive species); *normal* population in *pure water*. A zone may be missing. Such permanent populations which are indicative of environmental conditions may serve to identify the degree of water pollution.

2. *Modifications in quantitative composition*

The concept of biomass increasing with pollution is reviewed.

   On solid substrata

Average biomasses expressed in weight of dry organic matter (g per m²):

Pure water		Polluted water	
		populations composed of	
C. stricta	2226	*M. galloprovincialis*	1431
C. cf. *mediterranea*	232	*C.* cf. *officinalis*	253
C. crinita	1295		
H. scoparia	561	*Ulva lactuca*	338

Influencia de la contaminación sobre las poblaciones marinas de la región de Marsella

1. *Modificaciones en la composición específica de las poblaciones*

   Substratos sólidos

Agua pura	Agua contaminada
	poblaciones de
Mytilus galloprovincialis (en declives)	*M. galloprovincialis*
Cystoseira stricta (en rellanos)	
Corallina cf. *mediterranea*	*Corallina* cf. *officinalis*
Cystoseira crinita	
Halopteris scoparia	*Ulva lactuca*
Padina pavonia	

Los factores edáficos (especialmente el hidrodinamismo) desempeñan una función considerable.

   Substratos blandos

A partir de la fuente de contaminación se observa una sucesión de zonas concéntricas: *contaminación máxima*, azoica; *contaminada*, pobre en especies (5, cosmopolitas); *sub-normal* (alteración de las poblaciones naturales por desaparición de las especies sensibles); población *normal en agua pura*. Puede faltar una zona. Tales poblaciones permanentes que son indicadoras de las condiciones del medio, pueden servir para determinar el grado de contaminación de las aguas.

2. *Modificaciones en la composición cuantitativa*

Es preciso examinar el aumento de la biomasa con el de la contaminación.

   Sobre substratos sólidos

Biomasas medias expresadas en peso de materia orgánica seca (g/m²):

Agua pura		Agua contaminada	
		poblaciones de	
C. stricta	2 226	*M. galloprovincialis*	1431
C. cf. *mediterranea*	232	*C.* cf. *officinalis*	253
C. crinita	1 295		
H. scoparia	561	*Ulva lactuca*	338

* Station Marine d'Endoume, 13 Marseille, France.

In *port areas*, this biomass rises from 378 g/m² (the most polluted site) to 929 g/m² (least polluted site). The apparent increase in the (total) biomass may be due to the increase in organic limestone:

	Pure water	Polluted water
	populations composed of	
C. stricta	2383 g/m² *M. galloprovincialis*	
		22832 g/m²

Ratios of weight of organic limestone to weight of dry organic matter:

	Pure water	Polluted water
	populations composed of	
Mussel beds	10.81	15.95
Coralline populations	9.17	17.74

On loose substrata

Results are extremely variable according to the zone and locality studied

Zone of maximum pollution: zero biomass

Polluted zone (biomass expressed in g/m² of dry organic matter) from 1 to 5.5

Subnormal zone: 6.5 to (in exceptional cases) 61

Normal zone: 3 to 7.

En el *medio portuario*, esta biomasa pasa de 378 g/m² (la estación más contaminada) a 929 g/m² (la estación menos contaminada). El aumento aparente de la biomasa (total) puede deberse al crecimiento de materia calcárea orgánica:

	Agua pura	Agua contaminada
	poblaciones de	
C. stricta	2 383 g/m² *M. galloprovincialis*	
		22 832 g/m²

Relación entre peso de materia calcárea orgánica y peso de materia orgánica seca:

	Agua pura	Agua contaminada
	poblaciones de	
Criaderos de mejillones	10,81	15,95
Poblaciones de coralinas	9,17	17,74

Sobre substratos blandos

Los resultados son muy variables según las zonas y los lugares estudiados

Zona de contaminación máxima: biomasa nula

Zona contaminada (biomasa expresada en g/m² de materia orgánica seca) de 1 a 5,5

Zona subnormal: 6,5 a (excepcionalmente) 61

Zona normal: de 3 a 7

D E nombreux travaux ont été effectués, dans de nombreuses aires géographiques, sur les modifications des peuplements marins benthiques, entraînées par la pollution. Ces travaux ont été récemment résumés par Bellan-Santini (1968) et par Bellan et Pérès (1969). Ces travaux antérieurs traitent, pour la plupart, des modifications *qualitatives* dans les communautés. Les travaux d'ordre *quantitatif*, numéral ou pondéral, sont beaucoup plus rares. Aussi, insisterons-nous plus particulièrement sur les résultats que nous avons obtenus au cours de nos recherches *quantitatives*.

En ce qui concerne les substrats solides, les données essentielles proviennent de recherches sur les salissures biologiques (biofouling) dans les ports. Ces données ont l'inconvénient de mélanger l'action de deux facteurs: la pollution et l'installation de peuplements sur des surfaces artificielles nouvellement immergées.

Sur substrat solide, le prélèvement est effectué par grattage complet (jusqu'à la roche) du peuplement dont la strate élevée a été préalablement isolée du milieu à l'aide d'un sac en plastique. La surface grattée est un carré de 20 cm sur 20 cm de côté (1/25 de m²). Sur substrat meuble, les prélèvements (1/10 m², 5 dm³ de sédiment) ont été effectués avec une benne "Orange Peel". Les prélèvements sont traités de manière identique quel que soit le substrat d'origine: tamisage sur colonne de tamis, tri de tous les individus du macro-benthos (taille supérieure à 1,5 mm), déterminations, comptages des individus ou colonies, les pesées effectuées sont: poids humide formolé, poids après décalcification, poids après séchage à l'étuve à 120° jusqu'à poids constant. Par calcul, on obtient les poids humides, de calcaire organique et de matière organique sèche, décalcifiée.

Les résultats présentés ont été obtenus à partir de 160 prélèvements dans les peuplements de substrat solide (10 ou 20 prélèvements par peuplement; l'aspect saisonnier a été envisagé) et de 100 prélèvements dans les substrats meubles.

Modifications dans la composition des peuplements

Substrats solides

Les peuplements établis sur substrats solides sont largement tributaires des facteurs édaphiques, notamment de l'hydrodynamisme. Ceci est valable aussi bien pour les peuplements soumis à la pollution que pour ceux rencontrés en eau pure.

D'une manière générale, la pollution est marquée par un certain nombre de phénomènes:

(1) elle provoque la disparition de nombreuses espèces et la prolifération de quelques autres (Tableau 1);

(2) elle entraîne une certaine monotonie dans la répartition des peuplements;

(3) elle affecte surtout les niveaux superficiels, mais sur des surfaces considérables;

(4) de par le renouvellement généralement important des masses d'eau baignant ces substrats solides superficiels, les peuplements végétaux et animaux sont rarement complètement détruits.

Dans la région marseillaise, Bellan-Santini (1969) indique les modifications suivantes:

(1) Mode agité à très agité: en eau pure on trouve les peuplements à *Mytilus galloprovincialis* (sur les tombants) et à *Cystoseira stricta* (sur les replats); ils sont remplacés uniformément par une moulière à *M. galloprovincialis* en eau moyennement polluée, voire très polluée (ports).

(2) Mode peu agité: les peuplements à *Corallina mediterranea* et à *Cystoseira crinita* sont remplacés par le peuplement à *Corallina cf. officinalis* (cette coralline n'étant, peut-être, que la forme d'eau polluée de *C. mediterranea*).

(3) Mode calme: les peuplements à *Halopteris scoparia* et *Padina pavonia* sont remplacés par le peuplement à *C. cf. officinalis* sur lequel se surimpose l'aspect estival à *Ulva lactuca* qui devient dominant. Dans les enceintes portuaires calmes, le peuplement à *M. galloprovincialis* tend à s'étendre et à se mélanger au peuplement à *C. cf. officinalis*.

(4) Luminosité atténuée: on observe, en eau polluée, la disparition du peuplement à *Petroglossum nicaeense* (mode battu, lumière faible).

A partir d'une source de forte pollution, on observe (fig 1) une série de zones concentriques: une zone azoïque, une zone sans *Cystoseira stricta* (avec *M. galloprovincialis* et *C. cf. officinalis*), une zone où seulement quelques pieds de *C. stricta* survivent, on passe ensuite progressivement à la zone du peuplement à *Cystoseira stricta* (eau pure). La superficie occupée par

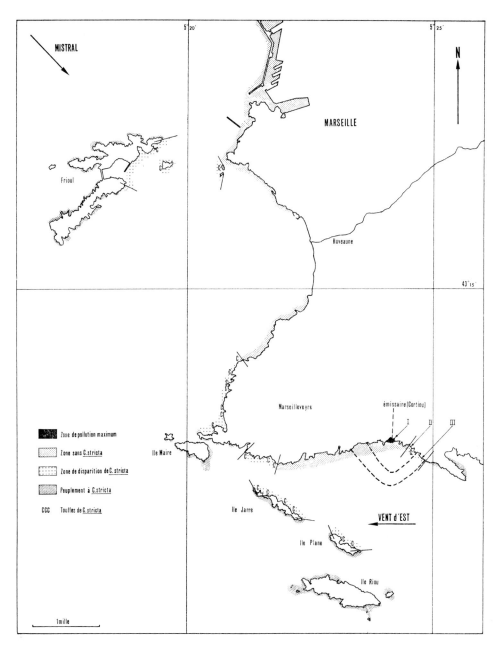

Fig. 1. Distribution de quelques peuplements soumis à la pollution dans la région marseillaise. Les peuplements de substrats meubles au large de l'émissaire (Cortiou) des égouts de la ville de Marseille sont indiqués en chiffres romains: I: Zone de pollution maximale, II: Zone polluée, III: Zone subnormale.

ces zones est fonction de la direction générale des vents et des courants tout autant que de l'importance de la source de pollution. Bellan-Santini a montré que ces zones étaient superposables à des concentrations définies de coliformes (utilisés comme "test" de niveau de pollution).

Substrats meubles

Les modifications des peuplements apparaissent plus tranchées que pour les substrats solides. L'action de la pollution se fait sentir non seulement en fonction des masses d'eau qui les baignent, mais aussi des débris qui se sédimentent sur le fond meuble (lequel peut résulter d'un ennoyage d'un substrat, solide à l'origine, plus ou moins plat). Le rôle de l'hydrodynamisme apparaît moins prépondérant.

Sur le plan bionomique, Bellan (1967, 1967a) considère qu'il n'y a pas de biocénose individualisée liée à la pollution, mais une série de faciès caractérisés chacun par la présence d'un très petit nombre d'espèces du macrobenthos (3 ou 4 au maximum) indicatrices de pollution. Ces espèces peuvent prendre un développement exubérant (jusqu'à 100.000 individus au m²) et exclusif (99,90 pour cent au minimum des individus récoltés). Ces mêmes espèces sont signalées au sein de biocénoses bien individualisées, soumises à des pollutions faibles (naturelles ou artificielles) ou se retrouvent dans des sédiments réduits. Dans la région marseillaise, ces espèces sont des polychètes: *Nereis caudata* (=*Neanthes arenaceodentata*), *Staurocephalus rudolphii* (=*Dorvillea rudolphi*), *Scolelepis fuliginosa* (=*Malacoceros fuliginosa*), *Audouinia tentaculata* (=*Cirriformia tentaculata*) et *Capitella capitata*.

On a typiquement, en partant à nouveau du foyer de pollution:

(1) une zone de pollution maximale azoïque (en ce qui concerne la macrofaune) et dépourvue de végétation macroscopique;

(2) une zone polluée à *Capitella capitata* et *Scolelepis fuliginosa* auxquelles viennent s'adjoindre, plus ou moins progressivement et localement, *Nereis caudata, Staurocephalus rudolphii* et *Audouinia tentaculata*;

(3) une zone subnormale caractérisée par les mollusques *Corbula gibba* et *Thyasira flexuosa* et une riche faunule de polychètes (plus de 45 espèces) à larges potentialités écologiques mais marquant des préférences nettes, dans l'ensemble, pour les sédiments riches en particules fines. Les espèces de polychètes les plus abondantes sont *Nematonereis unicornis, Hyalinoecia bilineata* (formes *brementi* et *fauveli*), *Lumbriconereis latreilli, Staurocephalus atlanticus, Aonides oxycephala, Thelepsavus costarum* et, surtout, *Heteromastus filiformis*. Les espèces indicatrices de pollution de la zone précédente ont, pratiquement, disparu. A l'ensemble de ces espèces, viennent s'adjoindre des éléments faunistiques caractéristiques des biocénoses que l'on devrait trouver au lieu considéré si le facteur pollution ne s'y faisait pas sentir. Dans la région marseillaise, on note essentiellement (et en fonction de la bathymétrie) soit un élément des sables vaseux de mode calme infralittoraux, soit des fonds détritiques côtiers circalittoraux;

(4) on passe, par l'intermédiaire d'une zone de transition plus ou moins marquée, au peuplement de la zone d'eau pure dans les conditions de milieu (profondeur et nature du sédiment principalement) présentes dans le secteur géographique étudié. Ces peuplements d'eau pure ont été, préalablement, définis par Picard (1965).

Il faut remarquer qu'une (ou deux) de ces zones successives peut manquer. Par exemple, si la source de pollution n'est pas très intense ou si un ouvrage (tel une digue) protège une zone d'eau pure.

Conclusions

Nous avons considéré que la délimitation des peuplements liés à la pollution et que la connaissance d'espèces indicatrices de niveaux déterminés de pollution permettraient de juger de l'importance de cette pollution en un lieu donné ainsi que des probabilités de voir s'installer et se développer tel ou tel peuplement, compte tenu des conditions édaphiques, climatiques et biotiques du milieu et, réciproquement, de prévoir les éventuels moyens de lutte contre la pollution. Le caractère pérennant des peuplements (à tout le moins la longueur du développement de leur cycle) fait qu'ils sont soustraits aux variations de faible durée dans le temps des conditions atmosphériques et hydrodynamiques des masses d'eau qui les baignent. De tels peuplements apparaissent donc comme les révélateurs fidèles des conditions moyennes du milieu. Il faut, bien entendu, connaître parfaitement les peuplements présents en eau pure.

Sur le plan de la recherche fondamentale, de telles études montrent que la pollution ne provoque pas le remplacement de biocénoses équilibrées par d'autres, mais se manifeste, par la destruction pure et simple d'une partie de la flore et de la faune, par la prolifération d'espèces résistantes ou adaptées et, lorsque la pollution devient trop intense, par la disparition de toute forme de vie macroscopique.

Sur le plan de la recherche appliquée, il suffira de reconnaître la présence de tel ou tel peuplement (ou d'espèces caractéristiques) pour en tirer des conclusions sur la pureté ou la pollution de l'eau. Par ailleurs, on peut proposer des méthodes pour limiter l'action de la pollution sur les peuplements benthiques. Ainsi, une simple digue pourra protéger un secteur non perturbé d'un autre très pollué. De même, il nous semble que des procédés tels que la dilacération (ou similaires) des matières rejetées est particulièrement néfaste et ne peut que conduire à l'accroissement des aires sous-marines soumises à la pollution.

Modifications dans la composition quantitative

Sur substrats solides

Le Tableau 1 montre que les "standing crops" exprimés en poids humides sont nettement supérieurs dans les peuplements en milieu pollué que dans leurs homologues d'eau pure. Par contre, les poids de matière organique sèche décalcifiée correspondants sont plus élevés (ou très voisins) dans les peuplements d'eau pure. La moulière en eau moyennement polluée fournit, en moyenne, 22.832 kg/m^2 de calcaire, celle en eau pure 13.211 kg/m^2 seulement. Le peuplement à *Cystoseira stricta*, remplacé en eau polluée par la moulière fournit 2 383 kg/m^2 de calcaire soit près de 10 fois moins (alors qu'il donne près de 2 fois plus de matière organique sèche). Si l'on compare les rapports: Poids de calcaire organique sur Poids de matière organique sèche de peuplements homologues présents en eau pure et en eau polluée, on remarque que la proportion de calcaire est plus grande dans les peuplements établis en eau moyennement polluée que dans ceux d'eau pure (les autres facteurs du milieu étant similaires):

Rapport $\dfrac{\text{Poids de calcaire}}{\text{Poids sec}}$	eau pure	eau polluée
Moulières	10,81	15,95
Corallines	9,17	17,74

La notion d'augmentation de la biomasse dans les milieux pollués pourrait n'être basée que sur une apparence liée à l'augmentation de la biomasse totale (exprimée en poids humide ou poids frais), elle-même sous la dépendance de l'accroissement de calcaire organique. En milieu pollué, sur substrat solide, au moins dans certains cas, la pollution favoriserait le métabolisme du calcaire organique plutôt que celui de la matière organique elle-même. A tout le moins, la production de matière organique ne serait pas nécessairement favorisée par la pollution. C'est ainsi qu'en milieu portuaire, nous avons observé que la biomasse exprimée en poids de matière organique sèche s'accroît de la zone la plus polluée du Vieux Port de Marseille vers la station la moins polluée (à la sortie du port):

	St. P-I (fond du port)	St. P-II (milieu)	St. P-III (sortie)
Poids matière organique sèche kg/m^2	0,378	0,883	0,929

Sur substrats meubles

Les résultats apparaissent moins homogènes que sur substrats solides (Tableau 2).

On note des zones dépourvues de toute trace de vie macroscopique où la biomasse du macro-peuplement est nulle (tout comme sur les substrats solides).

Dans la zone *polluée*, au large d'un égout, les biomasses sont extrêmement variables. Lorsque le peuplement ne comprend que des *S. fuliginosa* et des *C. capitata*, les biomasses sont relativement faibles, en moyenne 16 930 g/m² de matière organique sèche. Lorsque le peuplement s'enrichit d'espèces de plus grande taille (*N. caudata*, *A. tentaculata*) cette biomasse augmente: 54 710 g/m², alors que le nombre d'individus a diminué de moitié.

Dans les ports, la biomasse exprimée en poids de matière organique sèche est plus faible: 12 220 g/m², mais la présence de quelques mollusques *Tapes aureus* augmente la biomasse totale exprimée en poids humide (125 230 g/m²) en liaison, notamment, avec l'augmentation considérable des poids de calcaire (44 450 g/m²).

Dans la zone *subnormale*, au large de Cortiou, on note en moyenne 24 220 g/m² de matière organique sèche et 6 460 g/m² de calcaire organique. L'importance des polychètes décroît, par rapport à la zone précédente et passe de 99,6 pour cent à 77,5 pour cent, au bénéfice, essentiellement, des mollusques.

Dans les ports, il y a deux zones très distinctes: une superficielle (0–15 m) dont le peuplement, dominé par les *Tapes aureus* et *T. decussatus*, est très riche: 60 980 g/m² de matière organique sèche et 394 930 g/m² de calcaire organique; une, profonde (15–36 m) où la faune, voisine de celle de Cortiou, pauvre en *Tapes*, a fourni, en moyenne, 6 560 g/m² de matière organique sèche et 36 660 g/m² de calcaire.

La zone de transition de Cortiou a procuré 8 235 g/m² de matière organique et 16 796 g/m² de calcaire.

On a quelques données sur la biomasse (exprimée en poids de matière organique sèche) dans des biotopes d'eau pure voisins. Reys (1968), pour des fonds détritiques circalittoraux, donne des valeurs de 6 390 à 9 830 g/m²; Massé (comm. personnelle) trouve, pour des sables fins infralittoraux 2 924 g/m²; on n'a pas de données précises sur les biomasses des peuplements à *Tapes* dans des zones non polluées. Elles semblent, toutefois, élevées.

Evolution pondérale des populations en fonction de leur distance à une source de forte pollution (égout de Cortiou):

La fig 2 montre que: (a) le nombre maximum d'animaux par m² ne coïncide pas avec le poids maximum de matière organique sèche; (b) la pollution entraîne une augmentation de la matière organique, mais à une certaine distance (500 à 700 m) dans le cas présent. De part et d'autre, cette biomasse décroît et peut s'annuler; (c) l'on trouve, rapidement (1500 m environ) des bio-

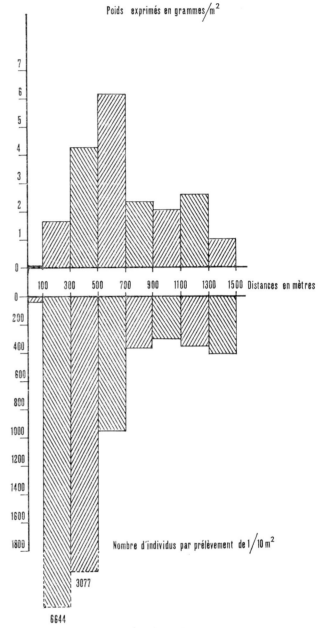

Fig. 2. Repartition quantitative de la faune des substrats meubles en fonction de la distance, au large de l'émissaire de Cortiou. Les poids sont exprimés en g/0,1 m² de matière organique sèche. Les nombres d'individus sont des moyennes par prélèvement (1/10 m²)

TABLEAU 1. ASPECTS QUANTITATIFS DE PEUPLEMENTS DE SUBSTRAT SOLIDE SOUMIS OU NON À LA POLLUTION

Peuplements à *C. stricta* (C.s), à *M. galloprovincialis* en eau pure (M.pr), à *C. mediterranea* (C.m), à *C. crinica* (C.c), à *H. scoparia* (H.s), *P. pavonia* (P.p), à *M. galloprovincialis*, en eau polluée (M.pl), à *C.cf. officinalis* (C.o), à *Ulva lactuca* (U.l)

	C.s	M.pr	C.m	C.c	H.s	P.p	M.pl	C.o	U.l
Moyennes annuelles des poids exprimés en *kg/m²* (P.H: Poids humide; P.C: Poids de calcaire; P.S: Poids de matière organique sèche)									
P.H	14,080	17,946	3,375	7,683	3,938	1,174	29,664	5,823	2,496
P.C	2,383	13,211	2,129	0,365	0,930	0,923	22,832	4,490	0,094
P.S	2,226	1,221	0,232	1,295	0,561	0,160	1,431	0,253	0,338
Pourcentage de calcaire									
	16,94	73,61	63,08	4,75	23,62	52,66	76,97	77,12	3,59
Nombre d'espèces recueillies (établi à partir de 10 prélèvements)									
	133	98	127	164	130	115	76	108	106

	Zone polluée			Zone subnormale		
	Cortiou		Ports	Cortiou	Ports	
	peuplements à *Scolelepis* et *Capitella*	Stations diverses			z. superficielle	z. profonde
P.H	99,02	205,75	125,23	101,54	744,42	66,32
P.C	0,51	2,76	44,45	6,46	394,93	36,66
P.S	16,93	54,71	12,22	24,22	60,98	6,56

Moyennes des poids de matière organique sèche exprimés en g/m^2.

masses identiques à celles reconnues dans des biotopes homologues, non soumis à la pollution.

Le problème de la fraction "calcaire organique" ou "calcaire de bioformation" de la biomasse totale (poids humide ou poids frais):

Ce problème est trop négligé. La fixation par les organismes de ce calcaire mobilise de l'énergie, distraite de l'élaboration de la matière organique et peut, comme on l'a vu, se faire aux dépens de cette dernière.

Il est curieux de noter que les poids de matière sèche totale [PSC = calcaire organique (PC) et matière organique sèche (PS)] de deux populations homologues peuvent être très voisins alors que les constituants sont très différents:

	PSC g/m^2	PC g/m^2	PS g/m^2
Z. polluée Cortiou	57,47	2,76	54,75
Z. polluée Port	56,67	44,45	12,22
Z. subnormale Cortiou	30,68	6,46	24,22
Z. subnormale Port	43,22	36,66	6,56

Le "métabolisme global" du peuplement n'est peut-être pas différent lorsqu'on compare des peuplements homologues mais, l'énergie dépensée peut l'être au profit d'une forme finalement "inutile": le calcaire organique.

Conclusion

La pollution provoque toujours un appauvrissement des peuplements benthiques par disparition de nombreuses espèces et même de peuplements équilibrés entiers. Les espèces qui remplacent celles qui ont disparu sont, le plus souvent, incapables d'être le noyau de peuplements eux-mêmes équilibrés. Ces regroupements d'espèces "indicatrices de pollution" sont souvent fugaces, facilement détruits.

L'accroissement occasionnel du nombre des individus cache, souvent, le fait que toute vie macroscopique a été détruite soit précédemment au même endroit, ou l'est tout à côté.

L'accroissement de la biomasse est souvent un leurre et peut masquer soit une augmentation du calcaire de bioformation, mobilisateur d'énergie au détriment de la matière organique vivante elle-même, soit la destruction de toute vie.

Bibliographie

BELLAN, G Pollution et peuplements benthiques sur substrat
1967 meuble dans la région de Marseille. Première partie. Le secteur de Cortiou. *Revue int. Océanogr. méd.*, 6–7:53–87.
BELLAN, G Pollution et peuplements benthiques sur substrat meuble
1967a dans la région de Marseille. Deuxième partie. L'ensemble portuaire marseillais. *Revue int. Océanogr. méd.*, 8:51–95.
BELLAN, G *et al.* Etat général des pollutions sur les côtes méditer-
1969 ranéennes de France. Document présenté au Congrès international "Acqua per il Domani", IV inchiesta internazionale "La Difesa del Mare", Milan.
BELLAN-SANTINI, D Influence des eaux polluées sur la faune et la
1966 flore marine benthique dans la région marseillaise. *Techqs Sci. munic.*, 61(7):285–92.
BELLAN-SANTINI, D Influence de la pollution sur les peuplements
1968 benthiques. *Revue int. Océanogr méd.*, 10:27–53.
BELLAN-SANTINI, D Contribution à l'étude des peuplements
1969 infralittoraux sur substrat rocheux (Etude qualitative et quantitative de la frange supérieure). *Recl. Trav. Stn mar. Endoume*, (63–47):5–294.
GILET, R Water pollution in Marseilles and its relation with fauna
1960 and flora. *In* Proceedings of the 1st International Conference on waste disposal in the Marine Environment, University of California, Berkeley, July 22–25, 1959, edited by E. A. Pearson. London, Pergamon Press, pp. 39–56.
MARION, A F Esquisse d'une topographie zoologique du golfe de
1883 Marseille. *Annls. Mus. Hist. nat. Marseille*, 1(1).
PICARD, J Recherches qualitatives sur les biocoenoses marines des
1965 substrats meubles dragables de la région marseillaise. *Recl. Trav. Stn. mar. Endoume*, Fasc. (52) Bull. (36):1–160.
REYS, J P Quelques données quantitatives sur les biocoenoses
1968 benthiques du golfe de Marseille. *Rapp. P.-v. Réun. Comn int. Explor. scient. Mer Méditerr.*, 19(2):121–3.

Problems and Approaches to Baseline Studies in Coastal Communities

J. R. Lewis*

Problèmes et méthodes des études de base sur les peuplements côtiers

Les études de base doivent fournir des données indiquant si les modifications futures des populations peuvent être imputées à des influences naturelles ou d'origine humaine. Une étude à long terme de la dynamique des populations d'invertébrés des rivages rocheux démontre la complexité des fluctuations naturelles et de leurs causes, et mène aux conclusions suivantes:
1. Les recensements sur place doivent tenir compte des variations résultant de l'hétérogénéité marquée du milieu local biologico-physique, des différences de longévité dans des conditions biologiques différentes, et de fluctuations qui sont saisonnières, annuelles ou fortement irrégulières.

Problemas de los estudios fundamentales sobre las comunidades costeras y sus soluciones

Los estudios fundamentales deben proporcionar datos de los cuales los cambios futuros de las poblaciones puedan ser atribuidos a influencias naturales o artificiales. Un estudio a largo plazo de la dinámica de las poblaciones de invertebrados de las costas rocosas demuestra la complejidad de las fluctuaciones naturales y sus causas, conduciendo a las conclusiones siguientes:
1. Los censos de campo deben comprender la variación procedente de una marcada heterogeneidad del medio biológico-físico local, de las diferencias de longevidad en diferentes condiciones biológicas y las fluctuaciones que sean estacionales, anuales o altamente irregulares.

* Wellcome Marine Laboratory, University of Leeds, Robin Hood's Bay, Yorkshire, England.

2. Pour être valable, l'interprétation des données numériques nécessite une compréhension de la structure des âges. Celle-ci dépend de la connaissance des facteurs suivants: (a) cycles de reproduction, taux de repeuplement, critères de détermination de l'âge, rapport taille/âge dans différentes conditions de croissance et mortalité; et (b) effets sur ces paramètres des influences biologiques (au sein de la communauté et à l'écart des prédateurs extérieurs) qui se font sentir dans chaque site en s'ajoutant à ceux des saisons précédentes.

3. Pour comprendre les causes de variation des taux de repeuplement, il faut disposer des éléments suivants: (a) comparaison annuelle des cycles gonadiques; (b) distinction entre les fluctuations très localisées et les fluctuations très largement répandues; et (c) renseignements sur le rapport croissance/mortalité, etc. au stade planctonique et/ou infralittorale (le cas échéant).

4. Le contrôle à une échelle géographique permettra de distinguer les influences climatiques et hydrographiques des facteurs spécifiquement locaux, mais pour recueillir de telles données, il faut un effort de coopération (peut-être dans le cadre international).

5. Les interactions entre les zones littorale, infralittorale et pélagique sont telles que, pour comprendre pleinement les fluctuations naturelles des populations littorales, et probablement de toutes les populations côtières, il convient d'inclure simultanément ces trois zones dans une étude globale à long terme.

Le document examine si cet "idéal" est réalisable et discute d'autres méthodes de portée plus restreinte, du point de vue des buts visés par les études de base.

2. Una interpretación razonada de los datos numéricos exige conocer la estructura por edades. Esto depende del conocimiento de: a) ciclos reproductivos, ritmos de repoblación, criterios de determinación de la edad, relación tamaño/edad en diferentes condiciones de crecimiento, y mortalidad; y b) el efecto sobre tales parámetros de las influencias biológicas (dentro de la comunidad y originadas por predadores externos) que actuaron en cada sector en las temporadas anteriores.

3. El conocimiento de las causas de la variación en los ritmos de repoblación exige: a) comparaciones anuales de ciclos de gónadas; b) distinción entre fluctuaciones muy locales y fluctuaciones muy amplias; y c) información sobre crecimiento/mortalidad, etc., durante las fases planctónicas o larvales en aguas de la zona intermareal (cuando sea aplicable).

4. La vigilancia a escala geográfica permitirá distinguir las influencias climáticas e hidrográficas de los factores muy locales, pero la obtención de tales datos exige un esfuerzo cooperativo (¿ internacional?).

5. Las interacciones entre las zonas litoral, intermareal y pelágica son tales que un conocimiento completo de las fluctuaciones naturales en todas las poblaciones litorales y probablemente costeras exige que se incluyan simultáneamente las tres zonas en un estudio amplio y a largo plazo.

Se examinan la practicabilidad de este "ideal" y de otras amplias soluciones en relación con las finalidades de los estudios fundamentales.

THE efficiency of pollution monitoring and of predictions about acceptable levels will finally be tested only in natural communities. The biological consequences of environmental change may take various forms and there is need to establish the "normal" range of growth or metabolic activities as precursors to detecting sublethal effects or for bio-assay purposes.

Personal awareness of the practical problems involved has emerged during a four-year study of rocky shore populations of *Balanus balanoides, Mytilus edulis, Patella vulgata, P. aspera, Thais lapillus* and *Fucus* spp. on the northeast coast of England. This project arose from the hypothesis that many anomalies in local distribution simply reflect irregular, unphased fluctuations among interacting species (Lewis, 1964). It was also known that among barnacles reproductive success has fluctuated in recent decades (Southward, 1967; Southward and Crisp, 1956), that settlement can be influenced by local winds (Barnes, 1956), and that survival for 10 to 15 years was possible in the absence of predation (personal observation). Assuming similar features occur among other littoral species, any one might have repercussions throughout entire communities.

Natural changes in littoral populations

The outcome of this study is concisely shown in figs 1, 2 & 3. Figure 1 shows the fluctuations that occurred in the four populations studied over the period. These were occasioned by natural factors. Figure 2 gives the percentage cover of *Mytilus* on five sites and fig 3, shows the density of Balanus, both over the four-year period: natural factors were wave exposure, storm effects, tide levels, availability of breeding areas as well as biological influences such as competition between barnacles, mussels and limpets under the varying seasonal conditions. Annual recruitment was not even but depended on natural conditions, also gonad conditions and climatic conditions.

Among examples of chain events were:

(a) The 1965 class of *Thais* ceased growing during winter and were eaten in great numbers by the sea bird *Calidrys maritima* (Feare, 1966). The following winter was mild, the 1966 class fed and grew continuously and became too big for *Calidrys* to eat. The increased numbers of larger, juvenile *Thais* (compared with 1965/66) ravaged the mussels and made space available for barnacle colonization.

(b) Extremely heavy *Mytilus* settlement (1967) destroyed all other species locally, but severe competition for space caused unstable hummock formation and eventual destruction by severe storms in March 1969. Space was now available for the barnacle settlement due in June. However, the mussels had excluded limpets since 1967 and had not themselves been destroyed until after the current limpet settlement period. Accordingly, diatoms and ephemeral algae developed abundantly and quickly, and barnacles failed to establish.

Because of the over-riding importance of local biological influences, most analyses, especially of standing populations, necessitate either data on the biological history of each site over several years, or at least practical experience of the species and types of interactions that can take place. Limited, random surveys without experience would yield little.

A base-line programme for coastal areas

These conclusions and experience gained in these investigations are embodied in the following programme:

1. Long-term, parallel studies of hard and soft bottoms, both littoral and sub-littoral, and of their overlying inshore waters in a single geographical area. The major objectives will be:

(a) To determine the principal interacting species and community structure of each habitat.

(b) To ascertain the scale and frequency of natural fluctuations and physical and biological causes thereof.

(c) To attempt to identify "key" species which might control the structure of communities, and which might lend themselves to long-term "key monitoring".

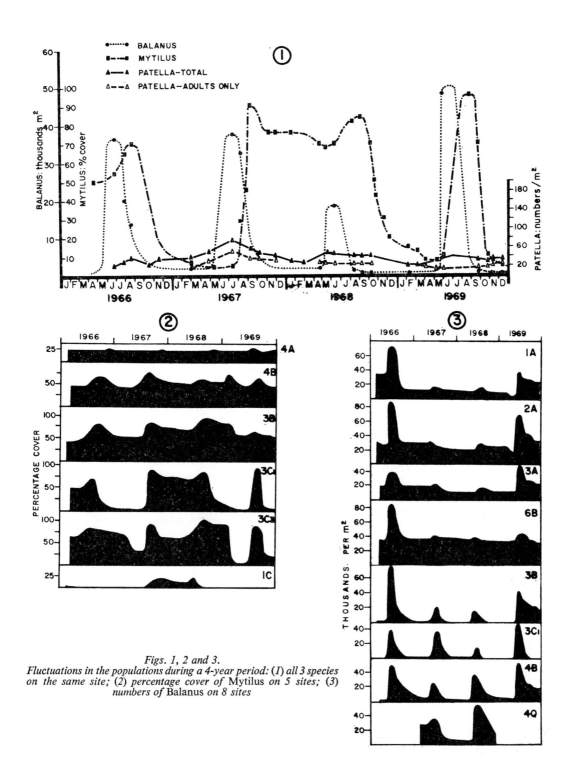

Figs. 1, 2 and 3.
Fluctuations in the populations during a 4-year period: (1) all 3 species on the same site; (2) percentage cover of Mytilus *on 5 sites; (3) numbers of* Balanus *on 8 sites*

Within each habitat the exact methods will vary with the species involved but required in all cases will be:

(i) Quantitative sampling techniques that allow for habitat heterogeneity, longevity differences and seasonal behavioural patterns.

(ii) Age determination methods and growth rates.

(iii) Influence of all associated species upon mortality and longevity.

(iv) Experimental manipulation of communities (where feasible) to assess the likely consequences of elimination or dramatic increases of particular species.

(v) Gonad cycles and relationships to environmental variables, with field data and experiments *in vitro* running parallel.

(vi) Larval phases: physical and biological influences on growth, mortality, dispersal and settlement.

2. Widescale monitoring of recruitment to permit distinction between the influence of local or widely operating factors. This aspect should cover the geographical range of each species with special attention at the limits. To this end attempts are being made in Britain to organize a country-wide network of collaborators and this approach, if successful, could well be developed on an international scale.

3. Supplementary surveys in other regions of the types

of populations under detailed long-term study at the base area.

4. Basic biological studies (i.e. reproduction, growth, mortality, ecological role) of species found suitable as pollution indicators.

5. Pollution gradient surveys to give comparative data on existing situations. Interpretation will often involve the prior knowledge gained under 1, 2 and 3.

6. Crash programmes by fully experienced staff for specific sites where new discharges are proposed, or have occurred unexpectedly. The value of this activity will increase as all other parts of the programme provide experience.

Conclusion

This approach to chronic pollution contends that without a massive expansion of ecological and reproductive data by simultaneous multidisciplinary studies not only will we be unable to detect significant long-term changes, but we will even remain unaware of the most suitable or important species and methods to build into a monitoring programme.

Against this total ideal approach it may be argued that it is impossibly difficult, too slow, too expensive, and can be circumvented by quicker methods. Apart from a complete ban of all possible pollutants at source, all other approaches ultimately depend upon biological data, frequently of the ecological type postulated. Predictions about safe levels, alone or in combination, can be adequately tested only in natural conditions where slight changes in competitive ability may have much more effect than expected. But this will not be detectable without prior data. Recruitment to some populations of *Echinocardium* and *Amphiura* has occurred only twice in 15 years (Buchanan, 1967) and has failed for five consecutive years in *Tellina* (McIntyre, personal comment). Data of this type is vital but clearly takes many years to obtain and perhaps longer to explain. If a species is particularly susceptible to certain substances, is its absence from field populations necessarily significant? Does its presence mean that there is no pollution, or that a

predator has proved more susceptible? Just how far can field data be interpreted solely by extrapolation of laboratory data? Perhaps we can bypass communities and concentrate upon a few "key" species but this designation depends upon an understanding of community structure. Whatever the role we must monitor the entire life cycle and so may be involved with planktonic phases and settlement stage predators.

In comparative studies of polluted and unpolluted sites can we be sure that they differ only in water quality? I find so much faunal variation within groups of unpolluted and apparently polluted sites that significant differences between them are slow to emerge.

The scale of simultaneous, interlocking investigation required is indeed the crux of our problems. This emerged in my very incomplete and relatively simple rocky shore investigation. It involved species that are easy to handle in the most accessible marine habitat, and led from individual species, through the local community to involve planktonic and subtidal spheres. Similar pathways are likely to arise with other benthic habitats, but there, with much less known initially and greater practical problems of quantitative study, the requirements of time and labour will be much greater. They must, however, be faced, for short-term, superficial surveys that ignore community dynamics may have immediate public relation value but will contribute little else.

References

BARNES, H   *Balanus balanoides* (L.) in the Firth of Clyde; the
1956   development and annual variation of the larval population and the causative factors. *J. Anim. Ecol.*, 25(1):72–84.

BUCHANAN, J B   Dispersal and demography of some infaunal
1967   echinoderm populations. *Symp. zool. Soc. Lond.*, 20:1–11.

FEARE, C J   The winter feeding of the purple sandpiper. *Br. Birds*.
1966   59:165–79.

LEWIS, J R   The ecology of rocky shores. London, English
1964   University Press, 323 p.

SEED, R   The ecology of *Mytilus edulis* L. on exposed rocky shores.
1969   2. Growth and mortality. *Oecologia*, 3:317–50.

SOUTHWARD, A J   Recent changes in abundance of intertidal
1967   barnacles in south-west England: a possible effect of climatic deterioration. *J. mar. biol. Ass. U.K.*, 47(1):81–95.

SOUTHWARD, A J and CRISP, D J   Fluctuations in the distribution
1956   and abundance of intertidal barnacles. *J. mar. biol. Ass. U.K.*, 35(1):211–30.

The Effects of Solid Waste Disposal on Benthic Communities in the New York Bight

*J. B. Pearce**

Les effets de l'évacuation de déchets solides sur les communautés benthiques dans la baie de New York

La baie de New York est polluée par le milliard de gallons d'effluents non épurés qu'y déversent quotidiennement l'Hudson et de plus petits voies d'eau. Une grande partie de cette zone se caractérise par l'absence de toute micro- ou macrofaune benthique et les communautés qui vivent à proximité sont très peu diversifiées et de faible biomasse. Nombre des *taxa* trouvés dans ces communautés marginales tolèrent de faibles taux de concentration d'oxygène et des taux élevés d'hydrogène sulfuré.

On a constaté que les gammaridés (amphipodes) benthiques étaient de bons indicateurs de la dégradation du milieu. Une corrélation inverse reliait la diversité et le nombre des amphipodes au degré de contamination. Les plus gros crustacés, notamment les crabes *Cancer irroratus* et *C. borealis*, et le homard *Homarus*

Efectos de la descarga de residuos sólidos en las comunidades bentónicas de la bahía de Nueva York

La bahía de Nueva York está contaminada por casi 4 500 millones de litros de aguas negras sin tratar que afluyen diariamente del río Hudson y de otros cursos más pequeños de agua. Gran parte de esta zona carece totalmente de microfauna o macrofauna bentónica y las comunidades que rodean a las zonas empobrecidas presentan una diversidad y una biomasa muy bajas. Se sabe que muchos de los *taxa* que se encuentran en estas comunidades marginales toleran valores bajos de oxígeno y valores elevados de sulfuro de hidrógeno.

Hemos encontrado que los anfípodos gamarídeos bentónicos son buenos indicadores de la deterioración del medio ambiente; tanto la diversidad de anfípodos como su número están en relación inversa con el grado de contaminación. Los crustáceos de mayor tamaño, incluidos los cangrejos *Cancer irroratus* y *C. borealis*, y el

* Sandy Hook Sport Fisheries Marine Laboratory, Highlands, New Jersey, U.S.A.

americanus, ont manifesté des signes d'affaiblissement et de maladie après migration dans les deux secteurs où les effluents sont déversés. Nous avons pu reproduire ces signes pathologiques dans notre milieu contrôlé en aquarium. La maladie semble résulter d'une obstruction des branchies et des cavités branchiales. Il s'ensuit de graves lésions sur la partie externe de l'exosquelette et les fines cuticules qui recouvrent les branchies. Dans la plupart des cas, les cellules hypodermiques sous-jacentes sont également atteintes.

Les sédiments des zones biologiquement appauvries se caractérisent par une très forte concentration des métaux lourds, y compris le cuivre, le chrome et le plomb, ainsi que des matières organiques et des produits pétrochimiques. Certains sédiments contiennent plus de 30 pour cent de matières organiques. Selon des renseignements préliminaires, il semble que les métaux lourds soient absorbés par certains vers polychètes. La décomposition bactérienne des matières organiques entraîne une réduction du taux d'oxygène dissous; pendant les mois d'été de 1969 et 1970, ce taux est tombé à moins de 2 ppm.

Le déversement de déchets solides pendant les quarante dernières années a gravement compromis l'écologie d'une partie notable de la baie de New York.

bogavante *Homarus americanus*, resultaron debilitados y enfermaron al migrar a estas zonas de evacuación. En nuestros acuarios de medio ambiente controlado hemos podido acelerar esta evolución patológica. La enfermedad comienza con la putrefacción de las branquias y las cavidades branquiales. Surgen después graves lesiones en el exoesqueleto y en las cutículas finas que cubren las branquias. En muchos casos resulta afectada también la hipodermis celular subyacente.

Los sedimentos de las zonas biológicamente empobrecidas se caracterizan por la presencia en proporciones considerables de metales pesados que incluyen cobre, cromo y plomo, así como de materiales petroquímicos y sustancias orgánicas. Algunos sedimentos contenían más del 30 por ciento de materia orgánica. La información preliminar indica que por lo menos algunos gusanos poliquetos absorben metales pesados. La descomposición bacteriana de la materia orgánica actúa reduciendo el contenido de oxígeno disuelto; durante los meses estivos de 1969 y 1970, el oxígeno se redujo a menos de 2 ppm.

La evacuación de desperdicios sólidos durante los cuarenta últimos años ha afectado gravemente la ecología de una parte notable de la bahía de Nueva York.

THE problem of wastes which originate in the New York Metropolitan area and are dumped at selected stations in the New York Bight and Long Island Sound has been reviewed (Gross, 1970). Approximately 9.6 million tons of waste solids per year were dumped during 1964 to 1968. From 1960 to 1963 eight million tons of waste were disposed of—an increase of about 4 per cent per year. This dumping represents the largest source of sediments entering the North Atlantic ocean from the North American continent.

In August 1968 Sandy Hook Sport Fisheries Marine Laboratory initiated a detailed biological and hydrographic survey of the New York Bight to determine what effects these dumping activities have had on marine resources.

Benthic samples were collected at 221 stations (fig 1) using the 0.1 m² Smith-McIntyre grab. Station positions were located with a Loran "C" unit. Dredge collections using the Sanders modified anchor dredge and a large shell dredge were made at selected stations to obtain larger semi-quantitative and qualitative samples.

Subsamples of sediment were removed from grab samples taken at selected stations. These were analysed quantitatively to compare the populations of meio- and microfauna present in uncontaminated sediments and sediments receiving dredge spoils and sewer sludge.

Zooplankton collections were made with 0.5 m diameter nets of 203 μ mesh equipped with flow meters (Tsurumi-Seiki Kosakusho Co., Yokohama) and were towed simultaneously at surface, mid-depth and bottom. Zooplankton samples were collected biweekly at stations indicated on fig 1. The feeding habits of demersal fish in relation to the distribution of dredge spoils and sewer sludge, were studied by making periodic collections with 40 and 60 ft otter trawls and analysing gut contents.

Sediments were analysed for grainsize distribution and heavy metals, pesticides, and petrochemicals. We also routinely measured temperature, salinity, and dissolved oxygen at each station where biological materials were collected. Phosphates and nitrates were measured at stations where zooplankton collections were made.

Hydrographic data were obtained through the use of *in situ* current meters (Geodyne Corp., model A-100) and thermographs (Geodyne, model A-119-4), sea bed and surface drifters, and temperature and salinity profiles.

Results

We found that areas heavily impinged upon by sewer sludge are devoid of normal benthic populations. These areas generally include the portions of the sewer sludge and dredge spoil disposal areas in which dried sediments contain more than 10 per cent organic matter (fig 2). Sediment samples collected from the sewer sludge beds are black and have odours characteristic of sewer sludge, hydrogen sulphide, and, occasionally, petrochemicals. We found these samples invariably contain large amounts of debris and artifacts of human origin and are contaminated with heavy metals (figs 3 and 4). During summer months the waters overlying the sediments of the dredge spoil and sewer sludge disposal areas have reduced dissolved oxygen (fig 5).

All these factors indicate a highly reduced environment inimical to normal benthic faunas. At the periphery of the sewer sludge dump area, the community is dominated by the burrowing anemone *Cerianthus americanus* which may be protected by its substantial tube. The associated fauna included rhynchocoel worms, representatives of several polychaete families and the protobranch bivalves *Yoldia limatula* and *Nucula proxima*. These bivalves are normally found in organic muddy sediments and might be expected to occur in highly organic sediments surrounding the sludge dumping grounds.

Sites to the west and northwest of the impoverished sludge dumping area (stations 21, 66 and A5) have similar grain size distribution and organic content (approximately 10 per cent dry weight) and a similar fauna.

Stations 7, 8 and 25 to the southwest of the sludge dumping area have a comparable grain size distribution but only 1.5–5 per cent organic matter. Polychaete worms Cossuridae and Paraonidae are rare or absent at stations 21, 66 and A5 but very abundant at stations 7 and 8. Conversely, nephtyid and large flabelligerid polychaetes are more abundant at the former than the latter stations. *Nucula* is found in large numbers at stations 21, 66 and A5 but lacking at 7 and 8. Station 25, at the periphery of the affected disposal area, has an intermediate fauna.

Stations to the northeast of the dumping area (including 13 and 31, about one nautical mile northeast) have similar sediments to the northwest stations, although a

Fig 1. *Location of numbered stations in New York Bight. Sandy Hook and New Jersey shorelines are shown along left margin and southern shore of Long Island, New York along upper margin. The Hudson River enters at upper left corner. Benthic samples were collected at stations indicated by large and small vive dots. Crosses indicate stations at which both benthic samples and zooplankton samples were collected. Small dots indicate stations located at grid intersections over the dredge spoil (designated by letters V–ZZ) and sewer sludge (A–I) disposal areas. The small dots are located ½ n mi apart. Ambrose Light is located at station 6*

few are somewhat coarser. The fauna of northeast stations is the same as at northwest stations but with one or two exceptions there are far fewer representatives of each species. Numerically, the spionid polychaete *Prionospio malmgreni* is the most abundant but *Cerianthus* makes the greatest contribution to the biomass. *Prionospio* is widespread to the northwest, west and southwest of the dumping grounds but is particularly abundant to the

northeast. This may reflect a preference for coarser substrates with a high organic content or more intensive settlement of larvae in the area.

Stations 3–4 mi east of the dumping ground (e.g. 63 and 65) have sediments with medium grain size and 2–3 per cent organic matter (dry weight). Many species found in other areas around the dumping ground are present here, but are in smaller numbers. *Cerianthus* and

[406]

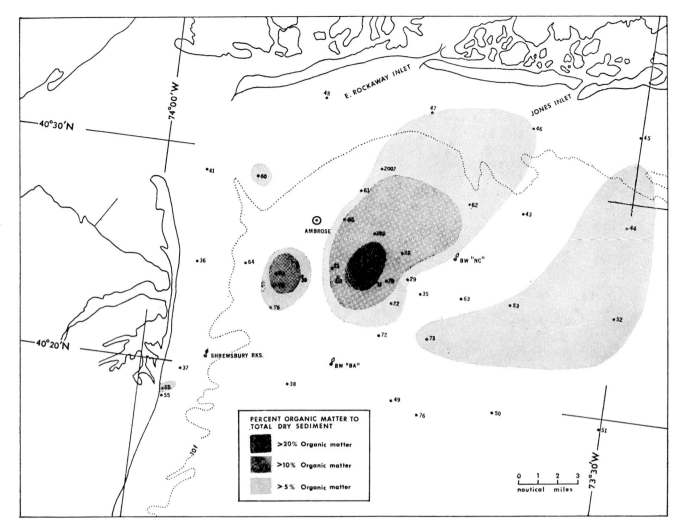

Fig 2. *Percentage organic matter in dry sediments collected at benthic sampling stations in New York Bight. Biologically impoverished bottoms roughly conform to the area characterized by greater than 10 per cent organic matter*

other groups dominant in fine organic sediments are absent. They are replaced by several species of bivalve mollusc, including large numbers of *Tellina agilis* which is generally found in clean sands.

Stations 9–10 mi northeast, east and southeast of the dumping area (e.g. 44, 52 and 53) have very coarse substrates but over 5 per cent organic matter which may be transported here from the dumping grounds by water currents. This organic enrichment does not increase the fauna and, in fact there are fewer individuals of species at stations 44 and 52 than at 63 and 65 where there is less than half the organic content.

Stations to the southeast of the dumping ground (18, 19 and 23) have a similar fauna to the more distant easterly stations but station 18 which is closest to the dumping grounds has some elements (e.g. *Cerianthus*) in common with stations to the northwest and southwest.

In summary, 14 mi² of sea bed is almost devoid of normal benthic macrofauna. Surrounding this is a belt of sporadically impoverished ground or with elements of the *Cerianthus* community which probably represents the indigenous fauna of the submerged Hudson Gorge area, the upper reaches of which are included in the sampling area. Coarser substrates to the south and southeast of the dumping grounds have sparser fauna in

variety and numbers, and probably have a smaller biomass and are less productive than the *Cerianthus* community. Increased organic matter in these coarser substrates does not appear to have a fertilizing effect and, indeed, Barber and Krieger (1970) have shown that organic matter originating from sewer sludge may inhibit phytoplankton productivity.

Our Smith-McIntyre grab samples to a depth of 15 cm but sludges have accumulated or affected sediments to a greater depth than this. Alpine Geophysical Associates, Inc., under a contract with the U.S. Corps of Engineers have furnished us with a series of deep cores (5–7 m) taken in or near the sludge disposal area. Cores 90 and 91 taken near our stations 83 and 27 respectively contained a limited amount of sludge debris.

Core 94 taken between our stations 11 and 70 had sludge contamination in the upper 3 ft, the remainder of the core was of compacted green sand much like some clays. At these stations, clean sand and sludge may be in alternate layers and we have noted that after a severe storm, sludge may be deposited where previously there was clean sand.

Phleger core samples have been taken from each Smith-McIntyre grab sample and examined for meiofauna. Samples from the centre of the sewer sludge and

[407]

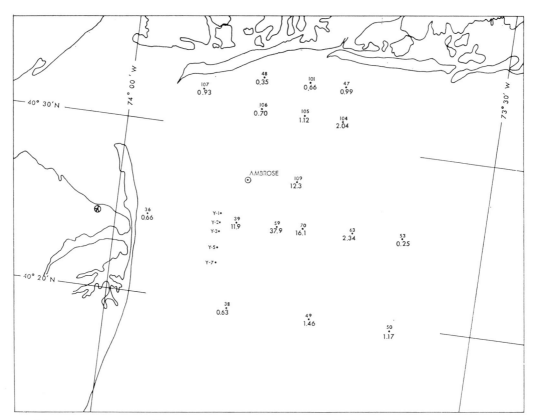

Fig 3. Amounts of chromium in parts per million at selected stations in the New York Bight. Station numbers are given above the dots which represent stations and the amounts of metal are given below the dots. Both copper and lead have a similar distribution with even higher values at most stations. Station 70 is located at the designated point of sewer sludge dumping; station 59 is at the approximate centre of the sludge bed

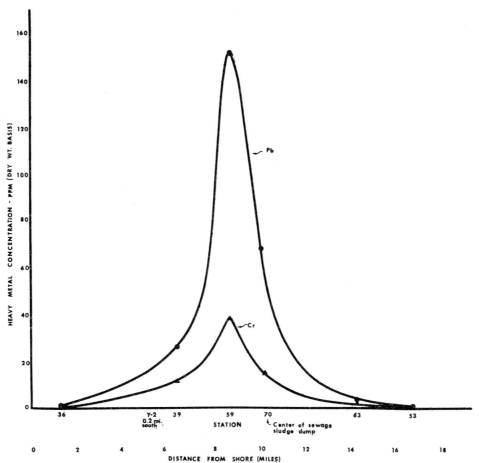

Fig 4. Distribution of lead and chromium in sediments at stations along a transect running east from the New Jersey shoreline through sewer sludge disposal (see fig 3 for location of stations)

[408]

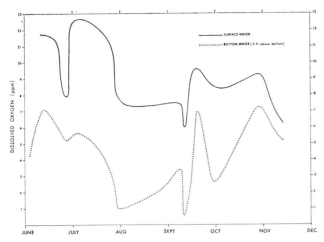

Fig 5. Dissolved oxygen values at station 59 (centre of sewer sludge bed) during the period 2 June to 12 November 1969. Water depth at station 59 is 37 m at high tide

dredge disposal areas are relatively lacking in meiofauna at all depths of sediment; samples from the peripheral areas have not yet been analysed.

Dredge spoil disposal area

We found the dredge spoil disposal area to be as seriously affected biologically as is the sewer sludge disposal area. The bottom sediments in the dredge spoil area were quite different from those of the sewer sludge disposal area. The cumulative curves of grain size distribution, for instance, are very irregular, and reflect the dumping of spoils of various sizes, from clays to gravels. When the dredge spoil sediments were washed through our screen stacks, we found fewer human artifacts (such as cellulose cigarette filters, band aids, and aluminum foil wrappers) than we did in the sewer sludge sediments.

We have, however, found levels of heavy metal contamination as high as those at the sewer sludge

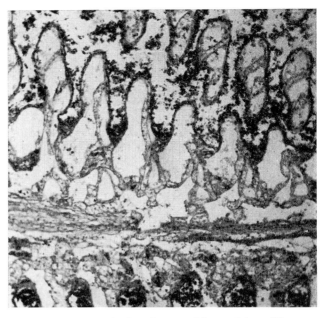

Fig 7. Photomicrograph of a gill removed from a lobster (Homarus americanus) which was kept on a sludge bed for six weeks. Notice the organic debris and silt between the filaments. These gills appeared black macroscopically; both crabs and lobsters held in aquaria with sludge-filled bottoms or collected from the sewer sludge disposal areas have gills similarly contaminated

disposal area, or higher, and also significant levels of pesticides belonging to the chlorinated hydrocarbon groups. Sediments from the dredge spoil disposal area always have a noticeable odour of petrochemicals, even when we find living benthic macrofauna in a particular sample. These sediments are oily in appearance and when they are washed through screens they make the decks, washing hoses, and apparatus oily. Preliminary analyses of sediments for hexane-extractable substances have revealed a large amount of these materials at the centre of the dumping area; quantities diminish in samples

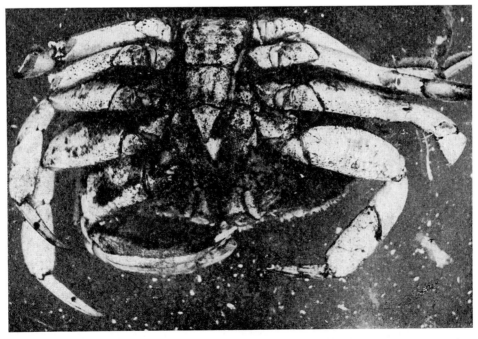

Fig 6. Ventral view of an adult male crab (Cancer irroratus) collected in the sewer sludge disposal area. Several necrotic lesions are visible on the thorax, abdomen and pereiopods. The gill chambers were partially filled with sludge sediments and the gills heavily fouled with sediments. Crabs in similar condition are frequently found dead on the sludge beds

taken at progressively increasing distances from the centre (station Y2). Over 6 mi² of bottom in the dredge spoil disposal area have been impoverished of normal macrofauna.

The benthic communities surrounding the dredge spoil disposal area reflect the diversity of bottom types which surround it. Dredge spoils are being dumped in an area between the shallow waters off Sandy Hook and the deeper waters of the Hudson Gorge. Stations 36 and 37 to the east of the dredge spoil area are representative of collection sites in 10 m of water along the northern New Jersey coast. The fauna is very limited and characterized by a few dominant bivalves including the surf clam (*Spisula solidissima*) and *Tellina*. The polychaete fauna is dominated by *Prionospio malmgreni* and *Spiophanes bombyx*, both of the family Spionidae. Sand dollars (*Echinarachnius parma*) are also very common in the collections and form the majority of the biomass in many samples. All of these forms are characteristically found in medium to coarse sand in high energy, wave swept, coastal environments.

Station 38 is representative of stations south and east of the dredge spoil disposal area. Although the sediment grain size distribution curve from this station is irregular, indicating an influence from limited waste disposal activities, the animal community is diverse and similar to faunas of the upper Hudson Gorge which are not substantially influenced by solid waste disposal activities. We found slightly over 6 per cent organic matter in sediments collected at this station.

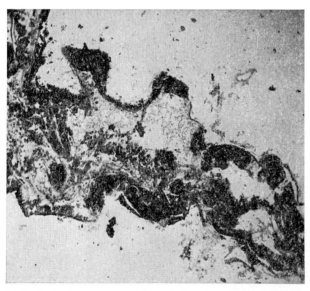

Fig 8. A gill filament removed from a lobster kept on a sludge bed for six weeks. The outer cuticle has become greatly thickened and distorted. The dark staining, thickened areas are not characteristic of control lobsters maintained on clean sediment or of lobsters and crabs collected from uncontaminated areas of the New York Bight

Partial analysis of samples collected from transect ZZ indicate benthic fauna very similar to station 38. There is, however, some evidence (from materials such as cigarette filters and petrochemical odours) which suggest a tendency for the movement of contaminated sediments into the Gorge itself. There is also some indication that contaminated sediments may have been moved westward from the sewer sludge disposal area into the Hudson Gorge.

Benthic crustaceans

Gammarid amphipods are extremely important in benthic communities. Many species form tubes which help stabilize the bottom sediments; these and other species are extremely important in the diet of reef- and bottom-dwelling finfishes. Fincham (1969) has suggested that the distribution of shallow-water Amphipoda may be correlated with pollution and pollution-related factors such as low levels of dissolved oxygen. Sanders (personal communication) has found a correlation between the distribution of benthic Amphipoda and oil pollution. We have identified 29 species of amphipods in 13 families associated with soft sediments in the New York Bight area. There are another 25 species which are associated with epibenthic communities in this area. Because of their ecological importance we have analysed their distribution in the study area.

As with other macrofaunal taxa, the Amphipoda are almost completely absent from the spoil and sludge disposal areas. Only two species, *Unciola irrorata* and *Monoculodes edwardsii*, occur in the impoverished portions of the dumping grounds; even then only an occasional single amphipod in a sample. Probably these isolates are carried to the area by water movements.

Compared with unpolluted portions of the New York Bight, those areas which are only marginally polluted contain few amphipods. This suggests that these small crustaceans may be more sensitive to the effects of pollution than *Cerianthus americanus*, many species of polychaete worms, and certain bivalve molluscs. For instance, whereas we found an average of 403 *Leptocheirus pinguis* at station 38 and 168 at station 7, both well south of the disposal areas, we found only two *Leptocheirus* at stations 8 and 9 which are marginally polluted and none at stations inside the heavily polluted sludge grounds.

We have observed that even though they are very abundant and widely distributed in the New York Bight, few living *Cancer irroratus* are collected in trawls or dredge hauls taken in the sewer sludge disposal area. During summer months, when these crabs migrate offshore, moribund crabs are frequently collected in dredge hauls taken in the sludge disposal area, invariably with extensive erosions of their exoskeleton (fig 6). Also, their gill chambers are filled with fine debris characteristic of sediments in the sludge disposal area (fig 7). Preliminary analyses of distribution of *Cancer* during the summer migration suggest that crabs may move onto the sludge beds where they perish.

We also found that crabs (*Cancer irroratus*) and lobsters (*Homarus americanus*) maintained on sewer sludge and dredge spoil sediments in well-aerated circulating aquaria developed anomalies similar to those found in crabs collected in the sewer sludge disposal areas. In both instances, erosions developed in the heavy exoskeleton which covers the outer portions of the carapace and appendages as well as in the thin cuticles which cover the gills (fig 8). These erosions developed within six weeks after initial exposure to the sludge sediments.

An even more significant finding is that during late May and June large numbers of crab megalopa larvae settle from the plankton and assume the benthic habitus throughout the New York Bight. The first crab stage often numbers 7/0.1 m² and is found in equal numbers on the dredge spoil and sewer sludge contaminated sediments as well as unpolluted environments. Later crab stages are not, however, found on contaminated sediments suggesting that larvae and juvenile stages die in these environments.

Tremendous numbers of crab and other larvae must therefore perish in the 20 mi² area contaminated by dredge spoils and sewer sludge.

Conclusions

The dumping of contaminated dredge spoils and sewer sludge in the New York Bight during 40 years has resulted in two separate areas becoming impoverished of normal benthic fauna; these two areas total 20 mi². Toxins associated with these wastes and low levels of dissolved oxygen at the sediment-water interface probably account for the lack of fauna. The two disposal areas are surrounded by benthic communities of low diversity and high dominance. The dominant organisms include taxa which are apparently resistant to the stress conditions associated with the dredge spoil and sewer sludge beds. Larger crustaceans exposed to sewer sludge and dredge spoils in the laboratory and field develop distinct pathological anomalies.

References

BARBER, R I and KRIEGER, D Growth of phytoplankton in waters
1970   from the New York City sludge dumping grounds. *In* Proceedings of the Thirty-third Annual Meeting American Society of Limnology and Oceanography, University of Rhode Island, Kingston, 25–29 August 1970 (Abstract).
FINCHAM, A A Amphipods of the shallow-water sand community
1969   in the northern Irish Sea. *J. mar. biol. Ass. U.K.*, 49(4): 1003–24.
GROSS, M G New York metropolitan region—a major sediment
1970   source. *Wat. Resour. Res.*, 6(3):927–31.

Acknowledgement

To the U.S. Army Corps of Engineers, Coastal Engineering Research Centre, for the funds for this study.

Waterfront Housing Developments: Their Effect on the Ecology of a Texas Estuarine Area*

W. L. Trent, E. J. Pullen and D. Moore†

La construction de logements sur le littoral: ses incidences sur l'écologie d'une zone estuarine au Texas

En 1969, on a effectué des études destinées à comparer l'écologie d'une zone estuarine naturelle (marais et baie) avec celle d'une zone estuarine adjacente modifiée par le creusement de canaux, le cloisonnement et l'assèchement. Dans chaque zone, on a étudié périodiquement par sondage, de mars à octobre, les facteurs hydrographiques, les poissons, les crustacés et les macro-invertébrés benthiques. On a mesuré les taux de fixation, de croissance et de mortalité des huîtres juvéniles (*Crassostrea virginica*) durant les mois de février à octobre et déterminé la productivité du phytoplancton du mois de juin au mois d'août.

Les taux de concentration de l'oxygène, des nitrites et de l'azote (méthode de Kjeldahl) étaient nettement plus élevés dans la zone naturelle, tandis que le taux de phosphate total était nettement plus élevé dans la zone modifiée. Bien que l'on ait dénombré au chalut 64 espèces de poissons et de crustacés, six d'entre elles représentaient 88,8 pour cent du total des prises. *Brevoortia patronus, Anchoa mitchilli* et *Micropogon undulatus* étaient les plus abondantes dans la zone modifiée: *Penaeus aztecus, P. setiferus* et *Leiostomus xanthurus* étaient les plus abondantes dans la zone naturelle. Le taux de fixation d'un essaim d'huîtres était 14 fois plus élevé dans le milieu naturel que dans la zone modifiée et le taux de croissance des jeunes huîtres était 1,8 fois plus rapide. La mortalité des huîtres était sensiblement plus élevée dans la zone modifiée pendant l'été.

La production brute de phytoplancton dans les eaux de surface atteignait en moyenne 2,24, 2,06 et 1,17 mg de carbone par litre et par jour dans la zone modifiée, le marais et la baie respectivement. Dans une partie de la zone modifiée, la production extrêmement forte de phytoplancton, responsable du faible taux d'oxygène dissous, a provoqué une réduction de l'abondance des poissons et des invertébrés pendant l'été.

Urbanizaciones a orillas del mar: sus efectos en la ecología de un estuario de Texas

Durante 1969 se hicieron estudios comparativos de la ecología de una zona de estuario natural (pantano y bahía) y la de otra adyacente alterada por la construcción de canales, muros de contención y rellenos. Desde marzo hasta octubre se hicieron en cada zona muestreos periódicos de factores hidrográficos, peces, crustáceos y macro-invertebrados bentónicos. De febrero a octubre se midieron los ritmos de fijación, crecimiento y mortandad de ostras juveniles (*Crassostrea virginica*) y de junio a agosto se determinó la productividad del fitoplancton.

Los índices de oxígeno, nitrito y nitrógeno (determinado por el método Kjeldahl) fueron sensiblemente mayores en la zona natural, en tanto que en la alterada fue bastante mayor el fósforo total. Si bien se recogieron con artes de arrastre 64 especies de peces y crustáceos, sólo seis de ellas constituían el 88,8 por ciento de la captura total. Las más abundantes especies en la zona alterada fueron *Brevooritia patronus, Anchoa mitchilli* y *Micropogon undulatus*, y en la zona natural, *Penaeus aztecus, P. setiferus* y *Leiostomus xanthurus*. Los ritmos de fijación de larvas de ostras y de crecimiento de ostras juveniles fueron 14 y 1,8 veces mayores respectivamente en la zona natural que en la alterada. La mortandad de las ostras era sensiblemente mayor en la zona alterada durante el verano.

La producción bruta de fitoplancton en las aguas superficiales alcanzó un promedio de 2,24, 2,06 y 1,17 mg de carbono por litro y por día en las zonas alterada, pantanosa y de bahía, respectivamente. En una parte de la zona alterada, la extraordinariamente elevada producción de fitoplancton que era motivo de que hubiera poco oxígeno disuelto, redujo la abundancia de peces e invertebrados durante el verano.

ESTUARIES along the United States coastline are being altered extensively by Federal, State and private institutions. More than 81,000 ha of shallow coastal bays (excluding marshes) in the Gulf of Mexico and South Atlantic areas have been altered over the past 20 years (Chapman, 1968). In Texas deepening of about 700 mi of Federal navigation channels has altered 5,265 ha of bay bottom and destroyed 2,830 ha of brackish marsh and the dredged spoil has filled 2,025 ha of shallow bay and covered 9,315 ha of brackish marsh. Presently, large areas of shallow bays and marshes are being developed for waterfront housing sites (fig 1). This involves dredging, bulkheading and filling. With expanding human populations and increased leisure time it is likely that the demand for these areas will increase.

When areas of shallow marsh and bay are deepened or filled with spoil major changes in the bayshore environment include: (1) reduction in acreage of shore zone and

---

* Contribution No. 311 from the National Marine Fisheries Service Biological Laboratory, Galveston, Texas 77550, U.S.A.
† National Marine Fisheries Service Biological Laboratory, Galveston, Texas 77550, U.S.A.

marsh vegetation, (2) changes in marsh drainage patterns and nutrient inputs and (3) changes in water depth and substrates. The effects of these changes on estuarine organisms are poorly understood.

Our studies compare natural and altered areas with respect to: (1) substrates, (2) selected hydrographic variables, (3) phytoplankton productivity, (4) relative abundance of benthic macro-invertebrates, fishes and crustaceans, and (5) the setting, growth and mortality rates of the American oyster (*Crassostrea virginica*).

Study area and methods

The study area, located in West Bay, Texas (near Galveston Bay system) included a natural marsh, an open bay area, and a canal area that was similar to the natural marsh before it was altered by channellization, bulkheading and filling (fig 1). The developed area, which originally comprised about 45 ha of emergent marsh vegetation, intertidal mud flats and subtidal water area was increased to about 32 ha of subtidal water area with a water volume of (mean low tide level) about 394,000 m³ compared to the original 184,000 m³.

Hydrographic measurements, fish and crustacean samples were taken the same day (between 10.00 h and 14.00 h) and night (between 22.00 h and 02.00 h) at 2-week intervals from 25 March to 21 October 1969, at ten stations (fig 1). Sediment samples were taken to determine the percentage composition of sand, silt and clay water. Samples were taken 30 cm above the bottom. Fishes and crustaceans were collected in a trawl that had a mouth opening of 0.6 m by 3 m and a stretched mesh of 28 mm in the body and 2.5 mm in the codend. At each station the trawl was towed over a distance of 200 m at 2 kn.

Primary productivity was determined twice each month at five stations (1, 2, 6, 7 and 10) in June, July and August using the light- and dark-bottle technique

described by Gaarder and Gran (1927). Water samples were taken 15 cm below the surface and incubated 24 h.

Benthic macro-invertebrates were sampled at 2-week intervals from 25 March to 21 October at six stations. Two stations (1 and 4) were in the canals, three (6, 7, 8) were in the natural marsh, and one (10) was in the open bay (fig 1). Cores of the substrate were taken with a metal cylinder 14 cm long and 9.6 cm in diameter. The number and volume of organisms, percentage of total carbon and the volume of detrital vegetation in the substrate were determined in each sample.

Oyster spatfall, growth, and mortality rates were monitored at stations 1 and 6 from February 1969 to February 1970. Asbestos plates were used to collect spatfall. Eight size groups of juvenile oysters were placed in trays at each station and total lengths of the oysters were determined every 4 weeks. Dead oysters were replaced every 2 weeks with oysters of similar size.

Substrate and hydrology

Substrates in the canals, marsh, and bay were distinctly different (fig 2). Sediments in the canals contained more silt and clay (41 per cent) than in the marsh (31 per cent). The lowest silt and clay contents occurred in the undredged bay area (17 per cent). In a similar study in Boca Ciega Bay, Florida, Taylor and Saloman (1968) found sediments in undredged areas averaged 94 per cent sand and shell and those in dredged canals averaged 92 per cent silt and clay. Values of total carbon in the sediments were highest in the canals and lowest in the bay, but the differences were not great. The amount of detrital vegetation on and in the substrates was more than twice as great in the marsh as in the canals. Detritus was almost absent in the bay.

Average temperature, salinity, total alkalinity and pH differed only slightly between areas (fig 3). The average and range of dissolved organic nitrogen were highest in the marsh. A major part of the nitrogen in the marsh may have originated from cattle grazing adjacent to this area. Total phosphorus was highest, and had the greatest range, in the canals and possibly originated in runoff from fertilized lawns and seepage from septic tanks.

Average values of nitrogen and phosphorus were much lower in West Bay than had been previously found in Clear Lake (Pullen, 1969); and in upper Galveston Bay (Pullen and Trent, 1969), probably because domestic and industrial wastes are not discharged into the middle area of West Bay. Clear Lake is a heavily populated area, with inadequate sewage treatment facilities and upper Galveston Bay receives domestic and industrial waste from the City of Houston via the Houston Ship Channel.

Average turbidity of bottom waters (Jackson turbidity unit—JTU) was highest in the bay, but the highest values were measured in the canals. In surface water samples, however, turbidities were over twice as high in the marsh and bay as in the canals. These observations agree with those reported by Taylor and Saloman (1968).

Average values of dissolved oxygen were highest in the bay, intermediate in the marsh, and lowest in the canals. During the summer, oxygen dropped to critically low levels (less than 0.2 ml/l) at the three stations (1, 2 and 3) farthest from the bay. The annual average at stations 1, 2 and 3 was 3.17 compared with 4.39 ml/l for stations 4 and 5 nearer the bay. Taylor and Saloman (1968) stated

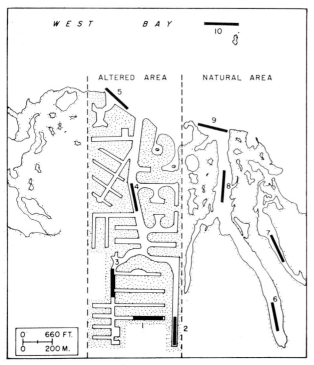

Fig 1. The Galveston Bay system showing the study area and station locations in West Bay, Texas

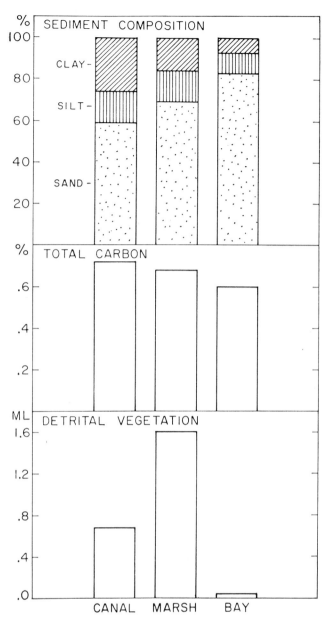

Fig 2. *Comparisons of sediment composition, total carbon, and detrital vegetation between the canals, marsh and bay*

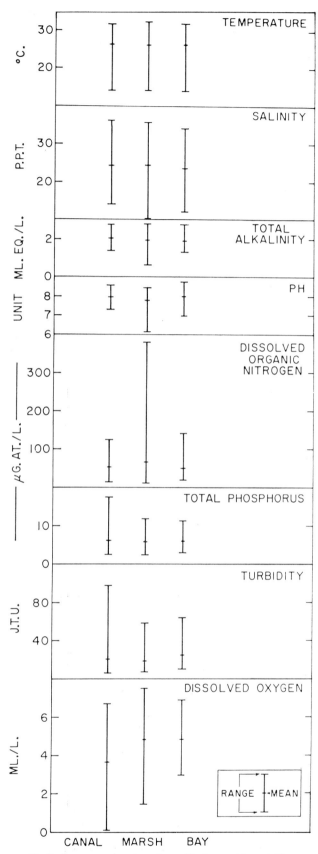

Fig 3. *Average and range values of hydrographic variables observed in the canals, marsh and bay*

that oxygen levels were reduced during the summer over the soft sediments of housing development canals.

Phytoplankton

Phytoplankton productivity in the West Bay area was high compared with other marine environments. Gross photosynthesis was 1.96 mg C/l day and respiration was 0.51 mg C/l day; the values ranged from 0.87 to 3.43 for photosynthesis and 0.23 to 1.19 for respiration. The average values were much higher than those reported by Williams (1966) for shallow estuaries along the Atlantic coast. At the 50 per cent insolation depth, his mean values were 0.50 mg C/l day for gross photosynthesis and 0.19 for respiration. The differences were even greater than indicated because, according to Williams, the rate of photosynthesis at the depth (about 50 cm in our area) of 50 per cent insolation is higher than at a depth of 15 cm. Net photosynthesis in the Atlantic coast estuaries was greater than most of the values from other marine environments (Williams, 1966).

Average gross photosynthesis rates for June, July and August ranged from 1.17 mg C/l day in the bay to 2.25 in the canals (fig 4). Average values at the canal stations were almost identical as were those at the two marsh

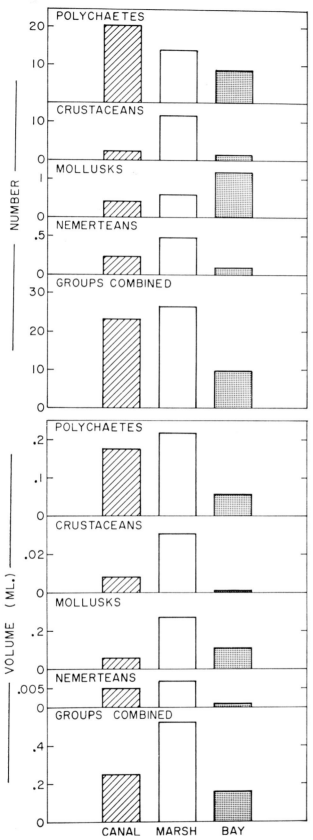

Fig 4. Gross and net photosynthesis, and respiration by area and date, and average values for all sampling dates

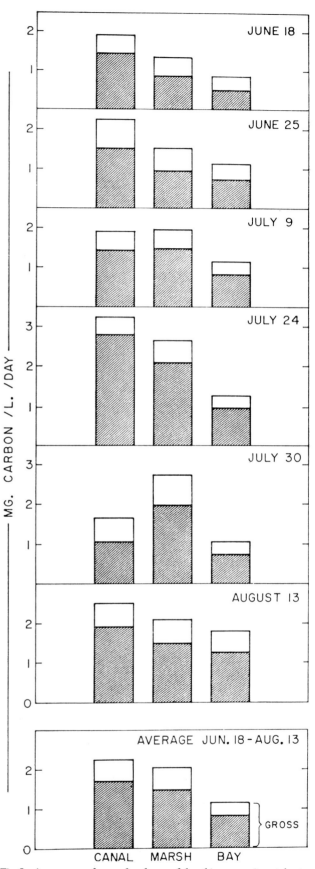

Fig 5. Average number and volume of benthic macro-invertebrates collected per 800 cm² of bottom material by area, taxonomic group, and for groups combined

stations. Average photosynthesis in the canals was slightly higher (8 per cent) than in the marsh and much higher (49 per cent) than in the bay. Taylor and Saloman (1968) reported that primary production of phytoplankton

did not differ consistently between development canals and open bay areas.

Benthic macro-invertebrates

Polychaetes comprised 66 per cent of the number and 44 per cent of the volume of benthic animals caught. Crustaceans were second in number (29 per cent) but lowest in volume (4 per cent). Molluscs were third in abundance (3 per cent), but second in volume (41 per cent). The volume of molluscan biomass was much lower than this indicates, however, because the volume includes shells. Nemerteans were lowest in number (1 per cent) and third in volume (11 per cent).

Benthic animals were slightly more abundant numerically and about twice as abundant volumetrically in the marsh as in the canals; they were least abundant in the bay (fig 5). The order of abundance varied, however, when individual groups were considered. Capitellidae were the dominant organisms caught (91 per cent of all polychaetes collected) and were most abundant in the marsh and canals where substrates were largely silt, clay and detritus. Individuals of this family burrow through the substrate and obtain their food by ingesting organic matter in the sand and mud (Barnes, 1963). Crustaceans, most of which belonged to the families Ampeliscidae and Corophiidae (99 per cent of all crustaceans collected and identified), were over three times as abundant in the marsh as in the other two areas. Molluscs, mostly the genera *Tellina*, *Tagelus* and *Mulinia* (95 per cent of all molluscs collected), were numerically most abundant in the bay although volumetrically, the marsh had the highest standing crop. Nemerteans were most abundant in the marsh and least abundant in the bay.

Oyster spatfall, growth and mortality

About 14 times more oyster spat attached to sampling plates in the marsh than in the canal during the 12-month period. On a 600 cm^2 surface area, 184 attached in the marsh and 13 in the canals. These rates were much lower than those observed by Hopkins (1931) in West Bay.

Juvenile oysters, 44 mm in average total initial length, achieved a yearly average increase in shell length of 52 mm in the marsh, 72 per cent faster than the 33 mm achieved in the canals. Growth in the marsh was similar to the average for Texas given by Hofstetter (1965) who estimated that it takes from 18 to 24 months for oysters to reach a length of 76 mm.

The annual rate of mortality averaged 91 per cent in the canals and 52 per cent in the marsh. In a nearby area (Louisiana) of similar climate, Mackin (1961) noted that the normal annual loss of oyster, 1 year and older, was between 50–70 per cent and might run as high as 90 per cent or as low as 30 per cent. On this basis the mortality of oysters in the marsh was a "low average mortality" whereas oyster mortality in the canals was slightly above the "high extreme mortality" observed in the Louisiana studies.

Fishes and crustaceans

Sixty-four species and 240,575 specimens of finfish and crustaceans were taken with the trawl. Of the 64 species, 54 occurred in the marsh, 52 in the canals and 44 in the bay. In terms of numbers of all species caught the marsh was the most productive area and the canals the second most productive. The average number of animals caught per tow was 951 in the marsh, 659 in the canals and 412 in the bay.

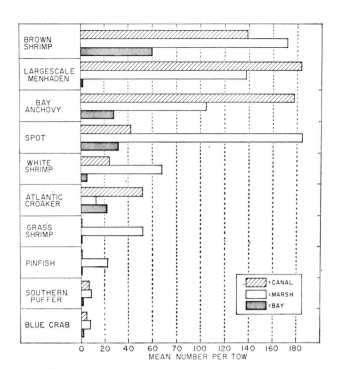

Fig 6. The 10 species caught in greatest abundance in the trawl, with comparisons of their abundance between canals, marsh and bay

The ten most abundant species represented 96 per cent of the total number of specimens (fig 6). Of the six most abundant species (89 per cent of the total catch), brown shrimp (*Penaeus aztecus*), white shrimp (*P. setiferus*), spot (*Leiostomus xanthurus*), large scale menhaden (*Brevoortia patronus*) and Atlantic croaker (*Micropogon undulatus*) are commercially valuable and the bay anchovy (*Anchoa mitchilli*) is important as food for commercial and sport fish species. The first three species were most abundant in the marsh and the last three were most abundant in the canals.

Brown shrimp, the most valuable commercial fishing species in the United States were more abundant in the canals and marsh, probably because of bottom type and food availability. Bottom sediments in these areas contained more silt and clay, vegetative material and total carbon than in the bay (fig 2). Juvenile brown shrimp feed mainly on detrital material and benthic organisms and prefer soft, muddy substrates with large amounts of detrital material (Williams, 1958). Benthic organisms were more abundant and phytoplankton productivity was higher, in the marsh and canals than in the open bay. Williams (1955) reported that stomachs of brown shrimp from estuarine areas along the Atlantic coast contained, in order of decreasing frequency of occurrence, masses of unrecognizable debris, chitin fragments of crustaceans, setae and jaws of annelids, plant fragments and sand.

Large scale menhaden and bay anchovy are plankton feeders during their juvenile stages. Menhaden feed predominantly on phytoplankton, and anchovy mostly on zooplankton (Darnell, 1958). The abundance of these fishes in the three areas was related, although not proportionately, to phytoplankton productivity in the areas (fig 4).

Juvenile spot were about four times more abundant in the marsh than in the canals and open bay. This is probably related to the high abundance of crustaceans

[415]

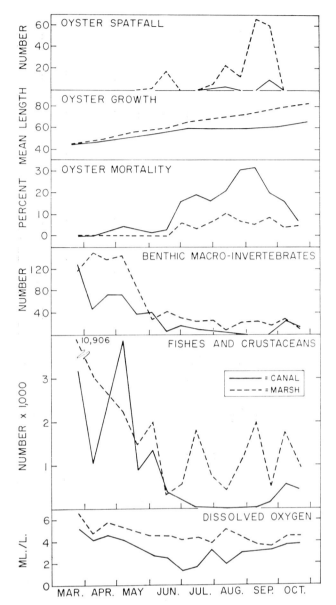

Fig 7. Oyster spatfall, growth and 2-week mortality rates, abundance of benthic organisms, total number of fish and crustaceans caught, and dissolved oxygen by date at stations 1 in the canal and 6 in the marsh

Seasonal relations

Seasonal relations between levels of dissolved oxygen, spatfall, growth and mortality rates of oysters, abundance of benthic organisms and fishes and crustaceans at stations 1 in the canal and 6 in the marsh are shown in fig 7. Oxygen levels were even more critical at station 1 than the figure indicates because: (1) the data were taken at a time of day that was not the most critical (about 06.00 h); and (2) in our regular schedule we did not sample during times of heavy plankton blooms. Other observations revealed, however, that plankton blooms sufficient to reduce oxygen levels to zero at night occurred at least four times at station 1 during June–August.

Poor oyster growth, high oyster mortality and low to nil standing crops of benthic organisms, fishes and crustaceans during June–September were probably directly or indirectly caused by low oxygen levels at station 1. Stations 2 and 3 in the area of the development farthest from the bay had low oxygen levels during the summer and a smaller than average standing crop of fishes and crustaceans. We think poor water circulation in parts of the development canals caused conditions favourable for high populations of phytoplankters.

Conclusions

In general, productivity was highest in the marsh, intermediate in the canals of the altered area and lowest in the open bay. Productivity in the canals would probably have been much higher if dissolved oxygen levels had been higher during the summer.

The standing crops of benthic organisms, fishes and crustaceans in the altered area were high and we are planning studies to determine the relative contributions of vegetative material by various primary producers. We know that phytoplankton, attached algae, and mud diatoms are produced in the altered area. In addition, submerged sea grasses and emergent vegetation (dominated by *Spartina alterniflora*) grow in the natural marsh. We do not know, however, whether the altered area is self-supporting in terms of vegetative productivity or derives much of the vegetative detritus from the natural marsh through tidal action. If the altered area is not self-supporting and if areas of marsh are developed in ways similar to the present, then biological productivity of the estuarine zone will be reduced in relation to the acres of marsh altered.

(fig 5) in the marsh area. In general, juvenile spot feed predominantly on planktonic and benthic micro-crustaceans (Gunter, 1945; Darnell, 1958).

White shrimp show a more distinct preference than brown shrimp for shallow water habitats characterized by muddy or peaty bottoms high in organic detritus and an abundance of marsh grasses (Weymouth, Lindner and Anderson, 1933; Williams, 1955; Loesch, 1965; Mock, 1967). These factors, along with those discussed previously for brown shrimp, are important in explaining the observed distribution of white shrimp.

Atlantic croakers were most abundant in the canals and least abundant in the marsh. Why they were so is difficult to explain. Juvenile croakers prefer soft substrates where they can obtain much of their food by digging for subsurface invertebrates and organic debris (Roelofs, 1954; Reid, 1955). This type of substrate was not present in the bay.

References

BARNES, R D   Invertebrates zoology. Philadelphia, W B Saunders
1963   Co., 632 p.
CHAPMAN, C   Channelization and spoiling in Gulf coast and South
1968   Atlantic estuaries. *In* Proceedings of the Marsh and Estuary Management Symposium, Louisiana State University, pp. 93–106.
DARNELL, R M   Food habits of fishes and larger invertebrates of
1958   Lake Pontchartrain, Louisiana, an estuarine community. *Publs Inst. Mar. Sci. Univ. Tex.*, (5):353–416.
GAARDER, T and GRAN, H H   Investigations of the production of
1927   plankton in the Oslo Fjord. *Rapp.P.-v. Réun. Cons. Perm. int. Explor. Mer.*, 42:1–48.
GUNTER, G   Studies on marine fishes of Texas. *Publs Inst. Mar. Sci.*
1945   *Univ. Tex.*, (1):1–190.
HOFSTETTER, R P   The Texas oyster fishery. *Bull. Tex. Parks Wildl.*
1965   *Dep.*, (40):1–39
HOPKINS, A E   Factors influencing the spawning and setting of
1931   oysters in Galveston Bay, Texas. *Bull. Bur. Fish., Wash.*, (47):57–83.
LOESCH, H   Distribution and growth of penaeid shrimp in Mobile
1965   Bay, Alabama. *Publs Inst. mar. Sci. Univ. Tex.*, (10):41–58.

MACKIN, J G   A method of estimation of mortality rates in
1961   oysters. *Proc. natn. Shellfish. Ass.*, 50(1959):41–51.
MOCK, C R   Natural and altered estuarine habitats of penaeid
1967   shrimp. *Proc. Gulf. Caribb. Fish. Inst.*, 19(1966):86–98.
PULLEN, E J   Hydrological conditions in Clear Lake, Texas, 1958–
1969   66. *Spec. scient. Rep. U.S. Fish Wildl. Serv. (Fish)* (578): 8 p.
PULLEN, E J and TRENT, L   Hydrographic observations from the
1969   Galveston Bay system, Texas, 1958–67. *Data Rep. U.S. Fish
Wildl. Serv. (microcards)* (31):1–151.
REID, G K   A summer study of the biology and ecology of East
1955   Bay, Texas. Part 2. The fish fauna of East Bay, the Gulf
beach, and summary. *Tex. J. Sci.*, 7:430–53.
ROELOFS, E W   Food studies of young sciaenid fishes *Micropogon*
1954   and *Leiostomus*, from North Carolina. *Copeia*, 1954 (2):
151–3.

TAYLOR, J L and SALOMAN, C H   Some effects of hydraulic dredg-
1968   ing and coastal development in Boca Ciega Bay, Florida.
Fishery Bull. U.S. Fish Wildl. Serv., 67:213–42.
WEYMOUTH, F W, LINDNER, M J and ANDERSON, W W   Prelim-
1933   inary report on the life history of the common shrimp,
Penaeus setiferus (Linn.). *Bull. Bur. Fish., Wash.*, (48):1–26.
WILLIAMS, A B   A contribution to the life histories of commercial
1955   shrimps (Penaeidae) in North Carolina. *Bull. mar. Sci. Gulf
Caribb.*, 5:117–46.
WILLIAMS, A B   Substrates as a factor in shrimp distribution.
1958   *Limnol. Oceanogr.*, 3:283–90.
WILLIAMS, R B   Annual phytoplanktonic production in a system of
1966   shallow temperate estuaries. *In* Some contemporary studies
in marine science, edited by Harold Barnes. London,
George Allen & Unwin Ltd., pp. 699–716.

A Trawl Study in an Area of Heavy Waste Discharge: Santa Monica Bay, California

*J. G. Carlisle, Jr.**

Etude de chalutage dans une zone de déversement abondant: baie de Santa Monica, Californie

Une étude de chalutage a été effectuée dans la baie de Santa Monica, importante zone de déversement des effluents, pendant les années 1958–1963. On espérait que les renseignements obtenus sur la fréquence et la répartition des poissons de fond dans cette zone mettraient en lumière les effets du déversement de matières usées sur leurs populations. Au total, on a effectué 705 traits de fond sur une profondeur allant de 18 à 183 mètres et on a capturé 112.799 poissons appartenant à 104 espèces, plus un certain nombre de juvéniles de scorpénidés non identifiés. Cinq espèces constituaient la majeure partie de ces prises. Le rendement des traits, assez élevé en 1958, très faible en 1959, s'est redressé en 1960, a atteint un sommet en 1961, puis a accusé une baisse légère en 1962 et 1963.

Il a été impossible de prouver que les fluctuations d'abondance du poisson, mesurées par les captures du chalutage, étaient causées par le déversement de matières usées dans la zone soumise à l'étude, et ne provenaient pas de causes naturelles. Les prises de poisson de sport entre 1958 et 1963 dans la baie de Santa Monica n'ont connu que de légères fluctuations. Sur les récifs artificiels de la baie de Santa Monica, on a constaté que les algues géantes (*Macrocystis*) sont incapables de survivre dans les zones affectées par les "retombées" d'éléments particulaires provenant des déchets de déversement, voire par la turbidité qui a la même origine.

Sur la base de cette étude, il est possible d'affirmer que le meilleur endroit pour déverser des effluents dans les eaux marines côtières se situe au-dessus des fonds les moins productifs, aussi loin vers le large que possible. Le choix du site devrait être effectué avant que le déversement ne soit autorisé. Il est prouvé que les courants qui entraînent des effluents sur les rochers du rivage et les récifs artificiels provoquent de graves dégâts et entravent la reproduction des algues géantes, voire d'autres organismes sessiles. Les décharges devraient être placées aussi loin que possible de ce type de milieu.

Estudio efectuádo con redes de arrastre en una zona sometida a intensa descarga de desechos: la Bahiá de Santa Mónica, en California

En la Bahía de Santa Mónica, donde se evacuan grandes cantidades de desechos, se ha realizado un estudio con redes de arrastre durante los años 1958–1963. Se esperaba que, recogiendo información con respecto a la presencia y distribución de peces de fondo en esa zona, sería posible determinar los efectos de la evacuación de desechos sobre las poblaciones. En total se realizaron 705 lances de fondo, en aguas de 60 a 600 pies de profundidad (18 a 183 m), y se capturaron 112.799 peces pertenecientes a 104 especies, más algunas formas juveniles de gallinetas no identificadas. El grueso de la captura lo constituyeron cinco especies. El rendimiento por lance, que fue moderadamente elevado en 1958 y extremadamente bajo en 1959, registró una recuperación en 1960, aumentó ulteriormente en 1961 y en 1962 y en 1963 registró una ligera disminución.

Ha sido imposible demostrar que las fluctuaciones en la abundancia, medidas por las capturas al arrastre, se debían a la descarga de desechos en la zona en estudio y no, sencillamente, a causas naturales. La pesca deportiva en la Bahía de Santa Mónica, entre los años 1958 y 1963, sólo ha registrado pequeñas fluctuaciones. Los arrecifes artificiales de la Bahía de Santa Mónica han demostrado que las algas gigantes (*Macrocystis*) no pueden sobrevivir en zonas afectadas por la "lluvia" de partículas procedentes de las descargas de desperdicios y tal vez incluso por la turbiedad causada por esas descargas.

Basándose en este estudio, se puede recomendar que, si es preciso descargar los desechos en las aguas costeras, el mejor lugar para ello es el fondo menos productivo que sea posible encontrar a cierta distancia de la costa. Antes de permitir la descarga, ha de determinarse dónde se encuentra esa zona. Cuando las corrientes transportan las aguas de descarga hasta llevarlas cerca de las rocas próximas a la costa y de los arrecifes artificiales, las algas gigantes y probablemente otros organismos sésiles, sufren serios daños y se produce una inhibición de la reproducción. Los desagües deben colocarse, pues, lo más lejos posible de estos lugares.

IN September 1957, the Bureau of Sanitation, City of Los Angeles, California, and the California Department of Fish and Game entered into an informal agreement to conduct a trawl study as part of a continuing surveillance programme for Santa Monica Bay. The purpose of this programme is to evaluate the effect on the marine environment of the waste discharges from the Hyperion Sewage Treatment Plant. This plant discharged effluent varying from 261 to 283 million gal/day during the 6 years of the study. Solids discharged by the sludge line ranged from 130 to 156 t/day.

The objective of this study was to determine changes in bottom fish populations of the Bay, which might result from the discharge of tremendous quantities of wastes. In the past, attempts to prove and document the long-term and subtle influence of major waste discharges on the open coastal biotope have invariably met with frustration and failure (Ludwig and Onodera, 1964). Six years of trawling on a quarterly basis from 1958 through 1963 provided a large quantity of data.

This paper deals principally with the trawl data for Santa Monica Bay. However, since they are pertinent to the overall problem of pollution of the area, work dealing with marine habitat development (Carlisle, Turner and Ebert, 1964) and the sport fish catch for the years 1958 through 1963 in Santa Monica Bay are also discussed.

Methods

Trawling operations were conducted from the *Prowler*, a 65-ft converted salmon troller used in the

* Marine Resources Operations, California Department of Fish and Game, Long Beach, California 90731, U.S.A.

Santa Monica Bay monitoring programme. Trawling equipment included a winch powered by the main engine and an A-frame on the fantail through which a single cable led over a block to the trawl net. The net was a 24-ft semi-balloon or tri-net with a body of 1½-in mesh, number 18 twine. Originally, a liner of ¼-in webbing was used to catch small invertebrates, but larger amounts of debris were retained. Therefore, we changed to a liner of ½-in mesh synthetic material which made little difference in fish catches.

A series of 39 regular stations in depths of 60 to 600 ft was trawled each quarter (fig 1). While difficulties in scheduling the boat and crew made it impossible always to adhere to the quarterly programme, it was followed as closely as possible.

Each trawl was for 10 min clocked from the time the net was on the bottom (and fishing) until the winch was started and net retrieving began. All fish caught in each haul were sacked and labelled by consecutive haul num-

ber. Shelled molluscs were counted or their numbers estimated and either returned to the water or preserved for later identification.

Surface and bottom temperatures and dissolved oxygen readings were taken at all stations. The monitoring programme also included chloride and coliform samples from some stations.

All fish taken in the trawl were taken back to the laboratory and examined. A record was kept listing haul number, date, depth of water from which taken, fish weight, length, and sex for each species. In large catches, subsamples were taken, and the remainder counted and their size range recorded. Notations were made on maturity, stomach contents parasites, and evidence of disease or abnormalities.

Discussion

During the study, 112,799 fish of 104 species plus unidentified juvenile rockfish were taken in 705 hauls

TABLE 1. FISHES TRAWLED IN SANTA MONICA BAY FROM THE *Prowler*, 1958–1963

Common name	Species	Number	Common name	Species	Number
1. Speckled sanddab	*Citharichthys stigmaeus*	30,581	54. Fringed sculpin	*Icelinus fimbriatus*	19
2. Yellowchin sculpin	*Icelinus quadriseriatus*	23,791	55. Pacific electric ray	*Torpedo californica*	18
3. California tonguefish	*Symphurus atricauda*	10,438	56. Rubberlip perch	*Rhacochilus toxotes*	16
4. Plainfin midshipman	*Porichthys notatus*	10,314	57. Cow rockfish	*Sebastodes levis*	16
5. Slender sole	*Lyopsetta exilis*	8,589	58. Thornback	*Platyrhinoidis triseriata*	14
6. English sole	*Parophrys vetulus*	4,178	59. Black perch	*Embiotoca jacksoni*	14
7. Pacific sanddab	*Citharichthys sordidus*	3,803	60. Ratfish	*Hydrolagus colliei*	13
8. Dover sole	*Microstomus pacificus*	3,274	61. Shortbelly rockfish	*Sebastodes jordani*	11
9. White Croaker	*Genyonemus lineatus*	2,138	62. Pile perch	*Rhacochilus vacca*	10
10. Pink seaperch	*Zalembius rosaceus*	2,036	63. California lizardfish	*Synodus lucioceps*	9
11. Blacktip poacher	*Xeneretmus latifrons*	1,466	64. Pacific pomano	*Palometa simillima*	9
12. Hornyhead turbot	*Pleuronichthys verticalis*	1,368	65. Roughback sculpin	*Chitonotus paugetensis*	9
13. Blackbelly eelpot	*Lycodopsis pacifica*	1,332	66. Dark-blotched rockfish	*Sebastodes crameri*	8
14. Stripetail rockfish	*Sebastodes saxicola*	1,249	67. Shovelnose guitarfish	*Rhinobatos productus*	8
15. Pigmy poacher	*Odontopyxis trispinosa*	1,169	68. Brown rockfish	*Sebastodes auriculatus*	8
16. Calico rockfish	*Sebastodes dalli*	842	69. Smooth stargazer	*Kethetostoma averruncus*	8
17. Queenfish	*Seriphus politus*	778	70. Starry rockfish	*Sebastodes constellatus*	7
18. Bigmouth sole	*Hippoglossina stomata*	584	71. Sandpaper skate	*Raja kincaidii*	6
19. Shortspine combfish	*Zaniolepis frenata*	509	72. Walleye surfperch	*Hyperprosopon argenteum*	6
20. Splitnose rockfish	*Sebastodes diploproa*	433	73. Kelp bass	*Paralabrax clathratus*	6
21. Greenspotted rockfish	*Sebastodes chlorostictus*	323	74. Sand bass	*Paralabrax nebulifer*	6
22. Bay Goby	*Lepidogobius lepidus*	264	75. Spotted turbot	*Pleuronichthus ritteri*	5
23. Rex sole	*Glyptocephalus zachirus*	253	76. Mexican rockfish	*Sebastodes macdonaldi*	5
24. Specklefin midshipman	*Porichthys myriaster*	244	77. Shortspine channel rockfish	*Sebastodes alascanus*	3
25. Bluespotted poacher	*Xeneretmus triacanthus*	216	78. Horn shark	*Heterodontus francisci*	3
26. Spotfin sculpin	*Icelinus tenuis*	207	79. California skate	*Raja inornata*	3
27. Longspine combfish	*Zaniolepis latipinnis*	204	80. Painted greenling	*Oxylebius pictus*	2
28. Shiner perch	*Cymatogaster aggregata*	178	81. Bearded ellpout	*Lyconema barbatum*	2
29. Curlfin turbot	*Pleuronichthys decurrens*	147	82. Pacific angel shark	*Squatina californica*	2
30. Pacific argentine	*Argentina sialis*	124	83. Bluebarred prickleback	*Plectobranchus evides*	2
31. Halfbanded rockfish	*Sebastodes semicinctus*	119	84. Smoothgum ellpout	*Aprodon cortezianus*	1
32. Fantail sole	*Xystreurys liolepis*	111	85. Red brotula	*Brosmophycis marginata*	1
33. Longfin sanddab	*Citharichthys xanthostigma*	111	86. Bocaccio	*Sebastodes paucispinis*	1
34. Pacific hake	*Merluccius productus*	106	87. Honeycomb rockfish	*Sebastodes umbrosus*	1
35. Basketweave cusk-eel	*Otophidium scrippsae*	105	88. Chilipepper	*Sebastodes goodei*	1
36. Vermilion rockfish	*Sebastodes miniatus*	89	89. Rainbow seaperch	*Hypsurus caryi*	1
37. Bluespot goby	*Coryphopterus nicholsi*	88	90. Sablefish	*Anoplopoma fimbria*	1
38. Pipefish	*Syngnathus* spp.	88	91. Cabezon	*Scorpaenichthys marmoratus*	1
39. California halibut	*Paralichthys californicus*	87	92. Giant kelpfish	*Heterostichus rostratus*	1
40. C-O turbot	*Pleuronichthys coenosus*	86	93. Northern spearnose poacher	*Agonopsis emmelane*	1
41. White seaperch	*Phanerodon furcatus*	80	94. Swell shark	*Cephaloscyllium uter*	1
42. Spotted cusk-eel	*Otophidium taylori*	63	95. Grey smoothhound	*Mustelus californicus*	1
43. Sculpin	*Scorpaena guttata*	55	96. Diamond turbot	*Hypsopsetta guttulata*	1
44. Scarcastic fringehead	*Neoclinus blanchardi*	49	97. Pacific ocean perch	*Sebastodes alutus*	1
45. Greenstriped rockfish	*Sebastodes elongatus*	46	98. Sharpchin rockfish	*Sebastodes sacentrus*	1
46. Bonehead sculpin	*Artedius notospilotus*	38	99. Longspine channel rockfish	*Sebastodes altivelis*	1
47. Petrale sole	*Eopsetta jordani*	30	100. Olive rockfish	*Sebastodes serranoides*	1
48. Slender sculpin	*Radulinus asprellus*	28	101. Threadfin sculpin	*Iclinus filamentosus*	1
49. Spiny dogfish	*Squalus acanthias*	26	102. Grunt sculpin	*Rhamphocottus richardsoni*	1
50. Unidentified juvenile rockfish	*Sebastodes* spp.	26	103. White barred blenny	*Poroclinus rothrocki*	1
51. Flag rockfish	*Sebastodes rubrivinctus*	23	104. Whitebelly rockfish	*Sebastodes vexillaris*	1
52. Southern spearnose poacher	*Agonopsis sterletus*	23			
53. Smooth ronguil	*Rathbunella hypoplectus*	21			

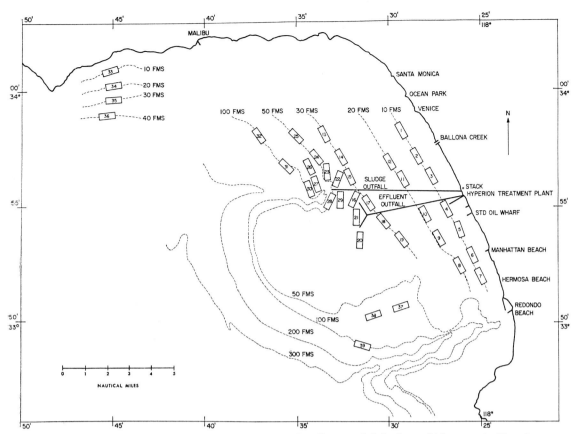

Fig 1. Santa Monica Bay trawl station and locations of the effluent and sludge lines which carry the Hyperion Treatment Plant waste discharge

(Table 1). Twenty species made up 96 per cent (108,872 individuals) of the total catch and were represented in all hauls. The other 84 species were represented by 3,922 individuals.

Five species of fish were taken in such numbers (83,713 or 74 per cent of the total catch) and with such consistency that evaluation of the catch data is based mainly on them: speckled sanddab, *Citharichthys stigmaeus*, yellowchin sculpin, *Icelinus quadriseriatus*, California tonguefish, *Symphurus atricauda*, plainfin midshipman, *Porichthys notatus*, and slender sole, *Lyopsetta exilis*.

When the catch-per-haul of the species is compared with the catch-per-haul of all fish taken on a yearly

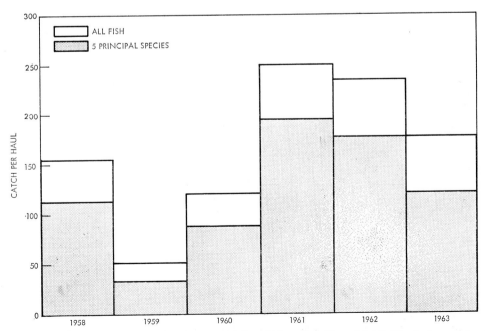

Fig 2. Catch per haul of the five principal species: speckled sanddab, yellowchin sculpin, California tonguefish, plainfin midshipman, and slender sole, compared with that of all species combined

[419]

basis, their dominance is clearly indicated as well as the trend from a moderately high yield in 1958, a severe drop in 1959, a build-up in 1960, a high for the period in 1961, a moderate drop in 1962 and again in 1963 (fig 2).

The speckled sanddab showed a similar population trend. A positive attraction of this species to the nutrient-enriched area near the sludge outfall (7 mi offshore) is evident during the last three years of the study, perhaps as the sludge accumulated. An increased catch of speckled sanddabs was evident at Stations 22, 28, and 29 when compared with the more remote Stations 23, 24, and 25. Speckled sanddabs avoided the effluent area.

Yellowchin sculpin were caught in fewer numbers than flatfish in 1958; they peaked in abundance in 1961, especially in the southern inshore area of the Bay. Yellowchin sculpin catches in general followed the catch trend for the Bay. This species also avoided the effluent area, stations 16, 17, and 21, being taken in considerably greater abundance at Stations 13, 14, and 15.

California tonguefish catches were relatively high in 1958, very low in 1959, and showed a gradual steady increase through 1962, followed by a moderate decrease in 1963. The same avoidance of the effluent area, but to a lesser extent, is exhibited by this species; with greater numbers being caught at Stations 13, 14, and 15 away from the effluent outfall, than at the nearer Stations 16, 17, and 21.

The next species in order of abundance, the plainfin midshipman, followed a somewhat different pattern, as a gradual increase in catch occurred from 1958 through 1961, without the usual 1959 decrease. A slight decrease occurred in 1962 and a greater decrease in 1963.

Catches of slender sole increased gradually from 1958 through 1962 with some drop off in 1963. No special effect of the sludge or effluent was noted on these two species.

There appears to be no valid reason for the fluctuations in the trawl catches, but it is evident from the data that some fishes tend to concentrate in enriched areas near sludge outfalls and others tend to avoid the effluent.

Other studies (Hartman, 1956) show that within a $\frac{1}{2}$ mi radius of the end of the old 1 ml Hyperion outfall (discontinued in 1962, solid loading removed in 1960) the faunal elements are limited both qualitatively and quantitatively and that the most abundant animals are polychaetes (Nereis procera, Capitella capitata, Dioptra ornata or Nothria elegans). Molluscs, echinoderms, and crustaceans appear to be usually low in abundance.

An examination of the surface and bottom temperature data taken during trawling operations revealed no correlation between the mean annual temperatures and the fluctuations in trawl catches.

It is impossible to show that fluctuations in overall abundance, as measured by trawl catches, are the result of large-scale waste discharges in the study area, and are not due to natural causes. Nevertheless our data indicate that some species tend to avoid the immediate effluent discharge area and other species appear to be attracted to the sludge area. This attraction or repulsion to sludge and effluent causes local changes, but do not show Santa Monica Bay to be changing to an area untenable to fish life as was once thought.

On the basis of this study it can be recommended that if wastes must be discharged into coastal marine waters, the best place for the discharge is over the least productive bottom possible, well offshore. This should be determined before discharge is permitted.

Effect of waste discharge on sport catch

When the sport catch for the years 1958 through 1963 for Santa Monica Bay is scrutinized, no effect of large-scale waste discharge is evident on the catch per angler day. The large fluctuations evident in the yearly comparisons of trawl catches are missing in the sport-catch data. Small fluctuations in catch per angler day occurred from year to year, but no trend is apparent, and there is nothing to suggest any effect of the waste discharge.

Effect of pollutants on algal growth

Although the direct effects of pollutants on large marine fish populations are difficult to measure, the effects of particulate matter deposited on the rocky nearshore environment appear all too obvious (North, 1964). Rocky areas and reefs, whether natural or artificial, may be subjected to a build-up of particulate matter which smothers the gametes or other microscopic products of plant reproduction to the extent that growth is almost completely inhibited. Turbidity of the water within the influence of large waste discharges may also cause serious problems of a major nature, and other constituents of the effluent may compound the problem.

Twenty old car bodies were placed off Paradise Cove during May 1958 in an area of normally healthy and plentiful giant kelp, Macrocystis, growth. By October 1958, giant kelp had established itself, by natural seeding, on the car body reef. Within one year a luxuriant kelp bed developed, undergoing the normal fluctuations between scarcity during warm water periods (above 19°C) and heavy reproduction and growth when water temperatures fell below that.

The kelp on the artificial reef remained plentiful from the time of the reef's placement until 1961 when the car bodies had deteriorated to a point where kelp holdfasts could no longer find attachment. The fish population reached a high of about 24,000 semi-residents but averaged approximately 11,000. Paradise Cove does not appear to be influenced very much by the effluent from the Hyperion plant.

A streetcar reef established in September 1958 off the Redondo-Palos Verdes Coast, about 22n mi south of the Paradise Cove reef, has supported a peak population of only about 3,000 semi-resident fish, with an average somewhat less than that. The lack of kelp could easily explain the difference. In this area, where deposition of particulate matter was related to sewage discharges, kelps of various species tried but invariably failed to gain a foothold (Carlisle, Turner and Ebert, 1964). Kelp and sand bass supported a good sport fishery until the reef finally became reduced to rubble by late September 1961; after which, the rubble continued to support a good population of sand bass for a period of about $4\frac{1}{2}$ years. The fish population undoubtedly would have been much larger had kelp been able to develop. Kelp also failed to gain a foothold on experimental artificial reefs composed of streetcars, car bodies, concrete shelters, and quarry rock or on large quarry rock reefs placed later (1960) off Malibu, Santa Monica, Hermosa Beach and Redondo Beach. The entire Palos Verdes Peninsula is now practically devoid of the kelp that was once so abundant.

That this particulate matter originates either at the Hyperion outfall, $8\frac{1}{2}$ mi north of the streetcar reef, or the Los Angeles County White Point outfall, $9\frac{1}{2}$ mi south, or at both, is obvious because the coastal surface current in Santa Monica Bay flows in a southerly direction about 85 per cent of the time (Paul Horrer, Marine Adviser, pers. comm.). This northerly reversal is in effect about 25 to 30 per cent of the time along the Palos Verdes Peninsula (Wayne Messick, Los Angeles County Sanitation District, pers. comm.), and this current pattern would bring particulate matter and other outfall constituents into the area of the streetcar reef.

References

BAXTER, J L   Fish surveillance in Santa Monica Bay, California.
1962   In Proceedings of the First National Coastal and Shallow Water Research Conference, edited by D. L. Gorsline. National Science Foundation and Office of Naval Research, Tallahassee, pp. 574–7.

California State Water Quality Control Board. An oceanographic
1965   and biological survey of the southern California mainland shelf. *Publs St. Wat. Qual. Cont. Bd Calif.*, (27):1–232.

CARLISLE, J G, Jr.   Results of a six-year trawl study in an area of
1969   heavy waste discharge: Santa Monica Bay, California. *Calif. Fish. Game*, 55(1):26–46.

CARLISLE, J G, Jr, TURNER, C H and EBERT, E E   Artificial habitat
1964   in the marine environment. *Fish Bull. Calif.*, (124):93 p.

HARTMAN, O   Results of investigations of pollution and its effects
1956   on benthonic population in Santa Monica Bay, California. Los Angeles, Allan Hancock Foundation, University South California, 23 p. (mimeo).

LUDWIG, H F and ONODERA, D   Scientific parameters of marine
1964   waste discharge. *Adv. Wat. Pollut. Res.*, 1(3):37–49.

NORTH, J W   Ecology of the rocky nearshore environment in
1964   southern California and possible influences of discharged wastes. *Adv. Wat. Pollut. Res.*, 1(3):247–62.

ULREY, A B and GREELEY, P O   A list of the marine fishes
1928   (Teleostei) of southern California with their distribution. *S. Calif. Acad. Sci.*, 27(1):1–53.

The Place of Ecological Monitoring in the Study of Pollution of the Marine Environment**

David J. Bellamy, *David M. John,†*
David J. Jones,‡ Alan Starkie§
and Alan Whittick‖

Rôle du controle écologique dans l'étude de la pollution du milieu marin

Les auteurs examinent l'utilisation des attributs généraux de l'écosystème constitué par "les forêts de varech" pour contrôler le milieu marin sur la côte est de la Grande-Bretagne pour deux gradients de pollution. Ils donnent également des données se rapportant à deux sites de référence.

Les principaux effets de la pollution sont les suivants: (1) diminution de luminosité sous l'effet des matières en suspension, réduisant la productivité des algues dominantes et le potentiel de l'écosystème; (2) réduction de la diversité et de la stabilité des populations d'invertébrés qui occupent les crampons du varech. Ceci semble être provoqué par l'interaction de deux facteurs: l'influence d'eaux usées riches en éléments énergétiques qui permettent aux organismes qui se nourrissent de matières en suspension de dominer l'habitat; l'action d'un autre facteur, peut-être les toxicoïdes, qui provoque des fluctuations considérables dans les populations de l'organisme dominant; (3) l'analyse des éléments de la chaîne trophodynamique indique que le taux de concentration des toxicoïdes est plus élevé dans les organismes provenant des sites de pollution. On n'a pas constaté d'accumulation massive de toxicoïdes dans la chaîne trophodynamique, ce qui indique soit l'existence de mécanismes régulateurs efficaces, soit la rupture de cette chaîne.

Les auteurs examinent dans quelle mesure les données provenant d'un système de contrôle *a posteriori* peuvent répondre à la question "de quoi est faite la pollution?"

Funcion de la vigilancia ecológica en el estudio de la contaminación del medio ambiente marino

El estudio expone el empleo de los grandes atributos del ecosistema del "bosque de algas gigantes" para vigilar el ambiente marino costero a lo largo de dos gradientes de contaminación en la costa oriental de Gran Bretaña. Se presentan también datos de dos lugares de referencia.

Los principales efectos de la contaminación son: (1) La atenuación de la luz por material en suspensión, lo que reduce la productividad de las algas dominantes y el potencial del ecosistema. (2) La reducción de la diversidad y estabilidad de las poblaciones de invertebrados en los lugares donde crecen las algas. Esto parece deberse a la interacción de dos factores principales: las aguas cloacles ricas en energía, que permiten a los organismos que se alimentan de materiales en suspensión dominar el habitat, y algún otro factor, quizás agentes toxicoideos, que producen violentas fluctuaciones en las poblaciones de los organismos dominantes. (3) El análisis de los miembros de la cadena alimentaria indica que los niveles de productos toxicoideos son más altos en los organismos procedentes de los lugares contaminados. No se encontró ninguna acumulación masiva de productos toxicoideos a través de la cadena alimentaria que señalase la existencia de mecanismos eficaces de regulación, o ruptura de la cadena alimentaria misma.

Se examina la importancia que tienen los datos de un sistema de vigilancia *a posteriori* para responder a la pregunta "¿qué es lo que constituye la contaminación?"

THE use of growth parameters of a particular plant species to monitor environmental factors has long been a tool of the terrestrial ecologist (e.g. Paterson, 1961).

This paper describes the use of gross ecosystem attributes to monitor and compare the inshore marine environment along two major pollution gradients on the east cost of Britain.

Table 1 gives details of the water quality of the sites at which studies were carried out along the pollution gradients. Wherever rock outcrops occur along this

coast the sublittoral zone is typified by "kelp forests" dominated by *Laminaria hyperborea* (Gunn.) Fosl. Preliminary investigations (Bellamy, John and Whittick, 1968; Bellamy, Whittick and Jones, 1969; and Starkie, 1970) have shown that attributes of the kelp forest can form a basis for comparative study.

Ecosystem attributes

Figure 1 summarizes performance data from two widely separated unpolluted sites, Petticoewick Lat. 55° 55' N, Long. 2° 09' W and Sennen Cove Lat. 50° 5' N, Long.

---

* Department of Botany, University of Durham, Durham City, U.K.
† Department of Botany, University of Ghana, Legon, Ghana.
‡ Department of Biology, Simon Frazer University, B.C., Canada.
§ Department of Forestry and Natural Resources, University of Edinburgh, Scotland.
‖ Department of Biological Sciences, Memorial University, St. Johns, Newfoundland.
** This work was done under grants from NERC, the British Sub Aqua Club and Durham University Expensive Equipment Fund.

TABLE 1. DETAILS OF STUDY SITES. ALL FIGURES ARE MEANS FOR ALL ANALYSES CARRIED OUT

Site number, name and position	Type of pollution	Faecal bacteria /100 mls	PO_4 µg/litre	NO_3 µg/litre	NH_4 µg/litre	Suspended solids mg/litre
1. Aberdeen 57° 8′ N 9° 05′ W	None	20	—	—	—	—
2. Forth Bridge 56° 3′ N 3° 30′ W	Estuarine industrial, domestic	2,000	6	—	8.2	—
3. Inch Mickery 56° 5′ N 3° 20′ W	Estuarine industrial, domestic	—	—	—	—	—
4. Inch Keith 56° 5′ N 3° 10′ W	Estuarine industrial, domestic	—	—	—	—	—
5. Dunbar 55° 59′ N 2° 30′ W	None	120	—	—	—	20
6. St. Abbs Petticoewick 55° 55′ N 2° 9′ W	None	17	0.4	2	2	12
7. Holy Island 55° 38′ N 1° 47′ W	None	20	1	8	1	—
8. Beadnall 55° 33′ N 1° 37′ W	None—some natural turbidity	40	2	11	3	40
9. Craster 55° 18′ N 1° 40′ W	Coal washings, some sewage	170	2	5	4	—
10. Newbiggin 55° 8′ N 1° 40′ W	Coal washings, some sewage	230	2	6	2	—
11. St. Mary's 55° 4′ N 1° 42′ W	Heavy pollution, industrial colliery, domestic	2,000	2	16	10	150
12. Souter 54° 48′ N 1° 21′ W	Heavy pollution, industrial, colliery, domestic	2,000	3	19	14	200
13. Seaham 54° 44′ N 1° 28′ W	Heavy pollution, industrial, colliery, domestic	2,000	5	23	27	—
14. Long Scar 54° 40′ N 1° 20′W	Heavy industry, chemical, works, domestic	2,000	72	140	37	70
15. Redcar 54° 37′ N 1° 04′ W	Heavy industry, chemical works, domestic	2,000	3	18	19	—
16. Robin Hoods Bay 54° 24′ N 0° 38′ W	Natural turbidity	10	2	19	21	50
17. Flamborough 54° 8′ N 0° 0′ W	Natural turbidity	—	2	12	5	140

5° 25′ W. The rapid attenuation of primary production over the lower depth range, although bare rock suitable for kelp growth is present in deeper water, indicates that the vertical extent of the kelp is controlled by lack of light, not substrate, at these sites.

The marked reduction of the depth range of the kelp at the polluted sites is at once obvious (fig 2). The fact that extensive rock platforms are present in deeper water at all sites, coupled with the fact that at each polluted site there is a marked reduction in individual performance over depth ranges which in unpolluted water would have little or no effect, again indicates light as being the limiting factor. The extended depth range coupled with the further drop in performance at sites 16 and 17, which have naturally turbid waters and no sewage nutrients, indicates that natural erosion being an intermittent phenomenon related to high tides and storms, allows longer periods with good light penetration, than the blanketing effects of pollutant in suspension; and there may be some interaction between the effect of reduction of light and the increase of nutrients along certain stretches of the polluted coast. Unfortunately, the pollution gradient is not perfect for this type of study, since no kelp forest has been found south of site 17 which is subject to neither pollution nor natural turbidity. Nevertheless, it seems safe to conclude that one of the main effects of pollution is a reduction in light penetration due to suspended material.

The fauna of kelp holdfasts

The fauna of holdfasts is trophically independent of the kelp in which it develops, being a detritus based eco-system, and can therefore be studied as a unit in its own right.

TABLE 2. DIVERSITY INDICES (MARGALEF 1968) FOR THE FAUNA FROM A 0–7 YEAR AGE RANGE OF KELP HOLDFASTS AT EACH SITE

Site No.	2	3	4	5	6	7	8	9	10	11	12	13	16	17
	1.6	2.3	4.2	3.9	4.1	3.8	3.6	3.8	3.7	2.2	2.1	3.8	3.7	

Polluted Estuarine	Unpolluted	Polluted	Natural Turbidity

Diversity which can be taken as a measure of maturity and stability (Margalef, 1968) is consistently greater in the unpolluted sites (Table 2). More detailed comparison of the composite data collected over three years at two sites (Table 3), shows that the ecosystems in the polluted waters are dominated by suspension feeders with a great reduction in all other trophic groups. This is exactly the situation found at all sites for the initial phases of development of the ecosystem, but is in marked contrast to the mature ecosystem from unpolluted water (Jones, 1970). Thus, it would appear that pollution is maintaining an immature ecosystem dominated by suspension feeders, (cf. the conclusions of Margalef, 1967).

Figure 3 shows the variation of the diversity of the fauna over the three years of study. The diversity of both fluctuate about a mean which is consistently higher in the unpolluted water. Both, therefore, exhibit instability, but the fluctuations of the less diverse polluted system must be regarded as more "serious" at the ecosystem level. This is borne out by Table 4, where data for the

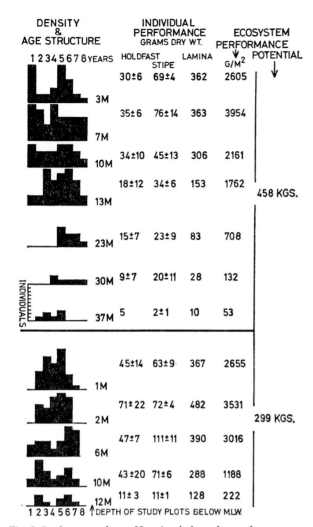

DENSITY & AGE STRUCTURE	INDIVIDUAL PERFORMANCE GRAMS DRY WT.		ECOSYSTEM PERFORMANCE	POTENTIAL
1 2 3 4 5 6 7 8 YEARS	HOLDFAST STIPE	LAMINA	G/M²	
3M	30±6 69±4	362	2605	
7M	35±6 76±14	363	3954	
10M	34±10 45±13	306	2161	
13M	18±12 34±6	153	1762	458 KGS.
23M	15±7 23±9	83	708	
30M	9±7 20±11	28	132	
37M	5 2±1	10	53	
1M	45±14 63±9	367	2655	
2M	71±22 72±4	482	3531	299 KGS.
6M	47±7 111±11	390	3016	
10M	43±20 71±6	288	1188	
12M	11±3 11±1	128	222	

1 2 3 4 5 6 7 8 ↑ DEPTH OF STUDY PLOTS BELOW MLW.

Fig. 1. Performance data of Laminaria hyperborea *from two unpolluted monitoring stations, top Sennen Cove, bottom Petticoewick. Individual performance (given as net annual production over a 7 year life span) is measured using peak and increment cropping techniques (Bellamy et al., 1968). Density and age structure is the mean of at least 40 quadrats in each case; ecosystem performance is based on this figure. Ecosystem potential is the total net annual production for a continuous metre wide strip of kelp forest with a gradient of 1 in 10 calculated from the composite data*

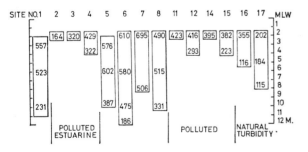

Fig 2. The depth range of the kelp forest at each site is shown together with the total individual performance of the kelp (calculated by adding the annual increments of stipe and holdfast biomass to the peak biomass of lamina for each age class 1 to 7 years), measured at the depths shown

total number of individuals of the two dominant species at the polluted site are shown. The populations of these two species contain very few mature individuals. These observations would point to the interaction of at least two factors, nutrient rich material in suspension favour-

ing suspension feeders, to the exclusion of practically all other organisms except their direct predators; and some factor, perhaps toxicoids, causing the observed fluctuations in the population of the dominants.

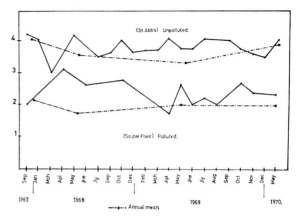

Fig 3. Margalef indices of diversity over four years at two study sites—one polluted, the other unpolluted

Chemical analysis for toxicoids

Table 5 gives the results of chemical analyses for copper and lead of *Laminaria hyperborea* (a primary producer), *Mytilus edulis L.*, the mussel (a suspension feeder), *Nucella lapillus L.*, the whelk and *Asterias rubens L.*, the common starfish (both direct predators of *Mytilus*), all collected from the pollution gradient.

In all cases the organisms collected from the polluted waters contain more toxicoids than those from unpolluted water. It is interesting to note that there is no evidence of a massive concentration of either metal through the food chain.

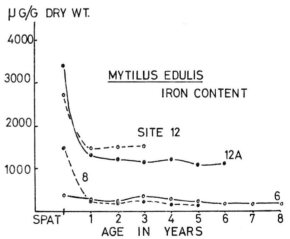

Fig. 4. Concentration of iron in the tissue of Mytilus edulis *(from four sites, 12 and 12A polluted, 8 and 6 unpolluted) plotted against age*

Study of the concentration of toxicoids with age and/or size class of the organism, shows that the concentration of all toxicoids within the tissues falls with increase in age, size. Figure 4 shows the result for iron in the mussel. The reaction is the same for all populations sampled, only the concentrations differ between populations.

However, as certain of the toxicoids do show significant concentration through the food chain at the unpolluted sites, it may be that in the polluted waters the

TABLE 3. TROPHIC ANALYSIS OF 1–7 YEAR ECOPERIOD, COMPOSITE RESULTS OF FOUR YEARS' SAMPLING AT TWO SITES, TOP PETTICOEWICK (UNPOLLUTED), BOTTOM SOUTER (POLLUTED). THE NUMBER OF INDIVIDUALS IN EACH TROPHIC GROUP EXPRESSED AS A PERCENTAGE OF THE TOTAL NUMBER OF INDIVIDUALS

| Year | 1967 | 1968 | | | | | | | | | | |
Month		1	3	4	5	6	7	9	10	12	Mean
Omnivores	44	27	43	—	37	—	45	22	28	23	32
Carnivores	16	19	14	—	11	—	2	23	12	24	15
Herbivores	6	2	3	—	2	—	2	0	1	6	2
Deposit feeders	.4	3	0	—	0	—	0	4	0	0	1
Suspension feeders	35	52	43	—	50	—	53	50	58	47	50
Omnivores	0	—	—	39	—	3	—	—	8	—	17
Carnivores	1	—	—	33	—	4	—	—	12	—	17
Herbivores	0	—	—	11	—	1	—	—	.9	—	4
Deposit feeders	0	—	—	6	—	0	—	—	0	—	2
Suspension feeders	98	—	—	11	—	91	—	—	79	—	60

| | 1969 | | | | | | | | | | | | 1970 | 1967–1970 |
	2	3	4	5	6	7	8	9	10	11	12	Mean		
O	30	37	40	39	39	21	—	38	41	41	20	35	50	40
C	12	23	23	19	8	21	—	15	9	12	30	17	13	15
H	4	2	6	2	2	2	—	3	2	1	2	3	6	4
DF	3	0	.4	.5	0	.6	—	0	0	.4	.6	.6	—	.5
SF	50	38	30	41	51	53	—	42	47	42	46	44	31	40
O	—	—	20	12	.2	.8	4	—	15	.8	—	8	.1	6
C	—	—	16	23	0	1	7	—	18	4	—	10	1	7
H	—	—	.5	1	0	.1	.3	—	1	.4	—	.5	.1	1
DF	—	—	.1	1	0	0	.3	—	3	.4	—	.7	.1	.7
SF	—	—	64	63	99	98	89	—	63	92	—	81	99	85

TABLE 4. FLUCTUATION IN THE NUMBER OF INDIVIDUALS OF THE TWO DOMINANT SUSPENSION FEEDERS AS A PERCENTAGE OF THE TOTAL POPULATION WITH TIME. A BEING THE DATE (MONTH AND YEAR). B. *Mytilus edulis* L. C. *Sabellaria spinulosa* L.

A.	9/67	4/68	6/68	10/68	4/69	5/69	6/69	7/69	8/69	10/69	5/70
B.	96.0	0	1	15.7	1.1	34.0	96.0	93.0	57.0	2.0	96.0
C.	0.3	0	89.4	58.3	48.0	0	2.0	3.0	22.0	56.0	1.0

food chain breaks down, the animals ceasing to feed on their normal source, but on energy-rich sewage. In this case all the organisms would represent the same trophic level as regards accumulation of toxicoids.

Much more work is needed before the relevance of these trends can be understood. However, if the right toxicoid food chain system could be found, the limits of pollution could be determined quite accurately.

TABLE 5. ANALYSES RESULTS OF 4 MEMBERS OF THE FOOD CHAIN COLLECTED AT 7 SITES FOR COPPER AND LEAD μg/g. DRY WEIGHT (STANDARD ERRORS). Tr. = TRACE.

| | Unpolluted | | | | | |
| Site No. | 5 | | 6 | | 8 | |
	Pb	Cu	Pb	Cu	Pb	Cu
Asterias rubens Carnivore	—	—	170 ± 16	10 ± 1	147 ± 10	Tr.
Nucella lapillus Carnivore	143 ± 16	22 ± 4	146 ± 5	10 ± 3	133 ± 10	Tr.
Mytilus edulis Suspension feeder	161 ± 7	7 ± 1	125 ± 21	8 ± 1	144 ± 6	Tr.
Laminaria hyperborea Primary producer	—	—	25 ± 2	1.9 ± 0.2	Tr.	Tr.

| | Polluted | | | | | | Naturally turbid | |
| Site No. | 12 | | 14 | | 15 | | 16 | |
	Pb	Cu	Pb	Cu	Pb	Cu	Pb	Cu
Asteiar rubens Carnivore	305 ± 90	21 ± 2	390 ± 78	20 ± 1	146 ± 36	13 ± 3	120 ± 18	16 ± 4
Nucella lapillus Carnivore	313 ± 124	20 ± 5	170	20	274 ± 50	34 ± 2	180 ± 25	24 ± 5
Mytilus edulis. Suspensiyn feeder	343 ± 99	13 ± 3	289 ± 38	19 ± 3	163 ± 9	16 ± 2	118 ± 8	12 ± 2
Laminaria hyperborea Primary producer	57 ± 8	1.7 ± 2	40	25	16 ± 6	22 ± 3	22 ± 2	12 ± 2

References

BELLAMY, D J, JOHN, D M and WHITTICK, A The kelp forest
1968   ecosystem as a "phytometer" in the study of pollution of
        the inshore environment. *In* Underwater Association
        Report, pp. 79–82.
BELLAMY, D J, WHITTICK, A and JONES, D J How to live with
1969   pollution. *Spectrum, Oxford*, 62:8–11.
JONES, D J Ecological studies on the fauna kelp holdfast. Thesis,
1970   University of Durham.

MARGALEF, R The food web in the pelagic environment. *Helgo-*
1967   *länder wiss. Meeresunters.*, 18: 548–59.
MARGALEF, R Perspectives in ecological theory. Chicago, Chicago
1968   University Press. 111 p.
PATERSON, S S Introduction to phyochrology in Norden. *Meddr*
1961   *St.SkogsforskInst.*, 50.
STARKIE, A An investigation of heavy metal ion concentrations in
1970   littoral and sublittoral marine ecosystems. Dissertation.
        University of Durham, 40 p.

Thermal Pollution of a Tropical Marine Estuary

R. G. Bader, M. A. Roessler
and A. Thorhaug*

La pollution thermique dans un estuaire tropical

La croissance et le déplacement constants des populations et le
progrès technologique font peser sur l'industrie des exigences qui
rendront nécessaire la construction de grandes centrales
électriques dans les régions côtières. Pour prévenir les effets nocifs
qui pourraient en résulter pour la production d'aliments marins,
pour les activités récréatives et pour d'autres usages, il convient de
veiller à l'emplacement choisi pour ces établissements et d'assurer
une élimination adéquate des effluents thermiques afin d'éviter que
l'équilibre du milieu marin ne soit compromis. On a donc entrepris
une étude pour déterminer les effets des effluents thermiques sur un
estuaire tropical. Il existe une étroite corrélation entre les résultats
des travaux effectués sur le terrain en matière d'hydrographie, de
chimie et d'écologie quantitative et les études de détail menées en
laboratoire sur les limites de létalité thermique pour certains
organismes qui font partie de la chaîne alimentaire et pour certaines
espèces d'importance commerciale. A une température soutenue
de 33°C, d'importantes macro-algues et herbes marines ont été
détruites. Après la disparition des végétaux, les formes animales
diminuent sous le double aspect du nombre et de la diversité. Un
léger rétablissement n'intervient que pendant les mois froids d'hiver.
Le déversement continu d'effluents thermiques exerce un effet
contraire sur le biota et dévalorise la région en tant que zone de
reproduction des poissons et aire de récréation pour l'homme. Un
surcroît de chaleur fait naître des problèmes critiques dans un
estuaire tropical, car les organismes y vivent dans des conditions
très proches des limites de tolérance de la température maximale.

Contaminación térmica de un estuario marino tropical

El constante crecimiento y movimiento de las poblaciones y el
desarrollo tecnológico de los países con las consiguientes demandas
industriales exigirá la construcción de centrales eléctricas enormes
en las zonas costeras. Para impedir los efectos perjudiciales sobre
la producción de alimentos marinos, el recreo y otras formas de
utilización, tales centrales deberán estar emplazadas favorablemente
y sus residuos caldeados deberán ser evacuados en forma que no
perturben el equilibrio marino. Se llevó a cabo un estudio para
averiguar los efectos de los residuos caldeados en un estuario
marino tropical. Los resultados de los estudios directos en el mar,
de la hidrología, química y ecología cuantitativa se correlacionan
estrechamente con los estudios detallados de laboratorio de los
límites térmicos letales de algunas especies importantes y de los
organismos de las cadenas alimentarias. A temperaturas
sostenidas de 33°C mueren importantes macroalgas y vegetales
marinos. Una vez que las plantas desaparecen disminuye la
diversidad y número de los animales. Solamente se observa una
ligera recuperación durante los meses fríos de invierno. La descarga
constante de desechos calentados perjudicará a la biota y disminuirá
el valor de la zona como lugar de cría de peces y como zona de
recreo. La adición de calor a un estuario en los trópicos determina
problemas críticos porque los organismos están ya viviendo
próximos a su máximo límite de tolerancia de temperatura.

INCREASED urbanization in all parts of the world,
coupled with a movement of population to coastal
regions will require the construction of fossil fuel and
nuclear power plants at situations where they can dispose
of their heated effluents, that is, chiefly in coastal and
estuarine waters. Tropical estuaries are nursery grounds
for penaeid shrimp, lobsters, crabs, clupeid fishes and
sciaenid fishes which constitute vast food supplies and
high value fisheries. Since many of these animals live
near their upper temperature limit, particular care must
be taken to site heated water discharges so that they do
not damage this resource. The effects of thermal additions
to a tropical estuary in Florida have been examined
near the site of the outfall from a new power plant.

Field studies

Circulation patterns, chemical variables and quantitative
estimates of the diversity and abundance of the biota
have been investigated for the past two years at 32
stations in the area of Turkey Point, Florida (fig 1),
water samples taken twice each month to measure
temperature, salinity and dissolved oxygen; monthly
measurements of copper and iron were also made.
Quantitative determinations of the flora were made in
12 areas every 2 weeks, and plankton samples were

collected at 8 stations with 30 µ and 300 µ mesh nets.
Monthly trawl samples were collected at 20 stations, and
8 additional locations were sampled quarterly (fig 1).
Trap, gill net, and slate-settling panels were used to
compare the abundance of animals within the effluent
canals with canals not receiving heated effluents.

Studies conducted prior to the opening of the Turkey
Point plant showed that the currents flowed NNE to
SSW on flood tides and SSW to NNE on ebb tides
(Anon., 1966). Since the opening of the major effluent
canal, there is a net flow to the NNE on all stages of the
tide except when strong NW winds cause a flow in an
easterly or southeasterly direction. The latter, according
to our observations, occurred less than 6 per cent of
the time.

The effect of the NNE flow has been to create a consistent temperature pattern (fig 2). With the current
discharge rate of 36 m³/sec, an area of 12–20 ha immediately off the canal mouth has temperatures raised
5°C. An area of approximately 50 ha is raised 4°C and an
area of 120 ha is raised 3°C above ambient Bay temperatures. Temperatures in unaffected shallow portions of
Biscayne Bay have ranged from a low of 9°C during a
cold spell in winter to 33°C at midday in summer. In
winter the temperature averaged about 17°C and in

* Rosenstiel School of Marine and Atmospheric Sciences, University of Miami, Miami, Florida 33149, U.S.A.

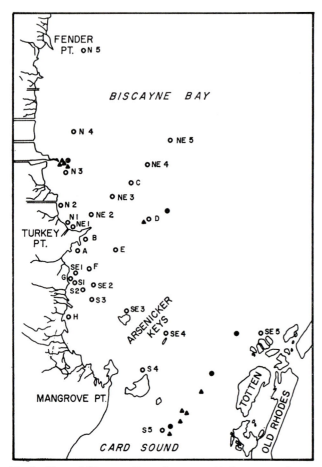

Fig 1. Map of Biscayne Bay adjacent to Turkey Point showing location of trawl sample stations

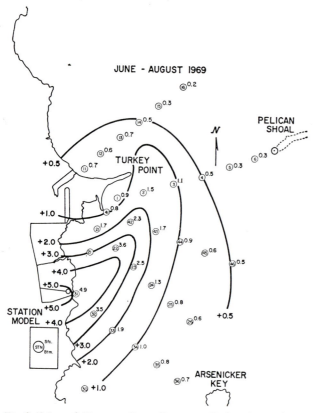

Fig 2. Map of Biscayne Bay adjacent to Turkey Point showing hydrographic stations and average increase in temperature

summer between 30 and 31°C. There is a 3°C diurnal fluctuation in temperature. The highest temperature at the outfall has exceeded 40°C but generally maintains a level between 35 and 37°C in summer. Temperature stratification occurs rarely and only during ebb tides when wind velocities are less than 8 km/h for a period of 24 h or more. Highest temperatures occur when high slack water occurs near noon: this is the period when industrial and airconditioning power demands are greatest (Bader, 1969; Bader and Tabb, 1970).

Salinity patterns in Biscayne Bay generally show the influence of land runoff. Salinities along the shore are generally low and fluctuate widely while those farther from the mainland are near 35‰ and remain relatively constant between 30 and 38‰. The power plant tends to draw some water from the north on flood tides and this water is usually of somewhat lower salinity than found south of Turkey Point. This causes reduced salinities in the area south of Turkey Point on flood tides.

The iron content of the effluent is higher than surrounding waters and organic flocculent material causes increased turbidity. Other nutrients and chemicals fall within the ranges expected in tropical estuaries.

Macro-algae and the turtle grass *Thalassia testudinum* have disappeared from an area of 20–25 ha off the mouth of the effluent canal. The typical flora has been replaced by a mat of blue green algae which does not provide food suitable for most estuarine animals (Roessler and Zieman, 1970). In an area about 50 ha algal depletion is severe, with some summer effects detectable in a 120-ha area. These are correlated with +5, +4 and +3°C isotherms respectively, above the average ambient summer temperature of about 30°C. Trawling has shown that virtually all animals have disappeared from the inner (+5°C) area. There are fewer species and fewer individuals of most species within the +4°C isotherm, but more molluscs and crustacea in the zone subtended by the +3°C isotherm. They feed on the detrital algae, but will leave the area when the food and shelter disappear.

Studies on fish populations sampled with traps and gill nets within the canal systems associated with the Turkey Point plant show that in summer there are fewer fish in the effluent canals than in adjacent control canals. In winter this is reversed. Settling of benthic fouling organisms including oysters is inhibited during summer in the effluent canals. In winter there is settlement of fouling organisms (Nugent, 1970).

Laboratory studies

To interpret the array of variables in a marine estuary, detailed laboratory studies must accompany field studies. Lethal temperature limits must be determined for the principal commercial species and the important members of the food chain. Organisms chosen for study were: the commercial pink shrimp, *Penaeus duorarum*; two shrimp important in the natural food chain, *Palaemonetes intermedius* and *Periclimenes americanus*; the green macro-algae *Penicillus* and *Dictyosphera* as well as the single cell alga *Valonia*, used often as an index of physiological properties of green algae.

The temperature controlling device was an aluminium bar fitted for glass tubes to contain the organisms, which was heated and cooled to ±0.01°C. One instrument ranged between 10 and 38°C in 1°C intervals (the natural extremes encountered in Biscayne Bay); a second instrument was set at 0.1°C intervals between 31 and 35°C in order to ascertain the exact death points. The temperature control for these investigations was the most precise ever used for studies of tropical benthic organisms. Also, the culture media and other conditions were optima developed by specialists for each species. With all other conditions held at optimal levels, temperature was the sole variable.

More than 5,000 specimens of five species of the marine chlorophyte *Valonia* showed that the mean limit for the thermal tolerance was 14.8°C minimum and 31.2°C maximum, with standard deviations of ±0.2°C. *Valonia*, although not representing a large percent of the biomass, is used conventionally as an index plant for green marine algal physiology. This is convenient because its large size simplifies some of the technical problems. Figure 3 shows a typical live–dead curve (represented by percent irreversible plasmolysis); each point represents 12 specimens of *Valonia macrophysa* from lower Biscayne Bay. Similar studies were done on cells from Bermuda (mean annual maximum temperature 22.6°C), Florida Keys (26.3°C), Dry Tortugas (27.0°C), Jamaica (27.4°C), Puerto Rico (28.5°C), and Curaçao (24.5°C). The lethal limits of *V. macrophysa* from these locations differed only in tenths of a degree C, which indicates little ability for physiological temperature adaptation. Two other species of *Valonia* from these localities showed similar results. Fig 4 indicates that some organisms exist in delicate thermal balance. The *Valonia* data agree with field work at Turkey Point.

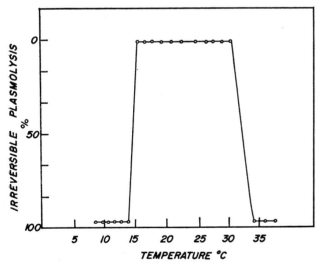

Fig 3. *Irreversible plasmolysis versus temperature for* Valonia macrophysa *after 3 days' exposure to the given temperature. Each point represents 12 cells*

A second green macro-algae, *Penicillus*, is important as shelter for small shrimp, molluscs, and other invertebrates. It is one of the three most abundant plants in the back reef environment of Biscayne Bay and the Florida Keys. It also contributes large quantities of calcium carbonate to the sediments. Figure 5 shows death after 5 days for *Penicillus capitatus*, the most common Bay species. The thermal limits of 15 and 31°C determined experimentally correlate well with those of *Valonia* as well as with field observations for *Penicillus*. The algae

Dictyosphera cavernosa from Biscayne Bay has upper limits between 33°C and 34°C, which also correlates well with field data.

Newly hatched, healthy nauplii (540 specimens) of the commercial pink shrimp were held at intervals of 1°C from

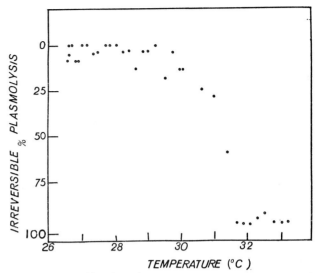

Fig. 4. *Irreversible plasmolysis versus temperature for* Valonia utricularis *after 5 days' exposure to the given temperature. Each point represents 25 cells*

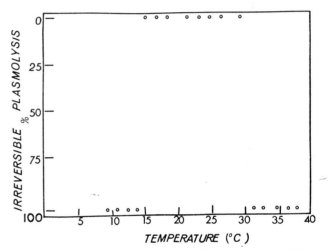

Fig 5. *Death versus temperature for* Penicillus capitatus *after 5 days exposure to the given temperature*

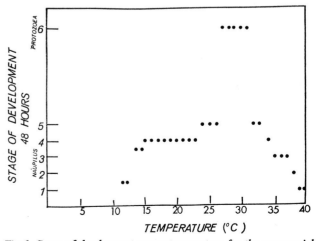

Fig 6. *Stage of development versus temperature for the commercial pink shrimp* Penaeus duorarum. *Each point represents 18 shrimp.*

[427]

11.0 to 38.3°C. The organisms were kept in filtered sea water at a salinity of 33‰ and aerated by vigorous bubbling (Idyll *et al.*, 1969). Figure 6 shows the results after 32 h; the shrimp between 25.0 to 26.0°C and 30.9 to 31.8°C were able to metamorphose to a protozoea. It became progressively more difficult to attain advanced nauplii stages below and above this temperature range. Ultimately those unable to metamorphose died because these nauplii do not feed. The adult *Penaeus duorarum* has been found in water up to 37°C. These results also correlate well with field and mariculture data.

Various shrimp that are important in the food chain were studied also. Death occurs to gravid female *Palaemonetes intermedius* and *Periclemenes americanus* near 36.5 and 34.2°C respectively after 2 days; at 37°C they both die in 2 min. *Palaemonetes intermedius* eggs cannot hatch above 34.6°C; this indicates that although adult shrimp may survive elevated temperatures, widespread areas of heated waters would not allow the life cycle to be completed.

Hatching of eggs from the commercial stone crab *Menippe mercenaria* was possible up to 34.5°C. Above 35.5°C the eggs ceased to hatch although heart-beat was very strong for an initial 12-h period.

Conclusions

The results of field and laboratory studies of the Biscayne Bay, Florida, tropical marine estuary clearly showed that sustained temperatures above 33°C can cause extensive mortalities of some of the most important macroalgae and sea grasses. This, in turn, may eliminate the major food source and shelter for a great number of herbivores and detritus feeders. These include many of the juvenile stages of commercial species such as shrimp, lobsters and fishes. The resulting loss of the carnivores can decrease the value of the area for fishing and recreation. In addition to the immediate losses of fish and invertebrates, the death of the algae and sea grasses which normally hold the sediments in place can permit erosion of the sediment with a resulting increased turbidity of contiguous waters. This process may have a detrimental effect on productivity, which will further contribute to the deterioration of estuarine areas.

Laboratory studies and field observations showed that important green benthic macroalgae have abrupt thermal limits between 31 and 33°C. Field observations demonstrated some temporary algal and grass recovery during winter months. Laboratory experiments using more than 6,000 specimens of algae from many parts of the tropics (acclimatized to temperatures ranging from 22.5 to 28.5°C) confirmed this by showing that the only adaptations possible were in the order of tenths of a degree. Laboratory studies further indicated that invertebrate adults, such as commercial and important food chain shrimp as well as crabs, have upper thermal limits of 33 to 37°C. Investigation of larval stages and eggs showed thermal limits near 31°C. The elimination of this important link in the growth of the shrimp and crab would not permit population regeneration.

Tropical estuarine organisms live close to their upper lethal temperatures in summer. Therefore, extremely careful planning is essential in developing estuaries to insure that technological progress does not cause a loss of the food and recreational properties essential to the population attracted to the area.

References

BADER, R G   An ecological study of South Biscayne Bay in the
1969   vicinity of Turkey Point. *In* Progress Report to Federal Water Pollution Control Administration. Coral Gables, Florida, Institute of Marine Science, University of Miami, 63 p.

BADER, R G and TABB, D C   An ecological study of South Biscayne
1970   Bay in the vicinity of Turkey Point. *In* Progress Report to U.S. Atomic Energy Commission. Coral Gables, Florida, Rosenstiel School Marine and Atmospherical Sciences, University of Miami, ML 70008a:81 p.

IDYLL, C P, *et al.*   Shrimp and Pompano Culture Facilities at the
1969   University of Miami. *Sea Grant Inf. Bull. Univ. Miami*, (2):1–18.

NUGENT, R   Some effects of heated power plant effluents on the
1970   macrofauna of a tropical mangrove habitat. Dissertation. Coral Gables, Florida, University of Miami, 125 p.

ROESSLER, M A and ZIEMAN, J C, JR.   The effects of thermal
1970   additions on the biota of Biscayne Bay, Florida. *Proc. Gulf Caribb. Fish. Inst.*, 22:136–45.

ZIEMAN, J C, JR.   The effects of a thermal effluent stress on the
1970   seagrasses and macroalgae in the vicinity of Turkey Point, Biscayne Bay, Florida. Dissertation. Coral Gables, Florida, University of Miami, 160 p.

ANON,   Oceanography. *In* Preliminary safety analysis report sub-
1966   mitted to U.S. Atomic Energy Commission by Florida Power and Light Co. Section 8, (1):1–15.

Acknowledgements

This work was supported by the United States Atomic Energy Commission and the Federal Water Quality Administration. The field work and laboratory studies summarized herein have been conducted by the authors and others at the School of Marine and Atmospheric Sciences, University of Miami.

Tissue Levels in Animals and Effects Caused by Chlorinated Hydrocarbon Insecticides, Chlorinated Biphenyls and Mercury in the Marine Environment along the Netherlands Coast

*J. H. Koeman and
H. van Genderen**

Taux de concentration dans les tissus animaux des insecticides à base d'hydrocarbures chlorés, des diphényles chlorés et du mercure et leurs effets sur le milieu marin du littoral des Pays-Bas

Les côtes néerlandaises sont soumises à une surveillance toxicologique depuis 1965. On a analysé des échantillons de mollusques, de poissons et d'oiseaux de mer pour suivre la distribution topographique de la contamination, noter ses effets sur la chaîne alimentaire et déterminer les rapports éventuels entre les résidus déposés dans les tissus et l'absorption de certains contaminants. Parmi les oiseaux de mer, on a constaté qu'une grande colonie de sternes (*Sterna*

Indices de concentración en los tejidos de los animales y efectos que causan los insecticidas a base de hidrocarburos clorurados, los bifenilos clorurados y el mercurio en el ambiente marino a lo largo de la costa de los Países Bajos

Desde 1965 se llevaron a cabo estudios toxicológicos a lo largo de la costa de los Países Bajos. Se han tomado muestras de moluscos, peces y aves marinas y se han analizado para determinar la distribución topográfica de la contaminación, observar su efecto en la cadena alimentaria y relacionar los residuos concentrados en los tejidos con los efectos que podrían deberse a la absorción

---

* Institute of Veterinary Pharmacology and Toxicology, University of Utrecht, Utrecht, Netherlands.

sandvicensis) vivant dans la mer des Wadden a présenté une forte mortalité en 1965. La cause pourrait en être attribuée à la pollution par les insecticides à base d'hydrocarbures chlorés. Dans la même région, la pollution semble menacer l'eider (*Somateria mollissima*), notamment la femelle, par suite de ses habitudes de reproduction.

On a trouvé par chromatographie en phase gazeuse associée avec la spectrométrie de masse, un grand nombre d'isomères des diphényles polychlorés dans les organismes marins. Les auteurs examinent l'influence de ces substances du point de vue toxicologique. On a déterminé les taux de concentration du mercure total et du mercure méthylique par les méthodes de l'activation neutronique et de la chromatographie en phase gazeuse. On a trouvé notamment des taux résiduels élevés chez certains eiders, mais il n'est pas encore possible de déterminer si les taux résiduels de mercure présentent des risques pour l'écosystème.

Les études de distribution ont révélé que la contamination provenait surtout du Rhin. On a pu fixer en partie l'origine de cette pollution par les insecticides à base d'hydrocarbures chlorés: elle provenait d'une usine de transformation située à proximité de l'embouchure du fleuve.

de los contaminantes. En 1965 se observó que entre las aves marinas, perecían muchos ejemplares de una gran colonia de golondrinas de mar (*Sterna sandvicensis*) en la costa de Wadden. La causa pudo atribuirse a la contaminación por insecticidas a base de hidrocarburos clorurados. En el mismo lugar el eider o pato del norte (*Somateria mollissima*) parece estar amenazado por la contaminación, particularmente la hembra debido a sus hábitos característicos de la reproducción.

En organismos marinos se encontraron por cromatografía gaseosa en combinación con espectrometría de masas un gran número de isómeros de bifenilos policlorurados. Se examinan las repercusiones toxicológicas de estas sustancias. Mediante métodos de activación de neutrones y de cromatografía gaseosa se determinaron los índices de mercurio total y metil mercurio. Se han encontrado importantes cantidades de residuos, entre otros, en algunos patos del norte. Todavía no es posible evaluar los niveles de residuos de mercurio que representan un peligro para el ecosistema.

Estudios de distribución han revelado que la contaminación procedía principalmente del río Rhin. La causada por los insecticidas a base de hidrocarburos clorurados se pudo comprobar que en parte tenía su origen en una fábrica situada en las cercanías de la desembocadura del río.

IN 1964 and 1965 during a survey to examine the degree of contamination of marine organisms with chlorinated hydrocarbon pesticides, relatively high levels of these compounds were found in molluscs, fish and birds. In the same period a heavy mortality was noticed among sandwich terns (*Sterna sandvicensis*) and eiders (*Somateria mollissima*), two species breeding on islands in the Dutch Wadden Sea. This coincidence made it urgent to consider the possible causal relationship between both phenomena. Dead, dying, as well as apparently healthy birds were collected and their tissues analysed. Residues were also measured in food organisms of terns and eiders. A monitoring programme was set up, using the mussel (*Mytilus edulis*) as indicator organism, to measure the distribution of the compounds along the Dutch coast. Continued observations were made in following years to see whether the effects observed were dependent upon changes in the degree of contamination of the environment.

Methods

The chlorinated hydrocarbons were analysed by gas chromatography (Koeman *et al.*, 1967). Confirmation of the identity of the compounds found was obtained by mass spectrometry (Koeman *et al.*, 1969). The polychlorinated biphenyls (PCB's) were measured quantitatively by peak height comparison of a peak with R_x 1.45 (dieldrin = 1) present in the wildlife extracts with the same peak in a standard solution prepared from a technical PCB mixture (Phenoclor DP 6). The pattern of the PCB peaks in the wildlife-extracts resembles that obtained from technical PCB mixtures containing 60 per cent chlorine (Phenoclor DP 6, Aroclor 1260 and Clophen A 60) (Koeman *et al.*, 1969). Mercury residues were measured by neutron-activation analysis in the Interuniversitair Reactor Instituut at Delft. Methyl mercury residues were estimated by the methods of Westöö (1967).

Qualitative composition of the residues in birds

The relative retention times and mass-spectrometric data of the peaks present in the chromatograms are listed in Table 1. Dieldrin and endrin are listed separately because of fractionation of the extracts during the preceding clean-up step. In fig 1 the corresponding chromatograms are represented. The profiles of the chromatograms obtained from different bird species in the marine environment are very similar.

Twenty-one compounds were detected, 1 fungicide hexachlorobenzene (HCB) (calculated quantitatively from 1969 onwards); 3 insecticides: telodrin, dieldrin and endrin; 1 metabolite of an insecticide p,p'-DDE (metabolite of p,p'-DDT); 15 PCB isomers (calculated quantitatively from 1967 onwards) and 1 unknown substance ($R_x = 0.71$; 376(8)).

Eider tissues and molluscs were analysed for mercury from 1968 onwards. As will be demonstrated, relatively high residues were found. The mercury residues consist partly of methyl-mercury compounds.

The sandwich tern

The sandwich tern is migratory and arrives at breeding sites between the end of April and beginning of June. One or two eggs are laid and hatch in 3 to 4 weeks. At about 4 weeks the juvenile terns are able to fly and normally the site is deserted by mid-August.

In July and August 1964 sandwich terns died with conspicuous symptoms on the island of Texel. They fell from the air uttering alarm calls and lay in spasmodic posture; death occurred 1–2 days afterwards. Tissue analyses of a few birds revealed high residues of dieldrin and endrin. Comparing the residues with published data relating residues and symptoms, we tentatively concluded the birds were poisoned (Koeman and van Genderen, 1966).

Since 1954 the sandwich tern had shown a remarkable decline from about 40,000 breeding pairs in 1954 to about 1,500 in 1964. In order to evaluate the effect of the environmental contamination on this decline, further studies were made in 1965 and 1966. Preliminary results of this study have been published (Koeman *et al.*, 1967a).

As a first result in 1965 it appeared that besides dieldrin and endrin also telodrin was present. Mortality was observed among adults, 4–6-week-old juvenile birds and newly hatched chicks; 15 adult birds died (about 1 per cent of total adults present). One bird was found alive showing heavy convulsions at 10 to 30 sec intervals. After 30 min the bird died.

In the beginning of August, 29 juvenile birds died suddenly at the age of 3 to 6 weeks (5 per cent of number hatched). Dead juvenile birds were also reported from places outside the breeding area. The birds were found

The relative retention times were measured by gas chromatography with electron-capture detection. The mass numbers of the compounds and the number of chlorine atoms per molecule, which is given in parentheses after each figure, were obtained by gas chromatography in conjunction with a mass-spectrometer. In both cases a column consisting of 10 per cent DC-200 on Gaschrom Q 80/100 mesh was used

Fraction I (hexane)

Mass-spectrometric data

R_x[a]	Sandwich tern	Eider	Identity
0.25	282(6)	282(6)	Hexachlorobenzene
0.40	256(3)	256(3)	PCB
0.48	290(4)	—	PCB
0.57	290(4)	—	PCB
0.61	408(8)	408(8)	Telodrin[b]
0.71	290(4)	290(4)	PCB
	376(8)	376(8)	Unknown
0.83	324(5)	324(5)	PCB
1.01	316(4)	316(4)	p,p'-DDE
	324(5)	324(5)	PCB
1.23	358(6)	358(6)	PCB
	324(5)	324(5)	PCB
1.45	358(6)	358(6)	PCB
1.69	358(6)	358(6)	PCB
1.96	392(7)	392(7)	PCB
2.27	392(7)	392(7)	PCB
2.67	392(7)	392(7)	PCB
3.20	392(7)	392(7)	PCB
3.58	426(8)	426(8)	PCB

Fraction II (10% diethyl ether in hexane)

1.00	378(6)	378(6)	Dieldrin[c]
1.21	378(6)	378(6)	Endrin[d]

[a] Relative retention time (dieldrin = 1.00).

[b] Telodrin = 1,3,4,5,6,7,8,8-octachloro-1,3,3a,4,7,7a-hexahydro-4,7-methanoisobenzofuran

[c] Dieldrin = 1,2,3,4,10,10-hexachloro-6,7-epoxy-1,4,4a,5,6,7,8,8a-octahydro-1,4-endo, exo-5,8-dimethanonaphthalene

[d] Endrin = 1,2,3,4,10,10-hexachloro-6,7-epoxy-1,4,4a,5,6,7,8,8a-octahydro-1,4-endo, endo-5,8-dimethanonaphthalene

dead in a conspicuous attitude, showing a marked opisthotonus. The wings were kept half-spread. On several occasions marks in the sand around the birds indicated vigorous movements before death.

Many chicks (about 12 per cent of number hatched) died in June/July during their first week of life. The chicks were found lying on their backs, squeaking and making trampling movements with their legs. Then suddenly convulsions appeared during which the legs were stretched. The head was bent backwards into a marked opisthotonus. Mortality among young chicks is normal in tern colonies, particularly when bad weather conditions prevail in hatching time. In 1965, however, weather conditions were good and there was no indication that food was insufficient. Pathological and parasitological examinations of 16 adult and juvenile birds found dead did not explain the mortality.

Results of tissue analyses of dead or dying birds are in Table 2. For comparison some adult and juvenile birds were killed. Residue analyses of these birds are presented in Table 3.

A comparison of the liver residues in adult birds found dead with those killed shows levels of telodrin, dieldrin and DDE were 4 to 5 times higher in the birds found dead ($P = 0.01$) (Rank sum test, Wilcoxon and Wilcox, 1964). The difference in the endrin residues was not significant

($P > 0.1$). In juvenile birds the residues in the liver were also markedly higher in the birds found dead. The differences in results of the total body analyses were much less and not significant ($P > 0.1$; adults and juveniles combined). The difference in the lipid content was highly significant and 6 to 8 times higher in the birds killed than in those found dead or dying. Considering the high concentrations of insecticides in the fat of the former (Table 3), it is likely that the high levels in the livers of the dead birds resulted from a redistribution of the insecticides originating from the fat deposits initially present.

The toxicological interpretation of the residues detected is complicated because different compounds are present. The studies of Keplinger and Deichmann (1967) with rats and mice indicated that most of the combinations of two or three pesticides induced additive effects in both species. This was found for instance for endrin and dieldrin. Endrin plus aldrin and endrin plus chlordane gave more than additive effects. In our observations with the chicken embryo assay technique telodrin and dieldrin demonstrated additive action. This was not affected by PCB's. Keplinger and Deichmann also found less than additive effects when endrin was combined with DDT. This is in agreement with Robinson (1969) who found a tendency for the dieldrin concentrations in tissues at death to be increased by simultaneous exposure to DDT-type compounds. Since residues of DDE found in terns are relatively low and even over-estimated because of interference by PCB's, it seems justifiable to suppose additive action for the compounds present.

From a number of studies it could be concluded that residues in the brain and liver of animals suspected of being killed by a chlorinated hydrocarbon pesticide can be used as diagnostic indicators for the cause of death. It is not within the scope of this paper to discuss the experimental evidence supporting this conclusion. It has been considered that in the terns telodrin, dieldrin and endrin were the compounds of major importance.

Tissue residues indicative of death by dieldrin poisoning can be derived from the experimental studies of Robinson *et al.* (1967) and Stickel *et al.* (1969) with Japanese quail and domestic pigeons. A comparable experiment was carried out with telodrin in our laboratory on Japanese quail. In this experiment 100 per cent mortality occurred between $2\frac{1}{2}$ and 65 days in two groups each composed of 5 male and 5 female birds, which were fed diets containing 2 and 10 mg/kg of telodrin respectively. In the brains and livers of the birds mean residue levels of 1.48 and 4.06 parts per million were found. For dieldrin, Robinson found levels of 17.4 in the brain and 40.0 in the livers of the birds given various treatments. Hence it can be concluded that based on a comparison of residues in tissues telodrin is 10 times more toxic than dieldrin. In a previous study (Koeman *et al.*, 1967) a factor of 17 was reported. This factor was obtained from a comparison of the tissue levels in domestic chicks which had been poisoned experimentally with telodrin and dieldrin. To avoid over-estimating telodrin toxicity for wild birds, the factor 10 is used to express the residues in the terns in equivalents of dieldrin toxicity. The equivalent values are given in the last column of Table 2. In this calculation endrin is considered to have the same toxicity as dieldrin, which

TABLE 2. RESIDUE LEVELS IN PPM (WET WEIGHT) IN SANDWICH TERNS FOUND DEAD OR DYING IN 1965 (MEAN AND RANGE)

Tissue	n	Telodrin	Dieldrin	Endrin	p,p'-DDE[a]	Dieldrin toxicity equivalent
A. *Adult birds*						
Liver	5	1.1 (0.77–1.6)	5.5 (4.7–7.2)	0.67 (0.50–0.80)	2.5 (1.9–3.6)	17.2 (12.9–24.0)
Brain	3	0.78 (0.50–1.1)	3.2 (2.8–3.4)	0.42 (0.30–0.60)	1.4 (1.1–2.1)	11.4 (8.1–15.0)
Total body (2.3% lipids)[b]	3	0.90 (0.80–1.1)	5.6 (4.3–7.5)	0.50 (0.37–0.70)	2.4 (1.9–3.7)	15.1 (12.7–19.2)
B. *Juvenile birds*						
Liver	8	0.86 (0.60–1.7)	4.7 (1.9–6.6)	0.43 (0.10–1.2)	1.7 (0.90–3.4)	13.7 (8.0–24.8)
Total body (1.9% lipids)[b]	12	0.34 (0.19–0.50)	2.2 (1.4–4.1)	0.26 (0.10–0.70)	0.77 (0.45–2.0)	5.9 (3.4–9.8)
C. *Chicks*						
Liver	6	2.3 (0.63–3.8)	5.6 (2.4–12)	0.47 (0.19–1.3)	5.9 (2.0–12)	29.1 (8.9–51)

[a] Interfered by PCB material, see Table 1
[b] Percentage of hexane extractable lipids

in fact is an under-estimate; p,p'-DDE was not taken into account. In experimental studies it was found that dead birds carrying liver residues larger than 15 to 20 ppm (Turtle *et al.*, 1963; Robinson, *et al.*, 1967) or brain residues larger than 5 to 10 ppm (Robinson *et al.*, 1967; Stickel *et al.*, 1969) are likely to have been killed by that cause. A comparison with the dieldrin equivalents in Table 2 shows that these approximate or exceed the levels found in the experiments. Hence, it is highly probable that most birds found dead or dying in 1965 were poisoned with these chlorinated hydrocarbon insecticides, of which telodrin played a dominant role.

Poisoning of terns occurred mainly in two stages: first in chicks shortly after hatching and second in juvenile birds when they started flying. The mortality shortly after hatching is very probably related with the resorption of the yolk sac. It could be shown in experiments with domestic chicks that a marked release of chlorinated hydrocarbon insecticides may occur during the first days after hatching (Koeman *et al.*, 1967a). The low lipid contents of the juvenile birds found dead suggest that their expenditure of energy increases when they start flying; at that time the birds are susceptible to the toxic action of chlorinated hydrocarbon pesticides released from the fat deposits.

The percentages of banded sandwich terns reported dead from places outside the colony during their first year of life were obtained from the Bird Banding Station at Arnhem, Netherlands. These data are presented in Table 4. Birds that were shot or captured are not included. Results show the percentage of yearling birds recaptured increased in 1956, 1957, and from 1962 through 1965. In these years 96 per cent of the birds were found in the months July, August and September; 49 per cent in the Netherlands, 26 per cent in France, 13 per cent in Germany and the remaining 12 per cent in Belgium, Great Britain, and Denmark. As can be seen in Table 4 there was no excessive mortality from 1966 onwards. In spite of careful observations in the colony no abnormal mortality was seen in 1966. Both in the eggs of terns as well as in the fish species on which terns feed the residues were lower in 1966 than in 1965. The cause of the decrease in the level of contamination will be discussed below.

In 1965 the total body residues in the terns found dead were compared with the levels present in fish collected in the area where the terns fish, in order to make an estimate of the storage ratio: concentration in the body/concentration in the food. Eleven samples, including 103 specimens of sprat (*Clupea sprattus*), juvenile herring (*Clupea*

TABLE 3. RESIDUE LEVELS IN PPM (WET WEIGHT) IN SANDWICH TERNS SHOT IN THE BREEDING COLONY IN 1965

Tissue	n	Telodrin	Dieldrin	Endrin	p,p'-DDE[a]	Dieldrin toxicity equivalent
A. *Adult birds*						
Fat	3	4.9 (3.5–5.9)	16 (12–28)	3.6 (3.3–4.7)	8.3 (5.2–13)	68 (50.3–91.7)
Liver	5	0.23 (0.07–0.50)	0.84 (0.48–2.0)	0.29 (0.13–0.80)	0.58 (0.30–1.3)	3.4 (1.3–7.8)
Total body (15.4% lipids)[b] n = 4	5	0.52 (0.30–0.90)	2.6 (1.7–4.3)	0.81 (0.61–1.0)	1.7 (0.73–3.0)	8.6 (5.3–14)
B. *Juvenile birds*						
Fat	2	1.3 (1.1–1.6)	6.0 (5.5–6.7)	1.8 (1.6–2.0)	3.0 (3.0–3.1)	21 (18–25)
Liver	3	0.070 (0.05–0.10)	0.31 (0.20–0.42)	0.12 (0.07–0.19)	0.14 (0.10–0.17)	1.1 (0.77–1.6)
Total body (15.1% lipids)[b]	4	0.11 (0.07–0.26)	1.1 (0.70–1.6)	0.24 (0.20–0.34)	0.28 (0.20–0.56)	2.4 (1.6–4.5)

[a] Interfered by PCB material, see Table 1
[b] Percentage of hexane extractable lipids

TABLE 4. BANDED JUVENILE SANDWICH TERNS FOUND DEAD OR
DYING EXPRESSED AS PERCENTAGE OF THE NUMBER OF BIRDS BANDED
PER YEAR
(birds found in the colony are not included)

Year	Number of chicks banded	Percentage found dead or dying within one year
1914–1954	5124	0.56
1955	682	0.73
1956	286	2.09
1957	308	2.26
1958	163	0.61
1959	731	0.68
1960	211	1.20
1961	1055	0.85
1962	384	2.60
1963	161	2.48
1964	855	4.91
1965	350	1.71
1966	304	0.32
1967	518	0.38
1968	759	0.52

harengus) and sandeel (*Ammodytes lanceolatus*), were analysed. The following ratios were calculated: telodrin 0.55/0.05 = 11; dieldrin 3.5/0.27 = 13 and endrin 0.36/0.14 = 2.5.

Observations on the eider

Since 1962 it had been noticed that at Vlieland, an island in the Dutch Wadden Sea, many adult female eiders died with convulsive symptoms at the end of the breeding season. In 1964 the number dying was estimated to be 650 (about 30 per cent of breeding females). In 1965 about 800 females died. One was sent for residue analysis. The results of these analyses are given in Table 5.

TABLE 5. TISSUE RESIDUES IN A FEMALE EIDER FOUND DEAD NEAR THE
ISLAND OF TEXEL ON 27 MAY 1965 (PPM WET WEIGHT)

	Telodrin	Dieldrin	Endrin	p,p'-DDE[a]	Dieldr. tox. equiv.
Liver	0.65	9.5	0.47	1.2	16.5
Brain	0.33	2.9	—	0.62	6.2
Breast muscle	0.30	2.3	0.29	0.13	5.6
Fat (mesenterial)	3.0	35.0	1.3	10.0	66.0

[a] See Table 2

We concluded it was highly probable the bird was poisoned by insecticides.

Only female birds died. In eiders only the female birds take care of the brood, and they consume little or no food during the incubation period. For that reason it was postulated that the pesticides were mobilized from the fat during incubation. To test this hypothesis blood samples were taken from breeding ducks in the

beginning of the breeding season in 1966. Samples of blood were taken also from females which were found dying at breeding period end (Table 6).

Results show concentration in the blood increases strongly in the incubation period, while body weight decreases by about 36 per cent.

Residues in the livers of dead and dying birds as well as in those which were shot in 1966, 1967 and 1968 are given in Table 7. The levels of telodrin, p,p'-DDE and PCB are much higher in the dead birds than in those shot. For endrin only a slight difference is found. The average dieldrin toxicity equivalents calculated for the dead birds approximate the above-mentioned values in experimentally poisoned birds. In the period from 1962 to 1965 the female eiders died when they were still breeding or when they had just finished breeding. From 1966 onwards the mortality generally occurred after the birds and the ducklings had reached the tidal flats bordering the island where they breed. It was also reported that before 1966 convulsions in dying birds were frequent, but from 1966 onwards only occasional. Many birds developed symptoms of convulsion and tremors, however, when they were handled for taking blood. The diseased birds hardly tried to escape on approach. The explanation for the differences in both the time of death and the degree of symptom development may be that the average rate of contamination of the marine environment with pesticides was higher before 1966. The bird analysed in 1965 (Table 5) still contained mesenterial fat deposits. None of the birds examined after 1965 had fat deposits. Obviously a further atrophy of adipose and other tissues was needed to mobilize a lethal dose of insecticides. After 1965 there was further complication by infection of the ducks with large numbers of trematodes in the intestines (Koeman et al., MS). The intestinal epithelia of most of the birds were damaged seriously by these trematodes and therefore it is likely that from 1966 on the chance of being poisoned was enhanced by the extremely poor physical condition induced by parasite infection. It is not likely, however, that the parasites were alone responsible—a few of the birds which died were infected to a low degree or not at all. Further Gerassimova and Baranova (1960) and Kulackova (1958), who studied the eider in Russia, do not report mortality in female ducks at the end of the incubation period, although 100 per cent of the population was infected by large numbers of trematodes.

The toxicological interpretation of the PCB residues is difficult. The levels found are considerably lower than those in livers of experimentally poisoned chickens (Vos and Koeman, 1970); so it is not likely that PCB's were a major cause of death. In seven eiders found dead or dying in 1969 the livers were analysed for mercury. The residues varied from 6.9 to 14.1 ppm, 22 per cent

TABLE 6. RESIDUE LEVELS IN BLOOD OF FEMALE EIDERS IN 1966 (μG/ML)

Period	n	Average body wt (kg)	Telodrin	Dieldrin	Endrin	p,p'-DDE[a]
3/5–4/5	18	2.2	<0.001–0.02	0.027 (0.01–0.04)	<0.005–0.01	<0.002–0.08
23/5–10/6	32	1.4	0.23 (0.11–0.40)	0.57 (0.37–1.2)	0.010–0.12	0.57 (0.22–0.98)

[a] See Table 2

[432]

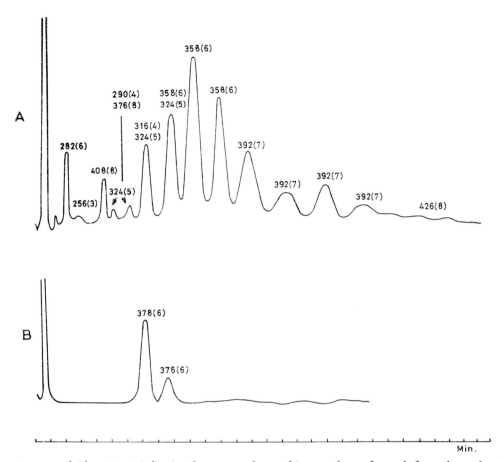

Fig 1. Gas chromatograms of eider extracts indicating the mass-numbers and in parentheses after each figure the number of chlorine atom per molecule: A. Fraction I (hexane); B. Fraction II (10% diethyl ether in hexane). Column: 10% DC 200 Gaschrom Q 8/100 mesh

being present as methyl mercury. Although probably not yet present in fatal concentrations, the levels of mercury and PCB's detected warrant further investigations concerning the possible sublethal effects on marine animals.

The reproductive success of eiders was very small in preceding years. Many nests were deserted before hatching, while many ducklings which hatched were lost because too few adult birds were left to care for them.

In 1965 it was observed, for instance (Swennen, personal communication) that a flock of 718 ducklings was accompanied by only 45 adult females, which implies a deficit of 135 adults. Many of the ducklings were lost to herring gulls. There are no indications, however, that the clutch size was smaller or that the fertility and hatchability of the eggs were markedly reduced. For the first time, the breeding success of the population seems normal again in 1970.

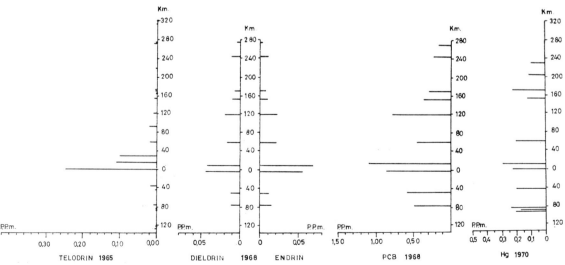

Fig 2. Residues of persistent compounds in mussels (Mytilus edulis) along the Dutch coast. Point 0 in the km scale indicates the site where the fraction of the Rhine water, which runs through the industrial area near Rotterdam, is entering into the sea. The distance of the sampling places from this point in both northern and southern directions is indicated on the km scale

Year	History	n	Telodrin	Dieldrin	Endrin	p,p′–DDE[a]	PCB	Dieldrin-toxicity equivalent
1966	Dead or dying	31	0.96 (0.47–1.7)	2.9 (1.1–4.8)	0.24 (0.08–0.42)	2.0 (0.94–3.7)	—	12.7 (5.9–22)
	Shot	5	0.06 (0.02–0.15)	0.37 (0.20–0.81)	0.12 (0.08–0.16)	0.26 (0.04–0.80)	—	1.1 (0.48–2.5)
1967	Dead or dying	10	1.08 (0.26–4.5)	2.2 (0.96–3.4)	0.18 (0.02–0.77)	2.59 (1.5–4.5)	63 (23–96)	13.2 (3.6–49)
	Shot	5	0.06 (0.02–0.12)	0.31 (0.14–0.69)	0.11 (0.07–0.14)	0.14 (0.06–0.29)	5.5 (2.1–12)	1.0 (0.41–2.0)
1968	Dead or dying	9	0.45 (0.16–0.72)	2.6 (0.80–5.0)	0.089 (0.010–0.31)	3.2 (1.1–5.2)	89 (67–134)	7.2 (2.4–12.5)
	Shot	3	0.11 (0.02–0.26)	0.27 (0.10–0.41)	0.066 (0.017–0.11)	0.17 (0.10–0.24)	5.6 (3.9–6.5)	1.4 (0.31–3.1)

[a] Interfered by PCB material, see Table 1

Distribution of persistent compounds along the coast

From 1965 to 1970 mussels (*Mytilus edulis*) were analysed to study the distribution of contamination along the coast and to trace the source. Figure 2 shows some of the results. The concentrations are set out on a kilometre scale indicating the distance from where Rhine water from the industrial area near Rotterdam enters the sea (indicated by 0). The insecticides and the PCB's obviously originate from this site. Probably most insecticidal material comes from near Rotterdam where chlorinated hydrocarbon pesticides are manufactured. Use of telodrin is not allowed in the Netherlands and the neighbouring countries. This compound was manufactured from 1961 to the end of 1965. Most of the endrin and dieldrin very probably also originated from this industry. In fish collected from up-stream parts of the Rhine no endrin could be detected, and the residues of dieldrin were relatively small. Considerable amounts PCB were present in these fish, however, which strongly suggests that most of the PCB's come from industrial areas in neighbouring countries. This probably also applies to mercury which is more evenly distributed than the chlorinated hydrocarbon compounds. The relatively high residues in the southern part of the study area further suggest that the River Schelde also contributes to the mercury contamination of the marine environment.

Concluding remarks

This study shows serious mortality can occur in marine birds from pollution with chlorinated hydrocarbon pesticides—particularly telodrin for terns and eiders. This causal relationship could only be proved from 1965 onwards. The reported deaths in eiders since 1962 and the increased percentages of recaptured juvenile terns from 1962 to 1966 (Table 4) coincide with the period that telodrin was manufactured. This suggests strongly that the increased mortality before 1965 was also mainly caused by this insecticide. The continued mortality in eiders from 1966 to 1969 is related to the long persistence of telodrin in the birds' tissues and further to the presence of large numbers of intestinal parasites which probably made them more susceptible to the toxic action of the compounds.

In spite of the relatively high endrin contamination of the lower trophic levels (the residues in mussels and fish being equal or sometimes even higher than those of dieldrin), relatively low concentrations are found in the tissues of terns and eiders. This shows that this insecticide, which is generally more toxic for warm-blooded animals than dieldrin, is less hazardous for the higher trophic levels.

Both in terns and eiders the reproductive physiology was not noticeably disturbed. A recent measurement of the egg shell thickness in the sandwich tern revealed that on average egg shells were 5 per cent thinner in 1970 (n = 35) than they were between 1932 and 1950 (n = 53) (P < 0.01; t–test).

For assessment of the biological meaning of this change further investigations are needed.

Finally it can be said that elaborate measures were made by the industry concerned, to avoid as much further contamination as possible.

References

GERASSIMOVA, T D and BARANOVA, Z M Ekologiia obyknovennoi
1960 gagi (*S. mollissima* L.) Kandalaksskom zapovednike. *Trudy Kandalaksskogo gos. zapovednika*, 3:8–90.

KEPLINGER, M L and DEICHMANN, W B Acute toxicity of
1967 combinations of pesticides. *Toxicol. appl. Pharmacol.*, 10:586–95.

KOEMAN, J H and VAN GENDEREN, H Some preliminary notes on
1966 residues of chlorinated hydrocarbon insecticides in birds and mammals in the Netherlands. *J. appl. Ecol.*, 3(Suppl.): 99–106.

KOEMAN, J H, OUDEJANS, R C H M and HUISMAN, E A Danger of
1967 chlorinated hydrocarbon insecticides in birds' eggs. *Nature, Lond.*, 215:1094–6.

KOEMAN, J H, *et al.* Insecticides as a factor in the mortality of the
1967a sandwich tern (*Sterna sandvicensis*). *Meded. Rijksfac. Landbouwwetensch. Gent*, 32:841–854.

KOEMAN, J H, TEN NOEVER DE BRAUW, M C and DE VOS, R H
1969 Chlorinated biphenyls in fish, mussels, and birds from the River Rhine and the Netherlands coastal area. *Nature, Lond.*, 221(5186):1126–8.

KULACKOVA, V G Ekologo-faunisticeskii obsor parazitofauny
1958 obyknovennoi gagi Kandalaksskogo Zaliva. *Trudy Kandalaksskogo gos. zapovednika*, 1:103–159.

ROBINSON, J, *et al.* Residues of dieldrin (HEOD) in the tissues of
1967 experimentally poisoned birds. *Life Sci.*, 6:1207–20.

ROBINSON, J Organochlorine insecticides and bird populations in
1969 Britain. *In* Chemical fallout, edited by M. W. Miller, and G. G. Berg, Springfield, Charles C. Thomas, pp. 113–69.

STICKEL, W H, STICKEL, L F and SPANN, J W Tissue residues of
1969 dieldrin in relation to mortality in birds and mammals. *In* Chemical fallout, edited by M. W. Miller and G. G. Berg, Springfield. Charles C. Thomas, pp. 174–204.

TURTLE, E E, *et al.* The effects on birds of certain chlorinated
1963 insecticides used as seed dressings. *J. Sci. Fd Agric.*, 14: 567–77.

Vos, J G and KOEMAN, J H Comparative toxicological study with
1970   polychlorinated biphenyls in chickens with special reference
        to porphyria, edema formation, liver necrosis, and tissue
        residues. *Toxicol. appl. Pharmacol.*

WESTÖÖ, G Determination of methyl mercury compounds in
1967   foodstuffs. 2. Determination of methyl mercury in fish, egg,
        meat and liver. *Acta chem. scand.*, 21:1790–800.

WILCOXON, F and WILCOX, R A Some rapid approximate
1964   statistical procedures. New York. Lederle Laboratories,
        60 p.

Subtle Effects of Pollution on Inshore Tropical Plankton

R. I. Clutter*

Les effets insidieux de la pollution sur le plancton des eaux côtières tropicales

Les eaux usées domestiques et les apports d'eau des fleuves fournissent des éléments nutritifs supplémentaires dans la baie de Kaneohe, aux îles Hawaii, surtout dans le secteur sud. Dans le secteur le plus productif (sud) de la baie, des modifications sont apparues dans le plancton au cours des quelques dernières années; ces modifications indiquent une évolution vers l'eutrophisation, la diminution de la diversité, la transformation de la structure écologique et le décroissement de la stabilité des populations de plancton avec de violentes fluctuations concomitantes dans les stocks permanents de quelques espèces. Aucune de ces modifications n'est vraiment souhaitable et quelques-unes sont mêmes définitivement indésirables.

Les données relatives à la situation actuelle de la baie fournissent un modèle pour prévoir les modifications spécifiques susceptibles de survenir dans les parties les moins productives de la baie si la tendance à l'eutrophisation se poursuit. On note une courbe décroissante allant de la haute productivité primaire et secondaire dans le secteur sud à la faible productivité du secteur nord. La diversité et, apparemment, la stabilité biotiques sont inversement proportionnelles à la productivité dans la baie. Le réseau des cours d'eau continuera à maintenir des différences sectorielles dans la baie même si la collectivité planctonienne globale est atteinte par la pollution. Toutefois, on peut s'attendre à ce que les zones relativement moins productrices régressent par une série d'étapes analogues ou homologues à la situation actuelle dans la zone qui est relativement la plus productrice.

Sutiles efectos de la contaminación sobre el plancton de aguas costeras tropicales

Las aguas de albañal y las de los ríos y arroyos proporcionan nutrientes complementarios a la Bahía de Kaneohe, Hawaii, fundamentalmente al sector sur. En este sector de la Bahía, que es el más productivo, en los últimos años han tenido lugar cambios en el plancton que indican que existe una tendencia hacia la eutrofización, menor diversidad, estructura del ecosistema alterada y menor estabilidad de la población planctónica con las consiguientes violentas fluctuaciones en las poblaciones permanentes de algunas especies. Sin duda alguna, ninguno de estos cambios es deseable, siendo algunos de ellos evidentemente indeseables.

El conocimiento del estado actual de la Bahía facilita un modelo para predecir los cambios específicos que pueden ocurrir en las partes menos productivas de ella, y en el caso de que continúe la tendencia hacia la eutrofización. Hay una transición de una productividad primaria y secundaria elevada en el sector sur a la más baja en el sector norte. La diversidad biótica y al parecer la estabilidad son inversamente proporcionales a la productividad de la Bahía, incluso aunque toda la comunidad planctónica se altere con la contaminación. Sin embargo, las zonas que relativamente producen menos es de esperar que disminuyan en una serie de fases y que sean análogas u homólogas a las condiciones actuales en la zona que relativamente produce más.

COMPARED with the immediate effects of acute pollution incidents the effects of chronic pollution often are subtle and difficult to demonstrate. These effects can be longer-lasting and broader in scope. They can pass undetected while causing ecosystems to become unstable and eventually irretrievably transformed.

There is evidence of subtle change in the plankton community of Kaneohe Bay on a moderately populated shore of the island of Oahu, Hawaii. The cause apparently is nutrients that enter the south sector of the bay in the treated effluents from two domestic sewers and from stream runoff, especially in the south sector where the human population is densest.

The purposes of this study were to define similarities and differences in the existing plankton populations in the various sectors of Kaneohe Bay so that it might be possible to predict the changes that might occur in less polluted areas should any significant change occur in environment, and to determine whether the present level of pollution has measurably affected the plankton community.

Technical details and full results are given in Clutter (1970).

Speculations

Continued increases in influx of nutrients and possibly of organic compounds to the Bay seem likely to increase the general level of plankton productivity. This will increase turbidity and reduce aesthetic quality as well

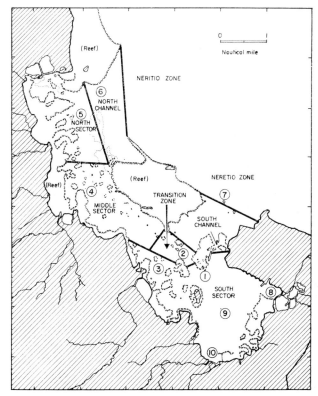

Fig 1. Chart of Kaneohe Bay, showing 1968–1969 sampling station positions (1–7, 9), sewer outfalls (8 and 10), and biographic divisions assigned from the results of this study

Consultant, Department of Fisheries FAO, Rome, Italy

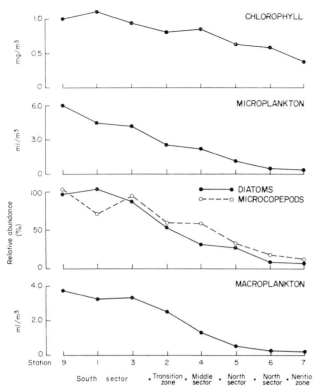

Fig 2. Mean standing stocks of chlorophyll-a, microplankton (0.06 mm net) settled volumes, relative numbers of diatom cells, relative numbers of microcopepods, and macroplankton (0.33 mm net) settled volumes at sampling stations in Kaneohe Bay, July 1968–May 1969

as cause reduced biological diversity, accompanied by reduced stability of the ecosystem. Increasing fluctuations in standing stocks will occur at all trophic levels, but be most pronounced among phytoplankton. Red tides can be expected to occur increasingly. Some outbreaks may involve species that are toxic or produce toxic by-products.

It is commonly assumed that increased fertilization of a body of water will increase the production of organisms that are useful to man, but this is not necessarily true. Enrichment may unbalance systems that have developed through long evolution. This may have unpredictable effects. Furthermore, the returns in fish or other desirable animal flesh may not increase in proportion to increased primary production. A considerable amount of fixed energy in the form of protein may shift out of the direct plant-to-man food chain into predators like chaetognaths, which are not preferred food for many of the pelagic fishes.

Present state of Kaneohe Bay plankton

There is a cline from higher standing stocks and productivity in the south sector to lower standing stocks and productivity in the north sector (fig 1). Water transparency is lower where productivity is higher.

Taxonomic diversity is inversely related to productivity, which conforms to the findings and prediction of Margalef (1963) The proportions of microflora (<0.06 mm) in the phytoplankton, microplankton (0.06 mm–0.33 mm) in the herbivores, and carnivores in the macroplankton (>0.33 mm) give clear evidence of differences in trophic structure between areas.

Concomitant with lower biological diversity, the plankton populations appear to fluctuate most violently in the higher productivity area (south sector). The relatively unproductive neritic zone has very small fluctuations. This conforms to the theory discussed by McArthur (1955), which contends that diversity confers stability because there are more interacting links in the food web.

The observed differences between the south sector and other parts of the Bay are maintained and partly caused by circulation patterns and intensified by stream runoff and sewage disposal caused by urbanization. The residence time of land-derived nutrients is longer in the south basin because it is deeper than the rest of the bay and has a relatively smaller exchange transport cross-section area. Circulation occurs as a self-contained gyre. The south basin volume is 27 per cent of the total bay volume, yet it apparently receives less than 10 per cent (8 per cent, according to Bathen, 1968) of the total exchange transport in and out of the bay.

Whatever the other consequences of increasing pollution may be, the circulation pattern will doubtless maintain differences among the sectors of the bay, but significant continued pollution would probably cause the productivity and trophic structure of the middle-sector plankton to become more like the present south sector.

Geographical distribution

Eight stations (1 to 7, and 9) were sampled 24 times during July 1968–May 1969, giving 192 microplankton (larger phytoplankton + microzooplankton) and 192 macroplankton (zooplankton only) samples. For mean standing stocks, by station, of chlorophyll-a, microplankton (0.06 mm net) settled volumes, phytoplankton cells, microcopepods (including nauplii), and macroplankton (0.33 mm net) settled volumes, see Fig 2.

These chlorophyll-a concentrations were determined from surface samples. Other samples taken at three discrete depths showed that the surface samples represented the whole water column with respect to geographical distribution. The surface chlorophyll-a concentrations/m³ were 70–80 per cent of the average values/m calculated for the whole water column.

The standing stocks of phytoplankton cells and microcopepods are shown as relative values; the maximum mean standing stock of phytoplankton (station 1) and copepods (station 9) are each assigned a value 100 per cent and the other station values are shown as percentages of their respective maximum.

Differences in standing stocks of plants and animals are not necessarily accurate indications of differences in primary and secondary productivity, but standing stocks can be used as indices of productivity when appropriate rates of energy fixation and exchange per unit standing stock are determined.

Primary productivity experiments demonstrated that productivity is related to the standing stock of chlorophyll-a, and to the standing stock of phytoplankton cells. On 17 July 1969 the standing stock of chlorophyll-a was 3.4 times as high in the south sector (station 9) as in the middle sector; and the primary productivity was 5.4 times as high. On 27 August 1969 the standing stock of chlorophyll-a was 3.6 times as high in the south sector as in the middle sector; and the primary productivity was 3.0 times as high.

Secondary productivity is necessarily directly related to standing stocks. The same general types and sizes of planktonic animals occur throughout the bay, but in different densities. Experiments have shown that the metabolism of each general type of animal is a function of size (age), and environmental temperature. Since the animals within taxa are about the same size throughout the bay, and the temperature is almost uniform it is assumed that growth and metabolism are direct functions of the standing stocks.

Taxonomic composition and diversity

The taxonomic composition and diversity of the phytoplankton in the bay are incompletely known because no study has included examination of the microflora (nannoplankton). Sixty-seven species of diatoms were observed in this survey. The same species of diatoms occur at one time or another in all sectors, but the diversity (number of species/number of individuals) is lowest in the south sector and becomes progressively higher northward.

The taxonomic composition of the microzooplankton was analysed for station 4 (middle sector) and station 9 (south sector) for four sampling dates. Fourteen microzooplankton taxa were distinguished and enumerated; some include several species. Six taxa were holoplankton (entire lifetime in the plankton) and eight were meroplankton (larvae of benthic molluscs, shrimps, etc.). The holoplankton constituted 97 per cent of the numbers at station 9, and 90 per cent at station 4. The diversity is low among the Kaneohe Bay microzooplankton, especially in the south sector.

Compared with the microzooplankton, the diversity of the macroplankton is high. The taxonomic composition of the macroplankton was analysed for eight stations. Thirty-four macroplankton taxa were distinguished and enumerated; some included several species. Fifteen taxa were holoplankton and nineteen were meroplankton. Holoplankton were much more abundant in the south sector; the meroplankton were almost uniformly distributed.

Biotic diversity can be expressed, simply, as the ratio of number of taxa/number of individuals. This is a useful index, but it does not take into account the distribution of numbers within taxa. The Shannon-Wiener function (MacArthur and MacArthur, 1961) is an index based on information theory that accounts for the frequency of occurrence of individuals in each taxon. The index was calculated from macroplankton counts for each of the inner bay sampling stations (1, 2, 3, 4, 5, 9) on four sampling dates. To determine the order of diversity in the different sections the results were treated in two ways. First, mean values of diversity index were calculated for each station and arranged by rank, from highest to lowest diversity: stations 5, 4, 2, 1, 3, 9. Highest diversity of macroplankton was in the northern sector, middle sector next, the transition zone intermediate, and the south sector least diverse.

Secondly the rank order, by station was determined for each date and these ranks were summed for each station over the four sampling dates. The sums of ranks

coincided with the rank of the mean values of the index. A Kendall 2-way analysis of variance on ranks (Tate and Clelland, 1957) showed that there was significant (p <0.01) concordance of rank by sampling station and the diversity of macroplankton increases progressively from the south to the north sector. This is inversely correlated with the decrease in standing stocks from south to north (fig 2).

Trophic structure

Microflora (nannoplankton) were not examined directly, but indirect evidence suggests that their proportions among the primary producers may be highest in the south sector. The ratio of mean chlorophyll-a, south sector/middle sector, was less than 1.4. The ratio of microzooplankton, south sector/middle sector was greater than 3. The ratio of numbers of *Oikopleura* (larvacean tunicates) in the south sector/middle-north sectors was 2.7, about the same as that for microzooplankton in the same sectors.

The microzooplankton probably feed on nannoplankton, and *Oikopleura*, the most abundant herbivore in the bay, have filtering apparatus that allows them to feed only on nannoplankton. To support the much larger populations of nannoplankton-feeders in the south sector, the numbers of nannoplankton/unit chlorophyll-a must be higher in the south sector than in the middle and north sectors. The same thing may be indicated by the limited ^{14}C productivity data, which showed a somewhat higher mean productivity/unit chlorophyll-a ratio in the south sector than in the middle sector. If a significant fraction of the nannoplankton were heterotrophs, the nannoplankton productivity could be larger without affecting the level of the chlorophyll-a standing stock.

Both the microplankton and macroplankton are most abundant in the south sector, and become progressively less abundant northward, and least abundant in the neritic zone (fig 2), but the ratios differ.

Among the microzooplankton, carnivores constituted only 0.07 per cent in south sector samples. Carnivores constituted from 2.6 per cent (station 7) to 52.2 per cent (station 1) of the macroplankton samples for the same period. For the period July–August 1968, the number of carnivores in the macroplankton samples declined regularly and markedly from the more productive south sector to the less productive areas. The herbivores were much more evenly distributed.

Among the macroplankton, the ratio of numbers of carnivores/herbivores was 0.8 in the south sector and 0.2 in the middle-north sectors. Among the microplankton, the ratios of carnivores/herbivores was very low (less than 0.001) and about the same in both parts of the bay. Among the herbivores, the ratio of numbers of microplankton/macroplankton was 196 in the south sector and 76 in the middle sector.

In less productive areas the small herbivores are fewer both in absolute numbers and in numbers relative to the numbers of macroplankton. The larger herbivores are more numerous per unit standing stock of plant material in less productive areas than in more productive areas. The numbers of carnivorous macroplankton apparently are related, indirectly, to the abundance of primary producers.

Variability

Moderate to large changes in standing stocks occur at all trophic levels in Kaneohe Bay plankton. The temporal variability of the average standing stocks of both phytoplankton and zooplankton in the northern part of the bay (stations 4, 5, 6) is less pronounced than the variability in the south sector.

As a measure of the amount of temporal variability at different specific locations in the bay, indices of dispersion (variance/mean) were calculated for each of the eight stations from the sampling data on settled volumes of microplankton and macroplankton obtained from July 1968 to May 1969. The index values for microplankton are greater than the values for macroplankton at all stations in the bay proper. This indicates that the microplankton populations are less stable. The index values for both categories of plankton were low for both the north channel (station 6) and the neritic zone (station 7). Significant variability seems to be present only within inner bay populations, which are more productive than the neritic populations.

Water transparency

A few light penetration measurements in open waters to 10 m depth were made with a submarine photometer in the middle and south sectors in 1969. The calculated extinction coefficients were: middle sector—0.204/m; south sector—0.322/m.

Changes in the plankton

The mean standing stocks of chlorophyll-a show no clearly demonstrable change over the past 5 to 9 years (Table 1). Values in the less productive area (represented by station 4) were about the same nine years ago. Values in the more productive south sector are now 20 per cent higher than nine years ago, but are somewhat lower than the estimate made five years ago. By comparison, transparency measurements made about the same time of year in 1969 as in 1963 (Piyakarnchana, 1965) give extinction coefficients that were 2.2 (middle sector) and 2.0 (south sector) times as high as those determined six years before. The differences in turbidity could be caused by phytoplankton, by suspended non-living organic matter, and by fine inorganic particles. The relative importance of these components in Kaneohe Bay is unknown. There had been no heavy stream runoff during the periods of measurement, therefore the turbidity may be indicative of the abundance of phytoplankton and detrital organic matter.

The chlorophyll-a concentration could stay about the same level, if this were actually the case, even if changes in primary producers occurred. The proportion of small,

metabolically active and perhaps partly heterotrophic flagellates, and consequently the gross primary productivity, may have increased without measurably affecting the average chlorophyll-a level. Possible increase in these nannoplankton is suggested by the relative decrease in the abundance of macrocopepods and the relative increase in the abundance of *Oikopleura* in the south sector over the past five years. The macrocopepods feed on the larger phytoplankton, primarily chain-forming diatoms, and *Oikopleura* are obligate feeders on nannoplankton, including naked flagellates. From this, I infer that the ratio of nannoplankton to diatoms may have increased in recent years. The photosynthetic nannoplankton may have increased as a result of increased input of inorganic nutrients and the heterotrophic nannoplankton may have increased as a result of increases in organic carbon compounds from sewage, stream runoff, and from the exudates of the other phytoplankton.

There is evidence of other changes in the phytoplankton. It is reported at the Hawaii Institute of Marine Biology that the average clarity and colour of the water in the bay have changed gradually over the last twenty years, especially in the south sector; I have shown some evidence of increased turbidity between 1963 and 1969. Also, "red tides" have occurred in recent years where they were not observed before.

The earliest detailed observation of a red tide was in October 1965 in the south sector. At least three red tides occurred in the south sector in November 1969. One of these persisted as a sharply defined patch about 100 m in diameter for about two weeks; it was composed almost entirely of a dinoflagellate, *Exuviella* sp., as was another red tide patch, located about 1,500 m away, in the vicinity of the Kaneohe municipal sewer outfall. Samples taken from this patch gave chlorophyll-a values as high as 12 mg/m³. Repeated outbreaks such as these can occur at random times and locations. Therefore, they are not often detected by routine sampling at fixed stations, such as in the 1968–1969 survey.

Zooplankton

The only information about the former state of microzooplankton in the bay is from Edmondson (1937). In 1931–1932 he took samples from surface waters only, but his data suggest that the microcopepods in the south and northern sectors were more nearly alike in 1931–1932 than in 1968–1969.

There is evidence of increase in the standing stock and changes in species composition of the macroplankton in the south sector (Table 2). The increased density implies increased primary productivity (perhaps increase in heterotrophic nannoplankton) because the water temperature, and therefore metabolism of the animals, has not changed. The changed species composition is evidence of changes in trophic structure and energy pathways. Some observations on the gut contents of chaetognaths, the dominant carnivores, support this conclusion.

The mean number of all species of macroplankton has apparently increased by 70 per cent over the past five years. Much of this increase was caused by the twofold increase in chaetognaths, the dominant species. The numbers of *Oikopleura* (larvacean tunicate), also

TABLE 1. COMPARISON OF MEAN CHLOROPHYLL-A AT THE SURFACE AT STATIONS 4 (MIDDLE SECTOR) AND 9 (SOUTH SECTOR): 1959–1960 (DATA FROM WORLD DATA CENTRE A IGY OCEANOGRAPHY REPORT NO. 4); 1963–1964 (DATA FROM PIYAKARNCHANA, 1965); 1968–1969 (DATA FROM THIS STUDY)

	1959–1960	1963–1964	1968–1969
Station 4			
Number of samples	6	no data	13
Mean chlorophyll-a (mg/m³)	0.78	no data	0.74
Station 9			
Number of samples	7	12	13
Mean chlorophyll-a (mg/m³)	0.86	1.11	1.04

increased dramatically in the past five years. This seems to represent a trend of continuing increase. In 1950, Hiatt (1951) classified *Oikopleura* as uncommon or rare, whereas it is now the second most abundant macroplankton. Conversely, the relative numbers of *Lucifer* (especially adults and advance juveniles) and the relative numbers of macrocopepods (larger than 0.3 mm), which feed on the larger phytoplankton, have both decreased.

Supporting evidence that some of the observed changes in relative abundance of macroplankton apparently have altered the trophic relationships is given by Peterson (1969). He shows that the gut contents of the chaetognaths he observed in 1968 were different from the gut contents observed by Piyakarnchana (1965) in 1963–1964. Among the herbivores fed on by chaetognaths in 1963–1964, macrocopepods constituted 74 per cent and *Oikopleura* constituted 22 per cent of the gut contents.

Among the herbivores observed in chaetognath guts in 1968, only 16 per cent were macrocopepods while 35 per cent were *Oikopleura*.

Series of macroplankton volumes/m³ were estimated during the periods indicated in Table 2 (1963–1964, 1966–1967, 1968–1969). These volumes were estimated by different methods in each of the three periods, therefore they cannot be used to assess changes in abundance, but they do provide evidence of variability within the sampling periods. They suggest that the variability in standing stocks is somewhat greater now than it has been in the past. This may be evidence that the population structure of the plankton community is becoming less stable.

TABLE 2. COMPARISON OF MACROPLANKTON STATISTICS FOR SOUTH SECTOR: 1963–1964 (PIYAKARNCHANA, 1965); 1966–1967 (CLUTTER AND MURPHY, UNPUBLISHED DATA); 1968–1969 (THIS STUDY).

	1963–1964	1966–1967	1968–1969
Mean number macroplankton/m³	706	—	1,217
Mean number sagitta/m³	215	424	447
Mean number *Oikopleura*/m³	176	—	263
Mean number *Lucifer* (all stages)/m³	79	—	77
Mean number *Lucifer* mastigopus & adults/m³	30	10	9
Mean number copepoda/m³	24	—	20
Mean number carnivores/m³	321	—	515
Mean number herbivores/m³	385	—	689
Per cent carnivores of total macroplankton	45%	—	42%
Mean number holoplankton/m³	481	—	867
Mean number meroplankton/m³	225	—	350
Per cent holoplankton of total macroplankton	68%	—	71%

Note: 1968–1969 carnivore-herbivore counts do not include eggs

References

BATHEN, K A descriptive study of the physical oceanography of
1968   Kaneohe Bay, Oahu, Hawaii. *Tech. Rep. Hawaii Inst. mar. Biol.*, (14):352 p.

CLUTTER, R I Plankton ecology. *In* Study of estuarine pollution in
1970   the State of Hawaii. 2. Kaneohe Bay study. *Tech. Rep. Wat. Resour. Res. Cent. Univ. Hawaii*, (31).

EDMONDSON, C H Quantitative studies of copepods in Hawaii with
1937   brief surveys in Fiji and Tahiti. *Occ. Pap. Bernice P. Bishop Mus.*, 13(12):131–46.

HIATT, R W Food and feeding habits of the nehu, *Stolephorus*
1951   *purpureus* Fowler. *Pacif. Sci.*, 5(4):347–58.

McARTHUR, R Fluctuations of animal populations, and a measure
1955   of community stability. *Ecology*, 36(3):533–6.

MACARTHUR, R H and MACARTHUR, J W On bird species
1961   diversity. *Ecology*, 42:594–8.

MARGALEF, R On certain unifying principles in ecology. *Am.*
1963   *Natst*, 97:357–74.

PETERSON, W T Species diversity and community structure of the
1969   macrozooplankton of Kaneohe Bay, Oahu, Hawaii. Unpbl. M.S. Thesis, University of Hawaii, 91 p.

PIYAKARNCHANA, T The plankton community in the southern part
1965   of Kaneohe Bay, Oahu, with special emphasis on the distribution, breeding season and population fluctuation of *Sagitta enflata* Grassi. Thesis, University of Hawaii, 227 p. (Unpubl.).

TATE, M W and CLELLAND, R C Nonparametric and shortcut
1957   statistics. Danville, Illinois, Interstate Printers and Publishers, 171 p.

Plankton in the North Atlantic—An Example of the Problems of Analysing Variability in the Environment

R. S. Glover, G. A. Robinson and J. M. Colebrook*

Le plancton dans l'Atlantique Nord: un exemple des problèmes posés par l'analyse de la variabilité du milieu

On peut détecter et mesurer la variabilité résultant de l'action de la pollution sur le milieu par comparaison avec le spectre de variation naturelle. Il convient d'intensifier la recherche sur la variabilité dans les mers et de la rattacher plus étroitement aux études atmosphériques.

Un exemple des problèmes qui se posent nous est donné par l'enquête effectuée dans l'Atlantique Nord et la mer du Nord par collecte permanente du plancton. Au cours des 22 dernières années, on a constaté que dans une partie de la région de nombreuses espèces accusaient progressivement une diminution quantitative alors que la biomasse du zooplancton se contractait; la saison d'activité biologique, à en juger par les stocks permanents à une profondeur d'échantillonnage de 10 m, semble s'être réduite.

Il est difficile d'établir un rapport entre ces modifications et certains facteurs causatifs par suite de l'insuffisance des observations sur le milieu. On ne sait pas exactement si les modifications de la température de l'eau de mer, ou les tendances du climat et des radiations solaires sont de caractère cyclique ou continu, si elles résultent de la turbidité croissante de l'atmosphère ou des variations de l'activité solaire. Des travaux expérimentaux ont montré que les résidus de pesticides peuvent faire baisser le taux de photo-

El plancton en el Atlántico Norte: un ejemplo de los problemas que plantea el análisis de la variabilidad del medio ambiente

La variabilidad inducida por la contaminación debe ser descubierta y medida comparándola con el espectro de la variación natural. Será necesario intensificar las investigaciones sobre variabilidad en los mares y correlacionarla más estrechamente con los estudios de la atmósfera.

Los problemas de que se trata resultan manifestos por el estudio realizado en el Atlántico Norte y en el Mar del Norte con el registrador continuo de plancton. Durante los 22 últimos años, en una parte de esta zona ha habido una disminución progresiva de la abundancia de muchas especies y de la biomasa del zooplancton; a juzgar por los efectivos de la población a la profundidad de 10 m en que se tomaban las muestras, la temporada anual de actividad biológica resulta acortada.

Es difícil relacionar estos cambios con los factores causantes dada la insuficiencia de observaciones ambientales. No está claro si los cambios de la temperatura del agua o de las tendencias en cuanto a clima y radiación solar, son cíclicos o constantes, ni si son el resultado de la mayor turbidez de la atmósfera o de variaciones en la energía radiada por el sol. Los trabajos experimentales han demostrado que los residuos de plaguicidas pueden reducir el ritmo

* Oceanographic Laboratory, Edinburgh, Scotland, U.K.

synthèse, mais on ne dispose pas de données à long terme concernant la présence de pesticides dans la haute mer. On ignore s'il y a eu des modifications de la répartition verticale du plancton, ou si ses taux de production et de remplacement ont changé. En l'absence d'enquêtes permettant la comparaison entre les régions benthiques et côtières, on ignore si l'évolution constatée dans le plancton reflète une tendance générale des peuplements marins.

Le progrès technique, par exemple dans le domaine des ordinateurs et des instruments de terrain, nous donnera la possibilité de résoudre nombre de ces difficultés.

de la fotosíntesis pero no se dispone de datos de un período prolongado en lo que se refiere a los plaguicidas en el mar abierto. No se sabe si se producen cambios en la distribución vertical del plancton o en las tasas de producción y de reposición. A falta de estudios comparables del bentos y de las regiones costeras, no es posible inferir, si lo que ocurre en el plancton, refleja una tendencia general en las comunidades marinas.

Los adelantos técnicos en lo que se refiere, por ejemplo, a calculadoras electrónicas e instrumentos utilizados en el mar, ofrecen la oportunidad de resolver muchas de estas dificultades.

THE immediate effects of pollution can often be recognized in the field and studies in the laboratory when the precise nature of the pollutant and its source are known; for example in oil spills and localized industrial effluents.

But it will always be difficult to detect and assess the effects of pollutants whose temporal and spatial gradients are not sharply defined and localized. These include those by-products of society and industry which may be dispersed via the atmosphere or drained into the sea at numerous points through river systems; for example, fertilizers, pesticides, dusts, carbon dioxide and heavy metals from vehicle exhausts and fungicides. Most of these substances are likely to be diluted quickly to sublethal levels in the natural environment, so that it will not only be difficult to detect their concentration gradients but, also, there may not be any dramatic and easily recognizable symptoms. Nevertheless, they may give rise to subtle effects in natural ecosystems which are exposed to them over very long periods of time. Moreover, pollution at a sublethal level may interact with "normal" natural stresses and so create effects which are out of all proportion to the component risks.

In dramatic pollution incidents as well as the long-term accumulation of "trace-pollutants", there is a possibility that populations and communities which are not themselves exposed to pollution may be affected by disturbance of the ecosystem; for example, by lethal effects or changes in productivity in a distant part of the food chain. Difficulties in extrapolating and predicting from experimental studies will arise because the effects of a given pollutant, as determined in the laboratory, will apply with varying force in different natural communities and in different locations; the effects may be more critical during some seasons or phases of life cycles than at others. These ecological problems are complicated by the dependence of equilibrium, in an ecosystem, on the interrelations of migratory as well as non-migratory animals. The effects of all but the most dramatic of pollution incidents will probably be small in relation to the scale of natural variations. However, natural changes tend to be cyclic whereas many pollutants are likely to have linear and cumulative effects. The superimposition of a small, but systematic, change on the natural cycles could have major effects on the productivity or composition of an ecosystem in the course of, say, a decade or two.

Because of these and similar difficulties in detecting and assessing the effects of pollution, it has been argued that there is an urgent need to advance our understanding of variability, and its causes, in natural ecosystems. In particular, it will be essential to establish ecological "base-lines" so that the effect of pollutants can be identified against natural variation.

But there has been relatively little research on

ecosystems. Fisheries research provides many examples of such studies but the data reflect the direct activities of man to such a large extent that natural sources of variability are often obscured. What is needed are assessments of variability in natural populations and communities which are not directly exploited by man.

The Continuous Plankton Recorder Survey provides an example of a long-term study of a small part of the marine ecosystem. Some of the data from this survey have been selected to illustrate just one of the kinds of variability which occur in unexploited populations and to suggest some of the problems that must be faced if the background of natural variability is to be studied as the basis for detecting, assessing and predicting the effects of pollutants.

It must be emphasized (a) that this is not a paper about pollution and (b) that the examples of environmental variation shown here are not to be interpreted as the sources of the changes in the plankton; they are given only to illustrate the problems of monitoring and analysing field data.

The Continuous Plankton Recorder Survey

Continuous Plankton Recorders (Hardy, 1939) are towed by merchant ships and ocean weather ships in the North Atlantic Ocean and the North Sea. They sample at a standard depth of 10 m; the material is analysed to provide numerical estimates of the abundance of the common species in alternate ten-mile sections of each tow. Sampling is repeated at monthly intervals along more than twenty standard routes. Details of the survey are given by Glover (1967).

The survey has formed the basis of analyses of many aspects of variation in the distribution, abundance and composition of the plankton. Colebrook (1965, 1969) has discussed the problems of such analyses and has explored the use of multi-variate statistics in dealing with the complex of interactions between the variability due to species, regions, months, years and parameters of the environment. In this paper we shall ignore all aspects of this complex except annual variation; indeed we shall deal with only a limited aspect of this, to demonstrate some selected long-term trends, without attempting the intensive statistical analysis which will be required if these trends are to be understood in the context of all the other sources of variation. This paper deals only with the eastern part of the survey which has been sampled consistently since 1948 (extension into the western North Atlantic was not started until 1961).

It is not necessary here to describe the standard methods of analysis of Plankton Recorder samples, except to say that the numbers of each organism in each ten-mile sample are transformed logarithmically [$y = \log 10 (x + 1.0)$]. These transformed estimates are then averaged to give an expression of the abundance of

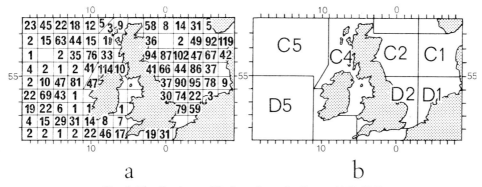

a   b

Fig. 1. The Continuous Plankton Recorder Survey 1948–1969
(a) Total numbers of analysed samples ($\times 10^{-1}$) in the standard rectangles of the eastern part of the survey
(b) The seven sub-areas for which data are presented in this paper

each organism in each standard rectangle (1° latitude by 2° longitude). For most purposes the rectangle means are averaged again to give the results for larger sub-areas which are designed to correspond, approximately, with the regional sub-division of fishery statistics. Although the logarithmic transformation has many merits in statistical analysis of the results (Colebrook, 1960), this standard treatment has the disadvantage that back-transformation of the processed data does not restore the rectangle or sub-area estimates to the original numbers per sample. For most purposes, therefore, the results must be considered as indices of abundance.

Figure 1a gives the numbers of analysed samples ($\times 10^{-1}$) in each of the rectangles of the area around the British Isles. Figure 1b shows the seven sub-areas which are the basis of all the results expressed in this paper. There was an observation for every month from January 1948 to December 1969 in all of these seven sub-areas.

Annual fluctuations in the plankton

Previous papers describing annual variations of plankton in the Recorder Survey have dealt with relatively short periods of about 12 years; see, for example, Colebrook and Robinson (1964). Glover (1967a, 1970), showed graphs of the abundance of 19 species of zooplankton during a period of 18 years; there appeared to have been a progressive decline in the abundance of many of these organisms. Robinson (1969) described trends in the phytoplankton and, especially, an apparently progressive delay in the date at which the phytoplankton blooms in the spring of each year.

It is now possible to bring these results up to date (figs 2 and 3), incorporating results for the last 22 years. Figure 2 shows the annual fluctuations in abundance of eleven zooplankton organisms, selected to typify the range of variation. The results for each sub-area were standardized (with a mean of zero and variance of one) so that (a) results from sub-areas with different levels of abundance could be combined and (b) results for species with different levels of abundance can be compared directly. In this way, the pattern of the fluctuations in abundance can be observed by eliminating differences in abundance between areas and species. Figure 2 serves to illustrate the wide range of variation, within and between species, but it is also apparent that there are certain consistent trends.

Some data for the first two organisms, *Pleuromamma*

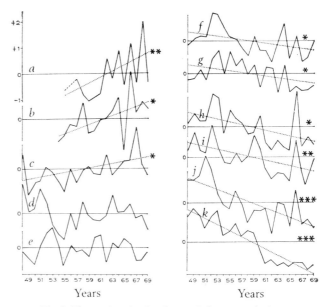

Fig 2. Fluctuations in abundance of eleven zooplankton organisms in the seven areas shown in fig 1(b)

The graphs show the average annual abundance of each species, standardized about a mean of zero; see scale, in standard deviation units, at top left. Calculated trend lines are drawn for those graphs which give a significant fit to a straight line (indicated by one, two or three asterisks for P = <5.0%, <1.0% and <0.1%, respectively).

The organisms are: a. *Pleuromamma borealis*; b. *Euchaeta norvegica*; c. *Acartia clausi*; d. *Temora longicornis*; e. *Clione limacina*; f. *Calanus helgolandicus* and *finmarchicus*, stages V and VI; g. *Metridia lucens*; h. *Candacia armata*; i. *Centropages typicus*; j. *Spiratella retroversa*; k. *Pseudocalanus* and *Paracalanus*, combined.

borealis and *Euchaeta norvegica*, are missing because these species were not counted and identified separately from their genera in the first six or seven years of the survey. Nevertheless, it looks as though these two, and *Acartia clausi*, have been tending to increase in abundance. The next two species (d, *Temora longicornis* and e, *Clione limacina*) do not show any marked trends unless it be the maintenance of their numbers about their long-term mean. The graph f, for stages V and VI of *Calanus* (*helgolandicus* and *finmarchicus* combined), shows a moderate decline in abundance. All the remaining species (g to k) give strong indications of a consistent decrease in abundance throughout the period of 22 years; this is most marked in the case of *Spiratella retroversa* and the combined genera *Pseudocalanus* and *Paracalanus*

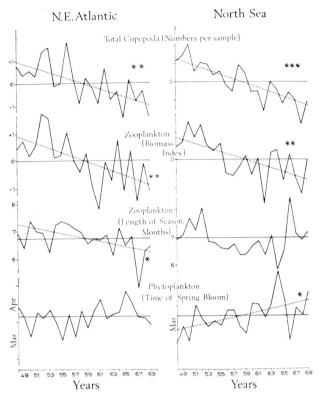

N.E. Atlantic   North Sea

Total Copepoda (Numbers per sample)

Zooplankton (Biomass Index)

Zooplankton (Length of Season, Months)

Phytoplankton (Time of Spring Bloom)

49 51 53 55 57 59 61 63 65 67 69   49 51 53 55 57 59 61 63 65 67 69
Years   Years

Fig. 3. Fluctuations in the plankton in the north-east Atlantic (sub-areas C4, C5 and D5 combined, see fig 1(b)) and the North Sea (sub-areas C1, C2, D1 and D2 combined) see text for detailed explanation. The results in the top two pairs of graphs are standardized about a mean of zero; the scales in standard deviation units being shown in the left-hand of each pair of graphs. The scales for the bottom two pairs of graphs are in months; plotted about the overall means for 22 years. Calculated trend lines are drawn for those graphs which give a significant fit to a straight line (indicated by one, two or three asterisks for P = <5.0%, <1.0% and <0.1%, respectively)

(which are not separated in routine analysis of the samples). In those cases where the fit to a straight line is significant at better than 5 per cent, the line and its significance, are indicated.

Figure 3 is an attempt to summarize these trends and to relate them to other indices of events in the plankton; the results for the three Atlantic areas (C4, C5 and D5) are shown separately from those for the North Sea (C1, C2, D1 and D2).

The numbers of copepods (all species combined) were standardized to permit the combination of data from different sub-areas. The graphs in fig 3 show a most dramatic decline with a highly significant fit to a straight line in the Atlantic as well as the North Sea.

Zooplankton biomass was calculated from the estimates of numerical abundance using the method developed by Glover (1968) to plot the geographical distribution of biomass. Briefly, the numbers of organisms in the six major groups in the zooplankton were multiplied by the net weights, using the best available estimates extracted from various published material. These were: 1.24 and 0.10 for Copepoda larger and smaller than *Calanus* stage IV, respectively, 0.8 to 14.5 for Euphausiidae (depending on the month), 3.0 for Hyperiidae, 1.3 for Chaetognatha and 0.12 for Gastropoda, Thecosomata. No attempt was made to allow for regional variations in weight nor, with the exception of Euphausiidae, for seasonal changes. Moreover, because

of the logarithmic transformations and the standardization of the data for each sub-area, mentioned above, the results cannot be expressed as biomass in terms of wet weight per unit volume of water sampled: they can only be regarded as a "biomass index". Despite the crudity of the method, fig 3 gives a clear indication of a systematic decline in the biomass of the standing stock of zooplankton at the standard sampling depth of 10 m. In the north-eastern Atlantic as well as the North Sea the fit to a straight line is significant at < 1 per cent. A more detailed analysis, which is not yet complete, suggests that the greater part of the fall in biomass arose from a decline in the abundance of small copepods, chiefly *Pseudocalanus* and *Paracalanus*.

Previous work in the Recorder Survey has often emphasized the importance of variations, between years and areas, in the timing of the seasonal cycle and its duration. Colebrook and Robinson (1965) used a statistical technique to estimate the "season duration" of the phytoplankton and copepods. For this paper we have used a different technique. The beginning and end of the "season" were calculated as the dates in the spring and autumn when the graphs of zooplankton biomass crossed the mean in each sub-area for the whole year; we assumed a linear rate of change between months. In this way differences in overall abundance, between years and sub-areas, were eliminated; the dates were calculated with regard to the shape of the seasonal curve in each year and area without regard to its height (or abundance). The results in fig 3 show that there has been considerable variation, between years, in the length of the season, measured from the standing stock of zooplankton biomass. Moreover, in the north-east Atlantic at least, the duration of the season appears to have become progressively shorter, amounting to a decrease of about 4½ weeks during the past 22 years, judging from the slope of the calculated straight line.

An event with major consequences for almost every aspect of the plankton is the blooming of the phytoplankton in the spring of each year. Visual estimates of the green coloration of the Recorder collecting silks can be used to provide a measure of abundance of phytoplankton (Robinson, 1970). From these estimates, the date of the spring bloom was calculated, in the same way as the season duration of the biomass; that is, the date when the seasonal graph crossed the mean in each sub-area and year. The resultant estimates, in fig 3, show a wide range of variation with a suggestion of a progressive delay in both regions, although the fit to a straight line is significant (at < 5 per cent) in the North Sea only. From these results, it might be estimated that the delay in the start of the phytoplankton bloom has amounted to about 3 weeks, judging by the slope of the calculated straight line for the North Sea data.

Obviously, further analysis of the data will be needed to define the extent and precise form of the changes in the plankton which are suggested by this limited selection from the results. The attempt to fit straight lines, for example, is probably an unjustifiable simplification and it will be necessary to investigate the variability of large numbers of species not presented here. An examination of the diatoms, which is under way, is revealing a very complicated pattern of variability but, again, there appears to have been a decline in abundance of some of the common species in parts of the survey area—these include *Rhizosolenia styliformis*, *Chaetoceros* (especially *Phaeoceros*), *Nitzschia delicatissima*, *Rhizosolenia imbricata* var. *shrubsolei* and *Dactyliosolen mediterraneus*.

Discussion

It would be premature to try to relate these trends in the plankton to changes in physical parameters. However, in order to illustrate the scale of the problem, we conclude with some examples of variability in one of the most critical aspects of the environment—radiant energy—which is too often ignored in spite of its great importance in regard to biological processes in general and photosynthesis in particular.

It is not easy to find appropriate records of radiation over the Atlantic and North Sea and such data as are available often reveal conflicting patterns of variation. Figure 4 shows four of the various direct and indirect estimates that might be used to describe solar radiation at the surface of the sea or on land in the northern hemisphere. The estimate for the two Ocean Weather Stations was calculated (using the equation developed by Black, 1956) from tables of low cloud, published by the U.K. Meteorological Office. These two stations lie at the western edge of sub-areas C5 and D5 [fig 1(b)]. Although there was considerable variability, between years, there is some indication of a progressive decrease in radiation with a significant fit (< 5 per cent) to a straight line. On the other hand, there is little evidence of a systematic trend in the "hours of bright sunshine" at Lerwick in the Shetlands (from Meteorological Office records) or direct measurements of radiation at Valencia on the west coast of Ireland (from the Irish Meteorological Office).

The last graph in fig 4 is re-drawn from fig 2a of Pivovarova (1968). Direct measurements of solar radiation were made at eight stations in the Soviet Union, at or near mid-day when there was no trace of cloud obscuring the sun's disc. Monthly means were calculated and, from these, the annual means and the long-term average. The results in fig 4, expressed as percentages of the long-term mean, show a sharp change in the pattern of the fluctuations at about the middle of the time-series; for the last twenty years, the recorded values of solar radiation have declined steadily. Pivovarova (1968) assumes that the solar constant has not changed and points out that there has been no systematic increase in water vapour in the atmosphere. She concludes that the decrease in radiation was the result of increases in aerosols in the atmosphere; this increased turbidity could have resulted from volcanic eruptions and the gradual pollution of the atmosphere from man's activities. Since there was only one major eruption during this period (in 1963 at Mount Agung in Indonesia), the inference is that the trend of the past twenty years is explained largely as the result of man's increased industrial activity. Pivovarova estimates that the attenuation of radiation due to aerosols increased about 30 per cent in the period 1953 to 1963. This could lead to a reduction of radiation by about 5 per cent.

Workers in other parts of the world have drawn attention to the progressive increase in the turbidity of the atmosphere; see, for example, Petersen and Bryson (1968) who based their conclusions on calculations of

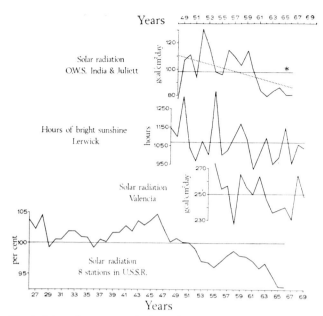

Fig 4. *Selected estimates of solar radiation at the surface of the sea and land in the northern hemisphere; see text for detailed explanation. The calculated trend line for solar radiation at the two Ocean Weather Stations fits a straight line (P < 1.0%). The observations at Lerwick, in the Shetland Isles, and Valencia, on the west coast of Ireland, do not fit a straight line at P = 5.0%. No attempt was made to calculate a linear trend line for the observations in the U.S.S.R. (redrawn from Pivovarova, 1968) because of the obvious change in the pattern of variation after 1946*

turbidity from measurements of solar radiation in Hawaii. They quote other workers who have measured major increases in the amount of atmospheric particulate matter over the U.S.A. and the Caucasus. McCormick and Ludwig (1967) report increases of 57 per cent and 88 per cent in the turbidity of the air above Washington and Davos, respectively, since the early part of this century; they suggest that global increases in turbidity during the past few decades may be responsible for the decrease in world-wide air temperatures, in spite of the apparent increase of CO_2 which would have the opposite effect. Professor R. A. Bryson, in a paper delivered at the 1968 National Meeting of the American Association for the Advancement of Science, argued that changes in air temperature in this century are related to the observed changes of CO_2 and atmospheric dusts. On the other hand, the Annual Report for 1969 of the Chief Inspector of Alkali Works in the U.K., as reported in *Nature*, vol. 227 (5261), dismisses these theories as mere speculation; he argues that there is no sign that those physical features of the earth which "undergo regular long-term periodicity in fluctuation" are being disturbed by man's efforts "which are puny compared with nature's".

Clearly, there are strong differences of opinion. Lamb (1969) sees the recent global decrease of air temperature as part of a much longer series of climatic oscillations and cycles with their origin in the fluctuating energy output of the sun. He estimates that the reduction of temperature over central England, since 1680, amounts to about 1°C but he points out that this is sufficient to shorten the average growing season, on the land, by about two weeks. Johnson *et al.* (1970) suggest that the fluctuations in temperature are cyclical, compounded of oscillations with periods of 78 and 181 years, related to solar activity. Superimposed on these long-term cycles are shorter period fluctuations of which the sun-spot cycle of 11 years is the best known.

Despite the conflicts in the evidence and its interpretation, there appear to have been changes in solar radiation that could, conceivably, contribute towards patterns of variation in the plankton of the kind described in this paper. However, it is far from clear whether the atmospheric changes are parts of cycles or continuing trends, and whether they are generated from outside the earth's atmosphere or are the products of man's activities. There has been very little research into the possible relationships, partly because there is a gross shortage of observations made with the required resolution and accuracy over a sufficient period of time; most observational programmes in the atmosphere have been directed largely towards the solution of purely meteorological problems.

The same inconclusive comments could be made, with varying force, with regard to all the other aspects of the marine environment which we have not considered here. For example, there is little understanding of the biogeographical effects of observed variations of sea temperature, although Rodewald (1967) has demonstrated some major trends in sea surface temperature during the past few decades (which, incidentally, he attributes partly to changes in the atmospheric pressure system over the North Atlantic). Although more research could be done, it will be severely restricted by the availability of past records which consist largely of surface temperatures taken by merchant ships and reported to the Meteorological Office with a low degree of resolution.

The situation is even worse if we consider variables that are not included in weather monitoring programmes. It would be quite impossible, for example, to compile a record of fluctuations of nutrients in time and space, during the past twenty years in the North Atlantic. The failure to make such records reflects the shortage of funds and man-power for research but it also arises from the technical difficulty of making measurements in the field on the required scale. However, the absence of such field records has the effect of sterilizing knowledge gained from laboratory experiments which have shown, for example, that light, temperature and nutrients are, indeed, critical to the fundamental biological processes of marine organisms.

The same argument applies to a consideration of pollutants. For example, in laboratory experiments, Wurster (1968) found that DDT reduced the rate of photosynthesis in four species of phytoplankton. Menzel, Anderson and Randtke (1970) showed that the effect varied considerably between species; thus, although "chlorinated hydrocarbons may not be universally toxic to all species, they may exert a dramatic effect on the succession and dominance of individual forms". Valuable laboratory studies of this kind are wasted unless we know whether there are chlorinated hydrocarbons in the seas and, if so, whether their concentration has varied in time and space. Moreover the results of short-term laboratory experiments (whether they be studies of fundamental physiology or of pollutants) cannot be extrapolated simply to conditions in the field. In this particular example, the depression of photosynthesis occurred at concentrations near or above the limit of solubility of DDT in water. But the experiment was not designed to measure the effects of long-term exposure to lower concentrations (and we wonder whether sufficient attention is being paid to regional and temporal variations in lipids which might hold pesticide residues in the sea water?).

Variations in the plankton of the open sea were used as the starting point in this paper but, because of the lack of comparable surveys of the coasts and inshore regions, we do not know whether there have been similar variations in the abundance, biomass or seasonal cycles of benthic organisms. Indeed, the validity of the apparent trends in the plankton is in question. Continuous Plankton Recorders sample at only one depth, 10 m, and we do not know whether there have been systematic changes in the vertical distribution of the plankton. It must be conceded, also, that the results described here refer only to standing stocks; we do not know whether there have been changes in the rate of production and turnover. In an attempt to overcome a few of these problems, the Edinburgh Oceanographic Laboratory is now engaged on an ambitious programme of design of a new Oceanographic Recorder which will undulate vertically as it is towed by merchant ships. In addition to sampling the plankton in the upper 75 or 100 m, it will contain a data logger and sensors for various physical and biophysical parameters (Glover, 1967, 1970).

Concluding remarks

We have used very simple examples, in this paper, designed to make what should be a simple and obvious point. The present state of knowledge of variability in the field and the present level of monitoring the natural environment are inadequate for the detection and identification of the sources of variation and, especially, for the separation of natural processes from all but the most obvious consequences of pollution incidents.

It follows from the arguments advanced in this paper that it is essential to implement field monitoring programmes designed to provide the basic data for environmental research as a whole; the problems of pollution cannot be considered in isolation from those of ecology in general. Theoretical and laboratory studies have shown the kinds of variables that should be measured but these are often ignored. For example, as Professor R. A. Bryson pointed out (at the 1968 National Meeting of the American Association for the Advancement of Science), although it has been established theoretically that the atmospheric dust load could affect many aspects of climate, "there are no systematic observations of dust densities and distribution, and apparently no plans for such observations contained in the Global Atmospheric Research Program of the World Weather Watch!"

It ought not to be necessary to emphasize the need to integrate environmental monitoring programmes so that, for example, atmospheric observations are designed to meet the requirements, not only of meteorologists, but also of biologists and, conversely, to ensure that plans for oceanographic monitoring should consider the needs of meteorologists. Indeed, as the examples in this paper have shown, the need to share monitoring programmes is only one aspect of the need to achieve a truly multidisciplinary approach to marine science. Dickson and Lee (1969) made the same point when they said, "It is clear therefore that changes in the atmospheric circulation over the North Atlantic have a dramatic response in the sea itself and that there are possible feedback effects since the ocean is seen to be actively transporting heat from one place to another and not acting merely as a reservoir. But the standard oceanographic observations made in the past only indicate the sort of things which might be happening: they are inadequate for use in any detailed analysis of the coupling between atmosphere and ocean".

Even if satisfactory monitoring can be achieved, there remain many problems in the development of methods for analysing the data. The calculation of a simple average is unlikely to reveal the complexities of environmental interaction; the data for the plankton, in this paper, have been subjected to several transformations (not described in full, for reasons of brevity). Data processing and analysis is another field in which meteorologists, oceanographers and biologists have much to gain from each other.

Until recently, arguments of this kind were no more than idealistic platitudes. However, during the past two or three decades, the problems of environmental science have assumed a new order of dimension and, at the same time, the developments of instrument technology have provided solutions to what, previously, were intractable problems on grounds of intellectual difficulty alone. Technology is now providing instruments to make observations, the data loggers to record them and the computers with which to analyse them.

Acknowledgements

The Continuous Plankton Recorder Survey would be impossible without the generous help of the captains and crews of merchant ships and weather ships of eight nations. The survey was financed by the Natural Environment Research Council and by Contracts N62558-3612 and F61052-67C-0091 between the Office of Research, Department of the U.S. Navy and the Scottish Marine Biological Association.

We received atmospheric data from many sources but we are particularly indebted to the Meteorological Offices of the U.K. and Ireland. We are grateful to our colleagues in the Edinburgh Laboratory who carry out the standard analysis of the Recorder samples as well as those who processed the data presented here.

References

COLEBROOK, J M Continuous plankton records: methods of
1960 analysis, 1950–1958. *Bull. mar. Ecol.*, 5:51–64.

COLEBROOK, J M On the analysis of variation in the plankton, the
1965 environment and the fisheries. *Spec. Publs int. Commn N.W. Atlant. Fish.*, (6):291–302.

COLEBROOK, J M Variability in the plankton. *Progr. Oceanogr.*,
1969 5:115–25.

COLEBROOK J M and ROBINSON, G A Continuous plankton
1964 records: annual variations of abundance of plankton, 1948–1960. *Bull. mar. Ecol.*, 6(3):52–69.

COLEBROOK J M and ROBINSON, G A Continuous plankton records:
1965 seasonal cycles of phytoplankton and copepods in the north-eastern Atlantic and the North Sea. *Bull. mar. Ecol.*, 6(5):123–40.

DICKSON, R and LEE, A J Atmospheric and marine climate
1969 fluctuations in the north Atlantic region. *Progr. Oceanogr.*, 5:55–65.

GLOVER, R S The continuous plankton recorder survey of the
1967 north Atlantic. *Symp. zool. Soc. Lond.*, 19:189–210.

GLOVER, R S Laboratory report: Oceanographic Laboratory,
1967a Edinburgh. *Rep. Scott. mar. biol. Ass.*, 1966–67:29–42.

GLOVER, R S Laboratory report: Oceanographic Laboratory,
1968 Edinburgh. *Rep. Scott. mar. biol. Ass.*, 1967–68: 42–58.

GLOVER, R S Synoptic oceanography; the work of the Edinburgh
1970 Oceanographic Laboratory. *J. underwat. Sci. Technol.*, 2(1):34–40.

HARDY, A C Ecological investigations with the continuous
1939 plankton recorder: object, plan and methods. *Hull Bull. mar. Ecol.*, 1(1):1–57.

JOHNSON, S J, *et al.* Climatic oscillations 1200–2000 A.D. *Nature,*
1970 *Lond.*, 227:482–3.

LAMB, H H The new look of climatology. *Nature, Lond.*, 223:
1969 1209–15.

McCORMICK, R A and LUDWIG, J H Climatic modification by
1967 atmospheric aerosols. *Science, N.Y.*, 156(3780):1358–9.

MENZEL, D W, ANDERSON, J and RANDTKE, A Marine phyto-
1970 plankton vary in their response to chlorinated hydro-carbons. *Science, N.Y.*, 167(3926):1724–6.

PETERSON, J T and BRYSON, R A Atmospheric aerosols:
1968 increased concentrations during the last decade. *Science, N.Y.*, 162:120–1.

PIVOVAROVA, Z I The long-term variation of intensity of solar
1968 radiation according to observations of actinometric stations. *Leningr. Glav. Geof. Obs.*, (233):17–37.

ROBINSON, G A Fluctuations in the timing of the spring outbreak
1969 of phytoplankton in the north-east Atlantic and the North Sea. *ICES, CM* 1969(L:17):2 p. (mimeo).

ROBINSON, G A Continuous plankton records: variation in the
1970 seasonal cycle of phytoplankton in the north Atlantic. *Bull. mar. Ecol.*, 6:33–45.

RODEWALD, M Beiträge zur Klimaschwankung im Meere 15.
1967 Beitrag Die Entwicklung der Oberflächentemperatur des Meeres im Bereich der nordatlantischen Wetterschiff-Stationen 1951 bis 1967. *Dt. Hydrogr. Z.* 20(6):269–75.

WURSTER, C F DDT reduces photosynthesis by marine phyto-
1968 plankton. *Science, N.Y.*, 159(3822):1474–5.

BLACK, J N The distribution of solar radiation over the earth's
1956 surface. *Arid Zone Res.*, (7):138–43.

Antagonism of the Native Microflora to Microbial Pollutants in the Sea†

R. Mitchell*

Antagonisme de la microflore naturelle envers les polluants microbiens dans la mer

La pollution provoque souvent un déséquilibre dans la communauté microbienne marine. Chaque jour, des millions de virus et de bactéries entériques sont déversés dans la mer par les décharges des égouts, causant un danger de maladie. L'écoulement des eaux d'égouts et le rejet d'eaux riches en nutriments ont également souvent pour résultat l'apparition de floraisons massives d'algues dans les eaux côtières. L'auteur examine le rôle de la microflore marine naturelle dans la destruction de ces polluants microbiens et a découvert qu'il existe un rapport direct entre l'ampleur des populations bactériennes indigènes et le taux de destruction d'*Escherichia coli* dans l'eau de mer naturelle. Il a isolé les micro-

Antagonismo entre la microflora nativa y la contaminación microbiana del mar

Ocurre con frecuencia que por efecto de la contaminación la comunidad microbiana marina se desequilibra. A diario entran en el mar procedentes de las descargas de aguas residuales millones de bacterias y virus entéricos que crean peligros de enfermedades. Además, las aguas cloacales y las de descarga contienen muchos nutrientes que dan por resultado floraciones de algas en aguas costeras. Se examina la función de la microflora marina nativa en la destrucción de estos contaminantes microbianos. Se ha encontrado una relación directa entre la magnitud de la población bacteriana indígena y la velocidad de destrucción de *Escherichia coli* en agua de mar natural. Se han aislado los microorganismos antagónicos y

* Laboratory of Applied Microbiology, Division of Engineering and Applied Physics, Harvard University, Cambridge, Massachusetts 02138, U.S.A.
† This work was supported in part by the U.S. Department of the Interior, Federal Water Pollution Control Administration.

organismes antagonistes qu'il décrit en détail. Il a également étudié en laboratoire le processus de destruction des virus dans la mer. Une fraction de la microflore naturelle participe à l'inactivation, ainsi qu'une composante chimique de l'eau de mer. Toutefois, le virus est souvent protégé contre ces deux modes d'inactivation par adsorption sur des surfaces microbiennes. La microflore marine naturelle exerce un effet régulateur sur la population d'algues. On a isolé un groupe de micro-organismes capables de provoquer la dégradation enzymatique des algues dans la mer. Certains indices indiquent que ces micro-organismes marins sont particulièrement actifs lorsque la productivité atteint un maximum. Les organismes antagonistes servent à détruire les communautés d'algues dans lesquelles un principe nutritif joue un rôle limitant et lorsque leur activité physiologique est affaiblie. L'auteur examine la possibilité d'utiliser les enseignements fournis par ce processus d'auto-épuration marine pour développer les moyens d'action contre la pollution.

se describen con pormenores. En este laboratorio también se han estudiado los procesos que regulan la destrucción de virus en el mar. En la inactivación participa una parte de la microflora nativa, junto con un compuesto químico del agua del mar, pero ocurre con frecuencia que los virus están protegidos de estos dos procesos de inactivación por adsorción a superficies microbianas. La población de algas regula la microflora marina nativa. Se ha aislado un grupo de microorganismos capaces de degradar enzimáticamente las algas en el mar. Se han obtenido pruebas de que la mayor actividad de estos microorganismos marinos ocurre cuando la productividad ha alcanzado un máximo. Los antagonistas sirven para destruir comunidades de algas en las que un elemento nutriente se ha convertido en factor limitante y se ha debilitado la actividad fisiológica. Se examina la posibilidad de emplear los conocimientos adquiridos en estos procesos de autopurificación marina para regular la contaminación más intensa.

MARINE microbial ecosystems frequently become unbalanced as a result of pollution. Millions of micro-organisms are disposed daily into the sea from sewage outfalls, posing a danger of disease. These micro-organisms include fungi, bacteria, and viruses pathogenic to man. Each year the amount of sewage entering the ocean from sewage outfalls increases. The sea has an inherent capacity to eradicate these foreign micro-organisms. However, as the volume of sewage entering the oceans increases, the self-purification processes become incapacitated and the marine ecosystem becomes overloaded and saturated by the microbial pollutants.

Surprisingly little is known about the processes by which these microbial pollutants are eradicated by self-purification processes in the sea. As early as 1936 ZoBell noted that natural, unsterilized sea water had a bactericidal effect on non-marine bacteria. Ketchum, Ayres and Vaccaro (1952), in a study of the destruction of Escherichia coli in estuaries, demonstrated that a biologically produced material was involved. On the other hand Pramer (1963) and his co-workers suggested that a physico-chemical fraction of the sea water, rather than an anti-bacterial agent, was responsible for the destruction of enteric bacteria in the sea. Krasil'nikova (1961) isolated a quantity of antibiotic producing micro-organisms from the ocean. However, she detected a minimal quantity of antibiotics produced when these organisms were grown in natural sea water. Thus it appears that antibiotic production is not of great importance in the destruction of enteric bacteria in the sea. The bactericidal and viricidal effect of marine micro-organisms, which do not produce antibiotics, on enteric bacteria and viruses has been investigated by the author and his co-workers and is now reported.

Methods

The sea water used in these studies was sampled at Woods Hole, Massachusetts, during June–August. The water was stored at 3°C for not longer than 4 weeks before use. Escherichia coli was counted on eosin methelene blue plates following incubation at 37°C. Counts of marine bacteria were made on Oppenheimer and ZoBell's (1952) medium following incubation at 25°C. Sea water was filtered through 0.45 μ pore size filters for the isolation of Bdellovibrio. These filterable bacteria and amoeba were plated on double layer plates. The lower layer consisted of sterile sea water with 1.5% agar added. The top layer was made of sterile sea water with a heavy suspension of washed E. coli added and solidi-

fied with 0.7% agar. Clearing zones caused by growth of lytic micro-organisms in the opaque E. coli layer appeared following 5–7 days incubation at 25°C. The virus used in these studies was the X-174 strain of bacteriophage. This virus is specific for E. coli strain C, and has been described by Sinsheimer (1959). Bacteriophage particles were counted by the double layer method described by Adams (1959). The sterile sea water used in all of these studies was sterilized by filtration through 0.22 μ pore size filters. Algae used in this study were grown under a light intensity of approximately 200 lux at 22°C.

Destruction of coliform bacteria

Initially we investigated the effect of the size of the marine microbial community on the rate and extent of destruction of Escherichia coli in sea water (Mitchell, Yankofsky and Jannasch, 1967). E. coli was added to sterile sea water containing different concentrations of the natural microbial population of the sea, which was obtained by centrifuging fresh sea water. Micro-organisms were returned to the sterile sea water at concentrations of 10/ml, 10^3/ml, and 10^5/ml. A direct relationship was found between the rate of degradation of E. coli in the sea water and the size of population of native marine micro-organisms. Our data showed that a fraction of the native marine microflora was responsible for the destruction of E. coli in the sea. Similar results were found repeatedly when natural sea water was sampled. Sea water containing high concentrations of micro-organisms had a much more effective self-purifying effect than water which contained small numbers of micro-organisms. Water samples from the deep ocean containing less than 10^3 bacteria/ml declined from 10^9 cells/ml to 10^7 cells/ml in 5 days. In contrast water samples from inshore areas containing populations of more than 10^6/ml caused a decline of E. coli population from 10^9 to 10^4/ml in 5 days.

This inactivation of coliform bacteria by a fraction of the marine microflora was investigated further. E. coli was added to natural sea water and the rate of decline was measured. The coliform population declined from the inoculum of 10^6/ml to 10^2/ml in 7 days. There was an initial lag period of 4 days. A subsequent inoculum of E. coli at 10^6/ml disappeared in 5 days, indicating either an enrichment of a specific antagonistic microflora or the inducement of specific degradative enzymes.

Samples of water were taken during these studies and analysed to determine whether a specific antagonistic microflora developed in the presence of E. coli.

Antagonistic micro-organisms were counted on double layer plates. The lower layer contained sea water solidified with 1.5% agar. A thin upper layer contained sea water solidified with 0.7% agar containing a heavy suspension of washed *E. coli* cells as the sole carbon source. Micro-organisms from sea water were inoculated to this medium. Occasional clearing zones, caused by lytic micro-organisms, were observed. However, as the enrichment of micro-organisms antagonistic to *E. coli* developed in response to the enteric bacteria the numbers of clearing zones increased, indicating the development of a specific antagonistic microflora in the water. Three groups of micro-organisms were isolated from these clearing zones, after 2 to 3 days (Mitchell, Yankofsky and Jannasch, 1967; Mitchell and Yankofsky, 1969): (1) bacteria which were capable of lysing *E. coli* to cell walls; (2) the obligately parasitic bacterium *Bdellovibrio bacteriovorus*, originally described by Stolp and Starr (1963); and (3) a marine amoeba, *Vexillifera*. The bacteria acted by lysing the *E. coli* cell walls with extracellular enzymes. The Bdellovibrios ranged in size from 0.2 to 0.5 microns. These bacteria used only Gram negative bacteria as their carbon source, and were incapable of growing on inert media. They differed from other Bdellovibrios in being halophilic. The lytic amoeba appeared to be the most effective of the three groups of micro-organisms in destroying *E. coli*. Pure cultures of the *Vexillifera* were capable of destroying 10^6/ml cells of *E. coli* in four days.

Inactivation of viruses

The role of the native microflora in destroying viruses in sea water was determined by adding the bacterial virus X-174 to sterile and to natural sea water as described by Mitchell and Jannasch (1969). The viruses survived well in autoclaved sea water. But in natural sea water the virus titer declined from 10^{12}/ml to 10^3/ml in 6 days (Table 1). The effect of the concentration of marine micro-organisms in the sea water on virus destruction was determined by adding different concentrations of marine micro-organisms to sterile sea water. When 10^6/ml marine bacteria were returned to the sterile sea water the virus concentration declined from 10^{12}/ml to 10^3/ml in 6 days. Addition of 10^3/ml marine bacteria to sterile sea water resulted in a decline from 10^{12} virus particles/ml to 10^8/ml in the same period.

Repeated attempts were made to isolate micro-organisms capable of degrading the viruses from the sea water. No antagonistic micro-organisms were isolated. It appears reasonable to conclude that an antagonistic microflora is responsible for the destruction of this virus in sea water, but either the concentration of the virus or the inability of conventional techniques to yield the antagonists is the cause of our failure to isolate these micro-organisms.

It appears that in addition to a microbiological inactivation of the virus in sea water, some chemical component of the water caused destruction of the virus. Sea water filtered through 0.45 µ filters displayed strong anti-viral activity. The virus concentration declined from 10^{11}/ml to 10^3/ml in 6 days. By comparison, sea water filtered through 0.22 micron filters displayed an even stronger anti-viral effect. Viruses were added to this sea water at 10^{11} cells/ml and no virus could be detected after four days of incubation.

The viruses could be protected against this microbiological or chemical inactivation by adsorption to colloids in the water. This was shown by adding ultraviolet irradiated cells to the sea water. The addition of these cells resulted in a total protection of the viruses from either microbiological or chemical inactivation. Sea water, to which ultraviolet irradiated cells had been added, was inoculated with the virus at 10^{13}/ml virus particles. After 8 days of incubation the concentration of virus particles had declined to 10^{11}/ml, indicating a strong protective effect of the dead bacterial cells.

Inactivation of fungi

The destruction of non-marine fungi in the sea by an antagonistic marine microflora was shown by Mitchell and Wirsen (1968). Mycelial mats of the fresh water fungus *Pythium debaryanum* were added to natural sea water in an attempt to determine the effect of the natural marine microflora on that fungus. The fungus was actively degraded by an antagonistic portion of the marine microflora. Mycelial mats added to natural sea water were degraded in 10 days. Typical data are shown in Table 2.

Cell-free filtrates of sea water were also capable of degrading the fungus. After repeated addition of the fungus to natural sea water, the water was filtered and the filtrate added to fresh mycelium. Activity against the fungus was detected by release of glucose into the medium. After 15 min of contact glucose began to be released into the medium. This activity was increased by concentration of the active factor. The factor was inactivated by boiling the sea water, indicating the presence of degradative enzymes in the sea water.

Enrichment cultures with *Pythium* mycelium and with mycelial cell walls resulted in the isolation of bacteria in large numbers which were capable of causing this rapid degradation of fungal mycelium in sea water. These bacteria, which were also capable of decomposing agar, and which were extremely common in the sea, acted on the fungus by producing extracellularly a cell wall degrading enzyme system. When 10^4/ml lytic bacteria were added to sterile sea water total degradation of the mycelium occurred within five days. A second inoculum of mycelium was degraded in three days.

A number of fungi related to *Pythium* were also tested for degradation in natural sea water. Mycelial mats of *Apodachlya saprolegnia, Achlya, Isoachlya* and *Thraustotheca* were added to natural sea water. All were degraded in less than 8 days. When added at 10^6/ml the pure culture of the lytic bacterium completely degraded all five fungi in less than five days. These data indicate that a specific antagonistic bacterial population capable of degrading fungi is present in the sea and is rapidly enriched by the addition of these foreign fungi to the sea water.

Inactivation of algae

The biodegradation of algae was tested by adding the marine diatom *Skeletonema* to sea water. The alga survived well in sterile sea water but decayed in 24 h in natural sea water in the dark. The rapidity of development of an antagonistic microflora is shown in Table 3. The population of diatoms declined very rapidly from

10^5/ml to 10^3/ml in 24 h. In the same period the bacterial population increased from 10^4/ml to 10^6/ml. The antagonist of this alga, isolated by enrichment culture, was identified as a species of *Pseudomonas*. This bacterium apparently lysed the algae in a similar manner to the fungal lysis, i.e. by enzymatically degrading a portion of the cell wall. Antagonistic activity was specific for the genus *Skeletonema* although all species of *Skeletonema* tested were degraded. No other genera of alga were degraded by this bacterium.

TABLE 1. DESTRUCTION OF VIRUSES PARTICLES BY MARINE MICRO-ORGANISMS

Time	Autoclaved sea water	Natural sea water
Days	No. virus/ml	No. virus/ml
0	5×10^{12}	5×10^{12}
2	3×10^{12}	2×10^5
4	3×10^{12}	8×10^5
6	2×10^{12}	7×10^3

TABLE 2. DISINTEGRATION OF A NON-MARINE FUNGUS BY THE NATIVE MARINE MICROFLORA

Time	Autoclaved sea water	Natural sea water
Days	Mycelial mat wt. mg	Mycelial mat wt. mg
0	92	90
3	99	78
5	86	72
7	83	60
10	80	43

TABLE 3. RELATIONSHIP BETWEEN THE RAPID DESTRUCTION OF THE DIATOM *Skeletonema* IN NATURAL SEA WATER STORED IN THE DARK AND THE ACTIVITY OF THE NATIVE MARINE MICROFLORA

Time	Diatom population	Bacterial population
Hours	No./ml	No./ml
0	3.5×10^5	1.0×10^4
3	2.0×10^5	1.0×10^5
7	1.0×10^4	3.0×10^6
12	9×10^4	5.0×10^6
24	3×10^3	6.0×10^6

Discussion

Our data demonstrate that enteric bacteria carried into the sea in sewage are rapidly destroyed by the native marine microflora. These results support the frequent observations that enteric bacteria are more rapidly destroyed in the summer than in the winter. The observation that two of the three groups of micro-organisms causing the destruction of enteric bacteria are obligately parasitic is of the utmost importance. Conventional culture techniques would not yield these micro-organisms because they require living bacteria as a substrate. It would be advisable in studies of this sort to always include a culture medium in which the microbial pollutant is the sole carbon source so as to detect obligate parasites.

The strong self-purifying effect of sea water is obviously the reason why so few cases of disease have been detected among people who swim in sea water polluted by domestic sewage. Enteric bacteria are incapable of surviving for a long time in the sea. The native microflora rapidly lyse the enteric bacteria.

Our studies have shown that an antagonistic marine microflora destroys non-marine fungi in the sea. These micro-organisms are primarily a group of bacteria which degrade the fungal cell walls. These data are important in assessing the danger of contamination from pollution of the sea with sewage containing yeasts and fungi pathogenic to man. Dabrowa *et al.* (1964) found a number of fungi pathogenic to man in sea water near sewage outfalls. Taysi and Van Uden (1964) failed to detect intestinal yeasts in temperate estuaries, and postulated that temperature is important in the destruction of yeasts carried into estuaries in sewage. Our data are supported by Buck *et al.* (1963), who showed that a marine *Pseudomonas* is capable of strongly inhibiting the pathogenic yeast *Cryptococcus*. It appears that pathogenic fungi carried into the sea in sewage are not a serious threat to man because of the destructive action of the native marine microflora.

By contrast, enteric viruses do pose a threat of disease when they enter the sea. These viruses are concentrated in shellfish and are responsible for epidemics of infectious hepatitis. Similarly, poliomyelitis virus and coxsackie virus are also probably transmitted to shellfish. Our data, showing the effect of the native microflora on the destruction of viruses in the sea, confirm those of Metcalf and Stiles (1967) who showed that viruses survived in sea water up to 30 days during the summer and up to 50 days in the winter. They also showed that domestic sewage had a protective effect on the virus. The data presented in this paper suggest that the longer survival rate of viruses in the winter is caused by a low microbiological activity in the sea compared with a much higher activity during the summer. Metcalf and Stiles (1967) observation of the protective action of domestic sewage supports our results, showing that when dead cells are added to sea water they have a protective effect on the virus, presumably by adsorption of virus onto the colloidal surfaces. It is apparent from these studies that viruses are not immediately and rapidly inactivated in sea water even though they are incapable of multiplying in the absence of a host. They seem capable of surviving by adsorbing onto colloidal particles. These particles are presumably ingested by shellfish, where the viruses are concentrated. In the absence of adsorption, the viruses are susceptible to both microbiological and chemical inactivation and are rapidly destroyed.

The algal population is frequently unbalanced by eutrophication of marine estuaries. Under these conditions algal blooms develop. The algal biomass is the result of the balance between productivity and degradation. Our data have shown that when productivity declines, degradation is extremely rapid: *Skeletonema* cultures can be lysed within 24 h by a native microflora. Carter and Lund (1948) have observed degradation of algal populations in lakes by microbial parasites. It seems that similar predation occurs in the sea. During the development of a bloom, productivity is much more rapid than degradation. However, at some point, when a nutrient becomes limiting, productivity declines and degradation becomes dominant. At this point a lytic microflora develops and the algal bloom is rapidly and completely destroyed. We have detected bacteria responsible for this destruction. However, it must be assumed that a wide variety of micro-organisms are capable of developing, depending on the nature of the algal substrate.

References

ADAMS, M R   Bacteriophages, New York, Interscience, 450 p.
1959

BUCK, J D, *et al.*   Inhibition of yeasts by a marine bacterium.
1963   *J. Bact.* 85(5):1132–5.

CARTER, H M and LUND, J W G   Studies on plankton parasites.
1948   1. *New Phytol.*, 47:238–61.

DABROWA, N, *et al.*   A survey of tide-washed coastal areas of
1964   Southern California for fungi pathogenic to man. *Mycopath. Mycol. appl.*, 24:137–50.

KETCHUM, B H, AYRES, J C and VACCARO, R F   Processes con-
1952   tributing to the decrease of coliform bacteria in a tidal estuary. *Ecology*, 33:247–58.

KRASIL'NIKOVA, E N   Antibiotic properties of microorganisms
1961   isolated from various depths of world oceans. *Mikrobiologiia*, 30:545–50.

METCALF, T C and STILES, W C   Survival of enteric viruses in
1967   estuary waters and shellfish. *In* Symposium on transmission of viruses by the water route, edited by G Berg, New York, Interscience Publishers, John Wiley and Sons, pp. 439–47.

MITCHELL, R and JANNASCH, H W   Processes controlling virus
1969   inactivation in sea water. *Environ. Sci. Technol.*, 3:941–3.

MITCHELL, R and WIRSEN, C   Lysis of non-marine fungi by
1968   marine micro-organisms. *J. gen. Microbiol.*, 52:335–45.

MITCHELL, R and YANKOFSKY, S   Implication of a marine
1969   *Amoeba* in the decline of *Escherichia coli* in seawater. *Envir. Sci. Technol.*, 3:574–6.

MITCHELL, R, YANKOFSKY, S and JANNASCH, H W   Lysis of
1967   *Escherichia coli* by marine micro-organisms. *Nature, Lond.*, 215:891–93.

OPPENHEIMER, C H and ZOBELL, C E   The growth and viability of
1952   sixty-three species of marine bacteria as influenced by hydrostatic pressure. *J. mar. Res.*, 11:10.

PRAMER, D, CARLUCCI, A F and SCARPINO, P V   The bactericidal
1963   action of seawater. *In* Symposium on marine microbiology, edited by C H Oppenheimer, Springfield, Ill., Charles C Thomas.

SINSHEIMER, R L   Purification and properties of bacteriophage
1959   X-174. *J. molec. Biol.*, 1:37–41.

STOLP, H and STARR, M P   *Bdellovibrio bacteriovorus* Gen.
1963   ETSP, N., a predatory ectoparasitic, and bacteriolytic microorganisms. *Antonie van Leeuwenhoek*, 29:217–48.

TAYSI, I and VAN UDEN, N   Occurrence and population density of
1964   yeast species in an estuarine marine area. *Limnol. Oceanogr.*, 9:42–5.

ZOBELL, C   Bactericidal action of seawater. *Proc. Soc. exp. Biol.*
1936   *Med.*, 34, 113–116.

Section 5

TECHNICAL ASPECTS OF MINIMISING POLLUTION AND COUNTERING ITS EFFECTS

Summary of Discussion

Waste treatment

It was noted that waste water treatment technology is available to meet any desired quality requirements including that for drinking water supply. Generally, the cost of waste water treatment is directly related to effluent quality. Thus, to ensure high quality of effluents, a realistic case must be made for the need for such quality and of the consequences if the desired standards are not met. More attention needs to be given to the control of objectionable pollutants at the source, especially those that are non-degradable, toxic, and of such character that they can be concentrated by several orders of magnitude in marine organisms. The most effective and economic solution to many industrial pollutants is by control at the source (within the plant) through good housekeeping practices, process modification, and product recovery and utilization. It was noted that in a particular plant (unbleached kraft pulping) the total plant discharge of BOD had been reduced by over 85 per cent through in-house process modifications. It must, however, be recognized that it is rarely possible to completely eliminate water discharge by in-house modifications, making some form of external treatment and waste discharge to the environment unavoidable. This is also true for modern municipal waste management systems where a high degree of waste water reclamation is practised. Here, too, there will be a residual waste stream—the so-called "non-reclaimables"—which is mostly salt, that must receive terminal treatment and discharge to the environment, generally reaching the ocean.

Waste disposal by submarine outlets

The theoretical bases for the design of effective open coastal submarine outfall dispersion systems have been verified in both small (<0.1 m³/s) and large (>15 m³/s) systems. Marine waste dispersion systems having outfall lengths from shore as great as 8 kms, and pipe diameters of 2 to 3 metres discharging at depths of 30 to 80 metres are capable of initial dilutions of waste with receiving water at the discharge source of at least 1:100 and often in the order of 1:150 or more. It is obvious that dilutions of this magnitude provide a substantial safety factor—equivalent to a high degree of treatment for non-conservative waste constituents in reducing the concentration of pollutants and subsequent effects in the immediate vicinity of the discharge. However, dilution alone is not an adequate disposal solution for materials that are conservative and that can concentrate in the biota, such as mercury and chlorinated hydrocarbons.

Differences in conditions in receiving water, such as deep open coasts with marked density structure (pycnoclines), as compared to shallow embayments and restricted seas without density gradations, require different concepts and mathematical models for the prediction of waste dilutions and concentrations. The proper design of marine waste disposal systems requires detailed consideration of the hydrographic characteristics of the disposal site, the water quality requirements and the optimum combination of treatment and dispersion to produce acceptable and safe waste concentrations in the immediate area of the discharge.

Effluent standards or criteria and receiving water criteria were considered and, while somewhat controversial, there appeared to be a consensus that some type of general receiving water criteria or objectives are needed and should be developed.

Waste disposal by dumping

Some concentrated wastes, such as inorganic acid solutions, can be more economically and safely handled by discharge from a moving ship (barge) at a great distance from shore and at considerable depth, rather than by discharge into conventional sewerage systems. Field studies of actual barge operations indicate that waste dilutions of >100 are attainable at the discharge location in the propeller stream of a ship moving at about 10 kts and equipped with a well designed and operated discharge system. Dilutions increased rapidly in the wake of the ship, reaching values of >1000 at a distance of 500 metres from the release point.

Studies conducted in areas where containers have been dumped in the ocean, sometimes in significant fishery areas, have indicated lack of integrity and inappropriate buoyancy in containers of such degree as to be a hazard and to cause gross economic loss to fishing operations. Faulty containers have been recovered in trawls—sometimes as many as eight in one haul.

The hazards of deep-ocean dumping of exotic chemicals including chlorinated hydrocarbons, were reviewed. Chemical "fingerprints" of discharged wastes have been traced over a broad expanse of the Atlantic in the water column and in the biota.

More attention needs to be given to evaluation of the short- and long-term effects of dumping concentrated waste of all types from ships, both with and without containers. Especial attention needs to be given to the conservative and exotic wastes, including chlorinated hydrocarbons and the polychlorinated biphenyls. It must be recognized that waste discharge from moving ships may be acceptable for contaminants that are non-conservative and where toxic concentrations can be eliminated by dilution alone. However, such operations are risky when used for conservative toxic wastes—especially those that can be accumulated or concentrated in the biota by chemical or biological processes. Technology should be developed for the economic destruction of toxic materials without discharge to the sea.

Oil pollution

It has been estimated that the influx of oil to the ocean from shipping and by accidents in ports is at least one million metric tons per year. In addition, oil from drilling accidents and land-based sources, such as municipal and industrial wastes, are substantial in tonnage and nuisance level. The total hydrocarbons discharged annually in the treated waste waters from the city and county of Los

Angeles, California, is of the same magnitude as the total oil released during the first year of the ill-famed Santa Barbara Channel oil accident. Similar quantities of hydrocarbons have been discharged each year, likely for decades, in the treated waste waters reaching San Francisco Bay. Considering the quantities involved and the general acceptability of these bodies of water with respect to oil it must be concluded that such hydrocarbons are probably degraded rapidly and are not very toxic to the marine biota.

Effects on marine organisms

Some participants expressed the opinion that crude oils and fractions thereof are more or less poisonous to all marine organisms. However, only very few data are available so far to show these effects on organisms as a function of the concentration, kinds of oil and the time of exposure of the organism. Such information is needed urgently if optimum progress is to be made in abatement of the oil pollution problem. Claims both about immediate toxic effects and the destructive action of oil and about long-term "low-level" effects must be documented.

Considerable disagreement exists as to whether or not the large-scale oil spills have been as damaging to the biota as has been claimed. Also the question of the significance of the nuisance and aesthetic problems with oil, as compared to the ecological effects, is most controversial. It is alleged by some that countermeasures such as the use of detergents and dispersants—even the so-called "non-toxic" types—are more harmful than if the oil were left to decay naturally. It is claimed that the use of such materials increases the rate of absorption of the entire oil mass into the marine organisms. There is a wide difference of opinion and need for additional study on the role and effects of dispersants in cleaning up oil spills.

Microbiological degradation

Microbiological degradation of hydrocarbons is well known; the oil concentration limits within which such degradation can occur are not. The conditions for optimum rates of oil degradation, as well as kinetic descriptions (rate constants), are not known. Much work is needed to assess the toxic-time-concentration function and the degradation kinetics so that proper control measures can be instituted to minimize the effects of oil pollution. Preliminary research indicates that it may be possible to increase the rate of *in situ* degradation of oils by the "seeding" of the oil slicks with oil-utilizing strains of yeast (such as *Candida* spp); however, it was noted that it may be necessary to add nutrients (such as N and P) and essential co-factors. The field application of such methods requires much additional study.

Combating accidental releases

Much work is under way on various methods to combat accidental spills. However, the success of the various methods used, as reported by different investigations, has been highly variable. It is apparent that this, too, is a subject needing much additional research and evaluation.

Instrumentation and monitoring

Only by substantial field use of available instruments and methods can the problem of accurate routine monitoring at expected pollution levels be solved. Development of instrumentation for continuous recording of chemical quality parameters is in its infancy, and the recording of biological parameters scarcely has been considered. Environmental monitoring efforts, particularly with respect to land-based waste input and ocean and, possibly atmosphere, should be coordinated.

Aquaculture

There are many examples of the utilization of the organic and nutrient content of domestic, and even of some industrial wastes, for the production of edible fish and shellfish. More attention should be given to the utilization of waste material and waste heat, for accelerated food production in both developing and developed countries. Average yields of commercially important fish of 147 kg/ha in the sewage-enriched Lake of Tunis is one good example of such utilization. Problems include aesthetic considerations as well as fish diseases resulting from the restricted and densely populated rearing facilities; it is believed both can be overcome by research and enlightened practices. Considering the rapidly increasing practice of re-using highly purified municipal waste water for all purposes including drinking, the aesthetic problem of fish production from wastes should be soluble by education. Aquaculture is an excellent example of a possible benefit from enlightened waste disposal practices—and such benefits may also accrue in the ocean if the treatment and disposal systems are properly designed and operated.

Concluding remarks by convenor

One of the most effective ways of reducing waste discharge is by recycling or re-use at the source. In industry this includes process modification and product recovery through good housekeeping practices; in municipal waste systems, it includes control of industrial effluents, separation of storm and cooling water, and treatment and re-use of domestic wastes wherever feasible and economic.

Modern technology is capable of treating municipal waste waters and some industrial wastes to as high a quality level as is required for drinking or any other use, the cost of treatment being generally proportional to the quality level.

Specific and statistically-defined water quality criteria are needed to ensure protection of the marine resources and to provide a basis for planning waste management systems.

Waste management systems include both treatment to remove pollutants, and dispersion (dilution) systems, to reduce the impact of the residual wastes on the marine environment. Dilutions of waste in excess of 100 are possible in most marine waste disposal situations. Dilution can be as good as treatment for non-conservative materials. However, dilution alone is not adequate for materials that are conservative, toxic, and concentrated in the marine biota. The only solutions for such materials are elimination at the source or treatment to remove a high fraction of the toxic residual materials.

There is inadequate information on the magnitude and effects of oil discharges and on the efficacy of methods to remove oil from waste waters and the environment. There is no doubt, however, that the quantities of oil discharged to the oceans must be markedly reduced.

The increasing practice of ship and barge discharge of exotic chemicals (including the chlorinated hydrocarbons) in both liquid and solid form (containers), should be examined critically in all respects to assess its short- and long-term effects.

There is inadequate information on the magnitude of pollutant discharges to the sea from all significant sources and on the current concentrations of pollutants.

Inadequate quantitative data are available on the present (to say nothing of the future) condition of the marine biota and such data are needed urgently if the technology of waste disposal is to be improved significantly.

The success of limited efforts in aquaculture indicates that one of the effective ways to utilize domestic and selected industrial wastes is to produce, under carefully controlled conditions, marketable food products from the energy and nutrients available in the waste stream. The results obtained in aquaculture are an indication of the possible benefits from carefully controlled waste disposal practices utilizing the ocean as the culture system, i.e. controlled enrichment and increased productivity by well designed systems for waste disposal.

Chemical and Biological Aspects of Waste Treatment   *W. W. Eckenfelder, Jr.**

Aspects chimiques et biologiques du traitement des déchets
Pendant la dernière décennie, des recherches considérables ont porté sur l'amélioration des procédés biologiques et chimiques de traitement des eaux usées et sur la mise au point de nouvelles techniques destinées à produire des effluents qui répondent à des normes plus élevées en ce qui concerne la qualité de l'eau. L'expansion démographique et la concentration industrielle dans les zones urbaines nécessitent l'intensification du traitement des eaux usées, l'accent devant porter sur le traitement combiné des eaux usées d'origine urbaine et des eaux résiduaries industrielles. Le présent document passe en revue les procédés chimiques et biologiques utilisés pour éliminer les polluants organiques et inorganiques dans les eaux usées. L'auteur examine la qualité de l'effluent obtenu grâce à ces méthodes et les aspects économiques du traitement. Il discute des nouvelles méthodes de traitement des eaux usées actuellement en cours d'élaboration.

Aspectos químicos y biológicos del tratamiento de desechos
Durante el decenio pasado se han investigado considerablemente los procedimientos necesarios para mejorar el tratamiento biológico y químico de las aguas residuales y para crear nuevos procedimientos encaminados a producir que las descargas se ajusten a normas de calidad más estrictas. El crecimiento demográfico y la concentración de la industria en las zonas urbanas exige un mayor tratamiento de las aguas residuales, debiendo concederse atención al tratamiento combinado de las aguas residuales municipales e industriales. Este documento examina los procedimientos químicos y biológicos empleados para eliminar los contaminantes orgánicos e inorgánicos de las aguas residuales. Se presentan datos sobre la calidad de las aguas obtenidas por estos procedimientos y los aspectos económicos del tratamiento. Se examinan nuevos procedimientos de tratamiento de aguas residuales que se investigan actualmente.

D URING the past decade there has been an increase in wastewater discharges to surface waters in virtually all parts of the world. This has been accompanied by an upgrading in water quality standards. To meet these considerations wastewater treatment technology has also rapidly improved. This paper summarizes the technology available today for the treatment of domestic and industrial wastewaters to meet various water quality criteria.

Characteristics of treatment processes

There are a number of pollutants in municipal and industrial wastewaters which are detrimental to the intended use of that water. Intended use in this context may include a potable water supply, recreational uses including bathing, fishing, etc., agricultural and industrial uses, navigation and preservation of aquatic life. The primary pollutants in wastewaters are summarized in Table 1. The levels to which these pollutants must be reduced to meet water quality criteria for intended use have been established (U.S. Dept. of Interior, 1968).

The type of treatment process to be employed depends upon the physical and chemical characteristics of the wastewater and the effluent quality desired. These considerations are broadly shown in Table 2. In any

TABLE 1. UNDESIRABLE CHARACTERISTICS OF INDUSTRIAL WASTES

1. Soluble organics: dissolved oxygen depletion in streams and estuaries.
2. Soluble organics that result and odours in water supplies, e.g. phenol.
3. Toxic materials and heavy metal ions; e.g., cyanide, Cu, and Zn.
4. Colour and turbidity: aesthetically undesirable; imposes increased loads on water-treatment plants; e.g., colour from pulp andpaper mills.
5. Nutrients (nitrogen and phosphorus): enhance eutrophication of lakes and ponded areas; critical in recreational areas.
6. Refractory materials, e.g., ABS: results in foaming in streams.
7. Oil and floating material: aesthetically undesirable.
8. Acids and alkalis.
9. Substances resulting in atmospheric odours, e.g., sulphides from tanneries.
10. Suspended solids: results in sludge banks in streams.
11. Temperature: thermal pollution resulting in depletion of dissolved oxygen (lowering of saturation value).

TABLE 2. PROCESSES APPLICABLE TO WASTEWATER TREATMENT

Pollutant	Processes
Biodegradable organics (BOD)	Aerobic biological (activated sludge), aerated lagoons, trickling filters, stabilization basins, anaerobic biological (lagoons, anaerobic contact), deep-well disposal
Suspended solids (SS)	Sedimentation, flotation, screening
Refractory organics (COD, TOC)	Carbon adsorption, deep-well disposal
Nitrogen	Maturation ponds, ammonia stripping, nitrification and denitrification, ion exchange
Phosphorus	Lime precipitation; Al or Fe precipitation, biological coprecipitation, ion exchange
Heavy metals	Ion exchange, chemical precipitation
Dissolved inorganic solids	Ion exchange, reverse osmosis, electro-dialysis

*Department of Environmental and Water Resources Engineering, Vanderbilt University, Nashville, Tennessee 37203, U.S.A.

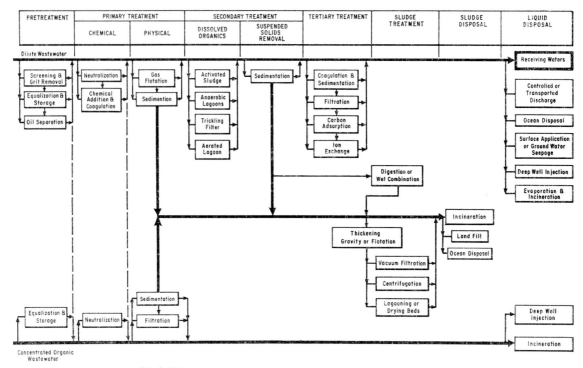

Fig 1. Wastewater treatment sequence/processes substitution diagram

category there will be a number of process alternatives, that selected will depend upon the effluent quality desired as well as such factors as land availability and cost. The various process alternatives are shown in fig 1 and are broken down into the general categories of primary, secondary, tertiary treatment. The effluent quality attainable for these various processes is shown in Table 3.

When considering industrial wastewaters, special attention must be given to the presence of heavy metals e.g., mercury, nickel, etc. These metals are commonly found in chemical and petrochemical wastewaters due to catalyst losses, from slimicides in pulp and paper mill wastewaters and in the metal treating and finishing industries. Pretreatment procedures are commonly employed for the removal of heavy metals prior to biological treatment. In an acclimated biological system, metals are also removed in the process and incorporated in the biomass. However, the metal concentration still remaining after treatment is sufficient to cause a gradual

build-up in concentration in fish and other aquatic life. This has been evidenced by the high concentration of mercury encountered in fish life subjected to pulp and paper mill effluents.

It is apparent, therefore, that modified pre- or post-treatment procedures will have to be employed for metal removal or the metal containing wastewaters be isolated for separate treatment.

Biological treatment

Biological processes are the most common methods of treatment for organic wastewaters. In an aerobic biological treatment process micro-organisms remove organic matter some of which is consumed for energy resulting in carbon dioxide and water and the remainder converted to biomass. The biomass is subsequently degraded by endogenous metabolism. The excess biomass not degraded must be disposed of as excess sludge. Increasing the concentration of biomass in the process will reduce the retention period required to remove the

TABLE 3. MAXIMUM EFFLUENT QUALITY ATTAINABLE FROM WASTE-TREATMENT PROCESSES

Process	BOD	COD	SS	N	P	TDS
Sedimentation, % removal	10–30	—	50–90	—	—	—
Flotation,[a] % removal	10–50	—	70–95	—	—	—
Activated sludge, mg/l	<25	[b]	<20	[c]	[c]	—
Aerated lagoons, mg/l	<50	—	>50	—	—	—
Anaerobic ponds, mg/l	>100	—	<100	—	—	—
Deep-well disposal	Total disposal of waste					
Carbon adsorption, mg/l	<2	<10	<1	—	—	—
Ammonia stripping, % removal	—	—	—	>95	—	—
Denitrification and nitrification, mg/l	<10	—	—	<5	—	—
Chemical precipitation, mg/l	—	—	<10	—	<1	—
Ion exchange, mg/l	—	—	<1	[d]	[d]	[d]

[a] Higher removals are attained when coagulating chemicals are used.
[b] $COD_{inf} - [BOD_u (Removed)/0.9]$.
[c] $N_{inf} - 0.12$ (excess biological sludge), lb; $P_{inf} - 0.026$ (excess biological sludge), lb.
[d] Depends on resin used, molecular state, and efficiency desired.

[454]

organic matter. The various aerobic processes employed today are combinations of retention period and biomass concentration to achieve a desired effluent quality. The characteristics of the various biological processes for the treatment of domestic sewage are summarized in Table 4.

TABLE 4. OPERATING CHARACTERISTICS OF AEROBIC BIOLOGICAL TREATMENT PROCESSES

Type of process	Biological solids concentration, mg/l	Aeration, detention period
Conventional activated sludge	2000–4000	2–6 h.
Extended aeration	3000–5000	18–24 h.
Aerated lagoons	100–300	3–8 days
Stabilization basins	<25	10–30 days
Pure oxygen activated sludge	7000–10000	<2 h.

Anaerobic processes yield end products of methane and carbon dioxide.

Lagoons are commonly employed as a low-cost method where land area is available. Lagoons are classified as anaerobic, facultative, aerobic or aerated. The differentiation between the first three classifications is primarily related to the organic loading expressed as lbs BOD/acre/day and to the depth of the basin which regulates the penetration of light to the basin. Table 5 shows the inter-relationships between these classifications.

At organic loadings above approximately 250 lbs of BOD/acre/day, the basin will be essentially anaerobic (Eckenfelder and O'Connor, 1961). The high rate of biological activity under these loading conditions prohibits the growth of algae.

If the loading is less than 250 lbs of BOD/acre/day algal growths produce oxygen and create an aerobic upper layer. This is defined as a facultative pond in which aerobic activity is evident in the upper layers and anaerobic activity in the bottom layers and the deposited sludge. Aerobic depths of up to three feet have been evidenced with organic loadings of 50 lbs of BOD/acre/day; with shallow depths (less than 1.5 ft) fully aerobic activity can be maintained by algal growth (Eckenfelder and O'Connor, 1961).

The aerated lagoon is commonly employed for the treatment of domestic and industrial wastewaters in the United States and Canada (Oswald, 1961; Randall and Bartsch, 1971). Mechanical aeration equipment is employed to provide oxygen and mixing in the basin.

In aerobic lagoons sufficient power is provided to maintain complete mixing conditions in the basin. Organisms synthesized from the wastes and suspended solids present are discharged with the effluent. In a facultative lagoon the power level is sufficiently low to permit suspended solids to deposit to the bottom of the basin where they undergo anaerobic decomposition. Many modern aerated lagoon systems employ combinations of the aerobic and facultative basins. The design and operating characteristics of lagoon systems are summarized in Table 6 (Porges, 1963).

Some of the problems associated with the operation of lagoon systems include odour development at high organic loadings and high effluent suspended solids particularly in aerated basins. High temperature dependency leads to reduced efficiency and high effluent BOD during the winter months (Gloyna, 1966)

The design variables for lagoons depending on the type of process, i.e., aerobic or anaerobic, include detention time, biological solids level and temperature. Anaerobic reaction rates are $\frac{1}{5}$ to $\frac{1}{10}$ that of aerobic systems, and hence longer retention periods (15 to 60 days) are usually employed in anaerobic lagoons. In both types the BOD removal is directly related to detention time and biological solids level. Increasing the biological solids level reduces the required retention period for a given effluent quality. Performance data for several lagoon systems are summarized in Table 6.

TABLE 5. DESIGN CRITERIA FOR STABILIZATION BASINS

	Aerobic[a]	Facultative	Anaerobic	Aerated
Depth, ft	0.6–1.0	3–8	8–15	8–15
metres	0.2–0.3	1–2.5	2.5–5	2.5–5
Detention, days	2–6	7–50	5–50	2–10
BOD loading				
lb/acre/day	100–200	20–50	250–4000	—
kg/ha/day	111–222	22–55	280–4500	
Per cent BOD removal	80–95	70–95	50–80	80–95
Algae concentration, mg/l	100	10–50	Nil	Nil

[a] Must be periodically mixed; velocity 1 to 1.5 ft/s.

TABLE 6. PERFORMANCE OF LAGOON SYSTEMS*

Waste	Type of pond	Depth, ft	Detention, days	Loading, lbs BOD/acre/day	BOD removal %
Canning	Facultative	5.8	37.5	139	98
Paper	Facultative	5.0	30.0	105	80
Dairy	Facultative	5.0	98.0	22	95
Sugar	Facultative	1.5	2.0	86	67
Canning	Anaerobic	6.0	15.0	392	51
Paper	Anaerobic	6.0	18.4	347	50

*Data from Porges (1963).

The activated sludge process is widely employed where land area is restricted and/or high effluent quality is desired. Solids liquid separation and sludge recycle permit high aeration solids levels to be maintained with corresponding short retention periods. Modern activated sludge plants employ completely mixed basins with high capacity mechanical surface aerators. The activated sludge process has been widely employed for the treatment of domestic and industrial wastewaters (Eckenfelder, 1970).

The conventional activated sludge process employs a single aeration basin (or several single pass basins) of either a plug flow or completely mixed design. Settled biological sludge from the clarifier is recycled to the head end of the aeration basin. Excess sludge must be removed from the process and disposed of. The contact-stabilization process employs a short detention time aeration contact basin followed by a clarifier in which the biological sludge is settled. It is then recycled to a sludge reaeration basin (stabilization tank) for several hours aeration prior to admixture with the raw wastewater. This process is particularly adaptable to wastewaters containing a major portion of the BOD in suspended or colloidal form such as municipal sewage (Eckenfelder, 1970).

The extended aeration process employs a long aeration period in order to oxidize most of the biological sludge produced in the process. Although it minimizes sludge handling and disposal because of the longer retention periods required, the extended aeration process is usually limited to relatively low daily volumes of wastewater. Recently, pure oxygen has been employed in the activated sludge process (U.S. Dept. of Interior, 1969).

The higher dissolved oxygen levels (>10 mg/l) result in a denser sludge permitting higher aeration solids levels and shorter aeration retention periods.

In a trickling filter, wastewaters are passed over a packing on which biological growths have been developed. BOD is progressively removed as the wastewater passes through the filter. Recent application of plastic packing permits use of deep columns with a high specific surface (30 ft^2/ft^3). Columns of up to 40 ft have been employed for the treatment of industrial wastewaters in the United States. In general, efficiency is lower than that attainable in the activated sludge process.

Tertiary treatment

Biological treatment will remove most of the biodegradable organic matter but not inorganic salts or refractory organics, such as those present in some textile wastes and kraft mill effluents. Some nitrogen and phosphorus is removed through synthesis of the biological cell mass. In the case of domestic sewage this accounts for about 20 per cent of that originally present in the wastewaters.

Tertiary treatment can be defined as the removal of pollutants not removed by conventional biological treatment processes. These include some suspended solids, BOD (usually less than 10 to 15 mg/l), refractory organics (usually reported as COD or TOC), nutrients (nitrogen and phosphorus) and inorganic salts. The various tertiary-treatment processes available are shown in Table 7. The effluent qualities attainable from the tertiary processes are shown in Table 3.

TABLE 7. TERTIARY-TREATMENT PROCESSES FOR THE REMOVAL OF SPECIFIC POLLUTANTS

Unit process	Major elements removed	Additional features
Filtration—sand, diatomite, or mixed media	Suspended solids	Removal of BOD, COD, PO$_4$ in suspended form
Filtration plus coagulation-mixed media	Suspended solids and phosphate, colour and turbidity	As above plus colloidal solids
Coagulation	Colour, turbidity and PO$_4$	Some COD and BOD removed
Air stripping	NH$_3$	High pH required
Nitrification and denitrification	Nitrogen	Ultimate BOD reduced
Carbon adsorption	COD or TOC	Reduction in colour and residual suspended solids
Ion exchange	PO$_4$, nitrogen, total dissolved solids	Resins selected for specific purposes
Reverse osmosis	Organics and inorganics	Pretreatment required to avoid membrane fouling; treatment or disposal of residue
Electrodialysis	Inorganic salts	Pretreatment required to avoid membrane fouling; treatment or disposal of residue

In order to attain effluent qualities superior to those encountered in conventional biological processes several physical-chemical treatment systems have recently been developed (Weber and Bloom, 1970). These involve chemical coagulation using alum, lime or ferric chloride as a coagulant for suspended and colloidal materials, e.g. phosphorus followed by multistage, counterflow activated carbon columns in which dissolved organics are removed. This process is capable of reducing the BOD in excess of 95 per cent and the phosphorus in excess of 90 per cent. Nitrogen removal can also be achieved by breakpoint chlorination in the carbon columns.

Discussion

Wastewater treatment technology can now achieve any desired effluent quality but since the cost of wastewater treatment is directly related to effluent quality, the establishment of realistic water quality requirements is an essential adjunct to defining environmental economics. One approach is the use of benefit/cost ratios, such as those developed by Stone and Friedland (1970) for various water uses in San Diego Bay. These workers used as a basis, primary, secondary and tertiary treatment, assuming that tertiary treatment would remove 93 per cent of the BOD and 90 per cent of the phosphorus. Secondary treatment would result in limited water contact sports and propagation of wild life. Secondary treatment yielded a benefit/cost ratio of 1.96 and tertiary treatment of 2.20. On the basis of these ratios tertiary treatment was justified.

One of the most difficult problems in augmenting programmes of this type is the political and economic aspect. Several approaches are being considered in the United States to offset these difficulties. In the Houston, Texas, area the Gulf Coast Waste Disposal Authority has been created to develop and operate combined and joint wastewater treatment facilities with costs allocated to the municipal and industrial users.

Heretofore the primary concern with effluent quality related to coliform organisms, aesthetic considerations such as floatables, oils etc., and dissolved oxygen deficits (BOD). In many cases, however, such as shellfish areas and aquatic reserves, these usual water quality parameters do not apply because they are non-specific as to the detrimental effect on aquatic life. In these cases, a species diversity index has been employed as related to either free floating or benthic organisms. In an increasing number of cases, fish bioassay tests are being employed as an effluent quality parameter.

References

CAMP, T R Water and its impurities. New York, Rienhold.
1964
ECKENFELDER, W W Industrial water pollution control. New York,
1966   McGraw Hill.
ECKENFELDER, W W Water quality engineering. New York,
1970   Barnes and Noble.

ECKENFELDER, W W and O'CONNOR, D J Biological waste treat-
1961   ment. Oxford, Pergamon Press.
GLOYNA, E F Basis for waste stabilization pond design. *Adv. Wat.*
1966   *Qual. Improv.*, 1.
McGAUHEY, P H Engineering management of water quality.
1968   New York, McGraw Hill.
McKEE, J D and WOLF, H W Water quality criteria. Pasadena,
1963   California, California Institute of Technology.
OSWALD, W Fundamental factors in waste stabilization pond
1961   design. *In* Advances in Biological Waste Treatment.
PORGES, R Performance of lagoon systems. *J. Wat. Pollut. Control*
1963   *Fed.*, 35(4):456.
RANDALL, C W and BARTSCH, E H Aerated lagoons—a state of
1971   the art. *J. Wat. Pollut. Control Fed.*, 43(4):699.
ROSS, R D Industrial waste disposal. New York, Rienhold.
1968
STONE, R and FRIEDLAND, H Estuarine clean water cost benefit
1970   studies. Paper presented at the 5th International Conference
    on Water Pollution Research, July 1970.
U.S. Department of the Interior, Federal Water Pollution Control
1968   Administration, Report of the Committee on water quality
    criteria. Washington, D.C., April 1968.
U.S. Department of the Interior, Federal Water Quality Adminis-
1969   stration Report, The application of pure oxygen in biological
    treatment.
WEBER, W and BLOOM, R Physical-chemical treatment of domestic
1970   sewage. *J. Wat. Pollut. Control Fed.*, 42(12).

Waste Water Treatment in Chemical Industries in the Federal Republic of Germany

*K. J. Bock**

Le traitement des eaux usées dans les industries chimiques en République fédérale d'Allemagne

L'implantation de grandes entreprises industrielles sur les côtes et les difficultés accrues qu'éprouvent les usines de l'intérieur pour éliminer leurs effluents ont abouti, ces dernières années, au rejet dans la mer de quantités croissantes de produits résiduaires solides ou liquides. Il a donc fallu prendre des mesures de protection, appliquées tout particulièrement aux zones côtières. L'industrie chimique fabrique un grand nombre de produits par des méthodes très différentes qui sont sujettes à des développements techniques rapides et posent des problèmes spéciaux en ce qui concerne l'évacuation des effluents. On démontre au moyen d'exemples que la plus grande partie de ces effluents peuvent être traités dans les établissements mêmes à un degré tel que les quantités à rejeter sont peu élevées. Les techniques modernes de fonctionnement, les méthodes d'épuration en usine, l'utilisation d'un équipement spécial pour l'élimination des effluents et des produits résiduaires ainsi que l'emploi d'incinérateurs spéciaux, tous ces éléments contribuent à des degrés divers à prévenir la pollution. Néanmoins, il subsiste un certain nombre de difficultés que l'on ne peut surmonter à l'heure actuelle qu'en recourant à des méthodes dont le coût n'est pas négligeable pour rejeter les effluents à la mer en des points choisis de manière adéquate.

Tratamiento de aguas residuales en las industrias químicas de la República Federal de Alemania

El emplazamiento de grandes establecimientos industriales en las costas y la creciente dificultad experimentada por las factorías continentales para dar salida a sus aguas residuales ha conducido en estos años a que vayan a parar al mar cada vez más desechos sólidos o líquidos. Esto ha exigido la adopción de medidas protectoras, principalmente aplicadas a las zonas costeras. La industria química fabrica gran número de productos distintos mediante procedimientos muy diferentes que están sujetos a una rápida evolución técnica y plantean determinados problemas con respecto a la evacuación de las aguas residuales. Se demuestra por medio de ejemplos que la mayor parte de las aguas residuales resultantes se pueden tratar en las fábricas de tal modo que sólo se hace necesario evacuar pequeñas cantidades. Contribuyen en varios grados a la prevención de la contaminación operaciones modernas, métodos internos de purificación, equipo especial para evacuar las aguas residuales y los desechos, así como incineradores especiales. Sin embargo, existen todavía algunas dificultades que en la actualidad sólo se pueden superar mediante el barato método de descargar las aguas residuales en lugares determinados de la costa.

IN recent years, industrialization has progressed rapidly along the coasts of Europe, especially near large ports. Increasing demands for environmental hygiene in the inland industrial centres and the more favourable conditions and easier disposal of waste water, exhaust air and refuse intensify the migration to the coastal areas.

During each chemical process, transformation into the desired end product is always incomplete. Waste water must be purified and residues or by-products in gaseous, liquid or solid form must be disposed of in such a way that they have no adverse effect on the environment. Some solid and liquid wastes are discharged or dumped into the sea but these represent only a small fraction of the total waste produced and a much larger fraction is treated at the factory. Examples of methods which are used by the chemical industry in the Federal

Republic of Germany for the treatment and purification of waste water are here given.

Industrial waste water disposal, especially in the chemical industry, differs in many respects from the disposal of household waste water. The composition of city waste water is almost the same everywhere and can be purified satisfactorily by well-known mechanical and biochemical methods. In the chemical industry many processes are used to manufacture inorganic and organic chemicals, fertilizers, plastics, pharmaceutical products, etc. Each process has its own specific characteristics and, accordingly, its very specific waste water problems; in addition, because of constant technological developments, manufacturing processes are subject to constant change. Purifying methods that have proved successful in one place can rarely be adopted elsewhere. For example, batch operations and continuous processes use different

* Chemische Werke Hüls A.G., Marl, Federal Republic of Germany.

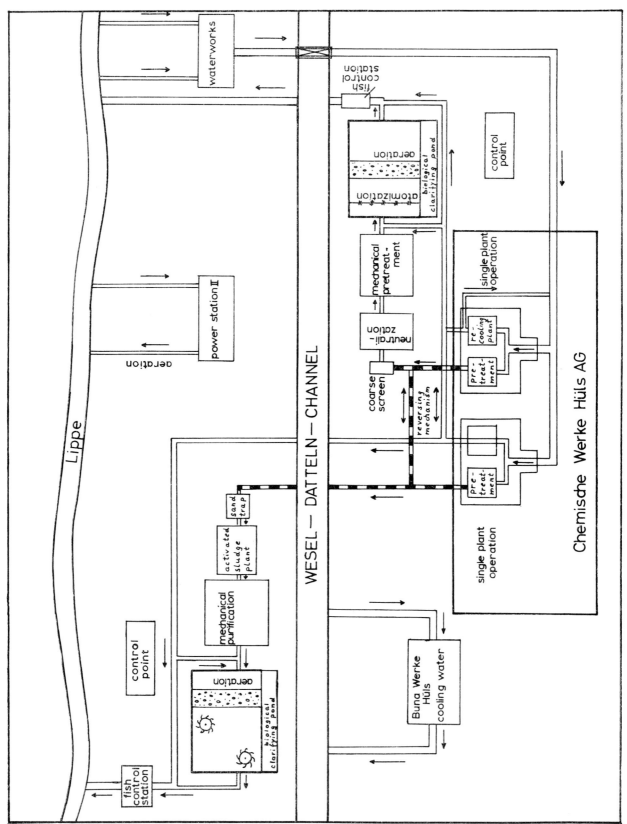

Fig 1. Water supply and waste water purifying plants of Chemische Werke Hüls AG, Marl.

amounts of waste water; as a result constant research and development work is necessary for effective purification.

The chemical industry needs considerable quantities of water. According to Government statistics, about 3×10^9 m³ of water were used in 1965. Of this 78.5 per cent was used for cooling purposes and was not contaminated but about 18 per cent, which had been in contact with the products, had to be treated. A basic condition for the economical purification of waste water is the separate collection and discharge of (1) clean cooling water, (2) rain water, and (3) contaminated waste water. Often the amount of water required for cooling can be reduced by circulating and re-cooling. Air cooling systems have often proved satisfactory substitutes for water cooling. Since the type and quantity of waste water depends on the manufacturing process, it is practical to consider the waste water problem during the planning and development of a production plant. Sometimes it is possible to considerably reduce the amount of waste water or even eliminate it completely.

The following are examples of methods used to purify waste water in industrial plants: Ethylene oxide is used for the manufacture of textile auxiliaries, glycol and its derivatives, ethanolamine and secondary products, as well as ethoxides. Ethylene oxide was formerly produced via chlorohydrin and for each ton of the finished product, waste water containing about 2.5 tons of contaminated calcium chloride was produced. Conversion of the process to direct oxidation completely eliminated this sludge. Before the remaining waste water is discharged into the sewer system, the glycol concentration is reduced to a small fraction of the former level.

Octanol (2-ethyl hexanol) and butanol are needed in considerable quantity as intermediate products for the manufacture of plasticizers. Also butanol is processed to such products as butyl acetate and butylene glycol. Formerly, octanol and butanol were produced in a duplexing process from acetylene via acetaldehyde, acetaldol, crotonaldehyde and butyraldehyde. During this several-stage process, waste water and sludge with a high organic contamination occurred and had to be purified. The bottoms obtained in the crotonaldehyde distillation process had to be separated from an extremely high amount of impurities in a special column of high-grade steel.

Conversion to the so-called oxosynthesis, whereby carbon monoxide and hydrogen are combined with propylene in the presence of cobalt catalysts thus directly producing butyraldehyde and butanol, greatly contributed to an improvement of the waste water situation. As a result of intensive research and development, waste water contaminated by cobalt compounds and organic matter, no longer occurs during the cobalt recovery process. A further improvement was the installation of a special plant which makes it possible to recycle the waste water, thereby reducing waste water contamination and fresh water needs.

During the manufacture of dyestuffs, many dyes are produced in aqueous solutions. Formerly the dyes were precipitated by the addition of considerable quantities of sodium chloride and Glauber salt and filtered off. The highly saline mother liquors were discharged as waste water into the canal. Dyestuffs are now being produced by evaporating the reaction solution in spray dryers whereby waste water is eliminated. However, the dyestuff contains impurities from the mother liquor so that for quality reasons this method cannot be employed for all dyestuffs.

The complete elimination of waste is exceptional and in most cases the waste water is subjected to preliminary purification. This may require special equipment; in some cases, basins are adequate for separating solids, oils, fats, polymerization residues, etc. In other cases, additional filters must be installed. Inorganic or organic products dissolved in waste water cause considerably greater difficulties. Often additional and costly equipment must be developed to achieve effective waste water purification. During the process of concentrating butadiene, for the manufacture of BUNA®, waste water containing copper is produced. To remove the copper the concentrated waste water is rendered alkaline and the copper is precipitated by steam injection. The remaining weak copper solution is purified by ion exchangers. The reclaimed copper oxide is passed on to the copper processing industry for recovery of the metal.

During the manufacture of ethyl benzene, styrene, and cumene, waste water containing benzene is obtained. The benzene can be separated from the waste water in a special plant by means of nitrogen and be recovered by treatment with activated carbon.

These examples show that by special measures, waste water contamination can be considerably reduced. The recovery of valuable products is possible only in exceptional cases. In most cases, the substances separated from the waste water are of no value and must either be disposed of at additional expense or be destroyed in special furnaces. Some of the waste materials may be disposed of in the sea. This is being done with diluted acids from various manufacturing processes since neutralization and discharge into inland waters would increase the salinity of the water to an intolerable level.

Even after successful pretreatment, some waste waters must be further treated before discharge. The treatment plants installed at various chemical plants are operated by different processes. At the Chemische Werke Hüls plant, approximately 110,000 m³/day of pretreated waste water is purified biologically in aerated ponds after the solids have been mechanically separated (fig 1). Farbwerke Hoechst have built a central clarifying plant with a daily capacity of 24,000 m³ of waste water; here the waste water is neutralized, mechanically cleaned and subsequently purified by means of activated sludge. Farbenfabriken Bayer have a plant under construction which is to serve for the mechanical and biochemical purification of 60,000 m³/day of chemical waste water and 70,000 m³/day of city waste water. Other companies have installed similar plants.

The cost of this type of waste water purification cannot easily be determined because changes in manufacturing processes as a result of waste water problems often cannot be separated from the building and operating costs. Only the costs for central clarifying plants can be specified. The construction of the clarifying plant of Farbwerke Hoechst, for instance, cost about 20 million DM, and the annual operating cost is 8 million DM. From 1948–1968, Chemische Werke Hüls spent about 80 million DM on clarifying plants, with an annual operating cost of 3–4 million DM. The costs for constructing the plants of Farbenfabriken Bayer and the Wupperverband (Wupper Association) are expected to reach a total of about 80 million DM, with annual operating costs estimated at 12 million DM.

These examples show that the chemical industry is making a significant contribution to keeping streams and oceans clean. But despite these achievements, some

waste water must be discharged into the sea to protect fresh water supplies from contamination. This dumping of waste water into the sea should be carried out under regulations established by marine research in co-operation with industry.

International conventions applicable in all countries should be established to prevent adverse effects on sea ecology.

Effects of Proposed Second Entrance on the Flushing Characteristics of San Diego Bay, California‡

H. B. Simmons and
*F. A. Herrmann, Jr.**

Effets d'une nouvelle voie d'accès projetée pour la baie de San Diego (Californie) sur le mode d'écoulement des eaux

On a construit un modèle hydraulique de la baie de San Diego (Californie) pour étudier les effets de l'ouverture d'une seconde voie d'accès à cette baie sur ses caractéristiques hydrauliques et sur le mode d'écoulement des eaux. Cette ouverture aurait pour objet de fournir un accès plus bref et plus sûr à l'océan, pour les navires de guerre comme pour les navires marchands et, si possible, d'améliorer le taux d'écoulement des eaux de la baie qui actuellement est lent.

Le modèle qui a été construit à une échelle linéaire de 1:500 horizontalement et de 1:100 verticalement était du type à fond fixe. Il a été soigneusement calculé pour reproduire exactement les courants prototypes observés, les courants de marée avec leur direction et leur vitesse et la dispersion des indicateurs colorants. Les eaux de la baie et celles des zones océaniques voisines étant essentiellement homogènes, il n'a pas été nécessaire de reproduire les effets de la salinité dans le modèle.

Des essais ont porté sur deux sites différents pour la seconde voie d'accès, à proximité de l'extrémité sud de la baie. On a constaté que la vitesse maximale du courant dans la moitié nord de la baie était généralement réduite d'environ 50 pour cent dans les deux cas. Les résultats des essais effectués avec des indicateurs colorants ont montré que l'un et l'autre plans permettraient d'améliorer substantiellement les caractéristiques globales de l'écoulement des eaux de la baie, l'ouverture la plus septentrionale permettant l'amélioration maximale de l'écoulement. Avec l'une et l'autre des deux ouvertures nouvelles proposées dans le modèle, il est apparu que le point nodal du courant d'entrée était quelque peu au nord du point nodal du courant de sortie, ce qui crée un schéma d'échange comportant une arrivée d'eau nette dans la baie par la voie d'accès actuelle et une sortie d'eau nette par la seconde ouverture proposée.

Efectos que podría tener la propuesta construcción de una segunda entrada en las características del flujo de aguas de la bahía de San Diego, California

Se construyó un modelo hidráulico de la bahía de San Diego, California, para estudiar los efectos que la propuesta construcción de una segunda entrada tendría en las características hidráulicas y de renovación del agua. La segunda entrada serviría para acortar y hacer más seguro el acceso el océano para barcos de guerra y mercantes y, de ser posible, aumentar la velocidad de renovación del agua de la bahía.

La maqueta se construyó a las escales lineales 1:500 horizontalmente y 1:100 verticalmente y era del tipo de lecho fijo. Se ajustó con cuidado para que reprodujera con exactitud las mareas prototipo observadas, las direcciones y velocidades de las corrientes creadas por las mareas y la dispersión de colorantes trazadores. Como son muy homogéneas las aguas dentro y fuera de la bahía, no fue necesario reproducir en el modelo los efectos que podría tener la salinidad.

Los ensayos se efectuaron colocando la segunda entrada en dos sitios distintos cerca del extremo sur de la bahía. Se observó que en ambos planos las velocidades máximas de la corriente en toda la parte norte de la bahía se reducían, de manera general, en cerca de un 50 por ciento. Los resultados de las pruebas con colorantes trazadores demostraron que ambos planes mejorarían sensiblemente las características generales de renovación del agua, y que la más septentrional de las dos entradas produciría la mayor renovación. Con cualquiera de las dos segundas entradas de la maqueta parecía desprenderse que el punto nodal de la marea entrante quedaba algo al norte del punto nodal de la vaciante, creándose así una circulación con una entrada neta en la bahía por la entrada actual y una salida neta por la segunda entrada propuesta.

SAN DIEGO BAY, California, lies about 26 km north of the United States–Mexico boundary. The bay is about 29 km long, has a maximum width of about 3.1 km, and covers approximately 90 sq km. The only connection between the bay and the ocean is Zuniga Channel (see fig 1). The fresh water inflow is relatively small—principally infrequent heavy local rainfall. For all practical purposes the water within the bay can be considered homogeneous. The flushing rate is extremely slow, there being no net outflow or vertical density gradients to establish a significant vertical circulation and thus accelerate flushing time.

The Pacific Ocean tides at the entrance to the bay are mixed, with two high and two low tides in each average diurnal period of about 25 hours; however, there is normally an appreciable diurnal inequality between the two tides. The normal tidal sequence is higher high water (HHW), lower low water (LLW), lower high water (LHW), and higher low water (HLW). Because of this, the stronger current velocities in the bay and Zuniga Channel occur during the ebb interval between HHW and LLW. The average range of tide at Ballast Point in Zuniga Channel is about 1.89 m, that in

the central portion of the bay is about 1.98 m, and that near the south end of the bay is about 2.10 m. Ebb and flood current velocities vary from a maximum of about 0.7 m per sec in Zuniga Channel to essentially zero at the south end of the bay.

The bay is used extensively by naval vessels and by recreational craft of various types and sizes. The locations of channels maintained for navigation in Zuniga Channel and throughout San Diego Bay are shown in fig 1. The channel depth is 12.80 m below mean lower low water (mllw) from the ocean to the vicinity of the Navy Pier, thence, 9.14 m to the vicinity of the South Bay gauge. As vessel size and traffic density increases, or new docking facilities are planned or constructed, requests are frequently made for increases in both depth and width and extent of navigation channels. San Diego Bay is not an area of heavy silting, as compared to most deep draught harbours in the United States, although some maintenance dredging must be done periodically in Zuniga Channel and at other locations where silting occurs.

For several reasons consideration is being given to the construction of a second navigation entrance to San

---

* U.S. Army Waterways Experiment Station, Vicksburg, Mississipi 39180, U.S.A.
‡ Investigation sponsored by the U.S. Army Engineer District, Los Angeles. Permission to publish appreciated.

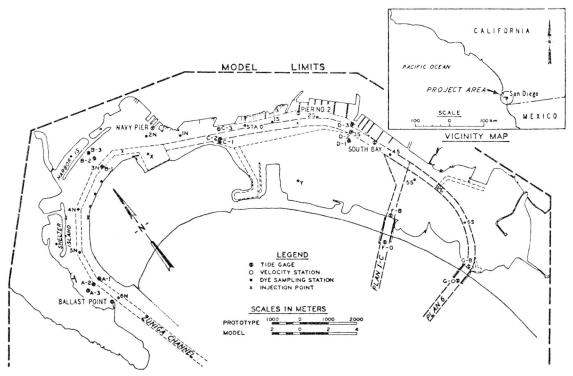

Fig 1. Location map

Diego Bay located somewhere along Silver Strand with access to the southern portion of the bay. Since the extent to which a second entrance would increase the flushing rate of San Diego Bay could not be computed reliably by available analytical methods, it was decided

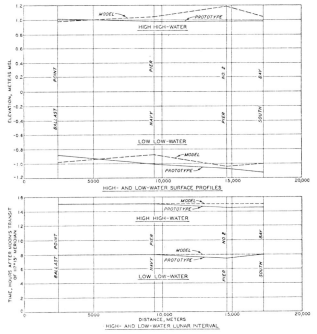

Fig 2. Verification of tides

to construct a physical hydraulic model of the bay in which several suggested locations for the second entrance could be tested and the effects of each one on the hydraulic and flushing characteristics of the bay investigated in detail.

The model

The San Diego Bay model reproduced about 280 sq km of prototype area, including the entire bay and offshore ocean areas to well beyond the −35 m contour. The limits of the model are shown in fig 1.

The model, which was approximately 35 m long and 40 m wide, was constructed to linear scale ratios of 1:500 horizontally and 1:100 vertically. From these basic ratios the following scale relations were computed using the Froudian relations: velocity 1:10, time 1:50, discharge 1:500,000, and volume 1:25,000,000. Since the waters within the bay and the adjacent ocean areas are essentially homogeneous, it was not necessary to reproduce salinity effects in the model. The model was of the fixed-bed type, although provisions were made to convert certain areas to movable bed if detailed studies of sedimentation and littoral transport were required. The navigation channels were moulded in movable blocks so that alterations could readily be made. The model was equipped to reproduce and measure all pertinent phenomena such as tidal elevations, current velocities, dye dispersion characteristics, waves, littoral current, and shoaling distribution.

Verification of model

The worth of any model study is wholly dependent on the proven ability of the model to produce with reasonable accuracy the results which can be expected to occur in the prototype under given conditions. It is essential, therefore, that the required similitude be established between model and prototype, and that all scale relations between the two be determined.

Verification of the San Diego Bay model consisted of two phases: (a) hydraulic verification, which ensured that tidal elevations and phases and current velocities

[461]

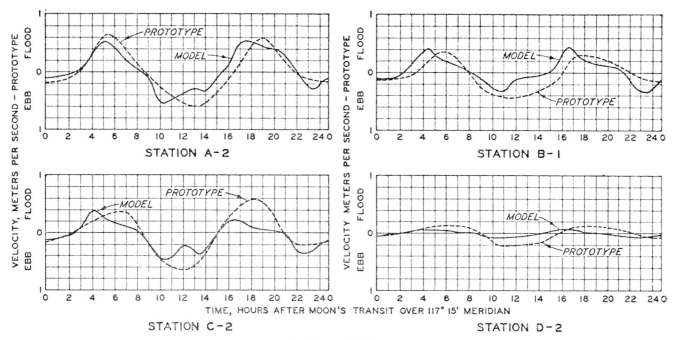

Fig 3. Verification of velocities

and directions were in proper agreement with the prototype, and (b) dye dispersion verification, which ensured that the dispersion patterns and flushing rate were adequately reproduced.

Hydraulic verification

Verification of tidal elevations and phases was accomplished by adjusting the tide generator to reproduce observed prototype tides at Ballast Point (fig 1), then adjusting the model roughness within the bay until reproduction of observed tides at the interior tide gauges was satisfactory. A comparison of model to prototype high- and low-water profiles and high- and low-water arrival times is shown in fig 2.

Following tidal verification, the model roughness was rearranged until a satisfactory distribution of current

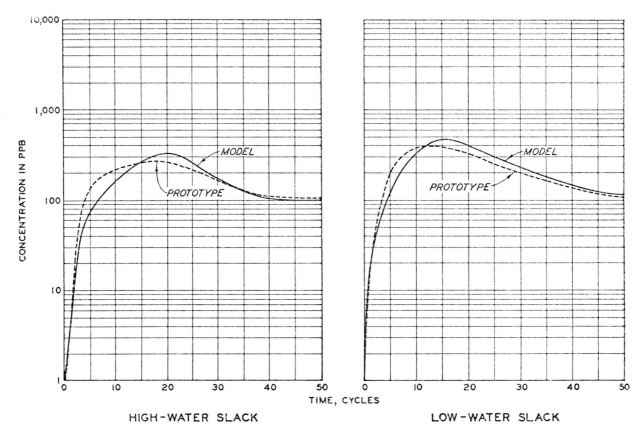

Fig 4. Verification of dye concentration history at station 1N

velocities, both laterally and in depth, was achieved. Prototype velocity measurements at each range were obtained on successive days. Since only one tide was reproduced in the model, the prototype velocity observations at three of the ranges did not correspond with the model tide. Thus, more emphasis was placed on the reproduction of the proper percentage distribution of flow, both vertically and laterally, than on the reproduction of the absolute magnitude and phasing of velocities. Comparisons of model to prototype mid-depth velocities at four stations located along the navigation channel are shown in fig 3.

Agreement between model and prototype hydraulic phenomena, as illustrated by results of tidal and velocity data, was considered satisfactory and sufficiently accurate to provide quantitative results on the effects of the proposed channel on the hydraulic regimen of the bay.

Dye dispersion verification

The Federal Water Pollution Control Administration (now the Federal Water Quality Administration) provided the results of a dye dispersion test in which a fluorescent dye with an initial concentration of 200,000,000 parts per billion (ppb) was injected continuously into the bay at a rate of 33 cc per min over a period of 15 tidal cycles (average duration 24.84 h each). Concentrations were measured throughout the bay over a period of 160 tidal cycles after the injection was started. Using the model discharge scale of 1:500,000 would have resulted in too small an input of dye to the model for accurate measurements. In addition, the extremely high initial concentration used in the field would have required that many of the model samples be diluted prior to analysis for dye concentration. It was thus necessary to increase the scale input of the dye to the model, as well as to reduce the initial concentration of the dye used in the model test. The model dye was injected at the corresponding location and over the corresponding period as was done in the field. After the model samples had been analysed, it was found that the time-concentration curves for each sampling station were almost identical in shape, and that by multiplying the observed prototype concentrations by a factor of 100 the magnitude of model and prototype concentration was in good agreement. No further adjustment of the model roughness was required in order to achieve a satisfactory reproduction of prototype dispersion characteristics.

Samples were withdrawn from the model at the 13 stations sampled during the prototype test (O, 1S, 1N, 2S, 2N, etc.) at times of slack currents following HHW and LLW and were subsequently analysed to determine dye concentration. A typical comparison of model and prototype time vs dye concentration curves is shown in fig 4. Profiles of model and prototype dye concentration along the bay at the end of 15 and 45 tidal cycles are presented in fig 5. On the basis of the excellent agreement between model and prototype dispersion characteristics it was considered that the model could be used to predict the effects of construction of the proposed second entrance.

Tests and results

Prior to tests of the improvement plans, a series of base tests was conducted to determine hydraulic and disper-

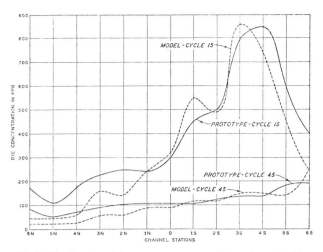

Fig 5. Verification of dye concentration profiles at the end of 15 and 45 tidal cycles

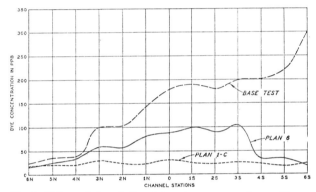

Fig 6. Effects of plans 1-C and 6 on dye concentration profiles at the end of 40 tidal cycles

sion characteristics throughout the bay for existing conditions. The same tide was reproduced in the model for the base tests as was used for the model verification, but several additional tide gauges, velocity stations, and dye dispersion sampling stations were established. In addition, mosaics were made of time exposure photographs of confetti floating on the water surface to establish surface velocities and current patterns for each prototype hour of a complete tidal cycle.

For base (and plan) conditions, dye was injected at the three locations along the navigation channel shown in fig 1, whereas only the central injection point was used during the verification tests. The overall rate of injection and total quantity of dye were, however, the same. The three-point injection was necessary to ensure that dye would be available to all portions of the bay after the second entrance was installed in the model.

Model conditions for the base (and plan) tests were as follows: tide range (HHW to LLW) of 1.9 m at Ballast Point; no fresh water inflow. The results of the base tests were used to evaluate the effects of the plans tested. Thus, any differences noted in the plan test results were attributable to the plan being tested and not to minor differences between model and prototype phenomena.

Two locations for the proposed second entrance were investigated in the model study. The location nearest to the south end of the bay was designated plan 6, while the more centrally located entrance was designated plan 1-C. For both plans the second entrance channel was 15.24 m deep at mllw and 3,048 m wide at the bottom. Inside the bay the channel narrowed to 2,438 m wide and extended northward to about the Navy Pier as shown in fig 1. Three jetties were located at the mouth of each proposed second entrance (fig 1) for the purpose of confining flow and impounding littoral drift material.

For both plans, the tide range in the southern portion of the bay was reduced by about 15 cm. Maximum velocities in the northern portion of the bay were generally reduced by 60 to 80 per cent, whereas in the southern portion they were relatively unchanged or showed increases in the immediate vicinity of the proposed second entrance. Furthermore, it was noted that a nodal point developed in the central portion of the bay. In this area there was very little horizontal water movement at any time during the tidal cycle. The current pattern photographs showed that the nodal point during the ebb tide was about 2,000 m south of that during the flood tide. Thus, a net southward circulation was created, with a net inflow through the existing entrance and a net outflow through either of the proposed second entrances.

Results of dye dispersion tests

Compared to the base test, plan 1-C caused reductions in maximum dye concentrations of about 80 per cent throughout the bay at low-water slack and a like reduction in the southern portion of the bay at high-water slack. Peak concentrations in the northern portion of the bay were reduced by about 30 per cent. Maximum concentrations were increased by about 50–60 per cent at both high- and low-water slack in the vicinity of the nodal point, but the subsequent decay rate was much faster than for base test conditions. Throughout the model the decay rate for the first 5–10 tidal cycles following the time of maximum concentration was much faster for plan 1-C than for the base test; later, however,

the opposite was true. Profiles of dye concentration along the bay for base and plan conditions at tidal cycle 40 are shown in fig 6. It can be seen that the dye concentrations for plan 1-C at that time are essentially uniform throughout the bay.

Plan 6 caused reductions in maximum dye concentrations of 40–60 per cent throughout the bay at low-water slack and by about 50 per cent in the southern portion of the bay at high-water slack. Maximum high-water slack concentrations, however, were increased by about 50 per cent in the northern portion of the bay, although the subsequent decay rate was much greater than for base conditions. In the later stages of the tests, the base test decay rate was much greater than that for plan 6. It can be seen in fig 6 that the greatest reductions in concentration at tidal cycle 40 occurred in the extreme southern end of the bay, in the vicinity of the plan 6 second entrance. Dye concentrations in the central reaches of the bay were intermediate between base and plan 1-C concentrations.

Conclusions

A satisfactory verification of both hydraulic and dispersion characteristics of the model was achieved. It was thus determined that the model was capable of providing quantitative results concerning the effects of the proposed improvement plans on the hydraulic and dispersion characteristics of the bay.

Both plans tested resulted in considerably reduced current velocities throughout most of the bay and an increased flushing rate. In this latter respect, plan 1-C was significantly more effective than plan 6. A net circulation, with net flow into the bay through the existing entrance was generated by both plans tested and is largely responsible for the significant increase in flushing rate.

Technical Aspects of Waste Disposal in the Sea through Submarine Outlets

H. B. Fischer and *N. H. Brooks*†

Les aspects techniques de l'évacuation des déchets dans la mer par décharges sous-marines

Les eaux d'égouts sont couramment évacuées dans les océans par des décharges sous-marines. Dans les systèmes simples, les eaux d'égouts sont déversées au fond de l'océan au moyen de conduits. Ces dernières années, l'usage de diffuseurs à orifices multiples déversant les effluents en une multitude de petits jets d'eau est devenu plus fréquent. Les jets d'eau se combinent pour former un vaste nuage au sein duquel la dilution atteint facilement le taux de 100:1. Une dilution supplémentaire se produit lorsque le nuage d'effluents s'écarte du diffuseur et se mélange avec les eaux de l'océan. S'il existe une stratification des eaux de réception en fonction de leur densité, on peut souvent concevoir un diffuseur qui maintienne le nuage de polluants sous la surface marine.

Le document examine les méthodes permettant de prévoir le taux de dilution d'un déversement sous-marin, ainsi que la diffusion du nuage d'effluents dans l'océan. On y joint des diagrammes de dilution pour les jets d'eau ascendants, ainsi que pour la diffusion du nuage d'effluents à mesure qu'il se déplace vers le rivage. Les auteurs exposent une méthode permettant de prévoir les effets de la stratification des eaux sur la montée du nuage de polluants.

Aspectos técnicos de la eliminación de aguas residuales en el mar mediante desagües submarinos

Las aguas negras contaminadas se eliminan de ordinario mediante desagües submarinos. Los desagües ordinarios descargan las aguas negras por el extremo de conductos situados sobre el fondo del mar. En los últimos años se ha extendido el empleo de difusores de boca múltiple, que descargan las aguas residuales en forma de numerosísimos chorros pequeños. El conjunto de los chorros forma una turbina de aguas residuales, con una dilución inicial que puede llegar fácilmente a 100:1. A medida que la turbina se aleja del difusor y se mezcla con las aguas oceánicas, se produce una dilución ulterior. Si en las aguas que reciben la descarga existe una estratificación por densidades, es posible, a menudo, proyectar el difusor de forma que la turbina permanezca por debajo de la superficie del mar.

En este documento se examinan los métodos empleados para predecir la dilución de un desagüe submarino y la difusión de la turbina en el océano. Se presentan diagramas de dilución teniendo en cuenta los chorros ascendentes y la difusión de la turbina en su avanzar hacia la costa. Se describe un método para predecir el efecto de la estratificación en la ascensión hacia la superficie de la turbina formada por los contaminantes.

* University of California at Berkeley, Berkeley, California 94720, U.S.A.
† W. M. Keck Laboratory of Hydraulics and Water Resources, California Institute of Technology, Pasadena, California 91109, U.S.A.

Les déversements d'eau chaude par les centrales électriques ont également une grande importance pour l'écologie marine. S'ils sont effectués par l'intermédiaire de décharges sous-marines, il est possible de leur appliquer les techniques suivies pour éliminer les eaux d'égouts. Cependant, l'eau chaude est souvent déversée à la surface par des canaux de décharge. Les auteurs passent en revue de récentes études sur la flottabilité de jets d'eau de surface et sur son application à la prévision de la répartition des températures dans les eaux de réception.

También las descargas de aguas calentadas de las centrales eléctricas revisten un importante significado para la ecología marina. Si estas se descargan mediante desagües submarinos, puede aplicarse la tecnología utilizada para la eliminación de las aguas negras. Pero a menudo las aguas calentadas se descargan directamente en la superficie del océano mediante canales de desagüe. En el documento se examinan los últimos estudios sobre chorros superficiales flotantes y su aplicación para predecir la distribución de la temperatura en las aguas que reciben la descarga.

OUTLETS for disposal of wastes at sea must be designed to obtain rapid initial dilution and avoid toxic effects on the marine environment in the immediate vicinity. Ocean currents are relied upon for further dilution and purification. For instance, the Hyperion outfall of the City of Los Angeles discharges approximately 18 cu metres per sec, whereas a current two kilometres wide and 30 metres deep moving at only 10 cm per sec would represent a flow of roughly 6,000 cu metres per sec. Thus if the Hyperion outfall discharge can be distributed to mix uniformly with the typical small current described, a dilution of 350 to 1 can be obtained. After this initial dilution, natural purification of sewage effluent is generally sufficient.

Where deep water and ocean currents are available near the coast, the optimum general plan for disposal of sewage is usually a primary treatment plant connected to a deep-water outfall. Once the sewage enters the ocean there are essentially three stages of turbulent mixing or diffusion:

(1) Initial jet mixing;
(2) Development of a homogeneous sewage field;
(3) Turbulent diffusion of sewage field as a whole due to natural oceanic turbulence.

In stage 1 mixing occurs between the jets of sewage from the diffuser and the surrounding ocean water. In this stage a dilution of as much as 100 to 1 is often obtained. Stage 2 is a transition from the rising jet, or line source, to a diffused cloud being carried by the ocean current; there is no special analysis for this stage. In stage 3, the cloud is carried away from the vicinity of the diffuser and is further diluted by oceanic diffusion. In this stage, sedimentation and bacterial die-away also reduce the concentration of sewage organisms. A total reduction of concentration of indicator organisms of approximately a factor of 10^5 is generally required, of which approximately 10^2 can usually be obtained by jet mixing and an additional 10 by subsequent oceanic mixing. An additional factor of 10^2 must be obtained either by chlorination of the effluent or by natural sedimentation and die-off during transit to the coast.

Since sewage has approximately the density of fresh water it tends to rise when discharged at the ocean bottom whereas ocean water is denser near the bottom than the surface. Since bottom water is entrained in the rising jet the jet often becomes heavier than the surface water, and never reaches the surface.

Design of diffusers for initial dilution

Improvement of sewage effluent dispersal can be accomplished by a multiple jet manifold or diffuser at the end of the outfall. Without such diffusers, other conditions being equal, much longer outfalls into deeper water are necessary. An effective and simple type of diffuser is one which distributes the outflow through many ports over a large are a with minimum head loss. The following discussion presumes a diffuser consisting of one long pipe, or several branching ones, with discharge ports at intervals along the pipes.

Hydraulic design of diffusers—The basic hydraulic requirements for such a multiple-port diffuser, as outlined by Rawn, Bowerman and Brooks (1961), are: (a) the flow distribution should be such that the discharge through each part is approximately uniform; (b) the flow velocity in all parts of the diffuser should be high enough to prevent deposition of particles remaining after primary sedimentation; (c) the design must allow for occasional cleaning; (d) all ports should flow full at all times to prevent intrusion of sea water; and (e) the total head loss in the system should be kept reasonably small. The outlet ports may be circular holes in the side of the pipe without nozzles or tubes or other projecting fittings. The hydraulic analysis in the diffuser is basically that of manifold flow.

Examples of the hydraulic design of outfalls have been given by Rawn, Bowerman and Brooks (1961), Municipality of Metropolitan Seattle (1965), Carollo (1970), and Vigander, Elder and Brooks (1970).

Diffuser loading

For comparison of diffuser pipe length of various outfalls, it is useful to compute the ratio of length in metres to design value of average daily dry-weather discharge in cms (cubic metres per second). Some values of diffuser loading in recent large designs are as follows (with year operation started):

Meters of diffuser/average dry-weather flow
County Sanitation Districts of Los Angeles County
(at Whites Point)
   90" outfall (1956)   730 m/6.7 cms = 110 m/cms
   120" outfall (1965)   1350 m/9.7 cms = 140 m/cms
City of Los Angeles at Hyperion
   144" outfall (1960)   2440 m/18 cms = 135 m/cms
Metropolitan Seattle at West Point
   96" outfall (1965)   180m/5.5 cms = 33 m/cms
County Sanitation Districts of Orange County
   120" outfall (1970)   1800 m/13 cms = 140 m/cms

Initial dilution in the rising jet

Sewage leaving a diffuser rises towards the ocean surface at the same time the jetting action entrains the surrounding water. The rate at which the jet is diluted has been studied experimentally and analytically by Rawn, Bowerman, and Brooks (1961), Abraham (1963, 1965), and Fan and Brooks (1966, 1969).

Dilution graphs—The dilution S_0 at the top of a rising plume formed by a horizontal jet (fig 1) is a function of the following variables (assuming no temperature gradients or currents):

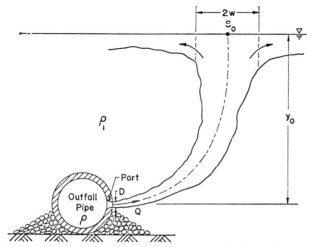

Fig 1. Definition sketch for rising jet of sewage discharged horizontally from a diffuser port into the ocean

y_0 = depth from surface to centre of discharge jet
D = diameter of jet at point of discharge
Q = jet discharge (of sewage effluent)
g' = $\Delta\rho/\rho \, g$ = apparent acceleration due to gravity where $\Delta\rho/\rho$ is the relative density difference, commonly about 0.026

The viscosity does not enter directly, because Reynolds numbers are so large that the flow is fully turbulent.

The four variables may be arranged into two dimensionless groups, namely y_0/D, and

$$F = \frac{Q}{\frac{\pi}{4} D^2 \sqrt{g'D}} = \frac{V}{\sqrt{g'D}} = \text{Froude number} \quad (2.1)$$

where V = nominal mean velocity of discharge. Therefore, the dilution S_0 is a function

$$S_0 = f(y_0/D, F). \quad (2.2)$$

The functional relation can best be shown graphically as in figs 2 and 3. Figure 2, from Rawn, Bowerman, and Brooks (1961), is based entirely on experiments by Rawn

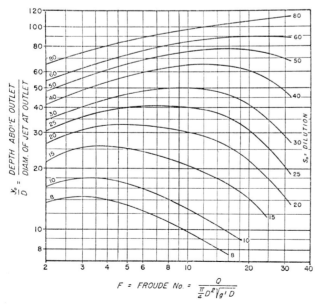

Fig 2. Dilution S_0 on the axis of rising plume at water surface, as function of y_0/D and F, for horizontal round jet into uniform environment (after Rawn, Bowerman and Brooks (1961))

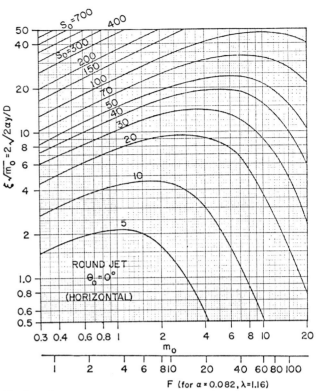

Fig 3. Dilution S_0 on the axis of rising plume, as function of dimensionless height of rise y/D and F, for horizontal round jet into uniform environment (after Fan and Brooks (1969)). Use ordinate $2\sqrt{2} \alpha y/D = 0.23 \, y/D$, based on $\alpha = 0.082$

and Palmer (1930), whereas fig 3 from Fan and Brooks (1966) covers a wider range of values, and is based on theoretical computer solutions. The theoretical analysis by Fan and Brooks (1966) is almost identical to Abraham's (1965), but is simpler and more readily extended to stratified environments.

For ordinates $2\sqrt{2} \alpha y/D$ (= 0.23 y/D for $\alpha = 0.082$ = the entrainment coefficient) greater then 50 (off the top of fig 3), the following asymptotic form for simple plumes in a uniform environment may be used for the centreline dilution:

$$S_0 = 0.092(y/D)^{5/3} F^{-2/3} = \frac{0.078 g'^{1/3} y^{5/3}}{Q^{2/3}} \quad (2.3)$$

Since $F \sim D^{-5/2}$ for given Q, S_0 is actually independent of D in this regime.

A comprehensive collection of numerical solutions to buoyant jet problems has been presented in graphical form by Fan and Brooks (1969) for round and two-dimensional turbulent jets in either uniform or linearly stratified environments. Results include the centreline dilutions, width of the plumes, and trajectories, for various angles of discharge ranging from horizontal to vertical.

The trajectory and jet widths are shown in fig 4 for horizontal round jets. In the ordinate and abscissa it is recommended that the entrainment coefficient $\alpha = 0.082$, or $2\sqrt{2}\alpha = 0.23$. The half-width w is considered to be the 2σ (i.e., two standard deviations of concentration distribution curve across the plume).

Calculations made from the dilution diagrams should not be considered highly accurate, as errors may be of the order of ± 10 to 20%, due to turbulent fluctuations, density stratification, and currents in the ambient fluid.

[466]

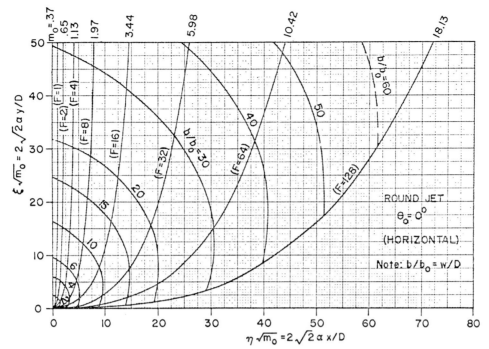

Fig 4. Trajectories and half-widths (w) for horizontal round jet into stationary uniform environment. Use $2\sqrt{2}\,\alpha = 0.23$, based on $\alpha = 0.082$; Froude numbers (F) are also based on $\alpha = 0.082$

The largest number of ports used on an ocean outfall (as far as is known to the authors) is 742, ranging in diameter from 5.1 to 9.2 cm on the fourth ocean outfall of the County Sanitation Districts of Los Angeles (completed in 1965). The overall length of this outfall is 3620 m, including 1350 m of diffuser at depths of 50 to 58 m.

For diffusers with many jets in a row, the effect is that of a line source or slot. For full details on slot jets, the reader is referred to Fan and Brooks (1969). For large depths the asymptotic formula for two-dimensional plumes in uniform environment is:

$$S_0 = 0.38 \frac{g'^{1/3} y}{q^{2/3}} \tag{2.4}$$

wherein q is the discharge per unit length of diffuser pipe, and S_0 is the centreline dilution.

Limiting height of rise of a buoyant plume

When it is desired to predict whether or not a sewage field will be submerged below the surface of the ocean when there is stable density stratification, the problem may be considered either as axisymmetric (point source case) or two-dimensional (line source case). Notation is as follows:

Q_0 = discharge from point source
q_0 = discharge per unit length from line source
ρ_d = density of fluid released at source
ρ_1 = density of ambient fluid at level of source
ρ_0 = density of ambient fluid at height y above source
$\dfrac{d\rho_0}{dy}$ = density gradient
g = acceleration due to gravity
$y_{\max}$ = maximum height of rise of plume

The formulae for the case of the point source are based on the analysis given by Morton, Taylor, and Turner (1956); using the same general procedure, Brooks and Koh (1965) extended the work for line sources.

The following simplifying assumptions are made:

(1) Source is a simple point or line source;
(2) Fluid is released from source with zero initial momentum*;
(3) Variations in ρ are small compared to ρ;
(4) Density gradient of the ambient fluid is constant (i.e., linear density profile);
(5) Plume is turbulent and rate of mixing around edges of plume at height y is proportional to a characteristic velocity and size at that height;
(6) Profiles of density deficiency and velocity are geometrically similar at all heights.

The analysis is based on three fundamental equations: continuity of volume flux and buoyancy flux, and momentum. Using the similarity assumption the authors integrate over the cross-sectional area of the plume at every height y and obtain three simultaneous ordinary differential equations for the three dependent variables:

$u(y)$ = upward velocity on the centreline of the plume
$\rho(y)$ = density on the centreline of the plume
$b(y)$ = nominal radius of the plume

The numerical integration by the authors yields graphs of these functions in dimensionless form.

Application to rising column of sewage effluent (or digested sludge) in sea water—Of special interest is the height of rise of a plume, found by setting the velocity of rise equal to zero, i.e.,

$$u(y_{\max}) = 0.$$

After the plume reaches this height it sinks back to a slightly lower position where its density is in balance with the ambient fluid (approximately at $0.8\, y_{\max}$). Because of

---

* Brooks and Koh (1965) have shown by analyses including initial momentum as an added parameter that this is a reasonable assumption for outfall problems. See also Fan and Brooks (1969).

the momentum of the plume during its rise it "over-shoots" this neutral position.

From the equations of Morton, Taylor, and Turner (1956) and an experimentally determined entrainment constant ($\alpha = 0.093$) it may be shown that:

$$y_{max}^4 = 198 \frac{Q_0 \sqrt{\rho_1}(\rho_1 - \rho_d)}{\sqrt{g}\left|\dfrac{d\rho_0}{dy}\right|^{3/2}}. \tag{2.5}$$

Equation 2.5 may be rewritten using σ, as frequently defined by oceanographers by the equation $\sigma = $ (specific gravity $-$ 1)$\cdot$1000, as

$$y_{max}^4 = 6340 \frac{Q_0(\sigma_1 - \sigma_d)}{\sqrt{g}\left|\dfrac{d\sigma_0}{dy}\right|^{3/2}}. \tag{2.5a}$$

Note that the σ's and the constant 6340 are dimensionless; any consistent set of units may be used for y, Q, and g.

The analysis gives the following value for the terminal dilution (centreline or minimum value):

$$S_t = 0.28 \frac{g^{1/8}(\sigma_1 - \sigma_d)^{3/4}}{Q_0^{1/4}\left|\dfrac{d\sigma_0}{dy}\right|^{5/8}}. \tag{2.6}$$

Line sources. Following the same method of analysis and approximation the maximum height of rise for a line source is

$$y_{max}^3 = 610 \frac{q_0(\sigma_1 - \sigma_d)}{\sqrt{g}\left|\dfrac{d\sigma_0}{dy}\right|^{3/2}}.$$

The corresponding terminal dilution is given by

$$S_t = 0.41 \frac{g^{1/6}(\sigma_1 - \sigma_d)^{2/3}}{q_0^{1/3}\left|\dfrac{d\sigma_0}{dy}\right|^{1/2}}. \tag{2.8}$$

Limitations of buoyant plume formulae

Besides the usual non-linearity of density gradients, which must be approximated by a linear assumption, some other limitations are that: (1) the discharge jet is not vertical with no momentum, but usually horizontal with appreciable initial lateral momentum; (2) ocean currents have not been included in the analysis; and (3) no allowance has been made for major changes in the overall environment due to waste discharges. However, limited field checks show that equation 2.5a works in spite of its apparent limitations.

Fan (1967) presents analytical solutions for horizontal buoyant jets in a stratified environment without current. With horizontal jet discharge the height of rise is reduced somewhat below that indicated for simple plumes; hence using the simple plume formula is on the conservative side. Also treated by Fan is the case of a vertical jet into a uniform current, which shows that even very weak currents can bend a rising plume and carry it long distances before it rises to the surface. The case of a buoyant jet in a stratified current has not yet been solved.

Prediction of coastal waste concentrations

When waste is discharged from a submarine outfall three processes reduce the waste concentration. The first is entrainment of ocean water in the rising jet. After the jet

has reached the surface, or its equilibrium position if it remains submerged, it is carried along by the current and further diluted by turbulent mixing. Finally, coliform concentrations are reduced by mortality and sedimentation. The reduction in coliform concentrations due to physical dilution by unpolluted ocean water and due to die-off are assumed to be entirely independent, and the resulting concentration is the product of the two effects.

Predicting the rate at which dilution will occur near the coast is among the most difficult problems faced by engineers. The rate of dilution depends on coastal currents, which are difficult either to predict or to measure adequately, and also on turbulent diffusion, a process which is ill understood. Theoretical and experimental studies have been made to describe turbulent diffusion in the open sea, where the boundaries are sufficiently far away not to affect the process, and in simple open channel flows, where the turbulence is entirely generated by boundary shear. Along the coast, however, where flow and turbulence are generated by tides, wind, and currents induced from the open sea it seems unlikely that any general analytical solution will be found adequate.

For the specific case in which a current carries a sewage cloud from the vicinity of a diffuser towards the coast, the coliform concentration to be expected at the coast can be estimated by the following procedure:

Let

c_0 = coliform concentration in the discharge from the outfall

Q = total discharge from the outfall

b = length of the outfall

S_a = average dilution obtained by initial jet mixing, normally about half the maximum obtained from a dilution diagram as explained in the previous section

If there is no current above the outfall a cloud of sewage will be formed which, in theory, will increase in thickness indefinitely. However, in all cases of practical interest a current will be flowing which will tend to carry off the cloud. If we assume that the cloud and the current move with the same velocity the initial thickness of the cloud can be determined by an equation of continuity of mass flux (assuming steady flow), in the form,

$$h = \frac{QS_a}{ub_n} \tag{3.1}$$

in which

h = initial thickness of sewage cloud

u = current velocity

b_n = length of outfall perpendicular to the current (see definition of notation in fig 5)

This equation applies so long as the current is sufficient to produce a cloud thickness significantly less than the total depth; if the cloud thickness is as much as half the depth the mechanics of the rising jet will be altered, and the computation of S_a must be revised.

Turbulent mixing will act to increase the dimensions of the cloud both horizontally and vertically, and to smooth out any differences in concentration which may have resulted from incomplete mixing of the flow from adjacent jets of the outfall. If the vertical depth of the

[468]

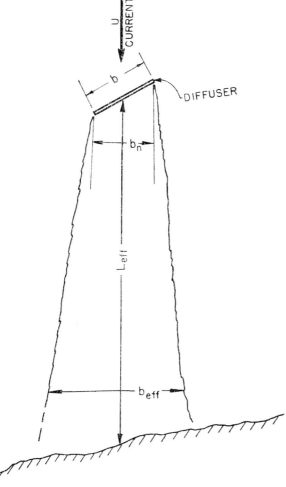

Fig. 5. Definition of symbols for a cloud of sewage being carried towards a coast

cloud, h, as given by equation 3.1, is greater than one-quarter the total depth, or if there is stratification, turbulent mixing may not greatly increase the depth during transit to the shore. If, on the other hand, equation 3.1 yields a small value of h this will probably be increased during the shoreward transit. The value of h which is important in predicting the coastal concentration is that which is found at the coast, denoted hereafter by h_{eff}.

Brooks (1960) has given a detailed method for estimating the increase in cloud width, based on Richardson's 4/3 law for turbulent mixing. Although Brooks' method has worked well in a number of practical applications, there is no assurance that the 4/3 law holds in coastal waters. Harremoes (1967), for instance, has described a number of experiments off the coasts of Denmark and Sweden in which the results were not appropriately described by the 4/3 law. A simpler approach which agrees equally well with the results of Harremoes' experiments, is to assume that the line marking the edge of the cloud diverges on each side from the current direction by an angle of approximately 5°. According to this approach the width of the cloud at the coast, denoted by b_{eff}, is given by (see fig 5)

$$b_{\mathrm{eff}} = b_n + 0.15\,L_{\mathrm{eff}} \qquad (3.2)$$

where L_{eff} = distance from diffuser to coast in direction of current.

If one assumes that within its boundaries the cloud is reasonably homogeneous equation 3.2 gives a dilution due to turbulent mixing not greatly different from the result of Brooks' analysis. Since the dilution factor obtained by this portion of the process is only about 2, its exact value is not of great importance. Whether in actual fact the boundaries of the cloud would be straight, or curved parabolically inward, as would result from a constant Fickian diffusion coefficient, or parabolically outward as given by the 4/3 law, probably depends on local conditions which cannot yet be predicted accurately.

While the cloud is in transit from the outfall to the coast coliform die-off is occurring at a rate assumed proportional to local concentration, i.e., a first order decay. Thus the concentration of coliforms at any time due to die-off alone is given by the formula

$$C = C_0\,e^{-kt} = C_0\,10^{-t/T_{90}} \qquad (3.3)$$

where t = time
 C_0 = concentration at $t = 0$
 T_{90} = time for a 90% decrease in concentration

Observed values of T_{90} in the ocean have been summarized by Gunnerson (1959); typical values are from 3 to 5 hours. This means that one order of magnitude of concentration decrease will occur for each 3–5 hours of transit time from outfall to coast, or in other words the concentration would be decreased by a factor of 100 after 6–10 hours, 1000 after 9–15 hours, etc.

The concentration of coliforms actually observed at the coast depends, as previously stated, on the product of reduction due to die-off and due to dilution. Combining equations 3.1, 3.2 and 3.3 yields

$$C_m = C_0\,\frac{Q}{u b_{\mathrm{eff}} h_{\mathrm{eff}}}\,10^{-L_{\mathrm{eff}}/T_{90}u} \qquad (3.4)$$

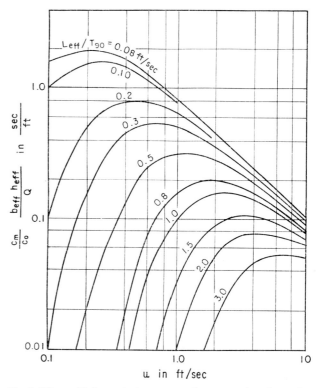

Fig 6. Effect of inflow velocity on coastal concentration, for various values of the decay parameter L_{eff}/T_{90}

in which C_m is the maximum concentration at the coast (assumed constant throughout the effective width of the cloud), and the other symbols are as previously defined. This equation requires on site measurements of current velocity and direction. Inspection of the equation shows that as the current increases reduction of concentration by dilution is increased, whereas reduction by die-off is decreased. Figure 6 is a plot of equation 3.4 for various values of the parameter L_{eff}/T_{90}. For each value of this parameter one velocity can be seen to yield the highest coastal concentration. It should be noted, however, that this is true only if h_{eff} does not depend on the current velocity. The original depth of the cloud, h, will certainly be smaller as u becomes greater, as shown by equation 3.1. However, it seems likely that the depth of the cloud on reaching the coast, h_{eff}, probably depends in large part on the ambient stratification of the receiving water, and is probably constant over a significant range of currents.

In practice a coastal concentration should be computed from equation 3.4 for each of a number of current directions and velocities, and the results weighted according to current observations obtained at the site. Judgement should also be applied to assure that the results are reasonable, particularly with respect to the use of equation 3.2 to predict the lateral spread of the cloud. Currents, particularly between the outfall and the coast, should be observed in as much detail as possible. In general, it seems likely that the analysis given above will produce conservative results, if only because it is physically impossible for a steady current to flow directly from the outfall all the way to the shore.

Disposal of warm water discharges

With the construction of many large electric generating stations along the coasts, considerable thought has been given to disposal of warm water from cooling systems. Tamai, Wiegel, and Tornberg (1969) have estimated that the usual rise in cooling water temperature is between 10° and 20°F, with an average usage of approximately 2.5 cubic metres per minute per megawatt of installed capacity.

Research into warm water discharges has for the most part been directed towards preventing recirculation of warm water into the cooling water intakes. Equally important is the effect of the warm water on the coastal environment ecology. There are two possible strategies towards disposal: (a) surface spreading which minimizes mixing between the warm and cool waters; or (b) use of jet mixing and sub-surface diffusers to obtain as much mixing as possible to reduce temperature difference.

The two approaches may thus be compared. With surface-spreading the rate of heat transfer to the atmosphere is maximized; by avoiding mixing, the volume of water which has a higher-than-normal temperature is minimized and recirculation to the intake is more easily avoided. Whereas with extensive mixing of the hot-water discharge into an ocean current, the temperature rise above the background level is minimized, although the volume of water affected by a temperature rise is greatly increased. Furthermore, the rate of heat loss is reduced, because the temperature differential at the air-water interface is reduced and the surface area is not likely to be increased by a proportional amount.

The specification of only the maximum allowable

temperature rise (Δt) or maximum temperature is probably not sufficient for regulating thermal pollution effects. For example, there may be instances where the ecological impact of a small surface-warm area is less than that for widespread diffusion of heat throughout an entire bay or estuary. Whether in a particular case warm water disposal should be by an undersea diffuser or by surface spreading depends on both engineering and ecological considerations. Because of its very great economy, surface water spreading has almost always been used in coastal installations. In inland installations, however, such as the Browns Ferry site of the Tennessee Valley Authority described by Harleman (1969) and Vigander, Elder and Brooks (1970), a maximum temperature rise specification sometimes requires use of a diffuser. If the mixing is not restricted by shallow depths, the mechanics of spreading from a diffuser are the same for warm water as for sewage, and the equations 2.1 to 2.8 can be applied without change.

Mechanics of spreading a surface warm water discharge

The spreading of a surface discharge may be divided into two zones: (1) a zone of jet entrainment, and (2) a zone in which the discharge is carried with the recipient water and in which the spreading is in part determined by the turbulence and convective patterns in the receiving water. As before, a short transition zone exists between the two. Almost all of the analytical and laboratory research to date has concerned the jet mixing zone; little is known about the interaction between a slightly buoyant surface layer and turbulence in the underlying flow.

The characteristics of a surface buoyant jet have been studied in the laboratory (Jen et al., 1964; Wiegel et al., 1966; Hayashi and Shuto, 1967; Tamai et al., 1969). The dimensionless number which primarily characterizes the flow pattern is the densimetric Froude number,

$$F = \frac{v}{\sqrt{g'D}} \qquad (4.1)$$

in which the notation is as given for 2.1 except that D is the hydraulic radius for flow in a rectangular outlet channel. Laboratory studies have been conducted for Froude numbers in the range 2.5 to 180. For higher Froude numbers the jet resembles a momentum jet with no density difference, but for lower Froude numbers buoyancy causes the jet to spread laterally as a surface flow and reduces the vertical mixing. Thus entrainment of the recipient water is greater for high Froude numbers, whereas a low Froude number discharge forms a surface warm water layer. The range of Froude numbers in prototype installations, as listed by Tamai, Wiegel and Tornberg (1969), range from 2.0 to 15. Field data on temperature distributions in the receiving water are, however, not yet adequate to yield a quantitative relationship between Froude number and flow pattern.

If a warm water surface layer is produced by the discharge, it will be carried on top of the receiving flow and will eventually be dissipated. In most instances much of the reduction in temperature is accomplished by mixing with cooler water; for instance, Hoopes, Zeller and Rohlich (1968) report that in warm water discharges into Lake Monona, Wisconsin, surface heat loss from the surface jet into the atmosphere was negligible compared to dilution of heat in the lake within the initial mixing zone.

Summary

Disposal of waste hot water may be accomplished with the intent either to obtain maximum or minimum mixing with the surroundings. Maximum mixing reduces the excess temperature by the greatest amount but affects the largest area of receiving water, whereas minimum mixing maintains the higher discharge temperature but confines the affected area. In most cases minimum mixing makes less likely recirculation of hot water into cooling water intakes and speeds up transfer of heat to the atmosphere. On the other hand, a smaller area of hotter water may be more damaging to the ecology of the receiving water than a larger area of less hot water.

Conclusions

During the past twenty years significant advances have been made in the technology of waste disposal at sea. Multi-port submarine diffusers can be used to obtain initial dilutions of the waste cloud of the order of 100 to 1. Furthermore, if a density stratification exists in the receiving water the diffuser can be designed so that the waste cloud always remains below the surface. These concepts have been used successfully in the design of a number of diffusers along the Pacific Coast of the United States, and the dilutions predicted by the analysis have generally been either achieved or bettered.

References

ABRAHAM, G Jet diffusion in stagnant ambient fluid. *Publs Delft*
1963 *Hydraul. Lab.*, (29).

ABRAHAM, G Horizontal jets in stagnant fluid of other density.
1965 *J. Hydraul. Div. Am. Soc. civ. Engrs*, 91(HY4):139–54 (Closing discussion, 93(HY1):63–8, Jan. 1967).

BROOKS, N H Diffusion of sewage effluent in an ocean current.
1960 *In* Proceedings First International Conference on Waste Disposal in the Marine Environment, University of California, 1959. New York, Pergamon Press.

BROOKS, N H and KOH, R C Y Discharge of sewage effluent from
1965 a line source into a stratified ocean. *Proc. Congr. int. Ass. Hydraul. Res.*, Pap.(2.19).

CAROLLO, J Engineers (Lafayette, Calif.). Final report on design
1970 of ocean outfall. No. 2 for County Sanitation Districts of Orange County, P.O. Box 8127, Fountain Valley, Calif.

FAN, L N Turbulent buoyant jets into stratified or flowing ambient
1967 fluids. *Rep. W. M. Keck Lab. Hydraul. Wat. Resourc. Calif. Inst. Tech.*, (KH-R-15).

FAN, L N and BROOKS, N H Discussion of horizontal jets in
1966 stagnant fluid of other density. *J. Hydraul. Div. Am. Soc. civ. Engrs*, 92(HY2):423–9.

FAN, L N and BROOKS, N H Numerical solutions of turbulent
1969 buoyant jet problems. *Rep. W. M. Keck Lab. Hydraul. Wat. Resour. Calif. Inst. Tech.*, (KH-R-18).

GUNNERSON, C G Sewage disposal in Santa Monica Bay,
1959 California. *Trans. Am. Soc. civ. Engrs*, 124:823–51.

HARLEMAN, D R F Mechanics of condenser water discharge from
1969 thermal-power plants. *In* Engineering aspects of thermal pollution, edited by F L Parker and P A Krenkel, Vanderbilt University Press.

HARREMOES, P Theoretical treatment of data on turbulent dis-
1967 persion related to disposal of industrial wastes. Reports Nos. 1 and 2 for Research Contract #402RB. Copenhagen, Danish Isotope Center.

HAYASHI, T and SHUTO, N Diffusion of warm water jets discharged
1967 horizontally at the water surface. *Proc. Congr. int. Ass. Hydraul. Res.*, 12, vol. 4:47–59.

HOOPES, J A, ZELLER, R W and ROHLICH, G A Heat dissipation
1968 and induced circulations from condenser cooling water discharges into Lake Monona. *Rep. Engng exp. Stn Univ. Wisc.*, (35):213 p.

JEN, Y, WIEGEL R L and MOBAREK, I Surface discharge of
1964 horizontal warm water jet. *Tech. Rep. Hydraul. Engng Lab., Univ. Calif.*, (3–3).

MORTON, B R, TAYLOR, G I and TURNER, J S Turbulent
1956 gravitational convection from maintained and instantaneous sources. *Proc. Roy. Soc. (A)*, 234:1–23.

Municipality of Metropolitan Seattle. Disposal of digested sludge
1965 to Puget Sound, the engineering and water quality aspects. Seattle, Washington.

RAWN, A M and PALMER, H K Predetermining the extent of a
1930 sewage field in sea water. *Trans. Am. Soc. civ. Engrs*, 94:1036.

RAWN, A M, BOWERMAN, F R and BROOKS, N H Diffusers for
1961 disposal of sewage in sea water. *Trans. Am. Soc. civ. Engrs*, 126(3):344–88.

TAMAI, N, WIEGEL, R L and TORNBERG, G F Horizontal surface
1969 discharge of warm water jets. *J. Power Div. Am. Soc. civ. Engrs*, 95(PO2):253–76.

VIGANDER, S, ELDER, R A and BROOKS, N H Internal hydraulics
1970 of thermal discharge diffusers. *J. Hydraul. Div. Am. Soc. civ. Engrs*, 96(HY1):509–27.

WIEGEL, R L, MOBAREK, I and JEN, Y Discharge of warm water
1966 jet over sloping bottom. *In* Modern trends in hydraulic engineering research, Golden Jubilee Symposia, Vol. 2. Poona, Central Water and Power Research Station.

Full-scale Experiments on Disposal of Waste Fluids into Propeller Stream of Ship

G. Abraham, W. D. Eysink,* G. C. van Dam, J. S. Sydow** and K. Müller***

Expériences en grandeur réelle sur le déversement de déchets liquides dans le sillage des navires

Un navire de décharge d'environ 1.000 tonnes, spécialement conçu à cet effet, déverse quotidiennement quelques 1.500 tonnes de déchets acides dans la mer du Nord, sur la côte des Pays-Bas. Pour obtenir un degré initial élevé de mélange, ces déchets sont déversés dans le sillage du bateau. Le taux de dilution obtenu a été déterminé par des mesures effectuées en mer pendant les voyages de routine du navire de décharge, la seule variante par rapport à la procédure normale étant l'addition de Rhodamine-B aux déchets acides pour faciliter les mesures. On a mesuré la dilution de la Rhodamine-B—et partant celle des déchets acides—jusqu'à une distance de 800 mètres sur l'arrière du navire de décharge. Le document communique les résultats de ces expériences en grandeur réelle, effectuées pendant l'été de 1969, avec des considérations théoriques et des expériences sur modèles.

Experimentos en gran escala sobre la eliminación de fluidos residuales en la corriente producida por la hélice de los barcos

Diariamente se arrojan al Mar del Norte a lo largo de la costa de los Países Bajos unas 1.500 toneladas de ácidos contra los residuos desde un barco de descarga de unas 1.000 toneladas, construido especialmente con este objeto. Para obtener un grado elevado de mezcla inicial, el ácido se echa en la corriente producida por la hélice del barco. Para determinar la dilución obtenida se efectuaron mediciones en el mar durante los viajes normales del barco de descarga, siendo la única variación del procedimiento de descarga normal el hecho de que se añadió Rodamina-B al ácido contra los residuos con fines de medición. La dilución de la Rodamina-B—y, por tanto, la dilución del ácido residual—se midió hasta una distancia de 800 m por detrás del barco. Se presentan los resultados de estas mediciones en gran escala, realizadas durante el verano de 1969, junto con consideraciones teóricas y experimentos modelo

ABOUT 1,500 tons of waste acid per day is discharged into the North Sea off the Netherlands coast from a specially designed vessel of about 1,000 tons. The waste acid is discharged into the propeller stream to obtain maximum initial dilution. To determine the dilution achieved, model investigations and measurements at sea during routine trips of the discharge vessel were made.

---

* Delft Hydraulics Laboratory, Delft, Netherlands.
** Directorate for Water Management and Water Research, The Hague, Netherlands.
*** Farbenfabriken Bayer, A.G., Leverkusen, Federal Republic of Germany.

As a ship progresses water is jetted away from the propeller. For a sea-going vessel of about 1,000 tons the discharge of the propeller flow is about 10 m³/sec. When a small quantity of waste acid is discharged into the propeller stream some initial dilution takes place close to its point of discharge. Given the discharge of the propeller stream, the initial dilution may be determined as

$$S_0 = \frac{Q_{pr} + Q_w}{Q_w} \qquad (1)$$

where S_0 is the rate of initial dilution; Q_{pr} is the initial discharge of propeller stream, as measured from a moving ship; and Q_w is the quantity of waste effluent discharged into propeller stream per unit time.

The water of the propeller stream has both kinetic energy due to the momentum supplied by the propeller and potential energy because the waste acid is denser (ρ_w about 1,250 kg/m³) than the surrounding sea water (ρ_a about 1,025 kg/m³). The kinetic and potential energy causes some further dilution of the waste acid. Accordingly, distinction must be made between the following conditions:

If waste effluent denser than sea water is discharged into homogeneous receiving sea water (case 1), the water of the propeller stream will sink until it reaches the bottom of the sea (fig 1A).

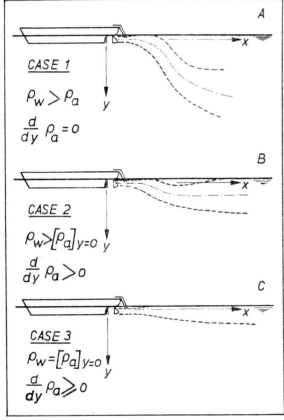

Fig 1. Further dilution patterns

If waste effluent denser than sea water is discharged into sea water having a vertical density gradient (case 2), the water of the propeller stream will sink until it reaches the level where it becomes as dense as the surrounding sea water (fig 1B). Due to the high degree of initial dilution of the waste acid (1 part of waste acid being mixed with about 100 parts of water in the propeller stream) and due to the further dilution caused by the momentum of the propeller stream, a difference in density in vertical direction in the order of 1/2 kg/m³ is sufficiently strong to keep the water of the propeller stream from sinking further.

If waste effluent of the same density as sea water is discharged into the propeller stream (case 3), the propeller stream remains near the surface of the sea (fig 1C).

The full-scale experiments described in this paper were performed under the conditions of case 2.

Knowing the initial discharge of the propeller stream and its initial diameter, the problem of determining the further dilution can be reduced to one of jet diffusion. In the first approximation an observer moving with the

Fig 2. Flexible discharge structure of Käthe H

same velocity as the ship may describe the propeller stream as a jet, issuing into water flowing with the same velocity as the ship. The further dilution may then be treated as the problem of a jet, characterized by the following parameters

$$D \approx D_{pr} \quad (2) \qquad\qquad U_0 \approx \frac{Q_{pr}}{D_{pr}^2} \quad (3)$$

$$U_a = U_s \quad (4) \qquad (\rho_0 - \rho_a) = \frac{1}{S_0}(\rho_w - \rho_a) \quad (5)$$

where D is the diameter of the jet at the nozzle; D_{pr} is the diameter of the propeller; U_0 is the velocity of the jet at the nozzle; U_s is the velocity of the ship; U_a is the velocity of the fluid into which jet issues, having same direction as U_0; ρ_0 is the density of the jet fluid at the nozzle; ρ_a is the density of the receiving fluid; and ρ_w is the density of the effluent.

Thus far the problem has been studied by considering the effects of the kinetic energy and the potential energy separately. When only the kinetic energy is studied the conditions are as in case 3 or case 2 provided the level where the propeller stream stops sinking is close to the water surface. When only the potential energy is studied the conditions approximate to case 1, provided that the depth of water is sufficiently great that when the waste reaches the bottom the effect of the kinetic energy can be neglected.

Jets having only kinetic energy in the direction of flow of the ambient fluid have been studied by, amongst others, Reichardt (1964) and Hinze (1959). These studies were made for jets that issue into an infinite space, in which the jet can entrain fluid from all sides. At great distances from the ship the diameter of the propeller stream becomes so great that the entrainment of ambient fluid is hampered because no fluid can be entrained through the water surface. In this case the further dilution in the propeller stream is about half as great as the dilution in the jet that issues into a infinite space.

By neglecting the effect of kinetic energy, the propeller stream can be treated as a buoyant thermal (Abraham and Hilberts, 1967).

Full-scale experiments

The tanker *Käthe H.* (about 1,000 tons) was used for the full-scale experiments. Waste acid can be pumped at a maximum rate of 0.2 m³/sec into the propeller stream through a flexible tube, which can be submerged in the water so that the waste acid is released below the water surface (see fig 2). The ship is designed to dispose of the acid while moving at about 12 kn.

For the experiments a solution of iron sulphate in sulphuric acid (S.G. range 1.18 to 1.28) was released. Known amounts of the fluorescent dye Rhodamine-B were added to the waste acid. By measuring the fluorescence of the mixture of Rhodamine-B and sea water, the dilution of the dye and therefore the acid can be determined. Preliminary investigations by Sydow, Minderhoud, and Hulshof (1969) showed that the fluorescence of Rhodamine-B in sea water was unaffected by the waste acid provided the dilution was greater than 1,000 to 1,500.

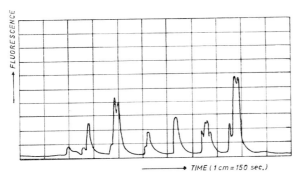

Fig 3. Example of fluorescence registration

Measurements were taken aboard a coaster of about 500 tons. The coaster towed a paravane and a rigid structure through the sea water. Flexible tubes with inlet openings at different depths, were attached to the suspension cable of the paravane and the rigid structure. These were connected to a pump aboard the measuring vessel. It was thus possible to take continuous samples of sea water at depths from 1 to 12 m.

Samples taken continuously at different depths were pumped through continuously registering fluorimeters

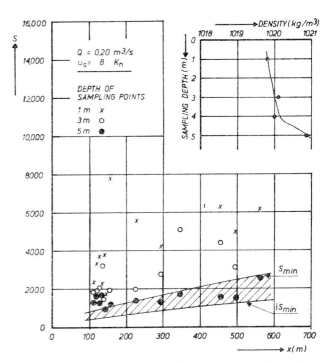

Fig 4. Dilution of acid in vertical plane of symmetry of propeller stream. Experiment 9

and conductivity meters. By means of thermometers and the conductivity meters the density of the sea water could be determined as a function of the distance from the water surface.

Both the measuring vessel and the discharge vessel were equipped to determine the positions of both ships relative to each other.

During an experimental run of about 1 h the two vessels sailed in a straight course at constant speed and the rate of discharge of the waste acid and Rhodamine-B

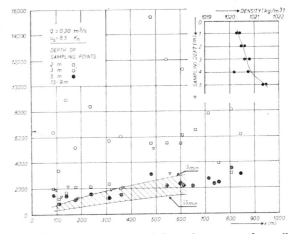

Fig 5. Dilution of acid in vertical plane of symmetry of propeller stream. Experiment 11

was kept as constant as possible. The distance between the ships was varied (more or less stepwise) during the experiments. The measuring vessel sailed in a zig-zag course, therefore the sampling tubes moved through the propeller stream at different depths perpendicular to its axis. Thus the fluorescence of the water pumped through the fluorimeters was greatest in samples taken when the inlet point of the sampling tube passed the vertical plane of symmetry of the propeller stream. From the maximum Rhodamine-B content that corresponded to the fluorescence observed during the specific crossing of the propeller stream, the dilution of the waste acid at the vertical plane of symmetry of the propeller stream could be determined for various distances between the ships and various depth levels.

While the dilution of the waste acid was thus measured all parameters relevant for a theoretical analysis were measured as well.

Results achieved

Figure 3 shows a typical example of the variation of the fluorescence in the water of the propeller stream obtained while the measuring vessel sailed on a zig-zag course. As explained above the dilution of the waste acid at the vertical plane of symmetry of the propeller stream was determined from the rhodamine B content corresponding to the peak values of the fluorescence shown in fig 3.

For each experiment the dilution at the vertical plane of symmetry was determined as a function of the horizontal distance behind the discharge vessel and the vertical distance from the water surface. Typical examples of the results obtained are given in figs 4 and 5. Each point represented in figs 4 and 5 corresponds to a peak value of the fluorescence as represented in fig 3. Figure 5 shows that the dilution rate of the acid at a depth of 3 m was greater than at depths of 2 m and 5 m. This does not seem to be realistic.

Figures 4 and 5 show that at a distance of about 5 m from the water surface the vertical variation of the density of the receiving sea water was sufficiently strong to keep the propeller stream from sinking further. This implies that the propeller stream sank such a small vertical distance that the effect of potential energy on further dilution could be neglected. Therefore, further

dilution observed during the experiments is compared with values calculated from the theory for jets having kinetic energy only. Jet theory allows calculation of S_{MIN}, i.e. the smallest rate of dilution, which occurs at the axis of the jet or at the axis of the propeller stream. Calculated values of S_{MIN} are indicated in figs 4 and 5 as a function of x, the distance behind the ship.

Close to the ship the observed minimum dilution tends to coincide with the minimum dilution calculated from theory. Farther from the ship the observed minimum dilution is approximately half of the calculated minimum dilution. This is due to the influence of the free water surface on entrainment, as explained earlier.

Conclusions

(1) The flow phenomenon within the propeller stream into which waste effluent is discharged must be classified as indicated in fig 1 and as described in text.

(2) The full-scale experiments described show that the dilution within the propeller stream can be determined theoretically by jet diffusion theory for flow patterns of case 3 (fig 1C) and of case 2 (fig 1B), provided that for case 2 the propeller stream sinks such a small vertical distance that the effect of potential energy on dilution of the waste effluent can be neglected.

References

ABRAHAM, G and HILBERTS, B Vermischung von Abfallsäure im
1967   Propellerstrahl eines Küstenmotorschiffes. *Publs Delft. Hydraulics Lab.*, (51).
HINZE, J O Turbulence—an introduction to its mechanism and
1959   theory. New York, McGraw-Hill, pp. 404–31.
REICHARDT, H Turbulente Strahlausbreitung in gleichgerichteter
1964   Grundströmung. *Forschr. IngWiss.*, 30(5):133–64.
SYDOW, J S, MINDERHOUD, J and HULSHOF, J E Invloed afvalzuur
1969   Titaanwitfabrieken op rhodamine B. *Rep. Rijkwaterst. Div. Math. Phys.*, (690219).

The Disposal of Containers with Industrial Waste into the North Sea: A Fisheries Problem

*G. Berge, R. Ljøen and K. H. Palmork**

Le rejet de récipients contenant des déchets industriels dans la mer du Nord: un problème pour la pêche
Des récipients contenant divers déchets de l'industrie chimique ont été repêchés par des chalutiers en mer du Nord. Ce problème est grave étant donné le grand nombre des récipients ainsi repêchés et la toxicité de la plupart des produits qu'ils contenaient. Les auteurs examinent l'utilisation de la mer du Nord pour ce genre d'évacuation.

Un problema para la pesca: la evacuación de recipientes con desechos industriales en el Mar del Norte
En el Mar del Norte los arrastreros están recogiendo envases con distintos contenidos de desechos procedentes de la industria química. La gravedad de este problema se demuestra por el gran número de recipientes recogidos y la toxidad de casi todos los productos contenidos en ellos. Se examina el uso que se hace del Mar del Norte al arrojar a sus aguas tales recipientes.

BECAUSE of increasing public concern over national resources, land tipping and incineration are not acceptable methods for disposing of many noxious industrial by-products, so an increasing amount of industrial waste is dumped in the sea—usually in international waters where no laws prohibit the action.

The nature of the waste often precludes pumping so iron drums are used similar to those used for radioactive waste. Often the containers are weighted by concrete to ensure sinking. However, the disposal sites are not always carefully selected and a considerable number of containers have been recovered by trawlers on

North Sea fishing grounds. Ten years ago such captures were infrequent but recently the captures became regular, even eight a day.

Five randomly selected containers were analysed. The containers were all 200 l oil drums, and contained two types of material:

(a) One drum contained a black, tarry substance, rich in nitrogen (nitriles and amides) and partly water soluble. The acute lethal doses for small flatfish were ~100 ppm. The chemical results indicated that the content originated from thermoplastic industry. It was possible, through

---

* Institute of Marine Research, Bergen, Norway.

[474]

examination of the different rocks and minerals (jaspis and chalk ooliths, Jurassic?) to trace, from the concrete in the bottom of the containers, the origin of these minerals (Main or Weser in Germany).

(b) Four drums contained a green greyish slurry. Gas chromatographic analysis of the contents using electron capture detection (ECD) revealed that there were more than twenty different halogenated hydrocarbons present (fig 2). The analyses (see fig 3) indicated that the waste originated from the petrochemical industry, e.g. PVC. 1.2-dichloropropane was lethal to flounder and pollock at 50 ppm.

According to reports by fishermen other types of waste have been recovered from the North Sea but these have not yet been analysed by our Institute.

A list of petrochemical factories in Northern Europe, including their annual waste production and methods of disposal, was obtained with the help of two newspaper journalists. One of the larger firms involved reported that probably 10,000 containers had been dumped. The procedure for such dumpings was as follows: The waste was taken by a contractor cooperating with a shipbroker who supplied the vessel for transport. An area north of 65°N in at least 2,000 m was commonly used (fig 1). The ship's captain subsequently reported back to the shipbroker enclosing a copy of the log. Because of the number of captures made on the fishing grounds south of this area, some ships evidently do not follow the instructions, or possibly some contractors actually use the North Sea area for dumping.

Discussion

Dumping in the North Sea and on the continental shelf off Norway is not advisable for several reasons. Most of the area is fished by bottom trawlers and the capture of containers definitely hampers fishing. About half of the captured containers were leaking and might be hazardous to fishermen. Destroyed gear, or catch, is a direct economic loss. Of greater concern is the eventual release

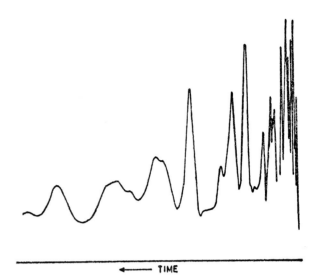

Fig 2. Gas chromatrogram of content of type b container (Electron capture detector)

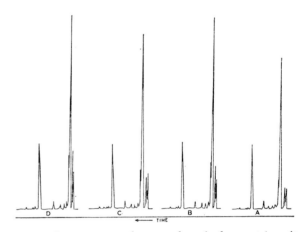

Fig 3. Gas chromatograms of contents from the four containers (type b) captured in different areas (Flame ionization detector)

of the contents into the bottom water. Important fishing resources might be damaged. The main current system can transport the compounds into Skagerrak (fig 1). By vertical mixing the compounds can reach the surface layers and the compounds could enter the important nursery areas for the fish populations of the North Sea and the Norwegian Sea, and damage the plankton and the fish larvae. The chlorinated aliphatic hydrocarbons are similar in fat webs to DDT and PCB, and might be concentrated in organisms. The effects of small concentrations in organisms are not well known, but some petrochemical waste products are known to be carcinogens (IMCO et al., 1970).

Conclusion

As a rule, the sensitivity of the larvae is much greater than that of the adults. Contamination of the fish might result in loads on the eggs which, as a consequence, could damage the populations through increased mortality at the larval stage.

Reference

IMCO/FAO/UNESCO/WMO/WHO/IAEA, Joint Group of Experts on the Scientific Aspects of Marine Pollution (GESAMP), Report of the second session, Paris, Unesco, 2–6 March 1970.
1970

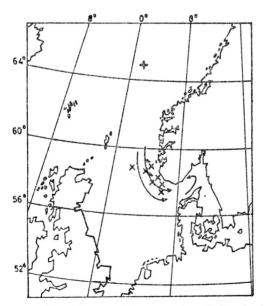

Fig 1. Localities of captured waste containers (×) and a known dumping site (+). Superimposed, the current system near the slope

[475]

Oil Contamination and the Living Resources of the Sea* *Max Blumer†*

La contamination par les hydrocarbures et les ressources biologiques de la mer

La quantité de pétrole brut et d'hydrocarbures déversés dans l'océan est estimée à 5–10 millions de tonnes par an. Dans l'environnement côtier où les ressources biologiques de la mer sont exposées à de très graves menaces, la pollution par les produits pétroliers vient s'ajouter à d'autres facteurs d'agression: eaux usées, insecticides, produits chimiques, surpêche et assèchement des terres humides.

Tous les hydrocarbures bruts sont des poisons, mais de nombreux produits de la distillation du pétrole sont encore plus dangereux car ils contiennent davantage de composés immédiatement toxiques. La toxicité à long terme peut affecter la flore et la faune marines que les épanchements ne tuent pas immédiatement, et des hydrocarbures peuvent s'accumuler dans les tissus des animaux marins, les rendant ainsi impropres à la consommation humaine. Le pétrole brut et les hydrocarbures sont cancérogènes pour les organismes marins et l'homme; même à très faible concentration, le pétrole peut perturber le processus vital de propagation des espèces marines.

Les fractions du pétrole qui présentent la toxicité aiguë la plus élevée sont solubles dans l'eau; de ce fait, il est souvent illusoire de chercher à éliminer les nappes d'hydrocarbures, sauf pour des raisons d'ordre esthétique. Le recours aux détergents, même "non toxiques", est dangereux car il expose les organismes marins à de plus fortes concentrations d'hydrocarbures solubles et toxiques; d'autre part, le pétrole est ainsi dispersé en gouttelettes que de nombreux organismes peuvent ingérer et retenir.

Les épanchements de pétrole finissent par être décomposés sous l'action naturelle des bactéries; toutefois, les fractions les plus toxiques sont celles qui disparaissent le plus lentement; en outre, les hydrocarbures sont très stables dans les sédiments et les organismes marins.

La contaminación por hidrocarburos y los recursos vivos del mar

Se calcula que anualmente se derraman en los océanos de 5 a 10 millones de toneladas de petróleo crudo y productos derivados. En el medio ambiente costero, donde más gravemente amenazados están los recursos vivos del mar, la contaminación de petróleo viene a sumarse a la producida por las descargas de aguas negras, de los insecticidas y de los productos químicos. También a este proceso, se agrega la acción de la sobrepesca y del rescate de las marismas.

Todos los petróleos crudos son venenosos, pero muchos destilados de petróleo crudo lo son aún más, dado que contienen proporciones más elevadas de compuestos inmediatamente tóxicos. La toxicidad a largo plazo puede dañar a la fauna y flora marinas que no perecen inmediatamente amenazadas a causa de los derrames de hidrocarburos, y el petróleo puede incorporarse en el organismo de los animales marinos, haciéndolos inadecuados para el consumo humano. El petróleo crudo y sus derivados pueden producir el cáncer en los organismos marinos y en el hombre. Incluso en concentraciones bajísimas, los hidrocarburos pueden interferir en los procesos que son vitales para la propagación de las especies marinas.

Los componentes más tóxicos de los hidrocarburos son hidrosolubles; por tanto, de poco sirve, si se exceptúa el punto de vista estético, eliminar las películas de hidrocarburos que flotan sobre el agua. El tratamiento con detergentes, incluso con detergentes "no tóxicos", es peligroso, porque expone a los organismos marinos a concentraciones más elevadas de hidrocarburos solubles y tóxicos y dispersa el petróleo en gotitas que pueden ser ingeridas por muchos organismos y acumularse en ellos.

La acción natural de las bacterias termina por descomponer los hidrocarburos presentes en el océano; de todas formas, las partes más tóxicas son las más lentas en desaparecer en razón de que los hidrocarburos son muy estables en los animales y sedimentos marinos.

THROUGHOUT history man has used the ocean and especially the coastal waters as a source of food and minerals, for shipping and for disposal of his wastes. The present annual world income from marine fishing is now roughly $8 billion. The world ocean freight bill is nearly twice that. In contrast, mineral recovery has a relatively small value; the world oil and gas production from the seabed is worth approximately half that of the fish catch, and all other mineral production adds only $250 million (Holt, 1969). It is not easy to place a figure on the value of the ocean for recreation or for waste disposal but a healthy ocean may well be essential for the survival of the human species.

It has been estimated that with presently available technology Puget Sound alone could produce annually 6 million pounds of oyster meat, equal in value to the entire present U.S. fish catch (Westley, 1967). Ryther (1969) states that the open sea is a biological desert, which produces a negligible fraction of the present fish catch and has little future potential. The coastal waters produce almost the entire shellfish crop and nearly half of the total fish crop; the remainder comes from regions of upwelling waters that occupy one-tenth of 1 per cent of the ocean surface. Oil and mineral resources and recreational and waste disposal areas are also almost all in the coastal regions of the ocean.

Marine resources—multiple stresses

Different uses of marine resources are often in conflict and are made and planned with little regard for the marine environment as a large interrelated ecosystem. Oil pollution is only one of many unrelated causes which contribute to the deterioration of the environment.

Additional stresses come from the loss of marshland, overfishing, persistent chemicals and domestic or industrial wastes. Many individual actions and even single large stresses can be tolerated but whether this remains true for the sum of the stresses imposed on the environment should be a matter of great concern. The wastes that now enter the ocean are similar to those that have already damaged the Great Lakes and other freshwater lakes and rivers. The ocean differs from the lakes principally in its size and time constant; changes may take much longer to become evident but equally restoration of a polluted ocean will also take a very long time. Lake Erie may or may not be restored within fifty years but a polluted ocean will remain irreversibly damaged for many generations.

Ketchum (1970a) has pointed out "that nature has a tremendous capacity to recover from the abuses of pollution, so long as the rate of addition does not exceed the rate of recovery of the environment. When this limit is exceeded, however, the deterioration of the environment is rapid and sometimes irreversible".

Oil pollution—extent

Oil pollution is the almost inevitable consequence of our dependence on an oil-based technology. Large catastrophes like that of the *Torrey Canyon*, the blow-outs at Santa Barbara and in the Gulf of Mexico get the attention of the public because of the obvious aesthetic damage and the harm to birds. Small and continuing spills and their far greater impact on less visible resources are less apparent to the public. It is estimated that 10,000 pollution incidents occur annually in U.S. waters alone and that oil pollution accounts for 7,500 of these. We have

* This is Contribution No. 2474 of the Woods Hole Oceanographic Institution. Supported by grants from Office of Naval Research, National Science Foundation, and Federal Water Pollution Control Administration.
† Woods Hole Oceanographic Institution, Woods Hole, Massachusetts 02543, U.S.A.

estimated that the present practices in tanker ballasting introduce about 3 million tons of petroleum into the ocean. The pumping of bilges by vessels other than tankers contributes another 500,000 tons. In addition, in-port losses from collisions and during loading and unloading contribute an estimated 1 million tons. (Blumer, 1970)

Oil enters the ocean from many other sources whose magnitude is much less readily assessed; among these are shipping accidents, losses during exploration (oil-based drilling mud), production (e.g. Santa Barbara, Gulf of Mexico), storage (submarine storage tanks) and in pipeline breaks, also spent marine lubricants and incompletely burned fuels. A major contribution may come from untreated domestic and industrial wastes; it is estimated that nearly 2 million tons of used lubricating oil is unaccounted for each year in the United States alone, a significant portion of this reaches our coastal waters (Anon., 1970; Murphy, 1970). Thus, the total annual oil influx to the ocean is probably between 5 and 10 million tons.

Oil—composition and persistence

Petroleum contains thousands of compounds and is one of the most complex natural materials. Different crude oils differ markedly in physical properties, such as gravity, viscosity and boiling point distribution. It is beyond the scope of this paper to describe the crude oil composition more than superficially (see reviews in: Eglinton and Murphy, 1969). Every crude oil contains the same homologous series of closely related compounds. Different crudes differ mainly in the relative contribution of the individual member of these series. Thus, low and high boiling saturated and aromatic hydrocarbons occur in every crude oil and though their numbers may go into thousands, individual members of these series have very similar chemical and biological properties. It follows that in their chemical, biological and toxicological properties crude oils are very similar.

Petroleum and petroleum hydrocarbons in the marine environment are remarkably stable. Hydrocarbons that are dissolved in the water column are eventually destroyed by bacterial attack, though it should be pointed out that the most toxic compounds are also the most refractory ones.

We have demonstrated that hydrocarbons that are ingested by marine organisms can pass through the wall of the digestive tract and can be retained for long time periods (Blumer, 1967; Blumer, Mullin and Guillard, 1970; Blumer, Souza and Sass, 1970). Analysis of the fat of oysters that had been polluted by a fuel oil spill showed that the amount and chemical composition of fuel oil hydrocarbons remained nearly unchanged even after six months in a clean aquarium (Blumer et al., 1970). Hydrocarbons spread through the marine food web in a manner similar to that of other persistent chemicals, e.g. DDT (Blumer, et al., 1969; Blumer, Mullin and Guillard, 1970).

Within marine sediments, hydrocarbons are protected from bacterial degradation, especially if the sediments are anaerobic. Thus in a spill of fuel oil in 1969 at West Falmouth, Massachusetts, U.S.A., oil was incorporated into the sediments of coastal waters, rivers, harbours and marshes. The oil was still present in the sediments one

year after the accident, and transport of oil laden sediment contaminated areas beyond those immediately affected by the spill (Blumer et al., 1970).

Oil—immediate toxicity

All crude oils and oil fractions except highly purified and pure materials are poisonous to all marine organisms. The wreck of the Tampico in Baja California, Mexico, "created a situation where a completely natural area was almost totally and suddenly destroyed. Among the dead species were lobsters, abalone, sea urchins, starfish, mussels, clams and hosts of smaller forms" (North, 1967). Similarly, the spill of fuel oil in West Falmouth, Massachusetts, U.S.A., has virtually extinguished life in a productive coastal and intertidal area (Hampson and Sanders, 1969). Toxicity is immediate and leads to death within minutes or hours (Wilber, 1969).

Three complex fractions are principally responsible for this immediate toxicity. It has been found that the low boiling saturated hydrocarbon fraction, which is rather readily soluble in sea water, produces at low concentration anaesthesia and narcosis and at greater concentration cell damage and death in a wide variety of lower animals. They may be especially damaging to the young forms of marine life (Goldacre, 1968). Benzene, toluene and xylene are acute poisons for man as well as for other organisms; naphthalene and phenanthrene are even more toxic to fishes than benzene, toluene and xylene. These low boiling aromatic hydrocarbons and substituted one, two and three ring hydrocarbons of similar toxicity are abundant in all oils and most oil products. Low boiling aromatics are more water soluble than the saturates and can kill marine organisms either by direct contact or in dilute solutions. Olefinic hydrocarbons, intermediate in structure and properties and probably in toxicity between saturated and aromatic hydrocarbons, are absent in crude oil but occur in many refining products, e.g. gasoline and cracked products.

Numerous other components of crude oils are toxic, among those named by Speers and Whitehead (1969) cresols, xylenols, naphthols, quinoline and substituted quinolines and pyridines and hydroxybenzoquinolines are of special concern because of their great toxicity and solubility in water.

It is unfortunate that statements are still circulated which disclaim this toxicity. Simpson (1968) claimed that "there is no evidence that oil spilt round the British Isles has ever killed any of these (mussels, cockles, winkles, oysters, shrimps, lobsters, crabs) shellfish". It was obvious when this statement was made that such animals were indeed killed by the accident of the Torrey Canyon as well as by earlier accidents. In addition, by its emphasizing only the effect on adult life forms such a statement implies that juvenile forms were also unaffected.

Oil and cancer

The higher boiling crude oil fractions are rich in multiring aromatic compounds. It was at one time thought that only a few of these compounds, mainly 3,4-benzpyrene, were capable of inducing cancer. R. A. Dean (1968) of British Petroleum Company stated "as far as I know, no 3,4-benzpyrene has been detected in any crude oil . . . it therefore seems that the risk to the health of a member of the public by spillage of oil at sea is probably far less than that which he normally encounters by eating the foods he enjoys". However, carcinogenic fractions containing 1,2-benzanthracene and alkylbenzanthracenes had been isolated (Carruthers *et al.*, 1967) from crude oil and it was known that "biological tests have shown that the extracts obtained from high-boiling fractions of the Kuwait oil . . . are carcinogenic".

In 1968, the year when Dean claimed the absence of the powerful carcinogen 3,4-benzopyrene in crude oil, this hydrocarbon was isolated in crude oil from Libya, Venezuela and the Persian Gulf (Graef and Winter, 1968). The amounts measured were between 450 and 1,800 milligrams per ton of crude oil. Thus, we do know that chemicals responsible for cancer in animals and man occur in petroleum.

According to Wilber (1969) "there is evidence that even a highly refined, diesel engine lubricating oil obtained from a naphthenic base crude oil, and lacking in substances ordinarily known to be carcinogenic, can induce tumours of the digestive tract of animals". Also, "Cutting oil is known to have carcinogenic potency". These references and a general knowledge of the composition of crude oils suggest that all crude oils and all oil products containing hydrocarbons boiling between 300 and 500°C should be viewed as potential cancer inducers.

This has severe implications for fisheries and human health. In our study of the West Falmouth oil spill (Blumer *et al.*, 1970; Blumer, Souza and Sass, 1970) we showed that oil from that spill was taken up by shellfish and incorporated in their body fat without fractionation of the hydrocarbons. In that incident an oil boiling between 170 and 370°C was involved; this boiling range overlaps that within which carcinogens have to be expected. Carcinogenic hydrocarbons can enter the chain leading to human food at an even lower level of the food chain; thus, it was shown by Doerr (1965) that intact plant roots can take up carcinogens like 3,4-benzopyrene from their growth medium.

Other questions suggest themselves: Floating masses of crude oil now cover all oceans and are being washed up on shores. We have shown that such lumps, even after considerable weathering still contain nearly the full range of hydrocarbons of the original crude oil, extending in boiling point as low as 100°C. Thus, such lumps contain some of the immediately toxic lower boiling hydrocarbons. In addition, they contain all the potentially carcinogenic material in the 300–500°C boiling fraction. The presence of oil lumps ("tar") or finely dispersed oil on recreational beaches may well constitute a severe public health hazard, through contaminated skin contact. The level of oil pollution encountered in many oceanic regions suggests that fisheries resources may often be contaminated with toxic petroleum derived hydrocarbons at levels that may constitute a public health hazard. Public health authorities should be urged to establish laboratories for continuous surveys of the pollution level encountered in commercial sea food.

Oil—destruction of fisheries resources

It has been said that "a review of the literature indicates that in deep water, whether in the open ocean or a mile or so offshore, no significant damage to marine life is encountered from even large oil spills because pelagic fish avoid the spill and few other marine species are present" (Little, 1969). The dead fish washed ashore after the West Falmouth oil spill were clearly unable to avoid the spill, nor will the fish fry in estuaries and marshes or the planktonic food organisms in the open ocean be able to avoid a large spill or the plume of toxic dissolved hydrocarbons descending from it. Unfortunately, investigations of the effects of major accidents (e.g. *Torrey Canyon*, Santa Barbara) have largely concentrated on the study of damage to adult fish or immediate reduction in fish catches. We should also consider the damage to the often more delicate juvenile forms and to the food organisms on which commercial fishes feed. Damage to these will not show up immediately nor will it necessarily be evident at the location of the accident. A large spill may lead to a gradual reduction of productivity over a large area. The combined effect of many such spills and other stresses, may lead to a reduction in fishing income which is difficult to trace to any single cause.

The so called "tainting" of fish and shellfish by oil spills has been recognized for many years, but it has only recently been realized that oil passes through the intestinal barrier and is incorporated and stabilized in the lipid pool of the organisms (Blumer *et al.*, 1970; Blumer, Souza and Sass, 1970). It has been widely assumed that fish and shellfish "tainted" by oil will again be fit for human consumption after a period from two weeks (Simpson, 1968) to several months (Little, 1969). Our experience suggests that this is highly improbable. The disappearance of an "oily smell" is no indication of whether a fish or shellfish has cleansed itself of the oil pollution. Only a small fraction of petroleum has a pronounced odour and loss of these compounds may occur while the more harmful high boiling, taste-and odourless carcinogens are retained. It has been reported that boiling or frying will remove the odour, but it will not affect the presence of polycyclic aromatic hydrocarbons.

Oil—Low level effects

Many biological processes which are important for the survival of marine organisms and which occupy key positions in their life processes are mediated by extremely low concentration of chemical messengers. We have demonstrated that marine predators are attracted to their prey by organic compounds at concentrations below the part per billion level (Whittle and Blumer, 1970). Such chemical attraction, or repulsion, plays a role in the finding of food, the escape from predators, in homing of many commercially important species of fishes, in the selection of habitats and in sex attraction. There is good reason to believe that pollution interferes with these processes in two ways: by blocking the taste receptors and by mimicking natural stimuli. Those crude oil fractions likely to interfere with such processes are the high boiling saturated and aromatic hydrocarbons and the full range of the olefinic hydrocarbons.

It has long been known that lobsters are attracted to crude oil distillate fractions, especially kerosene (Prudden, 1967; Anon., 1969) and this has been confirmed in the laboratory with purified hydrocarbon fractions derived from kerosene (Boylan, 1970). Thus an oil spill might attract lobsters away from their normal food and guide them towards the spill, where they may be severely contaminated or killed, indeed after the West Falmouth oil spill numerous dead lobsters were washed ashore. This is in direct contradiction to the opinion quoted above (Little, 1969) that marine animals will actively avoid oil spills.

Countermeasures

Compared to the size and numbers of accidents the present countermeasures are inadequate; however, a rapidly advancing technology is hopeful of developing techniques that will be effective in dealing even with very large spills under severe sea conditions. Although the gross aesthetic damage from oil spills may be avoided some time in the future, there is no reason to believe that existing or planned countermeasures will eliminate the biological impact of oil pollution.

The most acutely toxic oil fractions are soluble in sea water. Water currents will immediately spread the toxic plume and, if the accident occurs in inshore waters, the whole water column will be poisoned even if the bulk of the oil floats on the surface. Under storm conditions the oil will partly emulsify and will present a much larger surface area to the water; consequently, the toxic fraction will dissolve more rapidly and reach higher concentrations. From the point of view of avoiding the immediate biological effect of oil spills, countermeasures can be completely effective only if all of the oil is recovered immediately after the spill. The technology to achieve this goal does not exist.

Detergents and dispersants. The solvent-based detergents which did so much damage after the *Torrey Canyon* accident are in limited use only at present. However, so-called "non-toxic dispersants" have been developed. The term "non-toxic" is misleading, since although they may be non-toxic to a limited number of often quite resistant test organisms they are rarely tested in their effects upon a very wide spectrum of marine organisms including their juvenile forms. Moreover, in actual use the dispersant-oil mixtures are severely toxic, because the oil is toxic and bacterial degradation of "non-toxic" detergents may lead to toxic breakdown products.

A dispersant lowers the surface tension of the oil to a point where it will disperse in the form of small droplets. The break-up of the slick can then be aided by agitation, thus achieving the objective of those who wish to alleviate only the aesthetic damage. The recommendation to apply dispersants is often made in disregard of their ecological effects, since owing to the finer degree of dispersion, the immediately toxic fraction dissolves rapidly and reaches a higher concentration in the sea water than it would if natural dispersal were allowed. Long term poisons (e.g. the carcinogens) are made available to and are ingested by marine filter feeders, and can eventually return to man incorporated into the food he recovers from the ocean.

For these reasons the use of dispersants is only acceptable under special circumstances, e.g. extreme fire hazard from spillage or gasoline, as outlined in the contingency plan of the Federal Water Pollution Control Administration (1970).

Physical sinking. Sinking has been recommended: "the long-term effects on marine life will not be as disastrous as previously envisaged. Sinking of oil may result in the

mobile bottom dwellers moving to new locations for several years; however, conditions may return to normal as the oil decays" (Little, 1969). Again, these conclusions disregard our present knowledge of the effect of oil spills. Sunken oil will kill the bottom fauna before most mobile bottom dwellers have time to move away. Sessile forms of commercial importance (oysters, scallops, etc.) will be killed and some mobile organisms (lobsters) may be attracted into the direction of the spill where the exposure will contaminate or kill them. The persistent fraction of the oil which is not readily attacked by bacteria contains the long term poisons, e.g. the carcinogens. Exposure to these compounds may damage bottom organisms or render them unfit for human consumption.

Combustion. Burning the oil through the addition of wicks or oxidants appears more attractive than dispersion and sinking. However, it will be effective only if started immediately after a spill. For complete combustion, the entire spill must be covered by the combustion promoters, since burning will not extend to the untreated areas; in practice this may be impossible.

Mechanical containment and removal. Containment and removal appear ideal from the point of avoiding biological damage. Under severe weather conditions floating booms and barriers are ineffective. Booms were applied during the West Falmouth oil spill but the biological damage in the sealed-off harbours was severe and was probably caused by the oil in solution in sea water and in the form of wind-dispersed droplets.

Biological degradation. Hydrocarbons are naturally degraded by marine microorganisms. It is hoped to develop an oil removal technology based on bacterial seeding and fertilization of oil slicks but great obstacles and many unknowns stand in the way of the application of this principally attractive idea.

Bacteria are highly selective and complete degradation of a whole crude oil requires many different bacterial species (ZoBell, 1969). Bacterial oxidation of hydrocarbons produces many intermediates which may be more toxic than the hydrocarbons. Hydrocarbons and other compounds in crude oil may be bacteriostatic or bacteriocidal (ZoBell, 1969) and may therefore reduce the rate of degradation. The normal paraffin fraction of crude oil is most readily attacked by bacteria but is the least toxic, whereas the toxic aromatic hydrocarbons, especially the carcinogenic polynuclear aromatics, are not rapidly attacked (Blumer *et al.*, 1969, 1970; ZoBell, 1969). The complete oxidation of 1 gallon of crude oil requires all the dissolved oxygen in 320,000 gallons of sea water (ZoBell, 1969). Therefore, oxidation may be slow in areas where the oxygen content has been lowered by previous pollution and bacterial degradation may cause additional ecological damage through oxygen depletion.

Cost effectiveness. The high value of fisheries resources, which exceed that of the oil recovery from the sea, and the importance of marine proteins for human nutrition demand that the cost of direct and indirect ecological damage be considered. Unfortunately existing studies completely neglect these real values. A similarly one-sided approach would be a demand that all marine oil production and shipping be terminated, since it clearly interferes with fisheries interests.

Experience has shown that cleaning up a polluted aquatic environment is much more expensive than it would have been to keep the environment clean from the beginning (Ketchum, 1970). In terms of minimizing the environmental damage, spill prevention will produce far greater returns than clean-up.

Self control and law enforcement

The oil industry has an outstanding personnel and plant safety record. Oil refineries probably operate more safely than any other plants of equal production capacity. The industry has achieved this record through internal control because of the realization of the cost effectiveness of personnel safety. In the past the oil industry has not been fully aware of the substantiated toxicity of oil in the marine environment. It is hoped that the increasing recognition of the threat of oil pollution to marine resources will lead to an ecological safety record similar to the plant safety record.

Methods for the identification of oil spills by day and night through aerial spectroscopic surveys are becoming available (Swaby, 1969). Active tagging of oil in marine transit (Horowitz, 1969) should provide for simple and conclusive identification of spills. Even without active tagging, which depends on the willing cooperation of the ship owners and operators, each oil and oil product has its unique fingerprint. Fast and simple analytical techniques are available (e.g. capillary gas chromatography combined with mass spectrometry) that can qualitatively and quantitatively determine hundreds of different compounds in a spilled oil within a very short time. These techniques should be a great aid to more effective law enforcement. These techniques could be greatly supported if the oil industry would make available samples or analyses of those crude oils and products which are being transported across the sea.

CONCLUSIONS

The toxicity of crude oil and oil products to marine life and the danger of oil pollution to the marine ecology has been established in several independent ways:

(1) Studies of crude oil composition and isolation of compounds known to be toxic, e.g. low boiling aromatic hydrocarbons and the carcinogenic, high boiling polycyclic aromatics.
(2) Laboratory studies of the effect of oil and oil fractions on marine organisms.
(3) Field studies of the effect of oil spills on marine organisms in their normal habitat.

Pollution by crude oil and oil fractions damages the marine ecology through different effects:

(1) Direct kill of organisms through coating and asphyxiation (Arthur, 1968).
(2) Direct kill through contact poisoning of organisms.
(3) Direct kill through exposure to the water soluble toxic components of oil.
(4) Destruction of the more sensitive juvenile forms of organisms.
(5) Destruction of the food sources of higher species.
(6) Incorporation of sublethal amounts of oil and oil products into organisms resulting in reduced resistance to infection and other stresses (the principal cause of death in birds surviving the immediate exposure to oil (Beer, 1968)).

(7) Destruction of food values through the incorporation of oil and oil products, into fisheries resources.
(8) Incorporation of carcinogens into the marine food chain and human food sources.
(9) Low level effects that may interrupt any of the numerous events necessary for the propagation of marine species and for the survival of those species which stand higher in the marine food web.

Some oil products may be more poisonous than whole crude oils—thus, kerosene and No. 2 fuel oil are particularly rich in the low boiling water soluble poisons whereas higher boiling distillates are rich in carcinogenic hydrocarbons. Nevertheless all crude oils and distillates must be considered severe environmental poisons.

Crude oil and most products are persistent poisons; they enter the marine food chain, are stabilized in the lipids of marine organisms and are transferred from prey to predator. The most poisonous compounds of oil are the most persistent and since most of them do not normally occur in organisms natural pathways for their biodegradation are missing.

The presence of toxic and potentially carcinogenic hydrocarbons in fisheries products may constitute a public health hazard. Laboratories to measure routinely the contamination level should be organized to carry out continuous surveys of the safety of fisheries resources to public health.

Because of their low density, relative to sea water, crude oil and distillates should float but experience after the *Torrey Canyon* and West Falmouth oil spills has shown oil on the sea floor. Oil in sediments is not readily biodegraded; it can move with the sediments and can contaminate unpolluted areas long after an accident.

None of the presently used containment and recovery techniques prevents ecological damage and damage to fisheries products. Toxicity is evident immediately and the poisonous fraction will be carried in solution away from the accident, even if the surface spill is contained and recovered rapidly. Sinking and the use of detergents and dispersants, while cosmetically effective, are especially harmful since they introduce all the oil into the environment. The use of sinking agents and of dispersants should be most strongly discouraged.

Natural mechanisms for the degradation of oil in the sea exist; unfortunately, these are least effective for the most severely toxic components of oil. The breakdown products of oil and dispersants may also be toxic. Further, oil that has been incorporated into the lipids of marine organisms and into the sediments at the sea bottom and in estuaries and marshes is largely unavailable to the natural degradations. Poisoning of the bottom habitats and of the marine food web will therefore be more severe and more persistent than the poisoning of the water column itself.

The great persistence of oil and the existence of low level effects suggest the effects of oil pollution, especially in the coastal environment, must be considered in conjunction with other stresses. The combined impact of oil and oil products, chemicals, domestic sewage and municipal wastes, of the filling of wet lands, of dredging and of overfishing might lead to a deterioration of the coastal regions similar to that which we have brought about in the Great Lakes. Because of the much longer time scale of the oceans such a catastrophic deterioration would probably not likely be reversed for many generations.

References

ARTHUR, D R The biological problems of littoral pollution by oil
1968   and emulsifiers—a summing up. *Fld Stud.*, 2(Suppl.): 159–64.
BEER, J V Post-mortem findings in oiled auks during attempted
1968   rehabilitation. *Fld Stud.*, 2(Suppl.): 123–9.
BLUMER, M Hydrocarbons in digestive tract and liver of a basking
1967   shark. *Science. N.Y.*, 156(3773): 390–1.
BLUMER, M Scientific aspects of the oil spill problem. Paper presented
1970   at the NATO/CCMS Conference on Ocean Oil Spills,
        Brussels, November 2–6.
BLUMER, M, MULLIN, M M and GUILLARD, R R L A polyunsatu-
1970   rated hydrocarbon (3,6,9,12,15,18 heneicosahexaene) in the
        marine food web. *Mar. Biol.*, 6: 226–36.
BLUMER, M, SOUZA, G and SASS, J Hydrocarbon pollution of edible
1970   shellfish by an oil spill. (WHOI 70–1) Jan. 1970: 14p. Also
        issued in *Mar. Biol.*, 5 (3): 195–202.
BLUMER, M *et al.* Phytol-derived C$_{19}$ di- and triolefinic hydro-
1969   carbons in marine zooplankton and fishes. *Biochemistry*,
        8: 4067.
BLUMER, M *et al.* The West Falmouth oil spill. Woods Hole Oceano-
1970   graphic Institution, Ref. No. (70–44) (Unpubl. MS).
BOYLAN, D (Unpublished) observation reported in *Oceanology*,
1970   8(13): 146, 27 March 1970.
CARRUTHERS, W, STEWART, H N M and WATKINS, D A M 1,2-
1967   Benzanthracene derivatives in a Kuwait mineral oil.
        Nature, Lond., 213: 691–2.
DEAN, R A The chemistry of crude oils in relation to their spillage
1968   on the sea. *Fld Stud.*, 2(Suppl.): 1–6.
DOERR, R Alkaloid and benzopyrene uptake by intact plant roots.
1965   *Naturwissenschaften*, 52: 166.
EGLINTON, G and MURPHY, M T J (Eds.), Organic geochemistry.
1969   Berlin, Springer Verlag.
Federal Water Pollution Control Administration. Contingency plan
1970   for spills of oil and other hazardous materials in New
        England. (MS)
GOLDACRE, R J The effects of detergents and oils on the cell mem-
1968   brane. *Fld Stud.*, 2(Suppl.): 131–7.
GRAEF, W and WINTER, C 3,4–Benzpyren in Erdöl. *Arch. Hyg.*,
1968   152(4): 289–93.
HAMPSON, G R and SANDERS, H L Local oil spill. *Oceanus*, 15:
1969   8–10.

HOLT, S J The food resources of the ocean. *Scient. Am.*, 221(3):
1969   178–82, 187–94.
HOROWITZ, J Oil spills: comparison of several methods of identifi-
1969   cation. Paper presented to Joint API-FWCPA Conference
        on Prevention and Control of Oil Spills. Press release
        Dec. 17, 1969.
KETCHUM, B H Biological effects of pollution of estuaries and
1970   coastal waters. Boston, University Press (in press).
KETCHUM, B H Testimony before the Subcommittee on Public
1970a   Works, March 5, 1970. (Unpubl. MS).
LITTLE, A D, Inc. Combating pollution created by oil spills. *In*
1969   Report to the Department of Transportation, U.S. Coast
        Guard Methods. vol. 1: 713–86 (R) June 30, 1969.
MURPHY, T A Environmental effects of oil pollution. Paper
1970   presented to the session on oil pollution control, American
        Society of Civil Engineers, Boston, Mass., July 13.
NORTH, W J Tampico, a study of destruction and restoration.
1967   *Sea Front.*, 13: 212–7.
PRUDDEN, T M About lobsters. Freeport, Maine, Bond Wheelright
1967   Co.
RYTHER, J H Photosynthesis and Fish Production in the Sea.
1969   *Science*, N.Y., 166: 72–6.
SIMPSON, A C Oil, emulsifiers and commercial shellfish. *Fld Stud.*,
1968   2(Suppl.): 91–8.
SPEERS, G C and WHITEHEAD, E V Crude petroleum. *In* Organic
1969   geochemistry, edited by G Eglinton and M T J Murphy,
        Berlin, Springer, pp. 638–75.
SWABY, L G Remote sensing of oil slicks. Paper presented at Joint
1969   API-FWPCA Conference on Prevention and Control of
        Oil Spills. Press release, Dec. 17, 1969.
WESTLEY, R *In* Conference on pollution of the navigable waters of
1967   Puget Sound, the Strait of Juan de Fuca and their tributaries
        and estuaries. Seattle, Vol. 1: 174.
WHITTLE, K J and BLUMER, M A predator-prey relationship: sea
1970   stars-bivalves. The chemical basis of the response of
        Asterias vulgaris to *Crassostrea virginica*. Woods Hole,
        Oceanographic Institution, Ref. No. (70–20) (Unpubl. MS).
WILBER, C G The biological aspects of water pollution. Springfield,
1969   Ill., Ch.C. Thomas, 298 p.
ZOBELL, C E Microbial modification of crude oil in the sea. Paper
1969   presented at Joint API-FWPCA Conference on Prevention
        and Control of Oil Spills. Press release, Dec. 17, 1969.
ANON. Editorial. *Sea Secrets*, 13(11): 7.
1969
ANON. Final report of the Task Force on used oil disposal. New
1970   York, N.Y., American Petroleum Institute.

Mycological Degradation of Petroleum Products in Marine Environments†

S. P. Meyers and
D. G. Ahearn***

Dégradation mycologique des hydrocarbures dans les milieux marins

La proportion des levures et des champignons ressemblant aux levures dans les eaux des estuaries et marines dépasse rarement 500 unités viables/100 ml, encore que l'on observe des densités sensible-ment plus fortes dans des habitats o rganiquement enrichis. Selon les auteurs, les schémas de distribution des espèces pourraient servir d'indicateurs de la qualité de l'eau. Dans les sédiments des marais côtiers, on rencontre des concentrations aussi élevées que 9×10^4 unités de levures viables par cm³ de sédiments. La pollution des eaux ou des plages par des hydrocarbures peut modifier considérablement les biota des levures normales, et entraîner une augmentation parallèle du nombre des espèces hydrocarbonoclastes. Ces dernières, le plus souvent des souches de *Rhodotorula* et d'*Aureobasidium*, attaquent une grande variété d'hydrocarbures, notamment les alkanes en C$_{10}$ à C$_{18}$, le kérosène, le fuel pour moteurs diesel et le fuel pour trac-teurs. La respiration de populations mixtes de levures et bactéries choisies cultivées sur des fractions de pétrole brut est supérieure à celle des organismes pris isolément. Les auteurs suggèrent d'em-ployer des populations microbiennes mixtes pour combattre les épanchements d'hydrocarbures et en faciliter la décomposition, grâce à des mécanismes ou à des "systèmes" de dispersion comme la microencapsulation.

Degradación micológica del petróleo en medios marinos

Las levaduras y los hongos análogos a éstas de las aguas de estuarios y marinas raramente exceden de 500 unidades viables/100 ml, aunque se observan densidades sensiblemente mayores en habitats enriquecidos orgánicamente. Se propone la posibilidad de emplear las modalidades de distribución de las especies como indicadores de la calidad del agua. En los sedimentos de los pantanos costeros se encuentran concentraciones hasta de 9×10^4 unidades de levadura viables/cm³ sedimento. La contaminación de las aguas de las playas con petróleo puede dar por resultado una alteración radical de la biota de levaduras normales, con incremento simultáneo de los números de especies desdobladoras de hidrocarburos. Estas especies, comúnmente cepas de *Rhodotorula* y *Aureobasidium*, utilizan varios hidrocarburos, compredidos los alquenos de C$_{10}$–C$_{18}$, keroseno, y combustibles para motores diesel y para tractores. Las poblaciones mezcladas de levaduras y bacterias seleccionadas que han proliferado en fracciones de petróleo bruto, respiran más que los organismos individuales. Se propone el empleo de pob-laciones microbianas mixtas en el control y facilitación de la des-composición de petróleo derramado, empleando mecanismos o "sistemas" de dispersión, como la microencapsulación.

THE activity of yeasts and moulds in degradation of a variety of petroleum products is well docu-mented. Considerable literature is available on the ability of yeasts, especially *Candida* species, to convert paraffin hydrocarbons and other petroleum fractions into single cell protein. Komagata *et al.* (1964) found that 56 of 498 yeast strains (comprising 28 genera) utilized kerosene as a sole source of carbon and energy.

† Studies supported in part by Office of Sea Grant Programs and Office of Naval Research.
* Department of Food Science, Louisiana State University, Baton Rouge, Louisiana 70803, U.S.A.
** Department of Biology, Georgia State University, Atlanta, Georgia, U.S.A.

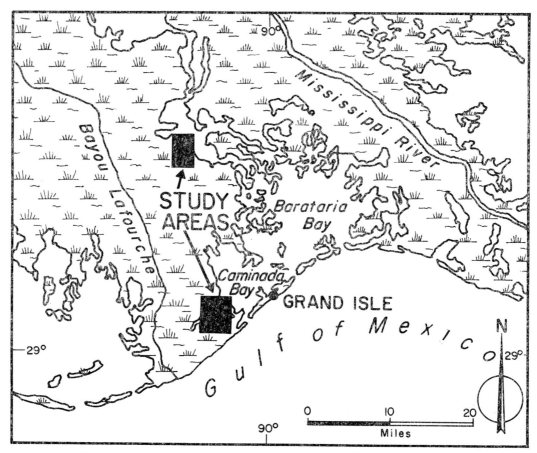

Fig 1. Location of study areas in the Barataria Bay marshland region, Louisiana

Scheda and Bos (1966) tested 1,200 yeast strains (244 species within 10 genera) for assimilation of n–decane, n–hexadecane, and kerosene. Variable response was noted, with the predominant utilizers belonging to the genera *Pichia, Debaryomyces, Candida,* and *Torulopsis.* Species of *Candida* have been isolated from microbiologically contaminated fuel in aircraft storage facilities; similarly, the ability of these isolates to utilize commercial kerosene fuels has been shown.

A few studies have been made on aspects of yeasts in natural petroleum-bearing environments (Iizuka and Komagata 1965; Iizuka and Goto 1965); these investigations, largely from Japan, involve microbiological studies of oil brines and oil fields. More recently, large concentrations of fungi have been reported from oil and natural gas deposits in Japan. Among the yeasts were representatives of *Rhodotorula, Candida,* and *Trichosporon.* However, other than the recent work of Turner and Ahearn

(1970, 1970a) and Ahearn *et al.* (1971), reports on activities of hydrocarbonoclastic yeasts in the marine/estuarine environment are notably lacking. Turner and Ahearn tabulated yeast populations in a petroleum waste lagoon and demonstrated the response of various species to oil intrusion into a watershed environment. Following discharge of large quantities of waste oils into the watershed, yeast densities, of a restricted spectrum of species, increased within 5 days from 20 to 150 viable units/ml to over 10^4 to 10^6/ml.

As part of our development of marine food resources at Louisiana State University, we are investigating biodegradative processes in the *Spartina* ecosystem, with particular emphasis on the ecological role of yeasts and filamentous fungi. Among the dominant vegetation in this marshland system is *Spartina alterniflora* (oyster grass), a "keystone" of estuarine productivity and a tremendous source of substrates and nutrients for

TABLE 1. UTILIZATION OF LABORATORY DISTILLATES OF LOUISIANA CRUDE OIL AT 4% (v/v) IN YEAST NITROGEN BASE[a]

Cult. no.	Initial crude		Laboratory distillates (°F)									
			Gasoline 65 – 430		Kerosene-jet fuel 430 – 550		Light gas oil 555 – 650		Heavy gas oil 650 – 900		Residue 900 +	
	FW[b]	SW	FW	SW	FW	SW	FW	SW	FW	SW	FW	SW
180	3	3	–	–	2	2	2	2	3	3	–	1
181	4	4	–	–	4	3	4	4	4	4	–	1
183	3	3	–	–	2	3	3	3	4	3	–	2
W12B	4	4	–	–	4	3	4	4	4	4	–	1
JK29	3	3	–	–	2	2	3	2	3	3	–	2
37–1	4	4	4	4	4	4	4	4	4	4	–	–
MIX°	4	4	–	–	4	4	4	4	4	4	–	–

[482]

TABLE 2. UTILIZATION OF REFINERY FRACTIONS OF LOUISIANA CRUDE OIL AT 4% (v/v) IN YEAST NITROGEN BASE[a]

Cult. no.	Initial crude		Virgin diesel oil 430 – –		Light catalytic heating oil – 650		Marine lube oil with additives (Tromar-65) 850 – 1100		Bunker oil (residual) 1050 +	
	FW[b]	SW	FW	SW	FW	SW	FW	SW	FW	SW
180	3	3	2	2	–	2	1	1	–	2
181	4	4	3	3	4	4	–	1	–	3
183	3	3	2	2	2	2	–	2	–	–
W12B	4	4	3	3	4	4	–	1	–	3
JK29	3	3	2	2	–	1	1	2	–	2
37–1	4	4	3	3	3	3	–	–	–	1
MIX[c]	4	4	4	4	4	4	–	–	–	2

[a] Growth recorded after 10 days incubation;
cultures showing no growth in all fractions: GM 119, GM 182, GM 179, FST 125, W-4-1, 76E, MENA
[b] FW, fresh water, SW, sea water
[c] Inoculated with, GM180, GM181, GM183, W-12-B and JK29

microbial transformation processes (Burkholder 1956). To date, we have demonstrated the occurrence of hydrocarbonoclastic yeasts, the response of these organisms to oil-permeated habitats, and their ability to utilize a range of petroleum breakdown products.

In the coastal marsh sediments in Barataria Bay, Louisiana, yeast concentrations as great as 9×10^4 viable units/cm³ of sediment may be encountered. This "standing crop" of yeasts is considerably greater than

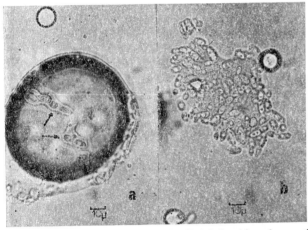

Fig 2. Development of yeast cells on oil globules. (a) early growth showing effect of the yeasts on surface tension (arrows); (b) late stage of yeast growth with complete entrapment of oil droplet by microbial development

that reported heretofore for estuarine regions. These wetlands (fig 1) are among the most productive estuarine regions in the world, supporting a tremendous economically important shrimp, blue crab, and oyster industry. The salt marsh is a nursery ground for menhaden and is a winter feeding area for duck, geese, and other bird-life. Possible spills of oil from the sizeable offshore drilling industry in Louisiana has continually posed a real and potential threat to the normal ecology of the area. Only recently, a serious oil disaster endangered the shrimp- and oyster-producing areas of Louisiana shoreline as well as the ecology of adjacent offshore regions.

Significant densities of various yeast taxa in the Louisiana marshland have suggested the need for meaningful field evaluation to study population dynamics of these organisms under a variety of stress and oil

spill conditions. Thus, we are attempting to establish guidelines to gauge the overall influence of oil spills on the indigenous microbial biota of Louisiana shellfish-producing areas and subsequent effects.

Our studies are directed along two interrelated lines: the hydrocarbonoclastic ability of yeasts and yeast-like fungi from a broad range of environments; and the response of the dominant mycota of the marshland biosphere to sudden intrusion of oil via spills, etc.

Experimental data

Over 700 yeasts from fresh water, estuarine, marine and terrestrial sources were examined in tube culture for ability to assimilate hexadecane and kerosene at 2 per cent (v/v) as sole sources of carbon. The basal medium used in the assimilation studies was the yeast nitrogen base of Wickerham (1951). A full description of methodology used is described elsewhere (see Turner and Ahearn 1970a; Ahearn et al. 1971).

Representatives of *Candida*, *Trichosporon*, *Rhodosporidium*, *Rhodotorula*, *Debaryomyces*, *Endomycopsis* and *Pichia* from all environments readily assimilated hexadecane and kerosene. Most rapid utilization was with strains of *Candida parapsilosis*, *C. tropicalis*, *C. guilliermondii*, *C. lipolytica* and *Rhodosporidium toruloides*. All of these organisms were originally isolated from oil-polluted habitats. It is noteworthy that strains of these same species from non-polluted regions were relatively poor in degree of hydrocarbon utilization. In the Gulf of Mexico area, hydrocarbonoclastic yeasts were markedly concentrated within regions of petroleum contamination. Except in regions of localized organic enrichment the extant yeast population in offshore waters seldom exceeds 10 viable units/100 ml. During the recent oil spill in the Gulf of Mexico, waters 1/2 mile from the oil well showed significantly increased numbers of yeasts, i.e., 500–1,088 viable units/100 ml.

The ability of selected yeasts to assimilate laboratory and refinery distillates of Louisiana crude is indicated in Tables 1 and 2. The heaviest and most rapid growth was obtained in 3 days with *Endomycopsis* (*Candida*) *lipolytica* Wickerham, 1970 (Culture 37-1). This isolate was the only yeast that grew rapidly on the gasoline fraction. In general, better growth occurred on the laboratory distillates than on the refinery fractions. However, regardless of the source of isolation of the organism, growth was more rapid when natural sea water was used as a diluent. This was particularly evident in tests of the residue of the laboratory distillates and bunker oil. The actual concentration of these two substrates varied from about 2 to 5 per cent (v/v), since their viscosity precluded precise manipulation. In several instances, yeasts, particularly marine strains, grew better on heavy gas oil than on the yeast nitrogen base medium with glucose.

It is noteworthy that certain *mixed* cultures utilized substrates that did not support growth of the individual strains alone. This confirms earlier observations of Miller and Johnson (1966). The response of the mixed culture to the residue fraction (Table 1) and bunker oil (Table 2) is less compared with several of the single isolates. However, the former did develop on sea water minimal media with 0.05 per cent $(NH_4)_2SO_4$ and the heavy gas oil and bunker oil. A number of isolates that failed to grow in 4 per cent (v/v) crude oil produced good growth in concentrations of <2 per cent (v/v) or, in mixed culture, grew at concentrations of >4 per cent.

Microscopic observations of the yeast/hydrocarbon systems (fig 2a, b) show development of the yeast cells over the periphery of the oil droplet, frequently affecting the surface tension of the globule.

It is significant that isolates FST 125 and MEN–A, representatives of the predominant species within the marshland sediments, did not utilize the petroleum fractions within 10 days at any concentration tested. Thus, it would appear that the primary yeast biota of the *Spartina* biosphere is not a hydrocarbonoclastic population.

Discussion

As a means of biodegradation of petroleum and petroleum fractions in the sea, various bacteria are being examined and have been proposed for destruction of petroleum effluents. Recently, Schwendinger (1968) suggested seeding of soil with *Cellulomonas* sp. to hasten oil decomposition. However, in spite of the wealth of excellent experimental laboratory data, hydrocarbonoclastic yeasts and moulds have not been examined intensively as *in situ* biodegradation agents. Biological treatment methods may be particularly desirable in estuaries, marshes and rivers where burning or sinking of the petroleum is not feasible. Ultimately, techniques may be developed to permit establishment of certain species as bioassay organisms for detection of petroleum contamination in marine and estuarine regions.

Use of yeasts to mediate oil decomposition has various advantages. The vegetative cells of yeasts are more resistant than those of bacteria to external stress conditions such as exposure to UV rays and to fluctuations in osmotic pressure and salinity. Present knowledge indicates that yeast strains effective in petroleum decomposition in fresh waters also function in sea water. This euryhaline adaptability is apparently lacking in many bacteria, especially in marine isolates that exhibit considerable lytic susceptibility. Ultimately, it may prove desirable to select hydrocarbonoclastic organisms that comprise the "normal" microbiota of the environment to reduce possible adverse effects of seeding large areas of petroleum-contaminated water surface with a particular organism or combination of organisms. This consideration is all the more important in shallow estuarine regions where artificially induced imbalance in the ecosystem may cause deleterious effects on the extant productivity. It is conceivable that addition of hydrocarbonoclastic yeasts (in view of their nutritional value to many filter-feeding organisms) to the environment may assist in restoration of the natural ecosystem. However, the potential toxicity of intermediate products of degradation needs thorough evaluation. Recently, Brown and Tischer (1969), in studies of decomposition of petroleum products, demonstrated the ichthyotoxicity of compounds produced microbiologically from oil.

The apparent inability of the dominant yeasts of the *Spartina* biosphere to utilize the various petroleum fractions tested is quite significant. Presumably, oil entering into this highly productive region would serve as an enrichment substrate for development of an extraneous hydrocarbonoclastic yeast biota. The effect of this latter population on the marshland productivity, and, concurrently, the effect of the petroleum on the prior-established dominant yeasts, warrants serious consideration.

The need to enhance nutrient supplies for accelerated microbial metabolism of oil slicks has been noted (Degler 1969). Undoubtedly, effective microbial degradation of petroleum in conjunction with a "seeding" programme will require addition of nutrients (especially nitrogen and phosphate sources) as well as possible essential co-factors. Similarly, the maintenance of viability and maximal activity of the organism in question is of paramount importance. In this regard, we are examining possible use of microcapsules for delivery of hydrocarbonoclastic microorganisms to the oil-polluted region in question. The microencapsulation procedure is well described by Sirine (1967, 1968) and Herbig (1967). In essence, the technique consists of enclosing substrates (chemicals, enzymes, microorganisms, etc.) within a carefully developed wall or shell to effect subsequent controlled release of the internal component(s). Through proper tailoring of the encapsulation system, microorganisms susceptible to environmental conditions can be protected and their release into the environment controlled.

References

AHEARN, D G, MEYERS, S P and STANDARD, P G   The role of
1971   yeasts in the decomposition of oils in marine environments. *Dev. Indust. Microbiol.* 12.

AHEARN, D G, ROTH, F J, JR. and MEYERS, S P   Ecology and
1968   characterization of yeasts from aquatic regions of South Florida. *Mar. Biol.*, 1(4):291–308.

BURKHOLDER, P R   Studies on the nutritive value of *Spartina* grass
1956   growing in the marsh areas of coastal Georgia. *Bull. Torrey bot. Club.*, 83:327–34.

BROWN, L R and TISCHER, R G   The decomposition of petroleum
1969   products in our natural waters. *Rep. Wat. Resour. Res. Inst. Miss. St. Univ.*, 1969:31 p.

DEGLER, S E   Oil pollution: problems and policies. BNA's En-
1969   vironmental Management Series, 142 p.

HERBIG, J A   Microencapsulation. Kirk-Othmer Encyclopedia of
1967   chemical technology (2nd Ed.), New York, Interscience Publishers, Vol. 13:436–56.

IIZUKA, H and GOTO, S   Microbiological studies on petroleum and
1965   natural gas. 8. Determination of red yeasts isolated from oil-brines and related materials. *J. gen. appl. Microbiol.*, 11:331–37.

IIZUKA, H and KOMAGATA, K   Microbiological studies on petrol-
1965   eum and natural gas. VI. Microflora of Niigata and Mobara gas-fields in Japan. *J. gen. appl. Microbiol.*, 11:103–14.

KOMAGATA, K, NAKASE, T and KATSUYA, N   Assimilation of
1964   hydrocarbons by yeasts. I. Preliminary screening. *J. gen. appl. Microbiol.*, 10:313–21.

MILLER, T and JOHNSON, M Utilization of normal alkanes by
1966   yeasts. *Biotechnol. Bioeng.*, 8:549–65.

SCHEDA, S and BOS, P Hydrocarbons as substrates for yeasts.
1966   *Nature, Lond.*, 211:660.

SCHWENDINGER, R B Reclamation of soil contaminated with oil.
1968   *J. Inst. Petrol.*, 54:182–97.

SIRINE, G Microencapsulation. *Stanford Res. Inst. J.*, 15:1–6.
1967

SIRINE, G Microencapsulation—a technique for limitless products
1968   *Fd Prod. Dev.*, (April–May).

TURNER, W E and AHEARN, D G Effects of oil pollution on
1970   populations of yeasts in fresh water. *Bacteriol. Proc.* 70:
G-24.

TURNER, W E and AHEARN, D G Ecology and physiology of yeasts
1970a   of an asphalt refinery and its watershed. *In* Recent trends
in yeast research, edited by D. G. Ahearn. *Spectrum,
Monogr. Ser. Arts Sci.*, 1:113–24.

VAN UDEN, N, FELL, J W and WOOD, E J Marine yeasts. *Adv.*
1968   *Microbiol. Sea*, 1:167–201.

WICKERHAM, L J Taxonomy of yeasts. *Tech. Bull. U.S. Dep. Agric.*,
1951 (1029):1–55.

Experiments on Combating Accidental Release of Oil

H. Hellmann and
H.-J. Marcinowski***

Expériences de lutte contre les déversements accidentels d'hydro-carbures

Expériences de lutte contre les hydrocarbures de surface au moyen d'émulsifiants et de dispersants (H. Hellmann)

L'auteur compare les résultats obtenus par l'utilisation d'émulsi-fiants et de dispersants pour provoquer la dissociation d'une nappe d'hydrocarbure sous forme de fines gouttelettes dispersées. Des essais de laboratoire ont été effectués pour déterminer quel est le produit le plus efficace et le plus économique. Trois produits se sont révélés satisfaisants. Afin d'examiner si l'on pourrait obtenir des effets analogues dans des conditions naturelles, on a déversé en mer 12 tonnes de pétrole brut que l'on a attaqué par pulvérisation d'un dispersant dilué dans l'eau. Etant donné la faible turbulence atmos-phérique et aquatique, l'effet dispersant n'a pu atteindre toute son ampleur. Néanmoins, le résultat a été acceptable, car, par dessus tout, il a été possible d'accélérer la vitesse de diffusion de l'hydro-carbure et l'on a pu prévenir l'émulsion eau-dans-huile (mousse de chocolat).

Brûlage des hydrocarbures de surface (H.-J. Marcinowski)

En juillet 1969, le Rijkswaterstaat néerlandais a effectué deux expériences de brûlage de pétrole brut léger d'Arabie altéré par les agents atmosphériques à la surface de la mer et sur les plages. On a utilisé un produit, mélange flottant de carbure associé à un composé de métal alcalin. Le premier composant produit au contact de l'eau une forte élévation de température qui favorise l'évaporation des composants du pétrole, formant en même temps de l'acétylène inflammable. Le mélange ainsi produit est allumé par le composé alcalin au contact de l'eau. La combustion ainsi déclenchée s'entre-tient d'elle-même par suite de la forte aspiration d'oxygène pro-voquée par les grandes flammes produites. Un pourcentage très élevé d'hydrocarbure de surface a été brûlé et, dans l'expérience effectuée à terre, les sables ne contenaient que de très petites quantités de matière organique.

Experimentos efectuados para combatir los derrames accidentales de hidrocarburos

Experiencias efectuadas con emulsionantes y dispersantes para combatir los hidrocarburos flotantes (H. Hellmann)

Se describe la diferencia que existe entre los emulsionantes y dispersantes como agentes para destruir las películas formadas por los hidrocarburos en gotas finas y dispersas. Para hallar el producto más eficaz y económico, se realizaron pruebas de laboratorio. Se revelaron satisfactorios tres productos. Con objeto de examinar si se pueden lograr efectos similares en condiciones naturales se arrojaron al mar 12 toneladas de petróleo bruto que se pulverizó con un dispersante previamente diluido en agua. A causa de la poca turbulencia del aire y del agua, no se pudo obtener por completo el efecto de la dispersión. Sin embargo, el resultado fue aceptable porque, sobre todo, se pudo acelerar la velocidad de extensión del petróleo e impedir la emulsión de éste en el agua.

Quema de hidrocarburos flotantes (H.-J. Marcinowski)

En julio de 1969, la Dutch Rijkswaterstaat efectuó dos experimentos para quemar petróleo crudo de Arabia, ligero y meteorizado, que flotaba en el mar o existía en las playas. El producto empleado fue una mezcla flotante de carburo junto con un compuesto de metal alcalino. En contacto con el agua, el primer componente produce un fuerte calor que favorece la evaporación de los componentes del petróleo, formando al mismo tiempo acetileno inflamable. La mezcla así producida arde al ponerse el compuesto alcalino en contacto con el agua. La ignición, iniciada de este modo continúa por sí misma debido a la mayor succión de oxígeno determinada por las grandes llamas producidas. Se quemó un porcentaje muy elevado del petróleo flotante, observándose que la arena subyacente en el experimento en tierra sólo contenía pequeñas cantidades de materia orgánica.

CHEMICAL liquids were used to emulsify oil after the accident suffered by tanker *Anne Mildred Brøvig* in Heligoland Bay in 1966. At the time, the full physico-chemical effect of these emulsifiers was not known. Even the techniques of control measures were still in a stage of improvisation. To avoid the pronounced toxic effect of the chemicals used on organic marine life, new and considerably less toxic dispersing agents were developed and intensive investigations initiated.

Dispersion generally means the scattering of a liquid in another liquid with which it will not mix, in this case oil droplets in water. If these droplets are invisible to the human eye, i.e. if their diameter is less than 10 to 20 microns, we have an emulsion (fine dispersion). If their diameter is larger than 20 microns, we speak of disper-sion in the narrow sense of the word (coarse dispersion).

The Bundesanstalt für Gewässerkunde, Koblenz, investigated to find out which commercially available chemical could be effectively used. Model tests were carried out under most realistic conditions working on the assumption that in practice an emulsifier (dispersing agent)/oil ratio of 1:10 is the highest that can be con-sidered, and that, for technical reasons, an intensive and permanent intermixture cannot be achieved by propeller action.

By these criteria Corexit 7664 (ESSO) was judged to be a dispersing agent, Hoe S 1708 (Farbwerke Hoechst) an emulsifier, and Fina Sol SC (Deutsche Fina), somewhere between the two. The laboratory tests were carried out with Kuwait crude oil (viscosity about 100 cSt at 20°C) in ocean water.

Experiences with emulsifiers and dispersants to deal with floating oil

In order to judge the full efficiency of the low toxicity dispersing agent Corexit 7664, a large-scale test was carried out in the German Bight in May 1969 under the direction of the Oelunfallausschuss (Federal Committee in Marine Pollution by Oil) in cooperation with the ESSO Chemie and in consultation with the Deutsches Hydrographisches Institut (DHI) and the Bundesanstalt für Gewässerkunde.

* Bundesanstalt für Gewässerkunde, Koblenz, Federal Republic of Germany.
** Stitchting Concawe, The Hague, Netherlands.

Three ships of the Wasser und Schiffahrtsamt Cuxhaven assembled at test position (54°30.2′N, 6°37.8′ E). To get a general idea of how and at what speed the oil would spread, only 1 ton of crude oil (Arabian light; density 0.851 at 15°C; viscosity 9.0 cSt. at 20°C) was dumped and left to spread freely (wind force 2 and water temperature 10°C). Shortly afterwards, another 11 tons (55 barrels) were dumped. After 22 min the crude oil patch covered an area of 100 m × 300 m with a mean thickness of 0.4 mm. As the patch was much thinner at the edges (about 1 micron), patch thickness in the central zones was at that time probably about 1 mm. Pollution control measures were then initiated. The ship used for the work, moved through the patch at a speed of 3 to 4 kn., followed by the two other ships, which were intended to effect the intermixture. ESSO provided and operated the spray equipment which consisted of two pipes 7 m long with an inside diameter of 1.5 m. The hole-type nozzles of 5 mm diameter were spaced at intervals of 230 mm. The admixture of Corexit to the water was effected by means of a regulatory valve, and the dose was later calculated to be 1.8 per cent. The output through the pipes was about 15,000 l/h. Altogether, a total of 1,030 l of dispersing agent was used.

During the test, samples were drawn continuously and evaluated according to the field methods of the Bundesanstalt für Gewässerkunde. Later, the oil patch which covered about 1 km² was circumnavigated. A paper (size DIN A4) dipped into a sample showed the dispersion achieved. The droplet diameter was several mm but was somewhat enlarged due to the suction effect of the paper. Non-dispersed oil would have formed a uniform brown layer on the paper while emulsified oil would have left no mark whatsoever.

The next day one ship returned to the test area. The drift which had occurred in the meantime was estimated by DHI. Very thin films of oil which were about to disintegrate were detected about 2 km south of the original discharge point. In addition, a brown layer 20 × 100 m in size and several mm thick was floating on the water. A sample taken showed that it consisted of a water-in-oil emulsion which was still comparatively unstable and had probably originated from an untreated portion of the oil. After brief treatment with Corexit applied by means of a deck swabbing hose, the layer split into a number of smaller patches which then expanded under the influence of the surface-active substance until they formed oil films. The test was then terminated.

In evaluating the results, several aspects must be taken into account:

(1) What would have become of the oil if it had not been treated? Relatively small quantities of oil spread on the water surface even without the addition of chemicals finally forms an oil film (about 0.1 to 0.3 microns), though without treatment this process takes much longer. While heavy seas and surf favour the dispersion of oil, they also facilitate the formation of a stable water-in-oil emulsion which may drift on the sea for days or even weeks.

(2) How does dispersed oil behave under the influence of current and wind? When the sea is smooth, a surface treated with Corexit is visually indistinguishable from an untreated surface. However,

under the influence of waves, wind and current the internal coherence is lost and the droplets formed are spread over comparatively large water surfaces and under the influence of Corexit form small films which disintegrate rapidly.

The effects of the dispersing agent are the acceleration of the oil spreading process and the prevention of formation of a water-in-oil emulsion with all its hydrologically unfavourable properties.

In the case of oil accidents, situations may arise in which a dispersion of the described type will be insufficient, for example in estuaries, especially in tidal regions where the oil will have to be dispersed under conditions of low water turbulence. Also, in the case of larger oil patches, emulsification is preferable to dispersion if the oil patch is drifting towards the coast.

Burning off of oil

Following the *Torrey Canyon* disaster attempts were made to burn the oil (Beynon 1968). When attempting to ignite crude oil by explosive or incendiary bombs, it must be borne in mind that the pressure wave caused by the bursting charge reinforces the tendency of the oil to spread and thus jeopardizes the preservation of a combustible mixture by forcing away the gas mixture and breaking the layer of oil. The use of bombs for fighting burning oil wells is based on this characteristic. A flame-thrower can operate similarly. Even when the thickness of the oil layer still allows it to flow back into the centre, this usually does not take place until the ignition energy in the epicentre is dissipated.

Also the rapid increase in the surface area of oil, due to spreading, leads to a very rapid evaporation of volatile components. In the case of medium- and high-boiling-point oil a thick layer may be left behind, frequently with too few volatile components to produce a combustible mixture. In the case of low-boiling-point oils, the layer thickness decreases so sharply that even with the presence of an ignitable air/hydrocarbon mixture the cooling effect of the underlying water is evident. As a result conventional techniques for igniting or burning away of solid or viscous substances have had little success. The use of granulated, oleophilic and buoyant absorbing agents, to produce wick effects, also appears to be ineffective owing to the difficulty of proper deposition and the breaking-up effect of the water movements.

A recent development seeks to overcome these drawbacks by mixing alkali metals with a carbide which will be decomposed by water, producing a buoyant mixture in such a way that in contact with water the carbide heats up the oil to such an extent that an ignitable mixture is formed, and in contact with water the alkali metal acts as a primer which ignites the original hydrocarbon/air mixture and the acetylene/air mixture formed by the exothermic reaction of the carbide with water (Scheidemandel 1968).

The heat of reaction released is not only enough to balance the loss of energy to the water, but heats up the overlying oil layer to such an extent that an ignitable mixture is formed which, if the oil layer is not too thin, leads to a combustible mixture. According to the manufacturer's data, layers of heavy fuel oil 2 mm thick can still be burned away at 0°C. In test apparatus, temperatures between 980 and 1,400°C were measured at a height of 2.5 cm to 5 cm above the burning oil layer.

In a test carried out in July 1969 off the Hook of Holland with the aid of the Rijkswaterstaat (Government Department of Public Works), using 10 tons of partly weathered Arabian Light Crude, the oil burned away almost completely. The blast of fire was strong enough to draw neighbouring oil layers to within the ignition range and to recombine pockets of fire into the large burning area.

The product, packed ready for operation, meets the requirements of the "Règlement International concernant le Transport des Marchandises Dangereuses par Chemin de Fer" (RID). The protection of the operating vessel can be improved by the development of a thrower. The question of operation from the air still requires investigation.

Tests carried out with storage tank residues and oil mixed with sand, on a fine-sand beach near the Hook of Holland, led to an almost complete burning away of the oil, leaving behind only a slightly brown greasy sand. However, the consumption of incendiary agents was greater, probably as a result of a poorer transfer of heat from the hydrolysis of the carbide.

The limitation of this promising method of destruction lies in the fact that:

(a) The requisite minimum thickness will often only be present when the tanker is still leaking and thus an ignition of the floating oil may lead to the extension of the fire to the tanker, and even to its explosion;

(b) The critical minimum thickness of 2 mm is quickly reached after the leakage has been stopped, especially with more easily volatile products. The period of preparation is therefore short; and

(c) The potential danger to the vessels used in the operation, or nearby vessels and shore installations, imposes further limits on possible applications.

Finally, it must be decided whether considerable short-duration air pollution is preferable to a long-lasting pollution of the sea and coastal areas. This decision is not dependent on product, but is a problem of the order of priority of the assets actually and/or potentially threatened.

References

BEYNON, L R   The *Torrey Canyon* disaster. Part I. Here's what the
1968   industry learned by experience. *Oil Gas int.*, 8 (1):28.
SCHEIDEMANDEL, A G   Hamburg (Owner), Swiss Patent No. 7573,
1968   24th May.

Eliminación de Petrlóeo en el Mar mediante Solidificación
*E. Castellanos**

Elimination of oil slicks in the ocean by means of solidification

Objective: A new process is proposed for eliminating oil slicks in the ocean by solidifying the oil through the use of paraffins having a high molecular weight and solid at room temperature.

Basis: The process is based on the transformation of the oil from a liquid to a solid state by using paraffin hydrocarbons as agglomerants. The mechanization triggering this change from liquid to solid state is determined by the change in temperature and the affinity of the paraffins for the hydrocarbons which persist in water.

Utilization of the method: Paraffin residues from the refinery with a melting point of between 50 and 60°C are sprayed in a melted state over the oil slick. The difference in temperature with respect to the water results in the solidification of the paraffin which, due to its affinity for hydrocarbons, totally absorbs the oil. The quantity of paraffin necessary to reach adhesion is 20 to 40 per cent. The resulting mass is inert, harmless to marine life and may be recuperated from the water with nets or raked off the surface.

Advantages: The solid product is completely recoverable, free of water, harmless to marine life, and large quantities of paraffin residues are available at low cost. Afterwards the water remains perfectly clean and free of toxic substances.

Application: Practical experiments were carried out on the ocean and the results confirmed by Spanish naval authorities. This process can be used for preventive treatment, for cleaning port waters and in case of shipwrecks.

Le nettoyage des mers par précipitation du pétrole

But: L'auteur propose un nouveau procédé pour éliminer le pétrole des mers par précipitation à l'état solide à la température ambiante au moyen de paraffines de poids moléculaire élevé.

Fondement de la méthode: La méthode repose sur le passage de l'huile minérale de l'état liquide à l'état solide par agglomération sous l'effet d'hydrocarbures paraffiniques. Le mécanisme qui régit cette transformation est déterminé par le changement de température et par l'affinité des paraffines pour les produits pétroliers qui se maintiennent dans l'eau.

Utilisation de la méthode: Des résidus paraffiniques du raffinage dont le point de fusion se situe entre 50 et 60°C sont projetés en état de fusion sur la nappe d'huile minérale. La différence de température entre la paraffine et l'eau provoque la précipitation de la paraffine avec absorption totale du pétrole pour lequel elle a une grande affinité. La quantité de paraffine nécessaire pour réaliser cette précipitation est de 20 à 40 pour cent. L'aggloméré obtenu est inerte, sans danger pour les organismes marins et peut être retiré de l'eau au moyen de filets ou de herses.

Avantages: Le produit solide obtenu est parfaitement manipulable, exempt d'eau, sans danger pour les organismes marins, et l'on peut disposer de grandes quantités de résidus paraffiniques à bas prix. L'eau reste parfaitement propre et exempte de substances toxiques.

Applications: Des essais pratiques ont été effectués en mer et homologués par les autorités navales espagnoles. Ce procédé est utile dans les cas de traitement préventif, pour le nettoyage des ports et lors des catastrophes maritimes.

LA dificultad de la recogida y eliminación del petróleo en el mar se debe a su naturaleza viscosa, a su fuerte dispersión superficial, al estado permanente de agitación en el mar y a su estructura molecular, totalmente opuesta a la del agua. Estos inconvenientes hacen que cualquier tipo de tratamiento sea caro por exigir un gran empleo de material con escaso rendimiento y sin lograr una perfecta eliminación. Por otro lado, cualquier procedimiento que no esté basado en la separación y extracción del petróleo del agua, o en su degradación o transformación, no conseguirá más que la solución parcial del problema, por cuanto la toxicidad del petróleo en el mar subsiste mientras el aceite continúe en el agua.

El tratamiento mediante disolventes, detergentes, dispersantes o cualesquiera otros productos tensoactivos puede conseguir la desaparición del petróleo de la superficie del mar al producir la emulsificación en

---

* Laboratorio Central, Compañía Arrendataria del Monopolio de Petróleos, S.A., Madrid 7, España.

su seno, pero esto supone añadir un nuevo agente contaminante de consecuencias tan nocivas como el petróleo. El empleo de materiales precipitantes que hunden el petróleo, aparte de razones de eficacia, supone simplemente el traslado del efecto tóxico de la superficie al fondo marino, con previsibles desplazamientos sobre estos fondos.

Nuevo procedimiento

Está basado en la transformación del estado líquido del aceite, pasándolo a estado sólido, mediante la utilización, como aglomerante, de hidrocarburos parafínicos de elevado peso molecular y sólidos a temperatura ambiente. El mecanismo que regula este cambio de estado viene determinado por la temperatura y por la afinidad de las parafinas hacia los hidrocarburos persistentes en el agua causantes de la contaminación. En efecto, la elevación de la temperatura por encima del punto de fusión de las parafinas utilizadas, que está situado entre 50° y 60°C las hace aptas para ser utilizadas en forma líquida, pudiendo proyectarse sobre la mancha de petróleo cuya temperatura es notablemente inferior. Esta diferencia produce el enfriamiento de la parafina que vuelve a solidificarse absorbiendo totalmente el aceite. Esta absorción es favorecida por la gran afinidad existente entre las parafinas y el petróleo ya que éstas son sustancias componentes del propio petróleo crudo. La solidificación del aceite en estas condiciones lo convierte en un producto inerte, exento de agua, que flota en el mar y que puede retirarse fácilmente mediante redes, rastrillos, etc.

En los ensayos y prácticas realizados se han utilizado, en condiciones ventajosas de precio y cantidad, residuos parafínicos procedentes de "bright-stock" de refinerías petrolíferas y con un contenido en aceites minerales, como impurezas, comprendido entre 15 y 20 por ciento.

Las ventajas del procedimiento son:

1. Gran afinidad de la parafina por el petróleo y ninguna por el agua.
2. Facilidad—mediante cambio térmico y absorción—para transformar el líquido del aceite, en producto sólido inocuo para la vida de animales y plantas marinas.
3. El material resultante, físicamente transformado y exento de agua, es fácilmente aprehensible.
4. Puede disponerse de grandes cantidades de este subproducto de petróleo, hoy sin aplicación, a buen precio.
5. Puede aplicarse perfectamente con independencia del estado del mar.
6. Es posible el aprovechamiento del material resultante para fines energéticos o industriales y en todo caso para su destrucción en tierra—sin posibilidades de contaminación del mar—mediante combustión.
7. El agua queda perfectamente limpia y libre de sustancias tóxicas.

Pruebas prácticas

En multitud de ensayos realizados a escala de laboratorio, se ha estudiado cualitativa y cuantitativamente este proceso, utilizándose una variada gama de productos petrolíferos en agua salina. Los resultados medios obtenidos sobre los más importantes productos son los siguientes:

Petróleo crudo solidificado con 25–30% de parafina
Fuel-oil „ „ „ 20–25% „ „
Gas-oil „ „ „ 50–60% „ „
Keroseno „ „ „ 60–70% „ „
Bencina „ „ „ 75–80% „ „

Han sido efectuadas dos pruebas prácticas en el mar, homologadas oficialmente por las autoridades navales españolas. La primera se celebró en el puerto de Cartagena (España) el 22.5.1967; la segunda en el puerto de Tarragona (España) el 29.11.1967. En ambos casos se trataron, sobre el mar, 150 kg de fuel-oil, obteniéndose su solidificación con unos 100 kg de parafina a pesar de que no se utilizaron los medios mecánicos más idóneos para la calefacción, impulsión y proyección de la parafina ni embarcación apropiada a este fin.

Aplicaciones

Este nuevo procedimiento tiene utilidad en los siguientes casos:

1. Tratamientos preventivos. En el caso de rotura de depósitos o tanques de petroleros. Puede evitarse el derrame del petróleo solidificándolo con parafina en el propio depósito.
2. Limpieza de puertos. Puede conseguirse mediante la utilización de una pequeña embarcación acondicionada para la fusión y lanzamiento de la parafina y dotada de un sistema de recogida del producto resultante. Esta es la solución para el caso frecuente de las contaminaciones que a diario se producen con motivo de las cargas o descargas de petroleros y en el abastecimiento de combustibles a otros barcos.
3. Catástrofes marítimas. Es el caso más importante y donde el tratamiento con parafina puede ser de más fácil aplicación. Para el tratamiento de la mancha de aceite puede utilizarse un petrolero de cabotaje de los que normalmente transportan "fuel-oil" y cuyos tanques pueden ser calentados. Estos tanques pueden ser llenados con parafina que podrá ser proyectada por las bombas del barco, sobre la mancha de petróleo. La recogida del petróleo solidificado puede ser realizada mediante redes de arrastre.

Contraindicaciones

Se han realizado numerosos ensayos para conocer las limitaciones que puede tener este procedimiento en su aplicación, por el envejecimiento producido en el petróleo después de largos períodos en el agua, sin que se haya observado pérdida de eficacia de la parafina. No ocurre lo mismo cuando el petróleo ha sido tratado previamente con productos tenso-activos. En este caso la parafina no absorbe el petróleo debido a la película liófila que exteriormente rodea cada partícula de aceite, impidiendo la acción absorbente del sistema. Este procedimiento no puede pues aplicarse en el caso de que el aceite mineral haya sido tratado previamente por agentes tenso-activos.

Costos

Los residuos parafinosos empleados son subproductos excedentes de refinerías, como antes se ha indicado, que en la actualidad tienen muy escasas aplicaciones industriales, por lo que normalmente se mezclan en pequeñas proporciones al "fuel-oil" para su combustión y por lo tanto su precio no puede ser muy superior al de este producto. En España puede calcularse una producción anual de estos residuos superior a las 50.000 t estimándose su valor en 40 dólares/t. La eliminación de una tonelada de petróleo en el mar viene a suponer unos 10 dólares aproximadamente en valor de la parafina. Este precio habrá de incrementarse con los gastos

normales de tratamiento muy similares a los de cualquier otro procedimiento.

Situación actual de estos trabajos

Resueltos ya los problemas de aplicación técnica que plantea el procedimiento, falta únicamente la construcción de un prototipo apto para la realización de todas estas operaciones en forma automática. La falta de financiamiento de este proyecto ha impedido, hasta ahora, la terminación del prototipo.

Global Monitoring of Atmospheric Pollution and its Possible Relation to a Global Marine Monitoring Scheme

*Erik Eriksson**

Le contrôle global de la pollution atmosphérique: ses rapports éventuels avec un plan global de contrôle de la pollution des mers

L'auteur de ce document insiste sur l'étroit contact et les interactions qui lient intimement l'océan et l'atmosphère. Un example classique à ce propos est le cas du gaz carbonique dont la concentration dans l'atmosphère dépend dans une large mesure de l'état physique et chimique des océans. A l'heure actuelle, il paraît vraisemblable qu'un tiers au moins du gaz carbonique fossile mis en liberté du fait des activités humaines pénètre dans l'océan en traversant la surface marine. Il est également possible que les océans contribuent à accroître la charge de l'atmosphère en composés du soufre par libération d'hydrogène sulfuré, notamment dans les régions côtières et surtout lorsque de grandes quantités d'effluents sont déversées dans des eaux peu profondes. Le rôle joué par le système océano-atmosphérique dans la diffusion globale rapide du DDT a été évoqué, mais on n'a pu vérifier cette hypothèse faute de données pertinentes. Il ressort de ce qui précède qu'à l'avenir il faudra établir des rapports étroits entre les systèmes de contrôle appliqués aux océans et à l'atmosphère.

L'auteur décrit en détail le réseau global de stations établi sous les auspices de l'OMM pour surveiller la pollution atmosphérique. Le recueil des données se fera en deux types de stations: des stations globales formant ligne de base et des stations consacrées aux conditions régionales. La première catégorie comporte environ dix stations choisies soigneusement pour représenter les conditions qui prévalent sur de très grandes zones. Celles de la deuxième catégorie doivent être représentatives des caractéristiques régionales de la pollution atmosphérique. Les éléments auxquels on peut attribuer une influence considérable sur le climat—c'est-à-dire le gaz carbonique et la turbidité atmosphérique—reçoivent une très haute priorité.

Les méthodes d'échantillonnage, de collecte et d'analyse des données doivent soit être standardisées, soit devenir entièrement comparables grâce à des programmes de routine conçus dans ce but. Les données recueillies seront publiées par accord spécial entre Etats Membres de l'OMM.

L'auteur souligne que, dans tout système de surveillance des mers, il faut accorder une attention spéciale aux éléments de la pollution atmosphérique globale qui présentent un intérêt particulier, afin d'aboutir à une coordination adéquate dans la collecte des données.

Inspección global de la contaminación atmosférica: su posible relación con un plan mundial de inspecciones marinas

En este estudio se pone de relieve el estrecho contacto y la íntima interacción que existe entre el océano y la atmósfera. El ejemplo clásico lo constituye el anhídrido carbónico cuya concentración en la atmósfera depende en gran medida del estado físico y químico de los océanos. En la actualidad parece probable que penetre en el océano a través de la superficie del mar por lo menos un tercio del anhídrido carbónico fósil producido por las actividades humanas. También es posible que los océanos contribuyan a cargar la atmósfera con compuestos sulfúricos al desprender sulfuro de hidrógeno, especialmente en las regiones costeras y en particular en aquellos lugares en los que pentran en aguas someras grandes cantidades de aguas residuales. Se ha sugerido la función del sistema océano-atmósfera en la difusión global rápida del DDT, pero ello no se puede determinar por falta de datos adecuados. Estas circunstancias indican la necesidad de una conexión muy estrecha entre los futuros sistemas de inspección para los océanos y la atmósfera.

Se describe detenidamente la red global de estaciones promovida por la OMM para vigilar la contaminación atmosférica. La recogida de datos tendrá lugar en dos clases de estaciones: estaciones globales y estaciones regionales. La primera clase contiene unas 10 estaciones, cuidadosamente elegidas para que representen las condiciones de zonas muy extensas. La segunda categoría debe representar las características regionales de la contaminación atmosférica. Se presta atención preferente a los elementos que se sospecha que influyen en el clima en un alto grado, por ejemplo, al anhídrido carbónico y a la turbidez de la atmósfera.

Los métodos de toma de muestras, recogida y análisis o bien se normalizarán o bien por medio de programas regulares de intercomparación se harán completamente comparables. Los datos se publicarán mediante acuerdos especiales con los Estados Miembros de la OMM dispuestos a cooperar.

Se pone de relieve que en todo plan de inspección marina, los elementos que tienen un especial interés en la contaminación atmosférica global deben recibir atención particular con objeto de lograr una adecuada coordinación en la recogida de los datos.

OVER the past few decades significant changes in the earth's environment have been the source of increasing concern among scientists and governments in all parts of the world. These changes, almost without exception, are attributable to the expanding requirements of our present day society and, almost without exception, have had the effect of degrading the environment from the standpoint of human activities.

Of major importance among the environmental changes brought about by the increasing trend toward urbanization and industrialization in many parts of the world has been the introduction of compounds, chemicals, and other forms of matter in such concentrations that they cause or may be suspected to cause damage to

human health or property. Substances present in such concentrations are commonly referred to as pollutants.

Like other substances in nature, pollutants are transported and dispersed throughout the environment. Some cease to be pollutants because of dispersion or dilution and some will be broken down into other forms which have no harmful effects. On the other hand some will accumulate in the soil, some in water with the ocean as their final destination, and some may reside in high concentrations in the atmosphere. It is possible in many cases from a knowledge of chemistry, biochemistry, and geochemistry and the circulation patterns of various chemicals in nature, to infer their final form and area of concentration.

*Meteorological Institute, University of Stockholm, Stockholm, Sweden.

To fully understand the impact of environmental pollution on human activities it is necessary to closely monitor the changes in quality and composition of the air, soil and water environments and to relate these changes to specific causes. One such operational monitoring system is being carried out by the World Meteorological Organization; its relation to a similar monitoring system for the oceanic environment will be examined.

WMO global monitoring of atmospheric pollution

The WMO Executive Committee at its twenty-first session set up a Panel of Experts on Meteorological Aspects of Air Pollution to advise on these matters. The Panel first met in May 1970 and proposed a global network of observation stations.

These stations should be classified into two groups:—

(a) Baseline air pollution stations
(b) Regional air pollution stations

The baseline air pollution stations should number only about ten and should monitor such pollutants as can be expected to influence climate on a global scale, in particular carbon-dioxide and atmospheric turbidity. Both properties are known to influence the radiative balance of the earth and hence also the mean temperature. Although their effects in these respects are not known in full quantitative detail as yet, the reported increase of these properties warrants a close watch in future on carbon-dioxide and atmospheric turbidity. Carbon-dioxide is released through combustion of fossil fuels and man's activity in general produces finely divided particle matter which enters the atmosphere, increasing its opaqueness or turbidity.

The baseline air pollution stations must be carefully selected, the principal criteria being that no development, industrial, agricultural or silvicultural, can be foreseen in an area of several thousand square miles in the region of the station. This is a rather restrictive criterion which will automatically limit the number of stations possible to polar continental areas, deserts and small ocean islands. Added to the monitoring programme of these stations is the chemistry of precipitation, since this may reveal slow trends in pollutants of particular interest. Examples of such are metals like lead and mercury and organic components such as DDT. Although these substances cannot be anticipated to have any direct effect on climate, they are of considerable interest from other points of view.

The regional air pollution station network which is expected to give information on the spread of air pollutants on a regional scale has a basic measuring programme on the chemistry of precipitation and atmospheric turbidity. In precipitation the major inorganic constituent will be measured on a monthly basis. Turbidity readings with suitable instruments are to be made daily if weather permits. The selection criteria for station sites are less stringent than for the baseline stations. It is sufficient that local pollution sources are avoided. The stations should represent the average conditions in the region. The minimum number of stations over the world should be about 100 but every country can add any number they consider to be relevant. The stations can, of course, be used for measuring such atmospheric trace gases as sulphur dioxide and nitrogen oxides but this is optional.

Operational details of the networks

To ensure completely comparable data it is necessary to prescribe in detail the local selection of station sites, of sample collection techniques, and of analytical or other measuring techniques. This is partly achieved through fairly detailed manuals of which one on stations siting criteria, precipitation sample collection procedures, turbidity measurement and formats for data reporting is in preparation. It is to be followed by a detailed manual on the collection and chemical analyses of air samples.

However, further precautions have to be taken to achieve comparable data and this is through an inter-comparison scheme which will operate routinely. In this way there will be a continuous check particularly on analytical methods.

In the present scheme the primary responsibility for the programme will rest with the national meteorological services, giving them, of course, full freedom with respect to the actual number of stations within the countries as well as with respect to the arrangements for chemical analyses. It is likely, however, that a considerable part of the chemical analyses will be concentrated in countries with adequate analytical facilities, an arrangement which has been tested in Europe for a number of years in connection with studies of the chemistry of precipitation.

Common interests in monitoring activity

It seems rather obvious that monitoring of atmospheric and oceanic pollution should be co-ordinated in one way or another. Looking first at the global aspects, the carbon-dioxide system is rather heavily influenced by the oceans, being a considerable reservoir for carbon-dioxide. This system has been discussed on many occasions and one can infer, at least semiquantitatively, how an excess of carbon-dioxide put into the atmosphere would be distributed between the atmosphere and the oceans when equilibrium is established. The transient characteristics of the system are, however, much less known. Present knowledge is mainly derived from the carbon-14 distribution in the oceans. Since carbon-14 in atmospheric carbon-dioxide is likely to be monitored in future, and the fossil fuel carbon-dioxide is devoid of this isotope, a similar scheme for oceans, which would yield a considerable amount of information, seems appropriate. One would, of course, also need continuous information on the carbon-14 content of the biosphere.

In oceanography there is a fear that pollution from terrestrial areas brought to the oceans by rivers will modify the oceanic environment. Also atmospheric pollution may be transferred directly into the oceans through the two-thirds of the earth's surface covered by water. Apart from such important substances as DDT and some heavy metals, a transfer of substantial amounts of sulphur dioxide to the oceans can perhaps affect the carbon-dioxide equilibrium conditions in surface water in future when the expected release of sulphur dioxide to the atmosphere becomes much larger than at present. Preliminary estimates of this effect can, of course, be made already. There is however, a possible reverse transport process of sulphur, in particular from coastal regions. This is a release of hydrogen sulphide from intertidal flats and shallow waters. Such a release can be suspected to be influenced by organic pollution in rivers discharging into these areas, which will increase the oxygen demand in the sediments. Sulphate in sea water can act as a suitable oxygen source when the dissolved oxygen is exhausted. This would lead to an increased hydrogen sulphide production. The properties of hydrogen sulphide are such that it would escape almost quantitatively into the atmosphere, thereby increasing the atmospheric burden of sulphur.

Another possible reverse transport process can perhaps explain the surprisingly uniform distribution of DDT in soils in areas where it was never applied. DDT is known to accumulate in lipoid substances. In the oceans there is a rather substantial synthesis of lipoids particularly in zoo-plankton. One can suspect that a great deal of these lipoids form surface films on the ocean surfaces and are consequently carried into the atmosphere during the production of sea salt particles. Once in the atmosphere, part of the fatty substances would be precipitated over land areas together with the sea salts, and the DDT enriched in these substances would thus be transferred from oceans to land. Although this may seem highly speculative there are pieces of evidence to support these ideas.

From the above discussion it seems advisable to arrange for a close co-operation between global networks for atmospheric and oceanic pollution, as well as with other networks of a similar kind—in particular networks for monitoring pollution of the biosphere. Only such co-operation can lead to a complete understanding of the long-term effects of man-made pollution on our environment.

Automation of Monitoring Equipment for Marine Pollution Studies   *R. D. Gafford**

L'automatisation de l'équipement de surveillance pour les études sur la pollution des mers

Pour mettre au point un système de surveillance automatique, il faut tout d'abord définir les agents de pollution des mers en termes strictement chimiques. Il est difficile de poser les limites d'une telle définition, car de nombreuses substances qui se trouvent dans la nature agissent comme polluants dans des conditions spéciales ou en concentrations élevées, et parce que certaines substances sont des polluants actifs à de très faibles concentrations.

Un schéma préliminaire de classement des agents de pollution des mers repose sur une combinaison de caractéristiques fonctionnelles et chimiques. Les trois catégories générales sont les métaux rares et les métaux lourds, les composés organiques, et les nutriments. On inclut cette dernière catégorie car, dans de nombreux cas, les quantités excessives de nutriments introduits par l'homme dans le milieu marin y engendrent des effets délétères.

On a déjà démontré l'application du traitement automatique aux méthodes d'analyse classiques pour certains éléments de chaque catégorie de pollution. Il en est ainsi de la spectrophotométrie d'émission et d'absorption atomique pour la recherche des métaux rares et des métaux lourds, de la colorimétrie pour les nutriments y compris les phosphates, les nitrates et d'autres, et de la chromatographie en phase gazeuse pour l'analyse des pesticides organiques et des hydrocarbures.

De gros problèmes sont posés par l'élaboration et la mise en œuvre de dispositifs d'échantillonnages robustes et non contaminants, par l'automatisation de l'échantillonnage et des procédures de concentration, ainsi que par l'établissement d'un contrôle de qualité adéquat des analyses automatiques.

Des techniques pour le traitement et la conservation des données applicables à la surveillance automatique de la pollution marine sont actuellement mises en œuvre. Les caractéristiques spécifiques de ces systèmes de données dépendront des types d'instruments utilisés par le système et des buts auxquels on destine ces données.

Automatización de un equipo de vigilancia para efectuar estudios sobre contaminación marina

La creación de un sistema de vigilancia automática requiere la definición inicial de los contaminantes marinos en estrictos términos químicos. Es difícil establecer los límites de tal definición porque muchas de las sustancias que aparecen naturalmente actúan como contaminantes en condiciones especiales o en altas concentraciones y porque algunas sustancias son contaminantes efectivos a concentraciones muy bajas.

Un esquema de clasificación preliminar de los contaminantes marinos se basa en una combinación de características funcionales y químicas. Las tres categorías generales son metales traza y pesados, compuestos orgánicos y nutrientes. Se incluye esta última categoría porque, en muchos casos, las cantidades excesivas de nutrientes introducidas por el hombre causan efectos nocivos en el ambiente marino.

Se ha demostrado ya la automatización de métodos clásicos analíticos para algunas sustancias químicas de cada clase de contaminación. Incluyen éstos la automatizacion de la espectrofotometría de emisión y absorción atómica para metales traza y pesados, de la colorimetría para nutrientes que incluyen fosfatos, nitratos y otros, y de la cromatografía de gas para el análisis de plaguicidas orgánicos e hidrocarburos.

Existen problemas importantes en el desarrollo y despliegue de difíciles sistemas de toma de muestras no contaminadas, en la automatización de procedimientos de toma de muestras y de concentración, y en la provisión de adecuados controles de la calidad de los análisis automáticos.

Son satisfactorias las técnicas de elaboración y conservación de datos aplicables a la vigilancia automática de la contaminación marina. Las específicas características de tales sistemas de datos dependerán de los tipos de instrumentos incluidos en el sistema y del uso a que se destinan estos datos.

BEFORE discussing the problems and characteristics of a particular instrumentation system, the instrumentation specialist wants to know, in general, four things. What is to be measured? At what level (absolute or concentration)? How accurately? And, under what conditions? Our first duty is to attempt a measurement-oriented definition of the problem.

This is not easy. It is difficult to arrive at a universally acceptable definition of marine pollution in terms of specific chemical elements, or ions. Similar ambiguity exists in attempts to provide answers to the other questions. Even universally defined marine pollutants are often present in concentrations which vary widely with time and location. Further, confusion may be generated by various units being used.

Accuracy specifications are most difficult to define if one is unsure how much is to be measured. Should accuracy be expressed in terms of the quantity measured, or in terms of the full-scale response of the instrument? Are different standards of accuracy required for research as opposed to surveillance or regulatory programmes?

Perhaps the least difficult question to answer is under

*Beckman Instruments, Inc., Fullerton, California 92634, U.S.A.

what conditions the measurements should be made. We feel that the ultimate requirement is for continuous real-time, on-line analysis in the marine environment under all climatic conditions. We recognize the great difficulty in achieving this ultimate objective, however, and that valid, useful information can be obtained under a more limited specification.

FAO's general definition of marine pollution was "Introduction by man of substances into the marine environment resulting in such deleterious effects as harm to living resources, hazards to human health, hindrance to marine activities including fishing, impairment of quality for use of sea water and reduction of amenities".

Pollution by radioactive sources is specifically excluded. A simple, short version might be "harmful substances".

If a substance is potentially harmful, it must be considered a pollutant. Theoretically every known stable element is present in ocean water to some level in some locations. Some of these materials are universally present and in one sense could not be considered pollutants. Yet many elements, when present either in higher than normal concentrations or in unusual chemical combinations, produce harmful effects. For example, the phytoplankton nutrients, especially phosphate and fixed nitrogen, are potential pollutants if, in such quantities they result in deleterious effects. So are advanced eutrophication, sludge deposition and reduction in dissolved oxygen concentrations.

In this paper, we will deal mainly with analytical problems associated with dissolved pollutants and, to some extent, colloidal material, recognizing that we have not attacked the entire problem.

Classification of pollutants

In classifying marine chemical pollutants, we have chosen three general categories based on a combination of chemical characteristics and ecological functions. They are (1) trace elements and heavy metals, (2) nutrients, and (3) organic compounds.

Trace elements and heavy metals

The ecologic functions or effects of these elements vary from most beneficial (i.e., essential) to most detrimental (i.e., toxic). These substances are included in a common category because they are all normally present in sea water in relatively low concentrations and because their detection and quantification can be accomplished by similar techniques. Specific delineation is difficult but a loose definition might be those elements which normally occur in concentrations of less than one part per million (mg/l) and which exhibit some known biological function or effect. We start by eliminating the fourteen elements which Goldberg (1965) has identified as normally occurring in sea water in concentrations greater than one part per million. These are O, H, Cl, Na, Mg, S, Ca, K, Br, C, Sr, B, Si and F. We will add to this list the fourth halogen (I) even though the nominal sea water concentration is only 0.6 mg/l.

The second category of substances which do not meet our loose definition are the dissolved inert gases, He, Ne, A, Kr, Xe and Rn since they are essentially biologically inactive. Of the elements remaining, we can eliminate nitrogen and phosphorus—as well as lanthanum and other rare earth elements, all of which occur in concentrations of 10^{-5} ppm or less.

The elements remaining for consideration are listed in Table 1 in the order of their atomic numbers. Some of these are classified as alkaline earths (A), trace metals (T), and heavy metals (H). Those which have been reported to be essential to some or all classes of marine organisms are placed in the trace element classification (Altman and Dittmer, 1966). (The physiological functions of Al, Ni, Se, Sr and V are not well understood.)

TABLE 1. CLASSIFICATION OF THE MINOR ELEMENTS WITH RESPECT TO MARINE POLLUTION. (A) ALKALINE EARTH METAL, (C) CONCENTRATED BY MARINE ORGANISMS, (H) HEAVY METAL, (T) TRACE ELEMENT, (X) EXCEPTIONAL TOXICITY

Element	Classification	Sea water concentration* (mg/litre)	Element	Classification	Sea water concentration* (mg/litre)
Li	(A)	0.17	Ru	—	—
Be	—	6×10^{-7}	Rh	—	—
Al	(T) (C)	0.01	Pd	—	—
Sc	—	4×10^{-5}	Ag	(H) (C)	0.0003†
Ti	(C)	1×10^{-3}	Cd	(H)	0.02
V	(T) (C)	0.002	In	—	—
Cr	(X) (C)	0.0005	Sn	(C)	8×10^{-4}
Mn	(T) (C)	0.002	Sb	(X)	0.0005
Fe	(T)	0.01	Te	(X)	0.001 (est)
Co	(T) (C)	0.001	I	—	0.06
Ni	(T) (C)	0.002	Cs	(A)	5×10^{-4}
Cu	(H) (T) (C)	0.003	Ba	(X) (C)	0.03
Zn	(T) (C)	0.01	Hf	—	—
Ga	—	3×10^{-5}	Ta	—	—
Ge	—	6×10^{-5}	W	—	1×10^{-4}
As	(X)	0.003	Re	—	—
Se	(T) (X)	0.0004	Os	—	—
Rb	(A)	>0.1	Ir	—	—
Y	—	3×10^{-4}	Pt	—	—
Zr	—	—	Au	—	4×10^{-6}
Nb	(C)	1×10^{-5}	Hg	(H)	0.00003
Mo	(T)	0.01	Tl	(X)	0.00001
Tc	—		Pb	(H) (C)	0.00003
			Bi	(X)	0.00002

*(Goldberg, 1965)

†(Brooks, personal communication)

The term "heavy metal", although widely used, appears to lack specific definition, but for purposes of this discussion is arbitrarily taken to include those elements which have the chemical characteristics of metals, which are known to have demonstrable toxic effects in low concentrations, which are precipitable as sulphide, and which have a density in excess of 8 g/cc. Zinc and tin are ruled out due to their relatively low toxicity and density. Of those remaining, at least one (Cu) meets the criteria of both heavy metal and trace element.

Of the remaining elements, some have been singled out for their special interest due either to the exceptional toxicity of the elements or some of their compounds (X) or to the fact that they have been reported to be concentrated to a very high degree by some specific marine organism (C). The biological significance of some of these concentrations is now known. Where the information is available, the normal sea water concentration is indicated in Table 1. Based on the relatively low concentration values, and the general biological insignificance, those elements which do not fall into one of the above classifications are not considered further.

Concentration range of interest

The concentration of the 28 elements of interest ranges from the one or two tenths of a part-per-million of the alkaline earth metals, Li and Rb, down to 1×10^{-5} ppm (Nb, Tl). Since they cannot be considered pollutants at the level of their natural occurrence in sea water, these lower levels could be considered ideal limits of detectability. Better criteria for sensitivity or concentration range would be toxic levels, as evidenced by research, or the concentrations observed in polluted waters.

Preliminary unpublished data on the concentration of some elements in sewage plant effluents were collected by the County Sanitation District of Los Angeles and by the Los Angeles City Sanitation Bureau and were tabulated by Norman Brooks (personal communication). Although based on infrequent or single samples some generalizations can be made; thirteen of our 28 elements are listed in Brooks' tabulation. Most appear in the sewage effluents in concentrations one or two orders of magnitude higher than in sea water, at least two (chromium and lead) appear to be concentrated by factors of several thousand.

The other factor to be considered is toxicity level. For example, mercury exhibits toxic effects on both animals and plants in concentrations well below 1 ppm. Zinc, on the other hand, is an essential micro-nutrient for some species, and may be highly concentrated in biological material. The chemical and biological activity of zinc places it generally in the same class as iron and in our classification is not listed as a toxic substance. Yet some oyster varieties are adversely affected by it, and concentrations of 10 ppm have been shown to reduce the photosynthetic rate of kelp (Clendenning and North, 1960).

Obviously, additional research is required to establish the toxic level of the heavy metals and trace elements for a variety of marine organisms. Meanwhile, a preliminary criterion may be arbitrarily accepted subject to modification as more information becomes available. For initial consideration, we set a reliable detectability level (signal level exceeds system noise level by a factor of 7) for all (H) and (X) class elements at ten times the reported

natural concentration in sea water, and for (T), (C) and (A) class elements at 100 times the sea water concentration. Upper limits of concentration are arbitrarily set at ten times the lower limit unless data are available to indicate a more appropriate value.

Definition of nutrients

The second general category is nutrients. Here we include ammonia, nitrite and nitrate; dissolved inorganic and total phosphorus; reactive silicate; sulphate and sulphide. Sulphide is considered because of its relationship to the nutrient form of sulphur sulphate. Other major constituents include chloride, sodium, magnesium, calcium, potassium, bromide, strontium, boron and fluorine. Large variations in the concentration of these components have little or no effect on primary productivity and they are therefore not considered further.

Inorganic carbon in its various forms (CO_2, HCO_3^-, H_2CO_3) is an important nutrient and contributes significantly to the buffering system in sea water, but it is seldom a limiting factor in productivity and certainly not a pollutant; therefore it is not included.

Concentration ranges of interest

Of Nitrogen's nine oxidation states, only four are of biological significance: ammonia, nitrite, nitrate and molecular nitrogen. The only marine organisms capable of fixing molecular nitrogen are some of the blue-green algae. Their part in the marine food chain is minimal so analysis of molecular nitrogen need not be considered further.

Fixed nitrogen (largely nitrate) is often the principal factor limiting primary production in the marine food chain. Nitrate may vary in concentration from 1 to 350 μg/l depending both on the location and season (Vacarro, 1965). Occasional values up to 500 μg/l have been reported. In the temperate coastal zones, the concentration is generally higher and more uniform in winter than summer. Concentrations below 5 μg/l are seldom observed anywhere in the oceans.

The distribution of nitrite is much more variable than that of nitrate, and its presence usually signifies the existence of environmental conditions of a transient or non-equilibrium nature. Since these conditions may indicate pollution the separate analysis of nitrite is desirable even though it is often included in nitrate analysis in routine surveys. Probably an adequate range for analysis is from 0.5 to 50 μg/l, although values in excess of 50 μg/l have been reported in a polluted estuary (Federal Water Pollution Control Administration, 1970).

The concentration of ammonia in the open ocean seldom exceeds 2 or 3 μg/l, but may reach 30 to 50 μg/l in coastal zones. Recently values of between 50 and over 1000 μg/l were reported in the Escambia Bay (Florida) estuary (FWPCA, 1970). Since the toxic level varies between 1 and 25 mg/l, depending on pH and dissolved oxygen content, the recommended range of concentration to be monitored is zero to 50 μg/l.

Phosphorus—Since the advent of the biodegradable detergents which incorporate phosphates as their principal constituent, the phosphorus content of nearshore waters has increased significantly. Ordinarily at least two phosphorus analyses are conducted on sea water samples, one for reactive phosphorus and one for total phosphorus. If the samples are filtered before analysis, a third and fourth analysis may be conducted on the particulate matter to obtain data on adsorbed phosphorus compounds and the phosphorus actually incorporated in the particulate matter. The concentration of phosphate phosphorus in the open ocean varies often being less than 1 μg/l in actively photosynthetic areas and ranging up to 40 or 50 μg/l in deep water (Armstrong, 1965). Total phosphorus may equal three or four times the concentration of phosphate phosphorus depending on oceanographic and climatic conditions and the amount of suspended organic matter in the sample. Estuarine or coastal zone concentrations of phosphate phosphorus may exceed 100 μg/l. The recommended range for monitoring is 0 to 100 μg/l phosphate phosphorus. For total phosphorus, a full scale range of 300 μg/l should be adequate for most purposes.

Silicon—As with the other nutrients, silicate concentration varies widely with location, depth and season. Except under very unusual circumstances, these variations in silicate concentration are due mainly to metabolic activity of the phytoplankton. The observed concentration of "reactive" silicate varies from almost zero in some locations at the surface to values of 3 or 4 thousand μg/l in deep water (Armstrong, 1965a). Particular interest attaches to the effect of silicate concentration on primary productivity especially in concentrations characteristic of the photic zone. Surface values in temperate and tropic latitudes seldom equal 500 μg/l and this is therefore taken as the upper limit of our range. The lower limit should be as close to zero as measurement technology allows.

Sulphate and sulphide—Sulphate occurs in a concentration of 2.71 g/kg. It is considered to be a "conservative" substance—that is, it is present in sea water in a constant ratio to chlorinity ($SO_4^{--}/Cl = 0.140$) (Culkin, 1965). Our interest in sulphate stems partly from the general observation that river waters are often characterized by sulphate to chlorinity ratios of 10 to 100 but more importantly from the fact that the ratio of sulphate to sulphide is a very good measure of the environment with respect to oxidation-reduction level.

The development and stabilization of anoxic zones in the Black Sea, Baltic Sea and in some Scandinavian fjords is generally marked by reduction of sulphate to sulphide. These conditions are abetted by excessive amounts of sewage and organic industrial wastes which represent large oxidation demands and the resultant reduction in dissolved oxygen concentration.

Sulphate should be monitored in the range of 2.5 to 3.0 parts per thousand (w/v). The precision of this measurement should be at least 0.01 parts per thousand. Sulphide levels up to 10 mg/l have been reported in the most anoxic regions of the Black Sea, but oxygen depletion is essentially complete at the level of 1.6 mg/l of hydrogen sulphide. This appears to be a good upper limit to the range which should nominally start at zero.

Organic compounds—Organic chemicals are derived from many sources, vary widely in composition and con-

centration and may have profound effects on the metabolism and chemical constitution of marine organisms. The three principal classes of compounds for pollution studies are hydrocarbons and their derivatives, organic pesticides, and phenolic compounds. In so far as they affect primary production of phytoplankton, biologically active compounds such as vitamins and vitamin precursors are also of interest.

The general problem of classifying the identifying organic compounds in sea water was reviewed by Duursma (1965). Reviewing almost 100 reports, he lists 91 identified specific dissolved organic substances. The major heading were carbohydrates, proteins and their derivatives, aliphatic carboxylic and hydroxy-carboxylic acids, biologically active compounds (mainly vitamins), humic acids, phenolic compounds and hydrocarbons.

The adverse biological effects of both the chlorinated hydrocarbons and organic phosphorous compounds in fresh and brackish waters are well documented. Specific biological effects of other organic pollutants are not so well defined. Some biological effects of the petroleum-based contaminants appear subtle and poorly understood. For example, fuel oil spills off Massachusetts may have caused hormonal disturbances in crab populations which led to their death by inducing the population to mate rather than hibernate in November 1969 (Boylan, 1970).

Except probably for three specific classes of compounds, our analytic approach will be qualitative rather than quantitative and no concentration criteria will be defined.

The exceptions are halogenated hydrocarbons which should be monitored in the range from 0.05 to 5 μg/l, organophosphorous compounds in the range from 1 to 10 μg/l and phenolic compounds in the range from 0.5 to 5 μg/l. The low level specification for the halogenated hydrocarbons is based on the significantly greater persistence and tendency toward biological concentration of the halogenated compounds. The ability to taint food and the documented carcinogenic character of the phenolics justifies the low concentration range assigned them.

ANALYTICAL METHODS

Techniques for analysing samples for both essential trace elements and for potentially dangerous heavy metals are similar.

In trace metal analysis, great care must be taken to avoid changes in the sample during collection and handling. Contamination may come from samplers, laboratory equipment, glassware, even airborne dust. Losses can occur by adsorption on sampling and handling vessels, by coagulation of colloidal complexes, and by metabolic activity in improperly stored samples. Errors in analysis may also be due to impurities in reagents.

Concentration methods

It will be necessary to concentrate some trace elements from large volumes of sample. Where the concentration is at the level of 1 in 10^9 or less, the most useful methods are co-precipitation and co-crystallization (Riley, 1965). At higher concentrations, solvent extraction procedures are effective and may be more rapid. Ferric hydroxide

TABLE 2. TRACE ELEMENTS WHICH HAVE BEEN EFFECTIVELY CONCENTRATED FROM SEA WATER BY CO-PRECIPITATION, CO-CRYSTALLIZATION AND SOLVENT EXTRACTION (Riley, 1965)

Elements	Co-precipitation	Co-crystallization	Solvent extraction	Elements	Co-precipitation	Co-crystallization	Solvent extraction
Ag	X	X	X	Mo	X		
Al	X			Nb	X		
As	X	X		Ni	X		X
Au	X	X		Os		X	
Ba		X		P	X		
Be	X			Pb	X		X
Bi	X		X	Ru		X	
Cd	X		X	Sb	X		
Ce		X		Sc	X		
Co	X	X	X	Se	X		
Cr	X	X		Sn		X	
Cu		X	X	Ta		X	
Fe		X		Th	X		
Ga	X	X		Ti	X		
Ge	X			Tl		X	X
Hf		X		U	X	X	
Hg	X	X	X	V	X		
In		X		W	X	X	
Ir		X		Zn	X	X	X
Li	X			Zr		X	
Mn	X	X	X	Y	X		

has been shown to be an excellent co-precipitant and when properly used can recover elements in concentrations as low as 20 millimicrograms per litre with efficiency of 98 to 99 per cent.

Co-crystallization is the process in which the trace element and the reagent form a complex which is more insoluble than the reagent alone. Thionalide is an example of this class of reagent. With it several researchers recovered 90 to 99 per cent of several trace elements (Riley, 1965). For a list of trace elements effectively concentrated see Table 2.

Solvent extraction involves the reaction of metals with a chelating agent which forms complexes soluble in organic solvents. The method can be made reasonably specific by adjustment of the pH of the sample. Further, since many chelated complexes so developed are strongly coloured, direct photometric analysis can often be made. A typical chelating agent is diphenylthiocarbazone (dithiozone). Since the advent of atomic absorption spectrophotometry (AAS), this method has been used to prepare quantitative concentrates for that type of analysis. A commonly used complexing agent is pyrrolidine dithiocarbamate, with methyl isobutyl ketone being a favoured extracting solvent. Also see Table 2.

Methods of analysis

There are several analytic methods available for quantitative and qualitative analysis of trace elements and heavy metals in sea water.

Neutron activation analysis is one of the most sensitive methods available for analysis of small samples. The amount of the element is determined by measuring the energy spectrum of radioactivity which is produced by bombardment of the sample with thermal neutrons. This method requires considerable skill, access to an atomic reactor and very expensive analytical instrumentation and it seems almost inconceivable that the method could be automated.

Isotope dilution

Isotope dilution involves the addition of a quantity of the element in question which has been artificially enriched in favour of a specific isotope. Following chemical isolation of the mixture analysis may be made in a mass spectrometer, or if the enriched isotope is radioactive, by simple radiation counting. The method is comparatively simple and accurate but would appear difficult to automate.

Emission spectrography

Perhaps the most promising method for rapid qualitative screening of sea water samples (or concentrates) is emission spectrography. In this, the sample is subjected to extreme thermal excitation and the resultant atomic emissions are analysed in a high resolution spectrometer.

Although the emission spectra generated are complex, they have been extremely well studied and can be used to simultaneously identify many elements in a single sample. Other advantages of emission spectrometry include the fact that most samples can be excited regardless of their composition, and that analyses may be conducted over concentrations ranging from trace levels to high percentages. The spectrograph is ordinarily recorded photographically but by arraying the required number of photodetectors along the dispersed spectrum at the proper locations, multiple element analyses can be conducted photometrically.

Flame emission photometry

In flame emission photometry the sample is excited by atomizing or spraying into a flame. The emission line of the element to be analysed is isolated by optical filters or monochromators and detected photometrically. This method is relatively insensitive compared to either arc emission spectrography or atomic absorption spectrophotometry. It is capable of significantly better accuracy than the emission spectrograph, however, for major constituents. It has been used to obtain accurate quantitative analyses of sodium, potassium, calcium, strontium, and lithium in sea water.

Atomic absorption spectrophotometry

In this (AAS), the sample is aspirated or sprayed into a long flame which is illuminated lengthwise with a hollow cathode lamp emitting very strong resonance spectra of the element to be analysed. Measurement of the absorption of this energy by the neutral atoms of the element in the flame results in a quantitative analysis for that element. A very large number of atomic vapours at ordinary pressures and moderate temperatures (up to a few thousand degrees) have absorption spectra consisting of very narrow, sharp lines. The lines are sufficiently intense to give good analytical sensitivity and narrow enough to provide extremely high specificity. These three factors—applicability to most elements, high sensitivity, and high specificity—are the major advantages.

The very high energy of the arc or spark source used in emission spectroscopy produces a multitude of spectral lines so that a monochromator of exceptionally high resolution is required to separate the analytical lines from all of the other lines. In contrast, the hollow cathode sources used in atomic absorption produce very strong single emissions which result in an instrument requiring relatively low resolution.

Some elements of interest in our scheme must be monitored at concentrations of less than one part per million. Atomic absorption is capable of providing this type of sensitivity for many of these elements. For purpose of discussion here, we have accepted as the limit of detection the fluctuational concentration limit (FCL) described by Ramírez-Muñoz (1966) and by Shifrin and Ramírez-Muñoz (1969). This is essentially the level at which the signal due to the presence of the element in the sample is equal to the noise level of the system. Obviously the detection limit (or FCL) sets the lower theoretical limit of the range which can be monitored by atomic absorption. In practice, acceptable accuracy is obtained at concentrations greater than seven times the detection limit.

Atomic absorption is rapid, accurate and direct, and can be used for quantitative analysis of many of the important trace metals in sea water at concentrations down to 1 microgram per litre. Various aspects of the atomic absorption method have been automated. These include automatic sampling and digital data recording. Sampling systems have been designed which provide forced-feed of sample as well as systems which employ aspirated samples. To our knowledge, none of these

sampling systems has actually been coupled to a continuous flow sea water sampler, but this appears to be a trivial plumbing problem, with due attention being required to preclude contamination.

Some of the hollow cathode lamps needed for AAS analysis have been designed with more than one element in them so that two, three or four elements can be analysed with a single lamp. If more than three or four elements are to be analysed in a single sample it is necessary to use more than one hollow cathode source.

A monochromator is needed to isolate the absorption lines of the various elements and the spectrum is scanned either by means of a rotating dispersion element (prism or grating) on a single phototube or by phototubes set at the properly selected wavelength. The multiple sources may be rotated on a turret or energy from lamps in fixed positions collected, collimated, and projected through the flame by an appropriate set of mirrors. An atomic absorption spectrophotometer was recently designed to process up to 90 samples per hour and analysed for up to ten separate elements simultaneously. It incorporated multiple sources, a fixed monochromator and multiple detectors.

Automatic AAS system

An outline of an automated analysis scheme is presented based on information shown in Table 3 which lists the elements of interest along with the nominal desired minimum concentrations. The detectability limit (Fluctuational Concentration Limit for each element by AAS) is shown and the practical lower limit for analysis (FCL × 7) is compared to the desired sensitivity. About half the elements (14) can theoretically be analysed without concentration and the other 13 require a concentration step before analysis. The sensitivity without concentration for Cr, Ni and Ag is marginal but the low limit criterion was established arbitrarily so the sligth loss in sensitivity for these elements is acceptable.

It is assumed that a single complexing-solvent extraction system can be devised to concentrate the 13 low level elements in one step, although this has not been demonstrated. The lamp configuration is shown in Table 4. Based on the currently available multiple element

TABLE 3. HEAVY METALS AND TRACE ELEMENTS—DESIRED LEVEL OF DETECTABILITY COMPARED TO REPORTED SENSITIVITY OF ATOMIC ABSORPTION SPECTROPHOTOMETRY. (C) ELEMENTS REQUIRING CONCENTRATION

Elements	Desired detectability (ppm)	Fluctional concentration limit (ppm)*	Lower limit of range (FCL × 7)	Elements	Desired detectability (ppm)	Fluctional concentration limit (ppm)*	Lower limit of range (FCL × 7)
Li	17.0	0.01	0.07	Nb (C)	0.001	15.0	105.0
Al	1.0	0.1	0.7	Mo	1.0	0.01	0.07
Ti	0.1	0.01	0.07	Ag	0.003	0.001	0.007
V	0.2	0.03	0.21	Cd	0.2	0.01	0.07
Cr	0.005	0.001	0.007	Sn (C)	0.08	1.4	9.8
Mn	0.2	0.005	0.035	Sb (C)	0.005	2.2	15.4
Fe	1.0	0.01	0.07	Te (C)	0.001	0.1	0.7
Co (C)	0.01	0.03	0.21	Cs (C)	0.05	0.06	0.42
Ni	0.2	0.04	0.28	Ba (C)	0.3	0.4	2.8
Cu	0.03	0.004	0.028	Hg (C)	0.0003	0.25	1.75
Zn	1.0	0.004	0.028	Tl (C)	0.0001	0.1	0.7
As (C)	0.03	5.0	35.0	Pb (C)	0.0003	0.04	0.28
Se (C)	0.004	1.4	9.8	Bi (C)	0.0002	1.6	11.2
Rb	10.0	0.01	0.07				

*(Shifrin and Ramírez-Muñoz, 1969)

lamps, the high level group would require eight separate lamps and the low level group ten separate lamps. These lamps could be arranged on a precision turret and spectral analysis would require multiple scans of the monochromator with sufficient dwell time on the 13 or 14 wavelengths of interest to allow collection and identification of the data. Alternatively, separate photo-detectors could be positioned at the appropriate locations along the dispersed spectrum in a fixed monochromator. This approach is less desirable from a cost viewpoint because the instrument would be much more complex. Since scanning time would not exceed more than two or three minutes at most, the former approach appears more desirable.

At least three potential problems require further study before the suggested approach could be considered practical. Ramírez-Muñoz (1970) has recently reported on interference by sodium in AAS analysis of several elements including Cu, Fe, Mn and Zn. His results indicate that the concentration of sodium chloride in sea water could materially affect the accuracy of the analyses unless corrections are made.

Mercury is too volatile to allow easy analysis by AAS at flame temperatures. Techniques involving cold sample cells have been developed for this element, but it is questionable whether they could be incorporated into an automated sequence in a practical way.

TABLE 4. HOLLOW CATHODE LAMP ARRAYS FOR LOW LEVEL AND HIGH LEVEL ELEMENT ANALYSIS BY AAS

| High level group | | Low level concentration group | |
Element	Analysis wavelength	Element	Analysis wavelength
Li	670.8	Pb	283.3
Al	309.2	Sn	286.3
Cr	357.9	Co	240.7
Mn	279.5	As	193.7
Ni	232.0	Se	196.0
Cu	324.7	Te	214.3
M	213.9	Nb	405.9
Ag	328.1	Sb	217.6
Ti	364.3	Cs	852.1
V	318.4	Ba	553.6
Rb	780.0	Hg	253.7
Mo	313.3	Tl	377.6
Cd	228.8	Bi	223.1
Wavelength range 213–780		Wavelength range 193–852	

Instrument sensitivity varies with each element and is also influenced by such instrumentation parameters as the characteristic energy levels of each hollow cathode lamp and the spectral response of the photodetectors. For example, a separate (red sensitive) photodetector must be used to collect data from the Cesium line at 852.1 nanometres. Automatic scaling and range switching of the photometric gain would therefore be required in any automated AAS system.

Despite the problems mentioned, it appears technically feasible to develop automated systems for continuous (repetitive) analysis by AAS of the trace elements and heavy metal components of sea water. The cycle time for an analysis of the type outlined would probably be somewhere between two and three minutes providing exceptionally high temporal resolution for this type of analysis.

NUTRIENTS

Absorption photometry

Another widely used technique for chemical analysis of sea water is absorption photometry, in which reaction of the sample with various reagents produces light absorbing compounds which are measured in a spectrophotometer or colorimeter. In a spectrophotometer, wavelength isolation is achieved by optical dispersion using either rotating optical gratings or prisms, whereas in a colorimeter changeable optical filters are used. Generally speaking the spectrophotometer is more versatile, more sensitive, can be adapted to a greater number of analyses, and is significantly more expensive. However, where many samples are to be analysed routinely for a few different substances the economy and ease of operation of the colorimeter is advantageous.

Because photometric analysis is used in many different applications in medicine, scientific and industrial research and in oceanography, much effort has been directed towards the development of automatic analysers. These instruments will automatically analyse a large number of samples for one or several constituents. They can perform most of the operations routinely accomplished by the laboratory chemist including accurate volumetric sampling and dilution, precipitation, filtration, incubation at controlled temperature and colorimetric analyses.

In marine chemistry the most important application for automated chemical analysis is in evaluating nutrients such as fixed nitrogen, phosphorus, silicon and sulphur.

Ammonia—Current procedures for ammonia analysis are generally based on oxidation by hypochlorite or hypobromite followed by a colour reaction and photometric analysis. Recently an automatic analysis method has been devised (Grasshoff, 1967).

Nitrite—It is often practicable to analyse nitrate and nitrite together. All present methods for nitrite analysis in sea water are based on the diazotization of an aromatic amine (usually sulphenilic acid) and coupling the product with an aromatic amine to form an intensely coloured dye. The method has been automated (Grasshoff, 1970).

Nitrate—Except under unusual circumstances, nitrate is the principal source of fixed nitrogen in ocean ecology. The concentration in sea water of nitrate nitrogen may vary from 1 to 500 μg per litre. Although several chemical analyses have been developed for nitrate, most are based on reduction of nitrate to nitrite followed by the azo dye procedure. Until recently, the reduction process was conducted using hydrazine as the reducing agent. To obtain adequate precision and reasonable yield of nitrite, it was necessary to conduct the reduction process for several hours, and the process was very temperature sensitive. Recently reduction methods employing packed columns of metallic granules (i.e., amalgamated cadmium) have overcome these problems. The method has been automated to cover the range from 5 to 100 μg nitrate nitrogen per litre (Brewer and Riley, 1965; Grasshoff, 1970; Henriksen, 1967).

Phosphorus—Most methods for analysing for soluble phosphate in sea water are based on reaction of the sample with a molybdenum reagent to produce a phosphomolybdate complex which is then reduced to phosphomolybdenum blue. This is measured colorimetrically. Total phosphorus is obtained by conducting the same analysis on a sample which has been digested with perchloric acid. Both analyses have been modified for automatic analysis. The apparatus used for the total phosphorus determination includes an ingenious continuous flow UV digestor (Grasshoff, 1967).

Silicon—Silicon is generally classed as a micro-nutrient because low concentrations (as silicate) limit the productivity of siliceous diatoms, a very important class of phytoplankton. "Reactive" silicate (silicic acid and very short chain polymers of silicic acid) is analysed by reacting with acidic molybdate reagent to produce silicomolybdate which is subsequently reduced to an intense blue complex. Reduction is accomplished by organic reductants such as Metol (p-methyl-aminophenol sulphate) in a reagent combined with acid sulphite. At least two procedures have been automated (Grasshoff, 1966; Henriksen, 1967).

Sulphur

Sulphate—The accepted method for the quantitative analysis of sulphate in sea water is the gravimetric method in which the sulphate is precipitated as the barium salt. No automatic method has been developed. An indirect method could be based on determination of the precipitated barium sulphate which would be converted to barium carbonate by boiling with sodium carbonate solution. The carbonate is soluble in hydrochloric acid to form $BaCl_2$ which can be analysed by atomic absorption spectrophotometry. All the chemical manipulation can be accomplished in present day automatic chemical analysers, and the ASS method is quite sensitive for barium (FCL $\simeq$ 0.5 g/l).

Sulphide—Both titrimetric (oxidation-reduction) and colorimetric methods for quantitative analysis of H_2S has been reported (Riley, 1965). All methods require careful handling of the sample to prevent spontaneous oxidation of the sulphide by contact with air. The colorimetric method reported by Riley (1965) results in formation of methylene blue. The method is linear to a concentration of about 2 mg/l H_2S. A modification of the method has been automated by Grasshoff and Chan (1970) and has a somewhat narrower linear range (to about 1 mg/l).

Organic compounds

Since the potential numbers and types of organic substances which might be found in any given sea water sample is almost infinite, a screening process may be required as the initial part of the programme. Subsequent automated analyses would then be based on information obtained during the screening phase. The samples may be subjected to such routine analyses as extinction coefficient in the far ultraviolet, ultraviolet fluorescence analysis, total organic carbon content, pH, oxidation reduction potential and CNH analysis.

The ultraviolet absorption and fluorescent spectra can be obtained easily and quickly with a variety of UV spectrophotometers. Total organic carbon analyses are based on a technique developed by the Dow Chemical Company, in which the organic carbon in a small water sample is pyrolysed to carbon dioxide, which is quantified in a specially designed infrared absorption analyser. A continuous sampling version of this instrument has recently been developed.

Analytical methods for specific identification and quantification will include paper chromatography (particularly applicable to carbohydrate compounds and derivatives) column chromatography and gas-liquid chromatography. The latter technique holds excellent promise for quantitative and qualitative analysis of a very large variety of organic substances. Many non-volatile compounds can be analysed by this method by forming volatile esters of the unknowns for chromatographic analysis.

Many organic substances are in such low concentration that a concentration procedure is required prior to analyses. Concentration techniques include flash evaporation at reduced pressure, freeze drying, solvent extraction and dialysis. Freeze drying is the method preferred for recovering the more volatile substances. Great care must be taken in all methods to avoid contamination or destruction of the organic compounds.

Hydrocarbons

The amount of crude and refined petroleum products annually dumped into the world's oceans has been estimated at over one million tons (Goldberg, personal communication). This is concentrated in shipping lanes, offshore oil production sites and in harbours, estuaries and coastal zones.

A fully developed instrument for monitoring the hydrocarbon content of sea water continuously is the process carbonaceous analyser. This is an automated version of the total organic carbon analyser. It automatically subjects a 20 microlitre sample from the flowing sample stream to combustion and CO_2 analysis once every five minutes. Output data may be collected and stored in either analogue or digital format. No correction for inorganic carbon is incorporated in the automatic version, so the output cannot be expressed specifically as total organic carbon. In addition, the instrument responds to *all* organic carbon rather than just to hydrocarbons.

An alternative approach is to employ a hydrogen flame ionization detector, which is relatively specific for hydrocarbons. When trace quantities of hydrocarbons are introduced into a hydrogen flame, a large number of ions are produced. Under properly controlled conditions, the number of ions is directly proportional to the number of carbon atoms in the hydrocarbons passing through the flame. The instrument is employed routinely in process control applications and in continuous monitoring of hydrocarbons as air pollutants.

To function in a water pollution monitoring system, a method for extracting and volatilizing (or aerosolizing) the sample without destroying the hydrocarbons would require development. A simple approach (used successfully in a gas chromatograph system for monitoring trace quantities of hydrocarbons in sea water) would be to sweep the sample with either air or the fuel used in the burner. Further research required to assure full success.

[498]

Pesticides

Of many procedures for sample preparation and quantitative analysis of chloro-organic pesticides, one of the best general reviews is that of Samuel and Hodges (1967). Methods developed by the U.S. Federal Water Quality Administration specifically for analysis in water samples are described by Breidenbach *et al.* (1964) and Kawahara *et al.* (1967). These include liquid-liquid extraction into organic solvent or by adsorption on activated carbon followed by drying and organic extraction. The solution is then subjected to thin layer chromatography and to gas-liquid chromatography incorporating an electron capture detector. Ultimate sensitivity of these methods range from 0.001 to 1.0 $\mu g/l$, depending on the compound.

Automation of the carbon absorption method seems highly unlikely due to difficulties in one-time extraction. The liquid-liquid extraction process is already classed as "semi-automatic" and shows promise as a possible method for full automation. Automation of gas chromatography has already been carried to a very high degree of sophistication and accuracies of 5 per cent have been achieved over periods of 20 days or more of continuous operation.

For many purposes, a quantitative determination of all organic chlorine will serve very well as an indication of pesticide pollution. Such a determination would respond equally to chlorine in the chlorinated herbicides, but their presence is probably as harmful as that of the pesticides. A method described by Gunther *et al.* (1965) appears amenable to complete automation. Samples containing from 0.03 to 1,000 μg of insecticide are completely burned in a small automatic combustion chamber. The gas evolved is quantitatively trapped in a continuously flowing aqueous medium. The concentration of chloride ion in the solution is monitored with a standard silver-silver chloride electrode. A technique for concentrating the pesticides from sea water would be required, but simple adsorption on a chlorine-free solid medium (activated carbon?) would be feasible, since the carrier will be completely consumed in the furnace. The combustion process requires seven minutes, suggesting a probable analysis rate of six determinations per hour which should be more than adequate.

Organic phosphorous compounds

As reviewed by Samuel and Hodges (1967) quantitative analytical methods for organophosphorus insecticide analysis include electron capture gas chromatography, gas chromatography using flame emission detectors and with induced vaporization detectors, and various chemical methods. An automatic method based on cholinesterase inhibition has been described by Voss (1970). Considerable processing of the sample is required and because of instability of the reagents daily blanks and calibration standards are required, in addition the method responds to any substance which inhibits cholinesterase.

As with halogenated compounds, it should suffice for present purposes, to detect all organic phosphorus rather than each compound. The most likely method for continuous monitoring of total organic phosphorus is the hydrogen flame emission photometer. When phosphorous compounds are decomposed in a hydrogen flame, a chemiluminescent reaction occurs, producing a specific wavelength of energy. This light energy is detected optically by a photomultiplier tube which senses a change in phosphorus concentration, then converts this to a proportional change in an electrical signal.

Much of the original effort in this technology was directed toward the detection of airborne pesticides. Recently, chemists at the U.S. Naval Applied Sciences Laboratory in New York have adapted the method to water analysis (McCallum, 1970). The technique has a claimed sensitivity of 100 $\mu g/l$, which is about two orders of magnitude less than required. However, other versions of the hydrogen flame photometer have been built which exhibited much greater sensitivity. The improved sensitivity is achieved by incorporating a second photometric channel which detects and suppresses the effects of sulphur emission in the flame.

Phenols

Phenols have been reported in sea water at levels of 1 to 3 $\mu g/l$ (Degens *et al.*, 1964). Manual methods for identification and quantification include solvent extraction followed by paper or column chromatography. Instrumental methods include measurement of differential absorption between alkaline and slightly acidic solutions at 289 nanometres and colorimetric analysis after reacting with 4-amino antipyrene (APHA, 1965). A fully automatic method based on oxidative coupling with 3-methyl 2-benzothiazolinone hydrozone has been reported by Håkon *et al.*, (1970). The method incorporates automated steam distillation for clean-up and can be run at twenty samples per hour. Detectability limits for a variety of substituted phenols range between 10 and 100 $\mu g/l$, which indicates a requirement for pre-extraction or concentration.

Sampling and sample processing

Problems encountered in obtaining, processing and storing sea water samples for chemical analysis have been extensively reviewed by Riley (1965) and by Strickland and Parsons (1960). Grasshoff (1967) has described a simple system for obtaining continuous samples at multiple depths for automatic shipboard analysis. The system was capable of analysing for six nutrients (NH_3, NO_2, NO_3, PO_4, total P and Si) and pH at five different depths twice per hour.

For the analysis of some trace elements and organic compounds, extraction and concentration procedures will be required. The lack of suitable procedures constitutes the most important deficiency in our ability to conduct continuous real time analysis of marine pollution. The area deserves the maximum possible support since no fully competent automatic pollution monitoring system can be deployed until these problems are solved.

Data processing

An adequate data acquisition and processing system is, of course, essential in a total programme.

The problems to be solved, however, do not appear particularly difficult. In fact, the technology for conversion of the analogue signals (which are generally characteristic of the instruments discussed) to digital format and for performing any required computations or transformations is fully developed. The principal problems are those of sample identification, time correlation and, perhaps, scaling the data over wide dynamic ranges.

Perhaps the most important problem in data processing is that of presentation and reporting. It seems likely that only such abstractions as trends and out-of-limits values should be displayed or printed. However, the true value of much of the information which is collected may not be realized at once, so provision should be made for economical storage of all data for some finite period. This is particularly important during the early, research-oriented phase of a programme when the significance of any information is least well understood.

The purpose of an automatic marine pollution system is two-fold. One is to provide data on the pollution environment so that the ecological effects of the various pollutants can be properly evaluated and quantified. The second purpose is to provide monitoring and alarm capability so that pollution events can be detected and corrective action taken. Related functions include provisions of permanent records for legal use and trend summaries of value in historical context. The data handling and storage system should be designed to provide the proper format and presentation to meet all of these requirements.

References

ALTMAN, P L and DITTMER, D S Nutrient elements in sea water.
1966   Table 169. *In* Environmental biology, Ohio WP AFB, Aerospace Medical Research Laboratory, pp. 511–21.
APHA Standard Methods for the Examination of Water and
1965   Waste Water. 12th ed. New York, American Public Health Association, Inc., p. 514.
ARMSTRONG, F A J Phosphorus. *In* Chemical oceanography,
1965   edited by J. P. Riley and G. Skirrow, London, Academic Press Ltd., vol. 1:323–64.
ARMSTRONG, F A J Silicon. *In* Chemical oceanography, edited by
1965a J. P. Riley and G. Skirrow, London, Academic Press Ltd., vol. 1:409–32.
BOYLAN, D (Unpublished) observation reported in *Oceanology*,
1970   8(13):146, 27 March 1970.
BREIDENBACH, A W, *et al.* The identification and measurement of
1964   chlorinated hydrocarbon pesticides in surface waters. *Publ. Hlth Serv. Publs, Wash.*, (1241).
BREWER, P G and RILEY, J P The automatic determination of
1965   nitrate in sea water. *Deep-Sea Res.*, 12(6):765–72.
CLENDENNING, K A and NORTH, W J Effects of wastes on the
1960   giant kelp, *Macrocystis pyrifera*. *In* Proceedings of the First International Conference on waste disposal in the marine environment, New York, Pergamon Press, pp. 82–91.
CULKIN, F The major constituents of sea water. *In* Chemical
1965   oceanography, edited by J. P. Riley and G. Skirrow, London, Academic Press Ltd., vol. 1:121–61.
DEGENS, E T, REUTER, J H and SHAW, K N F Biochemical
1964   compounds in offshore California sediments and sea waters. *Geochim. Cosmochim. Acta*, 28(1):45–67.

DUURSMA, E K The dissolved organic constituents of sea water.
1965   *In* Chemical oceanography, edited by J. P. Riley and G. Skirrow, London, Academic Press Ltd., vol. 1:433–77.
FAO Technical Conference on Marine Pollution and its effects
1969   on Living Resources and Fishing, Prospectus. Rome, FAO, (FRm/CMP/Circular 1):11
Federal Water Pollution Control Administration. Effects of
1970   pollution on water quality, Escambia River and Bay, Florida. U.S. Department of the Interior, Federal Water Pollution Control Administration, Southeast Water Laboratory, Athens, Georgia, U.S.A., January 1970.
GOLDBERG, E D Minor elements in sea water. *In* Chemical ocean-
1965   ography, edited by J. P. Riley and G. Skirrow, London, Academic Press Ltd., vol. 7:163–96.
GRASSHOFF, K Automatic determination of fluoride, phosphate
1966   and silicate in sea water. *In* Automation in analytical chemistry, New York, Mediad, Inc., pp. 304–7.
GRASSHOFF, K Results and possibilities of automated analysis of
1967   nutrients in sea water. *In* Automation in analytical chemistry, New York, Mediad, Inc., pp. 573–9.
GRASSHOFF, K A simultaneous multiple channel system for
1970   nutrient analysis in sea water with analog and digital data record. *In* Advances in automated analysis, New York, Mediad, Inc., pp. 133–45.
GRASSHOFF, K and CHAN K An automatic method for the
1970   determination of hydrogen sulfide in natural water. *In* Advances in automated analysis, New York, Mediad, Inc., pp. 147–50.
GUNTHER, F A, MILLER, T A and JENKINS, T E Continuous
1965   chloride-ion combustion method applied to determination of organochlorine insecticide residues. *Analyt. Chem.*, 37(11):1386–91.
HÅKON, O F, OTT, D E and GUNTHER, F A Automated colori-
1970   metric micro determination of phenols by oxidative coupling with 3-methyl-2-benzothiazolinone hydrozone. *In* Advances in automated analysis, New York, Mediad, Inc., pp. 21–7.
HENRIKSEN, A Application of the auto analyzer in routine water
1967   analysis. *In* Automation in analytical chemistry, New York, Mediad, Inc., pp. 658–72.
KAWAHARA, F K J, *et al.* Semi-automatic extraction of organic
1967   materials from water. *J. Wat. Pollut. Control Fed.*, 39(4):572–8.
McCALLUM, J D Instrument for tiny quantities. *Ind. Res.*, June:
1970   73–5.
RAMÍREZ-MUÑOZ, J Qualitative and quantiative sensitivity in
1966   flame photometry. *Talanta*, 13(1):87–101.
RAMÍREZ-MUÑOZ, J High sodium systems in atomic absorption
1970   flame photometry. *Analyt. Chem.*, 42:517–8.
RILEY, J P Analytical chemistry of sea water. *In* Chemical
1965   oceanography, edited by J. P. Riley and G. Skirrow, London, Academic Press Ltd., pp. 295–422.
SAMUEL, B L and HODGES, H K Screening methods for organo-
1967   chlorine and organophosphate insecticides in food and feeds. *In* Residue reviews, edited by F. A. Gunther, New York, Springer-Verlag, Vol. 17:35–72.
SHIFRIN, N and RAMÍREZ-MUÑOZ, J Reconsideration of criteria
1969   to determine practical fluctuational concentration limits in atomic-absorption flame photometry. *Appl. Spectrosc.*, 23(4):358–64.
SIEBURTH, J M The influence of algal antibiosis on the ecology of
1968   marine micro-organisms. *Adv. Microbiol. Sea*, 1:63–94.
STRICKLAND, J D H and PARSONS, T R A manual of sea water
1960   analysis. *Bull. Fish. Res. Bd Can.*, (125):185 p.
VACARRO, R F Inorganic nitrogen in sea water. *In* Chemical
1965   oceanography, edited by J. P. Riley and G. Skirrow, London, Academic Press Ltd., pp. 365–408.
VOSS, G Autoanalysis of insecticidal organophosphates and
1970   carbonates, based on cholinesterase inhibition. *In* Advances in automated analysis, New York, Mediad, Inc., pp. 11–20.

Multi-spectral Remote Sensing for Monitoring of Marine Pollution   *V. E. Noble*

La téléanalyse multispectrale pour le contrôle de la pollution des mers

Au cours des dix dernières années, on a procédé à des mesures multispectrales de la surface océanique au moyen de téléanalyseurs couvrant toute la gamme du spectre électromagnétique, depuis le proche ultra-violet en passant par la lumière visible et l'infra-rouge, jusqu'à la zone des micro-ondes. Des mesures, des observations et des expériences ont été effectuées à des hauteurs variables par des appareils placés à bord de bateaux et de satellite. On a utilisé des analyseurs allant du simple photomètre à lumière incidente au

Teledetección con espectro multiespectral para vigilar la contaminación de las aguas del mar

En el último decenio se han hecho mediciones de espectro múltiple de la superficie del océano con teledetectores que abarcan toda la gama del espectro electromagnético, desde el ultravioleta próximo, hasta la región de las microondas, pasando por los rayos visibles y los infrarrojos. Las mediciones, observaciones y experimentos se han realizado desde alturas muy diversas, desde detectores montados a bordo de embarcaciones hasta detectores instalados en satélites.

*U.S. Naval Oceanographic Office, Washington, D.C. 20390, U.S.A.

[500]

système "imageant" d'analyse multispectrale qui nécessite l'emploi des ordinateurs pour l'exposition et l'analyse des données.

Des démonstrations ont été faites pour mettre à l'épreuve les méthodes suivantes: établissement d'une liaison entre la couleur de l'eau et la concentration chlorophyllienne; relevé des nappes d'hydrocarbures par mesure dans les zones spectrales du proche ultra-violet, de l'infra-rouge thermique et des micro-ondes; relevé des courants côtiers par la photographie en couleurs et l'"imagerie" thermique. Les résultats de ces expériences ont montré que les moyens techniques actuels permettent d'utiliser des mesures multispectrales significatives dans les domaines de l'océanographie et des pêches. La mise au point de systèmes opérationnels conçus spécifiquement pour l'océanographie, tels que: le Multi-channel Ocean Color System (MOCS), le Thermal Infrared Mapping System (TIMS) et le All-Weather Ocean Dynamics System (AWODS), dépend maintenant de l'établissement d'étroits rapports entre le concepteur d'instruments et l'utilisateur d'une discipline pour la définition des besoins en matière de mesures.

Les résultats des expériences indiquent qu'il est maintenant possible de mettre au point des instruments spécifiques relativement simples pour dresser par voie aérienne la carte des nappes d'hydrocarbures, des "plumes" thermiques et des charges sédimentaires en vue d'applications spéciales. Les possibilités de contrôle de la pollution des mers se développeront dans la mesure où les premiers essais d'application permettront aux utilisateurs d'acquérir de l'expérience et d'obtenir des données de base pour la mise au point de modèles écologiques d'une complexité croissante. Grâce aux nouveaux modèles, l'utilisateur pourra réviser ses spécifications en matière de mesures et obtenir ainsi des instruments nouveaux grâce auxquels il sera mieux à même d'exercer un contrôle sur la pollution et de mener des recherches sur le milieu.

El grado de perfección de los detectores ha variado de un simple fotómetro de luz incidente a sistemas complejos de imágenes de espectro múltiple que hacen necesario el empleo de computadoras para la presentación visual y el análisis de los datos.

Se han realizado experimentos de demostración para comprobar la posibilidad de: poner en relación el color de las aguas con la concentración de clorofila; levantar mapas de las manchas oleosas mediante mediciones con las secciones próximas a los rayos ultravioleta, infrarroja térmica y de microondas del espectro; y levantar mapas de las corrientes costeras mediante fotografía en color y representación térmica. Los resultados de estos experimentos han mostrado que es posible técnicamente proceder a mediciones válidas de espectro múltiple en relación con la oceanografía y la pesca. La preparación de sistemas operacionales proyectados específicamente para la oceanografía, tales como el "Multi-channel Ocean Color System (MOCS)", el "Thermal Infrared Mapping System (TIMS)", y el "All-Weather Ocean Dynamics System (AWODS)", depende actualmente de una estrecha interacción entre el diseñador del instrumento y el especialista que ha de utilizarlo, para definir las necesidades de la medición.

Los resultados de los experimentos de demostración indican que en la actualidad es posible preparar instrumentos monovalentes sencillos para levantar mapas destinados a fines determinados, de manchas oleosas, "penachos" térmicos y capas de sedimentos. La capacidad de vigilancia de la contaminación de las aguas del mar irá aumentando, a medida que los primeros experimentos permitan a los especialistas que manejan los aparatos adquirir experiencia y reunir los datos básicos necesarios para la preparación de modelos ambientales más perfectos. Los nuevos modelos permitirán al usuario de estos instrumentos revisar sus especificaciones para las necesidades de medición y, de esa forma, se obtendrán nuevos y mejores instrumentos para vigilar la contaminación y estudiar el medio ambiente.

UNDER the direction of the U.S. Naval Oceanographic Office, Spacecraft Oceanography Project (SPOC), three remote sensing instrumentation systems are being planned for future NASA Earth Resource Technology Satellites (ERTS). These systems are the Multi-channel Ocean Color System (MOCS), the Thermal Infrared Mapping System (TIMS), and the All-Weather Ocean Dynamics System (AWODS), for the measurement of ocean colour, water temperature and sea surface roughness, respectively. The capability for accurate, remote measurement of water colour, water temperature and sea surface roughness from airborne and spacecraft platforms has been demonstrated. Requirements for measurement accuracies and sensitivities will be defined by the user groups.

It is important that such remote sensors as are now available be immediately used for monitoring marine pollution! For three reasons: First, remote sensors could rapidly map selected pollution indicators over large areas. Second, users in the oceanographic and fisheries sciences must gain experience in the reduction and analysis of remote sensing data, and develop analytic models for assessment, prediction, management and control of marine resources and pollution. Third, so that optimum design parameters may be established for operational oceanographic spacecraft systems to be used in the Earth Resources series of satellites.

It is the purpose of the present discussion to concentrate upon first-generation types of aircraft spectrophotometric instruments and measurement techniques that may begin to be used now for monitoring marine pollution. The results from the first experimental programmes will be used to develop improved instrumentation and analytic models for monitoring and control of marine pollution.

Choice of spectral region

For immediate application, it is recommended that instrumentation systems be limited to the near ultra-violet through thermal infra-red portions of the spectrum (0.32 through 13.0 µ)—this because of the relative technical simplicity of instrumentation in this portion of the spectrum, and because of the yet unresolved problems of interpreting microwave measurements.

Passive microwave measurements, for example, while offering the potential for all-weather, high altitude observations, are responsive to sea surface temperature, salinity, sea surface roughness, atmospheric water vapour, foam on the sea surface, and oil slicks on the sea surface (Paris, 1969; Aukland et al., 1969; Ross et al., 1970).

The effects of each of the environmental parameters upon microwave radiometric measurements is a function of the wave-length and polarization of the measurement. Thus, in order to obtain a truly all-weather microwave measurement capability, it will be necessary to design a multi-spectral system within the microwave portion of the spectrum. The design of such a system will depend on the results of applications measurements such as are now under consideration, and, therefore microwave techniques remain beyond the scope of this paper.

Valuable measurements may be made in the near ultraviolet and blue portions of the spectrum from low-altitude (300 m to 700 m) aircraft. At these wave-lengths, measurement problems increase as a function of flight altitude because of atmospheric scattering and absorption of the electromagnetic radiation, thereby increasing measurement noise.

Detection and mapping of oil slicks

Recent U.S. studies for the development of techniques for detection and mapping of oil slicks have been officially supported. These studies have included laboratory studies of the spectral reflectivity characteristics of various crude and refined oil/water/dispersant mixtures, analytic modelling for the prediction of spectral radiance curves, and preliminary, qualitative, flight measurement programmes to test measurement techniques. The results of two groups of measurements of the Santa Barbara oil slick during the spring of 1969 will be used as the basis for a proposed relatively simple remote sensing system specifically designed for oil slick mapping from light aircraft.

A Wide-range Imaging Spectrophotometer (WISP) and a Water Color Spectrometer were flown over the Santa Barbara oil slick on 29 January and 14 February 1969 by TRW Systems Group (Bailey, 1969). These flights were made during the early period of the slick, before the main leakage was stopped. Typical reflectivity curves obtained for a flight altitude of 3,000 ft (915 m) for a "heavy" oil film, "light" oil film, and local sea water are shown in fig 1. Normalized reflectivity curves for the same series of measurements are shown in fig 2. These measurements indicate a high reflectivity of the oil film in the red portion of the spectrum.

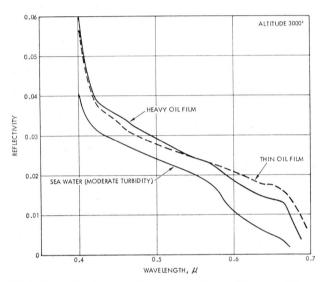

Fig 1. *Spectral reflectivity of sea water with or without an oil layer (after Bailey, 1969)*

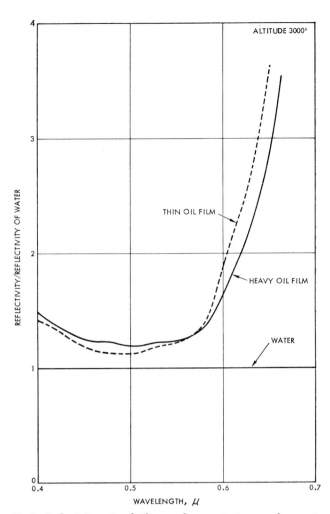

Fig 2. *Reflectivity ratio of oil-covered sea water to normal sea water (after Bailey, 1969)*

On 6 and 7 March 1969 (after the main leakage had been stopped), a series of flight measurements were made by the University of Michigan using a 17-channel multi-spectral scanner (Lowe and Hasell, 1969; Stewart *et al.*, 1970). These observations, which were made at a later time after the leakage rate had been reduced and significant quantities of dispersants had been used, showed a high contrast in the near ultra-violet portion of the spectrum (0.32 to 0.38 μ), and little or no contrast in the red and near infra-red portions of the spectrum (Horvath *et al.*, 1970). Simultaneous thermal infra-red (8.0 to 13.5 μ) imagery showed an apparent cooling at the thick portions of the oil slick.

Laboratory measurements of the optical properties of samples of the Santa Barbara oil and dispersants in sea water were carried out (Stewart *et al.*, 1970) to provide input data for an analytic reflectance model. These calculations showed that thin films of oil, and oil-dispersant mixtures (thicknesses from .01 mm to 0.5 mm) should exhibit positive contrasts with respect to the sea water surface in the spectral range below 0.4 μ. It is presumed that the high reflectivity measurements in the red portion of the spectrum as observed during the TRW field measurements were a consequence of the relatively great thickness of the oil slick during the early period of leakage.

The University of Michigan field measurements showed that the oil-dispersant slick was most visible in the UV region when the sun angle was low, i.e. during the early morning and late evening hours. This fact was not predicted by the analytic model, which assumed isotropic scattering.

It is therefore proposed that a simplified, prototype suite of instruments be used in a quasi-operational sense for the specific application of detection and mapping of oil slicks. This instrument suite would consist of three elements. The first item would be an aerial camera. Three possible camera selections deserve to be considered. The first camera to be considered is a 70 mm strip camera that would operate continuously during the flight mission. The primary function of the strip camera would be assistance in post-flight reconstruction of the actual flight tracks. It is essential that a time mark be recorded on the strip camera imagery. The strip camera imagery of the oil slick would be of secondary importance in mapping the oil slick.

The second camera selection to be considered is an aerial camera, preferably using 5 in film, that would be used for colour photography, with either standard aerial colour film, or with false-colour infra-red film. Colour photography, although presenting potential obstacles because of the difficulty of processing colour aerial film in the field, would be useful for mapping the larger features of the oil slick on the basis of the high red reflectivity of thick portions of the oil film as shown in the TRW measurements.

If quick-response requirements are imposed for an operational capability for detecting and mapping oil slicks in remote locations, a third type of camera system might be considered in order to avoid the practical problems of processing colour film in the field. This system is a multi-lens camera which provides the capability of simultaneously producing several images on a single piece of black-and-white film. The multi-lens camera uses a large film format, typically 9 in film. The exposure time is kept constant for all four images by the use of a single focal-plane shutter. The individual spectral bands are defined by optical filters used with each lens. The relative exposures between the four channels are established by independent diaphragm settings for each lens. Thus, the spectral sensitivities may be tailored for any given experiment, and, since the film is developed in a single chemical process, the relative film densities may be considered as a measure of the relative intensities in each of the spectral bands. The use of a single, large-film, multiple-lens camera has a technical advantage over the use of multiple small-film cameras (for example, four 70 mm cameras) because of the possibility of variations in shutter speed, film characteristics (aging, thermal history), and film processing between the individual units.

The second element of the instrumentation suite is an Airborne Radiation Thermometer (ART) designed to operate in the 8 to 13 portion of the spectrum. This device would be expected to give indications of the thick portions of the oil slick, and, from sea surface temperature patterns observed, to provide a means for estimation of probable circulation patterns in the area.

The third element of the instrumentation suite would be a simple radiometer designed to operate in the 0.32 μ to 0.38 μ portion of the spectrum. This device might be as simple as a telescope (of course capable of transmission in the near ultra-violet portion of the spectrum) with a field of view of 2°, a spectral filter, and a photo-multiplier tube. This device would be expected to give a significant output when viewing thin portions of the oil slick, or oil-dispersant mixtures—particularly during morning and evening hours when aerial cameras could not be used.

The first phase of a survey of possible oil slicks, would consist of a regular pattern of transects over the suspected area. These initial flights would probably be carried out during the middle portion of the day in order to have maximum light for visible observations and for photography of the area. If that survey yielded a probable indication of an oil slick, then the survey would go into the mapping phase.

It is expected that the use of a three-element sensor system (camera, ART and UV radiometer) will permit the detection and mapping of an oil slick in the visible, near ultra-violet, and thermal infra-red portions of the spectrum. The instruments described should provide the capability for conducting surveys with light aircraft in field locations where sophisticated field support data reduction equipment is not available.

Water colour measurements for pollution monitoring

The techniques and applications of water colour measurements for pollution monitoring depend on the user's definition of "pollution". Applications may range from monitoring algae blooms as indicators of accelerated eutrophication, through observation of sediment plumes to determine circulation and dissipation patterns, to direct measurements of obviously definable pollution phenomena, such as oil slicks. Therefore, before specific techniques for water colour measurements can be recommended, it is necessary that the intended user be able to specify the spectral sensitivity required for measurement, the spatial resolution required, and the frequency of observation required.

Properly speaking, water colour measurements are measurements of the spectral composition of light reflected from the water mass within the field of view of the sensor. Thus the apparent water colour is a function of both the reflectance of the water mass and the spectral composition of the illuminating radiation. Therefore, the apparent colour of a water mass will change as a function of time of day (shift to the red because of the low sun angle in early morning and evening hours) and cloud cover (shift to the blue because of the dominance of scattered skylight under a cloud).

Instrumentation and measurement techniques for water colour determinations are almost identical to those discussed for oil slick detection and mapping. The differences lie in the selection of spectral bandwidths to optimize the data for detection, identification and mapping of the oceanographic parameter selected for investigation. Spectral measurements of the incident illumination are necessary for accurate colorimetric measurements of the water mass. Examples of the results of preliminary measurements with a water colour spectrometer off Redondo Beach, California, on 7 March 1969 are shown in fig 3 (Bailey, 1969). These data show the shift of the reflectivity of the water mass from blue offshore water, through coastal water with an algae bloom, into nearshore water containing the "red tide" dino-flagellate (*Gonyaulax*, probably *G. polyedra*).

The results shown in fig 3 indicate that significant

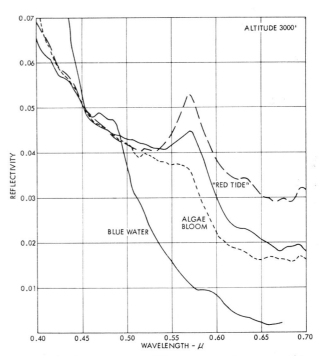

Fig 3. *Spectral reflectivity of different areas of sea water (after Bailey, 1969)*

[503]

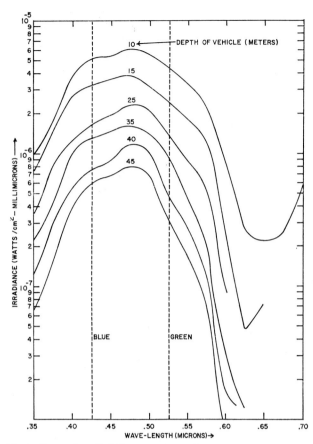

Fig 4. *Composite of full-corrected irradiance measurements as a function of vehicle depth* (Ben Franklin *experiment*) (*after Ross* et al., 1969)

amounts of water colour information may be expected in the blue (0.45 μ) and green (0.55 μ) portions of the spectrum. These indications are reinforced by spectral measurements of the downwelling illumination from the submersible *Ben Franklin* (Ross *et al.*, 1969). The

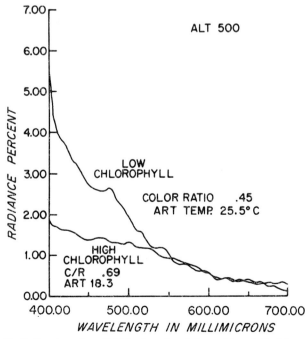

Fig 5. *Spectra of sunlight backscattered from chlorophyll-rich and chlorophyll-poor areas at 500 ft flight altitude* (*after Clarke* et al., 1969)

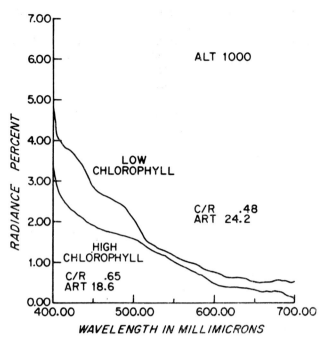

Fig 6. *Spectra of sunlight backscattered from chlorophyll-rich and chlorophyll-poor areas at 1,000 ft flight altitude* (*after Clarke* et al., 1969)

measurements from the *Ben Franklin* Gulf Stream Experiment, which was carried out in May 1969, are shown in fig 4. These data demonstrate that maximum penetration of the water column may be expected in the blue and green portions of the spectrum, thus indicating preferred spectral bands for bathymetric determinations.

Preliminary results from measurements of water colour over the Gulf Stream water (low chlorophyll) and fishing banks at 40°N, 68°W (high chlorophyll) made during August 1969 are shown in figs 5 and 6 (Clarke *et al.*, 1969). The mean low chlorophyll concentration of the Gulf Stream water mass was $0.07 \pm .02$ mg/m³. The mean chlorophyll concentration of the richer water mass was $0.22 \pm .04$ mg/m³. Clarke, Ewing and Lorenzen (1969) characterize the water colour spectrum by the colour ratio (C/R) which is defined as the ratio of the spectral intensities of the green (0.54 μ) to the blue (0.46 μ) components of the light reflected from the water surface. The increase of the colour ratio (C/R) for the water masses of high chlorophyll content is due to the increased green reflectance of the algal pigments. Additional measurements in this series of experiments showed, as expected, an increased blue component because of the effects of scattered skylight when measurements were made from higher altitudes (up to 3,000 m).

Pollution indices, such as the Clarke–Ewing–Lorenzen Color Ratio, may be used in operational surveys. In actual field operations, broad-band colour photography may be used for the observation of colour differences, and of colour quality (e.g. blue, green, red, or yellow) for descriptive mapping of the area under observation. Colour photography can be used in conjunction with ship surveys (for surface truth) to determine boundaries of regions of the water mass having similar characteristics. When useful indices can be defined, then water colour spectrometers, or multiple-lens cameras with specially selected spectral windows, may be used to map isopleths of constant values of the defined parameters. It is

[504]

essential that any colour measurement technique for marine application have a measurement capability well into the blue portion of the spectrum to maximize water penetration for bathymetric measurements and to measure colour changes associated with relatively low concentrations of phytoplankton.

Research in progress

During the first half of 1970, a series of 17 flight missions for the measurement of water colour was carried out under the direction of Dr. G. Ewing, Woods Hole Oceanographic Institution. These flights were co-ordinated with shipboard measurements of phytoplankton and of light attenuation in the water column. Water colour measurements were made over the Gulf of California, along the western coast of Mexico, in the Gulf of Panama, the Caribbean Sea and the Gulf of Mexico. The visible water colours ranged from bright blue, through various shades of green and brown, to the bright salmon pink of red tide patches at the head of the Gulf of California. The results, using a water colour spectrometer, are now being analysed.

A cooperative experiment under the sponsorship of the U.S. Coast Guard, for the comparison and evaluation of techniques for oil slick detection, was carried out off the California coast for a 6-week period in October 1970. Participants in this experiment were the University of Michigan, NASA, the Aerojet-General Corporation and the U.S. Naval Research Laboratory using multi-spectral scanners and colour photography, and microwave radiometry for the detection of oil slicks. The U.S. Naval Research Laboratory will continue experiments in mapping oil slicks with four-frequency side-looking airborne radar, as described by Guinard and Purves (1970) and measurements will be made of controlled oil plumes discharged from a surface support vessel.

Conclusion

Preliminary experiments have demonstrated the feasibility of using multi-spectral water colour parameters as potential indicators of defined marine pollution. An airborne instrumentation suite consisting of a multi-lens camera, thermal (8 µ to 13 µ) radiometer, and ultra-violet radiometer (0.32 µ to 0.38 µ) will provide a flexible capability for measuring water colour and detecting oil slicks from light aircraft with minimum requirements for ground support capability for data reduction. The operational efficiency of such a system will depend upon the user's definition of pollution, and upon the definition of measurable pollution indices.

It is recommended that such a system be fabricated and put into immediate use. Two of the spectral windows of the multi-spectral camera would be selected as 0.46 µ and 0.54 µ for determination of the Clarke–Ewing–Lorenzen Color Ratio. A third window would be selected in the region of 0.65 µ for detection of thick oil films and of suspended sediments in the water mass. A fourth window would be available for the user to select as required by the specific application. The thermal infra-red radiometer would be used for detection of the thick portion of oil slicks and for estimation of local circulation patterns from measurements of the sea surface temperature structure. The ultra-violet radiometer would be used to detect thin oil films and oil dispersant mixtures. Final specifications for such a system might need to be revised slightly on the basis of results from field experiments (October 1970 through December 1970) and data analyses.

Basic experiments are needed to define additional measurable parameters and to develop an all-weather microwave capability for monitoring marine pollution. A continuous evolution process, strongly dependent upon user interaction with the system designer, will be necessary for developing a reliable, sophisticated, capability for all-weather measurements of a wide range of variables definable as marine pollution.

References

AUKLAND, J C, CONWAY, W H and SAUNDERS, N K   Detection of
1969   oil slick pollution on water surfaces with microwave radiometer systems. *Proc. int. Symp. remote Sensing Envir.*, 6:789–96.

BAILEY, J S   Flight test of an ocean color measuring system. Final
1969   Report. Contract N62306–69–C–0024. TRW Systems Group, Sensor Systems Laboratory, Display and Imaging Department, April 1969.

CLARKE, G L, EWING, G C and LORENZEN, C J   Remote measure-
1969   ment of ocean color as an index of biological productivity. *Proc. int. Symp. remote Sensing Envir.*, 6:991–9.

GUINARD, N W and PURVES, C G   The remote sensing of oil slicks
1970   by radar. Washington, D.C. 20591, U.S. Coast Guard, Office of Research and Development. Project No. 714104/A/004.

HORVATH, R, MORGAN, W L and SPELLICY, R   Measurements
1970   program for oil slick characteristics. University of Michigan, Infrared and Optics Laboratory, Willow Run Laboratories, Institute of Science and Technology, Rep. (2766–7–F).

LOWE, D C and HASELL, P G   Multispectral sensing of oil pollution.
1969   *Proc. int. Symp. remote Sensing Envir.*, 6:765–75.

PARIS, J F   Microwave radiometry and its application to marine
1969   meteorology and oceanography. Texas A & M University, Department of Oceanography, Ref. No. (69–IT).

ROSS, D B, CARDONE, V and CONAWAY, J   Laser and microwave
1970   observations of sea surface condition for fetch limited 35 to 50 knot winds. *IEEE Trans. Geosci. Electron.*, GE–8 Special issue (4):476–87.

ROSS, D, FEDELCHAK, P and JENSEN, R   Experiments in oceano-
1969   graphic aerospace photography—I—*BEN FRANKLIN* spectral filter tests. Prepared for the U.S. Naval Oceanographic Office under Contract N62306–69–C–0072 by Philco, Ford Corporation, Palo Alto, California 94303.

STEWART, R, SPELLICY, R and POLCYN, F   Analysis of multispectral
1970   data of the Santa Barbara oil slick. University of Michigan, Infrared and Optics Laboratory, Willow Run Laboratories, Institute of Science and Technology, Rep. (3340–4–F).

The Constructive Use of Sewage, with Particular Reference to Fish Culture

G. H. Allen*

L'utilisation pratique des eaux usées, notamment en pisciculture

D'importantes récoltes de poisson proviennent de systèmes aquatiques fertilisés par des déchets d'origine animale, y compris ceux de l'homme. C'est là un fait qui est rarement mentionné dans la littérature scientifique des sociétés industrialisées d'Occident. Les idées occidentales relatives à la santé publique ont sans aucun doute fait obstacle à l'utilisation des eaux usées à cette fin. Cependant, l'usage des "bassins d'oxydation" par les ingénieurs sanitaires est largement répandu aux Etats-Unis et dans le monde. Ces bassins sont analogues aux nappes d'eaux naturelles qui sont fertilisées par le déversement d'eaux usées.

Les biologistes des pêches s'occupent depuis longtemps de l'utilisation des engrais dans les étangs artificiels ou les eaux naturelles pour accroître les récoltes aquatiques. Par conséquent, les spécialistes de la pisciculture et les ingénieurs sanitaires se trouvent en présence de problèmes qui se recoupent et dans leurs travaux ils ont recours à des connaissances analogues. Un tel chevauchement se produit actuellement entre ces deux domaines dans la littérature qu'il annonce la venue d'une approche interdisciplinaire face à des problèmes communs.

Il est possible d'utiliser les nutriments contenus dans les déchets des eaux usées pour cultiver des produits aquatiques récoltables au sein de systèmes aménagés dans des conditions hygiéniques. On peut tirer des avantages économiques et sociaux substantiels d'une exploitation de la thèse de Spilhaus "Les déchets sont simplement quelque substance utile que nous n'avons pas encore eu l'intelligence d'utiliser".

Aprovechamiento constructivo de las aguas residuales con especial referencia a la piscicultura

En el mundo se recoge una enorme cantidad de pescado en los sistemas acuáticos fertilizados con desperdicios de origen animal, incluyendo los del hombre. Este hecho se menciona raramente en la literatura científica de las sociedades occidentales industrializadas. Las actitudes que se adoptan en Occidente sobre la sanidad pública, sin duda alguna se han suavizado en relación con este aprovechamiento de las aguas residuales. No obstante, en los Estados Unidos y en el mundo los ingenieros sanitarios emplean en gran medida "estanques de oxidación", que son análogos a las aguas naturales fertilizadas por las descargas procedentes de las aguas de albañal.

Los biólogos de pesca se han ocupado durante mucho tiempo del empleo de los fertilizantes en estanques artificiales o aguas naturales para aumentar la recogida de peces. Por esto, los problemas de los piscicultores e ingenieros sanitarios se superponen extrayendo un conocimiento similar de su labor. El grado de superposición que ahora se da en la literatura sobre estas especialidades pronostica que a la solución de estos problemas se aplicarán muchas disciplinas.

Los nutrientes existentes en las aguas cloacales se pueden emplear para obtener productos acuáticos recolectables en sistemas sanitarios dirigidos. Se pueden lograr importantes beneficios económicos y sociales poniendo en ejecución la tesis de Spilhaus: "Los desechos son simplemente sustancias útiles que todavía no hemos sabido aprovechar".

THIS paper reviews current utilization of sewage in aquaculture, with particular emphasis on fish culture as developed in Germany, United States, Israel, and India.

One of the earliest papers on the subject comes from German literature of the late 19th century and reports on the treatment of city sewage in "sewage stabilization lagoons" which doubled as fish ponds. Hofer (1914) summarized these early German experiences and another major work on the subject was produced by Demoll (1926). Increasing industrialization with its concomitant wastes, usually highly toxic to fish life, tended to limit, if not entirely eliminate, the ability to rear fish in these purification systems, especially where there was insufficient water to mix with the raw or partly processed effluents. The same problem occurred in fish culture areas surrounding Calcutta (Bose, 1944).

In Germany, several sewage fish farm systems have persisted, however, including the unique system adjacent to Munich (Anon n. d.). Many factors have combined to make this system successful: (1) presence of wetlands available for ponds and lakes; (2) low-cost primary-treated sewage effluent from Munich; (3) high-quality diluting water from the Isar River; (4) good local market for carp; (5) purified water (summer—from fish ponds; winter—from Speicher See) immediately usable for power generation; (6) sport-fishing rights on Speicher See providing additional income; and (7) value of the system as a wildlife refuge. The Bavarian Biological Institute at Munich has published extensively on research conducted on the Munich Sewage fish pond complex (Liebmann, 1960; Reichenbach-Klinke, 1963; Kaufmann, 1958; Von Ammon and Stammer, 1958; by Von Schillinger from 1924 to 1949; Kisskalt and Ilzhofer, 1937).

The first interest in the United States seems to have coincided with the use of "oxidation ponds" or "sewage stabilization lagoons" as purification systems. Fitzgerald and Rohlich (1958) in evaluating stabilization pond literature considered the economics of oxidation ponds associated with the fertility of sewage pond water and muds. Despite this, few reports have mentioned the possibility of adding aquatic-related values into the systems, even though such units would be a non-consumptive use of water with much potential economic return (Tarzwell, 1966).

Agricultural re-use of waste water is reported to be practised in United States, Israel, South Africa, India, Japan, Hungary, Poland, and a few areas in South America. California has actively been pursuing waste water reclamation (Bush and Mulford, 1954; Merz, 1955, 1956). Deaner (1969) found approximately 200 reclamation facilities in California in 1968. Of various uses listed, irrigation is the most common in California (table 1). Use of reclaimed water for aquatic production is still in its infancy in California with most being used for fishing in recreational lakes. Scicluna-Spiteri (1969) also found that most of the treated sewage water in the world is used for crop irrigation. Bush and Mulford (1954) indicated that waste water, although about one-half of one per cent of California's total water resource, amounted to about ten per cent of the water from developed supplies and, hence, may be important locally. In Israel sewage is estimated to be about ten per cent of the total water potential of the country (Feinmesser, 1963).

A multi-purpose recreational system for using reclaimed sewage has been developed at Santee, California (Merrell et al. 1965, 1967). The most significant result has been the public approval gained. The United States government now appears to be increasing its unilateral aid to other countries for aquaculture in food production schemes (Bardach and Ryther, 1968; Ryther and Bardach, 1968). This increased interest was commented upon by Randal (1968) who believes that the lack of knowledge and manpower combined with public indifference will be serious obstacles to increased crops

* School of Natural Resources, Humboldt State College, Arcata, California 95521, U.S.A.

from aquaculture. A proposed cutback in aquaculture within the Bureau of Commercial Fisheries (Carter, 1970) is indicative of some ambiguity within U.S. Government. Worldwide, however, there will undoubtedly be increased research and development of waste water reclamation, and fishery uses will compete for such reclaimed water.

In the United States Huggins and Backmann (1969) used an existing terminal unit in a sewage treatment facility for catfish culture. An oxidation pond at Arcata, California was found to be marginal for fish life (Hansen, 1967), but the water when diluted was not (Allen and O'Brien, 1967); consequently, fish culture ponds are being constructed within the existing oxidation pond in which treated effluent water will be used to fertilize salt water for rearing salmonid fishes to migratory size. A third study (Odum and Chestnut, 1970, and collaborators) is a multi-disciplinary approach aimed at understanding the structure and functioning of ecosystems associated with estuarine and fresh waters enriched with treated effluents.

Fish production from sewage fertilized water

The increase in fish production by fertilizing waters is well documented (table 2) (Viosca, 1935; Bardach, 1968; Mihursky, 1967; Odum and Odum, 1966, tables 5–9). Recently Sreenivasan working in India, and Hepher in Israel, have expressed results as the yield of fish to man as a percentage of the estimated carbon produced at the lowest trophic level (Lindeman, 1942) (table 3). The highest yields were associated with the addition of

TABLE 1. WASTEWATER UTILIZATION IN CALIFORNIA FOR 1968 (Deaner, 1969)

Non-Agricultural Uses, Irrigation:
   Golf Course 24, Landscaping 14, Athletic and Parade
   Fields 5, Parks 3
                                     Total 46

Non-Irrigation:
   Ground Water Recharge 8, Cooling Water 4, Recreational and Storage Lakes 8, Miscellaneous 6
                                       Total 26

Agricultural Uses, Irrigation.:
   Pasture 63, "Fodder Crops" 10, Alfalfa 28.
                                     Total 101

Grains:
   Corn 15, Barley 14, "Grain" 4, Miscellaneous 7
                                     Total 40

Cotton 31
                                     Total 31

Fruit and Nut Trees:
   Peaches 4, Walnut 4, Orchard 3, Miscellaneous 7
                                     Total 18

   Other trees 2, Vineyard 9, Truck crops 6, Miscellaneous 3
                                     Total 20

GRAND TOTAL 282

TABLE 2. COMPARATIVE FISH PRODUCTION AS LBS PER ACRE PER ANNUM (from Mihursky, 1967, adapted from table 1 and 2 of Mortimer and Hickling, 1954).

Wild water	Lbs per acre	Pond culture	Lbs per acre
Swiss and German Alpine lakes	12.9	Germany	200–400
Fresh water of England and Wales	20.6	Yugoslavia	366
Lake Mendota, Wisconsin	22.0	Israel	2000
Lake Nakivali, Uganda	168.0	China	2800–6000
Lake Ktangiri, Tanganyika (Tanzania)	282.0		
Lake Waubesa, Wisconsin	400.0		

TABLE 3. ESTIMATES OF FISH YIELD IN VARIOUS AQUATIC ECOSYSTEMS EXPRESSED AS PERCENTAGE OF PRIMARY PRODUCTION OF CARBON HARVESTED AS FISH

System	Percentage	Author
Fort Moat (India) with some domestic sewage	2.3	Sreenivasan, 1968
K. C. Kulam (India) with some domestic sewage	5.4	Sreenivasan, 1968
Vellore Moat (India) with considerable domestic sewage	2.7	Sreenivasan, (1964 a)
Israel, fertilized ponds	2.3	Hepher (1962) as quoted in Sreenivasan, 1964
Vellore Moat (India) with considerable domestic sewage and mixed species	7.3	Sreenivasan, (1964 a)
Tidal marsh, wild waters	2.0	Odum as quoted in Sreenivasan, 1964
Small desert impoundment, wild water	0.98	McConnel (1936) as quoted in Sreenivasan, 1964

TABLE 4. MUSSEL PRODUCTION AS TONS PER ACRE PER ANNUM

Category of water	Tons per acre	Author
Natural water, relatively unfertilized, no culture	5	Yount
Spanish estuaries subjected to sewage fertilization and rope culture from boats	120	Bardach and Ryther, 1968; (see photographs in Pirie, 1967)

TABLE 5. CLASSIFICATION OF EXPERIMENTS ON FISH POPULATIONS BASED ON DEGREE OF CONTROL IMPOSED (from Backiel, 1968, Table 13.1).

Environment controlled or changed	Fish population controlled or changed		
	Fully	Partially	Not at all
Fully	1. Aquarium experiments	2. Aquarium experiments	3. Aquarium experiments
Partially	4. Aquarium and pond experiments; poisoning small lakes	5. Experiments in ponds and natural waters	6. Experiments in ponds and natural waters
Not at all	7. Experiments in ponds and natural waters	8. Experiments in natural waters	9. No experimenting

fertilizers plus supplemental feeding and the use of several species of fish, including at least one herbivore. Various marine aquatic invertebrates feeding directly on microscopic organisms show natural productions exceeding the best harvest from intensively managed fish ponds, and when cultured can provide exceedingly high harvests (table 4). Thus aquaculturists attempting to increase protein yields seek species adaptable to pond culture that eat plants, or animals only one step above the plants in the food chain (Birtwistle, 1931; Opuszynski, 1964; Nair, 1968).

In many cases, increasing the production of fish may be limited by consumer resistance. Thus elimination by breeding of features in species excellently adapted for pond culture (e.g. muscular bones in carp—Meske, 1968) would open new markets, and thus stimulate increased production. At present, however, attempts of

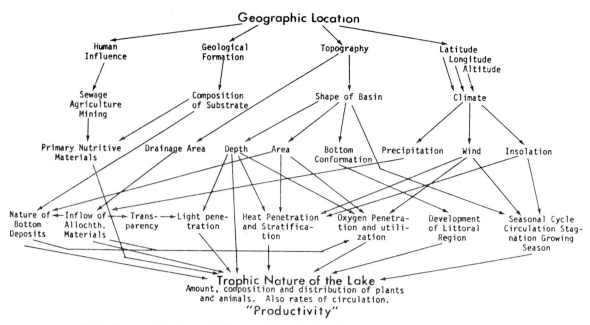

Fig 1. Factors affecting metabolism of a lake, developed by Mihursky (1967) as adapted from limnological literature

this nature lag behind similar work in agriculture (Riggs and Sneed, 1959; Donaldson, 1968; Bardach, 1968).

Fish raised in oxidation ponds readily acquire objectionable tastes and odours from their environment (Aschner *et al.*, 1967) and strict quality control prior to marketing is required. Cleansing of fish, or shellfish involves holding the animals in pure water for appropriate times. This is done for carp raised in the Munich system. Thus, biological competence and an understanding of fish processing, distributing, and marketing is required for good production.

Biology of sewage fish pond systems

The most important works summarizing all or major portions of the biology of fish ponds are: Hofer (1914–15), Demoll (1926), Fehlmann (1928), Kisskalt and Ilzhofer (1937), Hora (1944), Vaas (1948), Kaufmann (1958), Wolny (1962, 1962a), Liebmann (1960), Hickling (1962), and Reichenbach-Klinke (1963).

Hickling (1962) included many sections on use of sewage effluents for fish culture, drawing mainly from the important, now out-of-print work of Mortimer and Hickling (1954).

As originally outlined by Forbes (1925, see reprint in Keup *el al.*, 1967) the biology of sewage ponds is very complex (fig 1). Although organic production by plants is generally correlated with amounts of nutrients, the level of primary production is controlled by Liebig's "Law of the Minimum". A number of workers have considered the factors limiting organic production, e.g. Fogg (1966) on algae production; Oglesby and Edmonson (1966) on processes involved in eutrophication in lakes; Ludwig (Water Pollution Control Federation, 1964) on operation of stabilization ponds; and Neess (1949) on increasing fish production in ponds by fertilization. No species encounters in any given habitat the optimum conditions for all its functions (Darling, 1967) and achieving optimum conditions for fish production is the art and science of the fish biologist and fish culturist

(Schäperclaus, 1967; Pillay, 1965, 1967; Huet and Timmermans, 1970; Hora and Pillay, 1962).

Study needed

The possibilities for advancing general knowledge of aquatic ecosystems in the applied field of sewage use in fish ponds remains relatively unexploited. Neess (1949) concluded that "It is not only necessary to be familiar with beginning and ending quantities, but to attack intervening processes as well. This attack can be made through a deeper understanding of the place of such subjects as colloid chemistry, microbiology, biochemistry, and the thermodynamics of aquatic environments in pond fertilization." But in 1968, Backiel still was only able to say "There is a large literature concerning pond experiments in a great variety of languages. Unfortunately most pond experiments have not been conducted so as to shed light on the fundamental processes of production and the flow of energy through the fish population and ecosystem generally."

In ecological terms sewage fish pond systems are young highly productive but relatively unstable ecosystems (Odum, 1969). Older systems tend to be more complex, less productive, and more stable. Sewage fish pond systems lie half-way between controlled laboratory situations and highly eutrophic natural environments. This intermediate status is reflected in a classification of experiments on fish populations proposed by Backiel (1968) (table 5). The work by Huggins and Backmann (1969) plus that under way at Arcata, California, would represent experiments four and five in Backiel's scheme, while work under way in North Carolina (Odum and Chestnut, 1970) would represent experiments five and six.

The two most basic disciplines involved in sewage fish pond culture are sanitary engineering and aquaculture. Water pollution biology is of obvious applicability to sanitary engineering (Mackenthun, 1969; Hynes, 1966; Doudoroff and Katz, 1950), as is the use of ponds for treating waste waters (Fitzgerald and Rohlich, 1958;

Van Eck, 1959; Stander, *et al.*, 1970). The prevention and control of human disease (public health) would be of common concern to both sanitary engineer and aquaculturist. Oxidation ponds at Santee appeared effective in reducing virus (Askew *et al.*, 1965).

Literature on the primary plant production in aquatic systems is especially important (Ivlev, 1966; Ricker, 1968; Fogg, 1966). Applied studies in a number of areas provide basic information applicable to the biology of sewage-fertilized fish ponds. The literature on accelerated aging of lakes (eutrophication) is very pertinent (Stewart and Rohlich, 1967), as is the study of fjords, lochs, and brackish water impoundments which have been deliberately fertilized to increase fish production (Buljan, 1961—Yugoslavia; Bowers, 1966—North Sea area; Raymont, 1949—Scotland; Shelbourne, 1964—North Sea). Also relevant are those investigations of natural fish populations reacting to sewage effluents entering natural waters (Alabaster, 1959; Brinley, 1943; Allan *et al.*, 1958; Anon., 1946; Tsai, 1968). Space biology should not be overlooked since "adaptation of the techniques of spacecraft air and water pollution detection and quantitation promises to be a major spin-off of bioastronautics research to the environmental crisis of Earth" (Rolan, 1970).

Holm *et al.* (1969) suggested herbivores for the control of aquatic weeds (snails, beetles, wingless aquatic grasshoppers, four kinds of fish, three kinds of birds and four kinds of mammals) whereas Singh (1961) suggested that herbivorous fishes harvesting aquatic plants be used to increase food production for human populations. DeBont and DeBont Hers (1952) suggested that snail-eating fish be used for vector control in public health. The wedding of all such activities for water purification and aquaculture appears as a natural development.

Rodhe (1958) pointed out the importance of nanno-plankton assessment in productivity studies. Increasing attention is being directed toward the importance of the water-mud interface in controlling nutrient cycling (Mortimer, 1941, 1942; Lenhard, 1965; Deevey, 1970; Pravdic, 1970). Agricultural research on muds associated with rice culture will probably add significantly to the knowledge of pond mud chemistry. The importance of the air-water interface in aquatic systems is receiving more attention (Parker and Barsom, 1970).

Re-aeration of waters by mechanical means will have direct application to sewage fish pond biology (Bayley and Wyatt, 1961—polluted rivers; Hooper *et al.*, 1952, and Johnson, 1966—to improve fish production in small lakes; Bayley, 1963—engineering studies on aeration in sewage treatment). If aquaculture systems can be incorporated at the design stage of sewage treatment facilities, low-cost connections between sewage treatment aeration systems and aquaculture units would permit aeration during periods of oxygen depletion.

Resistance to re-use of sewage in fish culture

Pirie (1967) listed three parameters involving resistance to a suitable new food product, namely "(1) total; (2) quasi-logical, and (3) 'instant'. Total opposition is the denial of the problem. Quasi-logical opposition comes from economists, and 'instant opposition' arises because innovators are apt to irritate right-minded people, and enthusiasm invites skepticism. The innovators must, therefore, expect to run into trouble."

Our present level of technology in public health and medical knowledge in general, is such that objections by public health agencies to the re-use of sewage water can be answered with appropriate data. The experiences at Munich and Santee should be so utilized, as well as the emergency operations at Chanute, Kansas (Metzler *et al.*, 1958). In addition economic factors will favour aquaculture units in sewage treatment systems as public pressure for maintaining or improving natural waters and the need for animal protein increases.

Research and development possibilities

Biological treatment of wastes through the production of oxygen by algae in oxidation ponds has received considerable attention (Silva and Papenfuss, 1953). The objective is to eliminate nutrients by harvesting the algae, while the oxygen produced is available for use by bacteria in decomposition of the organic matter in the waste. Also, the algae apparently produce substances inhibitory to bacteria. Under aerobic conditions, however, nitrogen fixation occurs in some algae so that nitrogen may increase in treated water. The algal cells themselves contribute to the BOD load which must be met eventually in a receiving water. Thus more complex biological systems under semi-controlled conditions are worth exploring from both the waste water treatment viewpoint, and certainly from a fisheries point of view, since a product with sale value can be involved. Goldshmid (1970) noted that fish are an asset in the Israel National Water System of reservoirs by consuming biota deleterious to water quality, and that introduction of appropriate species is now being advocated.

Biological treatment systems tend to be geared to large-scale operations because urbanization and industrialization produce huge volumes of waste waters. More attention might be given to studying systems applicable to small volumes of water. Reichenbach-Klinke (1963) has written on the application of results at Munich to small and middle-sized towns. There are also possibilities for population units smaller than a small town. One private company is attempting to develop a trickling filter employing shredded redwood bark. Other products such as sawdust, fibrous matter of any kind, even the small plastic particles currently used in industrial packaging, might be successfully employed. Such a unit placed in series with a septic tank or some other primary treatment unit could provide an adequate effluent for small-scale watering of gardens and crops, fertilizing fish ponds, or maintaining water level in recreational lakes. Gibbons (1962) lists a large number of uses, including human nutrition, for cattails, a fact which might be incorporated into a pond scheme involving either fish production or recreational use.

Recently in the United States, the rearing of catfish in ponds has had a spectacular growth (U.S., 1970), and experiments are under way to utilize sewage effluents (Huggins and Backmann, 1969). A large food industry might find it profitable to organize such production from a series of small waste water treatment plants.

In agricultural communities throughout the world, health organizations have improved public sanitation by proper location and construction of community wells.

It is not a great technical advance to develop schemes for the collection of waste waters in oxidation ponds which could be planned and managed for fish production with the assistance of a fish culturist.

Undoubtedly further study of literature in allied fields will suggest new ideas for sewage fish culture. For example, Wurtz (1960) suggested (1) possible new species of fish for pond culture such as a cross between male *Coregonus albula* (native to Polish lakes) and female *Coregonus lavaretus*. These hybrids can live in shallow ponds with stagnant water and stand temperatures of up to 28°C. After one year they attain 50 g in weight and in the second year from 150 to 200 g; (2) a method for the more complete use of primary plant production with trial rearing of beavers such as the coypu *Myocastor coypus*—a South American aquatic rodent—which not only furnishes the very expensive nutria . . . but also browses the grasses found in ponds and converts them into organic matter and fertilizer very useful for the fish; and (3) an inexpensive analytical tool for productivity measurement: by immersing cotton twine into the water or mud and observing the rate of its decomposition.

Hynes (1966) suggested harvesting *Sphaerotilus* (a species associated with organically enriched waters) as food for other species, while Nikolsky (1963) discusses carp-duck combinations. Yount and Crossman (1966) in experimenting with water purification using water hyacinth found that turtles feed on this aquatic vegetation which clogs streams in areas where it has been introduced. Manatee control of water hyacinth is being investigated (Holm *et al.*, 1969; Allsopp, 1960) and their flesh might be utilized if controlled reproduction in captivity can be accomplished. *Daphnia-Lemna* system of biological treatment of sewage wastes (Ehrlich, 1966) suggests a fish-bird (carp-duck) component to obtain economic benefits beyond that of the primary value of water purification.

Experiments with freshwater mussels would suggest freshwater mussel culture in sewage effluents. In addition to rearing fishes used directly for food, it might be possible to rear short-lived, fast-growing species which could either be sold for bait (Dendy, 1966) or sold to feed species bringing higher prices such as trout reared in saltwater ponds (Jensen, 1967).

Low lying swampy lands, instead of being reclaimed at large expense, could be used as receiving waters for sewage effluents to enhance their role as refuges for aquatic life (e.g. part of Munich system). The greatest numbers and kinds of aquatic birds on north Humboldt Bay generally occur on the oxidation pond.

Daphnia from oxidation ponds have been captured by screening the outlet and selling them as fish food for aquarists in northern California. A similar source of *Daphnia* for young salmon and trout has been studied by DeWitt (1969). This principle could be used at any oxidation pond with fish rearing ponds in series or parallel. Dendy (1963) illustrated a method of concentrating aquatic organisms by suspending a series of fibrous mats in which aquatic organisms could be grown. The mats would be removed in rotation to adjacent fish ponds where the food could be shaken loose.

Although not compatible with their culture as food, aquatic organisms reared in terminal units of a sewage treatment facility could be used as monitors of undesirable concentrations of substances (methyl mercury, Abelson, 1970). This concept is used extensively in monitoring the radioactive discharges from nuclear power stations (Preston, 1967; Leet, 1969).

Tsai (1968) in reviewing the effects of sewage effluents on fish noted many reports of congregations of fish in the vicinity of sewage outfalls and that fish could live in undiluted sewage effluent if dissolved oxygen was high. This must have been known to fish culturists from practical experience in order for the basket culture of carp in highly fertilized but flowing natural waters to have developed (Vaas and Sachlan, 1956).

Conclusion

There is good reason to believe that worldwide there will be an increase in the use of the fertilizing potential of waste waters for fish and shellfish culture. Even in the United States, Tarzwell (1966) suggested "research should be carried out to develop methods by which micro-organisms could convert waste organic materials into useful products. Such processes would reduce the wastage of valuable materials, assist pollution abatement, and be an incentive to industry and municipalities to carry out more effective treatment of all waste waters. The development of such processes presents problems in applied ecology, physiology, biochemistry, engineering, and economics."

It seems likely that significant advances in this field will be made by that country or agency which can develop the necessary organization for a sustained programme. I also suspect that those who first control a substantial part of waste water by putting it to beneficial use will have established some sort of prior right, even if only a moral one. This would parallel the underlying philosophy of water right established by prior use as developed in the western portions of North America.

Economists are now studying methods by which the cost of water pollution can best be allocated (Kneese, 1968). However, the public might not wait for the development of the most efficient economic arrangements (Wildavsky, 1967; Hendee *et al.*, 1969). When public pressure is great enough, high quality domestic effluent could be placed in a storage reservoir for irrigation in which trout could flourish (Culp, 1969). Industrial concerns may be increasingly required to discharge pretreated wastes into domestic sewage systems to avoid malfunctionings of the central facility. As a result the quality of effluents from conventional treatment facilities could improve rapidly, and become more valuable for all types of re-use, including fish culture.

One of the more fundamental reasons for using sewage as a fertilizer is the long-term need for cheap food (Bardach, 1968). Selective breeding and diet research has resulted in food conversion ratios approaching 1:1 in salmon and trout culture, but the diets require high-cost, high quality fish protein ingredients. In the long run, aquaculture will have to compete with other consumers of animal protein (cattle, hogs, poultry, and in certain regions human populations).

Aquatic ecologists must involve themselves in environmental problems perhaps under new organizational arrangements. Viewing sewage effluents as valuable natural resources and as a scientific tool to understanding

[510

aquatic ecosystems should have aesthetic appeal to many scientists.

Qualified individuals should begin work in such applied fields as sewage utilization in food production. The intense concern of youth worldwide for relevant and environmentally oriented activities seems to offer a tremendous opportunity for channelling vigorous, idealistic people into a constructive endeavour.

Governmental and private agencies involved in public health, pollution control, food production, or any related area, must give more than moral support to those who do become interested in such work.

Appropriate agencies should organize a national conference, symposium, or at least a publication of invited papers, on the regional status of utilization of sewage effluents in food production, or any particular subdivision such as fish culture. Utilization of sewage effluents for food production should be given special attention at forthcoming symposiums on aquaculture, fish culture, pond culture, etc. Another approach might be to invite workers with appropriate specialities to organize the current state of knowledge in their field as applicable to sewage re-utilization schemes.

An appropriate international agency, or private foundation, could be solicited for funding an international programme of research and training in using waste waters in integrated agriculture-aquaculture systems, particularly by the use of unorthodox schemes for handling industrial or domestic effluents in food production.

References

ABELSON, P H Methylmercury. Editorial. *Science, N. Y.* 169(3942):
1970 237.

ALABASTER, J S The effect of a sewage effluent on the distribution
1959 of dissolved oxygen and fish in a stream. *J. Anim. Ecol.*, 28:
283–91.

ALLAN, I R H, HERBERT, D W M and ALABASTER, J S A field and
1958 laboratory investigation of fish in a sewage effluent. *Fishery
Invest., Lond.* (1), 6(2):1–76.

ALLEN, G H A preliminary bibliography on the utilization of
1969 sewage in fish culture. *FAO Fish. Circ.*, (308):15 p.

ALLEN, G H and O'BRIEN, P Preliminary experiments on the
1967 acclimatization of juvenile King salmon, *Oncorhynchus
tshawytscha*, to saline water mixed with sewage pond
effluent. *Calif. Fish Game*, 53(3):180–4.

ALLSOPP, W H L The manatee: ecology and use for weed control.
1960 *Nature, Lond.*, 188(4752):762.

ASCHNER, B, LAVENTER, C and CHORIN-KIRSCH, I Off-flavour in
1967 carp from ponds in coastal plain and the Galil. *Bamidgeh*,
19(1):23–5.

ASKEW, J B, et al. Microbiology of reclaimed water from sewage
1965 for recreational use. *Am. J. Publ. Hlth*, 55(3):453–62.

BACKIEL, T The experimental approach. *In* Methods for assess-
1968 ment of fish production in fresh waters, edited by W. E.
Ricker. *IBP Handbk.*, (3):246–51.

BARDACH, J E Aquaculture. Husbandry of aquatic animals can
1968 contribute increasingly to supplies of high-grade protein
food. *Science, N. Y.*, 161(3846):1098–106.

BARDACH, J E and RYTHER, J H The status and potential of
1968 aquaculture, particularly fish culture. Vol. 2, Pts 1 and 3.
Springfield, Virginia, Clearinghouse for Federal Scientific
and Technical Information, Doc. (PB. 177,768):255 p.

BAYLEY, R W Aeration by diffuse air bubbles: a cost analysis.
1963 *J. Proc. Inst. Sew. Purif.*, 1963(2):174–81.

BAYLEY, R W and WYATT, K Aeration of effluents by venturi
1961 tubes. *Wat. Waste Treatmt*, Nov./Dec.

BIRTWISTLE, W Rearing of carp in ponds, *Malay. agric. J.*, 19(8):
1931 372–83.

BOSE, P C Calcutta sewage and fish culture. *Proc. natn. Inst. Sci.
1944 India*, 10(4): 443–54.

BOWERS, A B Farming marine fish. *Sci. J., Lond.*, 2(6): 46–51.
1966

BRINLEY, F J Sewage, algae and fish. *Sewage Wks J.*, Easton Pa.,
1943 15(1):78–83.

BUSH, A F and MULFORD, S F Studies on "Waste water reclama-
1954 tion and utilization." *Publs St. Wat. Pollut. Control Bd
Calif.*, (9):82 p.

BULJAN, M Some results of fertilization experiments carried out in
1961 Yugoslav marine bays. *Proc. tech. Pap. gen. Fish. Coun.
Mediterr.*, (6):237–43.

CARTER, L J Fisheries Research: Rejuggling of priorities is
1970 assailed. *Science, N. Y.* 167: 1471–2.

CULP, R Water reclamation at South Tahoe. *Wat. Wastes Engng*,
1969 1969 (April):36–9.

CUTTING, C L Fish saving, a history of fish processing from an-
1955 cient to modern times. London, Leonard Hill (Books) Ltd.,
372 p.

DARLING, F F A wider environment of ecology and conservation.
1967 *Daedalus*, 96(4):1003–19.

DEANER, D G Directory of wastewater reclamation operations in
1969 California. Berkeley, Calif., California State Department of
Public Health, Bureau of Sanitary Engineering, 45 p.

DEBONT, A F and DEBONT HERS, M J Mollusc control and fish
1952 farming in Central Africa. *Nature, Lond.*, 170:323–4.

DEEVEY, E S In defense of mud. *Bull. ecol. Soc. Am.*, 51(1):5–8.
1970

DENDY, J S Living food for aquatic animals. *Turtox News*, 41(10):
1963 258–9.

DENDY, J S Farm ponds, *In* Limnology in North America, edited
1966 by D. G. Frey, Madison, University of Wisconsin Press, pp.
595–620.

DEMOLL, R Die Reinigung von Abwässern in Fischteichen.
1926 *Handb. Binnenfisch. Mitteleur.*, 6(2):222–62.

DEWITT, J W, JR. The pond, lagoon, bay, estuary, and impound-
1969 ment culture of anadromous and marine fishes, with emphasis on the culture of salmon and trout, along the Pacific
coast of the United States. U.S. Department of Commerce,
Economic Development Administration, Technical Assistance project, Feb., 1969: 36 p.

DONALDSON, L R Selective breeding of salmonoid fishes. *In*
1968 Marine aquaculture, edited by W. J. McNeil, Newport,
Oregon, Oregon State University Press, pp. 65–74.

DOUDOROFF, P and KATZ, M Critical review of literature on the
1950 toxicity of industrial wastes and their components to fish.
1. Alkalis, acids, and inorganic gases. *Sewage ind. wastes*,
22(11):1432–58.

EHRLICH, S Two experiments in the biological clarification of
1966 stabilization-pond effluents. *Hydrobiologia*, 27(1/2):70–80.

FEHLMANN, W Abwasserfischteiche. *Revue suisse Route*, (25/26):
1928 2–20.

FEINMESSER, A Survey of sewage utilization in Israel. Ministry of
1963 Agriculture, Water Commission, Water Utilization Division, 5 p.

FITZGERALD, G P and ROHLICH, G A An evaluation of stabiliza-
1958 tion pond literature. *Sewage ind. wastes*, 30(10):1213–24.

FOGG, G E Algal cultures and phytoplankton ecology. Madison,
1966 University of Wisconsin Press, 126 p.

FORBES, S A The lake as a microcosm. *Bull. Ill. nat. Hist. Surv.*,
1925 (15):537–50.

GIBBONS, E Stalking the wild asparagus. New York, David
1962 McKay Co., Inc. 303 p.

GOLDSHMID, Y Water quality management of the Israel National
1970 Water System. *In* Development in water quality research.
Proceedings of the Jerusalem International Conference on
Water Quality and Pollution Research, June 1969, edited by
H. I. Shuval, Ann Arbor, Humphrey Science Publishers,
pp. 3–17.

HANSEN, R J A study of some physical and chemical environment-
1967 al features of a large sewage oxidation pond. Arcata,
California, Humboldt State College, Thesis, 133 p.

HENDEE, J C, GALE, R P and HARRY, J Conservation, politics, and
1969 democracy. U.S. Department of Agriculture, U.S. Forest
Service (Reprint) 4 p.

HEPHER, B Primary production in fishponds and its application to
1962 fertilization experiments. *Limnol. Oceanogr.*, 7(2): 131–6.

HICKLING, C F Fish culture. London, Faber and Faber, 295 p.
1962

HOFER, H Teichdüngungsversuche. *Allg.FischZtg*, 39:139–44,
1914 166–70, 250–4, 326–33, 374–9. Also issued as *Ost. FischZtg*,
11:164–7, 179–81, 197–8.

HOFER, H Teichdüngungsversuche. *Allg.FischZtg*, 40: 251–6,
1915 266–70, 283–91. Also issued as *Ost. FischZtg*., 12:13–4,
70–2, 88–9.

HOLM, L G, WELDON, L W and BLACKBURN, R D Aquatic weeds.
1969 *Science, N. Y.*, 166(3906):699–709.

HOOPER, F F, BALL, R C and TANNER, H A An experiment in the
1952 artificial circulation of a small Michigan lake. *Trans. Am.
Fish. Soc.*, 82:222–41.

HORA, S L (ED.) Symposium on the utilization of sewage for fish
1944 culture. *Proc. natn. Inst. Sci. India*, 10(4):441–67.

HORA, S L and PILLAY, T V R   Handbook on fish culture in the
1962   Indo-Pacific Region. *FAO Fish. Biol. tech. Pap.*, (14): 203 p.

HUET, M and TIMMERMANS, J A   Traité de pisciculture. Brussels,
1970   Wyngaert, 718 p.

HUGGINS, T C and BACKMANN, R W   Production of channel catfish
1969   (*Ictalurus punctatus*) in tertiary treatment ponds. Ames, Iowa State University.

HYNES, H B N   The biology of polluted waters. Liverpool
1966   University Press. 202 p.

IVLEV, V S   (W. E. Ricker, Transl.) The biological productivity of
1966   waters. *J. Fish. Res. Bd Can.*, 23(11):1727–59. Translation of Biologicheskaia produktivnost' vodoemov, *Ush. sovrem. Biol.*, 19(1):98–120 (1945).

JENSEN, K W   Saltwater rearing of rainbow trout and salmon in
1967   Norway. *In* Symposium on feeding in trout and salmon culture, edited by J. L. Gaudet. *EIFAC tech. Pap.*, (3): 43–8.

JOHNSON, R C   The effect of artificial circulation on production of
1966   a thermally stratified lake. *Fish. Res. Pap. Wash. Dep. Fish.*, 2(4):5–15.

KAUFMANN, J   Chemische und biologische Untersuchungen an
1958   den Abwasserfischteichen von München. *Z. angew. Zool.*, 45(4):395–481.

KEUP, L E, INGRAM, W M and MACKENTHUM, K M   Biology of
1967   water pollution. A collection of selected papers on stream pollution, waste water, and water treatment. U.S. Department of the Interior, Federal Water Pollution Control Administration, (CWA-3):296 p.

KISSKALT, K und ILZHOFER, H   Die Reinigung von Abwasser in
1937   Fischteichen. *Arch. Hyg. Bakt.*, 118: 1–66.

KNEESE, A V   Why water pollution is economically unavoidable.
1968   *Trans-Action*, April:31–6.

KORMONDY, E J   Ecology and the environment of man. *Bio-*
1970   *Science*, 20(13):751–4.

LEET, W S   Accumulation of Zn-65 by oysters maintained in a
1969   discharge canal of a nuclear power plant. Thesis, Humboldt State College, 33 p.

LENHARD, G   Bottom deposits. A vital self-purification system in
1965   the degradation of polluting material in natural waters and in biological treatment of effluents. *Hydrobiologia*, 25: 404–11.

LIEBMANN, H   Biologie der Abwasserfischteiche. *In* Handbuch der
1960   Frischwasser-und Abwasser-Biologie, by H. Liebmann, München, Oldenbourg, Vol. 2:531–48.

LINDEMAN, R L   The trophic-dynamic aspect of ecology. *Ecology*,
1942   23(4):399–419.

MACKENTHUN, K M   The practice of water pollution biology. U.S.
1969   Department of the Interior, Federal Water Pollution Control Administration, Division of Technical Support, 281 p.

McCONNEL, W J   Primary productivity and fish harvest in a small
1936   desert impoundment. *Trans. Am. Fish. Soc.*, 92:1–12.

MERRELL, J C, JR, KATKO, A and PINTLER, H E   The Santee
1965   recreation project, Santee, California, Summary report, 1962–64. *Publ. Hlth, Serv. Publs, Wash.*, (999-WP-27): 69 p.

MERRELL, J C, *et al.*   Report on continued study of waste water
1956   reclamation and utilization. *Publs St. Wat. Pollut. Control Bd Calif.*, (15):89 p.

MERRELL, J C, *et, al.*   The Santee recreation project, Santee,
1967   California. Final Report. *Publ. Hlth Serv. Publs, Wash.*, (WP-20-7):165 p.

MERZ, R C   A survey of direct utilization of waste waters. *Publs*
1955   *St. Wat. Pollut. Control Bd Calif.*, (12):80 p.

MERZ, R C   Report on continued study of waste water reclama-
1956   tion and utilization. *Publs St. Wat. Pollut. Control Bd Calif.*, (15):89 p.

MESKE, CH   Breeding carp for reduced number of intermuscular
1968   bones, and growth of carp in aquaria. *Bamidgeh*, 20(4): 105–119.

METZLER, D F, *et al.*   Emergency use of reclaimed water for
1958   potable supply at Chanute, Kansas. *J. Am. Wat. Wks Ass.*, 50(8):1021.

MIHURSKY, J A   On possible constructive uses of thermal additions
1967   to estuaries. *BioScience*, 17(10):698–702.

MORTIMER, C H   The exchange of dissolved substances between
1941   mud and water in lakes. Pt. 2. *J. Ecol.*, 29:280–329.

MORTIMER, C H   The exchange of dissolved substances between
1942   mud and water in lakes. Pts 3 and 4. *J. Ecol.*, 30:147–201.

MORTIMER, C H and HICKLING, C F   Fertilizer in fishponds. A
1954   review and bibliography. *Fishery Publs colon. Off.*, 1(5): 155 p.

NAIR, K K   A preliminary bibliography of the grass carp, *Cteno-*
1968   *pharyngodon idella* Valenciennes. *FAO Fish. Circ.*, (302): 15 p.

NEESS, J C   Development and status of pond fertilization in central
1949   Europe. *Trans. Am. Fish. Soc.*, 76(1949):335–58.

NIKOLSKY, G V   The ecology of fishes. (L. Birkett, Transl.). New
1963   York, Academic Press, 352 p.

ODUM, E P   The strategy of ecosystem development. *Science, N.Y.*,
1969   164(3877):262–270.

ODUM, E P and ODUM, H T   Fundamentals of ecology. Phila-
1966   delphia, W. B. Sanders Co., 546 p.

ODUM, H T and CHESTNUT A F (Principal Investigators)   Studies
1970   on marine estuarine ecosystems developing with treated sewage wastes. Annual report for 1969–70 submitted to National Science Foundation, Sea Grants Projects Division, Grant #GH-18; N. Carolina Board of Science and Technology. Grant #180 and #232, by Institute of Marine Sciences, University of N. Carolina, 364 p.

OGLESBY, R T and EDMONDSON, W T   Control of eutrophication.
1966   *J. Wat. Pollut. Control Fedn*, 38(9): 1452–60.

OPUSZYNSKI, K   New possibilities of increasing fish production in
1964   pond cultures—the acclimatization of phytophagous fish. *Ekol. pol. (B)*, 10(3):201–14.

PARKER, B and BARSOM, G   Biological and chemical significance of
1970   surface microlayers in aquatic ecosystems. *BioScience*, 20(2):87–93.

PETER, W G III   Controlled fusion: a multifaceted approach to
1970   solving environmental degradation. *BioScience*, 20(12): 717–9.

PILLAY, T V R   A bibliography of brackish-water fish culture.
1965   *FAO Fish. Circ.*, (21):20 p.

PILLAY, T V R (ED.)   Proceedings of the FAO World symposium
1967–8   on warm-water pond fish culture, Rome, Italy, 18–25 May, 1966. *FAO Fish Rep.*, (44)Vol. 1–5:1600 p.

PIRIE, N W   Orthodox and unorthodox methods of meeting world
1967   food needs. *Scient. Am.*, 216(2):27–35.

POTTER, F M, JR.   Everyone wants to save the environment but no
1970   one knows quite what to do. *Center Mag.*, 3(2): 36–40.

PRAVDIC, V   Surface charge characterization of sea sediments.
1970   *Limnol. Oceanogr.*, 15(2): 230–3.

PRESTON, A   The concentration of ^{65}Zn in the flesh of oysters rela-
1967   ted to the discharge of cooling pond effluent from the C.E.G.B. nuclear power station at Bradwell-on-sea, Essex. *Proc. int. Symp. radioecol. Concent. Process.*, 1966:995–1004.

RANDAL, J   Aquaculture. *BioScience*, 18(10):979–80.
1968

RAYMONT, J E G   Further observations on changes in the bottom
1949   fauna of a fertilized sea loch. *J. mar. biol. Ass. U.K.*, 28(1): 9–19.

REICHENBACH-KLINKE, H   Abwässerfischteiche zur biologischen
1963   Nachreinigung der Abwasser kleinerer und mittlerer Gemeinden. *Münch. Beitr. Abwass.-Fisch.-Flussbiol.*, 10: 190–7.

RICKER, W E   Methods for assessment of fish production in fresh
1968   waters. *IBP Handbk*, (3):313 p.

RIGGS, C D and SNEED, K E   The effects of controlled spawning
1959   and genetic selection on the fish culture of the future. *Trans. Am. Fish. Soc.*, 88(1):53–7.

RODHE, W   The primary production in lakes: some results and
1958   restrictions of the ^{14}C method. *Rapp. P.-v. Reun. Cons. perm. int. Explor. Mer.*, 144: 122–8.

ROLAN, R B   Must we choose between men in space and mankind
1970   on earth? *BioScience*, 20(14):797–806.

RYTHER, J H and BARDACH, J E   The status and potential of
1968   aquaculture, particularly invertebrate and algae culture. Vol. 1, Parts 1 & 2. Springfield, Va., Clearinghouse for Federal Scientific and Technical Information U.S. Department of Commerce, National Bureau of Standards, Doc. (#PB 177,767): 261 p.

SCHÄPERCLAUS, W   Lehrbuch der Teichwirtschaft. Berlin, P. Parey,
1967   582 p.

SCICLUNA-SPITERI, A   The use of sewage water for irrigation.
1969   European Commission on Agriculture Working Party on Water Research and Irrigation, FAO Doc. LA:ECA/WR/ 69:44 p.

SHELBOURNE, J E   The artificial propagation of marine fish. *Adv.*
1964   *mar. Biol.*, 2:1–83.

SILVA, P C and PAPENFUSS, G F   A systematic study of the algae of
1953   sewage oxidation ponds. *Publs St. Wat. Pollut. Control Bd Calif.*, (7):36 p.

SINGH, R N   Role of blue-green algae in nitrogen economy of
1961   Indian agriculture. New Dehli, Indian Council of Agricultural Research, 175 p.

SREENIVASAN, A   Limnological features of and primary production
1964   in a polluted moat at Vellore, Madras State. *Envir. Hlth*, 6:237–47.

SREENIVASAN, A   The limnology, primary production, and fish
1964a   production in a tropical pond. *Limnol. Oceanogr.*, 9(3): 391–6.

SREENIVASAN, A   The limnology of and fish production in two
1968   ponds in Chinglepat (Madras). *Hydrobiologia*, 32(1–2): 131–44.

STANDER, G, et al., A guide to pond systems for wastewater
1970   purification. In Proceedings of the International Conference
        on Water Quality and Pollution Research, Jerusalem, June
        1969, edited by H. I. Shuval. Ann Arbor, Humphrey
        Science Publications, 312 p.
STEWART, K M and ROHLICH, G A Eutrophication—a review.
1967   Publs St. Wat. Qual. Control Bd Calif., (34):188 p.
TARZWELL, C M Sanitational limnology. In Limnology in North
1966   America, edited by D. G. Frey, Madison, University of
        Wisconsin Press, pp. 635–66.
TSAI, CHU-FA Effects of chlorinated sewage effluents on fishes in
1968   upper Patuxent River, Maryland, Chesapeake Sci., 9(2):
        83–93.
U.S. Department of the Interior, Bureau of Sport Fisheries and
1970   Wildlife, Report to the fish farmers. The status of warm-
        water fish farming and progress in fish farming research.
        Resour. Publs U.S. Bur. Sport Fish. Wildl., (83):124 p.
VAAS, K F Over het gebruik van visvijvers bij de reiniging van
1948   afvalwater in de tropen. Landbouw, Buitenz., 20:331–48.
        (English transln entitled "Notes on freshwater fish culture
        in domestic sewage").
VAAS, K F and SACHLAN, M Cultivation of common carp in
1956   running water in West Java, Proc. Indo-Pacif. Fish. Coun.,
        6(2–3):187–96.
VAN ECK, H Sewage stabilization ponds: a critical review. J. Proc.
1959   Inst. Sewage Purif., 1959(3): 320–34.
VIOSCA, P JR. Statistics on the productivity of inland waters: the
1935   master key to better fish culture. Trans. Am. Fish Soc.,
        65(1935):350–8.
VON AMMON, F K and STAMMER, H A Chemisch-biologische
1958   Untersuchungen zur Kartierung der Isar mit besonderer
        Berücksichtigung der Abwasserbeseitigung von München.
        Münch. Beitr. Abwass.-Fisch.-Flussbiol., 4:21–32.

VON SCHILLINGER, A Die Münchener Abwasserfischteichanlage in
1924   ihrem Zusammenhang mit dem Kraftwerkunternehmen der
        Mittleren Isar-AG. Allg. FischZtg, 49:231–4, 246–50.
VON SCHILLINGER, A Erfahrungen im Betrieb mit Abwasserfisch-
1949   teichen. Allg. FischZtg., 74.
Water Pollution Control Federation, Industry's idea clinic. Part 1.
1964   Stabilization ponds. J. Wat. Pollut. Control Fed., 36(8):
        931–48.
WHITE, L, JR. The historical roots of our ecologic crisis. In The
1970   environmental handbook, edited by G. DeBell, New York,
        Ballantine Books, pp. 12–30.
WILDAVSKY, A Aesthetic power or the triumph of the sensitive
1967   minority over the vulgar mass: a political analysis of the
        new economics. Daedalus, 19(4):1115–28.
WOLNY, P The use of purified town sewage for fish rearing.
1962   Roczn. Nauk rol. (B) 81(2):231–49.
WOLNY, P Fish rearing ponds filled with purified town sewage.
1962a   Gospod. rybna, 14(4):6–9.
WURTZ, A Fish culture in certain European countries. Stud. Rev.
1960   gen. Fish. Coun. Mediterr., (11):18–42.
YOUNT, J L and CROSSMAN, R A JR Causes and relief of hyper-
1966   eutrophication of lakes. Final report for Grant #00216-06.
        Vero Beach, Florida, Florida State Board of Health,
        Entomological Research Center. (Unpubl. Rep.):66 p.
ANON. Activated sludge with fins. Sewage Wks J., 18(6):1213–4.
1946
ANON. Power for Bavaria, München, Bayernwerk AG, Bayerische
n.d.   Landeselektrizitätsversorgung München (Public relations
        pamphlet).
ANON. Teichgut Birkenhof, München, Bayernwerk AG, Bayer-
n.d.   ische Landeselektrizitätsversorgung. München (Public
        relations pamphlet).

Thermal Pollution: Use of Deep, Cold, Nutrient-Rich Sea Water for Power Plant Cooling and Subsequent Aquaculture in Hawaii

K. Gundersen and P. Bienfang*

La pollution thermique: l'utilisation des eaux profondes, froides et riches en sels nutritifs pour le refroidissement des centrales électriques et les possibilités de leur utilisation en aquaculture à Hawaii

Etant donné l'accès facile aux eaux froides des fonds océaniques au large des îles Hawaii, les auteurs proposent que ces eaux soient utilisées pour le refroidissement des centrales électriques de l'Etat d'Hawaii au lieu de l'eau de surface utilisée actuellement. Cette méthode permettrait de réduire substantiellement le volume de l'eau nécessaire au refroidissement des centrales électriques en accroissant l'efficacité des condensateurs; du même coup, on éliminerait la pollution thermique des eaux côtières. De plus, par suite de la haute teneur en nutriments des eaux profondes, l'eau ainsi utilisée pour le refroidissement pourrait ensuite servir à alimenter un système d'échange aquatique où l'on pourrait pratiquer des cultures appropriées.

Afin de déterminer s'il serait possible de mettre sur pied une étude pilote fondée sur ces principes, des enquêtes ont été menées sur la productivité potentielle d'eaux prélevées au large de l'île d'Oahu (Hawaii). Les paramètres de l'enquête comportaient la fixation du carbone (méthode 14C), la production de pigment végétal et l'analyse des nitrates et des phosphates. Les organismes soumis aux essais ont été deux chlorophycées, une cyanophycée et des populations mixtes de phytoplancton.

La comparaison entre les eaux situées à 500 m de profondeur et les eaux côtières exposées aux conditions de lumière et de température qui prévalent à la surface a indiqué que la productivité des eaux profondes dépassait de manière significative celle des eaux de surface. La fixation de carbone par Dunaliella tertiolecta par 300 m de fond excédait cette même fixation dans les eaux de surface dans la proportion de 2,75, tandis que, par 600 m de fond, cette fixation était 16,78 fois plus élevée.

Contaminación térmica: empleo de las aguas profundas, frías y ricas en sales nutritivas para el enfriamiento de centrales eléctricas y la acuicultura posterior en Hawai

Por ser fácil la obtención de aguas oceánicas profundas y frías en las proximidades de las Islas Hawai, los autores han propuesto que se empleen para la refrigeración de las centrales eléctricas del Estado de Hawai, en vez del agua de la superficie como se hace actualmente. De esta manera se reduciría sensiblemente el volumen de agua de enfriamiento que necesitan las instalaciones, por aumentar el rendimiento de los condensadores y además se eliminaría la contaminación térmica de las aguas costeras. Asimismo, debido a las muchas materias nutritivas que contienen las aguas profundas, las de refrigeración descargadas se harían circular en un sistema de estanques para cultivar en ellos diversas especies apropiadas.

Como primera medida en la determinación de la posibilidad de iniciar un estudio-piloto basado en estos principios, se han efectuado investigaciones del valor productivo potencial de las aguas profundas obtenidas en las proximidades de la Isla de Oahu, Hawai. Los parámetros investigados comprendieron la fijación de carbono (método de C14), la producción de pigmentos de plantas y el análisis de nitratos y fosfatos. Como organismos de ensayo se emplearon dos clorofíceas, una cianofícea y poblaciones de fitoplancton mixtas.

En comparaciones hechas con agua obtenida a 500 m de profundidad y en las proximidades de la costa, en condiciones análogas de luz y temperatura inmediatamente debajo de la superficie, la productividad del agua profunda excedió a la de la superficie de una manera muy sensible. Dunaliella tertiolecta fijaba 2,75 veces más carbono a 300 m de profundidad que en la superficie, en tanto que a 600 m la fijación era de 16,78 veces mayor.

THERMAL pollution most commonly arises from the discharge of warm cooling water from electric power plants and is consequently a problem of concern to most developed countries. Recently, concern developed in Hawaii about potential damage to the invaluable coral reefs, beaches, and other seashore resources from thermal pollution.

At present Hawaii has five fossil fuelled power plants which use ocean surface water for cooling. A fifth plant is projected on the Island of Maui. These plants range

* Department of Microbiology, University of Hawaii, Honolulu, Hawaii 96822, U.S.A.

[513]

from 20,000 to 426,000 kW with a total maximum capacity of 943,000 kW. The corresponding cooling water requirement for all six plants is an average 3,500 m³ per minute with a discharge temperature of 5 to 6°C above ambient.

As yet, no adverse effects of thermal discharges have been observed although the State's water quality standards, regarding zones of mixing, are contravened in at least two cases. Expanding power requirements, increasing geometrically with population size, amplify the risk.

Two alternatives are: (1) that the warm water be discharged far offshore; and (2) that cooling water be taken from a depth where the initial temperature is low enough to yield a temperature of the discharge water not exceeding surface water temperature, i.e. 23 to 25°C. Hawaii has deep, cold water at a horizontal distance of only 2 to 10 km offshore and both alternatives are realistic possibilities.

Nitrate, phosphate, silicate and other plant nutrients all increase with depth in ocean waters (fig 1), but the surface water off the Hawaiian Islands is practically devoid of nutrients (Gundersen *et al.* 1970). If the second alternative for thermal pollution control were chosen the nutrient content of the discharged cooling water would have a stimulatory effect on productivity in the discharge area. If, instead of direct discharge to the ocean the nutrient-rich discharge water were circulated through a system of ponds there would be a great potential for mariculture. With this in mind investigations were made on the potential productivity value of ocean water from several depths off the Hawaiian Islands, particularly where deep water is close enough to shore and existing power plants to merit a pilot study.

Materials and methods

The organisms used were mixed plankton populations, endemic of Kaneohe Bay, Oahu, Hawaii, *Coccochloris stagnina* and an axenic culture of *Dunaliella tertiolecta*, both obtained from Dr. J. Caperon, University of Hawaii.

Stock cultures were grown in the complete phytoplankton medium of Jordan (1969) at 28°C at 5,500 lux. Transfers were made weekly into fresh medium.

Water was taken with conventional large-volume samplers from Kaneohe Bay (surface water) and from the Pacific Ocean off leeward Oahu, Hawaii (surface and deep water).

Nutrient content:	NO₃-N	PO₄-P	N/P
Kaneohe Bay, surface	3.1	0.64	4.9
Open ocean, 500 m	23.7	1.67	14.2
Open ocean, 700 m	41.3	2.38	17.4

The nutrient values of water used in the field experiments are shown in Table 1.

Carbon fixation was made according to the standard ^{14}C-method of Steemann Nielsen described in detail by Strickland and Parsons (1968). A $Na_2{}^{14}CO_3$-solution with a specific activity of 28.0 mCi/mM was used. Radioactivity of cell material was determined in a gas-flow Geiger Counter (Nuclear, Chicago). pH and alkalinity were determined with a Beckman Expandomatic pH meter according to Strickland and Parsons (1968). Chlorophyll *a* and other plant pigments were determined according to Strickland and Parsons (1968) using a Beckman, Model DBG, Spectrophotometer. Nitrate, nitrite, and phosphate were determined according to the methods described by Strickland and Parsons (1968) using a Bausch & Lomb, Spectronic 20, and a Beckman, Model DU, spectrophotometer respectively. Ammonium was determined by the phenolhypochlorite method of Solorzano (1969) using a Beckman, Model DU, spectrophotometer.

Results

In the first set of experiments a comparison was made of the amount of carbon fixed by samples of naturally occurring phytoplankton from Kaneohe Bay surface water using one volume of inoculum water in 9 volumes of filtered organism-free Bay surface water or similarly treated ocean water from 500 m depth. In general, the cultures were prepared in 300-ml BOD bottles to which 5 μCi $Na_2{}^{14}CO_3$ solution was added and the bottles incubated in a light cabinet at 3,750 lux and 28°C. After eight days 100-ml aliquots were filtered through membrane filters (Millipore HA) for determination of totally assimilated and retained carbon. In this type of experiment, counts per minute were taken as a measure of biomass which in the deep water amounted to approximately four times that produced in the surface water.

Experiments with pure cultures of *Dunaliella tertiolecta* and the blue-green alga, *Coccochloris stagnina*, were carried out in a somewhat different manner. The organisms were grown in complete medium in light, centrifuged, resuspended in sterile surface water and

TABLE 1. HYDROGRAPHIC DATA FROM STATIONS, CRUISE 7008, R/V *Teritu*, 1970, AND CARBON FIXATION BY *Dunaliella tertiolecta* IN WATER TAKEN FROM DIFFERENT DEPTHS. LIGHT INTENSITY 5,500 LUX

Station	Depth m	Temp. °C	PO₄–P	NH₄–N	NO₂–N	NO₃–N	Productivity C fixed mg/m³/hr	Fold increase deep: surface
					mg-at/m³			
A	0	26.5	0.25	0	0.020	0.12	9.38	—
	300	11.7	1.220	0.117	0.040	13.50	59.10	6.30
	600	5.83	2.877	0.227	0.018	35.00	165.01	17.69
B–2	0	26.5	0.099	0.267	0.030	0.15	13.12	—
	7	26.0	0.094	0.487	0.030	0.21	4.32	—
B–7	0	26.5	0.170	0	0.060	0.21	7.11	—
	285	19.1	0.481	0	0.098	3.80	8.48	1.19
B–11	0	26.5	0.094	0.371	0.008	0.13	6.22	—
	200	13.6	1.195	0.262	0.024	13.20	27.56	4.43
	0	26.5	0.124	0.227	0.020	0.13	15.57	—
C	300	11.7	1.057	0.098	0.020	12.78	43.01	2.76
	600	5.83	2.812	0.143	0.012	34.78	261.31	16.78

[514]

Organism	Days	Surface water		Deep water		Fold increase deep: surface	
		C-fixed cpm	Chl. a mg/m³	C-fixed cpm	Chl. a mg/m³	C-fixed	Chl. a
Mixed plankton	8	3,380	—	13,700	—	4.05	—
Dunaliella tertiolecta	2	5,436	—	43,535	—	8.02	—
	4	2,605	16.70	13,654	49.10	5.25	2.94
Co ccochloris stagnina	6	993	5.67	2,542	27.23	2.56	4.80
Nutrients in water							
Nitrate-N mg-at/m³		3.1		23.7		7.65	
Phosphate-P mg-at/m³		0.66		1.67		2.33	

starved in the dark, usually for 2 days. After another washing, standard inocula were suspended in sterile-filtered surface or deep water and dispersed in 125 ml bottles to which was added 1 μCi Na$_2$^{14}CO$_3$ solution. Chlorophyll and other plant pigments were determined in identical bottles without radiocarbon. The results, summarized in Table 2, showed that carbon fixation and chlorophyll synthesis in the deep water exceeded that in surface water from 2.5 to 8-fold. The nitrate-N and phosphate-P ratio for deep and surface waters indicates no distinct correlation of these parameters, although when a limiting factor did exist, it appeared to be the nitrate-N.

The filters used, composed partially of cellulose nitrate ester, were demonstrated to add as much as 0.25 μg-at nitrate-N per litre of filtered sea water.

Productivity ratios for filtered deep and surface water were thus confounded.

In a study of the dynamics of the rapid growth observed during the first three days of incubation, duplicate experiments were run with *Dunaliella tertiolecta* in which carbon fixation, pigment production, nitrate and phosphate uptake, pH, and alkalinity could be monitored continuously. Production was measured via chlorophyll determination and by carbon fixation. Kaneohe Bay surface water and 500 m ocean water were separately sterile-filtered, inoculated with starved (1 day) cells of *Dunaliella tertiolecta*, dispersed in 1 l sterile flasks and incubated at 5,500 lux on a light table. Samples were analysed at zero-time and at 8 h intervals over 72 h using 1 μCi Na$_2$^{14}CO$_3$ solution added to 100 ml aliquots of the cultures and continuing incubation under identical conditions for another 4 h. Plant pigments were determined by filtering aliquots directly from the culture flasks and the clear filtrate used for nitrate, phosphate and pH determinations.

The results are shown in fig 2. Within 16 h, there was little difference between the deep and surface water with respect to carbon fixed and chlorophyll *a* synthesized. At this time, however, as the nitrate content of the surface water had been reduced to approximately 1 mg-at/m³ and the phosphate to 0.3 mg-at/m³, carbon fixation was affected and fell rapidly for the remainder of the period. For the deep water, the events were similar but carbon

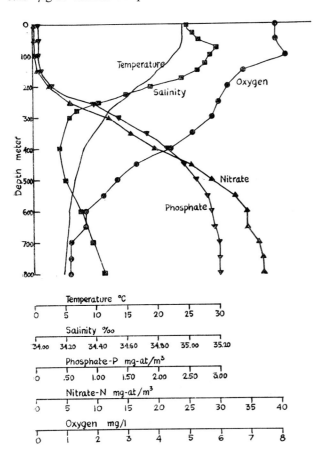

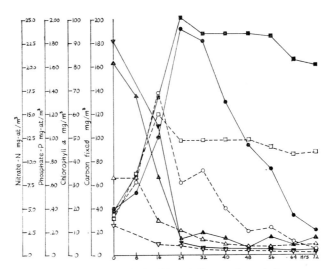

Fig 1. *Vertical distribution of temperature, salinity, oxygen, nitrate, and phosphate in the Pacific Ocean off the leeward Hawaiian Islands. Temperature and salinity from Charnell et al. (1967), other data from Gundersen et al. (1970)*

Fig 2. *Carbon fixation (○), chlorophyll* a *synthesis (□), nitrate (▽), and phosphate (△) uptake by* Dunaliella tertiolecta *grown in Kaneohe Bay surface water (open signs, dashed lines) and in 500-m ocean water (filled signs, solid lines). Light intensity 5,500 lux*

fixation, pigment synthesis, and nutrient uptake continued vigorously 8 h longer. The plant pigments (only chlorophyll *a* is shown in the figure) remained almost constant after the cessation of carbon fixation. Microscopy revealed a significant and increasing lysis of cells after the nutrients had been depleted but the chromatophores remained, by and large, intact explaining the unaltered level of chlorophyll after carbon fixation had ceased.

Carbon fixation compared

Since a comparison of deep ocean water with the polluted, phosphate rich water of Kaneohe Bay did not give a complete indication of the value of the deep water in stimulating plant production under surface conditions, an experiment was carried out using open ocean water, again using *Dunaliella tertiolecta* as a test organism.

Water was sampled with Mannon bottles at different depths at five different stations off the leeward coasts of the Hawaiian Islands. Aliquots of water were filtered through asbestos filters (to prevent nitrate leaching from membrane filters) and inoculated with *Dunaliella tertiolecta* to a final density of 2.3×10^5 cells per ml. Thereafter, 100 ml portions were distributed to 125 ml Pyrex bottles and pre-incubated for 48 h at 5,500 lux at 28 to 32°C. This pre-incubation demonstrated the increased potential of the deep water to support growth. After pre-incubation, 1 μCi $Na_2{}^{14}CO_3$ was added and the incubation continued for another 4 h period. The cultures were then filtered and carbon uptake determined in the prescribed manner. The results together with hydrographic data are given in Table 1.

As expected from the low phosphate and inorganic nitrogen values of the surface water at all stations investigated, the ratios obtained for deep to surface productivity were high. The nearshore B-stations illustrate that these significant relationships are valid for non-polluted, normally less productive coastal waters (Doty 1964, 1969; Doty and Capurro 1961; Doty and Oguri 1958) which are devoid of large quantities of terrestrial nutrient enrichment.

Discussion

Although not yet completed, the investigations have shown that a significant increase in growth is obtained when phytoplankton is placed in deep ocean water and exposed to surface light and temperature conditions. The production in deep water relative to open ocean surface water also showed a much higher value when compared to the land-influenced and phosphate-rich Kaneohe Bay water. Roels *et al.* (1970) observed similar high yields with mixed phytoplankton and cultures of different planktonic algae using deep, nutrient-rich water off St. Croix in the Virgin Islands.

Estimates of the potential yields in aquaculture systems utilizing discharged cooling water of deep origin can be based on this value. The maximum amount of carbon fixed in this study was well above 250 mg C/m³/h. Thus fixation rates under optimal field conditions which, over

a 10 h daylight period, would be approximately 2.5 g of carbon (dry weight)/m³/day. If we assume that the C:N:P composition of the phytoplankton is close to the value proposed by Fleming (1940), i.e. 106:16:1, the production of 2.5 g of cell carbon will require approximately 32 mg-at NO_3-N/m³, and 2 mg-at PO_4-P/m³. These are present at roughly the 600 m level off the Hawaiian Islands.

An average Hawaiian power plant at present uses some 400 m³ of cooling water per minute, with an intake water temperature of 25° and a discharge temperature of 31°C. If 600 m water of a temperature of 6°C were used, only 1/3 of this volume or some 200,000 m³ per day would be required. Assuming that about 80 per cent of the nutrients of this volume of water can be assimilated, a production of 400 kg of cell carbon, or 1,000 kg of organic matter (dry weight with 40 per cent C) will be produced per day at the primary level. In terms of secondary production at least 100 kg dry weight, or 500 kg wet weight, can be expected daily and some 182 t per year. Under field conditions, production might become even higher due to recycling of both nitrogen and phosphorus within the pond system.

The calculations presented above are more conservative than the calculations made by Roels *et al.* (1970) for Caribbean water. However, the results appear to be encouraging enough for continued studies of the potential of deep, nutrient-rich ocean water as a source of power plant cooling water and a medium for marine farming. This applies particularly to areas like the Hawaiian Islands where solar radiation and water temperature favour primary production.

References

CHARNELL, R L, AU, D W K and SECKEL, G R   The trade wind
1967   zone oceanography pilot study. Part 1. *Townsend Cromwell* cruises 1, 2 and 3, February to April, 1964. *Spec. scient. Rep. U.S. Fish Wildl. Serv. (Fish)*, (552):75 p.

DOTY, M S   Algal productivity of tropical Pacific as determined by
1964   isotope tracer techniques. *Tech. Rep. mat. Lab. Univ. Hawaii*, (1).

DOTY, M S   Primary productivity and plankton biological samp-
1969   ling program results from USCGS ship *Pioneer* 1964. *IIOE Cruise Rep.*, (4):19–90.

DOTY, M S and CAPURRO, L R   Productivity measurements in the
1961   world oceans. Part 1. *IGY Oceanogr. Rep.*, (4).

DOTY, M S and OGURI, M   Primary productivity patterns in
1958   enriched areas. *Proc. Pacif. Sci. Congr.*, 9 (16):94–7.

FLEMING, R H   The composition of plankton and units for report-
1940   ing populations and production. *Proc. Pacif. Sci. Congr.*, 6(3):535–40.

GUNDERSEN, K *et al.*   Some microbiological and chemical charac-
1970   teristics of the Pacific Ocean off the Leeward Hawaiian Islands. *Pacif. Sci.*

JORDAN, J B   A representative collection of Pacific coastal phyto-
1969   plankton. (Unpubl. MS).

ROELS, O A *et al.*   Artificial upwelling. Paper presented to the
1970   Second Annual Offshore Technology Conference, Texas, (OTC 1179).

SOLORZANO, L   Determination of ammonia in natural waters by
1969   the phenol-hypochlorite method. *Limnol. Oceanogr.*, 14:799–801.

STRICKLAND, J D H and PARSONS, T R   A practical handbook of
1968   seawater analysis. *Bull. Fish. Res. Bd Can.*, (167):311 p.

Possibilities for Constructive Use of Domestic Sewage (with an Example of the Lake of Tunis)

*J. Stirn**

Possibilités d'utilisation profitable des eaux usées domestiques (exemple du lac de Tunis)

Pour diverses raisons (esthétique, hygiène, pression des autorités et des milieux économiques pour protéger les activités touristiques), les ingénieurs sanitaires font en sorte que les effluents domestiques ne soient pas déversés dans les zones côtières. A cet effet, on utilise deux techniques: le traitement des déchets et leur dispersion par voie de décharges. Toutes deux sont coûteuses, sans être assez efficaces pour protéger entièrement les écosystèmes marins. A part quelques fertilisations limitées (insignifiantes pour l'exploitation halieutique) résultant du déversement d'eaux d'égout dans les eaux libres par de telles méthodes, ces techniques représentent une perte presque totale et permanente d'éléments nutritifs au sens large du terme. C'est pourquoi, envisageant de faire l'économie de la lutte contre la pollution, l'auteur préconise l'organisation d'expériences poussées en mariculture fondées sur la valeur nutritive des effluents domestiques; ces expériences auraient lieu notamment dans la Méditerranée pour des raisons qui lui sont propres (productivité naturelle faible, menace spécifique de la pollution à l'égard du tourisme qui constitue un important élément de l'économie nationale), et aussi à cause des possibilités offertes par cette mer (richesse des formations côtières—lagunes, baies, etc.).

L'auteur étaye ses théories sur les résultats d'expériences de laboratoire qui ont porté sur un cycle alimentaire artificiel dans un milieu fortement pollué, dont la production tertiaire pourrait avoir une valeur commerciale. Tous les organismes envisagés ont été isolés dans le lac extrêmement pollué de Tunis, où une expérience à grande échelle menée dans des conditions naturelles a fourni l'argument le plus fort à l'appui de la théorie mentionnée précédemment. Malgré une eutrophication non contrôlée qui provoque des désastres anoxiques accompagnées d'une mortalité massive du poisson, ce lac permet de réaliser des captures pouvant atteindre 460 tonnes de poisson par an (147 kg/ha/année) et il offre un potentiel de production de 100.000 tonnes d'ulves pour la nutrition et de 2.000 tonnes de gracilarias pour la fabrication de agar-agar.

Posibilidades de aprovechamiento constructivo de las aguas de descarga domésticas (con un ejemplo del lago de Túnez)

Por varias razones (estéticas, de presión social y económica para proteger el turismo) la ingeniería sanitaria se esfuerza por conservar las zonas costeras "limpias" de aguas de descarga domésticas. Con este fin se emplean dos técnicas: el tratamiento de los desechos y la dispersión de los mismos mediante desagües. Ambas son muy costosas, sin ser suficientemente eficaces para proteger como conviene los ecosistemas marinos. Prescindiendo de la limitada fertilización (insignificante para la explotación pesquera) causada por los residuos descargados en aguas abiertas por cualquiera de estos métodos, ambas técnicas representaban una pérdida casi total y permanente de sustancias nutritivas en *sensu lato*.

Por esto, teniendo en cuenta la necesidad de controlar la contaminación sin incurrir en grandes gastos, el autor aboga por la realización de experimentos avanzados de maricultura, basados en el valor nutritivo de las aguas de descarga domésticas, sobre todo en el Mediterráneo, por razones muy concretas (baja productividad natural, amenaza específica de la contaminación para el turismo, factor importante de la economía nacional) y porque ese mar proporciona determinadas condiciones específicas (abundancia de accidentes costeros adecuados: lagunas, bahías, etc.).

El autor defiende sus teorías con los resultados de algunos experimentos de laboratorio sobre un ciclo artificial de alimentación en un ambiente muy contaminado, cuya producción terciaria puede tener valor comercial. Todos los organismos examinados se obtuvieron del lago de Túnez, sujeto a una contaminación intensísima. Dicho lago se utilizó, además, en un experimento natural en gran escala, que aportó el argumento más fuerte en favor de la teoría mencionada. A pesar de su eutroficación incontrolada, que causa trastornos anóxicos, acompañada por una mortalidad masiva de los peces, el lago permite una captura media anual de 460 toneladas de pescado (147 kg/ha/año) y ofrece un potencial de 100.000 toneladas de ulvas para la alimentación y 2.000 toneladas de gracilarias para la preparación de agar-agar.

A POSSIBLE solution to the problems of domestic sewage pollution might be the development of a combined treatment and mariculture technology. This would not be a complete solution to the problem since for coastal mariculture an adequate geomorphological formation (bay, fjord, estuary, lagoon) must exist and be available for this purpose. There are in the Mediterranean an enormous number of such sites, located as a rule in economically passive areas (if we exclude tourism).

In discussing its possible use in mariculture the term sewage is used only in the sense of domestic waste, free of significant amounts of industrial toxic effluents. Phytoplankton refers to the total population of autotrophic unicellular algae, regardless of their size, i.e. it includes the nannoplanktonic fraction, without regard to its diversity (which in our case is usually very low). It must be stressed that because of diversity reduction due to pollution, only a small number of extremely adaptive species invade the changed ecosystem. From the synecological point of view, these species face little or no competition in space and nutrition and are not heavily consumed. In addition, new nutrients are continuously supplied due to the decomposition of the sewage-born organic matter.

In all cases where such ecosystems have been investigated, a significantly higher density of primary producers has been found than in comparable normal marine environments (Stirn, 1968). The same phenomenon developed in a coastal zone where an extensive perimeter bordering on open waters enables relatively fast dilution and dispersal of effluents. This is seen in Table 1 which shows phytoplankton densities during 1965 at two hydrologically similar stations in the North Adriatic. Station No. 18 is in clean waters off the Peninsula of Istra and Station No. 23 is in the Gulf of Trieste in slightly polluted waters.

High phytoplankton densities within polluted ecosystems depend on a few species. These are green flagellata (*Eutreptia*, *Chlamydomonas*) and coccal forms (*Nannochloris*), some dinophyts (*Prorocentrum*) and a very reduced number of diatom species (*Nitzschia seriata*, *Nitzschia closterium*, *Sceletonema costata*). Diatoms seldom dominate polluted pelagic communities and the use of classical phytoplanktological methods focused on diatoms and dinophyts may give even lower densities in polluted marine environments because these methods eliminate green nannoplankters which constitute the bulk of the total population.

The amount of nutrients in an area polluted by

TABLE 1. PHYTOPLANKTON DENSITY IN CELLS PER LITRE $\times 10^3$

Month	I	II	III	V	VI	VII	VIII	IX	XII
Station No. 18	8	6	301	2,222	873	11	23	254	340
Station No. 23	680	405	340	872	4,010	211	181	373	1,212

* Marine Biological Station of the University of Ljubljana, Portoroz, Yugoslavia.

sewage is always higher than in a normal marine environment. It may also be higher than the nutritional requirements of producers throughout the seasonal cycle. Other questions also arise: Are organic factors the only cause of phytoplankton blooms? Should vitamin requirements be taken into consideration? Furthermore, are there other unknown factors at work? The answer to the first question is certainly "no"; the second one may, in our particular case, be ignored as it is already known that our test species, *Nannochloris occulata*, does not require vitamins. Therefore, we shall consider the third question in more detail.

The author observed at various places in the North Adriatic and very clearly in the Lake of Tunis an anomalous situation showing that in spite of a relatively homogenous distribution of phosphates and nitrates, the differences in densities of green nannoplankters were usually very large (of the order of 1×10^9 cells per/1, particularly for *Nannochloris occulata*). The highest densities occurred within heavily but not extremely polluted areas. Table 2 gives the most important data for a characteristic case.

TABLE 2. PHYSICAL-CHEMICAL DATA AND DENSITIES OF *Nannochloris occulata* AT TWO STATIONS IN LAKE TUNIS. STATION NO. 6 IS IN THE HEAVILY POLLUTED PART AND STATION NO. 29 IS IN THE SLIGHTLY POLLUTED PART OF THE LAKE.

Data	Station No. 6		Station No. 29	
	July 23/66	January 7/67	July 26/66	January 7/67
Water temperature (°C)	29.40	12.00	30.00	12.00
Salinity	43.52	33.03	40.15	37.05
pH	8.10	8.40	9.30	8.00
Dissolved oxygen (ml/l)	6.00	7.30	9.10	9.80
Dissolved H_2S (mg/l)	1.47	0.08	0.00	0.00
Dissolved NH_3 (mg/l)	1.00	0.30	Trace	0.00
N-NO_2 (mg at/m³)	8.00	9.40	1.70	1.20
N-NO_3 (mg at/m³)	14.30	19.50	9.10	8.50
P-PO_4 (mg at/m³)	7.80	13.00	4.90	3.20
Nannochloris cells/l $\times 10^6$	1,200.00	18.00	3.50	0.30

These data suggest the existence of an unknown growth-promoting factor. Attempts were made to identify it by carrying out a number of cultivation experiments with *Nannochloris occulata* isolated from the Lake of Tunis and with clones kindly provided by Dr. M. Aubert. The following media were used:

Provasoli's medium (Provasoli, 1957) in lake water
Provasoli's medium in lake water and enriched with fresh sewage effluent (10 ml of sewage per/1)
Lake water enriched only with sewage (10 ml per/1)

The sewage effluent used for enrichment was collected at the outfall of the treatment plant. Its total organic matter was very high (872 mg/1), the total amount of inorganic phosphorus was 0.9 mg/1 and the total amount of inorganic nitrogen was 14.6 mg/1, of which 14.0 mg/1 was ammonia. There was no trace of oxygen but 9.2 mg/1 of H_2S. Cultures were inoculated with 1×10^6 cells/1 and kept for 12 days at 20° to 23°C and under natural light conditions.

Results of these experiments, expressed in densities of cells in billions per/1 in 12-day old cultures showed averages for Provasoli's medium of 1 billion, and for lake water enriched only with sewage 6 to 7 billion cells (about the same also for Provasoli's medium enriched

with sewage). These results suggest that there exists in sewage a factor which is not included in Provasoli's medium.

This factor may be related to plant hormonal activity, an assumption which has been made by some other authors (Bentley, 1959), (Duursma, 1965), (Provasoli, 1958). Sewage could be a source of these kinds of biologically active substances, as their basic compound indole is always present as the intermediate product of the anaerobic decomposition of proteins.

The influence of various phytohormones (auxin 2.4D, beta-indol-acetic acid, cytokinin and giberellic acid, all provided by Dr. M. Vardjan) was tested by adding them in various concentrations to the cultures in Provasoli's medium. The results are insignificant for all except beta-indol-acetic acid. At a concentration of 0.3 mg/1, this increased the densities of *Nannochloris* to three times those for Provasoli's medium alone, i.e. a result quite similar to that with sewage enrichment.

It is clear that in addition to inorganic sewage-born fertilizers biologically active substances are also present and these may strongly influence eutrophication of marine environments. This should also be considered from the standpoint of sanitary engineering because such substances may pass treatment plants or they may even be produced by the biota of a biological treatment plant. Even precipitation of inorganic nutrients (Føyn, 1965) may not eliminate them; thus even the most carefully treated effluent may still cause cultural eutrophication.

The utilization of sewage in mariculture seems to be possible because the basic food for higher trophic levels can be easily obtained using sewage-born fertilizers. There are also, as proved by the example of the Lake of Tunis, a number of secondary and tertiary producers, taking the food chain to organisms of potential commercial value (mostly mugilid fish and some molluscs). Among benthic primary producers there are enormous numbers of ulvas and gracilarias. These should be harvested in order to prevent intensive decomposition processes and they could possibly be used for food or in agriculture.

Eutrophication, i.e. secondary pollution, is a surplus of organic matter composed of undecomposed organic material from effluents and from increased productivity within the polluted ecosystem due to sewage-born fertilizers. The decomposition of these surplus organic materials can lead to complete consumption of oxygen, resulting in anaerobic decomposition processes, production of toxic compounds (H_2S, etc.), and causing periodical mass mortalities.

The Lake of Tunis, which is actually a shallow lagoon (average depth less than 1 m) with 30 km² of surface limited water exchange with the sea, receives sewage from an estimated equivalent of 300,000 inhabitants, most of it passing a relatively modern treatment plant before being discharged into the lake. Throughout all seasons the lake has, in addition to an extremely rich phytoplankton, an average standing crop of 1 kg/m² of ulvas in heavily polluted areas and 0.4 kg/m² of *Gracilaria confervoides* in slightly polluted parts. The lake's ecosystem also feeds very rich zoo-benthic populations, especially developed reefs and micro-atolls made by serpulid worms *Merceriella enigmatica*, *Hydroides uncinata* and *H. norvegica*. The fish stock in the lake is also quite rich and represents an important fisheries resource (Table 3) although once a year there is a mass mortality of fish.

TABLE 3. OFFICIAL FISH PRODUCTION IN LAKE TUNIS DURING 1968

Fish	Total catch in kg
Mugil ssp.	281,415
Anguilla anguilla	81,198
Chrysophrys aurata	45,245
Dicentrarchus labrax	17,086
Solea ssp	2,606
Various	4,976
Total catch 1968	432,526

The average yearly catch for the period from 1962 to 1968 has been calculated to be 147 kg/ha, which is close to the maximum yields known for the Mediterranean area (de Angelis, 1960). This yield is achieved without stocking with fry under conditions of uncontrolled eutrophication, often coupled with anaerobic conditions and fish mortality. The author therefore advocates the development of a combined sewage treatment-mariculture technology, controlling over-fertilization by regulatory mechanisms before the treated effluents enter a given water body and controlling eutrophication by consumption of a biomass surplus, using an artificial increase of consumers and harvesting of consumers as well as using benthic producers for food and industry. A preliminary scheme is here shown.

(1) Technical and administrative measures to ensure optimum effect of existing biological treatment plant.

(2) Improvements of existing treatment plant.

(3) Control of eutrophication by control of discharged volume of treated sewage and control of water exchange between lake and sea.

(4) Use of treated domestic sewage and consumption of surplus organic matter by farming molluscs and stocking detritophagous fish.

(5) Harvesting the lake: from green algae min 80,000 t/year (wet weight); from red algae min 2,000 t/year (wet weight) and fish and molluscs 3,000 t/year.

(6) Establishment of industries to give additional nutriment for cattle from green algae; an industry for agar-agar from red algae; and an industry for smoked fish.

References

BENTLEY, J A   Plant hormones in marine phytoplankton, zoo-
1959   plankton and seaweeds. *Prepr. int. Oceanogr. Congr. AAAS*, 1:910–1.
DE ANGELIS, R   Mediterranean brackish water lagoons and their
1960   exploitation. *Stud. Rev. gen. Fish. Coun. Mediterr.*, 12:1–41.
DUURSMA, E K   The dissolved organic constituents of sea water. *In*
1965   Chemical oceanography edited by J. P. Riley and G. Skirrow, London, Academic Press, vol. 1:433–75.
FØYN, E   Disposal of waste in the marine environment and the
1965   pollution of the sea. *Oceanogr. mar. biol.*, 3:95–114.
PROVASOLI, L   Effect of plant hormones on Ulva. *Biol. Bull. Mar.*
1958   *biol. Lab.*, Woods Hole, (114): 375–84.
PROVASOLI, L, *et al.*   The development of artificial media for
1957   marine algae. *Arch. mikrobiol.*, 25:392–428.
STIRN, J   The consequences of the increased sea bioproduction
1968   caused by organic pollution and possibilities for the protection. *Revue int. Océanogr. méd.*, 10:123–9.

The Use of Nutrients in the Enrichment of Sockeye Salmon Nursery Lakes (A Preliminary Report)

*T. R. Parsons, C. D. McAllister, R. J. LeBrasseur and W. E. Barraclough**

Utilisation des nutriments pour enrichir les lacs d'élevage du saumon rouge

On a montré que, dans certains cas, les nutriments libérés dans les lacs par les carcasses de poissons anadromes constituent une fraction substantielle du bilan nutritif total des lacs oligotrophes. Cependant, le rapport entre l'enrichissement nutritif des lacs par ce procédé et la nourriture dont disposent les jeunes poissons anadromes est complexe. L'abaissement constant par la pêche du nombre de poissons anadromes adultes peut avoir des conséquences graves en réduisant la fertilité naturelle de certains lacs d'élevage. Il est proposé de procéder à l'enrichissement artificiel de ces lacs au moyen de nutriments organiques ou inorganiques, et les auteurs examinent des études préliminaires consacrées à un grand lac (51 km²) producteur de saumon rouge où l'on a eu recours à l'utilisation constructive de "polluants".

El empleo de elementos nutrientes en el enriquecimiento de los lagos de cría de salmón "sockeye"

Se ha demostrado que, en algunos casos, la cantidad de elementos nutrientes que aportan a los lagos los cadáveres de los peces anádromos constituyen una fracción sensible del total de tales elementos en los lagos oligotróficos. Sin embargo, la relación entre el enriquecimiento de los lagos por este procedimiento y la disponibilidad de alimentos para los peces anádromos jóvenes es compleja. El agotamiento continuo de algunos lagos de cría por efecto de la pesca de peces anádromos adultos puede haber tenido graves consecuencias al reducir su fertilidad natural. Se propone el enriquecimiento artificial de estos lagos con elementos nutrientes orgánicos e inorgánicos y se discutan los estudios preliminares sobre un gran lago (51 km²) productor de salmón "sockeye" con referencia al empleo constructivo de "contaminantes".

CARCASSES of dead salmon have been shown in some cases to contribute an appreciable fraction of the total nutrient budget in oligotrophic lakes (Krokhin, 1967). However, the relationship between nutrient enrichment of lakes by this process and the availability of food to young anadromous fish is complex (Donaldson, 1967) and inefficient. This is because nutrients from salmon carcasses become available in the autumn and their availability for plankton in the spring when young fry hatch, is only made possible through elaborate recycling within the lake. The replacement of nutrients by artificial fertilization of sockeye salmon-producing lakes has been suggested by a number of authors (e.g. Foerster, 1968; Hasler, 1969) but to quote from a FAO report on lake fertilization (Gooch, 1967) ". . . no one has yet mounted an aggressive and sustained

---

* Environmental Research Group, Fisheries Research Board of Canada, Nanaimo, B.C., Canada.

study to elucidate the role played by artificial fertilization in aquatic productivity".

The possible need for nutrient enrichment of some salmon nursery lakes contrasts to current over-enrichment of many aquatic habitats caused by the disposal of sewage and other wastes. In the particular area of study reported on here, a city of approximately 20,000 inhabitants disposes of its sewage into the marine environment while in a 50-mi radius, several large sockeye salmon-producing lakes appear to be deficient in the same nutrients that are being dumped into the sea. In this presentation, the fertilization of a large sockeye salmon-producing lake is described in the context of the possible beneficial use of certain effluents which, when uncontrolled, are regarded as pollutants.

Natural characteristics of the lake

Great Central Lake is located on Vancouver Island close to Alberni. Its dimensions are: length, 34 km; area, 51 km²; volume, 7.7 km³; mean depth, 200 m. The surface is 83 m above sea level and yearly mean discharge is 6×10^6 m³/day (range 0.4×10^6 to 32×10^6).

The lake becomes thermally stratified to a depth of 10 to 20 m from April through November. Maximum surface temperatures of 20°C occur in July/August while deep water below the thermocline is 4°C. Ice cover occurs some years during winter.

The lake resembles a simple trough in which 60 per cent of the drainage area empties into the western 4 mi. A small hydro-electric plant, one-third of the way from the eastern outlet contributes between 15 to 50 per cent of the total outflow depending on season. Under calm or easterly winds, the upper 10 m in the western part of the lake flows west; this is generally reversed under strong westerly winds. A clockwise gyral circulation is often apparent in the vicinity of 125° 13'W, which represents about the centre point of a 3-mi² area in which nutrients have been introduced. Net transport in the upper 10 m range from 0 to 64×10^3, c.f.s. in either an easterly or westerly direction depending largely on winds.

From data collected during 1969, the annual primary productivity of the lake is estimated to be approximately 5 g C/m² (Stephens et al., MS 1970). Nutrient concentrations (nitrate and phosphate) are at all times very low but a seasonal cycle in productivity is apparent (fig 1b). The maximum standing stock of phytoplankton occurs at the bottom of the thermocline and nutrients are depleted in the surface layers from June to September. The increase in surface nutrients in September is associated with an increase in rainfall. Zooplankton samples collected during 1969 and the spring of 1970 generally weighed less than 0.5 g/m², but a measurable increase (to 1 g/m²) in the total biomass of zooplankton (fig 2b, 1969 data) occurred in the autumn.

Sockeye salmon fry are the principal planktivorous fish in the lake. During day fry are at an average depth of 70 m (maximum, 110 m) but during evening they rise to feed at a mean depth of 14 m. Other fish, such as juvenile coho salmon (*O. kisutch*), cut-throat trout (*Salmo clarki*), stickleback (*Gasterosteus aculeatus*) and sculpin (*Cottus asper*), are seldom caught in the midwater trawls since they are believed to feed in the littoral zone.

The number of adult sockeye spawning in the lake has varied over a 20-year period from 6,000 to 100,000. Based on these figures the number of fry spending one year is probably between 1×10^6 and 10×10^6. The size composition of migrant smolts is approximately 63 ± 10 mm, which contrasts with the 79 mm size of one-year-old smolts from Babine Lake (McDonald, 1969) on the mainland of British Columbia and which has a nutrient content and primary productivity at least two to three times the level found in Great Central Lake (Narver and Anderson, MS, 1968). For growth curves for fish in the two lakes see fig 3. The growth curve for Great Central Lake does not represent the true growth of fish during the period of smolt migration since the departure of larger fish tends to depress the estimate of normal spring growth until after June; however, it is apparent that under natural conditions, Babine Lake fish grow about twice as fast as Great Central Lake fish.

Nutrient enrichment

It was not our intention to identify any one nutrient as a limiting growth factor but rather to apply a broad nutrient spectrum. The reason for this is that the rate-

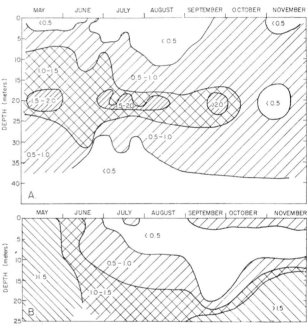

Fig 1. Great Central Lake (A) Chlorophyll a, (B) Nitrate (µg–at/l) —all data from 1969

limiting nutrients may differ with seasons and for practical application it would be excessively expensive to continue analysis and selection of nutrients for different times in the year. In preliminary studies on the addition of reagent grade chemicals to lake water incubated under continuous illumination, it was apparent that the greatest effect on primary productivity was obtained when a combination of nitrate, phosphate, trace elements, EDTA and vitamins were added. Since the cost of using these chemicals commercially in a large lake was prohibitive, further experiments had to be conducted with commercially available fertilizers considerably cheaper than reagent grade chemicals. The primary requirement was for a highly soluble fertilizer with an appropriate nitrogen to phosphorus ratio. For this purpose a combination of ammonium phosphate and ammonium nitrate

[520]

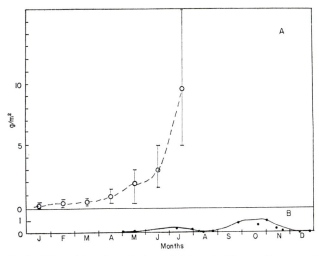

Fig. 2. Wet weight of zooplankton, 0–50 m. A (○) monthly mean of three or more observations during 1970 and range, (I); B (●) 1969 wet weights

was considered and the ratio was adjusted to give a N : P ratio of 10 : 1. The mixed fertilizer is known as 27-14-0 and it was found to be 94 per cent soluble under favourable laboratory conditions. In addition, the fertilizer contains traces of impurities, such as manganese, zinc and iron, which serve to further supplement nutrient content. The role of vitamins and EDTA in enhancing the primary productivity was not examined in detail but a cheap organic supplement (fish solubles—FS) was added in very small amounts (2 × 10⁻⁴ ml/100 ml) to give a similar effect.

Results of nutrient additions described are in fig 4. All experiments were carried out in duplicate by incubating samples of lake surface water at 14°C for three to four days following the addition of nutrients. ¹⁴C-

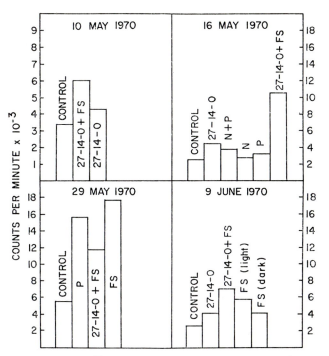

Fig. 4. Nutrient additions to 100 ml aliquots of Great Central Lake water. Additions as follows: 27-14-0, 3 µg-at N/l, 0.3 µg-at P/l; Phosphate (P) 0.3 µg-at P/l; Nitrate (N) 3 µg-at/l; Fish solubles (FS) 2 × 10⁻⁴ ml/100 ml

carbonate was then added and the samples were incubated for 24 h. Filtered residues were counted using a Nuclear-Chicago liquid scintillation counter. All incubations were carried out in the light with one exception. The total response of the water samples to nutrient enrichment was not consistent but an increase in ¹⁴C-carbonate uptake over the control was always encountered with any one addition of 27-14-0, fish solubles or phosphate. In

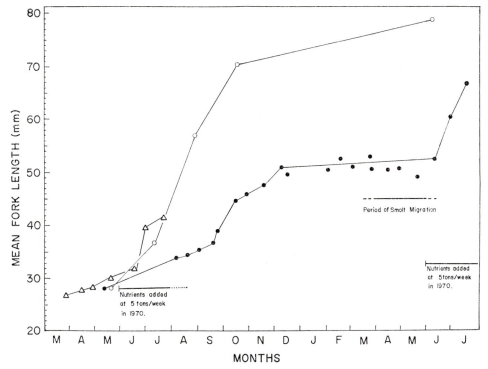

Fig. 3. Growth rate of Great Central Lake sockeye in 1969 (●—●), age "zero" to 1 year old; growth rate of sockeye in 1970 (△—△), age "zero" fish; and Babine Lake sockeye in 1966 (○—○), age "zero" to 1 year old (data from McDonald, 1969)

[521]

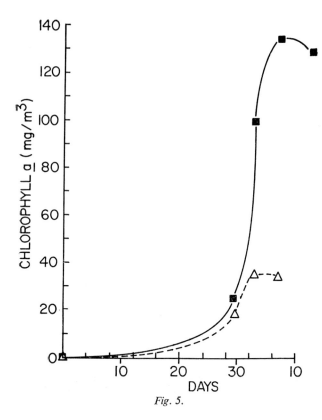

Fig. 5.

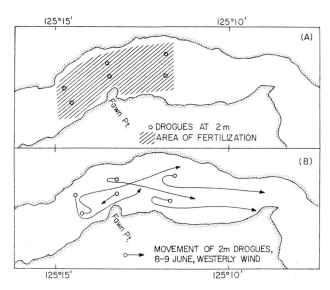

Fig 6. *Area of nutrient additions (A) and an example of water movement during 24 h (B).*

the examples shown, the combined effect of 27-14-0 and fish solubles was generally greater than separate additions while the effect of fish solubles in stimulating carbon dioxide uptake in the dark was apparent from the one exception on 9 June 1970.

The extent to which the nutrient mixture was capable of sustained support of algal growth is illustrated in fig 5. Two incubations were carried out; in the first a 0.25×10^{-5} dilution was made of the dissolved fertilizer concentrate; in the second a similar dilution was made at a level of 0.25×10^{-4}. The first solution analysed at 14.6 μg-at N/1 (NH_4^+); 13.1 μg-at N/1 (NO_3^-) and 3.0 μg-at P/1 (PO_4). Initial chlorophyll a concentration was 0.5 mg/m³ and the incubation conditions were continuous illumination under fluorescent lights at 14°C. In the low level nutrient addition, nitrate after 37 days was less than 0.5 μg-at/1 and the lack of nutrients was the probable reason for the termination of growth but in the high nutrient flask the level of nitrate after 37 days was 40 μg-at/1. However, assuming a chlorophyll a: carbon ratio of 30 in the latter flask, the total amount of particulate organic carbon synthesized would be ca 4.1 mgC/1. Since inorganic carbon in the lake water was only about 3.7 mg C/1, the maximum chlorophyll a:C ratio would have been 27, and it may be assumed that the culture was being growth-limited by an absence of inorganic carbon.

Addition of nutrients to Great Central Lake

The addition of nutrients to Great Central Lake was intended to increase production *but not to change the trophic relationships which lead to the food for the young sockeye salmon.* Thus it was considered important not to add excessive quantities of nutrients and in order to control the level of production the nutrients had to be added at frequent intervals. The effective quantity of

nutrients was estimated to be approximately 100 tons. (This was based on obtaining a 50 per cent increase in growth of 1×10^6 salmon fry assuming a transfer efficiency of 0.001; it is also roughly equivalent to the quantity of phosphorus in the carcasses of 10^6 adult salmon.)

Initial experiments on the introduction of nutrients were carried out in May using ca 2 tons of fertilizer in order to assess dispersal patterns. From 1 June dissolved nutrients were applied as 5 ton aliquots every seven days. If a single addition was mixed to a depth of 10 m the phosphate concentration over the entire lake would have

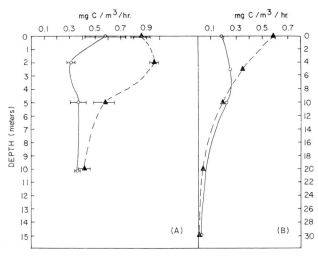

Fig. 7. In situ *primary productivity profiles in fertilized area* ($\blacktriangle$—$\blacktriangle$) *and unfertilized area* (O—O). *(A) mean of two productivity buoys and range* (—) *one day after fertilization, (B) three days after fertilization.*

been ca 0.02 μg-at P/1; a level approximately equivalent to the winter concentration of phosphate in the lake. In practice a 5-ton load of nutrients was distributed at 10 g a minute in the wake of a vessel and over an area of ca 3 mi² or ca 15 per cent of the lake surface; in addition, the depth of instantaneous mixing in the vessel's wake was determined to be about 5 m. Thus the immediate nutrient concentration in 15 per cent of the lake's surface waters was about 10 times the winter

Station number Date	1	2	3	4	5	6	7	8	9	10
5.6.70	.22	.35	.30	.45	.34	.30	.25	.35	.28	.31
9.6.70	.50	.57	.51	.26	.35	.27	.45	.33	.46	.54
16.6.70	.86	.55	1.00	.89	.76	.81	.66	.62	.52	.57

TABLE 2. AREAL CHLOROPHYLL a SURVEY, 0–14 m INTEGRATED SAMPLES

Station number	1	2	3	4	5	6	7	8	9	10
Chlorophyll a(mg/m³) 13.6.70	.74	.69	.66	.72	.89	.89	.96	.82	.69	.63
Chlorophyll a(mg/m³) 21.6.70	.82	1.20	.98	1.16	.98	1.17	1.11	1.05	1.22	.92

value of 0.02 μg-at P/l. However, the lateral movement of lake water rapidly dispersed the nutrients over an area which, depending on wind strength, could amount to a five-fold dispersion factor in less than a week. The exact location of the area repeatedly fertilized is shown in fig 6A and the movement of current drogues during the following 24 h period is shown in fig 6B.

The effect of two 5-ton loads of fertilizer on chlorophyll a concentration at ten stations covering two-thirds of the lake, is shown in table 1. It can be seen that the standing stock of primary producers increased by a factor of 2 to 3 following nutrient additions.

The standing stock of phytoplankton in the water column was determined at a further ten stations by using a hose to integrate the water from 0 to 14 m depth. These results which are given in table 2, show that the standing stock of phytoplankton had increased by ca 30 per cent between the second nutrient addition on 16 June 1970 and prior to the third addition on 22 June 1970. The actual increase in standing stock for the water column, based on the previous years' data was about 100 per cent on 21 June 1970. This increase was quite transient however and by July, following further nutrient additions, the standing stock of chlorophyll a decreased to <0.5 mg/m³ at the surface. Secchi disc values of 8 and 14 m were similar to values found in 1969 (Stephens *et al.*, MS 1970). However primary productivity continued to be enhanced by nutrient additions suggesting that the natural levels of nutrients in the lake were rate-limiting and that the increased primary production was cropped as it was produced.

In situ primary productivity profiles taken in connection with nutrient additions are shown in fig 7A. The results are the mean of two productivity buoys taken in a nutrient patch 24 h after the addition of fertilizer; the same measurements were made in an area 5 mi east of where the nutrients were being added. These results indicate that the primary productivity of the first 5 m was increased by a factor of 2 to 4 in the fertilized water, while for the water column to 10 m, primary production was increased by about 50 per cent in 24 h. The effect of fertilization was still apparent (fig 7B) three days after fertilization.

The principal species involved in the increase in primary production was *Cyclotella* sp. and a small (7 μ) unidentified flagellate. A decrease in silicate concentrations from around 46 μg-at/l to 42 μg-at/l in the surface waters was recorded during the diatom growth in early June. Ammonia and phosphate were taken up

within hours but nitrate could be detected up to 24 h later. This rapid uptake of nutrients was believed to be in part due to heterotrophic activity and as an indication of this the difference in plate counts of bacteria is shown in table 3 for two stations, one inside and the other outside the area of fertilization. Plate counts were made using Endo Broth medium and carrying out incubations for 48 h at room temperature.

TABLE 3. BACTERIAL PLATE COUNTS, 11.6.70.

	Colonies per plate	
Depth (m)	In fertilized area	Outside fertilized area
0	10,000	320
5	2,300	800
10	2,100	410
20	17	11
30	110	100

Secondary producers

Samples of zooplankton were collected every four days in 50 m vertical hauls using a 100 μ mesh net. Samples were analysed for species, life history stages and total biomass. Vertical and areal distributions of species were determined weekly.

The average standing stock of zooplankton following the start of fertilization in June are shown in fig 2A (1970 data). Zooplankton wet weight in July was 10 to 16 times the maximum observed in 1969. The abundance of principal zooplankton species are shown in table 4 for May and July, 1970. These data show that all species of zooplankton increased in number but that the greatest increase occurred in an unidentified colonial rotifer and in the number of *Holopedium*. However, since *Epischura* formed the principal food for young salmon during this period (table 5), the increase in *Epischura* recorded in table 4 must have occurred, in spite of heavy cropping by young salmon.

Growth of young sockeye salmon

Juvenile salmon were sampled at three-week intervals using a midwater trawl having a mouth dimension of 3×7 m. The growth rate of young sockeye salmon (fig 3) following intensive nutrient additions in June shows a sharp increase in both "zero" and one-year-old fry. The change in growth rate of the 1970 0 group, between the beginning and end of July, can be attributed to the further introduction of newly hatched fry during this period. In spite of this the total growth of the 0 group

[523]

TABLE 4. SUMMARY OF ZOOPLANKTON ABUNDANCE

	Rotifer	Cyclops	Epischura	Bosmina	Holopedium	Daphnia
May						
$\bar{m}$ No/1 (50 m vert.)[a]	0.32	2.9	0.12	1.28	0.16	0.02
Max. No/1 (Horiz. tow)[b]	1.2	9.0	0.27	4.0	0.7	0.16
d-max (m)[c]	10–25	15–30	0–5	5–15	0–15	15–20
July						
$\bar{m}$ No/1 (50 m vert.)[a]	9.6*	5.0	0.36	1.36	2.24	.08
Max. No/1 (Horiz. tow)[b]	10.0	12.0	1.0	5.0	6.0	0.3
d-max (m)[c]	0–10	20–40	0–5	0–5	0–12	20–30

[a] $\bar{m}$, mean counts based on 50 m vertical hauls.
[b] Max., counts based on horizontal net tows spaced at 2 to 4 m in upper 50 m
[c] d-max, depth range of maximum counts.
   For a count to be significantly different between May and July they should differ by at least a factor of 2
* July rotifers different species from May rotifers.

TABLE 5. FOOD ORGANISMS EATEN BY SOCKEYE SALMON FRY

	No. fish	Size range (mm)	No. fish for food analysis	Bosmina	Cyclops	Daphnia	Holopedium	Epischura	Chironomid larvae	Other insects	Total
August, 1969	37	24–53	12	1	27	2	135	329	29	10	533
% by number				0.2	5.0	0.4	25.3	61.6	5.4	2.1	100
Sept. to Nov. 1969	1,032	29–64	124	363	1,400	26	11,379	1,367	46	85	14,666
% by number				2.5	9.5	0.2	77.5	9.3	0.3	0.7	100
Dec. to Feb. 1969 and 1970	350	35–63	63	2,620	945		6,993	1,302	31	6	11,897
% by number				22.0	8.0		58.8	10.9	0.3	0	100
Mar. to May 1970	1,508	24–50	124	3,226	1,120	26	108	1,764	63	30	6,337
% by number				50.9	17.7	0.4	1.7	27.8	1.0	0.5	100
June and July 1970	2,983	25–53	277	1,689	812	1,284	2,428	12,175	43	17	18,448
% by number				9.2	4.4	7.0	13.2	66.0	0.2	0.1	100

□ principal food organism

by the end of July 1970 was as great as that encountered at the end of September during 1969.

Food organisms eaten by the salmon fry are shown for five periods of the year in 1969 and 1970 (table 5). *Bosmina* forms the principal food organism for salmon fry during the early spring, *Epischura* is taken during the summer and *Holopedium* is the principal food organism during the winter. The data are based on number of food organisms and illustrate a feeding pattern rather than the actual biomass of different food organisms consumed.

Conclusions

All three levels of production in an ultra-oligotrophic lake can be increased by small, frequently repeated additions of soluble fertilizers. The spectrum of food organisms before and after fertilization indicates that the productivity of the lake can be increased without appreciable changes in the pathways of production. The absolute increase in primary, secondary and tertiary production attained by July 1970, compared with 1969 data, may also have been influenced to some extent by other factors. For example, the formation of the thermocline was more rapid in the spring of 1970 than in 1969. A number of estimates of zooplankton wet weights in 1970 were increased to some extent by contamination with phytoplankton. It is unlikely, however, that such factors would cause the sustained order of magnitude

change in zooplankton standing stock and approximate doubling in the growth rate of salmon fry. The full effect of nutrient enrichment, together with its implications with respect to the beneficial use of certain pollutants now entering the sea in the same area, will be judged over a longer time period (up to five years) so that the initial favourable trend in these data can be more fully assessed. It should be emphasized that controlled addition of fertilizer appears to have caused a useful increase in production with no detectable decrease in water quality of the lake itself or the outlet river. No "blooms" were produced and the water remained clear as determined from Secchi disk readings. Nutrient levels in the river draining the lake remained at their pre-fertilization level and the river retained its weedless clarity. Thus the concern over eutrophication should not be allowed to obscure the possible benefits of controlled enrichment of natural waters.

References

DONALDSON, J R   The phosphorus budget of Iliamna Lake, Alaska, 1967   as related to the cyclic abundance of sockeye salmon. Thesis, University of Washington, 153 p.

FOERSTER, R E   The sockeye salmon, *Oncorhynchus nerka. Bull.* 1968   *Fish. Res. Bd Can.*, (162):422 p.

GOOCH, B C   Appraisal of North American fish culture fertiliza- 1967   tion studies. *FAO Fish. Rep.*, (44) vol. 3:13–26.

HASLER, A D   Cultural eutrophication is reversible. *BioScience,* 1969   19:425–31.

KROKHIN, E M Influence of the intensity of passage of the sockeye
1967 salmon *Oncorhynchus nerka* (Walb.) on the phosphate content of spawning lakes. Leningrad, Izdatel' stvo "Nauka", 15:26–31. Also issued as *Trans. Ser. Fish. Res. Ba Can.,* (1273).

McDONALD, J G Distribution, growth, and survival of sockeye fry
1969 (*Oncorhynchus nerka*) produced in natural and artificial stream environments. *J. Fish. Res. Bd Can.,* 26(2):229–67.

NARVER, D W and ANDERSON, B C Primary production and
1968 associated measurements in the Babine Lake area in 1966–67 and in Owikeno Lake in 1967. *Manuscr. Rep. Ser. Fish. Res. Bd Can.,* (1000):20 p.

STEPHENS, K, SHEEHAN, S and NEUMAN, R Chemical and physical
1970 limnological observations at Babine Lake, B.C., 1963 and 1969, and Great Central Lake, B.C., 1969. *Manuscr. Rep. Ser. Fish. Res. Bd Can.,* (1065).

Use of Potential Lagoon Pollutants to Produce Protein in the South Pacific

*G. L. Chan**

L'utilisation potentielle des polluants pour la production de protéines dans les lagons du Pacifique sud

Les îles des mers du Sud sont disséminées sur près d'un quart du globe, dans le plus grand des océans, et ne représentent guère qu'un pour cent de cette superficie aquatique. Par conséquent, l'élimination des petites quantités de polluants déversées dans cette région s'est toujours faite de manière naturelle grâce à cette vaste étendue d'eau limpide et non polluée. Les lagons aux eaux claires qui entourent les innombrables atolls et îles basses du Pacifique sud sont dotés d'une abondante vie marine qui, depuis toujours, est pour les insulaires une source d'aliments sains.

Toutefois, de nombreuses activités humaines résultant de l'exploitation commerciale de ces régions en ont altéré l'écologie avec des effets regrettables pour les peuplements de poissons et de mollusques. Le déversement d'excréments humains non traités dans ces nappes d'eau relativement immobiles est particulièrement dangereux, car il peut entraîner une réaction en chaîne nuisible à la biologie marine.

Il est possible de prévenir cette pollution en éliminant les déchets par les voies sanitaires et en utilisant tous les produits obtenus lors des cycles du carbone, de l'azote et de l'eau; grâce à l'action ininterrompue du soleil dans ces eaux tropicales, on pourrait en tirer une source importante de protéines pour la nourriture des animaux, de nutriments pour la pisciculture, ainsi qu'un effluent stabilisé riche en sels minéraux pouvant servir à l'horticulture.

Utilización de los contaminantes potenciales de las lagunas para producir proteínas en el sur del Pacífico

Las islas tropicales del Pacífico se extienden en el más grande de los océanos, cuya superficie es, aproximadamente, la cuarta parte del globo terráqueo, cubriendo apenas el 1 por ciento de la superficie acuática de la región. Por esa razón, una masa enorme de agua limpia e impoluta tenía siempre la capacidad de eliminar mediante procesos naturales las pequeñas cantidades de contaminantes que llegaban a ella. En las lagunas límpidas que bordean los innumerables atolones e islas bajas abunda la vida marina, que ha sido siempre una fuente de alimentos sanos para los isleños.

Muchas actividades humanas, derivadas de la explotación comercial de estos lugares, han trastornado la ecología, con efectos indeseables para las poblaciones de peces, moluscos y crustáceos. La más peligrosa es la descraga de excrementos humanos, sin tratarlos previamente, en estas masas de agua relativamente tranquilas, porque puede dar inicio a una reacción en cadena que influya negativamente en la vida marina. El tratamiento sanitario de estas aguas negras, aprovechando todos los productos obtenidos durante los ciclos del carbono, nitrógeno y agua, puede impedir esta contaminación y, gracias a la acción solar perenne en estas zonas tropicales, ofrecer una fuenta importante de proteínas para la alimentación animal, nutrientes para la piscicultura, y una aportación estabilizada de aguas ricas en sales minerales para la horticultura.

THE Pacific Ocean, is dotted with hundreds of small, isolated islands, no more than a few feet above the tides. The only fresh water resource on them is groundwater, strictly limited in volume. The islands suffer periodic droughts; their soils are deficient in many plant nutrients and they have a limited range of terrestrial animal life. For mankind they are a restricted environment, yet on them many generations of men have lived simple, healthy and contented lives in small communities. The coconut, taro, breadfruit, papaya, and rich, varied seafood from the clear lagoons provided them with a balanced diet, hunger being unknown.

Copra production upset the delicate ecological balance and food assumed a monetary value. The time spent in producing copra was no longer available for fishing or growing crops, and people resorted to canned and processed foods, which led to malnutrition and deteriorating dental health. As urban environment developed in some islands the discharge of human excreta into lagoons and shallow waters began.

Undesirable effects of excreta on water

For centuries, people have regarded the sea and any body of water as sewers, with disastrous results in many cases. But in the South Pacific, populations were small and scattered, and most of them defaecated in holes dug in the bush or other hiding places and covered them afterwards. This traditional pattern of excreta disposal was disrupted by Western influence. As a means of controlling worm infestation among the bare-footed islanders, latrines were built over the sea, polluting the lagoons and encouraging the growth of coarser forms of marine life while reducing edible fish and making shellfish potentially infectious. This practice deprived the soil of essential nutrients (nitrogen, etc.) and trace minerals, making the coral sands even less productive. Even the ubiquitous coconut palm must derive nutriment from the soil.

Some lagoons in the South Pacific show signs of gross pollution; with greenish brown slime on the white sand or coral reef. Both the flora and fauna are plentiful in unspoiled lagoons, but the ecology is fragile. Organic matter discharged into them will consume some of the dissolved oxygen, already limited because of the relatively small movement of the shallow and practically enclosed body of water.

There is no definite explanation why some lagoons are now short of fish and other seafoods, when from time immemorial they have provided the islanders with unlimited amounts of healthy food; there is only the suspicion that man has upset the ecology by indiscriminate pollution of the water with his own wastes.

In addition to this pollution, there is now the threat of industrial pollution. Wastes from fish and meat canning plants, dairies, and breweries are now discharged into the sea, and very soon a big oil refinery will add to this danger. Agricultural wastes also contribute to the in-

* South Pacific Commission, Noumea, New Caledonia.

creasing tide of pollution on the smaller islands. In many islands pigs roam freely; often they are kept enclosed on the beach to keep them out of the village. Their wastes are washed away with the tide, further polluting the lagoon.

Approach to the pollution problem

South Pacific peoples still have the opportunity to protect the ecosystems of their small islands and conserve their limited resources, mainly through the isolation of human and animal excreta on the land. Experience shows it cannot be done easily. People will listen to the germ theory of disease but they will do little about it. Many latrine programmes have been implemented, but the latrines are seldom used properly or kept clean, and many have been abandoned altogether. Even people with enough money to build proper toilets will seldom do so. If this sanitation programme is to succeed, there must be a new approach within the small community by making the people derive tangible benefit from it, without having to spend too much money. Therefore, local resources should be used as much as possible.

The main objects of this programme are to isolate both human and animal wastes to prevent them from contaminating the soil and the sea, particularly the lagoon,

and at the same time to use the end products to produce gas for cooking, algae for animal feed, nutrients for fish culture, and mineral salts and humus for vegetable gardening.

The requirements are: concrete receptacles, water supply, digester and gasholder, algae pond, fish pond, and irrigation canals.

A water-seal bowl and slab placed over a small concrete basin, costing about $10.00* is the most economical way to isolate human excreta, which is flushed by hand into a 4-inch diam pipe connected to a digester.

For pigs and chickens, a concrete slab will collect their wastes, which are washed daily into the digester as well. The cost of a slab, 15 × 20 ft, is about $20.00.

Water supply

In most islands, the only source of fresh water is rain collected from roofs in drums, etc. It is already inadequate for domestic use so an alternative must be found for flushing and cleaning. This can be obtained from groundwater by means of a well and a small hand pump, costing about $20.00. If this water is too brackish, and the effluent from the digester is to be used subsequently for

* Cost in Fiji; labour not included, as all the work can be done by local people.

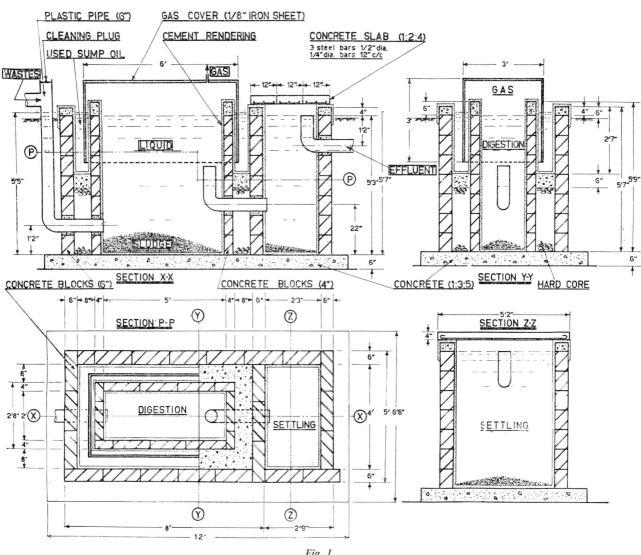

Fig. 1.

vegetable gardening, water from a solar still that produces fresh water from brackish or salt water can be mixed to reduce salinity. A small still of 50 ft², capable of producing 5 gal of fresh water per day, costs $50.00.

It would be more economical to abstract water from the fresh water lens that exists a few feet below ground on most atolls and low islands by means of one or more communal skimming wells and infiltration galleries. The water could be pumped by a solar engine to provide a reticulated supply to the whole village, to replace the individual wells and hand pumps. A solar engine, 3/4 hp, costs about $300.00, but running costs are minimal.

Digester and gasholder

The digester is similar to a simple septic tank, except that it is fitted with a gasholder (fig 1). It isolates the excreta in an anaerobic atmosphere so that quick liquefaction takes place, producing gas that can be used for cooking, and converts the organic matter in the excreta into organic salts. A digester of 600 gal capacity together with a small overflow compartment, suitable for ten pigs, costs about $150.00.

The gasholder is fitted with a $\frac{1}{2}$ in diameter plastic pipe that has a simple device to remove the water vapour before it is connected to a burner made of a small piece of perforated pipe. Made of 12-gauge iron sheet, the gasholder and the fitting cost about $50.00. It is painted black to absorb the maximum solar heat to promote more active digestion.

The production of gas for cooking and even small industries all the year round will eliminate unnecessary toil such as collecting firewood or husking coconuts, and make more time available for crop cultivation and handicrafts or other cottage industries.

Algae and fish ponds

The effluent from the digester flows into a narrow pond, excavated in the soil and lined with butyl rubber or fibreglass reinforced plastic sheet, where the action of light and warmth from the sun encourages algae to proliferate. In the process, oxygen is liberated to purify the effluent, oxidizing any remaining organic matter present and stabilizing the inorganic salts for use as fertilizer. The algae is removed and mixed with feed for the pigs and chickens. This provides a supplementary source of pro-

teins and vitamins for them. A pond 20 ft long, 3 ft wide and 3 ft deep costs about $50.00.

The purified effluent from the algae pond can in turn be fed into one or more ponds for fish culture. Some selected species of carp reproduce very well and are perfectly safe for human consumption as they feed on algae and protozoa, not excreta. In any case, any enteric microorganisms will have been killed by the digestion and oxidation processes. Fish from effluent ponds of this nature would be safer than several species of fish (obtainable in markets) which feed on raw sewage from sea outfalls. A pond built along similar lines as the algae pond costs about $0.30 per ft².

The overflow from the fish pond contains the end products of the decomposition of the organic matter, and these are mineral salts that are very good for growing vegetables. In some countries, up to ten crops are obtained in one year, by using such water for irrigation. Moreover, the sludge from the digester is removed about once a month with a small, inexpensive plastic pump and mixed with the sand to improve its physical fertility.

The effluent can be distributed through cheap irrigation canals, lined with butyl rubber or reinforced plastic, which cost about $0.30 per ft.

Pilot project

With the participation of the Community Development and Public Health Departments of one territory in the South Pacific, various specialists of the South Pacific Commission will be involved in a pilot project that embodies all the concepts outlined in this paper on a selected atoll. It is hoped that this will prove conclusively that the environment of an isolated island can be improved through the continuous development of three well-known cycles: carbon, nitrogen and water. But most important, it will control the pollution of the lagoon so as not to disturb the ecology, thus preserving the fauna and flora that make the lagoon the natural food reserve of the islands.

References

Chinese-American Joint Commission on Rural Reconstruction. The 1965   animal-methane-*Chlorella* cycle. Taipeh, Taiwan.
GOTASS, H B Composting. *Monogr. Ser. WHO*, (13). 1956

Engineering-Economic Approaches to the Management of Marine Water Quality

James A. Crutchfield[*]

L'application des systèmes à l'analyse économique de la lutte contre la pollution des mers

La possibilité d'une contamination à grande échelle du milieu marin à la suite du rejet délibéré ou non de déchets entre dans le domaine des réalités. L'attention du public a été attirée sur les incidents plus spectaculaires provoqués par des déversements importants d'hydrocarbures et par l'exploitation sans contrôle de gisements de pétrole et de gaz à proximité des côtes, mais l'agression insidieuse due à la capacité productrice des environnements voisins de la côte et des estuaires pose probablement un problème plus grave à l'échelle mondiale. Cependant, lorsqu'elle est utilisée sous un contrôle adéquat, la capacité d'assimilation de l'eau salée rend une multitude de services d'une immense valeur économique pour l'homme.

Enfoque sistemático del análisis económico de la lucha contra la contaminación del mar

Es una realidad la contaminación en gran escala del medio marino por la evacuación de desechos, organizada y no organizada. Han despertado el mayor interés entre el público los hechos más sensacionales relacionados con los grandes derramamientos de petróleo y las pérdidas incontroladas de éste y de gas de los yacimientos costeros, pero el lento agotamiento de la capacidad productiva del medio costero y de los estuarios es posiblemente mucho más inquietante en el plano mundial. Al mismo tiempo, y cuando se usa con las debidas precauciones, la capacidad asimilativa del agua del mar puede prestar a la humanidad una serie de servicios de enorme valor económico.

*Department of Economics, University of Washington, Seattle, Washington, U.S.A.

[527]

Le présent document élabore le cadre analytique d'un système technico-économique promettant d'évaluer les problèmes résultant de l'évacuation des déchets dans le milieu marin et de donner une mesure quantitative des variables pertinentes pour atteindre des objectifs spécifiques. On accorde une attention spéciale aux aspects dynamiques de la question, car les paramètres de la qualité de l'eau sont fonction de modifications démographiques constantes et de divers types d'effluents agricoles et industriels produits par une économie en expansion. Ainsi, l'analyse doit non seulement évaluer la réduction quantitative et qualitative des produits marins qui résulte de la pollution, mais également le coût de cette réduction. L'utilité opérationnelle d'un système de gestion destiné à utiliser la capacité d'assimilation des océans—système qui ne causerait qu'un tort minime aux autres productions économiques—requiert une évaluation soigneuse de la répartition des nuisances occasionnées aux diverses ressources marines par chaque cause de pollution. De tels systèmes doivent également tenir compte de la difficulté et du coût d'évaluations précises requises lorsque l'on traite de populations complexes et intriquées d'êtres vivants; ce qui souligne combien il faut être prudent lorsque l'on définit des normes minimales de sécurité pour l'évacuation des déchets.

Il est possible de concevoir des systèmes opérationnels qui ne nécessitent pas d'excessives entrées de données et permettent d'utiliser substantiellement les océans pour l'évacuation des effluents sans causer de dommages disproportionnées aux autres utilisateurs. De telles solutions ne sont possibles que si l'on exploite vigoureusement le fait que le montant total d'effluents produits peut être grandement réduit à l'aide de stimulants financiers de nature à faciliter la mise au point de méthodes de remplacement évitant tout préjudice aux autres utilisateurs de la mer. De la sorte, un programme adéquat de préservation de la productivité du milieu marin est conçu comme partie intégrante du problème de l'évaluation et de la répartition des coûts de l'élimination totale des effluents sur une base régionale et internationale.

El autor propone una manera de analizar un sistema económico técnico para evaluar los conflictos que surjan de la evacuación de desechos en el ambiente marino y para cuantificar las variables que sean de aplicación para alcanzar objetivos específicos. Se presta especial atención a los aspectos dinámicos del problema, porque los parámetros de calidad del agua quedan influidos por los cambios demográficos continuous y por las diversas clases de desechos agrícolas e industriales que producen una economía en crecimiento. Por esto, el análisis no solamente debe evaluar la reducción de la calidad y la cantidad de los productos del mar por efecto de la contaminación, sino también los costos de la supresión de ésta. La utilidad operacional de un sistema de ordenación destinada a utilizar la capacidad asimilativa de los océanos, con perjuicios mínimos para otras producciones económicas, exige una evaluación cuidadosa de la distribución de los daños ocasionados a recursos marinos específicos por cada una de las acciones de contaminación. Tales sistemas deben tener igualmente en cuenta las dificultades de necesidad y de los costos de evaluación exacta requerida cuando se trata de poblaciones complejas e intricadas de seres vivos. Esto implica la necesidad de proceder con cautela cuando se trata de fijar normas mínimas de seguridad para la evacuación de los desechos.

Pueden proyectarse sistemas de operación que no exijan muchos datos y que permitan un amplio uso de los océanos para la evacuación sin crear daños para los otros usuarios. Tales soluciones dependen de una explotación vigorosa del hecho de que la cantidad total de desechos generados puede reducirse mucho mediante estímulos financieros para encontrar métodos que no causen daños a otros usuarios del mar. Por esto un buen programa para salvaguardar la productividad del medio marino se considera como parte integrante del problema de evaluar y asignar los costos de toda la evacuación de desechos sobre una base regional e internacional.

THIS paper outlines briefly a framework for economic choice among feasible engineering techniques for minimizing detrimental effects of waste disposal in the marine environment.

A systematic approach to this vitally important worldwide problem, must be keyed to four basic concepts. First, waste generation is inevitable; the concentration of people in metropolitan areas near seashores is the heart of the problem. Second, the amount of waste generated is not a fixed quantity per ton of output or per head, but can be altered over a wide range, given the financial incentive to treat wastes and to make necessary adjustments in process and product. Third, salt water has the capacity to degrade substantial amounts of organic and inorganic material and can absorb large amounts of heat, properly introduced, without deleterious effects. Finally, the emphasis on the marine environment, should not obscure the fact that an efficient system of waste disposal must consider all possibilities, including disposal on land, in the air (through combustion) and through fresh and salt water. Efficient design of disposal systems and administrative machinery for pollution control must be conceived within this broader framework.

Water pollution, in the ocean or in fresh water, presents a classic case of the "externality problem" (i.e. the creation of costs in one sector as a result of activities in another sector). The flood of waste that now threatens bodies of salt water with the same fate that has overtaken so many rivers and lakes comes from households, industry and agriculture with varying identifiable points.

The problem of non-responsibility

The common element in all of these types of pollution is that those responsible, (householders, industrial plant managers, municipal sewage plant managers, or farmers) are not obligated to shoulder the economic or social costs created by the waste they discharge. In his own interest, each of these decision makers will choose that combination of production (or consumption) and associated waste disposal techniques that provides the largest net income (or, with householders, the minimum cost for waste disposal) within a given budget. If no cost accrues to him from pollution, he simply treats disposal in water as a free good. The essential problem in any systematic attack on pollution must be to "internalize" the decision-making processes, by forcing generators and dischargers of waste to account for the costs they impose on society and so seek alternatives that minimize *all* disposal costs.

The basic framework for static economic analysis of a water quality management system set forth below follows the pioneering work of Kneese. It is then modified to indicate the complications that arise in quantifying the essential parameters, and in adapting both the analytical framework and the operational control system to deal with growth and change.

The essential nature of the problem is best illustrated by imagining a unified decision-making authority which can intervene in all economic activity affecting water quality in a given area, and which has the power to adopt measures of control. Having such control, the authority will seek a system that minimizes the *total* cost of waste disposal, including costs of abatement or protection to itself and others. Abatement and protection measures include not only waste-treatment techniques but also a full range of potential adjustments in process and location by the offending firm, farm, or household.

The authority faces a series of functions relating water quality to control costs and pollution costs, and these functions, for simplicity, are assumed continuous. The control costs are, over some range, increasing functions of water quality (with the distinct possibility that the function may be stepped rather than continuous, because of changing techniques). Minimization of total cost occurs when: (1) the incremental cost of abatement is equal to that of pollution damage; and (2) for any given level of

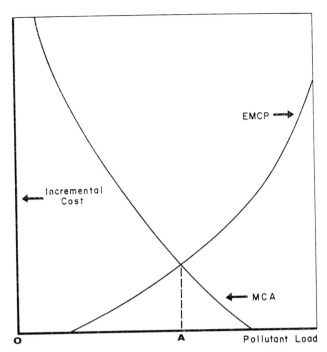

Fig 1. Optimal use of waste assimilative capacity
EMCP=sum of additional damages with increment of pollution
MCA=marginal costs of abatement

water quality, no change in the mix of abatement measures can further reduce costs. This idea is expressed in fig. 1.

The nature of the marginal costs here should be clarified, since the simple graphics conceal at each point a highly complex set of comparisons. The function, as drawn, implicitly assumes that all possible combinations of treatment and abatement are assessed to determine the minimum cost of achieving any given level of water quality. Thus, as the authority moves to better and better water quality, the actual combination of techniques employed is likely to change. It is virtually certain that the marginal cost of further improvement will rise at an increasing rate beyond some relatively low level of water quality.

Even a static model can and must be altered to take account of seasonal fluctuations in waste-carrying capacity—in effect, of regular shifts in the MCP function. While such changes are more important in the analysis of water quality in flowing streams, they may also be a factor in estuaries—either indirectly, through discharges from tributary rivers, or directly from variations in water temperature and mixing action.

"Pure" water v. controlled pollution

Important conclusions emerge even from this oversimple approach. First, it is essential to recognize that "pure" water is not an operational concept: it may be necessary to tolerate some degree of pollution, even in the marine environment, to avoid incurring even larger costs elsewhere. This simply recognizes that preservation of water quality is itself costly, and that the cost functions involved are likely to vary so that additional control may mean additional cost. It is also clear that the waste-absorbing capacity of salt water is a priceless asset. It would be senseless not to utilize that capacity, within

limits set by a system taking into account all costs of waste disposal and control.

Unfortunately, the simplicity of the model is more apparent than real: in designing operating systems, with real numbers in the functions, many appallingly difficult complications must be faced.

The most serious obstacle to the design of an optimal discharge system utilizing salt water is the obscure nature of potential damage. Some users are affected very little by polluted inshore or estuarine waters; sensitive ones will shift. However, an overwhelming proportion of the damage expected from degraded water is not borne by shore-dwellers but by life within the estuarine and inshore waters. For several reasons, including those listed below, it is almost impossible to quantify the incremental losses in marine life associated with incremental changes in water quality:

1. Most populations affected are subject to so many other influences on stocks and yields that it is difficult (often impossible) to isolate and measure the impact of one factor alone, except where the degradation is catastrophic.

2. Pollution damage to living marine populations may be indirect and/or cumulative: a desirable predator species may be completely unaffected by a particular contaminant, but one or more elements of its food web may be adversely affected. Similarly, a fish population may be completely unaffected by limited contact with one pollutant, but prolonged exposure may have serious effects on adults or their progeny, even though it simply cannot be quantified with any precision, even after decades of scientific work.

3. Assessment of damages resulting from degradation of water quality requires a substantial series of baseline studies. Very few estuaries or inshore areas can provide such benchmarks; yet without them there is simply no way of determining the magnitude of the damage inflicted by pollutants.

4. The nature of commercial fishing under open access conditions makes it peculiarly difficult to determine the net economic value of fish stocks whose productivity may be affected adversely by water pollution. Also, where recreational use of marine species provides a highly prized service, no satisfactory way has yet been devised to measure its net economic benefits. Thus, even where physical losses can be related to pollution, the economic valuation of the losses may be of dubious reliability.

Tangled interrelationships

These complications are compounded by the possibility of synergistic effects. The fact that we are presently unable to say much about this potentially explosive factor should not conceal its real threat. For example, moderate increases in water temperature in an estuary from any source, may produce either beneficial or harmful effects on fish and shellfish production. Direct relations are difficult enough to establish; but if thermal changes trigger adverse effects the results may prove far more serious in kind or degree than either alone would produce.

In North America this danger is accentuated by existing legal requirements which place the burden of proof largely on those who argue the possibility of irreversible damage rather than on those who propose to engage in activities which potentially cause pollution.

Most standard engineering-economic formulations of cost-minimizing waste systems assume implicitly that damage functions are monotonically increasing. In fact, however, the damage function for living marine populations is likely to be S-shaped and very steeply sloped. Up to some critical level, waste disposal has no discernible effect on the population; thereafter, even a small increase in waste loading causes severe damage; and beyond another critical point, no further damage results simply because the population has been destroyed. In incremental terms, this means that the *marginal* damage inflicted in the population may be zero above or below a comparatively narrow range of contamination, but within that range it may be very severe.

This fact carries elements of both hope and danger. On the one hand, a damage function of this type suggests that no sophisticated economic evaluation of engineering alternatives may be required. Establishing standards for receiving water above the level at which a critical problem could conceivably rise is not only simple and straight-forward, but it may well be economically efficient, particularly if the body of water is subject to increasingly heavy use by population and industry. The model then becomes much simpler: rigorous quality standards for receiving water constitute a constraint within which waste generation and waste discharge may be managed to minimize aggregate costs.

On the other hand, the very existence of such a narrow safety margin implies a serious threat to many valuable marine resources. Our simple model is essentially a formulation for steady-state or long-term equilibrium. In real life, damage to marine populations is likely to be much slower, less readily detectable, and more insidious. The shameful history of pollution of fresh water rivers and lakes attests that the damage may already be irreversible—at the very least, politically difficult and economically costly to remedy. In brief, the potential damage associated with marine pollution suggests the need for a strong public policy toward the use of the sea for waste disposal—in sharp contrast to the current casual attitude.

The economies of scale affect many types of waste treatment and disposal. By implication, our simple model suggests that an optimal waste disposal system can be defined and realized simply by imposing on the discharger the cost inflicted on others, thus forcing him to choose the most efficient way of minimizing all costs, which would then include a charge for disposal in water. While the techniques could obviously differ, the principle of assigning control costs to identifiable polluters is just as applicable in socialist as in market-oriented societies. But even with comprehensive schemes it might still be economical and useful to permit small individual disposal schemes to operate as hitherto.

Who is to pay the costs?

Obviously, these situations create problems of equity that must be resolved by imposing shared costs on all dischargers.

Simple models of the type developed above assume the possibility of identifying point sources of pollution, and so determining abatement procedures, and distribution of costs. Unfortunately, many combinations of pollutants make it impossible to pin responsibility on individuals, firms, or governments. Pesticides, siltation, temperature changes, and a host of other environmental modifications may produce significant repercussions on marine life in the ocean or estuarine areas adjacent to river mouths. Yet it is literally impossible to identify, much less assign legal responsibility to, the individual perpetrators.

This emphasizes the integral nature of the waste disposal problems. Degradation of environmental quality in the sea is inseparable from the processes that threaten the economic and social usefulness of land and air resources as well.

In designing and implementing measures to prevent marine pollution, one cannot ignore the critical importance of time lags. There is, first, a lag in recognizing the damage after its onset—particularly important where marine life is affected only after long exposure. In every country a second and disturbingly long lag occurs between recognition of a pollution problem and mobilization of efforts to deal with it. Finally, many pollutants are so persistent (or produce such lingering effects) that damage may continue long after taking corrective action.

It is all too clear that even vast increases in scientific effort coupled with overnight streamlining of governmental design and enforcement processes cannot shorten the lag sufficiently to protect many valuable living marine resources from damage—much of it irreversible. Standards for protecting ocean waters must be initiated well before need, must be more broadly based, and must be more severe than those currently applied. The costs of moving "too soon" and "too severely" may well be lower than the costs of painful rehabilitation if we wait for overt signs of danger.

CONCLUSIONS

Some conclusions from points discussed are too important to be ignored.

First, emphasis on the inherent complexity (and frequent instability) of all the parameters involved is intended not to counsel against intervention but merely as a warning that there are no magic solutions. Modelling is, after all, only a freehand sketching of reality, and the value of the rigorous logic imposed on the analyst must be compared with the cost of the data-devouring demands of the theoretical approach, as it becomes more complex. In short, models for water quality management will never substitute for good judgment based sometimes only on very sketchy information and some intuition. The issue is whether a systematic way of integrating socio-economic and engineering variables can *reduce* the burden on the decision maker.

Both economics and technology can offer a qualified "yes". Sensible testing of sensibly designed systems for water quality control can, even now, bring some order to what threatens to become a dangerously chaotic demand for instant solutions to dimly perceived problems.

Second, there exists a most serious need to devise political institutions with jurisdiction broad enough to permit full evaluation of alternative strategies for marine pollution control. Within individual nations, administrative responsibility over estuarine and inshore waters is typically so fragmented that no rational control of water quality can be expected. While it behoves all nations to continue research into causes and effects of marine pollution, it is even more imperative that they develop the political capacity to make effective use of knowledge already at hand.

It is equally essential to allocate the costs of abatement and treatment in ways that provide a positive incentive to reduce total social costs, and to seek new ways to reduce the amounts of waste generated. A case in point is the rather poor performance of the petroleum industry acting now only under pressure and threats.

To subsidize industrial and municipal agencies to abate pollution is economically costly on two grounds. First, the subsidy (relief or direct grant) is usually tied to specific control, such as treatment plants; thus no incentive appears for considering a change in plant location or process. Second, subsidy minimizes the incentive to strive, through research and adaptation, to attack the problem at its roots by reducing waste—which, after all, is the prime desideratum.

Section 6

EFFECTS OF POLLUTANTS ON QUALITY OF MARINE PRODUCTS AND EFFECTS ON FISHING

Status and suitability of microbiological detection methods

Several factors render marine bivalves particularly susceptible to pollution in estuarine and coastal water. These species live close to man's activities and their feeding mechanism results in the extraction and concentration of bacterial and viral pathogens, as well as toxic substances such as pesticides, biotoxins, radionuclides and heavy metals. The most important bacterial pathogens include *Salmonella* spp., *Shigella* spp., *Vibria* spp., and under certain conditions also *Clostridia* spp. The only documented virus infection appears to be that of infectious hepatitis.

There was reason to question the validity of a single bacteriological test to judge the suitability of shellfish for human consumption, water quality in the growing areas, the use of water for recreational purposes, and the use of water for fish processing. It was generally agreed that the coliform test was insufficient by itself; it needs to be supplemented by an *E. coli* test, and where appropriate, by a fecal streptococci test. However, it was noted that salmonella can be present in the absence of coliforms at considerable distances from sewer outlets. Some concern was expressed that bacteriological criteria may not accurately reflect the presence of viral contamination.

Depuration has been demonstrated to be a useful technique for cleansing shellfish of bacterial contamination when the bacterial load is moderate. However, shellfish growing areas must be closed when the bacterial load is heavy, and in this case, the only remedy is complete sewage treatment. No evidence was presented that depuration eliminated viral or chemical contaminants.

Marine protein concentrates

It was suggested that if intoxicants such as mercury, hydrocarbons, pesticides, heavy metals, etc., were protein bound, then these poisons would be passed on to man in even higher concentration than normally received by eating fresh fish. It was pointed out that FPC is intended as a protein supplement and that FPC would therefore essentially contain the same amount of the pollutant that might be ingested on a protein-unit basis as a result of eating fresh fish. Moreover, the protein of FPC would probably contain less intoxicants because of the extraction process. Hot, aqueous isopropanol is expected to remove some of the protein-bound pollutants. There was no laboratory evidence, however, to indicate to what extent this would be true in the broad spectrum of intoxicants present in the marine environment. The question was also raised as to whether or not FPC extraction procedures would remove natural-occurring marine biotoxins.

It was pointed out that there are no controlled studies presently available on that aspect of the problem. Although FPC might be used on only a supplementary basis in developed countries, there is the possibility that in developing countries it could conceivably be used on a larger-scale basis, in which case, toxic pollutants even at low levels over an extended period of time might be of concern.

It was agreed that further research is needed on the possible contamination of FPC by either marine biotoxins or by toxic industrial pollutants. Fish protein concentrate is expected to provide one means of removing pollutants from man's food.

Where pollution has affected fish resources

Toxic chemical pollutants are derived from many sources. Although an inefficient industrial technology was recognized as a major contributor, it was also pointed out that domestic sewage and waste from research laboratories, military installations, hospitals, and from a wide variety of governmental operations, agricultural practices, etc., are also significant contributors. In brief, the widespread indiscriminate use of poisons by the public must share in the responsibility for the overall influx of intoxicants into the marine environment.

Included within the general category of industrial intoxicants are such substances as heavy metals and other inorganics, petroleum, petrochemicals and other organics, and pesticides. Many of these materials have been found to be toxic to both marine organisms and man.

Natural-occurring marine biotoxins

Although a broad spectrum of marine biotoxins are know to occur in the marine environment, only paralytic shellfish poison, ciguatoxin, clupeotoxin, and possibly tetrotoxins, have been suggested as having a possible relationship to the overall pollution problem. The biogenesis of these biotoxins is presently unknown. However, field observations indicate that contaminants such as various metallic compounds, dumping of war materials, industrial wastes, ship wreckage, etc., may trigger toxicity cycles in marine organisms. The organisms affected include algae, shellfish and vertebrate fishes. Environmental disturbances such as storms, dredging, mining operations, blasting, volcanic eruptions, alteration in salinity, thermal pollution, etc., may cause serious ecological changes and thereby provide an environment conducive to the development of toxic organisms.

A correlation has been observed repeatedly between: (1) the availability of new surfaces in the reef environment; (2) the rapid growth of algae, and (3) the development of ciguatoxicity in normally edible species of reef fishes within the immediate vicinity. These problems are especially common in tropical insular areas of the world. The mechanisms involved in the ecological biogenesis of these poisons are unknown. The overall problem of marine biotoxicity in tropical insular areas is deemed to be of great economic importance to the development of fisheries in tropical insular countries.

Heavy metals and other inorganics

The inorganics considered as pollutants of the marine environment include the following: aluminium, arsenic, beryllium, fluoride, hydrogen acids, hydrogen sulphide, iron, lead, mercury, phosphorus, selenium, titanium, vanadium, zinc.

All of these substances are toxic to one extent or another. The degree of toxicity in any given amount of the substance varies from one species of organism to the next. These substances are of great public health concern because of their ever-increasing build-up and long-term persistence in the marine environment. In many instances, the portals of entry into the marine environment are unknown. Although mercury and lead are generally considered to be the most threatening pollutants, there are others which in the overall picture may be just as serious but are not readily detected because of their insidious toxic properties.

Heavy metals play a role in colour and possibly in flavour problems; copper was implicated in a green discoloration of oysters and iron in a green discoloration of cod fillets.

Petroleum

Oil pollution is primarily the consequence of our dependence upon petroleum-based technology. The rapid advances of technology have resulted in a steady increase in the use of petroleum for power, heating, paving and the production of a vast array of petrochemicals. It is estimated that annually there are about 7,500 oil pollution incidents in the United States alone. The total influx of oil to the ocean through shipping and by accidents in port is estimated to be at least 1 million metric tons per year. This figure does not include major catastrophes in production and at sea, unburned fuel, natural seepage, spent lubricants, and a significant hydrocarbon contribution entering the sea from land-based sources, e.g. municipal wastes, etc., and the total hydrocarbon influx may well be substantial in tonnage and nuisance value.

Crude petroleum is a complex mixture of natural products and includes many thousands of different compounds. Although crude oils differ markedly in their physical properties, the basic chemical, biological and toxicological properties of crude oil are quite similar. Petroleum and its hydrocarbons have been found to be remarkably stable in the marine environment. Although hydrocarbons that are dissolved in the water column are eventually destroyed by bacteria, the most toxic compounds are the ones that are the most refractory.

The immediate short-term effects of oil pollution are rather obvious. However, some of the most serious aspects of oil pollution deal with the low-level, long-term toxic effects, particularly on young forms of marine animals. The great complexity of the marine food chain and the stability of the hydrocarbons in marine organisms present a potentially dangerous situation which may adversely affect our fisheries resources. Hydrocarbons may not only be retained, but they may also be concentrated and become protein bound. Thus petroleum may contribute to the destruction of food values through the incorporation of oil and oil products into fisheries resources.

Reference was made to some of the recent activities of the Intergovernmental Maritime Consultative Organization (IMCO), concerned with efforts to minimize the discharge of oil from cargo vessels. Instances were cited where petroleum wastes were giving rise to odour and flavour problems in species which fed on contaminated marine organisms. Considerable concern was expressed about the carcinogenic nature of many hydrocarbons and other substances going into the environment. The serious socio-economic implications and effects on fishing of a recent spill of Bunker C oil in Canada were mentioned.

Petrochemicals and other organics

The production of chemicals derived from petroleum has become a highly diversified industry involving thousands of substances. At present, the chemical literature describes more than one million organic compounds, and this number is small when compared to the structural variations possible within even a few classes of organic compounds. Organic pollutants are characterized by their complexity. Petrochemical wastes include both inorganic and organic products. Some polynuclear aromatic hydrocarbon substances have been shown to be some of the most potent environmental carcinogens to test animals, and may probably be so also to others.

All organic compounds which are presently entering the marine environment as a result of technological activity may be classified as potential pollutants. Some of these compounds have been demonstrated to have deleterious effects on marine organisms, but the bulk of them have not. The task remains to test the toxicity of these additives and to develop suitable analytical methods for their isolation, detection and determination, as well as their identification in sea water.

Pesticides

An enormous number of man-made synthetic economic pesticides have been released into the marine environment. There are over 45,000 registered pesticidal formulations within the United States alone. These highly toxic substances have been released in such large quantities and so broadly that they are among the most widely distributed chemicals on this planet. Pesticides are sold and used primarily for specific purposes and they have the ability to exert biological effects on all living organisms. In addition to their toxic characteristics, some of these compounds have also displayed carcinogenic properties.

Pesticides have become of increasing concern to public health workers because of their uncontrolled ubiquitous distribution which now extends beyond their intended use. Probably the chief known hazard from residues of chlorinated hydrocarbon pesticides in the ocean is their ability to be concentrated by marine organisms. It is apparent that man is disturbing the delicate ecological balance of the marine ecosystem with some highly toxic substances having long-term residual properties, which are now entering the food web of marine organisms and man on a grand scale.

Many cases cited

Numerous examples were cited where toxic chemical wastes had seriously damaged coastal and estuarine fisheries. This damage was due to pollution by heavy metals, pesticides, petroleum, petrochemicals, etc. Extensive shellfish grounds have been closed in Canada, the United States, Spain, France, Germany, Japan and elsewhere because of chemical pollution. It is anticipated that additional shellfish grounds will be closed within the near future if the present rate of chemical pollution continues. It is estimated that about 25 per cent of the total shellfish grounds of North America have been closed because of pollution.

Excessive levels of mercury have been found in fish in North America, Sweden, Japan, Spain and elsewhere, which has resulted in the unsaleability of the fish. Excessive levels of lead, copper, and pesticidal residues have also been reported. Some of these high residues are now being found in pelagic fish species such as mackerel and tuna.

The impact of pollution in fisheries has not only resulted in a loss of many millions of dollars to fisheries, but has also produced serious social problems in some areas such as unemployment and a need to re-educate fishermen in other lines of endeavour.

Toxicological significance of pollutants

Industrial pollutants consisting of heavy metals and other inorganics, petroleum, petrochemicals and other organics, and a vast spectrum of pesticides are entering the marine environment on a massive scale. Most of these substances have varying degrees of toxicity to both marine organisms and man. Most have great public health importance and constitute a serious threat to the future utilization of fisheries resources. These poisons pose special problems in environmental toxicology because of their extreme persistence and bio-accumulation. Many of these materials are highly lipophilic and are not adversely affected by water. Some become protein bound within the body of marine organisms and thus enter the food web at all trophic levels including that of man.

Evidence of the toxicity of these industrial pollutants is now observed in the marine environment by the steady increase in cancerous growths, leukaemia, skin ulceration, tail deformities, genetic changes, and other disease conditions in fish and shellfish taken from heavily polluted areas in Southern California, Washington, New York,

the Baltic Sea, etc. A vast panorama of highly toxic and carcinogenic substances now persist in the marine environment.

Serious long-term effects

Some of the most serious aspects of the overall marine toxicological problems are those of low level, long-term effects. At present, acute toxicity, particularly lethality, is used as the prime indicator of toxicity and pollution. However, acute toxicity is an unreliable standard and a poor indicator of what may actually be taking place within the environment. Therefore, attention must be directed toward the detection of early or relatively low levels of injury. Consideration should also be given to the extrapolation of data on the toxic effects of marine animals and their application to man.

Permissible levels of intoxicants in the marine environment were discussed. It was pointed out that much of the data are of questionable significance and validity in the light of the long-term persistence and bioaccumulation of intoxicants in the marine environment. There is an almost complete lack of data on the synergistic effects of these poisons on both marine organisms and man.

Research is urgently needed to throw light on the acute and low level chronic effects of intoxicants on a wider range of marine organisms, i.e. plants, invertebrates, and vertebrates.

An ecotoxicological approach is required which will elucidate the synergistic effects of the poisons, and the resulting interactions between plant and animal populations, and their effects on fisheries resources. Studies on the role of pollutants in the biogenesis of marine biotoxins and their effects on fisheries resources, particularly in tropical latitudes, are needed.

It was clearly recognized by many of the participants that there is required a multi-disciplinary approach involving the basic biology, fisheries, chemical, toxicological and health sciences if these problems are to be effectively handled.

Effects of Pollutants on Quality of Marine Products and Effects on Fishing

D. R. Idler*

Effets des contaminants sur la qualité des produits de la mer et sur la pêche

Le présent document passe en revue les contaminants organiques et inorganiques provenant de diverses sources et examine leur nocivité pour les produits de la pêche. Les contaminants étudiés se répartissent entre trois catégories: (1) substances toxiques; (2) contaminants microbiologiques; (3) substances qui altèrent la qualité des produits de la pêche. L'auteur examine séparément les contaminants de chacune de ces catégories, donnant des précisions sur leur origine et leur mode de formation.

L'étude porte tout particulièrement sur les toxines qui affectent le système nerveux (ciguatera, etc.), puis aux substances toxiques d'origine industrielle (métaux, etc.) et aux contaminants provenant des industries pétrolières et pétrochimiques. Elle traite aussi des pesticides et de leurs effets toxiques sur l'écosystème marin. Un chapitre spécial est consacré aux effets de la contamination côtière sur l'industrie de la pêche au Canada; il indique les mesures législatives et les activités de surveillance nécessaires pour éviter la dégradation des produits manufacturés en liaison avec l'alimentation humaine. Pour terminer, l'auteur présente un plan d'action inter-institutions destiné à protéger les ressources halieutiques et l'industrie au moyen de mesures adéquates.

Efectos de los contaminantes sobre la calidad de los productos marinos y sobre la pesca

En el presente trabajo, se realiza una reseña sobre los contaminantes orgánicos e inorgánicos provenientes de diversas fuentes y se considera su acción nociva sobre los productos de la pesca. Los contaminantes estudiados se clasifican en tres categorías: (1) substancias intoxicantes; (2) contaminantes microbiológicos; y (3) substancias que alteran la calidad de los productos pesqueros. Se examinan separadamente los contaminantes de cada una de las categorías mencionadas, detallándose su procedencia y modalidad de formación.

Una atención especial se da a las toxinas que afectan el sistema nervioso (ciguatera, etc.), luego a los intoxicantes de origen industrial (metales, etc.) y a los contaminantes derivados de las industrias petrolera y petroquímica. También se examinan los pesticidas con sus consiguientes efectos tóxicos en el ecosistema marino. Un capítulo especial se dedica a los efectos de la contaminación costera en la industria pesquera del Canadá, destacándose las medidas legislativas y de vigilancia para evitar la degradación de los productos industrializados en relación con la alimentación humana. Finalmente, se presenta un plan de acción inter-institucional para la protección de los recursos pesqueros y de la industria con las medidas necesarias.

* Fisheries Research Board of Canada, Atlantic Regional Office, Halifax, Nova Scotia.

THIS review paper is primarily concerned with tainting, undesirable modification of texture or colour, induced infestation of microbiological organisms, parasites and toxicity of fish, and their by-products, such as FPC, due to pollutants. The topics discussed may be divided into three major categories: (1) intoxicants; (2) microbiological contaminants; and (3) substances altering the flavour, odour, texture, or colour of fisheries products.

INTOXICANTS

Toxic substances commonly encountered in the marine environment and associated with marine pollution fall into two major categories: (a) naturally occurring marine biotoxins; and (b) agricultural-industrial intoxicants.

(a) Naturally occurring marine biotoxins

Although a broad spectrum of marine biotoxins is known to occur in the marine environment, only paralytic shellfish poison, ciguatoxin, clupeotoxin and possibly tetrodotoxin have been suggested as having a possible link with pollution. The biogenesis of these biotoxins is presently unknown. However, field observations indicate that various metallic compounds, war materials, industrial wastes, ship wreckage, etc., may trigger toxicity cycles. Environmental disturbances, such as storms, dredging, mining operations, blasting, volcanic eruptions, alterations in salinity, thermal pollution, etc., may cause serious ecological changes and thereby provide an environment conducive to the development of toxic organisms. A correlation has repeatedly been observed between: (1) the availability of new surfaces in the reef environment; (2) the rapid growth of algae; and (3) the development of ciguatoxicity in normally edible species of reef fishes in the immediate area. These problems are especially common in tropical insular areas.

Ciguatera toxins affect the nervous system. The poison is not usually fatal but discomfort is severe and effects can last from days to years. The persons affected are sometimes sensitized so that even non-toxic fish cannot be consumed. The toxin is frequently observed only in fish from a narrowly defined area. Fish consumption is decreased in affected areas because of the risk factor.

De Sylva and Hine (1970) suggested that electrical power plants generate excess heat which favours the proliferation or organisms linked with ciguatera toxin and postulate that incidence of ciguatera poisoning may increase with the spread of such plants into tropical areas. Several outbreaks of ciguatera poisoning have been reported from an area near a plant south of Miami and they urge that great care should be taken to avoid sites where heated effluents may trigger production of ciguatera toxin.

The toxic nature of red tides is well known and Nitta (1970) suggests that these are affected by pollution, pointing out that red tides occur locally in Tokyo Bay, encouraged by eutrophication, even in winter. In Tokuyama Bay, nitrogen is supplied from industrial effluents and phosphorus is the limiting nutrient. Dredging caused a bloom in the Bay in 1957 and it is suggested that phosphorus was supplied by deoxidized bottom mud.

(b) Industrial intoxicants

Included within this general category are such substances as heavy metals and other inorganics, petroleum, petro-chemicals and other organics, and pesticides. Many of these have been found to be toxic to both marine organisms and man.

The ions considered as pollutants of the marine environment include the following heavy metals and other inorganics: aluminium, arsenic, beryllium, bismuth, cadmium, chromium, copper, cyanide, fluoride, hydrogen acids, hydrogen sulfide, iron, lead, mercury, phosphorus, selenium, titanium, vanadium, zinc. All of these substances are toxic but the degree of toxicity varies from one species of organism to the next. These chemicals are of great public health concern because of their increasing build-up and long-term persistence in the marine environment. Although these substances arise due to the activities of an enormously complex, and oft-times inefficient, industrial technology, in many instances the pathways to the marine environment are unknown. Mercury and lead are generally considered to be the most threatening pollutants but there are others which may be just as serious but are not readily detected because of their insidious toxic properties.

Metals also play a role in colour and possibly flavour problems. Nitta (1970), attributes a green discolouration of oysters to copper and/or zinc pollution either in solution or from sediments. The tissues contained concentrations as much as 100 times normal. The discolouration occurs in waters containing .01 ppm Cu or 0.1 ppm Zn. Above these concentrations the metals are toxic to the oyster.

An emerald green discolouration of the flesh has sometimes rendered yellow tail flounder fillets unsaleable. Studies at the Fisheries Research Board of Canada Laboratory in Newfoundland (unpublished) have shown that the pigmented portions of the flesh are high in iron. The discolouration is generally associated with the ventral surface of the fillet and is most probably caused by contact with sediments rich in iron. Heavy metals have been implicated in rancidity (Castell and Spears, 1968) and it would be of interest to know how metal containing wastes contribute to this problem either through physical contact or ingestion. Red colours are sometimes observed in the flesh of species such as cod but this is probably due to ingestion of invertebrates containing the pigment, astaxanthine, and not to pollution. However, it is conceivable that eutrophication of inshore waters would stimulate growth of red algae and laboratory experiments could readily resolve this question. When astaxanthine is linked to the appropriate protein it produces a green colouration which turns red on heating the flesh.

Petroleum

The rapid advances of technology have resulted in a great increase in the use of petroleum and it is estimated that annually there are about 7,500 oil pollution incidents in U.S. alone. The total influx of oil to the ocean through shipping and by accidents in port is at least 1 million metric tons a year. This figure does not include major catastrophes at sea, unburned fuel, natural seepage, spent lubricants, and that entering the sea from land-based sources, e.g. municipal wastes, etc. It is estimated that the total hydrocarbon influx may be between five and 10 million metric tons a year. Crude petroleum is a complex mixture of natural products and includes many thousands of compounds. Although crude oils differ markedly in their physical properties their basic chemical, biological and toxicological properties are similar. Blumer (1970) has pointed out that petroleum and hydrocarbons have been found to be reasonably stable in the marine environment. Although hydrocarbons that are dissolved in the water column are eventually destroyed by bacteria the most toxic compounds are the most refractory.

The immediate short-term effects of oil pollution are rather obvious but low-level long-term toxic effects, particularly on young forms of marine animals may be the most serious. The great complexity of the marine food chain and the stability of the hydrocarbons in marine organisms present a potentially dangerous situation which may adversely affect our fisheries resources.

Petrochemical and other organics

Present chemical literature describes more than one million organic compounds—this number is still small considering the structural variations possible within a few classes of organic compounds. Petrochemical wastes include both inorganic and organic products. Petrochemicals are considered some of the most important and potent environmental carcinogens. All organic compounds presently entering the marine environment as a result of technological activity may be classified as potential pollutants. Some have been demonstrated to have deleterious effects on marine organisms, although most have not. The task remains to test the toxicity of these additives and develop suitable analytical methods for the isolation, detection, determination and identification in sea water.

Petroleum wastes can contribute to odours found in fishery products. Nitta (1970) observes that fish held in Osaka Harbour become unmarketable and attributed this to ship discharge of oil rather than from shore.

The same author notes that grey mullet, in a reservoir which receives oil wastes from Iwakuni Airport, have odours. Sea-eel, squilla and flat fish taken up to several kilometres from Yokkaichi Harbour contained odours attributed to discharge from petroleum plants. Laboratory studies with two species suggest that the tolerance limit for oil in sea water lies between .01 and .001 ppm, in so far as odour is concerned. It is interesting that the treatment of oil with activated sludge permitted the concentration of oil to be increased to 0.1 ppm before odours were produced in fish flesh.

When oil was administered orally it was difficult to impart odour to the flesh and the author suggests that uptake from sediments is via the gills. Power plants contributed to the oil pollution problem in winter by causing fish to congregate in the warmer waters of the harbour.

Nitta (1970) also refers to bitter tastes in soft clams in a river estuary. It is thought that such bitter tasting substances are metabolites of aromatic compounds and arise from coal plants and other industries—10% of the clams were affected.

Not all fuel oil odours of marine fauna have their origin in petroleum wastes. The so-called "black-berry" odour of Labrador cod, and the fuel-oil odour of Pacific salmon and oysters are traced to dimethyl sulphide. This chemical has its origin in thetin (dimethyl carboxyethyl sulphonium chloride) which is a component of several algae. In the case of fish the vector is the pteropod, *Limacina helicina*. The problem is seasonal and occurs when the pteropod is abundant. It is possible that coastal eutrophication could contribute by encouraging growth of thetin containing algae.

Sidhu and associates (1970) report petroleum hydrocarbons of the type found in kerosene to be present in mullet. This affects several regions of Australia. Other studies suggest the residues may originate in waste water discharged by oil refineries. There is at present no simple method for separating tainted from untainted fish. In Canada, Krishnaswami has reported tainted fish attributed to hydrocarbon fractions (Krishnaswami and Kupchanko, 1969).

Pesticides

Many synthetic pesticides have been released into the marine environment. There are over 45,000 registered pesticidal formulations within the United States alone and they have been released in such large quantities and so broadly that they are among the most widely distributed chemicals on this planet. Pesticides are used primarily for specific purposes but they have the ability to exert biological effects on all living organisms; some have also displayed carcinogenic properties. Probably the chief known hazard from residues of chlorinated hydrocarbon pesticides in the ocean is their ability to be concentrated by marine organisms. So man is distributing the delicate ecological balance of the marine ecosystem with some highly toxic substances which have long-term residual properties that are now entering the food web on a grand scale.

The data of Jensen and co-workers quoted by Dybern (1970) illustrated the lipophyllic nature of halogenated pesticides as evidence by their accumulation in fatty and lean fish and within the animal on the same basis. The data were gathered between 1965–68 and cover fish, shellfish and seal for the Baltic area. Values for oil and flesh of seal were particularly high exceeding the 7 ppm level. This was sometimes also true, but less pronounced for five fish oils of commercial species. As would be expected pesticide concentrations in the flesh of these species were generally much lower usually 1/10 to 1/100th of the oil levels depending on whether the flesh was lean or fatty. The flesh of fish from the Baltic contained higher concentrations of all three organochlorine pollutants than did the flesh of the same species from Sweden's west coast.

COASTAL POLLUTION AND THE FISHING INDUSTRY

Sea water for fish processing plants

The problem in Canada and elsewhere is that estuarine and coastal waters have a limited capacity to absorb sewage. Blackwood (1970) noted that filleting plant with a capacity for 91,000 kg of round fish per 8 hours requires as much water as a city of 50,000 inhabitants.

Where the plant uses sea water or brackish water the site must be carefully selected to allow for contamination by the plant itself as well as industrial and sewage contamination, present and future. Far too frequently these conditions are not met and the cost to the Canadian taxpayer in the past ten years has amounted to twenty million dollars for assistance with pipeline extensions and water treatment in the Atlantic region.

The Regulations in Canada require that:

(1) Water for use in fish processing must be from a protected source and the distribution system must not present health hazards.

(2) There must be a most probable number (MPN) of coliform bacteria of two or less per 100 ml. Chlorination can be used to reach this level providing the MPN was less than 100 prior to treatment; if greater than 100 then acceptable purification must be used prior to chlorination.

(3) The water must contain no chemical or toxic substances which are injurious to health or otherwise adversely affect the acceptability of the product. Blackwood presents a case history of an east coast plant where it was found necessary to employ flocculation and sand filtration followed by chlorination or ozonization. The pretreatment facility for 4.6 kl/min was estimated at $140,000 with operating costs varying from 1 to 3 cents per min.

Canadian Guidelines require filtration of solids from fish plants. This is followed by chlorination or use of a tile disposal field when discharge will affect shellfish, recreational areas or a water supply for a food processing plant.

Several factors render marine bivalves particularly vulnerable to bacterial and viral pollution:

(1) The commercial species are largely estuarine, thereby living close to man's activities and maximally exposed to pollution.

(2) They are predominantly filter feeders. Their feeding mechanism results in the automatic ability to extract human pathogens (bacteria and viruses) from sewage pollution and also to extract molecules and ions from solution and concentrate them (e.g. pesticides, toxins, radio nuclides and heavy metals generally). This is a unique feature of an animal so low in trophic level.

(3) They are often eaten raw. This poses with (2) above special health hazards.

(4) They are largely sedentary and can avoid intermittent pollution only by relatively short-term closure of their shells.

(5) Their entire life history is normally confined to a severely restricted area, thereby maximizing exposure to pollution.

The problem

It was estimated in 1970 that at least 20–25% of shellfish beds in the Atlantic region of Canada were contaminated by sewage. The closure of clam beds was estimated at 20%. In 1961 25% of oyster beds in the Gulf of St. Lawrence were closed—this figure now probably increased. The criterion for closure is that water overlying growing areas shall not contain in excess of 70 coliforms per 100 ml.

Blackwood (1970) outlines the serious implications of closure to the commercial operation and the severe demands that such contamination places on the Canadian Fish Inspection Service. Dewling *et al.* (1970), describe the closure of an $18 million per year shellfishing industry in Raritan Bay. The closure was made necessary by an outbreak of infectious hepatitis resulting from faecal material in municipal sewage wastes.

Several outbreaks of infectious hepatitis, caused by the consumption of raw shellfish, are referred to by Metcalf and his associates (1970).

Wood (1970) points out that infectious hepatitis is the only well-documented virus infection associated with polluted shellfish. In non-temperate areas parasitic protozoa may be transmitted by shellfish. The most important bacterial pathogens include the organisms which cause typhoid fever and dysentery or produce pathogenic exotoxins. Wood points out that there is a risk from contaminated raw shucked shellfish being stored at temperatures which allow excessive multiplication of bacteria. The risk is less when they are stored alive. Approximately 20% of the food poisoning cases in Japan are attributed to storage and handling conditions which allow *Vibrio parahaemolyticus* to multiply.

Processing polluted shellfish

Wood has noted that Public Health Regulations permit the sale of shellfish from polluted areas after they have been: (1) sterilized by heat; (2) relayed in clean water; or (3) purified in an approved plant. However, Cole (1970) states that purification does not necessarily eliminate the risk from virus (e.g. hepatitis).

The ability of the sea to cleanse itself of bacterial and viral pollution is discussed (Shuval *et al.*, 1970). They detected enteroviruses 1,500 m from the sewage discharge while the coliform count has been reduced 10,000 times. They concluded that there was little or no virus mortality and suggested that laboratory experiments showing viral inactivation in seawater may not be extrapolated to the natural environment. They presented evidence that the rate of viral inactivation is much slower than the rate of coliform reduction in the laboratory. The message seems to be that coliform counts in the sea are not a good criterion of the level of viral contamination. Metcalf and his co-workers (1970) isolated coliphage and enteric virus from waste water plant effluents, shellfish, and waters in which shellfish were growing. They found shortcomings in trying to relate coliphage enumeration to viral contamination. Paoletti (1970) describes the value of shellfish in keeping the bacterial content of coastal waters and muds in check but he warns that shellfish are true concentrators and can be extremely harmful to man. Paoletti also describes the recent discoveries of two relatively new groups of organisms which contribute to the bactericidal power of seawater. Yoshpe-Purer and Shuval (1970) in commenting on the relatively rapid die-away of coliform bacteria in seawater note that pathogens such as salmonella can survive for long periods. They said "these findings cast new doubts on the validity of the universally accepted coliform standard for the safety of bathing waters". Their reports emphasize the point that there are seldom one or two all-inclusive indicators to describe a phenomenon as complex as shellfish sanitation let alone human health.

Fish protein concentrate

It is suggested by Halstead (1970) that if the poisons are protein-bound such as organic mercurial compounds, toxic hydrocarbons, and some of the heavy metals, then these poisons may be passed on to man in fish protein concentrate (FPC) in even higher concentrations than normally received by eating fresh fish. FPC is intended as a protein supplement and might possibly contain the same concentration of pollutant on a protein basis as the fresh fish from which it was prepared. It is more likely however that the protein concentrate will contain much less since hot aqueous isopropanol would probably extract some protein-bound pollutants. It should also be noted that the commonly used Halifax Process involves acidulation as well as high temperature. In any event the use of an FPC process to remove pollutants is one in which research should be encouraged. Certainly it is to be expected that fish protein concentrate provides one means of removing pollutants from man's food. Recalling the destructions of salmon from Lake Michigan because of pesticide contamination it is quite likely that such a lipid soluble pollutant would have been removed by isopropanol extraction. While it is true that metals and certain oil contaminants are bound to protein (Blumer, 1970) their levels may well, in my opinion, be reduced by extraction with hot acidified isopropanol.

Johannes (1970) observes that when coral reefs die they are rapidly colonized by filamentous algae which can increase the crop of herbivorous fish dramatically. Randall (1963) attributed the high standing crop of fish over artificial reefs to the absence of significant coral growth and an increase in benthic algae. However, it has been noted that fish feeding on algae growing on newly colonized surfaces may become poisonous from ciguatera toxin. It would be interesting to know if ciguatera toxin and other biotoxins such as the ones associated with red tide are removed during the preparation of FPC.

A case history of industrial coastal pollution

Some 275 fishermen in Placentia Bay, Newfoundland, Canada, fish principally for herring, cod and lobsters; and scallops and other species in smaller quantities. A phosphorous plant at Long Harbour was opened in December 1968—an investment of approximately $40 million and employing about 375 people. During February there were numerous fish mortalities around Placentia Bay in the direction of the prevailing currents. The fish involved were herring and they exhibited extensive haemorrhaging, particularly around the gills; hence the trouble was called "red herring". In addition to haemorrhaging, there was an almost complete haemolysis of the blood cells at death.

Some tests on cod were carried out in late April 1969. Portions of plant effluent were diluted and cod were placed in the tanks. Result was almost 100% mortality, but the fish did not discolour like the dead herring. Towards the end of April cod were found dead at Long Harbour and in May, divers discovered very large numbers of dead fish on the bottom—a high percentage being cod. The portion of Placentia Bay in which "red herring" were found was closed early in May and the plant ceased to operate. No "red herrings" reached the market and, to our knowledge, a portion of Long Harbour, not Placentia Bay, was the polluted area.

As soon as the problem was identified with pollution, samples of fish were tested by the authorities and the resources of the Ministry of Health and Welfare were mobilized by the Minister of Fisheries and Forestry, Jack Davis. D. R. Idler, then the Director, Halifax Laboratory, Fisheries Research Board, was instructed to negotiate with the plant owners and to give technical information. *Five* Federal Agencies, *six* Federal laboratories, *two* Provincial Government laboratories, *three* universities and almost every branch of the Federal Department of Fisheries and Forestry were involved with the problem; within two months the fishery was again in operation and the company back in production.

Early steps taken included: technical discussions with company officials to identify potential pollutants; a thorough search of scientific literature; bacteriological and virological studies to definitely rule out disease; sampling and analysis of bottom sediments near the plant since divers had reported the area to be a biological desert; laboratory bioassay tests with live fish to show that elemental phosphorus, bottom sediments and/or other effluent materials from the plant were toxic and could produce similar symptoms; tests to determine the lower limits of toxicity of these materials for many commercial fish species; and the development of analytical methods for these substances for quick and accurate determination, etc.

It was found finally that the principal potential pollutants were phosphorus in the colloidal state and fluosilicic acid. Cyanide and sulphur dioxide were present in sufficient quantity to be of concern. The fluosilicic acid was intended to be neutralized by dilution of the effluent with sufficient seawater to precipitate calcium and magnesium fluoride thereby producing an effluent of neutral pH. Unfortunately, fluosilicic acid requires a great deal of alkali for neutralization. This, coupled with the high fluoride content of the effluent, the large volume of the effluent, and possibly corrosion of the seawater pump resulted in incomplete neutralization. When the gaseous phosphorus was condensed to a liquid in water there were numerous impurities trapped with it, including silicon dioxide which assisted in keeping some phosphorus in suspension. The suspended and colloidal phosphorus was discharged in the plant effluent and proved to be extremely toxic to fish.

Antipollution measures

In a series of long meetings the control measures that would be required to eliminate release of toxic effluent to the harbour were discussed. The company had installed several centrifuges in the "phossy" water line in order to reduce the content of phosphorus. However, in the light of experimental evidence which was becoming available, it soon became evident that this was not enough to render the effluent harmless.

The company finally decided to retain and treat all effluents on their property. Since there was no natural site near the plant where the effluent could be pumped, ponds were excavated and constructed nearby with a total capacity of 3.8 million cubic feet.

There were two main effluent streams to be treated; the scrubbing water from the pelleting plant and the "phossy" water circulating through the phosphorus condensers. The first stream, originally very acid, was then treated by adding lime to precipitate the fluorides and neutralize the effluent. It was then pumped into a settling pond and ultimately re-used in the plant.

In the case of the "phossy" water it contained, in addition to elemental phosphorus, some cyanide and ammonia. After passing through the centrifuge where solid particles are removed, the "phossy" water was mixed with lime and passed through large tanks where the phosphorous settled out. The water then went to a pond and could be re-used.

Several fish kills took place after the plant was shut down because the mud on the bottom was toxic. This was dramatically emphasized when a stray school of herring spawned right in the effluent from the plant. Blood samples from these fish showed extensive haemolysis. This confirmed the laboratory studies which showed that the bottom sediments alone were then toxic to fish.

Removal of bottom sediments

Since phosphorous is quite stable under water, the bottom sediments near the effluent pipe posed a serious threat and so the company agreed to remove this deposit. It was decided that a suction dredge should pump mud and water to the retaining pond. The bulk of the solids was then "settled" and the water pumped to a treatment tank where charcoal, lime and alum were added to achieve clarification. After exposure to oxygen for 24 hours, the water flowed back into the harbour. The Fish Inspection Service ensured, throughout this crisis, that affected fish did not reach the consumer and on 30th June, formal notice was given that herring, cod and lobster were safe for consumption.

Thus in a period of two months the situation had changed from a crisis to a situation where employment of both the fishermen and plant personnel had been re-established.

Accidents can occur

Even with pollution control, poor housekeeping can result in "spills" which can do great damage before the fault is found and remedied. The risk of these "spills" would be far less if those charged with the responsibility of designing plants had pollution control specifically in mind.

Much pollution of the aquatic environment probably results from a failure to appreciate the difference between toxic substances in drinking water and food as contrasted with water in which fish live. The fact to be borne in mind is that water contains the air which fish breathe and this can no more contain chlorine, cyanide or similar substances than can the air which mammals breathe. The analogy is further complicated by the fact that many aquatic flora and fauna concentrate poisonous substances. This may affect the aquatic form directly or may affect the next organism in the food chain. By this process substances can be concentrated a millionfold from say the water to the first living form removed from the water. Jack Davis, Canadian Minister of Fisheries and Forestry, in stressing the sensitivity of fish to pollutants, declared that if we protect fish, we will provide an environment

where other animals can protect themselves: thus, fish can be the first line of defence against pollution.

It has been estimated that the closure and its consequences cost the company as much as $5 million— including damages to fishermen. Costs of research would increase this sum considerably. The cost of prevention would have been but a fraction of the cost of correction. There was no lack of adequate technology—simply lack of forethought.

Concluding remarks

The living aquatic resources of the world are being attacked by a formidable array of toxic substances. Many species of fish migrate over great distances and they can, and do, go to the pollutants. The rate at which they can cleanse themselves of pollutants, providing they survive, depends on their ability to metabolize and eliminate the foreign substances. There are all too many cases, involving the so-called cumulative toxins, for which this time can be very long indeed. Blumer (1970) has noted the extremely slow turnover of many fat soluble pollutants in aquatic life. He also observed that the disappearance of an "oily smell" from fish or shellfish is no evidence that the other substances which originally accompanied the odorous component have also gone. The implication is serious. Frequently the presence of objectionable odour, flavour or colour serve as a warning that a food product cannot be eaten without risk. Far better for the oily flavour to remain as a warning if there truly are deleterious components still present. The problem is even worse with pollutants such as pesticides and some heavy metals where there is no olefactory warning to the consumer.

Miettinen and associates (1970) report biological half-lives of the average order of a year for methyl mercury in molluscs. Bligh (1970) has reported that the biological half-life of mercury in pike flesh is of the order of 70 days. Juvenile salmon retain mercury in tissues for several months after exposure. The half-life of zinc for a marine mussel is of the order of 75 days. It is this slow metabolism of some petroleum wastes, heavy metals, chlorinated pesticides, etc. which render even non-resident species prone to the effects of the chemicals, hence the emphasis on searching for biodegradable substitutes. A further complicating factor is that large quantities of mercury, pesticides, elemental phosphorus, and many other pollutants are present in bottom sediments and are able to pollute aquatic resources for long after sources are eliminated.

The thought of a comprehensive worldwide monitoring system is particularly unattractive. It is true that monitoring can be both necessary and valuable in selected instances, but the capacity of chemists to come up with new compounds is almost without limit, as is the potential for our society to put them into the environment. If we become over enamoured with monitoring, the ends of history may be served but, partly due to the difficulty in interpreting pollutant levels in relation to effects on man, it is doubtful that we will do much to stop pollution. For the most part the answer must lie in research, education and legislation, and in scientists, in and out of government and industry, working to prevent rather than measure and record pollution.

References

BLACKWOOD, C M Canadian experience on sewage pollution of
1970 coastal waters: effect on fish plant water supplies. Paper presented to the FAO Technical Conference on Marine Pollution and its Effects on Living Resources and Fishing, Rome, Italy, 9–18 December 1970, FIR:MP/70/E-74:14 p.

BLIGH, E G Mercury in freshwater fish. *Fish. Can.*, 22(10):7–8.
1970

BLUMER, M Oil contamination and the living resources of the
1970 sea. Paper presented to the FAO Technical Conference on Marine Pollution and its Effects on Living Resources and Fishing, Rome, Italy, 9–18 December 1970, FIR:MP/70/R–1:11 p.

CASTELL, C H and SPEARS, D M Heavy metal ions and the
1968 development of rancidity in blended fish muscle. *J. Fish. Res. Bd Can.*, 25:639–56.

COLE, H A North Sea pollution. Paper presented to the FAO
1970 Technical Conference on Marine Pollution and its Effects on Living Resources and Fishing, Rome, Italy, 9–18 December 1970, FIR:MP/70/R-20:12 p.

DE SYLVA, D P and HINE, A E Ciguatera—marine fish poisoning:
1970 a possible consequence of thermal pollution in tropical seas? Paper presented to the FAO Technical Conference on Marine Pollution and its Effects on Living Resources and Fishing, Rome, Italy, 9–18 December 1970, FIR: MP/70, E–111.

DEWLING, R T, WALKER, K H and BREZENSKI, F T Effects of
1970 pollution: loss of an $18-million/year shellfishery. Paper presented to the FAO Technical Conference on Marine Pollution and its Effects on Living Resources and Fishing, Rome, Italy, 9–18 December 1970, FIR:MP/70/E-78:14 p.

DYBERN, B I Pollution in the Baltic. Paper presented to the FAO
1970 Technical Conference on Marine Pollution and its Effects on Living Resources and Fishing, Rome, Italy, 9–18 December 1970, FIR:MP/70/R-3: 17 p.

HALSTEAD, B W Toxicity of marine organisms caused by pollu-
1970 tants. Paper presented to the FAO Technical Conference on Marine Pollution and its Effects on Living Resources and Fishing, Rome, Italy, 9–18 December 1970, FIR:MP/70/R-6:21 p.

HUNT, D A Sanitary control of shellfish and marine pollution.
1970 Paper presented to the FAO Technical Conference on Marine Pollution and its Effects on Living Resources and Fishing, Rome, Italy, 9–18 December 1970, FIR:MP/70/E-87:6 p.

IDLER, D R Co-existence of a fishery and a major industry in
1969 Placentia Bay. *Chem. Can.*, 21(11).

JÄRVENPÄÄ, T, TILLANDER, M and MIETTINEN, J K Methyl-
1970 mercury: half-time of elimination in flounder, pike and eel. Paper presented to the FAO Technical Conference on Marine Pollution and its Effects on Living Resources and Fishing, Rome, Italy, 9–18 December 1970, FIR:MP/70/E-66:6 p.

JOHANNES, R E Coral reefs and pollution. Paper presented to the
1970 FAO Technical Conference on Marine Pollution and its Effects on Living Resources and Fishing, Rome, Italy, 9–18 December 1970, FIR:MP/70/R-14:15 p.

KECKES, S and MIETTINEN, J K Mercury as a marine pollutant.
1970 Paper presented to the FAO Technical Conference on Marine Pollution and its Effects on Living Resources and Fishing, Rome, Italy, 9–18 December 1970, FIR:MP/70/R-26:34 p.

KOEMAN, J H and VAN GENDEREN, H Tissue levels in animals
1970 and effects caused by chlorinated hydrocarbon insecticides, biphenyls and mercury in the marine environment along the Netherlands coast. Paper presented to the FAO Technical Conference on Marine Pollution and its Effects on Living Resources and Fishing, Rome, Italy, 9–18 December 1970, FIR:MP/70/E-21:12 p.

KRISHNASWAMI, S K and KUPCHANKO, E E Relationships between
1969 odour of petroleum refinery waste water and occurrence of "oily" taste-flavour in rainbow trout *Salmo gairdnerii*. *J. Wat. Pollut. Control Feb.*, 41:184–96.

METCALF, T G, VAUGHN, J M and STILES, W C The occurrence of
1970 human viruses and coliphage in marine waters and shellfish. Paper presented to the FAO Technical Conference on Marine Pollution and its Effects on Living Resources and Fishing, Rome, Italy, 9–18 December 1970, FIR:MP/70/E-24:9 p.

MIETTINEN, J K, HEYRAUD, M and KECKES, S Mercury as a
1970 hydrospheric pollutant. 2. Biological half-time of methyl mercury in four Mediterranean species: a fish, a crab and two molluscs. Paper presented to the FAO Technical Conference on Marine Pollution and its Effects on Living Resources and Fishing, Rome, Italy, 9–18 December 1970, FIR:MP/70/E-90:8 p.

NITTA, T Marine pollution in Japan. Paper presented to the FAO
1970 Technical Conference on Marine Pollution and its Effects on Living Resources and Fishing, Rome, Italy, 9–18 December 1970, FIR:MP/70/R-16:8 p.

PAOLETTI, A Phénomènes lytiques dans l'auto-épuration des eaux
1970 de mer. Document présenté à la Conférence Technique de la FAO sur la Pollution des Mers et ses Effets sur les Ressources Biologiques et la Pêche, Rome, Italie, 9–18 décembre 1970, FIR:MP/70/E-28:6 p.

RANDALL, J E An analysis of the fish populations of artificial and
1963 natural reefs in the Virgin Islands. *Caribb. J. Sci.*, 3(1):31–48.

SHUVAL, H I, *et al.* Virus inactivation in the marine environment.
1970 Paper presented to the FAO Technical Conference on Marine Pollution and its Effects on Living Resources and Fishing, Rome, Italy, 9–18 December 1970, FIR:MP/70/E-38:12 p.

SIDHU, G S, *et al.* Nature and effects of a kerosene-like taint in
1970 mullet (*Mugil cephalus*). Paper presented to the FAO Technical Conference on Marine Pollution and its Effects on Living Resources and Fishing, Rome, Italy, 9–18 December 1970, FIR:MP/70/E-39:9 p.

STOUT, V F, BEEZHOLD, F L and HOULE, C R DDT residue levels
1970 in some U.S. fishery products and the effectiveness of some treatments in reducing them. Paper presented to the FAO Technical Conference on Marine Pollution and its Effects on Living Resources and Fishing, Rome, Italy, 9–18 December 1970, FIR:MP/70/E-106:8 p.

SUESS, M J Polynuclear aromatic hydrocarbon pollution of the
1970 marine environment. Paper presented to the FAO Technical Conference on Marine Pollution and its Effects on Living Resources and Fishing, Rome, Italy, 9–18 December 1970, FIR:MP/70/E-42:5 p.

TILLANDER, M, MIETTINEN, J K and KOIVISTO, I Excretion rate
1970 of methyl mercury in the seal (*Pusa hispida*). Paper presented to the FAO Technical Conference on Marine Pollution and its Effects on Living Resources and Fishing, Rome, Italy, 9–18 December 1970, FIR:MP/70/E-67:4 p.

TRITES, R W The Gulf of St. Lawrence from a pollution view-
1970 point. Paper presented to the FAO Technical Conference on Marine Pollution and its Effects on Living Resources and Fishing, Rome, Italy, 9–18 December 1970, FIR:MP/70/R-18:23 p.

WOOD, P C The principles and methods employed for the sanitary
1970 control of molluscan shellfish. Paper presented to the FAO Technical Conference on Marine Pollution and its Effects on Living Resources and Fishing, Rome, Italy, 9–18 December 1970, FIR:MP/70/R-12:14 p.

YOSHPE-PURER, Y and SHUVAL, H I Salmonellae and bacterial
1970 indicator organisms in polluted coastal water and their hygienic significance. Paper presented to the FAO Technical Conference on Marine Pollution and its Effects on Living Resources and Fishing, Rome, Italy, 9–18 December 1970, FIR:MP/70/E-47:15 p.

Canadian Experience on Sewage Pollution of Coastal Waters: Effect on Fish Plant Water Supplies

C. M. Blackwood*

Expériences canadiennes sur la pollution des eaux côtières par les égouts: effet sur l'alimentation en eau des usines de traitement du poisson

Le déversement indiscriminé des eaux d'égout non traitées dans les eaux côtières a créé un sérieux problème de pollution. Même dans les zones côtières à faible peuplement humain les eaux voisines du rivage présentent des nombres moyennement élevés de coliformes. Dans l'est du Canada la pollution par les eaux d'égout a eu pour résultat la fermeture de plus de 50 pour cent des zones d'élevage de mollusques.

Une source appropriée et contrôlable d'une eau approuvée est une exigence dans le choix du lieu d'établissement d'une usine de traitement du poisson. L'alimentation en eau douce est souvent impossible et les industries de traitement du poisson installées ou en développement dépendent de l'eau de mer pour satisfaire les besoins s'élevant à environ 80 1/kg de filets traités. Au Canada toute l'eau de mer utilisée dans les usines de traitement du poisson doit satisfaire une norme de qualité bactériologique équivalent à celle de l'Organisation mondiale de la santé pour l'eau potable et un nombre croissant d'usines font face à de sérieuses difficultés pour obtenir une eau qui ne nécessite pas de pré-traitement. Les eaux de port dans les zones urbaines sont généralement sévèrement contaminées et le coût du traitement est prohibitif pour leur emploi dans les usines de traitement du poisson. L'emploi d'une eau de ville approuvée est recommandé partout où elle existe. Une omission importante est l'évacuation par les municipalités des eaux d'égout dans le milieu marin, sans prendre en considération les multiples usages des eaux côtières. Récemment les autorités des services des eaux ont été instituées pour la diminution et la prévention de la pollution.

Il est donné la description de l'état d'une zone côtière typique pour illustrer le type d'études d'hygiène et de bactériologie requises pour évaluer une source d'alimentation en eau des usines. Les normes canadiennes et les critères de traitement pour l'alimentation en eau d'usines de traitement du poisson sont décrits.

Experiencia canadiense sobre la contaminación por aguas cloacales en la región costera: efectos sobre el suministro de agua de las plantas de la industria del pescado

La indiscriminada descarga de residuos cloacales en las aguas costeras ha creado un problema serio de la contaminación. Incluso en las regiones costeras escasamente pobladas, se comprueba en las aguas del litoral una bastante alta densidad de gérmenes coliformes. La contaminación por aguas cloacales ha producido en el Canadá oriental la reducción de más del 50 por ciento de la superficie de los criaderos de moluscos.

En la elección del lugar para una planta fileteadora de pescado, una condición esencial es disponer de una fuente adecuada y controlada de aguas higiénicamente puras. Los suministros de agua dulce son a menudo inaprovechables y las industrias establecidas o en desarrollo para la elaboración de pescado necesitan agua de mar en una cantidad aproximada de 80 litros por kg de filetes elaborados. En el Canadá los suministros de agua de mar para las plantas fileteadoras deben tener una calidad bacteriológica equivalente al standard del agua potable requerido por la Organización Mundial de la Salud. Con el aumento del número de las plantas industriales se experimentan ya serias dificultades en los suministros de aguas sin tratar. Las aguas portuarias en las zonas urbanas están en general fuertemente contaminadas y los costos de tratamiento son prohibitivos para su uso en las plantas fileteadoras. Se recomienda el uso de aguas controladas por municipios siempre que estén disponibles. Un gran descuido en el problema de la contaminación ha sido la disposición de las municipalidades de drenar las aguas cloacales al ambiente marino, sin tener en cuenta el uso múltiple de las aguas costeras. Recientemente se han constituido las "Water Authorities" para prevenir y reducir la contaminación.

Se describe la situación de una región costera característica para ilustrar el tipo de los reconocimientos sanitarios y bacteriológicos necesario para evaluar la fuente del suministro con agua de una planta industrial. Además se describen los criterios canadienses sobre las condiciones standard y de tratamiento para el suministro de agua de las plantas industriales.

IN October 1970 the Government of Canada announced the intention of forming a new department of environmental control dealing with water, soil and air covering responsibilities for renewable resources including fish and forests.

Most industries and municipalities today realize that there are biological limits to what man can do to his surroundings and that in many coastal areas these have already been exceeded.

Recent experiences in Canada with oil spills, mercury, phosphorus and pesticide contaminants have made us aware of the gravity and magnitude of our problem. The fact that, during the past year, 12 per cent of Canada's important commercial freshwater fishery has been closed due to mercury pollution indicates the size of the problem. Recycling must become the order of the day. It is clear that new facilities will have to be built and new clean-up procedures developed and passed on to the public in higher prices.

Water quality requirements for fish plants

An adequate and controllable source of fresh or sea water is an essential requirement in selecting a site for a fish processing plant. Fresh water is often unavailable. A filleting plant with a processing capacity of 91,000 kg of round fish in an 8 h operation may use as much as 4,500 kl of water. This is equivalent to the needs of a city of 50,000 persons consuming 90 1 per person per day.

In many locations, procuring sea water of acceptable quality for fish processing merely involves laying a suction line and pumping the water. Initially there is little or no pollution. Water is used for fish washing, fluming round fish, fillets and offal, machinery cooling and other services. But a plant utilizing the sea both as a source of water and as a repository for sewage and plant effluent quickly contaminates its essential water supply. Protein waste products attract insects, rodents and birds. The problems are magnified a hundred-fold in instances where processing water is obtained from a harbour used for the disposal of human and processing wastes. Water suction lines may be extended so that the plant intake is moved to good quality water. This system can and does work in some instances but too often the initial error in site location is merely compounded.

Prior to 1946, only plants processing shellfish were required to use water of a quality suited for drinking. The requirement was extended in 1946 to include all fish canneries. Fish filleting plants participating in or complying with the Canadian Government Specifications Board's standards (1956) were required to use water of approved quality, having a most probable number (MPN) of coliform bacteria of 2 or less per 100 ml. This standard was made compulsory for all filleting plants in 1964 and for salt fish plants in 1967.

This created problems of compliance for many plants.

Water quality standards

To say that a water source is "safe" or "approved" does not necessarily signify that no risk is ever incurred in using it; only that the public health hazard incident upon its use is relatively low or minimal. The drinking water standard in Canada is employed as a guide in assessing

---

* Department of Fisheries and Forestry, Ottawa, Ontario, Canada.

the safety of water for use in food processing. The standard is based on three criteria:

    (a) source and its protection,
    (b) bacteriological quality, and
    (c) physical and chemical characteristics.

This means that the water must come from a protected source and be adequately treated to ensure safety, also the distribution system must be free from sanitary defects and possible health hazards. The bacteriological standard, requires in part, that a minimum number of water samples be collected at representative points in the source area and that organisms of the coliform group do not exceed limits specified in the standard. The standard applied to water for processing requires that the water shall have no chemical or toxic substances which may be injurious to health or which in any way may affect the acceptability of the fish product. Sources of water having an MPN of up to 100 coliform bacteria per 100 ml may be used after simple chlorination while those with counts in excess of 100 per 100 ml must undergo purification treatment prior to chlorination.

Policy guidelines in treatment of water

For the treatment of sea water, the Department of Fisheries and Forestry applies a policy similar to that used for municipal supplies. The differences lie in treatments and final use.

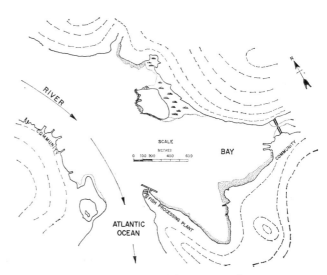

Fig 2. Location of fish processing plant

Prior to bacteriological assessment, the area is surveyed by a sanitary engineer for potential sources of contamination such as sewer outfalls, run-off, or proximity to wharves. Fish plants using sea water with mean coliform counts below 100 per 100 ml must chlorinate to a level of 5 ppm residual and with a minimum 3-min

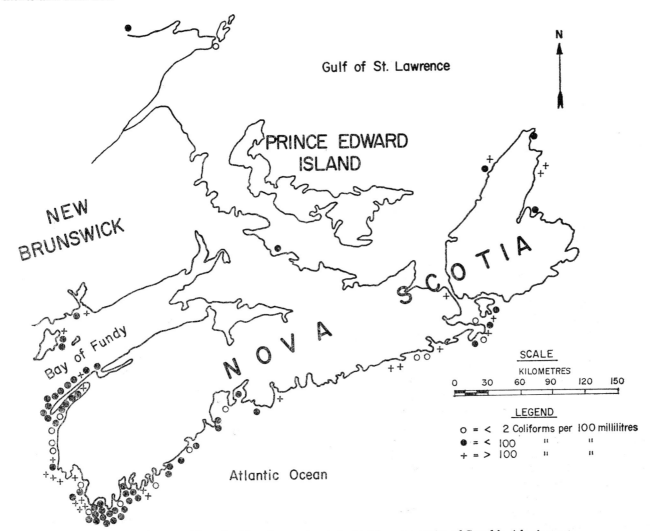

Fig 1. Typical mean coliform densities of sea water at plant intakes on a section of Canada's Atlantic coast

[543]

TABLE 1 MEAN COLIFORM DENSITIES PER 100 MILLILITRES OF SEA WATER AT PLANT INTAKES AND AFTER TREATMENT FOR A SECTION OF CANADA'S ATLANTIC COAST

Sampling Year	M.P.N. of Coliforms per 100 millilitres					
	Little or No Pollution <2	Slightly Polluted <100	Moderately Polluted 100 - 5000	Heavily Polluted > 5000	After Treatment	
					<2	>2
1967	8·1%	48·4%	43·5%	0%	92·3%	7·7%
1968	17·5%	47·4%	35·1%	0%	88·5%	11·5%
1969	7·1%	61·4%	31·4%	0%	100%	0·0%
1970	20·4%	57·1%	22·4%	0%	–	–

contact time. If the only available source yields water with a MPN coliform count in excess of 100 per 100 ml, pre-treatment must precede chlorination. Water with a MPN in excess of 5,000 coliform bacteria per 100 ml would never be approved regardless of treatment.

Sewage pollution of coastal waters

The cost of providing fish plants on the Atlantic coast with quality water has been over 20 million dollars and Governments have had to assist plants both financially and technologically. The quality of water for 1967 to 1970 is shown in Table 1 and fig 1 and reveals that the water quality at intake points is improving. This is directly attributable to changes in location of plant intake lines and, to a lesser extent, to pollution abatement.

Sanitary engineering and bacteriological assessment

This case history reveals a problem: plant employs up to 170 people and is located on a small peninsula in an estuary and discharges its wastes to the bay (fig 2).

Water surveys by bacteriologists and sanitary engineers were carried out in the river, bay area and seaward side of the peninsula to:

    (a) assess quantities of domestic and industrial wastes being discharged and their dispersion under various conditions of tide;

    (b) evaluate the effects on water quality of the discharge of existing plant waste;

    (c) determine water quality under varying tidal, weather and wind conditions; and

    (d) determine measures required to assure water of acceptable quality for use in fish processing.

The survey showed that several communities along the river were discharging untreated wastes directly into the river. The largest, a small town 18 km upstream and with a population of about 4,500, has a sewer system but no primary treatment. Other river communities with populations of 306, 138 and 258 have no sewers discharging directly, but liquid effluents from individual septic tanks find their way into the river.

The largest community discharges a small quantity of untreated industrial effluents to the river. A new relatively large industrial plant, which will soon be in operation, will treat its wastes. The fish plant discharges human and processing wastes on the seaward side of the peninsula, without even screening. Vessels unloading fish at the plant add to the general pollution of the water.

Surveys showed that the pollutants, both domestic and industrial have far less effect on water quality at the fish plant than had been expected. Dilution of the effluents, the current and direction of flow of the river and little or no vertical mixing all tend to minimize contamination of the sea water by the river in the plant area. However, it was apparent that the situation was borderline, and either a change in the river pattern, or an increase in pollutants could lead to a deterioration in water quality.

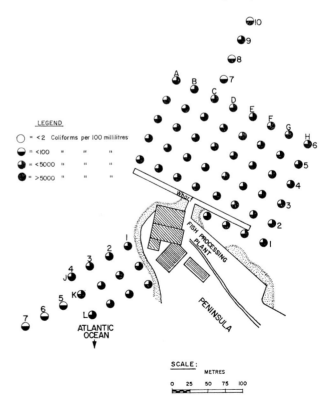

Fig 3. Mean coliform densities of sea water in vicinity of the fish processing plant.

Bacteriological assessment of water quality

To determine whether it was possible to obtain acceptable quality water close to the plant a bacteriological investigation was made. Water was sampled 0.9 m from the bottom at a network of stations (fig 3). At each, four samples were taken at low tide and three at high tide. Wave action, wind direction and velocity, rainfall and land run-off were noted. The mean MPN at the sampling stations were calculated for low tide because that coincided with the most severe pollution.

Survey data and engineering considerations showed the most suitable water intake was at station C7 and that the water should be chlorinated, and monitored weekly for chlorine residual levels and coliform density before and after chlorination—records of the water system, chlorine residual and coliform levels had to be kept.

Over a two-year period the recorded data and increasing flows of untreated domestic sewage prompted a further comprehensive bacteriological survey. This judged that chlorination alone was no longer sufficient and pre-treatment of the water would be required.

After an engineering study of applicable treatment systems a combination of flocculation and sand filtration was found to be the most economical and efficient, with ozone or chlorine recommended as final treatment. The required capital investment for the treatment plant with a capacity of 4.6 kl per minute was $140,000. The calculated operating costs varied from a low of 1 cent to 3 cents per 4.6 kl, depending on pollution density levels and quantities of water used.

Policy on disposal of sewage from fish plants

The guidelines developed for the sanitary disposal of plant sewage require that:

(a) all waste discharged by a fish plant to a water source must receive primary treatment (i.e. to remove solids).

(b) where the effluent from the primary treatment system could affect an open shellfish area; a recreational area; a water supply for a food processing plant; or be objectionable for any other valid reason, it shall, if practicable, be discharged to a tile disposal field constructed so as to meet the requirements of the health authorities having jurisdiction. If such a field is not practicable, then the wastes, after primary treatment, shall be chlorinated by passing through a chlorine contact chamber with a minimum retention time of 30 min and be discharged with a minimum chlorine residual of 5 ppm. The assimilative capacity of receiving water must be considered in respect of other possible water uses.

(c) the design and maintenance of the septic tanks and disposal fields shall conform to the requirements of the health authorities having jurisdiction.

(d) septic tank design shall be based on a usage of 92 l of water per person per day. All septic tanks shall hold at least 9,200 l.

(e) septic tank systems must be vented, either directly or through the plumbing in the building.

(f) all fixtures, wash basins and toilets must have individual traps,

(g) flush toilets using sea water are to be of marine type to minimize corrosion.

Sewage characteristics vary and only by on-the-spot inspection can a suitable method of disposal be designed. Because environmental conditions may change, sewage disposal systems need reassessment and inspection from time to time. Government inspection officers report any changes which may affect sewage disposal and water use to secure corrective action.

Pollution in shellfish-growing areas

Sewage pollution is directly responsible for the closure of many molluscan shellfish growing areas in Canada. The growing areas are surveyed on a continuing basis and the boundaries of "open" and "closed" areas are adjusted accordingly. Canadian authorities have adhered to the principle that only shellfish from "open" areas should be made available for use as food. No commercial use of molluscan shellstock from "closed" areas is permitted. The latter are growing areas wherein overlying waters have coliform densities in excess of MPN 70 per 100 ml.

To commercial operators, closure of areas means:
— an overall reduction in total supply of shellstock,
— greater competition in the procurement of available supplies,
— higher cost of procurement from distant sources,
— intermittent operations due to inconstancy of supply,
— reduced employment for local workers, and
— reduced profits due to lower volume of shellstock processed.

To control agencies, closure of areas necessitates:
— defining, surveying, monitoring and patrolling of "closed" areas to prevent the harvesting of contaminated shellstock,
— increased surveillance of harvesting, transport and utilization of shellstock to preclude any public health hazard presented by possible processing of illegally harvested shellstock, and
— costly surveys and resurveys of shellfish areas to ensure compliance with bacterial and public health engineering criteria.

Shellfish can accumulate and retain pathogenic organisms and toxic organic and inorganic substances present in the growing area waters.

While vast sums are being expended on programmes of pollution control and abatement, it appears to be technically and economically impracticable to try to achieve that degree of purity desirable in shellfish growing areas. This does not mean that consumers need to be denied a full range of gourmet shellfish products nor that a viable shellfish industry cannot be maintained. Rigid enforcement of effective pollution control and abatement can assure a containing flow of safe shellfish. Shellfish depuration could be utilized to maintain and increase production. Benefits that would accrue include:

— improved water quality and, hence, improved shellfish quality,
— control and rotational cropping of molluscan shellfish resources located in moderately contaminated "closed" areas,

- access by industry to such shellstock following depuration,
- utilization of otherwise unexploited, non-productive shellfish growing areas,
- increased volume and constancy of supply of raw material,
- increased opportunities for employment,
- reduction in costs associated with the enforcement, field patrol, and monitoring of conditions within shellfish harvesting areas,
- reduced public health hazards in that shellstock in "closed" areas would be reduced in volume because of controlled harvesting and depuration programmes.

Abatement and prevention of pollution

Public reaction to the overall effects of pollution has enabled governments to formulate and enforce policies in the abatement and prevention of pollution. The Department now has authority to approve water sources and the quality of water used for processing in fish plants. The Inspection Branch of the Department regularly tests fresh water and sea water before and after treatment to assess quality. When the analytical data warrant, corrective action is taken.

To assure the supply of adequate quantities of quality water, water authorities have been established to develop norms in water usage and waste disposal. Provinces have passed legislation to enable these authorities to enforce regulations. At the federal level, a new department concerned with environmental affairs is being established to control functions at the national level.

To tackle sewage pollution problems, we will have to set standards and these standards must be national and apply from coast to coast and to all municipalities and plants, both for old and new sewage installations.

A Kerosene-like Taint in Mullet (Mugil cephalus)*

G. S. Sidhu, G. L. Vale,
J. Shipton and K. E. Murray†

Nature et effets d'une odeur rappellant celle du kérosène chez le mulet (Mugil cephalus)

Les mulets (Mugil cephalus) pêchés au voisinage de Brisbane, Australie, et d'autres poissons capturés dans les eaux proches des raffineries de pétrole et des installations portuaires, présentent une odeur qui rappelle le kérosène. Les recherches ont montré que cette altération est différente de celle due au "pétrole", que l'on trouve parfois dans la morue (Gadus morhua) et dans le saumon kéta (Oncorhynchus keta), et qui est provoquée par la décomposition de la diméthyl-β-propiothétine en diméthylsulfure. On fait disparaître l'odeur kérosènique des tissus du poisson par extraction au moyen de solvants des graisses; elle est associée à la fraction hydrocarburée du lipide.

On a constaté, à l'examen du poisson altéré, une infiltration graisseuse du foie et, dans les filets, une teneur en lipides plus élevée que dans le produit non altéré. On a observé une proportion plus faible d'acides gras C16 et une proportion plus élevée d'acides gras C18 dans le poisson altéré que dans les lipides provenant des filets et du foie de poisson non altéré.

L'extraction de produits volatils par sublimation sous vide puis par concentration du sublimé a donné 2,7 ml d'hydrocarbures volatils pour 25 kg de poisson altéré, contre environ 10 μl pour le poisson non altéré. La chromatographie gaz-liquide (CGL) des hydrocarbures volatils et d'un échantillon commercial de kérosène a permis d'établir l'existence d'une analogie qualitative et quantitative étroite entre les principaux constituants ainsi qu'une analogie entre les descriptions olfactives correspondant aux sommets de périodes de rétention comparables. Les spectres dans l'infra-rouge et de résonance magnétique des protons ont confirmé la similitude de composition des deux substances et indiqué qu'elles étaient surtout constituées d'hydrocarbures paraffiniques avec peu de composantes aromatiques. L'identité de certains des principaux constituants contribuant à donner leur odeur aux deux substances a été pleinement établie par les spectres de masse identiques obtenus à partir du matériel capturé correspondant aux principaux sommets pour lesquels les descriptions olfactives rappellent le kérosène et par les temps de rétention analogues réalisés sur colonne capillaire CGL.

Les auteurs discutent les résultats obtenus du point de vue de la pollution provoquée par le déversement d'hydrocarbures, de leur pénétration dans les tissus du poisson et de leurs effets sur les prises effectuées en milieu marin pollué.

Naturaleza y efectos del sabor a kerosene en la lisa (Mugil cephalus)

La lisas (Mugil cephalus) que se pescan cerca de Brisbane, Australia y algunos otros peces que se capturan en aguas próximas a refinerías de petróleo e instalaciones portuarias tienen un sabor parecido al del kerosene. Las investigaciones han demostrado que este sabor es distinto del de "petróleo" que se nota algunas veces en el bacalao (Gadus morhua) y en el salmón "chum" (Oncorhynchus keta) y que lo causa la descomposición del dimetil-β-propiotetina en dimetilsulfuro. El sabor a kerosene se elimina del tejido muscular del cuerpo del pescado por extracción con solventes de la grasa y está relacionado con la fracción hidrocarburo del lípido.

En el pescado contaminado existía una infiltración grasosa de los hígados y un mayor contenido de lípidos en los filetes que en los del pescado sin sabor; se observó en él una menor proporción de ácidos grasos en C16 y una mayor en C18 que en los lípidos obtenidos a partir de filetes e hígados de pescado sin sabor.

La extracción de sustancias volátiles por sublimación al vacío y subsiguiente concentración del sublimado dio 2,7 ml de aceite volátil en 25 kg de pescado contaminado y alrededor de 10 μl en pescado sin aquel sabor. La cromatografía de gas líquido (CGL) del aceite volátil y de una muestra comercial de kerosene demostró que había una gran analogía cualitativa y cuantitativa de los principales componentes y similaridad de las descripciones del olor asignado a máximos de tiempo de retención comparables. Los espectros de resonancia magnética de protones e infrarrojos confirmaron la analogía de la composición de los dos materiales e indicaron que se trataba principalmente de hidrocarburos parafinados con un pequeño contenido de componentes aromáticos. En el material capturado se obtuvieron espectros de masa idénticos que correspondían a los principales máximos cuyas descripciones de olor a kerosene y tiempos de retención similares de una columna capilar de CGL demostraron plenamente la identidad de algunos de los principales constituyentes que contribuían al olor de los dos materiales.

Se discuten los resultados con respecto a la contaminación causada por la evacuación de hidrocarburos de petróleo, su entrada en los tejidos musculares de los peces y su efecto en los que se pescan en medios marinos contaminados.

IN Australia the annual catch of mullet, mostly *Mugil cephalus*, is about 12 million lb. Of the total wet fish sold in Brisbane mullet comprises 47 per cent. Immature mullet feed on detritus, small crustaceans, algae and plankton in coastal estuaries. When mature, the fish form schools and migrate northwards to warmer water to spawn.

Two types of taint, described as "earthy" and "kerosene", occur in Australian mullet. "Earthy" taint has been on record for a number of years (Thaysen, 1936;

---

*The Fish Board, Queensland, provided financial assistance.
†Commonwealth Scientific and Industrial Research Organization, Division of Food Preservation, Ryde, N.S.W., Australia.

and Kesteven, 1942) and is thought to be produced by an odoriferous species of *Actinomyces* (Morris *et al.*, 1963; Gerber and Lechevalier, 1965; Dougherty *et al.*, 1966). During the last few years a "kerosene" taint in mullet taken from Moreton Bay area and Brisbane River, Queensland, has been serious enough to affect commercial use of the resource. The taint has also been found in mullet and other fish caught in Corio Bay, Geelong, Victoria, some areas of Sydney Harbour and adjoining river estuaries, Botany Bay, N.S.W., and Cockburn Sound, Western Australia. In each area there are oil refineries and some docks, sewage outlets and heavy industry.

A taint described as "petroleum odour" in chum salmon (*Oncorhynchus keta*) from Pacific waters (Motohiro, 1962), "gunpowder" or "blackberry" in cod (*Gadus morhua*) from the Labrador coast, Canada, is reported caused by thermal decomposition of dimethyl-β-propriothetin (DMPT) to dimethyl sulphide (DMS), (Sipos and Ackman, 1964; Ackman *et al.*, 1966 and 1967). Investigations have been carried out for three years on the nature of "kerosene" taint and its effects on mullet.

Mullet samples analysed

Samples of mullet (whole frozen or fillets) from Queensland, and tainted fish from the Brisbane River, were considered against control supplies from northern New South Wales as well as tainted and controlled supplies from the area 60 mi north of Sydney. A panel of at least three persons examined the samples for physical appearance, odour and taste and the presence or absence of "kerosene" taint.

The samples were analysed for DMPT, total lipid, lipid fatty acid composition and liver glycogen by methods described by Vale *et al.* (1970). Lipids were fractionated by the method of Carroll (1961). Methods described by Shipton *et al.* (1969 and 1970) were used for the extraction and concentration of volatile constituents from fish fillets. 25 kg of tainted mullet fillets yielded 2.7 ml of volatile oil as compared to 10 μl from the same quantity of untainted fillets. The oil was examined by gas liquid chromatography, combined gas chromatography-mass spectrometry and infra-red and proton magnetic resonance spectrometry as described by Shipton *et al.* (1970).

Pieces of tissue, approximately 1 mm³, were excised from the livers of three tainted fish and two untainted fish. After fixing with glutaraldehyde and staining with 1 per cent uranylacetate, the pieces were embedded in "Araldite" epoxy resin (CIBA Ltd., Basle, Switzerland). Sections were stained with lead and examined under a Siemens "Elmiskop 1" electron microscope.

Radiotracer studies

Live mullet kept in sea water at 15° for two weeks were divided into three lots. Each lot was held for six days in three separate 135 l tanks, the first containing sea water, the second sea water with 4 ppm "Tween-60" (Chemical Materials Ltd., Glebe, N.S.W.), and the third sea water with 7.5 ppm commercial kerosene and 4 ppm "Tween-60". Half of the solution in each tank was replaced every two days. The fish were offered brine shrimp (*Artemia salina*) and a synthetic diet prepared according to Kelly *et al.* (1958) but rejected both diets.

After the experimental period, the fish were killed and their livers were removed. Each liver was sliced and 0.26 ± 0.01 g of the slices were placed in each of four flasks together with 1.9 ml frog Ringer solution (Dawson *et al.*, 1959) containing 70 mg/100 ml glucose, 5 μCi/0.33 mg glucose-C¹⁴(U) in 0.1 ml, and 4 μCi/12 μmole sodium acetate 1-C¹⁴ in 0.1 ml. Each flask was flushed with oxygen, stoppered and incubated for 1 h at 30°C. After incubation, slices from duplicate flasks were digested in 30 per cent KOH, the glycogen was isolated (Good *et al.*, 1933) and counted in a Packard Tricarb Liquid Scintillation Spectrometer Model 3003, using diotol as a scintillator. The lipid fatty acids from the slices in the remaining duplicate flasks were extracted with ether after saponification with alcoholic KOH and acidification. Fatty acid methyl esters were prepared using the method of Metcalfe and Schmitz (1961) and separated into saturated and unsaturated groups by column chromatography using acid-treated "Florisil" impregnated with silver nitrate (Willner, 1965), and the radioactivity of each group determined.

Results and discussion

The tainted and untainted mullet were indistinguishable by appearance or smell. Tainted fish fillets lacked the fresh "seaweedy" odour of untainted fillets and those from the heavily tainted fish showed some breakdown of muscle tissue or gaping. The stomach contents of the tainted fish showed no visible difference from those of the untainted fish. The taint became faintly apparent when the fillets were minced. When raw or cooked flesh was placed on the tongue and chewed, it gave an immediate and persistent impression of kerosene. Cooking made the taint more obvious but was only necessary with very slightly tainted fish. The dark meat and the fatty layers had a stronger taint than the white meat. The livers from the tainted fish were lighter in colour, larger, more friable and showed evidence of fatty infiltration.

Thin layer chromatography of extracts from both tainted and untainted mullet indicated the presence of DMPT. Quantitative estimation by GLC revealed that the levels of DMPT present (2 to 5 ppm) in the fillets from both the tainted and untainted fish were below those required to produce "petroleum" taint (Ackman *et al.*, 1966) suggesting that the thermal breakdown of DMPT was unlikely to be responsible for the taint in mullet.

TABLE 1. LIPID AND GLYCOGEN CONTENT OF MULLET TISSUE

1. Lipid per cent wet tissue

Liver

Tainted	16.0 ± 1.44	Difference between
Untainted	7.1 ± 0.63	means significant
	(17 d.f.)	at 1 per cent level

Fillets

Tainted	9.0 ± 0.44	Difference between
Untainted	4.5 ± 0.44	means significant
	(36 d.f.)	at 0.1 per cent level

2. Glycogen mg/100 g wet liver

	1969	*1970*
Tainted	980 ± 188	1630 ± 204
Untainted	320 ± 188	2500 ± 204
	n' = 7	n' = 6

TABLE 2. FATTY ACID COMPOSITION OF LIPID

Fatty acid	Liver lipid Tainted wt %	Liver lipid Untainted wt %	Sig. of difference	Fatty acid	Fillet lipid Tainted wt %	Fillet lipid Untainted wt %	Sig. of difference
C14:0	2.8 ± 0.2	3.0 ± 0.2	n.s.	C14:0	5.7 ± 0.5	9.3 ± 0.5	**
C15:0	1.3 ± 0.13	2.5 ± 0.53	n.s.	C15:0	1.4 ± 0.2	3.6 ± 0.8	*
C16:0	27.2 ± 0.75	30.9 ± 0.75	*	C16:0	23.2 ± 1.2	33.7 ± 1.2	***
C16:1	8.6 ± 0.6	10.9 ± 0.61	*	C16:1	11.8 ± 1.4	21.2 ± 0.8	**
C17:0	0.6 ± 0.2	0.7 ± 0.2	n.s.	C17:1	1.1 ± 0.2	3.3 ± 0.2	***
C17:1	0.0 ± 0.0	6.3 ± 1.2	***	C18:0	5.1 ± 0.3	3.6 ± 0.3	*
C18:0	7.2 ± 0.4	8.0 ± 0.4	n.s.	C18:1	35.2 ± 5.0	5.9 ± 0.7	***
C18:1	36.4 ± 2.2	11.6 ± 0.83	***	C18:2	3.9 ± 0.2	2.0 ± 0.2	***
C18:2	6.4 ± 0.4	2.4 ± 0.4	***	C18:3	1.8 ± 0.3	0.9 ± 0.1	*
C18:3	2.4 ± 0.24	0.9 ± 0.24	**	C20:1	1.7 ± 0.4	0.7 ± 0.1	n.s.
C20:1	0.6 ± 0.24	0.6 ± 0.24	n.s.	C20:2	0.8 ± 0.3	1.0 ± 0.3	n.s.
C20:3	0.7 ± 0.28	5.1 ± 0.28	***	C20:3	1.1 ± 0.3	3.4 ± 0.3	***
C20:4	0.4 ± 0.13	0.4 ± 0.13	n.s.	C20:5	3.7 ± 0.6	8.8 ± 0.6	***
C20:5	2.2 ± 0.56	6.9 ± 0.56	***	C22:5	0.9 ± 0.3	1.2 ± 0.3	n.s.
C22:3	0.3 ± 0.15	0.6 ± 0.15	n.s.	C22:6	1.9 ± 0.7	1.2 ± 0.5	n.s.
C22:5	0.8 ± 0.43	4.9 ± 0.98	**				
C22:6	1.5 ± 0.64	4.5 ± 1.64	n.s.				

In tainted samples $n' = 11$
In untainted samples $n' = 12$

In tainted and untainted samples $n' = 13$

* Significant at 5% level   ** Significant at 1% level   *** Significant at 0.1% level

TABLE 3. COMPOSITION OF LIPID FROM MULLET FILLETS
(per cent by weight)

	Tainted	Untainted
I Neutral lipids		
Hydrocarbons	0.6	0.0
Cholesterol esters	0.0	1.6
Triglycerides	87.7	81.7
Cholesterol	0.5	0.9
Diglycerides	1.4	2.4
Monoglycerides	0.0	0.2
F.F.A.	2.8	5.2
II Phospholipids	7.0	8.0

Livers and fillets from tainted and untainted fish, caught over a period of $2\frac{1}{2}$ years, were analysed for total lipid, and livers were analysed for glycogen content. The data are set out in Table 1. The considerable variations observed in lipid content of these tissues from one mullet to another were presumably due to variations in season, sex, stage of maturity and environment. Nevertheless, the lipid content of fillets and liver from the tainted fish was much higher than those from the untainted fish, indicating fatty infiltration in the tainted fish. In 1969, but not in 1970, livers from the tainted fish had a higher glycogen content than those from untainted fish.

Fatty acid composition of the liver and fillet lipids differed markedly between tainted and untainted fish (Table 2). There was a marked increase in C18 unsaturated fatty acids, mainly C18:1, and a corresponding decrease in C16 and C20 and higher fatty acids in the tainted fish. Fractionation of fillet lipids (Table 3) showed that in tainted and untainted fish triglycerides formed the major class. A small amount of hydrocarbons was present in the tainted fish but was not detected in untainted fish. Changes in the fatty acid composition of the triglycerides from the fillets were similar to those observed in the total lipid from the liver and fillets of tainted fish (Table 1).

Vivarium and radiotracer experiments

Mullet placed in a glass tank containing 5 ppm kerosene developed a mild kerosene taint in 24 h, confirming an earlier report by Nitta et al. (1965) that taints from water can be picked up through the gills by fish in a few hours. The fish kept for six days in sea water containing 7 ppm kerosene and 4 ppm "Tween-60" became heavily tainted.

The data obtained from the radiotracer studies are set out in Table 4. Fish in the control group, although starved for three weeks, maintained a higher level of liver glycogen. On the other hand, the remaining treatments resulted in a fall in liver glycogen, kerosene-treated fish showing the greatest fall. This ran parallel to the drop in activity of glycogen-synthesizing enzymes as measured by the incorporation of glucose-^{14}C (U) into glycogen. Fatty acid synthesis, was highest in the kerosene-treated group and lowest in the control, and was reflected in the levels of lipid in the livers of fish under the respective treatments. The high level of lipid in kerosene-affected liver was apparently maintained by the supply of intermediates for lipid synthesis by the breakdown of glycogen and other endogenous sources.

In actively feeding fish affected by petroleum hydrocarbons, fat accumulation in the liver does not seem to

TABLE 4. DATA FROM RADIOTRACER EXPERIMENT

	Control	"Tween-60"	"Tween-60" + Kerosene
Liver glycogen (mg/100 g)	2170	833	120
Label (C^{14}) incorporated into glycogen (10^{-5} xd/min/g liver)	31.3	7.3	0.2
Lipid content			
1. Liver (%)	0.8	1.5	4.5
2. Fillets (%)	5.4	10.0	11.3
Label (C^{14}) incorporated into fatty acids (10^{-5} xd/min/g)	2.1	3.5	5.5
Label in saturated fatty acids (%)	31	36	32
Label in unsaturated fatty acids (%)	69	64	68

[548]

arise from the mobilization of depot fatty acids and their deposition in liver as triglycerides. Fat accumulation could, however, arise from a higher rate of lipid synthesis in the liver from dietary and endogenously synthesized fatty acids and other intermediates for lipid synthesis.

Rats fed ethanol over long periods (24 days) deposited in the liver fat mainly derived from dietary or endogenously synthesized fatty acids (Lieber *et al.*, 1966). Ethanol, carbon tetrachloride and ethionine have been reported (Isselbacher and Greenberger, 1964; Maling *et al.*, 1963) to inhibit esterification of fatty acids to esters other than triglycerides in liver. A similar mechanism which partially inhibits fat transport and increases fat synthesis in mullet liver could result in fatty infiltration of the type observed.

Chemical nature of volatile oil

The volatile oil collected from the tainted mullet had a kerosene-like odour. When the components of this oil were separated on a 60-ft Carbowax 20 M column and sniffed, several had a kerosene-like odour and others had odours described as "turpentine", "oily", "aromatic" and "naphthalenic". When commercial kerosene was examined by GLC under identical conditions, components having comparable retention times were found to possess odours similar to those from the volatile mullet oil. Comparison of the chromatograms of the mullet oil and kerosene showed that, out of a total of 72 peaks, 59 were common and of these 18 were given odour descriptions appropriate to petroleum products. The infra-red and NMR spectra also showed a close similarity of the mullet oil and kerosene.

A rigorous comparison of the volatile oil from the tainted fish and kerosene was made by selecting six pairs of fractions (incorporating a major peak) having corresponding retention times and odour descriptions. These fractions were analysed by combined gas chromatography-mass spectrometry using a capillary column (500 ft × 0.030 in. silicone OV101). From their mass spectra and retention times it was possible to unequivocally identify in both the volatile oil and kerosene, n-tetradecane, naphthalene, 2-methylnaphthalene and from mass spectra alone to tentatively identify isopropylbenzene, 3-(2-methylphenyl) pentane, 2,6-dimethyl-1,2,-3,4-tetrahydronaphthalene. These results together clearly established the qualitative identity of the tainted mullet oil and a commercial kerosene.

General discussion

The evidence shows that petroleum hydrocarbons of the type found in kerosene can enter the tissues of mullet. Location of oil refineries close to areas in Australia from where reports of taint in fish have been received and observations in Japan (Nitta *et al.*, 1965) and Canada (Krishnaswami and Kupchanko, 1969) indicate that the hydrocarbons may originate from oil refineries. However, spillage by boats and additions through sewage from heavy industry cannot be entirely disregarded. It is unlikely that kerosene as such would be released in waste water from refineries but evaporation of lighter fractions from the polluted water and slow uptake by fish of heavier fractions and/or their removal through microbiological processes, may explain the observed presence of the hydrocarbon fraction resembling kerosene in the fish.

The solubility of these hydrocarbons range from 5 to 11 ppm in sea water and can be considerably enhanced by the presence of detergents. The Brisbane River water was found to contain 2 to 3 ppm detergents. Threshold levels of these hydrocarbons which would impart taint to rainbow trout (*Salmo gairdnerii*), Japanese mackerel and some other species of fish vary from 0.01 to 0.02 ppm. Bottom mud obtained from waters close to oil refineries could also impart the taint (Nitta *et al.*, 1965; Krishnaswami and Kupchanko, 1969). Mullet, being a detritus feeder with relatively high body fat, is likely to take up petroleum hydrocarbons more readily than other species of fish living in the same environment. During processing the effective exclusion of the tainted fish from the processing line could be very costly as there is at present no simple method of eliminating the tainted fish.

The long-term effects of hydrocarbons include changes in the lipid metabolism of mullet, with consequent accumulation of fat in the liver and other body tissues. Fatty infiltration of the liver has been observed in higher animals exposed to toxic levels of petroleum hydrocarbons (Lewin, 1932). Mullet seems to respond in a similar manner. Very little is known about the processes leading to fatty infiltration of the fish liver, but the higher deposition of fat by mullet seems to be a detoxifying response to the presence of hydrocarbons in the body tissues. The response is mediated through enhanced activity of fatty acid and lipid-synthesizing systems. These, coupled with a partial disturbance of the lipid transport mechanism from the liver, result in the accumulation of fat in the liver as well as the flesh of the fish. The main fatty acid synthesized seems to be C18:1. The effects of these long-term changes on the life cycle of the fish and the possibility of their reversing when the fish moves to untainted waters are not known.

It can be said that petroleum hydrocarbons can cause serious tainting problems and biochemical changes in fish. The scale of this pollution is likely to increase as more refineries and industries releasing hydrocarbons are established throughout the world. A satisfactory and permanent solution, therefore, requires the treatment of waste water and sewage to reduce hydrocarbon concentration to an acceptable level or preventive industrial action to keep hydrocarbons out of the effluent. The establishment of such an acceptable level, and effective methods of effluent treatment or preventive industrial action require investigation.

References

ACKMAN, R G, DALE, J and HINGLEY, J Deposition of dimethyl-β-
1966   propiothetin in Atlantic cod during feeding experiments. *J. Fish. Res. Bd Can.*, 23(4):487–97.
ACKMAN, R G, HINGLEY, J and MAY, A W Dimethyl-β-propio-
1967   thetin dimethyl sulphide in Labrador cod. *J. Fish. Res. Bd Can.*, 24(2):457–61.
CARROLL, K K Separation of lipid classes by chromatography on
1961   Florosil. *J. Lipid Res.*, 2(2):135–41.
DAWSON, R M C, *et al.* Mammalian and frog ringers. *In* Data for
1959   biochemical research, Oxford, Clarendon Press, p. 209.
DOUGHERTY, J D, CAMPBELL, R D and MORRIS, R L Actino-
1966   mycete: isolation and identification agent responsible for musty odors. *Science, N.Y.*, 152(3727):1372–3.
GERBER, N N and LECHEVALIER, H A Geosmin, an earthy-smelling
1965   substance isolated from Actinomycetes. *Appl. Microbiol.*, 13(6):935–8.

Good, C A, Kramer, H and Somogyi, M The determination of
1933   glycogen. *J. biol. Chem.*, 100(2):485–91.
Isselbacher, K J and Greenberger, N J Metabolic effects of
1964   alcohol on the liver. *New Engl. J. Med.*, 270:351–6, 402–10.
Kelly, P B, Reiser, R and Hood, D W The origin and metabol-
1958   ism of marine fatty acids: the effect of diet on the depot
        fats of *Mugil cephalus* (the common mullet). *J. Am. Oil
        Chem. Soc.*, 35(5):189–92.
Kesteven, G L Studies in the biology of Australian mullet.
1942   1. Account of the fishery and preliminary statement of the
        biology of *Mugil dobula* Gunther. *Bull. Coun. scient. ind.
        Res. Melb.*, (157):62.
Krishnaswami, S K and Kupchanko, E E Relationship between
1969   odour of petroleum refinery wastewater and occurrence of
        "oily" taste-flavour in rainbow trout *Salmo gairdnerii*.
        J. Wat. Pollut. Control Fed., 41:184–96.
Lewin, I E Zur Frage der pathologischen Veränderungen und der
1932   Funktionsfähigkeit des Reticuloendothelsystems bei Vergif-
        tung mit Benzindämpfen. *Arch. Gewerbepath. Gewerbehyg.*,
        3:340.
Lieber, C S, Spritz, N and DeCarli, L M Role of dietary, adipose
1966   and endogenously synthesized fatty acids in the patho-
        genesis of the alcoholic fatty liver. *J. clin. Invest.*, 45(1):
        51–61.
Maling, H M, Wakabayashi, M and Horning, M G Alterations
1963   in hepatic lipid biosynthetic pathways after ethanol,
        ethionine and carbon tetrachloride. *In* Advances in enzyme
        regulation, edited by G. Weber, New York, MacMillan Co.,
        Vol. 1:247–57.
Metcalfe, L D and Schmitz, A A The rapid preparation of fatty
1961   acid esters for gas chromatographic analysis. *Analyt. Chem.*,
        33(3):363–4.

Morris, R L, Dougherty, J D and Ronald, G W Chemical
1963   viewpoint on Actinomycetes metabolism products as taste
        and odour causes. *J. Am. Wat. Wks Ass.*, 55:1380–90.
Motohiro, T Studies on the petroleum odour in canned chum
1962   salmon. *Mem. Fac. Fish. Hokkaido Univ.*, 10(1):1–65.
Nitta, T, *et al.* Studies on the problems of offensive odour in
1965   fish caused by wastes from petroleum industries. *Bull.
        Tokai reg. Fish. Res. Lab.*, (42):23–37 (in Japanese).
Shipton, J, Whitfield, F B and Last, J H Extraction of volatile
1969   compounds from green peas (*Pisum sativum*). *J. agric. Fd
        Chem.*, 17(5):1113–8.
Shipton, J, *et al.* Studies in kerosine-like taint in mullet (*Mugil
1970   cephalus*). 2. Chemical nature of the volatile constituents.
        J. Sci. Fd Agric., 21(8):433–6.
Sipos, J C and Ackman, R G Association of dimethyl sulphide
1964   with the "blackberry" problem in cod from the Labrador
        area. *J. Fish. Res. Bd Can.*, 21(2):423–5.
Thaysen, A C The origin of an earthy or muddy taint in fish. 1.
1936   The nature and isolation of the taint. *Ann. appl. Biol.*, 23:
        99–104.
Vale, G L, *et al.* Studies on a kerosine-like taint in mullet (*Mugil
1970   cephalus*) 1. General nature of the taint. *J. Sci. Fd Agric.*
Willner, D Separation of fatty acid esters on acid-treated Florosil
1965   impregnated with silver nitrate. *Chem. Ind.*, 1965:1839–40.

Acknowledgements

To Mr. G. G. Coote for statistical analysis and to The Fish Board,
Queensland, for financial assistance in carrying out this investigation.

DDT Residue Levels in some U.S. Fishery Products and some Treatments in Reducing them

*V. F. Stout, F. L. Beezhold
and C. R. Houle**

**Niveaux résiduels de concentration en DDT dans certains produits
de la pêche aux Etats-Unis et efficacité de certaines méthodes de
traitement utilisées pour les réduire**

Aux Etats-Unis, on a examiné des résidus de pesticides trouvés dans
des poissons et des mollusques destinés à l'alimentation et provenant
des eaux situées au large de la côte du Pacifique, pour déterminer
dans quelle mesure ces produits étaient conformes aux normes de
tolérance intérimaires de 5 ppm établies par la Food and Drug
Administration des Etats-Unis pour le DDT total. On a également
examiné des saumons "sockeye" (*Oncorhynchus nerka*) capturés en
haute mer au sud des îles Aléoutiennes dans l'océan Pacifique.
Aucun des échantillons prélevés sur des produits capturés en haute
mer ou au large du Washington et de l'Oregon ne contenait plus de
0,5 ppm de DDT. Par contre, certains échantillons provenant des
eaux côtières de la Californie méridionale contenaient des taux
notables de résidus de DDT, pouvant aller jusqu'à 57 ppm.

Les efforts entrepris pour éliminer le DDT du chinchard
(*Trachurus symmetricus*) par modification des temps d'épuisement
et des températures n'ont pas donné de résultats. A la suite d'analyses
effectuées sur des chinchards on n'a pas constaté de variation
sensible des taux résiduels. Les plus gros poissons contenaient une
plus forte accumulation de DDT.

Il serait possible de réduire le taux de concentration des résidus
contenus dans le saumon "coho" (*Oncorhynchus kisutch*) des Grands
Lacs en éliminant la peau, les volets abdominaux et la graisse
contenue dans la partie dorsale médiane et les parties latérales du
poisson.

La transformation du poisson en farine ou en concentré protéique
fournit des produits riches en protéines dont le taux de concentration
en DDT est faible: 0,097–0,6 ppm dans la farine du menhaden
(*Brevoortia tyrannus* et *B. patronus*) et moins de 0,002 ppm dans le
concentré protéique de hareng du Pacifique (*Clupea harengus pallasi*).
Les huiles obtenues lors de la fabrication de la farine de menhaden
contiennent de 2,1 à 9,4 ppm de DDT. L'huile extraite du hareng du
Pacifique traité pour la fabrication de concentré protéique contient
0,28 ppm de DDT. On a étudié des méthodes permettant d'abaisser
la teneur en DDT de l'huile de poisson, hydrogénée ou non.

**Niveles residuales de concentración de DDT en algunos productos
pesqueros dé los Estados Unidos y eficacia de algunos métodos de
tratamiento para reducirlos**

Se han examinado residuos de plaguicidas en el pescado y mariscos
comestibles procedentes de las aguas de la costa del Pacífico de los
Estados Unidos, con objeto de determinar si cumplen la tolerancia
provisional de 5 ppm de DDT total que establecen las disposiciones
de la Food and Drug Administration de los Estados Unidos. Se
examinó igualmente el salmón "sockeye" (*Oncorhynchus nerka*)
capturado en alta mar en el Océano Pacífico, al sur de las Islas
Aleutianas. Ninguna de las muestras del mar abierto o de las aguas
costeras de Washington y Oregón contenían más de 0,5 ppm de
DDT. Sin embargo, algunas muestras de las aguas costeras de
California meridional contenían residuos considerables de DDT,
hasta 57 ppm.

Los esfuerzos hechos para eliminar el DDT del jurel (*Trachurus
symmetricus*) modificando los tiempos y las temperaturas de
tratamiento al vacío no tuvieron éxito. Los análisis de distintos
ejemplares de jurel indicaron amplias variaciones de residuos. En los
peces más grandes se descubrió una mayor acumulación de DDT.

El nivel de los residuos en el salmón "coho" (*Oncorhynchus
kisutch*) de los Grandes Lagos podría reducirse quitándole la piel
y la grasa de las paredes abdominales, de la línea lateral y la del
dorso del cuerpo.

Al convertir el pescado en harina o concentrados proteínicos
(CPP) se obtienen productos proteináceos con bajos niveles de
DDT: 0,097–0,6 ppm en la harina de lacha (*Brevoortia tyrannus* y
B. patronus) y menos de 0,002 ppm en CPP de arenque del Pacífico
(*Clupea harengus pallasi*). Los aceites obtenidos durante la fabricación de harina de lacha contenían 2,1–9,4 ppm de DDT. El aceite
extraido del arenque del Pacífico elaborado en CPP contenía
0,28 ppm de DDT. Se han examinado métodos para disminuir el
contenido de DDT del aceite sin hidrogenar e hidrogenado.

BECAUSE of the extensive use of DDT since World
War II, the ease with which it disperses, and the
stability of the chemical, it has become a widespread pollutant. The National Marine Fisheries Service
has been examining the position.

The aim on food fish and shellfish was to determine the

---

* National Marine Fisheries Service, Technological Laboratory, Seattle, Washington 98102, U.S.A.

residue levels in edible portions, and in those species with high residue levels to attempt to reduce the levels to meet the U.S. Food and Drug Administration interim tolerance for edible flesh, 5 ppm wet weight. Both marine and fresh water species were studied.

Marine species

We examined some thirty marine fish and shellfish of the Northeast Pacific Ocean. Although all are edible, several are used primarily for industrial or bait purposes or are not fished commercially.

None of the samples from the Pacific Ocean or Washington and Oregon coasts contained more than 0.5 ppm total DDT. These residues of less than 10 per cent of the current FDA limit for fish constitute some of the lowest levels found for fish in the United States. DDT usage depends on the intensity of farming, the weather, and the density of the population. Washington and Oregon have a short growing season, a winter cold enough to kill many pests, and a relatively low population density. Thus, relatively little DDT has been used in these areas, so the marine life contains little DDT.

In southern California, farming is intensive and the growing season is long. There are as many as three crops a year. Also, the density of population is high, and the winter mild. It is not surprising, therefore, that some fish from this area contain excessive residues of total DDT. We gave particular attention to jack mackerel, because it has been involved in FDA action.

During the fall of 1969, the FDA seized jack mackerel from southern California because some samples contained more than 5 ppm total DDT. Since that time, high levels of residues have been found in the livers and edible portion of other species of fish. The livers from one sample, for example, contained 1,026 ppm total DDT; fillets of the same fish contained 57 ppm. Although the relation between residues in the livers and flesh was not constant, the concentration of residues in the liver was always greater than that in the flesh. In fish with low levels of DDT, such as those from the coastal waters of Washington, the ratio of DDT in the liver to that in the flesh was greater than 100. When the concentration of DDT was high, the ratio decreased in an irregular, unpredictable way to as low as 5. Generally, it was in the range of 20–50.

Attempts to lower residue levels

Several experiments were done with jack mackerel to determine if simple precanning process alterations would reduce significantly the level of DDT in the flesh. These experiments show that no simple means exist for decreasing the residues of DDT in jack mackerel and still producing the normal canned product. As has been found with chicken, DDT is too closely associated with muscle components to be removed by conventional processing procedures.

The level of DDT in jack mackerel dropped early in 1970, so fishing was started again, but the threat of excessively high levels still continues. A cyclic rise and fall related to the season of the year or to the age of the fish may be suggested. In the fall, fish could be accumulating residues applied throughout the agricultural growing season.

Fresh water species

In spring, 1969, excessive residues of DDT in fresh water fish was brought to the attention of the general public by the seizure of Lake Michigan coho salmon containing up to 20 ppm DDT and up to 0.32 ppm dieldrin: the maximum permissible content of dieldrin in the edible portion is 0.30 ppm.

At about the same time excessive levels of DDT residues were being reported in chub (*Coregonus hoyi*).

The seriousness of the DDT residue problem in fresh water fish was shown by the data of the National Pesticide Monitoring Program of the U.S. Bureau of Sport Fisheries and Wildlife (Henderson *et al.*, 1969). In 12 of 44 lakes and rivers sampled, some or all of the fish contained more than 5 ppm DDT, and in 8 of these locations, the mean for all fish examined exceeded the tolerance level.

Work has been in progress at the National Marine Fisheries Service Laboratory at Ann Arbor, Michigan, to study methods of reducing the residues in local fish. A partial solution has been found for salmon (Reinert, 1969). Analyses of various portions showed (Table 1) that DDT is concentrated in areas of high fat content. Removal of the skin and the fatty deposits in the dorsal median, lateral line, and belly flap areas of fish with high residues would therefore reduce the contamination substantially.

TABLE 1. DISTRIBUTION OF DDT IN LAKE MICHIGAN COHO SALMON (FROM REINERT, 1969)

Sample	Total DDE + DDT[a] (ppm)
Whole fish	12.2
Whole eviscerated steak	14.9
Flesh (muscle)	5.7
Dorsal median fat	62.8
Laterial line fat	34.3
Abdominal adipose	92.3
Belly fat	69.7

[a] TDE is not found in significant amounts in Great Lakes fish

The situation with other fish is less readily improved. Trimming is practical only in a product of high value such as salmon. Furthermore, in many fatty fish such as chub, residues are so evenly distributed that trimming would be ineffective.

Fish protein concentrate

Studies of methods for production of fish protein concentrate (FPC) required information about the fate of the pesticides during processing (Stout and Beezhold, 1970). The lipophilicity of organochlorine pesticides suggested that DDT would be concentrated into the by-product oil. Table 2 indicates the validity of this inference by showing that FPC made from Pacific herring both by the isopropyl alcohol extraction method (Bureau of Commercial Fisheries, 1966) and by the aqueous fractionation method (Spinelli, 1969) contained a barely detectable amount of DDT. The largest concentration of pesticide occurred in the oil portion. Phospholipids are potent solubilizing agents for organic molecules, which may explain the concentration of the pesticides in Miscella 1 of the isopropyl alcohol extraction method. Fortunately, because of the fractionation of the DDT into Miscella 1, the edible-oil fraction contains a very low level of DDT, especially after being washed with water, which tends to remove any of the suspended phospholipids and associated compounds.

TABLE 2. FATE OF PESTICIDE RESIDUES DURING FPC PRODUCTION

Sample	Total DDT (ppm)	
	Isopropyl alcohol extraction	Aqueous fractionation
Whole raw Pacific herring	0.097	0.097
Separated raw herring[a]	0.084	0.084
FPC	<0.002	0.006
Oil		1.12
Miscella 1[b]	1.75	
Miscellas 2 + 3[c]	0.33	
Miscellas 2 + 3, washed with H_2O	0.28	

[a] Skin and bones removed
[b] Mainly phospholipids, not usually considered edible
[c] Triglycerides, potentially for food use

Industrial fishery products

Fish meal and its accompanying oil are used in animal feeds. The residues in meal contribute to the pesticide burden imposed on poultry and contribute to the load in foods derived from the animals.

Residue levels in menhaden (*Brevoortia tyrannus* and *B. patronus*) meal and oil are shown in Table 3. As with FPC, the meal contains less than one-tenth as much DDT as does the oil. Because meal may contain up to about 15 per cent oil, processes that reduce the oil content would also reduce the pesticide content. The Food and Drug Administration tolerance of 5 ppm for DDT residues in edible fish flesh does not apply to fish meal or oil.

TABLE 3. RANGE OF PESTICIDE RESIDUE LEVELS IN MENHADEN OIL AND MEAL IN 1969

States	No. of samples	Total DDT (ppm)	
		Oil	Meal
Middle Atlantic	6	4.4–9.4	0.29–0.61
Chesapeake	2	2.8–4.0	0.21–0.25
South Atlantic	12	2.1–7.2	0.10–0.32
Gulf of Mexico	2	2.2–2.3	0.10–0.11

The concentration of pesticides in oil suggests that removal of DDT from fish oil may become necessary. Deodorization of oil in the refining process of edible vegetable oil (Saha *et al.*, 1970; Smith *et al.*, 1968) removes almost all of the chlorinated hydrocarbon pesticides originally present. Unfortunately deodorization is too expensive for animal feed products so we tried to find simple methods to reduce the residue content of fish oil. Washing with cold or hot water, steam distillation at atmospheric pressure, or partitioning with 55 and 90 per cent (v/v) aqueous isopropyl alcohol, however, effected no change in the residue level.

Removal from fish oils

Table 4 presents data on the effectiveness of vacuum steam refining of two crude bleached Pacific herring oils containing closely similar concentrations of chlorinated pesticides. The two oils were subjected to the same conditions during the steam refining except for the temperature of distillation.

Although the rate of removal of pesticide residues is more rapid at the higher temperature, the quality of the fish oil is undoubtedly better retained at the lower temperature.

TABLE 4. PARTIAL REMOVAL OF PESTICIDE RESIDUES FROM TWO BLEACHED[a] PACIFIC HERRING OILS BY STEAM REFINING

Distillation temperature[b]	Pesticide	Pesticides (ppm)		
		Original oil	Refined oil	Condensate fraction
190 to 200°C	p,p'–DDE	0.73	0.47	40.3
	p,p'–TDE	0.59	0.53	21.3
	p,p'–DDT	0.37	BDL[c]	BDL
	Total	1.69	1.00	61.6
230 to 240°C	p,p'–DDE	0.82	0.49	34.1
	p,p'–TDE	0.70	0.62	24.5
	p,p'–DDT	0.44	BDL	BDL
	Total	1.96	1.11	58.6

[a] Bleached with activated clay (4 per cent weight of oil) for 30 min at 8 mm Hg and 80°C
[b] Deodorized for 3 hr at 8 mm Hg with 9 wt per cent steam based on the wt of the oil
[c] BDL: below detectable limit

The pesticide p,p'-DDE, a product of the natural breakdown of p,p'-DDT, is often the chlorinated pesticide found in highest concentrations in marine animals. Table 5 presents data on the effectiveness of removal of p,p'-DDE from menhaden oil containing added amounts of this substance. The table shows that the combination of hydrogenation and deodorization is effective in reducing the level of this pesticide. The deodorization process, did not, under the conditions used, reduce the pesticide level in menhaden oil at as high a rate nor to as great a degree as Smith *et al.* (1968) and Saha, Nielsen and Sumner (1970) found with vegetable oils.

TABLE 5. PARTIAL REMOVAL OF ARTIFICIALLY HIGH RESIDUES OF P,P'–DDE FROM BLEACHED MENHADEN OILS BY HYDROGENATION AND VACUUM STEAM DEODORIZATION

Sample	Residues of p,p'-DDE (ppm)		
	Trial 1	Trial 2	Trial 3
Original	63.1	23.4	12.0
Hydrogenated oil[a]	51.8	21.9	10.5
Deodorized oil[b]	28.7	16.8	8.2

[a] Melting point: 42°C
[b] Deodorized for 3 hours at 12 mm Hg and 190–200°C with 10 wt per cent steam based on the wt of the oil

The major difference in the application of deodorization to the removal of pesticides from fish oils, compared with vegetable oils, is the lower temperature that is nearly mandatory on fish oils to prevent thermal degradation. Because of this requirement the duration of the deodorization must be extended if a substantial reduction in the pesticide level is to be accomplished.

References

BURDICK, G E, *et al.* The accumulation of DDT in lake trout and
1964 the effect on reproduction. *Trans. Am. Fish. Soc.*, 93(2):127–36.
Bureau of Commercial Fisheries, Technological Laboratory,
1966 College Park, Maryland, Marine protein concentrate. *Fish. Leafl. U.S. Fish Wildl. Serv.*, (584):27 p.

CHANG, S S   Processing of fish oils. *In* Fish oils, edited by M. E.
1967   Stansby, Westport, Conn., Avi Publ. Co., pp. 206–21.

GOODING, C M B   Fate of chlorinated organic pesticide residues in
1966   the production of edible vegetable oil. *Chem. Ind.*, 334.

HENDERSON, C, JOHNSON, W L and INGLIS, A   Organochlorine
1969   insecticide residues in fish (National Pesticide Monitoring
Program). *Pestic. monitg J.*, 3(3):145–71.

REINERT, R E   Insecticides and the Great Lakes. *Limnos Magazine*
1969   (*U.S.*), 2(3):4–9.

RITCHEY, S J, YOUNG, R W and ESSARY, E O   Cooking methods
1969   and heating effects on DDT in chicken tissues. *J. Fd Sci.*,
34(6):569–71.

SAHA, J G, NIELSEN, M A and SUMNER, A K   Effect of commercial
1970   processing techniques on lindane- and DDT-^{14}C residues in
rapeseed oil. *J. agric. Fd Chem.*, 18(1):43–44.

SMITH, K J, *et al.*   Removal of chlorinated pesticides from crude
1968   vegetable oil by simulated commercial processing procedures. *J. Am. Oil Chem. Soc.*, 45(12):866–9.

SPINELLI, J   Preparation of fish protein concentrate and fish meal.
1969   U.S. Patent Application, Serial No. (779.117) Feb. 13.

STOUT, V F   Pesticide levels in fish of the Northeast Pacific. *Bull.*
1968   *environ. Contam. Toxic.*, 3(4):240–6.

STOUT, V F and BEEZHOLD, F L   The fate of DDT and its meta-
1970   bolites during the preparation of fish protein concentrate.
(Unpubl. MS).

Effects of Pollution: Loss of an $18 million/year Shellfishery

R. T. Dewling, K. H. Walker and
*F. T. Brezenski**

Les effets de la pollution: une pêcherie de mollusques perd 18 millions de dollars par an

Les pertes subies par la pêche des mollusques dans la Baie de Raritan, zone estuarine de 147 km² située entre l'Etat de New York et le New Jersey, sont directement imputables aux déversements d'effluents municipaux et industriels insuffisamment traités qui proviennent de ces deux Etats. Les effluents municipaux, d'origine fécale, ont contaminé les bancs à un tel point qu'une grave épidémie d'hépatite s'est produite. Les déversements d'effluents industriels ont avarié les bivalves et, dans certains secteurs, leur action sur le milieu a engendré une toxicité qui a pratiquement interrompu la multiplication de ces mollusques.

Afin de rendre aux eaux de la Baie de Raritan une qualité conforme à leur utilisation la plus favorable, à savoir la pêche des mollusques, la Federal Water Quality Administration a entrepris une enquête poussée d'une durée de 13 mois, ainsi qu'un programme de surveil-lance de 3 ans. L'étude comportait le prélèvement d'échantillons en profondeur et l'analyse d'effluents industriels et municipaux ainsi que celle des eaux de réception. A toutes ses phases elle comprenait des programmes chimiques, biologiques et microbiologiques. De nouvelles techniques et méthodes ont été élaborées pour analyser les polluants dans les eaux des estuaires. Les recherches micro-biologiques comportaient l'analyse et le calcul des taux de concen-tration des organismes indicateurs, et la détection des agents pathogènes dans la chair des mollusques ainsi que dans le milieu aquatique. En injectant des colorants dans les principales sources de pollution, on a démontré que la prévalence des organismes patho-gènes était plus marquée dans les zones qui reçoivent des déverse-ments directs ou sont sous l'influence des courants et des mouve-ments de marée.

Les données techniques et scientifiques recueillies ont contribué à la mise au point de recommandations destinées aux deux Etats concernant les mesures à prendre pour rendre leur qualité aux eaux de la Baie de Raritan. Parmi ces recommandations figuraient les éléments suivants: normes spécifiques pour le traitement minimum des effluents municipaux et industriels; limites minimales et maxi-males des tolérances autorisées pour certains éléments toxiques; dates prévues pour l'application du programme de protection.

La contaminación causa anualmente la pérdida de mariscos por valor de 18 millines de dólares

La causa directa de la pérdida de la industria marisquera de la Bahía de Raritan, un complejo estuarino de 57 millas cuadradas situado entre los Estados de Nueva York y Nueva Jersey, fueron las descargas, de ambos estados, de desechos municipales e in-dustriales insuficientemente tratados. Los desechos municipales, de origen fecal, contaminaron de tal manera los criaderos, que ocurrió un brote grave de hepatitis. Las descargas industriales contami-naron los bivalvos y en ciertos lugares produjeron un ambiente tóxico que impidió casi por completo la propagación de estos moluscos.

Para devolver a las aguas de la Bahía de Raritan una calidad en consonancia con el mejor uso al que se pueden poner—marisqueo—la Federal Water Quality Administration inició una intensiva investigación, y un programa de vigilancia de trece meses y tres años de duración, respectivamente, que consistía en analizar muestras, obtenidas a profundidad, de descargas industriales y municipales, así como de las aguas que las recibían. En todas las fases se incluyeron programas de estudios químicos, biológicos y micro-biológicos. Se idearon nuevas técnicas y métodos para analizar los contaminantes en el medio del estuario. Las investigaciones micro-biológicas comprendieron el análisis y la determinación de la concentración de organismos indicadores y la detección de patógenos en las carnes de los mariscos, así como en las aguas circunvecinas. Los colorantes inyectados en los principales focos de contaminación demostraron que los organismos patógenos abundaban más en los lugares que recibían descargas directas o quedaban influidos por las mareas y las corrientes.

Los datos técnicos y científicos recogidos contribuyeron a que en ambos estados se formularan recomendaciones con respecto a las medidas necesarias para devolver al agua de Raritan su buena calidad. Las recomendaciones comprendían estipulaciones concretas para el tratamiento mínimo de los desechos municipales e industriales; límites máximos y mínimos permisibles de ciertos elementos tóxicos; y fechas en que se esperaba poner en práctica el programa para remediar la situación.

A N epidemic of infectious hepatitis, traceable to the consumption of raw clams from Raritan Bay, led to closure of most of this estuary, and an intensive study of water quality and shellfish resources by the Federal Water Quality Administration.†

During the early 19th century, Raritan Bay supported a highly productive fin and shellfish industry; but over-fishing and increased municipal and industrial pollution, significantly reduced the resource.

Present commercial shellfish exploitation is limited to hard clams, *Mercenaria mercenaria*, and blue crabs. Although FWQA studies have shown that the standing crop of hard clams could provide an annual harvest worth $3.85 million if water quality were suitable, the present

harvest is worth only $40,000. The soft clam, *Mya arenaria*, resource, once significant, is estimated to have a potential commercial value of $18 million annually, assuming suitable water quality and market development, but is not at present exploited. The blue crab commercial fishery appears to be affected only slightly, if at all, by the degraded water quality; in fact, in 1960, the blue crab harvest was one of the largest on record.

The present annual commercial finfish harvest in the Bay is estimated at $200,000. With improved water quality along with more modern fishing techniques, its value could be doubled.

FWQA role

The Federal Water Pollution Control Act provides that pollution of interstate waters, which endangers the health

---

† Study started under direction of USPHS. Legislation transferred PHS activities to FWPCA, now known as FWQA.

* Edison Water Quality Laboratory, Federal Water Quality Administration, Edison, New Jersey, 08817, U.S.A.

or welfare of any person, is subject to abatement under prescribed procedures. Preliminary surveys and studies suggested pollution of Raritan Bay and adjacent waters was endangering health, so a conference was called. The aim was to review existing water quality problems, establish a basis for future action in order to restore shellfishing—and to give all concerned an opportunity to take any indicated remedial action under State and local law.

In August 1961, FWQA was requested to obtain scientific data. After several additional conferences, the final pollution abatement and water quality recommendations were presented for adoption in May 1967. Pending complete implementation of the technical recommendations by 1972, surveillance activities are being carried out.

Programme objectives

The planned aim was to develop scientific data necessary to establish effective abatement and control of pollution in the area (fig 1), which includes Lower Sandy Hook and Raritan Bays, a portion of the Narrows, Arthur Kill, the tidal reach of the Raritan River and other small tributaries. However, because the area is likely to be affected by wastes discharged in adjacent waters the control programme considered the study area as part of a system which includes Upper Bay, Kill Van Kull and Newark Bay.

Major pollutional loads to the area are presented in Table 1 and indicate the large demand placed upon these waters by treated and untreated municipal and industrial wastes. Raritan Bay and Arthur Kill receive directly more than 480 MGD (1,810,000 cu m/day) of wastes from a population exceeding 1.3 million people. These discharges represent a Biochemical Oxygen Demand (BOD) loading of 430,000 lb/day (194,000 kg/day). When discharges to Upper Bay and Raritan River are included, the total wastes volume approaches 1,500 MGD (5,675,000 cu m/day), which represents a BOD loading of greater than 1,300,000 lb/day (585,000 kg/day) from a population exceeding 5 million people.

Contamination by pollutants other than BOD from these same sources is also a significant problem with the bacteria from more than 900 MGD (3,405,000 cu m/day) of unchlorinated and raw municipal wastes from a population of 3.8 million persons. Such pollution constitutes a definite hazard to the health of persons in contact with these waters.

Nearly 75 per cent of the total wastes volume is from industry. This results in pollution by a variety of contaminants in addition to oxygen consuming material. Pollutants such as oil, phenol, phosphate and nitrogen

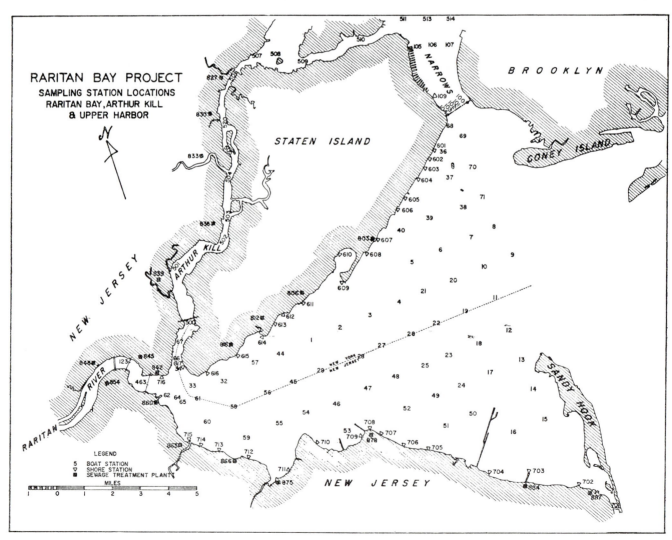

Fig 1. Sampling station locations: Raritan Bay, Arthur Kill and Upper Harbor

cause unsightly conditions, destruction of desirable aquatic life, tainting of fish and shellfish and eutrophication of the water.

Additional pollution results from the discharge of more than 1.0 billion gallons per day (3.785×10^{12} cu m/day) of cooling waters from power generating plants. Further contamination occurring in localized areas arises from wastes from recreational and commercial vessels. The overflow of sewage from combined storm-sanitary sewer systems also causes pollution.

Two-phase study programme

A sampling programme was designed which would permit an evaluation of the variations in water quality and long-term trends. An intensive programme with weekly samples at each station (fig 1) and mathematical analysis of the resulting parameter values, was run from August 1962 to September 1963. From September 1963 to May 1966 a surveillance programme, which involved collecting monthly samples at selected stations, was conducted to provide updated water quality data and to indicate any changes in water quality which might be implemented by the recommendations.

(3) Isolation of certain pathogenic bacteria from study area waters, sewage effluents and shellfish

(4) Simultaneous sampling of Raritan Bay, the Arthur Kill and waste treatment plant effluents emptying into these waters to assess the relationships between the waste loads and water quality

(5) Intensive bacteriological sampling of Raritan Bay and shoreline, entrant waters, and waste water treatment plants discharging to the Bay to determine bacterial densities

(6) Biological investigations designed to define the area of biological degradation, with particular emphasis on the benthic populations

(7) Chemical evaluation of existing water quality in the Bay and characterization of waste effluents, with particular emphasis on nutrients and oxygen-demanding components.

A number of special investigations have been undertaken to provide further data. These included an examination of water movement and dispersion patterns within Raritan Bay; an evaluation of the effects on water quality of combined sewer overflows; mathematical

TABLE 1 MUNICIPAL AND INDUSTRIAL WASTE LOADINGS[1]

Discharges to	Type source	Flow MGD	Loadings BOD	(lb/day) Suspended solids	Tributary population	Population equivalent (BOD) discharged
Raritan Bay	Municipal	72.1	182,500	40,560	507,800	1,069,200
	Industrial[2]	0.1	2,500			14,700
	Total[2]	72.2	185,000			1,083,900
Arthur Kill	Municipal	81.8	138,360	55,350	831,000	812,750
	Industrial[2]	367.3	104,640			615,000
	Total[2]	449.1	243,000			1,427,750
Raritan River	Municipal	2.0	1,605	845	20,365	9,430
	Industrial[2]	85.7	70,100			421,000
	Total[2]	87.7	71,705			430,430
Study Area	Municipal	155.9	322,465	96,755	1,359,165	1,891,380
	Industrial	453.1	177,240			1,050,700
	Total	609.0	499,705			2,942,080
Upper Bay	Municipal	915.9	808,510	645,100	3,815,100	4,758,400
	Industrial[3]	N.D.	N.D.	N.D.	N.D.	N.D.
	Total	915.9	808,510	645,100	3,815,100	4,758,400

NOTES: [1]Does not include additional wastes loadings from recreational and commercial vessels, or from stormwater overflow.
[2]Excludes flow from power generating industry.
[3]No data available.
MGD × 3,785 = cu m/day   lb/day × 0.45 = kg/day

The surveillance programme has continued since May 1966, although at a reduced scale (bimonthly chemical, biological and microbiological analyses of at least 40 stations). The programme has been expanded to include determination of mercury levels in fin and shellfish. Samples collected and analysed in August and September 1970 indicate that the mercury levels are below the recommended level of 0.5 mg/1, with concentrations varying from 0.025 mg/kg to 0.345 mg/kg in the shellfish.

Major activities undertaken during this investigation have included:

(1) Determination of bacteriological and chemical quality of the shellfish

(2) Determination of density and distribution of hard clams within the Bay area

analyses of the variations found in the chemical and bacteriological analysis of Bay water samples; and a study of the relationship between chlorination of waste-water treatment plant effluents and bacteriological densities in Raritan Bay. Results of these special investigations are in the FWQA Conference Proceedings (1967).

Shellfish resources

The shellfish resources of Raritan Bay have been studied on several occasions: Dr Julius Nelson's (1909) records on oyster production; Cumming, Purdy, Ritter (1916); and Cumming's (1917) exhaustive studies of pollution of growing areas. Cumming and his associates confined their studies to the effect of domestic waste on shellfish waters, while Dr Nelson (1916), during the same period, investigated the effect of industrial pollution in the form

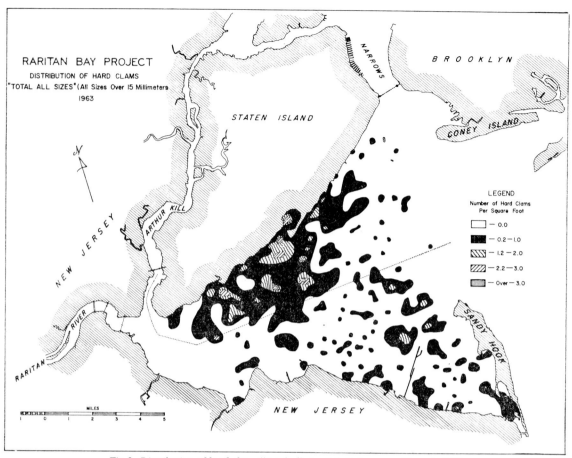

Fig 2. Distribution of hard clams "total all sizes" (all sizes over 15 mm), 1963

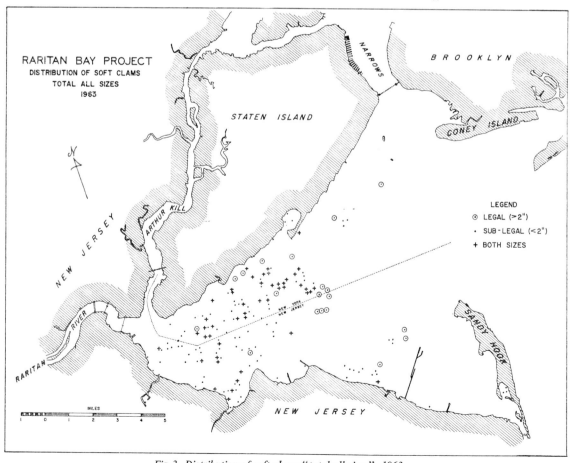

Fig 3. Distribution of soft clams "total all sizes", 1963

of metallic copper upon the oyster. Within a few years of Dr Nelson's studies, the oyster industry in Raritan Bay was virtually extinct due to this predicted effect of metallic copper on the environment (Nelson 1916).

Through public demand early this century, oysters were accorded a much higher rank than any other shellfish in Raritan Bay. According to Cumming, about 20,000 acres on the New York side of the Bay contained oysters, 8,000 of which were under private cultivation. New Jersey, in comparison, accounted for much less in total productivity. He also reported that "flats and foreshores have many extensive hard clam and soft clam producing areas". Sandy Hook was most notable. Shellfish growing and shipping during this period was one of the most important New Jersey industries with an annual oyster catch value of from 2 to 4 million dollars.

It has declined from major commercial importance prior to 1917, to a nonentity today. The soft shelled clam (*Mya arenaria*) although having declined considerably, has nevertheless managed to maintain a population in the western portion as well as in the Sandy Hook section of the Bay. Predictions relevant to the fate of the Northern Quahaug (*Mercenaria mercenaria*) fishery are more difficult to form, due to the area-confining aspects pollution has had on the fishery. Until the shellfish resource survey was completed, any estimates of the commercial shellfish resource in the whole Bay was questionable.

Density distribution: hard, soft clam

For density-distribution of hard clams in Raritan Bay see fig 2. This pattern is rather closely adhered to, but to a lesser degree, for each size category—over 66 mm, 47 to 66 mm, 15 to 46 mm. The general distribution patterns, particularly those for "sub-legals", are noticeably irregular and may be interpreted as being directly related to setting intensity and other factors such as current patterns, bottom sediments, or general hydrographic conditions.

The soft clam, *Mya arenaria*, proved more widely dispersed in the western sector than expected. Patches of sub-legal animals were also apparent in Sandy Hook Bay, as well as protected coves within Sandy Hook peninsula. The general distribution, illustrated in fig 3, suggests a rather evenly but widely scattered pattern of "sub-legal" size (< 5.08 cm) soft clams. "Legal" size shellfish were less abundant and appeared more or less confined to specific locations. The soft clam results shown are somewhat biased toward the smaller sizes because of ineffective sampling. Samples from some areas contained 90 per cent or more of the "sub-legal" size.

Microbiological investigations

Analyses were performed for both total and faecal coliform by both the MPN and MF procedures and for faecal streptococcus by the MF procedure. Figure 4 presents the mean MPN coliform count. High densities of coliforms were found in the vicinity of the Narrows and at the junction of the Arthur Kill and Raritan River. From these two sources coliforms appear to radiate out into the Bay. Those stations with the lowest mean count formed an apparent edge between the two radiating sources appearing as a straight band running from Princess Bay,

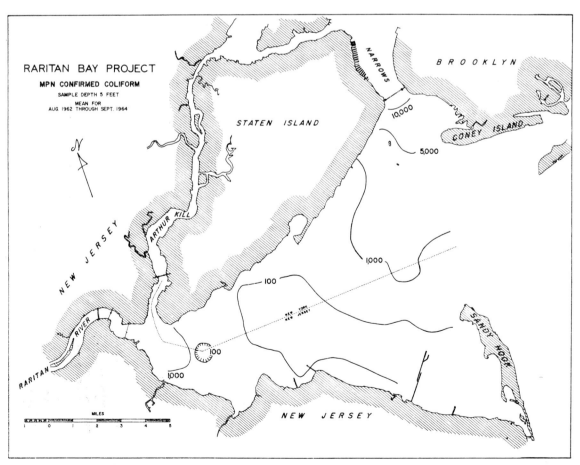

Fig 4. MPN confirmed coliform, sample depth 5 ft. Mean for August 1962 through September 1964

Staten Island, New York to Sandy Hook Bay, New Jersey. Geometric mean counts for MPN confirmed coliform ranged from 10,000/100 ml at the Narrows, and 7,000/100 ml at the mouth of the Raritan River to less than 50/100 ml in Sandy Hook Bay. The high faecal coliform densities and the ratio of faecal coliform to faecal streptococcus group organisms, strongly suggested that contamination resulted from human sources.

Bacteriological analyses were conducted on 391 shellfish samples from 50 stations throughout Raritan Bay. Samples from 12 of the 50 stations had geometric mean total coliform densities greater than 2,400 per 100 g of shellfish meat. The geometric mean faecal coliform density in shellfish taken from these same 12 stations ranged from 610 to 16,000 per 100 g. The presence of high total coliform densities appeared to show some correlation with water temperature. None of the shellfish taken from waters with temperatures less than 8.5°C had total coliform MPN's of 2,400 or more per 100 g. The 12 stations having geometric mean coliform densities greater than 2,400 per 100 g were located in the northerly sector of the Bay in an area extending generally south of Staten Island to and across the New York-New Jersey state line.

Salmonellae were isolated from clam meats collected at 14 of the 50 sampling stations but were not necessarily associated with high faecal coliform counts. A total of 23 *Salmonella* isolations were made with 13 serotypes identified. *Salmonella derby* was the predominant serotype and was isolated in shellfish from seven of the 14 stations. Stations which showed the presence of *Salmonella* in the clam meats covered two general areas, one of which corresponded with the location of high coliform counts in the clam meats as described above. The second area was located along the New York-New Jersey state line, in an area bounded roughly by Great Kills, Staten Island, New York and Keyport and Keansburg, New Jersey.

Shellfish chemical quality

The investigative programme was equally concerned with the chemical quality of the shellfish meats and overlying waters. In addition to the routine sanitary analyses both the shellfish and the water were analysed for selected trace metals, pesticides and certain organic materials.

Approximately 20 shellfish samples from some 400 collected from five areas within Raritan Bay were chosen, on the basis of shellfish concentrations and the prevailing currents, as being most indicative of the quality information desired.

In order to assess the analytical data properly it was necessary to develop a normal comparative pattern which could be used to evaluate the experimental results. Com-

parable data in the literature for these contaminants in shellfish were either non-existent, or out-dated. It was therefore decided to collect a representative group of "normal" shellfish samples from chemically and biologically clean areas and to analyse them for those compounds and metals under study. The resulting data which are shown in Table 2 were used as "normal" levels in evaluating the Raritan Bay analyses.

Selected data from this study are illustrated in figs 5, 6 and 7. The results of chemical analyses of shellfish can be summarized:

(a) Phenol values within the study areas were significantly elevated when compared with the normal values
(b) Results of mineral oil analyses suggest pollution by municipal and industrial waste discharges
(c) Copper levels, although somewhat higher than the baseline values, do not appear to indicate any gross contamination

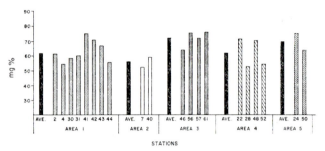

Fig 5. Phenol, mg/100 g tissue

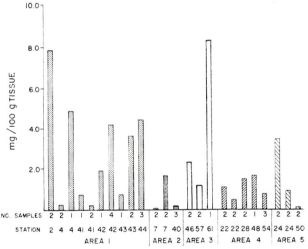

Fig 6. Mineral oil

TABLE 2. BASE LINE VALUES

Trace metals mg./kilo tissue	Phenols mg/100 g tissue	Mineral oils mg./100 g tissue	Pesticides ppm	
Cu 0– 5 mg.	35.2	0–4	Aldrin	0
Zn 40–60 mg.			Dieldrin	0
Pb 0–0.3 mg.			Lindane	0
Cr 0–0.2 mg.				
Ni 0–0.2 mg.				

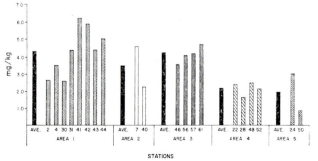

Fig 7. Lead, mg/kg tissue

(d) Lead appears to be a possible contaminant in all areas studied. The values are approximately tenfold higher than normal

(e) Nickel levels were similar to those for lead, inasmuch as they were also found to be at least ten times those of the normal baseline results

(f) Zinc values fall within the upper level of the normal range, and probably do not represent a contaminant source for shellfish in these areas.

(g) Chromium results, in general, are not indicative of a contaminant source within any of the five areas studied. Levels were found to be within the normal range with the exception of Areas 1 and 2. In these particular areas, the levels were not high enough to indicate chromium as a possible pollutant

(h) The fact that at least one of the three pesticides under study was detected in every area, does indicate that these materials are contributing to pollution within the areas studied

(i) The analytical results indicate that there is probably less contamination within Area 5 than in the sections surveyed

(j) All other sections vary as to the degree of contamination, but Areas 1 and 2 indicate pollution of greater significance for those materials studied.

Recommendations for remedial action

On the basis of these studies the following recommendations were made with a view to reclaiming the study area waters for maximum beneficial uses:

(1) Municipal treatment facilities should provide a minimum of 80 per cent removal of BOD and suspended solids at all times, including any four-hour period when the strength of the raw wastes might be expected to exceed average conditions. Effective year round disinfection (effluent coliform count of no greater than one per ml in more than 10 per cent of samples examined) shall be provided at all municipal plants discharging directly to these waters.

Unless existing orders specify even earlier completion dates all improvements are to be completed by 1972.

(2) Industrial plants shall improve practices for the segregation and treatment of wastes so as to effect maximum reduction of the following:
(a) Acids and alkalis
(b) Oil and tarry substances
(c) Phenolic and other compounds that contribute to taste, odour and tainting of fin and shellfish meat

(d) Nutrient materials, including nitrogenous and phosphorous compounds
(e) Suspended materials
(f) Toxic and highly coloured wastes
(g) Oxygen-requiring substances
(h) Heat
(i) Foam producing discharges
(j) Bacteria
(k) Wastes which detract from optimum use and enjoyment of receiving waters.

Industrial treatment facilities, to accomplish such reduction, must provide removals at least equivalent to those required for municipal treatment plants. Such facilities or reduction methods must be provided by 1972 unless existing orders specify even earlier compliance dates.

(3) Facilities and procedures for laboratory control be established at each treatment facility.

(4) State regulations be extended to require waste treatment facilities or holding tanks on all vessels and recreational boats using the area. If holding tanks are to be used, adequate dockside facilities should be required to ensure proper disposal of wastes.

(5) Investigate additional proposals to safeguard water quality in the study area. These studies are to include:
(a) Relocation of the main shipping channel through Raritan Bay to improve circulation characteristics
(b) Selection of areas for dredging for construction materials
(c) Suitable outfall locations for waste effluents to include possible trunk systems to divert effluents from the Arthur Kill.

Conferees, which include representatives from FWQA, the State of New York and New Jersey, and the Interstate Sanitation Commission meet every six months to review progress on the water quality improvements outlined above.

References

CUMMING, H S Investigation of the pollution of certain tidal waters
1917   of New Jersey, New York, and Delaware. *Publ. Hlth Bull., Wash.*, (86):150 p.
CUMMING, H S, PURDY, W C and RITTER, H C Investigation of
1916   pollution and sanitary conditions of the Potomac Watershed with special reference to self purification and sanitary condition of shellfish in the lower Potomac River. *Bull. hyg. Lab., Wash.*, (104):239 p.
NELSON, J Reports of Department of Biology, New Jersey. New
1909–16 Brunswick, New Jersey, Agricultural Experiment Station.
U.S. Federal Water Quality Administration, Proceedings of the
1967   conference on pollution of Raritan Bay and adjacent interstate waters, Third session. Vols. 1–3.

The authors acknowledge the contributions of Mr. Paul DeFalco, Jr., Regional Director, Southwest Region, FWQA, who was Project Director of the Raritan Bay Study, and Mr. Merrill S. Hohman, Director, Planning and Program Management, Northeast Region, FWQA, who was Chief of Planning and Evaluation for the investigation. Special acknowledgement is given to the USPHS, Northeast Shellfish Sanitation Research Center, Narragansett, Rhode Island, for assistance in chemical analyses and density-distribution studies.

The Principles and Methods Employed for the Sanitary Control of Molluscan Shellfish

P. C. Wood *

Principes et méthodes employés pour le contrôle sanitaire des coquillages

L'intensification de l'exploitation des coquillages dans les eaux côtières polluées pose des problèmes aux responsables du contrôle sanitaire des crustacés et mollusques. L'expansion de la pêche aux coquillages dans les pays en voie de développement et l'ampleur croissante du commerce international de ces produits indiquent qu'il est nécessaire de mettre en commun l'expérience acquise dans ce domaine et de mieux connaître les pratiques et les méthodes utilisées. Se servant d'exemples, l'auteur examine les fondements du contrôle sanitaire des mollusques en Europe et en Amérique du Nord, compte tenu des méthodes bactériologiques en usage, et souligne l'intérêt d'évaluer la qualité hygiénique des coquillages par analyse de l'eau de mer ou du tissu du mollusque. Il envisage les normes bactériologiques en vigueur, en étudiant notamment la valeur de plusieurs organismes indicateurs, particulièrement en ce qui concerne la transmission des virus, et attire l'attention sur l'importance pour la santé publique des organismes non fécaux d'origine marine. Enfin, il évoque les difficultés d'une application des normes en usage en Europe aux régions où la proportion d'animaux porteurs de virus est élevée.

Los principios y métodos empleados para la inspección sanitaria de moluscos

El aumento de la explotación de moluscos en aguas costeras contaminadas plantea problemas para los encargados de inspección sanitaria. La expansión de la extracción de moluscos en los países en desarrollo y el aumento del comercio internacional indican que es necesario efectuar un intercambio de experiencias en este sector y difundir mucho más los principios y métodos empleados. Se examina, citando ejemplos, la inspección sanitaria de moluscos en Europa y Norteamérica, comprendidos métodos bacteriológicos actuales y las ventajas de determinar la calidad higiénica de los mariscos analizando sus tejidos o el agua del mar. Se enumeran las actuales formas bacteriológicas comprendido el valor de varios organismos indicadores, con especial referencia a la transmisión de virus y se advierte de la importancia que pueden tener para la higiene pública los organismos no fecales de origen marino. Se mencionan las dificultades de aplicar normas actuales europeas a lugares en que abundan los vectores de enfermedades.

MOST commercially exploited molluscs come from coastal and estuarine waters, many of which are subject to pollution by sewage. As expansion of these resources continues more of these polluted areas will be utilized, and the presence of sewage organisms will require special attention, particularly where the shellfish are consumed in a raw or lightly preserved state. Methods for assessing the level of sewage contamination of molluscs and for controlling their exploitation from polluted areas have been developed, but these differ from country to country.

Between 1954 and 1963, many reports on this subject were made to the Shellfish Committee of the International Council for the Exploration of the Sea by member countries (Wood, 1963), and this has facilitated international trade within Europe. Elsewhere common methods have been adopted; for instance Canada, Japan and Korea have adopted U.S. methods (Felsing, 1964; Kelly *et al.*, 1968). A review of European regulations was presented to the General Fisheries Council of the Mediterranean by Coppini (1965), and recommendations were made for the international standardization of methods of examination and control.

This paper discusses the scientific basis of current European and North American methods of sanitary control, their limitations and how they may be applied elsewhere; national regulations are not considered except in so far as they relate to scientific aspects.

Objectives of sanitary control

In Europe and North America, the main risks associated with polluted shellfish are the presence of bacterial and virus pathogens. The most important bacterial pathogens are the *Salmonellae*, which include the organism causing typhoid fever, the *Shigella* species which can cause dysentery, and certain species of the *Clostridia* which can produce exotoxins pathogenic to man. A wide range of virus pathogens is present in both raw and treated sewage (Clarke *et al.*, 1964; Grabow, 1968) and several have been found in raw shellfish (Metcalf and Stiles, 1968).

At present, the only well-documented virus infection associated with shellfish is infective hepatitis (Mason and McLean, 1962). In parts of the world where parasitic protozoa and worms are common, it is possible that shellfish polluted by sewage may be a vector of transmission, through resistant cysts and eggs (Kabler, 1962).

The objective of sanitary control is to ensure that pathogenic organisms are either absent from the product or, present only in numbers that will not have any debilitating effects. When shellfish are heat-preserved, the risk from non-sporing organisms and viruses of sewage origin is negligible, except in canning, where there is risk of toxins developing from certain *Clostridia* during storage. Toxin production in canned shellfish can be overcome by ensuring efficient heat-processing. When lightly preserved or raw shellfish are stored under unsatisfactory conditions, there is a risk of toxin production by *Clostridium botulinum* type E (Cann *et al.*, 1965a; Johannsen, 1965; Graikoski, 1969; Presnell, Miescier and Hill, 1967). However, toxin production can be readily controlled by holding at temperatures low enough (Cann *et al.*, 1965). It is unlikely that heat-treated shellfish are vectors for transmitting parasitic protozoa or worms, but the more resistant cysts and eggs of the parasites may reach man in only lightly preserved shellfish (Cheng, 1965; Katoh, 1969).

Raw shellfish reach the consumer either in the live state, or removed from the shell (shucked), washed and held at a low temperature until they are eaten (Jensen, 1965). In both cases the major risk is from the transmission of the pathogenic non-sporing bacteria, viruses and other parasitic organisms. With shucked shellfish, and to a lesser extent with live shellfish, there is an additional risk of multiplication of bacteria if the ambient storage temperature is allowed to rise (Kelly, 1964; Wood, 1964). The bacterium *Vibrio parahaemolyticus* which is responsible for up to 20 per cent of all cases of food poisoning in Japan, is a free-living marine bacterium, widely distributed, which can affect man when shellfish and fish are stored or prepared under conditions which allow the organism to multiply (Liston *et al.*, 1969). The

---

* Ministry of Agriculture, Fisheries and Food, Fisheries Laboratory, Burnham-on-Crouch, Essex, England.

problem is essentially one of hygiene, and can be overcome by the shellfish being stored at a low temperature.

The basis for control

To ensure that sewage organisms do not reach the consumer, the sea water from which the shellfish are taken or the product, or both can be examined.

The use of sea water is attractive since sampling and examination of water samples is relatively simple, but at best this method can only be regarded as a very indirect way of assessing the quality of shellfish. There is no constant relationship between the bacterial content of filter-feeding molluscs and the water in which they rest, for many factors, including water temperature, suspended matter, and the presence of stimulating substances, can influence this relationship (Collier et al., 1953; Kelly et al., 1960; Wood, 1957, 1965, 1965a). The ability of molluscs to concentrate bacteria, particularly at high temperatures, implies that sea water of a very high standard is required if shellfish are to be of a high quality. Nevertheless, the examination of water is a simple way of monitoring pollution of an area, and of tracing the source.

If water quality is to be used as a guide to the acceptability of shellfish, further critical tests are needed to determine the water/shellfish pollution relationship over a wide range of conditions.

Selection of indicator organisms

To be satisfactory an indicator organism must always be present when pathogenic organisms are likely to be present, but not when they are absent. It must not multiply except when the pathogens do so. It must be excreted in much greater numbers than the pathogens, and must have a comparable resistance. To be acceptable, it should be capable of identification and enumeration by simple techniques in a reasonably short period. Many indicator organisms have been used, but the faecal streptococci, certain faecal anaerobes and members of the coliform group have received most attention.

The faecal streptococci—This group has found particular use in specialized situations such as in the examination of deep-frozen foodstuffs (Raj and Liston, 1961), because of its greater resistance to adverse conditions than the coliforms (Burman, 1961; Buttiaux, 1959; Deibel, 1964). It has been recommended that both E. coli and faecal streptococci should be used as indicators of pollution in oysters (Wilson and McClesky, 1951), and in drinking waters and food (Buttiaux, 1959). Slanetz, Bartley and Metcalf (1965), after comparing coliforms, E. coli and faecal streptococci as indices of pollution in sea water and shellfish, recommended that more attention should be directed towards the faecal streptococci as indicators. However, faecal streptococci in live shellfish undergo massive multiplication when stored at an ambient temperature of 11°C or more (Wood, 1965a) and more information is clearly required on the significance of these organisms as indicators of pollution.

The anaerobes—The anaerobe most widely used as an indicator is Clostridium welchii (perfringens) (Bonde, 1963; Ministry of Housing and Local Government, 1956). It is common in faeces, sewage, and soil (Willis, 1956),

but since its spores are extremely resistant to adverse conditions its significance as an indicator of non-sporing pathogenic bacteria is open to question. It has been suggested that the greater resistance of viruses in sea water, and the ability of certain strains of E. coli and coliforms to multiply outside the body, make Cl. welchii a useful indicator (Bonde, 1963). At the present time, Cl. welchii is not used by any national body for assessing the sanitary quality of shellfish or shellfish-growing waters.

The coliform group—This group has been used for many years, but because of the widespread distribution of some members of the group in natural unpolluted waters, opinion has shifted towards the use of E. coli because of its greater specificity (Ballentine and Kittrell, 1968; Geldreich, 1968). The coliform group is also of limited value as an indicator of faecal pollution because many of its members can multiply in shellfish held over a wide range of conditions (Kelly, 1964; Hoff et al., 1967, 1967a), whereas the behaviour of faecal coliforms in stored shellfish has been shown to be closely correlated with changes in the Salmonellae (Presnell and Kelly, 1961).

In considering which methods to employ for assessing E. coli a compromise has to be made between specificity, yield, ease of conducting the test and the time needed to complete it. Because of the perishable nature of shellfish, results are often needed within 24 hours, and the variability of their environment makes it necessary to examine a large number of samples. For routine work it is therefore usual to sacrifice specificity and yield for a simple method which will provide a result within 24 hours.

Other bacterial indicators—Because of doubts concerning the significance of the traditional bacterial indicators, it has been suggested that routine observations should be made for pathogens in sea water and other materials, using methods which are able to detect bacterial pathogens at concentrations too low to be of immediate public health significance (Gallagher and Spino, 1968; Grunnet and Nielsen, 1969). Whilst this would enable a direct assessment to be made of numbers of pathogens present, it will be necessary to assess with considerable caution the significance of small numbers of each type of pathogen. Such a judgement raises problems in cases where low levels of E. coli are normally permitted, as in shellfish.

The examination of shellfish

There are two basic methods of enumerating E. coli and coliforms in shellfish: that in which the organism is enumerated by colony count on solid media, and that which employs two or more series of dilution tubes in the so-called most probable number (MPN) technique. Direct enumeration of colonies is more accurate than the MPN method, but because the organisms are distributed according to the Poisson distribution its accuracy decreases substantially as the number of colonies is reduced. In the United Kingdom and in Holland, estimates of E. coli in shellfish are made using a modified McConkey agar incubated at 44 ± 0.2°C (Clegg and Sherwood, 1947; Reynolds and Wood, 1956). This method is reasonably specific (Clegg and Sherwood, 1939), gives a good yield, is economic in time and labour and produces a result in 18 hours. The yield and specificity are improved by pre-incubation for 2 hours at 37°C (Pretorius, 1961).

The dilution tube method is subject to wide sampling errors, the 95 per cent confidence limits of the 3 dilution 15 tube test which is in common use in North America being between one-third and three times the expected value (Ministry of Housing and Local Government, 1956; Swaroop, 1956). The method has the advantage of being able to detect relatively low densities of bacteria. Until recently the method in use in the United States, and in countries which export shellfish to America, was non-specific (Houser, 1965). The test has now been modified to enumerate only faecal coliforms (a group similar to, but not identical with, *E. coli*), using an elevated temperature of incubation (44.5 ± 0.2°C), and the result is now available in 24 hours (Kelly, 1964a).

In France an MPN estimate is made using a 2 dilution 10 tube test in phenol broth incubated at 44.5°C (Boury and Borde, 1957; Mazières, 1963). Although this method is reasonably specific for *E. coli*, it has a very wide sampling error, and takes 3 days to complete.

Where it is necessary to enumerate low densities of bacteria in shellfish, the size of the sample must be increased. Using the direct colony count, in roll-tubes with a solid medium, for all practical purposes the maximum size of sample is 5 ml, although with duplication this might be increased to 10 ml. Thus the maximum sensitivity of the method is 10 *E. coli*/100 ml, but with very large sampling errors. With the MPN technique, the sample size can be readily increased and it is possible, subject to the usual sampling errors, to detect 1 *E. coli*/100 ml. A high level of sensitivity without sacrifice of accuracy is obtained by the membrane filter technique, for after incubation a direct count can be made of the numbers of colonies of *E. coli*. For these reasons the method is widely used in water examination, but at present it is not applicable to the examination of shellfish, because of the difficulty of filtering extracts of tissue.

Current standards for shellfish

British standards are not laid down by law, but the recommendations of Sherwood and Thomson (1953) based upon the standards of the Fishmongers' Company (Knott, 1951) have been used for many years (Table 1.) However, during the last 10 years, an increasing proportion of shellfish has passed through some form of purification or tank storage, and current public health authority standards are considerably stricter (Wood, 1963). The tendency is now to accept shellfish which usually contain up to 200 *E. coli*/100 ml with occasional samples in the 200–500/100 ml range. It has been shown by Reynolds (1968) that if all the samples from a particular source are to come within sanitary grade I of Sherwood and Thomson (up to 500/100 ml) at least 90 per cent of the samples will contain less than 200 *E. coli*/ml.

The French standards for shellfish (Boury, 1962) are generally similar to the 1953 British values (Table 1). Shellfish from purification plants are expected to fall into Class I, with those falling into Classes II, III and IV being regarded as acceptable, suspicious or unacceptable, respectively. These standards are used as a guide for the assessment of quality, taking into account the topography and distribution of *E. coli* in waters of the growing area.

Faecal coliforms are now used in the United States as indicators (Kelly, 1964a), and shellfish are considered satisfactory if their density does not exceed 230/100 g. Coupled with this is a standard plate count (an estimate of non-specific bacteria including those of faecal origin) which should not exceed 500 000/g. Shellfish failing these standards are accepted on the condition that the second sample from the same source is satisfactory. The faecal coliforms enumerated by the American technique are similar to the *E. coli* enumerated by the roll-tube method, and there is now considerable similarity between the standards in use in America and the United Kingdom.

Examination of sea water from shellfish-growing areas

Most methods used for enumerating coliforms and *E. coli* in sea waters have been specifically devised for sea water. In England and Wales, most public health laboratories use the 15 tube 3 dilution MPN test for coliforms, with final confirmation of the *E. coli* present. In its shortest form the test takes 48 hours to complete (Ministry of Housing and Local Government, 1956). A few laboratories make direct colony counts at 44°C. Membrane filters have not been used extensively, although several techniques have been devised (American Public Health Association, 1965; Presnell, Arcisz and Kelly, 1954). More recently the author has employed a rapid method, with resuscitation, for enumerating *E. coli* on membrane filters using a modified Teepol lactose medium (Windle Taylor and Burman, 1964; Burman, pers. comm.).

The standard U.S. and French methods for examining sea water are almost identical with the methods used by each country for shellfish. Although techniques have been developed for the enumeration of faecal streptococci and anaerobes in sea water, there has been no widespread acceptance of either of these indicators.

Current standards for shellfish-growing waters

In the United States and other countries extensive use is made of the sanitary quality of water of the growing areas for ascertaining the quality of shellfish. Waters are classified as approved, restricted or prohibited on the basis of the coliform content of the water (MPN's of up to 70, 70–700 and greater than 700/100 ml respectively) (Houser, 1965). Shellfish from an approved area may be taken for direct human consumption. Because of the low specificity of coliforms, attempts have been made to establish standards for waters based upon faecal coliforms, and Beck (1964) proposed a median value of 7.8 faecal coliforms/100 ml, with not more than 10 per cent of the samples having an MPN of 33 or more for approved growing areas.

In France, the growing areas are classified (Mazières, 1965) as follows:

		E. coli/100 ml
Class I	Satisfactory	0
Class II	Acceptable	1–60
Class III	Suspicious	60–120
Class IV	Unfavourable	>120

These values are considered in conjunction with the topography, and the aim is to ensure there is no reduction of the sanitary quality of water in an area known to produce good quality shellfish.

Quality		British standard (Numbers of *E. coli*/100 ml by roll-tube)		French standard (Numbers of *E. coli*/ 100 ml in phenol broth)		U.S. standard (Numbers of faecal coliforms/100 g by MPN technique)
	Grade	Sherwood and Thomson (1953)	In current use (Wood, 1963)	Class	Boury (1962)	Kelly (1964)
Acceptable	I	Up to 500	90% of samples—up to 200 10% of samples—200–500	I	Less than 100	Up to 230
				II	Between 100 and less than 500	—
Suspicious	II	Greater than 500 not more than 1,500	—	III	Between 500 and less than 1,500	—
Unacceptable	III	More than 1,500	More than 500	IV	1,500 or more	Greater than 230

In the United Kingdom, where the *E. coli* content of shellfish is used as a basis for control, sea water is not examined on routine. Similar arrangements exist in Holland (Grijns, 1959).

Current standards on pathogenic viruses

The only known cases of virus infection caused by shellfish are those due to the consumption of grossly polluted molluscs (Mason and McLean, 1962) and it would appear that in practice the existing bacterial indicators and standards are satisfactory. However, it could be that the diagnostic methods at present available for the identification of virus infections are inadequate and the incubation period too long, for a causal relationship to be established.

Experimental studies suggest that members of the *E. coli*/coliform group are unsatisfactory as indicators of enteric viruses. Human enteric virus has been found in sea water containing less than 70 coliforms/100 ml and in up to 16 per cent of samples of shellfish judged to be of satisfactory quality by their faecal coliform content (less than 230/100 ml) (Metcalf and Stiles, 1968). Oysters retained enteric virus for 30 days when laid in an estuary where the water temperature was about 1°C and a significant reduction took place only when the water temperature rose to about 5°C (Metcalf and Stiles, 1967). There is clearly a need for work to assess the ecology of enteric viruses, with special reference to the use of existing bacterial standards; work is also needed to develop the use of virus indicators in shellfish and sea water. Consideration should be given to the value of *E. coli* bacteriophage as an indicator of virus pollution, for simple techniques already exist for its enumeration (Kott, 1966).

Existing European and North American indicators in other areas

Caution is required if bacterial indicators are to be used in controlling molluscan shellfish in areas where the climate and environmental conditions are very different. It may be necessary to use different organisms, since although *E. coli* appears to be universal in tropical areas it often has a higher temperature tolerance than in Europe. European and North American standards may be totally unrealistic where the carrier rate for certain diseases is higher and the ratio between pathogen and indicator in sewage is of a different order. In Europe, typhoid and paratyphoid were at one time the main diseases spread by shellfish, and indicators were developed accordingly; in other countries protozoan and metazoan parasites (Cheng, 1965) may assume greater significance and require the use of other indicators. Certain parasites have already been shown to be present or able to survive in raw shellfish. (Cheng and Burton, 1965, 1965a). Where standards are being set in countries with totally different environmental conditions it is unlikely that they can be derived mathematically, in which case it will be necessary to set an arbitrary standard.

Significance of non-faecal bacteria

There is evidence of wide variation in the numbers of bacteria of non-faecal origin in shellfish when fished (Wood, 1964), shucked and stored (Hoff *et al.*, 1967; Kelly, 1964). Some of these variations are associated with seasonal changes in the estuarine environment, others with the storage and handling conditions of the shellfish after leaving the water. In the United Kingdom, there is some evidence that mild gastro-enteritis following the consumption of raw shellfish is associated with non-faecal organisms (Wood, 1963). Psychrophilic marine bacteria may be the cause (Barrow and Miller, 1969). Another marine organism *Vibrio parahaemolyticus* has already been associated with food poisoning. The significance of marine bacteria present in shellfish, should be examined with a view to establishing a bacterial standard which would protect the consumer against these less clearly defined illnesses.

Conclusions

From this review it is apparent that several fields of research still require attention. The major advance in recent years has been the general acceptance of *E. coli* as the indicator of faecal pollution. However, the possibility of virus transmission by shellfish conforming to current bacterial standards cannot be excluded and there is clearly a case for a reassessment of the value of *E. coli* and the other bacterial indicators, such as the faecal streptococci and the anaerobes. Further examination of current standards may also be required.

There is increasing evidence that within the marine

and estuarine environment other organisms exist which may cause illness in man. This area of knowledge needs to be explored with a view to the creation of standards to protect the consumer.

References

American Public Health Association. Standard methods for the
1965 examination of water and wastewater. 12th ed. New York, American Public Health Association Inc., 769 p.

BALLENTINE, R K and KITTRELL, F W Observation of fecal coli-
1968 forms in several recent stream pollution studies. *In* Proceedings of the Symposium on Fecal Coliform Bacteria in Water and Wastewater, edited by A E Greenberg, sponsored by the Bureau of Sanitary Engineering, California State Department of Public Health.

BARROW, G I and MILLER, D C Marine bacteria in oysters purified
1969 for human consumption. *Lancet*, 1969 Aug. 23: 421–3.

BECK, W J Bacteriological criteria for shellfish growing areas.
1964 Report on collaborative study. *Proc. natn. Shellfish Sanit. Wkshop*, 5: 143–5.

BONDE, G J Bacterial indicators of water pollution: a study of
1963 quantitative estimation. Copenhagen, Teknisk Forlag, 422 p.

BOURY, M Appréciation de la qualité bactériologique des coquil-
1962 lages. *Sci. Pêche*, (103): 1–3.

BOURY, M and BORDE, J Méthodes d'examen bactériologique de
1957 l'eau de mer et des coquillages, essais comparatifs. *Sci. Pêche*, (51): 9 p.

BURMAN, N P Some observation on coli-aerogenes bacteria and
1961 streptococci in water. *J. appl. Bact.*, 24(3): 368–76.

BUTTIAUX, R The value of the association of *Escherichieae*—
1959 Group D Streptococci in the diagnosis of contamination in foods. *J. appl. Bact.*, 22(1): 153–8.

CANN, D C, *et al.* The growth and toxin production of *Clostridium*
1965 *botulinum*, type E in certain vacuum packed fish. *J. appl. Bact.*, 28(3): 431–6.

CANN, D C, *et al.* The incidence of *Clostridium botulinum*, type E
1965a in fish and bottom deposits in the North Sea and off the coast of Scandinavia. *J. appl. Bact.*, 28(3): 426–30.

CHENG, T C Parasitological problems associated with food pro-
1965 tection. *J. Environ. Hlth*, 28(1): 208–14.

CHENG, T C and BURTON, R W Relationship between *Bucephalus*
1965 sp. and *Crassostrea virginica*: histopathology and sites of infection. *Chesapeake Sci.*, 6: 3–16.

CHENG, T C and BURTON, R W The American oyster and clam as
1965a experimental intermediate hosts of *Angiostrongylus cantonensis*. *J. Parasit.*, 51(2): 296–7.

CLARKE, N A, *et al.* Human enteric viruses in water: source,
1964 survival and removability. Pergamon Press. *Adv. Wat. Pollut. Res.*, 1(2): 523–36.

CLEGG, L F L and SHERWOOD, H P Incubation at 44°C as a test for
1939 faecal coli. *J. Hyg., Camb.*, 39(4): 361–74.

CLEGG, L F L and SHERWOOD, H P The bacteriological examination
1947 of molluscan shellfish. *J. Hyg., Camb.*, 45: 504–21.

COLLIER, A, *et al.* The effect of dissolved organic substances on
1953 oysters. *Fishery Bull. U.S. Fish Wildl. Serv.*, 54(84): 167–85.

COPPINI, R Sanitary regulations for molluscs. *Stud. Rev. gen. Fish.*
1965 *Coun. Mediterr.*, 29:16 p.

DEIBEL, R H The Group D streptococci. *Bact. Rev.*, 28(3): 330–66.
1964

FELSING, W A Shellfish imports. *Proc. natn. Shellfish Sanit.*
1964 *Wkshop*, 5:136–9.

GALLAGHER, T P and SPINO, D F The significance of numbers of
1968 coliform bacteria as an indicator of enteric pathogens. *Wat. Res.*, 2:169–75.

GELDREICH, E E Fecal coliform concepts in stream pollution. *In*
1968 Proceedings of the Symposium on Fecal Coliform Bacteria in Water and Wastewater, edited by A E Greenberg, sponsored by the Bureau of Sanitary Engineering, California State Department of Public Health pp. 3–21.

GRABOW, W O K The virology of waste water treatment. *Wat. Res.*,
1968 2:675–701.

GRAIKOSKI, J T Seafoods and botulism. Paper presented to FAO
1969 Technical Conference on Fish Inspection and Quality Control, held at Halifax, Canada, 15–25/7/1969, (FE: FIC 69/0/61):15 p.

GRIJNS, A The bacteriological examination of shellfish by means
1959 of the roll-tube method of Clegg and Sherwood, modified by Reynolds and Wood. ICES C.M. 1959, Shellfish Committee, Doc. (9) (mimeo).

GRUNNET, K and NIELSEN, B B *Salmonella* types isolated from the
1969 Gulf of Aarhus compared with types from infected human beings, animals and feed products in Denmark. *Appl. Microbiol.*, 18(6): 985–90.

HOFF, J C, *et al.* Time-temperature effects on the bacteriological
1967 quality of stored shellfish. 1. Bacteriological changes in live shellfish: Pacific oysters (*Crassostrea gigas*), Olympia oysters (*Ostrea lurida*), Native littleneck clams (*Protothaca staminea*), and Manila clams (*Venerupis japonica*). *J. Fd Sci.*, 32(1): 121–4.

HOFF, J C, *et al.* Time-temperature effects on the bacteriological
1967a quality of stored shellfish. 2. Bacteriological changes in shucked Pacific oysters (*Crassostrea gigas*) and Olympia oysters (*Ostrea lurida*). *J. Fd Sci.*, 32(1): 125–9.

HOUSER, L S (Ed.) Sanitation of shellfish growing areas. National
1965 shellfish sanitation programme manual of operations. *Publ. Hlth Serv. Publs, Wash.*, (33)Pt. 1:32 p.

JENSEN, E T (Ed.) Sanitation of the harvesting and processing of
1965 shellfish. National shellfish sanitation programme manual of operations. *Publ. Hlth Serv. Publs, Wash.*, (33)Pt. 2:54 p.

JOHANNSEN, A *Clostridium botulinum* type E in foods and the
1965 environment generally. *J. appl. Bact.*, 28(1): 90–4.

KABLER, P W Reduction of pathogenic micro-organisms by
1962 sewage treatment. *Environ. Hlth*, 4(4): 258–68.

KATOH, T Fish and shellfish inspection at the Tokyo Central
1969 Wholesale Market with special reference to the sanitary quality assessment. Paper presented to FAO Technical Conference on Fish Inspection and Quality Control, held at Halifax, Canada, 15–25/7/1969 (FE: FIC/69/0/40):6 p.

KELLY, C B Time-temperature effect on bacteriological quality of
1964 stored oysters. *Proc. natn. Shellfish Sanit. Wkshop*, 5: 193–202.

KELLY, C B Interim bacteriological criteria for oysters. *Proc. natn.*
1964a *Shellfish Sanit. Wkshop*, 5:155–61.

KELLY, C B, *et al.*, Bacterial accumulation by the oyster *Cras-*
1960 *sostrea virginica* on the Gulf Coast. *Tech. Rep. U.S. Dept. Hlth Educ. Welf.*, (F60-4):45 p.

KELLY, C B, *et al.* Sanitary control of shellfish in Korea. Report of
1968 a study made for the Agency for International Development, U.S. Department of State, at the request of the Government of the Republic of Korea. U.S. Department of Health, Education and Welfare, 49 p.

KNOTT, F A Memorandum on the principles and standards
1951 employed by the Worshipful Company of Fishmongers in the bacteriological control of shellfish in the London markets. Presented to the Court of the Fishmongers' Company on 27 September 1951. London, Fishmongers' Company, 16 p.

KOTT, Y Estimation of low numbers of *Escherichia coli* bacteriophage
1966 by use of the Most Probable Number method. *Appl. Microbiol.*, 14(2):141–4.

LISTON, J, MATCHES, J R and BAROSS, J Survival and growth of
1969 pathogenic bacteria in sea foods. Paper presented to FAO Technical Conference on Fish Inspection and Quality Control, held at Halifax, Canada, 15–25/7/1969 (FE: FIC/69/0/31):7 p.

MASON, J O and McLEAN, W R Infectious hepatitis traced to the
1962 consumption of raw oysters. *Am. J. Hyg.*, 75(1):90–111.

MAZIÈRES, J Les coliformes dans les eaux marines et les huîtres.
1963 Application à l'hygiène ostréicole. *Rev. Trav. Inst. Pêch. marit.*, 27(Fasc.1):111 p.

MAZIÈRES, J Les germes-tests de contamination et l'appréciation de
1965 la qualité bactériologique des huîtres. *In* Pollutions marines par les micro-organismes et les produits pétroliers, Symposium de Monaco (avril 1964), Commission Internationale pour l'Exploration Scientifique de la Mer Méditerranée, Monaco, pp. 265–75.

METCALF, T G and STILES, W C Survival of enteric viruses in
1967 estuary waters and shellfish. *In* Symposium of transmission of viruses by the water route, edited by G Berg, New York, John Wiley-Interscience, pp. 439–47.

METCALF, T G and STILES, W C Enteroviruses within an estuarine
1968 environment. *Am. J. publ. Hlth*, 88(3):179–91.

Ministry of Housing and Local Government, The bacteriological
1956 examination of water supplies. *Rep. Publ. Hlth med. Subj. Lond.*, (71):52 p.

PRESNELL, M W, ARCISZ, W and KELLY, C B Comparison of the
1954 MF and MPN techniques in examining sea water. *Pub. Hlth Rep.*, 69(3):300–4.

PRESNELL, M W and KELLY, C B Bacteriological studies of com-
1961 mercial shellfish operations on the Gulf Coast. *Tech. Rep. U.S. Dep. Hlth Educ. Welf.*, (F61-9):20–6.

PRESNELL, M W, MIESCIER, J J and HILL, W F JR *Clostridium*
1967 *botulinum* in marine sediments and in the oyster *Crassostrea virginica* from Mobile Bay. *Appl. Microbiol.*, 15(3):668–9.

PRETORIUS, W A Investigations on the use of the roll tube method
1961 for counting *Escherichia coli* I in water. *J. appl. Bact.*, 24(2):212.

[564]

RAJ, H and LISTON, J Detection and enumeration of fecal indicator
1961   organisms in frozen sea foods. 1 *Escherichia coli*. *Appl.
        Microbiol.*, 9:171–4.

REYNOLDS, N Bacteriological standards for mussels. *Publ. Hlth
1968   Inspector*, Sept., 524–7.

REYNOLDS, N and WOOD, P C Improved techniques for the bac-
1956   teriological examination of molluscan shellfish. *J. appl.
        Bact.*, 19(1):20–5.

SHERWOOD, H P and THOMSON, S Bacteriological examination of
1953   shellfish as a basis for sanitary control. *Mon. Bull. Minist.
        Hlth*, (12):103–11.

SLANETZ, L W, BARTLEY, C H and METCALF, T G Correlation of
1965   coliform and fecal streptococcal indices with the presence
        of salmonella and enteric viruses in sea water and shellfish.
        Adv. wat. Pollut. Res., 2(3):27–35.

SWAROOP, S Estimation of bacterial density of water samples.
1956   Methods of attaining international comparability. *Bull.
        Wld Hlth Org.*, 14:1089–107.

WILLIS, A T Anaerobes as an index of faecal pollution in water.
1956   *J. appl. Bact.*, 19(1):105–7.

WILSON, T E and McCLESKEY, C S Indices of pollution in oysters.
1951   *Fd Res.*, 16(4):313.

WINDLE TAYLOR, E and BURMAN, N P The application of membrane
1964   filtration techniques to the bacteriological examination of
        water. *J. appl. Bact.*, 27(2):294–303.

WOOD, P C Factors affecting the pollution and self-purification of
1957   molluscan shellfish. *J. Cons. perm. int. Explor. Mer*,
        22(2):200–7.

WOOD, P C The sanitary control of molluscan shellfish. Some observa-
1963   tions on existing methods and their possible improvement.
        ICES C.M., 1963, Shellfish Committee, Doc. (24) (mimeo).

WOOD, P C Quantitative changes in the bacterial flora of air-stored
1964   oysters. ICES C.M., 1964, Shellfish Committee, Doc. (62)
        (mimeo).

WOOD, P C The effect of water temperature on the sanitary quality of
1965   *Ostrea edulis* and *Crassostrea angulata* held in polluted
        waters. *In* Pollutions marines par les micro-organismes et
        les produits pétroliers, Symposium de Monaco (avril 1964),
        Monaco, CIESMM, pp. 307–16.

WOOD, P C A preliminary appraisal of the use of faecal streptococci
1965a   in the sanitary control of shellfish. ICES C.M., 1965,
        Shellfish Committee, Doc. (94) (mimeo).

Sanitary Control of Shellfish and Marine Pollution

*D. A. Hunt**

Le contrôle sanitaire des mollusques et la pollution des mers

L'auteur examine le programme national de contrôle sanitaire des
mollusques, ses principes de base, ses problèmes et ses réalisations.
Ce programme a été mis sur pied en 1925 pour prévenir les maladies
associées aux mollusques en exerçant un contrôle sanitaire rigou-
reux sur la culture, la récolte, le traitement et la commercialisation
des mollusques (huîtres, clams et moules). Il s'agit d'une action
coopérative à laquelle participent 22 états des U.S.A., l'industrie
des mollusques et crustacés et la Food and Drug Administration du
Gouvernement fédéral.

Le Programme repose essentiellement sur le contrôle de la zone
de culture, considéré comme principal obstacle à la diffusion des
maladies transmises par les mollusques. L'augmentation de la
population et la croissance industrielle dans les régions côtières
s'accompagnent d'une pollution préjudiciable aux zones de conchyliculture. Cette pollution présente de multiples dangers pour la
santé, parmi lesquels les bactéries et les virus pathogènes, les
pesticides et les métaux lourds. La protection de la zone de culture
nécessite des enquêtes quantitatives complètes visant à déterminer
les sources présentes et potentielles de pollution, des études hydrographiques, ainsi que des analyses bactériologiques et chimiques
des eaux servant à la conchyliculture et des mollusques eux-mêmes.
Les zones de culture qui ne répondent pas aux critères approuvés
sont fermées et soumises à une inspection dont l'objet est de
prévenir la récolte de mollusques pollués.

On mène actuellement des recherches sur l'amélioration des
normes appliquées aux zones de culture pour les protéger contre la
pollution d'origine industrielle et domestique et en assurer la
purification sous contrôle. On a établi des normes bactériologiques
pour les eaux servant à la conchyliculture, pour la teneur des zones
de culture en pesticides chlorés, pour le taux de concentration des
radionucléides dans la chair des mollusques, ainsi que des normes
bactériologiques concernant ce même produit au niveau de la vente
en gros. On étudie actuellement des normes applicables aux zones
de culture pour le taux de concentration des métaux lourds dans
les chairs des mollusques.

La inspección sanitaria de los mariscos y la contaminación del mar

Se examinan el programa nacional de higiene de los mariscos, sus
fundamentos, sus problemas y lo que ha alcanzado hasta la fecha.
El programa se organizó en 1925 para impedir las enfermedades
relacionadas con los mariscos mediante un sistema de rigurosa
inspección higiénica del crecimiento, recolección, elaboración y
mercadeo de mariscos (ostras, almejas y mejillones). Se trata de un
programa en el que cooperan 22 estados de los E.U.A., la industria
marisquera y la Food and Drug Administration del Gobierno
Federal.

El tema principal del programa como barrera a la propagación de
las enfermedades transportadas por mariscos es una mayor protección de zonas. El aumento de la población y de la industria junto
a la costa y la contaminación que les acompaña tiene un efecto
perjudicial en los lugares en los que se crían los mariscos. Tal
contaminación presenta muchos peligros para la salud, que comprenden los creados por las bacterias y virus patógenos, los pesticidas y los metales pesados. El aumento de la protección de zonas
exige el estudio cuantitativo completo de los posibles focos e
intensidades de la contaminación actual y potencial, estudios
hidrográficos y análisis bacteriológicos y químicos de las aguas
donde se crían los mariscos. Los lugares de cría que no reúnan las
condiciones aprobadas, se cierran e inspeccionan para impedir la
recolección de mariscos contaminados.

Se efectúan investigaciones para reforzar las normas aplicables a
las zonas de cría para proteger de la contaminación industrial y
doméstica y para la purificación regulada. Se han fijado normas
bacteriológicas relativas a las aguas en las que se crían los mariscos,
a los lugares a los que llegan los plaguicidas clorados y radionúclidos que se encuentran en la carne de mariscos y normas bacteriológicas para esa carne en los mercados mayoristas. Se estudian en la
actualidad normas más rigurosas para los metales pesados
contenidos en la carne de mariscos.

THE effects of pollution on marine food resources,
specifically molluscan bivalves, have been of
major interest to the U.S. Public Health authorities.
Molluscan bivalves concentrate a wide variety of man's
waste products and are excellent indicators of industrial
and domestic pollution.

This paper describes in part, the operation of the National Shellfish Sanitation Program, particularly sanitary
control aspects and past and present problems caused by
marine pollution.

This programme was conceived as a combined effort
to establish a sanitary control mechanism for preventing
further shellfish-borne outbreaks of typhoid fever and
other enteric diseases of bacterial origin.

Membership in this State, Federal, industry cooperative triumvirate includes shellfish sanitation control
agencies in 22 States of the United States, Canada, and
the Hiroshima Prefecture of Japan, and shellfish industry
organizations. The Food and Drug Administration's
Shellfish Sanitation Branch is the Federal representative
and is responsible for administration. Operational
guidelines are developed at regional and national work-
shops and published in the Manual of Operations (U.S.
Public Health Service 1965, 1968, 1968a).

Although good sanitary practices during harvesting,
processing, storage, or in transit to the wholesale market
are important, sanitary control of shellfish-growing areas
forms the major part of the programme. This includes:

* Office of Compliance, Food and Drug Administration, Department of Health, Education and Welfare, Washington, D.C. 20204, U.S.A.

the determination of actual and potential sources and levels of biological, chemical and radiological pollutants; classification of the growing area; closure of areas not meeting approved growing area criteria; and the posting and patrolling of closed areas to prevent harvesting of polluted shellfish.

In establishing sanitary control of a specific growing area, several basic actions are taken. Information is collected on known pollution, tidal flushing, meteorological characteristics, and biological populations of the study area. This is followed by a comprehensive shoreline survey of the estuary to pinpoint actual and potential sources of both domestic and industrial pollution. In addition transects of the area are charted to determine positions for bacteriological, chemical, or physical sampling stations. Hydrographic studies provide valuable information on overall water quality and functional dynamics of the estuarine system. Areas subject to periodic blooms of toxic dinoflagellates such as *Gonyaulax* must be monitored to determine the levels of paralytic shellfish poison, or other marine biotoxins in shellfish meats. At the completion of the sanitary survey and associated studies, all data are tabulated and analysed. Those areas not meeting approved growing-area criteria must be delineated, properly posted, and patrolled by the official agency. In establishing the position of the closure line, bacteriological data should be correlated with known sources of pollution as determined by the shoreline survey. It is recommended that sampling be scheduled during the most unfavourable hydrographic and pollution conditions.

Development of standards criteria has been a continuing activity of the programme throughout its 45-year history. This includes recommended practices for growing-area waters, sanitary harvesting procedures, inplant sanitation, storage and shipping facilities, chlorinated pesticides, radionuclides, and bacteriological criteria at the wholesale market level. Pesticide and radionuclide standards are in the "Proceedings of the Sixth National Shellfish Sanitation Workshop". Standards criteria are continuously modified according to results of all parties interested.

Areas of responsibility

There is a distinct division of responsibilities between State shellfish sanitation agencies and the Federal agency responsible for the administration of the National Shellfish Sanitation Program. Each State food control agency has the responsibility of assuring that all components of the State programme comply with recommended procedures. The State also certifies that shellfish dealers conform to recommended operational procedures. The Food and Drug Administration annually evaluates each State or foreign programme, publishes a biweekly list of certified shellfish dealers and the Manual conducts training programmes and field investigations, and develops standards criteria. Shellfish shippers in States whose control agency fails to meet minimum programme requirements will not be placed on the list of dealers certified for interstate shipment of shellfish. Some State regulations prevent the sale of noncertified shellfish, and wholesale markets will not purchase the product. In this manner, control of the interstate shipment of shellfish is maintained without Federal legislative authority.

During the first 25 years of the programme, the primary health hazard was enteric bacterial pathogens. Harvesting shellfish from areas adjacent to towns and cities was known to be hazardous. Although Eijkman (1904) had proposed a more restrictive high temperature test for "coli bacteria" for determining faecal pollution in water, the "*B. coli*" indicating group and methodology defined in Standard Methods of Water Analysis (American Public Health Association, 1920) was the indicator group and bacteriological procedure in general use. Continued studies and modifications of Eijkman's procedure resulted in a proposal by Dr C. A. Perry (1939) that, ". . . *Escherichia coli* rather than the colon group should be the index of pollution for both shellfish and shellfish waters . . .". Two decades later, the present "faecal coliform" test using EC broth incubated at 44.5°C was accepted as recommended procedure for determining the sanitary quality of shellfish at the wholesale market level and is currently under consideration for growing water standards.

Population shifts from inland areas to coastal regions and the rapid development of industrial complexes on estuaries often adjacent to highly-productive shellfish areas, caused concern to those agencies attempting to maintain the sanitary quality of shellfish. New sewage treatment facilities were constructed and chlorination of treatment plant effluents became common practice, but both effluents were affecting shellfish-growing areas. In 1955, the first shellfish-associated epidemic of infectious hepatitis was reported in Sweden, soon to be followed by the Raritan Bay and Pascagoula, Mississippi outbreaks. and several smaller epidemics in the mid-Atlantic and New England States. Questions of major importance now confronted the programme. Were the current indicator group and growing area standards adequate to protect the consumer from viral diseases? Did chlorination kill the indicator while the virus survived to be concentrated in shellfish? What was the coliform/hepatitis virus relationship in shellfish areas contaminated by chlorinated sewage effluents? Did low bacterial counts give the control agency a false sense of security? It appears at this time that our current indicator group and bacteriological criteria for the growing area are providing adequate consumer protection. All shellfish-associated hepatitis outbreaks investigated by the Public Health Service, to date, have implicated heavily-polluted shellfish. Illegal harvesting from closed areas and a breakdown in control procedures at the local level are usually the cause of shellfish-associated epidemics. We have no evidence that shellfish harvested, processed, and shipped according to recommended procedures in the Manual have been implicated in the transmission of an infectious disease.

"Conditionally approved" areas

Improved sewage treatment and higher quality effluents stimulated the establishment of the "conditionally approved" classification for shellfish harvest areas adjacent to those plants producing high quality effluents. In case of a breakdown resulting in the release of raw or improperly treated sewage, the local control authority must be notified immediately and the "conditionally approved" area reclassified and closed to harvesting. The conditional area classification is also used on a seasonal or meteorological basis in some areas *following* high rainfall resulting in the release of raw sewage or pollution from excessive surface runoff.

Infectious hepatitis was not the only new shellfish-associated disease to be identified in the 1950's. More than 100 people were killed or severely disabled from consuming mercury-contaminated seafood products in Japan from 1953 to 1960. Although mercury was not thought to be a problem at that time the National Shellfish Sanitation Program was aware of the possible hazard of heavy metals in shellfish and surveys were undertaken to determine background levels in shellfish for such metals as copper, zinc, cadmium, chromium, lead and nickel. Uptake and depletion studies for selected shellfish species were also conducted which demonstrated that concentration factors varied both with shellfish species and metal. These studies further indicated that high metal levels would not be shown by shellfish in a controlled purification system within a commercially-feasible period of time. Guidelines for metals in shellfish-growing areas were proposed at the 1968 workshop, but rejected pending additional data. Studies are currently being conducted to determine metal levels and gradients of levels in shellfish from 14 estuaries on the Atlantic, Pacific and Gulf coasts. Chlorinated pesticides are also concentrated by shellfish to levels many times higher than those found in the water. Although the toxic effects of these chemicals to humans are still being evaluated, low levels in growing-area waters have proved to be toxic to a number of marine food species, especially in the larval stages. Growing area guidelines in shellfish meats for 11 pesticides and one herbicide were approved by the 1968 Shellfish Sanitation Workshop.

Depuration, or controlled purification of shellfish, appeared to be a means of recovering an otherwise unusable resource as well as a means of providing an additional safety barrier to consumers of shellfish harvested from approved growing areas. Studies on depuration included evaluation of ultra-violet sterilization systems, bacterial, viral and chemical depletion rates; rapid laboratory test procedures; and biological activity of shellfish during depuration. Although practised on a small scale in the United States some of the problems associated with depuration have yet to be solved. Even under optimum conditions the rate of biological activity for all shellfish in a batch being processed cannot be controlled. Hazardous levels of indicator organisms may still be found in shellfish which are not active during the depuration process, thus posing a hazard to the consumer if eaten raw.

Responsible for sea clams

With addition of the sea clam industry to the programme control of growing areas on the continental shelf beyond State jurisdictional limits was added to the Sanitation Branch's responsibilities.

The first investigation of offshore sewage sludge dump sites was conducted by the Northeast Technical Services Unit in 1966 in the general area off Sandy Hook, New Jersey, known as the New York Bight where sewage sludge, rubble, and acid are dumped. Our primary interest was in the sewage sludge dump site which was adjacent to sea clam beds along the Long Island, New York shore to the north and the New Jersey shore to the south. Sludge was being dumped from tankers at a rate of approximately 7,500 m³ per day. The 1969 harvest of surf clam meats was valued at $6¼ million illustrating the importance of this fishery to coastal economy. A study was also conducted at a second dump site off the mouth of the Delaware Bay in 1967. Investigations resulted in a circular area having a radius of 9.7 km from the centre of each of these two dump sites being closed to harvesting. Patrolled by the United States Coast Guard these areas are being monitored at regular intervals for changes in levels of bacteria and heavy metals.

Investigations of chemical contaminants in shellfish conducted by our research programme have been almost exclusively confined to chlorinated hydrocarbons and heavy metals. Carcinogenic hydrocarbons from oil spills and toxic detergents used to emulsify such spills are two more chemical contaminants which may be affecting consumers of shellfish.

The National Shellfish Sanitation Program and the Food and Drug Administration support the continued safe use of molluscan shellfish as a natural resource, actively encourage water quality and conservation programmes designed to combat marine and estuarine pollution, and vigorously oppose the disposal in marine and estuarine areas of any materials having a toxic or deleterious effect on marine food products. Furthermore, we believe that the comprehensive sanitary surveys of shellfish-growing areas followed by closure and patrol of areas not meeting approved criteria, is the primary line of defence in consumer protection from shellfish-borne disease, and that bacteriological and chemical sampling are essential adjuncts of, not substitutes for, the sanitary survey of the shoreline. The National Program recommends the multiple tube fermentation procedure and the coliform indicator group as defined in *Recommended Procedures for the Bacteriological Examination of Sea Water and Shellfish*, Third Edition, 1962, for bacteriological surveys of shellfish-growing areas.

Approximately 3,200,000 ha of shellfish-producing areas in the United States meet the "approved growing area" criteria, and approximately 800,000 ha of potential harvest areas are closed because of pollution and patrolled to prevent harvesting.

Depuration or controlled purification is an acceptable method of processing shellfish if proper procedures are followed and there is adequate supervision of the operation by the local food control agency. A varying percentage of any given lot of shellfish being processed in a controlled-purification system fail to be biologically active to the degree necessary for adequate cleansing within a commercially-feasible time period. There are also individual, species and seasonal differences in biological activity rates. Therefore, the health hazard to consumers of uncooked shellfish harvested from polluted waters and processed in a controlled purification system is roughly proportional to the degree of pollution of the harvest area. The degree of cleansing is markedly influenced by the level of bacteria in the shellfish when harvested. High levels of trace metals will not be depurated in a commercially-feasible time period. In the United States, shellfish harvested from areas failing to meet "approved" growing area criteria may not be sold in interstate commerce unless previously treated in a controlled-purification system.

[567]

References

American Public Health Association, Standard methods of water
1920   analysis. New York, N.Y., American Public Health
    Association.
EIJKMAN, C Die Gartingsprobe bei 46° als Hilfsmittel bei der
1904   Trinkwasseruntersuchung, *Zentbl. Bakt. ParasitKde (1)*,
    37:742.
PERRY, C A A summary of studies on pollution in shellfish.
1939   *Escherichia coli* versus the coliform group as an index of
    fecal pollution and the value of a modified Eijkman test.
    Food Res., 4(4):381–95.

U.S. Public Health Service, Shellfish Sanitation Branch. Shellfish
1965   sanitation program manual of operations. Part 1. Sanita-
    tion of shellfish-growing areas. *Publs publ. Hlth Serv.*,
    Wash., (33).
U.S. Public Health Service, Shellfish Sanitation Branch. Proposed
1968   Appendix B, Part 1, Shellfish sanitation manual of opera-
    tions. Interim guidelines for pesticides in shellfish. Appendix
    D. *Proc. natn. Shellfish sanit. Wkshop*, 1968:53–4.
U.S. Public Health Service, Shellfish Sanitation Branch. Proposed
1968a   Appendix C, Part 1. *Proc. natn. Shellfish sanit. Wkshop*,
    1968:54–8.

Polynuclear Aromatic Hydrocarbon Pollution of the Marine Environment†

M. J. Suess[*]

La contamination du milieu marin par les hydrocarbures aromatiques polynucléaires

La présence d'hydrocarbures aromatiques polynucléaires (HAP) en général, et notamment en benzo 3,4-pyrène (BP), dans les estuaires et les eaux côtières est maintenant bien établie. On a trouvé du BP, substance dont l'action cancérogène est reconnue, dans les sédiments, la flore et la faune des eaux des mers qu'elles soient polluées ou non. On estime que les HAP sont ingérés par la faune des eaux environnantes, mais que dans la flore ils peuvent résulter en partie de la synthèse endogène. La présence des HAP ayant été démontrée dans les denrées alimentaires, l'air et l'eau consommés par l'homme, on peut considérer qu'ils présentent un danger potentiel pour la santé. C'est pourquoi leur présence à des taux de concentration excessifs dans les ressources alimentaires des eaux côtières par suite de pollution est inopportune.

Contaminación del medio marino por hidrocarburos aromáticos polinucleares

Está demostrada la presencia de hidrocarburos aromáticos polinucleares (HAP) en general y de 3,4-benzopireno (BP) en particular en estuarios y aguas costeras. El BP es una conocida sustancia carcinógena que se ha encontrado en los sedimentos, la flora y la fauna de aguas de mar contaminadas y no contaminadas. Se considera que los HAP que se encuentran en las aguas que rodean a la fauna son ingeridos por ésta y que su presencia en la flora puede deberse parcialmente a síntesis endógena. Como se ha demostrado su presencia en los alimentos, agua y aire que consume el hombre, los HAP tienen que considerarse un posible peligro para su salud y por tanto su presencia en cantidades excesivas, debidas a la contaminación, en los recursos de alimentos de las aguas costeras es indeseable.

MANY polynuclear aromatic hydrocarbons (PAH) are present, at least in minute concentrations, in all marine waters, probably as a result of formation by aqueous flora. It has been postulated that phytoplankton may produce as much as 3 tons of aliphatic and aromatic hydrocarbon products per square kilometre of ocean per annum (Smith 1954). Moreover it has been reported that PAH are endogenically synthesized by bacteria (Knorr and Schenk 1968), fresh water algae (Borneff *et al.* 1968), and land plants (Gräf 1964, 1965; Gräf and Diehl 1966). Coastal waters also receive PAH from municipal and industrial effluents, atmospheric precipitation and fallout, runoff, and oil spillage in harbours and from ships.

PH IN THE MARINE ENVIRONMENT

The compound 3,4-benzpyrene (BP), a 5-ring PAH-type, is ubiquitous and one of the most potent PAH carcinogens (Falk *et al.* 1964). Consequently its presence has been thoroughly investigated in various samples of marine biota and sediments. These surveys demonstrated the widespread though non-uniform occurrence of BP in waters ranging from those off the unpopulated western coast of Greenland to the pollution-receiving waters off the highly populated coasts of the U.S.A., France and Italy (Table 1) (See also Suess 1970).

While, in certain locations, high BP concentrations were found in bottom sediments it did not necessarily occur in the corresponding biota (Mallet 1961). The high BP content in the sediments may be attributed in part to the activity of micro-organisms present, also in the Bay of Naples to volcanic ash (Bourcart and Mallet 1965). Concentrations are often highest in the upper sediment layers. In addition to BP, nine other PAH substances were found in bottom sediments; three of which are known carcinogens (Münch 1966).

PAH are not considered to be metabolic products of the fauna, but are ingested and possibly fixed and concentrated in process of feeding. This also explains the notable amounts of BP in a variety of fish and shellfish taken from both polluted and unpolluted areas (Table 1).

Edible French molluscs, and oysters from Alabama, sold on Paris markets were shown to contain high concentrations of PAH, including BP (Mallet and Schneider 1964; Perdriau 1964). Although there appears to be no information on PAH concentrations in edible seaweed (important in Japan), it is possible that PAH is synthesized by the seaweed, as it is by other flora, thus contributing to the overall human intake of PAH. Observations have been made on the western coast of the U.S.A. that fish developed papillomas from digging into mud (Russel and Kotin 1957). There is good reason to believe that the muds were contaminated with, among other things, carcinogenic PAH.

Carcinogenic effects

Experiments with test animals has indicated that repeated exposure to carcinogenic substances is more effective than an equivalent single exposure, and that repeated doses may overcome inhibiting responses of the host organism (Hueper and Conway 1964). Numerous experiments have confirmed the potential danger of continuous exposure to carcinogenic PAH introduced

---

* World Health Organisation, Regional Office for Europe, Copenhagen, Denmark.
† This article expresses the author's views and not necessarily those of the World Health Organization.

TABLE 1. SOME BP CONCENTRATIONS IN MARINE SAMPLES

Sea coast	Sample	BP concentration in µg/kg of dry sample	Reference
Virginia, U.S.A.	Oysters	2–6 (Carcinogenic PAH: 300) (Total PAH: 1,200)	Cahnmann and Kuratsune 1957

France (various)
—Channel
—Mediterranean
 	
—Toulon Harbour
	Crab, shrimps, oysters, mussels and molluscs
	Plankton
	Sediments:
	  0 to 0.5 m deep
	  1 and 2 m deep
	Sands and muds:
	  14 to 82 m deep
	Mussels
		Traces to 90
		400
		
		420–5,000
		26 and 16
		
		Traces to 1,500
		16–22
		(Total PAH: 1,100–3,400)
			Mallet, 1961
			Mallet and Sardou 1964
			
			Lalou et al. 1962
			
			Bourcart et al. 1961
			Greffard and Meury 1967

Italy—Bay of Naples	Sardine, mussels, and molluscs Algae and plankton Sand and muds: 2 to 120 m deep	65, 130 and 540, and 2 2 and 6–21 1–3,000	Bourcart and Mallet 1965
Greenland—west coast	Cod fish, mussels and molluscs Algae and plankton Sand: 0.2 m deep	15, 55 and 60 60 and 5 5	Mallet et al. 1963

into the gastro-intestinal tract by water and food, even at low concentrations. However, it can be assumed that a high percentage of the carcinogenic PAH or their metabolites will rapidly be excreted.

Laboratory investigations

BP was chosen for laboratory investigations, with the aim of studying its behaviour and fate in aqueous systems (Suess 1967). Experiments were conducted on the sorption of BP onto mineral surfaces, and its rate of photodegradation as affected by environmental factors was studied.

Under white fluorescent light the photodegradation of BP adsorbed onto calcite suspended in water followed a first-order rate equation with respect to BP surface concentration, the initial adsorbed amounts varying from 0.08 to 1.3 μg/g. In a typical reaction vessel with surface illumination of 0.31 mw/cm^2 and temperature of 21°C the rate constant was 0.009 per hour for the suspension deaerated with helium, and 0.020 and 0.027 per hour when saturated with air and oxygen, respectively. Degradation was not affected by varying the pH in the range of 7 to 10, nor by ionic strength up to 0.008. The degradation rate constant increased with temperature over the range of 5 to 31°C, leading to a calculated activation energy of 15.4 kcal/mole. The most likely mechanism in the photodegradation of BP in the presence of oxygen is the reaction of the latter with BP excited to a triplet state.

Summary and conclusion

Carcinogenic PAH are widely distributed, being found in sea waters off both heavily populated and unpopulated coasts. While the endogenic synthesis of flora appears to be a major source of PAH in the open seas, municipal and industrial waste water and atmospheric precipitation may further increase their concentrations along coasts.

Experimental results have indicated that mineral surfaces may be important in the transport of relatively insoluble BP and that such factors as temperature, oxygen concentration and, particularly light intensity, will affect its stability. An evaluation of the experimental results with respect to natural waters, leads to the conclusion that BP degradation will depend on water depth, as well as on daily and seasonal changes due to fluctuation of solar radiation, ambient temperature and dissolved oxygen. The degradation rate constant of BP will be higher in the upper layers of the water, where illumination, temperature and oxygen concentration are high, but will decrease with depth owing to reduced water clarity, reduced illumination and reduced temperature and oxygen concentration. It is also to be expected that BP in bottom deposits will degrade very slowly, if at all, owing to the lack of light and oxygen.

Food is one of the more important sources of PAH for man. In the marine environment, micro-organisms contain varying amounts of these substances which are accumulated in edible fish and shellfish. Although man is exposed to PAH from several sources (food, tobacco, air and water) no comprehensive study of their relative significance has yet been made, and it is not possible to estimate the contribution of seafood to the total human intake of PAH. Probably the consumption of carcinogen-containing seafood is not dangerous at present, but the combination of sources may well lead to a potential hazard, and the presence in food of any excess PAH carcinogens due to pollution is undesirable.

A comprehensive view dealing with PAH in the aquatic environment has recently been published (Andelman and Suess 1970). This reviews physico-chemical properties; origin, sources and carriers; concentrations in water, fauna, flora and sediments; the effect of water and waste water treatment on PAH removal; and health considerations.

References

ANDELMAN, J B and SUESS, M J Polynuclear aromatic hydro-
1970   carbons in the water environment. *Bull. Wld Hlth Org.*, 43(3):479–508.

BORNEFF, J et al. Experimental studies on the formation of poly-
1968   cyclic aromatic hydrocarbons in plants. Envir. Res., 2(1):
        22–9.

BOURCART, J and MALLET, L   Marine pollution of the shores of the
1965   central region of the Tyrrhenian Sea (Bay of Naples) by
        3,4-BP type polycyclic hydrocarbons. C. r. hebd. Séanc.
        Acad. Sci., Paris, 260(13):3729–34.

BOURCART, J, LALOU, C and MALLET, L   About the presence of
1961   3,4-BP type hydrocarbons in the coastal muds and the
        beach sands along the coast of Villefranche (Alpes-
        Maritimes). C. r. hebd. Séanc. Acad. Sci., Paris, 252:640–4.

CAHNMANN, H J and KURATSUNE, M   Determination of polycyclic
1957   aromatic hydrocarbons in oysters collected in polluted
        water. Analyt. Chem., 29(9):1312–7.

FALK, H L, KOTIN, P and MEHLER, A   Polycyclic hydrocarbons as
1964   carcinogens for man. Archs environ. Hlth., 8:721–30.

GRÄF, W   3,4-BP as growth factor in plants. Arch. Hyg. Bakt.,
1964   148(6):489–92.

GRÄF, W   About the natural occurrence and significance of car-
1965   cinogenic polycyclic aromatic hydrocarbons. Medsche Klin.,
        60(15):561–5.

GRÄF, W and DIEHL, H   About the natural background of car-
1966   cinogenic polycyclic aromates and its cause. Arch. Hyg.
        Bakt., 150(1/2):49–59.

GREFFARD, J and MEURY, J   Carcinogenic hydrocarbon pollution
1967   in Toulon harbour. Cah. océanogr., 19(6):457–68.

HUEPER, W C and CONWAY, W D   Chemical carcinogenesis and
1964   cancers. Springfield, Illinois, Charles C. Thomas, Publisher.

KNORR, M and SCHENK, D   About the question of the synthesis of
1968   polycyclic aromates by bacteria. Arch. Hyg. Bakt., 152
        (3):282–5.

LALOU, C, MALLET, L and HÉROS, M   Depth distribution of 3,4-BP
1962   in a core sample from the bay at Villefranche-sur-Mer.
        C.r. hebd. Séanc. Acad. Sci., Paris, 255(1):145–7.

MALLET, L   Investigations for 3,4-BP type polycyclic hydrocarbons
1961   in the fauna of marine environments (the Channel, Atlantic
        and Mediterranean). C. r. hebd. Séanc. Acad. Sci., Paris,
        253(1):168–70.

MALLET, L and SARDOU, J   Investigation on the presence of the
1964   polycyclic hydrocarbon 3,4-BP in the plankton environ-
        ment of the Bay of Villefranche (Alpes-Maritimes region).
        C. r. hebd. Séanc. Acad. Sci., Paris, 258(21):5264–7.

MALLET, L and SCHNEIDER, C   Presence of BP-type polycyclic
1964   hydrocarbons in geological and archeological levels. C. r.
        hebd. Séanc. Acad. Sci., Paris, 259(3):675–6.

MALLET, L, PERDRIAU, A and PERDRIAU, J   The extent of pollution
1963   of BP-type polycyclic hydrocarbons of the western region
        of the Arctic Ocean. C. r. hebd. Séanc. Acad. Sci., Paris,
        256(16):3487–9.

MÜNCH, H D   Carcinogenic materials in water. Z. ges. Hyg.,
1966   12(7):468–76.

PERDRIAU, J   Marine pollution by carcinogenic BP-type hydro-
1964   carbon—biological incidences. Part 2. Cah. océanogr.,
        16(3):205–29.

RUSSEL, F F and KOTIN, P   Squamous papilloma in the white
1957   croaker. Natn. Cancer Inst. J., 18(6):857–61.

SMITH, P V   Studies on origin of petroleum: Occurrence of hydro-
1954   carbons in recent sediments. Bull. Am. Ass. Petrol. Geol.,
        38(3):377–404.

SUESS, M J   The behaviour and fate of 3,4-Benzpyrene in aqueous
1967   systems. Pittsburgh, Pa., University of Pittsburgh Graduate
        School of Public Health, Doctoral Dissertation.

SUESS, M J   Presence of polycyclic aromatic hydrocarbons in
1970   coastal waters and their possible effects on human health.
        Arch. Hyg. Bakt., 154(1):1–7.

The Occurrence of Human Viruses and Coliphage in Marine Waters and Shellfish†

T. G. Metcalf,
J. M. Vaughn and W. C. Stiles*

Incidence des virus pathogènes de l'homme et des coliphages dans les eaux marines, les mollusques et les crustacés

On a fait une étude sur l'incidence des virus pathogènes de l'homme et des coliphages dans un estuaire soumis à la pollution par des eaux d'égouts domestiques non traitées. Des examens comparatifs ont porté sur la fréquence des entérovirus et des coliphages dans des échantillons d'huîtres, d'eau des couches superficielles et d'effluents d'usines de traitement déversés dans les affluents de l'estuaire. Les taux d'isolement (nombre d'organismes isolés/nombre d'échantillons soumis aux tests) pour les virus et les coliphages ont été établis par examen conjoint du matériel soumis aux tests. Les taux de survie (nombre de virus détectables par jour) pour chaque groupe ont été déterminés dans les eaux de l'estuaire. On a également dosé le taux de rétention (nombre de virus détectables par jour) d'huîtres déposées dans les eaux de l'estuaire et transformées expérimentalement en vecteurs des virus et des coliformes. On a établi les effets de la saison, de la température et du degré de pollution sur les taux d'isolement et de survie, et déterminé les taux de rétention des virus et des coliphages dans les huîtres.

Les données recueillies ont permis de tirer les conclusions suivantes: (1) dans tous les échantillons examinés, on a trouvé des coliphages plus fréquemment que des entérovirus; (2) l'incidence des entérovirus variait largement d'une semaine à l'autre pendant l'été, et d'une saison à l'autre; (3) les fluctuations hebdomadaires et saisonnières de l'incidence des coliphages n'étaient pas parallèles aux variations des entérovirus; (4) les entérovirus se maintenaient généralement plus longtemps que les coliphages dans l'estuaire.

A la suite de ces études, il ne semble pas possible d'utiliser les coliphages pour prévoir avec exactitude la présence de virus pathogènes de l'homme dans un estuaire pollué. Le document examine la portée de ces conclusions à l'égard des problèmes que pose à la santé publique l'examen des eaux, des mollusques et crustacés dans les estuaires.

Presencia de virus humanos y colifagos en las aguas del mar y en mariscos

Se realizó un estudio sobre la presencia de virus patógenos y colifagos en un estuario expuesto a la contaminación de aguas de descarga domésticas no tratadas previamente. Se realizaron exámenes comparativos de la frecuencia de enterovirus y colifagos en muestras de ostras, de aguas superficiales y de las aguas de descarga que una planta de tratamiento de desechos vierte en varios ríos tributarios del estuario. Se estableció la frecuencia numérica (número de ejemplares encontrados/número de muestras examinadas) de virus y colifagos, examinando conjuntamente los materiales experimentales. Se determinaron, además, los índices de supervivencia (días durante los cuales se encontraron virus detectables) de cada grupo en las aguas del estuario y, mediante experimentos realizados en lagunas ostreras, se determinó el índice de retención (días durante los cuales se recuperaron virus detectables) en ostras depositadas en aguas del estuario tras inocularles experimentalmente virus y colifagos. Se determinó la influencia de la estación, la temperatura y el grado de contaminación en la frecuencia numérica y el índice de supervivencia de virus y colifagos y en el índice de retención de las ostras.

Los datos obtenidos permitieron llegar a las conclusiones siguientes: (1) en todas las muestras examinadas, los colifagos resultaron más frecuentes que los enterovirus; (2) el número de enterovirus experimentó grandes fluctuaciones de una semana a otra, durante el verano, y de una estación a otra; (3) las fluctuaciones semales y estacionales del número de colifagos no correspondieron a las fluctuaciones de los enterovirus; (4) de ordinario, los enterovirus permanecieron más tiempo en el estuario que los colifagos.

Los resultados obtenidos indican que no es posible utilizar los colifagos para establecer con seguridad la presencia de virus humanos en un estuario contaminado.

Se examina el significado de estas conclusiones para diversos problemas de sanidad pública, entre ellos el examen sanitario de los mariscos y de las aguas del estuario.

POLLUTION of estuaries and the occasional occurrence of enteric viruses in sewage may cause human viral pathogens to appear in shellfish. Epidemiological studies identified several outbreaks of infectious hepatitis that were caused by consumption of raw shellfish (Roos 1956; Mason and McLean 1962; Dougherty and Altman 1962).

Successful isolation of human virus from shellfish or

* Department of Microbiology, University of New Hampshire, Durham, New Hampshire 03824, U.S.A.
† Work supported by research grant UI-00730-02, now EC-00299-02, from Environmental Control Administration, U.S. Public Health Service.

their habitat necessitates its separation from large volumes of water or homogenates of pooled shellfish tissues. Considerable interest has been shown in the possible use of coliphage as an indicator of enteric virus. Advantages cited for use of coliphage include ease and economy of laboratory procedures. A coliphage indicator is only acceptable if it constantly and accurately reflects the enteric virus status of shellfish and shellfish waters. This report describes the results of examinations for enteric virus and coliphage in waste treatment plant effluents, shellfish and waters polluted by domestic sewage; and the use of coliphage as an indicator for enteric virus is discussed.

Methods of investigation

Water examinations were conducted on preparations divided from either 100 ml grab samples or liquid expressed from gauze pads that had been immersed at collection sites for 3 days.

Fluids extracted from pads were clarified by low speed centrifugation and divided into two aliquots. One aliquot was treated overnight with chloroform and further divided for qualitative and quantitative examinations. Both analyses were carried out using four 0.5 ml replicate samples added to 2.5 ml of 0.7 per cent agar overlays containing 0.2 ml of the appropriate *Escherichia coli* host cell. After mixing the overlay was poured on an agar

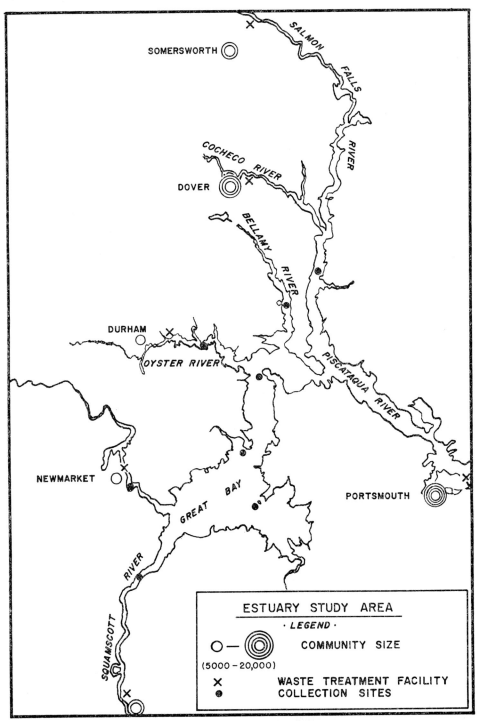

Fig 1. The estuary study area

[571]

basal layer and incubated at 36.5°C. Quantitative examinations were made on the initial chloroform treated supernates while qualitative examinations were performed on chloroform treated supernates following overnight enrichment with E. coli.

Coliphage isolations from grab samples were performed by the same methods following initial overnight chloroform treatment of the 100 ml sample and separation of a supernate upon centrifugation.

Samples for enteric virus examination were restricted to the second aliquot obtained from gauze pad extraction. This was inoculated into cell cultures following treatment and centrifugation to recover a concentrated sample. Details of the preparation procedure have been described (Metcalf and Stiles 1968). Three cell cultures were used routinely; primary human embryonic kidney, primary monkey kidney (African Green) and HEp II. Isolates were recognized by cytopathogenic effects (CPE) developing in test monolayers. A sample was considered negative only after two consecutive passages failed to reveal development of CPE in test monolayers. Viral isolates were identified by means of serum neutralization tests conducted on test tube monolayers.

Shellfish samples were obtained from homogenates representing pools of 10 or more oysters. Supernates from the homogenates were divided into two aliquots. One was used for coliphage isolations following the same processing method as for water and the other for enteric virus isolation using the method of sample preparation for water.

Survival of coliphage and enteric virus in water was defined as the length of time that coliphage or virus could be detected in water retained within seamless cellulose tubing submerged in estuary waters. Enteric virus assays for detectable virus were conducted in monolayers of a stable monkey kidney cell line, LLCNK$_2$. Fifty per cent infectivity endpoints (TCID$_{50}$) or plaque forming units (PFU) were determined.

Survival of coliphage and enteric virus in shellfish was determined by recovery from oysters following initial uptake of 100 to 1,000 or more PFU of each virus type and subsequent storage of the shellfish in estuary waters.

Results obtained

The estuary study area (fig 1) is an inlet of the North Atlantic which contains many oyster and clam beds and has a tidal area of 15,407 acres. Seven communities, with a total population in excess of 60,000, discharge effluents into the area.

Of nine treated effluent samples taken from one waste treatment plant (primary treatment plus chlorination) four were positive for coliphage and one was positive for enteric virus (Poliovirus I). Unlike enteric virus, coliphage shown to be present by grab sample isolates usually did not appear during examination of pad samples. Since pad samples yielded many coliphage isolates from estuary waters, failure to retain coliphage in or on pad surfaces could not be responsible for the negative results. This is probably due to inactivation of retained coliphage by residual chlorine (5 to 10 ppm during study interval). If this is so coliphage and enteric virus retained in pads react differently to residual chlorine in treated effluents.

Nine samples were taken of the raw sewage and all were positive for coliphage, with PFU per ml values ranging from 12 to 660. Comparison of PFU per ml values in raw sewage and treated effluents showed a 10 to 100-fold reduction of coliphage in the exiting effluents. Despite the reduction in PFU per ml, the number of coliphage PFU being discharged together with the greater number of coliphage isolations obtained, suggested that significantly greater numbers of coliphage than enteric virus were being released into the estuary.

Estuary waters and shellfish were examined for coliphage and enteric virus from June to September. Collection sites frequently were in "fringe" areas where coliform indices indicated that bacterial pollution fluctuated above

TABLE 1. COLIPHAGE AND ENTERIC VIRUS ISOLATES FROM WATER AND OYSTER SAMPLES

| Month | SAMPLES | | OYSTER | | WATER | | | |
| | | | | | Pad Sample | | Grab Sample | |
	Number tested	Per cent positive	Coliphage	Enteric virus	Coliphage	Enteric virus	Coliphage	Enteric virus
June	30	56	11	0	3	1	2	NT
July	83	35	10	0	15	0	4	NT
August	14	78	7	1	3	0	0	NT
September	3	66	2	0	0	0	0	NT
Total			30	1	21	1	6	

NT = No Test.

TABLE 2 THE ACCUMULATION BY OYSTERS OF COLIPHAGE AND ENTERIC VIRUS FROM SEA WATER

| Experiment | Description | COLIPHAGE | | ENTERIC VIRUS | |
		Sea water (PFU/ml)	Oyster (PFU/gm)	Sea water (PFU/ml)	Oyster (PFU/gm)
1	Sea water after addition of coliphage and virus	6.3×10^3		5×10^3	
	Coliphage and enteric virus accumulation by oysters		2×10^2		4×10^1
	Sea water after 24 h accumulation period	3×10^3		2.9×10^3	
2	Sea water after addition of coliphage and virus	1×10^4		8×10^3	
	Coliphage and enteric virus accumulation by oysters		7.1×10^3		2.4×10^2
	Sea water after 24 h accumulation period	2×10^3		3×10^3	

and below 70 coliform median MPN per 100 ml of estuary water. Fifty-seven coliphage and two enteric virus isolations were made from 130 examinations (Table 1).

Numerically, coliphage occurrence in oysters appeared to be uniform throughout the summer while the frequency of coliphage in the water fluctuated; notably higher isolation rates occurred in July. These findings were different from the pattern of enteric virus isolations made from the estuary over a period of 7 years. Enteric virus isolation usually showed significant increases in numbers as the summer progressed, with an isolation peak between July and September. The discrepancy between isolations from water and oysters indicates that coliphage is much more abundant than enteric virus in the "fringe" areas of the estuary.

Isolation of enteric viruses from grab samples is virtually impossible in "fringe" areas although coliphage were occasionally isolated from grab samples taken in "fringe" areas.

A controlled experiment on the accumulation of coliphage and enteric virus by oysters was done in running sea water to which coliphage and enteric virus had been added prior to introduction of oysters. All oysters used in this study were depurated before use. Results of the experiment are given in Table 2. Uptake of coliphage was 5 to 30 times greater than enteric virus, indicating a greater propensity of shellfish for coliphage. The results suggested a much greater chance of finding coliphage than enteric virus at most concentrations normally present in estuary waters. Thus, whether or not enteric virus was present numbers of coliphage isolations might well be high.

The influence of the E. coli strain used for demonstrating coliphage in estuary waters and shellfish was examined (Table 3). In experiment 1, only two E. coli strains were compared. A definite increase in the number of isolations from both oysters and sea water was obtained with E. coli B. One sample of oysters yielded an isolate with E. coli that was not obtained with E. coli B. The same result was obtained with one sea water sample. In all other samples of oysters and sea water, wherever an isolate was obtained with E. coli, an isolate was also obtained with E. coli B. In experiment 2 three E. coli strains were compared: Six isolations would not have been made without E. coli 9637. One isolation was obtained with E. coli B that was not found with E. coli 9637.

These examinations of oysters and estuary water indicated the presence of at least three qualitatively different coliphage reactivities with E. coli strains. The two major reactivities found were for E. coli B and E. coli 9637. Only two of nine samples reactive with E. coli 9637 were also reactive with E. coli B. The association of

E. coli 9637 coliphage reactivity with oysters only was in sharp contrast to the occurrence of E. coli B coliphage reactivity in oysters and sea water. More work on this subject is needed but on the basis of the data presented, it would appear desirable to use at least two E. coli host strains for coliphage isolations from an estuarine environment.

Survival analysis

Survival studies of coliphage and Coxsackievirus B-3 in estuary waters were carried out in separate dialysis tubes immersed in the estuary during the late winter and early spring. Water temperatures rose during the test interval from 1°C to 15°C, survival results are shown in fig. 2.

A slightly greater survival tendency for coliphage eventually translated into a longer survival time in terms of detectable activity. All detectable Coxsackievirus activity had ceased by day 42. Coliphage assay showed 500 PFU per ml remaining at 58 days. The survival characteristics of Coxsackievirus B-3 in estuary waters have been considered to be representative of the several virus groups recognized as enteric pathogens. On the basis of these findings coliphage survival potential can be considered at least equal to that shown by many enteric viruses.

The survival, or retention, of coliphage and Coxsackievirus B-3 in shellfish was also studied during the late winter and early spring, the results are shown in fig. 3. Comparative retention in oysters closely resembled comparative survival in estuary waters for coliphage and Coxsackievirus. Loss of original activity occurred at similar rates. Coliphage activity diminished at a slightly slower rate than that of Coxsackievirus. After 88 days no Coxsackievirus activity could be detected, but 15 PFU per ml of coliphage activity remained. The time that coliphage is retained by oysters appears to be comparable to retention by members of the enteric virus group.

Discussion and summary

Proposals for using coliphage to indicate the presence of enteric viruses of public health concern in water, stress as advantages, convenience, economy and quantitative determination (Guelin 1948; Carstens 1963). Berg (1969) shared the view that coliphage must always be present with viruses of human origin and in equal or greater numbers; coliphage must be at least as resistant to a marine environment as human viruses. The presence of coliphage in greater numbers than human viruses would confer a margin of safety when coliphage could not be demonstrated.

TABLE 3. COLIPHAGE ISOLATIONS FROM SEA WATER AND OYSTERS USING DIFFERENT *Escherichia coli* STRAINS

E. coli (UNH)*	OYSTERS E. coli B (ATCC 11303)†	E. coli (ATCC 9637)	E. coli (UNH)	SEA WATER E. coli B (ATCC 11303)	E. coli (ATCC 9637)
Experiment 1 3/22	15/22	—	3/9	6/9	—
Experiment 2 0/12	3/12	9/12	0/23	1/23	0/23

*Isolated from raw domestic sewage.
†American Type Culture Collection.

[573]

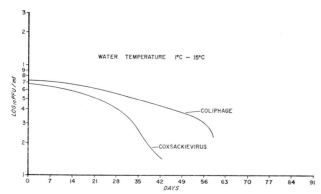

Fig 2. Survival of Coxsackievirus B-3 and coliphage in estuary waters during late winter and early spring months

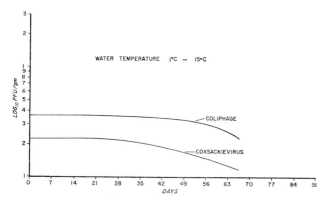

Fig 3. Survival of Coxsackievirus B-3 and coliphage in the oyster, Crassostrea virginica, immersed in estuary waters during late winter and early spring months

The present study fulfils the criteria advanced by Berg. Coliphage was constantly present in raw sewage entering a waste treatment plant in concentrations ranging from 12 to 853 PFU per ml during the winter months, when the enteric virus content had been previously shown to fluctuate between 0 and 100 per cent. Coliphage presence in treated effluents was sufficient to permit quantitative measurements when no enteric viruses could be detected. Frequent isolations of coliphage were made from estuary waters and shellfish that contained no detectable enteric virus. Coliphage enumeration, unlike enteric virus, occasionally could be measured directly from grab samples. Coliphage isolation from oysters demonstrated a greater accumulation potential for coliphage than for enteric virus.

Survival time of coliphage in estuary waters was slightly longer than that for Coxsackievirus. Retention of coliphage by dormant oysters during the winter months paralleled the retention shown for Coxsackievirus.

A potential difficulty in conducting coliphage determination was illustrated by the finding of two coliphage types in oysters that required different *E. coli* strains for isolation and had no cross reactivity. The existence of multiple coliphage types exhibiting differential sensitivity to several *E. coli* strains could impose constraints on effective use of coliphage indicators.

The realization of the criteria advanced by Berg for the acceptability of a coliphage indicator system should not be considered an automatic endorsement. Buttiaux (1958), for example, has questioned the value of faecal phage enumeration for estimating the faecal pollution of water.

The results of our study suggest that a coliphage indicator system for enteric viruses possesses inherent shortcomings and its capability of satisfactorily resolving the virus detection problem in marine waters is subject to question.

References

BERG, G Discussion of, The fate of viruses in a marine environment.
1969 *Adv. Wat. Pollut. Res.*, 4(3):833–4.
BUTTIAUX, R Surveillance et contrôle des eaux d'alimentation. La
1958 standardisation des méthodes d'analyse bactériologique de l'eau. *Revue Hyg. Méd. Soc.*, 6:170–92.
CARSTENS, E M J Bacteriophages and their possible use in sewage
1963 purification. *J. Proc. Inst. Sew. Purif. S. Afr.*, 1963(5):467–8.
DOUGHERTY, W J and ALTMAN, R Viral hepatitis in New Jersey.
1962 *Am. J. Med.*, 32:704–36.
GUELIN, A Etude quantitative de bactériophage de la mer. *Annls*
1948 *Inst. Pasteur, Paris*, 74:104–12.
MASON, J O and MCLEAN, W R Infectious hepatitis traced to the
1962 consumption of raw oysters. *Am. J. Hyg.*, 75:90–111.
METCALF, T G and STILES, W C Enteroviruses within an estuarine
1968 environment. *Am. J. Epidemiol.*, 88:379–91.
ROOS, B Hepatitis epidemic conveyed by oysters. *Svenska Lakartidn.*,
1956 *tidn.*, 53:989–1003.

Salmonellae and Bacterial Indicator Organisms in Polluted Coastal Water and their Hygienic Significance†

Y. Yoshpe-Purer* and H. I. Shuval**

Salmonelles et organismes bactériens indicateurs dans les eaux côtières polluées et leur signification du point de vue sanitaire

Une étude des schémas de dispersion et de dégradation de la pollution microbienne dans les eaux côtières a été effectuée pendant l'été et l'automne de 1968 et 1969 dans une zone voisine de Tel-Aviv, où les coliformes, salmonelles et entérovirus ont été utilisés comme indicateurs de la pollution microbienne. En 1968 on a mis au point une méthode nouvelle et précise pour isoler les salmonelles dans de grandes quantités d'eau de mer, et on a effectué des études comparatives à l'aide de deux autres méthodes—suspension de tampons "moore pads" dans l'eau de mer pendant 1 à 2 jours, et passage de l'eau de mer à travers des membranes filtrantes—afin de déterminer leur efficacité pour la détection des salmonelles. La méthode nouvelle, reposant sur la concentration du milieu d'enrichissement au lieu de l'échantillon, est moins fastidieuse et prend moins de temps que les autres, tout en étant plus efficace.

On a isolé 204 souches de salmonelles, appartenant à 35 sérotypes (sans inclure les divers types de phages). En 1968, 54 d'entre elles

Salmonelas y organismos bacterianos indicadores en aguas costeras contaminadas y su significado higiénico

Durante el verano y el otoño de 1968 y 1969 se realizó en una zona próxima a Tel Aviv un estudio de la dispersión y desvanecimiento de la contaminación microbiana en aguas costeras, empleando como indicadores de la contaminación salmonelas y enterovirus. En 1968 se puso a punto un método nuevo y más sensible para aislar salmonelas en grandes volúmenes de agua del mar y se realizaron estudios comparativos con otros dos métodos—"las almohadillas moore", suspendidas en agua del mar durante 1–2 días, y la filtración a través de filtros de membrana—para determinar su eficacia en la detección de salmonelas. El nuevo método, que se basa en la concentración del medio de enriquecimiento y no de la muestra es menos tedioso y dilatorio y más eficaz que los otros métodos.

Se aislaron 204 variedades de salmonelas pertenecientes a 35 serotipos (sin incluir los varios fagotipos). Cincuenta y cuatro de ellas se aislaron en 1968, utilizando todos los métodos citados a

---

  * A. Felix Public Health Laboratory, Ministry of Health, Tel Aviv, Israel.
 ** Environmental Health Laboratory, Hadassah Medical School, Hebrew University, Jerusalem, Israel.
  † Grants from the Israel National Council for Research and Development, and the National Communicable Disease Center, U.S. Public Health Service.

ont été isolées par toutes ces méthodes dans 37 échantillons positifs sur 115 soumis au test (32.2 pour cent de positifs), alors qu'en 1969, 150 ont été isolées par la seule méthode de concentration du milieu, dans 73 échantillons positifs sur 91 examinés (80 pour cent). En une seule journée, sur 22 échantillons prélevés à des distances différentes entre le point de déversement et un point éloigné de 2 km—parmi divers niveaux de pollution indiqués par les dénombrements de coliformes—20 étaient positifs (91 pour cent) et l'on a isolé sur ceux-ci 46 souches de salmonelles appartenant à 22 sérotypes. Dans un cas où 4 échantillons de 500 ml provenant du même lieu, prélevés chacun à une profondeur différente, ont été inoculés, on a isolé 10 souches appartenant à 8 sérotypes, dont la moitié provenait de la couche supérieure.

Les principaux sérotypes isolés étaient *S. paratyphi* B et *S. paratyphi* B var. *odense* appartenant à divers types de phages. D'autres sérotypes ont été isolés de une à neuf fois. Dans un certain nombre de cas, on a détecté des salmonelles dans des échantillons d'eau de mer d'une densité de coliformes inférieure à 2.400 par 100 ml ce qui est considéré comme une norme acceptable pour l'eau de baignade dans certains pays. Du point de vue sanitaire, on établit un rapport entre ces nombreux cas d'isolement de salmonelles dans les eaux côtières voisines des plages publiques et la dilution physique des eaux usées ainsi que l'indice des coliformes.

partir de 37 muestras positivas de las 115 ensayadas (32,3 por ciento positivas), mientras en 1969 se aislaron 150 utilizando sólo el método de concentración del medio y a partir de 73 muestras positivas, de las 91 examinadas (80 por ciento). En un sólo día se tomaron 22 muestras a diversas distancias del punto de salida de las aguas de descarga, hasta un máximo de dos kilómetros. De dichas muestras, que mostraban diferente contaminación, calculada por los recuentos de gérmenes coliformes, 20 resultaron positivas (91 por ciento) y de ellas se aislaron 46 variedades de salmonelas pertenecientes a 22 serotipos. En un caso, se inocularon cuatro muestras de 500 ml procedentes del mismo lugar pero tomadas todas a profundidades diversas y se aislaron 10 variedades, la mitad de ellas de la capa superior, pertenecientes a 8 serotipos.

Los serotipos predominantes fueron *S. paratyphi* B y *S. paratyphi* B var. *odense* de varios fagotipos. Otros serotipos se aislaron de 1 a 9 veces. En varios casos se encontraron salmonelas en muestras de agua de mar con densidad de gérmenes coliformes inferior a 2.400 por 100 ml, que en algunos países se considera como aceptable para el baño. El significado higiénico de todos estos aislamientos de salmonelas en aguas costeras próximas a playas públicas utilizadas para baños, se evalúa en relación con la dilución física de las aguas de desecho y el índice de gérmenes coliformes.

THE disposal of waste water into coastal waters without endangering the health of bathers requires an understanding of the natural purification capacity of the sea under various climatic conditions. Contamination of sea water by sewage organisms and their die-away rate in sea water in different conditions have been studied by numerous workers in various parts of the world, and good reviews are given by Orlob (1956) and Greenberg (1956).

Materials and methods

In 1968 samples were taken from four main points shown in fig 1 as A, B, C and D. Occasional samples were also taken at an additional station E.

At points B–D buoys were set up to establish sampling locations and to facilitate the attachment of pads which were left for 24–48 h. Weekly samples were taken between June and November and tested for coliforms (by two methods), faecal coliforms, enterococci (two methods) and salmonellae (three methods). Parallel samples were taken for virological analysis and are reported upon elsewhere (Shuval 1970).

In 1969 up to 100 daily samples were taken north and south of the outfall. These samples were taken as part of an extensive study on the dispersion and die-away patterns of coliforms in the sea, associated with radio isotope tracer experiments. All samples were examined for coliform density by the membrane filtration method and then pooled according to their distance from the outfall and examined for the presence of salmonellae by the TSS method.

The most probable number (MPN) of coliforms was determined according to Standard Methods (1965) using the multiple fermentation tube procedure. Presumptive tests with lactose broth were based on three tubes in each of up to seven decimal dilutions at 35°C. Confirmation was on Brilliant Green bile broth (Difco) or McConkey agar.

In the membrane filtration method (MF), which was used exclusively in 1969, the filters (Gelman-Metricel 0.45 μ) were placed on absorbant pads saturated with 2 ml of m-Endo broth MF (Difco) and incubated for 24 h at 35°C. Dark red colonies with metallic sheen were counted as coliforms. Doubtful colonies were confirmed by subculturing on McConkey agar (Difco).

Faecal coliforms were determined by the multiple fermentation tube procedure using EC medium (Difco) incubated at 44.5°C and reported as MPN (Standard Methods 1965).

For the determination of enterococci both the multiple fermentation tube and MF methods incubated at 35°C were used. In the former Azide broth was used for the presumptive test and EVA broth was used for confirmation. In the MF method m-Enterococcus agar (Difco) was used. Dark red colonies, or those with a dark red centre were counted. Selected colonies from filters as well as positive confirmed tubes were plated on blood agar for haemolysis and the Sherman test (growth at pH 9.6, 6.5 per cent NaCl at 45°C and 10°C) was applied to them.

During 1968 three different methods for the isolation of salmonellae were compared: the MF method, the pad

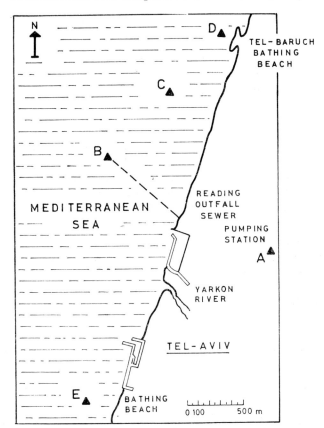

Fig 1. Sampling stations in the area of the Tel Aviv main sea outfall sewer

[575]

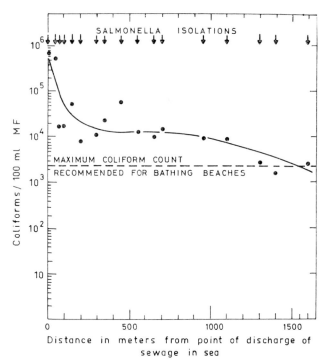

Fig 2. Salmonella isolations and coliform counts in sea as a function of distance from point of sewage discharge (12 November 1969)

method and the triple strength selenite (TSS) method developed in the course of this study (full details to be published elsewhere).

(1) MF method—One l portions of sea water or 50–100 ml portions of sewage were filtered through membrane filters (Gelman/Metricel 0.45 μ). The filter was then immersed in single strength Salenite broth (Difco)

(2) Pad method—Polyethylene sponges were immersed at the sampling stations and collected in new plastic bags after 24 or 48 h. The sponges were cut into small pieces and together with their expressed fluid contents placed in jars containing 400 ml single strength Selenite broth

(3) TSS method—The triple strength selenite method is based on using concentrated medium which is then diluted to normal strength with large volumes of the sewage or sea water sample to be tested. Thus 500 ml of inoculum added to 250 ml of TSS achieve a normal strength medium with all the organisms of the sample in it.

All cultures were incubated at 42°C. Plating was done on Brilliant Green agar after 24 and 48 or 72 h. Suspected colonies were picked and inoculated to slants of Kligler's triple sugar agar and tubes of peptone water with urea. Cultures showing typical appearance on Kligler's medium

and which showed no urease activity were tested with polyvalent salmonella antiserum and then typed with group antiserum and checked for fermentation of lactose, sucrose, salicin and dulcitol. Strains that fermented the first three materials but not dulcitol were sent for final typing to the Enterobacteriaceae Centre of the Government Central Laboratories in Jerusalem. Phage typing of the *S. paratyphi B* strains was done at the A. Felix Laboratory in Tel Aviv.

Results of field studies

A comparison of the MPN and MF methods for coliforms is given in Tables 1 and 2. The relationship between the various indicator organisms tested in 1968 is given in Table 3. For the purpose of comparison, only the MPN results were used.

In 1968, 54 strains of salmonella (14 serotypes) were isolated, most of them from sewage and very polluted sea water above the sewage outfall and at two additional points examined by pads only. Two strains were isolated from sampling point C and none from D. The percentage of salmonella isolations in this season is related to coliform and enterococci densities in Table 3.

During the summer of 1969 only ten samples of sewage and very contaminated sea water were examined for salmonellae. They were all positive and 20 strains were isolated. However, intensive weekly studies were conducted between the middle of October and the middle of November, and 155 strains belonging to 30 serotypes were isolated from the samples. For results of one of these surveys (12 November 1969) see fig 2. On that day three samples of sewage and 26 pooled samples of sea water were examined, all three sewage samples and 20 of the sea water samples were positive for salmonellae. Four of the positive samples were 1,000 to 1,600 m from the outfall, some of them showing coliform counts of 1,800/100 ml, i.e. below the limits of 2,400/100 ml, often recommended as maximum permissible for bathing beaches. *S. paratyphi B* was found in all sewage samples and 15 of the sea water samples. Two other dominant serotypes on that day were *S. give* and *S. london*, the latter not isolated from the population in that year. There was a wide difference between the serotypes isolated from sewage and sea water of the same day, as illustrated by results of one day given in Table 4. As many as five serotypes have been isolated from a single sample of sea water.

A comparison of the dominant serotypes isolated from sewage and sea water with those isolated from the population in Israel in 1968 is presented in Table 5.

Information on the isolation from the population is based on the Annual Report of the Ministry of Health Salmonella-typing Centre and the A. Felix Public Health Laboratory.

TABLE 1. COMPARISON OF COLIFORM COUNTS BY MPN AND MF METHODS IN SEWAGE AND SEA WATER

Sampling station	Type of sample	No. of samples	Coliforms/100 ml (logarithmic average) MPN	MF	Ratio MPN:MF	Distribution of MF counts in relation to MPN 95% confidence limits % within	% above	% below
A	Raw sewage	18	8.27×10^8	2.76×10^8	3:1	33	12	55
B	Heavily contaminated sea water	18	1.74×10^6	9.0×10^5	2:1	60	6	34
C + D	Lightly contaminated sea water	36	1.08×10^3	5.6×10^2	2:1	50	0	50

TABLE 2. RATIO BETWEEN COLIFORMS, FAECAL COLIFORMS AND ENTEROCOCCI IN SEWAGE AND SEA WATER OF VARYING DEGREES OF CONTAMINATION

Sampling station	Type of sample	No. of samples	Coliform:FC	Coliform: enterococci	FC: enterococci
A	Raw sewage	18	3:1	70:1	22:1
B	Heavily contaminated sea water	18	3:1	95:1	38:1
C + D	Lightly contaminated sea water	36	6:1	19:1	3:1

TABLE 3. SALMONELLAE ISOLATIONS IN SEWAGE AND SEA WATER AS RELATED TO COLIFORM AND ENTEROCOCCI CONCENTRATIONS

Sampling station	Type of sample	No. of samples	Coliforms/100ml MPN (median)	Enterococci/100ml MPN (median)	% positive for salmonellae
A	Raw sewage	26	4.6×10^8	2.4×10^7	47
B	Heavily contaminated sea water	35	2.4×10^6	4.6×10^4	47
C + D	Lightly contaminated sea water	35	460	36	6

TABLE 4. TYPICAL DISTRIBUTION OF SALMONELLA SEROTYPES ISOLATED FROM RAW SEWAGE AND IN THE SEA ON THE SAME DAY (5 NOVEMBER 1969)

Raw sewage	Sea = above outfall sewer	Sea = from 200 m–2,000 m from point of sewage discharge	
*S. para B (6)	*S. para B (5)	*S. para B (8)	S. reading (2)
*S. para B var. odense	*S. montevideo	*S. para B var. odense	S. blockley (2)
*S. emek	S. thompson	*S. emek	S. tennessee
*S. newport	S. siegburg	*S. newport	S. branderup
*S. montevideo	S. enteritidis	S. thompson	S. jerusalem (2)
S. give		S. 6:7:K (2)	S. kentucky
S. 4, 5, 12:0 form		S. stanley	S. typhimurium
		S. infantis	S. —:b:2
		S. sofia (2)	

* serotypes isolated both in sewage and in sea.
(n) number of repeat isolations.

TABLE 5. DISTRIBUTION OF SALMONELLA SEROTYPES ISOLATED FROM SEWAGE AND SEA WATER IN TEL AVIV IN 1968 AND 1969 COMPARED TO THE DOMINANT TYPES ISOLATED IN POPULATION 1968

Salmonella serotype	Population number of isolations	%	Sewage and sea water number of isolations	%
S. typhimurium	1,025	31.1	3	1.3
S. typhimurium var. copenhagen	485	14.7	0	0
S. blockley	201	6.1	3	1.3
S. haifa	143	4.3	0	0
S. sofia	137	4.1	3	1.3
S. branderup	118	3.6	7	3.1
S. infantis	98	3.0	8	3.5
S. emek	94	2.8	10	4.4
S. montevideo	91	2.8	5	2.2
S. newport	79	2.4	7	3.1
S. paratyphi B	44	1.3	83	36.2
S. jerusalem	16	0.5	8	3.5
S. paratyphi B var. odense	0	0	35	15.3
S. reading	0	0	10	4.4
S. give	0	0	10	4.4
S. london	0	0	4	1.7
S. kentucky	0	0	5	2.2
S. enteritidis	0	0	3	1.3
others	770[a]	23.3	25[b]	10.9
	3,301	100	229	100.1

[a] Distributed among 75 serotypes not including S. typhi and S. paratyphi A.
[b] Serotypes isolated in sewage or sea once or twice only.

Laboratory studies

The effect of sunlight, temperature, depth and storage conditions on the die-away rates of coliforms and salmonellae in sea water were studied under laboratory conditions.

Analysis of shallow and deep containers

White enamel containers were filled with 10 l of sea water (depth of water 20 cm) and seeded with 1 per cent raw sewage. A control for possible biological factors influencing die-away rates was set up by using boiled sea water. One set of containers was left in a room where the

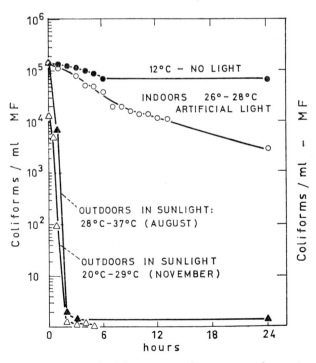

Fig 3. Die-away of coliforms in normal sea water under varying conditions of light and temperature; 1 per cent sewage added to sea water in shallow containers

[577]

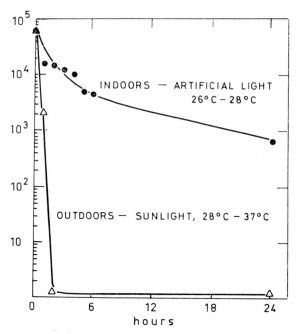

Fig 4. *Die-away of coliforms in sterilized sea water with 1 per cent sewage added, indoors and outdoors; shallow containers*

The results of coliform die-away studies in shallow containers are presented in figs 3 and 4. There was practically no die-away in 24 h in the refrigerated sample, while there was 1.5 to 2.0 log reduction in coliforms in 24 h in the samples of normal and sterilized sea water held indoors at 26° to 28°C. Samples held in direct sunlight outdoors decreased in coliform count by 4 to 5 logs within 1 to 2 h in both normal and sterilized sea water in August and also in November when the temperature outdoors rose to the same level as in August indoors.

The results of coliform die-away in the large deep opaque container exposed to daylight are shown in fig 5. It appears that, although less than in samples exposed to direct sunlight, there is a more rapid die-away of coliforms at the surface than at the 95 cm depth; salmonellae were isolated after 4 h at all depths.

The results of the study of die-away rates of various salmonellae serotypes is summarized in Table 6. These suggest that the difference in longevity is rather between individual strains and not serotypes, one of the *S. typhimurium* strains tested appeared more resistant than the *S. paratyphi B* strains.

Discussion

In a survey on beach pollution, Moore (1954) concluded that a large series of tests by a simple method is preferable to a limited investigation by more complex methods; also that the membrane filter method is one technique which might prove suitable. Comparison of our results of 1968 and 1969 confirms both statements.

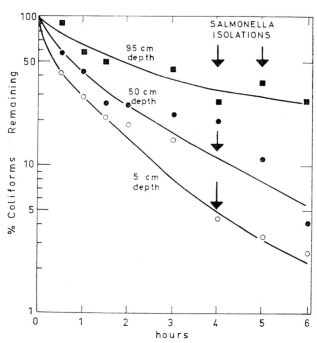

Fig 5. *Die-away of coliforms and salmonellae at varying depths in normal sea water exposed to daylight, 23°–26°C, 1 per cent sewage added to sea water in deep containers (average of two experiments)*

temperature was 26° to 28°C (August). The second set was placed outdoors in direct sunlight, where the temperature rose from 28° to 37°C during 6 h in August and from 20° to 29°C in November. A third set was kept at 12°C in the dark. Samples for coliform examinations by the MF method were drawn at half and one hour intervals after thorough mixing of the water.

One hundred l of sea water were placed in a special plastic tank of 1.20 m depth and 1,000 cm² surface area and seeded with 1 per cent sewage. The top layer of the container was exposed to very little direct sunlight due to the height of the container walls above the water surface and the low angle of the sun in November. Samples were drawn aseptically at half and one hour intervals, without disturbing the contents, from depths of 5, 50 and 95 cm. The rise in temperature was from 20.5° to 25°C. Enumeration of coliforms was done by the MF method and after 4 and 5 h they were tested for salmonellae.

Die-away rates of salmonellae

It became apparent from the field studies that the dominant salmonellae serotypes isolated from sea water in both 1968 and 1969 were *S. paratyphi B* varieties while 46 per cent of the serotypes isolated in the human population were *S. typhimurium* varieties which appeared in only 1.3 per cent of the sewage and sea water isolations. It was hypothesized that possibly *S. typhimurium* died more rapidly in sea water than *S. paratyphi B*. To test this 12 strains of salmonellae isolated from sewage and sea water, including several strains each of *S. typhimurium* and *S. paratyphi B*, were inoculated into sea water samples sterilized by membrane filtration. Parallel samples were inoculated into the same sea water enriched with 0.01 per cent peptone to simulate the nutrients in sewage contamination. The initial inoculum in each case was 10⁴ organisms per ml of sea water. The containers were held in the laboratory for three weeks at 20° ± 2°C. Samples were withdrawn daily to determine salmonellae counts.

Table 1 indicates that coliform counts obtained by the membrane filter method (MF) are somewhat lower than those of the MPN. On average only 50 per cent of the MF results fall within the 95 per cent confidence limits of the MPN—in sewage only 33 per cent and in sea water 50 to 60 per cent. Nevertheless, we decided to apply only the

TABLE 6. THE SURVIVAL OF VARIOUS SALMONELLA SEROTYPES IN FILTERED SEA WATER AT 20° ± 2°C — INITIAL INOCULUM 10⁴/ml

Salmonella serotype	Source	Number of days for decrease below 10/ml	Bacteria/ml after 20 days (0.01% peptone added)
1. S. paratyphi B	sea	10	10^3
2. S. paratyphi B	sea	8	10
3. S. paratyphi B var. odense	sea	> 15 < 21	5×10^3
4. S. typhimurium	sea	> 14 < 21	10^2
5. S. typhimurium	human	14	10^4
6. S. typhimurium	human	> 21	5×10^2
7. S. blockley	sewage	> 21	10^3
8. S. blockley	sewage	> 14 < 21	2×10^2
9. S. newport	sewage	> 14 < 21	10^3
10. S. newport	sewage	7	2.2×10^2
11. S. infantis	sewage	> 14 < 12	10^4
12. S. montevideo	sewage	8	10^3

MF method in 1969, when as many as 100 samples a day had to be examined for coliforms. The MF method was less time and space consuming and proved to have a high level of reproducibility (Slanetz et al. 1965).

In comparing the relationship between coliforms and faecal coliforms it appears that in raw sewage the ratio is 3:1 while it increased 6:1 in lightly contaminated sea water. This indicates a more rapid die-away of the faecal coliforms, which may reduce its usefulness as a reliable indicator organism in the sea. On the other hand, the coliforms: enterococci ratio was reduced from 70:1 to 19:1 under the same circumstances, confirming the previously reported value of this organism as an indicator of faecal contamination (Slanetz et al. 1965).

Altogether 229 strains of salmonellae belonging to 34 serotypes were isolated in sewage and in the sea—54 of them in 1968 (14 serotypes) and 175 in 1969 (30 serotypes). Ten of these serotypes appeared in both seasons. Salmonella typhi was not isolated since our plating medium (Brilliant Green agar) is not suitable for it and plating on Wilson/Blair medium was not done due to the extra work involved. Steiniger (1956) who studied the sea water in Barcelona and Palma de Mallorca isolated many strains of S. typhi where, together with S. java and S. bareilly, it was one of the three dominant types.

In 1968 most salmonellae were isolated from sewage or around the sewage outlet, while in 1969, 81 isolations (46 per cent) were from sea water 100 to 2,000 m from outfall. The reason for this discrepancy is that in 1968 only two points in the sea were examined. One of them, 1,500 m from outfall, yielded no salmonellae and the other, 750 m from the outfall, yielded only two isolations. In 1969 many samples of sea water were examined, and although many of them were taken at relatively short distances from the outfall, about 30 per cent of the positive samples were 650 to 2,000 m from the outfall.

Some outstanding features are illustrated in Table 5:

(1) The dominant type in sewage and the sea in both seasons was S. paratyphi B and S. paratyphi B var. odense—over 50 per cent. These findings agree with Moore et al. (1959).

(2) The most common types isolated from various sources in the population such as S. typhimurium, S. blockley and S. haifa, were only occasionally recovered from sewage and sea water.

(3) The wide variety of serotypes isolated in sewage and the sea, some of them rare in the population, others isolated from sewage only (S. london, S. westerstede, S cubana and others). Schmidt and Volker (1960) reported on a similar finding in Germany.

The number of serotypes isolated from sea water, which was taken several hours after the sewage samples, is greater than that found in the sewage, even when three sewage samples were examined. The study showed that on certain days there were one or more dominant serotypes—e.g. S. reading on 14 October 1969 and S. para B, S. london and S. give on 12 November 1969. On other days there were mostly single or double isolations of each serotype.

(1) There must be a rich source of S. paratyphi B in the sewage. These may be from undetected human carriers or from rats and other rodents which infect the sewage with a variety of phage types of these salmonellae.

(2) It might appear that the survival rate of S. paratyphi B in sewage and sea water is quite high. Moore (1954) found that they can be recovered from sea water contaminated in the laboratory (10⁴/ml) even after 2 months. Buttiaux and Zeurs (1954) state, however, that S. paratyphi B and S. enteritidis decrease to about 50 per cent in 44 h and that S. typhimurium is little affected by 24 h in sea water. Our results show no consistent or clear cut differential die-away rates for S. typhimurium as compared to S. paratyphi B strains, and differential survival rates in sea water do not appear to provide an explanation for this phenomenon.

(3) The wide variety of serotypes isolated from sea water but not from sewage indicate either that they originated from sewage of previous days or that they were not detected in the sewage due to overcrowding by S. paratyphi B or other microorganisms. Some of the rare types might have come from sources such as seagulls and migrating birds.

Infection of seagulls and other birds with salmonellae and their relation to contaminated water is reported by Steiniger (1953, 1954, 1956). Reitler (1955) described an outbreak of salmonellosis in migratory birds (starlings) in the Jordan valley caused by S. hessarek, a type endemic in Persia. These starlings also harboured S. montevideo and S. imeleagsidis encountered in Israel.

Figures 3 and 4 demonstrate the striking difference in decrease of coliforms indoors and in sunlight. The rapid decrease of coliforms outdoors appears to be due primarily to sunlight and not to the temperature, since in the outdoor experiment of November (fig 3) the coliform die-away was as rapid as in August, although the temperature was lower. This is also brought out by difference in various depths in tall opaque containers (fig 5). The very rapid die-away of coliforms in sterile sea water exposed to sunlight indicates that biological factors were not critical in this case. The effect of direct sunlight appears to be greater than indirect light. Gaarder and Sparck (cited by Orlob, 1956) considered the bactericidal effect of sunlight to contribute most significantly to destruction of bacteria in sea water in Norway and our experiments confirm this opinion. However, it is difficult to predict conditions at sea from laboratory experiments. ZoBell (1936) who immersed semipermeable tubes with bacteria in sea water found a 69 per cent die-away in 1 h and 97 per cent in 2 h. Our results in sunlight of over 99 per cent die-away in 1–1.5 h agree with these results. The findings of Weston and Edwards (1930) of a 90 per cent decrease in coliforms in 4 h and a 40 to 56 per cent decrease of total bacteria, agree with our results in the tall container, where a reduction of 97 per cent was reached in 6 h in the surface sample. Carpenter *et al.* (1938) state that 80 per cent of all sewage organisms indigenous to fresh sewage die within half an hour after contact with sea water. Nusbaum and Garver (1955) also reported high mortality rates of coliforms in laboratory experiments with raw San Diego Bay water, up to 90 per cent mortality in 1.4 days, but they are much lower than our values. However, in experiments with dialysis bags suspended in the bay they found large initial increase in the first 2 days, dropping to 13 per cent after 4 days.

Greenberg (1956), in a review of the literature on survival of enteric organisms in sea water, concludes that they can create a health hazard in estuaries, bays and beaches. Berger *et al.* (1963) in their review of the coliform standard for recreational waters cite a number of cases where bathing in sewage polluted water led to suspected cases of enteric disease, including ten cases of typhoid fever in Perth, Australia in 1962 which were, according to local health authorities, directly associated with bathing in sewage contaminated coastal waters. A study by the U.S. Public Health Service (Stevenson, 1953) provided some evidence that when MPN's were in the range of 2,400 coliforms per 100 ml, there was a significant increased incidence of illness in swimmers, especially gastroenteritis, and possibly upper respiratory.

Although in England four cases of paratyphoid due to sea bathing were recorded and 569 strains of salmonellae were isolated from over 40 polluted bathing beaches in 5 years, the Committee on Bathing Beach Contamination (Moore *et al.* 1959) conclude that:

(1) Bathing in sewage-polluted sea water carries only a negligible risk to health, even on beaches that are aesthetically very unsatisfactory.

(2) Isolation of pathogenic organisms is more important as evidence of an existing hazard in population from which sewage is derived than evidence of further risk of infection in bathers.

It is certainly not clear, in view of these contradictory opinions and the sparse epidemiological evidence, to what extent human disease does result from bathing in sewage-contaminated water. Our work strongly indicates that pathogens such as salmonellae, can survive for long periods in the sea, and due to the rapid die-away of coliforms salmonellae can be present even in coastal

waters of relatively low coliform densities ($< 2,400/100$ ml). Shuval (1970) also reports the isolation of enteroviruses in the low coliform count waters. These findings cast new doubts on the validity of the universally accepted coliform standard for the safety of bathing waters. Bathers who spend many hours in such coastal water may swallow some of it inadvertently but the risk of infection can only be surmised.

References

BERGER, B B *et al.* Senn, Coliform standards for recreational
1963   waters. *J. Sanit. Engng Div. Am. Soc. civ. Engrs*, 89:57–94.
BUTTIAUX, R and ZEURS, T Survival of salmonella in seawater.
1954   *J. Am. Wat. Wks. Ass.*, 16:82.
CARPENTER, L V, SETTER, L R and WEINBERG, M Chloramine
1938   treatment of sea water. *Am. J. Publ. Hlth*, 28:929.
GAMESON, A L H and SAXON, J R Field studies on effect of day-
1967   light on mortality of coliform bacteria. *Wat. Res.*, 1:279–95.
GILAT, Ch *et al.* The use of radioactive tracers in the study of the
1970   dispersion and inactivation of bacteria discharged into
        coastal waters. *In* Developments in Water Quality Research,
        Ann Arbor, Humphrey Science Publishers, 269 p.
GREENBERG, A E Survival of enteric organisms in sea water. *Publ.*
1956   *Hlth Rep., Wash.*, (71):77.
MOORE, B A survey of beach pollution at a seaside resort. *J. Hyg.*
1954   (*Lond.*), 52:71–86.
MOORE, B *et al.* Sewage contamination of coastal bathing waters
1959   in England and Wales. Bacteriological and epidemilogical
        study by the committee on bathing beach contamination of
        the Public Health Laboratory Service. *J. Hyg.* (*Lond.*), 57:
        435–72.
NUSBAUM, J and GARVER, R M Survival of coliform organisms in
1955   Pacific Ocean coastal waters. *Sewage ind. wastes*, 27:(12).
ORLOB, G T Viability of sewage bacteria in sea water. *Sewage ind.*
1956   *wastes*, 28:1147–67.
REITLER, R Salmonellosis in migratory birds. *Acta med. orient.*,
1955   14:52.
SCHMIDT, B and VOLKER, L Der Nachweis von Salmonellen in
1960   Abwasser als möglicher Massstab für die Seuchenlage einer
        Bevölkerung. *Zentbl. Bakt. ParasitKde*, 178:459–483.
SHUVAL, H Detection and control of enteroviruses in the water
1970   environment. *In* Developments in water quality research,
        Ann Arbor, Humphrey Science Publishers, 47 p.
SHUVAL, H, COHEN, N and YOSPHE-PURER, Y The dispersion of
1968   bacterial pollution along the Tel Aviv shore. *Revue int.*
        Oceanogr. Med., 9:107–21.
SLANETZ, L W and BARTLEY, C H Survival of fecal streptococci in
1965   sea water. *Hlth Lab. Sci.*, 2:145–8.
SLANETZ, L W, BARTLEY, C H and METCALF, T G Correlation of
1965   coliform and fecal streptococcal indices with the presence
        of salmonellae and enteri viruses in sea water and shellfish.
        Proc. int. Conf. Wat. Pollut. Res., 2(3):27–41. Also known
        as *Adv. Wat. Pollut. Res.*, 2(3):27–41.
Standard methods for the examination of water and waste-water.
1965   American Public Health Association. New York, 12th
        Edition.
STEVENSON, A H Studies of bathing water quality and health.
1953   *J. Am. Publ. Hlth Ass.*, 43:529.
STEINIGER, F Der Zyklus Vogel-Verunreinigtes Wasser in der
1953   Freilandbiologie der Salmonellen. *Zentbl. Bakt. Parasit-*
        Kde, 160:80.
STEINIGER, F Aus der Freilandbiologie der Salmonellen. *Dt. med.*
1954   *Wschr.*, 29/30:1118–20.
STEINIGER, F Zur Freilandbiologie der Salmonellen im Bereich des
1956   Westlichen Mittelmeers. *Zentbl. Bakt. ParasitKde*, 166:245.
WESTON, A D and EDWARDS, G P Pollution of Boston harbor.
1930   *Proc. Am. Soc. civ. Engnrs*, 65:383.
ZOBELL, C E Bactericidal action of sea water. *Proc. Soc. exp. biol.*
1936   *Med.*, 34:113.

Phénomènes Lytiques dans l'Auto-épuration des Eaux de Mer†

*A. Paoletti**

Lysis in the self-purification of sea water

Very recently two new groups of organisms, which are perhaps important in the self-purification of the sea, have been reported in external environments, namely, the *Bdellovibrios bacteriovorus* and lytic bacteria:

1. The *Bdellovibrios* (discovered in 1962) have been isolated in sea waters, especially along the coasts, in fresh water and sewage and on the soil; in the laboratory they are capable of lysing a large number of bacteria and of purifying polluted water; the explanation for this action very probably also lies in the external environment.
2. The lytic bacteria strains, which have been known for a long time (since 1907), were only isolated by us and other research workers almost simultaneously in 1969. They have been isolated in such great numbers in the marine environment and have shown themselves so active in the laboratory that it is to them that the bactericidal power of sea water—the cause of which is still debated—can be attributed.

The technique of isolation and the quantitative definition of these organisms is very simple and is being given in detail; their frequency, particularly in the polluted waters of the Bay of Naples, is adequately documented.

Fenómenos líticos en la autodepuración de las aguas del mar

Muy recientemente han sido señalados en el medio exterior dos nuevos grupos de organismos que tal vez desempeñen un papel importante en la autodepuración del mar: los *Bdellovibrios bacteriovorus* y las bacterias líticas.

1. Los *Bdellovibrios* (descubiertos en 1962) han sido aislados en las aguas del mar (sobre todo a lo largo de las costas), en aguas dulces y de alcantarillado, y en tierra. Son capaces de lisar en laboratorio gran cantidad de bacterias y de purificar las aguas contaminadas y es muy posible que realicen también esa acción en el medio exterior.
2. Las cepas de bacterias líticas, conocidas desde hace mucho tiempo (1907), fueron aisladas en gran número en el medio marino sólo en 1969 por nuestro grupo y casi simultáneamente por otros autores, y se han demostrado tan activas en laboratorio que se podría atribuirles el poder bactericida del agua del mar, cuya causa se discute aún.

La técnica de aislamiento, así como la definición cuantitativa de estos organismos, es muy sencilla y se expone en detalle; se documenta también abundantemente su frecuencia numérica, sobre todo en las aguas contaminadas de la Bahía de Nápoles.

LES facteurs biologiques d'auto-épuration des eaux de mer, jusqu'à présent pris en considération, sont les suivants (Paoletti, 1970):

(a) Pouvoir bactéricide;
(b) Substances antibiotiques;
(c) Facteurs lytiques: (a) Bactériophages;
                      (b) *Bdellovibrio bacteriovorus;*
                      (c) Souches bactériolytiques;
(d) Phénomènes prédateurs.

L'existence d'un *pouvoir bactéricide* de l'eau de mer est indiscutable et bien témoignée par de nombreuses recherches.

Selon certains chercheurs les *substances antibiotiques* isolées des bactéries, levures, mollusques, éponges, coraux, algues macro- et microscopiques sont les responsables du pouvoir bactéricide des eaux, tandis que d'autres affirment que l'isolement de substances anti-bactériennes et antivirales obtenu à partir de certains organismes marins ne justifie pas l'auto-épuration. La question est encore discutée et de nouvelles études sont

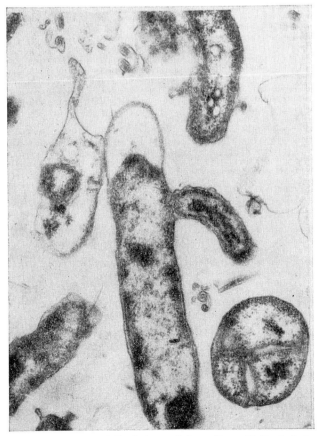

Fig. 1. Fixation de Bdellovibrio *sur* Salmonella typhi *Vi. Inclusion Epn., coupe et double coloration, microscope électronique 60.000 X. (amabilité de A. Guèlin et P. Lépine, Institut Pasteur, Paris)*

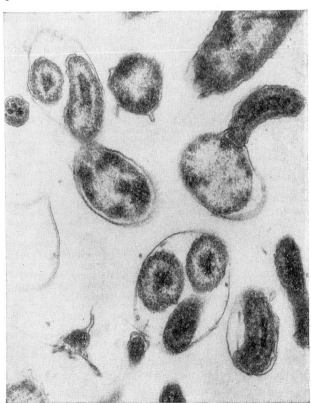

Fig. 2. Fixation et multiplication de Bdellovibrio *sur* Salmonella typhi *Vi. Inclusion Epn., coupe et double coloration, microscope électronique 60.000 X. (amabilité de A. Guèlin et P. Lépine, Institut Pasteur, Paris)*

† Recherches subsidiées par le Conseil National de Recherches, Institut des Recherches sur les Eaux, Italie.
* Facoltà delle Scienze della Università di Napoli, Naples, Italie.

nécessaires à ce sujet. En tous cas il semble utile de ne plus employer comme synonymes les mots pouvoir bactéricide et pouvoir antibiotique de l'eau de mer pour ne pas créer des incompréhensions et des désaccords.

Les bactériophages comme facteurs lytiques, après 40 ans d'études, ont perdu leur importance, au moins dans l'auto-épuration de la mer, à cause de constatations bien précises qu'il n'est pas possible de discuter ici.

Les phénomènes prédateurs ont une importance indiscutable et nous avons présenté à ce Congrès des expérimentations indiscutables à cet égard.

Les *Bdellovibrio bacteriovorus* et les souches bactériolytiques sont le sujet de ce rapport.

Bdellovibrio bacteriovorus

Parasites intracellulaires des bactéries Gram-négatives et, plus rarement, des bactéries Gram-positives, ces organismes pénètrent dans la cellule hôte et s'y multiplient jusqu'à la détruire (fig 1, 2, 3).

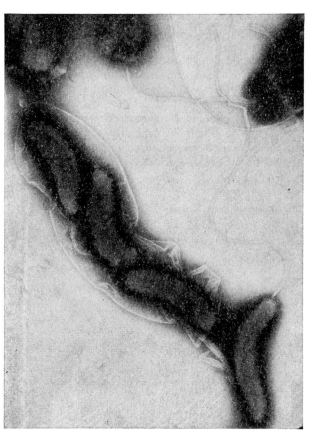

Fig. 3. *Multiplication de* Bdellovibrio *dans* Salmonella typhi *Vi. Coloration négative, microscope électronique 60.000 X. (amabilité de A. Guèlin et P. Lépine, Institut Pasteur, Paris)*

Ils sont facilement isolables dans les eaux polluées après double filtration par millipore (1,2 micron et 0,45 micron) et après culture en gélose molle ou en milieu liquide, avec substrat de bactéries sensibles. Le milieu de culture doit être très pauvre en substances nutritives et très riche en bactéries sensibles pour donner un phénomène visible sous forme de taches de lyse ou sous forme d'éclaircissement du substrat. La multiplication de *Bdellovibrio* et son isolement est possible aussi dans un milieu dépourvu de toute substance organique autre que celle des cellules bactériennes. Le phénomène

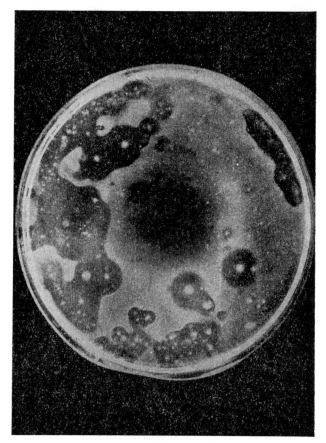

Fig. 4. *Taches de lyse par bactéries marines sur* Pseudomonas aeruginosa

lytique s'avère soit sur les bactéries vivantes soit sur celles tuées et à une température très variable (de 37 à 4°C); la lyse est progressive parce que les taches augmentent d'extension avec le temps et apparaissent en retard (5–30 jours), sauf dans le cas des souches adaptées par repiquage qui donnent une lyse très rapide.

Pendant les dernières deux années nous avons isolé les *Bdellovibrio* non seulement dans tous les échantillons d'eau de mer de la baie de Naples mais aussi dans les fleuves, les lacs, le sol et dans les eaux d'égout. Dans la mer ils sont présents tout le long des côtes en quantité variable. Ils semblent être d'autant plus nombreux que la quantité des bactéries dans les échantillons étudiés est élevée. Nous ne les avons pas encore cherchés dans la mer au large des côtes. Leur isolement est plus facile si l'on emploie *Salmonella typhi* Vi. et *Pseudomonas aeruginosa* comme bactéries-test. Ils n'ont pas une spécificité d'action et, au moins pour les souches que nous avons testées, résistent quelques mois dans les différentes eaux et à des températures variables (4–37°C). La fig 5 donne un schéma de la technique employée.

En dehors de l'intérêt que ces nouveaux microorganismes vont prendre pour la microbiologie générale, nous pensons qu'il est nécessaire de bien les étudier pour établir leur véritable rôle dans l'auto-épuration des eaux polluées.

Les souches bactériolytiques

Il est connu depuis longtemps que des bactéries et des champignons microscopiques sont capables de lyser d'autres micro-organismes (hétérolyse) et nos connais-

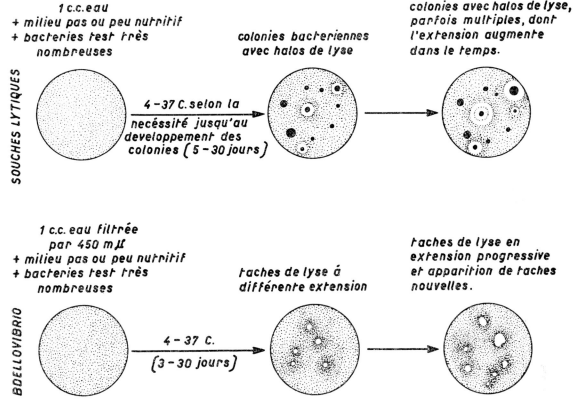

Fig. 5. Schéma de recherche des souches lytiques et de Bdellovibrio *dans un même échantillon d'eau*

sances sur le sujet ont été mises au point très récemment (Paoletti, 1970).

Ce genre d'action lytique est lié aux enzymes microbiens qui agissent sur les constituants pariétaux de la cellule bactérienne, avec des mécanismes suffisamment connus (Ghuysen, Tipper et Strominger, 1966; Tinelli, 1968).

La possibilité qu'une telle action lytique soit, au moins en partie, responsable des phénomènes d'autoépuration des eaux de mer a été considérée séparément et récemment par Mitchell, Yankofsky et Jannash (1967), par Mitchell et Morris (1969) et par nous-mêmes.

Nos études de ces dernières années mènent aux conclusions suivantes:

(1) Il est possible d'isoler dans l'eau de mer, avec des techniques très simples, des bactéries et champignons microscopiques qui manifestent une activité lytique remarquable contre d'autres micro-organismes tests.

(2) Le nombre de ces souches est extraordinairement élevé et le plus souvent proportionnel à la flore microbienne de l'échantillon d'eau. En général nous pouvons dire qu'avec la technique employée, 2 à 10 pour cent environ des bactéries de la flore totale sont douées d'action hétérolytique.

(3) Cette action apparaît aussi bien sur les bactéries-test vivantes que mortes, Gram-positives ou Gram-négatives.

(4) L'activité lytique est évidente si l'on emploie des milieux faiblement nutritifs ou, mieux encore, si dans le milieu de culture il y a seulement des cellules bactériennes sensibles comme substrat organique ajouté.

(5) Les halos de lyse apparaissent après 2–3 jours d'incubation aux températures adaptées pour la multiplication de la bactérie lytique (10–37°C) et aussi plus tard (5–15 jours) si les conditions entraînent un retard dans la multiplication de la bactérie isolée.

(6) Avec le temps, les halos de lyse s'étendent progressivement au fur et à mesure que l'enzyme se diffuse dans le substrat, jusqu'à intéresser la boîte de Pétri toute entière (fig 4).

(7) Il est possible d'observer fréquemment 2 ou 3 halos concentriques de lyse, dont l'interprétation a été discutée (Swertz, 1948–1949) (fig 5).

(8) Le phénomène lytique est absolument indépendant de l'activité antibiotique; néanmoins certaines souches bactériennes sont douées à la fois d'une activité lytique (halos de lyse) et antibiotique (halos d'inhibition), tandis que d'autres possèdent seulement une de ces activités.

(9) Comme les *Bdellovibrio*, les microbes hétérolytiques sont répandus dans tout le milieu extérieur (eaux de mer, des lacs, des fleuves; eaux d'égout; sol; air); il s'agit donc d'une activité qui est propre à un grand nombre de micro-organismes, d'une propriété très commune ou, peut-être, universelle.

(10) L'activité hétérolytique n'est pas spécifique parce que la même souche bactérienne peut agir sur plusieurs bactéries d'espèces différentes.

Les observations faites ici nous amènent à des conclusions vraiment suggestives et captivantes d'ordre général sur la survie des bactéries, c'est-à-dire que l'hétérolyse (en tant que phénomène très répandu) pourrait être une des causes universelles de la mort des micro-organismes dans le milieu extérieur; ce phénomène se manifeste dans les milieux où la diffusion des enzymes responsables est possible (liquides ou solidifiés avec gélose); il se réalise par action des micro-organismes en multiplication; il est plus actif dans les milieux pauvres en substances nutritives.

[583]

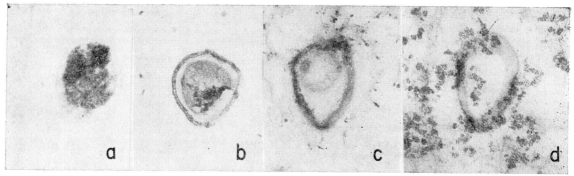

Fig. 6. De gauche à droite:—témoin et lyse progressive de Staphilococcus aureus *par action d'un enzyme microbien. L'altération progressive de la paroi bactérienne et la perte de la substance nucléaire sont évidentes. Inclusion Epn du halo lytique en gélose, coupe et double coloration, microscope électronique du "Laboratorio di Microscopia Elettronica dell'Istituto Anatomico Veterinario della Università di Napoli", 69.000 X.*

Il n'y a aucun doute que ces conditions se rencontrent dans les eaux de surface (surtout de la mer) où seule la multiplication des bactéries marines est possible; les bactéries terrestres et les entérobactéries surtout sont destinées à en subir les conséquences et à disparaître. Si les études ultérieures viennent confirmer qu'une telle activité hétérolytique des bactéries se manifeste dans la mer avec l'intensité indiscutable que nous avons observée en laboratoire, nous aurons un nouveau témoignage de l'origine biologique du pouvoir bactéricide de l'eau de mer.

Très récemment, au microscope électronique, nous avons observé une évidente et progressive altération de la paroi bactérienne et une perte de substance nucléaire (fig 6: a, b, c, d). En même temps nous poursuivons nos études pour établir le mécanisme d'action des enzymes bactériolytiques observés, pour en déterminer la nature chimique et reconnaître leur point d'attaque sur la paroi bactérienne.

Il faut souligner que les phénomènes lytiques étudiés sont répandus dans tous les milieux multimicrobiens. Pourtant nous ne sommes absolument pas autorisés à affirmer que la mer seule est capable d'auto-épuration, parce qu'il s'agit d'une propriété universelle très répandue dans les micro-organismes.

Bibliographie

GHUYSEN, J M, TIPPER, D J et STROMINGER, J L Enzymes that
1966   degrade bacterial cell walls. *In* Methods in enzymology, edited by Colowick and Kaplan, Vol. 8. Complex carbohydrates. New York, Academic Press.
MITCHELL, R et MORRIS, J C The fate of intestinal bacteria in the
1969   sea. *Adv. Wat. Pollut. Res.*, 4.
MITCHELL, R, YANKOFSKY, S et JANNASH, H W Lysis of *E. coli* by
1967   marine microorganisms. *Nature, Lond.*, 215:891.
PAOLETTI, A Facteurs biologiques d'auto-épuration des eaux de
1970   mer. IVème colloque international d'océanographie médicale. *Revue int. Océanogr. méd.*, 18–19: 33–68.
SWERTZ, L Etude de l'activité bactériolytique des actinomycètes et
1948–9 particulièrement des phénomènes de halo. *Revue belge Path. méd.*, 19(Suppl. 3–4):225.
TINELLI, R Le glycopeptide des parois bactériennes. *Bull. Inst.*
1968   *Pasteur, Paris*, (12):2508–38.

Toxicity of Marine Organisms caused by Pollutants*

B. W. Halstead†

La toxicité des organismes marins due aux polluants

La toxicité chez les organismes marins, qu'elle soit due à des causes naturelles ou à l'action d'agents polluants produits par l'homme, devient préoccupante. Environ 50 pour cent des disponibilités mondiales en poisson proviennent des régions de remontée des eaux côtières qui représentent environ 0,01 pour cent de la surface des océans; le reste se trouve en grande partie dans d'autres eaux littorales et dans quelques zones de haute mer où la fertilité est relativement élevée. Sur la base de la répartition géographique de la biomasse marine totale, il est patent que les zones les plus productives des océans sont concentrées surtout dans la région néritique. Cette région est particulièrement sensible à l'action destructive des polluants produits par l'homme, car elle est le lieu de l'activité la plus intense.

Une grande partie de la pollution est d'origine toxicologique: il s'agit des effets toxiques d'une substance chimique sur les systèmes cellulaires des organismes. Parmi les plus dangereux pour le milieu marin figurent les effets à long terme de faible degré.

Des preuves toujours plus nombreuses révèlent que les polluants peuvent déclencher des cycles de biotoxicité marine "d'incidence naturelle" dans les régions insulaires tropicales. Lorsque les composantes chimiques adéquates sont réunies dans le milieu, certains types de biotoxines sont produits dans les organismes marins.

Toxicidad de organismos marinos causada por contaminantes

El problema de la toxicidad en los organismos marinos debido a causas naturales o a contaminantes artificiales llega a un punto crítico. El 56 por ciento, aproximadamente, de todos los suministros de pescado del mundo se obtienen de zonas costeras con corrientes de surgencia que comprenden alrededor del 0,01 por ciento de la superficie del océano; la mitad restante se produce sobre todo en otras aguas costeras y en las pocas zonas próximas a la costa de alta fertilidad relativa. Basándose en la distribución geográfica de toda la biomasa marina, resulta que las partes más productivas de los océanos mundiales se concentran dentro de la región nerítica más que de la oceánica. La región nerítica es más susceptible a las fuerzas destructoras de los contaminantes artificiales porque es la de mayor actividad humana.

Una parte importante del problema de la contaminación es de carácter toxicológico y concierne a los efectos tóxicos de una sustancia química en el sistema celular del organismo. Algunos de los efectos más graves del medio ambiente marino son poco intensos, pero a largo plazo.

Una cantidad creciente de pruebas indica que los contaminantes pueden obstaculizar los ciclos de la biotoxicidad marina que surgen de un modo natural en las regiones insulares tropicales. Se producen biotoxinas cuando hay presente una mezcla adecuada de constitutivos químicos en el medio ambiente de ciertos tipos de organismos marinos.

---

* Prepared by Dr. Halstead in his capacity as a Consultant to the World Health Organization.
† International Biotoxicological Center, World Life Research Institute, Colton, California 92324, U.S.A.

La maladie de Minamata est devenue un exemple classique de l'action des polluants industriels sur la nourriture des organismes marins, aboutissant à des effets létaux chez l'homme. L'auteur examine la production d'un sel méthyl-mercurique toxique.

Les hydrocarbures bruts (huiles minérales) et raffinés sont toxiques pour tous les organismes marins et pour l'homme. L'empoisonnement peut être immédiat ou lent et durable. Les hydrocarbures peuvent être absorbés par les tissus des mollusques, ce qui rend ces derniers impropres à la commercialisation. Les hydrocarbures toxiques produisent des phénomènes d'anesthésie et de narcose, provoquent des lésions cellulaires et entraînent la mort chez divers animaux inférieurs. Les pétroles bruts contiennent des hydrocarbures cancérigènes. On ignore dans quelle mesure ces hydrocarbures toxiques s'associent aux protéines et passent dans l'alimentation de l'homme.

Les phénols et les crésols qui se trouvent dans les effluents industriels peuvent endommager la chair des poissons de consommation et détruire leur valeur commerciale. Les métaux et les colorants, même en très petites quantités, peuvent endommager ou décolorer la chair des organismes marins et leur faire perdre leur valeur commerciale. On ignore quels sont les effets de ces détériorations sur l'homme.

Des milliers de flétans, de maigres, de brèmes de mer, de soles, de limandes et d'autres poissons côtiers capturés sur le littoral de la Californie méridionale et de la baie de New-York au voisinage des décharges d'égouts ont présenté une fréquence anormalement élevée de tumeurs cancéreuses, d'ulcères cutanés, de déformations, d'émaciations et de transformations génétiques. On pense que ces troubles pathologiques pourraient être dus aux effets toxiques des polluants. On ignore quels sont exactement les agents étiologiques. Les répercussions de cette situation pour l'homme du point de vue de la santé publique causent des préoccupations croissantes.

L'auteur discute rapidement l'urgente nécessité de mettre sur pied les bases d'un système de communication et de récupération des renseignements fondamentaux dans les domaines de la biologie, de la chimie et de la pharmacotoxicologie.

La enfermedad "Minamata" se ha convertido en un ejemplo clásico del modo en que los contaminantes industriales pueden inmiscuirse en la cadena alimentaria de los organismos marinos y, por último, producir efectos letales en las personas. Se examina el mecanismo de la producción de una sal tóxica de metilo de mercurio.

El petróleo bruto y refinado son tóxicos para todos los organismos marinos y para el hombre. El envenenamiento puede ser inmediato o lento y de larga duración. El petróleo puede penetrar en los tejidos de los mariscos e inutilizarlos para su venta en el mercado. Los hidrocarburos tóxicos producen anestesia, narcosis, daños a las células y la muerte en varios animales inferiores. El petróleo bruto contiene hidrocarburos carcinógenos. No se conoce la medida en que estos hidrocarburos tóxicos pueden pasar a formar parte de las proteínas y penetrar en los alimentos del hombre.

Los fenoles y los cresoles en los desechos industriales pueden corromper la carne de los peces destinados a la alimentación y destruir su valor comercial. Los metales y los colorantes presentes, incluso en cantidades muy pequeñas pueden modificar el sabor y el color de la carne de los organismos marinos y destruir su valor comercial. No se conocen sus efectos en las personas.

Miles de hipoglosos, roncadores, meros, lenguados, limandas y otros peces costeros capturados a lo largo de la costa del sur de California y de la bahía de Nueva York, cerca de las salidas de las alcantarillas tenían una incidencia muy alarmante de excrecencias cancerosas, úlceras de la piel, deformaciones, emaciación y cambios genéticos. Se cree que estos disturbios patológicos se deben a los efectos tóxicos de los contaminantes. Se desconocen los agentes precisos que los causan. Son de creciente interés las posibles consecuencias higiénicas para el hombre.

Se examina brevemente la urgente necesidad que existe de crear un sistema base de datos fundamentales biológicos, químicos y farmacotoxicológicos, de localización de literatura, y de sistemas de comunicaciones.

THE problem of toxicity in marine organisms from natural causes or man-produced pollutants is coming into critical focus. The chain reaction of over-population, build-up of land and marine living resources, and the resulting impact on the health and economic welfare of mankind are matters of universal concern that require international action now. There is a lack of reliable analytical data as to precisely what toxic, chemical and biological interactions are taking place today. This paper does not review dose levels of chemical intoxicants but rather discusses how these chemicals affect the quality of marine products and possibly thereby human health. A useful critical review of the literature on the toxicity of industrial wastes has been published by Doudoroff and his associates (1950, 1951, 1953).

Toxicological effects of marine pollutants

Aside from mechanical and amenities aspects most pollution problems, in final analysis, are toxicological. In most instances one is concerned with the toxic effects of a chemical substance on a specific cellular system of an organism. These effects may be acute, dramatic, short-term, and readily discernible, but correction may be difficult, costly, and long term. Unfortunately some of the most serious problems facing us today are associated with chronic intoxications. The pathology may pass undetected until it is too late to correct or control the causative pollutants. This may result in destruction and build-up of toxic constituents in the environment long before detection. Increasing evidence shows this is taking place in many areas.

Gross pathological conditions in marine organisms are seldom observed in nature. Once the organism becomes ineffective in its activities it is usually consumed by predators or dies. Field investigations in California during 1956–1970 revealed that thousands of California halibut (*Paralichthys californicus*), caught in polluted areas near Long Beach, were dull-coloured, listless and had soft bodies (Young, 1964). Halibut caught in non-polluted areas were vigorous, brightly coloured, and firm fleshed. Spotfin croaker (*Roncador stearnsi*) taken in polluted areas suffered from exophthalmia. White seabass (*Cynoscion nobilis*) had exophthalmia and large skin ulcerations. Dover sole (*Microstomus pacificus*) from polluted areas had cancerous lesions about the mouth or body. Benign tumours or papillomas were found in white croakers (*Genyonemus lineatus*), tongue soles (*Symphurus atricauda*), cusk eels (*Otophidium scrippsae*), and Pacific sanddabs (*Citharichthys sordidus*). Similar cancerous growths were not observed in marine fishes taken from non-polluted areas during the same period. Lesions similar to those found on the white seabass were produced experimentally on killifish (*Fundulus parvipinnis*) by exposing them for 12 days to a 3–7 per cent solution of polluted water taken from the Dominguez Channel effluent.

Wellings, Chuinard and Bens (1965) discovered surface tumours in four species of flatfish (flathead sole *Hippoglossoides elassodon*, rex sole *Glyptocephalus zachirus*, English sole *Parophrys vetulus*, sand sole *Psettichthys melanostictus*) taken during 1962 to 1964 near Orcas Island, Puget Sound. Out of a total of 5,250 flathead sole 6.4 per cent had observable tumours, in addition 3 rex sole, 2 English sole, and 2 sand sole also had tumours. The authors suggested that carcinogenic substances from petroleum or other industrial wastes in the mud-sand substrate may have caused the tumours. Nigrelli, Ketchen and Ruggieri (1965) found skin tumours in sand sole, rock sole (*Lepidopsetta bilineata*), but not in English sole from northern Hecate Strait, British Columbia. Cooper and Keller (1969) examined 15,739 English sole from San Francisco Bay and found 12 per cent of these fish had as many as 33 cancerous tumours per fish, the highest incidence being found in fish taken in the north of the bay where petrochemical industrial waste concentrations were highest.

Cauliflower disease, a papillomatous growth on the skin of eels (*Anguilla vulgaris*) appears to have spread as an epidemic during 1944 to 1956 from Baltic waters to north-western Europe. The disease reached such proportions in 1946 that Danish fishermen complained it was disastrous because the eels could not be sold (Deys, 1969). Since 1956 the disease has spread to Holland and Belgium. Russell and Kotin (1957) found that 10 out of 353 white croakers taken about 2 mi from the outfall of a sewage-treatment plant in Santa Monica Bay, contained papillomas. No similar lesions were noted in 1,116 white croakers from non-polluted waters 50 mi away. In 1970 a trawl taken near the Santa Ana River outfall yielded 52 white croakers of which 27 had caudal fin abnormalities and 18 had lip cancers. Over the past three years the incidence of lip cancers has been between 15–18 per cent and the incidence is increasing. Large numbers of dover sole had large ulcerations on their undersides (Schnitger, 1970).

Leukaemia in oysters and other shellfish taken from polluted waters in Puget Sound, was reported by Smith (1970). In his document on pollution in the New York Bight, a number of studies indicated a high incidence of disease, cancer malformations, emaciation, and genetic changes among fish and shellfish found in the vicinity of disposal areas and sewage outfalls. There is also concern about the increasing numbers of outbreaks of human poisonings and illnesses associated with eating contaminated fish and shellfish, increasing outbreaks of red tide organisms, sea nettles, and other noxious plants and animals.

Naturally occurring marine biotoxins

Many "naturally occurring" marine biotoxins are presently recognized. These involve a broad spectrum of marine organisms ranging from protozoans to polar bears. One of these biotoxins iciguatoxin, is intimately involved with the food webs of a great variety of tropical and subtropical marine organisms (plants, invertebrates, and many kinds of shore fishes). The biology, pharmacology, toxicology and chemistry of ciguatera fish have been discussed elsewhere (Halstead and Courville, 1967).

The sinking of the MS *Southbank* in December 1964, reputedly caused outbreaks of ciguatera fish poisoning in the Line Islands, Central Pacific Ocean. The freighter was heavily laden with metals and other substances which may have helped trigger off the outbreak. According to all available documents poisonous fishes were never known to occur in the Line Islands even though they were found elsewhere in the tropical Pacific. Many common reef fishes previously eaten daily by both natives and salvage crews with impunity were found in August 1965 to be ciguatoxic and people who ate fish caught near the ship became violently ill (Halstead and Courville, 1967). Similar occurrences have been reported elsewhere in the past 100 years.

Marine organisms concentrate chemical substances from sea water or other organisms. Shellfish feed on plankton that feed on microorganisms and extract chemical nutrients from sea water. Shellfish that feed on

sufficient quantities of toxic dinoflagellates (*Gonyaulax*, etc.) become poisonous. A similar process apparently takes place in ciguatoxic fishes, clupeotoxic fishes, poisonous crabs, etc. The chain or web does not always begin with dinoflagellates, but may include any of a number of marine plants, invertebrates, or even herbivorous fishes. In the case of ciguatera, the sequence is believed to start with benthic marine algae which are ingested by herbivorous fishes, which in turn are eaten by carnivorous fishes. Each trophic level ingests the poisons that occur in the organisms they eat. When man eats a large toxic carnivorous reef fish he receives the accumulated poisons of all the lower trophic levels. With the advent of fish protein concentrate (FPC), man has added one more step by concentrating biological concentrators. If the toxins present are water- or alcohol-soluble, they are probably removed by the FPC process.* If the poisons are protein-bound, such as organic mercurial compounds, toxic hydrocarbons, and some of the heavy metals, then these poisons may be passed on to man in even greater concentration than normally received from fresh fish.

A recent survey of fish poisoning in the West Indies under FAO auspices (Halstead, 1970) revealed that ichtoyotoxism is an important public health problem in the Virgin–St. Kitts–St. Eustatius–Redondo island complex. A long list of commercially valuable fish species, involving such families as barracuda (Sphyraenidae), crevally or jacks (Carangidae), grouper (Serranidae), snapper (Lutjanidae), mackerels (Scombridae), and herrings (Clupeidae) had been repeatedly incriminated in human intoxications. Fish poisoning occurs in several forms in the West Indies: ciguatera, clupeoid, scombroid and puffer poisoning. In current morbidity and mortality rates, ciguatera is predominant and the greatest public health concern. Poisoning from the ingestion of violently toxic herring-like fishes does cause deaths but is relatively rare.

Fish poisoning is a major problem with various fisheries and public health implications in many tropical insular areas. It encourages the importation of canned fish into West Indian island areas where they already have an abundance of fresh fish. In some instances this adversely affects their balance of payments and so involves economic factors.

Heavy metals and other inorganics

This category includes the transition metals such as antimony, arsenic, cadmium, chromium, cobalt, copper, lead, mercury, nickel and zinc. All are relatively toxic and are readily concentrated by marine animals. These metals may enter the marine environment through industrial effluent and mining operations, and in some instances by aerial pollution, from burning fossil fuels. Important factors are their solubility and chemical or physical form in the effluent or sea water (Portmann, 1970). In most places no environmental monitoring operations for toxic heavy metals are presently in force.

* There is no published report available documenting that these poisons are removed by any FPC processes. Apparently no tests have been conducted on the toxicity of fish or other marine organisms used in the FPC process.

Antimony

Few of the salts of antimony are soluble, and when present they tend to precipitate from solution as Sb_2O_3 or Sb_2O_5. The element is seldom found in pure form, but occurs chiefly as stibnite, cervantite and valentinite. The probability that large concentrations will appear in the marine environment is not very great. Nevertheless it should be considered because it is reported to be lethal to humans in quantities as low as 97 mg (McKee and Wolf, 1963). Some marine organisms are able to concentrate antimony more than 300 times the concentration present in the surrounding water (known as concentration factor) (Noddack and Noddack, 1940). Portmann (1970) has suggested that antimony ore mining operations should be closely monitored to safeguard life—human and marine.

Arsenic

This element occurs in nature mainly as arsenides of true metals and as orpiment, AsS_3. It is usually associated with ores of copper, lead, zinc and tin. Arsenic may also appear in industrial effluents involving dyes, pesticides, chemicals, pharmaceuticals and smelting. The toxicity of arsenic to some types of marine animals is of the order of 1–10 ppm in the water; 100 mg of arsenic may cause severe poisoning in man and 130 mg has proved fatal (Browning, 1961). The concentration factor for arsenic for some marine animals may be as high as 3,300. In areas of high arsenic concentration shellfish can contain as much as 100 ppm. This may necessitate prohibiting the sale of shellfish and fishes from the area. Arsenic is a cumulative poison; the symptoms may not develop for several years. Arsenic has caused cancer of the skin, lung and liver and is suspect in cancer of the bladder in humans (Hueper, 1963).

Beryllium

According to Portmann (1970) the risk to humans of ingesting toxic doses of beryllium from marine organisms is minimal. Beryllium is suspected to cause cancer of the bone in humans (Dvizhkov, 1967; Hueper, 1963).

Cadmium

Concentration factors of more than 4,500 have been reported for cadmium for some marine organisms (Noddack and Noddack, 1940). Cadmium is lethal to certain marine life at levels of 1.01 to 1 ppm (McKee and Wolf, 1963). Ingestion of 35 mg of cadmium may produce serious intoxication in humans (Locket, 1957). The LD_{50} of cadmium chloride in rats is 88 mg/kg (Spector, 1955). Monitoring of the cadmium content of marine organisms may be required in future, since it may occur in certain industrial effluents.

Chromium

Some chromium compounds are extremely irritating, corrosive and carcinogenic (Goodman and Gilman, 1965; Roe and Carter, 1969; von Oettingen, 1958). Chromic acid and chromate salts are produced by the electroplating, dyeing, plating and chemical industries. Chromic oxide is said to have an LD_{50} of about 100 ppm for certain species of shrimp and fishes (Portmann, 1970). The lethal concentration of chromium varies with the valency state of the element and may range from 18 to more than 200 ppm. There is apparently a growing incidence of cancerous growths and skin ulcers on marine fishes taken near industrial outfall areas in the United States, and it is suspected that chromium compounds are contributing factors (Dvizhkov, 1967). Toxic chromium compounds should be monitored in the future.

Cobalt

The aqueous solutions of the salts of cobalt may cause skin lesions in humans (von Oettingen, 1958) and may cause similar lesions in fishes. Cobalt occurs in nature principally in admixture with other elements as an arsenide such as smaltide, $CoAs_2$, and cobaltite, $CoAsS$. It may appear in other forms in effluents of smelting, plating, dye, paint and other industries. Concentrations up to 10 ppm do not appear to be toxic to marine organisms. The concentration factor for marine organisms can be as high as 21,000 (Noddack and Noddack, 1940). Cobalt is suspected of producing cancers of connective tissue and lungs in humans (Hueper, 1963).

Copper

An enormous amount has been published concerning the toxicity of copper and its salts to marine organisms. Concentrations up to 1 ppm may occur in coastal waters near copper deposits. The concentration factor for copper among some marine organisms is about 7,500 (Noddack and Noddack, 1940). Oysters are especially susceptible and may accumulate copper, causing a green discoloration of the mantle and gills and giving them an unpleasant coppery flavour which ruins their market value (Galtsoff, 1932, 1964; Nelson, 1921). Copper is toxic to man in quantities of 100 mg (McKee and Wolf, 1963). However, copper poisoning as a result of eating copper-contaminated marine organisms is unlikely because their taste renders them unpalatable. Human taste threshold is about 5–7 ppm (Portmann, 1970).

Lead

Lead appears in industrial effluent in a variety of forms. The solubility of lead salts in sea water is about 1 ppm. The toxicity of lead for marine organisms has not been well established, but it appears to be in excess of 1 ppm for short term exposures for some animals (Portmann, 1970). Chronic lead poisoning in humans from direct contact or from contaminated food or water is common (Sollman, 1949). Lead is a slow-acting cumulative poison, and individual susceptibility varies greatly. Lead is also a potential urinary carcinogenic agent in man (Hueper, 1969). Some marine animals have shown concentration factors up to 1,400. The presence of lead in effluents entering the sea should be carefully monitored.

Mercury

The first reported human poisoning by mercury in seafoods occurred in Japan between 1953 and 1961. During this period more than 100 persons were poisoned. The intoxications were caused by eating shellfish, crabs and fishes from Minamata Bay, Kyushu, Japan. The marine organisms had been contaminated by mercury from the effluent of a vinyl chloride factory. The mercury compound in the seafood could not be extracted with organic acids or basic solvents. It was thought that a

[587]

simple alkyl mercury compound, which was later identified as methyl mercury, combined to form a protein complex that was released during ingestion and digestion. The victims developed a wide range of serious neurological disturbances, including blindness, deafness, stupor, coma, convulsions. About one-third of the people involved died (Halstead and Courville, 1967; McAlpine and Araki, 1958; Kurland *et al.*, 1960). Fish-eating birds and mammals were also affected. In 1965 about 35 cases of methyl mercury poisoning occurred among fishermen in Niigata, Japan (Hoshino *et al.*, 1966).

A mercury problem has been observed in Sweden and Finland since the 1950s. The main sources of mercury are fungicides and slimicides in the paper and pulp industries, seed dressings and fungicides in agriculture, the chlor-alkali industry, electrical installations, heating of ores and clays, burning of oil and coal containing small amounts of mercury, burning of paper and sweepings containing mercury, wastes from dental clinics and hospitals, and a variety of other industries (Ackefors, 1968). Investigations have shown that fishes and other aquatic organisms from lakes, rivers and coastal waters contain very high concentrations of mercury. Examples of the methyl mercury concentrations in marine fishes are as follows: flounder (*Pleuronectes flesus*) 50 to 860 ng/g,* plaice (*Pleuronectes platessa*) 71 to 3,100 ng/g, and cod (*Gadus morrhua*) 245 to 2,700 ng/g (Henriksson, 1968). Mercury levels in fish from non-contaminated areas of Sweden are said to be about 25 to 155 ng/g.

In 1967 Sweden established a practical residue limit of 1 mg of Hg/kg (Berglund, 1968). Fishes containing residues exceeding this limit are considered unfit for human consumption. There is considerable uncertainty concerning the safety factors in the Swedish mercurial residue limit. The Food and Agriculture Organization and World Health Organization have temporarily proposed practical residue limits at 0.02 to 0.05 mg of Hg/kg. In Japanese experiments, cats showed evidence of toxicity after consuming seafood containing a total of only 20 to 80 mg of mercury over a period of several weeks. One group of human victims of Minamata disease showed a mercury content in their organs as follows: liver 6–71 ppm, kidney 13–144 ppm, and brain 1–21 ppm (Kurland *et al.*, 1960). Data from studies on rat and man suggest that more than 90 per cent of methyl mercury obtained from food is absorbed (Ekman *et al.*, 1968). However, rats are able to tolerate higher blood levels of mercury than man before neurological disorders are evident (Gage, 1964). It is significant that a no-effect level has not been demonstrated for mercury (FAO/ WHO, 1967, 1967a).

Mercury poisoning is of considerable public health importance. It is believed that regardless of the source or original form of the mercury, about 90 per cent of it is transformed to highly toxic methyl mercury by marine microorganisms (Norén and Westöö, 1967). The intoxicant becomes involved in a cumulative sequence which may not destroy the vector organisms, but may be lethal to man as a result of bioaccumulation. There is evidence that toxic mercurial compounds may involve almost every major trophic level of the marine food chain.

Minamata disease is a prime example of knowledge of

* The units of measure are: 1 ppm = 1 mg/kg = 1,000 ng/g.

the ecology of marine organisms contributing to the understanding of intoxication. From Japanese and Swedish reports, it is apparent that the rapidity with which an area becomes dangerously contaminated is directly affected by its circulation. An area of closed or limited circulation such as a lake or bay becomes contaminated rapidly whereas in an open coastal area the process takes longer. Once established the contaminated protein food resources cannot be used for long periods of time. Moreover, because the mercurial compounds occur in the water, bottom mud and sand as well as in living organisms, the entire environment can be thoroughly contaminated. Mercurial contamination has recently been found in Lake St. Clair and Lake Erie. The Canadian Government has banned fishing from Lake Champlain, the Saskatchewan River, Lake Winnipeg, Winnipeg River and Cedar Lake; fishing has been banned in Lake Onondaga, and people are urged not to eat fish from the Niagara, Oswego and Seneca Rivers. Undoubtedly there are undetected high mercurial residues in many other areas.

Nickel

The toxicity of most nickel compounds to either marine animals or humans does not appear to be great, but even small concentrations may injure plants (Portmann, 1970). Human fatalities have been caused by nickel carbonyl in the production of nickel by the Mond process (von Oettingen, 1958). To what extent this contaminant may be dangerous is not known. Nickel is reported to have carcinogenic properties in laboratory animals (Hueper, 1963). Because of its injurious effects to plant life and possibly to man, nickel should be monitored in future.

Selenium

This non-metallic element exists in several allotropic forms and occurs in nature in the form of selenides. Selenium is used in the electronic industry, in the manufacture of glass, ceramics, rubber and chemicals. Selenium occurs naturally and may accumulate in the marine environment from soil run-off in addition to industrial effluent. Some selenium compounds cause serious injury to the kidneys, liver and heart muscle. Chronic intoxication has been shown to cause cancer of the liver in rats (DiPalma, 1965) and is suspected to cause cancer of the liver and thyroid in humans (Hueper, 1963). The degree to which selenium is becoming involved in marine organisms is not known, and it should be monitored.

Tin

Tin does not appear to be a threat to human health in the marine environment.

Uranium

The salts of uranium are corrosive, readily absorbed and highly toxic. Undoubtedly uranium will continue to receive considerable attention because of the radiological implications. It should be monitored regardless of the form in which it appears in industrial effluents.

Zinc

Although zinc is one of the more abundant toxic heavy metals, the oral toxicity in humans to most zinc compounds is relatively low. Concentrations of up to 0.4 mg/l have been reported in some estuarine waters. Concentrations at this level are lethal to mollusc larvae. Toxic levels for adult shellfish and fish are about 10 ppm (Portmann, 1968). The concentration factor for ^{65}Zn for some marine organisms may be as high as 100,000 (Silker, 1961) and for the stable isotope 32,500 (Noddack and Noddack, 1940). Zinc is almost always present to some degree in oysters, the concentration increasing during summer and decreasing in winter (Galtsoff, 1964). Since zinc is toxic to man only in high doses and tends to impart a blue–green colour to fish and shellfish, it is unlikely that acute intoxications would be produced. Long-term effects do not appear to be a serious threat. But high zinc concentrations may have a deleterious effect on larval fish and shellfish (Portmann, 1970). Zinc should be monitored because of its effects on marine life.

Petroleum

The rapid advances of technology have resulted in a steady increase in the use of petroleum. Large scale offshore oil drilling and transportation operations involve the movement of an estimated 1,000 million tons across the world ocean annually, with a yearly increase of 4 per cent. Through accidental or intentional discharge or leakage, petroleum and its products are among the major pollutants of our planet.

Crude oil is a complex mixture of natural products. It was believed that the low boiling saturated hydrocarbons were harmless to marine life, but as Blumer (1969, 1969a) has pointed out, low concentrations of these hydrocarbons produce anaesthesia and narcosis in a wide variety of lower animals. Higher concentrations may produce cell damage and death. Higher boiling saturated hydrocarbons may interfere with nutrition and communication processes in marine animals. The low boiling aromatics include such products as benzene, toluene and xylene, which are very toxic to man and other animals. The high boiling aromatic hydrocarbons are suspect as carcinogenic agents since crude oil contains alkylated 4- and 5-ring aromatic hydrocarbons similar to those in tobacco tar. Oil pollution has been suggested as a cause of tumour growths in soft clams (Hueper, 1963). Cahnmann and Kuratsune (1957) demonstrated the presence of polycyclic aromatic hydrocarbons in oysters taken from waters polluted with ship oil and industrial wastes. Up to 1 mg of polycyclic aromatic hydrocarbons has been detected per kg of shucked oysters harvested from moderately polluted waters (Callaghan, 1961). Some of the most serious aspects of oil pollution are low-level long-term toxic effects. Blumer (1967) and Blumer and Thomas (1965) found that once hydrocarbons are incorporated into a particular marine organism they are stable regardless of their structure and may pass through many members of the marine food chain without alteration. This stability is so great that it has been found to be a useful tool in tracing the food sources of the marine organisms. These hydrocarbons may also be concentrated in a similar manner to the chlorinated pesticides and some of the heavy metals, i.e. to the point where toxic levels may be reached. In the case of an oil spill, the natural dispersion of oil or treatment with detergents or dispersants produces oil droplets of a particle-size range which permits ingestion and assimilation by numerous marine organisms. Once the oil is assimilated, it passes through the food chain and eventually reaches organisms that are harvested for human consumption. This may result in a serious problem in the production of fish protein concentrates because these toxic carcinogenic hydrocarbons are protein bound and would not be eliminated by the routine FPC extraction process.

A less serious problem is tainting of fish and shellfish. This has immediate economic implications because the products lose their market value (Hawkes, 1961). There are reports of large numbers of oysters and cockles in certain beds along the French and English coasts being killed by crude or fuel oil, and many of the surviving shellfish have become tainted (Korringa, 1968; ZoBell, 1963). Some oil dispersants may be as toxic, or more so, than petroleum. At the present time there is no information available concerning their toxicological effects as they relate to human health. But some detergents may facilitate the penetration of carcinogens into tissues and thereby intensify and facilitate the action of co-existing carcinogens, or the detergents themselves may act as cocarcinogens (Eckardt, 1959; Hueper, 1963; Saffiotti, Shubik and Opdyke, 1962; Setälä et al., 1959).

Petrochemicals and other organics

The production of chemicals derived from petroleum has become a highly diversified industry involving thousands of substances. Some organic waste compounds can taint the flesh of marine organisms. Boëtius (1954) set out levels of the various phenols that cause tainting. Chlorophenol, for example, produces an unpleasant taste in fish at a concentration of 0.001 mg/l. Hueper (1963) in his review "Environmental carcinogenesis in man and animals" pointed out that petrochemicals are among the most important groups of environmental carcinogens. The occurrence of carcinomas and papillomas of the lips in croakers caught in Southern California waters, Russell and Kotin (1957), were attributed to the carcinogenically potent wastes released from a nearby oil refinery. Finkelstein (1960) suggested that the presence of papillomas in eels from the Baltic Sea might in part be attributed to an accumulation of carcinogenic substances from oil wastes.

Many industries use azo-dyes and such aromatic amino compounds as alpha-naphthylamine, beta-naphthylamine, xenylamine, anthracene, benzidine, and their derivatives, many of which are known to cause tumours of the urinary tract and particularly the bladder in humans (Hueper, 1963; Pauley, 1969).

Chloronitrobenzol has been found in river waters near New Orleans. This chemical is used in the manufacture of aromatic amino- and nitro-compounds. Carcinogenic bioassay performed on eluates of activated carbon from the industrial effluent of rubber plants into a mideastern United States river revealed that when these extracts were injected under the skin of mice they developed cancers of the subcutaneous tissue and bladder tumours (Middleton, Grant and Rosen, 1956).

Tumours have been produced in cuttlefish by implanting pellets of 1, 2, 5, 6-dibenzanthracene (Jacquemain et al., 1947). Benzene compounds have been found to produce tumours in molluscs (Jacquemain and Jullien,

[589]

1947) and prolonged exposure causes leukaemia in humans.

Investigations on possible carcinogenic agents in commercial fish feeds, which were believed to be causing severe epidemics of primary liver cancer in rainbow trout in American hatcheries, revealed that the liver of this species develops cancer when aromatic amines are added to an otherwise non-carcinogenic feed (Halver et al., 1962).

Nitrosamines are derived from dialkylamines when acted upon by nitrous oxide or nitric acid. These substances have been incriminated by the skin reactions of workers handling aniline dyes for cotton printing. Studies on over forty different nitrosamines by Druckrey et al., some of which were synthesized especially for this purpose, revealed that these amino compounds were extremely potent carcinogens in laboratory animals.

It is evident that an increasing number of potential carcinogens are entering the environment through the atmosphere or the discharge of effluents to streams and rivers or the ocean. Thus man is gradually building up enormous deposits of toxic and carcinogenic wastes which are contaminating and destroying our food supply and possibly also valuable potential pharmaceutical agents produced by marine organisms. Schmeer (1970) found that the New England marine clam (Mercenaria mercenaria) produced an anti-tumour substance under normal conditions but ceased to do so when the water became polluted.

Pesticides

In recent years a great many man-made pesticidal poisons have been released into the environment either by intent or accident. The term "pesticide" is generally used in the generic sense and may include insecticides herbicides, fungicides, rodenticides, helminthicides, miticides, molluscicides, amoebicides, algicides, fumigants and virucides. In addition there are related poisons such as defoliants, anti-sprouting and anti-maturation agents, soil sterilizers and water evaporation retardants. The United States production of pesticides for 1967 was in excess of 1 billion lb (American Chemical Society, 1969). The annual consumption of synthetic organic pesticides in the United States alone is estimated to be about 800 million lb per year. More than 300 organic pesticidal chemicals are now in use in the U.S.A. in more than 10,000 formulations. Fortunately many are short-lived or are not used in sufficient quantity to be considered serious hazards at present.

Pesticides are commercially valuable because of their toxic properties. Although they are used primarily for specific purposes they have the ability to exert biological effects on all living organisms. In addition to their toxic characteristics, some compounds display carcinogenic properties, and have become of increasing concern to public health because of their uncontrolled ubiquitous distribution which now extends far beyond their intended use. One analytical study identified dieldren, endrin, DDT and DDE, in that order of frequency of occurrence, in all major river basins in the United States. The study showed that the concentrations of these pesticides ranged from 4 to 118 ppt (Weaver et al., 1965). These pesticides have entered the sea and are consistently appearing in fisheries products for human consumption.

Chlorinated hydrocarbon pesticides may exert serious adverse effects directly on the plankton, resulting in a disturbance of the metabolism of the organism. Wurster (1968, 1968a) has shown that concentrations of DDT as low as a few parts per billion in water reduced photosynthesis in laboratory cultures of four species of coastal and oceanic phytoplankton representing four major classes of algae (Skeletonema costatum, Coccolithus huxleyi, Pyramimonas sp., Peridinium trochoideum). In the same study, photosynthesis was also reduced in a natural phytoplankton community. The toxicity to diatoms increased as cell concentration decreased. It was suggested that selective toxic stress by DDT on certain algae may alter the species composition of a natural phytoplankton community. A floral imbalance could favour species normally suppressed by others, producing population explosions and dominance of the planktonic community by one or a few species (Korringa, 1968). Other studies show that in some instances an undesirable planktonic community of highly toxic dinoflagellates could lead to poisonous shellfish, and possibly toxic fishes (Halstead and Courville, 1965, 1967). Thus a man-made toxin could lead to the growth of a population of naturally occurring toxic dinoflagellates which would be concentrated by shellfish and possibly even by fish. There is also the possibility that pesticides may, under proper circumstances, provide the missing chemical links to permit a ciguatoxic or clupeotoxic cycle to come into being.

The chlorinated hydrocarbons pose special problems in environmental toxicology because of their extreme persistence and bioaccumulation. They are highly lipophilic and may be found in solution in water and surface films and adsorbed on particulate matter. Predatory marine animals, many of great economic value, are especially susceptible to exposure since they accumulate and retain most of the pesticides ingested. These pesticides are finally deposited in the adipose tissue of the body. Fish may also acquire pesticides such as DDT through their gills and skin (American Chemical Society, 1969). Analyses conducted in California waters (Keith and Hunt, 1966; Modin, 1969) have shown oysters, mussels, and clams in estuaries to contain DDT, DDD, DDE, dieldrin and endrin in concentrations from 10 to 3,600 ppb. Offshore studies revealed 2,739 ppb of these compounds in king crab and 668 ppb in the eggs of a king salmon. Pesticidal residues of these compounds were as high as 591 ppb in the ova of prawn, flounder, halibut and sole. In the North Sea chlorinated hydrocarbons were found in cod, whiting, other gadoids, mackerel, sand eels, dabs, flounders, herring, sea trout, mussels, zooplankton, seals and porpoises (ICES, 1969). Pesticidal residues in codling from the Firth of Clyde had residues up to 0.5 ppm in the muscle tissue and concentrations up to 12.0 ppm in the liver. Extensive investigations carried out in the Baltic Sea (along the Swedish coasts) (ICES, 1970) showed DDT in concentrations up to 8.6 ppm in mussels, up to 37 ppm in fish (herring, plaice, cod and salmon), and up to 790 ppm in seals. The DDT concentrations in seals from the Baltic were up to 10 times higher than from North Sea seals and from Canadian seals. DDT has been found in Antarctic fishes in concentrations of 0.18 ppm (American Chemical Society, 1969).

The effects of pesticides upon shellfish, particularly oysters, is of considerable importance because of the concentration ability of these organisms and their economic value as food. Butler, Wilson and Rick (1960) found pesticides such as aldrin, chlordane, O-dichlorobenzene, DDD, DDT, dieldrin, endrin, heptachlor, rotenone, sevin and toxaphene at concentrations as low as one part in 100 million were able to inhibit the activity and growth of oysters after only 24 hours of exposure. Such pesticides as azinphos-methyl are lethal to brown shrimp in concentrations as low as 0.0003 ppm (Portmann and Connor, 1968). A study on the freshwater brook trout has shown that DDT in quantities as low as 20 ppb affects the learning ability of the fish within a period of 24 hours (Anderson and Prins, 1969).

The polychlorinated biphenyls (PCB's) have wide usage. These chlorinated compounds resemble organochlorine pesticides in that they are very persistent and are subject to bioaccumulation. Concentration of these materials may be very high in predatory marine animals, but there is little information on their effects on marine life.

Generally, the chlorinated hydrocarbons can cause toxic reactions in the liver and bone marrow of laboratory animals and man. They may cause liver necrosis and cirrhosis, anaemia, polycythemia, agranulocytosis, panmyelophthisis and maturation arrest in bone marrow (Hueper, 1963). In addition to the usual array of toxicological effects, many chlorinated hydrocarbons produce benign and malignant cancers and leukaemias in humans and laboratory animals. The chlorinated hydrocarbons tri-p-anisyl-chloro-ethylene and methoxychlor produce oestrogenic activity in experimental animals. There exists, therefore, a distinct possibility that these compounds, like all other oestrogens, could cause cancer in such organs as the breast, uterus, bladder, kidney, testis, and haematopoietic tissue. Some of the synthetic auxins (herbicides) of the chlorinated hydrocarbon type are also under suspicion as carcinogenic agents. The carbamates, which are widely employed as insecticides, herbicides, plasticizers, fish dope and medicine, have also been incriminated as cancer-producing agents in both experimental animals and man.

Discussion

The world's oceans (71 per cent of the earth's surface) are the depository of all major fresh water river systems. This is the final cesspool of most of man's activities. The most susceptible portions are the continental shelf and tropical insular shallow water regions. These produce our world fishery products and at best probably do not exceed 10 per cent of the total ocean area. Because of the extreme sensitivity of tropical reef organisms to all types of pollutants some of the more serious potential problems are in the tropical islands; but the insidious long-term, low-level effects are frequently not discerned until the situation has achieved catastrophic proportions. These problems are not receiving adequate attention.

The seriousness of the toxicological situation is due largely to the persistence in a salt-water medium and bioaccumulation of certain chemicals. Evidence of the destructive forces of environmental pollutants is seen in a steadily increasing number of cancerous growths, skin ulcerations, tail deformities, genetic changes, and other disease conditions in fish, and leukaemias and tumour growths in shellfish. In some instances these disease conditions have achieved epidemic proportions. The most seriously polluted geographical areas of the world are in the countries that have highly developed industrial technology.

The overall marine environmental toxicological picture involves "naturally occurring" marine biotoxins as well as man-made synthetic toxic substances. Some field evidence strongly suggests that man-made chemical pollutants may trigger biotoxic cycles either directly or indirectly by producing a planktonic imbalance favouring a population of toxic organisms. It is believed that tropical insular areas may be more susceptible to biotoxicological disturbances of this type than temperate regions.

Pollutants produced by industrial technology include a broad spectrum of heavy metals and other inorganics. The rapid advances of technology have resulted in steady increases in the drilling for and transportation of crude oil and in the development of an enormous petrochemical industry producing thousands of substances. Petrochemical wastes include both organic and inorganic products. The entire gamut of these industrial wastes eventually make their way into the marine environment. In recent years a large number of man-made pesticidal poisons have entered the marine environment in such large quantities that they have become some of the most widely distributed chemical agents on this planet.

Certain fundamental observations have emerged from the studies conducted thus far. Industrial prosperity has given rise to environmental destruction, primarily because of inefficient world technology. Most of the industrial wastes that are discharged into the world's ocean are of economic value. In most instances these toxic "wastes" command high prices when properly packaged. Yet these same elements or compounds are discharged into sewer outfalls and dumped into the sea.

The sea has been considered a limitless and bottomless receptacle which can absorb an unlimited quantity of anything toxic, but such is not the case. Our most valuable marine food resources and our greatest potential of biodynamic resources (new drugs from the sea) are limited to the shallow water continental shelves and tropical islands. The temperate continental regions have already been subjected to the destructive forces of pollution. Now temperate zone industrialists are beginning to invade tropical insular areas (Indonesia, Puerto Rico, and the U.S. Virgin Islands) for economic exploitation.

The ocean is a very dynamic and complex ecosystem which has as its basic life-support system a delicately balanced fluid medium—salt water. Man has now begun to destroy this chemical balance on a massive, ever-accelerating scale. The more people there are, the more industrial development, the more wastes produced, and the greater the destruction of the environment. When chemical pollutants are dumped into the ocean they enter a dynamic system. They are not only diluted and dispersed by storms, winds, tides, currents, upwellings, etc., but become intimately involved in the complexities of the biological food web of the sea. These toxic pollutants are not merely diluted, but may be re-

[591]

concentrated by the marine biomass. This fundamental principle of the bioeconomics of the sea is generally overlooked. This food web involves all living things in, on, or around the ocean. It is estimated that this includes about four-fifths of all living things. Marine animals pick up pollutants and also help to distribute them in their migratory activities. Microorganisms may act directly on various chemical constituents in the water. Animals feed on the plants and concentrate these chemical substances still further. Predatory animals concentrate the chemical constituents still further. If the final species is eaten by man, he receives the combined chemical concentrates at the end of the food chain.

When persistent toxic and carcinogenic pollutants enter the food web the pollution problem may become extremely serious and lethal to plants, animals and man. Most of these toxic agents affect all living things. Nutritional, communicative, reproductive, respiratory, genetic and a variety of other metabolic activities of the organisms may be seriously altered or destroyed. This may result in annihilation of some populations of organisms and increase in noxious groups. This results in an imbalance of the normal population pattern which can lead to serious problems in the ecosystem. Under more devastating conditions toxic substances build up in such excessive concentrations that a toxic and/or anoxic situation may develop.

Toxic substances may be stored in the fatty tissues or become protein bound. Thus when man is the final link in the food chain he receives a dosage which may be many thousands of times greater than that present in the environment. Also once a toxic substance enters a marine organism it may be metabolized into a different toxic substance. There are a large number of possible chemical and biological interactions; these are determined by the chemical substance, physical factors, the organism and the general ecology. At present very little is known concerning the exact physiopathological mechanisms involved. These matters should be seriously considered by those producing marine protein concentrates.

The broad spectrum of toxins and carcinogens with which man is polluting the marine environment are among the more lethal agents known to toxicologists. There is no longer any question that these materials are entering man's marine food resources. The question is how and to what extent these poisons affect marine plants, lower animals and human biological systems. At the moment there are no baselines of technical marine biological and chemical data. Our present toxicological and medical data are grossly inadequate; much being conflicting or questionable. The data having a direct bearing on the toxicological aspects of marine pollution are scattered through such widely divergent disciplines as chemistry, toxicology, fisheries, oceanography, marine biology, physiology, and a host of other subjects and are so divergent that only a computerized data retrieval system can provide the necessary cohesiveness.

In most instances there is not only a gross deficiency of adequate taxonomic floral and faunal studies, but there is also a deficiency of trained taxonomists capable of identifying the vast array of organisms involved. This taxonomic work is prerequisite to any ecological or toxicological work that is to be done.

More information needed

Information is urgently required on the toxic effects of pollutants on a broad phylogenetic spectrum of marine plants and animals, including their larval and adult stages in arctic, antarctic, temperate and tropical waters. Sensitivity to these various substances varies enormously from one plant or one animal to the next, and according to the stage of maturation of the organism. Pollutants at extremely low levels may be devastating to organisms during their larval periods. This could result in the loss of large populations of valuable food resources from a causative agent that is undetected because our bioassay methods are too insensitive. Most toxicological data now available are largely applicable to temperate zone organisms and are not pertinent to highly sensitive tropical reef organisms. The interpolation of toxicological data from one group of organisms to the next is a dangerous procedure and can lead to erroneous conclusions. The amount of work in this one segment alone is of staggering proportions. Controlled acute toxicity studies in a laboratory on goldfish and mice are not an answer to today's environmental toxicological problems. We need a shift in emphasis to low-level, long-term, chronic toxicity studies, on a wide variety of test organisms.

On the chemical side of the ledger no one knows what is being dumped into the sewer, nor the chemical interactions in the sewer, nor what comes out at the other end. There is no information available as to the chemistry of this witch's brew once it mixes with salt water and the manner and extent to which it is distributed. Before one can begin to intelligently predict what is taking place in the environment it is essential to know what inorganic and organic pollutants are present. This problem requires a new generation of rugged automated chemical sample analysers operating on a continuous basis over an extended period of time in broadly scattered areas of the marine environment. This could also necessitate a system coupled to long-range sensing equipment and a satellite communication programme.

There is a growing belief by some medical investigators that some of the chronic degenerative diseases of man (particularly neurological disorders and cancers) may be caused by environmental intoxicants. These pathological effects may be caused directly by the action of the poison, or indirectly through the inhibition of a chemical substance required for the metabolism of the organism. As Hueper (1963) has pointed out, man has made marked qualitative and quantitative changes in the environmental carcinogenic spectrum and on the cancer panorama. He states: "It should be obvious from these observations that a further indiscriminate, injudicious, careless and irresponsible continuation of the present practices might, or probably will, invite the occurrence of some uncontrollable disaster to man and animals taking the form of an acute and severe epidemic of cancer, resembling that which recently overtook the rainbow trout population of American fish hatcheries."

On this cancer epidemic in rainbow trout, it was found that almost 100 per cent of the rainbow trout in some American trout hatcheries developed cancer of the liver in fish that were over 3 years of age. Hueper and Payne (1961) of the Environmental Cancer Section, U.S. National Cancer Institute, reported: "The occurrence of this epidemic among edible fish subjected to an artificial nutritive regimen provides a serious warning of the possible future production of a similar cancer epidemic in the human population through an increasing contamination of the human environment with some of the many industry-related carcinogenic chemicals."

It is generally conceded that because of its highly developed technology the United States is the greatest producer of environmental pollutants in the world today. It is estimated that the United States dumps into the marine environment more than 47 million tons of wastes each year. This includes such products as dredging spoils, refuse, waste oil, industrial chemicals and sludges, and sewage sludge. On 6 May 1970 Dr. James B. Shrewsbury of Murray State University, Kentucky, testified before the U.S. Senate Committee on Public Works after having investigated the present state of water-quality surveillance system operations in the United States:

"I will make no attempt to mollify our step-by-step or cumulative alarm with what we found. It was our considered opinion that the concept of a water-quality surveillance system was nowhere in evidence and that actual surveillance was not merely inelegant but was trivial in the extreme."

National and international environmental monitoring systems are urgently required. At the risk of being classed as an alarmist, one cannot help but wonder if mankind is not heading for an eco-catastrophe such as Ehrlich (1969) has so graphically depicted.

References

ACKEFORS, H A survey of the mercury problem in Sweden with
1968   special reference to fish. Paper presented to Informal Meeting of a Group of Consultants on the Problems of Toxicants in Seafood, 18–22 November 1968 (W.P. 11) (Unpubl.).

AMERICAN CHEMICAL SOCIETY Cleaning our environment: the
1969   chemical basis for action. A report by the Subcommittee on Environmental Improvement, Committee on Chemistry and Public Affairs. Washington, D.C. American Chemical Society.

ANDERSON, J M and PRINS, H B Effect of sublethal DDT on a
1969   simple reflex in book trout. International Council for the Exploration of the Sea, C.M. 1969/E–19.

BERGLUND, F The problem of methylmercury in fish. Paper pre-
1968   sented at Informal Meeting of a Group of Consultants on the Problems of Toxicants in Seafood, 18–22 November 1968, (W.P. 10) (Unpubl.).

BLUMER, M Hydrocarbons in digestive tract and liver of a basking
1967   shark. Science, N.Y., 156(3773):390–1.

BLUMER, M Oil pollution of the ocean. Oceanus, 15(2):2–7.
1969

BLUMER, M Oil pollution of the ocean. In Oil on the sea, edited
1969a   by D. P. Hoult, New York, Plenum Press, pp. 5–13.

BLUMER, M and THOMAS, D W Zamene isomeric carbon-19 mono-
1965   olefins from marine zooplankton, fishes and mammals. Science, N.Y., 148(3668):370–1.

BOËTIUS, J Foul taste of fish and oysters caused by chlorophenol.
1954   Meddr Danm. Fisk-og. Havunders., 1(4): 8 p.

BOËTIUS, J Lethal action of mercuric chloride and phenyl
1960   mercuric acetate in fishes. Meddr Danm. Fisk-og. Havunders., 3(4):93–115.

BROWNING, E Toxicity of industrial metals. London, Butter-
1961   worth & Co.

BUTLER, P A, WILSON, A J, JR and RICK, A J Effect of pesticides
1960   on oysters. Proc. natn. Shellfish. Ass., 51:23–32.

CAHNMANN, H J and KURATSUNE, M Determination of polycyclic
1957   aromatic hydrocarbons in oysters collected in polluted water. Analyt. Chem., 29(9):1312–7.

CALLAGHAN, J International aspects of oil pollution. Trans.
1961   N. Am. Wildl. Nat. Res. Conf., 26:328–342.

CHAPMAN, W M Food from the sea. Van Camp Sea Food
1965   Company (Unpubl.).

COOPER, R C and KELLER, C A Epizootiology of papillomas in
1969   English sole, Parophrys vetulus. Natn. Cancer Inst. Monogr., (37):173–86.

DER MARDEROSIAN, A Marine pharmaceuticals. J. pharm. Sci.,
1969   58(1):1–33.

DEYS, B F Papillomas in the Atlantic eel, Anguilla vulgaris.
1969   Natn. Cancer Inst. Monogr., (31):187–94.

DIPALMA, J R Drill's pharmacology in medicine. New York,
1965   McGraw-Hill Book Co., 1488 p.

DOUDOROFF, P Bio-assay methods for the evaluation of acute
1951   toxicity of industrial wastes to fish. Sewage ind. Wastes, 23(11):1380–97.

DOUDOROFF, P and KATZ, M Critical review of literature on the
1950   toxicity of industrial wastes and their components to fish. 1. Alkalies, acids and inorganic gases. Sewage ind. Wastes, 22(11):1432–58.

DOUDOROFF, P and KATZ, M Critical review of literature on the
1953   toxicity of industrial wastes and their components to fish. 2. The metals as salts. Sewage ind. Wastes, 25(7):802–39.

DVIZHKOV, P P Glastomogenic effects of industrial metals and
1967   their compounds. Arkh. Patol., 29:3–11 (in Russian).

ECKARDT, R E Industrial carcinogens. New York, Grune and
1959   Stratton, Inc.

EHRLICH, P Eco-catastrophe. Ramparts, 1969:24–8.
1969

EKMAN, L et al. Metabolism and retention of methyl-203-
1968   mercury-nitrate in man. Nord med., 79:450–6 (in Swedish).

FAO/WHO Evaluation of some pesticide residues in food.
1967   Rome, FAO, (PL:CP/15):237 p. Geneva, WHO/Food Add./67.32:237 p.

FAO/WHO Pesticide residues in food. FAO Agric. Stud., (73):
1967a   19 p. Also issued as Tech. Rep. Ser. Wld Hlth Ord., (370):19 p.

FINKELSTEIN, E A Tumours of fish. Arkh. Patol., 22(9):56–61 (in
1960   Russian).

GAGE, J C Distribution and excretion of methyl and phenyl
1964   mercury salts. Br. J. ind. Med., 21:197–202.

GALTSOFF, P S Introduction of Japanese oysters into the United
1932   States. Fishery Circ. Bur. Fish. U.S., (12):16 p.

GALTSOFF, P S The American oyster Crassostrea virginica Gmelin.
1964   Fishery Bull. U.S. Fish Wildl. Serv., 64:480 p.

GOODMAN, L S and GILMAN, A (Eds.) The pharmacological basis
1965   of therapeutics. New York, Macmillan Co., 1785 p.

HALSTEAD, B W Marine biotoxins: a new source of medicinals.
1969   Lloydia, 32(4):484–88.

HALSTEAD, B W Results of a field survey on fish poisoning in the
1970   Virgin and Leeward Islands, 7–18 January 1970. Rome, Food and Agriculture Organization of the United Nations, (Unpubl.).

HALSTEAD, B W and COURVILLE, D A Poisonous and venomous
1965   marine animals of the world. Invertebrates. Washington, D.C., U.S. Government Printing Office, vol. 1:994 p.

HALSTEAD, B W and COURVILLE, D A Poisonous and venomous
1967   marine animals of the world. Vertebrates. Washington, D.C., U.S. Government Printing Office, vol. 2:1070 p.

HALVER, J E, JOHNSON, C L and ASHLEY, L M Dietary carcinogens
1962   induce fish hepatoma. Fedn Proc. Fedn Am. Socs exp. Biol., 21:390 (Abstr.).

HAWKES, A L A review of the nature and extent of damage caused
1961   by oil pollution at sea. Trans. N. Am. Wildl. Nat. Res. Conf., 26:343–55.

HENRIKSSON, R Ang. kvicksilversituationen i Öresund. Föredr.
1968   Oresundskomm. Sammanträde, 25:5.

HOSHINO, O et al. Quantitative determination of mercury in
1966   hair by activation analysis. J. hyg. chem. Soc. Japan, 12:94–9.

HUEPER, W C Environmental carcinogenesis in man and animals.
1963   Ann. N.Y. Acad. Sci., 108:961–1038.

HUEPER, W C Occupational and environmental cancers of
1969   the urinary system. New Haven, Yale University Press, 465 p.

HUEPER, W C and PAYNE, W W Observations on the occurrence
1961   of hepatomas in rainbow trout. J. natn. Cancer Inst., 27(5):1123–43.

HUEPER, W C and RUCHHOFT, C C Carcinogenic studies on ad-
1954   sorbates of industrially polluted raw and finished water supplies. Archs ind. Hyg. Occup. Med., 9:488–95.

ICES Report of the ICES Working Group on pollution of the
1969   North Sea. Coop. Res. Rep. int. Coun. Explor. Sea (A), (13):61 p.

ICES Report of the ICES Working Group on pollution of the
1970   Baltic Sea. Coop. Res. Rep. int. Coun. Explor. Sea (A), (15):86 p.

JACQUEMAIN, R and JULLIEN, A Réactions des céphalopodes aux
1947   injections intradermiques de substances polaires. C. r. hebd. Séanc. Acad. Sci., Paris, 225:698–9.

JACQUEMAIN, R, JULLIEN, A and NOEL, R Sur l'action de certains
1947 corps cancérigènes chez les céphalopodes. *C. r. hebd. Séanc. Acad. Sci., Paris*, 225:441–3.

KEITH, J O and HUNT, E G Levels of insecticide residues in fish
1966 and wildlife in California. *Trans. N. Am. Wildl. Nat. Res. Conf.*, (31):150–77.

KORRINGA, P Biological consequences of marine pollution with
1968 special reference to the North Sea fisheries. *Helgoländer wiss. Meeresunters.*, 17(1–4):126–40.

KURLAND, L T, FARO, S N and SIEDLER, H Minamata disease.
1960 *Wld Neurol.*, 1(5):370–95.

LOCKET, S Clinical toxicology. London, H. Kempton, 772 p.
1957

McALPINE, D and ARAKI, S Minamata disease: an unusual
1958 neurological disorder caused by contaminated fish. *Lancet*, 1958:629–31.

McKEE, J E and WOLF, H W Water quality criteria. *Publs St.*
1963 *Wat. Qual. Control Bd Calif.*, (3–A):548 p.

MENARD, H W and SMITH, S M Hypsometry of ocean basin
1966 provinces. *J. geophys. Res.*, 71(18):4305–25.

MODIN, J C Residues in fish, wildlife and estuaries. *Pestic. monitg*
1969 *J.*, 3(1):1–7.

NELSON, T C Some aspects of pollution as affecting oyster
1921 propagation. *Am. J. Publ. Hlth*, 11(6):498–501.

NIGRELLI, R F, KETCHEN, K S and RUGGIERI, G D Studies on
1965 virus diseases of fishes. *Zoologica, N.Y.*, 50(3):115–22.

NODDACK, I and NODDACK, W Die Häufigkeiten der Schwer-
1940 metalle im Meerestieren. *Ark. Zool.*, 32A(4):1–35.

NORÉN, K and WESTÖÖ, G Methylmercury in fish. *Var Föda*,
1967 1967:13–7 (in Swedish).

PAULEY, G B A critical review of neoplasia and tumor-like lesions
1969 to mollusks. *Natn. Cancer Inst. Monogr.*, (31):509–40.

PORTMANN, J E Progress report on a programme of insecticide
1968 analysis and toxicity-testing in relation to the marine environment. *Helgoländer wiss. Meeresunters.*, 17(1–4): 247–56.

PORTMANN, J E Marine pollution by mining operations with
1970 particular reference to possible metal-ore mining. Paper presented to IMCO/FAO/Unesco/WMO/WHO/IAEA Joint group of experts on the scientific aspects of marine pollution, held at Unesco headquarters, Paris, 2–6 March 1970. (GESAMP/20).

PORTMANN, J E and CONNOR, P M The toxicity of several oil-spill
1968 removers to some species of fish and shellfish. *Mar. Biol.*, 1(4):322–9.

ROE, F J C and CARTER, R L Chromium carcinogenesis: calcium
1969 chromate as a potent carcinogen for the subcutaneous tissues of the rat. *Br. J. Cancer*, 23:172–6.

RUSSELL, F E and KOTIN, P Squamous papilloma in the white
1957 croaker. *J. natn. Cancer Inst.*, 18(6):857–61.

RYTHER, J H Photosynthesis and fish production in the sea.
1969 *Science, N.Y.*, 166(3901):72–6.

SAFFIOTTI, U, SHUBIK, P and OPDYKE, D L Carcinogenic tests on
1962 alkylbenzenes and alkylbenzenes sulfonates. *Toxic. appl. Pharmac.*, 4:763–9.

SCHAEFER, M B The potential harvest of the sea. *Trans. Am. Fish.*
1965 *Soc.*, 94(2):123–8.

SCHMEER, A C Mercenene's effectiveness decreased when clams
1970 "stored" in polluted waters. (Personal communication, July 1, 1970).

SCHNITGER, R Summary background material for the white
1970 croaker and dover sole. (Personal correspondence, 10 April 1970.)

SETÄLÄ, H, et al. Mechanisms of experimental tumorigenesis.
1959 *J. natn. Cancer Inst.*, 23:925–77.

SHREWSBURY, J B Comments on technical capabilities for monitor-
1970 ing water quality, made to the Subcommittee on Air and Water Pollution of the Senate Committee on Public Works, 6 May 1970. (Unpubl.).

SILKER, W B Separation of radioactive zinc from reactor cooling
1961 water by an isotope exchange method. *Analyt. Chem.*, 33:233.

SMITH, R F Evaluation of influence of dumpings in New York
1970 Bight with a brief review of general ocean pollution problems. U.S. Dept. Interior Memorandum. (Unpubl.).

SOLLMAN, T A manual of pharmacology and its applications to
1949 therapeutics and toxicology. Philadelphia, W. B. Saunders Co., 1132 p.

SPECTOR, W S (Ed.) Handbook of toxicology. Vol. 1. *W.A.D.C.*
1955 *tech. Rep.*, (55–16):408 p.

THOMAS, E On the poison of fish. *Mem. Med. Soc., Lond.*, 5:94–
1799 111.

VON OETTINGEN, W F Poisoning; a guide to clinical diagnosis and
1958 treatment. Philadelphia, W. B. Saunders Co., 627 p.

WEAVER, L et al. Chlorinated hydrocarbon pesticides in major
1965 U.S. river basins. *Publ. Hlth Rep., Wash.*, (80):481.

WELLINGS, S R, CHUINARD, R G and BENS, M A comparative study
1965 of skin neoplasms in four species of pleuronectid fishes. *Ann. N.Y. Acad. Sci.*, 126:479–501.

WURSTER, C F, JR DDT reduces photosynthesis by marine phyto-
1968 plankton. *Science, N.Y.*, 159(3822):1474–5.

WURSTER, C F, JR DDT threatens ocean life, chemical balance of
1968a atmosphere. *UST*, 1968, June:2 p.

YOUNG, P H Some effects of sewer effluent on marine life. *Calif.*
1964 *Fish Game*, 50(1):33–41.

ZOBELL, C E The occurrence, effects, and fate of oil polluting the
1963 sea. *J. Air Wat. Pollut.*, 7:173–97.

Ciguatera—Marine Fish Poisoning—a Possible Consequence of Thermal Pollution in Tropical Seas?*

Donald P. de Sylva and Alden E. Hine†

Ciguatera—l'intoxication par le poisson de mer—serait-elle une conséquence de la pollution thermique dans les mers tropicales?

La toxine responsable des intoxications du type ciguatera ne provient pas d'une décomposition bactérienne et n'est pas non plus associée au cycle de reproduction des organismes toxiques. D'après les observations, l'affection serait provoquée par une ou plusieurs toxines transmises par l'intermédiaire de la chaîne alimentaire et tirant sans doute leur origine de plusieurs genres de cyano-chlorophycées, bien que l'on ait isolé des toxines sur certaines genres d'algues vertes.

Les cyano-chlorophycées prédominent dans les eaux dont la température est élevée dans les conditions naturelles ou par suite d'un processus écologique faisant suite à l'eutrophisation ou au réchauffement artificiel du milieu. Dans les eaux tropicales peu profondes, le réchauffement des eaux ambiantes par les effluents caloriques des centrales thermiques provoque un accroissement de la densité relative des cyano-chlorophycées jugées toxiques dans l'écosystème immédiatement adjacent à l'effluent. Il en est ainsi à proximité d'une centrale électrique située dans la baie de Biscayne méridionale, en Floride. On trouve dans cette région des poissons qui se nourrissent d'algues toxiques, ainsi que des espèces pré-datrices qui, à leur tour, les consomment. On estime que, tous autres

La ciguatera (intoxicación por peces marinos) es una posible consecuencia de la contaminación térmica de los mares tropicales

La ciguatera es causada en el hombre por la ingestión de algunos peces e invertebrados marinos recién capturados en aguas tropicales someras. La toxina no es un producto de la descomposición bacte-riana ni está relacionada con el ciclo del desove de los organismos tóxicos. Todos los elementos indican que se forman una o más toxinas que se transmiten por la cadena alimentaria y que probable-mente se originan en varios géneros de algas verdeazules, aunque también se han aislado a partir de varios géneros de algas verdes.

Las algas verdeazules predominan en aguas de temperaturas más elevadas naturalmente o por medio de sucesiones ecológicas, causadas por eutroficación o calentamiento artificial del ambiente. En los mares tropicales someros, el mayor calentamiento del agua ambiente, que tiene ya una temperatura relativamente elevada, podría causar por el efecto de las descargas de aguas calientes de las centrales termo-eléctricas un aumento del número relativo de algas verdeazules—probablemente tóxicas—en el ecosistema inmediatamente adyacente a la descarga. Se ha demostrado que esto ocurre cerca de las centrales generadoras de electricidad en el sur de la Bahía Biscayne, Florida. En estas aguas se encuentran especies ícticas que se alimentan de algas tóxicas y peces predadores

* Scientific Contribution No. 1346, University of Miami, Rosenstiel School of Marine and Atmospheric Sciences, University of Miami.
† Rosenstiel School of Marine and Atmospheric Sciences, University of Miami, Miami, Florida, U.S.A.

facteurs étant égaux, les flambées de ciguatera risquent de devenir plus fréquentes à mesure que la capacité énergétique des centrales électriques augmente dans la zone tropicale. Les auteurs recommandent la prudence lors du choix de sites destinés à accueillir de telles centrales sous le tropiques, où cette intoxication peut réellement affecter la santé des populations et pourrait agir sur le marché de tous les produits de la mer par la menace psychologique qu'elle fait peser sur eux.

que a su vez se alimentan de estos peces algívoros. Este hecho permite afirmar que, si los demás factores son iguales, los brotes de ciguatera pueden ocurrir con más frecuencia al aumentar la capacidad de las centrales eléctricas en las regiones tropicales. Se recomienda por tanto, proceder con cautela en la selección en esas partes del mundo de los emplazamientos para las centrales eléctricas, ya que la ciguatera puede causar perjuicios en la población o su efecto sicológico puede influir adversamente en el mercado de todos los productos alimenticios del mar.

THERMAL pollution has been the subject of extensive studies, and documentation of varying degrees of damage to the environment has been presented in Krenkel and Parker (1969). Because thermal pollution heretofore has not been evaluated in tropical marine waters, its potential or actual effects have only been documented in limited areas (Bader *et al.*, 1970). We propose that in shallow tropical seas, because electrical power stations generate excess heat which favours organisms linked with ciguatera, the incidence of ciguatera poisoning may increase.

Symptoms of ciguatera

Ciguatera from tropical seas was first reported by Martyr (1555). Its nature and occurrence are well documented by Russell (1965) and Halstead and Courville (1965, 1967). Symptoms of ciguatera in man include muscular weakness, tingling or numbness of the lips, hands, and feet, reversal of hot and cold sensations, nausea, joint and muscular pain, loss of voluntary muscular movement, and itching (Halstead and Courville, 1967; Randall, 1958; Russell, 1965). The toxins, which appear to be of several kinds, usually demonstrate anticholinesterase (Banner, 1967). Ciguatera poisoning, which can be fatal, causes severe discomfort in victims, lasting from several days to several years (Halstead and Courville, 1967), and even as long as 25 years (Gudger, 1930). Further, persons affected by toxic fishes became sensitized to subsequent meals of fish, whether or not the portions are toxic (Cooper, 1964; Randall, 1958).

Occurrence

Ciguatera occurs throughout shallow tropic seas, almost invariably being associated with coral reefs (Hiyama, 1943). Often one side of a small reef may harbour toxic fishes, while the same species are non-toxic on the other side (de Sylva, 1963; Mowbray, 1916). In addition, ciguatera may occur for several years and then cease, only to reappear, or may move to new areas.

Persons living in ciguatoxic regions tend to avoid a ready source of inexpensive protein. Clearly, an understanding of what factors cause these seafoods to be toxic on the one hand and the rendering of them edible on the other must have high priority. The presence of ciguatoxic organisms is doubtless more common than records indicate, because, where ciguatera is commom, those species or areas known to be toxic are simply avoided by man. Such avoidance often occurs in the Caribbean where animal protein shortage is not yet acute. In some Indo-Pacific islands the natives actually choose to suffer the consequences of toxic fishes because the taste of certain species is preferred (Cooper, 1964). Data on the etiology of poisonous fishes are scarce, because outbreaks of fish poisoning may go unrecorded unless medical attention is sought. Even though the symptoms of ciguatera poisoning are readily separable from bac-

terial food poisoning, most doctors and public health officers do not recognize ciguatera poisoning.

Theories

Poisoning is not caused by bacterial decomposition (Arcisz, 1950; Halstead and Courville, 1967) or the spawning cycle (Arcisz, 1950; de Sylva, 1963), nor is it related to any endogenous rhythm (Arcisz, 1950). There is no evidence that ciguatera is directly related to season, chemical or physical parameters of the marine environment, rainfall, or geographic location (Cooper, 1964; Helfrich *et al.*, 1968). There is convincing evidence that tropical fishes and invertebrates causing poisoning in man have acquired toxicity as a result of their food habits (Dawson *et al.*, 1955; de Sylva, 1956, 1963; Habekost *et al.*, 1955; Halstead and Courville, 1965, 1967; Helfrich *et al.*, 1968; Randall, 1958).

Role of food web

It is generally thought that the toxin(s) originate with benthic or, in some cases, planktonic algae which are eaten first by small browsing and grazing fishes and invertebrates, then by large predatory fishes such as barracudas, snappers, groupers, and jacks and finally by man (de Sylva, 1963).

Algae in the food web

The role of fresh water bluegreen algae as causal agents for death in cattle, waterfowl, and other animals has been summarized by Schwimmer and Schwimmer (1968). Certain toxic factors have been studied in detail (Bishop *et al.*, 1959, for *Microcystis aeruginosa*; Gorham *et al.*, 1964, for *Anabaena flos-aquae*; Louw, 1950, for *Microcystis toxica*). Zarnecki (1968) has also summarized evidence relating marine bluegreen algae to toxic (ciguatoxic) food fishes. Direct extraction of toxic factors from marine bluegreen algae (Banner, 1967, 1967a; Banner *et al.*, 1960) further argues for the relationship between ciguatera and ingested food matter.

Thompson *et al.* (1957) isolated toxic bacteria from the polysaccharide sheath of bluegreen algae. Viruses have also been isolated (Goryushin and Chaplinskaia, 1968). These findings suggest that bluegreen algae *per se* may not be wholly responsible for toxin production. Randall (1968) and others have suggested that a precursor of the toxic substances involved in ciguatera might occur in an alga which exhibits no toxic properties until it is metabolized by animals.

On coral reefs zooxanthellae live symbiotically with coral polyps. These may also be involved in toxin production through bacterial or viral symbiosis. In any event, algae appear to be indicators, if not the cause, of the ciguatera cycle.

Factors influencing algal populations

It has been suggested that the ciguatera cycle may begin with colonization of newly cleaned surfaces by pioneering bluegreen algae (Dawson, 1959; Dawson et al., 1955; Randall, 1958). Initially, these may or may not be of a toxic strain. New surfaces may result from natural storm damage, from dredging, filling, and blasting (Dawson, 1959), and from military activities or the dragging of heavy anchors (Randall, 1958).

The major factors favouring the occurrence and growth of bluegreen algae appear to be high concentrations of nutrients, man-made disturbances, and high temperatures. Euthrophication favouring growth of bluegreen algae is well documented (National Academy of Sciences, 1969), as is growth of bluegreen algae at higher temperatures (Cairns, 1956; Patrick, 1969; Wallace, 1955). Dawson (1959) showed that the potentially toxic alga *Lyngbya majuscula* became dominant on Palmyra Island from the combined effect of dredging and filling activities plus polluting wastes from the local human population. The extensive reef flat area, which had become isolated from ocean circulation by dredging and filling harboured only *Lyngbya*, which occurred at mid-day water temperatures of 30.5°C. Ecological succession in fresh water bluegreen algae was noted by Wallace (1955), Mihursky (1968), and Patrick (1969). Cairns (1956) found that in an unpolluted stream, diatoms flourished at 18–20°C, green algae at 30°C, and bluegreen algae at 35–40°C. Naylor (1965) concluded that bluegreen algae seemed to be indicators of extreme thermal pollution.

Role of thermal pollution

Electrical power production using fossil-fuelled or nuclear generators is presently concentrated in temperate climates (Guyol, 1969). As power requirements increase in the less-developed, tropical countries, where hydroelectric power is scarce, increasing numbers of steam-electric generating stations will be located along coastlines near population centres, where the greatest potential harm to the marine ecosystem may occur. An example of a power station in the tropics posing several environmental questions is Florida Power and Light's Turkey Point station in lower Biscayne Bay, Florida (Bader et al., 1970). The shallow embayments, mangrove islands, and adjacent coral reefs typify the warm, tropical environment, in which the faunal elements are already living, in summer, within a few degrees of their thermal death point (de Sylva, 1969).

Some damage to this ecosystem has already occurred (Bader et al., 1970; Hagan and Purkerson, 1970; Roessler and Zieman, 1970), and is expected to increase as the 864 MW generating capacity increases from the present two conventional fossil units, having a cooling water requirement of 35.9 m³/sec, to a total output of 2,384 MW and cooling water requirement of 120.3 m³/sec. With the planned nuclear units in operation the temperature increase through the generators will be 8–9°C, compared to 6–8°C at present. With a presently planned drainage canal southward into nearby Card Sound, a maximum cooling of 1°C is anticipated prior to discharge into the Bay proper.

Summer temperatures at the mouth of the present effluent canal at Turkey Point are now 36–37°C with a recorded maximum of 40.3°C in 1969. At the same time the average ambient temperatures of Biscayne Bay was 30.5°C. As of September, 1969, about 125 acres of bottom sediment adjacent to the effluent canal was colonized by bluegreen and filamentous-green algal mat (Roessler and Zieman, 1970). In the study area Zieman (1970) has shown that algal species numbers, algal species diversity, and bluegreen algal cover correspond well with temperature patterns. Zieman's work supports the findings of Cairns (1956) and Patrick (1969), who found that prolonged temperature increases above ambient favour dominance of bluegreen algae.

Hagan and Purkerson (1970), who studied the same area of Biscayne Bay, also reported that the exposed sediments near the mouth of the canal were colonized by green algae (*Boodleopsis pusilla*) and bluegreen microalgae, alternating with diatoms. Neither Zieman nor Hagan and Purkerson identified the bluegreen algae cover to genus, but the junior author of the present paper has found *Lyngbya majuscula* and *Schizothrix calcicola* to be year-round components of this community.

Marine algae presently implicated in ciguatera food webs include *Lyngbya majuscula* (Dawson et al., 1955; Habekost et al., 1955; Randall, 1968), *Schizothrix calcicola* (Banner, 1967a; Cooper, 1964), and *Calothrix crustacea* (Dawson et al., 1955; Randall, 1968). Of these, *Lyngbya majuscula* (first reported from Biscayne Bay by Humm (1964)) is especially prevalent, both in Biscayne Bay and in the adjacent reef areas. *Schizothrix* appears frequently as an epiphyte on *Thalassia testudinum* and coarse macro-algae. *Calothrix* is not common, and is usually found on bare rock surfaces. In addition, several species of common green algae *Caulerpa* have now been shown to contain transferable toxins (Doty and Aguilar-Santos, 1970). These are found rather generally throughout the Biscayne Bay area. Since potentially toxic algae are now present, both in Biscayne Bay and in the adjacent coral reef areas of the Pennekamp State Park, we may expect that further increases in the effluent temperature and volume (Roessler and Zieman, 1970) will increase the habitat area favourable for these suspect algae.

Any significant increase in extent of the bluegreen algal cover should logically result in increased use of these plants as food by local fish populations. Should a toxic algal strain develop, one would expect an increase in the incidence of toxic herbivorous fishes. Among the inshore and reef herbivores of the Caribbean, Randall (1965) includes: *Archosargus rhomboidalis; Abudefduf saxatilis; Microspathodon chrysurus; Pomacentrus leucostictus; P. planifrons; Scarus* spp.; *Blennius* spp.; *Centropyge argi; Acanthurus bahianus; A. chirurgus; A. coeruleus* and *Monocanthus ciliatus*. These herbivores occur commonly in the environs of Biscayne Bay and are, in turn, eaten by larger predatory sport and commercial fishes (de Sylva, 1963; Randall, 1965). Because several outbreaks of ciguatera poisoning have already been reported from the reefs of nearby Key Largo (de Sylva, 1963), reef areas contiguous to Biscayne Bay may be considered as potentially ciguatoxic.

Ciguatera is insidious because it is at present impossible to detect from field tests. The implication of elevated temperatures causing the spread of bluegreen algae requires that caution be used in selecting power station sites throughout tropical areas dependent upon fish protein as a basic food.

References

Arcisz, W Ciguatera: tropical fish poisoning. *Spec. scient. Rep.*
1950 *U.S. Fish Wildl. Serv.*, (27):23 p.

BADER, R G, ROESSLER, M A and THORHAUG, A Thermal pollution
1970 in a tropical marine estuary. Paper presented to the FAO
Technical Conference on Marine Pollution and its Effects
on Living Resources and Fishing, Rome, Italy, 9–18
December 1970, FIR:MP/70/E–4:9p.

BANNER, A H Poisonous marine animals, a synopsis. *J. forens. Sci.*,
1967 12(2):180–92.

BANNER, A H Marine toxins from the Pacific. I. Advances in the
1967a investigation of fish toxins. *In* Animal toxins. Oxford,
Pergamon Press, pp. 157–65.

BANNER, A H *et al.* Observations on ciguatera-type toxin in fish.
1960 *Ann. N.Y. Acad. Sci.*, 90(3):770–87.

BISHOP, C T, *et al.* Isolation and identification of the fast-death
1959 factor in *Microcystis aeruginosa* NRC-1. *Can. J. Biochem.
Physiol.*, 37:453.

CAIRNS, J JR Effects of increased temperatures on aquatic organisms.
1956 *Ind. Wastes*, 1(4):150–2.

COOPER, M J Ciguatera and other marine poisoning in the Gilbert
1964 Islands. *Pacif. Sci.* 13(4):411–40.

DAWSON, E Y Changes in Palmyra Atoll and its vegetation through
1959 activities of man, 1913–1958. *Pacif. Nat.*, 1(2):1–51.

DAWSON, E Y, ALEEM, A A and HALSTEAD, B W Marine algae from
1955 Palmyra Island with special reference to the feeding habits
and toxicology of reef fishes. *Occ. Pap. Allan Hancock Fdn*,
(17):39 p.

DE SYLVA, DONALD P Poisoning by barracuda and other fishes.
1956 *Spec. Serv. Bull. mar. Lab. Univ. Miami*, (13):9 p.

DE SYLVA, DONALD P Systematics and life history of the great
1963 barracuda, *Sphyraena barracuda* (Walbaum). *Stud. trop.
Oceanogr. Miami*, 1:1–179.

DE SYLVA, DONALD P Theoretical considerations of the effects of
1969 heated effluents on marine fishes. *In* Biological aspects of
thermal pollution, edited by P. A. Krenkel and F. L. Parker,
Nashville, Tennessee, Vanderbilt Univ. Press, pp. 229–93.

DOTY, M S and AGUILAR-SANTOS, G Transfer of toxic algal sub-
1970 stances in marine food chains. *Pacif. Sci.*, 24(3):351–5.

GORHAM, P R *et al.* Isolation and culture of toxic strains of
1964 *Anabaena flos-aquae* (Lyngb.) de Brebisson. *Verh. int.
Verein. theor. angew. Limnol.*, 15(2):796–804.

GORYUSHIN, V A and CHAPLINSKAIA, S M (Existence of viruses of
1968 bluegreen algae.) *Mikrobiol. Zh.*, 28(2):94–7. (in Russian).

GUDGER, E W Poisonous fishes and fish poisonings, with special
1930 reference to ciguatera in the West Indies. *Am. J. trop. Med.*,
10(3):43–55.

GUYOL, N B The world electric power industry. Berkeley, Univer-
1969 sity of California Press, 366 p.

HABEKOST, R C, FRASER, I M and HALSTEAD, B W Toxicology—
1955 observations on toxic marine algae. *J. Wash. Acad. Sci.*,
45(4):101–3.

HAGAN, J E and PURKERSON, L L Report on thermal pollution of
1970 intrastate waters, Biscayne Bay, Florida. Ft. Lauderdale,
Fla., U.S. Dept. Interior, Federal Water Pollution Control
Admin., Southeast Water Laboratory, Technical Services
Lower Florida Estuary Study, 44 p.

HALSTEAD, B W and COURVILLE, D A Poisonous and venomous
1965 marine animals of the world. Invertebrates. Washington,
D.C., U.S. Government Printing Office, vol. 1:994 p.

HALSTEAD, B W and COURVILLE, D A Poisonous and venomous
1967 marine animals of the world. Vertebrates. Washington,
D.C., U.S. Government Printing Office, vol. 2:1070 p.

HELFRICH, P, PIIAKARNCHANA, T and MILES, P S Ciguatera fish
1968 poisoning. 1. The ecology of ciguateric reef fishes in the
Line Islands. *Occ. Pap. Bernice P. Bishop Mus.*, 23(14):
305–82.

HIYAMA, Y Report of an investigation of poisonous fishes of the
1943 South Seas. *Spec. Publ. Nissen Fish. exp. Sta., Odawara
Brch*, 137 p. Issued also as *Spec. scient. Rep. U.S. Fish
Wildl. Serv.*, (25):188 p. (1950, without original plates).

HUMM, H J Epiphytes of the sea grass, *Thalassia testudinum*, in
1964 Florida. *Bull. mar. Sci. Gulf Caribb.*, 14(2):306–41.

KRENKEL, P A and PARKER, F L (Eds.) Biological aspects of thermal
1969 pollution. Nashville, Tennessee, Vanderbilt University
Press, 407 p.

LOUW, P G J The active constituent of the poisonous algae, *Micro-
1950 cystis toxica* Stephens. *S. Afr. ind. Chem.*, 4:62.

MARTYR, P De Orbe Novo, the eight decades of Peter Martyr
1555 d'Anghere. 372 (1912 English translation by F. A. McNutt,
New York, G. P. Putnam's Sons).

MIHURSKY, J A Effects of thermal discharges on the ecology of a
1968 waterway. *In* Proceedings Fifth Annual Environmental
Health Research Symposium. Sponsored by New York
State Department of Health, New York Water Pollution
Control Association, Albany, N.Y., 16 May, 1968, pp. 22–30.

MOWBRAY, L L Fish poisoning (ichthyotoxismus). *Bull. N.Y. zool.
1916 Soc.*, 19:1422–3.

National Academy of Sciences, U.S., Eutrophication: causes,
1969 consequences, correctives. Proceedings of an international
symposium. Washington, D.C., National Academy of
Sciences, 661 p.

NAYLOR, E Effects of heated effluents upon marine and estuarine
1965 organisms. *Adv. mar. Biol.*, 3:63–103.

PATRICK, R Some effects of temperature on freshwater algae. *In*
1969 Biological aspects of thermal pollution, edited by P. A.
Krenkel and F. L. Parker, Nashville, Tennessee, Vanderbilt
University Press, pp. 161–90.

RANDALL, J E A review of ciguatera tropical fish poisoning, with a
1958 tentative explanation of its cause. *Bull. mar. Sci. Gulf
Caribb.*, 8(3):236–67.

RANDALL, J E Food habits of reef fishes of the West Indies. *Stud.
1965 trop. Oceanogr. Miami*, 5:665–840.

RANDALL, J E Marine algae as a possible source of ciguatera toxins.
1968 *In* South Pacific Commission, Seminar on Ichthyosarco-
toxism, Noumea, New Caledonia, August 1968, p. 237
(Abstract).

ROESSLER, M S and ZIEMAN, J C The effects of thermal additions
1970 on the biota of southern Biscayne Bay, Florida. *Proc. Gulf
Caribb. Fish. Inst.*, 22:136–45.

RUSSELL, F E Marine toxins and venomous and poisonous marine
1965 animals. *Adv. mar. Biol.*, 3:255–384.

SCHWIMMER, M and SCHWIMMER, D Medical aspects of phycology.
1968 *In* Algae, man, and the environment, edited by D F
Jackson, Syracuse University Press, pp. 279–358.

THOMPSON, W K, LAING, A C and GRANT, G A Toxic algae. 4.
1957 Isolation of toxic bacterial contaminants. *Rep. Def. Res.
Kingston Lab. Can.*, (51):1–7.

WALLACE, N W The effect of temperature on the growth of some
1955 freshwater diatoms. *Notul. Nat.*, (280):1–11.

ZARNECKI, S Algae and fish relationships. *In* Algae, man, and the
1968 environment, edited by D. F. Jackson, Syracuse University
Press. pp. 459–77.

ZIEMAN, J C The effects of a thermal effluent stress on the sea-
1970 grasses and macro-algae in the vicinity of Turkey Point,
Biscayne Bay, Florida. Dissertation, University of Miami,
129 p.

Ciguatera et Intervention Humaine sur les Ecosystèmes Coralliens en Polynésie Française

R. A. Bagnis*

Ciguatera and man's influence on the coral ecosystems of French Polynesia

In many atolls of French Polynesia, outbreaks of ciguatera fish poisoning have been reported after what could be only physical modifications of the biotope had been observed on limited areas of coral reefs or lagoons. The author gives some results of a study concerning the conditions in which the disease has appeared and developed after important aggression on atoll benthos. In the light of this information he considers ciguatera fish poisoning as one aspect of the reaction of coral reef biota to different types of aggression to which it has been subjected.

Ciguatera y la influencia del hombre sobre los ecosistemas coralinos en la Polinesia francesa

La aparición de envenenamientos (ciguatera) por consumo de pescado ha sido relatada en muchos atolones de la Polinesia francesa, después de haberse producido cambios físicos dentro de una limitada área de los arrecifes coralinos o lagunas.

El autor presenta algunos resultados de un estudio sobre las condiciones en las cuales se ha manifestado y desarrollado la enfermedad después de importantes modificaciones en la comunidad bentónica de los atolones. A la luz de estas informaciones, se considera el envenenamiento (ciguatera) por consumo de pescado como un aspecto de la reacción biótica de los arrecifes coralinos a diferentes tipos de cambios violentos a los cuales han sido sometidos.

---

* Institut de Recherches Médicales Louis Malardé, Papeete, Tahiti, Polynésie, Française.

DANS la plupart des mers tropicales, le long de la ceinture corallienne, entre les 25èmes degrés de latitude nord et sud, sévit depuis fort longtemps une endémie souvent méconnue, désignée actuellement par l'épithète ciguatera. Il s'agit d'une forme d'ichthyosarcotoxisme caractérisée par un syndrome clinique polymorphe dont les éléments principaux sont : vomissements, diarrhée, troubles de la sensibilité, douleurs articulaires et musculaires, frilosité, fatigue marquée, troubles du rythme cardiaque, chute de la tension artérielle. L'évolution est en règle bénigne. Dans les formes graves s'installent parfois des paralysies ou des états de collapsus pouvant entraîner exceptionnellement la mort.

Toutes ces manifestations surviennent après la consommation de la chair de poissons ou plus rarement d'invertébrés en parfait état de fraîcheur, appartenant à des espèces benthiques multiples et variées, habituellement comestibles.

L'importance du phénomène est loin d'être négligeable puisqu'en 1966, 8 pour cent de la population de Tahiti et 12 pour cent de celle de Bora Bora ont été intoxiqués au moins une fois (Bagnis, 1969, 1969a). Dans un atoll des Iles Tuamotu le taux d'endémie a même atteint 60 pour cent à tel point que les habitants l'ont pratiquement déserté depuis. Fait plus grave : la ciguatera constitue un obstacle capital à l'exploitation des ressources lagunaires, en particulier au développement de l'aquaculture dans ces gigantesques viviers naturels que représentent les lagons des atolls du Pacifique.

Phénomène très ancien, puisque déjà décrit au 16ème siècle par les navigateurs qui découvrirent le Nouveau Monde, la ciguatera semble, a priori, pouvoir être difficilement liée par des relations de cause à effet à une pollution déterminée de type chimique, bactérien ou thermique.

Phénomène variable dans le temps et dans l'espace, mais parfois très localisé, la ciguatera paraît davantage en rapport avec une modification progressive ou soudaine de certaines conditions biomarines d'un lieu donné. Dans une telle éventualité, on est forcé de prendre en considération quelques faits empiriques patents, relevés en Polynésie Française, et dans lesquels l'intervention humaine directe ou indirecte sous ses diverses formes, a pu jouer un rôle déterminant dans la biogénèse d'une flambée de ciguatera, en modifiant les biotopes et biocénoses complexes et fragiles des écosystèmes coralliens considérés.

Données de l'empirisme recueillies en Polynésie Française

Eu égard aux notions précédemment évoquées, nous livrons à la réflexion, sans la moindre interprétation, quelques hypothèses formulées par les insulaires pour expliquer la vénénosité des poissons en certaines zones relativement limitées du récif ou du lagon :

Le naufrage d'un navire avec présence d'une épave plus ou moins immergée à Faaa (Tahiti), dans le chenal Raiatea-Tahaa (Iles Sous le Vent), Niau, Takaroa (Tuamotu),
Les décharges de boîtes de conserve métalliques à Vairao et Afaahiti (Tahiti),
L'immersion de ferraille à Faaa (Tahiti), Mangareva (Gambier), Faaoné (Tahiti),

L'immersion de blocs de ciment armé pour fixation de bouées d'amarrage ou de balisage à Tikehau, Fakarava (Tuamotu), Vairao (Tahiti), Mangareva (Gambier),
Les travaux de dragage ou d'élargissement des passes à Bora Bora (Iles Sous le Vent), Fakarava (Tuamotu), Faaoné (Tahiti),
Les constructions de digues ou wharfs à Hikueru, Tikehau (Tuamotu), Miva Oa (Marquises).

Effectivement, il est troublant de constater que dans ces cas les premiers poissons toxicophores sont apparus au voisinage immédiat des emplacements concernés par l'événement et dans les mois ou années qui ont suivi cet événement.

Cette influence souvent néfaste des corps étrangers introduits dans le milieu récifolagunaire a si bien sensibilisé la population de certains atolls déjà affectés par des antécédents du même ordre, qu'en 1966 à Hikueru nous avons vu les habitants retirer du lagon au prix d'efforts colossaux, l'épave d'un hydravion coulé accidentellement quelques semaines auparavant.

Données analogues notées dans d'autres régions du Pacifique

Les faits que nous venons d'énoncer ne sont pas nouveaux, puisque lors de la guerre du Pacifique certains atolls des Iles de la Ligne où avaient été immergées des quantités massives de matériel de guerre usagé, se sont révélées toxicogènes dans les années qui ont suivi (Halstead et Courville, 1967). Un exemple précis est fourni par l'épidémie de l'Ile de Fanning survenue un peu plus d'un an après de l'immersion (Ross, 1948). Dans la même archipel, l'Ile de Washington, où aucune immersion du même genre n'avait été effectuée, est restée indemne. Les premiers poissons vénéneux de cet atoll ont été pêchés, en 1968, dans les mois qui ont suivi l'échouage d'un navire sur le récif et dans l'environnement immédiat de l'épave (Helfrich et al., 1968).

Etude d'une flambée de ciguatera apparemment liée à une importante intervention humaine

A la lumière des données précédentes, il nous a paru indispensable de mieux appréhender les circonstances dans lesquelles une zone récifolagunaire non toxicogène pouvait abriter soudain des espèces ciguatérigènes à la suite de remaniements artificiels importants de son benthos.

L'atoll de Hao, dans les Tuamotu, nous en a fourni l'occasion. Cette île basse ouverte était l'une des rares de l'archipel à n'avoir jamais connu la ciguatera. A partir de 1965, la construction d'un port et d'un aérodrome dans la partie nord-est de l'anneau corallien, entre le village d'Otepa et la passe Kaki (fig 1), a entraîné un accroissement subit de la population. En même temps étaient mis en œuvre d'importants travaux de dragage et terrassement, de construction de quais et d'édification de digues. La passe fut également élargie et approfondie. Tous ces facteurs pouvaient donc concourir à l'éclosion d'une flambée d'ichthyosarcotoxisme (Bagnis, 1967, 1967a).

Effectivement, dès le 2e semestre 1966, un certain nombre de syndromes cliniques commencèrent à éveiller l'attention médicale et au cours de l'année 1967, la présence de ciguatera devint indiscutable.

Nous donnons ci-dessous brièvement les conclusions principales de l'enquête épidémiologique réalisée en 1968,

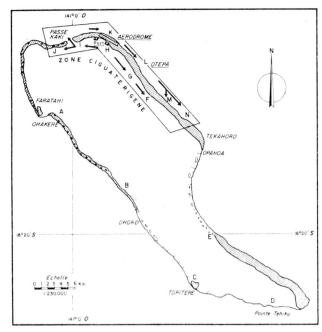

Fig. 1.

les données méthodologiques et les résultats détaillés, ayant fait l'objet d'une publication antérieure (Bagnis, 1969a).

(a) Les premiers poissons vénéneux ont été pêchés dans la zone de "beaching" des navires débarquant le matériel indispensable au début des travaux, zone H de la fig 1. Pendant quelques mois l'endémie ciguatoxique est restée limitée au voisinage immédiat de cette plage de débarquement. Ensuite l'extension dans le temps et dans l'espace, illustré fig 2, a suivi grossièrement l'ordre chronologique et topographique des remaniements du substrat marin déjà évoqués, avec un temps de latence que l'on peut estimer entre 1 et 2 ans suivant les cas.

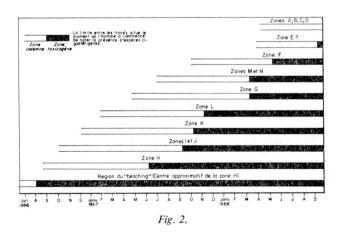

Fig. 2.

(b) Les premières intoxications ont été provoquées par des poissons herbivores, coralliphages ou détritivores tels les chirurgiens, perroquets, balistes, mulets. Les carnivores purs tels les loches, mérous, bec-de-cane et autres lutjans ne sont devenus toxicophores pour l'homme et les animaux qu'environ 18 mois plus tard, ce qui vient étayer la thèse généralement admise, d'une transmission des composés toxiques à travers la chaîne alimentaire.

(c) En janvier 1969, environ 30 mois après le début de la

TABLEAU 1

Famille de poissons en cause	Nombre de cas de ciguatera
Carangidae	48
Scaridae	47
Acanthuridae	30
Serranidae	28
Mullidae	22
Siganidae (Teuthidae)	19
Scombridae	15
Lutjanidae	14
Labridae	11
Balistidae	8
Lethrinidae	7
Priacanthidae	4
Mugilidae, Holocentridae	3
Sparidae	2
Kuhlidae	1

flambée, 35 espèces représentant 16 familles ont été signalées ciguatérigènes.

L'accroissement progressif du nombre des espèces vénéneuses est représenté fig 3.

(d) Sur le plan humain, pendant le même laps de temps, plus de 220 cas d'intoxications ont été officiellement recensés (fig 4).

Pour objectiver les résultats de notre enquête épidémiologique, nous avons procédé en septembre 1968,

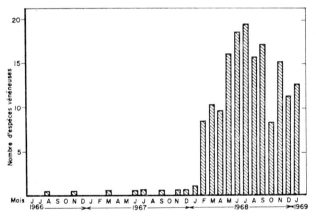

Fig. 3.

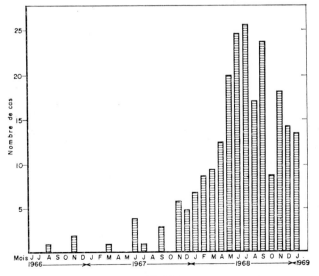

Fig. 4.

[599]

sur tout le pourtour de l'anneau corallien, à la capture d'environ 400 poissons appartenant à 32 espèces qui s'étaient avérées vénéneuses précédemment. Les résultats de l'expérimentation animale ont montré qu'environ 4 ans après le début de l'intervention humaine massive sur le substrat corallien, les espèces toxicogènes étaient uniquement cantonnées dans le voisinage des zones lagunaires et océaniques concernées par ladite intervention (F à N de la fig 1). La situation n'a toujours pas évolué depuis, puisque des poissons provenant des régions A à E de la fig 3 sont encore aujourd'hui couramment consommés sans dommage par la population.

Rôle éventuel des microagressions réitérées

La corrélation entre l'intervention de l'homme sur le milieu marin et la biogénèse de la vénénosité des poissons n'est pas toujours facile à mettre en évidence. Et pourtant, 2 ordres de faits laissent à penser que les massifs coralliens lagunaires et océaniques pourraient être extrêmement sensibles aux multiples microagressions localisées dont ils sont parfois l'objet de façon intempestive.

Le premier concerne les flambées de ciguatera survenues à Marutea-Sud, Takaroa, Hikueru, Raraka, Aratika dans les îles Tuamotu, 2 à 3 ans après une saison de plongée à l'huître perlière. Les poissons vénéneux, dans la plupart des cas, provenaient du secteur limité du lagon ouvert à la plongée. Or il faut savoir que lors d'une saison de plongée, un atoll peuplé habituellement d'une centaine d'habitants, voit débarquer 500 à 1 000 plongeurs, qui pendant plusieurs mois fouillent systématiquement chaque bloc corallien du secteur, blessent souvent les polypes pour en arracher les nacres solidement amarrées par leur byssus et laissent sur le fond sableux quantité de débris coralliens morts. La succession des saisons de plongée massive maintient l'endémicité ciguatérique à un certain niveau. A cet égard, il est intéressant de noter que dans les atolls de Raraka et Aratika, où aucune saison de plongée organisée ne s'est déroulée pendant une dizaine d'années, la plupart des poissons sont redevenus comestibles.

Le deuxième a trait à une observation faite dans les atolls fermés du nord-est de l'archipel des Tuamotu. Les espèces vénéneuses, peu nombreuses, sont localisées au voisinage immédiat des points de débarquement des baleinières de récif, sur le versant océanique des atolls.

Rôle favorisant des pluies importantes et répétées dans le déterminisme de certaines formes d'ichthyosarcotoxisme non ciguatériques

Les bénitiers de l'espèce *Tridacna maxima* sont consommés abondamment en Polynésie Française. Même ceux provenant de zones ciguatérigènes sont généralement comestibles. Or, entre les mois d'avril et juin 1964, les tridacnes de la passe de Bora Bora furent à l'origine d'une trentaine d'intoxications humaines graves dont deux mortelles. La symptomatologie rappelait par quelques aspects la maladie de Minamata (Bagnis, 1967, 1967a). Des études toxicologiques faites à Hawaii (Banner, 1967) sur des échantillons malheureusement prélevés bien après la flambée toxique ne permirent pas de trouver la trace d'un polluant chimique déterminé, bien que certains tissus tels les manteaux et les viscères se soient avérés toxiques pour l'animal.

Cet incident s'étant produit à la suite de pluies abondantes, il n'est pas déraisonnable de penser que ces précipitations ont pu déverser vers la passe unique de l'île une assez grande quantité de pesticides utilisés peu de temps auparavant pour protéger les cultures de pastèques et détruire certains crabes de sable.

Signalons que des tests de toxicité sur l'animal réalisés 5 ans après la flambée ne montrent plus la moindre trace de toxicité alors que dans la région de nombreux poissons, déjà vénéneux en 1964, le sont encore.

Commentaires

Dans ce dernier cas, l'action directe d'un polluant est probable, et elle doit être considérée comme une première alerte, étant donné la consommation abusive d'insecticides et pesticides qui est pratiquée dans de nombreuses îles.

En ce qui concerne la ciguatoxicité, la pollution directe ne paraît pas entrer en ligne de compte. Mais elle pourrait très bien intervenir de façon plus insidieuse et progressive à côté de nombreuses autres agressions que l'homme fait subir au milieu marin. En d'autres termes, la vénénosité des poissons benthiques pourrait être le facteur révélateur d'une réaction de certains biotopes coralliens d'eau peu profonde aux modifications entraînées par les attaques mécaniques ou chimiques dont ils sont l'objet, attaques qui créeraient des conditions peu propices au maintien de l'équilibre biologique des diverses populations benthiques. Il s'agirait d'une véritable réaction d'un écosystème vis-à-vis d'une agression, réaction qui se traduirait soit par la libération de zooxanthelles ou la prolifération de virus ou autres ultra-microorganismes commensaux toxicogènes. Ce phénomène immunologique se poursuivrait jusqu'à ce que l'écosystème corallien troublé se soit adapté aux nouvelles conditions ou jusqu'à ce que le facteur agresseur ait disparu ou ait été réduit suffisamment pour ne plus nuire à l'équilibre harmonieux entre les diverses biocénoses récifolagunaires.

Etant donné la menace que fait peser la ciguatera sur l'exploitation éventuelle des eaux insulaires tropicales, les réflexions précédentes, inspirées de données empiriques, devraient ouvrir la voie à une investigation plus poussée du milieu lagunaire.

Bibliographie

BAGNIS, R A Les empoisonnements par le poisson en Polynésie
1967 française: étude chimique et épidémiologique. *Revue Hyg. Méd. soc.*, 15(7):619–46.

BAGNIS, R A A propos de 30 cas d'intoxication par bénitiers
1967a dans une île de la Société. *Bull. Soc. Path. exot.*, 60:580–92.

BAGNIS, R A Mise au point sur les intoxications par poissons
1969 vénéneux du type ciguatera. *Presse méd.*, 77(2):59.

BAGNIS, R A Naissance et développement d'une flambée de
1969a ciguatera dans un atoll de l'archipel des Tuamotu. *Revue Corps Santé*, 10(6):783–95.

BANNER, A H Marine toxins from the Pacific. 1. Advances in the
1967 investigation of fish toxins. *In* Animal toxins. Oxford, Pergamon Press, pp. 157–65.

HALSTEAD, B W and COURVILLE, D A Poisonous and venomous
1967 marine animals of the world. Vol. 2. Vertebrates. Washington, D.C., U.S. Government Printing Office, 1070 p.

HELFRICH, P, PIYAKARNCHANA, T and MILES, P Ciguatera fish
1968 poisoning. 1. The ecology of ciguateric reef fishes in the Line Islands. 2. General patterns of development in the Pacific. *Occ. Pap. Bernice P. Bishop Mus.*, 23(14):305–82.

ROSS, S G Preliminary report on fish poisoning at Fanning Island
1948 (Central Pacific). *Med. J. Aust.*, 2(21):617–21.

Section 7

SCIENTIFIC BASIS FOR INTERNATIONAL LEGISLATIVE CONTROL OF MARINE POLLUTION IN THE INTERESTS OF MARINE RESOURCES AND FISHERIES

Summary of Discussion

International regulations for other pollutants, besides oil and nuclear substances, must be devised, particularly for those types of pollutants which are of high toxicity and are especially persistent. The Conference singled out mercury and persistent pesticides among pollutants requiring special attention.

As far as DDT and related pesticides were concerned, many participants favoured an outright ban, at least as soon as effective narrow-spectrum substitutes become available. Also the development of resistance by insect species, and the fact that predator species important for control often are destroyed by presently used pesticides, suggest the adoption of other means of pest control. Some participants from developing countries felt that in the absence of inexpensive substitutes, they would have to continue, at least for the time being, using DDT and related pesticides.

On mercury pollution, the relatively high concentration in the environment, as well as the lack of data on the sources, concentrations, and sinks in many areas that are likely to be affected, were noted. Special care had to be taken not to add further to mercury pollution. Some participants favoured an outright ban, although regulations would have to be introduced carefully to mitigate the impact on the large segments of industry dependent on its use. Much could be achieved by improving plant technology. Avoidance of mercury pollution is relatively simple and not expensive, especially for new plant installations.

The Conference felt more emphasis should be placed on the overall impact of pollution on the environment because of the interrelation of atmospheric and aquatic processes. As far as atmospheric pollution is concerned, its impact on the marine environment has not yet been scientifically determined, though it is known that several pollutants are carried in the atmosphere to the sea.

Factors influencing the selection of waste disposal areas include the type of pollutant (degree of toxicity and persistence), bottom topography, the type of containers and depth of water.

The criteria for the selection of disposal areas in marine waters must be provided to the policy makers by marine biologists on the basis of knowledge concerning evaluating the possible effects in each specific situation on the living resources and the marine ecosystem.

Effective control of pollution will involve the setting up of maximum permissible levels for pollutants, concentration in effluents, provisions for registering of dumping activities, monitoring levels of pollutants in the waters and in marine organisms and setting up of an efficient institutional framework for the anti-pollution fight. As a guide to industries, the drawing up and dissemination of principles for prevention of pollution from specific causes can be recommended.

The effectiveness of pollution control measures and the time lag in making them operational depends, *inter alia*, on administrative feasibility and cost factors. Ecological areas for the purpose of international management and control of pollution should be established, in the first place, with regard to fish stocks and local ocean currents. To become politically acceptable, economic and social factors also have to be considered in regional arrangements.

There are several documented examples demonstrating the causal link between pollutants and damage to fisheries and the marine environment. In many cases, however, it is difficult to assess the amount of damage or to establish the source responsible for the damaging pollutants.

Concluding remarks by Convener

Some international measures are now taken to control the pollution of the sea by oil and radioactive substances. Measures should also be taken against the continued introduction into the marine environment of certain other pollutants such as chlorinated hydrocarbons and mercury on account of their high toxicity and persistency.

Pollutants carried in the atmosphere contribute to marine pollution, but it is not known to what extent.

Regional integrated areas can be defined by using ecological criteria such as the distribution of fish stocks and the systems of ocean currents; and the recognition of such areas should form an initial step towards international control of marine pollution.

Policy makers require further information on the criteria for establishing regulations for the protection of resources threatened by dumping of waste materials into the ocean.

The Control of Marine Pollution and the Protection of Living Resources of the Sea

G. Moore*

A comparative study of international controls and national legislation and administration

La lutte contre la pollution marine et la protection des ressources biologiques de la mer

Etude comparative des mesures de lutte prévues par les législations et les administrations nationales

Ce document traite tout d'abord de la lutte contre la pollution des mers à l'échelon international, exposant l'état actuel du droit international et les activités des organismes internationaux. L'auteur analyse ensuite les principaux types de législation nationale traitant de la pollution des mers, en citant tout d'abord certains des problèmes les plus généraux que doit aborder cette législation, tels que l'étendue des eaux territoriales et les questions qui s'y rattachent. Certaines lois récemment adoptées qui ne se conforment pas à la juridiction traditionnelle sont examinées et les lacunes des structures administratives nationales sont exposées. La conclusion met l'accent sur l'insuffisance de la législation nationale, qui se confine traditionnellement aux limites des eaux territoriales, pour faire face à la pollution des mers et préconise le développement de l'action internationale et régionale qui devrait se substituer à l'extension unilatérale de la juridiction nationale.

Lucha contra la contaminacion de las aguas y protección de los recursos vivos del mar

Estudio comparativo de las medidas internacionales y de las legislaciones y administraciones nacionales

El autor empieza examinando la lucha internacional contra la contaminación de las aguas del mar, describiendo la situación actual del derecho internacional y las actividades de los organismos internacionales. A continuación, analiza los principales tipos de legislación nacional sobre contaminación de las aguas del mar, mencionando algunos de los problemas generales que dicha legislación tiene que abordar, tales como las cuestiones de sus objetivos y de ámbito territorial. Se examinan algunas leyes recientes que no responden a los conceptos jurisdiccionales tradicionales y se ponen de relieve las deficiencias de las estructuras administrativas nacionales. Para terminar, el autor señala la inadecuación de la legislatión nacional, confinada tradicionalmente dentro de sus propios límites territoriales para afrontar el problema de la contaminación de las aguas del mar, y propone, como alternativa a una extensión unilateral de la jurisdicción nacional, un sistema de control más internacional y regional.

UNTIL comparatively recently there were few conflicts between uses of the oceans. Navigation and fishing were not incompatible and the amount of waste disposed of into the sea was insignificant in comparison with its vastness. Within the last hundred years, however, the growth of technology has made increased production of food from the ocean possible and given more people time and money to enjoy the sea for recreation. It has, at the same time, intensified the use of the sea for waste disposal and increased incidental pollution from powered ships and mining. And there is now an awakening realization that these uses, unrestrained, are no longer completely compatible.

This paper examines the extent to which the international community and national communities faced with this conflict have acted to control those uses that threaten sources of food and recreation.

INTERNATIONAL MEASURES

International management of marine pollution to date has been concerned primarily with the control of pollution-causing activities occurring outside the territorial jurisdiction of the individual states. There, state powers of regulation, based solely on the tie of nationality, are clearly insufficient to provide any adequate system of control on their own. Until recently, international activity has been directed almost exclusively at the problems of oil pollution from ships and pollution by radio-active substances. Where attempts have been made to cover all forms of marine pollution, these have resulted in convention provisions so broad as to appear more as recommendations than as the basis of a comprehensive system of control.

In discussing international management, this paper will examine those international conventions that regulate activities causing marine pollution and the activities of international bodies in this field.

International Conventions: Pollution by oil

Oil pollution endangers fish and other living resources of the sea both directly and through its dispersal. The problem is widespread; Thor Heyerdahl on both of his recent *Ra* expeditions has reported to the United Nations alarming sightings of oil slicks in mid ocean stretching from horizon to horizon. It has also received a good deal of attention from the international community over the last half century and has been the subject of several conventions.

International Convention for the Prevention of Pollution of the Sea by Oil, London, 1954/1962/1969

A Preliminary Conference on Oil Pollution of Navigable Waters was convened in Washington in 1926. No convention came out of this conference, however, nor out of a subsequent movement in 1934, and it was not until 1954 that the first International Convention for the Prevention of Pollution of the Sea by Oil (London Convention) was signed. Apart from a few general provisions in the 1958 Geneva Conventions on the Law of the Sea, the international treaty law on marine oil pollution control is contained in this London Convention, as amended in 1962 and 1969, and in the two Brussels Conventions signed in November 1969.

As it now stands (the 1962 amendments entered into force in May 1967), the London Convention prohibits the discharge of oily wastes within certain prohibited zones from merchant ships, including tankers, over a certain size and prohibits entirely such discharges from new ships over 20,000 GT, except where the discharge results from damage or is dictated by safety considerations. The Convention also contains certain provisions relating to the availability of facilities for the reception of oily wastes in ports to provide an alternative to their discharge at sea and provisions requiring the installation of ship's equipment designed to prevent oil pollution. To facilitate the detection and prosecution of contraventions, each ship subject to the Convention is required to maintain an oil record book detailing certain operations capable of causing pollution and is required to hold the

---

* Legislation Branch, Legal Office, FAO, in collaboration with the Fishery Resources Division, FAO, Rome.

book available for inspection in the ports of any contracting party.

Although the London Convention, as amended in 1962, presents a good deal of progress in the control of oil pollution, the problem is still far from being solved. The provisions of the Convention cover only certain ships, only certain areas of the sea and only intentional discharges and the measures of enforcement are basically inefficacious. Some of these deficiencies have been remedied in the latest amendments to the Convention adopted in November 1969. These amendments have been communicated to the contracting governments and will enter into force after their acceptance by two-thirds of these governments. Under these amendments the basic concept of prohibited zones has been removed from the Convention. All discharges of oily wastes from ships covered by the Convention would be prohibited, except for slight discharges en route, clearly defined in terms of rate of flow and cargo carrying capacity. The amendments also include stronger provisions with regard to ship equipment and the operations to be detailed in oil record books. But the powers of enforcement are still left with the government of the state in which the offending ship is registered, which will have less interest in discovering and confirming violations than the coastal states affected by such violations. The present rather cumbersome system whereby one state notifies another of suspected violations by ships flying its flag, although supported by the oil record book system is still inadequate. Suggestions for improving the efficacy of these enforcement procedures range from the increase of the jurisdiction and powers of the coastal state to the establishment of international enforcement authorities. But no international solution has as yet been adopted. As a result, while numerous prosecutions have been initiated by states against ships infringing the provisions of the Convention within their own territorial waters, prosecutions by flag states following illicit discharges on the high seas have been rare.

The Brussels Conventions, 1969

Two Conventions were adopted in Brussels in November 1969 which sought to cover other gaps left by the London Convention. The first Convention, the International Convention relating to Intervention on the High Seas in Cases of Oil Pollution Casualties (the "Public Law Convention"), constitutes a significant departure from traditional concepts of the freedom of the high seas. Under its provisions, coastal states faced with a grave and imminent danger to their coastlines or related interests from pollution or threats of pollution of the sea by oil following a "maritime casualty" would be allowed to take such action on the high seas as may be necessary to avert or mitigate that danger (Art. 1).

A coastal state's "related interests", as defined in the second article of the Convention, include the conservation of living marine resources and wildlife, tourist attractions, the health of the coastal populations and maritime activities, including fishing upon which the livelihood of the persons concerned depends. Certain procedures of notification and consultation are envisaged by the Convention, although these may be waived in cases of extreme urgency (Art. 3).

The second Convention, the International Convention on Civil Liability for Oil Pollution Damage (the "Private Law Convention"), provides for the liability of a ship for all pollution damage caused in the territory or in the territorial waters of another contracting state by oil which has escaped or has been discharged from that ship. Negligence need not be proved, although the owner may be exempted from liability where the damage resulted from acts completely beyond his control. The Convention also contains certain provisions designed to ensure the financial capability of owners to settle prospective pollution damage claims. In general, the scheme of liability envisaged under the Private Law Convention goes somewhat further than a similar scheme proposed by the major oil transporting companies under their Tanker Owners' Voluntary Agreement Concerning Liability for Oil Pollution (Tovalop) signed earlier in the year.

In addition to adopting the two Conventions, the Brussels Conference also suggested the setting up of an International Compensation Fund and requested the Intergovernmental Maritime Consultative Organization (IMCO) to prepare a draft scheme of compensation for oil pollution damage, based on the existence of such a fund, to provide for the full and adequate compensation of all victims on the basis of strict liability.

The Geneva Conventions, 1958

The Geneva Conventions of 1958 added little to the existing 1954 Convention on oil pollution, although Article 24 of the Convention on the High Seas did provide that every state should draw up regulations to prevent the pollution of the sea by oil discharged not only from ships but also from pipelines, and pollution resulting from the exploitation and exploration of the seabed and its subsoil. Articles 5.1 and 5.7 of the Convention on the Continental Shelf also cover pollution by oil as well as other forms of pollution. "Any unjustifiable interference with navigation, fishing or the conservation of the living resources of the sea" as a result of exploration and exploitation operations is forbidden and coastal states are obliged to undertake "all appropriate measures for the protection of the living resources of the sea from harmful agents", in the safety zones around installations for the exploration and exploitation of the resources of the continental shelf.

Regional agreement for cooperation in dealing with pollution of the North Sea by oil

The *Torrey Canyon* disaster in 1967 emphasized the desirability of cooperation between countries threatened by a particular discharge or spillage of oil in averting or mitigating the danger of pollution damage. The Council of IMCO called for the establishment of international or regional schemes of cooperation, and the first of these, the Agreement for Cooperation in Dealing with Pollution of the North Sea by Oil, 9 June 1969, was set up in June 1969 by eight countries bordering the North Sea (Belgium, Denmark, France, the Federal Republic of Germany, the Netherlands, Norway, Sweden and the United Kingdom). Under this Agreement the North Sea is divided into administrative zones, in each of which one contracting party (or two or three in the case of zones of joint responsibility) is to be primarily responsible for initial assessments and action on oil pollution threats. The Agreement also provides for the exchange of information and reports on the existence of pollution threats, and for the making and handling of requests for assistance.

Pollution by radioactive substances

The problem of the pollution of the sea by radioactive substances is more recent than that of oil pollution, and has received correspondingly less treatment in international conventions and treaties.

A move by some states at the 1958 Conference on the Law of the Sea, in Geneva, to adopt a provision banning all discharges of radioactive substances into the sea proved unsuccessful. But one article (Art. 25) was adopted which committed every state "to take measures to prevent pollution of the seas from the dumping of radioactive waste, taking into account any standards and regulations which may be formulated by the competent international organizations".

The second paragraph of Article 25 is more broadly drafted and obliges states to "cooperate with the competent international organizations in taking measures for the prevention of pollution of the seas or air space above, resulting from any activities with radioactive materials or other harmful agents". This appears to cover not only the disposal of radioactive wastes but also the testing of civil or military nuclear devices, as well as pollution caused by harmful non-radioactive substances. The Conference urged those competent international organizations, in particular the International Atomic Energy Agency (IAEA), to press forward with their task of assisting states in the promulgation of standards and international regulations for the prevention of radioactive pollution that would adversely affect man and his marine resources.

A few treaties besides the 1958 Geneva Conventions have dealt with the problem of the radioactive contamination of the sea. In 1959 the Antarctic Treaty expressly prohibited nuclear explosions and the disposal of radioactive waste in the Antarctic (Art. V(i)). Four years later the Nuclear Test Ban Treaty outlawed test explosions anywhere in the atmosphere, in outer space and under water and this included territorial waters and the high seas. Certain aspects of the problem were also considered in the 1960 International Convention for the Safety of Life at Sea which dealt, *inter alia*, with the safety of crew, passengers, waterways and food and water resources from radiation from nuclear ships and the transport of radioactive substances as cargo. The Convention requires that the design and construction of nuclear reaction installations be examined and approved by the competent national authorities. A safety assessment must also be prepared for each ship and made available to the competent authorities of other countries which the ship intends to visit and a safety certificate issued. Radioactive substances carried as cargo, on the other hand, are treated as but one of nine classes of dangerous cargoes, all of which are subjected to various requirements with respect to packaging, labelling, storage and documentation.

Under the aegis of IAEA or the European Nuclear Energy Agency, several treaties have been signed during the last ten years* dealing with the question of civil

* Convention on Third Party Liability in the field on Nuclear Energy, Paris, 1960, supplemented and modified in 1963 and 1964; Vienna Convention on Civil Liability for Nuclear Damage, Vienna, 1963, and Convention on the Liability of Operators of Nuclear Ships, Brussels, 1962.

liability for damage caused by nuclear incidents. Although not specifically designed to prevent such incidents, such agreements clarifying liability stimulate improved safeguards for the prevention of nuclear incidents and the minimization of damage therefrom. Other agreements such as the Treaty on Prohibiting Emplacement of Nuclear Weapons on Seabed and Ocean Floor, signed in February 1971, may also have the effect of preventing incidents that might lead to widespread radioactive pollution damage.

Although little progress has been made so far towards a world-wide system for controlling marine pollution by radioactive substances, something more positive has been accomplished on a regional basis. The system set up between the EURATOM countries is a case in point. Under the EURATOM Treaty of 1959, it is the task of the Commission to establish basic standards, including maximum permissible levels of contamination, for the protection of workers and the general public. Each member state is bound to enact the legislation necessary to ensure compliance with the basic standards in its territory, but each has a free hand in the choice of the form and means of achieving those standards, although it is required to report to the Commission on the legislation and other measures adopted. The Commission, for its part, has the power to inspect control facilities and the right to make recommendations to the individual states, both on the measures adopted by them and on the general level of radioactivity in the atmosphere, soil and water. In cases of urgency, the Commission may direct a member state to take such measures as may be necessary to ensure that the basic standards are not exceeded and may stipulate a time limit for the completion of those measures. If the measures are not taken before the expiration of the time limit, the matter may be referred directly to the Court of Justice of the Community. An important aspect of the EURATOM system is the requirement that member states submit to the Commission detailed information on any plan for the disposal of radioactive waste at least six months before that plan is to be put into operation, so that the Commission may determine whether its operation is likely to involve radioactive contamination of the water, soil or atmosphere of any other state.

It has been suggested that the EURATOM system of control might serve as a model for an international system of control of radioactive pollution and possibly of other forms of marine, or even environmental, pollution. However, such a system of prior notification and recommendation might well prove unwieldy if merely expanded onto an international level. Such control is probably easier and more effective at the regional level, through regional agencies, although there would be a need for coordination of their activities on a world-wide scale.

Conventions for the control of pollution of the sea in general

Although not concerned specifically with the problem of marine pollution, Article 24 of the 1958 Geneva Convention on Territorial Waters and the Contiguous Zone allows the coastal state to exercise such control within the contiguous zone as may be necessary to "prevent infringement of its customs, fiscal, immigration or sanitary regulations within its territory or territorial sea".

It is certainly arguable that this provision would allow coastal states to exercise pollution control throughout the continuous zone. And in fact the latest U.S. legislation, which extends national pollution control jurisdiction to the outer limits of the contiguous zone, bases itself on this article. Article 5.7 of the Convention on the Continental Shelf and Article 25 of the Convention on the High Seas, are also drafted in general terms for the protection of the sea from pollution by "harmful agents".

None of the conventions establishing regional fishery bodies mentions water pollution specifically, although the problem has been discussed on occasions within those bodies and some studies have been undertaken. So far no positive action has been taken nor recommendations adopted for its control. The Convention on the Conduct of Fishing Operations in the North Atlantic of 1967, however, includes a provision prohibiting any fishing vessel from "dumping in the sea any article or substance which may interfere with fishing or obstruct or cause damage to fish, fishing gear or fishing vessels except in cases of *force majeure*" (Annex V, Rule 4).

Regional arrangements

Developments at the regional level have been of particular interest. In the U.S.A. the identity of interests of various states in the protection of their coasts and coastal waters from pollution has led to the establishment of interstate water pollution control commissions, such as the Interstate Sanitation Commission, which is charged with the task of controlling the level of pollution in the harbour and coastal waters of the New York metropolitan area. To this end the Commission has effected classifications of the waters in its area in terms of their main uses and has established effluent standards designed to protect those uses. At the international, as opposed to interstate level, regional arrangements and groupings are similarly taking shape, though the process is, of course, much less advanced. Within the framework of the Council for Mutual Economic Assistance (CMEA) for example, studies have been undertaken on the feasibility of establishing some regime to protect the Black Sea from pollution, while the North Sea and the Baltic Sea are the subjects of excellent reports recently published by the International Council for the Exploration of the Sea (ICES, 1969, 1970) examining the state of pollution and measures of control taken by each country bordering those seas. Similar studies are being undertaken by the General Fisheries Council for the Mediterranean (GFCM) with respect to the Mediterranean.

Already several bilateral agreements have been concluded that include general provisions on the pollution control of some particular marine region. The Agreement of 1971 between the U.S.S.R. and Iran on the Caspian is one example. More recently, in October 1971, a meeting was held in Oslo attended by member governments of the North East Atlantic Fisheries Commission to consider regional action for the prevention of pollution of the North East Atlantic region with special reference to the control of dumping operations. It appears likely that, as a result of this meeting, an agreement on the control of waste dumping practices in the area will be concluded in the fairly near future. Similar activity is taking place between the countries bordering the Mediterranean.

International product controls—European Agreement on detergents

The European Agreement on the restriction of the use of certain detergents in washing powder and cleaning products, opened for signature in Strasbourg in September 1968, presents an interesting example of international action to attack one root cause of the problem of pollution. By this Agreement the contracting parties bind themselves to take measures to prevent the marketing in their territories of washing powders, etc., containing detergents which are less than 80 per cent biodegradable. It is probable that this example will eventually be taken up by other international agreements regulating the manufacture and distribution of other products that cause pollution. Already the Commission of the European Economic Community, spurred by the recent unilateral actions of Germany, is considering the question of community agreement on the limitation of the amount of lead additives in petroleum fuels.

Activities of international agencies in the control of marine pollution

There are now many international organizations active in the area of marine pollution. Thus IAEA deals with marine pollution from the point of view of the specific pollutant radioactivity and has done much successful work to control pollution through the promulgation of safety standards. IMCO, on the other hand, is interested in marine pollution primarily in terms of one source—shipping—although it has recently shown interest in pollution from equipment other than ships operating in the marine environment. IMCO is holding a conference in 1973 on marine pollution control and has plans to eliminate all wilful and intentional pollution of the sea by oil and other noxious substances and to minimize accidental spills by 1975 or at least by the end of the decade. FAO and the various regional fisheries commissions view marine pollution primarily in terms of its effect on fishing and the living resources of the sea, although also in terms of the polluting effect of agricultural, fisheries and forestry industries. WHO, for its part, looks at pollution of the seas primarily from the standpoint of the health of human beings. At the regional level, some of the activities of the CMEA have already been mentioned. Within the European Economic Community there is also a general movement for the harmonization of water pollution control measures and standards throughout the Community, a movement occasioned primarily by the economic repercussions of pollution control measures. The movement is perhaps likely to be concerned particularly with community-wide controls over products that cause pollution, such as lead additives to petroleum and hard detergents.

On a general level, Unesco through its Intergovernmental Oceanographic Commission (IOC), has been involved with the scientific aspects of marine pollution, while many non-governmental organizations such as the International Union of Scientific Unions, the International Law Association and the Institute of International Law are concerned with both the scientific and the legal problems involved.

There is considerable overlap in the work of the various organizations. Recognizing this problem and to achieve a measure of coordination at the international level, several of the international agencies most concerned with marine pollution (IMCO, FAO, UNESCO, WMO, WHO, IAEA and the UN) have established a Joint Group of Experts on the Scientific Aspects of Marine Pollution (GESAMP) comprised of no more than 20 experts acting in their individual capacities but nominated by the sponsoring organizations to advise these organizations and member states on the scientific aspects of marine pollution. Three main sessions of GESAMP have been held to date. A further element of coordination is provided through the Administrative Committee on Coordination of the UN and its Sub-Committee on Marine Science and its Applications.

The UN itself also has had a broad interest in the problems of marine pollution in the past and has recently paid increasing attention to the problem. On the one hand, its concern with the human environment has led to the calling of a Conference on the Human Environment in 1972 at which marine pollution will be discussed as a major problem. Its concern with the peaceful uses of the sea bed, on the other hand, has expanded as a result of General Assembly Resolution 2750C and a general conference on the law of the sea is now to be held in 1973, to deal, *inter alia*, with the preservation of the marine environment.

NATIONAL MEASURES

Pollutants may enter the marine environment through direct discharges into the seas or indirectly, through pollution of inland waters that discharge into the seas or pollution of the atmosphere. The national legislation regulating each of these major categories of sources will be examined and a third body of legislation regulating the environmental effects of certain substances and imposing product controls will be considered separately.

Direct marine pollution

Under traditional international law, the waters of the world are divided into zones which determine the extent of the control that may be exercised by individual states. Inland waters fall entirely within the territorial jurisdiction of states and may be subjected to almost complete control. So may a state's territorial waters, subject, of course, to the right of innocent passage and other treaty and customary law obligations. A state may also exercise certain powers of a protective nature in the contiguous zone, but beyond that, it may exercise control only over its nationals and ships or aircraft registered in its territory and over activities connected with the exploration and exploitation of the natural resources of the continental shelf. Some states also claim limited powers of control within fishery conservation zones.

The bulk of the sea, therefore, falls outside the jurisdiction of any state. There any attempt by an individual state to exercise continuous control over all activities that might cause pollution would run counter to the principle of the freedom of the high seas. This fragmentary nature of national jurisdiction over the seas has emphasized the natural legal tendency of dealing with each problem separately as it arises. The result has been an equally fragmentary array of uncoordinated sectoral laws and departmental jurisdictions, as may be seen from the following discussion of the various types of legislation that exist.

Pollution from coastal establishments

In theory, the control of pollution from coastal and estuarine establishments, both urban and industrial, falls, in the majority of countries, within the same institutional and legal framework as inland water pollution control (as, for example, in Belgium, Bulgaria, Denmark, Finland, Ireland, Japan, the Netherlands, Norway and the U.S.A.). This equivalence of control is, in general, not undesirable. Inland and marine pollution are inextricably related and administratively it is simpler, fairer and more effective to subject industries and urban communities to the same forms of control wherever they are situated.

Two qualifications should be made in this connection, however. In the first place, it appears that in many cases the extension of the inland water pollution control system to coastal waters is, to a large extent, theoretical only and that the implementation of the relevant legislative provisions ends in practice at, or even before, estuarine areas. Secondly, the conditions of mixing and dilution and the uses to which waters are put and thus the quality required, differs substantially from coastal to inland waters. This latter aspect is recognized in the legislation of some countries, which, while maintaining the same administrative framework, expressly differentiates between the classifications and/or standards of water quality required for coastal and inland waters. Other countries do not appear to make such a differentiation.

In a few countries the systems of control over coastal pollution differ substantially from those over inland water pollution. In France, for example, while the general framework of inland water pollution control does extend to coastal waters, provision is also made for a separate permit system under the Prefects for discharges into the marine environment (Act No 64.21 of 1964, Article 4). In the U.K., while inland water and estuarine pollution falls under the jurisdiction of the River Authorities, coastal water pollution is the primary responsibility of the local Sea Fisheries Committees set up under the Sea Fisheries Regulations Act of 1888–1966. Proposals for the reform of coastal water pollution control in the U.K. are now under consideration. It is as yet unclear whether the powers will be transferred to the River Authorities or new, more powerful coastal water protection agencies set up.

Pollution from ships

(a) Oil: In most cases the legislation purporting to control the pollution of the sea by oil discharged from ships is merely the national implementation of the provisions of the 1954 London Convention on the prevention of the Pollution of the Sea by Oil, as amended in 1962, with only minor differences in such matters as organization and powers of inspection and enforcement and nature and extent of penalties.

Some legislation provides that ships causing oil pollution in contravention of the legislation may be charged with the cost of removing that pollution in addition to any normal fines imposed or that the proceeds of any fines imposed on offenders may be applied towards such cost. Under the new legislation now being introduced as a result of the Brussels Private Law Convention, ship-owners are made liable for all damage caused and the cost of measures taken as the result of discharges of oil both within national territory and territory within the jurisdiction of other Convention countries. The London Convention itself gives no standards for the nature and amount of penalties that the contracting governments should exact for offences, but merely requires that penalties for offences committed by ships outside the territorial waters of the state be adequate in severity to discourage any such unlawful discharge and shall not be less than the penalties imposed by that government for offences committed within its territorial waters (Art. VI(2)).

In some countries, the scope of the legislation has been extended beyond that of the London Convention to include oil pollution stemming from activities on the land or on the seabed underlying the territorial sea or internal waters of that country, mostly in cases where oil pollution had already been regulated by previous legislation. Such provisions are to be found in the oil pollution acts of, for example, the United Kingdom, New Zealand, Norway and the various states of Australia.

(b) Radioactive substances: Pollution of the sea by radioactive substances from ships is mostly governed by legislation dealing generally with the handling and disposal of radioactive wastes. In some cases, however, the problem of pollution by nuclear-powered ships or by ships carrying radioactive substances has been dealt with separately. The main provisions are often national implementations of the 1960 International Convention for the Safety of Life at Sea.

Thus in Australia, the competent Minister may issue a safety certificate, if satisfied on receipt of declaration of survey that a nuclear passenger or cargo ship has complied with the relevant construction, equipment and machinery regulations. Without this certificate no nuclear ship registered in Australia may set sail, nor may any ship, whether registered in Australia or elsewhere, enter Australia unless the master or owner of the ship has been informed officially that the Minister is satisfied as to the safety of that ship with regard to radiation and other nuclear hazards. The 1960 Safety Convention also deals with the carriage of radioactive substances as one species of dangerous goods under Chapter VII and lays down certain requirements regarding packing, marking, storage and documentation which are to be implemented by national legislation.

Waste dumping at sea

The pollution of the sea or territorial waters by the dumping of municipal or industrial wastes or surplus products, is not normally provided against specifically by national legislation. The basic laws controlling pollution of the territorial waters of the state, however, are normally drafted in sufficiently broad terms to cover this type of pollution when the dumping takes place within the limits of the territorial waters of the state.

Beyond the territorial waters zone such controls as exist are normally of a voluntary nature only (ICES, 1970; IMCO, 1969). A few countries, however, are now beginning to introduce compulsory systems of control over dumping beyond the limits of the territorial waters. In the USA, for example, the Army Corps of Engineers already exercises some control over dumping operations outside territorial waters, but only where the vessels in question come from or have passed through the waters of three clearly defined harbour areas (U.S.A. Supervisory Harbors Act, 1888). It is now the stated policy of the U.S.A. to extend this control to cover all dumping operations both within and outside the territorial waters. Similar recommendations have been made officially in the U.K. and compulsory systems are already in operation or are envisaged in the near future in the five Nordic countries. The Finnish Law on the Prevention of Pollution of the Sea of 1965 already prohibits pollution of the high seas and territorial waters of another state by Finnish ships. Norway has introduced Regulations in June 1971 covering the dumping of certain harmful substances into international waters by Norwegian ships.

In the Netherlands legislation has recently been passed requiring permits for the disposal of waste up to a certain distance beyond its territorial waters by Dutch ships and even by ships of other nationalities where the waste has been transported from or through the Netherlands. Similar provisions have been placed before the German Federal Parliament.

Pollution from operations on the continental shelf

The dangers of pollution from operations for the exploration and exploitation of the resources of the continental shelf have become increasingly apparent during the last few years, and the volume of national legislation has increased accordingly. There still appears to be, however, several states possessing and exploiting off-shore mineral resources which have enacted no, or insufficiently detailed, regulating legislation.

Although the operations creating the dangers of pollution often take place outside the limits of the territorial sea, the exclusive rights of the coastal state to control these activities on the continental shelf is recognized by Article 2 of the Geneva Convention on the Continental Shelf of 1958. Its obligation to take appropriate measures for the protection of the living resources of the sea from harmful agents is laid down in Article 5, paragraph 7, of the same Convention. General provisions implementing the terms of these articles are to be found in many of the national laws providing for the general regime of the continental shelf; for example, in the United Kingdom, the Federal Republic of Germany, Malta and New Zealand. The prohibitions, which may carry fines of up to £1,000 or imprisonment of up to 12 months, generally cover only oil pollution. The normal method of control, adopted by regulations under the main acts, appears to be the establishment of certain conditions to safeguard against pollution on the granting of licences for exploration or production. These conditions are either set out in the general regulations on petroleum drilling and the granting of licences (as in the case of Norway and Germany) or are incorporated into the licences themselves. Thus the U.K. Petroleum (Production) (Continental Shelf and Territorial Seas) Regulations of 1964 provide for model clauses to be inserted into each licence for exploration or production, obliging the licensee to avoid harmful methods of working and to take steps to prevent the escape of petroleum into any waters in the exploration or production area. Breach of the terms and conditions of the licence may lead to its revocation.

The degree of detail to be found in the regulatory provisions varies widely from very broad and general provisions to fairly detailed and stringent regulations, as in the case of the U.S.A. The vast majority of regulations, however, are far too general to be of great use. This is a particularly important failing in countries where lack of skilled personnel makes it difficult or impossible to exercise day-to-day technical control over safety aspects of individual drilling operations. It is also difficult to ensure that regulations, whatever their degree of detail, are strictly applied. In an effort to tighten up compliance in this area, the latest U.S.A. legislation provides that companies shall in future be liable without proof of fault for the cost of removal of oil pollution from oil drilling operations up to a maximum of $8 million.

Conditions established on the granting of authorization to commence exploration operations and seismic surveys normally contain restrictions on the use of explosives (ICES, 1969). In some states, such as the Netherlands, and in California, U.S.A., fisheries inspectors accompany exploration vessels to enforce such restrictions.

Legislation covering specific fields

Most countries exercise some control over coastal water pollution under legislative provisions drafted in terms of the protection of certain sectoral interests, such as fisheries, health and navigation. Such provisions are administered, normally without any measure of co-ordination, by the agencies, local or central, responsible for the various sectoral interests.

(a) Fisheries

Most countries possess broad legislative provisions prohibiting discharges harmful to fish. A typical example is Article 55 of the Mexican Law on Fishery of 1950, which prohibits "the discharge or allowing to flow into waters containing species of fish, substances toxic or harmful to those species of fish, except in cases of accident or *force majeure*". In Ireland, the Fisheries (Consolidation) Act, 1959, requires that any such discharge can only be made with the prior authorization of the competent fisheries authority. In other countries (e.g. the United Kingdom, Norway and Canada) fisheries authorities are empowered to make regulations for the more detailed control of activities causing pollution. Very often pollution control provisions in fisheries acts apply both to inland and coastal waters.

(b) Public health

Provinces or local communities are often required or empowered to adopt sanitary regulations regulating drainage and public health nuisances within these communities. Thus, the French model departmental health regulations prohibit certain discharges into the sea and restrict certain activities in the proximity of beaches and bathing places.

(c) Navigation

Many countries also possess legislative provisions for the control of marine pollution within the general acts or regulations for the management of ports and harbours (e.g. Bulgaria, Canada, Denmark, India, Japan, New Zealand and Norway). The scope of these acts and regulations is normally restricted to the harbour area with emphasis, naturally enough, on protection of navigation.

Some provision is also to be found in several countries prohibiting the causing of pollution detrimental to navigation outside of harbour areas, although normally within the limits of the territorial waters; for example, in Australia, Canada and the United Kingdom.

Indirect marine pollution

Inland water pollution

There is a good deal of variety among countries in the approaches taken to the control of inland water pollution. In some countries water pollution is not yet considered a problem and little or no control is exercised. In these countries water pollution laws either do not exist or are only broadly drafted provisions not yet implemented in practice. In other countries where water pollution problems have existed for some time, but where no comprehensive overhaul of the relevant legislation has yet been undertaken, water pollution control is often exercised under fragmented sectoral laws dealing with pollution from the point of view of the activities causing pollution, or from the point of view of activities or sectors affected by pollution. Pollution control provisions are thus to be found in forestry and mining laws, in health codes, in fisheries acts and the like.

In the last 25 years or so, a good many countries have introduced more comprehensive legislation, whose perspective is the water resource itself rather than particular sectoral interests and whose objective is the maintenance of the quality of that resource at a level that will satisfy the requirements of all users of the resource. Although the process is often by no means complete, the tendency of these modern laws is to consolidate regulatory powers in a single body or series of bodies within the framework of water resource management and/or environmental protection. In an increasing number of countries the river basin has been adopted as the basic unit for water pollution control, normally under the general policy guidance of a central body. The English River Authorities, the German Genossenschaften, the Belgian Water Purification Societies and the French River Basin Agencies are all examples of this trend.

The actual methods employed in the more recent water pollution control legislation vary considerably from country to country. But most countries make use of a combination of the following techniques.

(a) Permit requirements

Most of the more recent systems rely to some extent on a permit or prior authorization system, giving the pollution control agencies the power to prevent pollution from the outset rather than merely punishing offenders once the fact of pollution has been established. The actual point of time at which a permit is required varies from country to country. Some of the recent laws aim specifically to bring water pollution control considerations into the initial planning process for both polluter and planning authorities by requiring the approval of plans and specifications from the water quality point of view before any construction is commenced.

[609]

Authorizing agencies are normally empowered to impose conditions concerning the quality and quantity of effluents when granting permits. Such conditions may usually be revised or the permit withdrawn after a certain period of time, or should the nature or volume of the discharge be substantially changed, or new and unforeseen circumstances require the imposition of new conditions. Non-observance of the conditions imposed on the issuance of the permit may cause the permit to be withdrawn and/or render the persons contravening the conditions liable to fines or imprisonment.

In most new systems, the permit requirement applies to municipal domestic discharges as well as to industrial discharges. Sewerage authorities are also often given some control over the nature of industrial discharges into sewers under their control. In other cases, standards for such discharges are laid down in general regulations.

(b) Classification and standards

Whether or not a system of prior authorization for effluent discharge is relied upon as the principal measure for the control of water pollution it is necessary for the executive authority to have up-to-date information on the various uses to which the particular water resources are put or could be put and the standards of water quality required by these uses. Such information may be provided in some countries through the personal knowledge or connections of the members of the executive body or through a system of consultation with other bodies. In other countries, the procedure for acquiring these data is more centralized and depends on a system of classification or inventory of the water resources and/or the establishment of standards of water quality for those water resources.

Classification is normally based on the quality of waters, on the existing uses to which the particular waters are put, the potential uses to which they could be put or on a combination of all such factors.

The concept of classification has been shunned by many national water pollution control authorities as being too inflexible and tending to sanction increasing pollution of the lower categories of water courses. However, it would seem that the system has the advantage of facilitating planning on a national scale and that some such process of classification according to predominant use is, in fact, essential to allow for a rational policy of water quality management, whether or not the results of that process are eventually promulgated in legislation.

Most legislation which provides for the classification of waters also provides for the establishment of receiving water quality standards. For each category of water, standards of water quality which will meet the requirements of the particular use or uses are determined and promulgated.

The number of factors on which specifications are given in the standards so far promulgated varies considerably. They normally contain, however, specifications on pH values, oxygen content, coliform bacteria counts, suspended solids and oil, often, though not always, specifications on temperature and in some cases limits on permissible concentrations of certain toxic substances. Levels of contamination by radioisotopes are also sometimes included but are more often treated separately

in legislation dealing specifically with controls on the use and disposal of radioactive substances.

Some countries have steered away from the idea of promulgating nation-wide water quality standards, arguing that establishing such standards may lead to a decline in the quality of waters to the lower limits of the standards. Water quality standards do, however, have the advantage of ensuring some measure of uniformity of water pollution control measures throughout the state and of reducing the vulnerability of particular localities to pressure from individual polluters.

It is not entirely satisfactory to establish receiving water standards on their own. Where discharges are few and far between such standards may be sufficient. But where there are many discharges into the same body of water, the problems of proof and allocation of responsibility become a hindrance. For this reason receiving water standards are normally coupled with nation-wide effluent standards and/or a permit system in which individual effluent standards are set.

(c) Financial measures

The legislation of most countries provides in some way for financial incentives or assistance to municipalities or industrial establishments for water pollution control. This may take the form of national and federal government grants to municipal bodies for the construction of sewage treatment plants and the granting of tax and other financial concessions to industries which construct their own treatment plants. The U.S.A. legislation also provides for research, development and training grants.

One increasingly prevalent feature of inland water pollution control legislation is the effluent charge. The underlying principle of this charge is to ensure that the costs of pollution and, in particular, of treatment installations, are borne by those who cause the pollution and to make it economically desirable for them to reduce their polluting effects. The effluent charge has always been an essential tool of the German Genossenschaften and the principle has now been adopted in varying degrees of sophistication by most of the more recent legislation with respect to discharges into public sewers and discharges into open waters. The recent Hungarian legislation, for example, has an elaborate system of waste water charges which develop after a period of years into annually increasing fines.

Atmospheric pollution

It is only comparatively recently that the great importance of the atmosphere as a pathway for pollutants entering the marine environment has been recognized. GESAMP has now reported that "many marine pollutants reach the ocean in significant quantities via the atmosphere and, for some materials, this is the principal pathway" (IMCO/FAO/UNESCO/WMO/WHO/IAEA/UN, 1971). Even now one can only hazard vague guesses at the full extent of that importance. But enough is known to justify a brief mention of national legislation for the control of atmospheric pollution in the context of marine pollution control.

To a large extent serious control over atmospheric pollution is of recent vintage. For some of the pollutants which enter the seas in large quantities through the atmosphere, such as DDT and other pesticides, controls consist of product manufacture, registration, sale, and application regulations which will be discussed later in the general context of environmental product controls. Likewise with the problems of lead content in petroleum fuels. For the rest, such control as exists consists, broadly speaking, of emission controls for vehicles, controls on industrial emissions (through planning permission procedures and local or state-wide emission standards) and local controls (mostly through zoning ordinances) on emissions from domestic heating appliances. More recently, measures have included the establishment of air pollution districts and the formulation of contingency plans for air pollution emergencies.

In only a few countries have the air pollution effects of industrial activities been considered at the same time as their water polluting effects. Consequently, an increase in the effectiveness of atmospheric pollution control has in many cases resulted in corresponding increases in water pollution and vice versa. This interrelationship between air and water pollution control has been recognized in some recent institutional and legislative innovations. In Sweden, for example, the new Nature Protection Board controls air, water and land pollution, while in the U.S.A. the protection of all three media has become the responsibility of the Environmental Protection Agency. In many other countries Departments of the Environment have been set up or other co-ordinating machinery created.

Environmental pollution by specific pollutants

In addition to legislation controlling various sources of pollution or setting up resource oriented systems for the conservation of water resources, many countries now have an increasing body of legislation regulating the polluting effects on the environment of specific pollutants. Such legislation is concerned both with certain dangerous substances requiring specialized control, such as radioactive materials, and with the elimination or control of certain products or components of products that cause environmental pollution.

Radioactive substances

In some countries (e.g. Bulgaria, Finland, France and Poland) provisions on the pollution of sea water by radioactive substances are included within the general water pollution regulations. In view of the highly specialized nature of radioactive pollution and the activities which cause it, many countries prefer to treat pollution by radioactive substances separately within the framework of special legislation for the control of the use and disposal of radioactive substances. Thus the U.K. legislation has specifically withdrawn the question of radioactive pollution from the executive competence of the Sea Fisheries Committees and River Authorities and the Belgian legislation has similarly withdrawn pollution by radioactive substances from the purview of the basic water pollution control legislation and administration.

Only rarely does the legislation of a country deal directly with the disposal of radioactive wastes into the sea as opposed to other methods of disposal. The Finnish Law on the Prevention of the Pollution of the Sea of 5 March 1965, however, is one example. This law expressly prohibits the discharges of radioactive materials into the sea in such a way as to cause harm to "human beings, the environment of the living resources of the sea or to expose them to danger". Permission for the sinking of radioactive materials must in all cases be sought from the competent water rights court and they must in any case be disposed of at a depth of more than 2,000 m.

In most cases the legislation deals with the use and disposal of radioactive materials in general. Control of pollution is normally accomplished through a system of prior authorization for the construction of establishments or commencement of activities involving a significant amount of radioactive substances and for the disposal of radioactive wastes. This is often coupled with the publication of detailed standards of maximum permissible concentrations of radioactivity in water; any discharge violating these standards is prohibited.

Procedures for the consultation of the various interests in water resources before the granting of any authorization are sometimes laid down in the legislation. In those consultations those persons and bodies interested from a health aspect figure more prominently than any other uses and sometimes to the apparent exclusion of fishery interests.

Product controls

One of the earlier examples of controls over products or components of products that are a root cause of environmental pollution is the German Law on detergents of 1961. The law fixes a minimum biodegradability of 80 per cent for anionic detergents contained in washing and cleaning products before those products can be placed on the market by importers or manufacturers. Similar provisions have now been introduced or are contemplated in several other countries and an international agreement on this subject has been signed in Europe. This new control is of special interest both substantively and as a new concept of environmental pollution control whereby regulations are applied at an earlier point in the chain of causation.

Of similar interest are the new controls that are recently being introduced over pesticides. Thus, the use of DDT and certain other toxic and persistent wide spectrum pesticides has now been banned in several countries including Sweden, Denmark, Finland, Switzerland, the Ukranian S.S.R. and several states of the U.S.A. Stronger registration controls have also been placed on new types of pesticides and several countries expressly require pollution risk to be evaluated before a new pesticide can be registered. The use of pesticides is also subjected to stricter controls ranging from the requirement of clear instructions on containers to special licences for the use of the most dangerous products. This approach has been adopted in Sweden and contemplated in the U.S.A. Many countries also place restrictions on spraying near water bodies (UNECE, 1971; WHO, 1969).

Other instances of recent product controls of particular interest to marine pollution control include regulations fixing the maximum amount of lead that may be contained in petroleum products.

Some recent trends

The more recent legislation of relevance to marine pollution control has shown two general tendencies. The first has been a gradual extension of national jurisdiction over polluting activities beyond traditional territorial limits. In some cases this extension has been based on an international convention and in other cases it has been unilateral. One of the earliest extensions of national jurisdiction was in 1936 when Cuba prescribed penalties for the discharge of materials harmful to fisheries in coastal seas up to 5 mi from the coast; the width of her territorial sea was then, as now, only 3 mi. More recently, legislation has been passed, or will shortly be introduced, in several countries to implement the terms of the 1969 Brussels Public Law Convention which would empower the national authorities to take action against any vessel, whatever its nationality, outside the traditional limits of the territorial jurisdiction of the state, where the coastline or related interests of that state are threatened by oil pollution following upon a maritime casualty involving that vessel. The latter legislation is the direct implementation of an international agreement, but other recent or proposed legislation is not so directly linked to international cooperative activity. In the U.S.A., for example, the control of activities causing pollution has been pushed out as far as the outer limits of the contiguous zone (12 mi under the Geneva Convention). More ambitious legislation has been passed in Canada, extending the territorial scope of Canadian marine pollution control activities over large areas of the waters of the Arctic Ocean, well beyond the traditional limits of the Canadian territorial waters. A geographical extension of the powers of the coastal state to control pollution-causing activities is, of course, also inherent in the recent legislation claiming new territorial sea zones of up to 200 mi.

The second general tendency is the drawing together of legislative provisions relating to direct marine pollution, to water pollution as a whole, or to all environmental pollution. Where such administrative control was formerly scattered between many departments dealing with navigation, fisheries, agriculture, health, atomic energy, harbours, commerce and mines, for example, now overall control or at least co-ordination is being increasingly handed over to a single body. The Finnish law on the Prevention of Pollution of the Sea of 1965 is an example.* This covers pollution of the high seas and territorial waters of another state by Finnish ships or from Finnish territory and includes pollution from the continental shelf and from the disposal of radioactive material at sea. The recent Canadian Arctic Waters Pollution Prevention Act also draws together into one Act, now under the general administration of the new Department of the Environment, control over all direct sources of pollution affecting the entire Arctic environment. Thus, the Canadian Act deals not only with all pollution from ships but also with pollution from exploration and exploitation activities on the continental shelf and pollution from undertakings on the mainland and islands of Canada, including, apparently, from waste dumping. So far as undertakings on the mainland or islands and in the Arctic waters are concerned, the Act requires prior approval of plans and specifications of the proposed works from the point of view of their effect on the marine environment before construction can commence. It is, in effect, the first national legislation on

direct marine pollution that is truly aimed at, and drafted in terms of, the conservation of a particular region of the marine environment.

On a broader level, the Swedish Environmental Protection Act of 1969 covers all aspects of environmental pollution under a single Environment Protection Board, as does the Bulgarian Law on the Protection of the Air, Waters and Soil against Pollution of 1963. In the U.S.A. recent administrative reorganizations have drawn together all federal pollution control activities affecting the entire environment into a single Environmental Protection Agency. The agency is now responsible for water pollution, coastal air pollution control, solid waste management, radioactive contamination and pesticide regulation and residue standards. Similar reorganizations to ensure co-ordination of environmental protection activities are being discussed in other countries and several governments have recently set up Departments of the Environment for this purpose.

CONCLUSIONS

During the last few years both governments and public have come to realize that we cannot continue to use the seas (and, indeed, the whole environment) for the uncontrolled disposal of waste without endangering its usefulness as a source of food and recreation. The result has been a dramatic increase in both national legislation and international co-operation for the control of marine pollution. This much is encouraging. But a great deal more needs to be done at both levels. It is perhaps appropriate, therefore, to mention at this point some general considerations relevant to the future development of both national and international controls and to give some indications of the way in which those controls may evolve.

A movement towards the unilateral extension of national controls can be seen in some of the recent legislation. But this approach does suffer from serious disadvantages. Quite apart from encroachment on the principle of freedom of the high seas, including freedom of navigation and scientific study and the inevitable conflicts of jurisdiction, such unilateral extensions of national control would still only achieve an incomplete and uncoordinated patchwork of controls over what is, in fact, an international resource. The future, rather, would seem to lie in international action, or at least international coordination of national action.

The questions then arise, what form should that international action take, and at what level should it be taken? To what extent and in what circumstances should it be at the regional or global level? Should it confine itself to the construction and clarification of rules concerning liability for damage caused by pollution, require the passive registration and recording of certain discharges, or go further to construct regulatory machinery to restrict pollution causing activities on an international scale? Should any machinery be restricted to water quality management or must it expand to include other facets of environmental protection?

To answer these questions, lawyers and policy-makers must have sound and comprehensible information on all aspects of the problem, whether legal, economic, political, social or scientific. Some of the legal considerations have

---

* The law is only a framework law and it is not known whether implementing regulations have yet been made.

[612]

already been mentioned. Economic, political and social considerations, for their part, are too vast to be discussed here in depth. But a few more obvious points might perhaps be mentioned. In particular, the economic impact of water pollution control measures must be recognized and the fact that different physical, social and economic factors in different countries may result in different positions accorded to the pollution problem in the order of priorities of those countries. It should also be recognized that states most directly affected by particular sources of pollution have the most direct interest in controlling those sources and, conversely, that states may be less willing to submit to control over their activities by states not directly affected.

Some of these considerations in themselves suggest that at least where the formulation of relatively sophisticated resource oriented regulatory machinery is concerned, a primarily regional approach would be more acceptable and effective. Certainly, the problems in areas like the Baltic or the North Sea are not the same as those in the Pacific, whether from the physical, scientific, political, economic or social points of view. They may, therefore, call for different types of solution. Indeed, some regional groupings are already forming in areas where the problems of pollution are intense and where the concerned states have sufficient economic, political and social heritage in common to allow for effective co-operation, and more such regional groupings may be expected in the future. Other aspects of the problem of marine pollution, on the other hand, may well call for global action in addition to, or in place of, regional action. The formulation and clarification of international rules on liability for damage caused by pollution, the registration of dumping operations, the establishment of rules and machinery to prevent pollution from international shipping or the operation and exploitation of the natural resources of the seabed outside the limits of national jurisdiction are examples.

But before the optimum international regulatory machinery can be designed, lawyers and policy-makers will need answers generally agreed upon by the scientific community to some basic questions. They will need to know, for example, whether there are sound scientific and ecological reasons for tackling marine pollution on a regional basis, and if so, what the scientific community believes would be the optimum regions for control.

On a more general level, lawyers and policy-makers will need to know more about the nature and behaviour of pollutants—to know those classes of pollutants that are likely to create only local or regional problems of pollution, and those whose toxic effects may be so persistent and widespread as to require global control. They will also need to know more about the ways in which pollutants enter the marine environment and a clearer estimate, for example, of the importance of the effects of atmospheric pollution relative to other sources, or pathways, of pollution. More basically, they will need to know, in broad and understandable terms, just what the present state of marine pollution is, what the effect of that pollution is now on the various uses of the sea as a resource, including its use and potential use as a source of food, and a projection of what that effect may be in the future if not controlled. They will thus also need to know and identify not only the present uses to which the

sea resource is put but also some estimate of the potential uses to which it may be put in the future.

Not all these questions may be answerable as yet, and it may be that in most cases international action will have to be taken before the full picture is known—but two general considerations appear to be clear. The first is that any international regulatory machinery established, whether at the regional or the global level, will need to be flexible enough to allow for the continuous input into its decision-making processes of the rapidly increasing and changing scientific information, and flexible enough to act on that information. The classical method of treaty-making, relying on the one-time formulation of permanent and inflexible norms may not, perhaps, be adequate in this respect. More flexible machinery for the formulation and re-formulation of standards on the basis of changing scientific information may be necessary, just as it has proved necessary in the cases of international fishery agreements and the European Atomic Energy Community.

The second point is closely related to the first. In order to evaluate and act on the latest scientific information any international machinery must be designed and operated on a multi-disciplinary basis—not by lawyers and policy-makers alone but in concert with marine scientists, economists and other specialists concerned.

Although it is suggested that any efficient system of marine pollution control must ultimately depend upon some measures of international action and co-operation, this is not to deny the paramount importance of adequate national water pollution control machinery. On its own, it may not be sufficient to control marine pollution, but without adequate national machinery no international system can hope to function. Even without further international action, considerable improvement in the state of marine pollution could be achieved by the reform of legislation and administration. Some points for reform will be obvious from the preceding discussions of national control of marine pollution. But four points may perhaps be stressed again here. The first is that the problem of estuarine and coastal water pollution is inextricably linked with the problem of inland water pollution control, and that this link should be reflected, or at least recognized, in the national machinery. Secondly, although it may be true that coastal waters have a greater capacity than inland water for diluting and dispersing wastes, this capacity is still limited. Pollution control measures should therefore be applied in practice as rigorously to estuarine and coastal waters as to inland waters although the actual standards required may well be different. The third proposition is that pollution control measures are most effective the higher they are applied in the chain of causation. Such measures should therefore be aimed more and more at regulating the products and the processes which cause pollution and pollution control should be made, whether by economic or administrative methods, an integral part of both public and private planning processes. Finally, although the immediate regulations of various sources of marine pollution, such as shipping, seabed exploration and mining, coastal establishments, etc., may have to remain scattered between various ministries, some form of overall co-ordination of the various marine pollution control measures should be established to ensure a uniform policy.

References

ICES Reports of the ICES working group on pollution of the
1969 North Sea. *Coop. Res. Rep. int. Coun. Explor. Sea(A)*, (13):4.

ICES Report of the ICES working group on pollution of the Baltic
1970 Sea. *Coop. Res. Rep. int. Coun. Explor. Sea(A)*, (15):4–13.

IMCO Marine pollution. Questionnaire on pollution of the marine
1969 environment. OPS/Circ. 15/GESAMP/22: Anex II: pag. var.

IMCO/FAO/UNESCO/WMO/WHO/ISEA/UN Joint Group of
1971 Experts on the Scientific Aspects of Marine Pollution
(GESAMP), Rome, February 22–27 1971. Third Session
Report. *FAO Fish. Rep.*, (102):4.

JOHNSTON, D M The international law of fisheries: a framework for
1965 policy-oriented inquiries. New Haven, Yale University
Press, 554 p.

UN Economic and Social Council, Economic Commission for
1971 Europe, Committee on Water Problems, UNECE Water/
W.P. No. 38, 18 February 1971.

U.S. Department of the Interior. Federal Water Pollution Control
1968 Administration, Water quality criteria. Report of the
National Technical Advisory Committee to the Secretary of
the Interior. Washington, D.C., U.S. Government Printing
Office, 234 p.

WHO Comparative Health Legislation, Control of Pesticides. *Int.*
1969 *Digest Hlth Legis.*, 20(4).

Present Needs for Scientific Advice on Legislation on Pollution *J. Lopuski**

Les avis scientifiques nécessaires à l'élaboration d'une législation sur la pollution

Il ressort d'un examen général des lois et règlements existants sur la pollution des mers qu'il est nécessaire de mettre sur pied un programme en vue d'élaborer une législation internationale et de promouvoir l'action législatrice dans le cadre national. L'auteur formule des suggestions concernant le champ d'application d'accords mondiaux et régionaux. Des obstacles politiques s'opposent au contrôle international. Il convient d'envisager les intérêts spéciaux de certaines régions.

La pollution des mers est un problème nouveau, qui résulte du progrès technologique et ne pourra être résolu par des méthodes et des techniques juridiques orthodoxes. Il y a lieu d'envisager la législation comme un processus accompagnant le développement de la civilisation technique. Cela nécessite la collaboration entre les juristes et les experts des sciences de la mer.

Une motivation générale d'ordre scientifique est nécessaire à l'adoption de mesures législatives. Cette motivation comporterait la prévision des tendances à long terme de la pollution des mers, notamment dans les régions où elle sévit plus particulièrement. Il est temps de renoncer à voir dans cette pollution une "inconnue", même au risque de faire des erreurs. Les experts doivent procéder à l'identification des sources de ce phénomène et recommander des mesures de lutte et de prévention. Les limites de "pollution admissible" doivent être définies en termes scientifiques bien avant que ces limites soient atteintes. Un avis scientifique est nécessaire pour établir des méthodes juridiquement agréées afin de mesurer la concentration des polluants, et pour prendre les mesures législatives nécessaires à la mise en oeuvre des programmes de recherche et de contrôle.

Necesidades actuales de asesoramiento científico para promulgar legislación sobre contaminación

Un examen general de la legislación vigente sobre la contaminación de las aguas del mar indica la necessidad que existe de un programa de legislación internacional y de promover una legislación nacional. Se hacen indicaciones sobre el ámbito de aplicación de acuerdos mundiales y regionales. Obstáculos de tipo político se oponen al control internacional. Debe prestarse la debida consideración a los intereses especiales de ciertas regiones.

La contaminación de las aguas del mar es un problema nuevo creado por la creciente tecnología, no pudiendo resolverse por métodos y técnicas jurídicas ortodoxas. La legislación debe preverse como un proceso que acompaña a la evolución de la civilización técnica. Exige una colaboración entre juristas y talasólogos.

Se necesitan razones científicas de carácter general para adoptar medidas legislativas. Estas incluirían pronósticos de las tendencias a largo plazo en la contaminación de las aguas de mar, sobre todo en las regiones particularmente afectadas. Dicha contaminación dejaría de considerarse "desconocida", incluso a riesgo de cometer errores. La identificación de los orígenes de la contaminación deberán hacerla los especialistas, acompanándola con recomendaciones con respecto a las medidas necesarias para evitarla y luchar contra ella. Los límites de la "contaminación aceptable" deben definirse de un modo científico antes de que se alcancen. Se necesita asesoramiento científico con objeto de promulgar medidas legalmente reconocidas para medir la concentración de contaminantes, siendo necesario el asesoramiento de especialistas en relación con las medidas legislativas precisas para ejecutar programas de investigación y vigilancia.

O NE consequence of technological progress is environmental pollution. Marine scientists must determine the causes, effects, and present state of marine pollution and the best means of controlling it. Because legal measures are necessary to maintain social control over technological development, governments must establish the laws needed to control and prevent marine pollution.

Despite scientific research and warnings, little attempt has been made to control and prevent marine pollution by legal and administrative means. In this respect the law is lagging behind in performing its controlling function. This is true of international law as well as national legislation in maritime countries.

The Convention for the Prevention of Pollution of the Sea by Oil (1954, amended in 1962) deals only with the deliberate discharge of oil from ships. The preventive measures adopted are considered inadequate and amendments are now being considered which would allow only discharges meeting a requirement of 100 parts of oil to one million parts of water.

The two IMCO conventions adopted at the Brussels Diplomatic Conference in 1969 (the Convention relating to Intervention on the High Seas in cases of Oil Pollution Casualties and the Convention on Civil Liability for Oil Pollution Damage) deal only with the consequences of pollution incidents caused by oil from ships and therefore provide only limited prevention.

All these conventions deal solely with oil discharged from ships as a marine pollutant and oil is not considered the most harmful. Marine pollution resulting from the exploration or exploitation of the sea bottom and subsoil or from other sources (e.g. submarine pipelines) is not governed by any international rules.

Articles 24 and 25 of the Geneva Convention on the High Seas deals only with pollution by oil and radioactive wastes. It says that States should cooperate with the international organizations in taking measures for the prevention of marine or air pollution resulting from activities involving radioactive materials or other harmful agents. Unfortunately, these articles have formulated the obligations of the contracting States in such general terms that they are little more than declarations. The same is true of articles 5 (paragraph 7) of the Geneva Convention on the Continental Shelf. This states that in the safety zones to be established for the exploration and or exploitation of the Continental Shelf, the coastal State is obligated to take all appropriate measures in

* Sea Fisheries Institute, Gdynia, Poland.

order to ensure the protection of the living marine resources from harmful agents. It is obvious now that these conventions are inadequate for coping with the growing problems of oil pollution caused by offshore drilling or the dumping of harmful substances into the sea.

National legislation for preventing and controlling marine pollution is also ineffective in most cases. The States are primarily concerned with protecting their national inland waters against pollution. The pollution of their coastal sea waters is still considered marginal. In most cases the laws dealing with the protection of inland waters from pollution extend, at least in theory, to cover territorial waters as well. But seldom are those laws enforced as rigorously, if at all, where coastal waters are concerned. Too many administrators consider the sea a convenient receptacle for all forms of waste. They do not object to the discharge of harmful substances into the sea, provided it is done at a proper distance from the shore. Up to now, the national water administrations of the coastal states have given little consideration to the interests of the international community.

Complexity of the problem

Professor Despax, the French expert on the juridical problems of pollution control, states that the complexity of the problems of environmental pollution is discouraging. The international aspects when combined with the pollution problems of national inland waters, considerably increases this complexity.

Generally, the inland waters problem is not dealt with effectively until the administrations concerned are pressured by the public. Usually they wait too long, unable or unwilling to overcome the opposition of industries guided by economic factors and hostile towards preventive measures. Political questions are also involved: competition among individual states, political blocs, and relations between the developing and developed countries. Political issues may hinder the establishment of new legislation concerning the use of the seas which is necessary for effective international control and prevention of future marine pollution.

Need for identifying sources

Before control and prevention can be attempted, sources of pollution must first be distinguished. Pollutants may come into the sea from the atmosphere or with the water of rivers or streams discharging into the sea. They may come from sewers or the effluents of industrial plants near the shore. They may be emitted from ships at sea or the result of the exploration or exploitation of the sea-bed and subsoil, or from deliberate dumping of harmful substances, or from submarine pipelines.

It follows therefore that any international control and prevention of marine pollution must be partly interconnected with air, land, and fresh water pollution. The polluting substances present in the atmosphere may, depending on various factors, fall on the sea, on land or on inland waters; therefore, marine pollution is only one of the consequences of air pollution.

Where international control is concerned, it is difficult to imagine how marine pollution deriving from the atmosphere can be distinguished from air pollution. Marine pollution resulting from international rivers discharging into the sea is a consequence of the pollution of international inland waters. In both cases, the control and prevention of marine pollution cannot be clearly separated from the controlling and preventing air pollution or pollution of inland waters.

Internationally coordinated scientific research should cover all types and sources of marine pollution; but where preventive and controlling measures are involved, the principal sources of marine pollution should be distinguished.

It is suggested that the following activities of a purely marine character should be distinguished:—

> Shipping
> Exploration and exploitation of the sea-bed and the subsoil
> Dumping of radioactive materials and other harmful substances
> Submarine pipelines

The problem needs a philosophy

Modern technological advances have brought about rapid biological, economic, and social changes. To cope with them a critical appraisal of the basic principles and philosophy of the present laws is of primary importance. The traditional concepts of the international law of the sea are inadequate. The general evolutionary trend of the international law of the sea with respect to marine pollution seems to be towards a certain restriction of the State's right of unqualified sovereignty over its own ships on the high seas.

Another significant legal problem involves the basic concept of preventive measures and compensation. Who should bear the cost of such measures and with whom should the responsibility for their implementation rest: the general society or specifically those who cause the pollution?

It has been said that the problem of environmental pollution has been created by modern technology and therefore that technology should combat it. But this does not solve the problem of distributing responsibilities. An effective system of sanctions and incentives should be created which would help to eliminate pollution or at least keep it within permissible limits.

Nevertheless it seems that those who cause pollution should bear responsibility for preventing or rectifying it. This seems to agree with A. W. H. Needler's statement that the prevention of pollution will be one of the costs of industrial development. Therefore, industries which are the probable sources of marine pollution should bear the cost of preventive measures and those who have definitely caused pollution should be fully responsible for damage and the cost of clean-up operations.

Collaboration between lawmakers and scientists

Legislation on pollution should be envisaged as accompanying technical civilization. It necessitates collaboration between lawmakers and marine scientists.

To enable governments to perform their task, it is necessary to arouse public interest. Society needs more scientific information on the long-term consequences of marine pollution, particularly those consequences that will affect future generations and scientists should make a prognosis of effects on marine life for the next 10, 20 and 30 years.

Marine scientists should identify the principal sources of marine pollution and significant work has been done in this area by the Joint Group of Experts on the Scientific Aspects of Marine Pollution. They have indicated the major categories and principal sources of marine pollution, appraised their relative importance for the marine environment, and given some suggestions for control (IMCO/FAO, etc., 1970).

References

DESPAX, M   La pollution des eaux et ses problèmes juridiques.
1968   Paris, Librairies Techniques, 219 p.

IMCO/FAO/UNESCO/WMO/WHO/IAEA Joint Group of Ex-
1970   perts on the Scientific Aspects of Marine Pollution (GESAMP). Report of the second session held at UNESCO headquarters, Paris, from 2–6 March 1970. GESAMP II/11: pag. var.

JENKS, C W   The common law of mankind. New York, Praeger,
1958   456 p.

NEEDLER, A W H   Pollution prevention is costly. *Ceres F.A.O. Rev.*,
1970   3(3):34–8.

Legal Aspects of Sea Water Pollution Control

K. W. Cuperus*

Aspects juridiques de la lutte contre la pollution des mers

Depuis 1954, l'Association de droit international, organisation non gouvernementale de juristes internationaux créée en 1879, s'occupe des aspects juridiques de la lutte contre la pollution des eaux. En 1966, à sa conférence réunie à Helsinki, l'Association a adopté un ensemble de règles concernant les usages de l'eau des voies d'eau internationales, dont une partie traite de la lutte contre la pollution des eaux de surface. En 1967, un sous-comité de l'Association a entrepris l'étude de problèmes liés à la lutte contre la pollution des mers.

Comme première étape, on a élaboré des règlements provisoires concernant la répartition des responsabilités parmi les Etats pour les dommages causés par la "pollution continentale", c'est-à-dire la pollution résultant des activités humaines exercées dans la zone de compétence de l'Etat intéressé ou sur le plateau continental qui en dépend. Après l'élaboration de règlements relatifs à la "pollution continentale", le comité étudiera la possibilité d'établir des règlements analogues concernant la lutte contre la pollution en haute mer; des propositions ont été formulées à ce propos.

Aspectos jurídicos de la lucha contra la contaminàcion del agua del mar

Desde 1954, la Asociación Internacional de Juristas, que es una organización no gubernamental creada en 1879, se ha interesado en los aspectos jurídicos de la lucha contra la contaminación de las aguas. En 1966 celebró en Helsinki una conferencia en la que aprobó un reglamento sobre el empleo de las aguas de los ríos internacionales, parte del cual trata de la lucha contra la contaminación de las aguas de la superficie. En 1967, un Subcomité de la Asociación se encargó del estudio de los problemas relacionados con la lucha contra la contaminación de las aguas marítimas.

Como primera medida, se ha formulado un reglamento provisional relacionado con la responsabilidad interestatal por los perjuicios causados por la "contaminación continental", es decir, la producida por actividades humanes dentro de la jurisdicción territorial del estado o en su plataforma continental. Una vez que se apruebe el reglamento sobre "contaminación continental", el Comité estudiará otro análogo relativo a la lucha contra la contaminación en alta mar, para el que se han hecho ya algunas propuestas.

THE study of the legal aspects of pollution control of maritime waters was preceded by that of pollution of surface waters. This is not surprising. The unpleasantness caused by the latter form of pollution made itself felt earlier and more acutely than the particular problems and dangers caused by the pollution of sea water. During the age of sail, rivers carried few polluting agents into the sea, and only a few seafaring countries felt the need to introduce simple legislation prohibiting ships from emptying rubbish into harbours and navigable rivers. When sail gave way to coal and in turn to oil, and when industrialization on the continent gradually changed clean rivers into open sewer-systems, it became apparent that sea waters also needed to be protected, especially the coastal areas and enclosed seas where pollution first made its undesired appearance.

River and surface-water pollution

At the biennial Conference of the International Law Association (I.L.A.), held in 1954 in Edinburgh, a paper outlining the relevance of study and analysis of international water law was presented by the late Professor C. Eagleton. Subsequently the I.L.A. decided to set up a special Committee on the Uses of Waters of International Rivers.

A first report of the Committee, containing a statement of four basic principles, was introduced at the following Conference (Dubrovnik, 1956). On the basis of this report the Committee began the study of pollution of surface waters in 1958. In an interim report in 1964 the Committee confessed that this problem was the most difficult with which it was concerned: "Undoubtedly this difficulty results from the fact that pollution of waters is a major problem of our era. The conflict of interests is most direct and the application of the rule of law would, of course, not be completely pleasing to any interested state." In 1966 the Committee presented a recommendation to the Conference of the Association in Helsinki. The recommendation included 37 articles, the "Helsinki Rules", dealing with the use of international waters.

Marine pollution

The remark made in 1969 by the Committee in respect of the pollution of international surface waters is equally valid for the problems caused by the pollution of maritime waters. This subject is now being studied by a Committee of the Association, appointed immediately after the adoption of the Helsinki Rules and *inter alia* mandated: "to give further consideration to the subject of pollution of coastal areas and enclosed seas". A special Working Group was entrusted with this task.

The Working Group, which includes jurists from four continents, began its work in 1967. Although it limited its work in the first stage, in conformity with its mandate, to the study of the pollution problems in coastal areas and enclosed seas, the Committee soon concluded that this mandate needed to be enlarged. Legal rules needed to be expanded because modern technical development has been such that oceanic waters are now as much in

* Capelle a/d Ijssel, The Netherlands.

danger of becoming polluted by continental or non-continental sources, as coastal maritime regions.

Pollution from continental activities

The principal decision of the Working Group was to try to establish rules for both coastal and oceanic waters, but the Group felt that for various reasons, mainly practical, urgency rules dealing with the liability for damages caused by pollution originating on the continent should be elaborated first. The present main sources of maritime pollution (industrial effluents, oil, domestic wastes) can be traced to continental activities. Some provisional rules for this form of pollution have now been drafted. These rules, which conform largely with the Helsinki Rules on pollution of surface waters, will be reviewed by the Group at its next meeting.

In abbreviated form the rules are as follows:

(1) The term "sea water pollution" refers to any sea water pollution resulting from activities taking place within the territorial jurisdiction of a state or on its continental shelf. It shall include, *inter alia*, the discharge or introduction of substances that pollute the sea directly, or indirectly through rivers or other watercourses whether natural or artificial.

(2) To promote "equitable use", a state must take measures to prevent any new form of pollution and abate existing sea water pollution that may cause substantial injury to the interests of other states.

(3) Relevant factors with respect to "equitable use" include but are not limited to: The geography and hydrography of the area, the climatological conditions, quality (and composition) of receiving sea waters, the conservation of marine environment (flora and fauna) of the area, the resources of the seabed and sub-soil and their economic values for present and potential users, the recreation of the population dependent on the coastal area, the past, present and future use of sea water for waste disposal, taking into account the economic and social needs of populations of the coastal states, the comparative costs of alternative means for waste disposal, the avoidance of unnecessary waste disposal.

(4) In case of violation of rule 2, a state is required to compensate for the injury caused and/or must enter into negotiations in order to reach a settlement equitable under the circumstances.

Rules 2 and 4 restate in slightly different wording the basic principles of the Helsinki Rules concerning surface water pollution and are therefore thought to be acceptable. The concept and definition of "pollution caused by continental uses", however, is new and will be examined further. Rule 3 is adapted from the corresponding article in the Helsinki Rules. Technical advice will be required before its wording can be considered satisfactory.

Non-continental pollution

Assuming, as a working hypothesis, that the rules on "pollution caused by continental uses", are basically acceptable then the Working Group's attention will be focused on rules for "non-continental pollution"; this is pollution resulting from human activities originating outside the territorial jurisdiction of a state and its continental shelf. Eventually a distinction will be made

between pollution caused by shipping,[1] and by other activities (e.g. deep-sea mining).

As yet, rules on on-continental pollution have not been established by the Working Group, but in a preliminary exchange of views some tentative ideas have been expressed. These may be summarized as follows:

A. Rather than establish precise rules on inter-state liability for damages caused by "non-continental pollution", it is thought to be sufficient to adopt some general principles which refer to inter-state obligations resulting from a state's membership in the United Nations and other international organizations. The principles introduce an obligation to negotiate with neighbouring states on efficient means of pollution control, and to establish joint commissions when and where appropriate.

B. A functional set of rules is thought to be desirable. The following rules have been suggested:

(1) A state is responsible for any pollution originating from its own publicly-owned ships that cause substantial injury to the interests of other states.
(2) A state shall take all reasonable regulatory and control measures to prevent ships bearing its flag from creating pollution that would cause substantial injury to the interests of other states.
(3) A state shall take, within its territorial waters, all reasonable protective and control measures to prevent pollution from any source, as well as to abate, contain and remove such substances as would cause substantial injury to the interests of other states.

C. On certain points these suggested rules do not conform with the rules now formulated on continental pollution. They would therefore have to be revised and aligned with the continental rules if those were adopted. Moreover the question will have to be examined whether, to become really effective, the non-continental rules envisaged would need to be complemented by remedial rules, such are at present envisaged for continental pollution (see continental rule 4). This is of particular interest if in case of an accident involving polluting substances no adequate measures are taken to abate, contain and remove the substances which caused substantial injury to other states.

A third line of thought furthers the idea to draft rules for "non-continental" pollution caused by human activities, which are as much in conformity as possible with the rules on "continental" pollution, eventually making a distinction, as regards the liability for its injurious consequences, between pollution caused by accident or pollution caused otherwise.

In cases of "non-continental pollution" caused by human activities, rules based on this concept might be formulated as follows:

(1) A state shall take all reasonable measures to prevent or abate pollution caused by its nationals or ships registered in its territory so as to ensure that no substantial damage is caused to the interests of another state.

[1] At the International Conference on Marine Pollution Damage, held in Brussels in November 1969 under the auspices of IMCO, two conventions dealing with oil pollution were adopted. One related to intervention on the high seas in cases of oil pollution casualties and the other to civil liability for oil pollution damage.

(2) Non-compliance with this rule may result in claims for compensation by injured states, with the restriction however that

(3) A state which might reasonably be expected to co-operate in measures to prevent or abate pollution and fails to cooperate may forfeit its right to claim compensation for any damages eventually resulting therefrom.

The above outlined ideas on rules in respect to the control of "non-continental" pollution do little more than constitute a basis for discussion. They show to some extent the complex legal questions pertaining to pollution control of maritime waters. These questions will have to be solved in order to arrive at a coherent, equitable, and efficient set of rules which in the future should contribute to protect the sea against pollution and preserve marine flora and fauna both for the benefit and health of mankind.

Les Bases Scientifiques Nécessaires et Préalables à l'Adoption des Mesures Législatives contre la Pollution des Eaux de Mer

*E. Du Pontavice**

The scientific bases required prior to the adoption of legislative measures against pollution of the seas

A law which enforces expensive and difficult measures can be accepted and efficient only if its usefulness is understood, if the technical and scientific bases are accurate, and if it is uncontestable by the pressure groups (e.g. private and public industries, municipalities) which will be involved. It is essential that all the scientific and pseudo-scientific arguments which may be put forward by the interested parties should be dismissed on the basis of meticulous studies.

From the economic point of view, the cost of marine pollution and the measures to be taken against it should be evaluated. From the technical point of view, a list of pollutants should be compiled, their effects ascertained and the techniques of combating pollution evaluated. Finally, from the biological and medical points of view, it is necessary to study the autoepuration and antibiotic capacities attributed to the sea, as well as the microbial aspect of pollution.

Bases científicas previas y necesarias para la adopción de medidas legislativas contra la contaminación de las aguas del mar

Una ley que imponga medidas difíciles y costosas puede ser aceptada y eficaz cuando se ha comprendido su utilidad y cuando se funda en bases científicas y técnicas serias, difícilmente discutibles por los grupos de presión interesados (p.e., industrias públicas y privadas, municipalidades). Es preciso que todos los argumentos cientiíficos y pseudocientíficos presentados por las partes interesadas puedan ser rechazados basándose en estudios minuciosos.

Desde el punto de vista económico, el costo de la polución marina y de las medidas que se deban tomar para combatirla debe ser evaluado. Desde el punto de vista técnico, se debería hacer una lista de los productos contaminantes, averiguar sus efectos y evaluar las técnicas que combaten la contaminación. Finalmente, desde los puntos de vista biológico y médico, es preciso estudiar la autodepuración y capacidad antibiótica atribuida al mar, así como también el aspecto microbiano de la contaminación.

TOUTE action législative ou psychologique dans la lutte contre la pollution des eaux de mer exige une information exacte d'une part sur le niveau de la pollution et ses origines, et d'autre part sur la prévision de sa progression dans les années à venir. Cette prévision fait elle-même intervenir certaines hypothèses sur le développement de la production et de la consommation, l'évolution des techniques industrielles et agricoles, et les perspectives de lutte contre la pollution à l'échelle nationale et internationale.

Pour remplir pleinement son rôle, le législateur doit avoir une connaissance claire des problèmes de pollution et des conséquences directes comme des effets psychologiques que peut entraîner la mise en application des mesures qu'il est amené à prendre dans la lutte contre la pollution.

L'écologie nous enseigne que le milieu forme un tout et qu'il est dangereux d'agir sur une de ses composantes, sans savoir quel est son rôle et en quoi sa modification portera atteinte à l'ensemble. Lors du désastre du *Torrey Canyon* (Smith, 1968), on a critiqué l'emploi de bombes incendiaires contre l'épave, comme au contraire l'intervention tardive de l'aviation; de même on a fait valoir que le traitement par des détergents de la nappe de pétrole et des plages polluées avait eu, sur la faune et la flore, des effets pires que le pétrole lui-même. Ceci démontre que le législateur et les pouvoirs publics doivent être parfaitement informés des aspects techniques du problème.

A plus long terme, les mesures à prendre pour contrôler la pollution seront à la fois coûteuses et peu populaires. Le législateur devra donc s'assurer au préalable de leur efficacité. En outre il sera amené à faire, entre les diverses mesures idéalement les meilleures, un choix rationnel qui tienne compte de leur coût respectif. L'efficacité de son action sera là aussi fonction de son information.

Cette information doit être également étendue au public. Une connaissance lucide de la situation et des solutions possibles permet d'éviter de dramatiser les faits et de créer une psychose qui risque de décourager les efforts ou d'entraîner l'adoption de mesures disproportionnées aux fins à atteindre. Le public doit donc être objectivement informé de l'intérêt des efforts qu'on lui demande.

Enfin le législateur doit s'assurer de la valeur des mesures préconisées. Toute législation, qui serait fondée sur des postulats non démontrés ou qui serait élaborée à partir de propositions insuffisamment étayées, risquerait, non seulement d'être inefficace, mais surtout serait aisément critiquée par certains pollueurs—industries privées ou publiques, municipalités—qui n'ont, à court terme, aucun intérêt à ce que des mesures de protection contre la pollution soient appliquées; ce qui ne ferait que retarder l'application des mesures nécessaires.

C'est pourquoi nous pensons qu'un congrés ayant pour objet "les effets de la pollution des mers sur les ressources biologiques et la pêche" ne devrait pas se borner à des communications rendant compte d'expériences de chercheurs éminents; ceux-ci doivent

* Institut d'Administration des Entreprises, Université de Nantes, 44 Nantes, France.

[618]

profiter de leur réunion pour faire la synthèse des résultats acquis, exposer ceux-ci sous une forme accessible aux profanes et proposer des solutions, des actions concrètes aux pouvoirs publics.

Les forces de dégradation du monde où nous vivons sont maintenant telles que les chercheurs doivent être conscients de leur mission, de leur rôle. A partir des analyses disponibles à un moment donné, ils doivent présenter une synthèse provisoire. Le monde, s'il veut survivre, ne peut attendre que les travaux scientifiques soient achevés: il faut dès maintenant faire le point et proposer des solutions concrètes, que le législateur pourra traduire en textes législatifs ou réglementaires.

Voici ce qu'écrit à ce sujet l'éditorial du "Marine Pollution Bulletin" (Anon., 1970b) sous le titre "Ecological Monitoring":

" . . . The reasons for this are the complexity of the biological relationships which may be disturbed by an increased pollution load and the serious shortage of pertinent background information. We are paying now for the relative neglect of environmental biology in the past."

" . . . With other pollutants we are faced with the ridiculous task of searching for unknown compounds producing unknown effects in unknown animals, or alternatively with detecting environmental changes brought about by synergistic or cumulative effects of individually innocuous effluents." " . . . this is an area where legislation must often run ahead of scientific knowledge."

Voyons, dans cette optique, quelles sont les bases scientifiques nécessaires et préalables à l'adoption de mesures législatives.

Données d'ordre économique

Deux questions doivent être étudiées successivement: le coût de la pollution des eaux de mer, puis le coût des mesures à prendre contre cette pollution. Le bilan permettra en effet de chiffrer l'économie que chaque nation peut espérer réaliser par une lutte judicieuse; il est bien entendu que le contrôle de la pollution ne permet pas seulement d'obtenir des économies immédiatement mesurables, mais donne à l'environnement un équilibre et un agrément, dont les bienfaits indirects sur la santé mentale et sur la capacité de travail des individus sont considérables.

Le coût de la pollution des eaux de mer

Comme l'écrit Tendron (1970), expert auprès du Comité européen pour la sauvegarde de la nature et des ressource naturelles: "Tout compte fait, la dégradation de la nature finit par coûter beaucoup plus cher que le manque à gagner provisoire résultant de la non-exploitation. . . . Il y a des données qui ne sont pas "quantifiables", au moins dans l'immédiat: celles qui concernent la "qualité" de la vie dans nos civilisations industrielles; cette qualité dépend dans une large mesure de l'environnement dans lequel nous vivons; elle a d'autre part des répercussions certaines sur la santé, la démographie, l'équilibre psychologique. Il est difficile de chiffrer la rentabilité en "santé" d'une ville préservée de la pollution par un certain nombre d'investissements adéquats. Mais il est vraisemblable que les sommes épargnées grâce aux

maladies et aux fatigues évitées, l'emporteront sur celles qui auraient été dépensées dans la lutte contre la pollution". Il existe sur ce point une étude importante de Christ (1961).

On se rappelle que les assureurs du *Torrey Canyon* ont accepté de payer amiablement £15 000 000 à la France et la même somme à la Grande-Bretagne; en outre, les assureurs indemniseront directement les victimes de tout dommage causé matériellement par le contact direct du pétrole, à concurrence de £25 000, dont moitié aux ressortissants de chacun de ces Etats; en réalité, il s'agit d'une transaction parce que précisément le coût de la pollution était très difficile à calculer: coût d'une compagnie d'infanterie avec officiers et camions sur les plages françaises, coût d'un officier et de l'équipage d'un avion militaire anglais et des bombes incendiaires. Que l'on songe à ce qui se serait passé si le navire avait contenu des herbicides au lieu de pétrole: son naufrage aurait tué autant de plancton qu'il y en a dans toute la mer du Nord.

Nous avons voulu par ces simples exemples montrer que le coût de la pollution est considérable, qu'il est toutefois difficile à chiffrer; ce qui n'a du reste pas encore été tenté. Or c'est là une œuvre indispensable: si l'on veut entreprendre la lutte contre ce fléau, il faut d'abord savoir, non plus intuitivement, mais scientifiquement, ce qu'il coûte. C'est une œuvre que des économistes pourraient et devraient rapidement mener à bien dans les différents pays et les organismes internationaux intéressés comme l'ONU, la FAO, la COI/UNESCO et l'IMCO.

Le coût de la lutte contre la pollution

Le coût des mesures à prendre est un peu mieux connu, ne serait-ce que parce qu'il alimente l'argumentation des pollueurs; mais une étude systématique reste à faire.

En France, quelques entreprises ont fait leurs calculs: ainsi une raffinerie de pétrole non polluante coûte de 4 à 5 pour cent plus cher qu'une raffinerie ordinaire ("Lettre de l'Expansion", 15 juin, 1970). De même, en France où peu de villes côtières sont dotées de stations d'épuration et en tout cas de stations suffisantes, Nice, qui n'en possède point, va reporter l'émission de ses eaux d'égout à 2 500 m en mer après un simple traitement physique (dilacération, déshuilage, séparation des flottants), pour un coût total de 130 millions de francs; comme, chaque jour d'été, 130 000 baigneurs environ fréquentent les plages niçoises, la dépense paraît peu importante (Denuzière, 1970).

Pour l'Angleterre, on trouvera d'utiles indications dans l'article de MacNaughton (1961), président de l'Institution of Civil Engineers, Londres. Selon le Secrétaire d'Etat anglais pour le gouvernement local et la planification régionale, le coût économique de la lutte contre la pollution ne serait pas prohibitif dans son pays si on se contente d'un succès à 80 pour cent ("Observer", 22 mars 1970).

Le coût de la lutte pourrait être abaissé et son efficacité accrue par plusieurs moyens:

prêts à long terme aux collectivités locales (MacNaughton, 1961) ou aux industriels intéressés (Association Nationale pour la Protection des Eaux, 1970);

usage de l'information qui améliore le rendement des stations d'épuration (Marin, 1970);

création d'une institution publique centrale dans chaque Etat du type du Ministère de l'environnement créé en France, et au niveau international (Mansholt, 1970);

création d'un fonds européen de lutte, réunissant les six pays du Marché commun, ou bien les membres du Counseil de l'Europe (Lipkowski, 1970) ou encore les pays de l'OCDE.

Données d'ordre technique

Pour quelles raisons le juriste est-il amené à demander aux techniciens de préciser les effets toxiques des agents polluants? En droit interne ou en droit international, on estime généralement—ce n'est pas le lieu ici de discuter du bien-fondé de cette thèse—que la prétendue victime d'une pollution doit faire la preuve que la pollution incriminée existe, dire quelle est son origine et démontrer qu'il existe un rapport de cause à effet entre cette pollution et le dommage subi. Ainsi, en 1941 dans l'affaire de la Fonderie de Trail, le tribunal arbitral condamna le gouvernement Canadien envers celui des Etats-Unis parce que les dommages étaient "constatés par une preuve évidente et convaincante" (Bystricky, 1966). La victime, le plaignant, doit donc prouver que l'effluent ou le produit est toxique. Or il ne le pourra que si la science a mis à la disposition des experts, des informations et des instruments de mesure sûrs.

Il convient donc d'encourager les recherches scientifiques par des aides financières, des avantages fiscaux aux industries qui se livrent à ces recherches, des récompenses pour la mise au point de dispositifs efficaces de lutte contre les nuisances; il convient de mettre sur le plan interne à la disposition des collectivités locales, et sur le plan international, à la disposition des pays en voie de développement, des services d'assistance technique (Valiron, 1970; Anon., 1970a).

La Commission de défense contre la pollution des eaux du Congrès international juridique et scientifique réuni à Royan le 16 mai 1970, a conclu "qu'il convient de constituer un laboratoire national de recherches ayant pour objet l'étude des moyens techniques tendant à prévenir et combattre la pollution par le mazout des rivages nationaux; que la protection des plages et des ports de plaisance, contre la pollution par les eaux usées, doit faire l'objet d'études techniques approfondies, auxquelles seront conviés les autorités locales et les services administratifs compétents".

Liste et effets des produits polluants

Beaucoup reste à faire sur ce point évidemment fondamental. Ici, les aveux d'impuissance sont nombreux. L'éditorial du "Marine Pollution Bulletin" (Anon., 1970), intitulé "Oil at sea", note: "We are almost totally ignorant of the long-term effects of low level but chronic exposure of marine organisms to oil and emulsifiers. Whether or not this is a matter for concern will only be revealed by the kinds of research that are least in the public eye".

Mais, précisément, comme le reconnaît l'auteur de cet éditorial, faute de connaissances précises apportées par les savants, le public s'engoue de modes successives sur la pollution marine et, après avoir accusé le pétrole, met en accusation les pesticides.

Même incertitude pour les pollutions chimiques qui poseront à relativement court terme les problèmes les plus aigus: leur comportement physico-chimique en milieu marin n'est pas mieux connu que leur longévité ou leurs effets à faible concentration sur les êtres vivants. Il en va de même de l'interréaction entre l'eau et la vase, favorisée par la pollution organique (Vivier et Laurent, 1967).

Viel (1967) met également en lumière l'insuffisance de documentation statistique sur les pesticides. L'incertitude est particulièrement grande à l'égard du DDT; il semble que ses méfaits ne soient plus à établir (Le Gouriérec, 1970). Jusqu'en haute mer, le poisson s'en trouve imprégné (ICES, 1970). L'emploi du DDT a donc été interdit en Suède, en URSS, et partiellement au Danemark. Toutefois, aux Etats-Unis, si son usage est limité, la production (50 000 t dont 80 pour cent l'exportation) n'y a pas été réduite d'un kg; en effet, d'une part, la loi n'est pas entrée en application, d'autre part, le Ministère de l'agriculture continue à encourager son emploi ("Time", 1 juin 1970, USA). L'exemple de ce pays n'étant pas unique, il y a là une incohérence qui requiert une information scientifique incontestable: ou bien le DDT est un produit toxique pour les poissons qui s'en imprègnent et il doit alors être remplacé par les produits ou méthodes de substitution connus à l'heure actuelle, ou bien il est inoffensif et il faut alors rapporter les interdictions. Il existe à vrai dire un moyen terme: devenu inutile—grâce à des procédés et produits de substitution—dans les pays développés, il est possible que son utilisation reste indispensable pour juguler les conditions favorables à l'éclosion de certaines maladies de l'homme dans des pays en voie de développement; mais le DDT devrait avoir une aire d'utilisation limitée et de toute façon pourrait être employé moins massivement qu'aujourd'hui, même à l'intérieur de cette aire.

En France, à l'imitation de certains pays étrangers, on a interdit, avec un délai de mise en application d'un an le déversement des détergents dont le taux de bio-dégradabilité est inférieur à 80 pour cent (Décret 25 septembre 1970, *Journal Officiel*, 30 septembre 1970). Or, on a prétendu que le remède serait pire que le mal: les nouveaux détergents seraient plus toxiques pour le poisson que les précédents (Sibthorp, 1969). Ici encore, les savants ont à élucider un problème avant que le législateur puisse définir des normes.

Ce ne sont là que des exemples. Il faut ajouter à cette première incertitude relative au produit lui-même une seconde source d'incertitude: les polluants, isolés dans l'usage industriel, sont regroupés dans les eaux usées et se combinent, donnant des effets mal connus mais qui peuvent être synergiques et cumulatifs et avoir de ce fait des conséquences aussi redoutables qu'ignorées sur la vie du poisson et la santé de l'homme.

En outre, en mer, les effets doivent être étudiés en fonction de la dispersion des produits, elle-même fonction des vents et des courants ainsi que de la situation géographique: les effets sont différents selon qu'il s'agit d'une mer fermée ou d'une mer ouverte, d'un estuaire ou d'une côte rectiligne, de sable ou de rochers, d'une mer avec ou sans marée. Il faut également prendre en considération la profondeur de la mer, la nature du fond.

Définition de normes

Il faut tout d'abord trouver un langage commun entre le savant et le journaliste, ainsi que le public. C'est ce qu'a tenté Cowell (1970). Puis il faut établir des normes de nuisance, tâche à laquelle collaboreront notamment des biologistes. En effet, en l'absence de normes d'interdiction, l'industriel qui pollue sans scrupules concurrence victorieusement son collègue qui hésite à recourir à des agents toxiques ou à les libérer sans traitement préalable: il y a là une concurrence déloyale, contraire à la déontologie et pourtant légale.

Ces normes de nuisance devront être appliquées progressivement, sur cinq ans par exemple, afin de permettre aux industriels et aux municipalités de faire des prévisions à long terme et de développer en temps voulu des programmes de recherche pour améliorer leur fabrication ou trouver des substituts aux produits interdits. Dès lors que les pouvoirs publics auront édicté des normes à application progressive, une industrie peut prévoir un programme de recherche de produits nouveaux non polluants, qui lui conféreront une avance technologique sur ses concurrents.

Mais ces normes de nuisance dépendent précisément des travaux scientifiques. Toutefois, il n'est pas toujours possible d'attendre que ces travaux aient atteint leur perfection pour agir. Ainsi la loi française du 16 décembre 1964 relative au régime et à la répartition des eaux et à la lutte contre leur pollution (*Journal Officiel*, 18 décembre 1964 et rectificatif du 15 janvier et 6 février 1965) prévoit qu'un inventaire établissant le degré de pollution des eaux douces sera établi dans un délai de deux ans à compter du 16 décembre 1964; or cet inventaire n'était pas achevé en 1970. S'il faut donc créer des normes, définir des seuils de toxicité; on ne peut attendre d'avoir une connaissance parfaite des phénomènes pour agir: la norme sera perfectible au fur et à mesure de l'avancement des travaux scientifiques, mais il faut que les techniciens acceptent de présenter des conclusions provisoires, qui serviront de base à la définition des normes légales; celles-ci seront modifiables dans certaines conditions, comme l'est du reste toute œuvre législative: la règle de droit n'est pas une vérité scientifique, mais la solution la meilleure à un moment donné pour permettre aux hommes de vivre en société.

Surveillance et mesure de la pollution

Il convient également de surveiller la pollution, en particulier le long des côtes, par des expériences officielles, pouvant être contrôlées et comparées avec des mesures effectuées dans des conditions identiques mais où la pollution est moindre, près des îles par exemple. C'est ainsi qu'au congrès de Royan précité, Faugère, directeur du laboratoire municipal de Bordeaux, déclarait opérer dans la région de Bordeaux, un contrôle systématique de la pollution en mer sur un réseau comprenant une station tous les deux kilomètres carrés. Il convient également de déterminer la pollution dans les mers et océans, au fur et à mesure que l'on s'éloigne des sources terrestres de pollution.

Parmi les retombées aérospatiales dont le volume et les effets devraient être surveillés, on peut citer: les retombées radioactives des bombes nucléaires, celles des carburants utilisés par les avions et les fusées spatiales (Dost *et al.*, 1969) et les retombées de DDT (ICES, 1970).

Les problèmes posés par l'évacuation de déchets radioactifs dans les mers et les océans sont largement décrits dans le Rapport du Comité scientifique des Nations Unies pour l'étude des effets des radiations ionisantes (Nations Unies, 1962).

Prévention et lutte contre la pollution.

Si les accidents les plus spectaculaires sont jusqu'à présent survenus lors du transport par mer des produits pétroliers, le problème est beaucoup plus vaste. Avec le développement prodigieux du forage sous-marin se pose la question de la protection de la mer contre les accidents de forage et de stockage sous-marin des produits pétroliers. Cette question devrait être mise à l'étude sur le plan technique en vue de son approfondissement. En effet, à la conférence sur la prévention et la maîtrise du déversement de pétrole, tenue à New York en décembre 1969 sous le patronage de la Federal Water Pollution Control Administration, l'auteur du rapport de synthèse, appartenant à la Mobil Oil Corporation, concluait: "We still have a long way to go . . . the unknowns still outweigh the knowns" (Cywin, 1970; voir aussi Allen, 1969).

Dans la lutte contre la pollution, il est indispensable de pouvoir apprécier l'efficacité des techniques susceptibles d'être utilisées contre les agents polluants et leur innocuité pour la faune et la flore. Faut-il proscrire l'usage des détergents pour nettoyer une plage souillée par le pétrole? Quel produit est à la fois efficace dans la lutte contre l'agent de pollution et inoffensif pour le milieu? Le législateur et les pouvoirs publics, pour prévoir des plans de lutte contre la pollution accidentelle, ont besoin de connaissances scientifiquement établies sur ces points.

La lutte contre la pollution ne suppose pas seulement la connaissance des produits polluants et des conditions de leur utilisation (forage et stockage sous-marin), elle suppose également des connaissances en des domaines encore mal connus de l'écologie. Ainsi, pour le transport par mer du pétrole de l'Alaska, l'éditorial de "Marine Pollution Bulletin" intitulé "The vulnerable Arctic" déclarait: "Most knowledge of the biological consequences of marine pollution is dervied from studies in temperate waters. Information about these environments is woefully inadequate, but it is encyclopaedic compared with what we know about even the basic ecology of arctic and tropical waters, let alone the consequences of effluent disposal and accidental pollution in them".

Il faut souligner ici l'idée que le milieu écologique, l'environnement n'obéissent pas aux frontières—politiques—arbitrairement tracées par les hommes: la lutte anglaise contre la pollution du *Torrey Canyon* a modifié la composition chimique du produit pétrolier qui a dévasté la côte française. Ceci met en lumière la nécessité technique d'une coopération internationale en matière de lutte contre la pollution.

Pollution bactérienne et microbienne

Bien qu'on tende aujourd'hui à mettre l'accent sur les pollutions d'origine chimique, la pollution d'origine microbienne est importante (Brisou, 1968, 1969). Son étude devrait porter sur les effluents au voisinage des plages et des installations de conchyliculture et de mariculture.

Il existe, sur les questions des pouvoirs autoépurateur et antibiotique attribués à l'eau de mer, des controverses considérables tant en France qu'à l'étranger. Pour établir des normes de salubrité et des normes de baignade maritime, le législateur doit demander aux spécialistes de présenter des propositions concrètes et universellement admises.

L'étude de l'aspect quantitatif et qualitatif des maladies subies par les populations côtières sera d'autant plus difficile à mener que, d'une part la population estivale est d'origine disparate et se disperse les vacances achevées, et que d'autre la population autochtone, avec parfois la complicité du corps médical local, tait certaines maladies comme la fièvre typhoïde dont la révélation pourrait nuire à son activité (conchyliculture) en entraînant son interdiction ou du moins le renforcement de sa surveillance administrative. Une telle étude devrait considérer les variations dans le temps et dans l'espace des contaminations observées.

Tel est l'inventaire des problèmes à résoudre par les hommes de science, que le juriste présente comme condition préalable à l'adoption de mesures législatives dans la lutte contre la pollution du milieu marin.

Bibliographie

ALLEN, A A   Santa Barbara oil spill: statement presented to the
1969   U.S. Senate Interior Subcommittee on Minerals, Materials and Fuel.
Association Nationale pour la Protection des Eaux, L'épuration
1970   coûte-t-elle aux usines qui la pratiquent? *Eau pure*, février: 21.
BRISOU, J   La pollution microbienne, virale et parasitaire des eaux
1968   littorales et ses conséquences pour la santé publique. *Bull. Org. Mond. Santé*, 38(79):118.
BRISOU, J   La vie des microbes dans les mers et pollution, situation
1969   actuelle; perspectives. Document présenté à la Conférence de Naples, octobre 1969.
BYSTRICKY, R   La pollution des eaux de surface. *Revue droit*
1966   *contemp. Belg.*, (2):59.
CHRIST, W   L'évaluation des dommages économiques causés par
1961   la pollution des eaux. *In* Quelques aspects de la protection des eaux contre la pollution. *Publ. Hlth Pap. W.H.O.*, 13:93–106.
COWELL, E B   Possible pollution scale. *Mar. Pollut. Bull.*, 2(4):62.
1970

CYWIN, A   Prevention and control of oil spillage. *Mar. Pollut.*
1970   *Bull.*, 1(2):30–1.
DENUZIÈRE, M   Nice va dépenser 130 millions de francs pour
1970   enrayer la pollution de ses plages. *Le Monde*, 7 juin 1970. Paris.
DOST, F N, *et al.*   Studies on environmental pollution by missile
1969   propellants. U.S. Air Force Aerospace Medical Research Laboratory (AMRL. TR. 68.85):28 p.
ICES, Report of the ICES working group on pollution of the
1970   Baltic Sea. *Coop. Res. Rep. Int. Coun. Explor. Sea*, (A), (15):31–9.
LE GOURIÉREC, L   Rapport au Colloque consacré à la protection
1970   des populations contre les agressions de la vie moderne, novembre 1961, Paris (France). *Revue Administration*, France, No. spécial:126.
LIPKOWSKI, J DE   La défense de l'homme contre les pollutions,
1970   Colloque de Royan, *Le Monde*, 19 mai 1970:12, Paris.
MACNAUGHTON, G   La protection contre la pollution des eaux:
1961   aspects économiques et financiers. *In* Cahiers de santé publique précités, pp. 107–23.
MANSHOLT, S   Congrès sur "L'homme dans son environnement",
1970   à Rotterdam. *Le Figaro*, 26 mai 1970, Paris.
MARIN, G   "L'informatique au service des problèmes de l'eau",
1970   article sur le Colloque international organisé à Montpellier (France) sur l'initiative d'I.B.M., France, *Le Figaro*, 2 juin 1970:32, Paris.
MORIN, J Y   La pollution des cours d'eau au regard du droit
1967–8   international. Institut des Hautes Etudes Internationales, France, 1967–8, p. 32.
Nations Unies, Comité scientifique pour l'étude des effets des
1962   radiations ionisantes, A.G., 17e session, New York, 1962, sup. No. 16 (A/5216):399–400.
SIBTHORP, M M   Oceanic pollution; a survey and some suggestions
1969   for control. David Davies Memorial Institute of International Studies, p. 12.
SMITH, J E (Ed.)   *Torrey Canyon* pollution and marine life.
1968   A report by the Plymouth Laboratory of the Marine Biological Association of the United Kingdom. Cambridge University Press, 196 p.
TENDRON, G   Destruction de l'équilibre naturel: les réponses de
1970   l'écologie. *Preuves*, 1er trimestre:139, Paris.
VALIRON, F   Communication au Colloque sur la protection des
1970   particuliers contre les agressions de la vie moderne, novembre 1969. *Revue Administration*, No. spécial:263.
VIEL, G   L'usage agricole des pesticides et sa conséquence pour
1967   la pollution des eaux. *Bull. Tech. Inf. Minist. Agric. Fr.*, (224):845–50.
VIVIER, P et LAURENT, M   Interréactions vase-eau. La pollution
1967   des eaux et l'agriculture. *Bull. Tech. Inf. Minist. Agric. Fr.*, (224):775–82.
ANON   Oil at sea. *Mar. Pollut. Bull.*, 1(2):17.
1970
ANON   Protection des populations contre les agressions de la vie
1970a   moderne. *Revue Administration*, 1970:126, 268, 311.
ANON   Ecological monitoring. Editorial. *Mar. Pollut. Bull.*, 1(6): 81–2.
1970b

Information Requirements for Rational Decision-making in Control of Coastal and Estuarine Oil Pollution

*J. B. Glude**

Informations nécessaires à la prise de décisions en matière de pollution des côtes et des estuaires

Les épanchements importants d'hydrocarbures résultant d'accidents survenus aux pétroliers et de fuites des puits de pétrole, qui sont examinés dans le présent document, se sont produits dans des régions où la composition saisonnière des espèces aquatiques n'était pas entièrement connue. De même, on connaissait mal les effets des méthodes de lutte contre ces épanchements—par l'application de dispersants chimiques, de matériaux d'agglomération ou d'agents de précipitation—sur les espèces aquatiques ou sur leur habitat.

Dans ces conditions d'ignorance, il n'est pas surprenant que les organismes gouvernementaux ou les sociétés privées aient décidé d'utiliser des produits chimiques ou autres qui ont causé des dégâts considérables aux organismes marins. De plus, la nécessité d'agir d'urgence pour limiter la diffusion des hydrocarbures jointe à la forte réaction du public n'ont généralement pas permis de mener des recherches méthodiques pour réunir les renseignements dis-

La información requerida para tomar decisiones racionales en el control de la contaminación por petróleo en aguas costeras y estuarinas

Los principales derrames de hidrocarburos debidos a accidentes sufridos por petroleros u ocurridos en pozos situados en tierra han tenido lugar en zonas en las que no se conocía bien la composición estacional de las especies acuáticas. Del mismo modo, se desconocían también, en gran parte, los efectos de los procedimientos de control de los hidrocarburos, incluidos los de la aplicación de sustancias químicas dispersantes, sustancias aglomerantes o agentes de sumersión sobre las especies acuáticas o su habitat.

En estas condiciones de ignorancia no es sorprendente que los organismos gubernamentales o las compañías privadas decidieran usar sustancias químicas u otras materias que causaban amplias pérdidas de organismos marinos. Además, la urgencia de una acción inmediata para limitar la difusión de los hidrocarburos, y la intensa reacción pública impedían por lo general, que se investigara

---

* Northwest Region, National Marine Fisheries Service, National Oceanic and Atmospheric Administration, Seattle, Washington 98101, U.S.A.

ponibles concernant la composition des espèces ou les effets contraires qui peuvent résulter de l'application des diverses méthodes de lutte.

L'expérience acquise à la suite des importants épanchements d'hydrocarbures qui se sont produits récemment indique que l'on a besoin de deux éléments principaux pour pouvoir prendre des décisions opportunes et rationnelles sur le choix de méthodes et de procédures permettant de lutter contre ces épanchements ou de nettoyer le milieu marin :

1. Un "atlas des ressources" rassemblant les données disponibles sur la composition saisonnière des espèces dans les zones côtières et estuarines, notamment dans celles où des épanchements d'hydrocarbures ont le plus de chances de se produire, et fournissant les renseignements dont on dispose sur les mouvements de l'eau.
2. Un sommaire des connaissances relatives aux "effets biologiques" des divers matériaux chimiques ou physiques dont l'utilisation est prévue pour lutter contre les épanchements d'hydrocarbures sur les espèces aquatiques importantes dans chaque zone côtière ou estuarine.

Sans aucun doute, la compilation des séries de données qu'il est proposé de réunir pour l'atlas des ressources et à propos des effets biologiques mettront en lumière des lacunes que l'on pourra combler au moyen de programmes de recherche appropriés.

Lorsqu'ils auront à leur disposition des connaissances sur les espèces présumées présentes et sur leur tolérance aux divers produits employés pour lutter contre les hydrocarbures, les organismes ou les sociétés responsables de leur utilisation pourront prendre des décisions plus rationnelles en ce qui concerne les méthodes qu'il convient d'appliquer pour lutter contre les polluants en infligeant le minimum de dommages aux espèces marines. Le même type d'information servirait en outre à prendre des décisions contre la pollution d'autres matériaux.

ordenadamente la información existente relativa a la composición por especies o a los posibles efectos perjudiciales de los distintos métodos de control.

De la experiencia obtenida con los recientes e importantes derrames de hidrocarburos se ha llegado a la conclusión de que hacen falta dos principales sectores de conocimiento para permitir decisiones ordenadas y racionales sobre la selección de métodos y procedimientos para controlar o eliminar los derrames de hidrocarburos :

1. Un "atlas de recursos" en el que se resuman los conocimientos existentes sobre la composición estacional por especies en las zonas costeras y de los estuarios, especialmente de aquellas en las que son más probables los derrames de hidrocarburos y que faciliten información sobre los movimientos de las aguas.
2. Un resumen de los conocimientos que se poseen sobre los "efectos biológicos" que ejercen las distintas sustancias químicas o físicas—propuestas para controlar con ellas los derrames de hidrocarburos—sobre importantes especies acuáticas de cada zona costera o de los estuarios.

La compilación de las series propuestas del atlas de recursos y efectos biológicos es evidente que revelaría lagunas de conocimientos que podrían llenarse posteriormente por medio de adecuados programas de investigación.

Con el conocimiento inmediatamente existente de las especies que se supone que se hallan presentes y su tolerancia a las distintas sustancias que se emplean para eliminar los hidrocarburos, los organismos o compañías responsables podrían adoptar decisiones más racionales con respecto a los procedimientos adecuados para eliminar los contaminantes con un daño mínimo a las especies marinas. La misma clase de información ayudaría a tomar decisiones sobre la contaminación producida por otras sustancias peligrosas.

THE basis for selection of methods for control and clean-up of oil spills, both large and small, should include the following types of information in order that aquatic species and their environment can best be protected :

(1) The species and abundance of intertidal and subtidal animals and plants which are expected to be present in the area at the time of the spill.
(2) The effects of the type of oils spilled and proposed control methods on these forms and their environment.
(3) Tidal or oceanic currents in the area.
(4) Expected wind drift.

Even in many harbours the seasonal composition of aquatic populations was and still is imperfectly known. Under these conditions of ignorance, it is not surprising that government agencies or private companies decided to use chemicals which caused extensive mortality of marine organisms.

It is now known that the biological effects of the oil and the methods selected for control or clean-up must be based on the characteristics of the oil and to choose the best method of treatment the following three sources of information would facilitate timely, rational decisions on courses to be taken :

(1) A "resource atlas" summarizing available knowledge concerning seasonal composition of living species in coastal and estuarine areas, especially those in which oil spills are most likely.
(2) A series of maps, charts or tables providing information concerning water movements caused by tides, winds and river or oceanic currents for the coastline generally, but especially for those areas in which oil spills are most likely.
(3) A summary of knowledge concerning the "biological effects" of various kinds of petroleum products and the chemical or physical materials or

methods proposed for use in spill control upon important aquatic species and their environment in each coastal or estuarine area.

A resource atlas for the coasts of the United States might take the form of a series of charts and tables listing for each harbour or similar geographic unit the important aquatic species which would be expected to be present each month or each quarter during the year. For example, a chart for Elliott Bay, Seattle, Washington, for September would indicate the presence of adult chinook and coho salmon migrating upstream through Duwamish Waterway, and would include a caution "Do not use oil dispersant chemicals in this area during September". A similar chart for January would indicate the absence of salmon in Duwamish Waterway and would suggest that limited applications of dispersant chemicals would be acceptable.

Another suggestion by Mr. Keith Hay, American Petroleum Institute, is that resource information could be stored in a computer where it could provide a print-out of available information for each location and time of year.

A combination of the two systems might be the most usable since a printed resource atlas could be provided to government agencies, private companies, municipalities, port commissions, and firms specializing in clean-up of oil spills, for immediate reference. Storage of information in a computer bank would provide new information concerning each geographic unit.

Charts and tables of expected water movements by areas and seasons should accompany the resource atlas since this would provide the best means of predicting movement of oil masses following an accidental spill. An indication that water currents would probably transport oil toward an important shellfish producing area would suggest immediate containment and removal of oil or even the application of chemical dispersants at the scene of the spill.

[623]

Finally, a series of tables describing the biological effects of oils and the various chemical or physical materials proposed for the control of oil spills upon important aquatic species and their environment in each coastal or estuarine area would help the responsible agency or individual to select the materials which would be least harmful and still be reasonably effective and economical. With over 400 dispersant chemicals on the market, numerous gellants, absorbents and sinking agents, as well as several types of containment or removal systems, it would be difficult for the responsible agency or person to select the most appropriate procedure. However, with immediately available knowledge of the species expected to be present and their tolerance to the various oil and oil control materials, and information concerning water movements, the administering agency or company could make a rational decision with minimum damage to living marine species.

It is therefore proposed that government and the oil industry cooperate in the preparation and publication of a resource atlas eventually covering the entire coastline of the United States, but beginning with areas most susceptible to oil pollution. Secondly, that charts and tables describing water movements be prepared and published as a joint government-industry project to accompany the resource atlas series. Finally, that government and industry cooperate in sponsoring research to determine the biological effects of various petroleum products and chemical and physical oil control materials upon important aquatic species and their environment. This information should be published and widely distributed to agencies or companies responsible for oil pollution control and clean-up.

Much of the information needed is presently available through published articles and in the files of agencies, universities, and individual scientists. The first job would be to assemble this information. Further research would then be necessary to obtain missing information.

According to Sir Solly Zuckerman's report to the International Conference on Oil Pollution of the Sea in 1968, some efforts are under way in the United Kingdom to assemble information similar to that proposed in this paper. He stated that the Burnham-on-Crouch Laboratory is cooperating with the Nature Conservancy and the Marine Biological Association in preparing a series of charts and schedules showing coastal areas of especial importance to fisheries and nature conservation, which are particularly vulnerable to oil pollution and to toxic solvent emulsifiers. An extension of this concept to include a listing of species expected to be present during various seasons, together with hydrographic information and a table of biological effects of oils and proposed oil control methods, would improve the basis for decision-making regarding oil pollution control and clean-up in the United Kingdom. Eventually, similar information should be obtained and published for all coastlines of the world susceptible to accidental oil spills.

References

BLUMER, M   Testimony of Max Blumer before subcommittee
1970   on air and water pollution. Machios, Maine, Senate Committee on Public Works, 8 September 1970.

BLUMER, M, SASS, J and SOUZA, G   Hydrocarbon pollution of
1970   edible shellfish by an oil spill. *Mar. Biol.*, 5:195–202.

BLUMER, M, SOUZA, G and SASS, J   Hydrocarbon pollution of
1970   edible shellfish by an oil spill. WHOI (Ref. #70–1):14 p. (Unpubl. MS).

CORNER, E D S, SOUTHWARD, A J and SOUTHWARD, E C   Toxicity
1968   of oil spill removers ("Detergents") to marine life: an assessment using the intertidal barnacle (*Elminius modestus*). *J. mar. biol. Ass. U.K.*, 48(1):29–47.

DORRLER, J S   Limited oil spills in harbor areas. *In* Proceedings of
1969   the Joint Conference on Prevention and Control of Oil Spills, 15–17 December 1969, American Petroleum Institute, pp. 151–6.

GLUDE, J B   Observations on the effect of oil from the tanker
1968   *Ocean Eagle* and oil control operations on the fisheries of San Juan, Puerto Rico. Manuscript report to Director, Bureau of Commercial Fisheries, Washington, D.C., March 1968, 42 p.

GLUDE, J B and PETERS, J A   Observations on the effect of oil
1967   from the tanker *Torrey Canyon* and oil control measures on marine resources of Cornwall, England, and Brittany, France. Manuscript report to Director, Bureau of Commercial Fisheries, Washington D.C., June 1967, 10 p.

JONES, L G, *et al.*   Just how serious was the Santa Barbara oil
1969   spill? *Ocean Ind.*, 4(6):4 p.

NORTH, W J   *Tampico*, an experiment in destruction. *Sea Fish.*,
1967   13:212–7.

NORTH, W J, NEUSHUL, M and CLENDENNING, K A   Successive
1965   biological changes observed in a marine cove exposed to a large spillage of miheral oil. *In* Pollutions marines par les microorganismes et les produits pétroliers. Symposium de Monaco (avril 1964). CIESMM, pp. 335–54.

SMITH, J E (Ed.)   *Torrey Canyon*, pollution and marine
1968   life. A report by the Plymouth Laboratory of the Marine Biological Association of the United Kingdom, Cambridge, Cambridge University Press, 196 p.

TENDRON, G   Contamination of marine flora and fauna by oil
1969   and the biological consequences of the *Torrey Canyon* incident. *In* Report of Proceedings of the International Conference on Oil Pollution of the Sea, Rome, 7–9 October, 1968, pp. 114–21.

ZUCKERMAN, S   A scientific approach to the problems of oil
1969   pollution. *In* Report of Proceedings of the International Conference on Oil Pollution of the Sea, Rome, 7–9 October, 1968, pp. 148–59.

important
FAO books
on fisheries

POLLUTION–an international problem for fisheries

A comprehensive review of pollution of inland waters and oceans and its effect upon aquatic living resources.

Available in separate English, French and Spanish versions.

85 pages. **Price: $1.50, or £0.60 or FF7,50.**

FAO YEARBOOK OF FISHERIES STATISTICS–Trilingual

Contains statistical data on production and foreign trade in fishery products and in products derived from related industries, including tables that show imports and exports in terms of standardized statistical groups of fishery commodities. There are also statistics on world catches and landings, utilization and production of preserved and processed fishery commodities, and composition of fishing fleets.

The volume subtitled **Catches and Landings** covers data on quantities and values of fish caught and landed by countries, by species and by fishing areas. The volume subtitled **Fishery Commodities** covers disposition of catches and both production and international trade data by types of fishery commodities.

Volume 27	**Fishery Commodities**	1969 data	**$5·50 or £2·80 or FF19,25**
Volume 29	**Fishery Commodities**	1969 data	**$7·00 or £2·80 or FF35,00**
Volume 30	**Catches and Landings**	1970 data	**$9·00 or £3·60 or FF45,00**
Volume 31	**Fishery Commodities**	1970 data	**$9·00 or £3·60 or FF45,00**

The YEARBOOK is supplemented by a trilingual series of **Bulletins of Fishery Statistics,** issued from time to time. Limited supplies are available on request.

FAO FISHERIES INDEX

The INDEX covers publications and documents produced by the Department of Fisheries from 1945 to 1969. The biographical list contains some 4,500 entries from which the analytic index and the author index have been derived. Fisheries publications and documents issued after 1969 are included in the monthly **FAO Documentation–Current Bibliography.**

ATLAS OF THE LIVING RESOURCES OF THE SEAS

This work comprises 62 maps in three series. The first series of ten maps illustrates the geographical distribution and present state of exploitation of resources living in the world's oceans. A second series of seven maps provides some characteristic examples of migration in fish, while the third—and more detailed—series of 45 geographical maps presents in illustrated form the geographical and vertical distribution, as well as the abundance, of main stocks of pelagic and demersal (bottom-dwelling) fish and crustaceans in each ocean region.

Much of the material contained in the **Atlas of the Living Resources of the Seas** has never before appeared in published form. Thus, it is of inestimable practical value to scientific, engineering, management, academic, government, industry and commercial areas concerned with oceanology, as well as to those concerned with or interested in problems of environment and conservation.

Available in a trilingual (English, French and Spanish) edition.
Price: $12.00 or £4.80 or FF60,00.

For a free copy of the catalogue of **FAO Books in Print** write to:
DISTRIBUTION AND SALES SECTION, FOOD AND AGRICULTURE ORGANIZATION,
Via delle Terme di Caracalla, 00100 Rome, Italy.

Reference library on Modern Fisheries

Since 1953 the Department of Fisheries FAO has organized Congresses of experts on aspects of fisheries. From the resulting papers and discussions have emerged a number of important books which have distributed the knowledge thus assembled all over the world and contributed to the considerable development of new and old fisheries. The books are here listed, all are still in print by original or successive issues:

	Price £
Boats	
Fishing Boats of the World 1.	8·50
Fishing Boats of the World 2.	9·00
Fishing Boats of the World 3.	9·50
Gear	
Modern Fishing Gear of the World 1	8·50
Modern Fishing Gear of the World 2	8·50
Modern Fishing Gear of the World 3	14·50
FAO Catalogue of Fishing Gear Designs	5·00
Utilization and Processing	
Fish in Nutrition	7·50
The Technology of Fish Utilization	7·00
Fish Inspection and Quality Control	12·50
Freezing and Irradiation of Fish	11·50
Other Branches	
Fishing Ports and Markets	12·50
Mechanization of Small Fishing Craft	2·00
Fishing With Electricity	3·75
European Inland Water Fish	8·75
The Fish Resources of the Ocean	10·50
Marine Pollution and Sea Life	18·00
Coastal Aquaculture (to come)	

All the books are printed to a high standard and are well bound to stand long and sustained usage.

Fishing News (Books) Ltd.

23 Rosemount Avenue, West Byfleet
Surrey KT14 England

(A subsidiary of Arthur J Heighway Publications Ltd.
110 Fleet Street, London EC4A 2JL)

Expanded service through books

In addition to publishing the big books from FAO Congresses we have supplemented that service with a number of independently written works by competent writers. These are here listed under their varying categories:

Fishing Boats

	Price £
Commercial Fishing Methods	3·75
The Fishing Cadets Handbook	1·40
Introduction to Trawling	1·00
Fishing Boats and Equipment	2·25
Stability and Trim of Fishing Vessels	1·50
Refrigeration on Fishing Vessels	2·75
The Stern Trawler	8·50

Fishing Gear

	Price £
Trawlermen's Handbook	2·00
The Seine Net	4·25
Modern Deep Sea Trawling Gear	2·10
Underwater Observation Using Sonar	2·00
Sonar in Fisheries	2·20
How to Make and Set Nets	1·75
More Scottish Fishing Craft and Their Work	2·25
Fish Catching Methods of the World	5·50

Fish Production

Farming the Edge of the Sea	4·25
Tuna Distribution and Migration	2·10
Seafood Fishing for Amateur & Professional	3·00
A Living From Lobsters	1·25
The Stocks of Whales	2·80
The Lemon Sole	1·50
Inshore Fishing: its Skills Risks Rewards	2·65
The Fertile Sea	2·90
Fish and Shellfish Farming in Coastal Waters	7·50

Marketing and Miscellaneous

Marketing of Shellfish	4·50
Multilingual Dictionary of Fish and Fish Products	7·50
Inshore Craft of Britain in the Days of Sail & Oar (2 vols)	6·30
Japan's World Success in Fishing	3·25
Handy Medical Guide for Seafarers	1·50
Escape to Sea	1·25

For Postage add 5 per cent

Fishing News (Books) Ltd.

23 Rosemount Avenue, West Byfleet

Surrey KT14 England

(and at 110 Fleet Street, London EC4A 2JL)

Aquaculture in various types of water

With increasing interest in the breeding and cultivation of fish in fresh, brackish and coastal waters, an expanding range of books is called for to meet the commercial need for practical information. In addition to the range of books listed below a new journal entitled "Fish Farming International" will appear in 1973—first as a yearbook but with the intent of conversion to more frequent publication as and when needed. This list shows the range of material already available.

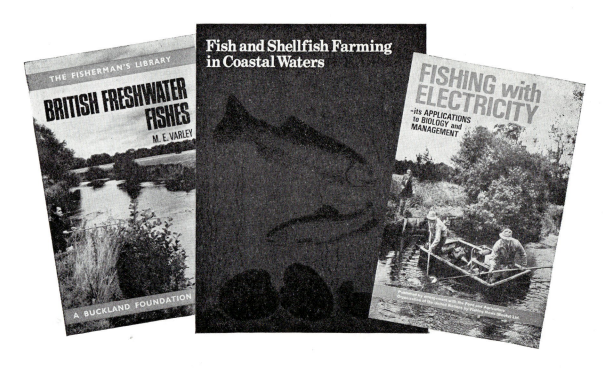

	Price £
Textbook of Fish Culture: Breeding and Cultivation of Fish *by Marcel Huet*	12·50
Fish and Shellfish Farming in Coastal Waters *by P H Milne*	7·50
European Inland Water Fish: A Multilingual Catalogue (*an FAO book*)	8·75
British Freshwater Fish *by Margaret E Varley*	1·80
Coastal Aquaculture edited by *T V R Pillay* (to come)	
Fishing With Electricity (*FAO and EIFAC experts*)	3·75
Farming the Edge of the Sea *by E S Iversen*	4·25
The Technique of Eel Culture *by Atsushi Usui and G Williamson* (to come)	

Fishing News (Books) Ltd.

23 Rosemount Avenue, West Byfleet
Surrey KT14 England

(a subsidiary of Arthur J Heighway Publications Ltd.
110 Fleet Street, London EC4A 2JL)